F THE ELEMENTS

(1971 values for atomic weights)

VIII			Ib	IIb	IIIa	IVa	Va	VIa	VIIa	O

Groups

Representative elements (nonmetals)

Noble gases

										2 He 4.0026
					5 B 10.81	6 C 12.011	7 N 14.0067	8 O 15.9994	9 F 18.9984	10 Ne 20.179
	Transition Metals				13 Al 26.9815	14 Si 28.086	15 P 30.9738	16 S 32.06	17 Cl 35.453	18 Ar 39.948
27 Co 8.9332	28 Ni 58 71	29 Cu 63 546	30 Zn 65.38		31 Ga 69.72	32 Ge 72.59	33 As 74.9216	34 Se 78.96	35 Br 79.904	36 Kr 83.80
45 Rh 02.905	46 Pd 106.4	47 Ag 107.868	48 Cd 112.40		49 In 114.82	50 Sn 118.69	51 Sb 121.75	52 Te 127.60	53 I 126.9045	54 Xe 131.30
77 Ir 192.22	78 Pt 195.09	79 Au 196.967	80 Hg 200.59		81 Tl 204.37	82 Pb 207.2	83 Bi 208.9804	84 Po (210)	85 At (210)	86 Rn (222)

Representative elements (metals)

Inner-transition metals

62 Sm 150.4	63 Eu 151.96	64 Gd 157.25	65 Tb 158.925	66 Dy 162.50	67 Ho 164.930	68 Er 167.26	69 Tm 168.934	70 Yb 173.04	71 Lu 174.97
94 Pu (242)	95 Am (243)	96 Cm (247)	97 Bk (247)	98 Cf (249)	99 Es (254)	100 Fm (253)	101 Md (256)	102 No (254)	103 Lr (257)

LANGE'S
HANDBOOK OF CHEMISTRY

LANGE'S
HANDBOOK OF
CHEMISTRY

Editor: JOHN A. DEAN

Professor of Chemistry
University of Tennessee (Knoxville)

Formerly Compiled and Edited by
NORBERT ADOLPH LANGE, Ph.D.

I enjoy sharing my books as I do my friends, asking only that you treat them well and see them safely home

Persinger

annesburg
hi
o

Library of Congress Cataloging in Publication Data

Lange, Norbert Adolph, 1892–1970
  Lange's Handbook of chemistry.

  1. Chemistry—Handbooks, manuals, etc.   I. Dean,
John Aurie, 1921–   ed. II. Title. III. Title:
Handbook of chemistry.
QD65.L36   1973   540   73-6553
ISBN 0-07-016190-9

1 2 3 4 5 6 7 8 9 0   COCO   7 6 5 4 3

*The editors for this book were Harold B. Crawford and
Ross J. Kepler, and its production was supervised
by Teresa F. Leaden. It was set in Fototronic Bodoni by
York Graphic Services, Inc.*

It was printed and bound by The Colonial Press Inc.

17236

*To those workers in science who through
their labors determined the values recorded herein,
this compilation is dedicated. Their devotion
to the search for the constants of nature
and the dissemination of this knowledge are the
foundations upon which rest the achievements
of applied science.*

CONTENTS

For the detailed contents of any section, consult the title page of that section. See also the alphabetical index in the back of this handbook.

PREFACE TO THE FIRST EDITION

This book is the result of a number of years' experience in the compiling and editing of data useful to chemists. In it an effort has been made to select material to meet the needs of chemists who cannot command the unlimited time available to the research specialist, or who lack the facilities of a large technical library which so often is not conveniently located at many manufacturing centers. If the information contained herein serves this purpose, the compiler will feel that he has accomplished a worthy task. Even the worker with the facilities of a comprehensive library may find this volume of value as a time-saver because of the many tables of numerical data which have been especially computed for this purpose.

Every effort has been made to select the most reliable information and to record it with accuracy. Many years of occupation with this type of work bring a realization of the opportunities for the occurrence of errors, and while every endeavor has been made to prevent them, yet it would be remarkable if the attempts towards this end had always been successful. In this connection it is desired to express appreciation to those who in the past have called attention to errors and it will be appreciated if this be done again with the present compilation for the publishers have given their assurance that no expense will be spared in making the necessary changes in subsequent printings.

It has been aimed to produce a compilation complete within the limits set by the economy of available space. One difficulty always at hand to the compiler of such a book is that he must decide what data are to be excluded in order to keep the volume from becoming unwieldy because of its size. He can hardly be expected to have an expert's knowledge of all branches of the science nor the intuition necessary to decide in all cases which particular value to record especially when many differing values are given in the literature for the same constant. If the expert in a particular field will judge the usefulness of this book by the data which it supplies to him from fields other than his specialty and not by the lack of highly specialized information in which only he and his co-workers are interested (and with which he is familiar and for which he would never have occasion to consult this compilation), then an estimate of its value to him will be apparent. However, if such specialists will call attention to missing data with which they are familiar and which they believe others less specialized will also need, then works of this type can be improved in succeeding editions.

Many of the gaps in this volume are caused by the lack of such information in the literature. It is hoped that to one of the most important classes of workers in chemistry, namely the teachers, the book will be of value not only as an aid in answering the most varied questions with which they are confronted by interested students, but also as an inspiration through what it suggests by the gaps and inconsistencies, challenging as they do the incentive to engage in the creative and experimental work necessary to supply the missing information.

While the principal value of the book is for the professional chemist or student of chemistry, it should also be of value to many people not especially educated as chemists. Workers in the natural sciences— physicists, mineralogists, biologists, pharmacists, engineers, patent attorneys, and librarians—are often called upon to solve problems dealing with the properties of chemical products or materials of construction. For such needs this compilation supplies helpful information and will serve not only as an economical substitute for the costly accumulation of a large library of monographs on specialized subjects, but also as a means of conserving the time required to search for information so widely scattered throughout the literature. For this reason especial care has been taken in compiling a comprehensive index and in furnishing cross references with many of the tables.

It is hoped that this book will be of the same usefulness to the worker in science as is the dictionary to the worker in literature, and that its resting place will be on the desk rather than on the book shelf.

N. A. Lange

Cleveland, Ohio
May 2, 1934

PREFACE
TO THE ELEVENTH EDITION

The new editor has assumed the task of data compilation from the late Dr. Lange, the man who initiated the *Handbook of Chemistry* almost four decades ago. It seems only fitting that his name should be embodied within the new title in recognition for his efforts on the ten preceding editions of this handbook.

Perhaps it would be simplest to begin by stating the ways in which this new edition has not been changed. It remains the one-volume source of factual information for chemists, both professionals and students—the first place in which to "look it up" on the spot. The aim is to provide sufficient data to satisfy all one's general needs.

The changes, however, are both numerous and significant. First of all, there is a basic change in organization. The handbook is now divided into sections—mathematics, general information and conversion tables, atomic and molecular structure, inorganic chemistry, analytical chemistry, electrochemistry, organic chemistry, spectroscopy, thermodynamic properties, physical properties, miscellaneous—and within these sections related groups of factual data are presented. This arrangement, plus the new sectional tables of contents, which provide a complete listing of items within each section, backed up by a thorough and extensively cross-indexed subject index, makes it possible to find the information quickly.

The following subject matter is offered for the first time:

Emission and absorption lines for arc, spark, and flame and atomic absorption—with sensitivities and/or detection limits

Formation constants of metal complexes with organic and inorganic ligands

Mass absorption coefficients of X-ray emission lines commonly used in X-ray absorption work, with coefficients for all elements

Statistical tables

Atomic electron affinities

Electronegativities of the elements

Spatial orientation of common hybrid bonds
Hammett and Taft substituent constants
Selectivity coefficients for ion-exchange resins
Cross contamination and separation factors in separation methods

Also new are self-instructional sections developed for certain areas: measurement of pH, use of statistics, separation methods, and X-ray methods.

Expanded coverage is provided in such important areas as:

Solubility products
Proton transfer reactions (acid dissociation constants) with 1200 entries for organic compounds and 150 for inorganic compounds
Electrode potentials of elements and their compounds listed by element
Bond energies and radii of atoms and ions
Reference electrodes
Reference pH buffers for water, deuterium oxide, and aqueous-organic systems
Approved symbols and abbreviations

Updating has increased the usefulness of such valuable tabulations as:

Physical properties of 4000 inorganic compounds
Nomenclature of inorganic compounds
Heats and free energies of formation, entropies, and heat capacities—incorporating the latest recommended values of the National Bureau of Standards
X-ray emission spectra and X-ray K and L absorption edges, given both as wavelengths and as energies in keV
Critical properties
Limiting equivalent ionic conductances in aqueous solution
Table of nuclides, now 100 pages extending through element 105
Ionization potentials of the elements

Finally, the mathematical section has been expanded from its rather restricted size in recent editions so as to include mathematical information commonly needed by an upper-division or graduate student, or professional, without recourse to other reference sources.

It is hoped that users of this and previous editions will continue to offer friendly criticism and suggestions, and call attention to errors.

John A. Dean

Knoxville, Tennessee

ACKNOWLEDGMENT

Grateful acknowledgment is hereby made of an indebtedness to those who have contributed to former editions and whose compilations continue in use in this edition. In particular, acknowledgment is made of the contribution of Dr. Joseph R. Peterson, who prepared the expanded Table of Nuclides; also that of Mr. Theodore C. Rains who supplied many of the atomic absorption sensitivities.

LANGE'S
HANDBOOK OF
CHEMISTRY

Section 1
MATHEMATICS

MATHEMATICAL TABLES

Table 1-1
SQUARES OF NUMBERS

From Baumeister and Marks, *Standard Handbook for Mechanical Engineers*, 7th ed., 1967, McGraw-Hill Book Company; by permission

N	0	1	2	3	4	5	6	7	8	9	Avg diff
1.00	1.000	1.002	1.004	1.006	1.008	1.010	1.012	1.014	1.016	1.018	2
1	1.020	1.022	1.024	1.026	1.028	1.030	1.032	1.034	1.036	1.038	
2	1.040	1.042	1.044	1.047	1.049	1.051	1.053	1.055	1.057	1.059	
3	1.061	1.063	1.065	1.067	1.069	1.071	1.073	1.075	1.077	1.080	
4	1.082	1.084	1.086	1.088	1.090	1.092	1.094	1.096	1.098	1.100	
1.05	1.102	1.105	1.107	1.109	1.111	1.113	1.115	1.117	1.119	1.121	
6	1.124	1.126	1.128	1.130	1.132	1.134	1.136	1.138	1.141	1.143	
7	1.145	1.147	1.149	1.151	1.153	1.156	1.158	1.160	1.162	1.164	
8	1.166	1.169	1.171	1.173	1.175	1.177	1.179	1.182	1.184	1.186	
9	1.188	1.190	1.192	1.195	1.197	1.199	1.201	1.203	1.206	1.208	
1.10	1.210	1.212	1.214	1.217	1.219	1.221	1.223	1.225	1.228	1.230	
1	1.232	1.234	1.237	1.239	1.241	1.243	1.245	1.248	1.250	1.252	
2	1.254	1.257	1.259	1.261	1.263	1.266	1.268	1.270	1.272	1.275	
3	1.277	1.279	1.281	1.284	1.286	1.288	1.290	1.293	1.295	1.297	
4	1.300	1.302	1.304	1.306	1.309	1.311	1.313	1.316	1.318	1.320	
1.15	1.322	1.325	1.327	1.329	1.332	1.334	1.336	1.339	1.341	1.343	
6	1.346	1.348	1.350	1.353	1.355	1.357	1.360	1.362	1.364	1.367	
7	1.369	1.371	1.374	1.376	1.378	1.381	1.383	1.385	1.388	1.390	
8	1.392	1.395	1.397	1.399	1.402	1.404	1.407	1.409	1.411	1.414	
9	1.416	1.418	1.421	1.423	1.426	1.428	1.430	1.433	1.435	1.438	
1.20	1.440	1.442	1.445	1.447	1.450	1.452	1.454	1.457	1.459	1.462	
1	1.464	1.467	1.469	1.471	1.474	1.476	1.479	1.481	1.484	1.486	
2	1.488	1.491	1.493	1.496	1.498	1.501	1.503	1.506	1.508	1.510	
3	1.513	1.515	1.518	1.520	1.523	1.525	1.528	1.530	1.533	1.535	
4	1.538	1.540	1.543	1.545	1.548	1.550	1.553	1.555	1.558	1.560	
1.25	1.562	1.565	1.568	1.570	1.573	1.575	1.578	1.580	1.583	1.585	3
6	1.588	1.590	1.593	1.595	1.598	1.600	1.603	1.605	1.608	1.610	
7	1.613	1.615	1.618	1.621	1.623	1.626	1.628	1.631	1.633	1.636	
8	1.638	1.641	1.644	1.646	1.649	1.651	1.654	1.656	1.659	1.662	
9	1.664	1.667	1.669	1.672	1.674	1.677	1.680	1.682	1.685	1.687	
1.30	1.690	1.693	1.695	1.698	1.700	1.703	1.706	1.708	1.711	1.713	
1	1.716	1.719	1.721	1.724	1.727	1.729	1.732	1.734	1.737	1.740	
2	1.742	1.745	1.748	1.750	1.753	1.756	1.758	1.761	1.764	1.766	
3	1.769	1.772	1.774	1.777	1.780	1.782	1.785	1.788	1.790	1.793	
4	1.796	1.798	1.801	1.804	1.806	1.809	1.812	1.814	1.817	1.820	
1.35	1.822	1.825	1.828	1.831	1.833	1.836	1.839	1.841	1.844	1.847	
6	1.850	1.852	1.855	1.858	1.860	1.863	1.866	1.869	1.871	1.874	
7	1.877	1.880	1.882	1.885	1.888	1.891	1.893	1.896	1.899	1.902	
8	1.904	1.907	1.910	1.913	1.915	1.918	1.921	1.924	1.927	1.929	
9	1.932	1.935	1.938	1.940	1.943	1.946	1.949	1.952	1.954	1.957	
1.40	1.960	1.963	1.966	1.968	1.971	1.974	1.977	1.980	1.982	1.985	
1	1.988	1.991	1.994	1.997	1.999	2.002	2.005	2.008	2.011	2.014	
2	2.016	2.019	2.022	2.025	2.028	2.031	2.033	2.036	2.039	2.042	
3	2.045	2.048	2.051	2.053	2.056	2.059	2.062	2.065	2.068	2.071	
4	2.074	2.076	2.079	2.082	2.085	2.088	2.091	2.094	2.097	2.100	
1.45	2.102	2.105	2.108	2.111	2.114	2.117	2.120	2.123	2.126	2.129	
6	2.132	2.135	2.137	2.140	2.143	2.146	2.149	2.152	2.155	2.158	
7	2.161	2.164	2.167	2.170	2.173	2.176	2.179	2.182	2.184	2.187	
8	2.190	2.193	2.196	2.199	2.202	2.205	2.208	2.211	2.214	2.217	
9	2.220	2.223	2.226	2.229	2.232	2.235	2.238	2.241	2.244	2.247	

Moving the decimal point ONE place in N requires moving it TWO places in body of table (see p. 1–6).

Table 1-1 (*Continued*)
SQUARES OF NUMBERS

N	0	1	2	3	4	5	6	7	8	9	Avg diff
1.50	2.250	2.253	2.256	2.259	2.262	2.265	2.268	2.271	2.274	2.277	3
1	2.280	2.283	2.286	2.289	2.292	2.295	2.298	2.301	2.304	2.307	
2	2.310	2.313	2.316	2.320	2.323	2.326	2.329	2.332	2.335	2.338	
3	2.341	2.344	2.347	2.350	2.353	2.356	2.359	2.362	2.365	2.369	
4	2.372	2.375	2.378	2.381	2.384	2.387	2.390	2.393	2.396	2.399	
1.55	2.402	2.406	2.409	2.412	2.415	2.418	2.421	2.424	2.427	2.430	
6	2.434	2.437	2.440	2.443	2.446	2.449	2.452	2.455	2.459	2.462	
7	2.465	2.468	2.471	2.474	2.477	2.481	2.484	2.487	2.490	2.493	
8	2.496	2.500	2.503	2.506	2.509	2.512	2.515	2.519	2.522	2.525	
9	2.528	2.531	2.534	2.538	2.541	2.544	2.547	2.550	2.554	2.557	
1.60	2.560	2.563	2.566	2.570	2.573	2.576	2.579	2.582	2.586	2.589	
1	2.592	2.595	2.599	2.602	2.605	2.608	2.611	2.615	2.618	2.621	
2	2.624	2.628	2.631	2.634	2.637	2.641	2.644	2.647	2.650	2.654	
3	2.657	2.660	2.663	2.667	2.670	2.673	2.676	2.680	2.683	2.686	
4	2.690	2.693	2.696	2.699	2.703	2.706	2.709	2.713	2.716	2.719	
1.65	2.722	2.726	2.729	2.732	2.736	2.739	2.742	2.746	2.749	2.752	
6	2.756	2.759	2.762	2.766	2.769	2.772	2.776	2.779	2.782	2.786	
7	2.789	2.792	2.796	2.799	2.802	2.806	2.809	2.812	2.816	2.819	
8	2.822	2.826	2.829	2.832	2.836	2.839	2.843	2.846	2.849	2.853	
9	2.856	2.859	2.863	2.866	2.870	2.873	2.876	2.880	2.883	2.887	
1.70	2.890	2.893	2.897	2.900	2.904	2.907	2.910	2.914	2.917	2.921	
1	2.924	2.928	2.931	2.934	2.938	2.941	2.945	2.948	2.952	2.955	
2	2.958	2.962	2.965	2.969	2.972	2.976	2.979	2.983	2.986	2.989	
3	2.993	2.996	3.000	3.003	3.007	3.010	3.014	3.017	3.021	3.024	
4	3.028	3.031	3.035	3.038	3.042	3.045	3.049	3.052	3.056	3.059	
1.75	3.062	3.066	3.070	3.073	3.077	3.080	3.084	3.087	3.091	3.094	4
6	3.098	3.101	3.105	3.108	3.112	3.115	3.119	3.122	3.126	3.129	
7	3.133	3.136	3.140	3.144	3.147	3.151	3.154	3.158	3.161	3.165	
8	3.168	3.172	3.176	3.179	3.183	3.186	3.190	3.193	3.197	3.201	
9	3.204	3.208	3.211	3.215	3.218	3.222	3.226	3.229	3.233	3.236	
1.80	3.240	3.244	3.247	3.251	3.254	3.258	3.262	3.265	3.269	3.272	
1	3.276	3.280	3.283	3.287	3.291	3.294	3.298	3.301	3.305	3.309	
2	3.312	3.316	3.320	3.323	3.327	3.331	3.334	3.338	3.342	3.345	
3	3.349	3.353	3.356	3.360	3.364	3.367	3.371	3.375	3.378	3.382	
4	3.386	3.389	3.393	3.397	3.400	3.404	3.408	3.411	3.415	3.419	
1.85	3.422	3.426	3.430	3.434	3.437	3.441	3.445	3.448	3.452	3.456	
6	3.460	3.463	3.467	3.471	3.474	3.478	3.482	3.486	3.489	3.493	
7	3.497	3.501	3.504	3.508	3.512	3.516	3.519	3.523	3.527	3.531	
8	3.534	3.538	3.542	3.546	3.549	3.553	3.557	3.561	3.565	3.568	
9	3.572	3.576	3.580	3.583	3.587	3.591	3.595	3.599	3.602	3.606	
1.90	3.610	3.614	3.618	3.621	3.625	3.629	3.633	3.637	3.640	3.644	
1	3.648	3.652	3.656	3.660	3.663	3.667	3.671	3.675	3.679	3.683	
2	3.686	3.690	3.694	3.698	3.702	3.706	3.709	3.713	3.717	3.721	
3	3.725	3.729	3.733	3.736	3.740	3.744	3.748	3.752	3.756	3.760	
4	3.764	3.767	3.771	3.775	3.779	3.783	3.787	3.791	3.795	3.799	
1.95	3.802	3.806	3.810	3.814	3.818	3.822	3.826	3.830	3.834	3.838	
6	3.842	3.846	3.849	3.853	3.857	3.861	3.865	3.869	3.873	3.877	
7	3.881	3.885	3.889	3.893	3.897	3.901	3.905	3.909	3.912	3.916	
8	3.920	3.924	3.928	3.932	3.936	3.940	3.944	3.948	3.952	3.956	
9	3.960	3.964	3.968	3.972	3.976	3.980	3.984	3.988	3.992	3.996	

$$\pi^2 = 9.86960 \qquad 1/\pi^2 = 0.101321 \qquad e^2 = 7.38906$$

Table 1-1 (*Continued*)
SQUARES OF NUMBERS

N	0	1	2	3	4	5	6	7	8	9	Avg diff
2.00	4.000	4.004	4.008	4.012	4.016	4.020	4.024	4.028	4.032	4.036	4
1	4.040	4.044	4.048	4.052	4.056	4.060	4.064	4.068	4.072	4.076	
2	4.080	4.084	4.088	4.093	4.097	4.101	4.105	4.109	4.113	4.117	
3	4.121	4.125	4.129	4.133	4.137	4.141	4.145	4.149	4.153	4.158	
4	4.162	4.166	4.170	4.174	4.178	4.182	4.186	4.190	4.194	4.198	
2.05	4.202	4.207	4.211	4.215	4.219	4.223	4.227	4.231	4.235	4.239	
6	4.244	4.248	4.252	4.256	4.260	4.264	4.268	4.272	4.277	4.281	
7	4.285	4.289	4.293	4.297	4.301	4.306	4.310	4.314	4.318	4.322	
8	4.326	4.331	4.335	4.339	4.343	4.347	4.351	4.356	4.360	4.364	
9	4.368	4.372	4.376	4.381	4.385	4.389	4.393	4.397	4.402	4.406	
2.10	4.410	4.414	4.418	4.423	4.427	4.431	4.435	4.439	4.444	4.448	
1	4.452	4.456	4.461	4.465	4.469	4.473	4.477	4.482	4.486	4.490	
2	4.494	4.499	4.503	4.507	4.511	4.516	4.520	4.524	4.528	4.533	
3	4.537	4.541	4.545	4.550	4.554	4.558	4.562	4.567	4.571	4.575	
4	4.580	4.584	4.588	4.592	4.597	4.601	4.605	4.610	4.614	4.618	
2.15	4.622	4.627	4.631	4.635	4.640	4.644	4.648	4.653	4.657	4.661	
6	4.666	4.670	4.674	4.679	4.683	4.687	4.692	4.696	4.700	4.705	
7	4.709	4.713	4.718	4.722	4.726	4.731	4.735	4.739	4.744	4.748	
8	4.752	4.757	4.761	4.765	4.770	4.774	4.779	4.783	4.787	4.792	
9	4.796	4.800	4.805	4.809	4.814	4.818	4.822	4.827	4.831	4.836	
2.20	4.840	4.844	4.849	4.853	4.858	4.862	4.866	4.871	4.875	4.880	
1	4.884	4.889	4.893	4.897	4.902	4.906	4.911	4.915	4.920	4.924	
2	4.928	4.933	4.937	4.942	4.946	4.951	4.955	4.960	4.964	4.968	
3	4.973	4.977	4.982	4.986	4.991	4.995	5.000	5.004	5.009	5.013	
4	5.018	5.022	5.027	5.031	5.036	5.040	5.045	5.049	5.054	5.058	
2.25	5.062	5.067	5.072	5.076	5.081	5.085	5.090	5.094	5.099	5.103	5
6	5.108	5.112	5.117	5.121	5.126	5.130	5.135	5.139	5.144	5.148	
7	5.153	5.157	5.162	5.167	5.171	5.176	5.180	5.185	5.189	5.194	
8	5.198	5.203	5.208	5.212	5.217	5.221	5.226	5.230	5.235	5.240	
9	5.244	5.249	5.253	5.258	5.262	5.267	5.272	5.276	5.281	5.285	
2.30	5.290	5.295	5.299	5.304	5.308	5.313	5.318	5.322	5.327	5.331	
1	5.336	5.341	5.345	5.350	5.355	5.359	5.364	5.368	5.373	5.378	
2	5.382	5.387	5.392	5.396	5.401	5.406	5.410	5.415	5.420	5.424	
3	5.429	5.434	5.438	5.443	5.448	5.452	5.457	5.462	5.466	5.471	
4	5.476	5.480	5.485	5.490	5.494	5.499	5.504	5.508	5.513	5.518	
2.35	5.522	5.527	5.532	5.537	5.541	5.546	5.551	5.555	5.560	5.565	
6	5.570	5.574	5.579	5.584	5.588	5.593	5.598	5.603	5.607	5.612	
7	5.617	5.622	5.626	5.631	5.636	5.641	5.645	5.650	5.655	5.660	
8	5.664	5.669	5.674	5.679	5.683	5.688	5.693	5.698	5.703	5.707	
9	5.712	5.717	5.722	5.726	5.731	5.736	5.741	5.746	5.750	5.755	
2.40	5.760	5.765	5.770	5.774	5.779	5.784	5.789	5.794	5.798	5.803	
1	5.808	5.813	5.818	5.823	5.827	5.832	5.837	5.842	5.847	5.852	
2	5.856	5.861	5.866	5.871	5.876	5.881	5.885	5.890	5.895	5.900	
3	5.905	5.910	5.915	5.919	5.924	5.929	5.934	5.939	5.944	5.949	
4	5.954	5.958	5.963	5.968	5.973	5.978	5.983	5.988	5.993	5.998	
2.45	6.002	6.007	6.012	6.017	6.022	6.027	6.032	6.037	6.042	6.047	
6	6.052	6.057	6.061	6.066	6.071	6.076	6.081	6.086	6.091	6.096	
7	6.101	6.106	6.111	6.116	6.121	6.126	6.131	6.136	6.140	6.145	
8	6.150	6.155	6.160	6.165	6.170	6.175	6.180	6.185	6.190	6.195	
9	6.200	6.205	6.210	6.215	6.220	6.225	6.230	6.235	6.240	6.245	

Moving the decimal point ONE place in *N* requires moving it TWO places in body of table (see p. 1–6).

Table 1-1 (*Continued*)
SQUARES OF NUMBERS

N	0	1	2	3	4	5	6	7	8	9	Avg diff
2.50	6.250	6.255	6.260	6.265	6.270	6.275	6.280	6.285	6.290	6.295	5
1	6.300	6.305	6.310	6.315	6.320	6.325	6.330	6.335	6.340	6.345	
2	6.350	6.355	6.360	6.366	6.371	6.376	6.381	6.386	6.391	6.396	
3	6.401	6.406	6.411	6.416	6.421	6.426	6.431	6.436	6.441	6.447	
4	6.452	6.457	6.462	6.467	6.472	6.477	6.482	6.487	6.492	6.497	
2.55	6.502	6.508	6.513	6.518	6.523	6.528	6.533	6.538	6.543	6.548	
6	6.554	6.559	6.564	6.569	6.574	6.579	6.584	6.589	6.595	6.600	
7	6.605	6.610	6.615	6.620	6.625	6.631	6.636	6.641	6.646	6.651	
8	6.656	6.662	6.667	6.672	6.677	6.682	6.687	6.693	6.698	6.703	
9	6.708	6.713	6.718	6.724	6.729	6.734	6.739	6.744	6.750	6.755	
2.60	6.760	6.765	6.770	6.776	6.781	6.786	6.791	6.796	6.802	6.807	
1	6.812	6.817	6.823	6.828	6.833	6.838	6.843	6.849	6.854	6.859	
2	6.864	6.870	6.875	6.880	6.885	6.891	6.896	6.901	6.906	6.912	
3	6.917	6.922	6.927	6.933	6.938	6.943	6.948	6.954	6.959	6.964	
4	6.970	6.975	6.980	6.985	6.991	6.996	7.001	7.007	7.012	7.017	
2.65	7.022	7.028	7.033	7.038	7.044	7.049	7.054	7.060	7.065	7.070	
6	7.076	7.081	7.086	7.092	7.097	7.102	7.108	7.113	7.118	7.124	
7	7.129	7.134	7.140	7.145	7.150	7.156	7.161	7.166	7.172	7.177	
8	7.182	7.188	7.193	7.198	7.204	7.209	7.215	7.220	7.225	7.231	
9	7.236	7.241	7.247	7.252	7.258	7.263	7.268	7.274	7.279	7.285	
2.70	7.290	7.295	7.301	7.306	7.312	7.317	7.322	7.328	7.333	7.339	
1	7.344	7.350	7.355	7.360	7.366	7.371	7.377	7.382	7.388	7.393	
2	7.398	7.404	7.409	7.415	7.420	7.426	7.431	7.437	7.442	7.447	
3	7.453	7.458	7.464	7.469	7.475	7.480	7.486	7.491	7.497	7.502	
4	7.508	7.513	7.519	7.524	7.530	7.535	7.541	7.546	7.552	7.557	
2.75	7.562	7.568	7.574	7.579	7.585	7.590	7.596	7.601	7.607	7.612	6
6	7.618	7.623	7.629	7.634	7.640	7.645	7.651	7.656	7.662	7.667	
7	7.673	7.678	7.684	7.690	7.695	7.701	7.706	7.712	7.717	7.723	
8	7.728	7.734	7.740	7.745	7.751	7.756	7.762	7.767	7.773	7.779	
9	7.784	7.790	7.795	7.801	7.806	7.812	7.818	7.823	7.829	7.834	
2.80	7.840	7.846	7.851	7.857	7.862	7.868	7.874	7.879	7.885	7.890	
1	7.896	7.902	7.907	7.913	7.919	7.924	7.930	7.935	7.941	7.947	
2	7.952	7.958	7.964	7.969	7.975	7.981	7.986	7.992	7.998	8.003	
3	8.009	8.015	8.020	8.026	8.032	8.037	8.043	8.049	8.054	8.060	
4	8.066	8.071	8.077	8.083	8.088	8.094	8.100	8.105	8.111	8.117	
2.85	8.122	8.128	8.134	8.140	8.145	8.151	8.157	8.162	8.168	8.174	
6	8.180	8.185	8.191	8.197	8.202	8.208	8.214	8.220	8.225	8.231	
7	8.237	8.243	8.248	8.254	8.260	8.266	8.271	8.277	8.283	8.289	
8	8.294	8.300	8.306	8.312	8.317	8.323	8.329	8.335	8.341	8.346	
9	8.352	8.358	8.364	8.369	8.375	8.381	8.387	8.393	8.398	8.404	
2.90	8.410	8.416	8.422	8.427	8.433	8.439	8.445	8.451	8.456	8.462	
1	8.468	8.474	8.480	8.486	8.491	8.497	8.503	8.509	8.515	8.521	
2	8.526	8.532	8.538	8.544	8.550	8.556	8.561	8.567	8.573	8.579	
3	8.585	8.591	8.597	8.602	8.608	8.614	8.620	8.626	8.632	8.638	
4	8.644	8.649	8.655	8.661	8.667	8.673	8.679	8.685	8.691	8.697	
2.95	8.702	8.708	8.714	8.720	8.726	8.732	8.738	8.744	8.750	8.756	
6	8.762	8.768	8.773	8.779	8.785	8.791	8.797	8.803	8.809	8.815	
7	8.821	8.827	8.833	8.839	8.845	8.851	8.857	8.863	8.868	8.874	
8	8.880	8.886	8.892	8.898	8.904	8.910	8.916	8.922	8.928	8.934	
9	8.940	8.946	8.952	8.958	8.964	8.970	8.976	8.982	8.988	8.994	

$$\pi^2 = 9.86960 \qquad 1/\pi^2 = 0.101321 \qquad e^2 = 7.38906$$

Table 1-1 (*Continued*)
SQUARES OF NUMBERS

N	0	1	2	3	4	5	6	7	8	9	Avg diff
3.00	9.000	9.006	9.012	9.018	9.024	9.030	9.036	9.042	9.048	9.054	6
1	9.060	9.066	9.072	9.078	9.084	9.090	9.096	9.102	9.108	9.114	
2	9.120	9.126	9.132	9.139	9.145	9.151	9.157	9.163	9.169	9.175	
3	9.181	9.187	9.193	9.199	9.205	9.211	9.217	9.223	9.229	9.236	
4	9.242	9.248	9.254	9.260	9.266	9.272	9.278	9.284	9.290	9.296	
3.05	9.302	9.309	9.315	9.321	9.327	9.333	9.339	9.345	9.351	9.357	
6	9.364	9.370	9.376	9.382	9.388	9.394	9.400	9.406	9.413	9.419	
7	9.425	9.431	9.437	9.443	9.449	9.456	9.462	9.468	9.474	9.480	
8	9.486	9.493	9.499	9.505	9.511	9.517	9.523	9.530	9.536	9.542	
9	9.548	9.554	9.560	9.567	9.573	9.579	9.585	9.591	9.598	9.604	
3.10	9.610	9.616	9.622	9.629	9.635	9.641	9.647	9.653	9.660	9.666	
1	9.672	9.678	9.685	9.691	9.697	9.703	9.709	9.716	9.722	9.728	
2	9.734	9.741	9.747	9.753	9.759	9.766	9.772	9.778	9.784	9.791	
3	9.797	9.803	9.809	9.816	9.822	9.828	9.834	9.841	9.847	9.853	
4	9.860	9.866	9.872	9.878	9.885	9.891	9.897	9.904	9.910	9.916	
3.15	9.922	9.929	9.935	9.941	9.948	9.954	9.960	9.967	9.973	9.979	
6	9.986	9.992	9.998	10.005							6
3.1							9.99	10.05	10.11	10.18	6
2	10.24	10.30	10.37	10.43	10.50	10.56	10.63	10.69	10.76	10.82	
3	10.89	10.96	11.02	11.09	11.16	11.22	11.29	11.36	11.42	11.49	7
4	11.56	11.63	11.70	11.76	11.83	11.90	11.97	12.04	12.11	12.18	
3.5	12.25	12.32	12.39	12.46	12.53	12.60	12.67	12.74	12.82	12.89	
6	12.96	13.03	13.10	13.18	13.25	13.32	13.40	13.47	13.54	13.62	
7	13.69	13.76	13.84	13.91	13.99	14.06	14.14	14.21	14.29	14.36	8
8	14.44	14.52	14.59	14.67	14.75	14.82	14.90	14.98	15.05	15.13	
9	15.21	15.29	15.37	15.44	15.52	15.60	15.68	15.76	15.84	15.92	
4.0	16.00	16.08	16.16	16.24	16.32	16.40	16.48	16.56	16.65	16.73	
1	16.81	16.89	16.97	17.06	17.14	17.22	17.31	17.39	17.47	17.56	
2	17.64	17.72	17.81	17.89	17.98	18.06	18.15	18.23	18.32	18.40	
3	18.49	18.58	18.66	18.75	18.84	18.92	19.01	19.10	19.18	19.27	9
4	19.36	19.45	19.54	19.62	19.71	19.80	19.89	19.98	20.07	20.16	
4.5	20.25	20.34	20.43	20.52	20.61	20.70	20.79	20.88	20.98	21.07	
6	21.16	21.25	21.34	21.44	21.53	21.62	21.72	21.81	21.90	22.00	
7	22.09	22.18	22.28	22.37	22.47	22.56	22.66	22.75	22.85	22.94	10
8	23.04	23.14	23.23	23.33	23.43	23.52	23.62	23.72	23.81	23.91	
9	24.01	24.11	24.21	24.30	24.40	24.50	24.60	24.70	24.80	24.90	

$$\pi^2 = 9.86960 \qquad (\pi/2)^2 = 2.46740 \qquad 1/\pi^2 = 0.101321$$

Explanation of Table of Squares (pp. 1–2 to 1–7).

This table gives the value of N^2 for values of N from 1 to 10, correct to four figures. (Interpolated values may be in error by 1 in the fourth figure.)

To find the square of a number N outside the range from 1 to 10, note that moving the decimal point ONE place in column N is equivalent to moving it TWO places in the body of the table. For example:
$$(3.217)^2 = 10.35 \qquad (0.03217)^2 = 0.001035 \qquad (3217)^2 = 10350000$$
This table can also be used inversely, to give square roots.

Table 1-1 (*Continued*)
SQUARES OF NUMBERS

N	0	1	2	3	4	5	6	7	8	9	Avg diff
5.0	25.00	25.10	25.20	25.30	25.40	25.50	25.60	25.70	25.81	25.91	10
1	26.01	26.11	26.21	26.32	26.42	26.52	26.63	26.73	26.83	26.94	
2	27.04	27.14	27.25	27.35	27.46	27.56	27.67	27.77	27.88	27.98	
3	28.09	28.20	28.30	28.41	28.52	28.62	28.73	28.84	28.94	29.05	11
4	29.16	29.27	29.38	29.48	29.59	29.70	29.81	29.92	30.03	30.14	
5.5	30.25	30.36	30.47	30.58	30.69	30.80	30.91	31.02	31.14	31.25	
6	31.36	31.47	31.58	31.70	31.81	31.92	32.04	32.15	32.26	32.38	
7	32.49	32.60	32.72	32.83	32.95	33.06	33.18	33.29	33.41	33.52	
8	33.64	33.76	33.87	33.99	34.11	34.22	34.34	34.46	34.57	34.69	12
9	34.81	34.93	35.05	35.16	35.28	35.40	35.52	35.64	35.76	35.88	
6.0	36.00	36.12	36.24	36.36	36.48	36.60	36.72	36.84	36.97	37.09	
1	37.21	37.33	37.45	37.58	37.70	37.82	37.95	38.07	38.19	38.32	
2	38.44	38.56	38.69	38.81	38.94	39.06	39.19	39.31	39.44	39.56	
3	39.69	39.82	39.94	40.07	40.20	40.32	40.45	40.58	40.70	40.83	13
4	40.96	41.09	41.22	41.34	41.47	41.60	41.73	41.86	41.99	42.12	
6.5	42.25	42.38	42.51	42.64	42.77	42.90	43.03	43.16	43.30	43.43	
6	43.56	43.69	43.82	43.96	44.09	44.22	44.36	44.49	44.62	44.76	
7	44.89	45.02	45.16	45.29	45.43	45.56	45.70	45.83	45.97	46.10	
8	46.24	46.38	46.51	46.65	46.79	46.92	47.06	47.20	47.33	47.47	14
9	47.61	47.75	47.89	48.02	48.16	48.30	48.44	48.58	48.72	48.86	
7.0	49.00	49.14	49.28	49.42	49.56	49.70	49.84	49.98	50.13	50.27	
1	50.41	50.55	50.69	50.84	50.98	51.12	51.27	51.41	51.55	51.70	
2	51.84	51.98	52.13	52.27	52.42	52.56	52.71	52.85	53.00	53.14	
3	53.29	53.44	53.58	53.73	53.88	54.02	54.17	54.32	54.46	54.61	15
4	54.76	54.91	55.06	55.20	55.35	55.50	55.65	55.80	55.95	56.10	
7.5	56.25	56.40	56.55	56.70	56.85	57.00	57.15	57.30	57.46	57.61	
6	57.76	57.91	58.06	58.22	58.37	58.52	58.68	58.83	58.98	59.14	
7	59.29	59.44	59.60	59.75	59.91	60.06	60.22	60.37	60.53	60.68	
8	60.84	61.00	61.15	61.31	61.47	61.62	61.78	61.94	62.09	62.25	16
9	62.41	62.57	62.73	62.88	63.04	63.20	63.36	63.52	63.68	63.84	
8.0	64.00	64.16	64.32	64.48	64.64	64.80	64.96	65.12	65.29	65.45	
1	65.61	65.77	65.93	66.10	66.26	66.42	66.59	66.75	66.91	67.08	
2	67.24	67.40	67.57	67.73	67.90	68.06	68.23	68.39	68.56	68.72	
3	68.89	69.06	69.22	69.39	69.56	69.72	69.89	70.06	70.22	70.39	17
4	70.56	70.73	70.90	71.06	71.23	71.40	71.57	71.74	71.91	72.08	
8.5	72.25	72.42	72.59	72.76	72.93	73.10	73.27	73.44	73.62	73.79	
6	73.96	74.13	74.30	74.48	74.65	74.82	75.00	75.17	75.34	75.52	
7	75.69	75.86	76.04	76.21	76.39	76.56	76.74	76.91	77.09	77.26	
8	77.44	77.62	77.79	77.97	78.15	78.32	78.50	78.68	78.85	79.03	18
9	79.21	79.39	79.57	79.74	79.92	80.10	80.28	80.46	80.64	80.82	
9.0	81.00	81.18	81.36	81.54	81.72	81.90	82.08	82.26	82.45	82.63	
1	82.81	82.99	83.17	83.36	83.54	83.72	83.91	84.09	84.27	84.46	
2	84.64	84.82	85.01	85.19	85.38	85.56	85.75	85.93	86.12	86.30	
3	86.49	86.68	86.86	87.05	87.24	87.42	87.61	87.80	87.98	88.17	19
4	88.36	88.55	88.74	88.92	89.11	89.30	89.49	89.68	89.87	90.06	
9.5	90.25	90.44	90.63	90.82	91.01	91.20	91.39	91.58	91.78	91.97	
6	92.16	92.35	92.54	92.74	92.93	93.12	93.32	93.51	93.70	93.90	
7	94.09	94.28	94.48	94.67	94.87	95.06	95.26	95.45	95.65	95.84	
8	96.04	96.24	96.43	96.63	96.83	97.02	97.22	97.42	97.61	97.81	20
9	98.01	98.21	98.41	98.60	98.80	99.00	99.20	99.40	99.60	99.80	
10.0	100.0										

Moving the decimal point ONE place in N requires moving it TWO places in body of table (see p. 1-6).

Table 1-2
CUBES OF NUMBERS

From Baumeister and Marks, *Standard Handbook for Mechanical Engineers,*
7th ed., 1967, McGraw-Hill Book Company; by permission

N	0	1	2	3	4	5	6	7	8	9	Avg diff
1.00	1.000	1.003	1.006	1.009	1.012	1.015	1.018	1.021	1.024	1.027	3
1	1.030	1.033	1.036	1.040	1.043	1.046	1.049	1.052	1.055	1.058	
2	1.061	1.064	1.067	1.071	1.074	1.077	1.080	1.083	1.086	1.090	
3	1.093	1.096	1.099	1.102	1.106	1.109	1.112	1.115	1.118	1.122	
4	1.125	1.128	1.131	1.135	1.138	1.141	1.144	1.148	1.151	1.154	
1.05	1.158	1.161	1.164	1.168	1.171	1.174	1.178	1.181	1.184	1.188	
6	1.191	1.194	1.198	1.201	1.205	1.208	1.211	1.215	1.218	1.222	
7	1.225	1.228	1.232	1.235	1.239	1.242	1.246	1.249	1.253	1.256	
8	1.260	1.263	1.267	1.270	1.274	1.277	1.281	1.284	1.288	1.291	4
9	1.295	1.299	1.302	1.306	1.309	1.313	1.317	1.320	1.324	1.327	
1.10	1.331	1.335	1.338	1.342	1.346	1.349	1.353	1.357	1.360	1.364	
1	1.368	1.371	1.375	1.379	1.382	1.386	1.390	1.394	1.397	1.401	
2	1.405	1.409	1.412	1.416	1.420	1.424	1.428	1.431	1.435	1.439	
3	1.443	1.447	1.451	1.454	1.458	1.462	1.466	1.470	1.474	1.478	
4	1.482	1.485	1.489	1.493	1.497	1.501	1.505	1.509	1.513	1.517	
1.15	1.521	1.525	1.529	1.533	1.537	1.541	1.545	1.549	1.553	1.557	
6	1.561	1.565	1.569	1.573	1.577	1.581	1.585	1.589	1.593	1.598	
7	1.602	1.606	1.610	1.614	1.618	1.622	1.626	1.631	1.635	1.639	
8	1.643	1.647	1.651	1.656	1.660	1.664	1.668	1.672	1.677	1.681	
9	1.685	1.689	1.694	1.698	1.702	1.706	1.711	1.715	1.719	1.724	
1.20	1.728	1.732	1.737	1.741	1.745	1.750	1.754	1.758	1.763	1.767	
1	1.772	1.776	1.780	1.785	1.789	1.794	1.798	1.802	1.807	1.811	
2	1.816	1.820	1.825	1.829	1.834	1.838	1.843	1.847	1.852	1.856	
3	1.861	1.865	1.870	1.875	1.879	1.884	1.888	1.893	1.897	1.902	
4	1.907	1.911	1.916	1.920	1.925	1.930	1.934	1.939	1.944	1.948	5
1.25	1.953	1.958	1.963	1.967	1.972	1.977	1.981	1.986	1.991	1.996	
6	2.000	2.005	2.010	2.015	2.019	2.024	2.029	2.034	2.039	2.044	
7	2.048	2.053	2.058	2.063	2.068	2.073	2.078	2.082	2.087	2.092	
8	2.097	2.102	2.107	2.112	2.117	2.122	2.127	2.132	2.137	2.142	
9	2.147	2.152	2.157	2.162	2.167	2.172	2.177	2.182	2.187	2.192	
1.30	2.197	2.202	2.207	2.212	2.217	2.222	2.228	2.233	2.238	2.243	
1	2.248	2.253	2.258	2.264	2.269	2.274	2.279	2.284	2.290	2.295	
2	2.300	2.305	2.310	2.316	2.321	2.326	2.331	2.337	2.342	2.347	
3	2.353	2.358	2.363	2.369	2.374	2.379	2.385	2.390	2.395	2.401	
4	2.406	2.411	2.417	2.422	2.428	2.433	2.439	2.444	2.449	2.455	
1.35	2.460	2.466	2.471	2.477	2.482	2.488	2.493	2.499	2.504	2.510	6
6	2.515	2.521	2.527	2.532	2.538	2.543	2.549	2.554	2.560	2.566	
7	2.571	2.577	2.583	2.588	2.594	2.600	2.605	2.611	2.617	2.622	
8	2.628	2.634	2.640	2.645	2.651	2.657	2.663	2.668	2.674	2.680	
9	2.686	2.691	2.697	2.703	2.709	2.715	2.721	2.726	2.732	2.738	
1.40	2.744	2.750	2.756	2.762	2.768	2.774	2.779	2.785	2.791	2.797	
1	2.803	2.809	2.815	2.821	2.827	2.833	2.839	2.845	2.851	2.857	
2	2.863	2.869	2.875	2.881	2.888	2.894	2.900	2.906	2.912	2.918	
3	2.924	2.930	2.936	2.943	2.949	2.955	2.961	2.967	2.974	2.980	
4	2.986	2.992	2.998	3.005	3.011	3.017	3.023	3.030	3.036	3.042	
1.45	3.049	3.055	3.061	3.068	3.074	3.080	3.087	3.093	3.099	3.106	
6	3.112	3.119	3.125	3.131	3.138	3.144	3.151	3.157	3.164	3.170	
7	3.177	3.183	3.190	3.196	3.203	3.209	3.216	3.222	3.229	3.235	
8	3.242	3.248	3.255	3.262	3.268	3.275	3.281	3.288	3.295	3.301	7
9	3.308	3.315	3.321	3.328	3.335	3.341	3.348	3.355	3.362	3.368	

Moving the decimal point ONE place in N requires moving it THREE places in body of table (see p. 1–10).

Table 1-2 (*Continued*)
CUBES OF NUMBERS

N	0	1	2	3	4	5	6	7	8	9	Avg diff
1.50	3.375	3.382	3.389	3.395	3.402	3.409	3.416	3.422	3.429	3.436	7
1	3.443	3.450	3.457	3.464	3.470	3.477	3.484	3.491	3.498	3.505	
2	3.512	3.519	3.526	3.533	3.540	3.547	3.554	3.561	3.568	3.575	
3	3.582	3.589	3.596	3.603	3.610	3.617	3.624	3.631	3.638	3.645	
4	3.652	3.659	3.667	3.674	3.681	3.688	3.695	3.702	3.709	3.717	
1.55	3.724	3.731	3.738	3.746	3.753	3.760	3.767	3.775	3.782	3.789	
6	3.796	3.804	3.811	3.818	3.826	3.833	3.840	3.848	3.855	3.863	
7	3.870	3.877	3.885	3.892	3.900	3.907	3.914	3.922	3.929	3.937	
8	3.944	3.952	3.959	3.967	3.974	3.982	3.989	3.997	4.005	4.012	8
9	4.020	4.027	4.035	4.042	4.050	4.058	4.065	4.073	4.081	4.088	
1.60	4.096	4.104	4.111	4.119	4.127	4.135	4.142	4.150	4.158	4.166	
1	4.173	4.181	4.189	4.197	4.204	4.212	4.220	4.228	4.236	4.244	
2	4.252	4.259	4.267	4.275	4.283	4.291	4.299	4.307	4.315	4.323	
3	4.331	4.339	4.347	4.355	4.363	4.371	4.379	4.387	4.395	4.403	
4	4.411	4.419	4.427	4.435	4.443	4.451	4.460	4.468	4.476	4.484	
1.65	4.492	4.500	4.508	4.517	4.525	4.533	4.541	4.550	4.558	4.566	
6	4.574	4.583	4.591	4.599	4.607	4.616	4.624	4.632	4.641	4.649	
7	4.657	4.666	4.674	4.683	4.691	4.699	4.708	4.716	4.725	4.733	
8	4.742	4.750	4.759	4.767	4.776	4.784	4.793	4.801	4.810	4.818	
9	4.827	4.835	4.844	4.853	4.861	4.870	4.878	4.887	4.896	4.904	9
1.70	4.913	4.922	4.930	4.939	4.948	4.956	4.965	4.974	4.983	4.991	
1	5.000	5.009	5.018	5.027	5.035	5.044	5.053	5.062	5.071	5.080	
2	5.088	5.097	5.106	5.115	5.124	5.133	5.142	5.151	5.160	5.169	
3	5.178	5.187	5.196	5.205	5.214	5.223	5.232	5.241	5.250	5.259	
4	5.268	5.277	5.286	5.295	5.304	5.314	5.323	5.332	5.341	5.350	
1.75	5.359	5.369	5.378	5.387	5.396	5.405	5.415	5.424	5.433	5.442	
6	5.452	5.461	5.470	5.480	5.489	5.498	5.508	5.517	5.526	5.536	
7	5.545	5.555	5.564	5.573	5.583	5.592	5.602	5.611	5.621	5.630	10
8	5.640	5.649	5.659	5.668	5.678	5.687	5.697	5.707	5.716	5.726	
9	5.735	5.745	5.755	5.764	5.774	5.784	5.793	5.803	5.813	5.822	
1.80	5.832	5.842	5.851	5.861	5.871	5.881	5.891	5.900	5.910	5.920	
1	5.930	5.940	5.949	5.959	5.969	5.979	5.989	5.999	6.009	6.019	
2	6.029	6.039	6.048	6.058	6.068	6.078	6.088	6.098	6.108	6.118	
3	6.128	6.139	6.149	6.159	6.169	6.179	6.189	6.199	6.209	6.219	
4	6.230	6.240	6.250	6.260	6.270	6.280	6.291	6.301	6.311	6.321	
1.85	6.332	6.342	6.352	6.362	6.373	6.383	6.393	6.404	6.414	6.424	
6	6.435	6.445	6.456	6.466	6.476	6.487	6.497	6.508	6.518	6.529	
7	6.539	6.550	6.560	6.571	6.581	6.592	6.602	6.613	6.623	6.634	11
8	6.645	6.655	6.666	6.677	6.687	6.698	6.708	6.719	6.730	6.741	
9	6.751	6.762	6.773	6.783	6.794	6.805	6.816	6.827	6.837	6.848	
1.90	6.859	6.870	6.881	6.892	6.902	6.913	6.924	6.935	6.946	6.957	
1	6.968	6.979	6.990	7.001	7.012	7.023	7.034	7.045	7.056	7.067	
2	7.078	7.089	7.100	7.111	7.122	7.133	7.144	7.156	7.167	7.178	
3	7.189	7.200	7.211	7.223	7.234	7.245	7.256	7.268	7.279	7.290	
4	7.301	7.313	7.324	7.335	7.347	7.358	7.369	7.381	7.392	7.403	
1.95	7.415	7.426	7.438	7.449	7.461	7.472	7.484	7.495	7.507	7.518	12
6	7.530	7.541	7.553	7.564	7.576	7.587	7.599	7.610	7.622	7.634	
7	7.645	7.657	7.669	7.680	7.692	7.704	7.715	7.727	7.739	7.751	
8	7.762	7.774	7.786	7.798	7.810	7.821	7.833	7.845	7.857	7.869	
9	7.881	7.892	7.904	7.916	7.928	7.940	7.952	7.964	7.976	7.988	

$$\pi^3 = 31.0063 \qquad 1/\pi^3 = 0.0322515 +$$

Table 1-2 (*Continued*)
CUBES OF NUMBERS

N	0	1	2	3	4	5	6	7	8	9	Avg diff
2.00	8.000	8.012	8.024	8.036	8.048	8.060	8.072	8.084	8.096	8.108	12
1	8.121	8.133	8.145	8.157	8.169	8.181	8.194	8.206	8.218	8.230	
2	8.242	8.255	8.267	8.279	8.291	8.304	8.316	8.328	8.341	8.353	
3	8.365	8.378	8.390	8.403	8.415	8.427	8.440	8.452	8.465	8.477	
4	8.490	8.502	8.515	8.527	8.540	8.552	8.565	8.577	8.590	8.603	
2.05	8.615	8.628	8.640	8.653	8.666	8.678	8.691	8.704	8.716	8.729	13
6	8.742	8.755	8.767	8.780	8.793	8.806	8.818	8.831	8.844	8.857	
7	8.870	8.883	8.895	8.908	8.921	8.934	8.947	8.960	8.973	8.986	
8	8.999	9.012	9.025	9.038	9.051	9.064	9.077	9.090	9.103	9.116	
9	9.129	9.142	9.156	9.169	9.182	9.195	9.208	9.221	9.235	9.248	
2.10	9.261	9.274	9.287	9.301	9.314	9.327	9.341	9.354	9.367	9.381	
1	9.394	9.407	9.421	9.434	9.447	9.461	9.474	9.488	9.501	9.515	
2	9.528	9.542	9.555	9.569	9.582	9.596	9.609	9.623	9.636	9.650	14
3	9.664	9.677	9.691	9.704	9.718	9.732	9.745	9.759	9.773	9.787	
4	9.800	9.814	9.828	9.842	9.855	9.869	9.883	9.897	9.911	9.925	
2.15	9.938	9.952	9.966	9.980	9.994	10.008					14
2.1						9.94	10.08	10.22	10.36	10.50	14
2	10.65	10.79	10.94	11.09	11.24	11.39	11.54	11.70	11.85	12.01	15
3	12.17	12.33	12.49	12.65	12.81	12.98	13.14	13.31	13.48	13.65	16
4	13.82	14.00	14.17	14.35	14.53	14.71	14.89	15.07	15.25	15.44	18
2.5	15.62	15.81	16.00	16.19	16.39	16.58	16.78	16.97	17.17	17.37	20
6	17.58	17.78	17.98	18.19	18.40	18.61	18.82	19.03	19.25	19.47	21
7	19.68	19.90	20.12	20.35	20.57	20.80	21.02	21.25	21.48	21.72	23
8	21.95	22.19	22.43	22.67	22.91	23.15	23.39	23.64	23.89	24.14	24
9	24.39	24.64	24.90	25.15	25.41	25.67	25.93	26.20	26.46	26.73	26
3.0	27.00	27.27	27.54	27.82	28.09	28.37	28.65	28.93	29.22	29.50	28
1	29.79	30.08	30.37	30.66	30.96	31.26	31.55	31.86	32.16	32.46	30
2	32.77	33.08	33.39	33.70	34.01	34.33	34.65	34.97	35.29	35.61	32
3	35.94	36.26	36.59	36.93	37.26	37.60	37.93	38.27	38.61	38.96	34
4	39.30	39.65	40.00	40.35	40.71	41.06	41.42	41.78	42.14	42.51	36
3.5	42.88	43.24	43.61	43.99	44.36	44.74	45.12	45.50	45.88	46.27	39
6	46.66	47.05	47.44	47.83	48.23	48.63	49.03	49.43	49.84	50.24	40
7	50.65	51.06	51.48	51.90	52.31	52.73	53.16	53.58	54.01	54.44	42
8	54.87	55.31	55.74	56.18	56.62	57.07	57.51	57.96	58.41	58.86	44
9	59.32	59.78	60.24	60.70	61.16	61.63	62.10	62.57	63.04	63.52	47
4.0	64.00	64.48	64.96	65.45	65.94	66.43	66.92	67.42	67.92	68.42	49
1	68.92	69.43	69.93	70.44	70.96	71.47	71.99	72.51	73.03	73.56	52
2	74.09	74.62	75.15	75.69	76.23	76.77	77.31	77.85	78.40	78.95	54
3	79.51	80.06	80.62	81.18	81.75	82.31	82.88	83.45	84.03	84.60	58
4	85.18	85.77	86.35	86.94	87.53	88.12	88.72	89.31	89.92	90.52	59
4.5	91.12	91.73	92.35	92.96	93.58	94.20	94.82	95.44	96.07	96.70	62
6	97.34	97.97	98.61	99.25	99.90	100.54					64
6						100.5	101.2	101.8	102.5	103.2	7
7	103.8	104.5	105.2	105.8	106.5	107.2	107.9	108.5	109.2	109.9	7
8	110.6	111.3	112.0	112.7	113.4	114.1	114.8	115.5	116.2	116.9	7
9	117.6	118.4	119.1	119.8	120.6	121.3	122.0	122.8	123.5	124.3	7

Explanation of Table of Cubes (pp. 1–8 to 1–11).

This table gives the value of N^3 for values of N from 1 to 10, correct to four figures. (Interpolated values may be in error by 1 in the fourth figure.)

To find the cube of a number N outside the range from 1 to 10, note that moving the decimal point ONE place in column N is equivalent to moving it THREE places in the body of the table. For example:

$$(4.852)^3 = 114.2 \qquad (0.4852)^3 = 0.1142 \qquad (485.2)^3 = 114200000$$

This table may also be used inversely, to give cube roots.

Table 1-2 (*Continued*)
CUBES OF NUMBERS

N	0	1	2	3	4	5	6	7	8	9	Avg diff
5.0	125.0	125.8	126.5	127.3	128.0	128.8	129.6	130.3	131.1	131.9	8
1	132.7	133.4	134.2	135.0	135.8	136.6	137.4	138.2	139.0	139.8	
2	140.6	141.4	142.2	143.1	143.9	144.7	145.5	146.4	147.2	148.0	
3	148.9	149.7	150.6	151.4	152.3	153.1	154.0	154.9	155.7	156.6	9
4	157.5	158.3	159.2	160.1	161.0	161.9	162.8	163.7	164.6	165.5	
5.5	166.4	167.3	168.2	169.1	170.0	171.0	171.9	172.8	173.7	174.7	
6	175.6	176.6	177.5	178.5	179.4	180.4	181.3	182.3	183.3	184.2	10
7	185.2	186.2	187.1	188.1	189.1	190.1	191.1	192.1	193.1	194.1	
8	195.1	196.1	197.1	198.2	199.2	200.2	201.2	202.3	203.3	204.3	
9	205.4	206.4	207.5	208.5	209.6	210.6	211.7	212.8	213.8	214.9	
6.0	216.0	217.1	218.2	219.3	220.3	221.4	222.5	223.6	224.8	225.9	11
1	227.0	228.1	229.2	230.3	231.5	232.6	233.7	234.9	236.0	237.2	
2	238.3	239.5	240.6	241.8	243.0	244.1	245.3	246.5	247.7	248.9	12
3	250.0	251.2	252.4	253.6	254.8	256.0	257.3	258.5	259.7	260.9	
4	262.1	263.4	264.6	265.8	267.1	268.3	269.6	270.8	272.1	273.4	
6.5	274.6	275.9	277.2	278.4	279.7	281.0	282.3	283.6	284.9	286.2	13
6	287.5	288.8	290.1	291.4	292.8	294.1	295.4	296.7	298.1	299.4	
7	300.8	302.1	303.5	304.8	306.2	307.5	308.9	310.3	311.7	313.0	14
8	314.4	315.8	317.2	318.6	320.0	321.4	322.8	324.2	325.7	327.1	
9	328.5	329.9	331.4	332.8	334.3	335.7	337.2	338.6	340.1	341.5	
7.0	343.0	344.5	345.9	347.4	348.9	350.4	351.9	353.4	354.9	356.4	15
1	357.9	359.4	360.9	362.5	364.0	365.5	367.1	368.6	370.1	371.7	
2	373.2	374.8	376.4	377.9	379.5	381.1	382.7	384.2	385.8	387.4	16
3	389.0	390.6	392.2	393.8	395.4	397.1	398.7	400.3	401.9	403.6	
4	405.2	406.9	408.5	410.2	411.8	413.5	415.2	416.8	418.5	420.2	17
7.5	421.9	423.6	425.3	427.0	428.7	430.4	432.1	433.8	435.5	437.2	
6	439.0	440.7	442.5	444.2	445.9	447.7	449.5	451.2	453.0	454.8	18
7	456.5	458.3	460.1	461.9	463.7	465.5	467.3	469.1	470.9	472.7	
8	474.6	476.4	478.2	480.0	481.9	483.7	485.6	487.4	489.3	491.2	
9	493.0	494.9	496.8	498.7	500.6	502.5	504.4	506.3	508.2	510.1	19
8.0	512.0	513.9	515.8	517.8	519.7	521.7	523.6	525.6	527.5	529.5	
1	531.4	533.4	535.4	537.4	539.4	541.3	543.3	545.3	547.3	549.4	20
2	551.4	553.4	555.4	557.4	559.5	561.5	563.6	565.6	567.7	569.7	
3	571.8	573.9	575.9	578.0	580.1	582.2	584.3	586.4	588.5	590.6	21
4	592.7	594.8	596.9	599.1	601.2	603.4	605.5	607.6	609.8	612.0	
8.5	614.1	616.3	618.5	620.7	622.8	625.0	627.2	629.4	631.6	633.8	22
6	636.1	638.3	640.5	642.7	645.0	647.2	649.5	651.7	654.0	656.2	
7	658.5	660.8	663.1	665.3	667.6	669.9	672.2	674.5	676.8	679.2	23
8	681.5	683.8	686.1	688.5	690.8	693.2	695.5	697.9	700.2	702.6	
9	705.0	707.3	709.7	712.1	714.5	716.9	719.3	721.7	724.2	726.6	24
9.0	729.0	731.4	733.9	736.3	738.8	741.2	743.7	746.1	748.6	751.1	25
1	753.6	756.1	758.6	761.0	763.6	766.1	768.6	771.1	773.6	776.2	
2	778.7	781.2	783.8	786.3	788.9	791.5	794.0	796.6	799.2	801.8	26
3	804.4	807.0	809.6	812.2	814.8	817.4	820.0	822.7	825.3	827.9	
4	830.6	833.2	835.9	838.6	841.2	843.9	846.6	849.3	852.0	854.7	27
9.5	857.4	860.1	862.8	865.5	868.3	871.0	873.7	876.5	879.2	882.0	
6	884.7	887.5	890.3	893.1	895.8	898.6	901.4	904.2	907.0	909.9	28
7	912.7	915.5	918.3	921.2	924.0	926.9	929.7	932.6	935.4	938.3	
8	941.2	944.1	947.0	949.9	952.8	955.7	958.6	961.5	964.4	967.4	29
9	970.3	973.2	976.2	979.1	982.1	985.1	988.0	991.0	994.0	997.0	
10.0	1000.0										

$$\pi^3 = 31.0063 \qquad 1/\pi^3 = 0.0322515+$$

Moving the decimal point ONE place in N requires moving it THREE places in body of table (see p. 1–10).

Table 1-3
SQUARE ROOTS OF NUMBERS

From Baumeister and Marks, *Standard Handbook for Mechanical Engineers,*
7th ed., 1967, McGraw-Hill Book Company; by permission

N	0	1	2	3	4	5	6	7	8	9	Avg diff
1.0	1.000	1.005	1.010	1.015	1.020	1.025	1.030	1.034	1.039	1.044	5
1	1.049	1.054	1.058	1.063	1.068	1.072	1.077	1.082	1.086	1.091	
2	1.095	1.100	1.105	1.109	1.114	1.118	1.122	1.127	1.131	1.136	4
3	1.140	1.145	1.149	1.153	1.158	1.162	1.166	1.170	1.175	1.179	
4	1.183	1.187	1.192	1.196	1.200	1.204	1.208	1.212	1.217	1.221	
1.5	1.225	1.229	1.233	1.237	1.241	1.245	1.249	1.253	1.257	1.261	
6	1.265	1.269	1.273	1.277	1.281	1.285	1.288	1.292	1.296	1.300	
7	1.304	1.308	1.311	1.315	1.319	1.323	1.327	1.330	1.334	1.338	
8	1.342	1.345	1.349	1.353	1.356	1.360	1.364	1.367	1.371	1.375	
9	1.378	1.382	1.386	1.389	1.393	1.396	1.400	1.404	1.407	1.411	
2.0	1.414	1.418	1.421	1.425	1.428	1.432	1.435	1.439	1.442	1.446	
1	1.449	1.453	1.456	1.459	1.463	1.466	1.470	1.473	1.476	1.480	3
2	1.483	1.487	1.490	1.493	1.497	1.500	1.503	1.507	1.510	1.513	
3	1.517	1.520	1.523	1.526	1.530	1.533	1.536	1.539	1.543	1.546	
4	1.549	1.552	1.556	1.559	1.562	1.565	1.568	1.572	1.575	1.578	
2.5	1.581	1.584	1.587	1.591	1.594	1.597	1.600	1.603	1.606	1.609	
6	1.612	1.616	1.619	1.622	1.625	1.628	1.631	1.634	1.637	1.640	
7	1.643	1.646	1.649	1.652	1.655	1.658	1.661	1.664	1.667	1.670	
8	1.673	1.676	1.679	1.682	1.685	1.688	1.691	1.694	1.697	1.700	
9	1.703	1.706	1.709	1.712	1.715	1.718	1.720	1.723	1.726	1.729	
3.0	1.732	1.735	1.738	1.741	1.744	1.746	1.749	1.752	1.755	1.758	
1	1.761	1.764	1.766	1.769	1.772	1.775	1.778	1.780	1.783	1.786	
2	1.789	1.792	1.794	1.797	1.800	1.803	1.806	1.808	1.811	1.814	
3	1.817	1.819	1.822	1.825	1.828	1.830	1.833	1.836	1.838	1.841	
4	1.844	1.847	1.849	1.852	1.855	1.857	1.860	1.863	1.865	1.868	
3.5	1.871	1.873	1.876	1.879	1.881	1.884	1.887	1.889	1.892	1.895	
6	1.897	1.900	1.903	1.905	1.908	1.910	1.913	1.916	1.918	1.921	
7	1.924	1.926	1.929	1.931	1.934	1.936	1.939	1.942	1.944	1.947	
8	1.949	1.952	1.954	1.957	1.960	1.962	1.965	1.967	1.970	1.972	
9	1.975	1.977	1.980	1.982	1.985	1.987	1.990	1.992	1.995	1.997	
4.0	2.000	2.002	2.005	2.007	2.010	2.012	2.015	2.017	2.020	2.022	
1	2.025	2.027	2.030	2.032	2.035	2.037	2.040	2.042	2.045	2.047	2
2	2.049	2.052	2.054	2.057	2.059	2.062	2.064	2.066	2.069	2.071	
3	2.074	2.076	2.078	2.081	2.083	2.086	2.088	2.090	2.093	2.095	
4	2.098	2.100	2.102	2.105	2.107	2.110	2.112	2.114	2.117	2.119	
4.5	2.121	2.124	2.126	2.128	2.131	2.133	2.135	2.138	2.140	2.142	
6	2.145	2.147	2.149	2.152	2.154	2.156	2.159	2.161	2.163	2.166	
7	2.168	2.170	2.173	2.175	2.177	2.179	2.182	2.184	2.186	2.189	
8	2.191	2.193	2.195	2.198	2.200	2.202	2.205	2.207	2.209	2.211	
9	2.214	2.216	2.218	2.220	2.223	2.225	2.227	2.229	2.232	2.234	

$$\sqrt{\pi} = 1.77245+ \qquad 1/\sqrt{\pi} = 0.56419 \qquad \sqrt{\pi/2} = 1.25331 \qquad \sqrt{e} = 1.64872$$

Explanation of Table of Square Roots (pp. 1-12 to 1-15).

This table gives the values of $\sqrt{N}$ for values of N from 1 to 100, correct to four figures. (Interpolated values may be in error by 1 in the fourth figure.)

To find the square root of a number N outside the range from 1 to 100, divide the digits of the number into blocks of two (beginning with the decimal point), and note that moving the decimal point TWO places in N is equivalent to moving it ONE place in the square root of N. For example:

$$\sqrt{2.718} = 1.648 \qquad \sqrt{271.8} = 16.48 \qquad \sqrt{0.0002718} = 0.01648$$
$$\sqrt{27.18} = 5.213 \qquad \sqrt{2718} = 52.13 \qquad \sqrt{0.002718} = 0.05213$$

Table 1-3 (*Continued*)
SQUARE ROOTS OF NUMBERS

N	0	1	2	3	4	5	6	7	8	9	Avg diff
5.0	2.236	2.238	2.241	2.243	2.245	2.247	2.249	2.252	2.254	2.256	2
1	2.258	2.261	2.263	2.265	2.267	2.269	2.272	2.274	2.276	2.278	
2	2.280	2.283	2.285	2.287	2.289	2.291	2.293	2.296	2.298	2.300	
3	2.302	2.304	2.307	2.309	2.311	2.313	2.315	2.317	2.319	2.322	
4	2.324	2.326	2.328	2.330	2.332	2.335	2.337	2.339	2.341	2.343	
5.5	2.345	2.347	2.349	2.352	2.354	2.356	2.358	2.360	2.362	2.364	
6	2.366	2.369	2.371	2.373	2.375	2.377	2.379	2.381	2.383	2.385	
7	2.387	2.390	2.392	2.394	2.396	2.398	2.400	2.402	2.404	2.406	
8	2.408	2.410	2.412	2.415	2.417	2.419	2.421	2.423	2.425	2.427	
9	2.429	2.431	2.433	2.435	2.437	2.439	2.441	2.443	2.445	2.447	
6.0	2.449	2.452	2.454	2.456	2.458	2.460	2.462	2.464	2.466	2.468	
1	2.470	2.472	2.474	2.476	2.478	2.480	2.482	2.484	2.486	2.488	
2	2.490	2.492	2.494	2.496	2.498	2.500	2.502	2.504	2.506	2.508	
3	2.510	2.512	2.514	2.516	2.518	2.520	2.522	2.524	2.526	2.528	
4	2.530	2.532	2.534	2.536	2.538	2.540	2.542	2.544	2.546	2.548	
6.5	2.550	2.551	2.553	2.555	2.557	2.559	2.561	2.563	2.565	2.567	
6	2.569	2.571	2.573	2.575	2.577	2.579	2.581	2.583	2.585	2.587	
7	2.588	2.590	2.592	2.594	2.596	2.598	2.600	2.602	2.604	2.606	
8	2.608	2.610	2.612	2.613	2.615	2.617	2.619	2.621	2.623	2.625	
9	2.627	2.629	2.631	2.632	2.634	2.636	2.638	2.640	2.642	2.644	
7.0	2.646	2.648	2.650	2.651	2.653	2.655	2.657	2.659	2.661	2.663	
1	2.665	2.666	2.668	2.670	2.672	2.674	2.676	2.678	2.680	2.681	
2	2.683	2.685	2.687	2.689	2.691	2.693	2.694	2.696	2.698	2.700	
3	2.702	2.704	2.706	2.707	2.709	2.711	2.713	2.715	2.717	2.718	
4	2.720	2.722	2.724	2.726	2.728	2.729	2.731	2.733	2.735	2.737	
7.5	2.739	2.740	2.742	2.744	2.746	2.748	2.750	2.751	2.753	2.755	
6	2.757	2.759	2.760	2.762	2.764	2.766	2.768	2.769	2.771	2.773	
7	2.775	2.777	2.778	2.780	2.782	2.784	2.786	2.787	2.789	2.791	
8	2.793	2.795	2.796	2.798	2.800	2.802	2.804	2.805	2.807	2.809	
9	2.811	2.812	2.814	2.816	2.818	2.820	2.821	2.823	2.825	2.827	
8.0	2.828	2.830	2.832	2.834	2.835	2.837	2.839	2.841	2.843	2.844	
1	2.846	2.848	2.850	2.851	2.853	2.855	2.857	2.858	2.860	2.862	
2	2.864	2.865	2.867	2.869	2.871	2.872	2.874	2.876	2.877	2.879	
3	2.881	2.883	2.884	2.886	2.888	2.890	2.891	2.893	2.895	2.897	
4	2.898	2.900	2.902	2.903	2.905	2.907	2.909	2.910	2.912	2.914	
8.5	2.915	2.917	2.919	2.921	2.922	2.924	2.926	2.927	2.929	2.931	
6	2.933	2.934	2.936	2.938	2.939	2.941	2.943	2.944	2.946	2.948	
7	2.950	2.951	2.953	2.955	2.956	2.958	2.960	2.961	2.963	2.965	
8	2.966	2.968	2.970	2.972	2.973	2.975	2.977	2.978	2.980	2.982	
9	2.983	2.985	2.987	2.988	2.990	2.992	2.993	2.995	2.997	2.998	
9.0	3.000	3.002	3.003	3.005	3.007	3.008	3.010	3.012	3.013	3.015	
1	3.017	3.018	3.020	3.022	3.023	3.025	3.027	3.028	3.030	3.032	
2	3.033	3.035	3.036	3.038	3.040	3.041	3.043	3.045	3.046	3.048	
3	3.050	3.051	3.053	3.055	3.056	3.058	3.059	3.061	3.063	3.064	
4	3.066	3.068	3.069	3.071	3.072	3.074	3.076	3.077	3.079	3.081	
9.5	3.082	3.084	3.085	3.087	3.089	3.090	3.092	3.094	3.095	3.097	
6	3.098	3.100	3.102	3.103	3.105	3.106	3.108	3.110	3.111	3.113	
7	3.114	3.116	3.118	3.119	3.121	3.122	3.124	3.126	3.127	3.129	
8	3.130	3.132	3.134	3.135	3.137	3.138	3.140	3.142	3.143	3.145	
9	3.146	3.148	3.150	3.151	3.153	3.154	3.156	3.158	3.159	3.161	

Moving the decimal point TWO places in *N* requires moving it ONE place in body of table (see p. 1–12).

Table 1-3 (*Continued*)
SQUARE ROOTS OF NUMBERS

N	0	1	2	3	4	5	6	7	8	9	Avg diff
10.	3.162	3.178	3.194	3.209	3.225	3.240	3.256	3.271	3.286	3.302	16
1.	3.317	3.332	3.347	3.362	3.376	3.391	3.406	3.421	3.435	3.450	15
2.	3.464	3.479	3.493	3.507	3.521	3.536	3.550	3.564	3.578	3.592	14
3.	3.606	3.619	3.633	3.647	3.661	3.674	3.688	3.701	3.715	3.728	
4.	3.742	3.755	3.768	3.782	3.795	3.808	3.821	3.834	3.847	3.860	13
15.	3.873	3.886	3.899	3.912	3.924	3.937	3.950	3.962	3.975	3.987	
6.	4.000	4.012	4.025	4.037	4.050	4.062	4.074	4.087	4.099	4.111	12
7.	4.123	4.135	4.147	4.159	4.171	4.183	4.195	4.207	4.219	4.231	
8.	4.243	4.254	4.266	4.278	4.290	4.301	4.313	4.324	4.336	4.347	
9.	4.359	4.370	4.382	4.393	4.405	4.416	4.427	4.438	4.450	4.461	11
20.	4.472	4.483	4.494	4.506	4.517	4.528	4.539	4.550	4.561	4.572	
1.	4.583	4.593	4.604	4.615	4.626	4.637	4.648	4.658	4.669	4.680	
2.	4.690	4.701	4.712	4.722	4.733	4.743	4.754	4.764	4.775	4.785	
3.	4.796	4.806	4.817	4.827	4.837	4.848	4.858	4.868	4.879	4.889	10
4.	4.899	4.909	4.919	4.930	4.940	4.950	4.960	4.970	4.980	4.990	
25.	5.000	5.010	5.020	5.030	5.040	5.050	5.060	5.070	5.079	5.089	
6.	5.099	5.109	5.119	5.128	5.138	5.148	5.158	5.167	5.177	5.187	
7.	5.196	5.206	5.215	5.225	5.235	5.244	5.254	5.263	5.273	5.282	
8.	5.292	5.301	5.310	5.320	5.329	5.339	5.348	5.357	5.367	5.376	9
9.	5.385	5.394	5.404	5.413	5.422	5.431	5.441	5.450	5.459	5.468	
30.	5.477	5.486	5.495	5.505	5.514	5.523	5.532	5.541	5.550	5.559	
1.	5.568	5.577	5.586	5.595	5.604	5.612	5.621	5.630	5.639	5.648	
2.	5.657	5.666	5.675	5.683	5.692	5.701	5.710	5.718	5.727	5.736	
3.	5.745	5.753	5.762	5.771	5.779	5.788	5.797	5.805	5.814	5.822	
4.	5.831	5.840	5.848	5.857	5.865	5.874	5.882	5.891	5.899	5.908	8
35.	5.916	5.925	5.933	5.941	5.950	5.958	5.967	5.975	5.983	5.992	
6.	6.000	6.008	6.017	6.025	6.033	6.042	6.050	6.058	6.066	6.075	
7.	6.083	6.091	6.099	6.107	6.116	6.124	6.132	6.140	6.148	6.156	
8.	6.164	6.173	6.181	6.189	6.197	6.205	6.213	6.221	6.229	6.237	
9.	6.245	6.253	6.261	6.269	6.277	6.285	6.293	6.301	6.309	6.317	
40.	6.325	6.332	6.340	6.348	6.356	6.364	6.372	6.380	6.387	6.395	
1.	6.403	6.411	6.419	6.427	6.434	6.442	6.450	6.458	6.465	6.473	
2.	6.481	6.488	6.496	6.504	6.512	6.519	6.527	6.535	6.542	6.550	
3.	6.557	6.565	6.573	6.580	6.588	6.595	6.603	6.611	6.618	6.626	
4.	6.633	6.641	6.648	6.656	6.663	6.671	6.678	6.686	6.693	6.701	
45.	6.708	6.716	6.723	6.731	6.738	6.745	6.753	6.760	6.768	6.775	7
6.	6.782	6.790	6.797	6.804	6.812	6.819	6.826	6.834	6.841	6.848	
7.	6.856	6.863	6.870	6.877	6.885	6.892	6.899	6.907	6.914	6.921	
8.	6.928	6.935	6.943	6.950	6.957	6.964	6.971	6.979	6.986	6.993	
9.	7.000	7.007	7.014	7.021	7.029	7.036	7.043	7.050	7.057	7.064	

Square Roots of Certain Fractions

N	$\sqrt{N}$	N	$\sqrt{N}$	N	$\sqrt{N}$	N	$\sqrt{N}$	N	$\sqrt{N}$	N	$\sqrt{N}$
1/2	0.7071	3/5	0.7746	4/7	0.7559	1/9	0.3333	5/12	0.6455	9/16	0.7500
1/3	0.5774	4/5	0.8944	5/7	0.8452	2/9	0.4714	7/12	0.7638	11/16	0.8292
2/3	0.8165	1/6	0.4082	6/7	0.9258	4/9	0.6667	11/12	0.9574	13/16	0.9014
1/4	0.5000	5/6	0.9129	1/8	0.3536	5/9	0.7454	1/16	0.2500	15/16	0.9682
3/4	0.8660	1/7	0.3780	3/8	0.6124	7/9	0.8819	3/16	0.4330	1/32	0.1768
1/5	0.4472	2/7	0.5345	5/8	0.7906	8/9	0.9428	5/16	0.5590	1/64	0.1250
2/5	0.6325	3/7	0.6547	7/8	0.9354	1/12	0.2887	7/16	0.6614	1/50	0.1414

Table 1-3 (*Continued*)
SQUARE ROOTS OF NUMBERS

N	0	1	2	3	4	5	6	7	8	9	Avg diff
50.	7.071	7.078	7.085	7.092	7.099	7.106	7.113	7.120	7.127	7.134	7
1.	7.141	7.148	7.155	7.162	7.169	7.176	7.183	7.190	7.197	7.204	
2.	7.211	7.218	7.225	7.232	7.239	7.246	7.253	7.259	7.266	7.273	
3.	7.280	7.287	7.294	7.301	7.308	7.314	7.321	7.328	7.335	7.342	
4.	7.348	7.355	7.362	7.369	7.376	7.382	7.389	7.396	7.403	7.409	
55.	7.416	7.423	7.430	7.436	7.443	7.450	7.457	7.463	7.470	7.477	
6.	7.483	7.490	7.497	7.503	7.510	7.517	7.523	7.530	7.537	7.543	
7.	7.550	7.556	7.563	7.570	7.576	7.583	7.589	7.596	7.603	7.609	
8.	7.616	7.622	7.629	7.635	7.642	7.649	7.655	7.662	7.668	7.675	
9.	7.681	7.688	7.694	7.701	7.707	7.714	7.720	7.727	7.733	7.740	6
60.	7.746	7.752	7.759	7.765	7.772	7.778	7.785	7.791	7.797	7.804	
1.	7.810	7.817	7.823	7.829	7.836	7.842	7.849	7.855	7.861	7.868	
2.	7.874	7.880	7.887	7.893	7.899	7.906	7.912	7.918	7.925	7.931	
3.	7.937	7.944	7.950	7.956	7.962	7.969	7.975	7.981	7.987	7.994	
4.	8.000	8.006	8.012	8.019	8.025	8.031	8.037	8.044	8.050	8.056	
65.	8.062	8.068	8.075	8.081	8.087	8.093	8.099	8.106	8.112	8.118	
6.	8.124	8.130	8.136	8.142	8.149	8.155	8.161	8.167	8.173	8.179	
7.	8.185	8.191	8.198	8.204	8.210	8.216	8.222	8.228	8.234	8.240	
8.	8.246	8.252	8.258	8.264	8.270	8.276	8.283	8.289	8.295	8.301	
9.	8.307	8.313	8.319	8.325	8.331	8.337	8.343	8.349	8.355	8.361	
70.	8.367	8.373	8.379	8.385	8.390	8.396	8.402	8.408	8.414	8.420	
1.	8.426	8.432	8.438	8.444	8.450	8.456	8.462	8.468	8.473	8.479	
2.	8.485	8.491	8.497	8.503	8.509	8.515	8.521	8.526	8.532	8.538	
3.	8.544	8.550	8.556	8.562	8.567	8.573	8.579	8.585	8.591	8.597	
4.	8.602	8.608	8.614	8.620	8.626	8.631	8.637	8.643	8.649	8.654	
75.	8.660	8.666	8.672	8.678	8.683	8.689	8.695	8.701	8.706	8.712	
6.	8.718	8.724	8.729	8.735	8.741	8.746	8.752	8.758	8.764	8.769	
7.	8.775	8.781	8.786	8.792	8.798	8.803	8.809	8.815	8.820	8.826	
8.	8.832	8.837	8.843	8.849	8.854	8.860	8.866	8.871	8.877	8.883	
9.	8.888	8.894	8.899	8.905	8.911	8.916	8.922	8.927	8.933	8.939	
80.	8.944	8.950	8.955	8.961	8.967	8.972	8.978	8.983	8.989	8.994	
1.	9.000	9.006	9.011	9.017	9.022	9.028	9.033	9.039	9.044	9.050	
2.	9.055	9.061	9.066	9.072	9.077	9.083	9.088	9.094	9.099	9.105	5
3.	9.110	9.116	9.121	9.127	9.132	9.138	9.143	9.149	9.154	9.160	
4.	9.165	9.171	9.176	9.182	9.187	9.192	9.198	9.203	9.209	9.214	
85.	9.220	9.225	9.230	9.236	9.241	9.247	9.252	9.257	9.263	9.268	
6.	9.274	9.279	9.284	9.290	9.295	9.301	9.306	9.311	9.317	9.322	
7.	9.327	9.333	9.338	9.343	9.349	9.354	9.359	9.365	9.370	9.375	
8.	9.381	9.386	9.391	9.397	9.402	9.407	9.413	9.418	9.423	9.429	
9.	9.434	9.439	9.445	9.450	9.455	9.460	9.466	9.471	9.476	9.482	
90.	9.487	9.492	9.497	9.503	9.508	9.513	9.518	9.524	9.529	9.534	
1.	9.539	9.545	9.550	9.555	9.560	9.566	9.571	9.576	9.581	9.586	
2.	9.592	9.597	9.602	9.607	9.612	9.618	9.623	9.628	9.633	9.638	
3.	9.644	9.649	9.654	9.659	9.664	9.670	9.675	9.680	9.685	9.690	
4.	9.695	9.701	9.706	9.711	9.716	9.721	9.726	9.731	9.737	9.742	
95.	9.747	9.752	9.757	9.762	9.767	9.772	9.778	9.783	9.788	9.793	
6.	9.798	9.803	9.808	9.813	9.818	9.823	9.829	9.834	9.839	9.844	
7.	9.849	9.854	9.859	9.864	9.869	9.874	9.879	9.884	9.889	9.894	
8.	9.899	9.905	9.910	9.915	9.920	9.925	9.930	9.935	9.940	9.945	
9.	9.950	9.955	9.960	9.965	9.970	9.975	9.980	9.985	9.990	9.995	

$$\sqrt{\pi} = 1.77245 + \qquad 1/\sqrt{\pi} = 0.56419 \qquad \sqrt{\pi/2} = 1.25331 \qquad \sqrt{e} = 1.64872$$

Moving the decimal point TWO places in N requires moving it ONE place in body of table (see p. 1–12).

Table 1-4
CUBE ROOTS OF NUMBERS

From Baumeister and Marks, *Standard Handbook for Mechanical Engineers,*
7th ed., 1967, McGraw-Hill Book Company; by permission

N	0	1	2	3	4	5	6	7	8	9	Avg diff
1.0	1.000	1.003	1.007	1.010	1.013	1.016	1.020	1.023	1.026	1.029	3
1	1.032	1.035	1.038	1.042	1.045	1.048	1.051	1.054	1.057	1.060	
2	1.063	1.066	1.069	1.071	1.074	1.077	1.080	1.083	1.086	1.089	
3	1.091	1.094	1.097	1.100	1.102	1.105	1.108	1.111	1.113	1.116	
4	1.119	1.121	1.124	1.127	1.129	1.132	1.134	1.137	1.140	1.142	
1.5	1.145	1.147	1.150	1.152	1.155	1.157	1.160	1.162	1.165	1.167	2
6	1.170	1.172	1.174	1.177	1.179	1.182	1.184	1.186	1.189	1.191	
7	1.193	1.196	1.198	1.200	1.203	1.205	1.207	1.210	1.212	1.214	
8	1.216	1.219	1.221	1.223	1.225	1.228	1.230	1.232	1.234	1.236	
9	1.239	1.241	1.243	1.245	1.247	1.249	1.251	1.254	1.256	1.258	
2.0	1.260	1.262	1.264	1.266	1.268	1.270	1.272	1.274	1.277	1.279	
1	1.281	1.283	1.285	1.287	1.289	1.291	1.293	1.295	1.297	1.299	
2	1.301	1.303	1.305	1.306	1.308	1.310	1.312	1.314	1.316	1.318	
3	1.320	1.322	1.324	1.326	1.328	1.330	1.331	1.333	1.335	1.337	
4	1.339	1.341	1.343	1.344	1.346	1.348	1.350	1.352	1.354	1.355	
2.5	1.357	1.359	1.361	1.363	1.364	1.366	1.368	1.370	1.372	1.373	
6	1.375	1.377	1.379	1.380	1.382	1.384	1.386	1.387	1.389	1.391	
7	1.392	1.394	1.396	1.398	1.399	1.401	1.403	1.404	1.406	1.408	
8	1.409	1.411	1.413	1.414	1.416	1.418	1.419	1.421	1.423	1.424	
9	1.426	1.428	1.429	1.431	1.433	1.434	1.436	1.437	1.439	1.441	
3.0	1.442	1.444	1.445	1.447	1.449	1.450	1.452	1.453	1.455	1.457	
1	1.458	1.460	1.461	1.463	1.464	1.466	1.467	1.469	1.471	1.472	
2	1.474	1.475	1.477	1.478	1.480	1.481	1.483	1.484	1.486	1.487	
3	1.489	1.490	1.492	1.493	1.495	1.496	1.498	1.499	1.501	1.502	
4	1.504	1.505	1.507	1.508	1.510	1.511	1.512	1.514	1.515	1.517	
3.5	1.518	1.520	1.521	1.523	1.524	1.525	1.527	1.528	1.530	1.531	
6	1.533	1.534	1.535	1.537	1.538	1.540	1.541	1.542	1.544	1.545	1
7	1.547	1.548	1.549	1.551	1.552	1.554	1.555	1.556	1.558	1.559	
8	1.560	1.562	1.563	1.565	1.566	1.567	1.569	1.570	1.571	1.573	
9	1.574	1.575	1.577	1.578	1.579	1.581	1.582	1.583	1.585	1.586	
4.0	1.587	1.589	1.590	1.591	1.593	1.594	1.595	1.597	1.598	1.599	
1	1.601	1.602	1.603	1.604	1.606	1.607	1.608	1.610	1.611	1.612	
2	1.613	1.615	1.616	1.617	1.619	1.620	1.621	1.622	1.624	1.625	
3	1.626	1.627	1.629	1.630	1.631	1.632	1.634	1.635	1.636	1.637	
4	1.639	1.640	1.641	1.642	1.644	1.645	1.646	1.647	1.649	1.650	
4.5	1.651	1.652	1.653	1.655	1.656	1.657	1.658	1.659	1.661	1.662	
6	1.663	1.664	1.666	1.667	1.668	1.669	1.670	1.671	1.673	1.674	
7	1.675	1.676	1.677	1.679	1.680	1.681	1.682	1.683	1.685	1.686	
8	1.687	1.688	1.689	1.690	1.692	1.693	1.694	1.695	1.696	1.697	
9	1.698	1.700	1.701	1.702	1.703	1.704	1.705	1.707	1.708	1.709	

$$\sqrt[3]{\pi} = 1.46459 \qquad 1/\sqrt[3]{\pi} = 0.682784$$

Explanation of Table of Cube Roots (pp. 1–16 to 1–21).

This table gives the values of $\sqrt[3]{N}$ for all values of N from 1 to 1000, correct to four figures. (Interpolated values may be in error by 1 in the fourth figure.)

To find the cube root of a number N outside the range from 1 to 1000, divide the digits of the number into blocks of three (beginning with the decimal point), and note that moving the decimal point THREE places in column N is equivalent to moving it ONE place in the cube root of N. For example:

$$\sqrt[3]{2.718} = 1.396 \qquad \sqrt[3]{2718} = 13.96 \qquad \sqrt[3]{0.000002718} = 0.01396$$
$$\sqrt[3]{27.18} = 3.007 \qquad \sqrt[3]{27180} = 30.07 \qquad \sqrt[3]{0.00002718} = 0.03007$$
$$\sqrt[3]{271.8} = 6.477 \qquad \sqrt[3]{271800} = 64.77 \qquad \sqrt[3]{0.0002718} = 0.06477$$

Table 1-4 (*Continued*)
CUBE ROOTS OF NUMBERS

N	0	1	2	3	4	5	6	7	8	9	Avg diff
5.0	1.710	1.711	1.712	1.713	1.715	1.716	1.717	1.718	1.719	1.720	1
1	1.721	1.722	1.724	1.725	1.726	1.727	1.728	1.729	1.730	1.731	
2	1.732	1.734	1.735	1.736	1.737	1.738	1.739	1.740	1.741	1.742	
3	1.744	1.745	1.746	1.747	1.748	1.749	1.750	1.751	1.752	1.753	
4	1.754	1.755	1.757	1.758	1.759	1.760	1.761	1.762	1.763	1.764	
5.5	1.765	1.766	1.767	1.768	1.769	1.771	1.772	1.773	1.774	1.775	
6	1.776	1.777	1.778	1.779	1.780	1.781	1.782	1.783	1.784	1.785	
7	1.786	1.787	1.788	1.789	1.790	1.792	1.793	1.794	1.795	1.796	
8	1.797	1.798	1.799	1.800	1.801	1.802	1.803	1.804	1.805	1.806	
9	1.807	1.808	1.809	1.810	1.811	1.812	1.813	1.814	1.815	1.816	
6.0	1.817	1.818	1.819	1.820	1.821	1.822	1.823	1.824	1.825	1.826	
1	1.827	1.828	1.829	1.830	1.831	1.832	1.833	1.834	1.835	1.836	
2	1.837	1.838	1.839	1.840	1.841	1.842	1.843	1.844	1.845	1.846	
3	1.847	1.848	1.849	1.850	1.851	1.852	1.853	1.854	1.855	1.856	
4	1.857	1.858	1.859	1.860	1.860	1.861	1.862	1.863	1.864	1.865	
6.5	1.866	1.867	1.868	1.869	1.870	1.871	1.872	1.873	1.874	1.875	
6	1.876	1.877	1.878	1.879	1.880	1.881	1.881	1.882	1.883	1.884	
7	1.885	1.886	1.887	1.888	1.889	1.890	1.891	1.892	1.893	1.894	
8	1.895	1.895	1.896	1.897	1.898	1.899	1.900	1.901	1.902	1.903	
9	1.904	1.905	1.906	1.907	1.907	1.908	1.909	1.910	1.911	1.912	
7.0	1.913	1.914	1.915	1.916	1.917	1.917	1.918	1.919	1.920	1.921	
1	1.922	1.923	1.924	1.925	1.926	1.926	1.927	1.928	1.929	1.930	
2	1.931	1.932	1.933	1.934	1.935	1.935	1.936	1.937	1.938	1.939	
3	1.940	1.941	1.942	1.943	1.943	1.944	1.945	1.946	1.947	1.948	
4	1.949	1.950	1.950	1.951	1.952	1.953	1.954	1.955	1.956	1.957	
7.5	1.957	1.958	1.959	1.960	1.961	1.962	1.963	1.964	1.964	1.965	
6	1.966	1.967	1.968	1.969	1.970	1.970	1.971	1.972	1.973	1.974	
7	1.975	1.976	1.976	1.977	1.978	1.979	1.980	1.981	1.981	1.982	
8	1.983	1.984	1.985	1.986	1.987	1.987	1.988	1.989	1.990	1.991	
9	1.992	1.992	1.993	1.994	1.995	1.996	1.997	1.997	1.998	1.999	
8.0	2.000	2.001	2.002	2.002	2.003	2.004	2.005	2.006	2.007	2.007	
1	2.008	2.009	2.010	2.011	2.012	2.012	2.013	2.014	2.015	2.016	
2	2.017	2.017	2.018	2.019	2.020	2.021	2.021	2.022	2.023	2.024	
3	2.025	2.026	2.026	2.027	2.028	2.029	2.030	2.030	2.031	2.032	
4	2.033	2.034	2.034	2.035	2.036	2.037	2.038	2.038	2.039	2.040	
8.5	2.041	2.042	2.042	2.043	2.044	2.045	2.046	2.046	2.047	2.048	
6	2.049	2.050	2.050	2.051	2.052	2.053	2.054	2.054	2.055	2.056	
7	2.057	2.057	2.058	2.059	2.060	2.061	2.061	2.062	2.063	2.064	
8	2.065	2.065	2.066	2.067	2.068	2.068	2.069	2.070	2.071	2.072	
9	2.072	2.073	2.074	2.075	2.075	2.076	2.077	2.078	2.079	2.079	
9.0	2.080	2.081	2.082	2.082	2.083	2.084	2.085	2.085	2.086	2.087	
1	2.088	2.089	2.089	2.090	2.091	2.092	2.092	2.093	2.094	2.095	
2	2.095	2.096	2.097	2.098	2.098	2.099	2.100	2.101	2.101	2.102	
3	2.103	2.104	2.104	2.105	2.106	2.107	2.107	2.108	2.109	2.110	
4	2.110	2.111	2.112	2.113	2.113	2.114	2.115	2.116	2.116	2.117	
9.5	2.118	2.119	2.119	2.120	2.121	2.122	2.122	2.123	2.124	2.125	
6	2.125	2.126	2.127	2.128	2.128	2.129	2.130	2.130	2.131	2.132	
7	2.133	2.133	2.134	2.135	2.136	2.136	2.137	2.138	2.139	2.139	
8	2.140	2.141	2.141	2.142	2.143	2.144	2.144	2.145	2.146	2.147	
9	2.147	2.148	2.149	2.149	2.150	2.151	2.152	2.152	2.153	2.154	

Moving the decimal point THREE places in N requires moving it ONE place in body of table (see p. 1–16).

Table 1-4 (*Continued*)
CUBE ROOTS OF NUMBERS

N	0	1	2	3	4	5	6	7	8	9	Avg diff
10.	2.154	2.162	2.169	2.176	2.183	2.190	2.197	2.204	2.210	2.217	7
1.	2.224	2.231	2.237	2.244	2.251	2.257	2.264	2.270	2.277	2.283	6
2.	2.289	2.296	2.302	2.308	2.315	2.321	2.327	2.333	2.339	2.345	
3.	2.351	2.357	2.363	2.369	2.375	2.381	2.387	2.393	2.399	2.404	
4.	2.410	2.416	2.422	2.427	2.433	2.438	2.444	2.450	2.455	2.461	
15.	2.466	2.472	2.477	2.483	2.488	2.493	2.499	2.504	2.509	2.515	5
6.	2.520	2.525	2.530	2.535	2.541	2.546	2.551	2.556	2.561	2.566	
7.	2.571	2.576	2.581	2.586	2.591	2.596	2.601	2.606	2.611	2.616	
8.	2.621	2.626	2.630	2.635	2.640	2.645	2.650	2.654	2.659	2.664	
9.	2.668	2.673	2.678	2.682	2.687	2.692	2.696	2.701	2.705	2.710	
20.	2.714	2.719	2.723	2.728	2.732	2.737	2.741	2.746	2.750	2.755	4
1.	2.759	2.763	2.768	2.772	2.776	2.781	2.785	2.789	2.794	2.798	
2.	2.802	2.806	2.811	2.815	2.819	2.823	2.827	2.831	2.836	2.840	
3.	2.844	2.848	2.852	2.856	2.860	2.864	2.868	2.872	2.876	2.880	
4.	2.884	2.888	2.892	2.896	2.900	2.904	2.908	2.912	2.916	2.920	
25.	2.924	2.928	2.932	2.936	2.940	2.943	2.947	2.951	2.955	2.959	
6.	2.962	2.966	2.970	2.974	2.978	2.981	2.985	2.989	2.993	2.996	
7.	3.000	3.004	3.007	3.011	3.015	3.018	3.022	3.026	3.029	3.033	
8.	3.037	3.040	3.044	3.047	3.051	3.055	3.058	3.062	3.065	3.069	
9.	3.072	3.076	3.079	3.083	3.086	3.090	3.093	3.097	3.100	3.104	
30.	3.107	3.111	3.114	3.118	3.121	3.124	3.128	3.131	3.135	3.138	3
1.	3.141	3.145	3.148	3.151	3.155	3.158	3.162	3.165	3.168	3.171	
2.	3.175	3.178	3.181	3.185	3.188	3.191	3.195	3.198	3.201	3.204	
3.	3.208	3.211	3.214	3.217	3.220	3.224	3.227	3.230	3.233	3.236	
4.	3.240	3.243	3.246	3.249	3.252	3.255	3.259	3.262	3.265	3.268	
35.	3.271	3.274	3.277	3.280	3.283	3.287	3.290	3.293	3.296	3.299	
6.	3.302	3.305	3.308	3.311	3.314	3.317	3.320	3.323	3.326	3.329	
7.	3.332	3.335	3.338	3.341	3.344	3.347	3.350	3.353	3.356	3.359	
8.	3.362	3.365	3.368	3.371	3.374	3.377	3.380	3.382	3.385	3.388	
9.	3.391	3.394	3.397	3.400	3.403	3.406	3.409	3.411	3.414	3.417	
40.	3.420	3.423	3.426	3.428	3.431	3.434	3.437	3.440	3.443	3.445	
1.	3.448	3.451	3.454	3.457	3.459	3.462	3.465	3.468	3.471	3.473	
2.	3.476	3.479	3.482	3.484	3.487	3.490	3.493	3.495	3.498	3.501	
3.	3.503	3.506	3.509	3.512	3.514	3.517	3.520	3.522	3.525	3.528	
4.	3.530	3.533	3.536	3.538	3.541	3.544	3.546	3.549	3.552	3.554	
45.	3.557	3.560	3.562	3.565	3.567	3.570	3.573	3.575	3.578	3.580	
6.	3.583	3.586	3.588	3.591	3.593	3.596	3.599	3.601	3.604	3.606	
7.	3.609	3.611	3.614	3.616	3.619	3.622	3.624	3.627	3.629	3.632	
8.	3.634	3.637	3.639	3.642	3.644	3.647	3.649	3.652	3.654	3.657	2
9.	3.659	3.662	3.664	3.667	3.669	3.672	3.674	3.677	3.679	3.682	

Cube Roots of Certain Fractions

N	$\sqrt[3]{N}$	N	$\sqrt[3]{N}$	N	$\sqrt[3]{N}$	N	$\sqrt[3]{N}$	N	$\sqrt[3]{N}$	N	$\sqrt[3]{N}$
$\frac{1}{2}$	.7937	$\frac{3}{5}$	.8434	$\frac{4}{7}$	.8298	$\frac{1}{9}$	.4807	$\frac{5}{12}$	.7469	$\frac{9}{16}$	.8255
$\frac{1}{3}$	.6934	$\frac{4}{5}$	.9283	$\frac{5}{7}$	.8939	$\frac{2}{9}$	.6057	$\frac{7}{12}$	.8355	$\frac{11}{16}$	.8826
$\frac{2}{3}$	.8736	$\frac{1}{6}$	.5503	$\frac{6}{7}$	.9499	$\frac{4}{9}$	.7631	$\frac{11}{12}$	.9714	$\frac{13}{16}$	.9331
$\frac{1}{4}$	.6300	$\frac{5}{6}$	.9410	$\frac{1}{8}$	.5000	$\frac{5}{9}$	.8221	$\frac{1}{16}$	.3969	$\frac{15}{16}$	.9787
$\frac{3}{4}$	.9086	$\frac{1}{7}$	.5228	$\frac{3}{8}$	.7211	$\frac{7}{9}$	.9196	$\frac{3}{16}$	.5724	$\frac{1}{32}$	.3150
$\frac{1}{5}$	.5848	$\frac{3}{7}$	.6586	$\frac{5}{8}$	.8550	$\frac{8}{9}$	.9615	$\frac{5}{16}$	.6786	$\frac{1}{64}$	.2500
$\frac{2}{5}$	.7368	$\frac{3}{7}$	.7539	$\frac{7}{8}$	.9565	$\frac{1}{12}$	.4368	$\frac{7}{16}$	.7591	$\frac{1}{50}$	.2714

Table 1-4 (*Continued*)
CUBE ROOTS OF NUMBERS

N	0	1	2	3	4	5	6	7	8	9	Avg Diff
50.	3.684	3.686	3.689	3.691	3.694	3.696	3.699	3.701	3.704	3.706	2
1.	3.708	3.711	3.713	3.716	3.718	3.721	3.723	3.725	3.728	3.730	
2.	3.733	3.735	3.737	3.740	3.742	3.744	3.747	3.749	3.752	3.754	
3.	3.756	3.759	3.761	3.763	3.766	3.768	3.770	3.773	3.775	3.777	
4.	3.780	3.782	3.784	3.787	3.789	3.791	3.794	3.796	3.798	3.801	
55.	3.803	3.805	3.808	3.810	3.812	3.814	3.817	3.819	3.821	3.824	
6.	3.826	3.828	3.830	3.833	3.835	3.837	3.839	3.842	3.844	3.846	
7.	3.849	3.851	3.853	3.855	3.857	3.860	3.862	3.864	3.866	3.869	
8.	3.871	3.873	3.875	3.878	3.880	3.882	3.884	3.886	3.889	3.891	
9.	3.893	3.895	3.897	3.900	3.902	3.904	3.906	3.908	3.911	3.913	
60.	3.915	3.917	3.919	3.921	3.924	3.926	3.928	3.930	3.932	3.934	
1.	3.936	3.939	3.941	3.943	3.945	3.947	3.949	3.951	3.954	3.956	
2.	3.958	3.960	3.962	3.964	3.966	3.968	3.971	3.973	3.975	3.977	
3.	3.979	3.981	3.983	3.985	3.987	3.990	3.992	3.994	3.996	3.998	
4.	4.000	4.002	4.004	4.006	4.008	4.010	4.012	4.015	4.017	4.019	
65.	4.021	4.023	4.025	4.027	4.029	4.031	4.033	4.035	4.037	4.039	
6.	4.041	4.043	4.045	4.047	4.049	4.051	4.053	4.055	4.058	4.060	
7.	4.062	4.064	4.066	4.068	4.070	4.072	4.074	4.076	4.078	4.080	
8.	4.082	4.084	4.086	4.088	4.090	4.092	4.094	4.096	4.098	4.100	
9.	4.102	4.104	4.106	4.108	4.109	4.111	4.113	4.115	4.117	4.119	
70.	4.121	4.123	4.125	4.127	4.129	4.131	4.133	4.135	4.137	4.139	
1.	4.141	4.143	4.145	4.147	4.149	4.151	4.152	4.154	4.156	4.158	
2.	4.160	4.162	4.164	4.166	4.168	4.170	4.172	4.174	4.176	4.177	
3.	4.179	4.181	4.183	4.185	4.187	4.189	4.191	4.193	4.195	4.196	
4.	4.198	4.200	4.202	4.204	4.206	4.208	4.210	4.212	4.213	4.215	
75.	4.217	4.219	4.221	4.223	4.225	4.227	4.228	4.230	4.232	4.234	
6.	4.236	4.238	4.240	4.241	4.243	4.245	4.247	4.249	4.251	4.252	
7.	4.254	4.256	4.258	4.260	4.262	4.264	4.265	4.267	4.269	4.271	
8.	4.273	4.274	4.276	4.278	4.280	4.282	4.284	4.285	4.287	4.289	
9.	4.291	4.293	4.294	4.296	4.298	4.300	4.302	4.303	4.305	4.307	
80.	4.309	4.311	4.312	4.314	4.316	4.318	4.320	4.321	4.323	4.325	
1.	4.327	4.329	4.330	4.332	4.334	4.336	4.337	4.339	4.341	4.343	
2.	4.344	4.346	4.348	4.350	4.352	4.353	4.355	4.357	4.359	4.360	
3.	4.362	4.364	4.366	4.367	4.369	4.371	4.373	4.374	4.376	4.378	
4.	4.380	4.381	4.383	4.385	4.386	4.388	4.390	4.392	4.393	4.395	
85.	4.397	4.399	4.400	4.402	4.404	4.405	4.407	4.409	4.411	4.412	
6.	4.414	4.416	4.417	4.419	4.421	4.423	4.424	4.426	4.428	4.429	
7.	4.431	4.433	4.434	4.436	4.438	4.440	4.441	4.443	4.445	4.446	
8.	4.448	4.450	4.451	4.453	4.455	4.456	4.458	4.460	4.461	4.463	
9.	4.465	4.466	4.468	4.470	4.471	4.473	4.475	4.476	4.478	4.480	
90.	4.481	4.483	4.485	4.486	4.488	4.490	4.491	4.493	4.495	4.496	
1.	4.498	4.500	4.501	4.503	4.505	4.506	4.508	4.509	4.511	4.513	
2.	4.514	4.516	4.518	4.519	4.521	4.523	4.524	4.526	4.527	4.529	
3.	4.531	4.532	4.534	4.536	4.537	4.539	4.540	4.542	4.544	4.545	
4.	4.547	4.548	4.550	4.552	4.553	4.555	4.556	4.558	4.560	4.561	
95.	4.563	4.565	4.566	4.568	4.569	4.571	4.572	4.574	4.576	4.577	
6.	4.579	4.580	4.582	4.584	4.585	4.587	4.588	4.590	4.592	4.593	
7.	4.595	4.596	4.598	4.599	4.601	4.603	4.604	4.606	4.607	4.609	
8.	4.610	4.612	4.614	4.615	4.617	4.618	4.620	4.621	4.623	4.625	
9.	4.626	4.628	4.629	4.631	4.632	4.634	4.635	4.637	4.638	4.640	

Moving the decimal point THREE places in N requires moving it ONE place in body of table (see p. 1–16).

Table 1-4 (*Continued*)
CUBE ROOTS OF NUMBERS

Cube roots of numbers from 100. to 499.

N	0.	1.	2.	3.	4.	5.	6.	7.	8.	9.	Avg diff
10	4.642	4.657	4.672	4.688	4.703	4.718	4.733	4.747	4.762	4.777	15
1	4.791	4.806	4.820	4.835	4.849	4.863	4.877	4.891	4.905	4.919	14
2	4.932	4.946	4.960	4.973	4.987	5.000	5.013	5.027	5.040	5.053	13
3	5.066	5.079	5.092	5.104	5.117	5.130	5.143	5.155	5.168	5.180	
4	5.192	5.205	5.217	5.229	5.241	5.254	5.266	5.278	5.290	5.301	12
15	5.313	5.325	5.337	5.348	5.360	5.372	5.383	5.395	5.406	5.418	
6	5.429	5.440	5.451	5.463	5.474	5.485	5.496	5.507	5.518	5.529	11
7	5.540	5.550	5.561	5.572	5.583	5.593	5.604	5.615	5.625	5.636	
8	5.646	5.657	5.667	5.677	5.688	5.698	5.708	5.718	5.729	5.739	10
9	5.749	5.759	5.769	5.779	5.789	5.799	5.809	5.819	5.828	5.838	
20	5.848	5.858	5.867	5.877	5.887	5.896	5.906	5.915	5.925	5.934	
1	5.944	5.953	5.963	5.972	5.981	5.991	6.000	6.009	6.018	6.028	9
2	6.037	6.046	6.055	6.064	6.073	6.082	6.091	6.100	6.109	6.118	
3	6.127	6.136	6.145	6.153	6.162	6.171	6.180	6.188	6.197	6.206	
4	6.214	6.223	6.232	6.240	6.249	6.257	6.266	6.274	6.283	6.291	
25	6.300	6.308	6.316	6.325	6.333	6.341	6.350	6.358	6.366	6.374	8
6	6.383	6.391	6.399	6.407	6.415	6.423	6.431	6.439	6.447	6.455	
7	6.463	6.471	6.479	6.487	6.495	6.503	6.511	6.519	6.527	6.534	
8	6.542	6.550	6.558	6.565	6.573	6.581	6.589	6.596	6.604	6.611	
9	6.619	6.627	6.634	6.642	6.649	6.657	6.664	6.672	6.679	6.687	
30	6.694	6.702	6.709	6.717	6.724	6.731	6.739	6.746	6.753	6.761	7
1	6.768	6.775	6.782	6.790	6.797	6.804	6.811	6.818	6.826	6.833	
2	6.840	6.847	6.854	6.861	6.868	6.875	6.882	6.889	6.896	6.903	
3	6.910	6.917	6.924	6.931	6.938	6.945	6.952	6.959	6.966	6.973	
4	6.980	6.986	6.993	7.000	7.007	7.014	7.020	7.027	7.034	7.041	
35	7.047	7.054	7.061	7.067	7.074	7.081	7.087	7.094	7.101	7.107	
6	7.114	7.120	7.127	7.133	7.140	7.147	7.153	7.160	7.166	7.173	6
7	7.179	7.186	7.192	7.198	7.205	7.211	7.218	7.224	7.230	7.237	
8	7.243	7.250	7.256	7.262	7.268	7.275	7.281	7.287	7.294	7.300	
9	7.306	7.312	7.319	7.325	7.331	7.337	7.343	7.350	7.356	7.362	
40	7.368	7.374	7.380	7.386	7.393	7.399	7.405	7.411	7.417	7.423	
1	7.429	7.435	7.441	7.447	7.453	7.459	7.465	7.471	7.477	7.483	
2	7.489	7.495	7.501	7.507	7.513	7.518	7.524	7.530	7.536	7.542	
3	7.548	7.554	7.560	7.565	7.571	7.577	7.583	7.589	7.594	7.600	
4	7.606	7.612	7.617	7.623	7.629	7.635	7.640	7.646	7.652	7.657	
45	7.663	7.669	7.674	7.680	7.686	7.691	7.697	7.703	7.708	7.714	5
6	7.719	7.725	7.731	7.736	7.742	7.747	7.753	7.758	7.764	7.769	
7	7.775	7.780	7.786	7.791	7.797	7.802	7.808	7.813	7.819	7.824	
8	7.830	7.835	7.841	7.846	7.851	7.857	7.862	7.868	7.873	7.878	
9	7.884	7.889	7.894	7.900	7.905	7.910	7.916	7.921	7.926	7.932	

Auxiliary Table of Two-thirds Powers and Three-halves Powers (see pp. 1–22 to 1–23)
(To assist in locating the decimal point)

N	$N^{2/3} (= \sqrt[3]{N^2})$	$N^{3/2} (= \sqrt{N^3})$	
.0001	.002154	.000001	For complete table of three-halves powers, see pp. 1–22 to 1–23. That table, used inversely, provides a complete table of two-thirds powers.
.001	.01	.00003162	
.01	.0464	.001	
.1	.2154	.03162278	
.1	1.	1.	
10.	4.64	31.62278	
100.	21.54	1000.	
1000.	100.	31622.78	
10000.	464.16	1000000.	

Table 1-4 (*Continued*)
CUBE ROOTS OF NUMBERS

N	0.	1.	2.	3.	4.	5.	6.	7.	8.	9.	Avg Diff
50	7.937	7.942	7.948	7.953	7.958	7.963	7.969	7.974	7.979	7.984	5
1	7.990	7.995	8.000	8.005	8.010	8.016	8.021	8.026	8.031	8.036	
2	8.041	8.047	8.052	8.057	8.062	8.067	8.072	8.077	8.082	8.088	
3	8.093	8.098	8.103	8.108	8.113	8.118	8.123	8.128	8.133	8.138	
4	8.143	8.148	8.153	8.158	8.163	8.168	8.173	8.178	8.183	8.188	
55	8.193	8.198	8.203	8.208	8.213	8.218	8.223	8.228	8.233	8.238	
6	8.243	8.247	8.252	8.257	8.262	8.267	8.272	8.277	8.282	8.286	
7	8.291	8.296	8.301	8.306	8.311	8.316	8.320	8.325	8.330	8.335	
8	8.340	8.344	8.349	8.354	8.359	8.363	8.368	8.373	8.378	8.382	
9	8.387	8.392	8.397	8.401	8.406	8.411	8.416	8.420	8.425	8.430	
60	8.434	8.439	8.444	8.448	8.453	8.458	8.462	8.467	8.472	8.476	
1	8.481	8.486	8.490	8.495	8.499	8.504	8.509	8.513	8.518	8.522	
2	8.527	8.532	8.536	8.541	8.545	8.550	8.554	8.559	8.564	8.568	
3	8.573	8.577	8.582	8.586	8.591	8.595	8.600	8.604	8.609	8.613	
4	8.618	8.622	8.627	8.631	8.636	8.640	8.645	8.649	8.653	8.658	
65	8.662	8.667	8.671	8.676	8.680	8.685	8.689	8.693	8.698	8.702	
6	8.707	8.711	8.715	8.720	8.724	8.729	8.733	8.737	8.742	8.746	
7	8.750	8.755	8.759	8.763	8.768	8.772	8.776	8.781	8.785	8.789	
8	8.794	8.798	8.802	8.807	8.811	8.815	8.819	8.824	8.828	8.832	
9	8.837	8.841	8.845	8.849	8.854	8.858	8.862	8.866	8.871	8.875	
70	8.879	8.883	8.887	8.892	8.896	8.900	8.904	8.909	8.913	8.917	
1	8.921	8.925	8.929	8.934	8.938	8.942	8.946	8.950	8.955	8.959	
2	8.963	8.967	8.971	8.975	8.979	8.984	8.988	8.992	8.996	9.000	
3	9.004	9.008	9.012	9.016	9.021	9.025	9.029	9.033	9.037	9.041	
4	9.045	9.049	9.053	9.057	9.061	9.065	9.069	9.073	9.078	9.082	
75	9.086	9.090	9.094	9.098	9.102	9.106	9.110	9.114	9.118	9.122	
6	9.126	9.130	9.134	9.138	9.142	9.146	9.150	9.154	9.158	9.162	
7	9.166	9.170	9.174	9.178	9.182	9.185	9.189	9.193	9.197	9.201	
8	9.205	9.209	9.213	9.217	9.221	9.225	9.229	9.233	9.237	9.240	
9	9.244	9.248	9.252	9.256	9.260	9.264	9.268	9.272	9.275	9.279	
80	9.283	9.287	9.291	9.295	9.299	9.302	9.306	9.310	9.314	9.318	
1	9.322	9.326	9.329	9.333	9.337	9.341	9.345	9.348	9.352	9.356	
2	9.360	9.364	9.368	9.371	9.375	9.379	9.383	9.386	9.390	9.394	
3	9.398	9.402	9.405	9.409	9.413	9.417	9.420	9.424	9.428	9.432	
4	9.435	9.439	9.443	9.447	9.450	9.454	9.458	9.462	9.465	9.469	
85	9.473	9.476	9.480	9.484	9.488	9.491	9.495	9.499	9.502	9.506	
6	9.510	9.513	9.517	9.521	9.524	9.528	9.532	9.535	9.539	9.543	
7	9.546	9.550	9.554	9.557	9.561	9.565	9.568	9.572	9.576	9.579	
8	9.583	9.586	9.590	9.594	9.597	9.601	9.605	9.608	9.612	9.615	
9	9.619	9.623	9.626	9.630	9.633	9.637	9.641	9.644	9.648	9.651	
90	9.655	9.658	9.662	9.666	9.669	9.673	9.676	9.680	9.683	9.687	
1	9.691	9.694	9.698	9.701	9.705	9.708	9.712	9.715	9.719	9.722	
2	9.726	9.729	9.733	9.736	9.740	9.743	9.747	9.750	9.754	9.758	
3	9.761	9.764	9.768	9.771	9.775	9.778	9.782	9.785	9.789	9.792	
4	9.796	9.799	9.803	9.806	9.810	9.813	9.817	9.820	9.824	9.827	
95	9.830	9.834	9.837	9.841	9.844	9.848	9.851	9.855	9.858	9.861	
6	9.865	9.868	9.872	9.875	9.879	9.882	9.885	9.889	9.892	9.896	
7	9.899	9.902	9.906	9.909	9.913	9.916	9.919	9.923	9.926	9.930	
8	9.933	9.936	9.940	9.943	9.946	9.950	9.953	9.956	9.960	9.963	
9	9.967	9.970	9.973	9.977	9.980	9.983	9.987	9.990	9.993	9.997	
100	10.00										

Cube roots of numbers from 500. to 1000.

Moving the decimal point THREE places in N requires moving it ONE place in body of table (see p. 1–16).

Table 1-5
THREE-HALVES POWERS OF NUMBERS*

From Baumeister and Marks, *Standard Handbook for Mechanical Engineers,*
7th ed., 1967, McGraw-Hill Book Company; by permission

N	0	1	2	3	4	5	6	7	8	9	Avg diff
1.	1.000	1.154	1.315	1.482	1.657	1.837	2.024	2.217	2.415	2.619	183
2.	2.828	3.043	3.263	3.488	3.718	3.953	4.192	4.437	4.685	4.939	237
3.	5.196	5.458	5.724	5.995	6.269	6.548	6.831	7.117	7.408	7.702	280
4.	8.000	8.302	8.607	8.917	9.230	9.546	9.866	10.190			313
4.								10.19	10.52	10.85	33
5.	11.18	11.52	11.86	12.20	12.55	12.90	13.25	13.61	13.97	14.33	35
6.	14.70	15.07	15.44	15.81	16.19	16.57	16.96	17.34	17.73	18.12	38
7.	18.52	18.92	19.32	19.72	20.13	20.54	20.95	21.37	21.78	22.20	41
8.	22.63	23.05	23.48	23.91	24.35	24.78	25.22	25.66	26.11	26.55	44
9.	27.00	27.45	27.90	28.36	28.82	29.28	29.74	30.21	30.68	31.15	46
10.	31.62	32.10	32.58	33.06	33.54	34.02	34.51	35.00	35.49	35.99	49
1.	36.48	36.98	37.48	37.99	38.49	39.00	39.51	40.02	40.53	41.05	51
2.	41.57	42.09	42.61	43.14	43.66	44.19	44.73	45.26	45.79	46.33	53
3.	46.87	47.41	47.96	48.50	49.05	49.60	50.15	50.71	51.26	51.82	55
4.	52.38	52.95	53.51	54.08	54.64	55.21	55.79	56.36	56.94	57.51	57
15.	58.09	58.68	59.26	59.85	60.43	61.02	61.62	62.21	62.80	63.40	59
6.	64.00	64.60	65.20	65.81	66.41	67.02	67.63	68.25	68.86	69.48	61
7.	70.09	70.71	71.33	71.96	72.58	73.21	73.84	74.47	75.10	75.73	63
8.	76.37	77.00	77.64	78.28	78.93	79.57	80.22	80.87	81.51	82.17	65
9.	82.82	83.47	84.13	84.79	85.45	86.11	86.77	87.44	88.10	88.77	66
20.	89.44	90.11	90.79	91.46	92.14	92.82	93.50	94.18	94.86	95.55	68
1.	96.23	96.92	97.61	98.30	99.00	99.69	100.38				69
1.							100.4	101.1	101.8	102.5	7
2.	103.2	103.9	104.6	105.3	106.0	106.7	107.4	108.2	108.9	109.6	7
3.	110.3	111.0	111.7	112.5	113.2	113.9	114.6	115.4	116.1	116.8	7
4.	117.6	118.3	119.0	119.8	120.5	121.3	122.0	122.8	123.5	124.3	7
25.	125.0	125.8	126.5	127.3	128.0	128.8	129.5	130.3	131.0	131.8	8
6.	132.6	133.3	134.1	134.9	135.6	136.4	137.2	138.0	138.7	139.5	8
7.	140.3	141.1	141.9	142.6	143.4	144.2	145.0	145.8	146.6	147.4	8
8.	148.2	149.0	149.8	150.5	151.3	152.1	153.0	153.8	154.6	155.4	8
9.	156.2	157.0	157.8	158.6	159.4	160.2	161.9	161.9	162.7	163.5	8
30.	164.3	165.1	166.0	166.8	167.6	168.4	169.3	170.1	170.9	171.8	8
1.	172.6	173.4	174.3	175.1	176.0	176.8	177.6	178.5	179.3	180.2	8
2.	181.0	181.9	182.7	183.6	184.4	185.3	186.1	187.0	187.9	188.7	9
3.	189.6	190.4	191.3	192.2	193.0	193.9	194.8	195.6	196.5	197.4	9
4.	198.3	199.1	200.0	200.9	201.8	202.6	203.5	204.4	205.3	206.2	9
35.	207.1	208.0	208.8	209.7	210.6	211.5	212.4	213.3	214.2	215.1	9
6.	216.0	216.9	217.8	218.7	219.6	220.5	221.4	222.3	223.2	224.2	9
7.	225.1	226.0	226.9	227.8	228.7	229.6	230.6	231.5	232.4	233.3	9
8.	234.2	235.2	236.1	237.0	238.0	238.9	239.8	240.8	241.7	242.6	9
9.	243.6	244.5	245.4	246.4	247.3	248.3	249.2	250.1	251.1	252.0	9
40.	253.0	253.9	254.9	255.8	256.8	257.7	258.7	259.7	260.6	261.6	10
1.	262.5	263.5	264.5	265.4	266.4	267.3	268.3	269.3	270.2	271.2	10
2.	272.2	273.2	274.1	275.1	276.1	277.1	278.0	279.0	280.0	281.0	10
3.	282.0	283.0	283.9	284.9	285.9	286.9	287.9	288.9	289.9	290.9	10
4.	291.9	292.9	293.9	294.9	295.9	296.9	297.9	298.9	299.9	300.9	10
45.	301.9	302.9	303.9	304.9	305.9	306.9	307.9	308.9	310.0	311.0	10
6.	312.0	313.0	314.0	315.0	316.1	317.1	318.1	319.1	320.2	321.2	10
7.	322.2	323.2	324.3	325.3	326.3	327.4	328.4	329.4	330.5	331.5	10
8.	332.6	333.6	334.6	335.7	336.7	337.8	338.8	339.9	340.9	342.0	10
9.	343.0	344.0	345.1	346.2	347.2	348.3	349.3	350.4	351.4	352.5	11

This table gives $N^{3/2}$ from $N = 1$ to $N = 100$. Moving the decimal point TWO places in N requires moving it THREE places in body of table. Thus:

$$(7.23)^{3/2} = 19.44 \qquad (723.)^{3/2} = 19440 \qquad (0.0723)^{3/2} = 0.01944$$
$$(72.3)^{3/2} = 614.8 \qquad (7230.)^{3/2} = 614800 \qquad (0.723)^{3/2} = 0.6148$$

Used inversely, table gives $M^{2/3}$ from $M = 1$ to $M = 1000$. Thus: $(0.6148)^{2/3} = 0.7230$.

*See also p. 1–20.

Table 1-5 (*Continued*)
THREE-HALVES POWERS OF NUMBERS

N	0	1	2	3	4	5	6	7	8	9	Avg diff
50.	353.6	354.6	355.7	356.7	357.8	358.9	359.9	361.0	362.1	363.1	11
1.	364.2	365.3	366.4	367.4	368.5	369.6	370.7	371.7	372.8	373.9	11
2.	375.0	376.1	377.1	378.2	379.3	380.4	381.5	382.6	383.7	384.8	11
3.	385.8	386.9	388.0	389.1	390.2	391.3	392.4	393.5	394.6	395.7	11
4.	396.8	397.9	399.0	400.1	401.2	402.3	403.4	404.6	405.7	406.8	11
55.	407.9	409.0	410.1	411.2	412.3	413.5	414.6	415.7	416.8	418.0	11
6.	419.1	420.2	421.3	422.4	423.6	424.7	425.8	426.9	428.1	429.2	11
7.	430.3	431.5	432.6	433.7	434.9	436.0	437.1	438.3	439.4	440.6	11
8.	441.7	442.9	444.0	445.1	446.3	447.4	448.6	449.7	450.9	452.0	11
9.	453.2	454.3	455.5	456.6	457.8	459.0	460.1	461.3	462.4	463.6	12
60.	464.8	465.9	467.1	468.2	469.4	470.6	471.7	472.9	474.1	475.3	12
1.	476.4	477.6	478.7	479.9	481.1	482.3	483.5	484.6	485.8	487.0	12
2.	488.2	489.4	490.6	491.7	492.9	494.1	495.3	496.5	497.7	498.9	12
3.	500.0	501.2	502.4	503.6	504.8	506.0	507.2	508.4	509.6	510.8	12
4.	512.0	513.2	514.4	515.6	516.8	518.0	519.2	520.4	521.6	522.8	12
65.	524.0	525.3	526.5	527.7	528.9	530.1	531.3	532.5	533.8	535.0	12
6.	536.2	537.4	538.6	539.8	541.1	542.3	543.5	544.7	546.0	547.2	12
7.	548.4	549.6	550.9	552.1	553.3	554.6	555.8	557.0	558.3	559.5	12
8.	560.7	562.0	563.2	564.5	565.7	566.9	568.2	569.4	570.7	571.9	12
9.	573.2	574.4	575.7	576.9	578.1	579.4	580.6	581.9	583.2	584.4	13
70.	585.7	586.9	588.2	589.4	590.7	591.9	593.2	594.5	595.7	597.0	13
1.	598.3	599.5	600.8	602.1	603.3	604.6	605.9	607.1	608.4	609.7	13
2.	610.9	612.2	613.5	614.8	616.0	617.3	618.6	619.9	621.2	622.4	13
3.	623.7	625.0	626.3	627.6	628.8	630.1	631.4	632.7	634.0	635.3	13
4.	636.6	637.9	639.2	640.4	641.7	643.0	644.3	645.6	646.9	648.2	13
75.	649.5	650.8	652.1	653.4	654.7	656.0	657.3	658.6	659.9	661.2	13
6.	662.6	663.9	665.2	666.5	667.8	669.1	670.4	671.7	673.0	674.4	13
7.	675.7	677.0	678.3	679.6	680.9	682.3	683.6	684.9	686.2	687.6	13
8.	688.9	690.2	691.5	692.9	694.2	695.5	696.8	698.2	699.5	700.8	13
9.	702.2	703.5	704.8	706.2	707.5	708.8	710.2	711.5	712.9	714.2	13
80.	715.5	716.9	718.2	719.6	720.9	722.3	723.6	725.0	726.3	727.7	13
1.	729.0	730.4	731.7	733.1	734.4	735.8	737.1	738.5	739.8	741.2	14
2.	742.5	743.9	745.3	746.6	748.0	749.3	750.7	752.1	753.4	754.8	14
3.	756.2	757.5	758.9	760.3	761.6	763.0	764.4	765.8	767.1	768.5	14
4.	769.9	771.2	772.6	774.0	775.4	776.8	778.1	779.5	780.9	782.3	14
85.	783.7	785.0	786.4	787.8	789.2	790.6	792.0	793.4	794.8	796.1	14
6.	797.5	798.9	800.3	801.7	803.1	804.5	805.9	807.3	808.7	810.1	14
7.	811.5	812.9	814.3	815.7	817.1	818.5	819.9	821.3	822.7	824.1	14
8.	825.5	826.9	828.3	829.7	831.1	832.6	834.0	835.4	836.8	838.2	14
9.	839.6	841.0	842.5	843.9	845.3	846.7	848.1	849.5	851.0	852.4	14
90.	853.8	855.2	856.7	858.1	859.5	860.9	862.4	863.8	865.2	866.7	14
1.	868.1	869.5	870.9	872.4	873.8	875.2	876.7	878.1	879.6	881.0	14
2.	882.4	883.9	885.3	886.8	888.2	889.6	891.1	892.5	894.0	895.4	14
3.	896.9	898.3	899.8	901.2	902.7	904.1	905.6	907.0	908.5	909.9	15
4.	911.4	912.8	914.3	915.7	917.2	918.6	920.1	921.6	923.0	924.5	15
95.	925.9	927.4	928.9	930.3	931.8	933.3	934.7	936.2	937.7	939.1	15
6.	940.6	942.1	943.5	945.0	946.5	948.0	949.4	950.9	952.4	953.9	15
7.	955.3	956.8	958.3	959.8	961.3	962.7	964.2	965.7	967.2	968.7	15
8.	970.2	971.6	973.1	974.6	976.1	977.6	979.1	980.6	982.1	983.5	15
9.	985.0	986.5	988.0	989.5	991.0	992.5	994.0	995.5	997.0	998.5	15
100.	1000.0										

Moving the decimal point TWO places in N requires moving it THREE places in body of table (see also auxiliary table on p. 1-20).

Table 1-6
RECIPROCALS OF NUMBERS

From Baumeister and Marks, *Standard Handbook for Mechanical Engineers*, 7th ed., 1967, McGraw-Hill Book Company; by permission

N	0	1	2	3	4	5	6	7	8	9	Avg diff
1.00		.9990	.9980	.9970	.9960	.9950	.9940	.9930	.9921	.9911	−10
1	.9901	.9891	.9881	.9872	.9862	.9852	.9843	.9833	.9823	.9814	
2	.9804	.9794	.9785	.9775	.9766	.9756	.9747	.9737	.9728	.9718	
3	.9709	.9699	.9690	.9681	.9671	.9662	.9653	.9643	.9634	.9625	−9
4	.9615	.9606	.9597	.9588	.9579	.9569	.9560	.9551	.9542	.9533	
1.05	.9524	.9515	.9506	.9497	.9488	.9479	.9470	.9461	.9452	.9443	
6	.9434	.9425	.9416	.9407	.9398	.9390	.9381	.9372	.9363	.9355	
7	.9346	.9337	.9328	.9320	.9311	.9302	.9294	.9285	.9276	.9268	
8	.9259	.9251	.9242	.9234	.9225	.9217	.9208	.9200	.9191	.9183	−8
9	.9174	.9166	.9158	.9149	.9141	.9132	.9124	.9116	.9107	.9099	
1.10	.9091	.9083	.9074	.9066	.9058	.9050	.9042	.9033	.9025	.9017	
1	.9009	.9001	.8993	.8985	.8977	.8969	.8961	.8953	.8945	.8937	
2	.8929	.8921	.8913	.8905	.8897	.8889	.8881	.8873	.8865	.8857	
3	.8850	.8842	.8834	.8826	.8818	.8811	.8803	.8795	.8787	.8780	
4	.8772	.8764	.8757	.8749	.8741	.8734	.8726	.8718	.8711	.8703	
1.15	.8696	.8688	.8681	.8673	.8666	.8658	.8651	.8643	.8636	.8628	
6	.8621	.8613	.8606	.8598	.8591	.8584	.8576	.8569	.8562	.8554	−7
7	.8547	.8540	.8532	.8525	.8518	.8511	.8503	.8496	.8489	.8482	
8	.8475	.8467	.8460	.8453	.8446	.8439	.8432	.8425	.8418	.8410	
9	.8403	.8396	.8389	.8382	.8375	.8368	.8361	.8354	.8347	.8340	
1.20	.8333	.8326	.8319	.8313	.8306	.8299	.8292	.8285	.8278	.8271	
1	.8264	.8258	.8251	.8244	.8237	.8230	.8224	.8217	.8210	.8203	
2	.8197	.8190	.8183	.8177	.8170	.8163	.8157	.8150	.8143	.8137	
3	.8130	.8123	.8117	.8110	.8104	.8097	.8091	.8084	.8078	.8071	−6
4	.8065	.8058	.8052	.8045	.8039	.8032	.8026	.8019	.8013	.8006	
1.25	.8000	.7994	.7987	.7981	.7974	.7968	.7962	.7955	.7949	.7943	
6	.7937	.7930	.7924	.7918	.7911	.7905	.7899	.7893	.7886	.7880	
7	.7874	.7868	.7862	.7855	.7849	.7843	.7837	.7831	.7825	.7819	
8	.7812	.7806	.7800	.7794	,7788	.7782	.7776	.7770	.7764	.7758	
9	.7752	.7746	.7740	.7734	.7728	.7722	.7716	.7710	.7704	.7698	
1.30	.7692	.7686	.7680	.7675	.7669	.7663	.7657	.7651	.7645	.7639	
1	.7634	.7628	.7622	.7616	.7610	.7605	.7599	.7593	.7587	.7582	
2	.7576	.7570	.7564	.7559	.7553	.7547	.7541	.7536	.7530	.7524	
3	.7519	.7513	.7508	.7502	.7496	.7491	.7485	.7479	.7474	.7468	
4	.7463	.7457	.7452	.7446	.7440	.7435	.7429	.7424	.7418	.7413	
1.35	.7407	.7402	.7396	.7391	.7386	.7380	.7375	.7369	.7364	.7358	−5
6	.7353	.7348	.7342	.7337	.7331	.7326	.7321	.7315	.7310	.7305	
7	.7299	.7294	.7289	.7283	.7278	.7273	.7267	.7262	.7257	.7252	
8	.7246	.7241	.7236	.7231	.7225	.7220	.7215	.7210	.7205	.7199	
9	.7194	.7189	.7184	.7179	.7174	.7168	.7163	.7158	.7153	.7148	
1.40	.7143	.7138	.7133	.7128	.7123	.7117	.7112	.7107	.7102	.7097	
1	.7092	.7087	.7082	.7077	.7072	.7067	.7062	.7057	.7052	.7047	
2	.7042	.7037	.7032	.7027	.7022	.7018	.7013	.7008	.7003	.6998	
3	.6993	.6988	.6983	.6978	.6974	.6969	.6964	.6959	.6954	.6949	
4	.6944	.6940	.6935	.6930	.6925	.6920	.6916	.6911	.6906	.6901	
1.45	.6897	.6892	.6887	.6882	.6878	.6873	.6868	.6863	.6859	.6854	
6	.6849	.6845	.6840	.6835	.6831	.6826	.6821	.6817	.6812	.6807	
7	.6803	.6798	.6793	.6789	.6784	.6780	.6775	.6770	.6766	.6761	
8	.6757	.6752	.6748	.6743	.6739	.6734	.6729	.6725	.6720	.6716	
9	.6711	.6707	.6702	.6698	.6693	.6689	.6684	.6680	.6676	.6671	

$1/\pi = 0.318310$ $1/e = 0.367879$

Moving the decimal point in either direction in N requires moving it in the OPPOSITE direction in body of table (see p. 1–26).

Table 1-6 (*Continued*)
RECIPROCALS OF NUMBERS

N	0	1	2	3	4	5	6	7	8	9	Avg Diff
1.50	.6667	.6662	.6658	.6653	.6649	.6645	.6640	.6636	.6631	.6627	−4
1	.6623	.6618	.6614	.6609	.6605	.6601	.6596	.6592	.6588	.6583	
2	.6579	.6575	.6570	.6566	.6562	.6557	.6553	.6549	.6545	.6540	
3	.6536	.6532	.6527	.6523	.6519	.6515	.6510	.6506	.6502	.6498	
4	.6494	.6489	.6485	.6481	.6477	.6472	.6468	.6464	.6460	.6456	
1.55	.6452	.6447	.6443	.6439	.6435	.6431	.6427	.6423	.6418	.6414	
6	.6410	.6406	.6402	.6398	.6394	.6390	.6386	.6382	.6378	.6373	
7	.6369	.6365	.6361	.6357	.6353	.6349	.6345	.6341	.6337	.6333	
8	.6329	.6325	.6321	.6317	.6313	.6309	.6305	.6301	.6297	.6293	
9	.6289	.6285	.6281	.6277	.6274	.6270	.6266	.6262	.6258	.6254	
1.60	.6250	.6246	.6242	.6238	.6234	.6231	.6227	.6223	.6219	.6215	
1	.6211	.6207	.6203	.6200	.6196	.6192	.6188	.6184	.6180	.6177	
2	.6173	.6169	.6165	.6161	.6158	.6154	.6150	.6146	.6143	.6139	
3	.6135	.6131	.6127	.6124	.6120	.6116	.6112	.6109	.6105	.6101	
4	.6098	.6094	.6090	.6086	.6083	.6079	.6075	.6072	.6068	.6064	
1.65	.6061	.6057	.6053	.6050	.6046	.6042	.6039	.6035	.6031	.6028	
6	.6024	.6020	.6017	.6013	.6010	.6006	.6002	.5999	.5995	.5992	
7	.5988	.5984	.5981	.5977	.5974	.5970	.5967	.5963	.5959	.5956	
8	.5952	.5949	.5945	.5942	.5938	.5935	.5931	.5928	.5924	.5921	
9	.5917	.5914	.5910	.5907	.5903	.5900	.5896	.5893	.5889	.5886	
1.70	.5882	.5879	.5875	.5872	.5869	.5865	.5862	.5858	.5855	.5851	−3
1	.5848	.5845	.5841	.5838	.5834	.5831	.5828	.5824	.5821	.5817	
2	.5814	.5811	.5807	.5804	.5800	.5797	.5794	.5790	.5787	.5784	
3	.5780	.5777	.5774	.5770	.5767	.5764	.5760	.5757	.5754	.5750	
4	.5747	.5744	.5741	.5737	.5734	.5731	.5727	.5724	.5721	.5718	
1.75	.5714	.5711	.5708	.5705	.5701	.5698	.5695	.5692	.5688	.5685	
6	.5682	.5679	.5675	.5672	.5669	.5666	.5663	.5659	.5656	.5653	
7	.5650	.5647	.5643	.5640	.5637	.5634	.5631	.5627	.5624	.5621	
8	.5618	.5615	.5612	.5609	.5605	.5602	.5599	.5596	.5593	.5590	
9	.5587	.5583	.5580	.5577	.5574	.5571	.5568	.5565	.5562	.5559	
1.80	.5556	.5552	.5549	.5546	.5543	.5540	.5537	.5534	.5531	.5528	
1	.5525	.5522	.5519	.5516	.5513	.5510	.5507	.5504	.5501	.5498	
2	.5495	.5491	.5488	.5485	.5482	.5479	.5476	.5473	.5470	.5467	
3	.5464	.5461	.5459	.5456	.5453	.5450	.5447	.5444	.5441	.5438	
4	.5435	.5432	.5429	.5426	.5423	.5420	.5417	.5414	.5411	.5408	
1.85	.5405	.5402	.5400	.5397	.5394	.5391	.5388	.5385	.5382	.5379	
6	.5376	.5373	.5371	.5368	.5365	.5362	.5359	.5356	.5353	.5350	
7	.5348	.5345	.5342	.5339	.5336	.5333	.5330	.5328	.5325	.5322	
8	.5319	.5316	.5313	.5311	.5308	.5305	.5302	.5299	.5297	.5294	
9	.5291	.5288	.5285	.5283	.5280	.5277	.5274	.5271	.5269	.5266	
1.90	.5263	.5260	.5258	.5255	.5252	.5249	.5247	.5244	.5241	.5238	
1	.5236	.5233	.5230	.5227	.5225	.5222	.5219	.5216	.5214	.5211	
2	.5208	.5206	.5203	.5200	.5198	.5195	.5192	.5189	.5187	.5184	
3	.5181	.5179	.5176	.5173	.5171	.5168	.5165	.5163	.5160	.5157	
4	.5155	.5152	.5149	.5147	.5144	.5141	.5139	.5136	.5133	.5131	
1.95	.5128	.5126	.5123	.5120	.5118	.5115	.5112	.5110	.5107	.5105	
6	.5102	.5099	.5097	.5094	.5092	.5089	.5086	.5084	.5081	.5079	
7	.5076	.5074	.5071	.5068	.5066	.5063	.5061	.5058	.5056	.5053	−2
8	.5051	.5048	.5045	.5043	.5040	.5038	.5035	.5033	.5030	.5028	
9	.5025	.5023	.5020	.5018	.5015	.5013	.5010	.5008	.5005	.5003	

Moving the decimal point in either direction in N requires moving it in the OPPOSITE direction in body of table (see p. 1–26).

Table 1-6 (*Continued*)
RECIPROCALS OF NUMBERS

N	0	1	2	3	4	5	6	7	8	9	Avg diff
2.0	.5000	.4975	.4950	.4926	.4902	.4878	.4854	.4831	.4808	.4785	−24
1	.4762	.4739	.4717	.4695	.4673	.4651	.4630	.4608	.4587	.4566	−21
2	.4545	.4525	.4505	.4484	.4464	.4444	.4425	.4405	.4386	.4367	−20
3	.4348	.4329	.4310	.4292	.4274	.4255	.4237	.4219	.4202	.4184	−18
4	.4167	.4149	.4132	.4115	.4098	.4082	.4065	.4049	.4032	.4016	−17
2.5	.4000	.3984	.3968	.3953	.3937	.3922	.3906	.3891	.3876	.3861	−15
6	.3846	.3831	.3817	.3802	.3788	.3774	.3759	.3745	.3731	.3717	−14
7	.3704	.3690	.3676	.3663	.3650	.3636	.3623	.3610	.3597	.3584	−13
8	.3571	.3559	.3546	.3534	.3521	.3509	.3497	.3484	.3472	.3460	−12
9	.3448	.3436	.3425	.3413	.3401	.3390	.3378	.3367	.3356	.3344	−12
3.0	.3333	.3322	.3311	.3300	.3289	.3279	.3268	.3257	.3247	.3236	−11
1	.3226	.3215	.3205	.3195	.3185	.3175	.3165	.3155	.3145	.3135	−10
2	.3125	.3115	.3106	.3096	.3086	.3077	.3067	.3058	.3049	.3040	−10
3	.3030	.3021	.3012	.3003	.2994	.2985	.2976	.2967	.2959	.2950	−9
4	.2941	.2933	.2924	.2915	.2907	.2899	.2890	.2882	.2874	.2865	−8
3.5	.2857	.2849	.2841	.2833	.2825	.2817	.2809	.2801	.2793	.2786	−8
6	.2778	.2770	.2762	.2755	.2747	.2740	.2732	.2725	.2717	.2710	−8
7	.2703	.2695	.2688	.2681	.2674	.2667	.2660	.2653	.2646	.2639	−7
8	.2632	.2625	.2618	.2611	.2604	.2597	.2591	.2584	.2577	.2571	−7
9	.2564	.2558	.2551	.2545	.2538	.2532	.2525	.2519	.2513	.2506	−6
4.0	.2500	.2494	.2488	.2481	.2475	.2469	.2463	.2457	.2451	.2445	−6
1	.2439	.2433	.2427	.2421	.2415	.2410	.2404	.2398	.2392	.2387	−6
2	.2381	.2375	.2370	.2364	.2358	.2353	.2347	.2342	.2336	.2331	−6
3	.2326	.2320	.2315	.2309	.2304	.2299	.2294	.2288	.2283	.2278	−5
4	.2273	.2268	.2262	.2257	.2252	.2247	.2242	.2237	.2232	.2227	−5
4.5	.2222	.2217	.2212	.2208	.2203	.2198	.2193	.2188	.2183	.2179	−5
6	.2174	.2169	.2165	.2160	.2155	.2151	.2146	.2141	.2137	.2132	−5
7	.2128	.2123	.2119	.2114	.2110	.2105	.2101	.2096	.2092	.2088	−4
8	.2083	.2079	.2075	.2070	.2066	.2062	.2058	.2053	.2049	.2045	−4
9	.2041	.2037	.2033	.2028	.2024	.2020	.2016	.2012	.2008	.2004	−4

$$1/\pi = 0.318310 \qquad 1/e = 0.367879$$

Explanation of Table of Reciprocals (pp. 1–24 to 1–27).

This table gives the values of $1/N$ for values of N from 1 to 10, correct to four figures. (Interpolated values may be in error by 1 in the fourth figure.)

To find the reciprocal of a number N outside the range from 1 to 10, note that moving the decimal point any number of places in either direction in column N is equivalent to moving it the same number of places in the OPPOSITE direction in the body of the table. For example:

$$\frac{1}{3.217} = 0.3109 \qquad \frac{1}{3217.} = 0.000\ 3109 \qquad \frac{1}{0.003217} = 310.9$$

Table 1-6 (*Continued*)
RECIPROCALS OF NUMBERS

N	0	1	2	3	4	5	6	7	8	9	Diff Avg
5.0	.2000	.1996	.1992	.1988	.1984	.1980	.1976	.1972	.1969	.1965	−4
.1	.1961	.1957	.1953	.1949	.1946	.1942	.1938	.1934	.1931	.1927	
.2	.1923	.1919	.1916	.1912	.1908	.1905	.1901	.1898	.1894	.1890	
.3	.1887	.1883	.1880	.1876	.1873	.1869	.1866	.1862	.1859	.1855	
.4	.1852	.1848	.1845	.1842	.1838	.1835	.1832	.1828	.1825	.1821	−3
5.5	.1818	.1815	.1812	.1808	.1805	.1802	.1799	.1795	.1792	.1789	
.6	.1786	.1783	.1779	.1776	.1773	.1770	.1767	.1764	.1761	.1757	
.7	.1754	.1751	.1748	.1745	.1742	.1739	.1736	.1733	.1730	.1727	
.8	.1724	.1721	.1718	.1715	.1712	.1709	.1706	.1704	.1701	.1698	
.9	.1695	.1692	.1689	.1686	.1684	.1681	.1678	.1675	.1672	.1669	
6.0	.1667	.1664	.1661	.1658	.1656	.1653	.1650	.1647	.1645	.1642	
.1	.1639	.1637	.1634	.1631	.1629	.1626	.1623	.1621	.1618	.1616	
.2	.1613	.1610	.1608	.1605	.1603	.1600	.1597	.1595	.1592	.1590	
.3	.1587	.1585	.1582	.1580	.1577	.1575	.1572	.1570	.1567	.1565	−2
.4	.1563	.1560	.1558	.1555	.1553	.1550	.1548	.1546	.1543	.1541	
6.5	.1538	.1536	.1534	.1531	.1529	.1527	.1524	.1522	.1520	.1517	
.6	.1515	.1513	.1511	.1508	.1506	.1504	.1502	.1499	.1497	.1495	
.7	.1493	.1490	.1488	.1486	.1484	.1481	.1479	.1477	.1475	.1473	
.8	.1471	.1468	.1466	.1464	.1462	.1460	.1458	.1456	.1453	.1451	
.9	.1449	.1447	.1445	.1443	.1441	.1439	.1437	.1435	.1433	.1431	
7.0	.1429	.1427	.1425	.1422	.1420	.1418	.1416	.1414	.1412	.1410	
.1	.1408	.1406	.1404	.1403	.1401	.1399	.1397	.1395	.1393	.1391	
.2	.1389	.1387	.1385	.1383	.1381	.1379	.1377	.1376	.1374	.1372	
.3	.1370	.1368	.1366	.1364	.1362	.1361	.1359	.1357	.1355	.1353	
.4	.1351	.1350	.1348	.1346	.1344	.1342	.1340	.1339	.1337	.1335	
7.5	.1333	.1332	.1330	.1328	.1326	.1325	.1323	.1321	.1319	.1318	
.6	.1316	.1314	.1312	.1311	.1309	.1307	.1305	.1304	.1302	.1300	
.7	.1299	.1297	.1295	.1294	.1292	.1290	.1289	.1287	.1285	.1284	
.8	.1282	.1280	.1279	.1277	.1276	.1274	.1272	.1271	.1269	.1267	
.9	.1266	.1264	.1263	.1261	.1259	.1258	.1256	.1255	.1253	.1252	
8.0	.1250	.1248	.1247	.1245	.1244	.1242	.1241	.1239	.1238	.1236	
.1	.1235	.1233	.1232	.1230	.1229	.1227	.1225	.1224	.1222	.1221	
.2	.1220	.1218	.1217	.1215	.1214	.1212	.1211	.1209	.1208	.1206	
.3	.1205	.1203	.1202	.1200	.1199	.1198	.1196	.1195	.1193	.1192	
.4	.1190	.1189	.1188	.1186	.1185	.1183	.1182	.1181	.1179	.1178	−1
8.5	.1176	.1175	.1174	.1172	.1171	.1170	.1168	.1167	.1166	.1164	
.6	.1163	.1161	.1160	.1159	.1157	.1156	.1155	.1153	.1152	.1151	
.7	.1149	.1148	.1147	.1145	.1144	.1143	.1142	.1140	.1139	.1138	
.8	.1136	.1135	.1134	.1133	.1131	.1130	.1129	.1127	.1126	.1125	
.9	.1124	.1122	.1121	.1120	.1119	.1117	.1116	.1115	.1114	.1112	
9.0	.1111	.1110	.1109	.1107	.1106	.1105	.1104	.1103	.1101	.1100	
.1	.1099	.1098	.1096	.1095	.1094	.1093	.1092	.1091	.1089	.1088	
.2	.1087	.1086	.1085	.1083	.1082	.1081	.1080	.1079	.1078	.1076	
.3	.1075	.1074	.1073	.1072	.1071	.1070	.1068	.1067	.1066	.1065	
.4	.1064	.1063	.1062	.1060	.1059	.1058	.1057	.1056	.1055	.1054	
9.5	.1053	.1052	.1050	.1049	.1048	.1047	.1046	.1045	.1044	.1043	
.6	.1042	.1041	.1040	.1038	.1037	.1036	.1035	.1034	.1033	.1032	
.7	.1031	.1030	.1029	.1028	.1027	.1026	.1025	.1024	.1022	.1021	
.8	.1020	.1019	.1018	.1017	.1016	.1015	.1014	.1013	.1012	.1011	
.9	.1010	.1009	.1008	.1007	.1006	.1005	.1004	.1003	.1002	.1001	

Moving the decimal point in either direction in N requires moving it in the OPPOSITE direction in body of table (see p. 1–26).

LOGARITHMS

Properties and Uses

Definition of Logarithm. The *logarithm* x of the number N to the base b is the exponent of the power to which b must be raised to give N. That is,

$$\log_b N = x \text{ or } b^x = N.$$

The number N is positive and b may be any positive number except 1.

Properties of Logarithms.

(**a.**) *The logarithm of a product is equal to the sum of the logarithms of the factors; thus,*

$$\log_b M \cdot N = \log_b M + \log_b N.$$

(**b.**) *The logarithm of a quotient is equal to the logarithm of the numerator minus the logarithm of the denominator; thus,*

$$\log_b \frac{M}{N} = \log_b M - \log_b N.$$

(**c.**) *The logarithm of a power of a number is equal to the logarithm of the base multiplied by the exponent of the power; thus,*

$$\log_b M^p = p \cdot \log_b M.$$

(**d.**) *The logarithm of a root of a number is equal to the logarithm of the number divided by the index of the root; thus,*

$$\log_b \sqrt[q]{M} = \frac{1}{q} \log_b M.$$

Other properties of logarithms:

$$\log_b b = 1. \qquad\qquad \log_b \sqrt[q]{M^p} = \frac{p}{q} \cdot \log_b M.$$

$$\log_b 1 = 0. \qquad\qquad \log_b N = \log_a N \cdot \log_b a = \frac{\log_a N}{\log_a b}$$

$$\log_b (b^N) = N. \qquad\qquad b^{\log_b N} = N.$$

Systems of Logarithms. There are two common systems of logarithms in use: (1) the *natural* (Napierian or hyperbolic) system which uses the base $e = 2.71828. . .$; (2) the *common* (Briggsian) system which uses the base 10.

We shall use the abbreviation $\log N \equiv \log_{10} N$ in this section.

Unless otherwise stated, tables of logarithms are always tables of common logarithms.

Characteristic of a Common Logarithm of a Number. Every real positive number has a real common logarithm such that if $a < b$, $\log a < \log b$. Neither zero nor any negative number has a real logarithm.

A common logarithm, in general, consists of an integer, which is called the *characteristic*, and a decimal (usually endless) which is called the *mantissa*. The characteristic of any number may be determined from the following rules.

Rule I. The characteristic of any number greater than 1 is one less than the number of digits before the decimal point.

Rule II.† The characteristic of a number less than 1 is found by subtracting from 9 the number of ciphers between the decimal point and the first significant digit, and writing −10 after the result.

Thus the characteristic of log 936 is 2; the characteristic of log 9.36 is 0; of log 0.936 is 9 −10; of log 0.00936 is 7 −10.

Mantissa of a Common Logarithm of a Number. An important consequence of the use of base 10 is that the mantissa of a number is independent of the position of the decimal point. Thus 93,600, 93.600, 0.000936, all have the same mantissa. Hence in Tables of Common Logarithms only mantissas are given. A five place table gives the values of the mantissa correct to five places of decimals.

To Find the Logarithm of a Given Number N. By means of Rules I and II determine the characteristic. Then use the Table to find mantissa.

To find mantissa when the given number (neglecting decimal point) consists of four, or less, digits (exclusive of ciphers at the beginning or end), look in the column marked N for the first three significant digits and pick the column headed by the fourth digit—the mantissa is the number appearing at the intersection of this row and column. Thus to find the logarithm of 64030, first note (by Rule I) that the characteristic is 4. Next, in the Table, find 640 in column marked N

---

† Some writers use a dash over the characteristic to indicate a negative value: for example,
$$\log .004657 = 7.66811 - 10 = \bar{3}.66811.$$

and opposite it in column 3 is the desired mantissa, .80638. Hence log 64030 = 4.80638.   Likewise, log 0.0064030 = 7.80638 − 10; log 0.64030 = 9.80638 −10.

Interpolation. The mantissa of a number of more than four significant figures can be found approximately by assuming that the mantissa varies directly as the number in the small interval not tabulated. Thus if N has five digits (significant), and f is the fifth digit of N, the mantissa of N is

$$m = m_1 + \frac{f}{10}\left(m_2 - m_1\right),$$

where m_1 is the mantissa corresponding to the first four digits of N, m_2 is the next larger mantissa in the table. $(m_2 - m_1)$ is called a *tabular difference*. The proportional part of the difference $m_2 - m_1$ is called the *correction*. These proportional parts are printed without zeros at the right-hand side of each page as an aid to mental multiplications.

For example, find log 64034. Here $f = 4$. From the table we see $m_1 = .80638$, $m_2 = .80645$, whence $m = .80638 + (4/10)\,(.00007)$, log 64034 = $4 + \bar{m} = 4.80641$.

To Find the Number N when its Logarithm is Known. (The number N whose logarithm is k is called the *anti logarithm* of k.)

Case 1. If the mantissa m is found exactly in the Table, join the figure at the top of the column containing m to the right of the figures in the column marked N and in the same row as m, and place the decimal point according to the characteristic of the logarithm.

Case 2. If the mantissa m is not found exactly in the table, interpolate as follows: find the next smaller mantissa m_1 to m; the first four significant digits of N correspond to the mantissa m_1, and the fifth digit f equals the nearest whole number to

$$f = 10\left(\frac{m - m_1}{m_2 - m_1}\right),$$

where m_2 is the next larger mantissa to m appearing in the table. Then locate the decimal point according to the characteristic.

The decimal point may be located by means of the following rules:

Rule III. *If the characteristic of the logarithm is positive (then the mantissa is not followed by −10), begin at the left, count digits one more than the characteristic, and place the decimal point to the right of the last digit counted.*

Rule IV. *If the characteristic is negative (then the mantissa will be preceded by an integer n and followed by* −10), *prefix* (9−n) *ciphers, and place the decimal point to the left of these ciphers.*

Illustrations of the Use of Logarithms.

Example 1. Given log x = 2.91089, find x. The mantissa .91089 appears in the table. Join the figure 5 which appears at the top of the column to the right of the number 814 in the column N, giving the number 8145. By Rule III, the decimal point is placed to the right of 4, thus giving x = 814.5.

Example 2. Given log x = 2.34917, find x. The mantissa m = .34917 does not appear in the table. The next smaller and next larger mantissas are m_1 and m_2,

$$m_1 = .34908, \quad m = .34917, \quad m_2 = .34928.$$

The first four digits of N, corresponding to m_1, are 2234 and the fifth digit is the nearest whole number (5) to

$$10 \left(\frac{m - m_1}{m_2 - m_1}\right) = 10 \left(\frac{.00009}{.00020}\right) = 4.5$$

By Rule III, we locate decimal point, thus giving x = 223.45.

Example 3. Find x = (396.21) (.004657) (21.21).
$$\log 396.21 = 2.59792$$
$$\log .004657 = 7.66811 -10$$
$$\log 21.210 = 1.32654 \qquad \text{(add)}$$
$$\log x \quad = 11.59257 -10, \quad x = 39.135.$$

Example 4. Find $x = \dfrac{396.21*}{24.3}$.

$$\log 396.21 = 2.59792$$
$$\log 24.3 \quad = 1.38561 \text{ (subtract)}$$
$$\log x \qquad = 1.21231, \quad x = 16.305.$$

Example 5. Find x = (3.5273)^4.
$$\log 3.5273 = 0.54745$$
$$4 \text{ (multiply)}$$
$$\log x \qquad = 2.18980, \quad x = 154.81.$$

---

*Some writers use *cologarithms*. The cologarithm of a number N is the negative of the logarithm of N; i. e., colog N = 10.00000 − log N − 10. Adding the cologarithm is equivalent to subtracting the logarithm. Thus in our example, colog 24.3 = (10.00000 − 1.38561) − 10 = 8.61439–10, log 396.21 − log 24.3 = log 396.21 + colog 24.3 = 2.59792 + 8.61439–10 = 11.21231–10 = anti-log 16.305.

Example 6.   Given log x = −2.23653, to find x. To convert this logarithm to one with a positive mantissa, add algebraically − 2.23653 to 10.00000–10. Thus

$$10.00000-10$$
$$\underline{-2.23653}$$
log x = 7.76347–10, hence x = 0.0058006.

Example 7.   Find $x = \sqrt[3]{.04657}$.

log x = $\frac{1}{3}$ log (.04657),
          = $\frac{1}{3}$(8.66811–10) = $\frac{1}{3}$(−1.33189),
log x = −0.44396 = 9.55604–10, x = 0.35978.

or

log x = $\frac{1}{3}$(8.66811–10) = $\frac{1}{3}$(28.66811–30)
log x = 9.55604–10, x = 0.35978.

Example 8.   Find $x = \dfrac{1}{21.210}$.

log x = log 1 − log 21.210
log 1   = 10.00000–10
log 21.210 = $\underline{1.32654}$ (subtract)
log x = 8.67346–10, x = 0.047148.

Table 1-7
COMMON LOGARITHMS OF NUMBERS
100–150

N.	0	1	2	3	4	5	6	7	8	9
100	00 000	043	087	130	173	217	260	303	346	389
101	432	475	518	561	604	647	689	732	775	817
102	860	903	945	988	*030	*072	*115	*157	*199	*242
103	01 284	326	368	410	452	494	536	578	620	662
104	703	745	787	828	870	912	953	995	*036	*078
105	02 119	160	202	243	284	325	366	407	449	490
106	531	572	612	653	694	735	776	816	857	898
107	938	979	*019	*060	*100	*141	*181	*222	*262	*302
108	03 342	383	423	463	503	543	583	623	663	703
109	743	782	822	862	902	941	981	*021	*060	*100
110	04 139	179	218	258	297	336	376	415	454	493
111	532	571	610	650	689	727	766	805	844	883
112	922	961	999	*038	*077	*115	*154	*192	*231	*269
113	05 308	346	385	423	461	500	538	576	614	652
114	690	729	767	805	843	881	918	956	994	*032
115	06 070	108	145	183	221	258	296	333	371	408
116	446	483	521	558	595	633	670	707	744	781
117	819	856	893	930	967	*004	*041	*078	*115	*151
118	07 188	225	262	298	335	372	408	445	482	518
119	555	591	628	664	700	737	773	809	846	882
120	918	954	990	*027	*063	*099	*135	*171	*207	*243
121	08 279	314	350	386	422	458	493	529	565	600
122	636	672	707	743	778	814	849	884	920	955
123	991	*026	*061	*096	*132	*167	*202	*237	*272	*307
124	09 342	377	412	447	482	517	552	587	621	656
125	691	726	760	795	830	864	899	934	968	*003
126	10 037	072	106	140	175	209	243	278	312	346
127	380	415	449	483	517	551	585	619	653	687
128	721	755	789	823	857	890	924	958	992	*025
129	11 059	093	126	160	193	227	261	294	327	361
130	394	428	461	494	528	561	594	628	661	694
131	727	760	793	826	860	893	926	959	992	*024
132	12 057	090	123	156	189	222	254	287	320	352
133	385	418	450	483	516	548	581	613	646	678
134	710	743	775	808	840	872	905	937	969	*001
135	13 033	066	098	130	162	194	226	258	290	322
136	354	386	418	450	481	513	545	577	609	640
137	672	704	735	767	799	830	862	893	925	956
138	988	*019	*051	*082	*114	*145	*176	*208	*239	*270
139	14 301	333	364	395	426	457	489	520	551	582
140	613	644	675	706	737	768	799	829	860	891
141	922	953	983	*014	*045	*076	*106	*137	*168	*198
142	15 229	259	290	320	351	381	412	442	473	503
143	534	564	594	625	655	685	715	746	776	806
144	836	866	897	927	957	987	*017	*047	*077	*107
145	16 137	167	197	227	256	286	316	346	376	406
146	435	465	495	524	554	584	613	643	673	702
147	732	761	791	820	850	879	909	938	967	997
148	17 026	056	085	114	143	173	202	231	260	289
149	319	348	377	406	435	464	493	522	551	580
150	609	638	667	696	725	754	782	811	840	869
N.	0	1	2	3	4	5	6	7	8	9

Proportional parts

	44	43	42
1	4.4	4.3	4.2
2	8.8	8.6	8.4
3	13.2	12.9	12.6
4	17.6	17.2	16.8
5	22.0	21.5	21.0
6	26.4	25.8	25.2
7	30.8	30.1	29.4
8	35.2	34.4	33.6
9	39.6	38.7	37.8

	41	40	39
1	4.1	4.0	3.9
2	8.2	8.0	7.8
3	12.3	12.0	11.7
4	16.4	16 0	15.6
5	20.5	20.0	19.5
6	24.6	24.0	23.4
7	28.7	28.0	27.3
8	32.8	32.0	31.2
9	36.9	36.0	35.1

	38	37	36
1	3.8	3.7	3.6
2	7.6	7.4	7.2
3	11.4	11.1	10.8
4	15.2	14.8	14.4
5	19.0	18.5	18.0
6	22.8	22.2	21.6
7	26.6	25.9	25.2
8	30.4	29.6	28.8
9	34.2	33.3	32.4

	35	34	33
1	3.5	3.4	3.3
2	7.0	6.8	6.6
3	10.5	10.2	9.9
4	14.0	13.6	13.2
5	17.5	17.0	16.5
6	21.0	20.4	19.8
7	24.5	23.8	23.1
8	28.0	27.2	26.4
9	31.5	30.6	29.7

	32	31	30
1	3.2	3.1	3.0
2	6.4	6.2	6.0
3	9.6	9.3	9.0
4	12.8	12.4	12.0
5	16.0	15.5	15.0
6	19.2	18.6	18.0
7	22.4	21.7	21.0
8	25.6	24.8	24.0
9	28.8	27.9	27.0

Proportional parts

.00 000–.17 869

Table 1-7 (*Continued*)
COMMON LOGARITHMS OF NUMBERS
150-200

N.	0	1	2	3	4	5	6	7	8	9
150	17 609	638	667	696	725	754	782	811	840	869
151	898	926	955	984	*013	*041	*070	*099	*127	*156
152	18 184	213	241	270	298	327	355	384	412	441
153	469	498	526	554	583	611	639	667	696	724
154	752	780	808	837	865	893	921	949	977	*005
155	19 033	061	089	117	145	173	201	229	257	285
156	312	340	368	396	424	451	479	507	535	562
157	590	618	645	673	700	728	756	783	811	838
158	866	893	921	948	976	*003	*030	*058	*085	*112
159	20 140	167	194	222	249	276	303	330	358	385
160	412	439	466	493	520	548	575	602	629	656
161	683	710	737	763	790	817	844	871	898	925
162	952	978	*005	*032	*059	*085	*112	*139	*165	*192
163	21 219	245	272	299	325	352	378	405	431	458
164	484	511	537	564	590	617	643	669	696	722
165	748	775	801	827	854	880	906	932	958	985
166	22 011	037	063	089	115	141	167	194	220	246
167	272	298	324	350	376	401	427	453	479	505
168	531	557	583	608	634	660	686	712	737	763
169	789	814	840	866	891	917	943	968	994	*019
170	23 045	070	096	121	147	172	198	223	249	274
171	300	325	350	376	401	426	452	477	502	528
172	553	578	603	629	654	679	704	729	754	779
173	805	830	855	880	905	930	955	980	*005	*030
174	24 055	080	105	130	155	180	204	229	254	279
175	304	329	353	378	403	428	452	477	502	527
176	551	576	601	625	650	674	699	724	748	773
177	797	822	846	871	895	920	944	969	993	*018
178	25 042	066	091	115	139	164	188	212	237	261
179	285	310	334	358	382	406	431	455	479	503
180	527	551	575	600	624	648	672	696	720	744
181	768	792	816	840	864	888	912	935	959	983
182	26 007	031	055	079	102	126	150	174	198	221
183	245	269	293	316	340	364	387	411	435	458
184	482	505	529	553	576	600	623	647	670	694
185	717	741	764	788	811	834	858	881	905	928
186	951	975	998	*021	*045	*068	*091	*114	*138	*161
187	27 184	207	231	254	277	300	323	346	370	393
188	416	439	462	485	508	531	554	577	600	623
189	646	669	692	715	738	761	784	807	830	852
190	875	898	921	944	967	989	*012	*035	*058	*081
191	28 103	126	149	171	194	217	240	262	285	307
192	330	353	375	398	421	443	466	488	511	533
193	556	578	601	623	646	668	691	713	735	758
194	780	803	825	847	870	892	914	937	959	981
195	29 003	026	048	070	092	115	137	159	181	203
196	226	248	270	292	314	336	358	380	403	425
197	447	469	491	513	535	557	579	601	623	645
198	667	688	710	732	754	776	798	820	842	863
199	885	907	929	951	973	994	*016	*038	*060	*081
200	30 103	125	146	168	190	211	233	255	276	298
N.	0	1	2	3	4	5	6	7	8	9

Proportional parts

	29	28
1	2.9	2.8
2	5.8	5.6
3	8.7	8.4
4	11.6	11.2
5	14.5	14.0
6	17.4	16.8
7	20.3	19.6
8	23.2	22.4
9	26.1	25.2

	27	26
1	2.7	2.6
2	5.4	5.2
3	8.1	7.8
4	10.8	10.4
5	13.5	13.0
6	16.2	15.6
7	18.9	18.2
8	21.6	20.8
9	24.3	23.4

	25
1	2.5
2	5.0
3	7.5
4	10.0
5	12.5
6	15.0
7	17.5
8	20.0
9	22.5

	24	23
1	2.4	2.3
2	4.8	4.6
3	7.2	6.9
4	9.6	9.2
5	12.0	11.5
6	14.4	13.8
7	16.8	16.1
8	19.2	18.4
9	21.6	20.7

	22	21
1	2.2	2.1
2	4.4	4.2
3	6.6	6.3
4	8.8	8.4
5	11.0	10.5
6	13.2	12.6
7	15.4	14.7
8	17.6	16.8
9	19.8	18.9

Proportional parts

.17 609-.30 298

Table 1-7 (*Continued*)
COMMON LOGARITHMS OF NUMBERS
200–250

N.	0	1	2	3	4	5	6	7	8	9
200	30 103	125	146	168	190	211	233	255	276	298
201	320	341	363	384	406	428	449	471	492	514
202	535	557	578	600	621	643	664	685	707	728
203	750	771	792	814	835	856	878	899	920	942
204	963	984	*006	*027	*048	*069	*091	*112	*133	*154
205	31 175	197	218	239	260	281	302	323	345	366
206	387	408	429	450	471	492	513	534	555	576
207	597	618	639	660	681	702	723	744	765	785
208	806	827	848	869	890	911	931	952	973	994
209	32 015	035	056	077	098	118	139	160	181	201
210	222	243	263	284	305	325	346	366	387	408
211	428	449	469	490	510	531	552	572	593	613
212	634	654	675	695	715	736	756	777	797	818
213	838	858	879	899	919	940	960	980	*001	*021
214	33 041	062	082	102	122	143	163	183	203	224
215	244	264	284	304	325	345	365	385	405	425
216	445	465	486	506	526	546	566	586	606	626
217	646	666	686	706	726	746	766	786	806	826
218	846	866	885	905	925	945	965	985	*005	*025
219	34 044	064	084	104	124	143	163	183	203	223
220	242	262	282	301	321	341	361	380	400	420
221	439	459	479	498	518	537	557	577	596	616
222	635	655	674	694	713	733	753	772	792	811
223	830	850	869	889	908	928	947	967	986	*005
224	35 025	044	064	083	102	122	141	160	180	199
225	218	238	257	276	295	315	334	353	372	392
226	411	430	449	468	488	507	526	545	564	583
227	603	622	641	660	679	698	717	736	755	774
228	793	813	832	851	870	889	908	927	946	965
229	984	*003	*021	*040	*059	*078	*097	*116	*135	*154
230	36 173	192	211	229	248	267	286	305	324	342
231	361	380	399	418	436	455	474	493	511	530
232	549	568	586	605	624	642	661	680	698	717
233	736	754	773	791	810	829	847	866	884	903
234	922	940	959	977	996	*014	*033	*051	*070	*088
235	37 107	125	144	162	181	199	218	236	254	273
236	291	310	328	346	365	383	401	420	438	457
237	475	493	511	530	548	566	585	603	621	639
238	658	676	694	712	731	749	767	785	803	822
239	840	858	876	894	912	931	949	967	985	*003
240	38 021	039	057	075	093	112	130	148	166	184
241	202	220	238	256	274	292	310	328	346	364
242	382	399	417	435	453	471	489	507	525	543
243	561	578	596	614	632	650	668	686	703	721
244	739	757	775	792	810	828	846	863	881	899
245	917	934	952	970	987	*005	*023	*041	*058	*076
246	39 094	111	129	146	164	182	199	217	235	252
247	270	287	305	322	340	358	375	393	410	428
248	445	463	480	498	515	533	550	568	585	602
249	620	637	655	672	690	707	724	742	759	777
250	794	811	829	846	863	881	898	915	933	950
N.	0	1	2	3	4	5	6	7	8	9

Proportional parts

	22	21
1	2.2	2.1
2	4.4	4.2
3	6.6	6.3
4	8.8	8.4
5	11.0	10.5
6	13.2	12.6
7	15.4	14.7
8	17.6	16.8
9	19.8	18.9

	20
1	2.0
2	4.0
3	6.0
4	8.0
5	10.0
6	12.0
7	14.0
8	16.0
9	18.0

	19
1	1.9
2	3.8
3	5.7
4	7.6
5	9.5
6	11.4
7	13.3
8	15.2
9	17.1

	18
1	1.8
2	3.6
3	5.4
4	7.2
5	9.0
6	10.8
7	12.6
8	14.4
9	16.2

	17
1	1.7
2	3.4
3	5.1
4	6.8
5	8.5
6	10.2
7	11.9
8	13.6
9	15.3

.30 103–.39 950

Table 1-7 (*Continued*)
COMMON LOGARITHMS OF NUMBERS
250–300

N.	0	1	2	3	4	5	6	7	8	9
250	39 794	811	829	846	863	881	898	915	933	950
251	967	985	*002	*019	*037	*054	*071	*088	*106	*123
252	40 140	157	175	192	209	226	243	261	278	295
253	312	329	346	364	381	398	415	432	449	466
254	483	500	518	535	552	569	586	603	620	637
255	654	671	688	705	722	739	756	773	790	807
256	824	841	858	875	892	909	926	943	960	976
257	993	*010	*027	*044	*061	*078	*095	*111	*128	*145
258	41 162	179	196	212	229	246	263	280	296	313
259	330	347	363	380	397	414	430	447	464	481
260	497	514	531	547	564	581	597	614	631	647
261	664	681	697	714	731	747	764	780	797	814
262	830	847	863	880	896	913	929	946	963	979
263	996	*012	*029	*045	*062	*078	*095	*111	*127	*144
264	42 160	177	193	210	226	243	259	275	292	308
265	325	341	357	374	390	406	423	439	455	472
266	488	504	521	537	553	570	586	602	619	635
267	651	667	684	700	716	732	749	765	781	797
268	813	830	846	862	878	894	911	927	943	959
269	975	991	*008	*024	*040	*056	*072	*088	*104	*120
270	43 136	152	169	185	201	217	233	249	265	281
271	297	313	329	345	361	377	393	409	425	441
272	457	473	489	505	521	537	553	569	584	600
273	616	632	648	664	680	696	712	727	743	759
274	775	791	807	823	838	854	870	886	902	917
275	933	949	965	981	996	*012	*028	*044	*059	*075
276	44 091	107	122	138	154	170	185	201	217	232
277	248	264	279	295	311	326	342	358	373	389
278	404	420	436	451	467	483	498	514	529	545
279	560	576	592	607	623	638	654	669	685	700
280	716	731	747	762	778	793	809	824	840	855
281	871	886	902	917	932	948	963	979	994	*010
282	45 025	040	056	071	086	102	117	133	148	163
283	179	194	209	225	240	255	271	286	301	317
284	332	347	362	378	393	408	423	439	454	469
285	484	500	515	530	545	561	576	591	606	621
286	637	652	667	682	697	712	728	743	758	773
287	788	803	818	834	849	864	879	894	909	924
288	939	954	969	984	*000	*015	*030	*045	*060	*075
289	46 090	105	120	135	150	165	180	195	210	225
290	240	255	270	285	300	315	330	345	359	374
291	389	404	419	434	449	464	479	494	509	523
292	538	553	568	583	598	613	627	642	657	672
293	687	702	716	731	746	761	776	790	805	820
294	835	850	864	879	894	909	923	938	953	967
295	982	997	*012	*026	*041	*056	*070	*085	*100	*114
296	47 129	144	159	173	188	202	217	232	246	261
297	276	290	305	319	334	349	363	378	392	407
298	422	436	451	465	480	494	509	524	538	553
299	567	582	596	611	625	640	654	669	683	698
300	712	727	741	756	770	784	799	813	828	842
N.	0	1	2	3	4	5	6	7	8	9

Proportional parts

18		17		16		15		14	
1	1.8	1	1.7	1	1.6	1	1.5	1	1.4
2	3.6	2	3.4	2	3.2	2	3.0	2	2.8
3	5.4	3	5.1	3	4.8	3	4.5	3	4.2
4	7.2	4	6.8	4	6.4	4	6.0	4	5.6
5	9.0	5	8.5	5	8.0	5	7.5	5	7.0
6	10.8	6	10.2	6	9.6	6	9.0	6	8.4
7	12.6	7	11.9	7	11.2	7	10.5	7	9.8
8	14.4	8	13.6	8	12.8	8	12.0	8	11.2
9	16.2	9	15.3	9	14.4	9	13.5	9	12.6

$\log e = 0.43429$

Proportional parts

.39 794–.47 842

Table 1-7 (*Continued*)
COMMON LOGARITHMS OF NUMBERS
300–350

N.	0	1	2	3	4	5	6	7	8	9
300	47 712	727	741	756	770	784	799	813	828	842
301	857	871	885	900	914	929	943	958	972	986
302	48 001	015	029	044	058	073	087	101	116	130
303	144	159	173	187	202	216	230	244	259	273
304	287	302	316	330	344	359	373	387	401	416
305	430	444	458	473	487	501	515	530	544	558
306	572	586	601	615	629	643	657	671	686	700
307	714	728	742	756	770	785	799	813	827	841
308	855	869	883	897	911	926	940	954	968	982
309	996	*010	*024	*038	*052	*066	*080	*094	*108	*122
310	49 136	150	164	178	192	206	220	234	248	262
311	276	290	304	318	332	346	360	374	388	402
312	415	429	443	457	471	485	499	513	527	541
313	554	568	582	596	610	624	638	651	665	679
314	693	707	721	734	748	762	776	790	803	817
315	831	845	859	872	886	900	914	927	941	955
316	969	982	996	*010	*024	*037	*051	*065	*079	*092
317	50 106	120	133	147	161	174	188	202	215	229
318	243	256	270	284	297	311	325	338	352	365
319	379	393	406	420	433	447	461	474	488	501
320	515	529	542	556	569	583	596	610	623	637
321	651	664	678	691	705	718	732	745	759	772
322	786	799	813	826	840	853	866	880	893	907
323	920	934	947	961	974	987	*001	*014	*028	*041
324	51 055	068	081	095	108	121	135	148	162	175
325	188	202	215	228	242	255	268	282	295	308
326	322	335	348	362	375	388	402	415	428	441
327	455	468	481	495	508	521	534	548	561	574
328	587	601	614	627	640	654	667	680	693	706
329	720	733	746	759	772	786	799	812	825	838
330	851	865	878	891	904	917	930	943	957	970
331	983	996	*009	*022	*035	*048	*061	*075	*088	*101
332	52 114	127	140	153	166	179	192	205	218	231
333	244	257	270	284	297	310	323	336	349	362
334	375	388	401	414	427	440	453	466	479	492
335	504	517	530	543	556	569	582	595	608	621
336	634	647	660	673	686	699	711	724	737	750
337	763	776	789	802	815	827	840	853	866	879
338	892	905	917	930	943	956	969	982	994	*007
339	53 020	033	046	058	071	084	097	110	122	135
340	148	161	173	186	199	212	224	237	250	263
341	275	288	301	314	326	339	352	364	377	390
342	403	415	428	441	453	466	479	491	504	517
343	529	542	555	567	580	593	605	618	631	643
344	656	668	681	694	706	719	732	744	757	769
345	782	794	807	820	832	845	857	870	882	895
346	908	920	933	945	958	970	983	995	*008	*020
347	54 033	045	058	070	083	095	108	120	133	145
348	158	170	183	195	208	220	233	245	258	270
349	283	295	307	320	332	345	357	370	382	394
350	407	419	432	444	456	469	481	494	506	518
N.	0	1	2	3	4	5	6	7	8	9

Proportional parts

15		14		13		12	
1	1.5	1	1.4	1	1.3	1	1.2
2	3.0	2	2.8	2	2.6	2	2.4
3	4.5	3	4.2	3	3.9	3	3.6
4	6.0	4	5.6	4	5.2	4	4.8
5	7.5	5	7.0	5	6.5	5	6.0
6	9.0	6	8.4	6	7.8	6	7.2
7	10.5	7	9.8	7	9.1	7	8.4
8	12.0	8	11.2	8	10.4	8	9.6
9	13.5	9	12.6	9	11.7	9	10.8

$\log \pi = 0.49715$

.47 712–.54 518

Table 1-7 (*Continued*)
COMMON LOGARITHMS OF NUMBERS
350–400

N.	0	1	2	3	4	5	6	7	8	9
350	54 407	419	432	444	456	469	481	494	506	518
351	531	543	555	568	580	593	605	617	630	642
352	654	667	679	691	704	716	728	741	753	765
353	777	790	802	814	827	839	851	864	876	888
354	900	913	925	937	949	962	974	986	998	*011
355	55 023	035	047	060	072	084	096	108	121	133
356	145	157	169	182	194	206	218	230	242	255
357	267	279	291	303	315	328	340	352	364	376
358	388	400	413	425	437	449	461	473	485	497
359	509	522	534	546	558	570	582	594	606	618
360	630	642	654	666	678	691	703	715	727	739
361	751	763	775	787	799	811	823	835	847	859
362	871	883	895	907	919	931	943	955	967	979
363	991	*003	*015	*027	*038	*050	*062	*074	*086	*098
364	56 110	122	134	146	158	170	182	194	205	217
365	229	241	253	265	277	289	301	312	324	336
366	348	360	372	384	396	407	419	431	443	455
367	467	478	490	502	514	526	538	549	561	573
368	585	597	608	620	632	644	656	667	679	691
369	703	714	726	738	750	761	773	785	797	808
370	820	832	844	855	867	879	891	902	914	926
371	937	949	961	972	984	996	*008	*019	*031	*043
372	57 054	066	078	089	101	113	124	136	148	159
373	171	183	194	206	217	229	241	252	264	276
374	287	299	310	322	334	345	357	368	380	392
375	403	415	426	438	449	461	473	484	496	507
376	519	530	542	553	565	576	588	600	611	623
377	634	646	657	669	680	692	703	715	726	738
378	749	761	772	784	795	807	818	830	841	852
379	864	875	887	898	910	921	933	944	955	967
380	978	990	*001	*013	*024	*035	*047	*058	*070	*081
381	58 092	104	115	127	138	149	161	172	184	195
382	206	218	229	240	252	263	274	286	297	309
383	320	331	343	354	365	377	388	399	410	422
384	433	444	456	467	478	490	501	512	524	535
385	546	557	569	580	591	602	614	625	636	647
386	659	670	681	692	704	715	726	737	749	760
387	771	782	794	805	816	827	838	850	861	872
388	883	894	906	917	928	939	950	961	973	984
389	995	*006	*017	*028	*040	*051	*062	*073	*084	*095
390	59 106	118	129	140	151	162	173	184	195	207
391	218	229	240	251	262	273	284	295	306	318
392	329	340	351	362	373	384	395	406	417	428
393	439	450	461	472	483	494	506	517	528	539
394	550	561	572	583	594	605	616	627	638	649
395	660	671	682	693	704	715	726	737	748	759
396	770	780	791	802	813	824	835	846	857	868
397	879	890	901	912	923	934	945	956	966	977
398	988	999	*010	*021	*032	*043	*054	*065	*076	*086
399	60 097	108	119	130	141	152	163	173	184	195
400	206	217	228	239	249	260	271	282	293	304
N.	0	1	2	3	4	5	6	7	8	9

Proportional parts

	13
1	1.3
2	2.6
3	3.9
4	5.2
5	6.5
6	7.8
7	9.1
8	10.4
9	11.7

	12
1	1.2
2	2.4
3	3.6
4	4.8
5	6.0
6	7.2
7	8.4
8	9.6
9	10.8

	11
1	1.1
2	2.2
3	3.3
4	4.4
5	5.5
6	6.6
7	7.7
8	8.8
9	9.9

	10
1	1.0
2	2.0
3	3.0
4	4.0
5	5.0
6	6.0
7	7.0
8	8.0
9	9.0

.54 407–.60 304

Table 1-7 (*Continued*)
COMMON LOGARITHMS OF NUMBERS
400–450

N.	0	1	2	3	4	5	6	7	8	9
400	60 206	217	228	239	249	260	271	282	293	304
401	314	325	336	347	358	369	379	390	401	412
402	423	433	444	455	466	477	487	498	509	520
403	531	541	552	563	574	584	595	606	617	627
404	638	649	660	670	681	692	703	713	724	735
405	746	756	767	778	788	799	810	821	831	842
406	853	863	874	885	895	906	917	927	938	949
407	959	970	981	991	*002	*013	*023	*034	*045	*055
408	61 066	077	087	098	109	119	130	140	151	162
409	172	183	194	204	215	225	236	247	257	268
410	278	289	300	310	321	331	342	352	363	374
411	384	395	405	416	426	437	448	458	469	479
412	490	500	511	521	532	542	553	563	574	584
413	595	606	616	627	637	648	658	669	679	690
414	700	711	721	731	742	752	763	773	784	794
415	805	815	826	836	847	857	868	878	888	899
416	909	920	930	941	951	962	972	982	993	*003
417	62 014	024	034	045	055	066	076	086	097	107
418	118	128	138	149	159	170	180	190	201	211
419	221	232	242	252	263	273	284	294	304	315
420	325	335	346	356	366	377	387	397	408	418
421	428	439	449	459	469	480	490	500	511	521
422	531	542	552	562	572	583	593	603	613	624
423	634	644	655	665	675	685	696	706	716	726
424	737	747	757	767	778	788	798	808	818	829
425	839	849	859	870	880	890	900	910	921	931
426	941	951	961	972	982	992	*002	*012	*022	*033
427	63 043	053	063	073	083	094	104	114	124	134
428	144	155	165	175	185	195	205	215	225	236
429	246	256	266	276	286	296	306	317	327	337
430	347	357	367	377	387	397	407	417	428	438
431	448	458	468	478	488	498	508	518	528	538
432	548	558	568	579	589	599	609	619	629	639
433	649	659	669	679	689	699	709	719	729	739
434	749	759	769	779	789	799	809	819	829	839
435	849	859	869	879	889	899	909	919	929	939
436	949	959	969	979	988	998	*008	*018	*028	*038
437	64 048	058	068	078	088	098	108	118	128	137
438	147	157	167	177	187	197	207	217	227	237
439	246	256	266	276	286	296	306	316	326	335
440	345	355	365	375	385	395	404	414	424	434
441	444	454	464	473	483	493	503	513	523	532
442	542	552	562	572	582	591	601	611	621	631
443	640	650	660	670	680	689	699	709	719	729
444	738	748	758	768	777	787	797	807	816	826
445	836	846	856	865	875	885	895	904	914	924
446	933	943	953	963	972	982	992	*002	*011	*021
447	65 031	040	050	060	070	079	089	099	108	118
448	128	137	147	157	167	176	186	196	205	215
449	225	234	244	254	263	273	283	292	302	312
450	321	331	341	350	360	369	379	389	398	408
N.	0	1	2	3	4	5	6	7	8	9

Proportional parts

	11		10		9
1	1.1	1	1.0	1	0.9
2	2.2	2	2.0	2	1.8
3	3.3	3	3.0	3	2.7
4	4.4	4	4.0	4	3.6
5	5.5	5	5.0	5	4.5
'6	6.6	6	6.0	6	5.4
7	7.7	7	7.0	7	6.3
8	8.8	8	8.0	8	7.2
9	9.9	9	9.0	9	8.1

.60 206–.65 408

Table 1-7 (*Continued*)
COMMON LOGARITHMS OF NUMBERS
450–500

N.	0	1	2	3	4	5	6	7	8	9
450	65 321	331	341	350	360	369	379	389	398	408
451	418	427	437	447	456	466	475	485	495	504
452	514	523	533	543	552	562	571	581	591	600
453	610	619	629	639	648	658	667	677	686	696
454	706	715	725	734	744	753	763	772	782	792
455	801	811	820	830	839	849	858	868	877	887
456	896	906	916	925	935	944	954	963	973	982
457	992	*001	*011	*020	*030	*039	*049	*058	*068	*077
458	66 087	096	106	115	124	134	143	153	162	172
459	181	191	200	210	219	229	238	247	257	266
460	276	285	295	304	314	323	332	342	351	361
461	370	380	389	398	408	417	427	436	445	455
462	464	474	483	492	502	511	521	530	539	549
463	558	567	577	586	596	605	614	624	633	642
464	652	661	671	680	689	699	708	717	727	736
465	745	755	764	773	783	792	801	811	820	829
466	839	848	857	867	876	885	894	904	913	922
467	932	941	950	960	969	978	987	997	*006	*015
468	67 025	034	043	052	062	071	080	089	099	108
469	117	127	136	145	154	164	173	182	191	201
470	210	219	228	237	247	256	265	274	284	293
471	302	311	321	330	339	348	357	367	376	385
472	394	403	413	422	431	440	449	459	468	477
473	486	495	504	514	523	532	541	550	560	569
474	578	587	596	605	614	624	633	642	651	660
475	669	679	688	697	706	715	724	733	742	752
476	761	770	779	788	797	806	815	825	834	843
477	852	861	870	879	888	897	906	916	925	934
478	943	952	961	970	979	988	997	*006	*015	*024
479	68 034	043	052	061	070	079	088	097	106	115
480	124	133	142	151	160	169	178	187	196	205
481	215	224	233	242	251	260	269	278	287	296
482	305	314	323	332	341	350	359	368	377	386
483	395	404	413	422	431	440	449	458	467	476
484	485	494	502	511	520	529	538	547	556	565
485	574	583	592	601	610	619	628	637	646	655
486	664	673	681	690	699	708	717	726	735	744
487	753	762	771	780	789	797	806	815	824	833
488	842	851	860	869	878	886	895	904	913	922
489	931	940	949	958	966	975	984	993	*002	*011
490	69 020	028	037	046	055	064	073	082	090	099
491	108	117	126	135	144	152	161	170	179	188
492	197	205	214	223	232	241	249	258	267	276
493	285	294	302	311	320	329	338	346	355	364
494	373	381	390	399	408	417	425	434	443	452
495	461	469	478	487	496	504	513	522	531	539
496	548	557	566	574	583	592	601	609	618	627
497	636	644	653	662	671	679	688	697	705	714
498	723	732	740	749	758	767	775	784	793	801
499	810	819	827	836	845	854	862	871	880	888
500	897	906	914	923	932	940	949	958	966	975
N.	0	1	2	3	4	5	6	7	8	9

Proportional parts

	10		9		8
1	1.0	1	0.9	1	0.8
2	2.0	2	1.8	2	1.6
3	3.0	3	2.7	3	2.4
4	4.0	4	3.6	4	3.2
5	5.0	5	4.5	5	4.0
6	6.0	6	5.4	6	4.8
7	7.0	7	6.3	7	5.6
8	8.0	8	7.2	8	6.4
9	9.0	9	8.1	9	7.2

.65 321–.69 975

Table 1-7 (*Continued*)
COMMON LOGARITHMS OF NUMBERS
500–550

N.	0	1	2	3	4	5	6	7	8	9
500	69 897	906	914	923	932	940	949	958	966	975
501	984	992	*001	*010	*018	*027	*036	*044	*053	*062
502	70 070	079	088	096	105	114	122	131	140	148
503	157	165	174	183	191	200	209	217	226	234
504	243	252	260	269	278	286	295	303	312	321
505	329	338	346	355	364	372	381	389	398	406
506	415	424	432	441	449	458	467	475	484	492
507	501	509	518	526	535	544	552	561	569	578
508	586	595	603	612	621	629	638	646	655	663
509	672	680	689	697	706	714	723	731	740	749
510	757	766	774	783	791	800	808	817	825	834
511	842	851	859	868	876	885	893	902	910	919
512	927	935	944	952	961	969	978	986	995	*003
513	71 012	020	029	037	046	054	063	071	079	088
514	096	105	113	122	130	139	147	155	164	172
515	181	189	198	206	214	223	231	240	248	257
516	265	273	282	290	299	307	315	324	332	341
517	349	357	366	374	383	391	399	408	416	425
518	433	441	450	458	466	475	483	492	500	508
519	517	525	533	542	550	559	567	575	584	592
520	600	609	617	625	634	642	650	659	667	675
521	684	692	700	709	717	725	734	742	750	759
522	767	775	784	792	800	809	817	825	834	842
523	850	858	867	875	883	892	900	908	917	925
524	933	941	950	958	966	975	983	991	999	*008
525	72 016	024	032	041	049	057	066	074	082	090
526	099	107	115	123	132	140	148	156	165	173
527	181	189	198	206	214	222	230	239	247	255
528	263	272	280	288	296	304	313	321	329	337
529	346	354	362	370	378	387	395	403	411	419
530	428	436	444	452	460	469	477	485	493	501
531	509	518	526	534	542	550	558	567	575	583
532	591	599	607	616	624	632	640	648	656	665
533	673	681	689	697	705	713	722	730	738	746
534	754	762	770	779	787	795	803	811	819	827
535	835	843	852	860	868	876	884	892	900	908
536	916	925	933	941	949	957	965	973	981	989
537	997	*006	*014	*022	*030	*038	*046	*054	*062	*070
538	73 078	086	094	102	111	119	127	135	143	151
539	159	167	175	183	191	199	207	215	223	231
540	239	247	255	263	272	280	288	296	304	312
541	320	328	336	344	352	360	368	376	384	392
542	400	408	416	424	432	440	448	456	464	472
543	480	488	496	504	512	520	528	536	544	552
544	560	568	576	584	592	600	608	616	624	632
545	640	648	656	664	672	679	687	695	703	711
546	719	727	735	743	751	759	767	775	783	791
547	799	807	815	823	830	838	846	854	862	870
548	878	886	894	902	910	918	926	933	941	949
549	957	965	973	981	989	997	*005	*013	*020	*028
550	74 036	044	052	060	068	076	084	092	099	107
N.	0	1	2	3	4	5	6	7	8	9

Proportional parts

	9
1	0.9
2	1.8
3	2.7
4	3.6
5	4.5
6	5.4
7	6.3
8	7.2
9	8.1

	8
1	0.8
2	1.6
3	2.4
4	3.2
5	4.0
6	4.8
7	5.6
8	6.4
9	7.2

	7
1	0.7
2	1.4
3	2.1
4	2.8
5	3.5
6	4.2
7	4.9
8	5.6
9	6.3

.69 897–.74 107

Table 1-7 (Continued)
COMMON LOGARITHMS OF NUMBERS
550-600

N.	0	1	2	3	4	5	6	7	8	9
550	74 036	044	052	060	068	076	084	092	099	107
551	115	123	131	139	147	155	162	170	178	186
552	194	202	210	218	225	233	241	249	257	265
553	273	280	288	296	304	312	320	327	335	343
554	351	359	367	374	382	390	398	406	414	421
555	429	437	445	453	461	468	476	484	492	500
556	507	515	523	531	539	547	554	562	570	578
557	586	593	601	609	617	624	632	640	648	656
558	663	671	679	687	695	702	710	718	726	733
559	741	749	757	764	772	780	788	796	803	811
560	819	827	834	842	850	858	865	873	881	889
561	896	904	912	920	927	935	943	950	958	966
562	974	981	989	997	*005	*012	*020	*028	*035	*043
563	75 051	059	066	074	082	089	097	105	113	120
564	128	136	143	151	159	166	174	182	189	197
565	205	213	220	228	236	243	251	259	266	274
566	282	289	297	305	312	320	328	335	343	351
567	358	366	374	381	389	397	404	412	420	427
568	435	442	450	458	465	473	481	488	496	504
569	511	519	526	534	542	549	557	565	572	580
570	587	595	603	610	618	626	633	641	648	656
571	664	671	679	686	694	702	709	717	724	732
572	740	747	755	762	770	778	785	793	800	808
573	815	823	831	838	846	853	861	868	876	884
574	891	899	906	914	921	929	937	944	952	959
575	967	974	982	989	997	*005	*012	*020	*027	*035
576	76 042	050	057	065	072	080	087	095	103	110
577	118	125	133	140	148	155	163	170	178	185
578	193	200	208	215	223	230	238	245	253	260
579	268	275	283	290	298	305	313	320	328	335
580	343	350	358	365	373	380	388	395	403	410
581	418	425	433	440	448	455	462	470	477	485
582	492	500	507	515	522	530	537	545	552	559
583	567	574	582	589	597	604	612	619	626	634
584	641	649	656	664	671	678	686	693	701	708
585	716	723	730	738	745	753	760	768	775	782
586	790	797	805	812	819	827	834	842	849	856
587	864	871	879	886	893	901	908	916	923	930
588	938	945	953	960	967	975	982	989	997	*004
589	77 012	019	026	034	041	048	056	063	070	078
590	085	093	100	107	115	122	129	137	144	151
591	159	166	173	181	188	195	203	210	217	225
592	232	240	247	254	262	269	276	283	291	298
593	305	313	320	327	335	342	349	357	364	371
594	379	386	393	401	408	415	422	430	437	444
595	452	459	466	474	481	488	495	503	510	517
596	525	532	539	546	554	561	568	576	583	590
597	597	605	612	619	627	634	641	648	656	663
598	670	677	685	692	699	706	714	721	728	735
599	743	750	757	764	772	779	786	793	801	808
600	815	822	830	837	844	851	859	866	873	880
N.	0	1	2	3	4	5	6	7	8	9

Proportional parts

8
1	0.8
2	1.6
3	2.4
4	3.2
5	4.0
6	4.8
7	5.6
8	6.4
9	7.2

7
1	0.7
2	1.4
3	2.1
4	2.8
5	3.5
6	4.2
7	4.9
8	5.6
9	6.3

.74 036-.77 880

Table 1-7 (*Continued*)
COMMON LOGARITHMS OF NUMBERS
600–650

N.	0	1	2	3	4	5	6	7	8	9	Proportional parts
600	77 815	822	830	837	844	851	859	866	873	880	
601	887	895	902	909	916	924	931	938	945	952	
602	960	967	974	981	988	996	*003	*010	*017	*025	
603	78 032	039	046	053	061	068	075	082	089	097	
604	104	111	118	125	132	140	147	154	161	168	
605	176	183	190	197	204	211	219	226	233	240	
606	247	254	262	269	276	283	290	297	305	312	
607	319	326	333	340	347	355	362	369	376	383	8
608	390	398	405	412	419	426	433	440	447	455	1 0.8
609	462	469	476	483	490	497	504	512	519	526	2 1.6
											3 2.4
610	533	540	547	554	561	569	576	583	590	597	4 3.2
611	604	611	618	625	633	640	647	654	661	668	5 4.0
612	675	682	689	696	704	711	718	725	732	739	6 4.8
613	746	753	760	767	774	781	789	796	803	810	7 5.6
614	817	824	831	838	845	852	859	866	873	880	8 6.4
											9 7.2
615	888	895	902	909	916	923	930	937	944	951	
616	958	965	972	979	986	993	*000	*007	*014	*021	
617	79 029	036	043	050	057	064	071	078	085	092	
618	099	106	113	120	127	134	141	148	155	162	
619	169	176	183	190	197	204	211	218	225	232	
620	239	246	253	260	267	274	281	288	295	302	
621	309	316	323	330	337	344	351	358	365	372	7
622	379	386	393	400	407	414	421	428	435	442	1 0.7
623	449	456	463	470	477	484	491	498	505	511	2 1.4
624	518	525	532	539	546	553	560	567	574	581	3 2.1
											4 2.8
625	588	595	602	609	616	623	630	637	644	650	5 3.5
626	657	664	671	678	685	692	699	706	713	720	6 4.2
627	727	734	741	748	754	761	768	775	782	789	7 4.9
628	796	803	810	817	824	831	837	844	851	858	8 5.6
629	865	872	879	886	893	900	906	913	920	927	9 6.3
630	934	941	948	955	962	969	975	982	989	996	
631	80 003	010	017	024	030	037	044	051	058	065	
632	072	079	085	092	099	106	113	120	127	134	
633	140	147	154	161	168	175	182	188	195	202	
634	209	216	223	229	236	243	250	257	264	271	
635	277	284	291	298	305	312	318	325	332	339	6
636	346	353	359	366	373	380	387	393	400	407	1 0.6
637	414	421	428	434	441	448	455	462	468	475	2 1.2
638	482	489	496	502	509	516	523	530	536	543	3 1.8
639	550	557	564	570	577	584	591	598	604	611	4 2.4
											5 3.0
640	618	625	632	638	645	652	659	665	672	679	6 3.6
641	686	693	699	706	713	720	726	733	740	747	7 4.2
642	754	760	767	774	781	787	794	801	808	814	8 4.8
643	821	828	835	841	848	855	862	868	875	882	9 5.4
644	889	895	902	909	916	922	929	936	943	949	
645	956	963	969	976	983	990	996	*003	*010	*017	
646	81 023	030	037	043	050	057	064	070	077	084	
647	090	097	104	111	117	124	131	137	144	151	
648	158	164	171	178	184	191	198	204	211	218	
649	224	231	238	245	251	258	265	271	278	285	
650	291	298	305	311	318	325	331	338	345	351	
N.	0	1	2	3	4	5	6	7	8	9	Proportional parts

.77 815–.81 351

Table 1-7 (*Continued*)
COMMON LOGARITHMS OF NUMBERS
650–700

N.	0	1	2	3	4	5	6	7	8	9	Proportional parts
650	81 291	298	305	311	318	325	331	338	345	351	
651	358	365	371	378	385	391	398	405	411	418	
652	425	431	438	445	451	458	465	471	478	485	
653	491	498	505	511	518	525	531	538	544	551	
654	558	564	571	578	584	591	598	604	611	617	
655	624	631	637	644	651	657	664	671	677	684	
656	690	697	704	710	717	723	730	737	743	750	
657	757	763	770	776	783	790	796	803	809	816	
658	823	829	836	842	849	856	862	869	875	882	
659	889	895	902	908	915	921	928	935	941	948	
660	954	961	968	974	981	987	994	*000	*007	*014	
661	82 020	027	033	040	046	053	060	066	073	079	
662	086	092	099	105	112	119	125	132	138	145	
663	151	158	164	171	178	184	191	197	204	210	
664	217	223	230	236	243	249	256	263	269	276	
665	282	289	295	302	308	315	321	328	334	341	
666	347	354	360	367	373	380	387	393	400	406	
667	413	419	426	432	439	445	452	458	465	471	
668	478	484	491	497	504	510	517	523	530	536	
669	543	549	556	562	569	575	582	588	595	601	
670	607	614	620	627	633	640	646	653	659	666	
671	672	679	685	692	698	705	711	718	724	730	
672	737	743	750	756	763	769	776	782	789	795	
673	802	808	814	821	827	834	840	847	853	860	
674	866	872	879	885	892	898	905	911	918	924	
675	930	937	943	950	956	963	969	975	982	988	
676	995	*001	*008	*014	*020	*027	*033	*040	*046	*052	
677	83 059	065	072	078	085	091	097	104	110	117	
678	123	129	136	142	149	155	161	168	174	181	
679	187	193	200	206	213	219	225	232	238	245	
680	251	257	264	270	276	283	289	296	302	308	
681	315	321	327	334	340	347	353	359	366	372	
682	378	385	391	398	404	410	417	423	429	436	
683	442	448	455	461	467	474	480	487	493	499	
684	506	512	518	525	531	537	544	550	556	563	
685	569	575	582	588	594	601	607	613	620	626	
686	632	639	645	651	658	664	670	677	683	689	
687	696	702	708	715	721	727	734	740	746	753	
688	759	765	771	778	784	790	797	803	809	816	
689	822	828	835	841	847	853	860	866	872	879	
690	885	891	897	904	910	916	923	929	935	942	
691	948	954	960	967	973	979	985	992	998	*004	
692	84 011	017	023	029	036	042	048	055	061	067	
693	073	080	086	092	098	105	111	117	123	130	
694	136	142	148	155	161	167	173	180	186	192	
695	198	205	211	217	223	230	236	242	248	255	
696	261	267	273	280	286	292	298	305	311	317	
697	323	330	336	342	348	354	361	367	373	379	
698	386	392	398	404	410	417	423	429	435	442	
699	448	454	460	466	473	479	485	491	497	504	
700	510	516	522	528	535	541	547	553	559	566	
N.	0	1	2	3	4	5	6	7	8	9	Proportional parts

Proportional parts:

	7
1	0.7
2	1.4
3	2.1
4	2.8
5	3.5
6	4.2
7	4.9
8	5.6
9	6 3

	6
1	0.6
2	1.2
3	1.8
4	2.4
5	3.0
6	3.6
7	4.2
8	4.8
9	5.4

.81 291–.84 566

Table 1-7 (*Continued*)
COMMON LOGARITHMS OF NUMBERS
700–750

N.	0	1	2	3	4	5	6	7	8	9	Proportional parts
700	84 510	516	522	528	535	541	547	553	559	566	
701	572	578	584	590	597	603	609	615	621	628	
702	634	640	646	652	658	665	671	677	683	689	
703	696	702	708	714	720	726	733	739	745	751	
704	757	763	770	776	782	788	794	800	807	813	
705	819	825	831	837	844	850	856	862	868	874	
706	880	887	893	899	905	911	917	924	930	936	
707	942	948	954	960	967	973	979	985	991	997	
708	85 003	009	016	022	028	034	040	046	052	058	
709	065	071	077	083	089	095	101	107	114	120	
710	126	132	138	144	150	156	163	169	175	181	
711	187	193	199	205	211	217	224	230	236	242	
712	248	254	260	266	272	278	285	291	297	303	
713	309	315	321	327	333	339	345	352	358	364	
714	370	376	382	388	394	400	406	412	418	425	
715	431	437	443	449	455	461	467	473	479	485	
716	491	497	503	509	516	522	528	534	540	546	
717	552	558	564	570	576	582	588	594	600	606	
718	612	618	625	631	637	643	649	655	661	667	
719	673	679	685	691	697	703	709	715	721	727	
720	733	739	745	751	757	763	769	775	781	788	
721	794	800	806	812	818	824	830	836	842	848	
722	854	860	866	872	878	884	890	896	902	908	
723	914	920	926	932	938	944	950	956	962	968	
724	974	980	986	992	998	*004	*010	*016	*022	*028	
725	86 034	040	046	052	058	064	070	076	082	088	
726	094	100	106	112	118	124	130	136	141	147	
727	153	159	165	171	177	183	189	195	201	207	
728	213	219	225	231	237	243	249	255	261	267	
729	273	279	285	291	297	303	308	314	320	326	
730	332	338	344	350	356	362	368	374	380	386	
731	392	398	404	410	415	421	427	433	439	445	
732	451	457	463	469	475	481	487	493	499	504	
733	510	516	522	528	534	540	546	552	558	564	
734	570	576	581	587	593	599	605	611	617	623	
735	629	635	641	646	652	658	664	670	676	682	
736	688	694	700	705	711	717	723	729	735	741	
737	747	753	759	764	770	776	782	788	794	800	
738	806	812	817	823	829	835	841	847	853	859	
739	864	870	876	882	888	894	900	906	911	917	
740	923	929	935	941	947	953	958	964	970	976	
741	982	988	994	999	*005	*011	*017	*023	*029	*035	
742	87 040	046	052	058	064	070	075	081	087	093	
743	099	105	111	116	122	128	134	140	146	151	
744	157	163	169	175	181	186	192	198	204	210	
745	216	221	227	233	239	245	251	256	262	268	
746	274	280	286	291	297	303	309	315	320	326	
747	332	338	344	349	355	361	367	373	379	384	
748	390	396	402	408	413	419	425	431	437	442	
749	448	454	460	466	471	477	483	489	495	500	
750	506	512	518	523	529	535	541	547	552	558	
N.	0	1	2	3	4	5	6	7	8	9	Proportional parts

Proportional parts:

7
1	0.7
2	1.4
3	2.1
4	2.8
5	3.5
6	4.2
7	4.9
8	5.6
9	6.3

6
1	0.6
2	1.2
3	1.8
4	2.4
5	3.0
6	3.6
7	4.2
8	4.8
9	5.4

5
1	0.5
2	1.0
3	1.5
4	2.0
5	2.5
6	3.0
7	3.5
8	4.0
9	4.5

.84 510–.87 558

Table 1-7 (*Continued*)
COMMON LOGARITHMS OF NUMBERS
750–800

N.	0	1	2	3	4	5	6	7	8	9	Proportional parts
750	87 506	512	518	523	529	535	541	547	552	558	
751	564	570	576	581	587	593	599	604	610	616	
752	622	628	633	639	645	651	656	662	668	674	
753	679	685	691	697	703	708	714	720	726	731	
754	737	743	749	754	760	766	772	777	783	789	
755	795	800	806	812	818	823	829	835	841	846	
756	852	858	864	869	875	881	887	892	898	904	
757	910	915	921	927	933	938	944	950	955	961	
758	967	973	978	984	990	996	*001	*007	*013	*018	
759	88 024	030	036	041	047	053	058	064	070	076	
760	081	087	093	098	104	110	116	121	127	133	**6**
761	138	144	150	156	161	167	173	178	184	190	1 0.6
762	195	201	207	213	218	224	230	235	241	247	2 1.2
763	252	258	264	270	275	281	287	292	298	304	3 1.8
764	309	315	321	326	332	338	343	349	355	360	4 2.4
765	366	372	377	383	389	395	400	406	412	417	5 3.0
766	423	429	434	440	446	451	457	463	468	474	6 3.6
767	480	485	491	497	502	508	513	519	525	530	7 4.2
768	536	542	547	553	559	564	570	576	581	587	8 4.8
769	593	598	604	610	615	621	627	632	638	643	9 5.4
770	649	655	660	666	672	677	683	689	694	700	
771	705	711	717	722	728	734	739	745	750	756	
772	762	767	773	779	784	790	795	801	807	812	
773	818	824	829	835	840	846	852	857	863	868	
774	874	880	885	891	897	902	908	913	919	925	
775	930	936	941	947	953	958	964	969	975	981	
776	986	992	997	*003	*009	*014	*020	*025	*031	*037	
777	89 042	048	053	059	064	070	076	081	087	092	
778	098	104	109	115	120	126	131	137	143	148	
779	154	159	165	170	176	182	187	193	198	204	
780	209	215	221	226	232	237	243	248	254	260	**5**
781	265	271	276	282	287	293	298	304	310	315	1 0.5
782	321	326	332	337	343	348	354	360	365	371	2 1.0
783	376	382	387	393	398	404	409	415	421	426	3 1.5
784	432	437	443	448	454	459	465	470	476	481	4 2.0
785	487	492	498	504	509	515	520	526	531	537	5 2.5
786	542	548	553	559	564	570	575	581	586	592	6 3.0
787	597	603	609	614	620	625	631	636	642	647	7 3.5
788	653	658	664	669	675	680	686	691	697	702	8 4.0
789	708	713	719	724	730	735	741	746	752	757	9 4.5
790	763	768	774	779	785	790	796	801	807	812	
791	818	823	829	834	840	845	851	856	862	867	
792	873	878	883	889	894	900	905	911	916	922	
793	927	933	938	944	949	955	960	966	971	977	
794	982	988	993	998	*004	*009	*015	*020	*026	*031	
795	90 037	042	048	053	059	064	069	075	080	086	
796	091	097	102	108	113	119	124	129	135	140	
797	146	151	157	162	168	173	179	184	189	195	
798	200	206	211	217	222	227	233	238	244	249	
799	255	260	266	271	276	282	287	293	298	304	
800	309	314	320	325	331	336	342	347	352	358	
N.	0	1	2	3	4	5	6	7	8	9	Proportional parts

.87 506–.90 358

Table 1-7 (*Continued*)
COMMON LOGARITHMS OF NUMBERS
800–850

N.	0	1	2	3	4	5	6	7	8	9	Proportional parts
800	90 309	314	320	325	331	336	342	347	352	358	
801	363	369	374	380	385	390	396	401	407	412	
802	417	423	428	434	439	445	450	455	461	466	
803	472	477	482	488	493	499	504	509	515	520	
804	526	531	536	542	547	553	558	563	569	574	
805	580	585	590	596	601	607	612	617	623	628	
806	634	639	644	650	655	660	666	671	677	682	
807	687	693	698	703	709	714	720	725	730	736	
808	741	747	752	757	763	768	773	779	784	789	
809	795	800	806	811	816	822	827	832	838	843	
810	849	854	859	865	870	875	881	886	891	897	6
811	902	907	913	918	924	929	934	940	945	950	1 0.6
812	956	961	966	972	977	982	988	993	998	*004	2 1.2
813	91 009	014	020	025	030	036	041	046	052	057	3 1.8
814	062	068	073	078	084	089	094	100	105	110	4 2.4 5 3.0
815	116	121	126	132	137	142	148	153	158	164	6 3.6
816	169	174	180	185	190	196	201	206	212	217	7 4.2
817	222	228	233	238	243	249	254	259	265	270	8 4.8
818	275	281	286	291	297	302	307	312	318	323	9 5.4
819	328	334	339	344	350	355	360	365	371	376	
820	381	387	392	397	403	408	413	418	424	429	
821	434	440	445	450	455	461	466	471	477	482	
822	487	492	498	503	508	514	519	524	529	535	
823	540	545	551	556	561	566	572	577	582	587	
824	593	598	603	609	614	619	624	630	635	640	
825	645	651	656	661	666	672	677	682	687	693	
826	698	703	709	714	719	724	730	735	740	745	
827	751	756	761	766	772	777	782	787	793	798	
828	803	808	814	819	824	829	834	840	845	850	
829	855	861	866	871	876	882	887	892	897	903	
830	908	913	918	924	929	934	939	944	950	955	5
831	960	965	971	976	981	986	991	997	*002	*007	1 0.5
832	92 012	018	023	028	033	038	044	049	054	059	2 1.0
833	065	070	075	080	085	091	096	101	106	111	3 1.5
834	117	122	127	132	137	143	148	153	158	163	4 2.0 5 2.5
835	169	174	179	184	189	195	200	205	210	215	6 3.0
836	221	226	231	236	241	247	252	257	262	267	7 3.5
837	273	278	283	288	293	298	304	309	314	319	8 4.0
838	324	330	335	340	345	350	355	361	366	371	9 4.5
839	376	381	387	392	397	402	407	412	418	423	
840	428	433	438	443	449	454	459	464	469	474	
841	480	485	490	495	500	505	511	516	521	526	
842	531	536	542	547	552	557	562	567	572	578	
843	583	588	593	598	603	609	614	619	624	629	
844	634	639	645	650	655	660	665	670	675	681	
845	686	691	696	701	706	711	716	722	727	732	
846	737	742	747	752	758	763	768	773	778	783	
847	788	793	799	804	809	814	819	824	829	834	
848	840	845	850	855	860	865	870	875	881	886	
849	891	896	901	906	911	916	921	927	932	937	
850	942	947	952	957	962	967	973	978	983	988	
N.	0	1	2	3	4	5	6	7	8	9	Proportional parts

.90 309–.92 988

Table 1-7 (*Continued*)
COMMON LOGARITHMS OF NUMBERS
850–900

N.	0	1	2	3	4	5	6	7	8	9	Proportional parts
850	92 942	947	952	957	962	967	973	978	983	988	
851	993	998	*003	*008	*013	*018	*024	*029	*034	*039	
852	93 044	049	054	059	064	069	075	080	085	090	
853	095	100	105	110	115	120	125	131	136	141	
854	146	151	156	161	166	171	176	181	186	192	
855	197	202	207	212	217	222	227	232	237	242	
856	247	252	258	263	268	273	278	283	288	293	
857	298	303	308	313	318	323	328	334	339	344	
858	349	354	359	364	369	374	379	384	389	394	6
859	399	404	409	414	420	425	430	435	440	445	1 0.6
860	450	455	460	465	470	475	480	485	490	495	2 1.2
861	500	505	510	515	520	526	531	536	541	546	3 1.8
862	551	556	561	566	571	576	581	586	591	596	4 2.4
863	601	606	611	616	621	626	631	636	641	646	5 3.0
864	651	656	661	666	671	676	682	687	692	697	6 3.6
865	702	707	712	717	722	727	732	737	742	747	7 4.2
866	752	757	762	767	772	777	782	787	792	797	8 4.8
867	802	807	812	817	822	827	832	837	842	847	9 5.4
868	852	857	862	867	872	877	882	887	892	897	
869	902	907	912	917	922	927	932	937	942	947	
870	94 952	957	962	967	972	977	982	987	992	997	
871	94 002	007	012	017	022	027	032	037	042	047	5
872	052	057	062	067	072	077	082	086	091	096	1 0.5
873	101	106	111	116	121	126	131	136	141	146	2 1.0
874	151	156	161	166	171	176	181	186	191	196	3 1.5
875	201	206	211	216	221	226	231	236	240	245	4 2.0
876	250	255	260	265	270	275	280	285	290	295	5 2.5
877	300	305	310	315	320	325	330	335	340	345	6 3.0
878	349	354	359	364	369	374	379	384	389	394	7 3.5
879	399	404	409	414	419	424	429	433	438	443	8 4.0
880	448	453	458	463	468	473	478	483	488	493	9 4.5
881	498	503	507	512	517	522	527	532	537	542	
882	547	552	557	562	567	571	576	581	586	591	
883	596	601	606	611	616	621	626	630	635	640	
884	645	650	655	660	665	670	675	680	685	689	
885	694	699	704	709	714	719	724	729	734	738	4
886	743	748	753	758	763	768	773	778	783	787	1 0.4
887	792	797	802	807	812	817	822	827	832	836	2 0.8
888	841	846	851	856	861	866	871	876	880	885	3 1.2
889	890	895	900	905	910	915	919	924	929	934	4 1.6
890	939	944	949	954	959	963	968	973	978	983	5 2.0
891	988	993	998	*002	*007	*012	*017	*022	*027	*032	6 2.4
892	95 036	041	046	051	056	061	066	071	075	080	7 2.8
893	085	090	095	100	105	109	114	119	124	129	8 3.2
894	134	139	143	148	153	158	163	168	173	177	9 3.6
895	182	187	192	197	202	207	211	216	221	226	
896	231	236	240	245	250	255	260	265	270	274	
897	279	284	289	294	299	303	308	313	318	323	
898	328	332	337	342	347	352	357	361	366	371	
899	376	381	386	390	395	400	405	410	415	419	
900	424	429	434	439	444	448	453	458	463	468	
N.	0	1	2	3	4	5	6	7	8	9	Proportional parts

.92 942–.95 468

Table 1-7 (*Continued*)
COMMON LOGARITHMS OF NUMBERS
900-950

N.	0	1	2	3	4	5	6	7	8	9	Proportional parts
900	95 424	429	434	439	444	448	453	458	463	468	
901	472	477	482	487	492	497	501	506	511	516	
902	521	525	530	535	540	545	550	554	559	564	
903	569	574	578	583	588	593	598	602	607	612	
904	617	622	626	631	636	641	646	650	655	660	
905	665	670	674	679	684	689	694	698	703	708	
906	713	718	722	727	732	737	742	746	751	756	
907	761	766	770	775	780	785	789	794	799	804	
908	809	813	818	823	828	832	837	842	847	852	
909	856	861	866	871	875	880	885	890	895	899	
910	904	909	914	918	923	928	933	938	942	947	
911	952	957	961	966	971	976	980	985	990	995	
912	999	*004	*009	*014	*019	*023	*028	*033	*038	*042	
913	96 047	052	057	061	066	071	076	080	085	090	
914	095	099	104	109	114	118	123	128	133	137	
915	142	147	152	156	161	166	171	175	180	185	
916	190	194	199	204	209	213	218	223	227	232	
917	237	242	246	251	256	261	265	270	275	280	
918	284	289	294	298	303	308	313	317	322	327	
919	332	336	341	346	350	355	360	365	369	374	
920	379	384	388	393	398	402	407	412	417	421	
921	426	431	435	440	445	450	454	459	464	468	
922	473	478	483	487	492	497	501	506	511	515	
923	520	525	530	534	539	544	548	553	558	562	
924	567	572	577	581	586	591	595	600	605	609	
925	614	619	624	628	633	638	642	647	652	656	
926	661	666	670	675	680	685	689	694	699	703	
927	708	713	717	722	727	731	736	741	745	750	
928	755	759	764	769	774	778	783	788	792	797	
929	802	806	811	816	820	825	830	834	839	844	
930	848	853	858	862	867	872	876	881	886	890	
931	895	900	904	909	914	918	923	928	932	937	
932	942	946	951	956	960	965	970	974	979	984	
933	988	993	997	*002	*007	*011	*016	*021	*025	*030	
934	97 035	039	044	049	053	058	063	067	072	077	
935	081	086	090	095	100	104	109	114	118	123	
936	128	132	137	142	146	151	155	160	165	169	
937	174	179	183	188	192	197	202	206	211	216	
938	220	225	230	234	239	243	248	253	257	262	
939	267	271	276	280	285	290	294	299	304	308	
940	313	317	322	327	331	336	340	345	350	354	
941	359	364	368	373	377	382	387	391	396	400	
942	405	410	414	419	424	428	433	437	442	447	
943	451	456	460	465	470	474	479	483	488	493	
944	497	502	506	511	516	520	525	529	534	539	
945	543	548	552	557	562	566	571	575	580	585	
946	589	594	598	603	607	612	617	621	626	630	
947	635	640	644	649	653	658	663	667	672	676	
948	681	685	690	695	699	704	708	713	717	722	
949	727	731	736	740	745	749	754	759	763	768	
950	772	777	782	786	791	795	800	804	809	813	
N.	0	1	2	3	4	5	6	7	8	9	Proportional parts

Proportional parts:

	5
1	0.5
2	1.0
3	1.5
4	2.0
5	2.5
6	3.0
7	3.5
8	4.0
9	4.5

	4
1	0.4
2	0.8
3	1.2
4	1.6
5	2.0
6	2.4
7	2.8
8	3.2
9	3.6

.95 424–.97 813

Table 1-7 (*Continued*)
COMMON LOGARITHMS OF NUMBERS
950–1000

N.	0	1	2	3	4	5	6	7	8	9	Proportional parts
950	97 772	777	782	786	791	795	800	804	809	813	
951	818	823	827	832	836	841	845	850	855	859	
952	864	868	873	877	882	886	891	896	900	905	
953	909	914	918	923	928	932	937	941	946	950	
954	955	959	964	968	973	978	982	987	991	996	
955	98 000	005	009	014	019	023	028	032	037	041	
956	046	050	055	059	064	068	073	078	082	087	
957	091	096	100	105	109	114	118	123	127	132	
958	137	141	146	150	155	159	164	168	173	177	
959	182	186	191	195	200	204	209	214	218	223	
960	227	232	236	241	245	250	254	259	263	268	
961	272	277	281	286	290	295	299	304	308	313	
962	318	322	327	331	336	340	345	349	354	358	
963	363	367	372	376	381	385	390	394	399	403	
964	408	412	417	421	426	430	435	439	444	448	
965	453	457	462	466	471	475	480	484	489	493	
966	498	502	507	511	516	520	525	529	534	538	
967	543	547	552	556	561	565	570	574	579	583	
968	588	592	597	601	605	610	614	619	623	628	
969	632	637	641	646	650	655	659	664	668	673	
970	677	682	686	691	695	700	704	709	713	717	
971	722	726	731	735	740	744	749	753	758	762	
972	767	771	776	780	784	789	793	798	802	807	
973	811	816	820	825	829	834	838	843	847	851	
974	856	860	865	869	874	878	883	887	892	896	
975	900	905	909	914	918	923	927	932	936	941	
976	945	949	954	958	963	967	972	976	981	985	
977	989	994	998	*003	*007	*012	*016	*021	*025	*029	
978	99 034	038	043	047	052	056	061	065	069	074	
979	078	083	087	092	096	100	105	109	114	118	
980	123	127	131	136	140	145	149	154	158	162	
981	167	171	176	180	185	189	193	198	202	207	
982	211	216	220	224	229	233	238	242	247	251	
983	255	260	264	269	273	277	282	286	291	295	
984	300	304	308	313	317	322	326	330	335	339	
985	344	348	352	357	361	366	370	374	379	383	
986	388	392	396	401	405	410	414	419	423	427	
987	432	436	441	445	449	454	458	463	467	471	
988	476	480	484	489	493	498	502	506	511	515	
989	520	524	528	533	537	542	546	550	555	559	
990	564	568	572	577	581	585	590	594	599	603	
991	607	612	616	621	625	629	634	638	642	647	
992	651	656	660	664	669	673	677	682	686	691	
993	695	699	704	708	712	717	721	726	730	734	
994	739	743	747	752	756	760	765	769	774	778	
995	782	787	791	795	800	804	808	813	817	822	
996	826	830	835	839	843	848	852	856	861	865	
997	870	874	878	883	887	891	896	900	904	909	
998	913	917	922	926	930	935	939	944	948	952	
999	957	961	965	970	974	978	983	987	991	996	
1000	00 000	004	009	013	017	022	026	030	035	039	
N.	0	1	2	3	4	5	6	7	8	9	Proportional parts

Proportional parts

5	
1	0.5
2	1.0
3	1.5
4	2.0
5	2.5
6	3.0
7	3.5
8	4.0
9	4.5

4	
1	0.4
2	0.8
3	1.2
4	1.6
5	2.0
6	2.4
7	2.8
8	3.2
9	3.6

.97 772–.99 996

Table 1-7 (*Continued*)
COMMON LOGARITHMS OF NUMBERS
1000–1050

N.	0	1	2	3	4	5	6	7	8	9	d.
1000	000 0000	0434	0869	1303	1737	2171	2605	3039	3473	3907	434
1001	4341	4775	5208	5642	6076	6510	6943	7377	7810	8244	434
1002	8677	9111	9544	9977	*0411	*0844	*1277	*1710	*2143	*2576	433
1003	001 3009	3442	3875	4308	4741	5174	5607	6039	6472	6905	433
1004	7337	7770	8202	8635	9067	9499	9932	*0364	*0796	*1228	432
1005	002 1661	2093	2525	2957	3389	3821	4253	4685	5116	5548	432
1006	5980	6411	6843	7275	7706	8138	8569	9001	9432	9863	431
1007	003 0295	0726	1157	1588	2019	2451	2882	3313	3744	4174	431
1008	4605	5036	5467	5898	6328	6759	7190	7620	8051	8481	431
1009	8912	9342	9772	*0203	*0633	*1063	*1493	*1924	*2354	*2784	430
1010	004 3214	3644	4074	4504	4933	5363	5793	6223	6652	7082	430
1011	7512	7941	8371	8800	9229	9659	*0088	*0517	*0947	*1376	429
1012	005 1805	2234	2663	3092	3521	3950	4379	4808	5237	5666	429
1013	6094	6523	6952	7380	7809	8238	8666	9094	9523	9951	429
1014	006 0380	0808	1236	1664	2092	2521	2949	3377	3805	4233	428
1015	4660	5088	5516	5944	6372	6799	7227	7655	8082	8510	428
1016	8937	9365	9792	*0219	*0647	*1074	*1501	*1928	*2355	*2782	427
1017	007 3210	3637	4064	4490	4917	5344	5771	6198	6624	7051	427
1018	7478	7904	8331	8757	9184	9610	*0037	*0463	*0889	*1316	426
1019	008 1742	2168	2594	3020	3446	3872	4298	4724	5150	5576	426
1020	6002	6427	6853	7279	7704	8130	8556	8981	9407	9832	426
1021	009 0257	0683	1108	1533	1959	2384	2809	3234	3659	4084	425
1022	4509	4934	5359	5784	6208	6633	7058	7483	7907	8332	425
1023	8756	9181	9605	*0030	*0454	*0878	*1303	*1727	*2151	*2575	424
1024	010 3000	3424	3848	4272	4696	5120	5544	5967	6391	6815	424
1025	7239	7662	8086	8510	8933	9357	9780	*0204	*0627	*1050	424
1026	011 1474	1897	2320	2743	3166	3590	4013	4436	4859	5282	423
1027	5704	6127	6550	6973	7396	7818	8241	8664	9086	9509	423
1028	9931	*0354	*0776	*1198	*1621	*2043	*2465	*2887	*3310	*3732	422
1029	012 4154	4576	4998	5420	5842	6264	6685	7107	7529	7951	422
1030	8372	8794	9215	9637	*0059	*0480	*0901	*1323	*1744	*2165	422
1031	013 2587	3008	3429	3850	4271	4692	5113	5534	5955	6376	421
1032	6797	7218	7639	8059	8480	8901	9321	9742	*0162	*0583	421
1033	014 1003	1424	1844	2264	2685	3105	3525	3945	4365	4785	420
1034	5205	5625	6045	6465	6885	7305	7725	8144	8564	8984	420
1035	9403	9823	*0243	*0662	*1082	*1501	*1920	*2340	*2759	*3178	420
1036	015 3598	4017	4436	4855	5274	5693	6112	6531	6950	7369	419
1037	7788	8206	8625	9044	9462	9881	*0300	*0718	*1137	*1555	419
1038	016 1974	2392	2810	3229	3647	4065	4483	4901	5319	5737	418
1039	6155	6573	6991	7409	7827	8245	8663	9080	9498	9916	418
1040	017 0333	0751	1168	1586	2003	2421	2838	3256	3673	4090	417
1041	4507	4924	5342	5759	6176	6593	7010	7427	7844	8260	417
1042	8677	9094	9511	9927	*0344	*0761	*1177	*1594	*2010	*2427	417
1043	018 2843	3259	3676	4092	4508	4925	5341	5757	6173	6589	416
1044	7005	7421	7837	8253	8669	9084	9500	9916	*0332	*0747	416
1045	019 1163	1578	1994	2410	2825	3240	3656	4071	4486	4902	415
1046	5317	5732	6147	6562	6977	7392	7807	8222	8637	9052	415
1047	9467	9882	*0296	*0711	*1126	*1540	*1955	*2369	*2784	*3198	415
1048	020 3613	4027	4442	4856	5270	5684	6099	6513	6927	7341	414
1049	7755	8169	8583	8997	9411	9824	*0238	*0652	*1066	*1479	414
1050	021 1893	2307	2720	3134	3547	3961	4374	4787	5201	5614	413
N.	0	1	2	3	4	5	6	7	8	9	d.

.000 0000–.021 5614

Table 1-7 (*Continued*)
COMMON LOGARITHMS OF NUMBERS
1050–1100

N.	0	1	2	3	4	5	6	7	8	9	d.
1050	021 1893	2307	2720	3134	3547	3961	4374	4787	5201	5614	413
1051	6027	6440	6854	7267	7680	8093	8506	8919	9332	9745	413
1052	022 0157	0570	0983	1396	1808	2221	2634	3046	3459	3871	413
1053	4284	4696	5109	5521	5933	6345	6758	7170	7582	7994	412
1054	8406	8818	9230	9642	*0054	*0466	*0878	*1289	*1701	*2113	412
1055	023 2525	2936	3348	3759	4171	4582	4994	5405	5817	6228	411
1056	6639	7050	7462	7873	8284	8695	9106	9517	9928	*0339	411
1057	024 0750	1161	1572	1982	2393	2804	3214	3625	4036	4446	411
1058	4857	5267	5678	6088	6498	6909	7319	7729	8139	8549	410
1059	8960	9370	9780	*0190	*0600	*1010	*1419	*1829	*2239	*2649	410
1060	025 3059	3468	3878	4288	4697	5107	5516	5926	6335	6744	410
1061	7154	7563	7972	8382	8791	9200	9609	*0018	*0427	*0836	409
1062	026 1245	1654	2063	2472	2881	3289	3698	4107	4515	4924	409
1063	5333	5741	6150	6558	6967	7375	7783	8192	8600	9008	408
1064	9416	9824	*0233	*0641	*1049	*1457	*1865	*2273	*2680	*3088	408
1065	027 3496	3904	4312	4719	5127	5535	5942	6350	6757	7165	408
1066	7572	7979	8387	8794	9201	9609	*0016	*0423	*0830	*1237	407
1067	028 1644	2051	2458	2865	3272	3679	4086	4492	4899	5306	407
1068	5713	6119	6526	6932	7339	7745	8152	8558	8964	9371	406
1069	9777	*0183	*0590	*0996	*1402	*1808	*2214	*2620	*3026	*3432	406
1070	029 3838	4244	4649	5055	5461	5867	6272	6678	7084	7489	406
1071	7895	8300	8706	9111	9516	9922	*0327	*0732	*1138	*1543	405
1072	030 1948	2353	2758	3163	3568	3973	4378	4783	5188	5592	405
1073	5997	6402	6807	7211	7616	8020	8425	8830	9234	9638	405
1074	031 0043	0447	0851	1256	1660	2064	2468	2872	3277	3681	404
1075	4085	4489	4893	5296	5700	6104	6508	6912	7315	7719	404
1076	8123	8526	8930	9333	9737	*0140	*0544	*0947	*1350	*1754	403
1077	032 2157	2560	2963	3367	3770	4173	4576	4979	5382	5785	403
1078	6188	6590	6993	7396	7799	8201	8604	9007	9409	9812	403
1079	033 0214	0617	1019	1422	1824	2226	2629	3031	3433	3835	402
1080	4238	4640	5042	5444	5846	6248	6650	7052	7453	7855	402
1081	8257	8659	9060	9462	9864	*0265	*0667	*1068	*1470	*1871	402
1082	034 2273	2674	3075	3477	3878	4279	4680	5081	5482	5884	401
1083	6285	6686	7087	7487	7888	8289	8690	9091	9491	9892	401
1084	035 0293	0693	1094	1495	1895	2296	2696	3096	3497	3897	400
1085	4297	4698	5098	5498	5898	6298	6698	7098	7498	7898	400
1086	8298	8698	9098	9498	9898	*0297	*0697	*1097	*1496	*1896	400
1087	036 2295	2695	3094	3494	3893	4293	4692	5091	5491	5890	399
1088	6289	6688	7087	7486	7885	8284	8683	9082	9481	9880	399
1089	037 0279	0678	1076	1475	1874	2272	2671	3070	3468	3867	399
1090	4265	4663	5062	5460	5858	6257	6655	7053	7451	7849	398
1091	8248	8646	9044	9442	9839	*0237	*0635	*1033	*1431	*1829	398
1092	038 2226	2624	3022	3419	3817	4214	4612	5009	5407	5804	398
1093	6202	6599	6996	7393	7791	8188	8585	8982	9379	9776	397
1094	039 0173	0570	0967	1364	1761	2158	2554	2951	3348	3745	397
1095	4141	4538	4934	5331	5727	6124	6520	6917	7313	7709	397
1096	8106	8502	8898	9294	9690	*0086	*0482	*0878	*1274	*1670	396
1097	040 2066	2462	2858	3254	3650	4045	4441	4837	5232	5628	396
1098	6023	6419	6814	7210	7605	8001	8396	8791	9187	9582	395
1099	9977	*0372	*0767	*1162	*1557	*1952	*2347	*2742	*3137	*3532	395
1100	041 3927	4322	4716	5111	5506	5900	6295	6690	7084	7479	395
N.	0	1	2	3	4	5	6	7	8	9	d.

.021 1893–.041 7479

Table 1-7 (*Continued*)
COMMON LOGARITHMS OF NUMBERS
1100–1150

N.	0	1	2	3	4	5	6	7	8	9	d.
1100	041 3927	4322	4716	5111	5506	5900	6295	6690	7084	7479	395
1101	7873	8268	8662	9056	9451	9845	*0239	*0633	*1028	*1422	394
1102	042 1816	2210	2604	2998	3392	3786	4180	4574	4968	5361	394
1103	5755	6149	6543	6936	7330	7723	8117	8510	8904	9297	394
1104	9691	*0004	*0477	*0871	*1264	*1657	*2050	*2444	*2837	*3230	393
1105	043 3623	4016	4409	4802	5195	5587	5980	6373	6766	7159	393
1106	7551	7944	8337	8729	9122	9514	9907	*0299	*0692	*1084	393
1107	044 1476	1869	2261	2653	3045	3437	3829	4222	4614	5006	392
1108	5398	5790	6181	6573	6965	7357	7749	8140	8532	8924	392
1109	9315	9707	*0099	*0490	*0882	*1273	*1664	*2056	*2447	*2839	392
1110	045 3230	3621	4012	4403	4795	5186	5577	5968	6359	6750	391
1111	7141	7531	7922	8313	8704	9095	9485	9876	*0267	*0657	391
1112	046 1048	1438	1829	2219	2610	3000	3391	3781	4171	4561	390
1113	4952	5342	5732	6122	6512	6902	7292	7682	8072	8462	390
1114	8852	9242	9632	*0021	*0411	*0801	*1190	*1580	*1970	*2359	390
1115	047 2749	3138	3528	3917	4306	4696	5085	5474	5864	6253	389
1116	6642	7031	7420	7809	8198	8587	8976	9365	9754	*0143	389
1117	048 0532	0921	1309	1698	2087	2475	2864	3253	3641	4030	389
1118	4418	4806	5195	5583	5972	6360	6748	7136	7525	7913	388
1119	8301	8689	9077	9465	9853	*0241	*0629	*1017	*1405	*1792	388
1120	049 2180	2568	2956	3343	3731	4119	4506	4894	5281	5669	388
1121	6056	6444	6831	7218	7606	7993	8380	8767	9154	9541	387
1122	9929	*0316	*0703	*1090	*1477	*1863	*2250	*2637	*3024	*3411	387
1123	050 3798	4184	4571	4958	5344	5731	6117	6504	6890	7277	387
1124	7663	8049	8436	8822	9208	9595	9981	*0367	*0753	*1139	386
1125	051 1525	1911	2297	2683	3069	3455	3841	4227	4612	4998	386
1126	5384	5770	6155	6541	6926	7312	7697	8083	8468	8854	386
1127	9239	9624	*0010	*0395	*0780	*1166	*1551	*1936	*2321	*2706	385
1128	052 3091	3476	3861	4246	4631	5016	5400	5785	6170	6555	385
1129	6939	7324	7709	8093	8478	8862	9247	9631	*0016	*0400	385
1130	053 0784	1169	1553	1937	2321	2706	3090	3474	3858	4242	384
1131	4626	5010	5394	5778	6162	6546	6929	7313	7697	8081	384
1132	8464	8848	9232	9615	9999	*0382	*0766	*1149	*1532	*1916	384
1133	054 2299	2682	3066	3449	3832	4215	4598	4981	5365	5748	383
1134	6131	6514	6896	7279	7662	8045	8428	8811	9193	9576	383
1135	9959	*0341	*0724	*1106	*1489	*1871	*2254	*2636	*3019	*3401	382
1136	055 3783	4166	4548	4930	5312	5694	6077	6459	6841	7223	382
1137	7605	7987	8369	8750	9132	9514	9896	*0278	*0659	*1041	382
1138	056 1423	1804	2186	2567	2949	3330	3712	4093	4475	4856	381
1139	5237	5619	6000	6381	6762	7143	7524	7905	8287	8668	381
1140	9049	9429	9810	*0191	*0572	*0953	*1334	*1714	*2095	*2476	381
1141	057 2856	3237	3618	3998	4379	4759	5140	5520	5900	6281	381
1142	6661	7041	7422	7802	8182	8562	8942	9322	9702	*0082	380
1143	058 0462	0842	1222	1602	1982	2362	2741	3121	3501	3881	380
1144	4260	4640	5019	5399	5778	6158	6537	6917	7296	7676	380
1145	8055	8434	8813	9193	9572	9951	*0330	*0709	*1088	*1467	379
1146	059 1846	2225	2604	2983	3362	3741	4119	4498	4877	5256	379
1147	5634	6013	6391	6770	7148	7527	7905	8284	8662	9041	379
1148	9419	9797	*0175	*0554	*0932	*1310	*1688	*2066	*2444	*2822	378
1149	060 3200	3578	3956	4334	4712	5090	5468	5845	6223	6601	378
1150	6978	7356	7734	8111	8489	8866	9244	9621	9999	*0376	378
N.	0	1	2	3	4	5	6	7	8	9	d.

.041 3927–.061 0376

Table 1-8
NATURAL SINES AND COSINES

From Baumeister and Marks, *Standard Handbook for Mechanical Engineers,*
7th ed., 1967, McGraw-Hill Book Company; by permission

Natural Sines at intervals of 0°.1, or 6'. (For 10' intervals, see pp. 1-60 to 1-64)

Deg	°.0 =(0')	°.1 (6')	°.2 (12')	°.3 (18')	°.4 (24')	°.5 (30')	°.6 (36')	°.7 (42')	°.8 (48')	°.9 (54')			Avg diff
											0.0000	90°	
0°	0.0000	0017	0035	0052	0070	0087	0105	0122	0140	0157	0175	89	17
1	0175	0192	0209	0227	0244	0262	0279	0297	0314	0332	0349	88	17
2	0349	0366	0384	0401	0419	0436	0454	0471	0488	0506	0523	87	17
3	0523	0541	0558	0576	0593	0610	0628	0645	0663	0680	0698	86	17
4	0698	0715	0732	0750	0767	0785	0802	0819	0837	0854	0.0872	85	17
5	0.0872	0889	0906	0924	0941	0958	0976	0993	1011	1028	1045	84	17
6	1045	1063	1080	1097	1115	1132	1149	1167	1184	1201	1219	83	17
7	1219	1236	1253	1271	1288	1305	1323	1340	1357	1374	1392	82	17
8	1392	1409	1426	1444	1461	1478	1495	1513	1530	1547	1564	81	17
9	1564	1582	1599	1616	1633	1650	1668	1685	1702	1719	0.1736	80°	17
10°	0.1736	1754	1771	1788	1805	1822	1840	1857	1874	1891	1908	79	17
11	1908	1925	1942	1959	1977	1994	2011	2028	2045	2062	2079	78	17
12	2079	2096	2113	2130	2147	2164	2181	2198	2215	2233	2250	77	17
13	2250	2267	2284	2300	2317	2334	2351	2368	2385	2402	2419	76	17
14	2419	2436	2453	2470	2487	2504	2521	2538	2554	2571	0.2588	75	17
15	0.2588	2605	2622	2639	2656	2672	2689	2706	2723	2740	2756	74	17
16	2756	2773	2790	2807	2823	2840	2857	2874	2890	2907	2924	73	17
17	2924	2940	2957	2974	2990	3007	3024	3040	3057	3074	3090	72	17
18	3090	3107	3123	3140	3156	3173	3190	3206	3223	3239	3256	71	17
19	3256	3272	3289	3305	3322	3338	3355	3371	3387	3404	0.3420	70°	16
20°	0.3420	3437	3453	3469	3486	3502	3518	3535	3551	3567	3584	69	16
21	3584	3600	3616	3633	3649	3665	3681	3697	3714	3730	3746	68	16
22	3746	3762	3778	3795	3811	3827	3843	3859	3875	3891	3907	67	16
23	3907	3923	3939	3955	3971	3987	4003	4019	4035	4051	4067	66	16
24	4067	4083	4099	4115	4131	4147	4163	4179	4195	4210	0.4226	65	16
25	0.4226	4242	4258	4274	4289	4305	4321	4337	4352	4368	4384	64	16
26	4384	4399	4415	4431	4446	4462	4478	4493	4509	4524	4540	63	16
27	4540	4555	4571	4586	4602	4617	4633	4648	4664	4679	4695	62	16
28	4695	4710	4726	4741	4756	4772	4787	4802	4818	4833	4848	61	15
29	4848	4863	4879	4894	4909	4924	4939	4955	4970	4985	0.5000	60°	15
30°	0.5000	5015	5030	5045	5060	5075	5090	5105	5120	5135	5150	59	15
31	5150	5165	5180	5195	5210	5225	5240	5255	5270	5284	5299	58	15
32	5299	5314	5329	5344	5358	5373	5388	5402	5417	5432	5446	57	15
33	5446	5461	5476	5490	5505	5519	5534	5548	5563	5577	5592	56	15
34	5592	5606	5621	5635	5650	5664	5678	5693	5707	5721	0.5736	55	14
35	0.5736	5750	5764	5779	5793	5807	5821	5835	5850	5864	5878	54	14
36	5878	5892	5906	5920	5934	5948	5962	5976	5990	6004	6018	53	14
37	6018	6032	6046	6060	6074	6088	6101	6115	6129	6143	6157	52	14
38	6157	6170	6184	6198	6211	6225	6239	6252	6266	6280	6293	51	14
39	6293	6307	6320	6334	6347	6361	6374	6388	6401	6414	0.6428	50°	13
40°	0.6428	6441	6455	6468	6481	6494	6508	6521	6534	6547	6561	49	13
41	6561	6574	6587	6600	6613	6626	6639	6652	6665	6678	6691	48	13
42	6691	6704	6717	6730	6743	6756	6769	6782	6794	6807	6820	47	13
43	6820	6833	6845	6858	6871	6884	6896	6909	6921	6934	6947	46	13
44	6947	6959	6972	6984	6997	7009	7022	7034	7046	7059	0.7071	45°	12
45°	0.7071												

	°.9 =(54')	°.8 (48')	°.7 (42')	°.6 (36')	°.5 (30')	°.4 (24')	°.3 (18')	°.2 (12')	°.1 (6')	°.0 (0')	Deg

Natural Cosines

Table 1-8 (*Continued*)
NATURAL SINES AND COSINES

Natural Sines at intervals of 0°.1, or 6'.   (For 10' intervals, see pp. 1-60 to 1-64)

Deg	°.0 =(0')	°.1 (6')	°.2 (12')	°.3 (18')	°.4 (24')	°.5 (30')	°.6 (36')	°.7 (42')	°.8 (48')	°.9 (54')			Avg diff
										0.7071	45°		
45°	0.7071	7083	7096	7108	7120	7133	7145	7157	7169	7181	7193	44	12
46	7193	7206	7218	7230	7242	7254	7266	7278	7290	7302	7314	43	12
47	7314	7325	7337	7349	7361	7373	7385	7396	7408	7420	7431	42	12
48	7431	7443	7455	7466	7478	7490	7501	7513	7524	7536	7547	41	12
49	7547	7559	7570	7581	7593	7604	7615	7627	7638	7649	0.7660	40°	11
50°	0.7660	7672	7683	7694	7705	7716	7727	7738	7749	7760	7771	39	11
51	7771	7782	7793	7804	7815	7826	7837	7848	7859	7869	7880	38	11
52	7880	7891	7902	7912	7923	7934	7944	7955	7965	7976	7986	37	11
53	7986	7997	8007	8018	8028	8039	8049	8059	8070	8080	8090	36	10
54	8090	8100	8111	8121	8131	8141	8151	8161	8171	8181	0.8192	35	10
55	0.8192	8202	8211	8221	8231	8241	8251	8261	8271	8281	8290	34	10
56	8290	8300	8310	8320	8329	8339	8348	8358	8368	8377	8387	33	10
57	8387	8396	8406	8415	8425	8434	8443	8453	8462	8471	8480	32	9
58	8480	8490	8499	8508	8517	8526	8536	8545	8554	8563	8572	31	9
59	8572	8581	8590	8599	8607	8616	8625	8634	8643	8652	0.8660	30°	9
60°	0.8660	8669	8678	8686	8695	8704	8712	8721	8729	8738	8746	29	9
61	8746	8755	8763	8771	8780	8788	8796	8805	8813	8821	8829	28	8
62	8829	8838	8846	8854	8862	8870	8878	8886	8894	8902	8910	27	8
63	8910	8918	8926	8934	8942	8949	8957	8965	8973	8980	8988	26	8
64	8988	8996	9003	9011	9018	9026	9033	9041	9048	9056	0.9063	25	7
65	0.9063	9070	9078	9085	9092	9100	9107	9114	9121	9128	9135	24	7
66	9135	9143	9150	9157	9164	9171	9178	9184	9191	9198	9205	23	7
67	9205	9212	9219	9225	9232	9239	9245	9252	9259	9265	9272	22	7
68	9272	9278	9285	9291	9298	9304	9311	9317	9323	9330	9336	21	6
69	9336	9342	9348	9354	9361	9367	9373	9379	9385	9391	0.9397	20°	6
70°	0.9397	9403	9409	9415	9421	9426	9432	9438	9444	9449	9455	19	6
71	9455	9461	9466	9472	9478	9483	9489	9494	9500	9505	9511	18	6
72	9511	9516	9521	9527	9532	9537	9542	9548	9553	9558	9563	17	5
73	9563	9568	9573	9578	9583	9588	9593	9598	9603	9608	9613	16	5
74	9613	9617	9622	9627	9632	9636	9641	9646	9650	9655	0.9659	15	5
75	0.9659	9664	9668	9673	9677	9681	9686	9690	9694	9699	9703	14	4
76	9703	9707	9711	9715	9720	9724	9728	9732	9736	9740	9744	13	4
77	9744	9748	9751	9755	9759	9763	9767	9770	9774	9778	9781	12	4
78	9781	9785	9789	9792	9796	9799	9803	9806	9810	9813	9816	11	3
79	9816	9820	9823	9826	9829	9833	9836	9839	9842	9845	0.9848	10°	3
80°	0.9848	9851	9854	9857	9860	9863	9866	9869	9871	9874	9877	9	3
81	9877	9880	9882	9885	9888	9890	9893	9895	9898	9900	9903	8	3
82	9903	9905	9907	9910	9912	9914	9917	9919	9921	9923	9925	7	2
83	9925	9928	9930	9932	9934	9936	9938	9940	9942	9943	9945	6	2
84	9945	9947	9949	9951	9952	9954	9956	9957	9959	9960	0.9962	5	2
85	0.9962	9963	9965	9966	9968	9969	9971	9972	9973	9974	9976	4	1
86	9976	9977	9978	9979	9980	9981	9982	9983	9984	9985	9986	3	1
87	9986	9987	9988	9989	9990	9990	9991	9992	9993	9993	9994	2	1
88	9994	9995	9995	9996	9996	9997	9997	9997	9998	9998	0.9998	1	0
89	0.9998	9999	9999	9999	9999	0000	0000	0000	0000	0000	1.0000	0°	0
90°	1.0000												

	°.9 =(54')	°.8 (48')	°.7 (42')	°.6 (36')	°.5 (30')	°.4 (24')	°.3 (18')	°.2 (12')	°.1 (6')	°.0 (0')	Deg

Natural Cosines

Table 1-9
NATURAL TANGENTS AND COTANGENTS

From Baumeister and Marks, *Standard Handbook for Mechanical Engineers,*
7th ed., 1967, McGraw-Hill Book Company; by permission

Natural Tangents at intervals of 0°.1, or 6′.   (For 10′ intervals, see pp. 1–60 to 1–64)

Deg	°.0 =(0′)	°.1 (6′)	°.2 (12′)	°.3 (18′)	°.4 (24′)	°.5 (30′)	°.6 (36′)	°.7 (42′)	°.8 (48′)	°.9 (54′)			Avg diff
											0.0000	90°	
0°	0.0000	0017	0035	0052	0070	0087	0105	0122	0140	0157	0175	89	17
1	0175	0192	0209	0227	0244	0262	0279	0297	0314	0332	0349	88	17
2	0349	0367	0384	0402	0419	0437	0454	0472	0489	0507	0524	87	17
3	0524	0542	0559	0577	0594	0612	0629	0647	0664	0682	0699	86	18
4	0699	0717	0734	0752	0769	0787	0805	0822	0840	0857	0.0875	85	18
5	0.0875	0892	0910	0928	0945	0963	0981	0998	1016	1033	1051	84	18
6	1051	1069	1086	1104	1122	1139	1157	1175	1192	1210	1228	83	18
7	1228	1246	1263	1281	1299	1317	1334	1352	1370	1388	1405	82	18
8	1405	1423	1441	1459	1477	1495	1512	1530	1548	1566	1584	81	18
9	1584	1602	1620	1638	1655	1673	1691	1709	1727	1745	0.1763	80°	18
10°	0.1763	1781	1799	1817	1835	1853	1871	1890	1908	1926	1944	79	18
11	1944	1962	1980	1998	2016	2035	2053	2071	2089	2107	2126	78	18
12	2126	2144	2162	2180	2199	2217	2235	2254	2272	2290	2309	77	18
13	2309	2327	2345	2364	2382	2401	2419	2438	2456	2475	2493	76	18
14	2493	2512	2530	2549	2568	2586	2605	2623	2642	2661	0.2679	75	19
15	0.2679	2698	2717	2736	2754	2773	2792	2811	2830	2849	2867	74	19
16	2867	2886	2905	2924	2943	2962	2981	3000	3019	3038	3057	73	19
17	3057	3076	3096	3115	3134	3153	3172	3191	3211	3230	3249	72	19
18	3249	3269	3288	3307	3327	3346	3365	3385	3404	3424	3443	71	19
19	3443	3463	3482	3502	3522	3541	3561	3581	3600	3620	0.3640	70°	20
20°	0.3640	3659	3679	3699	3719	3739	3759	3779	3799	3819	3839	69	20
21	3839	3859	3879	3899	3919	3939	3959	3979	4000	4020	4040	68	20
22	4040	4061	4081	4101	4122	4142	4163	4183	4204	4224	4245	67	21
23	4245	4265	4286	4307	4327	4348	4369	4390	4411	4431	4452	66	21
24	4452	4473	4494	4515	4536	4557	4578	4599	4621	4642	0.4663	65	21
25	0.4663	4684	4706	4727	4748	4770	4791	4813	4834	4856	4877	64	21
26	4877	4899	4921	4942	4964	4986	5008	5029	5051	5073	5095	63	22
27	5095	5117	5139	5161	5184	5206	5228	5250	5272	5295	5317	62	22
28	5317	5340	5362	5384	5407	5430	5452	5475	5498	5520	5543	61	23
29	5543	5566	5589	5612	5635	5658	5681	5704	5727	5750	0.5774	60°	23
30°	0.5774	5797	5820	5844	5867	5890	5914	5938	5961	5985	6009	59	24
31	6009	6032	6056	6080	6104	6128	6152	6176	6200	6224	6249	58	24
32	6249	6273	6297	6322	6346	6371	6395	6420	6445	6469	6494	57	25
33	6494	6519	6544	6569	6594	6619	6644	6669	6694	6720	6745	56	25
34	6745	6771	6796	6822	6847	6873	6899	6924	6950	6976	0.7002	55	26
35	0.7002	7028	7054	7080	7107	7133	7159	7186	7212	7239	7265	54	26
36	7265	7292	7319	7346	7373	7400	7427	7454	7481	7508	7536	53	27
37	7536	7563	7590	7618	7646	7673	7701	7729	7757	7785	7813	52	28
38	7813	7841	7869	7898	7926	7954	7983	8012	8040	8069	8098	51	28
39	8098	8127	8156	8185	8214	8243	8273	8302	8332	8361	0.8391	50°	29
40°	0.8391	8421	8451	8481	8511	8541	8571	8601	8632	8662	8693	49	30
41	8693	8724	8754	8785	8816	8847	8878	8910	8941	8972	9004	48	31
42	9004	9036	9067	9099	9131	9163	9195	9228	9260	9293	9325	47	32
43	9325	9358	9391	9424	9457	9490	9523	9556	9590	9623	0.9657	46	33
44	0.9657	9691	9725	9759	9793	9827	9861	9896	9930	9965	1.0000	45°	34
45°	1.0000												

	°.9 =(54′)	°.8 (48′)	°.7 (42′)	°.6 (36′)	°.5 (30′)	°.4 (24′)	°.3 (18′)	°.2 (12′)	°.1 (6′)	°.0 (0′)	Deg

Natural Cotangents

Table 1-9 (*Continued*)
NATURAL TANGENTS AND COTANGENTS

Natural Tangents at intervals of 0°.1, or 6′. (For 10′ intervals, see pp. 1-60 to 1-64)

Deg	°.0 =(0′)	°.1 (6′)	°.2 (12′)	°.3 (18′)	°.4 (24′)	°.5 (30′)	°.6 (36′)	°.7 (42′)	°.8 (48′)	°.9 (54′)			Avg diff
											1.0000	45°	
45°	1.0000	0035	0070	0105	0141	0176	0212	0247	0283	0319	0355	44	35
46	0355	0392	0428	0464	0501	0538	0575	0612	0649	0686	0724	43	37
47	0724	0761	0799	0837	0875	0913	0951	0990	1028	1067	1106	42	38
48	1106	1145	1184	1224	1263	1303	1343	1383	1423	1463	1504	41	40
49	1504	1544	1585	1626	1667	1708	1750	1792	1833	1875	1.1918	40°	41
50°	1.1918	1960	2002	2045	2088	2131	2174	2218	2261	2305	2349	39	43
51	2349	2393	2437	2482	2527	2572	2617	2662	2708	2753	2799	38	45
52	2799	2846	2892	2938	2985	3032	3079	3127	3175	3222	3270	37	47
53	3270	3319	3367	3416	3465	3514	3564	3613	3663	3713	3764	36	49
54	3764	3814	3865	3916	3968	4019	4071	4124	4176	4229	1.4281	35	52
55	1.4281	4335	4388	4442	4496	4550	4605	4659	4715	4770	4826	34	55
56	4826	4882	4938	4994	5051	5108	5166	5224	5282	5340	5399	33	57
57	5399	5458	5517	5577	5637	5697	5757	5818	5880	5941	6003	32	60
58	6003	6066	6128	6191	6255	6319	6383	6447	6512	6577	6643	31	64
59	1.6643	6709	6775	6842	6909	6977	7045	7113	7182	7251	1.7321	30°	67
60°	1.732	1.739	1.746	1.753	1.760	1.767	1.775	1.782	1.789	1.797	1.804	29	7
61	1.804	1.811	1.819	1.827	1.834	1.842	1.849	1.857	1.865	1.873	1.881	28	8
62	1.881	1.889	1.897	1.905	1.913	1.921	1.929	1.937	1.946	1.954	1.963	27	8
63	1.963	1.971	1.980	1.988	1.997	2.006	2.014	2.023	2.032	2.041	2.050	26	9
64	2.050	2.059	2.069	2.078	2.087	2.097	2.106	2.116	2.125	2.135	2.145	25	9
65	2.145	2.154	2.164	2.174	2.184	2.194	2.204	2.215	2.225	2.236	2.246	24	10
66	2.246	2.257	2.267	2.278	2.289	2.300	2.311	2.322	2.333	2.344	2.356	23	11
67	2.356	2.367	2.379	2.391	2.402	2.414	2.426	2.438	2.450	2.463	2.475	22	12
68	2.475	2.488	2.500	2.513	2.526	2.539	2.552	2.565	2.578	2.592	2.605	21	13
69	2.605	2.619	2.633	2.646	2.660	2.675	2.689	2.703	2.718	2.733	2.747	20°	14
70°	2.747	2.762	2.778	2.793	2.808	2.824	2.840	2.856	2.872	2.888	2.904	19	16
71	2.904	2.921	2.937	2.954	2.971	2.989	3.006	3.024	3.042	3.060	3.078	18	17
72	3.078	3.096	3.115	3.133	3.152	3.172	3.191	3.211	3.230	3.251	3.271	17	19
73	3.271	3.291	3.312	3.333	3.354	3.376	3.398	3.420	3.442	3.465	3.487	16	22
74	3.487	3.511	3.534	3.558	3.582	3.606	3.630	3.655	3.681	3.706	3.732	15	24
75	3.732	3.758	3.785	3.812	3.839	3.867	3.895	3.923	3.952	3.981	4.011	14	28
76	4.011	4.041	4.071	4.102	4.134	4.165	4.198	4.230	4.264	4.297	4.331	13	32
77	4.331	4.366	4.402	4.437	4.474	4.511	4.548	4.586	4.625	4.665	4.705	12	37
78	4.705	4.745	4.787	4.829	4.872	4.915	4.959	5.005	5.050	5.097	5.145	11	44
79	5.145	5.193	5.242	5.292	5.343	5.396	5.449	5.503	5.558	5.614	5.671	10°	53
80°	5.671	5.730	5.789	5.850	5.912	5.976	6.041	6.107	6.174	6.243	6.314	9	
81	6.314	6.386	6.460	6.535	6.612	6.691	6.772	6.855	6.940	7.026	7.115	8	
82	7.115	7.207	7.300	7.396	7.495	7.596	7.700	7.806	7.916	8.028	8.144	7	
83	8.144	8.264	8.386	8.513	8.643	8.777	8.915	9.058	9.205	9.357	9.514	6	
84	9.514	9.677	9.845	10.02	10.20	10.39	10.58	10.78	10.99	11.20	11.43	5	
85	11.43	11.66	11.91	12.16	12.43	12.71	13.00	13.30	13.62	13.95	14.30	4	
86	14.30	14.67	15.06	15.46	15.90	16.35	16.83	17.34	17.89	18.46	19.08	3	
87	19.08	19.74	20.45	21.20	22.02	22.90	23.86	24.90	26.03	27.27	28.64	2	
88	28.64	30.14	31.82	33.69	35.80	38.19	40.92	44.07	47.74	52.08	57.29	1	
89	57.29	63.66	71.62	81.85	95.49	114.6	143.2	191.0	286.5	573.0	∞	0°	
90°	∞												
	°.9 =(54′)	°.8 (48′)	°.7 (42′)	°.6 (36′)	°.5 (30′)	°.4 (24′)	°.3 (18′)	°.2 (12′)	°.1 (6′)	°.0 (0′)		Deg	

Natural Cotangents

Table 1-10
NATURAL SECANTS AND COSECANTS

From Baumeister and Marks, *Standard Handbook for Mechanical Engineers,*
7th ed., 1967, McGraw-Hill Book Company; by permission

Natural Secants at intervals of 0°.1, or 6'.

Deg	°.0 =(0')	°.1 (6')	°.2 (12')	°.3 (18')	°.4 (24')	°.5 (30')	°.6 (36')	°.7 (42')	°.8 (48')	°.9 (54')			Avg diff
											1.0000	90°	
0°	1.0000	0000	0000	0000	0000	0000	0001	0001	0001	0001	0002	89	0
1	0002	0002	0002	0003	0003	0003	0004	0004	0005	0006	0006	88	0
2	0006	0007	0007	0008	0009	0010	0010	0011	0012	0013	0014	87	1
3	0014	0015	0016	0017	0018	0019	0020	0021	0022	0023	0024	86	1
4	0024	0026	0027	0028	0030	0031	0032	0034	0035	0037	1.0038	85	1
5	1.0038	0040	0041	0043	0045	0046	0048	0050	0051	0053	0055	84	2
6	0055	0057	0059	0061	0063	0065	0067	0069	0071	0073	0075	83	2
7	0075	0077	0079	0082	0084	0086	0089	0091	0093	0096	0098	82	2
8	0098	0101	0103	0106	0108	0111	0114	0116	0119	0122	0125	81	3
9	0125	0127	0130	0133	0136	0139	0142	0145	0148	0151	1.0154	80°	3
10°	1.0154	0157	0161	0164	0167	0170	0174	0177	0180	0184	0187	79	3
11	0187	0191	0194	0198	0201	0205	0209	0212	0216	0220	0223	78	4
12	0223	0227	0231	0235	0239	0243	0247	0251	0255	0259	0263	77	4
13	0263	0267	0271	0276	0280	0284	0288	0293	0297	0302	0306	76	4
14	0306	0311	0315	0320	0324	0329	0334	0338	0343	0348	1.0353	75	5
15	1.0353	0358	0363	0367	0372	0377	0382	0388	0393	0398	0403	74	5
16	0403	0408	0413	0419	0424	0429	0435	0440	0446	0451	0457	73	5
17	0457	0463	0468	0474	0480	0485	0491	0497	0503	0509	0515	72	6
18	0515	0521	0527	0533	0539	0545	0551	0557	0564	0570	0576	71	6
19	0576	0583	0589	0595	0602	0608	0615	0622	0628	0635	1.0642	70°	7
20°	1.0642	0649	0655	0662	0669	0676	0683	0690	0697	0704	0711	69	7
21	0711	0719	0726	0733	0740	0748	0755	0763	0770	0778	0785	68	7
22	0785	0793	0801	0808	0816	0824	0832	0840	0848	0856	0864	67	8
23	0864	0872	0880	0888	0896	0904	0913	0921	0929	0938	0946	66	8
24	0946	0955	0963	0972	0981	0989	0998	1007	1016	1025	1.1034	65	9
25	1.1034	1043	1052	1061	1070	1079	1089	1098	1107	1117	1126	64	9
26	1126	1136	1145	1155	1164	1174	1184	1194	1203	1213	1223	63	10
27	1223	1233	1243	1253	1264	1274	1284	1294	1305	1315	1326	62	10
28	1326	1336	1347	1357	1368	1379	1390	1401	1412	1423	1434	61	11
29	1434	1445	1456	1467	1478	1490	1501	1512	1524	1535	1.1547	60°	11
30°	1.1547	1559	1570	1582	1594	1606	1618	1630	1642	1654	1666	59	12
31	1666	1679	1691	1703	1716	1728	1741	1753	1766	1779	1792	58	13
32	1792	1805	1818	1831	1844	1857	1870	1883	1897	1910	1924	57	13
33	1924	1937	1951	1964	1978	1992	2006	2020	2034	2048	2062	56	14
34	2062	2076	2091	2105	2120	2134	2149	2163	2178	2193	1.2208	55	15
35	1.2208	2223	2238	2253	2268	2283	2299	2314	2329	2345	2361	54	15
36	2361	2376	2392	2408	2424	2440	2456	2472	2489	2505	2521	53	16
37	2521	2538	2554	2571	2588	2605	2622	2639	2656	2673	2690	52	17
38	2690	2708	2725	2742	2760	2778	2796	2813	2831	2849	2868	51	18
39	2868	2886	2904	2923	2941	2960	2978	2997	3016	3035	1.3054	50°	19
40°	1.3054	3073	3093	3112	3131	3151	3171	3190	3210	3230	3250	49	20
41	3250	3270	3291	3311	3331	3352	3373	3393	3414	3435	3456	48	21
42	3456	3478	3499	3520	3542	3563	3585	3607	3629	3651	3673	47	22
43	3673	3696	3718	3741	3763	3786	3809	3832	3855	3878	3902	46	23
44	3902	3925	3949	3972	3996	4020	4044	4069	4093	4118	1.4142	45°	24
45°	1.4142												

	°.9 =(54')	°.8 (48')	°.7 (42')	°.6 (36')	°.5 (30')	°.4 (24')	°.3 (18')	°.2 (12')	°.1 (6')	°.0 (0')	Deg

Natural Cosecants

Table 1-10 (*Continued*)
NATURAL SECANTS AND COSECANTS
Natural Secants at intervals of 0°.1, or 6′.

Deg	°.0 =(0′)	°.1 (6′)	°.2 (12′)	°.3 (18′)	°.4 (24′)	°.5 (30′)	°.6 (36′)	°.7 (42′)	°.8 (48′)	°.9 (54′)			Avg diff
											1.4142	45°	
45°	1.4142	4167	4192	4217	4242	4267	4293	4318	4344	4370	4396	44	25
46	4396	4422	4448	4474	4501	4527	4554	4581	4608	4635	4663	43	27
47	4663	4690	4718	4746	4774	4802	4830	4859	4887	4916	4945	42	28
48	4945	4974	5003	5032	5062	5092	5121	5151	5182	5212	5243	41	30
49	5243	5273	5304	5335	5366	5398	5429	5461	5493	5525	1.5557	40°	31
50°	1.5557	5590	5622	5655	5688	5721	5755	5788	5822	5856	5890	39	33
51	5890	5925	5959	5994	6029	6064	6099	6135	6171	6207	6243	38	35
52	6243	6279	6316	6353	6390	6427	6464	6502	6540	6578	6616	37	37
53	6616	6655	6694	6733	6772	6812	6852	6892	6932	6972	7013	36	40
54	7013	7054	7095	7137	7179	7221	7263	7305	7348	7391	1.7434	35	42
55	1.7434	7478	7522	7566	7610	7655	7700	7745	7791	7837	7883	34	45
56	7883	7929	7976	8023	8070	8118	8166	8214	8263	8312	8361	33	48
57	8361	8410	8460	8510	8561	8612	8663	8714	8766	8818	8871	32	51
58	8871	8924	8977	9031	9084	9139	9194	9249	9304	9360	1.9416	31	54
59	1.9416	9473	9530	9587	9645	9703	9762	9821	9880	9940	2.0000	30°	58
60°	2.000	2.006	2.012	2.018	2.025	2.031	2.037	2.043	2.050	2.056	2.063	29	6
61	2.063	2.069	2.076	2.082	2.089	2.096	2.103	2.109	2.116	2.123	2.130	28	7
62	2.130	2.137	2.144	2.151	2.158	2.166	2.173	2.180	2.188	2.195	2.203	27	7
63	2.203	2.210	2.218	2.226	2.233	2.241	2.249	2.257	2.265	2.273	2.281	26	8
64	2.281	2.289	2.298	2.306	2.314	2.323	2.331	2.340	2.349	2.357	2.366	25	8
65	2.366	2.375	2.384	2.393	2.402	2.411	2.421	2.430	2.439	2.449	2.459	24	9
66	2.459	2.468	2.478	2.488	2.498	2.508	2.518	2.528	2.538	2.549	2.559	23	10
67	2.559	2.570	2.581	2.591	2.602	2.613	2.624	2.635	2.647	2.658	2.669	22	11
68	2.669	2.681	2.693	2.705	2.716	2.729	2.741	2.753	2.765	2.778	2.790	21	12
69	2.790	2.803	2.816	2.829	2.842	2.855	2.869	2.882	2.896	2.910	2.924	20°	13
70°	2.924	2.938	2.952	2.967	2.981	2.996	3.011	3.026	3.041	3.056	3.072	19	15
71	3.072	3.087	3.103	3.119	3.135	3.152	3.168	3.185	3.202	3.219	3.236	18	16
72	3.236	3.254	3.271	3.289	3.307	3.326	3.344	3.363	3.382	3.401	3.420	17	18
73	3.420	3.440	3.460	3.480	3.500	3.521	3.542	3.563	3.584	3.606	3.628	16	21
74	3.628	3.650	3.673	3.695	3.719	3.742	3.766	3.790	3.814	3.839	3.864	15	24
75	3.864	3.889	3.915	3.941	3.967	3.994	4.021	4.049	4.077	4.105	4.134	14	27
76	4.134	4.163	4.192	4.222	4.253	4.284	4.315	4.347	4.379	4.412	4.445	13	31
77	4.445	4.479	4.514	4.549	4.584	4.620	4.657	4.694	4.732	4.771	4.810	12	36
78	4.810	4.850	4.890	4.931	4.973	5.016	5.059	5.103	5.148	5.194	5.241	11	43
79	5.241	5.288	5.337	5.386	5.436	5.487	5.540	5.593	5.647	5.702	5.759	10°	52
80°	5.759	5.816	5.875	5.935	5.996	6.059	6.123	6.188	6.255	6.323	6.392	9	
81	6.392	6.464	6.537	6.611	6.687	6.765	6.845	6.927	7.011	7.097	7.185	8	
82	7.185	7.276	7.368	7.463	7.561	7.661	7.764	7.870	7.979	8.091	8.206	7	
83	8.206	8.324	8.446	8.571	8.700	8.834	8.971	9.113	9.259	9.411	9.567	6	
84	9.567	9.728	9.895	10.07	10.25	10.43	10.63	10.83	11.03	11.25	11.47	5	
85	11.47	11.71	11.95	12.20	12.47	12.75	13.03	13.34	13.65	13.99	14.34	4	
86	14.34	14.70	15.09	15.50	15.93	16.38	16.86	17.37	17.91	18.49	19.11	3	
87	19.11	19.77	20.47	21.23	22.04	22.93	23.88	24.92	26.05	27.29	28.65	2	
88	28.65	30.16	31.84	33.71	35.81	38.20	40.93	44.08	47.75	52.09	57.30	1	
89	57.30	63.66	71.62	81.85	95.49	114.6	143.2	191.0	286.5	573.0	∞	0°	
90°	∞												
	°.9 =(54′)	°.8 (48′)	°.7 (42′)	°.6 (36′)	°.5 (30′)	°.4 (24′)	°.3 (18′)	°.2 (12′)	°.1 (6′)	°.0 (0′)		Deg	

Natural Cosecants

Table 1-11
TRIGONOMETRIC FUNCTIONS
At Intervals of 10′
From Baumeister and Marks, *Standard Handbook for Mechanical Engineers*,
7th ed., 1967, McGraw-Hill Book Company; by permission
Annex—10 in columns marked *.   (For 0°.1 intervals, see pp. 1–54 to 1–59)

De-grees	Ra-dians	Sines Nat.	Sines Log *	Cosines Nat.	Cosines Log *	Tangents Nat.	Tangents Log *	Cotangents Nat.	Cotangents Log		
0° 00′	0.0000	.0000	∞	1.0000	0.0000	.0000	∞	∞	∞	1.5708	90° 00′
10	0.0029	.0029	7.4637	1.0000	.0000	.0029	7.4637	343.77	2.5363	1.5679	50
20	0.0058	.0058	.7648	1.0000	.0000	.0058	.7648	171.89	.2352	1.5650	40
30	0.0087	.0087	.9408	1.0000	.0000	.0087	.9409	114.59	.0591	1.5621	30
40	0.0116	.0116	8.0658	.9999	.0000	.0116	8.0658	85.940	1.9342	1.5592	20
50	0.0145	.0145	.1627	.9999	.0000	.0145	.1627	68.750	.8373	1.5563	10
1° 00′	0.0175	.0175	8.2419	.9998	9.9999	.0175	8.2419	57.290	1.7581	1.5533	89° 00′
10	0.0204	.0204	.3088	.9998	.9999	.0204	.3089	49.104	.6911	1.5504	50
20	0.0233	.0233	.3668	.9997	.9999	.0233	.3669	42.964	.6331	1.5475	40
30	0.0262	.0262	.4179	.9997	.9999	.0262	.4181	38.188	.5819	1.5446	30
40	0.0291	.0291	.4637	.9996	.9998	.0291	.4638	34.368	.5362	1.5417	20
50	0.0320	.0320	.5050	.9995	.9998	.0320	.5053	31.242	.4947	1.5388	10
2° 00′	0.0349	.0349	8.5428	.9994	9.9997	.0349	8.5431	28.636	1.4569	1.5359	88° 00′
10	0.0378	.0378	.5776	.9993	.9997	.0378	.5779	26.432	.4221	1.5330	50
20	0.0407	.0407	.6097	.9992	.9996	.0407	.6101	24.542	.3899	1.5301	40
30	0.0436	.0436	.6397	.9990	.9996	.0437	.6401	22.904	.3599	1.5272	30
40	0.0465	.0465	.6677	.9989	.9995	.0466	.6682	21.470	.3318	1.5243	20
50	0.0495	.0494	.6940	.9988	.9995	.0495	.6945	20.206	.3055	1.5213	10
3° 00′	0.0524	.0523	8.7188	.9986	9.9994	.0524	8.7194	19.081	1.2806	1.5184	87° 00′
10	0.0553	.0552	.7423	.9985	.9993	.0553	.7429	18.075	.2571	1.5155	50
20	0.0582	.0581	.7645	.9983	.9993	.0582	.7652	17.169	.2348	1.5126	40
30	0.0611	.0610	.7857	.9981	.9992	.0612	.7865	16.350	.2135	1.5097	30
40	0.0640	.0640	.8059	.9980	.9991	.0641	.8067	15.605	.1933	1.5068	20
50	0.0669	.0669	.8251	.9978	.9990	.0670	.8261	14.924	.1739	1.5039	10
4° 00′	0.0698	.0698	8.8436	.9976	9.9989	.0699	8.8446	14.301	1.1554	1.5010	86° 00′
10	0.0727	.0727	.8613	.9974	.9989	.0729	.8624	13.727	.1376	1.4981	50
20	0.0756	.0756	.8783	.9971	.9988	.0758	.8795	13.197	.1205	1.4952	40
30	0.0785	.0785	.8946	.9969	.9987	.0787	.8960	12.706	.1040	1.4923	30
40	0.0814	.0814	.9104	.9967	.9986	.0816	.9118	12.251	.0882	1.4893	20
50	0.0844	.0843	.9256	.9964	.9985	.0846	.9272	11.826	.0728	1.4864	10
5° 00′	0.0873	.0872	8.9403	.9962	9.9983	.0875	8.9420	11.430	1.0580	1.4835	85° 00′
10	0.0902	.0901	.9545	.9959	.9982	.0904	.9563	11.059	.0437	1.4806	50
20	0.0931	.0929	.9682	.9957	.9981	.0934	.9701	10.712	.0299	1.4777	40
30	0.0960	.0958	.9816	.9954	.9980	.0963	.9836	10.385	.0164	1.4748	30
40	0.0989	.0987	.9945	.9951	.9979	.0992	.9966	10.078	.0034	1.4719	20
50	0.1018	.1016	9.0070	.9948	.9977	.1022	9.0093	9.7882	0.9907	1.4690	10
6° 00′	0.1047	.1045	9.0192	.9945	9.9976	.1051	9.0216	9.5144	0.9784	1.4661	84° 00′
10	0.1076	.1074	.0311	.9942	.9975	.1080	.0336	9.2553	.9664	1.4632	50
20	0.1105	.1103	.0426	.9939	.9973	.1110	.0453	9.0098	.9547	1.4603	40
30	0.1134	.1132	.0539	.9936	.9972	.1139	.0567	8.7769	.9433	1.4574	30
40	0.1164	.1161	.0648	.9932	.9971	.1169	.0678	8.5555	.9322	1.4544	20
50	0.1193	.1190	.0755	.9929	.9969	.1198	.0786	8.3450	.9214	1.4515	10
7° 00′	0.1222	.1219	9.0859	.9925	9.9968	.1228	9.0891	8.1443	0.9109	1.4486	83° 00′
10	0.1251	.1248	.0961	.9922	.9966	.1257	.0995	7.9530	.9005	1.4457	50
20	0.1280	.1276	.1060	.9918	.9964	.1287	.1096	7.7704	.8904	1.4428	40
30	0.1309	.1305	.1157	.9914	.9963	.1317	.1194	7.5958	.8806	1.4399	30
40	0.1338	.1334	.1252	.9911	.9961	.1346	.1291	7.4287	.8709	1.4370	20
50	0.1367	.1363	.1345	.9907	.9959	.1376	.1385	7.2687	.8615	1.4341	10
8° 00′	0.1396	.1392	9.1436	.9903	9.9958	.1405	9.1478	7.1154	0.8522	1.4312	82° 00′
10	0.1425	.1421	.1525	.9899	.9956	.1435	.1569	6.9682	.8431	1.4283	50
20	0.1454	.1449	.1612	.9894	.9954	.1465	.1658	6.8269	.8342	1.4254	40
30	0.1484	.1478	.1697	.9890	.9952	.1495	.1745	6.6912	.8255	1.4224	30
40	0.1513	.1507	.1781	.9886	.9950	.1524	.1831	6.5606	.8169	1.4195	20
50	0.1542	.1536	.1863	.9881	.9948	.1554	.1915	6.4348	.8085	1.4166	10
9° 00′	0.1571	.1564	9.1943	.9877	9.9946	.1584	9.1997	6.3138	0.8003	1.4137	81° 00′
		Nat.	Log *	Nat.	Log *	Nat.	Log *	Nat.	Log		
		Cosines		Sines		Cotangents		Tangents		Ra-dians	De-grees

Table 1-11 (*Continued*)
TRIGONOMETRIC FUNCTIONS
Annex—10 in columns marked *.   (For 0°.1 intervals, see pp. 1-54 to 1-59)

De-grees	Ra-dians	Sines		Cosines		Tangents		Cotangents			
		Nat.	Log *	Nat.	Log *	Nat.	Log *	Nat.	Log		
9° 00′	0.1571	.1564	9.1943	.9877	9.9946	.1584	9.1997	6.3138	0.8003	1.4137	81° 00′
10	0.1600	.1593	.2022	.9872	.9944	.1614	.2078	6.1970	.7922	1.4108	50
20	0.1629	.1622	.2100	.9868	.9942	.1644	.2158	6.0844	.7842	1.4079	40
30	0.1658	.1650	.2176	.9863	.9940	.1673	.2236	5.9758	.7764	1.4050	30
40	0.1687	.1679	.2251	.9858	.9938	.1703	.2313	5.8708	.7687	1.4021	20
50	0.1716	.1708	.2324	.9853	.9936	.1733	.2389	5.7694	.7611	1.3992	10
10° 00′	0.1745	.1736	9.2397	.9848	9.9934	.1763	9.2463	5.6713	0.7537	1.3963	80° 00′
10	0.1774	.1765	.2468	.9843	.9931	.1793	.2536	5.5764	.7464	1.3934	50
20	0.1804	.1794	.2538	.9838	.9929	.1823	.2609	5.4845	.7391	1.3904	40
30	0.1833	.1822	.2606	.9833	.9927	.1853	.2680	5.3955	.7320	1.3875	30
40	0.1862	.1851	.2674	.9827	.9924	.1883	.2750	5.3093	.7250	1.3846	20
50	0.1891	.1880	.2740	.9822	.9922	.1914	.2819	5.2257	.7181	1.3817	10
11° 00′	0.1920	.1908	9.2806	.9816	9.9919	.1944	9.2887	5.1446	0.7113	1.3788	79° 00′
10	0.1949	.1937	.2870	.9811	.9917	.1974	.2953	5.0658	.7047	1.3759	50
20	0.1978	.1965	.2934	.9805	.9914	.2004	.3020	4.9894	.6980	1.3730	40
30	0.2007	.1994	.2997	.9799	.9912	.2035	.3085	4.9152	.6915	1.3701	30
40	0.2036	.2022	.3058	.9793	.9909	.2065	.3149	4.8430	.6851	1.3672	20
50	0.2065	.2051	.3119	.9787	.9907	.2095	.3212	4.7729	.6788	1.3643	10
12° 00′	0.2094	.2079	9.3179	.9781	9.9904	.2126	9.3275	4.7046	0.6725	1.3614	78° 00′
10	0.2123	.2108	.3238	.9775	.9901	.2156	.3336	4.6382	.6664	1.3584	50
20	0.2153	.2136	.3296	.9769	.9899	.2186	.3397	4.5736	.6603	1.3555	40
30	0.2182	.2164	.3353	.9763	.9896	.2217	.3458	4.5107	.6542	1.3526	30
40	0.2211	.2193	.3410	.9757	.9893	.2247	.3517	4.4494	.6483	1.3497	20
50	0.2240	.2221	.3466	.9750	.9890	.2278	.3576	4.3897	.6424	1.3468	10
13° 00′	0.2269	.2250	9.3521	.9744	9.9887	.2309	9.3634	4.3315	0.6366	1.3439	77° 00′
10	0.2298	.2278	.3575	.9737	.9884	.2339	.3691	4.2747	.6309	1.3410	50
20	0.2327	.2306	.3629	.9730	.9881	.2370	.3748	4.2193	.6252	1.3381	40
30	0.2356	.2334	.3682	.9724	.9878	.2401	.3804	4.1653	.6196	1.3352	30
40	0.2385	.2363	.3734	.9717	.9875	.2432	.3859	4.1126	.6141	1.3323	20
50	0.2414	.2391	.3786	.9710	.9872	.2462	.3914	4.0611	.6086	1.3294	10
14° 00′	0.2443	.2419	9.3837	.9703	9.9869	.2493	9.3968	4.0108	0.6032	1.3265	76° 00′
10	0.2473	.2447	.3887	.9696	.9866	.2524	.4021	3.9617	.5979	1.3235	50
20	0.2502	.2476	.3937	.9689	.9863	.2555	.4074	3.9136	.5926	1.3206	40
30	0.2531	.2504	.3986	.9681	.9859	.2586	.4127	3.8667	.5873	1.3177	30
40	0.2560	.2532	.4035	.9674	.9856	.2617	.4178	3.8208	.5822	1.3148	20
50	0.2589	.2560	.4083	.9667	.9853	.2648	.4230	3.7760	.5770	1.3119	10
15° 00′	0.2618	.2588	9.4130	.9659	9.9849	.2679	9.4281	3.7321	0.5719	1.3090	75° 00′
10	0.2647	.2616	.4177	.9652	.9846	.2711	.4331	3.6891	.5669	1.3061	50
20	0.2676	.2644	.4223	.9644	.9843	.2742	.4381	3.6470	.5619	1.3032	40
30	0.2705	.2672	.4269	.9636	.9839	.2773	.4430	3.6059	.5570	1.3003	30
40	0.2734	.2700	.4314	.9628	.9836	.2805	.4479	3.5656	.5521	1.2974	20
50	0.2763	.2728	.4359	.9621	.9832	.2836	.4527	3.5261	.5473	1.2945	10
16° 00′	0.2793	.2756	9.4403	.9613	9.9828	.2867	9.4575	3.4874	0.5425	1.2915	74° 00′
10	0.2822	.2784	.4447	.9605	.9825	.2899	.4622	3.4495	.5378	1.2886	50
20	0.2851	.2812	.4491	.9596	.9821	.2931	.4669	3.4124	.5331	1.2857	40
30	0.2880	.2840	.4533	.9588	.9817	.2962	.4716	3.3759	.5284	1.2828	30
40	0.2909	.2868	.4576	.9580	.9814	.2994	.4762	3.3402	.5238	1.2799	20
50	0.2938	.2896	.4618	.9572	.9810	.3026	.4808	3.3052	.5192	1.2770	10
17° 00′	0.2967	.2924	9.4659	.9563	9.9806	.3057	9.4853	3.2709	0.5147	1.2741	73° 00′
10	0.2996	.2952	.4700	.9555	.9802	.3089	.4898	3.2371	.5102	1.2712	50
20	0.3025	.2979	.4741	.9546	.9798	.3121	.4943	3.2041	.5057	1.2683	40
30	0.3054	.3007	.4781	.9537	.9794	.3153	.4987	3.1716	.5013	1.2654	30
40	0.3083	.3035	.4821	.9528	.9790	.3185	.5031	3.1397	.4969	1.2625	20
50	0.3113	.3062	.4861	.9520	.9786	.3217	.5075	3.1084	.4925	1.2595	10
18° 00′	0.3142	.3090	9.4900	.9511	9.9782	.3249	9.5118	3.0777	0.4882	1.2566	72° 00′
		Nat.	Log *	Nat.	Log *	Nat.	Log *	Nat.	Log		
		Cosines		Sines		Cotangents		Tangents		Ra-dians	De-grees

Table 1-11 (*Continued*)
TRIGONOMETRIC FUNCTIONS
Annex—10 in columns marked *.　　(For 0°.1 intervals, see pp. 1-54 to 1-59)

De-grees	Ra-dians	Sines		Cosines		Tangents		Cotangents			
		Nat.	Log *	Nat.	Log *	Nat.	Log *	Nat.	Log		
18° 00′	0.3142	.3090	9.4900	.9511	9.9782	.3249	9.5118	3.0777	0.4882	1.2566	72° 00′
10	0.3171	.3118	.4939	.9502	.9778	.3281	.5161	3.0475	.4839	1.2537	50
20	0.3200	.3145	.4977	.9492	.9774	.3314	.5203	3.0178	.4797	1.2508	40
30	0.3229	.3173	.5015	.9483	.9770	.3346	.5245	2.9887	.4755	1.2479	30
40	0.3258	.3201	.5052	.9474	.9765	.3378	.5287	2.9600	.4713	1.2450	20
50	0.3287	.3228	.5090	.9465	.9761	.3411	.5329	2.9319	.4671	1.2421	10
19° 00′	0.3316	.3256	9.5126	.9455	9.9757	.3443	9.5370	2.9042	0.4630	1.2392	71° 00′
10	0.3345	.3283	.5163	.9446	.9752	.3476	.5411	2.8770	.4589	1.2363	50
20	0.3374	.3311	.5199	.9436	.9748	.3508	.5451	2.8502	.4549	1.2334	40
30	0.3403	.3338	.5235	.9426	.9743	.3541	.5491	2.8239	.4509	1.2305	30
40	0.3432	.3365	.5270	.9417	.9739	.3574	.5531	2.7980	.4469	1.2275	20
50	0.3462	.3393	.5306	.9407	.9734	.3607	.5571	2.7725	.4429	1.2246	10
20° 00′	0.3491	.3420	9.5341	.9397	9.9730	.3640	9.5611	2.7475	0.4389	1.2217	70° 00′
10	0.3520	.3448	.5375	.9387	.9725	.3673	.5650	2.7228	.4350	1.2188	50
20	0.3549	.3475	.5409	.9377	.9721	.3706	.5689	2.6985	.4311	1.2159	40
30	0.3578	.3502	.5443	.9367	.9716	.3739	.5727	2.6746	.4273	1.2130	30
40	0.3607	.3529	.5477	.9356	.9711	.3772	.5766	2.6511	.4234	1.2101	20
50	0.3636	.3557	.5510	.9346	.9706	.3805	.5804	2.6279	.4196	1.2072	10
21° 00′	0.3665	.3584	9.5543	.9336	9.9702	.3839	9.5842	2.6051	0.4158	1.2043	69° 00′
10	0.3694	.3611	.5576	.9325	.9697	.3872	.5879	2.5826	.4121	1.2014	50
20	0.3723	.3638	.5609	.9315	.9692	.3906	.5917	2.5605	.4083	1.1985	40
30	0.3752	.3665	.5641	.9304	.9687	.3939	.5954	2.5386	.4046	1.1956	30
40	0.3782	.3692	.5673	.9293	.9682	.3973	.5991	2.5172	.4009	1.1926	20
50	0.3811	.3719	.5704	.9283	.9677	.4006	.6028	2.4960	.3972	1.1897	10
22° 00′	0.3840	.3746	9.5736	.9272	9.9672	.4040	9.6064	2.4751	0.3936	1.1868	68° 00′
10	0.3869	.3773	.5767	.9261	.9667	.4074	.6100	2.4545	.3900	1.1839	50
20	0.3898	.3800	.5798	.9250	.9661	.4108	.6136	2.4342	.3864	1.1810	40
30	0.3927	.3827	.5828	.9239	.9656	.4142	.6172	2.4142	.3828	1.1781	30
40	0.3956	.3854	.5859	.9228	.9651	.4176	.6208	2.3945	.3792	1.1752	20
50	0.3985	.3881	.5889	.9216	.9646	.4210	.6243	2.3750	.3757	1.1723	10
23° 00′	0.4014	.3907	9.5919	.9205	9.9640	.4245	9.6279	2.3559	0.3721	1.1694	67° 00′
10	0.4043	.3934	.5948	.9194	.9635	.4279	.6314	2.3369	.3686	1.1665	50
20	0.4072	.3961	.5978	.9182	.9629	.4314	.6348	2.3183	.3652	1.1636	40
30	0.4102	.3987	.6007	.9171	.9624	.4348	.6383	2.2998	.3617	1.1606	30
40	0.4131	.4014	.6036	.9159	.9618	.4383	.6417	2.2817	.3583	1.1577	20
50	0.4160	.4041	.6065	.9147	.9613	.4417	.6452	2.2637	.3548	1.1548	10
24° 00′	0.4189	.4067	9.6093	.9135	9.9607	.4452	9.6486	2.2460	0.3514	1.1519	66° 00′
10	0.4218	.4094	.6121	.9124	.9602	.4487	.6520	2.2286	.3480	1.1490	50
20	0.4247	.4120	.6149	.9112	.9596	.4522	.6553	2.2113	.3447	1.1461	40
30	0.4276	.4147	.6177	.9100	.9590	.4557	.6587	2.1943	.3413	1.1432	30
40	0.4305	.4173	.6205	.9088	.9584	.4592	.6620	2.1775	.3380	1.1403	20
50	0.4334	.4200	.6232	.9075	.9579	.4628	.6654	2.1609	.3346	1.1374	10
25° 00′	0.4363	.4226	9.6259	.9063	9.9573	.4663	9.6687	2.1445	0.3313	1.1345	65° 00′
10	0.4392	.4253	.6286	.9051	.9567	.4699	.6720	2.1283	.3280	1.1316	50
20	0.4422	.4279	.6313	.9038	.9561	.4734	.6752	2.1123	.3248	1.1286	40
30	0.4451	.4305	.6340	.9026	.9555	.4770	.6785	2.0965	.3215	1.1257	30
40	0.4480	.4331	.6366	.9013	.9549	.4806	.6817	2.0809	.3183	1.1228	20
50	0.4509	.4358	.6392	.9001	.9543	.4841	.6850	2.0655	.3150	1.1199	10
26° 00′	0.4538	.4384	9.6418	.8988	9.9537	.4877	9.6882	2.0503	0.3118	1.1170	64° 00′
10	0.4567	.4410	.6444	.8975	.9530	.4913	.6914	2.0353	.3086	1.1141	50
20	0.4596	.4436	.6470	.8962	.9524	.4950	.6946	2.0204	.3054	1.1112	40
30	0.4625	.4462	.6495	.8949	.9518	.4986	.6977	2.0057	.3023	1.1083	30
40	0.4654	.4488	.6521	.8936	.9512	.5022	.7009	1.9912	.2991	1.1054	20
50	0.4683	.4514	.6546	.8923	.9505	.5059	.7040	1.9768	.2960	1.1025	10
27° 00′	0.4712	.4540	9.6570	.8910	9.9499	.5095	9.7072	1.9626	0.2928	1.0996	63° 00′
		Nat.	Log *	Nat.	Log *	Nat.	Log *	Nat.	Log		
		Cosines		Sines		Cotangents		Tangents		Ra-dians	De-grees

Table 1-11 (*Continued*)
TRIGONOMETRIC FUNCTIONS
Annex—10 in columns marked *.   (For 0°.1 intervals, see pp. 1–54 to 1–59)

De-grees	Ra-dians	Sines		Cosines		Tangents		Cotangents			
		Nat.	Log *	Nat.	Log *	Nat.	Log *	Nat.	Log		
27° 00′	0.4712	.4540	9.6570	.8910	9.9499	.5095	9.7072	1.9626	0.2928	1.0996	63° 00′
10	0.4741	.4566	.6595	.8897	.9492	.5132	.7103	1.9486	.2897	1.0966	50
20	0.4771	.4592	.6620	.8884	.9486	.5169	.7134	1.9347	.2866	1.0937	40
30	0.4800	.4617	.6644	.8870	.9479	.5206	.7165	1.9210	.2835	1.0908	30
40	0.4829	.4643	.6668	.8857	.9473	.5243	.7196	1.9074	.2804	1.0879	20
50	0.4858	.4669	.6692	.8843	.9466	.5280	.7226	1.8940	.2774	1.0850	10
28° 00′	0.4887	.4695	9.6716	.8829	9.9459	.5317	9.7257	1.8807	0.2743	1.0821	62° 00′
10	0.4916	.4720	.6740	.8816	.9453	.5354	.7287	1.8676	.2713	1.0792	50
20	0.4945	.4746	.6763	.8802	.9446	.5392	.7317	1.8546	.2683	1.0763	40
30	0.4974	.4772	.6787	.8788	.9439	.5430	.7348	1.8418	.2652	1.0734	30
40	0.5003	.4797	.6810	.8774	.9432	.5467	.7378	1.8291	.2622	1.0705	20
50	0.5032	.4823	.6833	.8760	.9425	.5505	.7408	1.8165	.2592	1.0676	10
29° 00′	0.5061	.4848	9.6856	.8746	9.9418	.5543	9.7438	1.8040	0.2562	1.0647	61° 00′
10	0.5091	.4874	.6878	.8732	.9411	.5581	.7467	1.7917	.2533	1.0617	50
20	0.5120	.4899	.6901	.8718	.9404	.5619	.7497	1.7796	.2503	1.0588	40
30	0.5149	.4924	.6923	.8704	.9397	.5658	.7526	1.7675	.2474	1.0559	30
40	0.5178	.4950	.6946	.8689	.9390	.5696	.7556	1.7556	.2444	1.0530	20
50	0.5207	.4975	.6968	.8675	.9383	.5735	.7585	1.7437	.2415	1.0501	10
30° 00′	0.5236	.5000	9.6990	.8660	9.9375	.5774	9.7614	1.7321	0.2386	1.0472	60° 00′
10	0.5265	.5025	.7012	.8646	.9368	.5812	.7644	1.7205	.2356	1.0443	50
20	0.5294	.5050	.7033	.8631	.9361	.5851	.7673	1.7090	.2327	1.0414	40
30	0.5323	.5075	.7055	.8616	.9353	.5890	.7701	1.6977	.2299	1.0385	30
40	0.5352	.5100	.7076	.8601	.9346	.5930	.7730	1.6864	.2270	1.0356	20
50	0.5381	.5125	.7097	.8587	.9338	.5969	.7759	1.6753	.2241	1.0327	10
31° 00′	0.5411	.5150	9.7118	.8572	9.9331	.6009	9.7788	1.6643	0.2212	1.0297	59° 00′
10	0.5440	.5175	.7139	.8557	.9323	.6048	.7816	1.6534	.2184	1.0268	50
20	0.5469	.5200	.7160	.8542	.9315	.6088	.7845	1.6426	.2155	1.0239	40
30	0.5498	.5225	.7181	.8526	.9308	.6128	.7873	1.6319	.2127	1.0210	30
40	0.5527	.5250	.7201	.8511	.9300	.6168	.7902	1.6212	.2098	1.0181	20
50	0.5556	.5275	.7222	.8496	.9292	.6208	.7930	1.6107	.2070	1.0152	10
32° 00′	0.5585	.5299	9.7242	.8480	9.9284	.6249	9.7958	1.6003	0.2042	1.0123	58° 00′
10	0.5614	.5324	.7262	.8465	.9276	.6289	.7986	1.5900	.2014	1.0094	50
20	0.5643	.5348	.7282	.8450	.9268	.6330	.8014	1.5798	.1986	1.0065	40
30	0.5672	.5373	.7302	.8434	.9260	.6371	.8042	1.5697	.1958	1.0036	30
40	0.5701	.5398	.7322	.8418	.9252	.6412	.8070	1.5597	.1930	1.0007	20
50	0.5730	.5422	.7342	.8403	.9244	.6453	.8097	1.5497	.1903	0.9977	10
33° 00′	0.5760	.5446	9.7361	.8387	9.9236	.6494	9.8125	1.5399	0.1875	0.9948	57° 00′
10	0.5789	.5471	.7380	.8371	.9228	.6536	.8153	1.5301	.1847	0.9919	50
20	0.5818	.5495	.7400	.8355	.9219	.6577	.8180	1.5204	.1820	0.9890	40
30	0.5847	.5519	.7419	.8339	.9211	.6619	.8208	1.5108	.1792	0.9861	30
40	0.5876	.5544	.7438	.8323	.9203	.6661	.8235	1.5013	.1765	0.9832	20
50	0.5905	.5568	.7457	.8307	.9194	.6703	.8263	1.4919	.1737	0.9803	10
34° 00′	0.5934	.5592	9.7476	.8290	9.9186	.6745	9.8290	1.4826	0.1710	0.9774	56° 00′
10	0.5963	.5616	.7494	.8274	.9177	.6787	.8317	1.4733	.1683	0.9745	50
20	0.5992	.5640	.7513	.8258	.9169	.6830	.8344	1.4641	.1656	0.9716	40
30	0.6021	.5664	.7531	.8241	.9160	.6873	.8371	1.4550	.1629	0.9687	30
40	0.6050	.5688	.7550	.8225	.9151	.6916	.8398	1.4460	.1602	0.9657	20
50	0.6080	.5712	.7568	.8208	.9142	.6959	.8425	1.4370	.1575	0.9628	10
35° 00′	0.6109	.5736	9.7586	.8192	9.9134	.7002	9.8452	1.4281	0.1548	0.9599	55° 00′
10	0.6138	.5760	.7604	.8175	.9125	.7046	.8479	1.4193	.1521	0.9570	50
20	0.6167	.5783	.7622	.8158	.9116	.7089	.8506	1.4106	.1494	0.9541	40
30	0.6196	.5807	.7640	.8141	.9107	.7133	.8533	1.4019	.1467	0.9512	30
40	0.6225	.5831	.7657	.8124	.9098	.7177	.8559	1.3934	.1441	0.9483	20
50	0.6254	.5854	.7675	.8107	.9089	.7221	.8586	1.3848	.1414	0.9454	10
36° 00′	0.6283	.5878	9.7692	.8090	9.9080	.7265	9.8613	1.3764	0.1387	0.9425	54° 00′
		Nat.	Log *	Nat.	Log *	Nat.	Log *	Nat.	Log		
		Cosines		Sines		Cotangents		Tangents		Ra-dians	De-grees

Table 1-11 (*Continued*)
TRIGONOMETRIC FUNCTIONS
Annex—10 in columns marked *.   (For 0°.1 intervals, see pp. 1-54 to 1-59)

De-grees	Ra-dians	Sines		Cosines		Tangents		Cotangents			
		Nat.	Log *	Nat.	Log *	Nat.	Log *	Nat.	Log		
36° 00′	0.6283	.5878	9.7692	.8090	9.9080	.7265	9.8613	1.3764	0.1387	0.9425	54° 00′
10	0.6312	.5901	.7710	.8073	.9070	.7310	.8639	1.3680	.1361	0.9396	50
20	0.6341	.5925	.7727	.8056	.9061	.7355	.8666	1.3597	.1334	0.9367	40
30	0.6370	.5948	.7744	.8039	.9052	.7400	.8692	1.3514	.1308	0.9338	30
40	0.6400	.5972	.7761	.8021	.9042	.7445	.8718	1.3432	.1282	0.9308	20
50	0.6429	.5995	.7778	.8004	.9033	.7490	.8745	1.3351	.1255	0.9279	10
37° 00′	0.6458	.6018	9.7795	.7986	9.9023	.7536	9.8771	1.3270	0.1229	0.9250	53° 00′
10	0.6487	.6041	.7811	.7969	.9014	.7581	.8797	1.3190	.1203	0.9221	50
20	0.6516	.6065	.7828	.7951	.9004	.7627	.8824	1.3111	.1176	0.9192	40
30	0.6545	.6088	.7844	.7934	.8995	.7673	.8850	1.3032	.1150	0.9163	30
40	0.6574	.6111	.7861	.7916	.8985	.7720	.8876	1.2954	.1124	0.9134	20
50	0.6603	.6134	.7877	.7898	.8975	.7766	.8902	1.2876	.1098	0.9105	10
38° 00′	0.6632	.6157	9.7893	.7880	9.8965	.7813	9.8928	1.2799	0.1072	0.9076	52° 00′
10	0.6661	.6180	.7910	.7862	.8955	.7860	.8954	1.2723	.1046	0.9047	50
20	0.6690	.6202	.7926	.7844	.8945	.7907	.8980	1.2647	.1020	0.9018	40
30	0.6720	.6225	.7941	.7826	.8935	.7954	.9006	1.2572	.0994	0.8988	30
40	0.6749	.6248	.7957	.7808	.8925	.8002	.9032	1.2497	.0968	0.8959	20
50	0.6778	.6271	.7973	.7790	.8915	.8050	.9058	1.2423	.0942	0.8930	10
39° 00′	0.6807	.6293	9.7989	.7771	9.8905	.8098	9.9084	1.2349	0.0916	0.8901	51° 00′
10	0.6836	.6316	.8004	.7753	.8895	.8146	.9110	1.2276	.0890	0.8872	50
20	0.6865	.6338	.8020	.7735	.8884	.8195	.9135	1.2203	.0865	0.8843	40
30	0.6894	.6361	.8035	.7716	.8874	.8243	.9161	1.2131	.0839	0.8814	30
40	0.6923	.6383	.8050	.7698	.8864	.8292	.9187	1.2059	.0813	0.8785	20
50	0.6952	.6406	.8066	.7679	.8853	.8342	.9212	1.1988	.0788	0.8756	10
40° 00′	0.6981	.6428	9.8081	.7660	9.8843	.8391	9.9238	1.1918	0.0762	0.8727	50° 00′
10	0.7010	.6450	.8096	.7642	.8832	.8441	.9264	1.1847	.0736	0.8698	50
20	0.7039	.6472	.8111	.7623	.8821	.8491	.9289	1.1778	.0711	0.8668	40
30	0.7069	.6494	.8125	.7604	.8810	.8541	.9315	1.1708	.0685	0.8639	30
40	0.7098	.6517	.8140	.7585	.8800	.8591	.9341	1.1640	.0659	0.8610	20
50	0.7127	.6539	.8155	.7566	.8789	.8642	.9366	1.1571	.0634	0.8581	10
41° 00′	0.7156	.6561	9.8169	.7547	9.8778	.8693	9.9392	1.1504	0.0608	0.8552	49° 00′
10	0.7185	.6583	.8184	.7528	.8767	.8744	.9417	1.1436	.0583	0.8523	50
20	0.7214	.6604	.8198	.7509	.8756	.8796	.9443	1.1369	.0557	0.8494	40
30	0.7243	.6626	.8213	.7490	.8745	.8847	.9468	1.1303	.0532	0.8465	30
40	0.7272	.6648	.8227	.7470	.8733	.8899	.9494	1.1237	.0506	0.8436	20
50	0.7301	.6670	.8241	.7451	.8722	.8952	.9519	1.1171	.0481	0.8407	10
42° 00′	0.7330	.6691	9.8255	.7431	9.8711	.9004	9.9544	1.1106	0.0456	0.8378	48° 00′
10	0.7359	.6713	.8269	.7412	.8699	.9057	.9570	1.1041	.0430	0.8348	50
20	0.7389	.6734	.8283	.7392	.8688	.9110	.9595	1.0977	.0405	0.8319	40
30	0.7418	.6756	.8297	.7373	.8676	.9163	.9621	1.0913	.0379	0.8290	30
40	0.7447	.6777	.8311	.7353	.8665	.9217	.9646	1.0850	.0354	0.8261	20
50	0.7476	.6799	.8324	.7333	.8653	.9271	.9671	1.0786	.0329	0.8232	10
43° 00′	0.7505	.6820	9.8338	.7314	9.8641	.9325	9.9697	1.0724	0.0303	0.8203	47° 00′
10	0.7534	.6841	.8351	.7294	.8629	.9380	.9722	1.0661	.0278	0.8174	50
20	0.7563	.6862	.8365	.7274	.8618	.9435	.9747	1.0599	.0253	0.8145	40
30	0.7592	.6884	.8378	.7254	.8606	.9490	.9772	1.0538	.0228	0.8116	30
40	0.7621	.6905	.8391	.7234	.8594	.9545	.9798	1.0477	.0202	0.8087	20
50	0.7650	.6926	.8405	.7214	.8582	.9601	.9823	1.0416	.0177	0.8058	10
44° 00′	0.7679	.6947	9.8418	.7193	9.8569	.9657	9.9848	1.0355	0.0152	0.8029	46° 00′
10	0.7709	.6967	.8431	.7173	.8557	.9713	.9874	1.0295	.0126	0.7999	50
20	0.7738	.6988	.8444	.7153	.8545	.9770	.9899	1.0235	.0101	0.7970	40
30	0.7767	.7009	.8457	.7133	.8532	.9827	.9924	1.0176	.0076	0.7941	30
40	0.7796	.7030	.8469	.7112	.8520	.9884	.9949	1.0117	.0051	0.7912	20
50	0.7825	.7050	.8482	.7092	.8507	.9942	.9975	1.0058	.0025	0.7883	10
45° 00′	0.7854	.7071	9.8495	.7071	9.8495	1.0000	0.0000	1.0000	0.0000	0.7854	45° 00′
		Nat.	Log *	Nat.	Log *	Nat.	Log *	Nat.	Log		
		Cosines		Sines		Cotangents		Tangents		Ra-dians	De-grees

Table 1-12
EXPONENTIALS
e^n and e^{-n}

From Baumeister and Marks, *Standard Handbook for Mechanical Engineers,* 7th ed., 1967, McGraw-Hill Book Company; by permission

n	e^n	Diff	n	e^n	Diff	n	e^n	n	e^{-n}	Diff	n	e^{-n}	n	e^{-n}
0.00	1.000	10	0.50	1.649	16	1.0	2.718*	0.00	1.000	−10	0.50	.607	1.0	.368*
.01	1.010	10	.51	1.665	17	.1	3.004	.01	0.990	−10	.51	.600	.1	.333
.02	1.020	10	.52	1.682	17	.2	3.320	.02	.980	−10	.52	.595	.2	.301
.03	1.030	11	.53	1.699	17	.3	3.669	.03	.970	−9	.53	.589	.3	.273
.04	1.041	10	.54	1.716	17	.4	4.055	.04	.961	−10	.54	.583	.4	.247
0.05	1.051	11	0.55	1.733	18	1.5	4.482	0.05	.951	−9	0.55	.577	1.5	.223
.06	1.062	11	.56	1.751	17	.6	4.953	.06	.942	−10	.56	.571	.6	.202
.07	1.073	10	.57	1.768	18	.7	5.474	.07	.932	−9	.57	.566	.7	.183
.08	1.083	11	.58	1.786	18	.8	6.050	.08	.923	−9	.58	.560	.8	.165
.09	1.094	11	.59	1.804	18	.9	6.686	.09	.914	−9	.59	.554	.9	.150
0.10	1.105	11	0.60	1.822	18	2.0	7.389	0.10	.905	−9	0.60	.549	2.0	.135
.11	1.116	11	.61	1.840	19	.1	8.166	.11	.896	−9	.61	.543	.1	.122
.12	1.127	12	.62	1.859	19	.2	9.025	.12	.887	−9	.62	.538	.2	.111
.13	1.139	11	.63	1.878	18	.3	9.974	.13	.878	−9	.63	.533	.3	.100
.14	1.150	12	.64	1.896	20	.4	11.02	.14	.869	−8	.64	.527	.4	.0907
0.15	1.162	12	0.65	1.916	19	2.5	12.18	0.15	.861	−9	0.65	.522	2.5	.0821
.16	1.174	11	.66	1.935	19	.6	13.46	.16	.852	−8	.66	.517	.6	.0743
.17	1.185	12	.67	1.954	20	.7	14.88	.17	.844	−9	.67	.512	.7	.0672
.18	1.197	12	.68	1.974	20	.8	16.44	.18	.835	−8	.68	.507	.8	.0608
.19	1.209	12	.69	1.994	20	.9	18.17	.19	.827	−8	.69	.502	.9	.0550
0.20	1.221	13	0.70	2.014	20	3.0	20.09	0.20	.819	−8	0.70	.497	3.0	.0498
.21	1.234	12	.71	2.034	20	.1	22.20	.21	.811	−8	.71	.492	.1	.0450
.22	1.246	13	.72	2.054	21	.2	24.53	.22	.803	−8	.72	.487	.2	.0408
.23	1.259	12	.73	2.075	21	.3	27.11	.23	.795	−8	.73	.482	.3	.0369
.24	1.271	13	.74	2.096	21	.4	29.96	.24	.787	−8	.74	.477	.4	.0334
0.25	1.284	13	0.75	2.117	21	3.5	33.12	0.25	.779	−8	0.75	.472	3.5	.0302
.26	1.297	13	.76	2.138	22	.6	36.60	.26	.771	−8	.76	.468	.6	.0273
.27	1.310	13	.77	2.160	21	.7	40.45	.27	.763	−7	.77	.463	.7	.0247
.28	1.323	13	.78	2.181	22	.8	44.70	.28	.756	−8	.78	.458	.8	.0224
.29	1.336	14	.79	2.203	23	.9	49.40	.29	.748	−7	.79	.454	.9	.0202
0.30	1.350	13	0.80	2.226	22	4.0	54.60	0.30	.741	−8	0.80	.449	4.0	.0183
.31	1.363	14	.81	2.248	22	.1	60.34	.31	.733	−7	.81	.445	.1	.0166
.32	1.377	14	.82	2.270	23	.2	66.69	.32	.726	−7	.82	.440	.2	.0150
.33	1.391	14	.83	2.293	23	.3	73.70	.33	.719	−7	.83	.436	.3	.0136
.34	1.405	14	.84	2.316	24	.4	81.45	.34	.712	−7	.84	.432	.4	.0123
0.35	1.419	14	0.85	2.340	23	4.5	90.02	0.35	.705	−7	0.85	.427	4.5	.0111
.36	1.433	15	.86	2.363	24			.36	.698	−7	.86	.423		
.37	1.448	14	.87	2.387	24	5.0	148.4	.37	.691	−7	.87	.419	5.0	.00674
.38	1.462	15	.88	2.411	24	6.0	403.4	.38	.684	−7	.88	.415	6.0	.00248
.39	1.477	15	.89	2.435	25	7.0	1097.	.39	.677	−7	.89	.411	7.0	.000912
0.40	1.492	15	0.90	2.460	24	8.0	2981.	0.40	.670	−6	0.90	.407	8.0	.000335
.41	1.507	15	.91	2.484	25	9.0	8103.	.41	.664	−7	.91	.403	9.0	.000123
.42	1.522	15	.92	2.509	26	10.0	22026.	.42	.657	−6	.92	.399	10.0	.000045
.43	1.537	16	.93	2.535	25	$\pi/2$	4.810	.43	.651	−7	.93	.395	$\pi/2$	.208
.44	1.553	15	.94	2.560	26	$2\pi/2$	23.14	.44	.644	−6	.94	.391	$2\pi/2$	.0432
0.45	1.568	16	0.95	2.586	26	$3\pi/2$	111.3	0.45	.638	−7	0.95	.387	$3\pi/2$	.00898
.46	1.584	16	.96	2.612	26	$4\pi/2$	535.5	.46	.631	−6	.96	.383	$4\pi/2$	.00187
.47	1.600	16	.97	2.638	26	$5\pi/2$	2576.	.47	.625	−6	.97	.379	$5\pi/2$	.000388
.48	1.616	16	.98	2.664	27	$6\pi/2$	12392.	.48	.619	−6	.98	.375	$6\pi/2$	.000081
.49	1.632	17	.99	2.691	27	$7\pi/2$	59610.	.49	.613	−6	.99	.372	$7\pi/2$	.000017
						$8\pi/2$	286751.						$8\pi/2$	.000003
0.50	1.649		1.00	2.718				0.50	0.607		1.00	.368		

* Note: Do not interpolate in this column.

$e = 2.71828$ $1/e = 0.367879$ $\log_{10} e = 0.4343$ $1/(0.4343) = 2.3026$

$\log_{10}(0.4343) = \bar{1}.6378$ $\log_{10}(e^n) = n(0.4343)$

Table 1-13
NATURAL LOGARITHMS

From Baumeister and Marks, *Standard Handbook for Mechanical Engineers,*
7th ed., 1967, McGraw-Hill Book Company; by permission

	n	$n\,(2.3026)$	$n\,(0.6974-3)$
These two pages give the natural or Napierian logarithms (ln) of numbers between 1 and 10, correct to four places. Moving the decimal point n places to the right [or left] in the number is equivalent to adding n times 2.3026 [or n times $\overline{3}.6974$] to the logarithm. Base $e = 2.71828+$	1	2.3026	0.6974–3
	2	4.6052	0.3948–5
	3	6.9078	0.0922–7
	4	9.2103	0.7897–10
	5	11.5129	0.4871–12
	6	13.8155	0.1845–14
	7	16.1181	0.8819–17
	8	18.4207	0.5793–19
	9	20.7233	0.2767–21

Number	0	1	2	3	4	5	6	7	8	9	Avg diff
1.0	0.0000	0100	0198	0296	0392	0488	0583	0677	0770	0862	95
1.1	0953	1044	1133	1222	1310	1398	1484	1570	1655	1740	87
1.2	1823	1906	1989	2070	2151	2231	2311	2390	2469	2546	80
1.3	2624	2700	2776	2852	2927	3001	3075	3148	3221	3293	74
1.4	3365	3436	3507	3577	3646	3716	3784	3853	3920	3988	69
1.5	0.4055	4121	4187	4253	4318	4383	4447	4511	4574	4637	65
1.6	4700	4762	4824	4886	4947	5008	5068	5128	5188	5247	61
1.7	5306	5365	5423	5481	5539	5596	5653	5710	5766	5822	57
1.8	5878	5933	5988	6043	6098	6152	6206	6259	6313	6366	54
1.9	6419	6471	6523	6575	6627	6678	6729	6780	6831	6881	51
2.0	0.6931	6981	7031	7080	7129	7178	7227	7275	7324	7372	49
2.1	7419	7467	7514	7561	7608	7655	7701	7747	7793	7839	47
2.2	7885	7930	7975	8020	8065	8109	8154	8198	8242	8286	44
2.3	8329	8372	8416	8459	8502	8544	8587	8629	8671	8713	43
2.4	8755	8796	8838	8879	8920	8961	9002	9042	9083	9123	41
2.5	0.9163	9203	9243	9282	9322	9361	9400	9439	9478	9517	39
2.6	9555	9594	9632	9670	9708	9746	9783	9821	9858	9895	38
2.7	0.9933	9969	*0006	*0043	*0080	*0116	*0152	*0188	*0225	*0260	36
2.8	1.0296	0332	0367	0403	0438	0473	0508	0543	0578	0613	35
2.9	0647	0682	0716	0750	0784	0818	0852	0886	0919	0953	34
3.0	1.0986	1019	1053	1086	1119	1151	1184	1217	1249	1282	33
3.1	1314	1346	1378	1410	1442	1474	1506	1537	1569	1600	32
3.2	1632	1663	1694	1725	1756	1787	1817	1848	1878	1909	31
3.3	1939	1969	2000	2030	2060	2090	2119	2149	2179	2208	30
3.4	2238	2267	2296	2326	2355	2384	2413	2442	2470	2499	29
3.5	1.2528	2556	2585	2613	2641	2669	2698	2726	2754	2782	28
3.6	2809	2837	2865	2892	2920	2947	2975	3002	3029	3056	27
3.7	3083	3110	3137	3164	3191	3218	3244	3271	3297	3324	27
3.8	3350	3376	3403	3429	3455	3481	3507	3533	3558	3584	26
3.9	3610	3635	3661	3686	3712	3737	3762	3788	3813	3838	25
4.0	1.3863	3888	3913	3938	3962	3987	4012	4036	4061	4085	25
4.1	4110	4134	4159	4183	4207	4231	4255	4279	4303	4327	24
4.2	4351	4375	4398	4422	4446	4469	4493	4516	4540	4563	23
4.3	4586	4609	4633	4656	4679	4702	4725	4748	4770	4793	23
4.4	4816	4839	4861	4884	4907	4929	4951	4974	4996	5019	22
4.5	1.5041	5063	5085	5107	5129	5151	5173	5195	5217	5239	22
4.6	5261	5282	5304	5326	5347	5369	5390	5412	5433	5454	21
4.7	5476	5497	5518	5539	5560	5581	5602	5623	5644	5665	21
4.8	5686	5707	5728	5748	5769	5790	5810	5831	5851	5872	20
4.9	5892	5913	5933	5953	5974	5994	6014	6034	6054	6074	20

$$\ln x = (2.3026)\log_{10} x \qquad \log_{10} x = (0.4343)\ln x$$
where $2.3026 = \ln 10$ and $0.4343 = \log_{10} e$

Table 1-13 (*Continued*)
NATURAL LOGARITHMS

Num-ber	0	1	2	3	4	5	6	7	8	9	Avg diff
5.0	1.6094	6114	6134	6154	6174	6194	6214	6233	6253	6273	20
5.1	6292	6312	6332	6351	6371	6390	6409	6429	6448	6467	19
5.2	6487	6506	6525	6544	6563	6582	6601	6620	6639	6658	19
5.3	6677	6696	6715	6734	6752	6771	6790	6808	6827	6845	18
5.4	6864	6882	6901	6919	6938	6956	6974	6993	7011	7029	18
5.5	1.7047	7066	7084	7102	7120	7138	7156	7174	7192	7210	18
5.6	7228	7246	7263	7281	7299	7317	7334	7352	7370	7387	18
5.7	7405	7422	7440	7457	7475	7492	7509	7527	7544	7561	17
5.8	7579	7596	7613	7630	7647	7664	7681	7699	7716	7733	17
5.9	7750	7766	7783	7800	7817	7834	7851	7867	7884	7901	17
6.0	1.7918	7934	7951	7967	7984	8001	8017	8034	8050	8066	16
6.1	8083	8099	8116	8132	8148	8165	8181	8197	8213	8229	16
6.2	8245	8262	8278	8294	8310	8326	8342	8358	8374	8390	16
6.3	8405	8421	8437	8453	8469	8485	8500	8516	8532	8547	16
6.4	8563	8579	8594	8610	8625	8641	8656	8672	8687	8703	15
6.5	1.8718	8733	8749	8764	8779	8795	8810	8825	8840	8856	15
6.6	8871	8886	8901	8916	8931	8946	8961	8976	8991	9006	15
6.7	9021	9036	9051	9066	9081	9095	9110	9125	9140	9155	15
6.8	9169	9184	9199	9213	9228	9242	9257	9272	9286	9301	15
6.9	9315	9330	9344	9359	9373	9387	9402	9416	9430	9445	14
7.0	1.9459	9473	9488	9502	9516	9530	9544	9559	9573	9587	14
7.1	9601	9615	9629	9643	9657	9671	9685	9699	9713	9727	14
7.2	9741	9755	9769	9782	9796	9810	9824	9838	9851	9865	14
7.3	1.9879	9892	9906	9920	9933	9947	9961	9974	9988	*0001	13
7.4	2.0015	0028	0042	0055	0069	0082	0096	0109	0122	0136	13
7.5	2.0149	0162	0176	0189	0202	0215	0229	0242	0255	0268	13
7.6	0281	0295	0308	0321	0334	0347	0360	0373	0386	0399	13
7.7	0412	0425	0438	0451	0464	0477	0490	0503	0516	0528	13
7.8	0541	0554	0567	0580	0592	0605	0618	0631	0643	0656	13
7.9	0669	0681	0694	0707	0719	0732	0744	0757	0769	0782	12
8.0	2.0794	0807	0819	0832	0844	0857	0869	0882	0894	0906	12
8.1	0919	0931	0943	0956	0968	0980	0992	1005	1017	1029	12
8.2	1041	1054	1066	1078	1090	1102	1114	1126	1138	1150	12
8.3	1163	1175	1187	1199	1211	1223	1235	1247	1258	1270	12
8.4	1282	1294	1306	1318	1330	1342	1353	1365	1377	1389	12
8.5	2.1401	1412	1424	1436	1448	1459	1471	1483	1494	1506	12
8.6	1518	1529	1541	1552	1564	1576	1587	1599	1610	1622	12
8.7	1633	1645	1656	1668	1679	1691	1702	1713	1725	1736	11
8.8	1748	1759	1770	1782	1793	1804	1815	1827	1838	1849	11
8.9	1861	1872	1883	1894	1905	1917	1928	1939	1950	1961	11
9.0	2.1972	1983	1994	2006	2017	2028	2039	2050	2061	2072	11
9.1	2083	2094	2105	2116	2127	2138	2148	2159	2170	2181	11
9.2	2192	2203	2214	2225	2235	2246	2257	2268	2279	2289	11
9.3	2300	2311	2322	2332	2343	2354	2364	2375	2386	2396	11
9.4	2407	2418	2428	2439	2450	2460	2471	2481	2492	2502	11
9.5	2.2513	2523	2534	2544	2555	2565	2576	2586	2597	2607	10
9.6	2618	2628	2638	2649	2659	2670	2680	2690	2701	2711	10
9.7	2721	2732	2742	2752	2762	2773	2783	2793	2803	2814	10
9.8	2824	2834	2844	2854	2865	2875	2885	2895	2905	2915	10
9.9	2925	2935	2946	2956	2966	2976	2986	2996	3006	3016	10
10.0	2.3026										

Moving the decimal point n places to the right [or left] in the number requires adding n times 2.3026 [or n times (0.6074-3)] in the body of the table. See auxiliary table of multiples on top of the preceding page.

Table 1-14
DEGREES AND MINUTES EXPRESSED IN RADIANS

From Baumeister and Marks, *Standard Handbook for Mechanical Engineers,*
7th ed., 1967, McGraw-Hill Book Company; by permission

Degrees			Hundredths		Minutes
1° .0175	61° 1.0647	121° 2.1118	0°.01 .0002	0°.51 .0089	1' .0003
2 .0349	2 1.0821	2 2.1293	2 .0003	2 .0091	2' .0006
3 .0524	3 1.0996	3 2.1468	3 .0005	3 .0093	3' .0009
4 .0698	4 1.1170	4 2.1642	4 .0007	4 .0094	4' .0012
5° .0873	65° 1.1345	125° 2.1817	.05 .0009	.55 .0096	5' .0015
6 .1047	6 1.1519	6 2.1991	6 .0010	6 .0098	6' .0017
7 .1222	7 1.1694	7 2.2166	7 .0012	7 .0099	7' .0020
8 .1396	8 1.1868	8 2.2340	8 .0014	8 .0101	8' .0023
9 .1571	9 1.2043	9 2.2515	9 .0016	9 .0103	9' .0026
10° .1745	70° 1.2217	130° 2.2689	0°.10 .0017	0°.60 .0105	10' .0029
1 .1920	1 1.2392	1 2.2864	1 .0019	1 .0106	11' .0032
2 .2094	2 1.2566	2 2.3038	2 .0021	2 .0108	12' .0035
3 .2269	3 1.2741	3 2.3213	3 .0023	3 .0110	13' .0038
4 .2443	4 1.2915	4 2.3387	4 .0024	4 .0112	14' .0041
15° .2618	75° 1.3090	135° 2.3562	.15 .0026	.65 .0113	15' .0044
6 .2793	6 1.3265	6 2.3736	6 .0028	6 .0115	16' .0047
7 .2967	7 1.3439	7 2.3911	7 .0030	7 .0117	17' .0049
8 .3142	8 1.3614	8 2.4086	8 .0031	8 .0119	18' .0052
9 .3316	9 1.3788	9 2.4260	9 .0033	9 .0120	19' .0055
20° .3491	80° 1.3963	140° 2.4435	0°.20 .0035	0°.70 .0122	20' .0058
1 .3665	1 1.4137	1 2.4609	1 .0037	1 .0124	21' .0061
2 .3840	2 1.4312	2 2.4784	2 .0038	2 .0126	22' .0064
3 .4014	3 1.4486	3 2.4958	3 .0040	3 .0127	23' .0067
4 .4189	4 1.4661	4 2.5133	4 .0042	4 .0129	24' .0070
25° .4363	85° 1.4835	145° 2.5307	.25 .0044	.75 .0131	25' .0073
6 .4538	6 1.5010	6 2.5482	6 .0045	6 .0133	26' .0076
7 .4712	7 1.5184	7 2.5656	7 .0047	7 .0134	27' .0079
8 .4887	8 1.5359	8 2.5831	8 .0049	8 .0136	28' .0081
9 .5061	9 1.5533	9 2.6005	9 .0051	9 .0138	29' .0084
30° .5236	90° 1.5708	150° 2.6180	0°.30 .0052	0°.80 .0140	30' .0087
1 .5411	1 1.5882	1 2.6354	1 .0054	1 .0141	31' .0090
2 .5585	2 1.6057	2 2.6529	2 .0056	2 .0143	32' .0093
3 .5760	3 1.6232	3 2.6704	3 .0058	3 .0145	33' .0096
4 .5934	4 1.6406	4 2.6878	4 .0059	4 .0147	34' .0099
35° .6109	95° 1.6581	155° 2.7053	.35 .0061	.85 .0148	35' .0102
6 .6283	6 1.6755	6 2.7227	6 .0063	6 .0150	36' .0105
7 .6458	7 1.6930	7 2.7402	7 .0065	7 .0152	37' .0108
8 .6632	8 1.7104	8 2.7576	8 .0066	8 .0154	38' .0111
9 .6807	9 1.7279	9 2.7751	9 .0068	9 .0155	39' .0113
40° .6981	100° 1.7453	160° 2.7925	0°.40 .0070	0°.90 .0157	40' .0116
1 .7156	1 1.7628	1 2.8100	1 .0072	1 .0159	41' .0119
2 .7330	2 1.7802	2 2.8274	2 .0073	2 .0161	42' .0122
3 .7505	3 1.7977	3 2.8449	3 .0075	3 .0162	43' .0125
4 .7679	4 1.8151	4 2.8623	4 .0077	4 .0164	44' .0128
45° .7854	105° 1.8326	165° 2.8798	.45 .0079	.95 .0166	45' .0131
6 .8029	6 1.8500	6 2.8972	6 .0080	6 .0168	46' .0134
7 .8203	7 1.8675	7 2.9147	7 .0082	7 .0169	47' .0137
8 .8378	8 1.8850	8 2.9322	8 .0084	8 .0171	48' .0140
9 .8552	9 1.9024	9 2.9496	9 .0086	9 .0173	49' .0143
50° .8727	110° 1.9199	170° 2.9671	0°.50 .0087	1°.00 .0175	50' .0145
1 .8901	1 1.9373	1 2.9845			51' .0148
2 .9076	2 1.9548	2 3.0020			52' .0151
3 .9250	3 1.9722	3 3.0194			53' .0154
4 .9425	4 1.9897	4 3.0369			54' .0157
55° .9599	115° 2.0071	175° 3.0543			55' .0160
6 .9774	6 2.0246	6 3.0718			56' .0163
7 .9948	7 2.0420	7 3.0892			57' .0166
8 1.0123	8 2.0595	8 3.1067			58' .0169
9 1.0297	9 2.0769	9 3.1241			59' .0172
60° 1.0472	120° 2.0944	180° 3.1416			60' .0175

Arc 1° = 0.0174533   Arc 1' = 0.000290888   Arc 1'' = 0.00000484814
1 radian = 57°.295780 = 57° 17'.7468 = 57° 17' 44''.806

Table 1-14 (*Continued*)
RADIANS EXPRESSED IN DEGREES

0.01	0°.57	.64	36°.67	1.27	72°.77	1.90	108°.86	2.53	144°.96
2	1°.15	.65	37°.24	8	73°.34	1	109°.43	4	145°.53
3	1°.72	6	37°.82	9	73°.91	2	110°.01	2.55	146°.10
4	2°.29	7	38°.39	1.30	74°.48	3	110°.58	6	146°.68
.05	2°.86	8	38°.96	1	75°.06	4	111°.15	7	147°.25
6	3°.44	9	39°.53	2	75°.63	1.95	111°.73	8	147°.82
7	4°.01	.70	40°.11	3	76°.20	6	112°.30	9	148°.40
8	4°.58	1	40°.68	4	76°.78	7	112°.87	2.60	148°.97
9	5°.16	2	41°.25	1.35	77°.35	8	113°.45	1	149°.54
.10	5°.73	3	41°.83	6	77°.92	9	114°.02	2	150°.11
1	6°.30	4	42°.40	7	78°.50	2.00	114°.59	3	150°.69
2	6°.88	.75	42°.97	8	79°.07	1	115°.16	4	151°.26
3	7°.45	6	43°.54	9	79°.64	2	115°.74	2.65	151°.83
4	8°.02	7	44°.12	1.40	80°.21	3	116°.31	6	152°.41
.15	8°.59	8	44°.69	1	80°.79	4	116°.88	7	152°.98
6	9°.17	9	45°.26	2	81°.36	2.05	117°.46	8	153°.55
7	9°.74	.80	45°.84	3	81°.93	6	118°.03	9	154°.13
8	10°.31	1	46°.41	4	82°.51	7	118°.60	2.70	154°.70
9	10°.89	2	46°.98	1.45	83°.08	8	119°.18	1	155°.27
.20	11°.46	3	47°.56	6	83°.65	9	119°.75	2	155°.84
1	12°.03	4	48°.13	7	84°.22	2.10	120°.32	3	156°.42
2	12°.61	.85	48°.70	8	84°.80	1	120°.89	4	156°.99
3	13°.18	6	49°.27	9	85°.37	2	121°.47	2.75	157°.56
4	13°.75	7	49°.85	1.50	85°.94	3	122°.04	6	158°.14
.25	14°.32	8	50°.42	1	86°.52	4	122°.61	7	158°.71
6	14°.90	9	50°.99	2	87°.09	2.15	123°.19	8	159°.28
7	15°.47	.90	51°.57	3	87°.66	6	123°.76	9	159°.86
8	16°.04	1	52°.14	4	88°.24	7	124°.33	2.80	160°.43
9	16°.62	2	52°.71	1.55	88°.81	8	124°.90	1	161°.00
.30	17°.19	3	53°.29	6	89°.38	9	125°.48	2	161°.57
1	17°.76	4	53°.86	7	89°.95	2.20	126°.05	3	162°.15
2	18°.33	.95	54°.43	8	90°.53	1	126°.62	4	162°.72
3	18°.91	6	55°.00	9	91°.10	2	127°.20	2.85	163°.29
4	19°.48	7	55°.58	1.60	91°.67	3	127°.77	6	163°.87
.35	20°.05	8	56°.15	1	92°.25	4	128°.34	7	164°.44
6	20°.63	9	56°.72	2	92°.82	2.25	128°.92	8	165°.01
7	21°.20	1.00	57°.30	3	93°.39	6	129°.49	9	165°.58
8	21°.77	1	57°.87	4	93°.97	7	130°.06	2.90	166°.16
9	22°.35	2	58°.44	1.65	94°.54	8	130°.63	1	166°.73
.40	22°.92	3	59°.01	6	95°.11	9	131°.21	2	167°.30
1	23°.49	4	59°.59	7	95°.68	2.30	131°.78	3	167°.88
2	24°.06	1.05	60°.16	8	96°.26	1	132°.35	4	168°.45
3	24°.64	6	60°.73	9	96°.83	2	132°.93	2.95	169°.02
4	25°.21	7	61°.31	1.70	97°.40	3	133°.50	6	169°.60
.45	25°.78	8	61°.88	1	97°.98	4	134°.07	7	170°.17
6	26°.36	9	62°.45	2	98°.55	2.35	134°.65	8	170°.74
7	26°.93	1.10	63°.03	3	99°.12	6	135°.22	9	171°.31
8	27°.50	1	63°.60	4	99°.69	7	135°.79	3.00	171°.89
9	28°.07	2	64°.17	1.75	100°.27	8	136°.36	1	172°.46
.50	28°.65	3	64°.74	6	100°.84	9	136°.94	2	173°.03
1	29°.22	4	65°.32	7	101°.41	2.40	137°.51	3	173°.61
2	29°.79	1.15	65°.89	8	101°.99	1	138°.08	4	174°.18
3	30°.37	6	66°.46	9	102°.56	2	138°.66	3.05	174°.75
4	30°.94	7	67°.04	1.80	103°.13	3	139°.23	6	175°.33
.55	31°.51	8	67°.61	1	103°.71	4	139°.80	7	175°.90
6	32°.09	9	68°.18	2	104°.28	2.45	140°.37	8	176°.47
7	32°.66	1.20	68°.75	3	104°.85	6	140°.95	9	177°.04
8	33°.23	1	69°.33	4	105°.42	7	141°.52	3.10	177°.62
9	33°.80	2	69°.90	1.85	106°.00	8	142°.09	1	178°.19
.60	34°.38	3	70°.47	6	106°.57	9	142°.67	2	178°.76
1	34°.95	4	71°.05	7	107°.14	2.50	143°.24	3	179°.34
2	35°.52	1.25	71°.62	8	107°.72	1	143°.81	4	179°.91
3	36°.10	6	72°.19	9	108°.29	2	144°.39	3.15	180°.48

Interpolation

.0002	0°.01
04	.02
06	.03
08	.05
.0010	0°.06
12	.07
14	.08
16	.09
18	.10
.0020	0°.11
22	.13
24	.14
26	.15
28	.16
.0030	0°.17
32	.18
34	.19
36	.21
38	.22
.0040	0°.23
42	.24
44	.25
46	.26
48	.28
.0050	0°.29
52	.30
54	.31
56	.32
58	.33
.0060	0°.34
62	.36
64	.37
66	.38
68	.39
.0070	0°.40
72	.41
74	.42
76	.44
78	.45
.0080	0°.46
82	.47
84	.48
86	.49
88	.50
.0090	0°.52
92	.53
94	.54
96	.55
98	.56

Multiples of π

1	3.1416	180°
2	6.2832	360°
3	9.4248	540°
4	12.5664	720°
5	15.7080	900°
6	18.8496	1080°
7	21.9911	1260°
8	25.1327	1440°
9	28.2743	1620°
10	31.4159	1800°

SURFACE AREAS AND VOLUMES*

Let a, b, c, d, and s denote lengths, A denote areas, and V denote volumes.

Triangle.

$A = bh/2$, where b denotes the base and h the altitude.

Rectangle.

$A = ab$, where a and b denote the lengths of the sides.

Parallelogram (opposite sides parallel).

$A = ah = ab \sin \theta$, where a and b denote the sides, h the altitude and θ the angle between the sides.

Trapezoid (four sides, two parallel).

$A = \frac{1}{2}h(a + b)$, where a and b are the sides and h the altitude.

Regular Polygon of n Sides. (Fig. 1-1).

$A = \dfrac{1}{4} n\, a^2 \operatorname{ctn} \dfrac{180°}{n}$,   where a is length of side.

$R = \dfrac{a}{2} \csc \dfrac{180°}{n}$,   where R is radius of circumscribed circle.

$r = \dfrac{a}{2} \operatorname{ctn} \dfrac{180°}{n}$,   where r is radius of inscribed circle.

$\alpha = \dfrac{360°}{n} = \dfrac{2\pi}{n}$,   radians,

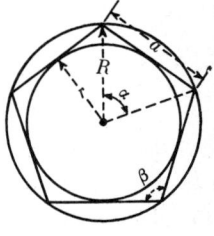

Fig. 1-1

$\beta = \left(\dfrac{n-2}{n}\right) \cdot 180° = \left(\dfrac{n-2}{n}\right)\pi$ radians where α and β are the angles indicated in Fig. 1-1.

$a = 2\,r \tan \dfrac{\alpha}{2} = 2R \sin \dfrac{\alpha}{2}.$

Circle. (Fig. 1-2).

Let C = circumference, S = length of arc subtended by θ,
R = radius, l = chord subtended by arc S,
D = diameter, h = rise,
A = area, θ = central angle in radians.

---

* Adapted by permission from Burington, "*Handbook of Mathematical Tables and Formulas,*" 3d ed., McGraw-Hill, New York (1959).

$$C = 2\pi R = \pi D, \quad \pi = 3.14159\cdots.$$

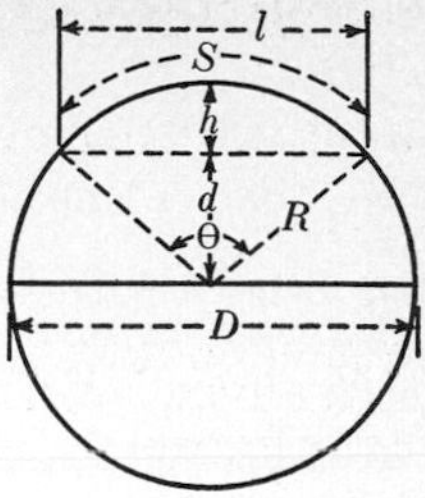

$$S = R\theta = \tfrac{1}{2}D\theta = D \cos^{-1}\frac{d}{R}.$$

$$l = 2\sqrt{R^2 - d^2} = 2R \sin\frac{\theta}{2} = 2d \tan\frac{\theta}{2}.$$

$$d = \tfrac{1}{2}\sqrt{4R^2 - l^2} = R \cos\frac{\theta}{2} = \tfrac{1}{2}l \operatorname{ctn}\frac{\theta}{2}.$$

$$h = R - d.$$

Fig. 1-2

$$\theta = \frac{S}{R} = \frac{2S}{D} = 2 \cos^{-1}\frac{d}{R} = 2 \tan^{-1}\frac{l}{2d} = 2 \sin^{-1}\frac{l}{D}.$$

A (circle) $= \pi R^2 = \tfrac{1}{4}\pi D^2.$

A (sector) $= \tfrac{1}{2}Rs = \tfrac{1}{2}R^2\theta.$

A (segment) $= A$ (sector) $- A$ (triangle) $= \tfrac{1}{2}R^2(\theta - \sin\theta)$

$$= R^2 \cos^{-1}\frac{(R - h)}{R} - (R - h)\sqrt{2Rh - h^2}.$$

Perimeter of a n-side regular polygon inscribed in a circle

$$= 2n\, R \sin\frac{\pi}{n}.$$

Area of inscribed polygon $= \tfrac{1}{2}nR^2 \sin\frac{2\pi}{n}.$

Perimeter of a n-side regular polygon circumscribed about a circle

$$= 2nR \tan\frac{\pi}{n}.$$

Area of circumscribed polygon $= nR^2 \tan\frac{\pi}{n}.$

Radius of circle inscribed in a triangle of sides a, b, and c is

$$r = \sqrt{\frac{(s - a)\,(s - b)\,(s - c)}{s}}, \quad s = \tfrac{1}{2}(a + b + c)\cdot$$

Radius of circle circumscribed about a triangle is

$$R = \frac{abc}{4\sqrt{s(s - a)\,(s - b)\,(s - c)}}.$$

Ellipse. (Fig. 1-3).

$A = \pi ab$, where a and b are lengths of semi-major and semi-minor axes respectively.

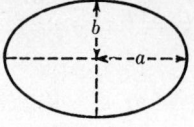

Fig. 1-3

Parabola. (Fig. 1-4).

$A = 2ld/3.$

Height of $d_1 = \dfrac{d}{l^2}\,(l^2 - l_1^2).$

Width of $l_1 = l\sqrt{\dfrac{d-d_1}{d}}.$

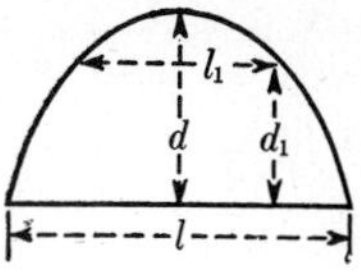

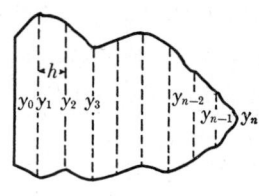

Fig. 1-4

Length of arc $= l\left[1 + \dfrac{2}{3}\left(\dfrac{2d}{l}\right)^2 - \dfrac{2}{5}\left(\dfrac{2d}{l}\right)^4 + \cdots\right].$

Area by Approximation. (Fig. 1-5). If $y_0, y_1, y_2, \cdots, y_n$ be the lengths of a series of equally spaced parallel chords, and if h be their distance apart, the area enclosed by boundary is given approximately by any one of the following formulae:

Fig. 1-5

$A_T = h\left[\tfrac{1}{2}(y_0 + y_n) + y_1 + y_2 + \cdots + y_{n-1}\right].$ (Trapezoidal Rule.)

$A_D = h\left[0.4(y_0 + y_n) + 1.1(y_1 + y_{n-1}) + y_2 + y_3 + \cdots + y_{n-2}\right].$
(Durand's Rule.)

$A_S = \tfrac{1}{3}h\left[(y_0 + y_n) + 4(y_1 + y_3 + \cdots + y_{n-1})\right.$
$\left. + 2(y_2 + y_4 + \cdots + y_{n-2})\right].$ (n even, Simpson's Rule.)

In general, A_S gives the most accurate approximation.

The greater the value of n, the greater the accuracy of approximation.

Cube.

$V = a^3;\quad a = a\sqrt{3};\quad$ total surface area $= 6a^2$, where a is length of side and d is length of diagonal.

Rectangular Parallelopiped.
$V = abc$; $d = \sqrt{a^2 + b^2 + c^2}$; total surface area $= 2\,(ab + bc + ca)$, where a, b, and c are the lengths of the sides and d is length of diagonal.

Prism or Cylinder.
$V =$ (area of base)·(altitude).
 Lateral area $=$ (perimeter of right section)·(lateral edge).

Pyramid or Cone.
$V = \frac{1}{3}$ (area of base)·(altitude).
 Lateral area of regular pyramid
$$= \tfrac{1}{2}\text{ (perimeter of base)}\cdot\text{(slant height)}.$$

Frustum of Pyramid or Cone.
$V = \frac{1}{3}(A_1 + A_2 + \sqrt{A_1 \cdot A_2})\,h$, where h is the altitude and A_1 and A_2 are the areas of the bases.
 Lateral area of a regular figure
$$= \tfrac{1}{2}\text{ (sum of perimeters of base)}\cdot\text{(slant height)}.$$

Prismoid.
$V = \dfrac{h}{6}(A_1 + A_2 + 4A_3)$, where $h =$ altitude, A_1 and A_2 are the areas of the bases, and A_3 is the area of the midsection parallel to bases.

Area of Surface and Volume of Regular Polyhedra of edge l.

Name	Type of Surface	Area of Surface	Volume
Tetrahedron	4 equilateral triangles	$1.73205l^2$	$0.11785l^3$
Hexahedron (cube)	6 squares	$6.00000l^2$	$1.00000l^3$
Octahedron	8 equilateral triangles	$3.46410l^2$	$0.47140l^3$
Dodecahedron	12 pentagons	$20.64578l^2$	$7.66312l^3$
Icosahedron	20 equilateral triangles	$8.66025l^2$	$2.18170l^3$

Sphere. (Fig. 1-6)

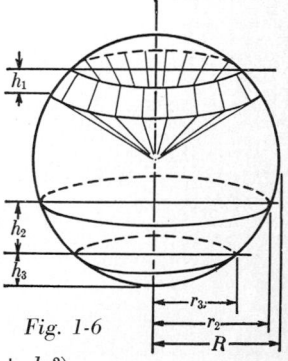

A (sphere) $= 4\pi R^2 = \pi D^2.$
A (zone) $= 2\pi R h_1 = \pi D h_1.$
V (sphere) $= \frac{4}{3}\pi R^3 = \frac{1}{6}\pi D^3.$
V (spherical sector) $= \frac{2}{3}\pi R^2 h_1 = \frac{1}{6}\pi D^2 h_1.$
V (spherical segment
 of one base) $= \frac{1}{6}\pi h_3\,(3r_3{}^2 + h_3{}^2).$
V (spherical segment
 of two bases) $= \frac{1}{6}\pi h_2\,(3r_3{}^2 + 3r_2{}^2 + h_2{}^2).$
A (lune) $= 2R^2\theta$, where θ is angle in radians of lune.

Fig. 1-6

Ellipsoid.

$V = \frac{4}{3}\pi abc$, where a, b, and c are the lengths of the semi-axes.

Torus. (Fig. 1-7). $V = 2\pi^2 R r^2$.

Area of surface $= S = 4\pi^2 R r$.

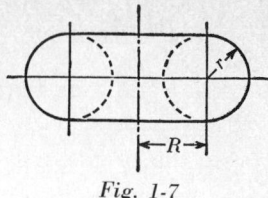

Fig. 1-7

Theorems of Pappus:

(a) If a plane area A be rotated about a line l in the plane of A and not cutting A, the volume of the solid generated is equal to the product of A and the distance traveled by the center of gravity of A.

(b) If a plane curve C be rotated about a line l in the plane of C and not cutting C, the area of the surface generated is equal to the product of the length of C and the distance traveled by the center of gravity of C.

Table 1-15
IMPORTANT CONSTANTS

N	Log₁₀ N	N	Log₁₀ N
$\pi = 3.14159265$	0.4971499	$\pi^2 = 9.86960440$	0.9942997
$2\pi = 6.28318531$	0.7981799	$\frac{1}{\pi^2} = 0.10132118$	9.0057003−10
$4\pi = 12.56637061$	1.0992099	$\sqrt{\pi} = 1.77245385$	0.2485749
$\frac{\pi}{2} = 1.57079633$	0.1961199	$\frac{1}{\sqrt{\pi}} = 0.56418958$	9.7514251−10
$\frac{\pi}{3} = 1.04719755$	0.0200286	$\sqrt{\frac{3}{\pi}} = 0.97720502$	9.9899857−10
$\frac{4\pi}{3} = 4.18879020$	0.6220886	$\sqrt{\frac{4}{\pi}} = 1.12837917$	0.0524551
$\frac{\pi}{4} = 0.78539816$	9.8950899−10	$\sqrt[3]{\pi} = 1.46459189$	0.1657166
$\frac{\pi}{6} = 0.52359878$	9.7189986−10	$\frac{1}{\sqrt[3]{\pi}} = 0.68278406$	9.8342834−10
$\frac{1}{\pi} = 0.31830989$	9.5028501−10	$\sqrt[3]{\pi^2} = 2.14502940$	0.3314332
$\frac{1}{2\pi} = 0.15915494$	9.2018201−10	$\sqrt[3]{\frac{3}{4\pi}} = 0.62035049$	9.7926371−10
$\frac{3}{\pi} = 0.95492966$	9.9799714−10	$\sqrt[3]{\frac{\pi}{6}} = 0.80599598$	9.9063329−10
$\frac{4}{\pi} = 1.27323954$	0.1049101		

Constant	Number	Log₁₀ of Number
Pi (π)	3.14159 26535 89793 23846	0.49714 98726 94133 85435
Napierian Base (e)	2.71828 18284 59045 23536	0.43429 448
$M = \log_{10}e$	0.43429 44819 03251 82765	9.63778 43113 00536 78912 −10
$1 \div M = \log_e 10$	2.30258 50929 94045 68402	0.36221 569
$180 \div \pi =$ degrees in 1 radian	57.2957 795	1.75812 263
$\pi \div 180 =$ radians in 1°	0.01745 329	8.24187 737 −10
$\pi \div 10800 =$ radians in 1'	0.00029 08882	6.46372 612 −10
$\pi \div 648000 =$ radians in 1''	0.00000 48481 36811 095	4.68557 487 −10

TRIGONOMETRIC FUNCTIONS OF AN ANGLE α.

Let α be any angle whose initial side lies on the positive x-axis and whose vertex is at the origin, and (x, y) be any point on the terminal side of the angle. (x is positive if measured along OX to the right, from the y-axis; and negative, if measured along OX' to the left from the y-axis. Likewise, y is positive if measured parallel to OY, and

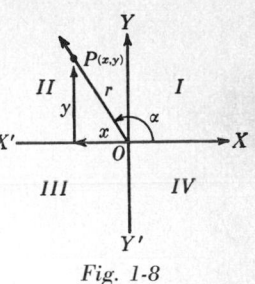

Fig. 1-8

negative if measured parallel to OY'.) Let r be the positive distance from the origin to the point. The trigonometric functions of an angle are defined as follows:

sine α	=	sin α	=	$\dfrac{y}{r}$.
cosine α	=	cos α	=	$\dfrac{x}{r}$.
tangent α	=	tan α	=	$\dfrac{y}{x}$.
cotangent α	=	ctn α = cot α	=	$\dfrac{x}{y}$.
secant α	=	sec α	=	$\dfrac{r}{x}$.
cosecant α	=	csc α	=	$\dfrac{r}{y}$.
exsecant α	=	exsec α	=	sec α -1 .
versine α	=	vers α	=	$1 - \cos α$.
coversine α	=	covers α	=	$1 - \sin α$.
haversine α	=	hav α	=	$\frac{1}{2}$ vers α .

Signs of the Functions.

Quadrant	sin	cos	tan	ctn	sec	csc
I..........	+	+	+	+	+	+
II	+	−	−	−	−	+
III.......	−	−	+	+	−	−
IV........	−	+	−	−	+	−

Relations between the Functions of a Single Angle*

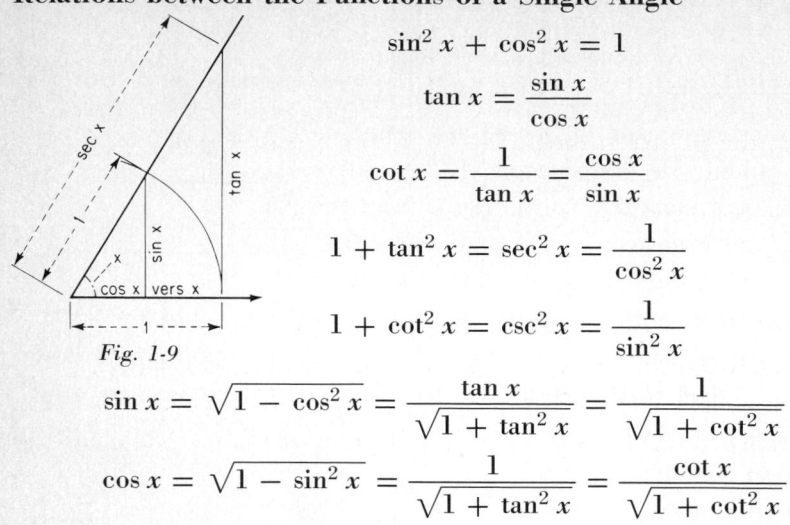

Fig. 1-9

$$\sin^2 x + \cos^2 x = 1$$

$$\tan x = \frac{\sin x}{\cos x}$$

$$\cot x = \frac{1}{\tan x} = \frac{\cos x}{\sin x}$$

$$1 + \tan^2 x = \sec^2 x = \frac{1}{\cos^2 x}$$

$$1 + \cot^2 x = \csc^2 x = \frac{1}{\sin^2 x}$$

$$\sin x = \sqrt{1 - \cos^2 x} = \frac{\tan x}{\sqrt{1 + \tan^2 x}} = \frac{1}{\sqrt{1 + \cot^2 x}}$$

$$\cos x = \sqrt{1 - \sin^2 x} = \frac{1}{\sqrt{1 + \tan^2 x}} = \frac{\cot x}{\sqrt{1 + \cot^2 x}}$$

Functions of Negative Angles. $\sin(-x) = -\sin x$; $\cos(-x) = \cos x$; $\tan(-x) = -\tan x$.

Functions of the Sum and Difference of Two Angles

$\sin(x + y) = \sin x \cos y + \cos x \sin y$.
$\cos(x + y) = \cos x \cos y - \sin x \sin y$.
$\tan(x + y) = (\tan x + \tan y)/(1 - \tan x \tan y)$.
$\cot(x + y) = (\cot x \cot y - 1)/(\cot x + \cot y)$.
$\sin(x - y) = \sin x \cos y - \cos x \sin y$.
$\cos(x - y) = \cos x \cos y + \sin x \sin y$.
$\tan(x - y) = (\tan x - \tan y)/(1 + \tan x \tan y)$.
$\cot(x - y) = (\cot x \cot y + 1)/(\cot y - \cot x)$.
$\sin x + \sin y = 2 \sin \frac{1}{2}(x + y) \cos \frac{1}{2}(x - y)$.
$\sin x - \sin y = 2 \cos \frac{1}{2}(x + y) \sin \frac{1}{2}(x - y)$.
$\cos x + \cos y = 2 \cos \frac{1}{2}(x + y) \cos \frac{1}{2}(x - y)$.
$\cos x - \cos y = -2 \sin \frac{1}{2}(x + y) \sin \frac{1}{2}(x - y)$.

$$\tan x + \tan y = \frac{\sin(x + y)}{\cos x \cos y}; \quad \cot x + \cot y = \frac{\sin(x + y)}{\sin x \sin y}.$$

$$\tan x - \tan y = \frac{\sin(x - y)}{\cos x \cos y}; \quad \cot x - \cot y = \frac{\sin(y - x)}{\sin x \sin y}.$$

$\sin^2 x - \sin^2 y = \cos^2 y - \cos^2 x = \sin(x + y)\sin(x - y)$.
$\cos^2 x - \sin^2 y = \cos^2 y - \sin^2 x = \cos(x + y)\cos(x - y)$.

* From Baumeister and Marks, *Standard Handbook for Mechanical Engineers*, 7th ed., 1967, McGraw-Hill Book Company; by permission.

$\sin (45° + x) = \cos (45° - x)$; $\tan (45° + x) = \cot (45° - x)$.
$\sin (45° - x) = \cos (45° + x)$; $\tan (45° - x) = \cot (45° + x)$.

In the following transformations, a and b are supposed to be positive, $c = \sqrt{a^2 + b^2}$, $A =$ the positive acute angle for which $\tan A = a/b$, and $B =$ the positive acute angle for which $\tan B = b/a$:

$a \cos x + b \sin x = c \sin (A + x) = c \cos (B - x)$.
$a \cos x - b \sin x = c \sin (A - x) = c \cos (B + x)$.

EXPANSION IN SERIES

From Baumeister and Marks, *Standard Handbook for Mechanical Engineers*, 7th ed., 1967, McGraw-Hill Book Company; by permission

The range of values of x for which each of the series is convergent is stated at the right of the series.

Exponential and Logarithmic Series

$$e^x = 1 + \frac{x}{1!} + \frac{x^2}{2!} + \frac{x^3}{3!} + \frac{x^4}{4!} + \cdots \qquad (-\infty < x < +\infty)$$

$$a^x = e^{mx} = 1 + \frac{m}{1!}x + \frac{m^2}{2!}x^2 + \frac{m^3}{3!}x^3 + \cdots$$

$$[a > 0, \; -\infty < x < +\infty]$$

where $m = \ln a = (2.3026)(\log_{10} a)$.

$$\ln (1 + x) = x - \frac{x^2}{2} + \frac{x^3}{3} - \frac{x^4}{4} + \frac{x^5}{5} \cdots \qquad [-1 < x < +1]$$

$$\ln (1 - x) = -x - \frac{x^2}{2} - \frac{x^3}{3} - \frac{x^4}{4} - \frac{x^5}{5} - \cdots \qquad [-1 < x < +1]$$

$$\ln \left(\frac{1 + x}{1 - x}\right) = 2\left(x + \frac{x^3}{3} + \frac{x^5}{5} + \frac{x^7}{7} + \cdots\right) \qquad [-1 < x < +1]$$

$$\ln \left(\frac{x + 1}{x - 1}\right) = 2\left(\frac{1}{x} + \frac{1}{3x^3} + \frac{1}{5x^5} + \frac{1}{7x^7} + \cdots\right)$$

$$[x < -1 \text{ or } +1 < x]$$

$$\ln x = 2\left[\frac{x - 1}{x + 1} + \frac{1}{3}\left(\frac{x - 1}{x + 1}\right)^3 + \frac{1}{5}\left(\frac{x - 1}{x + 1}\right)^5 + \cdots\right] \, [0 < x < \infty]$$

$$\ln (a + x) = \ln a + 2\left[\frac{x}{2a + x} + \frac{1}{3}\left(\frac{x}{2a + x}\right)^3 + \frac{1}{5}\left(\frac{x}{2a + x}\right)^5 + \cdots\right]$$

$$[0 < a < +\infty, \; -a < x < +\infty]$$

Series for the Trigonometric Functions. In the following formulas, *all angles must be expressed in radians.* If D = the number of degrees in the angle, and x = its radian measure, then $x = 0.017453D$.

$$\sin x = x - \frac{x^3}{3!} + \frac{x^5}{5!} - \frac{x^7}{7!} + \cdots \qquad [-\infty < x < +\infty]$$

$$\cos x = 1 - \frac{x^2}{2!} + \frac{x^4}{4!} - \frac{x^6}{6!} + \frac{x^8}{8!} - \cdots \qquad [-\infty < x < +\infty]$$

$$\tan x = x + \frac{x^3}{3} + \frac{2x^5}{15} + \frac{17x^7}{315} + \frac{62x^9}{2835} + \cdots$$
$$[-\pi/2 < x < +\pi/2]$$

$$\cot x = \frac{1}{x} - \frac{x}{3} - \frac{x^3}{45} - \frac{2x^5}{945} - \frac{x^7}{4725} - \cdots \qquad [-\pi < x < +\pi]$$

$$\sin^{-1} y = y + \frac{y^3}{6} + \frac{3y^5}{40} + \frac{5y^7}{112} + \cdots \qquad [-1 \leqq y \leqq +1]$$

$$\tan^{-1} y = y - \frac{y^3}{3} + \frac{y^5}{5} - \frac{y^7}{7} + \cdots \qquad [-1 \leqq y \leqq +1]$$

$$\cos^{-1} y = \tfrac{1}{2}\pi - \sin^{-1} y; \; \cot^{-1} y = \tfrac{1}{2}\pi - \tan^{-1} y.$$

STATISTICAL TABLES

STATISTICS

Raw data are collected observations that have not been organized numerically. An array is an arrangement of raw numerical data in ascending or descending order of magnitude. An *average* is a value that is typical or representative of a set of data. Several averages can be defined, the most common being the arithmetic mean (or briefly the mean), the median, the mode, and the geometric mean.

The *mean* of a set of N numbers, $x_1, x_2, x_3, \ldots, x_N$ is denoted by $\bar{x}$ and is defined as

$$\bar{x} = \frac{x_1 + x_2 + x_3 + \cdots + x_N}{N}$$

It is an estimate of the unknown true value μ of an infinite population.

The *median* of a set of numbers arranged in order of magnitude is the middle value or the arithmetic mean of the two middle values. The median allows inclusion of all data in a set without undue influence from outlying values; it is preferable to the mean for small sets of data.

The *mode* of a set of numbers is that value which occurs with the greatest frequency (the most common value). The mode may not exist, and even if it does exist it may not be unique. The empirical relation that exists between the mean, the mode, and the median for unimodal frequency curves which are moderately asymmetrical is

$$\text{Mean} - \text{mode} = 3(\text{mean} - \text{median})$$

The *geometric mean* of a set of N numbers is the Nth root of the product of the numbers:

$$\sqrt[N]{x_1 x_2 x_3 \ldots x_N}$$

Distribution of Measurements

When a large number of measurements are made, the individual measurements are not all identical and equal to the accepted value μ, which is the mean of an infinite population or universe of data, but are scattered about μ, owing to random error. If the magnitude of any single measurement is the abscissa and the relative frequencies (i.e., the probability) of occurrence of different-sized measurements are the ordinate, the smooth curve drawn through the points (Fig. 1-10) is the *normal distribution curve* (also called the *Gaussian distribution curve*, the *error curve*, or the *probability curve*). The term "error curve" arises when one considers the distribution of errors $(x - \mu)$ about the true value.

The breadth or spread of the curve indicates the precision of the measurements, and is determined by and related to the standard deviation, a relationship that is expressed in the equation for the normal curve,

$$Y = \frac{1}{\sigma\sqrt{2\pi}} e^{-1/2(x-\mu)^2/\sigma^2}$$

where σ is the standard deviation of the infinite population. The population mean μ expresses the magnitude of the quantity being measured; the standard deviation σ expresses the scatter. When $(x - \mu)/\sigma$ is replaced by the standardized variable z, then

$$Y = \frac{1}{\sqrt{2\pi}} e^{-1/2z^2}$$

The standardized variable measures the deviation from the population mean in units of standard deviation. Y is 0.399 for the most probable value, μ.

Table 1-16 lists the height of an ordinate (column Y) as a distance z from the mean, and the area under the normal curve (column A) at

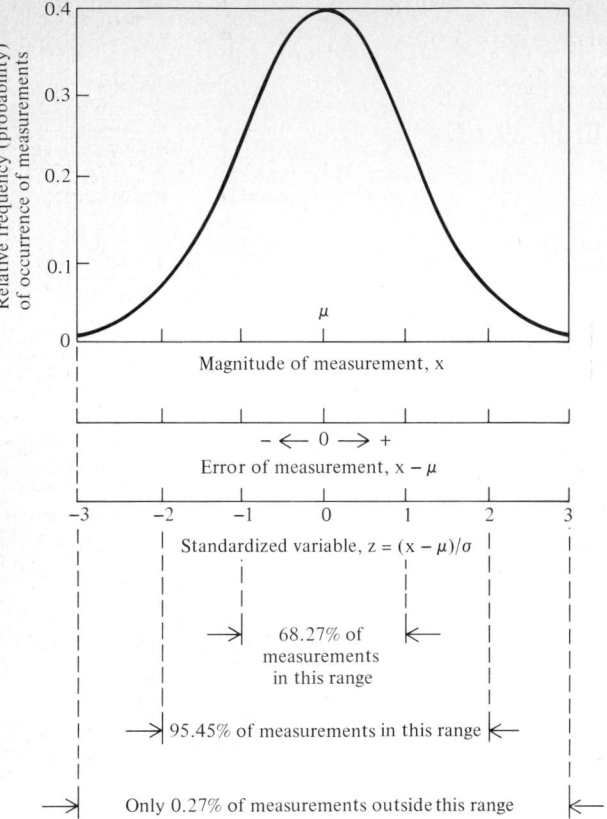

Fig. 1-10. The normal distribution curve.

a distance z from the mean, expressed as fractions of the total area, 1.000.

Example 1. The true value of a quantity is 30.00, and σ for the method of measurement is 0.30. What is the probability that a single measurement will have a deviation from the mean greater than 0.45; that is, what percentage of results will fall outside the range 30.00 ± 0.45?

$$z = \frac{x - \mu}{\sigma} = \frac{0.45}{0.30} = 1.5$$

From Table 1-16 the area under the normal curve from -1.5σ to $+1.5\sigma$ is 0.866, meaning that 86.6 per cent of the measurements will fall within the range 30.00 ± 0.45 and 13.4 per cent will lie outside this range. Half of these measurements, 6.8 per cent, will be less than 29.55; and a similar percentage will exceed 30.45. In actuality the uncertainty in

z is about 1 in 15; therefore, the value of z could lie between 1.4 and 1.6; the corresponding areas under the curve could lie between 84 and 89 per cent.

Example 2. In the foregoing example, what expected number of samples will be 29.40 or less from a total of 400 samples?

$$z = \frac{29.40 - 30.00}{0.30} = 2.0$$

The area under the negative portion of the normal curve corresponds to $0.500 - 0.477 = 0.023$. Only $0.023(400) = 9.2$ or nine samples will be 29.40 or less.

Example 3. If the mean value of 500 determinations is 151 and $\sigma = 15$, how many results lie between 120 and 155 (actually any value between 119.5 and 155.5)?

$$z = \frac{119.5 - 151}{15} = -2.10 \qquad \text{Area: } 0.482$$

$$z = \frac{155.5 - 151}{15} = 0.30$$

$$0.118$$

Total area: 0.600

$500(0.600) = 300$ results

Measures of Dispersion

Several ways may be used to characterize the spread or dispersion in the original data. The *range* is the difference between the largest value and the smallest value in a set of observations. However, almost always the most efficient quantity for characterizing variability is the *standard deviation* (also called the *root mean square*).

The standard deviation is the square root of the average squared difference between the individual observations and the population mean:

$$\sigma = \sqrt{\frac{\sum\limits_{i=1}^{N} (x_i - \mu)^2}{N}}$$

The standard deviation may be estimated by calculating s drawn from a sample set as follows:

$$s = \sqrt{\frac{\sum\limits_{i=1}^{N} (x_i - \bar{x})^2}{N - 1}} \quad \text{or} \quad s = \sqrt{\frac{x_1^2 + x_2^2 + \cdots - [(x_1 + x_2 + \cdots)^2]/N}{N - 1}}$$

where $x_i - \bar{x}$ represents the deviation of each number in the array from the arithmetic mean. Since two pieces of information, namely s and $\bar{x}$, have been extracted from the data, we are left with $N - 1$ independent data available for measurement of precision. The divisor is termed the *degrees of freedom*. If a relatively large sample of data corresponding to $N > 30$ is available, its mean can be taken as a measure of μ, and $s \simeq \sigma$.

So basic is the notion of a statistical *estimate* of a physical parameter that statisticians use Greek letters for the *parameters* and English letters for the estimates.

For many purposes, one uses the *variance* which for the sample is s^2, and for the entire population is σ^2. The variance s^2 of a finite sample is an unbiased estimate of σ^2, whereas the standard deviation s is not an unbiased estimate of σ.

When a series of observations can be logically arranged into k subgroups, the variance is calculated by summing the squares of the deviations for each subgroup, and then adding all the k sums and dividing by $N - k$ because one degree of freedom is lost in each subgroup. For two groups of observations consisting of N_A and N_B members, of standard deviations s_A and s_B, respectively, the variance is given by

$$s^2 = \frac{(N_A - 1)s_A^2 + (N_B - 1)s_B^2}{N_A + N_B - 2}$$

Another measure of dispersion is the *coefficient of variation*, which is merely the standard deviation expressed as a percentage of the arithmetic mean, viz: $\dfrac{100s}{\bar{x}}$. It is useful mainly to show whether the relative or the absolute spread of values is constant as the values are changed.

Theoretical Distributions and Tests of Significance

If the data contained only random (or chance) errors, the cumulative estimates $\bar{x}$ and s would gradually approach the limits μ and σ. The distribution of results would be normally distributed with mean μ and standard deviation σ. Were the true mean of the infinite population known, it would be expected that the averaged means for each group of data would also have some symmetrical type of distribution centered around μ. However, it would be expected that the dispersion or spread of this dispersion about the mean would depend on the sample size. The standard deviation of the distribution of means equals $\sigma/\sqrt{N}$. A distribution of this type is called the "Student's distribution"; and the corresponding test of significance, a measure of error between μ and $\bar{x}$, the t test. The Student t takes into account both the possible variation of the value of $\bar{x}$ from μ on the basis of the expected variance

$\sigma^2/\sqrt{N}$ and the reliability of using s in place of σ. The distribution of the statistic is

$$\pm t = \frac{\overline{x} - \mu}{s/\sqrt{N}} \qquad \text{or} \qquad \mu = \overline{x} \pm \frac{ts}{\sqrt{N}}$$

This distribution is symmetrical about zero, and its dispersion is a function of the degrees of freedom $N - 1$. Its limits are called *confidence limits*. The percentage probability that μ lies within this interval is called the *confidence level*. The *level of significance* or *error probability* (100 $-$ confidence level or $100 - \alpha$) is the per cent probability that μ will lie outside the confidence interval, and represents the chances of being incorrect in stating that μ lies within the confidence interval. Values of t are in Table 1-17 for any desired degrees of freedom and various confidence levels.

Statistical methods are frequently used to give a yes or no answer to a particular question concerning the significance of data. The answer is qualified by a confidence level indicating the degree of certainty of the answer. A common procedure is to set up a *null hypothesis*, which states that there is no significant difference between two sets of data or that a variable exerts no significant effect. Generally confidence levels of 95 and 99 per cent are chosen to express the probability that the answer is correct. These are also denoted as the 0.05 and 0.01 level of significance, respectively. When the hypothesis can be rejected at the 0.05 level of significance, but not at the 0.01 level, we can say that the sample results are probably significant. If, however, the hypothesis is also rejected at the 0.01 level, the results become highly significant. For a small number of samples we replace z, obtained from Table 1-16, by t from Table 1-17, and we replace σ by $[\sqrt{N/(N - 1)}]s$.

Example 4. In the past a method gave $\mu = 0.050$ per cent. A recent set of 10 results gave $\overline{x} = 0.053$ per cent and $s = 0.003$ per cent. Is everything satisfactory at a level of significance of 0.05? Of 0.01? We wish to decide between the hypotheses

H_0: $\mu = 0.050\%$ and the method is working properly; and
H_1: $\mu \neq 0.050\%$ and the method is not working properly.

A *two-tailed test* is required; that is, both tails on the distribution curve are involved:

$$t = \frac{0.053 - 0.050}{0.003} \sqrt{10 - 1} = -3.00$$

Enter Table 1-17 for 9 degrees of freedom under the column headed $t_{.975}$ for the 0.05 level of significance, and the column $t_{.995}$ for the 0.01 level of significance. At the 0.05 level, accept H_0 if t lies inside the interval $-t_{.975}$ to $t_{.975}$, that is, within -2.26 and 2.26; reject otherwise.

Since $t = -3.00$, we reject H_0. At the 0.01 level of significance, the corresponding interval is -3.25 to 3.25, which t lies within, indicating acceptance of H_0. Because we can reject H_0 at the 0.05 level but not at the 0.01 level of significance, we can say that the sample results are probably significant and that the method is not working properly.

Example 5. Six samples from a bulk chemical shipment averaged 77.50 per cent active ingredient with $s = 1.45$ per cent. The manufacturer claimed 80.00 per cent. Can his claim be supported? A one-tailed test is required.

$$t = \frac{77.50 - 80.00}{1.45} \sqrt{6 - 1} = -3.86$$

Since $t_{.95} = -2.01$, and $t_{.99} = -3.36$, the hypothesis is rejected at both the 0.05 and the 0.01 levels of significance. It is extremely unlikely that the claim is justified.

It is possible to compare the means of two relatively small sets of observations when the variances within the sets can be regarded as the same, as indicated by the F test. Also the t test can be applied to differences between pairs of observations. Perhaps only a single pair can be performed at one time, or possibly one wishes to compare two methods using samples of differing analytical content. It is still necessary that the two methods possess the same inherent standard deviation. An average difference $\bar{d}$ is calculated, and individual deviations from $\bar{d}$ are used to evaluate the variance of the differences.

Example 6. From the following data do the two methods actually give concordant results?

Sample	Method A	Method B	Difference
1	33.27	33.04	$d_1 = 0.23$
2	51.34	50.96	$d_2 = 0.38$
3	23.91	23.77	$d_3 = 0.14$
4	47.04	46.79	$d_4 = 0.25$
			$\bar{d} = 0.25$

$$s_d = \sqrt{\frac{\Sigma(d - \bar{d})^2}{N - 1}} = 0.099$$

$$t = \frac{0.25}{0.099} \sqrt{4 - 1} = 4.30$$

From Table 1-17, $t_{.975} = 3.18$ (at 95 per cent probability) and $t_{.995} = 5.84$ (at 99 per cent probability). The difference between the two methods is probably significant.

One can consider the distribution involving estimates of the true variance. With s_1^2 determined from a group of N_1 observations and s_2^2

from a second group of N_2 observations, the distribution of the ratio of the sample variances is given by the F test:

$$F = \frac{s_1^2}{s_2^2}$$

The larger variance is placed in the numerator. For example, the F test allows judgment regarding the existence of a significant difference in the precision between two sets of data or between two analysts. The hypothesis assumed is that both variances are indeed alike and a measure of the same σ.

Example 7. Suppose Analyst A made five observations and obtained a standard deviation of 0.06 whereas Analyst B with six observations obtained $s_B = 0.03$. The experimental variance ratio is

$$F = \frac{(0.06)^2}{(0.03)^2} = 4.00$$

From Table 1-19, with four degrees of freedom for A and five degrees of freedom for B, the value of F would exceed 5.19 five per cent of the time. Therefore, the null hypothesis is valid, and comparable skills are exhibited by the two analysts. As applied, the F test was one-tailed. The F test may also be applied as a two-tailed test in which the alternative to the null hypothesis is $\sigma_1^2 \neq \sigma_2^2$. This doubles the probability that the null hypothesis is invalid and has the effect of changing the confidence level, in the above example, from 95 to 90 per cent.

If improvement in precision is claimed for a set of measurements, the variance for the set against which comparison is being made should be placed in the numerator, regardless of magnitude. An experimental F smaller than unity indicates that the claim for improved precision cannot be supported.

As we have seen, for each group of samples a standard deviation can be calculated. These estimates of σ possess a distribution called the chi-square (χ^2) distribution:

$$\chi^2 = \frac{s^2}{\sigma^2/df}$$

The upper and lower confidence limits for the standard deviation are obtained by dividing $(N - 1)s^2$ by two numbers taken from Table 1-18.

Example 8. The variance obtained for 10 samples is $(0.65)^2$. How reliable is s^2 as an estimate of σ^2? σ^2 is known to be $(0.75)^2$.

$$\frac{s^2(N-1)}{\chi^2_{.975}} < \sigma^2 < \frac{s^2(N-1)}{\chi^2_{.025}}$$

$$\frac{(0.65)^2(10-1)}{19.02} < \sigma^2 < \frac{(0.65)^2(10-1)}{2.70}$$

$$0.20 < \sigma^2 < 1.43$$

Thus, only one time in 40 will $9s^2/\sigma^2$ be less than 2.70 by chance alone. Similarly, only one time in 40 will $9s^2/\sigma^2$ be greater than 19.02. Consequently, it is not unlikely that s^2 is a reliable estimate of σ^2. Stated differently:

Upper limit: $\sigma^2 = \dfrac{9s^2}{2.7} = 3.3s^2$

Lower Limit: $\sigma^2 = \dfrac{9s^2}{19.02} = 0.48s^2$

Ten measurements give an estimate of σ^2 that may be as much as 3.3 times or only about one-half the true variance.

Table 1-16
ORDINATES AND AREAS BETWEEN ABSCISSA VALUES
−z and +z OF THE NORMAL DISTRIBUTION CURVE

From Perry, Chilton, and Kirkpatrick, *Chemical Engineers' Handbook*, 4th ed., 1963, McGraw-Hill Book Company; by permission

z	X	Y	A	1−A
0	μ	0.399	0.0000	1.0000
±.05	$\mu\pm0.05\sigma$	.398	.0399	.9601
.10	$\mu\pm.10\sigma$	.397	.0797	.9203
.15	$\mu\pm.15\sigma$	.394	.1192	.8808
.20	$\mu\pm.20\sigma$	.391	.1585	.8415
.25	$\mu\pm.25\sigma$	.387	.1974	.8026
.30	$\mu\pm.30\sigma$	.381	.2358	.7642
.35	$\mu\pm.35\sigma$	.375	.2737	.7263
.40	$\mu\pm.40\sigma$	.368	.3108	.6892
.45	$\mu\pm.45\sigma$	.361	.3473	.6527
.50	$\mu\pm.50\sigma$	.352	.3829	.6171
.55	$\mu\pm.55\sigma$	.343	.4177	.5823
.60	$\mu\pm.60\sigma$	.333	.4515	.5485
.65	$\mu\pm.65\sigma$	.323	.4843	.5157
.70	$\mu\pm.70\sigma$	.312	.5161	.4839
.75	$\mu\pm.75\sigma$	.301	.5467	.4533
.80	$\mu\pm.80\sigma$	.290	.5763	.4237
.85	$\mu\pm.85\sigma$	.278	.6047	.3953
.90	$\mu\pm.90\sigma$	.266	.6319	.3681
.95	$\mu\pm.95\sigma$	.254	.6579	.3421
1.00	$\mu\pm1.00\sigma$	.242	.6827	.3173
1.05	$\mu\pm1.05\sigma$	.230	.7063	.2937
1.10	$\mu\pm1.10\sigma$	.218	.7287	.2713
1.15	$\mu\pm1.15\sigma$	.206	.7499	.2501
1.20	$\mu\pm1.20\sigma$	.194	.7699	.2301
1.25	$\mu\pm1.25\sigma$	.183	.7887	.2113
1.30	$\mu\pm1.30\sigma$	.171	.8064	.1936
1.35	$\mu\pm1.35\sigma$	.160	.8230	.1770
1.40	$\mu\pm1.40\sigma$	.150	.8385	.1615
1.45	$\mu\pm1.45\sigma$	.139	.8529	.1471
±1.50	$\mu\pm1.50\sigma$	.130	.8664	.1336
±0.000	μ	0.3989	0.0000	1.0000
.126	$\mu\pm0.126\sigma$	.3958	.1000	.9000
.253	$\mu\pm.253\sigma$	.3863	.2000	.8000
.385	$\mu\pm.385\sigma$	.3704	.3000	.7000
.524	$\mu\pm.524\sigma$	.3477	.4000	.6000
.674	$\mu\pm.674\sigma$	.3178	.5000	.5000
.842	$\mu\pm.842\sigma$	.2800	.6000	.4000

z	X	Y	A	1−A
±1.50	$\mu\pm1.50\sigma$	0.1295	0.8664	0.1336
±1.55	$\mu\pm1.55\sigma$	.1200	.8789	.1211
±1.60	$\mu\pm1.60\sigma$	.1109	.8904	.1096
±1.65	$\mu\pm1.65\sigma$	.1023	.9011	.0989
±1.70	$\mu\pm1.70\sigma$	.0940	.9109	.0891
±1.75	$\mu\pm1.75\sigma$	.0863	.9199	.0801
±1.80	$\mu\pm1.80\sigma$	.0790	.9281	.0719
±1.85	$\mu\pm1.85\sigma$	.0721	.9357	.0643
±1.90	$\mu\pm1.90\sigma$	.0656	.9426	.0574
±1.95	$\mu\pm1.95\sigma$	.0596	.9488	.0512
±2.00	$\mu\pm2.00\sigma$	.0540	.9545	.0455
±2.05	$\mu\pm2.05\sigma$	.0488	.9596	.0404
±2.10	$\mu\pm2.10\sigma$	.0440	.9643	.0357
±2.15	$\mu\pm2.15\sigma$	.0396	.9684	.0316
±2.20	$\mu\pm2.20\sigma$	.0355	.9722	.0278
±2.25	$\mu\pm2.25\sigma$	.0317	.9756	.0244
±2.30	$\mu\pm2.30\sigma$	.0283	.9786	.0214
±2.35	$\mu\pm2.35\sigma$	.0252	.9812	.0188
±2.40	$\mu\pm2.40\sigma$	.0224	.9836	.0164
±2.45	$\mu\pm2.45\sigma$	.0198	.9857	.0143
±2.50	$\mu\pm2.50\sigma$	.0175	.9876	.0124
±2.55	$\mu\pm2.55\sigma$	.0154	.9892	.0108
±2.60	$\mu\pm2.60\sigma$	.0136	.9907	.0093
±2.65	$\mu\pm2.65\sigma$	.0119	.9920	.0080
±2.70	$\mu\pm2.70\sigma$	.0104	.9931	.0069
±2.75	$\mu\pm2.75\sigma$	.0091	.9940	.0060
±2.80	$\mu\pm2.80\sigma$	.0079	.9949	.0051
±2.85	$\mu\pm2.85\sigma$	.0069	.9956	.0044
±2.90	$\mu\pm2.90\sigma$	.0060	.9963	.0037
±2.95	$\mu\pm2.95\sigma$	.0051	.9968	.0032
±3.00	$\mu\pm3.00\sigma$	.0044	.9973	.0027
±4.00	$\mu\pm4.00\sigma$	.0001	.99994	.00006
±5.00	$\mu\pm5.00\sigma$	.00001	.999994	.000006
±1.036	$\mu\pm1.036\sigma$	0.2331	0.7000	0.3000
±1.282	$\mu\pm1.282\sigma$	.1755	.8000	.2000
±1.645	$\mu\pm1.645\sigma$	.1031	.9000	.1000
±1.960	$\mu\pm1.960\sigma$	.0584	.9500	.0500
±2.576	$\mu\pm2.576\sigma$	.0145	.9900	.0100
±3.291	$\mu\pm3.291\sigma$	.0018	.9990	.0010
±3.891	$\mu\pm3.891\sigma$	.0002	.9999	.0001

Table 1-17
VALUES OF t

From Perry, Chilton, and Kirkpatrick, *Chemical Engineers' Handbook*, 4th
ed., 1963, McGraw-Hill Book Company; by permission

df	$t_{.60}$	$t_{.70}$	$t_{.80}$	$t_{.90}$	$t_{.95}$	$t_{.975}$	$t_{.99}$	$t_{.995}$
1	0.325	0.727	1.376	3.078	6.314	12.706	31.821	63.657
2	.289	.617	1.061	1.886	2.920	4.303	6.965	9.925
3	.277	.584	0.978	1.638	2.353	3.182	4.541	5.841
4	.271	.569	.941	1.533	2.132	2.776	3.747	4.604
5	.267	.559	.920	1.476	2.015	2.571	3.365	4.032
6	.265	.553	.906	1.440	1.943	2.447	3.143	3.707
7	.263	.549	.896	1.415	1.895	2.365	2.998	3.499
8	.262	.546	.889	1.397	1.860	2.306	2.896	3.355
9	.261	.543	.883	1.383	1.833	2.262	2.821	3.250
10	.260	.542	.879	1.372	1.812	2.228	2.764	3.169
11	.260	.540	.876	1.363	1.796	2.201	2.718	3.106
12	.259	.539	.873	1.356	1.782	2.179	2.681	3.055
13	.259	.538	.870	1.350	1.771	2.160	2.650	3.012
14	.258	.537	.868	1.345	1.761	2.145	2.624	2.977
15	.258	.536	.866	1.341	1.753	2.131	2.602	2.947
16	.258	.535	.865	1.337	1.746	2.120	2.583	2.921
17	.257	.534	.863	1.333	1.740	2.110	2.567	2.898
18	.257	.534	.862	1.330	1.734	2.101	2.552	2.878
19	.257	.533	.861	1.328	1.729	2.093	2.539	2.861
20	.257	.533	.860	1.325	1.725	2.086	2.528	2.845
21	.257	.532	.859	1.323	1.721	2.080	2.518	2.831
22	.256	.532	.858	1.321	1.717	2.074	2.508	2.819
23	.256	.532	.858	1.319	1.714	2.069	2.500	2.807
24	.256	.531	.857	1.318	1.711	2.064	2.492	2.797
25	.256	.531	.856	1.316	1.708	2.060	2.485	2.787
26	.256	.531	.856	1.315	1.706	2.056	2.479	2.779
27	.256	.531	.855	1.314	1.703	2.052	2.473	2.771
28	.256	.530	.855	1.313	1.701	2.048	2.467	2.763
29	.256	.530	.854	1.311	1.699	2.045	2.462	2.756
30	.256	.530	.854	1.310	1.697	2.042	2.457	2.750
40	.255	.529	.851	1.303	1.684	2.021	2.423	2.704
60	.254	.527	.848	1.296	1.671	2.000	2.390	2.660
120	.254	.526	.845	1.289	1.658	1.980	2.358	2.617
∞	.253	.524	.842	1.282	1.645	1.960	2.326	2.576
df	$-t_{.40}$	$-t_{.30}$	$-t_{.20}$	$-t_{.10}$	$-t_{.05}$	$-t_{.025}$	$-t_{.01}$	$-t_{.005}$

When the table is read from the foot, the tabled values are to be prefixed with
a negative sign. Interpolation should be performed using the reciprocals of
the degrees of freedom.

Table 1-18
PERCENTILES OF THE χ^2 DISTRIBUTION

From Perry, Chilton, and Kirkpatrick, *Chemical Engineers' Handbook*, 4th ed., 1963, McGraw-Hill Book Company; by permission

df	Per cent									
	0.5	1	2.5	5	10	90	95	97.5	99	99.5
1	0.000039	0.00016	0.00098	0.0039	0.0158	2.71	3.84	5.02	6.63	7.88
2	.0100	.0201	.0506	.1026	.2107	4.61	5.99	7.38	9.21	10.60
3	.0717	.115	.216	.352	.584	6.25	7.81	9.35	11.34	12.84
4	.207	.297	.484	.711	1.064	7.78	9.49	11.14	13.28	14.86
5	.412	.554	.831	1.15	1.61	9.24	11.07	12.83	15.09	16.75
6	.676	.872	1.24	1.64	2.20	10.64	12.59	14.45	16.81	18.55
7	.989	1.24	1.69	2.17	2.83	12.02	14.07	16.01	18.48	20.28
8	1.34	1.65	2.18	2.73	3.49	13.36	15.51	17.53	20.09	21.96
9	1.73	2.09	2.70	3.33	4.17	14.68	16.92	19.02	21.67	23.59
10	2.16	2.56	3.25	3.94	4.87	15.99	18.31	20.48	23.21	25.19
11	2.60	3.05	3.82	4.57	5.58	17.28	19.68	21.92	24.73	26.76
12	3.07	3.57	4.40	5.23	6.30	18.55	21.03	23.34	26.22	28.30
13	3.57	4.11	5.01	5.89	7.04	19.81	22.36	24.74	27.69	29.82
14	4.07	4.66	5.63	6.57	7.79	21.06	23.68	26.12	29.14	31.32
15	4.60	5.23	6.26	7.26	8.55	22.31	25.00	27.49	30.58	32.80
16	5.14	5.81	6.91	7.96	9.31	23.54	26.30	28.85	32.00	34.27
18	6.26	7.01	8.23	9.39	10.86	25.99	28.87	31.53	34.81	37.16
20	7.43	8.26	9.59	10.85	12.44	28.41	31.41	34.17	37.57	40.00
24	9.89	10.86	12.40	13.85	15.66	33.20	36.42	39.36	42.98	45.56
30	13.79	14.95	16.79	18.49	20.60	40.26	43.77	46.98	50.89	53.67
40	20.71	22.16	24.43	26.51	29.05	51.81	55.76	59.34	63.69	66.77
60	35.53	37.48	40.48	43.19	46.46	74.40	79.08	83.30	88.38	91.95
120	83.85	86.92	91.58	95.70	100.62	140.23	146.57	152.21	158.95	163.64

For large values of degrees of freedom the approximate formula

$$x_\alpha^2 = n \left(1 - \frac{2}{9n} + z_\alpha \sqrt{\frac{2}{9n}} \right)^3$$

where z_α is the normal deviate and n is the number of degrees of freedom, may be used. For example, $x_{.99}^2 = 60[1 - 0.00370 + 2.326(0.06086)]^3 = 60(1.1379)^3$ = 88.4 for the 99th percentile for 60 degrees of freedom.

Table 1-19
F DISTRIBUTION

From Perry, Chilton, and Kirkpatrick, *Chemical Engineers' Handbook*, 4th ed., 1963, McGraw-Hill Book Company; by permission

Upper 5% Points ($F_{.95}$)

Denominator df	\	\	\	\	\	\	\	\	\	\	\	\	\	\	\	\	\	\	\
Numerator df →	1	2	3	4	5	6	7	8	9	10	12	15	20	24	30	40	60	120	∞
1	161	200	216	225	230	234	237	239	241	242	244	246	248	249	250	251	252	253	254
2	18.5	19.0	19.2	19.2	19.3	19.3	19.4	19.4	19.4	19.4	19.4	19.4	19.4	19.5	19.5	19.5	19.5	19.5	19.5
3	10.1	9.55	9.28	9.12	9.01	8.94	8.89	8.85	8.81	8.79	8.74	8.70	8.66	8.64	8.62	8.59	8.57	8.55	8.53
4	7.71	6.94	6.59	6.39	6.26	6.16	6.09	6.04	6.00	5.96	5.91	5.86	5.80	5.77	5.75	5.72	5.69	5.66	5.63
5	6.61	5.79	5.41	5.19	5.05	4.95	4.88	4.82	4.77	4.74	4.68	4.62	4.56	4.53	4.50	4.46	4.43	4.40	4.37
6	5.99	5.14	4.76	4.53	4.39	4.28	4.21	4.15	4.10	4.06	4.00	3.94	3.87	3.84	3.81	3.77	3.74	3.70	3.67
7	5.59	4.74	4.35	4.12	3.97	3.87	3.79	3.73	3.68	3.64	3.57	3.51	3.44	3.41	3.38	3.34	3.30	3.27	3.23
8	5.32	4.46	4.07	3.84	3.69	3.58	3.50	3.44	3.39	3.35	3.28	3.22	3.15	3.12	3.08	3.04	3.01	2.97	2.93
9	5.12	4.26	3.86	3.63	3.48	3.37	3.29	3.23	3.18	3.14	3.07	3.01	2.94	2.90	2.86	2.83	2.79	2.75	2.71
10	4.96	4.10	3.71	3.48	3.33	3.22	3.14	3.07	3.02	2.98	2.91	2.85	2.77	2.74	2.70	2.66	2.62	2.58	2.54
11	4.84	3.98	3.59	3.36	3.20	3.09	3.01	2.95	2.90	2.85	2.79	2.72	2.65	2.61	2.57	2.53	2.49	2.45	2.40
12	4.75	3.89	3.49	3.26	3.11	3.00	2.91	2.85	2.80	2.75	2.69	2.62	2.54	2.51	2.47	2.43	2.38	2.34	2.30
13	4.67	3.81	3.41	3.18	3.03	2.92	2.83	2.77	2.71	2.67	2.60	2.53	2.46	2.42	2.38	2.34	2.30	2.25	2.21
14	4.60	3.74	3.34	3.11	2.96	2.85	2.76	2.70	2.65	2.60	2.53	2.46	2.39	2.35	2.31	2.27	2.22	2.18	2.13
15	4.54	3.68	3.29	3.06	2.90	2.79	2.71	2.64	2.59	2.54	2.48	2.40	2.33	2.29	2.25	2.20	2.16	2.11	2.07
16	4.49	3.63	3.24	3.01	2.85	2.74	2.66	2.59	2.54	2.49	2.42	2.35	2.28	2.24	2.19	2.15	2.11	2.06	2.01
17	4.45	3.59	3.20	2.96	2.81	2.70	2.61	2.55	2.49	2.45	2.38	2.31	2.23	2.19	2.15	2.10	2.06	2.01	1.96
18	4.41	3.55	3.16	2.93	2.77	2.66	2.58	2.51	2.46	2.41	2.34	2.27	2.19	2.15	2.11	2.06	2.02	1.97	1.92
19	4.38	3.52	3.13	2.90	2.74	2.63	2.54	2.48	2.42	2.38	2.31	2.23	2.16	2.11	2.07	2.03	1.98	1.93	1.88
20	4.35	3.49	3.10	2.87	2.71	2.60	2.51	2.45	2.39	2.35	2.28	2.20	2.12	2.08	2.04	1.99	1.95	1.90	1.84
21	4.32	3.47	3.07	2.84	2.68	2.57	2.49	2.42	2.37	2.32	2.25	2.18	2.10	2.05	2.01	1.96	1.92	1.87	1.81
22	4.30	3.44	3.05	2.82	2.66	2.55	2.46	2.40	2.34	2.30	2.23	2.15	2.07	2.03	1.98	1.94	1.89	1.84	1.78
23	4.28	3.42	3.03	2.80	2.64	2.53	2.44	2.37	2.32	2.27	2.20	2.13	2.05	2.01	1.96	1.91	1.86	1.81	1.76
24	4.26	3.40	3.01	2.78	2.62	2.51	2.42	2.36	2.30	2.25	2.18	2.11	2.03	1.98	1.94	1.89	1.84	1.79	1.73
25	4.24	3.39	2.99	2.76	2.60	2.49	2.40	2.34	2.28	2.24	2.16	2.09	2.01	1.96	1.92	1.87	1.82	1.77	1.71
30	4.17	3.32	2.92	2.69	2.53	2.42	2.33	2.27	2.21	2.16	2.09	2.01	1.93	1.89	1.84	1.79	1.74	1.68	1.62
40	4.08	3.23	2.84	2.61	2.45	2.34	2.25	2.18	2.12	2.08	2.00	1.92	1.84	1.79	1.74	1.69	1.64	1.58	1.51
60	4.00	3.15	2.76	2.53	2.37	2.25	2.17	2.10	2.04	1.99	1.92	1.84	1.75	1.70	1.65	1.59	1.53	1.47	1.39
120	3.92	3.07	2.68	2.45	2.29	2.18	2.09	2.02	1.96	1.91	1.83	1.75	1.66	1.61	1.55	1.50	1.43	1.35	1.25
∞	3.84	3.00	2.60	2.37	2.21	2.10	2.01	1.94	1.88	1.83	1.75	1.67	1.57	1.52	1.46	1.39	1.32	1.22	1.00

Degrees of freedom for numerator

Degrees of freedom for denominator

Table 1-19 (Continued)
F DISTRIBUTION
Upper 1% Points ($F_{.99}$)

Degrees of freedom for denominator	Degrees of freedom for numerator																		
	1	2	3	4	5	6	7	8	9	10	12	15	20	24	30	40	60	120	∞
1	4052	5000	5403	5625	5764	5859	5928	5982	6023	6056	6106	6157	6209	6235	6261	6287	6313	6339	6366
2	98.5	99.0	99.2	99.2	99.3	99.3	99.4	99.4	99.4	99.4	99.4	99.4	99.4	99.5	99.5	99.5	99.5	99.5	99.5
3	34.1	30.8	29.5	28.7	28.2	27.9	27.7	27.5	27.3	27.2	27.1	26.9	26.7	26.6	26.5	26.4	26.3	26.2	26.1
4	21.2	18.0	16.7	16.0	15.5	15.2	15.0	14.8	14.7	14.5	14.4	14.2	14.0	13.9	13.8	13.7	13.7	13.6	13.5
5	16.3	13.3	12.1	11.4	11.0	10.7	10.5	10.3	10.2	10.1	9.89	9.72	9.55	9.47	9.38	9.29	9.20	9.11	9.02
6	13.7	10.9	9.78	9.15	8.75	8.47	8.26	8.10	7.98	7.87	7.72	7.56	7.40	7.31	7.23	7.14	7.06	6.97	6.88
7	12.2	9.55	8.45	7.85	7.46	7.19	6.99	6.84	6.72	6.62	6.47	6.31	6.16	6.07	5.99	5.91	5.82	5.74	5.65
8	11.3	8.65	7.59	7.01	6.63	6.37	6.18	6.03	5.91	5.81	5.67	5.52	5.36	5.28	5.20	5.12	5.03	4.95	4.86
9	10.6	8.02	6.99	6.42	6.06	5.80	5.61	5.47	5.35	5.26	5.11	4.96	4.81	4.73	4.65	4.57	4.48	4.40	4.31
10	10.0	7.56	6.55	5.99	5.64	5.39	5.20	5.06	4.94	4.85	4.71	4.56	4.41	4.33	4.25	4.17	4.08	4.00	3.91
11	9.65	7.21	6.22	5.67	5.32	5.07	4.89	4.74	4.63	4.54	4.40	4.25	4.10	4.02	3.94	3.86	3.78	3.69	3.60
12	9.33	6.93	5.95	5.41	5.06	4.82	4.64	4.50	4.39	4.30	4.16	4.01	3.86	3.78	3.70	3.62	3.54	3.45	3.36
13	9.07	6.70	5.74	5.21	4.86	4.62	4.44	4.30	4.19	4.10	3.96	3.82	3.66	3.59	3.51	3.43	3.34	3.25	3.17
14	8.86	6.51	5.56	5.04	4.70	4.46	4.28	4.14	4.03	3.94	3.80	3.66	3.51	3.43	3.35	3.27	3.18	3.09	3.00
15	8.68	6.36	5.42	4.89	4.56	4.32	4.14	4.00	3.89	3.80	3.67	3.52	3.37	3.29	3.21	3.13	3.05	2.96	2.87
16	8.53	6.23	5.29	4.77	4.44	4.20	4.03	3.89	3.78	3.69	3.55	3.41	3.26	3.18	3.10	3.02	2.93	2.84	2.75
17	8.40	6.11	5.19	4.67	4.34	4.10	3.93	3.79	3.68	3.59	3.46	3.31	3.16	3.08	3.00	2.92	2.83	2.75	2.65
18	8.29	6.01	5.09	4.58	4.25	4.01	3.84	3.71	3.60	3.51	3.37	3.23	3.08	3.00	2.92	2.84	2.75	2.66	2.57
19	8.19	5.93	5.01	4.50	4.17	3.94	3.77	3.63	3.52	3.43	3.30	3.15	3.00	2.92	2.84	2.76	2.67	2.58	2.49
20	8.10	5.85	4.94	4.43	4.10	3.87	3.70	3.56	3.46	3.37	3.23	3.09	2.94	2.86	2.78	2.69	2.61	2.52	2.42
21	8.02	5.78	4.87	4.37	4.04	3.81	3.64	3.51	3.40	3.31	3.17	3.03	2.88	2.80	2.72	2.64	2.55	2.46	2.36
22	7.95	5.72	4.82	4.31	3.99	3.76	3.59	3.45	3.35	3.26	3.12	2.98	2.83	2.75	2.67	2.58	2.50	2.40	2.31
23	7.88	5.66	4.76	4.26	3.94	3.71	3.54	3.41	3.30	3.21	3.07	2.93	2.78	2.70	2.62	2.54	2.45	2.35	2.26
24	7.82	5.61	4.72	4.22	3.90	3.67	3.50	3.36	3.26	3.17	3.03	2.89	2.74	2.66	2.58	2.49	2.40	2.31	2.21
25	7.77	5.57	4.68	4.18	3.86	3.63	3.46	3.32	3.22	3.13	2.99	2.85	2.70	2.62	2.53	2.45	2.36	2.27	2.17
30	7.56	5.39	4.51	4.02	3.70	3.47	3.30	3.17	3.07	2.98	2.84	2.70	2.55	2.47	2.39	2.30	2.21	2.11	2.01
40	7.31	5.18	4.31	3.83	3.51	3.29	3.12	2.99	2.89	2.80	2.66	2.52	2.37	2.29	2.20	2.11	2.02	1.92	1.80
60	7.08	4.98	4.13	3.65	3.34	3.12	2.95	2.82	2.72	2.63	2.50	2.35	2.20	2.12	2.03	1.94	1.84	1.73	1.60
120	6.85	4.79	3.95	3.48	3.17	2.96	2.79	2.66	2.56	2.47	2.34	2.19	2.03	1.95	1.86	1.76	1.66	1.53	1.38
∞	6.63	4.61	3.78	3.32	3.02	2.80	2.64	2.51	2.41	2.32	2.18	2.04	1.88	1.79	1.70	1.59	1.47	1.32	1.00

Interpolation should be performed using reciprocals of the degrees of freedom.

Section 2

GENERAL INFORMATION
AND CONVERSION TABLES

Table 2-1
FUNDAMENTAL PHYSICAL CONSTANTS

Nat. Bur. Stand. Tech. News Bull., January, 1971; reprinted in *J. Chem. Educ.*, **48**, 569 (1971)

A. Defined Values (SI Base Units)

Meter (metre) . . . . m = 1 650 763.73 wavelengths in vacuum of the orange-red line of the spectrum of krypton-86.

Kilogram . . . . kg = mass of a cylinder of platinum-iridium alloy kept at Paris

Second . . . . .s = duration of 9 192 631 770 cycles of the radiation associated with a specified transition of the cesium atom

Ampere . . . . A = magnitude of the current that, when flowing through each of two long parallel wires separated by one meter in free space, results in a force between the two wires 2×10^{-7} newton for each meter of length

Kelvin (degree Kelvin) . . . . K = defined in the thermodynamic scale by assigning $273.16°K$ to the triple point of water (freezing point, $273.15°K = 0°C$)

Candela . . . . cd = luminous intensity of 1/600 000 of a square meter of a radiating cavity at the temperature of freezing platinum ($2042°K$)

Mole . . . . mol = amount of substance which contains as many specified entities (molecules, atoms, ions, electrons, photons, etc.) as there are atoms of carbon-12 in exactly 0.012 kg of that nuclide

B. Derived SI Units

Physical quantity	Name of SI Unit	Symbol	Definition
Force	newton	N	$kg \cdot m \cdot s^{-2}$
Pressure	pascal	Pa	$kg \cdot m^{-1} s^{-2}$ ($= N \cdot m^{-2}$)
Energy	joule	J	$kg \cdot m^2 \cdot s^{-2}$
Power	watt	W	$kg \cdot m^2 \cdot s^{-3}$ ($= J \cdot s^{-1}$)
Electric charge	coulomb	C	$A \cdot s$
Electric potential difference	volt	V	$kg \cdot m^2 \cdot s^{-3} \cdot A^{-1}$ ($= J \cdot A^{-1} \cdot s^{-1}$)
Electric resistance	ohm	Ω	$kg \cdot m^2 \cdot s^{-3} \cdot A^{-2}$ ($= V \cdot A^{-1}$)
Electric conductance	siemens	S	$kg^{-1} \cdot m^{-2} \cdot s^3 \cdot A^2$ ($= A \cdot V^{-1} = \Omega^{-1}$)
Electric capacitance	farad	F	$A^2 \cdot s^4 \cdot kg^{-1} \cdot m^{-2}$ ($= A \cdot s \cdot V^{-1}$)
Magnetic flux	weber	Wb	$kg \cdot m^2 \cdot s^{-2} \cdot A^{-1}$ ($= V \cdot s$)
Inductance	henry	H	$kg \cdot m^2 \cdot s^{-2} \cdot A^{-2}$ ($= V \cdot A^{-1} \cdot s$)
Magnetic flux density	tesla	T	$kg \cdot s^{-2} \cdot A^{-1}$ ($= V \cdot s \cdot m^{-2}$)
Luminous flux	lumen	lm	$cd \cdot sr$
Illumination	lux	lx	$cd \cdot sr \cdot m^{-2}$
Frequency	hertz	Hz	s^{-1} (cycle per second)
Length	Angstrom	Å	10^{-10} m
Area	barn	b	10^{-28} m^2
Force	dyne	dyn	10^{-5} N
Pressure	bar	bar	10^5 N $\cdot$ m^{-2}
Energy	erg	erg	10^{-7} J
Kinematic viscosity	stokes	St	10^{-4} m$^2 \cdot$ s^{-1}
Viscosity	poise	P	10^{-1} kg $\cdot$ m$^{-1} \cdot$ s^{-1}
Magnetic flux density	gauss	G	10^{-4} T
Absorbed dose (ionizing radiation)	rad	rd	10^{-2} J $\cdot$ kg^{-1}
Energy	electron volt	eV	$1.602\ 10 \times 10^{-19}$ J
Mass	unified atomic mass unit	u	$1.660\ 41 \times 10^{-27}$ kg
Electric dipole moment	debye	D	$3.335\ 64 \times 10^{-30}$ A $\cdot$ m $\cdot$ s
Distance (astronomical)	parsec	pc	$3.085\ 68 \times 10^{16}$ m

Table 2-1 (*Continued*)
FUNDAMENTAL PHYSICAL CONSTANTS

C. Adjusted Values of Constants

Speed of light in vacuum . . . . $c = (2.997\ 925 \pm 0.000\ 001) \times 10^8\ \text{m} \cdot \text{s}^{-1}$

Elementary charge . . . . $e = (1.602\ 192 \pm 0.000\ 007) \times 10^{-19}\ \text{C}$
$$= (4.803\ 25 \pm 0.000\ 02) \times 10^{-10}\ \text{cm}^{1/2}\text{g}^{1/2}\text{s}^{-1*}$$

Avogadro constant . . . . $N_A = (6.022\ 17 \pm 0.000\ 04) \times 10^{23}\ \text{mol}^{-1}$

Atomic mass unit . . . . $u = (1.660\ 53 \pm 0.000\ 01) \times 10^{-27}\ \text{kg}$

Electron rest mass . . . . $m_e = (9.109\ 56 \pm 0.000\ 05) \times 10^{-31}\ \text{kg}$

Proton rest mass . . . . $m_p = (1.672\ 61 \pm 0.000\ 01) \times 10^{-27}\ \text{kg}$

Faraday constant . . . . $F = (9.648\ 67 \pm 0.000\ 05) \times 10^4\ \text{C} \cdot \text{mol}^{-1}$

Planck constant . . . . $h = (6.626\ 20 \pm 0.000\ 05) \times 10^{-34}\ \text{J} \cdot \text{s}$

Fine structure constant . . . . $\alpha = (7.297\ 35 \pm 0.000\ 01) \times 10^{-3}$
$$1/\alpha = (1.370\ 36 \pm 0.000\ 02) \times 10^2$$
$$\alpha/2\pi = (1.161\ 41 \pm 0.000\ 01) \times 10^{-3}$$

Charge-to-mass ratio for electron . . . . $e/m_e = (1.758\ 803 \pm 0.000\ 005) \times 10^{11}\ \text{C} \cdot \text{kg}^{-1}$

Rydberg constant . . . . $R_\infty = (1.097\ 373\ 1 \pm 0.000\ 000\ 1) \times 10^7\ \text{m}^{-1}$

Gyromagnetic ratio of proton . . . . $\gamma_p = (2.675\ 197 \pm 0.000\ 008) \times 10^8\ \text{rad} \cdot \text{s}^{-1}\text{T}^{-1}$

(uncorrected for diagmagnetism, H_2O) . . . . . $\gamma_p = (2.675\ 127 \pm 0.000\ 008) \times 10^8\ \text{rad} \cdot \text{s}^{-1}\text{T}^{-1}$

Bohr magneton . . . . $\mu_B = (9.274\ 10 \pm 0.000\ 07) \times 10^{-24}\ \text{J} \cdot \text{T}^{-1}$

Gas constant . . . . $R = (8.314\ 3 \pm 0.000\ 3)\ \text{J} \cdot \text{K}^{-1}\ \text{mol}^{-1}$

Boltzmann constant . . . . $k = (1.380\ 62 \pm 0.000\ 06) \times 10^{-23}\ \text{J} \cdot \text{K}^{-1}$

First radiation constant $(8\pi hc)$ . . . . $c_1 = (4.992\ 58 \pm 0.000\ 04) \times 10^{-24}\ \text{J} \cdot \text{m}$

Second radiation constant . . . . $c_2 = (1.438\ 83 \pm 0.000\ 06) \times 10^{-2}\ \text{m} \cdot \text{K}$

Stefan-Boltzmann constant . . . . $\sigma = (5.669\ 61 \pm 0.000\ 96) \times 10^{-8}\ \text{W} \cdot \text{m}^{-2}\text{K}^{-4}$

Gravitational constant . . . . $G = (6.673 \pm 0.003) \times 10^{-11}\ \text{N} \cdot \text{m}^2\text{kg}^{-2}$

Neutron rest mass . . . . $m_n = (1.67482 \pm 0.00008) \times 10^{-24}\ \text{g}$
$$= (1.0086654 \pm 0.0000013) \times 10^0\ \text{u}$$

Quantum-charge ratio . . . . $h/e = (4.13556 \pm 0.00012) \times 10^{-7}\ \text{cm}^{3/2}\text{g}^{1/2}\text{s}^{-1*}$
$$= (1.37947 \pm 0.00004) \times 10^{-17}\ \text{cm}^{1/2}\text{g}^{1/2}\dagger$$

Compton wave length of electron . . . . $\lambda_C = (2.42621 \pm 0.00006) \times 10^{-10}\ \text{cm}$
$$\lambda_C/2\pi = (3.86144 \pm 0.00009) \times 10^{-11}\ \text{cm}$$

Compton wave length of proton . . . . $\lambda_{C,p} = (1.32140 \pm 0.00004) \times 10^{-13}\ \text{cm}$
$$\lambda_{C,p}/2\pi = (2.10307 \pm 0.00006) \times 10^{-14}\ \text{cm}$$

Bohr radius . . . . $a_0 = (5.29167 \pm 0.00007) \times 10^{-9}\ \text{cm}$

Electron radius . . . . $r_e = (2.81777 \pm 0.00011) \times 10^{-13}\ \text{cm}$
$$r_e^2 = (7.9398 \pm 0.0006) \times 10^{-26}\ \text{cm}^2$$

Thomson cross section . . . . $8\pi r_e^2/3 = (6.6516 \pm 0.0005) \times 10^{-25}\ \text{cm}^2$

Nuclear magneton . . . . $\mu_N = (5.0505 \pm 0.0004) \times 10^{-24}\ \text{erg}\ \text{G}^{-1*}$

Proton moment . . . . $\mu_p = (1.41049 \pm 0.00013) \times 10^{-23}\ \text{erg}\ \text{G}^{-1*}$
$$\mu_p/\mu_N = (2.79276 \pm 0.00007) \times 10^0$$

(uncorrected for diamagnetism, H_2O) . . . . $\mu_p'/\mu_N = (2.79268 \pm 0.00007) \times 10^0$

Anomalous electron moment corrn . . . . $(\mu_e/\mu_0) - 1 = (1.159615 \pm 0.000015) \times 10^{-3}$

Zeeman splitting constant . . . . $\mu_B/hc = (4.66858 \pm 0.00004) \times 10^{-5}\ \text{cm}^{-1}\ \text{G}^{-1*}$

Normal volume perfect gas . . . . $V_0 = (2.24136 \pm 0.00030) \times 10^4\ \text{cm}^3\ \text{mol}^{-1}$

Wien displacement constant . . . . $b = (2.8978 \pm 0.0004) \times 10^{-1}\ \text{cm}\ °\text{K}$

* Electromagnetic system.
† Electrostatic system.

Table 2-2
PHYSICAL AND CHEMICAL SYMBOLS AND TERMINOLOGY

Symbols separated by commas represent equivalent recommendations. Symbols preceded by three dots are alternatives to be used only when there is some reason for not using a symbol before the three dots. Symbols for physical and chemical quantities should be printed in *italic* type (and therefore must be underlined in typewritten material). Subscripts and superscripts which are themselves symbols for physical quantities should be italicized; all others should be in Roman type. References: Manual of Physico-chemical Symbols and Terminology (International Union of Pure and Applied Chemistry, *J. Am. Chem. Soc.*, **82**, 5517 (1960); Recommendations on Nomenclature and Presentation of Data in Gas Chromatography (IUPAC Division of Analytical Chemistry), *Pure Appl. Chem.*, **8**, 553 (1964).

A. Atomic and Molecular Quantities; Chemical Reactions

Quantity	Symbol	Quantity	Symbol
Affinity of a reaction $(-\Sigma \nu_B \mu_B)$	A	Molar concentration of substance B	c_B, $[B]$, $c(B)$
Atomic mass number	A	Molarity	M
Avogadro number	N_A	Mole fraction	$x \ldots X, y$
Binding energy of an electron in an atom	B_e	Molecular concentration	C
		Molecular mass	m
Bohr (or orbital) magneton $(= eh/4\pi mc)$	$\mu_B \ldots \mu B \ldots BM$	Neutron flux	ϕ
		Neutron rest mass	m_n
Boltzmann constant	k	Nuclear cross section	σ
Characteristic temperature	Θ	Nuclear magneton	$\mu_N \ldots \mu N$
Charge-to-mass ratio for electron	e/m_e	Number of molecules	N
		Normality	N
Collision number (collisions per unit volume and unit time)	Z	Number of moles	n
		Osmotic pressure	Π
		Planck constant	h
Concentration	$c(c_B)$	Proton moment	μ_p
Decay constant	λ	Proton rest mass	m_p
Degree of reaction (e.g., degree of dissociation)	α	Quantum-charge ratio	h/e
Diameter of molecule	$\sigma \ldots D$	Quantum number:	
		Azimuthal or orbital	l
Diffusion coefficient	D	Azimuthal or orbital, total	L
Dirac $\bar{h}$ [$= h/2\pi$]	$\bar{h}$		
Disintegration constant $[N = N_o e^{-\lambda t}]$	λ	Inner	j
		Inner, total	J
Electron radius	r_e	Magnetic	m
Electron rest mass	m_e	Magnetic, total	M
Equilibrium constant	K	Principal	n
Equivalent weight	M/Z	Rotational	R
Equilibrium quotient (or equilibrium product of molalities)	Q	Spin	s
		Spin, nuclear	I
		Spin, total	S
Extent of reaction $(dn_B = \nu_B d\xi)$	ξ	Rate constant	k
		Rate constant corresponding rate Z	z
Fine structure constant	α		
Force constant	k	Rotation, specific $(= \theta/lc)$	α
Gyromagnetic ratio	g, γ		
Loschmidt's number (2.68715 $\times 10^{19}$ molecules per cm³)	n_o	Rydberg constant for infinite mass	R_∞
		Rydberg constant for hydrogen	R_H
Mean free path	l		
Molality	m	Solubility product	K_{sp}

Quantity	Symbol	Quantity	Symbol
Statistical weight	$g \ldots p$	Transition probability of	B_{nm}
Surface concentration	Γ	induced or stimulated	
Symmetry number	σ	emission ($m \rightarrow n$	
Transition probability of	A_{nm}	energy level)	
spontaneous emission		Valence	z
($m \rightarrow n$ energy level)		Work function	ϕ
Transition probability of	B_{mn}		
absorption ($n \rightarrow m$			
energy level)			

B. Chromatography

Quantity	Symbol	Quantity	Symbol
Adjusted retention time	t'_R	Partition coefficient	K, K_d
Adjusted retention volume	V'_R	Peak resolution	R_{ij}
Average linear gas velocity in column	$\bar{u}$	Plate height	H
Capacity, volume	Q_v	Relative retention	r_{is}
Capacity, weight	Q_w	Relative retention ratio	α
Capacity or partition ratio	k	Retardation value	R
Column inlet pressure	p_i	Retardation factor based on relative	R_f
Column length	L	distances traveled	
Column outlet pressure	p_o	Retention distance on chromatogram	d_R
Column pressure gradient correction	j	Retention time	$t_R, \ t_{max}$
factor		Retention volume	$V_R, \ V_{max}$
Column temperature	T_c	Retention index (isothermal gas	I_x
Concentration at peak maximum	C_{max}	chromatography)	
Corrected retention time	t_R°	Retention index (linear programmed	I_{PT}
Corrected retention volume	V_R°	temperature gas chromatography)	
Density of liquid phase	ρ_L	Selectivity coefficient (equivalents	$E_H^{M/n}$
Distribution ratio (extraction)	D	of M replacing hydrogen ion)	
Diffusion coefficient in liquid film	D_f	Specific retention volume	V_g
Diffusion coefficient in mobile	D_M	Temperature, retention	T_R
phase		Theoretical plates (plate number)	n, N
Diffusion coefficient within resin	D_r	Volume, bed	V_b
bead		Volume distribution ratio	D_v
Diffusion coefficient in stationary	D_S	Volume, inner	V_i
phase		Volume of liquid phase in column	V_L
Elution volume (exclusion	V_e	Volume of mobile phase (carrier	$V_M(V_G)$
chromatography)		gas) in column	
Gas flow rate from column	F_c	Volume, outer or interstitial	V_o
Gas holdup time	t_M	Volume, resin or gel	V_r
Gas holdup volume	V_M	Volume, supporting solid in column	V_s
Gas/liquid volume ratio	β	Zone width at baseline	W
Height equivalent to a theoretical	H	Zone width at C_{max}/e or $0.368C_{max}$	W_e
plate		Zone width at peak height/2	$W_{1/2}$
Net retention volume	V_N		

C. Electrical Units

Quantity	Symbol	Equation (mks)	mks unit	cgs unit
Admittance (reciprocal impedance)	Y	$Y = I/E = \sqrt{G^2 + B^2}$	mho	abmho
Attenuation	$\ldots$		decibel	

Quantity	Symbol	Equation (mks)	mks unit	cgs unit
Capacitance	C	$C = Q/E$	farad	abfarad
Charge density, surface	σ_s			
Charge density, volume	σ_v			
Conductance	G	$G = R/Z^2 = \kappa A/l$	mho; siemens	abmho
Current, electric	$I \ldots i$	$I = E/R = E/Z = Q/t$	ampere	abampere
Current density	J			
Dielectric constant	ε			
Dielectric polarization	P			
Dipole moment	μ		debye	
Displacement	D			
Electric charge	e			
Electrokinetic potential	ζ			
Electromotive force (emf)	E	$E = IR = W/Q$	volt	abvolt
Energy	W	$W = eIt$	joule (watt-second)	erg
Frequency	f	$f = 1/t$	hertz (cycles per second)	hertz
Impedance	Z	$Z = E/I = \sqrt{R^2 + (X_L + X_C)^2}$	ohm	abohm
Inductance, mutual	M	$M = K\sqrt{L_1 L_2}$	henry	abhenry
Inductance, self	L	$L = -N(d\phi/dI)$	henry	abhenry
Moment, electric quadrupole	Q			
Period	T, t	$T = 1/f$	second	second
Permittivity, see dielectric constant				
Potential difference, see electromotive force				
Power, active	$P \ldots p$	$P = EI \cos \theta = dw/dt$	watt	abwatt
Power, apparent	$P \ldots p$	$P = EI$	volt-ampere	
Power, reactive	jQ	$jQ = EI \sin \theta$	var	abvar
Resistance	R	$R = E/I = \rho l/A$	ohm	abohm
Resistivity	ρ	$\rho = RA/l$	ohm-centimeter	abohm-centimeter
Susceptance	B	$B = X/Z^2$	mho	abmho
Power factor	pf	$pf = P/EI = \cos \theta$		
Quantity of electricity	Q	$Q = it = CE$	coulomb	abcoulomb
Reactance, capacitive	X_C	$X_C = 1/(2\pi f C)$	ohm	abohm
Reactance, inductive	X_L	$X_L = 2\pi f L$	ohm	abohm

D. Electrochemistry

Quantity	Symbol	Quantity	Symbol
Anodic current	i_a	Half-wave potential	$E_{1/2}$
Cathodic current	i_c	Faraday's constant (the faraday)	F, **F**
Charge number of an ion, plus or minus	$z_+ z_-$	Ionic conductance	λ
Degree of electrolytic dissociation	α	Ionic strength	$I \ldots \mu$
Diffusion current	i_d	Limiting current	i_l
Effective ionic radius	$\dot{a}$	Mobility	$\mu_+ \mu_-$
Electrolytic conductivity or specific conductance ($\kappa = l/\rho = l/RA$)	κ	Overpotential (or overvoltage)	η
		Residual current	i_r
Equivalent conductance of electrolyte or ion	Λ	Standard potential of half-reaction	$E°$
Half-neutralization potential	**HNP**	Transport number [transference number $= \mu_\pm/(\mu_+ + \mu_-)$]	$t_+ t_-$

E. Light

Quantity	Symbol	Quantity	Symbol
Absorbance (extinction) [$A = \log (1/T)$]	$A \dots E$	Planck's constant	h
Absorption (extinction) coefficient	κ	Planck's constant divided by 2π	$\hbar$
[$\kappa l c = \ln (1/T)$]		Quantity of light	Q
Absorption factor (fraction of incident radiant power which is absorbed	α	Radiancy or radiant flux density ($= d\Phi/dA$)	
Absorptivity (specific absorbance or decadic absorption coefficient)	a	Radiant power (dQ/dt)	Φ
Angle of diffraction	θ	Reflection factor (fraction of incident radiant power which is absorbed)	ρ
Angle of incidence	ϕ		
Angle of optical rotation	$\alpha \dots \theta$	Refraction index	n
Angle of reflection	r	Refractivity	r
Focal length	f	Transmission factor (fraction of incident radiant power which is transmitted)	τ
Illumination ($d\Phi/dS$)	E		
Luminance	L, B		
Luminous emittance	H	Transmittance ($T = I/I_o$)	T
Luminous flux	Φ	Velocity of light	c
Luminous intensity ($d\Phi/d\omega$)	I	Wavelength	λ
Molar absorptivity (molar extinction coefficient) [$\varepsilon l c = A$]	ε		

F. Magnetic Units

Quantity	Symbol	Equation (mks)	mks unit	cgs unit
Field strength	H	$H = \mathcal{F}/l = \mathbf{F}/m$	ampere-turn per meter	oersted
Flux	ϕ	$\phi = \mathcal{F}/\mathcal{R}$	weber	maxwell
Induction (flux density)	B	$B = \phi/A$	weber per centimeter squared (tesla)	gauss
Magnetization	M			
Magnetomotive force (magnetomotance)	$\mathcal{F}$	$\mathcal{F} = \phi\mathcal{R} = 0.4\pi NI$	ampere-turn	gilbert
Permeability, relative	μ_r	$\mu_r = B/H$	numeric	numeric
Pole strength	$\dots$		$mm'/\mu_r(4\pi)^2 10^{-7}$ weber	$mm'/\mu_r l^2 = \mathbf{F}$ maxwell
Reluctance (magnetic resistance)	$\mathcal{R}$	$\mathcal{R} = l/A\mu_r$	ampere-turn per weber	gilbert per maxwell
Susceptibility, specific	χ	$\chi = k/\rho$		
Susceptibility, volume	k	$k = M/H$		

$\mathbf{F}$ = force in dynes (cgs) or newton (mks) m = moment (loop)
A = sectional area in centimeters squared m' = moment (dipole)
N = number of turns

G. Mechanical and Related Quantities

Quantity	Symbol	Quantity	Symbol
Angle of contact	θ	Pressure	p, P
Compressibility $\left[-\dfrac{1}{V}\left(\dfrac{\partial V}{\partial P}\right)_T\right]$	κ	Shear modulus of elasticity	G
		Shear stress	τ
Compression modulus, bulk modulus	$1/\kappa, \mathbf{K}$	Surface tension	$\gamma \ldots \sigma$
Force	F	Traction	σ
Force due to gravity (weight)	$G \ldots W$	Viscosity, dynamic	η
Fluidity ($= 1/\eta$)	ϕ	Viscosity, kinematic ($= \eta/\rho$)	ν
Friction coefficient	f	Volume expansivity, coefficient of	β
Modulus (Young's) of elasticity	E, Y	volume expansion	
Moment of force	M		

H. Space, Time, Mass, and Related Quantities

Quantity	Symbol	Quantity	Symbol
Acceleration	a	Path, length of arc	s
Acceleration of free fall (gravity)	g	Period (any characteristic interval) [$= 1/f$]	T, τ
Angular frequency	ω	Plane angle	$\alpha, \beta, \gamma, \theta, \phi, \psi$
Angular velocity ($= 2\pi f$)	ω	Radius	r
Area	A, S	Rectangular coordinates	x, y, z
Density	ρ	Relative density	d
Diameter	d	Solid angle	ω
Frequency ($= 1/T$)	ν, f	Specific volume	ν
Height	h	Time	t
Length	l	Time constant	τ
Mass	m	Velocity	$v \ldots u, w$
Mass, reduced [$1/\mu =$ $(1/m_1 + 1/m_2)$]	μ	Volume	$V \ldots v$
		Wavelength	λ
Moment of inertia	I	Wavenumber	$\sigma, \nu, \bar{\nu}$

I. Thermodynamic and Related Quantities

Quantity	Symbol	Quantity	Symbol
Activity, absolute	λ	Heat capacity at constant pressure	$c_p \ldots C_p$
Activity, relative	a	Heat capacity at constant volume	$c_v \ldots C_v$
Activity coefficient	f, γ	Helmholtz energy (Gibb's ψ)	A
Amplitude of wave associated with moving particle	Ψ	[$= U - TS$]	
		Joule-Thompson (Kelvin) coefficient	μ
Boltzmann constant	k	[$= (\partial T/\partial P)_H$]	
Chemical potential	μ	Osmotic coefficient	g, ϕ
Energy, internal (Gibb's ε)	$E \ldots U$	Partition function	Q
Enthalpy, heat content (Gibb's χ)	H, Hf	Ratio of specific heats ($= c_p/c_v$)	γ, κ
[$H = U + PV$]		Temperature	θ, t
Entropy (Gibb's η)	S	Temperature, absolute or thermodynamic	T
Fugacity	f		
Gas constant ($= PV/nT$)	$R, \mathbf{R}$	Thermal conductivity	λ
Gibb's (free) energy [$= H - TS$]	G, Gf	Work	w, A
Heat	q, Q		

Table 2-3
INTERNATIONAL SYSTEM (SI) OF UNITS AND
CONVERSION FACTORS

The following coherent system was accepted as the preferred system of units by representatives of 36 participating countries, including the United States, at the Eleventh General Conference on Weights & Measures (1960). This system, known as the Système International d'Unites, is designated "SI" in all languages. Relations which are exact by National Bureau of Standards (U.S.) definitions are indicated by an asterisk(*). Relationships without an asterisk are either the results of physical measurements or are only approximate. Cf. E. A. Mechtly, *Sci. & Tech. Inform. Div., NASA, SP*-7012, Supt. of Documents, Washington, D.C. Price 20¢.

To convert from	To	Multiply by
Abampere	ampere*	10
Abcoulomb	coulomb*	10
Abfarad	farad*	1×10^9
Abhenry	henry*	1×10^{-9}
Abmho	mho	1×10^9
Abohm	ohm*	1×10^{-9}
Abvolt	volt*	1×10^{-8}
Acre	sq meter*	4046.8564224
Ampere (int 1948)	ampere	0.999835
Angstrom	meter*	1×10^{-10}
Are	sq meter	100
Astronomical unit	meter	1.49598×10^{11}
Atmosphere	newton/sq meter*	1.01325×10^5
Bar	newton/sq meter*	1.00×10^5
Barn	sq meter	1×10^{-28}
Barye	newton/sq meter*	0.100
Btu (int steam table)	joule	1055.04
" (mean)	joule	1055.87
" (thermochemical)	joule	1054.350264488888
" (39°F)	joule	1059.67
Bushel (U.S.)	cu cm*	35239.07016688
Cable	meter*	219.456
Caliber	meter*	0.000254
Calorie, gram (int steam table)	joule	4.18674
" " (mean)	joule	4.19002
" " (thermochemical)	joule*	4.184
" " (15°C)	joule	4.18580
Calorie, gram (20°C)	joule	4.18190
Carat (metric)	grams*	2.00×10^{-1}
Cm of mercury at 0°C	newton/sq meter	1333.22
Cm of water at 4°C	newton/sq meter	98.0638
Chain (engineers' or Ramden's)	meter*	30.48
" (surveyors' or Gunter's)	meter*	20.1168
Circular mil	sq meter	$5.0670748 \times 10^{-10}$
Cord	cu meter	3.6245563
Coulomb (int 1948)	coulomb	0.999835
Cubit	meter*	0.4572
Cup	cu cm*	236.5882365
Curie	disintegration/sec*	3.70×10^{10}
Day (mean solar)	second (mean solar)*	8.64×10^4
" (sidereal)	second (mean solar)	86164.090
Degree (angular)	radian	0.017453292519943
Denier (int)	kg/meter*	1.00×10^{-7}
Dram (avoir.)	gram*	1.7718451953125
" (troy or apoth.)	gram*	3.8879346
" (U.S. fluid)	cu cm*	3.6966911953125
Dyne	newton*	1.00×10^{-5}

Table 2-3 (*Continued*)
INTERNATIONAL (SI) CONVERSION FACTORS

To convert from	To	Multiply by
Electron volt	joule	1.60210×10^{-19}
Erg	joule*	1.00×10^{-7}
Farad (int 1948)	farad	0.999505
Faraday (based on carbon 12)	coulomb	96487.0
" (chemical)	coulomb	96495.7
" (physical)	coulomb	96521.9
Fathom	meter*	1.8288
Fermi	meter*	1.00×10^{-15}
Foot	meter*	0.3048
" (U.S. survey)	meter*	1200/3937
" (")	meter	0.3048006096
Foot of water at 39.2°F	newton/sq meter	2988.98
Foot-candle	lumen/sq meter	10.763910
Furlong	meter*	201.168
Gal	meter/sq sec*	1.00×10^{-2}
Gallon (British)	cu cm	4546.087
" (U.S. dry)	cu cm*	4404.88377086
" (U.S. liquid)	cu cm*	3785.411784
Gamma	tesla*	1.00×10^{-9}
Gauss	tesla*	1.00×10^{-4}
Gilbert	ampere turn	0.79577472
Gill (British)	cu cm	142.0652
" (U.S.)	cu cm	118.29412
Grad	degree (angular)*	9.00×10^{-1}
"	radian	0.015707963
Grain	gram*	0.06479891
Gram	kilogram*	1.00×10^{-3}
Hand	meter*	0.1016
Hectare	sq meter*	1.00×10^{4}
Henry (int 1948)	henry	1.000495
Hogshead (U.S.)	cu meter*	0.238480942392
Horsepower (550 ft-lbf/sec)	watt	745.69987
" (boiler)	watt	9809.50
" (electric)	watt*	746
" (metric)	watt	735.499
Horsepower (water)	watt	746.043
Hour (mean solar)	second (mean solar)*	3600
" (sidereal)	second (mean solar)	3590.1704
Hundredweight (long)	kilogram*	50.80234544
" (short)	kilogram*	45.359237
Inch	meter*	0.0254
Inch of mercury at 32°F	newton/sq meter	3386.389
" at 60°F	newton/sq meter	3376.85
Inch of water at 39.2°F	newton/sq meter	249.082
" at 60°F	newton/sq meter	248.84
Joule (int 1948)	joule	1.000165
Kayser	1/meter	1.00×10^{2}
Kilogram mass	kilogram*	1.00
" force, kgf	newton*	9.80665
Kip	newton*	4448.2216152605
Knot (int)	meter/second	0.5144444444
Lambert	candela/sq meter*	$(1/\pi) \times 10^{4}$
"	candela/sq meter	3183.0988
Langley	joule/sq meter*	4.184×10^{4}
lbf (pound force, avoir.)	newton*	4.4482216152605
lbm (pound mass, avoir.)	kilogram*	0.45359237
League (British, nautical)	meter	5559.552
" (int, nautical)	meter*	5556
" (statute)	meter*	4828.032

Table 2-3 (*Continued*)
INTERNATIONAL (SI) CONVERSION FACTORS

To convert from	To	Multiply by
Link (engineers' or Ramden's)	meter*	0.3048
" (surveyors' or Gunter's)	meter*	0.201168
Liter†	cu cm	1000.028
Lux	lumen/sq meter	1.00
Maxwell	weber*	1.00×10^{-8}
Meter	wave lengths of the orange-red line of Kr-86*	1.65076373×10^{6}
Micron	meter*	1.00×10^{-6}
Mil	meter*	2.54×10^{-5}
Mile (U.S., statute)	meter*	1609.344
" (British, nautical)	meter*	1853.184
" (int, nautical)	meter*	1852
" (U.S., nautical)	meter*	1852
Millibar	newton/sq meter*	1.00×10^{2}
Millimeter of mercury (0°C)	newton/sq meter	133.3224
Minute (angular)	radian	$2.90888208666 \times 10^{-4}$
Minute (mean solar)	second (mean solar)*	6.00×10
" (sidereal)	second (mean solar)*	59.836174
Month (mean calendar)	second (mean solar)	2.628×10^{6}
Oersted	ampere/meter	79.577472
Ohm (int 1948)	ohm	1.000495
Ounce force (avoir.)	newton	0.27801385
Ounce mass (avoir.)	gram*	28.349523125
" (troy or apoth.)	gram*	31.1034768
Ounce (U.S. fluid)	cu cm	29.5735295625
Pace	meter*	0.762
Parsec	meter	3.08374×10^{16}
Pascal	newton/sq meter*	1.00
Peck (U.S.)	cu meter	0.00880976754172
Pennyweight	gram*	1.55517384
Perch	meter*	5.0292
Phot	lumen/sq meter	1.00×10^{4}
Pica (printers')	cm*	0.42175176
Pint (U.S., dry)	cu cm*	550.6104713575
" (U.S., liquid)	cu cm*	473.176473
Point (printers')	cm	0.03514598
Poise	newton sec/sq meter*	1.00×10^{-1}
Pole	meter*	5.0292
Pound force, lbf (avoir.)	newton*	4.4482216152605
" mass, lbm (avoir.)	kilogram*	0.45359237
" " (troy or apoth.)	kilogram*	0.3732417216
Poundal	newton*	0.138254954376
Quart (U.S., dry)	cu cm*	1101.220942715
" (U.S., liquid)	cu cm	946.35295
Rad (radiation dose absorbed)	joule/kilogram*	1.00×10^{-2}
Rhe	sq meter/newton sec*	1.00×10
Rod	meter*	5.0292
Roentgen	coulomb/kilogram	2.57976×10^{-4}
Section	sq meter*	258998.8110336
Scruple (apoth.)	gram*	1.2959782
Shake	second	1.00×10^{-8}
Skein	meter*	109.728
Slug	kilogram	14.5939029
Span	meter*	0.2286
Statampere	ampere	3.335640×10^{-10}
Statcoulomb	coulomb	3.335640×10^{-10}
Statfarad	farad	1.112650×10^{-12}

† The 12th Gen. Conf. on Wts. & Measures, Paris (1964), defined the liter as equal to exactly 1000 cu cm.

Table 2-3 (*Continued*)
INTERNATIONAL (SI) CONVERSION FACTORS

To convert from	To	Multiply by
Stathenry	henry	8.987554×10^{11}
Statmho	mho	1.112650×10^{-12}
Statohm -	ohm - - - - - - - - - - - - - - - -	8.987554×10^{11}
Statvolt	volt	299.7925
Stere	cu meter*	1.00
Stilb	candela/sq meter	1.00×10^4
Stoke	sq meter/second*	1.00×10^{-4}
Tablespoon - - - - - - - - - - - - - - - - - -	cu cm* - - - - - - - - - - - -	14.78676478125
Teaspoon	cu cm*	4.92892159375
Ton (assay)	gram	29.166666
" (long)	kilogram*	1016.0469088
" (metric)	kilogram*	1.00×10^3
" (nuclear equiv of TNT) - - - - - - - - -	joule - - - - - - - - - - - - - -	4.20×10^9
" (register)	cu meter*	2.8316846592
" (short, 2,000 lb)	kilogram*	907.18474
Torr (0°C)	newton/sq meter	133.322
Township	sq kilometer	93.239572
Unit pole - - - - - - - - - - - - - - - - - - -	weber - - - - - - - - - - - -	1.256637×10^{-7}
Volt (int 1948)	volt	1.000330
Watt (int 1948)	watt	1.000165
Yard (U.S.)	meter*	0.9144
Year (calendar)	second (mean solar)*	3.1536×10^7
" (sidereal) - - - - - - - - - - - - - - -	second (mean solar) - - - - - - - -	3.1558150×10^7
" (tropical)	second (mean solar)	3.1556926×10^7
" 1900, tropical, Jan, day 0, hr 12	second (ephemeris)*	$3.15569259747 \times 10^7$
" light	meter	9.46055×10^{15}

Table 2-4
ABBREVIATIONS AND STANDARD LETTER SYMBOLS

Abampere	abamp	Beta particle	β
Absolute	abs	Body-centered cubic	bcc
Acetyl-	Ac	Boiling point	bp
Alcohol	alc	British thermal unit	btu
Alkaline	alk	Butyl-	Bu
Alpha particle	α	Calorie	cal
Alternating current	ac	Calorie, international	cal_{IT}
Ampere	A	steam table	
Amplification factor	μ	Calorie, thermochemical	cal_{th}
Anhydrous	anhyd	Candela	cd
Angstrom	Å, A	Centimeter-gram-second (system)	cgs
Approximate (*circa*)	*ca*	Chemically pure	CP
Aqueous	aq	*circa*	*ca*
Atmosphere	atm	Citrate	Cit
Atomic mass unit	amu	Confer (compare)	cf.
Atomic per cent	at.%	Counts per minute	cpm
Atomic weight	at. wt.	Coulomb	C
Average	av	Critical temperature	t_c
Barn	b	Crystalline	cryst
Barrel	bbl	Cubic	cub
Base of natural logarithms	*e*	Curie	Ci
[= 2.718..]		Cycles per second (hertz)	Hz

Table 2-4 (*Continued*)
ABBREVIATIONS AND STANDARD LETTER SYMBOLS

Day	d	Horsepower	hp
Decibel	db	Hour	hr, h
Decompose	dec	Hygroscopic	hygr
Density, critical	d_c	*ibidem* (in the same place)	*ibid*.
Degrees Baumé	°Bé	*id est* (that is)	i.e.
Degrees Celsius	°C	Ignition	ign
Degrees Fahrenheit	°F	Inch	in.
Degrees Kelvin	°K	Indices of a family of	(*hkl*)
Degrees Rankine	°R	crystallographic planes	
Degrees Réaumur	°Ré	Infrared	ir
Determine(d); detect	det(d)	Inorganic	inorg
Determination	detn	Inside diameter	i.d.
Deuteron	d	Insoluble	insol
Diameter	diam	Joule	J
Differential thermal analysis	DTA	Kilocalorie	kcal
Dilute	dil	Kilogram	kg
Dilution	diln	Kilogram-force	kg-f
Direct current	dc	Kilowatt-hour	kWh
Disintegrations per minute	dpm	Limit (math.)	lim
Distil(led)	dist(d)	Liquid	liq, *l*
Dropping mercury electrode	dme	Liter	l
Dyne	dyn	*loco citato* (in the place	*loc cit*.
Electromagnetic unit	emu	cited)	
Electromotive force	emf	Logarithm (common)	log
Electron	e^-, e	Logarithm (natural)	ln
Electron volt	eV	Lumen	lum, lm
Electrostatic unit	esu	Lux	lx
Entropy unit	eu, e.u.	Maximum	max
Equivalent weight	equiv wt	Maxwell	Mx
Especially	esp.	Meter-kilogram-second	mks
et alii (and others)	*et al*.	(system)	
et cetera	etc.	Melting point	mp
Ethyl-	Et	Metallic	met
Ethylenediamine	en	Metastable	*m*
Ethylenediaminetetraacetic acid	EDTA	Meter	m
exempli gratia (for example)	e.g.	Methyl-	Me
Experiment(al)	expt(l)	Micron (micrometer)	μm
Exponential	exp	Miller indices	(*hkl*)
Face-centered cubic	fcc	Milliequivalent	meq
Farad	F	Millimeters of mercury	mm Hg
Focal length	*f*	Millimicron (nanometer)	nm
Foot	ft	Millimole	mmole, mM
Formal (concentration)	*F*	Minimum	min
Freezing point	fp	Minute	m, min
Gallon	gal	Mixture	mixt
Gamma radiation	γ	Molal	*m*
Gas (physical state)	*g*	Molar	*M*
Gauss	G	Mole	mol
Gram	g	Molecular weight	mol wt
Gram-atom	g-atom	Mole per cent	mol %
Gram-calorie	g-cal	Monoclinic	mn
Gram-mole	g-mol	Naperian base	*e*
Gravimetric	grav	Negative	neg
Henry	H	Neutralization equivalent	neut equiv
Hexagonal	hex	Neutrino	ν

Table 2-4 (*Continued*)
ABBREVIATIONS AND STANDARD LETTER SYMBOLS

Newton	N	Saturated	satd
Normal (concentration)	*N*	Saturated calomel electrode	SCE
Normal hydrogen electrode	NHE	Second	s, sec
Nuclear magnetic resonance	nmr	Section	sect
Nuclear magneton	μ_N	Separation	sepn
Oersted	Oe	Slightly	sl
Ohm	Ω	Solid	s
opere citato (in the work cited)	*op cit.*	Soluble	sol
		Solution	soln
Optical speed	*f*/number	Solvent	solv
Organic	org	Specific gravity	sp gr
Orthorhombic	o-rh	Square	sq
Ounce	oz	Standard	std
Outer diameter	o.d.	Standard temperature and pressure	STP
Oxalate	Ox		
Oxidant	ox	Steradian	sr
Page(s)	p. (pp.)	Stokes	St
Parts per billion, volume	ng/ml	Substance	subst
Parts per billion, weight	ng/g	Symmetrical	sym
Parts per million, volume	μg/ml	Tartrate	Tart
Parts per million, weight	μg/g	Temperature	temp
Per cent	%	Temperature at boiling point	T_b
Phenyl-	Ph, ϕ	Temperature, critical	t_c
Photon	γ	Tesla	T
Physical	phys	Tetragonal	tetr
Poise	P	Thermogravimetric analysis	TGA
Positive	pos	Thin-layer chromatography	TLC
Positron	β^+	Tonne	t, ton
Pound	lb	Torr (mm of mercury)	Torr
Powder	pwd, powd	Transconductance	g_m
Precipitate(d)	ppt(d)	Transition	tr
Pressure, critical	p_c	Triclinic	tric
Propyl-	Pr	Trigonal	trig
Proton	p	Triton (tritium)	t
Proton magnetic resonance	pmr	Ultrahigh frequency	uhf
Pyridine	py	Ultraviolet	uv
Qualitative	qual	United States Pharmacopoeia	USP
Quantitative	quant	Vacuum	vac
Quantum (energy)	$h\nu$	Vapor pressure	vp
Quart	qt	Versus	*vs*
Radian	rad	Volt	V
Radio frequency	rf	Volt-ampere-reactive	var
Recrystallized	recryst	Volume	vol
Reductant	red	Volume per cent	vol %
Reference	ref	Volume per volume	v/v
Revolutions per minute	rpm	Watt	W
Rhombic	rh, rhb	Weber	Wb
Rhombohedral	rh-hed	Weight	wt
Roentgen	R	Weight per cent	wt %
Root-mean-square	rms	Weight per volume	w/v

Table 2-5
MATHEMATICAL SYMBOLS AND ABBREVIATIONS

Symbol	Meaning	Symbol	Meaning
$+$	Plus or Positive	sec	Secant
$-$	Minus or Negative	csc	Cosecant
$\pm$	{Plus or minus / Positive or negative	vers	Versed sine
		covers	Coversed sine
$\mp$	{Minus or plus / Negative or positive	exsec	Exsecant
$\times$ or $\cdot$	Multiplied by	$\sin^{-1}a$ or arc sin a	{Anti-sine a / Angle whose sine is a / Inverse sine a
$\div$ or $:$	Divided by	sinh	Hyperbolic sine
$=$ or $::$	Equals, as	cosh	Hyperbolic cosine
$\neq$, $\pm\!\!\!\!\!/$	Does not equal	tanh	Hyperbolic tangent
$\cong$	{Equals approximately / Congruent	$\sinh^{-1}a$ or arc sinh a	{Anti-hyperbolic sine a / Number whose hyperbolic sine is a
$>$	Greater than	$P\,(x,y)$	Rect. coörd. of point P
$<$	Less than	$P(r,\,\theta)$	Polar coörd. of point P
$\geqq$	Greater than or equal to	$f(x), F(x)$ or $\Phi\,(x)$	{Function of x
$\leqq$	Less than or equal to		
$\sim$	Similar to	Δy	Increment of y
$\therefore$	Therefore	$\doteq$ or $\rightarrow$	Approaches as a limit
$\sqrt{\ }$	Square root	Σ	Summation of
$\sqrt[n]{\ }$	nth root	∞	Infinity
a^n	nth power of a	dy	Differential of y
$\log$ or $\log_{10}$	{Common logarithm / Briggsian "	$\dfrac{dy}{dx}$ or $f'(x)$	Derivative of $y=f(x)$ with respect to x
$\ln$ or $\log_e$ or $\log$	{Natural logarithm / Hyperbolic " / Napierian "	$\dfrac{d^2y}{dx^2}$ or $f''(x)$	Second deriv. of $y=f(x)$ with respect to x
e or ϵ	Base (2.718) of natural system of logarithms	$\dfrac{d^n y}{dx^n}$ or $f^{(n)}(x)$	nth deriv. of $y=f(x)$ with respect to x
π	Pi (3.1416)	$\dfrac{\partial z}{\partial x}$	Partial derivative of z with respect to x
$\angle$	Angle		
$\perp$	Perpendicular to	$\dfrac{\partial^2 z}{\partial x\,\partial y}$	Second partial deriv. of z with respect to y and x
$\parallel$	Parallel to		
a°	a degrees (angle)	$\displaystyle\int$	Integral of
a'	{a minutes (angle) / a prime		
a''	{a seconds (angle) / a second / a double-prime	$\displaystyle\int_a^b$	Integral between the limits a and b
a'''	{a third / a triple-prime	j or i	{Imaginary quantity / $(\sqrt{-1})$, $i^2=-1$
a_n	a sub n	$x=a+jb$	Symbolic vector notation
sin	Sine	$n!=1\cdot2\cdot3\ \cdots\ n$	
cos	Cosine		
tan	Tangent		
cot or ctn	Cotangent		

Greek Alphabet

A	α	Alpha	N	ν	Nu
B	β	Beta	Ξ	ξ	Xi
Γ	γ	Gamma	O	o	Omicron
Δ	δ	Delta	Π	π	Pi
E	ϵ	Epsilon	P	ρ	Rho
Z	ζ	Zeta	Σ	σ	Sigma
H	η	Eta	T	τ	Tau
Θ	θ	Theta	Υ	υ	Upsilon
I	ι	Iota	Φ	ϕ	Phi
K	κ	Kappa	X	χ	Chi
Λ	λ	Lambda	Ψ	ψ	Psi
M	μ	Mu	Ω	ω	Omega

Prefixes for Naming Multiples and Submultiples of Units
For example: 10^{-9} gram is one nanogram or 1 ng.

Factor	Prefix	Symbol	Factor	Prefix	Symbol
10^{12}	tera	T	10^{-2}	centi	c
10^{9}	giga	G	10^{-3}	milli	m
10^{6}	mega	M	10^{-6}	micro	μ
10^{3}	hilo	k	10^{-9}	nano	n
10^{2}	hecto	h	10^{-12}	pico	p
10	deka	da	10^{-15}	femto	f
10^{-1}	deci	d	10^{-18}	atto	a

The prefix "myria" is sometimes used for 10^4 and "lakh" for 10^5.

Numerical Prefixes

½—Hemi	20—Eicosa	41—Hentetraconta
1—Mono	21—Henicosa	42—Dotetraconta
1½—Sesqui	22—Docosa	43—Tritetraconta
2—Di or Bi	23—Tricosa	44—Tetratetraconta
3—Tri	24—Tetracosa	45—Pentatetraconta
4—Tetra	25—Pentacosa	46—Hexatetraconta
5—Penta	26—Hexacosa	47—Heptatetraconta
6—Hexa	27—Heptacosa	48—Octatetraconta
7—Hepta	28—Octacosa	49—Nonatetraconta
8—Octa	29—Nonacosa	50—Pentaconta
9—Nona or Ennea	30—Triaconta	51—Henpentaconta
10—Deca	31—Hentriaconta	52—Dopentaconta
11—Undeca or Henadeca	32—Dotriaconta	53—Tripentaconta
12—Dodeca	33—Tritriaconta	54—Tetrapentaconta
13—Trideca	34—Tetratriaconta	55—Pentapentaconta
14—Tetradeca	35—Pentatriaconta	56—Hexapentaconta
15—Pentadeca	36—Hexatriaconta	57—Heptapentaconta
16—Hexadeca	37—Heptatriaconta	58—Octapentaconta
17—Heptadeca	38—Octatriaconta	59—Nonapentaconta
18—Octadeca	39—Nonatriaconta	60—Hexaconta
19—Nonadeca	40—Tetraconta	

CONVERSION OF THERMOMETER SCALES

The following abbreviations are used: F, degrees Fahrenheit; C, degrees Centigrade; R, degrees Reaumur; A.$_K$, degrees Kelvin; A.$_R$, degrees Rankine; Z, degrees on any scale; (f. p.)"Z", the freezing point of water on the Z scale; (b. p.) "Z", the boiling point of water on the Z scale. Cf. Dodds, *Chemical and Metallurgical Engineering*, Vol. 38, p. 476 (1931).

$$\frac{F-32}{180} = \frac{C}{100} = \frac{R}{80} = \frac{A_K - 273}{100} = \frac{A_R - 492}{180} = \frac{Z - (f.p.)"Z"}{(b.p.)"Z" - (f.p.)"Z"}$$

Examples: (1) To find the Fahrenheit temperature corresponding to –20°C:

$$\frac{F-32}{180} = \frac{C}{100} \text{ or } \frac{F-32}{180} = \frac{-20}{100} ; \text{ i.e., } -20°C = -4°F$$

(2) To find the Reaumur temperature corresponding to 20°F:

$$\frac{F-32}{180} = \frac{R}{80} \text{ or } \frac{20-32}{180} = \frac{R}{80} ; \text{ i.e., } 20°F = -5.33°R$$

(3) To find the correct temperature on a thermometer reading 80°C and which shows a reading of –0.30°C in melting ice and 99.0°C. in steam at 760mm pressure:

$$\frac{C}{100} = \frac{Z - (f.p.)"Z"}{(b.p.)"Z" - (f.p.)"Z"} = \frac{80 - (-0.30)}{99.0 - (0.30)} ; \text{ i.e., } C = 80.87° \text{ (corrected)}$$

The reading of any inaccurate thermometer can be corrected and converted in this manner into a reading on any of the other scales; thus, in the example (3) above, to convert into a Fahrenheit scale:

$$\frac{F-32}{180} = \frac{80 - (-0.30)}{99\,0 - (-0.30)} ; \text{ i.e., } F = 177.56° \text{ (corrected)}$$

Centigrade to Fahrenheit

Formula: $(°C \times 9/5) + 32 = °F$ Examples: (1) To convert 60°C to °F, $(60 \times 9/5) + 32 = 140°F$
(2) To convert −20°C to °F, $(-20 \times 9/5) + 32 = -4°F$
(See also special table)

Centigrade to Reaumur

Formula: $°C \times 4/5 = °R$ Examples: (1) To convert 40°C to °R, $40 \times 4/5 = 32°R$
(2) To convert −30°C to °R, $-30 \times 4/5 = -24°R$

Centigrade to Centigrade Absolute or Kelvin

Formula: $°C \div 273.1 = °K$ Examples: (1) To convert 20°C to °K, $20 + 273.1 = 293.1°K$
(2) To convert −20°C to °K, $-20 + 273.1 = 253.1°K$

Fahrenheit to Centigrade

Formula: $(°F - 32) \times 5/9 = °C$ Examples: (1) To convert 140°F to °C, $(140 - 32) \times 5/9 = 60°C$
(2) To convert −76°F to °C, $(-76 - 32) \times 5/9 = -60°C$
(See also special table)

Fahrenheit to Reaumur

Formula: $(°F - 32) \times 4/9 = °R$ Examples: (1) To convert 41°F to °R, $(41 - 32) \times 4/9 = 4°R$
(2) To convert −13°F to °R, $(-13 - 32) \times 4/9 = -20°R$

Fahrenheit to Fahrenheit Absolute or Rankine

Formula: $°F + 459.58 = °R'$ Examples: (1) To convert 20°F to °R', $20 + 459.58 = 479.58°R'$
(2) To convert −20°F to °R', $-20 + 459.58 = 439.58°R'$

Reaumur to Centigrade

Formula: $°R \times 5/4 = °C$ Examples: (1) To convert 32°R to °C, $32 \times 5/4 = 40°C$
(2) To convert −24°R to °C, $-24 \times 5/4 = -30°C$

Reaumur to Fahrenheit

Formula: $(°R \times 9/4) + 32 = °F$ Examples: (1) To convert 8°R to °F, $(8 \times 9/4) + 32 = 50°F$
(2) To convert −8°R to °F, $(-8 \times 9/4) + 32 = 14°F$
(3) To convert −20°R to °F, $(-20 \times 9/4) + 32 = -13°F$

Table 2-6
TEMPERATURE CONVERSION TABLE

The column of figures in bold and which is headed "Reading in °F. or °C. to be converted" refers to the temperature either in degrees Fahrenheit or Centigrade which it is desired to convert into the other scale. If converting from Fahrenheit degrees to Centigrade degrees the equivalent temperature will be found in the column headed "°C." while if converting from degrees Centigrade to degrees Fahrenheit, the equivalent temperature will be found in the column headed "°F." This arrangement is very similar to that of Sauveur and Boylston, copyrighted 1920 and is published with their permission.

°F.	Reading in °F. or °C. to be converted	°C.	°F.	Reading in °F. or °C. to be converted	°C.
....	−458	−272.22	....	−358	−216.67
....	−456	−271.11	....	−356	−215.56
....	−454	−270.00	....	−354	−214.44
....	−452	−268.89	....	−352	−213.33
....	−450	−267.78	....	−350	−212.22
....	−448	−266.67	....	−348	−211.11
....	−446	−265.56	....	−346	−210.00
....	−444	−264.44	....	−344	−208.89
....	−442	−263.33	....	−342	−207.78
....	−440	−262.22	....	−340	−206.67
....	−438	−261.11	....	−338	−205.56
....	−436	−260.00	....	−336	−204.44
....	−434	−258.89	....	−334	−203.33
....	−432	−257.78	....	−332	−202.22
....	−430	−256.67	....	−330	−201.11
....	−428	−255.56	....	−328	−200.00
....	−426	−254.44	....	−326	−198.89
....	−424	−253.33	......	−324	−197.78
....	−422	−252.22	....	−322	−196.67
....	−420	−251.11	....	−320	−195.56
....	−418	−250.00	....	−318	−194.44
....	−416	−248.89	....	−316	−193.33
....	−414	−247.78	....	−314	−192.22
....	−412	−246.67	....	−312	−191.11
....	−410	−245.56	....	−310	−190.00
....	−408	−244.44	....	−308	−188.89
....	−406	−243.33	....	−306	−187.78
....	−404	−242.22	....	−304	−186.67
....	−402	−241.11	....	−302	−185.56
....	−400	−240.00	....	−300	−184.44
....	−398	−238.89	....	−298	−183.33
....	−396	−237.78	....	−296	−182.22
....	−394	−236.67	....	−294	−181.11
....	−392	−235.56	....	−292	−180.00
....	−390	−234.44	....	−290	−178.89
....	−388	−233.33	....	−288	−177.78
....	−386	−232.22	....	−286	−176.67
....	−384	−231.11	....	−284	−175.56
....	−382	−230.00	....	−282	−174.44
....	−380	−228.89	....	−280	−173.33
....	−378	−227.78	....	−278	−172.22
....	−376	−226.67	....	−276	−171.11
....	−374	−225.56	....	−274	−170.00
....	−372	−224.44	−457.6	−272	−168.89
....	−370	−223.33	−454.0	−270	−167.78
....	−368	−222.22	−450.4	−268	−166.67
....	−366	−221.11	−446.8	−266	−165.56
....	−364	−220.00	−443.2	−264	−164.44
....	−362	−218.89	−439.6	−262	−163.33
....	−360	−217.78	−436.0	−260	−162.22

Table 2-6 (*Continued*)
TEMPERATURE CONVERSION TABLE

°F.	Reading in °F. or °C. to be converted	°C.	°F.	Reading in °F. or °C. to be converted	°C.
−432.4	−258	−161.11	−216.4	−138	−94.44
−428.8	−256	−160.00	−212.8	−136	−93.33
−425.2	−254	−158.89	−209.2	−134	−92.22
−421.6	−252	−157.78	−205.6	−132	−91.11
−418.0	−250	−156.67	−202.0	−130	−90.00
−414.4	−248	−155.56	−198.4	−128	−88.89
−410.8	−246	−154.44	−194.8	−126	−87.78
−407.2	−244	−153.33	−191.2	−124	−86.67
−403.6	−242	−152.22	−187.6	−122	−85.56
−400.0	−240	−151.11	−184.0	−120	−84.44
−396.4	−238	−150.00	−180.4	−118	−83.33
−392.8	−236	−148.89	−176.8	−116	−82.22
−389.2	−234	−147.78	−173.2	−114	−81.11
−385.6	−232	−146.67	−169.6	−112	−80.00
−382.0	−230	−145.56	−166.0	−110	−78.89
−378.4	−228	−144.44	−162.4	−108	−77.78
−374.8	−226	−143.33	−158.8	−106	−76.67
−371.2	−224	−142.22	−155.2	−104	−75.56
−367.6	−222	−141.11	−151.6	−102	−74.44
−364.0	−220	−140.00	−148.0	−100	−73.33
−360.4	−218	−138.89	−144.4	−98	−72.22
−356.8	−216	−137.78	−140.8	−96	−71.11
−353.2	−214	−136.67	−137.2	−94	−70.00
−349.6	−212	−135.56	−133.6	−92	−68.89
−346.0	−210	−134.44	−130.0	−90	−67.78
−342.4	−208	−133.33	−126.4	−88	−66.67
−338.8	−206	−132.22	−122.8	−86	−65.56
−335.2	−204	−131.11	−119.2	−84	−64.44
−331.6	−202	−130.00	−115.6	−82	−63.33
−328.0	−200	−128.89	−112.0	−80	−62.22
−324.4	−198	−127.78	−108.4	−78	−61.11
−320.8	−196	−126.67	−104.8	−76	−60.00
−317.2	−194	−125.56	−101.2	−74	−58.89
−313.6	−192	−124.44	−97.6	−72	−57.78
−310.0	−190	−123.33	−94.0	−70	−56.67
−306.4	−188	−122.22	−90.4	−68	−55.56
−302.8	−186	−121.11	−86.8	−66	−54.44
−299.2	−184	−120.00	−83.2	−64	−53.33
−295.6	−182	−118.89	−79.6	−62	−52.22
−292.0	−180	−117.78	−76.0	−60	−51.11
−288.4	−178	−116.67	−72.4	−58	−50.00
−284.8	−176	−115.56	−68.8	−56	−48.89
−281.2	−174	−114.44	−65.2	−54	−47.78
−277.6	−172	−113.33	−61.6	−52	−46.67
−274.0	−170	−112.22	−58.0	−50	−45.56
−270.4	−168	−111.11	−54.4	−48	−44.44
−266.8	−166	−110.00	−50.8	−46	−43.33
−263.2	−164	−108.89	−47.2	−44	−42.22
−259.6	−162	−107.78	−43.6	−42	−41.11
−256.0	−160	−106.67	−40.0	−40	−40.00
−252.4	−158	−105.56	−36.4	−38	−38.89
−248.8	−156	−104.44	−32.8	−36	−37.78
−245.2	−154	−103.33	−29.2	−34	−36.67
−241.6	−152	−102.22	−25.6	−32	−35.56
−238.0	−150	−101.11	−22.0	−30	−34.44
−234.4	−148	−100.00	−18.4	−28	−33.33
−230.8	−146	−98.89	−14.8	−26	−32.22
−227.2	−144	−97.78	−11.2	−24	−31.11
−223.6	−142	−96.67	−7.6	−22	−30.00
−220.0	−140	−95.56	−4.0	−20	−28.89

Table 2-6 (*Continued*)
TEMPERATURE CONVERSION TABLE

°F.	Reading in °F. or °C to be converted	°C.	°F.	Reading in °F. or °C. to be converted	°C.
−0.4	−18	−27.78	+116.6	+47	+8.33
+3.2	−16	−26.67	+118.4	+48	+8.89
+6.8	−14	−25.56	+120.2	+49	+9.44
+10.4	−12	−24.44	+122.0	+50	+10.00
+14.0	−10	−23.33	+123.8	+51	+10.56
+17.6	−8	−22.22	+125.6	+52	+11.11
+19.4	−7	−21.67	+127.4	+53	+11.67
+21.2	−6	−21.11	+129.2	+54	+12.22
+23.0	−5	−20.56	+131.0	+55	+12.78
+24.8	−4	−20.00	+132.8	+56	+13.33
+26.6	−3	−19.44	+134.6	+57	+13.89
+28.4	−2	−18.89	+136.4	+58	+14.44
+30.2	−1	−18.33	+138.2	+59	+15.00
+32.0	±0	−17.78	+140.0	+60	+15.56
+33.8	+1	−17.22	+141.8	+61	+16.11
+35.6	+2	−16.67	+143.6	+62	+16.67
+37.4	+3	−16.11	+145.4	+63	+17.22
+39.2	+4	−15.56	+147.2	+64	+17.78
+41.0	+5	−15.00	+149.0	+65	+18.33
+42.8	+6	−14.44	+150.8	+66	+18.89
+44.6	+7	−13.89	+152.6	+67	+19.44
+46.4	+8	−13.33	+154.4	+68	+20.00
+48.2	+9	−12.78	+156.2	+69	+20.56
+50.0	+10	−12.22	+158.0	+70	+21.11
+51.8	+11	−11.67	+159.8	+71	+21.67
+53.6	+12	−11.11	+161.6	+72	+22.22
+55.4	+13	−10.56	+163.4	+73	+22.78
+57.2	+14	−10.00	+165.2	+74	+23.33
+59.0	+15	−9.44	+167.0	+75	+23.89
+60.8	+16	−8.89	+168.8	+76	+24.44
+62.6	+17	−8.33	+170.6	+77	+25.00
+64.4	+18	−7.78	+172.4	+78	+25.56
+66.2	+19	−7.22	+174.2	+79	+26.11
+68.0	+20	−6.67	+176.0	+80	+26.67
+69.8	+21	−6.11	+177.8	+81	+27.22
+71.6	+22	−5.56	+179.6	+82	+27.78
+73.4	+23	−5.00	+181.4	+83	+28.33
+75.2	+24	−4.44	+183.2	+84	+28.89
+77.0	+25	−3.89	+185.0	+85	+29.44
+78.8	+26	−3.33	+186.8	+86	+30.00
+80.6	+27	−2.78	+188.6	+87	+30.56
+82.4	+28	−2.22	+190.4	+88	+31.11
+84.2	+29	−1.67	+192.2	+89	+31.67
+86.0	+30	−1.11	+194.0	+90	+32.22
+87.8	+31	−0.56	+195.8	+91	+32.78
+89.6	+32	±0.00	+197.6	+92	+33.33
+91.4	+33	+0.56	+199.4	+93	+33.89
+93.2	+34	+1.11	+201.2	+94	+34.44
+95.0	+35	+1.67	+203.0	+95	+35.00
+96.8	+36	+2.22	+204.8	+96	+35.56
+98.6	+37	+2.78	+206.6	+97	+36.11
+100.4	+38	+3.33	+208.4	+98	+36.67
+102.2	+39	+3.89	+210.2	+99	+37.22
+104.0	+40	+4.44	+212.0	+100	+37.78
+105.8	+41	+5.00	+213.8	+101	+38.33
+107.6	+42	+5.56	+215.6	+102	+38.89
+109.4	+43	+6.11	+217.4	+103	+39.44
+111.2	+44	+6.67	+219.2	+104	+40.00
+113.0	+45	+7.22	+221.0	+105	+40.56
+114.8	+46	+7.78	+222.8	+106	+41.11

Table 2-6 (*Continued*)
TEMPERATURE CONVERSION TABLE

°F.	Reading in °F. or °C. to be converted	°C.	°F.	Reading in °F. or °C. to be converted	°C.
+224.6	+107	+41.67	+332.6	+167	+75.00
+226.4	+108	+42.22	+334.4	+168	+75.56
+228.2	+109	+42.78	+336.2	+169	+76.11
+230.0	+110	+43.33	+338.0	+170	+76.67
+231.8	+111	+43.89	+339.8	+171	+77.22
+233.6	+112	+44.44	+341.6	+172	+77.78
+235.4	+113	+45.00	+343.4	+173	+78.33
+237.2	+114	+45.56	+345.2	+174	+78.89
+239.0	+115	+46.11	+347.0	+175	+79.44
+240.8	+116	+46.67	+348.8	+176	+80.00
+242.6	+117	+47.22	+350.6	+177	+80.56
+244.4	+118	+47.78	+352.4	+178	+81.11
+246.2	+119	+48.33	+354.2	+179	+81.67
+248.0	+120	+48.89	+356.0	+180	+82.22
+249.8	+121	+49.44	+357.8	+181	+82.78
+251.6	+122	+50.00	+359.6	+182	+83.33
+253.4	+123	+50.56	+361.4	+183	+83.89
+255.2	+124	+51.11	+363.2	+184	+84.44
+257.0	+125	+51.67	+365.0	+185	+85.00
+258.8	+126	+52.22	+366.8	+186	+85.56
+260.6	+127	+52.78	+368.6	+187	+86.11
+262.4	+128	+53.33	+370.4	+188	+86.67
+264.2	+129	+53.89	+372.2	+189	+87.22
+266.0	+130	+54.44	+374.0	+190	+87.78
+267.8	+131	+55.00	+375.8	+191	+88.33
+269.6	+132	+55.56	+377.6	+192	+88.89
+271.4	+133	+56.11	+379.4	+193	+89.44
+273.2	+134	+56.67	+381.2	+194	+90.00
+275.0	+135	+57.22	+383.0	+195	+90.56
+276.8	+136	+57.78	+384.8	+196	+91.11
+278.6	+137	+58.33	+386.6	+197	+91.67
+280.4	+138	+58.89	+388.4	+198	+92.22
+282.2	+139	+59.44	+390.2	+199	+92.78
+284.0	+140	+60.00	+392.0	+200	+93.33
+285.8	+141	+60.56	+393.8	+201	+93.89
+287.6	+142	+61.11	+395.6	+202	+94.44
+289.4	+143	+61.67	+397.4	+203	+95.00
+291.2	+144	+62.22	+399.2	+204	+95.56
+293.0	+145	+62.78	+401.0	+205	+96.11
+294.8	+146	+63.33	+402.8	+206	+96.67
+296.6	+147	+63.89	+404.6	+207	+97.22
+298.4	+148	+64.44	+406.4	+208	+97.78
+300.2	+149	+65.00	+408.2	+209	+98.33
+302.0	+150	+65.56	+410.0	+210	+98.89
+303.8	+151	+66.11	+411.8	+211	+99.44
+305.6	+152	+66.67	+413.6	+212	+100.00
+307.4	+153	+67.22	+415.4	+213	+100.56
+309.2	+154	+67.78	+417.2	+214	+101.11
+311.0	+155	+68.33	+419.0	+215	+101.67
+312.8	+156	+68.89	+420.8	+216	+102.22
+314.6	+157	+69.44	+422.6	+217	+102.78
+316.4	+158	+70.00	+424.4	+218	+103.33
+318.2	+159	+70.56	+426.2	+219	+103.89
+320.0	+160	+71.11	+428.0	+220	+104.44
+321.8	+161	+71.67	+431.6	+222	+105.56
+323.6	+162	+72.22	+435.2	+224	+106.67
+325.4	+163	+72.78	+438.8	+226	+107.78
+327.2	+164	+73.33	+442.4	+228	+108.89
+329.0	+165	+73.89	+446.0	+230	+110.00
+330.8	+166	+74.44	+449.6	+232	+111.11

Table 2-6 (*Continued*)
TEMPERATURE CONVERSION TABLE

°F.	Reading in °F. or °C. to be converted	°C.	°F.	Reading in °F. or °C. to be converted	°C.
+453.2	+234	+112.22	+669.2	+354	+178.89
+456.8	+236	+113.33	+672.8	+356	+180.00
+460.4	+238	+114.44	+676.4	+358	+181.11
+464.0	+240	+115.56	+680.0	+360	+182.22
+467.6	+242	+116.67	+683.6	+362	+183.33
+471.2	+244	+117.78	+687.2	+364	+184.44
+474.8	+246	+118.89	+690.8	+366	+185.56
+478.4	+248	+120.00	+694.4	+368	+186.67
+482.0	+250	+121.11	+698.0	+370	+187.78
+485.6	+252	+122.22	+701.6	+372	+188.89
+489.2	+254	+123.33	+705.2	+374	+190.00
+492.8	+256	+124.44	+708.8	+376	+191.11
+496.4	+258	+125.56	+712.4	+378	+192.22
+500.0	+260	+126.67	+716.0	+380	+193.33
+503.6	+262	+127.78	+719.6	+382	+194.44
+507.2	+264	+128.89	+723.2	+384	+195.56
+510.8	+266	+130.00	+726.8	+386	+196.67
+514.4	+268	+131.11	+730.4	+388	+197.78
+518.0	+270	+132.22	+734.0	+390	+198.89
+521.6	+272	+133.33	+737.6	+392	+200.00
+525.2	+274	+134.44	+741.2	+394	+201.11
+528.8	+276	+135.56	+744.8	+396	+202.22
+532.4	+278	+136.67	+748.4	+398	+203.33
+536.0	+280	+137.78	+752.0	+400	+204.44
+539.6	+282	+138.89	+755.6	+402	+205.56
+543.2	+284	+140.00	+759.2	+404	+206.67
+546.8	+286	+141.11	+762.8	+406	+207.78
+550.4	+288	+142.22	+766.4	+408	+208.89
+554.0	+290	+143.33	+770.0	+410	+210.00
+557.6	+292	+144.44	+773.6	+412	+211.11
+561.2	+294	+145.56	+777.2	+414	+212.22
+564.8	+296	+146.67	+780.8	+416	+213.33
+568.4	+298	+147.78	+784.4	+418	+214.44
+572.0	+300	+148.89	+788.0	+420	+215.56
+575.6	+302	+150.00	+791.6	+422	+216.67
+579.2	+304	+151.11	+795.2	+424	+217.78
+582.8	+306	+152.22	+798.8	+426	+218.89
+586.4	+308	+153.33	+802.4	+428	+220.00
+590.0	+310	+154.44	+806.0	+430	+221.11
+593.6	+312	+155.56	+809.6	+432	+222.22
+597.2	+314	+156.67	+813.2	+434	+223.33
+600.8	+316	+157.78	+816.8	+436	+224.14
+604.4	+318	+158.89	+820.4	+438	+225.56
+608.0	+320	+160.00	+824.0	+440	+226.67
+611.6	+322	+161.11	+827.6	+442	+227.78
+615.2	+324	+162.22	+831.2	+444	+228.89
+618.8	+326	+163.33	+834.8	+446	+230.00
+622.4	+328	+164.44	+838.4	+448	+231.11
+626.0	+330	+165.56	+842.0	+450	+232.22
+629.6	+332	+166.67	+845.6	+452	+233.33
+633.2	+334	+167.78	+849.2	+454	+234.44
+636.8	+336	+168.89	+852.8	+456	+235.56
+640.4	+338	+170.00	+856.4	+458	+236.67
+644.0	+340	+171.11	+860.0	+460	+237.78
+647.6	+342	+172.22	+863.6	+462	+238.89
+651.2	+344	+173.33	+867.2	+464	+240.00
+654.8	+346	+174.44	+870.8	+466	+241.11
+658.4	+348	+175.56	+874.4	+468	+242.22
+662.0	+350	+176.67	+878.0	+470	+243.33
+665.6	+352	+177.78	+881.6	+472	+244.44

Table 2-6 (*Continued*)
TEMPERATURE CONVERSION TABLE

°F.	Reading in °F. or °C. to be converted	°C.	°F.	Reading in °F. or °C. to be converted	°C.
+885.2	+474	+245.56	+1101.2	+594	+312.22
+888.8	+476	+246.67	+1104.8	+596	+313.33
+892.4	+478	+247.78	+1108.4	+598	+314.44
+896.0	+480	+248.89	+1112.0	+600	+315.56
+899.6	+482	+250.00	+1115.6	+602	+316.67
+903.2	+484	+251.11	+1119.2	+604	+317.78
+906.8	+486	+252.22	+1122.8	+606	+318.89
+910.4	+488	+253.33	+1126.4	+608	+320.00
+914.0	+490	+254.44	+1130.0	+610	+321.11
+917.6	+492	+255.56	+1133.6	+612	+322.22
+921.2	+494	+256.67	+1137.2	+614	+323.33
+924.8	+496	+257.78	+1140.8	+616	+324.44
+928.4	+498	+258.89	+1144.4	+618	+325.56
+932.0	+500	+260.00	+1148.0	+620	+326.67
+935.6	+502	+261.11	+1151.6	+622	+327.78
+939.2	+504	+262.22	+1155.2	+624	+328.89
+942.8	+506	+263.33	+1158.8	+626	+330.00
+946.4	+508	+264.44	+1162.4	+628	+331.11
+950.0	+510	+265.56	+1166.0	+630	+332.22
+953.6	+512	+266.67	+1169.6	+632	+333.33
+957.2	+514	+267.78	+1173.2	+634	+334.44
+960.8	+516	+268.89	+1176.8	+636	+335.56
+964.4	+518	+270.00	+1180.4	+638	+336.67
+968.0	+520	+271.11	+1184.0	+640	+337.78
+971.6	+522	+272.22	+1187.6	+642	+338.89
+975.2	+524	+273.33	+1191.2	+644	+340.00
+978.8	+526	+274.44	+1194.8	+646	+341.11
+982.4	+528	+275.56	+1198.4	+648	+342.22
+986.0	+530	+276.67	+1202.0	+650	+343.33
+989.6	+532	+277.78	+1205.6	+652	+344.44
+993.2	+534	+278.89	+1209.2	+654	+345.56
+996.8	+536	+280.00	+1212.8	+656	+346.67
+1000.4	+538	+281.11	+1216.4	+658	+347.78
+1004.0	+540	+282.22	+1220.0	+660	+348.89
+1007.6	+542	+283.33	+1223.6	+662	+350.00
+1011.2	+544	+284.44	+1227.2	+664	+351.11
+1014.8	+546	+285.56	+1230.8	+666	+352.22
+1018.4	+548	+286.67	+1234.4	+668	+353.33
+1022.0	+550	+287.78	+1238.0	+670	+354.44
+1025.6	+552	+288.89	+1241.6	+672	+355.56
+1029.2	+554	+290.00	+1245.2	+674	+356.67
+1032.8	+556	+291.11	+1248.8	+676	+357.78
+1036.4	+558	+292.22	+1252.4	+678	+358.89
+1040.0	+560	+293.33	+1256.0	+680	+360.00
+1043.6	+562	+294.44	+1259.6	+682	+361.11
+1047.2	+564	+295.56	+1263.2	+684	+362.22
+1050.8	+566	+296.67	+1266.8	+686	+363.33
+1054.4	+568	+297.78	+1270.4	+688	+364.44
+1058.0	+570	+298.89	+1274.0	+690	+365.56
+1061.6	+572	+300.00	+1277.6	+692	+366.67
+1065.2	+574	+301.11	+1281.2	+694	+367.78
+1068.8	+576	+302.22	+1284.8	+696	+368.89
+1072.4	+578	+303.33	+1288.4	+698	+370.00
+1076.0	+580	+304.44	+1292.0	+700	+371.11
+1079.6	+582	+305.56	+1295.6	+702	+372.22
+1083.2	+584	+306.67	+1299.2	+704	+373.33
+1086.8	+586	+307.78	+1302.8	+706	+374.44
+1090.4	+588	+308.89	+1306.4	+708	+375.56
+1094.0	+590	+310.00	+1310.0	+710	+376.67
+1097.6	+592	+311.11	+1313.6	+712	+377.78

Table 2-6 (*Continued*)
TEMPERATURE CONVERSION TABLE

°F.	Reading in °F. or °C. to be converted	°C.	°F.	Reading in °F. or °C. to be converted	°C.
+1317.2	+714	+378.89	+1533.2	+834	+445.56
+1320.8	+716	+380.00	+1536.8	+836	+446.67
+1324.4	+718	+381.11	+1540.4	+838	+447.78
+1328.0	+720	+382.22	+1544.0	+840	+448.89
+1331.6	+722	+383.33	+1547.6	+842	+450.00
+1335.2	+724	+384.44	+1551.2	+844	+451.11
+1338.8	+726	+385.56	+1554.8	+846	+452.22
+1342.4	+728	+386.67	+1558.4	+848	+453.33
+1346.0	+730	+387.78	+1562.0	+850	+454.44
+1349.6	+732	+388.89	+1565.6	+852	+455.56
+1353.2	+734	+390.00	+1569.2	+854	+456.67
+1356.8	+736	+391.11	+1572.8	+856	+457.78
+1360.4	+738	+392.22	+1576.4	+858	+458.89
+1364.0	+740	+393.33	+1580.0	+860	+460.00
+1367.6	+742	+394.44	+1583.6	+862	+461.11
+1371.2	+744	+395.56	+1587.2	+864	+462.22
+1374.8	+746	+396.67	+1590.8	+866	+463.33
+1378.4	+748	+397.78	+1594.4	+868	+464.44
+1382.0	+750	+398.89	+1598.0	+870	+465.56
+1385.6	+752	+400.00	+1601.6	+872	+466.67
+1389.2	+754	+401.11	+1605.2	+874	+467.78
+1392.8	+756	+402.22	+1608.8	+876	+468.89
+1396.4	+758	+403.33	+1612.4	+878	+470.00
+1400.0	+760	+404.44	+1616.0	+880	+471.11
+1403.6	+762	+405.56	+1619.6	+882	+472.22
+1407.2	+764	+406.67	+1623.2	+884	+473.33
+1410.8	+766	+407.78	+1626.8	+886	+474.44
+1414.4	+768	+408.89	+1630.4	+888	+475.56
+1418.0	+770	+410.00	+1634.0	+890	+476.67
+1421.6	+772	+411.11	+1637.6	+892	+477.78
+1425.2	+774	+412.22	+1641.2	+894	+478.89
+1428.8	+776	+413.33	+1644.8	+896	+480.00
+1432.4	+778	+414.44	+1648.4	+898	+481.11
+1436.0	+780	+415.56	+1652.0	+900	+482.22
+1439.6	+782	+416.67	+1655.6	+902	+483.33
+1443.2	+784	+417.78	+1659.2	+904	+484.44
+1446.8	+786	+418.89	+1662.8	+906	+485.56
+1450.4	+788	+420.00	+1666.4	+908	+486.67
+1454.0	+790	+421.11	+1670.0	+910	+487.78
+1457.6	+792	+422.22	+1673.6	+912	+488.89
+1461.2	+794	+423.33	+1677.2	+914	+490.00
+1464.8	+796	+424.44	+1680.8	+916	+491.11
+1468.4	+798	+425.56	+1684.4	+918	+492.22
+1472.0	+800	+426.67	+1688.0	+920	+493.33
+1475.6	+802	+427.78	+1691.6	+922	+494.44
+1479.2	+804	+428.89	+1695.2	+924	+495.56
+1482.8	+806	+430.00	+1698.8	+926	+496.67
+1486.4	+808	+431.11	+1702.4	+928	+497.78
+1490.0	+810	+432.22	+1706.0	+930	+498.89
+1493.6	+812	+433.33	+1709.6	+932	+500.00
+1497.2	+814	+434.44	+1713.2	+934	+501.11
+1500.8	+816	+435.56	+1716.8	+936	+502.22
+1504.4	+818	+436.67	+1720.4	+938	+503.33
+1508.0	+820	+437.78	+1724.0	+940	+504.44
+1511.6	+822	+438.89	+1727.6	+942	+505.56
+1515.2	+824	+440.00	+1731.2	+944	+506.67
+1518.8	+826	+441.11	+1734.8	+946	+507.78
+1522.4	+828	+442.22	+1738.4	+948	+508.89
+1526.0	+830	+443.33	+1742.0	+950	+510.00
+1529.6	+832	+444.44	+1745.6	+952	+511.11

Table 2-6 (*Continued*)
TEMPERATURE CONVERSION TABLE

°F.	Reading in °F. or °C. to be converted	°C.	°F.	Reading in °F. or °C. to be converted	°C.
+1749.2	+954	+512.22	+2498.0	+1370	+743.33
+1752.8	+956	+513.33	+2516.0	+1380	+748.89
+1756.4	+958	+514.44	+2534.0	+1390	+754.44
+1760.0	+960	+515.56	+2552.0	+1400	+760.00
+1763.6	+962	+516.67	+2570.0	+1410	+765.56
+1767.2	+964	+517.78	+2588.0	+1420	+771.11
+1770.8	+966	+518.89	+2606.0	+1430	+776.67
+1774.4	+968	+520.00	+2624.0	+1440	+782.22
+1778.0	+970	+521.11	+2642.0	+1450	+787.78
+1781.6	+972	+522.22	+2660.0	+1460	+793.33
+1785.2	+974	+523.33	+2678.0	+1470	+798.89
+1788.8	+976	+524.44	+2696.0	+1480	+804.44
+1792.4	+978	+525.56	+2714.0	+1490	+810.00
+1796.0	+980	+526.67	+2732.0	+1500	+815.56
+1799.6	+982	+527.78	+2750.0	+1510	+821.11
+1803.2	+984	+528.89	+2768.0	+1520	+826.67
+1806.8	+986	+530.00	+2786.0	+1530	+832.22
+1810.4	+988	+531.11	+2804.0	+1540	+837.78
+1814.0	+990	+532.22	+2822.0	+1550	+843.33
+1817.6	+992	+533.33	+2840.0	+1560	+848.89
+1821.2	+994	+534.44	+2858.0	+1570	+854.44
+1824.8	+996	+535.56	+2876.0	+1580	+860.00
+1828.4	+998	+536.67	+2894.0	+1590	+865.56
+1832.0	+1000	+537.78	+2912.0	+1600	+871.11
+1850.0	+1010	+543.33	+2930.0	+1610	+876.67
+1868.0	+1020	+548.89	+2948.0	+1620	+882.22
+1886.0	+1030	+554.44	+2966.0	+1630	+887.78
+1904.0	+1040	+560.00	+2984.0	+1640	+893.33
+1922.0	+1050	+565.56	+3002.0	+1650	+898.89
+1940.0	+1060	+571.11	+3020.0	+1660	+904.44
+1958.0	+1070	+576.67	+3038.0	+1670	+910.00
+1976.0	+1080	+582.22	+3056.0	+1680	+915.56
+1994.0	+1090	+587.78	+3074.0	+1690	+921.11
+2012.0	+1100	+593.33	+3092.0	+1700	+926.67
+2030.0	+1110	+598.89	+3110.0	+1710	+932.22
+2048.0	+1120	+604.44	+3128.0	+1720	+937.78
+2066.0	+1130	+610.00	+3146.0	+1730	+943.33
+2084.0	+1140	+615.56	+3164.0	+1740	+948.89
+2102.0	+1150	+621.11	+3182.0	+1750	+954.44
+2120.0	+1160	+626.67	+3200.0	+1760	+960.00
+2138.0	+1170	+632.22	+3218.0	+1770	+965.56
+2156.0	+1180	+637.78	+3236.0	+1780	+971.11
+2174.0	+1190	+643.33	+3254.0	+1790	+976.67
+2192.0	+1200	+648.89	+3272.0	+1800	+982.22
+2210.0	+1210	+654.44	+3290.0	+1810	+987.78
+2228.0	+1220	+660.00	+3308.0	+1820	+993.33
+2246.0	+1230	+665.56	+3326.0	+1830	+998.89
+2264.0	+1240	+671.11	+3344.0	+1840	+1004.4
+2282.0	+1250	+676.67	+3362.0	+1850	+1010.0
+2300.0	+1260	+682.22	+3380.0	+1860	+1015.6
+2318.0	+1270	+687.78	+3398.0	+1870	+1021.1
+2336.0	+1280	+693.33	+3416.0	+1880	+1026.7
+2354.0	+1290	+698.89	+3434.0	+1890	+1032.2
+2372.0	+1300	+704.44	+3452.0	+1900	+1037.8
+2390.0	+1310	+710.00	+3470.0	+1910	+1043.3
+2408.0	+1320	+715.56	+3488.0	+1920	+1048.9
+2426.0	+1330	+721.11	+3506.0	+1930	+1054.4
+2444.0	+1340	+726.67	+3524.0	+1940	+1060.0
+2462.0	+1350	+732.22	+3542.0	+1950	+1065.6
+2480.0	+1360	+737.78	+3560.0	+1960	+1071.1

Table 2-6 (*Continued*)
TEMPERATURE CONVERSION TABLE

°F.	Reading in °F. or °C. to be converted	°C.	°F.	Reading in °F. or °C. to be converted	°C.
+3578.0	+1970	+1076.7	+4604.0	+2540	+1393.3
+3596.0	+1980	+1082.2	+4622.0	+2550	+1398.9
+3614.0	+1990	+1087.8	+4640.0	+2560	+1404.4
+3632.0	+2000	+1093.3	+4658.0	+2570	+1410.0
+3650.0	+2010	+1098.9	+4676.0	+2580	+1415.6
+3668.0	+2020	+1104.4	+4694.0	+2590	+1421.1
+3686.0	+2030	+1110.0	+4712.0	+2600	+1426.7
+3704.0	+2040	+1115.6	+4730.0	+2610	+1432.2
+3722.0	+2050	+1121.1	+4748.0	+2620	+1437.8
+3740.0	+2060	+1126.7	+4766.0	+2630	+1443.3
+3758.0	+2070	+1132.2	+4784.0	+2640	+1448.9
+3776.0	+2080	+1137.8	+4802.0	+2650	+1454.4
+3794.0	+2090	+1143.3	+4820.0	+2660	+1460.0
+3812.0	+2100	+1148.9	+4838.0	+2670	+1465.6
+3830.0	+2110	+1154.4	+4856.0	+2680	+1471.1
+3848.0	+2120	+1160.0	+4874.0	+2690	+1476.7
+3866.0	+2130	+1165.6	+4892.0	+2700	+1482.2
+3884.0	+2140	+1171.1	+4910.0	+2710	+1487.8
+3902.0	+2150	+1176.7	+4928.0	+2720	+1493.3
+3920.0	+2160	+1182.2	+4946.0	+2730	+1498.9
+3938.0	+2170	+1187.8	+4964.0	+2740	+1504.4
+3956.0	+2180	+1193.3	+4982.0	+2750	+1510.0
+3974.0	+2190	+1198.9	+5000.0	+2760	+1515.6
+3992.0	+2200	+1204.4	+5018.0	+2770	+1521.1
+4010.0	+2210	+1210.0	+5036.0	+2780	+1526.7
+4028.0	+2220	+1215.6	+5054.0	+2790	+1532.2
+4046.0	+2230	+1221.1	+5072.0	+2800	+1537.8
+4064.0	+2240	+1226.7	+5090.0	+2810	+1543.3
+4082.0	+2250	+1232.2	+5108.0	+2820	+1548.9
+4100.0	+2260	+1237.8	+5126.0	+2830	+1554.4
+4118.0	+2270	+1243.3	+5144.0	+2840	+1560.0
+4136.0	+2280	+1248.9	+5162.0	+2850	+1565.6
+4154.0	+2290	+1254.4	+5180.0	+2860	+1571.1
+4172.0	+2300	+1260.0	+5198.0	+2870	+1576.7
+4190.0	+2310	+1265.6	+5216.0	+2880	+1582.2
+4208.0	+2320	+1271.1	+5234.0	+2890	+1587.8
+4226.0	+2330	+1276.7	+5252.0	+2900	+1593.3
+4244.0	+2340	+1282.2	+5270.0	+2910	+1598.9
+4262.0	+2350	+1287.8	+5288.0	+2920	+1604.4
+4280.0	+2360	+1293.3	+5306.0	+2930	+1610.0
+4298.0	+2370	+1298.9	+5324.0	+2940	+1615.6
+4316.0	+2380	+1304.4	+5342.0	+2950	+1621.1
+4334.0	+2390	+1310.0	+5360.0	+2960	+1626.7
+4352.0	+2400	+1315.6	+5378.0	+2970	+1632.2
+4370.0	+2410	+1321.1	+5396.0	+2980	+1637.8
+4388.0	+2420	+1326.7	+5414.0	+2990	+1643.3
+4406.0	+2430	+1332.2	+5432.0	+3000	+1648.9
+4424.0	+2440	+1337.8	+5450.0	+3010	+1654.4
+4442.0	+2450	+1343.3	+5468.0	+3020	+1660.0
+4460.0	+2460	+1348.9	+5486.0	+3030	+1665.6
+4478.0	+2470	+1354.4	+5504.0	+3040	+1671.1
+4496.0	+2480	+1360.0	+5522.0	+3050	+1676.7
+4514.0	+2490	+1365.6	+5540.0	+3060	+1682.2
+4532.0	+2500	+1371.1	+5558.0	+3070	+1687.8
+4550.0	+2510	+1376.7	+5576.0	+3080	+1693.3
+4568.0	+2520	+1382.2	+5594.0	+3090	+1698.9
+4586.0	+2530	+1387.8	+5612.0	+3100	+1704.4

Table 2-7
HYDROMETER CONVERSION TABLE

This table gives the relation between density (C.G.S.) and degrees on the Baumé and Twaddell scales. The Twaddell scale is never used for densities less than unity. See also section on *Hydrometers*.

Density	Degrees Baumé Bu. Stand. Scale	Degrees Baumé A.P.I. Scale	Density	Degrees Baume Bu. Stand. Scale	Degrees Baumé A.P.I. Scale
0.600	103.33	104.33	0.895	26.42	26.60
0.605	101.40	102.38	0.900	25.56	25.72
0.610	99.51	100.47	0.905	24.70	24.85
0.615	97.64	98.58	0.910	23.85	23.99
0.620	95.81	96.73	0.915	23.01	23.14
0.625	94.00	94.90	0.920	22.17	22.30
0.630	92.22	93.10	0.925	21.35	21.47
0.635	90.47	91.33	0.930	20.54	20.65
0.640	88.75	89.59	0.935	19.73	19.84
0.645	87.05	87.88	0.940	18.94	19.03
0.650	85.38	86.19	0.945	18.15	18.24
0.655	83.74	84.53	0.950	17.37	17.45
0.660	82.12	82.89	0.955	16.60	16.67
0.665	80.52	81.28	0.960	15.83	15.90
0.670	78.95	79.69	0.965	15.08	15.13
0.675	77.41	78.13	0.970	14.33	14.38
0.680	75.88	76.59	0.975	13.59	13.63
0.685	74.38	75.07	0.980	12.86	12.89
0.690	72.90	73.57	0.985	12.13	12.15
0.695	71.43	72.10	0.990	11.41	11.43
0.700	70.00	70.64	0.995	10.70	10.71
0.705	68.57	69.21	1.000	10.00	10.00
0.710	67.18	67.80			
0.715	65.80	66.40			
0.720	64.44	65.03			

DENSITIES GREATER THAN UNITY

Density	Degrees Baumé Bu. Stand. Scale	Degrees Twaddell
1.00	0.00	0
1.01	1.44	2
1.02	2.84	4
1.03	4.22	6
1.04	5.58	8
1.05	6.91	10
1.06	8.21	12
1.07	9.49	14
1.08	10.78	16
1.09	11.97	18
1.10	13.18	20
1.11	14.37	22
1.12	15.54	24
1.13	16.68	26
1.14	17.81	28
1.15	18.91	30
1.16	20.00	32
1.17	21.07	34
1.18	22.12	36
1.19	23.15	38
1.20	24.17	40
1.21	25.16	42
1.22	26.15	44
1.23	27.11	46
1.24	28.06	48
1.25	29.00	50
1.26	29.92	52
1.27	30.83	54
1.28	31.72	56
1.29	32.60	58
1.30	33.46	60
1.31	34.31	62

Continuation of left table (lower densities):

Density	Degrees Baumé Bu. Stand. Scale	Degrees Baumé A.P.I. Scale
0.725	63.10	63.67
0.730	61.78	62.34
0.735	60.48	61.02
0.740	59.19	59.72
0.745	57.92	58.43
0.750	56.67	57.17
0.755	55.43	55.92
0.760	54.21	54.68
0.765	53.01	53.47
0.770	51.82	52.27
0.775	50.65	51.08
0.780	49.49	49.91
0.785	48.34	48.75
0.790	47.22	47.61
0.795	46.10	46.49
0.800	45.00	45.38
0.805	43.91	44.28
0.810	42.84	43.19
0.815	41.78	42.12
0.820	40.73	41.06
0.825	39.70	40.02
0.830	38.68	38.98
0.835	37.66	37.96
0.840	36.67	36.95
0.845	35.68	35.96
0.850	34.71	34.97
0.855	33.74	34.00
0.860	32.79	33.03
0.865	31.85	32.08
0.870	30.92	31.14
0.875	30.00	30.21
0.880	29.09	29.30
0.885	28.19	28.39
0.890	27.30	27.49

Table 2-7 (*Continued*)
HYDROMETER CONVERSION TABLE

Density	Degrees Baumé Bu. Stand. Scale	Degrees Twaddell	Density	Degrees Baumé Bu. Stand. Scale	Degrees Twaddell
1.32	35.15	64	1.67	58.17	134
1.33	35.98	66	1.68	58.69	136
1.34	36.79	68	1.69	59.20	138
1.35	37.59	70	1.70	59.71	140
1.36	38.38	72	1.71	60.20	142
1.37	39.16	74	1.72	60.70	144
1.38	39.93	76	1.73	61.18	146
1.39	40.68	78	1.74	61.67	148
1.40	41.43	80	1.75	62.14	150
1.41	42.16	82	1.76	62.61	152
1.42	42.89	84	1.77	63.08	154
1.43	43.60	86	1.78	63.54	156
1.44	44.31	88	1.79	63.99	158
1.45	45.00	90	1.80	64.44	160
1.46	45.68	92	1.81	64.89	162
1.47	46.36	94	1.82	65.31	164
1.48	47.03	96	1.83	65.77	166
1.49	47.68	98	1.84	66.20	168
1.50	48.33	100	1.85	66.62	170
1.51	48.97	102	1.86	67.04	172
1.52	49.60	104	1.87	67.46	174
1.53	50.23	106	1.88	67.87	176
1.54	50.84	108	1.89	68.28	178
1.55	51.45	110	1.90	68.68	180
1.56	52.05	112	1.91	69.08	182
1.57	52.64	114	1.92	69.48	184
1.58	53.23	116	1.93	69.87	186
1.59	53.80	118	1.94	70.26	188
1.60	54.38	120	1.95	70.64	190
1.61	54.94	122	1.96	71.02	192
1.62	55.49	124	1.97	71.40	194
1.63	56.04	126	1.98	71.77	196
1.64	56.58	128	1.99	72.14	198
1.65	57.12	130	2.00	72.50	200
1.66	57.65	132			

BAROMETRY AND BAROMETRIC CORRECTIONS

In principle, the mercurial barometer balances a column of pure mercury against the weight of the atmosphere. The height of the column above the level of the mercury in the reservoir can be measured and serves as a direct index of atmospheric pressure. The space above the mercury in a barometer tube should be a Torricellian vacuum, perfect except for the practically negligible vapor pressure of mercury. The perfection of the vacuum is indicated by the sharpness of the click noted when the barometer tube is inclined. A barometer should be in a vertical position, suspended rather than fastened to a wall, and in a good light but not exposed to direct sunlight or too near a source of heat. The standard conditions for barometric measurements are 0°C., and gravity as at 45° latitude and sea level. There are numerous sources of error but corrections for most of these are readily applied. Some of the corrections are very small and their application may be questionable in view of the probably larger errors. The degree of consistency to be expected in careful measurements is about 0.005 inch with a 0.25 inch tube, increasing to 0.0015 inch with a tube 0.5 inch in diameter.

In reading a barometer of the Fortin type (the usual laboratory instrument for precision measurements) the procedure should be as follows: (1) Observe and record the temperature as indicated by the thermometer attached to the barometer. The temperature correction is very important and may be affected by heat from the observer's body. (2) Set the mercury in the reservoir at zero level, so that the point of the pin above the mercury just touches the surface, making a barely noticeable dimple therein. Tap the tube at the top and verify the zero setting. (3) Bring the vernier down until the view at the light background is cut off at the highest point of the meniscus. Record the reading.

The corrections to be made on the reading are as follow: (1) Temperature, to correct for the difference in thermal expansion of the mercury and the brass (or glass) to which the scale is attached.—This correction converts the reading into the value at 0°C. The brass scale table is applicable to the Fortin barometer. See tables. (2a) Latitude-gravity correction, and (2b) altitude-gravity correction, to compensate for differences in gravity, which would affect the height of the mercury column by variation in mass.—If local gravity is unknown, an approximate correction may be made from the tables. Local values of gravity are often subject to irregularities which lead to errors even when the corrections here provided are made. It is, therefore, advisable to determine the local value of gravity, from which the correction can be effected in the following manner:

$$Bt = Br + \left(\frac{g_1 - g_0}{g_0}\right) \times Br$$

in which Bt and Br are the true and the observed heights of the barometer respectively, g_0 is standard gravity (980,665 cm sec^{-2}), and g_1 is the local gravity. It may be noted that for most localities, g_1 is smaller than g_0, which makes the correction negative. These corrections compensate the reading to gravity at 45° latitude and sea level. (3) Correction for capillary depression of the level of the meniscus.—This varies with the tube diameter and actual height of the meniscus in a particular case. Some barometers are calibrated to allow for an average value of the latter and approximating the correction. See table. (4) Correction for vapor pressure of mercury.—This correction is usually negligible, being only 0.001 mm at 20°C. and 0.006 mm at 40°C. This correction is added. See table of vapor pressure of mercury.

All tables are given in both metric and English units. When a barometer reading is to be converted from one set of units to another, it is important that all corrections be made in the original units before the conversion is made.

The corrections above do not apply to an aneroid barometer. These instruments should be calibrated at regular intervals by checking them against a corrected mercurial barometer.

For records on weather maps, meteorologists customarily correct barometer readings to sea level, and some barometers may be calibrated accordingly. Such instruments are not suitable for laboratory use where true pressure under standard conditions is required. Scale corrections should be specified in the maker's instructions with the instrument, and are also indicated by the lack of correspondence between a gauge mark usually placed exactly 30 inches from the zero point and the 30 inch scale graduation

Table 2-8
BAROMETER TEMPERATURE CORRECTION—METRIC UNITS
A. Glass Scale

The values in the table below are to be subtracted from the observed readings to correct for the difference in the expansion of the mercury and the glass scale at different temperatures.

Temp. °C.	Observed barometer height in millimeters						
	700	730	740	750	760	770	800
	mm.	mm.	mm.	mm.	mm.	mm.	mm.
0	0.00	0.00	0.00	0.00	0.00	0.00	0.00
1	0.12	0.13	0.13	0.13	0.13	0.13	0.14
2	0.24	0.25	0.26	0.26	0.26	0.27	0.27
3	0.36	0.38	0.38	0.39	0.40	0.40	0.42
4	0.49	0.51	0.51	0.52	0.53	0.53	0.55
5	0.61	0.63	0.64	0.65	0.66	0.67	0.69
6	0.73	0.76	0.77	0.78	0.79	0.80	0.83
7	0.85	0.89	0.90	0.91	0.92	0.93	0.97
8	0.97	1.01	1.03	1.04	1.05	1.07	1.11
9	1.09	1.14	1.15	1.17	1.18	1.20	1.25
10	1.21	1.26	1.28	1.30	1.32	1.33	1.39
11	1.33	1.39	1.41	1.43	1.45	1.47	1.52
12	1.45	1.52	1.54	1.56	1.58	1.60	1.66
13	1.58	1.64	1.67	1.69	1.71	1.73	1.80
14	1.70	1.77	1.79	1.82	1.84	1.87	1.94
15	1.82	1.90	1.92	1.95	1.97	2.00	2.08
16	1.94	2.02	2.05	2.08	2.10	2.13	2.21
17	2.06	2.15	2.18	2.21	2.23	2.26	2.35
18	2.18	2.27	2.30	2.33	2.37	2.40	2.49
19	2.30	2.40	2.43	2.46	2.50	2.53	2.63
20	2.42	2.52	2.56	2.59	2.63	2.66	2.77
21	2.54	2.65	2.69	2.72	2.76	2.79	2.90
22	2.66	2.78	2.81	2.85	2.89	2.93	3.04
23	2.78	2.90	2.94	2.98	3.02	3.06	3.18
24	2.90	3.03	3.07	3.11	3.15	3.19	3.32
25	3.02	3.15	3.20	3.24	3.28	3.32	3.45
26	3.14	3.28	3.32	3.37	3.41	3.46	3.59
27	3.26	3.40	3.45	3.50	3.54	3.59	3.73
28	3.38	3.53	3.58	3.63	3.67	3.72	3.87
29	3.50	3.65	3.70	3.75	3.80	3.85	4.00
30	3.62	3.78	3.83	3.88	3.93	3.99	4.14
31	3.74	3.90	3.96	4.01	4.06	4.12	4.28
32	3.86	4.03	4.08	4.14	4.20	4.25	4.42
33	3.98	4.15	4.21	4.27	4.33	4.38	4.55
34	4.10	4.28	4.34	4.40	4.46	4.51	4.69
35	4.22	4.40	4.47	4.53	4.59	4.65	4.83

Table 2-8 (*Continued*)
BAROMETER TEMPERATURE CORRECTION—METRIC UNITS
B. Brass Scale

The values in the table below are to be subtracted from the observed readings to correct for the difference in the expansion of the mercury and the brass scale at different temperatures.

Temp. °C.	Observed barometer height in millimeters						
	640	650	660	670	680	690	700
	mm.	mm.	mm.	mm.	mm.	mm.	mm.
0	0.00	0.00	0.00	0.00	0.00	0.00	0.00
1	0.10	0.11	0.11	0.11	0.11	0.11	0.11
2	0.21	0.21	0.22	0.22	0.22	0.23	0.23
3	0.31	0.32	0.32	0.33	0.33	0.34	0.34
4	0.42	0.42	0.43	0.44	0.44	0.45	0.46
5	0.52	0.53	0.54	0.55	0.55	0.56	0.57
6	0.63	0.64	0.65	0.66	0.66	0.67	0.68
7	0.73	0.74	0.75	0.76	0.78	0.79	0.80
8	0.84	0.85	0.86	0.87	0.89	0.90	0.91
9	0.94	0.95	0.97	0.98	1.00	1.01	1.03
10	1.04	1.06	1.07	1.09	1.11	1.12	1.14
11	1.15	1.16	1.18	1.20	1.22	1.24	1.25
12	1.25	1.27	1.29	1.31	1.33	1.35	1.37
13	1.35	1.38	1.40	1.42	1.44	1.46	1.48
14	1.46	1.48	1.50	1.53	1.55	1.57	1.59
15	1.56	1.59	1.61	1.64	1.66	1.68	1.71
16	1.67	1.69	1.72	1.74	1.77	1.80	1.82
17	1.77	1.80	1.82	1.85	1.88	1.91	1.94
18	1.87	1.90	1.93	1.96	1.99	2.02	2.05
19	1.98	2.01	2.04	2.07	2.10	2.13	2.16
20	2.08	2.11	2.15	2.18	2.21	2.24	2.28
21	2.18	2.22	2.25	2.29	2.32	2.35	2.39
22	2.29	2.32	2.36	2.40	2.43	2.47	2.50
23	2.39	2.43	2.47	2.50	2.54	2.58	2.62
24	2.49	2.53	2.57	2.61	2.65	2.69	2.73
25	2.60	2.64	2.68	2.72	2.76	2.80	2.84
26	2.70	2.74	2.79	2.83	2.87	2.91	2.96
27	2.81	2.85	2.89	2.94	2.98	3.02	3.07
28	2.91	2.95	3.00	3.05	3.09	3.14	3.18
29	3.01	3.06	3.11	3.15	3.20	3.25	3.29
30	3.12	3.16	3.21	3.26	3.31	3.36	3.41
31	3.22	3.27	3.32	3.37	3.42	3.47	3.52
32	3.32	3.37	3.43	3.48	3.53	3.58	3.63
33	3.42	3.48	3.53	3.59	3.64	3.69	3.75
34	3.53	3.58	3.64	3.69	3.75	3.80	3.86
35	3.63	3.69	3.74	3.80	3.86	3.91	3.97

Table 2-8 (*Continued*)
BAROMETER TEMPERATURE CORRECTION—METRIC UNITS
B. Brass Scale

| Observed barometer height in millimeters | | | | | | | | Temp. °C. |
| 710 | 720 | 730 | 740 | 750 | 760 | 770 | 780 | |
mm.	mm.	mm.	mm.	mm.	mm.	mm.	mm.	
0.00	0.00	0.00	0.00	0.00	0.00	0.00	0.00	0
0.12	0.12	0.12	0.12	0.12	0.12	0.13	0.13	1
0.23	0.23	0.24	0.24	0.24	0.25	0.25	0.25	2
0.35	0.35	0.36	0.36	0.37	0.37	0.38	0.38	3
0.46	0.47	0.48	0.48	0.49	0.50	0.50	0.51	4
0.58	0.59	0.59	0.60	0.61	0.62	0.63	0.64	5
0.69	0.70	0.71	0.72	0.73	0.74	0.75	0.76	6
0.81	0.82	0.83	0.84	0.86	0.87	0.88	0.89	7
0.93	0.94	0.95	0.96	0.98	0.99	1.00	1.02	8
1.04	1.06	1.07	1.08	1.10	1.11	1.13	1.14	9
1.16	1.17	1.19	1.21	1.22	1.24	1.25	1.27	10
1.27	1.29	1.31	1.33	1.34	1.36	1.38	1.40	11
1.39	1.41	1.43	1.45	1.47	1.48	1.50	1.52	12
1.50	1.52	1.54	1.57	1.59	1.61	1.63	1.65	13
1.62	1.64	1.66	1.69	1.71	1.73	1.75	1.78	14
1.73	1.76	1.78	1.81	1.83	1.85	1.88	1.90	15
1.85	1.87	1.90	1.93	1.95	1.98	2.00	2.03	16
1.96	1.99	2.02	2.05	2.07	2.10	2.13	2.16	17
2.08	2.11	2.14	2.17	2.20	2.22	2.25	2.28	18
2.19	2.22	2.25	2.29	2.32	2.35	2.38	2.41	19
2.31	2.34	2.37	2.41	2.44	2.47	2.50	2.54	20
2.42	2.46	2.49	2.53	2.56	2.59	2.63	2.66	21
2.54	2.57	2.61	2.65	2.68	2.72	2.75	2.79	22
2.65	2.69	2.73	2.77	2.80	2.84	2.88	2.91	23
2.77	2.81	2.85	2.88	2.92	2.96	3.00	3.04	24
2.88	2.92	2.96	3.00	3.05	3.09	3.13	3.17	25
3.00	3.04	3.08	3.12	3.17	3.21	3.25	3.29	26
3.11	3.16	3.20	3.24	3.29	3.33	3.38	3.42	27
3.23	3.27	3.32	3.36	3.41	3.45	3.50	3.54	28
3.34	3.39	3.44	3.48	3.53	3.58	3.62	3.67	29
3.46	3.50	3.55	3.60	3.65	3.70	3.75	3.80	30
3.57	3.62	3.67	3.72	3.77	3.82	3.87	3.92	31
3.68	3.74	3.79	3.84	3.89	3.94	4.00	4.05	32
3.80	3.85	3.91	3.96	4.01	4.07	4.12	4.17	33
3.91	3.97	4.02	4.08	4.13	4.19	4.24	4.30	34
4.03	4.09	4.14	4.20	4.26	4.31	4.37	4.43	35

C. Correction of a Barometer for Capillarity
(*Smithsonian Tables*)

| Diameter of tube, millimeters | Height of meniscus in millimeters | | | | | | | |
| | 0.4 | 0.6 | 0.8 | 1.0 | 1.2 | 1.4 | 1.6 | 1.8 |
	Correction to be added in millimeters							
4	0.83	1.22	1.54	1.98	2.37	....	....	....
5	0.47	0.65	0.86	1.19	1.45	1.80	....	....
6	0.27	0.41	0.56	0.78	0.98	1.21	1.43	....
7	0.18	0.28	0.40	0.53	0.67	0.82	0.97	1.13
8	....	0.20	0.29	0.38	0.46	0.56	0.65	0.77
9	....	0.15	0.21	0.28	0.33	0.40	0.46	0.52
10	....	....	0.15	0.20	0.25	0.29	0.33	0.37
11	....	....	0.10	0.14	0.18	0.21	0.24	0.27
12	....	....	0.07	0.10	0.13	0.15	0.18	0.19
13	....	....	0.04	0.07	0.10	0.12	0.13	0.14

Table 2-9
BAROMETRIC LATITUDE-GRAVITY TABLE—METRIC UNITS
(*Smithsonian Tables*)

The values below are to be subtracted from the barometric reading for latitudes from 0 to 45° inclusive, and are to be added from 46 to 90°.

Deg. Lat.	Barometer readings, millimeters					
	680	700	720	740	760	780
	mm.	mm.	mm.	mm.	mm.	mm.
0	1.82	1.87	1.93	1.98	2.04	2.09
5	1.79	1.85	1.90	1.95	2.00	2.06
10	1.71	1.76	1.81	1.86	1.92	1.97
15	1.58	1.63	1.67	1.72	1.77	1.81
20	1.40	1.44	1.49	1.53	1.57	1.61
21	1.36	1.40	1.44	1.48	1.52	1.56
22	1.32	1.36	1.40	1.44	1.48	1.51
23	1.28	1.31	1.35	1.39	1.43	1.46
24	1.23	1.27	1.30	1.34	1.37	1.41
25	1.18	1.22	1.25	1.29	1.32	1.36
26	1.13	1.17	1.20	1.23	1.27	1.30
27	1.08	1.12	1.15	1.18	1.21	1.24
28	1.03	1.06	1.09	1.12	1.15	1.18
29	0.98	1.01	1.04	1.07	1.10	1.12
30	0.93	0.95	0.98	1.01	1.04	1.06
31	0.87	0.90	0.92	0.95	0.98	1.00
32	0.82	0.84	0.86	0.89	0.91	0.94
33	0.76	0.78	0.80	0.83	0.85	0.87
34	0.70	0.72	0.74	0.76	0.79	0.81
35	0.64	0.66	0.68	0.70	0.72	0.74
36	0.58	0.60	0.62	0.64	0.65	0.67
37	0.52	0.54	0.56	0.57	0.59	0.60
38	0.46	0.48	0.49	0.51	0.52	0.53
39	0.40	0.42	0.43	0.44	0.45	0.46
40	0.34	0.35	0.36	0.37	0.38	0.39
41	0.28	0.29	0.30	0.30	0.31	0.32
42	0.22	0.22	0.23	0.24	0.24	0.25
43	0.16	0.16	0.16	0.17	0.17	0.18
44	0.09	0.10	0.10	0.10	0.10	0.11
45	0.03	0.03	0.03	0.03	0.03	0.04
46	0.03	0.03	0.03	0.03	0.04	0.04
47	0.09	0.10	0.10	0.10	0.10	0.11
48	0.16	0.16	0.17	0.17	0.18	0.18
49	0.22	0.23	0.23	0.24	0.25	0.25
50	0.28	0.29	0.30	0.31	0.31	0.32
51	0.34	0.35	0.36	0.37	0.38	0.39
52	0.40	0.42	0.43	0.44	0.45	0.46
53	0.46	0.48	0.49	0.51	0.52	0.53
54	0.52	0.54	0.56	0.57	0.59	0.60
55	0.58	0.60	0.62	0.64	0.65	0.67
56	0.64	0.66	0.68	0.70	0.72	0.74
57	0.70	0.72	0.74	0.76	0.78	0.80
58	0.76	0.78	0.80	0.82	0.85	0.87
59	0.81	0.84	0.86	0.89	0.91	0.93
60	0.87	0.89	0.92	0.94	0.97	1.00
61	0.92	0.95	0.98	1.00	1.03	1.06
62	0.97	1.00	1.02	1.05	1.08	1.11
63	1.03	1.06	1.09	1.12	1.15	1.18
64	1.08	1.11	1.14	1.17	1.20	1.23
65	1.13	1.16	1.19	1.22	1.26	1.29
66	1.17	1.21	1.24	1.28	1.31	1.35
67	1.22	1.25	1.29	1.33	1.36	1.40
68	1.26	1.30	1.34	1.37	1.41	1.45
69	1.31	1.34	1.38	1.42	1.46	1.50
70	1.35	1.39	1.43	1.47	1.51	1.55
72	1.42	1.47	1.51	1.55	1.59	1.63
75	1.53	1.57	1.62	1.66	1.71	1.75
80	1.66	1.71	1.76	1.81	1.86	1.90
85	1.74	1.79	1.84	1.90	1.95	2.00
90	1.77	1.82	1.87	1.93	1.98	2.03

Table 2-9 (*Continued*)
BAROMETRIC CORRECTION FOR GRAVITY

The values in the table below are to be subtracted from the readings taken on a mercurial barometer to correct for the decrease in gravity with increase in altitude.

Metric Units

Height above sea-level meters	Observed barometer height in millimeters								
	400	450	500	550	600	650	700	750	800
	mm.	mm.	mm.	mm.	mm.	mm.	mm.	mm.	mm.
100	....	....	....	....	....	....	0.02	0.02	0.02
200	....	....	....	....	....	....	0.04	0.05	0.05
300	....	....	....	....	....	....	0.07	0.07	0.07
400	....	....	....	....	....	....	0.09	0.10	0.10
500	....	....	....	....	....	....	0.11	0.12	0.13
600	....	....	....	....	....	0.12	0.13	0.14	....
700	....	....	....	....	....	0.14	0.15	0.16	....
800	....	....	....	....	....	0.16	0.18	0.19	....
900	....	....	....	....	....	0.18	0.20	0.22	....
1000	....	....	....	0.18	0.19	0.20	0.22	0.24	....
1100	....	....	....	0.19	0.21	0.22	0.24	....	....
1200	....	....	....	0.21	0.23	0.24	0.26	....	....
1300	....	....	....	0.22	0.24	0.26	0.29	....	....
1400	....	....	....	0.24	0.26	0.28	0.31	....	....
1500	....	....	0.24	0.26	0.28	0.30	0.33	....	....
1600	....	....	0.25	0.28	0.30	0.32	....	....	....
1700	....	....	0.27	0.30	0.32	0.34	....	....	....
1800	....	....	0.28	0.31	0.34	0.36	....	....	....
1900	....	....	0.30	0.33	0.36	0.39	....	....	....
2000	....	0.28	0.31	0.34	0.38	0.41	....	....	....
2100	....	0.30	0.33	0.36	0.40	....	....	....	....
2200	....	0.31	0.35	0.38	0.41	....	....	....	....
2300	....	0.32	0.36	0.40	0.43	....	....	....	....
2400	....	0.34	0.38	0.42	0.45	....	....	....	....
2500	0.31	0.35	0.39	0.43	0.47	....	....	....	....
2600	0.33	0.37	0.41	....	....	....	....	....	....
2800	0.35	0.40	0.44	....	....	....	....	....	....
3000	0.38	0.42	0.47	....	....	....	....	....	....
3200	0.40	0.46	....	....	....	....	....	....	....
3400	0.43	0.48	....	....	....	....	....	....	....

English Units

Height above sea-level feet	Observed barometer height in inches								
	14	16	18	20	22	24	26	28	30
	in.	in.	in.	in.	in.	in.	in.	in.	in.
1000	.....	.....	.....	.....	.....	.....	0.003	0.003	0.003
2000	.....	.....	.....	.....	.....	0.004	0.005	0.005	0.006
3000	.....	.....	.....	.....	0.007	0.007	0.008	0.008	.....
4000	.....	.....	.....	.....	0.009	0.009	0.010	.....	.....
5000	.....	.....	.....	0.010	0.011	0.011	0.012	.....	.....
5500	.....	.....	.....	0.011	0.012	0.013	.....	.....	.....
6000	.....	.....	.....	0.011	0.013	0.014	.....	.....	.....
6500	.....	.....	0.011	0.012	0.014	0.015	.....	.....	.....
7000	.....	.....	0.012	0.013	0.015	0.016	.....	.....	.....
7500	.....	.....	0.013	0.014	0.016	0.017	.....	.....	.....
8000	.....	0.012	0.014	0.015	0.017	.....	.....	.....	.....
8500	.....	0.013	0.015	0.016	0.018	.....	.....	.....	.....
9000	0.012	0.014	0.016	0.017	0.019	.....	.....	.....	.....
9500	0.013	0.015	0.016	0.018	0.020	.....	.....	.....	.....
10000	0.013	0.015	0.017	0.019	.....	.....	.....	.....	.....
10500	0.014	0.016	0.018	0.020	.....	.....	.....	.....	.....
11000	0.015	0.017	0.019	0.021	.....	.....	.....	.....	.....
11500	0.015	0.018	0.020	.....	.....	.....	.....	.....	.....
12000	0.016	0.018	0.021	.....	.....	.....	.....	.....	.....
12500	0.017	0.019	0.021	.....	.....	.....	.....	.....	.....
13000	0.017	0.020	.....	.....	.....	.....	.....	.....	.....
13500	0.018	0.021	.....	.....	.....	.....	.....	.....	.....
14000	0.019	0.021	.....	.....	.....	.....	.....	.....	.....
14500	0.019	.....	.....	.....	.....	.....	.....	.....	.....
15000	0.020	.....	.....	.....	.....	.....	.....	.....	.....

Table 2-10
ATMOSPHERIC DATA

J. Chem. Ed., **31**, 115 (1954); *Phys. Rev.*, **88**, 1027 (1952); *Mon. Not. Roy. Astron. Soc.*, **112**, 101 (1952).

Altitude km	Temperature ($\mu = 28.966$) °K	Pressure mm Hg	Density g/cc	† Mean free path, cm
0	288.0	7.6×10^2	1.1×10^{-3}	8.6×10^{-6}
10	230.8	2.1×10^2	4.2×10^{-4}	2.1×10^{-5}
20	212.8	4.2×10^1	9.3×10^{-5}	9.7×10^{-5}
30	231.7	9.5×10^0	1.9×10^{-5}	4.8×10^{-4}
40	262.5	2.4×10^0	4.2×10^{-6}	2.2×10^{-3}
50	270.8	7.6×10^{-1}	1.2×10^{-6}	7.8×10^{-3}
60	252.8	2.2×10^{-1}	3.5×10^{-7}	2.6×10^{-2}
70	218.0	5.5×10^{-2}	9.7×10^{-8}	9.3×10^{-2}
80	205.0	1.1×10^{-2}	2.1×10^{-8}	4.3×10^{-1}
90	217.0	$2 \ \times 10^{-3}$	4.1×10^{-9}	2.1×10^0
100	240.0	$6 \ \times 10^{-4}$	8.6×10^{-10}	9.5×10^0
110	270.0	$2 \ \times 10^{-4}$	2.0×10^{-10}	3.8×10^1
120	330.0	$6 \ \times 10^{-5}$	5.6×10^{-11}	1.3×10^2
130	390.0	$2 \ \times 10^{-5}$	1.9×10^{-11}	3.7×10^2
140	447.0	$7 \ \times 10^{-6}$	7.6×10^{-12}	8.7×10^2
150	503.0	3.7×10^{-6}	3.4×10^{-12}	1.8×10^3
160	560.0	$2 \ \times 10^{-6}$	1.6×10^{-12}	3.6×10^3
180	676.9	$7 \ \times 10^{-7}$	4.8×10^{-13}	1.0×10^4
200	792.5	$3 \ \times 10^{-7}$	1.7×10^{-13}	3.0×10^4
220	906.6	1.4×10^{-7}	7.0×10^{-14}	8.7×10^4

† The average distance a particle travels between collisions with other particles; not the distance between particles.

Table 2-11
REDUCTION OF THE BAROMETER TO SEA LEVEL

A barometer located at an elevation above sea level will show a reading lower than a barometer at sea level by an amount approximately 2.5 millimeters (0.1 inch) for each 30.5 meters (100 feet) of elevation. A closer approximation can be made by reference to the following tables which take into account, (1) the effect of altitude of the station at which the barometer is read, (2) the mean temperature of the air column extending from the station down to sea level, (3) the latitude of the station at which the barometer is read and, (4) the reading of the barometer corrected for its temperature, a correction which is applied only to mercurial barometers since the aneroid barometers are compensated for temperature effects.

Example.—A barometer which has been corrected for its temperature read 650 mm. at a station whose altitude is 1350 meters above sea level and at a latitude of 30°. The mean temperature (outdoor temperature) at the station is 20°C.

Table A (metric units) gives for these conditions a temperature-altitude factor of. 135.2
The Latitude Factor Table gives for 135.2 at 30° lat. a correction of. . . . . +0.17
Therefore, the corrected value of the temperature-altitude factor is. 135.37

Entering Table B (metric units), with a temperature-altitude factor of 135.37 and a barometric reading of 650 mm. (corrected for temperature), the correction is found to be. 109.6
Accordingly the barometric reading reduced to sea level is 650 + 109.6 = 759.6 mm.

Latitude Factor—English or Metric Units

For latitudes 0°-45° add the latitude factor, for 45°-90° subtract the latitude factor, from the values obtained in Table A.

Temp.—Alt. Factor From Table A	Latitude				
	0°	10°	20°	30°	45°
50	0.1	0.1	0.1	0.1	0.0
100	0.3	0.3	0.2	0.1	0.0
150	0.4	0.4	0.3	0.2	0.0
200	0.5	0.5	0.4	0.3	0.0
250	0.7	0.6	0.5	0.3	0.0
300	0.8	0.8	0.6	0.4	0.0
350	0.9	0.9	0.7	0.5	0.0
	90°	80°	70°	60°	45°

Table 2-11 (*Continued*)
REDUCTION OF THE BAROMETER TO SEA LEVEL—METRIC UNITS
A. Values of the Temperature-Altitude Factor for
Use in Table B
(From *Smithsonian Meteorological Tables*, 3d Ed., 1907.)

Altitude in Meters	Mean Temperature of Air Column in Centigrade Degrees										
	−16°	−8°	−4°	0°	6°	10°	14°	18°	20°	22°	26°
10	1.2	1.1	1.1	1.1	1.1	1.0	1.0	1.0	1.0	1.0	1.0
50	5.8	5.6	5.5	5.4	5.3	5.2	5.1	5.0	5.0	5.0	4.9
100	11.5	11.2	11.0	10.8	10.6	10.4	10.3	10.1	10.0	9.9	9.8
150	17.3	16.7	16.5	16.2	15.9	15.6	15.4	15.1	15.0	14.9	14.7
200	23.0	22.3	22.0	21.6	21.1	20.8	20.5	20.2	20.0	19.9	19.6
250	28.8	27.9	27.5	27.0	26.4	26.0	25.6	25.2	25.0	24.9	24.5
300	34.5	33.5	33.0	32.5	31.7	31.2	30.7	30.3	30.1	29.8	29.4
350	40.3	39.0	38.5	37.9	37.0	36.4	35.9	35.3	35.1	34.8	34.3
400	46.0	44.6	43.9	43.3	42.3	41.6	41.0	40.4	40.1	39.8	39.2
450	51.8	51.3	49.4	48.7	47.6	46.8	46.1	45.4	45.1	44.8	44.1
500	57.5	55.8	54.9	54.1	52.9	52.0	51.2	50.5	50.1	49.7	49.0
550	63.3	61.4	60.4	59.5	58.1	57.2	56.4	55.5	55.1	54.7	53.9
600	69.0	66.9	65.9	64.9	63.4	62.4	61.5	60.6	60.1	59.7	58.8
650	74.8	72.5	71.4	70.3	68.7	67.6	66.6	65.6	65.1	64.6	63.7
700	80.6	78.1	76.9	75.7	74.0	72.9	71.7	70.7	70.1	69.6	68.6
750	86.3	83.7	82.4	81.1	79.3	78.1	76.9	75.7	75.1	74.6	73.5
800	92.1	89.2	87.9	86.5	84.6	83.3	82.0	80.8	80.1	79.6	78.4
850	97.8	94.8	93.4	92.0	89.8	88.5	87.1	85.8	85.2	84.5	83.3
900	103.6	100.4	98.9	97.4	95.1	93.7	92.2	90.8	90.2	89.5	88.2
950	109.3	106.0	104.4	102.8	100.4	98.9	97.4	95.9	95.2	94.5	93.1
1000	115.1	111.5	109.8	108.2	105.7	104.1	102.5	100.9	100.2	99.4	98.0
1050	120.8	117.1	115.3	113.6	111.0	109.3	107.6	106.0	105.2	104.4	102.9
1100	126.6	122.7	120.8	119.0	116.3	114.5	112.7	111.0	110.2	109.4	107.8
1150	132.3	128.3	126.3	124.4	121.6	119.7	117.9	116.1	115.2	114.4	112.7
1200	138.1	133.8	131.8	129.8	126.8	124.9	123.0	121.1	120.2	119.3	117.6
1250	143.8	139.4	137.3	135.2	132.1	130.1	128.1	126.2	125.2	124.3	122.5
1300	149.6	145.0	142.8	140.6	137.4	135.3	133.2	131.2	130.2	129.3	127.4
1350	155.3	150.6	148.3	146.0	142.7	140.5	138.4	136.3	135.2	134.2	132.3
1400	161.1	156.2	153.8	151.4	148.0	145.7	143.5	141.3	140.2	139.2	137.2
1450	166.8	161.7	159.3	156.8	153.3	150.9	148.6	146.4	145.3	144.2	142.1
1500	172.6	167.3	164.8	162.3	158.5	156.1	153.7	151.4	150.3	149.1	147.0
1550	178.3	172.9	170.2	167.7	163.8	161.3	158.8	156.4	155.3	154.1	151.8
1600	184.1	178.5	175.7	173.1	169.1	166.5	164.0	161.5	160.3	159.1	156.7
1650	189.8	184.0	181.2	178.5	174.4	171.7	169.1	166.5	165.3	164.1	161.6
1700	195.6	189.6	186.7	183.9	179.7	176.9	174.2	171.6	170.3	169.0	166.5
1750	201.4	195.2	192.2	189.3	185.0	182.1	179.3	176.6	175.3	174.0	171.4
1800	207.1	200.8	197.7	194.7	190.2	187.3	184.5	181.7	180.3	179.0	176.3
1850	212.9	206.3	203.2	200.1	195.5	192.5	189.6	186.7	185.3	183.9	181.2
1900	218.6	211.9	208.7	205.5	200.8	197.7	194.7	191.8	190.3	188.9	186.1
1950	224.4	217.5	214.2	210.9	206.1	202.9	199.8	196.8	195.3	193.9	191.0
2000	230.1	223.0	219.7	216.3	211.4	208.1	204.9	201.9	200.3	198.8	195.9
2050	235.9	228.6	225.1	221.7	216.7	213.3	210.1	206.9	205.3	203.8	200.8
2100	241.6	234.2	230.6	227.1	221.9	218.5	215.2	211.9	210.4	208.8	205.7
2150	247.4	239.8	236.1	232.5	227.2	223.7	220.3	217.0	215.4	213.8	210.6
2200	253.1	245.4	241.6	237.9	232.5	228.9	225.4	222.0	220.4	218.7	215.5
2250	258.9	250.9	247.1	243.4	237.8	234.1	230.6	227.1	225.4	223.7	220.4
2300	264.6	256.5	252.6	248.8	243.1	239.3	235.7	232.1	230.4	228.7	225.3
2350	270.4	262.1	258.1	254.2	248.3	244.5	240.8	237.2	235.4	233.6	230.2
2400	276.1	267.7	263.6	259.6	253.6	249.7	245.9	242.2	240.4	238.6	235.1
2450	281.9	273.2	269.1	265.0	258.9	254.9	251.0	247.3	245.4	243.6	240.0
2500	287.6	278.8	274.5	270.4	264.2	260.1	256.2	252.3	250.4	248.5	244.9
2550	293.4	284.4	280.0	275.8	269.5	265.3	261.3	257.3	255.4	253.5	249.8
2600	299.1	290.0	285.5	281.2	274.8	270.5	266.4	262.4	260.4	258.5	254.7
2650	304.9	295.5	291.0	286.6	280.0	275.7	271.5	267.4	265.4	263.4	259.6
2700	310.6	301.1	296.5	292.0	285.3	280.9	276.6	272.5	270.4	268.4	264.5
2750	316.4	306.7	302.0	297.4	290.6	286.1	281.8	277.5	275.4	273.4	269.4
2800	322.1	312.3	307.5	302.8	295.9	291.3	286.9	282.6	280.4	278.3	274.3
2850	327.9	317.8	313.0	308.2	301.2	296.5	292.0	287.6	285.4	283.3	279.2
2900	333.6	323.4	318.4	313.6	306.4	301.7	297.1	292.6	290.4	288.3	284.1
2950	339.4	329.0	323.9	319.0	311.7	306.9	302.2	297.7	295.5	293.3	289.0
3000	345.1	334.5	329.4	324.4	317.0	312.1	307.4	302.7	300.5	298.2	293.8

Table 2-11 (*Continued*)
REDUCTION OF THE BAROMETER TO SEA LEVEL—METRIC UNITS
B. Values in Millimeters to Be Added
(From *Smithsonian Meteorological Tables*, 3d Ed., 1907.)

Temp.—Alt. Factor	Barometer Reading in Millimeters						Temp.—Alt. Factor	Barometer Reading in Millimeters					
	790	770	750	730	710	690 / 670		630	610	590	570	550	530
1	0.9	0.9	0.9	0.8	0.8	0.8	200	163.1	157.9	152.8	147.6		
5	4.6	4.4	4.3	4.2	4.1	4.0	205	167.7	162.4	157.1	151.7		
10	9.1	8.9	8.7	8.5	8.2	8.0	210	172.3	166.8	161.4	155.9		
15	13.8	13.4	13.1	12.7	12.4	12.0	215	176.9	171.3	165.7	160.1	154.5	148.9
20	18.4	17.9	17.5	17.0	16.5	16.1	220		175.8	170.1	164.3	158.5	152.8
25		22.5	21.9	21.3	20.7	20.1	225		180.4	174.5	168.5	162.6	156.7
30		27.1	26.4	25.7	25.0	24.2	230		184.9	178.9	172.8	166.7	160.7
35		31.7	30.8	30.0	29.2	28.4	235		189.5	183.3	177.1	170.9	164.7
40		36.3	35.3	34.4	33.5	32.5 / 31.6	240		194.1	187.8	181.4	175.0	168.7
45			39.9	38.8	37.8	36.7 / 35.6	245		198.8	192.3	185.7	179.2	172.7

Temp.—Alt. Factor	750	730	710	690	670	650	630	Temp.—Alt. Factor	590	570	550	530	510
50	44.4	43.3	42.1	40.9	39.7			250	196.8	190.1	183.4	176.8	
55	49.0	47.7	46.4	45.1	43.8			255	201.3	194.5	187.7	180.8	
60	53.6	52.2	50.8	49.3	47.9			260	205.9	198.9	191.9	185.0	178.0
65	58.3	56.7	55.2	53.6	52.1			265	210.5	203.3	196.2	189.1	181.9
70		61.3	59.6	57.9	56.2			270	215.1	207.8	200.5	193.2	185.9
75		65.8	64.0	62.2	60.4			275	219.8	212.3	204.9	197.4	190.0
80		70.4	68.5	66.6	64.6	62.7	60.8	280		216.8	209.2	201.6	194.0
85		75.0	73.0	70.9	68.9	66.8	64.8	285		221.4	213.6	205.8	198.1
90			77.5	75.3	73.1	71.0	68.8	290		225.9	218.0	210.1	202.1
95			82.1	79.7	77.4	75.1	72.8	295		230.5	222.4	214.3	206.3

Temp.—Alt. Factor	710	690	670	650	630	610	Temp.—Alt. Factor	570	550	530	510	490
100	86.6	84.2	81.8	79.3	76.9		300	235.1	226.9	218.6	210.4	
105	91.2	88.7	86.1	83.5	81.0		305	239.8	231.4	223.0	214.6	206.1
110	95.9	93.2	90.5	87.8	85.1		310		235.9	227.3	218.7	210.1
115	100.5	97.7	94.8	92.0	89.2		315		240.4	231.7	222.9	214.2
120		102.2	99.3	96.3	93.3		320		245.0	236.1	227.2	218.3
125		106.8	103.7	100.6	97.5	94.4	325		249.6	240.5	231.4	222.4
130		111.4	108.2	104.9	101.7	98.5	330		254.2	244.9	235.7	226.5
135		116.0	112.7	109.3	105.9	102.6	335		258.8	249.4	240.0	230.6
140		120.7	117.2	113.7	110.2	106.7	340		263.5	253.9	244.4	234.8
145			121.7	118.1	114.5	110.8	345			258.4	248.7	238.9

Temp.—Alt. Factor	670	650	630	610	590	570
150	126.3	122.5	118.8	115.0		
155	130.9	127.0	123.1	119.2		
160	135.5	131.5	127.4	123.4		
165	140.2	136.0	131.8	127.6		
170		140.5	136.2	131.9	127.5	123.2
175		145.1	140.6	136.2	131.7	127.2
180		149.7	145.1	140.5	135.9	131.3
185		154.3	149.5	144.8	140.0	135.3
190		158.9	154.0	149.2	144.3	139.4
195			158.6	153.5	148.5	143.5

Table 2-12
VISCOSITY CONVERSION TABLE

Poise = c. g. s. unit of absolute viscosity. Centipoise = 0.01 poise.
Stoke = c. g. s. unit of kinematic viscosity. Centistoke = 0.01 stoke.
Centipoises = centistokes × density (at temperature under consideration).
Reyn (1 lb. sec. per sq. in.) = 69 × 10⁵ centipoises.

Centistokes to Saybolt, Redwood, and Engler Units

Cf. *Jour. Inst. Pet. Tech., Vol. 22, p. 21 (1936); Reports of A. S. T. M. Committee D-2, 1936 and 1937.*

The values of Saybolt Universal Viscosity at 100°F and at 210°F are taken directly from the comprehensive *ASTM Viscosity Tables, Special Technical Publication No. 43A* (1953) by permission of the publishers, American Society for Testing Materials, 1916 Race St., Philadelphia 3, Pa.

Centistokes	Saybolt Universal Viscosity at			Redwood Seconds at			Engler Degrees at all Temps.
	100°F.	130°F.	210°F.	70°F.	140°F.	200°F.	
2.0	32.62	32.68	32.85	30.2	31.0	31.2	1.14
3.0	36.03	36.10	36.28	32.7	33.5	33.7	1.22
4.0	39.14	39.22	39.41	35.3	36.0	36.3	1.31
5.0	42.35	42.43	42.65	37.9	38.5	38.9	1.40
6.0	45.56	45.65	45.88	40.5	41.0	41.5	1.48
7.0	48.77	48.86	49.11	43.2	43.7	44.2	1.56
8.0	52.09	52.19	52.45	46.0	46.4	46.9	1.65
9.0	55.50	55.61	55.89	48.9	49.1	49.7	1.75
10.0	58.91	59.02	59.32	51.7	52.0	52.6	1.84
11.0	62.43	62.55	62.86	54.8	55.0	55.6	1.93
12.0	66.04	66.17	66.50	57.9	58.1	58.8	2.02
14.0	73.57	73.71	74.09	64.4	64.6	65.3	2.22
16.0	81.30	81.46	81.87	71.0	71.4	72.2	2.43
18.0	89.44	89.61	90.06	77.9	78.5	79.4	2.64
20.0	97.77	97.96	98.45	85.0	85.8	86.9	2.87
22.0	106.4	106.6	107.1	92.4	93.3	94.5	3.10
24.0	115.0	115.2	115.8	99.9	100.9	102.2	3.34
26.0	123.7	123.9	124.5	107.5	108.6	110.0	3.58
28.0	132.5	132.8	133.4	115.3	116.5	118.0	3.82
30.0	141.3	141.6	142.3	123.1	124.4	126.0	4.07
32.0	150.2	150.5	151.2	131.0	132.3	134.1	4.32
34.0	159.2	159.5	160.3	138.9	140.2	142.2	4.57
36.0	168.2	168.5	169.4	146.9	148.2	150.3	4.83
38.0	177.3	177.6	178.5	155.0	156.2	158.3	5.08
40.0	186.3	186.7	187.6	163.0	164.3	166.7	5.34
42.0	195.3	195.7	196.7	171.0	172.3	175.0	5.59
44.0	204.4	204.8	205.9	179.1	180.4	183.3	5.85
46.0	213.7	214.1	215.2	187.1	188.5	191.7	6.11
48.0	222.9	223.3	224.5	195.2	196.6	200.0	6.37
50.0	232.1	232.5	233.8	203.3	204.7	208.3	6.63
60.0	278.3	278.8	280.2	243.5	245.3	250.0	7.90
70.0	324.4	325.0	326.7	283.9	286.0	291.7	9.21
80.0	370.8	371.5	373.4	323.9	326.6	333.4	10.53
90.0	417.1	417.9	420.0	364.4	367.4	375.0	11.84
100.0*	463.5	464.4	466.7	404.9	408.2	416.7	13.16

* At higher values use the same ratio as above for 100 centistokes; *e.g.*, 102 centistokes = 102 × 4.635 Saybolt seconds at 100 F.
  To obtain the Saybolt Universal viscosity equivalent to a kinematic viscosity determined at *t*°F., multiply the equivalent Saybolt Universal viscosity at 100°F. by 1 + (t − 100) 0.000064; *e.g.*, 10 centistokes at 210°F are equivalent to 58.91 × 1.0070, or 59.32 Saybolt Universal Viscosity at 210°F.

Table 2-13
CONVERSION OF WEIGHINGS IN AIR TO WEIGHINGS IN VACUO

If the mass of a substance in air is M_f, its density D_m, the density of weights used in making the weighing D_w, and the density* of air D_a, the true mass of the substance is vacuo, M_{vac} is

$$M_{vac} = M_f + D_a M_f \left(\frac{1}{D_m} - \frac{1}{D_w} \right)$$

For most purposes it is sufficient to assume a density of 8.4 for brass weights, and a density of 0.0012 for air under ordinary conditions. The equation then becomes

$$M_{vac} = M_f + 0.0012 M_f \left(\frac{1}{D_m} - \frac{1}{8.4} \right)$$

This table which follows gives the values of k (buoyancy reduction factor), which is the correction necessary because of the buoyant effect of the air upon the object weighed; the table is computed for air with the density of 0.0012; m is the weight in grams of the object when weighed in air; weight of object reduced to "in vacuo"

$$= m + \frac{km}{1000}$$

Density of object weighed	Buoyancy reduction factor, k			
	Brass weights, density =8.4	Pt or Pt-Ir weights, density =21.5	Al or quartz weights, density =2.7	Gold weights, density =17
0.2	5.89	5.98	5.58	5.97
0.3	3.87	3.96	3.56	3.95
0.4	2.87	2.95	2.55	2.94
0.5	2.26	2.35	1.95	2.34
0.6	1.86	1.95	1.55	1.93
0.7	1.57	1.66	1.26	1.65
0.75	1.46	1.55	1.15	1.53
0.80	1.36	1.45	1.05	1.43
0.82	1.32	1.41	1.01	1.39
0.84	1.29	1.37	0.98	1.36
0.86	1.25	1.34	0.94	1.33
0.88	1.22	1.31	0.91	1.29
0.90	1.19	1.28	0.88	1.26
0.92	1.16	1.25	0.85	1.24
0.94	1.13	1.22	0.82	1.21
0.96	1.11	1.20	0.80	1.18
0.98	1.08	1.17	0.77	1.16
1.00	1.06	1.15	0.75	1.13
1.02	1.03	1.12	0.72	1.11
1.04	1.01	1.10	0.70	1.08
1.06	0.99	1.08	0.68	1.06
1.08	0.97	1.06	0.66	1.04
1.10	0.95	1.04	0.64	1.02
1.12	0.93	1.02	0.62	1.00
1.14	0.91	1.00	0.60	0.98
1.16	0.89	0.98	0.58	0.96
1.18	0.87	0.96	0.56	0.95
1.20	0.86	0.95	0.55	0.93
1.25	0.82	0.91	0.51	0.89
1.30	0.78	0.87	0.47	0.85
1.35	0.75	0.83	0.44	0.82
1.40	0.71	0.80	0.40	0.79
1.50	0.66	0.74	0.35	0.73
1.6	0.61	0.69	0.30	0.68
1.7	0.56	0.65	0.25	0.64
1.8	0.52	0.61	0.21	0.60
1.9	0.49	0.58	0.18	0.56
2.0	0.46	0.54	0.15	0.53
2.2	0.40	0.49	0.09	0.48
2.4	0.36	0.44	0.05	0.43

*See special table: *Specific Gravity of Air.*

Table 2-13 (*Continued*)
CONVERSION OF WEIGHINGS IN AIR TO WEIGHINGS IN VACUO

Density of object weighed	Buoyancy reduction factor, k			
	Brass weights, density = 8.4	Pt or Pt-Ir weights, density = 21.5	Al or quartz weights density = 2.7	Gold weights, density = 17
2.6	0.32	0.41	0.01	0.39
2.8	0.29	0.37	−0.02	0.36
3.0	0.26	0.34	−0.05	0.33
3.5	0.20	0.29	−0.11	0.27
4	0.16	0.24	−0.15	0.23
5	0.10	0.18	−0.21	0.17
6	0.06	0.14	−0.25	0.13
7	0.03	0.12	−0.28	0.10
8	0.01	0.09	−0.30	0.08
9	−0.01	0.08	−0.32	0.06
10	−0.02	0.06	−0.33	0.05
12	−0.04	0.04	−0.35	0.03
14	−0.06	0.03	−0.37	0.02
16	−0.07	0.02	−0.38	0.00
18	−0.08	0.01	−0.39	0.00
20	−0.08	0.00	−0.39	−0.01
22	−0.09	0.00	−0.40	−0.02

TOLERANCES FOR VOLUMETRIC BURETS AND PIPETS

From Meites, *Handbook of Analytical Chemistry*, 1963, McGraw-Hill Book Company; by permission

Table 2-14
PERMISSIBLE DEVIATIONS FROM NOMINAL CAPACITY

Capacity, up to and including, ml	Deviations, ml, permitted for			
	Volumetric flasks, calibrated		Transfer pipets and burets	Measuring pipets
	To contain	To deliver		
1	±0.01	......	....	......
2	......	......	±0.006*	±0.01
3	0.015	......	....	......
5	0.02	......	0.01	0.02
10	0.02	±0.04	0.02	0.03
25	0.03	0.05	....	......
30	......	......	0.03	0.05
50	0.05	0.10	0.05	0.08
100	0.08	0.15	0.08†	0.15
200	0.10	0.20	0.10*	......
300	0.12	0.25	....	......
500	0.15	0.30	....	......
1000	0.30	0.50	....	......
2000	0.50	1.00	....	......
3000	0.75	1.50	....	......
4000	1.00	2.0	....	......
5000	1.2	2.4	....	......

* Applies to pipets only.
† ±0.10 ml. for burets.

Table 2-15
TOLERANCES FOR ANALYTICAL WEIGHTS
By Alan D. Westland with Fred E. Beamish

This table gives the individual and group tolerances established by the National Bureau of Standards (Washington, D.C.) for classes M, S, S-1, and P weights. Individual tolerances are "acceptance tolerances" for new weights. Group tolerances are defined by the National Bureau of Standards as follows: "The corrections of individual weights shall be such that no combination of weights that is intended to be used in a weighing shall differ from the sum of the nominal values by more than the amount listed under the group tolerances."

For class S-1 weights, two-thirds of the weights in a set must be within one-half of the individual tolerances given below. No group tolerances have been specified for class P weights. See *Natl. Bur. Standards Circ.* 547, sec. 1 (1954).

Denomination	Class M		Class S		Class S-1, individual tolerance, mg	Class P, individual tolerance, mg
	Individual tolerance, mg	Group tolerance, mg	Individual tolerance, mg	Group tolerance, mg		
100 g	0.50		0.25	None	1.0	2.0
50 g	0.25	None	0.12	specified	0.60	1.2
30 g	0.15	specified	0.074		0.45	0.90
20 g	0.10		0.074	0.154	0.35	0.70
10 g	0.050		0.074		0.25	0.50
5 g	0.034		0.054		0.18	0.36
3 g	0.034	0.065	0.054	0.105	0.15	0.14
2 g	0.034		0.054		0.13	0.26
1 g	0.034		0.054		0.10	0.20
500 mg	0.0054		0.025		0.080	0.16
300 mg	0.0054	0.0105	0.025	0.055	0.070	0.14
200 mg	0.0054		0.025		0.060	0.12
100 mg	0.0054		0.025		0.050	0.10
50 mg	0.0054		0.014		0.042	0.085
30 mg	0.0054	0.0105	0.014	0.034	0.038	0.076
20 mg	0.0054		0.014		0.035	0.070
10 mg	0.0054		0.014		0.030	0.060
5 mg	0.0054		0.014		0.028	0.055
3 mg	0.0054		0.014		0.026	0.052
2 mg	0.0054	0.0105	0.014	0.034	0.025	0.050
1 mg	0.0054		0.014		0.025	0.050
½ mg	0.0054		0.014		0.025	. . . . .

Table 2-16
FACTORS FOR SIMPLIFIED COMPUTATION OF VOLUME

The volume is determined by weighing the water, having a temperature of t°C, contained in or delivered by the apparatus at the same temperature. The weight of water, w grams, is obtained with brass weights in air having a density of 1.20 mg/ml.

For apparatus made of soft glass, the volume contained or delivered at 20°C is given by

$$v_{20} = w f_{20} \quad \text{ml}$$

where v_{20} is the volume at 20° and f_{20} is the factor (apparent specific volume) obtained from the table below for the temperature t at which the calibration is performed. The volume at any other temperature t' may then be obtained from

$$v' = v_{20}[1 + 0.00002(t' - 20)] \quad \text{ml}$$

For apparatus made of any other material, the volume contained or delivered at the temperature t is

$$v_t = w f_t \quad \text{ml}$$

where w is again the weight in air obtained with brass weights (in grams), and f_t is the factor given in the fourth column of the table for the temperature t. The volume at any temperature t' may then be obtained from

$$v_t' = v_t[1 + \beta(t' - t)] \quad \text{ml}$$

where β is the cubical coefficient of thermal expansion of the material from which the apparatus is made. Approximate values of β for some frequently encountered materials are given in Table 2-17.

Table 2-16 (*Continued*)
FACTORS FOR SIMPLIFIED COMPUTATION OF VOLUME

t, °C	f_{20}	$\log_{10} f_{20}$	f_t	$\log_{10} f_t$
0	1.0017	0.00074	1.0012	0.00052
1	16	70	11	49
2	15	67	11	47
3	15	65	11	46
4	15	63	11	46
5	1.0015	0.00063	1.0011	0.00046
6	15	63	11	47
7	15	63	11	49
8	15	64	12	51
9	15	66	13	54
10	1.0016	0.00069	1.0013	0.00058
11	17	72	14	62
12	17	75	15	67
13	18	79	17	72
14	19	84	18	78
15	1.0020	0.00089	1.0020	0.00084
16	22	095	21	091
17	23	101	23	098
18	25	108	24	106
19	27	115	26	114
20	1.0028	0.00123	1.0028	0.00123
21	30	130	30	132
22	32	139	33	142
23	34	149	35	152
24	37	159	38	163
25	1.0039	0.00168	1.0040	0.00174
26	41	178	43	185
27	44	189	45	197
28	46	200	48	209
29	49	211	51	222
30	1.0052	0.00224	1.0054	0.00235
31	54	236	57	248
32	57	248	60	262
33	60	261	64	276
34	63	275	67	290
35	1.0067	0.00289	1.0070	0.00305
36	71	307	74	319
37	74	320	77	335
38	77	333	81	350
39	81	350	85	367
40	1.0085	0.00368	1.0089	0.00383

Table 2-17
CUBICAL COEFFICIENTS OF THERMAL EXPANSION

This table lists values of β, the cubical coefficient of thermal expansion, taken from "Essentials of Quantitative Analysis," by Benedetti-Pichler, and from various other sources. The value of β represents the relative increase in volume for a change in temperature of $1°C$ at temperatures in the vicinity of $25°C$, and is equal to 3α, where α is the linear coefficient of thermal expansion. Data are given for the types of glass from which volumetric apparatus is most commonly made, and also for some other materials which have been or may be used in the fabrication of apparatus employed in analytical work.

Material	β
Glasses	
Alkali-resistant, Corning 728	1.90×10^{-5}
Geräteglas, Schott G20	1.47
Kimble KG-33 (borosilicate)	0.96
N-51A ("Resistant")	1.47
R-6 (soft)	2.79
Pyrex, Corning 744	0.96
Vitreous silica	0.15
Vycor, Corning 790	0.24
Metals	
Brass	*ca.* 5.5
Copper	5.0
Gold	4.3
Monel metal	4.0
Platinum	2.7
Silver	5.7
Stainless steel	*ca.* 5.3
Tantalum	*ca.* 2.0
Tungsten	1.3
Plastics and other materials	
Hard rubber	24×10^{-5}
Polyethylene	45–90
Polystyrene	18–24
Porcelain	*ca.* 1.2
Teflon (polytetrafluoroethylene)	16.5

Table 2-18
MOLAR EQUIVALENT OF ONE LITER OF GAS AT VARIOUS TEMPERATURES AND PRESSURES

The values in this table, which give the number of moles in one liter of gas, **are** based on the property of an "ideal" gas and were calculated by use of the formula:

$$\text{moles/liter} = \frac{P}{760} \times \frac{273}{T} \times \frac{1}{22.40}$$

where P is the pressure in millimeters of mercury and T is the temperature in degrees Absolute ($= t°C + 273$).

To convert to moles per cubic foot multiply the values in the table by 28.316.

Pressure mm of mercury	Temperature °C					
	10°	12°	14°	16°	18°	20°
655	0.03712	0.03686	0.03660	0.03634	0.03610	0.03585
660	3731	3714	3688	3662	3637	3612
665	3768	3742	3716	3690	3665	3640
670	3796	3770	3744	3718	3692	3667
675	3825	3798	3772	3745	3720	3695
680	0.03853	0.03826	0.03800	0.03773	0.03747	0.03694
685	3881	3854	3827	3801	3775	3749
690	3910	3882	3855	3829	3802	3776
695	3938	3910	3883	3856	3830	3804
700	3967	3939	3911	3884	3858	3831
702	0.03978	0.03950	0.03922	0.03895	0.03869	0.03842
704	3989	3961	3934	3906	3880	3853
706	4000	3972	3945	3917	3891	3864
708	4012	3984	3956	3929	3902	3875
710	4023	3995	3967	3940	3913	3886
712	0.04035	0.04006	0.03978	0.03951	0.03924	0.03897
714	4046	4018	3989	3962	3935	3908
716	4057	4029	4001	3973	3946	3919
718	4068	4040	4012	3984	3957	3930
720	4080	4051	4023	3995	3968	3941
722	0.04091	0.04063	0.04034	0.04006	0.03979	0.03952
724	4103	4074	4045	4017	3990	3963
726	4114	4085	4057	4028	4001	3973
728	4125	4096	4068	4040	4012	3984
730	4136	4108	4079	4051	4023	3995
732	0.04148	0.04119	0.04090	0.04062	0.04034	0.04006
734	4159	4130	4101	4073	4045	4017
736	4171	4141	4112	4084	4056	4028
738	4182	4153	4124	4095	4067	4039
740	4193	4164	4135	4106	4078	4050
742	0.04204	0.04175	0.04146	0.04117	0.04089	0.04061
744	4216	4186	4157	4128	4100	4072
746	4227	4198	4168	4139	4111	4083
748	4239	4209	4179	4151	4122	4094
750	4250	4220	4191	4162	4133	4105
752	0.04261	0.04231	0.04202	0.04173	0.04144	0.04116
754	4273	4243	4213	4184	4155	4127
756	4284	4254	4224	4195	4166	4138
758	4295	4265	4235	4206	4177	4149
760	4307	4276	4247	4217	4188	4160
762	0.04318	0.04287	0.04258	0.04228	0.04199	0.04171
764	4329	4299	4269	4239	4210	4181
766	4341	4310	4280	4250	4221	4192
768	4352	4321	4291	4262	4232	4203
770	4363	4333	4302	4273	4243	4214
772	0.04375	0.04344	0.04314	0.04284	0.04254	0.04225
774	4386	4355	4325	4295	4265	4236
776	4397	4366	4336	4306	4276	4247
778	4409	4378	4347	4317	4287	4258
780	4420	4389	4358	4328	4298	4269

Table 2-18 (*Continued*)
MOLAR EQUIVALENT OF ONE LITER OF GAS

Pressure mm of mercury	Temperature °C					
	22°	24°	26°	28°	30°	32°
655	0.03561	0.03537	0.03515	0.03490	0.03467	0.03444
660	3588	3564	3541	3516	3493	3470
665	3614	3591	3568	3543	3520	3496
670	3642	3618	3595	3569	3546	3523
675	3669	3645	3622	3596	3572	3549
680	0.03697	0.03672	0.03649	0.03623	0.03599	0.03575
685	3724	3699	3676	3649	3625	3602
690	3751	3726	3702	3676	3652	3628
695	3778	3753	3729	3703	3678	3654
700	3805	3780	3756	3729	3705	3680
702	0.03816	0.03790	0.03767	0.03740	0.03715	0.03691
704	3827	3801	3777	3750	3726	3701
706	3838	3812	3788	3761	3736	3712
708	3849	3823	3799	3772	3747	3722
710	3860	3834	3810	3783	3758	3733
712	0.03870	0.03844	0.03820	0.03793	0.03768	0.03744
714	3881	3855	3831	3804	3779	3754
716	3892	3866	3842	3815	3789	3765
718	3902	3877	3853	3825	3800	3775
720	3914	3888	3863	3836	3811	3786
722	0.03925	0.03898	0.03874	0.03847	0.03821	0.03796
724	3936	3909	3885	3857	3832	3807
726	3947	3920	3896	3868	3842	3817
728	3957	3931	3906	3878	3853	3828
730	3968	3941	3917	3889	3863	3838
732	0.03979	0.03952	0.03928	0.03900	0.03874	0.03849
734	3990	3963	3938	3910	3885	3859
736	4001	3974	3949	3921	3895	3870
738	4012	3985	3960	3932	3906	3880
740	4023	3995	3971	3942	3916	3891
742	0.04033	0.04006	0.03981	0.03953	0.03927	0.03901
744	4044	4017	3992	3964	3938	3912
746	4055	4028	4003	3974	3948	3922
748	4066	4039	4014	3985	3959	3933
750	4077	4049	4024	3996	3969	3943
752	0.04088	0.04060	0.04035	0.04006	0.03980	0.03954
754	4099	4071	4046	4017	3991	3964
756	4110	4082	4056	4028	4001	3975
758	4121	4093	4067	4038	4012	3985
760	4131	4103	4078	4049	4022	3996
762	0.04142	4114	4089	4060	4033	4006
764	4153	4125	4099	4070	4043	4017
766	4164	4136	4110	4081	4054	4027
768	4175	4147	4121	4092	4065	4038
770	4186	4158	4132	4102	4075	4048
772	0.04197	0.04168	0.04142	0.04113	0.04086	0.04059
774	4207	4179	4153	4124	4096	4070
776	4218	4190	4164	4134	4107	4080
778	4229	4201	4175	4145	4117	4091
780	4240	4211	4185	4155	4128	4101

Table 2-19
FACTORS FOR REDUCING GAS VOLUMES TO NORMAL TEMPERATURE AND PRESSURE (760 mm Hg)

Examples: (a) 20 ml of dry gas at 22°C and 730 mm = 20 × 0.8888 = 17.78 ml at 0°C and 760 mm. (b) 20 cc of a gas over water at 22° and 730 mm = 20 × (factor corrected for aqueous tension; i.e., 730 − 19.8 or 710.2 mm) = 20 ml of dry gas at 22° and 710.2 mm = 20 × 0.86475 = 17.30 ml at 0°C and 760 mm. Weight in milligrams of 1 ml of gas at N. T. P.: acetylene, 1.173; carbon dioxide, 1.9769; hydrogen, 0.0899; nitric oxide (NO), 1.3402; nitrogen, 1.25057; oxygen, 1.42904.

Pressure mm of mercury	Temperature °C							
	10°	11°	12°	13°	14°	15°	16°	17°
670	0.8504	0.8474	0.8445	0.8415	0.8386	0.8357	0.8328	0.8299
672	0.8530	0.8500	0.8470	0.8440	0.8411	0.8382	0.8353	0.8324
674	0.8555	0.8525	0.8495	0.8465	0.8436	0.8407	0.8377	0.8349
676	0.8580	0.8550	0.8520	0.8490	0.8461	0.8431	0.8402	0.8373
678	0.8606	0.8576	0.8545	0.8516	0.8486	0.8456	0.8427	0.8398
680	0.8631	0.8601	0.8571	0.8541	0.8511	0.8481	0.8452	0.8423
682	0.8657	0.8626	0.8596	0.8566	0.8536	0.8506	0.8477	0.8448
684	0.8682	0.8651	0.8621	0.8591	0.8561	0.8531	0.8502	0.8472
686	0.8707	0.8677	0.8646	0.8616	0.8586	0.8556	0.8527	0.8497
688	0.8733	0.8702	0.8672	0.8641	0.8611	0.8581	0.8551	0.8522
690	0.8758	0.8727	0.8697	0.8666	0.8636	0.8606	0.8576	0.8547
692	0.8784	0.8753	0.8722	0.8691	0.8661	0.8631	0.8601	0.8572
694	0.8809	0.8778	0.8747	0.3717	0.8686	0.8656	0.8626	0.8596
696	0.8834	0.8803	0.8772	0.8742	0.8711	0.8681	0.8651	0.8621
698	0.8860	0.8828	0.8798	0.8767	0.8736	0.8706	0.8676	0.8646
700	0.8885	0.8854	0.8823	0.8792	0.8761	0.8731	0.8700	0.8671
702	0.8910	0.8879	0.8848	0.8817	0.8786	0.8756	0.8725	0.8695
704	0.8936	0.8904	0.8873	0.8842	0.8811	0.8781	0.8750	0.8720
706	0.8961	0.8930	0.8898	0.8867	0.8836	0.8806	0.8775	0.8745
708	0.8987	0.8955	0.8924	0.8892	0.8861	0.8831	0.8800	0.8770
710	0.9012	0.8980	0.8949	0.8917	0.8886	0.8856	0.8825	0.8794
712	0.9037	0.9006	0.8974	0.8943	0.8911	0.8880	0.8850	0.8819
714	0.9063	0.9031	0.8999	0.8968	0.8936	0.8905	0.8875	0.8844
716	0.9088	0.9056	0.9024	0.8993	0.8961	0.8930	0.8899	0.8869
718	0.9114	0.9081	0.9050	0.9018	0.8987	0.8955	0.8924	0.8894
720	0.9139	0.9107	0.9075	0.9043	0.9012	0.8980	0.8949	0.8918
722	0.9164	0.9132	0.9100	0.9068	0.9037	0.9005	0.8974	0.8943
724	0.9190	0.9157	0.9125	0.9093	0.9062	0.9030	0.8999	0.8968
726	0.9215	0.9183	0.9151	0.9118	0.9087	0.9055	0.9024	0.8993
728	0.9241	0.9208	0.9176	0.9144	0.9112	0.9080	0.9049	0.9017
730	0.9266	0.9233	0.9201	0.9169	0.9137	0.9105	0.9073	0.9042
732	0.9291	0.9259	0.9226	0.9194	0.9162	0.9130	0.9098	0.9067
734	0.9317	0.9284	0.9251	0.9219	0.9187	0.9155	0.9123	0.9092
736	0.9342	0.9309	0.9277	0.9244	0.9212	0.9180	0.9148	0.9117
738	0.9368	0.9334	0.9302	0.9269	0.9237	0.9205	0.9173	0.9141
740	0.9393	0.9360	0.9327	0.9294	0.9262	0.9230	0.9198	0.9166
742	0.9418	0.9385	0.9352	0.9319	0.9287	0.9255	0.9223	0.9191
744	0.9444	0.9410	0.9377	0.9345	0.9312	0.9280	0.9248	0.9216
746	0.9469	0.9436	0.9403	0.9370	0.9337	0.9305	0.9272	0.9240
748	0.9494	0.9461	0.9428	0.9395	0.9362	0.9329	0.9297	0.9265
750	0.9520	0.9486	0.9453	0.9420	0.9387	0.9354	0.9322	0.9290
752	0.9545	0.9511	0.9478	0.9445	0.9412	0.9379	0.9347	0.9315
754	0.9571	0.9537	0.9504	0.9470	0.9437	0.9404	0.9372	0.9339
756	0.9596	0.9562	0.9529	0.9495	0.9462	0.9429	0.9397	0.9364
758	0.9621	0.9587	0.9554	0.9520	0.9487	0.9454	0.9422	0.9389
760	0.9647	0.9613	0.9579	0.9546	0.9512	0.9479	0.9446	0.9414
762	0.9672	0.9638	0.9604	0.9571	0.9537	0.9504	0.9471	0.9439
764	0.9698	0.9663	0.9630	0.9596	0.9562	0.9529	0.9496	0.9463
766	0.9723	0.9689	0.9655	0.9620	0.9587	0.9554	0.9521	0.9488
768	0.9748	0.9714	0.9680	0.9646	0.9612	0.9579	0.9546	0.9513
770	0.9774	0.9739	0.9705	0.9671	0.9637	0.9604	0.9571	0.9538
772	0.9799	0.9764	0.9730	0.9696	0.9632	0.9629	0.9596	0.9562
774	0.9825	0.9790	0.9756	0.9721	0.9687	0.9654	0.9620	0.9587
776	0.9850	0.9815	0.9781	0.9746	0.9712	0.9679	0.9645	0.9612
778	0.9875	0.9840	0.9806	0.9772	0.9737	0.9704	0.9670	0.9637
780	0.9901	0.9866	0.9831	0.9797	0.9763	0.9729	0.9695	0.9662
782	0.9926	0.9891	0.9856	0.9822	0.9788	0.9754	0.9720	0.9686
784	0.9952	0.9916	0.9882	0.9847	0.9813	0.9778	0.9745	0.9711
786	0.9977	0.9942	0.9907	0.9872	0.9838	0.9803	0.9770	0.9736
788	1.0002	0.9967	0.9932	0.9897	0.9863	0.9828	0.9794	0.9761

Table 2-19 (*Continued*)
FACTORS FOR REDUCING GAS VOLUMES TO NORMAL TEMPERATURE AND PRESSURE

Pressure mm of mercury	Temperature °C							
	18°	19°	20°	21°	22°	23°	24°	25°
670	0.8270	0.8242	0.8214	0.8186	0.8158	0.8131	0.8103	0.8076
672	0.8295	0.8267	0.8239	0.8211	0.8183	0.8155	0.8128	0.8100
674	0.8320	0.8291	0.8263	0.8235	0.8207	0.8179	0.8152	0.8124
676	0.8345	0.8316	0.8288	0.8259	0.8231	0.8204	0.8176	0.8149
678	0.8369	0.8341	0.8312	0.8284	0.8256	0.8228	0.8200	0.8173
680	0.8394	0.8365	0.8337	0.8308	0.8280	0.8252	0.8224	0.8197
682	0.8419	0.8390	0.8361	0.8333	0.8304	0.8276	0.8249	0.8221
684	0.8443	0.8414	0.8386	0.8357	0.8329	0.8301	0.8273	0.8245
686	0.8468	0.8439	0.8410	0.8382	0.8353	0.8325	0.8297	0.8269
688	0.8493	0.8464	0.8435	0.8406	0.8378	0.8349	0.8321	0.8293
690	0.8517	0.8488	0.8459	0.8430	0.8402	0.8373	0.8345	0.8317
692	0.8542	0.8513	0.8484	0.8455	0.8426	0.8398	0.8369	0.8341
694	0.8567	0.8537	0.8508	0.8479	0.8451	0.8422	0.8394	0.8366
696	0.8591	0.8562	0.8533	0.8504	0.8475	0.8446	0.8418	0.8390
698	0.8616	0.8587	0.8557	0.8528	0.8499	0.8471	0.8442	0.8414
700	0.8641	0.8611	0.8582	0.8553	0.8524	0.8495	0.8466	0.8438
702	0.8665	0.8636	0.8606	0.8577	0.8547	0.8519	0.8490	0.8462
704	0.8690	0.8660	0.8631	0.8602	0.8572	0.8543	0.8515	0.8486
706	0.8715	0.8685	0.8655	0.8626	0.8597	0.8568	0.8539	0.8510
708	0.8740	0.8710	0.8680	0.8650	0.8621	0.8592	0.8563	0.8534
710	0.8764	0.8734	0.8704	0.8675	0.8645	0.8616	0.8587	0.8558
712	0.8789	0.8759	0.8729	0.8699	0.8670	0.8640	0.8611	0.8582
714	0.8814	0.8783	p.8753	0.8724	0.8694	0.8665	0.8636	0.8607
716	0.8838	0.8808	0.8778	0.8748	0.8718	0.8689	0.8660	0.8631
718	0.8863	0.8833	0.8802	0.8773	0.8743	0.8713	0.8684	0.8655
720	0.8888	0.8857	0.8827	0.8797	0.8767	0.8738	0.8708	0.8679
722	0.8912	0.8882	0.8852	0.8821	0.8792	0.8762	0.8732	0.8703
724	0.8937	0.8906	0.8876	0.8846	0.8816	0.8786	0.8757	0.8727
726	0.8962	0.8931	0.8901	0.8870	0.8840	0.8810	0.8781	0.8751
728	0.8986	0.8956	0.8925	0.8895	0.8865	0.8835	0.8805	0.8775
730	0.9011	0.8980	0.8950	0.8919	0.8889	0.8859	0.8829	0.8799
732	0.9036	0.9005	0.8974	0.8944	0.8913	0.8883	0.8853	0.8824
734	0.9060	0.9029	0.8999	0.8968	0.8938	0.8907	0.8877	0.8848
736	0.9085	0.9054	0.9023	0.8992	0.8962	0.8932	0.8902	0.8872
738	0.9110	0.9079	0.9048	0.9017	0.8986	0.8956	0.8926	0.8896
740	0.9135	0.9103	0.9072	0.9041	0.9011	0.8980	0.8950	0.8920
742	0.9159	0.9128	0.9097	0.9066	0.9035	0.9005	0.8974	0.8944
744	0.9184	0.9153	0.9121	0.9090	0.9059	0.9029	0.8998	0.8968
746	0.9209	0.9177	0.9146	0.9115	0.9084	0.9053	0.9023	0.8992
748	0.9233	0.9202	0.9170	0.9139	0.9108	0.9077	0.9047	0.9016
750	0.9258	0.9226	0.9195	0.9164	0.9132	0.9102	0.9071	0.9041
752	0.9283	0.9251	0.9219	0.9188	0.9157	0.9126	0.9095	0.9065
754	0.9307	0.9276	0.9244	0.9212	0.9181	0.9150	0.9119	0.9089
756	0.9332	0.9300	0.9268	0.9237	0.9206	0.9174	0.9144	0.9113
758	0.9357	0.9325	0.9293	0.9261	0.9230	0.9199	0.9168	0.9137
760	0.9381	0.9349	0.9317	0.9286	0.9254	0.9223	0.9192	0.9161
762	0.9406	0.9374	0.9342	0.9310	0.9279	0.9247	0.9216	0.9185
764	0.9431	0.9399	0.9366	0.9335	0.9303	0.9272	0.9240	0.9209
766	0.9456	0.9423	0.9391	0.9359	0.9327	0.9296	0.9265	0.9233
768	0.9480	0.9448	0.9415	0.9383	0.9352	0.9320	0.9289	0.9258
770	0.9505	0.9472	0.9440	0.9408	0.9376	0.9344	0.9313	0.9282
772	0.9530	0.9497	0.9464	0.9432	0.9400	0.9369	0.9337	0.9306
774	0.9554	0.9522	0.9489	0.9457	0.9425	0.9393	0.9361	0.9330
776	0.9579	0.9546	0.9514	0.9481	0.9449	0.9417	0.9385	0.9354
778	0.9604	0.9571	0.9538	0.9506	0.9473	0.9441	0.9410	0.9378
780	0.9628	0.9595	0.9563	0.9530	0.9498	0.9466	0.9434	0.9402
782	0.9653	0.9620	0.9587	0.9555	0.9522	0.9490	0.9458	0.9426
784	0.9678	0.9645	0.9612	0.9579	0.9546	0.9514	0.9482	0.9450
786	0.9702	0.9669	0.9636	0.9603	0.9571	0.9538	0.9506	0.9474
788	0.9727	0.9694	0.9661	0.9628	0.9595	0.9563	0.9531	0.9499

Table 2-19 (*Continued*)
FACTORS FOR REDUCING GAS VOLUMES TO NORMAL
TEMPERATURE AND PRESSURE

Pressure mm of mercury	Temperature °C							
	26°	27°	28°	29°	30°	31°	32°	33°
670	0.8049	0.8022	0.7996	0.7969	0.7943	0.7917	0.7891	0.7865
672	0.8073	0.8046	0.8020	0.7993	0.7967	0.7940	0.7914	0.7889
674	0.8097	0.8070	0.8043	0.8017	0.7990	0.7964	0.7938	0.7912
676	0.8121	0.8094	0.8067	0.8041	0.8014	0.7988	0.7962	0.7936
678	0.8145	0.8118	0.8091	0.8064	0.8038	0.8011	0.7985	0.7959
680	0.8169	0.8142	0.8115	0.8088	0.8061	0.8035	0.8009	0.7982
682	0.8193	0.8166	0.8139	0.8112	0.8085	p.8059	0.8032	0.8006
684	0.8217	0.8190	0.8163	0.8136	0.8109	0.8082	0.8056	0.8029
686	0.8241	0.8214	0.8187	0.8160	0.8133	0.8106	0.8079	0.8053
688	0.8265	0.8238	0.8211	0.8183	0.8156	0.8129	0.8103	0.8076
690	0.8289	0.8262	0.8234	0.8207	0.8180	0.8153	0.8126	0.8100
692	0.8313	0.8286	0.8258	0.8231	0.8204	0.8177	0.8150	0.8123
694	0.8338	0.8310	0.8282	0.8255	0.8227	0.8200	0.8174	0.8147
696	0.8362	0.8334	0.8306	0.8278	0.8251	0.8224	0.8197	0.8170
698	0.8386	0.8358	0.8330	0.8302	0.8275	0.8248	0.8221	0.8194
700	0.8410	0.8382	0.8354	0.8326	0.8299	0.8271	0.8244	0.8217
702	0.8434	0.8406	0.8378	0.8350	0.8322	0.8295	0.8268	0.8241
704	0.8458	0.8429	0.8401	0.8374	0.8346	0.8319	0.8291	0.8264
706	0.8482	0.8453	0.8425	0.8397	0.8370	0.8342	0.8315	0.8288
708	0.8506	0.8477	0.8449	0.8421	0.8393	0.8366	0.8338	0.8311
710	0.8530	0.8501	0.8473	0.8445	0.8417	0.8389	0.8362	0.8335
712	0.8554	0.8525	0.8497	0.8469	0.8441	0.8413	0.8386	0.8358
714	0.8578	0.8549	0.8521	0.8493	0.8465	0.8437	0.8409	0.8382
716	0.8602	0.8573	0.8545	0.8516	0.8488	0.8460	0.8433	0.8405
718	0.8626	0.8597	0.8569	0.8540	0.8512	0.8484	0.8456	0.8429
720	0.8650	0.8621	0.8592	0.8564	0.8536	0.8508	0.8480	0.8452
722	0.8674	0.8645	0.8616	0.8588	0.8559	0.8531	0.8503	0.8475
724	0.8698	0.8669	0.8640	0.8612	0.8583	0.8555	0.8527	0.8499
726	0.8722	0.8693	0.8664	0.8635	0.8607	0.8579	0.8550	0.8522
728	0.8746	0.8717	0.8688	0.8659	0.8631	0.8602	0.8574	0.8546
730	0.8770	0.8741	0.8712	0.8683	0.8654	0.8626	0.8598	0.8569
732	0.8794	0.8765	0.8736	0.8707	0.8678	0.8649	0.8621	0.8593
734	0.8818	0.8789	0.8759	0.8730	0.8702	0.8673	0.8645	0.8616
736	0.8842	0.8813	0.8783	0.8754	0.8725	0.8697	0.8668	0.8640
738	0.8866	0.8837	0.8807	0.8778	0.8749	0.8720	0.8692	0.8663
740	0.8890	0.8861	0.8831	0.8802	0.8773	0.8744	0.8715	0.8687
742	0.8914	0.8884	0.8855	0.8826	0.8796	0.8768	0.8739	0.8710
744	0.8938	0.8908	0.8879	0.8849	0.8820	0.8791	0.8762	0.8734
746	0.8962	0.8932	0.8903	0.8873	0.8844	0.8815	0.8786	0.8757
748	0.8986	0.8956	0.8927	0.8897	0.8868	0.8838	0.8809	0.8781
750	0.9010	0.8980	0.8950	0.8921	0.8891	0.8862	0.8833	0.8804
752	0.9034	0.9004	0.8974	0.8945	0.8915	0.8886	0.8857	0.8828
754	0.9058	0.9028	0.8998	0.8968	0.8939	0.8909	0.8880	0.8851
756	0.9082	0.9052	0.9022	0.8992	0.8962	0.8933	0.8904	0.8875
758	0.9106	0.9076	0.9046	0.9016	0.8986	0.8957	0.8927	0.8898
760	0.9130	0.9100	0.9070	0.9040	0.9010	0.8980	0.8951	0.8922
762	0.9154	0.9124	0.9094	0.9064	0.9034	0.9004	0.8974	0.8945
764	0.9178	0.9148	0.9118	0.9087	0.9057	0.9028	0.8998	0.8969
766	0.9202	0.9172	0.9141	0.9111	0.9081	0.9051	0.9021	0.8992
768	0.9227	0.9196	0.9165	0.9135	0.9105	0.9075	0.9045	0.9015
770	0.9251	0.9220	0.9189	0.9159	0.9128	0.9098	0.9069	0.9039
772	0.9275	0.9244	0.9213	0.9182	0.9152	0.9122	0.9092	0.9062
774	0.9299	0.9268	0.9237	0.9206	0.9176	0.9146	0.9116	0.9086
776	0.9323	0.9292	0.9261	0.9230	0.9200	0.9169	0.9139	0.9109
778	0.9347	0.9316	0.9285	0.9254	0.9223	0.9193	0.9163	0.9133
780	0.9371	0.9340	0.9308	0.9278	0.9247	0.9217	0.9186	0.9156
782	0.9395	0.9363	0.9332	0.9301	0.9271	0.9240	0.9210	0.9180
784	0.9419	0.9387	0.9356	0.9325	0.9294	0.9264	0.9233	0.9203
786	0.9443	0.9411	0.9380	0.9349	0.9318	0.9287	0.9257	0.9227
788	0.9467	0.9435	0.9404	0.9373	0.9342	0.9311	0.9281	0.9250

Table 2-19 (*Continued*)
FACTORS FOR REDUCING GAS VOLUMES TO NORMAL TEMPERATURE AND PRESSURE

Pressure mm of mercury	Temperature °C.		
	34°	35°	36°
670	0.7839	0.7814	0.7789
672	0.7863	0.7837	0.7812
674	0.7886	0.7861	0.7835
676	0.7910	0.7884	0.7858
678	0.7933	0.7907	0.7882
680	0.7956	0.7931	0.7905
682	0.7980	0.7954	0.7928
684	0.8003	0.7977	0.7951
686	0.8027	0.8001	0.7975
688	0.8050	0.8024	0.7998
690	0.8073	0.8047	0.8021
692	0.8097	0.8071	0.8044
694	0.8120	0.8094	0.8068
696	0.8144	0.8117	0.8091
698	0.8167	0.8141	0.8114
700	0.8190	0.8164	0.8137
702	0.8214	0.8187	0.8161
704	0.8237	0.8211	0.8184
706	0.8261	0.8234	0.8207
708	0.8284	0.8257	0.8230
710	0.8307	0.8281	0.8254
712	0.8331	0.8304	0.8277
714	0.8354	0.8327	0.8300
716	0.8378	0.8350	0.8323
718	0.8401	0.8374	0.8347
720	0.8424	0.8397	0.8370
722	0.8448	0.8420	0.8393
724	0.8471	0.8444	0.8416
726	0.8495	0.8467	0.8440
728	0.8518	0.8490	0.8463
730	0.8541	0.8514	0.8486
732	0.8565	0.8537	0.8509
734	0.8588	0.8560	0.8533
736	0.8612	0.8584	0.8556
738	0.8635	0.8607	0.8579
740	0.8658	0.8630	0.8602
742	0.8682	0.8654	0.8626
744	0.8705	0.8677	0.8649
746	0.8729	0.8700	0.8672
748	0.8752	0.8724	0.8695
750	0.8775	0.8747	0.8719
752	0.8799	0.8770	0.8742
754	0.8822	0.8794	0.8765
756	0.8846	0.8817	0.8788
758	0.8869	0.8840	0.8812
760	0.8892	0.8864	0.8835
762	0.8916	0.8887	0.8858
764	0.8939	0.8910	0.8881
766	0.8963	0.8934	0.8905
768	0.8986	0.8957	0.8928
770	0.9009	0.8980	0.8951
772	0.9033	0.9004	0.8974
774	0.9056	0.9027	0.8998
776	0.9080	0.9050	0.9021
778	0.9103	0.9074	0.9044
780	0.9127	0.9097	0.9067
782	0.9150	0.9120	0.9091
784	0.9173	0.9144	0.9114
786	0.9197	0.9167	0.9137
788	0.9220	0.9190	0.9160

Aqueous Tension of Pure Water, Potassium Hydroxide Solutions and Saturated Salt Solution.

Deg. C.	10°	11°	12°	13°	14°	15°	16°	17°
H_2O	9.2	9.8	10.5	11.2	12.0	12.8	13.6	14.5
10KOH:100H_2O	8.6	9.2	9.8	10.5	11.2	11.9	12.7	13.6
20KOH:100H_2O	8.0	8.6	9.3	9.8	10.4	11.1	11.8	12.6
30KOH:100H_2O	7.3	7.8	8.3	8.9	9.5	10.1	10.8	11.5
40KOH:100H_2O	6.5	6.9	7.4	7.9	8.4	9.0	9.6	10.2
Satd. Aq. NaCl	6.9	7.4	7.9	8.5	9.1	9.7	10.3	11.0

Deg. C.	18°	19°	20°	21°	22°	23°	24°	25°
H_2O	15.5	16.5	17.5	18.7	19.8	21.1	22.4	23.8
10KOH:100H_2O	14.5	15.4	16.4	17.4	18.5	19.7	20.9	22.2
20KOH:100H_2O	13.4	14.3	15.2	16.2	17.2	18.3	19.5	20.7
30KOH:100H_2O	12.3	13.3	13.9	14.8	15.8	16.8	17.8	18.9
40KOH:100H_2O	10.9	11.6	12.4	13.2	14.0	14.9	15.8	16.8
Satd. Aq. NaCl	11.7	12.4	13.2	14.1	15.0	15.9	16.9	17.9

Table 2-20
VALUES OF ABSORBANCE FOR PER CENT ABSORPTION

To convert per cent absorption (% A) to absorbance, find the per cent absorption to the nearest whole digit in the left-hand column; read across to the column located under the tenth of a per cent desired, and read the value of absorbance. The value of absorbance corresponding to 26.8% absorption is thus 0.1355.

% A	.0	.1	.2	.3	.4	.5	.6	.7	.8	.9
0.0	.0000	.0004	.0009	.0013	.0017	.0022	.0026	.0031	.0035	.0039
1.0	.0044	.0048	.0052	.0057	.0061	.0066	.0070	.0074	.0079	.0083
2.0	.0088	.0092	.0097	.0101	.0106	.0110	.0114	.0119	.0123	.0128
3.0	.0132	.0137	.0141	.0146	.0150	.0155	.0159	.0164	.0168	.0173
4.0	.0177	.0182	.0186	.0191	.0195	.0200	.0205	.0209	.0214	.0218
5.0	.0223	.0227	.0232	.0236	.0241	.0246	.0250	.0255	.0259	.0264
6.0	.0269	.0273	.0278	.0283	.0287	.0292	.0297	.0301	.0306	.0311
7.0	.0315	.0320	.0325	.0329	.0334	.0339	.0343	.0348	.0353	.0357
8.0	.0362	.0367	.0372	.0376	.0381	.0386	.0391	.0395	.0400	.0405
9.0	.0410	.0414	.0419	.0424	.0429	.0434	.0438	.0443	.0448	.0453
10.0	.0458	.0462	.0467	.0472	.0477	.0482	.0487	.0491	.0496	.0501
11.0	.0506	.0511	.0516	.0521	.0526	.0531	.0535	.0540	.0545	.0550
12.0	.0555	.0560	.0565	.0570	.0575	.0580	.0585	.0590	.0595	.0600
13.0	.0605	.0610	.0615	.0620	.0625	.0630	.0635	.0640	.0645	.0650
14.0	.0655	.0660	.0665	.0670	.0675	.0680	.0685	.0691	.0696	.0701
15.0	.0706	.0711	.0716	.0721	.0726	.0731	.0737	.0742	.0747	.0752
16.0	.0757	.0762	.0768	.0773	.0778	.0783	.0788	.0794	.0799	.0804
17.0	.0809	.0814	.0820	.0825	.0830	.0835	.0841	.0846	.0851	.0857
18.0	.0862	.0867	.0872	.0878	.0883	.0888	.0894	.0899	.0904	.0910
19.0	.0915	.0921	.0926	.0931	.0937	.0942	.0947	.0953	.0958	.0964
20.0	.0969	.0975	.0980	.0985	.0991	.0996	.1002	.1007	.1013	.1018
21.0	.1024	.1029	.1035	.1040	.1046	.1051	.1057	.1062	.1068	.1073
22.0	.1079	.1085	.1090	.1096	.1101	.1107	.1113	.1118	.1124	.1129
23.0	.1135	.1141	.1146	.1152	.1158	.1163	.1169	.1175	.1180	.1186
24.0	.1192	.1198	.1203	.1209	.1215	.1221	.1226	.1232	.1238	.1244
25.0	.1249	.1255	.1261	.1267	.1273	.1278	.1284	.1290	.1296	.1302
26.0	.1308	.1314	.1319	.1325	.1331	.1337	.1343	.1349	.1355	.1361
27.0	.1367	.1373	.1379	.1385	.1391	.1397	.1403	.1409	.1415	.1421
28.0	.1427	.1433	.1439	.1445	.1451	.1457	.1463	.1469	.1475	.1481
29.0	.1487	.1494	.1500	.1506	.1512	.1518	.1524	.1530	.1537	.1543
30.0	.1549	.1555	.1561	.1568	.1574	.1580	.1586	.1593	.1599	.1605
31.0	.1612	.1618	.1624	.1630	.1637	.1643	.1649	.1656	.1662	.1669
32.0	.1675	.1681	.1688	.1694	.1701	.1707	.1713	.1720	.1726	.1733
33.0	.1739	.1746	.1752	.1759	.1765	.1772	.1778	.1785	.1791	.1798
34.0	.1805	.1811	.1818	.1824	.1831	.1838	.1844	.1851	.1858	.1864
35.0	.1871	.1878	.1884	.1891	.1898	.1904	.1911	.1918	.1925	.1931
36.0	.1938	.1945	.1952	.1959	.1965	.1972	.1979	.1986	.1993	.2000
37.0	.2007	.2013	.2020	.2027	.2034	.2041	.2048	.2055	.2062	.2069
38.0	.2076	.2083	.2090	.2097	.2104	.2111	.2118	.2125	.2132	.2140
39.0	.2147	.2154	.2161	.2168	.2175	.2182	.2190	.2197	.2204	.2211
40.0	.2218	.2226	.2233	.2240	.2248	.2255	.2262	.2269	.2277	.2284
41.0	.2291	.2299	.2306	.2314	.2321	.2328	.2336	.2343	.2351	.2358

Table 2-20 (*Continued*)
VALUES OF ABSORBANCE FOR PER CENT ABSORPTION

% A	.0	.1	.2	.3	.4	.5	.6	.7	.8	.9
42.0	.2366	.2373	.2381	.2388	.2396	.2403	.2411	.2418	.2426	.2434
43.0	.2441	.2449	.2457	.2464	.2472	.2480	.2487	.2495	.2503	.2510
44.0	.2518	.2526	.2534	.2541	.2549	.2557	.2565	.2573	.2581	.2588
45.0	.2596	.2604	.2612	.2620	.2628	.2636	.2644	.2652	.2660	.2668
46.0	.2676	.2684	.2692	.2700	.2708	.2716	.2725	.2733	.2741	.2749
47.0	.2757	.2765	.2774	.2782	.2790	.2798	.2807	.2815	.2823	.2832
48.0	.2840	.2848	.2857	.2865	.2874	.2882	.2890	.2899	.2907	.2916
49.0	.2924	.2933	.2941	.2950	.2958	.2967	.2976	.2984	.2993	.3002
50.0	.3010	.3019	.3028	.3036	.3045	.3054	.3063	.3072	.3080	.3089
51.0	.3098	.3107	.3116	.3125	.3134	.3143	.3152	.3161	.3170	.3179
52.0	.3188	.3197	.3206	.3215	.3224	.3233	.3242	.3251	.3261	.3270
53.0	.3279	.3288	.3298	.3307	.3316	.3325	.3335	.3344	.3354	.3363
54.0	.3372	.3382	.3391	.3401	.3410	.3420	.3429	.3439	.3449	.3458
55.0	.3468	.3478	.3487	.3497	.3507	.3516	.3526	.3536	.3546	.3556
56.0	.3565	.3575	.3585	.3595	.3605	.3615	.3625	.3635	.3645	.3655
57.0	.3665	.3675	.3686	.3696	.3706	.3716	.3726	.3737	.3747	.3757
58.0	.3768	.3778	.3788	.3799	.3809	.3820	.3830	.3840	.3851	.3862
59.0	.3872	.3883	.3893	.3904	.3915	.3925	.3936	.3947	.3958	.3969
60.0	.3979	.3990	.4001	.4012	.4023	.4034	.4045	.4056	.4067	.4078
61.0	.4089	.4101	.4112	.4123	.4134	.4145	.4157	.4168	.4179	.4191
62.0	.4202	.4214	.4225	.4237	.4248	.4260	.4271	.4283	.4295	.4306
63.0	.4318	.4330	.4342	.4353	.4365	.4377	.4389	.4401	.4413	.4425
64.0	.4437	.4449	.4461	.4473	.4485	.4498	.4510	.4522	.4535	.4547
65.0	.4559	.4572	.4584	.4597	.4609	.4622	.4634	.4647	.4660	.4672
66.0	.4685	.4698	.4711	.4724	.4737	.4750	.4763	.4776	.4789	.4802
67.0	.4815	.4828	.4841	.4855	.4868	.4881	.4895	.4908	.4921	.4935
68.0	.4948	.4962	.4976	.4989	.5003	.5017	.5031	.5045	.5058	.5072
69.0	.5086	.5100	.5114	.5129	.5143	.5157	.5171	.5186	.5200	.5214
70.0	.5229	.5243	.5258	.5272	.5287	.5302	.5317	.5331	.5346	.5361
71.0	.5376	.5391	.5406	.5421	.5436	.5452	.5467	.5482	.5498	.5513
72.0	.5528	.5544	.5560	.5575	.5591	.5607	.5622	.5638	.5654	.5670
73.0	.5686	.5702	.5719	.5735	.5751	.5768	.5784	.5800	.5817	.5834
74.0	.5850	.5867	.5884	.5901	.5918	.5935	.5952	.5969	.5986	.6003
75.0	.6021	.6038	.6055	.6073	.6091	.6108	.6126	.6144	.6162	.6180
76.0	.6198	.6216	.6234	.6253	.6271	.6289	.6308	.6326	.6345	.6364
77.0	.6383	.6402	.6421	.6440	.6459	.6478	.6498	.6517	.6536	.6556
78.0	.6576	.6596	.6615	.6635	.6655	.6676	.6696	.6716	.6737	.6757
79.0	.6778	.6799	.6819	.6840	.6861	.6882	.6904	.6925	.6946	.6968
80.0	.6990	.7011	.7033	.7055	.7077	.7100	.7122	.7144	.7167	.7190
81.0	.7212	.7235	.7258	.7282	.7305	.7328	.7352	.7375	.7399	.7423
82.0	.7447	.7471	.7496	.7520	.7545	.7570	.7595	.7620	.7645	.7670
83.0	.7696	.7721	.7747	.7773	.7799	.7825	.7852	.7878	.7905	.7932
84.0	.7959	.7986	.8013	.8041	.8069	.8097	.8125	.8153	.8182	.8210
85.0	.8239	.8268	.8297	.8327	.8356	.8386	.8416	.8447	.8477	.8508
86.0	.8539	.8570	.8601	.8633	.8665	.8697	.8729	.8761	.8794	.8827
87.0	.8861	.8894	.8928	.8962	.8996	.9031	.9066	.9101	.9136	.9172
88.0	.9208	.9245	.9281	.9318	.9355	.9393	.9431	.9469	.9508	.9547
89.0	.9586	.9626	.9666	.9706	.9747	.9788	.9830	.9872	.9914	.9957

Table 2-21
TRANSMITTANCE-ABSORBANCE CONVERSION TABLE

From Meites, *Handbook of Analytical Chemistry*, 1963, McGraw-Hill Book
Company; by permission

This table gives absorbance values to four significant figures corresponding to %
transmittance values which are given to three significant figures. The values of %
transmittance are given in the left-hand column and in the top row. For example, 8.4%
transmittance corresponds to an absorbance of 1.076.

Interpolation is facilitated and accuracy is maximized if the % transmittance is
between 1 and 10, by multiplying its value by 10, finding the absorbance corresponding
to the result, and adding 1. For example, to find the absorbance corresponding to 8.45%
transmittance, note that 84.5% transmittance corresponds to an absorbance of 0.0731,
so that 8.45% transmittance corresponds to an absorbance of 1.0731. For % transmittance values between 0.1 and 1, multiply by 100, find the absorbance corresponding
to the result, and add 2.

Conversely, to find the % transmittance corresponding to an absorbance between 1
and 2, subtract 1 from the absorbance, find the % transmittance corresponding to the
result, and divide by 10. For example, an absorbance of 1.219 can best be converted
to % transmittance by noting that an absorbance of 0.219 would correspond to 60.4%
transmittance; dividing this by 10 gives the desired value, 6.04% transmittance. For
absorbance values between 2 and 3, subtract 2 from the absorbance, find the %
transmittance corresponding to the result, and divide by 100.

% Trans-mittance	0.0	0.1	0.2	0.3	0.4	0.5	0.6	0.7	0.8	0.9
0	· · · · · ·	3.000	2.699	2.523	2.398	2.301	2.222	2.155	2.097	2.046
1	2.000	1.959	1.921	1.886	1.854	1.824	1.796	1.770	1.745	1.721
2	1.699	1.678	1.658	1.638	1.620	1.602	1.585	1.569	1.553	1.538
3	1.523	1.509	1.495	1.481	1.469	1.456	1.444	1.432	1.420	1.409
4	1.398	1.387	1.377	1.367	1.357	1.347	1.337	1.328	1.319	1.310
5	1.301	1.292	1.284	1.276	1.268	1.260	1.252	1.244	1.237	1.229
6	1.222	1.215	1.208	1.201	1.194	1.187	1.180	1.174	1.167	1.161
7	1.155	1.149	1.143	1.137	1.131	1.125	1.119	1.114	1.108	1.102
8	1.097	1.092	1.086	1.081	1.076	1.071	1.066	1.060	1.056	1.051
9	1.046	1.041	1.036	1.032	1.027	1.022	1.018	1.013	1.009	1.004
10	1.000	0.9957	0.9914	0.9872	0.9830	0.9788	0.9747	0.9706	0.9666	0.9626
11	0.9586	0.9547	0.9508	0.9469	0.9431	0.9393	0.9355	0.9318	0.9281	0.9245
12	0.9208	0.9172	0.9136	0.9101	0.9066	0.9031	0.8996	0.8962	0.8928	0.8894
13	0.8861	0.8827	0.8794	0.8761	0.8729	0.8697	0.8665	0.8633	0.8601	0.8570
14	0.8539	0.8508	0.8477	0.8447	0.8416	0.8386	0.8356	0.8327	0.8297	0.8268
15	0.8239	0.8210	0.8182	0.8153	0.8125	0.8097	0.8069	0.8041	0.8013	0.7986
16	0.7959	0.7932	0.7905	0.7878	0.7852	0.7825	0.7799	0.7773	0.7747	0.7721
17	0.7696	0.7670	0.7645	0.7620	0.7595	0.7570	0.7545	0.7520	0.7496	0.7471
18	0.7447	0.7423	0.7399	0.7375	0.7352	0.7328	0.7305	0.7282	0.7258	0.7235
19	0.7212	0.7190	0.7167	0.7144	0.7122	0.7100	0.7077	0.7055	0.7033	0.7011
20	0.6990	0.6968	0.6946	0.6925	0.6904	0.6882	0.6861	0.6840	0.6819	0.6799

Table 2-21 (*Continued*)
TRANSMITTANCE-ABSORBANCE CONVERSION TABLE

% Trans-mittance	0.0	0.1	0.2	0.3	0.4	0.5	0.6	0.7	0.8	0.9
21	0.6778	0.6757	0.6737	0.6716	0.6696	0.6676	0.6655	0.6635	0.6615	0.6596
22	0.6576	0.6556	0.6536	0.6517	0.6498	0.6478	0.6459	0.6440	0.6421	0.6402
23	0.6383	0.6364	0.6345	0.6326	0.6308	0.6289	0.6271	0.6253	0.6234	0.6216
24	0.6198	0.6180	0.6162	0.6144	0.6126	0.6108	0.6091	0.6073	0.6055	0.6038
25	0.6021	0.6003	0.5986	0.5969	0.5952	0.5935	0.5918	0.5901	0.5884	0.5867
26	0.5850	0.5834	0.5817	0.5800	0.5784	0.5766	0.5751	0.5735	0.5719	0.5702
27	0.5686	0.5670	0.5654	0.5638	0.5622	0.5607	0.5591	0.5575	0.5560	0.5544
28	0.5528	0.5513	0.5498	0.5482	0.5467	0.5452	0.5436	0.5421	0.5406	0.5391
29	0.5376	0.5361	0.5346	0.5331	0.5317	0.5302	0.5287	0.5272	0.5258	0.5243
30	0.5229	0.5214	0.5200	0.5186	0.5171	0.5157	0.5143	0.5129	0.5114	0.5100
31	0.5086	0.5072	0.5058	0.5045	0.5031	0.5017	0.5003	0.4989	0.4976	0.4962
32	0.4949	0.4935	0.4921	0.4908	0.4895	0.4881	0.4868	0.4855	0.4841	0.4828
33	0.4815	0.4802	0.4789	0.4776	0.4763	0.4750	0.4737	0.4724	0.4711	0.4698
34	0.4685	0.4672	0.4660	0.4647	0.4634	0.4622	0.4609	0.4597	0.4584	0.4572
35	0.4559	0.4547	0.4535	0.4522	0.4510	0.4498	0.4486	0.4473	0.4461	0.4449
36	0.4437	0.4425	0.4413	0.4401	0.4389	0.4377	0.4365	0.4353	0.4342	0.4330
37	0.4318	0.4306	0.4295	0.4283	0.4271	0.4260	0.4248	0.4237	0.4225	0.4214
38	0.4202	0.4191	0.4179	0.4168	0.4157	0.4145	0.4134	0.4123	0.4112	0.4101
39	0.4089	0.4078	0.4067	0.4056	0.4045	0.4034	0.4023	0.4012	0.4001	0.3989
40	0.3979	0.3969	0.3958	0.3947	0.3936	0.3925	0.3915	0.3904	0.3893	0.3883
41	0.3872	0.3862	0.3851	0.3840	0.3830	0.3820	0.3809	0.3799	0.3788	0.3778
42	0.3768	0.3757	0.3747	0.3737	0.3726	0.3716	0.3706	0.3696	0.3686	0.3675
43	0.3665	0.3655	0.3645	0.3635	0.3625	0.3615	0.3605	0.3595	0.3585	0.3575
44	0.3565	0.3556	0.3546	0.3536	0.3526	0.3516	0.3507	0.3497	0.3487	0.3478
45	0.3468	0.3458	0.3449	0.3439	0.3429	0.3420	0.3410	0.3401	0.3391	0.3382
46	0.3372	0.3363	0.3354	0.3344	0.3335	0.3325	0.3316	0.3307	0.3298	0.3288
47	0.3279	0.3270	0.3261	0.3251	0.3242	0.3233	0.3224	0.3215	0.3206	0.3197
48	0.3188	0.3179	0.3170	0.3161	0.3152	0.3143	0.3134	0.3125	0.3116	0.3107
49	0.3098	0.3089	0.3080	0.3072	0.3063	0.3054	0.3045	0.3036	0.3028	0.3019
50	0.3010	0.3002	0.2993	0.2984	0.2976	0.2967	0.2958	0.2950	0.2941	0.2933
51	0.2924	0.2916	0.2907	0.2899	0.2890	0.2882	0.2874	0.2865	0.2857	0.2848
52	0.2840	0.2832	0.2823	0.2815	0.2807	0.2798	0.2790	0.2782	0.2774	0.2765
53	0.2757	0.2749	0.2741	0.2733	0.2725	0.2716	0.2708	0.2700	0.2692	0.2684
54	0.2676	0.2668	0.2660	0.2652	0.2644	0.2636	0.2628	0.2620	0.2612	0.2604
55	0.2596	0.2588	0.2581	0.2573	0.2565	0.2557	0.2549	0.2541	0.2534	0.2526
56	0.2518	0.2510	0.2503	0.2495	0.2487	0.2480	0.2472	0.2464	0.2457	0.2449
57	0.2441	0.2434	0.2426	0.2418	0.2411	0.2403	0.2396	0.2388	0.2381	0.2373
58	0.2366	0.2358	0.2351	0.2343	0.2336	0.2328	0.2321	0.2314	0.2306	0.2299
59	0.2291	0.2284	0.2277	0.2269	0.2262	0.2255	0.2248	0.2240	0.2233	0.2226
60	0.2218	0.2211	0.2204	0.2197	0.2190	0.2182	0.2175	0.2168	0.2161	0.2154
61	0.2147	0.2140	0.2132	0.2125	0.2118	0.2111	0.2104	0.2097	0.2090	0.2083
62	0.2076	0.2069	0.2062	0.2055	0.2048	0.2041	0.2034	0.2027	0.2020	0.2013
63	0.2007	0.2000	0.1993	0.1986	0.1979	0.1972	0.1965	0.1959	0.1952	0.1945
64	0.1938	0.1931	0.1925	0.1918	0.1911	0.1904	0.1898	0.1891	0.1884	0.1878
65	0.1871	0.1864	0.1858	0.1851	0.1844	0.1838	0.1831	0.1824	0.1818	0.1811

Table 2-21 (*Continued*)
TRANSMITTANCE-ABSORBANCE CONVERSION TABLE

% Trans-mittance	0.0	0.1	0.2	0.3	0.4	0.5	0.6	0.7	0.8	0.9
66	0.1805	0.1798	0.1791	0.1785	0.1778	0.1772	0.1765	0.1759	0.1752	0.1746
67	0.1739	0.1733	0.1726	0.1720	0.1713	0.1707	0.1701	0.1694	0.1688	0.1681
68	0.1675	0.1669	0.1662	0.1656	0.1649	0.1643	0.1637	0.1630	0.1624	0.1618
69	0.1612	0.1605	0.1599	0.1593	0.1586	0.1580	0.1574	0.1568	0.1561	0.1555
70	0.1549	0.1543	0.1537	0.1530	0.1524	0.1518	0.1512	0.1506	0.1500	0.1494
71	0.1487	0.1481	0.1475	0.1469	0.1463	0.1457	0.1451	0.1445	0.1439	0.1433
72	0.1427	0.1421	0.1415	0.1409	0.1403	0.1397	0.1391	0.1385	0.1379	0.1373
73	0.1367	0.1361	0.1355	0.1349	0.1343	0.1337	0.1331	0.1325	0.1319	0.1314
74	0.1308	0.1302	0.1296	0.1290	0.1284	0.1278	0.1273	0.1267	0.1261	0.1255
75	0.1249	0.1244	0.1238	0.1232	0.1226	0.1221	0.1215	0.1209	0.1203	0.1198
76	0.1192	0.1186	0.1180	0.1175	0.1169	0.1163	0.1158	0.1152	0.1146	0.1141
77	0.1135	0.1129	0.1124	0.1118	0.1113	0.1107	0.1101	0.1096	0.1090	0.1085
78	0.1079	0.1073	0.1068	0.1062	0.1057	0.1051	0.1046	0.1040	0.1035	0.1029
79	0.1024	0.1018	0.1013	0.1007	0.1002	0.0996	0.0991	0.0985	0.0980	0.0975
80	0.0969	0.0964	0.0958	0.0953	0.0947	0.0942	0.0937	0.0931	0.0926	0.0921
81	0.0915	0.0910	0.0904	0.0899	0.0894	0.0888	0.0883	0.0878	0.0872	0.0867
82	0.0862	0.0857	0.0851	0.0846	0.0841	0.0835	0.0830	0.0825	0.0820	0.0814
83	0.0809	0.0804	0.0799	0.0794	0.0788	0.0783	0.0778	0.0773	0.0768	0.0762
84	0.0757	0.0752	0.0747	0.0742	0.0737	0.0731	0.0726	0.0721	0.0716	0.0711
85	0.0706	0.0701	0.0696	0.0691	0.0685	0.0680	0.0675	0.0670	0.0665	0.0660
86	0.0655	0.0650	0.0645	0.0640	0.0635	0.0630	0.0625	0.0620	0.0615	0.0610
87	0.0605	0.0600	0.0595	0.0590	0.0585	0.0580	0.0575	0.0570	0.0565	0.0560
88	0.0555	0.0550	0.0545	0.0540	0.0535	0.0531	0.0526	0.0521	0.0516	0.0511
89	0.0506	0.0501	0.0496	0.0491	0.0487	0.0482	0.0477	0.0472	0.0467	0.0462
90	0.0458	0.0453	0.0448	0.0443	0.0438	0.0434	0.0429	0.0424	0.0419	0.0414
91	0.0410	0.0405	0.0400	0.0395	0.0391	0.0386	0.0381	0.0376	0.0372	0.0367
92	0.0362	0.0357	0.0353	0.0348	0.0343	0.0339	0.0334	0.0329	0.0325	0.0320
93	0.0315	0.0311	0.0306	0.0301	0.0297	0.0292	0.0287	0.0283	0.0278	0.0273
94	0.0269	0.0264	0.0259	0.0255	0.0250	0.0246	0.0241	0.0237	0.0232	0.0227
95	0.0223	0.0218	0.0214	0.0209	0.0205	0.0200	0.0195	0.0191	0.0186	0.0182
96	0.0177	0.0173	0.0168	0.0164	0.0159	0.0155	0.0150	0.0146	0.0141	0.0137
97	0.0132	0.0128	0.0123	0.0119	0.0114	0.0110	0.0106	0.0101	0.0097	0.0092
98	0.0088	0.0083	0.0079	0.0074	0.0070	0.0066	0.0061	0.0057	0.0052	0.0048
99	0.0044	0.0039	0.0035	0.0031	0.0026	0.0022	0.0017	0.0013	0.0009	0.0004

Section 3

ATOMIC AND MOLECULAR STRUCTURE

Table 3-1
PHYSICAL PROPERTIES OF THE ELEMENTS
(1969 values for atomic weights)

Name	Symbol	At. No.	Electronic Configuration	Atomic Weight	Density, g/cc at 20°C	Melting Point, °C	Boiling Point, °C	Thermal Conductivity, cal/°C/cm/sec	Electrical Resistivity, microhms/cm
Actinium	Ac	89	$[Rn]6d7s^2$	(227)		1050	(3330)		2.6
Aluminum	Al	13	$[Ne]3s^23p$	26.9815	2.6984	660.2	2447	0.504	
Americium	Am	95	$[Rn]5f^77s^2$	(243)	13.67	>800	(2600)		39
Antimony	Sb	51	$[Kr]4d^{10}5s^25p^3$	121.75	6.684	630.5	1640	0.0538	35
Argon	Ar	18	$[Ne]3s^23p^6$	39.948	0.0017824	−189.38	−185.87	3.920×10^{-5}	
Arsenic	As	33	$[Ar]3d^{10}4s^24p^3$	74.9216	5.72 gray 2.026 yel 4.7 black	817 (28 atm)	613 subl		
Astatine	At	85	$[Xe]4f^{14}5d^{10}6s^26p^5$	(210)		302	334		
Barium	Ba	56	$[Xe]6s^2$	137.34	3.59	850	1537		60
Berkelium	Bk	97	$[Rn]5f^86d7s^2$	(245)					
Beryllium	Be	4	$[He]2s^2$	9.01218	1.86	1285	2970	0.38	12
Bismuth	Bi	83	$[Xe]4f^{14}5d^{10}6s^26p^3$	208.9806	9.80	271.3	1560	0.0177	110
Boron	B	5	$[He]2s^22p$	10.81	2.46	2074	3675		1.8×10^{12}
Bromine	Br	35	$[Ar]3d^{10}4s^24p^5$	79.904	3.119 lq	−7.08	58.76		7.8×10^{18} lq
Cadmium	Cd	48	$[Kr]4d^{10}5s^2$	112.40	8.642	320.9	767	0.264	5.9
Calcium	Ca	20	$[Ar]4s^2$	40.08	1.55	851	1487	0.3	4.5
Californium	Cf	98	$[Rn]5f^{10}7s^2$	(248)					
Carbon	C	6	$[He]2s^22p^2$	12.011	2.267 graphite 3.515 diamond	4000 (63 atm)	3850 subl	0.057	1375
Cerium	Ce	58	$[Xe]4f\ 5d\ 6s^2$	140.12	6.771	795	3470		71.6
Cesium	Cs	55	$[Xe]6s$	132.9055	1.8785^{15}	28.6	670		19
Chlorine	Cl	17	$[Ne]3s^23p^5$	35.453	0.00298 g	−101.0	−34.05		>10^9 lq
Chromium	Cr	24	$[Ar]3d^54s$	51.996	7.20	1900	2640	0.16	14
Cobalt	Co	27	$[Ar]3d^74s^2$	58.9332	8.9	1495	3550	0.165	8
Copper	Cu	29	$[Ar]3d^{10}4s$	63.546	8.92	1083	2582	0.989	1.6
Curium	Cm	96	$[Rn]5f^76d\ 7s^2$	(247)	13.51				
Dysprosium	Dy	66	$[Xe]4f^{10}6s^2$	162.50	8.536	1407	2600		90.9

3-2

Einsteinium	Es	99	$[Rn]5f^{11}7s^2$	(254)					
Erbium	Er	68	$[Xe]4f^{12}6s^2$	167.26	9.051	1495	2900		83
Europium	Eu	63	$[Xe]4f^{7}6s^2$	151.96	5.259	826	1440		83
Fermium	Fm	100	$[Rn]5f^{12}7s^2$	(253)					
Fluorine	F	9	$[He]2s^2 2p^5$	18.9984	0.001580 g	−219.62	−188.14		
Francium	Fr	87	$[Rn]7s$	(223)		(27)			
Gadolinium	Gd	64	$[Xe]4f^{7}5d6s^2$	157.25	7.895	1312	3000		143
Gallium	Ga	31	$[Ar]3d^{10}4s^2 4p$	69.72	5.907	29.75	1980		52
Germanium	Ge	32	$[Ar]3d^{10}4s^2 4p^2$	72.59	5.323	937	2830		89,000
Gold	Au	79	$[Xe]4f^{14}5d^{10}6s$	196.9665	19.3	1063	2707	0.700	2.4
Hafnium	Hf	72	$[Xe]4f^{14}5d^2 6s^2$	178.49	13.31	2225	~5200		32
Helium	He	2	$1s^2$	4.00260	0.17847 g (STP)	−272.2 (25 atm)	−268.935	33.90×10^{-5}	
Holmium	Ho	67	$[Xe]4f^{11}6s^2$	164.9303	8.803	1461	2600		90
Hydrogen	H	1	$1s$	1.0080	0.8887×10^{-4}	−259.20	−252.77		
Indium	In	49	$[Kr]4d^{10}5s^2 5p$	114.82	7.28	156.4	2050	0.057	8.5
Iodine	I	53	$[Kr]4d^{10}5s^2 5p^5$	126.9045	4.660 s	113.6	184.4		1.3×10^{15}
Iridium	Ir	77	$[Xe]4f^{14}5d^7 6s^2$	192.22	22.65	2448	4500	0.141	5.5
Iron	Fe	26	$[Ar]3d^6 4s^2$	55.847	7.86	1530	3000	0.18	10
Krypton	Kr	36	$[Ar]3d^{10}4s^2 4p^6$	83.80	0.003736	−157.2	−153.4	2.09×10^{-5}	
Lanthanum	La	57	$[Xe]5d6s^2$	138.9055	6.174	920	3470		56
Lawrencium	Lr	103	$[Rn]5f^{14}6d7s^2$	(257)					
Lead	Pb	82	$[Xe]4f^{14}5d^{10}6s^2 6p^2$	207.2	11.34	327.4	1751	0.083	21
Lithium	Li	3	$1s^2 2s$	6.941	0.535	179	1336	0.170°	8.6
Lutetium	Lu	71	$[Xe]4f^{14}5d6s^2$	174.97	9.842	1652	3330	0.376	67
Magnesium	Mg	12	$[Ne]3s^2$	24.305	1.74	650	1117		4.4
Manganese	Mn	25	$[Ar]3d^5 4s^2$	54.9380	7.30	1244	2120		
Mendelevium	Md	101	$[Rn]5f^{13}7s^2$	(256)					
Mercury	Hg	80	$[Xe]4f^{14}5d^{10}6s^2$	200.59	13.5939	−38.87	356.58	0.025	97 lq; 21 s
Molybdenum	Mo	42	$[Kr]4d^5 5s$	95.94	10.2	2625	4800	0.35	5
Neodymium	Nd	60	$[Xe]4f^4 6s^2$	144.24	7.004	1024	3030		79
Neon	Ne	10	$1s^2 2s^2 2p^6$	20.179	1.207 lq b.p.	−248.6	−246.1	11.00×10^{-5}	
Neptunium	Np	93	$[Rn]5f^4 6d7s^2$	237.0482	20.45	630			
Nickel	Ni	28	$[Ar]3d^8 4s^2$	58.71	8.90	1455	2840	0.140	6.8
Niobium	Nb	41	$[Kr]4d^4 5s$	92.9064	8.57	2468	5127		14

Table 3-1 (*Continued*)
PHYSICAL PROPERTIES OF THE ELEMENTS

Name	Symbol	At. No.	Electronic Configuration	Atomic Weight	Density, g/cc at 20°C	Melting Point, °C	Boiling Point, °C	Thermal Conductivity, cal/°C/cm/sec	Electrical Resistivity, microhms/cm
Nitrogen	N	7	$1s^22s^22p^3$	14.0067	0.001165 g	-209.97	-195.798		
Nobelium	No	102	$[Rn]5f^{14}7s^2$	(254)					
Osmium	Os	76	$[Xe]4f^{14}5d^66s^2$	190.2	22.61	2727	(4100)		9.5
Oxygen	O	8	$1s^22s^22p^4$	15.9994	0.001331 g	-218.787	-182.98		
Palladium	Pd	46	$[Kr]4d^{10}$	106.4	12.023	1552	2870	0.168	10.8
Phosphorus	P	15	$[Ne]3s^23p^3$	30.9738	1.828 white	44.2	280.3		10^{17}
					2.34 red	597	431 subl		
					2.699 black	610	453 subl		
Platinum	Pt	78	$[Xe]4f^{14}5d^96s$	195.09	21.45	1774	~3800	0.167	10.2
Plutonium	Pu	94	$[Rn]5f^67s^2$	239.05	19.82	638	3235		150
Polonium	Po	84	$[Xe]4f^{14}5d^{10}6s^26p^4$	210	9.20 cub	254	962		
					9.40 rh				
Potassium	K	19	$[Ar]4s$	39.102	0.87	63.5	758	0.232	6.6
Praseodymium	Pr	59	$[Xe]4f^36s^2$	140.9077	6.782	935	3130		68
Promethium	Pm	61	$[Xe]4f^56s^2$	(147)		(1027)	(2727)		
Protactinium	Pa	91	$[Rn]5f^26d7s^2$	231.0359	15.37	(1227)	(4027)		
Radium	Ra	88	$[Rn]7s^2$	226.0254	ca 6	ca 700	(1525)		
Radon	Rn	86	$[Xe]4f^{14}5d^{10}6s^26p^6$	(222)	4.4 lq bp	-71	-62		
Rhenium	Re	75	$[Xe]4f^{14}5d^56s^2$	186.2	21.04	3180	5885	0.14	19.3
Rhodium	Rh	45	$[Kr]4d^85s$	102.9055	12.41	1966	(3700)	0.210	5
Rubidium	Rb	37	$[Kr]5s$	85.4678	1.53	39.0	700		12.5
Ruthenium	Ru	44	$[Kr]4d^75s$	101.07	12.45	2430	3700		10
Samarium	Sm	62	$[Xe]4f^66s^2$	150.4	7.536	1052	1900		90
Scandium	Sc	21	$[Ar]3d4s^2$	44.9559	2.992	1397	2730		
Selenium	Se	34	$[Ar]3d^{10}4s^24p^4$	78.96	4.792 hex-rhb	217	685		1.2
					4.48 red, mn	170			
Silicon	Si	14	$[Ne]3s^23p^2$	28.086	2.33	1415	2680	0.20	10^5
Silver	Ag	47	$[Kr]4d^{10}5s$	107.868	10.50	960.15	2177	9.989	1.6
Sodium	Na	11	$[Ne]3s$	22.9898	0.97	97.8	883	0.317	4.4
Strontium	Sr	38	$[Kr]5s^2$	87.62	2.60	774	1366		24.8

Element	Symbol	At. no.	Electron configuration	At. weight	Density	m.p.	b.p.		
Sulfur	S	16	[Ne]$3s^23p^4$	32.06	2.08 α 1.96 β 1.92 γ	112.8 114.6 106.8	444.60	6.3×10^{-4}	2×10^{23}
Tantalum	Ta	73	[Xe]$4f^{14}5d^36s^2$	180.9479	16.60	2980	5425	0.130	13
Technetium	Tc	43	[Kr]$4d^55s^2$	98.9062	11.487	2200	(4700)		
Tellurium	Te	52	[Kr]$4d^{10}5s^25p^4$	127.60	6.24	450	994	0.014	$(5.8-33) \times 10^3$
Terbium	Tb	65	[Xe]$4f^96s^2$	158.9254	8.272	1356	2800		111
Thallium	Tl	81	[Xe]$4f^{14}5d^{10}6s^26p$	204.37	11.80	304	1470	0.093	18
Thorium	Th	90	[Rn]$6d^27s^2$	232.0381	11.71	1800	~4200		18
Thullium	Tm	69	[Xe]$4f^{13}6s^2$	168.9342	9.332	1545	1730		90
Tin	Sn	50	[Kr]$4d^{10}5s^25p^2$	118.69	7.28 white	231.89	2687	0.153	11.5
Titanium	Ti	22	[Ar]$3d^24s^2$	47.90	4.507 α 4.32 β	1672	3260		3
Tungsten (Wolfram)	W	74	[Xe]$4f^{14}5d^46s^2$	183.85	19.35	3415	5000	0.4	5.48
Uranium	U	92	[Rn]$5f^36d7s^2$	238.029	19.05	1132	3818		
Vanadium	V	23	[Ar]$3d^34s^2$	50.9414	6.1	1919	3400		59
Wolfram (see Tungsten)									
Xenon	Xe	54	[Kr]$4d^{10}5s^25p^6$	131.30	0.0058971^0	-111.8	-108.1	1.21×10^{-5}	29
Ytterbium	Yb	70	[Xe]$4f^{14}6s^2$	173.04	6.977	824	1430		53
Yttrium	Y	39	[Kr]$4d5s^2$	88.9059	4.478	1509	2930		5.9
Zinc	Zn	30	[Ar]$3d^{10}4s^2$	65.37	7.14	419.47	907	0.265	
Zirconium	Zr	40	[Kr]$4d^25s^2$	91.22	6.52^{30}	1855	4375		40

Table 3-2
IONIZATION POTENTIALS OF THE ELEMENTS
In Electron Volts

The minimum amount of energy required to remove the least strongly bound electron from a gaseous atom or ion is called the ionization energy or potential, and is expressed in electron volts (1 eV = 8065.73 cm^{-1}). In Table 3-2 the successive stages of ionization are indicated at the heading of each column: I, denoting first spectra (from neutral atoms)

$$M(\text{gas}) \longrightarrow M^+(\text{gas}) + e^-$$

II, second spectra (from singly ionized atoms), and so on for successive stages of ionization as the outermost electron is removed.

Reference: C. E. Moore, *National Standard Reference Data Series* **34**, U.S. Government Printing Office, Washington, D.C., 1970. This reference also lists ionization limits derived from the analysis of optical spectra and the literature references used for each spectrum.

At. No.	Element	Spectrum						
		I	II	III	IV	V	VI	VII
1	H	13.598						
2	He	24.587	54.416					
3	Li	5.392	75.638	122.451				
4	Be	9.322	18.211	153.893	217.713			
5	B	8.298	25.154	37.930	259.368	340.217		
6	C	11.260	24.383	47.887	64.492	392.077	489.981	
7	N	14.534	29.601	47.448	77.472	97.888	552.057	667.029
8	O	13.618	35.116	54.934	77.412	113.896	138.116	739.315
9	F	17.422	34.970	62.707	87.138	114.240	157.161	185.182
10	Ne	21.564	40.962	63.45	97.11	126.21	157.93	207.27
11	Na	5.139	47.286	71.64	98.91	138.39	172.15	208.47
12	Mg	7.646	15.035	80.143	109.24	141.26	186.50	224.94
13	Al	5.986	18.828	28.447	119.99	153.71	190.47	241.43
14	Si	8.151	16.345	33.492	45.141	166.77	205.05	246.52
15	P	10.486	19.725	30.18	51.37	65.023	220.43	263.22
16	S	10.360	23.33	34.83	47.30	72.68	88.049	280.93
17	Cl	12.967	23.81	39.61	53.46	67.8	97.03	114.193
18	Ar	15.759	27.629	40.74	59.81	75.02	91.007	124.319
19	K	4.341	31.625	45.72	60.91	82.66	100.0	117.56
20	Ca	6.113	11.871	50.908	67.10	84.41	108.78	127.7
21	Sc	6.54	12.80	24.76	73.47	91.66	111.1	138.0
22	Ti	6.82	13.58	27.491	43.266	99.22	119.36	140.8
23	V	6.74	14.65	29.310	46.707	65.23	128.12	150.17
24	Cr	6.766	16.50	30.96	49.1	69.3	90.56	161.1
25	Mn	7.435	15.640	33.667	51.2	72.4	95	119.27
26	Fe	7.870	16.18	30.651	54.8	75.0	99	125
27	Co	7.86	17.06	33.50	51.3	79.5	102	129
28	Ni	7.635	18.168	35.17	54.9	75.5	108	133
29	Cu	7.726	20.292	36.83	55.2	79.9	103	139
30	Zn	9.394	17.964	39.722	59.4	82.6	108	134
31	Ga	5.999	20.51	30.71	64			
32	Ge	7.899	15.934	34.22	45.71	93.5		
33	As	9.81	18.633	28.351	50.13	62.63	127.6	
34	Se	9.752	21.19	30.820	42.944	68.3	81.70	155.4

Table 3-2 (*Continued*)
IONIZATION POTENTIALS OF THE ELEMENTS
In Electron Volts

At. No.	Element	Spectrum						
		I	II	III	IV	V	VI	VII
35	Br	11.814	21.8	36	47.3	59.7	88.6	103.0
36	Kr	13.999	24.359	36.95	52.5	64.7	78.5	111.0
37	Rb	4.177	27.28	40	52.6	71.0	84.4	99.2
38	Sr	5.695	11.030	43.6	57	71.6	90.8	106
39	Y	6.38	12.24	20.52	61.8	77.0	93.0	116
40	Zr	6.84	13.13	22.99	34.34	81.5	(99)	
41	Nb	6.88	14.32	25.04	38.3	50.55	102.6	125
42	Mo	7.099	16.15	27.16	46.4	61.2	68	126.8
43	Tc	7.28	15.26	29.54				
44	Ru	7.37	16.76	28.47				
45	Rh	7.46	18.08	31.06				
46	Pd	8.34	19.43	32.93				
47	Ag	7.576	21.49	34.83				
48	Cd	8.993	16.908	37.48				
49	In	5.786	18.869	28.03	54			
50	Sn	7.344	14.632	30.502	40.734	72.28		
51	Sb	8.641	16.53	25.3	44.2	56	108	
52	Te	9.009	18.6	27.96	37.41	58.75	70.7	137
53	I	10.451	19.131	33				
54	Xe	12.130	21.21	32.1				
55	Cs	3.894	25.1	(35)				
56	Ba	5.212	10.004	(37)				
57	La	5.577	11.06	19.175				
58	Ce	5.47	10.85	20.20	36.72			
59	Pr	5.42	10.55	21.62	38.95	57.45		
60	Nd	5.49	10.72					
61	Pm	5.55	10.90					
62	Sm	5.63	11.07					
63	Eu	5.67	11.25					
64	Gd	6.14	12.1					
65	Tb	5.85	11.52					
66	Dy	5.93	11.67					
67	Ho	6.02	11.80					
68	Er	6.10	11.93					
69	Tm	6.18	12.05	23.71				
70	Yb	6.254	12.17	25.2				
71	Lu	5.426	13.9					
72	Hf	7.0	14.9	23.3	33.3			
73	Ta	7.89	(16)					
74	W	7.98	(18)					
75	Re	7.88	13.1	26.0	37.7			
76	Os	8.7	(17)					
77	Ir	9.1	(17)					
78	Pt	9.0	18.563					
79	Au	9.225	20.5					
80	Hg	10.437	18.756	34.2				
81	Tl	6.108	20.428	29.83				
82	Pb	7.416	15.032	31.937	42.32	68.8		
83	Bi	7.289	16.69	25.56	45.3	56.0	88.3	

Table 3-2 (*Continued*)
IONIZATION POTENTIALS OF THE ELEMENTS
In Electron Volts

At. No.	Element	Spectrum						
		I	II	III	IV	V	VI	VII
84	Po	8.42						
85	At							
86	Rn	10.748						
87	Fr							
88	Ra	5.279	10.147					
89	Ac	6.9	12.1					
90	Th	(6.95)	11.5	20.0	28.8			
91	Pa							
92	U	6.08	(14.7)					
93	Np							
94	Pu	5.8						
95	Am	6.0						

Table 3-3
ATOMIC ELECTRON AFFINITIES

When electrons are added to neutral atoms

$$X + e^- \longrightarrow X^-$$

energy may be released (negative value) or absorbed (positive value), and is called the electron affinity. Values of electron affinities are difficult to determine directly and are usually obtained from the Born-Haber cycle. They are expressed in electron volts.

Ion Produced	Electron Affinity, eV	Ion Produced	Electron Affinity, eV	Ion Produced	Electron Affinity, eV
F^-	−3.57	N^-	−0.05	Cu^-	−2.4
Cl^-	−3.7	P^-	−0.77	Ag^-	−2.5
Br^-	−3.53	As^-	−0.6	Au^-	−2.1
I^-	−3.06	Sb^-	−2.0	Hg^-	−1.54
		Bi^-	−0.7		
O^-	−1.47			Li^-	−0.82
O^{2-}	7.3*	C^-	−1.25	Na^-	−0.84
S^-	−2.07	Si^-	−1.5	K^-	−0.82
S^{2-}	4.0*			H^-	−0.752
Se^-	−1.7	B^-	−0.33	D^-	−0.754
Se^{2-}	4.6*	Al^-	−0.52		
Te^-	−2.2	Tl^-	−2.1	He^-	0.53
				Ne^-	0.57
		Be^-	0.19	Ar^-	1.0
		Mg^-	0.32		

* For addition of two electrons.

Table 3-4
ELECTRONEGATIVITIES OF THE ELEMENTS

According to Pauling, electronegativity χ is the relative attraction of an atom for the valence electrons in a covalent bond. It is proportional to the effective nuclear charge and inversely proportional to the covalent radius:

$$\chi = \frac{0.31(n + 1 \pm c)}{r} + 0.50$$

where n is the number of valence electrons, c is any formal valence charge on the atom and the sign before it corresponds to the sign of this charge, and r is the covalent radius. Because electronegativity is concerned with atoms in molecules rather than atoms in isolation, it is not possible to define precise electronegativity values. Pauling determined his set of values from bond energy data based on experimentally measured heats of dissociation and formation. Originally the element fluorine, whose atoms have the greatest attraction for electrons, was given an arbitrary electronegativity of 4.0. A revision of Pauling's values based on newer heat data assigns 3.90 to fluorine. Values given in Table 3-4 refer to the common oxidation states of the elements.

The greater the difference in electronegativity, the greater is the ionic character of the bond. A percentage of complete separation, or a "degree of ionic character" of the still covalent bond may be estimated from the approximate rule suggested by Gordy [*Discussions Faraday Soc.*, **19**, 23 (1955)]:

$$\text{Degree of ionic character} = \frac{|\chi_A - \chi_B|}{2}$$

for $|\chi_A - \chi_B| < 2$, and 100 per cent for $|\chi_A - \chi_B| > 2$. When $\Delta\chi = 0$ the bond is completely nonpolar. A more sophisticated expression was proposed by Hannay-Smyth [*J. Am. Chem. Soc.*, **68**, 171 (1946)]:

$$\text{Degree of ionic character} = 0.46|\chi_A - \chi_B| + 0.035(\chi_A - \chi_B)^2$$

The bond stretching force constant k (in units of 10^5 dynes cm^{-1}) can be estimated by the expression for stable molecules exhibiting their normal covalencies:

$$k = 1.67N\left(\frac{\chi_A\chi_B}{d^2}\right)^{3/4} + 0.30$$

where N is the bond order (i.e., the effective number of covalent or ionic bonds acting between the two atoms A and B) and d is the internuclear distance in angstroms.

Electronegativity is also proportional to the work function ϕ, which is the energy necessary to just remove an electron from the metal surface in thermoelectric or photoelectric emission:

$$\chi = 0.44\phi - 0.15$$

Element	χ	Element	χ	Element	χ	Element	χ
H	2.20	Na	0.9	Sc	1.3	Cu(I)	1.9
Li	1.0	Mg	1.2	Ti	1.5	Cu(II)	2.0
Be	1.5	Al	1.5	V	1.6	Zn	1.6
B	2.0	Si	1.90	Cr	1.6	Ga	1.6
C	2.60	P	2.15	Mn	1.5	Ge	1.90
N	3.05	S	2.60	Fe(II)	1.8	As	2.00
O	3.50	Cl	3.15	Fe(III)	1.9	Se	2.45
F	4.00*	K	0.8	Co	1.8	Br	2.85
	3.90†	Ca	1.0	Ni	1.8	Rb	0.8

* Arbitrary reference value.   † Recent value.

Table 3-4 (*Continued*)
ELECTRONEGATIVITIES OF THE ELEMENTS

Element	χ	Element	χ	Element	χ	Element	χ
Sr	1.0	Cd	1.7	Hf	1.3	Bi	1.9
Y	1.3	In	1.7	Ta	1.3	Po	2.0
Zr	1.6	Sn(II)	1.8	W	1.7	At	2.2
Nb	1.6	Sn(IV)	1.90	Re	1.9	Fr	0.65
Mo	1.8	Sb	2.05	Os	2.2	Ra	0.9
Tc	1.9	Te	2.30	Ir	2.2	Ac	1.1
Ru	2.2	I	2.65	Au	2.4	Th	1.3
Rh	2.2	Cs	0.7	Hg	1.9	Pa	1.5
Pd	2.2	Ba	0.9	Tl	1.8	U	1.7
Ag	1.9	La–Lu	1.1–1.2	Pb	1.8	Np-No	1.3

Table 3-5
NUCLEAR PROPERTIES OF THE ELEMENTS

In the following table the magnetic moment μ is in multiples of the nuclear magnetron $(eh/4\pi Mc)$, the spin I is in multiples of $h/2\pi$, and the electric quadrupole moment Q is in multiples of $e \times 10^{-24}$ cm^2.

Isotope	NMR Frequency, MHz		NMR Field Values, kG		Magnetic Moment μ	Spin I	Electric Quadrupole Moment Q
	at 14,092 G	at 23,490 G	at 60 MHz	at 90 MHz			
^1n	(29.167)*				−1.91315	−½	
^{1}H	60.000	100.00	14.09	21.06	2.79268	½	
^{2}H	9.2101	15.352	(45.5)†		0.85738	1	2.73×10^{-3}
^{3}H	(45.414)*				2.9788	½	
^{3}He	(32.435)*		18.49	27.75	−2.1274	−½	
^{6}Li	(6.265)*		(47.88)†		0.82192	1	4.6×10^{-4}
^{7}Li	23.317	38.867	36.26	54.39	3.2560	3/2	−0.03
^{9}Be	8.4318	14.055	(50.14)†		−1.1773	−3/2	0.052
^{10}B	6.4477	10.748			1.8005	3	0.074
^{11}B	19.250	32.087	43.92		2.6880	3/2	0.0355
^{13}C	15.086	25.147	56.05		0.70220	½	
^{14}N	4.3341	7.2246			0.40358	1	0.071
^{15}N	6.0796	10.134			−0.28304	−½	
^{17}O	8.134	13.56	(51.97)†		−1.8930	−5/2	−0.026
^{19}F	56.444	94.087	14.98	22.47	2.6273	½	
^{21}Ne	(3.363)*				−0.66176	−3/2	
^{22}Na	(4.434)*				1.746	3	
^{23}Na	15.870	26.454	53.28		2.2161	3/2	0.14–0.15
^{24}Na	(3.22)*				1.69	4	
^{25}Mg	(2.606)*				−0.85471	−5/2	
^{27}Al	15.634	26.060	54.08		3.6385	5/2	0.149
^{29}Si	11.919	19.867	(35.47)†		−0.55477	−½	
^{31}P	24.288	40.485	34.81	52.22	1.1305	½	
^{33}S	4.6016	7.6704			0.64274	3/2	−0.064
^{35}Cl	5.8788	9.7993			0.82091	3/2	−0.0789
^{37}Cl	4.893	8.156			0.68330	3/2	−0.0621
^{39}K	(1.987)*				0.39094	3/2	−0.07

* At 10,000 gauss.   † At 30 MHz.

Table 3-5 (*Continued*)
NUCLEAR PROPERTIES OF THE ELEMENTS

Isotope	NMR Frequency, MHz		NMR Field Values, kG		Magnetic Moment μ	Spin I	Electric Quadrupole Moment Q
	at 14,092 G	at 23,490 G	at 60 MHz	at 90 MHz			
^{41}K	(1.092)*				0.21488	$\frac{3}{2}$	
^{45}Sc	(10.344)*		58.00		4.7492	$\frac{7}{2}$	−0.22
^{47}Ti	(2.400)*				−0.78711	$-\frac{5}{2}$	
^{49}Ti	(2.401)*				−1.1022	$-\frac{7}{2}$	
^{51}V	15.77	26.29	53.60		5.1392	$\frac{7}{2}$	−0.04
^{55}Mn	14.798	24.667	56.85		3.4611	$\frac{5}{2}$	0.55
^{59}Co	14.168	23.617	59.39		4.6388	$\frac{7}{2}$	0.40
^{63}Cu	15.903	26.508	53.17		2.2206	$\frac{3}{2}$	−0.16
^{65}Cu	17.036	28.397	49.63		2.3790	$\frac{3}{2}$	−0.15
^{69}Ga	(10.219)*		58.71		2.0108	$\frac{3}{2}$	0.178
^{71}Ga	(12.984)*		46.21		2.5549	$\frac{3}{2}$	0.112
^{75}As	10.276	17.129	(41.14)†		1.4349	$\frac{3}{2}$	0.3
^{77}Se	11.44	19.07			0.5325	$\frac{1}{2}$	
^{79}Br	15.032	25.057	56.25		2.0991	$\frac{3}{2}$	0.33
^{81}Br	16.203	27.009	52.18		2.2626	$\frac{3}{2}$	0.28
^{83}Kr	(1.64)*				−0.96705	$\frac{9}{2}$	0.15
^{85}Rb	(4.111)*				1.3482	$\frac{5}{2}$	0.28
^{87}Rb	19.632	32.724			2.7414	$\frac{3}{2}$	0.14
^{89}Y	(2.086)*				−0.13682	$-\frac{1}{2}$	
^{91}Zr	(3.958)*				−1.298	$-\frac{5}{2}$	
^{93}Nb	14.666	24.446	57.65		6.1435	$\frac{9}{2}$	−0.2
^{95}Mo	(2.774)*				−0.9099	$-\frac{5}{2}$	
^{101}Ru	(2.1)*				−0.69	$-\frac{5}{2}$	
^{103}Rh	(1.340)*				−0.0879	$-\frac{1}{2}$	
^{105}Pd	(1.74)*				−0.57	$-\frac{5}{2}$	
^{107}Ag	(1.723)*				−0.1130	$-\frac{1}{2}$	
^{109}Ag	(1.981)*				−0.1299	$-\frac{1}{2}$	
^{111}Cd	(9.028)*		(33.23)†		−0.5922	$-\frac{1}{2}$	
^{113}Cd	(9.444)*		(31.77)†		−0.6195	$-\frac{1}{2}$	
^{115}In	(9.329)*		(32.16)†		5.5073	$\frac{9}{2}$	0.761; 1.16
^{117}Sn	21.375	35.630	39.55	59.33	−0.9949	$-\frac{1}{2}$	
^{119}Sn	22.363	37.276	37.81	56.71	−1.0409	$-\frac{1}{2}$	
^{121}Sb	14.358	23.934			3.3417	$\frac{5}{2}$	−0.53; −1.2
^{123}Sb	(5.518)*				2.5334	$\frac{7}{2}$	−0.68; −1.5
^{125}Te	(13.45)*				−0.8824	$-\frac{1}{2}$	
^{127}I	12.004	20.009	(35.21)†		2.7937	$\frac{5}{2}$	−0.69
^{129}Xe	(11.78)*				−0.77255	$-\frac{1}{2}$	
^{131}Xe	(3.490)*				0.68680	$\frac{3}{2}$	−0.12
^{133}Cs	7.8699	13.118	(53.71)†		2.5642	$\frac{7}{2}$	−0.004
^{137}Ba	(4.732)*				0.93107	$\frac{3}{2}$	
^{139}La	(6.014)*				2.7615	$\frac{7}{2}$	0.5
^{141}Pr	(11.95)*				3.92	$\frac{5}{2}$	0.054
^{143}Nd	(2.72)*				−1.25	$-\frac{7}{2}$	−0.57
^{145}Nd	(1.7)*				−0.78	$\frac{7}{2}$	−0.30
^{147}Sm	(1.5)*				−0.68	$\frac{7}{2}$	0.72
^{149}Sm	(1.2)*				−0.55	$\frac{7}{2}$	0.72
^{151}Eu	(10.49)*				3.441	$\frac{5}{2}$	1.2
^{153}Eu	(4.638)*				1.521	$\frac{5}{2}$	2.5
^{155}Gd	(1.2)*				−0.25	$\frac{3}{2}$	1.1
^{157}Gd	(1.7)*				−0.34	$\frac{3}{2}$	1.0

* At 10,000 gauss.   † At 30 MHz.

Table 3-5 (*Continued*)
NUCLEAR PROPERTIES OF THE ELEMENTS

Isotope	NMR Frequency, MHz		NMR Field Values, kG		Magnetic Moment μ	Spin I	Electric Quadrupole Moment Q
	at 14,092 G	at 23,490 G	at 60 MHz	at 90 MHz			
^{159}Tb	(7.72)*				1.52	$\frac{3}{2}$	
^{161}Dy	(1.2)*				−0.38	$\frac{5}{2}$	
^{163}Dy	(1.6)*				−0.53	$\frac{5}{2}$	
^{165}Ho	(7.22)*				3.31	$\frac{7}{2}$	2
^{167}Er	(1.04)*				0.48	$\frac{7}{2}$	(10)
^{169}Tm	(3.05)*				−0.20	$\frac{1}{2}$	
^{171}Yb	(7.51)*				0.4926	$\frac{1}{2}$	
^{173}Yb	(2.1)*				−0.677	$\frac{5}{2}$	3.9
^{175}Lu	(6.3)*				2.9	$\frac{7}{2}$	5.5
^{177}Hf	(1.3)*				0.61	$\frac{7}{2}$	3
^{179}Hf	(0.80)*				−0.47	$\frac{9}{2}$	3
^{181}Ta	(5.09)*				2.340	$\frac{7}{2}$	4.0
^{183}W	(1.75)*				0.115	$\frac{1}{2}$	
^{185}Re	(9.586)*				3.1437	$\frac{5}{2}$	2.8
^{187}Re	(9.684)*				3.1760	$\frac{5}{2}$	2.6
^{189}Os	(3.307)*				0.6507	$\frac{3}{2}$	2.0
^{191}Ir	(0.813)*				0.16	$\frac{3}{2}$	1.5
^{193}Ir	(0.86)*				0.17	$\frac{3}{2}$	1.5
^{195}Pt	12.90	21.50	(32.78)†		0.6004	$\frac{1}{2}$	
^{197}Au					0.1439	$\frac{3}{2}$	0.56
^{199}Hg	10.696	17.829			0.4979	$\frac{1}{2}$	
^{201}Hg	3.9597	6.6005			−0.5513	$−\frac{3}{2}$	0.50
^{203}Tl	34.289	57.156			1.5960	$\frac{1}{2}$	
^{205}Tl	34.624	57.715			1.6115	$\frac{1}{2}$	
^{207}Pb	12.553	20.924	(33.71)†		0.5837	$\frac{1}{2}$	
^{209}Bi	9.6414	16.071			4.0389	$\frac{9}{2}$	−0.4

* At 10,000 gauss.   † At 30 MHz.

Table 3-6
TABLE OF NUCLIDES

Introduction

The data presented here were compiled by J. R. Peterson from the *Table of Isotopes*, 6th ed. (C. M. Lederer, J. M. Hollander, and I. Perlman, John Wiley & Sons, New York, 1967), *Chart of the Nuclides*, 3d ed. (W. Seelmann-Eggebert, G. Pfennig, and H. Münzel, Gersbach und Sohn Verlag, Munich, 1968), the "1964 Atomic Mass Table" [J. H. E. Mattauch, W. Thiele, and A. H. Wapstra, *Nucl. Phys.*, **67**, 1 (1965)], and in the cases of the more recently discovered nuclides, both published and unpublished reports. Not all known nuclides were included, those omitted having usually very short half-lives.

Explanation of Column Headings

Nuclide. Each nuclide is identified by its atomic number Z, equal to the number of protons in the nucleus; the corresponding symbol for that element; and the mass number A, equal to the sum of the numbers of protons Z and neutrons N in the

Table 3-6 (*Continued*)
TABLE OF NUCLIDES

nucleus. Thus, $A = Z + N$, or $N = A - Z$. The m following the mass number (e.g., $^{58m}_{27}$Co) indicates an isomer of that nuclide. Isomers differ only in energy, the nuclide denoted by m (metastable) following its mass number being the higher-energy nuclide.

Mass Excess. The mass excess Δ is the difference between the actual mass M and the mass number A, i.e., $\Delta = M - A$. Thus, the actual mass M is obtained by adding the mass number A and the mass excess Δ. The mass excesses are expressed both in micro mass units (μmu; 1μmu $= 10^{-6}$ mu) and in million electron volts (MeV; 1 MeV $= 10^6$ eV) to facilitate mass and energy calculations. All values are based on the unified, or carbon-12, mass scale; i.e., the mass of one atom of carbon-12 is taken to be exactly 12 mu: thus Δ^{12}C $= 0$. For example, the mass of an atom of $^{34}_{16}$S would be $34 + (-0.032135) = 33.967865$ mu.

Half-life. For the radioactive nuclides this time period corresponds to that during which loss by disintegration of 50% of the nuclide occurs. The units of time are designated by year (y), day (d), hour (h), minute (m), and second (s).

Natural Abundance. The isotopic abundances listed are on an "atom per cent" basis for the stable nuclides present in naturally occurring elements in the earth's crust.

Thermal Neutron Absorption Cross Section.
The ease with which a given nuclide can absorb a thermal neutron (energy $\leq 1/40$ eV) and become a different nuclide is indicated by the cross section, given here in units of barns (1 barn $= 10^{-24}$ cm^2). If the mode of reaction is other than (n,γ), it is so indicated, e.g., (n,p) or (n,α), where $n =$ neutron, $p =$ proton, $\gamma =$ gamma ray, and $\alpha =$ alpha particle ($^4_2$He). In the case of the heaviest elements, neutron absorption leading to fission, i.e., $(n, \text{fission})$, is designated by (f) following the numerical value of the cross section.

Major Radiations. In this column are listed the principal mode(s) of decay and the energies of the emanating radiations in million electron volts (MeV). The gamma-ray (γ) intensities, where given, are given to the nearest whole percentage in parentheses following the numerical energy value for that particular γ. In most cases the radiations listed should be sufficient for identification of the particular nuclide.

The following designations are used: negatron (β^-), positron (β^+), conversion electron (e^-), gamma ray (γ), alpha particle (α), spontaneous fission (SF), neutron (n), and proton (p). More detailed information is available in the first source listed above (*Table of Isotopes*).

Nuclide		Mass Excess		Half-life y=yr, d=day h=hr, m=min s=sec	Natural Abundance, %	Thermal Neutron Absorption Cross Section σ, barns (mode if not n,γ)	Major Radiations, Energies in MeV and γ Intensities, %
Z Symbol	Mass Number	μmu	MeV				
$_0n$	1	8,665	8.0714	11.7 m	. . . .	. . . .	β^-, 0.78
$_1$H	1	7,825	7.2890	. . . .	99.985	0.332	. . . .
	2	14,102	13.1359	. . . .	0.015	0.0005	. . . .
	3	16,050	14.9500	12.26 y	. . . .	. . . .	β^-, 0.0186; no γ
$_2$He	3	16,030	14.9313	. . . .	1.3×10^{-4}	5,330(n,p)	. . . .
	4	2,603	2.4248	. . . .	~100	0	. . . .
	6	18,893	17.598	0.797 s	. . . .	. . . .	β^-, 3.508; no γ
	8	37,520	31.7	0.122 s	. . . .	. . . .	β^-, 9.7; γ, 0.98(88)
$_3$Li	6	15,125	14.088	. . . .	7.42	953(n,α)	. . . .

Table 3-6 (*Continued*)
TABLE OF NUCLIDES

Nuclide Z Symbol	Mass Number	Mass Excess μmu	Mass Excess MeV	Half-life y = yr, d = day h = hr, m = min s = sec	Natural Abundance, %	Thermal Neutron Absorption Cross Section σ, barns (mode if not n,γ)	Major-Radiations, Energies in MeV and γ Intensities, %
	7	16,004	14.907	....	92.58	0.037	....
	8	22,487	20.946	0.841 s	....	....	β−, 13; α, 1.6
	9	26,802	24.97	0.176 s	....	....	β−, 13.61; n, 0.76;
₄Be	6	19,717	18.37	~0.4 s	....	....	....
	7	16,929	15.769	53.6 d	....	54,000(n,p)	γ, 0.477(10)
	9	12,186	11.351	....	100	0.009	....
	10	13,534	12.607	2.5×10⁶ y	....	....	β−, 0.555; no γ
	11	21,666	20.18	13.6 s	....	....	β−, 11.5; γ, 2.14(32), 4.67(2), 5.85(2), 6.79(4), 7.99(2)
₅B	8	24,609	22.923	0.77 s	....	....	β+, 14.0; α, 1.6
	10	12,939	12.052	....	19.7	3,837(n,α)	....
	11	9,305	8.6677	....	80.3	0.005	....
₆C	9		29.0	0.127 s	....	....	β+, 3.5; p, 8.2, 1.1; α, 0.05, 1.6
	10	16,810	15.66	19.48 s	....	....	β+, 1.87; γ, 0.511, 0.717(100), 1.023(2)
	11	11,432	10.648	20.34 m	....	....	β+, 0.97; γ, 0.511
	12	0	0	....	98.892	0.0034	....
	13	3,354	3.125	....	1.108	0.0009	....
	14	3,242	3.0198	5730 y	....	....	β−, 0.156; no γ
	15	10,600	9.873	2.5 s	....	....	β−, 9.82, 4.51; γ, 5.299(68)
	16	14,700	13.69	0.74 s	....	....	....
₇N	13	5,738	5.345	9.96 m	....	....	β+, 1.20; γ, 0.511
	14	3,074	2.8637	....	99.635	1.81(n,p)	....
	15	108	0.100	....	0.365	2.4×10⁻⁵	....
	16	6,103	5.685	7.14 s	....	....	β−, 10.40, 4.27; γ, 2.75(1), 6.13(69), 7.11(5); α, 1.7
	17	8,450	7.87	....			β−, 8.68, 7.81, 4.1; γ, 0.87(3), 2.19(1); n, 0.40, 1.21, 1.81
	18		13.1	0.63 s	....	....	β−, 9.4; γ, 0.82(59), 1.65(59), 1.98(100), 2.47(41)
₈O	14	8,597	8.0080	70.91 s	....	....	β+, 4.12, 1.811; γ, 0.511, 2.312(99)
	15	3,070	2.860	123 s	....	....	β+, 1.74; γ,0.511
	16	−5,085	−4.7366	....	99.759	0.00018	....
	17	−867	−0.808	....	0.037	0.24(n,α)	....
	18	−840	−0.7824	....	0.204	0.00021	....
	19	3,578	3.333	29.1 s	....	....	β−, 4.60; γ, 0.197(97), 1.37(59)

Table 3-6 (*Continued*)
TABLE OF NUCLIDES

Nuclide		Mass Excess		Half-life y=yr, d=day h=hr, m=min s=sec	Natural Abundance, %	Thermal Neutron Absorption Cross Section σ, barns (mode if not n,γ)	Major Radiations, Energies in MeV and γ Intensities, %
Z Symbol	Mass Number	μmu	MeV				
	20	4,079	3.80	14 s	....	....	β^-, 2.75; γ, 1.06(100)
$_9$F	17	2,096	1.952	66.6 s	....	....	β^+, 1.74; γ, 0.511
	18	937	0.872	109.7 m	....	....	β^+, 0.635; γ, 0.511
	19	−1,595	−1.486	....	100	0.010	....
	20	−13	−0.012	11.56 s	....	....	β^-, 5.41; γ, 1.63(100)
	21	−49	−0.05	4.35 s	....	....	β^-, 5.4; γ, 0.350, 1.38
	22		4	4.0 s	....	....	β^-, 11; γ, 1.28(100), 2.06(67)
$_{10}$Ne	17		33.9	0.10 s	....	....	p, 4.59
	18	5,711	5.319	1.5 s	....	....	β^+, 3.42; γ, 0.511, 1.04(7)
	19	1,881	1.752	17.4 s	....	....	β^+, 2.22; γ, 0.511
	20	−7,560	−7.042	....	90.92	....	....
	21	−6,151	−5.730	....	0.257	....	....
	22	−8,615	−8.025	....	8.82	0.04	....
	23	−5,527	−5.148	37.6 s	....	....	β^-, 4.38; γ, 0.439(33), 1.64(1)
	24	−6,387	−5.95	3.38 m	....	....	β^-, 1.99; γ, 0.472(100), 0.88(8)
$_{11}$Na	20	8,880	8.3	0.39 s			β^+, 11.4; γ, 0.511, 1.63; α, 2.14
	21	−2,345	−2.19	23.0 s	....	....	β^+, 2.52; γ, 0.511, 0.350(2)
	22	−5,563	−5.182	2.62 y	....	....	β^+, 1.820, 0.545; γ, 0.511, 1.275(100)
	23	−10,229	−9.528	....	100	0.53	....
	24	−9,038	−8.418	14.96 h	....	....	β^-, 4.17, 1.389; γ, 1.369(100), 2.754(100)
	25	−10,045	−9.36	60 s	....	....	β^-, 3.83; γ, 0.39(14), 0.58(14), 0.98(15), 1.61(6)
	26	−8,260	−7.7	1.04 s	....	....	β^-, 6.7; γ, 1.82(100)
$_{12}$Mg	20		16	0.6 s	....	....	....
	21		10.9	0.121 s	....	....	p, 3.3, 3.8, 4.58, 6.14
	23	−5,875	−5.472	12.1 s	....	....	β^+, 3.03; γ, 0.44(9), 0.511
	24	−14,958	−13.933	....	78.60	0.03	....
	25	−14,161	−13.191	....	10.11	0.3	....
	26	−17,407	−16.214	....	11.29	0.027	
	27	−15,655	−14.583	9.46 m	....	<0.030	β^-, 1.75; γ, 0.18(1), 0.84(70), 1.013(30)

Table 3-6 (*Continued*)
TABLE OF NUCLIDES

| Nuclide | | Mass Excess | | Half-life | Natural | Thermal Neutron | Major Radiations, |
Z Symbol	Mass Number	μmu	MeV	y=yr, d=day h=hr, m=min s=sec	Abundance, %	Absorption Cross Section σ, barns (mode if not n,γ)	Energies in MeV and γ Intensities, %
	28	−16,125	−15.02	21.2 h	....	....	β⁻, 0.46; e⁻, 0.03; γ, 0.031(96), 0.40(30), 0.95(30), 1.35(70)
₁₃Al	24	100	−0.1	2.10 s	....	....	β⁺, 8.5; γ, 0.511, 1.368, 2.754, 4.2, 5.3, 7.1; α, 2
	25	−9,588	−8.93	7.24 s	....	....	β⁺, 3.24; γ, 0.511
	26	−13,109	−12.211	7.4×10⁵ y	....	....	β⁺, 1.17; γ, 0.511, 1.12(4), 1.81(100)
	26 m		−11.982	6.37 s	....	....	β⁺, 3.21; γ, 0.511
	27	−18,461	−17.196	....	100	0.235	....
	28	−18,095	−16.855	2.31 m	....	....	β⁻, 2.85; γ, 1.780(100)
	29	−19,558	−18.22	6.6 m	....	....	β⁻, 2.40; γ, 1.28(94), 2.43(6)
	30	−18,410	−17.2	3.3 s	....	....	β⁻, 5.0; γ, 2.23(61), 3.51(39)
₁₄Si	25		4.0	0.23 s	....	....	p, 3.34, 4.08, 4.68, 5.39
	26	−7,657	−7.13	2.1 s	....	....	β⁺, 3.83; γ, 0.511, 0.82(34)
	27	−13,297	−12.386	4.14 s	....	....	β⁺, 3.85; γ, 0.511
	28	−23,071	−21.490	....	92.18	0.08	....
	29	−23,504	−21.894	....	4.71	0.3	....
	30	−26,237	−24.439	....	3.12	0.11	....
	31	−24,651	−22.96	2.62 h	....	....	β⁻, 1.48; γ, 1.26
	32	−25,980	−24.08	~650 y	....	....	β⁻, 0.21; no γ
₁₅P	28	−8,220	−7.7	0.28 s	....	....	β⁺, 11.0; γ, 0.511, 1.780(75), 2.6, 4.44(10), 4.9, 6.1, 6.7, 7.0, 7.6(5)
	29	−18,192	−16.95	4.45 s	....	....	β⁺, 3.95; γ, 0.511, 1.28(1), 2.43
	30	−21,683	−20.20	2.50 m	....	....	β⁺, 3.24; γ, 0.511, 2.23(1)
	31	−26,235	−24.438	....	100	0.19	....
	32	−26,091	−24.303	14.28 d	....	....	β⁻, 1.710
	33	−28,272	−26.335	24.4 d	....	....	β⁻, 0.248; no γ
	34	−26,660	−24.8	12.4 s	....	....	β⁻, 5.1; γ, 2.13(25), 4.0
₁₆S	29		−2.9	0.19 s	....	....	p, 3.73, 5.40
	30	−15,127	−14.09	1.4 s	....	....	β⁺, 5.09, 4.42; γ, 0.511, 0.687(80)
	31	−20,389	−18.99	2.72 s	....	....	β⁺, 4.42; γ, 0.511, 1.27(1)
	32	−27,926	−26.013	....	95.0	....	....
	33	−28,538	−26.583	....	0.76	....	....
	34	−32,135	−29.934	....	4.22	0.27	....

Table 3-6 (*Continued*)
TABLE OF NUCLIDES

Nuclide		Mass Excess		Half-life y=yr, d=day h=hr, m=min s=sec	Natural Abundance, %	Thermal Neutron Absorption Cross Section σ, barns (mode if not n,γ)	Major Radiations, Energies in MeV and γ Intensities, %
Z Symbol	Mass Number	μmu	MeV				
	35	−30,969	−28.847	87.9 d	....	....	β−, 0.167; no γ
	36	−32,910	−30.655	....	0.014	0.14	....
	37	−28,990	−27.0	5.07 m	....	....	β−, 4.7, 1.6; γ, 3.09(90)
	38	−28,770	−26.8	2.87 h	....	....	β−, 3.0, 1.1; γ, 1.88(95)
$_{17}$Cl	32	−13,760	−12.8	0.306 s	....	....	β+, 9.9; γ, 0.511, 2.24(70), 4.29(7), 4.77(14)
	33	−22,560	−21.01	2.53 s	....	....	β+, 4.55; γ, 0.511, 2.9
	34	−26,250	−24.45	1.56 s	....	....	β+, 4.46; γ, 0.511
	34 *m*		−24.31	31.99 m	....	....	β+, 2.48; e−, 0.142; γ, 0.145(45), 0.511, 1.17(12), 2.12(38), 3.30(12)
	35	−31,149	−29.015	....	75.53	44	....
	36	−31,691	−29.520	3.08×10^5 y	....	100	β−, 0.714; γ, 0.511
	37	−34,102	−31.765	....	24.47	0.4	....
	38	−31,995	−29.80	37.29 m	....	....	β−, 4.91; γ, 1.60(38), 2.170(47)
	38 *m*		−29.13	0.74 s	....	....	γ, 0.66(100); e−, 0.66
	39	−31,992	−29.80	55.5 m	....	....	β−, 3.45, 2.18, 1.91; γ, 0.246(44), 1.27(50), 1.52(42)
	40	−29,600	−27.5	1.4 m	....	....	β−, 7.5; γ, 1.46, 2.83, 3.10, 5.8
$_{18}$Ar	33		−9.5	0.18 s	....	....	p, 3.16
	35	−24,746	−23.05	1.83 s	....	....	β+, 4.94; γ, 0.511, 1.22(5), 1.76(2)
	36	−32,456	−30.232	....	0.337	6	....
	37	−33,228	−30.951	35.1 d	....	....	Cl X-rays
	38	−37,272	−34.718	....	0.063	0.8	....
	39	−35,683	−33.24	269 y	....	....	β−, 0.565; no γ
	40	−37,616	−35.038	....	99.600	0.61	....
	41	−35,500	−33.061	1.83 h	....	0.5	β−, 2.49, 1.198; γ, 1.293(99)
	42	−36,952	−34.42	33 y	....	....	....
$_{19}$K	37	−26,635	−24.79	1.23 s	....	....	β+, 5.14; γ, 0.511, 2.79(2)
	38	−30,903	−28.79	7.71 m	....	....	β+, 2.68; γ, 0.511, 2.170(100)
	38 *m*		−28.66	0.95 s	....	....	β+, 5.0; γ, 0.511
	39	−36,290	−33.803	....	93.22	2.0	....
	40	−36,000	−33.533	1.26×10^9 y	0.118	70	β−, 1.314; β+, 0.483; γ, 1.460(11)

Table 3-6 (*Continued*)
TABLE OF NUCLIDES

Nuclide		Mass Excess		Half-life y = yr, d = day h = hr, m = min s = sec	Natural Abundance, %	Thermal Neutron Absorption Cross Section σ, barns (mode if not n,γ)	Major Radiations, Energies in MeV and γ Intensities, %
Z Symbol	Mass Number	μmu	MeV				
	41	−38,168	−35.552	. . . .	6.77	1.2	. . . .
	42	−37,594	−35.02	12.36 h	. . . .	. . . .	β⁻, 3.52; γ, 0.31, 1.524(18)
	43	−39,270	−36.58	22.4 h	. . . .	. . . .	β⁻, 1.82, 1.2, 0.83; γ, 0.220(3), 0.373(85), 0.39(18), 0.59(13), 0.619(81), 1.01(2)
	44	−37,960	−36.3	22.0 m			β⁻, 5.2; γ, 1.156(61), 1.74(8), 2.1(37), 2.6(7), 3.7(4)
	45	−39,320	−36.6	16.3 m	. . . .	. . . .	β⁻, 4.0, 2.1; γ, 0.175, 0.50, 0.95, 1.23, 1.71, 1.90, 2.10, 2.35, 2.60, 3.1
	47	−38, 910	−36.3	17.5 s	. . . .	. . . .	β⁻, 6.1, 4.1; γ, 2.0(84), 2.6(15)
₂₀Ca	37		−13.3	0.173 s	. . . .	. . . .	p, 3.10
	38	−23,280	−22	0.66 s	. . . .	. . . .	γ, 0.511. 3.5
	39	−29,309	−27.30	0.87 s	. . . .	. . . .	β⁺, 5.49; γ, 0.511
	40	−37,411	−34.848	. . . .	96.97	0.23	. . . .
	41	−37,725	−35.125	8×10⁴ y	. . . .	. . . .	K X-rays
	42	−41,375	−38.540	. . . .	0.64	42	. . . .
	43	−41,220	−38.396	. . . .	0.145	. . . .	. . . .
	44	−44,510	−41.460	. . . .	2.06	0.7	. . . .
	45	−43,811	−40.809	165 d	. . . .	. . . .	β⁻, 0.252
	46	−46,311	−43.14	. . . .	0.0033	0.3	. . . .
	47	−45, 462	−42.35	4.535 d	. . . .	. . . .	β⁻, 1.98, 0.67; γ, 0.49(5), 0.815(5), 1.308(74)
	48	−47,469	−44.22	>10¹⁸ y	0.185	1.1	. . . .
	49	−44,325	−41.29	8.8 m	. . . .	. . . .	β⁻, 1.95; γ, 3.10(89), 4.1(10)
	50		−41	9 s	. . . .	. . . .	γ, 0.072, 0.258
₂₁Sc	40	−22,430	−20.3	0.179 s	. . . .	. . . .	β⁺, 9.1; γ, 0.511, 3.75(100)
	41	−30,753	−28.63	0.60 s	. . . .	. . . .	β⁺, 5.47; γ, 0.511
	42	−34,505	−32.109	0.683 s	. . . .	. . . .	β⁺, 5.41; γ, 0.511
	42 m		−31.58	60.6 s	. . . .	. . . .	β⁺, 2.82; γ, 0.438(100), 0.511, 1.22(100), 1.52(100)
	43	−38,835	−36.17	3.92 h	. . . .	. . . .	β⁺, 1.20; γ, 0.375(22), 0.511
	44	−40,594	−37.81	3.92 h	. . . .	. . . .	β⁺, 1.47; γ, 0.511, 1.159(100)

Table 3-6 (*Continued*)
TABLE OF NUCLIDES

Z Symbol	Mass Number	Mass Excess μmu	Mass Excess MeV	Half-life y=yr, d=day h=hr, m=min s=sec	Natural Abundance, %	Thermal Neutron Absorption Cross Section σ, barns (mode if not n,γ)	Major Radiations, Energies in MeV and γ Intensities, %
	44 m		−37.54	2.44 d	....	....	γ, 0.271(86), 1.02(1), 1.14(3); e⁻, 0.267
	45	−44,081	−41.061	....	100	13	....
	46	−44,827	−41.756	83.9 d	....	....	β⁻, 1.48, 0.357; γ, 0.889(100), 1.120(100)
	46 m		−41.614	19.5 s	....	....	γ, 0.142; e⁻
	47	−47,587	−44.326	3.43 d	....	....	β⁻, 0.600; γ, 0.160(73)
	48	−47,779	−44.51	1.83 d	....	....	β⁻, 0.65; γ, 0.175(6), 0.983(100), 1.040(100), 1.314(100)
	49	−49,974	−46.55	57.5 m	....	....	β⁻, 2.01; γ, 1.76
	50	−48,270	−45.0	1.72 m	....	....	β⁻, 3.6; γ, 0.520(100), 1.12(100), 1.55(100)
	50 m		−44.7	0.35 s	....	....	γ, 0.258
$_{22}$Ti	43	−31,500	−29.3	0.56 s	....	....	β⁺, 5.8; γ, 0.511
	44	−40,428	−37.66	48 y	....	....	γ, 0.068(90), 0.078(98); e⁻, 0.065, 0.073
	45	−41,871	−39.002	3.09 h	....	....	β⁺, 1.04; γ, 0.718, 1.408
	46	−47,368	−44.123	....	7.99	0.6	....
	47	−48,232	−44.927	....	7.32	1.7	....
	48	−52,050	−48.483	....	73.99	8.0	....
	49	−52,130	−48.558	....	5.46	1.9	....
	50	−55,214	−51.431	....	5.25	0.14	....
	51	−53,397	−49.74	5.79 m	....	....	β⁻, 2.14; γ, 0.320(95), 0.605(2), 0.928(5)
$_{23}$V	46	−39,786	−37.069	0.426 s	....	....	β⁺, 6.03; γ, 0.511
	47	−45,101	−42.01	33 m	....	....	β⁺, 1.89; γ, 0.511, 1.80(1)
	48	−47,741	−44.470	16.0 d	....	....	β⁺, 0.696; γ, 0.511, 0.945(10), 0.983(100), 1.312(97), 2.241(3)
	49	−51,478	−47.950	330 d	....	....	Ti X-rays
	50	−52,836	−49.216	6×10¹⁵ y	0.25	130	γ, 0.783(30), 1.55(70)
	51	−56,039	−52.199	....	99.75	4.9	....
	52	−55,220	−51.44	3.75 m	....	....	β⁻, 2.47; γ, 1.434(100)

Table 3-6 (*Continued*)
TABLE OF NUCLIDES

Nuclide Z Sym-bol	Nuclide Mass Number	Mass Excess μmu	Mass Excess MeV	Half-life y=yr, d=day h=hr, m=min s=sec	Natural Abundance, %	Thermal Neutron Absorption Cross Section σ, barns (mode if not n,γ)	Major Radiations, Energies in MeV and γ Intensities, %
	53	−56,020	−51.8	2.0 m	....	....	β^-, 2.50; γ, 1.00(100)
	54	−53,280	−50	55 s	....	....	β^-, 3.3; γ, 0.84(100), 0.99(100), 2.21(100)
$_{24}$Cr	46			1.1 s	....	....	....
	48	−46,240	−43.1	23 h	....	....	γ, 0.116(98), 0.31(99); e^-, 0.111, 0.31
	49	−48,729	−45.39	41.9 m	....	....	β^+, 1.54; e^-, 0.058, 0.084, 0.148; γ, 0.063(14), 0.091(28), 0.153(13), 0.511
	50	−53,946	−50.249	....	4.31	17	....
	51	−55,232	−51.447	27.8 d	....	....	γ, 0.320(9); e^-, 0.315
	52	−59,487	−55.411	....	83.76	0.8	....
	53	−59,347	−55.281	....	9.55	18	....
	54	−61,119	−56.931	....	2.38	0.38	....
	55	−59,167	−55.11	3.52 m	....	....	β^-, 2.59
	56	−59,360	−55.3	5.9 m	....	....	β^-, 1.5; e^-, 0.020, 0.077; γ, 0.026, 0.083
$_{25}$Mn	50	−45,785	−42.648	0.286 s	....	....	β^+, 6.61; γ, 0.511
	50			2 m	....	....	γ, 0.511, 0.66(25), 0.783(100), 1.11(100), 1.28(25), 1.45(75)
	51	−51,810	−48.26	45.2 m	....	....	β^+, 2.17; γ, 0.511, 1.56, 2.03
	52	−54,432	−50.70	5.60 d	....	....	β^+, 0.575; γ, 0.511, 0.744(82), 0.935(84), 1.434(100)
	52 m		−50.32	21.1 m	....	....	β^+, 1.63; γ, 0.383(2), 0.511, 1.434(100)
	53	−58,705	−54.683	1.9×10⁶ y	....	170	Cr X-rays
	54	−59,638	−55.55	303 d	....	....	γ, 0.835(100); e^-, 0.829
	55	−61,950	−57.705	....	100	13.3	....
	56	−61,090	−56.904	2.576 h	....	....	β^-, 2.85; γ, 0.847(99), 1.811(29), 2.110(15)
	57	−61,700	−57.5	1.7 m	....	....	β^-, 2.55; γ, 0.122,

Table 3-6 (*Continued*)
TABLE OF NUCLIDES

| Nuclide | | Mass Excess | | Half-life | Natural | Thermal Neutron | Major Radiations, |
Z Symbol	Mass Number	μmu	MeV	y = yr, d = day h = hr, m = min s = sec	Abundance, %	Absorption Cross Section σ, barns (mode if not n,γ)	Energies in MeV and γ Intensities, %
							0.136, 0.22, 0.353, 0.692
	57			7 d	....	....	....
	58	−59,740	−56	1.1 m	....	....	γ, 0.36, 0.41, 0.52, 0.57, 0.82, 1.0, 1.25, 1.4, 1.6, 2.2, 2.8
$_{26}$Fe	52	−51,883	−48.33	8.2 h	....		β^+, 0.80; γ, 0.165(100), 0.511
	53	−54,428	−50.70	8.51 m	....		β^+, 3.0; γ, 0.38(32), 0.511
	54	−60,383	−56.246	....	5.84	2.9	....
	55	−61,701	−57.474	2.60 y	....	....	Mn X-rays
	56	−65,064	−60.605	....	91.68	2.7	....
	57	−64,602	−60.176	....	2.17	2.5	....
	58	−66,718	−62.147	....	0.31	1.1	....
	59	−65,122	−60.660	45.6 d	....	....	β^-, 1.57, 0.475; γ, 0.143(1), 0.192(3), 1.095(56), 1.292(44)
	60	−66,036	−61.51	3×10^5 y	....	....	....
	61	−63,480	−59	6.0 m	....	....	β^-, 2.8; γ, 0.13(11), 0.30(48), 1.03(98), 1.20(100)
$_{27}$Co	54	−51,525	−47.99	0.194 s	....	....	β^+, 7.23; γ, 0.511
	54			1.5 m	....	....	β^+, 4.3; γ, 0.41(100), 0.511, 1.14(100), 1.41(100)
	55	−57,987	−54.01	18.2 h	....	....	β^+, 1.50; γ, 0.480(12), 0.511, 0.930(80), 1.41(13)
	56	−60,153	−56.03	77.3 d	....	....	β^+, 1.49; γ, 0.511, 0.847(100), 1.04(15), 1.24(66), 1.76(15), 2.02(11), 2.60(17), 3.26(13)
	57	−63,704	−59.339	270 d	....	....	γ, 0.014(9), 0.122(87), 0.136(11), 0.692; e^-, 0.007, 0.013, 0.115, 0.129
	58	−64,239	−59.84	71.3 d	....	2,500	β^+, 0.474; γ, 0.511, 0.810(99),

Table 3-6 (*Continued*)
TABLE OF NUCLIDES

Nuclide		Mass Excess		Half-life y=yr, d=day h=hr, m=min s=sec	Natural Abundance, %	Thermal Neutron Absorption Cross Section σ, barns (mode if not n,γ)	Major Radiations, Energies in MeV and γ Intensities, %
Z Symbol	Mass Number	μmu	MeV				
	58 *m*		−59.81	9.2 h	....	1.4×10⁵	0.865(1), 1.67(1) e⁻, 0.017, 0.024
	59	−66,811	−62.233	....	100	19	....
	60	−66,187	−61.651	5.263 y	....	6	β⁻, 1.48, 0.314; γ, 1.173(100), 1.332(100)
	60 *m*		−61.593	10.47 m	....	100	β⁻, 1.55; e⁻, 0.051, 0.058; γ, 0.059(2), 1.33
	61	−67,560	−62.93	99.0 m	....	....	β⁻, 1.22; e⁻, 0.059; γ, 0.067(89)
	62	−66,054	−61.53	13.9 m	....	....	β⁻, 2.88; γ, 1.17(180), 1.47(20), 1.74(19), 2.03(7)
	62			1.5 m	....	....	....
	63	−66,470	−61.9	52 s	....	....	β⁻, 3.6; no γ
	63			1.40 h	....	....	....
	64			28 s	....	....	γ, 0.095
	64			7.8 m	....	....	....
	64			2.0 m	....	....	....
₂₈Ni	56	−57,884	−53.92	6.10 d	....	....	γ, 0.163(99), 0.276(31), 0.472(35), 0.748(48), 0.812(85), 1.56(14); e⁻, 0.155
	57	−60,231	−56.10	36.0 h	....	....	β⁺, 0.85; γ, 0.127(14), 0.511, 1.37(86), 1.89(14)
	58	−64,658	−60.23	....	67.76	4.4	....
	59	−65,658	−61.159	8×10⁴ y	....	....	Co X-rays
	60	−69,213	−64.471	....	26.16	2.6	....
	61	−68,944	−64.22	....	1.25	2	....
	62	−71,658	−66.75	....	3.66	15	....
	63	−70,336	−65.52	92 y	....	....	β⁻, 0.067; no γ
	64	−72,042	−67.11	....	1.16	1.5	....
	65	−69,928	−65.14	2.564 h	....	....	β⁻, 2.13; γ, 0.368(5), 1.115(16), 1.481(25)
	66	−70,915	−66.06	54.8 h	....	....	β⁻, 0.20; no γ
	67		−63.2	50 s	....	....	β⁻, 4.1; γ, 0.90(51), 1.26(15)
₂₉Cu	58	−55,459	−51.66	3.20 s	....	....	β⁺, 8.2; γ, 0.511
	58			9.5 m	....	....	....
	59	−60,504	−56.36	81.5 s	....	....	β⁺, 3.7; γ, 0.343(5), 0.463(5), 0.511,

Table 3-6 (*Continued*)
TABLE OF NUCLIDES

Nuclide		Mass Excess		Half-life	Natural	Thermal Neutron	Major Radiations,
Z Symbol	Mass Number	μmu	MeV	y = yr, d = day h = hr, m = min s = sec	Abundance, %	Absorption Cross Section σ, barns (mode if not *n,γ*)	Energies in MeV and γ Intensities, %
	60	−62,638	−58.35	23.4 m	....	....	0.872(9), 1.305(11), 1.70(1) β^+, 3.92, 3.00, 2.00; γ, 0.511, 0.85(15), 1.332(80), 1.76(52), 2.13(6), 2.64(5), 3.13(4), 2.52(2), 4.0(1)
	61	−66,543	−61.98	3.32 h	....	....	β^+, 1.22; e^-, 0.059; γ, 0.067(4), 0.284(12), 0.38(3), 0.511, 1.19(5)
	62	−67,434	−62.81	9.76 m	....	....	β^+, 2.91; γ, 0.511, 0.88, 1.17(1)
	63	−70,408	−65.583	....	69.1	4.5	....
	64	−70,241	−65.428	12.80 h	....	....	β^-, 0.573; β^+, 0.656; e^-, 1.33; γ, 0.511, 1.34(1)
	65	−72,214	−67.27	....	30.9	2.3	....
	66	−71,129	−66.26	5.10 m	....	130	β^-, 2.63; γ, 1.039(9)
	67	−72,241	−67.29	58.5 h	....	....	β^-. 0.57; e^-, 0.082, 0.091; γ, 0.092(23), 0.184(40)
	68	−70,230	−65.4	30 s	....	....	β^-, 3.5; γ, 0.80(17), 1.078(95), 1.24(3), 1.88(5)
$_{30}$Zn	60			2.1 m	....	....	....
	61	−60,750	−56.6	1.48 m	....	....	β^+, 4.4; γ, 0.48(11), 0.511, 0.98(3), 1.64(6)
	62	−65,620	−61.12	9.13 h	....	....	β^+, 0.66; e^-, 0.033; γ, 0.042(20), 0.51(47), 0.59(22)
	63	−66,794	−62.22	38.4 m	....	....	β^+, 2.34; γ, 0.511, 0.669(8), 0.962(6), 1.42(1)
	64	−70,855	−66.000	$>8 \times 10^{15}$ y	48.89	0.46	....
	65	−70,766	−65.92	245 d	....	....	β^+, 0.327; e^-, 1.106; γ, 0.511, 1.115(49)
	66	−73,948	−68.88	....	27.81	....	....
	67	−72,855	−67.86	....	4.11	....	....
	68	−75,143	−69.99	....	18.56	1.0	....
	69	−73,459	−68.43	57 m	....	....	β^-, 0.90; no γ
	69 *m*		−67.99	13.8 h	....	....	γ, 0.439(95); e^-, 0.429
	70	−74,666	−69.55	$>10^{15}$ y	0.62	0.10	....

Table 3-6 (*Continued*)
TABLE OF NUCLIDES

Nuclide		Mass Excess		Half-life y = yr, d = day h = hr, m = min s = sec	Natural Abundance, %	Thermal Neutron Absorption Cross Section σ, barns (mode if not n,γ)	Major Radiations, Energies in MeV and γ Intensities, %
Z Sym-bol	Mass Number	μmu	MeV				
	71	−72,490	−67.5	2.4 m	....	....	β^-, 2.61; γ, 0.120(1), 0.39(1), 0.510(13), 0.92(3), 1.12(1)
	71 *m*		−67.2	3.92 h	....	....	β^-, 1.46; γ, 0.13(9), 0.385(94), 0.495(75), 0.609(65), 0.76(5), 0.99(8), 1.11(4)
	72	−73,157	−68.14	46.5 h	....	....	β^-, 0.30; e^-, 0.005, 0.014; γ, 0.015(8), 0.046, 0.145(90), 0.192(10)
$_{31}$Ga	63	−60,890	−57	33 s	....	....	....
	64	−63,263	−58.93	2.6 m	....	....	β^+, 6.05, 2.8; γ, 0.511, 0.80(15), 0.992(43), 1.25(7), 1.38(14), 1.56(7), 1.78(5), 2.18(11), 2.34(9), 3.32(18)
	65	−67,267	−62.66	15.2 m	....	....	β^+, 2.24, 2.11; e^-, 0.044, 0.053, 0.105; γ, 0.054(8), 0.061(12), 0.115(55), 0.152(10), 0.206(4), 0.511, 0.75(10), 0.93(3)
	65			8.0 m	....	....	....
	66	−68,393	−63.71	9.45 h	....	....	β^+, 4.153; γ, 0.511, 0.828(5), 1.039(37), 1.91(3), 2.183(5), 2.748(25), 4.30(5)
	67	−71,784	−66.87	77.9 h	....	....	γ, 0.093(40), 0.184(24), 0.296(22), 0.388(7); e^-, 0.084, 0.092
	68	−72,008	−67.07	68.3 m	....	....	β^+, 1.90; γ, 0.511, 0.80, 1.078(4), 1.24, 1.87
	69	−74,426	−69.326	....	60.2	1.9	....
	70	−73,965	−68.90	21.1 m	....	....	β^-, 1.65; γ, 0.173, 1.040(1)
	71	−75,294	−70.135	....	39.8	5.0	....
	72	−73,628	−68.58	14.12 h	....	....	β^-, 3.15; γ, 0.601(8),

Table 3-6 (*Continued*)
TABLE OF NUCLIDES

Nuclide Z Symbol	Nuclide Mass Number	Mass Excess μmu	Mass Excess MeV	Half-life y=yr, d=day h=hr, m=min s=sec	Natural Abundance, %	Thermal Neutron Absorption Cross Section σ, barns (mode if not n,γ)	Major Radiations, Energies in MeV and γ Intensities, %
	73	−74,874	−69.74	4.9 h	....	....	0.630(27), 0.835(96), 0.894(10), 1.050(7), 1.465(4), 1.60(5), 1.860(5), 2.201(26), 2.50(20) β⁻, 1.19; e⁻, 0.012, 0.043, 0.053; γ, 0.054(9), 0.295(94), 0.74(6)
	74	−72,810	−67.8	8.0 m	....	....	β⁻, 2.5; γ, 0.50(11), 0.60(100), 0.87(9), 1.11(5), 1.20(8), 1.33(5), 1.46(8), 1.76(7), 2.35(45)
	75		−68.5	2.0 m	....	....	β⁻, 3.3; γ, 0.58(3)
	76			32 s	....	....	β⁻, 6; γ, 0.563, 0.96, 1.12
$_{32}$Ge	65	−60,400	−56	1.5 m	....	....	β⁺, 3.7; γ, 0.511, 0.67(3), 1.72(2)
	66	−65,200	−60.7	2.4 h	....	....	β⁺, 2.0, 1.3; γ, 0.046(37), 0.068(11), 0.114(22), 0.185(23), 0.245(7), 0.27(19), 0.30(6), 0.34(19), 0.38(48), 0.40(6), 0.47(19), 0.511
	67	−67,060	−62.5	18.7 m	....	....	β⁺, 3.1; γ, 0.170(105), 0.511, 0.84(4), 0.92(7), 1.48(5)
	68	−71,470	−67	275 d	....	....	Ga X-rays
	69	−72,037	−67.101	36 h	....	....	β⁺, 1.22; γ, 0.511, 0.573(13), 0.872(10), 1.107(28), 1.335(3)
	70	−75,749	−70.558	....	20.55	3.2	....
	71	−75,044	−69.90	11.4 d	....	....	Ga X-rays
	72	−77,918	−72.579	....	27.37	1.0	....
	73	−76,538	−71.293	....	7.67	14	....
	73 *m*		−71.226	0.53 s	....	....	γ, 0.054(9); e⁻, 0.012, 0.043,

Table 3-6 (*Continued*)
TABLE OF NUCLIDES

Nuclide		Mass Excess		Half-life y=yr, d=day h=hr, m=min s=sec	Natural Abundance, %	Thermal Neutron Absorption Cross Section σ, barns (mode if not n,γ)	Major Radiations, Energies in MeV and γ Intensities, %
Z Symbol	Mass Number	μmu	MeV				
	74	−78,819	−73.419	....	36.74	0.3	0.053 ....
	75	−77,117	−71.83	82 m	....	....	β^-, 1.19; γ, 0.066, 0.199(1), 0.265(11), 0.427, 0.477, 0.628
	75 m		−71.69	48 s	....	....	γ, 0.139(34); e^-, 0.128, 0.138
	76	−78,595	−73.209	>2×10¹⁶ y	7.67	0.1	....
	77	−76,400	−71.2	11.3 h	....	....	β^-, 2.2; e^-, 0.198, 0.253; γ, 0.21(61), 0.263(45), 0.368(15), 0.417(25), 0.563(18), 0.632(11), 0.73(14), 0.80(6), 0.93(5), 1.09(6)
	77 m		−71.0	54 s	....	....	β^-, 2.9; e^-, 0.148, 0.158; γ, 0.159(12), 0.215(21)
	78		−71.8	1.47 h	....	....	β^-, 0.71; γ, 0.277(94)
$_{33}$As	68			~7 m	....	....	....
	69	−67,850	−63.2	15 m	....	....	β^+, 2.9; γ, 0.23, 0.511
	70	−69,054	−64.32	52 m	....	....	β^+, 2.89, 2.14; γ, 0.511, 0.60(23), 0.67(25), 0.75(23), 0.91(17), 1.040(78), 1.12(23), 1.36(12), 1.42(10), 1.54(7), 1.71(22), 1.80(6), 2.03(19)
	71	−72,887	−67.89	62 h	....	....	β^+, 0.81; e^-, 0.012, 0.022, 0.164; γ, 0.175(90), 0.511
	72	−73,237	−68.22	26 h	....	....	β^+, 3.34, 2.50; e^-, 0.679; γ, 0.511, 0.630(8), 0.835(78)
	73	−76,139	−70.92	80.3 d	....	....	γ, 0.054(9); e^-, 0.012, 0.043, 0.053
	74	−76,067	−70.855	17.9 d	....	....	β^-, 1.36; β^+, 1.54,

Table 3-6 (*Continued*)
TABLE OF NUCLIDES

| Nuclide | | Mass Excess | | Half-life | Natural | Thermal Neutron | Major Radiations, |
Z Symbol	Mass Number	μmu	MeV	y = yr, d = day h = hr, m = min s = sec	Abundance, %	Absorption Cross Section σ, barns (mode if not n,γ)	Energies in MeV and γ Intensities, %
							0.95; γ, 0.511, 0.596(61), 0.635(14)
74 m			−70.572	8.0 s	....	....	γ, 0.283
75		−78,404	−73.031	....	100	4.5	....
76		−77,603	−72.29	26.4 h	....	....	β^-, 2.97; γ, 0.559(43), 0.657(6), 1.22(5), 1.44(1), 1.789, 2.10(1)
77		−79,354	−73.92	38.7 h	....	....	β^-, 0.68; γ, 0.086, 0.239(3), 0.522(1)
78		−78,100	−72.8	91 m	....	....	β^-, 4.1; γ, 0.614(42), 0.70(15), 0.83(8), 1.31(11)
78 m				6 m	....	....	....
79		−79,110	−73.7	9.0 m	....	....	β^-, 2.15; γ, 0.36(2), 0.43(2), 0.54(1), 0.73(1), 0.89(1)
80		−77,030	−71.8	15.3 s	....	....	β^-, 6.0; γ, 0.666(42), 0.8(1), 1.22(4), 1.64(4), 1.77(2)
81			−72.6	33 s	....	....	β^-, 3.8; no γ
85				0.43 s	....	....	....
$_{34}$Se	70			~44 m	....	....	....
	71	−68,160	−63.5	4.5 m	....	....	β^+, 3.4; γ, 0.16, 0.511
	72	−72,590	−68	8.4 d	....	....	γ, 0.046(59); e^-, 0.034, 0.044
	73	−73,186	−68.17	7.1 h	....	....	β^+, 1.30; e^-, 0.054, 0.064, 0.347; γ, 0.066(65), 0.359(99), 0.511
	73		−68.2	42 m	....	....	β^+, 1.7
	74	−77,524	−72.212	....	0.87	30	....
	75	−77,475	−72.166	120.4 d	....	....	γ, 0.066(1), 0.097(3), 0.121(17), 0.136(57), 0.265(60), 0.280(25), 0.401(12); e^-, 0.085, 0.095, 0.109, 0.124, 0.253

Table 3-6 (*Continued*)
TABLE OF NUCLIDES

Nuclide Z Sym-bol	Nuclide Mass Number	Mass Excess μmu	Mass Excess MeV	Half-life y=yr, d=day h=hr, m=min s=sec	Natural Abundance, %	Thermal Neutron Absorption Cross Section σ, barns (mode if not n,γ)	Major Radiations, Energies in MeV and γ Intensities, %
	76	−80,793	−75.26	....	9.02	63	....
	77	−80,089	−74.60	....	7.58	42	....
	77 m		−74.44	17.5 s	....	....	γ, 0.161(50); e⁻, 0.148, 0.160
	78	−82,686	−77.021	....	23.52	0.36	....
	79	−81,506	−75.921	≤6.5×10⁴ y	....	....	β⁻, 0.16; no γ
	79 m		−75.825	3.91 m	....	....	γ, 0.096(9); e⁻, 0.083, 0.095
	80	−83,473	−77.753	....	49.82	0.5	....
	81	−82,016	−76.40	18.6 m	....	....	β⁻, 1.58; γ, 0.030, 0.28(1), 0.56, 0.83
	81 m		−76.29	56.8 m	....	....	γ, 0.103(8); e⁻, 0.090, 0.102
	82	−83,293	−77.59	>10¹⁷ y	9.19	0.05	....
	83		−75.4	25 m	....	....	β⁻, 1.8; γ, 0.22(44), 0.36(69), 1.88(16), 2.29(9)
	83 m		−75.2	70 s	....	....	β⁻, 3.8; γ, 0.35(16), 0.65(20), 1.01(100), 2.02(40)
	84			3.3 m	....	....	....
	85			39 s	....	....	....
	87			16 s	....	....	....
₃₅Br	74	−70,220	−65	36 m	....	....	β⁺, 4.7; γ, 0.511
	75	−74,553	−69.44	1.7 h	....	....	β⁺, 1.70; γ, 0.285, 0.511, 0.62
	76	−75,820	−70.6	16.1 h	....	....	β⁺, 3.6; γ, 0.511, 0.559(63), 0.65(19), 0.75(6), 0.85(7), 1.21(13), 1.37(5), 1.47(7), 1.86(11), 2.10(7), 2.39(4), 2.78(5), 2.97(8), 3.57(2)
	77	−78,624	−73.24	57 h	....	....	β⁺, 0.34; e⁻, 0.229, 0.287, 0.508; γ, 0.24(30), 0.300(6), 0.52(24), 0.58(7), 0.75(2), 0.82(3), 1.00(1)
	77 m		−73.13	4.2 m	....	....	γ, 0.108; e⁻, 0.094, 0.106
	78	−78,850	−73.45	6.5 m	....	....	β⁺, 2.55; γ, 0.511,

Table 3-6 (*Continued*)
TABLE OF NUCLIDES

Nuclide		Mass Excess		Half-life y=yr, d=day h=hr, m=min s=sec	Natural Abundance, %	Thermal Neutron Absorption Cross Section σ, barns (mode if not n,γ)	Major Radiations, Energies in MeV and γ Intensities, %
Z Symbol	Mass Number	μmu	MeV				
							0.614(14)
79		−81,671	−76.075	....	50.52	8.5	....
79 m			−75.87	4.8 s	....	....	γ, 0.21
80		−81,464	−75.882	17.6 m	....	....	β⁻, 2.00; β⁺, 0.87; γ, 0.511, 0.618(7), 0.666(1)
80 m			−75.796	4.38 h	....	....	γ, 0.037(36); e⁻, 0.024, 0.036, 0.047
81		−83,708	−77.97	....	49.48	3	....
82		−83,198	−77.50	35.34 h	....	....	β⁻, 0.444; γ, 0.554(66), 0.619(41), 0.698(27), 0.777(83), 0.828(25), 1.044(29), 1.317(26), 1.475(17)
82 m			−77.45	6.05 m	....	....	γ, 0.046, 0.777, 1.475
83		−84,832	−79.02	2.41 h	....	....	β⁻, 0.93; γ, 0.530(1)
84		−83,450	−77.7	31.8 m	....	....	β⁻, 4.68; γ, 0.81(9), 0.88(51), 1.01(10), 1.21(4), 1.90(18), 2.47(8), 3.93(13)
84				6.0 m	....	....	β⁻, 1.9; γ, 0.44(68), 0.88(75), 1.46(75), 1.89(16)
85		−84,470	−78.7	3.00 m	....	....	β⁻, 2.5; no γ
86		−81,800	−76	54 s	....	....	β⁻, 7.1; γ, 1.29(12), 1.36(39), 1.56(100), 1.97(20), 2.34(20), 2.75(36)
87			−74.6	55.6 s	....	....	β⁻, 8.0, 2.6; n, 0.3; γ, 1.44(100), 1.85(18), 2.48(18),

Table 3-6 (*Continued*)
TABLE OF NUCLIDES

Nuclide		Mass Excess		Half-life y=yr, d=day h=hr, m=min s=sec	Natural Abundance, %	Thermal Neutron Absorption Cross Section σ, barns (mode if not n,γ)	Major Radiations, Energies in MeV and γ Intensities, %
Z Sym-bol	Mass Number	μmu	MeV				
							2.64(16), 2.98(25), 3.18(16), 3.80(11), 4.19(21), 4.8(17), 5.0(17), 5.2(12)
	88			15.5 s	....	....	γ, 0.76
	89			4.5 s	....	....	n, 0.5
	90			1.6 s	....	....	n
$_{36}$Kr	74	−66,900	−62	20 m	....	....	β^+, 3.1; γ, 0.511
	75	−69,080	−64	5.5 m	....	....	....
	76	−74,530	−69	14.8 h	....	....	γ, 0.039, 0.104, 0.135, 0.267, 0.316, 0.407, 0.452
	77	−75,520	−70.4	1.19 h	....	....	β^+, 1.86; e^-, 0.011, 0.023, 0.094, 0.106, 0.118, 0.136; γ, 0.024, 0.108, 0.131, 0.149, 0.665
	78	−79,597	−74.14	....	0.354	2	....
	79	−79,932	−74.46	34.92 h	....	....	β^+, 0.60; e^-, 0.031, 0.043, 0.123, 0.204, 0.248, 0.384; γ, 0.136(1), 0.261(9), 0.398(10), 0.511, 0.606(10), 0.836(2), 1.119(1), 1.336(1)
	79 *m*		−74.33	55 s	....	....	γ, 0.127; e^-, 0.113, 0.125
	80	−83,620	−77.89	....	2.27	15	....
	81	−83,390	−77.7	2.1×10^5 y	....	....	Br X-rays
	81 *m*		−77.5	13 s	....	....	γ, 0.190(65); e^-, 0.176, 0.188
	82	−86,518	−80.589	....	11.56	42	....
	83	−85,869	−79.985	....	11.55	180	
	83 *m*		−79.943	1.86 h	....	....	γ, 0.009(9); e^-, 0.007, 0.018, 0.031
	84	−88,497	−82.433	....	56.90	0.10	....
	85	−87,477	−81.48	10.76 y	....	<15	β^-, 0.67; γ, 0.514

Table 3-6 (*Continued*)
TABLE OF NUCLIDES

| Nuclide | | Mass Excess | | Half-life | Natural | Thermal Neutron | Major Radiations, |
Z Symbol	Mass Number	μmu	MeV	y = yr, d = day h = hr, m = min s = sec	Abundance, %	Absorption Cross Section σ, barns (mode if not n,γ)	Energies in MeV and γ Intensities, %
85 *m*			−81.18	4.4 h	....	....	β⁻, 0.82; e⁻, 0.134, 0.291; γ, 0.150(74), 0.305(13)
	86	−89,384	−83.259	....	17.37	0.06	....
	87	−86,635	−80.70	76 m	....	<600	β⁻, 3.8; γ, 0.403(84), 0.85(16), 2.57(35)
	88	−85,730	−79.9	2.80 h	....	....	β⁻, 2.8; e⁻, 0.013; γ, 0.028, 0.166(7), 0.191(35), 0.36(5), 0.85(23), 1.55(14), 2.19(18), 2.40(35)
	89	−83,400	−78	3.18 m	....	....	β⁻, 4.0; γ, 0.23(85), 0.36(28), 0.43(29), 0.51(42), 0.60(100), 0.74(32), 0.88(65), 1.12(45), 1.29(31), 1.51(88), 1.71(34), 1.93(10), 2.04(16), 2.23(10), 2.42(22), 2.57(10), 2.84(25)
	90	−80,280	−74.8	33 s	....	....	β⁻, 2.80; γ, 0.105(15), 0.120(65), 0.236(16), 0.495(12), 0.536(48), 1.11(48), 1.54(17), 1.79(11), 2.48(4)
	91			9.8 s	....	....	β⁻, 3.6; no γ
	92			3.0 s	....	....	β⁻
	93			2.0 s	....	....	β⁻
	94			1.4 s	....	....	β⁻
	97			∼1 s	....	....	β⁻

Table 3-6 (*Continued*)
TABLE OF NUCLIDES

Nuclide		Mass Excess		Half-life y = yr, d = day h = hr, m = min s = sec	Natural Abundance, %	Thermal Neutron Absorption Cross Section σ, barns (mode if not n,γ)	Major Radiations, Energies in MeV and γ Intensities, %
Z Sym- bol	Mass Number	μmu	MeV				
₃₇Rb	79			24 m	....	....	γ, 0.15(73), 0.19(29), 0.511
	80	−78,100	−73	34 s	....	....	β⁺, 4.1; γ, 0.511, 0.618(39)
	81	−80,980	−75.4	4.7 h	....	....	β⁺, 1.03; γ, 0.253, 0.450, 0.511, 1.10
	81 *m*		−75.3	31 m	....	....	β⁺, 1.4; e⁻, 0.071, 0.083
	82	−82,041	−76.42	1.25 m	....	....	β⁺, 3.15; γ, 0.511, 0.777(9)
	82 *m*		−76.14	6.3 h	....	....	β⁺, 0.78; γ, 0.511, 0.554(66), 0.619(41), 0.698(27), 0.777(83), 0.828(25), 1.044(29), 1.317(26), 1.475(17)
	83	−85,270	−79	83 d	....	....	γ, 0.53(93), 0.79(1); e⁻, 0.007, 0.52
	84	−85,619	−79.753	33.0 d	....	....	β⁺, 1.66; β⁻, 0.91; γ, 0.511, 0.88(74), 1.01(1), 1.90(1)
	84 *m*		−79.289	20 m	....	....	γ, 0.216(37), 0.250(65), 0.464(32); e⁻, 0.201, 0.214, 0.449
	85	−88,200	−82.16	....	72.15	0.9	....
	86	−88,807	−82.72	18.66 d	....	....	β⁻, 1.78; γ, 1.078(9)
	86 *m*		−82.16	1.02 m	....	....	γ, 0.56
	87	−90,814	−84.591	4.8×10¹⁰ y	27.85	0.12	β⁻, 0.274; no γ
	88	−88,730	−82.7	17.8 m	....	1.0	β⁻, 5.3; γ, 0.898(13), 1.863(21), 2.68(2)
	89	−88,350	−82.3	15.4 m	....	....	β⁻, 3.92, 2.9, 1.6; γ, 0.66(17), 1.05(75), 1.26(54), 2.20(14), 2.59(13)
	90	−85,180	−79.3	2.91 m	....	....	β⁻, 6.6; γ, 0.53(4), 0.83(61), 1.03(5),

Table 3-6 (*Continued*)
TABLE OF NUCLIDES

| Nuclide | | Mass Excess | | Half-life | Natural | Thermal Neutron | Major Radiations, |
Z Symbol	Mass Number	μmu	MeV	y=yr, d=day h=hr, m=min s=sec	Abundance, %	Absorption Cross Section σ, barns (mode if not n,γ)	Energies in MeV and γ Intensities, %
							1.11(7), 1.40(5), 1.70(3), 3.07(5), 3.34(15), 3.54(5), 4.13(11), 4.34(18), 4.60(5), 5.2(4)
	91	−83,930	−78	1.2 m	....	....	β−, 4.6
	91			14 m	....	....	β−
	92	−80,860	−75	5.3 s	....	....	....
	93			5.6 s	....	....	....
	94			2.9 s	....	....	....
	95			<2.5 s	....	....	....
₃₈Sr	80			1.7 h	....	....	γ, 0.58
	81			29 m	....	....	....
	82	−81,610	−76	25.0 d	....	....	Rb X-rays
	83	−82,800	−77	32.4 h	....	....	β+, 1.15; e−, 0.025, 0.040; γ, 0.040(24), 0.38(35), 0.511, 0.76(40), 1.16, 1.52
	84	−86,570	−80.638	....	0.56	0.8	....
	85	−87,011	−81.05	64.0 d	....	....	γ, 0.514(100); e−, 0.499
	85 m		−80.81	70 m	....	....	γ, 0.150(14), 0.231(85); e−, 0.005, 0.134, 0.215
	86	−90,715	−84.499	....	9.86	1.3	....
	87	−91,108	−84.865	....	7.02	....	....
	87 m		−84.477	2.83 h	....	....	γ, 0.388(80); e−, 0.372, 0.386
	88	−94,359	−87.89	....	82.56	0.006	....
	89	−92,558	−86.22	52.7 d	....	0.4	β−, 1.463; γ, 0.91
	89 m			10 d	....	....	....
	90	−92,253	−85.95	27.7 y	....	1	β−, 0.546; no γ
	91	−89,839	−83.68	9.67 h	....	....	β−, 2.67; γ, 0.645(15), 0.748(27), 0.93(3), 1.025(30), 1.413(5)
	92	−89,020	−82.9	2.71 h	....	....	β−, 1.5; 0.55; γ, 0.23(3), 0.44(4), 1.37(90)
	93	−85,290	−79.4	8.3 m	....	....	β−, 2.9; γ, 0.60, 0.8, 1.2
	94	−84,620	−78.8	1.35 m	....	....	β−, 2.1; γ, 1.42(100)
	95			0.8 m	....	....	β−

Table 3-6 (*Continued*)
TABLE OF NUCLIDES

Nuclide		Mass Excess		Half-life y=yr, d=day h=hr, m=min s=sec	Natural Abundance, %	Thermal Neutron Absorption Cross Section σ, barns (mode if not n,γ)	Major Radiations, Energies in MeV and γ Intensities, %
Z Symbol	Mass Number	μmu	MeV				
$_{39}$Y	82			12.3 m	....	....	....
	83			7.4 m	....	....	....
	84	−79,810	−74.3	43 m	....	....	$β^+$, 3.5; γ, 0.511, 0.590(15), 0.795(100), 0.982(100), 1.041(50), 1.27(9), 1.47(6)
	85	−83,511	−77.79	5.0 h	....	....	$β^+$, 2.24; e^-, 0.215; γ, 0.231(13), 0.511, 0.77(8), 2.16(9)
	85 m		−77.75	2.68 h	....	....	$β^+$, 1.54; γ, 0.51, 0.92(9)
	86	−85,054	−79.23	14.6 h	....	....	$β^+$, 3.15, 2.34; γ, 0.443(14), 0.511, 0.63(37), 0.704(14), 0.778(21), 0.836(7), 1.026(10), 1.077(82), 1.16(35), 1.857(18), 1.925(24)
	86 m		−79.01	48 m	....	....	γ, 0.208(94); e^-, 0.008
	87	−89,260	−83.2	80 h	....	....	$β^+$; γ, 0.483
	87 m		−82.8	14 h	....	....	$β^+$; e^-, 0.364, 0.379; γ, 0.381(74)
	88	−90,472	−84.27	108.1 d	....	....	$β^+$, 0.76; γ, 0.898(91), 1.836(100)
	89	−94,128	−87.678	....	100	1.3	....
	89 m		−86.77	16.1 s	....	....	γ, 0.91(99); e^-, 0.89
	90	−92,837	−86.50	64.0 h	....	....	$β^-$, 2.27; no γ
	90 m		−85.81	3.1 h	....	....	γ, 0.202(97), 0.482(91), 2.315; e^-, 0.185, 0.465
	91	−92,705	−86.35	58.8 d	....	1.4	$β^-$, 1.545; γ, 1.21
	91 m		−85.80	50.3 m	....	....	γ, 0.551(95); e^-, 0.534
	92	−91,074	−84.83	3.53 h	....	....	$β^-$, 3.63; γ, 0.448(2), 0.560(3), 0.934(14), 1.40(5), 1.83
	93	−90,448	−84.22	10.3 h	....	....	$β^-$, 2.89; γ,

Table 3-6 (*Continued*)
TABLE OF NUCLIDES

Nuclide		Mass Excess		Half-life y=yr, d=day h=hr, m=min s=sec	Natural Abundance, %	Thermal Neutron Absorption Cross Section σ, barns (mode if not n,γ)	Major Radiations, Energies in MeV and γ Intensities, %
Z Sym-bol	Mass Number	μmu	MeV				
	94	−88,320	−82.3	20.3 m	....	....	0.267(6), 0.67(1), 0.94(2), 1.42(1), 1.90(2), 2.18 β⁻, 5.0; γ, 0.56(6), 0.92(43), 1.13(5), 1.65(2), 1.90(2), 2.13(2), 2.57(2), 3.06(1), 3.53(1)
	95	−87,460	−81	10.9 m	....	....	γ, 1.30, 1.80
	96	−84,310	−79	2.3 m	....	....	β⁻, 3.5; γ, 0.7, 1.0, 1.5
$_{40}$Zr	86	−83,770	−78	16.5 h	....	....	γ, 0.028(20), 0.243(96), 0.612(5)
	87	−85,510	−79.7	1.6 h	....	....	β⁺, 2.10; γ, 0.511, 1.2, 2.2
	88	−89,940	−84	85 d	....	....	γ, 0.394(97); e⁻, 0.377
	89	−91,086	−84.85	78.4 h	....	....	β⁺, 0.90; e⁻, 0.89; γ, 0.511, 0.91(99), 1.71(1)
	89 m		−84.26	4.18 m	....	....	β⁺, 2.40, 0.89; e⁻, 0.570; γ, 0.588(87), 1.51(6)
	90	−95,300	−88.770	....	51.46	0.1	....
	90 m		−86.45	0.80 s	....		γ, 0.133(4), 2.18(14), 2.32(86); e⁻, 0.115, 0.130
	91	−94,358	−87.893	....	11.23	1	....
	92	−94,969	−88.462	....	17.11	0.2	....
	93	−93,550	−87.11	1.5×10^6 y	....	<4	β⁻, 0.060; no γ
	94	−93,687	−87.267	....	17.40	0.08	....
	95	−91,965	−85.663	65.5 d	....	....	β⁻, 0.89, 0.396; γ, 0.724(49), 0.756(49)
	96	−91,714	−85.430	$>3.6 \times 10^{17}$ y	2.80	0.05	....
	97	−89,034	−82.93	17.0 h	....	....	β⁻, 1.91; γ, 0.747(92)
	98	−88,040	−82	1 m	....	....	....
$_{41}$Nb	88	−82,210	−77	14 m	....	....	β⁺, 3.2; γ, 0.076, 0.141, 0.272, 0.399, 0.511, 0.671, 1.058, 1.083
	89	−86,920	−81.0	1.9 h	....	....	β⁺, 2.9; γ, 0.511, 1.626, 3.577, 3.838
	89 m		−80.2	42 m	....	....	β⁺, 3.1; e⁻, 0.570; γ,

Table 3-6 (*Continued*)
TABLE OF NUCLIDES

Nuclide		Mass Excess		Half-life y=yr, d=day h=hr, m=min s=sec	Natural Abundance, %	Thermal Neutron Absorption Cross Section σ, barns (mode if not n,γ)	Major Radiations, Energies in MeV and γ Intensities, %
Z Sym- bol	Mass Number	μmu	MeV				
	90	−88,741	−82.66	14.6 h	....	....	0.511, 0.588(93) β⁺, 1.50; e⁻, 0.115, 0.123; γ, 0.142(75), 0.511, 1.14(97), 2.18(14), 2.32(82)
	90 *m*		−82.54	24 s	....	....	γ, 0.122(71); e⁻, 0.104, 0.120
	91	−93,140	−86.8	long	....	....	....
	91 *m*		−86.6	64 d	....	....	γ, 0.104(1), 1.21(3); e⁻, 0.086, 0.102
	92	−92,789	−86.45	>350 y or <1 h	....	....	....
	92 *m*		−86.32	10.16 d	....	....	γ, 0.934(99)
	93	−93,618	−87.204	....	100	1	....
	93 *m*		−87.173	13.6 y	....	....	e⁻, 0.011, 0.028
	94	−92,697	−86.35	2.0×10⁴ y	....	~15	β⁻, 0.49; γ, 0.702(100), 0.871(100)
	94 *m*		−86.31	6.29 m	....	....	γ, 0.871; e⁻, 0.023, 0.039
	95	−93,168	−86.784	35.0 d	....	~7	β⁻, 0.160; γ, 0.765(100)
	95 *m*		−86.549	90 h	....	....	e⁻, 0.216
	96	−91,944	−85.64	23.35 h	....	....	β⁻, 0.7; γ, 0.459(28), 0.569(59), 0.778(97), 0.811(14), 0.851(22), 1.092(49), 1.200(21)
	97	−91,904	−85.61	72 m	....	....	β⁻, 1.27; γ, 0.665(98)
	97 *m*		−84.86	1.0 m	....	....	γ, 0.747(98); e⁻, 0.728
	98	−89,650	−83.5	51 m	....	....	β⁻, 3.1; γ, 0.330(9), 0.720(75), 0.787(100), 1.16(30), 1.44(10), 1.52(4), 1.68(10), 1.88(4), 1.93(8)
	98			<2 m	....	....	β⁻
	99	−88,950	−83	2.4 m	....	....	β⁻, 3.2; γ, 0.100(1), 0.260(1)
	99			10 s	....	....	β⁻
	100	−85,980	−80	3.0 m	....	....	γ, 0.140(10), 0.36(55), 0.45(40),

Table 3-6 (*Continued*)
TABLE OF NUCLIDES

Nuclide		Mass Excess		Half-life y=yr, d=day h=hr, m=min s=sec	Natural Abundance, %	Thermal Neutron Absorption Cross Section σ, barns (mode if not n,γ)	Major Radiations, Energies in MeV and γ Intensities, %
Z Symbol	Mass Number	μmu	MeV				
$_{42}$Mo	100		−80	11 m	. . . .	. . . .	0.53(100), 0.65, 2.2, 2.3, 2.65, 2.85 β⁻, 4.2, 3.5; γ, 0.535(100), 0.62(60), 1.04(10), 1.15(10), 1.47(5)
	88			27 m	. . . .	. . . .	β⁺, 2.5; γ, 0.511, 2.69
	89			7 m	. . . .	. . . .	β⁺, 4.9; γ, 0.511
	90	−86,060	−80.17	5.67 h	. . . .	. . . .	β⁺, 1.2; e⁻, 0.104, 0.120, 0.239, 0.255; γ, 0.122(71), 0.257(85), 0.445(9), 0.511, 0.945(10), 1.273(8), 1.389(4), 1.46(4)
	91	−88,350	−82.3	15.49 m	. . . .	. . . .	β⁺, 3.44; γ, 0.511
	91 m		−81.6	64 s	. . . .	. . . .	β⁺, 3.99, 2.78; e⁻, 0.638; γ, 0.511, 0.658(54), 1.21(22), 1.53(15)
	92	−93,190	−86.804	>4×10¹⁸ y	15.86	<0.3	. . . .
	93	−93,170	−86.79	>100 y	. . . .	. . . .	Nb X-rays
	93 m		−84.36	6.95 h	. . . .	. . . .	γ, 0.264(58), 0.685(100), 1.479(100); e⁻, 0.244, 0.261
	94	−94,910	−88.407	. . . .	9.12	. . . .	. . . .
	95	−94,161	−87.709	. . . .	15.70	14	. . . .
	96	−95,326	−88.794	. . . .	16.50	1	. . . .
	97	−93,979	−87.539	. . . .	9.45	2	. . . .
	98	−94,591	−88.110	. . . .	23.75	0.51	. . . .
	99	−92,280	−85.96	66.7 h	. . . .	. . . .	β⁻, 1.23; γ, 0.041(2), 0.181(7), 0.372(1), 0.740(12), 0.780(4)
	100	−92,525	−86.185	≥3×10¹⁷ y	9.62	0.2	. . . .
	101	−89,647	−83.50	14.6 m	. . . .	. . . .	β⁻, 2.23; e⁻, 0.170; γ, 0.191(25), 0.51(15), 0.59(21), 0.70(11), 0.89(15),

Table 3-6 (*Continued*)
TABLE OF NUCLIDES

Nuclide Z Sym-bol	Nuclide Mass Number	Mass Excess μmu	Mass Excess MeV	Half-life y=yr, d=day h=hr, m=min s=sec	Natural Abundance, %	Thermal Neutron Absorption Cross Section σ, barns (mode if not n,γ)	Major Radiations, Energies in MeV and γ Intensities, %
							1.02(25), 1.18(11), 1.38(9), 1.56(11), 2.08(16)
	102	−89,750	−84	11.5 m	....	....	β⁻, 1.2
	103			62 s	....	....	....
	104			1.1 m	....	....	β⁻, 4.8; γ, 0.070
	105			40 s	....	....	β⁻
₄₃Tc	92	−84,540	−78.8	4.4 m	....	....	β⁺, 4.1; γ, 0.090(20), 0.14(67), 0.24(30), 0.33(90), 0.511, 0.79(95), 1.54(100)
	93	−89,749	−83.60	2.75 h	....	....	β⁺, 0.80; γ, 0.511, 1.35(65), 1.49(33)
	93 m		−83.21	43 m	....	....	γ, 0.390(63), 2.66(18); e⁻, 0.369
	94	−90,337	−84.15	293 m	....	....	β⁺, 0.816; γ, 0.511, 0.702(100), 0.849(100), 0.871(100)
	94 m		−84.04	53 m	....	....	β⁺, 2.47; γ, 0.511, 0.871(91), 1.53(10), 1.87(9), 2.73(5), 3.20(2)
	95	−92,380	−86.05	20.0 h	....	....	γ, 0.768(82), 0.84(11), 1.06(4)
	95 m		−86.01	61 d	....	....	β⁺, 0.68; e⁻, 0.019, 0.036, 0.184; γ, 0.204(70), 0.584(36), 0.78(12), 0.823(9), 0.838(27), 1.042(4)
	96	−92,170	−85.9	4.35 d	....	....	γ, 0.32(5), 0.778(100), 0.81(84) 0.851(100), 1.12(16); e⁻, 0.30, 0.75, 0.79, 0.82
	96 m		−85.8	52 m	....	....	e⁻, 0.013, 0.032
	97	−93,660	−87	2.6×10⁶ y	....	....	Mo X-rays
	97 m		−87	91 d	....	....	e⁻, 0.075, 0.094
	98	−92,890	−86.5	1.5×10⁶ y	....	3	β⁻, 0.30; γ,

Table 3-6 (*Continued*)
TABLE OF NUCLIDES

Nuclide		Mass Excess		Half-life y = yr, d = day h = hr, m = min s = sec	Natural Abundance, %	Thermal Neutron Absorption Cross Section σ, barns (mode if not n,γ)	Major Radiations, Energies in MeV and γ Intensities, %
Z Symbol	Mass Number	μmu	MeV				
							0.66(100), 0.76(100)
	99	−93,751	−87.33	2.12×10⁵ y	....	22	β⁻, 0.292; no γ
	99 *m*		−87.18	6.049 h	....	....	γ, 0.140(90); e⁻, 0.001, 0.119
	100	−92,160	−85.9	15.8 s	....	....	β⁻, 3.38; γ, 0.540, 0.60, 0.71, 0.81, 0.89, 1.01, 1.31, 1.49, 1.8
	101	−92,674	−86.32	14.0 m	....	....	β⁻, 1.32; γ, 0.13(3), 0.307(91), 0.545(8)
	102	−90,820	−85	5 s	....	....	β⁻, 4.4
	103	−91,170	−84.9	50 s	....	....	β⁻, 2.2; γ, 0.135(17), 0.21(10), 0.35
	104	−88,290	−82.2	18 m	....	....	β⁻, 4.6; γ, 0.36, 0.53, 0.89, 1.15, 1.25, 1.37, 1.6, 1.9, 2.2, 2.7, 3.2, 3.4, 3.7, 4.0, 4.4, 4.7
	105	−88,670	−82.6	7.7 m	....	....	β⁻, 3.4; γ, 0.110
	106			37 s	....	....	....
	107			29 s	....	....	....
₄₄Ru	93			50 s	....	....	β⁺
	94			57 m	....	....	....
	95	−90,199	−84.02	1.65 h	....	....	β⁺, 1.33; γ, 0.340(70), 0.511, 0.625(13), 1.09(21), 1.43(5)
	96	−92,402	−86.07	....	5.46	0.2	....
	97	−92,370	−86	2.88 d	....	....	γ, 0.215(91), 0.324(8); e⁻, 0.194
	98	−94,711	−88.222	....	1.868	<8	....
	99	−94,065	−87.619	....	12.63	11	....
	100	−95,782	−89.219	....	12.53	10	....
	101	−94,423	−87.953	....	17.02	3	....
	102	−95,652	−89.098	....	31.6	1.4	....
	103	−93,694	−87.27	39.5 d	....	....	β⁻, 0.70, 0.21; γ, 0.497(88), 0.610(6)
	104	−94,570	−88.090	....	18.87	0.48	....
	105	−92,321	−86.00	4.44 h	....	0.2	β⁻, 1.87, 1.15; γ, 0.263(6), 0.317(11), 0.40(6), 0.475(20),

Table 3-6 (*Continued*)
TABLE OF NUCLIDES

Nuclide		Mass Excess		Half-life y = yr, d = day h = hr, m = min s = sec	Natural Abundance, %	Thermal Neutron Absorption Cross Section σ, barns (mode if not n,γ)	Major Radiations, Energies in MeV and γ Intensities, %
Z Symbol	Mass Number	μmu	MeV				
							0.67(16), 0.726(48)
	106	−92,678	−86.33	368 d	....	0.15	β⁻, 0.039; no γ
	107	−89,870	−83.7	4.2 m	....	....	β⁻, 3.2; γ, 0.195(14), 0.37, 0.48, 0.86(7), 0.93(4), 1.03(4), 1.29(4)
	108	−89,900	−84	4.5 m	....	....	β⁻, 1.3; γ, 0.165(28)
₄₅Rh	97	−88,620	−83	32 m	....	....	β⁺, 2.47; γ, 0.08, 0.187, 0.255, 0.420, 0.511, 0.86, 1.18, 1.57, 1.70, 1.96, 2.16
	97			1.0 m	....	....	γ, 0.75
	98	−90,200	−84.0	8.7 m	....	....	β⁺, 2.5; γ, 0.65(100)
	99		−85.57	16.1 d	....	....	β⁺, 1.03; γ, 0.090, 0.175, 0.31, 0.354, 0.444, 0.48, 0.511, 0.529
	99	−91,810	−85.52	4.7 h	....	....	β⁺, 0.74; γ, 0.34(70), 0.511, 0.62(20), 0.89, 1.26, 1.41
	100	−91,874	−85.58	20.8 h	....	....	β⁺, 2.62; e⁻, 0.516; γ, 0.444(8), 0.511, 0.540(88), 0.820(25), 1.11(13), 1.35(20), 1.55(23), 1.93(10), 2.37(39)
	101	−93,822	−87.39	3.0 y	....	....	γ, 0.127(88), 0.198(75), 0.325(11); e⁻, 0.105, 0.124, 0.176
	101 *m*		−87.24	4.4 d	....	....	γ, 0.307(83), 0.545(6); e⁻, 0.134, 0.154
	102	−93,158	−86.77	206 d	....	....	β⁻, 1.15; β⁺, 1.29; γ, 0.475(57), 0.511, 0.628(4), 1.103(3), 1.37(1), 1.57
	102			2.9 y	....	....	γ, 0.418(13), 0.475(95), 0.632(54),

Table 3-6 (*Continued*)
TABLE OF NUCLIDES

Nuclide		Mass Excess		Half-life	Natural	Thermal Neutron	Major Radiations,
Z Symbol	Mass Number	μmu	MeV	y=yr, d=day h=hr, m=min s=sec	Abundance, %	Absorption Cross Section σ, barns (mode if not n,γ)	Energies in MeV and γ Intensities, %
							0.698(41), 0.768(30), 1.05(41), 1.11(22)
	103	−94,489	−88.014	. . . .	100	144	. . . .
	103 *m*		−87.974	57.5 m	. . . .	. . . .	γ, 0.040; e^-, 0.017, 0.037
	104	−93,341	−86.95	43 s	. . . .	40	β^-, 2.44; γ, 0.56(2), 1.24
	104 *m*		−86.82	4.41 m	. . . .	800	γ, 0.051(47), 0.078(3), 0.097(3), 0.56, 0.77; e^-, 0.028, 0.054, 0.074
	105	−94,329	−87.87	35.88 h	. . . .	15,000	β^-, 0.568; γ, 0.306(5), 0.319(19)
	105 *m*		−87.74	45 s	. . . .	. . . .	γ, 0.129; e^-, 0.106, 0.126
	106	−92,721	−86.37	30 s	. . . .	. . . .	β^-, 3.54; γ, 0.512(21), 0.622(11), 1.05(2), 1.13(1), 1.55
	106		−86.3	130 m	. . . .	. . . .	β^-, 1.62, 1.1; γ, 0.220(18), 0.406(18), 0.451(35), 0.512(88), 0.616(29), 0.735(41), 0.82(35), 1.046(25), 1.128(12), 1.223(17), 1.56(18)
	107	−93,247	−86.86	21.7 m	. . . .	. . . .	β^-, 1.20; γ, 0.305(73), 0.390(11), 0.68(3)
	108	−91,300	−85	16.8 s	. . . .	. . . .	β^-, 4.5; γ, 0.434(43), 0.51(10), 0.62(22)
	109	−91,360	−85	<1 h	. . . .	. . . .	. . . .
	110	−88,900	−83	5 s	. . . .	. . . .	β^-, 5.5; γ, 0.374
$_{46}$Pd	98			17.5 m	. . . .	. . . .	γ, 0.132
	99	−87,730	−81.7	22 m	. . . .	. . . .	β^+, 2.0; γ, 0.140, 0.275, 0.420, 0.511, 0.67
	100	−91,230	−85	4.0 d	. . . .	. . . .	γ, 0.074(34), 0.084(49),

Table 3-6 (*Continued*)
TABLE OF NUCLIDES

Nuclide		Mass Excess		Half-life y = yr, d = day h = hr, m = min s = sec	Natural Abundance, %	Thermal Neutron Absorption Cross Section σ, barns (mode if not n,γ)	Major Radiations, Energies in MeV and γ Intensities, %
Z Symbol	Mass Number	μmu	MeV				
	101	−91,930	−85.40	8.4 h	....	....	0.126(16), 0.159(4); e⁻, 0.010, 0.019, 0.052, 0.061, 0.071, 0.081 γ, 0.270(8), 0.296(30), 0.511, 0.566(7), 0.590(24), 0.723(5), 0.993(2), 1.20(3), 1.30(3); β⁺, 0.78; e⁻, 0.021
	102	−94,391	−87.92	....	0.96	4.8	....
	103	−93,893	−87.46	17.0 d	....	....	γ, 0.297, 0.362, 0.498
	104	−95,989	−89.41	....	10.97	....	....
	105	−94,936	−88.43	....	22.2	....	....
	106	−96,521	−89.91	....	27.3	0.29	....
	107	−94,868	−88.368	~7×10⁶ y	....	....	β⁻, 0.04; no γ
	107 *m*		−88.16	21.3 s	....	....	γ, 0.21; e⁻, 0.19, 0.21
	108	−96,109	−89.52	....	26.7	12	....
	109	−94,046	−87.60	13.47 h	....	....	β⁻, 1.028; e⁻, 0.062, 0.084; γ, 0.088(5), 0.129, 0.31, 0.41, 0.60, 0.64
	109 *m*		−87.41	4.69 m	....	....	γ, 0.188(58); e⁻, 0.164, 0.185
	110	−94,836	−88.34	....	11.8	0.2	....
	111	−92,330	−86.0	22 m	....	....	β⁻, 2.2; γ, 0.38(5), 0.60(13), 0.81(1), 1.4(8)
	111 *m*		−85.8	5.5 h	....	....	β⁻, 2.0; e⁻, 0.148, 0.169; γ, 0.17
	112	−92,614	−86.27	21.0 h	....	....	β⁻, 0.28; γ, 0.019(20)
	113			1.4 m	....	....	no γ
	114			2.4 m	....	....	no γ
	115			45 s	....	....	....
₄₇Ag	102	−88,700	−83	15 m	....	....	....
	103	−91,110	−84.9	66 m	....	....	β⁺, 1.6; γ, 0.12(26), 0.15(23), 0.24(10), 0.27(34), 0.511, 1.01(10), 1.16(9), 1.28(13)
	103 *m*		−84.7	5.7 s	....	....	γ, 0.138

Table 3-6 (*Continued*)
TABLE OF NUCLIDES

Nuclide		Mass Excess		Half-life	Natural	Thermal Neutron	Major Radiations,
Z Sym-bol	Mass Number	μmu	MeV	y = yr, d = day h = hr, m = min s = sec	Abundance, %	Absorption Cross Section σ, barns (mode if not n,γ)	Energies in MeV and γ Intensities, %
	104	−91,404	−85.14	66 m	....	....	β^+, 0.99; e^-, 0.532, 0.743; γ, 0.511, 0.556(84), 0.764(48), 0.854(30), 1.34(8), 1.53(7), 1.62(8), 1.81(7)
	104 m		−85.12	29.8 m	....	....	β^+, 2.70; e^-, 0.532; γ, 0.511, 0.556(100)
	105	−93,540	−87	40 d	....	....	γ, 0.064(10), 0.280(32), 0.344(42), 0.443(10), 0.62–0.68(12), 1.088(2); e^-, 0.040, 0.060, 0.256, 0.320
	106	−93,339	−86.94	23.96 m	....	....	β^+, 1.96; γ, 0.511
	106 m		−86.6	8.5 d	....	....	γ, 0.221(9), 0.451(9), 0.512(86), 0.616(23), 0.717(31), 0.748(13), 0.80(41), 1.046(29), 1.128(9), 1.199(9), 1.528(15), 1.58(8), 1.83(3); e^-, 0.197, 0.382, 0.405, 0.426, 0.487, 0.508, 0.592, 0.693
	107	−94,906	−88.403	....	51.35	35	....
	107 m		−88.310	44.3 s	....	....	γ, 0.094(5); e^-, 0.068, 0.090
	108	−94,051	−87.61	2.42 m	....	....	β^-, 1.64; β^+, 0.90; γ, 0.434(1), 0.511, 0.615, 0.632(2)
	108 m		−87.50	>5 y	....	....	γ, 0.080(5), 0.434(89), 0.614(90), 0.722(90); e^-, 0.027
	109	−95,244	−88.717	....	48.65	89	....
	109 m		−88.630	39.2 s	....	....	γ, 0.088(5); e^-, 0.062, 0.084

Table 3-6 (*Continued*)
TABLE OF NUCLIDES

Nuclide		Mass Excess		Half-life y = yr, d = day h = hr, m = min s = sec	Natural Abundance, %	Thermal Neutron Absorption Cross Section σ, barns (mode if not n,γ)	Major Radiations, Energies in MeV and γ Intensities, %
Z Symbol	Mass Number	μmu	MeV				
	110	−93,905	−87.47	24.4 s	....	....	β⁻, 2.87; γ, 0.658(5)
	110 *m*		−87.35	255 d	....	80	β⁻, 1.5, 0.53, 0.087; e⁻, 0.090, 0.113; γ, 0.658(96), 0.68(16), 0.706(19), 0.764(23), 0.818(8), 0.885(71), 0.937(32), 1.384(21), 1.505(11)
	111	−94,684	−88.20	7.5 d	....	....	β⁻, 1.05; γ, 0.247(1), 0.342(6)
	111 *m*		−88.13	74 s	....	....	γ, 0.065
	112	−92,936	−86.57	3.14 h	....	....	β⁻, 3.94; γ, 0.617(41), 1.40(5), 1.63(3), 2.11(3), 2.55(2)
	113	−93,444	−87.04	5.3 h	....	....	β⁻, 2.0; γ, 0.12(10), 0.30(100), 0.58(5), 0.67(17), 0.88(4), 0.98(5), 1.18(4)
	113			1.2 m	....	....	β⁻, <2.0; γ, 0.14, 0.30, 0.39, 0.56, 0.70
	114	−91,700	−85.4	4.5 s	....	....	β⁻, 4.6; γ, 0.57
	115	−91,070	−84.8	20.0 m	....	....	β⁻, 3.2; γ, 0.14(12), 0.22(49), 0.28(13), 0.36(11), 0.42(7), 0.47(10), 0.64(4), 1.48(11), 1.66(8), 1.89(10), 2.12(13)
	116	−88,690	−83	2.5 m	....	....	β⁻, 5.0; γ, 0.52, 0.70
	117			1.1 m	....	....	....
₄₈Cd	103			10 m	....	....	γ, 0.22, 0.511, 0.63, 0.85
	104	−90,120	−84	57 m	....	....	γ, 0.084; e⁻, 0.041, 0.058, 0.080
	105	−90,530	−84	55 m	....	....	β⁺, 1.69; e⁻, 0.282, 0.295, 0.321, 0.408
	106	−93,537	−87.128	....	1.22	1	....
	107	−93,385	−86.99	6.49 h	....	....	β⁺, 0.302; γ, 0.511, 0.796, 0.829
	108	−95,813	−89.248	....	0.88	3	....
	109	−95,072	−88.55	453 d	....	....	γ, 0.088; e⁻, 0.062,

Table 3-6 (*Continued*)
TABLE OF NUCLIDES

Nuclide		Mass Excess		Half-life y = yr, d = day h = hr, m = min s = sec	Natural Abundance, %	Thermal Neutron Absorption Cross Section σ, barns (mode if not *n*,γ)	Major Radiations, Energies in MeV and γ Intensities, %
Z Symbol	Mass Number	μmu	MeV				
							0.084
	110	−96,988	−90.342	....	12.39	0.1	....
	111	−95,812	89.246	....	12.75	....	....
	111 *m*		−88.850	48.6 m	....	....	γ, 0.150(30), 0.247(94); *e*⁻, 0.123, 0.146
	112	−97,238	−90.575	....	24.07	0.03	....
	113	−95,592	−89.041	>1.3×10¹⁵ y	12.26	20,000	....
	113 *m*		−88.77	13.6 y	....	....	β⁻, 0.58; γ, 0.265
	114	−96,640	−90.018	....	28.86	1.1	....
	115	−94,569	−88.09	53.5 h	....	....	β⁻, 1.11; γ, 0.230(1), 0.262(2), 0.49(10), 0.53(26)
	115 *m*		−87.91	43 d	....	....	β⁻, 1.62; γ, 0.485, 0.935(2), 1.29(1)
	116	−95,238	−88.712	>10¹⁷ y	7.58	1.4	....
	117	−92,761	−86.41	2.4 h	....	....	β⁻, 2.23; *e*⁻, 0.286; γ, 0.089(7), 0.273(31), 0.314(16), 0.345(18), 0.434(13), 0.832(4), 0.880(3), 0.95(4), 1.052(5), 1.303(19), 1.577(17)
	117 *m*		−86.27	3.4 h	....	....	β⁻, 0.67; *e*⁻, 0.286; γ, 0.273(18), 0.314(8), 0.345(4), 0.434(4), 0.565(6), 0.715(4), 0.880(10), 1.065(9), 1.117(4), 1.24(11), 1.338(8), 1.408(8), 1.433(10), 1.562(6), 1.998(15), 2.319(3)
	118	−93,030	−87	49 m	....	....	β⁻
	119	−90,260	−84.1	2.7 m, 10 m	....	....	β⁻, 3.5
	121			12.8 ɘ	....	....	....
₄₉In	106	−86,560	−80.6	5.3 m	....	....	β⁺, 4.9; γ, 0.511,

Table 3-6 (*Continued*)
TABLE OF NUCLIDES

| Nuclide | | Mass Excess | | Half-life | Natural | Thermal Neutron | Major Radiations, |
Z Symbol	Mass Number	μmu	MeV	y=yr, d=day h=hr, m=min s=sec	Abundance, %	Absorption Cross Section σ, barns (mode if not n,γ)	Energies in MeV and γ Intensities, %
	107	−89,640	−83.5	33 m	....	....	0.63, 1.65, 1.85 β+, 2.2; γ, 0.22(46), 0.32, 0.511, 0.73, 0.84, 0.94, 1.05, 1.25
	108		−84.14	57 m	....	....	β+, 1.29; e−, 0.123, 0.147, 0.216, 0.238, 0.260, 0.606, 0.845; γ, 0.150, 0.175, 0.243, 0.511, 0.633, 0.872
	108	−90,280	−84.10	39 m	....	....	β+, 3.50; e−, 0.606; γ, 0.383, 0.511, 0.633, 0.842
	109	−92,904	−86.53	4.3 h	....	....	β+, 0.79; e−, 0.033, 0.056, 0.178, 0.201; γ, 0.205, 0.28, 0.35, 0.65, 0.91
	109 m₁		−85.87	1.3 m	....	....	γ, 0.658; e−, 0.630
	109 m₂		−84.42	0.20 s	....	....	γ, 0.17(12), 0.21(12), 0.40(20), 0.68(100), 1.04(20), 1.43(77)
	110	−92,769	−86.41	66 m	....	....	β+, 2.25; e−, 0.631; γ, 0.511, 0.658(95)
	110			4.9 h	....	....	γ, 0.66(160), 0.91(110); e−, 0.094, 0.558, 0.615, 0.631, 0.653, 0.680, 0.858, 0.910
	111	−94,640	−88.2	2.81 d	....	....	γ, 0.173(89), 0.247(94); e−, 0.146, 0.220, 0.243
	112	−94,456	−87.98	14.4 m	....	....	β−, 0.66; β+, 1.56; γ, 0.511, 0.617(6)
	112 m		−87.83	20.7 m	....	....	γ, 0.156(9); e−, 0.128, 0.152
	113	−95,911	−89.34	....	4.23	8	....
	113 m		−88.95	99.8 m	....	....	γ, 0.393(64); e−, 0.365, 0.389
	114	−95,095	−88.58	72 s	....	....	β−, 1.988; β+, 0.42; γ, 1.299
	114 m		−88.39	50.0 d	....	....	γ, 0.192(17),

Table 3-6 (*Continued*)
TABLE OF NUCLIDES

Nuclide		Mass Excess		Half-life y = yr, d = day h = hr, m = min s = sec	Natural Abundance, %	Thermal Neutron Absorption Cross Section σ, barns (mode if not n,γ)	Major Radiations, Energies in MeV and γ Intensities, %
Z Symbol	Mass Number	μmu	MeV				
							0.558(4), 0.724(4); e^-, 0.164, 0.188
	115	−96,129	−89.54	6×10^{14} y	95.77	154	....
	115 m		−89.21	4.50 h	....	....	β^-, 0.83; e^-, 0.308, 0.331; γ, 0.335(50)
	116	−94,683	−88.20	13.4 s	....	....	β^-, 3.3; γ, 0.434, 0.95, 1.293(1)
	116 m_1		−88.14	54.0 m	....	....	β^-, 1.00; γ, 0.138(3), 0.417(36), 0.819(17), 1.09(53), 1.293(80), 1.508(11), 2.111(20)
	116 m_2		−87.98	2.16 s	....	....	γ, 0.164; e^-, 0.138, 0.160
	117	−95,466	−88.93	45 m	....	....	β^-, 0.74; e^-, 0.132; γ, 0.158(87), 0.565(100)
	117 m		−88.61	1.93 h	....	....	β^-, 1.78; e^-, 0.286; γ, 0.158(14), 0.314(31)
	118	−93,890	−87.5	5.7 s	....	....	β^-, 4.2; γ, 1.230(15)
	118		−87.4	4.35 m	....	....	β^-, 2.0; γ, 0.69(41), 1.05(80), 1.230(97), 2.04(3)
	119	−94,010	−87.6	2.1 m	....	....	β^-, 1.6; γ, 0.82(95)
	119 m		−87.3	17.5 m	....	....	β^-, 2.7; γ, 0.024, 0.30, 0.91
	120	−92,000	−86	3.2 s	....	....	β^-, 5.6; γ, 1.711(15)
	120		−85.8	44 s	....	....	β^-, 3.1; γ, 0.090(12), 0.198(9), 0.71(12), 0.86(34), 0.94(12), 1.02(61), 1.171(100), 1.28(14), 1.47(6), 1.87(7), 2.01(6)
	121	−91,910	−86	30 s	....	....	γ, 0.94
	121		−86	3.1 m	....	....	β^-, 3.7
	122	−89,400	−83	8 s	....	....	β^-, 5; γ, 0.99, 1.14
	123	−89,430	−83	36 s	....	....	β^-, 4.6
	123		−03	10 s	....	....	γ, 1.1
	124	−86,800	−81	~3.6 s	....	....	β^-, 5; γ, 0.99(3),

Table 3-6 (*Continued*)
TABLE OF NUCLIDES

Nuclide		Mass Excess		Half-life y = yr, d = day h = hr, m = min s = sec	Natural Abundance, %	Thermal Neutron Absorption Cross Section σ, barns (mode if not n,γ)	Major Radiations, Energies in MeV and γ Intensities, %
Z Symbol	Mass Number	μmu	MeV				
$_{50}$Sn	108			9.2 m	....	....	1.13(10), 3.21(3) γ, 0.28, 0.42
	109			18.1 m	....	....	β^+, 1.6; e^-, 0.305, 0.491, 0.86, 1.09; γ, 0.335, 0.521, 0.89, 1.12
	110			4.0 h	....	....	γ, 0.283(95); e^-, 0.255
	111	−91,940	−85.6	35.0 m	....	....	β^+, 1.51; γ, 0.511, 0.75(1), 0.97(1), 1.14(2), 1.54(1), 1.59(1), 1.89(1), 2.11, 2.32
	112	−95,165	−88.64	....	0.95	0.9	....
	113	−94,813	−88.32	115 d	....	....	γ, 0.255(2)
	113 m		−88.24	20 m	....	....	γ, 0.079(1); e^-, 0.050, 0.075
	114	−97,227	−90.57	....	0.65	....	....
	115	−96,654	−90.03	....	0.34	....	....
	116	−98,255	−91.523	....	14.24	0.006	....
	117	−97,042	−90.392	....	7.57	....	....
	117 m		−90.075	14.0 d	....	....	γ, 0.158(87); e^-, 0.130, 0.155
	118	−98,394	−91.652	....	24.01	0.01	....
	119	−96,687	−90.062	....	8.58	....	....
	119 m		−89.973	∼250 d	....	....	γ, 0.024(16); e^-, 0.020, 0.026, 0.061
	120	−97,802	−91.100	....	32.97	0.14	....
	121	−95,773	−89.21	27.5 h	....	....	β^-, 0.383
	121 m		−89.14	76 y	....	....	β^-, 0.42; γ, 0.037
	122	−96,559	−89.943	....	4.71	0.2	....
	123	−94,262	−87.80	125 d	....	....	β^-, 1.42
	123 m		−87.78	39.5 m	....	....	β^-, 1.26; γ, 0.160
	124	−94,728	−88.237	$>2\times10^{17}$ y	5.98	0.1	....
	125	−92,254	−85.93	9.4 d	....	....	β^-, 2.34; γ, 0.342, 0.468, 0.811(2), 0.904(1), 1.068(4), 1.17, 1.41, 1.97(1), 2.23
	125 m		−85.91	9.5 m	....	....	β^-, 2.04; γ, 0.325(97)
	126	−92,360	−86	∼10^5 y	....	....	γ, 0.060, 0.067, 0.092
	127	−89,740	−84	2.05 h	....	....	β^-, 1.45; γ, 0.44, 0.49, 0.82, 1.10, 2.00, 2.32, 2.58, 2.68, 2.82

Table 3-6 (*Continued*)
TABLE OF NUCLIDES

| Nuclide | | Mass Excess | | Half-life | Natural | Thermal Neutron | Major Radiations, |
Z Symbol	Mass Number	μmu	MeV	y = yr, d = day h = hr, m = min s = sec	Abundance, %	Absorption Cross Section σ, barns (mode if not n,γ)	Energies in MeV and γ Intensities, %
	127		−83.5	4.1 m	....	....	β^-, 2.7; γ, 0.49(100)
	128	−89,530	−83.4	59 m	....	....	β^-, 0.80; γ, 0.044(7), 0.072(19), 0.50(61), 0.57(22)
	129			9 m	....	....	γ, 1.15
	129			1.0 h	....	....	....
	130			2.6 m	....	....	....
	131			3.4 m	....	....	....
	132			2.2 m	....	....	....
$_{51}$Sb	112			0.9 m	....	....	γ, 0.511, 1.27
	113	−90,014	−83.85	6.4 m	....	....	β^+, 2.42; γ, 0.32, 0.511, 0.6–0.9, 1.03, 1.2
	114	−90,490	−84.3	3.3 m	....	....	β^+, 2.7; γ, 0.9, 1.30
	115	−93,401	−87.00	31 m	....	....	β^+, 1.51; γ, 0.499(100), 0.511, 0.98(5), 1.24(5), 2.22(1)
	116	−93,370	−87.0	16 m	....	....	β^+, 2.3; γ, 0.511, 0.93(26), 1.293(85), 2.23(14)
	116 *m*		−86.5	60 m	....	....	β^+, 1.16; e^-, 0.070, 0.095, 0.111; γ, 0.099(30), 0.140(30), 0.406(36), 0.511, 0.545(68), 0.96(75), 1.06(27), 1.293(100)
	117	−95,088	−88.57	2.8 h	....	....	β^+, 0.57; γ, 0.158(87), 0.511
	118	−94,426	−87.96	3.5 m	....	....	β^+, 2.67; γ, 0.511, 0.83, 1.230(3)
	118 *m*		−87.77	5.1 h	....	....	γ, 0.041(29), 0.254(93), 1.049(100), 1.230(100); e^-, 0.012, 0.036, 0.223
	119	−96,065	−89.48	38.0 h	....	....	γ, 0.024(16); e^-, 0.020
	120	−94,919	−88.42	15.89 m	....	....	β^+, 1.70; γ, 0.511, 1.171(1)
	120			5.8 d	....	....	γ, 0.090(81), 0.200(88), 1.03(99),

Table 3-6 (*Continued*)
TABLE OF NUCLIDES

Nuclide		Mass Excess		Half-life y=yr, d=day h=hr, m=min s=sec	Natural Abundance, %	Thermal Neutron Absorption Cross Section σ, barns (mode if not n,γ)	Major Radiations, Energies in MeV and γ Intensities, %
Z Symbol	Mass Number	μmu	MeV				
							1.171(100); e^-, 0.061, 0.096, 0.171, 0.196
	121	−96,184	−89.593	···········	57.25	6	················
	122	−94,817	−88.32	2.80 d	·········	··············	β^-, 1.97; β^+, 0.56; γ, 0.564(66), 0.686(3), 1.14(1), 1.26(1)
	122 *m*		−88.16	4.2 m	·········	··············	γ, 0.061(50), 0.075(17); e^-, 0.021, 0.030, 0.045, 0.056, 0.071
	123	−95,787	−89.224	$>1.3\times10^{16}$ y	42.75	3.3	················
	124	−94,027	−87.58	60.4 d	·········	2000	β^-, 2.31; γ, 0.603(97), 0.644(7), 0.72(14), 0.967(2), 1.048(2), 1.31(3), 1.37(5), 1.45(2), 1.692(50), 2.088(7)
	124 m_1		−87.57	93 s	·········	··············	β^-, 1.19; e^-, 0.006, 0.009; γ, 0.505(20), 0.603(20), 0.644(20)
	124 m_2		−87.55	21 m	·········	··············	e^-, 0.021, 0.024; γ
	125	−94,768	−88.28	2.71 y	·········	<20	β^-, 0.61; e^-, 0.004, 0.030, 0.144, 0.395; γ, 0.176(6), 0.427(31), 0.463(10), 0.599(24), 0.634(11), 0.66(3)
	126			19.0 m	·········	··············	β^-, 1.9; γ, 0.41, 0.67
	126	−92,680	−86.3	12.5 d	·········	··············	β^-, 1.9; γ, 0.41, 0.69
	127	−93,073	−86.70	93 h	·········	··············	β^-, 1.5; γ, 0.060, 0.25, 0.41, 0.46, 0.68, 0.77, 0.92, 1.10, 1.34
	128	−90,930	−84.7	10.8 m	·········	··············	β^-, 2.6; γ, 0.320(83), 0.75(200), 1.07(4)
	128			8.6 h	···········	··············	β^-, 1; γ, 0.314, 0.53, 0.64, 0.75
	129	−90,740	−85	4.3 h	·········	··············	β^-, 1.87; γ, 0.073, 0.34, 0.460,

Table 3-6 (*Continued*)
TABLE OF NUCLIDES

Nuclide		Mass Excess		Half-life	Natural Abundance, %	Thermal Neutron Absorption Cross Section σ, barns (mode if not n,γ)	Major Radiations, Energies in MeV and γ Intensities, %
Z Symbol	Mass Number	μmu	MeV	y=yr, d=day h=hr, m=min s=sec			
	130	−87,960	−82	33 m	....	....	0.540, 0.81, 0.91, 1.04, 1.24 γ, 0.19, 0.33, 0.82, 0.94
	130		−82	7.1 m	....	....	γ, 0.20, 0.82, 1.03, 1.16
	131			26 m	....	....	γ, 0.64(37), 0.95(48)
	132			2.1 m	....	....	β⁻
	133			4.2 m	....	....	β⁻
	135			2 s	....	....	....
₅₂Te	107			2.2 s	....	....	α, 3.28
	108			5.3 s	....	....	α, 3.08; p, 2.6, 3.4, 3.7
	115		−82.5	6.0 m	....	....	β⁺, 2.8; γ, 0.511, 0.72(34), 0.96(6), 1.08(24), 1.28(32), 1.38(32), 1.58(6)
	116	−91,700	−85.4	2.50 h	....	....	γ, 0.094; e⁻, 0.063, 0.089
	117	−91,330	−85.1	61 m	....	....	β⁺, 1.81; γ, 0.511, 0.72(65), 0.93(6), 1.78(9)
	117			1.9 h	....	....	β⁺, 1.7
	118	−94,100	−88	6.00 d	....	....	Sb X-rays
	119	−93,602	−87.19	15.9 h	....	....	β⁺, 0.627; γ, 0.645(85), 0.70(11), 1.76(4)
	119 m		−86.9	4.68 d	....	....	γ, 0.153(62), 0.270(25), 0.92–1.14(36), 1.221(67), 2.09(4); e⁻, 0.122, 0.133, 0.148, 0.240, 0.266
	120	−95,977	−89.40	....	0.089	2.0	....
	121	−94,801	−88.31	17 d	....	....	γ, 0.508(18), 0.573(80); e⁻, 0.007, 0.033, 0.543
	121 m		−88.01	154 d	....	....	γ, 0.212(82), 1.10(3); e⁻, 0.007, 0.050, 0.077, 0.180
	122	−96,934	−90.29	....	2.46	2	....
	123	−95,723	−89.16	1.2×10¹³ y	0.87	400	Sb X-rays
	123 m		−88.92	117 d	....	....	γ, 0.159(84); e⁻, 0.057, 0.084,

Table 3-6 (*Continued*)
TABLE OF NUCLIDES

Nuclide		Mass Excess		Half-life y=yr, d=day h=hr, m=min s=sec	Natural Abundance, %	Thermal Neutron Absorption Cross Section σ, barns (mode if not n,γ)	Major Radiations, Energies in MeV and γ Intensities, %
Z Symbol	Mass Number	μmu	MeV				
	124	−97,158	−90.50	....	4.61	5	0.127 ....
	125	−95,582	−89.03	....	6.99	1.5	....
	125 m		−88.89	58 d	....	....	e^-, 0.004, 0.030, 0.078, 0.105; γ, 0.035(7), 0.110
	126	−96,678	−90.05	....	18.71	0.9	....
	127	−94,791	−88.30	9.4 h	....	....	β^-, 0.70; γ, 0.058, 0.21, 0.360, 0.417
	127 m		−88.21	109 d	....	....	γ, 0.059, 0.089, 0.67; e^-, 0.057, 0.084; β^-
	128	−95,524	−88.98	....	31.79	0.14	....
	129	−93,425	−87.02	68.7 m	....	....	β^-, 1.45; e^-, 0.022, 0.026; γ, 0.027(19), 0.275(2), 0.455(15), 0.81(1), 1.08(2)
	129 m		−86.92	34.1 d	....	....	β^-, 1.60; e^-, 0.074, 0.102; γ, 0.69(6)
	130	−93,762	−87.34	8×10^{20} y	34.49	0.2	....
	131	−91,425	−85.16	24.8 m	....	....	β^-, 2.14; e^-, 0.116, 0.144; γ, 0.150(68), 0.453(16), 0.493(5), 0.603(4), 0.95(3), 1.00(4), 1.147(6)
	131 m		−84.98	30 h	....	....	β^-, 2.46, 0.9; e^-, 0.048, 0.069, 0.149, 0.177; γ, 0.081(2), 0.102(5), 0.200(8), 0.241(8), 0.336(9), 0.78(60), 0.85(31), 1.127(13), 1.206(11), 1.629(3), 1.860(1), 1.965(2)
	132	−91,477	−85.21	77.7 h	....	....	β^-, 0.22; e^-, 0.020, 0.048, 0.197; γ, 0.053(17), 0.230(90)
	133			12.5 m	....	....	γ, 0.15, 0.31, 0.41, 0.73, 1.02, 1.33,

Table 3-6 (*Continued*)
TABLE OF NUCLIDES

| Nuclide | | Mass Excess | | Half-life | Natural | Thermal Neutron | Major Radiations, |
Z Symbol	Mass Number	μmu	MeV	y=yr, d=day h=hr, m=min s=sec	Abundance, %	Absorption Cross Section σ, barns (mode if not n,γ)	Energies in MeV and γ Intensities, %
	133 *m*			50 m	....	....	1.71, 1.85 β^-, 2.4; e^-, 0.303; γ, 0.31(21), 0.432(50), 0.47(22), 0.557(35), 0.63(18), 0.70(24), 0.754(85), 0.91(57), 1.01(10), 1.33, 1.71, 1.85
	134			42 m	....	....	γ, 0.08(13), 0.17(16), 0.204(21), 0.262(19)
	135			<2 m	....	....	β^-
$_{53}$I	117			7 m	....	....	β^+; γ, 0.16, 0.34, 0.522
	117			14.5 m	....	....	....
	118		−81	13.9 m	....	....	γ, 0.511, 0.55, 0.60, 1.15
	119			19.5 m	....	....	γ, 0.26, 0.511, 0.78
	120	−90,180	−83.8	1.35 h	....	....	β^+, 4.0; γ, 0.511, 0.56, 0.62, 1.52
	121	−92,270	−86.0	2.12 h	....	....	β^+, 1.2; γ, 0.212(90), 0.27(3), 0.32(6), 0.511
	122	−92,489	−86.15	3.5 m	....	....	β^+, 3.1; γ, 0.511, 0.564, 0.69, 0.78
	123	−94,270	−88	13.3 h	....	....	γ, 0.159(83); e^-, 0.127
	124	−93,754	−87.33	4.15 d	....	....	β^+, 2.14; γ, 0.511, 0.605(67), 0.644(12), 0.73(14), 1.37(3), 1.51(4), 1.69(14), 2.09(2), 2.26(2)
	125	−95,422	−88.88	60.2 d	....	900	γ, 0.035(7); e^-, 0.004, 0.030
	126	−94,369	−87.90	12.8 d	....	....	β^-, 1.25; β^+, 1.13; γ, 0.386(34), 0.667(33)
	127	−95,530	−88.984	....	100	6.4	....
	128	−94,162	−87.71	24.99 m	....	....	β^-, 2.12; γ, 0.441(14),

Table 3-6 (*Continued*)
TABLE OF NUCLIDES

Nuclide		Mass Excess		Half-life y=yr, d=day h=hr, m=min s=sec	Natural Abundance, %	Thermal Neutron Absorption Cross Section σ, barns (mode if not n,γ)	Major Radiations, Energies in MeV and γ Intensities, %
Z Sym-bol	Mass Number	μmu	MeV				
	129	−95,013	−88.50	1.7×10^7 y	....	28	0.528(1), 0.743, 0.969 β^-, 0.150; e^-, 0.005, 0.034; γ, 0.040(9)
	130	−93,324	−86.89	12.3 h	....	18	β^-, 1.7, 1.04; γ, 0.419(35), 0.538(99), 0.669(100), 0.743(87), 1.15(12)
	131	−93,873	−87.441	8.05 d	....	~0.7	β^-, 0.806, 0.606; e^-, 0.046, 0.330; γ, 0.080(3), 0.284(5), 0.364(82), 0.637(7), 0.723(2)
	132	−92,019	−85.71	2.26 h	....	....	β^-, 2.12; γ, 0.24(1), 0.52(20), 0.67(144), 0.773(89), 0.955(22), 1.14(6), 1.28(7), 1.40(14), 1.45(1), 1.91(1), 1.99(1)
	133	−92,250	−85.9	20.3 h	....	....	β^-, 1.27; γ, 0.53(90)
	134	−90,150	−84.0	52.0 m	....	....	β^-, 2.43; γ, 0.135(3), 0.41(8), 0.55(8), 0.61(18), 0.85(95), 0.89(65), 1.07(1), 1.15(10), 1.46(4), 1.62(5), 1.79(5)
	135	−89,980	−84	6.68 h	....	....	β^-, 1.4; γ, 0.42(7), 0.86(11), 1.04(9), 1.14(37), 1.28(34), 1.46(12), 1.72(19), 1.80(11)
	136	−85,260	−79.4	83 s	....	....	β^-, 7.0, 5.6; γ, 0.20(12), 0.27(18), 0.39(19), 1.32(95), 2.3(19), 2.63(10), 2.8(8), 3.2(5)
	137			22.0 s	....	....	β^-; n, 0.6 av.
	138			5.9 s	....	....	β^-; n
	139		‑	2.7 s	....	....	β^-; n

Table 3-6 (*Continued*)
TABLE OF NUCLIDES

Nuclide		Mass Excess		Half-life y=yr, d=day h=hr, m=min s=sec	Natural Abundance, %	Thermal Neutron Absorption Cross Section σ, barns (mode if not n,γ)	Major Radiations, Energies in MeV and γ Intensities, %
Z Symbol	Mass Number	μmu	MeV				
$_{54}$Xe	118			6 m	....	....	γ, 0.05, 0.511
	120			40 m	....	....	γ, 0.055, 0.073, 0.176, 0.76
	121	−88,200	−82.2	39 m	....	....	β^+, 2.8; γ, 0.080, 0.096, 0.132, 0.437, 0.511
	122			20.1 h	....	....	γ, 0.060, 0.090, 0.110, 0.148, 0.180, 0.345; e^-, 0.058, 0.116
	123	−91,270	−85	2.08 h	....	....	β^+, 1.51; e^-, 0.115, 0.144, 0.295; γ, 0.090, 0.110, 0.149, 0.178, 0.329, 0.511, 0.68, 0.90, 1.10
	124	−93,880	−87.5	....	0.096	110	....
	125	−93,380	−87	16.8 h	....	....	γ, 0.055, 0.188, 0.242; e^-, 0.022, 0.050, 0.154, 0.182, 0.209
	125 m			55 s	....	....	γ, 0.075, 0.111
	126	−95,712	−89.15	....	0.090	~2	....
	127	−94,780	−88.54	36.41 d	....	....	γ, 0.058(1), 0.145(4), 0.172(22), 0.203(65), 0.375(20); e^-, 0.024, 0.112, 0.139, 0.170, 0.198
	127 m			75 s	....	....	γ, 0.125, 0.175
	128	−96,460	−89.85	....	1.919	<5	....
	129	−95,216	−88.692	....	26.44	25	....
	129 m		−88.456	8.0 d	....	....	γ, 0.040(9), 0.197(6); e^-, 0.005, 0.034, 0.162, 0.191
	130	−96,491	−89.88	....	4.08	<5	....
	131	−94,915	−88.411	....	21.18	85	....
	131 m		−88.247	11.8 d	....	....	γ, 0.164(2); e^-, 0.129, 0.159
	132	−95,839	−89.272	....	26.89	<5	....
	133	−94,185	−87.73	5.270 d	....	190	β^-, 0.346; e^-, 0.045, 0.075; γ, 0.081(37)
	133 m		−87.50	2.26 d	....	....	γ, 0.233(14); e^-, 0.198, 0.227
	134	−94,603	−88.121	....	10.4	<5	....

Table 3-6 (*Continued*)
TABLE OF NUCLIDES

Z Symbol	Mass Number	Mass Excess μmu	Mass Excess MeV	Half-life y = yr, d = day h = hr, m = min s = sec	Natural Abundance, %	Thermal Neutron Absorption Cross Section σ, barns (mode if not n,γ)	Major Radiations, Energies in MeV and γ Intensities, %
	135	−92,980	−86.6	9.14 h	....	2.7×10⁶	β⁻, 0.92; e⁻, 0.214; γ, 0.250(91), 0.61(3)
	135 m		−86.1	15.6 m	....	....	γ, 0.527(80); e⁻, 0.493, 0.522
	136	−92,779	−86.42	....	8.87	0.15	....
	137	−88,900	−82.8	3.9 m	....	....	β⁻, 4.1; γ, 0.455(33)
	138	−86,190	−80.9	17.5 m	....	....	β⁻, 2.4; γ, 0.16(33), 0.26(100), 0.42(40), 0.51(8), 1.78(66), 2.02(58)
	139	−82,160	−76.5	43 s	....	....	γ, 0.18(41), 0.22(100), 0.30(57), 1.15(23)
	140			16.0 s	....	....	γ, 0.13
	141			1.7 s	....	....	β⁻
	142			~1.5 s	....	....	....
	143			1.0 s	....	....	β⁻
	144			~1 s	....	....	β⁻
₅₅Cs	123			8.0 m	....	....	β⁺
	125	−90,090	−84	45 m	....	....	β⁺, 2.05; e⁻, 0.077, 0.107; γ, 0.112, 0.511
	126	−90,560	−84.4	1.6 m	....	....	β⁺, 3.8; γ, 0.386(38), 0.511
	127	−92,520	−86.4	6.2 h	....		γ, 0.125(10), 0.406(72), 0.511; e⁻, 0.090, 0.119, 0.371; β⁺, 1.08
	128	−92,241	−85.92	3.8 m	....	....	β⁺, 2.9; e⁻, 0.407; γ, 0.441(27), 0.511, 0.528, 0.576, 0.97(1), 1.12(1)
	129	−94,040	−88	32.1 h	....	....	γ, 0.040(2), 0.280(3), 0.320(4), 0.375(48), 0.416(25), 0.550(5); e⁻, 0.005, 0.034, 0.057, 0.336, 0.376
	130	−93,280	−86.89	30 m	....	....	β⁺, 1.97; β⁻, 0.442; γ, 0.511
	131	−94,534	−88.06	9.70 d	....	....	Xe X-rays
	132	−93,607	−87.19	6.59 d	....	....	β⁺, 0.40; γ, 0.48(4), 0.668(99), 1.138(1), 1.320(1)
	133	−94,645	−88.16	....	100	28	....

Table 3-6 (*Continued*)
TABLE OF NUCLIDES

| Nuclide | | Mass Excess | | Half-life | Natural | Thermal Neutron | Major Radiations, |
Z Sym-bol	Mass Number	μmu	MeV	y = yr, d = day h = hr, m = min s = sec	Abundance, %	Absorption Cross Section σ, barns (mode if not n,γ)	Energies in MeV and γ Intensities, %
	134	−93,177	−86.79	2.046 y	....	136	β^-, 0.662; γ, 0.57(23), 0.605(98), 0.796(99), 1.038(1), 1.168(2), 1.365(3)
	134 *m*		−86.65	2.895 h	....	....	γ, 0.128(14); e^-, 0.005, 0.009, 0.092, 0.122; β^-, 0.55
	135	−94,230	−87.8	3.0×10⁶ y	....	8.7	β^-, 0.21; no γ
	135 *m*		−86.2	53 m	....	....	γ, 0.781(100), 0.840(96); e^-, 0.745, 0.775, 0.804
	136	−92,660	−86.6	13.7 d	....	....	β^-, 0.657, 0.341; e^-, 0.116, 0.126, 0.158, 0.302; γ, 0.067(11), 0.086(6), 0.16(36), 0.273(18), 0.340(53), 0.818(100), 1.05(82), 1.25(20)
	137	−93,230	−86.9	30.0 y	....	0.11	β^-, 1.176, 0.514; e^-, 0.624, 0.656; γ, 0.662(85)
	138	−89,200	−83.7	32.2 m	....	....	β^-, 3.40; γ, 0.463(23), 0.55(8), 1.01(25), 1.426(73), 2.21(18), 2.63(9)
	139	−87,100	−81.1	9.5 m	....	....	γ, 0.50, 0.63, 0.80, 1.28, 1.65, 1.90, 2.08
	140	−82,890	−77	66 s	....	....	γ, 0.59, 0.88, 1.14, 1.62, 1.85, 2.06, 2.32, 2.72, 3.15
	141			24 s	....	....	....
	142			2.3 s	....	....	....
	143			2.0 s	....	....	....
₅₆Ba	123			2.0 m	....	....	....
	125			6.5 m	....	....	....
	126			97 m	....	....	γ, 0.23(100), 0.70(33), 0.9
	127	−88,660	−83	10.0 m	....	....	β^+
	128	−91,490	−85	2.43 d	....	....	γ, 0.134, 0.278; e^-, 0.128, 0.242

Table 3-6 (*Continued*)
TABLE OF NUCLIDES

Nuclide Z Sym-bol	Mass Number	Mass Excess μmu	MeV	Half-life y=yr, d=day h=hr, m=min s=sec	Natural Abundance, %	Thermal Neutron Absorption Cross Section σ, barns (mode if not n,γ)	Major Radiations, Energies in MeV and γ Intensities, %
	129	−91,410	−85	2.61 h	....	....	β+, 1.42; e−, 0.017, 0.048, 0.093, 0.142, 0.171; γ, 0.129(26), 0.182(100), 0.21(65), 0.511, 1.45(42)
	130	−93,755	−87.33	....	0.101	8.8	....
	131	−93,284	−86.89	12.0 d	....	....	γ, 0.124(28), 0.216(19), 0.25(5), 0.373(13), 0.496(48), 0.60(3), 0.924(1), 1.048(1); e−, 0.019, 0.042, 0.049, 0.088, 0.097, 0.118, 0.180, 0.460
	131 m		−86.71	14.6 m	....	....	γ, 0.107(40); e−
	132	−94,880	−88.4	....	0.097	7	....
	133	−94,121	−87.67	7.2 y	....	....	γ, 0.080(36), 0.276(7), 0.302(14), 0.356(69), 0.382(8); e−, 0.045, 0.075, 0.266, 0.319
	133 m		−87.39	38.9 h	....	....	γ, 0.276(17); e−, 0.006, 0.011, 0.238, 0.270
	134	−95,388	−88.85	....	2.42	<4	....
	135	−94,450	−88.0	....	6.59	5	....
	135 m		−87.7	28.7 h	....	....	γ, 0.268(16); e−, 0.231, 0.262
	136	−95,700	−89.1	....	7.81	<1	....
	136 m		−87.1	0.32 s	....	....	γ, 0.164(40), 0.818(100), 1.05(100)
	137	−94,500	−88.0	....	11.32	4	....
	137 m		−87.4	2.554 m	....	....	γ, 0.662(89); e−, 0.624, 0.656
	138	−95,000	−88.5	....	71.66	0.4	....
	139	−91,400	−85.1	82.9 m	....	4	β−, 2.3; e−, 0.126, 0.159; γ, 0.166(23), 1.43
	140	−89,435	−83.31	12.80 d	....	<20	β−, 1.02; e−, 0.024, 0.029; γ, 0.030(11),

Table 3-6 (*Continued*)
TABLE OF NUCLIDES

Nuclide		Mass Excess		Half-life y=yr, d=day h=hr, m=min s=sec	Natural Abundance, %	Thermal Neutron Absorption Cross Section σ, barns (mode if not n.γ)	Major Radiations, Energies in MeV and γ Intensities, %
Z Symbol	Mass Number	μmu	MeV				
	141	−85,950	−80.1	18 m	....	....	0.163(6), 0.305(6), 0.438(5), 0.537(34) β⁻, 3.0; γ, 0.118(10), 0.193(100), 0.28(50), 0.35(20), 0.46(30), 0.64(20), 0.73(7), 0.86(6), 0.93(3), 1.19(8), 1.29(3), 1.42(4), 1.65(3)
	142	−83,650	−77.9	11 m	....	....	β⁻, 1.7; γ, 0.080(30), 0.26(100), 0.89(40), 0.97(15), 1.08(10), 1.20(35)
₅₇La	143			12 s	....	....	β⁻
	125			<1 m	....	....	....
	126			1.0 m	....	....	γ, 0.256, 0.511
	127			3.5 m	....	....	....
	128			4.2 m	....	....	γ, 0.279, 0.511
	129	−87,110	−81	10.0 m	....	....	....
	130	−87,740	−82	8.7 m	....	....	γ, 0.356, 0.45, 0.511, 0.55, 0.72, 0.81, 0.91, 1.01, 1.19, 1.45, 1.55
	131	−90,110	−83.9	56 m	....	....	β⁺, 1.94; e⁻, 0.078; γ, 0.115(23), 0.169(5), 0.214(8), 0.285(17), 0.364(20), 0.417(20), 0.455(8), 0.511, 0.597(7), 0.878(4)
	132	−89,700	−83.1	4.5 h	....	....	β⁺, 3.8; γ, 0.47, 0.511, 0.56, 0.66, 0.90, 1.03, 1.22, 1.58, 1.92
	133	−91,760	−85.5	4.0 h	....	....	γ, 0.511, 0.8; β⁺, 1.2; e⁻, 0.26
	134	−91,340	−85.1	6.8 m	....	....	β⁺, 2.7; γ, 0.511, 0.605(6)
	135	−93,110	−87.0	19.4 h	....	....	γ, 0.481(2), 0.588,

Table 3-6 (*Continued*)
TABLE OF NUCLIDES

Z Symbol	Mass Number	μmu	MeV	Half-life y=yr, d=day h=hr, m=min s=sec	Natural Abundance, %	Thermal Neutron Absorption Cross Section σ, barns (mode if not n,γ)	Major Radiations, Energies in MeV and γ Intensities, %
							0.87; e^-, 0.181, 0.444, 0.475
	136	−92,620	−86.3	9.5 m	. . . .	. . . .	β^+, 1.9; γ, 0.511, 0.818(3)
	137	−93,960	−88	6×10^4 y	. . . .	. . . .	Ba X-rays
	138	−93,090	−86.7	1.12×10^{11} y	0.089	. . . .	β^-, 0.21; γ, 0.81(30), 1.426(70)
	139	−93,860	−87.43	. . . .	99.911	8.9	. . . .
	140	−90,562	−84.36	40.22 h	. . . .	. . . .	β^-, 2.175, 1.69, 1.36; γ, 0.329(20), 0.487(40), 0.815(19), 0.923(10), 1.596(96), 2.53(3)
	141	−89,172	−83.06	3.87 h	. . . .	. . . .	β^-, 2.43; γ, 1.37(2)
	142	−86,020	−80.1	92.5 m	. . . .	. . . .	β^-, 4.51; γ, 0.65(48), 0.90(9), 1.01(5), 1.06(4), 1.55(5), 1.74(5), 1.91(9), 2.06(6), 2.41(15), 2.55(11), 2.99(5), 3.31(2), 3.65(2)
	143	−84,130	−78.4	14.0 m	. . . .	. . . .	β^-, 3.3; γ, 0.62(100), 0.80(44), 1.07(26), 1.17(57), 1.58(28), 1.98(35), 2.56(27)
$_{58}$Ce	129			~13 m	. . . .	. . . .	γ, 0.080, 0.32, 0.75
	130			30 m	. . . .	. . . .	γ, 0.13
	132	−88,410	−82	4.2 h	. . . .	. . . .	γ, 0.18
	133	−88,750	−83	6.3 h	. . . .	. . . .	β^+, 1.3; γ, 0.511, 1.8
	134	−91,190	−84.9	72.0 h	. . . .	. . . .	γ
	135	−90,860	−85	17.0 h	. . . .	. . . .	γ, 0.205(17), 0.265(100), 0.300(56), 0.39(10), 0.52(46), 0.59(98), 0.777(22), 0.821(22), 0.865(14), 0.901(10); e^-, 0.048, 0.078, 0.166, 0.225, 0.25; β^+, 0.81

Table 3-6 (*Continued*)
TABLE OF NUCLIDES

Nuclide		Mass Excess		Half-life y = yr, d = day h = hr, m = min s = sec	Natural Abundance, %	Thermal Neutron Absorption Cross Section σ, barns (mode if not *n,γ*)	Major Radiations, Energies in MeV and γ Intensities, %
Z Symbol	Mass Number	μmu	MeV				
	136	−92,900	−86.6	>2.9×10^{11} y	0.193	6.0	....
	137	−92,670	−86	9.0 h	....	....	γ, 0.446(2), 0.481, 0.698, 0.92; e⁻, 0.408
	137 *m*		−87	34.4 h	....	....	γ, 0.168, 0.255(11), 0.762, 0.825(1); e⁻, 0.214, 0.248
	138	−94,170	−87.7	....	0.250	1.0	....
	139	−93,570	−87.16	140 d	....	....	γ, 0.165(80); e⁻, 0.126, 0.159
	139 *m*		−86.41	54 s	....	....	γ, 0.746(93); e⁻, 0.706, 0.740
	140	−94,608	−88.13	....	88.48	0.6	....
	141	−91,781	−85.49	32.5 d	....	30	β⁻, 0.581; e⁻, 0.104, 0.139; γ, 0.145(48)
	142	−90,860	−84.63	>5×10^{16} y	11.07	1	....
	143	−87,673	−81.67	33 h	....	6	β⁻, 1.39; e⁻, 0.015, 0.051, 0.252; γ, 0.057(11), 0.293(46), 0.493(2), 0.668(7), 0.725(8), 0.88(1), 1.10(1)
	144	−86,409	−80.49	284 d	....	1.0	β⁻, 0.31; e⁻, 0.038, 0.092; γ, 0.080(2), 0.134(11)
	145	−82,730	−77	3.0 m	....	....	β⁻, 2.0; γ
	146	−81,330	−75.8	14 m	....	....	β⁻, 0.7; γ, 0.110(20), 0.142(42), 0.22(50), 0.27(12), 0.32(100)
	147			65 s	....	....	β⁻
	148			~43 s	....	....	β⁻
₅₉Pr	134			17 m	....	....	γ, 0.22, 0.30, 0.409, 0.511, 0.639, 0.96
	135			22 m	....	....	β⁺, 2.5; γ, 0.080, 0.22, 0.30, 0.511
	136			1.2 h	....	....	β⁺, 2.0; γ, 0.511
	137	−89,640	−84	1.5 h	....	....	β⁺, 1.7; γ, 0.511
	138	−89,540	−82.9	2.10 h	....	....	β⁺, 1.65; e⁻, 0.258, 0.292; γ, 0.298(77), 0.40(9), 0.511, 0.79(100), 1.04(100)

Table 3-6 (*Continued*)
TABLE OF NUCLIDES

Z Symbol	Mass Number	Mass Excess μmu	Mass Excess MeV	Half-life y=yr, d=day h=hr, m=min s=sec	Natural Abundance, %	Thermal Neutron Absorption Cross Section σ, barns (mode if not n,γ)	Major Radiations, Energies in MeV and γ Intensities, %
	139	−91,420	−85.0	4.5 h	....	....	β^+, 1.09; γ, 0.511, 1.35(1), 1.61
	140	−90,993	−84.78	3.39 m	....	....	β^+, 2.32; e^-, 1.862; γ, 0.511, 1.596
	141	−92,404	−86.07	>2×10¹⁶ y	100	12	....
	142	−90,022	−83.85	19.2 h	....	20	β^-, 2.16; γ, 1.57(4)
	143	−89,219	−83.11	13.59 d	....	89	β^-, 0.933; no γ
	144	−86,752	−80.81	17.27 m	....	....	β^-, 2.99; γ, 0.695(2), 1.487, 2.186(1)
	145	−85,524	−79.66	5.98 h	....	....	β^-, 1.80; γ, 0.072, 0.68, 0.75, 0.92, 0.98, 1.05, 1.16
	146	−82,410	−76.8	24.0 m	....	....	β^-, 3.7; γ, 0.455(77), 0.74(16), 0.78(15), 0.92(6), 1.37(6), 1.51(27), 1.72(4), 2.23(4), 2.39(3), 2.73(2)
	147	−81,200	−75.5	12.0 m	....	....	β^-, 2.1; γ, 0.078(17), 0.127(9), 0.32(47), 0.56(39), 0.61(10), 0.65(24), 1.26(11)
	148	−78,090	−72.9	2.0 m	....	....	β^-, 4.2; γ, 0.30
	149			2.3 m	....	....	β^-, 2.8; γ, 0.08, 0.155, 0.325, 0.36, 0.745
$_{60}$Nd	137			55 m	....	....	β^+, 3; e^-, 0.067
	138			22 m	....	....	β^+, 2.4
	139 *m*	−88,420	−82	5.5 h	....	....	β^+, 3.1; e^-, 0.072, 0.107, 0.189, 0.226; γ, 0.114(80), 0.327(50), 0.511(1400), 0.73(210), 0.82(70), 0.90(25), 0.983(70), 1.03(30), 1.24(20), 1.34(20), 1.48(10), 1.58(8), 2.05(10)

Table 3-6 (*Continued*)
TABLE OF NUCLIDES

Nuclide		Mass Excess		Half-life y = yr, d = day h = hr, m = min s = sec	Natural Abundance, %	Thermal Neutron Absorption Cross Section σ, barns (mode if not n,γ)	Major Radiations, Energies in MeV and γ Intensities, %
Z Sym-bol	Mass Number	μmu	MeV				
	140	−90,670	−84	3.3 d	....	....	Pr X-rays
	141	−90,472	−84.27	2.42 h	....	....	β^+, 0.79; γ, 0.145, 0.511, 1.14(2), 1.30(1)
	141 m		−83.52	64 s	....	....	γ, 0.755
	142	−92,337	−86.01	....	27.13	17	....
	143	−90,221	−84.04	....	12.20	330	....
	144	−89,961	−83.80	2.4×10^{15} y	23.87	5	α, 1.83
	145	−87,462	−81.47	....	8.29	50	....
	146	−86,914	−80.96	....	17.18	2	....
	147	−83,926	−78.18	11.06 d	....	....	β^-, 0.81; e^-, 0.046, 0.084; γ, 0.091(28), 0.319(3), 0.43(4), 0.533(13)
	148	−83,131	−77.44	....	5.72	4	....
	149	−79,878	−74.41	1.8 h	....	....	β^-, 1.5; e^-, 0.051, 0.068, 0.079, 0.090, 0.165, 0.195; γ, 0.114(18), 0.156(4), 0.210(27), 0.27(26), 0.327(5), 0.424(9), 0.541(10), 0.654(9)
	150	−79,085	−73.67	$>10^{16}$ y	5.60	1.5	....
	151	−76,230	−71.0	12 m	....	....	β^-, 2.0; e^-, 0.072; γ, 0.086(5), 0.118(40), 0.138(6), 0.174(10), 0.256(11), 0.425(5), 0.737(5), 0.797(3), 1.122(2), 1.180(9)
$_{61}$Pm	141	−86,590	−80.7	22 m	....	....	β^+, 2.6; γ, 0.195(13), 0.511
	142	−87,180	−81.2	40 s	....	....	β^+, 3.78; γ, 0.511
	143	−89,010	−82.9	0.73 y	....	....	γ, 0.742(47); e^-, 0.698
	144	−87,490	−82	0.96 y	....	....	γ, 0.474(45), 0.615(99), 0.695(99); e^-, 0.430, 0.571, 0.651

Table 3-6 (*Continued*)
TABLE OF NUCLIDES

Nuclide		Mass Excess		Half-life y=yr, d=day h=hr, m=min s=sec	Natural Abundance, %	Thermal Neutron Absorption Cross Section σ, barns (mode if not *n*,γ)	Major Radiations, Energies in MeV and γ Intensities, %
Z Sym-bol	Mass Number	μmu	MeV				
	145	−87,309	−81.33	17.7 y	....	....	γ, 0.067(1), 0.072(2); e^-, 0.023, 0.028, 0.061
	146	−85,368	−79.52	4.4 y	....	....	β^-, 0.78; γ, 0.453(65), 0.75(65)
	147	−84,892	−79.08	2.62 y	....	120	β^-, 0.224; no γ
	148	−82,579	−76.89	5.4 d	....	∼2000	β^-, 2.48; γ, 0.551(27), 0.914(15), 1.465(23)
	148 *m*		−76.75	41.8 d	....	30,000	β^-, 0.69; e^-, 0.031, 0.053, 0.091, 0.242, 0.503, 0.583; γ, 0.289(13), 0.413(17), 0.551(95), 0.630(87), 0.727(36), 0.916(21), 1.015(20)
	149	−81,670	−76.07	53.1 h	....	....	β^-, 1.07; γ, 0.286(2), 0.58, 0.85
	150	−79,040	−73.6	2.68 h	....	....	β^-, 3.05; γ, 0.334(71), 0.406(7), 0.71(8), 0.831(18), 0.88(12), 1.165(23), 1.33(22), 1.75(10), 1.96(3), 2.06(1), 2.53(1)
	151	−78,802	−73.40	27.8 h	....	....	β^-, 1.19; e^-, 0.003, 0.018, 0.053, 0.058; γ, 0.07(5), 0.10(7), 0.17(18), 0.24(5), 0.275(6), 0.340(21), 0.45(5), 0.66(3), 0.72(6)
	152	−76,490	−71	6.5 m	....	....	β^-, 2.2; γ, 0.122, 0.245
	153	−75,970	−70.8	5.5 m	....	....	β^-, 1.65; γ, 0.12, 0.18
	154			2.5 m	....	....	β^-, 2.5
$_{62}$Sm	142			73 m	....	....	γ, 0.15–0.35, 0.511
	143	−85,450	−79.6	9.0 m	....	....	γ, 0.511

Table 3-6 (*Continued*)
TABLE OF NUCLIDES

Nuclide		Mass Excess		Half-life y=yr, d=day h=hr, m=min s=sec	Natural Abundance, %	Thermal Neutron Absorption Cross Section σ, barns (mode if not n,γ)	Major Radiations, Energies in MeV and γ Intensities, %
Z Symbol	Mass Number	μmu	MeV				
	143 m		−78.8	64 s	....	....	γ, 0.748
	144	−88,011	−81.98	....	3.16	∼0.7	....
	145	−86,606	−80.67	340 d	....	∼100	γ, 0.061(13), 0.485; e−, 0.016, 0.054
	146	−87,008	−81.05	7×10⁷ y	<2×10⁻⁷	....	α, 2.46
	147	−85,133	−79.30	1.05×10¹¹ y	15.07	∼90	α, 2.23
	148	−85,209	−79.37	>2×10¹⁴ y	11.27	....	....
	149	−82,820	−77.15	>1×10¹⁵ y	13.82	41,500	....
	150	−82,724	−77.06	....	7.47	100	....
	151	−80,081	−74.59	∼87 y	....	15,000	β−, 0.076; e−, 0.014, 0.020; γ, 0.022(4)
	152	−80,244	−74.75	....	26.63	210	....
	153	−77,898	−72.56	46.8 h	....	....	β−, 0.80; e−, 0.022, 0.055, 0.062, 0.095, 0.101; γ, 0.070(5), 0.103(28), 0.41–0.64(1)
	154	−77,718	−72.39	....	22.53	5	....
	155	−75,299	−70.14	23.5 m	....	....	β−, 1.53; e−, 0.056, 0.097, 0.103; γ, 0.104(73), 0.246(4)
	156	−74,431	−69.33	9.4 h	....	....	β−, 0.72; e−, 0.014, 0.021, 0.030, 0.039; γ, 0.088(30), 0.166(10), 0.204(20), 0.25(5), 0.291(3)
	157			0.5 m	....	....	γ, 0.57
₆₃Eu	143			2.3 m	....	....	β+, 4.0; γ, 0.511
	144		−75.66	10.5 s	....	....	β+, 5.2; γ, 0.511
	145	−83,610	−77.9	5.9 d	....	....	γ, 0.53, 0.656(30), 0.766(10), 0.894(100), 1.66(16), 2.00(8); e−, 0.063, 0.103, 0.847
	146	−82,862	−77.18	4.59 d	....	....	γ, 0.511, 0.634(77), 0.666(12), 0.71(13), 0.749(100), 0.90(8), 1.058(7), 1.16(6), 1.298(6), 1.408(5), 1.535(8); β+,

Table 3-6 (*Continued*)
TABLE OF NUCLIDES

Nuclide		Mass Excess		Half-life y = yr, d = day h = hr, m = min s = sec	Natural Abundance, %	Thermal Neutron Absorption Cross Section σ, barns (mode if not n,γ)	Major Radiations, Energies in MeV and γ Intensities, %
Z Sym- bol	Mass Number	μmu	MeV				
	147	−83,200	−77.5	21.5 d	....	....	2.11, 1.47; e⁻, 0.586, 0.702 γ, 0.122(20), 0.198(24), 0.600(7), 0.680(11), 0.800(6), 0.957(9), 1.079(9), 1.25(1); e⁻, 0.030, 0.075, 0.114, 0.151; α, 2.91
	148	−81,890	−76.26	54 d	....	....	γ, 0.413(18), 0.551(120), 0.62(90), 0.72(18), 0.872(7), 0.917(5), 0.967(5), 1.033(7), 1.16(5), 1.345(8), 1.62(11); e⁻, 0.02–0.04, 0.51, 0.193, 0.366, 0.505, 0.544, 0.584; β⁺, 0.92; α, 2.63
	149	−82,000	−76	106 d	....	....	γ, 0.277(10), 0.328(10); e⁻, 0.015, 0.021, 0.230, 0.281
	150	−80,311	−74.81	12.55 h	....	....	β⁻, 1.01; β⁺, 1.24; γ, 0.334(4), 0.406(3), 0.511, 0.619, 0.713, 0.831(1), 0.921, 1.165, 1.224, 1.630, 1.964
	150			~5 y	....	....	γ, 0.334(96), 0.439(86), 0.584(60), 0.74(21), 1.049(9), 1.248(5), 1.347(4); e⁻, 0.287, 0.327, 0.392
	151	−80,162	−74.67	....	47.77	5,900	....
	152	−78,251	−72.89	12.7 y	....	5,000	β⁻, 1.48; e⁻, 0.075,

Table 3-6 (*Continued*)
TABLE OF NUCLIDES

Nuclide		Mass Excess		Half-life y = yr, d = day h = hr, m = min s = sec	Natural Abundance, %	Thermal Neutron Absorption Cross Section σ, barns (mode if not n,γ)	Major Radiations, Energies in MeV and γ Intensities, %
Z Sym-bol	Mass Number	μmu	MeV				
	152 m_1		−72.84	9.3 h	....	....	0.115, 0.120; β^+, 0.71; γ, 0.122(37), 0.245(8), 0.344(27), 0.779(14), 0.965(15), 1.087(12), 1.113(14), 1.408(22)
	152 m_2		−72.74	96 m	....	....	β^-, 1.88; e^-, 0.075, 0.115, 0.120; β^+, 0.89; γ, 0.122(8), 0.344(3), 0.842(13), 0.963(12), 1.315(1), 1.389(1)
	153	−78,758	−73.36	....	52.23	320	γ, 0.090(74); e^-, 0.010, 0.016, 0.032, 0.039 ....
	154	−76,947	−71.68	16 y	....	1,400	β^-, 1.85, 0.87; e^-, 0.073, 0.115, 0.122; γ, 0.123(38), 0.248(7), 0.593(6), 0.724(21), 0.759(5), 0.876(12), 1.00(31), 1.278(37)
	155	−77,070	−71.79	1.811 y	....	13,000	β^-, 0.25; e^-, 0.011, 0.017, 0.036, 0.054, 0.078, 0.082; γ, 0.087(32), 0.105(20)
	156	−75,198	−70.05	15.4 d	....	....	β^-, 2.45; e^-, 0.039, 0.081, 0.087; γ, 0.089(8), 0.646(7), 0.723(6), 0.812(9), 1.07(11), 1.15(14), 1.24(16), 1.97(7), 2.098(3), 2.19(5)
	157	−74,610	−69.43	15.1 h	....	....	β^-, 1.3; e^-, 0.004, 0.014, 0.046,

Table 3-6 (*Continued*)
TABLE OF NUCLIDES

Nuclide		Mass Excess		Half-life y = yr, d = day h = hr, m = min s = sec	Natural Abundance, %	Thermal Neutron Absorption Cross Section σ, barns (mode if not n,γ)	Major Radiations, Energies in MeV and γ Intensities, %
Z Sym- bol	Mass Number	μmu	MeV				
$_{64}$Gd	158	−72,060	−67.1	46 m	....	....	0.056; γ, 0.055(5), 0.064(27), 0.32(5), 0.37(14), 0.413(27), 0.477(5), 0.623(6) β⁻, 2.5; e⁻; γ, 0.080(100), 0.182, 0.52(25), 0.61(8), 0.95(95), 1.11(11), 1.19(16)
	159	−71,160	−66.02	18.1 m	....	....	β⁻, 2.6; γ, 0.07(42), 0.09(18), 0.15(14), 0.22(5), 0.67(21), 0.73(10), 0.8(11), 1.1(11), 1.5(5)
	160	−69,000	−64	∼2.5 m	....	....	β⁻, 3.6; no γ
	145			25 m	....	....	β⁺, 2.4; γ, 0.511, 0.80(9), 1.03(10), 1.75(100)
	146	−81,680	−76	50 d	....	....	γ, 0.078(30), 0.115(100), 0.155(45); e⁻, 0.066, 0.106
	147	−80,830	−75	35 h	....	....	γ, 0.229(150), 0.39(85), 0.64(70), 0.77(60), 0.932(60), 1.10(19); e⁻, 0.181, 0.221, 0.321, 0.348, 0.388
	148	−81,899	−76.29	84 y	....	....	α, 3.18
	149	−80,700	−75.2	9.5 d	....	....	γ, 0.150(48), 0.299(26), 0.347(25), 0.750(11), 0.790(10), 0.94(5); e⁻, 0.101, 0.142, 0.250, 0.298; α, 3.01
	150	−81,395	−75.82	2.1 × 10⁶ y	....	....	α, 2.73
	151	−79,730	−74	120 d	....	....	γ, 0.0216(3), 0.154(7), 0.175(3), 0.244(7),

Table 3-6 (*Continued*)
TABLE OF NUCLIDES

Nuclide Z Symbol	Nuclide Mass Number	Mass Excess μmu	Mass Excess MeV	Half-life y=yr, d=day h=hr, m=min s=sec	Natural Abundance, %	Thermal Neutron Absorption Cross Section σ, barns (mode if not n,γ)	Major Radiations, Energies in MeV and γ Intensities, %
							0.308(1); e⁻, 0.014, 0.020, 0.105, 0.127, 0.167; α, 2.60
	152	−80,206	−74.71	1.1×10¹⁴ y	0.20	<180	α, 2.1
	153	−78,497	−73.12	242 d	....	....	γ, 0.070(2), 0.099(55); e⁻, 0.021, 0.049, 0.065, 0.101
	154	−79,071	−73.65	....	2.15	....	....
	155	−77,336	−72.04	....	14.7	58,000	....
	156	−77,825	−72.49	....	20.47	....	....
	157	−75,975	−70.77	....	15.68	2.4×10⁵	....
	158	−75,822	−70.63	....	24.9	3.4	....
	159	−73,632	−68.59	18.0 h	....	....	β⁻, 0.95; e⁻, 0.006, 0.049, 0.056; γ, 0.058(3), 0.363(9)
	160	−72,885	−67.89	....	21.9	0.8	....
	161	−70,280	−65.5	3.6 m	....	....	β⁻, 1.6; e⁻, 0.005, 0.026, 0.049, 0.055, 0.263, 0.309; γ, 0.102(11), 0.284(8), 0.315(25), 0.361(66)
₆₅Tb	147			24 m	....	....	γ, 0.305, 0.511
	148	−75,870	−70.7	70 m	....	....	β⁺, 4.6; γ, 0.511, 0.78, 1.12
	149	−76,650	−71.4	4.10 h	....	....	γ, 0.16, 0.35; e⁻, 0.115, 0.127, 0.157, 0.301, 0.338, 0.587; α, 3.95
	149 m			4.3 m	....	....	α, 3.99
	150	−76,252	−71.03	3.1 h	....	....	β⁺, 3.6; γ, 0.511, 0.637(100), 0.93(35)
	151	−76,850	−71.6	18 h	....	....	γ, 0.108(35), 0.18(18), 0.252(35), 0.288(32), 0.40, 0.44, 0.48, 0.60, 0.72, 0.87; e⁻, 0.058, 0.100, 0.130, 0.202, 0.237; α, 3.42
	152	75,720	−70.5	17.4 h	....	....	β⁺, 2.82; e⁻, 0.221, 0.263, 0.294, 0.336, 0.382,

Table 3-6 (*Continued*)
TABLE OF NUCLIDES

Nuclide		Mass Excess		Half-life y = yr, d = day h = hr, m = min s = sec	Natural Abundance, %	Thermal Neutron Absorption Cross Section σ, barns (mode if not *n,γ*)	Major Radiations, Energies in MeV and γ Intensities, %
Z Sym-bol	Mass Number	μmu	MeV				
							0.536, 0.565, 0.607; γ, 0.271(13), 0.344(100), 0.411(6), 0.586(14), 0.779(14), 0.974(10), 1.12(10), 1.31(11), 1.60(7), 1.95(8), 2.40(9), 2.70(6)
	152			4.0 m	. . . .	. . . .	γ, 0.14, 0.23, 0.511
	153	−76,510	−71	55 h	. . . .	. . . .	γ, 0.083(11), 0.11(12), 0.17(9), 0.212(30), 0.250, 0.33, 0.88; *e*⁻, 0.012, 0.034, 0.037, 0.040, 0.044, 0.052, 0.057, 0.162
	154	−75,420	−70	21.0 h	. . . .	. . . .	γ, 0.123, 0.248, 0.30, 0.347, 0.53, 0.65; *e*⁻, 0.073, 0.115, 0.122, 0.198
	154		−70	8.5 h	. . . .	. . . .	γ, 0.123, 0.248, 0.53, 0.65; *e*⁻, 0.073, 0.115, 0.122, 0.198
	155	−76,370	−71	5.6 d	. . . .	. . . .	γ, 0.087(37), 0.105(25), 0.163(8), 0.180(8), 0.262(7), 0.368(4); *e*⁻, 0.011, 0.034, 0.053, 0.078, 0.110, 0.129, 0.210
	156	−75,250	−70	5.1 d	. . . .	. . . .	γ, 0.089(17), 0.199(40), 0.356(13), 0.535(70), 1.065(12), 1.16(17), 1.22(29), 1.42(15), 1.65(5), 1.85(4); *e*⁻,

Table 3-6 (*Continued*)
TABLE OF NUCLIDES

Nuclide		Mass Excess		Half-life y=yr, d=day h=hr, m=min s=sec	Natural Abundance, %	Thermal Neutron Absorption Cross Section σ, barns (mode if not n,γ)	Major Radiations, Energies in MeV and γ Intensities, %
Z Symbol	Mass Number	μmu	MeV				
	156 m		−70	5.5 h	....	....	0.039, 0.081, 0.087, 0.149 γ; e−, 0.036, 0.081
	157	−75,910	−70.71	1.5×10² y	....	....	Gd X-rays
	158	−74,536	−69.43	1.2×10³ y	....	....	β−, 0.85; e−, 0.029, 0.044, 0.072, 0.078, 0.092, 0.132; γ, 0.080(12), 0.182(10), 0.782(10), 0.95(69), 1.110(2), 1.190(2)
	158 m		−69.32	10.5 s	....	....	e−, 0.060, 0.102; γ, 0.110(1)
	159	−74,649	−69.53	>5×10¹⁶ y	100	46	....
	160	−72,854	−67.85	72.1 d	....	525	β−, 1.74, 0.86; e−, 0.033, 0.079, 0.085; γ, 0.087(12), 0.197(6), 0.299(30), 0.879(31), 0.966(31), 1.178(15), 1.272(7)
	161	−72,428	−67.47	6.9 d	....	....	β−, 0.59, 0.52; e−, 0.017, 0.040, 0.048; γ, 0.026(21), 0.049(19), 0.057(5), 0.075(10)
	162	−70,190	−65	7.48 m	....	....	γ, 0.040(17), 0.081(8), 0.140(6), 0.180(26), 0.258(100), 0.81(44), 0.89(54)
	162		−65	2.24 h	....	....	....
	163	−69,440	−64.7	6.5 h	....	....	β−, 1.65; γ, 0.025, 0.235, 0.330, 0.510
₆₆Dy	164	−66,720	−62	23 h	....	....	....
	149			10–20 m	....	....	
	150	−74,410	−69	7.2 m	....	....	γ, 0.39, 0.511; α, 4.23
	151	−73,750	−69	18.0 m	....	....	α, 4.06; γ, 0.145, 0.511

Table 3-6 (*Continued*)
TABLE OF NUCLIDES

Nuclide		Mass Excess		Half-life y=yr, d=day h=hr, m=min s=sec	Natural Abundance, %	Thermal Neutron Absorption Cross Section σ, barns (mode if not n,γ)	Major Radiations, Energies in MeV and γ Intensities, %
Z Symbol	Mass Number	μmu	MeV				
	152	−75,271	−70.11	2.41 h	....	....	γ, 0.257; α, 3.65
	153	−74,260	−69.2	6.4 h	....	....	γ, 0.08; e⁻, 0.029, 0.047, 0.072, 0.091, 0.192, 0.202; α, 3.48
	154	−75,650	−70.5	>10 y	....	....	α, 2.85
	154 m			13 h	....	....	α, 3.37
	155	−74,120	−69	10.2 h	....	....	γ, 0.227(68), 0.52(8), 0.65(5), 0.74(4), 0.91(5), 1.000(6), 1.091(5), 1.16(6), 1.250(4), 1.39(3), 1.45(4), 1.66(2); β⁺, 1.08, 0.85; e⁻, 0.013, 0.038, 0.057, 0.175
	156	−76,070	−70.9	>1×10¹⁸ y	0.0524	~3	....
	157	−74,730	−70	8.1 h	....	....	γ, 0.326(91); e⁻, 0.009, 0.031, 0.052, 0.074, 0.274
	158	−75,551	−70.37	....	0.0902	100	....
	159	−74,241	−69.15	144 d	....	....	γ, 0.058(4), 0.348; e⁻, 0.006, 0.049, 0.056
	160	−74,798	−69.67	....	2.294	....	....
	161	−73,055	−68.05	....	18.88	600	....
	162	−73,197	−68.18	....	25.53	140	....
	163	−71,245	−66.36	....	24.97	130	....
	164	−70,800	−65.95	....	28.18	2,000	....
	165	−68,184	−63.51	139.2 m	....	4,700	β⁻, 1.29; e⁻, 0.039, 0.085; γ, 0.095(4), 0.280(1), 0.361(1), 0.633(1), 0.716(1)
	165 m		−63.40	1.26 m	....	....	β⁻, 1.04, 0.89; e⁻, 0.054, 0.100, 0.106; γ, 0.108(3), 0.152, 0.362(1), 0.514(2)
	166	−67,193	−62.59	81.5 h	....	....	β⁻, 0.48, 0.40; e⁻, 0.019, 0.027, 0.046; γ, 0.082(12), 0.372(1), 0.426(1)
	167			4.4 m	....	....	....
₆₇Ho	150			~20 s	....	....	....
	151			35.6 s	....	....	α, 4.51

Table 3-6 (*Continued*)
TABLE OF NUCLIDES

Nuclide		Mass Excess		Half-life y = yr, d = day h = hr, m = min s = sec	Natural Abundance, %	Thermal Neutron Absorption Cross Section σ, barns (mode if not n,γ)	Major Radiations, Energies in MeV and γ Intensities, %
Z Symbol	Mass Number	μmu	MeV				
	151			42 s	....	....	α, 4.60
	152			52.3 s	....	....	α, **4.45**
	152	−68,440	−63.8	2.4 m	....	....	α, 4.38
	153	−69,730	−65.0	9 m	....	....	α, 3.92
	154	−69,740	−65	7 m	....	....	γ, 0.335, 0.511
	155			50 m	....	....	β^+, 2.1; γ, 0.092, 0.138, 0.511
	156			55 m	....	....	γ, 0.138(100), 0.266(99), 0.367(23), 0.511, 0.685, 0.89, 1.20, 1.41; e^-, 0.084, 0.130, 0.213; β^+, 2.9, 1.8
	157			14 m	....	....	γ, 0.087, 0.152, 0.190, 0.227, 0.511, 0.71, 0.86, 0.90, 1.20
	158	−71,210	−66.33	11.5 m	....	....	γ, 0.099, 0.218, 0.329, 0.412, 0.52, 0.647, 0.73, 0.86, 0.940, 1.21, 1.47, 1.6, 1.8, 2.05, 2.21, 2.87, 3.1; e^-, 0.045, 0.062, 0.091, 0.097, 0.164
	158 m		−66.26	29 m	....	....	γ, 0.099, 0.218, 0.32, 0.356, 0.412, 0.46, 0.52, 0.63, 0.73, 0.85, 0.95, 1.21, 1.47, 1.60, 1.80, 2.06, 2.20, 2.62; e^-, 0.029, 0.044, 0.072, 0.078, 0.092, 0.132; β^+, 1.32
	159	−72,310	−67	33 m	....	....	γ, 0.057, 0.080, 0.13, 0.18, 0.253, 0.309; e^-, 0.048, 0.071, 0.121, 0.198, 0.243, 0.256, 0.300
	159 m		−67	6.9 s	....	....	γ, 0.206; e^-, 0.150, 0.197
	160	−71,260	−66.4	25.6 m	....	....	see ^{160m}Ho
	160 m		−66.3	5.0 h	....	....	γ, 0.087(14), 0.197(20),

Table 3-6 (*Continued*)
TABLE OF NUCLIDES

Nuclide		Mass Excess		Half-life y=yr, d=day h=hr, m=min s=sec	Natural Abundance, %	Thermal Neutron Absorption Cross Section σ, barns (mode if not n,γ)	Major Radiations, Energies in MeV and γ Intensities, %
Z Sym-bol	Mass Number	μmu	MeV				
	161	−72,200	−67	2.4 h	....	....	0.539(5), 0.646(20), 0.729(50), 0.880(26), 0.965(37); e⁻, 0.033, 0.051, 0.058, 0.079, 0.085, 0.144, 0.188; β⁺, 1.9 γ, 0.026(23), 0.075(15), 0.157(1), 0.176(2); e⁻, 0.017, 0.024, 0.049, 0.069, 0.076
	161 m		−67	6.1 s	....	....	γ, 0.211(53); e⁻, 0.155, 0.202
	162	−70,878	−66.02	15 m	....	....	γ, 0.081(8), 0.511; β⁺, 1.10; e⁻, 0.27, 0.072, 0.079
	162 m		−65.92	68 m	....	....	γ, 0.081(10), 0.185(26), 0.283(12), 0.940(13), 1.224(24); e⁻, 0.027, 0.036, 0.048, 0.072, 0.079, 0.131, 0.177
	163	−71,234	−66.35	>10³ y	....	....	....
	163 m		−66.05	1.1 s	....	....	γ, 0.305; e⁻, 0.249, 0.296
	164	−69,610	−64.84	36.7 m	....	....	β⁻, 0.99; e⁻, 0.019, 0.034, 0.065, 0.071, 0.083, 0.089; γ, 0.073, 0.091
	165	−69,579	−64.81	>6×10¹⁶ y	100	64	....
	166	−67,711	−63.07	26.9 h	....	....	β⁻, 1.84; e⁻, 0.023, 0.072, 0.078; γ, 0.081(5), 1.380(1), 1.582, 1.663
	166 m		−63.06	1.2×10³ y	....	....	β⁻; e⁻, 0.023, 0.072, 0.078, 0.127, 0.175; γ, 0.081(12), 0.184(90),

Table 3-6 (*Continued*)
TABLE OF NUCLIDES

Nuclide		Mass Excess		Half-life	Natural	Thermal Neutron	Major Radiations,
Z Sym-bol	Mass Number	μmu	MeV	y = yr, d = day h = hr, m = min s = sec	Abundance, %	Absorption Cross Section σ, barns (mode if not n,γ)	Energies in MeV and γ Intensities, %
	167	−66,870	−62.3	3.1 h	....	....	0.280(30), 0.412(12), 0.532(12), 0.711(58), 0.810(60), 0.830(11) β⁻, 0.96; e⁻, 0.024, 0.048, 0.073, 0.150, 0.180, 0.199, 0.263; γ
	168	−64,070	−59.7	3.3 m	....	....	β⁻, 2.2; γ, 0.85
	169	−63,140	−58.8	4.8 m	....	....	β⁻, 1.95; γ, 0.15, 0.68, 0.76, 0.84, 0.92
₆₈Er	170	−59,930	−55.8	45 s	....	....	β⁻, 3.1; γ, 0.43
	152			10.7 s	....	....	α, 4.80
	153			36 s	....	....	α, 4.67
	154	−67,240	−63	5 m	....	....	α, 4.15
	157			~25 m	....	....	γ, 0.117, 0.386, 0.511, 1.32, 1.66, 1.82, 2.0
	158			2.3 h	....	....	γ, 0.072, 0.250, 0.315, 0.387, 0.511, 0.875, 0.906, 0.978; e⁻, 0.058, 0.065; β⁺, 0.8
	159			36 m	....	....	γ, 0.206, 0.37, 0.511, 0.62, 0.84, 1.20, 1.40, 1.80, 2.60; e⁻, 0.150, 0.197
	160			29.4 h	....	....	Ho X-rays
	161	−70,050	−65	3.1 h	....	....	γ, 0.211(9), 0.305(3), 0.592(8), 0.826(63), 1.17(8), 1.37(5), 1.66(2); e⁻, 0.059, 0.065, 0.155, 0.202; β⁺, 1.2
	162	−71,260	−66.4	....	0.136	2	....
	163	−69,935	−65.14	75.1 m	....	....	γ, 0.43, 1.10; β⁺, 0.19
	164	−70,713	−65.87	....	1.56	1.7	....
	165	−69,181	−64.44	10.34 h	....	....	Ho X-rays
	166	−69,693	−64.92	....	33.41	12	....
	167	−67,940	−63.29	....	22.94	700	....

Table 3-6 (*Continued*)
TABLE OF NUCLIDES

Nuclide		Mass Excess		Half-life y=yr, d=day h=hr, m=min s=sec	Natural Abundance, %	Thermal Neutron Absorption Cross Section σ, barns (mode if not n,γ)	Major Radiations, Energies in MeV and γ Intensities, %
Z Sym-bol	Mass Number	μmu	MeV				
	167 *m*		−63.08	2.3 s	....	....	γ, 0.208(43); e⁻, 0.150, 0.199
	168	−67,617	−62.98	....	27.07	2	....
	169	−65,390	−60.91	9.6 d	....	....	β⁻, 0.34; e⁻, 0.006; γ, 0.008
	170	−64,440	−60.0	....	14.88	9	....
	171	−61,870	−57.6	7.52 h	....	....	β⁻, 1.49, 1.06; e⁻, 0.004, 0.052, 0.065, 0.102, 0.115; γ, 0.112(25), 0.124(9), 0.296(28), 0.308(63)
	172	−60,670	−56.5	49.5 h	....	....	β⁻, 0.89, 0.37; e⁻, 0.010, 0.020, 0.049, 0.058, 0.348; γ, 0.407(40), 0.610(40)
₆₉Tm	153			1.6 s	....	....	α, 5.10
	154			3.0 s	....	....	α, 5.04
	154			5 s	....	....	α, 4.96
	161	−66,270	−62	32 m	....	....	γ, 0.084, 0.106, 0.112, 0.145, 0.172; e⁻, 0.027, 0.036, 0.050, 0.055, 0.065, 0.075, 0.089, 0.115
	162	−66,010	−61.5	77 m	....	....	γ, 0.102(20), 0.236(10)
	162		−61.5	22 m	....	....	β⁺, 3.82; e⁻, 0.045, 0.093, 0.100
	163	−67,498	−62.87	1.8 h	....	....	γ, 0.104(8), 0.17(1), 0.240(5), 0.29(3), 0.34(3); e⁻, 0.047, 0.095, 0.184; β⁺, 1.1
	164	−66,459	−61.91	2.0 m	....	....	γ, 0.091(4), 0.356, 0.361, 0.391, 0.511, 0.773, 0.862, 0.907, 0.930; β⁺, 2.94; e⁻, 0.034, 0.083, 0.089
	165	−67,460	−62.87	30.1 h	....	....	γ, 0.054, 0.113, 0.243(50), 0.297(35),

Table 3-6 (*Continued*)
TABLE OF NUCLIDES

Nuclide		Mass Excess		Half-life y = yr, d = day h = hr, m = min s = sec	Natural Abundance, %	Thermal Neutron Absorption Cross Section σ, barns (mode if not n,γ)	Major Radiations, Energies in MeV and γ Intensities, %
Z Sym- bol	Mass Number	μmu	MeV				
	166	−66,490	−61.88	7.7 h			0.34(10), 0.44(5), 0.70(2), 0.807(15), 1.13(5), 1.30(1); e^-, 0.038, 0.045, 0.052, 0.056, 0.068, 0.161, 0.185, 0.233, 0.240; β^+, 0.30 β^+, 1.94; e^-, 0.023, 0.072, 0.079, 0.127; γ, 0.081, 0.19, 0.215, 0.46, 0.60, 0.69, 0.78, 1.180, 1.277, 1.378, 1.873, 2.06
	167	−66,970	−62.13	9.6 d			γ, 0.057(4), 0.208(43), 0.532(2); e^-, 0.048, 0.150, 0.199
	168	−65,770	−61.27	85 d			γ, 0.080(11), 0.19(77), 0.448(27), 0.63(14), 0.73(40), 0.82(88), 0.917(4), 1.280(3); e^-, 0.022, 0.071, 0.077, 0.127, 0.141
	169	−65,755	−61.25	$>5\times10^{16}$ y	100	125	
	170	−63,940	−59.6	134 d		150	β^-, 0.97; e^-, 0.023, 0.075, 0.082; γ, 0.084(3)
	171	−63,470	−59.1	1.92 y			β^-, 0.097; e^-, 0.057, 0.065; γ, 0.067
	172	−61,620	−57.4	63.6 h			β^-, 1.88; γ, 0.079(5), 0.181(2), 0.91(1), 1.09(7), 1.39(7), 1.46(7), 1.53(6), 1.61(5)
	173	−60,520	−56.4	8.2 h			β^-, 1.3, 0.89; e^-, 0.008, 0.056, 0.064; γ, 0.066(1), 0.399(89),

Table 3-6 (*Continued*)
TABLE OF NUCLIDES

Nuclide		Mass Excess		Half-life	Natural	Thermal Neutron	Major Radiations,
Z Symbol	Mass Number	μmu	MeV	y=yr, d=day h=hr, m=min s=sec	Abundance, %	Absorption Cross Section σ, barns (mode if not n,γ)	Energies in MeV and γ Intensities, %
	174		−54.6	5.5 m	....	....	0.465(8) β−, 2.5; no γ
	174	−58,030	−54.1	5.2 m	....	....	β−, 1.2; e−, 0.015, 0.067, 0.074; γ, 0.176(67), 0.273(85), 0.366(93), 0.50(15),0.99(89)
	175	−56,170	−52.3	20 m	....	....	β−, 2.0; γ, 0.51
	176	−52,810	−49.2	1.5 m	....	....	β−, 4.2; no γ
$_{70}$Yb	154			0.39 s	....	....	α, 5.33
	155			1.6 s	....	....	α, 5.21
	162			~24 m	....	....	e−, 0.032, 0.039
	164			75 m	....	....	Tm X-rays
	165	−64,560	−60	10.5 m	....	....	....
	166	−66,150	−61.6	57.5 h	....	....	γ, 0.082(17); e−, 0.023, 0.072
	167	−64,870	−60.17	17.7 m	....	....	γ, 0.113(90), 0.176(15); e−, 0.047, 0.055, 0.096
	168	−65,840	−61.3	....	0.140	11,000	....
	169	−64,470	−60	31.8 d	....	....	γ, 0.063(45), 0.110(18), 0.131(11), 0.177(22), 0.198(35), 0.308(10); e−, 0.004–0.011, 0.034, 0.050, 0.053, 0.071, 0.100, 0.118, 0.121, 0.139
	169 m		−60	46 s	....	....	e−, 0.014, 0.022
	170	−64,980	−60.5	....	3.03	....	....
	171	−63,570	−59.2	....	14.31	....	....
	171 m		−59.1	<<8 d	....	....	γ, 0.019, 0.076; e−, 0.010, 0.017, 0.067, 0.074
	172	−63,640	−59.3	....	21.82	....	....
	173	−61,940	−57.7	....	16.13	....	....
	174	−61,260	−57.1	....	31.84	46	....
	175	−58,860	−54.8	101 h	....	....	β−, 0.466; γ, 0.114(2), 0.283(4), 0.396(6); e−, 0.051, 0.102, 0.112, 0.333
	176	−57,320	−53.4	....	12.73	7	....

Table 3-6 (*Continued*)
TABLE OF NUCLIDES

| Nuclide | | Mass Excess | | Half-life | Natural | Thermal Neutron | Major Radiations, |
| | | | | y = yr, d = day | Abundance, | Absorption Cross | Energies in MeV |
Z Symbol	Mass Number	μmu	MeV	h = hr, m = min s = sec	%	Section σ, barns (mode if not n,γ)	and γ Intensities, %
	176 m		−52.4	11.7 s	....	....	γ, 0.19, 0.29, 0.39
	177	−54,590	−50.8	1.9 h	....	....	β−, 1.40; γ, 0.122(3), 0.151(16), 1.080(5), 1.241(3); e−, 0.059, 0.075, 0.088, 0.110, 0.140
	177 m		−50.5	6.5 s	....	....	γ, 0.104(65), 0.228(13); e−, 0.043, 0.094, 0.167, 0.219
71Lu	156			0.23 s	....	....	α, 5.54
	156			0.5 s	....	....	α, 5.43
	167	−61,610	−57.1	54 m	....	....	γ, 0.030, 0.18–0.24, 0.278, 0.372, 0.402, 0.511; e−, 0.020, 0.028, 0.039, 0.069, 0.076, 0.152, 0.178; β+, 1.5
	168	−60,910	−57	7.1 m	....	....	γ, 0.087(7), 0.223, 0.71, 0.90(10), 0.99(13), 1.41, 1.81, 2.1; e−; β+, 1.2
	169	−62,040	−58	34 h	....	....	γ, 0.063, 0.111, 0.191, 0.577; e−, 0.010, 0.014, 0.022, 0.026, 0.050, 0.053, 0.060, 0.066, 0.077; β+, 1.2
	169 m		−58	2.7 m	....	....	e−, 0.019, 0.027
	170	−61,170	−57.1	2.05 d	....	....	γ, 0.084(13), 0.193, 0.24, 1.01, 1.03, 1.17, 1.27, 1.41, 2.03, 2.32, 2.67, 2.89, 3.09; e−, 0.023, 0.075, 0.082; β+, 2.4
	170 m		−57.0	0.7 s	....	....	e−, 0.036, 0.044
	171	−61,860	−58	8.3 d	....	....	γ, 0.019(20), 0.075(8), 0.668(14), 0.741(68), 0.842(7); e−, 0.010, 0.017,

Table 3-6 (*Continued*)
TABLE OF NUCLIDES

Nuclide		Mass Excess		Half-life y=yr, d=day h=hr, m=min s=sec	Natural Abundance, %	Thermal Neutron Absorption Cross Section σ, barns (mode if not n,γ)	Major Radiations, Energies in MeV and γ Intensities, %
Z Sym-bol	Mass Number	μmu	MeV				
	171 *m*		−58	76 s	....	....	0.057, 0.066, 0.074 γ, 0.071; e⁻, 0.061, 0.069
	172	−60,740	−57	6.70 d	....	....	γ, 0.079(13), 0.182(26), 0.81(21), 0.90(45), 1.09(60); e⁻, 0.017, 0.029, 0.069, 0.077, 0.081, 0.120
	172 *m*		−57	3.7 m	....	....	e⁻, 0.032, 0.040
	173	−61,200	−57.0	1.37 y	....	....	γ, 0.079(14), 0.101(7), 0.17(5), 0.272(18), 0.637(2); e⁻, 0.017, 0.039, 0.068, 0.077, 0.090
	174	−59,650	−55.6	3.6 y	....	....	γ, 0.076(6), 1.24(9); e⁻, 0.015, 0.067, 0.074
	174 *m*		−55.4	140 d	....	....	γ, 0.067, 0.176, 0.273, 0.994; e⁻, 0.004, 0.034, 0.050, 0.057
	175	−59,360	−55.3	>1×10¹⁷ y	97.40	18	....
	176	−57,340	−53.4	2.2×10¹⁰ y	2.60	....	β⁻, 0.43; e⁻, 0.023, 0.078, 0.086, 0.137; γ, 0.088(15), 0.202(85), 0.306(95)
	176 *m*		−53.1	3.69 h	....	....	β⁻, 1.31; e⁻, 0.023, 0.078, 0.086; γ, 0.088(10)
	177	−56,070	−52.2	6.74 d	....	....	β⁻, 0.497; γ, 0.113(3), 0.208(6); e⁻, 0.048, 0.103, 0.111, 0.143
	177 *m*		−51.3	155 d	....	....	γ, 0.105(13), 0.113(23), 0.128(17), 0.153(17), 0.174(13), 0.208(62), 0.228(37),

Table 3-6 (*Continued*)
TABLE OF NUCLIDES

Nuclide		Mass Excess		Half-life y = yr, d = day h = hr, m = min s = sec	Natural Abundance, %	Thermal Neutron Absorption Cross Section σ, barns (mode if not *n,γ*)	Major Radiations, Energies in MeV and γ Intensities, %
Z Symbol	Mass Number	μmu	MeV				
							0.281(14), 0.319(10), 0.327(18), 0.378(29), 0.414(17), 0.418(21); β^-; e^-, 0–0.47
	178	−53,700	−50.0	30 m	....	....	β^-, 2.25; no γ
	178		−49.6	22.0 m	....	....	β^-, 1.50; e^-, 0.023, 0.028, 0.077, 0.083, 0.091, 0.148, 0.204; γ, 0.089, 0.214, 0.326, 0.427
	178			5 m	....	....	β^-, 2.25; γ, 0.090, 0.22, 0.33, 0.43
	179	−52,530	−48.9	4.6 h	....	....	β^-, 1.35; γ, 0.213
	180	−49,630	−46.2	2.5 m	....	....	β^-, 3.3; no γ
$_{72}$Hf	157			0.12 s	....	....	α, 5.68
	158			3 s	....	....	α, 5.27
	168			22 m	....	....	γ, 0.129, 0.17
	169			1.5 h	....	....	γ, 0.115; β^+, 1.3
	170			12.2 h	....	....	γ, 0.120, 0.165, 0.99, 1.28, 1.65, 2.03, 2.36, 2.52, 2.94; e^-, 0.035, 0.057, 0.102, 0.145
	171			10.7 h	....	....	γ, 0.122, 0.188, 0.29, 0.34, 0.47, 0.66, 0.86, 1.07
	172			5 y	....	....	γ, 0.024(22), 0.082(10), 0.125(21); e^-, 0.014, 0.018, 0.032, 0.040, 0.063
	173			23.6 h	....	....	γ, 0.13(96), 0.162(5), 0.30(52), 0.55(1), 0.898(2), 1.04(1), 1.20; e^-, 0.060, 0.072, 0.076, 0.113, 0.127
	174	−59,640	−55.6	2.0×10^{15} y	0.163	400	....
	175	−58,390	−54.7	70 d	....	....	γ, 0.089(3), 0.343(85), 0.433(1); e^-, 0.026, 0.079,

Table 3-6 (*Continued*)
TABLE OF NUCLIDES

Nuclide		Mass Excess		Half-life	Natural Abundance, %	Thermal Neutron Absorption Cross Section σ, barns (mode if not *n*,γ)	Major Radiations, Energies in MeV and γ Intensities, %
Z Symbol	Mass Number	μmu	MeV	y = yr, d = day h = hr, m = min s = sec			
							0.280, 0.333
	176	−58,430	−54.4	....	5.21	<30	....
	177	−56,600	−52.7	....	18.56	370	....
	177 *m*		−51.4	1.1 s	....	....	γ, 0.105(17), 0.113(30), 0.128(21), 0.153(22), 0.174(16), 0.208(81), 0.228(48), 0.281(18), 0.327(23), 0.378(37), 0.418(27); *e*⁻, 0–0.47
	178	−56,120	−52.3	....	27.1	50	....
	178 *m*		−51.1	4.3 s	....	....	γ, 0.089(54), 0.093(14), 0.214(75), 0.326(94), 0.427(97); *e*⁻, 0.023, 0.028, 0.077, 0.083, 0.091, 0.148, 0.204
	179	−53,970	−50.3	....	13.75	65	....
	179 *m*		−49.9	18.6 s	....	....	γ, 0.217(94); *e*⁻, 0.096, 0.150
	180	−53,180	−49.5	....	35.22	10	....
	180 *m*		−48.4	5.5 h	....	....	γ, 0.058(48), 0.093(16), 0.215(82), 0.333(93), 0.444(80), 0.501(17); *e*⁻, 0.028, 0.047, 0.055, 0.083, 0.091, 0.150, 0.206, 0.267
	181	−50,895	−47.41	42.5 d	....	~40	β⁻, 0.41; *e*⁻, 0.066, 0.069, 0.122, 0.415; γ, 0.133(48), 0.346(13), 0.482(81)
	182	−49,300	−45.8	9×10⁶ y	....	....	β⁻; γ, 0.271(84)
	183	−46,170	−43.0	65 m	....	....	β⁻, 1.6; γ, 0.46(58), 0.82(100)
₇₃Ta	172			44 m	....	....	γ, 0.092, 0.208,

Table 3-6 (*Continued*)
TABLE OF NUCLIDES

Nuclide		Mass Excess		Half-life	Natural Abundance,	Thermal Neutron Absorption Cross Section σ, barns	Major Radiations, Energies in MeV and γ Intensities,
Z Symbol	Mass Number	μmu	MeV	y = yr, d = day h = hr, m = min s = sec	%	(mode if not n,γ)	%
	173			3.7 h	....	....	0.511 γ, 0.090, 0.170, 0.64, 1.00; e^-, 0.059, 0.069, 0.095, 0.107, 0.161
	174			1.2 h	....	....	γ, 0.091, 0.125, 0.160, 0.205, 0.280, 0.350, 0.511; e^-, 0.026, 0.081, 0.089
	175			10.5 h	....	....	γ, 0.08, 0.13, 0.21, 0.27, 0.35, 0.45, 0.60, 0.83, 1.2, 1.4, 1.7; e^-, 0.016, 0.039, 0.061, 0.070, 0.116, 0.202
	176		−51	8.0 h	....	....	γ, 0.088, 0.202; e^-, 0.023, 0.078, 0.086, 0.137
	177	−55,350	−51.6	56.6 h	....	....	γ, 0.113(6), 0.208(1), 0.425, 0.509, 0.746, 1.058; e^-, 0.048, 0.102, 0.111
	178	−54,070	−50.4	9.35 m	....	....	γ, 0.093(100), 0.511, 1.10(11), 1.18(4), 1.35(46), 1.45(9); β^+, 0.89; e^-, 0.028, 0.082
	178			2.1 h	....	....	γ, 0.089(54), 0.093(14), 0.214(75), 0.328(120), 0.427(97); e^-, 0.023, 0.028, 0.077, 0.083, 0.091, 0.148, 0.204
	179	−53,840	−50.2	∼600 d	....	....	Hf X-rays
	180	−52,456	−48.86	>1×10¹² y	0.0123	....	....
	180 *m*		−48.65	8.15 h	....	....	β^-, 0.71; e^-, 0.028, 0.083, 0.091; γ, 0.093(4), 0.103(1)
	181	−51,993	−48.43	....	99.9877	21	....
	182	−49,833	−46.35	115.1 d	....	8,000	β^-, 1.71, 0.522; e^-, 0.030, 0.044, 0.054, 0.073,

Table 3-6 (*Continued*)
TABLE OF NUCLIDES

Nuclide		Mass Excess		Half-life	Natural Abundance, %	Thermal Neutron Absorption Cross Section σ, barns (mode if not *n,γ*)	Major Radiations, Energies in MeV and γ Intensities, %
Z Symbol	Mass Number	μmu	MeV	y = yr, d = day h = hr, m = min s = sec			
	182 *m*		−45.84	16.5 m	....	....	0.089, 0.110; γ, 0.068(42), 0.100(14), 0.152(7), 0.222(8), 1.122(34), 1.189(16), 1.222(27), 1.231(13) γ, 0.147(40), 0.172(40), 0.184(20), 0.319(5), 0.356; *e*⁻, 0.080, 0.105, 0.117, 0.173
	183	−48,530	−45.20	5.0 d	....	....	β⁻, 0.62; γ, 0.046(5), 0.053(5), 0.099(7), 0.108(11), 0.161(17), 0.246(33), 0.30(11), 0.354(11); *e*⁻, 0.034–0.043, 0.050, 0.073, 0.088, 0.093, 0.177
	184	−46,020	−42.9	8.7 h	....	....	β⁻, 2.64, 1.76, 1.19; *e*⁻; γ, 0.111(21), 0.16(7), 0.21(7), 0.25(42), 0.30(24), 0.41(71), 0.53(19), 0.79(16), 0.90(49), 0.95(15), 1.16(12)
	185	−44,440	−41.3	50 m	....	....	β⁻, 1.7; γ, 0.075(5), 0.100(6), 0.175(60), 0.245(5)
	186	−41,590	−38.7	10.5 m	....	....	β⁻, 2.2; γ, 0.123(18), 0.20(74), 0.30(18), 0.41(15), 0.51(33), 0.61(33),

Table 3-6 (*Continued*)
TABLE OF NUCLIDES

Nuclide		Mass Excess		Half-life y=yr, d=day h=hr, m=min s=sec	Natural Abundance, %	Thermal Neutron Absorption Cross Section σ, barns (mode if not n.γ)	Major Radiations, Energies in MeV and γ Intensities, %
Z Symbol	Mass Number	μmu	MeV				
$_{74}$W							0.73(48), 0.94(11)
	173			16.5 m	....	....	e^-
	174			31 m	....	....	....
	175			34 m	....	....	γ, 0.26, 0.80, 1.3, 1.6
	176		−50	2.3 h	....	....	γ, 0.034, 0.100; e^-, 0.017, 0.023, 0.027, 0.033, 0.050, 0.083
	177		−50	135 m	....	....	γ, 0.20, 0.42, 0.62, 0.83, 1.00; e^-, 0.020, 0.028, 0.048, 0.059, 0.068, 0.075, 0.088, 0.119, 0.360
	178		−50	21.5 d	....	....	Ta X-rays
	179		−49	37.5 m	....	....	γ, 0.031(22); e^-, 0.020, 0.029
	179 m		−49	5.2 m	....	....	γ, 0.222; e^-, 0.152, 0.211
	180	−53,000	−49.37	>1.1×10^{15} y	0.135	<20	....
	181	−51,789	−48.24	140 d	....		γ, 0.006(1), 0.136, 0.152; e^-, 0.004, 0.006
	182	−51,699	−48.16	>2×10^{17} y	26.4	20	....
	183	−49,676	−46.27	>1.1×10^{17} y	14.4	11	....
	183 m		−45.96	5.3 s	....	....	γ, 0.046(8), 0.053(11), 0.099(9), 0.102(4), 0.108(19), 0.160(6); e^-, 0.034, 0.040
	184	−48,975	−45.62	....	30.6	2.1	....
	185	−46,481	−43.30	75 d	....	....	$β^-$, 0.429; no γ
	185 m		−42.93	1.62 m	....	....	γ, 0.075(8), 0.100(16), 0.13(70), 0.17(100)
	186	−45,560	−42.44	>6×10^{15} y	28.4	40	....
	187	−42,756	−39.83	23.9 h	....	~90	$β^-$, 1.31, 0.63; e^-, 0.063, 0.122; γ, 0.072(11), 0.134(9), 0.479(23), 0.552(5), 0.618(6),

Table 3-6 (*Continued*)
TABLE OF NUCLIDES

| Nuclide | | Mass Excess | | Half-life | Natural | Thermal Neutron | Major Radiations, |
Z Symbol	Mass Number	μmu	MeV	y = yr, d = day h = hr, m = min s = sec	Abundance, %	Absorption Cross Section σ, barns (mode if not n,γ)	Energies in MeV and γ Intensities, %
	188	−41,184	−38.44	69.4 d	....	....	0.686(27), 0.773(4) β⁻, 0.349; γ, 0.227, 0.290
	189		−35.3	11.5 m	....	....	β⁻, 2.5, 2.0; γ, 0.130(12), 0.178(13), 0.258(100), 0.417(96), 0.55(28), 0.86(20), 0.96(17)
$_{75}$Re	177		−47	17 m	....	....	....
	178			15 m	....	....	β⁺, 3.1
	179		−46	20 m	....	....	W X-rays
	180			2.4 m	....	....	β⁺, 1.1; γ, 0.11, 0.511, 0.88
	180			20 h	....	....	β⁺, 1.9
	181		−47	18 h	....	....	γ, 0.365; e⁻, 0.008, 0.040, 0.053, 0.296
	182	−48,628	−45.30	12.7 h	....	....	γ, 0.068, 0.100, 1.122, 1.189, 1.23, 2.01, 2.05; β⁺, 1.74; e⁻, 0.015, 0.031, 0.056, 0.089, 0.098
	182			64.0 h	....	....	γ, 0.068, 0.100, 0.15–0.36, 1.08, 1.112, 1.19, 1.22, 1.43; e⁻, 0.015, 0.031, 0.044, 0.061, 0.089, 0.098, 0.100, 0.122, 0.160, 0.187
	183	−48,740	−45	71 d	....	....	γ, 0.046, 0.053, 0.109, 0.209, 0.246, 0.292; e⁻, 0.030, 0.034, 0.040, 0.088, 0.093
	184	−47,220	−44	38 d	....	....	γ, 0.111, 0.78, 0.90; e⁻, 0.042, 0.100
	184 *m*		−44	169 d	....	....	γ, 0.111, 0.78, 0.90; e⁻, 0.035, 0.042, 0.073, 0.081, 0.100
	185	−46,941	−43.73	....	37.07	110	....

Table 3-6 (*Continued*)
TABLE OF NUCLIDES

Nuclide		Mass Excess		Half-life y=yr, d=day h=hr, m=min s=sec	Natural Abundance, %	Thermal Neutron Absorption Cross Section σ, barns (mode if not n,γ)	Major Radiations, Energies in MeV and γ Intensities, %
Z Symbol	Mass Number	μmu	MeV				
	186	−44,980	−41.9	88.9 h	....	....	β⁻, 1.07; e⁻, 0.063, 0.125; γ, 0.137(9), 0.632, 0.768
	187	−44,167	−41.14	4.3×10¹⁰ y	62.93	70	β⁻, 0.003
	188	−41,647	−38.79	16.7 h	....	<2	β⁻, 2.12; e⁻, 0.081, 0.143; γ, 0.155(10), 0.478(1), 0.633(1), 0.829, 0.932
	188 m		−38.62	18.7 m	....	....	γ, 0.092(5), 0.106(10); e⁻, 0.004, 0.013, 0.021, 0.034, 0.051, 0.061, 0.080, 0.093
	189	−40,630	−37.8	24.3 h	....	....	β⁻, 1.00; e⁻, 0.023, 0.028, 0.057, 0.074, 0.112, 0.143; γ, 0.150(4), 0.187(3), 0.218(10), 0.245(4)
	189			140 d	....	....	γ, 0.211, 0.57, 0.67
	190	−38,040	−35.4	2.8 m	....	....	β⁻, 1.6; γ, 0.191(10), 0.392(10), 0.57(10), 0.83(3)
	190 m			2.8 h	....	....	β⁻, 1.6; γ, 0.12, 0.19, 0.23, 0.38, 0.56, 0.82
	191		−34.6	9.8 m	....	....	β⁻, 1.8
	192			6 s	....	....	β⁻, 2.5; γ, 0.20, 0.29, 0.37, 0.48, 0.57
₇₆Os	181		−44	23 m	....	....	e⁻, 0.093, 0.101
	181			2.7 h	....	....	γ, 0.23
	182		−44	21.9 h	....	....	γ, 0.180(7), 0.263(1), 0.510(10); e⁻, 0.015, 0.025, 0.043, 0.052, 0.108, 0.438
	183		−43	12.0 h	....	....	γ, 0.114(27), 0.168(10), 0.236(5), 0.382(90), 0.48(9), 0.86(5), 1.44(1); e⁻, 0.043, 0.102
	183 m		−43	9.9 h	....	....	γ, 1.035(6),

Table 3-6 (*Continued*)
TABLE OF NUCLIDES

Nuclide		Mass Excess		Half-life y = yr, d = day h = hr, m = min s = sec	Natural Abundance, %	Thermal Neutron Absorption Cross Section σ, barns (mode if not n,γ)	Major Radiations, Energies in MeV and γ Intensities, %
Z Sym-bol	Mass Number	μmu	MeV				
							1.105(48); e^-, 0.055, 0.096, 0.158, 0.168
	184	−47,250	−44.0	··········	0.018	<200	················
	185	−45,887	−42.74	93.6 d	·········	··············	γ, 0.646(80), 0.875(14); e^-, 0.059, 0.091, 0.574, 0.634
	186	−46,130	−43.0	··········	1.59	··············	················
	187	−44,168	−41.14	··········	1.64	··············	················
	188	−43,919	−40.91	··········	13.3	··············	················
	189	−41,700	−38.8	··········	16.1	0.008	················
	189 m		−38.8	5.7 h	·········	··············	e^-, 0.019, 0.028
	190	−41,370	−38.5	··········	26.4	8.6	················
	190 m		−36.8	9.9 m	·········	··············	γ, 0.187(70), 0.361(94), 0.502(98), 0.616(99); e^-, 0.026, 0.036, 0.113, 0.175
	191	−39,030	−36.4	15.0 d	·········	··············	$β^-$, 0.143; e^-, 0.030, 0.042, 0.053, 0.116, 0.127; γ, 0.129(25)
	191 m		−36.3	13.0 h	·········	··············	e^-, 0.062, 0.072
	192	−38,550	−35.9	>10^{14} y	41.0	1.6	················
	193	−35,773	−33.32	31.5 h	·········	200	$β^-$, 1.13; e^-, 0.060, 0.070; γ, 0.139(3), 0.28(2), 0.322(1), 0.38(2), 0.460(4), 0.558(2)
	194	−34,771	−32.39	6.0 y	·········	··············	$β^-$, 0.053; e^-; γ, 0.043(10), 0.078
	195	−32,000	−30	6.5 m	·········	··············	$β^-$, 2
$_{77}$Ir	182		−39	15 m	·········	··············	γ, 0.133, 0.278, 0.510
	183			0.9 h	·········	··············	γ, 0.24
	184		−40	3.2 h	·········	··············	γ, 0.125(100), 0.267(200), 0.392(90), 0.51, 0.83, 0.96, 1.09
	185		−40	14 h	·········	··············	γ, 0.101, 0.254; e^-, 0.024, 0.034, 0.047, 0.085, 0.180
	186	−42,010	−39.1	15.8 h	·········	··············	γ, 0.137(45), 0.297(74), 0.434(35), 0.511,

Table 3-6 (*Continued*)
TABLE OF NUCLIDES

| Nuclide | | Mass Excess | | Half-life | Natural | Thermal Neutron | Major Radiations, |
Z Symbol	Mass Number	μmu	MeV	y=yr, d=day h=hr, m=min s=sec	Abundance, %	Absorption Cross Section σ, barns (mode if not n,γ)	Energies in MeV and γ Intensities, %
	186			1.7 h	....	....	0.64(9), 0.77(8), 1.60–1.75(4); β^+, 1.94; e^-, 0.063, 0.125, 0.135, 0.226 γ, 0.137, 0.295, 0.511, 0.630, 0.77, 0.99; β^+, 2.6; e^-, 0.063, 0.125
	187	−42,440	−40	10.5 h	....	....	γ, 0.18(45), 0.31(14), 0.41(100), 0.50(35), 0.61(45), 0.90(40), 0.98(50); e^-, 0.007, 0.013, 0.053, 0.063, 0.073, 0.104
	188	−40,878	−38.08	41.5 h	....	....	γ, 0.155(34), 0.478(16), 0.633(29), 0.829(7), 1.210(7), 1.717(4), 2.08(16), 2.217(13); β^+, 1.66; e^-, 0.081, 0.143
	189	−41,090	−38	13.3 d	....	....	γ, 0.245(18); e^-, 0.023, 0.046, 0.058, 0.067, 0.171
	190	−39,170	−36.5	11 d	....	....	γ, 0.187(51), 0.37(39), 0.40(39), 0.518(39), 0.56(72), 0.604(47); e^-, 0.113, 0.175
	190 m_1		−36.5	1.2 h	....	....	e^-, 0.015, 0.024
	190 m_2		−36.3	3.2 h	....	....	γ, 0.187(66), 0.361(88), 0.502(92), 0.616(93); e^-, 0.026, 0.036, 0.113, 0.175
	191	−39,360	−36.7	....	38.5	750	....

Table 3-6 (*Continued*)
TABLE OF NUCLIDES

Nuclide		Mass Excess		Half-life	Natural	Thermal Neutron	Major Radiations,
Z Symbol	Mass Number	μmu	MeV	y = yr, d = day h = hr, m = min s = sec	Abundance, %	Absorption Cross Section σ, barns (mode if not n,γ)	Energies in MeV and γ Intensities, %
	191 *m*		−36.5	4.9 s	....	....	γ, 0.129(25); e⁻, 0.030, 0.042, 0.053, 0.116, 0.127
	192	−37,300	−34.7	74.2 d	....	700	β⁻, 0.67; e⁻, 0.217, 0.230, 0.239, 0.390; γ, 0.296(29), 0.308(30), 0.317(81), 0.468(49), 0.589(4), 0.604(9), 0.612(6)
	192 *m₁*		−34.7	1.42 m	....	....	γ, 0.058, 0.317, 0.612; e⁻, 0.046, 0.056; β⁻, 1.5
	192 *m₂*		−34.6	>5 y	....	....	e⁻, 0.149, 0.158
	193	−36,988	−34.45	....	61.5	110	....
	193 *m*		−34.37	11.9 d	....	....	e⁻, 0.069, 0.078
	194	−34,875	−32.49	17.4 h	....	....	β⁻, 2.24; γ, 0.328(10), 0.64(1), 0.939, 1.16(1), 1.48(1), 1.7
	194 *m*			47 s	....	....	β⁻, 2.3; γ, 0.13, 0.32, 0.63
	195	−34,110	−31.8	4.2 h	....	....	β⁻, 1.0; γ, 0.10, 0.13, 0.33, 0.37, 0.43, 0.66
	196	−31,750	−29.23	120 m	....	....	β⁻, 0.95; γ, 0.100(33), 0.356(94), 0.39(95), 0.44(95), 0.522(99), 0.65(100)
	197	−30,510	−28.4	7 m	....	....	β⁻, 2.0; γ, 0.50
	198	−27,380	−25.5	50 s	....	....	β⁻, 3.6; γ, 0.78
₇₈Pt	173			short	....	....	α, 6.19
	174			0.7 s	....	....	α, 6.03
	175			2.1 s	....	....	α, 5.95
	176			6.0 s	....	....	α, 5.74
	177			6.6 s	....	....	α, 5.51
	178			21 s	....	....	α, 5.44
	179			33 s	....	....	α, 5.15
	180			50 s	....	....	α, 5.14
	181			51 s	....	....	α, 5.02
	182		−36	3.0 m	....	....	α, 4.84
	183			6.5 m	....	....	α, 4.73

Table 3-6 (*Continued*)
TABLE OF NUCLIDES

Nuclide		Mass Excess		Half-life y = yr, d = day h = hr, m = min s = sec	Natural Abundance, %	Thermal Neutron Absorption Cross Section σ, barns (mode if not n,γ)	Major Radiations, Energies in MeV and γ Intensities, %
Z Sym- bol	Mass Number	μmu	MeV				
	184			20 m	....	....	α, 4.50
	184			42 m	....	....	γ, 0.68, 1.72, 1.85
	185			1.2 h	....	....	γ, 0.035, 0.63, 1.56
	186			3.0 h	....	....	γ, 0.67; α, 4.23
	187			2.0 h	....	....	γ, 2.0
	188	−40,330	−37.6	10.2 d	....	....	γ, 0.140(22), 0.19(100), 0.38(15), 0.42(7); e⁻, 0.042, 0.111, 0.119; α, 3.93
	189	−39,390	−37	10.9 h	....	....	γ, 0.094(120), 0.114(61), 0.141(124), 0.187(137), 0.243(100), 0.31(96), 0.404(32), 0.56(230), 0.61(180), 0.722(156), 0.80(27); e⁻, 0.037, 0.058, 0.068, 0.082, 0.092, 0.168, 0.231, 0.241
	190	−40,050	−37.3	6.9×10^{11} y	0.0127	~150	α, 3.18
	191	−38,550	−36	3.00 d	....	....	γ, 0.096(1), 0.129(2), 0.175(1), 0.269(1), 0.36(5), 0.410(3), 0.457(1), 0.539(9), 0.624(1); e⁻, 0.020, 0.053, 0.069, 0.080
	192	−38,850	−36.2	$\sim 10^{15}$ y	0.78	<14	α
	193	−36,940	−34.41	<500 y	....	....	Ir X-rays
	193 m		−34.26	4.3 d	....	....	e⁻, 0.01, 0.057, 0.124, 0.133
	194	−37,275	−34.72	....	32.9	1.1	....
	195	−35,187	−32.78	....	33.8	27	....
	195 m		−32.52	4.1 d	....	....	γ, 0.099(11), 0.129(1); e⁻, 0.018, 0.028, 0.051, 0.085, 0.116, 0.126
	196	−35,033	−32.63	....	25.2	0.9	....
	197	−32,653	−30.42	18 h	....	....	β⁻, 0.670; e⁻,

Table 3-6 (*Continued*)
TABLE OF NUCLIDES

| Nuclide | | Mass Excess | | Half-life | Natural | Thermal Neutron | Major Radiations, |
Z Symbol	Mass Number	μmu	MeV	y=yr, d=day h=hr, m=min s=sec	Abundance, %	Absorption Cross Section σ, barns (mode if not n.γ)	Energies in MeV and γ Intensities, %
	197 *m*		−30.02	78 m	....	....	0.063, 0.074, 0.110; γ, 0.077(20), 0.191(6) γ, 0.279(3), 0.346(13); e−, 0.040, 0.050, 0.268, 0.332; β−, 0.737
	198	−32,105	−29.91	>10^{15} y	7.19	4	....
	199	−29,420	−27.40	31 m	....	~15	β−, 1.69; γ, 0.075, 0.197(9), 0.245(4), 0.32(8), 0.475(12), 0.540(24), 0.715(3), 0.790(2), 0.960(2)
	199 *m*		−26.98	14.1 s	....	....	γ, 0.393(90); e−, 0.018, 0.029, 0.315, 0.381
	200	−28,570	−27	11.5 h	....	....	....
	201	−25,230	−23.5	2.3 m	....	....	β−, 2.66; γ, 0.15, 0.23, 1.76
$_{79}$Au	177			1.4 s	....	....	α, 6.11
	178			2.7 s	....	....	α, 5.91
	179			7.1 s	....	....	α, 5.84
	181			10 s	....	....	α, 5.60, 5.47
	183			44 s	....	....	α, 5.34
	185			4.33 m	....	....	α, 5.07
	186			12 m	....	....	γ, 0.16, 0.22, 0.30, 0.40
	187			8 m	....	....	Pt X-rays
	188			8 m	....	....	γ, 0.25, 0.33, 0.63
	189			30 m	....	....	e−, 0.027, 0.036, 0.088, 0.137, 0.154, 0.166, 0.269
	190	−35,290	−33	39 m	....	....	γ, 0.29(100), 0.60(5); e−, 0.22, 0.29
	191	−36,450	−34	3.2 h	....	....	γ, 0.14(10), 0.30(60), 0.39(5), 0.48(4), 0.60(10); e−, 0.035, 0.046, 0.054, 0.080, 0.089
	192	−35,380	−33.0	4.1 h	....	....	γ, 0.137, 0.158, 0.296, 0.308, 0.317; e−, 0.032,

Table 3-6 (*Continued*)
TABLE OF NUCLIDES

Nuclide		Mass Excess		Half-life y=yr, d=day h=hr, m=min s=sec	Natural Abundance, %	Thermal Neutron Absorption Cross Section σ, barns (mode if not n,γ)	Major Radiations, Energies in MeV and γ Intensities, %
Z Sym- bol	Mass Number	μmu	MeV				
	193	−35,760	−33	15.8 h	. . . .	. . . .	0.143, 0.23, 0.30; β⁺, 2.2 γ, 0.114(5), 0.18(11), 0.26(9), 0.378(1), 0.440(3); e⁻, 0.034, 0.095, 0.108, 0.177
	193 m		−33	3.9 s	. . . .	. . . .	γ, 0.258(65); e⁻, 0.019, 0.030
	194	−34,582	−32.21	39.5 h	. . . .	. . . .	β⁺, 1.49; e⁻, 0.250, 0.315; γ, 0.294(12), 0.328(68), 1.469(8), 1.596(3), 1.887(4), 2.044(4)
	195	−34,949	−32.55	183 d	. . . .	. . . .	γ, 0.099(10), 0.129(1); e⁻, 0.018, 0.028, 0.085
	195 m		−32.23	30.6 s	. . . .	. . . .	γ, 0.261(77); e⁻, 0.044, 0.056, 0.180
	196	−33,445	−31.15	6.18 d	. . . .	. . . .	β⁻, 0.259; e⁻, 0.255, 0.277, 0.343; γ, 0.333(25), 0.356(94), 0.426(6), 1.091
	196 m		−30.56	9.7 h	. . . .	. . . .	γ, 0.148(42), 0.188(32), 0.285(5), 0.316(5); e⁻, 0.069, 0.081, 0.094, 0.108, 0.135, 0.160
	197	−33,459	−31.17	. . . .	100	98.8	. . . .
	197 m		−30.76	7.2 s	. . . .	. . . .	γ, 0.130(8), 0.279(75); e⁻, 0.050, 0.117, 0.127, 0.198, 0.265
	198	−31,769	−29.59	2.697 d	. . . .	26,000	β⁻, 0.962; e⁻, 0.329, 0.398; γ, 0.412(95), 0.676(1), 1.088
	199	−31,227	−29.09	3.15 d	. . . .	~30	β⁻, 0.46, 0.30; γ, 0.158(37),

Table 3-6 (*Continued*)
TABLE OF NUCLIDES

Nuclide		Mass Excess		Half-life y=yr, d=day h=hr, m=min s=sec	Natural Abundance, %	Thermal Neutron Absorption Cross Section σ, barns (mode if not n,γ)	Major Radiations, Energies in MeV and γ Intensities, %
Z Symbol	Mass Number	μmu	MeV				
$_{80}$Hg	200	−29,300	−27.3	48.4 m	....	....	0.208(8); e⁻, 0.075, 0.125, 0.145 β⁻, 2.2; γ, 0.368(24), 1.227(23), 1.593(1)
	201	−28,080	−26.2	26 m	....	....	β⁻, 1.5; γ, 0.53
	203	−24,870	−23	55 s	....	....	β⁻, 1.9; γ, 0.69
	185			50 s	....	....	....
	186			1.5 m	....	....	γ, 0.125, 0.27, 0.35, 0.44
	187			3 m	....	....	γ, 0.175, 0.255, 0.40
	188			3.7 m	....	....	γ, 0.14
	189			9.6 m	....	....	γ, 0.165, 0.24, 0.32, 0.50
	190		−31	20 m	....	....	γ, 0.14; e⁻, 0.015, 0.026, 0.049, 0.062, 0.076
	191			55 m	....	....	γ, 0.26; e⁻, 0.170, 0.191, 0.239
	192	−33,840	−32	4.8 h	....	....	γ, 0.114(10), 0.157(20), 0.274(100); e⁻, 0.017, 0.028, 0.034, 0.039, 0.077
	193	−33,250	−31	~6 h	....	....	γ, 0.187, 0.574, 0.762, 0.855, 1.04, 1.08; e⁻, 0.025, 0.035, 0.108, 0.174
	193 m		−31	10.0 h	....	....	γ, 0.218, 0.258, 0.574; e⁻, 0.020, 0.025, 0.029, 0.036, 0.087, 0.178, 0.243
	194	−34,210	−32.2	1.9 y	....	....	Au X-rays
	194 m			0.4 s	....	....	γ, 0.048, 0.134
	195	−33,380	−31	9.5 h	....	....	γ, 0.20, 0.261, 0.59, 0.780, 0.930, 1.110, 1.172; e⁻, 0.048, 0.058, 0.099
	195 m		−31	40.0 h	....	....	γ, 0.200(35), 0.261(20), 0.560(20); e⁻, 0.0014, 0.013,

Table 3-6 (*Continued*)
TABLE OF NUCLIDES

Z Symbol	Mass Number	Mass Excess μmu	Mass Excess MeV	Half-life y=yr, d=day h=hr, m=min s=sec	Natural Abundance, %	Thermal Neutron Absorption Cross Section σ, barns (mode if not n,γ)	Major Radiations, Energies in MeV and γ Intensities, %
	196	−34,180	−31.84	>1×10¹⁴ y	0.146	880	0.022, 0.034, 0.043, 0.048, 0.053, 0.058, 0.109, 0.120, 0.180
	197	−32,640	−30.75	65 h	....	....	γ, 0.077(18), 0.191(2), 0.268; e^-, 0.064, 0.074
	197 *m*		−30.45	24 h	....	....	γ, 0.134(42), 0.279(7); e^-, 0.051, 0.082, 0.120, 0.131, 0.152, 0.162
	198	−33,244	−30.97	....	10.02	0.02	....
	199	−31,721	−29.55	....	16.84	2,000	....
	199 *m*		−29.01	43 m	....	....	γ, 0.158(53), 0.375(15); e^-, 0.075, 0.144, 0.285, 0.354
	200	−31,673	−29.50	....	23.13	<50	....
	201	−29,692	−27.66	....	13.22	<50	....
	202	−29,358	−27.35	....	29.80	4	....
	203	−27,120	−25.26	46.9 d	....	....	β^-, 0.214; e^-, 0.194, 0.264, 0.275; γ, 0.279(77)
	204	−26,505	−24.69	....	6.85	0.4	....
	205	−23,790	−22.2	5.5 m	....	....	β^-, 1.7; γ, 0.205
	206	−22,487	−20.95	8.1 m	....	....	β^-; γ, 0.31
$_{81}$Tl	191			10 m	....	....	γ, 0.511
	192			11 m	....	....	γ, 0.424; e^-, 0.341
	193			23 m	....	....	γ, 0.158, 0.169, 0.178, 0.187, 0.208, 0.216, 0.238, 0.247, 0.511; e^-, 0.24
	193 *m*			2.1 m	....	....	γ, 0.365; e^-, 0.280
	194	−28,430	−26	33.0 m	....	....	γ, 0.427; e^-, 0.344
	194 *m*			32.8 m	....	....	γ, 0.097; e^-, 0.083
	195	−30,160	−28	1.16 h	....	....	e^-, 0.022, 0.034; β^+, 1.8
	195 *m*		−28	3.5 s	....	....	γ, 0.383(95); e^-, 0.084, 0.096
	196	−29,240	−27.2	1.84 h	....	....	γ, 0.426; e^-, 0.343
	196 *m*		−26.8	1.41 h	....	....	γ, 0.426; e^-, 0.071, 0.081, 0.107
	197	−30,200	−28.5	2.84 h	....	....	γ, 0.152, 0.426; e^-, 0.067, 0.137

Table 3-6 (*Continued*)
TABLE OF NUCLIDES

Z Symbol Mass Number	Mass Excess μmu	Mass Excess MeV	Half-life y=yr, d=day h=hr, m=min s=sec	Natural Abundance, %	Thermal Neutron Absorption Cross Section σ, barns (mode if not n,γ)	Major Radiations, Energies in MeV and γ Intensities, %
197 m		−27.9	0.54 s	....	....	γ, 0.222(40), 0.385(90); e⁻, 0.136, 0.207, 0.219, 0.300
198	−29,530	−27.5	5.3 h	....	....	γ, 0.412(90), 0.65(40), 1.20(21), 1.42(24), 2.01(15), 2.45(5), 2.78(2); β⁺, 2.4; e⁻, 0.111, 0.201, 0.317, 0.329
198 m		−27.0	1.87 h	....	....	γ, 0.283(30), 0.412(45), 0.586(35), 0.635(35); e⁻, 0.033, 0.046, 0.175, 0.197, 0.246
199	−30,540	−28.5	7.4 h	....	....	γ, 0.158(5), 0.208(12), 0.247(9), 0.455(14); e⁻, 0.035, 0.125, 0.161, 0.193
200	−29,038	−27.05	26.1 h	....	....	γ, 0.368(88), 0.579(10), 0.829(8), 1.21(35), 1.364(4), 1.410(2), 1.517(4); β⁺, 1.44, 1.07; e⁻, 0.285, 0.354
201	−29,250	−27.3	74 h	....	....	γ, 0.135(2), 0.167(8); e⁻, 0.016, 0.052, 0.084
202	−28,050	−26.13	12.0 d	....	....	γ, 0.439(95), 0.522, 0.961; e⁻, 0.356
203	−27,647	−25.75	....	29.50	11	....
204	−26,135	−24.34	3.81 y	....	....	β⁻, 0.766
205	−25,558	−23.81	....	70.50	0.11	....
206	−23,896	−22.26	4.19 m	....	....	β⁻, 1.52; no γ
207	−22,550	−21.01	4.79 m	....	....	β⁻, 1.44; γ, 0.897
207 m		−19.67	1.3 s	....	....	γ, 0.35, 1.00
208	−17,987	−16.76	3.10 m	....	....	β⁻, 1.80; e⁻, 0.187, 0.423, 0.495; γ, 0.511(23),

Table 3-6 (*Continued*)
TABLE OF NUCLIDES

Nuclide		Mass Excess		Half-life y = yr, d = day h = hr, m = min s = sec	Natural Abundance, %	Thermal Neutron Absorption Cross Section σ, barns (mode if not n,γ)	Major Radiations, Energies in MeV and γ Intensities, %
Z Sym- bol	Mass Number	μmu	MeV				
	209	−14,704	−13.65	2.2 m	....	....	0.583(86), 0.860(12), 2.614(100) β⁻, 1.99; e⁻, 0.03, 0.10; γ, 0.12(50), 0.45(100), 1.56(100)
	210	−9,946	−9.23	1.32 m	....	....	β⁻, 2.3; e⁻, 0.208, 0.28; γ, 0.296(80), 0.795(100), 1.08(19), 1.21(17), 1.31(21), 2.01(7), 2.09(5), 2.36(8), 2.43(9)
₈₂Pb	194			11 m	....	....	γ, 0.204
	195			17 m	....	....	γ, 0.39; e⁻, 0.084, 0.096, 0.30
	196	−26,200	−24	37 m	....	....	γ, 0.192, 0.240, 0.253, 0.367, 0.503; e⁻, 0.155, 0.168
	197	−25,910	−24	....	....	....	γ, 0.386
	197 m		−24	42 m	....	....	γ, 0.085, 0.222, 0.234, 0.386; e⁻, 0.069, 0.136, 0.146, 0.207, 0.219, 0.300
	198	−27,590	−26	2.4 h	....	....	γ, 0.117(3), 0.173(28), 0.259(8), 0.290(16), 0.38(40), 0.575(4), 0.649(2), 0.865(6); e⁻, 0.031, 0.088, 0.159, 0.172, 0.205, 0.270
	199	−27,140	−25	90 m	....	....	γ, 0.353(17), 0.367(80), 0.720(10); e⁻, 0.267; β⁺, 2.8
	199 m		−25	12.2 m	....	....	γ, 0.424(20); e⁻, 0.336, 0.409
	200	−28,030	−26	21.5 h	....	....	γ, 0.109, 0.146, 0.236, 0.26, 0.290, 0.450, 0.605; e⁻, 0.024,

Table 3-6 (*Continued*)
TABLE OF NUCLIDES

| Nuclide | | Mass Excess | | Half-life | Natural | Thermal Neutron | Major Radiations, |
Z Sym-bol	Mass Number	μmu	MeV	y = yr, d = day h = hr, m = min s = sec	Abundance, %	Absorption Cross Section σ, barns (mode if not n,γ)	Energies in MeV and γ Intensities, %
	201	−27,140	−25	9.4 h	....	....	0.06, 0.133, 0.150, 0.172, 0.183 γ, 0.330, 0.361, 0.406, 0.585, 0.766, 0.907, 0.946, 1.30, 1.40; e^-, 0.244, 0.275, 0.316; β^+, 0.55
	201 m		−25	61 s	....	....	γ, 0.629(51); e^-, 0.541, 0.614
	202	−27,997	−26.08	~3×10⁵ y	....	....	Tl X-rays
	202 m		−23.91	3.62 h	....	....	γ, 0.390(7), 0.422(90), 0.460(8), 0.490(10), 0.658(35), 0.787(45), 0.961(90); e^-, 0.115, 0.126, 0.302, 0.334, 0.699, 0.772
	203	−26,771	−24.94	52.1 h	....	....	γ, 0.279(81), 0.401(5), 0.680(1); e^-, 0.193, 0.264
	203 m		−24.11	6.1 s	....	....	γ, 0.825(70); e^-, 0.737, 0.810
	204	−26,956	−25.11	....	1.40	0.7	....
	204 m		−22.92	66.9 m	....	....	γ, 0.375(93), 0.90(189); e^-, 0.287, 0.360, 0.824, 0.897
	205	−25,520	−23.77	3.0×10⁷ y	....	....	Tl X-rays
	206	−25,532	−23.79	....	25.1	0.03	....
	207	−24,097	−22.45	....	21.7	0.72	....
	207 m		−20.81	0.80 s	....	....	γ, 0.570(98), 1.064(83); e^-, 0.482, 0.975, 1.048
	208	−23,350	−21.75	....	52.3	0.0005	....
	209	−18,918	−17.63	3.30 h	....	....	β^-, 0.635; no γ
	210	−15,813	−14.73	20.4 y	....	....	β^-, 0.061; e^-, 0.030, 0.043; γ, 0.047(4); α, 3.72
	211	−11,258	−10.46	36.1 m	....	....	β^-, 1.36; γ, 0.405(3), 0.427(2), 0.702, 0.766(1), 0.832(3)

Table 3-6 (*Continued*)
TABLE OF NUCLIDES

| Nuclide | | Mass Excess | | Half-life | Natural | Thermal Neutron | Major Radiations, |
Z Symbol	Mass Number	μmu	MeV	y=yr, d=day h=hr, m=min s=sec	Abundance, %	Absorption Cross Section σ, barns (mode if not n,γ)	Energies in MeV and γ Intensities, %
	212	−8,095	−7.55	10.64 h	....	....	β^-, 0.58; e^-, 0.148, 0.222; γ, 0.239(47), 0.300(3)
	213	−3,710	−3	10.2 m	....	....	....
	214	−234	−0.15	26.8 m	....	....	β^-, 1.03, 0.67; e^-, 0.037, 0.049; γ, 0.053(1), 0.242(4), 0.295(19), 0.352(36)
$_{83}$Bi	197?			8.0 m	....	....	α, 5.81
	≤198			1.7 m	....	....	α, 6.2
	199	−21,560	−20	24.4 m	....	....	α, 5.53
	200	−21,060	−20	35 m	....	....	....
	201	−22,630	−21	1.85 h	....	....	Pb X-rays
	201 m			52 m	....	....	α, 5.28
	202	−22,120	−21	95 m	....	....	γ, 0.422, 0.961
	203	−23,350	−21.8	11.8 h	....	....	β^+, 1.35; e^-, 0.045, 0.098, 0.112, 0.176, 0.737; γ, 0.186(6), 0.264(6), 0.381(9), 0.82(78), 1.034(16), 1.52(31), 1.87(35)
	204	−22,190	−21	11.2 h	....	....	γ, 0.21, 0.375, 0.671, 0.91, 0.98, 1.21; e^-, 0.063, 0.075, 0.087, 0.128, 0.133, 0.161, 0.201, 0.287, 0.360, 0.583, 0.811, 0.824, 0.897
	205	−22,618	−21.07	15.31 d	....	....	β^+, 0.98; e^-, 0.011, 0.023; γ, 0.26(3), 0.51(4), 0.57(14), 0.703(28), 0.911(4), 0.988(17), 1.044(8), 1.615(4), 1.766(27), 1.864(6), 1.906(2)
	206	−21,611	−20.18	6.243 d	....	....	γ, 0.184(21), 0.343(26), 0.398(10),

Table 3-6 (*Continued*)
TABLE OF NUCLIDES

Nuclide		Mass Excess		Half-life y=yr, d=day h=hr, m=min s=sec	Natural Abundance, %	Thermal Neutron Absorption Cross Section σ, barns (mode if not n,γ)	Major Radiations, Energies in MeV and γ Intensities, %
Z Sym- bol	Mass Number	μmu	MeV				
	207	−21,562	−20.04	30.2 y	....	....	0.497(18), 0.516(46), 0.538(34), 0.803(99), 0.880(72), 0.895(19), 1.019(8), 1.099(13), 1.596(8), 1.720(36); e^-, 0.096, 0.168, 0.255 γ, 0.570(98), 1.063(77), 1.771(9); e^-, 0.482, 0.975, 1.048
	208	−20,269	−18.88	3.68×10⁵ y	....	....	γ, 2.614(100)
	209	−19,606	−18.26	>2×10¹⁸ y	100	0.019	α(?), 3.0
	210	−15,813	−14.79	5.013 d	....	....	$β^-$, 1.160; α, 4.69, 4.65
	210 *m*		−14.52	~2.6×10⁶ y	....	....	α, 4.96, 4.92, 4.57; γ, 0.262(45), 0.30(23), 0.34, 0.61
	211	−12,700	−11.84	2.16 m	....	....	α, 6.62, 6.28; γ, 0.351(14); e^-, 0.265
	212	−8,721	−8.13	60.60 m	....	....	$β^-$, 2.25; e^-, 0.025, 0.036; α, 6.09, 6.05; γ, 0.040(2), 0.288(1), 0.46(1), 0.727(7), 0.785(1), 1.620(2)
	213	−5,683	−5.24	47 m	....	....	$β^-$, 1.39; γ, 0.437; α, 5.87
	214	−1,314	−1.19	19.7 m	....	....	$β^-$, 3.26; γ, 0.609(47), 0.769(5), 0.935(3), 1.120(17), 1.238(6), 1.378(5), 1.40(4), 1.509(2), 1.728(3), 1.764(17), 1.848(2), 2.117(1), 2.204(5),

Table 3-6 (*Continued*)
TABLE OF NUCLIDES

Nuclide		Mass Excess		Half-life y = yr, d = day h = hr, m = min s = sec	Natural Abundance, %	Thermal Neutron Absorption Cross Section σ, barns (mode if not n,γ)	Major Radiations, Energies in MeV and γ Intensities, %
Z Symbol	Mass Number	μmu	MeV				
							2.445(2); α, 5.51, 5.45
	215	1,830	1.7	7 m	. . . .	. . . .	β⁻
₈₄Po	193			short	. . . .	. . . .	α, 7.0
	194			0.5 s	. . . .	. . . .	α, 6.85
	195			3 s	. . . .	. . . .	α, 6.63
	195 m			1.4 s	. . . .	. . . .	α, 6.72
	196			6 s	. . . .	. . . .	α, 6.53
	197			54 s	. . . .	. . . .	α, 6.30
	197 m			25 s	. . . .	. . . .	α, 6.39
	198			1.7 m	. . . .	. . . .	α, 6.16
	199			5.0 m	. . . .	. . . .	α, 5.94
	199 m			4.2 m	. . . .	. . . .	α, 6.05
	200	−17,180	−16	10.5 m	. . . .	. . . .	α, 5.86
	201	−16,980	−16	15.1 m	. . . .	. . . .	α, 5.68
	201 m			8.9 m	. . . .	. . . .	α, 5.78
	202	−18,870	−18	45 m	. . . .	. . . .	α, 5.58
	203	−18,530	−17	42 m	. . . .	. . . .	α, 5.49
	204	−19,540	−18	3.6 h	. . . .	. . . .	α, 5.38
	205	−18,800	−18	1.8 h	. . . .	. . . .	α, 5.25
	206	−19,676	−18.33	8.8 d	. . . .	. . . .	γ, 0.286(35), 0.338(40), 0.51(100), 0.807(60), 1.02(85); e⁻, 0.045, 0.196, 0.248; α, 5.22
	207	−18,442	−17.14	5.7 h	. . . .	. . . .	γ, 0.25(5), 0.35(4), 0.41(13), 0.74(36), 0.95(84), 1.15(6), 1.37(4), 2.06(2); e⁻, 0.159, 0.255, 0.315, 0.652, 0.902; β⁺, 1.14; α, 5.11
	207 m		−15.75	2.8 s	. . . .	. . . .	γ, 0.26(42), 0.31(40), 0.82(100); e⁻, 0.22, 0.24
	208	−18,757	−17.47	2.93 y	. . . .	. . . .	α, 5.11; γ, 0.285, 0.60
	209	−17,574	−16.37	103 y	. . . .	. . . .	α, 4.88; γ, 0.261, 0.91(1); e⁻, 0.173
	210	−17,124	−15.95	138.40 d	. . . .	<0.03	α, 5.305; γ, 0.803
	211	−13,343	−12.43	0.52 s	. . . .	. . . .	α, 7.45; γ, 0.570(1), 0.90(1)
	211 m		−11.00	25 s	. . . .	. . . .	α, 8.88, 7.28; γ, 0.570(92),

Table 3-6 (*Continued*)
TABLE OF NUCLIDES

Nuclide Z Symbol	Nuclide Mass Number	Mass Excess μmu	Mass Excess MeV	Half-life y = yr, d = day h = hr, m = min s = sec	Natural Abundance, %	Thermal Neutron Absorption Cross Section σ, barns (mode if not n,γ)	Major Radiations, Energies in MeV and γ Intensities, %
	212 *m*		−7.44	45 s	....	....	1.063(77); e⁻ α, 11.65; γ, 0.57(2), 2.61(3)
	216	1,922	1.78	0.145 s	....	....	α, 6.78
	217	6,060	6	<10 s	....	....	α, 6.55
	218	8,930	8.38	3.05 m	....	....	α, 6.00
₈₅At	200			0.9 m	....	....	α, 6.47, 6.42
	201			1.5 m	....	....	α, 6.35
	202	−10,200	−10	3.0 m	....	....	α, 6.23, 6.12
	203	−12,290	−11	7.4 m	....	....	α, 6.09
	204	−11,940	−11	9.3 m	....	....	α, 5.95
	205	−13,560	−13	26.2 m	....	....	α, 5.90
	206	−13,210	−12	32.8 m	....	....	α, 5.70; γ, 0.068(10); e⁻, 0.052, 0.064
	207	−14,440	−13.41	1.8 h	....	....	α, 5.76
	208	−13,390	−12	1.6 h	....	....	γ, 0.18(25), 0.25, 0.66(100); α, 5.65
	209	−13,833	−12.89	5.5 h	....	....	γ, 0.195(23), 0.545(62), 0.780(94); e⁻, 0.076, 0.102, 0.178, 0.451, 0.686; α, 5.65
	210	−12,964	−12.12	8.3 h	....	....	γ, 0.245(79), 1.180(100), 1.436(29), 1.483(48), 1.599(14); e⁻, 0.023, 0.031, 0.043, 0.152, 0.229; α, 5.52, 5.44, 5.36
	211	−12,538	−11.64	7.21 h	....	....	α, 5.868; γ, 0.67
	212	−9,276	−8.64	0.30 s	....	....	α, 7.66, 7.60; e⁻, 0.047, 0.059
	212 *m*		−8.42	0.12 s	....	....	α, 7.88, 7.82; e⁻, 0.047, 0.059
	213	−6,930	−6.5	short	....	....	α, 9.2
	218	8,607	8.11	1.5–2.0 s	....	....	α, 6.70, 6.65
	219	11,290	10.5	0.9 m	....	....	α, 6.28
₈₆Rn	202			13 s	....	....	α, 6.64
	203			45 s	....	....	α, 6.50
	203 *m*			28 s	....	....	α, 6.55
	204	−7,700	−7	75 s	....	....	α, 6.42
	205	−7,440	−7	1.8 m	....	....	α, 6.26
	206	−9,420	−9	6.5 m	....	....	α, 6.26
	207	−9,240	−9	11 m	....	....	α, 6.15
	208	−10,210	−10	23 m	....	....	α, 6.15

Table 3-6 (*Continued*)
TABLE OF NUCLIDES

| Nuclide | | Mass Excess | | Half-life | Natural | Thermal Neutron | Major Radiations, |
Z Symbol	Mass Number	μmu	MeV	y=yr, d=day h=hr, m=min s=sec	Abundance, %	Absorption Cross Section σ, barns (mode if not n,γ)	Energies in MeV and γ Intensities, %
	209	−9,580	−9	30 m	....	....	α, 6.04
	210	−10,460	−9.74	2.42 h	....	....	α, 6.04
	211	−9,434	−8.75	15 h	....	....	α, 5.85, 5.78; γ, 0.445(29), 0.680(74), 0.865(18), 0.946(21), 1.13(23), 1.37(38); e⁻, 0.053, 0.065, 0.073, 0.153, 0.168, 0.200, 0.237, 0.349, 0.584, 0.665
	212	−9,293	−8.66	25 m	....	....	α, 6.27
	219	9,481	8.85	4.00 s	....	....	α, 6.82, 6.55, 6.42; γ, 0.272(9), 0.401(5); e⁻, 0.179, 0.255, 0.308
	220	11,401	10.61	55.3 s	....	<0.2	α, 6.29; γ, 0.55
	221	15,230	14	25 m	....	....	α, 6.0
	222	17,531	16.39	3.8229 d	....	0.7	α, 5.49; γ, 0.510
	223			43 m	....	....	....
	224			1.9 h	....	....	....
₈₇Fr	204			2.0 s	....	....	α, 7.03
	205			3.7 s	....	....	α, 6.92
	206	−160	−0	15.8 s	....	....	α, 6.80
	207	−2,270	−2	19 s	....	....	α, 6.78
	208	−2,050	−2	37 s	....	....	α, 6.66
	209	−3,680	−3	55 s	....	....	α, 6.66
	210	−3,430	−3	2.6 m	....	....	α, 6.56
	211	−4,670	−4.3	3.1 m	....	....	α, 6.56
	212	−3,770	−4	19.3 m	....	....	α, 6.42, 6.39, 6.35
	213	−3,816	−3.55	34 s	....	....	α, 6.78
	217	4,750	4.4	short	....	....	α, 8.3
	220	12,337	11.47	27.5 s	....	....	α, 6.68, 6.64
	221	14,183	13.27	4.8 m	....	....	α, 6.34, 6.12; γ, 0.218(14); e⁻, 0.122, 0.202
	222	17,630	16.34	14.8 m	....	....	....
	223	19,736	18.40	22 m	....	....	β⁻, 1.15; e⁻, 0.031, 0.045, 0.062, 0.075; γ, 0.050(40), 0.080(13), 0.234(4)
	224	23,590	22	<2 m	....	....	....
₈₈Ra	213	420	−0	2.7 m	....	....	α, 6.91

Table 3-6 (*Continued*)
TABLE OF NUCLIDES

Z Symbol	Mass Number	Mass Excess μmu	Mass Excess MeV	Half-life y = yr, d = day h = hr, m = min s = sec	Natural Abundance, %	Thermal Neutron Absorption Cross Section σ, barns (mode if not n,γ)	Major Radiations, Energies in MeV and γ Intensities, %
	221	13,892	12.96	30 s	....	....	α, 6.76, 6.67, 6.61, 6.59; γ, 0.091(4), 0.151(13), 0.175(2)
	222	15,376	14.32	38 s	....	....	α, 6.56; γ, 0.325(4), 0.473, 0.52, 0.85
	223	18,501	17.26	11.435 d	....	130	α, 5.75, 5.71, 5.61, 5.54; γ, 0.149(10), 0.270(10), 0.33(6); e^-, 0.024, 0.046, 0.056, 0.126, 0.136, 0.171
	224	20,218	18.82	3.64 d	....	12	α, 5.68, 5.45; γ, 0.241(4), 0.29, 0.41, 0.65; e^-, 0.144, 0.225
	225	23,528	22.01	14.8 d	....	....	β^-, 0.36; e^-, 0.021, 0.035; γ, 0.040(33)
	226	25,360	23.69	1,602 y	....	20	α, 4.78, 4.60; γ, 0.186(4), 0.26, 0.42, 0.61; e^-, 0.087, 0.170
	227	29,159	27.18	41.2 m	....	....	β^-, 1.31; e^-, 0.008, 0.023; γ, 0.291(4), 0.498(1)
	228	31,139	28.96	6.7 y	....	~36	β^-, 0.05; e^-, 0.005
	229			short	....	....	....
	230		35	1 h	....	....	β^-, 1.2
$_{89}$Ac	221	15,680	14.6	short	....	....	α, 7.6
	222	17,760	16.55	5.5 s	....	....	α, 7.00
	223	19,144	17.82	2.2 m	....	....	α, 6.66, 6.65, 6.57; γ, 0.082, 0.096
	224	21,690	20.21	2.9 h	....	....	γ, 0.132(28), 0.217(62); e^-, 0.067, 0.080; α, 6.20, 6.14, 6.04
	225	23,153	21.62	10.0 d	....	....	α, 5.83, 5.79, 5.73; γ, 0.099, 0.150, 0.187; e^-, 0.020, 0.032, 0.044, 0.081
	226	26,160	24.31	29 h	....	....	β^-, 1.2; e^-, 0.053, 0.067; γ, 0.158(32), 0.185(9), 0.230(47), 0.253(11); α

Table 3-6 (*Continued*)
TABLE OF NUCLIDES

Nuclide		Mass Excess		Half-life y = yr, d = day h = hr, m = min s = sec	Natural Abundance, %	Thermal Neutron Absorption Cross Section σ, barns (mode if not n,γ)	Major Radiations, Energies in MeV and γ Intensities, %
Z Sym- bol	Mass Number	μmu	MeV				
	227	27,753	25.87	21.6 y	....	830	β^-, 0.046; e^-, 0.005, 0.010; γ, 0.070, 0.166, 0.190; α, 4.95, 4.86
	228	31,080	28.91	6.13 h	....	....	β^-, 2.11; e^-, 0.040, 0.054, 0.110; γ, 0.34(15), 0.908(25), 0.96(20)
	229	32,800	31	66 m	....	....	....
	230	36,210	34	<1 m	....	....	β^-, 2.2
	231	38,550	35.9	15 m	....	....	β^-, 2.1; γ, 0.185, 0.28, 0.39, 0.71
$_{90}$Th	223		19.5	0.9 s	....	....	α, 7.56
	224		20.00	1.05 s	....	....	α, 7.18, 6.91; γ, 0.177(9), 0.235, 0.297, 0.410(1)
	225		22.30	8.0 m	....	....	α, 6.80, 6.75, 6.50, 6.48, 6.44; γ, 0.246(5), 0.322(27), 0.362(5), 0.45(1), 0.49(1)
	226		23.19	30.9 m	....	....	α, 6.34, 6.22; γ, 0.111(3), 0.131, 0.20, 0.242(1); e^-, 0.094, 0.107
	227		25.82	18.2 d	....	~1,500(f)	α, 6.04, 5.98, 5.76, 5.72; γ, 0.050(8), 0.237(15), 0.31(8); e^-, 0.013, 0.026, 0.044
	228		26.77	1.910 y	....	123; <0.3(f)	α, 5.43, 5.34; γ, 0.084(2), 0.132, 0.167, 0.214; e^-, 0.067, 0.080
	229	31,652	29.61	7,340 y	....	32(f)	α, 5.05, 4.97, 4.90, 4.84, 4.81; γ, 0.137(3), 0.20(10); e^-, 0.006–0.090
	230	33,087	30.87	8.0×10⁴ y	....	23; ≤0.001(f)	α, 4.68, 4.62; γ, 0.068(1), 0.142, 0.184, 0.253; e^-, 0.051, 0.064
	231	36,291	33.83	25.52 h	....	....	β^-, 0.30; e^-, 0.040, 0.054, 0.061; γ,

Table 3-6 (*Continued*)
TABLE OF NUCLIDES

Nuclide		Mass Excess		Half-life y = yr, d = day h = hr, m = min s = sec	Natural Abundance, %	Thermal Neutron Absorption Cross Section σ, barns (mode if not n,γ)	Major Radiations, Energies in MeV and γ Intensities, %
Z Sym- bol	Mass Number	μmu	MeV				
	232	38,124	35.47	1.41×10^10 y	100	7.4; <0.0002(f)	0.026(2), 0.084(10) α, 4.01, 3.95; e⁻, 0.042, 0.055
	233	41,469	38.76	22.12 m	·········	1,500; 15(f)	β⁻, 1.23; e⁻, 0.009, 0.024, 0.036, 0.051, 0.067, 0.082; γ, 0.029(2), 0.087(3), 0.171(1), 0.195, 0.453(1), 0.67, 0.895
	234	43,583	40.64	24.10 d	·········	1.8; <0.01(f)	β⁻, 0.191; e⁻, 0.012, 0.025, 0.072, 0.088; γ, 0.063(4), 0.093(4)
₉₁Pa	224			0.6 s	·········	··············	α
	225	26,230	25	0.8 s	·········	··············	·················
	226	27,810	25.96	1.8 m	·········	··············	α, 6.86, 6.82
	227	28,811	26.83	38.3 m	·········	··············	γ, 0.065(6), 0.110(2); α, 6.47, 6.42, 6.40, 6.36
	228	31,010	28.86	22 h	·········	··············	γ, 0.14(3), 0.20(9), 0.28(5), 0.33(18), 0.41(13), 0.46(32), 0.95(93), 1.57(7), 1.85(4); e⁻, 0.040, 0.054, 0.110; α, 6.11, 6.08, 6.03, 5.80
	229	32,022	29.88	1.5 d	·········	··············	e⁻, 0.023, 0.038; α, 5.67, 5.62, 5.58, 5.54
	230	34,433	32.17	17.7 d	·········	1,500(f)	β⁻, 0.41; e⁻, 0.034, 0.048; γ, 0.45(18), 0.51(8), 0.91(24), 0.954(50); α, 5.26–5.34
	231	35,877	33.44	3.25×10⁴ y	·········	200; 0.010(f)	α, 5.06, 5.02, 5.01, 4.95, 4.73; γ, 0.027(6), 0.29(6); e⁻, 0–0.10, 0.195, 0.323, 0.350
	232	38,612	35.95	1.31 d	·········	~760; ~700(f)	β⁻, 1.3, 0.32; e⁻, 0.028, 0.043, 0.091; γ, 0.107(5), 0.150(12), 0.39(9), 0.46(9),

Table 3-6 (*Continued*)
TABLE OF NUCLIDES

Nuclide		Mass Excess		Half-life y = yr, d = day h = hr, m = min s = sec	Natural Abundance, %	Thermal Neutron Absorption Cross Section σ, barns (mode if not n,γ)	Major Radiations, Energies in MeV and γ Intensities, %
Z Sym-bol	Mass Number	μmu	MeV				
	233	40,132	37.51	27.0 d	·········	22; <0.1(f)	0.57(8), 0.87(51), 0.971(40) β⁻, 0.568, 0.257; e⁻, 0.013, 0.023, 0.036, 0.054, 0.065, 0.185, 0.197, 0.291; γ, 0.31(44)
	234	43,298	40.38	6.75 h	·········	<5,000(f)	β⁻, 1.3, 1.13, 0.53; e⁻, 0.024, 0.039, 0.080, 0.095, 0.112; γ, 0.100(50), 0.126(26), 0.22(14), 0.36(13), 0.56(15), 0.70(24), 0.90(70), 1.08(12)
	234 *m*		40.45	1.175 m	·········	<500(f)	β⁻, 2.29; γ, 0.765, 1.001(1)
	235	45,420	42.3	23.7 m	·········	··············	β⁻, 1.4; no γ
	236	49,230	45	12 m	·········	··············	β⁻, 3.3
	237	51,080	47.7	39 m	·········	··············	β⁻, 2.3; γ, 0.090(50), 0.145(45), 0.205(55), 0.275(20), 0.330(40), 0.405(30), 0.46(100), 0.55(30), 0.59(25), 0.75(50), 0.80(45), 0.87(100), 0.92(100), 1.04(35), 1.32(10), 1.42(15)
₉₂U	227		29	1.3 m	·········	··············	α, 6.8
	228	31,387	29.23	9.1 m	·········	··············	α, 6.69, 6.60; γ, 0.152, 0.187, 0.246
	229	33,481	31.20	58 m	·········	··············	α, 6.36, 6.33, 6.30
	230	33,937	31.60	20.8 d	·········	25(f)	α, 5.89, 5.82; γ, 0.072, 0.156, 0.231; e⁻, 0.054, 0.068
	231	36,270	33.8	4.3 d	·········	~400(f)	γ, 0.026, 0.084(7),

Table 3-6 (*Continued*)
TABLE OF NUCLIDES

Nuclide Z Sym-bol	Nuclide Mass Number	Mass Excess μmu	Mass Excess MeV	Half-life y=yr, d=day h=hr, m=min s=sec	Natural Abundance, %	Thermal Neutron Absorption Cross Section σ, barns (mode if not n,γ)	Major Radiations, Energies in MeV and γ Intensities, %
	232	37,168	34.60	72 y	·········	78; 77(f)	0.218(1); e^-, 0.040, 0.054, 0.063; α, 5.46 α, 5.32, 5.27; γ, 0.058, 0.129, 0.270, 0.328; e^-, 0.040, 0.054
	233	39,522	36.94	1.62×10^5 y	·········	49; 524(f)	α, 4.82, 4.78; γ, 0.029(60), 0.042(310), 0.055(68), 0.097(100), 0.119(40), 0.146(35), 0.164(27), 0.22(45), 0.291(23), 0.32(43); e^-, 0.023, 0.038
	234	40,904	38.16	2.47×10^5 y	0.0057	95	α, 4.77, 4.72; γ, 0.053, 0.117, 0.48, 0.58
	235	43,915	40.93	7.1×10^8 y	0.7196	101; 577(f)	α, 4.58, 4.40, 4.37; γ, 0.143(11), 0.185(54), 0.204(5)
	235 m		40.93	26.1 m	·········	·············	e^-, ≤0.0001
	236	45,637	42.46	2.39×10^7 y	·········	6	α, 4.49, 4.44; e^-, 0.032, 0.045
	237	48,608	45.41	6.75 d	·········	·············	$β^-$, 0.248; e^-, 0.008, 0.011, 0.038, 0.089, 0.186; γ, 0.026(2), 0.060(36), 0.165(2), 0.208(23), 0.267(1), 0.332(1), 0.370
	238	50,770	47.33	4.51×10^9 y	99.276	2.73; <0.0005(f)	α, 4.20, 4.15; e^-, 0.030, 0.043
	239	54,300	50.60	23.54 m	·········	22; 14(f)	$β^-$, 1.29; e^-, 0.011, 0.023, 0.052, 0.069; γ, 0.044(4), 0.075(51)
	240	56,594	52.74	14.1 h	·········	·············	$β^-$, 0.36; e^-, 0.022, 0.038
₉₃Np	229			4.0 m	·········	·············	α, 6.89
	230	37,680	35.1	4.6 m	·········	·············	α, 6.66
	231	38,280	35.7	~50 m	·········	·············	α, 6.29

Table 3-6 (*Continued*)
TABLE OF NUCLIDES

Nuclide		Mass Excess		Half-life y=yr, d=day h=hr, m=min s=sec	Natural Abundance, %	Thermal Neutron Absorption Cross Section σ, barns (mode if not n,γ)	Major Radiations, Energies in MeV and γ Intensities, %
Z Symbol	Mass Number	μmu	MeV				
	232	39,860	37	~13 m	....	....	U X-rays
	233	40,670	38	35 m	....	....	α, 5.54; γ
	234	42,860	40.0	4.40 d	....	~900(f)	γ, 0.109, 0.23, 0.25, 0.45, 0.50, 0.75, 0.95, 1.21, 1.56; e⁻, 0.024, 0.039, 0.696; β⁺, 0.8
	235	44,049	41.05	410 d	....	....	U X-rays; α, 5.02
	236	46,624	43.41	22 h	....	....	β⁻, 0.52; e⁻, 0.025, 0.040; γ, 0.642, 0.688
	236			>5×10³ y	....	2,500(f)	....
	237	48,056	44.89	2.14×10⁶ y	....	170; 0.019(f)	α, 4.79, 4.77; γ, 0.030(14), 0.086(14), 0.145(1); e⁻, 0.009, 0.024, 0.036, 0.051, 0.067, 0.082
	238	50,896	47.47	2.10 d	....	1,600(f)	β⁻, 1.25; e⁻, 0.022, 0.039; γ, 1.01(42)
	239	52,924	49.32	2.346 d	....	35; <1(f)	β⁻, 0.713, 0.437; e⁻, 0.02–0.04, 0.048, 0.088, 0.106, 0.156; γ, 0.106(23), 0.209(4), 0.228(12), 0.278(14)
	240	56,080	52.2	63 m	....	....	β⁻, 0.89; γ, 0.16, 0.25, 0.44, 0.56, 0.60, 0.92, 1.00, 1.16
	240 m		52.3	7.3 m	....	....	β⁻, 2.16; e⁻, 0.022, 0.038; γ, 0.56(21), 0.60(13), 0.92(3), 1.5(3)
	241	58,200	54.3	16 m	....	....	β⁻, 1.4
	241			3.4 h	....	....	....
₉₄Pu	232	41,180	38.4	36 m	....	....	α, 6.59
	233	42,972	40.04	20 m	....	....	α, 6.31
	234	43,315	40.34	9.0 h	....	....	α, 6.20, 6.15; Np X-rays
	235	45,270	42.2	26 m	....	....	Np X-rays; α, 5.86
	236	46,071	42.90	2.85 y	....	170(f)	α, 5.77, 5.72; γ, 0.048, 0.109; e⁻, 0.028, 0.043
	237	48,298	45.12	45.6 d	....	2,500(f)	γ, 0.060(5); e⁻, 0.026, 0.032,

Table 3-6 (*Continued*)
TABLE OF NUCLIDES

Z Symbol	Mass Number	Mass Excess μmu	Mass Excess MeV	Half-life y=yr, d=day h=hr, m=min s=sec	Natural Abundance, %	Thermal Neutron Absorption Cross Section σ, barns (mode if not n,γ)	Major Radiations, Energies in MeV and γ Intensities, %
							0.038, 0.042, 0.056; α, 5.66, 5.37
	237 m		45.26	0.18 s	....	....	γ, 0.145(2); e⁻, 0.125, 0.140
	238	49,511	46.18	86.4 y	....	500; 16.8(f)	α, 5.50, 5.46; γ, 0.099, 0.150, 0.77; e⁻, 0.024, 0.039
	239	52,146	48.60	24,390 y	....	274; 741(f)	α, 5.16, 5.11; γ, 0.039, 0.052, 0.129, 0.375, 0.414, 0.65, 0.77; e⁻, 0.008, 0.019, 0.033, 0.047
	240	53,882	50.14	6,580 y	....	286; <0.08(f)	α, 5.17, 5.12; γ, 0.65; e⁻, 0.026, 0.040
	241	56,737	52.98	13.2 y	....	425; 950(f)	β⁻, 0.021; α, 4.90, 4.85; γ, 0.145
	242	58,725	54.74	3.79×10⁵ y	....	19; <0.2(f)	α, 4.90, 4.86
	243	61,972	57.77	4.98 h	....	170	β⁻, 0.58; e⁻, 0.019, 0.036; γ, 0.084(21), 0.381(1)
	244	64,100	59.83	~7.6×10⁷ y	....	1.8	α, 4.58; SF
	245	67,830	63	10.1 h	....	~260	β⁻
	246	70,090	65.3	10.85 d	....	....	β⁻, 0.33, 0.15; e⁻, 0.020, 0.038, 0.055, 0.156; γ, 0.044(30), 0.180(10), 0.224(25)
95Am	237	49,840	47	~1.3 h	....	....	α, 6.02
	238	51,940	48	1.9 h	....	....	γ, 0.36(12), 0.58(29), 0.98(80), 1.35(76)
	239	53,016	49.41	12.1 h	....	....	γ, 0.209(5), 0.228(18), 0.278(17); e⁻, 0.02–0.04, 0.048, 0.088, 0.106, 0.156; α, 5.78
	240	55,280	51	51.0 h	....	....	γ, 0.90(23), 1.00(77); e⁻, 0.022, 0.038, 0.079, 0.094
	241	56,714	52.96	433 y	....	700; 3.0(f)	α, 5.49, 5.44; γ, 0.060(36), 0.101,

Table 3-6 (*Continued*)
TABLE OF NUCLIDES

| Nuclide | | Mass Excess | | Half-life | Natural | Thermal Neutron | Major Radiations, |
Z Symbol	Mass Number	μmu	MeV	y = yr, d = day h = hr, m = min s = sec	Abundance, %	Absorption Cross Section σ, barns (mode if not n,γ)	Energies in MeV and γ Intensities, %
	242	59,502	55.48	16.01 h	....	2,900(*f*)	0.208, 0.335, 0.37, 0.663, 0.722; *e*⁻, 0.022, 0.038, 0.054 *β*⁻, 0.67; *e*⁻, 0.021, 0.037; Pu X-rays
	242 *m*		55.52	152 y	....	2,000; 6,000(*f*)	α, 5.21; *e*⁻, 0.028, 0.044; γ, 0.049, 0.087, 0.110, 0.163
	243	61,367	57.18	7.95 × 10³ y	....	74; <0.07(*f*)	α, 5.28, 5.23; γ, 0.044(4), 0.075(50); *e*⁻
	244	64,355	59.90	10.1 h	....	2,300(*f*)	*β*⁻, 0.387; *e*⁻, 0.020, 0.037, 0.077, 0.094; γ, 0.099(5), 0.154(19), 0.746(66), 0.900(25)
	244 *m*		60.02	26 m	....	....	*β*⁻, 1.50; *e*⁻, 0.020, 0.037; Cm X-rays
	245	66,340	61.93	2.07 h	....	....	*β*⁻, 0.91; *e*⁻, 0.125; γ, 0.253
	246	69,660	64.9	25.0 m	....	....	*β*⁻, 2.10, 1.60; γ, 0.799(29), 1.07(65)
₉₆Cm	247	72,090	67	24 m	....	....	*β*⁻; γ, 0.29, 0.23
	238	53,036	49.39	2.5 h	....	....	α, 6.51
	239	54,880	51	2.9 h	....	....	γ, 0.188
	240	55,545	51.72	26.8 d	....	....	α, 6.26, 6.25
	241	57,542	53.73	35 d	....	....	γ, 0.475(95), 0.60; *e*⁻, 0.123, 0.350; α, 5.94
	242	58,788	54.82	162.5 d	....	20; <5(*f*)	α, 6.12, 6.07; γ, 0.044, 0.102, 0.158, 0.58, 0.89; *e*⁻, 0.022, 0.039
	243	61,370	57.19	32 y	....	250; 660(*f*)	α, 6.06, 5.99, 5.79, 5.74; γ, 0.209(4), 0.228(12), 0.278(14); *e*⁻, 0.02–0.04, 0.048, 0.088, 0.106, 0.156
	244	62,821	58.47	18.099 y	....	15	α, 5.81, 5.77; γ, 0.043, 0.100, 0.150, 0.262, 0.59, 0.82; *e*⁻,

Table 3-6 (*Continued*)
TABLE OF NUCLIDES

Z Symbol	Mass Number	Mass Excess μmu	Mass Excess MeV	Half-life y=yr, d=day h=hr, m=min s=sec	Natural Abundance, %	Thermal Neutron Absorption Cross Section σ, barns (mode if not n,γ)	Major Radiations, Energies in MeV and γ Intensities, %
	245	65,371	61.02	8.3×10³ y	....	200; 1,900(*f*)	0.022, 0.038 α, 5.36, 5.31; γ, 0.13(5), 0.173(14)
	246	67,202	62.64	4.7×10³ y	....	15	α, 5.39, 5.34; γ
	247	70,280	65.56	1.6×10⁷ y	....	180	....
	248	72,220	67.43	3.8×10⁵ y	....	6	α, 5.08, 5.04; γ
	249	75,810	70.8	64 m	....	....	β⁻, 0.9
	250		73	1.7×10⁴ y	....	....	SF
₉₇Bk	243	62,965	58.70	4.6 h	....		α, 6.76, 6.72, 6.57, 6.54, 6.21; γ, 0.755, 0.84, 0.946
	244	65,170	61	4.4 h	....	....	α, 6.67, 6.62; γ, 0.145(7), 0.188(16), 0.218(100), 0.334(10), 0.490(14), 0.892(88), 0.922(17), 1.16(11)
	245	66,272	61.84	4.98 d	....	....	α, 6.36, 6.32, 6.15, 6.12, 5.89; γ, 0.253(31), 0.39(3); e⁻, 0.125
	246	68,770	64	1.8 d	....	....	γ, 0.800(40), 1.07(12)
	247	70,260	65.47	1.4×10³ y	....	....	α, 5.68, 5.52; γ, 0.084(40), 0.27(30)
	248	72,960	67.9	16 h	....	....	β⁻, 0.65; γ
	248			>9 y	....	....	....
	249	74,883	69.86	314 d	....	500	β⁻, 0.125; α, 5.42; γ, 0.32
	250	78,270	72.95	193.3 m	....	....	β⁻, 1.76, 0.73; e⁻, 0.019, 0.036; γ, 0.990(47), 1.032(39)
	251	80,810	75	57 m	....	....	β⁻, ~1.0, ~0.5; γ, 0.034, 0.14
₉₈Cf	242			3.4 m	....	....	α, 7.39; e⁻
	243	65,310	61	10.3 m	....	....	e⁻; α, 7.17, 7.06; γ, 0.12
	244	65,969	61.43	19.4 m	....	....	α, 7.21
	245	67,905	63.38	44 m	....	....	α, 7.12
	246	68,766	64.11	35.7 h	....	....	α, 6.76, 6.72; γ
	247	71,070	66	2.5 h	....	....	γ, 0.295(1), 0.417, 0.460; e⁻, 0.164
	248	72,262	67.26	350 d	....	....	α, 6.27, 6.22; γ

Table 3-6 (*Continued*)
TABLE OF NUCLIDES

Nuclide		Mass Excess		Half-life y=yr, d=day h=hr, m=min s=sec	Natural Abundance, %	Thermal Neutron Absorption Cross Section σ, barns (mode if not n,γ)	Major Radiations, Energies in MeV and γ Intensities, %
Z Sym-bol	Mass Number	μmu	MeV				
	249	74,749	69.74	352 y	....	270; 1,735(*f*)	α, 5.81; γ, 0.333(16), 0.388(72)
	250	76,384	71.19	13.1 y	....	1,500; <350(*f*)	α, 6.03, 5.99; e⁻, 0.023, 0.038; γ
	251	79,260	74.15	900 y	....	3,000; 3,000(*f*)	α, 5.85, 5.67; γ, 0.18
	252	81,500	76.05	2.646 y	....	30	α, 6.12, 6.08; e⁻, 0.022, 0.038; γ, SF
	253	85,020	79.3	17.6 d	....	....	β⁻, 0.27; α, 5.98
	254		81	60.5 d	....	<2	SF; α, 5.83, 5.79
₉₉Es	245	71,060	66	1.3 m	....	....	α, 7.70
	246	72,430	68	7.3 m	....	....	α, 7.33
	247	73,580	68	5.0 m	....	....	α, 7.33
	248	75,280	70	25 m	....	....	α, 6.88
	249	76,258	71.15	2 h	....	....	α, 6.77
	250	78,610	73	8 h	....	....	γ
	251	79,930	74.5	1.5 d	....	....	α, 6.48
	252	82,810	77.1	~140 d	....	....	α, 6.64, 6.58; γ, 0.074, 0.154, 0.198, 0.228, 0.278, 0.40(1)
	253	84,730	79.03	20.47 d	....	300	α, 6.64; e⁻, 0.017, 0.027, 0.035, 0.040; γ, 0.387, 0.429
	254	87,900	82.00	276 d	....	<40	α, 6.44; γ, 0.063(2), 0.27, 0.31, 0.39; e⁻, 0.011, 0.018, 0.030, 0.037
	254 *m*		82.10	39.3 h	....	....	β⁻, 1.13, 0.43; e⁻, 0.020, 0.038; γ, 0.65(31), 0.69(38)
	255		84	38.3 d	....	....	α, 6.31
₁₀₀Fm	245			4.2 s	....	....	α, 8.15
	246			1.2 s	....	....	α, 8.24; SF
	247			9.2 s	....	....	α, 8.18
	247			35 s	....	....	α, 7.93, 7.87
	248	77,092	72	38 s	....	....	α, 7.87, 7.83; SF
	249	79,140	73.8	~2.5 m	....	....	α, 7.53
	250	79,490	74.10	30 m	....	....	α, 7.44
	251	81,190	76	7 h	....	....	α, 6.90; γ, 0.41
	252	82,562	76.84	22.7 h	....	....	α, 7.06
	253	84,930	80	3 d	....	....	α, 6.96, 6.91
	254	86,839	80.93	3.24 h	....	....	α, 7.20, 7.16; γ; e⁻, 0.019, 0.036
	255	89,640	83.82	20.1 h	....	....	α, 7.03; γ, 0.059(1), 0.081(1); e⁻,

Table 3-6 (*Continued*)
TABLE OF NUCLIDES

| Nuclide | | Mass Excess | | Half-life | Natural | Thermal Neutron | Major Radiations, |
Z Symbol	Mass Number	μmu	MeV	y = yr, d = day h = hr, m = min s = sec	Abundance, %	Absorption Cross Section σ, barns (mode if not n.γ)	Energies in MeV and γ Intensities, %
							0.032, 0.05–0.07
	256		85.44	2.7 h	....	....	SF; α, 6.93
	257		88.6	80 d	....	....	α, 6.53; γ, 0.180(8), 0.242(10); e⁻, 0.037, 0.045, 0.055, 0.106
	258			380 μs	....	....	SF
₁₀₁Md	248			6 s	....	....	α, 8.32
	249			24 s	....	....	α, 8.03
	250			53 s	....	....	α, 7.73
	251			4.0 m	....	....	α, 7.53
	252	86,120	80	~8 m	....	....	....
	254			10 m	....	....	....
	255	90,550	84.4	27 m	....	....	α, 7.34
	256		86.9	77 m	....	....	α, 7.23, 7.17
	257		89	300 m	....	....	α, 7.08
	258			54 d	....	....	α, 6.79, 6.73
₁₀₂No	251			0.8 s	....	....	α, 8.68, 8.60
	252		83	2.4 s	....	....	α, 8.41; SF
	253	91,340	84	105 s	....	....	α, 8.01
	254	91,140	84.8	55 s	....	....	α, 8.10
	255	92,730	87	185 s	....	....	α, 8.11, 7.92, 7.76
	256		87.83	3.2 s	....	....	α, 8.43; SF
	257		90	26 s	....	....	α, 8.32, 8.27, 8.22
	258			≪1 s	....	....	SF
	259			~1 h	....	....	α, 7.50
₁₀₃Lr	253			0.5 s	....	....	α, 8.8
	254			20 s	....	....	....
	255			22 s	....	....	α, 8.37, 8.35
	256			31 s	....	....	α, 8.64, 8.52, 8.48, 8.43, 8.39, 8.32
	257			0.6 s	....	....	α, 8.87, 8.81
	258			4.2 s	....	....	α, 8.68, 8.65, 8.62, 8.59
	259			5.4 s	....	....	α, 8.45
	260			180 s	....	....	α, 8.03
₁₀₄Rf	257			4.5 s	....	....	α, 9.00, 8.95, 8.78, 8.70
	258?			~0.01 s	....	....	SF
	259			3 s	....	....	α, 8.86, 8.77
	260?			0.1 s	....	....	SF
	261			70 s	....	....	α, 8.40–8.25; SF
₁₀₅Ha	260			1.6 s	....	....	α, 9.14, 9.10, 9.06
	261			1.7 s	....	....	α, 8.93
	262			~40 s	....	....	α, 8.66–8.45
	263?			~60 s	....	....	α, 8.8

Rf = rutherfordium; Ha = hahnium; names and symbols not officially accepted by I.U.P.A.C.

Table 3-7
SPATIAL ORIENTATION OF COMMON HYBRID BONDS

On the assumption that the pairs of electrons in the valency shell of a bonded atom in a molecule are arranged in a definite way which depends on the number of electron pairs (coordination number), the geometrical arrangement or shape of molecules may be predicted. A multiple bond is regarded as equivalent to a single bond as far as molecular shape is concerned.

Coordination Number	Orbitals Hybridized	Geometrical Arrangement	Minimum Radius Ratio
2	sp dp	Linear	
	p^2 ds d^2	Bent (angular)	
3	sp^2 ds^2	Trigonal planar	0.155
	p^3 d^2p	Trigonal pyramidal	
	sp^2d p^2d^2	Square planar	
4	sp^3 d^3s	Tetrahedral	0.225
	d^4	Tetragonal pyramidal	
5	sp^3d d^3sp	Trigonal bipyramidal	0.155
6	d^2sp^3	Octahedral	0.414
	d^4sp	Trigonal prism	
7		One atom above the face of an octahedron, which is distorted chiefly by separating the atoms at the corners of this face.	0.592
8	d^4sp^3	Square antiprism (dodecahedral)	0.645
		Cube	0.732
9		Formed by adding atoms beyond each of the vertical faces of a right triangular prism.	0.732
12		Cube-octahedron	1.000

Table 3-8
CRYSTAL STRUCTURE

Unit cells of the different lattice types in each system are illustrated in Fig. 3-1.

System	Characteristics	Essential Symmetry	Axes in Unit Cell	Angles in Unit Cell
Cubic	Three axes equal and mutually perpendicular	Four threefold axes	$a = b = c$	$\alpha = \beta = \gamma = 90°$
Tetragonal	Two equal axes and one unequal axis mutually perpendicular	One fourfold axis	$a = b \neq c$	$\alpha = \beta = \gamma = 90°$
Orthorhombic (or rhombic)	Three unequal axes mutually perpendicular	Three mutually perpendicular twofold axes, or two planes intersecting in a twofold axis	$a \neq b \neq c$	$\alpha = \beta = \gamma = 90°$
Hexagonal or trigonal	Three equal axes inclined at 120° with a fourth axis unequal and perpendicular to the other three	One sixfold axis or one threefold axis	$a = b \neq c$ $a = b = c$	$\alpha = \beta = 90°;$ $\gamma = 120°$ $\alpha = \beta = \gamma \neq 90°$
Monoclinic	Two axes at an oblique angle with a third perpendicular to the other two	One twofold axis or one plane	$a \neq b \neq c$	$\alpha = \beta = 90°;$ $\gamma \neq 90°$
Triclinic	Three unequal axes intersecting obliquely	No planes or axes of symmetry	$a \neq b \neq c$	$\alpha \neq \beta \neq \gamma \neq 90°$
Rhombohedral	Two equal axes making equal angle with each other			

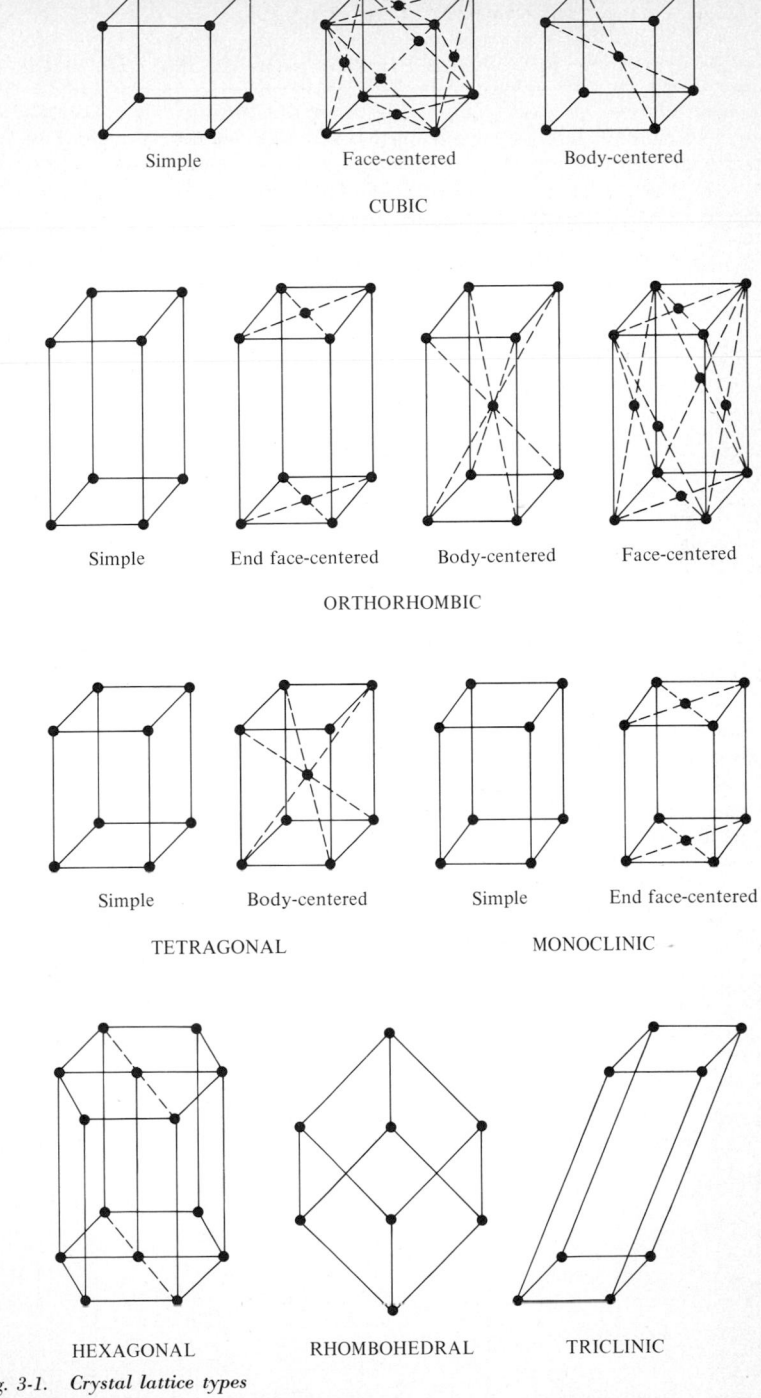

Fig. 3-1. *Crystal lattice types*

Table 3-9
RADII OF ATOMS AND IONS

The radius of an atom is the closest distance to which it will approach another atom of any size under the circumstances specified.

The *ionic radius* of an atom is the radius it exhibits in an ionic crystal like KCl in which the K^+ and Cl^- ions are packed together with their outermost shells of electrons in contact with each other. Values of the ionic radii are tabulated for a coordination number of 6. For structures with coordination numbers other than 6 the appropriate correction factors given below must be employed:

Coordination Number	Correction Factors	
	Ionic Radius	Metallic Radius
12	1.12	1.00
9	1.05	
8	1.03	0.98
6	1.00	0.96
4	0.94	0.88

The normal *covalent radius* is the effective distance from the center of the nucleus to the outer valence shell of that atom in a typical covalent or coordinate bond, assuming spherical symmetry. For example, it is half the distance between the centers of the two bromine atoms in the Br_2 molecule. The bond length contracts for multiple bonds; the double-bond radius is roughly 86 per cent of the length of a single-bond radius, and the triple bond is 78 per cent.

The *nonbonded* or *van der Waals radius* is the internuclear distance or radius of closest approach of an atom to another with which it forms no bond. For nonmetals the nonbonded radius is close to the anion radius.

One can also speak about the *radius of metal atoms* in a crystalline metal. A correction must be applied for coordination numbers other than 12 (see table above).

The *atomic volume*, a rough relative measure for the space occupied by the atoms in 1 gram atom of the element in question, is given by atomic weight/density.

Element	Crystal Ionic Radius		Covalent Radius, Å	Atomic Radius in Metals, Å	Van der Waals Radius, Å
	Charge	Radius, Å			
Actinium	+3	1.11		1.878	
Aluminum	+3	0.50	1.25	1.431	
Americium	+3	0.99		1.84	
	+4	0.89			
	+5	0.86			
	+6	0.80			
Antimony	−3	2.45	(s) 1.41		2.2
	+1	0.89	(d) 1.31		
	+3	0.9			
	+5	0.62			
Argon	+1 (gas)	1.54	1.74		1.91
Arsenic	−3	2.22	(s) 1.21	1.248	2.0
	+3	0.58	(d) 1.11		
	+5	0.47			

Table 3-9 (*Continued*)
RADII OF ATOMS AND IONS

Element	Crystal Ionic Radius		Covalent Radius, Å	Atomic Radius in Metals, Å	Van der Waals Radius, Å
	Charge	Radius, Å			
Astatine	−1	2.27			
	+5	0.57			
	+7	0.51			
Barium	+2	1.35	1.98	2.173	
Berkelium	+2	1.18			
	+3	0.98			
	+4	0.87			
Beryllium	−1	1.95	0.89	(α) 1.113	
	+2	0.31			
Bismuth	−3	2.13	1.52	1.547	
	+3	0.96			
	+5	0.74			
Boron	+1	0.35	(s) 0.88	0.83	2.08
	+3	0.20	(d) 0.76		
			(t) 0.68		
Bromine	−1	1.96	(s) 1.142		1.95
	+5	0.47	(d) 1.04		
	+7	0.39			
Cadmium	+1	1.14	1.41	1.489	
	+2	0.97			
Calcium	+2	0.99	1.74	(α) 1.973	
				(β) 1.939	
Californium	+2	1.17			
	+3	0.98			
	+4	0.86			
Carbon	−4	2.60	(s) 0.77		1.85
	+4	0.15	(d) 0.67		
			(t) 0.60		
Cerium	+3	1.03	1.646	1.825	
	+4	0.92			
Cesium	+1	1.69	2.35	2.654	2.62
Chlorine	−1	1.81	(s) 0.99		1.81
	+5	0.34	(d) 0.89		
	+7	0.26			
Chromium	+1	0.81	1.17	(α) 1.249	
	+2	0.84		(β) 1.305	
	+3	0.64			
	+4	0.56			
	+6	0.52			
Cobalt	+2	0.74	1.16	1.253	
	+3	0.63			
Copper	+1	0.96	1.17	1.278	
	+2	0.72			
Curium	+2	1.19			
	+3	0.99			
	+4	0.88			
Dysprosium	+3	0.91	1.589	1.773	
Einsteinium	+2	1.16			
	+3	0.98			
	+4	0.85			

Table 3-9 (*Continued*)
RADII OF ATOMS AND IONS

Element	Crystal Ionic Radius		Covalent Radius, Å	Atomic Radius in Metals, Å	Van der Waals Radius, Å
	Charge	Radius, Å			
Erbium	+3	0.88	1.567	1.757	
Europium	+2	1.12	1.850	2.042	
	+3	0.95			
Fermium	+2	1.15			
	+3	0.97			
	+4	0.84			
Fluorine	−1	1.36	(s) 0.64	0.717	1.35
	+7	0.07	(d) 0.54		
Francium	+1	1.76		2.7	
Gadolinium	+3	0.94	1.614	1.802	
Gallium	+3	0.62	1.25	1.221	
Germanium	−4	2.72	(s) 1.22	1.225	
	+2	0.70	(d) 1.12		
	+4	0.53			
Gold	+1	1.37	1.34	1.442	
	+2	1.05			
	+3	0.91			
Hafnium	+4	0.78	1.44	(α) 1.564	
Helium	+1 (gas)	0.93			1.22
Holmium	+3	0.89	1.580	1.766	
Hydrogen	−1	2.08	0.371		1.2
	+1	10^{-5}			
Indium	+1	1.32	1.50	1.626	
	+3	0.81			
Iodine	−1	2.16	(s) 1.333		2.15
	+5	0.62	(d) 1.23		
	+7	0.50			
Iridium	+2	0.89	1.26	1.357	
	+3	0.75			
	+4	0.64			
	+6	0.56			
Iron	+2	0.76	1.165	(α) 1.241	
	+3	0.64		(γ) 1.289	
				(δ) 1.27	
Krypton	+1 (gas)	1.69	1.89		1.98
Lanthanum	+3	1.06	1.690	1.877	
Lawrencium	+2	1.12			
	+3	0.94			
	+4	0.83			
Lead	−4	2.15	1.54	1.750	
	+2	1.20			
	+4	0.84			
Lithium	+1	0.68	1.23	1.52	
Lutetium	+3	0.85	1.557	1.734	
Magnesium	+2	0.65	1.36	1.60	
Manganese	+2	0.80	1.17	(α) 1.24	
	+3	0.62		(γ) 1.366	
	+4	0.54		(δ) 1.334	
	+7	0.46			
Mendelevium	+2	1.14			

Table 3-9 (*Continued*)
RADII OF ATOMS AND IONS

Element	Crystal Ionic Radius		Covalent Radius, Å	Atomic Radius in Metals, Å	Van der Waals Radius, Å
	Charge	Radius, Å			
	+3	0.96			
	+4	0.84			
Mercury	+1	1.27	1.44	1.60	
	+2	1.10			
Molybdenum	+4	0.66	1.29	1.362	
	+6	0.62			
Neodymium	+3	1.00	1.642	1.821	
Neon	+1 (gas)	1.12	1.31		1.60
Neptunium	+3	1.01		(α) 1.31	
	+4	0.92		(β) 1.38	
	+5	0.88		(γ) 1.52	
	+6	0.82			
Nickel	+2	0.72	1.15	1.246	
	+3	0.62			
Niobium	+4	0.74	1.34	1.429	
	+5	0.70			
Nitrogen	−3	1.71	(*s*) 0.70	[N_2 0.549]	1.54
	+1	0.25	(*d*) 0.60		
	+3	0.13	(*t*) 0.55		
	+5	0.11			
Nobelium	+2	1.13			
	+3	0.95			
	+4	0.83			
Osmium	+2	0.89	1.26	1.34	
	+3	0.81			
	+4	0.65			
	+6	0.6			
	+8	0.53			
Oxygen	−2	1.40	(*s*) 0.66	[O_2 0.603]	1.40
	−1	1.76	(*d*) 0.55		
	+1	0.22	(*t*) 0.51		
	+6	0.09			
Palladium	+2	0.86	1.28	1.376	
	+4	0.64			
Phosphorus	−3	2.12	(*s*) 1.10	black crys 1.08	1.9
	+3	0.42	(*d*) 1.00	yellow crys 0.93	
	+5	0.34	(*t*) 0.93	amorph red 1.15	
Platinum	+2	0.85	1.29	1.38	
	+4	0.70			
Plutonium	+3	1.00		(γ) 1.51	
	+4	0.90		(δ) 1.64	
	+5	0.87		(ϵ) 1.58	
	+6	0.81			
Polonium	−2	2.30	1.53	(α) 1.67	
	+4	0.65		(β) 1.68	
	+6	0.56			
Potassium	+1	1.33	2.025	2.272	2.31
Praseodymium	+3	1.01	1.648	1.828	
	+4	0.90			
Promethium	+3	0.98		1.810	

Table 3-9 (*Continued*)
RADII OF ATOMS AND IONS

Element	Crystal Ionic Radius		Covalent Radius, Å	Atomic Radius in Metals, Å	Van der Waals Radius, Å
	Charge	Radius, Å			
Protactinium	+3	1.05		1.606	
	+4	0.96			
	+5	0.90			
Radium	+2	1.40		2.20	
Radon			2.14		
Rhenium	+4	0.72	1.28	1.370	
	+6	0.61			
	+7	0.60			
Rhodium	+2	0.86	1.25	1.345	
	+3	0.75			
	+4	0.67			
Rubidium	+1	1.48	2.16	2.475	2.44
Ruthenium	+3	0.77	1.24	1.325	
	+4	0.63			
	+8	0.54			
Samarium	+2	1.11	1.66	1.802	
	+3	0.96			
Scandium	+3	0.81	1.44	1.606	
Selenium	−2	1.98	(s) 1.17		2.00
	−1	2.32	(d) 1.07		
	+1	0.66			
	+4	0.69			
	+6	0.42			
Silicon	−4	2.71	(s) 1.17		2.0
	−1	3.84	(d) 1.07		
	+1	0.65	(t) 1.00		
	+4	0.41			
Silver	+1	1.26	1.34	1.444	
	+2	0.97			
Sodium	+1	0.95	1.57	1.537	2.31
Strontium	+2	1.13	1.92	(α) 2.151	
				(β) 2.16	
				(γ) 2.10	
Sulfur	−2	1.84	(s) 1.04	[S_2 0.944]	1.85
	−1	2.19	(d) 0.94	[S_8 1.04]	
	+4	0.37	(t) 0.87		
	+6	0.29			
Tantalum	+5	0.7	1.34	1.43	
Technetium	+2	0.95		1.358	
	+4	0.72			
	+7	0.58			
Tellurium	−2	2.21	(s) 1.37	1.432	2.20
	−1	2.50	(d) 1.27		
	+4	0.81			
	+6	0.56			
Terbium	+3	0.92	1.592	1.782	
	+4	0.84			
Thallium	+1	1.44	1.55	(α) 1.704	
	+3	0.95		(β) 1.681	
Thorium	+3	1.08		(α) 1.798	

Table 3-9 (*Continued*)
RADII OF ATOMS AND IONS

Element	Crystal Ionic Radius		Covalent Radius, Å	Atomic Radius in Metals, Å	Van der Waals Radius, Å
	Charge	Radius, Å			
	+4	0.99		(β) 1.78	
Thullium	+3	0.87	1.562	1.746	
	+4	0.94			
Tin	−4	2.94	(s) 1.40	1.405	
	−1	3.70	(d) 1.30		
	+2	1.02			
	+4	0.71			
Titanium	+1	0.96	1.32	(α) 1.448	
	+2	0.90		(β) 1.432	
	+3	0.77			
	+4	0.68			
Tungsten	+4	0.68	1.30	1.370	
	+6	0.65			
Uranium	+3	1.03		(α) 1.385	
	+4	0.93		(β) 1.53	
	+5	0.87			
	+6	0.83			
Vanadium	+2	0.88	1.22	1.321	
	+3	0.74			
	+4	0.60			
	+5	0.59			
Xenon	+1 (gas)	1.90	2.09	2.18	
Ytterbium	+2	1.13	1.699	1.940	
	+3	0.86			
Yttrium	+3	0.93	1.62	1.81	
Zinc	+1	0.88	1.25	1.332	
	+2	0.74			
Zirconium	+1	1.09	1.45	1.60	
	+4	0.79			

(s) single bond; (d) double bond; (t) triple bond.

Table 3-10
BOND ENERGIES

The dissociation energy, D_0°, of a bond is the change in energy (in kcal) at absolute zero in the ideal gas state for the reaction: $A—B \longrightarrow A + B$ the products being in their ground states and the reactant in the zeroth vibrational level. The thermochemical bond energy E is defined as the standard enthalpy change per mole (in kcal at $25°C$) in the hypothetical ideal-gas reaction given above. When a molecule contains two or more equivalent bonds, $E(A—B)$ is taken as the average of the successive $A—B$ dissociation enthalpies.

Bond	D_0°	E	Bond	D_0°	E
Aluminum			Al—Cl	118 + 4	
Al—Br	99 ± 5		Al—F	158	
Al—C		61	Al—H	67	

Table 3-10 (*Continued*)
BOND ENERGIES

Bond	D_0°	E	Bond	D_0°	E
Aluminum (*cont.*)			Cadmium		
Al—I	90 ± 2		Cd—Cd	2.1	
Al—O	127–138		Cd—Br	76 ± 10	
Antimony			Cd—C		27
Sb—C		47	Cd—Cl	84?	
Sb—Cl		74	Cd—H	15.6	
Sb—F	92 ± 23		Cd—I	50?	
Sb—H		61	Cd—O	30 ± 10	
Sb—O	74 ± 9		Cd—Se	≤ 74	
Arsenic			Calcium		
As—As	91		Ca—Cl	≤ 64	
As—Br		58	Ca—F	< 73	
As—C	55	48	Ca—H	≤ 39	
As—Cl		70	Ca—I	58 ± 23	
As—F	111	111	Ca—O	100 ± 15	
As—H		70 ± 3	Ca—S	115 ± 12	
As—I		43	Carbon		
As—O	113 ± 2		C—B		89
Barium			C—Br	65	68
Ba—Br	65?		C—C		82.6
Ba—Cl	51 ± 11		C=C		145.8
Ba—F	69 ± 23		C≡C		199.6
Ba—H	42		C—Cl		81
Ba—O	130		C—F	121	116
Beryllium			C—H		98.7
Be—Br		89	C—I		52
Be—Cl		109	C—N		72.8
Be—F	92?		C=N		147
Be—H	53 ± 7		C≡N		212.6
Be—I		69	C—O		85.5
Be—O	124 ± 8		C=O (aldehydes)		176
Bismuth			(ketones)		179
Bi—Br		63 ± 3	C—P	138?	
Bi—C		31	C—S		65
Bi—Cl		67	C=S	166?	
Bi—F	74 ± 9		C—Si	51	63
Bi—H	58	47	(silicones)		78
Bi—O	85 ± 21		Cesium		
Boron			Cs—Cs	10.4	
B—Br	97	90	Cs—Br	99 ± 3	
B—C		89	Cs—Cl	101 ± 3	
B—Cl	117?	109	Cs—F	121 ± 8	
B—F	195?	154	Cs—H	42	
B—H	70		Cs—I	84 ± 4	
B—N	92?	106	Chlorine		
B—O	173–216	128	Cl—Cl	57.07	57.87
Bromine			Cl—F	60.5	
Br—Cl	52.5		Cl—H	102.2	103.1
Br—F	59.44		Cl—O	57	60
Br—H	86.5	87.5	Chromium		
Br—O	56 ± 3	48	Cr—Br	78 ± 6	

Table 3-10 (*Continued*)
BOND ENERGIES

Bond	D_0°	E	Bond	D_0°	E
Chromium (*cont.*)			**Indium**		
Cr—Cl	87 ± 6	91.3	In—Br	93 ± 3	63
Cr—I	68 ± 6	56	In—Cl	104 ± 3	75
Cr—O	105 ± 20		In—F	125 ± 8	
Cobalt			In—H		57
Co—Cl		86	In—I	78 ± 4	55
Copper			In—O	25 ± 5	
Cu—Cu	48 ± 3		**Iodine**		
Cu—Br	78 ± 6		I—I	35.55	36.06
Cu—Cl	88 ± 4		I—Br	41.90	
Cu—H	66		I—Cl	49.63	
Cu—I	76		I—F	66.2	
Cu—O	61		I—H	70.5	71.4
Cu—F	70 ± 3		I—O	44 ± 4	
Dysprosium			**Iridium**		
Dy—O		150	Ir—F		67?
Erbium			**Iron**		
Er—O		152	Fe—Cl	69 ± 46	81–95
Fluorine			Fe—O	95 ± 3	
F—F	36	37	**Krypton**		
F—H	134.0	135	Kr—F	12	
Gallium			**Lead**		
Ga—Ga	35		Pb—Br	58?	
Ga—Br	103 ± 4		Pb—C		31
Ga—Cl	115 ± 4		Pb—Cl	71 ± 6	
Ga—F	142		Pb—F	74 ± 9	
Ga—I	91 ± 7		Pb—H		49.0
Ga—O	58 ± 12		Pb—I	46?	
Germanium			Pb—O	98 ± 8	
Ge—Ge		37.6	Pb—S	75?	
Ge—Br	60 ± 7		Pb—Se	109?	
Ge—Cl		81	**Lithium**		
Ge—F	92 ± 23	111	Li—Li	25	
Ge—H		68.9	Li—Br	101 ± 3	
Ge—I		51	Li—Cl	115 ± 3	
Ge—O	157 ± 5		Li—F	137 ± 8	
Ge—S	131 ± 2		Li—H	~58	
Ge—Se	115	115	Li—I	81 ± 3	
Ge—Te		93	**Lutetium**		
Gold			Lu—O		163
Au—Au	52		**Magnesium**		
Au—Cl	65 ± 11		Mg—Br	57 ± 23	
Au—H	72		Mg—Cl	62 ± 16	
Hafnium			Mg—F	74 ± 16	
Hf—O		195 ± 12	Mg—H	46 ± 11	
Holmium			Mg—O	92–120	
Ho—O		150	**Manganese**		
Hydrogen			Mn—Cl	85 ± 2	
H—H	103.24	104.18	Mn—Br	74 ± 2	
H—D	104.0		Mn—F	81 ± 23	
D—D	105.0		Mn—H	55?	

Table 3-10 (*Continued*)
BOND ENERGIES

Bond	D_0°	E	Bond	D_0°	E
Manganese (*cont.*)			P—F		117
Mn—I	67 ± 3		P—H		79 ± 1
Mn—O	101 ± 6		P—I		44
Mercury			P≡N	~138	
Hg—Hg	3.2		P=O	122 ± 2	122
Hg—Br	72	44	P=S	82	
Hg—C	51	27	Potassium		
Hg—Cl	81 ± 2	54	K—K	11.8	
Hg—F	100	66	K—Br	91 ± 2	
Hg—H	8.6		K—Cl	101 ± 1	
Hg—S	≤65		K—F	118 ± 2	119 ± 2
Hg—Tl	0.7		K—H	43	
Molybdenum			K—I	77 ± 3	
Mo—I		89	Praseodymium		
Mo—O	140 ± 20		Pr—O		180
Neodymium			Rubidium		
Nd—O		170	Rb—H		119
Nickel			Samarium		
Ni—Br	85 ± 3		Sm—O		140
Ni—Cl	88 ± 5	87	Scandium		
Ni—H	~60		Sc—O		163
Ni—I	69 ± 5		Selenium		
Ni—O	≤99		Se—Se (Se₂)	65	
Niobium			(Se₆)		44
Nb—O	92 ± 13		Se—Cl		58
Nitrogen			Se—F	76	68
N—N	12.9	52	Se—H		66?
N=N	60	39	Se—O	81 ± 23	
N≡N	225.0	225.8	Silicon		
N—Cl	37	46	Si—Si		42.2
N—F		66.4	Si—Br	69 ± 14	74
N—H (NH)	~85		Si—Cl	76 ± 12	91
(NH₃)	102	93.4	Si—F		135
N—O		48	Si—H	74	76
N=O	162	151	Si—I		56
N—P	~138		Si—N	~104	
N—S	115?		Si—O	185 ± 7	
N—Si	~104		Si—S	147 ± 3	
Osmium			Si—Se	134 ± 6	
Os=O		123	Si—Te	122 ± 9	
Oxygen			Silver		
O—O (O₂)	117.96	119.1	Ag—Ag	39 ± 2	
(—O—O—)	32	47	Ag—Br	69 ± 7	
O—Cl	57	60	Ag—Cl	47	
O—F		45.3	Ag—H	58	
O—H (HOH)	117.5	110.6	Ag—O	32 ± 9	
(OH)	101.5		Sodium		
Phosphorus			Na—Na	17.3	
P—P (P₂)	116.0	116.7	Na—Br	88 ± 2	
(P₄)		51.3	Na—Cl	98.0	
P—Br		63.7	Na—F	107 ± 6	
P—Cl		78.5	Na—H	47	

Table 3-10 (*Continued*)
BOND ENERGIES

Bond	D_0°	E	Bond	D_0°	E
			Thullium		
Na—I	71 ± 2		Tm—O		138
Na—K	14.3		Tin		
Na—Rb	13.1		Sn—Sn	46	
Strontium			Sn—Br	46?	78 (SnBr$_2$)
Sr—Cl	58 ± 22				65 (SnBr$_4$)
Sr—F	62 ± 23		Sn—Cl	74?	76
Sr—H	38		Sn—H	<74	61.0 ± 0.7
Sr—I	46?		Sn—I		65
Sr—O	83 ± 5		Sn—O	132	
Sr—S	53?		Sn—S	110?	
Sulfur			Sn—Se	105?	
S—S (S$_2$)	83	84	Titanium		
(S$_8$)		54	Ti—Cl		120 (TiCl$_2$)
(—S—S—)		71 ± 2			109 (TiCl$_3$)
S—Br		52?			102 (TiCl$_4$)
S—Cl		61	Ti—O	167 ± 2	
S—F		76	Tungsten		
S—H	66–93	83	W—F		74
S—N	~115		W—O		160 ± 10
S—O	119		Uranium		
Tantalum			U—O		171 ± 10
Ta—O	193 ± 12		Vanadium		
Tellurium			V—Cl		92
Te—Te	53?		V—O	147	
Te—F		80	Xenon		
Te—H		57	Xe—F		31 ± 1
Te—O	63–79		Zinc		
Terbium			Zn—Cl	48?	
Tb—O		171	Zn—H	19.6 ± 0.5	
Thallium			Zn—I	41 ± 15	
Tl—Br	78 ± 2		Zn—O	≤92	
Tl—Cl	87 ± 2		Zn—S	98 ± 6	
Tl—F	109 ± 4		Zn—Te	49 ± 6	
Tl—H	46		Zirconium		
Tl—I	68 ± 3		Zr—Cl		116
Thorium			Zr—O	150 ± 35	183
Th—O		200 ± 12			

Table 3-11
HAMMETT AND TAFT SUBSTITUENT CONSTANTS

The Hammett sigma values σ_m and σ_p concern the group attached to an aromatic carbon atom in a system with which they can also engage in resonance interactions. The substituent constant σ_i is defined by the relation

$$\sigma_i = \log \frac{K_i}{K_o}$$

where K_o is the acidity constant of benzoic acid and K_i is the constant of the substituted benzoic acid.

The inductive sigma value of Taft, σ_I, performs a similar function with respect to aliphatic and alicyclic systems. It is determined by direct electrostatic and inductive effects of substituent groups when attached to a saturated carbon atom. The reference acid is acetic acid. Note that $\sigma_I = \sigma^*/6.23$. The σ_I constant for the group $-CH_2X$ is approximately 0.36 times σ_I for the group $-X$; useful when values for the former type group are not available.

Substituent	Hammett Constants		Taft Constant
	σ_m	σ_p	σ_I
$-AsO_3H^-$		-0.02	0.01
$-B(OH)_2$	0.01	0.45	
$-Br$	0.391	0.232	0.45
$-CH_2Br$			0.18
$-CH_3$	-0.069	-0.170	-0.05
$-C_2H_5$	-0.07	-0.12	-0.05
$n-C_3H_7-$			-0.03
$i-C_3H_7-$	-0.07	-0.10	-0.03
$n-C_4H_9-$	-0.07	-0.16	-0.04
$sec-C_4H_9-$		-0.12	-0.03
$iso-C_4H_9-$		-0.12	-0.03
$tert-C_4H_9-$	-0.10	-0.197	-0.07
$-CH_2-C(CH_3)_3$		-0.23	
$-C(CH_3)_2(C_2H_5)$		-0.19	
$n-C_5H_{11}-$			-0.04
n-Octyl-			-0.06
Cyclohexyl-			-0.02
$3,4-[CH_2]_3-$fused ring		-0.26	
$3,4-[CH_2]_4-$fused ring		-0.48	
$-CH=CH_2$			0.09
$-CH=C(CH_3)_2$			0.03
$trans-CH=CH(CH_3)$			0.05
$-CH_2-CH=CH_2$			0.00
$-CH=CH-C_6H_5$	0.14		
$-C\equiv CH$			0.35
$-CH_2-C\equiv CH$			0.13
$-C\equiv C-C_6H_5$	0.14	0.16	
$-C_6H_5$	0.06	-0.01	0.10
1-Naphthyl- (also 2-)			0.12
$-CH_2-C_6H_5$			0.04
$-CH_2-CH_2-C_6H_5$			-0.01
2-Furyl-			0.04
2-Indolyl-			-0.01

Table 3-11 (*Continued*)
HAMMETT AND TAFT SUBSTITUENT CONSTANTS

Substituent	Hammett Constants		Taft Constant
	σ_m	σ_p	σ_I
2-Thienyl-			0.21
2-Thienylmethyl-			0.05
—CHO	0.35	0.22	
—COCH$_3$	0.376	0.502	0.29
—COC$_6$H$_5$	0.34	0.36	0.35
—COCF$_3$	0.65		
—CONH$_2$	0.28	0.36	0.27
—CONHC$_6$H$_5$			0.25
—COO$^-$	0.07	0.12	0.17
—COOH	0.35	0.41	
—COOCH$_3$	0.39	0.385	0.34
—COOC$_2$H$_5$	0.37	0.45	0.34
—CH$_2$COOCH$_3$			0.17
—CH$_2$COOC$_2$H$_5$			0.13
—CH$_2$—CH$_2$—COOH	-0.03	-0.07	
—CH$_2$COO$^-$			0.01
—Cl	0.373	0.227	0.47
—CCl$_3$	0.40	0.46	
—CH$_2$Cl		0.18	0.15
—CH$_2$—CCl$_3$			0.14
—CH=CCl$_2$			0.16
—F	0.337	0.062	0.52
—CF$_3$	0.43	0.54	0.42
—CH$_2$—CF$_3$			0.14
—CHF$_2$			0.32
—Ge(CH$_3$)$_3$		0.00	
—I	0.352	0.18	0.39
—CH$_2$I			0.16
—IO$_2$	0.70	0.76	
—N$_2^+$	1.76	1.91	
—N$_3$	0.33	0.08	0.42
—NH$_2$	-0.038	-0.172	
—NH$_3^+$	1.13	1.70	0.60
—CH$_2$NH$_3^+$			0.36
—NH—NH$_2$	-0.02	-0.55	0.58
—CN	0.56	0.66	
—CH$_2$—CH			0.01
—NH—CH$_3$	-0.30	-0.84	0.18
—NH—C$_2$H$_5$	-0.24	-0.61·	
—NH-n-C$_4$H$_9$	-0.34	-0.51	
—NH(CH$_3$)$_2^+$			0.70
—N(CH$_3$)$_2$	-0.15	-0.44	
—NH—C$_6$H$_5$			0.27
—NH$_2$—CH$_3^+$	0.96		0.60
—NH$_2$—C$_2$H$_5^+$	0.96		0.60
N(CH$_3$)$_3^+$	0.86	0.02	0.73
—N=N—C$_6$H$_5$		0.64	
—N(CF$_3$)$_2$	0.45	0.53	

Table 3-11 (*Continued*)
HAMMETT AND TAFT SUBSTITUENT CONSTANTS

Substituent	Hammett Constants		Taft Constant
	σ_m	σ_p	σ_I
—NO		0.12	
—NO$_2$	0.710	0.778	
—NHOH	−0.04	−0.34	
—NHCOC$_2$H$_5$			0.25
—NHCOCH$_3$	0.25	0.04	0.28
—NHCOC$_6$H$_5$	0.22	0.08	0.27
—NHCHO			0.26
—NH—CONH$_2$			0.21
—CH$_2$—CONH$_2$			0.05
—CH$_2$—NH$_2$			0.08
—NH—SO$_2$—C$_6$H$_5$			0.32
—O$^-$	−0.2	−0.5	
—OH	0.121	−0.37	0.25
—OCH$_3$	0.115	−0.268	−0.25
—OC$_2$H$_5$	0.1	−0.24	0.27
—O-i-C$_3$H$_7$	0.1	−0.45	0.26
—O—C$_5$H$_{11}$	0.1	−0.34	
—O—C$_6$H$_5$	0.252	−0.320	0.39
—OCF$_3$	0.40	0.35	
3,4-O—CH$_2$—O—		−0.16	
—ONO$_2$			0.62
—O—N=C(CH$_3$)$_2$			0.29
—CH$_2$—OH	0.08	0.08	0.05
—CH$_2$—OCH$_3$			0.07
CH$_3$—CH(OH)—			0.02
C$_6$H$_5$—CH(OH)—			0.08
—CH$_2$—C(OH)(CH$_3$)$_2$			−0.04
—P(CH$_3$)$_2$	0.1	0.05	
—P(CH$_3$)$_3^+$	0.8	0.9	
—P(CF$_3$)$_2$	0.6	0.7	
—PO$_3$H$^-$	0.2	0.26	
—SH	0.25	0.15	0.26
—CH$_2$—SH			0.10
—SCH$_3$	0.23	0.22	0.25
—S(CH$_3$)$_2^+$	1.03	1.20	0.25
—SC$_2$H$_5$			0.25
—S-cyclohexyl			0.31
—SC$_6$H$_5$			0.30
—S—C(C$_6$H$_5$)$_3$			0.11
—S—CH$_2$—C$_6$H$_5$			0.25
—CH$_2$—S—CH$_2$—C$_6$H$_5$			0.06
—S—CF$_3$	0.40	0.50	
—S—CF$_5$	0.6	0.7	
—SCN		0.52	0.55
—S—COCH$_3$	0.39	0.44	
—S—CONH$_2$			0.33
—SOCH$_3$	0.52	0.49	
—SOC$_2$H$_5$			0.52

Table 3-11 (*Continued*)
HAMMETT AND TAFT SUBSTITUENT CONSTANTS

Substituent	Hammett Constants		Taft Constant
	σ_m	σ_p	σ_I
—SO$_2$CH$_3$	0.56	0.68	0.59
—SO$_2$C$_2$H$_5$			0.60
—SO$_2$C$_6$H$_5$			0.57
—SO$_2$CF$_3$	0.79	0.93	
—SO$_2$NH$_2$	0.55	0.62	
—SO$_3^-$	0.05	0.09	0.13
—SO$_3$H		0.50	
—SeCH$_3$	0.1	0.0	
—Se-cyclohexyl			0.38
—SeCN		0.66	0.58
—Si(CH$_3$)$_3$	−0.05	0.0	0.13
—Si(C$_2$H$_5$)$_3$		0.0	
—CH$_2$—Si(CH$_3$)$_3$	−0.19	−0.22	−0.05
—Sn(CH$_3$)$_3$		0.0	
—Sn(C$_2$H$_5$)$_3$		0.0	

Values of σ_m and σ_p are for most part from D. H. McDanial and H. C. Brown, *J. Org. Chem.*, **23,** 420 (1958). For electrically charged groups the Hammett constants may be particularly solvent-dependent.

Values of Taft σ_I constants are from M. Charton, *J. Org. Chem.*, **29,** 1222(1964).

Section 4

INORGANIC CHEMISTRY

NOMENCLATURE OF INORGANIC COMPOUNDS

Adopted by the 1957 Commission of the Nomenclature of Inorganic Chemistry of the International Union of Pure and Applied Chemistry (*IUPAC Nomenclature of Inorganic Chemistry*, 1957, Butterworths, London, 1959), the following short glossary is not intended to cover all the possible cases contained in the comprehensive report.

A. WRITING FORMULAS

1. The more electropositive constituent is placed first. Cations present should be arranged in order of increasing valence (except hydrogen); the cations of each valence group should be arranged in order of decreasing atomic number. *Examples:* KCl, NH_4MgPO_4, $NaNH_4HPO_4$.

2. For binary compounds between nonmetals, the sequence is: B, Si, C, Sb, As, P, N, H, Te, Se, S, At, I, Br, Cl, O, F. Place the constituent first which appears first in the sequence. *Examples:* $AsCl_3$, SbH_3, H_2Te, BrF_3, OF_2 and N_4S_4 (nitrogen sulfides, not sulfur nitrides).

For compounds containing three or more elements, the sequence should follow the order in which the atoms are actually bound in the molecule or ion.
Examples: $—SCN$, $—PH_2O_2$ (hypophosphite ion).
Exceptions: HNO_3, $HClO_4$, H_2SO_4, H_2CO_3, $HClO$.

3. The mass number, atomic number, number of atoms, and ionic charge of an element may be indicated by means of four indices placed around the symbol:

$$\begin{matrix} \text{mass number} \\ \text{atomic number} \end{matrix} \text{SYMBOL} \begin{matrix} \text{ionic charge} \\ \text{number of atoms} \end{matrix} \qquad {}^{15}_{7}N^{3-}_{2}$$

The atomic number, which is redundant, may be omitted in most cases. For multiple charges an Arabic superscript numeral should precede the plus or minus sign: Mg^{2+}, PO_4^{3-}.

4. A centered period is used to denote water of hydration, other solvates, and addition compounds; viz., $CuSO_4 \cdot 5H_2O$.

5. In the formula of a free radical the unshared electron may be indicated by a point in the middle position (or above the atom where it is localized):

$$H_3C \cdot \quad HO \cdot \quad (\text{or } \overset{\cdot}{N}H_3^+).$$

6. The prefixes *cis, trans, asym, cyclo, d(dextro), l(levo), dl, (meso)* should be connected with the chemical formula by a hyphen and should be italicized.

B. NAMING COMPOUNDS

1. Names and symbols for the elements are given in Table 3-1. Wolfram may be substituted for tungsten. A few Latin names are used

with affixes: cupr- (copper), aur- (gold), ferr- (iron), plumb- (lead), argent- (silver), and stann- (tin).

In forming a complete name the name of the electropositive constituent is left unmodified except when it is necessary to indicate the valency (see Stock System). *Examples:* $AlCl_3$, aluminum chloride; AgBr, silver bromide; MnO_2, manganese(IV) oxide.

2. Collective names include: halogens (F, Cl, Br, I, At); chalcogens (O, S, Se, Te, Po); alkali metals (Li to Fr); alkaline earth metals (Ca, Sr, Ba, Ra); lanthanides (Ce to Lu); actinides (Pa to Lr, or those whose $5f$ shell is being filled); lanthanum series (La to Lu); and actinium series (Ac to Lr).

3. The hydrogen isotopes are given special names: ^1H (protium), ^2H (deuterium), and ^3H (tritium); and one often encounters the symbols H, D, and T, respectively. Other isotopes are designated by mass numbers: boron-10 for ^{10}B.

4. Electronegative constituents which are monoatomic are named by stripping the name of the element back to the penultimate consonant (last syllable omitted) and then adding *-ide*. Examples: Cl^-, chloride; S^{2-}, sulfide; As^{3-}, arsenide; Si^{4-}, silicide. Exceptions: Bi^{3-}, bismuthide; H^-, hydride; Hg^{2-}, mercuride; N^{3-}, nitride; O^{2-}, oxide, P^{3-}, phosphide; Zn^{2-}, zincide.

5. *Pseudo-binary compounds.* A few common composite anions have names ending in *-ide.* These are

—OH, hydroxide	—NH$_2$, amide	—NH—NH$_2$, hydrazide
—CN, cyanide	—NH—, imide	—NHOH, hydroxylamide

Added to these anions are:

—O$_3$, ozonide	—I$_3$, triiodide	—S$_2$—, disulfide
—O—O—, peroxide	—N$_3$, azide	

but the compounds HS—[S]$_x$—SH are named as sulfanes: $x = 1$, trisulfane; $x = 2$, tetrasulfane; and so on. This resembles the nomenclature for boranes and silanes.

6. Compounds of hydrogen with the more electropositive elements are designated hydrides (NaH, sodium hydride), but the covalent compounds of hydrogen with the more electronegative elements are named as acids (H_2S, hydrogen sulfide). Certain hydrides possess trivial names: H_2O (water), NH_3 (ammonia), N_2H_4 (hydrazine), B_2H_6 (diborane, etc.), SiH_4 (silane), Si_2H_6 (disilane, etc.), PH_3 (phosphine), P_2H_4 (diphosphine), SbH_3 (stibine), AsH_3 (arsine).

7. If the electronegative constituent is polyatomic, it should be designated by the suffix *-ate.* Exceptions are listed under C-3. For K_2SiF_6 the name silicate applies irrespective of the oxidation state of the central atom (silicon) or the number or nature of the surrounding ligands (full

name: dipotassium hexafluorosilicate, usually shortened to potassium fluorosilicate). Similarly, Na_2SO_3 is sodium trioxosulfate, or is sodium sulfite, its trivial name.

8. *Stoichiometric proportions.* The proportions of the constituents in a formula may be denoted by Greek numerical prefixes: di, tri, tetra, penta, hexa, hepta, octa, ennea, undeca, and dodeca, joined to the name of the relevant element. The prefix *mono* can usually be omitted; occasionally *hemi* is used to denote $\frac{1}{2}$. Arabic numerals are used for numbers higher than 12.

When it is required to indicate the number of entire groups of atoms, the multiplicative numerals *bis, tris, tetrakis,* and so on, are used.

Stoichiometric proportions also may be indicated indirectly by the **Stock System.** Roman numerals representing the oxidation state of the element are placed in parentheses immediately following its name. This notation can be applied to both cations and anions.

Examples: PF_5, phosphorus pentafluoride; N_2O_3, dinitrogen trioxide; $Ca[PF_6]_2$, calcium *bis*(hexafluorophosphate); $CuCl_2$, copper(II) chloride; Fe_3O_4, iron(II,III) oxide; K_2PtCl_4, potassium chloroplatinate(II); K_2PtCl_6, potassium chloroplatinate(IV); Hg_2^{2+}, mercury(I) ion; Fe^{3+}, iron(III) ion.

9. In di- and polynuclear compounds a bridging group should be indicated by adding the Greek letter μ immediately before its name and separating this from the rest of the complex by a hyphen. Two or more bridging groups of the same kind are indicated by di-μ, etc. *Example:* $Be_4O(C_2H_3O_2)_6$ is named μ_4-oxohexa-μ-acetatotetraberyllium (where the sub 4 indicates that the oxygen bridges four atoms).

10. For chemical purposes polymorphs should be indicated by adding the crystal system after the name or formula: ZnS (cub) = zinc sulfide (cub) = zinc blende or sphalerite; ZnS (hex) = wurtzite.

11. Where there is more than one ion derived from one base, the ionic charges are indicated on their names: $N_2H_5^+$, hydrazinium(1+) ion and $N_2H_6^{2+}$, hydrazinium(2+) ion.

12. Names of *radicals* end in -yl and should be denoted by systematic names and formulas except where established practice warrants the retention of these special names:

OH	hydroxyl	CrO_2	chromyl	PO	phosphoryl
CO	carbonyl	UO_2	uranyl†	SO	sulfinyl (thionyl§)
ClO	chlorosyl*	VO	vanadyl	SO_2	sulfonyl (sulfuryl§)
ClO_2	chloryl*	NO	nitrosyl	S_2O_5	pyrosulfuryl
ClO_3	perchloryl*	NO_2	nitryl	SeO	seleninyl
				SeO_2	selenonyl

* Similarly for the other halides.
† Similarly for other actinides.
§ For use only with halides.

Thio, seleno and telluro prefixes may be used with radicals. Use of the Stock System extends the range: UO_2^{2+} uranyl(VI); UO_2^+ uranyl(V); vanadyl(V), (IV), and (III) all exist.

Examples: O_2PtF_6, oxygenyl hexafluoroplatinate(V); CS, thiocarbonyl.

C. ACIDS AND ANIONS

1. Binary acids with -*ide* anions should be named as hydrogen . . . .-*ide*; e.g., H_2S, hydrogen sulfide; HCN, hydrogen cyanide; HCl, hydrogen chloride.

Names such as hydrofluoric acid refer to aqueous solutions, and percentages such as 35 per cent hydrofluoric acid denote the weight/volume of hydrogen halide in the solution.

2. Anionic portions of polyatomic acids have endings -*ate* normally added to the penultimate constant of the name of the central atom (Latin name used if listed in B-1) with these exceptions: antimonate, bismuthate, carbonate, cobaltate, nickelate (also niccolate), nitrate, phosphate, tungstate, and zincate. *Examples:* H_4GeO_4, hydrogen germanate; H_2WO_4, hydrogen tungstate.

3. Acids given in the table retain their trivial names. Anions from these trivial names are formed by changing -*ous* acid to -*ite* and -*ic* acid to -*ate*. Ortho- and meta- distinguish acids of differing water content when distinction from other acids of the series is essential.

Hypo- is used as a prefix to denote a lower oxidation state.

Per- is used as a prefix to denote a higher oxidation state.

Trivial Names Retained for Acids
(alphabetically by central atom)

H_3AsO_4	arsenic acid	$HMnO_4$	permanganic acid
H_3AsO_3	arsenous acid	H_2MnO_4	manganic acid
H_3BO_3	orthoboric acid	HNO_3	nitric acid
HBO_2	metaboric acid	HNO_2	nitrous acid
$HBrO_3$	bromic acid	H_2NO_2	nitroxylic acid
$HBrO_2$	bromous acid	$H_2N_2O_2$	hyponitrous acid
$HBrO$	hypobromous acid	H_3PO_4	phosphoric acid (ortho-)
H_2CO_3	carbonic acid	$H_4P_2O_7$	diphosphoric acid or
$HOCN$	cyanic acid		pyrophosphoric acid
$HNCO$	isocyanic acid	$H_5P_3O_{10}$	triphosphoric acid
$HClO_4$	perchloric acid	HPO_3	metaphosphoric acid
$HClO_3$	chloric acid	H_2PHO_3	phosphorous acid
$HClO_2$	chlorous acid	$H_4P_2O_5$	diphosphorous or
$HClO$	hypochlorous acid		pyrophosphorous acid
H_2CrO_4	chromic acid	HPH_2O_2	hypophosphorous acid
$H_2Cr_2O_7$	dichromic acid	$HReO_4$	perrhenic acid
H_5IO_6	orthoperiodic acid ("para-")	H_2ReO_4	rhenic acid
HIO_4	periodic acid	H_2SO_4	sulfuric acid
HIO_3	iodic acid	$H_2S_2O_7$	disulfuric or pyrosulfuric acid
HIO	hypoiodous acid	$H_2S_2O_6$	dithionic acid

H_2SO_3	sulfurous acid	H_2SiO_3	metasilicic acid
$H_2S_2O_5$	disulfurous or pyrosulfurous acid	$HTcO_4$	pertechnetic acid
$H_2S_2O_4$	dithionous acid	H_2TcO_4	technetic acid
H_2SO_2	sulfoxylic acid	H_6TeO_6	orthotelluric acid
H_4SiO_4	orthosilicic acid		

4. **Peroxo-** is used as a prefix to denote the substitution of —O— by —O—O—; as $H_2S_2O_8$ (HO—SO_2—O—O—SO_2—OH), peroxo-disulfuric acid, or H_2SO_5, peroxosulfuric acid.

5. **Thio** (seleno and telluro) acids are derived from oxo acids by replacement of oxygen by sulfur (Se or Te) and are named by placing the prefix before the name of the acid. *Examples:* $H_2S_2O_3$, thiosulfuric acid; H_3AsS_4, tetrathioarsenic acid.

D. COORDINATION COMPOUNDS

1. To name a coordination complex, the names of the ligands are attached directly in front of the name of the central atom; and the oxidation number of the central atom is stated last in parentheses; or the proportions of the constituents are indicated by stoichiometric numerical prefixes.

The ligands must be cited in this order: (*a*) anionic ligands, (*b*) neutral ligands, and (*c*) cationic ligands.

2. Anionic ligands are cited in the following order: (*a*) H^-, (*b*) O^{2-}, (*c*) OH^-, (*d*) monoatomic anions (in the order given in A-2), (*e*) poly-atomic inorganic anions, and (*f*) organic anions in alphabetical order. Within category (*e*) the anions containing the smallest number of atoms should be cited first (CO_3^{2-} before CrO_4^{2-}) and two ions containing the same number of atoms should be cited in order of decreasing atomic number of the central atoms (CrO_4^{2-} before SO_4^{2-}).

Anionic ligands have names ending in *-o*. If the anion name ends in *-ide*, *-ite*, or *-ate*, the final *-e* is replaced by *-o*, giving *-ido, -ito,* and *-ato*, respectively. *Exceptions:* fluoro, chloro, bromo, iodo, oxo (O^{2-}), hydroxo (OH^-), peroxo (O_2^{2-}), thiolo (HS^-), thio (S^{2-}), disulfido (S_2^{2-}), cyano.

Organic radicals, when present as ligands, do not receive the terminal *-o*.

Examples: $[CoCl(NH_3)_5]^{2+}$ chloropentamminecobalt(III) ion; $Na[B(C_6H_5)_4]$ sodium tetraphenylborate.

3. Neutral and cationic ligands should be cited in the following order: (*a*) water (as aquo), (*b*) ammonia (as ammine), (*c*) other inorganic ligands as in A-2, (*d*) organic ligands in alphabetical order and used without change in name.

E. DOUBLE SALTS

1. Acid hydrogen salts are named by adding the word hydrogen; e.g. KH_2PO_4, potassium dihydrogen phosphate; $NaHCO_3$, sodium hydrogen carbonate.

2. Cations present in double salts are arranged as in A-1.

3. Anions should be arranged as in D-2.

Examples: $ZrOCl$, zirconium oxide chloride; $Al(OH)Cl_2$, aluminum hydroxide dichloride.

F. SOLVATES AND ADDITION COMPOUNDS

Naming involves merely speaking the compound. *Examples:* $CuSO_4 \cdot 5H_2O$ is copper(II) sulfate-5-water (or copper(II) sulfate penta-hydrate). Since some of the water is known to be coordinately bonded, it might be named tetraaquocopper(II) sulfate monohydrate for $[Cu(H_2O)_4]SO_4 \cdot H_2O$. $AlCl_3 \cdot 4C_2H_5OH$ is named aluminum chloride-4-ethanol (*not* ethanolate), and $CaCl_2 \cdot 8NH_3$ is calcium chloride-8-ammonia (*not* ammoniate).

G. ISOPOLY ANIONS

Where the composition and constitution of the structural units of isopoly acids and their salts are known, nomenclature follows the rules already given. When the constitution is not yet known, three methods may be used: (*a*) the composition, reduced to the simplest empirical formula, is indicated by Greek numerical prefixes; (*b*) the numbers of the individual atoms are placed in brackets after the simple name; (*c*) the ratio of basic anhydride to acidic anhydride is indicated by Arabic figures in parentheses. *Example:* $Na_2W_4O_{13}$, disodium tetratungstate or sodium tungstate(2.4.13); $Na_2O \cdot 4WO_3$, sodium(1:4)tungstate.

Table 4-1
PHYSICAL CONSTANTS OF INORGANIC COMPOUNDS

Names, while following the IUPAC Nomenclature, are often alphabeticized by the central atom to facilitate their location. Solvates are listed under the entry for the anhydrous salt. Acid salts follow the normal salt entry.

Formula Weights are based upon the International Atomic Weights of 1969 and are computed to the nearest hundredth. See Table 3-1 for additional significant figures in the atomic weights of the elements.

Refractive Index, unless otherwise specified, is given for the sodium line at 589.6 nm.

Density values are given at 25°C unless otherwise indicated by the superscript figure; thus, 2.487^{15} indicates a density of 2.487 for the substance at 15°C. For gases the values are given as grams per liter (g/l).

Melting Point is recorded in a certain case as 250 d and in some other cases as d 250, the distinction being made in this manner to indicate that the former is a melting

Table 4-1 (*Continued*)
PHYSICAL CONSTANTS OF INORGANIC COMPOUNDS

point with decomposition at 250°C, while in the latter decomposition only occurs at 250°C and higher temperatures. Where a value such as $-6H_2O$, 150 is given it indicates a loss of six moles of water per formula weight of the compound at a temperature of 150°C.

Boiling Point is given at atmospheric pressure (760 mm of mercury) unless otherwise indicated; thus 82^{15mm} indicates that the boiling point is 82°C when the pressure is 15 mm. Also, subl 550 indicates that the compound sublimes at 550°C.

Solubility is given in parts by weight (of the formula weight) per 100 parts by weight of the solvent (water unless otherwise specified) and at a temperature of 25°C. Other temperatures are indicated by the small superscript. The symbols of the common mineral acids represent aqueous solutions of these acids.

Synonyms and Mineral Names as well as alternate chemical names for compounds to be found in this table but under different names are listed in Table 4-2.

Abbreviations Used in the Table

a, acid	gelat, gelatin	sens, sensitive
abs, absolute	gly, glycerol	-sh (e.g., yelsh, yellowish)
ac a, acetic acid	grn, green	silv, silver
acet, acetone	h, hot	sl, slightly
al, 95% ethanol	hcp, hexagonal close-packed	slky, silky
alk, alkali (aq NaOH or KOH)	hex, hexagonal	sol, solid
amorp, amorphous	(Hg), mercury line at λ stated	soln, solution
anhyd, anhydrous	hygr, hygroscopic	subl, sublimes
aq, aqueous	i, insoluble	sulf, sulfides
aq reg, aqua regia	ign, ignites	tart, tartrate
atm, atmosphere	leaf, leaflets	tetr, tetragonal
bcc, body-centered cubic	(Li), lithium line at 671 nm	tr, transition
bct, body-centered tetragonal	liq, liquid	tric, triclinic
blk, black	lt, light	trig, trigonal
bl, blue	lust, lustrous	v, very
brn, brown	MeOH, methyl alcohol	vac, vacuum
bz, benzene	met, metal, metallic	viol, violently
cc, cubic centimeter	min, mineral	vlt, violet
chl, chloroform	mn, monoclinic	volat, volatile or volatilizes
col, colorless	nd, needles	wh, white
conc, concentrated	oct, octahedral	yel, yellow
cr, crystals or crystalline	ol, olive	
cub, cubic	or, orange	
d, decomposes	o-rh, orthorhombic	
deliq, deliquescent	oxid, oxidizing	
dil, dilute	pl, plates	
dk, dark	pois, poisonous	
effl, effloresces	ppt, precipitate	
eth, ether (ethyl)	pr, prisms	
EtOH, ethyl alcohol	purp, purple	
expl, explodes, explosive	pwd, powder	
fcc, face-centered cubic	py, pyridine	
fum, fuming	rh, rhombic	
fus, fusion, fuses	rh-hed, rhombohedral	
g, gas	s, soluble	
gel, gelatinous	satd, saturated	

Table 4-1 (*Continued*)
PHYSICAL CONSTANTS OF INORGANIC COMPOUNDS

Name	Formula	Formula Weight	Color, Crystalline Form, and Refractive Index	Density	Melting Point, °C	Boiling Point, °C	Solubility in 100 Parts
Actinium	Ac	(227)	silv-wh met, fcc	....	1050 ± 50	(3300)	d to Ac(OH)$_3$
bromide	AcBr$_3$	466.7	wh, hex	5.85	subl 800	....	s
chloride	AcCl$_3$	333.4	wh, hex	4.81	subl 960	....	....
fluoride	AcF$_3$	284.0	wh, hex	7.88	....	....	i
hydroxide	Ac(OH)$_3$	278.0	wh ppt	....	....	....	i
iodide	AcI$_3$	607.7	wh	....	subl $\sim$ 700	....	s
oxide	Ac$_2$O$_3$	502.0	wh, hex	9.19	....	....	i
oxide bromide	AcOBr	322.9	wh, tetr	7.9	....	....	....
oxide chloride	AcOCl	278.3	wh, tetr	9.70	....	....	....
oxide fluoride	AcOF	262.0	wh, cub	8.28	....	....	....
sulfide	Ac$_2$S$_3$	550.2	dk, cub	6.75	....	....	....
Aluminum	Al	26.98	silv, cub	2.70	660.2	2447	s HCl, H$_2$SO$_4$, alk
acetate	Al(C$_2$H$_3$O$_2$)$_3$	204.12	wh pwd	....	d	....	s
acetylacetonate	Al(C$_5$H$_7$O$_2$)$_3$	324.31	col, mn	1.27	193 subl	314	i; v s al, s eth, bz
arsenate	AlAsO$_4$	165.90	wh, trig, 1.596	3.25	....	....	i; sl s a
benzoate	Al(C$_7$H$_5$O$_2$)$_3$	390.33	wh cr pwd	....	....	....	v sl s
boride	AlB$_2$	48.60	red-br, hex	3.19	....	....	....
bromate	Al(BrO$_3$)$_3$·9H$_2$O	572.84	wh cr, hygr	....	62.3	d 100	s
bromide	AlBr$_3$	266.71	wh, rh, deliq	2.64	97.5	268	s
bromide	AlBr$_3$·6H$_2$O	374.80	wh cr, deliq	2.54	93	d 135	s
butoxide, tert-	Al(C$_4$H$_9$O)$_3$	246.33	wh cr	1.025	180 subl	....	v s org solv
carbide	Al$_4$C$_3$	143.96	yel-grn, hex, 2.70	2.36	d 2200	....	d to CH$_4$
chlorate	Al(ClO$_3$)$_3$·6H$_2$O	385.41	col, rh-hed, deliq	....	d	....	v s
chloride	AlCl$_3$	133.34	wh, hex, v deliq	2.44	194$^{5.2atm}$	183 subl	70 (viol); s eth, al
chloride	AlCl$_3$·6H$_2$O	241.43	wh, trig, deliq, 1.56	2.40	d 100	....	31, 50 al, s eth
chloride	AlCl$_3$·6NH$_3$	235.52	col cr, hygr	1.412	d	....	s
citrate	AlC$_6$H$_5$O$_7$	216.08	wh	....	....	....	0.5
ethoxide	Al(C$_2$H$_5$O)$_3$	162.14	wh cr	1.142	134	....	d; v sl s al, eth
fluoride	AlF$_3$	83.98	col, tric	2.88	1040	1200 subl	0.56; i a, alk

Table 4-1 (Continued)
PHYSICAL CONSTANTS OF INORGANIC COMPOUNDS

Name	Formula	Formula Weight	Color, Crystalline Form, and Refractive Index	Density	Melting Point, °C	Boiling Point, °C	Solubility in 100 Parts
Aluminum							
(fluellite)	$AlF_3 \cdot H_2O$	101.99	col, rh, 1.490	2.17	d	. . . .	sl s
fluorosilicate	$AlF_3 \cdot 3\frac{1}{2}H_2O$	147.03	wh cr pwd	1.914	anhyd 250	. . . .	sl s
(topaz)	$Al_2(SiF_6)_3$	480.19	wh, rh, 1.619, 1.620, 1.627	3.58	. . . .	. . . .	i
hydroxide	$Al(OH)_3$	78.00	wh, mn	2.42	$-H_2O$, 300	. . . .	i; s a, alk
iodide	AlI_3	407.69	brn cr	3.98	191	382	s d; s al, eth, CS_2
	$AlI_3 \cdot 6H_2O$	515.79	wh cr, hygr	2.63	d 185	. . . .	v s; s al, CS_2
isopropoxide	$Al(C_3H_7O)_3$	204.25	wh cr	1.0346	118.5	d 150	d; s al, bz, chl
nitrate	$Al(NO_3)_3 \cdot 9H_2O$	375.13	col, rh, deliq, 1.54	. . . .	73	. . . .	39; 100 al
nitride	AlN	40.99	wh, hex	3.26	. . . .	2000 subl	d; d a, alk
oxalate	$Al_2(C_2O_4)_3 \cdot 4H_2O$	390.08	wh pwd	. . . .	. . . .	. . . .	i; s a
oxide	Al_2O_3	101.96	col, hex, 1.768, 1.760	3.965	2045	2980	i; v sl s a, alk
α-, corundum	. . . .	. . . .	col, rh, 1.765	3.97	2015	2980	i; v sl s a, alk
γ-alumina	. . . .	. . . .	wh micr cr, 1.7	3.5–3.9	tr to α	. . . .	i; sl s a, alk
(boehmite)	$Al_2O_3 \cdot H_2O$	119.98	col, rh, 1.624	3.014	d 360	. . . .	i
(gibbsite)	$Al_2O_3 \cdot 3H_2O$	156.01	wh, mn, 1.577, 1.577	2.42	tr to boehmite	. . . .	i; s h a, alk
(bauxite)	$Al_2O_3 \cdot 2H_2O$	137.99	wh, rh, 1.571	2.55	d 360	. . . .	i
(bayerite)	. . . .	. . . .	wh micr cr, 1.583	2.53	tr to boehmite	. . . .	i; s h a, alk
perchlorate	$Al(ClO_4)_3$	325.34	col, hygr	2.209	. . . .	. . . .	$56.7^{14°}$
	$Al(ClO_4)_3 \cdot 6H_2O$	433.43	col, hygr	2.020	120.8	$-6H_2O$, 178	v s
palmitate (com'l)	$Al(OH)_2C_{16}H_{31}O_2$	316.41	wh	1.095	200	. . . .	i; s alk, hydrocarb
phenoxide	$Al(C_6H_5O)_3$	306.27	gray-wh	1.23	d 265	. . . .	d; s al, eth, chl
*meta*phosphate	$Al(PO_3)_3$	263.90	col, tetr	2.779	. . . .	. . . .	i; i a
*ortho*phosphate	$AlPO_4$	121.95	wh, rh, 1.546, 1.556	2.566	>1500	. . . .	i; s a, alk
propoxide	$Al(C_3H_7O)_3$	204.25	wh cr	$1.0578^{20°}$	106	. . . .	d; s al
selenide	Al_2Se_3	290.84	lt brn pwd, unstable	$3.43^{15°}$	. . . .	. . . .	d; d a

Name	Formula	Mol. wt.	Color, crystalline form, refractive index	Density	M.p., °C	B.p., °C	Solubility
silicate (andalusite, kyanite, sillimanite)	Al_2SiO_5	162.05	wh, rh, 1.66	3.247	tr 1545	......	i; d HF; s fus alk
(mullite)	$Al_6Si_2O_{13}$	426.05	col, rh, 1.638, 1.642, 1.653	3.156	1920	......	i; i a, HF
sodium fluoride (cryolite)	$AlF_3 \cdot 3NaF$	209.94	wh, mn, 1.3389	2.90	1000	......	......
stearate	$Al(C_{18}H_{35}O_2)_3$	877.42	wh pwd	1.010	103	......	i; s al, bz, alk
sulfate	$Al_2(SO_4)_3$	342.15	wh pwd, 1.47	2.71	d 770	......	27.8
(alunogenite)	$Al_2(SO_4)_3 \cdot 18H_2O$	666.43	col, mn, 1.474, 1.467	1.69^{17}	d 86.5	......	87°
sulfide	Al_2S_3	150.16	yel, hex	2.02^{13}	1100	1500 subl (in N_2)	d; s a
thallium sulfate	$AlTl(SO_4)_2 \cdot 12H_2O$	639.66	col, oct, 1.5011	2.325^{20}	91	......	$4.8°$; 65^{60}
Americium	Am	243	silv, hcp	13.67	>800	(2600)	s dil a
bromide	$AmBr_3$	482.9	wh, o-rh	6.83	......	subl 850	s
chloride	$AmCl_3$	349.5	pink, hex	5.78	......	subl 850	s
(III) fluoride	AmF_3	300.1	pink, hex	9.53	1395	......	i
(IV) fluoride	AmF_4	319.0	tan, mn	7.34	......	......	......
hydride, di-	AmH_2	245	cub	10.7	......	......	s
iodide	AmI_3	623.8	yel, hex	6.04	......	subl 900	s a
(III) oxide	Am_2O_3	534.3	red-brn, cub	11.68	......	......	s a
(IV) oxide	AmO_2	275.1	blk, cub	8.96	......	......	......
(III) oxide chloride	$AmOCl$	294.4	wh, tetr	......	......	......	......
Amidosulfuric acid	NH_2SO_3H	97.09	col, rh	2.126	200 d	d	14.68
Ammonia	NH_3	17.03	col g, 1.325^{16} liq 1.0003501 gas	0.7188^{20} g/l	-77.74	-33.43	89.9; 13.2 al
-d₃	ND_3	20.05	col g, 1.000347	0.8437^{20} g/l	-74.33	-31.05	......
Ammonium acetate	$NH_4C_2H_3O_2$	77.08	wh cr, hygr	1.17^{20}	114	d	$148^{4}°$; 7.9^{15} MeOH
aluminum chloride	NH_4AlCl_4	186.83	wh cr	......	304	......	s
aluminum sulfate	$NH_4Al(SO_4)_2$	237.14	col, hex	2.45^{20}	93.5	......	s; i al
aluminum sulfate	$NH_4Al(SO_4)_2 \cdot 12H_2O$	453.33	col, cub, 1.459	1.64	d	......	15
orthoarsenate	$(NH_4)_3AsO_4 \cdot 3H_2O$	247.08	wh, rh	......	d	......	sl s
orthoarsenate, di-H	$NH_4H_2AsO_4$	158.98	col, tetr, 1.577, 1.522	2.311^{9}	d	......	33.7°

4-11

Table 4-1 (*Continued*)
PHYSICAL CONSTANTS OF INORGANIC COMPOUNDS

Name	Formula	Formula Weight	Color, Crystalline Form, and Refractive Index	Density	Melting Point, °C	Boiling Point, °C	Solubility in 100 Parts
Ammonium							
*ortho*arsenate, mono-H	$(NH_4)_2HAsO_4$	176.00	col, mn	1.989	d	....	s
*meta*arsenate	NH_4AsO_2	124.96	col, rh, hygr	....	....	....	v s
azide	NH_4N_3	60.06	col pl	1.346	160	134 subl expl	20.2; i eth, bz
benzoate	$NH_4C_6H_5O_2$	139.16	col, rh	1.260	d 198	160 subl	$20^{15°}$
borate	$NH_4HB_4O_7 \cdot 3H_2O$	228.33	col cr	2.6	d	....	10
bromate	NH_4BrO_3	145.95	col, hex	....	expl	....	v s
bromide	NH_4Br	97.95	col, cub, hygr, 1.712	2.429	452 subl	....	97; s al, acet, eth
bromoplatinate(IV)	$(NH_4)_2PtBr_6$	710.62	red-brn, cub	4.265	d 145	....	0.6
bromoselenate(IV)	$(NH_4)_2SeBr_6$	594.49	red, oct	3.326	....	....	d; sl s eth
bromostannate(IV)	$(NH_4)_2SnBr_6$	634.22	col, cub	3.50	d	....	v s
cadmium chloride	$4NH_4Cl \cdot CdCl_2$	397.27	col, rh, 1.6038	2.01	....	....	s
calcium arsenate	$NH_4CaAsO_4 \cdot 6H_2O$	305.13	col, mn	$1.905^{15°}$	d 140	....	0.02; s NH_4Cl
calcium phosphate	$NH_4CaPO_4 \cdot 7H_2O$	279.20	col	$1.561^{15°}$	d	....	i; s a
carbamate	$NH_4NH_2CO_2$	78.07	col, rh	....	60 subl	....	v s
carbonate	$(NH_4)_2CO_3 \cdot H_2O$	114.10	col, cub	....	d <20	....	$100^{15°}$
carbonate, hydrogen	NH_4HCO_3	79.06	col, rh or mn, 1.423, 1.536, 1.555	1.58	d 35	....	$11.9^{0°}$
cerium(III) nitrate	$(NH_4)_2Ce(NO_3)_4 \cdot 4H_2O$	558.28	col, mn	....	74	....	318
cerium(III) sulfate	$(NH_4)_2SO_4 \cdot Ce_2(SO_4)_3 \cdot 8H_2O$	844.69	mn	2.523	$-6H_2O$, 100	$-8H_2O$, 150	$3.29^{50°}$ anhyd
chlorate	NH_4ClO_3	101.49	col, mn, nd	1.80	102 expl	....	28.7°
chloride	NH_4Cl	53.49	col, cub, 1.642	1.527	340 subl	....	29.7°
chloroaurate(III)	NH_4AuCl_4	356.82	yel, mn or rh	....	....	520	s; sl s al
chlorogallate	NH_4GaCl_4	229.57	wh cr	....	....	520	v s; s al
chloroiridate(IV)	$(NH_4)_2IrCl_6$	441.00	red-blk, cub	2.856	d	....	$0.69^{14°}$; s HCl

Name	Formula	Formula weight	Crystalline form, color, refractive index	Density	Melting point, °C	Boiling point, °C	Solubility
chlorosmate(IV)	$(NH_4)_2OsCl_6$	439.00	cub	2.93			sl s
chloropalladate(IV)	$(NH_4)_2PdCl_6$	355.20	red-brn, cub	2.418	d		s
chloropalladate(II)	$(NH_4)_2PdCl_4$	284.29	ol grn, tetr	2.17	d		0.7
chloroplatinate(IV)	$(NH_4)_2PtCl_6$	443.89	yel, cub, 1.8	3.065	d		s
chloroplatinate(II)	$(NH_4)_2PtCl_4$	372.98	red, rh	2.936	d 140		sl s; s a
chloroplumbate(IV)	$(NH_4)_2PbCl_6$	455.98	yel, cub	2.925	d 120		33^{15}
chlorostannate(IV)	$(NH_4)_2SnCl_6$	367.49	wh, cub	2.4	d		v s
chlorozincate	$(NH_4)_2ZnCl_4$	243.26	wh, rh, hygr	1.879	d 150		$40^{30°}$
chromate	$(NH_4)_2CrO_4$	152.08	yel, mn	$1.911^{2°}$	d 180		21.2
chromium(III) sulfate	$NH_4Cr(SO_4)_2\cdot12H_2O$	478.34	grn or vlt, cub, 1.4842	1.72	94, $-9H_2O$, 100		
citrate	$(NH_4)_3C_6H_5O_7$	243.22	wh cr, deliq		d		v s
citrate, hydrogen	$(NH_4)_2HC_6H_5O_7$	226.19	wh	1.48			100; sl s al
cobalt(II) selenate	$(NH_4)_2SeO_4\cdot CoSeO_4\cdot 6H_2O$	489.02	ruby-red, mn, 1.526, 1.532, 1.541	$2.228^{20°}$			$20.5^{20°}$
cobalt(II) sulfate	$(NH_4)_2SO_4\cdot CoSO_4\cdot 6H_2O$	395.23	ruby-red, mn, 1.490, 1.495, 1.503	1.902			$33.8°$; s al
copper(II) chloride	$(NH_4)_2CuCl_4\cdot 2H_2O$	277.46	bl, tetr, 1.744, 1.724	1.993			
copper(I) iodide	$NH_4CuI_2\cdot H_2O$	353.40	rh		d 110	40 subl	d; s NH_4I
cyanate	NH_4OCN	60.06	wh cr	$1.342^{20°}$	d 60		v s; sl s al
cyanide	NH_4CN	44.06	col, cub	$1.02^{100°}$	d 36		v s
cyanaurate(I)	$NH_4Au(CN)_2$	267.04	col, cub		d 100		vs; s al
cyanaurate(III)	$NH_4Au(CN)_4\cdot H_2O$	337.09	col	2.15	d 200		
dichromate	$(NH_4)_2Cr_2O_7$	252.06	or, mn	1.704	d 170		$30.8^{15°}$; s al
dithionate	$(NH_4)_2S_2O_6\cdot\frac{1}{2}H_2O$	205.21	col, mn		d 130		$135°$
ferricyanide	$(NH_4)_3Fe(CN)_6$	266.06	red, rh		d		v s
ferrocyanide	$(NH_4)_4Fe(CN)_6\cdot 3H_2O$	338.15	yel, mn	1.315	d		$100°$; s al
fluoride	NH_4F	37.04	col, hex, deliq	1.50	subl		v s; sl s al
fluoride, hydrogen	NH_4HF_2	57.04	rh or tetr, deliq, 1.390		125.6		
fluoroantimonate(III)	$(NH_4)_2SbF_5$	252.84	col, rh		d, subl		108
fluoroborate	NH_4BF_4	104.84	wh, rh	$1.87^{15°}$	subl		25^{16}
fluorogallate	$(NH_4)_3GaF_6$	237.83	wh, oct cr		d >250 to GaF_3		sl s

Table 4-1 (Continued)
PHYSICAL CONSTANTS OF INORGANIC COMPOUNDS

Name	Formula	Formula Weight	Color, Crystalline Form, and Refractive Index	Density	Melting Point, °C	Boiling Point, °C	Solubility in 100 Parts
Ammonium							
fluorogermanate	$(NH_4)_2GeF_6$	222.66	col, hex pr, 1.428, 1.425	2.564	......	......	s; i al
fluorophosphate, di-	$NH_4PO_2F_2$	119.01	col, rh	......	213	......	s; s al, acet
fluorophosphate, hexa-	NH_4PF_6	163.00	col pl	$2.180^{18°}$	d	......	s; s al, acet
fluorosilicate (cryptohalite)	$(NH_4)_2SiF_6$	178.14	α oct, 1.3696 β hex; col	α 2.011 β 2.152	d	......	$18.6^{17°}$
fluorosulfonate	NH_4FSO_3	117.10	wh nd	......	d 245	......	v s
	NH_4SO_3F	117.10	col nd	......	244.7	......	s; s MeOH
fluorozirconate	$(NH_4)_2ZrF_6$	241.29	hex	1.154	......	......	sl s
	$(NH_4)_3ZrF_7$	278.32	col, cub	1.433	......	......	sl s
formate	NH_4CHO_2	63.06	wh, mn, deliq	1.280	116	d 180	102°; s al, eth
gallium sulfate	$NH_4Ga(SO_4)_2 \cdot 12H_2O$	496.07	cub, 1.468	1.777	......	......	30.84
hydroxide	NH_4OH	35.05	in soln only	......	−77	......	s
iodate	NH_4IO_3	192.94	col, rh or mn	3.42	d 150	......	$2.06^{15°}$
iodide	NH_4I	144.94	col, cub, hygr, 1.7031	2.514	250 subl	......	$167^{15°}$
iodide, tri-	NH_4I_3	398.75	dk brn, rh	3.749	d 175	......	s d
iodoplatinate(IV)	$(NH_4)_2PtI_6$	992.59	blk, cub	4.61	......	......	i al
iridium(III) chloride	$(NH_4)_3IrCl_6 \cdot H_2O$	477.05	grn-blk, cub	......	d 350	......	$10.5^{20°}$
iron(III) chloride	$2NH_4Cl \cdot FeCl_3 \cdot H_2O$	485.93	red, rh, hygr, 1.78	1.99	234	......	v s
iron(III) fluoride	$3NH_4F \cdot FeF_3$	223.95	lt yel, oct	1.96	......	......	sl s
iron(II) selenate	$(NH_4)_2SeO_4 \cdot FeSO_4 \cdot 6H_2O$	485.93	lt grn, mn, 1.5226, 1.5260, 1.5334	$2.191^{20°}$	......	......	......
iron(III) sulfate	$(NH_4)_2SO_4 \cdot Fe_2(SO_4)_3$	532.02	wh, hex	$2.49^{22°}$	d 420	......	44.15
	$NH_4Fe(SO_4)_2 \cdot 12H_2O$	482.19	vlt, cub, 1.4854	1.71	39–41	$-12H_2O$, 230	124
iron(II) sulfate	$(NH_4)_2Fe(SO_4)_2 \cdot 6H_2O$	392.14	grn, mn, 1.487, 1.492, 1.499	$1.864^{20°}$	d 100	......	$26.9^{20°}$

Name	Formula	Mol. wt.	Crystalline form, color, refractive index	Density	M.p.	B.p.	Solubility
lactate	$NH_4C_3H_5O_3$	107.11	col-yelsh liq	$1.21^{15°}$	...	...	v s; v s al
magnesium arsenate	$NH_4MgAsO_4 \cdot 6H_2O$	289.36	col, tetr, 1.608	$1.932^{15°}$	...	d	$0.038^{20°}$; s a
magnesium chloride	$NH_4Cl \cdot MgCl_2 \cdot 6H_2O$	256.80	col, rh, deliq	1.456	$-2H_2O$, 100	d	16.7
magnesium chromate	$(NH_4)_2CrO_4 \cdot MgCrO_4 \cdot 6H_2O$	400.51	yel, mn, 1.636, 1.637, 1.653	1.84	...	d	v s
magnesium phosphate (guanite, struvite)	$NH_4MgPO_4 \cdot 6H_2O$	245.41	col, rh, 1.495, 1.496, 1.504	1.711	...	d	$0.023°$; v s dil a
magnesium selenate	$(NH_4)_2SeO_4 \cdot MgSeO_4 \cdot 6H_2O$	454.40	col, mn, 1.507, 1.509, 1.517	$2.058^{20°}$	...	...	sl s
magnesium sulfate (boussingaulite)	$(NH_4)_2SO_4 \cdot MgSO_4 \cdot 6H_2O$	360.60	col, mn, 1.472, 1.473, 1.479	1.723	>120	d 250	17.9°
l-malate, hydrogen	$NH_4HC_4H_4O_5$	151.08	col, rh	1.5	161	d	$32^{15°}$
manganese(II) phosphate	$NH_4MnPO_4 \cdot H_2O$	185.96	wh cr	...	...	...	0.0031
manganese(II) sulfate	$(NH_4)_2SO_4 \cdot MnSO_4 \cdot 6H_2O$	391.23	lt red, mn, 1.480, 1.484, 1.491	1.83	...	...	51.3
molybdate ("para-")	$(NH_4)_2MoO_4$	196.01	col, mn	2.276	d	...	s d; s a
"para-"	$(NH_4)_6Mo_7O_{24} \cdot 4H_2O$	1235.86	col-yelsh, mn	2.498	$-H_2O$, 90	d 190	43; s a, alk
nickel chloride	$NH_4Cl \cdot NiCl_2 \cdot 6H_2O$	291.20	grn, mn, deliq	1.654	...	...	v s
nickel sulfate	$(NH_4)_2SO_4 \cdot NiSO_4 \cdot 6H_2O$	395.00	dk blue-grn, mn, 1.495, 1.501	1.923	...	...	$10.4^{20°}$
nitrate	NH_4NO_3	80.04	col, rh; mn > 32.1°	1.725	169.6	...	$65^{17°}$
nitratocerate(IV)	$(NH_4)_2Ce(NO_3)_6$	548.23	yel-red, mn	...	...	...	142; s al
nitrite	NH_4NO_2	64.04	wh-yelsh cr	1.69	60 expl	30 subl vac	v s
oxalate	$(NH_4)_2C_2O_4 \cdot H_2O$	142.11	col, rh, 1.439, 1.546, 1.594	1.50	$d > 70$	...	2.5°
oxalate, hydrogen	$NH_4HC_2O_4 \cdot H_2O$	125.08	col, rh	1.556	$-H_2O$, 170	...	s
oxalatoferrate(III)	$(NH_4)_3Fe(C_2O_4)_3 \cdot 3H_2O$	428.07	grn, mn	1.78	d 165	...	42.7°
perchlorate	NH_4ClO_4	117.49	col, rh, 1.482	1.95	d	...	20.0
permanganate	NH_4MnO_4	136.97	purp, rh	$2.208^{10°}$	d 110	...	$7.9^{15°}$
peroxyborate	$NH_4BO_3 \cdot {}^1\!/_2H_2O$	85.86	wh cr	...	d	...	$1.55^{18°}$
peroxychromate	$(NH_4)_3CrO_8$	234.11	red-brn, cub	...	d 40	expl 50	sl s; expl H_2SO_4
peroxydisulfate	$(NH_4)_2S_2O_8$	228.18	col, mn, 1.498, 1.502, 1.587	1.982	d 120	expl 180	58.2°

Table 4-1 (Continued)
PHYSICAL CONSTANTS OF INORGANIC COMPOUNDS

Name	Formula	Formula Weight	Color, Crystalline Form, and Refractive Index	Density	Melting Point, °C	Boiling Point, °C	Solubility in 100 Parts
Ammonium							
periodate, meta-	NH_4IO_4	208.94	col, tetr	$3.056^{18°}$	expl	....	$2.7^{16°}$
periodate, para-	$(NH_4)_2H_3IO_6$	262.00	col, rh	2.85	....	....	....
perrhenate	NH_4ReO_4	268.24	wh, hex	3.97	d	....	$6.1^{20°}$
hypophosphate	$(NH_4)_2H_2P_2O_6$	196.04	col cr	....	170	....	7
orthophosphate	$(NH_4)_3PO_4 \cdot 3H_2O$	203.13	wh pr	....	$d > 150$	....	26.1
orthophosphate, di-H	$NH_4H_2PO_4$	115.03	col, tetr, 1.525, 1.479	$1.803^{19°}$	$d > 150$	....	$22.7°$
orthophosphate, mono-H	$(NH_4)_2HPO_4$	132.05	col, mn, 1.52	1.619	$d > 100$	....	$57^{10°}$
hypophosphite	$NH_4PH_2O_2$	83.03	rh	1.634	200	d 240	s
orthophosphite, di-H	$NH_4H_2PO_4$	99.03	col, mn	....	123	d 145	$171^{0°}$
picrate	$NH_4C_6H_2N_3O_7$	246.14	red or yel, rh	1.719	d	expl 423	$1.1^{20°}$
praseodymium sulfate	$(NH_4)_2SO_4 \cdot Pr_2(SO_4)_3 \cdot 8H_2O$	846.26	lt grn cr	$2.531^{17°}$	$-8H_2O$, 170	....	sl s
propionate	$NH_4C_3H_5O_2$	91.11	col, deliq	1.108	45	....	v s
rhodamide	NH_4CNS	76.12	col, mn, 1.685	1.305	147	....	$115°$
selenate	$(NH_4)_2SeO_4$	179.03	col, mn, 1.561, 1.563, 1.585	$2.194^{20°}$	d	....	$117°$
selenate, hydrogen	NH_4HSeO_4	162.00	rh, col	2.162	d	....	s
selenide	$(NH_4)_2Se$	115.04	col cr	....	d	....	....
sodium phosphate (microcosmic salt)	$NH_4NaHPO_4 \cdot 4H_2O$	209.07	col, mn, 1.439, 1.442, 1.469	1.574	d 97	....	s
succinate	$(NH_4)_2C_4H_4O_4$	152.15	col cr	1.37	....	....	s
sulfamate	$NH_4NH_2SO_3$	114.12	large pl, deliq	....	135	d 200	$167^{10°}$
sulfate (mascagnite)	$(NH_4)_2SO_4$	132.14	col, rh, 1.521, 1.523, 1.533	$1.769^{20°}$	d 230	....	$43^{20°}$
sulfate, hydrogen	NH_4HSO_4	115.11	col, rh, 1.473	1.78	146.9	d 350	100
sulfide	$(NH_4)_2S$	68.14	yel cr ($> -18°$), hygr	....	d	....	v s

Name	Formula	Mol. wt.	Color, crystalline form, etc.	Density	M.p., °C	B.p., °C	Solubility
sulfide, hydrogen	NH$_4$HS	51.11	wh, rh, 1.74	1.17	...	...	128°
sulfite	(NH$_4$)$_2$SO$_3$·H$_2$O	134.15	col, mn, 1.515	1.41	d 60	150 subl	32°
sulfite, hydrogen	NH$_4$HSO$_3$	99.10	rh pr, deliq	2.03	150 subl (in N$_2$)	...	v s
dl-tartrate	(NH$_4$)$_2$C$_4$H$_4$O$_6$	184.15	col, mn, α 1.55, β 1.581	1.601	d	...	58$^{15°}$
dl-tartrate, hydrogen	NH$_4$HC$_4$H$_4$O$_6$	167.12	col, mn, 1.561, 1.591	1.636	d 200	...	2.35$^{15°}$
tellurate	(NH$_4$)$_2$TeO$_4$	227.67	wh pwd	3.024	d	...	s
thioantimonate(V)	(NH$_4$)$_3$SbS$_4$·4H$_2$O	376.18	yel pr	...	d	...	71.2°
thiocarbamate	NH$_4$CS$_2$NH$_2$	110.20	yel cr	...	d 50	...	v s
thiocarbonate	(NH$_4$)$_2$CS$_3$	144.28	yel cr, hygr	...	subl	...	v s
thiocyanate	NH$_4$SCN	76.12	col, mn, deliq	1.305	149.6	d 170	128°
thiosulfate	(NH$_4$)$_2$S$_2$O$_3$	148.20	col, mn, hygr	1.679	d 100	...	2.15$^{15°}$
uranyl carbonate	(NH$_4$)$_4$UO$_2$(CO$_3$)$_3$·2H$_2$O	558.24	yel, mn	2.773	d 100	...	5.8$^{20°}$
uranyl pentafluoride	(NH$_4$)$_3$UO$_2$F$_5$	419.14	tetr, 1.495	3.186	subl	...	s
*meta*vanadate	NH$_4$VO$_3$	116.98	wh cr	2.326	d 130	...	0.44$^{18°}$
vanadium(III) sulfate	NH$_4$V(SO$_4$)$_2$·12H$_2$O	477.28	red to bl	1.687	49	...	28.5$^{20°}$
zinc sulfate	(NH$_4$)$_2$SO$_4$·ZnSO$_4$·6H$_2$O	401.66	wh, mn, 1.489, 1.493, 1.499	1.931	d	...	7° anhyd
Antimonic acid, ortho	H$_3$SbO$_4$	188.77	wh pwd	6.6	d	...	sl s
Antimony	Sb	121.75	silv-wh, rh-hed	6.684	630.5	1640	i; s hot conc H$_2$SO$_4$, aq reg
	Sb$_4$	...	yel	...	tr −90 to silv-form	...	
(III) bromide	SbBr$_3$	361.48	col, rh, hygr, 1.74	4.148	96.6	280	d; s HCl, HBr, CS$_2$
(III) chloride	SbCl$_3$	228.11	col, rh, deliq	3.06	73.4	223	v s; s al, bz, chl
(V) chloride	SbCl$_5$	299.02	col liq; trig bipy	2.336^{20}	5	140	d; s HCl, chl, tart
(III) fluoride	SbF$_3$	178.75	col, rh, deliq	4.379^{21}	292	319 subl	v s
(V) fluoride	SbF$_5$	216.74	col oily liq	2.993^{23}	7.0	150	s; s KF
hydride (stibine)	SbH$_3$	124.77	inflam gas	4.36^{15}	−91.5	−18.4	20 ml°; s CS$_2$
(III) iodide	SbI$_3$	502.46	yel-red, rh or mn	4.766^{22}	167	401	d; s al, KI, HCl
(V) iodide	SbI$_5$	756.27	dk brn	...	79	400	d
mercaptoacetamide	Sb(C$_2$H$_4$NOS)$_3$	392.12	wh cr	...	139	...	200
nitride	SbN	135.76	or pd	...	d	...	...

Table 4-1 (*Continued*)
PHYSICAL CONSTANTS OF INORGANIC COMPOUNDS

Name	Formula	Formula Weight	Color, Crystalline Form, and Refractive Index	Density	Melting Point, °C	Boiling Point, °C	Solubility in 100 Parts
Antimony							
(III) oxide	Sb_2O_3	291.50	wh, cub, 2.087	5.2	656	1425	v sl s; s KOH, HCl, tart
(senarmontite, valentinite)	....	....	wh, o-rh, 2.18, 2.35, 2.35	5.67	tr 567 to cub	....	....
oxide, tetr- (cervantite)	Sb_2O_4	307.50	wh pd, 2.00	5.82	$-O_2$, 930	....	v sl s
(V) oxide	Sb_2O_5	323.50	wh-yel pwd	3.8	$-O_2$, 380	....	v sl s; sl s KOH, HCl
(III) oxide chloride	SbOCl	173.20	wh, mn	5.01	d 170	....	i; s HCl, CS_2
(III) oxide sulfate	$Sb_2O_2SO_4$	371.56	wh	4.89	....	....	d; s H_2SO_4
(III) oxide potassium tartrate (tartar emetic)	$(SbO)KC_4H_4O_6 \cdot \frac{1}{2}H_2O$	333.93	col cr	2.6	$-H_2O$, 100	....	8.3; 6.7 gly
(III) selenide	Sb_2Se_3	480.38	gray, rh	....	611	....	v sl s; s conc HCl
(III) sulfate	$Sb_2(SO_4)_3$	531.68	col silky nd, deliq	$3.625°$	d	....	hyd; s a
(III) sulfide (stibnite)	Sb_2S_3	339.69	gray, hex	4.64	550	....	d; s H_2SO_4
precipitate	....	....	or-red amorp	4.15	....	....	....
(V) sulfide	Sb_2S_5	403.82	or pwd	4.120^{20}	$-S$, 135	....	i; s HCl, alk
d-tartrate	$Sb_2(C_4H_4O_6)_3 \cdot 6H_2O$	795.81	wh pwd	$6.50^{13°}$	....	....	s
telluride	Sb_2Te_3	626.30	gray		629	....	i; s HNO_3
Argon							
Ar		39.948	col g, 1.000284	1.7824 g/l, 1.400 liq 1.7840⁰	-189.38	-185.87	3.36^{20}cc
			col, fcc		c		
Arsenic							
As		74.9216	gray met, rh	5.72	817^{28atm}	subl 613	i; s HNO_3
			yel, cub	$2.026^{18°}$	d 358		i; s CS_2
bromide, tri-	$AsBr_3$	314.66	col, rh pr, deliq	$3.3972^{25°}_4$ sol, 3.3282 liq	31–33	221	hyd; s CS_2, HCl, HBr

Name	Formula	Formula wt	Color, crystalline form, sp. gr.	Density	m.p., °C	b.p., °C	Solubility
chloride, tri-	AsCl₃	181.28	col oily liq	2.17	-13	130	d; s al, eth, HCl
fluoride, tri-	AsF₃	131.92	col oily liq	3.01	-5.95	62.8	i; s HF, bz, eth, al
fluoride, penta-	AsF₅	169.91	col g	7.71 g/l	-79.8	-53.2	s; s al, eth, bz
hydride (arsine)	AsH₃	77.95	col g	2.695 g/l gas	-114	-55	28 ml; s chl, bz
iodide, di-	AsI₂	328.73	red pr	.....	124	380	d; s al, eth, chl, bz
iodide, tri-	AsI₃	455.64	red pl, hex	$4.38^{13°}$	146	403	$6^{25°}$; 5.2 CS₂
oxide, tri- (claudetite)	As₄O₆	395.68	col, o-rh	4.15	270	460	$1.2^{2°}$; s al, alk, HCl
(arsenolite)	.....	.....	col, cub, 1.755	4.15	315 subl	.....	$1.2^{2°}$; s al, alk, HCl
oxide, pent-	As₂O₅	229.84	wh, amorp, deliq	3.865	312	.....	$1.2^{2°}$; s al, alk, HCl
oxide chloride	AsOCl	126.37	brnsh	4.32	d 800	.....	39.7; s al, alk
oxide fluoride	AsOF	144.92	.....	.....	d	.....	.....
phosphide	AsP	105.90	blk cr	.....	-68.3	25.6	i; s HCl, H₂SO₄
selenide	As₂Se₃	386.72	br cr	4.75	subl d	.....	i; s alk
sulfide, tetra- (realgar)	As₄S₄	427.94	brn-red, mn, 2.68	$\alpha\ 3.506^{19°}$ $\beta\ 3.254^{19°}$	ca 360 tr 267	565	i; s K₂S, NaHCO₃
sulfide, tri- (orpiment)	As₂S₃	246.04	yel or red, mn	3.43	307 310	707	i; s alk, Na₂CO₃
sulfide, penta-	As₂S₅	310.16	yel	.....	subl d 500	.....	0.3 mg; s alk, alk sulf, HNO₃
Auric and **Aurous** salts, see under Gold							
Azine bromide	N₃Br	121.95	.....	.....	-45	-15	
chloride	N₃Cl	77.49	.....	.....	-100	.....	
fluoride	N₃F	61.04	grn-yel expl	.....	-154	-82	
(di-) d fluoro- (cis)	N₂F₂ (cis)	66.01	col g	.....	-195	-105.7	
(trans)	(trans)	.....	col g	.....	-172	-111.4	
Barium	Ba	137.34	silv met	3.59^{20}	850	1537	d ev H₂; s al
acetate	Ba(C₂H₃O₂)₂	255.43	col cr	2.468			72
	Ba(C₂H₃O₂)₂·H₂O	273.45	col, tric, 1.500	2.19	-H₂O, 150		76
amide	Ba(NH₂)₂	169.39	gray-wh cr		280		d
arsenate	Ba₃(AsO₄)₂	689.86	.....	5.10	1600		0.055; s a
arsenate, mono-H	BaHAsO₄·H₂O	295.28	col, mn, 1.635	1.93^{15}	-H₂O, 150		sl s; s HCl
arsenide	Ba₃As₂	561.86	brn	4.1^{15}			d

Table 4-1 (*Continued*)
PHYSICAL CONSTANTS OF INORGANIC COMPOUNDS

Name	Formula	Formula Weight	Color, Crystalline Form, and Refractive Index	Density	Melting Point, °C	Boiling Point, °C	Solubility in 100 Parts
Barium							
azide	$Ba(N_3)_2$	221.38	mn	2.936	$-N_2$, 120	expl	17[17°]
	$Ba(N_3)_2 \cdot H_2O$	239.40	tric, 1.7	....	expl	....	v s
boride	BaB_6	202.21	met-blk, cub	4.36[16°]	2270	....	i; s HNO_3
bromate	$Ba(BrO_3)_2 \cdot H_2O$	411.16	col, mn	3.99[18°]	d 260	....	0.65; s acet
bromide	$BaBr_2$	297.16	col cr, 1.75	4.781	847	d	110
	$BaBr_2 \cdot 2H_2O$	333.19	col, mn, 1.713, 1.727, 1.744	3.58	$-2H_2O$, 120; 880	....	151[20°]
bromoplatinate(IV)	$BaPtBr_6 \cdot 10H_2O$	992.04	mn	3.71	....	....	....
butyrate	$Ba(C_4H_7O_2)_2 \cdot 2H_2O$	347.57	col	....	....	....	35
carbide	BaC_2	161.36	gray, tetr	3.75	....	....	d to C_2H_2
carbonate (α)	$BaCO_3$	197.35	wh, hex	4.43	tr to α, 982	d 1360	0.002; s a, NH_4Cl
(β)							0.002; s a
carbonate (witherite) (γ)	$BaCO_3$	197.35	wh, rh, 1.529, 1.676, 1.677	4.43	to β, 811	d 1450	0.002
chlorate	$Ba(ClO_3)_2 \cdot H_2O$	322.26	col, mn, 1.5622, 1.577	3.18	$-H_2O$, 120; 414	$-O_2$, 250	33
chloride (α)	$BaCl_2$	208.25	col, mn, 1.7303, 1.7367, 1.7420	3.856	tr to cub, 925	1560	36
(β)			col, cub	3.917	963	1560	36
	$BaCl_2 \cdot 2H_2O$	244.28	col, mn, 1.629, 1.642, 1.658	3.097	$-2H_2O$, 113	....	36
chloroplatinate(IV)	$BaPtCl_6 \cdot 6H_2O$	653.24	or-yel, mn	2.868	$-5H_2O$, 70	d	s
chloroplatinate(II)	$BaPtCl_4 \cdot 3H_2O$	528.35	dk red	2.868	$-3H_2O$, 150	....	v s
chromate	$BaCrO_4$	253.33	yel, rh	4.498[15°]	....	....	0.0003[15°]
citrate	$Ba_3(C_6H_5O_7)_2 \cdot 7H_2O$	916.33	wh pwd	....	$-7H_2O$, 150	....	0.057
cyanide	$Ba(CN)_2$	189.38	wh cr pwd	....	....	....	80[14°]
cyanoplatinate(II)	$BaPt(CN)_4 \cdot 4H_2O$	508.56	(a) yel, mn, 1.6704	2.076	$-2H_2O$, 100	....	3[16°]
	....		(b) grn, rh	2.085			

Name	Formula	Formula wt	Color, crystalline form, refractive index	Density	M.P. °C	B.P. °C	Solubility
dichromate	$BaCr_2O_7$	353.33	red, mn				sl s; s h conc H_2SO_4
	$BaCr_2O_7 \cdot 2H_2O$	389.36	red-yel nd		$-2H_2O$, 120		d
dithionate	$BaS_2O_6 \cdot 2H_2O$	333.50	col, mn, 1.586, 1.595, 1.607	4.536^{14}	d 120		25^{20}
ferrocyanide	$Ba_2Fe(CN)_6 \cdot 6H_2O$	594.73	yel, mn	2.666	$-H_2O$, 40		$0.17^{15°}$
fluorogallate	$Ba_3(GaF_6)_2 \cdot H_2O$	797.46	wh cr	4.06	$-H_2O$, 230		i; s HF
fluorosilicate	$BaSiF_6$	279.42	col, rh	$4.29^{21°}$	d 300		0.026^{17}
formate	$Ba(CHO_2)_2$	227.38	col, rh, 1.573, 1.597, 1.636	3.21	d		30
gluconate	$Ba(C_6H_{11}O_7)_2 \cdot 3H_2O$	581.69	pr or rh leaf		$-3H_2O$, 120; d 120		$3.3^{15°}$
hydride	BaH_2	139.36	gray cr	$4.21^{0°}$	d > 1000		d
hydroxide	$Ba(OH)_2 \cdot 8H_2O$	315.48	col, mn, 1.471, 1.502, 1.50	$2.18^{16°}$	78		3.9
hypochlorite	$Ba(ClO)_2 \cdot 2H_2O$	276.28	col cr		d		
hyponitrite	$BaN_2O_2 \cdot 4H_2O$	269.41	wh cr pwd				
iodate	$Ba(IO_3)_2$	487.15	col, mn	2.742	d		0.033; s HCl
	$Ba(IO_3)_2 \cdot H_2O$	505.17	col, mn	4.998	$-H_2O$, 200		v sl s
iodide	BaI_2	391.15	col cr	$4.657^{18°}$	740		$170^{0°}$
	$BaI_2 \cdot 2H_2O$	427.18	col, rh, deliq	5.15	$-2H_2O$, 540		v s
laurate	$Ba(C_{12}H_{23}O_2)_2$	535.97	wh leaf cr	5.15	260		$0.008^{15°}$
l-malate	$BaC_4H_4O_5$	269.41	wh				1.2
malonate	$BaC_3H_2O_4 \cdot H_2O$	257.40	col				0.21
manganate	$BaMnO_4$	256.28	gray-grn, hex	4.85			v sl s
molybdate	$BaMoO_4$	297.28	wh pwd	4.65	1480		0.0058
nitrate (nitrobarite)	$Ba(NO_3)_2$	261.35	col, cub, 1.572	3.24	575	d	9.0
nitride	Ba_3N_2	440.03	yel-brn	4.783		1000 vac	d
nitrite	$Ba(NO_2)_2$	229.35	col, hex	3.23	220 d		$67.5^{20°}$
	$Ba(NO_2)_2 \cdot H_2O$	247.37	col-yelsh, hex	$3.173^{20°}$	115		$63^{20°}$
oxalate	BaC_2O_4	225.36	col cr	2.658	400 d		0.009
oxide	BaO	153.34	col, cub, 1.98	5.72	1920	ca 2000	$3.48^{20°}$
perchlorate	$Ba(ClO_4)_2$	336.24	col, hex	3.2	505		198
	$Ba(ClO_4)_2 \cdot 3H_2O$	390.29	col, hex, 1.533	2.74	d 400		198
permanganate	$Ba(MnO_4)_2$	375.21	brn-vlt cr	3.77	d 200		$62^{11°}$
peroxide	BaO_2	169.34	wh-gray pwd	4.96	450	$-O_2$, 800	$1.5^{0°}$
	$BaO_2 \cdot 8H_2O$	313.46	col, hex	2.292	$-8H_2O$, 100		0.17

Table 4-1 (Continued)
PHYSICAL CONSTANTS OF INORGANIC COMPOUNDS

Name	Formula	Formula Weight	Color, Crystalline Form, and Refractive Index	Density	Melting Point, °C	Boiling Point, °C	Solubility in 100 Parts
Barium							
hypophosphate	$BaPO_3$	216.31	col nd	....	....	....	sl s
orthophosphate	$Ba_3(PO_4)_2$	601.96	wh, cub	4.1^{16}	....	....	i; s a
orthophosphate, di-H	$Ba(H_2PO_4)_2$	331.31	col, tric	2.9^4	....	....	d; s a
orthophosphate, mono-H	$BaHPO_4$	233.32	wh, rh, 1.635, 1.617	4.165^{15}	d 410	....	0.01; s a
pyrophosphate	$Ba_2P_2O_7$	448.62	wh, rh	3.9^{20}	....	....	0.01; s a
hypophosphite	$Ba(PH_2O_2)_2 \cdot H_2O$	285.33	wh, mn	2.90^{17}	d 100–150	....	30^{15}
propionate	$Ba(C_3H_5O_2)_2 \cdot H_2O$	301.50	col, rh, β 1.518	....	d 300	....	36.5
selenate	$BaSeO_4$	280.30	wh, rh	4.75	d	....	0.012
selenide	$BaSe$	216.30	wh, cub, 2.268	5.02	1830	....	d
silicate	$BaSiO_3$	213.42	col, rh, 1.673, 1.674, 1.678	4.399	1600	....	i; s HCl
	$BaSiO_3 \cdot 6H_2O$	351.52	col, rh, 1.542, 1.548, 1.548	2.59	....	....	0.17
stearate	$Ba(C_{18}H_{35}O_2)_2$	704.13	wh pwd	....	....	....	0.004^{15}
succinate	$BaC_4H_4O_4$	253.37	wh pwd	....	....	....	0.42
sulfate (barite)	$BaSO_4$	233.40	wh, rh, 1.637, 1.638, 1.649	4.50^{15}	1580	tr 1150 mn	0.0002
sulfide	BaS	169.40	col, cub, 2.155	4.25^{15}	1200	....	7.9 d
sulfide, hydrogen	$Ba(HS)_2 \cdot 4H_2O$	275.56	yel, rh	....	d 50	....	s
sulfide, tetra-	$BaS_4 \cdot H_2O$	283.61	red or yel, rh	2.988	d 300	....	41^{15}
sulfide, tri-	BaS_3	233.53	yel-grn cr	....	d 554	....	s
sulfite	$BaSO_3$	217.40	col, hex	....	d	....	0.02
tartrate	$BaC_4H_4O_6 \cdot H_2O$	303.52	wh cr	2.980^{20}	d 200	....	0.026^{18}
tellurate	$BaTeO_4 \cdot 3H_2O$	383.07	wh	4.2^{20}	....	....	sl s; s HCl
telluride	$BaTe$	264.94	yel-wh, cub, 2.440	5.13	d	....	d a
thiocarbonate	$BaCS_3$	245.54	yel, hex	....	d	....	1.08^0
thiocyanate	$Ba(SCN)_2 \cdot 2H_2O$	289.53	col nd, deliq	2.28^{18}	$-H_2O$, 160	....	4.3^{20}
thiosulfate	BaS_2O_3	249.47	wh, rh	....	d 220	....	0.2
	$BaS_2O_3 \cdot H_2O$	267.48	wh, rh	3.5^{18}	d 100	....	0.21^{20}

Name	Formula	Formula weight	Color, crystalline form, refractive index	Density	Melting point, °C	Boiling point, °C	Solubility
titanate	$BaTiO_3$	233.24	tetr and hex, 2.40	tetr 6.017 hex 5.806	......	......	......
tungstate	$BaWO_4$	385.19	col, tetr	5.04	860	......	sl s; d a
divanadate(V)	$Ba_2V_2O_7$	488.56	wh cr	......	......	......	......
Berkelium	Bk	(245)	wh cr	......	......	......	......
Beryllium	Be	9.012	dk gray, hcp	1.86	1285	2970	s a, alk, NH_4HF_2
acetate	$Be(C_2H_3O_2)_2$	127.10	col	1.364°	d 294	......	s
acetate, basic	$3Be(C_2H_3O_2)_2 \cdot BeO$	406.32	wh oct	......	283	330	......
alumirate (chrysoberyl)	$BeAl_2O_4$	126.97	col, rh, 1.747, 1.748, 1.757	3.76	1870	......	......
aluminum silicate (beryl)	$Be_3Al_2(SiO_3)_6$	537.51	col, hex, 1.580, 1.547	2.66	1410	......	......
aluminum silicate (euclase)	$Be_2Al_2(SiO_4)_2(OH)_2$	290.17	col, mn, 1.652, 1.655, 1.671	3.1	......	......	......
ammonium arsenate	$BeNH_4AsO_4 \cdot 1.5H_2O$	193.02	......	......	$-H_2O$, 250 $-NH_3$, 400	d 1200 to BeO	0.0012
ammonium phosphate	$BeNH_4PO_4 \cdot H_2O$	140.04	wh, tetr	......	$-H_2O$, 250 $-NH_3$, 400	>700 to $Be_2P_2O_7$	......
azide	$Be(N_3)_2$	93.05	wh	......	......	......	......
boronate	$Be(BH_4)_2$	38.69	tetr	0.604	......	......	i; s tetrahydrofuran
*ortho*borate, basic (hambergite)	$Be_2(OH)BO_3$	93.84	col, rh, 1.560, 1.591, 1.631	2.35	......	......	......
bromide	$BeBr_2$	168.83	wh nd, deliq	3.465²⁵°	488	subl 473	s
carbide	Be_2C	30.04	yel, hex	1.905¹⁵°	>2100 d	......	s a
carbonate	$BeCO_3 \cdot 4H_2O$	141.08	wh, hex	......	......	......	0.36°; hyd
chloride	$BeCl_2$	79.92	wh, o-rh, deliq	1.899²⁵°	410	488 subl readily	42
fluoride	BeF_2	47.01	wh, deliq	1.986²⁵°	544	1160	v s but slow
hydride	BeH_2	11.03	wh cr	......	d 125	......	d
iodide	BeI_2	262.82	col nd, tetr	4.2	510 subl before melting	590	hyd; s al, eth, CS_2
nitrate	$Be(NO_3)_2 \cdot 4H_2O$	205.08	wh cr, deliq	1.557	60.5	$-H_2O$, 142	51.6

Table 4-1 (*Continued*)
PHYSICAL CONSTANTS OF INORGANIC COMPOUNDS

Name	Formula	Formula Weight	Color, Crystalline Form, and Refractive Index	Density	Melting Point, °C	Boiling Point, °C	Solubility in 100 Parts
Beryllium							
nitride	Be_3N_2	55.05	col, cub	. . . .	2200	. . . .	d h H_2O, alk
oxalate	$BeC_2O_4 \cdot 3H_2O$	151.08	col, o-rh	. . . .	$-2H_2O$, 50	d 225 to BeO	24.8
oxide (bromellite)	BeO	25.01	wh, hex, 1.719, 1.733	3.01	2570	4260	s conc H_2SO_4, fus alk
perchlorate	$Be(ClO_4)_2 \cdot 4H_2O$	279.97	wh cr	. . . .	. . . .	. . . .	59.5
phosphide	Be_3P_2	88.96	brn, cub	2.24	. . . .	. . . .	s
*ortho*phosphate	$Be_2(PO_4)_3 \cdot 3H_2O$	271.03	wh	. . . .	$-H_2O$, 100	. . . .	
selenate	$BeSeO_4 \cdot 4H_2O$	224.03	col, rh, 1.466, 1.501, 1.503	2.03	$-4H_2O$, 213	d 560	49
silicate (bertrandite)	$Be_4Si_2O_7(OH)_2$	238.23	col, rh, 1.591, 1.605, 1.604	2.6	. . . .	. . . .	. . . .
*ortho*silicate (phenazite)	Be_2SiO_4	110.11	col, tric, 1.654, 1.670	3.0	. . . .	. . . .	. . . .
stearate	$Be(C_{18}H_{35}O_2)_2$	575.97	wh waxy	. . . .	45	. . . .	i; s eth, CCl_4
sulfate	$BeSO_4$	105.07	col	2.443	d 550	. . . .	i
	$BeSO_4 \cdot 4H_2O$	177.14	col, tetr, 1.472, 1.440	1.713^{11}	$-4H_2O$, 270	d 580	42.5
sulfide	BeS	41.08	wh	2.36	. . . .	. . . .	i; s HNO_3
Bismuth							
	Bi	208.98	silv-wh, rh	9.80	271.3	1560	i; s h H_2SO_4, HNO_3
acetate	$Bi(C_2H_3O_2)_3$	386.12	wh cr	. . . .	d	. . . .	i; s acet a
*ortho*arsenate	$BiAsO_4$	347.90	wh, mn	7.14	. . . .	. . . .	i; sl s h conc HNO_3
bromide, tri-	$BiBr_3$	448.71	yel cr, hygr	5.72^{25}	218	460	hyd; s HCl, HBr, eth
chloride, tri-	$BiCl_3$	315.34	wh cr, deliq	4.6	232	447	d; s a al, eth, acet
citrate	$BiC_6H_5O_7$	398.08	wh cr	3.458	d	. . . .	sl s; s a, aq NH_3
fluoride, tri-	BiF_3	265.98	gray cr, cub 1.74	5.32^{20}	727	(1027)	i; s KF, inorg a
fluoride, penta-	BiF_5	303.98	bct	5.4	<160	230	. . . .
hydride (bismuthine)	BiH_3	212.00	v unstable liq	. . . .		22	d; s a
hydroxide	$Bi(OH)_3$	260.00	wh amorp pwd	4.36	$-H_2O$, 100	. . . .	

Name	Formula	Formula wt.	Color, crystalline form	Sp. gr.	m.p., °C	b.p., °C	Solubility
iodate	$Bi(IO_3)_3$	733.69	wh	......	......	......	i
iodide, tri-	BiI_3	589.69	blk leaf, hex	5.8	408	subl 439	i; s al, HCl, HI
lactate, *dl-*	$Bi(C_6H_4O_6)_3 \cdot 7H_2O$	512.22	pr nd	......	......	d 500	14.4
molybdate	$Bi_2(MoO_4)_3$	897.78	yel-wh, tetr nd	6.07	643	......	i; v s a
nitrate	$Bi(NO_3)_3 \cdot 5H_2O$	485.07	col, tric	2.83	d 30	......	d; s a
oxalate	$Bi_2(C_2O_4)_3 \cdot 7H_2O$	803.73	wh pwd	......	$-6H_2O$, 130	$-5H_2O$, 80	d; s inorg a
oxide, tri-	Bi_2O_3	495.96	yel, rh	8.76	820	(1887)	i; s a
oxide, pent-	Bi_2O_5	497.96	brn	5.10	$-O$, 150	$-2O$, 357	i; s KOH
*ortho*phosphate	$BiPO_4$	303.95	wh, mn	$6.323^{15°}$	d	d	i; s HCl
selenide, tri- (guanajuatite)	Bi_2Se_3	654.84	blk, rh	6.82	710	d	i
silicate (eulytite)	$2Bi_2O_3 \cdot 3SiO_2$	1112.17	yel, cub, 2.05	6.11	......	......	i; s HCl
sulfate	$Bi_2(SO_4)_3$	706.14	wh nd, hygr	$5.08^{15°}$	d 405	......	d; s a
sulfide, mono-	BiS	241.04	dk gray pwd	7.6–7.8	680 (inert)	......	i; s HNO_3
sulfide, tri-	Bi_2S_3	514.15	dk br, rh	7.39	d 685	......	i; s a, alk
tartrate	$Bi_2(C_4H_4O_6)_3 \cdot 3H_2O$	970.27	wh pwd	$2.595^{25°}$	$-3H_2O$, 105	......	......
tellurate (montanite)	$Bi_2TeO_6 \cdot 2H_2O$	677.59	......	3.79	......	......	......
telluride, tri-	Bi_2Te_3	800.76	gray, rh-hed	7.7^{20}_{4}	573	......	d HNO_3
vanadate	$Bi_2O_3 \cdot V_2O_5$	647.84	red-grn, rh	6.25	d	d	i; s a
Bismuthyl bromide	$BiOBr$	304.89	col, cub	$8.08^{15°}$	......	......	i; s a
carbonate	$(BiO)_2CO_3$	509.97	wh pwd	6.86	d	d	i; s a
chloride	$BiOCl$	260.43	wh pwd	$7.72^{15°}$	......	......	i; s a
dichromate	$(BiO)_2Cr_2O_7$	665.94	yel or red cr	......	d	......	i; s a
fluoride	$BiOF$	243.98	wh cr or pwd	7.5^{20}_{20}	d	......	i; s a
iodide	$BiOI$	351.88	red cr, tetr	7.922	d	......	i; s a
nitrate	$BiONO_3 \cdot H_2O$	305.00	hex leaf	$4.928^{18°}$	$-H_2O$, 105	$-HNO_3$, 260	i; s a
perchlorate	$BiOClO_4 \cdot H_2O$	342.44	wh, rh	0.588 sol	......	......	s
Borane, di-	B_2H_6	27.67	col g, hex sol	0.652 sol	−165.5	−92.5	s NH_4OH, conc H_2SO_4
bromo- (di-)	B_2H_5Br	106.67	col g	$0.56^{-35°}$ liq	−104	~10	hyd
tetra-	B_4H_{10}	53.32	col g, mn sol	0.725 sol	−120	18	sl s d; s bz

Table 4-1 (Continued)
PHYSICAL CONSTANTS OF INORGANIC COMPOUNDS

Name	Formula	Formula Weight	Color, Crystalline Form, and Refractive Index	Density	Melting Point, °C	Boiling Point, °C	Solubility in 100 Parts
Borane							
penta- (dihydro-)	B_5H_{11}	65.14	mn	0.745 sol	-123	63	....
hexa-	B_6H_{10}	74.95	col liq, o-rh sol	0.860 sol; 0.69° liq	-65.1	82.2	hyd
nona- (dihydro-)	B_9H_{15}	112.41	col liq, mn sol		2.7	....	....
deca-	$B_{10}H_{14}$	122.22	wh, mn	$0.78^{100°}$ liq; 0.948 sol	99.7	213	sl s; v s CS_2; s bz, al, eth
deca- (dihydro-)	$B_{10}H_{16}$	124.23	wh	0.84 sol	....	....	....
octadeca-	$B_{18}H_{22}$	216.76	col, o-rh	1.012	....	....	....
eicosa- (dihydro-)	$B_{20}H_{26}$	238.41	col, tetr	1.130	196–199	....	sl d
Borazine	$H_6B_3N_3$ (ring: HB–NH, HN–BH, HB–NH with B and N)	80.50	col liq	0.86	-58	55	
Boric acid, meta-	HBO_2	43.82	wh, cub, 1.619	2.486	236	d 300 to B_2O_3	v sl s
ortho-	H_3BO_3	61.83	col, tric, 1.337, 1.461, 1.462	$1.435^{15°}$	169 tr to HBO_2	....	$6.35^{30°}$
tetra-	$H_2B_4O_7$	157.26	wh pwd or vitr	....	....	....	s
fluoro-	HBF_4	87.81	col liq	....	....	d 130	v s
Borinoaminoborine	B_2H_7N	42.68	col liq	....	-66.5	76.2	v sl s HNO_3
Boron	B	10.81	brn, amorp, or rh	2.46	2074	3675	s min a
arsenate	$BAsO_4$	149.73	wh, tetr, 1.681, 1.690	3.64	subl ~700	....	
bromide, tri-	BBr_3	250.54	col fum liq, 1.5312	$2.6954^{0°}$	46.0	91	d
(di-) tetrabromide	B_2Br_4	341.24	unstable liq	....	0.5–1.5	....	
bromidediiodide	$BBrI_2$	344.53	col liq	....	....	180	d

Name	Formula	Formula wt	Color, crystalline form, refractive index	Density	Melting point	Boiling point	Solubility
dibromideiodide	BBr$_2$I	297.53	col liq	....	....	125	d
(tetra-) carbide	B$_4$C	55.26	blk, rh-hed	2.52	2350	>3500	i a; s fus alk
chloride, tri-	BCl$_3$	117.17	col fum liq, 1.4195^6 (α H$_2$)	1.3730^0	-107	12.4	d
(di-) *tetra*chloride	B$_2$Cl$_4$	163.43	col liq	....	-92	65.5 d	s conc H$_2$SO$_4$
fluoride, tri-	BF$_3$	67.81	col g	0.00299	-128.7	-99	s eth, dioxane
fluoride, ·2H$_2$O	BF$_3$·2H$_2$O	103.84	col liq	1.6316^{20}	6	....	s
fluoride, ammonia	BF$_3$·NH$_3$	84.84	....	1.86	163	-34	s
(di-) *tetra*fluoride	B$_2$F$_4$	97.62	col, mn	1.92	-56	....	....
hydrides, see Borane							
iodide, tri-	BI$_3$	391.52	col pl, hygr	3.35$^{50°}$	49	210	d; v s bz, CS$_2$, CCl$_4$
nitride	BN	24.82	wh slippery, hex diamond-like, cub	2.25 3.47	subl ~3000	....	sl s h a
oxide	B$_2$O$_3$	69.62	wh, rh, 1.64, 1.61 col glass, 1.485	2.46	460	ca 1860	sl s; 1.1^0
phosphide	BP	41.78	maroon pwd	1.812	ign 200	....	i
selenide	B$_2$Se$_3$	258.50	yel-gray pwd	....	....	....	d
(tri-) silicide	B$_3$Si	60.52	blk, rh	2.52	....	....	sl s HNO$_3$
(hexa-) silicide	B$_6$Si	92.95	blk cr	2.47	....	....	s HNO$_3$
sulfide, tri-	B$_2$S$_3$	117.81	wh	1.55	310	200 subl	d
sulfide, penta-	B$_2$S$_5$	181.94	col, tetr	1.85	390	....	d
Borotungstic acid	H$_5$BW$_{12}$O$_{40}$·30H$_2$O	3402.49	tetr	3	45-51	....	s
Bromauric acid	HAuBr$_4$·5H$_2$O	607.69	red-brn cr	....	27	....	v s
Bromic acid	HBrO$_3$	128.92	known in soln only, col	....	....	d 100	v s
Bromine	Br$_2$	159.81	red-brn fum liq, 1.654	3.119	-7.08	58.76	3.56^{20}; v s al, eth, chl, CS$_2$
azide	BrN$_3$	121.93	or-red, liq, expl	....	ca -45	expl	s eth, KI
chloride	BrCl	115.36	red liq or g	....	-54	ca 5	s d; s eth, CS$_2$
cyanide	BrCN	105.93	col, rh	1.863^{62}	52	61.6	....
fluoride, mono-	BrF	98.91	dk red	....	-33	20	....
fluoride, tri-	BrF$_3$	136.90	straw, 1.4536	2.803	8.77	125.75	d viol; d alk
fluoride, penta-	BrF$_5$	174.90	col fum liq, 1.3529	2.4634	-60.6	40.9	d; s HF
(I) fluorosulfate	BrSO$_3$F	178.98	dk red liq	2.60	....	120.5	d viol
(III) fluorosulfate	Br(SO$_3$F)$_3$	377.11	or, hygr	....	59 vac	....	d viol

Table 4-1 (*Continued*)
PHYSICAL CONSTANTS OF INORGANIC COMPOUNDS

Name	Formula	Formula Weight	Color, Crystalline Form, and Refractive Index	Density	Melting Point, °C	Boiling Point, °C	Solubility in 100 Parts
Bromine							
hydrate	$Br_2 \cdot 8H_2O$	303.95	red, oct	1.49	d 3.84	....	3.86^{mp}
nitrate	$BrNO_3$	141.91	yel liq	....	−42	d 0.0	s CCl_4, $CFCl_3$
nitrate dinitrogen pentoxide	$BrNO_3 \cdot N_2O_5$ or $[Br(NO_3)_3]$	249.92	wh or lt yel flakes	....	44–45	subl 25 vac	s CCl_4, $CFCl_3$
oxide, mon-	Br_2O	175.82	dk brn	....	−17.5 d	d	d; s d CCl_4
oxide, di-	BrO_2	111.91	lt yel cr	....	d −40	....	s CCl_4
(tri-) octoxide polymer	$(Br_3O_8)_n$	....	wh	....	−40 d	....	....
sulfanes, *see* Sulfur bromides							
Bromoplatinic(IV) acid	$H_2PtBr_6 \cdot 9H_2O$	838.70	red, mn, deliq	....	d −100	....	v s
Bromous acid, *hypo*	$HBrO$	96.92	known in soln only	....	....	....	s
Bromyl fluoride	BrO_2F	130.91	col cr	....	−9.0	d 20	....
nitrate	BrO_2NO_3	173.91	or	....	d −60	....	....
Cadmium	Cd	112.40	silv-wh, hex	8.642	320.9	767	i; s a
acetate	$Cd(C_2H_3O_2)_2 \cdot 3H_2O$	284.55	col, mn	2.01	$-H_2O$, 130	....	v s
	$Cd(C_2H_3O_2)_2$	230.50	col	2.341	256	d	v s
amide	$Cd(NH_2)_2$	144.45	....	3.05	d 120	....	....
ammonium chloride	$CdCl_2 \cdot NH_4Cl$	236.79	col, rh	2.93	289	....	$33.51^{6°}$
ammonium sulfate	$Cd(NH_4)_2(SO_4)_2 \cdot 6H_2O$	448.69	col, mn	$2.061^{20°}$	$-H_2O$, 100	....	s
arsenate, hydrogen	$CdHAsO_4 \cdot H_2O$	270.34	....	$4.164^{15°}$	>120	....	....
arsenide	Cd_3As_2	487.04	dk gray, cub	$6.21^{15°}$	721	....	i; s HNO_3
borate	$Cd(BO_2)_2 \cdot H_2O$	248.03	wh, rh	3.758	d	....	$125^{17°}$
borotungstate	$Cd_5(BW_{12}O_{40}) \cdot 18H_2O$	6600.25	yel, tric	....	75	....	1250
bromate	$Cd(BrO_3)_2 \cdot H_2O$	386.23	col, rh	3.758	d	....	125
bromide	$CdBr_2$	272.22	lt yel	5.192	570–585	863	117; $26^{15°}$ al
	$CdBr_2 \cdot 4H_2O$	344.28	wh nd, effl	....	....	....	v s
carbonate	$CdCO_3$	172.41	wh, trig	$4.258°$	d 500	....	i; s a, NH_4, KCN

		Mol wt	Color, crystalline form, index of refraction	Density	Melting point, °C	Boiling point, °C	Solubility
chlorate	$Cd(ClO_3)_2 \cdot 2H_2O$	315.33	col pr, deliq	2.28^{18}	80	.....	298°
chloride	$CdCl_2$	183.32	col, hex	4.047	570	960	140^{20}
	$CdCl_2 \cdot 2\tfrac{1}{2}H_2O$	228.35	col, mn, 1.6513	3.327	$-H_2O$, 170 d	.....	168^{20}
chloroplatinate(IV)	$CdPtCl_6 \cdot 3H_2O$	574.25	yel, trig	2.882	.....	.....	s
chromite	$CdCr_2O_4$	280.39	dk grn, cub	5.79^{17}	.....	.....	i; i a
cyanide	$Cd(CN)_2$	164.44	col cr	.....	d 200	.....	1.7^{15}
dithionate	$CdS_2O_6 \cdot 6H_2O$	380.62	col, tric	2.272^{20}	d	.....	.....
fluoride	CdF_2	150.40	wh, cub, 1.56	6.64	1100	1760	4.35
fluorogallate	$[Cd(H_2O)_6]GaF_5(H_2O)$	403.22	col cr, 1.45	2.79	$-5H_2O$, 110	.....	v s
fluorosilicate	$CdSiF_6 \cdot 6H_2O$	362.57	col, hex	.....	d	.....	s
formate	$Cd(CHO_2)_2 \cdot 2H_2O$	238.47	col, mn	2.44	d	.....	v s
hydroxide	$Cd(OH)_2$	146.11	wh, trig or amorp	4.79^{15}	d 300	.....	0.00026; s a
iodate	$Cd(IO_3)_2$	462.21	wh cr	6.43	d	.....	s
iodide	CdI_2	366.21	grn-yel pwd	5.670^{30}	387	796	86.2
molybdate	$CdMoO_4$	272.34	yel pl	5.347	.....	.....	sl s; s a, KCN, NH_4OH
nitrate	$Cd(NO_3)_2$	236.41	col	.....	350	.....	109°
	$Cd(NO_3)_2 \cdot 4H_2O$	200.42	wh pr nd, hygr	2.455^{17}	59.4	132	215
oxalate	CdC_2O_4	308.47	col cr	3.32^{18}	d 340	.....	i; s a
oxide	CdO	128.40	brn, amorp; brn, cub, 2.49^{Li}	6.95; 8.15	d 900	subl 1500	i; s a
permanganate	$Cd(MnO_4)_2 \cdot 6H_2O$	458.36	....	2.81	d 95	.....	v s
*ortho*phosphate	$Cd_3(PO_4)_2$	527.14	col, amor	.....	1500	.....	i; s a
pyrophosphate	$Cd_2P_2O_7$	398.74	wh cr leaf	4.965^{15}	>red heat	.....	sl s; s a
*ortho*phosphate, di-H	$Cd(H_2PO_4)_2 \cdot 2H_2O$	342.41	col, tric	2.741^{5}	d 100	.....	i; s a
phosphide	Cd_3P_2	399.15	grn, tetr	5.60	700	.....	i; s d HCl
potassium cyanide	$K_2[Cd(CN)_4]$	294.68	col, oct	1.847	.....	.....	33.3
potassium sulfate	$K_2Cd(SO_4)_2 \cdot 2H_2O$	322.69	col, tric	2.922^{16}	.....	.....	42.91^{6}
selenate	$CdSeO_4 \cdot 2H_2O$	291.39	col, rh	3.63	$-2H_2O$, 170	.....	v s
selenide	$CdSe$	191.36	grn-brn, hex	5.81^{15}	>1350	.....	i; d a
silicate	$CdSiO_3$	188.48	col, rh, 1.739	4.93	1240	.....	v sl s
sulfate	$CdSO_4$	208.46	wh, rh	4.691^{20}	1000	.....	75°
	$CdSO_4 \cdot H_2O$	226.48	col, mn	3.79^{20}	tr 108	.....	s
	$CdSO_4 \cdot \tfrac{8}{3}H_2O$	256.50	col, mn, 1.565	3.09	tr 41.5	.....	113°
sulfide (greenockite)	CdS	144.46	yel-or, hex, 2.506, 2.529	4.82	.....	subl 980 in N_2	i; s a

Table 4-1 (Continued)
PHYSICAL CONSTANTS OF INORGANIC COMPOUNDS

Name	Formula	Formula Weight	Color, Crystalline Form, and Refractive Index	Density	Melting Point, °C	Boiling Point, °C	Solubility in 100 Parts
Cadmium							
sulfite	$CdSO_3$	192.46	col cr	....	d	....	sl s; s a
telluride	$CdTe$	240.00	blk, cub	6.20^{15}	1041	....	i; i a; d HNO_3
tungstate	$CdWO_4$	360.25	yel cr	....	....	....	0.05; s NH_4OH
Calcium							
	Ca	40.08	silv-wh, fcc	1.55	851	1487	d; s a
acetate	$Ca(C_2H_3O_2)_2$	158.17	col cr, 1.55, 1.56, 1.57	....	d	....	$37.4°$
	$Ca(C_2H_3O_2)_2 \cdot 2H_2O$	194.21	col nd	....	$-H_2O$, 84	....	34.7^{20}
aluminate	$CaAl_2O_4$	158.04	wh, mn, tric, or rh, 1.643, 1.665, 1.663	2.981	1600	....	d; s HCl
	$Ca_3Al_2O_6$	270.20	wh, cub, 1.710	3.038	d 1535	....	i; s a
	$Ca_3Al_2O_6 \cdot 3H_2O$	378.29	col, oct, 1.603	2.52^{20}	d 700–800	....	d
aluminum silicate	$2CaO \cdot Al_2O_3 \cdot SiO_2$	274.20	col, tetr, 1.669, 1.658	3.048	1590	....	d a
(anorthite)	$CaAl_2Si_2O_8$	278.21	wh, tric, 1.5832	2.765	1551	....	....
orthoarsenate	$Ca_3(AsO_4)_2$	398.08	col amorp pwd	3.620	....	....	0.013
(haidingerite)	$2CaO \cdot As_2O_5 \cdot 3H_2O$	396.04	col, rh, 1.590, 1.602, 1.638	2.967	....	....	....
arsenide	Ca_3As_2	270.08	red cr	3.031	d	....	d; d a; s h HNO_3
azide	$Ca(N_3)_2$	124.12	col, rh, hygr	....	expl 144	....	$38.1°$
benzoate	$Ca(C_7H_5O_2)_2 \cdot 3H_2O$	336.36	col, rh	1.436	$-3H_2O$, 110	....	$2.7°$
borate, meta-	$Ca(BO_2)_2$	125.70	col, rh, 1.550, 1.660, 1.680	....	1154	....	sl s; s a
	$Ca(BO_2)_2 \cdot 6H_2O$	233.79	col, tetr, 1.520, 1.502	1.88	....	....	0.25
tetraborate	CaB_4O_7	195.32	....	....	986	....	....
boride	CaB_6	104.95	blk, cub	2.32^{20}	2235	....	i; s HNO_3
bromate	$Ca(BrO_3)_2 \cdot H_2O$	313.91	col, mn	3.329	$-H_2O$, 180	....	230
bromide	$CaBr_2$	199.90	col, rh, deliq	3.353	730, sl d	810	143; s al, acet

Name	Formula	Mol. wt.	Color, crystalline form, index of refraction	Density	Melting point	Boiling point	Solubility
	$CaBr_2 \cdot 6H_2O$	307.99	col, hex	2.295	38.2	149	222; s al, acet
carbide	CaC_2	64.10	col, tetr, 1.75	2.22	......	2300	d
carbonate (aragonite)	$CaCO_3$	100.09	col, rh, 1.530, 1.681, 1.685	2.930	tr 520 to calcite	d 900	0.0015; s a
(calcite)	$CaCO_3$	100.09	col, rh or hex, 1.6583, 1.4864	2.710^{18}	......	d 900	0.0014; s a
	$CaCO_3 \cdot 6H_2O$	208.18	col, mn, 1.460, 1.535, 1.545	1.771^{0}	......	......	......
chlorate	$Ca(ClO_3)_2$	206.99	wh cr, hygr	......	340	......	s; s al, acet
	$Ca(ClO_3)_2 \cdot 2H_2O$	243.01	yelsh, rh or mn, deliq	2.711	$-H_2O$, 100	......	230
chloride	$CaCl_2$	110.99	col, cub, deliq, 1.52	2.15	782	>1600	74.5; s al, acet
	$CaCl_2 \cdot H_2O$	129.00	col cr, deliq	......	260	......	76.8°
	$CaCl_2 \cdot 2H_2O$	147.02	col cr	......	......	......	97.7°
	$CaCl_2 \cdot 6H_2O$	219.08	col, trig, deliq, 1.417, 1.393	1.71	29.9; $-6H_2O$, 200	......	279°
chloride fluoride orthophosphate	$3Ca_3(PO_4)_2 \cdot CaClF$	1025.08	col cr, 1.634, 1.631	3.14	1270	......	v sl s
chlorite	$Ca(ClO_2)_2$	174.98	wh, cub	2.71	......	......	d
hypochlorite	$Ca(ClO)_2$	142.98	wh pwd, 1.545, 1.69	2.35	d 100	......	27.8
	$Ca(ClO)_2 \cdot 3H_2O$	197.03	wh, tetr, 1.535, 1.63	2.1	$-3H_2O$, 60	......	20.4; s a
chromate	$CaCrO_4 \cdot 2H_2O$	192.09	yel, mn	......	$-2H_2O$, 200	......	i a; s fus K_2CO_3
chromite	$CaCr_2O_4$	208.07	ol grn, cub	4.8^{18}	2090	......	0.22
citrate	$Ca_3(C_6H_5O_7)_2 \cdot 4H_2O$	570.51	wh nd	......	$-4H_2O$, 120	......	d ev NH_3
cyanamide	$CaCN_2$	80.10	col, hex	......	1300 subl	......	d
cyanide	$Ca(CN)_2$	92.12	wh pwd	......	d 350	......	160°
dithionate	$CaS_2O_6 \cdot 4H_2O$	272.27	col, trig, 1.5496	2.176	......	......	v s
ferricyanide	$Ca_3[Fe(CN)_6]_2 \cdot 12H_2O$	760.42	red nd, deliq	......	......	......	......
ferrite	$CaFe_2O_4$	215.77	dk red, rh, 2.58, 2.43 (Na)	5.08	1250	......	i; v sl s a
ferrocyanide	$Ca_2[Fe(CN)_6] \cdot 12H_2O$	490.28	yel, tric, 1.570, 1.582, 1.596	1.68	d	......	86.8
fluoride (fluorite)	CaF_2	78.08	col, cub, 1.434	3.180	1418	ca 2500	0.002; sl s a
fluorosilicate	$CaSiF_6$	182.16	col, tetr	2.66^{18}	......	......	sl s; s HF, HCl
	$CaSiF_6 \cdot 2H_2O$	218.19	col, tetr	2.254	......	......	sl s; s HF, HCl

Table 4-1 (*Continued*)
PHYSICAL CONSTANTS OF INORGANIC COMPOUNDS

Name	Formula	Formula Weight	Color, Crystalline Form, and Refractive Index	Density	Melting Point, °C	Boiling Point, °C	Solubility in 100 Parts
Calcium							
formate	$Ca(CHO_2)_2$	130.12	col, rh, 1.510, 1.514, 1.578	2.015	d	....	16.6
d-gluconate	$Ca(C_6H_{11}O_7)_2 \cdot H_2O$	448.40	wh cr pwd	....	$-H_2O$, 120	....	3.3^{15}
glycerophosphate	$CaC_3H_5(OH)_2PO_4$	210.16	wh cr pwd, hygr	....	d 170	....	2; i al
hydride	CaH_2	42.10	wh, rh	1.902	816 (in H_2)	....	d
hydroxide	$Ca(OH)_2$	74.09	col, hex, 1.574, 1.545	2.24	$-H_2O$, 580	d	0.19^0; s a
hyponitrite	$CaN_2O_2 \cdot 4H_2O$	172.15	wh cr	1.834	d 320	....	d dil a
iodate (lautarite)	$Ca(IO_3)_2$	389.89	col, mn	4.519^{15}	d 540	....	0.24; s HNO_3
	$Ca(IO_3)_2 \cdot 6H_2O$	497.98	col, rh	....	d 35	....	0.13; s HNO_3
iodide	CaI_2	293.89	yelsh-wh, hex, deliq	3.956	740	ca 1100	209^{20}; s al, acet
	$CaI_2 \cdot 6H_2O$	401.98	yel, hex nd	2.55	d 42	160	757^0; s al, acet
iron(III) aluminate (celite)	$4CaO \cdot Fe_2O_3 \cdot Al_2O_3$	485.97	brn, rh, 1.98, 2.05, 2.08	3.77	1420	....	....
lactate	$Ca(C_3H_5O_3)_2 \cdot 5H_2O$	308.30	wh nd, effl	....	$-3H_2O$, 100	....	5.4^{15}
laurate	$Ca(C_{12}H_{23}O_2)_2 \cdot H_2O$	456.73	wh nd, effl	....	182	....	0.004^{15}
magnesium carbonate (dolomite)	$Ca[Mg(CO_3)_2]$	184.41	col, trig, 1.6817, 1.5026	2.872	d 730–760	....	0.032^{18}; s HCl
magnesium silicate (diopside)	$CaO \cdot MgO \cdot 2SiO_2$	216.56	col, mn, 1.665, 1.672, 1.695	3.275	1390	....	i; i a
(mervinite)	$3CaO \cdot MgO \cdot SiO_2$	328.72	col to lt grn, mn, 1.708, 1.711, 1.718	3.150	....	....	....
molybdate (pawellite)	$CaMoO_4$	200.01	col, tetr, 1.967, 1.978	4.38–4.53	....	....	i; s a
nitrate	$Ca(NO_3)_2$	164.09	col, cub, hygr	2.504^{18}	561	....	129
	$Ca(NO_3)_2 \cdot 3H_2O$	218.14	col, tric	....	51.1	....	....
	$Ca(NO_3)_2 \cdot 4H_2O$	236.15	col, mn, deliq, 1.465, 1.498, 1.504	α 1.896 β 1.82	α 42.7 β 39.7	d 132	660^{30}

Name	Formula	Mol wt	Color, crystalline form, refractive index	Density	m.p.	b.p.	Solubility
nitride	Ca$_3$N$_2$	148.25	brn, hex	2.63^{17}	1195	. . . .	d; s dil a
nitrite	Ca(NO$_2$)$_2$·H$_2$O	150.11	col-yelsh, hex, deliq	2.2334°	−H$_2$O, 100	. . . .	84.5
	Ca(NO$_2$)$_2$·4H$_2$O	204.15	col, tetr	1.674°	−2H$_2$O, 44	. . . .	74.9°
oleate	Ca(C$_{18}$H$_{33}$O$_2$)$_2$	603.01	wh waxy cr	. . . .	83–84	. . . .	0.04
oxalate	CaC$_2$O$_4$	128.10	col, cub	2.24°	d	. . . .	i; s a
	CaC$_2$O$_4$·H$_2$O	146.12	col	2.2	−H$_2$O, 200	. . . .	i; s a
oxide (lime, calcia)	CaO	56.08	col, cub, 1.838	3.4	2590	2850	0.13^{10}; s a
palmitate	Ca(C$_{16}$H$_{31}$O$_2$)$_2$	550.93	wh fatty pwd	. . . .	. . . .	. . . .	0.003
perchlorate	Ca(ClO$_4$)$_2$	238.98	col cr	2.651	d 270	. . . .	188
permanganate	Ca(MnO$_4$)$_2$·5H$_2$O	368.03	purp cr	2.4	d	. . . .	338
peroxide	CaO$_2$	72.08	wh, tetr, 1.895	2.92	d 275	. . . .	sl s; s a
	CaO$_2$·8H$_2$O	216.20	wh, tetr	1.70	−8H$_2$O, 200	exp 275	sl s; s a
*meta*phosphate	Ca(PO$_3$)$_2$	198.02	col, 1.588, 1.595	2.82	975	. . . .	i; i a
*ortho*phosphate (whitlockite)	Ca$_3$(PO$_4$)$_2$	310.18	wh amorp pwd, 1.629, 1.626	3.14	1670	. . . .	0.002; s a
*ortho*phosphate, mono-H (brushite)	CaHPO$_4$·2H$_2$O	172.09	wh, tric, 1.5576, 1.5457, 1.5392	2.306^{16}	−H$_2$O, 109	. . . .	0.032; s a
*ortho*phosphate, di-H	Ca(H$_2$PO$_4$)$_2$·H$_2$O	252.07	col, tric, deliq, 1.5292, 1.5176, 1.4392	2.220^{15}	−H$_2$O, 109	d 203	1.8^{30}
*pyro*phosphate	Ca$_2$P$_2$O$_7$	254.10	col, biax, 1.585, 1.604	3.09	1230	. . . .	i; s a
	Ca$_2$P$_2$O$_7$·5H$_2$O	344.18	col, mn, 1.539, 1.545, 1.551	2.25	. . . .	. . . .	sl s; s a
phosphide	Ca$_3$P$_2$	182.19	gray lumps	2.51	ca 1600	. . . .	d; s a
hypophosphite	Ca(PH$_2$O$_2$)$_2$	170.06	wh-gray, mn	. . . .	d	. . . .	15.4
plumbate	Ca$_2$PbO$_4$	351.35	red-brn cr	5.71	d	. . . .	i; s a
salicylate	Ca(C$_7$H$_5$O$_3$)$_2$·2H$_2$O	350.34	wh, oct	. . . .	−2H$_2$O, 120	. . . .	2.3
selenate	CaSeO$_4$	183.04	col	2.88	. . . .	. . . .	7.94^4
	CaSeO$_4$·2H$_2$O	219.07	col, mn	2.68	. . . .	. . . .	. . . .
selenide	CaSe	119.04	cub, 2.274	3.57	. . . .	. . . .	. . . .
*meta*silicate (α) (pseudowollastonite)	CaSiO$_3$	116.16	col, mn, 1.610, 1.611, 1.664	2.905	1540	. . . .	0.0096^{17}; s HCl

Table 4-1 (Continued)
PHYSICAL CONSTANTS OF INORGANIC COMPOUNDS

Name	Formula	Formula Weight	Color, Crystalline Form, and Refractive Index	Density	Melting Point, °C	Boiling Point, °C	Solubility in 100 Parts
Calcium							
metasilicate (β) (wollastonite)	$CaSiO_3$	116.16	col, mn, 1.616, 1.629, 1.631	2.5	tr 1200	......	......
orthosilicate(I)	Ca_2SiO_4	172.24	col, mn, 1.717, 1.735	3.27	2130	......	......
(II)	......	......	col, rh, 1.717, 1.735	3.28	tr to (I) 1420	......	......
(III)	......	......	col, mn, 1.642, 1.645, 1.654	2.97	tr to (II) 675	......	......
(tri-) silicate (alite)	Ca_3SiO_5	228.32	col, mn, α 1.718, β 1.724	1900	......	......	i; s a, alk
silicide	$CaSi_2$	96.25	wh cr pwd	2.5	......	......	......
stearate	$Ca(C_{18}H_{35}O_2)_2$	607.04	col, 1.460, 1.540, 1.610	......	179–180	......	$0.004^{15°}$
succinate	$CaC_4H_4O_4 \cdot 3H_2O$	212.22	......	......	......	......	$0.19^{10°}$
sulfate (anhydrite)	$CaSO_4$	136.14	col, rh, 1.569, 1.575, 1.613	2.960	1360 d	tr to mn 1193	0.20; s a
sulfate (soluble anhydrite)	......	......	col, hex or tric, 1.505, 1.548	2.61	tr to rh, >200	......	......
(plaster of Paris)	$CaSO_4 \cdot \frac{1}{2}H_2O$	145.15	wh pwd	......	$-H_2O$, 163	......	$0.3^{20°}$; s a, gly
(gypsum)	$CaSO_4 \cdot 2H_2O$	172.17	col, mn, 1.521, 1.523, 1.530	2.32	$-1.5H_2O$, 128	$-2H_2O$, 163	0.24; s a, gly
sulfide (oldhamite)	CaS	72.14	col, cub, 2.137	2.5	d ~2400	......	$0.02^{15°}$ d; d a
sulfide, hydrogen	$Ca(HS)_2 \cdot 6H_2O$	214.32	col pr	......	d 15–18	......	v s
sulfite	$CaSO_3 \cdot 2H_2O$	156.17	col, hex	......	$-2H_2O$, 100	......	0.004; s a
d-tartrate	$CaC_4H_4O_6 \cdot 4H_2O$	260.21	col, rh, 1.525, 1.535, 1.550	......	d	......	$0.069^{38°}$; sl s al
dl-tartrate	......	......	tric pwd or nd	......	$-4H_2O$, 200	......	$0.0078^{38°}$; s HCl
meso-tartrate	$CaC_4H_4O_6 \cdot 3H_2O$	242.20	wh mn or tric pr	......	$-3H_2O$, 170	......	$0.28^{18°}$
telluride	$CaTe$	167.68	cub, 2.51, 2.58	4.873	......	......	......

Name	Formula	Formula wt	Color, crystalline form, refractive index	Density	Melting point	Boiling point	Solubility
tellurite	$CaTeO_3$	215.68	wh fl	....	>960	....	sl s; s a
thiocyanate	$Ca(SCN)_2 \cdot 3H_2O$	210.29	wh cr, deliq	....	d	....	150; v s al
thiosulfate	$CaS_2O_3 \cdot 6H_2O$	260.30	wh, tric	1.872	d	....	$100^{3\cdot}$; s al
metat tanate (perovskite)	$CaTiO_3$	135.98	col, cub, rh, β 2.34	4.10	1975	....	i
tungstate	$CaWO_4$	287.93	wh, tetr, 1.9263, 1.9107	$6.062^{20\circ}$	....	....	i
(scheelite)	....	....	col, tetr, 1.918, 1.934	6.06	....	....	0.2; s NH_4Cl
metatungstate	$Ca_3H_4[H_2(W_2O_7)_6] \cdot 27H_2O$	3500.96	col, tric	....	$-7H_2O$, 105	d	d a
zirconate	$CaZrO_3$	179.30	col, mn	4.78	2550	190	s al, eth, chl
Californium	Cf	(248)	silv met	....	....	227	....
chloride	$CfCl_3$	354	grn, hex	5.88	....	210	i
Carbon							
C diamond	12.011		col, cub, 2.4173	3.5153	4000^{63atm}	subl 3850	i
C graphite	12.011		blk, hex	2.267	....	....	i
C amorphous	12.011		blk	1.8–2.1	subl 3650	....	i
bromide, *tetra-* (di-) *tetrabromide*	CBr_4	331.65	col, mn (tr pt 47°)	3.42	90.1	190	s al, eth, chl
(di-) *hexabromide*	C_2Br_4	343.66	rh, 1.740, 1.847, 1.863	....	57.5	227	s al, eth
	C_2Br_6	503.48	......	3.823	148 d	210	s CS_2
chloride, *tetra-*	CCl_4	153.82	col liq, 1.4601	1.5867	−22.9	76.7	s al, eth, chl
(di-) *tetrachloride*	C_2Cl_4	165.83	col liq, 1.5055	$1.6311^{15\circ}$	−22.4	120.8	s al, eth
(di-) *hexachloride*	C_2Cl_6	236.74	col	2.091	subl 187	−128.0	s al, eth, oils
fluoride, *tetra-*	CF_4	88.00	col g	$1.96^{-184\circ}$	−183.7	−161.49	sl s
hydride or methane	CH_4	16.04	col g	$0.415^{-164\circ}$	−182.48	−128.0	s bz
iodide, *tetra-*	CI_4	519.63	dk red, cub	4.34	d 171	−161.49	s al, bz, eth, MeOH
oxide, mon-	CO	28.01	col g, 1.000335	0.001250 gas 0.793 liq	−205.00	−191.45	s al, bz, ac a, Cu_2Cl_2
oxide, di-	CO_2	44.01	col g, 1.000449	0.0019750 gas $1.56^{-79\circ}$ sol	$-56.57^{t.p.}$	−78.477 subl	31 cc^{15} (0.033M)
(tri-) *dioxide*	C_3O_2	68.03	col liq, 1.4538	1.11	$-112.18^{63.5atm}$	6.4	d acet, eth, CS_2, CCl_4
selenide, di-	CSe_2	169.93	yel liq, 1.845	$2.6626^{25\circ}$	−45.5	125	s acet, eth, CS_2, CCl_4

Table 4-1 (Continued)
PHYSICAL CONSTANTS OF INORGANIC COMPOUNDS

Name	Formula	Formula Weight	Color, Crystalline Form, and Refractive Index	Density	Melting Point, °C	Boiling Point, °C	Solubility in 100 Parts
Carbon							
selenide sulfide	CSeS	123.04	yel oily liq	1.9874	−85	84.5	s CS_2
sulfide	CS	44.08	red pwd	1.66	d 200	......	s eth, CS_2
sulfide, di-	CS_2	76.14	col liq, inflam, 1.6295	1.261^{22}	−111.6	46.3	0.22^{22}; s al, eth
(tri-) disulfide	C_3S_2	100.16	red liq	1.274	−0.5	d 90	s bz, CS_2
sulfide telluride	CTeS	171.68	yel-red liq	2.9^{-50}	−54	d −54	......
Carbonic acid	H_2CO_3	62.03	known in soln only	......	......	......	......
Carbonyl bromide	$COBr_2$	187.83	col liq	......	......	60	d; s bz, tol, ac a
chloride	$COCl_2$	98.92	col g; phosgene	1.392	−127.8	7.6	d
fluoride	COF_2	66.01	col g	1.139^{-114}	−114.0	−83.3	d; s $COCl_2$
selenide	COSe	106.97	col g	1.812 liq	−124.4	−21.7	s al, CS_2
sulfide	COS	60.07	col g	0.001073^{0}	−138.2	−50.2	s a
Cerium	Ce	140.12	gray-met, fcc	6.771	795	3470	20^{15}
(III) acetate	$Ce(C_2H_3O_2)_3$	317.26	col	......	d 308	......	i
boride, tetra-	CeB_4	183.36	tetr	5.74	2190	......	......
boride, hexa-	CeB_6	204.98	bl-vlt, cub	4.69	......	d	s
bromate	$Ce(BrO_3)_3 \cdot 9H_2O$	685.98	redsh-wh, hex	......	49	......	v s
bromide	$CeBr_3$	379.84	col, hygr, hex	5.18	733	1560	d; s.a
carbide	CeC_2	164.14	red, hex	5.23	......	......	s a
carbonate	$Ce_2(CO_3)_3 \cdot 5H_2O$	550.37	wh crys	3.92	......	......	v s
chloride	$CeCl_3$	246.48	wh, hex, deliq	6.16	817	1730	s H_2SO_4
fluoride, tri-	CeF_3	197.12	wh, hex	4.80	1430	2327	......
fluoride, tetra-	CeF_4	216.12	wh, mn	5.43	~650	d	......
hydride, di-	CeH_2	142.13	fcc	5.5	ign	......	s a
hydride, tri-	CeH_3	143.14	dk bl, amorp	......	......	......	s a
(III) hydroxide	$Ce(OH)_3$	191.14	wh gel	......	......	......	......
(IV) hydroxide	$Ce(OH)_4$	208.15	yel gel; $[CeO_2 \cdot 2H_2O]$	......	......	......	......
iodate	$Ce(IO_3)_4$	839.73	yel	......	......	......	0.015

Name	Formula	Mol wt	Color, crystalline form, etc.	Density	Melting point	Boiling point	Solubility
iodide, di-	CeI_2	393.93	dk bronze; $[Ce^{3+}(e^-)(I^-)_2]$		808		s
iodide, tri-	CeI_3	520.83	yel, rh	4.83	766	1400	v s
molybdate	$Ce_2(MoO_4)_3$	760.05	yel, tetr, 2.019, 2.007		973		v s
nitrate	$Ce(NO_3)_3 \cdot 6H_2O$	434.23	col, deliq		$-3H_2O$, 150	d 200	s HCl, H_2SO_4
oxalate	$Ce_2(C_2O_4)_3 \cdot 10H_2O$	724.46	wh cr		d		s H_2SO_4
oxide, sesqui-	Ce_2O_3	328.24	gray-grn, hex	6.86	1692		s dil a
oxide chloride	CeOCl	191.57	purp leaf				
(di-)dioxide sulfide	Ce_2O_2S	344.30	brn-maroon, hex	6.00	1950		i a
(III) metaphosphate	$Ce(PO_3)_3$	377.04	micr nd	3.272			s a
(III) orthophosphate	$CePO_4$	235.09	red, mn or yel, rh	5.22			s a
(III) selenate	$Ce_2(SeO_4)_3$	709.11	rh	4.456			$39.6^{30°}$
selenide, sesqui-	Ce_2Se_3	517.12	bl-blk	6.53			
silicide	$CeSi_2$	196.29		5.67			i
(III) sulfate	$Ce_2(SO_4)_3$	568.42	wh, mn, hygr	3.912	d 1000		$10.1^{0°}$
(III) sulfate	$Ce_2(SO_4)_3 \cdot 5H_2O$	658.50	wh, mn	3.17			$3.95^{0°}$
(III) sulfate	$Ce_2(SO_4)_3 \cdot 9H_2O$	730.56	wh nd, hex	2.831			$11.9^{15°}$
(IV) sulfate	$Ce(SO_4)_2$	332.24	yel cr	3.91	d 195		hyd; s dil H_2SO_4
sulfide, di-	CeS	172.18	gold, cub	5.88	2100		
sulfide, sesqui-	Ce_2S_3	376.43	red-brn pwd	$5.020^{11°}$	d 2100 vac		s dil a
(tri-) tetrasulfide	Ce_2S_4	548.60	blk, cub	5.3	2050		
tungstate	$Ce_2(WO_4)_3$	1023.78	yel, tetr	6.77	1089		
Cesium	Cs	132.91	silv met, hex	$1.8785^{15°}$	28.6	670	d; s a
acetate	$CsC_2H_3O_2$	191.95	col, deliq		194		v s
aluminum sulfate	$CsAl(SO_4)_2 \cdot 12H_2O$	568.19	col, cub, 1.4587	1.97	117		42^{100}
amide	$CsNH_2$	148.93	wh nd	3.44	262		d
azide	CsN_3	174.93	col nd, deliq		310		v s
boranate	$CsBH_4$	147.75	wh, fcc, 1.498	2.404			v s
bromate	$CsBrO_3$	260.81	wh, hex, 2.2	$4.109^{16°}$	420 d		3.66
bromide	CsBr	212.81	col, bcc, 1.6984	4.44	636	1300	124.3
bromide, tri-	$CsBr_3$	372.63	rh		180		
dibromide chloride	$CsBr_2Cl$	328.18	yel-red, rh		191	$-Br_2$, 150	s; d al
bromide chloride iodide	CsBrClI	375.17	yel-red, rh		235	d 290	s; s al
dibromide iodide	$CsIBr_2$	419.63		4.25	248	d 320	$4.61^{20°}$
carborate	Cs_2CO_3	325.82	col cr, deliq		d 610		v s; $11^{20°}$ al

Table 4-1 (Continued)
PHYSICAL CONSTANTS OF INORGANIC COMPOUNDS

Name	Formula	Formula Weight	Color, Crystalline Form, and Refractive Index	Density	Melting Point, °C	Boiling Point, °C	Solubility in 100 Parts
Cesium							
carbonate, hydrogen	$CsHCO_3$	193.92	col, rh	....	$-H_2O$, 175	....	v s; s al
chlorate	$CsClO_3$	216.36	wh cr	3.57	....	....	$6.28^{20°}$; s al
dichloride bromide	$CsBrCl_2$	283.72	glossy yel, rh	....	205	....	s; d al
chloride	$CsCl$	168.36	col, cub, deliq, 1.6418	3.988	646	1295	65.8
chloroaurate(III)	$CsAuCl_4$	471.68	yel, mn	....	....	....	9.0
dichloride iodide	$CsICl_2$	330.74	or, trig	3.86	230	d 290	s; s al
chloroplatinate(IV)	Cs_2PtCl_6	673.62	yel, cub	4.197	d 570	....	$0.0047^{0°}$
chlorostannate(IV)	Cs_2SnCl_6	597.22	wh, cub	3.33	....	....	....
chromate	Cs_2CrO_4	381.80	yel, rh	4.237	....	....	$71.4^{13°}$
chromium sulfate	$Cs[Cr(H_2O)_6](SO_4)_2 \cdot 6H_2O$	593.21	vlt cr	2.064	116	....	9.4
cyanide	$CsCN$	158.92	wh cr	2.93	....	....	v s
fluoride	CsF	151.90	wh, cub, deliq, 1.478	4.115	682	1250	v s
fluoride, hydrogen	$CsHF_2$	171.91	wh nd, deliq	....	160	....	v s
fluoroborate	$CsBF_4$	219.71	col, rh, 1.350	3.20	550 d	....	1.6
fluorogermanate	Cs_2GeF_6	452.39	oct	4.10	....	....	sl s; sl s a
fluorosilicate	Cs_2SiF_6	407.89	wh, cub	$3.372^{17°}$	....	....	60^{17}
formate	$CsCHO_2$	172.92	col	$1.0169^{21°}$	....	....	v s
gallium sulfate	$CsGa(SO_4)_2 \cdot 12H_2O$	610.93	col, cub, 1.46495	3.42	....	....	1.21
hydride	CsH	133.91	wh, cub	....	d	....	d; d a
hydrogen carbide	$CsHC_2$	157.94	....	....	300	....	d
hydroxide	$CsOH$	149.91	lt yel, deliq	3.675	272.3	....	v s
iodate	$CsIO_3$	307.81	wh, mn	4.85	....	....	2.6
iodide	CsI	259.81	wh, rh, deliq, 1.7876	4.510	621	1280	$44^{0°}$
iodide, tri-	CsI_3	513.62	blk, rh	4.47	207.5	....	sl s; s al
iodotetrachloride	$CsICl_4$	401.62	pale or nd	$3.374^{-10°}$	228	d	sl s

Name	Formula	Mol wt	Crystalline form, color, index of refraction	Density	mp, °C	bp / transition, °C	Solubility
iron(II) sulfate	$Cs_2SO_4\cdot FeSO_4\cdot 6H_2O$	621.87	lt grn, mn, 1.500, 1.504, 1.509	2.79^{20}	ca 70	......	101 anhyd
iron(III) sulfate	$CsFe(SO_4)_2\cdot 12H_2O$	597.06	lt vlt cr, 1.484	2.061^{20}	ca 90	......	s
magnesium sulfate	$Cs_2SO_4\cdot MgSO_4\cdot 6H_2O$	590.34	col, mn, 1.486, 1.452	2.676^{20}	......	......	......
mercury(II) bromide	$CsBr\cdot 2HgBr_2$	933.63	rh	......	......	......	0.81^{17}; sl s al
mercury(II) chloride	$CsCl\cdot HgCl_2$	439.85	col, cub or rh, 1.792	......	......	d 849	1.44^{17}
nitrate	$CsNO_3$	194.91	col, hex or cub, 1.55, 1.56	3.685	414	......	9.16°
nitrate, HNO_3 adduct	$CsNO_3\cdot HNO_3$	257.92	oct	......	100	......	......
nitrite	$CsNO_2$	178.91	yel cr	......	......	......	v s
oxalate	$Cs_2C_2O_4$	353.82	......	3.230^{15}	......	......	v s
oxide	Cs_2O	281.81	or nd	4.25	d 400 (490 in N_2)	......	v s
oxide, tri-	Cs_2O_3	313.81	choc brn, cub	4.25	400	......	d; s a
perchlorate	$CsClO_4$	232.36	col, rh at 219, cub, 1.4752, 1.4788, 1.4804	3.327^{4}	d 224	to CsCl 575	1.9
periodate	$CsIO_4$	323.81	wh, rh	4.259^{15}	......	......	2.15^{15}
permanganate	$CsMnO_4$	251.84	purp	3.507	d 320	......	0.097^{1}
peroxide	Cs_2O_2	297.81	lt yel nd	4.25	400	$-O_2$, 650	s
perrhenate	$CsReO_4$	383.11	wh cr	4.76	616	......	0.287^{30}
phthalate, hydrogen	$CsHC_8H_4O_4$	298.03	col, rh	2.178	......	......	......
polonium(IV) chloride	Cs_2PoCl_6	688.53	cub, 1.86	3.82	......	......	......
rhodium(III) sulfate	$CsRh(SO_4)_2\cdot 12H_2O$	644.12	yel, oct	2.238	111	......	sl s
selenate	Cs_2SeO_4	408.77	col, rh, deliq, 1.595, 1.506, 1.596	4.4528^{20}	......	......	244^{12}
sulfate	Cs_2SO_4	361.87	col, rh or hex, 1.560, 1.564, 1.566	4.243	995	tr to hex, 600	64^{20}
sulfate, hydrogen	$CsHSO_4$	229.97	col, rh	3.352^{16}	d	......	s
sulfide	$Cs_2S\cdot 4H_2O$	369.94	wh cr, deliq	......	......	......	v s
sulfide, di-	Cs_2S_2	329.94	dk red, amorp	......	460	>800	......
sulfide, penta-	Cs_2S_5	426.13	......	2.806^{15}	210	......	......
sulfide, tri-	Cs_2S_3	362.00	yel leaf	......	217	780	......
tartrate, hydrogen	$CsHC_4H_4O_6$	281.99	wh, rh	......	......	......	9.7

Table 4-1 (*Continued*)
PHYSICAL CONSTANTS OF INORGANIC COMPOUNDS

Name	Formula	Formula Weight	Color, Crystalline Form, and Refractive Index	Density	Melting Point, °C	Boiling Point, °C	Solubility in 100 Parts
Cesium							
l-tartrate	$Cs_2C_4H_4O_6$	413.88	col, trig	$3.03^{14°}$	....	....	v s d
12-tungstanosilicate	$Cs_8SiW_{12}O_{42} \cdot 12H_2O$	3969.50	wh cr	....	....	....	$0.005^{20°}$
vanadium(III) sulfate	$CsV(SO_4)_2 \cdot 12H_2O$	592.15	red, cub, 1.4780	$2.033^{20°}$	82	d 300	$0.46^{10°}$
Chloramine	NH_2Cl	51.48	yel liq	....	−66	....	s; s al, eth
Chloric acid	$HClO_3 \cdot 7H_2O$	210.57	known only in soln	$1.282^{14°}$	<−20	d 40	v s
Chlorine	Cl_2	70.91	grn-yel g	$2.98^{20°}$ g/l, 1.57^{bp} liq	−101.0	−34.05	3.26 g (310 cc[10°])
azide	ClN_3	77.48	col g, expl	....	....	....	sl s; d alk
fluoride, mono-	ClF	54.56	col g, 1.000494 (Hg 5461)	1.62^{bp}	−155.6	−90	....
fluoride, tri-	ClF_3	92.45	col g, 1.000633 (Hg 5461)	1.825^{bp}	−76.3	11.75	....
fluoride, penta-	ClF_5	120.45	col g	$1.79^{20°}$	−84	−13.1	....
fluorosulfate	$Cl(OSO_2F)$	134.51	lt yel, rh	....	45.1	....	....
hydrate	$Cl_2 \cdot 8H_2O$	215.03	yel-brn g; red-brn liq, expl	1.23	d 9.6	3.5	i
(di-) oxide	Cl_2O	86.91	yel-grn g; red-brn liq;	3.02^{bp}	−120.6	3.5	3.5 g (200 cc)
oxide, di-	ClO_2	67.45	red-yel sol	$1.642^{20°}$	−59	10.9	200 cc[4°]
oxide, tri-	$(ClO_3)_2$	166.90	....	$1.95^{20°}$	3.5	203	....
(di-) oxide, hept-	Cl_2O_7	182.90	col oily liq, expl	$1.82^{20°}$	−91.5	84	d
Chlorodifluoroamine	$NClF_2$	87.45	col g	....	−183 to −196	−67	....
(Di-) **Chlorofluoroamine**	NCl_2F	103.91	col g	....	....	−3	....
Chlorosulfonic acid	$ClSO_3H$	116.52	col fum liq, 1.437	1.787	−80	158	d to H_2SO_4 + HCl

		Mol. wt.	Color, crystalline form	Density	M.P. °C	B.P. °C	Solubility
Chlorous acid	$HClO_2$	68.46	col, known only in soln	....	....	....	d
Chloryl fluoride	ClO_2F	86.45	g, fumes in moisture	....	−115	−5.65	d
perchlorate	ClO_2ClO_4	182.90	g, or sol; dk red liq	2.023	3.5	203	d
Chromium	Cr	52.00	steel gray, bcc	7.20	1900	2640	i; s dil HCl, H_2SO_4
(II) acetate	$Cr(C_2H_3O_2)_2$	170.09	red cr	....	....	....	sl s
(III) acetate	$Cr(C_2H_3O_2)_3 \cdot H_2O$	247.15	gray-grn pwd	....	....	....	s
arsenide	$CrAs$	126.92	gray, hex	$6.35^{16°}$	....	....	i; i a
boride	CrB	62.81	silv cr	6.17	....	....	i; s fus Na_2O_2
(II) bromide	$CrBr_2$	211.81	wh, mn	4.36	842/4	....	s
(III) bromide	$CrBr_3$	291.72	grn, rh-hed	4.25	....	....	i; v s al
	$[Cr(H_2O)_6]Br_3$	399.81	blsh-gray to vlt	5.41^{17}	....	....	v s
(tri-) carbide, di-	Cr_3C_2	180.02	gray, rh	6.68	1890	3800	i
carbonyl	$Cr(CO)_6$	220.06	col, o-rh	1.77	subl, d 110	210 expl	i; i al, eth
(II) chloride	$CrCl_2$	122.90	wh, rh, hygr	2.878	820	1300	v s
(III) chloride	$CrCl_3$	158.35	vlt, mn	$2.76^{15°}$	1150	947 subl	i; i al, acet, eth
(IV) chloride	$CrCl_4$	193.81	brn g	....	....	....	....
(II) chloride, hydrate	$[Cr(H_2O)_4]Cl_2 \cdot 2H_2O$	266.45	vlt, mn	1.76	83	....	58.5
(II) fluoride	CrF_2	89.99	grn, mn	4.11	894	....	sl s; s hot HCl
(III) fluoride	CrF_3	108.99	grn, rh-hed	3.8	1400	....	i; s HF
(IV) fluoride	CrF_4	127.99	grn	....	−28	sub 100	....
(V) fluoride	CrF_5	147.00	crimson, o-rh	....	30	117	....
(VI) fluoride	CrF_6	166.00	lemon-yel	....	d −100	....	....
(II) hydroxide	$Cr(OH)_2$	86.01	yel-brn	....	d	....	d; s a
(III) hydroxide	$Cr(OH)_3$	103.02	grn or bl, gel	....	d	....	i; s a
(II) iodide	CrI_2	305.80	red-brn, mn	5.196	868	$-I_2$, vac 350	i a, alk
(III) icdide	CrI_3	432.71	blk, hex	4.915	>600	....	s
(III) nitrate	$Cr(NO_3)_3 \cdot 9H_2O$	400.15	purp, mn	....	60	....	....
nitride, mono-	CrN	66.00	cub or amorp	5.9	1500	....	i; s HCl
(II) oxalate	$CrC_2O_4 \cdot H_2O$	158.03	yel cr pwd	2.468	−H_2O, 120 tr to grn	....	sl s; s dil a
(III) oxalate	$Cr_2(C_2O_4)_3 \cdot 6H_2O$	476.14	red, amorp, hygr	....	....	....	s; v s (red) al; i (grn) al
(II) oxide	CrO	68.00	blk pwd	....	....	d 100	i

Table 4-1 (Continued)
PHYSICAL CONSTANTS OF INORGANIC COMPOUNDS

Name	Formula	Formula Weight	Color, Crystalline Form, and Refractive Index	Density	Melting Point, °C	Boiling Point, °C	Solubility in 100 Parts
Chromium							
(III) oxide	Cr_2O_3	151.99	grn, hex, 2.551	5.21	2280	3000	i
	$Cr_2O_3 \cdot xH_2O$	varies	vlt amorp or bl-gray-grn gel	....	....	....	i; s a, alk
(IV) oxide	CrO_2	84.00	brn-blk pwd	....	$-O_2$, 300	....	i; s HNO_3
(VI) oxide	CrO_3	99.99	red, rh, deliq	2.70	198	d	166[15]; s al, eth
oxidetetrafluoride	$CrOF_4$	143.99	dk red, mn	....	55	....	hyd
2,4-pentanedione	$Cr(C_5H_7O_2)_3$	349.33	....	....	216	340	i; s org solv
(III) orthophosphate	$CrPO_4 \cdot 2H_2O$	183.00	vlt cr	2.42	....	....	sl s
	$CrPO_4 \cdot 6H_2O$	255.06	vlt, tric, 1.568, 1.591, 1.699	$2.121^{14°}$	100	....	
(III) pyrophosphate	$Cr_4(P_2O_7)_3$	729.81	lt grn, mn	3.2	....	....	i; s alk
phosphide, mono-	CrP	82.97	gray-blk cr	$5.7^{15°}$	....	....	i; s HNO_3, HF
(tri-) phosphide	Cr_3P	186.97	gray-blk, tetr	6.1	1515	....	i
silicide	Cr_3Si_2	212.17	gray, tetr	$5.5^{0°}$	....	....	i; s HCl, HF
(II) sulfate	$CrSO_4 \cdot 7H_2O$	274.17	blue cr	....	....	....	$12.4^{0°}$
(III) sulfate	$Cr_2(SO_4)_3$	392.18	vlt or red pwd	3.012	....	....	i; i a
	$Cr_2(SO_4)_3 \cdot 15H_2O$	662.41	vlt, amorp	$1.867^{17°}$	100	$-10H_2O$, 100	s
	$Cr_2(SO_4)_3 \cdot 18H_2O$	716.45	bl-vlt, oct, 1.564	1.7	$-12H_2O$, 100	....	120
(II) sulfide (daubrelite)	CrS	84.06	blk, hex	4.85	1550	....	i; v s a
(III) sulfide	Cr_2S_3	200.18	brn-blk pwd	$3.77^{10°}$	$-S$, 1350	....	i; s HNO_3
(III) sulfite	$Cr_2(SO_3)_3$	344.18	grnsh-wh	2.2	d	....	
(II) tartrate	$CrC_4H_4O_6$	200.07	bl pwd	2.33	....	....	i; sl s a
Chromium complexes							
hexaamminechromium(III) chloride	$[Cr(NH_3)_6]Cl_3 \cdot H_2O$	278.55	yel cr	1.585	....	....	s
chloropentammine chromium(III) chloride	$[Cr(NH_3)_5Cl]Cl_2$	243.51	red, oct	1.696	....	....	$0.65^{16°}$; i HCl

Name	Formula	Mol. wt.	Color, crystalline form, optical properties	Density	Melting point, °C	Boiling point, °C	Solubility
hexaureachromium(III) perrhenate	[Cr(CON₂H₄)₆](ReO₄)₃	1162.92	grn nd	2.652	....	....	1.786
Chromyl bromide	CrO₂Br₂	243.80	dk vlt nd, hygr	....	d −70	....	d; s eth
chloride	CrO₂Cl₂	154.90	dk red liq, fum	....	−96.5	117	d; s eth
fluoride	CrO₂F₂	121.99	red-vlt cr	....	31.6	....	i; s a
Cobalt	Co	58.93	silv-gray, cub	8.9	1495	3550	s; s al
(II) acetate	Co(C₂H₃O₂)₂·4H₂O	249.08	red-vlt, mn, deliq, 1.542	1.705¹⁹°	−4H₂O, 140	....	i; s a
(III) acetate	Co(C₂H₃O₂)₃	236.07	grn, oct	....	d 100	....	hyd; s a
(II) *ortho*arsenate (erythrite)	Co₃(AsO₄)₂·8H₂O	598.75	red-vlt, mn, 1.626, 1.661, 1.669	3.178¹⁵°	d	....	i; s dil a, NH₄OH
arsen c sulfide (cobaltite)	CoAsS	156.92	redsh-gray	6.2–6.3	d	....	....
arsenide	Co₂As	192.79	cr pwd	8.28	950	....	i; s HNO₃
boride	CoB	69.74	pr	7.25¹⁸°	....	....	d; s HNO₃
(II) bromate	Co(BrO₃)₂·6H₂O	422.84	red, cub	2.55	....	....	45.5¹⁷
(II) bromide	CoBr₂	218.75	grn, hcp, deliq	4.909	678 (in N₂)	....	67⁹°
	CoBr₂·6H₂O	326.84	red-vlt pr, deliq	2.46	−6H₂O, 130	....	v s
bromoplatinate(IV)	CoPtBr₆·12H₂O	949.66	trig	2.762	....	....	....
carbonate (sphrerocobaltite)	CoCO₃	118.94	red, trig, 1.855, 1.60	4.13	d	....	i; s a
carbonyl, tetra-(dicobalt octa-carbonyl)	[Co(CO)₄]₂	341.95	or cr	1.73¹⁸°	51	d 52	i; s CS₂, eth
carbonyl, tri-(tetracobalt dodecacarbonyl)	Co₄(CO)₁₂ or [Co(CO)₃]₄	571.86	blk cr	....	d 60	....	sl s; s bz
chlorate	Co(ClO₃)₂·6H₂O	333.93	red, cub, deliq, 1.55	1.92	50	d 100	v s
chloride	CoCl₂	129.84	blue, hex, hygr	3.356	740	1053	45⁷°; 54 al; 8 ac
	CoCl₂·2H₂O	165.87	red-vlt, mn or tric, 1.625, 1.671, 1.67	2.477	....	....	s
	CoCl₂·6H₂O	237.93	red, mn	1.924	86; −6H₂O, 110	....	76⁰°
chloroplatinate(IV)	CoPtCl₆·6H₂O	574.83	trig,	2.699	d	....	....

Table 4-1 (Continued)
PHYSICAL CONSTANTS OF INORGANIC COMPOUNDS

Name	Formula	Formula Weight	Color, Crystalline Form, and Refractive Index	Density	Melting Point, °C	Boiling Point, °C	Solubility in 100 Parts
Cobalt							
chlorostannate(IV)	$CoSnCl_6 \cdot 6H_2O$	498.43	rh or trig	......	d 100	......	......
chromate	$CoCrO_4$	174.93	gray-blk cr	......	d	......	i; s a
citrate	$Co_3(C_6H_5O_7)_2 \cdot 2H_2O$	591.04	rose-red	......	$-2H_2O$, 150	......	0.8
cyanide	$Co(CN)_2$	110.99	buff	1.872	d 300	d 300	i; s KCN
	$Co(CN)_2 \cdot 2H_2O$	147.00	bl-vlt pwd	......	$-2H_2O$, 280	......	i; s KCN
ferricyanide	$Co_3[Fe(CN)_6]_2$	600.71	red nd	......	......	......	i; s NH_4OH
(II) fluoride	CoF_2	96.93	pink, tetr	4.46	1127	1400	1.5
	$CoF_2 \cdot 4H_2O$	168.99	red, rh	2.192	d 200	......	s
(III) fluoride	CoF_3	115.93	brn, rh-hed, hygr	3.88	......	......	i d
fluorosilicate	$CoSiF_6 \cdot 6H_2O$	309.10	pink, trig, 1.382, 1.387	2.113^{19}	......	......	$118^{21°}$
formate	$Co(CHO_2)_2 \cdot 2H_2O$	185.00	red cr	$2.129^{22°}$	$-2H_2O$, 140	d 175	$5.03^{20°}$
(II) hydroxide	$Co(OH)_2$	92.95	rose-red, rh	3.597^{15}	d	......	i; s a
(III) hydroxide	$Co(OH)_3$	109.96	blk-brn	4.46	$-H_2O$, 100	d	i; s a
iodate	$Co(IO_3)_2$	408.74	bl-vlt nd	4.931	d 200	......	vsl s
	$Co(IO_3)_2 \cdot 6H_2O$	516.83	red, oct	$3.689^{21°}$	d 61	......	s
iodide	CoI_2	312.74	blk, hex, hygr	5.68	505 d	570 vac	v s
	$CoI_2 \cdot 6H_2O$	420.83	red-brn, hex, hygr	2.90	$-6H_2O$, 130	......	s
nitrate	$Co(NO_3)_2 \cdot 6H_2O$	291.04	red, mn, 1.52	1.87	55 d	......	v s
nitrosyl tricarbonyl	$Co(NO)(CO)_3$	172.97	cherry-red liq	......	-1.05	48.6	i; s al, eth, bz
oxalate	CoC_2O_4	146.95	wh or pinksh	3.021	d 250	......	vsl s
	$CoC_2O_4 \cdot 2H_2O$	182.98	pink cr	......	$-H_2O$, 190	......	i; s a
(II) oxide	CoO	74.93	grn-brn, fcc	6.45	1935	......	i; s a
(III) oxide	Co_2O_3	165.86	blk-gray, hex or rh	5.18	d 895	......	i; s a
(II,III) oxide	Co_3O_4	240.80	blk, cub	6.07	tr to CoO, 900–950	......	i; v sl s a
perchlorate	$Co(ClO_4)_2$	257.83	red nd, 1.510, 1.490	3.327	d 182	......	$100^{0°}$
	$Co(ClO_4)_2 \cdot 6H_2O$	365.93	red, oct, deliq, 1.51	......	......	......	v s

Name	Formula	Formula wt	Color, crystalline form, refractive index	Density	Melting point, °C	Boiling point, °C	Solubility
*ortho*phosphate	$Co_3(PO_4)_2$	366.74	redsh cr	2.587	......	......	i; s H_3PO_4, NH_4OH
phosphate, mono-	$Co_3(PO_4)_2 \cdot 8H_2O$	510.87	redsh pwd	2.769	$-8H_2O$, 200	......	sl s; s min a
phosphide, mono-	CoP	89.91	gray, o-rh	6.24	......	......	i
phosphide, tri-	CoP_3	151.85	gray	4.26	......	......	i
(di-) phosphide	Co_2P	148.84	gray nd, o-rh	6.4^{15}	1386	......	i; s HNO_3
selenate	$Co(SeO_4)_2 \cdot 5H_2O$	291.97	ruby-red, tric	2.512	d	......	v s
	$Co(SeO_4)_2 \cdot 6H_2O$	309.98	red, mn, 1.5225	2.25^{17}	......	......	s
	$Co(SeO_4)_2 \cdot 7H_2O$	328.00	red, mn	2.135	......	......	
selenide	$CoSe$	137.89	yel, hex	7.65	red heat	......	i; s HNO_3
*ortho*silicate	Co_2SiO_4	209.95	vlt, rh	4.63	1345	......	i; s HCl
silicide, mono-	$CoSi$	87.03	rh	....	1395	......	s HCl
silicide, di-	$CoSi_2$	115.11	rh	5.3	1277	......	d a
(di-) silicide	Co_2Si	145.95	gray cr	7.28^0	1327	......	36.2^{20}
*ortho*stannate(IV)	Co_2SnO_4	300.55	grnsh-blue, cub	3.71	d 735	......	d a
sulfate	$CoSO_4 \cdot H_2O$	173.01	red cr, 1.603, 1.639, 1.683	3.075	d	......	s
	$CoSO_4 \cdot 6H_2O$	263.09	red, mn, 1.531, 1.549, 1.552	2.019^{15}	$-2H_2O$, 95	$-7H_2O$, 420	
(bieberite)	$CoSO_4 \cdot 7H_2O$	281.10	pink, mn, 1.477, 1.483, 1.489	1.948	96.8	......	62
sulfide (sycoporite)	CoS	91.00	redsh-silv, oct	5.45^{18}	......	......	i; sl s a
sulfide, sesqui-	Co_2S_3	214.06	blk cr	4.8	......	......	i; d a
sulfide, di-	CoS_2	123.06	blk, cub	4.269	......	......	i; s HNO_3
(tri-) *tetra*sulfide (linneite)	Co_3S_4	305.06	dk gray, cub	4.86	d 480	......	i; s hot conc a
sulfite	$CoSO_3 \cdot 5H_2O$	229.07	red	....	......	......	i; s a
tartrate	$CoC_4H_4O_6$	207.01	redsh, mn	....	......	......	sl s; s a
thiocyanate	$Co(SCN)_2 \cdot 3H_2O$	229.14	vlt, rh	....	$-3H_2O$, 105	......	s; s al, eth, MeOH
titanate	Co_2TiO_4	229.76	grnsh-blk, cub	5.1	......	......	i; s conc HCl
tungstate	$CoWO_4$	306.78	bl-grn, mn	8.42	......	......	i; s hot conc a
Cobalt complexes							
hexamminecobalt(II) chloride	$[Co(NH_3)_6]Cl_2$	232.02	rose, oct	1.497	d	d	d; s NH_4OH
hexamminecobalt(III) chloride	$[Co(NH_3)_6]Cl_3$	267.46	red, mn, 1.710		$-NH_3$, 215	......	5.9^{10}

Table 4-1 (*Continued*)
PHYSICAL CONSTANTS OF INORGANIC COMPOUNDS

Name	Formula	Formula Weight	Color, Crystalline Form, and Refractive Index	Density	Melting Point, °C	Boiling Point, °C	Solubility in 100 Parts
Cobalt complexes							
hexamminecobalt(III) nitrate	[Co(NH₃)₆](NO₃)₃	347.13	yel, tetr	1.804	....	....	1.7
hexamminecobalt(II) sulfate	[Co(NH₃)₆]SO₄	257.18	pink pwd	1.654	d 116	....	d; v s dil NH₄OH
hexamminecobalt(III) sulfate	[Co(NH₃)₆]₂(SO₄)₃·5H₂O	700.50	dk yel, mn	1.797 anhyd	−5H₂O, 150	....	$1.4^{17°}$
Copper	Cu	63.55	redsh met, fcc	8.92	1083	2582	i; s HNO₃, hot H₂SO₄
acetate (neutral verdigris)	Cu(C₂H₃O₂)₂·H₂O	199.65	dk grn pwd, 1.545, 1.550	1.882	115	d 240	7.2; 7.1 al
acetate meta-arsenate (Paris green)	Cu(C₂H₃O₂)₂·3Cu(AsO₂)₂	1013.77	emerald grn pwd	....	....	....	i; s a, NH₄OH
acetylide	Cu₂C₂	151.10	red, amorp, expl	....	expl	....	v sl s; s a, KCN
ammine azide	Cu(NH₃)₂(N₃)₂	181.64	dk grn cr, expl	....	d 100	expl 200	i; d a
ammine nitrate	[Cu(NH₃)₄](NO₃)₂	255.67	dk bl, oct	1.91	expl 210	....	s
ammine sulfate	[Cu(NH₃)₄]SO₄·H₂O	245.74	dk bl, rh	1.79	d	....	$18^{21°}$
antimonide	Cu₃Sb	312.37	gray	8.51	687	....	....
*ortho*arsenate	Cu₃(AsO₄)₂·4H₂O	540.52	blsh-grn	....	....	....	i; s a, NH₄OH
*ortho*arsenate, di-H	Cu₅H₂(AsO₄)₄·2H₂O	911.42	bl	....	....	....	i; s a, NH₄OH
arsenide	Cu₅As₂	467.54	bl, oct	7.56	....	....	i; s a
(tri-)arsenide (domeykite)	Cu₃As	265.54	hex	8.0	830	....	i; s a
(I) azide	CuN₃	105.56	col cr, v expl	3.26	....	....	i; s NH₄Cl
borate	Cu(BO₂)₂	149.16	blsh-grn pwd	3.859	....	....	s
boride	Cu₃B₂	212.24	yel	8.116	....	....	....
bromate	Cu(BrO₃)₂·6H₂O	427.45	bl-grn, cub	2.583	d 180	−6H₂O, 200	v s
(I) bromide	CuBr	143.45	wh, cub, 2.116	4.98	492	1345	v sl s; s a
(II) bromide	CuBr₂	223.31	blk, mn, deliq	4.77	498	....	v s; s al, acet, pyr

Name	Formula	Mol. wt.	Color, crystalline form, refractive index	Density	Melting point, °C	Boiling point, °C	Solubility
(I) carbonate	Cu_2CO_3	187.09	yel	4.40	d	.....	i; s a
(II) carbonate, basic (malachite)	$CuCO_3 \cdot Cu(OH)_2$	221.11	dk grn, mn, 1.655, 1.875, 1.909	4.0	d 200	.....	i; s a
(II) carbonate, basic (azurite, chessylite)	$2CuCO_3 \cdot Cu(OH)_2$	344.65	blue, mn, 1.730, 1.758, 1.838	3.88	d 220	.....	i; s NH_4OH, h Na_2CO_3
chlorate	$Cu(ClO_3)_2 \cdot 6H_2O$	338.53	grn, cub, deliq	.....	65	d 100	v s; s al, acet
(I) chloride (nantokite)	$CuCl$	98.99	wh, cub, 1.93	4.14	430	1490	0.006; s HCl, NH_4OH
(II) chloride (eriochalcite)	$CuCl_2$	134.44	brn-yel pwd, hygr	3.386	620	993 d to CuCl	70°
(II) chloride, basic	$CuCl_2 \cdot Cu(OH)_2$	232.00	yel-grn, hex	3.78	$-H_2O$, 250	d red heat	.....
(II) chloride	$CuCl_2 \cdot 2H_2O$	170.47	bl-grn, rh, deliq, 1.644, 1.683, 1.732	2.54	$-2H_2O$, 100	d	110°
(I) chromite	$Cu_2Cr_2O_4$	295.07	gray-blk, cub	5.24^{20}	.....	.....	i; s HNO_3
(II) citrate	$2Cu_2C_6H_4O_7 \cdot 5H_2O$	720.43	blsh-grn pwd	.....	$-H_2O$, 100	.....	i; s a
(I) cyanide	$CuCN$	89.56	wh, mn	2.92	473 (in N_2)	d	i; s HCl, KCN, NH_4OH
(II) cyanide	$Cu(CN)_2$	115.58	yel-grn pwd	.....	d	.....	i; s a, alk, KCN
(II) dichromate	$CuCr_2O_7 \cdot 2H_2O$	315.56	blk cr, deliq	2.283	$-2H_2O$, 100	d	v s
(II) ferricyanide	$Cu_2Fe(CN)_6$	402.57	brn-red	.....	.....	.....	i; s NH_4OH
(II) ferricyanide	$Cu_3[Fe(CN)_6]_2 \cdot 14H_2O$	866.74	yel-grn	.....	.....	.....	i; s NH_4OH
(I) fluoride	CuF	82.54	red cr	.....	908	.....	i; s HCl, HF
(II) fluoride	CuF_2	101.54	wh, mn, hygr	4.23	785	subl 1100	$4.7^{20°}$; s a
(II) fluoride	$CuF_2 \cdot 2H_2O$	137.57	blue, mn	2.93	d	.....	4.7
(I) fluorosilicate	Cu_2SiF_6	269.16	red pwd	.....	d	d to SiF_4	$d^{100°}$
(II) fluorosilicate	$CuSiF_6 \cdot 4H_2O$	277.68	mn	2.158	.....	.....	42.8
(II) fluorosilicate	$CuSiF_6 \cdot 6H_2O$	313.71	bl, rh, deliq, 1.409, 1.408	2.207	.....	.....	$232^{17°}$
formate	$Cu(CHO_2)_2$	153.55	bl, mn	1.831	.....	.....	12.5
	$Cu(CHO_2)_2 \cdot 4H_2O$	225.61	bl cr	1.81	$-H_2O$, 130	.....	6.2
hydride	CuH	64.55	red-brn	6.39	d	.....	i; d HCl
hydroxide	$Cu(OH)_2$	97.56	bl-grn gel pwd	3.368	d	.....	i; s a
iodate	$Cu(IO_3)_2$	413.35	grn, mn	5.241^{15}	d	.....	$0.14^{15°}$; s dil a

Table 4-1 (*Continued*)
PHYSICAL CONSTANTS OF INORGANIC COMPOUNDS

Name	Formula	Formula Weight	Color, Crystalline Form, and Refractive Index	Density	Melting Point, °C	Boiling Point, °C	Solubility in 100 Parts
Copper							
(I) iodide (marshite)	CuI	190.44	beige, cub, 2.346	5.62	605	1290	i; s dil HCl, KI, KCN
lactate	Cu(C$_3$H$_5$O$_3$)$_2$·2H$_2$O	277.71	dk bl, mn	.....	.....	.....	16.7
nitrate	Cu(NO$_3$)$_2$·3H$_2$O	241.60	bl cr, deliq	2.32	114.5	d 170	138°
nitrate	Cu(NO$_3$)$_2$·6H$_2$O	295.64	bl cr, deliq	2.074	.....	.....	v s
nitride	Cu$_3$N	204.63	dk grn pwd	5.84	d 300	.....	d a
(II) oxalate	CuC$_2$O$_4$·$^1/_2$H$_2$O	160.57	blsh-wh	.....	.....	.....	0.002; s NH$_4$OH
(I) oxide (cuprite)	Cu$_2$O	143.08	red, cub, 2.705	6.0	1235	$-$O$_2$, 1800	i; s HCl
(II) oxide (tenorite)	CuO	79.54	blk, mn, β 2.63	6.3–6.5	1325	.....	i; s a
perchlorate	Cu(ClO$_4$)$_2$	262.43	mn, 1.495, 1.505, 1.522	2.225$^{23°}$	82.3	.....	s
	Cu(ClO$_4$)$_2$·6H$_2$O	370.53	lt bl, tric, deliq, 1.505	2.225	82	d 120	v s
paraperiodate	Cu$_5$HIO$_6$	350.00	grn cr pwd	.....	d 110	.....	i; s HNO$_3$
peroxide	CuO$_2$·H$_2$O	113.55	brnsh-blk cr	.....	d 60	.....	i; s dil a
(II) *orthophosphate*	Cu$_3$(PO$_4$)$_2$·3H$_2$O	434.61	bl, rh	.....	d	.....	i; s a
phosphide, di-	CuP$_2$	135.51	gray-blk	4.20	.....	.....	i
(di-) phosphide	Cu$_2$P	158.07	gray-met	6.4	.....	.....	i
(tri-) phosphide	Cu$_3$P	221.59	gray-blk, hex	7.15	1023	.....	i; s HNO$_3$
(II) pyridine chloride	Cu(py)$_2$Cl$_2$	316.67	blsh-grn, mn, 1.60, 1.75	1.76	d 263	.....	s
(II) salicylate	Cu(C$_7$H$_5$O$_3$)$_2$·4H$_2$O	409.83	bl-grn nd	.....	.....	.....	v s
(II) selenate	CuSeO$_4$·5H$_2$O	296.57	bl, tric, 1.56	2.559	$-$5H$_2$O, 150	.....	25.7$^{15°}$
(I) selenide	Cu$_2$Se	206.04	blk, cub	6.749$^{30°}$	1113	.....	d HCl
(II) selenide	CuSe	142.50	grn-blk, hex	5.99	d red heat	.....	i; s hot HNO$_3$
selenite	CuSeO$_3$·2H$_2$O	226.53	bl-grn, rh	3.31	$-$H$_2$O, 100	.....	i
silicide	Cu$_4$Si	282.25	wh met	7.53	850	.....	d HNO$_3$
(I) sulfate	Cu$_2$SO$_4$	223.14	gray pwd, 1.724, 1.733, 1.739	.....	.....	.....	d; s conc a

Name	Formula	Mol. wt.	Color, crystalline form, refractive index	Density	M.P., °C	B.P., °C	Solubility
(II) sulfate (chalcanthite)	$CuSO_4 \cdot 5H_2O$	249.68	bl, tric, 1.514, 1.537, 1.543	2.284	$-5H_2O$, 150	....	31.6°
(II) sulfate, basic (brochantite)	$CuSO_4 \cdot 3Cu(OH)_3$	452.27	grn, mn, 1.728,	3.78	d 300	....	i; s a
(I) sulfide (chalcocite)	Cu_2S	159.14	blk, rh	5.6	1100	....	i; s HNO_3
(II) sulfide (covellite)	CuS	95.60	blk mn or hex, 1.45	4.6	d 220	....	i; s HNO_3, hot HCl
(I) sulfite	$Cu_2SO_3 \cdot H_2O$	225.16	wh, hex	$3.83^{15°}$	d	....	sl s; s HCl
(II) tartrate	$CuC_4H_4O_6$	211.61	lt bl pwd			....	v sl s; s a, alk
(I) telluride	Cu_2Te	254.68	bl-blk, oct	7.27		....	i; s Br_2 water
telluride (rickardite)	Cu_4Te_3	636.96	purp, tetr	7.54		....	....
tellurite	$CuTeO_3$	239.10	bl glass			....	i; s conc HCl
(I) thiocyanate	CuSCN	121.62	wh	2.843	1084	....	i; s NH_4OH
(II) thiocyanate	$Cu(SCN)_2$	179.70	blk		d 100	....	d; s a
tungstate	$CuWO_4 \cdot 2H_2O$	347.42	lt-grn, oct		red heat	....	$0.1^{15°}$; s NH_4OH
xanthate	$Cu(C_3H_5OS_2)_2$	305.94	yel		d	....	i; s NH_4OH
Curium	Cm	247	silv, hcp	13.51		....	s a
bromide	$CmBr_3$	486.7	wh, rh	6.87	400	....	s
chloride	$CmCl_3$	353.3	wh, hex	5.81	500	....	s
(III) fluoride	CmF_3	304	wh, hex	9.70	1406	....	i
(IV) fluoride	CmF_4	323	green-tan, mn	7.49		....	s
iodide	CmI_3	627.7	wh, hex	6.37		....	s
(III) oxide	Cm_2O_3	542	wh			....	s a
Cyanic acid, iso	HOCN	43.04	liq	1.14^{20}	d 100	....	s d; s eth, bz
Cyanogen	$(CN)_2$	52.04	col g	2.335 g/l, 0.9577^{bp} liq	-27.9	-20.7	450 cc^{20}; 230 cc al
Deuterium	D_2	4.032	col g	2	-252.89	-248.24	sl s
bromide	DBr	81.92	col g, 1.000569	3.39^{20} g/l	-87.48	-66.6	....
chloride	DCl	37.47	g, 1.000413		-114.66	-84.75	....
fluoride	DF	21.02	col g		-83.6	18.65	....
iodide	DI	128.92	col g		-51.87	-35.7	....
oxide (heavy water)	D_2O	20.03	col liq or hex, 1.33844	1.1045	3.82	101.43	....

Table 4-1 (Continued)
PHYSICAL CONSTANTS OF INORGANIC COMPOUNDS

Name	Formula	Formula Weight	Color, Crystalline Form, and Refractive Index	Density	Melting Point, °C	Boiling Point, °C	Solubility in 100 Parts
Deuterium							
selenide	D_2Se	82.99	col g	....	−66.92	....	....
sulfide	D_2S	36.09	col g	....	−86.02	....	....
Dysprosium	Dy	162.50	met, hcp	8.536	1407	2600	s a
acetate	$Dy(C_2H_3O_2)_3 \cdot 4H_2O$	411.64	yel nd	....	d 120	....	s
bromate	$Dy(BrO_3)_3 \cdot 9H_2O$	708.36	yel nd, hex	....	78	−6H₂O, 110	v s
bromide	$DyBr_3$	402.23	lt yel, rh-hed	4.78	880	1480	s
chloride, di-	$DyCl_2$	233.41	blk, rh	....	721 d	....	
chloride, tri-	$DyCl_3$	268.85	wh, mn	3.67°	647	1530	s
chromate	$Dy_2(CrO_4)_3 \cdot 10H_2O$	853.13	yel cr	....	−3.5H₂O, 150	d	1.00
fluoride	DyF_3	219.50	lt grn, hex	7.465	1154	2230	s H₂SO₄
hydroxide	$Dy(OH)_3$	213.52	wh, hex	....	−H₂O, 210	d 310 to Dy₂O₃	s a
iodide	DyI_3	543.21	dk grn, hex	....	978	1320	s
nitrate	$Dy(NO_3)_3 \cdot 5H_2O$	438.58	yel cr	....	88.6	....	s
oxalate	$Dy_2(C_2O_4)_3 \cdot 10H_2O$	769.21	wh pwd	....	−H₂O, 40	....	s dil a
oxide	Dy_2O_3	373.00	wh, bcc	8.15	2340	....	s a
*ortho*phosphate	$DyPO_4 \cdot 5H_2O$	347.55	yel-wh pwd	....	−5H₂O, 200	....	s dil a
selenate	$Dy_2(SeO_4)_3 \cdot 8H_2O$	898.00	yel nd	....	−8H₂O, 200	....	v s
sulfate	$Dy_2(SO_4)_3 \cdot 8H_2O$	757.31	wh	....	−8H₂O, 360	d red heat	5.07
sulfide	Dy_2S_3	421.18	yel or brn	....	1490	....	
Erbium	Er	167.26	gray, hcp	9.051	1495	2900	s a
acetate	$Er(C_2H_3O_2)_3 \cdot 4H_2O$	416.48	wh, tric	2.114	....	....	
boride, hexa-	ErB_6	232.12	bl, cub	4.61	....	....	
bromide	$ErBr_3$	406.97	vit-rose, rh-hed	4.93	923	1460	s

		Formula wt.	Color, crystalline form	Density	Melting point	Boiling point	Solubility
chloride	ErBr$_3$·9H$_2$O	569.15	vl-rose				s
	ErCl$_3$	273.62	vlt-rose, mn		774	1500	s
	ErCl$_3$·6H$_2$O	381.73	pink cr, deliq				s
fluoride	ErF$_3$	224.28	pink, hex	7.814	1140	2230	s H$_2$SO$_4$
hydroxide	Er(OH)$_3$	218.28	rose, hex		−H$_2$O, 200	d 315 to Er$_2$O$_3$	s a
iodide	ErI$_3$	547.99	vlt-red, hex		1015	1280	s
nitrate	Er(NO$_3$)$_3$·5H$_2$O	443.37	red cr		−4H$_2$O, 130		s
oxalate	Er$_2$(C$_2$O$_4$)$_3$·10H$_2$O	778.77	redsh pwd		d 575		i
selenide	Er$_2$Se$_3$	571.40	brn	6.96			
sulfate	Er$_2$(SO$_4$)$_3$	622.74	wh pwd, hygr	3.678	d 630		43°
	Er$_2$(SO$_4$)$_3$·8H$_2$O	766.87	rose-red, mn	3.217	−8H$_2$O, 110		16.0
sulfide	Er$_2$S$_3$	430.70	bright ochre		1730		
Europium	Eu	151.96	gray, bcc	5.259	826	1440	s a
(II) bromide	EuBr$_2$	311.78	wh		683	1880	s
(III) bromide	EuBr$_3$	391.69	rose, rh	5.40	702	d	s
(II) chloride	EuCl$_2$	222.87	wh, rh		731	2027	s
(III) chloride	EuCl$_3$	258.32	yel nd, hex	4.89	623 d	d	i
(II) fluoride	EuF$_2$	189.96	dk yel, cub	6.495	1416	2427	s H$_2$SO$_4$
(III) fluoride	EuF$_3$	208.96	wh, hex	6.793	1276	2277	
hydride, di-	EuH$_2$	153.98					
(III) hydroxide	Eu(OH)$_3$	202.96					s a
(II) iodide	EuI$_2$	405.77	ol grn, mn	5.50	580	1577	s
(III) iodide	EuI$_3$	532.68			d		s
(III) nitrate	Eu(NO$_3$)$_3$·6H$_2$O	446.07	col		85 (sealed tube)		s a
(III) oxide	Eu$_2$O$_3$	351.92	lt pink, mn	7.27	2050		s a
(tri-)tetroxide bromide	Eu$_3$O$_4$Br	599.78	rhb	6.80			
(II) sulfate	EuSO$_4$	248.02	col, o-rh	4.989		d 1600	i
(III) sulfate	Eu$_2$(SO$_4$)$_3$·8H$_2$O	736.23	lt pink		−8H$_2$O, 375		2.56
(II) sulfide	EuS	184.02	blk, cub	(5.75)	ign		
Ferricyanic acid	H$_3$Fe(CN)$_6$	214.98	grn-brn nd, deliq		d		s
Ferrocyanic acid	H$_4$Fe(CN)$_6$	215.99	wh nd		d		s

Table 4-1 (*Continued*)
PHYSICAL CONSTANTS OF INORGANIC COMPOUNDS

Name	Formula	Formula Weight	Color, Crystalline Form, and Refractive Index	Density	Melting Point, °C	Boiling Point, °C	Solubility in 100 Parts
Fluorine	F_2	38.00	yel-grn g, 1.000190	1.580 g/l	−219.62	−188.14	d to HF + O_3
amide	FNH_2	25.00	......	......	......	−77 subl	......
fluorosulfate	$F(SO_3F)$	118.06	col g	$1.78^{-74°}$ liq	−158.5	−31.3	......
nitride	FN_3	60.92	grn-yel g	......	−154	−82	d
perchlorate	$F(ClO_4)$	102.45	col g, expl	4.93 g/l	−167.3	−15.9	......
(Di-)Fluoroamine	HNF_2	53.01	unstable g	......	−116	−23.6	......
Fluoroboric acid	HBF_4	87.81	col liq	1.818	d 130	......	v s
Fluorophosphonic acid	H_2PO_3F	99.99	col oily liq	......	−80	......	v s
(Di-)Fluorophosphonic acid	$H_2PO_2F_2$	102.99	col fum liq	1.583	−96.5	115.9	s
HexaFluorophosphoric acid	HPF_6	145.97	col fum liq	ca 1.65 (65%)	......	......	s
Fluorosilicic acid	$H_2SiF_6 \cdot xH_2O$	......	col fum liq, 1.3465 (61% soln)	1.463 (61%)	d	......	s
	$H_2SiF_6 \cdot 2H_2O$	180.12	wh cr, fum, deliq	......	d	......	s
Fluorosulfonic acid	HSO_3F	100.07	col liq	$1.743^{15°}$	−87.3	165.5	s a
Gadolinium	Gd	157.25	met, hcp	7.895	1312	3000	s a
acetate	$Gd(C_2H_3O_2)_3 \cdot 4H_2O$	406.45	col, tric	1.611	......	......	11.6
boride, hexa-	GdB_6	222.11	bl, cub	4.65	......	......	......
bromide	$GdBr_3$	396.96	col, rh-hed	4.57	770	1490	s
	$GdBr_3 \cdot 6H_2O$	505.07	col, rh	$2.844^{16°}$	......	......	s
chloride	$GdCl_3$	263.61	wh, hex, hygr	4.52	602	1580	s
	$GdCl_3 \cdot 6H_2O$	371.70	wh pr, deliq	$2.424^{0°}$	......	......	s
fluoride	GdF_3	214.25	wh, hex	7.047	1231	2277	s H_2SO_4
hydride, di-	GdH_2	159.27	wh, hex	7.08	......	......	s a
hydroxide	$Gd(OH)_3$	308.27	wh, hex	......	−H_2O, 210	d 310 to Gd_2O_3	......
iodide, di-	GdI_2	411.06	bronze [$Gd^{3+}(e^-)(I^-)_2$]	......	831	......	......

iodide, tri-	GdI₃	537.96	yel, hex	....	925	1340	s
nitrate	Gd(NO₃)₃·6H₂O	451.36	col, tric, deliq	2.322	91	....	v s
oxalate	Gd₂(C₂O₄)₃·10H₂O	758.71	col, mn	....	-6H₂O, 110	....	s HNO₃
oxide	Gd₂O₃	362.50	wh, mn, hygr	7.64	2330	....	s a
selenate	Gd₂(SeO₄)₃·8H₂O	887.50	pearly, mn	3.309	-8H₂O, 130	....	s
sulfate	Gd₂(SO₄)₃	602.68	col	4.139¹⁵°	d 500	d 600	3.98°
	Gd₂(SO₄)₃·8H₂O	746.81	col, mn	3.010¹⁵°	....	....	2.89
sulfide	Gd₂S₃	410.69	yel-brn, hygr	3.8	1885	1980	s a
Gallium	Ga	69.72	gray-blk, o-rh	5.907 sol; 6.0948 liq	29.75	....	s a
antimonide	GaSb	191.47	cub	3.9	712	....	s HCl
arsenide	GaAs	144.64	dk-gray, cub	3.4	1238	....	s HCl
(I) tetrabromogallate (III)	Ga(GaBr₄)	459.06	col	3.46	166.7	....	hyd
(III) bromide	GaBr₃	309.45	wh	3.69	124	280	v s
(III) bromide, ammine	GaBr₃·NH₃	326.48	wh	3.112	124	....	d
(III) chloride	GaCl₃	176.03	wh nd, hygr	2.47	77.75	201.2	d; s bz, CCl₄, CS₂
(I) tetrachlorogallate (III)	Ga(GaCl₄)	281.25	lt gray, o-rh, hygr	2.417	170.5	d 200	d; 4.6 bz
ethyl, tri-	Ga(C₂H₅)₃	146.90	....	1.058³⁰°	-82.3	142.8	i conc HCl
ferrocyanide	Ga₄[Fe(CN)₆]₃	914.74	....	....	d	....	0.002; s HF
fluoride	GaF₃	126.72	wh, rh-hed	4.47	subl 950	~1000	s coord solv
hydride (digallane)	Ga₂H₆	145.49	volat liq	....	-21.4	139 d	....
hydroxide	Ga(OH)₃	120.74	....	....	-H₂O, 300	....	s a, alk
(III) iodide	GaI₃	450.43	col nd	4.15	212	346	v s
(I) tetraiodogallate (III)	Ga(GaI₄)	647.06	....	....	211	d 250	....
methyl, tri-	Ga(CH₃)₃	114.84	....	1.151¹⁵°	-15.7	55.8	....
nitrate	Ga(NO₃)₃·xH₂O	....	wh cr, deliq	....	d 110	d 200 to Ga₂O₃	v s
nitride	GaN	83.73	dk-gray pwd	6.1	subl 800	....	s h conc H₂SO₄ or alk
oxalate	Ga₂(C₂O₄)₃·4H₂O	475.56	wh cr, hygr	....	-4H₂O, 180	d 200	0.4
(I) oxide	Ga₂O	155.44	blk-brn pwd	4.77	>660	subl >500	s a, alk
(III) oxide	Ga₂O₃	187.44	α: wh, rh; tr to β 600; β: rh or mn	6.44; 5.88	1900	....	s alk

Table 4-1 (Continued)
PHYSICAL CONSTANTS OF INORGANIC COMPOUNDS

Name	Formula	Formula Weight	Color, Crystalline Form, and Refractive Index	Density	Melting Point, °C	Boiling Point, °C	Solubility in 100 Parts
Gallium							
(III) oxide, hydrate	$Ga_2O_3 \cdot H_2O$	205.45	wh cr, o-rh, 1.84	5.2	400, tr to Ga_2O_3	......	......
oxide chloride	GaOCl	121.17	wh nd	......	......	......	i
perchlorate	$Ga(ClO_4)_3 \cdot 6H_2O$	476.16	......	......	d 175	......	v s
phosphide	GaP	100.69	or trans cr, cub	......	1350	......	57.5
selenate	$Ga_2(SeO_4)_3 \cdot 16H_2O$	856.56	col, mn or tric	......	......	......	......
(I) selenide	Ga_2Se	218.40	bl	5.02	......	......	......
(II) selenide	GaSe	148.68	dk-red-brn leaf	5.03	960	d	......
(III) selenide	Ga_2Se_3	376.32	redsh-bl, brittle	4.92	>1020	......	......
sulfate	$Ga_2(SO_4)_3$	427.62	wh pwd	......	d 690	......	v s
	$Ga_2(SO_4)_3 \cdot 18H_2O$	751.90	oct cr	......	......	......	v s
(I) sulfide	Ga_2S	171.50	dk gray-grn, hex	4.18	d 800 vac	......	s a, alk
(II) sulfide	GaS	101.78	yel cr	3.80	970	d	......
(III) sulfide	Ga_2S_3	235.63	yel cr	3.65	d >950	......	hyd; s a, alk
(tetra-) *pentasulfide*	Ga_2S_5	439.18	......	3.82	d >1200	......	......
telluride	GaTe	197.32	blk soft cr	5.44	824	......	......
telluride, sesqui-	Ga_2Te_3	522.24	blk brittle cr	5.57	790	......	......
Germane	GeH_4	76.62	col g	$1.523^{-142°}$	-165	-88.5	sl s h HCl
bromo-	GeH_3Br	155.52	col liq	$2.34^{30°}$	-32	52	d
*di*bromo-	GeH_2Br_2	234.42	col liq	$2.80^{0°}$	-15	89	d
*tri*bromo-	$GeHBr_3$	313.36	col liq	......	-24.5	d	d
chloro-	GeH_3Cl	111.07	col liq	$1.75^{-25°}$	-52	28	d
*di*chloro-	GeH_2Cl_2	145.51	col liq	$1.90^{-68°}$	-68.0	69.5	d
*tri*chloro-	$GeHCl_3$	179.96	col liq	1.93	-71	75.2	hyd; s eth
chlorotrifluoro-	$GeClF_3$	165.04	col g	......	-66.2	-20.6	d; s abs al
*di*chloro*di*fluoro-	$GeCl_2F_2$	181.49	col g	......	-51.8	-2.8	d; s abs al
*tri*chlorofluoro-	$GeCl_3F$	197.95	col liq	......	-49.8	37.5	d; s abs al
cyano-	GeH_3CN	101.63	......	......	45-47	......	......

Name	Formula	Mol. wt.	Color, crystalline form, index of refraction	Density	Melting point, °C	Boiling point, °C	Solubility
silyl-	GeH₃SiH₃	106.72	.....	.....	.....	7.0	.....
Germane, di-	Ge₂H₆	151.23	col liq	$1.98^{-109°}$	−119.7	29	i; s CCl_4
Germane, tri-	Ge₃H₈	225.83	col liq	$2.20^{-105.6°}$	−109	110.5	.....
Germane, tetra-	Ge₄H₁₀	300.44	col liq	.....	105.6	176.9	.....
Germane, penta-	Ge₅H₁₂	374.04	.....	.....	.....	234	.....
Germanium	Ge	72.59	gray, cub	5.323	937	2830	s h H_2SO_4, aq reg
(II) bromide	GeBr₂	232.41	col nd	.....	122	d	s a
(IV) bromide	GeBr₄	392.23	col, oct, 1.6269	3.132	26.0	186	hyd; s eth, bz
(II) chloride	GeCl₂	143.50	wh cr	.....	d	.....	hyd
(IV) chloride	GeCl₄	214.40	col fum liq	1.88	−49.5	83.1	hyd; s dil HCl, al, eth
(di-) *hexachloride*	Ge₂Cl₆	358.30	col cr	.....	40–42	subl 25	hyd
(II) fluoride	GeF₂	110.59	wh cr, hygr	.....	110	d <160	hyd; s dil HCl
(IV) fluoride	GeF₄	148.58	col g, garlic odor	$2.46^{-37°}$	−36.6	.....	s
fluoride	GeF₄·3H₂O	202.63	col cr, hygr	.....	d	.....	.....
hydride, see under Germane							
(II) hydroxide	Ge(OH)₂	106.61	yel, easily oxidized	.....	−H₂O, 650	.....	s $HClO_4$
(II) iodide	GeI₂	326.40	yel pl, hex	5.37	subl 240 vac	.....	hyd; s conc HI
(IV) iodide	GeI₄	580.21	or, cub, deliq	4.322	146	subl $110^{0.7mm}$	sl d; s bz, MeOH, CS₂
(tri-) *dinitride*	Ge₃N₂	245.78	blk	.....	subl 650	.....	i h a; i h alk
(tri-) *tetranitride*	Ge₃N₄	273.80	lt brn inert pwd	5.25	d >1000	.....	i; s Cl₂ water
(II) oxide	GeO	88.59	blk nd, 1.607	.....	oxid >550	subl >700 (inert atm)	i; s Cl₂ water
(IV) oxide (soluble)	GeO₂	104.59	col, hex, 1.650	4.228	1115	1200	0.4
(IV) oxide (insoluble)	GeO₂	104.59	col, tetr	6.239	1086	.....	i
oxidedichloride	GeOCl₂	159.50	col liq	.....	−56	d >20	d; i
(II) phosphite, hydrogen	GeHPO₃	152.57	wh cr	.....	d >230	.....	s HX acids
(II) selenide	GeSe	151.55	brn-blk, tetr	5.30	667	.....	.....
(IV) selenide	GeSe₂	230.51	yel, rh	4.56	707	d	sl s alk
(IV) sulfate	Ge(SO₄)₂	264.71	wh	3.92	d >200	.....	.....

Table 4-1 (*Continued*)
PHYSICAL CONSTANTS OF INORGANIC COMPOUNDS

Name	Formula	Formula Weight	Color, Crystalline Form, and Refractive Index	Density	Melting Point, °C	Boiling Point, °C	Solubility in 100 Parts
Germanium							
(II) sulfide	GeS	104.65	gray-blk, cub	3.54	530 in H_2	d >350 to GeO_2	s a, alk, polysulf
(IV) sulfide	GeS_2	136.72	wh	.....	~800	~850	0.45; s NaOH
Gold	Au	196.97	yel met, fcc	19.3	1063	2707	saq reg, KCN, h H_2SO_4
(I) bromide	AuBr	276.88	yel-gray pwd	7.90	d 115	.....	i; d a: s KCN
(III) bromide	$AuBr_3$	436.69	gray pwd or brn cr	.....	97.5	$-Br$, 160	s eth, al
(I) chloride	AuCl	232.42	yel, o-rh	7.4	d 170 to $AuCl_3$	.....	s HCl, HBr
(III) chloride	$AuCl_3$	303.33	red, mn	3.9	d 420	subl 265	68
(I) cyanide	AuCN	222.98	lt yel pwd	7.12	d	.....	s KCN, NH_4OH
(III) cyanide	$Au(CN)_3 \cdot 3H_2O$	329.07	col, hygr	.....	d 50	.....	v s
(III) fluoride	AuF_3	253.96	or-yel, hex	.....	subl 300 vac	d 500	.....
(I) iodide	AuI	323.87	grnsh-yel, tetr	8.25	d 120	.....	s KI
(III) iodide	AuI_3	577.68	dk grn	.....	.....	.....	s KI
(III) nitrate, hydrogen	$H[Au(NO_3)_4] \cdot 3H_2O$	500.04	yel, tric	2.84	d 72	.....	s d; s HNO_3
(III) oxide	Au_2O_3	441.93	.....	8.12	$-O$, 160	$-3O$, 250	i; s HCl, HNO_3
phosphide	Au_2P_3	486.86	gray pwd	4.65^{22}	d	.....	i a
selenide	Au_2Se_3	630.81	.....	.....	d 240	.....	s aq reg, KCN
(I) sulfide	Au_2S	426.00	brn-blk pwd	8.754	d 197	.....	i; s Na_2S
(III) sulfide	Au_2S_3	490.13	brn-blk pwd	8.2-9.3	d 472	.....	i
telluride (krennerite)	$AuTe_2$	452.16	yel: rh, mn, tric	.....	.....	.....	s HF
Hafnium	Hf	178.49	met, hcp	13.31	2225 ± 25	~5200	.....
boride, di-	HfB_2	200.11	.....	.....	3335	.....	.....
boronate	$Hf(BH_4)_4$	237.85	.....	.....	29.0	118	s HF
(III) bromide	$HfBr_3$	418.20	bl-blk	.....	d 350	.....	hyd
(IV) bromide	$HfBr_4$	498.13	wh, cub	.....	425	subl 322	v s

Name	Formula	Formula wt.	Color, crystalline form, index of refraction	Density	Melting point	Boiling point	Solubility
carbide	HfC	190.54	....	12.20	4160	....	s MeOH, acet
chloride	HfCl$_4$	320.30	wh	....	434	subl 319	....
deuteride	HfD$_2$	182.52	tetr	11.68	....	....	....
fluoride	HfF$_4$	254.48	wh, mn, 1.56	....	....	....	....
hydride	HfH$_2$	180.51	tetr	11.37	....	....	....
iodide	HfI$_4$	686.11	yel-or, cub	....	449	400 subl vac	....
nitride	HfN	192.50	yel-brn, cub	....	3307	....	i
(IV) oxide	HfO$_2$	210.49	wh, cub	9.68	2812	(5400)	s
oxide dichloride	HfOCl$_2\cdot$8H$_2$O	409.52	col	....	....	....	s
Helium	He	4.003	col g, 1.000035	0.17847° g/l, 0.129 liq	-272.2^{25atm}	-268.935	0.861$^{20°}$ cc
Holmium	Ho	164.93	met, hcp	8.803	1461	2600	s a
bromide	HoBr$_3$	404.66	lt yel, rh-hed	4.86	919	1470	s
chloride	HoCl$_3$	271.29	lt yel, mn	....	718	1510	s H$_2$SO$_4$
fluoride	HoF$_3$	221.93	pink, hex	7.644	1143	2230	s
iodide	HoI$_3$	545.64	lt yel, hex	....	994	1300	s
oxalate	Ho$_2$(C$_2$O$_4$)$_3\cdot$10H$_2$O	774.10	lt tan	....	$-$H$_2$O, 40	d	s a
oxide	Ho$_2$O$_3$	377.86	lt yel, bcc	....	....	....	s a
sulfate	Ho$_2$(SO$_4$)$_3\cdot$8H$_2$O	762.23	lt yel cr	....	....	....	8.18
Hydrazine	H$_2$NNH$_2$	32.04	col, hygr, 1.4710	1.00833$^{20°}$	1.53	113.5	v s
hydrate	H$_2$NNH$_2\cdot$H$_2$O	50.16	col, 1.428	1.038	-51.7	119.4	v s
Hydrazinium							
(1+)azide	N$_2$H$_5$N$_3$	75.07	wh pr, deliq, 1.53 1.76	....	75.4	....	v s
(1+)bromide	N$_2$H$_5$Br	112.96	....	....	86.5	....	s
(2+)dibromide	N$_2$H$_6$Br$_2$	192.87	....	....	195	....	s
(1+)chloride	N$_2$H$_5$Cl	68.51	wh nd	....	92.6	d 240	v s
(2+)dichloride	N$_2$H$_6$Cl$_2$	104.97	col, oct	1.4226$^{20°}$	198	d 200	v s
(1+)chlorate	N$_2$H$_5$ClO$_3$	116.50	cr	....	80	....	s
(di-)(1+)fluoro-germanate	(N$_2$H$_5$)$_2$GeF$_6$	252.80	mn pr, 1.452, 1.460, 1.76	2.406	....	....	s
(2+)fluorosilicate	N$_2$H$_6$SiF$_6$	176.14	cr	....	d 186	....	v s
(2+)formate	N$_2$H$_6$(CO$_2$H)$_2$	124.10	....	....	128	....	s
(1+)iodide	N$_2$H$_5$I	159.98	col pr	....	127	....	s

Table 4-1 (Continued)
PHYSICAL CONSTANTS OF INORGANIC COMPOUNDS

Name	Formula	Formula Weight	Color, Crystalline Form, and Refractive Index	Density	Melting Point, °C	Boiling Point, °C	Solubility in 100 Parts
Hydrazinium							
(2+)diiodide hydrate	$N_2H_6I_2 \cdot 2H_2O$	323.90	cr	......	65	......	s
(1+)nitrate	$N_2H_5NO_3$	95.06	col dimorph nd	......	α 70.71 β 62.09	subl 140	175[10°]
(1+)dinitrate	$N_2H_6(NO_3)_2$	158.07	col cr	......	104	......	v s
(1+)oxalate	$(N_2H_5)_2(C_2O_4)$	154.14	wh nd	......	148	......	200[35°]
(1+)perchlorate	$N_2H_5ClO_4$	132.51	crys, expl	1.939[15°]	137	d 145	d, s al
	$N_2H_5ClO_4 \cdot {}^{1}\!/_2 H_2O$	141.51			85		d, s al
(1+)phosphate	$N_2H_5H_2PO_4$	130.05	cr, hygr		82		v s
(1+)phosphite	$N_2H_5H_2PO_3$	114.05	......		36		v s
(2+)selenate	$N_2H_6SeO_4$	177.02	col pwd, unstable		expl		v sl s
(1+)sulfate	$(N_2H_5)_2SO_4$	162.18	col cr, hygr		85		202
(2+)sulfate	$N_2H_6SO_4$	130.13	col, rh	1.37	254	d	3.415[20°]
tartrate	$(N_2H_5)_2(C_4H_6O_6)$	182.13	col cr, $[\alpha]_4^{20} + 22.5$		183		6.0°
(1+)trioxosulfate	$N_2H_5SO_3$	113.11	col		217.5		......
Hydrogen	H_2	2.016	col g	0.0899 g/l 0.07099[bp] liq	−259.20	−252.77	1.9 cc
(see also Deuterium)							
azide	HN_3	43.03	col liq, v expl	1.126	−80	37	v s
bromide	HBr	80.92	col g, 1.0005688	3.388[20°] g/l 2.160[bp] liq	−86.86	−66.72	70°
	$HBr \cdot 2H_2O$	116.95	wh cr, col liq	2.11[−15°]	−11.3		s
	$HBr \cdot 3H_2O$	134.96	col liq, 60% HBr		d −49.6		s
	$HBr \cdot 4H_2O$	152.98	col liq, 53% HBr		−57.9		s
(const, boiling)	48% HBr + H_2O	......	......	1.49	−11	126	v s

Name	Formula	Mol. wt.	Color, state, etc.	Density	M.p., °C	B.p., °C	Solubility
(const. boiling)	$HBr\cdot 6H_2O$	189.01	col liq, 43% HBr	....	d −88.2	....	s
chloride	HCl	36.46	col g, 1.000415	1.526^{20} g/l; 1.187^{bp} liq	−114.19	−85.03	v s $45°$
	$HCl\cdot H_2O$	54.48	col liq	1.48	−15.35	....	v s
	$HCl\cdot 2H_2O$	72.49	col liq	$1.461^{18°}$	−17.7	d	v s
	$HCl\cdot 3H_2O$	90.51	col liq	....	−24.9	d	v s
(const boiling)	20.24% HCl + H_2O	....		....	....	110	v s
cyanide	HCN	27.03	col, pois, liq or g, 1.2675 (liq)$^{10°}$	0.901 g/l; $0.699^{22°}$ liq	−14	26	v s
deuteride	HD	3.02	col g	....	−256.56	−251.03	....
fluoride	HF	20.01	fum col liq or g, g 1.90	0.957^{bp} liq; $0.922^{0°}$ g/l	−83.37	19.52	v s
(const. boiling)	35.35% HF + H_2O	....	col fum liq	....	....	....	
	$HF\cdot H_2O$	38.03	col fum liq	....	−35.4	120	v s
iodide	HI	127.91	col g, pale yel liq, $1.466^{16°}$ liq, g 1.000853	5.37^{20} g/l; 2.799^{bp} liq	−50.80	−35.6	v s $70°$
(const. boiling)	57% HI + H_2O	....	pale yel liq	1.70^{15}	....	127	v s
	$HI\cdot 2H_2O$	163.94	col liq	....	−43	....	v s
	$HI\cdot 3H_2O$	181.96	col liq	....	−48	....	v s
	$HI\cdot 4H_2O$	199.97	col liq	....	−36.5	....	v s
oxide (water)	H_2O	18.02	col liq, hex cr, liq 1.333, sol 1.309, 1.313	$1.000^{4°}$	0.000	100.000	
peroxide	H_2O_2	34.01	col liq, 1.414; 30% wt soln, 1.4084	1.4649^{0}; 1.4482	−89; −0.43	151.2; 150.2	v s; s al, eth
phosphide (phosphine)	H_3P	34.00	col g or liq, inflam, liq 1.317	1.529 g/l; 0.765^{bp} liq	−133.81	−87.78	26 cc^{17}; s al, eth

Table 4-1 (Continued)
PHYSICAL CONSTANTS OF INORGANIC COMPOUNDS

Name	Formula	Formula Weight	Color, Crystalline Form, and Refractive Index	Density	Melting Point, °C	Boiling Point, °C	Solubility in 100 Parts
Hydrogen							
phosphide, di-(diphosphine)	H_4P_2	65.98	col liq	1.012	-99	51.7	i; s al
selenide	H_2Se	80.98	col g or liq	2.12^{bp} liq	-65.73	-42	$9.5^{20°}$ cc
sulfide	H_2S	34.08	col g, inflam	1.1906 g/l	-85.60	-60.75	$0.102M$; $9.5^{20°}$ cc al, s CS_2
sulfide, di-	H_2S_2	66.14	yel oil, 1.631	$1.334^{20°}$	-89.7	70.7	d; s bz, eth, CS_2
sulfide, tri-(trisulfane)	H_2S_3	98.21	yel oil, 1.729	$1.491^{20°}$	-53	170	s bz, eth, CS_2
sulfide, tetra-(tetrasulfane)	H_2S_4	130.27	yel liq, 1.791	$1.582^{20°}$	-85	240	....
sulfide, penta-(pentasulfane)	H_2S_5	162.34	yel oil, 1.836	$1.644^{20°}$	-50	285	s bz, eth, CS_2
telluride	H_2Te	129.62	col g or yel nd	2.650^{bp} $2.57^{-20°}$ liq	-46	-2	d; s al, alk
Hydroxylamine	NH_2OH	33.03	wh nd or col liq, deliq	1.332	33.1	58^{22mm}	s; s al, MeOH
acetate	$NH_2OH \cdot HC_2H_3O_2$	93.08	col cr	....	87	subl 90	v s
bromide	$NH_2OH \cdot HBr$	113.95	wh, mn	$2.35^{22°}$	....	....	v s
chloride	$NH_2OH \cdot HCl$	69.49	col, mn	$1.680^{20°}$	150.5	d	$83^{17°}$; $4.4^{20°}$ al
iodide	$NH_2OH \cdot HI$	160.94	col nd, hygr	....	83	....	v s
sulfate	$(NH_2OH)_2 \cdot H_2SO_4$	164.14	col, mn	....	d 170	....	$68.^{20°}$ s a
Indium	In	114.82	silv-wh, tetr	7.28	156.4	2050	s a
antimonide	InSb	236.57	cr	....	535	....	....
arsenide	InAs	189.74	met cr	....	943	....	....
(I) bromide	InBr	194.73	red-brn	4.98	220	928	d; s a
(II) bromide	$InBr_2$	274.64	lt yel	4.22	235	subl 632	d; s a

Name	Formula	Mol. wt.	Color, crystalline form	Density	M.P. °C	B.P. °C	Solubility
(III) bromide	$InBr_3$	354.55	wh nd, deliq	4.75	436	subl	v s
(I) chloride	$InCl$	150.27	yel	4.19	......	......	d; s a
			red, deliq	4.18	225	608	d
(II) chloride	$InCl_2$	185.73	wh, rh, deliq	3.655	235	560	v s
(III) chloride	$InCl_3$	221.18	col pl, deliq	4.0	586	subl 600	v s
fluoride	InF_3	171.82	col	4.39	1170	......	0.040
	$InF_3 \cdot 3H_2O$	225.86	col	......	$-3H_2O$, 100	......	8.49
hydride	$[InH_3]_x$	......	wh polymer	......	d 80	......	......
hydroxide	$In(OH)_3$	165.84	wh ppt	......	$-H_2O$, 150	......	s a
iodate	$In(IO_3)_3$	639.53	wh	......	d	......	0.067; s dil min a
(I) iodide	InI	241.72	red-brn	5.31	351	700	s dil a
(II) iodide	InI_2	368.63	lt yel, hygr	4.71	212	......	s a
(III) iodide	InI_3	495.53	col	4.69	210	......	unstable; s a
methyl, tri-	$In(CH_3)_3$	159.93	col pl	1.568	88.4	135.8	d; s acet, bz
nitrate	$In(NO_3)_3 \cdot 3H_2O$	354.88	blk	......	d 100	......	v s
(I) oxide	In_2O	245.64	gray-wh	6.99	subl 565 vac	sl volat 900	s HCl
(II) oxide	InO	130.81		......	......	......	s a
(III) oxide	In_2O_3	277.64	lt yel, amorp, trig	7.179	......	d 200	s a
perchlorate	$In(ClO_4)_3 \cdot 8H_2O$	557.29	col, deliq	......	~800	......	v s
phosphide	InP	145.79	met, brittle	......	1070	......	v sl s min a
selenate	$In_2(SeO_4)_3 \cdot 10H_2O$	838.67	cr, deliq	......	......	......	v s
(III) selenide	In_2Se_3	466.52	blk	5.67	890	......	s
sulfate	$In_2(SO_4)_3$	517.83	lt gray, mn, hygr	3.438	......	......	s
	$In_2(SO_4)_3 \cdot 9H_2O$	679.96	wh pwd, hygr	3.44	d 250	......	v s
(I) sulfide	In_2S	261.70	blk or yel	5.87	653	......	s HCl, HNO_3
(II) sulfide	InS	146.88	wine red	5.18	692	d 850	s a
(III) sulfide	In_2S_3	325.83	yel or red	4.9	1050	......	s a
(II) telluride	$InTe$	242.42	dk met, shiny	6.29	696	......	s HNO_3
(III) telluride	In_2Te_3	612.44	bl brittle cr	5.78	667	......	......
Iodic acid	HIO_3	175.91	wh, rh	$4.629^{0°}$	d 110 to H_5IO_6	d 195 to I_2O_5	$310^{16°}$
Iodine	I_2	253.809	vlt, rh, 3.34	$4.660^{20°}$	113.6	184.4	$0.029^{20°}$; s al, bz, eth, chl, CS_2, CCl_4

Table 4-1 (*Continued*)
PHYSICAL CONSTANTS OF INORGANIC COMPOUNDS

Name	Formula	Formula Weight	Color, Crystalline Form, and Refractive Index	Density	Melting Point, °C	Boiling Point, °C	Solubility in 100 Parts
Iodine							
bromide	IBr	206.81	vlt-bl	4.4157^{0}; $3.76^{42°}$	42	116–119	s d
chloride, mono-	ICl	162.36	α ruby-red nd, cub; β red-brn pl, rh	3.20; 3.66	27.3; 13.9	97.8; 97	d; s al, eth, CS_2; d; s al, eth
chloride, tri-	ICl_3	233.26	yel nd, deliq	3.202	101 d		d; s al, eth, bz, CCl_4, ac a.
cyanide	ICN	152.92	wh cr				sl s; s al, eth
fluoride, penta-	IF_5	221.90	col liq	3.252	8.5	102	d
fluoride, hepta	IF_7	259.89	col g, liq, cr	2.86^{6} g/l	4.5	5.5	
fluorosulfate	ISO_3F	125.97			51.5		
tris-fluorosulfate	$I(SO_3F)_3$	424.10			32.2	$d\ 114^{3mm}$	
(tetra-) oxide, ennea-	I_4O_9	651.61	pale yel, hygr		d >75		187^{13}
oxide, penta-	I_2O_5	333.81	wh cr, trimetric	$4.799^{25°}$	d 275		d; s H_2SO_4
Iodosyl iodate	$IOIO_3$	158.90	yel sol	$4.21^{10°}$	d 130		
Iridium	Ir	192.22	silv met, fcc	$22.65^{20°}$	2448	4500	s oxid flux, aq reg
bromide, tri-	$IrBr_3$	431.92	yel cr				
bromide, tetra-	$IrBr_4$	511.84	blk, deliq				
(di-)octacarbonyl	$Ir_2(CO)_8$	608.48	grn-yel cr		subl 160 (in CO_2)		s CCl_4
carbonyl	$[Ir(CO)_3]_x$		yel, rh-hed		d 120		i CCl_4
chloride, tri-	$IrCl_3$	298.56	α: red, mn; β: red, o-rh		d >775		i a alk
fluoride, tri-	IrF_3	249.19	rh-hed		d >250		i
fluoride, penta-	IrF_5	287.19	yel cr, mn		104.5		hyd
fluoride, hexa-	IrF_6	306.19	yel cr, cub	6.0	44	53.6	d
iodide, tri-	IrI_3	572.91	grn		d		sl s
	$IrI_3 \cdot 3H_2O$	626.96	yel				s
(III) oxide	Ir_2O_3	432.40	blk		–O, >400		s a

Name	Formula	Mol. wt.	Color and form	Density	M.p., °C	B.p., °C	Solubility
(IV) oxide	IrO_2	224.20	blk, tetr	3.15	d 1100	.....	s HCl
	$IrO_2 \cdot 2H_2O$	260.23	dk bl cr	.....	$-2H_2O$, 350	.....	s HCl
selenide	$IrSe_2$	350.12	dk gray pwd	.....	d 600	.....	s sl aq reg
sulfate	$Ir_2(SO_4)_3 \cdot xH_2O$	.....	choc brn	.....	d	.....	s HNO_3
sulfide, di-	IrS_2	256.33	brn-blk	8.43	d 300	.....	i; s aq reg
sulfide, sesqui-	Ir_2S_3	480.59	brn-blk	9.64	d	.....	i; s K_2S, HNO_3
telluride	$IrTe_2$	575.00	dk gray cr	9.5	.....	.....	i; s hot aq reg
Iron	Fe	55.85	silv met, bcc	7.86	1530	3000	i; s a
(II) acetate	$Fe(C_2H_3O_2)_2 \cdot 4H_2O$	246.00	lt grn, mn	.....	d	.....	v s
(III) acetylacetonate	$Fe(C_5H_7O_2)_3$	353.18	red, rh	1.33	184	.....	sl s; s al, bz, chl
(II) *ortho*arsenate	$Fe_3(AsO_4)_2 \cdot 6H_2O$	553.47	grn amorp pwd	.....	d	.....	i; s HCl
(III) *ortho*arsenate (scorodite)	$FeAsO_4 \cdot 2H_2O$	230.80	grn, rh, 1.765, 1.774, 1.797	3.18	d	.....	i; s HCl
arsenide	FeAs	130.77	wh	7.83	1020	.....	v sl s
arsenide, di- (arsenoferrite)	$FeAs_2$	205.69	silv-gray, cub	7.4	990	.....	i; sl s HNO_3
boride	FeB	66.66	gray cr	$7.15^{18°}$	.....	.....	i; s HNO_3, h conc H_2SO_4
(II) bromide	$FeBr_2$	215.67	grn-yel, hex	4.636	d 690	.....	109^{10}
(III) bromide	$FeBr_3$	295.57	dk red-brn, rh, deliq	.....	subl	.....	s
	$FeBr_3 \cdot 6H_2O$	403.68	dk grn	.....	27	d	v s
carbide (cementite)	Fe_3C	179.55	gray, cub	7.694	1840	.....	i; s a
(II) carbonate (siderite)	$FeCO_3$	115.85	gray, trig, 1.875, 1.633	3.8	d	.....	0.0067; s a
carbonyl, ennea-	$Fe_2(CO)_9$	363.79	yel met, hex	2.09	d 100	.....	i; d HNO_3
carbonyl, penta-	$Fe(CO)_5$	195.00	viscous yel liq	1.49	-21	103	i; s al, eth, bz, alk, conc H_2SO_4
carbonyl, tetra-	$[Fe(CO)_4]_3$	503.67	dk grn, mn	$1.996^{18°}$	d 140	.....	i; s org solv, hot H_2SO_4, conc HNO_3
(II) chloride (lawrencite)	$FeCl_2$	126.75	grn to yel, hex, deliq, 1.567	3.16	670	subl	$64^{10°}$
	$FeCl_2 \cdot 2H_2O$	162.78	grn, mn	2.358	.....	.....	.....
	$FeCl_2 \cdot 4H_2O$	198.81	bl-grn, mn, deliq	1.93	.....	.....	$160^{10°}$
(III) chloride (molysite)	$FeCl_3$	162.21	blk-brn, hex	2.898	306	d 315	$740°$

Table 4-1 (Continued)
PHYSICAL CONSTANTS OF INORGANIC COMPOUNDS

Name	Formula	Formula Weight	Color, Crystalline Form, and Refractive Index	Density	Melting Point, °C	Boiling Point, °C	Solubility in 100 Parts
Iron	$FeCl_3 \cdot 6H_2O$	270.30	brnsh-yel, deliq	....	37	280	$91.6^{20°}$
(II) chloroplatinate(IV)	$FePtCl_6 \cdot 6H_2O$	571.75	yel, hex	2.714	d	....	v s
(II) chromite	$FeCr_2O_4$	223.84	brn-blk, cub	$4.97^{20°}$	....	....	i; sl s a
(III) dichromate	$Fe_2(Cr_2O_7)_3$	759.66	red-brn granules	....	....	....	s
(II) ferrocyanide	$Fe_2[Fe(CN)_6]$	323.65	bl-wh, amorp	1.601	d 100	....	i
(III) ferrocyanide	$Fe_4[Fe(CN)_6]_3$	859.25	dk bl cr	....	d	....	i; s HCl
(II) fluoride	FeF_2	93.84	wh, tetr	4.09	1020	....	sl s; s a
	$FeF_2 \cdot 4H_2O$	165.91	wh, rh	2.095	d	....	sl s; s a
	$FeF_2 \cdot 8H_2O$	237.97	brn-bl	....	$-8H_2O, 100$	....	sl s; s a
(III) fluoride	FeF_3	112.84	grn, rh	3.52	>1000	....	sl s; s a
(II) fluorosilicate	$FeSiF_6 \cdot 6H_2O$	306.01	col, trig, 1.361, 1.385	1.961	....	....	128
(III) formate	$Fe(CHO_2)_3 \cdot H_2O$	208.92	red cr pwd	....	....	....	s
(II) hydroxide	$Fe(OH)_2$	89.86	lt grn, hex or wh amorp	....	....	....	s a
(III) hydroxide (goethite)	$FeO(OH)$	88.85	brn-blk, rh, 2.260, 2.394, 2.400	4.28	$-H_2O, 136$	....	s HCl
(III) iodate	$Fe(IO_3)_3$	580.55	yel-grn pwd	$4.80^{20°}$	d 130	....	sl s
(II) iodide	FeI_2	309.66	gray, hex, hygr	5.315	red heat	....	v s
	$FeI_2 \cdot 4H_2O$	381.72	gray-blk cr, deliq	2.873	d 90	....	v s
(II) nitrate	$Fe(NO_3)_2 \cdot 6H_2O$	287.95	grn, rh	....	60.5	....	$83^{20°}$
(III) nitrate	$Fe(NO_3)_3 \cdot 6H_2O$	349.95	lt vlt, cub	....	35	d	v s
	$Fe(NO_3)_3 \cdot 9H_2O$	404.02	lt vlt, mn, deliq	1.684	47	d 125	v s
nitride	Fe_3N	125.70	gray	6.35	d 200	....	i; s HCl, H_2SO_4
nitrosyl carbonyl	$Fe(NO_2)_2(CO)_2$	171.88	dk red cr	1.56	18.5	d 50	s org solv
(II) oxalate	$FeC_2O_4 \cdot 2H_2O$	179.90	lt yel, rh	2.28	d 190	....	0.022, s a
(III) oxalate	$Fe_2(C_2O_4)_3 \cdot 5H_2O$	465.83	yel micro cr pwd	....	d 100	....	v s
(II) oxide (wuestite)	FeO	71.85	blk, cub, 2.32	5.7	1420	....	i; s a

(III) oxide (hematite)	Fe_2O_3	159.69	red-brn to blk, trig, 3.01, 2.94(Li)	5.24	1565		i; s HCl, H_2SO_4
oxide (magnetite)	Fe_3O_4	231.54	blk, cub, 2.42	5.1	1538		i; s a
oxide chloride	$FeOCl$	107.30	brn, rh	3.1	d 400		98°
(II) perchlorate	$Fe(ClO_4)_2 \cdot 6H_2O$	362.84	grn, 1.493, 1.478		d 100		i; s a
(II) orthophosphate (vivianite)	$Fe_3(PO_4)_2 \cdot 8H_2O$	501.61	wh-bl, min, 1.579, 1.603, 1.633	2.58			
(III) orthophosphate	$FePO_4 \cdot 2H_2O$	186.85	pink, mn	2.74	d		i; s HCl, H_2SO_4
(III) pyrophosphate	$Fe_4(P_2O_7)_3 \cdot 9H_2O$	907.36	yel-wh pwd				i; s a, alk
phosphide, mono- (di-) phosphide	FeP	86.82	gray, o-rh	6.07	1370		i; s aq reg, HNO_3 + HF
(di-) phosphide	Fe_2P	142.67	bl-gray	6.77			i; d HCl
(di-) d.phosphide	Fe_2P_2	173.64	gray, o-rh	4.95	1100		i
(tri-) phosphide	Fe_3P	198.51	gray, tetr	7.11	1150		i
(II) metasilicate (gruenerite)	$FeSiO_3$	131.93	gray-grn, rh, 1.672, 1.697, 1.717	3.5			
(II) orthosilicate (fayalite)	Fe_2SiO_4	203.78	col, rh	4.34	1500		i; d HCl
silicide	$FeSi$	83.93	yel-gray, oct	6.1			i; i a
(I) sulfate (szomolnikite)	$FeSO_4 \cdot H_2O$	169.96	grayish-wh, mn	2.970			sl s
	$FeSO_4 \cdot 4H_2O$	224.01	grn, mn, 1.533, 1.535	2.23–2.29			
(siderotil)	$FeSO_4 \cdot 5H_2O$	242.02	wh, tric, 1.526, 1.536, 1.542	2.2	$-5H_2O$, 300		s
(III) sulfate (coquimbite)	$Fe_2(SO_4)_3$	399.87	yel, rh, hygr, 1.814	$3.097^{18°}$	d 480		sl s
	$Fe_2(SO_4)_3 \cdot 9H_2O$	562.01	yel, rh, deliq, 1.552, 1.558	2.1	$-7H_2O$, 175		440
sulfide, di- (pyrite)	FeS_2	119.98	yel, cub	5.0	1171		i; d dil a
(marcasite)	FeS_2	119.98	yel, rh	4.87	tr 450	d	i; d HNO_3
(II) sulfide (troilite)	FeS	87.91	blk-brn, hex	4.74	1195	d	i; s dil a
(III) sulfide	Fe_2S_3	207.87	yel-grn	4.3	d		i; d a
(II) sulfite	$FeSO_3 \cdot 3H_2O$	189.96	grnsh-wh cr		d 250		v sl s

Table 4-1 (*Continued*)
PHYSICAL CONSTANTS OF INORGANIC COMPOUNDS

Name	Formula	Formula Weight	Color, Crystalline Form, and Refractive Index	Density	Melting Point, °C	Boiling Point, °C	Solubility in 100 Parts
Iron							
tantalate (tapiolite)	$Fe(TaO_3)_2$	513.73	lt brn, tetr, 2.27, 2.42(Li)	7.33	....	....	....
(II) thiocyanate	$Fe(SCN)_2 \cdot 3H_2O$	226.06	grn, rh	....	d	....	v s
(III) thiocyanate	$Fe(SCN)_3$	230.09	red-blk, cub, deliq	....	....	....	v s
(II) thiosulfate	$FeS_2O_3 \cdot 5H_2O$	258.05	grn cr, deliq	....	....	....	v s
tungstate (ferberite)	$FeWO_4$	303.69	tetr, 2.40(Li)	6.64	....	....	i; s a
vanadate, meta-	$Fe(VO_3)_2$	352.67	graysh-brn pwd	....	....	....	i; s a
Krypton	Kr	83.80	col g, 1.000427	3.736 g/l, 2.413^{bp} liq	−157.2	−153.4	5.94$^{20°}$ cc
fluoride, di-	KrF_2	122.80	col, tetr	....	subl −60	....	s anhy HF, SbF_5
Lanthanum	La	138.91	whsh met, hex	6.174	920	3470	s HCl
acetate	$La(C_2H_3O_2)_3 \cdot 1.5H_2O$	343.07	col cr	....	....	....	16.9
boride, hexa-	LaB_6	203.78	bl-vlt, cub	4.61	2210	d	i
bromate	$La(BrO_3)_3 \cdot 9H_2O$	684.77	col pr, hex	....	37.5	−7H₂O, 100	28.5$^{15°}$
bromide	$LaBr_3$	378.62	col, hex	5.07	789	1580	s
carbide	LaC_2	162.93	yel	(5.35)	1800–2100	....	d; s H_2SO_4
carbonate	$La_2(CO_3)_3 \cdot 8H_2O$	601.97	wh, rh	2.65	....	....	s a
chloride	$LaCl_3$	245.27	wh, hex, deliq	3.818	862	1750	v s
	$LaCl_3 \cdot 7H_2O$	371.38	wh, tric, hygr	....	−H₂O, 91	....	v s
fluoride	LaF_3	195.91	wh, hex	4.49	1493	2327	....
hydride, di-	LaH_2	140.92	fcc	5.14	....	....	....
hydride, tri-	LaH_3	141.93	bl-blk amorp	5.26	....	....	....
hydroxide	$La(OH)_3$	189.93	wh, hex	....	−H₂O, 260	>380 oxide	s a
iodate	$La(IO_3)_3$	663.62	col cr	....	....	....	1.7
iodide, di-	LaI_2	392.71	blk [$La^{3+}(e^-)(I^-)_2$]	....	820	....	....

name	formula	formula wt	crystalline form, color	density	m.p.	b.p.	solubility
iodide, tri-	LaI$_3$	519.62	gray, rh, hygr	5.63	766	1405	s
molybdate	La$_2$(MoO$_4$)$_3$	757.63	wh, tetr	4.77^{16}	1181	....	0.0018; s HCl
nitrate	La(NO$_3$)$_3$·6H$_2$O	433.02	wh, tric, deliq	....	40	d 126	151
oxalate	La$_2$(C$_2$O$_4$)$_3$·10H$_2$O	722.04	wh	....	d	....	0.0002; s a
oxide	La$_2$O$_3$	325.82	wh, hcp	6.48	2315	4200	s a
(di-)d^{i}oxide sulfide	La$_2$O$_2$S	341.88	yel-wh, hex	5.77	1940	....	....
selenide	La$_2$Se$_3$	514.70	brick red	6.32	....	....	3.0
sulfate	La$_2$(SO$_4$)$_3$	566.00	wh, hygr	3.60	....	....	3.8
	La$_2$(SO$_4$)$_3$·8H$_2$O	710.12	col, hex, 1.564	2.821	−H$_2$O, 400	d 500	
sulfide, mono-	LaS	170.97	gold, cub	5.75	1970	....	d; s a
sulfide, sesqui-	La$_2$S$_3$	374.01	yel-wh, cub	4.99	2080 (γ) 1915 (β)	....	d; s a
Lead	Pb	207.2	silv-blsh met, fcc	11.34 11.288 (Ra-Pb) 11.2960 (U-Pb)	327.4	1751	i; s HNO$_3$, h conc H$_2$SO$_4$
(II) acetate (sugar of lead)	Pb(C$_2$H$_3$O$_2$)$_2$	325.28	wh cr	3.25^{20}	280	....	44.3^{20}
	Pb(C$_2$H$_3$O$_2$)$_2$·3H$_2$O	379.33	wh, mn	2.55	−H$_2$O, 75	d 200	45.6^{15}
	Pb(C$_2$H$_3$O$_2$)$_2$·10H$_2$O	505.44	wh, rh	1.69	22	....	s
(IV) acetate	Pb(C$_2$H$_3$O$_2$)$_4$	443.37	col, mn	2.228^{17}	175	....	d; s chl; d al
dianti monate(V)	Pb$_2$Sb$_2$O$_7$	769.88	dk-yel pwd	6.72	....	....	i; v sl s HCl
orthoantimonate (mcnimolite)	Pb$_3$(SbO$_4$)$_2$	993.07	or-yel pwd	6.58^{20}	....	....	i; v sl s HCl
metaarsenate	Pb(AsO$_3$)$_2$	453.03	hex, col	6.42^{15}	....	....	d; s HNO$_3$
orthoarsenate	Pb$_3$(AsO$_4$)$_2$	899.41	wh cr, v pois	7.80	1040	....	s HNO$_3$
orthoarsenate, mono-H (schultenite)	PbHAsO$_4$	347.12	col, mn	5.79	d 720	....	s HNO$_3$, alk
orthoarsenate, di-H	Pb(H$_2$AsO$_4$)$_2$	489.06	col, tric, 1.74, 1.82	4.46^{15}	d 140	....	s HNO$_3$
diarsenate(V)	Pb$_2$As$_2$O$_7$	676.22	col, rh	6.85^{15}	802	....	i; s HCl, HNO$_3$
metaarsenite	Pb(AsO$_2$)$_2$	421.03	wh pwd	5.85	....	....	i; s HNO$_3$
azide	Pb(N$_3$)$_2$	291.23	col nd or pwd	....	expl 350	....	0.023^{18}; v s ac a
borate	Pb(BO$_2$)$_2$·H$_2$O	310.82	wh cr pwd	5.598 anhyd	−H$_2$O, 160	....	i; s a

Table 4-1 (Continued)
PHYSICAL CONSTANTS OF INORGANIC COMPOUNDS

Name	Formula	Formula Weight	Color, Crystalline Form, and Refractive Index	Density	Melting Point, °C	Boiling Point, °C	Solubility in 100 Parts
Lead							
bromate	Pb(BrO$_3$)$_2$·H$_2$O	481.02	col, mn	5.53	d 180	....	1.38^{20}
bromide	PbBr$_2$	367.01	wh, rh	6.67	360–373	918	0.84^{20}; s a
carbonate (cerussite)	PbCO$_3$	267.20	col, rh, 1.804, 2.076, 2.078	6.6	d 340	....	i; s a, alk
carbonate, basic (hydrocerussite)	2PbCO$_3$·Pb(OH)$_2$	775.60	wh, hex	6.14	d 400	....	i; s HNO$_3$
chlorate	Pb(ClO$_3$)$_2$	374.09	wh, mn, deliq	3.89	d 230	....	v s
	Pb(ClO$_3$)$_2$·H$_2$O	392.11	wh, mn, deliq	4.037	d 110	....	151^{18}
chloride (cotunite)	PbCl$_2$	278.10	wh, rh, 2.199, 2.217, 2.260	5.85	500	950	0.99^{20}; s NH$_4$ salts
(IV) chloride	PbCl$_4$	349.00	yel oily fum liq	3.18^0	−15	expl 105	d; s conc HCl
chlorite	Pb(ClO$_2$)$_2$	342.09	yel, mn	....	expl 126	....	0.095^{20}; s KOH
chromate (crocoite)	PbCrO$_4$	323.18	yel, mn, 2.31 2.37(Li), 2.66	6.12^{15}	844	d	i; s a
chromate, basic	PbCrO$_4$·PbO	546.37	red cr pwd	6.63	....	....	i; s a
citrate	Pb$_3$(C$_6$H$_5$O$_7$)$_3$·3H$_2$O	1053.82	wh cr pwd	....	....	....	s
cyanate	Pb(OCN)$_2$	291.22	wh nd	....	d	....	i
cyanide	Pb(CN)$_2$	259.23	yelsh-wh pwd	....	....	....	sl s; s KCN
dichromate	PbCr$_2$O$_7$	423.18	red cr	....	....	....	d; s a, alk
dithionate	PbS$_2$O$_6$·4H$_2$O	439.38	col, trig, 1.635, 1.653	3.22	d	....	115^{20}
ferricyanide	Pb$_3$[Fe(CN)$_6$]$_2$·5H$_2$O	1135.55	brn-blk, mn	....	−H$_2$O, 110 d	....	sl s, s alk, HNO$_3$
ferrite	PbFe$_2$O$_4$	382.88	hex	....	1530 d	....	....
ferrocyanide	Pb$_2$Fe(CN)$_6$·3H$_2$O	680.38	yelsh-wh pwd	....	−H$_2$O, 100	....	i
fluoride	PbF$_2$	245.19	col, rh	8.24	855	1290	0.064^{20}; s HNO$_3$
fluoroborate	Pb(BF$_4$)$_2$	380.80	col cr	....	....	....	d
fluorochloride (matlockite)	PbFCl	261.64	wh, tetr, 2.145, 2.006	7.05	601	....	0.0325

4-68

Name	Formula	Mol. wt.	Crystalline form, color, index of refraction	Density	M.p., °C	B.p., °C	Solubility
formate	$Pb(CHO_2)_2$	297.23	wh, rh, 1.789, 1.852	4.63	d 190	.....	$1.6^{20°}$
hydride	PbH_2	209.21	gray pwd	.....	d	.....	.....
hydroxide	$Pb(OH)_2$	241.20	wh, amorp	.....	d 145	.....	0.0155^{20}; s a, alk
iodate	$Pb(IO_3)_2$	557.00	wh	6.155^{20}	d 300	.....	0.003
iodide	PbI_2	461.00	yel, hex	6.16	400	954	0.063^{20}; s alk, KI
iodide, basic	$PbI_2 \cdot PbO \cdot H_2O$	702.20	rh	6.83^{20}	d 100	.....	i; s a, KOH
molybdate (sulfenite)	$PbMoO_4$	367.13	col–lt yel, tetr	6.92	1065	.....	.....
nitrate	$Pb(NO_3)_2$	331.20	col, cub or mn, 1.782	4.53^{20}	d 200	.....	56^{20}
nitrate, basic	$Pb(NO_3)(OH)$	286.20	wh, rh	5.93	d 180	.....	19.4^{19}
oxalate	PbC_2O_4	295.21	wh pwd	5.28	d 300	.....	i; s HNO_3
oxide, mon- (litharge)	PbO	223.19	yel, tetr	9.53	888	.....	0.0017^{20}; s HNO_3, alk
oxide, mon- (massicot)	PbO	223.19	yel, rh, 2.51, 2.61(Li), 2.71	8.0	886	1472	0.0023; s HNO_3, alk
oxide, mon-	$2PbO \cdot H_2O$	464.39	pwd, cub, or amorp	7.592	d 145	.....	0.014; s HCl
oxide, di- (plattnerite)	PbO_2	239.19	blk, tetr	9.375	$-O_2,\ 752$	.....	i; d a
oxide, sesqui-	Pb_2O_3	462.38	or-yel pwd, amorp	.....	d 370	.....	i; s HCl
oxide (red lead)	Pb_3O_4	685.57	red cr or pwd	9.1	$-O_2,\ 830$	.....	i; s HNO_3
oxide chloride (fiedlerite)	$2PbCl_2 \cdot PbO \cdot H_2O$	797.40	mn, 1.816, 2.102, 2.026	5.88^{20}	d 150	.....	.....
oxide chloride (laurionite)	$PbCl_2 \cdot PbO \cdot H_2O$	519.29	rh	6.24	d 142	.....	i; s alk
oxide chloride (mendipite)	$PbCl_2 \cdot 2PbO$	724.47	yel, rh, 2.24, 2.27, 2.31	7.08	693	.....	.....
oxide chloride (paralaurionite)	$PbCl_2 \cdot PbO \cdot H_2O$	519.29	wh, mn, 2.146	6.05^{15}	d 150	.....	.....
perchlorate	$Pb(ClO_4)_2 \cdot 3H_2O$	460.14	wh, rh,	2.6	d 88	.....	v s
periodate	$PbHIO_6$	415.10	wh cr	.....	d 130	.....	i; s dil HNO_3
metaphosphate	$Pb(PO_3)_2$	365.13	col cr	.....	800	.....	v sl s
orthophosphate	$Pb_3(PO_4)_2$	811.51	wh pwd, hex, 1.970, 1.936	6.9–7.3	1014	.....	i; s HNO_3
orthophosphate, mono-H	$PbHPO_4$	303.17	rh	5.661^{15}	d	.....	i; s HNO_3
phosphide, penta-	PbP_5	362.06	blk, inflam	.....	d 400 vac	.....	d; d dil a
orthophosphite, mono-H	$PbHPO_3$	287.17	wh pwd	.....	d	.....	i; s HNO_3

Table 4-1 (*Continued*)
PHYSICAL CONSTANTS OF INORGANIC COMPOUNDS

Name	Formula	Formula Weight	Color, Crystalline Form, and Refractive Index	Density	Melting Point, °C	Boiling Point, °C	Solubility in 100 Parts
Lead							
picrate	$Pb(C_6H_2N_3O_7)_2 \cdot 2H_2O$	681.45	yel nd	$2.831^{20°}$	$-H_2O$, 130	expl	$0.88^{15°}$
pyrophosphate	$Pb_2P_2O_7$	588.32	wh, rh	$5.8^{20°}$	824	....	i; s HNO_3, KOH
selenate	$PbSeO_4$	350.17	wh, rh	$6.37^{20°}$	d	....	i; s conc a
selenide (clausthalite)	$PbSe$	286.15	gray, cub	$8.10^{15°}$	1065	....	i; s HNO_3
*meta*silicate (alamosite)	$PbSiO_3$	283.27	wh, mn	6.49	766	....	i; d a
(di-)*di*silicate (barysilite)	$Pb_2Si_2O_7$	582.55	wh, trig, 2.070, 2.050	6.707	....	....	i
stearate	$Pb(C_{18}H_{35}O_2)_2$	774.15	wh pwd	....	115.7	....	$0.05^{35°}$
sulfate (anglesite)	$PbSO_4$	303.25	wh, mn or rh, 1.877, 1.822, 1.894	6.2	1170	....	0.004
sulfate, basic (lanarkite)	$PbSO_4 \cdot PbO$	526.44	wh, mn, 1.93, 1.99, 2.02	6.92	977	....	0.004
sulfide (galena)	PbS	239.25	bl met, cub, 3.921	7.5	1112	....	$0.012^{20°}$; s a
sulfite	$PbSO_3$	287.25	wh pwd	....	d	....	i; s HNO_3
tartrate	$PbC_4H_4O_6$	355.26	wh cr pwd	$2.53^{19°}$	d	....	0.0025; s HNO_3, KOH
thiosulfate	PbS_2O_3	319.32	wh cr	5.18	d	....	0.03
titanate	$PbTiO_3$	303.09	yel, rh py	7.52	....	....	i
telluride (altaite)	$PbTe$	334.79	wh, cub	$8.164^{20°}$	917	....	i
thiocyanate	$Pb(SCN)_2$	323.35	wh, mn	3.82	d 190	....	0.05; s KSCN, HNO_3
tungstate (stolzite)	$PbWO_4$	455.04	col, tetr, 2.269, 2.182	8.23	....	....	i; s KOH
tungstate (raspite)	$PbWO_4$	455.04	col, mn, 2.27, 2.27, 2.30	....	1125	....	0.03; d a
vanadate, meta-	$Pb(VO_3)_2$	405.07	yel pwd	....	....	....	sl s; s dil HNO_3
Lithium	Li	6.94	silv met, bcc	$0.535^{20°}$	179	1336	d; s a
acetate	$LiC_2H_3O_2 \cdot 2H_2O$	102.01	wh, rh	....	70	d	v s

Name	Formula	Formula weight	Crystalline form, color, refractive index	Density	Melting point, °C	Boiling point, °C	Solubility
aluminate	$LiAlO_2$	65.92	wh pwd	2.554	>1625	......	......
amide	$LiNH_2$	22.96	col, cub	1.178^{18}	374	430	d
antimonide	Li_3Sb	142.57	......	3.21^{17}	>950	......	......
arsenate, ortho-	Li_3AsO_4	159.74	......	3.07	......	......	......
borarate	$LiBH_4$	21.78	wh, o-rh	0.666	278	......	0.9^{0}
borate, meta-	$LiBO_2$	49.75	wh, tric	......	843	......	......
bromate	$LiBrO_3$	134.85	col, rh	......	......	......	164
bromide	$LiBr$	86.85	wh, fcc, deliq, 1.784	3.464	552	1265–1310	v s
	$LiBr \cdot 2H_2O$	122.88	wh pr	......	44	......	d
carbide	Li_2C_2	37.90	wh cr	1.65^{18}	......	......	......
carbonate	Li_2CO_3	73.89	col, mn, 1.567	2.11^{0}	618	d	1.32^{20}
carbonate, hydrogen	$LiHCO_3$	67.96	in soln only	......	......	......	......
chlorate	$LiClO_3$	90.39	col	......	129	d 270	75^{18}
chloride	$LiCl$	42.39	wh, cub, deliq, 1.662	2.068	610	1360	78
chloroaurate(III)	$LiAuCl_4$	345.72	or-red, hex	......	......	......	58
chloroplatinate(IV)	$Li_2PtCl_6 \cdot 6H_2O$	529.78	yel, rh	......	$-6H_2O$, 180	......	52^{20}
chromate	$Li_2CrO_4 \cdot 2H_2O$	165.90	wh cr	......	$-2H_2O$, 130	......	61.5^{15}
citrate	$Li_3C_6H_5O_7 \cdot 4H_2O$	281.98	......	......	d	......	......
deuteride	LiD	8.96	......	0.881	686	......	56^{30}
dichromate	$Li_2Cr_2O_7 \cdot 2H_2O$	265.90	yel-red, deliq	2.34^{30}	$-2H_2O$, 110	......	s
dithionate	$Li_2S_2O_6 \cdot 2H_2O$	210.03	col, rh, 1.5602	2.158	d	......	......
fluorice	LiF	25.94	wh, cub, 1.3915	2.295^{21}	847	1670	0.26^{18}
fluorosilicate	$Li_2SiF_6 \cdot 2H_2O$	191.99	col, mn	2.3^{12}	$-2H_2O$, 100	d	73^{17}
fluorosulfate	$LiSO_3F$	106.00	wh	......	360	......	49.2^{20}
formate	$LiHCO_2 \cdot H_2O$	69.97	col, rh	1.46	$-H_2O$, 94	......	d
hydride	LiH	7.95	col, bcc	0.780	692	......	d
*tetra*hydridoaluminate	$LiAlH_4$	37.95	col	0.917	......	......	......
hydroxide	$LiOH$	23.95	wh cr	2.54	$-H_2O$, 445	925	12.7^{0}
	$LiOH \cdot H_2O$	41.96	col, mn	1.83^{20}	......	d	22.31^{10}
iodide	LiI	133.84	cub, deliq, 1.955	4.061	446	1190	165^{20}; v s al
	$LiI \cdot 3H_2O$	187.89	wh, mn	3.5	73; $-3H_2O$, 300	......	v s
nitrate	$LiNO_3$	68.94	col, trig, 1.735	2.38	261	......	41^{20}
nitride	Li_3N	34.82	blk	......	849	......	d

Table 4-1 (Continued)
PHYSICAL CONSTANTS OF INORGANIC COMPOUNDS

Name	Formula	Formula Weight	Color, Crystalline Form, and Refractive Index	Density	Melting Point, °C	Boiling Point, °C	Solubility in 100 Parts
Lithium							
nitrite	$LiNO_2 \cdot H_2O$	70.96	wh nd	1.615^0	<100		125^0
oxalate	$Li_2C_2O_4$	101.90	col cr	2.121^{17}	d		8^{20}
oxide	Li_2O	29.88	col, 1.644	2.013	1730	subl 1000	forms LiOH
perchlorate	$LiClO_4$	106.39	col, deliq	2.43	236	d 470	37.8
	$LiClO_4 \cdot 3H_2O$	160.44	col, hex	1.84	95; −3H₂O, 130	d 470	s
permanganate	$LiMnO_4$	125.87	purp cr	2.06	d 190		29^{16}
peroxide	Li_2O_2	45.88	wh				
phosphate	Li_3PO_4	115.79	wh, rh	2.537^{18}	837		0.034^{18}
phosphate, di-H	$Li_3PO_4 \cdot 12H_2O$	331.97	wh, trig	1.645	100		
phosphate, di-H	LiH_2PO_4	103.93	col	2.461	>100		i
phosphide	Li_2P_5	200.02	red-brn		~650		
salicylate	$LiC_7H_5O_3$	144.06	col		d		128
selenide	Li_2Se	92.84	cub	2.91			57.7
silicate, meta-	Li_2SiO_3	89.96	col, rh, 1.548	2.52	1200		
silicate, o tho-	Li_4SiO_4	119.84	col, rh, 1.60	2.28	1256		
silicide	Li_6Si_2	97.81	bl cr	1.12	ign in air	d 500	
sulfate	Li_2SO_4	109.94	col, mn, 1.465	2.22	860		34.5^0
sulfate	$Li_2SO_4 \cdot H_2O$	127.95	col, mn, 1.477	2.06	−H₂O, 130		43.6^0
sulfate, hydrogen	$LiHSO_4$	104.01	col pr	2.123^{13}	171		
sulfide	Li_2S	45.94	wh or yel-red, amorp	1.68			
sulfite	Li_2SO_3	93.94	wh		455 d		
telluride	Li_2Te	141.46	cub	3.24			
Lutetium	Lu	174.97	met, hcp	9.842	1652	3330	
bromide	$LuBr_3$	414.70	col, rh-hed	5.17	1025	1410	s a
chloride	$LuCl_3$	281.33	wh, mn	3.98	925	1480	s
fluoride	LuF_3	231.97	wh, hex	8.332	1182	2230	s H_2SO_4
iodide	LuI_3	555.68	brn, hex		1050	1210	s

Name	Formula	Formula wt	Crystalline form, color, and refractive index	Density	Melting point, °C	Boiling point, °C	Solubility
oxalate	$Lu_2(C_2O_4)_3 \cdot 6H_2O$	722.09	wh cr	......	$-H_2O$, 50	......	s a
oxide	Lu_2O_3	397.94	wh, bcc	9.42	......	......	s a
sulfate	$Lu_2(SO_4)_3 \cdot 8H_2O$	782.25	col cr	......	......	1117	47.3
Magnesium	Mg	24.31	silv met, hex	1.74^{20}	650		i; s a
acetate	$Mg(C_2H_3O_2)_2$	142.40	wh	1.42	323 d	......	v s
	$Mg(C_2H_3O_2)_2 \cdot 4H_2O$	214.46	wh, mn, 1.491	1.454	80	......	v s
aluminate (spinel)	$MgAl_2O_4$	142.27	col, cub, 1.723	3.6	2135	......	i; v sl s HCl
ammonium arsenate	$MgNH_4AsO_4 \cdot 6H_2O$	289.36	wh, rh or tetr	1.932^{15}	d	......	0.04^{0}; s a
ammonium phosphate (struvite, guanite)	$MgNH_4PO_4 \cdot 6H_2O$	245.41	col, rh, 1.496	1.715	d 100	......	0.023^{0}
antimonide	Mg_3Sb_2	316.44	met, hex	4.088	961	......	i
orthoarsenate (hoernesite)	$Mg_3(AsO_4)_2 \cdot 8H_2O$	494.90	wh, mn	2.60	......	......	d
orthoarsenate, mono-H (roesslerite)	$MgHAsO_4 \cdot 7H_2O$	290.35	wh, mn	1.943^{15}	$-5H_2O$, 100	......	......
arsenide	Mg_3As_2	222.78	brn-red, cub	3.148	800	......	d; d dil a
orthoarsenite	$Mg_3(AsO_3)_2$	318.78	wh	......	d	......	s
bismuthide	Mg_3Bi_2	490.90	met, hex	5.945	823	......	i; s a
metaborate	$Mg(BO_2)_2 \cdot 8H_2O$	254.06	col, tetr, 1.565, 1.575	2.30	......	......	i; s min a
orthoborate	$Mg_3(BO_3)_2$	190.55	col, rh, 1.6527, 1.6537, 1.6548	2.99^{21}	......	......	......
diborate (ascharite)	$Mg_2B_2O_4 \cdot H_2O$	168.26	o-rh, 1.54	2.6–2.7	......	d	......
bromate	$Mg(BrO_3)_2 \cdot 6H_2O$	388.22	col, cub, 1.5136	2.29	$-6H_2O$, 200	......	71.5
bromide	$MgBr_2$	184.13	wh, hex, deliq	3.72	695	......	96
bromide	$MgBr_2 \cdot 6H_2O$	292.22	col, hex, deliq	2.00	160 d	......	v s
bromoplatinate(IV)	$MgPtBr_6 \cdot 12H_2O$	915.04	trig	2.802	......	......	......
carbonate (magnesite)	$MgCO_3$	84.32	wh, trig, 1.717, 1.515	2.958	d 400	......	0.01; s a
(nesquehonite)	$MgCO_3 \cdot 3H_2O$	138.37	col, rh, 1.501	1.852	$-H_2O$, 100	......	0.16; s a
(lansfordite)	$MgCO_3 \cdot 5H_2O$	174.40	wh, mn, 1.456, 1.476, 1.502	1.73	d in air	......	0.18^{8}; s HCl
carbonate, basic (artinite)	$MgCO_3 \cdot Mg(OH)_2 \cdot 3H_2O$	196.69	wh, rh, 1.489, 1.534, 1.557	2.02^{20}	......	......	......
(hydromagnesite)	$3MgCO_3 \cdot Mg(OH)_2 \cdot 3H_2O$	365.35	wh, rh, 1.527, 1.530, 1.540	2.16	d	......	0.04; s a

Table 4-1 (*Continued*)
PHYSICAL CONSTANTS OF INORGANIC COMPOUNDS

Name	Formula	Formula Weight	Color, Crystalline Form, and Refractive Index	Density	Melting Point, °C	Boiling Point, °C	Solubility in 100 Parts
Magnesium							
chlorate	$Mg(ClO_3)_2 \cdot 6H_2O$	299.31	wh, rh, deliq	1.80	d 120	....	v s
chloride (chloro-magnesite)	$MgCl_2$	95.22	col, hex, 1.675, 1.59	2.325	714	1412	35.5
(bischofite)	$MgCl_2 \cdot 6H_2O$	203.31	wh, mn, deliq, 1.495, 1.507, 1.528	1.569	118 d	....	v s
chloropalladate(IV)	$MgPdCl_6 \cdot 6H_2O$	451.52	wh, hex	2.12	d	....	....
chloroplatinate(IV)	$MgPtCl_6 \cdot 6H_2O$	540.21	yel, trig	2.437	$-H_2O$, 180	d	....
chromate	$MgCrO_4 \cdot 7H_2O$	266.41	yel, rh, 1.5500, 1.521, 1.568	1.695	$-3H_2O$, 120	....	v s
*di*chromate(III)	$MgCr_2O_4$	192.30	dk grn or red, cub	4.6^{20}	....	....	i; s conc H_2SO_4
citrate, mono-H	$MgHC_6H_5O_7 \cdot 5H_2O$	304.50	wh	....	....	....	20
ferrite	$MgFeO_4$	200.00	blk, oct, 2.35	4.44-4.60	1750	....	i; s conc HCl
ferrocyanide	$Mg_2Fe(CN)_6 \cdot 12H_2O$	476.76	pale yel	....	d ca 200	....	33
fluoride (sellaite)	MgF_2	62.31	col, tetr, 1.378, 1.390; 3.14	....	1263	2230	0.013; s HNO_3
fluorosilicate	$MgSiF_6 \cdot 6H_2O$	274.48	wh, trig	1.788	d 120	....	64^{17}
formate	$Mg(CHO_2)_2 \cdot 2H_2O$	150.38	col, rh	....	$-2H_2O$, 100	....	17^0
germanide	Mg_2Ge	121.21	....	....	1115	....	....
hydride	MgH_2	26.33	wh, tetr	1.45	d 280 vac	....	d viol
hydroxide (brucite)	$Mg(OH)_2$	58.33	col, hex, 1.559, 1.580	2.36	$-H_2O$, 350	....	i; s a
iodate	$Mg(IO_3)_2 \cdot 4H_2O$	446.18	wh, mn	3.3^{13}	$-4H_2O$, 210	d	10.2
iodide	MgI_2	278.12	wh, hex, deliq	4.43	d >700	....	148^{18}
	$MgI_2 \cdot 8H_2O$	422.24	wh pwd, deliq	....	d 41	....	81^{20}
molybdate	$MgMoO_4$	184.25	wh, rh, tric	2.208	....	....	13.7
nitrate	$Mg(NO_3)_2 \cdot 6H_2O$	256.41	col, mn, deliq	1.6363	129 d	....	125
nitride	Mg_3N_2	100.95	grn-yel pwd	2.712	d 800	subl 700 vac	d; s a

	Formula	Formula wt	Color, crystalline form, refractive index	Density	Melting point, °C	Boiling point, °C	Solubility
nitrite	$Mg(NO_2)_2 \cdot 3H_2O$	170.37	wh pr, hygr		d 100		s
oxalate	$MgC_2O_4 \cdot 2H_2O$	148.36	wh pwd	2.45	d 150		0.07; s a
oxide (magnesia, periclase)	MgO	40.31	col, cub, 1.736	3.58	2800	3580	i; s a
perchlorate	$Mg(ClO_4)_2$	223.21	wh, deliq	2.60	d 251		49.6
	$Mg(ClO_4)_2 \cdot 6H_2O$	299.31	wh, rh, deliq, 1.482, 1.458	1.970	185		v s
permanganate	$Mg(MnO_4)_2 \cdot 6H_2O$	370.27	dk purp nd, deliq	2.18	d		v s
orthophosphate	$Mg_3(PO_4)_2$	262.88	col, rh		1184		i
	$Mg_3(PO_4)_2 \cdot 4H_2O$	334.97	col, mn	$1.64^{15°}$	$-8H_2O$, 400		0.02; s a
(bobierite)	$Mg_3(PO_4)_2 \cdot 8H_2O$	407.00	wh, mn, 1.510, 1.520, 1.543	$2.195^{15°}$			i; s HNO_3, citrate
orthophosphate, mono-H (newberyite)	$MgHPO_4 \cdot 3H_2O$	174.34	wh, rh, 1.514, 1.518, 1.533	$2.123^{15°}$	$-H_2O$, 205	d 550	sl s; s a
	$MgHPO_4 \cdot 7H_2O$	246.40	wh, mn nd	$1.729^{15°}$	$-4H_2O$, 100	d 550	0.3; s a
phosphide	Mg_3P_2	134.88	yel-grn, cub	2.055			d; s min a
phosphite, hypo-	$Mg(PH_2O_2)_2 \cdot 6H_2O$	262.38	wh, ditetr	$1.591^{12°}$	$-6H_2O$, 180		20
pyrophosphate	$Mg_2P_2O_7$	222.57	col, mn, 1.602, 1.604, 1.615	2.559	1383		i; s a
selenate	$MgSeO_4 \cdot 6H_2O$	275.36	col, mn, 1.468, 1.489, 1.491	1.928			v s
selenide	$MgSe_2$	103.27	lt gray pwd, 2.44	4.21			d; d a
silicate, meta- (clinoenstatite)	$MgSiO_3$	100.40	wh, mn, 1.651, 1.660	3.192	d 1557		i
silicate, ortho- (forsterite)	Mg_2SiO_4	140.71	wh, o-rh, 1.65, 1.66, 1.67	3.21	1910		i; d hot HCl
silicide	Mg_2Si	76.71	bl, cub	1.94	1102		i; s a
sulfate	$MgSO_4$	120.37	col, rh, 1.56	2.66	d 1124		$26^{0°}$
(kieserite)	$MgSO_4 \cdot H_2O$	138.39	col, mn, 1.523, 1.535	2.445			s
(epsomite)	$MgSO_4 \cdot 7H_2O$	246.48	col, rh or mn, 1.433	1.68	$-7H_2O$, 200		$71^{20°}$
sulfide	MgS	56.38	pale red-brn, cub, 2.271	2.84	d >2000		d; s a
sulfite	$MgSO_3 \cdot 6H_2O$	212.47	wh, rh or hex, 1.511, 1.464 (hex)	1.725	$-6H_2O$, 200	d	66
d-tartrate	$MgC_4H_4O_6 \cdot 5H_2O$	262.46	wh, rh	1.67	$-5H_2O$, 200		$0.8^{16°}$; s min a

Table 4-1 (Continued)
PHYSICAL CONSTANTS OF INORGANIC COMPOUNDS

Name	Formula	Formula Weight	Color, Crystalline Form, and Refractive Index	Density	Melting Point, °C	Boiling Point, °C	Solubility in 100 Parts
Magnesium							
telluride	MgTe	151.91	wh, hex cr	3.86	. . . .	. . . .	d; d a
thiosulfate	$MgS_2O_3 \cdot 6H_2O$	244.53	col, rh pr	1.818	$-3H_2O$, 170	d	v s
tungstate	$MgWO_4$	272.16	col, mn	5.66	. . . .	. . . .	i; d a
Manganese							
acetate	Mn	54.94	gray met, cub or tetr	7.30	1244	2120	d; s a
	$Mn(C_2H_3O_2)_2$	173.02	brn cr	1.74	. . . .	. . . .	s d
	$Mn(C_2H_3O_2)_2 \cdot 4H_2O$	245.08	pale red, mn	1.589	. . . .	. . . .	s
arsenide	MnAs	129.86	blk, hex	6.18	d 400	. . . .	i; s HCl
arsenide, di-	$MnAs_2$	184.80			1400	. . . .	i; s aq reg
boride	MnB	65.75	cr pwd	$6.2^{15°}$	. . . .	. . . .	d; s a
boride, di-	MnB_2	76.56	gray-vlt cr	6.9	. . . .	. . . .	$1277°$
bromide	$MnBr_2 \cdot 4H_2O$	214.76	rose cr	4.39	698	. . . .	v s
carbide	Mn_3C	286.82	rose, mn, deliq	. . . .	d 64	. . . .	d; s a
carbonate	$MnCO_3$	176.83	blk, tetr	6.89^{17}	1520	. . . .	i; s dil a
(rodochrosite)		114.95	rose, rh	3.125	d	. . . .	
carbonyl	$Mn_2(CO)_{10}$	389.98	gold, mn	. . . .	155	. . . .	. . . .
chloride (scacchite)	$MnCl_2$	125.84	pink, hex, deliq	2.977	652	1225	72
	$MnCl_2 \cdot 4H_2O$	197.91	rose, mn, deliq	2.01	$-4H_2O$, 198	. . . .	$151°$
(III) chloride	$MnCl_3$	161.30	brn or grn-blk cr	. . . .	d -40	. . . .	s abs al
chloroplatinate(IV)	$MnPtCl_6 \cdot 6H_2O$	570.84	rose, trig	2.692	d	. . . .	i; i a
dichromate(III)	$MnCr_2O_4$	222.93	gray-blk, cub	$4.97^{20°}$	. . . .	. . . .	i; i a
dithionate	MnS_2O_6	215.06	rose, tric	1.757	. . . .	. . . .	s
fluoride	MnF_2	92.93	red, tetr or pwd	3.98	920	. . . .	$0.66^{40°}$; s a
(III) fluoride	MnF_3	111.93	purp, mn	3.54	d	. . . .	d; s a
(IV) fluoride	MnF_4	140.93	blue, hygr	. . . .	d 20	. . . .	. . . .
fluorogallate	$MnGaF_5 \cdot 7H_2O$	345.76	pink, o-rh, 1.45	2.22	d 230	. . . .	v s
fluorosilicate	$MnSiF_6 \cdot 6H_2O$	305.11	rose, hex pr, 1.357, 1.374	1.903	d	. . . .	140^{18}

Name	Formula	Formula wt	Color, crystalline form, properties	Density	MP, °C	BP, °C	Solubility
formate	Mn(CHO₂)₂·2H₂O	181.00	rose, rh	1.953	d	....	s
hydroxide (pyrochroite)	Mn(OH)₂	88.95	wh-pink, trig, 1.723, 1.681	3.2581³°	d	....	i; s a
(III) hydroxide (marganite)	MnO(OH)	87.94	brn-blk, rh, 2.24, 2.24, 2.53(Li)	4.2–4.4	d	....	i; s HCl, hot H₂SO₄
(IV) hydroxide	MnO(OH)₂	104.95	blk-brn, amorp	2.58	....	....	i; s HCl
iodide	MnI₂	308.75	pink, hex, deliq, brn in air	5.0¹	613 vac; d ca 80	....	s
	MnI₂·4H₂O	380.81	rose, mn, deliq	....	d	....	s
iodoplatinate(IV)	MnPtI₆·9H₂O	1173.59	trig	3.604	d	....	v s
nitrate	Mn(NO₃)₂·4H₂O	251.01	lt rose, mn	1.82	26	129	i; s a
oxalate	MnC₂O₄	142.96	wh cr pwd	2.43²²°	d 150	....	i; s a
	MnC₂O₄·2H₂O	178.98	pink, oct cr or pwd	....	−2H₂O, 100	d	0.031; s a
oxide (manganosite)	MnO	70.94	grn, cub, 2.16	5.44	....	....	i; s a
(III) oxide (braunite)	Mn₂O₃	157.87	blk, cub	4.50	−O₂, 940	....	i; s a
(IV) oxide (pyrolusite)	MnO₂	86.94	blk, rh or pwd	5.026	−O₂, 530	....	i; s HCl
(VII) oxide	Mn₂O₇	221.87	dk red oil, hygr, expl	2.396²⁰°	ca −20	ca 25	v s
(VI) oxide	MnO₃	102.94	redsh, deliq	....	d	....	s
(III,IV) oxide (hausmannite)	Mn₃O₄	228.81	blk, tetr, 2.46, 2.15(Li)	4.856	1705	....	i; s HCl
trioxide chloride	MnO₃Cl	138.29	grn-vlt g	....	....	....	
trioxide fluoride	MnO₃F	121.84	dk grn liq	....	−38	−50; expl 20	s
perchlorate	Mn(ClO₄)₂·6H₂O	361.95	pink, hex, 1.492	....	d 165	ca 60	s
(II) orthophosphate	Mn₃(PO₄)₂·3H₂O	408.80	rose, rh, 1.651, 1.656, 1.683	3.102	....	....	
orthophosphate, mono-H	MnPO₄·3H₂O	204.97	red rh or pink pwd, 1.656	....	....	....	sl s; s a
orthophosphate, di-H	Mn(H₂PO₄)₂·2H₂O	284.94	brn-pink, mn, 1.695, 1.704, 1.710	3.707	−H₂O, 100	....	s
pyrophosphate	Mn₂P₂O₇	283.82	dk gray, o-rh	....	1196	....	i; s a
phosphide	MnP	85.91	dk gray, o-rh	5.34	1150	....	i; sl s HNO₃
(di-) phosphide	Mn₂P	140.85	gray, hex	6.32	1327	....	i; s a
(tri-) phosphide, di-	Mn₃P₂	226.76	blk pwd	5.12¹⁸°	1095	....	i; sl s HNO₃

Table 4-1 (*Continued*)
PHYSICAL CONSTANTS OF INORGANIC COMPOUNDS

Name	Formula	Formula Weight	Color, Crystalline Form, and Refractive Index	Density	Melting Point, °C	Boiling Point, °C	Solubility in 100 Parts
Manganese							
*ortho*phosphite, mono-H	$MnHPO_3 \cdot H_2O$	152.93	redsh	......	$-H_2O$, 200	......	sl s
selenate	$MnSeO_4 \cdot 2H_2O$	233.93	pink, rh	2.95–3.01	......	......	s
	$MnSeO_4 \cdot 5H_2O$	287.97	pink, trig	2.33–2.39	......	......	s
selenide	MnSe	133.90	gray, cub	5.55^{15}	......	......	i; d a
selenite	$MnSeO_3 \cdot 2H_2O$	217.93	mn cr	......	......	......	v sl s
silicate, meta- (rhodonite, hermannite)	$MnSiO_3$	131.02	red, tric, 1.733, 1.740, 1.744	3.72	1325	......	i; i HCl
silicide	MnSi	83.02	tetrahed	5.90^{15}	1280	......	i; s HF
silicide, di-	$MnSi_2$	111.11	gray, oct	5.24^{13}	......	......	i; s HF, alk
(di-) silicide	Mn_2Si	137.96	quadr pr	6.20^{15}	1315	......	i; s HCl, alk
sulfate	$MnSO_4$	151.00	redsh	3.25	700	d 850	52^5
(szmikite)	$MnSO_4 \cdot H_2O$	169.01	pink, mn, 1.562, 1.595, 1.632	2.95	d 117	......	v s
	$MnSO_4 \cdot 4H_2O$	223.06	pink, mn or rh, effl, 1.508, 1.522	2.107	d 27	......	v s
	$MnSO_4 \cdot 5H_2O$	241.08	rose, tric, 1.495, 1.508, 1.514	2.103^{15}	d 26	......	v s
	$MnSO_4 \cdot 7H_2O$	277.11	red, mn or rh	2.09	$-7H_2O$, 280	......	v s
(III) sulfate	$Mn_2(SO_4)_3$	398.06	grn cr, hex, deliq	3.24	d 160	......	d; s HCl
sulfide (alabandite)	MnS	87.00	grn cub or pink amorp, 2.70(Li)	3.99	d	......	i; s a
sulfide, di- (hauerite)	MnS_2	119.07	blk cub, 2.69(Li)	3.463	d	......	i; d HCl
tantalate	$Mn(TaO_3)_2$	512.83	blk, rh, 2.22, 2.25, 2.29	7.03	......	......	......
thiocyanate	$Mn(SCN)_2 \cdot 3H_2O$	225.16	pink, deliq	......	$-3H_2O$, 160	......	s
titanate (pyrophanite)	$MnTiO_3$	150.84	yel, trig, 2.481, 2.210	4.54	1360	......	......

Mercury

Mercury (quicksilver)	Hg	200.59	silv-wh met, liq	13.5939^{20}	−38.87	356.58	i; s HNO_3
(I) acetate	$Hg_2(C_2H_3O_2)_2$	519.27	col pl	......	d	......	0.75^{12}; s HNO_3, H_2SO_4
(II) acetate	$Hg(C_2H_3O_2)_2$	318.76	wh pwd	3.270	d	......	$25^{10°}$; s al, ac a
(II) acetylide	$3HgC_2 \cdot H_2O$	691.85	wh pwd	5.3	expl	......	i; i al
(II) amidobromide	$Hg(NH_2)Br$	296.52	wh pwd	......	d	......	d; s NH_4OH
(II) amidechloride	$Hg(NH_2)Cl$	252.07	wh pwd	5.70	infusible	......	0.14; d a
(II) orthoarsenate	$Hg_3(AsO_4)_2$	879.61	yel	......		......	v sl s; s HCl, HNO_3
(I) azide	$Hg_2(N_3)_2$	485.22	wh cr	......	expl, d by light	......	0.025
(I) bromate	$Hg_2(BrO_3)_2$	656.99	wh cr	......	d	......	sl s HNO_3
(II) bromate	$Hg(BrO_3)_2 \cdot 2H_2O$	492.44	wh, rh	......	d 130–140	......	0.08; s HCl, HNO_3
(I) bromide	Hg_2Br_2	561.00	wh-yel, tetr	7.307	subl 350	......	i; s a
(II) bromide	$HgBr_2$	360.41	col, rh	6.109	236–244	319–335	1.1; $15°$ al, s MeOH
bromide iodide	HgBrI	407.40	yel, rh	......	229	360	i; s al, eth
(I) carbonate	Hg_2CO_3	461.19	yel-brn cr	......	d 130	......	i; s dil HNO_3
(II) carbonate, basic	$HgCO_3 \cdot 2HgO$	693.78	brn-red	......		......	i; s HNO_3
(I) chlorate	$Hg_2(ClO_3)_2$	568.08	wh, rh	6.409	d 250	......	s
(II) chlorate	$Hg(ClO_3)_2$	367.49	wh nd	4.998	d	......	25
(I) chloride (calomel)	Hg_2Cl_2	472.09	wh, tetr, 1.973, 2.656	7.150	subl 383; 525 d	......	0.00020; s aq reg
(II) chloride (corrosive sublimate)	$HgCl_2$	271.50	col rh, or wh pwd, 1.859	5.44	276	302	6.9^{20}; 33 al; 4 eth
(I) chromate	Hg_2CrO_4	517.17	red nd or pwd	......	d	......	v sl s; s HCl, HNO_3
(II) chromate	$HgCrO_4$	316.58	red, rh	......	d	......	sl s, d; d a
(II) cyanide	$Hg(CN)_2$	252.63	col, tetr, or wh pwd	3.996	d	d	9.3^{14}; $25^{20°}$ MeOH; 8 al
(I) fluoride	Hg_2F_2	439.18	yel, cub	$8.73^{15°}$	570		s HNO_3
(II) fluoride	HgF_2	238.59	col, cub	$8.95^{15°}$	d 645	650	d; s HF, dil HNO_3
(I) fluorosilicate	$Hg_2SiF_6 \cdot 2H_2O$	579.29	col pr	2.134	d easily	......	sl s; i HCl
(II) fluorosilicate	$HgSiF_6 \cdot 6H_2O$	450.76	col, rh, deliq	......	d	......	....
(II) formate	$Hg_2(CHO_2)_2$	491.22	col scales	......	d	......	0.4^{17}
(II) fulminate	$Hg(CNO)_2$	284.62	wh, cub	4.42	expl	......	sl s; s al, NH_4OH
(I) iodate	$Hg_2(IO_3)_2$	750.99	yelsh	......	d 250	......	i; s HCl, conc HNO_3
(II) iodate	$Hg(IO_3)_2$	550.48	yel, tetr or amorp pwd	......	subl 140	d 250	i; s HCl
(I) iodide	Hg_2I_2	654.99	yel, tetr or amorp pwd	7.70	d 290		v sl s; s KI

Table 4-1 (*Continued*)
PHYSICAL CONSTANTS OF INORGANIC COMPOUNDS

Name	Formula	Formula Weight	Color, Crystalline Form, and Refractive Index	Density	Melting Point, °C	Boiling Point, °C	Solubility in 100 Parts
Mercury							
(II) iodide (α)	HgI_2	454.90	red, tetr	6.36	tr 127	.....	0.01; 3.2 acet; 2.2 al
(β)			yel, rh cr or pwd	6.094	259	354	v sl s; s eth, KI, $Na_2S_2O_3$
(I) nitrate	$Hg_2(NO_3)_2 \cdot 2H_2O$	561.22	col, mn, effl	4.79ᵃ	70	.....	d; s dil HNO_3
(II) nitrate	$Hg(NO_3)_2 \cdot \frac{1}{2}H_2O$	333.61	wh-yelsh cr or pwd, deliq	4.39	79	d	v s; s acet
(I) nitrite	$Hg_2(NO_2)_2$	493.19	yel	7.33	d 100	.....	d
nitride	Hg_3N_2	629.78	brn pwd	.....	expl	.....	d; d a, s NH_4OH
(I) oxalate	$Hg_2C_2O_4$	489.20	.....	.....	.....	.....	i; sl s HNO_3
(II) oxalate	HgC_2O_4	288.61	.....	.....	d	.....	0.011; s HCl
(I) oxide	Hg_2O	417.18	brn-blk pwd	9.8	$-O_2$, 100	.....	i; s HNO_3
(II) oxide (montroydite)	HgO	216.59	yel or red, rh, 2.37, 2.5, 2.65	11.14	d 300	.....	0.005; s a
(II) selenide (tiemannite)	$HgSe$	279.55	gray pl	8.266	subl vac	.....	i; s aq reg
(I) sulfate	Hg_2SO_4	497.24	col, mn	7.56	d	d	0.06; s HNO_3, H_2SO_4
(II) sulfate	$HgSO_4$	296.65	col, rh or wh pwd	6.47	d	.....	d; s a
(II) sulfate, basic	$HgSO_4 \cdot 2HgO$	729.83	lemon yel pwd	6.44	.....	.....	0.003; s a
(I) sulfide	Hg_2S	433.24	blk	.....	d	.....	i
(II) sulfide (cinnabar, vermillion) (α)	HgS	232.65	red, hex or pwd, 2.854, 3.201	8.10	subl 583	.....	i; s aq reg, Na_2S
(metacinnabar) (β)	.....		blk, cub, or amorp	7.73	583 subl	.....	i; s aq reg, Na_2S
(I) thiocyanate	$Hg_2(SCN)_2$	517.34	wh pwd	.....	d	.....	i; s HCl, KCNS
(II) thiocyanate	$Hg(SCN)_2$	316.75	yel, amorp	.....	d	.....	0.07; s HCl, KCNS
(I) tungstate	Hg_2WO_4	649.03	yel	.....	d	.....	i; d a
(II) tungstate	$HgWO_4$	448.44	yel	.....	d	.....	i; d a
Molybdenum	Mo	95.94	silv-wh, bcc	10.2	2625 ± 50	4800	s hot conc H_2SO_4, HNO_3, or aq reg

Name	Formula	Formula wt	Color, crystalline form	Density	m.p., °C	b.p., °C	Solubility
boride	MoB	106.75	tetr	8.65	2100		
boride, di-	MoB$_2$	117.56	rh	7.12			
(di-) boride	Mo$_2$B	202.69	tetr	9.26			s alk
(II) bromide	MoBr$_2$	255.76	or, amorp	4.88$^{18°}$	high		i; d alk
(III) bromide	MoBr$_3$	335.67	dk grn nd		(977 subl)		v s
(IV) bromide	MoBr$_4$	415.58	blk nd, deliq		d		
carbide	MoC	107.95	gray, hex	8.20	2692		sl s HNO$_3$, HF, hot H$_2$SO$_4$, HCl
(di-) carbide	Mo$_2$C	203.89	wh, hex pr	8.9	2687		sl s HNO$_3$, HF, hot H$_2$SO$_4$, HCl
carbonyl	Mo(CO)$_6$	264.00	col, rh	1.96	subl	156.4	s bz
(II) chloride	MoCl$_2$	166.85	yel	3.714	d 530		s a, alk, NH$_4$OH
(III) chloride	MoCl$_3$	202.30	red-brn, mn	3.578	(1027)		s conc HNO$_3$ or H$_2$SO$_4$
(IV) chloride	MoCl$_4$	237.75	brn pwd, deliq		d		d; s conc min a
(V) chloride	MoCl$_5$	273.21	blk, mn, deliq	2.928	194.4	268	hyd; s conc min a
dichloridetrifluoride	MoCl$_2$F$_3$	223.85	or		d >90		hyd
(III) fluoride	MoF$_3$	152.94	tan, rh-hed		d 600		
(V) fluoride	MoF$_5$	190.94	yel, mn	3.44	64	214	hyd; s alk, NH$_4$OH
(VI) fluoride	MoF$_6$	209.93	wh, bcc or o-rh	rh 3.27$^{-46°}$ cub 2.88^{5}	17.4	34.0	0.2; sl s HCl; s H$_2$O$_2$
hydroxide	Mo(OH)$_3$	146.96	blk pwd		d		
(II) iodide	MoI$_2$	349.75	brn pwd	5.278	(927)		i; v sl s a
(III) iodide	MoI$_3$	476.65	blk		d 100		i
(IV) iodide	MoI$_4$	603.56	blk cr				i
(III) oxide	Mo$_2$O$_3$	239.88	blk				
(IV) oxide	MoO$_2$	127.94	gray, tetr or mn	6.47			sl s h conc H$_2$SO$_4$
(V) oxide	Mo$_2$O$_5$	271.88	vlt-blk pwd				s h H$_2$SO$_4$, h HCl
(VI) oxide (molybdite)	MoO$_3$	143.94	wh-yel, rh	4.692	795	subl 1155	s a, alk sulf, NH$_4$OH
(V) oxide trichloride	MoOCl$_3$	218.30	grn cr		295	d >215	d; s HCl
(VI) oxide tetrachloride	MoOCl$_4$	253.75	dk grn		102	197	hyd; s conc HCl
(VI) oxide tetrafluoride	MoOF$_4$	187.93	wh, deliq	3.001	97.2 (tr pt)	186	s
dioxide dibromide	MoO$_2$Br$_2$	287.76	purp-brn cr, deliq				s

Table 4-1 (Continued)
PHYSICAL CONSTANTS OF INORGANIC COMPOUNDS

Name	Formula	Formula Weight	Color, Crystalline Form, and Refractive Index	Density	Melting Point, °C	Boiling Point, °C	Solubility in 100 Parts
Molybdenum							
dioxide dichloride	MoO_2Cl_2	198.84	wh, mn	3.31^{17}	175	250	hyd; s conc HCl
dioxide difluoride	MoO_2F_2	165.94	wh, hygr	3.494	subl 270	. . . . . .	v s
*meta*phosphate	$Mo(PO_3)_6$	569.77	yel pwd	3.28^0	. . . . . .	. . . . . .	sl s h aq reg
phosphide	MoP	126.93	gray-grn, hex	6.58	. . . . . .	. . . . . .	i
phosphide, di- (tri-) phosphide	MoP_2	157.89	. . . . . .	5.35	. . . . . .	. . . . . .	i
	Mo_3P	318.79	gray, tetr	9.07	. . . . . .	. . . . . .	i
silicide, di-	$MoSi_2$	152.11	grey met, tetr	6.31	1185	. . . . . .	s HNO_3 + HF
sulfide, di- (molybdenite)	MoS_2	160.07	blk, hex	4.80^{14}	. . . . . .	subl 450	s h H_2SO_4, aq reg, HNO_3
sulfide, sesqui-	Mo_2S_3	288.07	grey nd	5.91^{15}	d 1100	(1200)	d h HNO_3
sulfide, penta-	Mo_2S_5	352.18	dk brn pwd	. . . . . .	d	d	s NH_4OH, alk, sulf
sulfide, tri-	MoS_3	192.13	blk	. . . . . .	d	. . . . . .	sl s; s alk sulf, alk
Molybdic acid	$H_2MoO_4 \cdot H_2O$	179.97	yel, mn	3.124^{15}	$-H_2O$, 70	d	0.133^{18}; s alk
	H_2MoO_4	161.95	wh or sl yelsh, hex	3.112	$-H_2O$, 70	. . . . . .	sl s; s alk, NH_4OH
Molybdic arsenic acid	$6MoO_3 \cdot As_2O_5 \cdot 18H_2O$	1475.75	col, trig	2.493^{20}	$-15H_2O$, 150	. . . . . .	v s
Molybdic phosphoric acid	$H_7[P(Mo_2O_7)_6] \cdot 28H_2O$	2365.71	yel, oct	2.53	78	. . . . . .	d
Molybdic silicic acid	$H_8[Si(Mo_2O_7)_6] \cdot 28H_2O$	2363.83	yel, tetr	. . . . . .	45	d 100	v s
Neodymium	Nd	144.24	silv-wh, hex	7.004	1024	3030	s hot H_2O, a
acetate	$Nd(C_2H_3O_2)_3 \cdot H_2O$	339.39	. . . . . .	. . . . . .	. . . . . .	. . . . . .	26.2
boride, hexa-	NdB_6	209.10	bl, cub	4.68	. . . . . .	. . . . . .	s
bromate	$Nd(BrO_3)_3 \cdot 9H_2O$	690.10	red, hex	. . . . . .	66.7	$-9H_2O$, 150	151
bromide	$NdBr_3$	383.97	red-vlt, rh	5.35	682	1537	sl s
carbide	NdC_2	168.26	yel leaf, hex	5.15	d	. . . . . .	d; s dil a
carbonate	$Nd_2(CO_3)_3 \cdot xH_2O$	. . . . . .	lt red	. . . . . .	. . . . . .	. . . . . .	s a
(II) chloride	$NdCl_2$	215.15	grn, rh	. . . . . .	841	. . . . . .	. . . . . .
(III) chloride	$NdCl_3$	250.60	rose, hygr, hex	4.134	760	1690	97^{13}
	$NdCl_3 \cdot 6H_2O$	358.69	pink	. . . . . .	d 124	. . . . . .	246^{13}
fluoride	NdF_3	201.24	lilac, hex	. . . . . .	1374	2327	s H_2SO_4

Name	Formula	M.W.	Color, crystalline form	Density	M.P.	B.P.	Solubility
hydride, di-	NdH_2	146.26		5.94			
hydride, tri-	NdH_3	147.26	indigo bl, amorp				s a
hydroxide	$Nd(OH)_3$	234.26	bl, hex		$-H_2O$, 210	>320, Nd_2O_3	s a
(II) iodide	NdI_2	398.05	red-vlt		562		s
(III) iodide	NdI_3	524.95	lt grn, rh		784	1370	153^{25^o}
molybdate	$Nd_2(MoO_4)_3$	768.29	tetr, 2.005	5.14	1176		s a
nitrate	$Nd(NO_3)_3 \cdot 6H_2O$	438.35	tric				s a
oxalate	$Nd_2(C_2O_4)_3 \cdot 10H_2O$	732.69	rose cr				
oxide	Nd_2O_3	336.48	lt bl, hex	7.28	~1900		7.0
(di-) dioxide sulfide	Nd_2O_2S	352.54	blsh-wh, hex	6.22	1990		
selenide	Nd_2Se_3	525.36	viol-blk	6.67			
sulfate	$Nd_2(SO_4)_3 \cdot 8H_2O$	720.79	red, mn, 1.41, 1.55, 1.56	2.85	1176		s dil a
(II) sulfide	NdS	176.30	gold, cub	6.24	2140		1.05^{20^o} cc
(III) sulfide	Nd_2S_3	384.67	yel-grn pwd	5.38	2010		
Neon	Ne	20.18	col g, 1.000067	0.8999^o g/l, 1.207^{bp} liq	-248.6	-246.1	
Neptunium	Np	237.05	silv, o-rh	20.45	630		s HCl
(III) bromide	$NpBr_3$	476.76	grn, hex (α), o-rh (β)	6.62 (α)	467	subl 800	s
(IV) bromide	$NpBr_4$	556.67	red-brn, mn		464	subl 500	
(III) chlcride	$NpCl_3$	343.41	grn, hex	5.58	~800		s
(IV) chloride	$NpCl_4$	378.86	red-brn, tetr	4.92	538	subl 500	
(III) fluoride	NpF_3	294.05	purple, hex	8.95	1425		i
(IV) fluoride	NpF_4	313.04	grn, mn	6.80			i
(VI) fluoride	NpF_6	351.04	or-brn, o-rh	5.00	53	55.2	d
hydride, di-	NpH_2	239.06	cub				
hydride, tri-	NpH_3	240.07	hex				
(III) iodide	NpI_3	617.76	brn, rh	6.92	770	subl 800	
(IV) oxide	NpO_2	269.05	apple grn, cub	11.11	d 500		s conc a
(tri-) octaoxide	Np_3O_8	839.14	brn, cub		subl 500 d		s HNO_3
Neptunyl(IV) dichloride	$NpOCl_2$	323.96	orange, rh				

Table 4-1 (Continued)
PHYSICAL CONSTANTS OF INORGANIC COMPOUNDS

Name	Formula	Formula Weight	Color, Crystalline Form, and Refractive Index	Density	Melting Point, °C	Boiling Point, °C	Solubility in 100 Parts
Neptunyl(V) trifluoride	$NpOF_3 \cdot 2H_2O$	346.07	grn	....	....	....	....
Neptunyl(VI) difluoride	NpO_2F_2	307.05	pink, rh-hed	6.41	....	2840	s
Nickel	Ni	58.71	silv met, cub	8.90	1455		i; s dil HNO_3
acetate	$Ni(C_2H_3O_2)_2$	176.80	grn pr	1.798	d	....	16.6
	$Ni(C_2H_3O_2)_2 \cdot 4H_2O$	248.86	grn pr	1.744	d	....	16
hexaammine bromide	$[Ni(NH_3)_6]Br_2$	320.71	vlt pwd	1.837	....	....	v s
hexaammine chloride	$[Ni(NH_3)_6]Cl_2$	231.80	blsh, cub	1.468	....	....	s
hexaammine iodide	$[Ni(NH_3)_6]I_2$	414.70	lt bl, cub	2.101	d	....	d
hexaammine nitrate	$[Ni(NH_3)_6](NO_3)_2$	284.90	bl, oct or cub	....	....	....	4.46
ammonium chloride	$NiCl_2 \cdot NH_4Cl \cdot 6H_2O$	291.20	grn, mn, deliq	1.645	....	....	150
ammonium sulfate	$NiSO_4 \cdot (NH_4)_2SO_4 \cdot 6H_2O$	395.00	bl-grn, mn, 1.5007	1.923	....	....	2.54°
antimonide (breithauptite)	NiSb	180.46	lt copper red, hex	7.54	1158	d 1400	....
arsenate (xanthiosite)	$Ni_3(AsO_4)_2$	453.97	yel, amorp	4.982	....	....	i; s a
arsenate	$Ni_3(AsO_4)_2 \cdot 8H_2O$	598.09	yelsh-grn, pwd	4.98	....	....	i; s a
arsenide (niccolite)	NiAs	133.63	hex	7.57°	968	....	i; s aq reg
arsenide, di-	Ni_3As_2	325.97	met, tetr	7.86°	1000	....	....
benzenesulfonate	$Ni(C_6H_5SO_3)_2 \cdot 6H_2O$	481.10	grn, mn	1.628	$-H_2O$	d	14.3^{18}
boride	NiB	69.52	pr	7.39^{18}	....	....	d; s HNO_3
bromate	$Ni(BrO_3)_2 \cdot 6H_2O$	422.62	grn, mn	2.60	d	....	28
bromide	$NiBr_2$	218.53	yel-brn, deliq	5.098	963	subl	112°
bromide	$NiBr_2 \cdot 3H_2O$	272.57	yelsh-grn nd, deliq	....	$-3H_2O$, 300	....	v s
bromoplatinate(IV)	$NiPtBr_6 \cdot 6H_2O$	841.35	trig	3.715	....	....	....
carbide	Ni_3C	188.14	dk gray pwd	7.957	....	....	....
carbonate	$NiCO_3$	118.72	lt grn, rh	....	d	....	0.009; s a
carbonate, basic (zaratite)	$NiCO_3 \cdot 2Ni(OH)_2 \cdot 4H_2O(?)$	....	grn, cub, 1.56–1.61	2.6	....	....	i; s h dil HCl
carbonyl	$Ni(CO)_4$	170.75	col, inflam	1.31	−25	43	0.018^{20}; s aq reg, al eth, bz

Name	Formula	Formula weight	Crystalline form, color, refractive index	Density	Melting point, °C	Boiling point, °C	Solubility
chlorate	$Ni(ClO_3)_2 \cdot 6H_2O$	333.70	dk red	2.07	d 80	......	0.9
chloride	$NiCl_2$	129.62	yel, deliq	3.55	1030	subl 973	64^{20}
chloride	$NiCl_2 \cdot 6H_2O$	237.70	grn, mn, deliq, 1.57	......	......	......	v s
chloropalladate(IV)	$NiPdCl_6 \cdot 6H_2O$	485.92	hex	2.353	......	......	......
chloroplatinate(IV)	$NiPtCl_6 \cdot 6H_2O$	574.61	trig	2.798	......	......	......
cyanide	$Ni(CN)_2$	110.75	yel-brn	......	......	......	i; s KCN
	$Ni(CN)_2 \cdot 4H_2O$	182.81	lt grn pwd	......	−4H₂O, 200	......	i; s KCN, NH₄OH, alk
dimethylglyoxime	$Ni(HC_4H_6N_2O_2)_2$	288.94	scarlet cr	......	subl 250	d	i; s a, abs al
dithionate	$NiS_2O_6 \cdot 6H_2O$	326.93	grn, tric	1.908	d	......	......
ferrocyanide	$Ni_2Fe(CN)_6 \cdot xH_2O$	......	grn-wh	1.89	......	......	i; s KCN, NH₄OH
fluoride	NiF_2	96.71	grn, tetr	4.63	1450	subl	4
fluorosilicate	$NiSiF_6 \cdot 6H_2O$	308.88	grn, trig, 1.391, 1.407	2.134	d	......	s
formate	$Ni(CHO_2)_2 \cdot 2H_2O$	184.78	grn cr	2.154	d	......	......
hydroxide	$Ni(OH)_2$	92.72	grn cr or amorp	......	d 230	......	0.013; s a, NH₄OH
iodate	$Ni(IO_3)_2$	408.52	yel nd	5.07	......	......	1.1
	$Ni(IO_3)_2 \cdot 4H_2O$	480.59	yel, hex	......	d 100	......	1.4
iodide	NiI_2	312.52	blk, hex, deliq	5.834	797	136.7	124°
nitrate	$Ni(NO_3)_2 \cdot 6H_2O$	290.81	grn, mn, deliq	2.05	56.7	......	238°
oxalate	$NiC_2O_4 \cdot 2H_2O$	182.76	lt grn pwd	......	......	......	i; s a
oxide (bunsenite)	NiO	74.71	grn-blk, cub, 2.182	6.67	1990	......	i; s a
perchlorate	$Ni(ClO_4)_2 \cdot 6H_2O$	365.70	grn, hex, 1.518, 1.498	......	140	......	225°; s al, chl, acet
orthophosphate	$Ni_3(PO_4)_2 \cdot 8H_2O$	510.20	lt grn	3.93 anhyd	d	......	i; s a
pyrophosphate	$Ni_2P_2O_7 \cdot xH_2O$	......	grn	......	......	......	i; s a, NH₄OH
phosphide, di-	NiP_2	120.66	gray, o-rh	4.62	......	......	......
phosphide, tri-	NiP_3	151.63	gray, hex	4.19	......	......	s HNO₃ + HF
(di-) phosphide	Ni_2P	148.39	gray, hex	7.2	1112	......	......
(tri-) phosphide	Ni_3P	207.10	gray, tetr	7.7	975	......	i; s HNO₃
(tri-) *diphosphide*	Ni_3P_2	238.08	dk grn-blk	5.99	......	......	......
(penta-) *diphosphide*	Ni_5P_2	355.50	gray nd	7.5	1185	......	......
(hexa-) *pentaphosphide*	Ni_6P_5	414.21	gray cr	5.85	1185	......	......
hypophosphite	$Ni(PH_2O_2)_2 \cdot 6H_2O$	296.78	grn	1.82^{20}	d 100	......	s

Table 4-1 (*Continued*)
PHYSICAL CONSTANTS OF INORGANIC COMPOUNDS

Name	Formula	Formula Weight	Color, Crystalline Form, and Refractive Index	Density	Melting Point, °C	Boiling Point, °C	Solubility in 100 Parts
Nickel							
tetrapyridinenickel fluorosilicate	$[Ni(C_5H_5N)_4]SiF_6$	517.20	bl-grn, rh	2.307	....	....	....
selenate	$NiSeO_4 \cdot 6H_2O$	309.76	grn, tetr. 1.5393	2.314	....	....	s
selenide	$NiSe$	137.67	wh or gray, cub	8.46	red heat	....	i; s HNO_3
silicide	Ni_2Si	145.51	....	7.21^{17}	1309	....	i; a
sulfate	$NiSO_4$	154.78	yel, cub	3.68	d 848	....	$29.3°$
	$NiSO_4 \cdot 6H_2O$	262.86	α: bl, tetr; β: grn, mn, 1.511, 1.487	2.07	tr; 53.3	$-6H_2O$, 103	$62.5°$
(morenosite)	$NiSO_4 \cdot 7H_2O$	280.88	grn, rh, 1.467, 1.489, 1.492	1.948	99	$-7H_2O$, 103	$76^{20°}$
sulfide (millerite)	NiS	90.77	blk, trig or amorp	5.3–5.65	797	....	i; s HNO_3, KHS
(tri-) sulfide, di- (heazlewoodite)	Ni_3S_2	240.26	lt yel bronze lust	5.82	790	....	i; s HNO_3
(tri-) sulfide, tetra- (poydomite)	Ni_3S_4	304.39	gray-blk, cub	4.7	....	....	i; s HNO_3
sulfite	$NiSO_3 \cdot 6H_2O$	246.86	grn, tetr	....	....	....	i; s HCl
Niobium							
	Nb	92.906	steel-gray, bcc	8.57	2468	5127	s fus alk
boride, di-	NbB_2	114.53	hex	6.97	3050	....	....
(III) bromide	$NbBr_3$	332.62	blk, o-rh	....	subl 400 vac	....	....
(IV) bromide	$NbBr_4$	412.53	dk, o-rh	4.65	....	....	s dil HCl
(V) bromide	$NbBr_5$	492.46	vlt-blk, o-rh, hygr	4.36	268	362	d; s alc, ethyl bromide
carbide	NbC	104.92	blk, cub	7.6	3500	....	s HNO_3, HF
(III) chloride	$NbCl_3$	199.27	blk cr	$3.61^{20°}$	d 100	....	i org, dil min a
(IV) chloride	$NbCl_4$	234.72	vlt-blk, o-rh or mn	3.23	d	....	hyd; s HCl
(V) chloride	$NbCl_5$	270.17	yel, mn, hygr	2.84	205	248	s HCl, CCl_4

Name	Formula	Formula wt	Crystalline form, color	Density	Melting point	Boiling point	Solubility
(V) tetrachloride fluoride	$NbCl_4F$	253.72	yel, hygr	......	......	......	......
(IV) fluoride	NbF_4	168.90	blk, tetr, hygr	......	d 400	......	d; s alc
(V) fluoride	NbF_5	187.90	wh, mn	$2.70^{80°}$	80.0	234.9	
(di-) pentafluoride	Nb_2F_5	280.82	cub	4.91	d 700	......	s HF + HNO_3
hydride	NbH	93.91	gray pwd	6.6	infus	infus	
(III) iodide	NbI_3	473.62	hex	......	d 513	......	
(IV) iodide	NbI_4	600.53	gray nd, o-rh	......	d 450	......	
(V) iocide	NbI_5	727.43	bronze, mn, hyd in air	5.11	d 280	......	hyd
nitride	NbN	106.91	blk, cub	7.3	d 2050	......	s HNO_3 + HF
oxalate, hydrogen	$Nb(HC_2O_4)_5$	538.05	col, mn	......	......	......	d; s $H_2C_2O_4$
(II) oxide	NbO	108.91	blk, cub	7.30	......	......	s a, alk
(III) oxide	Nb_2O_3	233.81	bl-blk	......	1780	......	
(IV) oxide	NbO_2	124.90	blk	5.9	......	......	sl s alk; i a
(V) oxide	Nb_2O_5	265.81	wh, rh	4.47	1460	......	s HF, alk
oxide tribromide	$NbOBr_3$	348.63	yel	......	subl	......	hyd; s a
oxide trichloride	$NbOCl_3$	215.26	wh, tetr	3.27	d 270	......	hyd
dioxide fluoride	NbO_2F	143.90	cub	......	201	......	
Nitramide	NH_2NO_2	62.03	unstable weak a	......	d 72	......	d
Nitric acid	HNO_3	63.01	col liq, 1.3970	1.5027	-41.60	83	v s
(const boiling)	69% HNO_3 + H_2O	......	col liq	$1.41^{20°}$	......	120.5	v s
Nitrogen	N_2	28.01	col g, 1.0002779	$0.0011652^{20°}$ $0.808^{-195°}$	-209.97	-195.798	$1.52^{20°}$ cc
	$^{15}N_2$	30.00	col g, 1.000298	$0.00125^{20°}$	-209.952	-195.73	
bromide, tri-	NBr_3	253.73	red-vlt, expl	......	expl -67	......	
chloride, tri-	NCl_3	120.37	yel oily liq, rh	$1.653^{20°}$	-27	71	s bz, CS_2, CCl_4
fluoride, tri-	NF_3	71.00	col g, 1.0004492	$0.00296^{20°}$	-206.78	-129.06	i
fluoride, di-	NF_2	52.00	......	$1.537^{-129°}$	......	-125	
iodide, tri-	NI_3	394.72	blk, expl	......	expl	......	
iodide, (tri-) ammonia	$NI_3 \cdot NH_3$	411.75	blk, rh, shock sens	......	d	......	s KCNS, $Na_2S_2O_3$
(di-)oxide	N_2O	44.01	col g, 1.0004732	$1.8433^{20°}$ g/l	-90.82	-88.48	130° cc

Table 4-1 (Continued)
PHYSICAL CONSTANTS OF INORGANIC COMPOUNDS

Name	Formula	Formula Weight	Color, Crystalline Form, and Refractive Index	Density	Melting Point, °C	Boiling Point, °C	Solubility in 100 Parts
Nitrogen							
oxide, mono-	NO	30.01	col g, 1.0002743	$1.2488^{20°}$ g/l $1.2906^{-151°}$	-163.64	-151.77	7° cc
(di-)oxide, tri-	N_2O_3	76.01	bl liq red-brn g, bl liq	$1.447^{2°}$	-111	2 (d 3.5)	s eth
(di-)oxide, tetr-	N_2O_4	92.01	col (brn as NO_2)	$1.447^{20°}$	-11.20	21.15 d	s HNO_3, H_2SO_4, CS_2, $CHCl_3$
(di-)oxide, pent-	N_2O_5	108.01	col, hex or rh	$1.63^{18°}$	40.7	32.3 subl	s
oxide, tri-	NO_3	62.00	blsh g		d 20		s eth
(di-)sulfide, di-	N_2S_2	92.14	col, polymer, bl-blk				s org
(di-)sulfide, tetra-	N_2S_4	156.26	red-brn liq, gray sol	$1.71^{20°}$	23	30 vac	s CS_2
(di-)sulfide, penta-	N_2S_5	188.33	red liq	$1.901^{20°}$	10	d	s org
(tetra-)sulfide, tetra-	N_4S_4	184.28	or-yel, mn, 2.046	$2.24^{18°}$	180	185	s org
(di-)sulfur difluoride	N_2SF_2	98.07			-108	-11.1	
Nitrosyl							
azide	NON_3	72.02				1.5	
bromide	$NOBr$	109.92	dk brn g or liq		-55.5	19	hyd
bromide, tri-	$NOBr_3$	269.73		2.637	-40	32	
chloride	$NOCl$	65.46	or-yel g, red sol	$1.592^{-6.4°}$	-64.5	-6.4	hyd
fluoride	NOF	49.00	col	0.002788^{20}	-132.5	-59.9	d
fluoroborate	$NOBF_4$	116.81	col, rh, hygr	2.185	subl 250^{10mm}		d
fluorosulfate	$NOSO_3F$	129.07			140		
hydrogen selenate	$NOHSeO_4$	173.97			85 d		d, s H_2SO_4
hydrogen sulfate	$NOHSO_4$	127.08	col, rh		73.5		d
perchlorate	$NOClO_4 \cdot H_2O$	147.47	rh, deliq	2.169	d 100		d
(di-) trisulfate	$(NO)_2S_3O_{10}$	286.19			143		
(di-) pyrosulfate	$(NO)_2S_2O_7$	206.13			233	357	

	Formula	Mol. wt.	Color, crystalline form	Density	M.p., °C	B.p., °C	Solubility
sulfonyl fluoride	$NOSO_2F$	113.07			8		hyd
sulfuric anhydride	$(NOSO_3)_2O$	236.14	tetr		217	360	
Nitrous acid	HNO_2	47.01	lt bl, known only in soln		d		s
hypo-	$H_2N_2O_2$	62.03	wh,		expl		s
Nitryl							
chloride	NO_2Cl	81.46	col	$1.41^{-16°}$	-145	-15.3	d
fluoride	NO_2F	65.00	col	$0.0027^{20°}$	-166	-72.4	d
fluorosulfate	NO_2SO_3F	145.07			200		
hydrogen disulfate	$NO_2HS_2O_7$	223.14		2.1788	106		
hypofluorite	NO_2OF	71.00	col g, toxic	$1.507^{-181°}$	-181	-45.9	
nitrosyl trisulfate	$(NO_2)(NO)S_3O_{10}$	332.19			126		
(di-)trisulfate	$(NO_2)_2S_3O_{10}$	302.20			128	200	
Osmium	Os	190.2	gray-bl met, hcp	$22.61^{20°}$	2727	(4100)	s molten alk, oxid flukes
(di-)carbonyl, enneacarbonyl	$Os_2(CO)_9$	632.49	yel cr		subl		s NaOH
tricarbonyl dichloride	$Os(CO)_3Cl_2$	345.14	col pr		d 280		
(II) chloride	$OsCl_2$	261.11	dk brn, deliq		d	d	s al, eth, HNO_3
(III) chloride	$OsCl_3$	296.56	dk gray, cub		350 subl	d >450	hyd, s HNO_3
(IV) c-loride	$OsCl_4$	332.01	blk nd		d >350		i, hyd alk
(IV) fluoride	OsF_4	266.19	yel pwd		230		s
(V) fluoride	OsF_5	247.19	bl-gray, mn		70	225.7	hyd
(VI) fluoride	OsF_6	304.19	grn-yel, bcc		33.2	47.0	
(IV) iodide	OsI_4	697.82	vlt-blk, hygr	7.91			v s
(IV) oxide	OsO_2	222.20	blk, tetr	$11.37^{21°}$			s HCl, HF
	OsO_2	222.20	brn				i a
(VI) oxide	OsO_3	238.20	known only as g				
(VIII) oxide	OsO_4	254.20	pale yel tetr	4.91	40.6	130	$7.24^{25°}$; 250/100 g CCl_4
oxide tetrachloride	$OsOCl_4$	348.01	col nd; dk yel vapor		32	~200	hyd
(V) oxide pentafluoride	$OsOF_5$	263.19	emerald grn, o-rh		59.8	100.5	
(VIII) trioxide difluoride	OsO_3F_2	276.20	or, hex		171		hyd

Table 4-1 (*Continued*)
PHYSICAL CONSTANTS OF INORGANIC COMPOUNDS

Name	Formula	Formula Weight	Color, Crystalline Form, and Refractive Index	Density	Melting Point, °C	Boiling Point, °C	Solubility in 100 Parts
Osmium							
sulfide, di-	OsS_2	254.33	blk, cub	9.47	d	....	sl s aq reg
sulfide, tetra-	OsS_4	318.46	brn-blk	....	d	....	s dil HNO_3
sulfite	$OsSO_3$	270.26	bl-blk	....	d	....	s dil HNO_3
telluride	$OsTe_2$	445.40	gray-blk cr	....	~600	....	s dil HNO_3
Oxygen	O_2	32.00	col g, 1.0002713	0.001331[20] gas 1.118[bp] liq	−218.787	−182.98	0.0013M (1 atm 25°C)
difluoride	OF_2	54.00	col pois g	0.00226[20]	−223.8	−144.9	6.8° cc
(di-) difluoride	O_2F_2	70.00	pale brn g	1.45[bp] liq	−163.5	−56 d	....
(tri-) difluoride	O_3F_2	86.00	d red liq	1.80[bp] liq	−190	d −157	....
(tetra-) difluoride	O_4F_2	102.00	red brn sol	....	....	d −183	....
Oxygenyl hexafluoro platinate(V)	O_2PtF_6	275.09	d red, cub	....	219 d	....	hyd
Ozone	O_3	48.00	blue expl liq	1.46[bp] liq 0.001998[20] gas	−192.5	−111.3	0.494 cc/cc (0°C)
Palladium	Pd	106.4	silv-wh met, fcc	12.023	1552	2870	s hot HNO_3, H_2SO_4
(II) bromide	$PdBr_2$	266.22	red-brn, mn	5.173[16°]	d	....	d, s HBr
(II) chloride	$PdCl_2$	177.31	red, o-rh	4.018°	680	d	v s
	$PdCl_2 \cdot 2H_2O$	213.34	brn pr, deliq	....	d	....	
(II) cyanide	$Pd(CN)_2$	158.44	yel-wh	....	d	....	s KCN, aq NH_3
(II) fluoride	PdF_2	144.40	vlt, tetr	5.80	volat	d red heat	s HF
(II) *hexafluoro*-palladate(IV)	$Pd[PdF_6]$	326.80	blk, rh	5.06	d	....	d, s HF
hydride	Pd_2H	213.81	silv met	10.76	d	....	
(II) iodide	PdI_2	360.21	blk pwd	6.003[18°]	d >350	....	s KI
(II) nitrate	$Pd(NO_3)_2$	230.41	brn-yel, rh, deliq	....	d	....	s hyd; s HNO_3

Name	Formula	Mol. wt.	Color, crystalline form, etc.	Density	M.p., °C	B.p., °C	Solubility
(II) oxide	PdO	122.40	grn-blk	8.70[20]	870	......	i aq reg
(IV) oxide hydrate	PdO₂·xH₂O	......	dull red	......	d	......	s a, alk
(II) selenate	PdSeO₄	249.36	dk brn-red, rh, deliq	6.5	d red heat	......	v s
selenide	PdSe	185.36	dk gray	......	<960	......	s aq reg
selenide, di-	PdSe₂	264.32	ol gray, hex	......	......	......	s aq reg
silicide	PdSi	134.49	cr	7.31[15]	......	......	i
(II) sulfide	PdS	138.46	brn-blk, tetr	6.6	d 950	......	sl s HNO₃, aq reg
sulfide, di-	PdS₂	170.53	dk brn cr	4.75	d	......	s aq reg, NH₄S
(II) sulfate	PdSO₄·2H₂O	238.50	red-brn cr, deliq	......	d	......	v s
telluride, di-	PdTe₂	361.60	silv cr, hex	......	d	......	s HNO₃
Perchloric acid	HClO₄	100.46	col nd, expl and shock sens	1.767[20]	50	expl 110	v s
	HClO₄·H₂O	118.47	col oily liq, 1.382	1.7756[20]	−112	d	v s
	HClO₄·2H₂O	136.49	col, comm 72% liq	1.65	−17.5	200	v s
Perchloryl fluoride	ClO₃F	102.45	col g	1.392 g/l	−147.7	−46.7	......
Periodic acid, meta-	HIO₄	191.91	col cr	......	subl 110	d 138	v s
para-	H₅IO₆	227.94	col pr	......	130	d 140	113
Peroxydisulfuric acid	H₂S₂O₈	194.14	col cr, hygr	......	d 60	......	v s
Phosphazine, hexabromocyclotri-	(PNBr₂)₃	614.40	o-rh	3.18	192	......	......
hexachlorocyclotri-	(PNCl₂)₃	347.67	rh	1.98	114	256.5	s eth, bz, CS₂, dioxane
octachlorocyclotetra-	(PNCl₂)₄	463.56	tetrahed	2.18	123.5	328.5	s eth, bz, CS₂, dioxane
decachlorocyclopenta-	(PNCl₂)₅	579.45	......	2.02	41.3	223[13mm]	s eth, bz, CS₂, dioxane
dodecachlorocyclohexa-	(PNCl₂)₆	695.34	tric	1.96	90–91	261[13mm]	s eth, bz, CS₂, dioxane
tetradecachlorocyclohepta-	(PNCl₂)₇	811.23	......	1.89	8–12	291[13mm]	s eth, bz, CS₂, dioxane
hexafluorocyclotri-	(PNF₂)₃	248.94	o-rh	2.34	27.8	50.9	s eth, bz, CS₂, dioxane

Structure (hexachlorocyclotri-phosphazine): a P₃N₃ ring with Cl substituents on the phosphorus atoms.

Table 4-1 (Continued)
PHYSICAL CONSTANTS OF INORGANIC COMPOUNDS

Name	Formula	Formula Weight	Color, Crystalline Form, and Refractive Index	Density	Melting Point, °C	Boiling Point, °C	Solubility in 100 Parts
Phospham	PN_2H	60.00	wh, amorp	....	infus	....	i; s conc H_2SO_4
Phosphamic acid	$PONH_2(OH)_2$	97.01	wh, cub	....	d	....	v s
Phosphonium							
bromide	PH_4Br	114.91	col, cub	2.464 g/l gas	−32 subl	38.8^{794mm}	d
chloride	PH_4Cl	70.46	col, cub	....	50 subl	....	d
iodide	PH_4I	161.91	col, tetr, deliq	2.86	18.5	80 ; 61.8 subl	d
Phosphoric acid							
difluoro-	$H_2PO_2F_2$	102.99	col fum liq	1.583	−96.5	115.9	d
hypo-	$H_4P_2O_6$	162.01	col, rh, deliq	....	70	d 100	d
meta-	HPO_3	79.98	col, vitreous, deliq	2.2–2.5	subl	....	s
monofluoro-	H_2PO_3F	99.99	col oily liq	1.818	−80	....	....
ortho-	H_3PO_4	98.00	col liq, 85% comm	1.88	42.3	$-H_2O$, 213	v s
pyro-	$H_4P_2O_7$	177.98	col nd or liq, hygr	....	61	....	s
Phosphorous acid							
hypo-	$H(H_2PO_2)$	66.00	col oily liq or leaf	$1.493^{19°}$	26.5	d 50	s
meta-	HPO_2	63.98	col feathery cr	....	....	....	d
ortho-	H_3PO_3	82.00	col-yelsh cr, deliq	$1.651^{21°}$	70.1	d 200	s
pyro-	$H_4P_2O_5$	145.98	col nd	....	38	d 120	d
Phosphorus, white	P_4	....	wh waxy, cub, 2.1463	1.828	44.2	280.3	i; 880^{10} CS_2
red amorp	....	....	....	2.16	595	subl 398	s
black amorp	....	....	....	2.25	d 550	subl 431	i
red	P_4	123.90	red (vit), tric	2.34	597	subl 431	i
black	P_4	123.90	blk, o-rh	2.699	610	subl 453	i
bromide, tri-	PBr_3	270.72	col fum liq, 1.6903	$2.8902^{20°}$	−40.5	173.2	d; s eth, chl, CS_2
bromide, penta-	PBr_5	430.55	red-yel nd, rh, $[PBr_4^+Br^-]$	$3.46^{20°}$	....	103.7	d; s bz, CS_2, CCl_4
bromide difluoride	$PBrF_2$	148.88	col g	....	−133.8	−16.1	d

Name	Formula	Mol. wt.	Crystalline form, color, etc.	Density	M.p., °C	B.p., °C	Solubility
dibromidefluoride	PBr_2F	209.79	col liq	2.127^{20}	−115.0	78.3	d
dibromidetrifluoride	PBr_2F_3	247.79	liq	.....	−20	ca 106 d	.....
tetrabromide fluoride	PBr_4F	369.61	.....	.....	87	.....	.....
chloride, tri-	PCl_3	137.33	col liq, 1.5151	1.575^{20}	−92	76.1	d
chloride, penta-	PCl_5	208.24	grn-wh, tetrahed $[PCl_4^+PCl_6^-]$	2.119^{20}	160	subl 160	.....
(di-)tetrachloride	P_2Cl_4	203.75	col oily liq	.....	−28	180 d	.....
chloride dicyanate	$PCl(OCN)_2$	150.45	col liq	1.50	−50	134.6	.....
dichloride thiocyanate	$PCl_2(SCN)$	159.95	col liq	1.54	−76	148	.....
chloride difluoride	$PClF_2$	104.42	col g	.....	−164.8	−47.3	.....
dichloride fluoride	PCl_2F	120.88	col g	.....	−144.0	13.85	.....
dichloride trifluoride	PCl_2F_3	158.88	col g	5.4 g/l	.....	−8 (d 70)	.....
tetrachloride fluoride	PCl_4F	191.79	cr $[PCl_4^+F^-]$	.....	177	175 subl	.....
cyanate	$P(OCN)_3$	157.02	col	1.439	−2	169.3	d; v s eth
cyanide	$P(CN)_3$	109.03	wh nd	.....	subl 130	.....	d; s al
fluoride, tri-	PF_3	87.97	col g	3.65^{20} g/l	−151.5	−101.2	d
fluoride, penta-	PF_5	125.97	col g	5.295^{20} g/l	−93.8	−84.5	.....
iodide, tri-	PI_3	411.68	dk red, hex, deliq	4.18	61.5	d 200	d; s CS_2
(di-)tetraiodide	P_2I_4	569.57	or, tric	.....	124.5	330	d; s CS_2
(tetra-) oxide, hexa-	P_4O_6	219.90	wh, waxy or mn, deliq, 1.540	2.136^{20}	23.8	175.3	d; s bz, chl, eth
(di-)oxide, tetr-	P_2O_4	125.95	col, rh, deliq	2.539^{20}	100	180 vac	d
(tetra-)oxide, deca-	P_4O_{10}	283.88	wh: hex clII, 1.47; o-rh clI, 1.545; o-rh cl, 1.529	2.28; 3.0; 3.05	420; 565; 580	subl 358.9; 605 subl; 360–400	d; s H_2SO_4
(tetra-)selenide, tri-	P_4Se_3	360.78	or-red cr	1.31	242	.....	.....
(di-)selenide, penta-	P_2Se_5	456.75	blk-red nd	.....	d	.....	.....
(tetra-)trisulfide	P_4S_3	220.09	yel, rh	2.03^{17}	174	408	d; s CCl_4
(tetra-)pentasulfide	P_4S_5	284.20	yel cr	.....	170 d	.....	i; $100^{17°}$ CS_2
(tetra-)heptasulfide	P_4S_7	348.33	col pr	2.19^{17}	308	523	sl s CS_2
(tetra-)decasulfide	P_4S_{10}	444.54	gray-yel, tric, deliq	2.03	288	514	i; s alk
thiocyanate	$P(SCN)_3$	205.22	.....	1.625^{18}	ca −4	265	d; s al, eth, bz, CS_2
Phosphoryl bromide	$POBr_3$	286.70	col pl, tetrahed	3.25	51.3	192.0	d; s CS_2, bz, eth, chl

Table 4-1 (*Continued*)
PHYSICAL CONSTANTS OF INORGANIC COMPOUNDS

Name	Formula	Formula Weight	Color, Crystalline Form, and Refractive Index	Density	Melting Point, °C	Boiling Point, °C	Solubility in 100 Parts
Phosphoryl							
*di*bromide chloride	$POBr_2Cl$	242.27	col cr	2.450^{50}	31	165	d
bromide *di*chloride	$POBrCl_2$	197.79	col liq, 1.5091	2.1167	13	136	d
*di*bromide fluoride	$POBr_2F$	225.81	col liq	2.0363^{20}	−117.2	110.1	....
bromide *di*fluoride	$POBrF_2$	154.89	col liq	2.4842^{20}	−84.8	31.9	....
bromide chloride fluoride	$POBrClF$	181.33	col liq	1.872^{20}	....	79	....
chloride	$POCl_3$	153.33	col fum liq; 1.4866	1.6766^{20}	1.17	105.5	d
chloride *di*fluoride	$POClF_2$	120.42	col	1.656^{0}	−96.4	3.1	....
*di*chloride fluoride	$POCl_2F$	136.88	col liq	1.5497^{20}	−80.1	52.90	....
cyanate	$PO(OCN)_3$	173.02	col liq	1.570	5.0	193.1	....
fluoride	POF_3	103.98	col g	4.69 g/l	−39.5 subl	−39.1	d
nitride	PON	60.98	wh amorph	....	....	....	i
thiocyanate	$PO(SCN)_3$	221.16	col liq	1.484	13.8	300.1	....
Platinic acid							
hexachloro-	$H_2PtCl_6 \cdot 6H_2O$	517.92	red-brn cr, deliq	2.431	60	....	v s
hexahydroxyl-	$H_2Pt(OH)_6$	299.15	yel nd	....	−2H₂O, 100	−3H₂O, 120	i; s a, alk
hexaiodo-	$H_2PtI_6 \cdot 6H_2O$	1120.67	blk-red, deliq	....	....	....	d
Platinum	Pt	195.09	silv met, fcc	21.45^{20}	1774	~3800	s aq reg, fus alk
arsenide	$PtAs_2$	344.93	sperrylith, gray, cub	11.8	d >800	....	v sl s a
(II) *di*bromide	$PtBr_2$	354.91	brn	6.65	d 250	....	s HBr, KBr
(IV) *tetra*bromide	$PtBr_4$	514.73	dk red	5.69	d 180	....	0.41; v s al, eth, HBr
carbonyl bromide	$[Pt_2(CO)_2]Br_4$	765.84	lt red nd, hygr	5.115	d 180	....	s d, s CCl₄, bz
carbonyl dichloride	$Pt(CO)Cl_2$	294.00	yel nd	4.2346	195	240 subl	d, s conc HCl, H₂SO₄, al
*di*carbonyl *di*chloride	$Pt(CO)_2Cl_2$	322.02	lt yel nd	3.4882	142	−CO, 210	d, s CCl₄
*di*carbonyl *tetra*chloride	$[Pt_2(CO)_2]Cl_4$	588.01	yel-or nd	4.235	195	240 subl	d

Name	Formula	Mol. wt.	Color, crystalline form, refractive index	Density	M.p., °C	B.p., °C	Solubility
tricarbonyl *tetra-*chloride	Pt₂(CO)₃Cl₄	616.02	yel-or nd		130	d 250	d, s h CCl₄
carbonyl *di*iodide	Pt(CO)I₂	476.91	red cr	5.257	d 140		s d, al; s bz
carbonyl sulfide	Pt(CO)S	255.16	brn-blk		d 300		d alk al
(II) chloride	PtCl₂	266.00	ol grn, hex	6.05	d 581		s HCl, aq NH₃
(III) chloride	PtCl₃	301.45	grn-blk	5.256	435		sl s; s h HCl
(IV) chloride	PtCl₄	336.90	red-brn, cub	4.303	d 370		$58.7^{25°}$
(IV) chloride	PtCl₄·5H₂O	426.98	red, mn	2.43	−H₂O, 100		v s
(II) cyanide	Pt(CN)₂	247.13	yel-brn cr				s KCN
(II) fluoride	PtF₂	233.09	yel-grn				i
(IV) fluoride	PtF₄	271.08	yel-brn, mn		d red heat		sl hyd; s a, alk
*hexa*fluoride	PtF₆	309.08	dk red, o-rh	3.826 liq	61.3	69.1	
*penta*fluoride	PtF₅	290.08	dk red		80	disprop	
(II) hydroxide	Pt(OH)₂	229.10	blk		d 500		s HCl, HBr, alk
(II) iodide	PtI₂	448.90	blk pwd	6.403	d 230		s HI
(III) iodide	PtI₃	575.80	blk, cub	7.414			disprop; s KI
(IV) iodide	PtI₄	702.71	blk cr (or brn amorp)	6.064	d >25		s d; s al, acet, alk, KI
(II) oxide	PtO	211.09	vlt-blk	14.9^{15}	d 550		s HCl
(IV) oxide	PtO₂	227.09	blk	10.2	450		i aq reg
(II,IV) oxide	Pt₃O₄	649.27			d		i aq reg
(VI) oxide	PtO₃	243.09	red-brn pwd				s HCl, H₂SO₄
oxide *tetra*fluoride	PtOF₄	286.08	red		260	150 subl	
phosphide	PtP₂	257.04	steel gray, cub	9.25	~1500		v sl s aq reg
pyrophosphate	PtP₂O₇	369.03	grn-yel	4.85	d 600		v sl s
*di*selenide	PtSe₂	353.01	gray cr	7.65	d dry		s aq reg
*tri*selenide	PtSe₃	431.97	bl flakes	7.15	d 140		s aq reg
(IV) sulfate	Pt(SO₄)₂·4H₂O	459.27	yel pl				s
sulfide, mono-(cooperite)	PtS	227.15	blk, tetr	10.04	d		i
sulfide, di-	PtS₂	259.22	blk-brn pwd	7.66	d 225		i; s HCl, HNO₃
sulfide, sesqui-	Pt₂S₃	486.37	gray	5.52	d		sl s aq reg
telluride, di-	PtTe₂	450.29	gray, hex		1200–1300		sl s Na₂S
Platinum complexes							
(II) *tetrammine* chloride	[Pt(NH₃)₄]Cl₂·H₂O	352.13	col, tetr, 1.672, 1.667	2.737	250		

Table 4-1 (Continued)
PHYSICAL CONSTANTS OF INORGANIC COMPOUNDS

Name	Formula	Formula Weight	Color, Crystalline Form, and Refractive Index	Density	Melting Point, °C	Boiling Point, °C	Solubility in 100 Parts
Platinum complexes							
(IV) tetrachloro-diammine, cis-trans-	[Pt(NH$_3$)$_2$]Cl$_4$	370.96	....	3.3	200–216	....	....
	....	....	or-yel, rh or hex pl or nd	....	....	....	....
Plutonium	Pu	239.05	silv-wh, mn	19.82	638	3235	s a
(III) arsenate	PuAsO$_4$	377.97	lt grn, mn, 1.66	7.97	....	....	i
arsenide	PuAs	313.05	met, fcc	10.39	d >2000?	d >1300	s
(III) bromide	PuBr$_3$	478.79	bl-grn, rh, hygr	6.69	681	....	s h HNO$_3$ + NaF
carbide, mono-	PuC	251.06	blk, fcc	13.6	1650 d	....	....
carbide, sesqui-	Pu$_2$C$_3$	514.15	dull blk, bcc	12.70	2050 d	....	
carbide, di-	PuC$_2$	263.07	blk, tetr	....	2250 d	....	
(III) chloride	PuCl$_3$	345.42	emerald grn, hex, hygr	5.70	760	1767	v s a
(III) chloride 6H$_2$O	PuCl$_3$·6H$_2$O	451.53	grn, deliq	....	....	....	s
(IV) chloride	PuCl$_4$	380.87	exist (?)	....	....	....	
(III) fluoride	PuF$_3$	296.06	purp, hex	9.32	1425	d >2000	hyd
(IV) fluoride	PuF$_4$	315.05	lt brn-pink, mn	7.00	1037 d	....	i
(VI) fluoride	PuF$_6$	353.05	yel-brn, o-rh	4.86	51.59	62.16	....
hydride, di-	PuH$_2$	241.08	blk, fcc	10.40	~727	....	
hydride, tri-	PuH$_3$	242.08	blk, hex	9.61	~327	....	
(III) hydroxide	Pu(OH)$_3$·xH$_2$O	....	bl	....	....	....	s a
(IV) hydroxide	Pu(OH)$_4$·xH$_2$O	....	grn-gray polymer	....	....	....	0.45 mg
(III) iodate	Pu(IO$_3$)$_3$	667.76	tan	....	−H$_2$O, 150	d >475 to PuO$_2$	v sl s
(IV) iodate	Pu(IO$_4$)$_4$·xH$_2$O	....	pink amorp	....	....	....	....
(III) iodide	PuI$_3$	619.77	bl-grn, o-rh	6.92	~777	d >1000	....
(IV) nitrate	Pu(NO$_3$)$_4$·5H$_2$O	578.16	grn, o-rh, 1.554	2.90	−H$_2$O, 40	180, d to PuO$_2$(NO$_3$)$_2$	s
nitride	PuN	253.07	brn-blk, fcc	14.25	2570 (in H$_2$)	....	hyd; s dil HCl, H$_2$SO$_4$

Name	Formula	Formula weight	Color, crystalline form, refractive index	Density	Melting point, °C	Boiling point / decomposition, °C	Solubility
(III) oxalate	Pu₂(C₂O₄)₃·10H₂O	......	grn	......	−10H₂O, 225	d >300 to PuO₂	4 mg
(IV) oxalate	Pu(C₂O₄)₂·6H₂O	515.19	yel-grn	13.9	−6H₂O, 135	d >150	3.1 mg Pu
(II) oxide	PuO	255.05	blk, fcc	10.2	1900	......	......
(III) oxide (α)	Pu₂O₃	526.12	blk, bcc	11.47	2085 (in He)	d 2800	......
(β)	Pu₂O₃	526.12	silv met, hex	11.46	2390 (in He)	......	......
(IV) oxide	PuO₂	271.05	grn-brn, fcc	9.07	......	......	......
(III) oxide bromide	PuOBr	334.97	dk grn, tetr	8.81	......	......	s dil a
(III) oxide chloride	PuOCl	290.51	bl-grn, tetr	9.76	1637	......	s dil a
(III) oxide fluoride	PuOF	274.05	met, fcc	8.46	......	......	s dil H₂SO₄
(III) oxide iodide	PuOI	381.96	grn, tetr	7.55	......	......	......
(III) phosphate	PuPO₄	334.02	blue, mn, 1.86	6.27	d 1000	......	0.4 (in 0.8M H₃PO₄)
	PuPO₄·¹⁄₂H₂O	343.03	blue, hex, 1.76	4.03	−H₂O, 950	......	......
(IV) metaphosphate	Pu(PO₃)₄	475.97	pink-tan, o-rh, 1.65	4.37	d 1000 to PuPO₄	......	......
(IV) pyrophosphate	PuP₂O₇	412.99	col, cub, 1.68	......	high	......	......
phosphide	PuP	270.02	blk, fcc	9.85	......	......	......
silicide	PuSi	267.14	met, o-rh	10.15	1576 (in N₂)	d 700 (air)	......
silicide, di-	PuSi₂	295.23	met, bct	9.08	1638 (in N₂)	d 700 (air)	......
(tri-)silicide, penta-	Pu₃Si₅	618.54	met, hex	8.96	1646 (in N₂)	d 700 (air)	......
(penta-)silicide, tri-	Pu₅Si₃	1279.53	met, tetr	11.98	1377 (in N₂)	d 700 (air)	4.2
(III) sulfate	Pu₂(SO₄)₃·7H₂O	882.40	vit cr	......	......	......	s
(IV) sulfate	Pu(SO₄)₂	431.18	pink, hygr	......	......	......	s
(IV) sulfate	Pu(SO₄)₂·4H₂O	503.24	lt pink to red, prim std	......	......	......	......
sulfide, mono-	PuS	271.12	yel-brn, cub	10.6	2350	d >1350	......
sulfide, sesqui-	Pu₂S₃	574.30	blk	9.95	1725 vac	......	......
Plutonyl							
carbonate	PuO₂CO₃	331.06	red	......	......	d >130	......
chloride	PuO₂Cl₂·6H₂O	350.06	yel-grn	......	−H₂O, 130	......	35
fluoride	PuO₂F₂	309.05	wh, rh-hed	6.50	......	d >230	s
nitrate	PuO₂(NO₃)₂·6H₂O	513.17	grn, o-rh, 1.554	......	expl >180	......	......
oxalate	PuO₂C₂O₄·3H₂O	413.12	red	......	......	......	s (NH₄)₂C₂O₄, (NH₄)₂CO₃

Table 4-1 (Continued)
PHYSICAL CONSTANTS OF INORGANIC COMPOUNDS

Name	Formula	Formula Weight	Color, Crystalline Form, and Refractive Index	Density	Melting Point, °C	Boiling Point, °C	Solubility in 100 Parts
Plutonyl							
sulfate	$PuO_2SO_4 \cdot xH_2O$	....	vlt	....	....	....	....
Polonium	Po	210	silv-gray: α cub β rh	9.20 9.40	254	962	sl s; s dil min a
(II) bromide	$PoBr_2$	369.8	purp-brn	....	270 d (in N_2)	subl 110$^{0.03mm}$	s HBr
(IV) bromide	$PoBr_4$	529.6	bright red, fcc, hygr	....	324 (in Br_2 vapor)	360^{200mm}	sl hyd; s HBr, alc, acet
(II) chloride	$PoCl_2$	280.9	dk red, o-rh	6.50	subl 190 d (in Cl_2 vapor)	....	s dil HNO_3
(IV) chloride	$PoCl_4$	351.8	yel, mn, hygr	....	300	390	sl hyd; s $SOCl_2$
(IV) hydroxide	$Po(OH)_4$	278.0	lt yel, fcc	....			s alk
(IV) iodate	$Po(IO_3)_4$	909.6	wh	....	d 350		s 2N HCl
(IV) iodide	PoI_4	717.6	blk	....	subl 200 d (in N_2)		sl s alk, s acet
(II) oxide	PoO	226.0	blk	....	oxidized		....
(IV) oxide	PoO_2	242.0	yel fcc (also tetr)	~9	885 subl	d 500 vac	....
(II) selenite	$PoSeO_3$	337.0	bright red	....	d 25 to PoO		....
(IV) sulfate	$Po(SO_4)_2$	402.2	purp	....	d 550		v s dil HCl
(II) sulfite	$PoSO_3$	290.1	bright red	....	d 25 to PoO		....
(II) sulfide	PoS	242.1	....	....		d 275 vac	s aq reg
Potassium	K	39.10	silv met, cub	0.87 0.83^{mp} liq	63.5	758	d to KOH; s a
acetate	$KC_2H_3O_2$	98.15	wh pwd	1.57	292	....	v s

Name	Formula	Mol wt	Color, crystalline form, refractive index	Density	M.p., °C	B.p., °C	Solubility
aluminate	$K_2Al_2O_4 \cdot 3H_2O$	250.21	col cr				s
amide	KNH_2	55.12	yel-grn	1.64	338	subl 400	d
antimonate, meta-	$KSbO_3$	208.85	col cr				i; s h KOH
antimonide	K_3Sb	239.06	yel-grn		812 (in N_2)	d in air	d
antimony tartrate	$KSbC_4H_4O_6 \cdot \frac{1}{2}H_2O$	333.93	col, rh, 1.620, 1.636, 1.638	2.607	$-H_2O$, 100		$5.3^{0°}$, 6.67 gly
arsenate, ortho-	K_3AsO_4	256.23	col cr, deliq		1310		19
arsenate, mono-H	$K_2HAsO_4 \cdot H_2O$	236.15	col, trig		$-H_2O$, 110		s
arsenate, di-H	KH_2AsO_4	180.04	col, tetr, 1.5674	2.867	288		$19^{6°}$
arsenite, meta-	$KAsO_2$	146.02	wh pwd, hygr				s
aurate(III)	$KAuO_2 \cdot 3H_2O$	322.11	yel nd		d		v s
azide	KN_3	81.12	col, tetr	2.04	350 vac		$47^{11°}$
borane, di-	$K_2B_2H_6$	105.87	wh, cub, 1.493	1.18	subl 400		s
borane, penta-	$K_2B_5H_9$	141.32	wh pwd		d 180		s d
borate, meta-	KBO_2	81.91	col, hex, 1.526, 1.450		950		$71^{30°}$
borate peroxo-	$KBO_3 \cdot \frac{1}{2}H_2O$	106.92	wh cr		$-O_2$, 100	d 150	$1.22^{0°}$
borate tetra-	$K_2B_4O_7 \cdot 8H_2O$	377.57	col, mn	1.74	d		$26.7^{30°}$
bromate	$KBrO_3$	167.01	col, trig	anhyd 3.27^{18}	434		8.14
bromide	KBr	119.01	col, fcc, 1.5594	2.75	748–760	1380	67
bromoaurate	$K[AuBr_4]$	555.71	brn, mn		d 120		sl s
	$K[AuBr_4] \cdot 2H_2O$	591.74	dk brn, rh	4.08			$19.5^{15°}$
bromoplatinate(II)	K_2PtBr_4	592.93	brn, rh	4.66			v s
bromoplatinate(IV)	$K_2[PtBr_6]$	752.75	dk red-brn, cub	3.783	d 400		$2.0^{20°}$
bromostannate(IV)	$K_2[SnBr_6]$	676.35	wh cr	1.846			33
cadmium cyanide	$K_2[Cd(CN)_4]$	294.68	col, oct	2.29	450	d	113
carbonate	K_2CO_3	138.21	wh pwd, deliq, 1.531	2.043	891		147
	$K_2CO_3 \cdot 2H_2O$	174.24	col, mn, deliq, 1.380, 1.432, 1.573		$-H_2O$, 130		
	$2K_2CO_3 \cdot 3H_2O$	330.47	col, mn, 1.380, 1.482, 1.573	2.043			129
carbonate, hydrogen	$KHCO_3$	100.12	col, mn, 1.482	2.17	d 100–200		36.1
chlorate	$KClO_3$	122.55	col, mn, 1.5167	2.32	368	d 400	$7.1^{20°}$

Table 4-1 (*Continued*)
PHYSICAL CONSTANTS OF INORGANIC COMPOUNDS

Name	Formula	Formula Weight	Color, Crystalline Form, and Refractive Index	Density	Melting Point, °C	Boiling Point, °C	Solubility in 100 Parts
Potassium							
chloride (sylvite)	KCl	74.56	col, cub, 1.4904	1.988	772	1411	25.5
chloroaurate(III)	K[AuCl$_4$]	377.88	yel, mn	3.75	d 357	....	61.8$^{20°}$
chlorochromate(VI)	K[CrO$_3$Cl]	174.55	red, mn	2.497	d	....	s d; s a
chloroiridate(IV)	K$_2$[IrCl$_6$]	483.12	blk, cub	3.546	d	....	1.3$^{20°}$
chloronitrosyl ruthenate(II)	K$_2$[Ru(NO)Cl$_5$]	386.55	dk red, rh	....	d	....	12
chloroosmate(III)	K$_3$[OsCl$_6$]·3H$_2$O	574.27	dk red cr	....	−3H$_2$O, 150	....	v s
chloroosmate(IV)	K$_2$[OsCl$_6$]	481.12	red, cub	d	d	....	sl s
chloropalladate(II)	K$_2$[PdCl$_4$]	326.42	red-brn, tetr	2.67	d 105	....	s
chloropalladate(IV)	K$_2$[PdCl$_6$]	397.32	red, cub	2.738	d	....	sl s d; sl s HCl
chloroplatinate(II)	K$_2$[PtCl$_4$]	415.11	red-brn, tetr, 1.64, 1.67	3.38	d	....	0.9$^{16°}$
chloroplatinate(IV)	K$_2$[PtCl$_6$]	486.01	yel, cub, 1.83	3.499	d 250	....	0.48$^{12°}$
chloroplumbate(IV)	K$_2$[PbCl$_6$]	498.11	lt yel, cub	....	d 190	....	d; s h HCl
chlororhenate(IV)	K$_2$[ReCl$_6$]	477.12	yel-grn, oct	3.34	....	....	0.8
*penta*chlororhenate(V)	K$_2$[ReOCl$_5$]	457.67	grn, hex pl	....	....	....	d; s a
chlororhodate(III)	K$_3$[RhCl$_6$]·3H$_2$O	486.98	red, tric	3.291	d	....	d
chlororhodate(IV)	K$_2$[RhCl$_6$]	358.37	blk, cub	....	d	....	s d
chlororuthenate(IV)	K$_2$[RuCl$_6$]	391.99	red, rh	....	d	....	sl s
chlorostannate(IV)	K$_2$[SnCl$_6$]	409.63	col, cub, 1.657	2.71	....	....	s
chlorotellurate(IV)	K$_2$[TeCl$_6$]	418.52	pale yel, oct	....	....	....	d
chromate (tarapacaite)	K$_2$CrO$_4$	194.20	yel, rh, 1.7261	2.732$^{18°}$	975	....	36$^{20°}$
chromate, peroxo-	K$_3$CrO$_8$	297.30	brn-red, cub	....	d 170	....	sl s
chromium(III) sulfate	K[Cr(SO$_4$)$_2$]·12H$_2$O	499.41	vlt-red, cub, 1.4814	1.826	89; −12H$_2$O, 400	....	24.4
citrate	K$_3$C$_6$H$_5$O$_7$·H$_2$O	324.42	col cr	1.98	d 230	....	167$^{15°}$
cobalt sulfate	K$_2$SO$_4$·CoSO$_4$·6H$_2$O	437.36	red pr, mn, 1.481, 1.487, 1.500	2.218	....	....	25.5$^{0°}$

Name	Formula	Mol. wt.	Crystalline form, color, refractive index	Density	M.p. (°C)	B.p. (°C)	Solubility
cyanate	KOCN	81.12	wh, tetr	2.048	d 700–900	....	s
cyanide	KCN	65.12	wh, cub, deliq, 1.410	1.52^{16}	634.5	....	50
cyanoaurate(I)	K[Au(CN)$_2$]	288.11	col, rh	3.45	d 200	....	14.3
cyanoaurate(III)	K[Au(CN)$_4$]·1.5H$_2$O	367.16	col pl	....	....	....	s
cyanoargentate(I)	K[Ag(CN)$_2$]	199.01	col, 1.625	2.36	....	....	25^{30}
cyanocobaltate(II)	K$_4$[Co(CN)$_6$]	371.45	red-brn cr, deliq	2.039	....	....	v s
cyanocobaltate(III)	K$_3$[Co(CN)$_6$]	332.35	wh-yel, mn pr	1.878	....	....	sl s; s a
cyanomanganate(III)	K$_3$[Mn(CN)$_6$]	328.35	red, rh, 1.553, 1.555, 1.571(Li)	....	....	....	s
cyanomolybdate(IV)	K$_4$[Mo(CN)$_8$]·2H$_2$O	496.52	yel, mn	2.337	−H$_2$O, 110	....	v s
cyanonickelate(II)	K$_2$[Ni(CN)$_4$]·H$_2$O	259.00	red-yel, mn	1.875^{11}	−H$_2$O, 100	....	s
cyanoplatinate(II)	K$_2$[Pt(CN)$_4$]·3H$_2$O	431.41	lt yel, rh, deliq	2.455^{16}	−3H$_2$O, 100	d >400	sl s
cyanotungstate(IV)	K$_4$[W(CN)$_8$]·2H$_2$O	584.43	lt yel-grn cr pwd	1.989	−2H$_2$O, 114	....	130^{18}
dichromate	K$_2$Cr$_2$O$_7$	294.19	red, tric	2.676	398	d 500	4.9^0
dithionate	K$_2$S$_2$O$_6$	238.33	col, trig, 1.4550	2.278	d	....	6.06^{16}
ferricyanide	K$_3$[Fe(CN)$_6$]	329.26	red, mn pr, 1.5689	1.84	d	....	33^4
ferrocyanide	K$_4$[Fe(CN)$_6$]·3H$_2$O	422.41	yel, mn, 1.5772	1.853^{17}	−3H$_2$O, 70	d	28^{12}
fluoride	KF	58.10	col, cub, deliq, 1.352	2.48	857	1500	92^{18}
	KF·2H$_2$O	94.13	col, mn pr, 1.352	2.454	41	....	v s
fluoride, hydrogen	KHF$_2$	78.11	col, cub, deliq	2.37	d ca 225	d	41^{21}
fluoroberyllate	K$_2$BeF$_6$	163.21	col, rh	....	red heat	....	2^{20}
fluoroborate	KBF$_4$	125.91	col, cub or rh, 1.324, 1.325, 1.325	2.498^{20}	d 350	....	0.44^{20}
fluorogermanate	K$_2$GeF$_6$	264.78	wh, hex	....	730	835	180^{18}
fluorohafniate	K$_2$HfF$_6$	370.68	wh, mn	....	....	....	3.1^{20}
pentafluoroniobate(V)	K$_2$[NbOF$_5$]·H$_2$O	300.12	col, mn pl or leaf	....	....	....	7.69
fluorophosphate	KPF$_6$	184.07	col	....	ca 575	....	9.3
fluorosilicate (hieratite)	K$_2$SiF$_6$	220.25	col, cub or hex, 1.3991	hex 3.08 cub 2.666^{17}	d	d	hex 0.12^{18} cub 6.9^{20}
fluorostannate(IV)	K$_2$SnF$_6$·H$_2$O	328.90	wh, mn	3.053	....	....	6.7^{18}
fluorosulfate	KSO$_3$F	138.16	wh	....	311.1	....	s
fluorotantalate(V)	K$_2$TaF$_7$	392.14	col, rh	4.5–5.2	....	....	sl s

Table 4-1 (Continued)
PHYSICAL CONSTANTS OF INORGANIC COMPOUNDS

Name	Formula	Formula Weight	Color, Crystalline Form, and Refractive Index	Density	Melting Point, °C	Boiling Point, °C	Solubility in 100 Parts
Potassium							
fluorotitanate(IV)	$K_2TiF_6 \cdot H_2O$	258.11	wh, mn	2.992	$-H_2O$, 100	......	1.3^{20}
tetrafluoro-tungstate(VI)	$K_2[WO_2F_4] \cdot H_2O$	388.06	col, mn	......	$-H_2O$, red heat	......	6^{17}
fluorozirconate	K_2ZrF_6	283.41	col, mn, 1.465	3.58	......	......	0.78^0
formate	$KHCO_2$	84.12	col, rh	1.91	167.5	d	v s
gallium sulfate	$KGa(SO_4)_2 \cdot 12H_2O$	517.13	col cr	1.895	......	......	s
germanate, meta-	K_2GeO_3	198.79	wh cr	3.40^{21}	823	......	s
hydride	KH	40.11	wh nd, cub, 1.453	1.43	d	......	d
tetrahydridoborate	KBH_4	53.94	wh, cub, 1.494	1.178	d 500	......	19.3^{20}
hydroxide	KOH	56.11	wh, rh, deliq	2.044	360	1320	112^{20}
hypochlorite	KClO	90.55	exists only in soln	......	d	......	v s
hypophosphite	KPH_2O_2	104.09	col, hex, deliq	......	ign	......	200^{24}
iodate	KIO_3	214.00	col, mn	3.89	560	......	4.7^0
bis-iodate, hydrogen	$KH(IO_3)_2$	389.92	col, rh or mn	......	......	......	1.33^{15}
iodide	KI	166.01	wh, cub, 1.6670	3.13	677	~1330	144^{20}
iodide, tri-	KI_3	419.92	dk bl, mn, deliq	3.498	45	d 225	v s
iodoaurate(III)	$KAuI_4$	743.69	blk lust cr	......	d 150	......	s d
iodomercurate(II)	K_2HgI_4	786.41	yel cr, deliq	......	......	......	v s
iodoplatinate(IV)	K_2PtI_6	1034.72	blk, cub	4.96	......	......	s
manganate(VI)	K_2MnO_4	197.14	grn, rh	......	d 190	......	d; s KOH
molybdate	K_2MoO_4	238.14	wh pwd, deliq	2.91^{18}	919	d 1400	184
nitrate (saltpeter)	KNO_3	111.11	col, rh or trig, 1.335, 1.5056, 1.5064	2.109^{16}	tr to trig, 129; mp 334	d 400	32^{30}
nitride	K_2N	131.31	grnsh-blk	......	d	d	d
nitrite	KNO_2	85.11	wh-yelsh pr, deliq	1.915	440	d	v s
hexanitrito-cobaltate(III)	$K_3[Co(NO_2)_6] \cdot 1.5H_2O$	479.30	yel, tetr	......	d 200	......	0.089^{17}

Name	Formula	Mol. wt.	Crystalline form, color	Density	Melting point	Boiling point	Solubility
oxalate	$K_2C_2O_4 \cdot H_2O$	184.24	wh, mn, 1.4410, 1.485, 1.550	2.127[4°]	$-H_2O$, 100	d	33[16°]
oxalate, hydrogen	KHC_2O_4	128.11	col, mn, 1.382, 1.553, 1.573	2.044	d	....	2.5
bis-oxalate, hydrogen	$KH_3(C_2O_4)_2 \cdot 2H_2O$	254.20	col, tric, 1.415, 1.536, 1.560	1.836	d	....	1.8[13°]
oxalatoferrate(III)	$K_3[Fe(C_2O_4)_3] \cdot 3H_2O$	491.25	grn, mn	....	$-3H_2O$, 100	d 230	4.7[0°]
oxide	K_2O	94.20	wh, cub	2.32[20°]	>760	....	d to KOH
oxide, "super"	KO_2	71.10	yel, cub leaf	2.14	380	d	v s d
perchlorate	$KClO_4$	138.55	col, rh, 1.4737	2.5298	610	d 653	1.99
periodate	KIO_4	230.00	col, tetr, 1.6205	3.618	582 d	....	0.66[13°]
permanganate	$KMnO_4$	158.04	purp, rh	2.703	d 240	....	6[20°]
peroxide	K_2O_2	110.20	wh, amorp, deliq	....	d	d	d
perrhenate	$KReO_4$	289.30	wh, tetr, 1.643	4.38	555	1370	1.21[20°]
phosphate	K_3PO_4	212.28	wh, rh	2.564[17°]	1340	....	193
phosphate, mono-H	K_2HPO_4	174.18	wh, deliq	....	d	....	33
phosphate, di-H	KH_2PO_4	136.09	col, tetr, deliq, 1.5095	2.338	256 d	....	33
phosphate, meta-	KPO_3	118.07	col, 1.458, 1.487	2.393	807	....	v sl s
phosphate, pyro-	$K_4P_2O_7 \cdot 3H_2O$	384.40	col, deliq	2.33	$-3H_2O$, 300	....	s
phthalate, hydrogen	$KHC_8H_4O_4$	204.23	wh cr, rh	1.636	d	....	10.2
picrate	$KC_6H_2N_3O_7$	267.20	yelsh or grnsh, rh, 1.527, 1.903, 1.952	1.852	expl 310	....	0.5[15°]
platinate(IV)	$K_2PtO_3 \cdot 3H_2O$	375.34	yel, rh	....	d	....	s
ruthenate(VI)	$K_2RuO_4 \cdot H_2O$	261.29	blk, rh	....	$-H_2O$, 200	....	v s
selenate	K_2SeO_4	221.16	col, rh, 1.535, 1.539, 1.545	3.066[20°]	....	....	110°
selenide	K_2Se	157.16	wh, cub, hygr	2.29	....	....	s d
selenite	K_2SeO_3	205.16	wh, deliq	....	d 875	....	s
selenocyanate	$KSeCN$	144.08	nd, deliq	....	d 100	....	s
selenocyanato-platinate(IV)	$K_2[Pt(SeCN)_6]$	903.16	rh	3.378[13°]	d 80	....	....
silicate, meta-	K_2SiO_3	154.29	col, rh, 1.530	....	976	....	s
silicate, meta-tetrasilicate	$K_2Si_4O_9 \cdot H_2O$	352.56	col, rh, 1.530	2.417	d 400	....	s
sodium carbonate	$KNaCO_3 \cdot 6H_2O$	230.19	col, mn	1.633	$-6H_2O$, 100	....	185[15°]

Table 4-1 (*Continued*)
PHYSICAL CONSTANTS OF INORGANIC COMPOUNDS

Name	Formula	Formula Weight	Color, Crystalline Form, and Refractive Index	Density	Melting Point, °C	Boiling Point, °C	Solubility in 100 Parts
Potassium							
sodium nitritocobaltate(III)	$K_2Na[Co(NO_2)_6]\cdot H_2O$	454.18	yel	1.633	d 135	....	0.07
sodium tartrate (Rochelle salt)	$KNaC_4H_4O_6\cdot 4H_2O$	282.23	col, rh, 1.492, 1.493, 1.496	1.790	70–80	$-4H_2O$, 215	66
stannate(IV)	$K_2SnO_3\cdot 3H_2O$	298.94	col, trig	3.197	$-3H_2O$, 140	....	110^{20}
sulfane, tri-	K_2S_3	174.40	yel-brn	....	252	....	s
sulfane, tetra-	K_2S_4	206.46	red-brn	....	145	d 850	s
sulfane, penta-	K_2S_5	238.52	or cr, hygr	....	206	....	v s
sulfate (arcanite)	K_2SO_4	174.27	col, rh, 1.494	2.662	1074	1670	12
sulfate, hydrogen (mercallite)	$KHSO_4$	136.17	col, rh, deliq, 1.480	2.322	214	d	36°
sulfate, peroxodi-	$K_2S_2O_8$	270.33	col, tric, 1.461, 1.467, 1.566	2.477	d	....	5.3^{20}
sulfate, pyro-	$K_2S_2O_7$	254.33	col nd	2.512	300 d	....	s
sulfide	K_2S	110.27	brn, deliq, cub	$1.805^{14°}$	840	....	s
sulfide, hydrogen	KHS	72.17	yel, rh, deliq	1.69	455	....	d
sulfide, di-	K_2S_2	142.33	red-yel	....	471	....	s
sulfite	$K_2SO_3\cdot 2H_2O$	194.30	wh, rh	....	d	....	100
sulfite, hydrogen	$KHSO_3$	120.17	wh, mn	....	d 190	....	$45.5^{15°}$
tartrate	$K_2C_4H_4O_6\cdot \frac{1}{2}H_2O$	235.28	col, mn, 1.526	1.98	$-H_2O$, 155	d 200	$150^{14°}$
tartrate, hydrogen	$KHC_4H_4O_6$	188.18	col, rh	1.956		....	0.520^{20}
tellurate, ortho-	$K_2H_4TeO_6\cdot 3H_2O$	359.88	col, rh, deliq	....	$-O_2$, 300	....	sl s
telluride	K_2Te	205.80	col, cub, hygr	2.51	....	....	s d
thioantimonate(V)	$2K_3SbS_4\cdot 9H_2O$	896.76	yel cr	....	d	....	300°
thioarsenate(V)	K_3AsS_4	320.48	wh cr, deliq	....	d	....	v s
thiocarbonate	K_2CS_3	186.41	yel-brn, deliq	....	d	....	v s
thiocyanate	$KSCN$	97.18	col, rh pr, deliq	$1.886^{14°}$	173.2	d 500	217^{20}
thionate, tri-	$K_2S_3O_6$	270.39	col, rh, 1.475, 1.480, 1.487	2.304	d 30–40	....	s

Name	Formula	Formula wt	Color, crystalline form	Density	M.P., °C	B.P., °C	Sol. cold water	Sol. other reagents
thiostannate(IV)	$K_2SnS_3 \cdot 3H_2O$	347.13	dk brn oil	1.847^{18}	$-3H_2O$, 100			s
thiosulfate	$K_2S_2O_3$	190.33	col, cub	......	d 400		96°	
tungstate	$K_2WO_4 \cdot 2H_2O$	362.08	col, mn, deliq	3.113	921		51.5	
uranyl acetate	$KUO_2(C_2H_3O_2)_3 \cdot H_2O$	504.28	yel, tetr	3.296^{15}	$-H_2O$, 275			s
uranyl carbonate	$2K_2CO_3 \cdot UO_2CO_3$	606.46	yel, hex	......	$-CO_2$,		$7.4^{15°}$	
uranyl sulfate	$K_2SO_4 \cdot UO_2SO_4 \cdot 2H_2O$	576.39	yel, mn	3.363^{19}	$-2H_2O$, 120			s
vanadium sulfate	$KV(SO_4)_2 \cdot 12H_2O$	498.35	vlt, cub; liq 20°	1.783^{20}	$-12H_2O$, 230			v s
xanthate, ethyl-	KC_2H_5OCSS	160.30	pale yel cr or pwd	1.558^{22}	d 200			v s
Praseodymium								
	Pr	140.907	pale yel, hex	6.782	935	3130		s hot H_2O, a
acetate	$Pr(C_2H_3O_2)_3 \cdot 3H_2O$	372.09	grn nd	......				v s
bromate	$Pr(BrO_3)_3 \cdot 9H_2O$	686.77	grn, hex		56.5	$-7H_2O$, 170		v s
bromide	$PrBr_3$	380.63	yel-grn, hex	5.26	691	1547		d sl s
carbide	PrC_2	164.93	yel cr	5.10	d			s dil a
carbonate	$Pr_2(CO_3)_3 \cdot 8H_2O$	605.96	grn slky pl		$-6H_2O$, 100			s a
chloride	$PrCl_3$	247.27	pale grn nd, hex	4.02	786	1710		v s
	$PrCl_3 \cdot 7H_2O$	373.37	grn, tric	2.25	115			v s
(III) fluoride	PrF_3	197.91	grn, hex	4.94	1395	2327		s H_2SO_4
(IV) fluoride	PrF_4	215.91	wh, mn	5.5				
hydride, tri-	PrH_3	143.92	grn amorp					
hydroxide	$Pr(OH)_3$	191.93	grn, hex		$-H_2O$, 220	d 340 to Pr_2O_3		s a
iodide, di-	PrI_2	394.72	bronze [$Pr^{3+}(e^-)(I^-)_2$]		758			
iodide, tri-	PrI_3	521.62	rh		737	1380		s
molybdate	$Pr_2(MoO_4)_3$	761.63	grn, tetr	4.84	1030			
nitrate	$Pr(NO_3)_3 \cdot 6H_2O$	339.03	grn nd					
oxalate	$Pr_2(C_2O_4)_3 \cdot 10H_2O$	726.03	lt grn cr					s a
(III) oxide	Pr_2O_3	329.81	lt grn, hex	7.07	d			s a
(IV) oxide	PrO_2	172.91	bl-blk, bcc	6.82	>350 tr to Pr_6O_{11}			
oxide	Pr_6O_{11}	1021.44	bcc					
selenate	$Pr_2(SeO_4)_3$	710.69	......	4.30^{15}			36°	
selenide	Pr_2Se_3	478.69	carmine red	6.64				
sulfate	$Pr_2(SO_4)_3$	570.00	lt grn	3.72			12.7	
	$Pr_2(SO_4)_3 \cdot 5H_2O$	660.08	pr, mn	3.176^{16}			$1.85^{85°}$	
	$Pr_2(SO_4)_3 \cdot 8H_2O$	714.12	grn, mn, 1.540	2.827^{13}			17.4	
sulfide	Pr_2S_3	378.01	brn pwd	5.024	1795			s dil a

Table 4-1 (Continued)
PHYSICAL CONSTANTS OF INORGANIC COMPOUNDS

Name	Formula	Formula Weight	Color, Crystalline Form, and Refractive Index	Density	Melting Point, °C	Boiling Point, °C	Solubility in 100 Parts
Promethium	Pm	(147)	....	....	(1227)	(4027)	....
bromide	PmBr₃	386.7	pink, rh	5.38	737	1667	s
chloride	PmCl₃	153.4	pale blue, hex	....	737	1670	s
fluoride	PmF₃	204	purpsh-pink, hex	....	1410	2330	s
iodide	PmI₃	527.7	....	....	800	1370	s
nitrate	Pm(NO₃)₃	333	pink	....	....	....	s
Protoactinium	Pa	231.04	gray-met, bct	15.37	(1227)	(4027)	....
(IV) bromide	PaBr₄	550.66	red, tetr	....	subl >500 vac	....	hyd; s alc, fum HNO₃
(V) bromide	PaBr₅	630.56	dk red, mn (β)	....	subl 250	....	s HCl
carbide	PaC	243.05	blk, fcc	....	>1900	....	....
(IV) chloride	PaCl₄	372.85	grn-yel, tetr	4.72	subl 400	....	d; s CH₃CN, fum HNO₃
(V) chloride	PaCl₅	408.31	yel, hygr, mn	3.74	306	420	i
(IV) fluoride	PaF₄	307.03	brn, mn	6.36	....	....	....
(V) fluoride	PaF₅	325.04	wh, tetr	6.28	....	subl >500 vac	....
(di-) nonafluoride	Pa₂F₉	632.08	blk, bcc	6.83	d	....	....
hydride	PaH₃	234.04	blk, cub	....	....	....	hyd; s CH₃CN
(III) iodide	PaI₃	611.76	blk, o-rh	....	....	....	....
(IV) iodide	PaI₄	738.66	dk grn	....	subl ~500	....	....
(V) iodide	PaI₅	865.56	blk, hygr, o-rh	....	....	....	hyd; s CH₃CN
(II) oxide	PaO	247.04	cub	....	....	....	....
(IV) oxide	PaO₂	263.04	blk, cub	....	>1550	....	....
(V) oxide	Pa₂O₅	542.14	wh, cub	....	....	....	s dil HF
(IV) oxide dibromide	PaOBr₂	406.85	or	....	d 550 vac	....	....
(V) oxide tribromide	PaOBr₃	486.75	grn-yel, mn	....	d 500 vac	....	....
(IV) oxide dichloride	PaOCl₂	317.95	grn-yel	....	d 550 vac	....	....
(di-)(V) oxide octa-chloride	Pa₂OCl₈	761.70	....	....	d 250	....	....

		Mol. wt.	Color, crystalline form	Density	M.p., °C	B.p., °C	Solubility
(IV) oxide diiodide	PaOI₂	500.85	burgundy				
(V) oxide triiodide	PaOI₃	627.75	dk brn				
(V) dioxide fluoride	PaO₂F	282.04	wh		d 450 vac		
(V) dioxide iodide	PaO₂I	389.94	brn, hex		d 450		
(di-)(V) oxide octafluoride	Pa₂OF₈	619.08	wh, bcc, hygr		d 850 vac		hyd; s CH₃CN
(IV) oxide sulfide	PaOS	279.10	yel, tetr				
Pyrophosphoryl							
chloride, tetra-	P₂O₃Cl₄	251.76	col fum liq	1.813^{20}	102	215	d; s a
(di-) decachloride	P₄O₄Cl₁₀	542.42	cr	2.057^{20}		37	s
(di-) tetrasulfide	P₄O₆S₄	343.76	tetr, deliq			295	i; s a
Pyrosulfuryl chloride	ClSO₂OSO₂Cl	215.03	col liq, 1.937	1.818^{11}	−37.5	152.5	s
chlor de fluoride	S₂O₅ClF	198.57	col liq	1.797^{20}	−65	100.1	d; d a
fluoride, di-	S₂O₅F₂	182.12	col liq	$1.75°$	−58	51	
Radium	Ra	226.03	silv-wh met	ca 6	ca 700	(1525)	d; s a
brom de	RaBr₂	385.82	col-yelsh, mn	5.79	728	820	s
	RaBr₂·2H₂O	421.85	wh, mn		−2H₂O, 100	820	s
carbcnate	RaCO₃	286.01	wh, mn				i; s a
chloride	RaCl₂	296.91	wh-yelsh, mn	4.91	1000		s
	RaCl₂·2H₂O	332.94	wh, mn		−2H₂O, 100		sl s
iodate	Ra(IO₃)₂	575.81	wh				$0.018°$
nitrate	Ra(NO₃)₂	350.01	wh, cr				13.9^{20}
sulfate	RaSO₄	332.06	wh, rh				i; i a
Radon	Rn	(222)	col g	$9.73°$ g/l; 4.4^{bp} liq	−71	−62	23 cc²⁰
Rhenium	Re	186.2	silv met, hcp	21.04	3180	5885	s HNO₃; sl s hot H₂SO₄
(III) bromide	ReBr₃	425.93	grn-blk		subl 500		v s
(V) bromide	ReBr₅	585.72	dk bl		~28	d	
(di-)carbonyl, deca-	Re₂(CO)₁₀	652.51	col, cub		d 250		
(III) chloride	ReCl₃	292.56	red		subl 500		s
(IV) chloride	ReCl₄	328.01	blk				hyd
(V) cl·loride	ReCl₅	363.47	brn-blk, mn		~260	~330	s HCl
(VI) chloride	ReCl₆	398.92	dk grn	4.9	25		hyd

Table 4-1 (Continued)
PHYSICAL CONSTANTS OF INORGANIC COMPOUNDS

Name	Formula	Formula Weight	Color, Crystalline Form, and Refractive Index	Density	Melting Point, °C	Boiling Point, °C	Solubility in 100 Parts
Rhenium							
(VI) chloride pentafluoride	$ReClF_5$	316.65	red	....	−2	d	....
(IV) fluoride	ReF_4	262.19	lt bl, tetr	5.38	124.5	795	hyd
(V) fluoride	ReF_5	280.20	yel-grn, o-rh	....	48	d 140	d
(VI) fluoride	ReF_6	300.19	lt yel, bcc, v hygr	3.58 liq	18.5	33.8	disprop; s HF
(VII) fluoride	ReF_7	319.19	yel, cub	3.65^{52} liq	48.3	73.7	hyd
(II) iodide	ReI	313.10	cub	....	....	....	....
(IV) iodide	ReI_4	693.82	amorp, hygr	....	....	....	hyd; s acet
(IV) oxide	ReO_2	218.20	blk, mn	11.4	d 1000	....	s HNO_3; alk H_2O_2
(IV) oxide	$ReO_2 \cdot 2H_2O$	254.23	....	....	−2H_2O, 250	....	....
(VI) oxide	ReO_3	234.20	red, cub	6.9–7.4	d 300	....	s HNO_3, alk H_2O_2
(VII) oxide	Re_2O_7	484.40	yel, hex, hygr	6.1	296	362	v s
oxide *tribromide*	$ReOBr_3$	441.91	blk	....	....	....	....
oxide *tetrabromide*	$ReOBr_4$	521.82	blue	....	(low)	d	....
oxide *trichloride*	$ReOCl_3$	308.56	brn	....	....	....	....
oxide *tetrachloride*	$ReOCl_4$	344.01	brn cr	3.309^{34}	34	225	hyd; s CCl_4
oxide *trifluoride*	$ReOF_3$	259.20	blk, tetr, hygr	....	....	....	s
oxide *tetrafluoride*	$ReOF_4$	278.19	blue, mn	4.032	108	171	d
oxide *pentafluoride*	$ReOF_5$	297.19	creamy cr or col liq	....	34.5	73.0	hyd
dioxide trifluoride	ReO_2F_3	275.19	wh	....	95	185 est	hyd
trioxide bromide	ReO_3Br	314.11	col	....	d	....	....
trioxide chloride	ReO_3Cl	269.65	col liq	....	4.5	131	hyd
trioxide fluoride	ReO_3F	253.20	yel	....	147	~164	hyd
phosphide	ReP	217.17	gray, cub	12.0	....	....	....
phosphide, di-	ReP_2	248.14	gray, cub	8.83	....	....	....
phosphide, tri-	ReP_3	279.11	gray	7.30	....	....	....
sulfide, di-	ReS_2	250.33	blk, hex	7.5	d 1000	....	s HNO_3
(di-)sulfide, hepta-	Re_2S_7	596.85	blk, tetr	4.866	d 460	....	s HNO_3; $H_2O_2 + NH_3$

Rhodium

Name	Formula	Formula weight	Crystalline form, color, refractive index	Density	Melting point	Boiling point	Solubility
	Rh	102.91	gray-wh, fcc	$12.41^{20°}$	1966	(3700)	s $KHSO_4$ melt
(0) carbonyl, octa-	$Rh_2(CO)_8$	429.90	or		76 d		
carbonyl hydride	$HRh(CO)_4$	215.96	yel liq		−10 d	800 subl	i a, aq reg
(III) chloride	$RhCl_3$	209.26	red, mn		d 450		v s
	$RhCl_3 \cdot xH_2O$		dk red, deliq		d 100		i a, alk
(III) fluoride	RhF_3	159.90	red, rh-hed	5.38	600 subl		hyd
(IV) fluoride	RhF_4	178.90	red-purp				hyd viol
(V) fluoride	RhF_5	197.90	dk red, mn		95.5		
(VI) fluoride	RhF_6	216.90	yel, cub		70	d	
(III) nitrate	$Rh(NO_3)_3 \cdot 2H_2O$	324.93	red, deliq				v s
oxide, sesqui-	Rh_2O_3	253.81	bl-gray, rh-hed	8.20	d 1100		i aq reg, KOH
	$Rh_2O_3 \cdot 5H_2O$	343.88	yel		d		s a
oxide, di-	RhO_2	134.90	brn				i a, alk
	$RhO_2 \cdot 2H_2O$	170.93	ol grn		d		s HCl, alk
(III) perchlorate	$[Rh(H_2O)_6](ClO_4)_3$	527.37	yel nd, fcc				s
phosphide	Rh_2P	236.79	steel gray, cub	9.13	~1500		
phosphide, tri-	RhP_3	195.82	steel gray	5.16			
(III) sulfate	$Rh_2(SO_4)_3 \cdot 4H_2O$	566.05	red				s
sulfide	RhS	134.97	gray-blk	d	d		i aq reg
sulfide, sesqui-	Rh_2S_3	302.00	blk	6.40	d		i aq reg
(III) sulfite	$Rh_2(SO_3)_3 \cdot 6H_2O$	554.09	yel cr		d		s

Rubidium

Name	Formula	Formula weight	Crystalline form, color, refractive index	Density	Melting point	Boiling point	Solubility
	Rb	85.47	silv-wh, bcc	$1.53^{20°}$, $1.475^{39°}$ liq	39.0	700	d; s a
acetate	$RbC_2H_3O_2$	144.52	col, hygr		246		$86^{45°}$
aluminum sulfate	$RbAl(SO_4)_2 \cdot 12H_2O$	520.76	col, cub, 1.457, 1.45232, 1.46618	$1.867^{0°}$	99		$2.6^{20°}$
amide	$RbNH_2$	115.49	col	2.58	309		
azide	RbN_3	127.49	col nd or pl	2.7876	d ca 310		v s
boranate	$RbBH_4$	100.31	wh, fcc	1.919			2.93
bromate	$RbBrO_3$	213.37	wh, hex, 2.2	3.68	430 d		100
bromide	$RbBr$	165.37	wh, fcc, 1.5528	3.36	682	1345	hyd
hexabromotitanate	Rb_2TiBr_6	698.26	dk red, fcc		672		v s; $0.74^{19°}$ al
carbonate	Rb_2CO_3	230.95	wh, deliq		837		$110^{20°}$
carbonate, hydrogen	$RbHCO_3$	146.49	wh, rh		d 175	d 900	

Table 4-1 (*Continued*)
PHYSICAL CONSTANTS OF INORGANIC COMPOUNDS

Name	Formula	Formula Weight	Color, Crystalline Form, and Refractive Index	Density	Melting Point, °C	Boiling Point, °C	Solubility in 100 Parts
Rubidium							
chlorate	$RbClO_3$	168.92	wh, trimetric	3.19	....	....	5.1[20]
chloride	RbCl	120.92	wh, cub, 1.4936	2.76	714	1383	48.5
hexachloroplatinate(IV)	Rb_2PtCl_6	578.75	yel, cub	....	d	....	0.028[20]
hexachlorotitanate(IV)	Rb_2TiCl_6	431.56	yel, fcc	3.94[18]	686	....	....
chromate	Rb_2CrO_4	286.93	yel, rh, 1.71	3.518	980	....	42.4[20]
chromium sulfate	$RbCr(SO_4)_2 \cdot 12H_2O$	545.77	vlt, cub, 1.482	1.946	107	....	43.4
cobalt sulfate	$Rb_2SO_4 \cdot CoSO_4 \cdot 6H_2O$	449.22	ruby red, mn, 1.486 1.491, 1.501	2.56[15]	....	....	9.3
copper sulfate	$Rb_2SO_4 \cdot CuSO_4 \cdot 6H_2O$	534.70	lt bl, mn, 1.489	2.57	....	....	10.28
cyanide	RbCN	111.49	col cr pwd	2.32	....	....	s
dichromate	$Rb_2Cr_2O_7$	386.93	red tric or yel mn	3.125 (tric) 3.021 (mn)	400	d 650	4.96[18] (tric), 5.41[8] (mn)
gallium sulfate	$RbGa(SO_4)_2 \cdot 12H_2O$	563.50	col cr, 1.46579	1.962	....	....	s
fluoride	RbF	104.47	wh, cub, 1.396	3.557	765	1410	75
fluoroborate	$RbBF_4$	172.27	wh, rh, 1.333	2.820[20]	590	d 500	0.6[17]
hexafluorosilicate	Rb_2SiF_6	313.02	wh, cub	3.332	....	....	0.16[20]
hydride	RbH	86.49	wh, bcc	2.595	d 300	....	d
hydroxide	RbOH	102.48	wh, deliq	3.203[11]	300	....	180[15], v s al
iodate	$RbIO_3$	260.37	wh, mn or cub	4.56	d	....	1.96
iodide	RbI	212.37	col, cub	3.55	642	1304	60[17]
iodide, tri-	RbI_3	466.18	blk, rh	4.03[22]	190	....	s
iron(II) selenate	$Rb_2SeO_4 \cdot FeSeO_4 \cdot 6H_2O$	620.79	bl-grn, mn, 1.513, 1.520, 1.532	2.819	....	....	s
iron(III) selenate	$RbFe(SeO_4)_2 \cdot 12H_2O$	643.42	cub, 1.507	2.311[11]	45	−12H₂O, 100	s
iron(II) sulfate	$Rb_2SO_4 \cdot FeSO_4 \cdot 6H_2O$	527.00	grn, mn, 1.4815, 1.4874, 1.4977	2.516	d 60	....	24.2 anhyd
iron(III) sulfate	$RbFe(SO_4)_2 \cdot 12H_2O$	549.62	cub, 1.4823	1.91–1.95	48–53	....	4.5[6]

Name	Formula	Formula wt.	Color, crystalline form, refractive index	Density	Melting point, °C	Boiling point, °C	Solubility
magnesium sulfate	$Rb_2SO_4 \cdot MgSO_4 \cdot 6H_2O$	495.47	col, mn, 1.467, 1.469, 1.478	2.386	...	...	20.2 anhyd
nitrate	$RbNO_3$	147.47	col, hygr, 1.51, 1.52, 1.524	3.11	tr cub 161 mp 310 hex tric, rh 219	...	44.3[16]
oxide	Rb_2O	186.94	yel, cub	3.72	477 d	...	s
oxide, di-	RbO_2	117.47	dk or	...	412	...	s
oxide, sesqui-	Rb_2O_3	218.94	blk, cub	3.53⁰	489	d 606	1.26
perchlorate	$RbClO_4$	184.92	col, rh, 1.4701	2.9	281	...	0.5[13]
periodate	$RbIO_4$	276.37	col, tetr	3.918[16]	...	...	1.06[20]
permanganate	$RbMnO_4$	204.41	vlt cr	3.235[10]	d 295	...	d
peroxide	Rb_2O_2	202.94	yel, cub	3.65⁰	600	d 1011	0.47[30]
perrhenate	$RbReO_4$	335.68	wh cr	4.73	598	...	s
phosphate, di-H	RbH_2PO_4	182.47	wh cr	...	840	...	...
praseodymium nitrate	$2RbNO_3 \cdot Pr(NO_3)_3 \cdot 4H_2O$	693.93	grnsh, mn, hygr	2.50	63.5	...	...
selenate	Rb_2SeO_4	313.90	col, rh	3.90	...	...	159[12]
sulfate	Rb_2SO_4	267.00	col, rh, 1.5133	3.613[20]	1074	...	32.5[20]
sulfate, hydrogen	$RbHSO_4$	182.54	col, rh, 1.473	2.892[16]	d red heat	...	s
sulfide	Rb_2S	203.00	red, cub, deliq	2.912	530 d vac	...	v s
sulfide, di-	Rb_2S_2	235.07	dk red	...	420	volat >850	...
sulfide, tri-	Rb_2S_3	267.13	redsh yel	...	213	...	d; s 70% al
sulfide, penta-	Rb_2S_5	331.26	red, rh, deliq	2.618[15]	225	...	v s
tartrate (dl)	$Rb_2C_4H_4O_6$	319.01	col, trig	2.658[20]	...	...	1.18
tartrate, hydrogen	$RbHC_4H_4O_6$	234.55	trimetric pr	2.282	201 d	d 300	...
vanadium sulfate	$RbV(SO_4)_2 \cdot 12H_2O$	544.72	yel, cub, 1.4689	1.915[20]	64; −12H₂O, 230	...	2.56[10]
Ruthenium	Ru	101.07	silv met, hcp	12.45[20]	2430	3700	s fus alk, oxid fluxes
(III) bromide	$RuBr_3$	340.80	dk brn pwd, hex	...	−22	...	s al, bz
carboryl, penta-	$Ru(CO)_5$	241.12	col liq	...	...	...	...
(di-) nonacarbonyl	$Ru_2(CO)_9$	454.23	...	...	d >500	...	s HCl
(III) chloride	$RuCl_3$	207.43	brn, hex, deliq	3.11	d	...	s al
(IV) chloride	$RuCl_4 \cdot 5H_2O$	332.96	red-brn cr, hygr	...	d	...	...
fluoride, tri-	RuF_3	158.06	rh-hed	...	...	...	...
fluoride, tetra-	RuF_4	177.06	yel	...	≥280	...	viol hyd

Table 4-1 (*Continued*)
PHYSICAL CONSTANTS OF INORGANIC COMPOUNDS

Name	Formula	Formula Weight	Color, Crystalline Form, and Refractive Index	Density	Melting Point, °C	Boiling Point, °C	Solubility in 100 Parts
Ruthenium							
fluoride, penta-	RuF_5	196.06	dk grn, mn	3.82	86.5	227	d
fluoride, hexa-	RuF_6	215.06	dk brn, cub	....	54.0	d	d
(III) hydroxide	$Ru(OH)_3$	152.09	blk pwd	....	....	....	s a
(III) iodide	RuI_3	481.78	bl, hex	....	d 590	....	s fus alk
oxide, di-	RuO_2	133.07	bl, tetr	6.97	d	....	....
oxide, tetra-	RuO_4	165.07	yel nd, rh, volat	3.29	25.4	d 108	2.03^{20}, v s CCl_4
oxide tetrafluoride	$RuOF_4$	193.06	pale grn cr	....	115	184	s HNO_3 + HF
silicide	$RuSi$	129.16	met pr	$5.40^{4°}$	....	....	s fus alk
sulfide, di-	RuS_2	165.20	blk, cub	6.14°	d >1000	....	s a
Samarium	Sm	150.4	wh-gray, rh	7.536	1052	1900	15
acetate	$Sm(C_2H_3O_2)_3 \cdot 3H_2O$	381.53	yel, hex	1.94	....	....	114
bromate	$Sm(BrO_3)_3 \cdot 9H_2O$	696.21	....	....	75	$-9H_2O$, 150	s d
(II) bromide	$SmBr_2$	310.17	brn	5.1	700	1880	....
(III) bromide	$SmBr_3$	390.06	yel, rh	5.40	640	d 900 to $SmBr_2$ vac	d; s a
	$SmBr_3 \cdot 6H_2O$	498.17	yel, deliq	2.97	....	....	s d
carbide	SmC_2	174.37	yel, hex	5.86	....	....	....
(II) chloride	$SmCl_2$	221.26	red-brn, o-rh	4.56	848	2030	....
(III) chloride	$SmCl_3$	256.71	lt yel, hex, hygr	4.46	682	d	92.4^{10}
	$SmCl_3 \cdot 6H_2O$	364.80	grn-yel, tric, hygr	2.383	$-5H_2O$, 110	....	....
chromate	$Sm_2(CrO_4)_3 \cdot 8H_2O$	792.80	yel	....	....	....	0.043
(II) fluoride	SmF_2	188.35	yel, cub	6.643	1417	2427	....
(III) fluoride	SmF_3	207.35	wh, hex	6.52	1306	2427	s H_2SO_4
hydride, di-	SmH_2	152.37	....	....	....	>325 Sm_2O_3	....
(III) hydroxide	$Sm(OH)_3$	201.37	lt yel, hex	....	$-H_2O$, 220	....	s a
(II) iodide	SmI_2	404.16	dk grn, mn	....	520	1583	hyd
(III) iodide	SmI_3	531.06	or-yel, hex	....	850 d	....	s
molybdate	$Sm_2(MoO_4)_3$	780.51	vlt, rh oct	5.36	....	....	....

Name	Formula	Formula weight	Color, crystalline form	Density	m.p., °C	b.p., °C	Solubility
nitrate	Sm(NO₃)₃·6H₂O	444.46	lt yel, tric	2.375	78	....	v s
oxalate	Sm₂(C₂O₄)₃·10H₂O	744.91	wh cr	....	....	....	s H₂SO₄
oxide	Sm₂O₃	348.70	yel-tan, mn	7.74	2300 ± 50	....	s a
(tri-) tetroxide bromide	Sm₃O₄Br	595.30	rh	6.64	1980	....	....
(di-) dioxide sulfide	Sm₂O₂S	364.76	lt tan, hex	6.90	d 1100	....	i
(di-) dioxide sulfate	Sm₂O₂SO₄	428.76	yel pwd	....	....	....	....
selenide	Sm₂Se₃	537.58	gray-blk	7.10	....	....	2.7
sulfate	Sm₂(SO₄)₃·8H₂O	733.01	lt yel, mn, 1.543, 1.552, 1.563	2.930	−8H₂O, 450	....	....
sulfide sesqui-	Sm₂S₃	396.89	yel-pink	5.729	1780	s dil a	d ev H₂
Scandium	Sc	44.956	silv met, hcp	2.992	1397	2730	s
bromide	ScBr₃	284.68	wh, rh	3.93	948	....	s alk, hot Na₂CO₃
	ScBr₃·6H₂O	392.78	col cr	....	−4.5H₂O, 120	....	s
carbonate	Sc₂(CO₃)₃·12H₂O	486.13	bulky wh ppt	....	....	subl 909	s a (if not ign)
chloride	ScCl₃	151.32	wh, rh, deliq	2.39	957	....	v s
fluoride	ScF₃	101.96	wh, rh	2.5	1515	....	v s NH₄F, s alk F
hydroxide	Sc(OH)₃	95.98	cub & hemihed	....	−H₂O, 550	....	s a
iodide	ScI₃	425.67	wh, hex	....	920	d 220 to oxide	....
nitrate	Sc(NO₃)₃·4H₂O	303.03	col pr, deliq	....	d 120 to oxidenitrate	....	s
nitride	ScN	58.96	bcc	....	2650	....	....
oxalate	Sc₂(C₂O₄)₃·6H₂O	462.07	wh	....	−4H₂O, 50	d 635 to oxide	0.006; s alk, oxalates
oxide	Sc₂O₃	137.91	wh pwd, bcc	3.864	....	....	s a (if not ign)
sulfate	Sc₂(SO₄)₃	378.10	col cr	2.579	1100, Sc₂O₃, 250	....	28.5
	Sc₂(SO₄)₃·5H₂O	468.17	col cr	2.519	−5H₂O, 250	−SO₂, 550	54.6
sulfide	Sc₂S₃	186.09	yel	....	....	....	d hot H₂O; s a
Selenic acid	H₂SeO₄	144.97	wh, hex pr, hygr	$3.004^{15°}$	58	d 260	v s
	H₂SeO₄·H₂O	162.99	wh nd	$2.627^{15°}$; $2.356^{15°}$ liq	26	205	v s
Selenious acid	H₂SeO₃	128.97	col, hex, deliq	$3.004^{15°}$	d 70	....	167^{20}
Selenium	Se	78.96	gray, hex, rh; red, mn, Se α; red pr, mn, Se β	$4.792^{20°}$; $4.48^{20°}$; 4.5	217; 170	685	s CS₂
(di-) dibromide	Se₂Br₂	317.74	dk red, oily, hygr, 1.96^{Li}	$3.604^{15°}$	144–180	225–230	s CS₂, CHCl₃

Table 4-1 (Continued)
PHYSICAL CONSTANTS OF INORGANIC COMPOUNDS

Name	Formula	Formula Weight	Color, Crystalline Form, and Refractive Index	Density	Melting Point, °C	Boiling Point, °C	Solubility in 100 Parts
Selenium							
bromide, di-	$SeBr_2$	238.78	dk red g	....	....	d 227	....
bromide, tetra-	$SeBr_4$	398.62	yel cr	4.029	d 75	115 subl	d; s HBr
(di-) dichloride	Se_2Cl_2	228.83	yel-brn liq, 1.5993	2.789^{20}	−85	d 127	s CS_2, $CHCl_3$
chloride, di-	$SeCl_2$	149.87	g	....	....	....	....
chloride, tetra-	$SeCl_4$	220.77	col cr	....	305	d 196 subl	....
fluoride, tetra-	SeF_4	154.95	col liq	2.732^{20}	−9.5	105	....
fluoride, hexa-	SeF_6	192.95	col g	2.108^{-10}	−34.8	−46.3 subl	....
(penta)fluoro- hypofluorite	$F_5Se(OF)$	208.95	col irritating g	....	−54	−29	d
oxide, di-	SeO_2	110.96	wh nd, tetr	3.954^{16}	340	315 subl	s H_2O, alc
oxide, tri-	SeO_3	126.96	wh, cub, hygr	3.6	118	d 180	s
oxide dibromide	$SeOBr_2$	254.79	yel-red nd	3.38^{50}	41.6	217 d	hyd
oxide dichloride	$SeOCl_2$	165.87	col, 1.6516	2.422^{7}	10.8	175.5	hyd
oxide difluoride	$SeOF_2$	132.96	....	2.801^{20}	15	126	hyd
dioxide difluoride	SeO_2F_2	148.96	....	3.05^{-73}	−99.5	−8.4	....
Silane, bromo-	SiH_3Br	111.02	col g, expl in air	1.72^{-80}	−94	1.9	d
dibromo-	SiH_2Br_2	189.92	col liq, inflam	2.17^{0}	−70.1	66	d
tribromo-	$SiHBr_3$	268.82	col liq, inflam	2.7^{17}	−73	109	d
bromotrichloro-	$SiBrCl_3$	214.35	col liq	1.826	−62	80.3	d
dibromodichloro-	$SiBr_2Cl_2$	258.81	col liq	2.172	−45.5	104	d
tribromochloro-	$SiBr_3Cl$	303.27	col liq	2.497	−21	127	d
chloro-	SiH_3Cl	66.56	col g	0.003033	−118.1	−30.4	....
dichloro-	SiH_2Cl_2	101.01	g	0.004599	−122	8.3	d
trichloro-	$SiHCl_3$	135.45	col liq	1.34	−126.5	33	d; s bz, chl, CS_2
chlorotrifluoro-	$SiClF_3$	120.53	g	0.005455	−138	−70.0	d
dichlorodifluoro-	$SiCl_2F_2$	136.99	g	0.0062784	−144	−31.7	d
trichloroiodo-	$SiCl_3I$	261.35	col liq	....	>−60	113.5	d
trifluoro-	$SiHF_3$	86.09	col g	0.00386	−131.4	ca −95	d; s tol

Name	Formula	Mol. wt.	Color, crystalline form	Density	M.P.	B.P.	Solubility
iodo-	SiH_3I	158.01	col liq	$2.035^{15°}$	−57.0	45.5	d
triiodo-	$SiHI_3$	409.81	red liq	3.314	8	220	d; s bz, CS_2
Silicic acid, meta-	H_2SiO_3	78.10	col, amorp	....	d	....	i; s HF, h alk
Silicon	Si	28.086	gray met, cub	2.33	1415	2680	s HF + HNO_3
acetate	$Si(C_2H_3O_2)_4$	264.27	col cr, hygr	....	subl 110	148^{6mm}	sl s acet, bz
brom de, tetra-	$SiBr_4$	347.72	col oily liq, tetrahed sol	2.7715	5.4	154	hyd
(di-) *hexabromide*	Si_2Br_6	535.62	wh, rh	....	95	240	d; s CS_2
(tri-) *octabromide*	Si_3Br_8	723.6	col cr	....	43.3	....	....
(tetra-) *decabromide*	Si_4Br_{10}	911.52	col cr	....	64.6	....	d; s bz, CS_2
dibromide sulfide	$SiBr_2S$	219.97	col pl	....	93	150^{18mm}	s fus alk
carbide	SiC	40.10	blk, hex or cub	3.217	subl 2700	d	hyd; s bz, eth, CCl_4
chloride, tetra-	$SiCl_4$	169.90	col fum liq, tetrahed sol	$1.48^{20°}$	−68	57.6	
(di-) *hexachloride*	Si_2Cl_6	268.89	col liq	$1.5624^{15°}$	−2.5	147	hyd
(penta-) *dodecachloride*	Si_5Cl_{12}	565.87	wh	....	345	....	....
dichloride sulfide	$SiCl_2S$	131.06	col pr	....	75	92^{23mm}	d; s bz, CS_2, CCl_4
trichloride hydrogen sulfide	$SiCl_3HS$	167.52	col liq	1.45	....	96–100	d
cyanate	$Si(OCN)_4$	196.15	col	1.414^{20}	34.5	247.2	d
cyanate, iso-	$Si(NCO)_4$	196.15	col cr	1.434	26	185.6	hyd; s bz, CCl_4, CS_2
fluoride, tetra-	SiF_4	104.08	col g	0.00469	-90.3^{1318mm}	subl −95.5	hyd; s HF
(di-) *hexafluoride*	Si_2F_6	170.16	col g	0.00776	-18^{780mm}	−19	i
hydride	SiH_4	32.12	col g	$0.68^{-185°}$	−184.7	−111.9	
see also Silane							
diimide	$Si(NH)_2$	58.12	wh pwd	....	d 900	....	d
iodide, tetra-	SiI_4	535.70	wh, cub	4.198	120.5	288	d; 2.2^{27} CS_2
(di-) *hexaiodide*	Si_2I_6	817.60	wh, hex	....	250	subl 150 vac	d; 19 CS_2
nitride	Si_3N_4	140.28	lt gray amorp pwd	3.44	1900	....	i; s HF
(II) oxide	SiO	44.09	wh, cub	2.13	>1702	1880	s HF + HNO_3
(IV) oxide (cristobalite)	SiO_2	60.08	col, cub, 1.487, 1.484	2.32	1713	2230	s HF
(lechatelierite)	SiO_2	60.08	col, 1.4588	2.19	....	....	s HF

Table 4-1 (*Continued*)

PHYSICAL CONSTANTS OF INORGANIC COMPOUNDS

Name	Formula	Formula Weight	Color, Crystalline Form, and Refractive Index	Density	Melting Point, °C	Boiling Point, °C	Solubility in 100 Parts
Silicon							
(IV) oxide (opal)	SiO_2	60.08	col, amorp, 1.41–1.46	2.17–2.20	>1600	....	s HF
(tridymite)	SiO_2	60.08	col, rh, 1.469, 1.470, 1.471	2.26	1703	2230	s HF
(quartz)	SiO_2	60.08	col, hex, 1.544, 1.553	2.635–2.660	1610	2230	s HF
(di-) oxide hexachloride	Si_2OCl_6	284.89	col liq	....	28.1	137	d; v s CS_2, CCl_4
(di-) oxide hexafluoride	Si_2OF_6	186.18	col g	1.358 liq	−48	−23.3	d
phosphide	SiP	59.06	lt yel pwd	2.4	....	....	hyd
sulfide	SiS	60.15	yel nd	$1.853^{15°}$	subl 940	....	d alk
sulfide, di-	SiS_2	92.21	wh nd, rh	2.02	subl 1090	....	s alk
thiocyanate, *iso*	$Si(NCS)_4$	260.40	wh pr	....	143.8	314.2	d
Silicotungstic acid	$SiO_2 \cdot 12WO_3 \cdot 26H_2O$	3310.66	wh-yelsh cr, deliq	....	....	....	v s
Silicyl oxide	$(SiH_3)_2O$	78.22	col g	3.491 g/l	−144	−15.2	v s l s
Silver	Ag	107.87	wh met, fcc	10.50	960.15	2177	i; s HNO_3, h H_2SO_4
acetate	$AgC_2H_3O_2$	166.92	wh pl	$3.259^{15°}$	d	....	1.04; s dil HNO_3
acetylide	Ag_2C_2	239.76	wh	....	expl	....	i; s a
arsenate, ortho-	Ag_3AsO_4	462.53	dk red, cub	6.657	d	....	i; s NH_4OH
arsenite, ortho-	Ag_3AsO_3	446.53	yel pwd	....	d 150	....	i; s NH_4OH, HNO_3
azide	AgN_3	149.89	wh, rh pr, expl	....	252	297	i; s KCN, dil HNO_3
bromate	$AgBrO_3$	235.78	col, tetr, 1.874, 1.920	5.206	d	....	0.2; s NH_4OH
bromide (bromyrite)	$AgBr$	187.78	pale yel, 2.253	6.473	432	1778	i; s KCN, $Na_2S_2O_3$
carbonate	Ag_2CO_3	275.75	yel pwd	6.077	d 170	....	i; s HNO_3, NH_4OH
chlorate	$AgClO_3$	191.32	wh, tetr	$4.430^{20°}$	230	d 270	$10^{15°}$
chloride (cerargyrite)	$AgCl$	143.32	wh, cub, 2.071	5.56	455	1818	i; s NH_4OH, KCN, $S_2O_3^{2-}$

Name	Formula	Formula weight	Color, crystalline form, and refractive index	M.p., °C	B.p., °C	Density	Solubility
chlorite	$AgClO_2$	175.32	yel cr	105 expl	.....	.....	0.45
chromate	Ag_2CrO_4	331.73	red, mn	d	.....	5.625	i; s NH_4OH, KCN
cyanate	AgOCN	149.89	col	d 320	.....	4.00	sl s; s HNO_3, NH_4OH
cyanide	AgCN	133.84	wh, hex	d	.....	3.95	i; s HNO_3, KCN, $S_2O_3^{2-}$
dichromate	$Ag_2Cr_2O_7$	413.73	red, tric	.....	.....	4.770	i; s a, NH_4OH, KCN
dithionate	$Ag_2S_2O_6 \cdot 2H_2O$	411.90	wh, rh, 1.66	.....	.....	3.61	.....
ferricyanide	$Ag_3Fe(CN)_6$	535.56	.....	.....	.....	.....	i; s NH_4OH
ferrocyanide	$Ag_4Fe(CN)_6 \cdot H_2O$	661.45	wh	.....	.....	.....	i; s KCN
fluoride	AgF	126.87	yel, cub, deliq	435	ca 1159	5.852^{16}	v s
(II) fluoride	AgF_2	145.87	brn, o-rh	690	d 700	4.57	d
(di-) fluoride	Ag_2F	234.74	yel, hex	d 90	.....	8.57	d
fulminate	$Ag_2C_2N_2O_2$	299.77	nd	expl	.....	.....	0.075^{13}; s NH_4OH
hyponitrite	$Ag_2N_2O_2$	275.75	yel	d 110	.....	5.75^{30}	v sl s; d a
iodate	$AgIO_3$	282.77	col, rh	>200	.....	5.525^{17}	0.003; s HNO_3, NH_4OH
iodide (α) (iodyrite)	AgI	234.77	yel, hex, 2.21, 2.22	tr 146 to β	d	5.683^{30}	i; s KCN, $Na_2S_2O_3$
(β)	......	......	or, cub	555.5	1506	6.010^{15}	.....
iodomercurate (α)	Ag_2HgI_4	923.95	yel, tetr	tr 50.7 to β	.....	6.02	i; s KI, KOH
(β)	......	......	red, cub	d 158	d 444	5.90	.....
nitrate	$AgNO_3$	169.87	col, rh, 1.729, 1.744, 1.788	208.5	.....	4.352^{19}	v s
nitrite	$AgNO_2$	153.88	wh, rh	d 140	.....	4.453	0.16^0; s ac a, HNO_3
oxalate	$Ag_2C_2O_4$	303.76	col cr	expl 140	.....	5.029^4	i; s a
oxide	Ag_2O	231.74	brn-blk, cub	d 300	.....	7.143^{17}	i; s a
perchlorate	$AgClO_4$	207.32	wh cr, deliq	d 486	.....	2.806	v s
periodate	$AgIO_4$	298.77	or-yel, tetr	d 180	.....	5.57	d, s HNO_3
permanganate	$AgMnO_4$	226.81	dk vlt, mn	d	.....	4.27	1.6
peroxide	Ag_2O_2	247.74	gray-blk, cub	d 100	.....	7.44	i; s HNO_3, H_2SO_4
perrhenate	$AgReO_4$	358.07	wh, tetr or rh	430	.....	7.05	0.32^{20}
phosphate, meta-	$AgPO_3$	186.84	wh amorp	ca 482	.....	6.37	i; s HNO_3
phosphate, ortho-	Ag_3PO_4	418.58	yel, cub	849	.....	6.370	i; s a
phosphate, ortho-, mono-H	Ag_2HPO_4	311.75	wh, trig	d 110	.....	1.8036	.....
phosphate, pyro-	$Ag_4P_2O_7$	605.42	wh	585	.....	5.306^{18}	i; s a
propionate	$AgC_3H_5O_2$	180.94	wh leaf or nd	.....	.....	2.687	0.84^{20}
selenate	Ag_2SeO_4	358.73	wh, o-rh	.....	.....	5.72	0.12^{20}

Table 4-1 (*Continued*)
PHYSICAL CONSTANTS OF INORGANIC COMPOUNDS

Name	Formula	Formula Weight	Color, Crystalline Form, and Refractive Index	Density	Melting Point, °C	Boiling Point, °C	Solubility in 100 Parts
Silver							
selenide (naumannite)	Ag_2Se	294.70	gray pl, cub	7.96	880	d	i; s hot HNO_3
sulfate	Ag_2SO_4	311.80	wh, rh, 1.7583, 1.7748, 1.7852	$5.45^{30°}$	652	d 1085	$0.570^{0°}$; s a
sulfide (acanthite) (argentite)	Ag_2S	247.80	gray-blk, rh	7.326	tr 175	d	i; s HNO_3, H_2SO_4
sulfite	Ag_2SO_3	295.80	wh cr	....	d 100	....	v sl s; s a
d-tartrate	$Ag_2C_4H_4O_6$	363.81	wh scales	$3.423^{15°}$	d	....	$0.21^{18°}$; s a
telluride (hessite)	Ag_2Te	343.34	gray, cub	8.5	955	....	i; s KCN, NH_4OH
thioantimonite (pyrargyrite)	Ag_3SbS_3	541.55	red, trig, 3.084, 2.881(Li)	5.76	486	....	i; s HNO_3
thioarsenite (proustite)	Ag_3AsS_3	494.72	red, trig, 3.088, 2.792	5.49	490	....	i; s HNO_3
thiocyanate	AgSCN	165.95	col cr	....	d	....	i; s NH_4OH
thiosulfate	$Ag_2S_2O_3$	327.87	wh cr	....	d	....	sl s; s $Na_2S_2O_3$
tungstate	Ag_2WO_4	463.59	lt yel cr	....	....	883	$0.05^{15°}$; s HNO_3
Sodium	Na	22.99	silv met, cub	0.97; liq 0.93^{100}	97.8	883	d, forms NaOH
acetate	$NaC_2H_3O_2$	82.03	wh, mn, 1.464	1.528	324	....	$46.5^{20°}$
	$NaC_2H_3O_2 \cdot 3H_2O$	136.08	wh, mn	1.45	58	$-3H_2O$, 120	v s
aluminate	$NaAlO_2$	81.97	amorp	....	1650	....	s
amide	$NaNH_2$	39.01	ol grn	....	210	400	d
ammonium phosphate, meta-	$NaNH_4HPO_4 \cdot 4H_2O$	209.07	col, mn	1.574	79 d	....	16.7
antimonate, meta-	$2NaSbO_3 \cdot 7H_2O$	511.58	cub	....	....	....	$0.031^{12.3°}$
antimonate, pyro-	$Na_2H_2Sb_2O_7 \cdot H_2O$	421.51	col cr	....	....	....	$0.03^{12.3°}$
antimonide	Na_3Sb	190.72	deep bl	....	856	....	d
antimonite, meta-	$NaSbO_2 \cdot 3H_2O$	230.78	col, rh	2.864	d	....	d
arsenate	$Na_3AsO_4 \cdot 12H_2O$	424.07	hex, 1.4589	1.759	86.3	....	$26.7^{17°}$

Name	Formula	Mol. wt.	Crystalline form, color, index of refraction	Density	M.P., °C	B.P., °C	Solubility
arsenate, di-H	$NaH_2AsO_4 \cdot H_2O$	181.94	rh, 1.5535	2.535	d 100	....	s
arsenate, mono-H	$Na_2HAsO_4 \cdot 7H_2O$	312.01	col, mn, 1.4658	1.871	125	$-7H_2O$, 100	$61^{15°}$
	$Na_2HAsO_4 \cdot 12H_2O$	402.09	mn, 1.4496	1.72	28	$-12H_2O$, 100	$5.59^{0.1°}$
arsenite, mono-H	Na_2HAsO_3	169.91	col	1.87	....	....	s
arsenite, meta-	$NaAsO_2$	129.91	wh-gray pwd, hygr	....	....	....	s
aurate(I), thio-	$NaAuS \cdot 4H_2O$	324.08	col, mn	....	d	....	
azide	NaN_3	65.01	wh, hex	$1.85^{25°}$	....	....	39°
benzoate	$NaC_7H_5O_2$	144.11	col cr	....	....	....	$62.5^{25°}$
bismuthate	$NaBiO_3$	279.97	yel-brn pwd	....	....	....	i d a
bismuthide	Na_3Bi	277.95	blue vlt	....	766	....	d
borate, meta-	$NaBO_2$	65.80	hex pr	....	966	1400	$26^{20°}$
	$NaBO_2 \cdot 4H_2O$	137.86	mn	....	57	....	
borate, tetra-	$Na_2B_4O_7$	201.22	....	2.367	741	....	1.3°
	$Na_2B_4O_7 \cdot 5H_2O$	291.30	col, rh, 1.461	1.815	....	....	$22^{62°}$ (anhyd)
(borax)	$Na_2B_4O_7 \cdot 10H_2O$	381.37	wh, mn, 1.4694	1.73	75	$-10H_2O$, 200	$1.3^{0.5}$ (anhyd)
bromate	$NaBrO_3$	150.90	col, cub	$3.339^{17.5°}$	381	....	27.5°
bromide	$NaBr$	102.90	col, cub, 1.6412	$3.205^{17.5°}$	755	1390	$90^{20°}$
	$NaBr \cdot 2H_2O$	138.93	col, mn	2.176	50.7	....	79.5° (anhyd)
bromoplatinate	$Na_2PtBr_6 \cdot 6H_2O$	828.62	dark red, tric	3.323	....	....	v s
carbide	Na_2C_2	70.00	brn pwd	$1.575^{15°}$	d forms Na_2CO_3	....	d
carbonate (soda ash)	Na_2CO_3	105.99	wh pwd, 1.535	2.533	851	d	7.1°
	$Na_2CO_3 \cdot H_2O$	124.00	wh, rh, 1.506–1.509	1.55	$-H_2O$, 100	....	s
	$Na_2CO_3 \cdot 7H_2O$	232.10	rh or trig	1.51	d 35.1	....	s
(sal soda)	$Na_2CO_3 \cdot 10H_2O$	286.14	wh, mn, 1.425	1.46	32	....	21.5°
carbonate, sesqui- (trona)	$Na_3H(CO_3)_2 \cdot 2H_2O$	226.03	wh, mn, 1.5073	2.112	d	....	13°
carbonate, hydrogen	$NaHCO_3$	84.01	wh, mn, 1.500	2.20	$-CO_2$, 270	d 350	6.9°
chlorate	$NaClO_3$	106.44	wh, cub, or trig, 1.5151	$2.490^{15°}$	248	....	79°
chloroaurate(III)	$NaAuCl_4 \cdot 2H_2O$	397.80	or-yel, rh, 1.6±	....	d	d	$150^{10°}$
chloride	$NaCl$	58.44	col, cub, 1.5443	2.164^{20}_{4}	808	1473	35.7°

Table 4-1 (*Continued*)
PHYSICAL CONSTANTS OF INORGANIC COMPOUNDS

Name	Formula	Formula Weight	Color, Crystalline Form, and Refractive Index	Density	Melting Point, °C	Boiling Point, °C	Solubility in 100 Parts
Sodium							
chloroiridate(IV)	$Na_2IrCl_6·6H_2O$	558.99	red, tric	......	d	d	$34^{15°}$
chloroplatinate(IV)	$Na_2PtCl_6·6H_2O$	561.88	red, tric	2.50	$-6H_2O$, 100	......	v s
chloroplatinate(II)	$Na_2PtCl_4·4H_2O$	454.94	red	......	100 d	......	$32°$
chromate	Na_2CrO_4	161.97	yel, rh	2.723	792	......	v s
	$Na_2CrO_4·10H_2O$	342.13	yel., deliq, mn	1.483	19.9	......	$77^{25°}$
citrate	$Na_3C_6H_5O_7·2H_2O$	294.10	wh	......	$-2H_2O$, 150	......	$91^{25°}$
	$2Na_3C_6H_5O_7·11H_2O$	714.31	wh, rh	$1.857^{23.5}_{4}$	$-11H_2O$, 150	d.	......
cobaltinitrite	$Na_3Co(NO_2)_6$	403.94	yel to yel-brn cr pwd	......	......	......	s d
cyanide	NaCN	49.01	wh, cub, 1.452	......	563.7	1496	$48^{10°}$
dichromate	$Na_2Cr_2O_7·2H_2O$	298.00	red, mn, 1.6994	$2.34^{25°}$	$-2H_2O$, 84.6; 356 (anhyd)	d 400	$236°$
dithionate	$Na_2S_2O_6·2H_2O$	242.13	col, rh, 1.4953	2.189	......	......	$47.6^{16°}$
ethylate	$NaOC_2H_5$	68.05	wh	......	......	......	d
ferricyanide	$Na_3Fe(CN)_6·H_2O$	298.94	red, deliq.	......	......	......	$18.9°$
ferric oxalate	$Na_3Fe(C_2O_4)_3·5\frac{1}{2}H_2O$	487.96	grn, mn	$1.973^{17.5°}$	$-4H_2O$, 100	$-5\frac{1}{2}H_2O$, 200	$32.5°$
ferrate(III)	$Na_2Fe_2O_4$	221.67	......	......	......	......	d
ferrocyanide	$Na_4Fe(CN)_6·10H_2O$	484.07	yel, mn	1.458	......	......	$17.9^{20°}$ (anhyd)
fluoroborate	$NaBF_4$	109.79	wh, rh	$2.47^{20°}$	384 sl d	d	$108^{26.5°}$
fluoride (villiaumite)	NaF	41.99	tetr, 1.3258, wh	2.79	992	1704	$4°$
fluoride, hydrogen	$NaHF_2$	61.99	col cr	......	d	......	$3.7^{20°}$
formate	$NaHCO_2$	68.01	wh, mn	1.919	255	......	$44°$
hydride	NaH	24.00	silv nd, 1.470	1.396	d 800	......	d

	Formula	Mol. wt.	Color, crystalline form, index of refraction	Density	m.p., °C	b.p., °C	Solubility in cold water
hydrosulfide	$NaSH \cdot 2H_2O$	92.09	col, deliq, nd	…	d	…	s
	$NaSH \cdot 3H_2O$	110.11	rh deliq	…	22	d	s
hydrosulfite	$Na_2S_2O_4 \cdot 2H_2O$	210.14	col cr	…	d	…	22^{20}
hydroxide	$NaOH$	40.00	wh, deliq	2.130	328	1390	108
	$NaOH \cdot 3\frac{1}{2}H_2O$	103.05	mn	…	15.5	…	s
hypochlorite	$NaOCl$	74.44	pale yel, in soln only	…	d	…	26°
hyponitrite	$Na_2(NO)_2$	105.99	…	2.466^{30}	300, d	…	s
hypophosphate	$Na_4P_2O_6 \cdot 10H_2O$	430.06	mn	1.832	−$6H_2O$, 100; 250 (anhyd)	…	3.25^{25}
hypophosphate, di-H	$Na_2H_2P_2O_6 \cdot 6H_2O$	314.03	col, mn	1.849	…	…	2.2
hypophosphite	$NaH_2PO_2 \cdot H_2O$	105.99	col, mn	…	d	…	$100^{25°}$
iodate	$NaIO_3$	197.89	col, rh	3.62	−H_2O, 200	…	2.5°
iodide	NaI	149.89	col, cub, 1.7745	3.667°	661	1300	158.7°
	$NaI \cdot 2H_2O$	185.92	col, mn	2.448	…	…	v s
iodoplatinate(IV)	$Na_2PtI_6 \cdot 6H_2O$	1110.59	mn (?)	3.707	d	…	v s
lactate	$NaC_3H_5O_3$	112.06	col, amorp	…	17	…	s
manganate(VI)	$Na_2MnO_4 \cdot 10H_2O$	345.07	grn, mn	…	…	…	v s
metabisulfite	$Na_2S_2O_5$	190.10	wh pwd	…	d, −SO_2	…	d
methylate	$NaOCH_3$	54.02	wh pwd	…	…	…	…
molybdate	Na_2MoO_4	205.92	wh	liq, 2.590^{1026}	687	…	44°
	$Na_2MoO_4 \cdot 2H_2O$	241.95	pl	3.28	…	…	56.2°
molybdate, di-	$Na_2Mo_2O_7$	349.86	nd	…	612	…	sl s
molybdate, tri-	$Na_2Mo_3O_{10} \cdot 7H_2O$	619.90	nd	…	…	…	$16^{30°}$
molybdate, tetra-	$Na_2Mo_4O_{13} \cdot 6H_2O$	745.82	nd	…	−$6H_2O$, 120	…	$28.4^{21°}$
molybdate, octa-	$Na_2Mo_8O_{25} \cdot 4H_2O$	1285.55	pwd	…	…	…	i
molybdate, deca-	$Na_2Mo_{10}O_{31} \cdot 12H_2O$	1717.55	wh cr	…	…	…	sl s
nitrate (soda niter)	$NaNO_3$	84.99	col, trig, 1.5874	2.257	308	d 380	73°
nitride	Na_3N	82.98	gray	…	d	…	d
nitrite	$NaNO_2$	69.00	pale yel, rh	2.168°	271	d 320	72.1°
nitroprusside	$Na_2Fe(CN)_5NO \cdot 2H_2O$	297.95	red, rh	1.72	…	…	$40^{16°}$
oleate	$NaOOCC_{17}H_{33}$	304.45	wh pwd	…	…	…	10
oxalate	$Na_2C_2O_4$	134.00	wh cr	2.27	…	…	$3.7^{20°}$
oxalate, hydrogen	$NaHC_2O_4 \cdot H_2O$	130.03	tric	…	…	…	$1.7^{15°}$

Table 4-1 (*Continued*)

PHYSICAL CONSTANTS OF INORGANIC COMPOUNDS

Name	Formula	Formula Weight	Color, Crystalline Form, and Refractive Index	Density	Melting Point, °C	Boiling Point, °C	Solubility in 100 Parts
Sodium							
oxide	Na_2O	61.98	wh, deliq	2.27	subl 917	....	forms NaOH
peroxyborate	$NaBO_3 \cdot H_2O$	99.81	wh pwd	....	d 40	....	sl s
	$NaBO_2 \cdot 3H_2O \cdot H_2O_2$	153.86	transparent, mn	....	d >60	....	$3.9^{15°}$
perchlorate	$NaClO_4$	122.44	rh, 1.4617	2.499	482 d	....	$170°$
	$NaClO_4 \cdot H_2O$	140.46	hex	2.02	d 130	....	$66°$
perchromate	Na_3CrO_8	248.96	or pl	....	d 115	....	$400°$
periodate	$NaIO_4$	213.89	tetr	$3.865^{16°}$	d 300	....	$4.6°$
	$NaIO_4 \cdot 3H_2O$	267.94	trig	3.219^{18}_{4}	d	....	s
permanganate	$NaMnO_4 \cdot 3H_2O$	195.94	purp, deliq	2.46	d 170	....	v s
peroxide	Na_2O_2	77.98	yel-wh pwd	2.805	d	....	s d
	$Na_2O_2 \cdot 8H_2O$	222.10	wh, hex	....	d 30	....	s d
perrhenate	$NaReO_4$	273.19	col pl	5.24	300	....	$33^{20°}$
perruthenate(VII)	$NaRuO_4 \cdot H_2O$	206.07	blk cr	....	....	....	sl s
persulfate	$Na_2S_2O_8$	238.10	wh cr	....	....	d	s
peruranate	$Na_2UO_5 \cdot 5H_2O$	454.08	red cr	....	d 100	....	d
phosphate, di-H	$NaH_2PO_4 \cdot H_2O$	137.99	col, rh, 1.4852	2.040	$-H_2O$, 100	d 200	$71°$
	$NaH_2PO_4 \cdot 2H_2O$	156.01	col, rh, 1.4629	1.91	60	....	$91.1°$
phosphate, hydrogen	Na_2HPO_4	141.96	wh, hygr	1.679	d	d	$1.63°$
	$Na_2HPO_4 \cdot 7H_2O$	268.07	col, mn, 1.4424	....	....	....	$185^{40°}$
	$Na_2HPO_4 \cdot 12H_2O$	358.14	col, mn, 1.4361	1.52	34.6	$-12H_2O$, 180	$4.3°$
phosphate	Na_3PO_4	163.94	wh	$2.537^{17.5°}$	1340	....	$11^{20°}$
	$Na_3PO_4 \cdot 12H_2O$	380.12	wh, trig, 1.4458	1.62	73.4	$-11H_2O^{100°}$	$28.3^{15°}$
phosphate, meta-(glass)	$NaPO_3$	101.96	col, hygr, 1.485	$2.484^{20°}$	cr >300	....	v s
(sol)	$NaPO_3$	101.96	col, 1.478	2.476	627.6	....	21
(insol)	$NaPO_3$	101.96	col, 1.510	....	tr 500	....	i

Name	Formula	Formula wt	Crystalline form, color, refractive index	Density	Melting point	Dehydration	Solubility
phosphate, pyro-	Na₄P₂O₇	265.90	wh	2.45	988	.....	2.26°
	Na₄P₂O₇·10H₂O	446.06	mn, 1.4525	1.82	d	.....	5.4°
di-H	Na₂H₂P₂O₇	221.94	col, mn, 1.510	1.862	d 220	.....	4.5°
	Na₂H₂P₂O₇·6H₂O	330.03	col, mn, 1.4645	1.848	.....	.....	6.9°
orthophosphite, mono-H	Na₂HPO₃·5H₂O	216.04	deliq, rh	.....	53	$-5H_2O$, 120	400°
orthophosphite, di-H	2NaH₂PO₃·5H₂O	298.03	mn pr	.....	42	$-5H_2O$, 100	56°
platinate(IV)	Na₂PtO₃·3H₂O	343.11	yel	.....	.....	$-3H_2O$, 150	s
potassium tartrate	NaKC₄H₄O₆·4H₂O	282.23	rh, 1.493	1.790	70–80	$-4H_2O$, 215	26°
salicylate	NaC₇H₅O₃	160.11	wh cr	.....	.....	.....	111¹⁵°
selenate	Na₂SeO₄	188.94	rh	3.098	.....	.....	83.4³⁵°
	Na₂SeO₄·10H₂O	369.09	mn	1.58	.....	.....	29.80°
selenide	Na₂Se	124.94	deliq cr	2.625¹⁰	>875	d	d
silicate, meta-	Na₂SiO₃	122.06	col, rh, 1.520	.....	1088	.....	s
	Na₂SiO₃·5H₂O	212.14	wh, tric, 1.456	1.749	72.2	$-5H_2O$, 100	v s
silicate, ortho-	Na₄SiO₄	184.04	col, hex, 1.530	.....	1018	.....	s
silicate, tetra-	Na₂Si₄O₉	302.32	amorp	.....	.....	.....	s
silicate, hexafluoro-	Na₂SiF₆	188.06	wh, hex, 1.312	2.679	d	.....	0.44°
stannate(IV)	Na₂SnO₃·3H₂O	266.71	hex tablets	.....	d 140	.....	50°
stearate	NaOOCC₁₇H₃₅	306.47	wh soap	.....	.....	.....	sl s
sulfate (thenardite)	Na₂SO₄	142.04	col, rh, 1.477	2.664	tr 234	d	16°
(metathenardite)	Na₂SO₄	142.04	col	2.697²⁵°	884	.....	48.84°
	Na₂SO₄·7H₂O	268.15	tetr	.....	.....	.....	44.9°
(Glauber's salt)	Na₂SO₄·10H₂O	322.19	col, mn, 1.396	1.464	32.4	$-10H_2O$, 100	36¹⁵°
sulfate, hydrogen	NaHSO₄	120.06	col, tric	2.742	>315	d, $-H_2O$	50°
	NaHSO₄·H₂O	138.07	col, deliq, mn	.....	58.5	d	100
sulfide, mono-	Na₂S	78.04	pink or wh, amorp	1.856	1180	d	15.4¹⁰°
	Na₂S·9H₂O	240.18	col, deliq, cr	.....	50	d	125²⁵°
sulfide, tetra-	Na₂S₄	174.24	yel, cub	.....	275	.....	s
sulfide, penta-	Na₂S₅	206.30	yel	.....	251.8	.....	s
sulfite	Na₂SO₃	126.04	hex pr, 1.565	2.633$^{15}_{4}$	d	.....	13.9°
	Na₂SO₃·7H₂O	252.15	mn	1.561	$-7H_2O$, 33.4	d	34.7²°
sulfite, hydrogen	NaHSO₃	104.06	col, mn, 1.526	1.48	d	.....	s
tartrate	Na₂C₄H₄O₆·2H₂O	230.08	rh	1.818	.....	.....	29°
tartrate, hydrogen	NaHC₄H₄O₆·H₂O	190.09	wh cr	.....	.....	.....	11

Table 4-1 (Continued)
PHYSICAL CONSTANTS OF INORGANIC COMPOUNDS

Name	Formula	Formula Weight	Color, Crystalline Form, and Refractive Index	Density	Melting Point, °C	Boiling Point, °C	Solubility in 100 Parts
Sodium							
thioantimonate	$Na_3SbS_4 \cdot 9H_2O$	481.11	yel, cub	1.839	....	....	21.6°
thioarsenate	$Na_3AsS_4 \cdot 8H_2O$	416.29	mn, 1.6802	....	d	....	s
	$2Na_3AsS_4 \cdot 15H_2O$	814.56	yel or col, mn	....	d	....	v s
thiocarbonate	$Na_2CS_3 \cdot H_2O$	172.20	yel	....	d	....	s
thiocyanate	$NaSCN$	81.07	delq, rh, 1.625±	....	287	....	$110^{10°}$
thioplatinate	$Na_4Pt_3S_6$	869.61	red, rh	....	d	....	i
thiosulfate	$Na_2S_2O_3$	158.11	mn	1.667	....	....	50°
(hypo)	$Na_2S_2O_3 \cdot 5H_2O$	248.18	mn pr, 1.5079	1.685	d 48.0	....	74.7°
tungstate	Na_2WO_4	293.83	wh, rh	4.179	692	....	57.58°
	$Na_2WO_4 \cdot 2H_2O$	329.86	wh, rh	3.245	$-2H_2O$, 100	....	88°
tungstate, hepta-	$Na_6W_7O_{24} \cdot 16H_2O$	2097.12	wh, tric	$3.987^{14°}$	$-16H_2O$, 300	....	8
uranate(VI)	Na_2UO_4	348.01	yel	....	....	....	i
vanadate	$Na_3VO_4 \cdot 16H_2O$	472.15	col nd	....	866 (anhyd)	....	v s
vanadate(V), di-	$Na_4V_2O_7$	305.84	hex	....	654	....	s
Strontium							
	Sr	87.62	silv-wh, fcc	$2.6^{20°}$	774	1366	d; s a
acetate	$Sr(C_2H_3O_2)_2$	205.71	wh cr	2.099	d	....	36.9
orthoarsenate, mono-H	$SrHAsO_4 \cdot H_2O$	245.56	wh, rh nd	$3.606^{15°}$ 4.035 anhyd	$-H_2O$, 125	....	$0.28^{16°}$; s a
boride, hexa-	SrB_6	152.49	blk, cub	$3.39^{15°}$	2235	....	i; s HNO_3
bromate	$Sr(BrO_3)_2 \cdot H_2O$	361.45	col-yelsh, mn, hygr	3.773	$-H_2O$, 120	d 240	43
bromide	$SrBr_2$	247.44	wh, hex, hygr, 1.575	4.216	643	d	$100^{20°}$
	$SrBr_2 \cdot 6H_2O$	355.53	col, hex, hygr	2.386	$-6H_2O$, 180	....	v s; $63^{20°}$ al
carbide	SrC_2	111.64	blk, tetr	3.2	>1700	....	d; d a
carbonate (strontianite)	$SrCO_3$	147.63	col, rh, 1.516, 1.664, 1.666	3.70	$-CO_2$, 1340	....	i; s a
chlorate	$Sr(ClO_3)_2$	254.52	col, rh, 1.516, 1.605, 1.626	3.152	d 120	....	v s

Name	Formula	Formula wt.	Crystalline form, color, index of refraction	Density	Melting point	Boiling point	Solubility
chloride	$SrCl_2$	158.53	col, cub, 1.650	3.052	875	1250	35.8
	$SrCl_2 \cdot 6H_2O$	266.62	col, trig, 1.536, 1.487	1.93	$-6H_2O$, 100	....	v s
chloride fluoride	$SrCl_2 \cdot SrF_2$	284.14	col, tetr, 1.651, 1.627	4.18	962	....	d; s conc HCl, HNO$_3$
chromate	$SrCrO_4$	203.61	yel, mn	3.895^{15}	....	....	0.12^{15}; s HCl, HNO$_3$
cyanide	$Sr(CN)_2 \cdot 4H_2O$	211.72	wh, rh, deliq	....	d	....	v s
ferrocyanide	$Sr_2Fe(CN)_6 \cdot 15H_2O$	657.42	yel, mn	....	....	....	50
fluoride	SrF_2	125.62	col, cub, 1.442	4.24	1450	2490	0.011, s hot HCl
fluorosilicate	$SrSiF_6 \cdot 2H_2O$	265.73	col, mn	2.99^{18}	d	....	3.2^{15}
formate	$Sr(CHO_2)_2$	177.66	col, rh, 1.559, 1.547, 1.598	2.693	71.9	....	9.1°
	$Sr(CHO_2)_2 \cdot 2H_2O$	213.69	col, rh, 1.484, 1.521, 1.538	2.25	d	....	12
hydride	SrH_2	89.65	rh, o-rh, hygr	3.269	d >1000	....	d
hydroxide	$Sr(OH)_2$	121.63	wh, deliq	3.625	375 (in H$_2$)	$-H_2O$, 710	0.41°; s a
	$Sr(OH)_2 \cdot 8H_2O$	265.76	col, tetr, deliq, 1.499, 1.476	1.90	$-8H_2O$, 100	....	0.90°; s a
hyponitrite	$SrN_2O_2 \cdot 5H_2O$	237.71	wh nd	2.173	....	....	v sl s
iodate	$Sr(IO_3)_2$	437.43	wh, tric	5.045^{15}	....	....	0.03^{15}
iodide	SrI_2	341.43	col pl	4.549	515	....	v s
	$SrI_2 \cdot 6H_2O$	449.52	col-yelsh, hex, deliq	2.672	d 90	d	v s
molybdate	$SrMoO_4$	247.56	col, tetr, 1.91	4.54	d	....	0.01; s a
nitrate	$Sr(NO_3)_2$	211.63	col, cub	2.986	645	1100 tr to SrO	71^{18}
	$Sr(NO_3)_2 \cdot 4H_2O$	283.69	col, mn	2.2	$-4H_2O$, 100	....	v s
nitride	Sr_3N_2	290.87	....	....	1030	....	d; s HCl
nitrite	$Sr(NO_2)_2 \cdot H_2O$	197.65	col, hex, 1.588	2.408°	$-H_2O$, 100	d 240	59°
oxalate	$SrC_2O_4 \cdot H_2O$	193.64	col cr	....	$-H_2O$, 150	....	i; s HCl, HNO$_3$
oxide	SrO	103.62	gray-wh, cub, 1.810	3.98–4.61	2460	....	0.69^{20}; s HCl
perchlorate	$Sr(ClO_4)_2$	286.52	col cr, hygr	....	d 175	....	v s
permanganate	$Sr(MnO_4)_2 \cdot 3H_2O$	379.54	vlt, cub	2.75	....	....	v s
peroxide	SrO_2	119.62	wh pwd	4.56	d 215	....	0.018^{20}; d h H$_2$O
	$SrO_2 \cdot 8H_2O$	263.74	col cr	1.951	$-8H_2O$, 100	....	0.02
orthophosphate, mono-H	$SrHPO_4$	183.60	col, rh	3.544^{15}	d	....	i; s a

Table 4-1 (*Continued*)
PHYSICAL CONSTANTS OF INORGANIC COMPOUNDS

Name	Formula	Formula Weight	Color, Crystalline Form, and Refractive Index	Density	Melting Point, °C	Boiling Point, °C	Solubility in 100 Parts
Strontium							
phosphide	Sr_3P_2	324.81	brn cr	2.68	....	....	hyd
selenate	$SrSeO_4$	230.58	col, rh	4.23	....	....	i; s hot HCl
selenide	SrSe	166.58	wh, cub, 2.220	4.38	....	....	i; s HCl
*meta*silicate	$SrSiO_3$	163.70	col, mn, 1.599, 1.637	3.65	1580	....	i
*ortho*silicate	Sr_2SiO_4	267.32	col, mn, 1.728, 1.732, 1.758	3.84	>1750	....	....
sulfate (celestite)	$SrSO_4$	183.68	col, rh, 1.622, 1.624, 1.631	3.96	1605	....	$0.01^{0°}$; sl s a
sulfide	SrS	119.68	lt gray, cub, 2.107	3.70^{15}	>2000	....	i; d a
sulfide, hydrogen	$Sr(HS)_2$	153.76	col, cub, 2.107	....	d	....	s
sulfite	$SrSO_3$	167.68	col cr	....	d	....	i; s a
tartrate	$SrC_4H_4O_6 \cdot 4H_2O$	307.75	wh, mn	1.966	....	....	$0.11^°$; s HCl
telluride	SrTe	215.22	wh, cub, 2.408	4.83	....	....	....
thiocyanate	$Sr(SCN)_2 \cdot 3H_2O$	257.83	wh, deliq	....	$-3H_2O$, 100	d 160	v s
thiosulfate	$SrS_2O_3 \cdot 5H_2O$	289.82	col, mn nd	$2.17^{17°}$	$-4H_2O$, 100	....	$2.5^{13°}$
tungstate	$SrWO_4$	335.47	col, tetr	6.187	d	....	$0.14^{15°}$
Sulfamic acid	NH_2SO_3H	97.09	col, rh	2.126	200 d	d	14.7
Sulfamide	$SO_2(NH_2)_2$	96.11	col, rh pl	1.807	91.5	d 250	s
Sulfanes, *see* Hydrogen sulfide, tri-, etc.							
Sulfinyl bromide	$SOBr_2$	207.88	red-yel liq	2.67	-49.5	139.7	d
chloride	$SOCl_2$	118.97	col liq	$1.656^{15°}$	-101	76	hyd
chlorofluoride	SOClF	102.51	col g	$1.576^{0°}$	-139	12.2	d
fluoride	SOF_2	86.06	col g	$3.0^{-44°}$	-110	-43.8	d; s bz, eth, chl, acet
imide	SONH	63.06	col g	....	-85	....	....
Sulfonyl *di*amide	$SO_2(NH_2)_2$	96.09	col cr	....	93	....	....
bromofluoride	SO_2BrF	162.98	col liq	$2.17^°$	-86.5	40	d

Name	Formula	Form/color	Mol wt	Density	mp	bp	Solubility
chloride	SO_2Cl_2	col liq, 1.444	134.97	$1.708^{0°}$	-54.1	69.1	d; s bz, ac a
chlorofluoride	SO_2ClF	col g	118.51	$1.623^{0°}$ g/l	-124.7	7.1	d
fluoride	SO_2F_2	col g	102.06	3.72^{20} g/l	-120	-55	$10^{0°}$; s al, CCl_4
(fluoro-) hypofluorite	$FSO_2(OF)$	pale yel-grn liq	118.06	....	0	d 50	....
(peroxydi-) difluoride	$S_2O_6F_2$	col liq	198.12	....	-55.4	67.1	....
(peroxydi-) tetrafluoride	$S_2O_5F_4$	....	220.12	....	-95	d -20	....
(tri-) chloride	$S_3O_8Cl_2$	col liq	295.09	1.90^{20}	18.7	61^{3mm}	....
(tri-) fluoride	$S_3O_8F_2$	col liq	262.18	1.83	-67.2	120	....
Sulfur (α)	S_8	yel, o-rh, 1.940	256.48	2.08^{20}	112.8 tr to β	....	i; 23° CS_2; s al, bz
(β)	S_8	yel, mn, 2.017	256.48	1.96^{20}	114.6 (natural)	444.60	i; s org solv
(γ)	S_8	yel, mn, 2.216	256.48	1.92	106.8	444.60	i
(di-) bromide, di-	S_2Br_2	dk red oily liq, 1.73	223.95	2.63^{20}	-46	d 90	d
(tri-) bromide, di-	S_3Br_2	red visc liq	256.01	2.52^{20}	....	....	....
bromide pentafluoride	$SBrF_5$	....	206.97	....	-79	3.1	hyd
(di-) chloride, di-	S_2Cl_2	or-yel liq	135.03	$1.709^{0°}$	-77	138	hyd
chloride, di-	SCl_2	red liq	102.97	$1.622^{15°}$	-122	59.5	hyd
chloride, tetra-	SCl_4	pale yel pwd	173.88	....	-30	d -15	hyd
chloride pentafluoride	$SClF_5$	....	162.51	....	-64	-21	hyd
(di-) fluoride, di-	S_2F_2	col liq	102.13	$1.5^{-100°}$	-133	15	d
fluoride, di-	SF_2	col	70.06	....	-35	....	d
fluoride, tetra-	SF_4	col g	108.06	$1.919^{-73°}$	-121	-38	d
fluoride, hexa-	SF_6	col g	146.05	$1.88^{-51°}$	-51	subl 63.8	sl s; s al, KOH
(di-) fluoride, deca-	S_2F_{10}	col liq	254.11	$2.08^{0°}$	-52.7	30	d fus KOH
(difluoroamino-) pentafluoride	F_2NSF_5	col g [dec 80°]	179.07	....	....	-17.5	....
(di-) decafluoride dioxide	$S_2O_2F_{10}$	col liq	286.11	1.82^{20}	-95	49	i
(pentafluoro-) hypofluorite	$F_5S(OF)$	pale yel liq	162.06	....	-86.0	-35.1	....
cyclo (hepta-) imide	S_7NH	col, rh	239.45	2.01^{20}	112.5	....	i; s org solv
cyclo (hexa-) diimide-1,3	$S_6(NH)_2$	col cr	222.38	....	130	....	....

Table 4-1 (Continued)
PHYSICAL CONSTANTS OF INORGANIC COMPOUNDS

Name	Formula	Formula Weight	Color, Crystalline Form, and Refractive Index	Density	Melting Point, °C	Boiling Point, °C	Solubility in 100 Parts
Sulfur							
cyclo(hexa-)diimide-1,4	$S_6(NH)_2$	222.38	col cr	....	133	....	....
cyclo(hexa-)diimide-1,5	$S_6(NH)_2$	222.38	col cr	....	155	....	....
cyclo (tetra-) imide, tetra-	$S_4(NH)_4$	188.26	col cr	....	145	....	s acet, al, py
(di-) nitride, di-	S_2N_2	92.14	col; polymer, bl-blk	....	....	....	s org solv
(tetra-) nitride, di-	S_4N_2	156.26	red-brn liq, gray sol	1.71^{20}	23	30 vac	i; s org solv
(tetra-) nitride, tetra-	S_4N_4	184.28	or-yel, mn, 2.046	2.24^{18}	180	185 (160 expl)	s org solv
oxide, di-	SO_2	64.06	col g, 1.000613	2.716^{20} g/l $1.46^{-10°}$ liq	−72.7	−10.02	$3937^{20°}$ cc
oxide, tri- (γ)	SO_3	80.06	col g or liq, 1.410	1.9225^{20}	16.86	44.75	forms fum H_2SO_4
oxide, sesqui-	S_2O_3	112.13	blue liq, hygr	1.663^{20}	d 20	....	d; s fum H_2SO_4
(di-) oxide tetrachloride	S_2OCl_4	221.94	dk red liq	1.656^{0}	....	60	d
Sulfuric acid	H_2SO_4	98.08	col liq, 1.4297	1.8318^{20}	10.3	330	v s
	$H_2SO_4 \cdot H_2O$	116.09	col liq or mn cr, 1.438	1.788	8.48	290	v s
	$H_2SO_4 \cdot 2H_2O$	134.11	col liq, 1.405	1.650^{0}	−39.47	167	v s
−d_2	D_2SO_4	100.09	col liq, 1.454	1.8620^{20}	14.35	....	v s
bromo-	$HOSO_2Br$	240.90	col liq, hygr	....	−6 to −8	d	hyd
chloro-	$HOSO_2Cl$	116.52	col liq	1.784^{0}	−80	152	d viol
fluoro-	$HOSO_2F$	100.07	col fum liq	1.726	−88.98	162.6	d slowly
peroxodi-	$H_2S_2O_8$	194.14	col cr, hygr	....	d 65	....	d; s eth, al, H_2SO_4
peroxo- (Caro's acid)	H_2SO_5	114.08	col	....	d 45	....	d; s H_3PO_4
pyro-	$H_2S_2O_7$	178.14	col cr, hygr	1.9^{20}	35	d	d

Name	Formula	Mol. wt	Crystalline form	Density	Melting point	Boiling point	Solubility
Sulfurous acid	H$_2$SO$_3$	82.08	in soln only	1.03	......	......	s
Tantalum	Ta	180.948	gray-blk, bcc	16.60	2980	5425	s HF, fus alk
boride, di-	TaB$_2$	202.57	gray	11.15	3200	......	d
(III) bromide	TaBr$_3$	420.66	......	......	d 220	d	......
(IV) bromide	TaBr$_4$	500.57	brn-blk cr, o-rh	......	392	......	d; s eth, abs alc
(V) bromide	TaBr$_5$	580.49	bronze, o-rh, hygr	4.989	280	348	sl s HF, H$_2$SO$_4$
carbide	TaC	192.96	blk, cub	13.9	3825	5500	s h min a
(III) chloride	TaCl$_3$	287.31	blk cr	5.03^{20}	d 340	......	hyd
(IV) chloride	TaCl$_4$	322.76	blk, o-rh or mn	4.35	d	......	s inert org solv
(V) chloride	TaCl$_5$	358.21	wh, hygr	3.76	216.5	234	s HF
(V) fluoride	TaF$_5$	275.94	wh, mn	3.880^{95}	95–97	229	hyd; s eth
(V) iodide	TaI$_5$	815.47	blk-brn pwd, o-rh	5.80^{26}	496	543	sl s aq reg
nitride	TaN	194.95	bronze or blk, hex	16.30	3090	......	i a
(IV) oxide	TaO$_2$	212.95	dk-gray pwd	......	oxidizes	......	s fus KHSO$_4$, HF
(V) oxide	Ta$_2$O$_5$	441.89	col, rh	8.2	1800	......	......
oxide tribromide	TaOBr$_3$	436.66	lt yel	......	d	......	......
dioxide fluoride	TaO$_2$F	231.92	cub	......	d 850	......	......
phosphide, mono-	TaP	211.92	blk pwd	10.9	......	......	......
phosphide, di-	TaP$_2$	242.90	brn cr	8.41	......	......	sl s HNO$_3$
sulfide, di-	TaS$_2$	245.08	blk pwd	......	>1300	......	s HNO$_3$, conc H$_2$SO$_4$
Technetium	Tc	98.906	hcp	11.487	2200	(4700)	s HCl
carbide	TcC	110.92	fcc	11.5	......	......	hyd
(IV) chloride	TcCl$_4$	240.72	red, o-rh	......	subl 300	......	hyd
(VI) chloride	TcCl$_6$	311.63	volat unstable grn sol	......	d 25 to TcCl$_4$	......	s HCl
(V) fluoride	TcF$_5$	193.91	yel, o-rh	......	50	d 60	s a, alk
(VI) fluoride	TcF$_6$	212.91	yel, cub	......	33.4	55.3	s
(IV) oxide	TcO$_2$	130.91	blk	6.9	subl 1000	......	......
(VII) oxide	Tc$_2$O$_7$	309.81	lt yel, hygr	......	119.5	310.6	......
oxide tribromide	TcOBr$_3$	354.62	brn	......	subl 400 (in Br$_2$ gas)	......	......
oxide tetrafluoride	TcOF$_4$	190.91	blue, mn	......	134	sl d	......
trioxide chloride	TcO$_3$Cl	182.36	col liq	......	18.3	~100	hyd
trioxide fluoride	TcO$_3$F	165.91	yel cr or liq	......	......	......	hyd

Table 4-1 (Continued)
PHYSICAL CONSTANTS OF INORGANIC COMPOUNDS

Name	Formula	Formula Weight	Color, Crystalline Form, and Refractive Index	Density	Melting Point, °C	Boiling Point, °C	Solubility in 100 Parts
Technetium							
(IV) sulfide	TcS_2	163.03	....	....	....	d 1000	....
(VII) sulfide	Tc_2S_7	422.23	dk brn	....	subl 100	d	s H_2O_2 + NH_3
Telluric acid	$H_2TeO_4 \cdot 2H_2O$ or H_6TeO_6 or $Te(OH)_6$	229.64	col, cub	3.20	$-2H_2O$, 120; 136 (sealed tube)	....	s
Tellurium	Te	127.60	silv, hex-rh	6.24^{20}	450	994	....
bromide, di-	$TeBr_2$	287.44	blk-grn sol	....	210	339	d
bromide, tetra-	$TeBr_4$	447.24	red-yel cr	4.31^{15}	~380	~420	hyd
(IV) dibromide dichloride	$TeBr_2Cl_2$	357.73	yel pwd, hygr	....	292	415	hyd
(IV) dibromide diiodide	$TeBr_2I_2$	541.25	red cr	....	325	d 420	hyd
chloride, di-	$TeCl_2$	198.51	blk-grn sol	....	208	328	....
chloride, tetra-	$TeCl_4$	269.41	col cr	2.56^{232}	224.1	388	....
fluoride, tetra-	TeF_4	203.60	col sol	....	129.6	....	....
fluoride, hexa-	TeF_6	241.59	col g	....	-37.8	-38.2 subl	hyd
(di-)fluoride, deca-	Te_2F_{10}	317.58	....	2.852^{20}	-33.7	59.0	....
iodide, tetra-	TeI_4	635.22	blk cr	5.76	259	....	....
oxide, di-	TeO_2	159.60	tellurite, col, tetr-octahed (artificial) rh	....	732.6	....	s HCl, HF, HNO_3
oxide, tri-	TeO_3	175.60	yel pwd	6.04	d 400	....	s conc alk
oxide, pent-	Te_2O_5	335.20	....	5.075	d 450	....	s 30% KOH
Tellurous acid	$H_2TeO_3(?)$	177.61	wh, indef, not isolated	4.14	d 40	....	d; s a, alk
Terbium	Tb	158.925	silv-gray, hcp	8.272	1356	2800	s a
bromide	$TbBr_3$	398.65	wh, rh	4.67	828	1490	s

Name	Formula	Formula wt	Color, crystalline form, index	Density	Melting point, °C	Boiling point, °C	Solubility
chloride	$TbCl_3$	265.28	wh, rh	4.35	582	1550	v s
	$TbCl_3 \cdot 6H_2O$	373.38	col pr, deliq				v s
fluoride, tri-	TbF_3	215.92	wh, hex	7.236	1172	2280	s H_2SO_4
fluoride, tetra-	TbF_4	234.92	wh, mn	5.88			s
iodide	TbI_3	539.64	wh, hex		957	1330	s
nitrate	$Tb(NO_3)_3 \cdot 6H_2O$	453.03	col, mn		893		
oxalate	$Tb_2(C_2O_4)_3 \cdot 10H_2O$	762.06	wh cr	2.60	$-H_2O$, 40		s a
oxide, sesqui-	Tb_2O_3	365.85	wh, bcc				s hot conc a
(tetra-;heptoxide)	Tb_4O_7	747.69	blk				
sulfate	$Tb_2(SO_4)_3 \cdot 8H_2O$	750.16	wh cr		$-8H_2O$, 360		3.56
Thallium	Tl	204.37	bl-wh met, hcp	11.80	304	1470	s HNO_3, H_2SO_4
(I) acetate	$TlC_2H_3O_2$	263.42	slky wh nd, deliq	$3.765^{137°}$	131		v s
(I) aluminum sulfate	$TlAl(SO_4)_2 \cdot 12H_2O$	639.66	oct, 1.488	2.306	91		11.8
(I) azide	TlN_3	246.39	yel, tetr		330 vac		0.3
(I) boronate	$TlBH_4$	219.21	wh, fcc				i, sl d
(I) bromate	$TlBrO_3$	332.28	col nd				0.35
(I) bromide	$TlBr$	284.28	yel-wh, cub	7.54		815	0.05; s al
(III) bromide	$TlBr_3$	444.10	yel, deliq, unstable		45.6		s; v s al
(I) tetrabromo-thallate(III)	$Tl[TlBr_4]$	728.38	yel nd		d		d
(I) carbonate	Tl_2CO_3	468.75	col, mn	7.11	273		4.03^{15}
(I) chlorate	$TlClO_3$	287.82	nd	5.047^{9}			2^{0}
(I) chloride	$TlCl$	239.82	wh, 2.247	7.004^{30}	430	720	0.3
(III) chloride	$TlCl_3$	310.73	wh, hex, hygr		33	-Cl, 40	v s
(III) chloride	$TlCl_3 \cdot 4H_2O$	382.79	col nd	$5.76^{17°}$	37	$-4H_2O$, 100	86
(I) hexachloro-platinate(IV)	$Tl_2[PtCl_6]$	816.55	lt or cr				$0.0064^{15°}$
(I) chromate	Tl_2CrO_4	524.73	yel				0.03; sl s a
(I) chromium sulfate	$Tl[Cr(H_2O)_6](SO_4)_2 \cdot 6H_2O$	664.67	vlt cr	2.394	92		164
(I) cyanate	$TlOCN$	246.39	col nd	5.487^{20}			s
(I) cyanide	$TlCN$	230.39		6.523			$16.8^{28°}$
(I) dichromate	$Tl_2Cr_2O_7$	624.73	red	5.572^{20}			i; s a
(I) dithionate	$Tl_2S_2O_6$	568.86	wh, mn	3.493^{20}	d		41.8
(I) ethylate	$TlOC_2H_5$	249.43	liq, 1.6714		-3		sl s al, s eth
(I) ferrocyanide	$Tl_4[Fe(CN)_6] \cdot 2H_2O$	1065.46	yel, tric	4.641		d 130	0.37

Table 4-1 (*Continued*)
PHYSICAL CONSTANTS OF INORGANIC COMPOUNDS

Name	Formula	Formula Weight	Color, Crystalline Form, and Refractive Index	Density	Melting Point, °C	Boiling Point, °C	Solubility in 100 Parts
Thallium							
(I) *pentafluoro-*gallate(III)	$Tl_2[GaF_5]\cdot H_2O$	591.47	col, o-rh	6.44	....	....	s
(I) *hexafluoro*silicate	$Tl_2SiF_6\cdot 2H_2O$	586.85	col, hex	5.72	....	....	v s
(I) fluoride	TlF	223.37	col, cub	8.23°	327	655	78[15]
(III) fluoride	TlF_3	261.37	ol grn	8.36	d 550	....	d
(I) formate	$TlCHO_2$	249.39	col nd, hygr	4.967[104]	101	....	500[10]
(I) hydroxide	$TlOH$	221.38	lt yel nd	....	d 139	....	26⁰
(I) iodate	$TlIO_3$	379.27	wh nd	....	....	....	0.058
(I) iodide	TlI	331.27	(α) yel, rh (β) red, cub	7.29 / 7.09[15]	tr β, 170 / 440	823	0.006
(I) iodide, tri-	$Tl(I_3)$	585.08	blk lust, rh	....	d	....	s
(I) iron(III) sulfate	$TlFe(SO_4)_2\cdot 12H_2O$	668.52	pink, oct, 1.524[17]	2.351[15]	....	....	36 anhyd
(I) magnesium sulfate	$Tl_2Mg(SO_4)_2\cdot 6H_2O$	733.26	wh cr, 1.5660, 1.5836, 1.5900	3.573[20]	−6H₂O, 40	....	....
methoxide	$TlOCH_3$	235.40	wh	....	d >120	....	s d: 1.7 MeOH; 3.2 bz
molybdate	Tl_2MoO_4	568.68	wh pwd	....	....	....	s alk carb, HF, NH₄OH
(I) nitrate	$TlNO_3$	266.37	col, rh; tr 60° trig; tr 143.5° cub	5.556	206	430	9.55
(III) nitrate	$Tl(NO_3)_3$	390.38	col cr	....	....	....	s
(III) nitrate	$Tl(NO_3)_3\cdot 3H_2O$	444.43	col, rh, deliq	....	....	....	
(I) nitrite	$TlNO_2$	250.38	yel cr	....	182	....	32.1[25]
oxalate	$Tl_2(C_2O_4)$	496.76	mn pr	6.31	....	....	1.5[15]
bis-oxalate, tri-hydrogen	$TlH_3(C_2O_4)_2\cdot 2H_2O$	419.46	tric, 1.5097, 1.6319, 1.6538	2.992[17]	d 100	....	76.9
(I) oxide	Tl_2O	424.74	blk, deliq	9.52[16]	300	1080	v s d; s a
(III) oxide	Tl_2O_3	456.74	col, amorp or hex	10.19[22]; hex; 9.65 amorp	717	−O₂, 875	s a

Name	Formula	Mol wt	Crystalline form, color, refractive index	Density	M.p., °C	B.p., °C	Solubility
perchlorate	TlClO$_4$	303.82	col, rh	4.89	501	d	20.5^{30}
*ortho*phosphate	Tl$_3$PO$_4$	708.08	col nd	6.89^{10}	....	....	0.5^{15}
*ortho*phosphate, di-H	TlH$_2$PO$_4$	301.36	mn	4.726	ca 190	....	40
pyrophosphate	Tl$_4$P$_2$O$_7$	991.42	mn pr	6.786	>120	....	2.13^{10}
selenate	Tl$_2$SeO$_4$	551.70	rh, 1.949, 1.959, 1.964	....	>400	....	....
selenide	Tl$_2$Se	487.70	gray leaf	9.05^{25}	340	....	s a
(I) sulfite	Tl$_2$SO$_4$	504.80	col, rh, 1.860, 1.867, 1.885	6.77	632	d	4.87
(I) sulfate, hydrogen	TlHSO$_4$	301.44	pr nd	....	120 d	....	d; s dil H$_2$SO$_4$
(III) sulfate	Tl$_2$(SO$_4$)$_3$·7H$_2$O	823.03	col leaf	....	−6H$_2$O, 220	oxidizes	0.02; s a
(I) sulfide	Tl$_2$S	440.80	bl-blk, tetr	8.0	448.5	d	s h H$_2$SO$_4$
(III) sulfide	Tl$_2$S$_3$	504.93	bl-blk, amorp	....	260 (in N$_2$)	....	....
(di-) *pentasulfide*	Tl$_2$S$_5$	569.04	brick red pwd	....	127	....	....
tartrate (*dl*)	Tl$_2$C$_4$H$_4$O$_6$	556.81	mn	4.659	d 165	....	13.3^{15}
metatellurate	Tl$_2$TeO$_4$	600.34	wh ppt	6.760^{18}	red heat	....	sl s
thiocyanate	TlSCN	262.45	col, tetr	4.956	260	....	0.32
*iso*thiocyanate	TlCNS	262.45	glossy leaf, rh	4.954^{30}	d low temp	....	0.39
thiosulfate	Tl$_2$S$_2$O$_3$	520.87	wh rh	....	d 130	....	sl s
metavanadate	TlVO$_3$	303.41	gray cr	6.09^{17}	424	....	0.87^{11}
(tetra-) divanadate(V)	Tl$_4$V$_2$O$_7$	1031.36	lt yel	8.21	454	....	0.2^{14}
Thiazyl chloride	NSCl	81.52	yel-grn g	....	polymerizes	....	v s bz, CS$_2$, CCl$_4$
(*cyclo*tri-) chloride, tri-	N$_3$S$_3$Cl$_3$	244.55	yel nd, hygr	2.09	162.5	polymerizes	....
fluoride	NSF	65.06	col g	....	−7.9	4.6	hyd
fluoride, tri-	NSF$_3$	103.06	col g	....	−80.8	−23.1	v s bz, CS$_2$, CCl$_4$
(*cyclo*tri-) fluoride, tri-	N$_3$S$_3$F$_3$	195.19	col cr, hygr	....	74.2	92.5	v s bz, CS$_2$, CCl$_4$
(*cyclo*tetra-) fluoride, tetra-	N$_4$S$_4$F$_4$	260.26	col nd, tetr	2.326^{20}	153	d 128	0.34 CCl$_4$
Thiocarbonyl chloride, di-	CSCl$_2$	114.98	red-yel liq, 1.5442	1.509^{15}	....	73.5	d; s eth
chloride, tetra-	CSCl$_4$	185.89	yel	1.712^{13}	146	....	v s; v s bz, al, eth
Thiocyanic acid, iso-	HNCS	59.09	col	....	ca −110	polymer- izes, −90	

Table 4-1 (*Continued*)
PHYSICAL CONSTANTS OF INORGANIC COMPOUNDS

Name	Formula	Formula Weight	Color, Crystalline Form, and Refractive Index	Density	Melting Point, °C	Boiling Point, °C	Solubility in 100 Parts
Thiocyanogen	$(SCN)_2$	116.16	liq or yel sol	....	ca −2	....	d; s al, eth, CS_2, CCl_4
Thiophosphoryl amide	$PS(NH_2)_3$	111.11	yel-wh amorp	$1.71^{3°}$	d 200	....	sl s
bromide	$PSBr_3$	302.78	yel fine cr, cub	$2.85^{17°}$	38.2	206 d	s; s eth, CS_2
bromide dichloride	$PSBrCl_2$	213.87	col oily liq, 1.6107	$2.037^{20°}$	....	d 154	
bromide(di-) chloride	$PSBr_2Cl$	258.33	lt yel oily liq, 1.6660	$2.407^{20°}$	....	d 170	
bromide difluoride	$PSBrF_2$	170.95	col liq	$1.940°$	−136.9	35.5	
bromide(di-) fluoride	$PSBr_2F$	241.87	col liq	$2.390°$	−75.2	125.3	
chloride	$PsCl_3$	169.40	col g, fum liq	1.668	−35	125	sl d; s CS_2
chloride difluoride	$PSClF_2$	136.48	col g	....	−155.2	6.3	
chloride(di-) fluoride	$PSCl_2F$	152.94	col liq	....	−96.0	64.7	
cyanate, iso-	$PS(NCO)_3$	237.22	col liq	1.538	8.8	215	
fluoride	PSF_3	120.03	col g	....	−148.8	−52.2	hyd
Thiosulfinyl fluoride	$S=SF_2$	102.13	col g	....	−165	−10.6	
Thorium	Th	232.038	gray, fcc	11.71	1800	~4200	s HCl, H_2SO_4, aq reg
boride, tetra-	ThB_4	275.28	tetr pr	$7.5^{15°}$	....	....	s HCl, HNO_3, h H_2SO_4
boride, hexa-	ThB_6	296.92	vlt-blk, cub	$6.4^{15°}$	2195	....	s HNO_3
boronate	$Th(BH_4)_4$	291.40	tetr	2.50	203	....	....
bromide	$ThBr_4$	551.67	wh, rh, deliq	5.55	725	subl 610	v s
carbide	ThC_2	256.06	yel, tetr	$8.96^{18°}$	2655	....	v sl s conc a
carbonate	$Th(CO_3)_2$	352.06	....	....	....	....	s a
chloride	$ThCl_4$	373.85	wh, tetr, deliq	4.60	770	d 928	v s
fluoride	ThF_4	308.03	wh, mn	5.71	1110	....	sl d dil HCl, H_2SO_4
	$ThF_4 \cdot 4H_2O$	380.09	cr	....	$-H_2O$, 100	$-2H_2O$, 140	0.017
hydride, di-	ThH_2	234.05	bct	9.20	....	....	....
hydride, pentadeca-	Th_4H_{15}	943.27	cub	8.24	....	....	....

Name	Formula	Mol. wt.	Color, crystalline form, refractive index	Density	Melting point, °C	Boiling point, °C	Solubility
hydroxide	$Th(OH)_4$	300.02	wh gelat		d		s a
iodate	$Th(IO_3)_4$	931.65	wh		d >300	>674, ThO_2	s dil H_2SO_4
iodide, di-	ThI_2	485.85	α blk, hex; β golden, hex		d 550		hyd
iodide, tri-	ThI_3	612.75	blk				d
iodide, tetra-	ThI_4	739.66	yel-wh, mn	6.00	566	837	s hyd
nitrate	$Th(NO_3)_4$	480.06	col pl, deliq		d >216	>630, ThO_2	v s
	$Th(NO_3)_4 \cdot 4H_2O$	552.12	col cr		d >216		v s
nitride	ThN	246.04			2630		d; s HCl
(tri-) nitride, tetra-	Th_3N_4	752.14	dk brn pwd				
oxalate	$Th(C_2O_4)_2$	408.08	wh cr	$4.637^{16°}$	d 350	>610, ThO_2	0.0017; sl s a
oxide (thorianite)	ThO_2	264.04	wh, cub	9.86	3050	4400	s hot H_2SO_4
oxide dichloride	$ThOCl_2$	318.94	wh, rh		d 550		v s
oxide diiodide	$ThOI_2$	501.85	wh, hygr	6.44°			s aq reg
oxide sulfide	$ThOS$	280.10	yel cr				
2,4-pentanedione	$Th(C_5H_7O_2)_4$	628.48	col cr	$4.08^{16°}$	subl 171	265^{10mm}	v s al, chl; s eth
metaphosphate	$Th(PO_3)_4$	547.93	col rh pr				s HCl
orthophosphate	$Th_3(PO_4)_4 \cdot 4H_2O$	1148.06	wh gelat		anhyd >540		i
pyrophosphate	ThP_2O_7	411.98					
phosphide	ThP	263.01	gray-blk, cub	8.81			
(tri-) phosphide, tetra-	Th_3P_4	820.01	gray-blk, cub	8.44			
selenate	$Th(SeO_4)_2 \cdot 9H_2O$	680.09	col, mn	3.026	−8H_2O, 200	d 1500	0.5°
orthosilicate	$ThSiO_4$	324.12	col, tetr, 1.80, 1.81	$6.82^{16°}$			v sl s; i a
silicide, di-	$ThSi_2$	288.21	blk, tetr	$7.96^{16°}$			s hot HCl
sulfate	$Th(SO_4)_2$	424.16	wh cr, hygr	$4.225^{17°}$			9.41
	$Th(SO_4)_2 \cdot 4H_2O$	496.22	wh nd	2.77	−4H_2O, 400		1.57
	$Th(SO_4)_2 \cdot 9H_2O$	586.31	wh, mn		−9H_2O, 400		
sulfide, di-	ThS_2	296.17	brn-blk,	7.30	1925 vac		s hot aq reg
Thulium	Tm	168.934	silv-wh, hcp	9.332	1545	1730	s a
bromide, di-	$TmBr_2$	239.84			619		
bromide, tri-	$TmBr_3$	408.66	wh, rh-hed	5.02	952	1440	s
chloride	$TmCl_3$	275.29	lt yel, mn		824	1490	s
	$TmCl_3 \cdot 7H_2O$	401.40	grn cr, deliq				v s

Table 4-1 (*Continued*)

PHYSICAL CONSTANTS OF INORGANIC COMPOUNDS

Name	Formula	Formula Weight	Color, Crystalline Form, and Refractive Index	Density	Melting Point, °C	Boiling Point, °C	Solubility in 100 Parts
Thulium							
fluoride	TmF_3	225.93	wh, hex	7.971	1158	2230	s H_2SO_4
iodide, di-	TmI_2	422.74	blk, hex	....	756	....	....
iodide, tri-	TmI_3	549.65	yel, hex	....	1021	1260	s
nitrate	$Tm(NO_3)_3 \cdot 4H_2O$	499.08	grn, deliq	....	....	....	s alk, oxal
oxalate	$Tm_2(C_2O_4)_3 \cdot 6H_2O$	710.02	wh-grn	8.6	$-H_2O$, 50	....	sl s a
oxide	Tm_2O_3	385.87	lt grn, bcc	7.28	231.89	....	i; s HCl, H_2SO_4, alk
Tin	Sn	118.69	silv-wh, tetr gray, cub	5.75	stable −161 to 13.2	2687	
(II) acetate	$Sn(C_2H_3O_2)_2$	236.78	yelsh pwd	....	182	240	d; s dil HCl
(II) bromide	$SnBr_2$	278.51	pale yel, rh	4.9–5.1	216	620	$85°$; s al, eth, acet
(IV) bromide	$SnBr_4$	438.33	col, rh py	$3.35^{33°}$	32	203	s d; s acet
bromide dichloride	$SnBrCl_3$	304.96	col liq	$2.51^{13°}$	−31	d 191	$50^{30°}$
*di*bromide dichloride	$SnBr_2Cl_2$	349.41	....	$2.82^{13°}$	−20	....	d
*tri*bromide chloride	$SnBr_3Cl$	393.87	liq	$3.12^{13°}$	1	....	....
*di*bromide diiodide	$SnBr_2I_2$	532.22	or-red, hex pl	$3.631^{15°}$	50	225	s
(II) chloride	$SnCl_2$	189.60	wh, rh	3.95	227	605	$84°$; s al, eth, acet, et acet, py
	$SnCl_2 \cdot 2H_2O$	225.63	wh, mn	$2.710^{16°}$	37.7	d	d; s al, eth, acet, glac ac a
(IV) chloride	$SnCl_4$	260.50	col liq; cub cr	2.226 liq	−33.3	114.1	s; s eth
*di*chloride diiodide	$SnCl_2I_2$	443.40	red liq	$3.287^{15°}$	....	297	s; s bz, chl, CS_2
(IV) chloride nitrosyl chloride	$SnCl_4 \cdot 2NOCl$	391.42	pale yel, oct cr	2.60	180	d	d
(IV) chromate	$Sn(CrO_4)_2$	350.68	brn-yel cr pwd	....	d		s
(II) fluoride (fluoristan)	SnF_2	156.69	wh, mn	....	....	....	s

Name	Formula	Mol. wt.	Color, crystalline form, refractive index	Density	Melting point	Boiling point	Solubility
(IV) fluoride	SnF_4	194.68	wh, mn, hygr	4.780^{19}	705 subl	.....	v s
hydride	SnH_4	122.72	col g	.....	−150	−52	s conc alk, H_2SO_4
(II) iodide	SnI_2	372.50	yelsh-red, mn nd	5.285	320	717	0.98^{20}; s dil HCl
(IV) iodide	SnI_4	626.31	or-red, cub, 2.106	4.473^{0}	144.5	364.5	s; hyd hot water
(II) nitrate	$Sn(NO_3)_2 \cdot 20H_2O$	603.07	col leaf	.....	−20	.....	d
(IV) nitrate	$Sn(NO_3)_4$	366.71	silky nd	.....	d 50	.....	hyd
(II) oxide	SnO	134.69	blk, cub	6.446^{0}	oxid, 300	.....	i; s a, alk
(IV) oxide (cassiterite)	SnO_2	150.69	wh, tetr, 1.997, 2.093	6.95	1625	subl 1900	i; d alk
(IV) oxide, hydrate	"Stannic acid" H_2SnO_3 or $SnO_2 \cdot xH_2O$	.....	amorp or gel	.....	.....	.....	i; s a, alk; after ign i a, s alk
(II) metaphosphate	$Sn(PO_3)_2$	276.63	wh amorp	3.380^{23}	.....	.....	.....
(II) orthophosphate	$Sn_3(PO_4)_2$	546.01	wh amorp	3.823^{17}	.....	.....	i; d a
(II) orthophosphate, di-H	$Sn(H_2PO_4)_2$	312.66	wh, rh cr	3.167^{23}	d	.....	.....
(II) pyrophosphate	$Sn_2P_2O_7$	411.32	wh amorp pwd	4.009^{16}	.....	.....	s conc a
phosphide	SnP	149.66	silv-wh	6.56	d	.....	i; s HCl
phosphide, tri-	SnP_3	211.61	silv-wh	4.10^{0}	530	.....	i; s HCl
(tri-) tetraphosphide	Sn_3P_4	479.96	silv-wh	.....	562	.....	.....
(tetra-) phosphide, tri-	Sn_4P_3	567.68	silv-wh	5.181	550	.....	d HCl
(II) selenide	$SnSe$	197.65	steel gray cr	6.179^{0}	861	.....	i; d a
(II) sulfate	$SnSO_4$	214.75	wh-yelsh cr pwd	.....	−SO_2, 300	.....	33
(IV) sulfate	$Sn(SO_4)_2 \cdot 2H_2O$	346.84	wh, hex, deliq	.....	.....	.....	v s
(II) sulfide	SnS	150.75	gray-blk, cub, mn	5.22	882	1230	i; d HCl, alk
(IV) sulfide (mosaic gold)	SnS_2	182.82	golden yel, hex	4.5	d 600	.....	i; d alk sulf, aq reg
(II) telluride	$SnTe$	246.29	gray cr	6.48	780	.....	i; d alk sulf
(IV) telluride	$SnTe_2$	373.89	blk flocc	.....	.....	.....	i; d dil a, alk
Titanium	Ti	47.90	silv, hcp	4.507	1672	3260	s h min a, s HF
boride, di-	TiB_2	69.52	hex	4.50	2920	.....	d
(II) bromide	$TiBr_2$	207.72	blk, hex	4.31	>400 d	.....	.....
(III) bromide	$TiBr_3$	287.63	gray-blk, rh-hed	4.24	d 350	.....	.....

Table 4-1 (Continued)
PHYSICAL CONSTANTS OF INORGANIC COMPOUNDS

Name	Formula	Formula Weight	Color, Crystalline Form, and Refractive Index	Density	Melting Point, °C	Boiling Point, °C	Solubility in 100 Parts
Titanium							
(IV) bromide	TiBr₃·6H₂O	395.72	red-vlt, deliq, mn	....	115	d 400	s
	TiBr₄	367.54	amber yel, mn	3.33 sol 2.59⁴⁰ liq	39.0	233	hyd; 287 alc
carbide	TiC	59.91	met-gray, cub	4.938	3150	....	s HNO₃ + HF
(II) chloride	TiCl₂	118.81	blk, hex	3.13	1025	(1500)	....
(III) chloride	TiCl₃	154.26	vlt, hex	2.71	subl 425 vac	subl 700 d 450	s
(IV) chloride	TiCl₄	189.71	col, 1.612	1.70	−23.5	136	hyd
deuteride	TiD₂	51.94	fcc	3.940	....	....	....
(III) fluoride	TiF₃	104.89	bl cr, rh-hed	3.00	d 100	subl 930 vac	i alc, dil a
(IV) fluoride	TiF₄	123.89	wh, v hygr	2.80	....	284	s
hydride	TiH₂	49.92	gray pwd, fcc	3.752	d 400	....	s HF, conc HCl
(II) iodide	TiI₂	301.70	bl-brn, hex	4.99	480 subl	d	s
(III) iodide	TiI₃	428.60	dk vlt, nd, hex	4.95	d 300	....	s dry nonpolar solv
(IV) iodide	TiI₄	555.50	red-yel, cub	4.40	150	377	s aq reg, alk
nitride	TiN	61.91	yel-brn pwd, bcc	5.21	2950 (in N₂)	d 1800 vac	s
(III) oxalate	Ti₂(C₂O₄)₃·10H₂O	540.01	yel	....	....	....	s
(II) oxide	TiO	63.90	bronze, cub or mn	4.88	1750	>3000	s dil H₂SO₄
(III) oxide	Ti₂O₃	143.80	vlt-blk, trig	4.49	1839	....	s H₂SO₄
(IV) oxide (rutile)	TiO₂	79.90	wh, tetr	4.27	1840	....	s H₂SO₄, fus K₂S₂O₇
(anatase)	....	......	wh, tetr	3.90	change to rutile		
(brookite)	....	......	wh, o-rh	4.13	~700 d		
(III) oxide chloride	TiOCl	99.35	bronze, o-rh	3.14			s HNO₃ + HF

Name	Formula	Mol wt	Color, form	Density	M.p., °C	B.p., °C	Solubility
oxide dibromide	$TiOBr_2$	223.71	yel, hygr	....	d 150	....	....
oxide dichloride	$TiOCl_2$	134.81	lt yel, cub, hygr	2.44	d 180	....	i
oxide difluoride	$TiOF_2$	101.90	yel, cub	3.09	high	....	....
oxide diiodide	$TiOI_2$	317.72	dk brn, amorp, hygr	....	d 125 to TiOI	....	hyd
oxide sulfate	$TiOSO_4$	159.96	wh pwd	....	....	....	s; d h water
phosphide	$TiP_{0.95}$	......	gray, hex	3.95	....	....	i a
(III) sulfate	$Ti_2(SO_4)_3$	383.98	grn pwd	....	....	....	s conc H_2SO_4
sulfide, mono-	TiS	79.96	bronze	4.05	1927	....	s conc H_2SO_4
sulfide, sesqui-	Ti_2S_3	191.99	gray-blk	3.584	....	....	s conc H_2SO_4
sulfide, di-	TiS_2	112.03	dk bronze, hex	3.27	d	....	s h H_2SO_4
sulfide, tri-	TiS_3	144.08	bl-blk, polysulfide	3.22	....	....	s h H_2SO_4
(di-) sulfide	Ti_2S	127.86	gray met, brittle	4.80	....	....	s HCl
W	W	183.85	gray-blk, bcc	19.35	3415 ± 20	5000	s HNO_3 + HF, fus NaOH + $NaNO_3$
Tungsten							
arsenide, di-	WAs_2	333.69	blk cr	6.9	d red heat	....	d hot HNO_3 or H_2SO_4
boride, di-	WB_2	205.47	silv, oct	10.77	(3195)	....	s aq reg
(II) bromide	WB_2	343.67	grn-brn, hygr	....	subl 600 (in N_2)	....	i
(III) bromide	WBr_3	423.56	blk pwd	....	d 80	....	i; sl s polar solv
(V) bromide	WBr_5	583.40	blk, grn irrid, hygr	....	286	392	hyd, s chl, eth, alk
(VI) bromide	WBr_6	663.30	bl-blk lust nd	6.9	309	d 200	hyd; s eth, CS_2
carbide	WC	195.86	blk, hex	15.63^{15}	3140	(6000)	s HNO_3 + HF, aq reg
(di-) carbide	W_2C	379.71	blk, hex	17.15	2860	(6000)	s aq reg
carbonyl, hexa-	$W(CO)_6$	351.91	col, rh	2.65	....	175	s fum HNO_3
(II) chloride	WCl_2	254.76	gray, amorp	5.436	....	....	d
(III) chloride	WCl_3	325.66	blk, deliq	4.624	d 450	subl 507	d to W_2O_5
(V) chloride	WCl_5	361.12	blk, deliq	3.875	230	286	d to W_2O_5
(VI) chloride	WCl_6	396.57	bl-blk	$2.721^{282°}$ liq	281.5	348	hyd; s CCl_4, CS_2
chloride pentafluoride	$WClF_5$	314.30	yel liq	....	-33.7	d 25	....
(IV) fluoride	WF_4	259.85	gray	....	d 800	....	hyd; s alk
(VI) fluoride	WF_6	297.84	wh, bcc or o-rh	3.44 liq 0.0129 gas	2.0	17.1	d
nitride, di-	WN_2	211.86	brn, cub	....	>400 vac	....	d

Table 4-1 (*Continued*)
PHYSICAL CONSTANTS OF INORGANIC COMPOUNDS

Name	Formula	Formula Weight	Color, Crystalline Form, and Refractive Index	Density	Melting Point, °C	Boiling Point, °C	Solubility in 100 Parts
Tungsten							
(IV) oxide	WO_2	215.85	brn, cub	12.11	1550 (in N_2)	subl 800	s a, KOH
(V) oxide	W_2O_5	447.70	bl-vlt, tric	....	subl 800	(1530)	i
(VI) oxide	WO_3	231.85	yel, rh	7.16	1473	....	s hot alk; sl s HF
oxide *tetrabromide*	$WOBr_4$	519.49	blk nd, bct	....	322.4	331	hyd
oxide *tetrachloride*	$WOCl_4$	341.66	red-or, bct	3.22^{209} liq	209	227	hyd
oxide *tetrafluoride*	WOF_4	275.85	col	....	110	185.9	....
dioxide *dibromide*	WO_2Br_2	375.67	purp-brn	....	d 200–300	....	hyd
dioxide *dichloride*	WO_2Cl_2	286.76	wh, o-rh	....	subl 260 d	....	hyd; s conc HCl
dioxide *difluoride*	WO_2F_2	253.85	wh	....	d 200–300	....	....
phosphide	WP	214.82	gray pr	8.5	d	....	s HNO_3 + HF, aq reg
phosphide, di-	WP_2	245.80	blk cr	5.8	d	....	s HNO_3 + HF, aq reg
(di-) phosphide	W_2P	398.67	dk gray pr	5.21	d	....	s fus Na_2CO_3 + $NaNO_3$
silicide, di-	WSi_2	240.02	bl-gray, tetr	9.4	>900	....	s HNO_3 + HF
sulfide, di-	WS_2	247.98	dk-gray, hex	7.5^{10}	d 1250	....	s HNO_3 + HF
sulfide, tri-	WS_3	280.04	choc brn,	....	....	....	sl s; s alk
Tungstic acid, meta-	$H_2W_4O_{13} \cdot 9H_2O$	1107.55	....	3.93	d 50	....	88.6^{22}
ortho-	H_2WO_4	249.86	yel pwd, 2.24	5.5	$-H_2O$, 100	1473	i; s alk, HF, NH_4OH
Uranium	U	238.03	silv, o-rh	19.05	1132	3818	s a
boride, di-	UB_2	259.65	hex	12.70	2365	....	i
boronate	UBH_4	252.87	tetr	2.67	....	....	....
(III) bromide	UBr_3	477.76	brn-red nd, hex, hygr	6.53	730	....	s
(IV) bromide	UBr_4	557.67	brn, mn, deliq	5.55	519	765	v s
(V) bromide	UBr_5	637.55	brn	....	....	....	s alc, acet, eth, acet
carbide, di-	UC_2	262.05	met	11.28^{16}	2350	4370	d; s dil a
(III) chloride	UCl_3	344.39	red nd, hex, hygr	5.51	835	....	s
(IV) chloride	UCl_4	379.84	dk grn, tetr, hygr	4.87	590	792	v s

Name	Formula	Mol. wt.	Color, crystalline form	Density	m.p.	b.p.	Solubility
(V) chloride	UCl₅	415.30	red-brn, mn	....	d 100	....	d; s CS₂, SOCl₂
(VI) chloride	UCl₆	450.75	grn-blk, hex	3.56	177.5	subl 75 vac	hyd; s chl, CCl₄
(III) fluoride	UF₃	295.03	blk, hex	8.95	d 1000	subl 1000	sl d; v sl s a
(IV) fluoride	UF₄	314.02	grn, mn	6.70	960	subl 1000	s conc a, alk
(VI) fluoride	UF₆	352.02	wh, rh, deliq	5.06	64	subl 56.8	d; s chl, CCl₄
(di-) nonafluoride	U₂F₉	646.06	blk, bcc	7.06	d	....	i
hydride	UH₃	241.05	blk-brn, cub (β)	10.95	....	....	....
(III) iodide	UI₃	618.74	blk, rh	6.76	766	....	....
(IV) iodide	UI₄	745.65	blk	5.6¹⁵°	506 (I₂ atm)	760	s
nitride	UN	252.04	brn pwd	14.31	2630	....	i a
(di-) trinitride	U₂N₃	518.08	....	11.2	d	....	....
(IV) oxide	UO₂	270.03	brn-blk, rh or cub	10.96	2500	....	s HNO₃; conc H₂SO₄
(tri-) octaoxide (pitchblende)	U₃O₈	842.09	ol grn-blk	8.30	d 1300 to UO₂	....	s HNO₃; H₂SO₄
(VI) oxide	UO₃	286.03	yel-red pwd	7.29	d	....	s HNO₃, HCl
(III) oxide chloride	UOCl	289.48	red, tetr	8.78	....	....	....
(IV) oxide dichloride	UOCl₂	324.94	grn	....	d 550	....	s d
(V) oxide trichloride	UOCl₃	360.40	red-brn	....	d 700	....	s alc
peroxide	UO₄·2H₂O	338.06	pale yel, hygr	10.2	d 115	....	i; d HCl
phosphide	UP	269.00	gray-blk, cub	8.57	....	....	....
phosphide, di-	UP₂	299.98	gray-blk	10.0	....	....	....
(tri-) tetraphosphide	U₃P₄	837.99	gray-blk, cub	....	....	....	....
(IV) sulfate	U(SO₄)₂·4H₂O	502.21	grn, rh	....	−4H₂O, 300	−9H₂O, red heat	23¹¹°
	U(SO₄)₂·9H₂O	592.29	grnsh, mn	....	−7H₂O, 230	....	11.3¹⁸°
sulfide	US	270.09	blk amorp pwd	10.87	>2000	....	i a
sulfide, di-	US₂	302.16	gray-blk, tetr	7.96	>1100	oxidizes	s conc HCl
sulfide, sesqui-	U₂S₃	572.25	gray-blk, rh	....	ign	....	conc HNO₃
Uranyl acetate	UO₂(C₂H₃O₂)₂·2H₂O	422.13	yel, rh	2.893¹⁵°	−2H₂O, 110	d 275	7.694¹⁵°
bromide	UO₂Br₂	429.85	dk brn	....	d 500 (N₂)	....	s hyd
chloride	UO₂Cl₂	340.93	yel, deliq, rh	5.43	578	d 900	320
fluoride	UO₂F₂	308.03	pale yel, rh-hed, hygr	6.37	d 300	....	v s

Table 4-1 (*Continued*)
PHYSICAL CONSTANTS OF INORGANIC COMPOUNDS

Name	Formula	Formula Weight	Color, Crystalline Form, and Refractive Index	Density	Melting Point, °C	Boiling Point, °C	Solubility in 100 Parts
Uranyl							
formate	$UO_2(CHO_2)_2 \cdot H_2O$	378.08	yel, oct	3.695	$-H_2O$, 110	......	7.2^{15}
iodate	$UO_2(IO_3)_2$	619.83	yel, rh	5.2	d 250	......	s
iodate	$UO_2(IO_3)_2 \cdot H_2O$	637.85	α prismatic	5.220^{18}	......	......	0.105
iodide	UO_2I_2	523.84	red, deliq	......	d in air	......	s al, eth, bz
nitrate	$UO_2(NO_3)_2 \cdot 6H_2O$	502.13	yel, rh, deliq, 1.4967	2.807^{13}	60.2, d 100	118	v s
oxalate	$UO_2C_2O_4 \cdot 3H_2O$	412.09	yel cr	......	$-H_2O$, 110	......	0.8^{14}
perchlorate	$UO_2(ClO_4)_2 \cdot 6H_2O$	577.02	yel, rh, deliq	......	90	......	s
phosphate, mono-H	$UO_2HPO_4 \cdot 4H_2O$	438.07	yel pl, tetr	......	......	......	i; s HNO_3, aq Na_2CO_3
potassium carbonate	$UO_2CO_3 \cdot 2K_2CO_3$	606.46	yel cr	......	$-CO_2$, 300	......	7.4^{15}
sulfate	$UO_2SO_4 \cdot 3H_2O$	420.14	yel-grn cr	3.28^{16}	d 100	......	21
sulfide	UO_2S	302.09	brn-blk, tetr	......	d 40	......	sl s; s dil a
Vanadic acid, meta-	HVO_3	99.95	yel	......	......	......	i; s a, alk
tetra-	$H_2V_4O_{11}$	381.78	brn, amorp	......	......	......	i; s a, alk, NH_4OH
Vanadium	V	50.941	lt gray, bcc, 3.0282	6.1	1919	3400	s HF, H_2SO_4, HNO_3
arsenide	VAs	125.86	gray	6.28	1345 vac	......	......
(di-) arsenide	V_2As	176.80	silv-gray	6.39	2400	......	......
boride, di-	VB_2	72.56	hex	5.10	~800 subl	......	v s
(II) bromide	VBr_2	210.76	or-brn, hex, hygr	4.58	d 400	......	s
(III) bromide	VBr_3	290.67	dk gray, hex, deliq	4.84	......	......	s
	$VBr_3 \cdot 6H_2O$	398.77	grn	......	>−23 d	......	s
(IV) bromide	VBr_4	370.58	magenta	......	......	......	s HNO_3, conc H_2SO_4
carbide	VC	62.95	blk, bcc	5.77	2830	3900	s d; s alc, eth
(II) chloride	VCl_2	121.85	lt grn, hex, hygr	3.09	910 subl	......	s
(III) chloride	VCl_3	157.30	red-vlt, rh-hed, hygr	2.87	d 425	......	hyd; s nonpolar solv
(IV) chloride	VCl_4	192.75	dk brn-redsh liq	1.82	−26	153	s
(II) fluoride	VF_2	88.94	bl nd, tetr	3.96	......	......	sl s; i org solv
(III) fluoride	VF_3	107.94	yel-grn pwd, rh-hed	3.36	~1400	subl 800	sl s; i org solv

Name	Formula	Mol. wt.	Color, form	Density	M.p.	B.p.	Solubility
(IV) fluoride	VF₃·3H₂O	161.98	dk grn, rh		−3H₂O 100		s
(IV) fluoride	VF₄	126.93	grn pwd, hex	3.15	subl 100 d		s
(V) fluoride	VF₅	145.93	wh, rh	2.502 liq	19.5	48	s ligroin, chl, acet, al
(II) iodide	VI₂	304.74	vlt-rose, hex	5.47	850 subl		s h HNO₃, alk
(III) iodide	VI₃	431.64	brn-blk, hygr	4.2	d 270		v s
	VI₃·6H₂O	539.75	grn cr, deliq		d		v s
nitride	VN	64.95	gray-vlt, bcc	6.04	2050 d		s HCl
(II) oxide	VO	66.94	gray, bcc	5.55	950 d		s HNO₃, HF, alk
(III) oxide	V₂O₃	149.88	gray-blk	4.84	1970		s a, alk
(IV) oxide	VO₂	82.94	dk bl	4.34	1640		0.05; s a, alk
(V) oxide	V₂O₅	181.88	red, mn	3.34	658	d 1750	s tetrahydrofuran, acetoacetic acid
dioxide chloride	VO₂Cl	118.39	or, hygr	2.29		d 150	hyd
dioxide fluoride	VO₂F	101.94	brn		>300		hyd
phosphide	VP	81.91	gray, hex	4.7			s HF
silicide, di-	VSi₂	107.11	met pr	4.42			s HF
(di-) s licide	V₂Si	129.97	silv-wh pr	$5.48^{17°}$			
su fate	VSO₄·7H₂O	273.11	vlt, mn		d in air		s h H₂SO₄, HNO₃
(II) su fide	VS	83.01	brn-blk	4.51	d		s alk sulf; sl s a
(III) sulfide	V₂S₃	198.08	grn-blk	4.72	d 600		s HNO₃, alk, alk sulf
(V) sulfide	V₂S₅	262.20	blk-grn	3.0	d		s alk
sulfide, tetra-	VS₄	179.18	blk pwd, patronite	2.80	d 500	s	sl s; s acet
Vanadyl (III) bromide	VOBr	146.85	vlt	4.00	d 480		s
(IV) bromide	VOBr₂	226.76	brn pwd, deliq		d 180		s
(V) bromide	VOBr₃	306.67	dk red liq	2.932	−59	170	s
(III) chloride	VOCl	102.39	brn, o-rh	3.45	620 d vac		i; v s HNO₃
(IV) chloride	VOCl₂	137.85	grn, deliq	$2.88^{13°}$	d 300		s
(V) chloride	VOCl₃	173.30	yel liq	1.830	−79	127	i; s CCl₄, hydrocarbons
(IV) fluoride	VOF₂	104.94	yel	$3.396^{18°}$	d		i; sl s acet
(V) fluoride	VOF₃	123.94	lt yel, hygr	2.459	110 subl	(480)	v s
(IV) sulfate	VOSO₄	163.00	bl				
Water, see Hydrogen oxide							

Table 4-1 (*Continued*)
PHYSICAL CONSTANTS OF INORGANIC COMPOUNDS

Name	Formula	Formula Weight	Color, Crystalline Form, and Refractive Index	Density	Melting Point, °C	Boiling Point, °C	Solubility in 100 Parts
Xenon	Xe	131.30	col g, 1.000702	5.8971^{0} g/l 3.057^{bp} liq	-111.8	-108.1	10.81^{20} cc
fluoride, di-	XeF_2	169.30	col, bct	4.32	140	$254.6mm$	2.5^{0}
fluoride, tetra-	XeF_4	207.30	col, mn	4.04	~114	20^{3mm}	i; s F_3CCOOH
fluoride, hexa-	XeF_6	245.30	col, mn	....	46	25^{29mm}	....
oxide, tri-	XeO_3	179.30	col, o-rh, hygr, expl	4.55	d 40	....	d
oxide, tetr-	XeO_4	195.30	col g, expl	....	d -40	....	hyd
oxide tetrafluoride	$XeOF_4$	223.30	col liq	3.168^{0}	-41 to -28	....	s a
*di*oxide difluoride	XeO_2F_2	201.30	col cr	....	30.8	....	v s
Ytterbium	Yb	173.04	fcc	6.977	824	1430	....
acetate	$Yb(C_2H_3O_2)_3 \cdot 4H_2O$	422.24	hex	2.09	$-4H_2O$, 100	....	s a
boride, hexa-	YbB_6	237.90	blk, cub	4.37	....	....	....
(II) bromide	$YbBr_2$	332.86	grn	5.91	613	1830	v s
(III) bromide	$YbBr_3$	412.77	col, rh-hed	5.10	940	d	s
(II) chloride	$YbCl_2$	243.95	grn, rh	5.08	702	1930	s
(III) chloride	$YbCl_3$	279.40	wh, mn	....	865	....	s
(III) chloride	$YbCl_3 \cdot 6H_2O$	387.49	grn, rh, deliq	....	$-6H_2O$, 180	....	v s
(II) fluoride	YbF_2	211.04	cub	....	1407	2380	i
(III) fluoride	YbF_3	230.04	wh, hex	8.168	1157	2230	s H_2SO_4
(III) hydroxide	$Yb(OH)_3$	224.06	....	....	$-H_2O$, 190	d 290 to Yb_2O_3	s a
(II) iodide	YbI_2	426.85	blk, hex	5.40	772	1330	s dil a
(III) iodide	YbI_3	553.75	yel, hex	....	700	d	s
oxalate	$Yb_2(C_2O_4)_3 \cdot 10H_2O$	790.29	col cr	2.644	....	....	sls dil a
oxide	Yb_2O_3	394.08	wh, bcc	9.18	....	....	s hot dil a
selenate	$Yb_2(SeO_4)_3 \cdot 8H_2O$	919.08	col hex	3.30	....	....	s d
selenide	Yb_2Se_3	582.96	vlt-blk	7.33	....	....	....

Name	Formula	Color, crystalline form	Mol. wt.	Density	Melting point	Boiling point	Solubility
sulfate	$Yb_2(SO_4)_3$	col	634.26	3.793	d 900	.....	44.20^0
	$Yb_2(SO_4)_3 \cdot 8H_2O$	pr	778.39	3.286	.....	2930	34.8
Yttrium	Y	gray, hcp	88.906	4.478	1509	.....	s hot water
acetate	$Y(C_2H_3O_2)_3 \cdot 4H_2O$	col, tric	338.10	.....	.....	.....	9.03
boride	YB_6	bl, cub	153.77	3.72	.....	.....	.....
bromate	$Y(BrO_3)_3 \cdot 9H_2O$	hex pr, deliq	634.76	.....	74	$-6H_2O$, 100	168
bromide	YBr_3	col, rh, deliq	328.63	.....	913	1470	83^{30}
carbide	YC_2	yel	112.93	4.13	.....	.....	d
carbonate	$Y_2(CO_3)_3 \cdot 2H_2O$	wh-redsh pwd	393.86	.....	$-H_2O$, 100	$-2H_2O$, 130	s a; s Na_2CO_3
chloride	YCl_3	wh, mn	195.26	2.67	720	1510	79
	$YCl_3 \cdot 6H_2O$	wh-redsh, rh, deliq	303.36	2.18	162	d	75.4
fluoride	YF_3	wh, o-rh	145.90	5.069	1152	2230	d conc a
hydroxide	$Y(OH)_3$	wh gelat, hex	139.93	.....	.....	anhyd 856	s a
iodide	YI_3	wh, hex, deliq	469.62	.....	965	1307	v s
molybdate	$Y_2(MoO_4)_3 \cdot 4H_2O$	yel pl, tetr	729.68	4.79	.....	.....	.....
nitrate	$Y(NO_3)_3 \cdot 6H_2O$	col-redsh, tric, deliq	383.01	2.68	$-3H_2O$, 100	.....	142
nitride	YN	cub	102.91	5.60	2670	.....	s HNO_3, $Na_2C_2O_4$
oxalate	$Y_2(C_2O_4)_3 \cdot 9H_2O$	wh, mn	604.01	.....	$-H_2O$, 45	anhyd 410	s a
oxide	Y_2O_3	wh, bcc, 1.92	225.81	4.84	2417	~4300	s a
(di-) dioxide sulfide	Y_2O_2S	gray, hex	241.87	4.86	2120	.....	s a
selenide	Y_2Se_3	gray-blk	414.69	5.13	.....	.....	.....
sulfate	$Y_2(SO_4)_3$	wh pwd, hygr	465.99	2.52	d >1000	.....	5.02
	$Y_2(SO_4)_3 \cdot 8H_2O$	col-yelsh, mn pr, 1.543	610.12	2.558	anhyd >400	.....	9.76
sulfide	YS	red, cub	120.97	4.51	2040	.....	s a
sulfide, sesqui-	Y_2S_3	lemon, mn	273.99	3.91	1600	.....	s a
Zinc	Zn	blsh-wh met, hex	65.37	7.14	419.47	907	i; s a, alk
acetate	$Zn(C_2H_3O_2)_2$	col, mn	183.46	1.84	d 200	.....	30^{20}
	$Zn(C_2H_3O_2)_2 \cdot 2H_2O$	col, mn	219.49	1.735	237	.....	31^{20}
aluminate (gahnite)	$ZnAl_2O_4$	grn, cub, 1.78	183.33	4.58	.....	.....	i; i a, sl s alk
amide	$Zn(NH_2)_2$	wh pwd, amorp	97.42	2.13	d 200 vac	.....	d; i al, eth
antimonide	Zn_3Sb_2	silv-wh, rh pr	439.61	6.33	570	.....	d
*ortho*arsenate (koettigite)	$Zn_3(AsO_4)_2 \cdot 8H_2O$	col, mn, 1.662, 1.683, 1.717	618.08	3.309^{15}	$-H_2O$, 100	.....	i; s HNO_3, alk

Table 4-1 (*Continued*)
PHYSICAL CONSTANTS OF INORGANIC COMPOUNDS

Name	Formula	Formula Weight	Color, Crystalline Form, and Refractive Index	Density	Melting Point, °C	Boiling Point, °C	Solubility in 100 Parts
Zinc							
orthoarsenate, basic (adamite)	$Zn_3(AsO_4)_2 \cdot Zn(OH)_2$	573.34	col, rh	4.475^{15}	d 250	....	....
orthoarsenate, mono-H	$ZnHAsO_4 \cdot 4H_2O$	277.36	wh, rh	....	$-H_2O$, 327	....	d
arsenide	Zn_3As_2	345.95	gray met, tetr	5.528	1015	....	i; d a
borate	$3ZnO \cdot 2B_2O_3$	383.35	wh tric cr, or amorp pwd	cr 4.22 pwd 3.64	980	....	cr: i HCl amorp: s HCl
bromate	$Zn(BrO_3)_2 \cdot 6H_2O$	429.28	wh, cub, 1.5452	2.566	100	$-6H_2O$, 200	v s
bromide	$ZnBr_2$	225.19	col, rh, hygr, 1.5452	4.201	390	690 d	v s
carbonate (smithsonite)	$ZnCO_3$	125.39	col, trig, 1.818, 1.618	4.398	$-CO_2$, 300	....	i; s a, alk
chlorate	$Zn(ClO_3)_2 \cdot 4H_2O$	304.33	col-yelsh, cub, deliq	2.15	d 60	d	v s
chloride	$ZnCl_2$	136.28	wh, hex, deliq, 1.681, 1.713	2.91	270	732	v s
chloroplatinate(IV)	$ZnPtCl_6 \cdot 6H_2O$	581.27	yel, trig, hygr	2.717^{12}	d 160	....	v s
chromate	$ZnCrO_4$	181.36	lemon yel pr	3.40	....	....	i; s a
chromite	$ZnCr_2O_4$	233.36	dk grn to blk, cub	5.30^{15}	....	....	i; s alk, KCN
cyanide	$Zn(CN)_2$	117.41	col, rh	1.852	d 800	....	v s
dichromate	$ZnCr_2O_7 \cdot 3H_2O$	335.40	redsh-brn cr or or-yel pwd, hygr	....	....	....	v s
ferrate(III) (ferrite)	$ZnFe_2O_4$	241.06	blk, oct	5.33^{20}	1590	....	s conc HCl
ferrocyanide	$Zn_2Fe(CN)_6$	342.69	wh pwd	1.85	....	....	i; s excess alk
fluoride	ZnF_2	103.37	col, mn or tric	4.95	872	ca 1500 to ZnO, 3000	1.6^{20}
	$ZnF_2 \cdot 4H_2O$	175.43	col, rh	2.255	$-4H_2O$, 100	....	1.6^{18}
fluorosilicate	$ZnSiF_6 \cdot 6H_2O$	315.54	col, hex pr, 1.3824, 1.3956	2.104	d 100	....	v s
formate	$Zn(CHO_2)_2$	155.41	col cr	2.368	d	d	3.8
	$Zn(CHO_2)_2 \cdot 2H_2O$	191.44	wh, mn, 1.513, 1.526, 1.566	2.207^{20}	$-2H_2O$, 140	....	5.2^{20}

Name	Formula	Formula wt	Color, crystalline form, index of refraction	Density	M.P., °C	B.P., °C	Solubility
*di*gallate(III)	$ZnGa_2O_4$	268.81	wh cr, 1.74	6.15 est	<800	....	i; s dil a
hydroxide	$Zn(OH)_2$	99.38	col, rh	3.053	d 125	....	v sl s; s a, alk
iodate	$Zn(IO_3)_2$	415.18	wh, nd	5.0625^5	d	....	0.87; s alk, HNO_3
	$Zn(IO_3)_2 \cdot 2H_2O$	451.21	wh cr pwd	4.223	$-H_2O$, 200	....	0.88; s HNO_3, NH_4OH
iodide	ZnI_2	319.18	col, hex	4.7364	446	d 624	v s
nitrate	$Zn(NO_3)_2 \cdot 6H_2O$	297.47	col, tetr	2.065^{14}	36.4	$-6H_2O$, 105–131	v s
nitride	Zn_3N_2	224.12	gray	6.22	....	....	d; s HCl
oxalate	$ZnC_2O_4 \cdot 2H_2O$	189.42	wh pwd	3.28	d 100	....	i; s a, alk
oxide (zincite)	ZnO	81.37	wh, hex, 2.008, 2.029	5.606	1975	....	i; s a, alk
perchlorate	$Zn(ClO_4)_2 \cdot 6H_2O$	372.36	wh, hex, deliq, 1.508, 1.480	2.252	106	d 200	s
permanganate	$Zn(MnO_4)_2 \cdot 6H_2O$	411.33	vlt-blk, deliq	2.47	$-H_2O$, 100	....	33.3
peroxide	$ZnO_2 \cdot \frac{1}{2}H_2O$	106.38	yelsh pwd	3.00	$-O_2$, vac	....	sl s d
*ortho*phosphate	$Zn_3(PO_4)_2$	386.05	col, rh	3.998^{15}	900	....	i; s a, NH_4OH
	$Zn_3(PO_4)_2 \cdot 8H_2O$	530.18	col, rh pl	3.109^{15}	....	....	i; s a, alk
(hopeite)	$Zn_3(PO_4)_2 \cdot 4H_2O$	458.11	col, rh, 1.572, 1.591, 1.59	3.04	....	....	i; s a
*pyro*phosphate	$Zn_2P_2O_7$	304.68	wh pwd	3.75^{23}	....	....	i; s a, alk
phosphide, di-	ZnP_2	127.32	or-red, cub	3.0	~700	....	....
(tri-) phosphide, di-	Zn_3P_2	258.06	dk gray, tetr	4.7	>420	1100 subl (in H_2)	d; s HNO_3, (viol) dil a
selenate	$ZnSeO_4 \cdot 5H_2O$	298.40	wh, tric	2.591^{20}	d 50	....	s
selenide	$ZnSe$	144.33	yelsh-redsh, cub, 2.89	5.3	>1100	....	i; s a
silicate, meta-silicate	$ZnSiO_3$	141.45	col, rh	3.42	1437	....	i; i a
(hemimorphite)	$2ZnO \cdot SiO_2 \cdot H_2O$	240.84	col, rh or trig, 1.614, 1.617, 1.636	3.45	....	....	i
*orthos*ilicate (willemite)	Zn_2SiO_4	222.82	col, trig, 1.694, 1.723	4.103	1509	....	i; s a
stearate	$Zn(C_{18}H_{35}O_2)_2$	632.33	lt pwd	....	130	....	i; i al, eth
sulfate (zinkosite)	$ZnSO_4$	161.43	col, rh, 1.658, 1.669, 1.670	3.54	d 600	....	s
(gos arite)	$ZnSO_4 \cdot 6H_2O$	269.52	col, mn or tetr	2.072^{15}	$-H_2O$, 70	....	s
	$ZnSO_4 \cdot 7H_2O$	287.54	col, rh, effl, 1.457, 1.480, 1.484	1.957	100	$-7H_2O$, 320	96^{20}

Table 4-1 (*Continued*)
PHYSICAL CONSTANTS OF INORGANIC COMPOUNDS

Name	Formula	Formula Weight	Color, Crystalline Form, and Refractive Index	Density	Melting Point, °C	Boiling Point, °C	Solubility in 100 Parts
Zinc							
sulfide (α) (wurzite)	ZnS	97.43	col, hex, 2.356, 2.378	3.98	subl 1185	......	i; v s a
(β) (sphalerite)	ZnS	97.43	col, cub, 2.368	4.102	tr 1020 to α	......	i; v s a
	$ZnS \cdot H_2O$	115.45	yelsh-wh pwd	3.98	1049	......	i; s a
sulfite	$ZnSO_3 \cdot 2H_2O$	181.46	wh cr pwd	......	$-2H_2O$, 100	d 200	0.16; s a
tellurate	Zn_3TeO_6	419.71	wh granules	......	......	......	i; s a
telluride	$ZnTe$	192.97	red, cub, 3.56	$6.34^{15°}$	1239	......	d; s dil a
thiocyanate	$Zn(SCN)_2$	181.53	wh pwd, deliq	......	......	......	s
Zinc complexes							
diammine chloride	$Zn(NH_3)_2Cl_2$	170.34	col, rh, 1.625, 1.590	2.10	210.8	d 271	d
tetrammine perrhenate	$Zn(NH_3)_4(ReO_4)_2$	633.89	wh, cub cr	3.608	......	......	i; 0.18 conc NH_4OH
Zirconium	Zr	91.22	wh, hcp	$6.52^{30°}$	1855	4375	s aq reg, hot conc H_2SO_4
boranate, tetra-	$Zr(BH_4)_4$	150.58	......	......	28.7	123	......
boride, di-	ZrB_2	112.84	......	6.085	3265	......	......
bromide, di-	$ZrBr_2$	251.04	blk pwd	......	ign	......	d
bromide, tri-	$ZrBr_3$	330.95	bl-blk, hex	......	d 300	......	d
bromide, tetra-	$ZrBr_4$	410.86	wh, cub, deliq	......	450	subl 357	......
carbide	ZrC	103.23	gray, cub	6.73	3500	5100	sl s conc H_2SO_4
chloride, di-	$ZrCl_2$	162.13	blk pwd	$3.6^{18°}$	d 350	......	d
chloride, tri-	$ZrCl_3$	197.58	bl-blk, hex or tetr	$3.00^{15°}$	d 350	......	d
chloride, tetra-	$ZrCl_4$	233.03	wh cr	$2.803^{15°}$	438	311 subl	s
fluoride, di-	ZrF_2	129.22	blk, o-rh	......	ign	......	......
fluoride, tetra-	ZrF_4	167.21	wh, hex, 1.59	4.43	subl 600	......	1.39
hydride, di-	ZrH_2	93.24	gray-glk, bct	5.61	......	......	......
(deuteride)	ZrD_2	94.25	tetr	5.75	......	......	......

name	formula	mol wt	color, form	density	mp	bp	solubility
hydroxide	Zr(OH)$_4$	159.25	wh amorp pwd	3.25	$-$2H$_2$O, 500	....	s min a
iodide, tri-	ZrI$_3$	471.93	bl-blk, hex	....	d 325	subl 420	s d; s eth
iodide, tetra-	ZrI$_4$	598.84	or-red to yel pwd, hygr	....	500	....	
nitrate	Zr(NO$_3$)$_4$·5H$_2$O	429.32	col cr, deliq	....	....	....	v s
nitride	ZrN	105.23	yel-brn cr	7.3	2980	....	sl s inorg a
(IV) oxide (zirconia)	ZrO$_2$	123.22	col, mn,	5.6	2715	....	s H$_2$SO$_4$, HF
oxide chloride	ZrOCl$_2$·8H$_2$O	322.25	wh nd, tetr, effl	....	$-$8H$_2$O, 210	....	s
oxide iodide	ZrOI$_2$·8H$_2$O	505.15	col nd, hygr	....	d	....	v s
oxide sulfide	ZrOS	139.28	yel pwd	4.87	ign in air	....	i
phosphide	ZrP	122.20	silv-gray, hex	....	....	....	
phosphide, di-	ZrP$_2$	153.17	gray	4.77	....	....	s hot conc H$_2$SO$_4$
selenate	Zr(SeO$_4$)$_2$·4H$_2$O	449.65	hex	....	....	....	s
selenite	Zr(SeO$_3$)$_2$	345.14	wh cr	4.3	d 400	....	s sl H$_2$SO$_4$
orthosilicate	ZrSiO$_4$	183.30	tetr	4.56	2550	....	i aq reg. alk
silicide, di-	ZrSi$_2$	147.39	gray, rh	4.88	....	....	s HF
sulfate	Zr(SO$_4$)$_2$	283.35	wh, hygr	$3.22^{16°}$	d 410	....	s
	Zr(SO$_4$)$_2$·4H$_2$O	355.41	wh, rh	$3.22^{16°}$	$-$3H$_2$O, 150	....	v s
sulfide	ZrS$_2$	155.35	gray, hex	3.87	~1550	....	i

Table 4-2
SYNONYMS AND MINERAL NAMES

Acanthite, see Silver sulfide
Adamite, see Zinc orthoarsenate, basic
Agate, see Silicon dioxide-x-water
Alabandite, see Manganese sulfide
Alamosite, see Lead meta-silicate
Albite, see Aluminum sodium silicate
Alite, see Calcium silicate
Altaite, see Lead telluride
Alumina, see Aluminum oxide
Alundum, see Aluminum oxide
Alunogenite, see Aluminum sulfate-18-water
Anatase, see Titanium dioxide
Andalusite, see Aluminum silicate
Anglesite, see Lead sulfate
Anhydrite, see Calcium sulfate
Anhydrone, see Magnesium perchlorate
Anorthite, see Calcium aluminum sulfate
Aragonite, see Calcium carbonate
Arcanite, see Potassium sulfate
Argentite, see Silver sulfide
Argol, see Potassium hydrogen tartrate
Arkansite (brookite), see Titanium dioxide
Arsenoferrite, see Iron diarsenide
Arsenolite, see Arsenic trioxide
Arsine, see Arsenic hydride
Artinite, see Magnesium carbonate, basic
Ascharite, see Magnesium diborate
Auric and aurous, see under Gold
Azoimide, see Hydrogen azide
Azurite, see Copper(II) carbonate, basic

Baddeleyite, see Zirconium dioxide
Baking soda, see Sodium hydrogen carbonate
Barite (barytes), see Barium sulfate
Barysilite, see Lead orthosilicate
Bauxite, see Aluminum oxide-2-water
Bayerite, see Aluminum oxide-3-water
Beryl, see Beryllium aluminum silicate
Bertrandite, see Beryllium hydroxide disilicate
Bieberite, see Cobalt sulfate-7-water
Bischofite, see Magnesium chloride
Bismuthine, see Bismuth hydride
Bleaching powder, see Calcium hypochlo-
  rite-3-water
Bleaching solution, see Sodium hypochlorite
Blue copperas, see Copper sulfate-5-water
Bobierite, see Magnesium orthophosphate
Boehmite, see Aluminum oxide-1-water
Boracic acid, see Boric acid
Borax, see Sodium tetraborate-10-water
Boussingaulite, see Ammonium magnesium
  sulfate-6-water
Braunite, see Manganese(III) oxide

Breithauptite, see Nickel antimonide
Brimstone, see Sulfur
Brochantite, see Copper sulfate, basic
Bromellite, see Beryllium oxide
Bromyrite, see Silver bromide
Brookite, see Titanium dioxide
Brucite, see Magnesium hydroxide
Brushite, see Calcium hydrogen phosphate
Bunsenite, see Nickel oxide

Caesium, see under Cesium
Calamine, see Zinc carbonate
Calcia, see Calcium oxide
Calcite, see Calcium carbonate
Calomel, see Mercury(I) chloride
Carnallite, see Magnesium potassium chloride
Caro's acid, see Sulfuric acid, peroxo-
Cassiopeium, see Lutetium
Cassiterite, see Tin(IV) oxide
Caustic potash, see Potassium hydroxide
Caustic soda, see Sodium hydroxide
Celestite, see Strontium sulfate
Celite, see Calcium iron(III) aluminate
Cementite, see Iron carbide
Cerargyrite, see Silver chloride
Cerussite, see Lead carbonate
Cervantite, see Antimony tetroxide
Chalcanthite, see Copper sulfate-5-water
Chalcocite, see Copper(I) sulfide
Chalk, see Calcium carbonate
Chessylite, see Copper(II) carbonate, basic
Chile nitre, see Sodium nitrate
Chile saltpeter, see Sodium nitrate
Chloromagnesite, see Magnesium chloride
Chrysoberyl, see Beryllium dialuminate
Cinnabar, see Mercury(II) sulfide
Cinnabar, Austrian, see Lead chromate, basic
Claudetite, see Arsenic(III) oxide
Clausthalite, see Lead selenide
Clinoenstatite, see Magnesium metasilicate
Cobaltite, see Cobalt arsenic sulfide
Columbium, see under Niobium
Cooperite, see Platinum(II) sulfide
Coquimbite, see Iron(III) sulfate
Corrosive sublimate, see Mercury(II) chloride
Corundum, see Aluminum oxide
Cotunite, see Lead Chloride
Covellite, see Copper(II) sulfide
Cream of tartar, see Potassium hydrogen tar-
  trate
Cristobalite, see Silicon dioxide
Crocoite, see Lead chromate
Cryolite, see Aluminum sodium fluoride

Table 4-2 (*Continued*)
SYNONYMS AND MINERAL NAMES

Cryptohalite, see Ammonium fluorosilicate
Cupric and cuprous, see under Copper
Cuprite, see Copper(I) oxide

Dakin's solution, see Sodium hypochlorite
Daubrelite, see Chromium(II) sulfide
Dehydrite, see Magnesium perchlorate
Dental gas, see Nitrogen(I) oxide
Diamond, see Carbon
Diaspore, see Aluminum oxide-1-water
Digallane, see Gallium hydride
Diopside, see Calcium magnesium silicate
Diphosphine, see Hydrogen diphosphide
Diuretic salt, see Potassium acetate
Dolomite, see Calcium magnesium carbonate
Domeykite, see Copper arsenide
Dry Ice, see Carbon dioxide (solid)

Epsomite, see Magnesium sulfate-7-water
Epsom salts, see Magnesium sulfate-7-water
Eriochalcite, see Copper(II) chloride
Erythrite, see Cobalt arsenate-8-water
Euclase, see Beryllium aluminum hydroxide
   sulfate
Eulytite, see Bismuth silicate

Fayalite, see Iron(II) orthosilicate
Ferberite, see Iron(II) tungstate
Ferric and ferrous salts, see under Iron
Ferrite, see Zinc diferrate
Fiedlerite, see Lead chloride, basic
Fluellite, see Aluminum fluoride-1-water
Fluorine oxide, see Oxygen difluoride
Fluoristan, see Tin(II) fluoride
Fluorite, see Calcium fluoride
Fluorspar, see Calcium fluoride
Forsterite, see Magnesium orthosilicate
Freezing salt, see Sodium chloride
Fulminating mercury, see Mercury fulminate

Gahnite, see Zinc dialuminate
Galena, see Lead sulfide
Gibbsite, see Aluminum oxide-3-water
Glauber's salt, see Sodium sulfate-10-water
Goethite, see Iron(III) oxide hydroxide
Goslarite, see Zinc sulfate
Graham's salt, see Sodium metaphosphate
Graphite, see Carbon
Greenockite, see Cadmium sulfide
Guanajuatite, see Bismuth selenide
Guanite, see Ammonium magnesium phos-
   phate-6-water
Gruenerite, see Iron(II) metasilicate

Gypsum, see Calcium sulfate-2-water

Haidingerite, see Calcium arsenate
Halite, see Sodium chloride
Hambergite, see Beryllium hydroxide borate
Hauerite, see Manganese disulfide
Hausmannite, see Manganese(II,IV) oxide
Heavy hydrogen, see Deuterium
Heavy water, see Deuterium oxide
Heazlewoodite, see Nickel (tri-) disulfide
Hematite, see Iron(III) oxide
Hemimorphite, see Zinc silicate
Hermannite, see Manganese silicate
Hessite, see Silver telluride
Hieratite, see Potassium fluorosilicate
Hoernesite, see Magnesium arsenate
Hopeite, see Zinc phosphate
Hydroazoic acid, see Hydrogen azide
Hydrocerussite, see Lead carbonate, basic
Hydrophilite, see Calcium chloride
Hydrosulfite, see Sodium disulfate(III) or So-
   dium hyposulfite
Hypo (photographic), see Sodium thiosulfate-
   5-water

Ice, see Hydrogen oxide (solid)
Iceland spar, see Calcium carbonate
Iodyrite, see Silver iodide

Jeweler's borax, see Sodium tetraborate-10-
   water
Jeweler's rouge, see Iron(III) oxide

Kalinite, see Aluminum potassium sulfate
Kernite, see Sodium tetraborate
Kieserite, see Magnesium sulfate-1-water
Kleinite, see Mercury(II) oxide chloride
Koettigite, see Zinc orthoarsenate
Krennerite, see Gold ditelluride
Kyanite, see Dialuminum silicate

Lanarkite, see Lead sulfate, basic
Lansfordite, see Magnesium carbonate
Laughing gas, see Nitrogen(I) oxide
Laurionite, see Lead chloride, basic
Lawrencite, see Iron(II) chloride
Lautarite, see Calcium iodate
Lechatelierite, see Silicon dioxide
Lime, see Calcium oxide
Linneite, see Tricobalt tetrasulfide
Litharge, see Lead(II) oxide
Lithium aluminum hydride, see Lithium tetra-
   hydridoaluminate

Table 4-2 (*Continued*)
SYNONYMS AND MINERAL NAMES

Lodestone, see Iron(II,III) oxide
Lunar caustic, see Silver nitrate
Lye, see Sodium hydroxide

Magnesia, see Magnesium oxide
Magnesite, see Magnesium carbonate
Magnetite, see Iron (II, III) oxide
Malachite, see Copper carbonate, basic
Manganite, see Manganese(III) oxide hydroxide
Manganosite, see Manganese(II) oxide
Marcasite, see Iron disulfide
Marshite, see Copper(I) iodide
Mascagnite, see Ammonium sulfate
Massicotite, see Lead oxide
Matlockite, see Lead chloride fluoride
Mendipite, see Lead chloride, basic
Mercuric and mercurous, see under Mercury
Mervinite, see Calcium magnesium silicate
Metacinnabar, see Mercury(II) sulfide
Metastannic acid, see Tin(IV) oxide hydrate
Microcline, see Aluminum potassium silicate
Microcosmic salt, see Ammonium sodium hydrogen phosphate-4-water
Millerite, see Nickel sulfide
Mohr's salt, see Iron(II) ammonium sulfate-7-water
Moissanite, see Silicon carbide
Molybdenite, see Molybdenum disulfide
Molybdite, see Molybdenum trioxide
Molysite, see Iron(II) carbonate
Monazite, see Cerium(III) phosphate
Monimolite, see Lead antimonate(V)
Montanite, see Bismuth tellurate
Montroydrite, see Mercury(II) oxide
Morenosite, see Nickel sulfate
Mosaic gold, see Tin disulfide
Mullite, see Aluminum silicate
Muriatic acid, see Hydrogen chloride, aqueous solutions
Muscovite, see Aluminum potassium silicate

Nantokite, see Copper(I) chloride
Natron, see Sodium carbonate
Naumannite, see Silver selenide
Nesquehonite, see Magnesium carbonate
Neutral verdigris, see Copper(II) acetate
Newberyite, see Magnesium hydrogen phosphate
Niccolite, see Nickel arsenide
Niter, see Potassium nitrate
Nitric oxide, see Nitrogen monoxide
Nitrobarite, see Barium nitrate

Oldhamite, see Calcium sulfide
Oleum, see Sulfuric acid, fuming
Opal, see Silicon dioxide
Orpiment, see Arsenic trisulfide
Orthoclase, see Potassium aluminum silicate
Oxygen powder, see Sodium peroxide

Parahopeite, see Zinc orthophosphate
Paralaurionite, see Lead chloride, basic
Paris green, see Copper arsenate acetate
Pawellite, see Calcium molybdate
Pearl ash, see Potassium carbonate
Perborax, see Sodium peroxoborate
Periclase, see Magnesium oxide
Perovskite, see Calcium titanate
Peroxodisulfuric acid, see Sulfuric acid, peroxodi-
Peroxosulfuric acid, see Sulfuric acid, peroxo-
Phenazite, see Beryllium silicate
Phosgene, see Carbonyl chloride
Phosphine, see Hydrogen phosphide
Pickling acid, see Sulfuric acid
Picromerite, see Magnesium potassium sulfate
Pitchblende, see Uranium (tri-) octaoxide
Plaster of Paris, see Calcium sulfate hemihydrate
Plattnerite, see Lead(IV) oxide
Polianite, see Manganese(IV) oxide
Polishing powder, see Silicon dioxide
Polydymite, see Nickel (tri-) tetrasulfide
Potash, see Potassium carbonate
Potassium acid phthalate, see Potassium hydrogen phthalate
Poydomite, see Nickel (tri-) tetrasulfide
Proustite, see Silver thioarsenate(III)
Prussic acid, see Hydrogen cyanide
Pseudowollastonite, see Calcium silicate
Pyrargyrite, see Silver thioarsenate(III)
Pyrite, see Iron disulfide
Pyrochroite, see Manganese(II) hydroxide
Pyrolusite, see Manganese(IV) oxide
Pyrophanite, see Manganese titanate
Pyrosulfuric acid, see Sulfuric acid, pyro-

Quartz, see Silicon dioxide
Quicksilver, see Mercury metal

Raspite, see Lead tungstate
Realgar, see Arsenic (tetra-) tetrasulfide
Red lead, see Lead (II,IV) oxide
Rhodochrosite, see Manganese carbonate
Rhodonite, see Manganese silicate
Richardite, see Copper telluride

Table 4-2 (*Continued*)
SYNONYMS AND MINERAL NAMES

Rochelle salt, see Sodium potassium tartrate
Rock crystal, see Silicon dioxide
Rodochrosite, see Manganese carbonate
Roesslerite, see Magnesium arsenate
Rutile, see Titanium dioxide

Sal soda, see Sodium carbonate-10-water
Saltpeter, see Potassium nitrate
Scacchite, see Manganese chloride
Scheelite, see Calcium tungstate
Schultenite, see Lead hydrogen arsenate
Scorodite, see Iron(III) arsenate
Sellaite, see Magnesium fluoride
Senarmontite, see Antimony(III) oxide
Siderite, see Iron(II) carbonate
Siderotil, see Iron(II) sulfate
Silica, see Silicon dioxide
Sillimanite, see Aluminum silicate
Smithsonite, see Zinc carbonate
Soda ash, Sodium carbonate
Spelter, see Zinc metal
Sphalerite, see Zinc sulfide
Sphercocobalitite, see Cobalt(II) carbonate
Spinel, see Magnesium aluminate
Stannic and stannous, see under Tin
Stibine, see Antimony hydride
Stibnite, see Antimony trisulfide
Stolzite, see Lead tungstate
Strengite, see Iron(III) phosphate
Strontianite, see Strontium carbonate
Struvite, see Ammonium magnesium phosphate-6-water
Sugar of lead, see Lead acetate
Sulfenite, see Lead molybdate
Sulfurated lime, see Calcium sulfide
Sulfuretted hydrogen, see Hydrogen sulfide
Sulphur, see Sulfur
Sycoporite, see Cobalt sulfide
Sylvite, see Potassium chloride
Szmikite, see Manganese(II) sulfate
Szomolnikite, see Iron(II) sulfate

Talc, see Magnesium silicate
Tapiolite, see Iron(II) tantalate

Tarapacaite, see Potassium chromate
Tartar emetic, see Antimony potassium oxide tartrate hemi-hydrate
Tellurite, see Tellurium dioxide
Tenorite, see Copper(II) oxide
Thenardite, see Sodium sulfate
Thionyl, see Sulfinyl
Thorianite, see Thorium dioxide
Tiemannite, see Mercury selenide
Topaz, see Aluminum fluorosilicate
Tridymite, see Silicon dioxide
Troilite, see Iron(II) sulfide
Trona, see Sodium carbonate hydrogen carbonate-2-water
Tschermigite, see Aluminum ammonium sulfate
Tungstenite, see Tungsten disulfide
Tungstite, see Tungstic acid

Uraninite, see Uranium(IV) oxide

Valentinite, see Antimony(III) oxide
Vermilion, see Mercury(II) sulfide
Villiaumite, see Sodium fluoride
Vivianite, see Iron(III) phosphate

Washing soda, see Sodium carbonate-10-water
White lead, see Lead carbonate, basic
Whitlockite, see Calcium orthophosphate
Willemite, see Zinc silicate
Witherite, see Barium carbonate
Wollastonite, see Calcium silicate
Wuestite, see Iron(II) oxide
Wolfenite, see Lead molybdate
Wurzite, see Zinc sulfide

Xantheosite, see Nickel orthoarsenate

Zaratite, see Nickel carbonate, basic
Zincite, see Zinc oxide
Zincosite, see Zinc sulfate
Zirconia, see Zirconium dioxide

Section 5

ANALYTICAL CHEMISTRY

ACTIVITY COEFFICIENTS

Although it is not possible to measure an individual ionic activity coefficient, f_i, it may be estimated from the following equation of the Debye-Hückel theory:

$$-\log f_i = \frac{A z_i^2 \sqrt{I}}{1 + B \mathring{a} \sqrt{I}}$$

where I is the ionic strength of the medium, and $\mathring{a}$ is the ion-size parameter—the effective ionic radius (Table 5-2). The values of A and B vary with the temperature and dielectric constant of the solvent; values from 0 to 100°C for aqueous medium ($\mathring{a}$ in angstrom units) are listed in Table 5-3. Corresponding values of A and B for unit weight of solvent (when employing molality) can be obtained by multiplying the corresponding values for unit volume (molarity units) by the square root of the density of water at the appropriate temperature.

The ionic strength can be estimated from the summation of the product molarity times ionic charge squared for all the ionic species present in the solution, i.e., $I = 0.5(c_1 z_1^2 + c_2 z_2^2 + \cdots + c_i z_i^2)$.

Values for the activity coefficients of ions in water at 25°C are given in Table 5-1 in terms of their effective ionic radii.

At moderate ionic strengths a considerable improvement is effected by subtracting a term bI from the Debye-Hückel expression; b is an adjustable parameter which is 0.2 for water at 25°C. Table 5-4 gives the values of the ionic activity coefficients (for z_i from 1 to 6) with $\mathring{a}$ taken to be 4.6Å.

Table 5-1
INDIVIDUAL ACTIVITY COEFFICIENTS
OF IONS IN WATER AT 25°C

Effective Ionic Radii $\mathring{a}$ (in Å)	f_i at Ionic Strength of				
	0.001	0.005	0.01	0.05	0.1
Univalent Ions					
9	0.967	0.933	0.914	0.86	0.83
8	0.966	0.931	0.912	0.85	0.82
7	0.965	0.930	0.909	0.845	0.81
6	0.965	0.929	0.907	0.835	0.80
5	0.964	0.928	0.904	0.83	0.79
4	0.964	0.928	0.902	0.82	0.775
3.5	0.964	0.926	0.900	0.81	0.76
3	0.964	0.925	0.899	0.805	0.755
2.5	0.964	0.924	0.898	0.80	0.75
Divalent Ions					
8	0.872	0.755	0.69	0.52	0.45
7	0.872	0.755	0.685	0.50	0.425
6	0.870	0.749	0.675	0.485	0.405
5	0.868	0.744	0.67	0.465	0.38
4.5	0.868	0.741	0.663	0.45	0.36
4	0.867	0.740	0.660	0.445	0.355
Trivalent Ions					
6	0.731	0.52	0.415	0.195	0.13
5	0.728	0.51	0.405	0.18	0.115
4	0.725	0.505	0.395	0.16	0.095
Tetravalent Ions					
11	0.588	0.35	0.255	0.10	0.065
5	0.57	0.31	0.20	0.048	0.021
Pentavalent Ions					
9	0.43	0.18	0.105	0.020	0.009

Table 5-2

APPROXIMATE EFFECTIVE IONIC RADII IN AQUEOUS SOLUTIONS AT 25°C

$\mathring{a}$ (in Å)	Inorganic Ions
2.5	Rb^+, Cs^+, NH_4^+, Tl^+, Ag^+
3	K^+, Cl^-, Br^-, I^-, CN^-, NO_2^-, NO_3^-
3.5	OH^-, F^-, SCN^-, OCN^-, HS^-, ClO_3^-, ClO_4^-, BrO_3^-, IO_4^-, MnO_4^-
4	Na^+, $CdCl^+$, Hg_2^{2+}, ClO_2^-, IO_3^-, HCO_3^-, $H_2PO_4^-$, HSO_3^-, $H_2AsO_4^-$, SO_4^{2-}, $S_2O_3^{2-}$, $S_2O_8^{2-}$, SeO_4^{2-}, CrO_4^{2-}, HPO_4^{2-}, $S_2O_6^{2-}$, PO_4^{3-}, $Fe(CN)_6^{3-}$, $Cr(NH_3)_6^{3+}$, $Co(NH_3)_6^{3+}$, $Co(NH_3)_5H_2O^{3+}$
4.5	Pb^{2+}, CO_3^{2-}, SO_3^{2-}, MoO_4^{2-}, $Co(NH_3)_5Cl^{2+}$, $Fe(CN)_5NO^{2-}$
5	Sr^{2+}, Ba^{2+}, Ra^{2+}, Cd^{2+}, Hg^{2+}, S^{2-}, $S_2O_4^{2-}$, WO_4^{2-}, $Fe(CN)_6^{4-}$
6	Li^+, Ca^{2+}, Cu^{2+}, Zn^{2+}, Sn^{2+}, Mn^{2+}, Fe^{2+}, Ni^{2+}, Co^{2+}, $Co(en)_3^{3+}$, $Co(S_2O_3)(CN)_5^{4-}$
8	Mg^{2+}, Be^{2+}
9	H^+, Al^{3+}, Fe^{3+}, Cr^{3+}, Sc^{3+}, Y^{3+}, La^{3+}, In^{3+}, Ce^{3+}, Pr^{3+}, Nd^{3+}, Sm^{3+}, $Co(SO_3)_2(CN)_4^{5-}$
11	Th^{4+}, Zr^{4+}, Ce^{4+}, Sn^{4+}

$\mathring{a}$ (in Å)	Organic Ions
3.5	$HCOO^-$, H_2Cit^-, $CH_3NH_3^+$, $(CH_3)_2NH_2^+$
4	$H_3N^+CH_2COOH$, $(CH_3)_3NH^+$, $C_2H_5NH_3^+$
4.5	CH_3COO^-, $ClCH_2COO^-$, $(CH_3)_4N^+$, $(C_2H_5)_2NH_2^+$, $H_2NCH_2COO^-$, oxalate^{2-}, $HCit^{2-}$
5	Cl_2CHCOO^-, Cl_3COO^-, $(C_2H_5)_3NH^+$, $C_3H_7NH_3^+$, Cit^{3-}, succinate^{2-}, malonate^{2-}, tartrate^{2-}
6	benzoate$^-$, hydroxybenzoate$^-$, chlorobenzoate$^-$, phenyl-acetate$^-$, vinylacetate$^-$, $(CH_3)_2C{=}CHCOO^-$, $(C_2H_5)_4N^+$, $(C_3H_7)_2NH_2^+$, phthalate^{2-}, glutarate^{2-}, adipate^{2-}
7	trinitrophenolate$^-$, $(C_3H_7)_3NH^+$, methoxybenzoate$^-$, pimelate^{2-}, suberate^{2-}, Congo red anion^{2-}
8	$(C_6H_5)_2CHCOO^-$, $(C_3H_7)_4N^+$

Table 5-3
CONSTANTS OF THE DEBYE-HÜCKEL EQUATION
FROM 0 TO 100°C

$$-\log f_i = \frac{A z_i^2 \sqrt{I}}{1 + B \mathring{a} \sqrt{I}}$$

Temp., °C	Unit Volume of Solvent		Temp., °C	Unit Volume of Solvent	
	A	B		A	B
0	0.4918	0.3248	55	0.5432	0.3358
5	0.4952	0.3256	60	0.5494	0.3371
10	0.4989	0.3264	65	0.5558	0.3384
15	0.5028	0.3273	70	0.5625	0.3397
20	0.5070	0.3282	75	0.5695	0.3411
25	0.5115	0.3291	80	0.5767	0.3426
30	0.5161	0.3301	85	0.5842	0.3440
35	0.5211	0.3312	90	0.5920	0.3456
40	0.5262	0.3323	95	0.6001	0.3471
45	0.5317	0.3334	100	0.6086	0.3488
50	0.5373	0.3346			

The values for unit weight of solvent (molality scale) can be obtained by multiplying the corresponding values for unit volume by the square root of the density of water at the appropriate temperature.

Table 5-4
INDIVIDUAL IONIC ACTIVITY COEFFICIENTS
AT HIGHER IONIC STRENGTHS AT 25°C

The values were calculated from the modified Debye-Hückel equation utilizing the modifications proposed by Robinson and by Guggenheim and Bates:

$$-\frac{\log f_i}{z_i^2} = \frac{0.511I}{1 + 1.5I} - 0.2I$$

where I is the ionic strength and $\mathring{a}$ is assumed to be 4.6 Å.

I	$-\dfrac{\log_{10} f_i}{z_i^2}$	f_i for $z_i =$					
		1	2	3	4	5	6
0.05	0.0756	0.840	0.498	0.209	0.0617	0.0129	0.00190
0.1	0.0896	0.814	0.438	0.156	0.0369	0.00576	0.000595
0.2	0.0968	0.800	0.410	0.138	0.0283	0.00380	0.000328
0.3	0.0936	0.806	0.422	0.144	0.0318	0.00457	0.000427
0.4	0.0858	0.821	0.454	0.169	0.0424	0.00716	0.000815
0.5	0.0753	0.841	0.500	0.210	0.0624	0.0131	0.00195
0.6	0.0631	0.865	0.559	0.270₅	0.0978	0.0265	0.00535
0.7	0.0496	0.892	0.633	0.358	0.161	0.0575₅	0.0164
0.8	0.0352	0.922	0.723	0.482	0.273	0.132	0.0541
0.9	0.0201	0.955	0.831	0.659	0.477	0.314	0.189
1.0	0.0044	0.900	0.960	0.913	0.850	0.776	0.694

EQUILIBRIUM CONSTANTS

Table 5-5
IONIC PRODUCT CONSTANT OF WATER

This table gives values of pK_w, where K_w is the ionic activity product constant of water, from 0 to 100°C. The values of pK_w from 0 to 60° were given by Harned and Robinson, *Trans. Faraday Soc.*, **36**, 973 (1940). Values from 65 to 100° were calculated from Eq. (1-5a) given by Harned and Robinson.

Temp., °C.	pK_w	Temp., °C.	pK_w	Temp., °C.	pK_w	Temp., °C.	pK_w
0	14.944	30	13.833	60	13.017	90	12.42
5	14.734	35	13.680	65	12.90	95	12.34
10	14.535	40	13.535	70	12.80	100	12.26
15	14.346	45	13.396	75	12.69	...	.....
20	14.167	50	13.262	80	12.60	...	.....
25	13.996	55	13.137	85	12.51	...	.....

Table 5-6
SOLUBILITY PRODUCTS

The data refer to various temperatures between 18 and 25°C, and were primarily compiled from values cited by Bjerrum, Schwarzenbach, and Sillen, *Stability Constants of Metal Complexes*, Part II, Chemical Society, London, 1958.

Substance	pK_{sp}	K_{sp}	Substance	pK_{sp}	K_{sp}
Actinium			$BaCrO_4$	9.93	1.2×10^{-10}
Ac(OH)$_3$	15	1×10^{-15}	$Ba_2[Fe(CN)_6] \cdot 6H_2O$	7.5	3.2×10^{-8}
Aluminum			BaF_2	5.98	1.0×10^{-6}
AlAsO$_4$	15.8	1.6×10^{-16}	$BaSiF_6$	6	1×10^{-6}
cupferrate, AlL$_3$	18.64	2.3×10^{-19}	$Ba(IO_3)_2 \cdot 2H_2O$	8.82	1.5×10^{-9}
Al(OH)$_3$ amorphous	32.9	1.3×10^{-33}	$Ba(OH)_2$	2.3	5×10^{-3}
AlPO$_4$	18.24	6.3×10^{-19}	$Ba(MnO_4)_2$	9.61	2.5×10^{-10}
8-quinolinolate, AlL$_3$	29.00	1.00×10^{-29}	$BaMoO_4$	7.40	4.0×10^{-8}
Al$_2$S$_3$	6.7	2×10^{-7}	$Ba(NbO_3)_2$	16.50	3.2×10^{-17}
Al$_2$Se$_3$	24.4	4×10^{-25}	$Ba(NO_3)_2$	2.35	4.5×10^{-3}
Americium			BaC_2O_4	6.79	1.6×10^{-7}
Am(OH)$_3$	19.57	2.7×10^{-20}	$BaC_2O_4 \cdot H_2O$	7.64	2.3×10^{-8}
Am(OH)$_4$	56	1×10^{-56}	$BaHPO_4$	6.5	3.2×10^{-7}
Ammonium			$Ba_3(PO_4)_2$	22.47	3.4×10^{-23}
NH$_4$UO$_2$AsO$_4$	23.77	1.7×10^{-24}	$Ba_2P_2O_7$	10.5	3.2×10^{-11}
Arsenic			$BaHPO_4 \cdot 0.5H_2O$	3	1×10^{-3}
As$_2$S$_3$ + 4H$_2$O →	21.68	2.1×10^{-22}	8-quinolinolate, BaL$_2$	8.3	5.0×10^{-9}
2HAsO$_2$ + 3H$_2$S			$Ba(ReO_4)_2$	1.28	5.2×10^{-2}
Barium			$BaSeO_4$	7.46	3.5×10^{-8}
Ba$_3$(AsO$_4$)$_2$	50.11	8.0×10^{-51}	$BaSO_4$	9.96	1.1×10^{-10}
Ba(BrO$_3$)$_2$	5.50	3.2×10^{-6}	$BaSO_3$	6.1	8×10^{-7}
BaCO$_3$	8.29	5.1×10^{-9}	BaS_2O_3	4.79	1.6×10^{-5}
BaCO$_3$ + CO$_2$ +	4.35	4.5×10^{-5}	Beryllium		
H$_2$O → Ba^{2+} +			$BeCO_3 \cdot 4H_2O$	3	1×10^{-3}
2HCO$_3^-$					

Table 5-6 (*Continued*)
SOLUBILITY PRODUCTS

Substance	pK_{sp}	K_{sp}	Substance	pK_{sp}	K_{sp}
Be(OH)$_2$ amorphous	21.8	1.6×10^{-22}	CaSeO$_3$	5.53	8.0×10^{-6}
Be(OH)$_2$ + OH$^-$ →	2.50	3.2×10^{-3}	CaSiO$_3$	7.60	2.5×10^{-8}
HBeO$_2^-$ + H$_2$O			CaSO$_4$	5.04	9.1×10^{-6}
BeMoO$_4$	1.5	3.2×10^{-2}	CaSO$_3$	7.17	6.8×10^{-8}
Be(NbO$_3$)$_2$	15.92	1.2×10^{-16}	tartrate dihydrate	6.11	7.7×10^{-7}
Bismuth			CaWO$_4$	8.06	8.7×10^{-9}
BiAsO$_4$	9.36	4.4×10^{-10}	Cerium		
cupferrate	27.22	6.0×10^{-28}	CeF$_3$	15.1	8×10^{-16}
Bi(OH)$_3$	30.4	4×10^{-31}	Ce(OH)$_3$	19.8	1.6×10^{-20}
BiI$_3$	18.09	8.1×10^{-19}	Ce(IO$_3$)$_3$	9.50	3.2×10^{-10}
BiPO$_4$	22.89	1.3×10^{-23}	Ce(IO$_3$)$_4$	16.3	5×10^{-17}
Bi$_2$S$_3$	97	1×10^{-97}	CeO$_2$	36.1	8×10^{-37}
BiOBr	6.52	3.0×10^{-7}	Ce$_2$(C$_2$O$_4$)$_3 \cdot$ 9H$_2$O	25.5	3.2×10^{-26}
BiOCl	30.75	1.8×10^{-31}	CePO$_4$	23	1×10^{-23}
BiOOH	9.4	4×10^{-10}	Ce$_2$(SeO$_3$)$_3$	24.43	3.7×10^{-25}
BiO(NO$_2$)	6.31	4.9×10^{-7}	Ce$_2$S$_3$	10.22	6.0×10^{-11}
BiO(NO$_3$)	2.55	2.82×10^{-3}	(III) tartrate	19.0	1×10^{-19}
BiOSCN	6.80	1.6×10^{-7}	Cesium		
Cadmium			CsBrO$_3$	1.7	5×10^{-2}
anthranilate, CdL$_2$	8.27	5.4×10^{-9}	CsClO$_3$	1.4	4×10^{-2}
Cd$_3$(AsO$_4$)$_2$	32.66	2.2×10^{-33}	Cs$_2$[PtCl$_6$]	7.5	3.2×10^{-8}
[Cd(NH$_3$)$_6$](BF$_4$)$_2$	5.7	2×10^{-6}	Cs$_3$[Co(NO$_2$)$_6$]	15.24	5.7×10^{-16}
benzoate $\cdot$ 2H$_2$O	2.7	2×10^{-3}	Cs[BF$_4$]	4.7	5×10^{-5}
Cd(BO$_2$)$_2$	8.64	2.3×10^{-9}	Cs$_2$[PtF$_6$]	5.62	2.4×10^{-6}
CdCO$_3$	11.28	5.2×10^{-12}	Cs$_2$[SiF$_6$]	4.90	1.3×10^{-5}
Cd(CN)$_2$	8.0	1.0×10^{-8}	CsClO$_4$	2.4	4×10^{-3}
Cd$_2$[Fe(CN)$_6$]	16.49	3.2×10^{-17}	CsIO$_4$	2.36	4.3×10^{-3}
Cd(OH)$_2$ fresh	13.6	2.5×10^{-14}	CsMnO$_4$	4.08	8.2×10^{-5}
CdC$_2$O$_4 \cdot$ 3H$_2$O	7.04	9.1×10^{-8}	CsReO$_4$	3.40	4.0×10^{-4}
Cd$_3$(PO$_4$)$_2$	32.6	2.5×10^{-33}	Chromium(II)		
quinaldate, CdL$_2$	12.3	5.0×10^{-13}	Cr(OH)$_2$	15.7	2×10^{-16}
CdS	26.1	8.0×10^{-27}	Chromium(III)		
CdWO$_4$	5.7	2×10^{-6}	CrAsO$_4$	20.11	7.7×10^{-21}
Calcium			CrF$_3$	10.18	6.6×10^{-11}
Ca$_3$(AsO$_4$)$_2$	18.17	6.8×10^{-19}	Cr(NH$_3$)$_6$(BF$_4$)$_3$	4.21	6.2×10^{-5}
acetate $\cdot$ 3H$_2$O	2.4	4×10^{-3}	Cr(OH)$_3$	30.2	6.3×10^{-31}
benzoate $\cdot$ 3H$_2$O	2.4	4×10^{-3}	Cr(NH$_3$)$_6$(ReO$_4$)$_3$	11.11	7.7×10^{-12}
CaCO$_3$	8.54	2.8×10^{-9}	CrPO$_4 \cdot$ 4H$_2$O green	22.62	2.4×10^{-23}
CaCO$_3$ calcite	8.35	4.5×10^{-9}	violet	17.00	1.0×10^{-17}
CaCO$_3$ aragonite	8.22	6.0×10^{-9}	Cobalt		
CaCrO$_4$	3.15	7.1×10^{-4}	anthranilate, CoL$_2$	9.68	2.1×10^{-10}
CaF$_2$	10.57	2.7×10^{-11}	Co$_3$(AsO$_4$)$_2$	28.12	7.6×10^{-29}
Ca[SiF$_6$]	3.09	8.1×10^{-4}	CoCO$_3$	12.84	1.4×10^{-13}
Ca(OH)$_2$	5.26	5.5×10^{-6}	Co$_2$[Fe(CN)$_6$]	14.74	1.8×10^{-15}
Ca(IO$_3$)$_2 \cdot$ 6H$_2$O	6.15	7.1×10^{-7}	Co(NH$_3$)$_6$(BF$_4$)$_2$	5.4	4×10^{-6}
Ca[Mg(CO$_3$)$_2$] dolomite	11	1×10^{-11}	Co(OH)$_2$ fresh	14.8	1.6×10^{-15}
CaMoO$_4$	7.38	4.2×10^{-8}	Co(OH)$_3$	43.8	1.6×10^{-44}
Ca(NbO$_3$)$_2$	17.06	8.7×10^{-18}	Co(IO$_3$)$_2$	4.0	1.0×10^{-4}
CaC$_2$O$_4 \cdot$ H$_2$O	8.4	4×10^{-9}	quinaldate, CoL$_2$	10.8	1.6×10^{-11}
CaHPO$_4$	7.0	1×10^{-7}	Co[Hg(SCN)$_4$]	5.82	1.5×10^{-6}
Ca$_3$(PO$_4$)$_2$	28.70	2.0×10^{-29}	α-CoS	20.4	4.0×10^{-21}
8-quinolinolate, CaL$_2$	11.12	7.6×10^{-12}	β-CoS	24.7	2.0×10^{-25}
CaSeO$_4$	3.09	8.1×10^{-4}	8-quinolinolate, CoL$_2$	24.8	1.6×10^{-25}

Table 5-6 (*Continued*)
SOLUBILITY PRODUCTS

Substance	pK_{sp}	K_{sp}	Substance	pK_{sp}	K_{sp}
$CoHPO_4$	6.7	2×10^{-7}	AuI_3	46	1×10^{-46}
$Co_3(PO_4)_2$	34.7	2×10^{-35}	$Au_2(C_2O_4)_3$	10	1×10^{-10}
$CoSeO_3$	6.8	1.6×10^{-7}	Hafnium		
Copper(I)			$Hf(OH)_3$	25.4	4.0×10^{-26}
CuN_3	8.31	4.9×10^{-9}	Holmium		
$Cu[B(C_6H_5)_4]$			$Ho(OH)_3$	22.3	5.0×10^{-23}
tetraphenylborate	8.0	1×10^{-8}	Indium		
$CuBr$	8.28	5.3×10^{-9}	$In_4[Fe(CN)_6]_3$	43.72	1.9×10^{-44}
$CuCl$	5.92	1.2×10^{-6}	$In(OH)_3$	33.2	6.3×10^{-34}
$CuCN$	19.49	3.2×10^{-20}	quinolinolate, InL_3	31.34	4.6×10^{-32}
CuI	11.96	1.1×10^{-12}	In_2S_3	73.24	5.7×10^{-74}
$CuOH$	14.0	1×10^{-14}	$In_2(SeO_3)_3$	32.6	4.0×10^{-33}
Cu_2S	47.6	2.5×10^{-48}	Iron(II)		
$CuSCN$	14.32	4.8×10^{-15}	$FeCO_3$	10.50	3.2×10^{-11}
Copper(II)			$Fe(OH)_2$	15.1	8.0×10^{-16}
anthranilate, CuL_2	13.22	6.0×10^{-14}	$FeC_2O_4 \cdot 2H_2O$	6.5	3.2×10^{-7}
$Cu_3(AsO_4)_2$	35.12	7.6×10^{-36}	FeS	17.2	6.3×10^{-18}
$Cu(N_3)_2$	9.2	6.3×10^{-10}	Iron(III)		
$CuCO_3$	9.86	1.4×10^{-10}	$FeAsO_4$	20.24	5.7×10^{-21}
$CuCrO_4$	5.44	3.6×10^{-6}	$Fe_4[Fe(CN)_6]_3$	40.52	3.3×10^{-41}
$Cu_2[Fe(CN)_6]$	15.89	1.3×10^{-16}	$Fe(OH)_3$	37.4	4×10^{-38}
$Cu(IO_3)_2$	7.13	7.4×10^{-8}	$FePO_4$	21.89	1.3×10^{-22}
$Cu(OH)_2$	19.66	2.2×10^{-20}	quinaldate, FeL_3	16.9	1.3×10^{-17}
CuC_2O_4	7.64	2.3×10^{-8}	$Fe_2(SeO_3)_3$	30.7	2.0×10^{-31}
$Cu_3(PO_4)_2$	36.9	1.3×10^{-37}	Lanthanum		
$Cu_2P_2O_7$	15.08	8.3×10^{-16}	$La(BrO_3)_3 \cdot 9H_2O$	2.5	3.2×10^{-3}
quinaldate, CuL_2	16.8	1.6×10^{-17}	$La(OH)_3$	18.7	2.0×10^{-19}
8-quinolinolate, CuL_2	29.7	2.0×10^{-30}	$La(IO_3)_3$	11.21	6.1×10^{-12}
CuS	35.2	6.3×10^{-36}	$La_2(MoO_4)_3$	20.4	4×10^{-21}
$CuSeO_3$	7.68	2.1×10^{-8}	$La_2(C_2O_4)_3 \cdot 9H_2O$	26.60	2.5×10^{-27}
$CuWO_4$	5	1×10^{-5}	$LaPO_4$	22.43	3.7×10^{-23}
Dysprosium			$La_2(SO_4)_3$	4.5	3.2×10^{-5}
$Dy_2(CrO_4)_3 \cdot 10H_2O$	8	1×10^{-8}	La_2S_3	12.70	2.0×10^{-13}
$Dy(OH)_3$	21.85	1.4×10^{-22}	$La_2(WO_4)_3 \cdot 3H_2O$	3.90	1.3×10^{-4}
Erbium			Lead		
$Er(OH)_3$	23.39	4.1×10^{-24}	acetate	2.75	1.8×10^{-3}
Europium			anthranilate, PbL_2	9.81	1.6×10^{-10}
$Eu(OH)_3$	23.05	8.9×10^{-24}	$Pb_3(AsO_4)_2$	35.39	4.0×10^{-36}
Gadolinium			$Pb(N_3)_2$	8.59	2.5×10^{-9}
$Gd(HCO_3)_3$	1.7	2×10^{-2}	$Pb(BO_2)_2$	10.78	1.6×10^{-11}
$Gd(OH)_3$	22.74	1.8×10^{-23}	$PbBr_2$	4.41	4.0×10^{-5}
Gallium			$Pb(BrO_3)_2$	1.70	2.0×10^{-2}
$Ga_4[Fe(CN)_6]_3$	33.82	1.5×10^{-34}	$PbCO_3$	13.13	7.4×10^{-14}
$Ga(OH)_3$	35.15	7.0×10^{-36}	$PbCl_2$	4.79	1.6×10^{-5}
8-quinolinolate, GaL_3	32.06	8.7×10^{-33}	$PbClF$	8.62	2.4×10^{-9}
Germanium			$PbCrO_4$	12.55	2.8×10^{-13}
GeO_2	57.0	1.0×10^{-57}	$Pb(ClO_2)_2$	8.4	4×10^{-9}
Gold(I)			$Pb_2[Fe(CN)_6]$	14.46	3.5×10^{-15}
$AuCl$	12.7	2.0×10^{-13}	PbF_2	7.57	2.7×10^{-8}
AuI	22.8	1.6×10^{-23}	$PbFI$	8.07	8.5×10^{-9}
Gold(III)			$Pb(OH)_2$	14.93	1.2×10^{-15}
$AuCl_3$	24.5	3.2×10^{-25}	$PbOHBr$	14.70	2.0×10^{-15}
$Au(OH)_3$	45.26	5.5×10^{-46}	$PbOHCl$	13.7	2×10^{-14}

Table 5-6 (*Continued*)
SOLUBILITY PRODUCTS

Substance	pK_{sp}	K_{sp}	Substance	pK_{sp}	K_{sp}
PbOHNO$_3$	3.55	2.8×10^{-4}	Hg$_2$CO$_3$	16.05	8.9×10^{-17}
PbI$_2$	8.15	7.1×10^{-9}	Hg$_2$(CN)$_2$	39.3	5×10^{-40}
Pb(IO$_3$)$_2$	12.49	3.2×10^{-13}	Hg$_2$Cl$_2$	17.88	1.3×10^{-18}
PbMoO$_4$	13.0	1.0×10^{-13}	Hg$_2$CrO$_4$	8.70	2.0×10^{-9}
Pb(NbO$_3$)$_2$	16.62	2.4×10^{-17}	(Hg$_2$)$_3$[Fe(CN)$_6$]$_2$	20.07	8.5×10^{-21}
PbC$_2$O$_4$	9.32	4.8×10^{-10}	Hg$_2$(OH)$_2$	23.7	2.0×10^{-24}
PbHPO$_4$	9.90	1.3×10^{-10}	Hg$_2$(IO$_3$)$_2$	13.71	2.0×10^{-14}
Pb$_3$(PO$_4$)$_2$	42.10	8.0×10^{-43}	Hg$_2$I$_2$	28.35	4.5×10^{-29}
PbHPO$_3$	6.24	5.8×10^{-7}	Hg$_2$C$_2$O$_4$	12.7	2.0×10^{-13}
quinaldate, PbL$_2$	10.6	2.5×10^{-11}	Hg$_2$HPO$_4$	12.40	4.0×10^{-13}
PbSeO$_4$	6.84	1.4×10^{-7}	quinaldate, Hg$_2$L$_2$	17.9	1.3×10^{-18}
PbSeO$_3$	11.5	3.2×10^{-12}	Hg$_2$SeO$_3$	14.2	8.4×10^{-15}
PbSO$_4$	7.79	1.6×10^{-8}	Hg$_2$SO$_4$	6.13	7.4×10^{-7}
PbS	27.9	8.0×10^{-28}	Hg$_2$SO$_3$	27.0	1.0×10^{-27}
Pb(SCN)$_2$	4.70	2.0×10^{-5}	Hg$_2$S	47.0	1.0×10^{-47}
PbS$_2$O$_3$	6.40	4.0×10^{-7}	Hg$_2$(SCN)$_2$	19.7	2.0×10^{-20}
PbWO$_4$	6.35	4.5×10^{-7}	Hg$_2$WO$_4$	16.96	1.1×10^{-17}
Lead(IV)			Mercury(II)		
Pb(OH)$_4$	65.5	3.2×10^{-66}	Hg(OH)$_2$	25.52	3.0×10^{-26}
Lithium			Hg(IO$_3$)$_2$	12.5	3.2×10^{-13}
Li$_2$CO$_3$	1.60	2.5×10^{-2}	1,10-phenanthroline	24.70	2.0×10^{-25}
LiF	2.42	3.8×10^{-3}	quinaldate, HgL$_2$	16.8	1.6×10^{-17}
Li$_3$PO$_4$	8.5	3.2×10^{-9}	HgSeO$_3$	13.82	1.5×10^{-14}
LiUO$_2$AsO$_4$	18.82	1.5×10^{-19}	HgS red	52.4	4×10^{-53}
Lutetium			HgS black	51.8	1.6×10^{-52}
Lu(OH)$_3$	23.72	1.9×10^{-24}	Neodymium		
Magnesium			Nd(OH)$_3$	21.49	3.2×10^{-22}
MgNH$_4$PO$_4$	12.6	2.5×10^{-13}	Neptunium		
Mg$_3$(AsO$_4$)$_2$	19.68	2.1×10^{-20}	NpO$_2$(OH)$_2$	21.6	2.5×10^{-22}
MgCO$_3$	7.46	3.5×10^{-8}	Nickel		
MgCO$_3 \cdot$ 3H$_2$O	4.67	2.1×10^{-5}	[Ni(NH$_3$)$_6$][ReO$_4$]$_2$	3.29	5.1×10^{-4}
MgF$_2$	8.19	6.5×10^{-9}	anthranilate, NiL$_2$	9.09	8.1×10^{-10}
Mg(OH)$_2$	10.74	1.8×10^{-11}	Ni$_3$(AsO$_4$)$_2$	25.51	3.1×10^{-26}
Mg(IO$_3$)$_2 \cdot$ 4H$_2$O	2.5	3.2×10^{-3}	NiCO$_3$	8.18	6.6×10^{-9}
Mg(NbO$_3$)$_2$	16.64	2.3×10^{-17}	Ni$_2$(CN)$_4 \rightarrow$ Ni^{2+} +	8.77	1.7×10^{-9}
Mg$_3$(PO$_4$)$_2$	23–27	10^{-23} to 10^{-27}	Ni(CN)$_4^{2-}$		
8-quinolinolate, MgL$_2$	15.4	4.0×10^{-16}	Ni$_2$[Fe(CN)$_6$]	14.89	1.3×10^{-15}
MgSeO$_3$	4.89	1.3×10^{-5}	[Ni(N$_2$H$_4$)$_3$]SO$_4$	13.15	7.1×10^{-14}
MgSO$_3$	2.5	3.2×10^{-3}	Ni(OH)$_2$ fresh	14.7	2.0×10^{-15}
Manganese			Ni(IO$_3$)$_2$	7.85	1.4×10^{-8}
anthranilate, MnL$_2$	6.75	1.8×10^{-7}	NiC$_2$O$_4$	9.4	4×10^{-10}
Mn$_3$(AsO$_4$)$_2$	28.72	1.9×10^{-29}	Ni$_3$(PO$_4$)$_2$	30.3	5×10^{-31}
MnCO$_3$	10.74	1.8×10^{-11}	Ni$_2$P$_2$O$_7$	12.77	1.7×10^{-13}
Mn$_2$[Fe(CN)$_6$]	12.10	8.0×10^{-13}	8-quinolinolate, NiL$_2$	26.1	8×10^{-27}
Mn(OH)$_2$	12.72	1.9×10^{-13}	quinaldate, NiL$_2$	10.1	8×10^{-11}
MnC$_2$O$_4 \cdot$ 2H$_2$O	14.96	1.1×10^{-15}	NiSeO$_3$	5.0	1.0×10^{-5}
8-quinolinolate, MnL$_2$	21.7	2.0×10^{-22}	α-NiS	18.5	3.2×10^{-19}
MgSeO$_3$	6.9	1.3×10^{-7}	β-NiS	24.0	1.0×10^{-24}
MnS amorphous	9.6	2.5×10^{-10}	γ-NiS	25.7	2.0×10^{-26}
crystalline	12.6	2.5×10^{-13}	Palladium		
Mercury(I)			Pd(OH)$_2$	31.0	1.0×10^{-31}
Hg$_2$(N$_3$)$_2$	9.15	7.1×10^{-10}	Pd(OH)$_4$	70.2	6.3×10^{-71}
Hg$_2$Br$_2$	22.24	5.6×10^{-23}	quinaldate, PdL$_2$	12.9	1.3×10^{-13}

Table 5-6 (*Continued*)
SOLUBILITY PRODUCTS

Substance	pK_{sp}	K_{sp}	Substance	pK_{sp}	K_{sp}
Platinum			$AgBrO_3$	4.28	5.3×10^{-5}
$PtBr_4$	40.5	3.2×10^{-41}	AgBr	12.30	5.0×10^{-13}
$Pt(OH)_2$	35	1×10^{-35}	Ag_2CO_3	11.09	8.1×10^{-12}
Plutonium			$AgClO_2$	3.7	2.0×10^{-4}
PuO_2CO_3	12.77	1.7×10^{-13}	AgCl	9.75	1.8×10^{-10}
PuF_3	15.6	2.5×10^{-16}	Ag_2CrO_4	11.95	1.1×10^{-12}
PuF_4	19.2	6.3×10^{-20}	$Ag_3[Co(NO_2)_6]$	20.07	8.5×10^{-21}
$Pu(OH)_3$	19.7	2.0×10^{-20}	cyanamide, Ag_2CN_2	10.14	7.2×10^{-11}
$Pu(OH)_4$	55	1×10^{-55}	AgOCN	6.64	2.3×10^{-7}
$PuO_2(OH)$	9.3	5×10^{-10}	AgCN	15.92	1.2×10^{-16}
$PuO_2(OH)_2$	24.7	2×10^{-25}	$Ag_2Cr_2O_7$	6.70	2.0×10^{-7}
$Pu(IO_3)_4$	12.3	5×10^{-13}	dicyanimide, $AgN(CN)_2$	8.85	1.4×10^{-9}
$Pu(HPO_4)_2 \cdot xH_2O$	27.7	2×10^{-28}	$Ag_4[Fe(CN)_6]$	40.81	1.6×10^{-41}
Polonium			AgOH	7.71	2.0×10^{-8}
PoS	28.26	5.5×10^{-29}	$Ag_2N_2O_2$	18.89	1.3×10^{-19}
Potassium			$AgIO_3$	7.52	3.0×10^{-8}
$K_2[PdCl_6]$	5.22	6.0×10^{-6}	AgI	16.08	8.3×10^{-17}
$K_2[PtCl_6]$	4.96	1.1×10^{-5}	Ag_2MoO_4	11.55	2.8×10^{-12}
$K_2[PtBr_6]$	4.2	6.3×10^{-5}	$AgNO_2$	3.22	6.0×10^{-4}
$K_2[PtF_6]$	4.54	2.9×10^{-5}	$Ag_2C_2O_4$	10.46	3.4×10^{-11}
K_2SiF_6	6.06	8.7×10^{-7}	Ag_3PO_4	15.84	1.4×10^{-16}
K_2ZrF_6	3.3	5×10^{-4}	quinaldate, AgL	17.9	1.3×10^{-18}
KIO_4	3.08	8.3×10^{-4}	$AgReO_4$	4.10	8.0×10^{-5}
$K_2Na[Co(NO_2)_6] \cdot H_2O$	10.66	2.2×10^{-11}	Ag_2SeO_3	15.00	1.0×10^{-15}
$K[B(C_6H_5)_4]$	7.65	2.2×10^{-8}	Ag_2SeO_4	7.25	5.7×10^{-8}
KUO_2AsO_4	22.60	2.5×10^{-23}	AgSeCN	15.40	4.0×10^{-16}
$K_4[UO_2(CO_3)_3]$	4.2	6.3×10^{-5}	Ag_2SO_4	4.84	1.4×10^{-5}
Praseodymium			Ag_2SO_3	13.82	1.5×10^{-14}
$Pr(OH)_3$	21.17	6.8×10^{-22}	Ag_2S	49.2	6.3×10^{-50}
Promethium			AgSCN	12.00	1.0×10^{-12}
$Pm(OH)_3$	21	1×10^{-21}	$AgVO_3$	6.3	5×10^{-7}
Radium			Ag_2WO_4	11.26	5.5×10^{-12}
$Ra(IO_3)_2$	9.06	8.7×10^{-10}	Sodium		
$RaSO_4$	10.37	4.2×10^{-11}	$Na[Sb(OH)_6]$	7.4	4.0×10^{-8}
Rhodium			Na_3AlF_6	9.39	4.0×10^{-10}
$Rh(OH)_3$	23	1×10^{-23}	$NaK_2[Co(NO_2)_6]$	10.66	2.2×10^{-11}
Rubidium			$Na(NH_4)_2[Co(NO_2)_6]$	11.4	4×10^{-12}
$Rb_3[Co(NO_2)_6]$	14.83	1.5×10^{-15}	$NaUO_2AsO_4$	21.87	1.3×10^{-22}
$Rb_2[PtCl_6]$	7.2	6.3×10^{-8}	Strontium		
$Rb_2[PtF_6]$	6.12	7.7×10^{-7}	$Sr_3(AsO_4)_2$	18.09	8.1×10^{-19}
$Rb_2[SiF_6]$	6.3	5.0×10^{-7}	$SrCO_3$	9.96	1.1×10^{-10}
$RbClO_4$	2.60	2.5×10^{-3}	$SrCrO_4$	4.65	2.2×10^{-5}
$RbIO_4$	3.26	5.5×10^{-4}	SrF_2	8.61	2.5×10^{-9}
Ruthenium			$Sr(IO_3)_2$	6.48	3.3×10^{-7}
$Ru(OH)_3$	36	1×10^{-36}	$SrMoO_4$	6.7	2×10^{-7}
Samarium			$Sr(NbO_3)_2$	17.38	4.2×10^{-18}
$Sm(OH)_3$	22.08	8.3×10^{-23}	$SrC_2O_4 \cdot H_2O$	6.80	1.6×10^{-7}
Scandium			$Sr_3(PO_4)_2$	27.39	4.0×10^{-28}
ScF_3	17.37	4.2×10^{-18}	8-quinolinolate, SrL_2	9.3	5×10^{-10}
$Sc(OH)_3$	30.1	8.0×10^{-31}	$SrSeO_3$	5.74	1.8×10^{-6}
Silver			$SrSeO_4$	3.09	8.1×10^{-4}
AgN_3	8.54	2.8×10^{-9}	$SrSO_3$	7.4	4×10^{-8}
Ag_3AsO_4	22.0	1.0×10^{-22}	$SrSO_4$	6.49	3.2×10^{-7}

Table 5-6 (*Continued*)
SOLUBILITY PRODUCTS

Substance	pK_{sp}	K_{sp}	Substance	pK_{sp}	K_{sp}
$SrWO_4$	9.77	1.7×10^{-10}	Uranium		
Terbium			UO_2HAsO_4	10.50	3.2×10^{-11}
$Tb(OH)_3$	21.70	2.0×10^{-22}	UO_2CO_3	11.73	1.8×10^{-12}
Tellurium			$(UO_2)_2[Fe(CN)_6]$	13.15	7.1×10^{-14}
$Te(OH)_4$	53.52	3.0×10^{-54}	$UF_4 \cdot 2.5H_2O$	21.24	5.7×10^{-22}
Thallium(I)			$UO_2(OH)_2$	21.95	1.1×10^{-22}
TlN_3	3.66	2.2×10^{-4}	$UO_2(IO_3)_2 \cdot H_2O$	7.5	3.2×10^{-8}
$TlBr$	5.47	3.4×10^{-6}	$UO_2C_2O_4 \cdot 3H_2O$	3.7	2×10^{-4}
$TlBrO_3$	4.07	8.5×10^{-5}	$(UO_2)_3(PO_4)_2$	46.7	2.0×10^{-47}
$Tl_2[PtCl_6]$	11.4	4.0×10^{-12}	UO_2HPO_4	10.67	2.1×10^{-11}
$TlCl$	3.76	1.7×10^{-4}	UO_2SO_3	8.59	2.6×10^{-9}
Tl_2CrO_4	12.00	1.0×10^{-12}	$UO_2(SCN)_2$	3.4	4×10^{-4}
$Tl_4[Fe(CN)_6] \cdot 2H_2O$	9.3	5×10^{-10}	Vanadium		
$TlIO_3$	5.51	3.1×10^{-6}	$VO(OH)_2$	22.13	5.9×10^{-23}
TlI	7.19	6.5×10^{-8}	$(VO)_3PO_4$	24.1	8×10^{-25}
$Tl_2C_2O_4$	3.7	2×10^{-4}	Ytterbium		
Tl_2SeO_3	38.7	2×10^{-39}	$Yt(OH)_3$	23.6	2.5×10^{-24}
Tl_2SeO_4	4.00	1.0×10^{-4}	Yttrium		
Tl_2S	20.3	5.0×10^{-21}	YF_3	12.14	6.6×10^{-13}
$TlSCN$	3.77	1.7×10^{-4}	$Y(OH)_3$	22.1	8.0×10^{-23}
Thallium(III)			$Y_2(C_2O_4)_3$	28.28	5.3×10^{-29}
$Tl(OH)_3$	45.20	6.3×10^{-46}	Zinc		
8-quinolinolate, TlL_3	32.4	4.0×10^{-33}	anthranilate, ZnL_2	9.23	5.9×10^{-10}
Thorium			$Zn_3(AsO_4)_2$	27.89	1.3×10^{-28}
$ThF_4 \cdot 4H_2O + 2H^+ \rightarrow$	7.23	5.9×10^{-6}	$Zn(BO_2)_2 \cdot H_2O$	10.18	6.6×10^{-11}
$ThF_2^{2+} + 2HF + 4H_2O$			$ZnCO_3$	10.84	1.4×10^{-11}
$Th(OH)_4$	44.4	4.0×10^{-45}	$Zn_2[Fe(CN)_6]$	15.39	4.0×10^{-16}
$Th(C_2O_4)_2$	22	1×10^{-22}	$Zn(IO_3)_2$	7.7	2.0×10^{-8}
$Th_3(PO_4)_4$	78.6	2.5×10^{-79}	$Zn(OH)_2$	16.92	1.2×10^{-17}
$Th(HPO_4)_2$	20	1×10^{-20}	ZnC_2O_4	7.56	2.7×10^{-8}
$Th(IO_3)_4$	14.6	2.5×10^{-15}	$Zn_3(PO_4)_2$	32.04	9.0×10^{-33}
Thullium			quinaldate, ZnL_2	13.8	1.6×10^{-14}
$Tm(OH)_3$	23.48	3.3×10^{-24}	8-quinolinolate, ZnL_2	24.3	5.0×10^{-25}
Tin			$ZnSeO_3$	6.59	2.6×10^{-7}
$Sn(OH)_2$	27.85	1.4×10^{-28}	α-ZnS	23.8	1.6×10^{-24}
$Sn(OH)_4$	56	1×10^{-56}	β-ZnS	21.6	2.5×10^{-22}
SnS	25.0	1.0×10^{-25}	$Zn[Hg(SCN)_4]$	6.66	2.2×10^{-7}
Titanium			Zirconium		
$Ti(OH)_3$	40	1×10^{-40}	$ZrO(OH)_2$	48.2	6.3×10^{-49}
$TiO(OH)_2$	29	1×10^{-29}	$Zr_3(PO_4)_4$	132	1×10^{-132}

PROTON-TRANSFER REACTIONS

The pK_a values listed in Tables 5-7 and 5-8 are the negative (decadic) logarithms of the acidic dissociation constant, i.e., $-\log_{10}K_a = pK_a$. For the general proton-transfer reaction

$$HB \rightleftharpoons H^+ + B$$

the acidic dissociation constant is formulated as follows:

$$K_a = \frac{[H^+][B]}{[HB]}$$

The most common charge types for the acid (HB) and its conjugate base (B) are as follows:

$CH_3COOH \rightleftharpoons H^+ + CH_3COO^-$ (acetic acid, acetate ion)
$HSO_4^- \rightleftharpoons H^+ + SO_4^{2-}$ (hydrogen sulfate ion, sulfate ion)
$NH_4^+ \rightleftharpoons H^+ + NH_3$ (ammonium ion, ammonia)

Acids that have more than one acidic hydrogen ionize in steps, as shown for phosphoric acid:

$H_3PO_4 \rightleftharpoons H^+ + H_2PO_4^-$ $pK_1 = 2.12$ $K_1 = 7.6 \times 10^{-3}$
$H_2PO_4^- \rightleftharpoons H^+ + HPO_4^{2-}$ $pK_2 = 7.20$ $K_2 = 6.3 \times 10^{-8}$
$HPO_4^{2-} \rightleftharpoons H^+ + PO_4^{3-}$ $pK_3 = 12.36$ $K_3 = 4.4 \times 10^{-13}$

If the basic dissociation constant K_b for the equilibrium such as

$$NH_3 + H_2O \rightleftharpoons NH_4^+ + OH^-$$

is required, pK_b may be calculated from the relationship

$$pK_b = pK_w - pK_a$$

where $K_w = [H^+][OH^-]$ is the ionic product of water and $pK_w = pH + pOH$ (see Table 5-5). Thus, for ammonia

$$pK_b = 14.00 - 9.24 = 4.76$$

or
$$K_b = 1.7 \times 10^{-5}$$

Table 5-7
PROTON-TRANSFER REACTIONS OF INORGANIC MATERIALS IN WATER AT 25°C

Substance	Formula	pK_1	pK_2
Antimonic acid	$HSb(OH)_6$	2.55	
Alumina	H_3AlO_3	11.2	
Amidophosphoric acid	$H_2NP(OH)_2$ [in $1M$ Me$_4$NBr]	3.3	8.28
Ammonium ion	NH_4^+	9.24	
Arsenic acid	H_3AsO_4	2.20	6.98
		pK_3 11.50	
Arsenous acid	$HAsO_2$ or $HAs(OH)_4$	9.22	
Boric acid	H_3BO_3	9.24	12.74
		pK_3 13.80	
Boric acid (tetra-)	$H_2B_4O_7$	4	9
Carbonic acid	$CO_2 + H_2O$	6.38*	10.25
Chloric acid	$HClO_3$	−2.7	

* $pK_1 = 3.76$ without including dehydration constant.

Table 5-7 (*Continued*)
PROTON-TRANSFER REACTIONS OF INORGANIC
MATERIALS IN WATER AT 25°C

Substance	Formula	pK_1	pK_2
Chlorous acid	$HClO_2$	1.96	
Chlorosulfonic acid	$HOSO_2Cl$	-10.43	
Chromic acid	H_2CrO_4	-0.98	6.50
Cyanic acid	$HOCN$	3.46	
Diamidophosphoric acid	$(NH_2)_2POH$	4.83	
Diimidotriphosphoric acid	$H_5P_3O_8(NH)_2$	0	2
		pK_3 3.94	pK_4 7.74
		pK_5 9.95	
Deuterium oxide	D_2O	14.80	
Dithionic acid	$H_2S_2O_6$	0.2	3.4
Dithionous acid	$H_2S_2O_4$	2.45	
Ferrocyanic acid	$H_2[Fe(CN)_6]^{2-}$	pK_3 2.22	pK_4 4.17
Germanic acid	H_2GeO_3	8.5–9.5	12.72
Hydrazinium ion	$^+H_3NNH_3^+$	-0.88	7.99
Hydrazine monosulfonic acid	H_2NNHSO_3H	3.85	
Hydrazoic acid	HN_3	4.72	
Hydrobromic acid	HBr	-9	
Hydrochloric acid	HCl	-6.1	
Hydrochloroferric acid	$HFeCl_4$	2.7	
Hydrocyanic acid	HCN	9.21	
Hydrofluoric acid	HF	3.18	
Hydrogen peroxide	H_2O_2	11.65	
Hydrogen selenide	H_2Se	3.89	11.0
Hydrogen sulfide	H_2S	6.88	14.15
Hydrogen telluride	H_2Te	2.64	11
Hydroiodic acid	HI	-9.5	
Hypobromous acid	$HBrO$	8.62	
Hypochlorous acid	$HClO$	7.50	
Hypoiodous acid	HIO	10.64	
Hyponitrous acid	$HON{=}NOH$	6.95	10.84
Hypophosphorous acid	HPH_2O_2	1.1	
Hydroxylammonium ion	$HONH_3^+$	5.96	
Imidodiphosphoric acid	$(HO)_2PONHPO(OH)_2$	2	2.85
		pK_3 7.08	pK_4 9.72
Iodic acid	HIO_3	0.77	
Manganic acid	H_2MnO_4		10.15
Methyltrioxoarsenic acid	$CH_3AsO_3H_2$	3.61	8.24
Molybdic acid	H_2MoO_4		3.75
Nitramide	NH_2NO_2	6.58	
Nitric acid	HNO_3	-1.34	
Nitrous acid	HNO_2	3.29	
α-Oxyhyponitric acid	$H_2N_2O_3$	2.51	9.70
para-Periodic acid	H_5IO_6	1.55	8.27

Table 5-7 (*Continued*)
PROTON-TRANSFER REACTIONS OF INORGANIC MATERIALS IN WATER AT 25°C

Substance	Formula	pK_1	pK_2
Perchloric acid	$HClO_4$	−7.3	
Permanganic acid	$HMnO_4$	−2.25	
Peroxomonosulfuric acid	H_2SO_5		9.3
Perrhenic acid	$HReO_4$	−1.25	
Pertechnic acid	$HTcO_4$	0.3	
Phosphonium ion	PH_4^+	0	
Phosphoric acid	H_3PO_4	2.12	7.20
		pK_3 12.36	
pyro-Phosphoric acid	$H_4P_2O_7$	1.52	2.36
		pK_3 6.60	pK_4 9.25
Phosphorous acid	H_3PO_3	1.30	6.6
Selenic acid	H_2SeO_4	−3	2.05
Selenous acid	H_2SeO_3	2.57	6.60
Silicic acid	H_2SiO_3	9.77	11.80
Sulfamic acid	$HOSO_2NH_2$	0.988	
Sulfuric acid	H_2SO_4	−3	1.92
Sulfurous acid+	$SO_2 + H_2O$	1.90	7.20
Telluric acid	H_6TeO_6	7.61	11.00
Tellurous acid	H_2TeO_3	2.57	7.74
Tetrametaphosphoric acid	$H_4P_4O_{12}$	pK_4 2.74	
Thiocyanic acid	$HSCN$	0.85	
Thiosulfuric acid	$H_2S_2O_3$	0.60	1.4–1.7
Triphosphoric acid	$H_5P_3O_{10}$	pK_3 2.30	pK_4 6.50
		pK_5 9.24	
Water	H_2O	15.74	

† Including dehydration constant.

Table 5-8
PROTON-TRANSFER REACTIONS OF ORGANIC MATERIALS IN WATER AT 25°C

Substance	pK_1	pK_2	pK_3	pK_4
Abietic acid	7.62			
Acetamide (protonated cation)	1.40			
Acetamidobenzoic acid: ortho-	3.63			
meta-	4.07			
para-	4.28			
Acetamidoglycine (protonated cation) (20°)	7.7			
N-(2-Acetamido)iminodiacetic acid (20°)	6.62			
Acetanilide (protonated cation)	0.4			
Acethydrazidinium ion	3.24			
Acetic acid	4.76			
Acetidinium ion	11.29			
Acetoacetic acid (18°)	3.58			
Acetone dicarboxylic acid	3.10			
Acetoxime	12.42			

Table 5-8 (*Continued*)
PROTON-TRANSFER REACTIONS OF ORGANIC MATERIALS IN WATER AT 25°C

Substance	pK_1	pK_2	pK_3	pK_4
Acetoxybenzoic acid: ortho-	3.48			
meta-	4.00			
para-	4.38			
Acetylacetic acid (18°)	3.58			
Acetylacetone	8.95			
Acetyl-α-alanine	3.72			
Acetyl-β-alanine	4.45			
β-Acetylamino-n-propionic acid	4.45			
Acetyl-α-amino-n-butyric acid	3.72			
Acetylbenzoic acid: ortho-	4.13			
meta-	3.83			
para-	3.70			
2-Acetylcyclohexanone	14.1			
N-Acetylcysteine (30°)	9.52			
Acetylene dicarboxylic acid	1.75	4.40		
N-Acetylglycine	3.67			
N-Acetylguanidinium ion	8.23			
Acetyl-L-histidine	7.08			
Acetylhydroxamic acid (20°)	9.40			
4-Acetyl-β-mercaptoisoleucine (30°)	10.30			
N-Acetyl-2-mercaptoethylamine	9.92			
2-Acetyl-1-naphthol (30°)	13.40			
N-Acetylpenicillamine (30°)	9.90			
Acetylphenol: ortho-	9.19			
para-	8.05			
2-Acetylpyridinium ion	3.18			
Acetylsalicylic acid	4.56			
3-Acetylthiophenol (49% aq EtOH)	6.93			
4-Acetylthiophenol (49% aq EtOH)	5.93			
Aconitine (protonated cation)	8.11			
Acridinium ion	5.60			
Acrylic acid	4.26			
Adenine	4.17	9.75		
Adenine-N-oxide	2.69	8.49		
Adenosinediphosphoric acid	3.99	6.35		
Adenosine-3'-phosphoric acid	3.63	5.80		
Adenosine-5'-phosphoric acid	3.81	6.21		
Adenosinetriphosphoric acid	4.05	6.50		
Adipamic acid	4.63			
Adipic acid	4.43	5.41		
α-Alanine (protonated cation)	2.34	9.87		
β-Alanine (protonated cation)	3.55	10.23		
Alanylglycine	8.17			
Alizarin black SN	5.79	12.8	>14	
Alizarin-3-sulfonic acid	5.54	11.01		
Allantoin	8.96			
Allothreonine	2.11	9.10		
Alloxanic acid	6.64			
Allylacetic acid	4.68			
Allylammonium ion	9.69			
β-Allylpropionic acid	4.72			

Table 5-8 (*Continued*)
PROTON-TRANSFER REACTIONS OF ORGANIC MATERIALS IN WATER AT 25°C

Substance	pK_1	pK_2	pK_3	pK_4
3-Amidotetrazolinium ion	3.95			
α-Aminoacetic acid (*see* Glycine)				
Aminobenzoic acid: ortho-	2.05	4.95		
meta-	3.07	4.74		
para-	2.38	4.89		
4-Aminobenzophenone (protonated cation)	2.17			
Aminobenzosulfonic acid: ortho-	2.48			
meta-	3.73			
para-	3.24			
α-Amino-*i*-butyric acid	2.36	10.21		
α-Amino-*n*-butyric acid	2.29	9.83		
γ-Amino-*n*-butyric acid	4.03	10.56		
ε-Amino-*n*-caproic acid	4.37	10.80		
1-Aminocycloheptanecarboxylic acid	2.59	10.46		
1,1-Aminocyclohexanecarboxylic acid	2.65	10.03		
2-Amino-4,5-dimethylphenol	5.28	10.40		
Aminoethane-β-sulfonic acid (20°)	~1.5	9.08		
2-Aminoethanol-1-phosphoric acid		5.84		
2-[2-(2-Aminoethyl)aminoethyl]pyridine H⁺	3.50	6.59	9.51	
2-Aminoethyldihydrogen phosphate	10.31			
2-Aminoethyldihydrogen sulfate	9.13			
2-Aminoethylphosphoric acid	2.45	7.0	10.8	
2,2'-Aminoethylpyrrolidine	6.56	9.74		
1-Aminoethylsulfonic acid	−0.33	9.06		
2-Amino-2'-hydroxydiethylsulfide	9.27			
4-Aminoisoxazolidine-3-one	7.4			
Aminomalonic acid	3.32	9.83		
Aminomethylphosphoric acid	2.35	5.9	10.8	
2-Amino-2-methylpropan-1,3-diol	8.79			
α-Amino-α-methylpropionic acid	2.48	9.9		
2-Aminomethylpyridinium ion	8.65			
Aminomethylsulfonic acid		5.75		
5-Amino-*n*-pentylsulfonic acid		10.95		
2-Aminophenol (protonated cation)	9.28	9.72		
3-Aminophenol (protonated cation)	9.83	9.87		
4-Aminophenol (protonated cation)	8.50	10.30		
o-Aminophenylphosphoric acid	4.10	7.29		
β-Aminopropionic acid	3.55	10.24		
2-Aminopyridine-1-oxide	2.58	16.74		
2-Aminopyridinium ion	6.86			
3-Aminopyridinium ion	5.98			
4-Aminopyridinium ion	9.11			
8-Aminoquinaldinium ion	4.86			
8-Aminoquinolinium ion	4.04			
4-Aminosalicylic acid	1.70	3.90		
5-Aminosalicylic acid	2.74	5.84		
4-Aminothiophenol (40% aq EtOH)	7.95			
o-Aminothiophenol	<2	7.90		
α-Amino-*n*-valeric acid	2.32	9.81		
5-Amino-*n*-valeric acid	4.27	10.77		
n-Amylammonium ion (also *iso*-)	10.60			
Angelic acid	4.30			

Table 5-8 (*Continued*)
PROTON-TRANSFER REACTIONS OF ORGANIC MATERIALS
IN WATER AT 25°C

Substance	pK_1	pK_2	pK_3	pK_4
Anilinediacetic acid (20°)	2.40	4.98		
Aniline-3-sulfonic acid	0.39	3.72		
Aniline-4-sulfonic acid	0.58	3.12		
Anilinium ion	4.60			
Anisic acid: ortho-	4.09			
meta-	4.08			
para-	4.49			
Anisidinium ion: ortho-	4.49			
meta-	4.20			
para-	4.29			
β-Anisylpropionic acid: ortho-	4.80			
meta-	4.65			
para-	4.69			
Anthracenecarboxylic acid −1	3.68			
−2	4.18			
−9	3.65			
Anthranilic acid	2.05	4.95		
Anthraquinonecarboxylic acid −1 (20°)	3.37			
−2 (20°)	3.42			
9,10-Anthraquinone monoxime	9.78			
Apomorphine (protonated cation)	7.00			
Arginine	2.02	$9.04(NH_3^+)$		
Arsenazo (pK_5 10.5; pK_6 12.0)	0.1	1.2	2.7	7.9
Ascorbic acid	4.30	11.82		
Asparagine (protonated cation)	2.02	8.80		
Aspartic acid	1.99	3.86	9.8	
Aureomycine	3.30	7.44	9.27	
Azelaic acid	4.53	5.40		
Aziridinium ion	8.04			
Barbituric acid	4.04			
Benzamide	(13–14)			
Benzene-1-carboxylic-2-phosphoric acid	1.71	3.78	9.17	
Benzene-1-carboxylic-3-phosphoric acid	1.55	4.03	7.03	
Benzene-1-carboxylic-4-phosphoric acid	1.50	3.95	6.89	
Benzenediazonium ion	11.08			
Benzenehexacarboxylic acid	1.40	2.19	3.31	4.78
(pK_5 5.89; pK_6 6.96)				
Benzenepentacarboxylic acid (pK_5 6.46)	1.80	2.73	3.96	5.25
Benzenesulfinic acid	1.50			
Benzene-1,2,3,5-tetracarboxylic acid	2.38	3.51	4.44	5.81
Benzene-1,2,3,4-tetracarboxylic acid	2.05	3.25	4.73	6.21
Benzene-1,2,4,5-tetracarboxylic acid	1.92	2.87	4.49	5.63
Benzene-1,2,3-tricarboxylic acid	2.80	4.20	5.87	
Benzene-1,2,4-tricarboxylic acid	2.52	3.84	5.20	
Benzene-1,3,5-tricarboxylic acid	2.12	3.89	5.20	
Benzhydroxamic acid (20°)	8.89			
Benzidinium ion	3.63	4.70		
Benzil-α-dioxime	12.0			
Benzilic acid	3.04			
Benzimidazolinium ion (ortho-)	5.53	12.3		
Benzoic acid	4.21			
Benzoic hydrazide	3.03	12.45		

Table 5-8 (*Continued*)
PROTON-TRANSFER REACTIONS OF ORGANIC MATERIALS IN WATER AT 25°C

Substance	pK_1	pK_2	pK_3	pK_4
Benzoquinone monoxime	6.20			
Benzosulfonic acid	0.70			
Benzoylacetone	8.23			
Benzoylammonium ion	9.34			
o-Benzoylbenzoic acid	3.54			
Benzoylglutamic acid	3.49	4.99		
Benzoylpyruvic acid	6.40	12.10		
3-Benzoyl-1,1,1-trifluoroacetone	6.35			
Benztriazolinium ion	1.6			
Benzylammonium ion	9.35			
Benzylmercaptan	10.70			
Benzylpyrrolidinium ion	9.51			
Benzylsuccinic acid	4.11	5.65		
Betaine (protonated cation) (20°)	12.16			
Biguanide	2.96	11.51		
2,2'-Bipyridylinium ion	4.35			
Bromoacetic acid	2.90			
Bromoanilinium ion: ortho-	2.60			
meta-	3.51			
para-	3.91			
Bromobenzoic acid: ortho-	2.85			
meta-	3.81			
para-	4.00			
α-Bromobutyric acid	2.97			
o-Bromo-(*trans*)-cinnamic acid	4.41			
m-Bromomandelic acid	3.13			
2-Bromo-6-nitrobenzoic acid	1.37			
Bromophenol: ortho-	8.40			
meta-	8.85			
para-	9.24			
Bromophenoxyacetic acid: ortho-	3.12			
meta-	3.09			
para-	3.13			
α-Bromopropionic acid	2.97			
β-Bromopropionic acid	4.00			
γ-Bromopropionic acid	4.58			
2-Bromopyridinium ion	0.90			
Bromosuccinic acid	2.55	4.41		
Brucine (protonated cation)	2.3	7.95		
1,4-Butanediammonium ion	9.35	10.82		
2,3-Butanediammonium ion	6.91	10.00		
But-3-ene-1-oic acid	4.68			
i-Butylammonium ion	10.41			
n-Butylammonium ion	10.61			
sec-Butylammonium ion	10.56			
tert-Butylammonium ion	10.45			
2-tert-Butylanilinium ion	3.78			
N-tert-Butylanilinium ion	7.10			
n-Butylarsonic acid	4.23	8.91		
t-Butylbenzoic acid: ortho-	3.57			
meta-	4.18			
para-	4.40			
N-Butylethylenediammonium ion	7.53	10.30		

Table 5-8 (*Continued*)
PROTON-TRANSFER REACTIONS OF ORGANIC MATERIALS
IN WATER AT 25°C

Substance	pK_1	pK_2	pK_3	pK_4
t-Butylhydroperoxide	12.80			
p-*tert*-Butylphenylacetic acid	4.42			
n-Butylphosphinic acid	3.41			
t-Butylphosphinic acid	4.24			
2-t-Butylpyridinium ion	5.76			
But-2-yn-1-oic acid	2.66			
But-2-yn-1,4-dioic acid	1.75	4.40		
i-Butyric acid	4.85			
n-Butyric acid	4.83			
Cacodylic acid	1.56			
Caffeine (40°)	10.4			
Calcein: <4, 5.4, 9.0, 10.5, >12				
Calmagite	8.14	12.35		
Camphoric acid	4.57	5.10		
n-Caproic acid (also *iso*-)	4.88			
n-Caprylic acid	4.90			
N-Carbamoylacetic acid	3.64			
N-Carbamoylalanine (*dl*-)	3.89			
N-Carbamoylglycine	3.88			
Carbamylmethylammonium ion	7.93			
β-Carboxymethylaminopropionic acid	3.61	9.46		
Catechol	9.45			
Chloranilic acid	1.09	2.42		
Chloroacetic acid	2.86			
Chloroanilinium ion: ortho-	3.64			
meta-	3.34			
para-	3.99			
Chlorobenzoic acid: ortho-	2.94			
meta-	3.82			
para-	3.99			
2-Chloro-n-butyric acid	2.86			
2-Chloro-iso-butyric acid	2.98			
3-Chloro-n-butyric acid	4.05			
4-Chloro-n-butyric acid	4.50			
Chloro-(*trans*)-cinnamic acid: ortho-	4.23			
meta-	4.29			
para-	4.41			
α-Chlorocrotonic acid	3.14			
β-Chlorocrotonic acid	3.84			
4-Chloro-2,6-dinitrophenol	2.97			
α-Chloroisocrotonic acid	2.80			
β-Chloroisocrotonic acid	4.02			
Chloromethylphosphoric acid	1.40	6.30		
2-Chloro-3-nitrobenzoic acid	2.02			
2-Chloro-4-nitrobenzoic acid	1.96			
2-Chloro-5-nitrobenzoic acid	2.17			
2-Chloro-6-nitrobenzoic acid	1.34			
3-Chloro-6-nitrobenzoic acid	1.85			
Chlorophenol: ortho-	8.48			
meta-	9.02			
para-	9.37			

Table 5-8 (*Continued*)
PROTON-TRANSFER REACTIONS OF ORGANIC MATERIALS IN WATER AT 25°C

Substance	pK_1	pK_2	pK_3	pK_4
Chlorophenoxyacetic acid: ortho-	3.05			
meta-	3.07			
para-	3.10			
2-Chlorophenylacetic acid	4.07			
3-Chlorophenylacetic acid	4.14			
4-Chlorophenylacetic acid	4.19			
2-Chlorophenylphosphoric acid	1.63	6.98		
3-Chlorophenylphosphoric acid	1.55	6.65		
4-Chlorophenylphosphoric acid	1.66	6.75		
3-(o-Chlorophenyl)propionic acid	4.58			
3-(m-Chlorophenyl)propionic acid	4.59			
3-(p-Chlorophenyl)propionic acid	4.61			
4-Chlorophthalic acid	1.60			
2-Chloropropionic acid	2.80			
3-Chloropropionic acid	4.00			
2-Chloropyridinium ion	0.72			
3-Chloropyridinium ion	2.84			
Chlorotetracycline	3.30	7.44	9.27	
4-Chlorothiophenol	5.9			
N-Chloro-p-tolylsulfonamide	4.54			
Cholamine chloride (20°)	7.1			
Choline (protonated cation)	8.94			
Chrome Azurol S	2.45	4.86	11.47	
Chrome Dark Blue	7.56	9.3	12.4	
Cinchonidine (protonated cation)	3.92	8.20		
Cinchonine (protonated cation) (15°)	4.04	8.15		
cis-Cinnamic acid	3.88			
trans-Cinnamic acid	4.44			
Citraconic acid	2.29	6.15		
Citric acid	3.13	4.76	6.40	
1-Citrullin	2.43	9.41		
Cocaine (protonated cation)	8.41			
Codeine (protonated cation)	7.95			
Colchicine (protonated cation)	1.65			
Coniine (protonated cation)	11.0			
Creatine (protonated cation) (40°)	3.28			
Creatinine (protonated cation)	3.57			
Cresol: ortho-	10.26			
meta-	10.00			
para-	10.26			
trans-Crotonic acid	4.69			
cis-Crotonic acid (18°)	4.41			
Cuminic acid	4.35			
Cyanamide	10.27			
N-Cyanoacetamide	4			
Cyanoacethydrazide	2.34	11.17		
Cyanoacetic acid	2.46			
Cyanoanilinium ion: meta-	2.76			
para-	1.74			
Cyanobenzoic acid: ortho-	3.14			
meta-	3.60			
para-	3.55			

Table 5-8 (*Continued*)
PROTON-TRANSFER REACTIONS OF ORGANIC MATERIALS IN WATER AT 25°C

Substance	pK_1	pK_2	pK_3	pK_4
2-Cyano-*i*-butyric acid	2.42			
4-Cyano-*n*-butyric acid	4.44			
2-Cyanoethylammonium ion	7.7			
Cyanomethylammonium ion	5.34			
Cyanophenol: meta-	8.61			
para-	7.95			
Cyanophenoxyacetic acid: ortho-	2.98			
meta-	3.03			
para-	2.93			
2-Cyanopropionic acid	2.37			
3-Cyanopropionic acid	3.99			
3-Cyanopyridinium ion	1.45			
Cyanuric acid	6.78			
trans-Cyclobutanecarboxylic acid	4.78			
cis-Cyclobutane-1,2-dicarboxylic acid	3.90	5.89		
trans-Cyclobutane-1,2-dicarboxylic acid	3.79	5.61		
cis-Cyclobutane-1,3-dicarboxylic acid	4.04	5.31		
trans-Cyclobutane-1,3-dicarboxylic acid	3.81	5.28		
Cyclohexanecarboxylic acid	4.90			
Cyclohexane-1,1-dicarboxylic acid	3.45	4.11		
cis-Cyclohexanediammonium ion	6.43	9.93		
trans-Cyclohexanediammonium ion	6.34	9.74		
Cyclohexane-1,2-*trans*-diamine-*NNN'N''*-tetraacetic acid	2.4	3.5	6.12	11.58
cis-Cyclohexane-1,2-dicarboxylic acid (20°)	4.23	6.77		
trans-Cyclohexane-1,2-dicarboxylic acid (18°)	4.18	5.93		
cis-Cyclohexane-1,3-dicarboxylic acid (16°)	4.10	5.46		
trans-Cyclohexane-1,3-dicarboxylic acid (19°)	4.31	5.73		
trans-Cyclohexane-1,4-dicarboxylic acid (16°)	4.18	5.42		
Cyclohexanonimine	9.15			
cis,cis-Cyclohexane-1,3,5-triamine	6.9	8.7	10.4	
cis-Cyclohex-4-en-1,2-dicarboxylic acid (20°)	3.89	6.79		
trans-Cyclohex-4-en-1,2-dicarboxylic acid (20°)	3.95	5.81		
Cyclohexylacetic acid	4.80			
Cyclohexylammonium ion	10.64			
4-Cyclohexyl-*n*-butyric acid	4.95			
Cyclohexyl-1,1-diacetic acid	3.49	6.96		
cis-Cyclohexyl-1,2-diacetic acid (20°)	4.42	5.45		
trans-Cyclohexyl-1,2-diacetic acid (20°)	4.38	5.42		
3-Cyclohexylpropionic acid	4.91			
Cyclopentanecarboxylic acid	4.99			
Cyclopentane-1,1-dicarboxylic acid	3.23	4.08		
cis-Cyclopentane-1,2-dicarboxylic acid	4.43	6.57		
trans-Cyclopentane-1,2-dicarboxylic acid	3.96	5.85		
cis-Cyclopentane-1,3-dicarboxylic acid	4.26	5.51		
trans-Cyclopentane-1,3-dicarboxylic acid	4.32	5.42		
Cyclopentyl-1,1-diacetic acid	3.80	6.77		
Cyclopropane-1,1-dicarboxylic acid	1.82	5.43		
cis-Cyclopropane-1,2-dicarboxylic acid	3.33	6.47		
trans-Cyclopropane-1,2-dicarboxylic acid	3.65	5.13		
Cumene hydroperoxide	12.60			
Cysteine (protonated cation)	1.96	8.36	10.28(SH)	
Cystine (protonated cation)	1.65	2.26	7.85	9.85

Table 5-8 (*Continued*)
PROTON-TRANSFER REACTIONS OF ORGANIC MATERIALS IN WATER AT 25°C

Substance	pK_1	pK_2	pK_3	pK_4
Dehydroascorbic acid (20°)	3.21	7.92	10.3	
Diacetylacetone	7.42			
1,3-Diamino-2-aminomethylpropane	6.44	8.56	10.38	
3,5-Diaminobenzoic acid	5.30			
α,γ-Diaminobutyric acid	1.85	8.24	10.40	
2,2'-Diaminodiethylsulfide (30°)	8.84	9.64		
3,3'-Diaminodipropylammonium ion (30°)	8.02	9.70	10.70	
Di-(2-aminoethyl)ether	8.62	9.59		
N,N'-Di(2-aminoethyl)ethylenediammonium ion (20°)	3.32	6.67	9.20	9.92
Diaminoglyoxime	2.95			
1,3-Diamino-2-propanol (protonated cation) 20°	7.93	9.69		
Di-*iso*-Amylammonium ion	10.89			
α,α-Dibenzylsuccinic acid	3.96	6.66		
α,α-Dibromopropionic acid	1.48			
α,β-Dibromopropionic acid	2.17			
Dibromosuccinic acid	1.47	2.80		
Di-*i*-butylammonium ion	10.59			
Dichloroacetic acid	1.30			
Dichloroacetylacetic acid	2.11			
Dichlorohydroquinone	7.30	10.0		
2,3-Dichlorophenol	7.44			
2,4-Dichlorophenol	7.85			
2,6-Dichlorophenol	6.78			
Dichloromethylphosphoric acid	1.14	5.61		
3,6-Dichlorophthalic acid	1.46			
Di-2-cyanoethylammonium ion	5.14			
Didodecylammonium ion	10.99			
Diethanolammonium ion	8.88			
3-(Diethoxyphosphinyl)phenol	8.66			
4-(Diethoxyphosphinyl)phenol	8.28			
3-(Diethoxyphosphonyl)benzoic acid	3.65			
4-(Diethoxyphosphonyl)benzoic acid	3.60			
Diethylacetic acid	4.74			
Diethylammonium ion	10.93			
N-Diethylanilinium ion	6.56			
5,5-Diethylbarbituric acid	7.98	13.31		
N,N-Diethylbenzylammonium ion	9.48			
Diethylbiguanide (protonated cation) (30°)	2.53	11.68		
N,N'-Diethylethylenediammonium ion	7.70	10.46		
Diethylenetriamine	4.42	9.21	10.02	
Diethylenetriamine-N,N,N',N'',N''-pentaacetic acid (pK_5 10.58)	1.80	2.55	4.33	8.60
β,β-Diethylglutaric acid	3.62	7.12		
N,N-Diethylglycine	2.04	10.47		
Diethylmalonic acid	2.15	7.29		
Diethylsuccinic acid: (*meso-*)	3.63	6.46		
(*rac-*)	3.54	6.59		
N-Diethyl-*o*-toluidinium ion	7.18			
Diglycolic acid	2.96			
Diguanidinium ion	12.8			

Table 5-8 (*Continued*)
PROTON-TRANSFER REACTIONS OF ORGANIC MATERIALS
IN WATER AT 25°C

Substance	pK_1	pK_2	pK_3	pK_4
Dihexylammonium ion	11.0			
Dihydroresorcinol	5.26			
3,4-Dihydroxyalanine (protonated cation)	2.32	8.68	9.87	
2,3-Dihydroxybenzoic acid	2.94	11.76	>13	
2,4-Dihydroxybenzoic acid	3.29	8.98	>13	
2,5-Dihydroxybenzoic acid	2.97	10.50	>12	
2,6-Dihydroxybenzoic acid	1.30			
3,4-Dihydroxybenzoic acid	4.48	8.67	11.74	
3,5-Dihydroxybenzoic acid	4.04			
Dihydroxybenzoquinone	2.71	5.18		
Dihydroxyfumaric acid	1.10			
Dihydroxymaleic acid	1.15			
Dihydroxymalic acid	1.92			
2,4-Dihydroxyoxazolidine (protonated cation)	6.11			
2,4-Dihydroxypteridine (protonated cation)	<1.3	7.92		
Dihydroxytartaric acid	1.95	4.00		
3,5-Diiodotyrosine (protonated cation)	2.12	6.48	7.82	
2,3-Dimercaptopropan-1-ol (BAL)	8.62	10.58		
α,β-Dimercaptosuccinic acid	2.71	3.48	8.89	10.79
2,6-Dimethoxybenzoic acid	3.44			
cis-Dimethylacrylic acid	4.30			
trans-Dimethylacrylic acid	5.00			
β,β-Dimethylacrylic acid	5.12			
Dimethylaminoantipyrine (protonated cation)	4.84			
Dimethylammonium ion	10.77			
2,3-Dimethylanilinium ion	4.70			
2,4-Dimethylanilinium ion	4.89			
2,5-Dimethylanilinium ion	4.53			
2,6-Dimethylanilinium ion	3.95			
3,4-Dimethylanilinium ion	5.17			
N,N-Dimethylanilinium ion	5.21			
1,3-Dimethylbarbituric acid	4.68			
2,3-Dimethylbenzoic acid	3.74			
2,4-Dimethylbenzoic acid	4.18			
2,5-Dimethylbenzoic acid	3.98			
2,6-Dimethylbenzoic acid	3.25			
3,4-Dimethylbenzoic acid	4.41			
3,5-Dimethylbenzoic acid	4.30			
N,N-Dimethylbenzylammonium ion	9.02			
Dimethylbiguanide (protonated cation)	2.77	11.52		
2,2'-Dimethyl-*n*-butyric acid (18°)	5.03			
2,6-Dimethyl-4-cyano-phenol	8.27			
3,5-Dimethyl-4-cyano-phenol	8.21			
N,N-Dimethylethylenediamine-*N,N*-diacetic acid	6.63	9.53		
N,N-Dimethylethylenediamine-*N,N'*-diacetic acid	7.40	10.16		
N,N-Dimethylethylenediamine-*N',N'*-diacetic acid	5.99	9.97		
β,β-Dimethylglutaric acid	3.70	6.34		
N,N-Dimethylglycine (protonated cation)	2.15	9.94		
Dimethylglyoxime	10.60			
2,2-Dimethylhexanedione-3,5	10.01			
5,5-Dimethylhydantoin	9.19			

Table 5-8 (*Continued*)
PROTON-TRANSFER REACTIONS OF ORGANIC MATERIALS
IN WATER AT 25°C

Substance	pK_1	pK_2	pK_3	pK_4
2,4-Dimethyl-8-hydroxyquinoline	6.20	10.60		
3,4-Dimethyl-8-hydroxyquinoline	5.80	10.05		
Dimethylimidazolium ion	8.52			
Dimethylmalic acid	3.17	6.06		
Dimethylmalonic acid	3.17	6.06		
α,α-Dimethyloxaloacetic acid	1.77	4.62		
2,3-Dimethylphenol	10.50			
2,4-Dimethylphenol	10.58			
2,5-Dimethylphenol	10.22			
3,4-Dimethylphenol	10.32			
3,5-Dimethylphenol	10.15			
Dimethylphosphoric acid	1.29			
2,3-Dimethylquinoline	4.94			
2,6-Dimethylquinoline	5.46			
Dimethylsuccinic acid: (*meso-*)	3.77	5.93		
(*rac-*)	3.94	6.20		
N,N-Dimethyltoluidinium ion: ortho-	5.86			
para-	7.24			
Dinicotinic acid	2.80			
2,3-Dinitrobenzoic acid	1.85			
2,4-Dinitrobenzoic acid	1.43			
2,5-Dinitrobenzoic acid	1.62			
2,6-Dinitrobenzoic acid	1.14			
3,4-Dinitrobenzoic acid	2.82			
3,5-Dinitrobenzoic acid	2.82			
Dinitro-*o*-cresol	4.35			
2,4-Dinitrophenol	4.09			
2,5-Dinitrophenol	5.22			
2,6-Dinitrophenol	3.71			
3,4-Dinitrophenol	5.42			
2,4-Dinitrophenylacetic acid	3.50			
3,5-Dinitrotoluic acid	2.97			
Diphenylacetic acid	3.94			
α,α-Diphenyladipic acid	4.17	5.40		
β,β'-Diphenyladipic acid	4.22	5.19		
Diphenylammonium ion	0.9			
α,α-Diphenylglutaric acid	3.91	5.38		
1,3-Diphenylguanidinium ion	10.12			
Diphenylketimine	6.82			
α,α-Diphenylsuccinic acid	3.05	7.3		
α,α'-Diphenylsuccinic acid (*meso-*)	3.48			
Diphenylthiocarbazone	4.50	15		
Dipicrylammonium ion	5.42			
Di-*n*-propylammonium ion	10.91			
Di-*n*-propylmalonic acid	2.04	7.51		
Dithiodiacetic acid	3.07	4.20		
1,4-Dithioerythritol	9.5			
Dithiooxamide	10.89			
1,4-Dithiothreitol	8.9			
Emetine (protonated cation)	7.36	8.23		
ψ-Ephedrinium ion	9.71			

Table 5-8 (*Continued*)
PROTON-TRANSFER REACTIONS OF ORGANIC MATERIALS
IN WATER AT 25°C

Substance	pK_1	pK_2	pK_3	pK_4
Eriochrome black T	6.3	11.55		
Ethane-1,2-dithiol	8.96	10.54		
Ethanolammonium ion	9.50			
Ethoxyacetic acid (18°)	3.65			
Ethoxyanilinium ion: ortho-	4.47			
meta-	4.17			
para-	5.25			
Ethoxybenzoic acid: ortho-	4.21			
meta-	4.17			
para-	4.80			
Ethoxycarbonylacetic acid	3.55			
Ethoxycarbonylethylammonium ion	9.13			
2-Ethoxyethanethiol	9.38			
Ethylacetoacetate	10.68			
Ethylammonium ion	10.63			
N-Ethylanilinium ion	5.11			
Ethylarsonic acid	3.89	8.35		
Ethylbenzoic acid: ortho-	3.79			
para-	4.35			
Ethylbiguanide (protonated cation)	2.09	11.47		
Ethylenebiguanide (protonated cation) (30°)	1.74	2.88	11.34	11.76
Ethylene-N,N-diacetic acid	5.58	11.05		
Ethylene-N,N'-diacetic acid	6.42	9.46		
Ethylenediaminetetraacetic acid	6.27	10.95		
Ethylenediammonium ion	6.85	9.93		
trans-Ethyleneoxidedicarboxylic acid	1.93	3.25		
cis-Ethyleneoxidedicarboxylic acid	1.93	3.92		
N-Ethylethylenediammonium ion	7.63	10.56		
β-Ethylglutaric acid	4.28	5.33		
N-Ethylglycine (protonated cation)	2.30	10.10		
Ethylhydroperoxide	11.80			
Ethylmalonic acid	2.90	5.55		
Ethylmercaptan (20°)	10.50			
N-Ethylmercaptoacetamide	8.14			
Ethylmercaptoacetate	7.95			
Ethyl-3-mercaptopropionate	9.48			
N,N-Ethylmethylammonium ion	4.23			
Ethylnitroacetate	5.85			
3-Ethylpentanedione-2,4	11.34			
p-Ethylphenylacetic acid	4.37			
Ethylphosphinic acid	3.29			
Ethylphosphoric acid	2.43	8.05		
Ethyl-n-propylmalonic acid	3.14	7.43		
Ethylpyrrolidine	10.43			
S-Ethylthioacetic acid	5.06			
N-Ethyl-o-toluidinium ion	4.92			
Eugenol	10.0			
Fluoroacetic acid	2.59			
Fluoroanilinium ion: meta-	3.39			
para-	4.53			

Table 5-8 (*Continued*)
PROTON-TRANSFER REACTIONS OF ORGANIC MATERIALS IN WATER AT 25°C

Substance	pK_1	pK_2	pK_3	pK_4
Fluorobenzoic acid: ortho-	3.27			
meta-	3.87			
para-	4.14			
m-Fluoromandelic acid	4.15			
Fluorophenol: ortho-	8.81			
meta-	9.28			
para-	9.95			
Fluorophenoxyacetic acid: ortho-	3.08			
meta-	3.08			
para-	3.13			
p-Fluorophenylacetic acid	4.25			
Fluorophenylalanine: ortho-	2.12	9.01		
meta-	2.10	8.98		
para-	2.18	9.05		
o-Fluorophenylphosphoric acid	1.64	6.80		
2-Fluoropyridinium ion	−0.44			
Folic acid (Pteroylglutamic acid)	8.26			
Formic acid	3.75			
N-Formylglycine	3.43			
Formylphenol: ortho-	8.37			
meta-	9.02			
para-	7.62			
Fumaric acid (*trans-*)	3.02	4.38		
Furancarboxylic acid	3.15			
2-Furoic acid	3.16			
Galactose-*l*-phosphoric acid	1.00	6.17		
Gallic acid	4.34	8.85		
Glucoascorbic acid	4.26	11.58		
Gluconic acid	3.86			
Glucose (18°)	12.43			
Glucose-*l*-phosphate	6.50			
Glutaconic acid (*trans*)	3.77	6.08		
Glutamic acid (protonated cation)	2.30	4.28	9.67	
Glutamic acid-α-ethyl half ester	3.85	7.84		
Glutamic acid-γ-ethyl half ester	2.15	9.19		
Glutammonium ion	2.17	9.13		
Glutaramic acid	4.60			
Glutaric acid	4.34	5.42		
Glutarimide	11.43			
Glutathione (30°)	8.56			
Glyceric acid	3.64			
Glycerol	14.15			
Glycero-2-phosphoric acid	1.32	6.50		
Glycine (protonated cation)	2.35	9.78		
Glycine amide	7.99			
Glycine hydroxamic acid	7.10	9.10		
Glycol	14.22			
Glycollic acid	3.82			
Glycylalanine (protonated cation)	3.15	8.17		
Glycylglycine (protonated cation)	3.15	8.25		

Table 5-8 (*Continued*)
PROTON-TRANSFER REACTIONS OF ORGANIC MATERIALS
IN WATER AT 25°C

Substance	pK_1	pK_2	pK_3	pK_4
Glycylsarcosine	8.77			
Glycylserine	2.91			
Glyoxalic acid	3.30			
Glyoxaline (protonated cation)	7.03			
Guanine (protonated cation)	3.3	9.2	12.3	
Guanylethyl urea	3.20	11.10		
Guanylurea	1.80	8.20		
Hemimellitic acid	2.80	4.20	5.87	
Heptafluoro-*n*-butyric acid	0.17			
Heptanoic acid	4.89			
Hexahydrobenzoic acid	4.89			
Hexamethylenediammonium ion	9.83	10.93		
i-Hexanoic acid	4.84			
n-Hexanoic acid	4.86			
Hex-2-en-3-oic acid	4.72			
Hex-3-en-4-oic acid	4.58			
Hex-4-en-5-oic acid (*also* Hex-5-en-6-oic acid)	4.74			
Hippuric acid	3.65			
Histamine (protonated cation)	6.14	9.85		
Histidine (protonated cation)	1.82	6.12(Imid)	9.17(NH$_3^+$)	
Hydantoin	9.12			
Hydrastine (protonated cation)	6.23			
Hydrazine-*N,N*-diacetic acid	<0.1	2.8	3.8	
Hydrazine-*N,N'*-diacetic acid	2.40	3.12	7.32	
Hydrazinium ion (20°)	−0.88	8.48		
Hydroquinone	10.0	12.0		
Hydroxyacetophenone: meta-	9.19			
para-	8.05			
Hydroxybenzoic acid: ortho-	3.00	12.38		
meta-	4.08	9.85		
para-	4.58	9.23		
o-Hydroxybiphenyl	9.55			
2-Hydroxy-5-bromobenzoic acid	2.61			
2-Hydroxybutyric acid (30°)	3.65			
3-Hydroxybutyric acid (30°)	4.41			
4-Hydroxybutyric acid (30°)	4.71			
2-Hydroxy-5-chlorobenzoic acid	2.63			
2-Hydroxy-6-chlorobenzoic acid	2.63			
trans-Hydroxycinnamic acid: ortho-	4.61			
meta-	4.40			
N-(2-Hydroxyethyl)biguanide	2.8	11.53		
N'-(2-Hydroxyethyl)ethylenediamine-*N,N,N'*-triacetic acid	2.39	5.37	9.93	
N-(2-Hydroxyethyl)ethylenediammonium ion	7.21	10.12		
N,N-bis(2-Hydroxyethyl)glycine (protonated cation)	8.35			
N-(2-Hydroxyethyl)iminodiacetic acid	2.2	8.73		
β-Hydroxyethylmercaptan	9.43			
N-(2-Hydroxyethyl)piperazine-*N'*-2-ethane sulfonic acid (20°)	7.55			
β-Hydroxyglutamic acid	2.32	4.23	9.56	

Table 5-8 (*Continued*)
PROTON-TRANSFER REACTIONS OF ORGANIC MATERIALS
IN WATER AT 25°C

Substance	pK_1	pK_2	pK_3	pK_4
Hydroxylammonium ion	5.96			
2-Hydroxy-1-methoxybenzylammonium ion	8.89	10.52		
3-Hydroxy-2-methoxybenzylammonium ion	8.94	10.52		
1-Hydroxy-2-methoxybenzylammonium ion	8.70	10.52		
2-Hydroxy-3-methylbenzoic acid	2.99			
2-Hydroxy-4-methylbenzoic acid	3.17			
2-Hydroxy-5-methylbenzoic acid	4.08			
2-Hydroxy-6-methylbenzoic acid	3.32			
N-*tris*(Hydroxymethyl)methyl-2-aminoethane sulfonic acid (20°)	7.5			
2,2-*bis*(Hydroxymethyl)2,2′,2″-nitrilotriethanol	6.46			
N-*tris*(Hydroxymethyl)methylglycine (protonated cation) (20°)	8.15			
Hydroxymethylphosphoric acid	1.91	7.15		
8-Hydroxy-5-methylquinoline-5-sulfonic acid		4.80	9.30	
2-Hydroxy-3-nitrobenzoic acid	1.87			
2-Hydroxy-4-nitrobenzoic acid	2.23			
2-Hydroxy-5-nitrobenzoic acid	2.12			
2-Hydroxy-6-nitrobenzoic acid	2.24			
Hydroxyproline (protonated cation)	1.82	9.66		
β-Hydroxypropionic acid	3.73			
4-Hydroxypteridine (protonated cation)	$<$1.3	7.89		
2-Hydroxypyridine (protonated cation) (20°)		11.62		
3-Hydroxypyridine (protonated cation) (20°)		8.72		
4-Hydroxypyridine (protonated cation) (20°)		11.09		
2-Hydroxypyrimidine (protonated cation) (20°)	2.24	9.17		
4-Hydroxypyrimidine (protonated cation) (20°)	1.85	8.59		
8-Hydroxyquinazolinium ion	3.41	8.65		
2-Hydroxyquinolinium ion (20°)	−0.31	11.74		
3-Hydroxyquinolinium ion (20°)	4.30	8.06		
4-Hydroxyquinolinium ion (20°)	2.27	11.25		
5-Hydroxyquinolinium ion (20°)	5.20	8.54		
6-Hydroxyquinolinium ion (20°)	5.17	8.88		
7-Hydroxyquinolinium ion (20°)	5.48	8.85		
8-Hydroxyquinolinium ion (20°)	4.91	9.81		
8-Hydroxyquinoline-5-sulfonic acid	1.3	4.11	8.75	
Hydroxytetracycline	3.27	7.32	9.11	
Hydroxyuracil	8.64			
Hypoxanthine	1.98	8.94		
Imidazolinium ion	7.03			
Iminodiacetic acid	2.54	9.12		
Indanol-4	10.32			
Iodoacetic acid	3.18			
Iodoanilinium ion: ortho-	2.60			
meta-	3.61			
para-	3.78			
Iodobenzoic acid: ortho-	2.86			
meta-	3.85			
para-	3.93			
7-Iodo-8-hydroxyquinoline-5-sulfonic acid	2.51	7.42		

Table 5-8 (*Continued*)
PROTON-TRANSFER REACTIONS OF ORGANIC MATERIALS
IN WATER AT 25°C

Substance	pK_1	pK_2	pK_3	pK_4
m-Iodomandelic acid	3.17			
Iodophenol: ortho-	8.46			
meta-	8.82			
para-	9.20			
Iodophenoxyacetic acid: ortho-	3.17			
meta-	3.13			
para-	3.16			
Iodophenylacetic acid (ortho- and meta-)	4.16			
2-Iodopropionic acid	3.11			
3-Iodopropionic acid	4.05			
2-Iodopyridinium ion	1.82			
Isocrotonic acid	4.44			
Isoleucine (protonated cation)	2.32	9.76		
Isophthalic acid	3.62	4.60		
Isothiocyanatoacetic acid	6.62			
Itaconic acid	3.85	5.45		
δ-Ketovaleric acid	4.72			
Kojic acid	7.75			
Lactic acid	3.86			
i-Leucine (protonated cation)	2.32	9.76		
Leucine (protonated cation)	2.33	9.74		
Levulinic acid	4.59			
Lutidine (protonated cation)	2.15			
Lysine (protonated cation)	2.18	8.95(α-NH$_3^+$)	10.53(ε-NH$_3^+$)	
Maleic acid	1.94	6.22		
Malic acid	3.40	5.05		
Malonamic acid	3.64			
Malonic acid	2.86	5.70		
Mandelic acid	3.41			
Mannose (18°)	13.57			
Mellitic acid (pK_5 5.89; pK_6 6.96)	1.40	2.19	3.31	4.78
Mellophanic acid	2.05	3.25	4.73	6.21
Mercaptoacetic acid	3.60	10.55		
2-Mercaptobutyric acid	3.53			
Mercaptodiacetic acid	3.32	4.29		
2-Mercaptoethanesulfonic acid (20°)		9.5		
2-Mercaptoethylammonium ion	8.27	10.53		
2-Mercaptopropionic acid	4.32	10.20		
2-Mercaptoquinoline (20°)	−1.44	10.21		
3-Mercaptoquinoline (20°)	2.33	6.13		
4-Mercaptoquinoline (20°)	0.77	8.83		
Mercaptosuccinic acid	3.30	4.94	10.94	
Mesaconic acid	3.09	4.75		
Mesitylenic acid	4.32			
Metanilic acid	0.39	3.72		
Methionine (protonated cation)	2.28	9.2		
Methoxyacetic acid	3.57			
2-Methoxyanilinium ion	4.49			
3-Methoxyanilinium ion	4.20			
4-Methoxyanilinium ion	5.29			
2-Methoxybenzoic acid	4.09			

Table 5-8 (*Continued*)
PROTON-TRANSFER REACTIONS OF ORGANIC MATERIALS IN WATER AT 25°C

Substance	pK_1	pK_2	pK_3	pK_4
3-Methoxybenzoic acid	4.08			
4-Methoxybenzoic acid	4.49			
N,N-Methoxybenzylammonium ion	9.68			
2-Methoxycarbonylanilinium ion	2.23			
3-Methoxycarbonylanilinium ion	3.64			
4-Methoxycarbonylanilinium ion	2.38			
Methoxycarbonylmethylammonium ion	7.66			
2-Methoxycinnamic acid (*trans-*)	4.46			
3-Methoxycinnamic acid (*trans-*)	4.38			
4-Methoxycinnamic acid (*trans-*)	4.54			
Methoxyethylammonium ion	9.45			
2-Methoxyphenol	9.98			
3-Methoxyphenol	9.65			
4-Methoxyphenol	10.21			
2-Methylacrylic acid (18°)	4.66			
N-Methylalanine (protonated cation)	2.22	10.19		
2-(*N*-Methylamino)benzoic acid	1.93	5.34		
3-(*N*-Methylamino)benzoic acid		5.10		
4-(*N*-Methylamino)benzoic acid		5.05		
Methylaminodiacetic acid	2.15	10.00		
Methylammonium ion	10.62			
N-Methylanilinium ion	4.85			
Methylarsonic acid (18°)	3.41	8.18		
1-Methylbarbituric acid	4.35			
Methylbiguanidinium ion	3.00	11.44		
2-Methyl-2-butanethiol	11.35			
trans-2-Methyl-*n*-but-1-ene-1-oic acid	5.13			
4-Methylcarboxyphenol	8.47			
trans-2-Methylcinnamic acid	4.50			
trans-3-Methylcinnamic acid	4.44			
trans-4-Methylcinnamic acid	4.56			
1-Methylcyclohexane carboxylic acid	5.13			
cis-2-Methylcyclohexane carboxylic acid	5.03			
trans-2-Methylcyclohexane carboxylic acid	5.73			
cis-3-Methylcyclohexane carboxylic acid	4.88			
trans-3-Methylcyclohexane carboxylic acid	5.02			
cis-4-Methylcyclohexane carboxylic acid	4.88			
trans-4-Methylcyclohexane carboxylic acid	5.03			
3-Methylcyclohexyl-1,1-diacetic acid	3.49	6.10		
4-Methylcyclohexyl-1,1-diacetic acid	3.49	6.10		
Methyldiethylammonium ion	10.43			
2,2'-Methylene-*bis*(4-chlorophenol)	7.6	11.5		
2,2'-Methylene-*bis*(4,6-dichlorophenol)	5.6	10.65		
2,2'-Methylene-*bis*(3,4,6-trichlorophenol)		10.1		
Methylethylacetic acid (18°)	4.81			
Methylethylacrylic acid (*cis-* and *trans-*)	5.15			
N-Methylethylenediammonium ion (20°)	6.86	10.15		
Methylethyl ketoxime	12.45			
β-Methylglutaric acid	4.24	5.41		
N-Methylglycine (protonated cation)	2.35	10.18		
N-Methylguanidinium ion	13.4			
2-Methylhexanedione-3,5	9.43			
3-Methylhistamine (protonated cation)	5.80	9.90		

Table 5-8 (*Continued*)
PROTON-TRANSFER REACTIONS OF ORGANIC MATERIALS IN WATER AT 25°C

Substance	pK_1	pK_2	pK_3	pK_4
2-Methyl-8-hydroxyquinolinium ion	5.53	10.71		
4-Methyl-8-hydroxyquinolinium ion	5.55	10.00		
1-Methylimidazole (protonated cation)	7.06			
4-Methylimidazole (protonated cation)	7.45			
N-Methyliminodiacetic acid	2.15	10.09		
S-Methylisothiourea	9.83			
O-Methylisourea	9.72			
Methylmalonic acid	3.07	5.87		
Methylmercaptan	10.70			
3-Methylmercaptophenol	9.53			
4-Methylmercaptophenol	9.53			
N-Methylmorpholinium ion	7.13			
2-Methylnaphthoic acid-1	3.11			
N-Methyl-1-naphthylammonium ion	3.70			
2-Methyl-6-nitrobenzoic acid	1.87			
1-Methyl-2-nitroterephthalic acid	3.11			
4-Methyl-2-nitroterephthalic acid	1.82			
3-Methylpentanedione-2,4	10.87			
Methylphosphinic acid	3.08			
Methylphosphoric acid	2.38	7.74		
Methylphthalic acid: ortho-	3.18			
meta- (or *iso*-)	3.89			
Methylpiperidine	11.02			
2-Methyl-2-propanethiol	11.2			
2-Methyl-*n*-propylacetic acid (18°)	4.79			
2-Methylpyridinium ion	5.97			
3-Methylpyridinium ion	5.68			
4-Methylpyridinium ion	6.02			
Methyl-2-pyridyl ketoxime	9.97			
N-Methylpyrrolidinium ion	10.18			
2-Methylquinolinium ion	5.42			
5-Methylquinolinium ion	4.62			
Methylsuccinic acid	4.13	5.64		
Methylsulfonylacetic acid	2.36			
Methylsulfonylanilinium ion: ortho-				
meta-	2.68			
para-	1.48			
3-Methylsulfonylbenzoic acid	3.52			
4-Methylsulfonylbenzoic acid	3.64			
Methylsulfonylphenol: meta-	8.40			
para-	7.83			
Methylthioacetic acid	3.72			
Methylthioanilinium ion: meta-	4.05			
para-	4.40			
Methylthioglycolic acid	7.68			
3-(*S*-Methylthio)phenol (*also* 4-)	9.53			
Methyluracil	9.52			
2-Methyl-*n*-valeric acid (18°)	4.79			
1-Methylxanthine	7.70	12.0		
3-Methylxanthine	8.10	11.3		
7-Methylxanthine	8.33	~13		
9-Methylxanthine	6.25			
Morphine (protonated cation)	7.87			

Table 5-8 (*Continued*)
PROTON-TRANSFER REACTIONS OF ORGANIC MATERIALS IN WATER AT 25°C

Substance	pK_1	pK_2	pK_3	pK_4
Morpholinium ion	8.70			
2-(*N*-Morpholino)ethanesulfonic acid (20°)	6.15			
Murexide	0.0	9.20	10.50	
1-Naphthalenesulfonic acid	0.57			
1-Naphthoic acid	3.70			
2-Naphthoic acid	4.16			
1-Naphthol (20°)	9.30			
2-Naphthol (20°)	9.57			
Naphthoquinone monoxime	8.01			
1-Naphthylacetic acid	4.24			
2-Naphthylacetic acid	4.26			
1-Naphthylammonium ion	3.92			
2-Naphthylammonium ion	4.11			
1-Naphthylarsonic acid	3.66	8.66		
Narceine (protonated cation)	3.3			
Narcotine (protonated cation)	6.18			
Nicotine (protonated cation)	3.15	7.87		
Nicotinic acid	4.82	11.0		
iso-Nicotinic acid	4.84	10.22		
Nicotinic acid amide	3.33			
iso-Nicotinic acid methyl ester	3.26			
Nitrilotriacetic acid	1.65	2.95	10.28	
Nitroacetic acid	1.68			
Nitroanilinium ion: ortho-	−0.26			
meta-	2.46			
para-	0.99			
Nitrobenzoic acid: ortho-	2.17			
meta-	3.49			
para-	3.44			
2-Nitro-4-chlorophenol	6.48			
Nitrocinnamic acid (*trans*-): ortho-	4.15			
meta-	4.12			
para-	4.05			
Nitroethane	8.44			
2-Nitrohydroquinone	7.63	10.06		
N-Nitroiminodiacetic acid	2.21	3.33		
Nitromethane	10.21			
Nitrophenol: ortho-	7.23			
meta-	8.40			
para-	7.15			
Nitrophenylacetic acid: ortho-	4.00			
meta-	3.97			
para-	3.85			
m-Nitrophenylphosphoric acid	1.30	6.27		
p-Nitrophenylphosphoric acid	1.24	6.23		
β-(*o*-Nitrophenyl)propionic acid	4.50			
β-(*p*-Nitrophenyl)propionic acid	4.47			
3-Nitrophthalic acid	1.88			
4-Nitrophthalic acid	2.11			
2-Nitropropionic acid	3.79			
N-Nitrosoiminodiacetic acid	2.28	3.38		
Nitroterephthalic acid	1.73			

Table 5-8 (*Continued*)
PROTON-TRANSFER REACTIONS OF ORGANIC MATERIALS IN WATER AT 25°C

Substance	pK_1	pK_2	pK_3	pK_4
4-Nitro-o-toluic acid	1.86			
Nitrourea	4.15			
Nonanoic acid	4.95			
Norleucine (protonated cation)	2.34	9.83		
Norvaline (protonated cation)	2.29	9.70		
Novocaine (protonated cation)	8.85			
Octanoic acid	4.89			
Octylammonium ion	10.65			
Ornithine (protonated cation)	1.94	8.65	10.76	
Orthanilic acid	2.46			
Oxaloacetic acid	2.56	4.37		
Oxalic acid	1.27	4.27		
Papaverine (protonated cation)	5.90			
Parabanic acid	6.10			
Pelargonic acid	4.95			
Pentenedioic acid	3.77	6.08		
Pent-1-en-oic acid	4.70			
Pent-2-en-oic acid	4.52			
Pent-3-en-oic acid	4.72			
Pent-4-en-oic acid	4.72			
Peracetic acid	8.20			
1,10-Phenanthrolinium ion	4.86			
Phenazinium ion	1.2			
Phenetidine (protonated cation): ortho-	4.47			
meta-	4.17			
para-	5.34			
Phenol	9.99			
Phenol-3-phosphoric acid	1.78	7.03	10.2	
Phenol-4-phosphoric acid	1.99	7.25	9.9	
p-Phenolsulfonic acid	......	9.05		
Phenoxyacetic acid	3.17			
Phenoxybenzoic acid: ortho-	3.53			
meta-	3.95			
para-	4.52			
Phenylacetic acid	4.31			
Phenylalanine (protonated cation)	2.16	9.13		
Phenylanilinium ion: ortho-	3.78			
meta-	4.18			
para-	4.27			
Phenylarsonic acid	3.46	8.48		
Phenylboric acid	8.86			
o-Phenylbenzoic acid	3.46			
α-Phenyl-α-benzylsuccinic acid	3.69	6.47		
γ-Phenylbutyric acid	4.76			
Phenylenediammonium ion: ortho-	<2	4.47		
meta-	2.65	4.88		
para-	3.29	6.08		
β-Phenylethylammonium ion	9.83			
β-Phenylethylboric acid	10.0			
Phenylguanidinium ion	10.77			
Phenylhydrazinium ion	5.20			

Table 5-8 (*Continued*)
PROTON-TRANSFER REACTIONS OF ORGANIC MATERIALS IN WATER AT 25°C

Substance	pK_1	pK_2	pK_3	pK_4
α-Phenyl-α-hydroxypropionic acid	3.53			
β-Phenyl-β-hydroxypropionic acid	4.40			
Phenylmalonic acid	2.58	5.03		
Phenylphenol: ortho-	9.97			
meta-	9.63			
para-	9.55			
Phenylphosphinic acid	2.1			
Phenylphosphoric acid	1.83	7.07		
α-Phenylpropionic acid	4.38			
β-Phenylpropionic acid	4.64			
γ-Phenylpropylammonium ion	10.39			
Phenylselenic acid	4.79			
Phenylsuccinic acid	3.78	5.55		
Phloroglucinol	8.68			
Phosphoserine (protonated cation)	2.08	5.65	9.74	
Phthalamic acid	3.79			
Phthalazinium ion	3.47			
Phthalic acid: ortho-	2.95	5.41		
meta-	3.62	4.60		
para-	3.54	4.46		
Phthalic acid monoamide	3.76			
Phthalimide	9.90			
Physostigmine (protonated cation)	1.76	7.88		
Picolinic acid	3.4	11.0		
Picolinic acid methylester	2.21			
2-Picolinium ion	6.48			
3-Picolinium ion (*and* 4-)	6.00			
Picric acid	0.29			
Pilocarpine (protonated cation)	1.3	6.85		
Pimelic acid	4.50	5.42		
Piperazine (protonated cation)	5.68	9.82		
Piperazine-2-carboxylic acid	1.5	5.41	9.53	
Piperidinium ion	11.12			
Piperine (protonated cation)	0.0			
Pivalic acid	5.05			
Polyacrylic acid	5.08			
Prehnitic acid	2.38	3.51	4.44	5.81
L-Proline (protonated cation)	1.95	10.64		
1,2-Propanediammonium ion	7.10	9.97		
1,3-Propanediammonium ion (20°)	8.64	10.62		
2-Propanethiol	10.86			
1,2,3-Propanetriammonium ion	3.72	7.95	9.59	
1,2,3-Propanetricarboxylic acid	3.50	4.63	5.95	
Propenylacetic acid	4.51			
Propiolic acid	1.84			
Propionic acid	4.87			
N-n-Propionylglycine (protonated cation)	3.72			
n-Propoxybenzoic acid: ortho- and meta-	4.22			
para-	4.78			
iso-Propylacrylic acid	4.70			
iso-Propylammonium ion	10.53			
N-iso-Propylanilinium ion	5.50			

Table 5-8 (*Continued*)
PROTON-TRANSFER REACTIONS OF ORGANIC MATERIALS
IN WATER AT 25°C

Substance	pK_1	pK_2	pK_3	pK_4
n-Propylarsonic acid	4.21	9.09		
iso-Propylbenzoic acid: ortho-	3.64			
para-	4.36			
β-n-Propylglutaric acid	4.31	5.39		
N-n-Propylglycine (protonated cation)	2.38	10.03		
n-Propylmalonic acid (also iso-)	2.97	5.84		
p-iso-Propylphenylacetic acid	4.39			
n-Propylphosphinic acid	3.46			
iso-Propylphosphinic acid	3.56			
n-Propylphosphoric acid	2.49	8.18		
iso-Propylphosphoric acid	2.66	8.44		
Pteroylglutamic acid	8.26			
Purine (protonated cation)	2.52	8.92		
Pyrazinium ion	0.6			
Pyrazolinium ion	2.53			
Pyridazinium ion	2.33			
Pyridine-2-carboxylic acid	1.06	5.37		
Pyridine-3-carboxylic acid	2.07	4.73		
Pyridine-4-carboxylic acid	1.70	4.89		
Pyridine-2,3-dicarboxylic acid	2.36	7.08		
Pyridine-2,4-dicarboxylic acid	2.23	7.02		
Pyridine-2,6-dicarboxylic acid	2.16	6.92		
2-Pyridinium ion	3.28			
3-Pyridinium ion	4.88			
4-Pyridinium ion	6.62			
Pyridinium-N-oxide ion	0.79			
Pyrimidinium ion	1.30			
Pyrocatechol	9.45	12.08		
Pyrocatechol Violet	7.82	9.76	11.73	
Pyrogallol	9.03	12.63		
Pyromellitic acid	1.92	2.87	4.49	5.63
Pyromucic acid	5.15			
Pyroracomic acid	2.49			
Pyroxilidinium ion	11.11			
1-Pyrrolecarboxylic acid	4.45			
2-Pyrrolecarboxylic acid	4.45			
Pyrrolinium ion	−0.27			
Pyruvic acid	2.49			
Quinidine (protonated cation)	4.0	8.54		
Quinine (protonated cation)	4.11	8.0		
Quinol	9.96			
Quinolinium ion	4.88			
Quinoxalinium ion	0.72			
Reductic acid (20°)	4.72			
Resorcinol	9.44	12.32		
β-Resorcylic acid	3.29	8.98	>13	
Rubeanic acid	10.89			
Saccharin	11.68			
Salicylaldehyde	8.34			

Table 5-8 (*Continued*)
PROTON-TRANSFER REACTIONS OF ORGANIC MATERIALS IN WATER AT 25°C

Substance	pK_1	pK_2	pK_3	pK_4
Salicylaldoxime	1.37	9.18	12.11	
Salicylamide	8.36			
Salicylic acid	3.00	12.38		
Sarcosine (protonated cation)	2.25	10.20		
Sarcosine amide (protonated cation)	8.35			
Sarcosine dimethylamide (protonated cation)	8.86			
Sarcosine methylamide (protonated cation)	8.28			
Sarcosylsarcosine (protonated cation)	2.92	9.15		
Sarcosylserine (protonated cation)	3.17	8.63		
Semicarbazide (protonated cation)	3.43			
Serine (protonated cation)	2.20	9.21		
Serine methyl ester (protonated cation)	7.10			
Solanine (protonated cation)	7.34			
Sorbic acid	4.77			
Sorbose (18°)	13.57			
Sparteine (protonated cation)	4.49	11.76		
Strychnine (protonated cation)	2.3	8.0		
Suberic acid	2.52	5.41		
Succinamic acid	4.54			
Succinic acid	4.21	5.64		
Succinimide	9.62			
Sulfamic acid	0.99			
β-(p-Sulfaminophenyl)-alanine (protonated cation)	1.99	8.64	10.26	
Sulfamoylbenzoic acid: meta-	3.54			
para-	3.47			
p-Sulfamoylphenylphosphoric acid	1.42	6.38	10.0	
Sulfanilic acid	0.58	3.12		
Sulfoacetic acid	4.07			
Sulfobenzoic acid: meta-	3.78			
para-	3.72			
Sulfophenol: meta-	0.39	9.07		
para-	0.58	8.70		
α-Sulfopropionic acid	1.99			
β-Sulfopropionic acid	3.62			
5-Sulfosalicylic acid	2.49	12.00		
Sylvic acid	7.62			
Tartaric acid: (*meso-*)	3.22	4.81		
(*d-*)	3.04	4.37		
Tartronic acid	2.37	4.74		
Taurine (protonated cation) (20°)	~1.5	9.08		
Terephthalic acid	3.54	4.46		
Terramycine	3.10	7.26	9.11	
Tetracycline	3.35	7.82	9.57	
Tetraethylenepentamine (protonated cation)	2.66	4.70	8.10	9.05; pK_5 9.54
Tetralol-1	10.28			
Tetralol-2	10.48			
Tetramethylenediammonium ion	9.22	10.75		
N,N,N',N'-Tetramethylethylenediammonium ion	6.56	10.13		
N,N'-Tetramethyl-p-phenylenediammonium ion	2.20	6.35		

Table 5-8 (*Continued*)
PROTON-TRANSFER REACTIONS OF ORGANIC MATERIALS
IN WATER AT 25°C

Substance	pK_1	pK_2	pK_3	pK_4
Tetrolic acid	2.66			
2-Thenoic acid	3.53			
3-Thenoic acid	4.10			
Thebaine (protonated cation)	7.95			
Theobromic acid	8 to 10			
Throbromine (protonated cation)	0.68	7.89		
Theophylline	8.80			
Thiazolinium ion	2.53			
Thioacetic acid	3.33			
Thiocyanatoacetic acid	2.58			
Thiodiacetic acid (18°)	3.30	4.50		
Thiodiglycollic acid	3.32	4.29		
Thioglycollic acid	3.60	10.55		
1-Thionylcarboxylic acid	3.53			
2-Thionylcarboxylic acid	4.10			
Thiophenecarboxylic acids (see Thenoic acid)				
Thiophenol	6.50			
Thiourea (protonated cation)	2.03			
Thorin	3.7	8.3	11.8	
Threonine (protonated cation)	2.09	9.10		
Tiglic acid	5.00			
Tiron	7.66	12.6		
2-Toluenethiol	6.64			
3-Toluenethiol	6.58			
4-Toluenethiol	6.52			
p-Toluenesulfinic acid	1.7			
Toluhydroquinone	10.03	11.62		
Toluic acid: ortho-	3.91			
meta-	4.27			
para-	4.37			
Toluidinium ion: ortho-	4.39			
meta-	4.68			
para-	5.09			
o-Tolylacetic acid (also p-) (18°)	4.36			
Tolylphosphoric acid: ortho-	2.10	7.68		
meta-	1.88	7.44		
para-	1.84	7.33		
Triacetylmethane	5.81			
1,2,3-Triaminopropane (protonated cation)	3.80	8.03	9.67	
2,2′,2″-Triaminotriethylamine (protonated cation)	8.64	9.67	10.37	
2,4,6-Tribromobenzoic acid	1.41			
Tri-i-butylammonium ion	10.42			
Tricarballylic acid	3.50	4.63	5.95	
Trichloroacetic acid	0.64			
3,3,3-Trichlorolactic acid	2.34			
Trichloromethylphosphoric acid	1.63	4.81		
Trichlorophenol	6.00			
Triethanolammonium ion	7.76			
Triethylammonium ion	10.72			
Triethylenediammonium ion	4.18	8.19		
Triethylenetetrammonium ion (20°)	3.32	6.67	9.20	9.92

Table 5-8 (*Continued*)
PROTON-TRANSFER REACTIONS OF ORGANIC MATERIALS IN WATER AT 25°C

Substance	pK_1	pK_2	pK_3	pK_4
Trifluoroacetic acid	0.25			
γ,γ,γ-Trifluoro-*n*-butyric acid	4.16			
Trifluoromethylanilinium ion: meta-	3.5			
para-	2.6			
β,β,β-Trifluoropropionic acid	3.06			
1,1,1-Trifluoro-3-(2-thenoyl)acetone	5.70			
2,4,6-Trihydroxybenzoic acid	1.68			
3,4,5-Trihydroxybenzoic acid	4.34	8.85		
Trimellitic acid	2.52	3.84	5.20	
Trimesic acid	2.12	3.89	5.20	
Trimethylacetic acid	5.08			
Trimethylammonium ion	9.80			
2,4,6-Trimethylanilinium ion	4.38			
2,4,6-Trimethylbenzoic acid	3.44			
Trimethylenediammonium ion	10.45			
2,4,6,-Trimethylphenol	10.88			
2,4,6-Trinitrobenzoic acid	0.65			
2,2,2-Trinitroethanol	2.36			
2,4,6-Trinitrophenol	0.80			
Triphenylacetic acid	3.96			
Tripropylammonium ion	10.70			
Trishydroxymethylamineomethane (protonated cation)	8.08			
Tryptophane (protonated cation)	2.38	9.39		
Tyrosine (protonated cation)	2.20	9.11	10.07(OH)	
Uric acid	5.40	5.53		
i-Valeric acid	4.78			
n-Valeric acid	4.84			
t-Valeric acid	5.08			
Valine (protonated cation)	2.29	9.74		
Vanillic acid	4.52			
Vanillin	7.40			
Veratrine (protonated cation)	8.85			
Veronal	7.43			
Vinylacetic acid	4.34			
Vinylmethylammonium ion	9.69			
Violuric acid	4.57			
Xahthene (protonated cation) (40°)	0.68			
Xanthine	7.53	11.63		
Xylenol orange (pK_5 10.46; pK_6 12.28)	......	2.58	3.23	6.37
Zincon	4	7.85	15	

Table 5-9
TEMPERATURE DEPENDENCE OF SELECTED EQUILIBRIUM CONSTANTS IN AQUEOUS SOLUTIONS

Substance	Temperature, °C									
	0°	5°	10°	15°	20°	25°	30°	35°	40°	50°
Acetic acid, pK_a	4.78	4.77	4.76	4.76	4.76	4.76	4.76	4.76	4.77	4.79
Ammonium ion, pK_a	10.081	9.904	.....	9.564	9.400	9.245	9.093	8.947	8.805	8.539
Boric acid, pK_a	.....	9.439	9.380	9.327	9.280	9.237	9.198	9.164	9.132	9.081
Carbon dioxide + water, pK_1	6.58	.....	6.46	6.42	6.38	6.35	.....	.....	6.30	6.28
pK_2	10.63	.....	10.49	10.43	.....	10.33	.....	.....	10.22	10.17
Hydrogen cyanide, pK_a	.....	.....	9.63	9.49	9.36	9.21	9.11	8.99	8.88	.....
Hydrogen peroxide, pK_1	.....	.....	.....	11.86	11.75	11.65	11.55	11.45	.....	.....
Hydrogen sulfide, pK_1	.....	7.33	7.25	7.15	7.05	6.96	6.89	6.82	6.75	6.62
pK_2	.....	13.5	.....	13.2	.....	12.90	12.75	12.6	.....	.....
Lead sulfate, pK_{sp}	8.01	.....	.....	7.87	.....	7.80	.....	7.73	.....	7.63
Mercury(I) chloride, pK_{sp}	.....	.....	18.65	18.48	18.27	17.88	.....	16.79	.....	.....
Phosphoric acid, pK_1	2.056	2.073	2.088	2.107	2.127	2.148	2.171	.....	2.224	2.277
pK_2	7.314	7.281	7.254	7.231	7.213	7.199	7.190	.....	7.178	7.18
Silver bromide, pK_{sp}	.....	13.33	.....	12.83	12.57	12.30	12.07	11.83	11.61	11.19
Silver chloride, pK_{sp}	.....	10.595	.....	10.152	.....	9.749	.....	9.381	9.21	8.88
Silver sulfate, pK_{sp}	.....	.....	.....	.....	.....	4.835	.....	.....	.....	4.62
Sulfuric acid, pK_2	1.76	1.79	1.83	1.88	1.94	1.99	2.05	2.12	2.19	2.35

ACIDITIES OF VARIOUS COMPOUNDS IN NONAQUEOUS SOLVENTS

The properties of common solvents employed in nonaqueous titrations of acids and bases are given in Table 5-10; these properties include the dielectric constant, the autoprotolysis constant $(-\log K_s)$, and the approximate potential span in millivolts for acid-base titrations. Figure 5-1 graphically displays the approximate potential ranges in various solvents; however, the basic end is truncated by the nature of the titrant employed—tetrabutylammonium hydroxide. All values refer to $0.01M$ solutions. These limiting potentials are subject to wide variation with changes in titrant concentration.

Only limited data pertaining to acidity constants in nonaqueous solvents are available. These are summarized in Table 5-11. For computing the approximate half-neutralization potential (HNP) values from pK_a values in aqueous solutions (Tables 5-7 and 5-8), a useful equation is

$$HNP = a - b(pK_a)$$

The constants a and b in the equation are given in Table 5-12 for the solvents listed in the first column. The classes of compounds for which the expression is applicable are denoted by the numbers appearing in the last column. The values of HNP computed for compounds of classes 1 and 2 are referred to diphenylguanidine ($pK_a = 10.12$; assigned a HNP value of 0.0 mV); those for compounds of classes 3 and 4 are referred to benzoic acid ($pK_a = 4.21$; assigned a HNP value of 0.0 mV).

The acidity function H_0 is not affected by the dielectric constant and allows a quantitative comparison of acidity in different solvents. H_0 is defined by

$$H_0 = (pK_{HB^+})_{water} + \log (C_B/C_{HB^+})_{solvent}$$

where K_{HB^+} refers to the dissociation constant of the protonated cation or cation acid with reference to water as the standard state. Table 5-13 gives H_0 for mineral acids in aqueous solutions; all values are referred to the value of $pK_{HB^+} = 0.99$ for p-nitroaniline. Values for weak bases are given by M. A. Paul and F. A. Long, *Chem. Rev.*, **57**, 1 (1957).

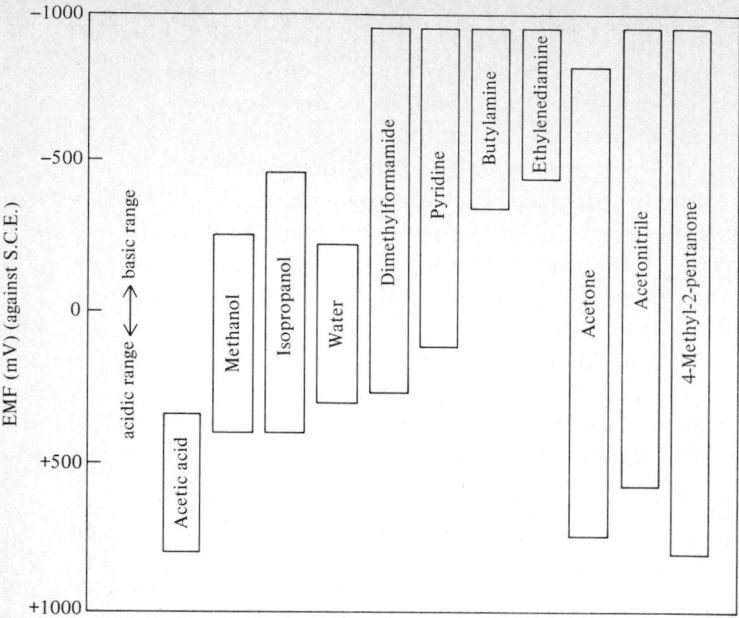

Fig. 5-1. *Approximate potential ranges in nonaqueous solvents*

Table 5-10
PROPERTIES OF COMMON ACID-BASE SOLVENTS

Solvent	Potential Span, mV	$-\log K_s$	Dielectric Constant, 25°C
Acetic acid	400	14.5	6.1(20°)
Acetic anhydride	800	14.5	20.7(20°)
Acetone	1600		20.7
Acetonitrile	1600	26.5	36.0(20°)
Ammonia (at −50°C)		33	22(−33°)
n-Butanol	900		17.1
n-Butylamine	500		5.3(20°)
Chlorobenzene	1500		5.62
N,N-Dimethylformamide	1300	18.0	37.6
Dimethylsulfoxide		17.3	46.6
Ethanol	800	19.1	24.3
Ethanolamine		5.1	37.7
Ethyl acetate	1500		6.02
Ethylenediamine	500	15.3	14.2(20°)
Formic acid	200	6.2	58.5
Methanol	800	16.7	32.6
4-Methyl-2-pentanone (methyl isobutyl ketone)	1600		13.1(20°)
Nitromethane	1000		35.8(30°)
iso-Propanol	900		18.3
Pyridine	1000		12.3
Sulfuric acid		3.85	101
Water	800	14.0	78.3

Table 5-11
EQUILIBRIUM CONSTANTS OF PROTON-TRANSFER REACTIONS IN NONAQUEOUS SOLVENTS

Acid	Methanol	Ethanol	
Acetic acid	9.52	10.32	11.4[a]
p-Aminobenzoic acid	10.25		
Ammonium ion	10.7		6.40[b]
Anilinium ion	6.0	5.70	
Benzoic acid		10.72	10.0[a]
Bromocresol purple	11.3		
Bromocresol green	9.8	10.65	
Bromophenol blue	8.9	9.5	
Bromothymol blue	12.4	13.2	
di-n-Butylammonium ion			10.3[a]
o-Chloroanilinium ion	3.4		
Cyanoacetic acid		7.49	
2,5-Dichloroanilinium ion			9.48[b]
Dimethylaminoazobenzene			6.32[b]
N,N'-Dimethylanilinium ion		4.37	
Formic acid		9.15	
Hydrobromic acid			5.5[c]
Hydrochloric acid			8.55[b], 8.9[c]
Methyl orange	3.8	3.4	
Methyl red (acid range)	4.1	3.55	
(alkaline range)	9.2	10.45	
Methyl yellow	3.4	3.55	
Neutral red	8.2	8.2	
o-Nitrobenzoic acid	7.6		
m-Nitrobenzoic acid	8.3		
p-Nitrobenzoic acid	8.4		
Perchloric acid			4.87[b]
Phenol	14.0		
Phenol red	12.8		
Phthalic acid, pK_2	11.65		
Picric acid	3.8	3.8	8.9[c]
Pyridinium ion			6.1[b]
Succinic acid, pK_2	11.4		
Salicylic acid	8.7	7.9	
Stearic acid	10.0		
Sulfuric acid, pK_1			7.24[b,c]
Tartaric acid, pK_2	9.9		
p-Toluenesulfonic acid			8.44[b]
p-Toluidinium ion		6.24	
Thymol blue (alkaline range)	14.0	15.2	
(acid range)	4.7	5.35	
Tropeoline 00	2.2	2.3	
Thymolbenzein (acid range)	3.5		
(alkaline range)	13.1		
Tribenzylammonium ion			5.40[b]
Urea (protonated cation)			6.96[b]
Veronal	12.6		

[a]Dimethylsulfoxide. [b]Glacial acetic acid. [c]Acetonitrile.

Table 5-12
CONSTANTS FOR EQUATION TO COMPUTE APPROXIMATE HNP VALUES

Solvent	Constant		Applicable to Class
	a	b	
Acetic acid*	253	48.5	1,2
Acetic anhydride	520	52.3	1,2
Acetonitrile	805	77.5	1
	665	69.4	2
N,N-Dimethylformamide	601	144	3
	198	37.9	4
Ethyl acetate	664	63.5	1
	730	84.7	2
4-Methyl-2-pentanone	616	145	3
	904	122	4
Nitromethane	784	75.1	1
	635	76.9	2
Pyridine†	646	156	3
	1126	160	4

* All bases with pK_a values above 4.76 are leveled.
† All phenols with pK_a values below 4.1 are leveled.
Class 1. Monofunctional amines.
Class 2. Amides, amines containing oxygen functions, diamines, heterocyclic bases, and ureas.
Class 3. Aliphatic and m- and p-substituted aromatic carboxylic acids.
Class 4. Hydroxyaromatic compounds.

Table 5-13
H_0 FOR MINERAL ACIDS IN AQUEOUS SOLUTIONS AT 25°C

Acid Concentration, moles per liter	Acid						
	$HClO_4$	HBr	H_2SO_4	HCl	H_3PO_4	HNO_3	HF
0.1		+0.98	+0.83	+0.98	+1.45	+0.98	
0.25		+0.44	+0.55	+1.15	+0.55		
0.5	+0.18	+0.20	+0.13	+0.20	+0.97	+0.21	
1.0	−0.22	−0.20	−0.26	−0.20	+0.63	−0.18	+1.20
2.0	−0.78	−0.71	−0.84	−0.69	+0.24	−0.67	+0.91
3.0	−1.23	−1.11	−1.38	−1.05	−0.08	−1.02	+0.60
4.0	−1.72	−1.50	−1.85	−1.40	−0.37	−1.32	+0.40
5.0	−2.23	−1.93	−2.28	−1.76	−0.69	−1.57	+0.28
6.0	−2.84	−2.38	−2.76	−2.12	−1.04	−1.79	+0.15
7.0	−3.61	−2.85	−3.32	−2.50	−1.45	−1.99	+0.02
8.0	−4.33	−3.34	−3.87	−2.86	−1.85		−0.11
9.0	−5.05	−3.89	−4.40	−3.22	−2.22		−0.24
10.0	−5.79	−4.44	−4.89	−3.59	−2.59		−0.36

Formation Constants of Metal Complexes

Each value listed in Tables 5-14 and 5-15 is the logarithm of the overall formation constant for the cumulative binding of a ligand L to the central metal cation M, viz:

	Cumulative Formation Constant	Stepwise Stability Constants
$M + L = ML$	K_1	k_1
$M + 2L = ML_2$	K_2	$k_1 k_2$
. . . .		
$M + nL = ML_n$	K_n	$k_1 k_2 \cdots k_n$

As an example, the entries in Table 5-14 for the zinc ammine complexes represent these equilibria:

$$Zn^{2+} + NH_3 = Zn(NH_3)^{2+} \qquad K_1 = \frac{[Zn(NH_3)^{2+}]}{[Zn^{2+}][NH_3]}$$

$$Zn^{2+} + 2NH_3 = Zn(NH_3)_2^{2+} \qquad K_2 = \frac{[Zn(NH_3)_2^{2+}]}{[Zn^{2+}][NH_3]^2}$$

$$Zn^{2+} + 3NH_3 = Zn(NH_3)_3^{2+} \qquad K_3 = \frac{[Zn(NH_3)_3^{2+}]}{[Zn^{2+}][NH_3]^3}$$

$$Zn^{2+} + 4NH_3 = Zn(NH_3)_4^{2+} \qquad K_4 = \frac{[Zn(NH_3)_4^{2+}]}{[Zn^{2+}][NH_3]^4}$$

If the stepwise stability or formation constants of the reactions are desired, for the first step $\log K_1 = \log k_1 = 2.37$. For the second and succeeding steps the equilibria and corresponding constants are as follows:

$Zn(NH_3)^{2+} + NH_3 = Zn(NH_3)_2^{2+}$	$\log k_2 = \log K_2 - \log K_1 = 2.44$
$Zn(NH_3)_2^{2+} + NH_3 = Zn(NH_3)_3^{2+}$	$\log k_3 = \log K_3 - \log K_2 = 3.50$
$Zn(NH_3)_3^{2+} + NH_3 = Zn(NH_3)_4^{2+}$	$\log k_4 = \log K_4 - \log K_3 = 2.15$

The reverse of the association or formation reactions would represent the dissociation or instability constant for the systems, i.e., $-\log K_f = \log K_{instab}$.

The data in the tables generally refer to temperatures of about 20 to 25 °C. Most of the values in Table 5-14 refer to zero ionic strength, but those in Table 5-15 often refer to a finite ionic strength.

Table 5-14

CUMULATIVE FORMATION CONSTANTS FOR METAL COMPLEXES WITH INORGANIC LIGANDS

	$\log K_1$	$\log K_2$	$\log K_3$	$\log K_4$	$\log K_5$	$\log K_6$
Ammonia						
Cadmium	2.65	4.75	6.19	7.12	6.80	5.14
Cobalt(II)	2.11	3.74	4.79	5.55	5.73	5.11
Cobalt(III)	6.7	14.0	20.1	25.7	30.8	35.2
Copper(I)	5.93	10.86				
Copper(II)	4.31	7.98	11.02	13.32	12.86	
Iron(II)	1.4	2.2				
Manganese(II)	0.8	1.3				
Mercury(II)	8.8	17.5	18.5	19.28		
Nickel	2.80	5.04	6.77	7.96	8.71	8.74
Platinum(II)						35.3
Silver(I)	3.24	7.05				
Zinc	2.37	4.81	7.31	9.46		
Bromide						
Astatine	2.51 [AtBr]					
Bismuth(III)	4.30	5.55	5.89	7.82		9.70
Bromine	1.24 [Br$_3^-$]					
Cadmium	1.75	2.34	3.32	3.70		
Cerium(III)	0.42					
Copper(I)		5.89				
Copper(II)	0.30					
Gold(I)		12.46				
Indium	1.30	1.88				
Iodine	2.64 [IBr]					
Iron(III)	−0.30	−0.50				
Lead	1.2	1.9		1.1		
Mercury(II)	9.05	17.32	19.74	21.00		
Palladium(II)				13.1		
Platinum(II)				20.5		
Rhodium(III)		14.3	16.3	17.6	18.4	17.2
Scandium	2.08	3.08				
Silver(I)	4.38	7.33	8.00	8.73		
Thallium(I)	0.93					
Thallium(III)	9.7	16.6	21.2	23.9	29.2	31.6
Tin(II)	1.11	1.81	1.46			
Uranium(IV)	0.18					
Yttrium	1.32					
Chloride						
Americium(III)	1.17					
Antimony(III)	2.26	3.49	4.18	4.72		
Bismuth(III)	2.44	4.7	5.0	5.6		
Cadmium	1.95	2.50	2.60	2.80		
Cerium(III)	0.48					
Copper(I)		5.5	5.7			
Copper(II)	0.1	−0.6				
Curium(III)	1.17					
Gold(III)		9.8				
Indium	1.42	2.23	3.23			
Iron(II)	0.36					
Iron(III)	1.48	2.13	1.99	0.01		
Lead	1.62	2.44	1.70	1.60		
Manganese(II)	0.96					

Table 5-14 (*Continued*)

CUMULATIVE FORMATION CONSTANTS FOR METAL COMPLEXES WITH INORGANIC LIGANDS

	$\log K_1$	$\log K_2$	$\log K_3$	$\log K_4$	$\log K_5$	$\log K_6$
Mercury(II)	6.74	13.22	14.07	15.07		
Palladium(II)	6.1	10.7	13.1	15.7		
Platinum(II)		11.5	14.5	16.0		
Plutonium(III)	1.17					
Silver(I)	3.04	5.04		5.30		
Thallium(I)	0.52					
Thallium(III)	8.14	13.60	15.78	18.00		
Thorium	1.38	0.38				
Tin(II)	1.51	2.24	2.03	1.48		
Tin(IV)						4
Uranium(IV)	0.8					
Uranium(VI)	0.22					
Zinc	0.43	0.61	0.53	0.20		
Zirconium	0.9	1.3	1.5	1.2		
Cyanide						
Cadmium	5.48	10.60	15.23	18.78		
Copper(I)		24.0	28.59	30.30		
Gold(I)		38.3				
Iron(II)						35
Iron(III)						42
Mercury(II)				41.4		
Nickel				31.3		
Silver(I)		21.1	21.7	20.6		
Zinc				16.7		
Fluoride						
Aluminum	6.10	11.15	15.00	17.75	19.37	19.84
Beryllium	5.1	8.8	12.6			
Cerium(III)	3.20					
Chromium(III)	4.41	7.81	10.29			
Gadolinium	3.46					
Gallium	5.08					
Indium	3.70	6.25	8.60	9.70		
Iron(III)	5.28	9.30	12.06			
Lanthanum	2.77					
Magnesium	1.30					
Manganese(II)	5.48					
Plutonium(III)	6.77					
Scandium						17.3
Thallium(I)	0.1					
Thallium(III) [TlO$^+$]	6.44					
Thorium	7.65	13.46	17.97			
Titanium(IV) [TiO^{2+}]	5.4	9.8	13.7	18.0		
Uranium(VI)	4.59	7.93	10.47	11.84		
Yttrium	4.81	8.54	12.14			
Zirconium	8.80	16.12	21.94			
Hydroxide						
Aluminum	9.27			33.03		
Antimony(III)		24.3	36.7	38.3		
Arsenic (as AsO$^+$)	14.33	18.73	20.60	21.20		
Beryllium	9.7	14.0	15.2			
Bismuth(III)	12.7	15.8		35.2		
Cadmium	4.17	8.33	9.02	8.62		

Table 5-14 (*Continued*)
CUMULATIVE FORMATION CONSTANTS FOR METAL COMPLEXES WITH INORGANIC LIGANDS

	$\log K_1$	$\log K_2$	$\log K_3$	$\log K_4$	$\log K_5$	$\log K_6$
Cerium(III)	4.6					
Cerium(IV)	13.28	26.46				
Chromium(III)	10.1	17.8		29.9		
Copper(II)	7.0	13.68	17.00	18.5		
Dysprosium	5.2					
Erbium(III)	5.4					
Gadolinium	4.6					
Gallium	11.0	21.7		34.3	38.0	40.3
Indium	9.9	19.8		28.7		
Iodine	9.49	11.24				
Iron(II)	5.56	9.77	9.67	8.58		
Iron(III)	11.87	21.17	29.67			
Lanthanum	3.3					
Lead(II)	7.82	10.85	14.58			61.0
Lutetium	6.6					
Magnesium	2.58					
Manganese(II)	3.90		8.3			
Neodymium	5.5					
Nickel	4.97	8.55	11.33			
Praseodymium	4.30					
Plutonium(III)	7.0					
Plutonium(IV)	12.39					
Plutonium (as PuO_2^{2+})	8.3	16.6	20.9			
Samarium(III)	4.8					
Scandium	8.9					
Tellurium(IV)			41.6	53.0	64.8	72.0
Thallium(III)	12.86	25.37				
Titanium(III)	12.71					
Uranium(IV)	13.3				41.2	
Uranium(VI) [as UO_2^{2+}]	9.5	22.80		32.4		
Vanadium(III)	11.1	21.6				
Vanadium(IV) [as VO^{2+}]	8.6		[25.8 for $V_2O_4(OH)^-$]			
Vanadium(V) [as VO^{3+}]		25.2		46.2	58.5	
Yttrium	5.0					
Zinc	4.40	11.30	14.14	17.66		
Zirconium	14.3	28.3	41.9	55.3		
Iodide						
Bismuth	3.63			14.95	16.80	18.80
Cadmium	2.10	3.43	4.49	5.41		
Copper(I)		8.85				
Indium	1.00	2.26				
Iodine	2.89	5.79				
Iron(III)	1.88					
Lead	2.00	3.15	3.92	4.47		
Mercury(II)	12.87	23.82	27.60	29.83		
Silver	6.58	11.74	13.68			
Thallium(I)	0.72	0.90	1.08			
Thallium(III)	11.41	20.88	27.60	31.82		
Iodate						
Barium	1.05					
Calcium	0.89					
Magnesium	0.72					

Table 5-14 (*Continued*)
CUMULATIVE FORMATION CONSTANTS FOR METAL COMPLEXES WITH INORGANIC LIGANDS

	$\log K_1$	$\log K_2$	$\log K_3$	$\log K_4$	$\log K_5$	$\log K_6$
Strontium	1.00					
Thorium	2.88	4.79	7.15			
Nitrate						
Barium	0.92					
Beryllium	1.62					
Bismuth(III)	1.26					
Cadmium	0.40					
Calcium	0.28					
Cerium(III)	1.04	2.55				
Curium(III)	0.57					
Hafnium	0.92	2.43	4.32	6.40	8.48	10.29
Iron(III)	1.0					
Lanthanum	0.26	0.69	1.27			
Lead	1.18					
Mercury(II)	0.35					
Neodymium	0.52	1.18				
Neptunium(IV)	0.38					
Plutonium(III)	0.77	1.93	3.09			
Plutonium(IV)	0.54					
Strontium	0.82					
Thallium(I)	0.33					
Thallium(III)	0.92					
Thorium	0.78	1.89	2.89	3.63		
Uranium(IV)	0.20	0.37				
Uranium(VI)	0.34	0.45				
Ytterbium	0.45	1.30	2.42			
Zirconium(as ZrO^{2+})		1.91		3.54		
Pyrophosphate						
Barium	4.6					
Calcium	4.6					
Cadmium	5.6					
Copper(II)	6.7	9.0				
Lead		5.3				
Magnesium	5.7					
Nickel	5.8	7.4				
Strontium	4.7					
Yttrium		9.7				
Zirconium		6.5				
Sulfate						
Cerium(III)	3.40					
Erbium	3.58					
Gadolinium	3.66					
Holmium	3.58					
Indium	1.78	1.88	2.36			
Iron(III)	2.03	2.98				
Lanthanum	3.64					
Neodymium	3.64					
Nickel	2.4					
Plutonium(IV)	3.66					
Praseodymium	3.62					
Samarium	3.66					
Thorium	3.32	5.50				

Table 5-14 (*Continued*)
CUMULATIVE FORMATION CONSTANTS FOR METAL COMPLEXES WITH INORGANIC LIGANDS

	$\log K_1$	$\log K_2$	$\log K_3$	$\log K_4$	$\log K_5$	$\log K_6$
Uranium(IV)	3.24	5.42				
Uranium(VI)	1.70	2.45	3.30			
Yttrium	3.47					
Ytterbium	3.58					
Zirconium	3.79	6.64	7.77			
Sulfite						
Copper(I)	7.5	8.5	9.2			
Mercury(II)		22.66				
Silver	5.30	7.35				
Thiocyanate						
Bismuth	1.15	2.26	3.41	4.23		
Cadmium	1.39	1.98	2.58	3.6		
Chromium(III)	1.87	2.98				
Cobalt(II)	−0.04	−0.70	0	3.00		
Copper(I)	12.11	5.18				
Gold(I)		23		42		
Indium	2.58	3.00	4.63			
Iron(III)	2.95	3.36				
Mercury(II)		17.47		21.23		
Nickel	1.18	1.64	1.81			
Ruthenium(III)	1.78					
Silver		7.57	9.08	10.08		
Thallium(I)	0.80					
Uranium(IV)	1.49	2.11				
Uranium(VI)	0.76	0.74	1.18			
Vanadium(III)	2.0					
Vanadium(IV)	0.92					
Zinc	1.62					
Thiosulfate						
Cadmium	3.92	6.44				
Copper(I)	10.27	12.22	13.84			
Iron(III)	2.10					
Lead		5.13	6.35			
Mercury(II)		29.44	31.90	33.24		
Silver	8.82	13.46				

Table 5-15
CUMULATIVE FORMATION CONSTANTS FOR METAL COMPLEXES
WITH ORGANIC LIGANDS

Temperature is 25°C, and ionic strengths are approaching zero unless indicated otherwise: (*a*) At 20°C, (*b*) at 30°C, (*c*) 0.1*M* uni-univalent salt, (*d*) 1.0*M* uni-univalent salt, (*e*) 2.0*M* uni-univalent salt present.

	$\log K_1$	$\log K_2$	$\log K_3$	$\log K_4$
Acetate				
Ag(I)	0.73	0.64		
Ba(II)	0.41			
Ca(II)	0.6			
Cd(II)	1.5	2.3	2.4	
Ce(III)	1.68	2.69	3.13	3.18
Co(II)	1.5	1.9		
Cr(III)	1.80	4.72		
Cu(II) *a*	2.16	3.20		
Fe(II) *c*	3.2	6.1	8.3	
Fe(III) *a,d*	3.2			
In(III)	3.50	5.95	7.90	9.08
Hg(II)		8.43		
La(III) *a,e*	1.56	2.48	2.98	2.95
Mg(II)	0.8			
Mn(II)	9.84	2.06		
Ni(II)	1.12	1.81		
Pb(II)	2.52	4.0	6.4	8.5
Rare earths *a,e*	1.6–1.9	2.8–3.0	3.3–3.7	
Sr(II)	0.44			
Tl(III)				15.4
UO$_2$(II) *a,e*	2.38	4.36	6.34	
Y(III) *a,e*	1.53	2.65	3.38	
Zn(II)	1.5			
Acetylacetone				
Al(III) *b*	8.6	15.5		
Be(II)	7.8	14.5		
Cd(II)	3.84	6.66		
Ce(III)	5.30	9.27	12.65	
Cr(II)	5.9	11.7		
Co(II)	5.40	9.54		
Cu(II)	8.27	16.34		
Dy(III) *b*	6.03	10.70	14.04	
Er(III) *b*	5.99	10.67	14.09	
Eu(III) *b*	5.87	10.35	13.64	
Fe(II)	5.07	8.67		
Fe(III)	11.4	22.1	26.7	
Ga(III)	9.5	17.9	23.6	
Gd(III) *b*	5.90	10.38	13.79	
Hf(IV)	8.7	15.4	21.8	28.1
Ho(III)	6.05	10.73	14.13	
In(III)	8.0	15.1		
La(III) *b*	5.1	8.90	11.90	
Lu(III) *b*	6.23	11.00	13.63	
Mg(II)	3.65	6.27		
Mn(II)	4.24	7.35		
Mn(III)			3.86	
Nd(III)	5.6	9.9	13.1	
Ni(II) *a*	6.06	10.77	13.09	
Pd(II) *b*	16.2	27.1		
Pr(III) *b*	6.4	9.5	12.5	
Pu(IV) *c*	10.5	19.7	28.1	34.1
Sc(III) *b*	8.0	15.2		

Table 5-15 (*Continued*)
CUMULATIVE FORMATION CONSTANTS FOR METAL COMPLEXES WITH ORGANIC LIGANDS

	$\log K_1$	$\log K_2$	$\log K_3$	$\log K_4$
Sm(III) b	5.9	10.4		
Tb(III) b	6.02	10.63	14.04	
Th(IV)	8.8	16.2	22.5	26.7
Tm(IV) b	6.09	10.85	14.33	
U(IV) a,c	8.6	17.0	23.4	29.5
UO$_2$(II) b	7.74	14.19		
VO(II)	8.68	15.79		
V(II)	5.4	10.2	14.7	
Y(III) b	6.4	11.1	13.9	
Yb(III) b	6.18	11.04	13.64	
Zn(II) b	4.98	8.81		
Zr(IV)	8.4	16.0	23.2	30.1
Alizarin red				
Cr(VI)	4.7			
Cu(II)	4.1			
Hf(IV)		10.4		
Mo(VI)		9.6		
Pb(II)	6.0			
Th(IV)		8.24		
UO$_2$(II)	4.22			
V(V)		8.6		
W(VI)		7.8		
Aurintricarboxylic acid				
Be(II)	4.54			
Cu(II)	4.1	8.81		
Fe(III)	4.68			
Th(IV)	5.04			
UO$_2$(II)	4.77			
Arsenazo				
Hf(IV)	10.07			
Zr(IV)	12.95			
Benzoylacetone (75% dioxane)				
Ba(II)		9.4		
Be(II)	12.59	24.01		
Cd(II)	7.79	14.36		
Ce(III)	10.09	19.42	27.04	
Co(II)	9.42	17.83		
Cu(II)	12.05	23.01		
La(III)	6.33	11.66	16.78	
Mg(II)	7.69	14.09		
Mn(II)	8.66	15.78		
Ni(II)	9.58	18.00		
Pb(II)	8.84	16.35		
Pr(III)	7.02	13.62	18.74	
UO$_2$(II)	12.15	23.27		
Y(III)	8.24	14.98	20.57	
Zn(II)	9.62	17.90		
Calmagite				
Ca	6.05			
Mg	8.05			

Table 5-15 (*Continued*)
CUMULATIVE FORMATION CONSTANTS FOR METAL COMPLEXES WITH ORGANIC LIGANDS

	Complex of HL^{2-} Anion		Complex of L^{3-} Anion		Complex of H_2L^-
	$\log K_1$	$\log K_2$	$\log K_1$	$\log K_2$	$\log K_3$
Citric acid					
Ag	7.1				
Al	7.0		20.0		
Ba	2.98				
Be	4.52				
Ca	4.68				
Cd	3.98		11.3		
Ce(III)		6.18		9.65	3.2
Co(II)	4.8		12.5		
Cu(II)	4.35		14.2		
Eu(III)		6.46		9.80	
Fe(II)	3.08		15.5		
Fe(III)	12.5		25.0		
La		6.97		9.45	6.22
Mg	3.29				
Mn(II)	3.67				
Nd(III)		6.32		9.70	
Ni	5.11		14.3		
Pb	6.50				
Pr					3.4
Ra	2.36				
Sr	2.8				
Tl(I)	1.04				
UO_2	8.5	10.8			
Y					3.6
Yb				8	
Zn	4.71		11.4		

	$\log K_1$	$\log K_2$	$\log K_3$	
1,2-Diaminocyclohexane-*N,N,N',N'*-tetraacetic acid				
Al *c*	17.63			
Ba *c*	8.64			
Ca *c*	12.3			
Cd *c*	19.88			
Ce(III) *c*	16.76			
Co(II) *c*	19.57			
Cu(II) *c*	21.95			
Dy(III) *c*	19.69			
Er(III) *c*	20.20			
Eu(III) *c*	18.77			
Fe(III) *c*	27.48			
Ga *c*	22.91			
Gd *c*	18.80			
Hg(II) *c*	24.4			
Ho *c*	19.89			
La *c*	16.35			
Lu *c*	21.51			
Mg *c*	10.41			
Mn(II) *c*	17.43			
Nd *c*	17.69			
Ni *c*	19.4			
Pb *c*	20.33			
Pr *c*	17.23			

Table 5-15 (*Continued*)
CUMULATIVE FORMATION CONSTANTS FOR METAL COMPLEXES WITH ORGANIC LIGANDS

	$\log K_1$	$\log K_2$	$\log K_3$	$\log K_4$
Sm(III) *c*	18.63			
Sr *c*	8.92			
Tb *c*	19.30			
Tm *c*	20.46			
VO(II) *c*	19.40			
Y *c*	19.41			
Yb *c*	20.80			
Zn *c*	18.6			
Dibenzoylmethane (75% dioxane)				
Ba	6.10	11.50		
Be	13.62	26.03		
Ca	7.17	13.55		
Cd	8.67	16.63		
Ce(III)	10.99	21.53	30.38	
Co(II)	10.35	20.05		
Cu(II)	12.98	24.98		
Cs	3.42			
Fe(II)	11.15	21.50		
K	3.67			
Li	5.95			
Mg	8.54	16.21		
Mn(II)	9.32	17.79		
Na	4.18			
Ni	10.83	20.72		
Pb	9.75	18.79		
Rb	3.52			
Sr	6.40	12.10		
Zn	10.23	19.65		

	$\log K_1$	$\log K_2$	$\log K_3$	$\log K_f$ [MHL]
4,5-Dihydroxybenzene-1,3-disulfonic acid (Tiron)				
Al	19.02	31.10	33.5	
Ba	4.10			14.6
Ca	5.80			14.8
Cd *d*	7.69	13.29		
Ce(III)		3.75		
Co(II) *d*	8.19	14.41		15.7
Cu(II) *d*	12.76	23.73		18.1
Fe(III) *a,c*	20.7	35.9	46.9	22.6
La	12.9			18.6 [La(OH)L]
Mg *a,c*	6.86			14.6
Mn(II) *c*	8.6			
Ni *a,c*	8.56	14.90		15.6
Pb *d*	11.95	18.28		
Sr *c*	4.55			
UO$_2$(II) *c*	15.90			
VO(II)	15.88			
Zn *d*	9.00	16.91		15.9

	$\log K_1$	$\log K_2$	$\log K_f$ [M$_2$L$_3$]
2,3-Dimercaptopropan-1-ol (BAL)			
Fe(II)	15.8		
Fe(III)	30.6 [Fe(OH)L]		28
Mn(II)	5.23	10.43	
Ni		22.78	
Zn	13.48	23.3	40.6

Table 5-15 (*Continued*)
CUMULATIVE FORMATION CONSTANTS FOR METAL COMPLEXES WITH ORGANIC LIGANDS

	$\log K_1$	$\log K_2$	$\log K_3$	$\log K_4$
Dimethylglyoxime (50% dioxane)				
Cd	5.7	10.7		
Co(II)	9.80	18.94		
Cu(II)	12.00	33.44		
Fe(II)		7.25		
La	6.6	12.5		
Ni	11.16			
Pb	7.3			
Zn	7.7	13.9		
2,2′-Dipyridyl				
Ag	3.65	7.15		
Cd	4.26	7.81	10.47	
Co(II)	5.73	11.57	17.59	
Cr(II)	4.5	10.5	14.0	
Cu(I)		14.2		
Cu(II)	8.0	13.60	17.08	
Fe(II)	4.36	8.0	17.45	
Hg(II)	9.64	16.74	19.54	
Mg	0.5			
Mn(II) *d*	4.06	7.84	11.47	
Ni	6.80	13.26	18.46	
Pb	3.0			
Ti(III)			25.28	
V(II)	4.9	9.6	13.1	
Zn	5.30	9.83	13.63	
Eriochrome Black T				
Ca	5.4			
Mg	7.0			
Zn	13.5	20.6		
Ethanolamine				
Ag	3.29	6.92		
Cu(II)		6.68		16.48
Hg(II)	8.51	17.32		
Ethylenediamine				
Ag	4.70	7.70		
Cd *a*	5.47	10.09	12.09	
Co(II)	5.91	10.64	13.94	
Co(III)	18.7	34.9	48.69	
Cr(II)	5.15	9.19		
Cu(I)		10.8		
Cu(II)	10.67	20.00	21.0	
Fe(II)	4.34	7.65	9.70	
Hg(II)	14.3	23.3		
Mg	0.37			
Mn(II)	2.73	4.79	5.67	
Ni	7.52	13.84	18.33	
Pd(II)		26.90		
V(II)	4.6	7.5	8.8	
Zn	5.77	10.83	14.11	
Ethylenediamine-*N,N,N′,N′*-tetraacetic acid				
Ag	7.32			
Al	16.11			
Am(III)	18.18			
Ba	7.78			
Be	9.3			
Bi	22.8			
Ca	11.0			
Cd	16.4			
Ce(III)	16.80			

Table 5-15 (*Continued*)
CUMULATIVE FORMATION CONSTANTS FOR METAL COMPLEXES WITH ORGANIC LIGANDS

	$\log K_1$	$\log K_2$	$\log K_3$	$\log K_4$
Cf(III)	19.09			
Cm(III)	18.45			
Co(II)	16.31			
Co(III)	36			
Cr(II)	13.6			
Cr(III)	23			
Cu(II)	18.7			
Dy	18.0			
Er	18.15			
Eu(III)	17.99			
Fe(II)	14.33			
Fe(III)	24.23			
Ga	20.25			
Gd	17.2			
Hg(II)	21.80			
Ho	18.1			
In	24.95			
La	16.34			
Li	2.79			
Lu	19.83			
Mg	8.64			
Mn(II)	13.8			
Mo(V)	6.36			
Na	1.66			
Nd	16.6			
Ni	18.56			
Pb	18.3			
Pd(II)	18.5			
Pm(III)	17.45			
Pr	16.55			
Pu(III)	18.12			
Pu(IV)	17.66			
Pu(VI)	17.66			
Ra	7.4			
Sc	23.1			
Sm	16.43			
Sn(II)	22.1			
Sr	8.80			
Tb	17.6			
Th	23.2			
Ti(III)	21.3			
TiO(II)	17.3			
Tl(III)	22.5			
Tm	19.49			
U(IV)	17.50			
V(II)	12.70			
V(III)	25.9			
VO(II)	18.0			
V(V)	18.05			
Y	18.32			
Yb	18.70			
Zn	16.4			
Zr	19.40			
Glycine				
Ag	3.41	6.89		
Ba	0.77			
Be		4.95		

Table 5-15 (*Continued*)
CUMULATIVE FORMATION CONSTANTS FOR METAL COMPLEXES WITH ORGANIC LIGANDS

	$\log K_1$	$\log K_2$	$\log K_3$	$\log K_4$
Ca	1.38			
Cd	4.74	8.60		
Co(II)	5.23	9.25	10.76	
Cu(II)	8.60	15.54	16.27	
Dy		12.2		
Er		12.7		
Fe(II) *a*	4.3	7.8		
Fe(III) *a,d*	10.0			
Gd		11.9		
Hg(II)	10.3	19.2		
La		11.2		
Mg	3.44	6.46		
Mn(II)	3.6	6.6		
Ni	6.18	11.14	15	
Pb	5.47	8.92		
Pd(II)	9.12	17.55		
Pr		11.5		
Sm		11.7		
Sr	0.91			
Y		12.5		
Yb		13.0		
Zn	5.52	9.96		
***N*'-(2-Hydroxyethyl)ethylenediamine-*N,N,N*'-triacetic acid**				
Ba *c*	5.54			
Ca *c*	8.43			
Cd *c*	13.0			
Ce(III) *c*	14.11			
Co(II) *c*	14.4			
Cu(II) *c*	17.40			
Dy *c*	15.30			
Er *c*	15.42			
Eu(III) *c*	15.35			
Fe(II) *c*	11.6			
Fe(III) *c*	19.8			
Gd *c*	15.22			
Hg(II) *c*	20.1			
Ho *c*	15.32			
La *c*	13.46			
Lu *c*	15.88			
Mg *c*	5.78			
Mn(II) *c*	10.7			
Nd *c*	14.86			
Ni *c*	17.0			
Pb *c*	15.5			
Pr *c*	14.61			
Sm *c*	15.28			
Sr *c*	6.92			
Tb *c*	15.32			
Th *c*	18.5			
Tm *c*	15.59			
Y *c*	14.65			
Yb *c*	15.88			
Zn *c*	14.5			
8-Hydroxy-2-methylquinoline (50% dioxane)				
Cd	9.00	9.00	16.60	
Ce(III)	7.71			
Co(II)	9.63	18.50		

Table 5-15 (*Continued*)
CUMULATIVE FORMATION CONSTANTS FOR METAL COMPLEXES WITH ORGANIC LIGANDS

	$\log K_1$	$\log K_2$	$\log K_3$	$\log K_4$
Cu(II)	12.48	24.00		
Fe(II)	8.75	17.10		
Mg	5.24	9.64		
Mn(II)	7.44	13.99		
Ni	9.41	17.76		
Pb	10.30	18.50		
UO$_2$(II)	9.4	17		
Zn	9.82	18.72		
8-Hydroxyquinoline-5-sulfonic acid				
Ba	2.31			
Ca	3.52			
Cd	7.70	14.20		
Ce(III)	6.05	11.05	14.95	
Co(II)	8.11	15.05	20.41	
Cu(II)	11.92	21.87		
Er	7.16	13.34	18.56	
Fe(II)	8.4	15.7	21.75	
Fe(III)	11.6	22.8	35.65	
Gd	6.64	12.37	17.27	
La	5.63	10.13	13.83	
Mg	4.79	8.19		
Mn(II)	5.67	10.72		
Nd	6.3	11.6	16.0	
Ni	9.57	18.27	22.9	
Pb	8.53	16.13		
Pr	6.17	11.37	15.67	
Sm	6.58	12.28	17.04	
Sr	2.75			
Th	9.56	18.29	25.92	32.04
UO$_2$(II)	8.52	15.67		
Zn	8.65	16.15		
Lactic acid				
Ba	0.64			
Ca	1.42			
Cd	1.70			
Ce(III) *a,c*	2.76	4.73	5.96	
Co(II)	1.90			
Cu(II)	3.02	4.85		
Er	2.77	5.11	6.70	
Eu(III)	2.53	4.60	5.88	
Fe(III)	7.1			
Gd	2.53	4.63	5.91	
Ho	2.71	4.97	6.55	
La *a,c*	2.60	4.34	5.64	
Li	0.20			
Mg	1.37			
Mn(II)	1.43			
Nd	2.47	4.37	5.60	
Ni	2.22			
Pb	2.40	3.80		
Pr *a,c*	2.85	4.90	6.10	
Rare earths *a,c*	2.8–3.0	4.9–5.4	6.1–7.8	
Sm	2.56	4.58	5.90	
Sr	0.98			
Tb	2.61	4.73	6.01	
Y	2.53	4.70	6.12	
Yb	2.85	5.27	7.96	
Zn	2.20	3.75		

Table 5-15 (*Continued*)
CUMULATIVE FORMATION CONSTANTS FOR METAL COMPLEXES WITH ORGANIC LIGANDS

	$\log K_1$	$\log K_2$	$\log K_3$	$\log K_4$
Nitrilotriacetic acid				
Al	>10			
Ba *a*	5.88			
Ca	7.60	11.61		
Cd *c*	9.80	15.2		
Ce(III) *c*	10.83	18.67		
Co(II) *c*	10.38	14.5		
Cr(III)	>10			
Cu(II) *c*	13.10			
Dy *c*	11.74	21.15		
Er *c*	12.03	21.29		
Eu(III) *c*	11.52	20.70		
Fe(II) *c*	8.84			
Fe(III) *c*	15.87	24.32		
Gd *c*	11.54	20.80		
Hg(II)	12.7			
Ho *c*	11.90	21.25		
In	15			
La *c*	10.36	17.60		
Li *a*	3.28			
Lu *c*	12.49	21.91		
Mg *c*	5.36	10.2		
Mn(II)	8.60	11.1		
Na	2.15			
Nd *c*	11.26	19.73		
Ni	11.26	16.0		
Pb *a,c*	11.8			
Pr *c*	11.07	19.25		
Sm(III) *c*	11.53	20.53		
Sr	6.73			
Tb *c*	11.59	20.97		
Tl(I)	3.44			
Th *c*	12.4			
Tm *c*	12.22	21.45		
Y *c*	11.48	20.43		
Yb *c*	12.40	21.69		
Zn *c*	10.45	13.45		
Zr *c*	20.8			
1-Nitroso-2-naphthol (75% dioxane)				
Ag	7.74			
Cd	6.18	11.38		
Co(II)	10.67	22.81		
Cu(II)	12.52	23.37		
Mg	6.2	10.60		
Nd	9.5	17.7	25.6	
Ni	10.75	21.29	28.09	
Pb	9.73	17.31		
Pr	9.04	17.06	23.85	
Th *c*	8.50	16.13	24.03	30.29
Y	9.02	17.74	25.04	
Zn	9.32	17.02		
Zr	3.6			
Oxalate				
Ag	2.41			
Al	7.26	13.0	16.3	
Am(III)		9.8		[Am(HL)$_4^-$ 11.0]
Ba	2.31			
Be	4.90			

Table 5-15 (*Continued*)
CUMULATIVE FORMATION CONSTANTS FOR METAL COMPLEXES
WITH ORGANIC LIGANDS

	$\log K_1$	$\log K_2$	$\log K_3$	$\log K_4$
Ca	3.0			
Cd	3.52	5.77		
Ce(III)	6.52	10.5	11.3	
Co(II)	4.79	6.7	9.7	
Co(III)			~20	
Cu(II)	6.16	8.5		
Er	4.82	8.21	10.03	
Fe(II)	2.9	4.52	5.22	
Fe(III)	9.4	16.2	20.2	
Gd	7.04			
Hg(II)		6.98		
Mg	3.43	4.38		
Mn(II)	3.97	5.80		
Mn(III) e	9.98	16.57	19.42	
Mo(III)	3.38			
Mo(VI)				[$MoO_3(L)^{2-}$ 13.0]
Nd	7.21	11.5	>14	
Ni	5.3	7.64	~8.5	
NpO_2(II)	3.30	7.07		
Pb		6.54		
Pu(III)	9.31	18.70	28	
Pu(IV)	8.74	16.91	23.39	27.50
PuO_2(II)		11.4		
Sr	2.54			
Th				24.48
TiO(II)	2.67			
Tl(I)	2.03			
UO_2(II)		10.57		
VO(II)		9.80		
V(II)	~2.7			
Y	6.52	10.10	11.47	
Yb	7.30	11.7	>14	
Zn	4.89	7.60	8.15	
Zr	9.80	17.14	20.86	21.15
1,10-Phenanthroline				
Ag	5.02	12.07		
Ca	0.7			
Cd	5.93	10.53	14.31	
Co(II)	7.25	13.95	19.90	
Cu(II)	9.08	15.76	20.94	
Fe(II)	5.85	11.45	21.3	
Fe(III)	6.5	11.4	23.5	
Hg(II)		19.65	23.35	
Mg	1.2			
Mn(II)	3.88	7.04	10.11	
Ni	8.80	17.10	24.80	
Pb	4.65	7.5	9	
VO(II)	5.47	9.69		
Zn	6.55	12.35	17.55	
Phthalic acid				
Ba	2.33			
Ca	2.43			
Cd	2.5			
Co(II)	1.81	4.51		
Cu(II)	3.46	4.83		
La		7.74		
Ni	2.14			
Pb d	3.4			

Table 5-15 (*Continued*)
CUMULATIVE FORMATION CONSTANTS FOR METAL COMPLEXES WITH ORGANIC LIGANDS

	$\log K_1$	$\log K_2$	$\log K_3$	$\log K_4$
UO$_2$(II)	4.38			
Zn	2.2			
Piperidine				
Ag	3.30	6.48		
Hg(II)	8.70	17.44		
Pt(II)			$\log K_5$ 5.7	$\log K_6$ 8.2
Propylene-1,2-diamine				
Cd *b,c*	5.42	9.97	12.12	
Co(II) *d*	6.41	11.47	14.72	
Cu(II) *c*	10.78	20.06		
Hg(II) *c*		23.53	23.25	
Ni *d*	7.43	13.62	17.89	
Zn *b,c*	5.89	10.87	12.57	
Pyridine				
Ag	1.97	4.35		
Cd	1.40	1.95	2.27	2.50
Co(II)	1.14	1.54		
Cu(I)		3.34	4.51	5.44
				$\log K_6$ 6.89
Cu(II)	2.59	4.33	5.93	6.54
			$\log K_5$ 7.00	$\log K_6$ 10.2
Fe(II)	0.71			
Hg(II)	5.1	10.0	10.4	
Mn(II)	1.92	2.77	3.37	3.50
VO(II)	−1.70			
Zn	1.41	1.11	1.61	1.93
Pyridine-2,6-dicarboxylic acid				
Ba *a,d*	3.46			
Ca *a,d*	4.6	7.2		
Cd *a,d*	5.7	10.0		
Ce(III) *a,d*	8.34	14.42	18.80	
Co(II) *a,d*	7.0	12.5		
Cu(II) *a,d*	9.14	16.52		
Dy *a,d*	8.69	16.19	22.14	
Er *a,d*	8.77	16.39	22.14	
Eu(III) *a,d*	8.84	15.98	21.00	
Fe(II) *a,d*	5.71	10.36		
Fe(III) *a,d*	10.91	17.13		
Gd *a,d*	8.74	16.06	21.83	
Ho *a,d*	8.72	16.23	22.08	
La *a,d*	7.98	13.79	18.06	
Lu *a,d*	9.03	16.80	21.48	
Hg(II) *a,d*	20.28			
Mg *a,d*	2.7			
Mn(II) *a,d*	5.01	8.49		
Nd *a,d*	8.78	15.60	20.66	
Ni *a,d*	6.95	13.50		
Pb *a,d*	8.70	10.60		
Pr *a,d*	8.63	15.10	19.94	
Sm *a,d*	8.86	15.88	21.23	
Sr *a,d*	3.89			
Tb *a,d*	8.68	16.11	22.03	
Tm *a,d*	8.83	16.54	22.04	
Y *a,d*	8.46	15.73	21.34	
Yb *a,d*	8.85	16.61	21.83	
Zn *a,d*	6.35	11.88		

Table 5-15 (*Continued*)
CUMULATIVE FORMATION CONSTANTS FOR METAL COMPLEXES WITH ORGANIC LIGANDS

	$\log K_1$	$\log K_2$	$\log K_3$	$\log K_4$
1-(2-Pyridylazo)-2-naphthol (PAN)				
Co(II)	>12			
Cu(II)	16			
Mn(II)	8.5	16.4		
Ni	12.7	25.3		
Tl(III)	2.29			
Zn	11.2	21.7		

	$\log K_f$ [ML]	$\log K_f$ [MHL]	$\log K_f$ [M(HL)$_2$]
4-(2-Pyridylazo)resorcinal (PAR)			
Co(II)		>12	
Cu(II)	10.3		
Mn(II)		9.7	18.9
Ni		13.2	26.0
Sc	4.8		
Tl(III)	4.23		
Zn		12.4	23.5

	$\log K_f$ [ML]	$\log K_f$ [M$_2$L]	$\log K_f$ [MHL]
Pyrocatechol-3,5-disulfonate (Pyrocatechol Violet)			
Al	19.13	4.95	
Bi	27.07	5.25	
Cd	8.13		5.86
Co(II)	9.01		6.53
Cu(II)	16.47		11.18
Ga	22.18	4.65	
In	18.10	4.81	
Mg	4.42	4.6	3.66
Mn(II)	7.13		5.36
Ni	9.35	4.38	6.85
Pb	13.25		10.19
Th	23.36	4.42	
Zn	10.41	6.21	7.21
Zr	27.40	4.18	

	$\log K_1$	$\log K_2$	$\log K_3$	$\log K_4$
8-Quinolinol				
Ba	2.07			
Be	3.36			
Ca (75% dioxane)	7.3	13.2		
Cd	7.2	13.4		
Ce(III) (50% dioxane)	9.15	17.13		
Co(II)	9.1	17.2		
Cu(II)	12.2	23.4		
Fe(II)	8.58	16.93	22.23	
Fe(III)	12.3	23.6	33.9	
La	5.85	16.95		
Mg (50% dioxane)	6.38	11.81		
Mn(II) (50% dioxane)	8.28	15.45		
Ni (50% dioxane)	11.44	21.38		
Pb (50% dioxane)	10.61	18.70		

Table 5-15 (*Continued*)
CUMULATIVE FORMATION CONSTANTS FOR METAL COMPLEXES WITH ORGANIC LIGANDS

	$\log K_1$	$\log K_2$	$\log K_3$	$\log K_4$
Sm	6.84		19.50	
Sr	2.89	6.08		
Th	10.45	20.40	29.85	38.80
UO$_2$(II) (50% dioxane)	11.25	20.89		
V(II)	12.8	23.6		
VO(II)	10.97	20.19		
Y	8.15	14.90	20.25	
Zn (50% dioxane)	9.96	18.86		

	$\log K_f$ [MHL$^+$]	$\log K_f$ [M(HL)$_2$]
Salicylaldoxime		
Ba	0.53	3.72
Be	<7	
Ca	0.92	3.72
Cd	<4.4	
Co(II)		8.13
Cu(II)		8.13
Mg	0.64	4.10
Ni		3.77
Sr		3.77
Zn	<5.2	

	$\log K_1$	$\log K_2$	$\log K_3$	$\log K_4$
Salicylic acid				
Al	14.11			
Be	17.4			
Cd	5.55			
Ce(III)	2.66			
Co(II)	6.72	11.42		
Cr(II)	8.4	15.3		
Cu(II)	10.60	18.45		
Fe(II)	6.55	11.25		
Fe(III) *a,c*	16.48	28.12	36.80	
La	2.64			
Mg (75% dioxane)	4.7			
Mn(II)	5.90	9.80		
Nd	2.70			
Ni	6.95	11.75		
Pr	2.68			
Th	4.25	7.60	10.05	11.60
TiO(II)	6.09			
UO$_2$(II)	13.4			
V(II)	6.3			
Zn	6.85			
Succinic acid				
Ba	2.08			
Be	3.08			
Ca	2.0			
Cd	2.2			
Co(II)	2.22			
Cu(II)	3.33			
Fe(III)	7.49			
Hg(II)		7.28		
La	3.96			

Table 5-15 (*Continued*)
CUMULATIVE FORMATION CONSTANTS FOR METAL COMPLEXES WITH ORGANIC LIGANDS

	$\log K_1$	$\log K_2$	$\log K_3$	$\log K_4$
Mg	1.20			
Mn(II)	2.26			
Nd	8.1			
Ni	2.36			
Pb	2.8			
Ra	1.0			
Sr	1.06			
Zn	1.6			
5-Sulfosalicylic acid				
Al *c*	13.20	22.83	28.89	
Be *c*	11.71	20.81		
Cd *c*	16.68	29.08		
Co(II) *c*	6.13	9.82		
Cr(II) *c*	7.1	12.9		
Cr(III) *c*	9.56			
Cu(II) *c*	9.52	16.45		
Fe(II) *c*	5.90			
Fe(III) *c*	14.64	25.18	32.12	
La *c*	9.11			
Mn(II) *c*	5.24	8.24		
NbO(III) *c*	4.0	7.7		
Ni *c*	6.42	10.24		
UO_2(II) *c*	11.14	19.20		
Zn *c*	6.05	10.65		
Tartaric acid				
Ba		1.62		
Bi			8.30	
Ca	2.98	9.01		
Cd	2.8			
Co(II)	2.1			
Cu(II)	3.2	5.11	4.78	6.51
				$\log K_f$ 19.14 [Cu(OH)$_2$L^{2-}]
Eu(III)	4.98	8.11		
Fe(III)	7.49			
La	3.06			
Mg		1.36		
Nd	9.0			
Pb	3.78		4.7	$\log K_f$ 14.1 [Pb(OH)$_2$L^{2-}]
Ra	1.24			
Sr	1.60			
Zn	2.68	8.32		
Thioglycolic acid				
Ce(III) *a,c*	1.99	3.03		
Co(II)	5.84	12.15		
Fe(II)		10.92		
Hg(II)		43.82		
La *a,c*	1.98	2.98		
Mn(II)	4.38	7.56		
Pb	8.5			
Ni	6.98	13.53		
Rare earths *a,c*	1.9–2.1	3.0–3.3		
Y *a,c*	1.91	3.19		
Zn	7.86	15.04		
Thiourea				
Ag	7.4	13.1		
Bi				$\log K_6$ 11.9
Cd	0.6	1.6	2.6	4.6

Table 5-15 (*Continued*)
CUMULATIVE FORMATION CONSTANTS FOR METAL COMPLEXES WITH ORGANIC LIGANDS

	$\log K_1$	$\log K_2$	$\log K_3$	$\log K_4$
Cu(I)			13	15.4
Hg(II)		22.1	24.7	26.8
Pb	1.4	3.1	4.7	8.3
Ru(III)	1.21		0.72	
Thoron				
Th		10.15		
Triethanolamine				
Ag	2.30	3.64		
Co(II)	1.73			
Cu(II)	4.30			
Hg(II)	6.90	13.08		
Ni	2.7			
Zn	2.00			
Triethylenetetramine (Trien)				
Ag	7.7			
Cd	10.75	13.9		
Co(II)	11.0			
Cu(II)	20.4			
Fe(II)	7.8			
Fe(III)	21.9			
Hg(II)	25.26			
Mn(II)	4.9			
Ni	14.0			
Pb	10.4			
Zn	11.9			
1,1,1-Trifluoro-3-2'-Thenoylacetone (TTA)				
Ba		10.6		
Cu(II)	6.55	13.0		
Fe(III)	6.9			
Ni	10.0			
Pr	9.53			
Pu(III)	9.53			
Pu(IV)	8.0			
Th	8.1			
U(IV)	7.2			
Zr	3.03 [as ZrL^{3+}]			
Xylenol orange				
Bi	5.52			
Fe(III)	5.70			
Hf	6.50			
Tl(III)	4.90			
Zn	6.15			
Zr	7.60			
Zincon				
Zn	13.1			

MEASUREMENT OF pH

ELECTROMETRIC MEASUREMENT OF pH

The pH value is defined for an aqueous solution in an operational (arbitrary but reproducible) manner according to the Bates-Guggenheim convention:

$$pH_x = pH_s + \frac{E_x - E_s}{2.3026RT/F}$$

where R is the gas constant per mole, T is the temperature on the absolute scale, and F is the faraday. The pH_x of the unknown medium is calculated from that of an accepted standard (pH_s) and the measured difference in the emf (E) of the electrode combination when the standard solution is removed from the cell and replaced by the unknown. The double vertical line marks a liquid junction. Electrodes as fabricated exhibit variations in the reproducibility of the reference electrode, in the liquid-junction potential, and, with glass electrodes, in the asymmetry potential. These differences are all eliminated in the standardizing procedure with standard reference pH buffers. (See R. G. Bates, *Determination of pH, Theory and Practice*, Wiley, New York, 1964.)

Electrode reversible to hydrogen ions	Standard reference buffer or unknown solution	Salt bridge (KCl, 3.5M or saturated)	Reference electrode

An electrometric pH-measurement system consists of (1) pH-responsive electrode, (2) reference electrode, and (3) potential-measuring device—some form of high-impedance electronic voltmeter for glass-electrode combinations and this or a potentiometer arrangement for other pH-responsive electrodes. Electronic pH meters are simply voltmeters with scale divisions in pH units which are equivalent to the values of 2.3026RT/F (in mV) per pH unit. Values of this function at several temperatures are given in Table 5-16. There is no compensation incorporated in the meter for the changes in pH of the test solution as a function of temperature. Reliability of an indicator–reference electrode combination must be ascertained by standardization of the pH meter with one standard buffer and checking the pH response by immersing the combination in a second and different reference buffer.

The temperature compensator on a pH meter varies the instrument definition of a pH unit from 54.20 mV at 0° to perhaps 66.10 mV at 60°C. This permits one to measure the pH of the sample (and reference buffer standard) at its actual temperature and thus avoid error due to dissociation equilibria and to junction potentials which have significant temperature coefficients.

Table 5-16
VALUES OF $2.3026RT/F$ AT SEVERAL TEMPERATURES
In Millivolts

$t\,^\circ C$	Value	$t\,^\circ C$	Value	$t\,^\circ C$	Value	$t\,^\circ C$	Value
0	54.197	25	59.157	50	64.118	80	70.070
5	55.189	30	60.149	55	65.110	85	71.062
10	56.181	35	61.141	60	66.102	90	72.054
15	57.173	38	61.737	65	67.094	95	73.046
18	57.767	40	62.133	70	68.086	100	74.038
20	58.165	45	63.126	75	69.078		

Report of the National Academy of Sciences: National Research Council Committee of Fundamental Constants, 1963.

Hydrogen Electrode. In theory the hydrogen gas electrode is the only perfect hydrogen ion electrode, and corrections must be applied for the imperfect response of other electrodes under specified conditions. The potential of a hydrogen electrode varies with the partial pressure of hydrogen gas in equilibrium with the platinum surface, which is in turn dependent on the barometric pressure and the vapor pressure of water. Customarily the observed emf's of cells with hydrogen electrodes are corrected to a partial pressure of hydrogen of 760 mm of Hg. The barometric corrections, in millivolts to be added to the observed reading, are listed in Table 5-17.

Quinhydrone Electrode. The quinhydrone electrode is formed by dipping a bright platinum or gold electrode, at least 1 cm^2 in area, into a solution saturated with quinhydrone (solubility, 3.63 g per liter), an equimolecular mixture of benzoquinone and hydroquinone (dissociation constant, $K_Q = 0.259$ at 25°C). The metal should be in contact with some of the solid quinhydrone in the cell. Commercial quinhydrone should be recrystallized from water at 70°C and dried at room temperature in subdued light. Standard potentials of the quinhydrone electrode are given in Table 5-18.

Antimony Electrode. This metal–metal oxide electrode consists of high-purity electrolytic antimony cast in stick form. An invisible coating of the very slightly soluble oxide always seems to be present as a surface film. Since the electrode potential differs from one electrode to another, and since the slope of the emf-pH curve is not rectilinear, it is necessary to standardize each electrode with solutions of known pH. It is a rugged electrode; accuracy is ±0.2 pH unit. The electrode potential is affected by dissolved oxygen, by oxidizing agents, and by reducing agents, and there is a marked sensitivity to complexing agents, notably fluoride ion, certain amino acids, and the anions of hydroxy acids. Ions of metals more electropositive than antimony in the emf series of the elements cannot be tolerated.

Table 5-17
BAROMETRIC PRESSURE CORRECTIONS FOR THE
HYDROGEN ELECTRODE FROM 0 TO 60°C

The first column gives the total barometric pressure. This includes the partial pressure of water vapor above the solution. It has been assumed, in computing the values in the body of the table, that this is equal to the vapor pressure of pure water at the temperatures specified.

Barometric pressure, mm Hg	Correction, mV., at $t =$								
	0°C	10°C	20°C	25°C	30°C	35°C	40°C	50°C	60°C
720	0.71	0.82	0.99	1.13	1.30	1.52	1.81	2.67	4.12
725	0.63	0.73	0.91	1.03	1.20	1.42	1.71	2.56	3.99
730	0.55	0.65	0.82	0.94	1.11	1.32	1.61	2.45	3.87
735	0.47	0.56	0.73	0.85	1.02	1.23	1.51	2.34	3.74
740	0.39	0.48	0.64	0.76	0.92	1.13	1.41	2.23	3.62
745	0.31	0.39	0.55	0.67	0.83	1.04	1.31	2.12	3.50
750	0.23	0.31	0.47	0.58	0.74	0.94	1.21	2.02	3.38
755	0.15	0.23	0.38	0.49	0.64	0.85	1.12	1.91	3.25
760	0.07	0.15	0.30	0.41	0.56	0.76	1.02	1.81	3.14
765	−0.01	0.07	0.21	0.32	0.47	0.67	0.92	1.70	3.02
770	−0.08	−0.01	0.13	0.24	0.38	0.57	0.83	1.60	2.91

Table 5-18
STANDARD POTENTIALS OF THE QUINHYDRONE ELECTRODE

$t°C$	$E°$, volts	$t°C$	$E°$, volts
0	0.71798	25	0.69976
5	0.71437	30	0.69607
10	0.71073	35	0.69237
15	0.70709	40	0.68865
20	0.70343		

REFERENCE ELECTRODES

To be suitable as a reference electrode, the potential of the half-cell must not be significantly altered if a small current passes through it, the resistance must not be too great, it should be easily assembled, and the components should be stable in contact with the atmosphere. The standard potential of the half-cell includes the liquid junction potential of the cell:

Pt; $H_2(g)$, $H^+(a = 1)$ ∥ KCl solution, reference electrode

Potentials of reference electrodes are given in Table 5-19; the values quoted may vary as much as 2 or 3 mV because of unknown liquid-junction potentials, and the care with which the reference electrodes are prepared and aged.

The calomel electrode is widely used. It comprises a nonattackable element, such as platinum, in contact with mercury and a potassium chloride solution of known concentration that is saturated with respect to mercurous chloride (calomel). Although it is relatively simple to prepare, the rather large temperature coefficient of a saturated calomel electrode (abbreviated SCE) necessitates maintenance of constant temperature in precise work. Electrodes prepared with unsaturated potassium chloride solutions possess smaller temperature coefficients, but these electrodes must be protected from evaporation. Calomel electrodes are not stable above 80°C.

Table 5-19
POTENTIALS OF REFERENCE ELECTRODES IN VOLTS
AS A FUNCTION OF TEMPERATURE
Liquid-junction Potential Included

Temp., °C	0.1M KCl Calomel*	1.0M KCl Calomel*	3.5M KCl Calomel*	Satd. KCl Calomel*	1.0M KCl Ag/AgCl†	1.0M KBr Ag/AgBr‡	1.0M KI Ag/AgI§
0	0.3367	0.2883		0.25918	0.23655	0.08128	−0.14637
5					0.23413	0.07961	−0.14719
10	0.3362	0.2868	0.2556	0.25387	0.23142	0.07773	−0.14822
15	0.3361			0.2511	0.22857	0.07572	−0.14942
20	0.3358	0.2844	0.2520	0.24775	0.22557	0.07349	−0.15081
25	0.3356	0.2830	0.2501	0.24453	0.22234	0.07106	−0.15244
30	0.3354	0.2815	0.2481	0.24118	0.21904	0.06856	−0.15405
35	0.3351			0.2376	0.21565	0.06585	−0.15590
38	0.3350		0.2448	0.2355			
40	0.3345	0.2782	0.2439	0.23449	0.21208	0.06310	−0.15788
45					0.20835	0.06012	−0.15998
50	0.3315	0.2745		0.22737	0.20449	0.05704	−0.16219
55					0.20056		
60	0.3248	0.2702		0.2235	0.19649		
70					0.18782		
80				0.2083	0.1787		
90					0.1695	0.0251	

*Bates et al., *J. Research Natl. Bur. Standards,* **45,** 418 (1950).
†Bates and Bower, *J. Research Natl. Bur. Standards,* **53,** 283 (1954).
‡Hetzer, Robinson and Bates, *J. Phys. Chem.,* **66,** 1423 (1962).
§Hetzer, Robinson and Bates, *J. Phys. Chem.,* **68,** 1929 (1964).

The silver halide electrode consists of metallic silver coated with the silver halide, and immersed in an appropriate halide solution of known concentration. These electrodes find use in systems where the presence of mercury cannot be tolerated, and at temperatures exceeding 80°C. Values at elevated temperatures follow:

Temp., °C	125	150	175	200	225	250	275
1.0M KCl Ag/AgCl*	0.1130	0.1032	0.0708	0.0348	−0.0051	−0.054	−0.090
1.0M KBr Ag/AgBr†	−0.0048	−0.0312	−0.0612	−0.0951			

* Greeley et al., *J. Phys. Chem.*, **64,** 652 (1960).
† Towns et al., *J. Phys. Chem.*, **64,** 1861 (1960).

The values of several additional reference electrodes at 25°C are listed:

Ag/AgCl, satd. KCl	0.198
Ag/AgCl, 0.1M KCl	0.288
Hg/HgO, 1.0M NaOH	0.140
Hg/HgO, 0.1M NaOH	0.165
Hg/Hg$_2$SO$_4$, satd. K$_2$SO$_4$ (22°C)	0.658
Hg/Hg$_2$SO$_4$, satd. KCl	0.655

STANDARD REFERENCE pH BUFFER SOLUTIONS

The assigned values of pH$_s$, according to the Bates-Guggenheim convention [*Pure Applied Chem.* **1,** 163 (1960)], for the primary standard solutions prepared from salts issued by the National Bureau of Standards (U.S.) are given in Table 5-20. These are smoothed values. The ionic strength of these reference solutions is 0.1 or less. Strictly speaking the NBS scale uses a molality concentration system; however, values are given in molarity units for convenience.

As a result of a variable liquid-junction potential, the measured pH may be expected to differ seriously from the pa$_H$ determined from cells without a liquid junction in solutions of high acidity or high alkalinity. Merely to affirm the proper functioning of the glass electrode at the extreme ends of the pH scale, two secondary standards are included in Table 5-20. In addition, values for a 0.1m solution of HCl are given to extend the pH scale up to 275°C [see R. S. Greeley, *Anal. Chem.* **32,** 1717 (1960)]:

t, °C:	25	60	90	125	150	175	200	225–275
pH:	1.10	1.11	1.12	1.13	1.14	1.15	1.16	1.2

Uncertainties in the values are ±0.03 pH unit from 25 to 90°C, ±0.05 pH unit from 125 to 200°C, and ±0.1 pH unit from 225 to 275°C.

Table 5-20
NATIONAL BUREAU OF STANDARDS (U.S.) REFERENCE pH BUFFER SOLUTIONS

Temperature, °C	Secondary Standard, 0.05M K tetroxalate	KH Tartrate (satd. at 25°C)	0.05M KH$_2$ Citrate	0.05M KH Phthalate	0.025M KH$_2$PO$_4$, 0.025M Na$_2$HPO$_4$	0.0087M KH$_2$PO$_4$, 0.0302M Na$_2$HPO$_4$	0.01M Na$_2$B$_4$O$_7$	0.025M NaHCO$_3$, 0.025M Na$_2$CO$_3$	Secondary Standard, Ca(OH)$_2$ (satd. at 25°C)
0	1.666		3.860	4.003	6.984	7.534	9.464	10.317	13.423
5	1.668		3.840	3.999	6.951	7.500	9.395	10.245	13.207
10	1.670		3.820	3.998	6.923	7.472	9.332	10.179	13.003
15	1.672		3.802	3.999	6.900	7.448	9.276	10.118	12.810
20	1.675		3.788	4.002	6.881	7.429	9.225	10.062	12.627
25	1.679	3.557	3.776	4.008	6.865	7.413	9.180	10.012	12.454
30	1.683	3.552	3.766	4.015	6.853	7.400	9.139	9.966	12.289
35	1.688	3.549	3.759	4.024	6.844	7.389	9.102	9.925	12.133
38	1.691	3.548		4.030	6.840	7.384	9.081		12.043
40	1.694	3.547	3.753	4.035	6.838	7.380	9.068	9.889	11.984
45	1.700	3.547		4.047	6.834	7.373	9.038		11.841
50	1.707	3.549	3.749	4.060	6.833	7.367	9.011	9.828	11.705
55	1.715	3.554		4.075	6.834		8.985		11.574
60	1.723	3.560		4.091	6.836		8.962		11.449
70	1.743	3.580		4.126	6.845		8.921		
80	1.766	3.609		4.164	6.859		8.885		
90	1.792	3.650		4.205	6.877		8.850		
95	1.806	3.674		4.227	6.886		8.833		
Dilution value, $\Delta pH_{1/2}$	+0.186	+0.049	0.024	+0.052	+0.080	+0.07	+0.01	0.079	−0.28

From R. G. Bates, J. Research Natl. Bur. Standards (U.S.), **66A**, 179 (1962) and B. R. Staples and R. G. Bates, ibid., **73A**, 37 (1969).

The buffer value for the NBS reference pH buffer solutions is given below:

	KH Tartrate	0.05M KH$_2$ Citrate	0.05M KH Phthalate	0.025M KH$_2$PO$_4$, 0.025M Na$_2$HPO$_4$	0.0087M KH$_2$PO$_4$, 0.0302M Na$_2$HPO$_4$	0.01M Na$_2$B$_4$O$_7$	0.025M NaHCO$_3$, 0.025M Na$_2$CO$_3$
Buffer value, β	0.027	0.034	0.016	0.029	0.016	0.020	0.029

For the secondary pH reference standards, the buffer value is 0.070 for potassium tetroxalate and 0.09 for calcium hydroxide.

To prepare the standard pH buffer solutions recommended by the National Bureau of Standards (U.S.), the indicated weights of the pure materials in Table 5-21 should be dissolved in water of specific conductivity not greater than 5 micromhos. The tartrate, phthalate, and phosphates can be dried for 2 hours at 110°C before use. Potassium tetroxalate and calcium hydroxide need not be dried. Fresh-looking crystals of borax should be used. Before use, excess solid potassium hydrogen tartrate and calcium hydroxide must be removed. Buffer solutions pH 6 or above should be stored in plastic containers and should be protected from carbon dioxide with soda-lime traps. The solutions should be replaced within two to three weeks, or sooner if formation of mold is noticed. A crystal of thymol may be added as preservative.

The British Standards Institution scale makes no direct attempt to assign a pH to any buffer solution. A primary standard, consisting of a 0.05M potassium hydrogen phthalate solution, is defined as having a pH of exactly 4.000 at 15°C; all other standard buffer solutions are referred to this (Table 5-22). The defining equation for the pH of the phthalate standard at a series of temperatures is

$$pH = 4.000 + 0.5\left(\frac{t-15}{100}\right)^2$$

for temperatures between 0 and 55°C, and by

$$pH = 4.000 + 0.5\left(\frac{t-15}{100}\right)^2 - \frac{t-55}{500}$$

for temperatures between 55 and 95°C, where t is the temperature in degrees centigrade.

The preparation of the buffer solutions of Clark and Lubs plus supplementary systems is given in Table 5-23; these solutions offer a means of preparing aqueous buffers in 0.2 pH intervals.

Table 5-21
COMPOSITIONS OF STANDARD pH BUFFER SOLUTIONS, NATIONAL BUREAU OF STANDARDS (U.S.)
Air Weight of Material per Liter of Buffer Solution

Standard	Weight, g
$KH_3(C_2O_4)_2 \cdot 2H_2O$, 0.05$M$	12.61
Potassium hydrogen tartrate, about 0.034M	Saturated at 25°C
Potassium hydrogen phthalate, 0.05M	10.12
Phosphate:	
$\quad KH_2PO_4$, 0.025M	3.39
$\quad Na_2HPO_4$, 0.025M	3.53
Phosphate:	
$\quad KH_2PO_4$, 0.008665M	1.179
$\quad Na_2HPO_4$, 0.03032M	4.30
$Na_2B_4O_7 \cdot 10H_2O$, 0.01M	3.80
Carbonate:	
$\quad NaHCO_3$, 0.025M	2.10
$\quad Na_2CO_3$, 0.025M	2.65
$Ca(OH)_2$, about 0.0203M	Saturated at 25°C

Table 5-22
pH VALUES OF BUFFERS BY THE BRITISH STANDARDS METHOD

The 0.05M KH phthalate solution is defined as having a pH of exactly 4.000 at 15°C, and all other standards are referred to this.

Temperature, °C	0.05M K tetroxalate	KH Tartrate (satd. at 25°C)	0.05M KH Phthalate	0.1M HOAc, 0.1M NaOAc	0.025M KH_2PO_4, 0.025M Na_2HPO_4	0.01M Borax	0.025M $NaHCO_3$, 0.025M Na_2CO_3
0	1.639		4.011	4.684	6.973	9.464	10.284
5	1.642		4.005	4.665	6.953	9.395	10.220
10	1.643		4.001	4.656	6.916	9.333	10.158
15	1.645		4.000	4.660	6.893	9.277	10.101
20	1.646		4.001	4.646	6.873	9.227	10.046
25	1.647	3.556	4.005	4.644	6.856	9.181	9.995
30	1.648	3.550	4.011	4.644	6.844	9.141	9.948
35	1.651	3.547	4.020	4.648	6.843	9.106	9.906
40	1.653	3.546	4.031	4.655	6.827	9.074	9.869
45	1.658	3.549	4.045	4.663	6.825	9.047	9.837
50	1.665	3.554	4.061	4.674	6.826	9.023	9.811
55	1.672	3.563	4.080	4.689	6.830	9.002	
60	1.672	3.564	4.091	4.695	6.827	8.974	

Table 5-23
COMPOSITION AND pH VALUES OF BUFFER SOLUTIONS

Values based on the conventional activity pH scale as defined by the National Bureau of Standards (U.S.) and pertain to a temperature of 25°C. Ref: Bower and Bates, *J. Research Natl. Bur. Standards* (U.S.), **55**, 197 (1955) and Bates and Bower, *Anal. Chem.*, **28**, 1322 (1956). Buffer value is denoted by column headed β.

25 ml 0.2M KCl + x ml 0.2M HCl, Diluted to 100 ml			50 ml 0.1M KH Phthalate + x ml 0.1M HCl, Diluted to 100 ml			50 ml 0.1M KH Phthalate + x ml 0.1M NaOH, Diluted to 100 ml		
pH	x	β	pH	x	β	pH	x	β
1.00	67.0	0.31	2.20	49.5		4.20	3.0	0.017
1.20	42.5	0.34	2.40	42.2	0.036	4.40	6.6	0.020
1.40	26.6	0.19	2.60	35.4	0.033	4.60	11.1	0.025
1.60	16.2	0.077	2.80	28.9	0.032	4.80	16.5	0.029
1.80	10.2	0.049	3.00	22.3	0.030	5.00	22.6	0.031
2.00	6.5	0.030	3.20	15.7	0.026	5.20	28.8	0.030
2.20	3.9	0.022	3.40	10.4	0.023	5.40	34.1	0.025
			3.60	6.3	0.018	5.60	38.8	0.020
			3.80	2.9	0.015	5.80	42.3	0.015

50 ml 0.1M KH$_2$PO$_4$ + x ml 0.1M NaOH, Diluted to 100 ml			50 ml 0.1M Tris(hydroxymethyl)aminomethane + x ml of 0.1M HCl, Diluted to 100 ml $\Delta pH/\Delta t \simeq -0.028$ $I = 0.001x$			50 ml of a Mixture 0.1M with Respect to Both KCl and H$_3$BO$_3$ + x ml 0.1M NaOH, Diluted to 100 ml		
pH	x	β	pH	x	β	pH	x	β
5.80	3.6		7.00	46.6		8.00	3.9	
6.00	5.6	0.010	7.20	44.7	0.012	8.20	6.0	0.011
6.20	8.1	0.015	7.40	42.0	0.015	8.40	8.6	0.015
6.40	11.6	0.021	7.60	38.5	0.018	8.60	11.8	0.018
6.60	16.4	0.027	7.80	34.5	0.023	8.80	15.8	0.022
6.80	22.4	0.033	8.00	29.2	0.029	9.00	20.8	0.027
7.00	29.1	0.031	8.20	22.9	0.031	9.20	26.4	0.029
7.20	34.7	0.025	8.40	17.2	0.026	9.40	32.1	0.027
7.40	39.1	0.020	8.60	12.4	0.022	9.60	36.9	0.022
7.60	42.4	0.013	8.80	8.5	0.016	9.80	40.6	0.016
7.80	44.5	0.009	9.00	5.7		10.00	43.7	0.014
8.00	46.1					10.20	46.2	

50 ml 0.025M Borax, + x ml 0.1M HCl, Diluted to 100 ml $\Delta pH/\Delta t \simeq -0.008$ $I = 0.025$			50 ml 0.025M Borax, + x ml 0.1M NaOH, Diluted to 100 ml $\Delta pH/\Delta t \simeq -0.008$ $I = 0.001(25 + x)$			50 ml 0.05M NaHCO$_3$ + x ml 0.1M NaOH, Diluted to 100 ml $\Delta pH/\Delta t \simeq -0.009$ $I = 0.001(25 + 2x)$		
pH	x	β	pH	x	β	pH	x	β
8.00	20.5		9.20	0.9		9.60	5.0	
8.20	19.7	0.010	9.40	3.6	0.026	9.80	6.2	0.014
8.40	16.6	0.012	9.60	11.1	0.022	10.00	10.7	0.016
8.60	13.5	0.018	9.80	15.0	0.018	10.20	13.8	0.015

Table 5-23 (*Continued*)
COMPOSITION AND pH VALUES OF BUFFER SOLUTIONS

50 ml 0.025M Borax, + x ml 0.1M HCl, Diluted to 100 ml ΔpH/$\Delta t \simeq -0.008$ $I = 0.025$			50 ml 0.025M Borax + x ml 0.1M NaOH, Diluted to 100 ml ΔpH/$\Delta t \simeq -0.008$ $I = 0.001(25 + x)$			50 ml 0.05M NaHCO$_3$ + x ml 0.1M NaOH, Diluted to 100 ml ΔpH/$\Delta t \simeq -0.009$ $I = 0.001(25 + 2x)$		
pH	x	β	pH	x	β	pH	x	β
8.80	9.4	0.023	10.00	18.3	0.014	10.40	16.5	0.013
9.00	4.6	0.026	10.20	20.5	0.009	10.60	19.1	0.012
9.10	2.0		10.40	22.1	0.007	10.80	21.2	0.009
			10.60	23.3	0.005	11.00	22.7	

50 ml 0.05M Na$_2$HPO$_4$ + x ml 0.1M NaOH, Diluted to 100 ml ΔpH/$\Delta t \simeq -0.025$ $I = 0.001(77 + 2x)$			25 ml 0.2M KCl + x ml 0.2M NaOH, Diluted to 100 ml ΔpH/$\Delta t \simeq -0.033$ $I = 0.001(50 + 2x)$		
pH	x	β	pH	x	β
11.00	4.1	0.009	12.00	6.0	0.028
11.20	6.3	0.012	12.20	10.2	0.048
11.40	9.1	0.017	12.40	16.2	0.076
11.60	13.5	0.026	12.60	25.6	0.12
11.80	19.4	0.034	12.80	41.2	0.21
11.90	23.0	0.037	13.00	66.0	0.30

Standards for pH Measurement of Blood and Biological Media. Blood is a well-buffered medium. In addition to the NBS phosphate standard of 0.025M (pH$_s$ = 6.840 at 38°C), another reference solution containing the same salts, but in the molal ratio 1:4, has an ionic strength of 0.13. It is prepared by dissolving 1.360 g of KH$_2$PO$_4$ and 5.677 g of Na$_2$HPO$_4$ (air weights) in carbon dioxide–free water to make 1 liter of solution. The pH$_s$ is 7.416 $\pm$ 0.004 at 37.5 and 38°C.

The compositions and pH$_s$ values of *tris*(hydroxymethyl)aminomethane, covering the pH range 7.0 to 8.9, are listed in Table 7-23.

The phosphate-succinate system gives the values of pH$_s$ shown below:

$\dfrac{\text{Molality}}{\text{KH}_2\text{PO}_4} = \dfrac{\text{Molality}}{\text{Na}_2\text{HC}_6\text{H}_5\text{O}_7}$	pH$_s$	Δ(pHs)/Δt
0.005	6.251	-0.00086 deg^{-1}
0.010	6.197	-0.00071
0.015	6.162	
0.020	6.131	
0.025	6.109	-0.0004

BUFFER SOLUTIONS OTHER THAN STANDARDS

The range of the buffering effect of a single weak acid group is approximately one pH unit on either side of pK_a. The ranges of some useful buffer systems are collected in Table 5-24. After all the components have been brought together, the pH of the resulting solution should be determined with reference to standard reference solutions at the temperature to be employed. Buffer components should be compatible with other components in the system under study; this is particularly significant for buffers employed in biological studies. Check the tables of formation constants to ascertain whether metal-binding character exists.

When there are two or more acid groups per molecule, or a mixture is composed of several overlapping acids, the useful range is larger. Universal buffer solutions consist of a mixture of acid groups which overlap such that successive pK_a values differ by two pH units or less. The Prideaux-Ward mixture comprises phosphate, phenyl acetate, and borate plus HCl, and covers the range from 2 to 12 pH units. McIlvaine's buffer is a mixture of citric acid and KH_2PO_4 that covers the range from pH 2.2 to 8.0. The Britton-Robinson system consists of acetic acid, phosphoric acid, and boric acid plus NaOH, and covers the range from pH 4.0 to 11.5. A mixture composed of Na_2CO_3, NaH_2PO_4, citric acid, and 2-amino-2-methyl-1,3-propanediol covers the range from pH 2.2 to 11.0.

General directions for the preparation of buffer solutions of varying pH but fixed ionic strength are given by Bates (*Determination of pH, Theory and Practice*, pp. 121–122, Wiley, New York, 1964). Preparation of McIlvaine buffered solutions at ionic strengths of 0.5 and 1.0 and Britton-Robinson solutions of constant ionic strength have been described by Elving, Markowitz, and Rosenthal, *Anal. Chem.*, **28**, 1179 (1956), and Frugoni, *Gazz. Chim. Ital.*, **87**, 403 (1957), respectively.

Table 5-24
pH RANGES OF BUFFER SOLUTIONS FOR CONTROL PURPOSES

Materials	pH range
Glycine and HCl	1.0 to 3.7
Citrate and HCl	1.3 to 4.7
p-Toluenesulfonate and p-toluenesulfonic acid	1.1 to 3.3
Formate and HCl	2.8 to 4.6
Succinic acid and borax	3.0 to 5.8
Phenyl acetate and HCl	3.5 to 5.0
Acetate and acetic acid	3.7 to 5.6
Succinate and succinic acid	4.8 to 6.3
2-(N-Morpholino)ethanesulfonic acid, "MES," and NaOH	5.2 to 7.1

Table 5-24 (*Continued*)
pH RANGES OF BUFFER SOLUTIONS FOR CONTROL PURPOSES

Materials	pH range
2,2-*Bis*(hydroxymethyl)-2,2′,2″-nitrilotriethanol and HCl	5.8 to 7.2
KH$_2$PO$_4$ and borax	5.8 to 9.2
N-*Tris*(hydroxymethyl)methyl-2-aminoethanesulfonic acid, "TES," and NaOH	6.8 to 8.2
KH$_2$PO$_4$ and Na$_2$HPO$_4$	6.1 to 7.5
N-2-Hydroxyethylpiperazine-N′-2-ethanesulfonic acid, "HEPES," and NaOH	6.9 to 8.3
Triethanolamine and HCl	6.9 to 8.5
Diethylbarbiturate (Veronal) and HCl	7.0 to 8.5
Tris(hydroxymethyl)aminomethane, "Tris," and HCl	7.2 to 9.0
N-*Tris*(hydroxymethyl)methylglycine, "Tricine," and HCl	7.2 to 9.0
N,N-*Bis*(2-hydroxyethyl)glycine, "Bicine," and HCl	7.4 to 9.2
Borax and HCl	7.6 to 8.9
Glycine and NaOH	8.2 to 10.1
Ammonia (aqueous) and NH$_4$Cl	8.3 to 9.2
Ethanolamine and HCl	8.6 to 10.4
Borax and NaOH	9.4 to 11.1
Carbonate and hydrogen carbonate	9.2 to 11.0
Na$_2$HPO$_4$ and NaOH	11.0 to 12.0

ACIDITY IN OTHER SOLVENT MEDIA

In nonaqueous and mixed solvents, the operational pH scale is defined as

$$pH_x^* = pH_s^* + \frac{(E_x - E_s)F}{2.3026RT}$$

where standard values of pH_s^* are assigned to buffer solutions, as shown in Table 5-25. Selected values of pH* for buffer solutions in methanol-water and ethanol-water solvents are gathered together in Table 5-26. Meaningful applications of pH* data in cells with liquid junction depends to a very large extent on using reference electrodes in which the salt-bridge solution employs the same solvent as that of the buffer.

Acidity standards for deuterium oxide (heavy water) can be used for measurements of pD. Assignment of pD_s values (Table 5-27) is analogous to those used for assigning pH_s values with the exception that the ultimate reference basis for a scale of deuterium ion activity is the deuterium gas electrode, by convention, taken to be zero at all temperatures. There appears to be a constant difference of 0.45 ± 0.03 between the operational pH, determined with the glass electrode in heavy water solutions with reference to aqueous standards, and the presumed or expected pD in these same solutions.

Table 5-25
STANDARD REFERENCE VALUES pH$_s^*$ FOR THE MEASUREMENT OF ACIDITY IN 50 WEIGHT PER CENT METHANOL-WATER

Temperature, °C	0.02m HOAc, 0.02m NaOAc, 0.02m NaCl	0.02m NaHSuc, 0.02m NaCl	0.02m KH$_2$PO$_4$, 0.02m Na$_2$HPO$_4$, 0.02m NaCl
10	5.560	5.806	7.937
15	5.549	5.786	7.916
20	5.543	5.770	7.898
25	5.540	5.757	7.884
30	5.540	5.748	7.872
35	5.543	5.743	7.863
40	5.550	5.741	7.858

OAc = acetate   Suc = succinate

Reference: R. G. Bates, *Anal. Chem.*, **40**(6), 35A (1968).

Table 5-26
pH* VALUES FOR BUFFER SOLUTIONS IN ALCOHOL-WATER SOLVENTS AT 25°C
Liquid-junction Potential not Included

Solvent Composition (weight per cent alcohol)	0.01M H$_2$C$_2$O$_4$, 0.01M NH$_4$HC$_2$O$_4$	0.01M H$_2$Suc, 0.01M LiHSuc	0.01M HSal, 0.01M NaSal
Methanol-Water Solvents			
0	2.15	4.12	
10	2.19	4.30	
20	2.25	4.48	
30	2.30	4.67	
40	2.38	4.87	
50	2.47	5.07	
60	2.58	5.30	
70	2.76	5.57	
80	3.13	6.01	
90	3.73	6.73	
92	3.90	6.92	
94	4.10	7.13	
96	4.39	7.43	
98	4.84	7.89	
99	5.20	8.23	
100	5.79	8.75	7.53
Ethanol-Water Solvents			
0	2.15	4.12	
30	2.32	4.70	
50	2.51	5.07	
71.9	2.98	5.71	
100			8.32

Suc = succinate   Sal = salicylate

Table 5-27
STANDARD REFERENCE VALUES pD_s FOR THE MEASUREMENT OF ACIDITY IN HEAVY WATER

Temperature, °C	0.05M KD$_2$ citrate	0.025M KD$_2$PO$_4$ + 0.025M Na$_2$DPO$_4$	0.025M NaDCO$_3$ + 0.025M Na$_2$CO$_3$
5	4.378	7.539	10.998
10	4.352	7.504	10.924
15	4.329	7.475	10.855
20	4.310	7.449	10.793
25	4.293	7.428	10.736
30	4.279	7.411	10.685
35	4.268	7.397	10.638
40	4.260	7.387	10.597
45	4.253	7.381	10.560
50	4.250	7.377	10.527

References: R. G. Bates, *Anal. Chem.*, **40**(6), 36A (1968); M. Paabo and R. G. Bates, *Anal. Chem.*, **41**, 283 (1969).

ION ACTIVITY STANDARDS

An approach, similar to the operational definition of pH values, can be applied to the problem of measuring the activities of other ions in solution. Standard reference values for pNa, pCl, pCa, and pF have been suggested for use in standardizing ion-selective electrodes:

Suggested Reference Standard Values, pI ($-\log$ [I]), at 25°C

Material	Molality (mole kg^{-1})	pNa	pCa	pCl	pF
NaCl	0.001	3.015		3.015	
	0.01	2.044		2.044	
	0.1	1.108		1.110	
	1.0	0.160		0.204	
NaF	0.001	3.015			3.015
	0.01	2.044			2.048
	0.1	1.108			1.124
CaCl$_2$	0.000333		3.537	3.191	
	0.00333		2.653	2.220	
	0.0333		1.887	1.286	
	0.333		1.105	0.381	

From R. G. Bates and M. Alfenaar in R. A. Durst (Ed.), "Ion-selective Electrodes," *National Bureau of Standards Spec. Publ.* **314,** Washington, 1969.

Table 5-28
INDICATORS FOR AQUEOUS ACID-BASE TITRATIONS AND COLORIMETRIC DETERMINATION OF pH

This table lists some selected indicators. The pH range or transition interval given in the third column may vary appreciably from one observer to another, and in addition it is affected by ionic strength, temperature, and illumination; consequently only approximate values can be given. They should be considered to refer to solutions having low ionic strengths and a temperature of about 25°C. In the fourth column the $pK_a (-\log K_a)$ of the indicator as determined spectrophotometrically is listed. In the fifth column the wavelength of maximum absorption is given first for the acidic and then for the basic form of the indicator, and the same order is followed in giving the colors in the sixth column. The abbreviations used to describe the colors of the two forms of the indicator are as follows:

B	blue	G	green	P	purple	V	violet
Br	brown						
C	colorless	O	orange	R	red	Y	yellow

Indicator	Chemical Name	pH Range	pK_a	λ_{max}, nm	Color Change
Cresol red (acid range)	o-Cresolsulfonephthalein	0.2 to 1.8			R–Y
Cresol purple (acid range)	m-Cresolsulfonephthalein	1.2 to 2.8	1.51	533, ⋯	R–Y
Thymol blue (acid range)	Thymolsulfonephthalein	1.2 to 2.8	1.65	544, 430	R–Y
Tropeolin 00	Diphenylamino-p-benzene sodium sulfonate	1.3 to 3.2	2.0	527, ⋯	R–Y
2,6-Dinitrophenol	2,6-Dinitrophenol	2.4 to 4.0	3.69		C–Y
2,4-Dinitrophenol	2,4-Dinitrophenol	2.5 to 4.3	3.90		C–Y
Methyl yellow	Dimethylaminoazobenzene	2.9 to 4.0	3.3	508, ⋯	R–Y
Methyl orange	Dimethylaminoazobenzene sodium sulfonate	3.1 to 4.4	3.40	522, 464	R–O
Bromophenol blue	Tetrabromophenolsulfonephthalein	3.0 to 4.6	3.85	436, 592	Y–BV
Bromocresol green	Tetrabromo-m-cresolsulfonephthalein	4.0 to 5.6	4.68	444, 617	Y–B
Methyl red	o-Carboxybenzeneazodimethylaniline	4.4 to 6.2	4.95	530, 427	R–Y
Chlorophenol red	Dichlorophenolsulfonephthalein	5.4 to 6.8	6.0	⋯, 573	Y–R
Bromocresol purple	Dibromo-o-cresolsulfonephthalein	5.2 to 6.8	6.3	433, 591	Y–P
Bromophenol red	Dibromophenolsulfonephthalein	5.2 to 6.8		⋯, 574	Y–R
p-Nitrophenol	p-Nitrophenol	5.3 to 7.6	7.15	320, 405	C–Y
Bromothymol blue	Dibromothymolsulfonephthalein	6.2 to 7.6	7.1	433, 617	Y–B
Neutral red	Aminodimethylaminotoluphenazonium chloride	6.8 to 8.0	7.4		R-Y
Phenol red	Phenolsulfonephthalein	6.4 to 8.0	7.9	433, 558	Y–R
m-Nitrophenol	m-Nitrophenol	6.4 to 8.8	8.3	⋯, 570	C–Y
Cresol red	o-Cresolsulfonephthalein	7.2 to 8.8	8.2	434, 572	Y–R
m-Cresol purple	m-Cresolsulfonephthalein	7.6 to 9.2	8.32	⋯, 580	Y–P
Thymol blue	Thymolsulfonephthalein	8.0 to 9.6	8.9	430, 596	Y–B
Phenolphthalein	Phenolphthalein	8.0 to 10.0	9.4	⋯, 553	C–R
α-Naphtholbenzein	α-Naphtholbenzein	9.0 to 11.0			Y–B
Thymolphthalein	Thymolphthalein	9.4 to 10.6	10.0	⋯, 598	C–B
Alizarin Yellow R	5-(p-Nitrophenylazo)-salicylic acid, Na salt	10.0 to 12.0	11.16		Y–V
Tropeolin 0	p-Sulfobenzeneazoresorcinol	11.0 to 13.0			Y–O Br
Nitramine	2,4,6-Trinitrophenylmethylnitroamine	10.8 to 13.0			C–O Br

Salt Errors of Indicators. The salt error in the colorimetric measurement of pH arises from the fact that the color of the indicator depends on the ratio of the concentrations of the conjugate acid and base forms, whereas it is the ratio of their activities which is fixed by the pH of the solution. Therefore, differences in the activity coefficients of the indicator forms, arising from differences in total ionic strength of the solutions and the indicator charge type, can cause difference in apparent color and pH. The salt error is given by the expression:

$$\text{Salt error} = \log\left(\frac{f_I}{f_{HI}}\right)_x - \log\left(\frac{f_I}{f_{HI}}\right)_s$$

where x and s are the unknown and reference solutions, respectively. An indicator acid can be an uncharged molecule (HI), an anion (HI$^-$), or a protonated cation (HI$^+$). Table 5-29 gives the values of the salt error terms, f_I/f_{HI}, for indicators of the three charge types at various ionic strengths, as calculated from the expanded Debye-Hückel equation.

Table 5-29
SALT ERRORS OF INDICATORS

Ionic strength	$\log(f_I/f_{HI^+})$	$\log(f_{I^-}/f_{HI})$	$\log(f_{I^-}/f_{HI^-})$
0.01	+0.042	−0.042	−0.13
0.02	+0.056	−0.056	−0.17
0.03	+0.064	−0.064	−0.19
0.04	+0.071	−0.071	−0.21
0.05	+0.076	−0.076	−0.23
0.06	+0.080	−0.080	−0.24
0.08	+0.086	−0.086	−0.26
0.10	+0.090	−0.090	−0.27
0.12	+0.093	−0.093	−0.28
0.14	+0.095	−0.095	−0.29
0.16	+0.096	−0.096	−0.29
0.18	+0.097	−0.097	−0.29
0.20	+0.097	−0.097	−0.29
0.25	+0.096	−0.096	−0.29
0.30	+0.094	−0.094	−0.28
0.40	+0.086	−0.086	−0.26
0.50	+0.076	−0.076	−0.23

Table 5-30
MIXED INDICATORS

Mixed indicators give sharp color changes and are especially useful in titrating to a given titration exponent (p_I). Methods for the preparation of solutions of indicators will be found in the section *LABORATORY SOLUTIONS*.

The information given in this table is from the two-volume work *VOLUMETRIC ANALYSIS* by Kolthoff and Stenger, published by Interscience Publishers, Inc., New York, 1942 and 1947, and reproduced with their permission.

* Keep in a dark bottle.   † Excellent indicator.

Composition of Indicator Solution		p_I	Color		Notes
			Acid	Alkaline	
1 part 0.1% methyl yellow in alc. 1 part 0.1% methylene blue in alc.	*	3.25	Blue-violet	Green	Still green at p_H 3.4, blue-violet at 3.2; †
1 part 0.1% hexamethoxytriphenyl carbinol in alc. 1 part 0.1% methyl green in alc.	*	4.0	Violet	Green	Color is blue-violet at p_H 4.0
1 part 0.1% methyl orange in aq. 1 part 0.25% indigo carmine in aq.	*	4.1	Violet	Green	Good indicator, especially in artificial light
1 part 0.1% methyl orange in aq. 1 part 0.1% aniline blue in aq.		4.3	Violet	Green	
1 part 0.1% bromcresol green sodium salt in aq. 1 part 0.02% methyl orange in aq.		4.3	Orange	Blue-green	Yellow at p_H 3.5, greenish yellow at 4.0, weakly green at 4.3
3 parts 0.1% bromcresol green in alc. 1 part 0.2% methyl red in alc.		5.1	Wine-red	Green	Very sharp color change; †
1 part 0.2% methyl red in alc. 1 part 0.1% methylene blue in alc.	*	5.4	Red-violet	Green	Color is red-violet at p_H 5.2, a dirty blue at 5.4, and a dirty green at 5.6
1 part 0.1% chlorphenol red sodium salt in aq. 1 part 0.1% aniline blue in water		5.8	Green	Violet	Pale violet at p_H 5.8

Table 5-30 (*Continued*)
MIXED INDICATORS

Composition of Indicator Solution	pI	Color — Acid	Color — Alkaline	Notes
1 part 0.1% bromcresol green sodium salt in aq. 1 part 0.1% chlorphenol red sodium salt in aq.	6.1	Yellow-green	Blue-violet	Blue-green at pH 5.4, blue at 5.8, blue with a touch of violet at 6.0, blue-violet at 6.2
1 part 0.1% bromcresol purple sodium salt in aq. 1 part 0.1% bromthymol blue sodium salt in aq.	6.7	Yellow	Violet-blue	Yellow-violet at pH 6.2, violet at 6.6, blue-violet at 6.8
2 parts 0.1% bromthymol blue sodium salt in aq. 1 part 0.1% azolitmin in aq.	6.9	Violet	Blue	
1 part 0.1% neutral red in alc. 1 part 0.1% methylene blue in alc. *	7.0	Violet-blue	Green	Violet blue at pH 7.0; †
1 part 0.1% neutral red in alc. 1 part 0.1% bromthymol blue in alc.	7.2	Rose	Green	Dirty green at pH 7.4, pale rose at 7.2, clear rose at 7.0
2 parts 0.1% cyanine in 50% alc. 1 part 0.1% phenol red in 50% alc.	7.3	Yellow	Violet	Orange at pH 7.2, beautiful violet at 7.4, color fades on standing
1 part 0.1% bromthymol blue sodium salt in aq. 1 part 0.1% phenol red sodium salt in aq.	7.5	Yellow	Violet	Dirty green at pH 7.2, pale violet at 7.4, strong violet at 7.6; †
1 part 0.1% cresol red sodium salt in aq. 3 parts 0.1% thymol blue sodium salt in aq.	8.3	Yellow	Violet	Rose at pH 8.2, distinctly violet at 8.4; †
2 parts 0.1% α-naphtholphthalein in alc. 1 part 0.1% cresol red in alc.	8.3	Pale rose	Violet	Pale violet at pH 8.2, strong violet at 8.4

Table 5-30 (*Continued*)
MIXED INDICATORS

Composition of Indicator Solution	p_I	Color		Notes
		Acid	Alkaline	
1 part 0.1% α-naphtholphthalein in alc. 3 parts 0.1% phenolphthalein in alc.	8.9	Pale rose	Violet	Pale green at p_H 8.6, violet at 9.0
1 part 0.1% phenolphthalein in alc. 2 parts 0.1% methyl green in alc. *	8.9	Green	Violet	Pale blue at p_H 8.8, violet at 9.0
1 part 0.1% thymol blue in 50% alc. 3 parts 0.1% phenolphthalein in 50% alc.	9.0	Yellow	Violet	From yellow thru green to violet; †
1 part 0.1% phenolphthalein in alc. 1 part 0.1% thymolphthalein in alc.	9.9	Colorless	Violet	Rose at p_H 9.6, violet at 10; sharp color change
1 part 0.1% phenolphthalein in alc. 2 parts 0.2% Nile blue in alc.	10.0	Blue	Red	Violet at p_H 10; †
2 parts 0.1% thymolphthalein in alc. 1 part 0.1% alizarin yellow in alc.	10.2	Yellow	Violet	Sharp color change
2 parts 0.2% Nile blue in aq. 1 part 0.1% alizarin yellow in alc.	10.8	Green	Red-brown	

Table 5-31
FLUORESCENT INDICATORS

Organic Chem. Bulletin, 29, No. 4, Eastman Kodak Co., Rochester, N.Y. (1957).

Fluorescent indicators have the advantage over other indicators in that they may be used in titrating very turbid or strongly colored liquids.

Abbreviations: B, blue; C, colorless; D, dark; G, green; L, light; O, orange; V, violet; W, weak; Y, yellow.

Name	pH	Color acid → base	Name	pH	Color acid → base
4-Methylumbelliferone	0.0– 2.0	G-WB	3,6-Dihydroxyphthalo-		
3,6-Dihydroxyphthalimide	0.0– 2.5	G-YG	nitrile	6.0– 8.0	B-G
Benzoflavine	0.3– 1.7	Y-G	2-Naphthol-6-sulfonic acid		
Ethoxyacridone	1.2– 3.2	G-B	Na salt	6.0– 8.0	C-B
3,6-Tetramethyldiamino-			Brilliant Diazo Yellow	6.5– 7.5	C-B
xanthone	1.2– 3.4	G-B	Thioflavin	6.5– 7.6	C-G
Esculin	1.5– 2.0	C-B	4-Methylumbelliferone	6.5– 8.0	WB-B
Eosin Yellowish	2.0– 3.5	C-Y	Umbelliferone	6.5– 8.0	O-B
7-Amino-1,3-naphthalene-			Orcinaurine	6.5– 8.0	C-G
disulfonic acid	2.0– 4.0	WB-B	2-Naphthol	7.0– 8.5	WB-B
1-Hydroxy-2-naphthoic			2-Naphthol-6,8-disulfonic		
acid	2.5– 3.5	C-B	acid diK salt	7.0– 8.5	WB-B
Salicylic acid	2.5– 4.0	C-B	Morin	7.0– 8.5	WG-G
2,′4′,5′,7′-Tetrabromo-			1-Naphthol	7.0– 9.0	C-BG
fluorescein Na salt	2.5– 4.5	C-G	2-Naphthol-3,6-disulfonic		
2-Naphthylamine	2.8– 4.4	C-V	acid diNa salt	7.0– 9.0	WB-B
Erythrosin	3.0– 4.0	C-B	Harmine	7.2– 8.9	B-Y
1-Naphthylamine	3.0– 4.5	C-B	trans-o-Hydroxycinnamic		
3-Hydroxy-2-naphthoic			acid	7.2– 9.0	C-G
acid	3.0– 6.8	B-G	1-Naphthol-4-sulfonic acid		
o-Phenylenediamine	3.1– 4.4	G-C	Na salt	8.0– 9.0	DB-LB
p-Phenylenediamine	3.1– 4.4	C-OY	1-Naphthol-2-sulfonic acid		
5-Aminosalicylic acid	3.1– 4.4	C-G	Na salt	8.0– 9.0	DB-LB
o-Methoxybenzaldehyde	3.1– 4.4	C-G	Coumarin	8.0– 9.5	WG-G
Phloxine	3.4– 5.0	C-Y	Acridine Orange	8.0–10.0	WYG-Y
Fluorescein	4.0– 5.0	WG-G	Naphthol AS	8.2–10.3	C-YG
Quinic acid	4.0– 5.0	Y-B	Ethoxyphenylnaphtha-		
4,5-Dihydroxy-2,7-naph-			stilbazonium chloride	9.0–11.0	G-O
thalenedisulfonic acid			6,7-Dimethoxyisoquino-		
diNa salt	4.0– 6.0	WB-B	line-1-carboxylic acid	9.5–11.0	Y-B
2′,7′-Dichlorofluorescein	4.0– 6.0	WG-G	Eosin BN	10.5–14.0	C-Y
Resorufin	4.0– 6.0	C-O	1-Naphthylamine	12.0–13.0	B-WB
β-Methyl esculetin	4.0– 6.2	C-B	7-Amino-1,3-naphthalene-		
Acridine	4.5– 6.0	G-B	disulfonic acid	12.0–13.0	B-WO
3,6-Dihydroxyxanthone	5.4– 7.6	C-BG	Cotarmine	12 0–13.0	Y-C
3,6-Dihydroxyphthalic			β-Naphthionic acid	12 0–13.0	B-V
acid	5.8– 8.2	B-G	4-Amino-1-naphthalene-		
Quinine	5.9– 6.1	B-V	sulfonic acid	12.0–14.0	B-G
5-Amino-2,3-dihydro-1,4-					
phthalazinedione	6.0– 7.0	B-WB			

SEPARATION METHODS

The extent of cross-contamination in column chromatographic methods can be estimated from the experimental elution curves (Fig. 5-2). Let us assume that the effluent from the column is divided into two portions to give products of equal percentage purity, i.e., $\eta_1 = \eta_2$ where $\eta_1 = \Delta m_2/m_1$ and $\eta_2 = \Delta m_1/m_2$. The ratio $\Delta m/m$ represents a chosen degree of purity. When the products are unequal in amount, the fractional impurity must be multiplied by the factor: $(A_1^2 + A_2^2)/2A_1A_2$, where A_1 and A_2 are the areas under the elution curves.

The relative peak retention ratio, or separation factor,

$$\alpha = \frac{V_{max,2}}{V_{max,1}} = \frac{(V_R')_2}{(V_R')_1}$$

is related to the volume distribution ratio or the molar distribution coefficient in ion-exchange chromatography by

$$\alpha = \frac{(D_v)_2 + \beta}{(D_v)_1 + \beta} = \frac{(K_d)_2 + 1}{(K_d)_1 + 1}$$

where the volume distribution ratio D_v is defined as

$$D_v = \frac{\text{amount of ion in 1 ml of resin bed}}{\text{amount of ion in 1 ml of interstitial volume}} = \frac{[M]_r/V_b}{[M]/V_0}$$

The volume distribution ratio is related to the weight distribution coefficient by

$$D_v = K_d\left(\frac{V_0}{V_b}\right) = K_d\beta$$

where V_0 is the liquid volume in the interstices between the individual resin heads and V_b is the total bed volume. The molar (or weight) distribution coefficient is given by

$$K_d = \frac{[M]_r}{[M]} = \frac{\text{amount of ion on resin per gram of resin}}{\text{amount of ion in solution per milliliter of solution}}$$

The interrelation of product purity, relative peak retention ratio (or separation factor), and number of effective plates is shown in Figs. 5-3 and 5-4.

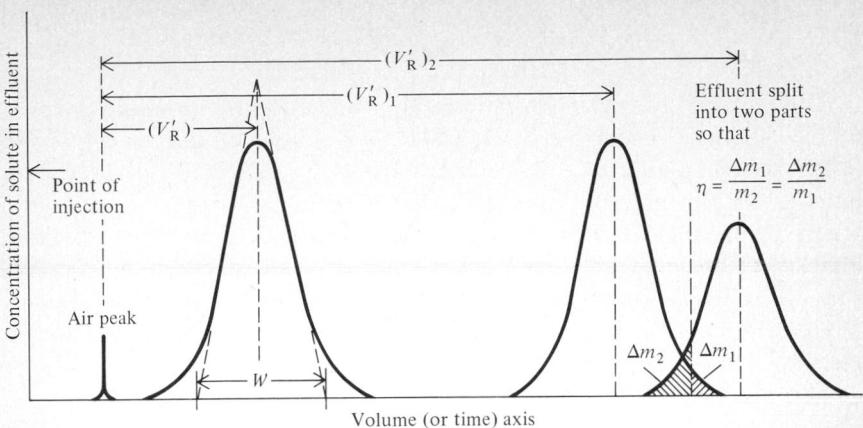

Fig. 5-2. *Idealized elution peaks showing method of handling overlapping bands.*

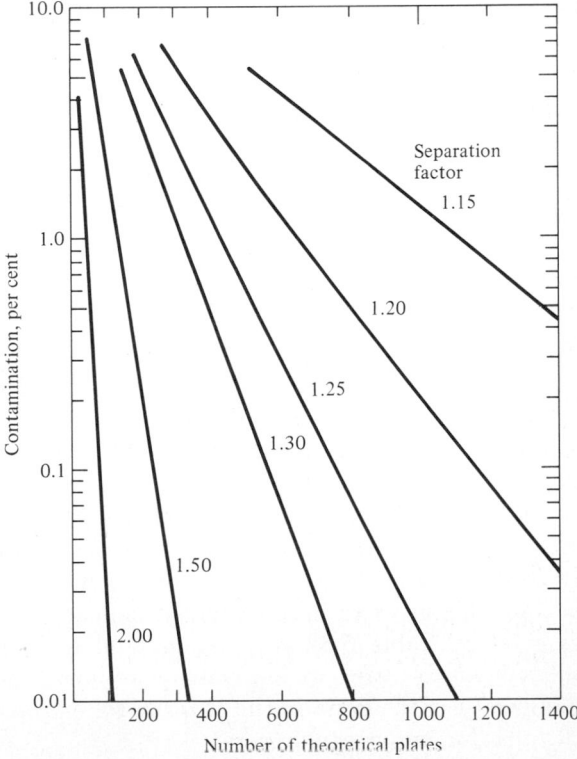

Fig. 5-3. *Cross-contamination, $100 \, \Delta m_2/m_1$, for different separation factors, as a function of number of theoretical plates (after Glueckauf), for case that $m_1 = m_2$.*

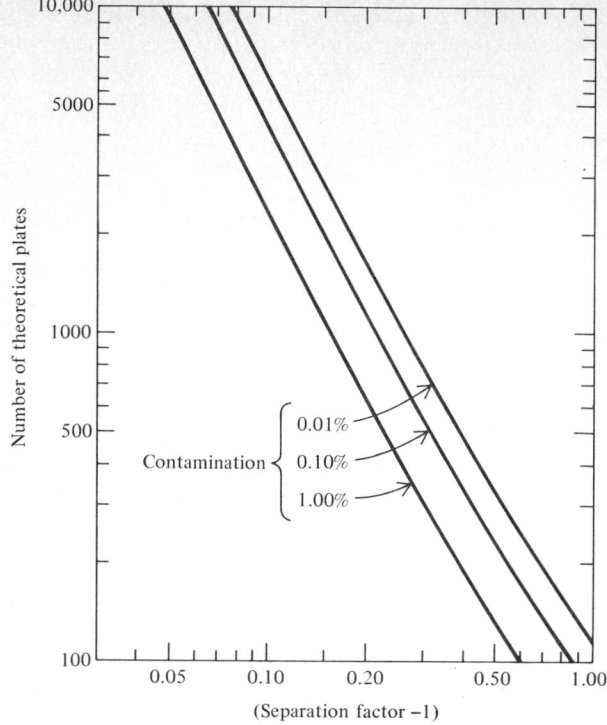

Fig. 5-4. Theoretical plates needed to achieve desired degrees of purity. (After Glueckauf.)

Selectivity Coefficients in Ion Exchange

At equilibrium the concentration of ions in solution and in the ion-exchange resin is related through the *selectivity coefficient,* $E_H^{M/n}$ (also denoted by the name concentration exchange constant) as follows:

$$E_H^{M/n} = \frac{[M^{n+}]_r[H^+]}{[H^+]_r[M^{n+}]}$$

where the subscript r refers to the resin phase. The selectivity coefficient expresses the selectivity of the resin for M^{n+} ions in preference to hydrogen ions from a solution containing an equal concentration of each (and for a specified degree of resin loading). A collection of selectivity coefficients is given in Table 5-32 for cations, and in Table 5-33 for anions. Tabulated values refer to equivalents of ion adsorbed from 1 ml of solution per 1 g of dry resin in the H form, for cation exchangers, and in the Cl form for anion exchangers.

From the selectivity coefficients listed in the tables, one is able to calculate the selectivity coefficient for the exchange between any pair of ions. For the exchange between a calcium salt solution and a Dowex 50 × 8 cation-exchange resin with 8 percent cross-linking loaded with sodium ions:

$$E_{Na}^{Ca/2} = \frac{E_{H}^{Ca/2}}{E_{H}^{Na}} = \frac{1.80}{1.56} = 1.15$$

or, expressed on a molar basis to account for the unequal charges of the ions:

$$E_{2Na}^{Ca} = \frac{[Ca^{2+}]_r[Na^+]^2}{[Ca^{2+}][Na^+]_r^2} = (1.15)^2 = 1.32$$

Table 5-32

SELECTIVITY COEFFICIENTS $E_{H}^{M/n}$ FOR SOME
METAL IONS ON DIFFERENTLY
CROSS-LINKED DOWEX 50

	4% DVB	8% DVB	16% DVB
Univalent ions			
Li	0.76	0.79	0.68
H	1.00	1.00	1.00
Na	1.20	1.56	1.61
NH_4	1.44	2.01	2.27
K	1.72	2.28	3.06
Rb	1.86	2.49	3.14
Cs	2.02	2.56	3.17
Ag	3.58	6.70	15.6
Bivalent ions			
UO_2	0.79	0.85	1.05
Mg	0.99	1.15	1.10
Zn	1.05	1.21	1.18
Co	1.08	1.31	1.19
Cu	1.10	1.35	1.40
Cd	1.13	1.36	1.55
Ni	1.16	1.37	1.27
Mn	1.15	1.43	1.54
Ca	1.39	1.80	2.28
Sr	1.57	2.27	3.16
Pb	2.20	3.46	5.65
Ba	2.50	4.02	6.52
Trivalent ions			
Cr	1.6	2.0	2.5
Ce	1.9	2.8	4.1
La	1.9	2.8	4.1

Based on data of Bonner et al., J. Phys. Chem. **61,** 326 (1957); ibid. **62,** 250 (1958).

Table 5-33
SELECTIVITY COEFFICIENTS E_{Cl}^{X} FOR SOME ANIONS ON DOWEX RESINS

	Dowex 1	Dowex 2
Hydroxide	0.09	0.65
Fluoride	0.09	0.13
Aminoacetate	0.10	0.10
Acetate	0.17	0.18
Formate	0.22	0.22
Dihydrogen phosphate	0.25	0.34
Bicarbonate	0.32	0.53
Chloride	1.00	1.00
Bromate	...	1.01
Bisulfite	1.3	1.3
Cyanide	1.6	1.3
Bromide	2.8	2.3
Nitrate	3.8	3.3
Benzene sulfonate	...	4.0
Bisulfate	4.1	6.1
Phenoxide	5.2	8.7
Iodide	8.7	7.3
p-Toluene sulfonate	...	13.7
Thiocyanate	...	18.5
Perchlorate	...	32
Salicylate	32.2	28

From S. Peterson, *Ann. N. Y. Acad. Sci.* **57,** 144 (1954).

CHEMISTRY 345 LABORATORY PLANNING

DO WE HAVE THE MATERIALS FOR SECTION 9 p764 IODIMETRIC MOA?
WE SHOULD PREPARE SEVERAL ONE LITER LOTS OF THE KI3 AND LET THEM
STANDARDIZE AGAINST BOTH THEIR THIOSULFATE WHICH WAS STANDARDIZED
AGAINST KIO3 AND THE BARIUM THIOSULFATE HYDRATE OR ARSEN IC OXIDE
THIS SHOULD BE THE NEXT EXPERIMENT.

EXPERIMENT 5A EDTA FOR CALCIUM++

EXPERIMENT 18.1 & 18.2 & 18.3

FOR 18.3 I WOULD LIKE TO RUN A SIMULATANEOUS TITRATION OF A
BUFFER MAKING MEASUREMENTS BOTH COLORIMETRIC AND GLASS ELECTRODE.

EXPERIMENT 19 PERHAPS THIS SHOULD BE A COMPARISON AGAINST A
VISUAL SET OF STANDARDS SINCE WE PROBABLY DO NOT HAVE A
FLUORIMETER. PLEASE CHECK TO SEE IF WE DO HAVE A TURNER OR OTHER
FLUORESCENCE INSTRUMENT ON HAND.

EXPERIMENT 14.5 THIS SHOULD BE SCHEDULED AFTER THEY HAVE USED THE
GLASS ELECTRODE IN 18.3. WE SHOULD ALSO FIND APPROPRIATE
INDICATORS FOR THE SAME EXPERIMENT AND HAVE THEM IN HAND TO DO
COMPARITIVE WORK.

WE SHOULD WORK OUT A FEW GAS CHROMATOGRAPHIC METHODS AND LINE UP
EQUIPMENT FOR THE WORK. VINCE, WE SHOULD ALS DO SOME WORK EVEN IF

Section 6

ELECTROCHEMISTRY

SINGLE ELECTRODE POTENTIALS AT 25°C

It is convenient to consider that the total emf of a cell is the sum of two "single electrode potentials." That is,

$$E_{cell} = E_{ox} + E_{red} \tag{1}$$

where E_{ox} is the single electrode potential of the electrode forming the negative pole of the cell, and E_{red} is the single electrode potential of the electrode forming the positive pole of the cell. An oxidation process always occurs at the negative pole of a primary cell, whereas a reduction process occurs at the positive pole. If a particular substance is more easily oxidized than hydrogen, its E_{ox} is assigned a positive value and its E_{red} a negative value. If a substance is not oxidized as easily as hydrogen, its E_{ox} is negative and E_{red} is positive in sign.

The values given in the table below are E_{red}° values, where E° is the single electrode potential when each substance involved in the electrode reaction is at unit "activity." Values of E_{ox}° may be obtained merely by changing the sign of the E_{red}° values.

If the half-reaction involves H^+ as product or reactant, the solution is acid. Half-reactions involving OH^-, NH_3, CN^-, CO_3^{2-}, or S^{2-} are reactions occurring in an alkaline solution.

Not all the E° values in this table were determined by emf measurements. Some were calculated from free energy data. For details, see W. M. Latimer, *Oxidation Potentials*, 2d ed., Prentice-Hall, Englewood Cliffs, N.J., 1952.

One substance on the left side of each half-reaction is an oxidizing agent, and one substance on the right side of each half-reaction is a reducing agent. As one proceeds downward in the table the oxidizing agents decrease in strength, and the reducing agents increase in strength. In general, if two half-reactions are represented by the following equations:

$$\text{ox. agent}_1 + ne = \text{red. agent}_1$$
$$\text{ox. agent}_2 + ne = \text{red. agent}_2$$

and the second equation occurs lower in the table than the first equation, then a reaction may occur between ox. agent$_1$ and red. agent$_2$, whereas no reaction is possible between ox. agent$_2$ and red. agent$_1$. For example, considering the half-reactions

$$Cu^{2+} + I^- + e = CuI \qquad (E_{red}^{\circ} = 0.86 \text{ volt})$$
$$I_2 + 2e = 2I^- \qquad (E_{red}^{\circ} = 0.535 \text{ volt})$$

from the rule that Cu^{2+} should react with I^-, but no reaction should occur between I_2 and CuI.

To obtain the overall reaction between two half-reactions, the half-reaction which is lower in the table is reversed. Each equation is multiplied through by the proper numbers so that the number of electrons lost in the reversed upper equation is equal to the number of electrons gained in the lower half-reaction. The two half-reactions are then added. For the reaction between Cu^{2+} and I^- the overall reaction is obtained as follows:

$$2I^- = I_2 + 2e \qquad \text{(lower half-reaction reversed)}$$
$$\underline{2Cu^{2+} + 2I^- + 2e = 2CuI} \qquad \text{(reaction multiplied by 2)}$$
$$2Cu^{2+} + 4I^- = I_2 + 2CuI \qquad \text{(overall reaction)}$$

If two or more reactions between two substances are possible, usually the reaction involving half-reactions which are farthest apart in the table will occur first.

To calculate the single electrode potential, E_{red}, when the concentration of ions involved in the electrode reaction is other than unit activity, the following equation is used:

$$E_{red} = E°_{red} - \frac{0.00019841T}{n} \log_{10} \frac{(a_p)^x}{(a_r)^y} \qquad (2)$$

where n number of electrons (e) involved in the half-reaction; a_p activity of products, a_r activity of reactants; and a and b are the coefficients of products and reactants, respectively, in the half-reaction *written as a reduction reaction*; T is the absolute temperature; the value of $0.00019841T$ at $25°C$ is 0.05916 (see Table 5-16).

Examples. For the cell:

$$^{(-)}Zn, \ Zn^{2+}(a = 1) \qquad Cu^{2+}(a = 1), \ Cu,^{(+)}$$
$$E°_{red} = -0.763 \qquad E°_{red} = 0.337$$
or
$$E°_{ox} = 0.763$$

$$E_{cell} = E°_{ox} + E°_{red} = 0.763 + 0.337 = 1.000 \text{ volts}$$

To calculate the emf and overall cell reaction for the cell:

$$^{(-)}(Pt)Sn^{2+}(a = 0.2) \qquad Mn^{2+}(a = 0.3)^{(+)}$$
$$Sn^{4+}(a = 0.8) \qquad MnO_4^-(a = 0.9)(Pt)$$
$$H^+(a = 0.5)$$
$$E°_{red} = 0.15 \qquad E°_{red} = 1.51$$
or
$$E°_{ox} = -0.15$$

The upper half of the two reactions, involving Sn, is reversed to give the oxidation half-reaction

$$Sn^{2+} = Sn^{4+} + 2e$$

$$E_{ox} = -0.15 - \frac{0.05916}{5} \log_{10} \frac{0.8}{0.2} = -0.15 - 0.018 = -0.168$$

The MnO_4^- half-reaction is written as a reduction reaction:

$$MnO_4^- + 8H + 5e = Mn^{2+} + 4H_2O$$

$$E_{red} = 1.51 - \frac{0.05916}{5}\log_{10}\frac{0.3}{(0.9)(0.5)^8} = 1.51 - 0.023 = 1.487$$

$$E_{cell} = E_{ox} + E_{red} = -0.168 + 1.487 = 1.319 \text{ volts}$$

$$5Sn^{2+} = 5Sn^{4+} + 10e \qquad \text{(ox. reaction multiplied by 5)}$$
$$\underline{2MnO_4^- + 16H^+ + 10e} \qquad \text{(red. reaction multiplied by 2)}$$
$$5Sn^{2+} + 2MnO_4^- + 16H^+ = 5Sn^{4+} + 2Mn^{2+} + 8H_2O \qquad \text{(complete reaction)}$$

Table 6-1
POTENTIALS OF SELECTED HALF-REACTIONS AT 25°C

A summary of oxidation-reduction half-reactions arranged in order of decreasing oxidation strength and useful for selecting reagent systems.

Half-reaction		$E°$, volts
$F_2(g) + 2H^+ + 2e^-$	$= 2HF$	3.06
$O_3 + 2H^+ + 2e^-$	$= O_2 + H_2O$	2.07
$S_2O_8^{2-} + 2e^-$	$= 2SO_4^{2-}$	2.01
$Ag^{2+} + e^-$	$= Ag^+$	2.00
$H_2O_2 + 2H^+ + 2e^-$	$= 2H_2O$	1.77
$MnO_4^- + 4H^+ + 3e^-$	$= MnO_2(s) + 2H_2O$	1.70
$Ce(IV) + e^-$	$= Ce(III)(\text{in } 1M \text{ } HClO_4)$	1.61
$H_5IO_6 + H^+ + 2e^-$	$= IO_3^- + 3H_2O$	1.6
$Bi_2O_4(\text{bismuthate}) + 4H^+ + 2e^-$	$= 2BiO^+ + 2H_2O$	1.59
$BrO_3^- + 6H^+ + 5e^-$	$= \frac{1}{2}Br_2 + 3H_2O$	1.52
$MnO_4^- + 8H^+ + 5e^-$	$= Mn^{2+} + 4H_2O$	1.51
$PbO_2 + 4H^+ + 2e^-$	$= Pb^{2+} + 2H_2O$	1.455
$Cl_2 + 2e^-$	$= 2Cl^-$	1.36
$Cr_2O_7^{2-} + 14H^+ + 6e^-$	$= 2Cr^{3+} + 7H_2O$	1.33
$MnO_2(s) + 4H^+ + 2e^-$	$= Mn^{2+} + 2H_2O$	1.23
$O_2(g) + 4H^+ + 4e^-$	$= 2H_2O$	1.229
$IO_3^- + 6H^+ + 5e^-$	$= \frac{1}{2}I_2 + 3H_2O$	1.20
$Br_2(\text{liq}) + 2e^-$	$= 2Br^-$	1.065
$ICl_2^- + e^-$	$= \frac{1}{2}I_2 + 2Cl^-$	1.06
$VO_2^+ + 2H^+ + e^-$	$= VO^{2+} + H_2O$	1.00
$HNO_2 + H^+ + e^-$	$= NO(g) + H_2O$	1.00
$NO_3^- + 3H^+ + 2e^-$	$= HNO_2 + H_2O$	0.94
$2Hg^{2+} + 2e^-$	$= Hg_2^{2+}$	0.92
$Cu^{2+} + I^- + e^-$	$= CuI$	0.86
$Ag^+ + e^-$	$= Ag$	0.799
$Hg_2^{2+} + 2e^-$	$= 2Hg$	0.79
$Fe(III) + e^-$	$= Fe^{2+}$	0.771
$O_2(g) + 2H^+ + 2e^-$	$= H_2O_2$	0.682
$2HgCl_2 + 2e^-$	$= Hg_2Cl_2(s) + 2Cl^-$	0.63
$Hg_2SO_4(s) + 2e^-$	$= 2Hg + SO_4^{2-}$	0.615
$H_3AsO_4 + 2H^+ + 2e^-$	$= HAsO_2 + 2H_2O$	0.581
$Sb_2O_5 + 6H^+ + 4e^-$	$= 2SbO^+ + 3H_2O$	0.559

Table 6-1 (*Continued*)
POTENTIALS OF SELECTED HALF-REACTIONS AT 25°C

Half-reaction		$E°$, volts
$I_3^- + 2e^-$	$= 3I^-$	0.545
$Cu^+ + e^-$	$= Cu$	0.52
$VO^{2+} + 2H^+ + e^-$	$= V^{3+} + H_2O$	0.337
$Fe(CN)_6^{3-} + e^-$	$= Fe(CN)_6^{4-}$	0.36
$Cu^{2+} + 2e^-$	$= Cu$	0.337
$UO_2^{2+} + 4H^+ + 2e^-$	$= U^{4+} + 2H_2O$	0.334
$BiO^+ + 2H^+ + 3e^-$	$= Bi + H_2O$	0.32
$Hg_2Cl_2(s) + 2e^-$	$= 2Hg + 2Cl^-$	0.2676
$AgCl(s) + e^-$	$= Ag + Cl^-$	0.2223
$SbO^+ + 2H^+ + 3e^-$	$= Sb + H_2O$	0.212
$CuCl_3^{2-} + e^-$	$= Cu + 3Cl^-$	0.178
$SO_4^{2-} + 4H^+ + 2e^-$	$= SO_2(aq) + 2H_2O$	0.17
$Sn^{4+} + 2e^-$	$= Sn^{2+}$	0.154
$S + 2H^+ + 2e^-$	$= H_2S(g)$	0.141
$TiO^{2+} + 2H^+ + e^-$	$= Ti^{3+} + H_2O$	0.10
$S_4O_6^{2-} + 2e^-$	$= 2S_2O_3^{2-}$	0.08
$AgBr(s) + e^-$	$= Ag + Br^-$	0.071
$2H^+ + 2e^-$	$= H_2$	0.0000
$Pb^{2+} + 2e^-$	$= Pb$	−0.126
$Sn^{2+} + 2e^-$	$= Sn$	−0.136
$AgI(s) + e^-$	$= Ag + I^-$	−0.152
$Mo^{3+} + 3e^-$	$= Mo$	ca −0.2
$N_2 + 5H^+ + 4e^-$	$= H_2NNH_3^+$	−0.23
$Ni^{2+} + 2e^-$	$= Ni$	−0.246
$V^{3+} + e^-$	$= V^{2+}$	−0.255
$Co^{2+} + 2e^-$	$= Co$	−0.277
$Ag(CN)_2^- + e^-$	$= Ag + 2CN^-$	−0.31
$Cd^{2+} + 2e^-$	$= Cd$	−0.403
$Cr^{3+} + e^-$	$= Cr^{2+}$	−0.41
$Fe^{2+} + 2e^-$	$= Fe$	−0.440
$2CO_2 + 2H^+ + 2e^-$	$= H_2C_2O_4$	−0.49
$H_3PO_3 + 2H^+ + 2e^-$	$= H_3PO_2 + H_2O$	−0.50
$U^{4+} + e^-$	$= U^{3+}$	−0.61
$Zn^{2+} + 2e^-$	$= Zn$	−0.763
$Cr^{2+} + 2e^-$	$= Cr$	−0.91
$Mn^{2+} + 2e^-$	$= Mn$	−1.18
$Zr^{4+} + 4e^-$	$= Zr$	−1.53
$Ti^{3+} + 3e^-$	$= Ti$	−1.63
$Al^{3+} + 3e^-$	$= Al$	−1.66
$Th^{4+} + 4e^-$	$= Th$	−1.90
$Mg^{2+} + 2e^-$	$= Mg$	−2.37
$La^{3+} + 3e^-$	$= La$	−2.52
$Na^+ + e^-$	$= Na$	−2.714
$Ca^{2+} + 2e^-$	$= Ca$	−2.87
$Sr^{2+} + 2e^-$	$= Sr$	−2.89
$K^+ + e^-$	$= K$	−2.925
$Li^+ + e^-$	$= Li$	−3.045

Table 6-2
POTENTIALS OF THE ELEMENTS AND THEIR COMPOUNDS AT 25°C

Standard potentials are denoted by 0 in the column headed Solution Composition. All other values are formal potentials.

Half-reaction	Standard or Formal Potential	Solution Composition
Actinium		
$Ac^{3+} + 3e^- = Ac$	ca -2.6	
Aluminum		
$Al^{3+} + 3e^- = Al$	-1.66	0
$AlF_6^{3-} + 3e^- = Al + 6F^-$	-2.07	0
$H_2AlO_2^- + H_2O + 3e^- = Al + 4OH^-$	-2.35	0
Americium		
$AmO_2^+ + 4H^+ + 2e^- = Am^{3+} + 2H_2O$	1.83	0
$AmO_2^{2+} + e^- = AmO_2^+$	1.64	$1M$ $HClO_4$
$AmO_2^+ + 4H^+ + e^- = Am^{4+} + 2H_2O$	1.26	0
$Am^{4+} + e^- = Am^{3+}$	2.4	0
$Am^{3+} + 3e^- = Am$	-2.38	0
Antimony		
$Sb(OH)_6^- + 2e^- = SbO_2^- + 2OH^- + 2H_2O$	-0.428	$3M$ NaOH
$SbO_3^- + H_2O + 2e^- = SbO_2^- + 2OH^-$	-0.589	$10M$ NaOH
$Sb(V) + 2e^- = Sb(III)$	0.82	$6M$ HCl
$Sb_2O_5 + 6H^+ + 4e^- = 2SbO^+ + 3H_2O$	0.58	0
$SbO_2^- + 2H_2O + 3e^- = Sb + 4OH^-$	-0.675	$10M$ KOH
$Sb_2O_3 + 6H^+ + 6e^- = 2Sb + 3H_2O$	0.152	0
$SbO^+ + 2H^+ + 3e^- = Sb + H_2O$	0.212	0
$Sb + 3H^+ + 3e^- = SbH_3$	-0.51	0
$Sb + 3H_2O + 3e^- = SbH_3 + 3OH^-$	-1.34	$1M$ NaOH
Arsenic		
$H_3AsO_4 + 2H^+ + 2e^- = HAsO_2 + 2H_2O$	0.559	0
$AsO_4^{3-} + 2H_2O + 2e^- = AsO_2^- + 4OH^-$	-0.67	0
$HAsO_2 + 3H^+ + 3e^- = As + 2H_2O$	0.248	0
$AsO_2^- + 2H_2O + 3e^- = As + 4OH^-$	0.68	0
$As + 3H^+ + 3e^- = AsH_3$	-0.38	0
$As + 3H_2O + 3e^- = AsH_3 + 3OH^-$	-1.21	0
Astatine		
$HAtO_3 + 4H^+ + 4e^- = HAtO + 2H_2O$	ca 1.4	0
$HAtO + H^+ + e^- = \frac{1}{2}At_2 + H_2O$	ca 0.7	0
$At_2 + 2e^- = 2At^-$	ca 0.2	0
Barium		
$Ba^{2+} + 2e^- = Ba$	-2.90	0
Berkelium		
$Bk(IV) + e^- = Bk(III)$	ca 1.6	0
Beryllium		
$Be^{2+} + 2e^- = Be$	-1.85	0
$Be_2O_3^{2-} + 3H_2O + 4e^- = 2Be + 6OH^-$	-2.62	0
Bismuth		
Bi_2O_4 (bismuthate) $+ 4H^+ + 2e^- = 2BiO^+ + 2H_2O$	1.59	
$BiO^+ + 2H^+ + 3e^- = Bi + H_2O$	0.32	0
$BiOCl + 2H^+ + 3e^- = Bi + H_2O + Cl^-$	0.160	0
$Bi_2O_3 + 3H_2O + 6e^- = 2Bi + 6OH^-$	-0.46	0
$Bi + 3H^+ + 3e^- = BiH_3$	ca -0.8	0

Table 6-2 (*Continued*)
POTENTIALS OF THE ELEMENTS AND THEIR COMPOUNDS AT 25°C

Half-reaction	Standard or Formal Potential	Solution Composition
Boron		
$H_3BO_3 + 3H^+ + 3e^- = B + 3H_2O$	−0.87	0
$H_2BO_3^- + H_2O + 3e^- = B + 4OH^-$	−1.79	0
Bromine		
$BrO_3^- + 6H^+ + 6e^- = Br^- + 3H_2O$	1.44	0
$BrO_3^- + 6H^+ + 5e^- = \frac{1}{2}Br_2 + 3H_2O$	1.52	0
$HBrO + H^+ + e^- = \frac{1}{2}Br_2 + H_2O$	1.59	0
$BrO^- + H_2O + 2e^- = Br^- + 2OH^-$	0.76	0
$Br_3^- + 2e^- = 3Br^-$	1.05	0
$Br_2(aq) + 2e^- = 2Br^-$	1.087	0
$Br_2(liq) + 2e^- = 2Br^-$	1.065	0
Cadmium		
$Cd^{2+} + 2e^- = Cd$	−0.403	0
$Cd^{2+} + 2e^- = Cd(Hg)$	−0.352	0
$Cd(NH_3)_4^{2+} + 2e^- = Cd + 4NH_3$	−0.61	0
$Cd(CN)_4^{2-} + 2e^- = Cd + 4CN^-$	−1.09	0
Calcium		
$Ca^{2+} + 2e^- = Ca$	−2.87	0
Carbon		
$\frac{1}{2}C_2N_2 + H^+ + e^- = HCN$	0.37	0
$HCNO + H^+ + e^- = \frac{1}{2}C_2N_2 + H_2O$	0.33	0
$2CO_2 + 2H^+ + 2e^- = H_2C_2O_4$	−0.49	0
$HCHO + 2H_2O + 2e^- = CH_3OH + 2OH^-$	−0.59	
$CNO^- + H_2O + 2e^- = CN^- + 2OH^-$	−0.97	0
Cerium		
$Ce(IV) + e^- = Ce(III)$	1.74	$1M$ $HClO_4$
	1.61	$1M$ HNO_3
	1.44	$0.5M$ H_2SO_4
	1.28	$1M$ HCl
$Ce^{3+} + 3e^- = Ce$	−2.48	0
Cesium		
$Cs^+ + e^- = Cs$	−2.923	0
Chlorine		
$ClO_4^- + 2H^+ + 2e^- = ClO_3^- + H_2O$	1.19	0
$ClO_4^- + 8H^+ + 7e^- = \frac{1}{2}Cl_2 + 4H_2O$	1.34	
$ClO_3^- + 6H^+ + 5e^- = \frac{1}{2}Cl_2 + 3H_2O$	1.47	0
$ClO_3^- + 6H^+ + 6e^- = Cl^- + 3H_2O$	1.45	
$ClO_3^- + 2H^+ + 3e^- = ClO_2 + H_2O$	1.15	0
$HClO_2 + 2H^+ + 2e^- = HClO + H_2O$	1.64	0
$HClO + H^+ + 2e^- = Cl^- + H_2O$	1.49	0
$HClO + H^+ + e^- = \frac{1}{2}Cl_2 + H_2O$	1.63	
$ClO^- + H_2O + 2e^- = Cl^- + 2OH^-$	0.89	0
$ClO_2(g) + e^- = ClO_2^-$	0.954	0
$Cl_2 + 2e^- = 2Cl^-$	1.3595	0
Chromium		
$Cr_2O_7^{2-} + 14H^+ + 6e^- = 2Cr^{3+} + 7H_2O$	1.33	0
	1.15	$4M$ H_2SO_4
	0.92	$0.1M$ H_2SO_4
	1.025	$1M$ $HClO_4$

Table 6-2 (*Continued*)
POTENTIALS OF THE ELEMENTS AND THEIR COMPOUNDS AT 25°C

Half-reaction	Standard or Formal Potential	Solution Composition
	0.84	0.1M HClO$_4$
	1.08	3M HCl
	0.93	0.1M HCl
$CrO_4^{2-} + 2H_2O + 3e^- = CrO_2^- + 4OH^-$	−0.12	1M NaOH
$Cr^{3+} + e^- = Cr^{2+}$	−0.41	0
$Cr(CN)_6^{3-} + e^- = Cr(CN)_6^{4-}$	−1.13	1M KCN
$Cr^{2+} + 2e^- = Cr$	−0.91	0
Cobalt		
$Co(III) + e^- = Co(II)$	1.84	3M HNO$_3$
$Co(OH)_3 + e^- = Co(OH)_2 + OH^-$	0.17	0
$Co(NH_3)_6^{3+} + e^- = Co(NH_3)_6^{2+}$	0.1	0
$Co(en)_3^{3+} + e^- = Co(en)_3^{2+}$ [en = ethylenediamine]	−0.2	0.1M en + 0.1M KNO$_3$
$Co(CN)_6^{3-} + e^- = Co(CN)_6^{3-}$	−0.8	
$Co^{2+} + 2e^- = Co$	−0.277	0
$Co(NH_3)_6^{2+} + 2e^- = Co + 6NH_3$	−0.422	
Copper		
$Cu^{2+} + 2e^- = Cu$	0.337	0
$Cu^{2+} + e^- = Cu^+$	0.159	0
$Cu^{2+} + I^- + e^- = CuI$	0.86	0
$Cu^{2+} + 2CN^- + e^- = Cu(CN)_2^-$	*ca* 1.12	0
$Cu(CN)_3^- + e^- = Cu + 3CN^-$	−1.0	7M KCN
$Cu(NH_3)_4^{2+} + e^- = Cu(NH_3)_2^+ + 2NH_3$	−0.01	1M NH$_3$ + 1M NH$_4^+$
$Cu(en)_2^{2+} + e^- = Cu(en)^+ + en$ (= ethylenediamine)	−0.35	
$CuCl_3^{2-} + e^- = Cu + 3Cl^-$	0.178	1M HCl
$Cu(NH_3)_2^+ + e^- = Cu + 2NH_3$	−0.12	1M NH$_3$ + 1M NH$_4^+$
$Cu(CN)_2^- + e^- = Cu + 2CN^-$	*ca* −0.43	0
$Cu(EDTA)^{2-} + 2e = Cu + EDTA^{-4}$	0.13	0.1M EDTA, pH 4–5
$Cu^+ + e^- = Cu$	0.52	0
$Cu_2O + H_2O + 2e^- = 2Cu + 2OH^-$	−0.361	
Dysprosium		
$Dy^{3+} + 3e^- = Dy$	−2.35	0
Erbium		
$Er^{3+} + 3e^- = Er$	−2.30	0
Europium		
$Eu^{3+} + e^- = Eu^{2+}$	−0.43	1M KCl
$Eu^{3+} + 3e^- = Eu$	−2.41	0
$Eu(EDTA)(III) + e^- = Eu(EDTA)(II)$	−0.92	0.1M EDTA, pH 6–8
Fluorine		
$F_2 + 2H^+ + 2e^- = 2HF$	3.06	0
$F_2 + 2e^- = 2F^-$	2.87	0
$F_2O + 2H^+ + 4e^- = H_2O + 2F^-$	2.1	0
Francium		
$Fr^+ + e^- = Fr$	*ca* −2.9	0
Gadolinium		
$Gd^{3+} + 3e^- = Gd$	−2.40	0
Gallium		
$Ga^{3+} + 3e^- = Ga$	−0.56	0
$Ga(OH)_4^- + 3e^- = Ga + 4OH^-$	−1.26	
Germanium		
GeO_2(s, hex) $+ 2H^+ + 2e^- = GeO$ (s, brown) $+ H_2O$	−0.118	0

Table 6-2 (*Continued*)
POTENTIALS OF THE ELEMENTS AND THEIR COMPOUNDS AT 25°C

Half-reaction	Standard or Formal Potential	Solution Composition
$H_2GeO_3 + 4H^+ + 4e^- = Ge + 3H_2O$	-0.131	0
$H_2GeO_3 + 4H^+ + 2e^- = Ge^{2+} + 3H_2O$	ca -0.3	0
$HGeO_3^- + 2H_2O + 4e^- = Ge + 5OH^-$	-1.0	
$Ge^{2+} + 2e^- = Ge$	0.23	0
$Ge + 4H^+ + 4e^- = GeH_4$	ca -0.3	0
Gold		
$Au(III) + 2e^- = Au(I)$	ca 1.41	0
$Au(III) + 3e^- = Au$	1.50	0
$AuCl_4^- + 2e^- = AuCl_2^- + 2Cl^-$	0.93	$1M$ HCl
$AuBr_4^- + 2e^- = AuBr_2^- + 2Br^-$	0.805	$1M$ HBr
$Au(OH)_3 + 3H^+ + 3e^- = Au + 3H_2O$	1.36	0
$AuCl_4^- + 3e^- = Au + 4Cl^-$	1.00	0
$AuBr_4^- + 3e^- = Au + 4Br^-$	0.86	0
$Au(CN)_2^- + e^- = Au + 2CN^-$	0.611	0
$AuCl_2^- + e^- = Au + 2Cl^-$	1.15	$1M$ Cl$^-$
$AuBr_2^- + e^- = Au + 2Br^-$	0.963	0
$Au(SCN)_2^- + e^- = Au + 2SCN^-$	0.69	
Hafnium		
$Hf(IV) + 4e^- = Hf$	-1.70	0
Holmium		
$Ho^{3+} + 3e^- = Ho$	-2.32	0
Hydrogen		
$2H^+ + 2e^- = H_2$	0.000	0
$2H_2O + 2e^- = H_2 + 2OH^-$	-0.828	0
$\frac{1}{2}H_2 + e^- = H^-$	-2.25	
$2D^+ + 2e^- = D_2$	-0.0034	
Indium		
$In^{3+} + 3e^- = In$	-0.345	0
$In(OH)_3 + 3e^- = In + 3OH^-$	-1.0	
$In^{3+} + 2e^- = In^+$	-0.40	dil H_2SO_4
Iodine		
$H_5IO_6 + H^+ + 2e^- = IO_3^- + 3H_2O$	1.6	0
$IO_3^- + 5H^+ + 4e^- = HIO + 2H_2O$	1.14	
$IO_3^- + 6H^+ + 5e^- = \frac{1}{2}I_2 + 3H_2O$	1.20	0
$HIO + H^+ + 2e^- = I^- + H_2O$	0.99	0
$HIO + H^+ + e^- = \frac{1}{2}I_2 + H_2O$	1.45	
$ICl_2^- + e^- = \frac{1}{2}I_2 + 2Cl^-$	1.06	0
$IBr_2^- + e^- = \frac{1}{2}I_2 + 2Br^-$	0.87	0
$2ICN + 2H^+ + 2e^- = I_2 + 2HCN$	0.63	0
$I_3^- + 2e^- = 3I^-$	0.5446	$0.5M$ H_2SO_4
$I_2(aq) + 2e^- = 2I^-$	0.6276	$0.5M$ H_2SO_4
$I_2(s) + 2e^- = 2I^-$	0.5345	0
Iridium		
$IrCl_6^{2-} + 4e^- = Ir + 6Cl^-$	0.835	0
$IrO_2 + 4H^+ + 4e^- = Ir + 2H_2O$	ca 0.93	0
$Ir^{3+} + 3e^- = Ir$	1.15	0
$IrCl_6^{2-} + e^- = IrCl_6^{3-}$	1.026	$1M$ HCl
$IrBr_6^{2-} + e^- = IrBr_6^{3-}$	0.99	0
Iron		
$Fe(III) + e^- = Fe(II)$	0.771	0

Table 6-2 (*Continued*)
POTENTIALS OF THE ELEMENTS AND THEIR COMPOUNDS AT 25°C

Half-reaction	Standard or Formal Potential	Solution Composition
	0.767	1M HClO$_4$
	0.746	1M HNO$_3$
	0.674	0.5M H$_2$SO$_4$
	0.64	5M HCl
	0.46	2M H$_3$PO$_4$
	−0.68	10M NaOH
Fe(CN)$_6^{3-}$ + e^- = Fe(CN)$_6^{4-}$	0.36	0
	0.71	1M HCl
Fe(EDTA)$^-$ + e^- = Fe(EDTA)$^{2-}$	0.12	0.1M EDTA, pH 4–6
FeO$_4^{2-}$ + 2H$_2$O + 3e^- = FeO$_2^-$ + 4OH$^-$	0.55	10M NaOH
Fe^{2+} + 2e^- = Fe	−0.440	0
Lanthanum		
La^{3+} + 3e^- = La	−2.52	0
Lead		
PbO$_2$ + SO$_4^{-2}$ + 4H$^+$ + 2e^- = PbSO$_4$ + 2H$_2$O	1.685	0
PbO$_2$ + 4H$^+$ + 2e^- = Pb^{2+} + 2H$_2$O	1.455	0
PbO$_2$ + H$_2$O + 2e^- = PbO + 2OH$^-$	0.28	0
Pb^{2+} + 2e^- = Pb	$\begin{cases} -0.126 \\ -0.32 \end{cases}$	0 1M NaOAc
HPbO$_2^-$ + H$_2$O + 2e^- = Pb + 3OH$^-$	−0.54	
PbSO$_4$ + 2e^- = Pb + SO$_4^{2-}$	−0.3553	0
PbF$_2$ + 2e^- = Pb + 2F$^-$	−0.350	0
PbBr$_2$ + 2e^- = Pb + 2Br$^-$	−0.280	
Lithium		
Li$^+$ + e^- = Li	−3.045	0
Lutetium		
Lu^{3+} + 3e^- = Lu	−2.25	0
Magnesium		
Mg^{2+} + 2e^- = Mg	−2.37	0
Mg(OH)$_2$ + 2e^- = Mg + 2OH$^-$	−2.69	0
Manganese		
MnO$_4^-$ + e^- = MnO$_4^{-2}$	0.564	0
MnO$_4^-$ + 4H$^+$ + 3e^- = MnO$_2$ + 2H$_2$O	1.695	0
MnO$_4^-$ + 8H$^+$ + 5e^- = Mn^{2+} + 4H$_2$O	1.51	0
MnO$_4^-$ + 2H$_2$O + 3e^- = MnO$_2$ + 4OH$^-$	0.588	0
MnO$_2$ + 4H$^+$ + 2e^- = Mn^{2+} + 2H$_2$O	1.23	0
Mn(III) + e^- = Mn(II)	1.488	7.5M H$_2$SO$_4$
Mn(CN)$_6^{3-}$ + e^- = Mn(CN)$_6^{4-}$	−0.24	1.5M NaCN
Mn(CN)$_6^{4-}$ + e^- = Mn(CN)$_6^{5-}$	−1.056	1.5M NaCN
Mn(OH)$_3$ + e^- = Mn(OH)$_2$ + OH$^-$	0.1	0
Mn^{2+} + 2e^- = Mn	−1.182	0
Mn(OH)$_2$ + 2e^- = Mn + 2OH$^-$	−1.55	0
Mercury		
2Hg^{2+} + 2e^- = Hg$_2^{2+}$	0.920	1M HClO$_4$
2HgCl$_2$ + 2e = Hg$_2$Cl$_2$ + 2Cl$^-$	0.63	
Hg^{2+} + 2e^- = Hg	0.854	0
Hg$_2^{2+}$ + 2e^- = 2Hg	0.793	0
HgO + 2H$^+$ + 2e^- = Hg + H$_2$O	0.926	0
HgO + H$_2$O + 2e^- = Hg + 2OH$^-$	0.098	0
Hg(CN)$_4^{2-}$ + 2e^- = Hg + 4CN$^-$	−0.37	

Table 6-2 (*Continued*)
POTENTIALS OF THE ELEMENTS AND THEIR COMPOUNDS AT 25°C

Half-reaction	Standard or Formal Potential	Solution Composition
$HgCl_4^{2-} + 2e^- = Hg + 4Cl^-$	0.48	
$Hg_2Cl_2(s) + 2e^- = 2Hg + 2Cl^-$	0.2676	0
$Hg_2Br_2 + 2e^- = 2Hg + 2Br^-$	0.1395	0
$Hg_2I_2 + 2e^- = 2Hg + 2I^-$	−0.0405	0
$Hg_2SO_4 + 2e^- = 2Hg + SO_4^{2-}$	0.6151	0
Molybdenum		
$H_2MoO_4(aq) + 2H^+ + e^- = MoO_2^+ + 2H_2O$	0.48	0
$Mo(VI) + e^- = Mo(V)$	0.53	2M HCl
	0.7	8M HCl
$MoO_2^+ + 4H^+ + 2e^- = Mo^{3+} + 2H_2O$	−0.01	0.5M H_2SO_4
$Mo(V) + 2e^- = Mo(III)$ (green)	−0.25	2M HCl
(red)	0.11	2M HCl
$Mo(CN)_8^{3-} + e^- = Mo(CN)_8^{4-}$	0.73	0.25M KCl
$MoO_2^{2+} + 2H^+ + e^- = MoO^{3+} + H_2O$	0.48	0
$Mo(III) + 3e^- = Mo$	*ca* −0.2	0
Neodymium		
$Nd^{3+} + 3e^- = Nd$	−2.44	0
Neptunium		
$NpO_2^{2+} + e^- = NpO_2^+$	1.153	1M HCl
$NpO_2^+ + 4H^+ + e^- = Np^{4+} + 2H_2O$	0.74	1M HNO_3
$Np^{4+} + e^- = Np^{3+}$	0.140	1M $HClO_4$
$Np^{3+} + 3e^- = Np$	−1.86	1M HCl
Nickel		
$NiO_2 + 2H_2O + 2e^- = Ni(OH)_2 + 2OH^-$	0.49	0
$NiO_2 + 4H^+ + 2e^- = Ni^{2+} + 2H_2O$	1.68	0
$Ni(CN)_4^{2-} + e^- = Ni(CN)_4^{3-}$	−0.82	1M KCN
$Ni^{2+} + 2e^- = Ni$	−0.246	0
$Ni(NH_3)_6^{2+} + 2e^- = Ni + 6NH_3$	−0.48	
Niobium		
$NbO^{3+} + 2H^+ + 2e^- = Nb^{3+} + H_2O$	−0.37	2–6M HCl
$Nb(V) + e^- = Nb(IV)$	−0.21	12M HCl
$Nb^{3+} + 3e^- = Nb$	*ca* −1.1	0
Nitrogen		
$NO_3^- + 3H^+ + 2e^- = HNO_2 + H_2O$	0.94	0
$NO_3^- + 2H^+ + e^- = NO_2 + H_2O$	0.80	0
$NO_3^- + 4H^+ + 3e^- = NO + 2H_2O$	0.96	0
$NO_3^- + H_2O + 2e^- = NO_2^- + 2OH^-$	0.01	
$NO_2 + H^+ + e^- = HNO_2$	1.07	0
$NO_2 + 2H^+ + 2e^- = NO + H_2O$	1.03	0
$HNO_2 + H^+ + e^- = NO + H_2O$	1.00	0
$HONH_3^+ + 2H^+ + 2e^- = NH_4^+ + H_2O$	1.35	0
$2HONH_3^+ + H^+ + 2e^- = N_2H_5^+ + 2H_2O$	1.46	
$H_2NNH_3^+ + 3H^+ + 2e^- = 2NH_4^+$	1.27	0
$N_2 + 2H_2O + 4H^+ + 2e^- = 2HONH_3^+$	−1.87	0
$N_2 + 5H^+ + 4e^- = H_2NNH_3^+$	−0.23	0
$3N_2 + 2H^+ + 2e^- = 2HN_3$	−3.1	0
Osmium		
$OsO_4 + 4H^+ + 4e^- = OsO_2 + 2H_2O$	0.964	0
$HOsO_5^- + 2e^- = OsO_4^{2-} + OH^-$	0.3	0

Table 6-2 (*Continued*)
POTENTIALS OF THE ELEMENTS AND THEIR COMPOUNDS AT 25°C

Half-reaction	Standard or Formal Potential	Solution Composition
$OsCl_6^{2-} + e^- = OsCl_6^{3-}$	0.85	1M HCl
$OsBr_6^{2-} + e^- = OsBr_6^{3-}$	0.349	2M HBr
$OsCl_6^{2-} + e^- = Os^{3+} + 6Cl^-$	0.4	
$OsCl_6^{3-} + 3e^- = Os + 6Cl^-$	0.71	1M HCl
$Os^{2+} + 2e^- = Os$	0.85	0
Oxygen		
$O_3 + 2H^+ + 2e^- = O_2 + H_2O$	2.07	0
$O_3 + 2H_2O + 2e^- = O_2 + 2OH^-$	1.24	
$O_2 + 4H^+ + 4e^- = 2H_2O$	1.229	0
$O_2 + 2H^+ + 2e^- = H_2O_2$	0.682	0
$O_2 + H_2O + 2e^- = HO_2^- + OH^-$	−0.076	0
$H_2O_2 + 2H^+ + 2e^- = 2H_2O$	1.77	0
$H_2O_2 + 2e^- = 2OH^-$	0.88	
$O_2 + 2H_2O + 4e^- = 4OH^-$	0.41	1M NaOH
Palladium		
$PdO_3 + H_2O + 2e^- = PdO_2 + 2OH^-$	1.22	0
$PdO_2 + H_2O + 2e^- = PdO + 2OH^-$	ca 0.73	0
$PdCl_6^{2-} + 2e^- = PdCl_4^{2-} + 2Cl^-$	1.288	1M HCl
$PdBr_6^{2-} + 2e^- = PdBr_4^{2-} + 2Br^-$	0.994	1M KBr
$PdCl_4^{2-} + 2e^- = Pd + 4Cl^-$	0.62	1M HCl
$PdBr_4^{2-} + 2e^- = Pd + 4Br^-$	0.6	0
$PdO + H_2O + 2e^- = Pd + 2OH^-$	0.07	
$Pd^{2+} + 2e^- = Pd$	0.987	4M HClO$_4$
Phosphorus		
$H_3PO_4 + 2H^+ + 2e^- = H_3PO_3 + H_2O$	−0.276	0
$H_3PO_3 + 2H^+ + 2e^- = HPH_2O_2 + H_2O$	−0.50	0
Platinum		
$PtO_4^{2-} + 4H_2O + 2e^- = Pt(OH)_6^{2-} + 2OH^-$	ca 0.4	0
$PtBr_6^{2-} + 2e^- = PtBr_4^{2-} + 2Br^-$	0.64	1M KBr
$PtCl_6^{2-} + 2e^- = PtCl_4^{2-} + 2Cl^-$	0.76	1M NaCl
$PtI_6^{2-} + 2e^- = PtI_4^{2-} + 2I^-$	0.395	1M KI
$PtBr_4^{2-} + 2e^- = Pt + 4Br^-$	0.58	0
$PtCl_4^{2-} + 2e^- = Pt + 4Cl^-$	0.73	0
$Pt(OH)_2 + 2H^+ + 2e^- = Pt + 2H_2O$	0.98	0
$Pt^{2+} + 2e^- = Pt$	ca 1.2	0
Plutonium		
$PuO_2^{2+} + e^- = PuO_2^+$	0.913	1M HClO$_4$
$PuO_2^{2+} + 4H^+ + 3e^- = Pu^{3+} + 2H_2O$	1.023	1M HClO$_4$
$PuO_2^{2+} + 4H^+ + 2e^- = Pu^{4+} + 2H_2O$	1.043	1M HClO$_4$
$PuO_2^+ + 4H^+ + e^- = Pu^{4+} + 2H_2O$	1.172	1M HClO$_4$
Pu(IV) + e^- = Pu(III)	0.970	1M HCl
	0.982	1M HClO$_4$
	0.92	1M HNO$_3$
	0.75	1M H$_2$SO$_4$
	0.50	1M HF
	0.59	0.6M H$_3$PO$_4$ + 1M HCl
	0.40	1M HOAc + 1M NaOAc
$Pu^{3+} + 3e^- = Pu$	2.02	0
Polonium		
$PoO_2 + 4H^+ + 2e^- = Po^{2+} + 2H_2O$	0.8	1M HNO$_3$

Table 6-2 (*Continued*)
POTENTIALS OF THE ELEMENTS AND THEIR COMPOUNDS AT 25°C

Half-reaction	Standard or Formal Potential	Solution Composition
$Po(IV) + 2e^- = Po(II)$	0.72	$1M$ HCl
$Po^{2+} + 2e^- = Po$	0.60	$1M$ HNO_3
Potassium		
$K^+ + e^- = K$	−2.925	0
Praseodymium		
$Pr(IV) + e^- = Pr(III)$	2.86	0
$Pr^{3+} + 3e^- = Pr$	−2.47	0
Promethium		
$Pm^{3+} + 3e^- = Pm$	−2.42	0
Protoactinium		
$PaO_2^+ + 4H^+ + 5e^- = Pa + 2H_2O$	*ca* −1.0	0
$Pa(V) + e^- = Pa(IV)$	−0.3	$6M$ HCl
Radium		
$Ra^{2+} + 2e^- = Ra$	−2.92	0
Rhenium		
$ReO_4^- + 2H^+ + e^- = ReO_3 + H_2O$	0.768	0
$ReO_4^- + 4H^+ + 3e^- = ReO_2 + 2H_2O$	0.510	0
$ReO_4^- + 2H_2O + 3e^- = ReO_2 + 4OH^-$	−0.594	0
$ReO_4^- + 8H^+ + 7e^- = Re + 4H_2O$	0.367	0
$ReO_3 + 2H^+ + 2e^- = ReO_2 + H_2O$	0.40	0
$ReO_2 + 4H^+ + 4e^- = Re + 2H_2O$	0.260	0
$ReCl_6^{2-} + e^- = ReCl_6^{3-} + 2Cl^-$	0.25	$1M$ HCl
$Re + e^- = Re^-$	−0.23	0.4–$2M$ H_2SO_4
Rhodium		
$RhO_4^{2-} + 6H^+ + 2e^- = RhO^{2+} + 3H_2O$	1.46	$0.1M$ H_2SO_4
$RhO^{2+} + 2H^+ + e^- = Rh^{3+} + H_2O$	1.435	$0.5M$ H_2SO_4
$Rh(VI) + 3e^- = Rh(III)$	1.48	$1M$ $HClO_4$
$Rh(III) + 3e^- = Rh$	*ca* 0.8	0
$RhCl_6^{2-} + e^- = RhCl_6^{3-}$	*ca* 1.2	0
$RhCl_6^{3-} + 3e^- = Rh + 6Cl^-$	0.44	0
$Rh^{2+} + e^- = Rh^+$	*ca* 0.6	0
$Rh^+ + e^- = Rh$	*ca* 0.6	0
Rubidium		
$Rb^+ + e^- = Rb$	−2.925	0
Ruthenium		
$RuO_4 + e^- = RuO_4^-$	1.00	0
$RuO_4^- + e^- = RuO_4^{2-}$	0.59	0
$RuO_4 + 8H^+ + 4e^- = Ru(IV) + 4H_2O$	1.40	$1M$ $HClO_4$
$Ru(IV) + e^- = Ru(III)$	0.908	$0.5M$ HCl
$Ru(III) + e^- = Ru(II)$	0.084	1.5–$6.8M$ HCl
$Ru(CN)_6^{3-} + e^- = Ru(CN)_6^{4-}$	0.86	$0.05M$ H_2SO_4
$RuCl_5^{2-} + 3e^- = Ru + 5Cl^-$	0.4	
Samarium		
$Sm^{3+} + 3e^- = Sm$	−2.41	0
$Sm^{3+} + e^- = Sm^{2+}$	−1.35	0
Scandium		
$Sc^{3+} + 3e^- = Sc$	−2.08	0
Selenium		
$SeO_4^{2-} + 4H^+ + 2e^- = H_2SeO_3 + H_2O$	1.15	0
$SeO_4^{2-} + H_2O + 2e^- = SeO_3^{2-} + 2OH^-$	0.05	0

Table 6-2 (*Continued*)
POTENTIALS OF THE ELEMENTS AND THEIR COMPOUNDS AT 25°C

Half-reaction	Standard or Formal Potential	Solution Composition
$H_2SeO_3 + 4H^+ + 4e^- = Se + 3H_2O$	0.740	0
$SeO_4^{2-} + 3H_2O + 4e^- = Se + 6OH^-$	−0.366	0
$Se + 2H^+ + 2e^- = H_2Se$	−0.40	0
$Se + 2e^- = Se^{2-}$	−0.92	0
Silicon		
$SiO_2(s) + 4H^+ + 4e^- = Si + 2H_2O$	−0.86	0
$SiF_6^{2-} + 4e^- = Si + 6F^-$	−1.2	0
$Si + 4H^+ + 4e^- = SiH_4(g)$	0.102	0
Silver		
$AgO^+ + 2H^+ + e^- = Ag^{2+} + H_2O$	ca 2.1	4M HNO$_3$
$Ag^{2+} + e^- = Ag^+$	1.927	4M HNO$_3$
	2.000	4M HClO$_4$
$Ag_2O_3 + H_2O + 2e^- = 2AgO + 2OH^-$	0.74	0
$AgO + H^+ + e^- = \frac{1}{2}Ag_2O + \frac{1}{2}H_2O$	1.41	0
$Ag^+ + e^- = Ag$	0.7995	0
$Ag_2O + 2H^+ + 2e^- = 2Ag + H_2O$	1.17	0
$Ag_2SO_4 + 2e^- = 2Ag + SO_4^{2-}$	0.653	0
$Ag_2CrO_4 + 2e^- = 2Ag + CrO_4^{2-}$	0.447	0
$Ag_2O + H_2O + 2e^- = 2Ag + 2OH^-$	0.342	0
$Ag(NH_3)_2^+ + e^- = Ag + 2NH_3$	0.373	0
$AgCl + e^- = Ag + Cl^-$	0.2223	0
$AgBr + e^- = Ag + Br^-$	0.0713	0
$AgCN + e^- = Ag + CN^-$	−0.017	
$AgI + e^- = Ag + I^-$	−0.152	0
$AgSCN + e^- = Ag + SCN^-$	0.09	
$Ag(CN)_2^- + e^- = Ag + 2CN^-$	−0.31	
$Ag_2S + 2e^- = 2Ag + S^{2-}$	−0.69	0
$AgN_3 + e^- = Ag + N_3^-$	0.293	0
Sodium		
$Na^+ + e^- = Na$	−2.714	0
Strontium		
$Sr^{2+} + 2e^- = Sr$	−2.89	0
Sulfur		
$S_2O_8^{2-} + 2e^- = 2SO_4^{2-}$	2.01	0
$SO_4^{2-} + 4H^+ + 2e^- = SO_2(aq) + H_2O$	0.17	0
$SO_4^{2-} + H_2O + 2e^- = SO_3^{2-} + 2OH^-$	−0.93	
$4SO_2(aq) + 4H^+ + 6e^- = S_4O_6^{2-} + 2H_2O$	0.51	
$2SO_2(aq) + 2H^+ + 4e^- = S_2O_3^{2-} + H_2O$	0.40	0
$2SO_3^{2-} + 3H_2O + 4e^- = S_2O_3^{2-} + 6OH^-$	−0.58	0
$S_4O_6^{2-} + 2e^- = 2S_2O_3^{2-}$	0.08	0
$SO_3^{2-} + 3H_2O + 4e^- = S + 6OH^-$	−0.66	0
$S_2^{2-} + 2e^- = 2S^{2-}$	−0.48	0
$S + 2H^+ + 2e^- = H_2S(g)$	0.141	0
$S + 2e^- = S^{2-}$	−0.48	
$(SCN)_2 + 2e^- = 2SCN^-$	0.77	
Tantalum		
$Ta_2O_5 + 10H^+ + 10e^- = 2Ta + 5H_2O$	−0.81	0
Technetium		
$TcO_4^- + 4H^+ + 3e^- = TcO_2 + 4H_2O$	0.738	0
$TcO_2 + 4H^+ + 2e^- = Tc^{2+} + 2H_2O$	ca 0.6	0

Table 6-2 (*Continued*)
POTENTIALS OF THE ELEMENTS AND THEIR COMPOUNDS AT 25°C

Half-reaction	Standard or Formal Potential	Solution Composition
$TcO_4^- + 2H^+ + e^- = TcO_3 + H_2O$	0.7	
$TcO_4^- + 8H^+ + 7e^- = Tc + 4H_2O$	0.472	0
$TcO_4^- + 4H_2O + 7e^- = Tc + 8OH^-$	−0.311	0
$TcO_3 + 2H^+ + 2e^- = TcO_2 + H_2O$	0.8	0
$TcO_2 + 4H^+ + 4e^- = Tc + 2H_2O$	0.272	0
Tellurium		
$H_6TeO_6 + 2H^+ + 2e^- = TeO_2 + 4H_2O$	1.02	0
$TeO_4^{2-} + H_2O + 2e^- = TeO_3^{2-} + 2OH^-$	*ca* 0.4	0
$TeOOH^+ + 3H^+ + 4e^- = Te + 2H_2O$	0.559	0
$TeO_2(s) + 4H^+ + 4e^- = Te + 2H_2O$	0.529	0
$TeCl_6^{2-} + 4e^- = Te + 6Cl^-$	0.646	0
$TeO_3^{2-} + 3H_2O + 4e^- = Te + 6OH^-$	−0.57	0
$Te + 2H^+ + 2e^- = H_2Te(g)$	0.72	0
$Te + 2e^- = Te^{2-}$	−1.14	0
Terbium		
$Tb^{3+} + 3e^- = Tb$	−2.39	0
Thallium		
$Tl(III) + 2e^- = Tl(I)$	1.26	$1M$ $HClO_4$
	1.22	0.5–5M H_2SO_4 or HNO_3
	0.77	0.5–1M HCl
$Tl(OH)_3 + 2e^- = Tl^+ + 3OH^-$	−0.05	0
$Tl^+ + e^- = Tl$	−0.3360	0
$TlOH + e^- = Tl + OH^-$	−0.345	
$TlCl + e^- = Tl + Cl^-$	−0.557	0
$TlBr + e^- = Tl + Br^-$	0.658	0
$TlI + e^- = Tl + I^-$	−0.766	0
Thorium		
$Th^{4+} + 4e^- = Th$	−1.90	0
Thulium		
$Tm^{3+} + 3e^- = Tm$	−2.28	0
Tin		
$Sn^{4+} + 2e^- = Sn^{2+}$	0.154	0
$SnCl_6^{2-} + 2e^- = SnCl_4^{2-} + 2Cl^-$	0.14	$1M$ HCl
$Sn(OH)_6^{2-} + 2e^- = HSnO_2^- + H_2O + 3OH^-$	−0.93	
$Sn^{2+} + 2e^- = Sn$	−0.136	0
$SnCl_4^{2-} + 2e^- = Sn + 4Cl^-$	−0.19	$1M$ HCl
$HSnO_2^- + H_2O + 2e^- = Sn + 3OH^-$	−0.91	
Titanium		
$TiO^{2+} + 2H^+ + e^- = Ti^{3+} + H_2O$	0.1	0
$TiOCl^+ + 2H^+ + 3Cl^- + e^- = TiCl_4^- + H_2O$	−0.09	$1M$ HCl
	+0.24	6M HCl
$TiF_6^{2-} + 4e^- = Ti + 6F^-$	−1.24	
$Ti(IV) + e^- = Ti(III)$	0.092	2M H_2SO_4
	0.130	4M H_2SO_4
	−0.05	1M H_3PO_4
$Ti^{3+} + e^- = Ti^{2+}$	−0.37	0
$Ti^{2+} + 2e^- = Ti$	−1.63	0
Tungsten		
$WO_2Cl_3^- + 2H^+ + 2Cl^- + e^- = WOCl_5^{2-} + H_2O$	0.26	12M HCl
$2WO_3 + 2H^+ + 2e^- = W_2O_5 + H_2O$	−0.03	

Table 6-2 (*Continued*)
POTENTIALS OF THE ELEMENTS AND THEIR COMPOUNDS AT 25°C

Half-reaction	Standard or Formal Potential	Solution Composition
$WO_3 + 6H^+ + 6e^- = W + 3H_2O$	-0.09	0
$WO_4^{2-} + 4H_2O + 6e^- = W + 8OH^-$	-1.05	0
$W_2O_5 + 2H^+ + 2e^- = 2WO_2 + H_2O$	-0.043	
$WOCl_5^{2-} + 2H^+ + 2e^- = WCl_5^{2-}(red) + H_2O$	-0.2	12M HCl
$2WOCl_5^{2-} + 4H^+ + 4e^- = W_2Cl_9^{3-}(green) + 2H_2O + Cl^-$	$+0.1$	12M HCl
$W(CN)_8^{3-} + e^- = W(CN)_8^{4-}$	0.457	0
$WO_2 + 4H^+ + 4e^- = W + 2H_2O$	-0.12	
$W(V) + e^- = W(IV)$	-0.3	12M HCl
Uranium		
$UO_2^{2+} + e^- = UO_2^+$	0.052	0
$UO_2^{2+} + 4H^+ + 2e^- = U^{4+} + 2H_2O$	0.334	0
$UO_2^+ + 4H^+ + e^- = U^{4+} + 2H_2O$	0.55	0
$U^{4+} + e^- = U^{3+}$	-0.61	0
$U^{3+} + 3e^- = U$	-1.80	0
Vanadium		
$VO_2^+ + 2H^+ + e^- = VO^{2+} + H_2O$	1.000	0
$VO^{2+} + 2H^+ + e^- = V^{3+} + H_2O$	0.337	0
$V^{3+} + e^- = V^{2+}$	-0.255	0
$V^{2+} + 2e^- = V$	-1.18	0
Ytterbium		
$Yb^{3+} + e^- = Yb^{2+}$	-1.15	0.1M NH$_4$Cl
$Yb^{3+} + 3e^- = Yb$	-2.27	0
Yttrium		
$Y^{3+} + 3e^- = Y$	-2.37	0
Xenon		
$H_4XeO_6 + 2H^+ + 2e^- = XeO_3 + 3H_2O$	$ca\ -2.3$	
$HXeO_6^{3-} + 2H_2O + e^- = HXeO_4 + 4OH^-$	$ca\ -0.9$	
$XeO_3 + 6H^+ + 6e^- = Xe + 3H_2O$	$ca\ -1.8$	
$HXeO_4 + 3H_2O + 7e^- = Xe + 7OH^-$	$ca\ -0.96$	
$XeO_3 + 6H^+ + 2F^- + 4e^- = XeF_2 + 3H_2O$	$ca\ -1.6$	
$XeF_2 + 2e^- = Xe + 2F^-$	$ca\ -2.2$	
Zinc		
$Zn^{2+} + 2e^- = Zn$	-0.763	0
$ZnO_2^{2-} + 2H_2O + 2e^- = Zn + 4OH^-$	-1.216	
$Zn(NH_3)_4^{2+} + 2e^- = Zn + 4NH_3$	-1.04	
$Zn(CN)_4^{2-} + 2e^- = Zn + 4CN^-$	-1.26	
Zirconium		
$Zr(IV) + 4e^- = Zr$	-1.53	0

Table 6-3
SELECTED LIST OF OXIDATION-REDUCTION INDICATORS

In the column headed π_0 are given the potentials, referred to the normal hydrogen electrode, at which the indicators are "half-changed"; these potentials are those of solutions one normal in hydrogen ions.

The information given in this table is from the two-volume work VOLUMETRIC ANALYSIS by Kolthoff and Stenger, published by Interscience Publishers, Inc., New York, 1942 and 1947, and reproduced with their permission.

* Distinct color change 1.20.   † Color change 1.31.

Indicator	π_0 at $pH = 0$	Color Change (Ox)	Color Change (Red)
Safranine T	0.24	red	colorless
Neutral red	0.24	red	colorless
Indigo monosulfonate	0.26	blue	colorless
Phenosafranine	0.28	red	colorless
Indigo tetrasulfonate	0.36	blue	colorless
Nile blue	0.41	blue	colorless
Methylene blue	0.53	green-blue	colorless
1-Naphthol-2-sulfonic acid indophenol	0.54	red	colorless
2,6-Dibromophenol indophenol	0.67	blue	colorless
Diphenylamine (diphenylbenzidine)	0.76 ± 0.1	violet	colorless
Diphenylamine sulfonic acid	0.85	red-violet	colorless
Erioglaucin A	1.0	red	green
Setoglaucin O	1.06	pale red	yellow-green
p-Nitrodiphenylamine	1.06	violet	colorless
o, m'-Diphenylamine dicarboxylic acid	1.12	blue-violet	colorless
o, o'-Diphenylamine dicarboxylic acid	1.26	blue-violet	colorless
o-Phenanthroline ferrous complex	1.14 *	pale blue	red
Nitro-o-phenanthroline ferrous complex	1.25 †	pale blue	violet-red

Table 6-4
OVERPOTENTIALS FOR COMMON ELECTRODE REACTIONS AT 25°C

The overpotential is defined as the difference between the actual potential of an electrode at a given current density and the reversible electrode potential for the reaction.

Electrode	Current Density, A/cm²					
	0.001	0.01	0.1	0.5	1.0	5.0
	Overpotential, volts					
Liberation of H_2 from $1M$ H_2SO_4						
Ag	0.097	0.13	0.3		0.48	0.69
Al	0.3	0.83	1.00		1.29	
Au	0.017		0.1		0.24	0.33
Bi	0.39	0.4			0.78	0.98
Cd		1.13	1.22		1.25	
Co		0.2				
Cr		0.4				
Cu			0.35		0.48	0.55
Fe		0.56	0.82		1.29	
Graphite	0.002		0.32		0.60	0.73

6-17

Table 6-4 (*Continued*)
OVERPOTENTIALS FOR COMMON ELECTRODE
REACTIONS AT 25°C

Electrode	Current Density, A/cm^2					
	0.001	0.01	0.1	0.5	1.0	5.0
	Overpotential, volts					
Liberation of H$_2$ from 1M H$_2$SO$_4$						
Hg	0.8	0.93	1.03		1.07	
Ir	0.0026	0.2				
Ni	0.14	0.3			0.56	0.71
Pb	0.40	0.4			0.52	1.06
Pd	0	0.04				
Pt (smooth)	0.0000	0.16	0.29		0.68	
Pt (platinized)	0.0000	0.030	0.041		0.048	0.051
Sb		0.4				
Sn		0.5	1.2			
Ta		0.39	0.4			
Zn	0.48	0.75	1.06		1.23	
Liberation of O$_2$ from 1M KOH						
Ag	0.58	0.73	0.98		1.13	
Au	0.67	0.96	1.24		1.63	
Cu	0.42	0.58	0.66		0.79	
Graphite	0.53	0.90	1.09		1.24	
Ni	0.35	0.52	0.73		0.85	
Pt (smooth)	0.72	0.85	1.28		1.49	
Pt (platinized)	0.40	0.52	0.64		0.77	
Liberation of Cl$_2$ from saturated NaCl solution						
Graphite			0.25	0.42	0.53	
Platinized Pt	0.006		0.026	0.05		
Smooth Pt	0.008	0.03	0.054	0.161	0.236	
Liberation of Br$_2$ from saturated NaBr solution						
Graphite		0.002	0.027	0.16	0.33	
Platinized Pt		0.002	0.012	0.069	0.21	
Smooth Pt		0.002	0.006*	0.26	0.38†	
Liberation of I$_2$ from saturated NaI solution						
Graphite	0.002	0.014	0.097			
Platinized Pt		0.006	0.032		0.196	
Smooth Pt		0.003	0.03	0.12	0.22	

* At 0.23 A/cm^2.   † At 0.72 A/cm^2.

The overpotential required for the evolution of O$_2$ from dilute solutions of HClO$_4$, HNO$_3$, H$_3$PO$_4$ or H$_2$SO$_4$ onto smooth platinum electrodes is approximately 0.5 V.

Table 6-5
POLAROGRAPHY

The values in the column headed $E_{\frac{1}{2}}$ are the half-wave potentials in units of volts referred to the saturated calomel electrode. The values in the column headed I_d are for $i_d/C \cdot m_{2/3} \cdot t^{1/6}$, where i_d is the diffusion current in microamperes, C is the concentration in millimoles per liter, m is the rate of flow of mercury in milligrams per second, and t is the drop time in seconds. The term "too positive" means that the diffusion current starts from zero applied emf and that the potential involved is more positive than the oxidation potential of mercury.

Half-wave Potentials of Inorganic Ions at 25°C

Ion Electrode Reaction	Electrolyte	$E_{1/2}$	I_d
$Ag^+ \rightarrow Ag$	KNO_3	*	.....
$Al^{3+} \rightarrow Al$	$0.05\,N\,BaCl_2$	-1.75	.....
$As^{3+} \rightarrow As$	$1\,N\,H_2SO_4$-0.01% gelatin†	-0.7	8.4
$As \rightarrow AsH_3$		-1.0	
$As^{3+} \rightarrow As^{5+}$	$0.5\,N\,KOH$-0.025% gelatin	-0.26	3.82
$Au^{3+} \rightarrow Au^+$	$0.1\,M\,KCN$†	*	.....
		-1.4	.....
$Ba^{2+} \leftrightarrow Ba(Hg)$	$0.1\,M\,(C_2H_5)_4NI$-50% C_2H_5OH	-1.875	2.91
$Be^{2+} \rightarrow Be$	$0.05\,M\,CH_3CO_2K$	-1.8‡	.....
$Bi^{3+} \rightarrow Bi$	$1\,M\,HCl$-0.01% gelatin	-0.09	.....
$Bi^{3+} \rightarrow Bi$	$0.5\,Na$ tartrate (ph 4.5)	-0.29	.....
$Ca^{2+} \rightarrow Ca(Hg)$	$0.1\,M\,(C_2H_5)_4NI$-50% C_2H_5OH-3.15 × $10^{-5}\,M\,Ba^{2+}$	-2.12	2.87
$Cd^{2+} \rightarrow Cd$	$0.1\,M\,KCl$-0.01% gelatin	-0.599	3.51
$Cd^{2+} \rightarrow Cd$	$1\,M\,HCl$-0.01% gelatin	-0.64	3.58
$Cd^{2+} \rightarrow Cd$	$0.5\,M\,Na$ tartrate-0.01% gelatin (pH 9)	-0.64	2.34
$Cd^{2+} \rightarrow Cd$	$1\,M\,NH_3$-1 $M\,NH_4Cl$-0.01% gelatin	-0.81	3.68
$Cd^{2+} \rightarrow Cd$	$1\,M\,KCN$	-1.18	.....
$Co^{2+} \rightarrow Co$	$0.05\,M\,K_2SO_4$	-1.428	.....
$Co^{3+} \rightarrow Co$	$1\,M\,NH_4Cl$-1 $M\,NH_3$-gelatin†	-0.5	.....
$Co^{2+} \rightarrow Co(Hg)$		-1.3	.....
$Co^{2+} \rightarrow Co$	$0.1\,M$ pyridine-0.1 M pyridinium chloride	-1.07	.....
$Co^{2+} \rightarrow Co$	$1\,M\,KSCN$	-1.03	.....
$Co^{2+} \rightarrow Co^+$	$1\,M\,KCN$	-1.2	.....
$Cr^{3+} \rightarrow Cr^{2+}$	$0.1\,M\,KCl$-0.01% gelatin†	-0.91	.....
$Cr^{2+} \rightarrow Cr$		-1.47	.....
$Cr^{3+} \rightarrow Cr^{2+}$	$0.1\,M$ pyridine-0.1 M pyridinium chloride	-0.95	.....
$Cr^{2+} \rightarrow Cr^{3+}$	$1\,M\,KCl$	-0.40	1.54
$Cr^{2+} \rightarrow Cr^{3+}$	$1\,M\,KSCN$	-0.80	1.64
$Cr^{6+} \rightarrow Cr^{3+}$	$1\,M\,NaOH$	-0.85	.....
$Cr^{6+} \rightarrow Cr^{3+}$	$0.1\,M\,KCl$§	$-0.3, -1.0$	.....
$Cr^{3+} \rightarrow Cr^{2+}$		-1.5	.....
$Cr^{2+} \rightarrow Cr$		-1.7	.....
$Cs^+ \rightarrow Cs$	$0.1\,M\,(C_2H_5)_4NOH$-50% C_2H_5OH	-2.05	.....
$Cu^{2+} \rightarrow Cu$	$0.1\,M\,KCl\,(HCl)$-0.01% gelatin	$+0.04$	3.23
$Cu^{2+} \rightarrow Cu$	$1\,M\,NaOH$-0.01% gelatin	-0.42	2.91
$Cu^{2+} \rightarrow Cu$	$0.5\,M\,Na$ tartrate-0.01% gelatin (pH 9)	-0.12	2.24
$Cu^{2+} \rightarrow Cu^+$	$1\,M\,NH_3$-1 $M\,NH_4Cl$†	-0.24	3.75
$Cu^+ \rightarrow Cu$		-0.50	.....
$Cu^{2+} \rightarrow Cu^+$	$1\,M\,K_2C_2O_4$ (pH 5.7 to 10)	-0.27	.....
$Cu^{2+} \leftrightarrow Cu$	$0.1\,M\,Na_4P_2O_7$-0.2 $M\,CH_3CO_2Na$ (pH 4.5)	-0.085	.....
$Eu^{3+} \rightarrow Eu^{2+}$	$0.1\,M\,NH_4Cl$	-0.671	.....

* Too positive.
† One run; two waves.
‡ Poorly defined.   § One run; three waves.

Table 6-5 (*Continued*)
POLAROGRAPHY

Ion Electrode Reaction	Electrolyte	$E_{1/2}$	I_d
$Fe^{3+} \to Fe^{2+}$	$1\ M\ NH_4ClO_4$†	{ *	.....
$Fe^{2+} \to Fe$		{ -1.44^a	.....
$Fe^{3+} \to Fe^{2+}$	$0.5\ M$ Na citrate-0.005% gelatin (pH 6)	-0.222	0.93
$Fe^{3+} \to Fe^{2+}$	$0.5\ M$ Na citrate-0.005% gelatin (pH 5.8)†	-0.21	1.11
$Fe^{2+} \to Fe$		-1.53	.....
$Fe^{3+} \to Fe^{2+}$	$0.2\ M\ Na_2C_2O_4$ (pH 5.25)	-0.245	1.50
$Fe^{2+} \to Fe$	$1\ M$ NaOH	-1.46	.....
$Ga^{3+} \to Ga$	$0.1\ M$ KCl	-1.15^b	.....
$Gd^{3+} \to Gd$	$0.1\ M$ LiCl-0.01% gelatin	-1.77	3.7
$Ge^{2+} \to Ge$	$6\ M$ HCl	-0.45	.....
$Hg^{2+} \to Hg^+$	$0.1\ N$ HNO$_3$	*	.....
$IO_4^- \to IO_3^-$	pH 12	-0.08	.....
$IO_3^- \to I^-$	$0.1\ M$ Na citrate (pH 5.95)	-0.650	.....
$In^{3+} \to In$	$0.1\ M$ KCl-gelatin	-0.561	.....
$K^+ \leftrightarrow K(Hg)$	$0.1\ M\ (C_2H_5)_4NOH$-50% C_2H_5OH	-2.10	1.69
$Li^+ \leftrightarrow Li(Hg)$	$0.1\ M\ (C_2H_5)_4NOH$-50% C_2H_5OH	-2.31	1.16
$Mg^{2+} \to Mg(Hg)$	$0.1\ M\ (CH_3)_4NCl$	-2.2‡	.....
$Mn^{2+} \to Mn$	$1\ M$ KCl	-1.51^c	.....
$Mn^{2+} \to Mn$	$1\ M\ NH_4Cl$-$1\ M\ NH_3$-0.005% gelatin	-1.65	.....
$Mn^{2+} \to Mn$	$1.5\ M$ KCN	-1.33	.....
$Mn^{2+} \to Mn$	$1\ M$ KSCN-0.01% gelatin	-1.553^c	.....
$Mo^{6+} \to Mo^{5+}$	$0.3\ M$ HCl†	{ -0.26	.....
$Mo^{5+} \to Mo^{3+}$		{ -0.63	.....
$Mo^{6+} \to Mo^{5+}$	$10\ M\ H_2SO_4$†	{ *	3.99
$Mo^{5+} \to Mo^{3+}$		{ -0.13	.....
$Mo^{6+} \to Mo^{5+}$	$0.03\ M\ Na_2HPO_4$-$0.1\ M$ citric acid-$0.1\ M$ KCl†	{ -0.23	.....
$Mo^{5+} \to Mo^{3+}$		{ -0.58	.....
$NO_3^- \to NH_2OH$	$0.01\ M$ LaCl	-1.3 to -1.5	.....
$NC_3^- \to {}_{1/2}N_2$	$0.1\ M$ KCl-$0.01\ M$ HCl-$2 \times 10^{-4}\ M\ UO_2Cl_2$	-0.98^d	13.95
$NO_2^- \to {}_{1/2}N_2$	$0.1\ M$ KCl-$0.01\ M$ HCl-$2 \times 10^{-4}\ M\ UO_2(O_2CCH_3)_2$	-1.0	7.45
$Na^+ \leftrightarrow Na(Hg)$	$0.1\ M\ (C_2H_5)_4NOH$-50% C_2H_5OH	-2.07	1.40
$Nb^{5+} \to Nb^{3+}$	$0.9\ M$ HNO$_3$	-0.76	.....
$Nd^{3+} \to Nd$	$0.1\ M\ (CH_3)_4NI$-$0.02\ M\ H_2SO_4$-0.1% gelatin	-1.82	4.40
$Ni^{2+} \to Ni$	$0.01\ M$ KCl	-1.1	.....
$Ni^{2+} \to Ni$	$1\ M$ KSCN	-0.70	.....
$Ni^{2+} \to Ni$	$1\ M$ KCl-$0.5\ M$ pyridine	-0.78	.....
$Ni^{2+} \to Ni$	$1\ M\ NH_3$-$1\ M\ NH_4Cl$	-1.09	3.56
$Ni^{2+} \to Ni$	$1\ M$ KCN-0.01% gelatin	-1.36	.....
$Ni^+ \to Ni^{2+}$	$1\ M$ KCN	-0.80	.....
$O_2 \to H_2O_2$	pH 1 to 10 with max. suppressor†	{ -0.05	12.3
$H_2O_2 \to H_2O$		{ -0.94	.....
$Os^{8+} \to Os^{6+}$	Saturated $Ca(OH)_2$§	{ *	.....
$Os^{6+} \to Os^{5+}$		{ -0.41	.....
$Os^{5+} \to Os^{3+}$		{ -1.16	.....
$Pb^{2+} \to Pb$	$0.1\ M$ KCl-0.01% gelatin	-0.396	3.80
$Pb^{2+} \to Pb$	$1\ M$ HNO$_3$-0.01% gelatin	-0.405	3.67
$Pb^{2+} \to Pb$	$1\ M$ NaOH-0.01% gelatin	-0.755	3.39
$Pb^{2+} \to Pb$	$0.5\ M$ Na tartrate (pH 9)	-0.50	2.30
$Pb^{2+} \to Pb$	$1\ M$ KCN	-0.72	.....
$Pd^{2+} \to Pd$	$1\ M$ KCl-$1\ M$ pyridine	-0.36^f	.....
$Pd^{2+} \to Pd$	$1\ M$ KCN	-1.77	.....

* Too positive.   †One run; two waves. ^{a} 0.01 M. ^{b} 0.005 M.
‡ Poorly defined. ^{c} Two seconds.
^{d} $4 \times 10^{-4}\ M$.   § One run; three waves. ^{f} 0.005 M

Table 6-5 (*Continued*)
POLAROGRAPHY

Ion Electrode Reaction	Electrolyte	$E_{1/2}$	I_d
$Pd^{2+} \rightarrow Pd$	1 M NH_3-1 M NH_4Cl	-0.80^f	.....
$Pr^{3+} \rightarrow Pr$	0.1 M LiCl-0.01% gelatin	-1.75^g	3.59
$Ra^{2+} \rightarrow Ra$	0.02 M KCl	-1.84	.....
$Rb^+ \rightarrow Rb$	0.1 M $(C_2H_5)_4NOH$-50% C_2H_5OH	-1.99	.....
$Re^{7+} \rightarrow Re^{4+}$	4 M $HClO_4$	-0.39^a	6.69
$Re^{7+} \rightarrow Re^-$	2 M KCl†	$\left\{ \begin{array}{c} -1.41 \\ -1.70 \end{array} \right.$	17.9^b
Catalytic wave^c			.....
$Re^{3+} \rightarrow Re^{2+}$	2 M $HClO_4$†	$\left\{ \begin{array}{c} -0.28 \\ -0.46 \end{array} \right.$	.....
$Re^{2+} \rightarrow Re$			.....
$Rh^{3+} \rightarrow Rh^{2+}$	1 M KCN	-1.47	.....
$Rh^{3+} \rightarrow Rh^{2+}$	1 M NH_4Cl	-0.93	.....
$Rh^{3+} \rightarrow Rh^{2+}$	1 M Pyridine-1 M KCl	-0.41	.....
$Ru^{4+} \rightarrow Ru^{3+}$	1 M $HClO_4$§	$\left\{ \begin{array}{c} *, +0.21 \\ -0.34 \end{array} \right.$	1.53
$Ru^{3+} \rightarrow Ru^{2+}$			1.38
$HSO_3^- \rightarrow HSO_2$	pH 6†	$\left\{ \begin{array}{c} -0.67 \\ -1.23 \end{array} \right.$	.....
$S_2O_4^{2-} \rightarrow S_2O_3^{2-}$	#		.....
$H_2SO_3 \leftrightarrow H_2SO_2$	0.1 M HNO_3	-0.37	.....
$S_2O_4^{2-} \leftrightarrow 2SO_3^{2-}$	0.5 M $(NH_4)_2HPO_4$-1 M NH_3-0.01% gelatin	-0.43	4.09
$S \rightarrow H_2S$	CH_3OH-pyridine-pyridinium hydrochloride (pH 6)	-0.50	.....
$Sb^{3+} \rightarrow Sb$	1 N H_2SO_4-0.01% gelatin	-0.32	.....
$Sb^{3+} \rightarrow Sb$	1 N NaOH-0.01% gelatin	-1.26	.....
$Sb^{3+} \rightarrow Sb$	0.5 M Na tartrate-0.1 M NaOH-0.01% gelatin	-1.32	.....
$Sb^{3+} \rightarrow Sb^{5+}$	1 M KOH	-0.45	.....
$Sc^{3+} \rightarrow Sc$	0.1 M LiCl-HCl (0.25 of Sc^{3+} concn.)	-1.80	.....
$Se^{4+} \rightarrow Se^{2-}$	1 M NH_4Cl-0.1 to 1 M NH_3-0.003% gelatin (pH 8.0)	-1.44	.....
$Se^{4+} \rightarrow Se^{2-}$	1 M NH_4Cl-0.1 to 1 M NH_3-0.003% gelatin (pH 9.5)	-1.54	.....
$Sm^{3+} \rightarrow Sm^{2+}$	0.1 M $(CH_3)_4NI$-0.001 N H_2SO_4-0.01% gelatin†	$\left\{ \begin{array}{c} -1.80 \\ ..... \end{array} \right.$	3.85^d
$Sm^{2+} \rightarrow Sm$			8.20^d
$Sn^{2+} \rightarrow Sn$	1 M HCl-0.01% gelatin	-0.47	4.07
$Sn^{2+} \rightarrow Sn$	1 M NaOH-0.01% gelatin†	$\left\{ \begin{array}{c} -1.22 \\ -0.73 \end{array} \right.$	3.45
$Sn^{2+} \rightarrow Sn^{4+}$			3.45
$Sn^{2+} \rightarrow Sn$	0.5 M Na tartrate-0.01% gelatin-0.1 M NaOH†	-0.71	2.86
$Sn^{2+} \rightarrow Sn^{4+}$		-1.16	2.86
$Sn^{4+} \rightarrow Sn^{2+}$	4 M NH_4Cl-1 M HCl-0.005% gelatin†	$\left\{ \begin{array}{c} -0.25 \\ -0.52 \end{array} \right.$	2.84
$Sn^{2+} \rightarrow Sn$			3.49
$Sr^{2+} \leftrightarrow Sr(Hg)$	0.1 M $(C_2H_5)_4NI$-50% C_2H_5OH	-2.06	3.17
$Te^{4+} \rightarrow Te$	1 M NH_4Cl-NH_3-0.003% gelatin (pH 8.4)	-0.63	.....
$Te^{4+} \rightarrow Te$	1 M NH_4Cl-NH_3-0.003% gelatin (pH 9.4)	-0.68	.....
$Te^{4+} \rightarrow Te^{2-}$	1 M NaOH-0.003% gelatin	-1.19	9.75
$Te^{2-} \leftrightarrow Te$	1 M HCl-0.003% gelatin	-0.72	.....
$Ti^{4+} \rightarrow Ti^{2+}$	0.1 M HCl-0.005% gelatin	-0.81	1.56
$Ti^{2+} \rightarrow Ti^{4+}$	0.01 M HCl	-0.14	.....
$Ti^{4+} \rightarrow Ti^{2+}$	0.1 M $K_2C_2O_4$-1 M H_2SO_4-0.005% gelatin	-0.173	1.25
$Ti^{4+} \rightarrow Ti^{2+}$	0.4 M Na tartrate-0.005% gelatin (pH 6.9)	-1.32	1.22
$Ti^{4+} \rightarrow Ti^{2+}$	0.4 M Na citrate-0.005% gelatin (pH 5.7)	-0.90	1.02
$Tl^+ \rightarrow Tl$	0.1 M KNO_3 or 0.1 M KCl, etc.	-0.460	2.13
$U^{6+} \leftrightarrow U^{5+}$	0.1 M KCl-0.01 M HCl-2 $\times$ 10^{-4}% thymol†	$\left\{ \begin{array}{c} -0.18 \\ -0.92 \end{array} \right.$	1.51
$U^{5+} \rightarrow U^{3+}$			3.20
$U^{4+} \rightarrow U^{3+}$	0.1 M $HClO_4$	-0.862	.....

^{f} 0.005 M. $\quad$ ^{g} 0.0025 M.
^{a} 0.001 M. $\quad$ † One run; two waves. $\quad$ ^{b} 0.0003 M.
^{c} $Re^- + 2H^+ \rightarrow Re^+ + H_2$; $Re^+ + 2e \rightarrow Re^-$.
§ One run; three waves. $\quad$ * Too positive.
# $2HSO_2 \rightarrow 2H^+ + S_2O_4^{2-}$.
^{d} 0.001 M
† One run; two waves.

Table 6-5 (*Continued*)
POLAROGRAPHY

Ion Electrode Reaction	Electrolyte	$E_{1/2}$	I_d
$/^{5+} \to V^{4+}$	0.05 M H$_2$SO$_4$-0.005% gelatin†	*	1.65
$V^{4+} \to V^{2+}$		−0.98	3.31
$V^{5+} \to V^{4+}$	1 M NH$_3$-1 M NH$_4$Cl-0.005% gelatin†	−0.97	4.72
$V^{4+} \to V^{2+}$		−1.26	.....
$V^{5+} \to V^{4+}$	1 M K$_2$C$_2$O$_4$ (pH 4.6)	*	1.86
$V^{4+} \to V^{2+}$		−1.33	3.74
$V^{4+} \to V^{2+}$	1 M NH$_3$-1 M NH$_4$Cl-0.08 M Na$_2$SO$_3$-0.005%	−1.28	1.82
$V^{4+} \to V^{5+}$	gelatin†	−0.32	0.94
$V^{4+} \to V^{5+}$	1 M NaOH-0.08 M Na$_2$SO$_3$	−0.432	1.47
$V^{3+} \leftrightarrow V^{2+}$	0.5 M H$_2$SO$_4$-0.005% gelatin	−0.55	1.41
$V^{3+} \to V^{5+}$	0.5 M KHCO$_3$-0.5 M Na$_2$CO$_3$ (pH 9.4)	−0.337	2.80
$V^{2+} \leftrightarrow V^{3+}$	0.5 M H$_2$SO$_4$	−0.508	1.74
$V^{2+} \leftrightarrow V^{3+}$	1 M K$_2$C$_2$O$_4$	−1.091	1.43
$V^{2+} \leftrightarrow V^{3+}$	1 M Na acetate buffer pH 5.4†	−0.89	1.09
$V^{3+} \to V^{5+}$		−0.11	3.36
$W^{6+} \to W^{5+}$	8 M HCl†	*	4.70
$W^{5+} \to W^{3+}$		−0.62	.....
$W^{5+} \to W^{3+}$	12 M HCl	−0.56	2.53
$Yb^{3+} \to Yb^{2+}$	0.1 M NH$_4$Cl	−1.169	.....
$Yb^{2+} \to Yb^{+}$	0.01 M LiCl	−2.05	.....
$Zn^{2+} \to Zn$	0.1 M KCl-0.01% gelatin	−0.995	3.42
$Zn^{2+} \to Zn$	1 M NaOH-0.01% gelatin	−1.53	3.14
$Zn^{2+} \to Zn$	0.5 M Na tartrate-0.01% gelatin (pH 9)	−1.15	2.30
$Zn^{2+} \to Zn$	1 M NH$_3$-1 M NH$_4$Cl-0.01% gelatin	−1.33	3.82
$Zr^{4+} \to Zr$	0.1 M KCl (pH 3)	−1.65	.....

* Too positive. † One run; two waves.

Anodic Depolarization Potentials of Inorganic Anions

Ion	Concn.	Electrolyte	$E_{1/2}$ (S.C.E.)
$2Cl^- + 2Hg \leftrightarrow Hg_2Cl_2$	0.001 M	0.1 M KNO$_3$	+0.25
$2Br^- + 2Hg \leftrightarrow Hg_2Br_2$	0.001 M	0.1 M KNO$_3$	+0.1
$2I^- + 2Hg \leftrightarrow Hg_2I_2$	0.001 M	0.1 M KNO$_3$	−0.05
$S^{2-} + Hg \leftrightarrow HgS$	0.001 M	0.1 M NaOH	−0.76
$Se^{2-} + Hg \leftrightarrow HgSe$	0.00013 M	0.5 M Na$_2$CO$_3$-0.003% gelatin (pH 10.7)	−0.86
$2S_2O_3^{2-} + Hg \leftrightarrow Hg(S_2O_3)_2^{2-}$	0.001 M	0.1 M KNO$_3$	−0.15
$SO_3^{2-} + Hg \leftrightarrow Hg(SO_3)_2^{2-}$	0.001 M	0.1 M KNO$_3$	−0.007
$2CN^- + Hg \leftrightarrow Hg(CN)_2$	0.0005 M	0.1 M NaOH	⧣
$2SCN^- + Hg \leftrightarrow Hg(SCN)_2$	0.001 M	0.1 M KNO$_3$	+0.18
$2OH^- + Hg \leftrightarrow Hg(OH)_2$	0.001 M	0.1 M KNO$_3$	+0.080

⧣ Starts at −0.45.

Table 6-5 (*Continued*)
POLAROGRAPHY

Half-wave Potentials of Organic Compounds at 25°C

Anodic waves are indicated in the half-wave column. The value in the column headed n indicates the number of electrons involved in the reduction or oxidation. Measurements made at temperatures other than 25°C are indicated by including the temperature value.

Compound	Electrolyte	n	$E_{1/2}$ (S.C.E.)
Aliphatic hydrocarbons:			
Allene	$0.05\ M\ (C_2H_5)_4$ NBr-75% dioxane	2	−2.29
1,3-Butadiene	$0.05\ M\ (C_2H_5)_4$NBr-75% dioxane	2	−2.59
Cyclooctatetraene	$0.1\ M\ (CH_3)_4$NOH-50% C_2H_5OH	2	−1.51
Diacetylene	$0.05\ M\ (C_2H_5)_4$NBr-75% dioxane	2	−2.27
Dimethylfulvene	$0.175\ M\ (C_4H_9)_4$NI-75% dioxane	2	−1.89
Vinylacetylene	$0.05\ M\ (C_2H_5)_4$NBr-75% dioxane	2	−2.40
Aromatic hydrocarbons:			
Acenaphthene	$0.175\ M\ (C_4H_9)_4$NI-75% dioxane	2	−2.58
Anthracene	$0.175\ M\ (C_4H_9)_4$NI-75% dioxane	2	−1.94
1,2-Benzanthracene	$0.175\ M\ (C_4H_9)_4$NI-75% dioxane	2	−2.03
		2	−2.54
Biphenyl	$0.175\ M\ (C_4H_9)_4$NI-75% dioxane	2	−2.70
1,2-Dihydronaphthalene	$0.175\ M\ (C_4H_9)_4$NI-75% dioxane	2	−2.57
Diphenylacetylene	$0.175\ M\ (C_4H_9)_4$NI-75% dioxane	2	−2.20
1,4-Diphenylbuta-1,3-diene	$0.175\ M\ (C_4H_9)_4$NI-75% dioxane	2	−1.98
1,1-Diphenylethylene	$0.175\ M\ (C_4H_9)_4$NI-75% dioxane	2	−2.14
Fluorene	$0.175\ M\ (C_4H_9)_4$NI-75% dioxane	2	−2.65
β-Methylstyrene	$0.175\ M\ (C_4H_9)_4$NI-75% dioxane	2	−2.54
Naphthalene	$0.175\ M\ (C_4H_9)_4$NI-75% dioxane	2	−2.50
Phenanthrene	$0.175\ M\ (C_4H_9)_4$NI-75% dioxane	2	−2.46
		2	−2.71
Phenylacetylene	$0.175\ M\ (C_4H_9)_4$NI-75% dioxane	2	−2.37
3-Phenylindene	$0.175\ M\ (C_4H_9)_4$NI-75% dioxane	2	−2.33
Stilbene	$0.175\ (C_4H_9)_4$NI-75% dioxane	2	−2.26
Styrene	$0.175\ (C_4H_9)4$NI-75% dioxane	2	−2.35
Tetraphenylethylene	$0.175\ M\ (C_4H_9)_4$NI-75% dioxane	2	−2.05
Aldehydes:			
Acetaldehyde	$0.6\ M$ LiOH-$0.07\ M$ LiCl (pH 12.7)	2	−1.89
Acrolein	pH 4.5	...	−1.36
Benzaldehyde	pH 1.2-50% C_2H_5OH	1	−0.94
	pH 11.3-50% C_2H_5OH	2	−1.44
Crotonaldehyde	pH 1.3-50% dioxane	...	−0.92
Formaldehyde	$0.05\ M$ KOH-$0.1\ M$ KCl	2	−1.59
Furfural	pH 3.9	...	−1.06
	pH 7.6	...	−1.38
Glucose ($0.25\ M$)	pH 7	...	−1.55
Glycolaldehyde	$0.1\ M$ NaOH	2	−1.70
Glyoxal	pH 3.4	...	−1.41
p-Hydroxybenzaldehyde	pH 1.81-50% C_2H_5OH	1	−1.16
	pH 11.98-50% C_2H_5OH	2	−1.85
o-Methoxybenzaldehyde	pH 1.81	1	−1.03
	pH 11.98	2	−1.53
p-Methoxybenzaldehyde	pH 1.81	1	−1.07
	pH 11.98	2	−1.60
Methyl glyoxal	pH 4.5	...	−0.83
Propienaldehyde	$0.1\ M$ LiOH	2	−1.93

Table 6-5 (*Continued*)
POLAROGRAPHY

Compound	Electrolyte	n	$E_{1/2}$ (S.C.E.)
Salicylaldehyde	pH 1.81	1	−1.02
	pH 11.98	2	−1.63
Ketones and derivatives:			
Acetone	$0.05 M$ $(C_2H_5)_4NI$-75% dioxane	2	−2.46
	$2.5 M$ NH_3-$2.5 M$ $(NH_4)_2SO_4$ (pH 9.3)	...	−1.52
Acetophenone	pH 1.3-50% C_2H_5OH	1	−1.12
	pH 8.6-50% C_2H_5OH	2	−1.62
Aurin (70°C)	pH 7-30% C_2H_5OH	...	−0.76
		...	−1.20
Benzalacetone	pH 1.3-50% C_2H_5OH	1	−0.72
	pH 8.6-50% C_2H_5OH	2	−1.27
Benzalacetophenone	pH 8.6-50% C_2H_5OH	2	−1.10
		2	−1.63
Benzil	pH 1.3-50% C_2H_5OH	2	−0.27
	pH 11.3-50% C_2H_5OH	2	−0.75
Benzoin	pH 1.3-50% C_2H_5OH	...	−0.90
	pH 11.3-50% C_2H_5OH	...	−1.51
Benzophenone	pH 1.3-50% C_2H_5OH	1	−0.94
	pH 11.3-50% C_2H_5OH	2	−1.42
Biacetyl	0.1 M HCl	...	−0.84
Colchceine	pH 6.80-50% C_2H_5OH	...	−1.42
Colchicine	pH 6.80-H_2O	...	−1.40
Cyclohexanone	$0.05 M$ $(C_2H_5)_4NI$-75% dioxane	2	−2.45
Dibenzoylethylene (*trans*)	pH 1.3-50% C_2H_5OH	...	−0.12
	pH 11.3-50% C_2H_5OH	...	−0.57
		...	−1.52
Dibenzoylethylene (*cis*)	pH 1.3-50% C_2H_5OH	...	−0.30
	pH 11.3-50% C_2H_5OH	...	−0.62
		...	−1.65
Dibenzoylmethane	pH 1.3-50% C_2H_5OH	...	−0.59
	pH 11.3-50% C_2H_5OH	...	−1.30
		...	−1.62
Fructose	0.02 M LiCl	...	−1.76
Girard derivatives of aliphatic ketones	pH 8.2	2	−1.52
o-Hydroxyacetophenone	0.1 M NH_4Cl-50% C_2H_5OH	...	−1.36
p-Hydroxyacetophenone	0.1 M NH_4Cl-50% C_2H_5OH	...	−1.45
Mesityl oxide	pH 1.3-50% C_2H_5OH	...	−1.014
	pH 11.3-50% C_2H_5OH	...	−1.604
Methyl vinyl ketone	0.1 M KCl	...	−1.42
Santonin	pH 2.5	...	−1.1
	pH 8	...	−1.6
Acids and derivatives:			
Acrylonitrile	$0.05 M$ $(C_2H_5)_4NI$	2	−1.94
Ascorbic acid	pH 3.4 (phosphate buffer-1.5% H_3PO_3)	...	+0.17†
Bromoacetic acid	pH 1.1	2	−0.54
α-Bromopropionic acid	pH 2.0	2	−0.39
Crotonic acid	$0.05 M$ $(C_2H_5)_4NI$-75% dioxane	2	−1.94
Dibromoacetic acid	pH 1.1	2	−0.03
		2	−0.59
Dichloroacetic acid	pH 8.19	2	−1.57
Diethyl fumarate	pH 3.97	2	−0.84
Diethyl maleate	pH 3.98	2	−0.95

† Anodic.

Table 6-5 (*Continued*)
POLAROGRAPHY

Compound	Electrolyte	n	$E_{1/2}$ (S.C.E.)
Ethyl chloroacetate (0°C)	pH 6.8 to 10.4	2	−1.50
Ethyl dichloroacetate (0°C)	pH 6.8 to 10.4	2	−0.86
		2	−1.50
Ethyl trichloroacetate (0°C)	pH 6.8 to 10.4	2	−0.22
		2	−0.86
		2	−1.51
Fumaric acid	pH 8.2	2	−1.60
Iodoacetic acid	pH 1.1	2	−0.16
Maleic acid	pH 8.2	2	−1.36
Methacrylonitrile	0.1 M $(C_2H_5)_4$NBr	2	−2.07
Phenolphthalein	pH 3.5-25% C_2H_5OH	2	−0.11
	pH 10.06-50% C_2H_5OH	...	−1.01
		...	−1.33
Phthalide	0.1 M $(C_4H_9)_4$NI-50% dioxane	2	−2.03
Pyruvic acid	pH 3	...	−0.86
	pH 7	...	−1.30
		...	−1.57
Trichloroacetic acid	pH 8.19	2	−0.84
		2	−1.57
Halogen compounds:			
Allyl bromide	0.05 M $(C_2H_5)_4$NBr-75% dioxane	2	−1.29
Allyl chloride	0.05 M $(C_2H_5)_4$NBr-75% dioxane	2	−1.91
Benzal chloride	0.05 M $(C_2H_5)_4$NBr-75% dioxane	4	−1.81
Benzotrichloride	0.05 M $(C_2H_5)_4$NBr-75% dioxane	2	−0.68
		2	−1.65
		2	−2.00
Benzyl chloride	0.05 M $(C_2H_5)_4$NBr-75% dioxane	2	−1.94
Bromobenzene	0.05 M $(C_2H_5)_4$NBr-75% dioxane	2	−2.32
Bromoform	0.05 M $(C_2H_5)_4$NBr-75% dioxane	2	−0.64
		4	−1.51
n-Butyl bromide	0.05 M $(C_2H_5)_4$NBr-75% dioxane	2	−2.27
Carbon tetrabromide	0.05 M $(C_2H_5)_4$NBr-75% dioxane	2	−0.3
		2	−0.75
		4	−1.49
Carbon tetrachloride	0.05 M $(C_2H_5)_4$NBr-75% dioxane	2	−0.78
		2	−1.71
Chloroform	0.05 M $(C_2H_5)_4$NBr-75% dioxane	2	−1.6
p-Dibromobenzene	0.05 M $(C_2H_5)_4$NBr-75% dioxane	2	−2.10
o-Dichlorobenzene	0.05 M $(C_2H_5)_4$NBr-75% dioxane	2	−2.51
m-Dichlorobenzene	0.05 M $(C_2H_5)_4$NBr-75% dioxane	2	−2.48
p-Dichlorobenzene	0.05 M $(C_2H_5)_4$NBr-75% dioxane	2	−2.49
Ethyl bromide	0.05 M $(C_2H_5)_4$NBr-75% dioxane	2	−2.08
Ethyl iodide	0.05 M $(C_2H_5)_4$NBr-75% dioxane	2	−1.67
γ-Hexachlorocyclohexane	0.1 M $(C_2H_5)_4$NI-80% C_2H_5OH	...	−1.65
	1% KI-50% C_2H_5OH	...	−1.35
Iodobenzene	0.05 M $(C_2H_5)_4$NBr-75% dioxane	2	−1.62
		2	−1.09
		2	−1.50
Methyl bromide	0.05 M $(C_2H_5)_4$NBr-75% dioxane	2	−1.63
Methyl chloride	0.05 M $(C_2H_5)_4$NBr-75% dioxane	2	−2.23
Methyl iodide	0.05 M $(C_2H_5)_4$NBr-75% dioxane	2	−1.63
Methylene bromide	0.05 M $(C_2H_5)_4$NBr-75% dioxane	4	−1.48
Methylene chloride	0.05 M $(C_2H_5)_4$NBr-75% dioxane	4	−1.6
Methylene iodide	0.05 M $(C_2H_5)_4$NBr-75% dioxane	2	−1.12
		2	−1.53

Table 6-5 (*Continued*)
POLAROGRAPHY

Compound	Electrolyte	n	$E_{1/2}$ (S.C.E.)
Nitro compounds and their reduction products:			
Azobenzene (*trans*)	pH 6.26-10% C_2H_5OH	2	−0.33
Azoxybenzene	pH 6.3-20% C_2H_5OH	4	−0.63
Benzenediazonium chloride	0.1 M HCl	1	−0.18
		3	−0.67
m-Dinitrobenzene	pH 1.7-8% C_2H_5OH	4	−0.12
		4	−0.26
Methyl *o*-nitrobenzoate	pH 2-10% C_2H_5OH	4	−0.25
		2	−0.74
Methyl *m*-nitrobenzoate	pH 2-10% C_2H_5OH	4	−0.240
		2	−0.68
Methyl *p*-nitrobenzoate	pH 2-10% C_2H_5OH	4	−0.200
		2	−0.73
p-Nitroaniline	pH 2	6	−0.36
o-Nitroanisole	pH 2-10% C_2H_5OH	4	−0.29
		2	−0.58
	pH 10-10% C_2H_5OH	4	−0.76
m-Nitroanisole	pH 2-10% C_2H_5OH	4	−0.28
		2	−0.69
	pH 10-10% C_2H_5OH	4	−0.71
p-Nitroanisole	pH 2-10% C_2H_5OH	4	−0.35
		2	−0.64
	pH 10-10% C_2H_5OH	4	−0.80
m-Nitrobenzaldehyde	pH 2-10% C_2H_5OH	4	−0.28
		2	−1.20
Nitrobenzene	pH 2-8% C_2H_5OH	4	−0.30
		2	−0.75
o-Nitrobenzoic acid	pH 2-10% C_2H_5OH	4	−0.23
		2	−0.73
m-Nitrobenzoic acid	pH 2-10% C_2H_5OH	4	−0.20
		2	−0.70
p-Nitrobenzoic acid	pH 2-10% C_2H_5OH	4	−0.17
		2	−0.74
Nitromethane	pH 3.3	4	−0.83
	pH 11.9	4	−0.90
o-Nitrophenol	pH 2-10% C_2H_5OH	4	−0.23
	pH 10-8% C_2H_5OH	6	−0.80
m-Nitrophenol	pH 2-8% C_2H_5OH	4	−0.37
		2	−1.10
	pH 10-8% C_2H_5OH	4	−0.76
p-Nitrophenol	pH 2-8% C_2H_5OH	4	−0.35
	pH 10-8% C_2H_5OH	2	−0.92
		4	−1.33
o-Nitrotoluene	pH 1-80% dioxane	4	−0.26
		2	−0.64
m-Nitrotoluene	pH 1-80% dioxane	4	−0.22
		2	−0.71
p-Nitrotoluene	pH 1-80% dioxane	4	−0.24
		2	−0.71
Nitrosobenzene	pH 7-10% C_2H_5OH	2	−0.11
α-Nitroso-β-naphthol	pH 5.9	4	−0.11
N-Nitrosophenylhydroxylamine	pH 2	6	−0.84
Phenylhydroxylamine	pH 7-10% C_2H_5OH	2	−0.09
Tetranitromethane	pH 12	...	−0.41

Table 6-5 (*Continued*)
POLAROGRAPHY

Compound	Electrolyte	n	$E_{1/2}$ (S.C.E.)
Trinitrotoluene	pH 0.5-8% C_2H_5OH	4	−0.01
		4	−0.08
		4	−0.14
Sulfur compounds:			
Cysteine	0.1 M $HClO_4$	1‡	−0.05
Dithiodiglycolic acid	pH 3.00-0.002% gelatin	2	−0.37
Thioglycolic acid	pH 6.75	1‡	−0.38
Peroxides:			
Benzoyl peroxide	0.3 M LiCl-50% CH_3OH-50%C_6H_6	...	0.00
Cumene hydroperoxide	0.3 M LiCl-50% CH_3OH-50% C_6H_6	...	−0.68
Quinones (Reversible system):			
Anthraquinone	pH 7.4-40% dioxane	2	−0.54
Benzoquinone	pH 5.40-50% CH_3OH	2	+0.146
2,3-Dimethylnaphthoquinone	pH 5.40-50% CH_3OH	2	−0.216
Duroquinone	pH 5.40-50% CH_3OH	2	−0.093
2-Methyl-1,4-naphthoquinone	pH 6.24-75% C_2H_5OH	2	−0.17
Toluquinone	pH 5.40-50% CH_3OH	2	+0.090
Heterocyclic compounds containing oxygen:			
Flavanone	pH 7.5-50% $(CH_3)_2CHOH$	...	−1.37
Flavone	pH 7.5-50% $(CH_3)_2CHOH$	...	−1.26
		...	−1.44
2,5,7,8-Tetramethyl-6-hydroxy-chroman	pH 3.56-50% CH_3OH	2	+0.230†
2,4,6,7-Tetramethyl-5-hydroxy-coumaran	pH 3.56-50% CH_3OH	2	+0.219†
Heterocyclic compounds containing nitrogen:			
Acridine	pH 8.29-60% C_2H_5OH	1	−0.80
		1	−1.48
8-Hydroxyquinoline	pH 10	...	−1.39
		...	−1.61
Nicotinamide	pH 8.7	...	−1.56
Iso-Nicotinic acid hydrazide	pH 1.5	...	−0.52
		...	−0.70
Picolinic acid	pH 7	...	−1.17
Pyridine	pH 2.9	...	−1.49
Quinaldinic acid	pH 4-8% C_2H_5OH-gelatin	...	−0.86
		...	−1.19
Quinoline	pH 6.51-50% C_2H_5OH-0.04% gelatin	2	−1.23
Quinoline-8-carboxylic acid	pH 9	...	−1.11
		...	−1.75
Saccharin	pH 7.0	...	−1.77
Thiamin	pH 7.2	...	−1.30

‡ HgSR formation.
† Anodic.

Table 6-6
VOLTAIC CELLS AND BATTERIES

Name of Cell	Electrochemical system	Nominal emf (volts)	Remarks
Conventional cells			
Air (oxygen) "depolarized":			
(a) Alkaline	$Zn(Hg)|KOH$ resp $NaOH(aq)|O_2$ [C]	1.4	Wet or "dry"
(b) Leclanché type	$Zn(Hg)|NH_4Cl + ZnCl_2(aq)|O_2$ [C(MnO_2)]	1.4	Wet or "dry"
Alkaline MnO_2	$Zn(Hg)|KOH(aq)|MnO_2$ [C]	1.6	Prim and sec (dry type)
Clark—Standard	$Zn(Hg)|ZnSO_4(aq) \| Hg_2SO_4(aq)|Hg$	1.4328[10]	2-Fluid
Daniell	$Zn(Hg)|ZnSO_4(aq) \| CuSO_4(aq)|Cu$	1.08	2-Fluid prototype
Drumm	$Zn(Hg)|KOH + K_2Zn_2O_3(aq)|NiO\text{-}OH$	1.8	Secondary
Edison	$Fe|KOH + LiOH(aq)|NiO\text{-}OH$	1.5	Secondary
Halogen (chlorine)	$Zn|ZnCl_2(aq)|Cl_2$ [C]	2.05	2.85 v with Mg anode
Lelande-Chaperon	$Zn(Hg)|NaOH(aq)|CuO$ [Fe]	0.95	....
Leclanché wet	$Zn(Hg)|NH_4Cl(aq)|MnO_2$ [C]	1.5	Prototype
Leclanché dry	$Zn(Hg)|NH_4Cl + ZnCl_2(aq)|MnO_2$ [C]	1.6	"Dry battery" of commerce
Magnesium, Leclanché type	$Mg|MgBr_2(aq)|MnO_2$ [C]	1.9	Dry cell
Magnesium, reserve cell:			
(a) AgCl	$Mg|MgCl_2(aq)|AgCl$ [Ag]	1.8	Water-activated
(b) Cu_2Cl_2	$Mg|MgCl_2(aq)|Cu_2Cl_2$ [Cu]	1.75	Water-activated
Main (Reynier)	$Zn(Hg)|H_2SO_4(aq)|PbO_2$ [Pb]	2.5	Dry type
Mercuric oxide (Ruben-Mallory)	$Zn(Hg)|KOH + K_2Zn_2O_3(aq)|HgO$ [C]	1.35	See battery
Nickel-cadmium	$Cd|KOH(aq)|NiO\text{-}OH$	1.3	Reserve cell
Perchloric acid	$Pb|HClO_4(aq)|PbO_2$ [Pb]	2.3	....
Poggendorff (dichromate)	$Zn(Hg)|K_2Cr_2O_7 + H_2SO_4(aq)|C$	2.0	See battery
Silver-cadmium	$Cd|KOH(aq)|Ag_2O_2$ [Ag]	1.4	

Table 6-6 (Continued)
VOLTAIC CELLS AND BATTERIES

Name of Cell	Electrochemical system	Nominal emf (volts)	Remarks
Conventional cells (cont.)			
Silver chloride	$Zn(Hg)|NH_4Cl(aq)|AgCl$ [Ag]	1.02	....
Silver–zinc	$Zn(Hg)|KOH(aq)|Ag_2O_2$ [Ag]	1.8	Sec battery
	$Zn(Hg)|KOH(aq)|Ag_2O$ [Ag]	1.6	Prim or sec
Volta	$Zn|acid, alk, or salt(aq)|Ag$ or Cu	0.9–1.0	First battery (wet or dry)
Weston—Standard	$Cd(Hg)|CdSO_4(aq) \parallel Hg_2SO_4(aq)|Hg$	$1.01827^{20°}$	2-Fluid
Zamboni	$Zn|paper|MnO_2$ [Ag or Au]	0.7–0.9	High-voltage dry battery of minimal current
Continuous-feed and fuel cells*			
Amalgam	[Fe] $Na(Hg)|NaOH(aq)|O_2$ [C or Ni(Ag)]	2.05	....
Hydrogen-oxygen:			
(a) Low temperature	[C or Ni] $H_2|KOH(aq)|O_2$ [C or Ni(Ag)]	1.1	20–90°C
(b) Intermediate temperature	[Ni] $H_2|KOH(aq)|O_2$ [Ni]	1.05	200–240°C (>400 psi)
(c) High temperature	$\begin{bmatrix} Porous \\ metals \end{bmatrix} H_2|Fused\ car-\\ bonates|O_2 \begin{bmatrix} Porous \\ metals \end{bmatrix}$	1.0	400–700°C
Ion-exchange membrane	[Pt] $H_2|Resin|O_2$ [Pt]	1.05	20–80°C

* Cells in early development stages not included; e.g., alcohol, hydrocarbons, redox, etc.

Table 6-7
LIMITING EQUIVALENT IONIC CONDUCTANCES
IN AQUEOUS SOLUTIONS AT 25°C

Except for H^+ (0.0139) and OH^- (0.018), a temperature coefficient of 0.02 deg^{-1} is applicable to the cations and anions.

Ion	Limiting Equivalent Ionic Conductance, mho-cm^2/equivalent	Ion	Limiting Equivalent Ionic Conductance, mho-cm^2/equivalent
Inorganic Cations		**Inorganic Anions**	
Ag^+	61.9	$Au(CN)_2^-$	50
Al^{3+}	61	$Au(CN)_4^-$	36
Ba^{2+}	63.9	$B(C_6H_5)_4^-$	21
Be^{2+}	45	Br^-	78.1
Ca^{2+}	59.5	Br_3^-	43
Cd^{2+}	54	BrO_3^-	55.8
Ce^{3+}	70	Cl^-	76.35
Co^{2+}	53	ClO_2^-	52
$Co(NH_3)_6^{3+}$	100	ClO_3^-	64.6
$Co(en)_3^{3+}$	74.7	ClO_4^-	67.9
Cr^{3+}	67	CN^-	78
Cs^+	77.3	CO_3^{2-}	72
Cu^{2+}	55	$Co(CN)_6^{3-}$	98.9
D^+ (deuterium)	213.7 (18°)	CrO_4^{2-}	85
Dy^{3+}	65.7	F^-	54.4
Er^{3+}	66.0	$Fe(CN)_6^{4-}$	111
Eu^{3+}	67.9	$Fe(CN)_6^{3-}$	101
Fe^{2+}	54	$H_2AsO_4^-$	34
Fe^{3+}	68	HCO_3^-	44.5
Gd^{3+}	67.4	HF_2^-	75
H^+	349.82	HPO_4^{2-}	57
Hg^{2+}	53	$H_2PO_4^-$	33
Ho^{3+}	66.3	$H_2PO_2^-$	46
K^+	73.5	HS^-	65
La^{3+}	69.6	HSO_3^-	50
Li^+	38.69	HSO_4^-	50
Mg^{2+}	53.06	$H_2SbO_4^-$	31
Mn^{2+}	53.5	I^-	76.8
NH_4^+	73.5	IO_3^-	40.5
$N_2H_5^+$	59	IO_4^-	54.5
Na^+	50.11	$N(CN)_2^-$	54.5
Nd^{3+}	69.6	NO_2^-	71.8
Ni^{2+}	50	NO_3^-	71.4
Pb^{2+}	71	$NH_2SO_3^-$	48.6
Pr^{3+}	69.6	N_3^-	69
Ra^{2+}	66.8	OCN^-	64.6
Rb^+	77.8	OH^-	198.6
Sc^{3+}	64.7	PF_6^-	56.9
Sm^{3+}	68.5	PO_3F^{2-}	63.3
Sr^{2+}	59.46	PO_4^{3-}	69.0
Tl^+	76	$P_2O_7^{4-}$	81.4
Tm^{3+}	65.5	$P_3O_9^{3-}$	83.6
UO_2^{2+}	32	$P_3O_{10}^{5-}$	109
Y^{3+}	62	ReO_4^-	54.7
Yb^{3+}	65.2	SCN^-	66
Zn^{2+}	52.8	$SeCN^-$	64.7

Table 6-7 (*Continued*)
LIMITING EQUIVALENT IONIC CONDUCTANCES
IN AQUEOUS SOLUTIONS AT 25°C

	Limiting Equivalent Ionic Conductance, mho-cm^2/equivalent		Limiting Equivalent Ionic Conductance, mho-cm^2/equivalent
Inorganic Anions		Bromobenzoate	30
SeO_4^{2-}	75.7	*n*-Butyrate	32.6
SO_3^{2-}	79.9	Chloroacetate	39.7
SO_4^{2-}	80.0	Chlorobenzoate	33
$S_2O_3^{2-}$	85.0	Citrate^{3-}	70.2
$S_2O_4^{2-}$	66.5	α-Crotonate	33.2
$S_2O_6^{2-}$	93	Cyanoacetate	41.8
$S_2O_8^{2-}$	86	Cyclohexane carboxylate	28.7
WO_4^{2-}	69	Cyclopropane-1,1-dicarboxylate^{2-}	53.4
Organic Cations		Decyl sulfonate	26
i-Butylammonium	38	Dichloroacetate	38.3
n-Decylpyridinium	29.5	Diethyl barbiturate^{2-}	26.3
Diethylammonium	42.0	Dihydrogen citrate	30
Dimethylammonium	51.5	Dimethyl malonate^{2-}	49.4
Dipropylammonium	30.1	3,5-Dinitrobenzoate	28.3
n-Dodecylammonium	23.8	Dodecyl sulfonate	24
Ethylammonium	47.2	Ethyl malonate	49.3
Ethyltrimethylammonium	40.5	Ethyl sulfonate	39.6
Methylammonium	58.3	Fluorobenzoate	33
Histadyl	23.0	Formate	54.6
Piperidinium	37.2	$HC_2O_4^-$	40.2
Propylammonium	40.8	Lactate	38.8
Pyrilammonium	24.3	Malonate^{2-}	63.5
Tetra-*n*-butylammonium	19.1	Methyl sulfonate	48.8
Tetraethylammonium	33.0	$C_2O_4^{2-}$	74.2
Tetramethylammonium	45.3	Octyl sulfonate	29
Tetra-*n*-propylammonium	23.5	Phenyl acetate	30.6
Triethylammonium	34.3	Picrate	30.2
Triethylsulfonium	36.1	Propionate	35.8
Trimethylammonium	46.6	Propyl sulfonate	37.1
Trimethylsulfonium	51.4	Salicylate	36
Tripropylammonium	26.1	Suberate^{2-}	36
Organic Anions		Succinate^{2-}	58.8
Acetate	40.9	Sulfonate	43.1
p-Anisate	29.0	Tartrate^{2-}	64
Azelate^{2-}	40.6	Trichloroacetate	36.6
Benzoate	32.4		

Table 6-8
SPECIFIC CONDUCTANCES OF STANDARD
POTASSIUM CHLORIDE SOLUTIONS

The values of κ, the specific conductance, are corrected for the conductivity of water. The cell constant θ of a conductance cell can be obtained from the equation

$$\theta = \frac{\kappa R R_{\text{solv}}}{R_{\text{solv}} - R}$$

where R is the resistance measured when the cell is filled with a solution of the stated composition in the table below, and R_{solv} is the resistance measured when the cell is filled with solvent at the same temperature.

Grams KCl/100 g Solution (in vacuo)	Specific Conductance (Int. ohm^{-1} cm^{-1}) at		
	0°C	18°C	25°C
71.1352	0.06518	0.09784	0.11134
7.41913	0.007138	0.011167	0.012856
0.745263*	0.0007736	0.0012205	0.0014088

* Virtually 0.0100M.
From the data of Jones and Bradshaw, *J. Am. Chem. Soc.*, **55,** 1780 (1933).

Table 6-9
ELECTRICAL CONDUCTIVITY OF VARIOUS PURE LIQUIDS

Liquid	Temp. °C.	mhos/cm or ohm^{-1} cm^{-1}	Liquid	Temp. °C.	mhos/cm or ohm^{-1} cm^{-1}
Acetaldehyde	15	1.7×10^{-6}	Benzylamine	25	$<1.7 \times 10^{-8}$
Acetamide	100	$<4.3 \times 10^{-5}$	Benzyl benzoate	25	$<1 \times 10^{-9}$
Acetic acid	0	5×10^{-9}	Bromine	17.2	1.3×10^{-13}
	25	1.12×10^{-8}	Bromobenzene	25	$<2 \times 10^{-11}$
Acetic anhydride	0	1×10^{-6}	Bromoform	25	$<2 \times 10^{-8}$
	25	4.8×10^{-7}	iso-Butyl alcohol	25	8×10^{-8}
Acetone	18	2×10^{-8}			
	25	6×10^{-8}	Capronitrile	25	3.7×10^{-6}
Acetonitrile	20	7×10^{-6}	Carbon disulfide	1	7.8×10^{-18}
Acetophenone	25	6×10^{-9}	Carbon tetrachloride	18	4×10^{-18}
Acetyl bromide	25	2.4×10^{-6}	Chlorine	-70	$<1 \times 10^{-16}$
Acetyl chloride	25	4×10^{-7}	Chloroacetic acid	60	1.4×10^{-6}
Alizarin	233	$1.45 \times 10^{-6}(?)$	m-Chloroaniline	25	5×10^{-8}
Allyl alcohol	25	7×10^{-6}	Chloroform	25	$<2 \times 10^{-8}$
Ammonia	-79	1.3×10^{-7}	Chlorohydrin	25	5×10^{-7}
Aniline	25	2.4×10^{-8}	m-Cresol	25	$<1.7 \times 10^{-8}$
Anthracene	230	3×10^{-10}	Cyanogen	...	$<7 \times 10^{-9}$
Arsenic tribromide	35	1.5×10^{-6}	Cymene	25	$<2 \times 10^{-8}$
Arsenic trichloride	25	1.2×10^{-6}			
			Dichloroacetic acid	25	7×10^{-8}
			Dichlorohydrin	25	1.2×10^{-5}
Benzaldehyde	25	1.5×10^{-7}	Diethylamine	-33.5	2.2×10^{-9}
Benzene	...	7.6×10^{-8}	Diethyl carbonate	25	1.7×10^{-8}
Benzoic acid	125	3×10^{-9}	Diethyl oxalate	25	7.6×10^{-7}
Benzonitrile	25	5×10^{-8}	Diethyl sulfate	25	2.6×10^{-7}
Benzyl alcohol	25	1.8×10^{-6}	Dimethyl sulfate	0	1.6×10^{-7}

Table 6-9 (*Continued*)
Electrical Conductivity of Various Pure Liquids

Liquid	Temp. °C.	mhos/cm or ohm^{-1} cm^{-1}	Liquid	Temp. °C.	mhos/cm or ohm^{-1} cm^{-1}
Epichlorohydrin	25	3.4×10^{-8}	Nitromethane	18	6×10^{-7}
Ethyl acetate	25	$<1 \times 10^{-9}$	o- or m-Nitrotoluene	25	$<2 \times 10^{-7}$
Ethyl acetoacetate	25	4×10^{-8}	Nonane	25	$<1.7 \times 10^{-8}$
Ethyl alcohol	25	1.35×10^{-9}	Oleic acid	15	$<2 \times 10^{-10}$
Ethylamine	0	4×10^{-7}			
Ethyl benzoate	25	$<1 \times 10^{-9}$	Pentane	19.5	$<2 \times 10^{-10}$
Ethyl bromide	25	$<2 \times 10^{-8}$	Petroleum	...	3×10^{-13}
Ethylene bromide	19	$<2 \times 10^{-10}$	Phenetole	25	$<1.7 \times 10^{-8}$
Ethylene chloride	25	3×10^{-8}	Phenol	25	$<1.7 \times 10^{-8}$
Ethyl ether	25	$<4 \times 10^{-13}$	Phenyl isothiocyanate	25	1.4×10^{-6}
Ethylidene chloride	25	$<1.7 \times 10^{-8}$	Phosgene	25	7×10^{-9}
Ethyl iodide	25	$<2 \times 10^{-8}$	Phosphorus	25	4×10^{-7}
Ethyl isothiocyanate	25	1.26×10^{-7}	Phosphorus oxychloride	25	2.2×10^{-6}
Ethyl nitrate	25	5.3×10^{-7}			
Ethyl thiocyanate	25	1.2×10^{-6}	Pinene	23	$<2 \times 10^{-10}$
Eugenol	25	$<1.7 \times 10^{-8}$	Piperidine	25	$<2 \times 10^{-7}$
			Propionaldehyde	25	8.5×10^{-7}
Formamide	25	4×10^{-6}	Propionic acid	25	$<1 \times 10^{-9}$
Formic acid	18	5.6×10^{-5}	Propionitrile	25	$<1 \times 10^{-7}$
	25	6.4×10^{-5}	n-Propyl alcohol	18	5×10^{-8}
Furfural	25	1.5×10^{-6}		25	2×10^{-8}
			iso-Propyl alcohol	25	3.5×10^{-6}
Gallium	30	36,800	n-Propyl bromide	25	$<2 \times 10^{-8}$
Glycerol	25	6.4×10^{-8}	Pyridine	18	5.3×10^{-8}
Glycol	25	3×10^{-7}			
Guaiacol	25	2.8×10^{-7}	Quinoline	25	2.2×10^{-8}
Heptane	...	$<1 \times 10^{-13}$	Salicylaldehyde	25	1.6×10^{-7}
Hexane	18	$<1 \times 10^{-18}$	Stearic acid	80	$<4 \times 10^{-13}$
Hydrogen bromide	−80	8×10^{-9}	Sulfonyl chloride, SOCl$_2$	25	2×10^{-6}
Hydrogen chloride	−96	1×10^{-8}			
Hydrogen cyanide	0	3.3×10^{-6}	Sulfur	115	1×10^{-12}
Hydrogen iodide	B.P.	2×10^{-7}		130	5×10^{-11}
Hydrogen sulfide	B.P.	1×10^{-11}		440	1.2×10^{-7}
			Sulfur dioxide	35	1.5×10^{-8}
Iodine	110	1.3×10^{-10}	Sulfuric acid	25	1×10^{-2}
			Sulfuryl chloride, SO$_2$Cl$_2$	25	3×10^{-8}
Kerosene	25	$<1.7 \times 10^{-8}$			
			Toluene	...	$<1 \times 10^{-14}$
Mercury	0	10,629.6	o-Toluidine	25	$<2 \times 10^{-6}$
Methyl acetate	25	3.4×10^{-6}	p-Toluidine	100	6.2×10^{-8}
Methyl alcohol	18	4.4×10^{-7}	Trichloroacetic acid	25	3×10^{-9}
Methyl ethyl ketone	25	1×10^{-7}	Trimethylamine	−33.5	2.2×10^{-10}
Methyl iodide	25	$<2 \times 10^{-8}$	Turpentine	...	2×10^{-13}
Methyl nitrate	25	4.5×10^{-6}			
Methyl thiocyanate	25	1.5×10^{-6}	iso-Valeric acid	80	$<4 \times 10^{-13}$
Naphthalene	82	4×10^{-10}	Water	18	4×10^{-8}
Nitrobenzene	0	5×10^{-9}	Xylene	...	$<1 \times 10^{-15}$

Section 7

ORGANIC CHEMISTRY

NOMENCLATURE

SHORT GLOSSARY OF ORGANIC CHEMICAL NOMENCLATURE

The following rules for naming organic compounds and the examples given in explanation are not intended to cover all of the possible cases. For a more comprehensive and detailed description see Beilstein: *Handbuch der Organischen Chemie*, 4th edition, and Patterson: *Definitive Report of the Commission on the Reform of the Nomenclature of Organic Chemistry, Jour. Am. Chem. Soc. 55, 3905* (1933); *News Edition, Ind. Eng. Chem. 14, 486* (1936).

HYDROCARBONS—The saturated open-chain hydrocarbons (C_nH_{2n+2}) have names ending in *ane*. The first four are: *methane* (CH_4); *ethane* (C_2H_6); *propane* (C_3H_8); and, *butane* (C_4H_{10}). Beginning with *pentane* (C_5H_{12}) the prefix (*e.g. penta*) indicates the number of carbon atoms; for example, octane, C_8H_{18}; nonatriacontane, $C_{39}H_{80}$. See special table of *Numerical Prefixes*. Saturated monocyclic hydrocarbons (C_nH_{2n}) have the corresponding names of the open-chain hydrocarbons (C_nH_{2n+2}) preceded by the prefix *cyclo;* for example, cyclopropane (C_3H_6); cyclopentane (C_5H_{10}). The straight (unbranched) chains are termed *normal* (abbreviated: *n*); the grouping $(CH_3)_2CH-$ in a chain is termed *iso;* for example, $CH_3(CH_2)_2CH_3$ is *n*-butane; $(CH_3)_2CHCH_3$ is *iso*-butane. Branched-chain hydrocarbons are often named as derivatives of some relatively simple hydrocarbon by stating what radicals* are attached to the simple hydrocarbon; for example, $CH_3CH_2CH_2CH_3$, which is *n*-butane, may also be named: methyl-ethyl-methane since a methyl radical (CH_3) and an ethyl radical (C_2H_5) are replacing the two hydrogen atoms of the original methane molecule, the carbon atom of which is printed in bold face type. Similarly $(CH_3)_2CHCH_3$, which is *iso*-butane, would be named trimethyl-methane, the carbon atom printed in bold face type being the methane carbon atom corresponding to this name. In this system of naming, a compound might be given any one of several names depending upon which carbon atom is selected; for example, $CH_3CH_2CH_2CH_2CH_3$ might be given any one of the following names: diethyl-methane; methyl-propyl-methane; butyl-methane.

In 1892 an international congress of chemists assembled in Geneva, Switzerland, to devise a uniform and scientific system of nomenclature. This *Geneva system* with modifications proposed at various times is the modern *systematic naming*. According to this system a hydrocarbon is named as a derivative of the normal compound corresponding to the longest chain of carbon atoms in the molecule. The word *normal* is usually omitted, but implied, so that the simple unqualified systematic name always means the compound with the normal constitution; for example, *n*-butane, $CH_3CH_2CH_2CH_3$, is simply butane; iso-butane, $(CH_3)_2CHCH_3$, is 2-methyl-propane. The numeral 2 placed before** the word *methyl* indicates that the methyl radical is attached to the second carbon atom in the longest chain which, in this example, is of three carbons in length and, therefore, a propane. In cases where several different side chains are attached the order of stating them is either according to increasing complexity (e.g., methyl, then ethyl, then propyl, followed by butyl, etc.) or, according to alphabetical order (e.g., butyl, then ethyl, then methyl, followed by propyl, etc.) Where branching occurs in a side chain and for other cases of ambiguity, or if a simpler name would result, then a chain is selected as the fundamental unit which admits of the maximum of substitutions. A detailed discussion of this and other exceptions caused by very complicated structures is beyond the scope of this exposition and the reader is referred to the references given below the title of this section.

UNSATURATED HYDROCARBONS AND UNSATURATION—Hydrocarbons with one double bond (C_nH_{2n}) are named by adding the suffix *ene* to the name of the corresponding alkyl radical (C_nH_{2n+1})*; for example, $CH_2:CH_2$ is ethylene; $CH_3CH:CH_2$ is propylene. Just as the saturated hydrocarbons may be named as

---

*See Radicals in this section and also a special section Names and Formulas of Organic Radicals.

**The numbering is such that the lowest numbers are obtained; e.g., $CH_3CH_2CH_2CH(CH_3)_2$ is 2-methylpentane and not 4-methylpentane. Usage varies as to the position in the word for the numerals not only in the case of the hydrocarbons but also with other types such as alcohols, aldehydes, ketones, etc. Some writers place them before, some after, the parts of the name to which they refer. In Beilstein a combination is used wherein numbers placed after are enclosed in parenthesis and those placed before are not; e.g., primary isobutyl alcohol is named: 2-methyl-butanol-(1).

ORGANIC CHEMICAL NOMENCLATURE

derivatives of methane, so these C_nH_{2n} hydrocarbons may be named as derivatives of ethylene; for example, $CH_3CH:CHCH_3$ which is a butylene, may also be named *sym*-dimethyl-ethylene since two methyl radicals (CH_3) are replacing two hydrogen atoms of the original ethylene molecule, the carbon atoms of which are printed in bold face type. Similarly $(CH_3)_2C:CH_2$, also a butylene, would be named *uns*-dimethyl-ethylene. The italicized prefixes *sym* and *uns* are abbreviations for *symmetrical* and *unsymmetrical*, respectively, and refer to the symmetrical or unsymmetrical manner in which the two methyl radicals are attached to the ethylene residue.

In the systematic naming, the longest chain is selected as in the case of the saturated hydrocarbons and the compound is named by replacing the *ane* ending of the corresponding saturated hydrocarbon with the ending *ene;* for example, $CH_3CH:CHCH_3$ is 2-butene. The numeral 2 indicates that the double bond is between the second and third carbon atoms. Similarly $(CH_3)_2C:CH_2$ is 2-methyl-1-propene** or 2-methyl-propene-(1), because a methyl group is on the second carbon atom, and the double bond is between the first and second carbon atoms in a chain of three carbon atoms. Where more than one double bond occurs in the molecule a numerical prefix is added to the ending *ene;* for example, isoprene, $CH_2:C(CH_3)CH:CH_2$ is 2-methyl-1,3-butadiene because the longest chain contains four carbon atoms with a methyl radical (CH_3) on the second carbon atom and double bonds between the first and second carbon atoms, and between the third and fourth carbon atoms of the four-carbon chain. Similarly benzene would be named cyclohexa-1,3,5-triene.

Hydrocarbons with one triple bond (C_nH_{2n-2}) may be named as derivatives of acetylene ($CH:CH$); for example, $CH_3C:CH$ which is allylene, may also be named methyl-acetylene. In the systematic naming the same scheme is used as with the double bond compounds except that the ending becomes *yne* or *ine*. Thus, allylene is 1-propyne or 1-propine. The Greek letter Δ is often used to indicate unsaturation; the numbers following it indicate the positions of the unsaturated linkings.

Unsaturation in compounds other than the hydrocarbons, such as alcohols, aldehydes, ketones, etc., is indicated in the same general manner by use of *ene, diene, ine*, etc.; for example, allyl alcohol, $CH_2:CHCH_2OH$, is Δ^1-propenol-(3)***.

ALCOHOLS—Alcohols are commonly named from the radical to which the hydroxyl radical (OH) is attached; for example, CH_3OH is methyl alcohol; $C_6H_5CH_2OH$ is benzyl alcohol. As in the case of the hydrocarbons, a complex alcohol may be named as the derivative of some relatively simple alcohol. When alcohols are named as derivatives of methyl alcohol the latter is commonly given the name of *carbinol;* for example, $(CH_3)_2CHOH$ is dimethyl carbinol because two of the hydrogen atoms in methyl alcohol have been replaced by two methyl radicals; similarly benzyl alcohol, $C_6H_5CH_2OH$, is phenyl carbinol. The adjectives *primary, secondary*, and *tertiary* are often used in connection with alcohols and refer to alcohol groupings which are of the type $-CH_2OH$, $=CHOH$, $\equiv COH$, respectively; for example, $C_3H_7CH_2OH$ is a primary butyl alcohol; $CH_3CH(OH)C_2H_5$ is a secondary butyl alcohol; $(CH_3)_3COH$ is a tertiary butyl alcohol.

In the systematic naming the hydroxyl radical is indicated by the ending *ol* and its position of attachment by an arabic numeral; for example, $C_2H_5CH_2CH_2OH$ is 1-butanol; $(CH_3)_3COH$ is 2-methyl-2-propanol because a methyl radical and an hydroxyl radical are both attached to the second carbon atom in a chain of three carbon atoms.** Two, three, etc., hydroxyl radicals in a molecule are indicated by the endings *diol, triol*, etc.; for example, glycol, CH_2OHCH_2OH is 1,2-ethandiol.

ETHERS—Ethers are most commonly named by adding the word *ether* to the name of the radicals connected by the ether oxygen atom; for example, $(C_2H_5)_2O$ is ethyl ether or diethyl ether; $CH_3OC_2H_5$ is methyl-ethyl ether. Often the name is such as to designate the presence of an alkoxy (OR) group; for example, $CH_3OC_6H_5$ is methoxy benzene.

ALDEHYDES—Aldehydes are most frequently named by dropping the ending *ic* from the name of the acid to which they are directly oxidizable and adding the suffix *aldehyde;* for example, CH_3CHO is named acetaldehyde because it is directly

***In this case the final vowel *e* of the ending *ene* has been dropped from the name because it is followed by the vowel *o* of the ending *ol* which is used to indicate the alcoholic hydroxyl radical.

ORGANIC CHEMICAL NOMENCLATURE

oxidized to acetic acid, CH_3CO_2H; similarly C_6H_5CHO is named benzaldehyde to show its relationship to benzoic acid. Sometimes aldehydes are named in a manner to show a relationship to some simpler aldehyde such as formaldehyde, $HCHO$; for example, CH_3CHO is methyl formaldehyde because the methyl radical has replaced the hydrogen atom in formaldehyde; similarly C_6H_5CHO is named phenyl formaldehyde. In the systematic naming the ending *al* indicates an aldehyde radical (CHO); for example, CH_3CHO is ethanal; $(CH_3)_2CHCHO$ is 2-methyl-1-propanal.**

KETONES—Ketones are most often named by adding the word *ketone* to the names of the radicals connected to the CO radical; for example, $(CH_3)_2CO$ is dimethyl ketone; $CH_3COCH(CH_3)_2$ is methyl-isopropyl ketone. The systematic names of the ketones differ from those of the alcohols and the aldehydes (*q. v.*) only in the termination *one* which is added to the name of the hydrocarbon; for example, $(CH_3)_2CO$ is propanone; $CH_3COCH(CH_3)_2$ is 2-methyl-butanone-(3).**

ACIDS—The common names of the normal carboxylic acids are derived from words having some relation to their natural occurrence or their properties; for example, formic, from ants (*Formicidae*); butyric, from butter; tannic, from the tanning property. Sometimes acids are named in a manner to show a relationship to some simpler acid such as acetic, CH_3CO_2H; for example, pivalic acid, $(CH_3)_3CCO_2H$, is tri-methyl-acetic acid because three methyl radicals (CH_3) are replacing the three hydrogen atoms in acetic acid. The systematic names are derived by adding the suffix *oic* and the word *acid* to the root of the name for the hydrocarbon of the *same number of carbon atoms;* for example, isovaleric acid, $(CH_3)_2CHCH_2CO_2H$, is 2-methyl-butanoic-4-acid; succinic acid, $(CH_2CO_2H)_2$, is butan-1,4-dioic acid. In compounds where such a naming would be awkward the carboxyl radical (CO_2H) is considered as a substituting group and the acid named accordingly; for example, citric acid, $HO_2CCH_2C(OH)(CO_2H)CH_2CO_2H$, is 2-hydroxy-propan-1,2,3-tricarboxylic acid.**

ESTERS AND SALTS—To the name of the alcoholic (or phenolic) radical, or the metal, is added the name of the acid modified to end in *ate;* for example, the ester of ethyl (C_2H_5) alcohol and acetic acid, $CH_3CO_2C_2H_5$, is ethyl acetate; the sodium salt of acetic acid, CH_3CO_2Na, is sodium acetate. To emphasize the ester structure they are often named as follows: acetic acid ethyl ester for ethyl acetate; malonic acid diethyl ester for diethyl malonate, $CH_2(CO_2C_2H_5)_2$.

ACID DERIVATIVES—The most common types in this classification are the anhydrides $(RCO)_2O$; halides $ROCl$; amides $RCONH_2$; amidoximes $RC(:NOH)NH_2$; amidines $RC(:NH)NH_2$; imides $R(CO)_2NH$; nitriles RCN. They are named by adding the endings *anhydride, chloride* (or other *halide*), *amide, amidoxime, amidine, imide, nitrile*, respectively, to the name of the acid or acid radical* (e.g., acetyl, acet, or aceto for the acid radical CH_3CO); for example, acetic anhydride, $(CH_3CO)_2O$; acetyl chloride, CH_3COCl; acetamide, CH_3CONH_2; acetamidoxime, $CH_3C(:NOH)NH_2$; acetamidine, $CH_3C(:NH)NH_2$; acetonitrile, CH_3CN; succinimide, $(CH_2CO)_2NH$. In compounds where such a naming would be awkward the modified carboxyl group is considered as a substituting group and named carbonamide, carbonamidine, carbonitrile, etc.

AMINES—Amines are most often named by adding the suffix *amine* to the name of the radical (or radicals) replacing one (or more) hydrogen atoms in ammonia (NH_3); for example, CH_3NH_2 is methylamine; $CH_3NHC_2H_5$ is methyl-ethylamine. Sometimes amines are named in a manner to show a relationship to some other simpler (or well-known) amine; for example, $C_6H_5NHCH_3$ is methylaniline; $C_6H_5N(CH_3)_2$ is named dimethylaniline to show the relationship to aniline, $C_6H_5NH_2$. In the systematic names the amino (NH_2) or modified amino (NHR, and NR_2) group is considered as a substituting group of the hydrocarbon; for example, $(CH_3)_2CHNH_2$ is 2-aminopropane; $(CH_3)_2CHCH_2NHCH_3$ is 2-methyl-1-methylamino-propane.

RADICALS—Alkyl radicals (C_nH_{2n+1}) are univalent groups derived by removal of one hydrogen atom from saturated hydrocarbons. They are named by replacing the ending *ane* of the corresponding hydrocarbon with the ending *yl;* for example, C_2H_5- is ethyl derived from ethane (C_2H_6); $CH_3CH_2CH_2-$ is propyl; $(CH_3)_2CH-$ is isopropyl.

ORGANIC CHEMICAL NOMENCLATURE

Univalent radicals derived from unsaturated aliphatic hydrocarbons have the endings *enyl; ynyl; dienyl;* etc.; for example, CH_2:CH- commonly called *vinyl* is also ethenyl; CH:C- is *ethynyl;* CH_2:CH-CH:CH- is *buta-1,3-dienyl.* In the aromatic series the corresponding *aryl* radicals are named in a similar manner; for example, C_6H_5- is *phenyl;* C_6H_4 = is phenylene; $CH_3C_6H_4$- is tolyl.

Bivalent radicals derived by removal of two hydrogen atoms from the *same carbon atom* are sometimes named in a manner similar to the corresponding aldehyde; for example, CH_3CH = is acetal; C_6H_5CH = is benzal. In the systematic naming the ending *ylidene* is used to replace the ending *ane* of the corresponding saturated hydrocarbon; for example, CH_3CH = is ethylidene; somewhat analogous is C_6H_5CH = when named benzylidene.

Acid radicals, derived by removal of OH from the carboxyl group ($COOH$), are named by replacing the ending *ic* of the acid name with the ending *yl;* for example, CH_3CO- is acetyl (from acetic) or ethanoyl (from ethanoic).

See also the special table NAMES AND FORMULAS OF ORGANIC RADICALS.

RING COMPOUNDS—Positions of substituting groups on the various rings are usually indicated by a system of numbering. In the benzene series of compounds disubstitution positions are often indicated by the terms *ortho* (abbreviated *o*) for the 1,2-positions; *meta* (abbreviated *m*) for the 1,3-positions; *para* (abbreviated *p*) for the 1,4-positions. In cases of trisubstitution the terms *vicinal* (*vic*), *unsymmetrical* (*uns*) and *symmetrical* (*sym*) are often employed to indicate the positions 1,2,3; 1,3,4; and 1,3,5, respectively. In the naphthalene series of compounds, especially in the older literature, disubstitution positions are sometimes indicated by the terms *ortho, meta,* and *para* with the same significance as in the benzene series, and by the terms *ana* for the 1,5-positions; *epi* for the 1,6-positions; *kata* for the 1,7-positions; *peri* for the 1,8-positions; *amphi* for the 2,6-positions; and *pros* for the 2,7-positions. See also the special table ORGANIC RING SYSTEMS.

PREFIXES AND SUFFIXES

-al, a suffix indicating the presence of an *aldehyde group* (-CHO), as in chlor*al,* Cl_3C·CHO; or pentenal, $CH_3(CH_2)_3$·CHO.*

allo-, a prefix indicating the compound to be a close relative, an isomer, or a variety of the compound to the name of which it is prefixed, as *allo*caffeine. In cases of ethylenic isomerism it is prefixed to the name of the more stable form into which the compound can be converted on heating; thus, fumaric acid is *allo*maleic acid since it is the stable isomer formed on heating maleic acid.

alpha-, beta-, gamma-, etc., usually written as letters of the Greek alphabet (α, β, γ, etc.) to indicate the first, second, third, etc., positions of substitution respectively, as in α-aminopropionic acid for CH_3·$CH(NH_2)$·CO_2H where the carbon atom attached to the CO_2H group is considered the α-carbon atom. See also the definition below for *omega* (ω). See also the numbering in the section ORGANIC RING SYSTEMS.

amphi-, a prefix used with disubstitution products of naphthalene to indicate the 2,6-positions. It is also used in naming certain stereoisomers of dioximes; thus,

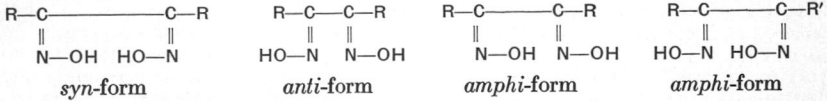

| syn-form | anti-form | amphi-form | amphi-form |

-ane, a suffix usually indicating a saturated (paraffin) hydrocarbon, C_nH_{2n+2}, as in pent*ane,* C_5H_{12}.* It is also used to indicate saturated parent compounds in the cyclic series, as camph*ane,* $C_{10}H_{18}$; or diox*ane,* $C_4H_8O_2$.

anti-, a term used in certain cases of stereoisomerism to indicate an opposed arrangement of groups or atoms, as in *anti*tartaric (mesotartaric) acid. It is often

* See this on preceding pages.

ORGANIC CHEMICAL NOMENCLATURE

employed to distinguish stereoisomers in the case of oximes and hydrazones. See also *amphi* and *syn.*

$$C_6H_5{-}C{-}C_6H_4\cdot CH_3$$
$$\|$$
$$HO{-}N$$
syn-Phenyl tolyl ketoxime

$$C_6H_5{-}C{-}C_6H_4\cdot CH_3$$
$$\|$$
$$N{-}OH$$
anti-Phenyl tolyl ketoxime

apo-, a prefix indicating a compound which is related to, or formed from, the compound to the name of which it is prefixed, as *apo*morphine.

-ase, a suffix used in forming the name of an enzyme. Usually the suffix is added to the name, or a part of the name, of a substance upon which the enzyme acts; e.g., malt*ase.*

-ate, a suffix used to denote an ester or a salt; e.g., ethyl acet*ate;* or sodium acet*ate.* *

cis, a term used in cases of *cis-trans* (i.e., ethylenic) stereoisomerism to indicate that certain elements or groups are on the *same side* of the molecule; e.g.,

$$H{-}C{-}CO_2H$$
$$\|$$
$$H{-}C{-}CO_2H$$
cis-form

$$H{-}C{-}CO_2H$$
$$\|$$
$$HO_2C{-}C{-}H$$
trans-form

d, an abbreviation for *dextro* and usually indicating that a compound is *dextro-rotatory.* Exceptions to this meaning are to be noted in certain cases of nomenclature; e.g., *d*-fructose, although levorotatory, is so named because of its relation in configuration to *d*-glucose and *d*-mannose; conversely, *l*-fructose, similarly related to *l*-glucose and *l*-mannose, is dextrorotatory.

-diene, a suffix indicating the *unsaturation* of two ethylenic (i.e., two double bond) linkages; e.g., a diolefin, C_nH_{2n}, as penta*diene*, $CH_3\cdot CH{:}CH\cdot CH{:}CH_2.$ *

-diine, or **-diyne,** a suffix indicating the *unsaturation* of two acetylenic (i.e., two triple bond) linkages; e.g., a diacetylene, $C_nH_{2n\text{-}6}$, as buta*diyne*, $CH{:}C\cdot C{:}CH.$ *

-dioic, a suffix indicating a *dicarboxylic* (two -CO_2H) acid; e.g., pentan*dioic* acid, $HO_2C\cdot (CH_2)_3\cdot CO_2H.$ *

-diol, a suffix indicating the presence of *two alcohol* (two hydroxyl) groups; e.g., pentan*diol,* $HO\cdot CH_2\cdot (CH_2)_3\cdot CH_2OH.$ *

dl, an abbreviation for *dextro-levo* indicating a substance composed of equal parts of the dextro- and levo-optical isomers of a compound.

-ene, a suffix indicating the *unsaturation* of one ethylenic (i.e., one double bond) linkage; e.g., an olefin, C_nH_{2n}, as pent*ene*, $CH_2{:}CH\cdot C_3H_7.$ * It is often used as a suffix in naming aromatic hydrocarbons; e.g., benz*ene*, tolu*ene*, naphthal*ene*, etc.

epi-, a prefix used with disubstitution products of naphthalene to indicate the 1,6-positions. See *RING COMPOUNDS* preceding. It is also prefixed to the name of an aldose, or related compound, to indicate an isomer that differs from the aldose only in the arrangement of the group about the α-carbon atom (the carbon atom next to the aldehyde group); thus, mannose is also called *epi*glucose. It is also used as a prefix to indicate a *bridge* or intramolecular connection, as in *epi*chlorohydrin, $O\cdot CH_2\cdot CH\cdot CH_2Cl.$
⌐_____⌐

eso-, a prefix indicating that substitution groups are *attached to the ring,* as *eso-*trinitromesitylene for the 2,4,6-trinitro derivative of mesitylene. See also *exo-.*

exo-, a prefix indicating that substitution groups are *attached to the side chain,* as *exo*-nitrotoluene for phenyl-nitromethane, $C_6H_5\cdot CH_2\cdot NO_2.$ See also *omega-* and *eso-.*

homo-, a prefix indicating a homologue of the compound to the name of which it is prefixed and usually differing by having one more CH_2 in the formula; e.g., *homo*-salicylic acid, $CH_3\cdot C_6H_3(OH)CO_2H.$

i or in, an abbreviation for *inactive* and indicating that the substance exhibits no activity towards polarized light. See also *dl.*

-ic, see **-oic.**

---

* See this on preceding pages.

ORGANIC CHEMICAL NOMENCLATURE

-in, a suffix used in forming the names of fats (i.e., the glycerides or esters of glycerol), the suffix being used to replace the *ic* ending of the name of the fatty acid forming the glyceride; e.g., tributyr*in*, $(C_3H_7CO)_3C_3H_5O_3$. It is often used as a suffix in the names of proteins and of glucosides; e.g., album*in* and amygdal*in*, resp. See also *i* above.

-ine, a suffix often used in forming the names of basic compounds such as amines, alkaloids, and amino acids; e.g., anil*ine*, quin*ine*, and glyc*ine*, resp. See also *-yne*.

-ite, a suffix used in various ways; thus, for esters of the *ous* acids, e.g., amyl nitr*ite*, $C_5H_{11}O \cdot NO$; for polyhydric alcohols, e.g., mann*ite* (mannitol), $C_6H_8(OH)_6$; for explosives, e.g., cord*ite*, a smokeless powder.

l, an abbreviation for *l(a)evo* and indicating that a compound is levorotatory. See also *d.*

N-, a letter used in the names of certain nitrogen compounds to indicate that the group to which the letter is prefixed has replaced a hydrogen directly attached to a nitrogen atom; e.g., *N*-methylpyrrole, $C_4H_4{:}N \cdot CH_3$. Sometimes the letter is erroneously used to indicate a *normal* compound, as *N*-butane, $CH_3 \cdot CH_2 \cdot CH_2 \cdot CH_3$, instead of a lower case letter for this meaning, as *n*-butane.

N. F., an abbreviation for the publication *National Formulary* which is issued by the *American Pharmaceutical Association* and quoted in Federal and State laws as having equal authority with the *United States Pharmacopoeia*. See also *U. S. P.*

nor-, a prefix indicating the parent compound from which another may be regarded as derived, as *nor*camphor (of which camphor is the 7-methyl derivative). The prefix has also been used to indicate a *normal* compound which is isomeric with the one to the name of which it is prefixed, as *nor*valine which is α-amino-*n*-valeric acid, whereas valine is α-amino-*iso*-valeric acid.

O-, a letter used in names of certain oxygen compounds to indicate that the group to which the letter is prefixed has replaced a hydrogen directly attached to an oxygen atom; e.g., *O*-ethyl phenol (phenetole), $C_2H_5 \cdot O \cdot C_6H_5$.

-oic, a suffix indicating a carboxylic (CO_2H) *acid;* e.g., pentan*oic* acid.* Often the suffix *-ic* is used in forming the name of an acid; e.g., acet*ic* acid.

-ol, a suffix indicating an *hydroxyl group* (OH) as in an alcohol,* e.g., pentan*ol;* or in a phenol, e.g., resorcin*ol*,* $C_6H_4(OH)_2$.

-ole, a suffix used in various ways; thus, for phenolic ethers, e.g., phenet*ole*, $C_2H_5 \cdot O \cdot C_6H_5$; for certain aldehydes, e.g., oenanth*ole*, $CH_3(CH_2)_5 \cdot CHO$; for five-membered rings, usually heterocyclic, e.g., pyrr*ole*, C_4H_5N, or oxaz*ole*, C_3H_3ON.

omega-, a prefix indicating a substitution group *on the last carbon* of a side chain, as *omega*-nitrostyrene for $C_6H_5 \cdot CH{:}CH \cdot NO_2$. It is most often indicated by the Greek letter omega (ω), as ω-nitrostyrene.

-one, a suffix indicating the presence of a *ketone group* (:CO), as in propan*one*,* $CH_3 \cdot CO \cdot CH_3$; or acetophen*one*, $CH_3 \cdot CO \cdot C_6H_5$.

-ose, a suffix usually indicating a *carbohydrate;* e.g., gluc*ose*, $C_6H_{12}O_6$; cellul*ose*, $(C_6H_{10}O_5)_x$. It is also used in naming the primary alteration or hydrolytic products of proteins; e.g., album*ose*.

par(a)-, a prefix or an adjective used to indicate that the substance is related in some way, as a polymer, an isomer, etc., of the compound to whose name it is attached; e.g., *par*aldehyde, *para*lactic acid. See also its significance when used in showing position of substituents in naming *RING COMPOUNDS.**

poly-, a prefix indicating a polymer of the compound to the name of which it is prefixed, as *poly*styrene, $(C_8H_8)_x$.

prim., an abbreviation for *primary* and indicating a replacement to the first degree; e.g., a primary amine $(R \cdot NH_2)$ where only one of the three hydrogens in ammonia has been replaced by a group (R) of atoms. It is frequently used in the designation of certain *alcohols.**

pseudo-, a prefix indicating an isomerism with, or some other relation to, the compound to the name of which it is prefixed, as *pseudo*tropine. It is sometimes indicated by the Greek letter psi (ψ); e.g., ψ-tropine.

---

* See this on preceding pages.

ORGANIC CHEMICAL NOMENCLATURE

S-, a letter used in the names of certain sulfur compounds to indicate that the group to which the letter is prefixed has replaced a hydrogen directly attached to a sulfur atom; e.g., S-methyl thioglycolic acid, $CH_3 \cdot S \cdot CH_2 \cdot CO_2H$.

sec., an abbreviation for *secondary* and indicating a replacement to the second degree; e.g., a secondary amine (R_2:NH) where two of the hydrogens in ammonia have been replaced by two groups (R_2) of atoms. It is frequently used in the designation of certain *alcohols.**

sym., an abbreviation for *symmetrical* and used to indicate a symmetrical arrangement of substitution on a parent compound. See also under *RING COMPOUNDS.**

syn-, a term used in certain cases of isomerism to indicate that certain groups or atoms are on the same side of the molecule. See also *amphi* and *anti*.

Benz-*syn*-aldoxime   Benz-*anti*-aldoxime

tert., an abbreviation for *tertiary* and indicating a replacement to the third degree; e.g., a tertiary amine (R_3:N) where three of the hydrogens in ammonia have been replaced by three groups (R_3) of atoms. It is frequently used in the designation of certain *alcohols.**

-thiol, a suffix indicating a mercaptan or thio-alcohol (-SH) group; e.g., pentan-*thiol*, $C_5H_{11} \cdot SH$.

trans, a term used in cases of cis-trans stereoisomerism. See also *cis*.

uns., an abbreviation for *unsymmetrical* and used to indicate an unsymmetrical arrangement of substituents on a parent compound. See also under *RING COMPOUNDS.**

U. S. P., an abbreviation for *United States Pharmacopoeia* and indicating that a compound conforms to the standards of purity, and other requirements, set by this publication which has been recognized in the United States as a standard in administering laws relating to the Federal Food, Drug and Cosmetic Act (1936, 1939). The Roman numeral often employed with this abbreviation indicates the revision to which reference is being made; e.g., *U. S. P. (XIII)* refers to the 13th revision.

vic., an abbreviation for *vicinal*. See under *RING COMPOUNDS.**

x-, a symbol for *unknown*. The unknown number of monomers composing a polymeric compound is often indicated by using this symbol as a subscript in the formula; e.g., starch, $(C_5H_{10}O_5)_x$. It is sometimes used to indicate the unknown position of attachment of a substituting group; e.g., in 3,x,3',x'-tetranitro-4,4'-diphenol where the positions of substitution of two of the four nitro groups are unknown.

-yl, a suffix used in forming the names of radicals, usually univalent radicals; e.g., alkyl, C_nH_{2n+1}; acyl, $R \cdot CO-$.

-ylene, a suffix used in forming the names of unsaturated hydrocarbons *; e.g., ethyl*ene*, $CH_2:CH_2$. It is frequently used in forming the names of bivalent radicals with the free valences on different carbon atoms; e.g., phenyl*ene*, $-C_6H_4-$.

-yne, a suffix indicating *unsaturation* * of one acetylenic (i.e., one triple bond) linkage; e.g., pentyne, $CH:CH \cdot C_3H_7$. In this sense modern nomenclature prefers this suffix to that of *-ine*.

* See this on preceding pages.

ORGANIC RING SYSTEMS

The rings given in the following pages are numbered as in Chemical Abstracts beginning with Volume 31 (1937). In all but a few cases (especially anthracene, phenanthrene and purine) the numberings are in accordance with the "Proposed International Rules for Numbering Organic Ring Systems"; cf. Patterson: Jour. Am. Chem. Soc. 47, 543-61 (1925). For a more complete listing of ring systems see the annual indexes of Chemical Abstracts commencing with the year 1916 and each of the decennial indexes; and, Richter: Lexikon der Kohlenstoff-Verbindungen, 3d Edition (1910), Volume I, p. 14, published by Leopold Voss, Hamburg and Leipzig; Patterson and Capell: The Ring Index (1940), Reinhold, N. Y.

When numbering positions in the case of substitution derivatives of benzoic acid, aniline, phenol, toluene, etc., the characteristic radical of each of these substances (i. e., CO_2H, NH_2, OH, CH_3, etc., respectively) is regarded as in position 1.

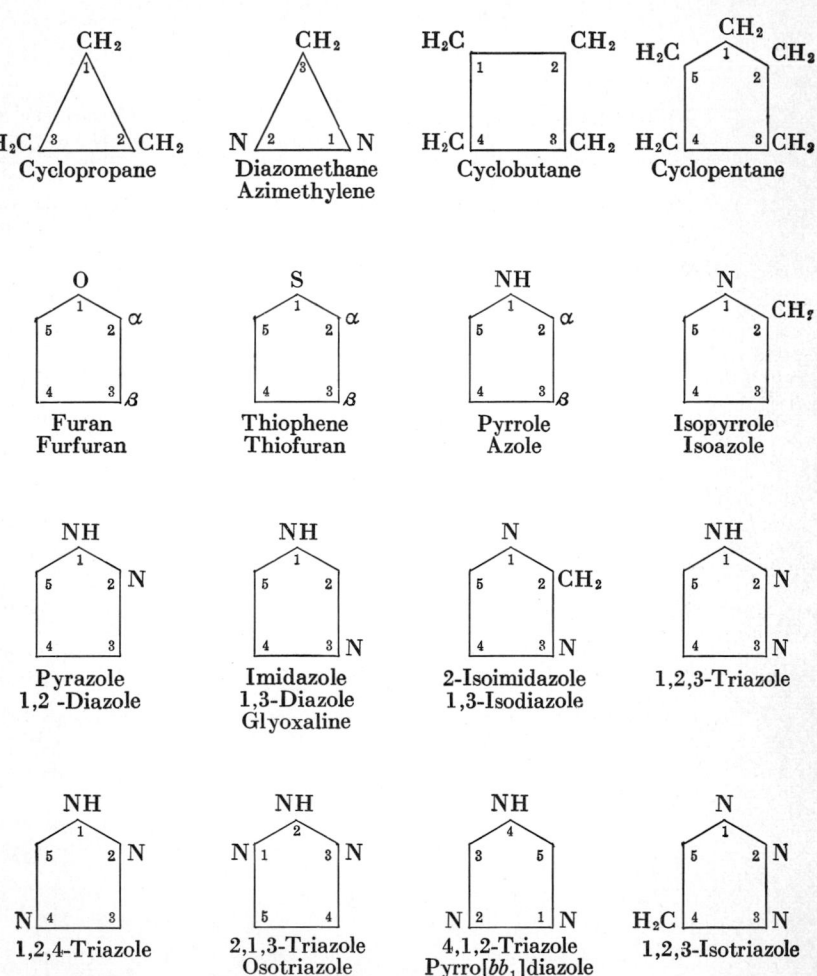

ORGANIC RING SYSTEMS

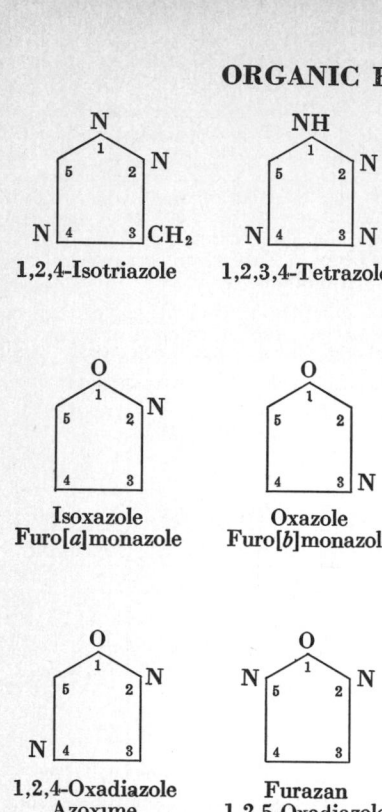

1,2,4-Isotriazole

1,2,3,4-Tetrazole

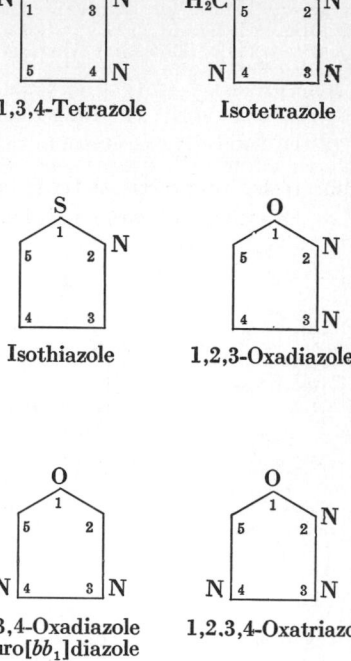

2,1,3,4-Tetrazole

Isotetrazole

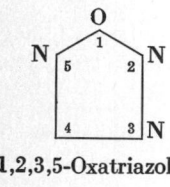

Isoxazole
Furo[a]monazole

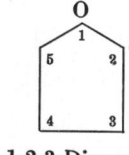

Oxazole
Furo[b]monazole

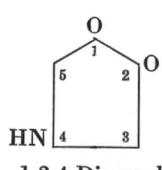

Isothiazole

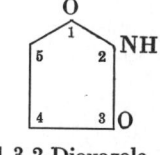

1,2,3-Oxadiazole

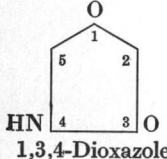

1,2,4-Oxadiazole
Azoxime
Furo[ab₁]diazole

Furazan
1,2,5-Oxadiazole

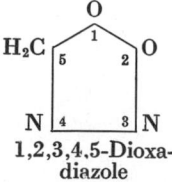

1,3,4-Oxadiazole
Furo[bb₁]diazole

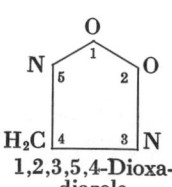

1,2.3,4-Oxatriazole

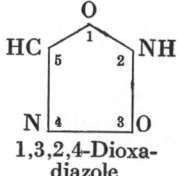

1,2,3,5-Oxatriazole

1,2,3-Dioxazole

1,2,4-Dioxazole

1,3,2-Dioxazole

1,3,4-Dioxazole

**1,2,3,4,5-Dioxa-
diazole**

**1,2,3,5,4-Dioxa-
diazole**

**1,3,2,4-Dioxa-
diazole**

ORGANIC RING SYSTEMS

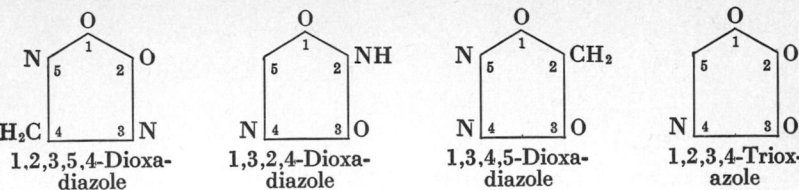

1,2,3,5,4-Dioxa-diazole   **1,3,2,4-Dioxa-diazole**   **1,3,4,5-Dioxa-diazole**   **1,2,3,4-Triox-azole**

Sulfur replacing oxygen in these five-membered rings forms the corresponding thiazoles; thiatriazoles; dithiazoles; dithiadiazoles; and trithiazoles.

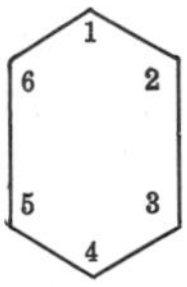

Benzene 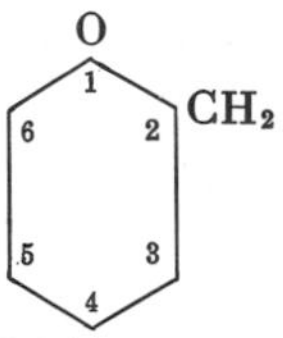   **1,2-Pyran** 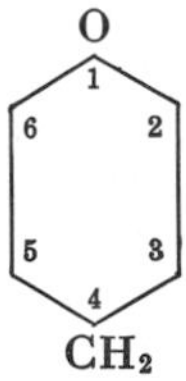   **1,4-Pyran** 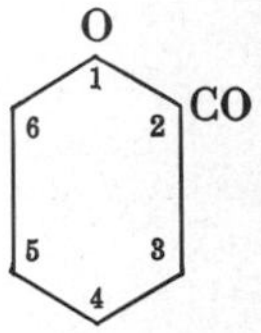   **1,2-Pyrone**
α-Pyrone
Coumalin

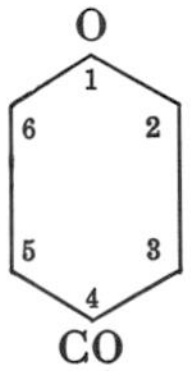

1,4-Pyrone
γ-Pyrone   **1,2-Thiapyran** 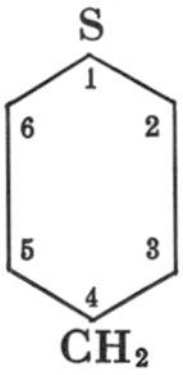   **1,4-Thiapyran**
Phenanthi-
ophene   **s-Trioxane**
1,3,5-Trioxane

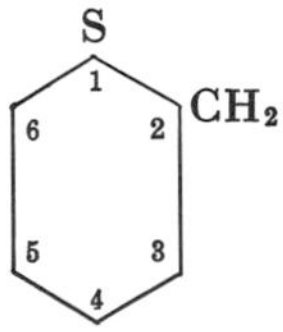

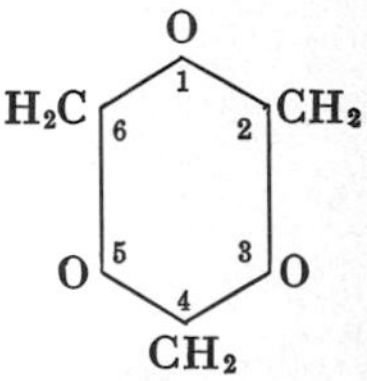

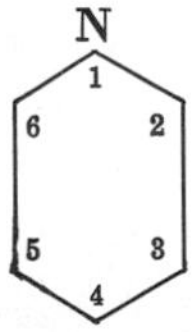

Pyridine
Azine   **Pyridazine**
1,2-Diazine
Orthodiazine   **Pyrimidine**
1,3-Diazine
Metadiazine
Miazine   **Pyrazine**
1,4-Diazine
Paradiazine
Piazine

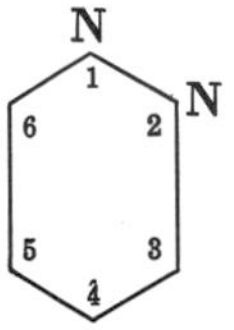

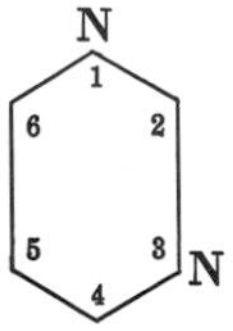

ORGANIC RING SYSTEMS

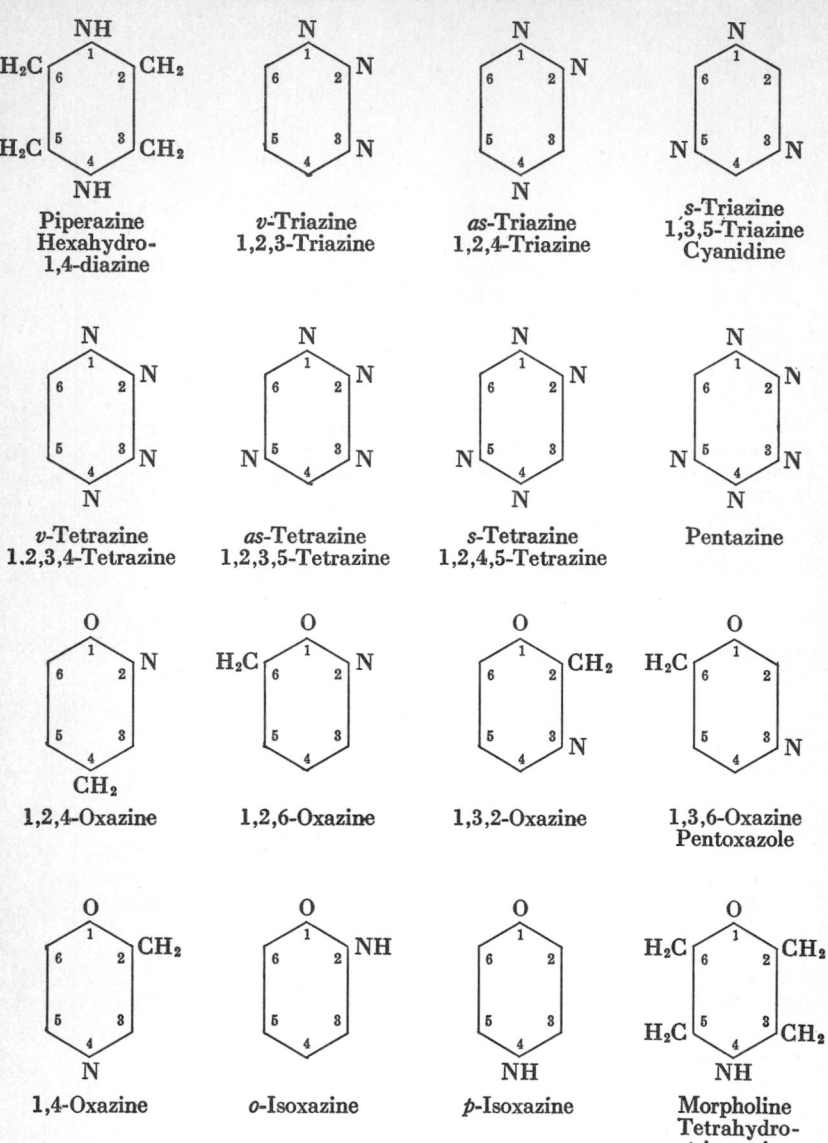

Piperazine
Hexahydro-
1,4-diazine

v-Triazine
1,2,3-Triazine

as-Triazine
1,2,4-Triazine

s-Triazine
1,3,5-Triazine
Cyanidine

v-Tetrazine
1.2,3,4-Tetrazine

as-Tetrazine
1,2,3,5-Tetrazine

s-Tetrazine
1,2,4,5-Tetrazine

Pentazine

1,2,4-Oxazine

1,2,6-Oxazine

1,3,2-Oxazine

1,3,6-Oxazine
Pentoxazole

1,4-Oxazine

o-Isoxazine

p-Isoxazine

Morpholine
Tetrahydro-
p-isoxazine

In a similar manner are derived:- oxadiazine; oxatriazine; oxatetrazine; dioxazine; dioxadiazine; dioxatriazine; trioxazine; trioxadiazine; tetroxazine. Sulfur
replacing oxygen in these rings forms the corresponding thia derivatives.

ORGANIC RING SYSTEMS

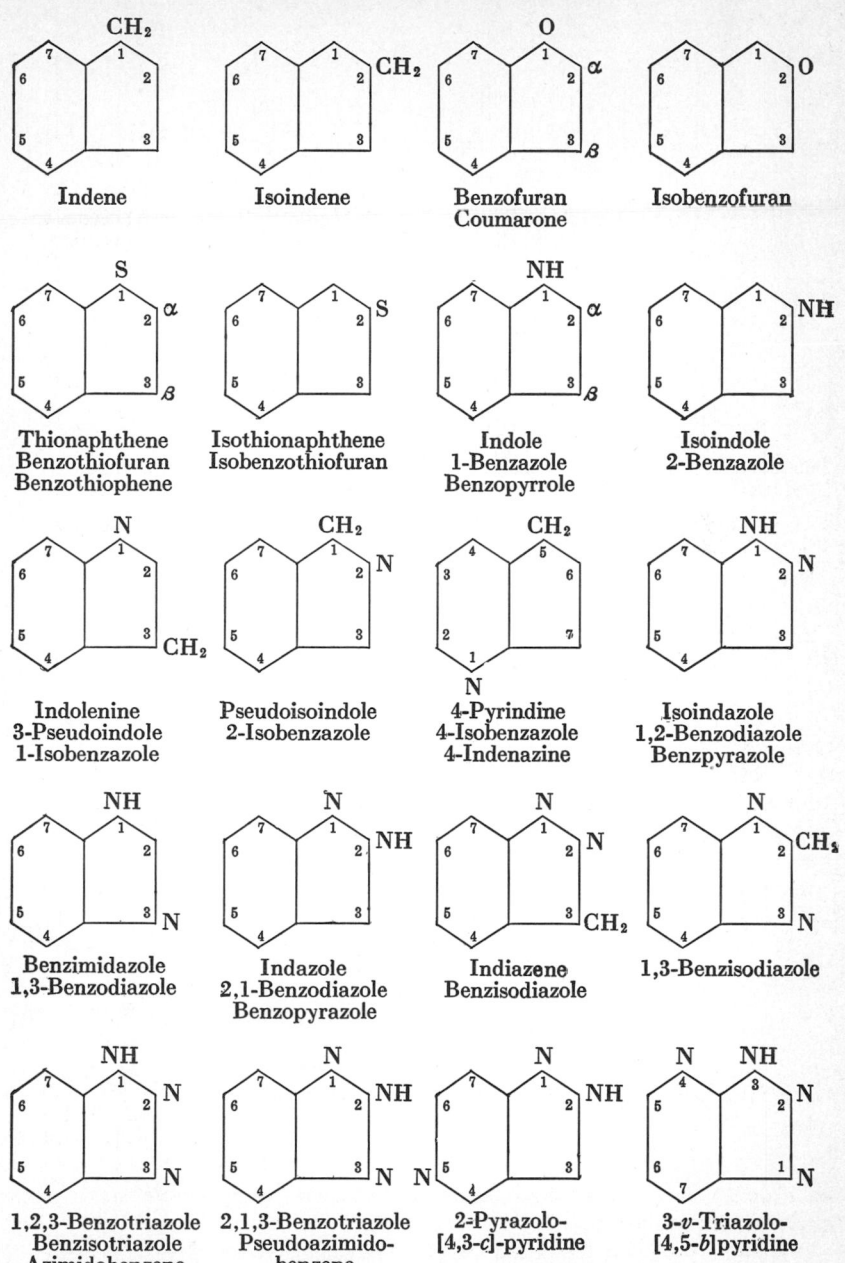

Indene

Isoindene

Benzofuran
Coumarone

Isobenzofuran

Thionaphthene
Benzothiofuran
Benzothiophene

Isothionaphthene
Isobenzothiofuran

Indole
1-Benzazole
Benzopyrrole

Isoindole
2-Benzazole

Indolenine
3-Pseudoindole
1-Isobenzazole

Pseudoisoindole
2-Isobenzazole

4-Pyrindine
4-Isobenzazole
4-Indenazine

Isoindazole
1,2-Benzodiazole
Benzpyrazole

Benzimidazole
1,3-Benzodiazole

Indazole
2,1-Benzodiazole
Benzopyrazole

Indiazene
Benzisodiazole

1,3-Benzisodiazole

1,2,3-Benzotriazole
Benzisotriazole
Azimidobenzene

2,1,3-Benzotriazole
Pseudoazimido-
benzene

2-Pyrazolo-
[4,3-c]-pyridine

3-v-Triazolo-
[4,5-b]pyridine

ORGANIC RING SYSTEMS

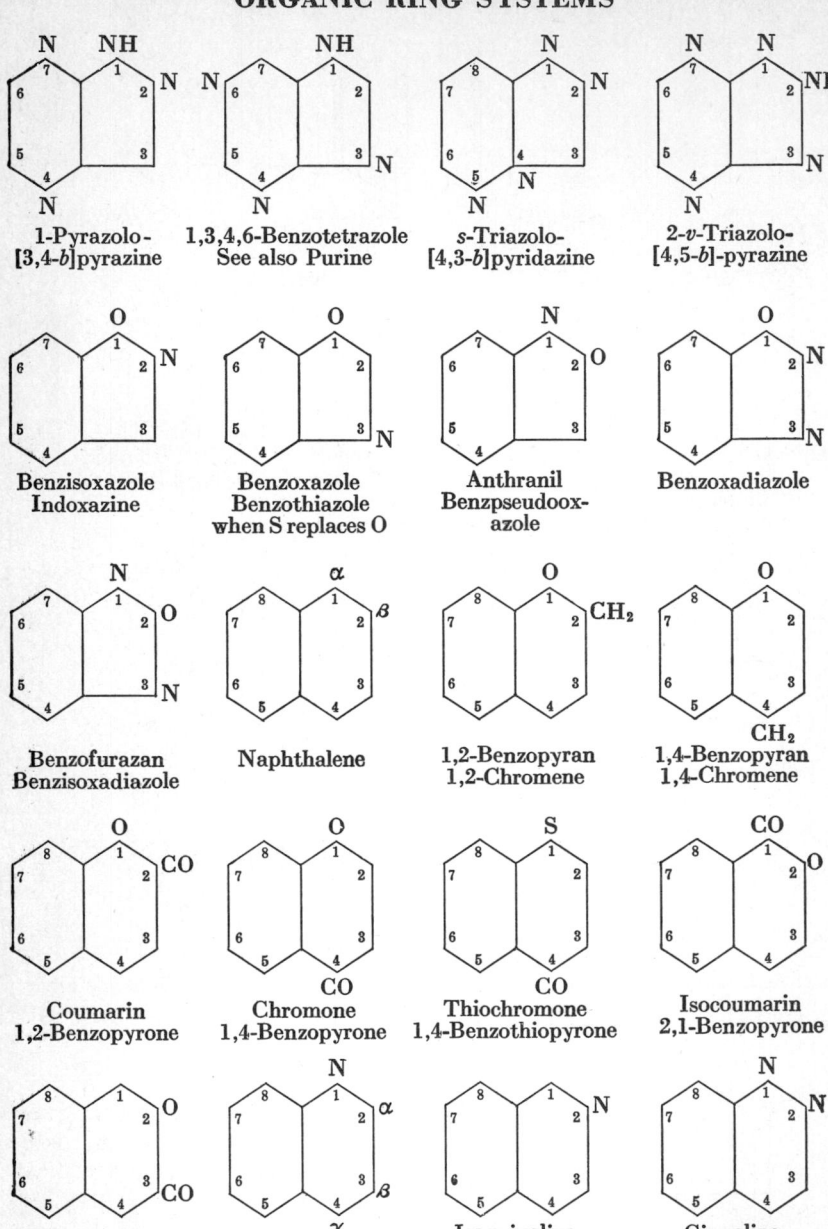

1-Pyrazolo-
[3,4-b]pyrazine

1,3,4,6-Benzotetrazole
See also Purine

s-Triazolo-
[4,3-b]pyridazine

2-v-Triazolo-
[4,5-b]-pyrazine

Benzisoxazole
Indoxazine

Benzoxazole
Benzothiazole
when S replaces O

Anthranil
Benzpseudoox-
azole

Benzoxadiazole

Benzofurazan
Benzisoxadiazole

Naphthalene

1,2-Benzopyran
1,2-Chromene

1,4-Benzopyran
1,4-Chromene

Coumarin
1,2-Benzopyrone

Chromone
1,4-Benzopyrone

Thiochromone
1,4-Benzothiopyrone

Isocoumarin
2,1-Benzopyrone

2,3-Benzopyrone

Quinoline
1-Benzazine

Isoquinoline
2-Benzazine
Leucoline

Cinnoline
1,2-Bezodiazine
α-Phenodiazine

ORGANIC RING SYSTEMS

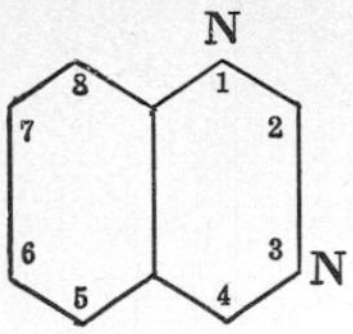

Quinazoline
1,3-Benzodiazine
Phenmiazine

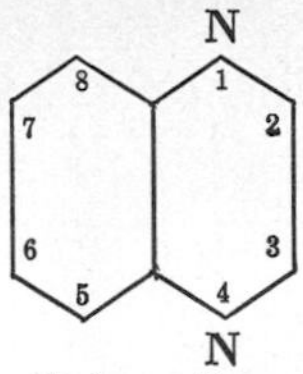

Quinoxaline
1,4-Benzodiazine
Phenpiazine

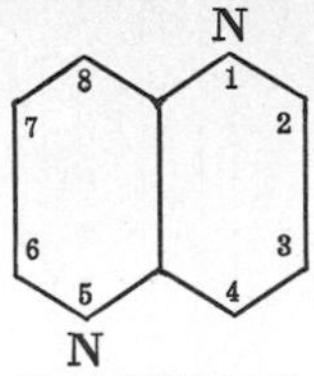

Pyrido[3,2-*b*]-
pyridine

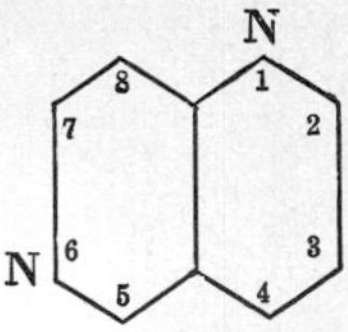

Pyrido[4,3-*b*]-
pyridine

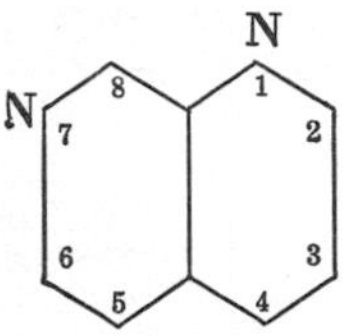

Pyrido[3,4-*b*]-
pyridine

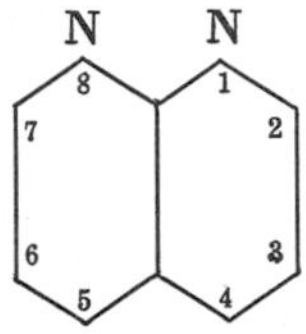

Naphthyridine

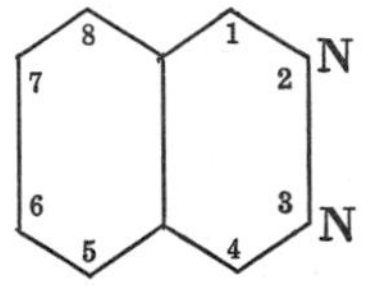

Phthalazine
2,3-Benzodiazine
β-Phenodiazine

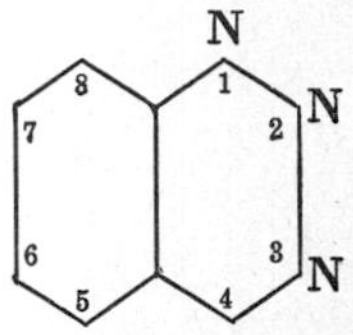

1,2,3-Benzotriazine
Phenotriazine

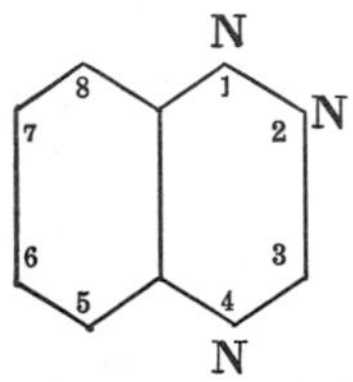

1,2,4-Benzotriazine

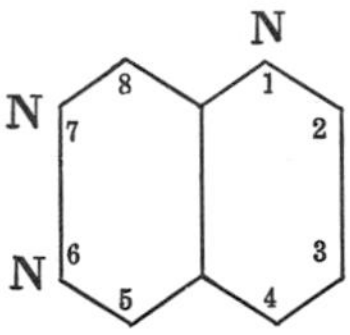

Pyrido[2,3-*d*]-
pyridazine

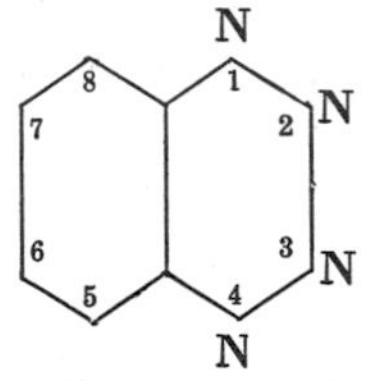

1,2,3,4-Benzotetrazine

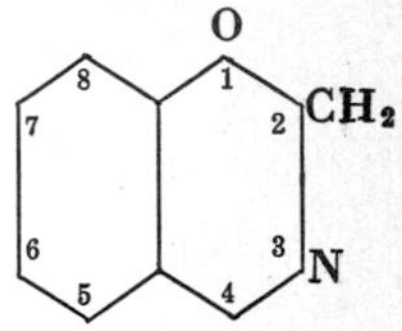

1,3,2-Benzoxazine

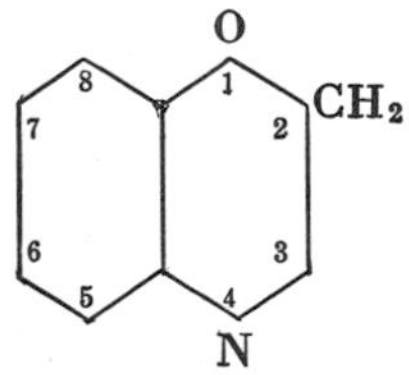

1,4,2-Benzoxazine

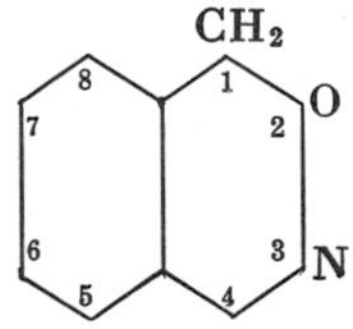

2,3,1-Benzoxazine

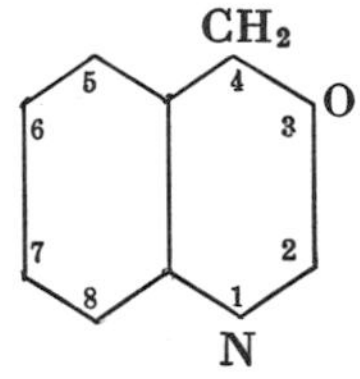

3,1,4-Benzoxazine

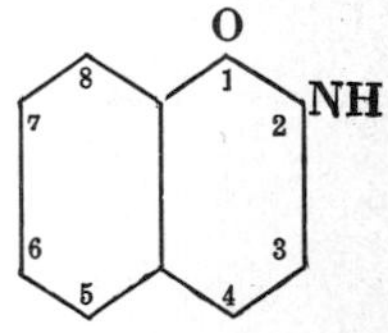

1,2-Benzisoxazine

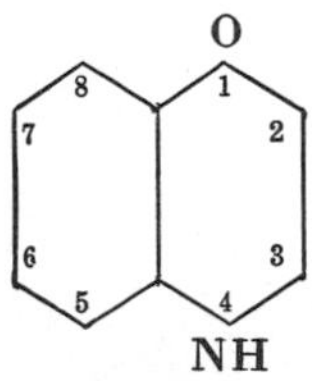

1,4-Benzisoxazine

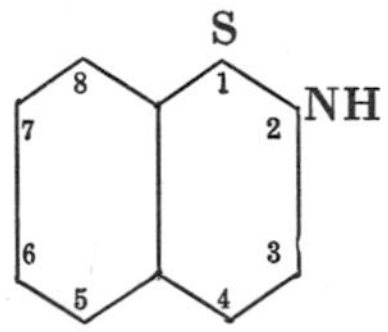

1,2-Benzisothiazine
Also a 1.4-Form

In a similar manner are derived: **benz-**oxadiazine; benzoxatriazine; etc.; **benzo-**dioxazine; etc.; benzotrioxazine; etc.

ORGANIC RING SYSTEMS

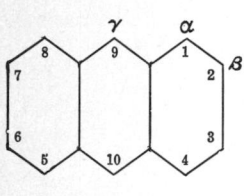

Acenaphthene

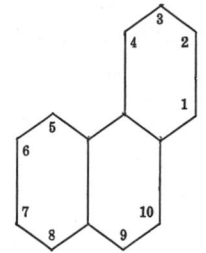

Fluorene
Diphenylenemethane

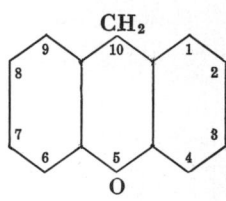

Carbazole
Dibenzopyrrole
Diphenyleneimide

Anthracene

Phenanthrene

Xanthene

Thianthrene

Acridine

Phenazine

ORGANIC RING SYSTEMS

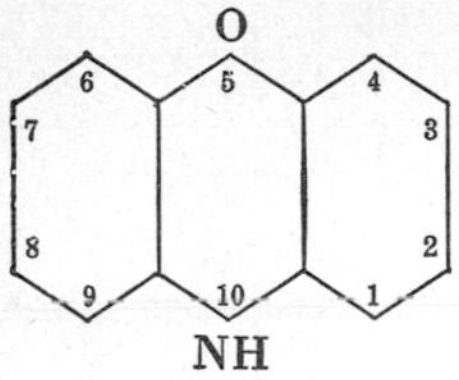

Phenoxazine
S replacing O = Phenothiazine

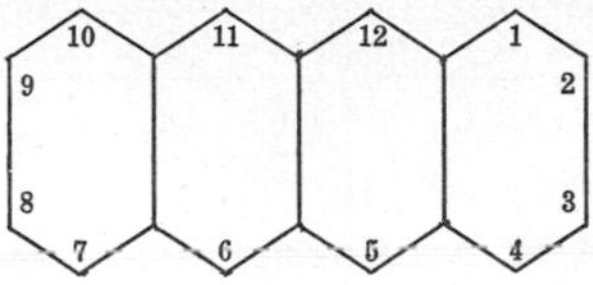

Naphthacene

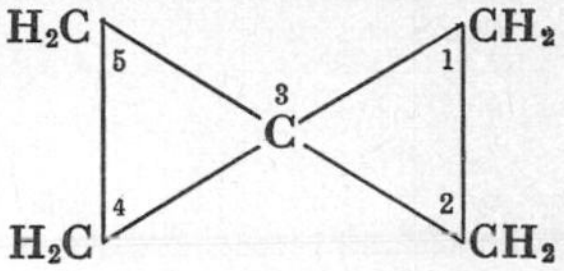

Spiropentane
Rings with but one
atom in common are known
as "spiro compounds".

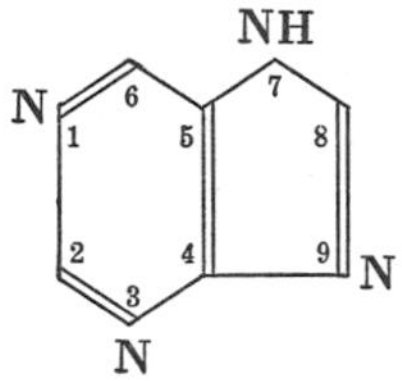

Purine

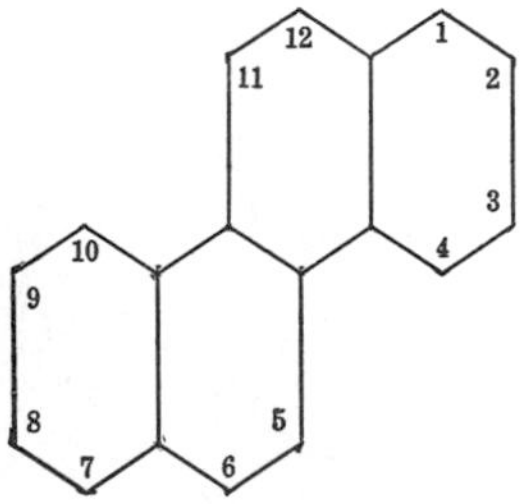

Chrysene

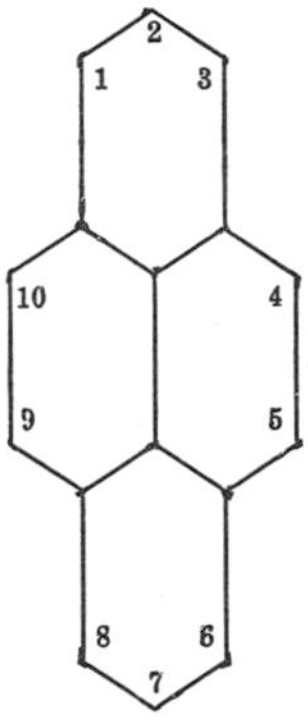

Pyrene

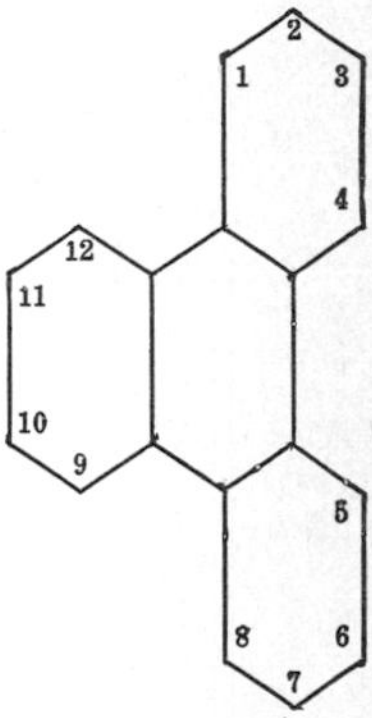

Triphenylene

7-17

Table 7-1
NAMES AND FORMULAS OF ORGANIC RADICALS

For more comprehensive lists see the various lists of radicals given in the subject indexes of the annual and decennial indexes of *Chemical Abstracts*.

Name	Formula	Name	Formula
Acenaphthenyl (from acenaphthene)	$C_{12}H_9-$	Antipyrinyl (from antipyrine)	$CO \cdot N(C_6H_5) \cdot N(CH_3) \cdot$ $\underline{\quad\quad}$ $C(CH_3):C-$
Acenaphthenylene	$-C_{12}H_8-$		
1-Acenaphthenylidene	$C_{12}H_8:$	Antipyroyl (from antipyric acid)	$CO \cdot N(C_6H_5) \cdot N(CH_3) \cdot$ $\underline{\quad\quad}$ $C(CH_3):C \cdot CO-$
Acetamido	$CH_3CO \cdot NH-$		
Acetenyl	$CH\colon C-$		
Acetimido	$CH_3C(:NH)-$ or $CH_3CON:$	Antipyryl	$CO \cdot N(C_6H_5) \cdot N(CH_3) \cdot$ $\underline{\quad\quad}$ $C(CH_3):C-$
Acetimidoyl	$CH_3C(:NH)-$		
Acetoacetyl	CH_3COCH_2CO-		
Acetonyl	CH_3COCH_2-	Arginyl	$H_2NC(:NH) \cdot NH(CH_2)_3 \cdot$ $CH(NH_2)CO-$
Acetonylidene	$CH_3COCH:$		
Acetoxy	$CH_3CO \cdot O-$	Arseno	$-As:As-$
Acetyl	CH_3CO-	Arsenoso	$O:As-$
Acetylene	$:CH \cdot CH:$	Arsinico	$(HO)O As:$
Acetylimino	$CH_3CON:$	Arsino	H_2As-
Aci-nitro	see nitro	Arso	O_2As-
Acridanyl	$C_{13}H_{10}N-$	Arsono (from arsonic acid)	$(HO)_2OAs-$
Acridinyl (from acridine)	$C_{13}H_8N-$		
Acryloyl	$CH_2:CHCO-$	Arsylene	$HAs:$
Acrylyl	$CH_2:CHCO-$	Asaryl	$(CH_3O)_3C_6H_2-$ *(2,4,5)*
Adipoyl	$-CO(CH_2)_4CO-$	Asparaginyl	$H_2NCOCH_2CH(NH_2) \cdot$ $CO-$
Adipyl	$-CO(CH_2)_4CO-$		
Alanyl	$CH_3CH(NH_2)CO-$	Asparagyl	$H_2NCOCH_2CH(NH_2) \cdot$ $CO-$
β-Alanyl	$H_2N(CH_2)_2CO-$		
Aldo (in generic sense)	$O:$	Aspartoyl	$-COCH_2CH(NH_2)CO-$
Allyl	$CH_2:CHCH_2-$	Aspartyl	$-COCH_2CH(NH_2)CO-$
β-Allyl	$CH_2:C(CH_3)-$	α-Aspartyl	$HO_2CCH_2CH(NH_2)CO-$
Allylidene	$CH_2:CHCH:$	β-Aspartyl	$HO_2CCH(NH_2)CH_2CO-$
Allyloxy	$CH_2:CHCH_2O-$	Atropoyl	$C_6H_5C(:CH_2)CO-$
Amidino	$H_2N \cdot C(:NH)-$	Auri	$Au\colon$
Amido	H_2N- (in acid groups)	Auro	$Au-$
Amidoxalyl	$H_2N \cdot CO \cdot CO-$	Azelaoyl	$-CO(CH_2)_7CO-$
Amino	H_2N-	Azido	$N:N \cdot N-$ $\underline{\quad\quad}$
Amoxy	$CH_3(CH_2)_3CH_2O-$		
Amyl	$CH_3(CH_2)_3CH_2-$	Azimino (azimido)	$-N:N \cdot NH-$
Amylidene	$CH_3(CH_2)_3CH:$	Azino	$:N:N:$
Anilino	C_6H_5NH-	Azo	$-N:N-$
Anisal	$CH_3O \cdot C_6H_4CH:$ *(p)*	Azoxy	$-N(O)N-$
Anisidino	$CH_3 \cdot C_6H_4 \cdot NH_2-$ *(p)*	Benzal	$C_6H_5CH:$
Anisoyl	$CH_3O \cdot C_6H_4CO-$ *(p)*	Benzamido	C_6H_5CONH-
Anisyl	$CH_3O \cdot C_6H_4-$ or $CH_3 \cdot C_6H_4CH_2-$	Benzenesulfinyl	C_6H_5SO-
		Benzenesulfonamido	$C_6H_5SO_2NH-$
		Benzenesulfonyl	$C_6H_5SO_2-$
Anisylidene	$CH_3O \cdot C_6H_4CH:$	Benzenyl	$C_6H_5C\colon$
Anthraniloyl	$H_2N \cdot C_6H_4CO-$ *(o)*	Benzhydryl	$(C_6H_5)_2CH-$
Anthranoyl	$H_2N \cdot C_6H_4CO-$ *(o)*	Benzhydrylidene	$(C_6H_5)_2C:$
Anthraquinonyl (from anthraquinone)	$C_{14}H_7O_2-$ (2 isomers)	Benzidino (from benzidine)	$(p-H_2N \cdot C_6H_4)C_6H_4NH-$ *(p)*
Anthraquinonylene	$-C_{14}H_6O_2-$	Benziloyl	$(C_6H_5)_2C(OH)CO-$
Anthryl (from anthracene)	$C_{14}H_9-$ (5 isomers)	Benzimidazolyl (from benzimidazole)	$C_7H_5N_2-$
Anthrylene (from anthracene)	$-C_{14}H_8-$ (11 isomers)		
Antimono	$-Sb:Sb-$		

Table 7-1 (*Continued*)
NAMES AND FORMULAS OF ORGANIC RADICALS

Name	Formula	Name	Formula
Benzimido	$C_6H_5C(:NH)-$ or	Butyryl	$CH_3(CH_2)_2CO-$
	$C_6H_5CON:$	Cacodyl	$(CH_3)_2As-$
Benzimidoyl	$C_6H_5C(:NH)-$	Camphanyl (from	$C_{10}H_{17}-$ (3 isomers)
Benzofuranyl (from	C_8H_5O-	camphane)	
benzofuran)		Camphoroyl (from	$C_{10}H_{14}O_2:$
Benzofuryl	C_8H_5O-	camphoric acid)	
Benzohydrylidene	$(C_6H_5)_2C:$	Camphoryl (from	$C_{10}H_{15}O-$
Benzopyranyl (from	C_9H_7O- (*2-α*, etc.)	camphor)	
benzopyran)		Camphorylidene (from	$C_{10}H_{14}O:$
Benzoquinonyl	$C_6H_3O_2-$ (*o* or *p*)	camphor)	
Benzoquinonylene	$-C_6H_2O_2-$ (*o* or *p*)	Caprinoyl	$CH_3(CH_2)_8CO-$
Benzothienyl	C_8H_5S-	Caproyl	$CH_3(CH_2)_4CO-$
Benzoxazinyl	C_8H_6NO-	Capryl	$CH_3(CH_2)_8CO-$
Benzoxazolyl (from	C_7H_4NO-	Capryloyl	$CH_3(CH_2)_6CO-$
benzoxazole)		Caprylyl	$CH_3(CH_2)_6CO-$
Benzoxy	$C_6H_5CO\cdot O-$	Carbamido	$H_2N\cdot CO\cdot NH-$
Benzoyl	C_6H_5CO-	Carbamoyl	$H_2N\cdot CO-$
Benzoylene	$-C_6H_4CO-$	Carbamyl	$H_2N\cdot CO-$
Benzoylimino	$C_6H_5CO\cdot N:$	Carbanilino	$C_6H_5NH\cdot CO-$
Benzoyloxy	$C_6H_5CO\cdot O-$	Carbaniloyl	$C_6H_5NH\cdot CO$
Benzyl	$C_6H_5CH_2-$	Carbazolyl	$C_{12}H_8N-$ (5 isomers)
Benzylidene	$C_6H_5CH:$	Carbazyl	$C_{12}H_8N-$ (5 isomers)
Benzylidyne	$C_6H_5C:$	Carbethoxy	$C_2H_5O\cdot CO-$
Benzyloxy	$C_6H_5CH_2\cdot O-$	Carbomethoxy	$CH_3O\cdot CO-$
Benzylthio	$C_6H_5CH_2S-$	Carbonyl	$OC:$
Biphenylene	$-C_6H_4\cdot C_6H_4-$	Carbonyldioxy	$-O\cdot CO\cdot O-$
Biphenylenebisazo	$-N:NC_6H_4\cdot C_6H_4N:N-$	Carboxy	$HO\cdot CO-$
Biphenylyl	$C_6H_5\cdot C_6H_4-$	Carbyl	$-C-$
Biphenylylcarbonyl	$C_6H_5\cdot C_6H_4CO-$	Carvacryl	$(CH_3)[(CH_3)_2CH]:$
Biphenylylene	$-C_6H_4\cdot C_6H_4-$		C_6H_3- (*1,4;2*)
Biphenylylenebisazo	$-N:NC_6H_4\cdot C_6H_4N:N-$	Caryl (from carane)	$C_{10}H_{17}-$
Biphenylyloxy	$C_6H_5\cdot C_6H_4\cdot O-$	Cetyl	$CH_3(CH_2)_{14}CH_2-$
Bismuthino	H_2Bi-	Chaulmoogroyl	$C_6H_7(CH_2)_{12}CO-$
Bornyl (from borneol)	$C_{10}H_{17}-$	Chaulmoogryl	$C_6H_7(CH_2)_{12}CH_2-$
Boryl	$O:B-$	Chloro	$Cl-$
Bromo	$Br-$	Chloroformyl	$Cl\cdot CO-$
Butadienyl	$CH_2:CHCH:CH-$	Chloromercuri	$ClHg-$
	(*1,3*-form shown)	Chromanyl	C_9H_9O-
1-Butenyl	$CH_3CH_2CH:CH-$	Chrysenyl	$C_{18}H_{11}-$
2-Butenyl	$CH_3CH:CHCH_2-$	Cinnamal	$C_6H_5CH:CHCH:$
3-Butenyl	$CH_2:CH(CH_2)_2-$	Cinnamenyl	$C_6H_5CH:CH-$
2-Butenylene	$-CH_2CH:CHCH_2-$	Cinnamoyl	$C_6H_5CH:CHCO-$
Butenylidene	$CH_3CH:CHCH:$	Cinnamyl	$C_6H_5CH:CHCH_2-$
	(*2*-form shown)	Cinnamylidene	$C_6H_5CH:CHCH:$
Butenylidyne	$CH_3CH:CHC:$	Citraconoyl	$-CO\cdot C(CH_3):CH\cdot CO-$
	(*2*-form shown)		(*cis*)
Butoxy	$CH_3(CH_2)_3O-$	Cresotoyl (from cresotinic	$HO(CH_3)C_6H_3CO-$
sec-Butoxy	$C_2H_5CH(CH_3)O-$	acid)	
tert-Butoxy	$(CH_3)_3C\cdot O-$	Cresoxy	$CH_3\cdot C_6H_4\cdot O-$
Butyl	$CH_3(CH_2)_3-$		(*o*, *m* or *p*)
sec-Butyl	$C_2H_5CH(CH_3)-$	Cresyl	$(HO)(CH_3)C_6H_3-$ or
tert-Butyl	$(CH_3)_3C-$		$CH_3\cdot C_6H_4-$
1,4-Butylene	$-CH_2(CH_2)_2CH_2-$	Cresylene	$-C_6H_3(CH_3)-$
Butylidene	$CH_3(CH_2)_2CH:$		(6 isomers)
sec-Butylidene	$C_2H_5(CH_3)C:$	Crotonoyl	$CH_3CH:CHCO-$
Butylidyne	$CH_3(CH_2)_2C:$	Crotonyl	$CH_3CH:CHCO-$
Butynylene	$-CH_2C:CCH_2-$	Crotyl	$CH_3CH:CHCH_2-$
	(*2*-form shown)		

Table 7-1 (*Continued*)
NAMES AND FORMULAS OF ORGANIC RADICALS

Name	Formula	Name	Formula
Cumal	$(CH_3)_2CH \cdot C_6H_4 \cdot CH:$ (*p*)	Cysteinyl	$HS \cdot CH_2CH(NH_2)CO-$
Cumenyl	$(CH_3)_2CH \cdot C_6H_4-$ (*o, m or p*)	Cystyl	$[-COCH(NH_2)CH_2S-]$
		Decanedioyl	$-CO(CH_2)_8CO-$
Cumidino	$(CH_3)_2CH \cdot C_6H_4 \cdot NH-$ (*p*)	Decanoyl	$CH_3(CH_2)_8CO-$
		Decyl	$CH_3(CH_2)_8CH_2-$
Cuminal	$(CH_3)_2CH \cdot C_6H_4 \cdot CH:$ (*p*)	Desyl	$(C_6H_5)(C_6H_5CO)CH-$
Cuminyl	$(CH_3)_2CH \cdot C_6H_4 \cdot CH_2-$ (*p*)	Diacetylamino	$(CH_3CO)_2N-$
		Diazo	$-N:N-$ or $N:N:$
Cuminylidene	$(CH_3)_2CH \cdot C_6H_4 \cdot CH:$ (*p*)	Diazoamino	$-N:N \cdot NH-$
Cumoyl	$(CH_3)_2CH \cdot C_6H_4 \cdot CO-$ (*p*)	Diazonium	$\overset{+}{N}(N:)-$
		Dimethyoxyphenethyl	$(CH_3O)_2C_6H_3CH_2CH_2-$
Cumyl	$(CH_3)_2CH \cdot C_6H_4-$ (*o, m or p*)	Dimethoxyphenylacetyl	$(CH_3O)_2C_6H_3CH_2CO-$
Cyanato	$N:CO-$	Dimethylamino	$(CH_3)_2N-$
Cyano	$N:C-$	Dimethylarsino	$(CH_3)_2As-$
Cyclobutyl	$CH_2CH_2CH_2CH-$ ⌞____⌟	Dimethylbenzoyl	$(CH_3)_2C_6H_3CO-$
Cycloheptyl	$CH_2(CH_2)_5CH-$ ⌞____⌟	Dimethylbenzyl	$(CH_3)_2C_6H_3CH_2-$
Cyclohexadienyl	$CH_2(CH:CH)_2CH-$ ⌞____⌟ (2,4-form shown)	Diphenylmethyl	$(C_6H_5)_2CH-$
		Diphenylmethylene	$(C_6H_5)_2C:$
		Disilanoxy	$H_3Si \cdot SiH_2 \cdot O-$
Cyclohexadienylene	$-C_6H_6-$	Disilanyl	$H_3Si \cdot SiH_2-$
Cyclohexadienylidene	$CH:CHCH_2CH:CHC:$ ⌞____⌟ (2,5-form shown)	Disilanylamino	$H_3Si \cdot SiH_2 \cdot NH-$
		Disilanylene	$-SiH_2 \cdot SiH_2-$
		Disylanylthio	$H_3Si \cdot SiH_2 \cdot S-$
Cyclohexasilanyl	$SiH_2(SiH_2)_4SiH-$ ⌞____⌟	Disilazanoxy	$H_3Si \cdot NH \cdot SiH_2 \cdot O-$
		Disilazanyl	$H_2Si \cdot NH \cdot SiH_2-$
Cyclohexenyl (from cyclohexene)	C_6H_9- (3 isomers)	Disilazanylamino	$H_3Si \cdot NH \cdot SiH_2 \cdot NH-$
		Disiloxanoxy	$H_3Si \cdot O \cdot SiH_2 \cdot O-$
Cyclohexenylene	$-C_6H_8-$	Disiloxanyl	$H_3Si \cdot O \cdot SiH_2-$
Cyclohexenylidene	$CH_2(CH_2)_2CH:CHC:$ ⌞____⌟ (2-form shown)	Disiloxanylamino	$H_3Si \cdot O \cdot SiH_2 \cdot NH-$
		Disiloxanylene	$-SiH_2 \cdot O \cdot SiH_2-$
		Disiloxanylthio	$H_3Si \cdot O \cdot SiH_2 \cdot S-$
		Disilthianoxy	$H_3Si \cdot S \cdot SiH_2 \cdot O-$
Cyclohexyl (from cyclohexane)	$C_6H_{11}-$	Disilthianyl	$H_3Si \cdot S \cdot SiH_2-$
Cyclohexylene	$-C_6H_{10}-$	Disilthianylthio	$H_3Si \cdot S \cdot SiH_2 \cdot S-$
Cyclohexylidene	$CH_2(CH_2)_4C:$ ⌞____⌟	Disilyldisilanyl	$(H_3Si)_3Si-$
		Dithio	$-S \cdot S-$
Cyclopentadienyl	C_5H_5-	Docosyl	$CH_3(CH_2)_{20}CH_2-$
Cyclopentadienylidene	$C_5H_4:$	Dodecanoyl	$CH_3(CH_2)_{10}CO-$
Cyclopentenyl (from cyclopentene)	C_5H_7-	Dodecyl	$CH_3(CH_2)_{10}CH_2-$
		Dotriacontyl	$CH_3(CH_2)_{30}CH_2-$
Cyclopentylidene	$CH_2CH_2CH:CHC:$ ⌞____⌟ (2-form shown)	Duryl	$(CH_3)_4C_6H-$ (2,3,5,6)
		Durylene	$(CH_3)_4C_6:$ (2,3,5,6)
		Eicosyl	$CH_3(CH_2)_{18}CH_2-$
Cyclopentyl (from cyclopentane)	C_5H_9-	Enanthoyl	$CH_3(CH_2)_5CO-$
Cyclopentylene	$-C_5H_8-$	Enanthyl	$CH_3(CH_2)_5CO$
Cyclopentylidene	$CH_2(CH_2)_3C:$ ⌞____⌟	Epidioxy	$-O \cdot O-$ (to different atoms or radicals already united in some other way)
Cyclopropyl	CH_2CH_2CH- ⌞____⌟	Epithio	$-S-$ (to different atoms or radicals already united in some other way)
Cymyl (from cymene)	$C_{10}H_{13}-$ (ring attachment)	Epoxy	$-O-$ (to different atoms or radicals already united in some other way)

Table 7-1 (*Continued*)
NAMES AND FORMULAS OF ORGANIC RADICALS

Name	Formula	Name	Formula
Ethanediylidene	:CH·CH:	Glutaminyl	$H_2NCOCH_2CH_2CH\cdot$
Ethene	$-CH_2\cdot CH_2-$		$(NH_2)CO-$
Ethenyl	$CH_3C\vdots$ or $CH_2:CH-$	Glutamoyl	$-OC\cdot CHNH_2\cdot(CH_2)_2\cdot CO-$
Ethenylene	$-CH:CH-$	α-Glutamyl	$HO_2C(CH_2)_2CH(NH_2)\cdot$
Ethenylidene	$CH_2:C:$		$CO-$
Ethinyl	$CH\vdots C-$	γ-Glutamyl	$HO_2CCH(NH_2)(CH_2)_2\cdot$
Ethoxalyl	$C_2H_5O\cdot OC\cdot CO-$		$CO-$
Ethoxy	C_2H_5O-	Glutamyl	see also glutamoyl
Ethoxycarbonyl	$C_2H_5O\cdot CO-$	Glutaryl	$-CO(CH_2)_3CO-$
Ethoxyphenyl	$C_2H_5O\cdot C_6H_4-$	Glyceroyl	$HOCH_2CH(OH)CO-$
Ethyl	CH_3CH_2-	Glyceryl	$-CH_2(CH-)CH_2-$
Ethylamino	C_2H_5NH-	Glycoloyl	$HOCH_2CO-$
Ethylene	$-CH_2\cdot CH_2-$	Glycolyl	$HOCH_2CO-$
Ethylenedioxy	$-O(CH_2)_2O-$	Glycyl	H_2NCH_2CO-
Ethylidene	$CH_3CH:$	Glyoxalinyl or glyoxalyl	$C_3H_3N_2-$ (4 isomers)
Ethylidyne	$CH_3C\vdots$	Glyoxyloyl	$OCH\cdot CO-$
Ethylthio	C_2H_5S-	Glyoxylyl	$OCH\cdot CO-$
Ethynyl	$CH\vdots C-$	Guaiacyl	$CH_3O\cdot C_6H_4-$ (*o*)
Ethynylene	$-C\vdots C-$	Guanidino	$H_2NC(:NH)NH-$
Fenchyl	$C_{10}H_{17}-$	Guanyl	$H_2NC(:NH)-$
Fluorenyl	$C_{13}H_9-$ (5 isomers)	Hendecyl	$CH_3(CH_2)_9CH_2-$
Fluorenylidene	$C_{13}H_8:$	Heneicosyl	$CH_3(CH_2)_{19}CH_2-$
Fluoro	$F-$	Hentriacontyl	$CH_3(CH_2)_{29}CH_2-$
Formamido	$H\cdot CO\cdot NH-$	Heptacosyl	$CH_3(CH_2)_{25}CH_2-$
Formazyl	$(C_6H_5N:N\cdot)(C_6H_5\cdot$	Heptadecanoyl	$CH_3(CH_2)_{15}CO-$
	$NHN:)C-$	Heptadecyl	$CH_3(CH_2)_{15}CH_2-$
Formimidoyl	$HC(:NH)-$	Heptanamido	$CH_3(CH_2)_5CONH-$
Formyl	$H\cdot CO-$	Heptanedioyl	$-CO(CH_2)_5CO-$
Formyloxy	$H\cdot CO\cdot O-$	Heptanoyl	$CH_3(CH_2)_5CO-$
Fumaraniloyl	$C_6H_5NHCOCH:CHCO-$	Heptyl	$CH_3(CH_2)_5CH_2-$
	(*trans*)	Hexacontyl	$CH_3(CH_2)_{58}CH_2-$
Fumaroyl	$-COCH:CHCO-$	Hexacosyl	$CH_3(CH_2)_{24}CH_2-$
	(*trans*)	Hexadecanoyl	$CH_3(CH_2)_{14}CO-$
Furfural	O·CH:CHCH:CCH:	Hexadecyl	$CH_3(CH_2)_{14}CH_2-$
	└─────────┘	Hexamethylene	$-CH_2(CH_2)_4CH_2-$
	(2-isomer only)	Hexanedioyl	$-CO(CH_2)_4CO-$
Furfuryl	O·CH:CH·CH:CCH₂-	Hexanoyl	$CH_3(CH_2)_4CO-$
	└─────────┘	Hexyl	$CH_3(CH_2)_4CH_2-$
	(2-isomer only)	Hexylidene	$CH_3(CH_2)_4CH:$
Furfurylidene	O·CH:CHCH:CCH:	Hexylidyne	$CH_3(CH_2)_4C\vdots$
	└─────────┘	Hippuroyl	$C_6H_5CONH\cdot CH_2CO-$
	(2-isomer only)	Hippuryl	$C_6H_5CONHCH_2CO-$
Furoyl	CH:CH·O·CH:CCO-	Histidyl	$N_2C_3H_3CH_2CH(NH_2)\cdot$
	└─────────┘		$CO-$
	(3-form shown)	Homopiperonyl	$(CH_2O_2)C_6H_3CH_2CH_2-$
Furyl	C_4H_3O- (2 isomers)		(*3,4*)
Furylidene	CH:CH·O·CH₂C:	Homoveratroyl	$(CH_3O)_2C_6H_3CH_2CO-$
	└─────────┘		(*3,4*)
	(3(2*H*)-form shown)	Homoveratryl	$(CH_3O)_2C_6H_3CH_2CH_2-$
Furylmethyl	CH:CH·O·CH·CCH₂-		(*3,4*)
	└─────────┘	Hydantoyl	$H_2NCONHCH_2CO-$
	(3)	Hydnocarpoyl	$C_6H_7(CH_2)_{10}CO-$
Galloyl	$(HO)_3C_6H_2CO-$ (*3,4,5*)	Hydnocarpyl	$C_6H_7(CH_2)_{10}CH_2-$
Geranyl (from geraniol)	$C_{10}H_{17}-$	Hydratropoyl	$C_6H_5CH(CH_3)CO-$
Germyl	H_3Ge-	Hydrazi	$-NH\cdot NH-$ (to same
Glucosyl	$C_6H_{11}O_5-$		atom)
Glucosyloxy	$C_6H_{11}O_6-$	Hydrazino	$H_2N\cdot NH-$

Table 7-1 (*Continued*)
NAMES AND FORMULAS OF ORGANIC RADICALS

Name	Formula	Name	Formula
Hydrazo	$-HN \cdot NH-$ (to different atoms)	Isonitro	$(HO)ON:$
		Isonitroso	$HO \cdot N:$
Hydrazono	$H_2N \cdot N:$	1-Isopentenyl	$(CH_3)_2CHCH:CH-$
Hydrocinnamoyl	$C_6H_5CH_2CH_2CO-$	Isopentylidene	$(CH_3)_2CHCH_2CH:$
Hydroperoxy	$HO \cdot O-$	Isopentylidyne	$(CH_3)_2CHCH_2C\vdots$
Hydroxy	$HO-$	Isopentyloxy	$(CH_3)_2CHCH_2CH_2O-$
Hydroxyamino	$HO \cdot NH-$	Isophthalal	$:HC \cdot C_6H_4 \cdot CH:$ (*m*)
Hydroxyimino	$HO \cdot N:$	Isophthaloyl	$-CO \cdot C_6H_4 \cdot CO-$ (*m*)
Hydroxyl	$HO-$	Isophthalylidene	$:HC \cdot C_6H_4 \cdot CH:$ (*m*)
Hydroxymethyl	$HOCH_2-$	Isopropenyl	$CH_2:C(CH_3)-$
Hydroxyphosphinyl	$(HO)HP(:O)-$	Isopropoxy	$(CH_3)_2CH \cdot O-$
ar-Hydroxytolyl	$(HO)(CH_3)C_6H_3-$	Isopropyl	$(CH_3)_2CH-$
-idene (added to any radical usually means a double bond at point of attachment)		Isopropylbenzoyl	$(CH_3)_2CH \cdot C_6H_4 \cdot CO-$
		ar-Isopropylbenzyl	$(CH_3)_2CH \cdot C_6H_4 \cdot CH_2-$
		ar-Isopropylbenzylidene	$(CH_3)_2CH \cdot C_6H_4 \cdot CH:$
		Isopropylidene	$(CH_3)_2C:$
Imidazolidinyl	$C_3H_7N_2$	Isoquinolyl (from isoquinoline)	C_9H_6N- (9 isomers)
Imidazolidyl	$C_3H_7N_2-$		
Imidazolinyl	$C_3H_5N_2-$	Isothiocyanato	$S:C:N-$
Imidazolyl (from imidazole)	$C_3H_3N_2-$ (4 isomers)	Isothiocyano	$S:C:N-$
		Isovaleryl	$(CH_3)_2CHCH_2CO-$
Imido	$HN:$	Isoxazolyl (from isoxazole)	C_3H_2NO- (5 isomers)
Imino	$HN:$		
Indanyl (from indan)	C_9H_9- (4 isomers)	Keto (in specific sense)	$O:$
Indazolyl	$C_7H_5N_2-$	Keto (in generic sense)	$O:$ (to same atom)
Indenyl (from indene)	C_9H_7- (7 isomers)	Lactoyl	$CH_3CH(OH)CO-$
Indolinyl	C_8H_8N-	Lauroyl	$CH_3(CH_2)_{10}CO-$
Indolinylidene	$CH_2 \cdot NH \cdot C_6H_4C:$ (*o*)	Leucyl	$(CH_3)_2CHCH_2 \cdot$ $CH(NH_2)CO-$
	⌞_____⌟		
	(*3*-form shown)	Linalyl	$C_{10}H_{17}-$
Indolyl (from indole)	C_8H_6N-	Lysyl	$H_2N(CH_2)_4CH(NH_2)CO-$
Indyl	C_8H_6N-	Maleoyl	$-COCH:CHCO-$ (*cis*)
Iodo	$I-$	Malonyl	$-CO \cdot CH_2 \cdot CO-$
Iodoso	$OI-$	Maloyl	$-COCH(OH)CH_2CO-$
Iodoxy	O_2I-	Mandeloyl	$C_6H_5CH(OH) \cdot CO-$
Isoallyl	$CH_3CH:CH-$	Menthyl (from menthane)	$CH_3CH(CH_2)_2CH \cdot$
Isoamoxy	$(CH_3)_2CHCH_2CH_2O-$		⌞_____
Isoamyl	$(CH_3)_2CHCH_2CH_2-$		$(iso\text{-}C_3H_7)CH_2CH-$
Isoamylidene	$(CH_3)_2CHCH_2CH:$		_____⌟
Isobornyl	$C_{10}H_{17}-$	Mercapto	$HS-$
Isobutenyl	$(CH_3)_2C:CH-$	Mercuri	$-Hg-$
Isobutoxy	$(CH_3)_2CHCH_2O-$	Mesaconoyl	$-COC(CH_3):CHCO-$
Isobutyl	$(CH_3)_2CHCH_2-$		(*trans*)
Isobutylidene	$(CH_3)_2CHCH:$	Mesidino	$(CH_3)_3C_6H_2 \cdot NH-$
Isobutylidyne	$(CH_3)_2CHC\vdots$		(*2,4,6*)
Isobutyryl	$(CH_3)_2CHCO-$	Mesityl	$(CH_3)_3C_6H_2-$ (*2,4,6*)
Isocyanato	$O:C:N-$	α-Mesityl	$(CH_3)_2C_6H_3 \cdot CH_2-$ (*3,5*)
Isocyano	$C:N-$	Mesityloxy	$(CH_3)_3C_6H_2 \cdot O-$ (*2,4,6*)
Isodiazo	$-HN \cdot N:$ (to the same atom)	Mesoxalyl	$-CO \cdot CO \cdot CO-$
		Mesyl	CH_3SO_2-
Isohexyl	$(CH_3)_2CH(CH_2)_2CH_2-$	Metanilamido	$H_2N \cdot C_6H_4 \cdot SO_2NH-$ (*m*)
Isohexylidene	$(CH_3)_2CH(CH_2)_2CH:$	Metanilyl	$H_2N \cdot C_6H_4 \cdot SO_2-$ (*m*)
Isohexylidyne	$(CH_3)_2CH(CH_2)_2C\vdots$	Methacryloyl	$CH_2:C(CH_3) \cdot CO-$
Isoindolinyl	C_8H_8N-	Methallyl	$CH_2:C(CH_3) \cdot CH_2-$
Isoindolyl (from isoindole)	C_8H_6N- (4 isomers)	Methanedisulfonyl	$CH_2(SO_2)_2:$
		Methene	$CH_2:$
Isoleucyl	$C_2H_5CH(CH_3) \cdot$ $CH(NH_2)CO-$	Methenyl	$CH\vdots$

Table 7-1 (*Continued*)
NAMES AND FORMULAS OF ORGANIC RADICALS

Name	Formula	Name	Formula
Methionyl	$CH_2S(CH_2)_2CH(NH_2)\cdot$ $CO-$ or $CH_2(SO_2)_2$:	Nonacosyl	$CH_3(CH_2)_{27}CH_2-$
		Nonadecyl	$CH_3(CH_2)_{17}CH_2-$
Methoxalyl	$CH_3O\cdot CO\cdot CO-$	Nonanedioyl	$-CO(CH_2)_7CO-$
Methoxy	CH_3O-	Nonanoyl	$CH_3(CH_2)_7CO-$
ar-Methoxyanilino	$CH_3O\cdot C_6H_4\cdot NH-$ (o, m or p)	Nonyl	$CH_3(CH_2)_7CH_2-$
Methoxybenzoyl	$CH_3O\cdot C_6H_4\cdot CO-$ (o, m or p)	Norbornyl (from norbornane)	$C_7H_{11}-$
		Norcamphanyl	$C_7H_{11}-$
ar-Methoxybenzyl	$CH_3O\cdot C_6H_4\cdot CH_2-$	Norcaryl (from norcarane)	$C_7H_{11}-$
ar-Methoxybenzylidene	$CH_3O\cdot C_6H_4\cdot CH$:		
Methoxycarbonyl	$CH_3O\cdot CO-$	Norleucyl	$CH_3(CH_2)_3CH(NH_2)CO-$
Methoxyphenyl	$CH_3O\cdot C_6H_4-$	Norpinyl (from norpinane)	$C_7H_{11}-$
Methyl	CH_3-		
Methylbenzyl	$CH_3\cdot C_6H_4\cdot CH_2-$ (o, m or p)	Octacosyl	$CH_3(CH_2)_{26}CH_2-$
		Octadecanoyl	$CH_3(CH_2)_{16}CO-$
α-Methylbenzyl	$C_6H_5\cdot CH(CH_3)-$	Octadecyl	$CH_3(CH_2)_{16}CH_2-$
Methylene	CH_2:	Octanedioyl	$-CO(CH_2)_6CO-$
Methylenedioxy	$-O\cdot CH_2\cdot O-$	Octanoyl	$CH_3(CH_2)_6CO-$
Methylidyne	HC:	Octyl	$CH_3(CH_2)_6CH_2-$
Methylol	$HOCH_2-$	Oenanthyl	$CH_3(CH_2)_5CO-$
Methylphenylene	$-C_6H_3(CH_3)-$ (6 isomers)	Oleoyl	$CH_3(CH_2)_7CH:CH\cdot$ $(CH_2)_7CO-$ (*cis*)
Methylsulfonyl	CH_3SO_2-	Ornithyl	$CH_2(NH_2)(CH_2)_2\cdot$ $CH(NH_2)CO-$
Methylthio	CH_3S-		
Morpholino	$CH_2CH_2\cdot O\cdot (CH_2)_2\cdot N-$ [____] (*4-position only*)	Oxalyl	$-OC\cdot CO-$
		Oxamido	$H_2NCO\cdot CONH-$
		Oxamoyl	$H_2NCO\cdot CO-$
Morpholinyl	$NH(CH_2)_2OCH_2CH-$ [____] (*3-form shown*)	Oxamyl	$H_2NCO\cdot CO-$
		Oxazinyl	C_4H_4NO-
		Oxazolidinyl	C_3H_6NO-
Myristoyl	$CH_3(CH_2)_{12}CO-$	Oxazolinyl	C_3H_4NO-
Naphthal	$C_{10}H_7CH$:	Oxazolyl	C_3H_2NO-
Naphthalimido (from naphthalimide)	$C_{10}H_6(CO)_2N-$ (*1,8*)	Oximido	$HO\cdot N$:
		Oxo (in specific sense)	O:
Naphthenyl	$C_{10}H_7C$:	Oxotrimethylene	$-CH_2COCH_2-$ (*2*)
Naphthionyl	$H_2N\cdot C_{10}H_6\cdot SO_2-$ (*4,1*)	Oxy	$-O-$ (used as a connective; see also epoxy and oxo)
Naphthothienyl	$C_{12}H_7S-$		
Naphthoxy	$C_{10}H_7O-$		
Naphthoyl	$C_{10}H_7CO-$	Palmitoyl	$CH_3(CH_2)_{14}CO-$
Naphthoyloxy	$C_{10}H_7CO\cdot O-$	Pelargonoyl	$CH_3(CH_2)_7CO-$
Naphthyl	$C_{10}H_7-$	Pelargonyl	$CH_3(CH_2)_7CO-$
Naphthylene	$-C_{10}H_6-$	Pentacontyl	$CH_3(CH_2)_{48}CH_2-$
Naphthylidene	$C_6H_4CH_2CH:CH\cdot C$: [____] (*1(4H)-form shown*)	Pentacosyl	$CH_3(CH_2)_{23}CH_2-$
		Pentadecanoyl	$CH_3(CH_2)_{13}CO-$
		Pentadecyl	$CH_3(CH_2)_{13}CH_2-$
Naphthylmethylene	$C_{10}H_7CH$:	Pentamethylene	$-CH_2(CH_2)_3CH_2-$
Naphthylmethylidyne	$C_{10}H_7C$:	Pentazolyl	$N:N\cdot:N\cdot N-$ [____]
Naphthyloxy	$C_{10}H_7O-$		
Neopentanetetryl	$C(CH_2-)_4$	Pentenyl	$CH_3CH_2CH:CHCH_2-$ (*2-form shown*)
Neopentyl	$(CH_3)_3CCH_2-$		
Neryl	$C_{10}H_{17}-$	Pentyl	$CH_3(CH_2)_3CH_2-$
Nitramino	$O_2N\cdot NH-$	tert-Pentyl	$CH_3CH_2C(CH_3)_2-$
Nitrilo	N:	Pentylidene	$CH_3(CH_2)_3CH$:
Nitro	O_2N-	Pentylidyne	$CH_3(CH_2)_3C$:
acī-Nitro	$(HO)ON$:	Pentyloxy	$CH_3(CH_2)_3CH_2O-$
Nitrosamino	$ON\cdot NH-$	Perimidinyl (from perimidine)	$C_{11}H_7N_2-$ (8 isomers)
Nitrosimino	$ON\cdot N$:		
Nitroso	$ON-$	Perseleno	$Se:Se$:

Table 7-1 (*Continued*)
NAMES AND FORMULAS OF ORGANIC RADICALS

Name	Formula	Name	Formula
Perthio	S:S: (replacing O only)	Phthalidylidene	$C_6H_4CO \cdot O \cdot C$: (*o*)
Phenacyl	$C_6H_5COCH_2-$		
Phenacylidene	C_6H_5COCH:	Phthalimido	$C_6H_4(CO)_2N-$ (*o*)
Phenanthridinyl	$C_{13}H_8N-$	Phthaloyl	$-OC \cdot C_6H_4 \cdot CO-$ (*o*)
Phenanthryl (from	$C_{14}H_9-$ (5 isomers)	Phthalylidene	:HCC_6H_4CH: (*o*)
phenanthrene)		Phytyl	$C_{20}H_{39}-$
Phenanthrylene	$C_{14}H_8$: (several isomers)	Picryl	$(NO_2)_3C_6H_2-$ (*2,4,6*)
Phenazinyl	$C_{12}H_7N_2-$	Pimeloyl	$-CO(CH_2)_5CO-$
Phenenyl	C_6H_3: (*s, as* or *v*)	Pinanyl (from pinane)	$C_{10}H_{17}-$
Phenethyl	$C_6H_5CH_2CH_2-$	Pinanylene	$-C_{10}H_{16}-$
Phenetidino	$C_2H_5O \cdot C_6H_4NH-$	Pinanylidene	$C_{10}H_{16}$:
	(*o, m* or *p*)	Piperidino (*1*-position	$CH_2CH_2CH_2CH_2CH_2N-$
Phenetyl	$C_2H_5O \cdot C_6H_4-$	only)	
Phenoxy	C_6H_5O-	Piperidyl	$C_5H_{10}N-$ (*2-, 3-* or *4-*)
Phenyl	C_6H_5-	Piperidylidene	C_5H_9N:
N-Phenylacetamido	$CH_3CON(C_6H_5)-$	Piperonyl	$(CH_2O_2):C_6H_3CH_2-$
Phenylacetyl	$C_6H_5CH_2CO-$		(*3,4*)
Phenylazo	$C_6H_5N:N-$	Piperonylidene	$(CH_2O_2):C_6H_3CH$: (*3,4*)
Phenylcarbamido	$C_6H_5NH \cdot CO \cdot NH-$ (*3*)	Piperonyloyl	$(CH_2O_2):C_6H_3CO$ (*3,4*)
Phenylcarbamoyl	C_6H_5NHCO-	Pivaloyl (from pivalic	$(CH_3)_3C \cdot CO-$
Phenylene	$-C_6H_4-$ (*o, m* or *p*)	acid)	
Phenylenebisazo	$-N:NC_6H_4N:N-$	Pivalyl	$(CH_3)_3C \cdot CO-$
	(*o, m* or *p*)	Plumbyl	H_3Pb-
Phenylenedimethylene	$-H_2C \cdot C_6H_4 \cdot CH_2-$	Prolyl (from proline)	$HN(CH_2)_3CH \cdot CO-$
Phenylenedimethylidyne	:$HC \cdot C_6H_4 \cdot CH$:		
Phenylethylene	$-CH(C_6H_5)CH_2-$	Propanetriyl	$-CH_2(CH-)CH_2-$
Phenylidene	$CH:CHCH_2CH:CHC$:	Propargyl	$CH:CCH_2-$
		Propenyl	$CH_3CH:CH-$
	(*2,5*-form shown)	Propenylidene	$CH_3CH:C$:
Phenylimino	C_6H_5N:	Propenylidene (*2*)	$CH_2:CHCH-$
α-Phenylphenacyl	$(C_6H_5)(C_6H_5CO)CH-$	Propioloyl	$CH:CCO-$
Phenylpropyl	$C_6H_5CH_2CH_2CH_2-$ (*3*)	Propiolyl	$CH:CCO-$
Phenylsulfamoyl	$C_6H_5NHSO_2-$	Propionamido	CH_3CH_2CONH-
Phenylsulfamyl	$C_6H_5NHSO_2-$	Propionyl	CH_3CH_2CO-
Phenylsulfinyl	C_6H_5SO-	Propionyloxy	$CH_3CH_2CO \cdot O-$
Phenylsulfonamido	$C_6H_4SO_2NH-$	Propoxy	$CH_3CH_2CH_2O-$
Phenylsulfonyl	$C_6H_5SO_2-$	Propyl	$CH_3CH_2CH_2-$ (*n*)
Phenylureido	$C_6H_5NH \cdot CO \cdot NH-$ (*3*)	*sec*-Propyl	$(CH_3)_2CH-$
Pharseno	$-P:As-$	Propylene	$-CH(CH_3)CH_2-$
Phosphazo	$-P:N-$	Propylidene	CH_3CH_2CH:
Phosphinico	(HO)OP- (only as	Propylidyne	CH_3CH_2C:
	doubling radical)	Propynyl	$CH_3C:C-$ (*1*)
Phosphinidene	HP:	Propynyl	$CH:CCH_2-$ (*2*)
Phosphinidyne	P:	Protocatechuoyl	$(HO)_2C_6H_3CO-$ (*3,4*)
Phosphino	H_2P-	Pseudoallyl	$CH_2:C(CH_3)-$
Phosphinothioyl	$H_2P(:S)-$	Pseudocumidino	$(CH_3)_3C_6H_2NH-$
Phosphinyl	$H_2P(:O)-$		(*2,4,5*)
Phosphinylidene	HP(:O):	*as*-Pseudocumyl	$(CH_3)_3C_6H_2$ (*2,3,5*)
Phosphinylidyne	O:P:	*s*-Pseudocumyl	$(CH_3)_3C_6H_2-$ (*2,4,5*)
Phospho	O_2P-	*v*-Pseudocumyl	$(CH_3)_3C_6H_2-$ (*2,3,6*)
Phosphono	$(HO)_2OP-$	Pseudoindolyl (from	C_8H_6N-
Phosphoro	$-P:P-$	pseudoindole)	
Phosphoroso	OP-	Pteridinyl (from	$C_6H_3N_4-$
Phthalal	:CHC_6H_4CH: (*o*)	pteridine)	
Phthalamoyl	$H_2NCOC_6H_4CO-$ (*o*)	Pteroyl	$C_{14}H_{11}N_6O_2-$
Phthalazinyl	$C_8H_5N_2-$	Pyranyl	C_5H_5O-
Phthalidyl	$C_6H_4CO \cdot O \cdot CH-$ (*o*)	Pyrazinyl	$C_4H_9N_2-$
		Pyrazolidinyl	$C_3H_7N_2-$

Table 7-1 (*Continued*)
NAMES AND FORMULAS OF ORGANIC RADICALS

Name	Formula	Name	Formula
Pyrazolidyl	$C_3H_7N_2-$	Stearoyl	$CH_3(CH_2)_{16}CO-$
Pyrazolinyl	$C_3H_5N_2-$	Stibarseno	$-Sb:As-$
Pyrazolyl	$C_3H_3N_2-$ (4 isomers)	Stibinico	$(HO)OSb:$
Pyrenyl	$C_{16}H_9-$	Stibino	H_2Sb-
Pyridazinyl	$C_4H_3N_2-$	Stibo	O_2Sb-
Pyridinediyl	$-C:CHCH:CH\cdot N:C-$	Stibono	$(HO)_2OSb-$
		Stiboso	$OSb-$
	(*2,3*-form shown)	Stibyl	H_2Sb-
Pyridyl (from pyridine)	C_5H_4N- (3 isomers)	Stibylene	$HSb:$
Pyridylidene	$CH:CH\cdot NH\cdot CH:CH\cdot C:$	Styrene	$-CH(C_6H_5)CH_2-$
		Styrolene	$-CH(C_6H_5)CH_2-$
	(*4(1)*)	Styryl	$C_6H_5CH:CH-$
Pyrimidinyl (from pyrimidine)	$C_4H_3N_2-$	Suberoyl	$-CO(CH_2)_6CO-$
Pyromucyl	$O\cdot CH:CH\cdot CH:C\cdot CO-$	Succinamoyl	$H_2NCOCH_2CH_2CO-$
		Succinamyl	$H_2NCOCH_2CH_2CO-$
		Succinimido	$CO\cdot CH_2\cdot CH_2CO\cdot N-$
Pyrrolidinyl (from pyrrolidine)	C_4H_8N- (3 isomers)		
Pyrrolidyl	C_4H_8N- (3 isomers)	Succinyl	$-COCH_2CH_2CO-$
Pyrrolinyl	C_4H_6N-	Sulfamino	HO_3SNH-
Pyrrolyl	C_4H_4N-	Sulfamoyl	H_2NSO_2-
Pyrrolylcarbonyl	C_4H_3NCO-	Sulfamyl	H_2NSO_2-
Pyrroyl	C_4H_3NCO-	Sulfanilamido	$H_2N\cdot C_6H_4\cdot SO_2NH-$ (*p*)
Pyrryl	C_4H_4N-	Sulfanilyl	$H_2N\cdot C_6H_4\cdot SO_2-$ (*p*)
Pyrruvoyl	$CH_3CO\cdot CO-$	Sulfhydryl	$HS-$
Quinazolinyl (from quinazoline)	$C_8H_5N_2-$	Sulfino	$(HO)OS-$
		Sulfinyl	$-SO-$
Quinolyl (from quinoline)	C_9H_6N-	Sulfo	$(HO)O_2S-$
Quinonyl	$C_6H_3O_2-$ (*o* or *p*)	Sulfoamino	HO_3SNH-
Quinoxalinyl (from quinoxaline)	$C_8H_5N_2-$	Sulfonamido	$-SO_2NH-$ (only as a doubling radical)
Quinuclidinyl	$C_7H_{12}N-$	Sulfonyl	$-SO_2-$
Salicyl	$HO\cdot C_6H_4\cdot CH_2-$ (*o*)	Sulfuryl	$-SO_2-$
Salicylidene	$HO\cdot C_6H_4\cdot CH:$ (*o*)	Tartronoyl	$-COCH(OH)CO-$
Salicyloyl	$HO\cdot C_6H_4\cdot CO-$ (*o*)	Tauryl	$H_2NCH_2CH_2SO_2-$
Sabacoyl	$-CO(CH_2)_8CO-$	Telluro	$-Te-$
Selenino	$(HO)OSe-$	Terephthalal	$:HCC_6H_4CH:$ (*p*)
Seleninyl	$OSe:$	Terephthaloyl	$-COC_6H_4CO-$ (*p*)
Seleno	$-Se-$	Terephthalylidene	$:HCC_6H_4CH:$ (*p*)
Selenocyanato	$N:CSe-$	Terphenylyl (from terphenyl)	$C_{18}H_{13}-$
Selenono	$(HO)O_2Se-$		
Selenonyl	$O_2Se:$	Terphenylylene (from terphenyl)	$-C_{18}H_{12}-$
Selenyl	$HSe-$		
Semicarbazido	$H_2NCONHNH-$	Tetracontyl	$CH_3(CH_2)_{38}CH_2-$
Semicarbazono	$H_2NCONHN:$	Tetracosyl	$CH_3(CH_2)_{22}CH_2-$
Senecicyl (from senecioic acid)	$(CH_3)_2C:CHCO-$	Tetradecanoyl	$CH_3(CH_2)_{12}CO-$
		Tetradecyl	$CH_3(CH_2)_{12}CH_2-$
Seryl	$HOCH_2CH(NH_2)CO-$	Tetramethylene	$-CH_2(CH_2)_2CH_2-$
Siloxy	$H_3Si\cdot O-$	Tetramethylphenyl	$(CH_3)_4C_6H-$
Silyl	H_3Si-	Tetramethylphenylene	$(CH_3)_4C_6:$
Silylamino	$H_3Si\cdot NH-$	*1*-Tetrazeno	$H_2N\cdot NH\cdot N:N-$
Silyldisilanyl	$(H_3Si)_2SiH-$	Tetrazolyl (from tetrazole)	$N:N\cdot N:N\cdot CH-$
Silylene	$H_2Si:$		
Silylidyne	$HSi\vdots$		(2 isomers)
Silylthio	$H_3Si\cdot S-$	Thenoyl	$S\cdot CH:CHCH:C\cdot CO-$
Stannono	$(HO)OSn-$		
Stannyl	H_3Sn-		(*2*-form shown)
Stannylene	$H_2Sn:$	Thenyl	$C_4H_3SCH_2-$
		Thenylidene	$C_4H_3SCH:$

Table 7-1 (*Continued*)
NAMES AND FORMULAS OF ORGANIC RADICALS

Name	Formula	Name	Formula
Thiazinyl	C_4H_4NS-	Triazinyl (from triazine)	$C_3H_2N_3-$
Thiazolidinyl	C_3H_6NS-	Triazolidinyl	$C_2H_5N_3-$
Thiazolidyl	C_3H_6NS-	Triazolyl (from triazole)	$C_2H_2N_3-$
Thiazolinyl	C_3H_4NS-	Tricosyl	$CH_3(CH_2)_{21}CH_2-$
Thiazolyl (from thiazole)	C_3H_2NS- (3 isomers)	Tridecanoyl	$CH_3(CH_2)_{11}CO-$
Thienyl (from thiophene)	C_4H_3S- (2 isomers)	Tridecyl	$CH_3(CH_2)_{11}CH_2-$
Thio	$-S-$	Trimethoxyphenyl	$(CH_3O)_3C_6H_2-$
Thiocarbamoyl	H_2NCS-	Trimethylanilino	$(CH_3)_3C_6H_2NH-$
Thiocarbamyl	H_2NCS-	Trimethylene	$-CH_2CH_2CH_2-$
Thiocarbonyl	$SC:$	Trimethylphenyl	$(CH_3)_3C_6H_2-$
Thiocyanato	$N\!:\!CS-$	Triphenylmethyl	$(C_6H_5)_3C-$
Thiocyano	$N\!:\!CS-$	Triphenylsilyl	$(C_6H_5)_3Si-$
Thioformyl	$S:CH-$	Trisilanyl	$H_3Si\cdot SiH_2\cdot SiH_2-$
Thiohydroxy	$HS-$	Trisilanylene	$-SiH_2\cdot SiH_2\cdot SiH_2-$
Thiomorpholino	$CH_2CH_2\cdot S\cdot CH_2CH_2\cdot N-$ ⌞_____⌟ (*4*-position only)	Tritriacontyl	$CH_3(CH_2)_{31}CH_2-$
		Trityl	$(C_6H_5)_3C-$
		Tropoyl	$C_6H_5CH(CH_2OH)CO-$
Thiomorpholinyl (2- or 3-)	$NH\cdot CH_2CH_2\cdot S\cdot CH_2CH-$ ⌞_____⌟ (*3*-form shown)	Tryptophyl (from tryptophane)	$C_8H_6NCH_2CH(NH_2)CO-$
		Tyrosyl (from tyrosine)	$HO\cdot C_6H_4\cdot CH_2\cdot CH\cdot$ $(NH_2)CO-$ (*p*)
Thionyl	$-SO-$		
Thioxo	$S:$ (replacing 2H in $:CH_2$)	Undecanoyl	$CH_3(CH_2)_9CO-$
		Undecyl	$CH_3(CH_2)_9CH_2-$
Thiuram	H_2NCS-	Uramino	$H_2NCONH-$
Threonyl	$CH_3CH(OH)CH(NH_2)\cdot$ $CO-$	Ureido	$H_2NCONH-$
		Ureylene	$-NHCONH-$
Thujyl (from thujane)	$C_{10}H_{17}-$	Valeryl	$CH_3(CH_2)_3CO-$
Thymyl (from thymol)	$HC:C(CH_3)\cdot CH:CH\cdot$ ⌞_____⌟ $C(CH(CH_3)_2):C-$	Valyl (from valine)	$(CH_3)_2CHCH(NH_2)CO-$
		Vanillal	$(CH_3O)(OH)C_6H_5\cdot CH:$ (*3,4*)
Thyronyl	$(p\text{-}HOC_6H_4O)C_6H_4CH_2\cdot$ $CH(NH_2)CO-$ (*p*)	Vanilloyl	$(CH_3O)(HO)C_6H_3\cdot CO-$ (*3,4*)
Toloxy	$CH_3\cdot C_6H_4\cdot O-$ (*o, m* or *p*)	Vanillyl	$(CH_3O)(HO)C_6H_3\cdot CH_2-$ (*3,4*)
Toluenesulfonyl	$C_7H_7SO_2-$	Vanillylidene	$(CH_3O)(HO)C_6H_3\cdot CH:$ (*3,4*)
Toluidino	$CH_3\cdot C_6H_4\cdot NH-$ (*o, m* or *p*)	Veratral	$(CH_3O)_2C_6H_3\cdot CH:$ (*3,4*)
Toluoyl	$CH_3\cdot C_6H_4\cdot CO-$ (*o, m* or *p*)	Veratroyl	$(CH_3O)_2C_6H_3\cdot CO-$ (*3,4*)
		o-Veratroyl	$(CH_3O)_2C_6H_3\cdot CO-$ (*2,3*)
Toluyl	$CH_3\cdot C_6H_4\cdot CO-$ (*o, m* or *p*)	Veratryl	$(CH_3O)_2C_6H_3\cdot CH_2-$ (*3,4*)
Tolyl	$CH_3\cdot C_6H_4-$ (*o, m* or *p*)	Veratrylidene	$(CH_3O)_2C_6H_3\cdot CH:$ (*3,4*)
α-Tolyl	$C_6H_5CH_2-$		
Tolylene	$-C_6H_3(CH_3)-$ (6 isomers)	Vinyl	$CH_2:CH-$
		Vinylene	$-CH:CH-$
α-Tolylene	$C_6H_5CH:$	Vinylidene	$CH_2:C:$
Tolyloxy	$CH_3\cdot C_6H_4\cdot O-$ (*o, m* or *p*)	Xanthenyl	$C_{13}H_9O-$ (6 isomers)
		Xanthyl (from xanthene)	$C_{13}H_9O-$ (6 isomers)
Tolylsulfonyl	$CH_3\cdot C_6H_4\cdot SO_2-$ (*o, m* or *p*)	Xenyl	$C_6H_5\cdot C_6H_4-$
		Xylidino	$(CH_3)_2C_6H_3NH-$
Tosyl	$CH_3\cdot C_6H_4\cdot SO_2-$ (*o, m* or *p*)	Xyloyl (from xylic acid)	$(CH_3)_2C_6H_3CO-$ (7 isomers)
Triacontyl	$CH_3(CH_2)_{28}CH_2-$	Xylyl	$(CH_3)_2C_6H_3-$
Triazeno	$H_2N\cdot N:N-$	Xylylene	$-H_2C\cdot C_6H_4\cdot CH_2-$

PHYSICAL PROPERTIES OF PURE SUBSTANCES

Table 7-2
FORMULA INDEX FOR ORGANIC COMPOUNDS

For finding compounds listed below see Table 7-4, Physical Constants of Organic Compounds.

The system used in this formula index is essentially the same as that employed in *Richter's Lexikon der Kohlenstoff Verbindungen*. The succession of the elements combined with carbon is as follows: H, O, N, S, F, Cl, Br, I. All the other elements are placed in alphabetical order according to their symbols. The arrangement depends (1) on the number of carbon atoms, (2) on the number of the other elements, which in addition to carbon, are contained in the compounds, (3) on the kind of elements, which in addition to carbon are contained in the molecule, in accordance with the order as stated above, and (4) on the number of atoms of each element which, in addition to carbon, is contained in the compound Thus 2 III means two carbon atoms combined with three other elements; *e.g.*, $C_2H_2O_2F_2$ 2239, is a compound with 2 carbon atoms and 3 other elements, and it is the 2239th compound in the table to which this formula index is a part.

1 I

CH_4 4061
CO 1239
CO_2 1237
CN 1589
CS_2 1238
CF_4 1244
CCl_4 1243
CBr_4 1242
CI_4 1245

1 II

CHN 3714
CHF_3 3287
$CHCl_3$ 1477
$CHBr_3$ 1016
CHI_3 3887
CH_2O 3288; 3290
CH_2O_2 3301
CH_2N_2 1577; 1772
CH_2N_4 5870
CH_2S_3 6349
CH_2F_2 4433
CH_2Cl_2 4429
CH_2Br_2 4428
CH_2I_2 4434
CH_3N_3 4134
CH_3F 4240
CH_3Cl 4183
CH_3Br 4147
CH_3I 4287
CH_4O 4100
CH_4N_2 3297
CH_4S 4298
CH_5N 4105
CH_5N_3 3488
CH_5As 4132
CH_5P 4341
CH_6N_2 4269
CH_6N_4 322
COS 1248
$COCl_2$ 1247
$COBr_2$ 1246
CO_8N_4 5853

$CNCl$ 1591
$CNBr$ 1590
CNI 1592
$CSCl_2$ 5916
$CSCl_4$ 5089
$CFCl_3$ 3286
CF_2Cl_2 2246
$CClBr_3$ 1472
CCl_2Br_2 1983
CCl_2I_2 1986
CCl_3Br 6143
CCl_3I 6157
$CBrI_3$ 1004

1 III

$CHON$ 1576; 1579
CHO_2Na 3302
CHO_2Tl 3303
CHO_6N_3 6300
$CHNS$ 5897
$CHFCl_2$ 3271
$CHClBr_2$ 1361
$CHClI_2$ 1363
$CHCl_2Br$ 1970
$CHCl_2I$ 2006
$CHBrI_2$ 928
$CHBr_2I$ 1849
$CH_2O_3N_2$ 4323
$CH_2O_4N_2$ 2636
CH_2FCl 3267
CH_2ClBr 1344
CH_2ClI 1393
CH_2BrI 950
CH_3ON 3294; 3296
CH_3OCl 4281
CH_3OAs 4131
CH_3O_2N 1228; 4316; 4765
CH_3O_3N 4315
$CH_3O_3N_3$ 4863
CH_3NS 5901
CH_3NS_2 2819
CH_3Cl_2As 4214
CH_4ON_2 3298; 3307; 6386
$CH_4O_2N_2$ 3842; 4314

$CH_4O_2N_4$ 4743
CH_4O_2Si 5615
CH_4O_3S 4382
CH_4O_4S 4384
$CH_4O_6S_2$ 4432
CH_4N_2S 5920
CH_4N_2Se 5608
CH_5ON 4279
CH_5ON_3 5609
CH_5O_3As 4130
$CH_5O_4N_3$ 6389
CH_5N_3S 5919
CH_6ON_4 1236
$CH_6O_3N_4$ 3492
CH_6NCl 4106
CH_6N_3Cl 3491
CO_2NCl_3 4697
CO_2NBr_3 4688
CO_6N_3Br 1005
CO_6N_3I 3908

1 IV

CHO_2NBr_2 1854
$CHFClBr$ 3266
CH_2ONCl 1229
CH_2O_2NBr 966
$CH_2O_5S_2Na_2$ 3293
CH_3O_2SCl 4383
CH_3O_3SNa 3295
CH_3O_4SNa 3292
CH_5ON_2Cl 6388
CH_6ONCl 4280
CH_6ON_3Cl 5610
$CH_8O_4N_2S$ 4270
$CH_8O_4N_4S$ 324

2 I

C_2H_2 130
C_2H_4 3192
C_2H_6 2904
C_2N_2 1588
C_2Cl_4 5741
C_2Cl_6 3564
C_2Br_2 1796

Table 7-2 (*Continued*)
FORMULA INDEX FOR ORGANIC COMPOUNDS

C_2Br_4 5721
C_2Br_6 3562
C_2I_2 2382
C_2I_4 5812

2 II

C_2HCl 1305
C_2HCl_3 6155
C_2HCl_5 5054
C_2HBr 891
C_2HBr_3 6105
C_2HBr_5 5050
C_2HI_5 5072
C_2H_2O 3933
$C_2H_2O_2$ 3478
$C_2H_2O_3$ 3480
$C_2H_2O_4$ 5006-7
$C_2H_2N_2$ 3715
$C_2H_2N_4$ 5869
$C_2H_2F_2$ 2250
$C_2H_2Cl_2$ 135-6; 1998
$C_2H_2Cl_4$ 5739-40
$C_2H_2Br_2$ 131-3
$C_2H_2Br_4$ 5719-20
$C_2H_2I_2$ 137-8
C_2H_3N 49; 4195
$C_2H_3N_3$ 6080
C_2H_3F 6439
C_2H_3Cl 6435
$C_2H_3Cl_3$ 6151-2
C_2H_3Br 6434
$C_2H_3Br_3$ 6103
C_2H_3I 6442
$C_2H_3I_3$ 6237
C_2H_4O 5; 3210; 6429
$C_2H_4O_2$ 36; 3470; 4242
$C_2H_4O_3$ 3468
$C_2H_4N_4$ 2080
$C_2H_4N_6$ 1761
$C_2H_4S_2$ 2818
$C_2H_4F_2$ 2249
$C_2H_4Cl_2$ 1993-4
$C_2H_4Br_2$ 1836-7
$C_2H_4I_2$ 2390-1
C_2H_5N 6430
C_2H_5F 3046
C_2H_5Cl 3015
C_2H_5Br 2989
C_2H_5I 3077
C_2H_5Na 5622
C_2H_6O 2478; 2946
$C_2H_6O_2$ 3444
$C_2H_6N_2$ 15
C_2H_6S 2560; 3083
$C_2H_6S_2$ 3209; 4222
C_2H_6Cd 1186
C_2H_6Hg 4044
C_2H_6Se 2553; 3071
C_2H_6Te 2566

C_2H_6Zn 6504
C_2H_7N 2429; 2950
$C_2H_7N_3$ 4258
C_2H_7As 2456; 2970
C_2H_7P 2529; 3135
$C_2H_8N_2$ 2494-5; 3064; 3196-9
C_2OCl_4 6129
$C_2O_2Cl_2$ 5011
$C_2O_2Cl_4$ 6159
$C_2O_4Na_2$ 5008
$C_2O_6N_4$ 6288
C_2N_2S 1593
C_2FCl_3 3285
C_2FCl_5 3276
C_2FBr_5 3275
$C_2F_2Cl_2$ 2245
$C_2F_2Cl_4$ 2253-4
$C_2F_2Br_4$ 2252
C_2F_3Cl 6203
$C_2F_3Cl_3$ 6206
$C_2F_3Br_3$ 6205
C_2ClBr_5 1423
$C_2Cl_2Br_4$ 2073
$C_2Cl_3Br_3$ 6170
$C_2Cl_4Br_2$ 5736-7

2 III

C_2HOCl_3 1281; 1931
C_2HOBr_3 876
C_2HO_2N 5012
$C_2HO_2Cl_3$ 6128
$C_2HO_2Br_3$ 6088
$C_2HO_2I_3$ 6232
C_2HFCl_4 3278
C_2HFBr_4 3277
$C_2HF_2Br_3$ 2255
$C_2HF_3Br_2$ 6204
C_2HClBr_4 1465
$C_2HCl_2Br_3$ 2075-6
$C_2HCl_3Br_2$ 6150
$C_2H_2OCl_2$ 1304; 1926
$C_2H_2OBr_2$ 890
$C_2H_2O_2F_2$ 2239
$C_2H_2O_2Cl_2$ 1928
$C_2H_2O_2Br_2$ 1795
$C_2H_2O_2I_2$ 2381
C_2H_2NCl 1399
C_2H_2NBr 954
C_2H_2NI 3890
$C_2H_2N_2S_3$ 5898
$C_2H_2FCl_3$ 3284
$C_2H_2FBr_3$ 3282-3
$C_2H_2F_2Cl_2$ 2244
$C_2H_2F_2Br_2$ 2242-3
$C_2H_2F_3Cl$ 6202
$C_2H_2ClBr_3$ 1471
$C_2H_2Cl_2Br_2$ 1980-2
$C_2H_2Cl_3Br$ 6142
$C_2H_2Cl_3As$ 1473

C_2H_3ON 3474; 4194
C_2H_3OF 89
C_2H_3OCl 81; 1293
$C_2H_3OCl_3$ 6154
C_2H_3OBr 75
$C_2H_3OBr_3$ 6104
C_2H_3OI 92
$C_2H_3O_2N_3$ 6078
$C_2H_3O_2F$ 3255
$C_2H_3O_2Cl$ 1299; 4187
$C_2H_3O_2Cl_3$ 1286
$C_2H_3O_2Br$ 883
$C_2H_3O_2Br_3$ 877
$C_2H_3O_2I$ 3874
$C_2H_3O_3N$ 5016
$C_2H_3O_4N$ 105; 4592
$C_2H_3O_6N_3$ 6299
C_2H_3NS 4387-8
$C_2H_3FBr_2$ 3268-9
$C_2H_3F_2Cl$ 2241
$C_2H_3F_2Br$ 2240
$C_2H_3F_2I$ 2251
$C_2H_3ClBr_2$ 1359-60
$C_2H_3Cl_2Br$ 1968-9
$C_2H_3Cl_2I$ 2005
C_2H_4OS 5885
$C_2H_4OCl_2$ 1997; 2008
$C_2H_4O_2N_2$ 2256; 3479; **5009**
$C_2H_4O_2N_4$ 603; 2851
$C_2H_4O_2S$ 5905
$C_2H_4O_3N_2$ 3105
$C_2H_4O_4N_2$ 2630; 3463
$C_2H_4O_5S$ 5675
$C_2H_4O_6N_2$ 3462
$C_2H_4N_2S_2$ 2821
$C_2H_4N_6Cl_2$ 1949
C_2H_4ClBr 1342-3
C_2H_4ClI 1390-1
C_2H_4BrI 948-9
C_2H_5ON 10; 14
C_2H_5OCl 1400; 3076; 3194
C_2H_5OBr 3193
C_2H_5OI 3208
$C_2H_5O_2N$ 3097; 3437; 3472; 4175; 4733
$C_2H_5O_2N_3$ 859; 5017
$C_2H_5O_3N$ 3096; 4735
C_2H_5NS 5882
C_2H_5SNa 5624
$C_2H_5Cl_2As$ 3040
$C_2H_6ON_2$ 2513; 3438; 4414
$C_2H_6ON_4$ 2081
C_2H_6OS 2563; 5904
$C_2H_6O_2N_4$ 3689
$C_2H_6O_2S$ 2562; 2905
$C_2H_6O_3S$ 2561; 2906
$C_2H_6O_4S$ 2559; 3069
$C_2H_6O_6S_2$ 3206
$C_2H_6N_2S$ 4390-1

Table 7-2 (*Continued*)
FORMULA INDEX FOR ORGANIC COMPOUNDS

$C_2H_6N_4S$ 3494
C_2H_6ClAs 1179
$C_2H_6Cl_3As$ 1181
C_2H_7ON 8; 2907; 3074-5
$C_2H_7O_2As$ 1182
$C_2H_7O_3As$ 2971
$C_2H_7O_4N_3$ 4415
$C_2H_7N_2Cl$ 16
$C_2H_8O_3N_4$ 323; 4259
C_2H_8NCl 2430; 2952
C_2H_8NBr 2951
$C_2H_9N_2Cl$ 2496
$C_2H_{10}ON_2$ 3199
$C_2H_{10}N_2Cl_2$ 3198
$C_2H_{10}N_2Br_2$ 3197
$C_2O_2N_2Ag_2$ 5620
$C_2O_2N_2Hg$ 4040

2 IV

$C_2HFCl_2Br_2$ 3270
$C_2H_2ONCl_3$ 6126
$C_2H_2ONBr_3$ 6086
$C_2H_3ONCl_2$ 1927; 2032
$C_2H_3ONBr_2$ 1793-4
$C_2H_3O_2NCl_2$ 2027
C_2H_4ONCl 1294-5; 1422
$C_2H_4ONCl_3$ 1283
C_2H_4ONBr 879
$C_2H_4O_2NCl$ 1416-7
$C_2H_4O_2NBr$ 965
$C_2H_5O_2SCl$ 3155
$C_2H_5O_3SCl$ 3022
$C_2H_5O_4SNa$ 13
$C_2H_7O_3NS$ 5696
$C_2H_{12}O_4N_6S$ 3493
$C_2H_{14}O_4N_8S$ 325

3 I

C_3H_4 183; 225
C_3H_6 1621; 5459
C_3H_8 5338
C_3O_2 1240
C_3S_2 1241
C_3Cl_8 4936

3 II

C_3HCl_7 3506
C_3H_2O 5349
$C_3H_2O_2$ 5348
$C_3H_2N_2$ 4006
$C_3H_2Br_2$ 1871
C_3H_3N 151
C_3H_3Cl 5343
C_3H_3Br 5342
C_3H_3I 3903; 5344
C_3H_4O 146; 226; 5341
$C_3H_4O_2$ 149
$C_3H_4O_3$ 5525
$C_3H_4O_4$ 4002; 4329

$C_3H_4O_5$ 5695
$C_3H_4O_6$ 4055
$C_3H_4N_2$ 3851; 5478
$C_3H_4Cl_2$ 2058-61
$C_3H_4Br_2$ 1869-70
C_3H_5N 3029; 5363
C_3H_5Cl 202; 5345-6
$C_3H_5Cl_3$ 6166-7
C_3H_5Br 199; 988-90
$C_3H_5Br_3$ 6112
C_3H_5I 208
C_3H_6O 56; 193; 5352; 5468-9
$C_3H_6O_2$ 55; 3048; 3436; 3454;
3942; 4087; 5351
$C_3H_6O_3$ 2292; 2468; 3289;
3380; 3456; 3682;
3939-40; 4063; 4255
$C_3H_6O_4$ 3379
$C_3H_6N_2$ 5479
$C_3H_6N_6$ 4019
C_3H_6S 210
$C_3H_6S_3$ 3309
$C_3H_6Cl_2$ 2050-3
$C_3H_6Br_2$ 1863-6
$C_3H_6I_2$ 2397-9
C_3H_7N 194
C_3H_7F 5406-7
C_3H_7Cl 5400-1
C_3H_7Br 5388-9
C_3H_7I 5417-8
C_3H_8O 4230; 5368-9
$C_3H_8O_2$ 3460; 4424; 5465; 6286
$C_3H_8O_3$ 3381
C_3H_8S 4238; 5424-5
$C_3H_8S_2$ 6284
C_3H_8Se 5448
C_3H_9N 4226; 5371-2; 6248
$C_3H_9N_3$ 2491
C_3H_9Al 231
C_3H_9As 6252
C_3H_9B 872
C_3H_9Bi 856
C_3H_9P 6269
C_3H_9Sb 6274
$C_3H_{10}N_2$ 5464; 6282
$C_3N_3Cl_3$ 6172

3 III

C_3HOBr_5 5048.1
$C_3H_2OCl_4$ 5728-9
$C_3H_2O_3N_2$ 5036
$C_3H_2O_3Br_2$ 1875
$C_3H_2O_4Br_2$ 1851
$C_3H_2O_4Ca$ 4003
C_3H_3ON 5527
$C_3H_3O_2N$ 1583
$C_3H_3O_2Cl$ 1306-7
$C_3H_3O_2Cl_3$ 4403
$C_3H_3O_3N_3$ 1594; 3313

$C_3H_3O_3Cl_3$ 6158
$C_3H_3O_4N$ 4762
$C_3H_3O_4Cl$ 1397
$C_3H_3O_4Br$ 952
C_3H_3NS 5878
$C_3H_3NCl_2$ 2056
$C_3H_3N_3S_3$ 5899
$C_3H_4ON_2$ 1581; 5480
$C_3H_4OCl_2$ 1444; 1929-30
$C_3H_4OBr_2$ 982
$C_3H_4O_2N_2$ 3679
$C_3H_4O_2Cl_2$ 1382; 2054-5
$C_3H_4O_2Br_2$ 1867-8
$C_3H_4O_3N_2$ 182
$C_3H_4O_4N_2$ 5013
$C_3H_4N_2S$ 377
C_3H_5ON 148; 3027-8; 3195;
3944
$C_3H_5ON_5$ 396
C_3H_5OF 5361
C_3H_5OCl 1301; 2879; 5360
$C_3H_5OCl_3$ 6168
C_3H_5OBr 886; 892; 5359
C_3H_5OI 2881; 5362
$C_3H_5O_2N$ 3927; 4829; 4873;
5526
$C_3H_5O_2Cl$ 1398; 1442-3; 3019;
4184
$C_3H_5O_2Br$ 955; 980-1; 4148
$C_3H_5O_2I$ 3901-2
$C_3H_5O_3N_3$ 5014
$C_3H_5O_4N$ 328
$C_3H_5O_6N_3$ 3429
$C_3H_5O_9N_3$ 3428
C_3H_5NS 3158-9
$C_3H_6ON_2$ 3212; 3216
$C_3H_6OS_2$ 6448
$C_3H_6OCl_2$ 3393-4
$C_3H_6OBr_2$ 3388-9
$C_3H_6OI_2$ 3401
$C_3H_6O_2N_2$ 123; 4004
$C_3H_6O_2S$ 5908; 5910
$C_3H_6O_3N_2$ 3678
$C_3H_6O_4N_2$ 2658-9; 4864
$C_3H_6O_7N_2$ 3405-6
C_3H_6ClBr 6279
C_3H_7ON 61; 235; 3047; 4085;
5353; 5356
$C_3H_7ON_3$ 12
C_3H_7OCl 4186; 5461-2; 6280
C_3H_7OBr 5460; 6278
C_3H_7OI 6287
$C_3H_7O_2N$ 158-60; 3291; 3943;
4107; 4827-8; 5428;
5600; 6392
$C_3H_7O_2N_3$ 3442
$C_3H_7O_2Cl$ 3395-6
$C_3H_7O_2I$ 3402
$C_3H_7O_3N$ 5427; 5611-4

Table 7-2 (*Continued*)
FORMULA INDEX FOR ORGANIC COMPOUNDS

C₃H₇O₃As 197
C₃H₇O₃Na 5623
C₃H₇O₅N 3407-8
C₃H₇NS 5917
C₃H₇NS₂ 3157
C₃H₇N₆Cl 4020
C₃H₈ON₂ 2572-3; 3183
C₃H₈O₂N₂ 5429
C₃H₈O₂S 4239; 5903
C₃H₈N₂S 2568; 3162
C₃H₉ON 4111; 5339; 5416
C₃H₉O₂N 3385
C₃H₉O₂N₃ 3489
C₃H₉O₃As 5376
C₃H₉O₃B 6262
C₃H₉O₄P 6268
C₃H₉O₆P 3411
C₃H₁₀ON₂ 3384
C₃H₁₀NCl 4227; 6249
C₃H₁₂O₃N₆ 3490

3 IV

C₃H₂ONCl₃ 1284
C₃H₃ONS₂ 5560
C₃H₃O₂NS 2369
C₃H₄ON₂S 5907
C₃H₄O₂NCl₃ 1285
C₃H₅O₂NCl₂ 2031
C₃H₅O₃NCl₂ 2028
C₃H₆ON₂S 122
C₃H₆O₂NCl 1420-1
C₃H₆NS₂Na 2477
C₃H₇ONS 5921; 6450
C₃H₇O₂NS 1625
C₃H₇O₄SNa 64
C₃H₇O₆PNa₂ 3412
C₃H₈O₂NCl 5601

3 V

C₃H₈O₂NSCl 1626

4 I

C₄H₂ 1679
C₄H₆ 1018-9; 1156-7
C₄H₈ 1147-9; 1595; 4210
C₄H₁₀ 1021-2

4 II

C₄H₂O₃ 3996
C₄H₂O₄ 134
C₄H₄O 3327
C₄H₄O₂ 5871
C₄H₄O₃ 5657
C₄H₄O₄ 3314; 3475; 3995
C₄H₄N₂ 5477; 5482; 5513; 5661
C₄H₄S 5923
C₄H₅N 204-5; 1563; 5521
C₄H₅N₃ 3853

C₄H₅Cl 1345
C₄H₆O 1559; 2499; 4058; 4348; 6436
C₄H₆O₂ 207; 1557-8; 1667; 3304; 4095-6; 5655; 6426-7
C₄H₆O₃ 37-8; 3934; 4366; 6440
C₄H₆O₄ 106; 2521; 3066; 3455; 5652-3
C₄H₆O₅ 2261; 3473; 3997-8; 4000
C₄H₆O₆ 5688; 5690-1
C₄H₆O₈ 2367
C₄H₆N₂ 4257
C₄H₆S 2852
C₄H₇N 1175-6; 5404-5; 5524
C₄H₇Cl 1566-7; 4060
C₄H₇Br₃ 6100-1
C₄H₇I 1568
C₄H₈O 1155; 1160; 1163; 1565; 4059; 4104; 4232; 4422; 5773; 6437-8
C₄H₈O₂ 43; 169; 1166-7; 2237; 2935; 3207; 4349; 5408-9; 6285
C₄H₈O₃ 2912; 3060; 3397; 3446; 3753-7; 4076; 4229; 4294
C₄H₈O₄ 4253
C₄H₈N₂ 9; 4256
C₄H₈S 3190
C₄H₈S₂ 2228
C₄H₈Cl₂ 1971-6
C₄H₈Br₂ 1821-5
C₄H₈I₂ 2389
C₄H₉N 4102; 5523
C₄H₉F 1085
C₄H₉Cl 1072-5
C₄H₉Br 1057-60
C₄H₉I 1095-8
C₄H₁₀O 1034-7; 2911; 4353-4
C₄H₁₀O₂ 1154; 1265; 2196; 2329-32; 2415; 3061; 3459
C₄H₁₀O₃ 3404; 3453; 4077
C₄H₁₀O₄ 2894
C₄H₁₀N₂ 2225-6
C₄H₁₀S 1103-5; 2214
C₄H₁₀S₂ 2166
C₄H₁₀Be 845
C₄H₁₀Cd 1185
C₄H₁₀Hg 4043
C₄H₁₀Se 2207
C₄H₁₀Se₂ 2208
C₄H₁₀Sn 5949
C₄H₁₀Te 2219
C₄H₁₀Zn 6503
C₄H₁₁N 1039-42; 2109; 2482; 4350

C₄H₁₁P 2197
C₄H₁₂N₂ 2178; 5474
C₄H₁₂As₂ 1178
C₄H₁₂Pb 3958
C₄H₁₂Si 5617
C₄H₁₂Sn 5951
C₄H₁₃N₃ 2238
C₄OCl₁₀ 5088

4 III

C₄HNI₄ 5817
C₄H₂O₂Cl₂ 3316
C₄H₂O₃Cl₂ 4442
C₄H₂O₃Br₂ 4441
C₄H₂O₄N₂ 184-6
C₄H₂O₄Cl₂ 2007
C₄H₂O₄Br₂ 1840; 1850
C₄H₂SI₂ 2400
C₄H₃OBr 942
C₄H₃O₃N 4740
C₄H₃O₄N₃ 4862; 6445
C₄H₃O₄Cl 1384; 1396
C₄H₃O₄Br 941; 951
C₄H₃O₅N₃ 4632
C₄H₄O₂N₂ 6383
C₄H₄O₂Cl₂ 5658
C₄H₄O₃N₂ 621
C₄H₄O₃N₄ 1774
C₄H₄O₃Cl₂ 1300
C₄H₄O₄N₂ 1686
C₄H₄O₄Br₂ 1877-8
C₄H₄O₄Na₂ 5654
C₄H₄O₅N₂ 186-7
C₄H₄O₆Ca 5689
C₄H₄O₈Na₂ 2368
C₄H₄O₁₀N₂ 4842
C₄H₄N₂S₂ 3211
C₄H₅ON 2880
C₄H₅OCl 1562
C₄H₅OCl₃ 1158
C₄H₅O₂N 1587; 3032; 4196; 5660
C₄H₅O₂N₃ 386
C₄H₅O₂Cl 1358
C₄H₅O₂Cl₃ 3179; 6153
C₄H₅O₂Br 925
C₄H₅O₂Br₃ 3178
C₄H₅O₃N₃ 6384
C₄H₅O₄Cl 1464
C₄H₅O₄Br 997
C₄H₅NS 219-20; 378
C₄H₆ON₂ 4360
C₄H₆O₂N₂ 3035; 3315; 3439; 4267-8
C₄H₆O₂N₄ 139
C₄H₆O₂S₂ 85
C₄H₆O₂Cl₂ 1448; 1978-9; 1988; 3039
C₄H₆O₂Br₂ 1826-7; 3036

Table 7-2 (*Continued*)
FORMULA INDEX FOR ORGANIC COMPOUNDS

$C_4H_6O_3N_4$ 181
$C_4H_6O_4S$ 5911
$C_4H_6O_6Ca$ 3469
C_4H_6NCl 1351
$C_4H_6N_2S$ 329
$C_4H_6O_{12}N_4$ 2895
C_4H_7ON 57; 1560; 3758; 3936
$C_4H_7ON_3$ 1540
C_4H_7OCl 1173-4; 1401; 4223
$C_4H_7OCl_3$ 6145
$C_4H_7OBr_3$ 6102
$C_4H_7O_2N$ 1662; 4888
$C_4H_7O_2N_3$ 3177
$C_4H_7O_2Cl$ 1347-50; 1377; 3016; 5402-3
$C_4H_7O_2Cl_3$ 1159; 1282
$C_4H_7O_2Br$ 917-8; 936; 2990
$C_4H_7O_2I$ 3078; 3886
$C_4H_7O_3N$ 90; 4880; 5015; 5651
$C_4H_7O_3N_3$ 74
$C_4H_7O_3Cl$ 4075
$C_4H_7O_4N$ 589-90; 3098; 3852
$C_4H_7O_5N$ 5693
C_4H_7NS 5450-1
$C_4H_8ON_2$ 223
C_4H_8OS 3156
$C_4H_8OS_2$ 3087
$C_4H_8OCl_2$ 1977; 1995-6
$C_4H_8O_2N_2$ 2490; 2522-3; 4091; 5656
$C_4H_8O_2S$ 3160-1
$C_4H_8O_3N_2$ 588; 2947; 3476; 4001; 4891
$C_4H_8O_4N_2$ 5692
$C_4H_8O_4N_6$ 5005
$C_4H_8O_7N_2$ 2230
$C_4H_8N_2S$ 221
$C_4H_8SCl_2$ 1985
C_4H_9ON 1161; 1168-9; 2235; 2416; 2933; 4233
$C_4H_9ON_3$ 63; 5354-5
C_4H_9OCl 1151-3; 1362
C_4H_9OBr 939
$C_4H_9O_2N$ 86; 277-80; 1108-9; 3057; 3059; 4118; 4416; 4690-3; 5398
$C_4H_9O_2N_3$ 1539
$C_4H_9O_2Cl$ 2260; 4254
$C_4H_9O_3N$ 1107; 4694-5; 5932-4
$C_4H_9O_4N$ 4689
$C_4H_9O_5N$ 4859
C_4H_9NS 5884
$C_4H_{10}ON_2$ 2191; 5453; 6276
$C_4H_{10}OS$ 2217
$C_4H_{10}O_2S$ 2216; 5900
$C_4H_{10}O_3S$ 2215; 2567
$C_4H_{10}O_4S$ 2213
$C_4H_{10}O_6N_4$ 6390-1
$C_4H_{10}S_2Hg$ 4051

$C_4H_{11}ON$ 1023-4; 2443; 2957
$C_4H_{11}O_2N$ 1020; 2090
$C_4H_{11}O_3N$ 6230
$C_4H_{11}O_3As$ 1048
$C_4H_{11}O_4P$ 2198
$C_4H_{12}ON_2$ 3771
$C_4H_{12}OAs_2$ 1183
$C_4H_{12}NCl$ 2110; 5822
$C_4H_{12}NBr$ 5821
$C_4H_{12}NI$ 5826
$C_4H_{12}SAs_2$ 1180
$C_4H_{13}ON$ 5823-5
$C_4H_{14}N_2Cl_2$ 5475

4 IV

$C_4H_2O_4N_3Na$ 6446
$C_4H_2O_3N_2Cl_2$ 1951
$C_4H_2O_3N_2Br_2$ 1810
$C_4H_3O_2NS$ 4843
$C_4H_4O_2NCl$ 5659
$C_4H_4O_2N_2S$ 5887
C_4H_6ONCl 3764
$C_4H_7O_2NCl_2$ 171
$C_4H_7O_3N_2I$ 179
$C_4H_{10}O_2NCl$ 3058
$C_4H_{14}O_4N_4S_3$ 4391
$C_4H_{14}O_6N_8S$ 2082
$C_4H_{16}O_4N_6S$ 4260

4 V

$C_4H_3O_4SAuNa_2$ 4447
$C_4H_{16}O_4N_8Cl_2Ca$ 6387

5 I

C_5H_6 1618; 6415
C_5H_8 4154-5; 4168; 5064-5; 5086-7
C_5H_{10} 498-502; 1619; 4198
C_5H_{12} 5083-5

5 II

$C_5H_2O_5$ 1556
$C_5H_4O_2$ 3317-8; 5518
$C_5H_4O_3$ 1511; 3339-40
$C_5H_4O_4$ 140
$C_5H_4N_4$ 5473
C_5H_5N 5483
C_5H_6O 4245
$C_5H_6O_2$ 3137; 3332; 5340; 6428
$C_5H_6O_4$ 1510; 3370; 3928; 4243; 6425
$C_5H_6O_5$ 59
$C_5H_6O_8$ 1659
$C_5H_6N_2$ 360-2; 3377; 5463
C_5H_6S 4389
C_5H_7N 192; 4362-4
C_5H_8O 1620; 2427; 3136; 3213
$C_5H_8O_2$ 68; 189-90; 2424-6; 2942-3; 3846; 3968; 4193

$C_5H_8O_3$ 67; 1672; 3150; 3967; 4088
$C_5H_8O_4$ 2502-3; 3082; 3376; 4236; 4430; 5520
$C_5H_8O_5$ 1514; 3398; 3773-4; 3815; 3929
$C_5H_8O_7$ 6228-9
$C_5H_8N_2$ 2534
C_5H_9N 1083; 4365; 6246; 6405; 6413
$C_5H_9N_3$ 3667
$C_5H_{10}O$ 2180; 2949; 3189; 4103; 4224; 4355; 4357; 6243; 6400; 6407
$C_5H_{10}O_2$ 1086; 1088-9; 3138; 4169-71; 5081; 5365-6; 5775; 6245; 6277; 6399; 6406
$C_5H_{10}O_3$ 2138; 3063; 3079; 3843-5; 4182; 5413; 5467
$C_5H_{10}O_4$ 3056; 3382
$C_5H_{10}O_5$ 570; 3991-2; 5564; 6480
$C_5H_{10}O_6$ 573
$C_5H_{10}N_2$ 2143
$C_5H_{10}Cl_2$ 2033-41
$C_5H_{10}Br_2$ 1856-8
$C_5H_{10}I_2$ 2396
$C_5H_{11}N$ 5322
$C_5H_{11}Cl$ 442-8
$C_5H_{11}Br$ 427-31
$C_5H_{11}I$ 459-63
$C_5H_{12}O$ 404-11; 3142-3; 4161-2
$C_5H_{12}O_2$ 503-4; 2172
$C_5H_{12}O_3$ 3403; 4177; 5069
$C_5H_{12}O_4$ 5066
$C_5H_{12}O_5$ 156; 572; 6478
$C_5H_{12}S$ 467-9; 4167
$C_5H_{13}N$ 412-8; 4156-7; 4215
$C_5H_{14}N_2$ 1184

5 III

C_5H_3ON 3345
$C_5H_3O_2Cl$ 3344
$C_5H_3O_3Cl$ 1385
$C_5H_3O_3Br$ 943
$C_5H_3O_5N$ 4741
$C_5H_3NBr_2$ 1872-4
C_5H_4OS 5925
$C_5H_4O_2N_4$ 6449
$C_5H_4O_2S$ 5926-7
$C_5H_4O_3N_4$ 6393
C_5H_4NCl 1450-2
C_5H_4NBr 991-2
C_5H_5ON 3812-4
C_5H_5OCl 3335
$C_5H_5O_2N$ 2362-3; 3322-3; 3341-2; 5522

Table 7-2 (*Continued*)
FORMULA INDEX FOR ORGANIC COMPOUNDS

$C_5H_5O_3N$ 6231
C_5H_6OS 3337; 5924
$C_5H_6O_2N_2$ 4410-3
C_5H_6NCl 5484
C_5H_7ON 3333
$C_5H_7O_2N$ 3030
$C_5H_7NCl_2$ 5485
$C_5H_8O_2N_2$ 66
$C_5H_8O_2Cl_2$ 45
$C_5H_8O_3N_2$ 1678
$C_5H_8O_3Cl_2$ 1984
C_5H_9OCl 6404; 6412
$C_5H_9O_2N$ 1674; 4890; 5335
$C_5H_9O_2Cl$ 1077-8; 1445-7; 3020-1
$C_5H_9O_2Br$ 983-4; 1006-7; 2995-6
$C_5H_9O_3N$ 3808-9
$C_5H_9O_3Cl$ 2923
$C_5H_9O_4N$ 3371-2
$C_5H_9O_5N$ 3772
C_5H_9NS 1131-6
$C_5H_{10}ON_2$ 4898
$C_5H_{10}OS_2$ 2163-4
$C_5H_{10}O_3N_2$ 3374-5; 4886
$C_5H_{10}N_2S_2$ 1249
$C_5H_{11}ON$ 2173; 4356; 4358; 6244; 6401-2; 6408-9
$C_5H_{11}ON_3$ 1164; 4234
$C_5H_{11}O_2N$ 387-9; 391-5; 473-4; 846; 1067-8; 3184; 4790-3
$C_5H_{11}O_3N$ 472; 2922
$C_5H_{11}O_4N$ 4789
$C_5H_{11}NS$ 6275
$C_5H_{11}NS_2$ 2162
$C_5H_{11}N_3Cl_2$ 3668
$C_5H_{12}ON_2$ 1138-9; 2222-3; 5845
$C_5H_{12}O_2N_2$ 5000
$C_5H_{12}N_2S$ 2220
$C_5H_{13}ON$ 4584
$C_5H_{13}O_2N$ 243; 2444
$C_5H_{14}N_2S_2$ 2449
$C_5H_{15}O_2N$ 1488

5 IV

$C_5H_5O_3NS$ 5505
$C_5H_6O_2N_2S$ 121
$C_5H_8O_3NCl_3$ 1288
$C_5H_{10}O_4NCl$ 3373
$C_5H_{11}O_2NS$ 4062
$C_5H_{12}O_2NCl$ 390; 847; 2945
$C_5H_{14}ONCl$ 1489

6 I

C_6H_6 656; 2768
C_6H_8 1599; 1600

C_6H_{10} 1609; 2769; 3573-4
C_6H_{12} 1601; 3655-61; 4209
C_6H_{14} 3603-7
C_6O_6 1604
C_6Cl_6 3563
C_6Br_6 3561
C_6I_6 3589

6 II

C_6HCl_5 5053
C_6HBr_5 5049
C_6HI_5 5071
$C_6H_2Cl_4$ 5733-5
$C_6H_2Br_4$ 5715-6
$C_6H_2I_4$ 5809-11
$C_6H_3Cl_3$ 6134-6
$C_6H_3Br_3$ 6095-7
$C_6H_3I_3$ 6233-5
$C_6H_4O_2$ 5541-2
$C_6H_4O_4$ 2364
$C_6H_4O_6$ 5808
$C_6H_4Cl_2$ 1952-4
$C_6H_4Br_2$ 1811-3
$C_6H_4I_2$ 2386-8
$C_6H_5N_3$ 252; 6079
C_6H_5F 3258
C_6H_5Cl 1324
C_6H_5Br 902
C_6H_5I 3879
C_6H_6O 5118
$C_6H_6O_2$ 2312-4; 4244
$C_6H_6O_3$ 150; 4246; 6216-9
$C_6H_6O_4$ 4443; 5802-3
$C_6H_6O_6$ 141; 3588
C_6H_6S 5915
$C_6H_6S_2$ 2820; 2822
$C_6H_6Cl_6$ 668-9
$C_6H_6Br_6$ 666
C_6H_6Se 5607
C_6H_7N 513; 5300-2
C_6H_7P 5219
C_6H_8O 2488
$C_6H_8O_2$ 2285; 5625; 5798
$C_6H_8O_4$ 209; 2487; 2501; 3945
$C_6H_8O_6$ 96; 586-7; 6123
$C_6H_8O_7$ 1515-6; 5578
$C_6H_8N_2$ 155; 2533; 5196; 5256; 5258; 5260
$C_6H_8N_4$ 4425
C_6H_8S 5928-31
C_6H_9N 2535-8; 3149
$C_6H_9N_3$ 6059-60; 6062
$C_6H_9N_{11}$ 4018
$C_6H_{10}O$ 191; 206; 1606; 4053; 4225
$C_6H_{10}O_2$ 65; 216; 1564; 3024-6; 3086
$C_6H_{10}O_3$ 2937; 3417; 3765-6; 4296; 5357

$C_6H_{10}O_4$ 153; 2193; 2554-8; 2940; 3154; 3214; 3445; 3926; 4235; 4251-2; 5422-3
$C_6H_{10}O_5$ 622; 1267; 1661; 2500; 3443; 3872; 3948; 3966; 3969; 5630
$C_6H_{10}O_6$ 2564-5; 3070; 3361; 3368
$C_6H_{10}O_8$ 4438-40; 5577
$C_6H_{10}N_4$ 5079
$C_6H_{10}S$ 1684
$C_6H_{10}S_3$ 222
$C_6H_{11}N$ 453; 1217-8; 1680; 2104
$C_6H_{11}N_5$ 5048
$C_6H_{11}Cl$ 1613
$C_6H_{11}Br$ 1612
$C_6H_{11}I$ 1614
$C_6H_{12}O$ 1212; 1605; 2099; 3144-5; 3614-5; 4163-6
$C_6H_{12}O_2$ 454-5; 1029-32; 1208-11; 1663; 2101; 2481; 3006-7; 4404; 4420-1; 5350; 5441-2
$C_6H_{12}O_3$ 7; 1266; 3073; 3759-60; 5419-20
$C_6H_{12}O_4$ 2924; 3413
$C_6H_{12}O_5$ 1608; 3312; 4423; 5558; 5561
$C_6H_{12}O_6$ 1602-3; 3616-20
$C_6H_{12}O_7$ 3359; 3946
$C_6H_{12}N_2$ 3681; 3932
$C_6H_{12}N_4$ 3599
$C_6H_{12}S_3$ 5879-81
$C_6H_{12}Cl_2$ 2001-2
$C_6H_{12}Br_2$ 1842-7
$C_6H_{12}I_2$ 2393
$C_6H_{13}N$ 326; 4344-7
$C_6H_{13}Cl$ 3644-7
$C_6H_{13}Br$ 3641-3
$C_6H_{13}I$ 3650-1
$C_6H_{14}O$ 2783-6; 3000-2; 3624-38; 4120-1
$C_6H_{14}O_2$ 4; 1071; 2139; 2346; 3598; 3662; 4332; 5312-3
$C_6H_{14}O_3$ 1232; 2808; 3471
$C_6H_{14}O_4$ 6199
$C_6H_{14}O_5$ 2259; 5557
$C_6H_{14}O_6$ 3608-9; 3611
$C_6H_{14}S$ 2799-800; 3005; 3652
$C_6H_{14}S_2$ 2782
$C_6H_{14}Hg$ 4046
$C_6H_{14}Zn$ 6505-6
$C_6H_{15}N$ 2772-3; 3639-40; 6180
$C_6H_{15}Al$ 230
$C_6H_{15}As$ 6183
$C_6H_{15}B$ 871

Table 7-2 (*Continued*)
FORMULA INDEX FOR ORGANIC COMPOUNDS

$C_6H_{15}Bi$ 855
$C_6H_{15}P$ 6189
$C_6H_{15}Sb$ 6198
$C_6H_{16}N_2$ 3596
$C_6H_{16}Si$ 6196
$C_6H_{18}N_4$ 6201
$C_6H_{26}O_8$ 5313
C_6OCl_6 3565-6
$C_6O_2Cl_4$ 5746
$C_6O_2Br_4$ 5726

6 III

C_6HOCl_5 5057
C_6HOBr_5 5051
$C_6HO_2Cl_3$ 6169
$C_6H_2OCl_4$ 5743
$C_6H_2O_2Cl_2$ 2070-1
$C_6H_2O_2Cl_4$ 5742
$C_6H_2O_8N_2$ 4586
$C_6H_2O_9N_4$ 5857
$C_6H_2NCl_3$ 5052
$C_6H_3OCl_3$ 6162-4
$C_6H_3OBr_3$ 6108
$C_6H_3OI_3$ 6238
$C_6H_3O_2Cl_3$ 6156
$C_6H_3O_2Br_3$ 6114
$C_6H_3O_6N_3$ 6292-4
$C_6H_3O_7N_3$ 6307-10
$C_6H_3O_8N_3$ 6312
$C_6H_3NCl_4$ 5730-2
$C_6H_3NBr_4$ 5714
$C_6H_4OCl_2$ 2042
$C_6H_4OBr_2$ 1859-62
$C_6H_4O_2Cl_2$ 2003-4
$C_6H_4O_3N_2$ 4896
$C_6H_4O_4N_2$ 2602-4
$C_6H_4O_5N_2$ 2646-7; 2649-52
$C_6H_4O_6N_2$ 2660-1
$C_6H_4O_6N_4$ 6289
$C_6H_4NCl_3$ 6130-3
$C_6H_4NBr_3$ 6089-93
$C_6H_4N_2Cl_2$ 5544
C_6H_4FCl 3263-5
C_6H_4FBr 3262
C_6H_4FI 3272
C_6H_4ClBr 1340-1
C_6H_4ClI 1387-9
C_6H_4BrI 945-7
C_6H_5ON 4876
C_6H_5OCl 1427-9
C_6H_5OBr 972-4
C_6H_5OI 3898-900; 3910
C_6H_5ONa 5119
$C_6H_5O_2N$ 4646; 5486; 5488; 5496; 5546
$C_6H_5O_2Cl$ 1386; 1460
$C_6H_5O_2Br$ 944
$C_6H_5O_2I$ 3912

$C_6H_5O_3N$ 3791-2; 4799; 4802-3; 4900
$C_6H_5O_3N_3$ 1767
$C_6H_5O_4N$ 4836
$C_6H_5O_4N_3$ 2587-92
$C_6H_5O_5N_3$ 5303
$C_6H_5O_6N_5$ 6311
$C_6H_5NCl_2$ 1932-7
$C_6H_5NBr_2$ 1798-802
$C_6H_5NI_2$ 2385
$C_6H_5N_2Cl$ 1765
$C_6H_5N_2Br_3$ 1768
C_6H_5ClHg 4037
$C_6H_5Cl_2P$ 5270
$C_6H_5Cl_3Si$ 5618
$C_6H_6ON_2$ 4875; 5493
C_6H_6OS 5909
$C_6H_6O_2N_2$ 4618-20; 5212; 5545
$C_6H_6O_2S$ 672
$C_6H_6O_2Si$ 5616
$C_6H_6O_3N_2$ 4610-6
$C_6H_6O_3N_4$ 4417-9
$C_6H_6O_3S$ 675
$C_6H_6O_4N_4$ 2657
$C_6H_6O_4S$ 5123-5
$C_6H_6O_5N_2$ 5490
$C_6H_6O_6S_2$ 664
$C_6H_6O_9S_3$ 686
C_6H_6NF 3256
C_6H_6NCl 1309-11
C_6H_6NBr 893-5
C_6H_6NI 3875-7
$C_6H_6N_2Cl_2$ 2044; 2048
$C_6H_6N_3Cl_9$ 1287
C_6H_7ON 113; 345; 347; 349; 4082; 5200-1
$C_6H_7ON_3$ 4897
$C_6H_7O_2N$ 372
$C_6H_7O_2N_3$ 4813; 4815
$C_6H_7O_3As$ 5158
$C_6H_7O_4As$ 3797
$C_6H_7N_2Cl$ 1433
$C_6H_7N_2Cl_3$ 2049
$C_6H_7N_2Br$ 975
$C_6H_8ON_2$ 1723; 1725-7
$C_6H_8O_2Cl_2$ 154
$C_6H_8O_3N_2$ 521
$C_6H_8O_5Na_2$ 2268
$C_6H_8O_{18}N_6$ 4011
C_6H_8NCl 520
C_6H_8NBr 519
$C_6H_9ON_3$ 6068
$C_6H_9O_2N_3$ 1574; 3669
$C_6H_9O_2Cl_3$ 6144
$C_6H_9O_3N$ 6055
$C_6H_9O_3N_3$ 4197; 5267
$C_6H_9O_3Cl$ 3017-8
$C_6H_9O_3Br$ 2991
$C_6H_9O_4N$ 88

$C_6H_9N_2Cl$ 5197
$C_6H_9N_2Cl_3$ 1434
$C_6H_{10}O_2Br_2$ 3037
$C_6H_{10}O_2S_2$ 2165
$C_6H_{10}O_6Ca$ 3941
$C_6H_{10}N_2Cl_2$ 5257; 5259; 5261
$C_6H_{11}ON$ 1607
$C_6H_{11}OCl$ 1216
$C_6H_{11}O_2N$ 2956
$C_6H_{11}O_2Cl$ 449-50; 1076
$C_6H_{11}O_2Cl_3$ 6124-5
$C_6H_{11}O_2Br$ 920; 2992-3
$C_6H_{11}O_3N$ 1111
$C_6H_{11}O_4N_3$ 1517; 2262
$C_6H_{11}NS$ 490-3
$C_6H_{11}N_3Cl_2$ 6061
$C_6H_{11}N_3Cl_4$ 6067
$C_6H_{12}OCl_2$ 1991-2
$C_6H_{12}O_2N_2$ 152; 2194
$C_6H_{12}O_2Cl_2$ 1925; 6207
$C_6H_{12}O_3N_2$ 3477
$C_6H_{12}O_4N_2$ 2631
$C_6H_{12}N_2S_4$ 5843
$C_6H_{12}N_3Cl_3$ 6063
$C_6H_{13}ON$ 1213; 1664; 2100
$C_6H_{13}ON_3$ 2181; 4359
$C_6H_{13}O_2N$ 282-7; 439-41; 2236; 3496; 3653; 4749-51; 5454
$C_6H_{13}O_2Cl$ 1292
$C_6H_{13}O_3N_3$ 1521
$C_6H_{13}O_5N$ 3311
$C_6H_{13}O_6N$ 3369
$C_6H_{13}NS_2$ 5875
$C_6H_{14}ON_2$ 494-5; 2790-1
$C_6H_{14}O_2N_2$ 3986; 3988
$C_6H_{14}O_2N_4$ 575
$C_6H_{14}O_2S$ 2802-3
$C_6H_{14}O_3S$ 2801
$C_6H_{14}O_4S$ 2798
$C_6H_{15}ON$ 2112
$C_6H_{15}OP$ 6190
$C_6H_{15}O_2N$ 2767
$C_6H_{15}O_3N$ 6178
$C_6H_{15}O_3Al$ 229
$C_6H_{15}O_3B$ 6186
$C_6H_{15}O_3P$ 6192
$C_6H_{15}O_4P$ 6188
$C_6H_{15}SP$ 6191
$C_6H_{16}OSi$ 6194
$C_6H_{16}NCl$ 6182
$C_6H_{16}NBr$ 6181

6 IV

$C_6H_2ONCl_3$ 2072
$C_6H_2O_6N_3Cl$ 5308
$C_6H_3O_2NCl_2$ 2022-6
$C_6H_3O_2SCl_3$ 1956
$C_6H_3O_3NCl_2$ 2029-30

Table 7-2 (*Continued*)
FORMULA INDEX FOR ORGANIC COMPOUNDS

Table 7-2 (*Continued*)
FORMULA INDEX FOR ORGANIC COMPOUNDS

$C_7H_4O_5N_2$ 2600-1
$C_7H_4O_5Br_2$ 1841
$C_7H_4O_6N_2$ 2607-11
$C_7H_4O_7N_2$ 2662
C_7H_4NCl 1330
C_7H_4NBr 909
C_7H_5ON 544; 3748; 5177
$C_7H_5ON_3$ 653
C_7H_5OF 757
C_7H_5OCl 755; 1318-20
$C_7H_5OCl_3$ 6146-9
C_7H_5OBr 753
$C_7H_5OBr_3$ 6094
C_7H_5OI 762
$C_7H_5O_2N_3$ 4662
$C_7H_5O_2F$ 3259-61
$C_7H_5O_2Cl$ 1327-9; 1461
$C_7H_5O_2Br$ 906-8
$C_7H_5O_2I$ 3880-2
$C_7H_5O_2Na$ 712
$C_7H_5O_3N$ 4637; 4639-40; 4877-9
$C_7H_5O_3Br$ 993
$C_7H_5O_3I$ 3904; 3911
$C_7H_5O_3Na$ 3744
$C_7H_5O_4N$ 4663-4; 4666; 5497-502
$C_7H_5O_5N$ 4753-60
$C_7H_5O_6N_3$ 2594; 6313-7
$C_7H_5O_7N_3$ 6290; 6298
$C_7H_5O_8N_3$ 6306
$C_7H_5O_8N_5$ 5873
C_7H_5NS 5241
$C_7H_5NS_2$ 4033
$C_7H_5NCl_2$ 5168
$C_7H_6ON_2$ 709
$C_7H_6OBr_2$ 1831
C_7H_6OS 5891
$C_7H_6OS_2$ 2823
$C_7H_6O_2S$ 5918
$C_7H_6O_3N_2$ 4642-4
$C_7H_6O_4N_2$ 2666-71; 4604-9
$C_7H_6O_5N_2$ 2593; 2616-7
$C_7H_6O_5S$ 5676-8
$C_7H_6O_6S$ 5681
$C_7H_6N_2S$ 273
C_7H_6ClBr 913-4; 1337-8
C_7H_7ON 254-6; 643-4; 713; 3299; 4903-5
C_7H_7OF 3257
C_7H_7OCl 1314; 1354-7
C_7H_7OBr 897-8
C_7H_7OI 3878
$C_7H_7O_2N$ 264-6; 710; 3348-9; 3736; 3738-40; 4844-6; 4881; 5167; 5213
$C_7H_7O_2N_3$ 4739
$C_7H_7O_3N$ 374-6; 4623-5; 4678-80; 4710

$C_7H_7O_3N_3$ 4658-60
$C_7H_7O_4N$ 3352; 4742
$C_7H_7O_5As$ 651
$C_7H_7O_5P$ 730
C_7H_7NS 5902
$C_7H_7N_3S$ 734
C_7H_7ClHg 4039
$C_7H_8ON_2$ 257-61; 688; 760; 3308; 4887; 5253
$C_7H_8O_2N_2$ 1700-5; 3684; 4767-9; 4849-58
$C_7H_8O_2S$ 5964
$C_7H_8O_3N$ 352
$C_7H_8O_3N_2$ 4601-3
$C_7H_8O_3N_4$ 1250
$C_7H_8O_3S$ 5966-8
$C_7H_8O_6S_2$ 5962
C_7H_8NCl 1469-70
C_7H_8NBr 1002-3
$C_7H_8N_2S$ 5244
C_7H_9ON 292-3; 295-9; 534-6; 810-1; 4115; 6028-30
$C_7H_9ON_3$ 5236-7
$C_7H_9O_2N_3$ 6064-6
$C_7H_9O_3N$ 1584
$C_7H_9O_3As$ 787
$C_7H_9N_3S$ 5243
$C_7H_{10}ON_2$ 1687
$C_7H_{10}O_4Br_2$ 2155
$C_7H_{10}NCl$ 5996; 5998; 6000
$C_7H_{11}O_4Br$ 2133
$C_7H_{11}O_6N$ 4763
$C_7H_{11}N_2Cl$ 6023; 6025; 6027
$C_7H_{12}O_3Cl_2$ 2057
$C_7H_{12}N_2Cl_2$ 4337; 6004; 6006; 6008
$C_7H_{13}ON$ 112
$C_7H_{13}O_2Br$ 2997-8
$C_7H_{14}N_3Cl_3$ 6070
$C_7H_{15}ON$ 2931; 3522-3; 4124
$C_7H_{15}ON_3$ 5314
$C_7H_{15}O_2N$ 1140-1; 3551; 4744-8
$C_7H_{16}ON_2$ 2806-7
$C_7H_{16}O_4S_2$ 5683
$C_7H_{17}ON$ 2117-8
$C_7H_{17}O_2N$ 2113

7 IV

$C_7H_3O_5N_2Cl$ 2613
$C_7H_3O_6N_2Cl$ 1370-1
$C_7H_3O_6N_2Br$ 932
C_7H_4ONCl 1333
C_7H_4OClBr 911
C_7H_4OClI 3883
$C_7H_4O_3NCl$ 4673-5
$C_7H_4O_4NBr$ 964
$C_7H_4O_4NNa$ 4665
C_7H_4NSCl 1332

$C_7H_5O_2NCl_2$ 4635-6
$C_7H_5O_2NBr_2$ 1804; 4634
$C_7H_5O_2NI_2$ 2383
$C_7H_5O_3NS$ 5579
$C_7H_5O_3NHg$ 4057
$C_7H_5O_4N_2Cl$ 2615
$C_7H_5O_6SNa$ 5682
C_7H_6ONCl 1321-3
C_7H_6ONNa 3300
$C_7H_6O_2NCl$ 1315; 4682-4
$C_7H_6O_2NBr$ 4681
$C_7H_6O_6S_2Na_2$ 5963
$C_7H_7O_2SCl$ 5974-5
$C_7H_7O_2SNa$ 5965
$C_7H_7O_3SNa$ 5969
$C_7H_7O_4NS$ 5671-3
$C_7H_7O_5NS$ 4847
C_7H_8ONCl 1312
$C_7H_9ONCl_2$ 1313
$C_7H_9O_2NS$ 5970-1
$C_7H_9O_3NS$ 379-83
$C_7H_9O_4N_2As$ 1230
$C_7H_{10}O_2N_2Cl_2$ 1705
$C_7H_{10}O_2N_4S$ 5666
$C_7H_{12}ON_2Cl_2$ 1688-9
$C_7H_{12}O_2N_3Cl_3$ 6066
$C_7H_{12}O_4N_2S$ 6009
$C_7H_{14}ONBr$ 4580
$C_7H_{14}O_4N_2S_2$ 2854
$C_7H_{16}ONBr$ 6264
$C_7H_{16}O_2NCl$ 83
$C_7H_{16}O_2NBr$ 82

7 V

$C_7H_4O_3NSNa$ 5580
$C_7H_6O_4NSCl$ 4848
$C_7H_7O_2NSCl_2$ 2074

7 VI

$C_7H_7O_2NSClNa$ 1290

8 I

C_8H_6 5146
C_8H_8 5646
C_8H_{10} 2973; 6454; 6457; 6462
C_8H_{12} 2289-91
C_8H_{14} 1530; 2493; 4991
C_8H_{16} 1617; 1923; 2471-3; 4984-90
C_8H_{18} 4944-58

8 II

$C_8H_4O_3$ 5282
$C_8H_4N_2$ 5288-9
$C_8H_5Cl_5$ 5055
C_8H_6O 1538
$C_8H_6O_2$ 5278-80; 5290
$C_8H_6O_3$ 162-4; 758; 5323
$C_8H_6O_4$ 165-8; 5274; 5276-7; 5327

Table 7-2 (*Continued*)
FORMULA INDEX FOR ORGANIC COMPOUNDS

$C_8H_6O_5$ 3802-7
$C_8H_6O_6$ 3719
$C_8H_6N_2$ 5534; 5548
C_8H_6S 5912
C_8H_7N 800; 3864; 5992-4
C_8H_7Cl 1462-3
C_8H_7Br 994-6
C_8H_8O 51; 5139; 5648; 5985-7; 6443-4
$C_8H_8O_2$ 531-2; 754; 807; 2549-51; 3319; 3721-3; 4138; 4379; 5141-2; 5982-4
$C_8H_8O_3$ 114; 3777; 3794-6; 3832-41; 4008; 4067-9; 4248; 4274; 4380; 5130; 5325; 5550; 6418; 6421
$C_8H_8O_4$ 1656; 2280-3; 3673; 4247; 5001; 6211; 6417
$C_8H_8O_5$ 4249
$C_8H_8N_2$ 274; 799; 4136-7
$C_8H_8Cl_2$ 6494-6
$C_8H_8Br_2$ 1882; 5647; 6491-3
C_8H_9N 218; 639
C_8H_9Cl 1378-80; 1474-6; 5191; 6488-90
C_8H_9Br 937-8; 1010-5; 6485-7
C_8H_9I 3909
$C_8H_{10}O$ 3114-6; 4145; 4399-401; 5111; 5186-7; 6016-8; 6465-70
$C_8H_{10}O_2$ 530; 1543; 2375-80; 2406-8; 2927-9; 4072; 5136; 6479; 6497-9
$C_8H_{10}O_3$ 1561; 3238; 3338; 5411; 5515; 6422
$C_8H_{10}O_4$ 1683; 3399; 5796
$C_8H_{10}O_8$ 5662
$C_8H_{10}N_2$ 11
$C_8H_{10}S$ 3132; 4402
$C_8H_{11}N$ 161; 319-21; 1524-6; 2450; 2965; 4396-8; 5188-9; 5443-5; 6471-5; 6477; 6482-4
$C_8H_{12}O_2$ 2475
$C_8H_{12}O_3$ 3033
$C_8H_{12}O_4$ 2174; 2183; 3034; 3583-6; 5702
$C_8H_{12}O_5$ 2192
$C_8H_{12}N_2$ 305-7; 2527; 3127-8; 3597; 6500-2
$C_8H_{13}N$ 1126
$C_8H_{14}O$ 2403; 3141; 4262
$C_8H_{14}O_2$ 78; 224; 1081-2; 1610; 5650
$C_8H_{14}O_3$ 1033; 1170-1; 2423; 2938; 5421; 5783

$C_8H_{14}O_4$ 465-6; 2211-2; 2224; 2428; 2792-3; 2944; 3465; 5649; 5779; 5842
$C_8H_{14}O_5$ 2182; 2229
$C_8H_{14}O_6$ 2218
$C_8H_{15}N$ 1226
$C_8H_{16}O$ 195; 1222; 2963-4; 3062; 4266; 5391
$C_8H_{16}O_2$ 483-5; 1062-4; 1150; 1177; 1220-1; 2470; 2770; 3009; 3548; 3621-3; 4263; 5315; 5455-7
$C_8H_{16}O_3$ 464; 1094; 3761-2
$C_8H_{16}O_4$ 6; 170; 1233; 2925
$C_8H_{17}F$ 4978
$C_8H_{17}Cl$ 4975-6
$C_8H_{17}Br$ 4973-4
$C_8H_{17}I$ 4980
$C_8H_{18}O$ 1898-9; 4264; 4961-9
$C_8H_{18}O_2$ 4943
$C_8H_{18}O_3$ 2232-3; 2936; 3387
$C_8H_{18}O_5$ 5768
$C_8H_{18}S$ 1914-6
$C_8H_{18}S_2$ 1897
$C_8H_{18}Hg$ 4042
$C_8H_{19}N$ 1884-6; 4970-2
$C_8H_{20}N_2$ 5832
$C_8H_{20}As_2$ 579
$C_8H_{20}Ge$ 3357
$C_8H_{20}Pb$ 3957
$C_8H_{20}Si$ 5619
$C_8H_{20}Sn$ 5950
$C_8H_{23}N_5$ 5770
$C_8O_3Br_4$ 5725
$C_8O_3I_4$ 5816

8 III

$C_8H_2O_4Cl_4$ 5745
$C_8H_3O_5N$ 4823
$C_8H_4O_2Cl_2$ 5283-6
$C_8H_4O_4N_2$ 4761; 4825-6
$C_8H_4O_6N_4$ 2614
$C_8H_4N_2Cl_2$ 2062
C_8H_5ON 756
$C_8H_5O_2N$ 645; 1585; 3919; 5287
$C_8H_5O_2Br_3$ 6109
$C_8H_5O_3N$ 3922
$C_8H_5O_4N$ 4824
$C_8H_5O_4Cl$ 1441
$C_8H_5O_4Br$ 978-9
$C_8H_5O_4Na$ 5275
$C_8H_5O_5N$ 4676
$C_8H_5O_6N$ 4819-22; 5506-9
$C_8H_6ON_2$ 3850; 3935
$C_8H_6OBr_2$ 967
$C_8H_6O_2N_2$ 779; 3923; 4685-7

$C_8H_6O_6N_2$ 2653
$C_8H_6O_7N_2$ 2632
$C_8H_6O_8N_4$ 188
C_8N_6NCl 1586
C_8H_7ON 2908; 3870; 4010; 5021; 6019-21
C_8H_7OCl 1302-3; 5145
C_8H_7OBr 887-8
$C_8H_7O_2N$ 2686; 4838-41; 4874
$C_8H_7O_2N_3$ 359
$C_8H_7O_2Cl$ 4071; 4185; 5169
$C_8H_7O_2Br$ 4149-51; 5166
$C_8H_7O_2I$ 4288-90
$C_8H_7O_3N$ 3921; 4594-5; 5018; 5272
$C_8H_7O_3Cl$ 1432
$C_8H_7O_3Na$ 4009
$C_8H_7O_4N$ 4317-9; 4808-9; 6398
$C_8H_7O_5N_3$ 2586
$C_8H_7O_6N_3$ 6323-5
C_8H_7NS 838; 2909; 6041-3
$C_8H_7NS_2$ 4299
$C_8H_8O_2N_2$ 778; 5281
$C_8H_8O_2Hg$ 4036
$C_8H_8O_3N_2$ 4588-90; 6385
$C_8H_8O_4N_2$ 2673-81; 4812
$C_8H_8O_5N_2$ 2645
$C_8H_8O_6N_6$ 4445
C_8H_9ON 17; 53; 236-9; 4135; 4241; 5140; 5988-90
C_8H_9OCl 1424-6
C_8H_9OBr 969-71
C_8H_9OI 3896-7
$C_8H_9O_2N$ 33-5; 356-7; 795; 3095; 4108-10; 4736-7; 4865-70; 5131; 5194
$C_8H_9O_3N$ 3800; 4112-3 4796-8
$C_8H_9O_5N$ 351
C_8H_9NS 5240; 5883
$C_8H_9N_2Cl$ 275
$C_8H_{10}ON_2$ 107-8; 111; 844; 4883; 4885; 6048-50
$C_8H_{10}O_2N_2$ 358; 4720-2
$C_8H_{10}O_2N_4$ 1569
$C_8H_{10}O_2S$ 3133
$C_8H_{10}O_3N_2$ 4795
$C_8H_{10}O_3S$ 2974; 4395; 6455; 6458; 6463
$C_8H_{10}O_4Br_2$ 2154
$C_8H_{10}NBr$ 929
$C_8H_{10}N_2S$ 839-40; 6045-7
$C_8H_{10}O_{10}Ca$ 3999
$C_8H_{11}ON$ 1544; 2445-6; 2930; 2958-60; 3799; 5105-7
$C_8H_{11}O_2N$ 514; 2404
$C_8H_{11}O_3N$ 5510
$C_8H_{11}O_4Cl$ 2140-1

Table 7-2 (*Continued*)
FORMULA INDEX FOR ORGANIC COMPOUNDS

$C_8H_{11}O_4Br$ 2132
$C_8H_{11}O_6Cl_3$ 1289
$C_8H_{11}N_3S$ 6044
$C_8H_{12}O_3N_2$ 2124
$C_8H_{12}O_4Br_2$ 2156
$C_8H_{14}O_5N_4$ 6208
$C_8H_{15}OCl$ 1225
$C_8H_{15}O_2N$ 79; 4889
$C_8H_{15}O_2Br$ 2994
$C_8H_{15}O_5N$ 1665
$C_8H_{16}O_3N_2$ 3965
$C_8H_{16}O_3Cl_2$ 5771
$C_8H_{16}O_4N_2$ 3215
$C_8H_{17}ON$ 1223
$C_8H_{17}O_2N$ 288; 4437; 4786-8; 4982
$C_8H_{17}O_2Br$ 1026
$C_8H_{17}O_3N$ 4981
$C_8H_{18}O_2S$ 1918
$C_8H_{18}O_3S$ 1917
$C_8H_{18}O_4S$ 1913
$C_8H_{18}O_4S_2$ 6326
$C_8H_{19}O_2N$ 1084
$C_8H_{20}OSi$ 6195
$C_8H_{20}O_4Si$ 3152
$C_8H_{20}NCl$ 5755
$C_8H_{20}NBr$ 5754
$C_8H_{20}NI$ 5757
$C_8H_{21}ON$ 5756
$C_8H_{21}O_5N$ 5753

8 IV

C_8H_4ONCl 3920
$C_8H_4O_2N_2Cl_2$ 1967
$C_8H_4O_2N_2Br_2$ 1820
$C_8H_5ON_2Cl$ 1394
$C_8H_6ONCl_3$ 6127
$C_8H_6ONBr_3$ 6087
$C_8H_6O_2N_2S$ 3325
$C_8H_7O_2NBr_2$ 4212
C_8H_8ONCl 1296-8
C_8H_8ONBr 880-2
C_8H_8ONI 3873
C_8H_8ONNa 5621
$C_8H_8ON_2S$ 774
$C_8H_9O_2SCl$ 6461
$C_8H_9O_3SNa$ 6456; 6459; 6464
$C_8H_{10}ONCl$ 240
$C_8H_{10}O_4NAs$ 577
$C_8H_{10}O_5NAs$ 128
$C_8H_{11}ON_2Cl$ 109
$C_8H_{11}O_2NS$ 5979
$C_8H_{11}O_3NS$ 2451-2; 2966
$C_8H_{11}O_3N_2Na$ 2125
$C_8H_{11}O_4N_2As$ 582
$C_8H_{11}N_2SCl$ 841
$C_8H_{12}ONCl$ 5109
$C_8H_{12}O_2N_2Cl_2$ 1728
$C_8H_{12}O_3NCl$ 5511

$C_8H_{14}O_4N_2S$ 308
$C_8H_{18}O_2NCl$ 100
$C_8H_{18}O_2NBr$ 99
$C_8H_{18}O_2NI$ 101

8 V

$C_8H_8O_4NSNa$ 21
$C_8H_9O_4NAsNa$ 578
$C_8H_9O_4N_2AsNa_2$ 583
$C_8H_{10}O_3NSNa$ 2967

9 I

C_9H_8 3855
C_9H_{10} 3695; 5223-5
C_9H_{12} 3166-8; 5378-9; 6253-5
C_9H_{16} 2492; 4325
C_9H_{18} 3578; 3581; 4930
C_9H_{20} 4910-7

9 II

$C_9H_6O_2$ 1490; 1534-5; 3854; 5220
$C_9H_6O_3$ 6369
$C_9H_6O_4$ 1628; 2898; 6239
$C_9H_6O_5$ 5291
$C_9H_6O_6$ 683-5
$C_9H_6O_7$ 5128
C_9H_7N 5538; 5540
C_9H_8O 1501; 3698-9
$C_9H_8O_2$ 72; 591; 1497-8; 1500
$C_9H_8O_3$ 69; 70-1; 738; 1531-3; 1537
$C_9H_8O_4$ 73; 115; 759; 2333-5; 3674; 6397
$C_9H_8O_5$ 5583
$C_9H_8N_2$ 365-71
$C_9H_9Br_3$ 6106
C_9H_9N 2284; 3709; 4282-5
$C_9H_{10}O$ 211; 214; 1507; 3129; 3706; 4089; 4146; 6431-3
$C_9H_{10}O_2$ 781; 1552-3; 2458-63; 2914; 2975-8; 3683; 3704; 3810-1; 4065; 4335; 4392-4; 5132; 5221; 6010-2; 6441
$C_9H_{10}O_3$ 2405; 2915-7; 3053; 3072; 3151; 3188; 3481; 3711-3; 3776; 4300-1; 6361; 6424
$C_9H_{10}O_4$ 3052; 3230; 3466; 4078; 6423
$C_9H_{10}O_5$ 3321; 5686
$C_9H_{11}N$ 196; 5797
$C_9H_{11}Br$ 953; 985-6; 5222
$C_9H_{12}O$ 2526; 2986; 3122-3; 3173-5; 3705; 5430-5; 5438; 6265-6
$C_9H_{12}O_2$ 3449; 4054

$C_9H_{12}O_3$ 147; 1090; 3334; 3410; 5266; 5517
$C_9H_{12}N_2$ 62
$C_9H_{12}S$ 3176
$C_9H_{13}N$ 654; 2465; 2569-71; 3170-2; 4228; 5040-1; 5374-5; 6250-1
$C_9H_{14}O$ 5268-9
$C_9H_{14}O_3$ 2948
$C_9H_{14}O_4$ 2142; 2175; 2179; 2186
$C_9H_{14}O_5$ 87; 2103
$C_9H_{14}O_6$ 1201-2; 3418
$C_9H_{14}O_7$ 6263
$C_9H_{14}N_2$ 597
$C_9H_{16}O_2$ 201
$C_9H_{16}O_3$ 1101; 3139-40; 5777
$C_9H_{16}O_4$ 596; 2161; 2169; 2176; 2789
$C_9H_{16}O_6$ 60
$C_9H_{17}N$ 1633; 5047
$C_9H_{18}O$ 1900-1; 4265; 5043
$C_9H_{18}O_2$ 432-5; 1142-5; 3108; 3526; 4174; 4979; 5042; 5396
$C_9H_{18}O_3$ 1892-4
$C_9H_{18}O_5$ 2094
$C_9H_{18}N_2$ 1895
$C_9H_{19}N$ 4114
$C_9H_{19}Cl$ 4928
$C_9H_{19}Br$ 4927
$C_9H_{19}I$ 4929
$C_9H_{20}O$ 4918-25
$C_9H_{20}O_3$ 6193
$C_9H_{20}O_4$ 3012
$C_9H_{21}N$ 4926; 6348

9 III

$C_9H_5O_2Cl$ 1353
$C_9H_5O_4N$ 4817-8
$C_9H_5NCl_2$ 2063-9
$C_9H_6O_2N_2$ 4831-5
$C_9H_6O_2Br_2$ 1829
C_9H_6NCl 1453-9
C_9H_7ON 742; 3820-6
C_9H_7OCl 1505
$C_9H_7O_2N$ 4291-3
$C_9H_7O_2Br$ 921-4
$C_9H_7O_2Na$ 1499
$C_9H_7O_3N$ 3871
$C_9H_7O_4N$ 4707-9
$C_9H_7O_4Br$ 76
$C_9H_7O_4I$ 93-4
$C_9H_7O_4Li$ 117
$C_9H_7O_4Na$ 119
$C_9H_8ON_2$ 1582
$C_9H_8O_2N_2$ 4141-2
$C_9H_8O_2Br_2$ 1830
$C_9H_8O_3I_2$ 3041

Table 7-2 (*Continued*)
FORMULA INDEX FOR ORGANIC COMPOUNDS

C₉H₈O₅N₂ 4752
C₉H₈O₆N₂ 3043
C₉H₉ON 1502; 3703
C₉H₉OCl 3708
C₉H₉OBr 4333
C₉H₉O₂N 289-91; 4899; 5586
C₉H₉O₂Cl 797
C₉H₉O₃N 125-7; 3664
C₉H₉O₄N 2539; 3099-101; 4677; 5195
C₉H₉O₆N₃ 6318-20
C₉H₉O₇N₃ 2672
C₉H₁₀O₃N₂ 4597; 4766
C₉H₁₀O₄N₂ 2635; 4591
C₉H₁₀O₅N₂ 4860
C₉H₁₁ON 26-8; 780; 2435-6; 2457; 2972; 3305; 3707; 4086; 5358
C₉H₁₁OBr 987
C₉H₁₁O₂N 18-20; 97-8; 2440-1; 2953-5; 4302; 4711-3; 4764; 5148-50; 5254
C₉H₁₁O₃N 6367
C₉H₁₁O₄N 2361
C₉H₁₂ON₂ 330; 3134
C₉H₁₂O₂N₂ 5112
C₉H₁₂O₃S 3169
C₉H₁₃O₂N 2882
C₉H₁₃O₃N 157; 5512
C₉H₁₄O₃N₄ 1251
C₉H₁₄NI 6267
C₉H₁₆O₂N₂ 217; 4906
C₉H₁₇ON 2107; 6056
C₉H₁₇OCl 5046
C₉H₁₇O₅N 5033-4
C₉H₁₇N₄I 3600
C₉H₁₉ON 5044-5; 6414
C₉H₁₉O₂N 4784-5
C₉H₂₀ON₂ 1922; 5767
C₉H₂₀O₄S₂ 5872
C₉H₂₀N₂S 1921
C₉H₂₁O₃N 6347
C₉H₂₂N₂S₂ 2120

9 IV

C₉H₅ONBr₂ 1848
C₉H₇O₄NS 3828
C₉H₈O₃NNa 3666
C₉H₉O₂N₃S₂ 5670
C₉H₉O₃NCl₂ 2077
C₉H₉O₃NBr₂ 1881
C₉H₉O₃NI₂ 2401-2
C₉H₁₀ONBr 889
C₉H₁₀O₂N₂S 5242
C₉H₁₁O₃SCl 1383
C₉H₁₃O₂NS 5977-8
C₉H₁₄O₂NCl 4582
C₉H₁₄O₃NCl 1523

C₉H₁₅O₂N₃S 2890
C₉H₁₅O₃N₂Br 84
C₉H₂₄ON₂I₂ 3591

9 V

C₉H₅ONClI 1392
C₉H₆O₄NSI 3888
C₉H₉O₂SHgNa 4052

10 I

C₁₀H₈ 4458
C₁₀H₁₀ 2278; 3120
C₁₀H₁₂ 2089; 5786
C₁₀H₁₄ 1049-52; 1622-4; 2126-8; 2483-6; 3582; 4351-2; 5827-9
C₁₀H₁₆ 870; 1189; 3232; 3972-3; 5092-3; 5318; 5576; 5684; 5704-5; 5709
C₁₀H₁₈ 1188; 1258; 1630-2; 2514-5; 4023; 5317
C₁₀H₂₀ 1650-3; 3579
C₁₀H₂₂ 1635-8

10 II

C₁₀H₅Cl₃ 6160-1
C₁₀H₆O₂ 4520-1; 4523
C₁₀H₆O₃ 3786-9
C₁₀H₆O₄ 1536; 3328; 4477-8
C₁₀H₆O₅ 3343
C₁₀H₆O₈ 680-2
C₁₀H₆Cl₂ 2009-18
C₁₀H₆Br₂ 1852
C₁₀H₇F 3273
C₁₀H₇Cl 1402-3
C₁₀H₇Br 956-7
C₁₀H₇I 3891-2
C₁₀H₈O 4489-90
C₁₀H₈O₂ 2348-57; 4192
C₁₀H₈O₃ 743; 4406
C₁₀H₈O₄ 510; 638; 1504; 3336; 3346; 5604
C₁₀H₈N₂ 2810-3; 6051-3
C₁₀H₈S 5913-4
C₁₀H₈Cl₄ 4473
C₁₀H₉N 4367; 4369-70; 4372; 4374-5; 4529-30; 5229
C₁₀H₁₀O 626
C₁₀H₁₀O₂ 198; 741; 4191; 5175-6; 5581-2
C₁₀H₁₀O₃ 637; 771; 2980; 4139; 5094
C₁₀H₁₀O₄ 763; 815; 2530-2; 3067; 3236-7; 3716; 4017; 4090; 5239; 5552
C₁₀H₁₀O₅ 837; 4993
C₁₀H₁₀O₆ 3502
C₁₀H₁₀N₂ 2809; 4563; 4565; 4571-9

C₁₀H₁₁Cl 4188
C₁₀H₁₁Br 4152
C₁₀H₁₂O 511; 2901; 2987; 5377; 5439-40; 5787-90
C₁₀H₁₂O₂ 829; 1280; 3118-9; 3163-5; 3218; 3223; 4126; 5185; 5380-5; 5831; 5943; 6256-61; 6481
C₁₀H₁₂O₃ 814; 1529; 2968; 5412; 5415; 5446-7; 6419
C₁₀H₁₂O₄ 3187; 6226-7
C₁₀H₁₂O₅ 3414; 6240-2
C₁₀H₁₂N₂ 6363
C₁₀H₁₃N 3930; 5792-5
C₁₀H₁₃Cl 1572
C₁₀H₁₃Br 915; 926
C₁₀H₁₄O 1116; 1119-20; 1255; 1261; 1571; 2466; 5436-7; 5938
C₁₀H₁₄O₂ 1070; 2091-3; 5937
C₁₀H₁₄O₃ 457; 1200
C₁₀H₁₄O₄ 2188; 3400
C₁₀H₁₅N 1046-7; 1256; 2121; 5818-20; 5946
C₁₀H₁₆O 1191; 1257; 1512; 2275; 3233; 5320; 5472; 5935-6
C₁₀H₁₆O₂ 585
C₁₀H₁₆O₄ 1197-9; 2108; 2788
C₁₀H₁₆O₅ 2105
C₁₀H₁₆O₈ 1673
C₁₀H₁₆N₂ 303; 5606; 5839
C₁₀H₁₆Cl₂ 1192-4
C₁₀H₁₇Cl 864; 5319
C₁₀H₁₈O 860; 1260; 1518; 2274; 2516-8; 3217; 3234-5; 3974; 4026; 5471; 5706-8
C₁₀H₁₈O₂ 1190
C₁₀H₁₈O₃ 2102; 6403; 6410
C₁₀H₁₈O₄ 1905-6; 2106; 2187; 2203-4; 2796-7; 3450; 3550; 5605
C₁₀H₁₈O₆ 2804-5; 6200
C₁₀H₁₉N 862; 1203-4; 1207
C₁₀H₂₀O 1206; 1259; 1519; 4024-5; 4326; 5414
C₁₀H₂₀O₂ 496-7; 1065; 1205; 1219; 3010; 3553; 4331; 4959-60; 5703
C₁₀H₂₀O₄ 1069
C₁₀H₂₁N 2145; 4029
C₁₀H₂₁Cl 1646
C₁₀H₂₁I 1647
C₁₀H₂₂O 1640-4; 1740-1; 2519-20; 3106
C₁₀H₂₂O₂ 1634; 1654

Table 7-2 (*Continued*)
FORMULA INDEX FOR ORGANIC COMPOUNDS

$C_{10}H_{22}O_3$ 5410; 5703
$C_{10}H_{22}O_5$ 5769
$C_{10}H_{22}S$ 1751-2
$C_{10}H_{22}S_2$ 1753-4
$C_{10}H_{23}N$ 1645; 1734-6

10 III

$C_{10}H_4O_2Cl_2$ 2020
$C_{10}H_4O_8N_4$ 5854-6
$C_{10}H_5OCl_5$ 5056
$C_{10}H_5OBr_3$ 6107
$C_{10}H_5O_6N_3$ 6301-4
$C_{10}H_5O_7N_3$ 6305
$C_{10}H_5O_{10}N$ 5503-4
$C_{10}H_6OCl_2$ 2019
$C_{10}H_6OBr_2$ 1853
$C_{10}H_6O_2N_2$ 5514
$C_{10}H_6O_3S$ 4524
$C_{10}H_6O_4N_2$ 2637-40
$C_{10}H_6O_4Cl_4$ 3068
$C_{10}H_6O_5N_2$ 2641-3
$C_{10}H_6NBr_3$ 6113
$C_{10}H_7OCl$ 1404-6
$C_{10}H_7OBr$ 958-9
$C_{10}H_7O_2N$ 4772-3; 4892-4; 5533
$C_{10}H_7O_3N$ 95; 3937; 4775-9
$C_{10}H_8ON_2$ 4895
$C_{10}H_8O_2N_2$ 4780-3; 4830
$C_{10}H_8O_2S$ 4465-6
$C_{10}H_8O_3S$ 4470-1
$C_{10}H_8O_4N_2$ 3329-30
$C_{10}H_8O_4S$ 4498; 4502-5; 4507; 4510
$C_{10}H_8O_5N_4$ 5305
$C_{10}H_8O_6S_2$ 4461-4
$C_{10}H_8O_7S_2$ 4491
$C_{10}H_8O_8N_2$ 2633
$C_{10}H_8O_8S_2$ 2359
$C_{10}H_8NCl$ 1395
$C_{10}H_8NI$ 3889
$C_{10}H_9ON$ 91; 332-4; 336; 3816-9; 4083; 4179-80; 4567
$C_{10}H_9O_2N$ 3865
$C_{10}H_9O_4N$ 4320-2
$C_{10}H_9O_6N$ 2512
$C_{10}H_{10}ON_2$ 5202
$C_{10}H_{10}O_2N_2$ 2414; 2464; 2981-2
$C_{10}H_{10}O_2Br_2$ 4213
$C_{10}H_{10}O_5N_2$ 4598
$C_{10}H_{10}O_8N_6$ 1541
$C_{10}H_{10}NCl$ 4531-2
$C_{10}H_{11}ON$ 3111; 5134
$C_{10}H_{11}OCl$ 2524
$C_{10}H_{11}O_2N$ 40; 1670
$C_{10}H_{11}O_2N_3$ 4581
$C_{10}H_{11}O_2Br$ 3121
$C_{10}H_{11}O_3N$ 110; 129; 744-5; 1669; 3110

$C_{10}H_{11}N_2Cl$ 4564; 4566
$C_{10}H_{12}ON_2$ 215
$C_{10}H_{12}O_2N_2$ 1676-7
$C_{10}H_{12}O_3N_2$ 1681; 4734; 4871
$C_{10}H_{12}O_4N_2$ 4596
$C_{10}H_{12}O_5N_2$ 2665
$C_{10}H_{13}ON$ 102-4; 1172; 2417-22; 2913; 2934; 5203; 5386; 5874
$C_{10}H_{13}OCl$ 1346; 5939
$C_{10}H_{13}OBr$ 916
$C_{10}H_{13}OI$ 5940
$C_{10}H_{13}O_2N$ 23-5; 2448; 3125; 4714; 4902; 5373
$C_{10}H_{13}O_2Cl$ 5135
$C_{10}H_{13}O_3N$ 4405
$C_{10}H_{13}O_4N_3$ 2619
$C_{10}H_{13}N_2Cl$ 6364
$C_{10}H_{14}ON_2$ 318; 4882
$C_{10}H_{14}OBr_2$ 1828
$C_{10}H_{14}O_2N_2$ 4717-9; 5115
$C_{10}H_{14}O_3N_2$ 5370
$C_{10}H_{14}O_3S$ 5452
$C_{10}H_{14}O_4S$ 5942
$C_{10}H_{14}NBr$ 927
$C_{10}H_{15}ON$ 1262; 2115-6; 2877-8; 3124; 3675
$C_{10}H_{15}OCl$ 1352
$C_{10}H_{15}OBr$ 919
$C_{10}H_{15}O_2N$ 5181; 6476
$C_{10}H_{15}O_3N$ 2345; 4696
$C_{10}H_{15}O_3N_2$ 2776; 2999
$C_{10}H_{16}O_4S$ 1196
$C_{10}H_{16}O_4Br_2$ 2781
$C_{10}H_{17}ON$ 281; 1195
$C_{10}H_{17}N_2Cl$ 304
$C_{10}H_{18}O_5N_4$ 6209
$C_{10}H_{18}N_2Cl_2$ 5840
$C_{10}H_{21}O_2N$ 1649; 4715-6
$C_{10}H_{21}O_3N$ 1648
$C_{10}H_{22}O_2S$ 1756
$C_{10}H_{22}O_3S$ 1755
$C_{10}H_{23}ON$ 1887

10 IV

$C_{10}H_5O_5N_2Na$ 2642
$C_{10}H_5O_5SNa$ 4522
$C_{10}H_6O_2NCl$ 1418-9
$C_{10}H_6O_4SNa$ 4512
$C_{10}H_6O_7S_2Ca$ 4492
$C_{10}H_6O_7S_2K_2$ 4493; 4495
$C_{10}H_6O_7S_2Na_2$ 4494; 4496
$C_{10}H_6O_8N_2S$ 3240
$C_{10}H_7O_2NS$ 4475
$C_{10}H_7O_2SCl$ 4468-9
$C_{10}H_7O_3SNa$ 4472
$C_{10}H_7O_4SK$ 4500; 4509
$C_{10}H_7O_4SNa$ 4501; 4511-2

$C_{10}H_8ON_2S$ 641
$C_{10}H_8O_2NBr$ 940
$C_{10}H_8O_2N_2S$ 773; 5589
$C_{10}H_9O_2NS$ 4467
$C_{10}H_9O_2NCl_2$ 42
$C_{10}H_9O_3NS$ 4541; 4543-53
$C_{10}H_9O_4NS$ 337-40
$C_{10}H_9O_6NS_2$ 4533-5; 4537
$C_{10}H_9O_7NS_2$ 342; 344
$C_{10}H_9O_9NS_3$ 4554
$C_{10}H_{10}ONCl$ 335
$C_{10}H_{10}O_2NCl$ 41
$C_{10}H_{10}O_2NBr$ 884
$C_{10}H_{10}O_2N_4S$ 5665; 5667
$C_{10}H_{10}O_4N_2S$ 5680
$C_{10}H_{13}O_3N_2Br$ 5390
$C_{10}H_{13}O_3SCl$ 1449
$C_{10}H_{15}O_2N_2Cl$ 5116
$C_{10}H_{15}O_3NS$ 2122
$C_{10}H_{16}O_3N_2S$ 853
$C_{10}H_{17}O_6N_3S$ 3378

10 V

$C_{10}H_4O_8NS_3Na_3$ 4476
$C_{10}H_4O_8N_2SNa_2$ 3241
$C_{10}H_5O_8NS_2Na_2$ 4901
$C_{10}H_6O_9NS_3Na_3$ 4555
$C_{10}H_7O_6NS_2Na_2$ 4540
$C_{10}H_7O_7NS_2Na_2$ 343
$C_{10}H_8O_3NSNa$ 4542
$C_{10}H_8O_4NSNa$ 341
$C_{10}H_8O_6NS_2K$ 4536; 4538
$C_{10}H_8O_6NS_2Na$ 4539

11 I

$C_{11}H_{10}$ 4304-5
$C_{11}H_{16}$ 421-4; 2221; 4158-60; 5074
$C_{11}H_{20}$ 6382
$C_{11}H_{22}$ 6373-4
$C_{11}H_{24}$ 6370

11 II

$C_{11}H_6O_{10}$ 671
$C_{11}H_7N$ 4487-8
$C_{11}H_8O$ 4482-3
$C_{11}H_8O_2$ 4306; 4480-1
$C_{11}H_8O_3$ 3778-85; 4275
$C_{11}H_9N$ 5226-8
$C_{11}H_{10}O$ 4311-2
$C_{11}H_{10}O_2$ 3131
$C_{11}H_{10}O_4$ 3696-7; 3971
$C_{11}H_{11}N$ 832; 2540-1; 2544; 2546-8; 4309-10; 6035-7
$C_{11}H_{12}O_2$ 52; 528; 3023; 5151-2
$C_{11}H_{12}O_3$ 739; 6416
$C_{11}H_{12}O_4$ 636; 834; 2941; 3225
$C_{11}H_{14}O$ 1117-8; 6247

Table 7-2 (*Continued*)
FORMULA INDEX FOR ORGANIC COMPOUNDS

$C_{11}H_{14}O_2$ 794; 1053-4; 3065; 3222; 3224; 5347
$C_{11}H_{14}O_3$ 1091-3; 5944; 6507
$C_{11}H_{16}O$ 476; 478-9; 1056; 1079-80; 1115; 5078
$C_{11}H_{16}O_3$ 2231
$C_{11}H_{17}N$ 419-20; 4216-8; 5073
$C_{11}H_{18}O_2$ 3356
$C_{11}H_{18}N_2$ 480
$C_{11}H_{20}O_2$ 6376
$C_{11}H_{20}O_3$ 5778
$C_{11}H_{20}O_4$ 1904; 2134-6; 2158; 2202; 6283
$C_{11}H_{21}N$ 6381
$C_{11}H_{22}O$ 1743-4; 3107; 4324; 6375; 6378
$C_{11}H_{22}O_2$ 436-7; 3113; 4172; 5397; 6377
$C_{11}H_{22}O_3$ 1737-8
$C_{11}H_{22}O_6$ 4386
$C_{11}H_{24}O$ 6371-2
$C_{11}H_{24}O_3$ 3390

11 III

$C_{11}H_7ON$ 4558-9
$C_{11}H_7OCl$ 4486
$C_{11}H_7O_4N$ 4774
$C_{11}H_7NS$ 4570
$C_{11}H_9ON$ 4484-5
$C_{11}H_9O_2N$ 4308; 5528-32
$C_{11}H_9O_3N$ 4278; 5537
$C_{11}H_{10}ON_2$ 3324
$C_{11}H_{11}ON$ 3013
$C_{11}H_{11}O_2N$ 3867
$C_{11}H_{11}O_4N$ 3102-4
$C_{11}H_{12}ON_2$ 565; 2497
$C_{11}H_{12}O_2N_2$ 3126; 6365
$C_{11}H_{12}NI$ 5539
$C_{11}H_{13}O_4N_2$ 29; 2205
$C_{11}H_{13}O_6N_3$ 6296
$C_{11}H_{15}ON$ 2111; 5364; 6411
$C_{11}H_{15}O_2N$ 1043; 1045
$C_{11}H_{15}O_3N$ 3949; 4064
$C_{11}H_{16}O_3N_2$ 1038
$C_{11}H_{16}O_3S$ 1137
$C_{11}H_{16}O_4N_2$ 571; 5565
$C_{11}H_{18}O_3N_2$ 505
$C_{11}H_{19}ON_3$ 1513
$C_{11}H_{19}N_2Cl$ 302
$C_{11}H_{23}ON$ 6379-80
$C_{11}H_{23}O_2N$ 3038; 4861
$C_{11}H_{25}ON$ 1888-9

11 IV

$C_{11}H_{11}ON_2Br$ 901
$C_{11}H_{11}O_2N_3S$ 5668
$C_{11}H_{15}O_3N_2Br$ 1061
$C_{11}H_{17}O_2NS$ 5973
$C_{11}H_{17}O_3N_2Na$ 506; 4583

$C_{11}H_{18}O_3N_2S$ 854

11 V

$C_{11}H_{10}O_2N_3SNa$ 5669

12 I

$C_{12}H_8$ 3
$C_{12}H_{10}$ 2; 2695
$C_{12}H_{12}$ 2507-9; 3089-90
$C_{12}H_{16}$ 5179
$C_{12}H_{18}$ 2777-8; 3590; 4119; 6184-5
$C_{12}H_{22}$ 2083; 2864
$C_{12}H_{24}$ 2863; 6122
$C_{12}H_{26}$ 2856-7
$C_{12}Cl_{10}$ 1629

12 II

$C_{12}H_6O_3$ 4474
$C_{12}H_6O_{12}$ 667
$C_{12}H_6Cl_4$ 5738
$C_{12}H_8O$ 2763
$C_{12}H_8O_4$ 4459-60; 5037
$C_{12}H_8N_2$ 5101-4
$C_{12}H_8S_2$ 5877
$C_{12}H_8F_2$ 2247-8
$C_{12}H_8Cl_2$ 1989-90
$C_{12}H_8Br_2$ 1833
$C_{12}H_9N$ 1231
$C_{12}H_9Cl$ 1291; 1373-5
$C_{12}H_9Br$ 878; 933-4
$C_{12}H_9I$ 3884-5
$C_{12}H_{10}O$ 2719; 4313; 5216-8
$C_{12}H_{10}O_2$ 48; 2338-43; 3770; 4526-8
$C_{12}H_{10}O_4$ 2817; 5321; 5535
$C_{12}H_{10}N_2$ 599; 2279
$C_{12}H_{10}S$ 2749
$C_{12}H_{10}O_2$ 2715
$C_{12}H_{10}I_2$ 2729
$C_{12}H_{10}$ Hg 4047
$C_{12}H_{10}P_2$ 5271
$C_{12}H_{10}Se$ 2742
$C_{12}H_{11}N$ 830-1; 2698; 5153-5
$C_{12}H_{11}N_3$ 247-9; 1762
$C_{12}H_{12}O$ 3093-4
$C_{12}H_{12}O_2$ 203
$C_{12}H_{12}O_6$ 3775; 5516
$C_{12}H_{12}O_{12}$ 3580
$C_{12}H_{12}N_2$ 313-4; 1710-4; 2725; 2727; 3685
$C_{12}H_{12}N_4$ 1696-9
$C_{12}H_{13}N$ 2505-6; 3091-2; 6270-3
$C_{12}H_{13}N_3$ 262; 1717-8
$C_{12}H_{13}N_5$ 6058
$C_{12}H_{14}O_3$ 3219; 5776
$C_{12}H_{14}O_4$ 569; 1122; 2199-201; 5784

$C_{12}H_{14}N_4$ 1722; 2269-70
$C_{12}H_{16}O$ 481-2; 1615-6; 4181
$C_{12}H_{16}O_2$ 425; 1113-4; 5075; 5945
$C_{12}H_{16}O_3$ 458; 486-7; 584; 1106; 3117; 3487
$C_{12}H_{16}O_8$ 1270
$C_{12}H_{18}O$ 426; 452; 477; 4066
$C_{12}H_{18}O_2$ 1112; 3654
$C_{12}H_{18}O_3$ 5265
$C_{12}H_{18}O_6$ 2146-51
$C_{12}H_{18}O_8$ 2152
$C_{12}H_{19}N$ 2775
$C_{12}H_{20}O_2$ 861; 3354; 3975; 5710
$C_{12}H_{20}O_4$ 1903
$C_{12}H_{20}O_6$ 1666; 2095; 3432
$C_{12}H_{20}O_7$ 6187
$C_{12}H_{22}O_2$ 1520; 4027-8; 4407
$C_{12}H_{22}O_3$ 1214
$C_{12}H_{22}O_4$ 1745; 1910-2; 2171; 2210; 2552; 2771
$C_{12}H_{22}O_5$ 1902
$C_{12}H_{22}O_6$ 1919-20; 2410
$C_{12}H_{22}O_{11}$ 3947; 4007; 5663; 6054
$C_{12}H_{23}N$ 2085; 3956
$C_{12}H_{24}O$ 3952
$C_{12}H_{24}O_2$ 1066; 1639; 3008; 3950
$C_{12}H_{24}O_3$ 1162; 1165
$C_{12}H_{25}Cl$ 2861
$C_{12}H_{25}Br$ 2860
$C_{12}H_{26}O$ 2859
$C_{12}H_{27}N$ 6115-6
$C_{12}H_{30}Sn_2$ 5953

12 III

$C_{12}H_5O_{12}N_7$ 2764
$C_{12}H_6O_8N_4$ 5847-9
$C_{12}H_6O_9N_4$ 5851
$C_{12}H_6O_{10}N_4$ 5846
$C_{12}H_7O_3N$ 4732; 5554
$C_{12}H_7O_4N$ 5551
$C_{12}H_8OS$ 5129
$C_{12}H_8OBr_2$ 1834
$C_{12}H_8O_4N_2$ 2620-3
$C_{12}H_8O_5N_2$ 2627-8
$C_{12}H_8O_5N_4$ 2598
$C_{12}H_8O_9N_4$ 5492
$C_{12}H_9OCl$ 1436-8
$C_{12}H_9OBr$ 885; 935; 977
$C_{12}H_9O_2N$ 3869; 4587; 4723-5
$C_{12}H_9O_3N$ 4729-31
$C_{12}H_9O_4N_3$ 2624-6; 4650-1
$C_{12}H_9O_4N_5$ 2618
$C_{12}H_9O_5N_3$ 2634
$C_{12}H_9O_8N_5$ 5495
$C_{12}H_9NS$ 5138

Table 7-2 (*Continued*)
FORMULA INDEX FOR ORGANIC COMPOUNDS

$C_{12}H_9N_3S$ 5922
$C_{12}H_{10}ON_2$ 615; 2733; 3727-9; 4884
$C_{12}H_{10}OS$ 2752
$C_{12}H_{10}O_2N_2$ 609-11; 662; 4726-7
$C_{12}H_{10}O_2S$ 2750
$C_{12}H_{10}O_2S_2$ 2716
$C_{12}H_{10}O_3N_2$ 4593
$C_{12}H_{10}N_2Cl_2$ 1957
$C_{12}H_{10}ClAs$ 2712
$C_{12}H_{10}Cl_2Se$ 2743
$C_{12}H_{11}ON$ 46-7; 3767-9
$C_{12}H_{11}OI$ 2728
$C_{12}H_{11}O_2N$ 31-2; 3031
$C_{12}H_{11}O_2N_3$ 4661
$C_{12}H_{11}O_4P$ 2735-6
$C_{12}H_{12}ON_2$ 1720
$C_{12}H_{12}O_3N_2$ 5113
$C_{12}H_{12}NCl$ 2699
$C_{12}H_{12}N_2S$ 5886
$C_{12}H_{12}N_2Cl_4$ 1958
$C_{12}H_{12}N_3Cl$ 250
$C_{12}H_{13}O_2N$ 3866
$C_{12}H_{13}O_3N$ 5476
$C_{12}H_{13}O_6N$ 2190
$C_{12}H_{13}N_2Cl$ 315; 2726
$C_{12}H_{13}N_4Cl$ 1699
$C_{12}H_{14}O_2N_2$ 2098; 2129
$C_{12}H_{14}O_8N_4$ 232
$C_{12}H_{14}NI$ 4368; 4371; 4373
$C_{12}H_{14}N_2Cl_2$ 1715
$C_{12}H_{15}ON$ 770
$C_{12}H_{15}O_2N$ 39
$C_{12}H_{15}O_6N_3$ 6297
$C_{12}H_{16}OCl_2$ 1381
$C_{12}H_{16}O_3N_2$ 3231; 5091
$C_{12}H_{16}O_8N_2$ 2528
$C_{12}H_{17}ON$ 77; 1215
$C_{12}H_{17}O_4N$ 2474
$C_{12}H_{17}O_5N$ 2116
$C_{12}H_{17}O_7N$ 5110
$C_{12}H_{18}O_4N_2$ 5559
$C_{12}H_{18}O_5N_2$ 3366-7
$C_{12}H_{19}ON$ 2114
$C_{12}H_{20}O_4Br_2$ 1896
$C_{12}H_{22}O_{14}Ca$ 3360
$C_{12}H_{23}OCl$ 3955
$C_{12}H_{25}ON$ 3953
$C_{12}H_{27}O_2N$ 3951
$C_{12}H_{27}O_3B$ 6117
$C_{12}H_{27}O_4P$ 6121
$C_{12}H_{28}O_4N_2$ 3200
$C_{12}H_{28}NI$ 5866
$C_{12}H_{30}OSi_2$ 6197
$C_{12}H_{30}N_6Cl_2$ 5685

12 IV

$C_{12}H_4O_{12}N_6S$ 2765

$C_{12}H_6O_3SCl_4$ 2043
$C_{12}H_6O_8N_4S_2$ 5850
$C_{12}H_8ON_2Cl_2$ 1950
$C_{12}H_8O_2SBr_2$ 1835
$C_{12}H_8O_3SCl_2$ 5180
$C_{12}H_8O_4N_2S_2$ 2655-6
$C_{12}H_9O_3SCl$ 5170
$C_{12}H_9O_3SNa$ 2751
$C_{12}H_{10}O_4N_2S$ 3730
$C_{12}H_{10}O_5N_2S$ 2307; 4728
$C_{12}H_{11}ONS$ 5906
$C_{12}H_{11}O_3NS$ 2701
$C_{12}H_{11}O_3N_2Na$ 5114
$C_{12}H_{12}ON_2S_2$ 2437
$C_{12}H_{12}O_3N_2S$ 696
$C_{12}H_{12}O_6N_2S_2$ 695
$C_{12}H_{13}O_4NS$ 2700
$C_{12}H_{14}ON_4S$ 5893
$C_{12}H_{14}O_4N_2S$ 1716
$C_{12}H_{16}O_2NCl$ 2413
$C_{12}H_{16}O_4N_2S$ 522
$C_{12}H_{19}O_2N_2Br$ 5470
$C_{12}H_{24}O_6N_6Cu$ 1522

12 V

$C_{12}H_6O_4N_2S_2Cl_2$ 1987
$C_{12}H_6O_8S_2I_4Zn$ 5629
$C_{12}H_9O_4N_2SNa$ 3731
$C_{12}H_9O_5N_2SNa$ 2308
$C_{12}H_{14}O_2N_2Cl_2As_2$ 5592
$C_{12}H_{14}O_2N_5SCl$ 5337
$C_{12}H_{18}ON_4SCl_2$ 5876
$C_{12}H_{19}O_3N_2Na$ 5002

13 I

$C_{13}H_{10}$ 3246
$C_{13}H_{12}$ 2731; 5245-7
$C_{13}H_{26}$ 6175
$C_{13}H_{28}$ 6173

13 II

$C_{13}H_8O$ 3248
$C_{13}H_8O_2$ 6451
$C_{13}H_8O_4$ 3227
$C_{13}H_8O_5$ 2891
$C_{13}H_9N$ 142; 4517-9
$C_{13}H_{10}O$ 724; 3247; 6447
$C_{13}H_{10}O_2$ 3320; 3749; 5159-62; 6453
$C_{13}H_{10}O_3$ 2321-8; 2711; 5133; 5235
$C_{13}H_{10}O_4$ 4276; 6224-5
$C_{13}H_{10}O_6$ 5070
$C_{13}H_{10}N_2$ 690; 1234-5; 4334
$C_{13}H_{10}Cl_2$ 726
$C_{13}H_{11}N$ 632; 2271
$C_{13}H_{12}O$ 691; 823-4; 4073-4
$C_{13}H_{12}O_2$ 2344; 3718
$C_{13}H_{12}N_2$ 646; 2721

$C_{13}H_{13}N$ 693; 786; 4219-20
$C_{13}H_{13}N_3$ 2723; 4211
$C_{13}H_{14}O_3$ 625
$C_{13}H_{14}O_4$ 740
$C_{13}H_{14}O_6$ 5584
$C_{13}H_{14}N_2$ 826; 1721; 4271-3; 4431
$C_{13}H_{15}N$ 5426
$C_{13}H_{16}O_3$ 2983
$C_{13}H_{18}O$ 1611
$C_{13}H_{18}O_3$ 3483
$C_{13}H_{18}O_4$ 1028
$C_{13}H_{20}O$ 3663; 3913-5
$C_{13}H_{20}O_8$ 5067
$C_{13}H_{24}O_4$ 2123; 2168
$C_{13}H_{25}N$ 2862
$C_{13}H_{26}O_2$ 2267; 4408
$C_{13}H_{26}O_2$ 438; 3181; 4295; 6176
$C_{13}H_{28}O$ 6118; 6174
$C_{13}H_{28}O_3$ 1087; 3386
$C_{13}H_{28}O_4$ 5399

13 III

$C_{13}H_7O_3Br_3$ 6110
$C_{13}H_8OCl_2$ 1965-6
$C_{13}H_8O_5N_2$ 2612
$C_{13}H_8O_8N_4$ 5852
$C_{13}H_9ON$ 145; 2714; 3249; 4525
$C_{13}H_9OCl$ 1331
$C_{13}H_9O_2N$ 4738
$C_{13}H_9O_3N$ 4670-2
$C_{13}H_9O_5N$ 4837
$C_{13}H_9O_5N_3$ 4652
$C_{13}H_9NS$ 687; 2755
$C_{13}H_{10}O_3N_2$ 4645
$C_{13}H_{10}O_4N_2$ 2629
$C_{13}H_{10}NCl$ 4519
$C_{13}H_{11}ON$ 270-2; 631; 715; 728; 3306
$C_{13}H_{11}O_2N$ 3741; 5157
$C_{13}H_{11}O_2N_3$ 4638
$C_{13}H_{11}O_4N_3$ 4649
$C_{13}H_{11}NS$ 5890
$C_{13}H_{12}ON_2$ 657-8; 767-8; 828; 1706-8; 2759-60; 3737
$C_{13}H_{12}ON_4$ 2709
$C_{13}H_{12}O_2N_4$ 4771
$C_{13}H_{12}O_2S$ 5252
$C_{13}H_{12}O_3S$ 5248
$C_{13}H_{12}O_5N_6$ 2654
$C_{13}H_{12}N_2S$ 5892
$C_{13}H_{12}N_4S$ 2753
$C_{13}H_{13}ON$ 22; 784; 5791
$C_{13}H_{13}ON_3$ 246; 2744-7
$C_{13}H_{14}ON_4$ 2710
$C_{13}H_{14}O_3N_2$ 5336

Table 7-2 (*Continued*)
FORMULA INDEX FOR ORGANIC COMPOUNDS

$C_{13}H_{14}N_4S$ 2754
$C_{13}H_{15}ON$ 2479
$C_{13}H_{15}N_2Cl$ 827
$C_{13}H_{16}NI$ 2543; 2545
$C_{13}H_{17}ON_3$ 363
$C_{13}H_{18}O_3N_4$ 3602
$C_{13}H_{19}ON$ 2774
$C_{13}H_{20}O_2N_2$ 4934-5
$C_{13}H_{25}O_2Br$ 4153
$C_{13}H_{27}ON$ 6177

13 IV

$C_{13}H_8O_5N_3Na$ 4653
$C_{13}H_9O_3SBr_3$ 6111
$C_{13}H_{10}ONCl$ 2708
$C_{13}H_{10}O_3SCl_2$ 2047
$C_{13}H_{11}O_3SCl$ 1439-40
$C_{13}H_{13}O_2NS$ 5972
$C_{13}H_{13}O_4N_3S$ 5328
$C_{13}H_{14}ONCl$ 785
$C_{13}H_{15}O_3N_2Cl_3$ 567
$C_{13}H_{18}O_2NBr$ 1008
$C_{13}H_{21}O_2N_2Cl$ 4934

13 V

$C_{13}H_{16}O_6NHgNa$ 5591

13 VI

$C_{13}H_{13}O_4N_2SAs_2Na$ 5593

14 I

$C_{14}H_{10}$ 538; 5095; 5954
$C_{14}H_{12}$ 2272; 5642-3
$C_{14}H_{14}$ 842-3; 2717-8; 2825-9
$C_{14}H_{16}$ 3576
$C_{14}H_{22}$ 5758-9
$C_{14}H_{26}$ 5752
$C_{14}H_{28}$ 5751
$C_{14}H_{30}$ 5748

14 II

$C_{14}H_6O_8$ 2872
$C_{14}H_8O_2$ 548-50; 5097
$C_{14}H_8O_3$ 3724-6
$C_{14}H_8O_4$ 2297-304; 3747
$C_{14}H_8O_5$ 6212-5
$C_{14}H_8O_6$ 5799-801
$C_{14}H_8O_8$ 5575
$C_{14}H_8Cl_2$ 1939-40
$C_{14}H_8Br_2$ 1803
$C_{14}H_9Br$ 968
$C_{14}H_{10}O$ 545; 561-2; 2730; 5098-100
$C_{14}H_{10}O_2$ 697; 2293-4; 2360; 5019; 5096
$C_{14}H_{10}O_3$ 714; 750-2; 2295-6
$C_{14}H_{10}O_4$ 766; 2688-91
$C_{14}H_{10}O_5$ 5585
$C_{14}H_{10}O_9$ 5687

$C_{14}H_{11}N$ 542-3; 4092-4; 4514-6
$C_{14}H_{12}O$ 1660; 2273; 5144; 5249-51
$C_{14}H_{12}O_2$ 717; 789-92; 2697; 4070; 6013-5
$C_{14}H_{12}O_3$ 694; 699; 809; 833; 3482; 6038-40
$C_{14}H_{12}O_4$ 2347; 2358; 3486
$C_{14}H_{12}N_2$ 633; 5182
$C_{14}H_{13}N$ 3011
$C_{14}H_{14}O$ 806; 2918; 4221; 5164
$C_{14}H_{14}O_2$ 2079; 2693; 3202; 3701-2
$C_{14}H_{14}O_3$ 3931
$C_{14}H_{14}O_8$ 5841
$C_{14}H_{14}N_2$ 54; 612-4; 1729-31; 2696
$C_{14}H_{14}S$ 1791; 2842
$C_{14}H_{14}S_2$ 805; 2836; 3204
$C_{14}H_{14}Hg$ 4048-50
$C_{14}H_{15}N$ 1783; 2830-2; 3044; 4144
$C_{14}H_{15}N_3$ 309-12; 1764; 2431-2
$C_{14}H_{15}P$ 3045
$C_{14}H_{16}N_2$ 1719; 2720; 3692-4; 5955; 5959
$C_{14}H_{17}N$ 2189
$C_{14}H_{18}O$ 451
$C_{14}H_{18}O_4$ 2130; 2794-5; 4231
$C_{14}H_{18}O_6$ 2412
$C_{14}H_{18}O_9$ 1269
$C_{14}H_{20}O_6$ 5781
$C_{14}H_{22}O$ 3918
$C_{14}H_{22}O_4$ 2087
$C_{14}H_{22}O_6$ 5785
$C_{14}H_{22}O_8$ 5764
$C_{14}H_{23}N$ 1891
$C_{14}H_{24}O_2$ 3355
$C_{14}H_{24}N_2$ 300
$C_{14}H_{26}O_3$ 3524
$C_{14}H_{26}O_4$ 1748-50; 1883; 2167; 2177; 2206; 3451
$C_{14}H_{26}O_6$ 1757; 2920
$C_{14}H_{27}N$ 4456
$C_{14}H_{28}O$ 3182; 4451
$C_{14}H_{28}O_2$ 2858; 3080; 3552; 4450
$C_{14}H_{29}Br$ 4457
$C_{14}H_{30}O$ 2264; 5750
$C_{14}H_{30}S$ 2266

14 III

$C_{14}H_4O_{12}N_4$ 1491
$C_{14}H_6O_2Cl_2$ 1941-8
$C_{14}H_6O_2Br_2$ 1805-9
$C_{14}H_6O_6N_2$ 2595-7; 2644
$C_{14}H_7O_2Cl$ 1316-7
$C_{14}H_7O_2Br$ 899-900
$C_{14}H_7O_4N$ 4627-8; 4794

$C_{14}H_7O_6N$ 4599-600
$C_{14}H_8O_5S$ 559
$C_{14}H_8O_7S$ 177
$C_{14}H_8O_8S_2$ 556
$C_{14}H_9O_2N$ 244-5; 4626; **5273**
$C_{14}H_9O_3N$ 172-3; 327
$C_{14}H_9O_4N$ 241-2
$C_{14}H_{10}O_2N_2$ 1690-5; 5163
$C_{14}H_{10}O_2S_2$ 1779
$C_{14}H_{10}O_4N_2$ 600-2; 2663-4
$C_{14}H_{10}O_5N_2$ 616-8
$C_{14}H_{10}O_6Ca$ 3743
$C_{14}H_{11}ON$ 80
$C_{14}H_{11}OCl$ 5215
$C_{14}H_{11}OBr$ 5214
$C_{14}H_{11}O_2N$ 703-5; 1776; **5587**
$C_{14}H_{11}O_3N$ 746-8
$C_{14}H_{12}O_2N_2$ 700-2; 5010
$C_{14}H_{12}O_3N_2$ 546
$C_{14}H_{12}O_4N_2$ 3686-8
$C_{14}H_{13}ON$ 84.1; 630; **775-7** 5143
$C_{14}H_{13}OCl$ 1376
$C_{14}H_{13}O_2N$ 720-1; 825
$C_{14}H_{14}ON_2$ 533; 4176
$C_{14}H_{14}OS$ 836; 2919
$C_{14}H_{14}O_2N_2$ 598
$C_{14}H_{14}O_2S$ 835; 2843
$C_{14}H_{14}O_3S$ 1554
$C_{14}H_{14}O_4S_2$ 3203
$C_{14}H_{15}ON$ 1786; 5137
$C_{14}H_{16}O_2N_2$ 1759
$C_{14}H_{16}O_6N_2$ 352; 5201
$C_{14}H_{16}N_3Cl$ 2432
$C_{14}H_{17}O_6N$ 3856
$C_{14}H_{18}ON_6$ 1709
$C_{14}H_{18}N_2Cl_2$ 5956
$C_{14}H_{19}O_8N$ 5108
$C_{14}H_{19}O_9Cl$ 44
$C_{14}H_{21}O_4N$ 2159
$C_{14}H_{23}ON_3$ 3916-7
$C_{14}H_{26}N_2Cl_2$ 301
$C_{14}H_{27}OCl$ 4455
$C_{14}H_{29}ON_3$ 4409
$C_{14}H_{29}O_2N$ 4452

14 IV

$C_{14}H_6O_8S_2Na_2$ 553-5; 557
$C_{14}H_7O_5SNa$ 558; 560
$C_{14}H_7O_7NS$ 4630-1
$C_{14}H_7O_7SNa$ 178
$C_{14}H_{10}OClBr$ 1435
$C_{14}H_{13}O_3SCl$ 4189
$C_{14}H_{14}O_6N_2S_2$ 1732
$C_{14}H_{15}O_2NS$ 5980-1
$C_{14}H_{16}O_2N_3As$ 2433
$C_{14}H_{16}O_3N_3As$ 2434
$C_{14}H_{16}O_6N_2S_2$ 5958
$C_{14}H_{17}O_4N_3S_2$ 6368

Table 7-2 (*Continued*)
FORMULA INDEX FOR ORGANIC COMPOUNDS

$C_{14}H_{18}O_4N_2S$ 5957
$C_{14}H_{20}O_6N_2S$ 294; 4116-7
$C_{14}H_{22}O_2NCl$ 5644
$C_{14}H_{22}O_4N_4S$ 4339
$C_{14}H_{23}O_2N_2Cl$ 6366

14 V

$C_{14}H_{14}O_3N_3SNa$ 4328

14 VI

$C_{14}H_{14}O_8N_2S_2As_2Na_2$ 5674

15 I

$C_{15}H_{12}$ 4127-8
$C_{15}H_{16}$ 2985
$C_{15}H_{18}$ 470
$C_{15}H_{24}$ 1263
$C_{15}H_{32}$ 5060

15 II

$C_{15}H_8O_4$ 551-2
$C_{15}H_8O_6$ 176; 2305-6
$C_{15}H_{10}O_2$ 539-41; 640; 3243; 4129
$C_{15}H_{10}O_3$ 2758
$C_{15}H_{10}O_4$ 1494-5; 4101
$C_{15}H_{10}O_5$ 727; 769; 772; 2874
$C_{15}H_{10}O_6$ 5805
$C_{15}H_{10}O_7$ 5806-7
$C_{15}H_{11}N$ 5230-3
$C_{15}H_{12}O$ 627
$C_{15}H_{12}O_2$ 1781; 3763; 5172-4
$C_{15}H_{12}O_3$ 1492; 2258; 4140; 6002
$C_{15}H_{12}O_4$ 120; 4143
$C_{15}H_{13}N$ 2510-1
$C_{15}H_{13}N_3$ 5199
$C_{15}H_{14}O$ 1787; 2838
$C_{15}H_{14}O_2$ 2739; 5190
$C_{15}H_{14}O_3$ 2409; 2833-5
$C_{15}H_{14}O_4$ 4307
$C_{15}H_{14}O_5$ 3484; 5264 5594
$C_{15}H_{16}O$ 1573
$C_{15}H_{16}O_2$ 2694
$C_{15}H_{16}O_3$ 3409
$C_{15}H_{16}O_9$ 2899
$C_{15}H_{17}N$ 2984
$C_{15}H_{17}N_3$ 2837
$C_{15}H_{18}O$ 471
$C_{15}H_{18}O_3$ 5598
$C_{15}H_{20}O_2$ 3497
$C_{15}H_{20}O_4$ 2170; 5597; 5599
$C_{15}H_{20}N_4$ 5713
$C_{15}H_{22}O_{10}$ 4385
$C_{15}H_{24}O$ 5595
$C_{15}H_{24}O_8$ 5765
$C_{15}H_{26}O_2$ 867; 869
$C_{15}H_{26}O_6$ 3421

$C_{15}H_{28}O_2$ 4032
$C_{15}H_{28}O_4$ 2157
$C_{15}H_{30}O$ 2265
$C_{15}H_{30}O_2$ 4303
$C_{15}H_{31}Br$ 5063
$C_{15}H_{32}O$ 5062
$C_{15}H_{33}N$ 6074-6

15 III

$C_{15}H_9O_4N$ 4770
$C_{15}H_{11}O_3N$ 4633
$C_{15}H_{11}O_4N$ 4641
$C_{15}H_{12}ON_2$ 2722
$C_{15}H_{12}OBr_2$ 628-9
$C_{15}H_{12}O_2N_2$ 2724
$C_{15}H_{12}O_3N_2$ 3326
$C_{15}H_{12}O_5N_2$ 3353
$C_{15}H_{12}N_2S$ 317
$C_{15}H_{13}ON$ 30
$C_{15}H_{14}O_2N_2$ 4005
$C_{15}H_{15}ON$ 812; 2442
$C_{15}H_{15}O_2N$ 2761; 5588
$C_{15}H_{15}O_2N_3$ 4376
$C_{15}H_{16}ON_2$ 2467; 2847-9
$C_{15}H_{16}N_2S$ 2844-6
$C_{15}H_{17}N_4Cl$ 4585
$C_{15}H_{25}O_2Br$ 863
$C_{15}H_{31}ON$ 5059
$C_{15}H_{33}O_3B$ 6077

15 IV

$C_{15}H_6O_5NCl$ 4629
$C_{15}H_{11}O_4NI_4$ 5947
$C_{15}H_{11}O_4N_3S$ 679
$C_{15}H_{14}ON_2Cl$ 114
$C_{15}H_{14}O_2N_3Na$ 4377-8
$C_{15}H_{20}ONI$ 2480
$C_{15}H_{25}O_2NS$ 5976
$C_{15}H_{25}O_2N_2Cl$ 5032

16 I

$C_{16}H_{10}$ 3245; 5481
$C_{16}H_{12}$ 5204-5
$C_{16}H_{14}$ 2453-5; 2525; 2969
$C_{16}H_{16}$ 2277
$C_{16}H_{26}$ 1739; 5068
$C_{16}H_{30}$ 3571-2
$C_{16}H_{32}$ 3570
$C_{16}H_{34}$ 3569

16 II

$C_{16}H_{10}N_2$ 4513
$C_{16}H_{12}O_3$ 5324
$C_{16}H_{12}O_5$ 874
$C_{16}H_{12}O_6$ 3498
$C_{16}H_{12}O_7$ 5556
$C_{16}H_{13}N$ 5206-7
$C_{16}H_{13}N_3$ 660-1
$C_{16}H_{14}O_2$ 798

$C_{16}H_{14}O_3$ 718; 2257; 2979; 3485; 5991
$C_{16}H_{14}O_42748$; 3447
$C_{16}H_{14}O_5$ 875
$C_{16}H_{14}O_6$ 3500; 3559; 3672; 5326
$C_{16}H_{14}N_2$ 3242
$C_{16}H_{16}O_2$ 719; 1782
$C_{16}H_{16}O_5$ 180
$C_{16}H_{16}O_6$ 1264
$C_{16}H_{18}O$ 1121
$C_{16}H_{18}O_2$ 3448
$C_{16}H_{18}O_3$ 2692
$C_{16}H_{18}O_4$ 3457
$C_{16}H_{18}O_6$ 3710
$C_{16}H_{18}O_{10}$ 3310
$C_{16}H_{18}N_2$ 2738
$C_{16}H_{19}N$ 2988
$C_{16}H_{20}O_{10}$ 1268
$C_{16}H_{20}N_2$ 5830
$C_{16}H_{21}N_3$ 5833
$C_{16}H_{22}O_4$ 1908
$C_{16}H_{22}O_6$ 2096
$C_{16}H_{22}O_8$ 1528
$C_{16}H_{22}O_{11}$ 3362-3
$C_{16}H_{24}O_4$ 2086
$C_{16}H_{26}O_2$ 1742
$C_{16}H_{28}O_2$ 3680; 5029
$C_{16}H_{30}O$ 4446
$C_{16}H_{30}O_2$ 3849
$C_{16}H_{30}O_3$ 1224
$C_{16}H_{31}N$ 5028
$C_{16}H_{32}O$ 5023
$C_{16}H_{32}O_2$ 2263; 3088; 5022; 5749
$C_{16}H_{32}O_4$ 2234
$C_{16}H_{33}Br$ 1276
$C_{16}H_{33}I$ 1277
$C_{16}H_{34}O$ 1275; 2684
$C_{16}H_{34}O_3$ 456

16 III

$C_{16}H_{10}O_2N_2$ 3857; 3862
$C_{16}H_{10}O_4N_2$ 850
$C_{16}H_{11}O_2N$ 5234
$C_{16}H_{11}O_3N_3$ 4648
$C_{16}H_{12}O_2N_2$ 3863
$C_{16}H_{13}O_{15}N_9$ 1542
$C_{16}H_{14}O_3N_2$ 1509
$C_{16}H_{15}O_3N$ 3848
$C_{16}H_{16}O_2N_2$ 1671; 1780
$C_{16}H_{17}ON_3$ 124
$C_{16}H_{17}O_2N$ 5165
$C_{16}H_{18}O_2N_2$ 606-8
$C_{16}H_{18}O_3N_2$ 851
$C_{16}H_{20}N_3Cl$ 253
$C_{16}H_{27}O_5N_3$ 737
$C_{16}H_{31}OCl$ 5027
$C_{16}H_{33}ON$ 5024-5
$C_{16}H_{36}NI$ 5727

Table 7-2 *(Continued)*
FORMULA INDEX FOR ORGANIC COMPOUNDS

16 IV

$C_{16}H_9O_2NI_2$ 2384
$C_{16}H_{10}O_5N_2S$ 3861
$C_{16}H_{10}O_8N_2S_2$ 3860
$C_{16}H_{11}O_{10}N_3S$ 5491
$C_{16}H_{12}O_9N_4S$ 5494
$C_{16}H_{13}O_3NS$ 5208
$C_{16}H_{13}O_3N_3S$ 331
$C_{16}H_{18}O_6S_2Ca$ 6460
$C_{16}H_{18}N_3SCl$ 4427
$C_{16}H_{19}ON_3S$ 4426
$C_{16}H_{24}O_2NCl$ 764; 2119
$C_{16}H_{24}O_6N_2S$ 2447
$C_{16}H_{27}O_2N_2Cl$ 276; 736

16 V

$C_{16}H_8O_8N_2S_2Na_2$ 3858
$C_{16}H_9O_9N_4S_2Na_3$ 5694
$C_{16}H_9O_{10}N_3S_2Na_2$ 4647
$C_{16}H_{11}O_4N_2SNa$ 4994–5
$C_{16}H_{14}O_4N_2As_2Na_2$ 581

17 I

$C_{17}H_{14}$ 817–8
$C_{17}H_{28}$ 3180
$C_{17}H_{36}$ 3508

17 II

$C_{17}H_{11}N$ 547
$C_{17}H_{12}O$ 5210-1
$C_{17}H_{12}O_2$ 4556-7
$C_{17}H_{12}O_3$ 4568-9
$C_{17}H_{14}O$ 819-20; 1506; 1775; 5209
$C_{17}H_{14}N_2$ 689
$C_{17}H_{15}N$ 6031-4
$C_{17}H_{15}N_3$ 5961
$C_{17}H_{16}O_2$ 5192
$C_{17}H_{16}O_3$ 3220; 5387
$C_{17}H_{16}O_6$ 4277
$C_{17}H_{16}O_7$ 3229
$C_{17}H_{19}N_3$ 143
$C_{17}H_{21}N_3$ 592
$C_{17}H_{22}O_3$ 866
$C_{17}H_{22}N_2$ 5834
$C_{17}H_{24}O_2$ 4030; 4977
$C_{17}H_{24}O_3$ 4031
$C_{17}H_{28}O_4$ 868
$C_{17}H_{32}O_3$ 5780
$C_{17}H_{34}O$ 2685; 4015
$C_{17}H_{34}O_2$ 4014; 4330; 5061
$C_{17}H_{36}O$ 3509

17 III

$C_{17}H_9O_4N$ 174
$C_{17}H_{11}O_5N$ 3801
$C_{17}H_{12}O_5N_2$ 2599
$C_{17}H_{13}ON$ 765
$C_{17}H_{20}ON_2$ 2137; 4436

$C_{17}H_{20}O_3N_2$ 2687
$C_{17}H_{20}O_5N_2$ 5117
$C_{17}H_{20}O_6N_4$ 5563
$C_{17}H_{22}ON_2$ 4435

17 IV

$C_{17}H_{19}ON_2Cl$ 5519
$C_{17}H_{24}O_3N_3Cl_3$ 364

17 V

$C_{17}H_{11}O_{10}NS_2Na_2$ 175

18 I

$C_{18}H_{12}$ 650; 1493
$C_{18}H_{14}$ 2703-5
$C_{18}H_{18}$ 5555
$C_{18}H_{20}$ 2476
$C_{18}H_{30}$ 3575; 5867
$C_{18}H_{34}$ 4942
$C_{18}H_{36}$ 4941
$C_{18}H_{38}$ 4938

18 II

$C_{18}H_{10}O_2$ 1496
$C_{18}H_{12}O_4$ 1658
$C_{18}H_{12}N_2$ 2814-6
$C_{18}H_{14}O$ 821
$C_{18}H_{14}O_3$ 1503
$C_{18}H_{14}O_4$ 3790; 5020
$C_{18}H_{15}N$ 6328
$C_{18}H_{15}N_3$ 659
$C_{18}H_{15}As$ 580
$C_{18}H_{15}Bi$ 857
$C_{18}H_{15}P$ 6342
$C_{18}H_{15}Sb$ 563
$C_{18}H_{16}O_2$ 1508
$C_{18}H_{16}O_4$ 1785; 1788; 3924-5
$C_{18}H_{16}O_7$ 6394-5
$C_{18}H_{16}N_2$ 6336
$C_{18}H_{17}N_3$ 2438-9
$C_{18}H_{18}O_2$ 2883-4; 2886-8
$C_{18}H_{18}O_3$ 1055
$C_{18}H_{18}O_4$ 1790; 2921
$C_{18}H_{18}O_6$ 1792
$C_{18}H_{20}O_2$ 2209
$C_{18}H_{22}O_2$ 2903
$C_{18}H_{23}N_3$ 4133
$C_{18}H_{24}O_2$ 2900
$C_{18}H_{24}O_3$ 2902
$C_{18}H_{26}O_4$ 1746-7
$C_{18}H_{26}O_{12}$ 3610; 3612
$C_{18}H_{28}O_4$ 2873
$C_{18}H_{30}O_2$ 3977
$C_{18}H_{30}O_4$ 2084
$C_{18}H_{32}O_2$ 1279; 2871; 3976; 5639
$C_{18}H_{32}O_3$ 5569
$C_{18}H_{32}O_4$ 5641
$C_{18}H_{32}O_6$ 3434-5

$C_{18}H_{32}O_7$ 6119
$C_{18}H_{32}O_{16}$ 5549
$C_{18}H_{34}O_2$ 2870; 3495; 4992
$C_{18}H_{34}O_3$ 5566; 5568
$C_{18}H_{34}O_4$ 1909; 3452
$C_{18}H_{34}O_6$ 2097
$C_{18}H_{35}N$ 5638
$C_{18}H_{36}O$ 5633
$C_{18}H_{36}O_2$ 1274; 2683; 3112; 5631
$C_{18}H_{36}O_3$ 3829-31
$C_{18}H_{36}O_4$ 2365-6
$C_{18}H_{37}I$ 4940
$C_{18}H_{38}O$ 4939
$C_{18}H_{39}N$ 6210

18 III

$C_{18}H_{10}O_6N_2$ 3859
$C_{18}H_{12}ON_2$ 670
$C_{18}H_{12}O_4N_2$ 3205
$C_{18}H_{14}O_2N_4$ 1778
$C_{18}H_{14}O_8Ca$ 116
$C_{18}H_{14}O_8Mg$ 118
$C_{18}H_{15}O_3P$ 6343
$C_{18}H_{15}O_4P$ 6341
$C_{18}H_{15}ClSn$ 5948
$C_{18}H_{15}Cl_2Bi$ 858
$C_{18}H_{15}Cl_2Sb$ 564
$C_{18}H_{16}O_2N_2$ 508; 5171
$C_{18}H_{17}O_3Na$ 5120
$C_{18}H_{18}O_4N_2$ 5590
$C_{18}H_{22}O_4N_4$ 5193
$C_{18}H_{27}O_3N$ 1227
$C_{18}H_{33}O_3Na$ 5567
$C_{18}H_{35}OCl$ 5637
$C_{18}H_{37}ON$ 5634
$C_{18}H_{39}O_2N$ 5632

18 IV

$C_{18}H_{12}O_4Cl_3P$ 6165
$C_{18}H_{13}O_4Cl_2P$ 2046
$C_{18}H_{14}O_4ClP$ 2713
$C_{18}H_{15}OS_3P$ 6350
$C_{18}H_{15}O_3SP$ 6346
$C_{18}H_{16}O_6N_2S$ 3827
$C_{18}H_{16}O_6N_2Ca$ 3665
$C_{18}H_{23}O_2N_2Cl$ 2910
$C_{18}H_{28}O_4N_2S$ 655
$C_{18}H_{32}O_5N_2S$ 5031
$C_{18}H_{32}O_{10}N_2Ca$ 5035

18 V

$C_{18}H_{14}O_3N_3SNa$ 4056; 4996

19 I

$C_{19}H_{15}$ 6339
$C_{19}H_{16}$ 802-3; 6337
$C_{19}H_{40}$ 4908

Table 7-2 (*Continued*)
FORMULA INDEX FOR ORGANIC COMPOUNDS

19 II

$C_{19}H_{12}O_2$ 4479
$C_{19}H_{13}N$ 5147
$C_{19}H_{14}O$ 2741
$C_{19}H_{14}O_2$ 652; 5255
$C_{19}H_{14}O_3$ 593
$C_{19}H_{14}O_5$ 2766
$C_{19}H_{15}N$ 725
$C_{19}H_{15}Cl$ 6333
$C_{19}H_{15}Br$ 6330
$C_{19}H_{16}O$ 6331
$C_{19}H_{16}O_3$ 3963
$C_{19}H_{16}N_2$ 729
$C_{19}H_{17}N$ 384-5; 804
$C_{19}H_{17}N_3$ 6334-5
$C_{19}H_{18}O$ 2824
$C_{19}H_{18}O_3$ 1758; 3221
$C_{19}H_{18}N_2$ 1733
$C_{19}H_{19}N_3$ 6071-3
$C_{19}H_{22}O_3$ 5004
$C_{19}H_{28}O_2$ 1657; 5711-2
$C_{19}H_{30}O_2$ 509
$C_{19}H_{36}O_2$ 4327
$C_{19}H_{38}O$ 2682; 4261
$C_{19}H_{38}O_2$ 4381; 4909

19 III

$C_{19}H_{13}O_6N_3$ 6322
$C_{19}H_{13}O_7N_3$ 6321
$C_{19}H_{14}O_5S$ 5127
$C_{19}H_{15}O_2N$ 212
$C_{19}H_{17}O_2N$ 3130
$C_{19}H_{18}N_3Cl$ 5039
$C_{19}H_{19}ON_3$ 5038
$C_{19}H_{20}O_4N_2$ 568
$C_{19}H_{21}O_5N$ 6420
$C_{19}H_{24}O_2Br_2$ 865

19 IV

$C_{19}H_{10}O_5SBr_4$ 5724
$C_{19}H_{10}O_5SI_4$ 5815
$C_{19}H_{10}O_{13}N_4S$ 5858
$C_{19}H_{14}ONCl$ 3980
$C_{19}H_{16}O_2NCl$ 213

20 I

$C_{20}H_{14}$ 2574-5; 5156
$C_{20}H_{18}$ 2756-7
$C_{20}H_{38}$ 5293
$C_{20}H_{40}$ 5297
$C_{20}H_{42}$ 1017; 2868; 5294

20 II

$C_{20}H_{12}O_3$ 3244
$C_{20}H_{12}O_5$ 3250; 3720
$C_{20}H_{14}O$ 4560-2
$C_{20}H_{14}O_2$ 2336-7; 5292
$C_{20}H_{14}O_4$ 2737; 5121; 5553
$C_{20}H_{14}O_5$ 3254
$C_{20}H_{14}N_2$ 604-5

$C_{20}H_{14}S$ 2583-4
$C_{20}H_{14}Hg$ 4045
$C_{20}H_{15}N$ 2576
$C_{20}H_{15}N_3$ 251; 1763
$C_{20}H_{16}O_2$ 6327; 6338
$C_{20}H_{16}O_3$ 5573
$C_{20}H_{16}O_4$ 5122
$C_{20}H_{16}N_2$ 3690-1
$C_{20}H_{16}N_4$ 4872
$C_{20}H_{18}O$ 6332
$C_{20}H_{18}O_2$ 529; 3717
$C_{20}H_{19}N$ 1784
$C_{20}H_{20}N_2$ 2839-40
$C_{20}H_{21}N_3$ 3962
$C_{20}H_{24}O_4$ 1555
$C_{20}H_{26}O_3$ 2209
$C_{20}H_{26}O_4$ 2088
$C_{20}H_{30}O_2$ 1
$C_{20}H_{30}O_6$ 1027
$C_{20}H_{32}N_2$ 3961
$C_{20}H_{36}O_2$ 3014
$C_{20}H_{38}O_2$ 3109
$C_{20}H_{40}O$ 5298
$C_{20}H_{40}O_2$ 574; 3153
$C_{20}H_{42}O$ 2869; 5295

20 III

$C_{20}H_8O_5Br_4$ 2875-6
$C_{20}H_8O_5I_4$ 2896-7
$C_{20}H_{10}O_3Cl_2$ 3252
$C_{20}H_{10}O_4Cl_4$ 5744
$C_{20}H_{10}O_4Br_4$ 5722
$C_{20}H_{10}O_4I_4$ 5813
$C_{20}H_{10}O_5Cl_2$ 1999; 2000
$C_{20}H_{10}O_5Br_2$ 1838-9
$C_{20}H_{10}O_5I_2$ 2392
$C_{20}H_{10}O_5Na_2$ 3251
$C_{20}H_{14}ON_2$ 619-20
$C_{20}H_{18}ON_2$ 722-3
$C_{20}H_{20}O_5N_2$ 566
$C_{20}H_{20}N_3Cl$ 5571
$C_{20}H_{21}ON_3$ 5570
$C_{20}H_{22}O_4N_2$ 5307
$C_{20}H_{33}ON$ 4454
$C_{20}H_{41}ONa$ 5296

20 IV

$C_{20}H_6O_5Br_4Na_2$ 2876
$C_{20}H_6O_5I_4Na_2$ 2897
$C_{20}H_8O_4Br_4Na_2$ 5723
$C_{20}H_8O_4I_4Na_2$ 5814
$C_{20}H_{14}O_8S_2Ca$ 4499; 4506; 4508
$C_{20}H_{30}O_2N_3Cl$ 1025
$C_{20}H_{32}O_6N_2S$ 3676-7

20 V

$C_{20}H_8O_6Br_2HgNa_2$ 4041
$C_{20}H_8O_{10}S_2Br_4Na_2$ 998
$C_{20}H_{11}O_{10}N_2S_3Na_3$ 3732

21 I

$C_{21}H_{16}$ 822; 2580-2; 4190
$C_{21}H_{20}$ 5183
$C_{21}H_{42}$ 3504
$C_{21}H_{44}$ 3503

21 II

$C_{21}H_{14}O$ 2577-9
$C_{21}H_{16}N_2$ 3978
$C_{21}H_{18}O$ 2078; 2740
$C_{21}H_{18}N_2$ 233; 3700
$C_{21}H_{20}O_6$ 1575
$C_{21}H_{20}O_9$ 227
$C_{21}H_{21}N$ 6084
$C_{21}H_{21}N_3$ 512
$C_{21}H_{24}O_4$ 2153
$C_{21}H_{30}O_2$ 5333-4
$C_{21}H_{30}N_2$ 5761
$C_{21}H_{32}O_2$ 4084
$C_{21}H_{32}O_{11}$ 3365
$C_{21}H_{36}O_2$ 5329-32
$C_{21}H_{38}O_6$ 3423
$C_{21}H_{40}O_3$ 5782
$C_{21}H_{42}O_4$ 3415-6

21 III

$C_{21}H_{15}ON$ 698
$C_{21}H_{15}O_3N$ 6081
$C_{21}H_{16}ON_2$ 2585
$C_{21}H_{18}O_5S$ 1550-1
$C_{21}H_{20}O_4N_2$ 228
$C_{21}H_{21}O_3P$ 6354-5
$C_{21}H_{21}O_4P$ 6351-3
$C_{21}H_{24}O_5N_2$ 2779
$C_{21}H_{28}ON_2$ 5760

21 IV

$C_{21}H_{14}O_5SBr_4$ 5718
$C_{21}H_{16}O_5SBr_2$ 1832
$C_{21}H_{21}O_3SP$ 6356-8

22 I

$C_{22}H_{14}$ 1777; 5299
$C_{22}H_{46}$ 2855

22 II

$C_{22}H_{15}N_3$ 5572
$C_{22}H_{16}O_3$ 6082-3
$C_{22}H_{16}C_5$ 4999
$C_{22}H_{18}O_4$ 1548; 1789; 2841
$C_{22}H_{20}O_4$ 1549
$C_{22}H_{22}O_8$ 5306
$C_{22}H_{28}O_5$ 5596
$C_{22}H_{32}O_3$ 507
$C_{22}H_{34}O_2$ 2932
$C_{22}H_{40}O_2$ 624
$C_{22}H_{42}O_2$ 873; 1110; 2892
$C_{22}H_{42}O_3$ 1127-8
$C_{22}H_{42}O_4$ 1924; 3201
$C_{22}H_{44}O$ 2893

Table 7-2 (*Continued*)
FORMULA INDEX FOR ORGANIC COMPOUNDS

$C_{22}H_{44}O_2$ 623; 1129-30
$C_{22}H_{44}O_4$ 2309

22 III

$C_{22}H_{16}ON_4$ 5664
$C_{22}H_{21}O_2N$ 5603
$C_{22}H_{23}O_4P$ 2707
$C_{22}H_{40}O_4N_8$ 3601

22 IV

$C_{22}H_{24}O_4N_2S$ 2542

23 I

$C_{23}H_{48}$6171

23 II

$C_{23}H_{17}N$ 50
$C_{23}H_{18}O$ 2732
$C_{23}H_{22}O_6$ 1655; 5574
$C_{23}H_{22}O_7$ 5699
$C_{23}H_{26}N_2$ 3964
$C_{23}H_{27}N_3$ 5844
$C_{23}H_{32}O_6$ 5645
$C_{23}H_{44}O_2$ 475
$C_{23}H_{46}O$ 2850
$C_{23}H_{46}O_2$ 488-9

23 III

$C_{23}H_{23}N_2I$ 2144
$C_{23}H_{25}O_2N_3$ 5838
$C_{23}H_{25}N_2Cl$ $(ZnCl_2)$ 3993
$C_{23}H_{26}ON_2$ 5835

23 IV

$C_{23}H_{26}O_3N_3Cl$ 1760

24 I

$C_{24}H_{18}$ 6329
$C_{24}H_{50}$ 5747

24 II

$C_{24}H_{16}O_7$ 3253
$C_{24}H_{20}O_6$ 3420
$C_{24}H_{20}N_2$ 2706; 5863
$C_{24}H_{20}Pb$ 3959
$C_{24}H_{20}Sn$ 5952
$C_{24}H_{28}O_4$ 4931-2
$C_{24}H_{28}N_2$ 5837
$C_{24}H_{40}O_2$ 5238
$C_{24}H_{40}O_4$ 1479
$C_{24}H_{40}O_5$ 1487
$C_{24}H_{44}O_6$ 3425
$C_{24}H_{46}O_3$ 3954
$C_{24}H_{48}O_2$ 3970

24 III

$C_{24}H_{19}O_4P$ 2762
$C_{24}H_{19}O_5N$ 1675
$C_{24}H_{20}BrAs$ 5859

$C_{24}H_{21}O_3N_3$ 801
$C_{24}H_{25}ON_3$ 749
$C_{24}H_{28}ON_2$ 5836
$C_{24}H_{28}N_3Cl$ 5077
$C_{24}H_{29}ON_3$ 5076
$C_{24}H_{41}ON$ 5636

24 IV

$C_{24}H_{14}O_2N_2S_2$ 3868
$C_{24}H_{26}O_4N_4S$ 316

24 V

$C_{24}H_{20}O_6N_2S_2Ba$ 2702

25 I

$C_{25}H_{20}$ 5864
$C_{25}H_{52}$ 5058

25 II

$C_{25}H_{21}N_3$ 5862
$C_{25}H_{22}O_3$ 2885
$C_{25}H_{26}O_9$ 3226
$C_{25}H_{30}O_4$ 4933
$C_{25}H_{31}N_3$ 3595
$C_{25}H_{50}O_2$ 3847

25 III

$C_{25}H_{20}ON_2$ 5865
$C_{25}H_{25}N_2I$ 5316
$C_{25}H_{27}N_2I$ 5003
$C_{25}H_{30}N_3Cl$ 3593
$C_{25}H_{31}ON_3$ 3592

25 IV

$C_{25}H_{29}ON_2I$ 2144

26 I

$C_{26}H_{18}$ 2734
$C_{26}H_{20}$ 5861
$C_{26}H_{22}$ 5860
$C_{26}H_{52}$ 1271
$C_{26}H_{54}$ 3567-8

26 II

$C_{26}H_{20}O$ 731
$C_{26}H_{22}O$ 692
$C_{26}H_{22}O_2$ 732; 2853
$C_{26}H_{22}N_4$ 706-7
$C_{26}H_{42}O_{11}$ 3364
$C_{26}H_{50}O_4$ 3458
$C_{26}H_{52}O_2$ 1272; 4983
$C_{26}H_{54}O$ 1273

26 III

$C_{26}H_{31}O_4P$ 1907
$C_{26}H_{33}N_3Cl_2$ 3594
$C_{26}H_{43}O_6N$ 3440

26 IV

$C_{26}H_{42}O_6NNa$ 3441
$C_{26}H_{45}O_7NS$ 5697

26 V

$C_{26}H_{44}O_7NSNa$ 5698

27 I

$C_{27}H_{18}$ 6362
$C_{27}H_{56}$ 3507

27 II

$C_{27}H_{26}O_7$ 6085
$C_{27}H_{36}O_4$ 595
$C_{27}H_{38}O_4$ 594
$C_{27}H_{44}O_3$ 5602
$C_{27}H_{44}O_4$ 1478
$C_{27}H_{46}O$ 1480-2
$C_{27}H_{48}O$ 2276
$C_{27}H_{50}O_6$ 3424

27 III

$C_{27}H_{34}ON_2$ 5762

27 IV

$C_{27}H_{28}O_5SBr_2$ 1879
$C_{27}H_{34}O_4N_2S$ 5763

28 I

$C_{28}H_{58}$ 4937

28 II

$C_{28}H_{18}O_4$ 4497
$C_{28}H_{20}N_2$ 234
$C_{28}H_{30}O_4$ 5941
$C_{28}H_{34}O_{15}$ 3560
$C_{28}H_{44}O$ 1187; 2889; **3979**
$C_{28}H_{54}O_3$ 4453
$C_{28}H_{54}O_4$ 2227

28 III

$C_{28}H_{24}O_7N_2$ 4997
$C_{28}H_{33}O_{11}N_5$ 1044

28 IV

$C_{28}H_{31}O_3N_2Cl$ 5766

29 I

$C_{29}H_{60}$ 4907

29 II

$C_{29}H_{48}O_2$ 1483

29 III

$C_{29}H_{35}N_2I$ 1580

30 I

$C_{30}H_{60}$ 4021
$C_{30}H_{62}$ 6057; 4021

Table 7-2 (*Continued*)

FORMULA INDEX FOR ORGANIC COMPOUNDS

30 II

$C_{30}H_{48}O_2$ 4444
$C_{30}H_{48}O_3$ 849
$C_{30}H_{50}O_2$ 848; 1486
$C_{30}H_{58}O_4$ 3461
$C_{30}H_{62}O$ 4449

30 III

$C_{30}H_{23}O_4P$ 5184
$C_{30}H_{39}O_4P$ 6120

31 I

$C_{31}H_{64}$ 3505

31 II

$C_{31}H_{46}O_2$ 4342
$C_{31}H_{62}O$ 5030
$C_{31}H_{62}O_2$ 4022
$C_{31}H_{64}O$ 4449

31 III

$C_{31}H_{43}ON_3$ 3191

32 I

$C_{32}H_{66}$ 2866

32 II

$C_{32}H_{52}O_2$ 2867
$C_{32}H_{62}O_3$ 5026
$C_{32}H_{64}O_2$ 1278

32 III

$C_{32}H_{36}O_6N_4$ 851
$C_{32}H_{36}O_8N_4$ 852

32 V

$C_{32}H_{22}O_6N_6S_2Na_2$ 1527

33 I

$C_{33}H_{68}$ 6359

33 II

$C_{33}H_{62}O_6$ 3422
$C_{33}H_{66}O_2$ 4448

34 I

$C_{34}H_{70}$ 5868

34 II

$C_{34}H_{50}O_2$ 1484-5
$C_{34}H_{66}O_4$ 3464

34 IV

$C_{34}H_{33}O_5N_4Fe$ 3499

34 V

$C_{34}H_{26}O_6N_6S_2Na_2$ 733
$C_{34}H_{26}O_{14}N_6S_4Na_4$ 3228
$C_{34}H_{32}O_4N_4ClFe$ 3501

35 I

$C_{35}H_{72}$ 5080

35 II

$C_{35}H_{28}O_2$ 647-8
$C_{35}H_{40}O_{12}$ 3239
$C_{35}H_{52}O_4$ 4343
$C_{35}H_{70}O$ 5640

36 I

$C_{36}H_{74}$ 3613

36 II

$C_{36}H_{70}O_3$ 5635

36 III

$C_{36}H_{27}O_4P$ 6360
$C_{36}H_{43}O_4P$ 1146

36 IV

$C_{36}H_{62}O_8N_4S$ 1890

37 V

$C_{37}H_{27}O_9N_3S_3Na_2$ 6345

38 II

$C_{38}H_{30}O_2$ 6340
$C_{38}H_{74}O_4$ 3467

38 III

$C_{38}H_{32}N_3Cl$ 516

39 II

$C_{39}H_{74}O_6$ 3426

40 I

$C_{40}H_{56}$ 1252-4; 3985

40 II

$C_{40}H_{50}O_2$ 5562
$C_{40}H_{56}O$ 1570
$C_{40}H_{56}O_2$ 6452

42 IV

$C_{42}H_{84}O_9NP$ 3960

45 II

$C_{45}H_{86}O_6$ 3427

51 II

$C_{51}H_{98}O_6$ 3431

52 III

$C_{52}H_{54}O_{12}N_4$ 3994

57 II

$C_{57}H_{104}O_6$ 3430
$C_{57}H_{110}O_6$ 3433

63 II

$C_{63}H_{122}O_6$ 3419

69 IV

$C_{69}H_{75}N_6Cl_7Zn_2$ 3993

76 IV

$C_{76}H_{64}O_4N_6S$ 6344

Table 7-3
MELTING POINTS OF ORGANIC COMPOUNDS

The compounds below are arranged in the order of the ascending values of the melting points, an arrangement intended to serve as an aid in the identification of organic compounds. The values given may be uncertain by one or more units in the last figure and consequently are not to be used as fixed points for calibration. See special table for *Calibration of Thermometers and Thermocouples*. The numbers listed under a temperature-interval heading refer to the compounds with corresponding numerical listings in the table *Physical Constants of Organic Compounds*. Thus, under the temperature interval 186° to 190° C. is listed *No. 31* which is *acetamino-α-naphthol* m. p. 187°C.

Below −140° C.

498; 499; 872; 1022; 1147; 1149; 1239; 1772; 2246; 2904; 3192; 3574; 3604; 3656; 3933; 4061; 4154; 4209; 5084; 5338; 5459; 6203; 6435; 6439

−130° to −139° C.

183; 202; 500; 501; 502; 1018; 1021; 1072; 1074; 1248; 2478; 3015; 3517; 3655; 3659; 3660; 3661; 3957; 5083; 5346; 5425; 6434

−120° to −129° C.

5; 193; 990; 1148; 1156; 1621; 2783; 2911; 3083; 3271; 3515; 3606; 4155; 4199; 4298; 4957; 4958; 5083; 5368; 5400; 6248

−110° to −119° C.

81; 199; 406; 988; 1035; 1057; 1058; 1059; 1103; 1175; 2245; 2415; 2911; 2946; 2989; 3210; 3280; 3513; 3514; 3516; 3518; 3554; 3556; 3607; 3657; 4154; 4161; 4917; 4946; 4950; 5388; 5401; 5424; 5950; 6180; 6404; 6438

−100° to −109° C.

225; 415; 1019; 1036; 1040; 1095; 1096; 1148.1; 1238; 1240; 1247; 1400; 1609; 1680; 1885; 1916; 1923; 2778; 2782; 3062; 3077; 3096; 3631; 4238; 4424; 4827; 4914; 4916; 4947; 4948; 4949; 4956; 4984; 5086; 5087; 5372; 5468

−90° to −99° C.

56; 75; 208; 427; 442; 455; 871; 1031; 1089; 1097; 1142; 1160; 1174; 1599; 1600; 1619; 1898; 1993; 2208; 2214; 2429; 2499; 2973; 3006; 3167; 3185; 3186; 3288; 3357; 3512; 3578; 3603; 3605; 3973; 4087; 4100; 4105; 4147; 4169; 4183; 4242; 4357; 4420; 4429; 4733; 4828; 4953; 4960; 4986; 5345; 5360; 5363; 5365; 5378; 5379; 5392; 5408; 5417; 5418; 5458; 5960; 6348; 6400; 6405; 6426

−80° to −89° C.

130; 135; 146; 151; 204; 220; 459; 1041; 1049; 1064; 1104; 1123; 1173; 1216; 1246; 1378; 1579; 1618; 1684; 2099; 2127; 2473; 2484; 2493; 2560; 2935; 2950; 3007; 3019; 3142; 3158; 3166; 3279; 3285; 3460; 3558; 4058; 4088; 4164; 4170; 4171; 4232; 4269; 4349; 4422; 4735; 4911; 5352; 5369; 5371; 5389; 5418; 6121

−70° to −79° C.

37; 161; 397; 404; 421; 432; 447; 454; 483; 989; 1029; 1034; 1050; 1104; 1125; 1217; 1265; 1379; 1622; 1636; 1825; 1886; 1914; 2051; 2240; 2472; 2769; 2942; 3048; 3049; 3138; 3279; 3418; 3421; 3449; 3474; 3557; 3658; 4123; 4156; 4355; 4965; 4991; 5300; 5366; 5441; 5455; 6155

−60° to −69° C.

89; 1042; 1083; 1141; 1163; 1266; 1380; 1477; 1559; 1623; 1624; 1740; 1821 1884; 1980; 2172; 2232; 2483; 2485; 2502; 2569; 2773; 2784; 2965; 3009; 3029; 3108; 3168; 3194; 3266; 3622; 3628; 3648; 3708; 4182; 4262; 4287; 4352; 4697; 4910; 4985; 5046; 5773

−50° to −59° C.

132; 136; 412; 724; 781; 886; 1039; 1051; 1052; 1132; 1171; 1183; 1237; 1281; 1344; 1563; 1595; 1620; 1631; 1638; 1864; 1870; 1970; 2133; 2184; 2242; 2407; 2443; 2471; 2486; 2792; 3269; 3459; 3624; 4053; 4125; 4163; 4166; 4214; 4289; 4387; 4392; 4428; 4944; 5318; 5380; 5710; 6159; 6255; 6399; 6403; 6407; 6448

−40° to −49° C.

49; 319; 994; 1167; 1179; 1180; 1214; 1218; 1301; 1324; 1467; 1606; 1613; 1663; 1735; 1997; 2128; 2138; 2180; 2184; 2193; 2233; 2749; 2772; 2937; 3010; 3113; 3114; 3195; 3257; 3258; 3263; 3521; 3637; 3944; 4174; 4195; 4694; 4980; 5146; 5317; 5357; 5483; 6254; 6255; 6285; 6457; 6504

−30° to −39° C.

537; 796; 843; 902; 1000; 1069; 1098; 1157; 1209; 1466; 1565; 1588; 1632; 1635; 1679; 1824; 1857; 1865; 1883; 1973; 1994; 2053; 2109; 2121; 2126; 2195; 2238; 2772; 2786; 2863; 2975; 3003; 3026; 3047; 3084; 3318; 3406; 3445; 3527; 3529; 3530; 3543; 3637; 4122; 4184; 4383; 4384; 4711; 4737; 4919; 4959; 4962; 5111; 5466; 5646; 5739; 5786; 5923; 5997; 6122; 6152; 6198; 6206; 6207; 6262; 6349; 6382; 6399; 6406; 6427

−20° to −29° C.

68; 91; 211; 306; 470; 642; 735; 800; 916; 926; 933; 948; 986; 999; 1014; 1028; 1053; 1075; 1140; 1152; 1162; 1185; 1233; 1243; 1340; 1346; 1361; 1377; 1381; 1402; 1442; 1459; 1462; 1474; 1475; 1635; 1783; 1823; 1880; 1905; 1910; 1953; 1981; 2106; 2176; 2195; 2202; 2211; 2213; 2416; 2559; 2718; 2771; 2879; 3008; 3016; 3030; 3035; 3057; 3084; 3089; 3265; 3272; 3377; 3393; 3411; 3423; 3424; 3462; 3519; 3544; 3683; 3846; 3879; 3958; 3968; 4025; 4089; 4196; 4201; 4237; 4265; 4282; 4372; 4407; 4430; 4729; 4736; 4765; 4891; 4920; 4964; 5054; 5068; 5105; 5224; 5241; 5351; 5375; 5429; 5828; 5895; 5993; 5995; 6017; 6042; 6103; 6116; 6119; 6143; 6167; 6253; 6282; 6370; 6454; 6503

−10° to −19° C.

55; 57; 137; 190; 230; 409; 493; 535; 544; 635; 716; 782; 845; 915; 936; 1012; 1060; 1256; 1310; 1314; 1343; 1391; 1473; 1538; 1596; 1952; 2101; 2109; 2123; 2183; 2230; 2399; 2481; 2507; 2517; 2636; 2796; 2831; 2906; 3080; 3090; 3109; 3163; 3170; 3223; 3272; 3444; 3453; 3463; 3520; 3547; 3629; 3638; 3639; 3647; 3705; 3714; 3756; 3846; 3955; 4010; 4080; 4126; 4138; 4172; 4181; 4229; 4265; 4266; 4304; 4311; 4403; 4788; 4844; 4961; 4973; 5011; 5085; 5249; 5341; 5538; 5741; 5751; 5885; 5900; 5905; 5992; 5995; 6157; 6166; 6471

0° to −9° C.

65; 68; 133; 231; 321; 482; 513; 735; 753; 755; 793; 917; 946; 953; 980; 996; 1011; 1023; 1166; 1178; 1186; 1208; 1221; 1224; 1255; 1309; 1334; 1363; 1387; 1501; 1591; 1748; 1812; 1900; 1907; 1928; 2046; 2075; 2097; 2174; 2235; 2239; 2468; 2707; 2713; 2762; 2768; 2856; 2864; 3017; 3115; 3159;

Table 7-3 (*Continued*)
MELTING POINTS OF ORGANIC COMPOUNDS

3161; 3256; 3262; 3273; 3344; 3430; 3597; 3733; 3855; 3976; 4026; 4200; 4219; 4257; 4317; 4367; 4380; 4455; 4839; 4844; 4918; 5269; 5322; 5414; 5482; 5719; 5720; 5827; 6150; 6173; 6199; 6375; 6378; 6482

1° to 10° C.

7; 155; 531; 532; 534; 643; 656; 753; 755; 762; 807; 945; 956; 971; 972; 995; 1015; 1016; 1092; 1136; 1184; 1255; 1333; 1387; 1427; 1436; 1468; 1470; 1511; 1543; 1601; 1640; 1811; 1837; 1858; 1928; 2083; 2097; 2174; 2206; 2210; 2237; 2389; 2393; 2396; 2428; 2450; 2468; 2501; 2547; 2742; 2827; 2907; 2914; 2943; 2968; 3067; 3088; 3093; 3095; 3151; 3196; 3199; 3217; 3218; 3233; 3296; 3301; 3424; 3428; 3504; 3570; 3627; 3956; 4063; 4097; 4108; 4144; 4295; 4370; 4434; 4457; 4623; 4646; 4688; 4796; 4868; 5060; 5061; 5107; 5179; 5302; 5568; 5582; 5643; 5748; 5752; 5897; 5910; 5974; 6243

11° to 20° C.

10; 36; 43; 51; 85; 86; 149; 569; 678; 894; 898; 973; 1003; 1005; 1112; 1129; 1136; 1220; 1276; 1302; 1318; 1319; 1336; 1346; 1348; 1349; 1383; 1546; 1558; 1617; 1645; 1743; 1985; 2085; 2089; 2092; 2147; 2153; 2188; 2200; 2217; 2218; 2224; 2234; 2237; 2331; 2428; 2533; 2554; 2825; 2907; 2909; 2914; 2915; 2939; 2953; 2956; 3023; 3088; 3107; 3214; 3336; 3381; 3428; 3478; 3524; 3528; 3569; 3571; 3572; 3940; 3967; 4096; 4241; 4302; 4303; 4324; 4326; 4369; 4382; 4456; 4673; 4679; 4713; 4838; 4845; 4865; 4887; 4941; 4992; 5027; 5042; 5061; 5063; 5136; 5196; 5221; 5283; 5348; 5433; 5439; 5463; 5513; 5525; 5581; 5658; 5749; 5758; 5759; 5797; 5853; 5894; 5908; 5910; 6078; 6112; 6118; 6135; 6154; 6178; 6201; 6246; 6349; 6371; 6372; 6462; 6477; 6484

21° to 30° C.

51; 62; 106; 452; 471; 481; 488; 489; 507; 511; 530; 569; 724; 789; 808; 813; 823; 861; 949; 980; 1001; 1024; 1037; 1079; 1129; 1130; 1274; 1277; 1332; 1339; 1365; 1367; 1413; 1425; 1426; 1469; 1545; 1584; 1605; 1849; 1919; 1932; 1956; 1983; 1989; 2085; 2090; 2147; 2165; 2227; 2254; 2263; 2278; 2386; 2406; 2466; 2526; 2552; 2719; 2731; 2802; 2822; 2859; 2914; 2927; 2944; 3053; 3099; 3112; 3129; 3219; 3274; 3336; 3422; 3452; 3506; 3508; 3511; 3577; 3643; 3883; 3897; 4006; 4071; 4073; 4079; 4083; 4085; 4108; 4146; 4248; 4330; 4395; 4408; 4451; 4679; 4693; 4740; 4840; 4866; 4938; 4942; 5154; 5186; 5196; 5268; 5353; 5431; 5432; 5434; 5474; 5513; 5540; 5548; 5637; 5772; 5797; 5820; 5891; 5901; 5994; 6043; 6052; 6174; 6178; 6200; 6349; 6376; 6377; 6443; 6469; 6485

31° to 40° C.

73; 79; 212; 243; 254; 531; 615; 643; 653; 756; 786; 798; 818; 831; 832; 893; 927; 930; 969; 973; 1024; 1119; 1205; 1350; 1365; 1373; 1412; 1428; 1443; 1453; 1455; 1457; 1458; 1465; 1505; 1507; 1545; 1547; 1670; 1756; 1787; 1859; 1935; 2009; 2022; 2055; 2084; 2089; 2096; 2148; 2190; 2265; 2267; 2332; 2387; 2400; 2498; 2552; 2615; 2683; 2725; 2803; 2868; 2877; 2892; 2893; 2930; 2956; 3050; 3094; 3124; 3153; 3169; 3182; 3201; 3234; 3255; 3319; 3383; 3422; 3426; 3451; 3466; 3488; 3503; 3577; 3843; 3849; 3876; 3883; 3894; 3898; 3899; 3907; 3934; 3936; 3967; 4006; 4015; 4028; 4065; 4150; 4191; 4239; 4305; 4314; 4329; 4330; 4336; 4374; 4381; 4388; 4394; 4487; 4624; 4637; 4674; 4703; 4710; 4723; 4750; 4767; 4797; 4808; 4848; 4908; 4940; 4983; 5023; 5028; 5166; 5312; 5313; 5323; 5353; 5518; 5633; 5706; 5707; 5708; 5750; 5755; 5795; 5909; 5912; 5991; 6016; 6038; 6040;

6056; 6165; 6174; 6245; 6357; 6440; 6450; 6461; 6487; 6495; 6507

41° to 50° C.

10; 147; 274; 307; 329; 563; 626; 632; 636; 675; 714; 724; 761; 770; 799; 866; 883; 887; 888; 911; 914; 918; 960; 1007; 1119; 1121; 1181; 1189; 1229; 1242; 1275; 1282; 1286; 1299; 1300; 1320; 1338; 1355; 1367; 1413; 1429; 1443; 1456; 1457; 1500; 1508; 1535; 1577; 1578; 1615; 1633; 1659; 1668; 1711; 1789; 1790; 1791; 1792; 1795; 1798; 1918; 1930; 1934; 1937; 2013; 2025; 2042; 2055; 2087; 2091; 2111; 2201; 2226; 2387; 2400; 2425; 2457; 2506; 2510; 2564; 2669; 2712; 2716; 2725; 2767; 2791; 2821; 2823; 2834; 2836; 2855; 2871; 2977; 3025; 3026; 3031; 3100; 3102; 3116; 3119; 3133; 3157; 3187; 3235; 3291; 3299; 3384; 3402; 3426; 3427; 3431; 3488; 3565; 3596; 3598; 3698; 3704; 3753; 3754; 3901; 3950; 3951; 3952; 3954; 4024; 4090; 4279; 4301; 4333; 4446; 4486; 4506; 4526; 4529; 4580; 4636; 4637; 4676; 4682; 4683; 4712; 4762; 4764; 4799; 4843; 4905; 5029; 5062; 5094; 5118; 5123; 5134; 5137; 5153; 5155; 5169; 5204; 5212; 5235; 5247; 5265; 5285; 5312; 5313; 5364; 5416; 5485; 5517; 5534; 5638; 5639; 5728; 5733; 5756; 5761; 5874; 5875; 5896; 5917; 5943; 5973; 5999; 6028; 6095; 6128; 6148; 6168; 6171; 6176; 6190; 6244; 6273; 6288; 6341; 6356; 6392; 6424; 6450; 6468; 6472; 6498; 6502; 6507

51° to 55° C.

88; 102; 267; 268; 410; 563; 568; 612; 613; 802; 823; 877; 878; 888; 916; 929; 1067; 1229; 1278; 1286; 1303; 1354; 1357; 1367; 1376; 1472; 1590; 1615; 1764; 1792; 1834; 1860; 1868; 1937; 1954; 2023; 2098; 2253; 2476; 2521; 2527; 2539; 2549; 2659; 2665; 2668; 2685; 2698; 2717; 2736; 2870; 2921; 2934; 3031; 3291; 3307; 3321; 3347; 3408; 3410; 3431; 3433; 3458; 3821; 3864; 3892; 3893; 3908; 3939; 3951; 4030; 4081; 4090; 4140; 4175; 4261; 4299; 4313; 4333; 4453; 4506; 4517; 4556; 4559; 4580; 4625; 4762; 4807; 4846; 4853; 4897; 4904; 4935; 4998; 5055; 5058; 5074; 5155; 5183; 5238; 5251; 5254; 5266; 5325; 5383; 5477; 5515; 5566; 5569; 5661; 5734; 5747; 5756; 5839; 5938; 6014; 6128; 6134; 6162; 6168; 6190; 6346; 6401; 6494

56° to 60° C.

11; 53; 61; 102; 247; 248; 360; 439; 536; 580; 624; 625; 632; 741; 780; 811; 817; 821; 828; 834; 928; 930; 957; 961; 1002; 1043; 1070; 1165; 1271; 1282; 1299; 1354; 1368; 1369; 1376; 1389; 1395; 1403; 1418; 1490; 1500; 1586; 1593; 1611; 1660; 1802; 1827; 1831; 1860; 2160; 2323; 2346; 2348; 2408; 2544; 2669; 2670; 2674; 2682; 2703; 2714; 2755; 2756; 2833; 2842; 2969; 2979; 3054; 3075; 3101; 3125; 3407; 3415; 3427; 3482; 3507; 3522; 3567; 3662; 3680; 3699; 3736; 3759; 3765; 3856; 3875; 3908; 3996; 4014; 4087; 4261; 4272; 4283; 4285; 4313; 4379; 4450; 4556; 4559; 4570; 4639; 4721; 4731; 4772; 4789; 4798; 4811; 4841; 4889; 4890; 4935; 4939; 4998; 5050; 5079; 5129; 5181; 5183; 5192; 5203; 5216; 5251; 5278; 5382; 5398; 5515; 5541; 5702; 5721; 5790; 5824; 5939; 5954; 6011; 6018; 6022; 6051; 6056; 6128; 6299; 6347; 6441

61° to 65° C.

27; 61; 330; 361; 440; 566; 627; 663; 675; 719; 741; 743; 758; 780; 873; 895; 912; 968; 974; 981; 1045; 1068; 1227; 1283; 1284; 1299; 1306; 1404; 1418; 1422; 1433; 1583; 1586; 1768; 1785; 1828; 1867; 1868; 1933; 1978; 2010; 2014; 2026; 2188; 2252; 2348; 2374; 2413; 2426; 2524; 2546; 2564; 2568; 2650; 2666; 2669; 2670; 2694; 2715; 2720; 2756; 2869; 2895; 2922; 2959; 3054; 3289; 3295; 3431; 3433; 3438; 3461; 3468; 3486; 3509; 3568; 3576;

Table 7-3 (*Continued*)
MELTING POINTS OF ORGANIC COMPOUNDS

3699; 3759; 3875; 4014; 4021; 4070; 4098; 4272; 4323; 4343; 4431; 4483; 4635; 4657; 4686; 4712; 4721; 4724; 4864; 4907; 4937; 4943; 5022; 5026; 5044; 5175; 5206; 5229; 5231; 5258; 5290; 5398; 5435; 5541; 5823; 5954; 5978; 6007; 6011; 6026; 6136; 6147; 6161; 6163; 6270; 6272; 6294; 6379; 6419; 6458; 6466; 6497

66° to 70° C.

14; 27; 44; 103; 281; 314; 367; 371; 567; 584; 599; 657; 691; 704; 758; 767; 879; 1017; 1093; 1279; 1291; 1311; 1326; 1341; 1356; 1404; 1406; 1437; 1497; 1534; 1583; 1672; 1687; 1725; 1766; 1784; 1921; 1965; 2011; 2026; 2064; 2072; 2088; 2120; 2131; 2152; 2156; 2292; 2370; 2405; 2424; 2440; 2512; 2531; 2540; 2600; 2613; 2667; 2692; 2695; 2733; 2735; 2752; 2758; 2810; 2813; 2850; 2866; 2869; 2929; 2972; 2976; 3011; 3110; 3204; 3220; 3410; 3433; 3438; 3495; 3498; 3505; 3654; 3751; 3761; 3767; 3769; 3877; 3927; 4323; 4468; 4488; 4510; 4527; 4576; 4654; 4705; 4768; 4873; 4876; 4909; 5088; 5102; 5159; 5164; 5478; 5541; 5631; 5676; 5743; 5788; 5818; 5940; 5975; 6001; 6026; 6029; 6057; 6130; 6149; 6161; 6164; 6250; 6262; 6270; 6290; 6467; 6476

71° to 75° C.

44; 108; 138; 256; 281; 314; 369; 370; 528; 805; 865; 879; 905; 931; 1122; 1170; 1262; 1297; 1311; 1406; 1430; 1437; 1529; 1557; 1573; 1634; 1672; 1781; 1784; 1882; 1920; 1936; 1979; 2024; 2093; 2120; 2216; 2223; 2292; 2414; 2436; 2551; 2676; 2737; 2752; 2757; 2761; 2866; 2871; 2918; 2972; 3055; 3130; 3204; 3220; 3306; 3323; 3343; 3419; 3420; 3433; 3447; 3464; 3496; 3613; 3825; 3943; 4180; 4312; 4320; 4448; 4618; 4673; 4675; 4678; 4684; 4705; 4726; 4739; 4831; 4869; 4903; 4906; 5048.1; 5080; 5159; 5210; 5217; 5290; 5373; 5631; 5632; 5635; 5647; 5819; 5868; 5883; 6015; 6039; 6145; 6265; 6266; 6276; 6359; 6379; 6465; 6470

76° to 80° C.

19; 23; 104; 171; 221; 260; 281; 309; 313; 574; 623; 649; 688; 742; 759; 819; 857; 865; 879; 885; 905; 919; 931; 1112; 1159; 1273; 1331; 1375; 1430; 1438; 1662; 1723; 1781; 1801; 1861; 1979; 2107; 2115; 2137; 2220; 2273; 2382; 2404; 2446; 2448; 2619; 2676; 2711; 2807; 2832; 3103; 3130; 3188; 3400; 3409; 3414; 3420; 3467; 3468; 3496; 3497; 3613; 3757; 3760; 3825; 3831; 3847; 3853; 4051; 4151; 4221; 4251; 4318; 4458; 4469; 4567; 4574; 4595; 4677; 4719; 4726; 4739; 4773; 4888; 5048.1; 5102; 5142; 5210; 5217; 5228; 5286; 5356; 5373; 5764; 5798; 5829; 5868; 5871; 5883; 5979; 6032; 6055; 6104; 6132; 6145; 6146; 6177; 6226; 6263; 6276; 6326; 6342; 6353; 6428; 6492; 6501

81° to 85° C.

14; 19; 40; 148; 170; 171; 223; 281; 364; 366; 385; 465; 659; 672; 718; 742; 754; 803; 824; 885; 940; 959; 1272; 1307; 1312; 1414; 1418; 1481; 1710; 1781; 1782; 1800; 1801; 1852; 1862; 1876; 1986; 2043; 2066; 2069; 2074; 2086; 2382; 2391; 2404; 2445; 2453; 2496; 2673; 2677; 2708; 2945; 3242; 3248; 3409; 3415; 3484; 3711; 3727; 3768; 3830; 3870; 3874; 3902; 3970; 4135; 4143; 4192; 4221; 4454; 4471; 4515; 4561; 4568; 4595; 4634; 4685; 4739; 4836; 4857; 4882; 5030; 5053; 5067; 5152; 5180; 5200; 5211; 5308; 5347; 5484; 5543; 5675; 5738; 5872; 5881; 5914; 5964; 5977; 6037; 6091; 6109; 6125; 6145; 6177; 6316; 6337; 6418; 6500

86° to 90° C.

18; 258; 269; 271; 281; 377; 441; 573; 658; 754; 784; 795; 804; 812; 881; 934; 963; 1025; 1111; 1242; 1296; 1364; 1366; 1374; 1460; 1607; 1782; 1813; 1826; 2015; 2069; 2150; 2442; 2479; 2489; 2565; 2603; 2616; 2635; 2645; 2679; 2704; 2708; 2763; 2945; 3070; 3105; 3211; 3221; 3322; 3475; 3477; 3484; 3611; 3667; 3715; 3750; 3790; 3844; 3851; 3889; 3926; 4022; 4049; 4074; 4092; 4115; 4127; 4211; 4212; 4220; 4252; 4273; 4449; 4470; 4592; 4680; 4695; 4701; 4739; 4834; 4883; 5024; 5053; 5059; 5170; 5176; 5209; 5230; 5279; 5284; 5347; 5354; 5423; 5543; 5549; 5564; 5640; 5641; 5675; 5696; 5702; 5731; 5732; 5834; 5964; 6003; 6010; 6091; 6094; 6096; 6145; 6271; 6332; 6420; 6463

91° to 95° C.

2; 3; 22; 42; 281; 333; 366; 369; 390; 466; 476; 494; 607; 633; 697; 750; 812; 822; 925; 934; 947; 977; 1172; 1192; 1215; 1242; 1280; 1330; 1352; 1486; 1510; 1544; 1604; 1721; 1872; 1875; 2067; 2248; 2385; 2442; 2581; 2582; 2593; 2623; 2632; 2671; 2678; 2679; 2838; 2955; 3043; 3068; 3070; 3183; 3221; 3303; 3322; 3470; 3523; 3550; 3611; 3616; 3722; 3766; 3774; 3790; 3814; 3829; 3900; 3964; 4022; 4099; 4131; 4189; 4211; 4284; 4471; 4510; 4516; 4518; 4569; 4588; 4597; 4671; 4680; 4695; 4813; 4852; 5101; 5115; 5170; 5182; 5248; 5415; 5449; 5588; 5590; 5603; 5616; 5636; 5678; 5760; 5834; 5837; 5980; 6012; 6030; 6031; 6033; 6084; 6090; 6145; 6191; 6234; 6237; 6337; 6358; 6491

96° to 100° C.

8; 11; 24; 48; 120; 162; 228; 237; 252; 281; 310; 318; 333; 640; 686; 721; 743; 820; 880; 925; 977; 1025; 1120; 1138; 1213; 1280; 1325; 1358; 1461; 1486; 1516; 1580; 1604; 1762; 1794; 1829; 1890; 1927; 2065; 2309; 2338; 2366; 2385; 2421; 2581; 2617; 2632; 2693; 2740; 2820; 2960; 2971; 3111; 3134; 3154; 3202; 3240; 3243; 3276; 3324; 3376; 3412; 3470; 3480; 3523; 3611; 3612; 3616; 3722; 3748; 3772; 3773; 3775; 3962; 3998; 4064; 4067; 4319; 4435; 4465; 4489; 4571; 4597; 4681; 4802; 4849; 4855; 5045; 5081; 5095; 5101; 5115; 5130; 5240; 5415; 5422; 5555; 5609; 5677; 5715; 5760; 5791; 5869; 5989; 6005; 6033; 6053; 6054; 6089; 6108; 6131; 6133; 6145; 6226; 6240; 6296; 6308; 6314; 6446; 6447

101° to 105° C.

11; 48; 54; 81.1; 156; 209; 281; 572; 661; 686; 738; 840; 904; 1102; 1213; 1223; 1287; 1288; 1409; 1506; 1580; 1713; 1733; 1955; 2044; 2063; 2068; 2081; 2276; 2285; 2312; 2340; 2371; 2487; 2508; 2523; 2577; 2649; 2806; 2853; 3206; 3298; 3303; 3348; 3351; 3611; 3616; 3700; 3707; 3931; 3942; 3953; 3964; 4017; 4049; 4086; 4176; 4256; 4271; 4291; 4385; 4414; 4452; 4466; 4562; 4587; 4596; 4670; 4696; 4742; 4777; 4783; 4850; 4945; 5007; 5115; 5131; 5151; 5205; 5256; 5309; 5310; 5358; 5512; 5584; 5594; 5836; 5879; 5890; 5968; 5972; 5982; 6034; 6059; 6060; 6089; 6120; 6258; 6296; 6315; 6380; 6446; 6447; 6496

106° to 110° C.

26; 41; 54; 74; 84; 142; 233; 238; 246; 259; 270; 281; 328; 363; 368; 565; 591; 596; 628; 658; 692; 723; 766; 791; 809; 879; 967; 975; 1020; 1044; 1190; 1223; 1287; 1295; 1386; 1410; 1460; 1656; 1666; 1713; 1733; 1874; 2012; 2019; 2129; 2225; 2269; 2272; 2285; 2313; 2340; 2509; 2534; 2562; 2572; 2580; 2583; 2605; 2841; 2952; 2958; 3180; 3245; 3359; 3566; 3723; 3734; 3770; 3812; 3916; 3991; 4048; 4068; 4113; 4276; 4306; 4359; 4438; 4503;

Table 7-3 (*Continued*)
MELTING POINTS OF ORGANIC COMPOUNDS

4504; 4510; 4557; 4560; 4594; 4613; 4640; 4667; 4806; 4835; 4850; 4854; 4894; 4998; 5020; 5025; 5207; 5232; 5453; 5497; 5634; 5836; 5886; 5921; 5959; 5983; 6046; 6258; 6296; 6297; 6298; 6402

111° to 115° C.

17; 69; 142; 215; 246; 565; 687; 698; 705; 725; 760; 763; 790; 809; 870; 909; 944; 952; 1004; 1020; 1168; 1187; 1396; 1483; 1666; 1713; 1775; 1853; 1874; 2018; 2129; 2222; 2228; 2313; 2367; 2375; 2509; 2627; 2647; 2676; 2811; 2835; 2841; 2970; 3072; 3082; 3162; 3246; 3362; 3587; 3611; 3712; 3728; 3776; 3885; 4011; 4094; 4109; 4137; 4213; 4290; 4307; 4359; 4503; 4510; 4521; 4530; 4613; 4619; 4698; 4725; 4795; 4803; 4899; 5015; 5133; 5160; 5232; 5280; 5292; 5520; 5542; 5831; 5882; 5884; 5906; 5942; 5948; 5983; 6046; 6111; 6114; 6275; 6296; 6301; 6313; 6333; 6350; 6360; 6411; 6421; 6422; 6499

116° to 120° C.

384; 518; 533; 645; 725; 771; 790; 889; 921; 1168; 1187; 1193; 1195; 1285; 1294; 1405; 1407; 1419; 1499; 1514; 1581; 1698; 1713; 1873; 2007; 2016; 2047; 2062; 2071; 2247; 2373; 2375; 2431; 2461; 2586; 2602; 2675; 2878; 3072; 3246; 3259; 3326; 3472; 3494; 3610; 3617; 3675; 3728; 3735; 3887; 3949; 3979; 3992; 4008; 4055; 4094; 4213; 4334; 4390; 4503; 4510; 4521; 4525; 4563; 4572; 4575; 4591; 4601; 4641; 4658; 4661; 4668; 4687; 4699; 4706; 4734; 4858; 5004; 5021; 5099; 5101; 5143; 5201; 5280; 5305; 5385; 5542; 5549; 5553; 5626; 5652; 5657; 5681; 5714; 5730; 5796; 5833; 5867; 5882; 5884; 5942; 5981; 6080; 6092; 6093; 6097; 6218; 6233; 6275; 6291; 6307; 6309; 6361; 6363; 6421; 6499

121° to 125° C.

272; 347; 533; 660; 674; 711; 815; 901; 921; 1164; 1294; 1665; 1730; 1786; 1850; 2029; 2030; 2048; 2071; 2185; 2322; 2342; 2349; 2372; 2467; 2550; 2579; 2601; 2620; 2652; 2681; 2700; 2748; 2828; 2845; 3215; 3225; 3254; 3259; 3260; 3339; 3702; 3716; 3718; 3752; 3815; 3866; 3945; 4047; 4235; 4321; 4344; 4441; 4471; 4490; 4507; 4520; 4565; 4581; 4593; 4603; 4658; 4676; 4753; 5078; 5099; 5144; 5201; 5214; 5237; 5252; 5324; 5334; 5526; 5546; 5557; 5577; 5642; 5652; 5660; 5697; 5880; 5961; 6080; 6086; 6158; 6302; 6310; 6323; 6391; 6453; 6479

126° to 130° C.

20; 111; 197; 249; 273; 311; 312; 365; 542; 619; 644; 657; 674; 713; 750; 874; 943; 951; 958; 962; 1061; 1164; 1169; 1435; 1665; 1709; 1714; 1717; 1758; 2324; 2347; 2369; 2388; 2418; 2449; 2459; 2511; 2558; 2579; 2587; 2750; 2894; 2999; 3200; 3215; 3254; 3473; 3485; 3575; 3694; 3713; 3717; 3777; 3795; 3813; 3854; 3993; 3995; 3997; 4002; 4227; 4361; 4409; 4415; 4425; 4442; 4520; 4581; 4615; 4707; 4756; 4775; 4781; 4847; 4851; 4874; 4896; 5194; 5202; 5215; 5282; 5333; 5376; 5544; 5546; 5558; 5577; 5652; 5660; 5662; 5679; 5683; 5697; 5825; 5862; 5873; 5926; 5944; 5955; 6113; 6137; 6228; 6256; 6292; 6328; 6416

131° to 135° C.

25; 47; 50; 96; 115; 233; 257; 262; 273; 299; 542; 606; 710; 717; 836; 874; 922; 924; 1061; 1194; 1322; 1397; 1498; 1503; 1513; 1616; 1696; 1706; 1719; 1759; 1779; 1957; 2017; 2159; 2309; 2310; 2347; 2417; 2460; 2474; 2578; 2633; 2651; 2696; 2727; 2916; 3041; 3212; 3340; 3359; 3363; 3619; 3685; 3694; 3741; 3749; 3854; 3867; 3993; 3995; 4002; 4012; 4093; 4133; 4227; 4274; 4292; 4415; 4446;

4523; 4528; 4615; 4700; 4727; 4833; 4856; 5282; 5319; 5493; 5605; 5625; 5652; 5825; 5862; 6088; 6156; 6160; 6216; 6225; 6335; 6386; 6396; 6409

136° to 140° C.

59; 96; 101; 115; 257; 529; 592; 637; 703; 717; 729; 874; 884; 951; 1038; 1139; 1482; 1503; 1657; 1701; 1733; 1829; 1851; 1962; 2119; 2181; 2310; 2339; 2345; 2352; 2354; 2365; 2375; 2419; 2525; 2532; 2578; 2589; 2590; 2633; 2641; 2721; 2727; 2732; 2798; 2839; 2916; 3240; 3346; 3367; 3369; 3370; 3457; 3500; 3601; 3701; 3738; 3842; 3916; 3966; 4460; 4582; 4602; 4610; 4664; 4672; 4771; 4856; 5173; 5220; 5260; 5370; 5384; 5486; 5649; 5735; 5809; 5817; 5857; 5902; 5927; 5937; 5971; 6002; 6217; 6224; 6239; 6317; 6325; 6409

141° to 145° C.

58; 82; 95; 213; 264; 298; 512; 564; 598; 614; 689; 728; 775; 825; 858; 951; 1139; 1321; 1327; 1502; 1701; 1722; 1855; 1962; 1966; 2003; 2257; 2321; 2345; 2360; 2409; 2422; 2458; 2557; 2574; 2628; 2637; 2646; 2721; 2776; 2798; 2873; 3058; 3104; 3162; 3231; 3240; 3308; 3312; 3313; 3341; 3350; 3367; 3380; 3691; 3737; 3794; 3832; 3966; 4000; 4048; 4112; 4113; 4234; 4268; 4270; 4371; 4460; 4513; 4582; 4611; 4633; 4643; 4661; 4664; 4779; 4780; 4782; 4807; 4824; 4884; 5167; 5291; 5326; 5384; 5550; 5561; 5649; 5676; 5677; 5727; 5817; 5841; 5937; 6044; 6049; 6058; 6072; 6126; 6172; 6217; 6224; 6241; 6334; 6336; 6339; 6388; 6493

146° to 150° C.

34; 97; 99; 180; 213; 227; 229; 295; 375; 506; 561; 699; 727; 745; 775; 835; 837; 841; 844; 906; 1287; 1432; 1441; 1480; 1484; 1502; 1592; 1657; 1669; 1776; 1814; 1817; 1841; 1922; 1990; 2078; 2144; 2261; 2318; 2325; 2344; 2470; 2475; 2660; 2680; 2697; 2723; 2754; 2798; 3240; 3290; 3308; 3600; 3618; 3794; 3796; 3802; 3811; 3814; 3917; 3946; 3960; 3971; 4036; 4054; 4234; 4259; 4280; 4460; 4467; 4598; 4620; 4626; 4663; 4669; 4689; 4704; 4753; 4778; 4832; 4895; 4900; 4993; 5018; 5032; 5096; 5253; 5271; 5272; 5481; 5550; 5585; 5659; 5810; 5817; 5843; 5863; 5911; 5988; 6044; 6072; 6073; 6232; 6259; 6304; 6339; 6449; 6493

151° to 155° C.

28; 153; 229; 367; 375; 495; 505; 506; 561; 571; 598; 646; 720; 722; 727; 745; 792; 835; 907; 1008; 1188; 1236; 1287; 1441; 1464; 1480; 1515; 1669; 1678; 1816; 1950; 1961; 2002; 2070; 2144; 2280; 2311; 2318; 2584; 2591; 2739; 2747; 2754; 2798; 3216; 3247; 3290; 3361; 3440; 3673; 3729; 3762; 3763; 3787; 3802; 3839; 3960; 4034; 4423; 4479; 4524; 4589; 4616; 4638; 4645; 4659; 4766; 4769; 4809; 4837; 4895; 5032; 5073; 5100; 5156; 5244; 5355; 5386; 5476; 5532; 5565; 5586; 5695; 5711; 5838; 5844; 5870; 5892; 6045; 6107; 6229; 6261; 6330; 6385; 6389; 6480

156° to 160° C.

46; 93; 229; 289; 292; 362; 389; 495; 505; 510; 545; 570; 608; 629; 646; 650; 668; 677; 722; 777; 792; 801; 923; 997; 1009; 1048; 1328; 1371; 1560; 1718; 1763; 1793; 1950; 2001; 2070; 2256; 2318; 2341; 2344; 2355; 2358; 2574; 2592; 2624; 2653; 2709; 2798; 2843; 2844; 2889; 3290; 3366; 3678; 3729; 3742; 3778; 3848; 3869; 3872; 3960; 4001; 4063; 4267; 4277; 4322; 4423; 4440; 4479; 4480; 4579; 4738; 4815; 4817; 4863; 4881; 4902; 4934; 4936; 5031; 5056; 5113; 5140; 5158; 5161; 5242; 5255; 5314; 5476; 5510; 5533; 5539; 5559; 5651; 5688; 5695; 5700; 5877; 5970; 6238; 6387

Table 7-3 (*Continued*)
MELTING POINTS OF ORGANIC COMPOUNDS

161° to 165° C.

12; 60; 107; 140; 163; 292; 324; 336; 527; 545; 662; 694; 702; 715; 737; 744; 751; 765; 773; 839; 862; 864; 993; 997; 1201; 1316; 1353; 1513; 1542; 1574; 1833; 1960; 2079; 2258; 2270; 2318; 2326; 2341; 2462; 2480; 2579; 2610; 2738; 2746; 2798; 2880; 2889; 3059; 3305; 3328; 3617; 3620; 3678; 3692; 3703; 3740; 3799; 3836; 3848; 3865; 3880; 3928; 3960; 4130; 4250; 4322; 4480; 4614; 4722; 4752; 4776; 4820; 4823; 4830; 4859; 4893; 4902; 5037; 5161; 5174; 5218; 5288; 5480; 5516; 5536; 5574; 5671; 5700; 5803; 5835; 5847; 5918; 6071; 6123; 6138; 6139; 6140; 6306; 6331; 6338

166° to 170° C.

35; 76; 94; 122; 163; 185; 186; 527; 545; 547; 587; 620; 662; 702; 708; 736; 744; 768; 774; 787; 827; 854; 882; 993; 1202; 1235; 1423; 1542; 1570; 1726; 1727; 1877; 1959; 2004; 2209; 2270; 2314; 2318; 2430; 2432; 2462; 2463; 2464; 2639; 2640; 2741; 2746; 2798; 2867; 3013; 3229; 3230; 3236; 3328; 3329; 3342; 3349; 3562; 3590; 3609; 3617; 3739; 3797; 3833; 3928; 3960; 4004; 4250; 4308; 4439; 4502; 4752; 4759; 4859; 4880; 5098; 5104; 5110; 5113; 5239; 5303; 5560; 5597; 5598; 5601; 5645; 5663; 5671; 5691; 5847; 5888; 6102; 6123; 6169; 6242; 6257; 6264; 6321; 6329; 6390; 6488; 6490

171° to 175° C.

66; 70; 100; 110; 121; 251; 265; 291; 293; 297; 323; 345; 587; 609; 695; 764; 785; 1199; 1230; 1290; 1298; 1479; 1504; 1513; 1569; 1626; 1655; 1658; 1681; 1707; 1815; 1835; 1847; 2209; 2350; 2420; 2430; 2432; 2439; 2576; 2604; 2640; 2662; 2710; 2779; 2814; 2824; 2900; 3230; 3674; 3739; 3797; 3837; 3850; 3895; 3960; 3985; 4136; 4197; 4275; 4373; 4410; 4436; 4439; 4650; 4871; 4875; 5071; 5091; 5103; 5104; 5112; 5113; 5171; 5172; 5236; 5332; 5535; 5597; 5601; 5602; 5610; 5644; 5645; 5663; 5693; 5701; 5852; 6102; 6211; 6230; 6329; 6390; 6425; 6451; 6488; 6490

176° to 180° C.

16; 121; 134; 166; 291; 294; 603; 648; 695; 731; 749; 841; 978; 1191; 1254; 1290; 1298; 1323; 1385; 1392; 1523; 1626; 1729; 1760; 2073; 2281; 2296; 2350; 2353; 2420; 2607; 2608; 2686; 2745; 2814; 2816; 2837; 2846; 3203; 3275; 3305; 3358; 3479; 3595; 3674; 3827; 3834; 3835; 3841; 3850; 3856; 3920; 3960; 4033; 4091; 4110; 4129; 4136; 4190; 4197; 4376; 4444; 4475; 4632; 4642; 4757; 5001; 5332; 5336; 5390; 5470; 5663; 5716; 5920; 5984; 6050; 6102; 6170; 6179; 6227; 6312; 6423

181° to 185° C.

1; 21; 124; 125; 143; 240; 261; 290; 349; 509; 548; 603; 631; 731; 732; 746; 850; 932; 941; 1253; 1528; 1575; 1676; 1742; 1842; 1964; 2573; 2629; 2656; 2729; 2743; 2766; 2840; 3239; 3244; 3261; 3358; 3502; 3835; 3873; 3920; 3960; 4069; 4110; 4376; 4406; 4473; 4481; 4607; 4628; 4732; 4741; 4760; 4818; 5072; 5138; 5147; 5157; 5331; 5578; 5663; 5865; 5919; 5920; 6050; 6235; 6320; 6324; 6340; 6423

186° to 190° C.

29; 31; 63; 124; 125; 240; 266; 349; 521; 548; 603; 604; 683; 732; 899; 941; 964; 1198; 1252; 1289; 1536; 1676; 1700; 1720; 1845; 1942; 2357; 2413; 2575; 2588; 2612; 2760; 2882; 2910; 3330; 3358; 3375; 3378; 3502; 3564; 3608; 3664; 3710; 3782; 3786; 3881; 3960; 4038; 4286; 4293; 4406; 4573; 4760; 4818; 4872; 5006; 5057; 5090; 5108; 5307; 5497; 5570; 5587; 5626; 5653; 5663; 5666; 5812;

5842; 5848; 6047; 6048; 6098; 6110; 6137; 6289; 6305; 6311; 6340; 6452

191° to 195° C.

30; 63; 141; 179; 217; 228; 278; 388; 521; 586; 595; 603; 638; 752; 849; 851; 859; 1196; 1384; 1485; 1495; 1532; 1700; 1797; 1847; 2020; 2021; 2082; 2124; 2301; 2316; 2335; 2363; 2503; 2528; 2598; 2634; 2643; 2655; 2744; 2815; 2917; 3249; 3378; 3583; 3710; 3724; 3782; 3788; 3823; 3960; 3989; 4249; 4297; 4485; 4577; 4818; 4826; 4892; 5005; 5199; 5274; 5307; 5490; 5495; 5522; 5583; 5664; 5668; 5685; 5692; 5830; 5842; 5851; 5854; 6048; 6321; 6368; 6395; 6452

196° to 200° C.

71; 160; 228; 276; 357; 388; 520; 562; 603; 638; 734; 1182; 1196; 1370; 1582; 1702; 1830; 1948; 2299; 2316; 2317; 2319; 2327; 2335; 2528; 2621; 2634; 2657; 2917; 2982; 3126; 3226; 3371; 3490; 3588; 3677; 3696; 3747; 3820; 3861; 3904; 3919; 3960; 4249; 4278; 4644; 4651; 4708; 4933; 5069; 5070; 5243; 5307; 5572; 5608; 5614; 5687; 5699; 5736; 5842; 5851; 5899; 5907; 6395

201° to 205° C.

33; 71; 279; 280; 284; 336; 388; 546; 577; 603; 605; 610; 706; 900; 1197; 1492; 1509; 1521; 1944; 1963; 2299; 2315; 2335; 2402; 2453; 2609; 2611; 2865; 3253; 3373; 3499; 3676; 3686; 3730; 3745; 3747; 3789; 3803; 3821; 3911; 3919; 3947; 4243; 4484; 4604; 4612; 4644; 4662; 4708; 5039; 5193; 5243; 5273; 5307; 5571; 5604; 5608; 5676; 5686; 5690; 5713; 5736; 5757; 5855; 5858; 5907; 6127; 6141; 6394

206° to 210° C.

90; 157; 317; 327; 388; 508; 541; 549; 630; 651; 701; 706; 860; 1308; 1317; 1492; 1517; 1521; 1531; 1533; 1703; 1805; 1940; 1941; 1948; 2080; 2295; 2328; 2334; 2335; 2447; 2556; 2663; 2887; 3780; 3826; 3838; 3840; 3925; 4128; 4438; 4608; 4649; 4660; 4877; 5013; 5075; 5097; 5307; 5487; 5511; 5690; 5717; 5860; 6054; 6081; 6082; 6219; 6220; 6295; 6319; 6417

211° to 215° C.

157; 388; 549; 594; 611; 630; 666; 681; 706; 778; 1315; 1394; 1517; 1521; 1548; 1686; 1948; 1987; 2279; 2282; 2335; 2401; 2522; 2626; 2661; 2663; 2705; 2981; 3238; 3373; 3476; 3492; 3584; 3746; 3784; 3819; 3935; 4438; 4590; 4649; 4774; 4816; 4825; 5016; 5075; 5307; 5487; 5600; 5802; 5860; 6082; 6219; 6260; 6295; 6322; 6366

216° to 220° C.

119; 123; 275; 351; 393; 538; 549; 647; 681; 684; 1521; 1541; 1549; 1585; 1819; 1951; 2337; 2356; 2497; 2522; 2625; 2638; 2663; 2690; 2848; 2981; 3309; 3476; 3584; 3592; 3679; 3684; 3746; 3783; 4360; 4578; 4590; 4825; 5162; 5195; 5234; 5281; 5307; 5321; 5335; 5473; 5487; 5504; 5562; 5712; 5802; 5813; 5846; 5849; 5893; 5996; 6082; 6219; 6221; 6231; 6260; 6295; 6303

221° to 225° C.

232; 263; 358; 576; 681; 707; 1200; 1521; 1778; 1803; 2294; 2542; 2545; 2618; 2625; 2687; 2885; 2887; 2900; 3237; 3372; 3672; 3679; 3684; 3697; 3822; 3852; 3923; 3988; 4005; 4035; 4112; 4179; 4812; 4819; 5017; 5051; 5289; 5335; 5487; 5492; 5579; 5712; 5813; 5846; 5861; 5889; 5893; 5933; 6061; 6106; 6208; 6231; 6236; 6369; 6445

Table 7-3 (*Continued*)
MELTING POINTS OF ORGANIC COMPOUNDS

226° to 230° C.

152; 348; 576; 588; 671; 681; 707; 847; 853; 1250; 1521; 1704; 1731; 1840; 2110; 2305; 2394; 2542; 2543; 2545; 2618; 2687; 2688; 2765; 3237; 3489; 3559; 3563; 3720; 3817; 3852; 3959; 4005; 4106; 4627; 4629; 4754; 4755; 4758; 4812; 4878; 5051; 5306; 5327; 5492; 5500; 5508; 5579; 5612; 5889; 5893; 5932; 5933; 5952; 5998; 6083; 6106; 6231; 6267; 6369

231° to 235° C.

32; 167; 181; 374; 515; 576; 588; 681; 700; 707; 853; 1250; 1496; 1506; 1608; 1704; 1804; 1810; 1818; 2305; 2383; 2394; 2622; 3205; 3437; 3563; 3720; 3726; 3779; 3817; 3818; 3824; 3852; 3990; 4035; 4039; 4368; 4411; 4648; 4754; 4999; 5052; 5109; 5122; 5150; 5274; 5330; 5488; 5528; 5645; 5742; 5821; 5826; 5887; 5947; 6083; 6087; 6099; 6223; 6318

236° to 240° C.

32; 98; 181; 543; 575; 600; 618; 671; 680; 681; 700; 1250; 1496; 1704; 1708; 1804; 2320; 2333; 2437; 2441; 2490; 2599; 2759; 3227; 3437; 3443; 3668; 3779; 3800; 3824; 3852; 3922; 3924; 3990; 4050; 4260; 4609; 4666; 4860; 5122; 5197; 5287; 5329; 5499; 5528; 5545; 5672; 5742; 6239

241° to 245° C.

98; 129; 168; 234; 539; 618; 621; 680; 1231; 1329; 1675; 1697; 1809; 1832; 1881; 2333; 2437; 2455; 2490; 2706; 2764; 3668; 3670; 3800; 3804; 3965; 4045; 4142; 4391; 4459; 4599; 4666; 4707; 5036; 5197; 5507; 5529; 5545; 5623; 5656; 5696; 6000

246° to 250° C.

6; 126; 165; 173; 234; 616; 621; 670; 680; 747; 875; 1489; 1780; 1881; 2077; 2262; 2333; 2437; 2490; 2734; 2764; 3309; 3325; 3498; 3560; 3668; 3671; 3980; 4019; 4101; 4117; 4743; 4761; 4879; 5010; 5337; 5491; 5498; 5506; 5529; 5611; 5613; 5934; 6181

251° to 260° C.

127; 244; 335; 356; 539; 589; 641; 848; 908; 1251; 1493; 1522; 1540; 1602; 1603; 1627; 1628; 1705; 1780; 1878; 1943; 2077; 2333; 2351; 2362; 2437; 2454; 2691; 2847; 2874; 2884; 2886; 2888; 2903; 3252; 3254; 3302; 3325; 3374; 3560; 3671; 3689; 3781; 3791; 3888; 4019; 4037; 4101; 4117; 4405; 4497; 4498; 4532; 4652; 4794; 4821; 4932; 5264; 5337; 5501; 5531; 5589; 5665; 5667; 5678; 5745; 5811; 5856; 5876; 5934; 5941; 6182; 6213; 6364

261° to 270° C.

174; 288; 355; 589; 682; 1478; 1690; 1692; 1777; 1808; 1939; 1946; 2122; 2293; 2298; 2343; 2362; 2437; 2451; 2452; 2594; 2597; 2654; 2849; 2851; 2891; 2898; 2903; 3315; 3442; 3785; 3808; 3809; 3816; 3882; 3888; 4141; 4412; 4544; 4605; 4606; 4770; 4822; 5066; 5121; 5494; 5509; 5514; 5531; 5724; 6209; 6327

271° to 280° C.

109; 174; 286; 287; 355; 359; 525; 540; 582; 589; 590; 748; 1494; 2300; 2343; 2361; 2395; 2614; 2883; 2888; 2898; 2902; 3439; 3442; 3591; 3690; 3828; 3978; 4412; 4474; 4478; 4606; 5148; 5149; 5316; 5489; 5537; 5673; 5722; 5724; 5725; 5801; 5866; 6249; 6398

281° to 290° C.

109; 164; 277; 352; 355; 359; 519; 525; 540; 550; 552; 589; 667; 1555; 1807; 1839; 1947; 2297; 2302; 2491; 2664; 2883; 2890; 2902; 2932; 3314; 3669; 3801; 3806; 3937; 4062; 4600; 4709; 5149; 5163; 5198; 5530; 5563; 5722; 5746; 5806; 5859; 5864; 5866; 6365; 6397; 6398

291° to 300° C.

139; 158; 159; 282; 285; 354; 359; 387; 391; 525; 551; 690; 772; 846; 979; 1539; 1838; 1967; 2336; 2384; 2585; 2664; 2724; 2854; 2890; 2966; 3250; 3586; 4443; 5049; 5475; 5726; 6367; 6397

301° to 310° C.

139; 176; 241; 245; 283; 350; 354; 394; 593; 669; 709; 730; 1677; 1694; 1820; 1838; 2644; 2817; 2854; 3586; 3725; 3792; 3805; 4118; 4931; 5556; 5573; 5744; 6212

311° to 320° C.

392; 395; 617; 669; 709; 1691; 1693; 1694; 2596; 2854; 3561; 4118; 5496; 5669; 5680; 5807; 5816; 6367

321° to 330° C.

602; 1671; 1693; 1695; 2306; 2722; 2854; 4413; 5502; 5816; 6205

331° to 340° C.

601; 602; 1629; 1671; 1695; 2303; 2304; 2595; 2854; 3589; 3807; 6383

341° to 360° C.

145; 601; 779; 2689; 2854; 3589; 4335; 5276; 5956

361° to 380° C.

685; 1594; 4417; 4418; 4562; 5299; 5800; 5805; 6214; 6215; 6362

Above 380° C.

676; 3857; 5009; 5277; 5822; 6384

Table 7-4

PHYSICAL CONSTANTS OF ORGANIC COMPOUNDS

See also special tables on Fats and Oils, Alkaloids, Glucosides, Resins, Nomenclature, Organic Radicals, and Common or Trade Names of Chemicals.

Names of the compounds in the table below are arranged alphabetically. The names were selected in such a manner as to bring isomers and closely related compounds in close sequence to enable a convenient means of comparison. In general, names with prefixes such as *iso* or *tertiary* are placed with the normal compounds; thus, *isopropyl* will be found listed under *propyl*. Indented words are suffixes of the word in bold face type immediately preceding; thus, compound No. 8 is acetaldehyde ammonia. No compound is listed more than once in the table and each compound is given a number.

Synonyms are listed in two ways. In the column headed "Synonym" will be found a different name for the same compound. At the bottom of each page is an alphabetical listing for compounds which are to be found in the main body of the table but under a different name; the number following the name refers to the numerical place of this compound in the table; thus, *acetaldazine 9* indicates that this compound is the 9th compound in the table where it will be found listed under the name *acetaldehyde azine*. Alkaloids which are listed in another table are indicated by *alkd.*; thus, *quinine, cf. alkd.*, means that this compound does not appear in the table below but will be found in a special table of alkaloids. Similarly, glucosides are indicated by *glcde.*; thus, *absinthin, cf. glcde.* indicates that the compound is listed in the table of glucosides.

Formulas are presented in a semi-structural form. To indicate the position of substituents on aromatic rings, position numbers are given in the same sequence as they appear in the formula; thus, in the formula "$HO \cdot C_6H_3(CHO) \cdot CO_2H$ (2;5,1)", the CO_2H group is at position 1, the CHO at 5 and the OH at 2.

Beil. Ref. In the column so headed will be found the reference to the volume and page numbers of the 4th edition of Beilstein: *Handbuch der Organischen Chemie* (published by Springer, Berlin). Where the Roman numeral, which indicates the volume number, is preceded by an asterisk, reference is made to the first supplementary volumes. Thus, "*I-23", indicates that the compound will be found listed in the supplement to volume 1 and on page 23.

Formula Weights are based upon the International Atomic Weights of 1961 and are computed to the nearest hundredth.

Crystalline Form and Color. In addition to the crystalline form and color, the solvent used in purification is often given; thus, "rhb./al." indicates that rhombic

Abbreviations used in the table

a., acid
abs., absolute
ac., acetic acid
Ac, acetyl (i. e. CH_3CO)
act., acetone
al., ethyl alcohol
alk., alkali (i. e. aqueous NaOH or KOH)
alkd., an alkaloid; see alkaloids
Am, amyl (i. e. C_5H_{11})
amor., amorphous
anh., anhydrous
anhyd., anhydride
art., artificial
atm., atmosphere
aq., aqueous; water

Beil. Ref., reference to 4th ed. Beilstein
b., blue
bl., black
b. p., boiling point
brn., brown
bz., benzene, C_6H_6
c., cold
carb., carbonate
chl., chloroform, $CHCl_3$
col., colorless or white
conc., concentrated
cr., crystals or crystalline
d, dextro-rotatory
d., decomposes or decomposed
delq., deliquescent

dil., dilute
diss., dissociates
dl, inactive (i. e. 50% d and 50% l)
off., effloresces
et., ether ($C_2H_5OC_2H_5$)
Et, ethyl (C_2H_5)
expl., explodes
fl., flakes
gly., glycerol
gn., green
gr., gray
h., hot
hex., hexagonal
hyg., hygroscopic
i., insoluble

No.	Name	Synonym	Formula	Beil. Ref.	Formula Weight
1	**Abietic acid**	sylvic acid	$C_{20}H_{30}O_2$	....	302.46
2	**Acenaphthene**	naphthylene ethylene	$C_{10}H_6(CH_2)_2$	V-586	154.21
3	**Acenaphthylene**	....	$C_{12}H_8$	V-625	152.20

Table 7-4 (*Continued*)
PHYSICAL CONSTANTS OF ORGANIC COMPOUNDS

crystals were obtained when the compound was crystallized from alcohol. Where no color is stated it may generally be inferred that the material is colorless.

Specific Gravity values are given at room temperatures (15 to 20°C.) unless otherwise indicated by the small figures which follow the value; thus, "$1.069^{95°}_{35}$" indicates a specific gravity of 1.069 for the substance at 95°C. referred to water at 95°C. Specific gravity of gases with the notation "(A)" or "(D)" indicate the density of the gas referred to (A) air = 1 or (D) hydrogen = 1 respectively. See also special tables.

Melting Point is recorded in a certain case as "82 d." and in some other case as "d. 82", the distinction being made in this manner to indicate that the former is a melting point with decomposition at 82°C., while in the latter decomposition only occurs at 82°C. Where a value such as "$-2H_2O$, 82" is given it indicates loss of two moles of water per formula weight of the compound at a temperature of 82°C.

Boiling Point is given at atmospheric pressure (760 mm of mercury) unless otherwise indicated; thus, "82^{15mm}" indicates the boiling point is 82°C. when the pressure is 15 mm.

Solubility is given in parts by weight (of the formula shown at the extreme left) per 100 parts by weight of the solvent; the small exponent indicates the temperature in degrees C. In the case of gases the solubility is often expressed in some manner as "$5^{10°}$ cc." which indicates that at 10°C., 5 cc. of the gas are soluble in 100 grams of the solvent.

References. The information given in the table has been collected mainly from the following sources:—

Beilstein: *Handbuch der Organischen Chemie*, 4th edition, Published by Springer, Berlin.

The Merck Index, 5th edition, Published by Merck and Co., Inc., Rahway, N. J.

Heilbron: *Dictionary of Organic Compounds*, Published by Oxford Press, New York.

Tables Annuelles Internationales de Constants et Donnes Numeriques. Published by McGraw-Hill Book Co., New York.

International Critical Tables, Vol. 1. Published by McGraw-Hill Book Co., New York.

Seidell: *Solubilities of Inorganic and Organic Compounds.* Published by Van Nostrand and Co., New York.

ign., ignites	*p*, para position	*tert*, tertiary
K_2CO_3, aqueous potassium carbonate	pa., pale	tet., tetragonal
l, levo-rotatory	pd., powder	tri., triclinic
lf., leaves or leaflets	pet., petroleum ether	trig., trigonal
lg., ligroin	Ph, phenyl (C_6H_5)	*uns*, unsymmetrical
lq., liquid	pl. plates	v., very
lt., light	pr. prisms	vl., violet
m, meta position	pyr., pyridine	v. s., very soluble
Me, methyl (CH_3)	r., red	v. sl. s., very slightly soluble
Me al., methyl alcohol	rhb., rhombic	vac., vacuo or vacuum
met., metallic	s., soluble	wh., white
mn., monoclinic	sc., scales	yel., yellow
m. p., melting point	*sec*, secondary	∞, soluble in all proportions;
n, normal	silv., silvery	i. e., miscible
NaOAc, aqueous sodium acetate	sl., slight or slightly	>, greater than
nd., needles	soln., solution	<, less than
o, ortho position	subl., sublimes	42±, about or near to 42
or., orange	*sym*, symmetrical	$-3H_2O$, 100, loses 3 molecules
	syr., syrup	of water at 100° C.

No.	Crystalline Form and Color	Specific Gravity	Melting Point °C.	Boiling Point °C.	Solubility in 100 Parts		
					Water	Alcohol	Ether
1	lf.	....	182	....	i.	v. s.	v. s.
2	rhb./al.	$1.069^{95°}_{35}$	95	278-9	i.; $26^{18°}$bz.	s. h. $31^{18°}$ac.	s. chl.
3	rhb.	$0.899^{16°}_{4}$	93.5-4.5	265-75(sl.d.)	i.	v. s.	v. s.

Absinthin, cf. glcde   Acecoline 83
Accelerine 4883

Table 7-4 (*Continued*)
PHYSICAL CONSTANTS OF ORGANIC COMPOUNDS

No.	Name	Synonym	Formula	Beil. Ref.	Formula Weight
4	**Acetal**	acetaldehyde-diethylacetal	$CH_3 \cdot CH(OC_2H_5)_2$	I-603	118.18
5	**Acetaldehyde**	ethanal	$CH_3 \cdot CHO$	I-594	44.05
6	**Acetaldehyde, meta-**	metaldehyde	$(C_2H_4O)_4$	I-602	176.21
7	**Acetaldehyde, par-**	paraldehyde	$(C_2H_4O)_3$	XIX-385	132.16
8	ammonia	....	$(CH_3 \cdot CHOH \cdot NH_2)_3$	XXVI-7	183.25
9	azine	acetaldazine	$CH_3CH:N \cdot N:CHCH_3$	I-609	84.12
10	oxime	acetaldoxime	$CH_3CH:NOH$	I-608	59.07
11	phenylhydrazone	....	$C_6H_5NHN:CHCH_3$	XV-127	134.18
12	semicarbazone	....	$CH_3CH:NNHCONH_2$	III-101	101.11
13	sodium bisulfite	....	$CH_3CHO \cdot NaHSO_3 \cdot \frac{1}{2}H_2O$	I-605	157.12
14	**Acetamide**	ethanamide	$CH_3 \cdot CO \cdot NH_2$	II-175	59.07
15	**Acetamidine**	....	$CH_3 \cdot C(NH)NH_2$	II-185	58.08
16	hydrochloride	....	$CH_3 \cdot C(NH)NH_2 \cdot HCl$	II-185	94.54
17	**Acet-**anilide	antifebrin	$C_6H_5NH \cdot COCH_3$	XII-237	135.17
18	anisidide (*o*)	acetyl-*o*-anisidine	$CH_3O \cdot C_6H_4NH \cdot COCH_3$	XIII-371	165.19
19	anisidide (*m*)	acetyl-*m*-anisidine	$CH_3O \cdot C_6H_4NH \cdot COCH_3$	XIII-416	165.19
20	anisidide (*p*)	methacetin; *p*-meth-oxy-acetanilide	$CH_3O \cdot C_6H_4NH \cdot COCH_3$	XIII-461	165.19
21	metanilic acid (Na)	....	$CH_3CO \cdot NH \cdot C_6H_4 \cdot SO_3Na \cdot 2H_2O$ (1,3)	XIV-691	273.24
22	methyl-*α*-naph-thylamide	....	$CH_3CO \cdot (CH_3)N \cdot C_{10}H_7$	XII-1231	199.25
23	*o*-phenetidide	*o*-ethoxyacetanilide	$CH_3CONH \cdot C_6H_4 \cdot OC_2H_5$	XIII-371	179.22
24	*m*-phenetidide	acetyl-*m*-phenetidine	$CH_3CONH \cdot C_6H_4 \cdot OC_2H_5$	XIII-416	179.22
25	*p*-phenetidide	phenacetin	$CH_3CONH \cdot C_6H_4 \cdot OC_2H_5$	XIII-461	179.22
26	toluidide (*o*)	N-tolylacetamide	$CH_3C_6H_4NH \cdot COCH_3$	XII-792	149.19
27	toluidide (*m*)	N-tolylacetamide	$CH_3C_6H_4NH \cdot COCH_3$	XII-860	149.19
28	toluidide (*p*)	N-tolylacetamide	$CH_3C_6H_4NH \cdot COCH_3$	XII-920	149.19
29	**Acetamino-**ethyl-salicylic acid (5;2,1)	benzacetin	$CH_3CONH \cdot C_6H_3 \cdot (OC_2H_5)CO_2H$	XIV-583	223.23
30	fluorene(2)	....	$C_{13}H_9NH \cdot COCH_3$	XII-1331	223.28
31	*α*-naphthol (4,1)	naphthacetol	$CH_3CO \cdot NH \cdot C_{10}H_6OH$	XIII-669	201.23
32	*β*-naphthol (1,2)	....	$CH_3CO \cdot NH \cdot C_{10}H_6OH$	XIII-679	201.23
33	phenol (*o*)	*o*-hydroxyacetanilide	$CH_3CONHC_6H_4OH$	XIII-370	151.17
34	phenol (*m*)	acetyl-aminophenol	$CH_3CONHC_6H_4OH$	XIII-415	151.17
35	phenol (*p*)	....	$CH_3CONHC_6H_4OH$	XIII-460	151.17
36	**Acetic** acid	ethanoic acid	$CH_3 \cdot COOH$	II-96	60.05
37	anhydride	....	$(CH_3CO)_2O$	II-166	102.09
38	**Acetoacetic** acid	acetyl acetic acid	$CH_3CO \cdot CH_2CO_2H$	III-630	102.09
39	*β*-anil	Et *β*-anilinocroton-ate	$CH_3C(:NC_6H_5) \cdot CH_2 \cdot CO_2C_2H_5$	XII-518	205.26

Table 7-4 (*Continued*)
PHYSICAL CONSTANTS OF ORGANIC COMPOUNDS

No.	Crystalline Form and Color	Specific Gravity	Melting Point °C.	Boiling Point °C.	Solubility in 100 Parts		
					Water	Alcohol	Ether
4	lq.	$0.821\frac{22}{4}°$	....	102.2	$6^{25°}$	∞	∞
5	col. lq.	$0.783\frac{18}{4}°$	−123.5	20.2	∞	∞	∞
6	nd.	....	246.2*	....	i.; s. chl.	2 h.; s. bz.	$0.5^{35°}$; v. sl. s. h.
7	col. cr.	$0.994\frac{20}{4}°$	12.6	124.4^{752mm}	$12^{13°}$ $6^{100°}$	∞; ∞ chl.	∞
8	col. cr.	....	97	100-10(sl.d.)	v. s.	v. s.	sl. s.
9	lq.	$0.832^{17°}$	....	95-6	....	....	....
10	nd. or lq.	$0.965\frac{20}{4}°$	47(13)	114-5	s.	∞	∞
11	col. nd.	....	(α)98-101 (β)57	$236-7^{20mm}$ $133-6^{20mm}$	....	....	s. pet.
12	nd. / aq. or al.	....	162-3	....	$3^{17°}$	s.	....
13	col. nd.	....	....	....	s.; d. a.	i.	i.
14	col. cr.	1.159	81(69.4)	221.2	s.	s.	v. sl. s.
15	in aq. soln.	....	....	....	d. h.	....	....
16	nd./al.	....	177	....	s.	v. s.	i.
17	rhb./al.	$1.214°$	113-4	305	$0.53^{6°}$; $3.5^{80°}$	$21^{20°}$; $46^{60°}$	$72^{5°}$
18	cr./aq.	....	87-8	303-5	v. s. h.	$55.3^{21°}$	v. s. ac.
19	cr./aq.	....	80-1	....	....	$80^{21°}$	....
20	pl./aq.	....	127	....	$0.2^{15°}$; $8^{100°}$	$12.7^{21°}$; s. act.	s. chl.; s. dil. alk.
21	nd./aq.	....	184-5	....	v. s. h.	v. sl. s.	i.
22	pr./aq.	....	94-5	....	v. sl. s.	s.	s.
23	lf./aq. al.	....	79	>250	i.	s.	....
24	lf./aq.	....	96-7	....	sl. s.	s.	....
25	col. mn.	....	134-5	d.	$0.7^{20°}$	7.4 c.; 40 h.	$1.6^{25°}$; s. gly.
26	rhb.	$1.168^{15°}$	110	296	$0.8^{619°}$	s.; s. bz.	s.; s. chl.
27	mn./aq.	$1.141^{15°}$	65.5	303	$0.44^{13°}$	v. s.	v. s.
28	rhb. or mn.	$1.212^{15°}$	153	306-7	$0.09^{22°}$	$10.2^{25°}$ abs.	s.; i. lg.
29	col. nd./aq.	....	189-90	....	sl. s.	s.	....
30	cr./50% ac.	....	194-5	....	....	....	....
31	nd./al.	....	187	....	s. h.	s; s. NH_4OH	s. Na_2CO_3
32	lf./aq. al.	....	235.5	subl. sl. d.	v. s. aq. NaOH	s; s. bz.	s.; s. h. ac.
33	lf./aq. al.	....	203	....	s. h.	s.	s. KOH
34	nd./aq.	....	148-9	....	s.	s.; sl. s. chl.	sl. s.; sl. s. bz.
35	mn./al.	$1.293^{21°}$	168-9	....	v. s. h.	v. s.	sl. s.
36	col. lq.	$1.049\frac{20}{4}°$	16.6	118.1	∞	∞	∞
37	col. lq.	$1.081\frac{20}{4}°$	−73	140.0	12 c.; d. h.	∞; d. h.	∞
38	col. oil	....	36-7	d. <100	∞	∞	s.
39	oil	....	....	d. 240	i.	s.	s.

* In sealed tube; subl. 112-6; partly depolymerized.
Acet-methylamide 4085
Acet-oxime 61
Acet-phenetidide 23-5
Acetamino-, cf. also acetylamino.

Acetic ester 2935
Acetic ether 2935
Acetin 3382
Aceto-acetic ester 2937

Table 7-4 (*Continued*)
PHYSICAL CONSTANTS OF ORGANIC COMPOUNDS

No.	Name	Synonym	Formula	Beil. Ref.	Formula Weight
	Acetoacetic				
40	anilide	acetoacetanilide	$CH_3COCH_2CONH \cdot C_6H_5$	XII-518	177.20
41	2-chloroanilide	*o*-chloroaceto-acet-anilide	$CH_3CO \cdot CH_2CO \cdot NH \cdot C_6H_4Cl$	*XII-300	211.65
42	2,5-dichloroanilide	....	$CH_3CO \cdot CH_2CO \cdot NH \cdot C_6H_3Cl_2$	....	246.09
43	**Acetoin**	methylacetyl carbinol	$CH_3CHOH \cdot CO \cdot CH_3$	I-827	88.11
44	**Aceto-**chloroglucose	tetraacetyl-α-gluco-syl chloride	$Cl \cdot CH(CHO_2CCH_3)_3 \cdot$ $\overline{\ \ \ \ \ \ \ \ \ \ \ \ }$ $CH \cdot (CH_2O_2CCH_3) \cdot O$ $\overline{\ \ \ \ \ \ \ \ \ \ \ \ \ \ \ \ }$	....	366.76
45	dichlorohydrin	dichloropropyl acetate (β,γ)	$CH_3CO_2CH_2CHCl \cdot CH_2Cl$	II-129	171.02
46	naphthalide (α)	acetyl naphthylamine	$C_{10}H_7NH \cdot COCH_3$	XII-1230	185.23
47	naphthalide (β)	....	$C_{10}H_7NH \cdot COCH_3$	XII-1284	185.23
48	naphthol (1,2)	hydroxy-aceto-naphthone	$HO \cdot C_{10}H_6 \cdot COCH_3$	VIII-149	186.21
49	nitrile	methyl cyanide	$CH_3 \cdot CN$	II-183	41.05
50	phenine	....	$C_{23}H_{17}N$	....	307.40
51	phenone	methyl-phenyl ketone	$CH_3 \cdot CO \cdot C_6H_5$	VII-271	120.15
52	phenone acetone	phenacyl acetone	$C_6H_5 \cdot CO \cdot C_2H_4 \cdot COCH_3$	VII-687	176.22
53	phenone oxime	....	$C_6H_5(CH_3)C:NOH$	VII-278	135.17
54	phenone phenyl-hydrazone	....	$C_6H_5NH \cdot N:C(CH_3) \cdot C_6H_5$	XV-139	210.28
55	**Acetol**	acetyl carbinol	$CH_3CO \cdot CH_2OH$	I-821	74.08
56	**Acetone**	propanone	$CH_3 \cdot CO \cdot CH_3$	I-635	58.08
57	cyanohydrin	....	$(CH_3)_2C(OH) \cdot CN$	III-316	85.11
58	diacetic acid	....	$CO:(C_2H_4 \cdot CO_2H)_2$	III-804	174.15
59	dicarboxylic acid	β-ketoglutaric acid	$CO:(CH_2CO_2H)_2$	III-789	146.10
60	glucose, mono	glucose diMe-ketal	$(CH_3)_2C:(HCO)_2 \cdot$ $\overline{\ \ \ \ \ \ \ \ \ }$ $(CHOH)_2 \cdot CH \cdot$ $\overline{\ \ \ \ \ \ \ \ \ \ \ }$ $(CH_2OH) \cdot O$ $\overline{\ \ \ \ \ \ \ \ \ }$	XXXI-153	220.22
61	oxime	acetoxime	$(CH_3)_2C:NOH$	I-649	73.10
62	phenylhydrazone	....	$C_6H_5NH \cdot N:C(CH_3)_2$	XV-129	148.21
63	semicarbazone	....	$(CH_3)_2C:N \cdot NH \cdot CONH_2$	III-101	115.14
64	sodium bisulfite	....	$(CH_3)_2CO \cdot NaHSO_3$	I-649	162.14
65	**Acetonyl** acetone	....	$(CH_3 \cdot CO \cdot CH_2)_2$	I-788	114.15
66	urea	dimethyl hydantoin	$(CH_3)_2:$ $CNHCONHCO$ $\overline{\ \ \ \ \ \ \ \ \ \ \ \ \ \ }$	XXIV-289	128.13
67	**Acetoxy acetone**	acetol acetate	$CH_3CO_2CH_2COCH_3$	II-155	116.12
68	**Acetyl-**acetone	diacetylmethane	$(CH_3CO)_2CH_2$	I-777	100.12
69	benzoic acid (*o*)	acetophenone car-boxylic acid	$CH_3CO \cdot C_6H_4 \cdot CO_2H$	X-690	164.16
70	benzoic acid (*m*)	....	$CH_3CO \cdot C_6H_4 \cdot CO_2H$	X-694	164.16
71	benzoic acid (*p*)	....	$CH_3CO \cdot C_6H_4 \cdot CO_2H$	X-694	164.16
72	benzoyl	Me-Ph-diketone	$CH_3CO \cdot CO \cdot C_6H_5$	VII-677	148.16

Table 7-4 (*Continued*)
PHYSICAL CONSTANTS OF ORGANIC COMPOUNDS

No.	Crystalline Form and Color	Specific Gravity	Melting Point °C.	Boiling Point °C.	Solubility in 100 Parts		
					Water	Alcohol	Ether
40	lf.	$1.103^{85°}$	85	d.	sl. s.; s. alk.	s.	s.; s. h. bz.
41	col. nd.	$1.190^{107°}$	107	d.	i.	s.	i.; i. lg.
42	col. nd.	....	94-5	....	i.	sl. s.	i.
43	lq.	$1.011^{\frac{15°}{15}}$	15	148	∞	v. s.	sl. s.; i. lg.
44	col. mn. nd./et.-lg.	....	70-1 (d.)	....	d. h.; i. pet.	s.; s. act.	s.; s. CCl_4
45	col. lq.	$1.168^{15°}$	....	$191-2^{755mm}$	....	....	....
46	cr./al.	....	159-60	....	s. h.	$4^{25°}$	....
47	lf./al.	....	134	....	s. h.	s. h.	....
48	yel. nd./lg.	....	103(98)	325(sl. d.)	i.; v. s. bz.	v. sl. s. c.	s. ac.; s. CS_2
49	col. lq.	$0.783^{20°}_{4}$	−44.9	81.6	∞	∞	∞
50	nd./al.	....	135	....	....	s.	....
51	cr. lf.	$1.028^{20°}_{4}$	19.6	202.0	i.	s.	s.
52	yel. oil	>1	....	162^{12}(sl. d.)	sl. s.	....	i. alk.
53	nd./aq.	....	59	d.	....	v. s.	v. s.
54	col. nd.	....	105-6	....	sl. s.	sl. s. c.	s.
55	col. lq.	$1.082^{20°}_{20}$	−17	145-6	∞	∞	∞
56	col. lq.	$0.791^{20°}_{4}$	−94.8	56.2	∞	∞	∞
57	col. lq.	$0.932^{19°}$	−19	82^{23mm}	v. s.	v. s.	v. s.; v. sl. s. pet.
58	rhb./aq.	....	142-3	....	s. h.	s.	sl. s.; i. bz.
59	nd./et.	....	138	....	v. s.	v. s.	sl. s.; i. bz.
60	col. mn. nd./aq.	....	162	....	s.	s.; s. h. EtOAc	I.
61	pr.	$0.97^{20°}_{20}$	60-1	136.3	v. s.	v. s.	v. s.; s. lg.
62	oil	....	26.6	163^{50mm}	s. c. dil. a.	....	....
63	nd./aq.	....	190-1(d.)	....	s. c.	sl. s. c.	i.
64	col. lf.	....	....	....	s.; d. a.	sl. s.	i.
65	col. lq.	$0.974^{20°}_{4}$	ca. −6	194^{754mm}	∞; i. KOH	∞	∞
66	tri. pr./al.	....	175	subl.	s.	s.	s.
67	col. oil	$1.075^{20°}_{4}$	....	174-5	v. s.	v. s.	v. s.
68	lq.	$0.976^{20°}_{4}$	−23.2(−9)	139^{746mm}	12.5	∞	∞; ∞ chl.
69	cr./aq.	....	114-5	....	s. h.	....	....
70	nd./aq.	....	172	....	s. h.	s.; sl. s. chl.	s.; sl. s. bz.
71	nd./h. aq.	....	200-5(d.)	....	s. h.	sl. s.	sl. s.; i. lg.
72	yel. oil	$1.101^{20°}_{4}$	....	216-8	$0.3^{20°}$	....	....

Aceto-xylidide 2417-22
Acetozone 73
Acetol acetate 67
Acetol phenyl ether 5132
Acetone azine 3932

Acetone bromoform 6102
Acetone chloride 2053
Acetone chloroform 6145
Acetone chloroform acetate 6144
Acetone diethylsulfone 5683

Acetone oxalic ester 2939
Acetonic acid 3757
Acetonyl-acetophenone 52
Acetonyl-amine 235
Acetonyl anisole 4126

Table 7-4 (*Continued*)
PHYSICAL CONSTANTS OF ORGANIC COMPOUNDS

No.	Name	Synonym	Formula	Beil. Ref.	Formula Weight
	Acetyl				
73	benzoyl peroxide	acetozone	$C_6H_5CO \cdot O_2 \cdot COCH_3$	IX-179	180.16
74	biuret	. . . .	$CH_3CO(NHCO)_2NH_2$	III-72	145.12
75	bromide	ethanoyl bromide	$CH_3CO \cdot Br$	II-174	122.95
76	5-bromosalicylic acid (2;5,1)	5-bromoaspirin	$CH_3CO_2 \cdot C_6H_3(Br) \cdot CO_2H$	X-108	259.06
77	*n*-butylaniline	. . . .	$CH_3CO \cdot N(C_4H_9) \cdot C_6H_5$	XII-247	191.28
78	caproyl	octandione-2,3	$CH_3(CH_2)_4CO \cdot COCH_3$	I-795	142.20
79	caproyl oxime	isonitroso-*n*-amyl-ketone	$C_8H_{15}O_2N$	I-795	157.21
80	carbazole (*N*)	. . . .	$CH_3CO \cdot NC_{12}H_8$	XX-436	209.25
81	chloride	ethanoyl chloride	$CH_3CO \cdot Cl$	II-173	78.50
82	choline bromide	pragmoline	$(CH_3)_3N(Br)(CH_2)_2 \cdot O_2C \cdot CH_3$	*IV-428	226.12
83	choline chloride	acecoline	$(CH_3)_3N(Cl)C_2H_4 \cdot O_2C \cdot CH_3$	IV-281	181.66
84	diethylbromo-acetyl-urea (*N,N'*)	abasin; acetylcar-bromal	$(C_2H_5)_2CBr \cdot CO \cdot NH \cdot CO \cdot NH \cdot COCH_3$	*III-30	279.14
84.1	diphenylamine	*N*-diPh acetamide	$(C_6H_5)_2N \cdot COCH_3$	XII-247	211.27
85	disulfide	diacetyl disulfide	$(CH_3CO)_2S_2$	II-232	150.22
86	ethanolamine (*N*)	. . . .	$HO(CH_2)_2NH \cdot COCH_3$	. . . .	103.12
87	ethylmalonate	. . . .	$CH_3CO \cdot CH \cdot (CO_2C_2H_5)_2$	III-796	202.21
88	ethyloxamate	. . . .	$(CH_3CO)NHC_2O_2 \cdot OC_2H_5$	II-545	159.14
89	fluoride	ethanoyl fluoride	$CH_3CO \cdot F$	II-172	62.04
90	glycine	aceturic acid	$CH_3CO \cdot NH \cdot CH_2 \cdot CO_2H$	IV-354	117.11
91	indole (1)(*N*)	. . . .	$CH_3CO \cdot NC_8H_6$	XX-309	159.19
92	iodide	ethanoyl iodide	$CH_3CO \cdot I$	II-174	169.95
93	iodosalicylic acid	4-iodoaspirin (2;4,1)	$CH_3CO_2C_6H_3(I)CO_2H$	*X-49	306.06
94	iodosalicylic acid	5-iodoaspirin (2;5,1)	$CH_3CO_2C_6H_3(I)CO_2H$	*X-49	306.06
95	isatin (*N*)	. . . .	$CH_3CO \cdot N(CO)_2C_6H_4$	XXI-447	189.17
96	malic acid	. . . .	$C_2H_2O_2 \cdot C_2H_3(CO_2H)_2$	III-429	176.13
97	*o*-methylamino-phenol	. . . .	$CH_3CO \cdot N(CH_3)C_6H_4 \cdot OH$	XIII-372	165.19
98	*p*-methylamino-phenol	. . . .	$CH_3CO \cdot N(CH_3)C_6H_4 \cdot OH$	XIII-466	165.19
99	β-methylcholine bromide	mecholyl bromide	$(CH_3)_3N(Br)CH_2 \cdot CH(CH_3)O_2C \cdot CH_3$	. . . .	240.15
100	β-methylcholine chloride	mecholyl; mecholin	$(CH_3)_3N(Cl)CH_2 \cdot CH(CH_3)O_2C \cdot CH_3$	. . . .	195.69
101	β-methylcholine iodide	mecholyl iodide	$(CH_3)_3N(I)CH_2 \cdot CH(CH_3)O_2C \cdot CH_3$	. . . .	287.14
102	methyl-*o*-toluidine	*N*-Me-acet-*o*-toluidide	$CH_3CO \cdot N(CH_3)C_6H_4 \cdot CH_3$	XII-793	163.22
103	methyl-*m*-toluidine	*N*-Me-acet-*m*-toluidide	$CH_3CO \cdot N(CH_3)C_6H_4 \cdot CH_3$	XII-861	163.22
104	methyl-*p*-toluidine	*N*-Me-acet-*p*-toluidide	$CH_3CO \cdot N(CH_3)C_6H_4 \cdot CH_3$	XII-922	163.22
105	nitrate	. . . .	$CH_3CO_2NO_2$	II-171	105.05
106	peroxide	diacetyl peroxide	$(CH_3CO)_2O_2$	II-170	118.09
107	*p*-phenylenedi-amine	amino-acetanilide (*p*)	$C_2H_3O \cdot NHC_6H_4NH_2$	XIII-94	150.18

Acetoxy-acetophenone 5094
Aceturic acid 90
Acetyl, cf. also aceto or acet.
Acetyl-acetic acid 38

Acetyl-anisidine 18–20
Acetyl-anisole 4065
Acetyl-anthranilic acid (*N*) 125
Acetyl-arsanilic acid 577

Acetyl-benzoin 718
Acetyl-benzoylaconine, cf. alkd.
Acetyl-benzylamine 780
Acetyl-carbinol 55

Table 7-4 (*Continued*)
PHYSICAL CONSTANTS OF ORGANIC COMPOUNDS

No.	Crystalline Form and Color	Specific Gravity	Melting Point °C.	Boiling Point °C.	Solubility in 100 Parts		
					Water	Alcohol	Ether
73	col. nd./lg.	....	37-9	130^{19mm}	$0.06^{25°}$ d.	s. oils	s.; s. chl.
74	nd.	....	107		s. bz.	v. s.	s.
75	col. lq.	$1.663\frac{16°}{4}$	−96.5	76^{750mm}	d.	d.; ∞ bz.	∞; ∞ chl.
76	cr./al.	....	168	....	i.; i. bz.	$16^{25°}$; i. CCl_4	$10^{25°}$
77	lq.	....	....	$273-57^{18mm}$	....	....	....
78	lq.	....	....	$172-3^{733mm}$	....	....	....
79	cr.	....	39	139^{16mm}	....	....	....
80	nd./aq.	....	69	>360 d.	v. sl. s. h.	v. s.	v. s.; s. bz.
81	col. lq.	$1.105\frac{20°}{4}$	−112.0	51-2	d.	d.; ∞ bz.	∞; ∞ chl.
82	hyg. pr./ abs. al.	....	143	....	v. s.; d. h.	s.	i.
83	hyg. col. pr.	....	....	....	v. s.; d. h.	s.	i.
84	col. cr./aq.	....	108-9	....	sl. s.	s.	s. EtOAc
84.1	rhb./aq.	....	103	subl.	sl. s.	s.	sl. s.
85	cr.	....	20	d.	i.; s. CS_2	v. s.	v. s.
86	col. syr.	$1.122\frac{20°}{20}$	15.8	d.	sl. s.	s.	s.
87	lq.	$1.083\frac{26°}{5}$	....	120^{17mm}	s. dil. alk.	....	....
88	nd.	....	54		d. h.	s.	s.
89	lg. or gas	$0.993^{20°}$	<−60	20.8^{770mm}	5; d.	∞; ∞ bz.	∞
90	cr./aq.	....	206	....	$2.7^{15°}$	s. abs.	i.
91	lq.	$1.071^{0°}$	<−20	239	v. sl. s.	v. s.	v. s.; v. s. bz.
92	col. lq.	$1.98^{17°}$	....	108	d.	d.	s.
93	col. cr.	....	156	....	i.	s.	s.
94	col. cr.	....	166	....	i.	s.	s.
95	yel. nd./bz.	....	141	....	sl. s. c.; d. h.	s.	d. h. HCl
96	col. cr. pd.	....	132-9 d.	....	s.; d. h.	....	i. bz.
97	nd./MeOH	....	150	....	sl. s.	s.	
98	sc./aq.	....	240-1	....	v. sl. s.	v. s.	s.
99	col. delq. cr.	....	148-9	....	v. s.	70	i.; d. alk.
100	col. delq. cr./al.	....	172-3	....	v. s.	v. s.	i.; d. alk.
101	col. cr.	....	138-9	....	v. s.	75	i.
102	col. cr.	....	55-6	260	....	s.	....
103	col. cr.	....	66	....	....	....	....
104	lf./al.-et.	....	80	283	....	s.	s. h. lg.
105	col. lq.	$1.24^{15°}$	expl.	227^{mm}	....	....	....
106	col. cr.	....	30	$63^{21°}$; expl.	sl. s. c.; d.	....	....
107	nd./aq.	....	162	....	s. h.	v. s.	v. s.

Table 7-4 (*Continued*)
PHYSICAL CONSTANTS OF ORGANIC COMPOUNDS

No.	Name	Synonym	Formula	Beil. Ref.	Formula Weight
	Acetyl				
108	*m*-phenylenedi-amine	aminoacetanilide (*m*)	$C_2H_3O \cdot NHC_6H_4NH_2$	XIII-45	150.18
109	*m*-phenylene diamine HCl	....	$C_2H_3ONHC_6H_4NH_2 \cdot$ HCl	XIII-45	186.64
110	phenylglycine	*N*-Ph-aceturic acid	$C_6H_5N(COCH_3)CH_2 \cdot CO_2H$	XII-476	193.20
111	phenylhydrazine (*β*)	hydracetin	$CH_3CO \cdot NH \cdot NHC_6H_5$	XV-241	150.18
112	piperidine	....	$CH_3CO \cdot NC_5H_{10}$	XX-45	127.19
113	pyrrol (*N*)	....	$CH_3CO \cdot NC_4H_4$	XX-165	109.13
114	resorcinol (1,3)	euresol; eurisol	$CH_3CO_2 \cdot C_6H_4 \cdot OH$	VI-816	152.15
115	*o*-salicylic acid	aspirin	$CH_3CO_2 \cdot C_6H_4 \cdot CO_2H$	X-67	180.16
116	*o*-salicylic acid (Ca)	aspirin soluble	$(C_9H_7O_4)_2Ca \cdot 2H_2O$	*X-29	434.42
117	*o*-salicylic acid (Li)	litmopyrin	$C_9H_7O_4Li$	X-68	186.09
118	*o*-salicylic acid (Mg)	magnespirin	$(C_9H_7O_4)_2Mg$†	*X-29	382.62
119	*o*-salicylic acid (Na)	hydropyrin	$C_9H_7O_4Na$	X-68	202.14
120	salol	spiroform; vesipyrin	$CH_3CO_2 \cdot C_6H_4 \cdot CO_2C_6H_5$	X-79	256.26
121	2-thiohydantoin	....	$CH_3CO \cdot N \cdot CH_2 \cdot$ $\mid\_\_\_\_\_$ $CONH \cdot CS$ $\mid\_\_\_\_\_$	*XXIV-293	158.18
122	thiourea	....	$CH_3CO \cdot NHCSNH_2$	III-191	118.16
123	urea	....	$CH_3CO \cdot NHCONH_2$	III-61	102.09
124	**Acetylamino-** azotoluene	azodermin (4-NH_2; 3,2'-diMe)	$CH_3 \cdot C_6H_4 \cdot N:N \cdot C_6H_3 \cdot$ $(CH_3) \cdot NH(COCH_3)$	....	267.33
125	benzoic acid (*o*)	*N*-acetyl-anthranilic acid	$CH_3CO \cdot NH \cdot C_6H_4 \cdot CO_2H$	XIV-337	179.18
126	benzoic acid (*m*)	....	$CH_3CO \cdot NH \cdot C_6H_4 \cdot CO_2H$	XIV-396	179.18
127	benzoic acid (*p*)	acetaminobenzoic acid	$CH_3CO \cdot NH \cdot C_6H_4 \cdot CO_2H$	XIV-432	179.18
128	hydroxyphenyl-arsonic acid (3;4,1)	stovarsol; spirocid	$CH_3CONH \cdot C_6H_3(OH) \cdot AsO_3H_2$	....	275.09
129	phenylglycine (*p*)	....	$CH_3CONH \cdot C_6H_4 \cdot CH_2CO_2H$	....	193.20
130	**Acetylene**	ethyne; ethine	$CH:CH$	I-228	26.04
131	dibromide (1,1)	α,α-dibromoethylene	$CH_2:CBr_2$	I-190	185.86
132	dibromide (*cis*)	1,2-dibromoethene	$CHBr:CHBr$	I-190	185.86
133	dibromide (*trans*)	1,2-dibromoethene	$CHBr:CHBr$	I-190	185.86
134	dicarboxylic acid	butynedioic acid	$(C \cdot CO_2H)_2$	II-801	114.06
135	dichloride (*cis*)	1,2-dichloroethene	$CHCl:CHCl$	I-187	96.94
136	dichloride (*trans*)	dioform	$CHCl:CHCl$	I-187	96.94
137	diiodide (*cis*)	1,2-diiodoethene	$CHI:CHI$	I-194	279.85
138	diiodide (*trans*)	1,2-diiodoethene	$CHI:CHI$	I-194	279.85
139	urea	glycoluril	$C_2H_2(CON_2H_2)_2$	XXVI-441	142.12

† Also cr. $+ 3H_2O$, and $+ 4H_2O$.
Acetyl-propiophenone 52
Acetyl-propylaniline 5364

Acetyl-resorcinol 5550
Acetyl-thymol 5945

Table 7-4 (*Continued*)
PHYSICAL CONSTANTS OF ORGANIC COMPOUNDS

No.	Crystalline Form and Color	Specific Gravity	Melting Point °C.	Boiling Point °C.	Solubility in 100 Parts		
					Water	Alcohol	Ether
108	pl./bz.	....	softens >70	d. >87	s.; s. act.	s.; sl. s. bz.	s.; i. lg.
109	pl./al.	....	280±	....	s.; i. bz.	sl. s.	i.; i. lg.
110	lf./aq.	....	190-2	....	v. sl. s.	s.; s. ac.	v. sl. s.
111	pr.	....	130-2	....	s. h.	s.	sl. s.
112	lq.	$1.011^{9°}$	....	226-7	∞	s.	....
113	lq.	....	....	181-2	i.	....	d. HCl
114	yel. oil	....	....	283 d.	i.; s. alk.	∞; ∞ bz.	∞ chl.
115	col. nd./aq.	....	135-6	....	$1^{37°}$	v. sl. s. bz.	$5^{20°}$
116	cr./aq. al.	....	....	....	16	1.4	i.
117	wh. hyg. pd.	....	....	....	100	25	i.
118	wh. pd.	....	....	....	s.	sl. s.	i.
119	cr./act.-et.	....	218 sl. d.	....	v. s.	v. s.	sl. s. act.
120	col. cr./ abs. al.	....	97-8	198^{11mm}	i.	s.	s.
121	cr./al.	....	175-6	....	i.	sl. s.	i.
122	pr./aq.	....	166-7	....	s. h.	s.	sl. s.
123	nd./al.	....	218	....	$2^{15°}$; v. s. h.	10 h.	....
124	r. nd./al.	....	185-6	....	i.	s. chl.	s.; s. oils
125	rhb./ac.	....	184-6	....	s. h.	s.	s.; s. bz.
126	nd./al.	....	248-50	subl.	v. sl. s. h.	s. h.	v. sl. s.
127	nd.	....	256.5 d.	....	sl. s.	s.	....
128	wh. pd.	....	....	....	sl. s.	s. alk.	....
129	col. fl.	....	241-2	....	i.	sl. s.	sl. s.
130	col. gas	lq. $0.613^{-80°}$ s. $0.730^{-85°}$ (A) 0.906	-81.5^{891mm}	-84^{760mm}	$100\ cc^{18°}$ cf. special table	$600\ cc^{18°}$	$2500\ cc^{15°}$ act.; 30,000 cc act. 12 atm.
131	col. lq.	$2.178^{\frac{21°}{4}}$	....	$91-2^{754mm}$	i.	v. s.	v. s.
132	col. lq.	$2.285^{\frac{17.5°}{4}}$	-53	110^{754mm}	i.	v. s.	v. s.
133	col. lq.	$2.267^{\frac{17.5°}{4}}$	-6.5	108	i.	v. s.	v. s.
134	pl./et.	....	179-80	....	v. s.	v. s.	v. s.
135	col. lq.	$1.282^{\frac{20°}{4}}$	-80.5	60.3	$0.35^{20°}$	∞	∞
136	col. lq.	$1.255^{\frac{20°}{4}}$	-50	47.7	$0.63^{20°}$	∞	∞
137	col. lq.	$3.063^{20°}$	-13.8	188 d.	i.	s.	s.
138	mn. pr.	$3.303^{21°}$	73	192	i.	s.	s.
139	nd./aq.	....	d. 300±	....	$0.117°$; $1.5^{100°}$	i.; s. conc. HCl	>1, h. NH₄OH

Table 7-4 (*Continued*)
PHYSICAL CONSTANTS OF ORGANIC COMPOUNDS

No.	Name	Synonym	Formula	Beil. Ref.	Formula Weight
140	**Aconic acid**	....	CH$_2$·CO·O·CH:C· $\quad$|_____| $\quad$ CO$_2$H	XVIII-395	128.09
141	**Aconitic acid**	equisetic acid; citridic acid	C$_3$H$_3$(CO$_2$H)$_3$	II-849	174.11
142	**Acridine**	....	C$_6$H$_4$·CH·C$_6$H$_4$N $\quad$|_____|	XX-459	179.22
143	orange (3,6)	....	NC$_{13}$H$_7$[N(CH$_3$)$_2$]$_2$	XXII-487	265.36
144	red (2,7)	....	CH$_3$NH·C$_{13}$H$_7$O: N(Cl)CH$_3$	....	273.74
145	**Acridone**	dihydroketo-acridine	C$_6$H$_4$·CO·C$_6$H$_4$·NH $\quad$|_____|	XXI-335	195.22
146	**Acrolein**	acrylic aldehyde	CH$_2$:CH·CHO	I-725	56.06
147	**Acrolein** (*meta*)	metacrolein	(C$_3$H$_4$O)$_3$	I-727	168.19
148	**Acrylamide**	acrylic amide	CH$_2$:CH·CONH$_2$	II-400	71.08
149	**Acrylic** acid	propenoic acid	CH$_2$:CH·CO$_2$H	II-397	72.06
150	anhydride	....	(CH$_2$:CH·CO)$_2$O	II-400	126.11
151	nitrile	vinyl cyanide	CH$_2$:CH·CN	II-400	53.06
152	**Adipamide**	adipic diamide	(·CH$_2$CH$_2$CONH$_2$)$_2$	II-653	144.17
153	**Adipic** acid	hexandioic acid	(·CH$_2$CH$_2$CO$_2$H)$_2$	II-649	146.14
154	**Adipyl** chloride	....	(·CH$_2$CH$_2$CO·Cl)$_2$	II-653	183.04
155	dinitrile	tetramethylene dicyanide	(·CH$_2$CH$_2$CN)$_2$	II-653	108.14
156	**Adonitol**	adonite	HOCH$_2$(CHOH)$_3$· CH$_2$OH	I-530	152.15
157	**Adrenaline** (*l*)(3,4,1)	*l*-suprarenine	C$_6$H$_3$(OH)$_2$(CHOH· CH$_2$NHCH$_3$)	XIII-830	183.21
158	**Alanine** (α) (*l* +)	α-aminopropionic acid	H$_2$N·CH(CH$_3$)·CO$_2$H	IV-387	89.09
159	**Alanine** (α) (*dl*)	....	H$_2$N·CH(CH$_3$)·CO$_2$H	IV-387	89.09
160	**Alanine** (β)	β-aminopropionic acid	H$_2$N·CH$_2$·CH$_2$·CO$_2$H	IV-401	89.09
161	**Aldehydin** (2,5)	2-Me-5-Et-pyridine	(CH$_3$)C$_5$H$_3$N(C$_2$H$_5$)	XX-248	121.18
162	**Aldehydo-**benzoic acid (*o*)	phthalic aldehyde	HO$_2$C·C$_6$H$_4$·CHO	X-666	150.14
163	benzoic acid (*m*)	isophthalic aldehyde	HO$_2$C·C$_6$H$_4$·CHO	X-671	150.14
164	benzoic acid (*p*)	terephthalic aldehyde	HO$_2$C·C$_6$H$_4$·CHO	X-671	150 14
165	*o*-hydroxybenzoic acid	(2;5,1)	HO·C$_6$H$_3$(CHO)· CO$_2$H	X-953	166 13
166	*o*-hydroxybenzoic acid	(2;3,1)	HO·C$_6$H$_3$(CHO)· CO$_2$H·1H$_2$O	X-952	184.15
167	*m*-hydroxybenzoic acid	(3;4,1)	HO·C$_6$H$_3$(CHO)· CO$_2$H	X-954	166.13
168	*p*-hydroxybenzoic acid	(4;3,1) formyl sal-icylic acid (5)	HO·C$_6$H$_3$(CHO)· CO$_2$H	X-953	166.13
169	**Aldol**	2-hydroxybutyralde-hyde	CH$_3$·CHOH·CH$_2$· CHO	I-824	88.11
170	**Aldol** par-	paraldol	(C$_4$H$_8$O$_2$)$_2$	I-825	176.21
171	**Aleudrin**	α,α'-diCl-*iso*Pr-carbamate	H$_2$N·CO$_2$·CH(CH$_2$· Cl)$_2$	III-29	172.01

Achilleic acid 141
Acid Cβ 4534
Acid F 4510
Acid G 4495-6
Acid H 342
Acid I 337
Acid J 337
Acid L 4503

Acid NW 4502
Acid R 4491
Acid S 339 or 344
Acid Sα 4533
Acid α S 4533
Acid β 559
Acid βC 4534
Acid βγ 4537

Acid γ 338
Acid δ 4533 or 4553
Acid λ 4544
Acid ρ 553
Acid χ 554
Acid 1:2:4 340
Acid of sugar 5006-7
Acid amido G 4537

Acid amino H 4555
Acid yellow D 4996
Acid yellow S 3241
Acidogen nitrate 6389
Acidol 847; also cf. alkd.
Acidulin 3373
Acocantherin, cf. glcde.
Acoin 1760

Table 7-4 (*Continued*)
PHYSICAL CONSTANTS OF ORGANIC COMPOUNDS

No.	Crystalline Form and Color	Specific Gravity	Melting Point °C.	Boiling Point °C.	Solubility in 100 Parts		
					Water	Alcohol	Ether
140	rhb./aq.	....	164	....	$18^{15°}$	....	s.
141	lf. or nd./aq.	....	192 d.	....	$33^{15°}$	$50^{12°}$, 88% al.	v. sl. s.
142	rhb./aq. al.	....	110-1	346	sl. s. h.	s.	s.; s. CS_2
143	yel. nd./aq. al.	....	181	....	s.; s. dil. a.	s.; s. act.	v. sl. s. pet.
144	bronze pd.	....	....	....	sl. s.	s.	i.
145	yel. nd./al.	....	354	....	i. aq. KOH; s. h. HCl	s. h.; s. h. ac.	i.; i. bz.; i. chl.
146	col. lq.	$0.841^{\frac{20°}{4}}$	−87.7	52.5	40	s.	s.
147	nd.	....	45-50	sl. diss.	v. sl. s. h.	s.	s.
148	lf./bz.	$1.122^{30°}$	84-5	125^{25mm}	204	74	v. s.
149	col. lq.	$1.062^{\frac{16°}{4}}$	13.0	141-2	∞; ∞ bz.	∞; ∞ chl.	∞; ∞ act.
150	col. lq.	$1.094^{0°}$	....	97^{35mm}	....	....	....
151	col. lq.	$0.806^{\frac{20°}{4}}$	−82	77.3	s.	....	....
152	cr. pd.	....	226-7	....	$0.4^{12°}$	....	....
153	mn. pr.	$1.360^{\frac{25°}{4}}$	151-3	265^{10mm}	$1.4^{15°}$	v. s.	$0.6^{15°}$
154	col. lq.	....	....	$125-8^{11mm}$	d. h.	d. h.	....
155	col. oil	$0.951^{\frac{19°}{19}}$	1	295	v. sl. s.	s.	v. sl. s.
156	pr./aq.	....	102	....	v. s.	s. h.	i.; i. lg.
157	col. cr. pd.	....	d. 207-11	....	$0.027^{20°}$; s. alk.	v. sl. s.; s. a.	i.; i. chl.
158	col. rhb./aq.	....	297 d.	....	$20.5^{45°}$	v. sl. s. abs.	i.
159	col. nd./aq.	....	295 d.	subl. >200	$21.7^{17°}$; $32^{75°}$	0.2, 80% c. al.	i.; i. act.
160	col. rhb.	....	196 d.	....	v. s.	v. sl. s. abs.	i.
161	lq.	$0.918^{23°}$	−70.3	178.3	i.; s. aq. a.	s.	s.;s. H_2SO_4
162	mn.	1.404	97-8	....	v. s.	v. s.	v. s.
163	nd.	....	164-6	....	....	....	....
164	nd./aq.	....	285	....	sl. s. h.	s. chl.	s.
165	nd.	....	248-9	....	$0.04^{25°}$ $0.7^{100°}$	s. h.; s. alk.	s.; i. chl.
166	nd./aq.	....	−H_2O, 100; m. anh. 179	....	$0.06^{25°}$; $6^{100°}$	s.	s. alk.
167	nd.	....	234	....	sl. s. h.	s.	s.
168	pr./aq.	....	243-4	subl.	sl. s.	s.	s.
169	col. lq.	$1.103^{\frac{20°}{4}}$	....	83^{20mm}; d. 85	∞	∞	s.
170	tri.	....	82	in vac. 90-100 (diss.)	v. s.	$25^{25°}$, 99% al.	$5^{23°}$
171	col. cr. pd.	....	80-2	....	sl.s.; s.bz.	s.; s. chl.	s.; s. oil

Table 7-4 (*Continued*)
PHYSICAL CONSTANTS OF ORGANIC COMPOUNDS

No.	Name	Synonym	Formula	Beil. Ref.	Formula Weight
172	**Alizarin** amide (β) (1,2)	. . . .	$C_6H_4(CO)_2C_6H_2 \cdot (OH)NH_2$	XIV-267	239.23
173	amide (α) (2,1)	. . . .	$C_6H_4(CO)_2C_6H_2 \cdot (OH)NH_2$	XIV-275	239.23
174	blue	diOH-anthraqui-none-quinoline	$C_6H_4(CO)_2C_9H_3N : (OH)_2$	XXI-632	291.27
175	blue S	. . . .	$C_{17}H_9O_4N \cdot 2NaHSO_3$	XXI-633	499.39
176	carboxylic acid (β)	. . . .	$(HO)_2C_6H_2(CO)_2 : C_6H_3 \cdot CO_2H$	X-1035	284.23
177	3-sulfonic acid (1,2;3)	. . . .	$(HO)_2C_{14}H_5O_2 \cdot SO_3H$	XI-355	320.28
178	3-sulfonic acid (Na)	alizarin S; alizarin carmine	$(HO)_2C_{14}H_5O_2 \cdot SO_3Na \cdot H_2O$	XI-355	360.28
179	**Aljodan**	iodoethyl allo-phanate	$IC_2H_4 \cdot CO_2NHCONH_2$	**III-56	258.02
180	**Alkannin**	. . . .	$C_{16}H_{16}O_5$	**VIII-544	288.30
181	**Allantoin** †	glyoxyldiureide	$C_4H_6O_3N_4$	XXV-474	158.12
182	**Allanturic acid**	lantanuric acid	$NH_2 \cdot CO \cdot N:CH \cdot CO_2H$	XXV-475	116.08
183	**Allene**	propadiene	$CH_2:C:CH_2$	I-248	40.07
184	**Alloxan**	mesoxalylurea	$HN \cdot (CO)_3 \cdot NH \cdot CO \cdot \overline{}$ 4H_2O	XXIV-500	214.13
185	**Alloxan**	. . . .	$HN \cdot (CO)_3 \cdot NH \cdot CO$	XXIV-500	142.07
186	**Alloxan** (mono-hydrate)	. . . .	$CO \cdot NHCONHCO \cdot C : (OH)_2$	XXIV-500	160.09
187	**Alloxanic acid**	. . . .	$C_4H_4O_5N_2$	III-772	160.09
188	**Alloxantin**	uroxin	$C_8H_6O_8N_4 \cdot 2H_2O$	XXVI-556	322.19
189	**Allyl** acetate	. . . .	$CH_3CO_2 \cdot C_3H_5$	II-136	100.12
190	acetic acid	pentenoic acid	$CH_2:CH \cdot C_2H_4 \cdot CO_2H$	II-425	100.12
191	acetone	. . . .	$C_3H_5 \cdot CH_2 \cdot CO \cdot CH_3$	I-734	98.15
192	acetonitrile	. . . .	$C_3H_5 \cdot CH_2CN$	II-426	81.12
193	alcohol	propen-1-ol-3	$CH_2:CH \cdot CH_2OH$	I-436	58.08
194	amine	. . . .	$CH_2:CH \cdot CH_2 \cdot NH_2$	IV-205	57.10
195	*iso*-amyl ether	. . . .	$C_3H_5 \cdot O \cdot C_5H_{11}$	I-438	128.22
196	aniline	. . . .	$C_3H_5 \cdot NH \cdot C_6H_5$	XII-170	133.19
197	arsonic acid	. . . .	$CH_2:CH \cdot CH_2 \cdot AsO(OH)_2$	**IV-998	166.01
198	benzoate	. . . .	$C_6H_5 \cdot CO_2 \cdot C_3H_5$	IX-114	162.19
199	bromide	3-bromo-propene-1	$CH_2:CH \cdot CH_2Br$	I-201	120.98
200	butyrate	. . . .	$CH_3(CH_2)_2CO_2 \cdot C_3H_5$	II-272	128.17
201	caproate (n)	. . . .	$CH_3(CH_2)_4CO_2 \cdot C_3H_5$	**II-285	156.23
202	chloride	3-chloro-propene-1	$CH_2:CH \cdot CH_2Cl$	I-198	76.53
203	cinnamate	. . . .	$C_6H_5 \cdot (CH)_2CO_2 \cdot C_3H_5$	*IX-230	188.23
204	cyanide	vinyl-acetonitrile	$CH_2:CH \cdot CH_2 \cdot CN$	II-408	67.09
205	*iso*-cyanide	allyl carbylamine	$CH_2:CH \cdot CH_2 \cdot NC$	IV-208	67.09
206	ether	diallyl ether	$(CH_2:CH \cdot CH_2)_2O$	I-438	98.15
207	formate	. . . .	$HCO_2 \cdot CH_2 \cdot CH:CH_2$	II-23	86.09
208	iodide	3-iodo-propene-1	$CH_2:CH \cdot CH_2I$	I-202	167.98
209	malonic acid	. . . .	$C_3H_5 \cdot CH(CO_2H)_2$	II-776	144.13
210	mercaptan	. . . .	$CH_2:CH \cdot CH_2 \cdot SH$	I-440	74.15
211	phenol (p)	chavicol	$CH_2:CH \cdot CH_2 \cdot C_6H_4 \cdot OH$	VI-571	134.18

† Cf. also Alkaloids table.

Alival 3402	Alizarin S 177	Alizarin orange 4599	Alizarin yellow A 6224
Alizarin 2297	Alizarin bordeaux 5801	Alizarin β-quinoline 174	Alizarin yellow C 6211
Alizarin (leuco) 2295	Alizarin carmine 177	Alizarin red R 178	Alizarin yellow R 4653

Table 7-4 (*Continued*)
PHYSICAL CONSTANTS OF ORGANIC COMPOUNDS

No.	Crystalline Form and Color	Specific Gravity	Melting Point °C.	Boiling Point °C.	Solubility in 100 Parts		
					Water	Alcohol	Ether
172	br. nd./al.	....	....	....	i.	s.	s.; sl. s. NH$_4$OH
173	br. nd./al.	....	250	....	s. alk.	s.	s.; s. alk. carb.
174	b. met. nd./bz.	....	270±	subl.	i.; s. h. bz.	sl. s.; s. ac.	sl. s.
175	r. br. cr.	....	....	....	v. s.; d. h.	sl. s.	i.
176	or. nd./ PhNO$_2$	....	305	subl.	s. aq. NaOAc	sl. s.	v. sl. s.
177	or. yel. cr.	....	....	....	s.	s.	i.
178	or. yel. pd.	....	....	....	s.	s.	....
179	col. pd.	....	192±d.	....	sl. s.	s.; s. act.	v. sl. s. bz.
180	br. met.	....	149	....	s. alk.	v. sl. s.	v. sl. s.
181	nd./h. aq.	....	235-6	....	$0.76^{22°}$; 3.3 h.	sl. s.; s. NaOH	i.
182	hyg. pd.	....	....	....	delq.	i.	d. h. alk.
183	gas	....	−146	−32	....	....	....
184	rhb. pr. /aq.; eff. −3H$_2$O	....	−4H$_2$O, 150	d. 170	s.	s.	....
185	rhb. yel. hyg.	....	d. 170	....	v. s.; s. ac.	v. s.; sl. s. pet.	i.; sl. s. chl.
186	tri./aq.	....	d. 170	....	v. s.	v. s.	i.
187	tri.	....	d.	....	v. s.	16	s.
188	rhb.	....	−2H$_2$O, 120	d. 253-5	$0.3^{25°}$; 6h.	v. sl. s.	v. sl. s.
189	col. lq.	0.928^{20}_{4}	....	104^{762mm}	i.	∞	∞
190	col. lq.	0.984^{18}_{4}	<−18	187-9	sl. s.	s.	s.
191	col. lq.	0.847^{16}_{4}	....	129^{748mm}	i.	....	....
192	lq.	$1.180^{13°}$	....	140	i.	....	....
193	lq.	0.855^{15}_{4}	−129	97.1	∞	∞	∞
194	col. lq.	0.761^{22}_{4}	....	56.5^{756mm}	∞	∞	∞
195	col. lq.	....	....	120	i.	∞	∞
196	yel. oil	$0.982^{25°}$	....	$217-9^{736mm}$	i.	s.	s.
197	col. cr./aq.	....	129-30	....	v.s.h.; d.h.a.	s.	....
198	yel. lq.	1.058^{15}_{15}	....	230	i.	s.	s.
199	lq.	1.398^{20}_{4}	−119.4	$70-17^{53mm}$	i.	∞	∞
200	col. lq.	....	....	143^{772mm}	i.	s.	s.
201	pa. yel. lq.	....	....	186-8	i.	s.	s.
202	col. lq.	0.938^{20}_{4}	−134.5	45.1	<0.1	∞	∞
203	col. lq.	1.052^{25}_{25}	....	$150-21^{5mm}$	i.	s.	v. s.
204	col. lq.	$0.837^{16°}$	−86.8	118.4	....	s.	....
205	lq.	$0.797^{17°}$	....	106	sl. s.	s.	∞
206	lq.	0.826^{20}_{4}	....	94.3	0.3	∞	∞
207	lq.	$0.948^{18°}$	....	83.6^{768mm}	i.	s.	....
208	yel. lq.	1.848^{12}_{12}	−99.3	102^{744mm}	i.	∞	∞
209	cr./et.	....	102-5	d.−CO$_2$,180	s.	s.	s.; s. h. bz.
210	lq.	$0.925^{23°}_{4}$	....	67-8	....	∞	∞
211	lq.	$1.023^{19.4°}_{4}$	<−25	237	∞ pet.	∞ ; ∞ chl.	∞

Table 7-4 (*Continued*)
PHYSICAL CONSTANTS OF ORGANIC COMPOUNDS

No.	Name	Synonym	Formula	Beil. Ref.	Formula Weight
	Allyl				
212	phenylcinchonate	atoquinol (2,4)	$C_9H_5N(C_6H_5)CO_2 \cdot$ C_3H_5	....	289.34
213	phenylcinchonate HCl	....	$C_{19}H_{15}O_2N \cdot HCl$	....	325.80
214	phenyl ether	....	$C_3H_5 \cdot O \cdot C_6H_5$	VI-144	134.18
215	phenyl urea (*N,N'*)	....	$C_3H_5NH \cdot CO \cdot$ NHC_6H_5	XII-350	176.22
216	propionate	....	$C_2H_5 \cdot CO_2 \cdot C_3H_5$	II-241	114.15
217	*iso*-propylacetyl-carbamide	sedormid	$(C_3H_7)(C_3H_5):CH \cdot$ $CO \cdot NHCONH_2$	**III-53	184.24
218	pyridine (2)	....	$C_3H_5 \cdot C_5H_4N$	....	119.17
219	thiocyanate	....	$C_3H_5 \cdot SCN$	III-177	99.16
220	*iso*-thiocyanate	mustard oil	$C_3H_5 \cdot N:C:S$	IV-214	99.16
221	thiourea	thiosinamine	$C_3H_5 \cdot NH \cdot CS \cdot NH_2$	IV-211	116.19
222	trisulfide	....	$(C_3H_5)_2S_3$	I-441	178.34
223	urea	....	$C_3H_5 \cdot NH \cdot CO \cdot NH_2$	IV-209	100.12
224	*iso*-valerate	....	$(CH_3)_2CH \cdot CH_2 \cdot$ $CO_2C_3H_5$	II-313	142.20
225	**Allylene**	propyne; propine	$CH_3 \cdot C:CH$	I-246	40.07
226	oxide	....	$CH_3 \cdot C:CH \cdot O$	XVII-20	56.06
227	**Aloin**	....	$C_{21}H_{20}O_9$	....	416.39
228	**Alstonine**	....	$C_{21}H_{20}O_4N_2 \cdot 3\frac{1}{2}H_2O$	....	427.46
229	**Aluminum** ethoxide	....	$Al(OC_2H_5)_3$	I-313	162.17
230	triethyl	....	$Al(C_2H_5)_3$	IV-643	114.17
231	trimethyl	....	$Al(CH_3)_3$	IV-643	72.09
232	**Amalinic acid**	tetraMe-alloxantin	$(CH_3)_4C_8H_2O_8N_4$	XXVI-559	342.27
233	**Amarine**	triphenyl-imina-zoline	$C_6H_5CHCH(C_6H_5) \cdot$ $N:C(C_6H_5) \cdot NH \cdot$ $\frac{1}{2}H_2O$	XXIII-304	307.40
234	**Amaron**	tetraphenylpyrazine	$C_{28}H_{20}N_2$	XXIII-343	384.49
235	**Amino**-acetone	acetonyl-amine	$NH_2 \cdot CH_2 \cdot CO \cdot CH_3$	IV-314	73.10
236	acetophenone (*o*)	....	$NH_2 \cdot C_6H_4 \cdot CO \cdot CH_3$	XIV-41	135.17
237	acetophenone (*m*)	....	$NH_2 \cdot C_6H_4 \cdot CO \cdot CH_3$	XIV-45	135.17
238	acetophenone (*p*)	....	$NH_2 \cdot C_6H_4 \cdot CO \cdot CH_3$	XIV-46	135.17
239	acetophenone (*ω*)	phenacylamine	$C_6H_5 \cdot CO \cdot CH_2 \cdot NH_2$	XIV-49	135.17
240	acetophenone HCl (*ω*)	....	$C_8H_7O \cdot NH_2 \cdot HCl$	XIV-49	171.63
241	alizarin (3)	*β*-aminoalizarin	$C_6H_4(CO)_2C_6H \cdot$ $(OH)_2NH_2$	XIV-285	255.23
242	alizarin (4)	*α*-aminoalizarin	$C_{14}H_5O_2(OH)_2NH_2$	XIV-286	255.23
243	amylene glycol	2-NH$_2$-2-Et-1,3-propandiol	$C_2H_5 \cdot C(NH_2):$ $(CH_2OH)_2$	....	119.16
244	anthraquinone (1)	....	$C_6H_4(CO)_2C_6H_3NH_2$	XIV-177	223.23
245	anthraquinone (2)	....	$C_6H_4(CO)_2C_6H_3NH_2$	XIV-191	223.23
246	2-azo-5-anisole	benzene-azo-*o*-anis-idine (5;2,1)	$C_6H_5 \cdot N:N \cdot C_6H_3(NH_2) \cdot$ OCH_3	XVI-396	227.27
247	azobenzene (*o*)	....	$NH_2 \cdot C_6H_4 \cdot N_2 \cdot C_6H_5$	XVI-303	197.24
248	azobenzene (*m*)	....	$NH_2 \cdot C_6H_4 \cdot N_2 \cdot C_6H_5$	XVI-304	197.24

Allyl sulfide 1684
Allylen dichloride 2059
Almond oil 642
Alpha S acid 4533
Alphogen 5662

Alphol 4568
Alphozone 5662
Alurate 5370
Alypin 736-7
Amatol, a mixt. of 6316 and ammonium nitrate.

Table 7-4 (*Continued*)
PHYSICAL CONSTANTS OF ORGANIC COMPOUNDS

No.	Crystalline Form and Color	Specific Gravity	Melting Point °C.	Boiling Point °C.	Solubility in 100 Parts		
					Water	Alcohol	Ether
212	yel. nd./al.	....	36	260^{15mm}	i.; s. oil	8; s. bz.	v. s.
213	yel. nd./al.	....	145-7	....	d.	s. h.	i.
214	col. oil	$0.986\frac{15°}{15}$	....	191.7	i.	....	....
215	nd./bz.	....	114.5-5.5	....	s. bz.	....	....
216	col. lq.	....	....	124.5^{774mm}	....	....	....
217	nd./al.	....	194	....	0.03c., 0.5 h.	10	1.3
218	lq.	$0.959^{0°}$	....	189-90	....	....	....
219	oil	$1.056^{15°}$	....	161	v. sl. s.	v. s.	v. s.
220	col. oil	$1.024\frac{15°}{4}$	−102.5	152	0.2	∞	∞
221	col. pr.	$1.219\frac{20°}{20}$	77-8	....	$3^{0°}$; s. h.	s.; i. bz.	v. sl. s.
222	lq.	$1.085^{15°}$	....	$112\text{-}22^{16mm}$	....	....	....
223	nd./al.	....	85	....	v. s.	v. s.	v. sl. s.
224	lq.	....	....	155	∞	....	∞
225	gas	$0.660^{\frac{-13°}{4}}$	−102.7	−23.2	v. sl. s.	v. s.	$3000cc.^{16°}$
226	lq.	....	....	62-3	sl. s.	....	i. aq. K_2CO_3
227	yel. pr./al.	....	147.9	....	sl. s. h.	sl. s. h.	v. s. alk.
228	br. amor.	....	<100; 195 ±(anh.)	....	sl. s.	s.	v. sl. s.
229	pd.	$1.142\frac{20°}{0}$	150-60	$200\text{-}5^{10mm}$	d.	i.; sl. s. bz.	v. sl. s.
230	col. lq.	....	<−18	194	d.	d. by air	....
231	col. lq.	....	0	129-30	d.	d. by air	....
232	cr./aq.	....	221 d.	....	sl. s. h.	v. sl. s.	s. alk.
233	cr./al.	....	106; 131-3 (anh.)	d. 198 (anh.)	i.	s.	s.
234	red nd./ac.	....	245-6	subl.	i.; v. s. h. bz.	sl. s. h.; s. chl.	sl. s. h.
235	only salts are known	....	....	....	....	....	....
236	yel. oil	....	....	250 (sl. d.)	i.	s.	s.
237	lf./aq.	....	99.5	289-90	....	....	....
238	cr./aq.	....	106	293-5	s. h.	s.	s.
239	unstable	....	....	....	....	....	....
240	cr.	....	184-8	....	s.	....	i.
241	red pr./ac.	....	>300	subl. d.	s. KOH	sl. s.	sl. s. HCl
242	bl. nd./al.	....	....	....	s. alk.	s.	....
243	col. cr.	$1.099\frac{20°}{20}$	37.5-8.5	$152\text{-}3^{10mm}$	∞	∞	v. sl. s.
244	red nd.	....	256	subl.	i.; s. HCl	s.; s. bz.	s.; s. chl.
245	red nd./al.	....	302	subl.	i.	s.	i.; s. bz.
246	red br. pl./PhMe	....	110.5-1.5	....	s. ac.	s.; s. bz.	s.
247	red pr./al.	....	59	....	i.	v. s.	v. s.
248	or. nd./lg.	....	56-7	....	s. bz.	s.	s.; s. chl.

Table 7-4 (*Continued*)
PHYSICAL CONSTANTS OF ORGANIC COMPOUNDS

No.	Name	Synonym	Formula	Beil. Ref.	Formula Weight
	Amino				
249	azobenzene (*p*)	aniline yellow	$NH_2 \cdot C_6H_4 \cdot N_2 \cdot C_6H_5$	XVI-307	197.24
250	azobenzene HCl (*p*)	....	$C_{12}H_9N_2(NH_2) \cdot HCl$	XVI-310	233.70
251	α-azonaphthalene	azodinaphthylamine	$NH_2 \cdot C_{10}H_6 \cdot N_2 \cdot C_{10}H_7$	XVI-365	297.36
252	azophenylene	benzotriazole	$C_6H_4NH \cdot N:N$ $\rule{1.2cm}{0.4pt}$	XXVI-38	119.13
253	azoxylene HCl	....	$(CH_3)_4C_{12}H_5(NH_2)N_2 \cdot$ HCl	....	289.81
254	benzaldehyde (*o*)	....	$NH_2 \cdot C_6H_4 \cdot CHO$	XIV-21	121.14
255	benzaldehyde (*m*)	....	$NH_2 \cdot C_6H_4 \cdot CHO$	XIV-28	121.14
256	benzaldehyde (*p*)	....	$NH_2 \cdot C_6H_4 \cdot CHO$	XIV-29	121.14
257	benzaldoxime (*o*)	....	$NH_2 \cdot C_6H_4 \cdot CH:NOH$	XIV-24	136.15
258	benzaldoxime (*m*)	....	$NH_2 \cdot C_6H_4 \cdot CH:NOH$	XIV-28	136.15
259	benzamide (*o*)	anthranilamide	$NH_2 \cdot C_6H_4 \cdot CO \cdot NH_2$	XIV-320	136.15
260	benzamide (*m*)	....	$NH_2 \cdot C_6H_4 \cdot CO \cdot NH_2 \cdot$ H_2O	XIV-390	154.17
261	benzamide (*p*)	....	$C_7H_8ON_2 \cdot \frac{1}{4}H_2O$	XIV-425	140.66
262	benzidine (2,4;4')	triamino-diphenyl	$(NH_2)_2C_6H_3C_6H_4 \cdot$ NH_2	XIII-306	199.26
263	benzimidazole (2)	*o*-phenylene-guanidine	$C_6H_4 \cdot NH \cdot C(NH_2):N$ $\rule{1.2cm}{0.4pt}$	XXIV-116	133.15
264	benzoic acid (*o*)	anthranilic acid	$NH_2 \cdot C_6H_4 \cdot CO_2H$	XIV-310	137.14
265	benzoic acid (*m*)	....	$NH_2 \cdot C_6H_4 \cdot CO_2H$	XIV-383	137.14
266	benzoic acid (*p*)	aminodracylic acid	$NH_2 \cdot C_6H_4 \cdot CO_2H$	XIV-418	137.14
267	benzonitrile (*o*)	anthranilic nitrile	$NH_2 \cdot C_6H_4 \cdot CN$	XIV-322	118.14
268	benzonitrile (*m*)	....	$NH_2 \cdot C_6H_4 \cdot CN$	XIV-391	118.14
269	benzonitrile (*p*)	....	$NH_2 \cdot C_6H_4 \cdot CN$	XIV-425	118.14
270	benzophenone (*o*)	....	$C_6H_5 \cdot CO \cdot C_6H_4 \cdot NH_2$	XIV-76	197.24
271	benzophenone (*m*)	....	$C_6H_5 \cdot CO \cdot C_6H_4 \cdot NH_2$	XIV-81	197.24
272	benzophenone (*p*)	....	$C_6H_5 \cdot CO \cdot C_6H_4 \cdot NH_2$	XIV-81	197.24
273	benzothiazole (2)	....	$C_6H_4 \cdot N:C(NH_2) \cdot S$ $\rule{1.4cm}{0.4pt}$	XXVII-182	150.20
274	benzyl-cyanide (*p*)	aminophenyl-aceto-nitrile	$H_2N \cdot C_6H_4 \cdot CH_2 \cdot CN$	XIV-457	132.17
275	benzyl-cyanide HCl (*p*)	....	$H_2N \cdot C_6H_4 \cdot CH_2 \cdot CN \cdot$ HCl	XIV-457	168.63
276	benzoyl-2,2-di-methyl-3-di-ethylaminopro-panol HCl (*p*)	larocaine	$H_2N \cdot C_6H_4 \cdot CO_2 \cdot CH_2 \cdot$ $C(CH_3)_2 \cdot CH_2 \cdot$ $N(C_2H_5)_2 \cdot HCl$	....	314.86
277	butyric acid (α) (*dl*)	....	$C_2H_5 \cdot CHNH_2 \cdot CO_2H$	IV-408	103.12
278	butyric acid (β) (*dl*)	....	$CH_3 \cdot CHNH_2 \cdot CH_2 \cdot$ CO_2H	IV-412	103.12
279	butyric acid (γ)	piperidinic acid	$NH_2 \cdot (CH_2)_3 \cdot CO_2H$	IV-413	103.12
280	*iso*-butyric acid (α)	....	$(CH_3)_2CNH_2 \cdot CO_2H$	IV-414	103.12
281	camphor (3) (α)	....	$C_{10}H_{15}O \cdot NH_2$	XIV-10	167.25
282	*n*-caproic acid (α) (*dl*)	....	$CH_3(CH_2)_3CHNH_2 \cdot$ CO_2H	IV-433	131.18
283	*n*-caproic acid (α)	norleucine (*l* +)	$CH_3(CH_2)_3CHNH_2 \cdot$ CO_2H	IV-432	131.18
284	*n*-caproic acid (ε)	....	$H_2N(CH_2)_5CO_2H$	IV-434	131.18

Amino-azotoluene 309-12
Amino-barbituric acid 6384
Amino-benzene 513

Amino-benzene sulfonic acid 523-5
Amino-butyl-guanidine, cf. alko.

Table 7-4 (*Continued*)
PHYSICAL CONSTANTS OF ORGANIC COMPOUNDS

No.	Crystalline Form and Color	Specific Gravity	Melting Point °C.	Boiling Point °C.	Solubility in 100 Parts		
					Water	Alcohol	Ether
249	yel. mn.	....	126-7	225^{120mm}	sl. s. h.	s. h.	s.
250	steel b.*	....	....	....	s.	....	....
251	br. nd.	....	175	....	sl. s. bz.	sl. s.	sl. s.
252	col. nd./bz.	....	98.5	....	i.	s.	s. bz.
253	br. pd.	....	....	....	i.	s.	sl. s.
254	lf.	....	39-40	d.	v. sl. s.	v. s.	v. s.
255	unstable	....	....	....	....	....	....
256	lf./aq.	....	71.5	....	s.	....	....
257	nd./bz.	....	135-6	subl.	v. sl. s.	s.	s.
258	nd./bz.	....	88	....	v. sl. s. lg.	s.	s.
259	lf./chl.	....	108	300 ± (sl.d.)	v. s. h.	v. s.	sl. s.; i. bz.
260	mn./aq.	....	78-9	$-H_2O > 100$	s.	s.	s.
261	pa. yel. cr.	....	182.9 (anh.)	$-\frac{1}{4}H_2O$, 170	sl. s.	....	....
262	nd.	....	134	....	....	....	....
263	lf./aq.	....	222-4	....	s.; s. alk.; s. dil. a.	s.; s. act.; v. sl. s. bz.	v. sl. s.
264	col. rhb.	....	146-7	subl.	0.35$^{14°}$	**10.7$^{10°}$	16$^{7°}$
265	nd./aq.	1.5114$^{°}$	173-4	....	**0.6$^{15°}$	**2.2$^{10°}$	1.8$^{6°}$
266	mn./pr.	....	187-8	....	0.31$^{3°}$	**11.3$^{10°}$	8.2$^{6°}$
267	nd./CS$_2$	....	51	267-8^{777mm}	v. sl. s.	s.	s.
268	nd./aq. al.	....	53	288-90	s. h.	v. s.	v. s.
269	mn. pr.	....	86	d.	v. s. h.	v. s.	v. s.; i. HCl
270	yel. lf./al.	....	106-10	....	....	s.	s.
271	yel. lf./aq.	....	86-7	....	sl. s.	s.	s.
272	lf./aq. al.	....	124	....	v. sl. s. c.	v. s.	2$^{25°}$
273	lf./aq.	....	130-2	d.	v. sl. s.; s. conc. a.	s.; s. chl.	s.
274	lf./aq.	....	46	312	sl. s. h.	s.	s.
275	cr.	....	217-20	....	....	v. sl. s. c.	....
276	col. cr.	....	196-7	....	33.3	10	sl. s. c.
277	lf.	....	285 d.	subl. > 300	33 c.	0.2 h.	i.
278	cr.	....	193-4	....	100	i. abs. al.	i. abs. et.
279	lf. or nd.	....	203 d.	....	v. s.	i.	i.
280	pl. or pr./aq.	....	203 d.	subl. 280	v. s.	sl. s.	i.
281	waxy cr.	....	70-110	245-6	i.	s.	s.
282	col. lf./aq.	....	300 ± †	....	1.25	....	....
283	pl. or lf./aq.	....	301	subl. > 275	1.74$^{230°}$	v. sl. s.	....
284	lf./MeOH-et.	....	202-3	softens > 190	v. s.	i.	sl. s. MeOH

* Also a red-black crystalline form.   † Sealed tube.
** In 90% alcohol.

Table 7-4 (*Continued*)
PHYSICAL CONSTANTS OF ORGANIC COMPOUNDS

No.	Name	Synonym	Formula	Beil. Ref.	Formula Weight
285	**Amino** *iso*-caproic (α)	leucine (*l* −)	$(CH_3)_2CHCH_2\cdot$ $CH(NH_2):CO_2H$	IV-437	131.18
286	caproic acid (α)	isoleucine (*l* +)	$(C_2H_5)(CH_3)CH\cdot$ $CHNH_2CO_2H$	IV-454	131.18
287	caproic acid (α)	allo-isoleucine (*d* −)	$(C_2H_5)(CH_3)CH\cdot$ $CHNH_2CO_2H$	IV-457	131.18
288	caprylic acid (α) (*dl*)	. . . .	$CH_3(CH_2)_5CHNH_2\cdot$ CO_2H	IV-461	159.23
289	cinnamic acid (*o*)	. . . .	$H_2N\cdot C_8H_6\cdot CO_2H$	XIV-517	163.18
290	cinnamic acid (*m*)	. . . .	$H_2N\cdot C_8H_6\cdot CO_2H$	XIV-520	163.18
291	cinnamic acid (*p*)	. . . .	$H_2N\cdot C_8H_6\cdot CO_2H$	XIV-521	163.18
292	*o*-cresol (4;1,2)	amino-hydroxy-toluene	$NH_2\cdot C_6H_3\cdot CH_3(OH)$	XIII-574	123.16
293	*o*-cresol (5;1,2)	. . . .	$NH_2\cdot C_6H_3CH_3\cdot (OH)$	XIII-576	123.16
294	*o*-cresol H_2SO_4	(5;1,2)	$(C_7H_9ON)_2\cdot H_2SO_4$	XIII-576	344.39
295	*m*-cresol (2;1,3)	. . . .	$NH_2\cdot C_6H_3\cdot CH_3(OH)$	XIII-589	123.16
296	*m*-cresol (4;1,3)	. . . .	$NH_2\cdot C_6H_3\cdot CH_3(OH)$	XIII-590	123.16
297	*m*-cresol (6;1,3)	. . . .	$NH_2\cdot C_6H_3\cdot CH_3(OH)$	XIII-593	123.16
298	*p*-cresol (2;1,4)	. . . .	$NH_2\cdot C_6H_3\cdot CH_3(OH)$	XIII-598	123.16
299	*p*-cresol (3;1,4)	. . . .	$NH_2\cdot C_6H_3\cdot CH_3(OH)$	XIII-601	123.16
300	dibutylaniline (*p*)	. . . .	$(C_4H_9)_2N\cdot C_6H_4\cdot NH_2$	*XIII-23	220.36
301	dibutylaniline (*p*)	HCl	$C_{14}H_{24}N_2\cdot 2HCl$	*XIII-23	293.28
302	diethylamino-toluene (mono) HCl	(5;1,2)	$(C_2H_5)_2N\cdot C_6H_3\cdot$ $(CH_3)NH_2\cdot HCl$	. . . .	214.74
303	diethylaniline (*p*)	diEt-phenylene-diNH_2	$(C_2H_5)_2N\cdot C_6H_4\cdot NH_2$	XIII-75	164.25
304	diethylaniline (*p*)	(mono-HCl)	$C_{10}H_{16}N_2\cdot HCl$	. . . .	200.71
305	dimethylaniline (*o*)	diMe-*o*-phenylene-diamine (*N,N*)	$(CH_3)_2N\cdot C_6H_4\cdot NH_2$	XIII-15	136.20
306	dimethylaniline (*m*)	. . . .	$(CH_3)_2N\cdot C_6H_4\cdot NH_2$	XIII-38	136.20
307	dimethylaniline (*p*)	. . . .	$(CH_3)_2N\cdot C_6H_4\cdot NH_2$	XIII-72	136.20
308	dimethylaniline (*p*)	H_2SO_4	$C_8H_{12}N_2\cdot H_2SO_4$	XIII-73	234.28
309	2,3′-dimethyl-azobenzene (4)	amino-azo-toluene (3′;1,2,4)	$CH_3\cdot C_6H_4\cdot N_2\cdot C_6H_3\cdot$ $(CH_3)NH_2$	XVI-348	225.30
310	2,3′-dimethyl-azobenzene (4′)	(2;1,3′,4′)	$CH_3\cdot C_6H_4\cdot N_2\cdot C_6H_3\cdot$ $(CH_3)NH_2$	XVI-344	225.30
311	2,4′-dimethyl-azobenzene (4)	(4′;1,2,4)	$CH_3\cdot C_6H_4\cdot N_2\cdot C_6H_3\cdot$ $(CH_3)NH_2$	XVI-348	225.30
312	3,4′-dimethyl-azobenzene (4)	(4′;1,3,4)	$CH_3\cdot C_6H_4\cdot N_2\cdot C_6H_3\cdot$ $(CH_3)NH_2$	XVI-345	225.30
313	diphenylamine (*o*)	*N*-phenyl-*o*-phenylenediamine	$NH_2\cdot C_6H_4\cdot NH\cdot C_6H_5$	XIII-16	184.24
314	diphenylamine (*p*)	. . . .	$NH_2\cdot C_6H_4\cdot NH\cdot C_6H_5$	XIII-76	184.24
315	diphenylamine (*p*)	HCl	$C_{12}H_{10}N\cdot NH_2\cdot HCl$	*XIII-23	220.70
316	diphenylamine (*p*)	H_2SO_4	$(C_{12}H_{10}N\cdot NH_2)_2\cdot$ H_2SO_4	XIII-78	466.56
317	4-(*p*-diphenyl)-thiazole (2)	. . . .	$C_6H_5\cdot C_6H_4\cdot C:CH\cdot$ $\underline{\qquad\mid}$ $S\cdot C(NH_2):N$ $\overline{\qquad\qquad\mid}$	. . . .	252.34
318	ethylacetanilide (*p*)	. . . .	$NH_2C_6H_4N(C_2H_5)\cdot$ $COCH_3$	. . . .	178.24

Table 7-4 (*Continued*)
PHYSICAL CONSTANTS OF ORGANIC COMPOUNDS

No.	Crystalline Form and Color	Specific Gravity	Melting Point °C.	Boiling Point °C.	Solubility in 100 Parts		
					Water	Alcohol	Ether
285	lf./aq. al.	$1.293\frac{18°}{4}$	293-5 d.*	subl.	$2^{17°}$; 6.6 h.	$0.07^{17°}$ abs.	i.; 10.9 ac.
286	rhb./80% al.	....	280 d.*	subl. >280	$3.9^{15.5°}$; $6.1^{75°}$	sl. s. h.; s. h. ac.	i.; s. h. gly.
287	lf. or rods	....	280 d.*	....	$2.6^{20°}$	i. c.	s. h. ac.
288	lf./aq.	....	263-4; subl.	....	sl. s. c.; 0.6 h.	v. sl. s.	v. sl. s.
289	yel. nd.	....	158-9 d.	....	s. h.	s.	s.
290	pa. yel. nd./al.	....	181-2	....	sl. s. h.	s.	s.
291	yel. nd.	....	175-6 d.	....	s. h.	s.	s.
292	col. lf./aq.	....	159-61	subl.	s. h.	v. s.	v. s.
293	lf./bz.	....	174-5	subl.	v. sl. s.	s.	s.; sl. s. bz.
294	....	....	178-9	....	s.	i.	i.
295	lf.	....	148-50	subl.	....	....	....
296	lf.	....	....	....	....	....	....
297	cr./bz.	....	174 d.	....	....	....	....
298	cr./aq.	....	144.5	subl.	sl. s. c.	....	....
299	rhb./bz.	....	135	subl.	i. c.	s.; sl. s. bz.	s.; s. chl.
300	lq.	....	....	....	....	....	....
301	nd./al.	....	....	....	s.	s.	i.
302	col. nd.	....	....	....	s.	sl. s.	i.
303	lq.	....	22-4	260-2	i.	s.	s.
304	col. nd.	....	....	....	s.	s.	i.
305	oil	....	....	$217.5^{751.5mm}$	sl. s.	s.	s.
306	oil	$0.995^{25°}$	<−20	$268-70^{740mm}$	sl. s.	s.	s.
307	col. nd.	$1.041\frac{15°}{15}$	41	262.3	s. c.	v. s.	s.; s. chl.
308	lf.	....	....	....	v. s.	....	....
309	yel. nd.	....	80	....	i.	s.	....
310	yel. mn./al.	....	100	....	i.	s.	s.; s. chl.
311	yel. lf./al.	....	127	....	i.	s.	sl. s. lg.
312	yel. nd./lg.	....	127-8	....	i.	s.	sl. s. lg.
313	nd./aq.	....	79-80	....	s. chl.	s. act.	s. bz.; sl s. lg.
314	nd./aq. al.	....	66-7**	354 (in H_2)	sl. s.	s. abs. al.	s.
315	grn. pd.	....	....	....	s. h.	s. h.	i.
316	gr. pd.	....	....	....	v. sl. s.	sl. s.	i.
317	col. pd.	....	206-8	....	i.	s. h.	i.
318	col. fl.	....	98-9	....	i.	v. s.	i.

* Sealed tube.
** Cryst./pet. m.p. 75°.
Amino-dracylic acid 266

Amino-ethane 2950
Amino-ethyl alcohol 8, 2907

Table 7-4 (*Continued*)
PHYSICAL CONSTANTS OF ORGANIC COMPOUNDS

No.	Name	Synonym	Formula	Beil. Ref.	Formula Weight
319	**Amino** ethylbenzene (*o*)	*o*-Et-aniline	$NH_2 \cdot C_6H_4 \cdot C_2H_5$	XII-1089	121.18
320	ethylbenzene (*m*)	....	$NH_2 \cdot C_6H_4 \cdot C_2H_5$	XII-1090	121.18
321	ethylbenzene (*p*)	....	$NH_2 \cdot C_6H_4 \cdot C_2H_5$	XII-1090	121.18
322	guanidine	guanyl hydrazine	$H_2N \cdot C(NH) \cdot NH \cdot NH_2$	III-117	74.09
323	guanidine H_2CO_3	(bicarbonate)	$CH_6N_4 \cdot H_2CO_3$	III-118	136.11
324	guanidine H_2SO_4	(acid sulfate)	$CH_6N_4 \cdot H_2SO_4$	III-118	172.16
325	guanidine H_2SO_4	....	$(CH_6N_4)_2 \cdot H_2SO_4 \cdot H_2O$	....	264.26
326	hexahydrobenzene	cyclohexylamine	$H_6C_6H_5 \cdot NH_2$	XII-5	99.19
327	1-hydroxyanthra-quinone (4)	quinizarinamide	$C_6H_4(CO)_2C_6H_2 \cdot (OH)NH_2$	XIV-268	239.23
328	malonic acid	....	$NH_2 \cdot CH(CO_2H)_2$	IV-469	119.08
329	4-methylthiazole (2)	....	$CH_3 \cdot C\!:\!CH \cdot S \cdot \underset{\underline{}}{\overset{\displaystyle |\underline{}}{C(NH_2)\!:\!N}}$	XXVII-159	114.17
330	methylacetanilide (*p*)	....	$NH_2 \cdot C_6H_4 \cdot N(CH_3) \cdot COCH_3$	*XIII-30	164.21
331	naphthalene-4-azobenzene-*p*-sulfonic acid (1)	(1,4;4')	$NH_2 \cdot C_{10}H_6 \cdot N\!:\!N \cdot C_6H_4 \cdot SO_3H$	XVI-367	327.36
332	α-naphthol (4,1)	4-amino-naphthol-1	$NH_2 \cdot C_{10}H_6 \cdot OH$	XIII-667	159.19
333	α-naphthol (8,1)	8-amino-naphthol-1	$NH_2 \cdot C_{10}H_6 \cdot OH$	XIII-672	159.19
334	β-naphthol (1,2)	1-amino-naphthol-2	$NH_2 \cdot C_{10}H_6 \cdot OH$	XIII-676	159.19
335	β-naphthol (1,2) HCl	HCl	$NH_2 \cdot C_{10}H_6 \cdot OH \cdot HCl$	XIII-677	195.65
336	β-naphthol (7,2)	7-amino-naphthol-2	$NH_2 \cdot C_{10}H_6 \cdot OH$	XIII-684	159.19
337	1-naphthol-3-sul-fonic acid (6) (6;1,3)	J acid	$NH_2 \cdot C_{10}H_5(OH) \cdot SO_3H$	XIV-823	239.25
338	1-naphthol-3-sul-fonic acid (7) (7;1,3)	gamma acid	$NH_2 \cdot C_{10}H_5(OH) \cdot SO_3H$	XIV-828	239.25
339	1-naphthol-5-sul-fonic acid (8) (8;1,5)	S acid	$NH_2 \cdot C_{10}H_5(OH) \cdot SO_3H$	XIV-835	239.25
340	2-naphthol-4-sul-fonic acid (1) (1;2,4)	1:2:4-acid	$NH_2 \cdot C_{10}H_5(OH) \cdot SO_3H \cdot \frac{1}{2}H_2O$	XIV-846	248.26
341	2-naphthol-6-sul-fonic acid (1) Na	eikonogen (1;2,6)	$NH_2 \cdot C_{10}H_5(OH) \cdot SO_3Na \cdot 2\frac{1}{2}H_2O$	XIV-848	306.27
342	1-naphthol-3,6-disulfonic acid (8;1,3,6)	H acid	$NH_2 \cdot C_{10}H_4(OH)\!:\!(SO_3H)_2$	XIV-840	319.31
343	naphthol-disulfonic acid Na	H acid Na salt	$C_{10}H_7O_7NS_2Na_2 \cdot 1\frac{1}{2}H_2O$	XIV-840	390.30
344	1-naphthol-5,7-disulfonic acid (8;1,5,7)	2S acid	$NH_2 \cdot C_{10}H_4(OH)\!:\!(SO_3H)_2$	XIV-845	319.31
345	phenol (*o*)	2-aminophenol	$NH_2 \cdot C_6H_4 \cdot OH$	XIII-354	109.13
346	phenol (*o*) HCl	....	$HO \cdot C_6H_4 \cdot NH_2 \cdot HCl$	XIII-358	145.59
347	phenol (*m*)	3-aminophenol	$NH_2 \cdot C_6H_4 \cdot OH$	XIII-401	109.13
348	phenol (*m*) HCl	....	$HO \cdot C_6H_4 \cdot NH_2 \cdot HCl$	XIII-403	145.59
349	phenol (*p*)	*p*-hydroxyaniline	$NH_2 \cdot C_6H_4 \cdot OH$	XIII-427	109.13
350	phenol (*p*) HCl	....	$HO \cdot C_6H_4 \cdot NH_2 \cdot HCl$	XIII-434	145.59

Amino-ethyl-indole 6363
Amino-ethyl propandiol 243
Amino-form 3599
Amino G acid 4537
Amino-glutaric acid 3371-2

Amino H acid 4555
Amino-hydrocinnamic acid 5149-50
Amino-hydroxybenzoic acid 374-6
Amino-hydroxybutyric acid 5932-4
Amino-hydroxypurine, cf. alkd.

Table 7-4 (*Continued*)
PHYSICAL CONSTANTS OF ORGANIC COMPOUNDS

No.	Crystalline Form and Color	Specific Gravity	Melting Point °C.	Boiling Point °C.	Solubility in 100 Parts		
					Water	Alcohol	Ether
319	lq.	$0.983^{22°}$	−45	214.3	i.	∞	∞
320	lq.	$0.990^{0°}$	....	$214\text{-}5^{764mm}$	....	....	....
321	col. oil	$0.976^{22°}$	−5	217.4	i.	∞	∞
322	cr.	....	d.	....	s.	s.	i.
323	cr.	....	172 d.	....	i. c.; d. h.	....	....
324	large pl.	....	161	....	s.	....	....
325	....	....	207 ±	....	s.	....	....
326	col. lq.	$0.865^{2.0°}_{0}$	....	134	s.	s.	....
327	red-violet pd.	....	207-8	....	s. conc. HCl	s.	s. bz.
328	cr.	....	109 d.	....	s. c.	sl. s.	....
329	hyg. cr.	....	42*	231-2 (sl.d.)	v. s.	v. s.	v. s.
330	nd./pet.	....	63		....	....	....
331	vl. nd.	....	....	....	i.	v. sl. s.	....
332	nd.	....	....	....	sl. s.	s.	s.
333	cr./bz.-lg.	....	95-7 d.	....	s. h.	s. alk.	s. HCl
334	lf./et.	....	....	....	v. sl. s. h.	....	sl. s.
335	nd.	....	255 d.	....	....	6.6 h.	....
336	nd./al.	....	201 (163)	....	sl. s.	s.	s.
337	....	....	....	....	sl. s.	....	....
338	nd.	....	....	....	0.4 h.	....	....
339	nd.	....	....	....	sl. s.	i.	s. alk.
340	nd.	....	....	....	v. sl. s.	i.	i.; i. bz.
341	col. pd.	....	....	....	s.	i.	i.
342	col. cr.	....	....	....	sl. s.	sl. s.	s. alk.
343	col. cr.	....	....	....	$0.2^{20°}$; $2.4^{60°}$	....	....
344	cr.	....	....	....	s.	....	s. alk.
345	col. nd.	....	173	subl.	$1.7^{0°}$	$4.3^{0°}$	v. s.
346	col. cr.	....	....	208d.	$80^{0°}$	$40^{0°}$	....
347	pr./toluene	....	122-3	....	$2.6^{0°}$	s.	sl. s.
348	pr./aq.	....	229	....	....	....	....
349	lf.	....	184-6 d.	subl. sl. d.	$1.1^{0°}$	$4.6^{0°}$, abs.	i. bz.
350	col. pr.	....	306 d.	....	$71^{0°}$	$10^{0°}$, abs.	....

* Cryst. + 1 HCl, m. 100-2°.
Amino-hydroxytoluene 292-9
Amino-menthane 4029
Amino-methane 4105
Amino-methylaniline 4336

Amino-methylanisole 1544
Amino-naphthalene 4529-30
Amino-nitrophenol 4610-6
Amino-nonane 4926
Amino-octane 4970-2

Amino-oxamide 5017
Amino-pentane, cf. amylamine.
Amino-phenetole 5105

Table 7-4 (*Continued*)
PHYSICAL CONSTANTS OF ORGANIC COMPOUNDS

No.	Name	Synonym	Formula	Beil. Ref.	Formula Weight
351	**Amino** phenol (*p*) $H_2C_2O_4$	acid oxalate	$HO \cdot C_6H_4 \cdot NH_2 \cdot$ $H_2C_2O_4$	*XIII-144	199.16
352	phenol (*p*) $H_2C_2O_4$	oxalate	$(HOC_6H_4NH_2)_2 \cdot$ $H_2C_2O_4$	*XIII-144	308.29
353	phenol sulfonic acid	(1,2,4)	$C_6H_3(OH)(NH_2)SO_3H \cdot$ $\frac{1}{2}H_2O$	XIV-814	198.20
354	phenol sulfonic acid	(1,4,2)	$C_6H_3(OH)(NH_2)SO_3H$	XIV-806	189.19
355	phenol sulfonic acid	(1,4,3)	$C_6H_3(OH)(NH_2)SO_3H$	XIV-812	189.19
356	phenylacetic acid (*α*) (*dl*)	....	$C_6H_5 \cdot CH(NH_2) \cdot CO_2H$	XIV-460	151.17
357	phenylacetic acid (*p*)	*p*-amino-toluic acid	$H_2N \cdot C_6H_4 \cdot CH_2CO_2H$	XIV-456	151.17
358	phenylglycine (*p*)	*N*-(4-aminophenyl)- glycine	$H_2N \cdot C_6H_4 \cdot NH \cdot CH_2 \cdot$ CO_2H	XIII-105	166.18
359	phthalhydrazide (3)	luminol	$H_2N \cdot C_6H_3 \cdot CO \cdot NH \cdot$ $\underline{\qquad\qquad}$ $NH \cdot CO$ $\underline{\qquad\qquad}$	*XXV-698	177.16
360	pyridine (2) (*α*)	....	$N \cdot CH(CH)_3 \cdot C \cdot NH_2$ $\underline{\qquad\qquad\qquad}$	XXII-428	94.12
361	pyridine (3) (*β*)	....	$NH_2 \cdot C_5H_4N$	XXII-431	94.12
362	pyridine (4) (*γ*)	....	$NH_2 \cdot C_5H_4N$	XXII-433	94.12
363	pyrine	3-keto-1,5-diMe- 4-dimethylamino- 2-Ph-2,3-dihy- dropyrazole	$(CH_3)_2N \cdot C : C(CH_3) \cdot$ $\underline{\qquad}$ $N(CH_3) \cdot N(C_6H_5) \cdot$ $\underline{\qquad\qquad\qquad}$ CO $\underline{\qquad}$	....	231.30
364	pyrine-butyl- chloral-hydrate	trigemin	$C_{13}H_{17}ON_3 \cdot C_4H_7O_2Cl_3$	XXV-453	424.76
365	quinoline (2) (*α*)	....	$NH_2 \cdot C_9H_6N$	XXII-443	144.18
366	quinoline (3) (*β*)	....	$NH_2 \cdot C_9H_6N$	*XXII-638	144.18
367	quinoline (4) (*γ*)	....	$NH_2 \cdot C_9H_6N \cdot H_2O$	XXII-444	162.19
368	quinoline (5)	....	$NH_2 \cdot C_6H_3 \cdot CH : CH \cdot$ $\underline{\qquad\qquad}$ $CH : N$ $\underline{\qquad}$	XXII-445	144.18
369	quinoline (6)	....	$NH_2 \cdot C_9H_6N \cdot 2H_2O$	XXII-447	180.21
370	quinoline (7)	....	$NH_2 \cdot C_9H_6N + aq.$	XXII-450	144.18
371	quinoline (8)	....	$NH_2 \cdot C_9H_6N$	XXII-450	144.18
372	resorcinol (2)	base unknown	....	....	125.13
373	resorcinol (2) HCl	(2;1,3)	$NH_2 \cdot C_6H_3(OH)_2 \cdot HCl$	XIII-782	161.59
374	salicylic acid (3)	3-NH_2-2-OH- benzoic acid	$C_6H_3(CO_2H)(OH) \cdot$ NH_2	XIV-577	153.14
375	salicylic acid (4)	....	$C_7H_4O_2(OH)NH_2$	XIV-579	153.14
376	salicylic acid (5)	....	$C_7H_4O_2(OH)NH_2$	XIV-579	153.14
377	thiazol (2)	....	$NH_2 \cdot C : N \cdot CH : CH \cdot S$ $\underline{\qquad\qquad\qquad}$	XXVII-155	100.14
378	thiophene (*α*)	thiophenine	$NH_2C_4H_3S$	XVII-248	99.16
379	toluene sulfonic acid (1,2,3)	toluidine sulfonic acid	$C_6H_3(CH_3)(NH_2) \cdot$ SO_3H	XIV-723	187.22

Amino-phenyl-acetonitrile 274
Amino-phenyl-lepidine 3242
Amino-phenyl-propionic acid 5148
Amino-propandiol 3385

Amino-propane 5371-2
Amino-propionic acid 158-60
Amino-*iso*-propyl alcohol 5339
Amino-propylene-glycol 3385

Table 7-4 (*Continued*)
PHYSICAL CONSTANTS OF ORGANIC COMPOUNDS

No.	Crystalline Form and Color	Specific Gravity	Melting Point °C.	Boiling Point °C.	Solubility in 100 Parts		
					Water	Alcohol	Ether
351	vl. cr./aq.	....	220 d.	....	sl. s.	....	....
352	col. fl.	....	290 d.	....	i.	sl. s. alc.	i.
353	col. cr.	....	....	....	114°	....	....
354	nd.+aq.	....	d. >300	....	0.0714°	i.	i.
355	nd.+aq.	....	d. >270-85	....	214°	i.	i.; s. alk.
356	pr./aq. al.	....	256	subl.256-65	v. sl. s.	v. sl. s.	v. sl. s.; s. alk.
357	col. lf./aq.	....	199-200 d.	....	s. h.	s.	s. alk.
358	lf./aq.	....	222-3 d.	....	sl. s.	....	....
359	yel. nd.	....	280-300	....	i.; s. h. ac.	sl. s.	sl. s.
360	lf./lg.	....	56	204	s.	v. s.	s.; sl. s. lg.
361	lf./bz. lg.	....	64	250-2	v. s.	v. s.	v. s.; i. lg.
362	nd./bz.	....	158	....	s.; s. alk.	s.; sl. s. bz.	sl. s.
363	col. cr.	....	107-9	....	5.5	75	10; 10 bz.
364	col. cr./bz	....	85-6	....	1.5	50	10; sl. s. bz.
365	lf./aq.	....	129	....	v. s. h.	s.; sl. s. bz.	s.; sl. s. lg.
366	cr.	....	94(84)	....	s.	s.	....
367	nd./aq.	....	69-70*	−H₂O, 100	s. h.	s.; s. chl.	v. sl. s. CS₂
368	nd./al.	....	109-10	310	sl. s.	s.	s.; i. lg.
369	cr./aq.	....	73.5**	subl.	sl. s.	s.	s.; s. NH₄OH
370	yel. nd./aq.	....	74-5	....	....	....	....
371	yel. nd./al.	....	70	....	s. h.	....	....
372	....	....	....	....	....	....	....
373	gr. pd.	....	....	....	v. s.	s.	i.
374	cr.	....	235 d.	....	....	v. sl. s.	....
375	cr. pd.	....	150-1	....	v. s.	v. s.	sl. s.
376	nd.	....	d. 260-80	....	sl. s. h.	i.; s. HCl	s. CS₂
377	yel. pl./al.	....	90	d.	sl. s.	sl. s.; s. h.	sl. s.
378	oil	....	....	77-9¹¹ᵐᵐ	v. s.	v. s.	i.
379	nd.	....	....	....	0.97¹¹°; v. s. h.	....	....

* Anh. m.p. 154°.
** Anh. m.p. 93-4°.
Amino-purine, cf. alkd.
Amino Schaeffer's acid 4552

Amino-succinic acid 589-90
Amino-thiophenylimine 5922
Amino-toluene (ω) 783

Table 7-4 (*Continued*)
PHYSICAL CONSTANTS OF ORGANIC COMPOUNDS

No.	Name	Synonym	Formula	Beil. Ref.	Formula Weight
380	**Amino** toluene sulfonic acid	(1,2,4)	$C_6H_3(CH_3)(NH_2)\cdot SO_3H\cdot H_2O$	XIV-728	205.23
381	toluene sulfonic acid	(1,4,2)	$C_6H_3(CH_3)(NH_2)\cdot SO_3H\cdot H_2O$	XIV-720	205.23
382	toluene sulfonic acid	(1,4,3)	$C_6H_3(CH_3)(NH_2)\cdot SO_3H\cdot\frac{1}{2}H_2O$	XIV-723	196.23
383	toluene sulfonic acid	(1,2,5)	$C_6H_3(CH_3)(NH_2)\cdot SO_3H\cdot H_2O$	XIV-726	205.23
384	triphenylmethane (*m*)	benzhydryl aniline	$(C_6H_5)_2CH\cdot C_6H_4NH_2$	XII-1342	259.35
385	triphenylmethane	(*p*)	$(C_6H_5)_2CH\cdot C_6H_4NH_2$	XII-1342	259.35
386	uracil (5)	....	$\begin{array}{l}H_2N\cdot C{:}CH\cdot NH\cdot CO\cdot \\ \underline{\quad\quad\quad\quad\quad\quad}\mid \\ \quad\quad\quad NH\cdot CO \\ \underline{\mid\quad} \end{array}$	XXIV-463	127.10
387	valeric acid (α)	....	$C_2H_5\cdot CH_2\cdot CHNH_2\cdot CO_2H$	IV-416	**117.15**
388	valeric acid (γ) (*dl*)	....	$CH_3\cdot CHNH_2\cdot (CH_2)_2\cdot CO_2H$	IV-418	**117.15**
389	valeric acid (δ)	....	$NH_2\cdot (CH_2)_4\cdot CO_2H$	IV-418	**117.15**
390	valeric acid HCl (δ)	....	$NH_2(CH_2)_4CO_2H\cdot HCl$	IV-419	153.61
391	*iso*-valeric acid (α) (*dl*)	α-valine	$(CH_3)_2CH\cdot CHNH_2\cdot CO_2H$	IV-430	**117.15**
392	*iso*-valeric acid (α) (*l*+)	valine	$(CH_3)_2CH\cdot CHNH_2\cdot CO_2H$	IV-427	**117.15**
393	*iso*-valeric acid (β)	β-valine	$(CH_3)_2CNH_2\cdot CH_2\cdot CO_2H$	IV-426	**117.15**
394	valeric acid (α)	*dl-iso*-valine	$C_2H_5\cdot C(CH_3)(NH_2)\cdot CO_2H$	IV-425	**117.15**
395	valeric acid (α)	*iso*-valine (*l*+)	$C_2H_5\cdot C(CH_3)(NH_2)\cdot CO_2H\cdot H_2O$	*IV-513	**117.15**
396	**Ammelin**	triuretdiamidine	$(CN)_3(NH_2)_2OH$	XXVI-244	127.11
397	**Amyl** acetate (*n*)	....	$CH_3\cdot CO_2\cdot C_5H_{11}$	II-131	130.19
398	acetate (*iso*)	common amyl acetate	$CH_3CO_2\cdot CH_2CH_2CH{:}(CH_3)_2$	II-132	130.19
399	acetate	β-Me-Bu-acetate	$CH_3CO_2\cdot CH_2\cdot CH\cdot (CH_3)\cdot C_2H_5$	II-132	130.19
400	acetate (*sec*)	α-Me-Bu-acetate	$CH_3CO_2\cdot CH(CH_3\text{·}CH_2\cdot C_2H_5$	II-131	130.19
401	acetate (*sec*)	diEt-carbinol acetate	$CH_3CO_2\cdot CH(C_2H_5)_2$	II-131	130.19
402	acetate (*tert*)	....	$CH_3CO_2\cdot C(CH_3)_2\cdot C_2H_5$	II-132	130.19
403	acetic acid (*iso*)	....	$(CH_3)_2{:}C_4H_7\cdot CO_2H$	II-342	130.19
404	alcohol (*n*)	pentanol-1	$CH_3\cdot (CH_2)_3\cdot CH_2OH$	I-383	88.15
405	alcohol (*sec. n*)	pentanol-2	$C_2H_5\cdot CH_2\cdot CHOH\cdot CH_3$	I-384	88.15
406	alcohol (*prim. iso*)	2-methyl-butanol-4	$(CH_3)_2CH\cdot CH_2\cdot CH_2OH$	I-392	88.15
407	alcohol (*sec. iso*)	2-methyl-butanol-3	$(CH_3)_2CH\cdot CHOH\cdot CH_3$	I-391	88.15
408	alcohol	pentanol-3	$(C_2H_5)_2{:}CHOH$	I-385	88.15
409	alcohol (*tert*)	2-methyl-butanol-2	$(CH_3)_2COH\cdot C_2H_5$	I-388	88.15
410	alcohol	2,2-diMe-propanol-1	$(CH_3)_3C\cdot CH_2OH$	I-406	88.15
411	alcohol (*d*)	active amyl alcohol	$C_2H_5\cdot CH(CH_3)\cdot CH_2OH$	I-385	88.15

Amino-toluic acid 357
Amphetamine 654

Amphotropin 3601
Amygdalin, cf. glcde.

Amyl, cf. also diamyl.

Table 7-4 (*Continued*)
PHYSICAL CONSTANTS OF ORGANIC COMPOUNDS

No.	Crystalline Form and Color	Specific Gravity	Melting Point °C.	Boiling Point °C.	Solubility in 100 Parts		
					Water	Alcohol	Ether
380	nd.	....	....	....	$0.9^{11°}$	i.	....
381	mn.	....	d.	....	$0.5^{20°}$	i.	....
382	nd.	....	....	....	0.47	....	....
383	tri./aq.	....	$-H_2O$, 120	....	$2.7^{11°}$ (anh.)	i.	....
384	nd./et.	....	120	....	....	....	....
385	pr./et.	....	83-4	$248^{12mm} \pm$	i.	s. bz.	....
386	nd./aq.	....	d.	subl. d.	0.05 c.; 1.6 h.	s. a.; s. alk.	s. NH_4OH
387	nd./aq.	....	291.5	subl.	$10^{15°}$; v. s. h.	sl. s.	i.
388	cr.	....	193-214 d.	....	v. s.	v. sl. s.	i.; i. bz.
389	lf.	....	157-8 d.	....	v. s.	v. sl. s. abs.	i.
390	pl. or pr.	....	90-2	....	s.	s.	i.
391	lf./al.	....	298 d.	subl.	$7^{25°}$	v. sl. s.	v. sl. s.
392	lf./aq. al.	....	315*	subl. d.	$9.1^{16.5°}$	v. sl. s.	v. sl. s.
393	cr./al.+et.	....	217	subl. >180	v. s.	v. sl. s.	i.
394	rhb.+aq./aq. al.	....	307.5**	subl. $300 \pm$	$36^{20°}$	$0.6^{15°}$	6 h.
395	nd./aq. al.	....	315	subl.	s.	sl. s.	....
396	nd./aq.	....	....	....	$0.008^{23°}$	i.	i.; s. HCl
397	col. lq.	$0.879^{20°}_{20°}$	-70.8	148.4^{737mm}	v. sl. s.	∞	∞
398	col. lq.	$0.876^{15°}_{4}$	....	142^{757mm}	$0.25^{15°}$	∞	∞
399	col. lq.	$0.880^{12.5°}$	....	141-2	v. sl. s.	∞	∞
400	col. lq.	$0.922^{0°}$	....	133.5	sl. s.	∞	∞
401	col. lq.	$0.871^{20°}_{4}$	....	133	sl. s.	∞	∞
402	col. lq.	$0.874^{19°}$	....	124.5^{749mm}	v. sl. s.	∞	∞
403	col. oil	$0.912^{18°}_{18°}$	....	216^{762mm}	sl. s. h.	∞	∞
404	col. lq.	$0.818^{15°}_{4}$	-78.9	138.1	$2.7^{22°}$	∞	∞
405	col. lq.	$0.813^{15°}_{4}$	....	119.9	$4^{20°}$	∞	∞
406	col. lq.	$0.813^{15°}_{4}$	-117.2	132.0	$2^{14°}$	∞	∞
407	col. lq.	$0.825^{15°}_{4}$	....	113-4	$2.8^{30°}$	∞	∞
408	col. lq.	$0.813^{15°}_{4}$	....	116.1	$5.5^{30°}$	∞	∞
409	col. lq.	$0.813^{15°}_{4}$	-9	102.4	sl. s.	s.	s.
410	cr.	....	52-3	113-4	sl. s.	∞	∞
411	col. lq.	$0.816^{20°}_{4}$	....	128	$3.6^{30°}$	∞	∞

* Sealed tube.
** Sealed tube; loses aq. at 100°.

Table 7-4 (*Continued*)

PHYSICAL CONSTANTS OF ORGANIC COMPOUNDS

No.	Name	Synonym	Formula	Beil. Ref.	Formula Weight
412	**Amyl** amine (*n*)	....	$CH_3(CH_2)_4NH_2$	IV-175	87.17
413	amine (*sec. n*)	....	$(C_3H_7)(CH_3):CHNH_2$	IV-177	87.17
414	amine (*iso*)	....	$(CH_3)_2CH(CH_2)_2NH_2$	IV-180	87.17
415	amine (*tert*)	....	$(C_2H_5)(CH_3)_2C \cdot NH_2$	IV-179	87.17
416	amine	1-NH_2-2-Me-butane	$C_2H_5CH(CH_3) \cdot CH_2NH_2$	IV-178	87.17
417	amine	3-aminopentane	$(C_2H_5)_2:CH \cdot NH_2$	IV-178	87.17
418	amine	3-NH_2-2-Me-butane	$(CH_3)_2CH \cdot CH \cdot (CH_3)NH_2$	IV-179	87.17
419	aniline (*iso*)	....	$C_6H_5 \cdot NH \cdot C_5H_{11}$	XII-169	163.26
420	aniline (*p*) (*tert*)	....	$(C_2H_5)(CH_3)_2C \cdot C_6H_4 \cdot NH_2$	XII-1179	163.26
421	benzene (*n*)	1-phenylpentane	$C_5H_{11} \cdot C_6H_5$	V-434	148.25
422	benzene (*iso*)	....	$C_5H_{11} \cdot C_6H_5$	V-434	148.25
423	benzene (*tert*)	diMe-Et-Ph-methane	$C_5H_{11} \cdot C_6H_5$	V-436	148.25
424	benzene (*sec*)	diEt-Ph-methane	$(C_2H_5)_2CH \cdot C_6H_5$	V-436	148.25
425	benzoate (*iso*)	....	$C_6H_5 \cdot CO_2 \cdot C_5H_{11}$	IX-113	192.26
426	benzyl ether (*iso*)	....	$C_6H_5 \cdot CH_2 \cdot O \cdot C_5H_{11}$	VI-431	178.28
427	bromide (*n*)	1-bromopentane	$CH_3 \cdot (CH_2)_3 \cdot CH_2Br$	I-131	151.05
428	bromide (*iso*)	4-Br-2-Me-butane	$(CH_3)_2CH(CH_2)_2Br$	I-136	151.05
429	bromide (*tert*)	2-Br-2-Me-butane	$(CH_3)_2C(Br) \cdot C_2H_5$	I-136	151.05
430	bromide	1-Br-2, 2-diMe-propane	$(CH_3)_3C \cdot CH_2Br$	I-141	151.05
431	bromide	1-Br-2-Me-butane	$C_2H_5 \cdot CH(CH_3) \cdot CH_2Br$	I-136	151.05
432	*n*-butyrate (*n*)	....	$C_2H_5CH_2CO_2(CH_2)_4 \cdot CH_3$	II-271	158.24
433	*n*-butyrate (*iso*)	....	$C_2H_5CH_2CO_2 \cdot C_5H_{11}$	II-271	158.24
434	*n*-butyrate (*tert*)	....	$C_3H_7CO_2 \cdot C(CH_3)_2 \cdot C_2H_5$	II-271	158.24
435	*iso*-butyrate (*iso*)	....	$(CH_3)_2CHCO_2 \cdot C_5H_{11}$	II-291	158.24
436	*n*-caproate (*n*)	....	$C_5H_{11}CO_2 \cdot C_5H_{11}$	II-323	186.30
437	*n*-caproate (*iso*)	....	$C_5H_{11}CO_2 \cdot (CH_2)_2 \cdot CH:(CH_3)_2$	....	186.30
438	*n*-caprylate (*n*)	....	$C_7H_{15}CO_2(CH_2)_4 \cdot CH_3$	....	214.35
439	carbamate (*n*)	....	$H_2N \cdot CO_2 \cdot C_5H_{11}$	....	131.18
440	carbamate (*iso*)	....	$NH_2 \cdot CO_2 \cdot C_5H_{11}$	III-30	131.18
441	carbamate (*tert*)	aponal	$H_2N \cdot CO_2 \cdot C(CH_3)_2 \cdot C_2H_5$	*III-14	131.18
442	chloride (*n*)	1-chloropentane	$CH_3 \cdot (CH_2)_3 \cdot CH_2Cl$	I-130	106.60
443	chloride (*sec*)	2-chloropentane	$C_3H_7 \cdot CHCl \cdot CH_3$	I-131	106.60
444	chloride (*sec*)	3-chloropentane	$(C_2H_5)_2CHCl$	I-131	106.60
445	chloride (*iso*)	4-Cl-2-Me-butane	$(CH_3)_2CH(CH_2)_2Cl$	I-135	106.60
446	chloride (*sec. iso*)	3-Cl-2-Me-butane	$(CH_3)_2CH \cdot CHCl \cdot CH_3$	I-135	106.60
447	chloride (*tert*)	2-Cl-2-Me-butane	$(CH_3)_2CCl \cdot C_2H_5$	I-134	106.60
448	chloride	1-Cl-2-Me-butane	$(CH_3)(C_2H_5)CH \cdot CH_2Cl$	I-134	106.60
449	chloroformate (*n*)	*n*-amyl chloro-carbonate	$C_2H_5(CH_2)_3O \cdot COCl$	....	150.61
450	chloroformate (*iso*)	*iso*-amyl chloro-carbonate	$(CH_3)_2CH(CH_2)_2 \cdot O \cdot COCl$	III-12	150.61

Amyl borate 6077   Amyl carbinol (*iso*) 3634

Table 7-4 (*Continued*)
PHYSICAL CONSTANTS OF ORGANIC COMPOUNDS

No.	Crystalline Form and Color	Specific Gravity	Melting Point °C.	Boiling Point °C.	Solubility in 100 Parts		
					Water	Alcohol	Ether
412	col. lq.	$0.766^{19°}$	−55	103-4	s.	s.	s.
413	col. lq.	$0.749^{20°}_4$	....	91-2	∞	∞	∞
414	col. lq.	$0.751^{18°}_4$	....	95	∞	∞	∞
415	col. lq.	$0.731^{25°}_4$	−105	77-8	∞	∞; ∞ chl.	∞
416	col. lq.	$0.755^{18°}$	....	95-6	∞	∞	∞
417	col. lq.	$0.749^{20°}_4$	....	90-1	∞	∞	∞
418	col. lq.	$0.757^{18.5°}$	....	83-4	∞	∞	∞
419	lq.	$0.928^{15°}_4$	....	254.5	....	....	....
420	col. lq.	....	....	259-62	i.	s.	s.
421	lq.	$0.858^{20°}_4$	−78.3	205.3	i.	∞	∞
422	lq.	$0.859^{20°}_4$	....	198-9	i.	∞	∞
423	lq.	$0.867^{20°}_4$	....	189-90	i.	∞	∞
424	col. lq.	$0.876^{15°}_4$	....	187^{753mm}	i.	s.	s.
425	col. lq.	$0.992^{14°}_{14}$	....	260.7^{746mm}	i.	∞	∞
426	col. lq.	....	....	237^{748mm}	i.	s.	s.
427	col. lq.	$1.218^{20°}_4$	−95	129.7	i.	s.	....
428	col. lq.	$1.203^{20°}_4$	−111.9	120.7	i.	s.	s.
429	lq.	$1.216^{19°}_0$	....	108^{765mm}	i.	s.	s.
430	col. lq.	$1.260^{20}_4{}^{4°}$	....	$89\text{-}91^{749mm}$	i.	s.	s.
431	lq.	$1.221^{20°}_4$	....	121^{755mm}	i.	s.	s.
432	col. lq.	$0.871^{15°}_4$	−73.2	186.4	$0.05^{50°}$	∞	∞
433	col. lq.	$0.866^{19°}_{18}$	....	178.6	i.	∞	∞
434	col. lq.	$0.865^{14.5°}_0$	....	164	sl. s.	∞	∞
435	lq.	$0.876^{0°}_4$	....	168.8	i.	s.	s.
436	col. lq.	$0.861^{25°}_4$	−47	226.2	....	....	....
437	col. lq.	....	....	$94\text{-}6^{10mm}$	i.	s.	....
438	col. lq.	$0.856^{25°}_4$	−34.5	260.2	i.	s.	....
439	col. pl.	....	56-7.5	....	i.	s.	s.
440	nd./aq.	$0.944^{70.6°}_4$	64	220	s. h.	s.	s.
441	nd./aq. al.	....	86	....	sl. s.	s.	s.; sl. s. pet.
442	col. lq.	$0.878^{20°}_4$	−99	108.4	i.	s.	s.
443	lq.	$0.870^{20°}_4$	....	96.7	i.	s.	s.
444	col. lq.	$0.895^{21°}$	....	97.3	i.	∞	∞
445	col. lq.	$0.893^{20°}_4$	....	99.7^{758mm}	i.	s.	∞
446	lq.	$0.883^{0°}$	....	91^{753mm}	i.	s.	s.
447	lq.	$0.871^{20°}_4$	−72.9	85.7	i.	s.	s.
448	lq.	$0.881^{17.5°}$	....	98-9	i.	s.	s.
449	col. lq.	....	....	$43\text{-}57^{mm}$	....	....	....
450	lq.	$1.028^{18.5°}_{15}$	....	155-6	i.	∞	∞

Amyl carbonate 1737-8   Amyl chlorocarbonate 449-50

Table 7-4 (Continued)
PHYSICAL CONSTANTS OF ORGANIC COMPOUNDS

No.	Name	Synonym	Formula	Beil. Ref.	Formula Weight
451	**Amyl** cinnamaldehyde (n) (α)	....	$C_6H_5 \cdot CH:C(C_5H_{11}) \cdot CHO$	**VII-310	202.30
452	m-cresol (n)	....	$C_5H_{11} \cdot C_6H_3(OH)CH_3$	....	178.28
453	iso-cyanide (iso)	iso-caproic iso-nitrile	$(CH_3)_2CH(CH_2)_2 \cdot NC$	IV-184	97.16
454	formate (n)	....	$HCO_2 \cdot C_5H_{11}$	II-22	116.16
455	formate (iso)	....	$HCO_2 \cdot C_5H_{11}$	II-22	116.16
456	formate, ortho (n)	....	$HC(OC_5H_{11})_3$	II-22	274.45
457	furoate (iso)	iso-Am pyromucate	$C_4H_3O \cdot CO_2 \cdot C_5H_{11}$	....	182.22
458	β-furylacrylate (n)	....	$C_4H_3O \cdot CH:CH \cdot CO_2C_5H_{11}$	....	208.26
459	iodide (n)	1-iodopentane	$CH_3 \cdot (CH_2)_3 \cdot CH_2I$	I-133	198.05
460	iodide (iso)	4-I-2-Me-butane	$(CH_3)_2CH \cdot CH_2 \cdot CH_2I$	I-138	198.05
461	iodide (sec n)	2-iodopentane	$C_3H_7 \cdot CHI \cdot CH_3$	I-133	198.05
462	iodide (tert)	2-I-2-Me-butane	$(CH_3)_2CI \cdot C_2H_5$	I-138	198.05
463	iodide	1-I-2-Me-butane	$C_2H_5 \cdot CH(CH_3) \cdot CH_2I$	I-138	198.05
464	d-lactate (dl) (−)	....	$C_2H_5OCO_2 \cdot CH_2 \cdot CH(CH_3)C_2H_5$	III-265	160.21
465	malonic acid (n)	....	$C_5H_{11}CH(CO_2H)_2$	II-695	174.20
466	malonic acid (iso)	....	$C_5H_{11}CH(CO_2H)_2$	II-700	174.20
467	mercaptan (n)	pentanthiol-1	$CH_3 \cdot (CH_2)_3 \cdot CH_2 \cdot SH$	I-384	104.22
468	mercaptan (n)	pentanthiol-3	$(C_2H_5)_2CH \cdot SH$	*I-194	104.22
469	mercaptan (iso)	2-Me-butanthiol-4	$(CH_3)_2CH(CH_2)_2SH$	I-405	104.22
470	naphthalene (iso) (β)	....	$C_5H_{11} \cdot C_{10}H_7$	V-574	198.31
471	β-naphthyl ether (iso)	....	$C_5H_{11} \cdot O \cdot C_{10}H_7$	VI-642	214.31
472	nitrate (iso)	....	$(CH_3)_2CH \cdot CH_2 \cdot CH_2 \cdot O \cdot NO_2$	I-403	133.15
473	nitrite (n)	....	$C_5H_{11} \cdot O \cdot NO$	I-384	117.15
474	nitrite (iso)	....	$(CH_3)_2CH \cdot CH_2 \cdot CH_2 \cdot O \cdot NO$	I-402	117.15
475	oleate (iso)	....	$C_8H_{17}CH:CHC_7H_{14} \cdot CO_2 \cdot C_5H_{11}$	II-467	352.61
476	phenol (tert) (p)	pentaphen	$C_5H_{11} \cdot C_6H_4 \cdot OH$	VI-548	164.25
477	phenol methyl ether (tert) (p)	....	$(C_2H_5)(CH_3)_2C \cdot C_6H_4 \cdot O \cdot CH_3$	VI-549	178.28
478	phenyl ether (n)	....	$CH_3(CH_2)_4 \cdot O \cdot C_6H_5$	*VI-82	164.25
479	phenyl ether (iso)	....	$C_5H_{11} \cdot O \cdot C_6H_5$	VI-143	164.25
480	α-phenylhydrazine	(iso) (α)	$C_6H_5 \cdot N(NH_2)C_5H_{11}$	XV-121	178.28
481	phenyl ketone (n)	caprophenone	$C_5H_{11} \cdot CO \cdot C_6H_5$	VII-333	176.26
482	phenyl ketone (iso)	iso-caprophenone	$C_5H_{11} \cdot CO \cdot C_6H_5$	VII-334	176.26
483	propionate (n)	....	$C_2H_5 \cdot CO_2 \cdot C_5H_{11}$	**II-221	144.22
484	propionate (iso)	....	$C_2H_5 \cdot CO_2 \cdot C_5H_{11}$	II-241	144.22
485	propionate (act)	....	$C_2H_5 \cdot CO_2 \cdot C_5H_{11}$	II-241	144.22
486	salicylate (n)	....	$HO \cdot C_6H_4 \cdot CO_2 \cdot C_5H_{11}$	....	208.26
487	salicylate (iso)	....	$HO \cdot C_6H_4 \cdot CO_2 \cdot C_5H_{11}$	X-76	208.26
488	stearate (iso)	....	$CH_3(CH_2)_{16}CO_2 \cdot C_5H_{11}$	II-380	354.62
489	stearate	....	$C_{17}H_{35}CO_2 \cdot CH_2 \cdot CH:(CH_3)C_2H_5$	II-380	354.62
490	thiocyanate (iso)	....	$(CH_3)_2CH(CH_2)_2 \cdot S \cdot CN$	III-177	129.23
491	iso-thiocyanate (n)	amyl mustard oil	$CH_3(CH_2)_4 \cdot N:C:S$	IV-176	129.23

Amyl cyanide 1217-8, 2104
Amyl disulfide 1753-4
Amyl ether 1740-1
Amyl-ethyl-barbituric acid 505

Amyl mustard oil 491-3
Amyl oxalate 1745
Amyl phthalate 1746-7

Table 7-4 (*Continued*)
PHYSICAL CONSTANTS OF ORGANIC COMPOUNDS

No.	Crystalline Form and Color	Specific Gravity	Melting Point °C.	Boiling Point °C.	Solubility in 100 Parts		
					Water	Alcohol	Ether
451	yel. lq.	$0.971\frac{20°}{20°}$	....	174-5^{20mm}	....	....	....
452	col.	....	24-5	137-9^{15mm}	i.	s.	s.
453	lq.	....	....	137-9	i.	s.	s.
454	lq.	0.902$^{20°}$	−73.5	132	v. sl. s.	∞	∞
455	lq.	$0.882\frac{20°}{4}$	−93.5	123.5	0.3$^{22°}$	∞	∞
456	lq.	0.864$^{23°}$	....	265-7 dec.	....	....	....
457	col. lq.	....	....	135-7^{25mm}	i.	∞	
458	col. lq.	$1.032\frac{20°}{4}$	....	119^{4mm}	i.	....	....
459	lq.	$1.510\frac{20°}{4}$	−86	157.0	i.	s.	∞
460	lq.	$1.515\frac{18°}{4}$	....	147^{765mm}	i.	∞	∞
461	lq.	$1.507\frac{17°}{4}$	....	144-5	i.	∞	∞
462	lq.	$1.471\frac{19°}{15}$	....	127^{765mm}	i.	∞	∞
463	lq.	$1.524\frac{20°}{4}$	....	148^{760mm}	i.	∞	∞
464	col. lq.	$0.971\frac{20°}{4}$	....	114-5^{36mm}	v. sl. s.	s.	s.
465	col. pr.	....	82	d. 140	v. s.	v. s.	v. s.
466	nd./aq.	....	95d.	d.	v. s.	v. s.	v. s.
467	lq.	0.857$^{20°}$	....	126^{767mm}	i.	∞	∞
468	col. lq.	....	....	105	i.	∞	∞
469	lq.	$0.835\frac{20°}{4}$	....	120	i.	∞	∞
470	col. lq.	0.973$^{0°}$	<−21	288-92	i.	s.	s.
471	col. lf.	$1.016\frac{12°}{4}$	26.5	323-6	i.	s.	s.
472	lq.	0.996$^{21.7°}$	....	147-8	v. sl. s.	i.	i.
473	pa. yel. lq.	0.853$^{20°}$	....	104^{761mm}	sl. s.	∞	∞
474	lq.	$0.872\frac{20°}{4}$	....	99	sl. s.	∞	∞
475	col. lq.	0.897$^{15°}$	....	223-4^{10mm}	i.	s.	v. s.
476	cr.	....	93	265-7	sl. s.	s.	s.
477	col. lq.	....	....	216-7	i.	s.	s.
478	col. lq.	....	....	111^{17mm}	i.	s.	s.
479	col. oil	$0.920\frac{22°}{4}$	....	224-5	i.	s.	s.
480	lq.	0.968$^{15°}$	....	260-2	....	....	....
481	lf.	$0.958\frac{25°}{4}$	24.7	265.2	....	....	....
482	lq.	$0.962\frac{15°}{4}$	−2	240-2^{720mm}	i.	v. s.	v. s.
483	lq.	$0.876\frac{15°}{4}$	−73.1	168.7	i.	∞	∞
484	col. lq.	$0.870\frac{20°}{4}$	....	160.2	0.1$^{25°}$	∞	∞
485	col. lq.	$0.866\frac{20°}{4}$	....	58^{16mm}	v. sl. s.	∞	∞
486	lq.	1.065$^{15°}$	....	265	i.	∞	∞
487	lq.	$1.045\frac{25°}{25}$	....	276-7^{743mm}	0.004$^{22°}$	∞	∞ ; ∞ chl.
488	col. pl.	$0.855\frac{20°}{4}$	23	185-90^{1mm}	i.	sl. s.	s.
489	col. lq.	$0.855\frac{20°}{4}$	21-2	....	i.	s.	s.
490	pa. yel. oil	0.905	....	197	i.	s.	s.
491	lq.	....	....	193.4	v. sl. s.	v. s.	v. s.

Table 7-4 (*Continued*)
PHYSICAL CONSTANTS OF ORGANIC COMPOUNDS

No.	Name	Synonym	Formula	Beil. Ref.	Formula Weight
492	**Amyl** *iso*-thiocyanate (*iso*)	amyl mustard oil	$(CH_3)_2CH(CH_2)_2\cdot N:C:S$	IV-186	129.23
493	*iso*-thiocyanate (*tert*)	amyl mustard oil	$C_5H_{11}N:C:S$	IV-179	129.23
494	urea (*iso*)	....	$C_5H_{11}\cdot NH\cdot CO\cdot NH_2$	IV-185	130.19
495	urea (*tert*)	....	$C_5H_{11}\cdot NH\cdot CO\cdot NH_2$	IV-179	130.19
496	*iso*-valerate (*iso*)	....	$C_4H_9\cdot CO_2\cdot C_5H_{11}$	II-312	172.27
497	*iso*-valerate (*tert*)	valamin	$C_4H_9\cdot CO_2\cdot C_5H_{11}$	II-312	172.27
498	**Amylene** (*n*)	pentene-1	$C_2H_5\cdot CH_2\cdot CH:CH_2$	I-210	70.14
499	**Amylene** (*iso*)	2-methyl-butene-3	$(CH_3)_2CH\cdot CH:CH_2$	I-213	70.14
500	**Amylene** (α)	2-methyl-butene-1	$(C_2H_5)(CH_3)C:CH_2$	I-211	70.14
501	**Amylene** (β)(*cis*)	pentene-2	$C_2H_5\cdot CH:CH\cdot CH_3$	I-210	70.14
501.1	**Amylene** (β)(*trans*)	pentene-2	$C_2H_5\cdot CH:CH\cdot CH_3$	I-210	70.14
502	**Amylene** (β) (*iso*)	2-methyl-butene-2	$(CH_3)_2:C:CH\cdot CH_3$	I-211	70.14
503	glycol (1,4)	γ-pentylene glycol	$C_5H_{10}(OH)_2$	I-480	104.15
504	glycol (2,4)	*iso*-amylene alcohol	$C_5H_{10}(OH)_2$	I-483	104.15
505	**Amytal**	*iso*-Am-Et-barbituric acid	$(C_2H_5)(C_5H_{11}):C\cdot CO\cdot$ ⌐___ $NH\cdot CO\cdot NH\cdot CO$ ___⌐	....	226.28
506	sodium	....	$C_{11}H_{17}O_3N_2Na$	....	248.26
507	**Anacardic acid**	....	$HO\cdot C_{21}H_{30}\cdot CO_2H$	X-327	344.50
508	**Analgen**	8-EtO-5-benzoyl-amino-quinoline	$C_7H_5ONH\cdot C_6H_2\cdot (OC_2H_5):C_3H_3N$	XXII-503	292.34
509	**Androsterone**	3-*trans*-hydroxy-17-keto-androstane	$C_{19}H_{30}O_2$	....	290.45
510	**Anemonin**	pulsatilla camphor	$C_{10}H_8O_4$	....	192.17
511	**Anethole**	*p*-propenyl anisole	$CH_3\cdot CH:CH\cdot C_6H_4\cdot OCH_3$	VI-566	148.21
512	**Anhydroformaldaniline**	methylene aniline	$(CH_2NC_6H_5)_3$	XXVI-3	315.42
513	**Aniline**	aminobenzene	$C_6H_5\cdot NH_2$	XII-59	93.13
514	acetate	....	$C_6H_5NH_2\cdot C_2H_4O_2$	XII-118	153.18
515	arsonic acid (*p*)	arsanilic acid	$H_2N\cdot C_6H_4\cdot AsO_3H_2$	XVI-878	217.06
516	blue	spirit blue	$C_{38}H_{32}N_3\cdot Cl$	XIII-768	566.15
517	disulfonic acid	(2,5)	$NH_2\cdot C_6H_3:(SO_3H)_2\cdot 4H_2O$	XIV-780	325.32
518	disulfonic acid	(2,4)	$NH_2\cdot C_6H_3:(SO_3H)_2\cdot 2H_2O$	XIV-778	289.28
519	hydrobromide	....	$C_6H_5NH_2\cdot HBr$	XII-116	174.05
520	hydrochloride	....	$C_6H_5NH_2\cdot HCl$	XII-116	129.59
521	nitrate	....	$C_6H_5NH_2\cdot HNO_3$	XII-116	156.14
522	sulfate	....	$(C_6H_5NH_2)_2\cdot H_2SO_4$	XII-117	284.34
523	*o*-sulfonic acid	orthanilic acid	$NH_2\cdot C_6H_4\cdot SO_3H$†	XIV-681	173.19
524	*m*-sulfonic acid	metanilic acid	$NH_2\cdot C_6H_4\cdot SO_3H$‡	XIV-688	173.19
525	*p*-sulfonic acid	sulfanilic acid	$NH_2\cdot C_6H_4\cdot SO_3H$§	XIV-695	173.19
526	*p*-sulfonic Na	....	$NH_2\cdot C_6H_4\cdot SO_3Na\cdot 2H_2O$	XIV-698	231.20
527	*p*-sulfonic amide	sulfanilamide	$NH_2\cdot C_6H_4\cdot SO_2\cdot NH_2$	XIV-698	172.21

† Also crysts. $+\frac{1}{2}H_2O$.
‡ Also crysts. $+1\frac{1}{2}H_2O$.
§ Also crysts. $+1H_2O$.
Amyl toluene 4119
Amylene alcohol 504, 4103

Amylene chloride 2034
Amylocaine 5644
Anabasine, cf. alkd.
Analgesine 565
Andirine 4405

Anemone camphor 510
Anesthesine 2955
Angelic acid 2425
Angeline 4405
Anhaline, cf. alkd.

Anhalonine, cf. alkd.
Aniline salt 520
Aniline violet 3593
Aniline yellow 249
Anilino-acetic acid 5194

Table 7-4 (*Continued*)
PHYSICAL CONSTANTS OF ORGANIC COMPOUNDS

No.	Crystalline Form and Color	Specific Gravity	Melting Point °C.	Boiling Point °C.	Solubility in 100 Parts		
					Water	Alcohol	Ether
492	yel. lq.	$0.942^{17°}$	....	183-4	v. sl. s.	v. s.	v. s.
493	lq.	....	<−10	166^{770mm}	v. sl. s.	v. 3.	v. s.
494	cr.	....	92-3	....	v. sl. s.	....	....
495	mn./aq.	....	152-8	....	$1.3^{27°}$	....	....
496	col. lq.	$0.858^{20°}_{15}$	....	194	v. sl. s.	∞	∞
497	col. lq.	$0.861^{14°}_{0}$	....	173-4	sl. s.	s.	∞ oils
498	lq.	$0.641^{20°}$	−165.2	30.0	i.	∞	∞
499	col. lq.	$0.627^{20°}$	−168.5	20.1	*	∞	∞
500	col. lq.	$0.650^{20°}$	−137.6	31.1	**	∞	∞
501	col. lq.	$0.656^{20°}$	−151.4	37.1	v. sl. s.	∞	∞
501.1	col. lq.	$0.648^{20°}$	−140.2	36.4	v. sl. s.	∞	∞
502	col. lq.	$0.662^{20°}$	−133.8	38.5	**	s.	∞
503	oil	$0.996^{17°}_{4}$	....	$219-20^{13mm}$	∞	∞	∞ chl.; i. lg.
504	syrup	$0.989^{20°}$	....	202-3	....	....	....
505	col. cr.	....	154-6	....	sl. s.	s.	s.; s. alk.
506	hyg. pd.	....	150-5	....	v. s.	100	i.
507	cr.	....	26	....	i.	s.	s.
508	yel. nd./al.	....	210	....	i.	v. sl. s. c.	s. a.
509	col. nd./ 65% al.	....	184-5	....	i.; s. MeOH	s.; s. CCl$_4$	s.; s. bz.
510	yel.-wh. cr.	....	157-8	....	sl. s. h.	s. h.	i.
511	lf./al.	$0.991^{20°}_{20}$	22.5	235.3	v. sl. s.	∞ abs. al.	∞
512	pr./al.	....	143	185	i.	sl. s.	s.; s. bz.
513	col. oil	$1.022^{20°}_{4}$	−6.1	184.4	$3.6^{18°}$	∞	∞
514	col. cr.	1.07	....	d., −H$_2$O	∞	∞	....
515	nd./aq.	....	232	d. −H$_2$O, 150	v. s. h.	v. s. h.	i.; i. bz.
516	bl.-brn. cr.	....	....	....	i.	sl. s.	i.
517	cr./aq.	....	....	....	v. s.	v. s.	....
518	nd./aq.	....	d. >120	....	v. s.	v. s.	i.; s. alk.
519	cr. pd.	....	286	....	s.	s.	....
520	lf. or nd.	$1.222^{4°}$	198	245	$88.4^{15°}$; $107^{25°}$	s.	i.
521	rhb.	$1.356^{4°}$	d. 190	....	s.	s.	sl. s.
522	lf./al.	$1.377^{4°}$	d.	....	$51^{4°}$	sl. s.	i.
523	col. cr.	....	d.	....	$1.5^{15°}$	v. sl. s.	v. sl. s.
524	col. nd.	....	d.	....	$2^{15°}$	v. sl. s.	v. sl. s.
525	col. cr.	....	d. 280-300	....	$0.8^{10°}$; $6.6^{100°}$	v. sl. s.	v. sl. s.
526	rhb.	....	....	....	v. s.	s. h.	i.
527	col. lf./ aq. al.	....	165-6	....	0.8 c.; v. s. h.	3 c.	s.; 20 act.

* Insoluble in a mixt. of 2H$_2$SO$_4$ + 1H$_2$O.
** Very soluble in mixt. of 2H$_2$SO$_4$ + 1H$_2$O.
Anilino-azobenzene 659
Anilino-benzoic acid 5157
Anilino-ethyl alcohol 2930

Anilino-glyoxylic acid 3921
Anilino-phenol 3767-9
Aniluvitonic acid 5529
Animal starch 3443
Anisal-benzylamine 812

Anise camphor 511
Anisic acid 4069
Anisoyl-anisole 2409
Anisoyl chloride 4071

Table 7-4 (*Continued*)
PHYSICAL CONSTANTS OF ORGANIC COMPOUNDS

No.	Name	Synonym	Formula	Beil. Ref.	Formula Weight
528	**Anis**-alacetone (*p*)	MeO-benzalacetone	$CH_3O \cdot C_6H_4 \cdot CH:CH \cdot CO \cdot CH_3$	VIII-131	176.22
529	alcinnamalacetone	(*α,α'*)	$C_6H_5 \cdot (CH:CH)_2 \cdot CO \cdot CH:CH \cdot C_6H_4 \cdot OCH_3$	VIII-208	290.37
530	alcohol	*p*-MeO-benzyl alcohol	$CH_3O \cdot C_6H_4 \cdot CH_2OH$	VI-897	138.17
531	aldehyde (*o*)	MeO-benzaldehyde	$CH_3O \cdot C_6H_4 \cdot CHO$	VIII-43	136.15
532	aldehyde (*p*)	....	$CH_3O \cdot C_6H_4 \cdot CHO$	VIII-67	136.15
533	aldehyde phenyl-hydrazone (*p*)	....	$C_6H_5NH \cdot N:CH \cdot C_6H_4 \cdot OCH_3$	XV-192	226.28
534	**Anisidine** (*o*)	2-amino-anisole	$NH_2 \cdot C_6H_4 \cdot OCH_3$	XIII-358	123.16
535	**Anisidine** (*m*)	MeO-aniline (*m*)	$NH_2 \cdot C_6H_4 \cdot OCH_3$	XIII-404	123.16
536	**Anisidine** (*p*)	4-amino-anisole	$NH_2 \cdot C_6H_4 \cdot OCH_3$	XIII-435	123.16
537	**Anisole**	methyl-phenyl ether	$CH_3 \cdot O \cdot C_6H_5$	VI-138	108.14
538	**Anthracene**	....	$(C_6H_4CH)_2$	V-657	178.24
539	carboxylic acid (1) (*β*)	....	$C_6H_4:C_2H(CO_2H): C_6H_4$	IX-704	222.25
540	carboxylic acid (2) (*γ*)	....	$C_6H_4:(CH)_2:C_6H_3 \cdot CO_2H$	IX-705	222.25
541	carboxylic acid (9) (*α*)	....	$C_6H_4:(CH)_2:C_6H_3 \cdot CO_2H$	IX-705	222.25
542	**Anthramine** (*α*)	*α*-amino-anthracene	$C_{14}H_9 \cdot NH_2$	XII-1335	193.25
543	**Anthramine** (*β*)	*β*-amino-anthracene	$C_{14}H_9 \cdot NH_2$	XII-1335	193.25
544	**Anthranil**	....	$O \cdot N:C_6H_4:CH$	XXVII-39	119.12
545	**Anthranol**	*γ*-hydroxy anthra-cene; anthrone	$(C_6H_4:)_2COHCH$ $(C_6H_4:)_2COCH_2$	VII-473	194.24
546	**Anthranoyl anthra-nilic acid**	....	$NH_2 \cdot C_6H_4 \cdot CO \cdot NHC_6H_4 \cdot CO_2H$	XIV-358	256.26
547	**Anthraquinoline** (*β*)	....	$C_{17}H_{11}N$	XX-506	229.28
548	**Anthraquinone** (1,2)	....	$C_{10}H_6(CO)_2CH:CH$	VII-780	208.22
549	**Anthraquinone** (1,4)	....	$C_{10}H_6(COCH)_2$	VII-781	208.22
550	**Anthraquinone** (9,10)	anthraquinone	$(C_6H_4)_2(CO)_2$	VII-781	208.22
551	carboxylic acid (*α*)	....	$C_6H_4(CO)_2C_6H_3 \cdot CO_2H$	X-834	252.23
552	carboxylic acid (*β*)	....	$C_6H_4(CO)_2C_6H_3 \cdot CO_2H$	X-835	252.23
553	disulfonate Na₂ (1,5)	*ρ*-anthraquinone disulfonate	$C_{14}H_6O_2(SO_3Na)_2 \cdot 5H_2O$	XI-340	502.38
554	disulfonate Na₂ (1,8)	*χ*-anthraquinone disulfonate	$C_{14}H_6O_2(SO_3Na)_2 \cdot 4H_2O$	XI-341	484.37
555	disulfonate Na₂ (2,6)	....	$C_{14}H_6O_2(SO_3Na)_2 \cdot 7H_2O$	*XI-84	538.41
556	disulfonic acid (2,7)	....	$C_{14}H_6O_2(SO_3H)_2$	*XI-84	368.34
557	disulfonate Na₂ (2,7)	....	$C_{14}H_6O_2(SO_3Na)_2 \cdot 4H_2O$	*XI-84	484.37
558	sulfonate Na (1)	....	$C_{14}H_7O_2 \cdot SO_3Na$	XI-336	310.26
559	sulfonic acid (2)	*β*-sulfoanthra-quinone	$C_{14}H_7O_2 \cdot SO_3H \cdot 3H_2O$	XI-337	342.33

Anisyl methyl ketone 4065
Anisylidene, cf. anisal.
Annidalin 5940
Anol 1605
Anon 1606

Anthracene-hexahydride 3576
Anthrachryson 5800
Anthraflavic acid 2303-4
Anthragallol 6212

Anthrahydroquinone 5019
Anthrahydroquinone diacetate 5020
Anthralin 2296
Anthranilamide 259

Anthranilic acid 264
Anthranilic nitrile 267
Anthrapurpurin 6215
Anthrarobin 2295
Anthrarufin 2300

Table 7-4 (*Continued*)
PHYSICAL CONSTANTS OF ORGANIC COMPOUNDS

No.	Crystalline Form and Color	Specific Gravity	Melting Point °C.	Boiling Point °C.	Solubility in 100 Parts		
					Water	Alcohol	Ether
528	lf./et.	....	73-4	....	i.; s. bz·	v. s.	v. s.; s. ac.
529	yel. lf./al. or CS_2	....	138-9	....	i.; s. chl.	sl. s.; s. bz.	sl. s.; sl. s. CCl_4
530	nd.	$1.113\frac{15}{15}°$	25	258.8	i.	s.	s.
531	col. pr.	$1.135\frac{15}{15}°$	35-8 (3)	243-4	i.; sl. s. bz.	sl. s	v. s.
532	col. oil	$1.123\frac{20}{4}°$	2.5	247-8	v. sl. s.	∞	∞
533	col. nd.	....	120-1	....	s. h. bz.	s. h.	s.
534	col. lq.	$1.098\frac{15}{15}°$	5.2	225	v. sl. s.	∞	∞
535	oil	$1.096\frac{20}{4}°$	<-12	251	v. sl. s.	s.	s.; s. a.
536	pl./aq.	$1.089\frac{55}{55}°$	57.2	243	s. h.	s.	s.
537	col. lq.	$0.990\frac{22}{2}°$	-37.4	153.8	i.	s.	s.
538	col mn.	$1.252\frac{27}{4}°$	217	339.9	i.	1.9 (20°), abs. al.	12.2 (100°), toluene
539	yel. pr./al.	....	245 (260)	subl.	i.	s.	s.; sl. s. bz.
540	yel. lf./al.	....	280±	subl.	i.	sl. s.	i. bz.; i. CS_2
541	lt. yel./al.	....	206 (-CO_2)	....	v. sl. s. h.	s.	....
542	yel. nd./al.	....	130±	....	i. HCl	s.	....
543	yel. lf./al.	....	238	subl. 93 (9mm)	i.	sl. s.	sl. s.
544	col. oil	$1.187\frac{15}{4}°$	<-18	d. >215	sl. s. h.	s.	s.; s. c. HCl
545	col. rhb.	....	160-70 d.	....	i.	v. s. h. bz.	s. h. alk.
546	nd./aq. al.	....	202-4	d.	v. sl. s. c.	3, in 50% h. al.	0.3, h. bz.
547	col. lf.	....	170	446	i.	s.	s.
548	or. nd./al.	....	185-90 d.	....	v. sl. s.	3	s. h. chl.
549	yel. nd./al.	....	206-18 d.	....	....	..........•	....
550	yel. rhb.	$1.438\frac{20}{4}°$	286	379-81	i.	0.05 (18°); 2.25 h.	v. sl. s.
551	yel. nd./aq.	....	293.4	....	sl. s. h.	....	....
552	yel. nd./al.	....	290	....	s. act.	v. sl. s. abs. al.	i.
553	yel. lf.	....	....	....	v. s.	i.	i.
554	yel. pr.	....	....	....	sl. s.	....	....
555	col. cr. darken in light	....	....	....	3.9 (20°); 18 (100°)	....	....
556	yel. cr.	....	....	....	v. s.	s.	i.; i. bz.
557	cr.	....	....	....	30.5 (20°); 13 (100°)	v. sl. s.	i.; i. bz.
558	yel. lf.	....	....	....	0.5 c.; s. h.	i.; v. sl. s. act.	i.
559	col. lf.	....	....	....	v. s. c.	v. s.	i.

Anthrone 545
Antiarin, cf. glcde.
Antifebrin 17
Antimony triethyl 6198
Antimony trimethyl 6274

Antipyrene salicylate 5590
Antisepsin 882
Antodin 3410
Antodyne 3410
Aphrodine, cf. alkd.

Apiin, cf. glcde.
Apocupreine, cf. alkd.
Apocynin 3776
Apomorphine, cf. alkd.
Aponal 1738

Apoquinine, cf. alkd.
Aposafranone 670
Apothesine 2119
Apple oil 496
Apyron 118

Table 7-4 (*Continued*)
PHYSICAL CONSTANTS OF ORGANIC COMPOUNDS

No.	Name	Synonym	Formula	Beil. Ref.	Formula Weight
560	**Anthraquinone** sulfonate Na (2)	"silver salt"	$C_{14}H_7O_2 \cdot SO_3Na$	XI-337	310.26
561	**Anthrol** (α)	1-hydroxy-anthracene	$C_6H_4(CH)_2C_6H_3OH$	VI-702	194.24
562	**Anthrol** (β)	2-anthrol	$C_6H_4(CH)_2C_6H_3OH$	VI-702	194.24
563	**Antimony triphenyl**	triphenyl stibine	$Sb(C_6H_5)_3$	XVI-891	353.07
564	dichloride	tri-Ph-stibine-di-Cl	$(C_6H_5)_3SbCl_2$	XVI-893	423.98
565	**Antipyrine**	1-Ph-2, 3-diMe-pyrazolone-5	$C_{11}H_{12}ON_2$	XXIV-27	188.23
566	acetylsalicylate	pyrosal	$C_{20}H_{20}O_5N_2$	XXIV-32	368.39
567	chloralhydrate	hypnal; chloral-antipyrine	$Cl_3C \cdot CH(OH)_2 \cdot$ $C_{11}H_{12}ON_2$	XXIV-31	353.64
568	mandelate	tussol	$C_{11}H_{12}ON_2 \cdot C_8H_8O_3$	. . . .	340.38
569	**Apiole**	1-allyl-2, 5-diMeO-3, 4-methylenedi-oxybenzene	$C_{12}H_{14}O_4$	XIX-87	222.24
570	**Arabinose** (α) (*d* or *l*)	. . . .	$C_4H_9O_4 \cdot CHO$	I-859	150.13
571	phenylhydrazone (*l*)	. . . .	$C_6H_5NH \cdot N:C_5H_{10}O_4$	XV-215	240.26
572	**Arabitol** (*d*)	pentanpentol	$C_5H_7(OH)_5$	I-531	152.15
573	**Arabonic acid**	. . . .	$HO \cdot CH_2(CHOH)_3 \cdot$ CO_2H	III-473	166.13
574	**Arachidic acid**	eicosanoic acid	$CH_3(CH_2)_{18} \cdot CO_2H$	II-389	312.54
575	**Arginine** (*l*+)	guanidine-amino-valeric acid	$H_2N \cdot C(:NH) \cdot NH \cdot$ $(CH_2)_3 \cdot CH(NH_2) \cdot$ CO_2H	IV-420	174.20
576	hydrochloride (*d*)	. . . .	$C_6H_{14}O_2N_4 \cdot HCl$	IV-423	210.67
577	**Arsacetin** (*p*)	acetyl arsanilic acid	$CH_3CO \cdot NH \cdot C_6H_4 \cdot$ $AsO(OH)_2$	XVI-880	259.09
578	Na salt	. . . .	$C_8H_9O_4NAsNa \cdot 4H_2O$	XVI-880	353.14
579	**Arsenic** diethyl	ethyl cacodyl	$[(C_2H_5)_2As]_2$	IV-616	266.09
580	triphenyl	triphenylarsine	$As(C_6H_5)_3$	XVI-828	306.24
581	**Arseno-phenyl-glycine** Na	. . . .	$(As \cdot C_6H_4 \cdot NH \cdot CH_2 \cdot$ $CO_2Na)_2$	. . . .	494.12
582	**Arsono-phenyl-glycinamide** (4)	. . . .	$H_2N \cdot CO \cdot CH_2 \cdot NH \cdot$ $C_6H_4 \cdot AsO(OH)_2$	*XVI-470	274.11
583	Na salt	tryparsamide	$C_2H_5ON_2 \cdot C_6H_4 \cdot$ $AsO(ONa)_2 \cdot 3H_3O$	*XVI-470	372.12
584	**Asarone**	propenyl-2,4,5-tri-MeO-benzene	$CH_3CH:CH \cdot$ $C_6H_2(OCH_3)_3$	VI-1129	208.26
585	**Ascaridole**	. . . .	$CH_3 \cdot C_6H_6O_2 \cdot CH:$ $(CH_3)_2$	XIX-17	168.24
586	**Ascorbic acid** (*l*+)	vitamin C; redoxon; cevitamic acid	$O \cdot CO \cdot C(OH):C(OH) \cdot$ $\underline{\quad\quad\quad\quad\quad}$ $CH \cdot CHOH \cdot CH_2OH$ $\rfloor$	. . . .	176.13
587	*iso*-**Ascorbic acid** (*d*)	. . . .	$C_6H_8O_6$	. . . .	176.13
588	**Asparagine** (*l*)	. . . .	$HO_2C \cdot CH(NH_2) \cdot$ $CH_2 \cdot CONH_2$	IV-476	132.12
589	**Aspartic acid** (*l*–)	amino-succinic acid	$HO_2C \cdot CH(NH_2) \cdot$ $CH_2 \cdot CO_2H$	IV-472	133.10
590	**Aspartic acid** (*dl*)	amino-succinic acid	$C_4H_7O_4N$	IV-483	133.10
591	**Atropic acid**	α-phenyl acrylic acid	$C_6H_5C(:CH_2) \cdot CO_2H$	IX-610	148.16

Table 7-4 (*Continued*)
PHYSICAL CONSTANTS OF ORGANIC COMPOUNDS

No.	Crystalline Form and Color	Specific Gravity	Melting Point °C.	Boiling Point °C.	Solubility in 100 Parts		
					Water	Alcohol	Ether
560	silver lf.	....	....	....	0.8 c.; 21 h.	i.	i.
561	nd./al.	....	150-3	....	i.	v. s.	v. s.
562	nd./aq.	....	d. 200	....	i.	v. s.	v. s.
563	tri. pr./pot.	$1.5^{12°}$	53-4.5	>360 d.	i.; v. s. bz.	sl. s.	v. s.
564	col. nd.	....	143	....	v. sl. s.	i. c.	i.; s. bz.
565	mn./aq.	$1.088^{113°}_{4}$	113 (109)	319^{174mm}	100	100	sl. s.
566	col. pd.	....	64-5	....	$1^{25°}$; $4^{80°}$	$126^{25°}$	$13^{25°}$; s. bz.
567	col. rhb.	....	67-8	....	7.9^{14}	33	sl. s.
568	col. cr.	....	52-5	....	6.6	30	5
569	col. nd.	$1.02^{20°}_{4}$	30*	294	i.; s. bz.	s.; s. act.	s.; s. oil
570	rhb. pr.	$1.585^{20°}_{4}$	158.5-9.5	....	46^{00}	$0.5^{9°}$, 90% al.	i.
571	col. nd.	....	151-3	....	$1.2^{15°}$	$1.3^{15°}$, abs.	v. sl. s.
572	pr. or warts	....	102-3	....	v. s.	$2^{12°}$, 90% al.	....
573	cr.	....	89	d., $-H_2O$	s.	....	....
574	col. lf.	....	77	328 sl. d.	i.	v. s. h. abs.	v. s.
575	pr./66% al.	....	238 d.	....	$15^{21°}$	sl. s.	i.
576	col. pl.	....	222-35 d.	softens 218	v. s.	....	....
577	lf.	....	>200	....	s. Na_2CO_3	v. sl. s. dil. HCl	....
578	wh. cr. pd.	....	....	185-90	$10^{20°}$	$33^{50°}$	....
579	lq.	>1	ign. in air	....	i.	v. s.	v. s.
580	pl./bz.	1.306	59-60	>360(CO$_2$)	i.; i. HCl	s.; v. s. bz.	v. s.
581	yel. pd.	....	d. in air	....	s.	....	....
582	pl./aq.	....	darkens 280	....	v. s. h.; s. h. ac.	v. sl. s. h.; i. c. chl.	i. c. act.
583	wh. pd.	....	....	....	50	sl. s.	i.; i. chl.
584	mn. nd./aq.	$1.165^{18°}$	67	296	sl. s. h.; s. ac.	$7^{18°}$, $165^{50°}$, 60% al.	s.; s. chl.
585	col. lq.	$0.999^{20°}_{20}$	expl. 250	115^{15mm}	i.	s.	expl. with a.
586	col. mn.	....	192	....	33; 1 gly.	4; i. pet.; i. fat	i.; i. bz.
587	col. cr.	....	168-71	....	v. s.	s.; i. act.	v. sl. s.
588	rhb.	$1.543^{15°}_{4}$	227-35**	235 d.	$3.1^{28°}$	i. c. abs. al.	s. NH^3
589	rhb.	....	251-83 d.	....	$0.5^{16°}$	i.	i.
590	mn. pr.	$1.663^{13°}_{13}$	271**	d. 290	$2.71^{75°}$	i.	i.
591	mn. nd./aq	....	106-7	267 d.	0.13 c.	s.; s. chl.	s.; s. CS$_2$

* Also modifications, m.p. 27.5 and 18-9.   ** Sealed tube.

Table 7-4 (*Continued*)
PHYSICAL CONSTANTS OF ORGANIC COMPOUNDS

No.	Name	Synonym	Formula	Beil. Ref.	Formula Weight
592	**Auramine**	4,4′-dimethylamino-benzophenonimide	$[(CH_3)_2NC_6H_4]_2C:NH$	XIV-91	267.38
593	**Aurin**	4′,4″-diOH-fuchsone	$(HOC_6H_4)_2C:C_6H_4:O$	VIII-361	290.32
594	**Azafrin**	....	$C_{27}H_{38}O_4$	XXX-115	426.60
595	**Azafrinone**	....	$C_{27}H_{36}O_4$	XXX-116	424.59
596	**Azelaic** acid	nonandioic acid	$(CH_2)_7(CO_2H)_2$	II-707	188.23
597	nitrile	....	$(CH_2)_7(CN)_2$	II-709	150.23
598	**Azo**-anisole (2,2′)	diMeO-azobenzene	$(CH_3O \cdot C_6H_4N:)_2$	XVI-92	242.28
599	benzene	diphenyldiimide	$C_6H_5 \cdot N:N \cdot C_6H_5$	XVI-8	182.23
600	benzoic acid (2,2′)	....	$(HO_2C \cdot C_6H_4 \cdot N:)_2$	XVI-228	270.25
601	benzoic acid (3,3′)	....	$(HO_2C \cdot C_6H_4 \cdot N:)_2 \cdot \frac{1}{2}H_2O$	XVI-233	279.25
602	benzoic acid (4,4′)	....	$(HO_2C \cdot C_6H_4 \cdot N:)_2$	XVI-236	270.25
603	dicarbonamide	................-	$(NH_2 \cdot CO \cdot N:)_2$	III-123	116.08
604	naphthalene (α,α′)	....	$C_{10}H_7 \cdot N:N \cdot C_{10}H_7$	XVI-79	282.35
605	naphthalene (β,β′)	....	$C_{10}H_7 \cdot N:N \cdot C_{10}H_7$	XVI-80	282.35
606	phenetole (2,2′)	....	$(C_2H_5O \cdot C_6H_4 \cdot N:)_2$	XVI-92	270.33
607	phenetole (3,3′)	....	$(C_2H_5O \cdot C_6H_4 \cdot N:)_2$	XVI-95	270.33
608	phenetole (4,4′)	....	$(C_2H_5O \cdot C_6H_4 \cdot N:)_2$	XVI-113	270.33
609	phenol (2,2′)	dihydroxy-azobenzene	$(HO \cdot C_6H_4 \cdot N:)_2$	XVI-91	214.23
610	phenol (3,3′)	....	$(HO \cdot C_6H_4 \cdot N:)_2$	XVI-95	214.23
611	phenol (4,4′)	....	$(HO \cdot C_6H_4 \cdot N:)_2$	XVI-110	214.23
612	toluene (2,2′)	dimethylazobenzene	$(CH_3 \cdot C_6H_4 \cdot N:)_2$	XVI-61	210.28
613	toluene (3,3′)	....	$(CH_3 \cdot C_6H_4 \cdot N:)_2$	XVI-64	210.28
614	toluene (4,4′)	....	$(CH_3 \cdot C_6H_4 \cdot N:)_2$	XVI-67	210.28
615	**Azoxy**-benzene	....	$C_6H_5 \cdot N(:O):N \cdot C_6H_5$	XVI-622	198.23
616	benzoic acid (2,2′)	....	$(HO_2C \cdot C_6H_4)_2:N_2O$	XVI-644	286.25
617	benzoic acid (3,3′)	....	$(HO_2C \cdot C_6H_4)_2:N_2O$	XVI-646	286.25
618	benzoic acid (4,4′)	....	$(HO_2C \cdot C_6H_4)_2:N_2O$	XVI-647	286.25
619	naphthalene (α,α′)	....	$(C_{10}H_7N)_2O$	XVI-633	298.35
620	naphthalene (β,β′)	....	$(C_{10}H_7N)_2O$	XVI-633	298.35
621	**Barbituric acid**	malonyl urea	$CH_2CONHCONHCO \cdot$ $\overline{}$ $2H_2O$	XXIV-467	164.12
622	**Bassorine**	....	$C_6H_{10}O_5$	....	162.14
623	**Behenic acid**	docosanoic acid	$CH_3 \cdot (CH_2)_{20} \cdot CO_2H$	II-391	340.59
624	**Behenolic acid**	docosinoic acid	$C_8H_{17}C \vdots C(CH_2)_{11} \cdot$ CO_2H	II-497	336.58
625	**Benzal** acetoacetic ester	....	$C_2H_3O \cdot C(C_7H_6) \cdot$ $CO_2C_2H_5$	X-731	218.25
626	acetone	Me-cinnamyl ketone	$C_6H_5 \cdot C_2H_2 \cdot CO \cdot CH_3$	VII-364	146.19
627	acetophenone	chalcone	$C_6H_5 \cdot C_2H_2 \cdot CO \cdot C_6H_5$	VII-479	208.26
628	acetophenone dibromide	....	$C_6H_5 \cdot (CHBr)_2 \cdot$ COC_6H_5	VII-445	368.08
629	acetophenone dibromide	high m. p. form	$C_6H_5 \cdot (CHBr)_2 \cdot CO \cdot$ C_6H_5	VII-445	368.08
630	amino-2-cresol (5)	(5; 1, 2)	$C_6H_5CH:N \cdot$ $C_6H_3(CH_3) \cdot OH)$	....	211.27
631	aminophenol (*p*)	....	$C_6H_5CH:N \cdot C_6H_4 \cdot OH$	XIII-453	197.24
632	aniline	benzaldehyde anil	$C_6H_5CH:N \cdot C_6H_5$	XII-195	181.24

Table 7-4 (*Continued*)
PHYSICAL CONSTANTS OF ORGANIC COMPOUNDS

No.	Crystalline Form and Color	Specific Gravity	Melting Point °C.	Boiling Point °C.	Solubility in 100 Parts		
					Water	Alcohol	Ether
592	col./al.	....	136	....	i.	$7^{20°}$	$2.3^{20°}$
593	rhb. red	....	308-10 d.	....	i.; i. bz.	s.	s.; s. alk.
594	or. nd./bz.	....	212-4	....	i.; s. alk.	sl. s.	i.; sl. s. bz.
595	or. cr./act.	....	191	....	....	....	....
596	lf. or nd.	$1.029^{20°}_4$	106.5	286.5^{100mm}	0.2 c.; ∞ h.	v. s.	$2.7^{15°}$
597	oil	$0.941^{0°}$	....	$195-6^{19mm}$	i.	v. s.	v. s.
598	or. pr./ MeOH	....	153 (141)	....	i.; s. act.	s.; s. chl.	s.; s. bz.
599	or. mn.	$1.203^{20°}_4$	68	297	i.	$4.2^{20°}$	$12^{20°}$ lg.
600	yel. nd./al.	....	237 d.	....	sl. s. h.	v. s. h.	i. bz.
601	yel. nd./ac.	....	340±	d.	v. sl. s.	0.2 h.	sl. s.
602	red nd./ac.	....	d. 330±	....	v. sl. s.	v. sl. s.	v. sl. s.
603	or. red pd.	....	d. 180-200	....	v. sl. s. h.	i.	d. h. HCl
604	red nd./ac.	....	190	subl. >190	i.; s. ac.	v. sl. s.	v. s. bz.
605	red pr./chl.	....	204	subl. 210	i.	sl. s.	s. bz.
606	red pr./al.	....	131	240 d.	i.; s. HCl	s.	s.
607	yel. pr./al.	....	91	d.	i.; i. HCl	s.	s.
608	or. lf.	....	159-60	d.	v. s. chl.	s. h.	v. s.
609	yel. lf.	....	171-2	subl.	i.	0.3 c.	s.; s. KOH
610	brn./aq. al.	....	205	....	v. sl. s.	s. h.	s. Na_2CO_3
611	brn. tri.	....	215 d.	....	sl. s.	s.	s.; s. bz.
612	red mn./et.	....	55	....	i.; s. bz.	6.03^{15}	$147.7^{17°}$
613	or. rhb.	....	54-5	....	i.	sl. s.	s.
614	or. nd./lg.	....	144-5	....	i.	sl. s.	s.; s. lg.
615	yel. rhb.	$1.248^{20°}_{20}$	36	d.	i.	$11.4^{15°}$	$43.5^{15°}$ lg.
616	brn./al.	....	246-8 d.	....	v. sl. s. h.	s. h.	s. h.
617	nd. or lf.	....	320 d.	....	i.	sl. s.	sl. s.
618	yel. armor.	....	d. 240±	....	i.	i.	s. C_5H_5N
619	yel.rhb./al.	....	127	....	i.	s.	....
620	yel.rhb./al.	....	167-8	....	i.	....	....
621	col. pr./aq.	....	d. 245±	....	s. h.	sl. s.	s.; s. HCl
622	wh. amor.	....	....	....	v. sl. s.	i.; d. h. a.	i. $(NH_4)_2S$
623	col. nd.	....	80	306^{60mm}	i.	sl. s.	sl. s.
624	nd.	....	57.5	....	i.	v. s.	v. s.
625	rhb./al.	....	59-60	295-7 sl. d.	v. s. chl.	sl. s. c.	sl. s. c.
626	pl.	$1.035^{20°}_{20}$	41-2	260-2	s. H_2SO_4	s.; s. chl.	s.; s. bz.
627	lt. yel. rhb.	$1.071^{62°}_4$	62	345-8 sl. d.	i.	sl. s.	v. s.
628	nd./al.	....	108-9	....	....	$1^{30°}$; s. h.	....
629	....	....	156-7	....	....	$0.16^{30°}$	....
630	brn. fl.	....	210-11	....	i.	sl. s.	i.
631	lf./aq. al.	....	185-6	....	....	v. s.	....
632	yel. cr.	....	56 (48)	300±	i.	s.	s.

Table 7-4 (*Continued*)
PHYSICAL CONSTANTS OF ORGANIC COMPOUNDS

No.	Name	Synonym	Formula	Beil. Ref.	Formula Weight
633	**Benzal** azine	dibenzal hydrazine	$(C_6H_5CH:N\cdot)_2$	VII-225	208.27
634	bromide	benzylidene bromide	$C_6H_5\cdot CH:Br_2$	V-308	249.94
635	chloride	benzylidene chloride	$C_6H_5\cdot CH:Cl_2$	V-297	161.03
636	diacetate	benzylidene diacetate	$C_6H_5CH:(O_2CCH_3)_2$	VII-210	208.22
637	lactic acid	phenyl-hydroxy-crotonic acid	$C_6H_5\cdot CH:CH\cdot CHOH\cdot CO_2H$	X-308	178.19
638	malonic acid	....	$C_6H_5\cdot CH:C(CO_2H)_2$	IX-891	192.17
639	methylamine	....	$C_6H_5\cdot CH:N\cdot CH_3$	VII-213	119.17
640	phthalide	benzylidene phthalide	$C_7H_6:C\cdot C_6H_4\cdot CO\cdot O$	XVII-376	222.25
641	2-thiohydantoin (5)	....	$CO\cdot NH\cdot CS\cdot NH\cdot C:C_7H_6$	XXIV-400	204.25
642	**Benzaldehyde**	art. almond oil	$C_6H_5\cdot CH:O$	VII-174	106.13
643	oxime (*α*; *syn*)	....	$C_6H_5CH:NOH$	VII-218	121.14
644	oxime (*β*; *anti*)	....	$C_6H_5CH:NOH$	VII-221	121.14
645	oxime carboxylic anhydride (*o*)	benzoxazinone	$C_6H_4\cdot CH:N\cdot O\cdot CO$	XXVII-198	147.13
646	phenylhydrazone	....	$C_6H_5NH\cdot N:CH\cdot C_6H_5$	XV-134	196.25
647	**Benzamaron**	....	$C_6H_5CH[CH(C_6H_5)\cdot COC_6H_5]_2$	VII-849	480.61
648	*iso*-**Benzamaron**	....	$C_{35}H_{28}O_2$	VII-849	480.61
649	**Benzamidine**	benzenylamidine	$C_6H_5C(:NH)\cdot NH_2$	IX-280	120.16
650	**Benzanthracene** (1,2)	naphthanthracene	$C_{18}H_{12}$	V-718	228.30
651	**Benzarsonic acid** (*p*)	....	$(HO)_2OAs\cdot C_6H_4\cdot CO_2H$	XVI-876	246.05
652	**Benzaurine**	....	$C_6H_5C(:C_6H_4O)\cdot C_6H_4OH$	VI-1145	274.32
653	**Benzazide**	benzoyl azide	$C_6H_5CO\cdot N_3$	IX-332	147.14
654	**Benzedrine**	*dl*-desoxy-nor-ephedrine	$C_6H_5CH_2CH(NH_2)\cdot CH_3$	XII-1145	135.21
655	sulfate	....	$(C_9H_{13}N)_2\cdot H_2SO_4$	....	368.50
656	**Benzene**	benzol	C_6H_6	V-179	78.11
657	azo-*o*-cresol (5)	(1;4,3)	$C_6H_5\cdot N_2\cdot C_6H_3(OH)\cdot CH_3$	XVI-130	212.25
658	azo-*m*-cresol (6)	(1;4,2)	$C_6H_5\cdot N_2\cdot C_6H_3(OH)\cdot CH_3$	XVI-134	212.25
659	azodiphenylamine (*p*)	4-anilino-azobenzene	$C_6H_5\cdot N_2\cdot C_6H_4NHC_6H_5$	XVI-314	273.34
660	azo-*α*-naphthyl-amine (4)	....	$C_6H_5\cdot N_2\cdot C_{10}H_6\cdot NH_2$	XVI-361	247.30
661	azo-*β*-naphthyl-amine (1)	Yellow AB	$C_6H_5\cdot N_2\cdot C_{10}H_6\cdot NH_2$	XVI-369	247.30
662	azoresorcinol	(1;2,4); Sudan G	$C_6H_5\cdot N_2\cdot C_6H_3(OH)_2$	XVI-180	214.23
663	disulfone chloride	(1,3)	$C_6H_4:(SO_2Cl)_2$	XI-200	275.14
664	disulfonic acid (*m*)	....	$C_6H_4\cdot(SO_3H)_2\cdot 2\frac{1}{2}H_2O$	XI-199	283.28
665	disulfonic Na (*m*)	....	$C_6H_4(SO_3Na)_2\cdot 4H_2O$	XI-199	354.27
666	hexabromide (*α, trans*)	hexabromo-cyclohexane	$C_6H_6Br_6$	V-25	557.57
667	hexacarboxylic acid	mellitic acid	$C_6(CO_2H)_6$	IX-1008	342.17
668	hexachloride (*α, trans*)	hexachloro-cyclohexane	$C_6H_6Cl_6$	V-23	290.83

Benzaldehyde-anil 632
Benzaldehyde-cyanohydrin 4010
Benz-amide 713
Benz-aminobenzoic acid 746-8
Benz-anilide 715

Benz-triazole 252
Benzene-azimide 252
Benzene azoanisidine 246
Benzene azodimethylaniline 2431

Table 7-4 (*Continued*)
PHYSICAL CONSTANTS OF ORGANIC COMPOUNDS

No.	Crystalline Form and Color	Specific Gravity	Melting Point °C.	Boiling Point °C.	Solubility in 100 Parts		
					Water	Alcohol	Ether
633	yel. pr.	....	93	....	i. c.	s. h.	s.
634	oil	$1.51^{15°}$	....	140^{20mm}	i.	∞	∞
635	col. lq.	$1.295^{16°}$	−16.1	214	i.	∞	∞
636	pl./et.	....	45-6	154^{20mm}	s. dil. alk.	v. s.	v. s.
637	nd./aq.	....	137±	....	s. h.	i. bz., CS_2	sl. s.; i. lg.
638	cr./et.-CS_2	....	195-6 d.	....	v. s. h.	s.	sl. s.
639	col. lq.	....	....	180±	....	....	....
640	mn./al.	....	98-9	....	i. h.	v. s. h.	....
641	yel. nd./al.	....	264-5	....	....	....	....
642	col. lq.	$1.046^{\frac{20}{4}°}$	−26*	179.1	0.3	∞	∞
643	pr.	$1.111^{\frac{20}{4}°}$	35 (5)	117.5^{14mm}	sl. s.	v. s. bz.	v. s.
644	nd./et.	....	128-30	....	s. h.	v. sl. s. bz.	....
645	cr./bz.	....	d. 120	....	....	....	....
646	col.** mn.	....	155-6	....	s. bz.	s. h.	sl. s.
647	cr.	....	218-9	....	$1.6^{12°}$, bz.	....	....
648	cr.	....	179-80	....	$4.1^{12°}$, bz.	....	....
649	cr.	....	80	....	s.	v. s.	sl. s.
650	lf./al.-ac.	....	159-60	subl.	....	sl. s. h. .	s. bz.
651	cr./aq.	....	d.-H_2O, 210	....	v. sl. s.	i.	i. ac.
652	r. pd.	....	....	....	i.; s. alk.	s.	s.
653	col. pl./act.	....	32	expl.	i.	sl. s.	s.
654	col. lq.	0.93	....	203	sl. s.; s. a.	s.; s. EtOAc	s.; s. chl.
655	wh. pd.	....	....	....	10	sl. s.	i.
656	col. lq.	$0.879^{\frac{20}{4}°}$	5.5	80.1	$0.07^{22°}$	∞ abs. al.	∞
657	yel. nd./al.	....	129-30	....	v. s. h.; s. alk.	s.; s. chl.	s.; s. bz.
658	yel. nd./lg.	....	109†	....	....	s.; s. chl.	s.; s. bz.
659	yel. pr./et. al.	....	86-8	....	v. s. lig.	v. s.	v. s.
660	r. nd./al.	....	123	....	s. bz.	s.	s.
661	r. pl./al.	....	102-4	....	i.; s. oil	s.; s. CCl_4	s. ac.
662	r. nd./aq. al.	....	170‡	....	i.; s. alk.	v. s.	v. s.; v. s. bz.
663	mn./et.	....	63	210.7^{20mm}	d. 130°	s. h. d.	s.
664	delq. cr.	....	....	....	s.	....	....
665	col. nd.	....	....	....	s.	....	....
666	mn. pr.	....	212	....	....	sl. s.	sl. s.
667	nd./al.	....	286-8§	d.	v. s.	v. s.	s. h. H_2SO_4
668	mn. pr.	....	157-8	d. >158	$4.4^{15°}$, chl.	s.	$6.5^{18°}$, bz.

* Solidifies −56°.
** Pink on exposure to light.
† Crysts. +1H_2O, m. 90°.
‡ Cryst. +⅓H_2O, m. 161°.
§ Sealed tube.

Benzene azonitromethane 4739
Benzene azophenol 3727-9
Benzene diazoanilide 1762
Benzene diazonium, cf. diazobenzene.

Table 7-4 (*Continued*)
PHYSICAL CONSTANTS OF ORGANIC COMPOUNDS

No.	Name	Synonym	Formula	Beil. Ref.	Formula Weight
669	**Benzene** hexachloride (β, *cis*)	....	$C_6H_6Cl_6$	V-23	290.83
670	indone	safranone	$C_{18}H_{12}ON_2$	XXIII-413	272.31
671	pentacarboxylic acid	....	$C_6H(CO_2H)_5 \cdot 5H_2O$	IX-1006	388.24
672	sulfinic acid	....	$C_6H_5 \cdot SO_2H$	XI-2	142.18
673	sulfinic Na	....	$C_6H_5 \cdot SO_2Na \cdot 2H_2O$	XI-6	200.19
674	sulfohydroxamic acid	Piloty's acid	$C_6H_5SO_2 \cdot NHOH$	XI-51	173.19
675	sulfonic acid	....	$C_6H_5 \cdot SO_3H$	XI-26	158.18
676	sulfonic Na	....	$C_6H_5SO_3Na \cdot H_2O$	XI-28	198.18
677	sulfonic amide	benzenesulfonamide	$C_6H_5 \cdot SO_2 \cdot NH_2$	XI-39	157.19
678	sulfonic chloride	benzenesulfonyl chloride	$C_6H_5 \cdot SO_2 \cdot Cl$	XI-34	176.62
679	1-sulfonic acid-(4-azo-5)-8-hydroxy-quinoline	sulfenazoxine	$HO_3S \cdot C_6H_4 \cdot N_2 \cdot C_9H_5N(OH)$	XXII-584	329.34
680	tetracarboxylic acid (1,2,3,4)	prehnitic acid	$C_6H_2(CO_2H)_4 \cdot 2H_2O$	IX-997	290.19
681	tetracarboxylic acid (1,2,3,5)	mellophanic acid	$C_6H_2(CO_2H)_4$	IX-997	254.15
682	tetracarboxylic acid (1,2,4,5)	pyromellitic acid	$C_6H_2(CO_2H)_4 \cdot 2H_2O$	IX-997	290.19
683	tricarboxylic acid (1,2,3)	hemimellitic acid	$C_6H_3(CO_2H)_3 \cdot 2H_2O$	IX-976	246.18
684	tricarboxylic acid (1,2,4)	trimellitic acid	$C_6H_3(CO_2H)_3$	IX-977	210.14
685	tricarboxylic acid (1,3,5)	trimesic acid	$C_6H_3(CO_2H)_3$	IX-978	210.14
686	trisulfonic acid	(1,3,5)	$C_6H_3(SO_3H)_3 +$ aq.	XI-227	318.30
687	**Benzenyl**-aminothiophenol	μ-phenylbenzothiazole	$C_6H_4 \cdot N:C(C_6H_5) \cdot S$	XXVII-74	211.29
688	aminoxime	....	$C_6H_5C(NOH)NH_2$	IX-304	136.15
689	naphthylamidine	α-naphthyl-benzamidine	$C_6H_5C(:NH)NH \cdot C_{10}H_7$	XII-1233	246.31
690	phenylenediamine (1,2)	α-phenylbenzimidazole	$C_6H_5C:N \cdot C_6H_4 \cdot NH$	XXIII-230	194.24
691	**Benzhydrol**	diphenyl carbinol	$(C_6H_5)_2CHOH$	VI-678	184.24
692	ether	....	$[(C_6H_5)_2CH]_2O$	VI-679	350.46
693	**Benzhydryl** amine	....	$(C_6H_5)_2CH \cdot NH_2$	XII-1323	183.26
694	benzoic acid (*p*)	....	$C_6H_5CH(OH) \cdot C_6H_4CO_2H$	X-346	228.25
695	**Benzidine** disulfonic acid (*o,o'*)	....	$[C_6H_3(NH_2) \cdot SO_3H]_2$	XIV-794	344.37
696	monosulfonic acid	(4;4',3)	$H_2N \cdot C_6H_4 \cdot C_6H_3(NH_2) \cdot SO_3H$	XIV-770	264.31
697	**Benzil**	dibenzoyl	$(C_6H_5 \cdot CO)_2$	VII-747	210.23
698	**Benzilam**	triphenyloxazole	$C_6H_5C:C(C_6H_5) \cdot O \cdot C(C_6H_5):N$	XXVII-88	297.36
699	**Benzilic acid**	diphenyl glycolic acid	$(C_6H_5)_2C(OH) \cdot CO_2H$	X-342	228.25

Benzene hexahydride 1601
Benzene sulfonamide 677
Benzene sulfonyl chloride 678
Benzene tetrahydride 1609

Table 7-4 (*Continued*)
PHYSICAL CONSTANTS OF ORGANIC COMPOUNDS

No.	Crystalline Form and Color	Specific Gravity	Melting Point °C.	Boiling Point °C.	Solubility in 100 Parts		
					Water	Alcohol	Ether
669	col. cr./xylene	$1.89^{19°}$	310-2	....	$0.13^{20°}$, chl.	v. sl. s.; $1^{22°}$bz.	v. sl. s. ac.
670	brn. met.	....	248-9	....	sl. s.	v. s.	i. alk.
671	rhb.	....	228-30; 238 (anh.)	....	v. s. h.	s.	sl. s.; i. bz.
672	pr./aq.	....	83-4	d.>100	v. s. h.; sl. s. c.	v. s.	v. s.
673	lf./aq.	....	....	....	s.	....	....
674	rhb. pl./aq.	....	126±	d.>m.p.	s. h.	s.; v. sl. s. bz.	s.; s. act.
675	col. nd.	....	65-6*	d.	v. s.	v. s.	i.; sl. s. bz.
676	nd./aq. al.	....	450 d.	....	$60^{30°}$	sl. s. h.	....
677	mn. nd./aq.	....	156	....	$0.43^{16°}$	v. s.	v. s.
678	cr.	$1.384^{\frac{15}{15}°}$	16.5-17.5	251.5	i.; d. h.	v. s.; d. h.	s.
679	or. nd./aq. al.	....	....	....	....	....	....
680	pr./aq.	....	237-50 d.	....	v. s.	....	sl. s.
681	cr./aq.	....	215-38 d.	....	v. s.	....	....
382	tri./aq.	....	269-71 d.	....	(anh.) $1.416°$	v. s.	....
683	pl./aq.	....	190 d.	....	$3.2^{19°}$; v. s. h.	....	s.
684	nd./aq.	....	216-8 d.	....	sl. s.	....	sl. s.
685	pr./aq.	....	375-80	subl.	$2.8^{22.5°}$	v. s.	s.
686	hyg. nd.	....	d.>100	....	s.	....	....
687	nd./al.	....	115	360± (sl. d.)	i.; s. HCl	sl. s.	s.; s. CS_2
688	mn./aq.	....	79-80	....	sl. s. c.	s.; s. bz.	s.; i. lg.
689	pl./aq.	....	141	....	i.	s.	s.
690	nd./aq.	....	291	....	sl. s.	s.	sl. s. chl., bz.
691	nd./lg.	....	68-9	$297-8^{48mm}$	$0.05^{20°}$	v. s.	v. s.
692	mn./bz.	....	109-10	267^{15mm}	....	sl. s. h.	v. s. bz.
693	lq.	$1.064^{\frac{21.5°}{0}}$	....	$301-2^{746mm}$	....	....	....
694	nd./aq.	....	164-5	d.	s. h.	s.	s.; sl. s. chl.
695	pr./aq. (+3H_2O)	....	d.>175	....	$0.08^{25°}$	i.	i.
696	lf.	....	....	....	v. sl. s. h.	v. sl. s.	v. sl. s.
697	trig. pr.	$1.23^{15°}$	95	346-8 (sl. d.)	i.	v. s.	v. s.
698	pr./al. et.	....	115	....	v. sl. s.	sl. s. c.	v. s.; s. h. ac.
699	nd./aq.	....	150	....	s. h.	s.	s.; s. H_2SO_4

* Cryst. +1.5H_2O, m.p. 43-4.
Benzenyl-amidine 649
Benzhydryl-aniline 384-5

Benzidine 1713-4
Benzidine (β) 1711

Table 7-4 (*Continued*)
PHYSICAL CONSTANTS OF ORGANIC COMPOUNDS

No.	Name	Synonym	Formula	Beil. Ref.	Formula Weight
700	**Benzil**-oxime, di- (*syn*)	α-benzildioxime	$(C_6H_5C:NOH)_2$	VII-760	240.26
701	oxime, di-(*anti*)	β-benzildioxime	$(C_6H_5C:NOH)_2$	VII-761	240.26
702	oxime, di- (*amphi*)	γ-benzildioxime	$(C_6H_5C:NOH)_2 \cdot C_2H_5OH$	VII-763	286.33
703	oxime, mono- (α)	....	$C_6H_5CO \cdot C(:NOH) \cdot C_6H_5$	VII-757	225.25
704	oxime, mono- (β)	....	$C_{14}H_{11}O_2N \cdot \frac{1}{2}C_6H_6$	VII-758	264.31
705	oxime, mono- (β)	....	$C_{14}H_{11}O_2N$	VII-758	225.25
706	osazone (α) (*syn*)	....	$(C_6H_5C)_2 \cdot (N \cdot NHC_6H_5)_2$	XV-173	390.49
707	osazone (β) (*anti*)	....	$(C_6H_5C)_2 \cdot (N \cdot NHC_6H_5)_2$	XV-174	390.49
708	**Benzimidazole** (*o*)	....	$C_6H_4 \cdot N:CH \cdot NH$ ⎿_____⏌	XXIII-131	118.14
709	**Benzimidazolone** (*o*)	*o*-phenylene urea	$C_6H_4 \cdot NH \cdot CO \cdot NH$ ⎿_____⏌	XXIV-116	134.14
710	**Benzohydroxamic acid**	*N*-benzoyl hydroxyl-amine	$C_6H_5CO \cdot NHOH$	IX-301	137.14
711	**Benzoic** acid	....	$C_6H_5 \cdot CO_2H$	IX-92	122.12
712	Na salt	sodium benzoate	$C_6H_5 \cdot CO_2Na \cdot H_2O$	IX-107	162.12
713	amide	benzamide	$C_6H_5CO \cdot NH_2$	IX-195	121.14
714	anhydride	....	$(C_6H_5CO)_2O$	IX-164	226.23
715	anilide	benzanilide; *N*-Ph-benzamide	$C_6H_5NH \cdot COC_6H_5$	XII-262	197.24
716	nitrile	phenyl cyanide	$C_6H_5 \cdot CN$	IX-275	103.12
717	**Benzoin** (*dl*)	....	$C_6H_5 \cdot CHOH \cdot CO \cdot C_6H_5$	VIII-166	212.25
718	acetate	acetyl benzoin	$C_6H_5CO \cdot CH(C_6H_5) \cdot O_2C \cdot CH_3$	VIII-174	254.29
719	ethyl ether	....	$C_6H_5CH(OC_2H_5) \cdot CO \cdot C_6H_5$	VIII-174	240.30
720	oxime (α)	cupron	$C_{13}H_{12}O:C:NOH$	VIII-175	227.27
721	oxime (β)	....	$C_{13}H_{12}O:C:NOH$	VIII-175	227.27
722	phenylhydrazone (α)	....	$C_6H_5NHN:C(C_6H_5) \cdot CHOHC_6H_5$	XV-200	302.38
723	phenylhydrazone (β)	....	$C_6H_5NHN:C(C_6H_5) \cdot CHOHC_6H_5$	XV-200	302.38
724	**Benzophenone**	diphenyl ketone	$(C_6H_5)_2CO$	VII-410	182.22
725	anil	....	$(C_6H_5)_2C:N \cdot C_6H_5$	XII-201	257.34
726	chloride	diPh-diCl-methane	$(C_6H_5)_2CCl_2$	V-590	237.13
727	dicarboxylic acid (2,2′)	....	$(HO_2C \cdot C_6H_4)_2:CO$	X-881	270.24
728	oxime	diphenyl ketoxime	$(C_6H_5)_2C:NOH$	VII-416	197.24
729	phenylhydrazone	....	$C_6H_5NHN:C(C_6H_5)_2$	XV-148	272.35
730	**Benzo**-phosphinic acid (*p*)	....	$HO_2C \cdot C_6H_4 \cdot PO(OH)_2$	XVI-820	202.10

Benznaphthalide 765   Benzoic acid sulfamide 5671-3

Table 7-4 (*Continued*)
PHYSICAL CONSTANTS OF ORGANIC COMPOUNDS

No.	Crystalline Form and Color	Specific Gravity	Melting Point °C.	Boiling Point °C.	Solubility in 100 Parts		
					Water	Alcohol	Ether
700	lf.	....	235-7 d.	....	i.; s. NaOH	$0.05^{17°}$	v. sl. s.
701	cr.	....	206-7 d.	....	sl. s. h.	$15.3^{17°}$	v. s.
702	nd./al.	....	anh. 164-6	....	i.	$>15.3^{17°}$	v. s.
703	lf./aq. al.	....	138-40	d. 200	i.	s. c.	s.
704	nd./bz.	....	70	....	v. sl. s.	v. s.	v. s.
705	....	....	113-4	....	....	....	....
706	yel. nd.	....	205-15	....	$1.7^{19°}$ act.	sl. s. c.	s.
707	nd.	....	225-35	....	$2.4^{19°}$ act.	sl. s.	sl. s.
708	rhb./al.	....	170	>360	sl. s.; s. a.	v. s.	sl. s.; s. alk.
709	lf./aq.	....	310-2	subl. >270	sl. s. h.	s.; sl. s. bz.	i. aq. a.
710	rhb.	....	131-2	expl.	$2.25^{6°}$	v. s.	sl. s.; i. bz.
711	mn. pr.	$1.316\frac{28°}{4}$	122.4; subl. > 100	250.0	$0.21^{17.5°}$; $2.27^{5°}$	$46.6^{15°}$, abs. al.	$66^{15°}$
712	col. cr.	....	$-H_2O$, 120	....	$61^{25°}$; $77^{100°}$	$2.3^{25°}$; $8.3^{78°}$	....
713	col. pr.	1.341	130	290	$1.35^{25°}$	$17.0^{25°}$, abs. al.	sl. s.
714	rhb./et.	$1.199\frac{15°}{4}$	42	360	i.; s. act.	s.; s. bz.	s.; s. chl.
715	lf./al.	$1.314°$	163	$117-9^{10mm}$	i.	$43^{0°}$, abs. al.	sl. s.
716	col. lq.	$1.001\frac{25°}{4}$	-13.8	191.1	$1^{100°}$	∞	∞
717	mn.	....	133-7	$343-4^{768mm}$	v. sl. s. h.	s. h.; s. act.	sl. s.
718	pr./et. or al.	....	83	....	....	v. s.	v. s.
719	nd./lg.	....	62-3	....	v. s. bz.	v. s.	v. s.
720	pr./bz.	....	151-2	....	sl. s.	s.	s. NH_4OH
721	pr.	....	99	....	....	....	s.
722	col. nd.	....	155-8	....	i.	$2.2^{20°}$	....
723	col. nd.	....	106	....	....	$8.8^{20°}$	s.
724	col. rhb.	(α) $1.083^{54°}$ (β) $1.108^{23°}$	(α) 48.1 (β) 26.5 (γ) 45.8 (δ) -51	305.9	i.	$6.5^{15°}$; s. chl.	$15^{13°}$
725	yel. rhb./ et.	....	117 (112)	356-8	s. bz., CS_2	sl. s.	sl. s.
726	col. lq.	$1.235^{19°}$	....	305 d.	d.	d.	s. bz.
727	cr.	....	150-5 d.	$-H_2O >160°$	i.	s.	s.
728	cr./lg.	....	143-4	....	v. sl. s.	v. s. act.	v. s.
729	col. nd.	....	137-8	....	.. . ...	sl. s. h.	....
730	nd./aq.	....	>300	d. >300	s.	sl. s.	sl. s. HCl

Benzophenone oxide 6451

Table 7-4 (*Continued*)
PHYSICAL CONSTANTS OF ORGANIC COMPOUNDS

No.	Name	Synonym	Formula	Beil. Ref.	Formula Weight
731	**Benzo-**pinacoline (β)	phenyl-trityl ketone	$(C_6H_5)_3C\cdot CO\cdot C_6H_5$	VII-544	348.45
732	pinacone	....	$[(C_6H_5)_2C\cdot OH]_2$	VI-1058	366.46
733	purpurin 4B	ditolyl-bis-(azo-naphthionic acid)	$[C_{10}H_5(NH_2)(SO_3Na)\cdot N:N\cdot C_6H_3(CH_3)\cdot]_2$	*XVI-342	724.73
734	thiazylhydrazine (2)	....	$C_6H_4N:C(N_2H_3)\cdot S\cdot \rule{1.5cm}{0.4pt}$	....	165.22
735	trichloride	phenyl chloroform	$C_6H_5CCl_3$	V-300	195.48
736	**Benzoxyl-2-di-methyl-amino-methyl-1-di-methyl-amino-butane (2) HCl**	alypin hydro-chloride	$[(CH_3)_2N\cdot CH_2]_2:C: (C_2H_5)(O_2C\cdot C_6H_5)\cdot HCl$	IX-175	314.86
737	nitrate	alypin nitrate	$C_{16}H_{26}O_2N_2\cdot HNO_3$	*IX-92	341.41
738	**Benzoyl** acetic acid	....	$C_6H_5\cdot CO\cdot CH_2\cdot CO_2H$	X-672	164.16
739	acetic ester	ethyl benzoyl-acetate	$C_6H_5\cdot CO\cdot CH_2\cdot CO_2\cdot C_2H_5$	X-674	192.22
740	acetoacetic ester	....	$C_2H_3O\cdot CH(C_7H_5O)\cdot CO_2C_2H_5$	X-817	234.25
741	acetone	....	$C_6H_5CO\cdot CH_2\cdot CO\cdot CH_3$	VII-680	162.19
742	acetonitrile	ω-cyanoacetophenone	$C_6H_5\cdot CO\cdot CH_2\cdot CN$	X-680	145.16
743	acrylic acid (β)	....	$C_7H_5O\cdot CH:CH\cdot CO_2H\cdot H_2O$	X-726	194.19
744	alanine (*dl*)	....	$C_6H_5CO\cdot NH\cdot CH(CH_3)\cdot CO_2H$	IX-248	193.20
745	alanine (*l*)	....	$C_{10}H_{11}O_3N$	IX-248	193.20
746	aminobenzoic acid (*o*)	N-benzoyl-an-thranilic acid	$C_7H_5O\cdot NH\cdot C_6H_4\cdot CO_2H$	XIV-340	241.25
747	aminobenzoic acid (*m*)	benzaminobenzoic acid	$C_7H_5O\cdot NH\cdot C_6H_4\cdot CO_2H$	XIV-397	241.25
748	aminobenzoic acid (*p*)	....	$C_7H_5O\cdot NH\cdot C_6H_4\cdot CO_2H$	XIV-433	241.25
749	auramine	....	$[(CH_3)_2N\cdot C_6H_4]_2C: N\cdot COC_6H_5$	XIV-95	371.49
750	benzoic acid (*o*)	....	$C_6H_5CO\cdot C_6H_4\cdot CO_2H\cdot H_2O$	X-747	244.25
751	benzoic acid (*m*)	....	$C_6H_5CO\cdot C_6H_4\cdot CO_2H$	X-752	226.23
752	benzoic acid (*p*)	....	$C_6H_5CO\cdot C_6H_4\cdot CO_2H$	X-753	226.23
753	bromide	....	$C_6H_5\cdot CO\cdot Br$	IX-195	185.03
754	carbinol	phenacyl alcohol	$C_6H_5\cdot CO\cdot CH_2OH$	VIII-90	136.15
755	chloride	....	$C_6H_5\cdot CO\cdot Cl$	IX-182	140.57
756	cyanide	....	$C_6H_5\cdot CO\cdot CN$	X-659	131.14
757	fluoride	....	$C_6H_5\cdot CO\cdot F$	IX-181	124.12
758	formic acid	phenylglyoxylic acid	$C_6H_5\cdot CO\cdot CO_2H$	X-654	150.14
759	glycolic acid	....	$C_7H_5O\cdot OCH_2\cdot CO_2H$	IX-167	180.16
760	hydrazine	....	$C_6H_5CO\cdot NH\cdot NH_2$	IX-319	136.15
761	hydrogen peroxide	....	$C_6H_5CO_2\cdot OH$	IX-178	138.12
762	iodide	....	$C_6H_5\cdot CO\cdot I$	IX-195	232.02
763	lactic acid (*dl*)	....	$CH_3\cdot CH(O_2C_7H_5)\cdot CO_2H$	IX-167	194.19
764	γ-(2-methylpiper-idino)-propanol HCl	neothesin	$C_{16}H_{23}O_2N\cdot HCl$	....	297.83

Table 7-4 (*Continued*)
PHYSICAL CONSTANTS OF ORGANIC COMPOUNDS

No.	Crystalline Form and Color	Specific Gravity	Melting Point °C.	Boiling Point °C.	Solubility in 100 Parts Water	Alcohol	Ether
731	nd./al.	....	179-81	....	i.	v. sl. s. c.; 1.2 h.	s.
732	mn.	....	185-6 d.	....	....	2.5 h.	v. s.
733	cr.	....	....	....	s.	....	....
734	pa. yel. cr.	....	197-8	....	i.	s. h.	i.
735	col. lq.	$1.380^{14°}$	−4.75	220.7	i.; d.	s.; s. bz.	s.
736	cr./act.	....	169 d.	....	v. s.; s. act.	s.; s. chl.	i.
737	col. pd.	....	163	....	v. s.	v. s.	v. sl. s.
738	nd./bz.	....	103-4 d.	....	s. h.	s.	s.
739	col. lq.	$1.111^{25°}_{4}$	....	$165\text{-}7^{20mm}$	v. sl. s.	∞	∞
740	oil	....	....	$175\text{-}6^{12mm}$ (sl. d.)	....	....	....
741	pr.	$1.090^{60°}_{60}$	60-1	260-2 (sl. d.)	s. h.	v. s.	v. s.
742	pr./aq.	....	80.5	....	sl. s. c.	s.	s.; v. s. bz.
743	cr.	....	65; anh. 99	....	sl. s. c.	s.	s.
744	lf./et.	....	165-6	....	0.4 c.	v. s.	v. sl. s.
745	pl./aq.	....	150-1	....	$1.1^{20°}$	v. s.	v. sl. s.
746	nd./al.	....	181-2	....	i.	s.	s.
747	pr./al.	....	248	....	sl. s.	s.	sl. s.
748	nd./al.	....	278	....	sl. s. h.	s.	s.; s. ac.
749	yel. nd./al.	....	179	....	....	$0.16^{16°}$	$0.93^{16°}$ bz.
750	tri./aq.	....	93-4; anh. 128	....	sl. s.	....	....
751	lf./aq. al.	....	161-2	subl.	sl. s. h.	s.	s.
752	mn.	....	194	subl.	v. sl. s. c.	s.	s.
753	lq.	$1.570^{20°}_{4}$	5.5-7.5	218-9	d.	d. h.	∞
754	hex. pl./al.	1.013	85-6	$118\text{-}20^{11mm}$	sl. s. h.	v. s.	v. s.
755	col. lq.	$1.212^{20°}_{4}$	−0.6	197.9	d.; s. bz.	d. h.	∞; s. CS_2
756	cr. pl.	....	33-4	208-10	i.	....	....
757	lq.	>1	....	161.5^{45mm}	sl. d. h.	d. h.	s.
758	pr.	....	65-6	148^{6mm}	s.	i. CS_2	s.
759	pr.	....	79	....	sl. s. c.; d. h.	s.	s.
760	pl./aq.	....	112.0-2.5	d.	s.	s.; sl. s. bz.	sl. s.
761	lf./lg.	....	41-3	$97\text{-}100^{13mm}$	i.	s.	s.
762	nd.	....	3	135^{25mm}	d. h.	d. h.	s.
763	pl.	....	112	....	0.25 c.; d. h.	v. s.	v. s.
764	wh. cr. pd.	....	171-3	....	100	s.; s. chl.	i.

Benzoyl-anisole 4070
Benzoyl-anthranilic acid (*N*) 746
Benzoyl-azide 653
Penzoyl-ecgonine, cf. alkd.
Benzoyl-eugenol 3220
Benzoyl-formaldoxime 4874
Benzoyl-glucose, cf. glcd.
Benzoyl-glycine 3664
Benzoyl-guaiacol 3482
Benzoyl-hydroquinone 2322
Benzoyl-hydroxylamine (*N*) 710
Benzoyl-naphthalene 5210-1

Table 7-4 (*Continued*)
PHYSICAL CONSTANTS OF ORGANIC COMPOUNDS

No.	Name	Synonym	Formula	Beil. Ref.	Formula Weight
	Benzoyl				
765	α-naphthylamine	benznaphthalide	$C_6H_5CO\cdot NH\cdot C_{10}H_7$	XII-1233	247.30
766	peroxide	....	$(C_6H_5CO)_2O_2$	IX-179	242.23
767	phenylhydrazine (α, α')	....	$C_6H_5N(COC_6H_5)\cdot$ NH_2	XV-250	212.25
768	phenylhydrazine	(α, β)	$C_6H_5NH\cdot NHCOC_6H_5$	XV-255	212.25
769	phthalic acid (1;2,3)	....	$C_6H_5CO\cdot C_6H_3\cdot$ $(CO_2H)_2\cdot H_2O$	X-880	288.26
770	piperidine (*N*)	....	$C_7H_5O\cdot N\cdot CH_2(CH_2)_4$	XX-46	189.26
771	propionic acid (β)	....	$C_7H_5O\cdot CH_2\cdot CH_2\cdot CO_2H$	X-696	178.19
772	terephthalic acid	(1;2,5)	$C_{13}H_8O(CO_2H)_2$	X-881	270.24
773	2-thiohydantoin (1)	....	$CH_2\cdot CO\cdot NH\cdot CS\cdot N\cdot$ C_7H_5O	*XXIV-294	220.25
774	thiourea	....	$C_7H_5O\cdot NH\cdot CS\cdot NH_2$	IX-219	180.23
775	toluidide (*o*)	*N*-tolylbenzamide (*o*)	$C_7H_5O\cdot NH\cdot C_6H_4\cdot CH_3$	XII-795	211.27
776	toluidide (*m*)	....	$C_7H_5O\cdot NH\cdot C_6H_4\cdot CH_3$	XII-861	211.27
777	toluidide (*p*)	....	$C_7H_5O\cdot NH\cdot C_6H_4\cdot CH_3$	XII-926	211.27
778	urea (*N*)	....	$C_6H_5CO\cdot NHCONH_2$	IX-215	164.17
779	**Benzoylene urea**	2,4-diketotetrahydro-quinazoline	$C_6H_4\cdot NH\cdot CO\cdot NH\cdot CO$	XXIV-373	162.15
780	**Benzyl** acetamide (*N*)	acetyl-benzylamine	$CH_3CO\cdot NH\cdot$ $CH_2C_6H_5$	XII-1044	149.19
781	acetate	....	$C_6H_5CH_2O\cdot COCH_3$	VI-435	150.18
782	alcohol	phenyl carbinol	$C_6H_5\cdot CH_2OH$	VI-428	108.14
783	amine	ω-aminotoluene	$C_6H_5\cdot CH_2\cdot NH_2$	XII-1013	107.16
784	aminophenol (*p*)	....	$C_6H_5\cdot CH_2\cdot NH\cdot$ C_6H_4OH	XIII-448	199.25
785	aminophenol HCl	(*p*)	$C_{13}H_{13}ON\cdot HCl\cdot H_2O$	XIII-448	253.73
786	aniline	phenyl-benzylamine	$C_6H_5CH_2\cdot NH\cdot C_6H_5$	XII-1023	183.26
787	arsonic acid	....	$C_6H_5CH_2\cdot AsO(OH)_2$	XVI-872	216.07
788	azide	....	$C_6H_5\cdot CH_2\cdot N_3$	V-350	133.15
789	benzoate	....	$C_6H_5CO_2\cdot CH_2\cdot C_6H_5$	IX-121	212.25
790	benzoic acid (*o*)	....	$C_6H_5\cdot CH_2\cdot C_6H_4\cdot CO_2H$	IX-676	212.25
791	benzoic acid (*m*)	....	$C_6H_5\cdot CH_2\cdot C_6H_4\cdot CO_2H$	IX-676	212.25
792	benzoic acid (*p*)	....	$C_6H_5\cdot CH_2\cdot C_6H_4\cdot CO_2H$	IX-677	212.25
793	bromide	ω-bromotoluene	$C_6H_5\cdot CH_2\cdot Br$	V-306	171.04
794	butyrate	....	$C_2H_5CH_2CO_2CH_2\cdot$ C_6H_5	VI-436	178.23
795	carbamate	....	$NH_2\cdot CO_2\cdot CH_2C_6H_5$	VI-437	151.17
796	chloride	ω-chlorotoluene	$C_6H_5\cdot CH_2\cdot Cl$	V-292	126.59
797	chloroacetate	....	$ClCH_2CO_2\cdot CH_2C_6H_5$	VI-435	184.62
798	cinnamate	cinnamein	$C_8H_7\cdot CO_2\cdot C_7H_7$	IX-584	238.29
799	cyanamide	....	$C_6H_5\cdot CH_2\cdot NH\cdot CN$	XII-1051	132.17
800	cyanide	phenylacetonitrile	$C_6H_5\cdot CH_2\cdot CN$	IX-441	117.15
801	cyanurate (*iso*)	....	$(C_6H_5CH_2N\cdot CO)_3$	XXVI-255	399.45
802	diphenyl (*o*)	....	$C_6H_5CH_2\cdot C_6H_4\cdot C_6H_5$	V-708	244.34
803	diphenyl (*p*)	....	$C_6H_5CH_2\cdot C_6H_4\cdot C_6H_5$	V-708	244.34
804	diphenylamine	diphenyl-benzylamine	$(C_6H_5)_2N\cdot CH_2C_6H_5$	XII-1033	259.35
805	disulfide	dibenzyl-disulfide	$(C_6H_5\cdot CH_2)_2S_2$	VI-465	246.40

Benzoyl-nitromethane 4594
Benzoyl-persulfide 1779
Benzoyl-pseudotropine, cf. alkd.
Benzoyl-pyrogallol 6224

Benzoyl-resorcin 2321
Benzoyl-salicin, cf. glcde.
Benzoyl-sulfimide 5579
Benzoyl-*pseudo*-tropine, cf. alkd.

Table 7-4 (*Continued*)
PHYSICAL CONSTANTS OF ORGANIC COMPOUNDS

No.	Crystalline Form and Color	Specific Gravity	Melting Point °C.	Boiling Point °C.	Solubility in 100 Parts		
					Water	Alcohol	Ether
765	nd./aq. al.	....	161-2	....	s. ac.	v. sl. s. abs.	....
766	rhb./et.	....	108 d.	expl.	i.	sl. s. c.; s. h.	s.; s. bz.
767	nd./aq.	....	70	....	sl. s. c.	v. s.	v. s.
768	pr./al.	....	168	....	sl. s. h.	s. h.; s. chl.	sl. s.
769	nd./aq.	....	$-H_2O$, 100	$-2H_2O >100$	s. h.	s.	v. sl. s. bz.
770	col. tri.	....	48	320-1	i.	s.	s.
771	nd./aq.	....	116	sl. d.	s. h.	s.	s.; i. lg.
772	cr./al.	....	>290	....	i.	s.	s.
773	cr./al.	....	165 d.	....	....	....	....
774	pr./aq. al.	....	169-70	....	sl. s. c.	s.	i.
775	rhb.	$1.205^{15°}$	145-6	....	sl. s. h.	s.	....
776	mn./aq. al.	$1.170^{15°}$	....	....	....	$13^{15°}$ abs.	....
777	rhb./al.	$1.202^{15°}$	157-8	232	i.	$4.2^{18°}$ abs.	....
778	nd./al.	....	214-5	....	s. h.	1, c,; 4, h.	i.
779	col. nd.	....	353-4	....	$0.013^{23°}$	v. sl. s.	s. alk.
780	lf./et.	....	60-1	>300	sl. s. pet.	v. s.	v. s.
781	col. lq.	$1.057^{17°}$	-51.5	213.5	i.	∞	∞
782	col. lq.	$1.045\frac{20°}{4}$	-15.2	205.4	$4^{17°}$	∞; ∞ chl.	∞
783	lq.	$0.982\frac{20°}{4}$	....	184.5	∞	∞	∞
784	lf.	....	89	....	v. sl. s.	v. s.	v. s. bz.; s. NaOH
785	pr./aq.	....	172 (anh.)	....	s. h.	s.	i.
786	mn. pr.	$1.065\frac{25°}{25}$	37-8	306^{759mm}	i.	s. h. Me al.	s.; s. chl.
787	nd.	....	167	d.	$0.34^{23°}$; $3.5^{97°}$	$0.87^{23°}$; $5.9^{70°}$	....
788	oil	$1.066^{25°}$	....	108^{23mm}	i.	∞	∞
789	nd. or lf.	1.122 (lq.)	21	323-4	i.; ∞ chl.	∞; s. oil	∞
790	nd./aq. al.	....	114-7	subl.	sl. s.	s.; s. chl.	s.; s. bz.
791	lf./aq. al.	....	107-8	subl.	sl. s.	s.; s. chl.	s.
792	lf./aq. al.	....	155-7	subl.	sl. s.	s.; s. chl.	s.
793	col. lq.	$1.443^{17°}$	-4	198-9	i.; sl. d.	∞	∞
794	col. lq.	$1.016\frac{16°}{18}$	....	238-40	i.	v. s.	v. s.
795	cr.	....	86	d.	v. sl. s.	s.	s.
796	col. lq.	$1.100\frac{20°}{20}$	-39.2	179.4	i.	∞; ∞ chl.	∞
797	col. oil	$1.222\frac{4°}{4}$	....	147.5^{9mm}	....	....	....
798	pr.	....	39	244^{25mm}	i.	s.	s.
799	pl./et.	....	43	....	i.	v. s.	v. s.
800	col. lq.	$1.018\frac{20°}{4}$	-23.8	233-4	i.	∞	∞
801	nd./al.	....	157	>320	i. c.	s.	sl. s.
802	mm.	....	54	$283-7^{110mm}$	v. s. bz.	s.	v. s.
803	lf.	$1.171\frac{0°}{4}$	85	$285-6^{110mm}$	v. s. bz.	sl. s.	v. s.
804	nd.	....	86-7	....	v. sl. s.	s. h.	s.
805	lf./al.	....	71-2	d. >270	v. sl. s.	s. h.	s.; s. bz.

Benzyl, cf. also dibenzyl.
Benzyl-acetic acid 3704
Benzyl-acetoacetic ester 2983
Benzyl-carbamide 844

Benzyl-carbinol 5187
Benzyl-cellosolve 3449
Benzyl citrate 6085

Table 7-4 (*Continued*)
PHYSICAL CONSTANTS OF ORGANIC COMPOUNDS

No.	Name	Synonym	Formula	Beil. Ref.	Formula Weight
806	**Benzyl** ether	dibenzyl ether	$(C_6H_5 \cdot CH_2)_2O$	VI-434	198.27
807	formate	....	$HCO_2 \cdot CH_2 \cdot C_6H_5$	VI-435	136.15
808	hydrazine	....	$C_6H_5 \cdot CH_2 \cdot NH \cdot NH_2$	XV-531	122.17
809	hydroxybenzoate (*p*)	....	$HO \cdot C_6H_4 \cdot CO_2 \cdot C_7H_7$	....	228.25
810	hydroxylamine (α)	....	$C_6H_5 \cdot CH_2 \cdot O \cdot NH_2$	VI-440	123.16
811	hydroxylamine (β)	....	$C_6H_5 \cdot CH_2 \cdot NH \cdot OH$	XV-17	123.16
812	imino-(4-methoxy-phenyl) methane	anisalbenzylamine	$C_6H_5CH_2N{:}CH \cdot C_6H_4 \cdot OCH_3$	XII-1043	225.29
813	iodide	ω-iodotoluene	$C_6H_5 \cdot CH_2 \cdot I$	V-314	218.04
814	lactate	....	$CH_3CHOHCO_2 \cdot C_7H_7$	**VI-420	180.21
815	malonic acid	....	$C_6H_5CH_2 \cdot CH(CO_2H)_2$	IX-868	194.19
816	mercaptan	....	$C_6H_5 \cdot CH_2 \cdot SH$	VI-453	124.21
817	naphthalene (α)	Ph-naphthylmethane	$C_6H_5 \cdot CH_2 \cdot C_{10}H_7$	V-689	218.30
818	naphthalene (β)	Ph-naphthylmethane	$C_6H_5 \cdot CH_2 \cdot C_{10}H_7$	V-690	218.30
819	α-naphthyl ether	....	$C_6H_5 \cdot CH_2 \cdot O \cdot C_{10}H_7$	....	234.30
820	β-naphthyl ether	....	$C_6H_5 \cdot CH_2 \cdot O \cdot C_{10}H_7$	VI-642	234.30
821	α-naphthyl ketone	....	$C_6H_5 \cdot CH_2 \cdot CO \cdot C_{10}H_7$	VII-512	246.31
822	phenanthracene (9)	....	$C_7H_7 \cdot C_6H_3{:}C_2H_2{:}C_6H_4$	*V-359	268.36
823	phenol (*o*)	HO-diPh-methane	$C_6H_5 \cdot CH_2 \cdot C_6H_4 \cdot OH$	VI-675	184.24
824	phenol (*p*)	4-hydroxy-ditane	$C_6H_5 \cdot CH_2 \cdot C_6H_4 \cdot OH$	VI-675	184.24
825	phenyl carbamate (*p*)	butolan	$C_7H_7 \cdot C_6H_4O_2C \cdot NH_2$	*VI-325	227.27
826	phenylhydrazine (α,α)	....	$(C_7H_7)(C_6H_5)N \cdot NH_2$	XV-532	198.27
827	phenylhydrazine HCl	(α,α)	$C_{13}H_{12}N \cdot NH_2 \cdot HCl$	XV-533	234.73
828	phenylnitrosamine	nitroso-benzyl-aniline	$C_6H_5CH_2 \cdot N(NO) \cdot C_6H_5$	XII-1071	212.25
829	propionate	....	$C_2H_5CO_2 \cdot CH_2C_6H_5$	VI-436	164.21
830	pyridine (2)(α)	....	$C_6H_5 \cdot CH_2 \cdot C_5H_4N$	XX-425	169.23
831	pyridine (3)(β)	....	$C_6H_5 \cdot CH_2 \cdot C_5H_4N$	XX-426	169.23
832	pyrrole (*N*)	....	$C_4H_4N \cdot CH_2 \cdot C_6H_5$	XX-164	157.22
833	salicylate	....	$HOC_6H_4CO_2 \cdot CH_2C_6H_5$	X-80	228.25
834	succinate (mono)	....	$C_7H_7O_2C(CH_2)_2 \cdot CO_2H$	VI-436	208.22
835	sulfone	dibenzyl sulfone	$(C_6H_5 \cdot CH_2)_2SO_2$	VI-456	246.33
836	sulfoxide	....	$(C_6H_5 \cdot CH_2)_2SO$	VI-456	230.33
837	tartronic acid	....	$C_7H_7 \cdot C(OH)(CO_2H)_2$	X-515	210.19
838	*iso*-thiocyanate	benzyl mustard oil	$C_6H_5 \cdot CH_2 \cdot N{:}CS$	XII-1059	149.22
839	thiourea	....	$C_6H_5CH_2 \cdot NH \cdot CS \cdot NH_2$	XII-1051	166.25
840	*iso*-thiourea	....	$C_7H_7 \cdot S \cdot C({:}NH) \cdot NH_2$	VI-461	166.25
841	*iso*-thiourea HCl	....	$C_8H_{10}N_2S \cdot HCl$	VI-461	202.71
842	toluene (*m*)	phenyl-tolyl-methane	$C_7H_7 \cdot C_6H_4 \cdot CH_3$	V-607	182.27
843	toluene (*p*)	....	$C_7H_7 \cdot C_6H_4 \cdot CH_3$	V-607	182.27
844	urea	benzyl carbamide	$C_6H_5CH_2 \cdot NH \cdot CO \cdot NH_2$	XII-1050	150.18
845	**Beryllium** diethyl	....	$(C_2H_5)_2Be$	IV-645	67.14
846	**Betaine**†	trimethyl glycine; oxyneurine	$(CH_3)_3N \cdot CH_2 \cdot CO \cdot O$	IV-346	117.15
347	hydrochloride	acidol; lycine	$C_5H_{11}O_2N \cdot HCl$	IV-348	153.61
848	**Betulin**	....	$C_{30}H_{50}O_2$	**VI-937	442.73

† See also alkaloids table.
Benzyl-ethyl-aniline 2984
Benzyl-ethyl ether 2986
Benzyl fumarate 1785
Benzyl maleate 1788

Benzyl-methyl ether 4145
Benzyl mustard oil 838
Benzyl phthalate 1789
Benzyl succinate 1790
Benzyl sulfide 1791

Table 7-4 (*Continued*)
PHYSICAL CONSTANTS OF ORGANIC COMPOUNDS

No.	Crystalline Form and Color	Specific Gravity	Melting Point °C.	Boiling Point °C.	Solubility in 100 Parts		
					Water	Alcohol	Ether
806	lq.	$1.036^{16°}$	1.5–3.5	295–8	i.	s. h.	s.
807	col. lq.	$1.081^{23°}$	3.6	202–37^{47mm}	i.	s.	∞
808	col. oil	....	26	135^{29mm}	∞	∞	∞
809	wh. pd.	....	110–2	....	i.	v. s.	s.
810	oil	....	....	118–9^{30mm}	....	....	....
811	nd./pet.	....	57–8	123^{50mm}	s.	....	....
812	col. cr./pet.	....	89–91	....	....	....	....
813	cr.	$1.734^{25°}$	24.1	93^{10mm}	i.	s.	s.
814	col. lq.	....	....	136–8^{10mm}	....	....	....
815	cr./et.	....	121	d. 180	s.	s.	s.; s. lı. bz.
816	col. lq.	$1.058^{20°}$	....	194–5	....	....	....
817	lf./al.	$1.165^{0°}$	58–9	350	s. chl., bz.	1.6 c., 3 h.	50 c.
818	mn. pr.	$1.176^{0°}$	35.5	350	....	v. s. h.	v. s. bz.
819	col. cr.	....	76–7	....	i.	s. h.	s.
820	lf./al.	....	98–9	....	s. chl., bz.	s.	s.
821	pl./al.	....	57	....	i.	s.	s.
822	lf./al.	....	91–2	....	i.	....	....
823	col. oil	....	52(21)	312	v. s. h.	s.	s.; s. alk.
824	nd./al.	....	84	320–2	s. h.	s.	s.; s. alk.
825	wh. cr./al.	....	144	....	sl. s.	s. h.	s. bz.
826	col. oil	....	....	216–8^{38mm}	....	....	....
827	nd./aq.	....	167–70	....	....	....	....
828	yel. nd./al	....	58	....	s. chl., lg.	s.	s.
829	lq.	$1.036^{\frac{16.5°}{17.5}}$	....	220–2	i.	....	....
830	lq.	$1.054^{\frac{20°}{0}}$	....	276^{742mm}	i.	v. s.	v. s.
831	lq.	$1.061^{\frac{20°}{0}}$	34	287^{742mm}	i.	v. s.	v. s.
832	....	....	<37.5	246–7	i.	s.	s.
833	col. oil	$1.176^{\frac{19.5°}{15}}$	23–5	208^{26mm}	v. sl. s.	∞	∞
834	cr.	....	59	....	i.	s.	s.
835	nd./al. + bz.	....	150–1	290(sl. d.)	i.	sl. s.	s. bz.
836	lf./al.	....	133–4	d. 210	s. h.	v. s.	v. s.
837	pr.	....	147 d.	....	s.	s.	s.
838	lq.	$1.125^{\frac{15°}{4}}$	....	243	i.	....	s.
839	pr./aq.	....	162–4	....	i. c.	1.5 c.	....
840	nd./bz.	....	103–4 d.	....	s. chl., bz.	s.	s.
841	pl./HCl	....	176(148)	....	s.	s.	i.
842	lq.	$0.997^{17.5°}$	....	275^{747mm}	....	s.	s.
843	lq.	$0.994^{18°}$	–30	285–6	v. s. chl.	v. s.	v. s.
844	nd./al.	....	147–8	d. 200	$1.74^{45°}$	$3.12^{3°}$, act.	$0.05^{22.5°}$
845	col. lq.	....	–11–3	185–8	d.	d.	i.
846	pr./al.	....	293 d.	....	$157^{19°}$	$8.6^{18°}$	i.
847	col. mn./al.	....	236–7 d.	....	60	0.6	i.; i. chl.
848	nd./al.	....	258	subl. sl. d.	i.; i. CS_2	0.7, c.; 4.27, h.	0.4, c.; 3, h.

Table 7-4 (*Continued*)
PHYSICAL CONSTANTS OF ORGANIC COMPOUNDS

No.	Name	Synonym	Formula	Beil. Ref.	Formula Weight
849	**Betulinic acid**	....	$C_{30}H_{48}O_3$	**VI-939	456.72
850	**Bilifuscin**	bile pigment	$C_{16}H_{10}O_4N_2$	....	294.27
851	**Bilirubin**	principal bile pigment	$(C_{16}H_{18}O_3N_2)_2$	....	572.67
852	**Biliverdin**	bile pigment	$C_{32}H_{36}O_8N_4$	....	604.67
853	**Biotin**	vitamin H	$C_{10}H_{16}O_3N_2S$	....	244.31
854	methyl ester	....	$C_{11}H_{18}O_3N_2S$	....	258.34
855	**Bismuth** triethyl	triethyl bismuthine	$(C_2H_5)_3Bi$	IV-622	296.17
856	trimethyl	trimethyl bismuthine	$(CH_3)_3Bi$	IV-622	254.09
857	triphenyl	triphenyl bismuthine	$(C_6H_5)_3Bi$	XVI-898	440.30
858	triphenyl dichloride	triphenylbismuthine dichloride	$(C_6H_5)_3Bi:Cl_2$	XVI-899	511.21
859	**Biuret**	allophanamide	$NH(CONH_2)_2$†	III-70	103.08
860	**Borneol** (*d* or *l*)	....	$C_{10}H_{17}OH$	VI-75	154.25
861	**Bornyl** acetate (*d*)	....	$C_{10}H_{17}O\cdot CO\cdot CH_3$	VI-78	196.29
862	amine (*d*)	....	$C_{10}H_{17}\cdot NH_2$	XII-45	153.27
863	α-bromo-*iso*-valerate (*d*)	brovalol; valisan; eubornyl	$(CH_3)_2CH\cdot CHBr\cdot CO_2C_{10}H_{17}$	VI-79	317.27
864	chloride‡	*iso*-bornyl chloride	$C_{10}H_{17}Cl$	V-97	172.70
865	dibromo-dihydrocinnamate	adamon	$C_6H_5(CHBr)_2CO_2\cdot C_{10}H_{17}$	*IX-202	444.22
866	salicylate (*d*)	salit	$HO\cdot C_6H_4\cdot CO_2\cdot C_{10}H_{17}$	X-76	274.36
867	*iso*-valerate (*d*)	bornyval	$(CH_3)_2CHCH_2CO_2\cdot C_{10}H_{17}$	VI-79	238.37
868	*iso*-valeryl-glycollate	neobornyval	$(CH_3)_2CHCH_2CO_2\cdot CH_2\cdot CO_2\cdot C_{10}H_{17}$	....	296.41
869	*iso*-**Bornyl**-*n*-valerate (*d*)	gynoval	$CH_3(CH_2)_3CO_2\cdot C_{10}H_{17}$	VI-88	238.37
870	**Bornylene** (*l*)	....	$C_{10}H_{16}$	V-155	136.24
871	**Boron** triethyl	....	$(C_2H_5)_3B$	IV-641	98.00
872	trimethyl	....	$(CH_3)_3B$	IV-641	55.92
873	**Brassidic acid**	....	$C_{21}H_{41}\cdot CO_2H$	II-474	338.58
874	**Brazilein**	....	$C_{16}H_{12}O_5\cdot H_2O$	XVIII-194	302.29
875	**Brazilin**	....	$C_{16}H_{14}O_5$	XVII-194	286.29
876	**Bromal**	tribromo-acetaldehyde	$Br_3C\cdot CH:O$	I-626	280.76
877	hydrate	....	$Br_3C\cdot CH(OH)_2$	I-626	298.77
878	**Bromo-**acenapththene (5)	....	$C_{12}H_9Br$	V-587	233.11
879	acetamide (*N*)	acetbromamide	$CH_3\cdot CO\cdot NHBr\cdot H_2O$	II-181	155.98
880	acetanilide (*o*)	....	$Br\cdot C_6H_4\cdot NH\cdot CO\cdot CH_3$	XII-632	214.07
881	acetanilide (*m*)	....	$Br\cdot C_6H_4\cdot NH\cdot CO\cdot CH_3$	XII-634	214.07
882	acetanilide (*p*)	asepsin	$Br\cdot C_6H_4\cdot NH\cdot CO\cdot CH_3$	XII-642	214.07
883	acetic acid	....	$Br\cdot CH_2\cdot CO_2H$	II-213	138.95
884	acetoacetanilide (α)	....	$CH_3CO\cdot CHBr\cdot CO\cdot NH\cdot C_6H_5$	XII-519	256.11
885	aceto-β-naphthone (ω)	....	$C_{10}H_7\cdot CO\cdot CH_2Br$	**VII-338	249.11
886	acetone	....	$Br\cdot CH_2\cdot CO\cdot CH_3$	I-657	136.98
887	acetophenone (ω)	phenacyl bromide	$C_6H_5CO\cdot CH_2\cdot Br$	VII-283	199.05

† Cryst. $+1H_2O$/aq.; $-H_2O$, 110°.   ‡ See also pinene hydrochloride.
Bi-, cf. also di.
Bile acid 1479
Bile pigments 850-2
Bilineurine 1488

Biloptin 2384
Bindschedler's green 5833
Biphenin 1722
Biphenyl 2695

Bismark brown 6058
Bistriazo-ethane 1761
Biuret-amidine 2081
Biuret base 6209

Bixin 4933
Blue cross gas 2712
Bonoform 5739
Bordeaux DH 3732

Table 7-4 (*Continued*)
PHYSICAL CONSTANTS OF ORGANIC COMPOUNDS

No.	Crystalline Form and Color	Specific Gravity	Melting Point °C.	Boiling Point °C.	Solubility in 100 Parts		
					Water	Alcohol	Ether
849	wh. pd.	....	295-7	....	v. sl. s.	s.	....
850	br. pd.	....	183	....	v. sl. s.	s.; v. s. alk.	v. sl. s.
851	or. pd.	....	darkens with d.	....	i.; s. alk.; s. bz.	v. sl. s.; s. CS_2	v. sl. s.; s. chl.
852	gn. pd.	....	....	....	i.; s. alk.	s.; s. bz.; s. CS_2	v. sl. s.; i. chl.
853	nd.	....	230-2	....	s. NaOH	....	....
854	cr./al. et.	....	166-7	....	i.	s.; s. Me al.	s.
855	col. oil	1.82	....	$*107^{79mm}$	i.	v. s.	v. s.
856	col. oil	$2.30^{18°}$	....	*110	i.	v. s.	v. s.
857	mn./al.	$1.952^{15°}_{4}$	77-8	242^{14mm}	i.; v. s. chl.	v. sl. s.	s.; s. act.
858	pr./chl. al.	....	141.5	....	v. s. bz.	v. sl. s.	v. sl. s.
859	nd./al.	....	192-3 d.	....	$1.2^{50°}$; $45^{106°}$	s.	....
860	col. cr.	$1.011^{20°}_{4}$	208.6	215.0	v. sl. s.	v. s.; s. bz.	v. s.
861	rhb./pet.	$0.991^{15°}$	29	226-7	i.	s.	s.
862	col. cr.	....	163	200	i.	v. s.	v. s.
863	col. oil	1.18	....	163	i.	s.	s.; s. chl.
864	col. cr.	....	161.5	....	i.	v. sl. s. c.	v. s.
865	col. cr./al.	....	75±	....	i.; s. h. chl.	s. h.	s. h.
866	cr.	....	44-5	$230-5^{50mm}$	i.; ∞ oil	∞ ; ∞ chl.	∞
867	col. lq.	$0.951^{20°}$	....	255-60	i.	s.	s.
868	col. oil	1.03	....	283-5 d.	i.; s. oil	v. s.; v. s. bz.	v. s.
869	col. lq.	$0.953^{18°}_{4}$	....	$143-5^{18mm}$	v. sl. s.; s. act.	s.; s. chl.	s.; s. bz.
870	col. cr.	....	113	146^{740mm}	i.	s. Me al.	s. toluene
871	col. lq.	$0.696^{23°}$	-92.9	95	v. sl. s.	d. by air	....
872	col. gas	....	-160	-20	v. sl. s.	v. s.	v. s.
873	lf./al.	$0.859^{57°}$	61-2	282^{30mm}	$0.7^{25°}$	v. sl. s. c.	sl. s.
874	rhd. red	....	softens 130-40	....	sl. s. h.	s. alk.	s. d. H_2SO_4
875	cr./abs. al.	....	250	....	s.; s. alk.	s.	s.
876	yel. lq.	$2.665^{25°}_{4}$	....	174 d.	forms hydr.	s.	s.
877	cr.	$2.566^{40°}_{4}$	53.5	....	s.	s.; s. chl.	s.; s. glyc.
878	yel. cr.	$1.437^{55°}$	51-2	336.4	i.	s.	v. s.
879	pl. + H_2O	....	70-80	m. anh. 108	s.	s.	v. s.
880	nd./al.	....	99	....	i.	s.	s.
881	nd./aq. al.	....	87.5	....	....	v. s.	v. s.
882	mn. pr.	1.717	166-7	....	sl. s. h.	s.; s. bz.	s. chl.
883	pl. or rhb.	$1.934^{50°}_{50}$	49-50	208	$∞^{25°}$	$∞^{25°}$	$∞^{25°}$
884	col. lf./al.	....	138 d.	....	v. sl. s.; s. alk.	v. s.	sl. s.; sl. s. chl.
885	red fl.	....	84	....	i.	sl. s.	v. s.
886	lq.	$1.634^{23°}$	-54	136.5^{725mm}	v. sl. s.	v. s.	v. s.
887	rhb.	$1.647^{20°}_{4}$	50	119	i.; s. bz.	v. s.	v. s.

* Explodes when heated in air.

Bordeaux S 3732	Brilliant green base 5762	Brom cresol purple 1832	Brom thymol blue 1879
Bornyval 867	Bromacetol 1866	Bromisoval 1009	Bromelia 3094
Bourbonal 3188	Bromanil 5726	Brom phenol blue 5724	Brometone 6102
Brilliant cotton blue 6345	Brom cresol green 5718	Brom-tetragnost 5723	Bromo-acetonitrile 954

Table 7-4 (*Continued*)
PHYSICAL CONSTANTS OF ORGANIC COMPOUNDS

No.	Name	Synonym	Formula	Beil. Ref.	Formula Weight
888	**Bromo** acetophenone (*p*)	Me-*p*-Br-phenyl ketone	Br·C$_6$H$_4$·CO·CH$_3$	VII-283	199.05
889	acettoluidide	3;1,4	Br·C$_6$H$_3$(CH$_3$)·NH·CO·CH$_3$	XII-991	228.10
890	acetyl bromide	. . . .	Br·CH$_2$·CO·Br	II-215	201.86
891	acetylene	bromo-ethyne	CH⫶C·Br(or C:CHBr)	I-245	104.94
892	allyl alcohol (*β*)	. . . .	CH$_2$:CBr·CH$_2$OH	I-439	136.98
893	aniline (*o*)	. . . .	Br·C$_6$H$_4$·NH$_2$	XII-631	172.03
894	aniline (*m*)	. . . .	Br·C$_6$H$_4$·NH$_2$	XII-633	172.03
895	aniline (*p*)	. . . .	Br·C$_6$H$_4$·NH$_2$	XII-636	172.03
896	aniline (*p*) HCl	. . . .	Br·C$_6$H$_4$·NH$_2$·HCl	XII-637	208.49
897	anisole (*o*)	. . . .	CH$_3$O·C$_6$H$_4$·Br	VI-197	187.04
898	anisole (*p*)	. . . .	CH$_3$O·C$_6$H$_4$·Br	VI-199	187.04
899	anthraquinone (1)	. . . .	C$_6$H$_4$:(CO)$_2$:C$_6$H$_3$Br	VII-789	287.12
900	anthraquinone (2)	. . . .	C$_6$H$_4$:(CO)$_2$:C$_6$H$_3$Br	VII-789	287.12
901	antipyrine (*p*)	bromopyrine	BrC$_6$H$_4$·C$_5$H$_7$ON$_2$	XXIV-33	267.13
902	benzene	phenyl bromide	C$_6$H$_5$Br	V-206	157.02
903	benzene sulfonic acid (*o*)	. . . .	Br·C$_6$H$_4$·SO$_3$H	XI-56	237.08
904	benzene sulfonic acid (*p*)	. . . .	Br·C$_6$H$_4$·SO$_3$H	XI-57	237.08
905	benzene sulfonyl chloride (*p*)	. . . .	Br·C$_6$H$_4$·SO$_2$Cl	XI-57	255.52
906	benzoic acid (*o*)	. . . .	Br·C$_6$H$_4$·CO$_2$H	IX-347	201.03
907	benzoic acid (*m*)	. . . .	Br·C$_6$H$_4$·CO$_2$H	IX-349	201.03
908	benzoic acid (*p*)	. . . .	Br·C$_6$H$_4$·CO$_2$H	IX-351	201.03
909	benzonitrile (*p*)	*p*-Br-phenyl cyanide	Br·C$_6$H$_4$·CN	IX-354	182.03
910	benzoyl bromide (*m*)	. . . .	Br·C$_6$H$_4$·CO·Br	. . . .	263.93
911	benzoyl chloride (*p*)	. . . .	Br·C$_6$H$_4$·CO·Cl	IX-353	219.47
912	benzyl bromide (*p*)	. . . .	Br·C$_6$H$_4$·CH$_2$Br	V-308	249.94
913	benzyl chloride (*o*)	ω-Cl-2-Br-toluene	Br·C$_6$H$_4$·CH$_2$Cl	*V-155	205.49
914	benzyl chloride (*p*)	. . . .	Br·C$_6$H$_4$·CH$_2$Cl	V-307	205.49
915	*iso*-butyl benzene (*p*)	. . . .	Br·C$_6$H$_4$·C$_4$H$_9$	V-415	213.12
916	*tert*-butyl phenol	(4;2,1)	(CH$_3$)$_3$C·C$_6$H$_3$(Br)OH	VI-525	229.12
917	butyric acid (*α*)(*dl*)	. . . .	C$_2$H$_5$·CHBr·CO$_2$H	II-281	167.01
918	*iso*-butyric acid (*α*)	. . . .	(CH$_3$)$_2$CBr.CO$_2$H	II-295	167.01
919	camphor (3)(*d*)	α-bromocamphor	C$_8$H$_{14}$·CO·CHBr	VII-120	231.15
920	*n*-caproic acid (*α*)	. . . .	CH$_3$(CH$_2$)$_3$·CHBr·CO$_2$H	II-325	195.06
921	cinnamic acid (*α*) (*cis*)	bromo-allocinnamic acid	C$_6$H$_5$·CH:CBr·CO$_2$H	IX-600	227.06
922	cinnamic acid (*α*) (*trans*)	. . . .	C$_6$H$_5$·CH:CBr·CO$_2$H	IX-599	227.06
923	cinnamic acid (*β*) (*cis*)	bromo-allocinnamic acid	C$_6$H$_5$·CBr:CH·CO$_2$H	IX-598	227.06

Bromo-acetylamino-toluene 889　　　　Bromo-allyl bromide 1869
Bromo-allocinnamic acid 921, 923　　　Bromo-allylene 5342

Table 7-4 (*Continued*)
PHYSICAL CONSTANTS OF ORGANIC COMPOUNDS

No.	Crystalline Form and Color	Specific Gravity	Melting Point °C.	Boiling Point °C.	Solubility in 100 Parts Water	Alcohol	Ether
888	col. lf./al.	....	50-1	255^{736mm}	i.; s. bz.	s.; s. CS_2	v. s.; s. ac.
889	nd./bz.	....	116-7	....	....	....	....
890	lq.	$2.317\frac{21.5°}{21.5}$	....	149-50	d.	d.	....
891	poisonous gas	....	....	-2	5000-6000 cc.$^{15°}$, CH_2CBr_2	s.	....
892	lq.	$1.6^{15°}$	....	153-4	....	....	....
893	cr.	....	31-2	229	i.	v. s.	s.
894	cr.	$1.579\frac{20.4°}{4}$	18.5	251	i.	s.	s.
895	rhb.	$1.80^{15-20°}$	63-4	....	i. c.	v. s.	v. s.
896	mn. pr.	....	....	....	s.	s.	i.
897	oil	....	....	218-21	....	s.	....
898	cr.	$1.4949°$	13-14	215	....	s.	....
899	yel. nd./bz.	....	188	subl.	s. H_2SO_4	....	....
900	cr./am. al.	....	204-5	....	....	....	....
901	col. mn./ aq.	....	122	$300^{9mm} \pm$	s. h.	s.; s. chl.	sl. s.
902	col. lq.	$1.495\frac{20°}{4}$	-30.6	156.2	i.	s.	∞; ∞ chl.
903	delq. nd.	....	....	....	v. s.	v. s.	i.
904	delq. nd.	....	102-3	155^{25mm}	s.	s.	i.
905	tri. pr./et.	....	75-6	153^{15mm}	d. h.	d. h.	....
906	nd./aq.	....	148-50	subl.	$0.18^{25°}$	s.	s.
907	nd.	....	156-8	>280	$0.04^{25°}$	s.	s.
908	mn. pr.	....	251-3	....	sl. s. h.	s.	s.
909	nd./aq. or al.	....	112-3	....	s. h.	s.	v. s.
910	col. lq.	....	....	$118-22^{8mm}$	....	....	....
911	nd./pet.	....	42	245-7 sl. d.	d.; v. s. bz.	v. s. lg.	s.
912	nd./al.	....	61	....	v. sl. s.	s. h.	s.
913	col. lq.	....	....	$124-6^{20mm}$	i.	v. s.	v. s.
914	nd./al.	....	41	236	i.	v. s. h.	v. s.
915	lq.	....	<-18	$232-3^{739mm}$	....	....	....
916	lq.	$1.338\frac{25°}{25}$	<-20*	$109-29^{5mm}$	i.	∞; ∞ Me al.	∞ bz., act.
917	col. oil	$1.567\frac{20°}{20}$	-4	$127-8^{25mm}$	6.6	s.	s.
918	pl.	$1.523\frac{60°}{60}$	48	198-200	forms oil	s.	s.
919	cr.	$1.449\frac{20°}{4}$	77-8	274(sl. d.)	i.	$20^{26°}$	v. s.
920	lq.	....	....	$128-31^{10mm}$	s.	s.	s.
921	rhb./aq.	....	120-1	$110^{0.6mm}$	s. h.	s.	s. bz.
922	nd./aq.	....	131-2	$121^{0.6mm}$	v. sl. s. h.	∞	∞
923	mn./al.	....	160	$110^{0.6mm}$	sl. s. h.; sl. d.	sl. s. c.	s.; s. h. bz.

* Crysts. $+1 H_2O$, m.p. 51-2°.
Bromo-aminotoluene 1002-3
Bromo-aspirin 76

Bromo-butane 1057-60

Table 7-4 (*Continued*)
PHYSICAL CONSTANTS OF ORGANIC COMPOUNDS

No.	Name	Synonym	Formula	Beil. Ref.	Formula Weight
924	**Bromo** cinnamic acid (β) (*trans*)	....	$C_6H_5 \cdot CBr : CH \cdot CO_2H$	IX-597	227.06
925	crotonic acid (β)	....	$CH_3 \cdot C(Br) : CH \cdot CO_2H$	II-419	164.99
926	cymene (2)	(4;2,1)	$(CH_3)_2CH \cdot C_6H_3(Br) \cdot CH_3$	V-423	213.12
927	diethylaniline (p)	....	$Br \cdot C_6H_4 \cdot N(C_2H_5)_2$	XII-638	228.15
928	diiodo-methane	....	$BrCHI_2$	I-72	346.74
929	dimethylaniline (p)	....	$Br \cdot C_6H_4 \cdot N(CH_3)_2$	XII-637	200.08
930	dinitrobenzene	(4;1,2)	$Br \cdot C_6H_3(NO_2)_2$	V-266	247.01
931	dinitrobenzene	(4;1,3)	$Br \cdot C_6H_3(NO_2)_2$	V-266	247.01
932	dinitrobenzoic acid	(4;3,5,1)	$Br \cdot C_6H_2(NO_2)_2CO_2H$	IX-416	291.02
933	diphenyl (o)	....	$Br \cdot C_6H_4 \cdot C_6H_5$	V-580	233.11
934	diphenyl (p)	....	$Br \cdot C_6H_4 \cdot C_6H_5$	V-580	233.11
935	diphenyl ether (p)	....	$Br \cdot C_6H_4 \cdot O \cdot C_6H_5$	*VI-105	249.11
936	ethyl acetate (β)	....	$CH_3CO_2 \cdot CH_2 \cdot CH_2Br$	II-128	167.01
937	ethyl benzene (β)	Ph-Et-bromide	$C_6H_5 \cdot CH_2 \cdot CH_2Br$	V-356	185.07
938	ethyl benzene (α)	Ph-Et-bromide	$C_6H_5 \cdot CH(Br)CH_3$	V-355	185.07
939	ethyl ethyl ether	β-Br-diethyl ether	$BrCH_2CH_2 \cdot O \cdot C_2H_5$	I-338	153.02
940	ethyl phthalimide (β)(N)	phthalimino-Et-bromide	$C_6H_4(CO)_2N \cdot CH_2 \cdot CH_2Br$	XXI-461	254.09
941	fumaric acid	....	$CH : CBr(CO_2H)_2$	II-745	194.98
942	furan (2)	....	$Br \cdot C_4H_3O$	XVII-27	146.98
943	furoic acid (3)	....	$Br \cdot C_4H_2O \cdot CO_2H$	XVIII-284	190.99
944	hydroquinone	....	$Br \cdot C_6H_3(OH)_2$	VI-852	189.01
945	iodobenzene (o)	....	$Br \cdot C_6H_4 \cdot I$	V-223	282.91
946	iodobenzene (m)	....	$Br \cdot C_6H_4 \cdot I$	V-223	282.91
947	iodobenzene (p)	....	$Br \cdot C_6H_4 \cdot I$	V-223	282.91
948	iodo-ethane (1,1)	....	$CH_3 \cdot CHBrI$	I-98	234.87
949	iodo-ethane (1,2)	....	$BrCH_2 \cdot CH_2I$	I-98	234.87
950	iodo-methane	....	$BrCH_2I$	I-71	220.84
951	maleic acid	....	$CH : CBr(CO_2H)_2$	II-754	194.98
952	malonic acid	....	$BrCH(CO_2H)_2$	II-594	182.96
953	mesitylene	(2;1,3,5)	$Br \cdot C_6H_2(CH_3)_3$	V-408	199.10
954	methyl cyanide	bromo-acetonitrile	$BrCH_2 \cdot CN$	II-216	119.95
955	methyl acetate	....	$CH_3 \cdot CO_2 \cdot CH_2Br$	II-152	152.98
956	naphthalene (α)	α-naphthyl bromide	$C_{10}H_7Br$	V-547	207.08
957	naphthalene (β)	β-naphthyl bromide	$C_{10}H_7Br$	V-548	207.08
958	α-naphthol (4,1)	....	$Br \cdot C_{10}H_6 \cdot OH$	VI-613	223.08
959	β-naphthol (1,2)	....	$Br \cdot C_{10}H_6 \cdot OH$	VI-650	223.08
960	nitrobenzene (o)	....	$Br \cdot C_6H_4 \cdot NO_2$	V-247	202.01
961	nitrobenzene (m)	....	$Br \cdot C_6H_4 \cdot NO_2$	V-248	202.01
962	nitrobenzene (p)	....	$Br \cdot C_6H_4 \cdot NO_2$	V-248	202.01
963	3-nitrobenzene-1-sulfonic acid (4)	(4;3,1)	$Br \cdot C_6H_3(NO_2) \cdot SO_3H$	XI-74	282.08
964	3-nitrobenzoic acid (2)	(2;3,1)	$Br \cdot C_6H_3(NO_2) \cdot CO_2H$	IX-406	246.02
965	nitroethane (1,1)	....	$CH_3 \cdot CHBr \cdot NO_2$	I-101	153.97
966	nitromethane	....	$Br \cdot CH_2 \cdot NO_2$	I-77	139.94

Table 7-4 (*Continued*)
PHYSICAL CONSTANTS OF ORGANIC COMPOUNDS

No.	Crystalline Form and Color	Specific Gravity	Melting Point °C.	Boiling Point °C.	Solubility in 100 Parts		
					Water	Alcohol	Ether
924	nd./aq.	....	134-5	$122^{0.6mm}$	sl. s. h.	s.	s. h. bz.
925	nd./lg.	....	95-7	....	sl. s. c.	v. s.; v. s. bz.	v. s.; v. s. CS_2
926	lq.	1.253^{25}_{25}	<−20	233-5	i.; $\infty^{25°}$ act.	$50^{25°}$ Me al.	∞; ∞ bz.
927	yel. red/ac.	....	33	270	i.	v.s.	v.s.
928	yel. cr.	....	60	110^{25mm}	sl. s. pet.	....	soln. d. in light
929	lf./al.	....	55	264	i.	v. s.	v. s.
930	mn. pr.	$1.801^{60°}$	59.5(34.8)	....	i.	s. h.	s.
931	yel. cr.	....	75.3	....	....	v. s. h.	....
932	pr./aq. H_2SO_4	....	181	....	sl. s.	v. s.	v. s.
933	lq.	....	<−20	296-8	i.	s.	v. s.
934	cr./al.	....	90-1	310	i.; $100^{25°}$ bz.	s.; $3^{25°}$ Me al.	$34^{25°}$
935	lq.	$1.449^{13°}$	17-18	305	....	....	....
936	col. lq.	$1.514^{20°}_4$	−13.8	162-3	i.	∞	∞
937	lq.	....	....	$217\text{-}8^{734mm}$	....	....	....
938	lq.	$1.311^{23°}$	....	200-10 d.	i.	s.	s.
939	col. lq.	$1.357^{20°}_4$	....	127-8	sl. s.	∞	∞
940	nd./al.	....	82-3	d.	....	....	....
941	lf./aq.	....	185-6	d.	v. s.	....	....
942	lq.	1.650	....	101-2	i.	s.	....
943	nd./aq.	....	128-9	....	$1.3^{20°}$	s.; v. sl. s. CS_2	s.; v. sl. s. lg.
944	lf./pet.	....	113	subl.	v. s.	v. s.	v. s.
945	col. lq.	$2.257^{25°}_4$	5.0(2.1)	257.4^{754mm}	i.	v. sl. s.	v. sl. s. ac.
946	col. lq.	....	−9.3	252^{754mm}	i.	v. sl. s.	v. sl. s. ac.
947	pl./al. et.	....	91-2	251.5^{754mm}	i.	v. sl. s. c.	sl. s.
948	lq.	$2.452^{16°}$	<−20	142-3	....	....	....
949	long nd.	$2.516^{29°}$	28	163	....	v. s. h.	....
950	lq.	$2.926^{17°}$	....	138-40	....	....	....
951	nd. or pr.	....	128*	d.	v. s.	v. s.	v. s.
952	nd./et.	....	112-3 d.	....	....	v. s.	v. s.
953	lq.	$1.319^{10°}$	−1	225-30	....	....	....
954	yel. oil	1.771	....	148-50	....	....	s.
955	col. lq.	$1.195^{14°}$	....	$130\text{-}3^{750mm}$	i.; sl. d.	s.	s.
956	col. oil	$1.482^{20°}_4$	6.1	281.2	i.; ∞ bz.	∞ abs. al.	∞
957	lf./al.	$1.605^{0°}$	59	281-2	i.; s. bz.	$6^{20°}$, 92% al.	v. s.; v. s. ehl.
958	nd./aq. al.	....	127-8	....	....	....	....
959	rhb. pr.	....	83-4	d. 130	....	....	....
960	yel. cr.	$1.623^{80°}_4$	43	261	i.	v. s.	s.
961	rhb.	$1.704^{20°}_4$	56.4	256-7	i.	s.	s.
962	tri.	$1.938^{0°}_4$	126-7	255-6	i.	1.4 c.	s.
963	yel. nd.	....	87-8	....	s.	s.	i.
964	cr./aq. al.	....	186-8	....	....	....	....
965	lq.	....	....	146-7	i.	....	....
966	lq.	....	....	152.5^{765mm}	i.	s. alk.	....

* Heated slowly, m.p. 136-8°; rapidly, m.p. 140-1°.

Bromo-methane 4147	Bromo-nitrotoluene 4681	Bromo-pentadecane 5063
Bromo-methylacetophenone 4333	Bromo-nonane 4927	Bromo-pentane 427-31
Bromo-methyl-*p*-tolyl ketone 4333	Bromo-octane 4973-4	

Table 7-4 *(Continued)*
PHYSICAL CONSTANTS OF ORGANIC COMPOUNDS

No.	Name	Synonym	Formula	Beil. Ref.	Formula Weight
967	**Bromo** phenacyl bromide *(p)*	diBr-acetophenone	$Br \cdot C_6H_4 \cdot CO \cdot CH_2Br$	VII-285	277.95
968	phenanthrene (9)	....	$C_{14}H_9Br$	V-671	257.14
969	phenetole *(β)*	*β*-PhO-ethyl bromide	$Br \cdot CH_2CH_2 \cdot O \cdot C_6H_5$	VI-142	201.07
970	phenetole *(o)*	....	$Br \cdot C_6H_4 \cdot O \cdot C_2H_5$	VI-197	201.07
971	phenetole *(p)*	....	$Br \cdot C_6H_4 \cdot O \cdot C_2H_5$	VI-199	201.07
972	phenol *(o)*	....	$Br \cdot C_6H_4 \cdot OH$	VI-197	173.02
973	phenol *(m)*	....	$Br \cdot C_6H_4 \cdot OH$	VI-198	173.02
974	phenol *(p)*	....	$Br \cdot C_6H_4 \cdot OH$	VI-198	173.02
975	phenylhydrazine *(p)*	....	$Br \cdot C_6H_4 \cdot NH \cdot NH_2$	XV-434	187.05
976	phenylhydrazine HCl	*(p)*	$Br \cdot C_6H_4 \cdot N_2H_3 \cdot HCl$	XV-435	223.51
977	phenylphenol	(1;3,4)	$C_6H_5 \cdot C_6H_3(Br)OH$	**VI-625	249.11
978	phthalic acid	(3;1,2)	$Br \cdot C_6H_3(CO_2H)_2$	IX-821	245.04
979	(tere)-phthalic acid	(2;1,4)	$Br \cdot C_6H_3(CO_2H)_2$	IX-848	245.04
980	propionic acid *(α)*	*(dl)*	$CH_3 \cdot CHBr \cdot CO_2H$	II-254	152.98
981	propionic acid *(β)*	....	$BrCH_2 \cdot CH_2 \cdot CO_2H$	II-256	152.98
982	propionyl bromide *(α)*	....	$CH_3 \cdot CHBr \cdot CO \cdot Br$	II-256	215.88
983	*n*-propyl acetate *(β)*	....	$CH_3CO_2 \cdot CH_2 \cdot CHBr \cdot CH_3$	....	181.04
984	*n*-propyl acetate *(γ)*	....	$CH_3CO_2 \cdot (CH_2)_3Br$	**II-139'	181.04
985	*n*-propyl benzene *(p)*	....	$Br \cdot C_6H_4 \cdot (CH_2)_2CH_3$	V-391	199.10
986	*iso*-propyl benzene *(p)*	*p*-bromocumene	$Br \cdot C_6H_4 \cdot CH(CH_3)_2$	V-395	199.10
987	*n*-propyl phenyl ether *(γ)*	....	$C_6H_5 \cdot O \cdot (CH_2)_3Br$	VI-142	215.10
988	propylene (1)*(α)*	1-bromo-propene-1	$CH_3 \cdot CH:CHBr$	I-200	120.98
989	propylene (1)*(β)*	1-bromo-propene-1	$CH_3 \cdot CH:CHBr$	I-200	120.98
990	propylene (2)	2-bromo-propene-1	$CH_3 \cdot CBr:CH_2$	I-200	120.98
991	pyridine (2)	*α*-bromo-pyridine	$Br \cdot C_5H_4N$	XX-233	158.01
992	pyridine (3)	*β*-bromo-pyridine	$Br \cdot C_5H_4N$	XX-233	158.01
993	salicylic acid (5)	(5;2,1)	$Br \cdot C_6H_3(OH)CO_2H$	X-107	217.03
994	styrene *(α)*	....	$C_6H_5 \cdot CBr:CH_2$	V-477	183.05
995	styrene *(ω)*	*isomer No. 1*	$C_6H_5 \cdot CH:CHBr$	V-477	183.05
996	styrene *(ω)*	*isomer No. 2*	$C_6H_5 \cdot CH:CHBr$	V-477	183.05
997	succinic acid *(dl)*	....	$(\cdot CH_2CHBr \cdot)(CO_2H)_2$	II-621	196.99
998	sulfalein	di-Na-phenol-tetra-bromo-phthalein-disulfonate	$Br_4C_6 \cdot CO \cdot O \cdot C:$ $\underset{(C_6H_3OHSO_3Na)_2}{\mid\underline{\qquad}\mid}$	....	838.02
999	toluene *(o)*	*o*-tolyl bromide	$Br \cdot C_6H_4 \cdot CH_3$	V-304	171.04
1000	toluene *(m)*	....	$Br \cdot C_6H_4 \cdot CH_3$	V-305	171.04
1001	toluene *(p)*	....	$Br \cdot C_6H_4 \cdot CH_3$	V-305	171.04
1002	*o*-toluidine (5)	(5;2,1)	$Br \cdot C_6H_3(NH_2)CH_3$	XII-838	186.06
1003	*p*-toluidine (3)	(3;4,1)	$Br \cdot C_6H_3(NH_2)CH_3$	XII-991	186.06
1004	triiodomethane	....	$BrCl_3$	I-74	472.63
1005	trinitromethane	....	$BrC(NO_2)_3$	I-79	229.94

Bromo-phenylacetophenone 5214   Bromo-phenyl-ethyl ether 970-1   Bromo-propane 5388-9
Bromo-phenyl-cyanide 909   Bromo-picrin 4688   Bromo-propyl alcohol 5460, 6278

Table 7-4 (*Continued*)
PHYSICAL CONSTANTS OF ORGANIC COMPOUNDS

No.	Crystalline Form and Color	Specific Gravity	Melting Point °C.	Boiling Point °C.	Solubility in 100 Parts		
					Water	Alcohol	Ether
967	nd.	....	110-2	....	i.	s. h.	s.
968	pr./al.	1.409$\frac{10°}{4}$	63	>360; subl.	v. s. ac.	....	v. s. CS$_2$
969	cr.	....	35	240-50 d.	i.	s.	s.
970	lq.	....	....	218-22	....	s.	s.
971	lq.	....	10-12	227-33	....	....	....
972	col. lq.	1.553$^{80°}$	5.6	194-5	s.; s. alk.	s.; ∞ chl.	∞
973	cr.	....	32-3(18)	236-7	s. alk.	s.	s.
974	tet. cr.	1.588$^{80°}$	63.5	238	1.4$^{15°}$	v. s.; 88$^{25°}$ act.	v. s.; 40$^{25°}$ bz.
975	nd./al.	....	106-7	....	s. bz.	s.	s.
976	cr./aq.	....	....	....	s. h.	....	....
977	cr.	....	94-6	....	i.; 32$^{25°}$bz.	125$^{25°}$	v. s.
978	nd./aq.	....	178.5 (-H$_2$O)	....	s.	s.	s.; v. sl. s. chl.
979	nd./al.	....	299	....	0.18$^{24°}$; s. h.	s.	i.; i. bz.
980	pr.	1.700$\frac{20°}{4}$	25.7(-3.9)	205.5 sl. d.	v. s.	v. s.	v. s.
981	pl.	....	62.5	140-245^{mm}	v. s.	v. s.	v. s.
982	lq.	2.061$\frac{18°}{4}$	....	152-4	....	....	....
983	col. lq.	....	....	51-3^{9mm}	....	....	....
984	col. lq.	....	....	88-90^{22mm}	....	....	....
985	col. lq.	....	....	220	....	....	....
986	lq.	1.365$\frac{22°}{4}$	<-20	216-8	....	....	....
987	col. lq.	1.365$\frac{16°}{16}$	....	211-2^{200mm}	....	....	....
988	lq.	1.434$\frac{15.8°}{4}$	-113	59-60	....	....	....
989	lq.	1.417$\frac{15.8°}{4}$	-76.5	63.3	....	....	....
990	lq.	1.397$\frac{15.8°}{4}$	-125	48.4	....	....	....
991	lq.	1.657$^{15°}$	....	193-4	sl. s.	....	....
992	col. lq.	1.632$^{10°}$	....	173-4^{758mm}	sl. s.	s.	s.
993	nd./aq. al.	....	167-9	subl. >100	0.38$^{0°}$	85$^{25°}$	70$^{25°}$
994	oil	1.406$\frac{20°}{4}$	-43.5	160^{75mm}	....	....	....
995	lq.	1.422$\frac{20°}{4}$	7	221 sl. d.	i.	∞	∞
996	lq.	1.427$\frac{20°}{4}$	-7.5	108^{26mm}	i.	∞	∞
997	col.	2.073	160-1	....	19$^{15°}$	s.	....
998	wh. pd.	....	....	....	s.	i.	i. act.
999	col. lq.	1.422$\frac{20°}{4}$	-28.1	181.5	i.	s.; ∞ bz.	∞$^{25°}$
1000	col. lq.	1.410$\frac{20°}{4}$	-39.8	183.7	i.	s.; ∞ bz.	s.
1001	cr./al.	1.390$\frac{20°}{4}$	26.7	185.0	i.	s.	∞$^{25°}$
1002	rhb./al.	....	58-9	240	v. sl. s.	s.	s.
1003	lf.	1.5$^{20°}$	16-8	240	i.	s.	s.
1004	cr.	....	113	....	....	....	....
1005	lq.	2.044$\frac{15°}{4}$	17-8	55-6^{12mm}	0.4$^{25°}$	....	....

Bromo-propylene 199   Bromo-pyrine 901
Bromo-propyne 5342   Bromo-toluene (ω) 793

Table 7-4 *(Continued)*
PHYSICAL CONSTANTS OF ORGANIC COMPOUNDS

No.	Name	Synonym	Formula	Beil. Ref.	Formula Weight
1006	**Bromo** *n*-valeric acid (α)	....	$CH_3(CH_2)_2CHBr\cdot CO_2H$	II-302	181.04
1007	*iso*-valeric acid	(α)(*dl*)	$(CH_3)_2CHCHBr\cdot CO_2H$	II-317	181.04
1008	*iso*-valeryl-*p*-phenetidine (α) (*d*)	phenoval	$(CH_3)_2CHCHBr\cdot CO\cdot NH\cdot C_6H_4\cdot O\cdot C_2H_5$	*XIII-163	300.20
1009	*iso*-valeryl urea (α)	bromural; bromisoval	$(CH_3)_2CHCHBr\cdot CO\cdot NH\cdot CO\cdot NH_2$	III-63	223.08
1010	xylene (3;1,2)	bromo-*o*-xylene (3)	$Br\cdot C_6H_3(CH_3)_2$	V-365	185.07
1011	xylene (4;1,2)	bromo-*o*-xylene (4)	$Br\cdot C_6H_3(CH_3)_2$	V-365	185.07
1012	xylene (2;1,3)	bromo-*m*-xylene (2)	$Br\cdot C_6H_3(CH_3)_2$	V-374	185.07
1013	xylene (4;1,3)	bromo-*m*-xylene (4)	$Br\cdot C_6H_3(CH_3)_2$	V-374	185.07
1014	xylene (5;1,3)	bromo-*m*-xylene (5)	$Br\cdot C_6H_3(CH_3)_2$	V-374	185.07
1015	xylene (2;1,4)	bromo-*p*-xylene (2)	$Br\cdot C_6H_3(CH_3)_2$	V-385	185.07
1016	**Bromoform**	tribromo-methane	$CHBr_3$	I-68	252.75
1017	**Bryonane**	laurane	$C_{20}H_{42}$	I-174	282.56
1018	**Butadiene** (1,2)	methyl-allene	$CH_3\cdot CH:C:CH_2$	I-249	54.09
1019	**Butadiene** (1,3)	erythrene	$CH_2:CH\cdot CH:CH_2$	I-249	54.09
1020	**Butandiolamine**	2-NH$_2$-2-Me-pro-pandiol-1,3	$CH_3\cdot C(NH_2):(CH_2OH)_2$	IV-303	105.14
1021	**Butane** (*n*)	diethyl	$CH_3\cdot CH_2\cdot CH_2\cdot CH_3$	I-118	58.12
1022	**Butane** (*iso*)	trimethyl-methane	$(CH_3)_3CH$	I-124	58.12
1023	**Butanolamine**	2-NH$_2$-butanol-1	$C_2H_5\cdot CH(NH_2)\cdot CH_2OH$	IV-291	89.14
1024	*iso*-**Butanolamine**	2-NH$_2$-2-Me-propanol-1	$(CH_3)_2C(NH_2)\cdot CH_2OH$	....	89.14
1025	**Butoxy**-cinchoninic acid diethyleth-ylene-diamide HCl	nupercaine; percaine	$C_4H_9O\cdot C_9H_5N\cdot CO\cdot NH(CH_2)_2N:(C_2H_5)_2\cdot HCl$	....	379.93
1026	ethoxyethyl bro-mide (2)(β)	....	$C_4H_9O\cdot (CH_2)_2\cdot O\cdot (CH_2)_2Br$	....	225.13
1027	ethyl phthalate	....	$C_6H_4(CO_2CH_2CH_2\cdot O\cdot C_4H_9)_2$	....	366.46
1028	ethyl salicylate (β)(*n*)	....	$HOC_6H_4CO_2\cdot CH_2\cdot CH_2\cdot O\cdot C_4H_9$	....	238.29
1029	**Butyl** acetate (*n*)	....	$CH_3CO_2\cdot CH_2CH_2\cdot C_2H_5$	II-130	116.16
1030	acetate (*sec*)	....	$CH_3CO_2\cdot CH(CH_3)\cdot C_2H_5$	II-131	116.16
1031	acetate (*iso*)	....	$CH_3CO_2\cdot CH_2CH:(CH_3)_2$	II-131	116.16
1032	acetate (*tert*)	....	$CH_3CO_2\cdot C(CH_3)_3$	II-131	116.16
1033	acetoacetate (*n*)		$CH_3COCH_2CO_2C_4H_9$	....	158.20
1034	alcohol (*n*)	butanol-1	$C_2H_5\cdot CH_2\cdot CH_2OH$	I-367	74.12
1035	alcohol (*sec*)	butanol-2	$C_2H_5\cdot CHOH\cdot CH_3$	I-371	74.12
1036	alcohol (*iso*)	2-methyl-propanol-1	$(CH_3)_2CH\cdot CH_2OH$	I-373	74.12
1037	alcohol (*tert*)	2-methyl-propanol-2	$(CH_3)_3COH$	I-379	74.12
1038	allylbarbituric acid (*iso*)	sandoptal	$C_4H_9(C_3H_5):C_4H_2O_3N_2$	....	224.26
1039	amine (*n*)	....	$C_2H_5\cdot CH_2\cdot CH_2\cdot NH_2$	IV-156	73.14

Table 7-4 (*Continued*)
PHYSICAL CONSTANTS OF ORGANIC COMPOUNDS

No.	Crystalline Form and Color	Specific Gravity	Melting Point °C.	Boiling Point °C.	Solubility in 100 Parts		
					Water	Alcohol	Ether
1006	lq.	....	....	$126-30^{27mm}$	sl. s.	v. s.	s.
1007	col. pr./et.	....	44	230 sl. d.	1.4 c.	v. s.	s.
1008	col. cr./aq. al	....	151	....	i.; sl. s. gly.	s. h.; v. sl. s. bz.	sl. s.; sl. s. chl.
1009	lf./toluene	....	160	subl.	2 c.; s. h.	s.	s.; s. alk.
1010	col. lq.	$1.365^{20°}_4$	....	213-4	i.	....	....
1011	col. lq.	$1.369^{15°}_{15}$	−0.2	214.5	i.	....	....
1012	col. lq.	....	<−10	206±	i.	....	....
1013	col. lq.	....	....	205-7	i.	....	....
1014	col. lq.	$1.362^{20°}$	<−20	204	....	....	....
1015	lf. or pl.	$1.356^{20°}_4$	9-10	205.5^{755mm}	....	....	....
1016	col. lq.	$2.890^{20°}_4$	8.3	149.6	0.1 c.	∞; ∞ bz.	∞; ∞ chl.
1017	nd.	....	69	400	v. s. pet.	v. s. h. abs.	v. s. bz.
1018	lq.	$0.652^{20°}$	−136.3	10.3	i.	∞	∞
1019	col. gas	$0.621^{20°}_4$	−108.9	−4.41	i.	∞	∞
1020	cr.	....	109-11	$151-2^{10mm}$	$250^{20°}$	$25^{20°}$	0.1
1021	col. gas	$0.579^{20°}$ (lq.)	−138.3	−0.50	$15\frac{17°}{772}$ cc.	$1883\frac{17°}{775}$ cc.	$2980\frac{18°}{773}$ cc.
1022	col. gas	$0.557^{20°}$ (lq.)	−159.6	−11.7	$13\frac{17°}{772}$ cc.	$1320\frac{17°}{775}$ cc.	$2790\frac{18°}{773}$ cc.
1023	lq.	$0.944^{20°}_{20}$	−2	174-8	∞	∞	∞
1024	col. cr.	$0.934^{20°}_{20}$	30-1	165	∞	∞	∞
1025	wh. cr.	....	90*	....	200; s. chl.	s.; sl. s. bz.	i.; s. act.
1026	col. lq.	....	....	$107-9^{12mm}$	....	....	....
1027	col. lq.	....	....	$189-91^{2mm}$	....	....	....
1028	col. lq.	$1.077^{25°}_{25}$	<−20	$186-93^{23mm}$	i.; ∞ bz.	∞; ∞ chl.	∞
1029	col. lq.	$0.882^{20°}$	−73.5	126.1	0.7	∞	∞
1030	col. lq.	$0.865^{25°}_4$	....	111.5-2.5^{744mm}	i.	∞	∞
1031	col. lq.	$0.871^{20°}_4$	−98.9	118	$0.6^{25°}$	∞	∞
1032	col. lq.	$0.866^{20°}_4$	....	$95-6^{750mm}$	i.	∞	∞
1033	col. lq.	....	....	$98-100^{16mm}$	i.	s.	s.
1034	col. lq.	$0.810^{20°}_4$	−89.8	118.0	$9^{15°}$	∞	∞
1035	col. lq.	$0.808^{20°}_4$	−114.7	99.5	$12.5^{20°}$	∞	∞
1036	col. lq.	$0.802^{20°}_4$	−108	108.1	$10^{15°}$	∞	∞
1037	lq. or rhb.	$0.779^{26°}$	25.6	82.6	∞	∞	∞
1038	wh. cr. pd.	....	138-9	....	sl. s.	s.; s. chl.	s.; s. act.
1039	col. lq.	$0.739^{25°}_4$	−50	77.8	∞	∞; ∞ gly.	∞

* Free base, m. 97-8°; s. al., aq., bz.; i. et.

Butenoic acid 1557-8, 6427
Butenol 1565, 4422
Butenyl chloride (*iso*) 4060
Butesin 1043

Butine, cf. butyne
Butoben 1093
Butolan 825
Butoxy-benzene 1116

Butter yellow 2431 or 2438
Butyl, cf. also dibutyl.
Butyl adipate 1883

Table 7-4 (*Continued*)
PHYSICAL CONSTANTS OF ORGANIC COMPOUNDS

No.	Name	Synonym	Formula	Beil. Ref.	Formula Weight
1040	**Butyl** amine (*sec*)	....	$(C_2H_5)(CH_3){:}CH{\cdot}NH_2$	IV-160	73.14
1041	amine (*iso*)	....	$(CH_3)_2CH{\cdot}CH_2{\cdot}NH_2$	IV-163	73.14
1042	amine (*tert*)	....	$(CH_3)_3C{\cdot}NH_2$	IV-173	73.14
1043	*p*-aminobenzoate (*n*)	butesin	$H_2N{\cdot}C_6H_4{\cdot}CO_2C_4H_9$	....	193.25
1044	aminobenzoate picrate	butesin picrate	$(C_{11}H_{15}O_2N)_2{\cdot}$ $C_6H_3O_7N_3$	....	615.60
1045	*p*-aminobenzoate (*iso*)	cycloform	$H_2N{\cdot}C_6H_4{\cdot}CO_2C_4H_9$	*XIV-567	193.25
1046	aniline (*n*)	....	$C_6H_5{\cdot}NH{\cdot}C_4H_9$	XII-168	149.24
1047	aniline (*iso*)	....	$C_6H_5{\cdot}NH{\cdot}C_4H_9$	XII-168	149.24
1048	arsonic acid (*n*)	....	$C_4H_9{\cdot}AsO(OH)_2$	**IV-997	182.05
1049	benzene (*n*)	....	$C_6H_5{\cdot}C_4H_9$	V-413	134.22
1050	benzene (*sec*)	....	$C_6H_5{\cdot}C_4H_9$	V-414	134.22
1051	benzene (*iso*)	....	$C_6H_5{\cdot}C_4H_9$	V-414	134.22
1052	benzene (*tert*)	triMe-Ph-methane	$C_6H_5{\cdot}C(CH_3)_3$	V-415	134.22
1053	benzoate (*n*)	....	$C_6H_5CO_2{\cdot}C_4H_9$	IX-112	178.23
1054	benzoate (*iso*)	....	$C_6H_5CO_2{\cdot}C_4H_9$	IX-113	178.23
1055	*o*-benzoylbenzoate (*n*)	....	$C_6H_5{\cdot}CO{\cdot}C_6H_4{\cdot}CO_2{\cdot}$ C_4H_9	....	282.34
1056	benzyl ether (*n*)	....	$C_4H_9{\cdot}O{\cdot}CH_2C_6H_5$	**VI-410	164.25
1057	bromide (*n*)	1-bromo-butane	$C_2H_5{\cdot}CH_2{\cdot}CH_2Br$	I-119	137.03
1058	bromide (*sec*)	2-bromo-butane	$C_2H_5{\cdot}CHBr{\cdot}CH_3$	I-119	137.03
1059	bromide (*iso*)	1-Br-2-Me-propane	$(CH_3)_2CH{\cdot}CH_2Br$	I-126	137.03
1060	bromide (*tert*)	2-Br-2-Me-propane	$(CH_3)_3CBr$	I-127	137.03
1061	*β*-bromoallyl barbituric acid	pernoston	$(CH_2{:}CBr{\cdot}CH_2){\cdot}$ $(C_4H_9){:}C_4H_2O_3N_2$	....	303.16
1062	*n*-butyrate (*n*)	....	$C_3H_7CO_2{\cdot}C_4H_9$	II-271	144.22
1063	*n*-butyrate (*iso*)	....	$C_3H_7CO_2{\cdot}C_4H_9$	II-271	144.22
1064	*iso*-butyrate (*iso*)	....	$(CH_3)_2CHCO_2{\cdot}C_4H_9$	II-291	144.22
1065	caproate	....	$C_5H_{11}CO_2{\cdot}C_4H_9$	II-323	172.27
1066	caprylate	....	$C_7H_{15}CO_2{\cdot}C_4H_9$	II-348	200.32
1067	carbamate (*n*)	....	$NH_2{\cdot}CO_2{\cdot}C_4H_9$	*III-14	117.15
1068	carbamate (*iso*)	....	$NH_2{\cdot}CO_2{\cdot}C_4H_9$	III-29	117.15
1069	carbitol acetate	*β*-BuO-*β*-EtO-ethyl-acetate	$CH_3CO(OC_2H_4)_2{\cdot}O{\cdot}$ C_4H_9	....	204.27
1070	catechol (*p*)(*tert*)	(4;1,2)	$(CH_3)_3C{\cdot}C_6H_3(OH)_2$	....	166.22
1071	cellosolve (*n*)	2-BuO-ethanol-1	$C_4H_9{\cdot}O{\cdot}CH_2CH_2OH$	**I-519	118.18
1072	chloride (*n*)	1-chloro-butane	$C_2H_5{\cdot}CH_2{\cdot}CH_2Cl$	I-118	92.57
1073	chloride (*sec*)	2-chloro-butane	$C_2H_5{\cdot}CHCl{\cdot}CH_3$	I-119	92.57
1074	chloride (*iso*)	1-Cl-2-Me-propane	$(CH_3)_2CH{\cdot}CH_2Cl$	I-124	92.57
1075	chloride (*tert*)	2-Cl-2-Me-propane	$(CH_3)_3C{\cdot}Cl$	I-125	92.57
1076	chloroacetate (*n*)	....	$Cl{\cdot}CH_2{\cdot}CO_2{\cdot}C_4H_9$	II-198	150.61
1077	chloroformate (*n*)	*n*-Bu-chlorocarbonate	$Cl{\cdot}CO_2{\cdot}C_4H_9$	**III-11	136.58
1078	chloroformate (*iso*)	*iso*-Bu-chlorocarbonate	$Cl{\cdot}CO_2{\cdot}C_4H_9$	III-12	136.58
1079	*o*-cresol (*p*)(*tert*)	2-Me-4-*tert*-Bu-phenol	$(CH_3)_3C{\cdot}C_6H_3(CH_3){\cdot}$ OH	VI-550	164.25
1080	*o*-cresyl ether (*n*)	*n*-Bu-*o*-tolyl ether	$CH_3{\cdot}C_6H_4{\cdot}O{\cdot}C_4H_9$	VI-353	164.25
1081	crotonate (*n*)	....	$CH_3CH{:}CHCO_2{\cdot}C_4H_9$	....	142.20
1082	crotonate (*iso*)	....	$CH_3CH{:}CHCO_2{\cdot}C_4H_9$	....	142.20
1083	*iso*-cyanide (*iso*)	*iso*-Bu-carbylamine	$(CH_3)_2CH{\cdot}CH_2{\cdot}NC$	IV-167	83.13

Butyl borate 6117
Butyl carbamide 1138-9
Butyl carbinol (*n*) 404

Butyl carbinol (*iso*) 406
Butyl carbinol (*sec*) 411

Butyl carbinol (*tert*) 410
Butyl carbitol 2232

Table 7-4 (*Continued*)
PHYSICAL CONSTANTS OF ORGANIC COMPOUNDS

No.	Crystalline Form and Color	Specific Gravity	Melting Point °C.	Boiling Point °C.	Solubility in 100 Parts		
					Water	Alcohol	Ether
1040	col. lq.	$0.724^{20°}_{4}$	−104	66^{772mm}	∞	∞	∞
1041	col. lq.	$0.732^{20°}_{4}$	−85	68-9	∞	∞	∞
1042	col. lq.	$0.698^{18°}_{4}$	−67.5	45.2	....	∞	....
1043	wh. cr. pd.	....	58-9	$173-4^{8mm}$	0.013	s.; s. chl.	s.; s. oil
1044	yel. amor. pd./bz.	....	109-10	....	0.05	s.; s. chl.	s.; s. bz.
1045	nd./lg.		65	....	0.022	s.; s. act.	s.; s. bz.
1046	lq.	....	....	235^{720mm}	i.	v. s.	v. s.
1047	oil	$0.940^{20°}_{4}$	....	231-2	$0.011^{15°}$	v. s.	v. s.
1048	col. fl.	....	160-2	....	s.	s.	i.
1049	lq.	$0.860^{20°}$	−88.0	183.3	i.	s.	s.
1050	lq.	$0.862^{20°}$	−75.5	173.3	i.	s.	s.
1051	lq.	$0.853^{20°}$	−51.5	172.8	i.	s.	s.
1052	col. lq.	$0.867^{20°}_{4}$	−57.9	169.1	i.	s.	s.
1053	col. oil	$1.005^{25°}_{25°}$	−22	248.5-9.5	i.	s.	s.
1054	col. oil	$0.997^{25°}_{25°}$	....	241.5	i.	∞	∞
1055	col. lq.	....	....	$229-32^{15mm}$	....	....	....
1056	col. lq.	$0.931^{10°}_{4}$	....	220.5^{744mm}	i.	∞	∞
1057	lq.	$1.275^{20°}_{4}$	−112.4	101.6	$0.06^{16°}$	∞	∞
1058	lq.	$1.261^{20°}_{4}$	−112.1	91.3	i.	....	....
1059	lq.	$1.264^{20°}_{4}$	−117.4	91.4	$0.06^{18°}$	∞	∞
1060	lq.	$1.220^{20°}_{4}$	−16.2	73.3; d. 210	$0.06^{18°}$	∞	∞
1061	wh. cr. pd.	....	130-3	....	v. sl. s.	s.; s. alk.	s.
1062	col. lq.	$0.872^{20°}_{20°}$	....	165.7^{736mm}	i.	∞	∞
1063	col. lq.	$0.863^{18°}_{4}$	....	156.9	i.	∞	∞
1064	col. lq.	$0.875^{0°}_{4}$	−80.7	148.7	i.	∞	∞
1065	col. lq.	$0.862^{25°}_{4}$	−64.3	207.7	i.	....	....
1066	col. lq.	$0.858^{25°}_{4}$	−43.0	245.0	....	....	....
1067	wh. fl.	....	53-4	203-4d.	i.	s.	v. s.
1068	col. lf.	$0.956^{76°}_{4}$	65	206-7	i.	s.	s.
1069	col. lq.	$0.981^{20°}_{20°}$	−32.2	246.8	$6.5^{20°}$	....	....
1070	col. cr.	$1.049^{60°}_{25}$	56-7	285	$0.2^{80°}$	s.; v. s. act.	$240^{25°}$
1071	col. lq.	$0.903^{20°}_{4}$	....	171.2	∞	∞	∞
1072	col. lq.	$0.878^{20°}_{4}$	−123.1	78.5	$0.07^{12.5°}$	∞	∞
1073	col. lq.	$0.873^{20°}_{4}$	−131.3	68.3	i.	∞	∞
1074	col. lq.	$0.884^{15°}$	−131.2	68.9	i.	∞	∞
1075	lq.	$0.847^{15°}$	−27.1	50.7	i.	∞	∞
1076	col. lq.	$1.081^{15°}$	....	181-3	....	....	....
1077	col. lq.	$1.074^{25°}_{4}$	....	140-5	d.	d.	∞
1078	lq.	$1.045^{18.5°}_{15}$	....	128.8	slow d.	slow d.; ∞ chl.	∞ ; ∞ bz.
1079	lq.	$0.969^{30°}_{25}$	27	$137-8^{25mm}$	i.; ∞ CCl_4	∞ Me al.	∞ ; ∞ bz.
1080	col. lq.	$0.944^{0°}_{0}$	....	223	....	....	....
1081	lt. yel. lq.	....	....	175-80	....	....	....
1082	lt. yel. oil	....	....	170-2	....	....	....
1083	lq.	$0.787^{4°}$	<−60	110.5	sl. s.	s.	s.

Butyl carbonate 1892-4
Butyl chlorocarbonate 1077-8

Butyl citrate 6119
Butyl cyanide (*n*) 6405

Butyl cyanide (*iso*) 6413
Butyl cyanide (*tert*) 6246

Table 7-4 (*Continued*)

PHYSICAL CONSTANTS OF ORGANIC COMPOUNDS

No.	Name	Synonym	Formula	Beil. Ref.	Formula Weight
	Butyl				
1084	diethanolamine (*n*)	....	$C_4H_9 \cdot N(CH_2CH_2OH)_2$	IV-285	161.25
1085	fluoride (*iso*)	1-F-2-Me-propane	$(CH_3)_2CH \cdot CH_2F$	I-124	76.11
1086	formate (*n*)	....	$HCO_2 \cdot CH_2CH_2C_2H_5$	II-21	102.13
1087	formate, ortho (*n*)	triBu-orthoformate	$HC(O \cdot CH_2CH_2C_2H_5)_3$	....	232.37
1088	formate (*sec*)	....	$HCO_2CH(CH_3) \cdot C_2H_5$	**II-30	102.13
1089	formate (*iso*)	....	$HCO_2 \cdot CH_2 \cdot CH(CH_3)_2$	II-21	102.13
1090	furoate (*n*)	....	$C_4H_3O \cdot CO_2 \cdot C_4H_9$	....	168.19
1091	β-furylacrylate (*n*)	....	$C_4H_3O \cdot CH:CH \cdot CO_2 \cdot$ C_4H_9	....	194.23
1092	(*o*)-hydroxy-benzoate	*n*-Bu-salicylate	$HO \cdot C_6H_4 \cdot CO_2 \cdot C_4H_9$	....	194.23
1093	*p*-hydroxy-benzoate (*n*)	butoben	$HO \cdot C_6H_4 \cdot CO_2 \cdot C_4H_9$	....	194.23
1094	α-hydroxy-*iso*-butyrate (*n*)	....	$(CH_3)_2C(OH) \cdot CO_2 \cdot$ C_4H_9	....	160.21
1095	iodide (*n*)	1-iodo-butane	$C_2H_5 \cdot CH_2 \cdot CH_2I$	I-123	184.02
1096	iodide (*sec*)	2-iodo-butane	$C_2H_5 \cdot CHI \cdot CH_3$	I-123	184.02
1097	iodide (*iso*)	1-iodo-2-Me-propane	$(CH_3)_2CH \cdot CH_2I$	I-128	184.02
1098	iodide (*tert*)	2-iodo-2-Me-propane	$(CH_3)_3CI$	I-129	184.02
1099	lactate (*n*)	....	$CH_3CHOHCO_2 \cdot C_4H_9$	....	146.19
1100	lactate (*iso*)	....	$CH_3CHOHCO_2 \cdot C_4H_9$	**III-188	146.19
1101	levulinate (*n*)	....	$CH_3CO(CH_2)_2CO_2 \cdot$ C_4H_9	**III-207	172.23
1102	malonic acid	....	$C_4H_9 \cdot CH(CO_2H)_2$	II-673	160.17
1103	mercaptan (*n*)	butanthiol-1	$C_2H_5(CH_2)_2 \cdot SH$	I-370	90.19
1104	mercaptan (*iso*)	2-Me-propanthiol-1	$(CH_3)_2CH \cdot CH_2 \cdot SH$	I-378	90.19
1105	mercaptan (*tert*)	....	$(CH_3)_3C \cdot SH$	I-383	90.19
1106	*o*-methoxybenzoate (*n*)	....	$CH_3O \cdot C_6H_4 \cdot CO_2 \cdot$ C_4H_9	....	208.26
1107	nitrate (*iso*)	....	$C_4H_9 \cdot O \cdot NO_2$	I-377	119.12
1108	nitrite (*n*)	....	$C_4H_9 \cdot O \cdot NO$	I-369	103.12
1109	nitrite (*iso*)	....	$(CH_3)_2CHCH_2 \cdot O \cdot NO$	I-377	103.12
1110	oleate (*n*)	....	$C_{17}H_{33}CO_2 \cdot C_4H_9$	**II-439	338.58
1111	oxamate (*n*)	....	$H_2N \cdot CO \cdot CO_2 \cdot C_4H_9$	....	145.16
1112	phenoxy-ethanol	(β)(*p-tert*)	$(CH_3)_3C \cdot C_6H_4 \cdot O \cdot$ $CH_2 \cdot CH_2OH$	....	194.28
1113	phenylacetate (*n*)	....	$C_6H_5 \cdot CH_2 \cdot CO_2 \cdot C_4H_9$	....	192.26
1114	phenylacetate (*iso*)	....	$C_6H_5 \cdot CH_2 \cdot CO_2 \cdot C_4H_9$	IX-435	192.26
1115	phenyl carbinol (*n*)	....	$C_6H_5 \cdot CHOH \cdot C_4H_9$	**II-504	164.25
1116	phenyl ether (*n*)	*n*-BuO-benzene	$C_4H_9 \cdot O \cdot C_6H_5$	VI-143	150.22
1117	phenyl ketone (*n*)	valerophenone	$C_4H_9 \cdot CO \cdot C_6H_5$	VII-327	162.23
1118	phenyl ketone (*iso*)	*iso*-valerophenone	$C_4H_9 \cdot CO \cdot C_6H_5$	VII-329	162.23
1119	phenol (*m*)(*tert*)	....	$(CH_3)_3C \cdot C_6H_4 \cdot OH$	....	150.22
1120	phenol (*p*)(*tert*)	....	$(CH_3)_3C \cdot C_6H_4 \cdot OH$	VI-524	150.22
1121	*o*-phenylphenol (*tert*)	(1;2,5)	$C_6H_5 \cdot C_6H_3(OH) \cdot$ $C(CH_3)_3$	....	226.32
1122	phthalate (*n*) (mono)	....	$HO_2C \cdot C_6H_4 \cdot CO_2 \cdot$ C_4H_9	**IX-586	222.24
1123	propionate (*n*)	....	$C_2H_5 \cdot CO_2 \cdot C_4H_9$	II-241	130.19
1124	propionate (*sec*)	....	$C_2H_5 \cdot CO_2 \cdot C_4H_9$	II-241	130.19
1125	propionate (*iso*)	....	$C_2H_5 \cdot CO_2 \cdot C_4H_9$	II-241	130.19

Butyl dibromosuccinate 1896
Butyl disulfide 1897

Butyl ether 1898-9
Butyl ethylene 3655, 3657

Butyl malate 1902
Butyl maleate **1903**

Table 7-4 (*Continued*)
PHYSICAL CONSTANTS OF ORGANIC COMPOUNDS

No.	Crystalline Form and Color	Specific Gravity	Melting Point °C.	Boiling Point °C.	Solubility in 100 Parts		
					Water	Alcohol	Ether
1084	col. lq.	$0.968^{20°}_{4}$	....	273-574^{1mm}	∞	∞	∞ ; ∞ bz.
1085	gas	....	....	16±	....	....	....
1086	lq.	$0.911^{0°}$	....	106.9	v. sl. s.	∞	∞
1087	lq.	$0.869^{20°}_{4}$	....	245-7	....	....	....
1088	lq.	$0.882^{20°}_{4}$	....	97	sl. s.	∞	∞
1089	lq.	$0.885^{20°}_{4}$	−95.3	98.2	1.1$^{22°}$	∞	∞
1090	col. lq.	$1.056^{20°}_{4}$	....	118-20^{25mm}	i.	∞	∞
1091	col. lq.	$1.048^{20°}_{4}$	....	121^{5mm}	i.	s.	....
1092	col. lq.	....	5.9	259-60	....	....	....
1093	wh. pd.	....	68-9	....	0.02	s.	s.; s. chl.
1094	col. lq.	....	....	64-5^{5mm}	....	....	....
1095	lq.	$1.615^{20°}_{4}$	−103.0	130.4	i.	∞	∞
1096	lq.	$1.597^{20°}$	−104	120.0	i.	∞	∞
1097	lq.	$1.603^{20°}_{4}$	−93.5	121.0	i.	∞	∞
1098	lq.	$1.570^{18;5°}_{15}$	−34	99	i.	∞	∞
1099	col. lq.	0.968	....	75-6^{6mm}	sl. s.	∞	∞
1100	col. lq.	$0.964^{30°}_{4}$	....	60-2^{7mm}	....	....	....
1101	lq.	$0.974^{20°}_{4}$	....	237.8	....	....	....
1102	pr./aq.	....	101.5	d. 150	v. s.	v. s.	v. s.
1103	col. lq.	$0.837^{25°}_{4}$	−116	97-8	sl. s.	v. s.	v. s.
1104	lq.	$0.836^{20°}_{4}$	<−79	88	v. sl. s.	s.	s.
1105	lq.	....	....	65-7	....	....	....
1106	col. lq.	....	....	185-6^{20mm}	....	....	....
1107	lq.	$1.015^{20°}_{4}$	....	122.9	i.	∞	∞
1108	lq.	$0.911^{0°}$	....	77-9	....	∞	∞
1109	lq.	$0.870^{20°}_{20}$	....	67-8	sl. s.; d.	∞	....
1110	lq.	$0.868^{25°}$	....	227-8^{15mm}	i.	s.	s.
1111	wh. nd.	....	87-8	....	i.	s. h.	sl. s.
1112	col. lq.	$1.014^{25°}_{25}$	13	147-56^{8mm}	i.; ∞ act.	∞ Me al.	∞ bz.
1113	col. lq.	....	....	128-32^{18mm}	....	....	....
1114	col. lq.	....	....	247	....	....	....
1115	col. lq.	$0.967^{20°}_{20}$	....	128-30^{8mm}	....	....	....
1116	col. lq.	$0.930^{20°}_{4}$	....	210.3	i.	s.	s.
1117	lq.	....	....	248.5	i.	s.	s.
1118	lq.	$0.993^{17.5°}$	....	227-8^{720mm}	i.	∞	∞
1119	col. cr.	....	40.6	240	....	....	....
1120	nd./aq.	$0.908^{11.3°}_{4}$	99	236-8	s.	s.	s.
1121	col. cr.	$1.022^{25°}_{25}$	50	196-9^{25mm}	i.; v. s. bz.	v. s. Me al.; v. s. act.	v. s.; v. s. CCl$_4$
1122	col. cr.	....	73-7	....	s. h.	s.	v. s.
1123	col. lq.	$0.883^{15°}$	−89.6	146	i.	∞	∞
1124	col. lq.	$0.866^{20°}_{4}$	....	132.0-2.5	i.	∞	∞
1125	col. lq.	$0.888^{0°}_{4}$	−71	136.8	i.	∞	∞

Butyl malonate 1904
Butyl mustard oil 1133-6
Butyl oxalate 1905-6
Butyl phosphate 6121
Butyl phthalate 1908

Table 7-4 (*Continued*)
PHYSICAL CONSTANTS OF ORGANIC COMPOUNDS

No.	Name	Synonym	Formula	Beil. Ref.	Formula Weight
	Butyl				
1126	pyrrole (*n*)(*N*)	....	$C_4H_4N\cdot C_4H_9$	....	123.20
1127	ricinoleate (*n*)	....	$HO\cdot C_{17}H_{32}CO_2\cdot C_4H_9$	III-388	354.58
1128	ricinoleate (*iso*)	....	$HO\cdot C_{17}H_{32}CO_2\cdot C_4H_9$	III-388	354.58
1129	stearate (*n*)	....	$C_{17}H_{35}CO_2\cdot C_4H_9$	**II-352	340.59
1130	stearate (*iso*)	....	$C_{17}H_{35}CO_2\cdot C_4H_9$	*II-173	340.59
1131	thiocyanate (*n*)	....	$C_2H_5CH_2CH_2\cdot S\cdot CN$	**III-122	115.20
1132	thiocyanate (*iso*)	....	$(CH_3)_2CHCH_2\cdot S\cdot CN$	III-177	115.20
1133	*iso*-thiocyanate (*n*)	butyl mustard oil	$C_2H_5CH_2CH_2\cdot N:CS$	IV-158	115.20
1134	*iso*-thiocyanate	(*sec*)(*d*)	$C_4H_9\cdot N:CS$	IV-161	115.20
1135	*iso*-thiocyanate (*iso*)	*iso*-Bu mustard oil	$(CH_3)_2CHCH_2\cdot N:CS$	IV-171	115.20
1136	*iso*-thiocyanate (*tert*)	....	$(CH_3)_3C\cdot N:CS$	IV-175	115.20
1137	*p*-toluene sulfonate (*n*)	....	$CH_3\cdot C_6H_4\cdot SO_2\cdot O\cdot C_4H_9$	**XI-46	228.31
1138	urea (*n*)(*N*)	*n*-Bu carbamide	$C_4H_9\cdot NH\cdot CO\cdot NH_2$	*IV-371	116.16
1139	urea (*iso*)(*N*)	....	$C_4H_9\cdot NH\cdot CO\cdot NH_2$	IV-168	116.16
1140	urethane (*n*)(*N*)	Et *N*-*n*-Bu carbamate	$C_4H_9\cdot NH\cdot CO_2\cdot C_2H_5$	IV-158	145.20
1141	urethane (*iso*)	....	$C_4H_9\cdot NH\cdot CO_2\cdot C_2H_5$	IV-168	145.20
1142	*n*-valerate (*n*)	....	$CH_3(CH_2)_3CO_2\cdot C_4H_9$	II-301	158.24
1143	*iso*-valerate (*n*)	....	$(CH_3)_2CHCH_2CO_2\cdot C_4H_9$	**II-275	158.24
1144	*iso*-valerate (*sec*)	....	$(CH_3)_2CHCH_2CO_2\cdot C_4H_9$	II-312	158.24
1145	*iso*-valerate (*iso*)	....	$C_4H_9CO_2\cdot C_4H_9$	II-312	158.24
1146	2-xenyl-di (*p*-*tert*-butyl-phenyl) phosphate (5-*tert*)	phosphen 11	$(CH_3)_3C\cdot C_{12}H_8\cdot O\cdot PO[O\cdot C_6H_4\cdot C(CH_3)_3]_2$	....	570.72
1147	**Butylene** (*α*)	butene-1	$C_2H_5\cdot CH:CH_2$	I-203	56.11
1148	**Butylene** (*β*)(*cis*)	butene-2	$CH_3\cdot CH:CH\cdot CH_3$	I-205	56.11
1148.1	**Butylene** (*β*)(*trans*)	butene-2	$CH_3\cdot CH:CH\cdot CH_3$	I-205	56.11
1149	**Butylene** (*γ*) (*iso*)	2-methyl-propene-1	$(CH_3)_2C:CH_2$	I-207	56.11
1150	*iso*-butyracetal (*iso*)	....	$(CH_3)_2CH\cdot CH\cdot O$ — $C(CH_3)_2\cdot CH_2\cdot O$	....	144.22
1151	chlorohydrin (*β, γ*)	3-chloro-butanol-2	$CH_3CHCl\cdot CHOH\cdot CH_3$	I-373	108.57
1152	chlorohydrin (*α*) (*iso*)	1-chloro-2-Me-propanol-2	$(CH_3)_2C(OH)\cdot CH_2Cl$	I-382	108.57
1153	chlorohydrin (*β*) (*iso*)	2-chloro-2-Me-propanol-1	$(CH_3)_2CCl\cdot CH_2OH$	I-378	108.57
1154	glycol (*iso*)	2-Me-propandiol-1,2	$(CH_3)_2COH\cdot CH_2OH$	I-480	90.12
1155	oxide (*iso*)	*α, α* di Me-ethylene oxide	$(CH_3)_2C\cdot CH_2\cdot O$	XVII-11	72.11
1156	**Butyne**-1	ethyl-acetylene	$C_2H_5\cdot C:CH$	I-249	54.09
1157	**Butyne**-2	crotonylene	$CH_3\cdot C:C\cdot CH_3$	I-249	54.09
1158	**Butyrchloral**	triCl-butaldehyde	$CH_3\cdot CHCl\cdot CCl_2\cdot CHO$	I-664	175.44
1159	hydrate	butylchloral hydrate	$CH_3\cdot CHCl\cdot CCl_2\cdot CH(OH)_2$	I-664	193.46
1160	**Butyraldehyde** (*n*)	butanal	$C_2H_5\cdot CH_2\cdot CHO$	I-662	72.11
1161	oxime (*n*)	butyraldoxime	$C_2H_5\cdot CH_2\cdot CH:NOH$	I-663	87.12

Butyl salicylate 1092
Butyl sebacate 1909
Butyl succinate 1910-2

Butyl sulfate 1913
Butyl sulfide 1914-6
Butyl sulfite 1917

Butyl sulfone 1918
Butyl tartrate 1919-20
Butyl-toluene 4158-60

Table 7-4 (*Continued*)
PHYSICAL CONSTANTS OF ORGANIC COMPOUNDS

No.	Crystalline Form and Color	Specific Gravity	Melting Point °C.	Boiling Point °C.	Solubility in 100 Parts		
					Water	Alcohol	Ether
1126	yel. lq.	....	....	$91\text{-}2^{55mm}$	....	....	....
1127	lq.	$0.906^{22°}$	....	275^{13mm}	i.	....	s.
1128	lq.	$0.903^{22°}$	....	262^{9mm}	i.	s.	s.
1129	col. lq.	$0.855\text{-}8^{\frac{2.5}{5}°}$	27.5(19.5)	$220\text{-}5^{25mm}$	$0.3^{29°}$	s.	s.
1130	waxy	....	25	....	i.	....	....
1131	col. lq.	$0.956^{25°}$	....	185-6	i.	s.	s.
1132	col. lq.	....	−59	175.4	i.	∞	....
1133	lq.	$0.956^{11.2°}$	....	165^{724mm}	i.	s.	s.
1134	lq.	$0.943^{\frac{20}{4}°}$	....	159-63	i.	s.	s.
1135	lq.	$0.964^{\frac{1.4}{4}°}$	....	162	i.	s.	s.
1136	lq.	$0.919^{10°}$	10.5	140^{770mm}	i.	s.	s.
1137	lq.	$1.120^{\frac{20}{4}°}$	....	$174\text{-}5^{10mm}$	....		....
1138	col. nd./bz.	....	96	....	s.	s.	s.
1139	nd./act.	....	140.5-1.5	....	sl. s. act.	sl. s. bz.	v. sl. s.
1140	col. lq.	$0.951^{15°}$	−22	202-3	....	s.	s.
1141	col. lq.	$0.943^{\frac{20}{4}°}$	<−65	$95\text{-}6^{15mm}$	i.	s.	....
1142	lq.	$0.870^{\frac{15}{4}°}$	−93	186	v. sl. s.	∞	∞
1143	lq.	$0.862^{\frac{25}{4}°}$	....	168.8	....	....	....
1144	col. lq.	$0.848^{\frac{20}{4}°}$	....	$163\text{-}4^{752mm}$	i.	∞	∞
1145	col. lq.	$0.874^{\frac{0}{4}°}$	....	168.7	i.	∞	∞
1146	resin	$1.07^{\frac{6.0}{4}°}$	....	$300\text{-}25^{5mm}$	i.; ∞ CCl_4	∞ ; ∞ act.	∞ bz.
1147	col. gas	lq. $0.60^{20°}$	−185.4	−6.3	i.	v. s.	v. s.
1148	col. gas	lq. $0.62^{20°}$	−138.9	3.7	....	....	....
1148.1	col. gas	lq. $0.60^{20°}$	−105.6	0.88			
1149	col. gas	lq. $0.59^{20°}$	−140.4	−6.9	s. H_2SO_4	....	....
1150	lq.	$0.890^{\frac{20}{4}°}$	....	138	<1	....	....
1151	col. lq.	$1.105^{20°}$	....	136-7.5	$6.6^{20°}$	....	....
1152	col. lq.	$1.061^{\frac{20}{4}°}$	−20	127-9	sl. s.	s. HCl	....
1153	col. lq.	....	....	132-3 sl. d.	s. HCl	....	....
1154	lq.	$0.994^{\frac{20}{4}°}$	....	178.6	∞	....	....
1155	lq.	$0.805^{\frac{20}{4}°}$	....	52.4	5.8	s.	s.
1156	col. lq.	$0.65^{20°}$	−125.8	8.7	i.	s.	s.
1157	col. lq.	$0.693^{200°}$	−32.2	27.0	i.	....	....
1158	lq.	$1.396^{\frac{20}{4}°}$	....	$164\text{-}5^{750mm}$	....	....	....
1159	rhb.	$1.694^{\frac{4}{5}}$	78 sl. d.	d.	s. h.	v. s.	....
1160	col. lq.	$0.817^{\frac{20}{4}°}$	−99	75.7	4	∞	∞
1161	lq.	....	....	152^{715mm}	....	....	....

Table 7-4 (*Continued*)
PHYSICAL CONSTANTS OF ORGANIC COMPOUNDS

No.	Name	Synonym	Formula	Beil. Ref.	Formula Weight
	Butyraldehyde				
1162	trimer	para-butyraldehyde	$(C_4H_8O)_3$	*XIX-807	216.32
1163	**Butyraldehyde** (*iso*)	2-Me-propanal	$(CH_3)_2CH \cdot CHO$	I-671	72.11
1164	semicarbazone (*iso*)	....	$C_3H_7 \cdot CH:N_3CH_3O$	III-103	129.16
1165	trimer	(para)	$(C_4H_8O)_3$	XIX-390	216.32
1166	**Butyric** acid (*n*)	butanoic acid	$C_2H_5 \cdot CH_2 \cdot CO_2H$	II-264	88.11
1167	acid (*iso*)	2-Me-propanoic acid	$(CH_3)_2CH \cdot CO_2H$	II-288	88.11
1168	amide (*n*)	*n*-butyramide	$C_2H_5 \cdot CH_2 \cdot CO \cdot NH_2$	II-275	87.12
1169	amide (*iso*)	*iso*-butyramide	$(CH_3)_2CH \cdot CO \cdot NH_2$	II-293	87.12
1170	anhydride (*n*)	....	$(C_2H_5 \cdot CH_2 \cdot CO)_2O$	II-274	158.20
1171	anhydride (*iso*)	....	$[(CH_3)_2CH \cdot CO]_2O$	II-292	158.20
1172	anilide (*n*)	*n*-butyranilide	$C_3H_7CO \cdot NHC_6H_5$	XII-252	163.22
1173	chloride (*n*)	butyryl chloride	$C_2H_5 \cdot CH_2 \cdot CO \cdot Cl$	II-274	106.55
1174	chloride (*iso*)	....	$(CH_3)_2CH \cdot CO \cdot Cl$	II-293	106.55
1175	nitrile (*n*)	*n*-butyronitrile	$C_2H_5 \cdot CH_2 \cdot CN$	II-275	69.11
1176	nitrile (*iso*)	*iso*-propyl cyanide	$(CH_3)_2CH \cdot CN$	II-294	69.11
1177	**Butyroin**	....	$C_3H_7 \cdot CHOH \cdot CO \cdot C_3H_7$	I-840	144.22
1178	**Cacodyl**	....	$(CH_3)_2:As \cdot As:(CH_3)_2$	IV-615	209.98
1179	chloride	di Me-chloroarsine	$(CH_3)_2AsCl$	IV-607	140.45
1180	sulfide	....	$[(CH_3)_2As]_2S$	IV-608	242.05
1181	trichloride	....	$(CH_3)_2AsCl_3$	IV-612	211.35
1182	**Cacodylic** acid	....	$(CH_3)_2AsO \cdot OH$	IV-610	138.00
1183	oxide	....	$[(CH_3)_2As]_2O$	IV-608	225.98
1184	**Cadaverine**	pentamethylene diamine	$H_2N \cdot (CH_2)_5 \cdot NH_2$	IV-266	102.18
1185	**Cadmium** diethyl	....	$(C_2H_5)_2Cd$	IV-677	170.52
1186	dimethyl	....	$(CH_3)_2Cd$	IV-677	142.47
1187	**Calciferol**	vitamin D_2	$C_{28}H_{43}OH$	....	396.66
1188	**Camphane**	....	$C_{10}H_{18}$	V-93	138.25
1189	**Camphene** (*dl*)	....	$C_{10}H_{16}$	V-156	136.24
1190	**Campholic acid**	(*d* or *l*)	$C_{10}H_{18}O_2$	IX-34	170.25
1191	**Camphor** (*d*)	....	$C_{10}H_{16}O$	VII-101	152.24
1192	dichloride (3)(α)	chlorocamphor	$C_{10}H_{16}Cl_2$	VII-117	207.15
1193	dichloride (3)(α')	....	$C_{10}H_{16}Cl_2$	VII-117	207.15
1194	dichloride (3)(β)	....	$C_{10}H_{16}Cl_2$	VII-117	207.15
1195	oxime	....	$C_9H_{16}C:NOH$	VII-112	167.25
1196	sulfonic acid	(10 or 6) Reychler	$C_{10}H_{15}O \cdot SO_3H$	XI-315	232.30
1197	**Camphoric** acid (*dl*)	paracamphoric acid	$C_8H_{14}(CO_2H)_2$	IX-760	200.24
1198	acid (*d*)	....	$C_8H_{14}(CO_2H)_2$	IX-745	200.24
1199	acid (*iso*)(*l*)	....	$C_8H_{14}(CO_2H)_2$	IX-762	200.24
1200	anhydride (*dl*)	....	$C_{10}H_{14}O_3$	XVII-459	182.22
1201	**Camphoronic acid** (*l*)	....	$C_9H_{14}O_6$	II-837	218.21
1202	acid (*iso*)	....	$C_9H_{14}O_6$	II-835	218.21
1203	**Camphyl** amine (α)	....	$C_8H_{13} \cdot CH_2 \cdot CH_2 \cdot NH_2$	XII-40	153.27
1204	amine (β)	....	$C_8H_{13} \cdot CH_2 \cdot CH_2 \cdot NH_2$	XII-40	153.27
1205	**Capric** acid	decanoic acid	$CH_3 \cdot (CH_2)_8 \cdot CO_2H$	II-355	172.27
1206	aldehyde	decanal	$CH_3 \cdot (CH_2)_8 \cdot CH:O$	I-711	156.27
1207	nitrile	nonyl cyanide	$CH_3 \cdot (CH_2)_8 \cdot CN$	II-356	153.27
1208	**Caproic** acid (*n*)	hexanoic acid	$CH_3 \cdot (CH_2)_4 \cdot CO_2H$	II-321	116.16

Butyramide 1168-9
Butyranilide 1172
Butyrolactam 3936

Butyrophenone 5439-40
Butyrone 2786
Butyryl chloride 1173-4

C β acid 4534
Cacodyl hydride 2456
Caffeic acid 2335

Table 7-4 (*Continued*)
PHYSICAL CONSTANTS OF ORGANIC COMPOUNDS

No.	Crystalline Form and Color	Specific Gravity	Melting Point °C.	Boiling Point °C.	Solubility in 100 Parts		
					Water	Alcohol	Ether
1162	oil	0.918	<-20	$115\text{-}7^{17mm}$	....	....	....
1163	col. lq.	0.794^{20}_{4}	-65.9	$63\text{-}4^{757mm}$	112^{0}	∞ ; ∞ chl.	∞; ∞ bz.
1164	lf.	....	$125.5\text{-}6.0$	....	s. bz.	s.	i. pet.
1165	nd./al.	....	$59\text{-}60$	195 sl. d.	i.	s.	v. s.
1166	col. lq.	0.958^{20}_{4}	-5.5	164.1	∞	∞	∞
1167	col. lq.	0.949^{20}_{4}	-47	154.7	20^{20}	∞ ; ∞ chl.	∞
1168	rhb.	1.032	$115\text{-}6$	216	16.3^{15}	s.	sl. s.
1169	mn. pl.	1.013	$129\text{-}30$	$216\text{-}20$	v. s.	s.	sl. s.
1170	col. lq.	0.968^{20}_{20}	-75	199.5^{760mm}	d.	d.	∞
1171	col. lq.	0.950^{25}_{4}	-53.5	181.5^{734mm}	d.	d.	∞
1172	mn. pr.	1.134	92	189^{15mm}	i.	s.	s.
1173	col. lq.	1.028^{20}_{4}	-89	$101\text{-}2$	d.	d.	s.
1174	col. lq.	1.017^{20}_{4}	-90	92	d.	d.	s.
1175	col. lq.	0.795^{15}_{4}	-111.9	117.9	....	....	....
1176	col. lq.	0.775^{15}_{4}	-71.5	103.9	sl. s.	v. s.	v. s.
1177	lq.	$0.911^{16.7}_{4}$	....	$180\text{-}90$	....	....	....
1178	col. oil	1.447^{15}	-5	163	v. sl. s.	s.	s.
1179	col. lq.	1.505^{12}_{4}	<-45	106.5	i.	∞	i.
1180	oil	....	<-40	211	v. sl. s.	s.	s.
1181	cr./et.	....	50 d.	....	d.; s. CS_2	d. abs. al.	s. abs. et.
1182	tri.	....	200	....	82.9^{22}	19.5^{15}	i. abs. et.
1183	col. lq.	1.486^{15}	-57	150	sl. s.	s.	s.
1184	syrup	0.873^{25}_{4}	9	$178\text{-}80$	v. s.	v. s.	sl. s.
1185	col. lq.	1.653^{22}_{4}	-21	$64^{19.5mm}$	....	....	∞
1186	col. lq.	1.985^{1}_{4}	-4.5	105.5^{758mm}	d.	∞	∞
1187	cr./act.	....	$115\text{-}7$	....	i.; s. act.	s.; s. chl.	s.
1188	pr.	....	$152\text{-}3$	160^{763mm}	i.	s. h.	s.
1189	cr.	0.822^{78}_{0}	50	$159\text{-}60$	i.	s.	s.
1190	mn./80% al.	....	$106\text{-}7$	$255\text{-}60$	0.02^{19}	64.5^{15}, 80% al.	s.
1191	trig.	0.999^{8}_{0}	179.5	207.4	0.1	120^{12}	v. s.
1192	mn.	....	$93\text{-}4$	$244\text{-}7$ el. d	sl. s. h.	v. s. h.	v. s.
1193	....	....	117	....	220^{18}	....	
1194	pr./al.	....	132.5	....	v. s. bz.	s.	v. s. ac.
1195	mn./al.	1.011^{16}	$119\text{-}20$	$249\text{-}54$ sl. d.	i.	v. s.	v. s.
1196	pr./ac.	....	$195\text{-}6$	....	v. s.	sl. s. ac.	v. sl. s.
1197	mn.	1.228	$202\text{-}3$	....	0.8^{25}; 10^{100}	s.	
1198	mn.	1.186	187	....	0.6^{12}	s.	i. chl.
1199	cr./aq. al.	1.243	172	....	0.34^{20}	47.5^{20}, abs. al.	....
1200	nd./al.	....	$221\text{-}3$	270	25 chl.	1.5	4
1201	nd./aq.	....	$164\text{-}5$	$195\text{-}210^{13mm}$	12.5^{16}	75.8^{16}, abs. al.	7.4^{16}, abs. et.
1202	tri.	....	170	....	s.	s.	s.
1203	lq.	$0.874^{17.8}_{4}$	....	$194\text{-}6$	....	....	....
1204	lq.	0.870^{20}_{20}	....	$205.5\text{-}6.5$	....	....	....
1205	col. nd.	0.888^{35}_{4}	31.4	268.70	0.003^{15}	s.; s. chl.	s.; s. bz.
1206	lq.	0.828^{15}	....	$207\text{-}9^{755mm}$	....	....	....
1207	col. lq.	0.823^{15}_{4}	-14.5	$235\text{-}7$	....	....	....
1208	oily lq.	0.931^{15}_{4}	-4.0	205.4	1.10^{20}	s.	s.

Table 7-4 (Continued)
PHYSICAL CONSTANTS OF ORGANIC COMPOUNDS

No.	Name	Synonym	Formula	Beil. Ref.	Formula Weight
1209	**Caproic** acid (iso)	2-Me-pentanoic-5 acid	$(CH_3)_2CH \cdot (CH_2)_2 \cdot CO_2H$	II–327	116.16
1210	acid	3-Me-pentanoic-1 acid	$C_2H_5 \cdot CH(CH_3) \cdot CH_2 \cdot CO_2H$	II–331	116.16
1211	acid	2-Me-pentanoic-1 acid	$C_2H_5 \cdot CH_2 \cdot CH(CH_3) \cdot CO_2H$	II–326	116.16
1212	aldehyde (n)	hexanal	$CH_3 \cdot (CH_2)_4 \cdot CH{:}O$	I–688	100.16
1213	amide	caproamide	$C_5H_{11} \cdot CO \cdot NH_2$	II–324	115.18
1214	anhydride (n)	....	$(C_5H_{11}CO)_2O$	II–324	214.31
1215	anilide (n)	n-caproanilide	$C_5H_{11}CO \cdot NH \cdot C_6H_5$	XII–255	191.28
1216	chloride (n)	n-caproyl chloride	$CH_3 \cdot (CH_2)_4 \cdot CO \cdot Cl$	II–324	134.61
1217	nitrile (n)	n-amyl cyanide	$CH_3 \cdot (CH_2)_4 \cdot CN$	II–324	97.16
1218	nitrile (iso)	iso-amyl cyanide	$(CH_3)_2CH(CH_2)_2CN$	II–329	97.16
1219	**Capryl** acetate†	(sec)-octyl acetate	$CH_3CO_2 \cdot C_8H_{17}$	II–134	172.27
1220	**Caprylic** acid (n)	octanoic acid	$CH_3 \cdot (CH_2)_6 \cdot CO_2H$	II–347	144.22
1221	acid	2-Et-hexanoic acid	$CH_3(CH_2)_3CH(C_2H_5) \cdot CO_2H$	II–349	144.22
1222	aldehyde (n)	octaldehyde	$CH_3 \cdot (CH_2)_6 \cdot CH{:}O$	I–704	128.22
1223	amide (n)	caprylamide	$C_7H_{15}CO \cdot NH_2$	II–349	143.23
1224	anhydride (n)	....	$(C_7H_{15}CO)_2O$	II–348	270.42
1225	chloride (n)	octanoyl chloride	$CH_3 \cdot (CH_2)_6 \cdot CO \cdot Cl$	II–348	162.66
1226	nitrile (n)	heptyl cyanide	$CH_3 \cdot (CH_2)_6 \cdot CN$	II–349	125.22
1227	**Capsaicine**	(2;1,4)	$CH_3O \cdot C_6H_3(OH) \cdot CH_2 \cdot NH \cdot CO \cdot C_9H_{17}$	*XIII–322	305.42
1228	**Carbamic** acid	known only in derivs.	$H_2N \cdot CO \cdot OH$	III–20	61.04
1229	chloride	urea chloride	$H_2N \cdot CO \cdot Cl$	III–31	79.49
1230	**Carbasone**	p-carbamylamino-phenyl-arsonic acid	$H_2NCONH\ C_6H_4 \cdot AsO{:}(OH)_2$	XVI–880	260.08
1231	**Carbazole**	diphenyleneim-ine; dibenzopyrrole	$C_6H_4 \cdot NH \cdot C_6H_4$	XX–433	167.21
1232	**Carbitol**	diethylene glycol mono-Et ether	$HO \cdot (CH_2)_2 \cdot O \cdot (CH_2)_2 \cdot O \cdot C_2H_5$	**I–520	134.18
1233	acetate	....	$CH_3CO_2 \cdot C_6H_{13}O_2$	**II–155	176.21
1234	**Carbodiphenyl-imide**	(α)	$C_6H_5 \cdot N{:}C{:}N \cdot C_6H_5$	XII–449	194.24
1235	(β)	....	$C_6H_5 \cdot N{:}C{:}N \cdot C_6H_5$	XII–450	194.24
1236	**Carbohydrazide**	carbazide	$(H_2N \cdot NH)_2{:}CO$	III–121	90.09
1237	**Carbon** dioxide	carbonic anhydride	CO_2	III–4	44.01
1238	disulfide	....	CS_2	III–197	76.14
1239	monoxide	....	CO	I–720	28.01
1240	suboxide	dioxo-allene	$O{:}C{:}C{:}C{:}O$	I–805	68.03
1241	subsulfide	....	$S{:}C{:}C{:}C{:}S$	III–207	100.16
1242	tetrabromide	tetrabromo-methane	CBr_4	I–69	331.65
1243	tetrachloride	tetrachloro-methane	CCl_4	I–65	153.82

† See also Nos. 4959–60.

Caproamide 1213	Caprone 1743	Caprylamide 1223
Caproanilide 1215	Caprophenone (iso) 481–2	Caprylamine 4970–2
Caproic isonitrile 453	Capryl alcohol 4962	Caprylene 4984
		Caprylidene 4991

Caprylone 2265
Carbamic nitrile 1577
Carbamide 6386
Carbamides, cf. ureas

Table 7-4 (*Continued*)
PHYSICAL CONSTANTS OF ORGANIC COMPOUNDS

No.	Crystalline Form and Color	Specific Gravity	Melting Point °C.	Boiling Point °C.	Solubility in 100 Parts		
					Water	Alcohol	Ether
1209	col. oil	$0.923\frac{20}{4}°$	−33	199.5	v. sl. s.	s.	s.
1210	col. lq.	$0.9241^{9°}$	....	197–8	....	....	....
1211	col. lq.	$0.928\frac{18}{0}°$	....	193^{748mm}	$0.57^{17°}$	....	....
1212	col. lq.	$0.834\frac{20}{20}°$	....	128.6	$0.5^{20°}$	....	....
1213	pl.	0.999	100–1	....	s. h.	s.	s.
1214	col. oil	$0.928^{17°}$	−40.6	241–3 sl. d.	....	s.	....
1215	pr./al.	1.112	95	....		v. s.	v. s.
1216	col. lq.	$0.981\frac{15}{4}°$	−87.3	152.6	d.	d.; s. chl.	s.
1217	col. lq.	$0.809\frac{15}{4}°$	−79.4	164.1	i.	v. s.	v. s.
1218	lq.	$0.804\frac{20}{4}°$	−51.1	155.5	i.	∞	∞
1219	col. lq.	$0.863\frac{14}{4}°$	....	$194–5^{744mm}$	i.	s.	s.
1220	col. lf.	$0.908\frac{20}{4}°$	16.3	239.3	$0.25^{100°}$; $0.07^{15°}$	s.; s. chl.; s. pet.	s.; s. CS_2
1221	lq.	$0.908\frac{20}{20}°$	<0	226.9	$0.25^{20°}$	....	....
1222	lq.	$0.821^{20°}$	....	167–70	....	....	....
1223	col. lf.	....	105–10	>200 d.	$0.45^{100°}$	s.	s.
1224	lq.	$0.907\frac{18}{4}°$	−1	280–5	....	s.	....
1225	col. lq.	$0.953\frac{15}{4}°$	−61.0	195.6	d.	d.	s.
1226	lq.	$0.817\frac{15}{4}°$	−45.6	204.6			
1227	cr./aq. al.	....	64–5	>210, in vac.	i. c.; s. bz.	s.; s. alk.	s.; s. chl.
1228	....	....	....	....	....	....	....
1229	col. lq.	....	50±	61–2 d.	d.	d.	....
1230	col. cr.	....	174	....	i.	s. alk.	i. HCl
1231	lf. or pl.	....	246–7	354.8	i.; s. bz.; s. chl.	$0.92^{14°}$; 3.88 h.	sl. s.; s. pet.
1232	col. lq.	$0.990\frac{20}{20}°$	....	201.9	∞	v. s.	s.
1233	col. lq.	$1.011\frac{20}{20}°$	−25+	217.7	∞	∞	∞
1234	syrup	....	....	330–1	v. s. bz.	....	....
1235	cr.	....	168–70	....	v. sl. s. bz.	v. sl. s	v. sl. s.
1236	nd./aq. al.	....	154	....	i. bz.	i. chl.	i.
1237	col. gas	1.53 (A); $1.101^{-37°}$ (lq.); $1.56^{-79°}$ solid	$-56^{5.2atm.}$	subl. −78.5	$179.7^{0°}$ cc. See special table	s. a.	s. alk.
1238	col. lq.	$1.263\frac{20}{4}°$; 2.63(A)	−111.5	46.3	$0.2^{0°}$; $0.0145^{0°}$	∞	∞
1239	poison. gas	$0.814^{-\frac{195}{4}°}$; 0.968(A)	−207	−191.5	$0.0044^{0°}$; $3.5^{0°}$ cc.	s.	s. Cu_2Cl_2
1240	gas	$1.114^{0°}$	−107	776^{1mm}	d.	....	s.
1241	red lq.	....	....	in vac. 60–70*	v. s. bz.	v. s.	v. s.
1242	col. mn.	3.42	α48.4 β90.1	189.5	i.	s.	s.
1243	col. lq.	$1.594\frac{20}{4}°$	−22.96	76.8	$0.097^{0°}$; $0.08^{20°}$	∞	∞

* Partial polymerization.

Table 7-4 (*Continued*)
PHYSICAL CONSTANTS OF ORGANIC COMPOUNDS

No.	Name	Synonym	Formula	Beil. Ref.	Formula Weight
1244	**Carbon** tetrafluoride	tetrafluoro-methane	CF_4	I-59	88.00
1245	tetraiodide	tetraiodo-methane	CI_4	I-74	519.63
1246	**Carbonyl** bromide	....	$O{:}C{:}Br_2$	III-20	187.83
1247	chloride	phosgene	$O{:}C{:}Cl_2$	III-13	98.92
1248	sulfide	carbon oxysulfide	$O{:}C{:}S$	III-131	60.07
1249	**Carbothiaidine**	....	$NH_2CS{\cdot}S{\cdot}N(CHCH_3)_2$	XXVI-9	162.28
1250	**Carnine**	....	$C_7H_8O_3N_4{\cdot}H_2O$	....	214.18
1251	**Carnosine**	$N(\alpha){-}\beta{-}$alanyl-histidine	$N_2C_3H_3{\cdot}CH_2{\cdot}CH{\cdot}(CO_2H){\cdot}NHC_3H_6ON$	XXV-516	226.24
1252	**Carotene** (α)	provitamin A	$C_{40}H_{56}$	XXX-91	536.89
1253	**Carotene** (β)	provitamin A	$C_{40}H_{56}$	XXX-87	536.89
1254	**Carotene** (γ)(*dl*)	provitamin A	$C_{40}H_{56}$	XXX-92	536.89
1255	**Carvacrol**	2-Me-5-*iso*-Pr-phenol	$(CH_3)(C_3H_7)C_6H_3{\cdot}OH$	VI-527	150.22
1256	**Carvacrylamine** (2;1,4)	2-NH_2-*p*-cymene	$H_2N{\cdot}C_6H_3(CH_3){\cdot}C_3H_7$	XII-1171	149.24
1257	**Carvenone** (*dl*)	....	$C_{10}H_{16}O$	VII-78	152.24
1258	**Carvo**-menthene (*d*)	*p*-menthene-1	$C_{10}H_{18}$	V-84	138.25
1259	menthol	....	$C_{10}H_{19}OH$	VI-26	156.27
1260	menthone	....	$C_{10}H_{18}O$	VII-34	154.25
1261	**Carvone** (*d*)	carvol	$C_{10}H_{14}O$	VII-153	150.22
1262	oxime	carvoxime	$C_{10}H_{14}{:}NOH$	VII-156	165.24
1263	**Cedrene**	....	$C_{15}H_{24}$	V-461	204.36
1264	**Cedriret**	cörulignone	$[(CH_3O)_2C_6H_2{\cdot}O{\cdot}]_2$	VIII-537	304.30
1265	**Cellosolve**	2-ethoxy-ethanol-1	$C_2H_5O(CH_2)_2OH$	I-467	90.12
1266	acetate	....	$CH_3CO_2C_4H_9O$	II-141	132.16
1267	**Cellulose**	....	$(C_6H_{10}O_5)x$	....	162.14
1270	acetate, tri-	....	$C_6H_7O_2(O_2CCH_3)_3$	....	288.26
1271	**Cerotene**	....	$C_{26}H_{52}$	I-227	364.70
1272	**Cerotic acid**	....	$C_{25}H_{51}{\cdot}CO_2H$	II-394	396.70
1273	**Ceryl alcohol**	....	$C_{26}H_{53}{\cdot}OH$	I-432	382.72
1274	**Cetyl** acetate	*n*-hexadecyl acetate	$CH_3CO_2{\cdot}C_{16}H_{33}$	II-136	284.49
1275	alcohol	hexadecanol	$CH_3(CH_2)_{14}CH_2OH$	I-429	242.45
1276	bromide	hexadecyl bromide	$CH_3(CH_2)_{14}CH_2Br$	I-172	305.35
1277	iodide (*n*)	1-iodohexadecane	$CH_3(CH_2)_{14}CH_2I$	I-172	352.35
1278	palmitate	....	$C_{15}H_{31}CO_2{\cdot}C_{16}H_{33}$	II-373	480.87
1279	**Chaulmoogric acid**	....	$C_5H_7{\cdot}(CH_2)_{12}{\cdot}CO_2H$	IX-80	280.45
1280	**Chavibetol** (*iso*) (5;1,2)	propenyl guaiacol	$CH_3{\cdot}CH{:}CH{\cdot}C_6H_3{\cdot}(OH)OCH_3$	VI-956	164.21
1281	**Chloral**	trichloro-acetalde-hyde	$Cl_3C{\cdot}CH{:}O$	I-616	147.39
1282	alcoholate	....	$Cl_3C{\cdot}CH(OH){\cdot}OC_2H_5$	I-621	193.46
1283	ammonia	....	$Cl_3C{\cdot}CH(OH)NH_2$	I-624	164.42
1284	cyanohydrine	tri-Cl-lactic nitrile	$Cl_3C{\cdot}CHOH{\cdot}CN$	III-288	174.41
1285	formamide	chlor(al)amide	$Cl_3C{\cdot}CHOH{\cdot}NH{\cdot}CO{\cdot}H$	II-27	192.43

Table 7-4 (*Continued*)
PHYSICAL CONSTANTS OF ORGANIC COMPOUNDS

No.	Crystalline Form and Color	Specific Gravity	Melting Point °C.	Boiling Point °C.	Solubility in 100 Parts		
					Water	Alcohol	Ether
1244	gas	$1.96^{-184°}$	−184	−128	sl. s.	....	....
1245	red, cubic	$4.32^{20.2°}$	d.	subl. in vac. 90–100	i.; d. h.	d. h.	s.
1246	col. lq.	$2.45^{15°}$	−80	64–5	v. sl. d.	....	....
1247	poison. gas	$1.371^{20°}_{4}$	−127.9	7.6	v. sl.; sl. d.	v. s. bz.	v. s. ac.
1248	col. gas	$2.105(A)$ 1.24^{-87}	−138.2	-50.2^{760mm}	$80^{14°}$ cc.	s.	s.
1249	cr./al.	....	....	....	i.	s. h.	i.; s. a.
1250	cr.	....	d. 230–9	....	s. h.	i.	i.
1251	col. nd./al.	....	254 d.	....	$32.3^{25°}$	sl. s.	....
1252	red cr./pet.	$1.000^{20°}_{20}$	187	....	i.; v. s. chl.	sl. s.; s.bz.	s.; s. CS_2
1253	red/bz.-al.	$1.000^{20°}_{20}$	184	....	i.; v. s. chl.	sl. s.; s. bz.	s.; s. CS_2
1254	red pr.	....	177–8	....	i.; s. chl.	s. bz.	s. CS_2
1255	col. lq.	$0.977^{20°}_{4}$	0.5	237.9	v. sl. s.	∞	∞
1256	oil	$0.994^{20°}$	−16	241	v. sl. s.	s.	s.
1257	lq.	$0.926^{20°}_{4}$	....	233	i.	....	....
1258	lq.	$0.829^{20°}$	....	174–6	....	....	....
1259	oil	$0.908^{20°}_{4}$	....	220	....	....	....
1260	oil	....	....	222–3	....	....	....
1261	col. lq.	$0.961^{20°}_{4}$	....	230^{755mm}	i.	∞; 20, 60% al.	∞
1262	mn. pl./al.	$1.016^{73°}$	71–2	....	s.	s.	....
1263	col. lq.	$0.939^{15°}$	....	$262-3^{750mm}$	....	....	....
1264	b. bl. nd.	....	d.	....	i.	s. phenol	s. H_2SO_4
1265	col. lq.	$0.931^{20°}_{4}$	−70	135.1	∞	∞; ∞ chl.	∞; ∞ act.
1266	col. lq.	$0.975^{20°}_{4}$	−61.7	156.3	22	∞	∞
1267	amor.	1.3–1.4	....	....	i.; *	i.	i.
1270	fl.	....	57–8	....	i.	s. ac.	i.; s. chl.
1271	cr.	....	57–8	....	....	....	....
1272	col. cr.	$0.836^{72°}_{4}$	82.5	....	i.	v. s. h.	s. h.
1273	cr.	....	80	....	i.	s.	....
1274	nd.	$0.859^{25°}_{4}$	18.5	$200-1^{15mm}$	i.	v. sl. s. c.	....
1275	lf.	$0.815^{55°}_{4}$	49.2	189.5^{15mm}	i.	s.	s.; s. chl.
1276	col. lq.	....	16.5–8	187	....	....	....
1277	lf.	$1.123^{20°}_{4}$	22–3	211^{15mm}	i.	s.	s.
1278	lf.	$0.832^{20°}_{4}$	53–4	....	i.; v. s. bz.	$0.05^{22°}$	$21^{22°}$
1279	lf./al.	....	68	$247-8^{20mm}$	i.	v. sl. s.	s.; s. chl.
1280	cr.	....	95–6	147^{19mm}	....	....	....
1281	col. lq.	$1.505^{25°}_{4}$	−57±	97.6^{768mm}	v. s.	∞	∞
1282	nd.	$1.143^{40°}$	56–7 (50)	115–6	s.	s.	s.
1283	col. nd.	....	63–4	100±, d.	v. sl. s.	s.; s. bz.	v. s.
1284	col. cr./aq.	....	61	215–20 sl. d.	s.	s.	s.
1285	col. cr.	....	116–8	d.	5; d. h.	40; d. h.	v. s.

*Soluble in $Cu(NH_3)_4(OH)_2$.
Cerulignone 1264
Cetane 3569
Cetylene 3572
Cevadine, cf. alkd.

Cevitamic acid 586
CG (choky) gas 1247
Chalkone 627
Chaulmestrol 3014
Chavicol 211

Chelerythrine, cf. alkd.
Chelidonine, cf. alkd.
Chi acid 554
Chicago acid 344
Chinicine, cf. alkd.

Chinidine, cf. alkd.
Chinosol 3827
Chinotine, cf. alkd.
Chinovin, cf. glcde.
Chloral-amide 1285

Table 7-4 (*Continued*)
PHYSICAL CONSTANTS OF ORGANIC COMPOUNDS

No.	Name	Synonym	Formula	Beil. Ref.	Formula Weight
1286	**Chloral** hydrate	....	$Cl_3C \cdot CH(OH)_2$	I-619	165.40
1287	imide (trimer)	tri-Cl-ethylidene-imide	$(Cl_3C \cdot CH:NH)_3$	XXVI-9	439.21
1288	urethane	ural(ine)	$C_2H_2OCl_3 \cdot NHCO_2 \cdot C_2H_5$	III-24	236.48
1289	**Chloralose** (α)	glucochloral	$Cl_3C \cdot CH:O_2:C_6H_7O \cdot (OH)_3$	XXXI-151	309.53
1290	**Chloramine T**	chlorazone (1,4)	$CH_3 \cdot C_6H_4 \cdot S:O \cdot (ONa):NCl \cdot H_2O$	XI-107	245.66
1291	**Chloro-**acenaphthene	(3)	$Cl \cdot C_{10}H_5(CH_2)_2$	*V-276	188.66
1292	acetal	....	$ClCH_2 \cdot CH(OC_2H_5)_2$	I-611	152.62
1293	acetaldehyde	2-chloro-ethanal	$ClCH_2 \cdot CH:O$	I-610	78.50
1294	acetamide	....	$ClCH_2 \cdot CO \cdot NH_2$	II-199	93.51
1295	acetamide (***N***)	acetochloramide	$CH_3 \cdot CO \cdot NHCl$	II-181	93.51
1296	acetanilide (*o*)	....	$Cl \cdot C_6H_4 \cdot NH \cdot COCH_3$	XII-599	169.61
1297	acetanilide (*m*)	....	$Cl \cdot C_6H_4 \cdot NH \cdot COCH_3$	XII-604	169.61
1298	acetanilide (*p*)	....	$Cl \cdot C_6H_4 \cdot NH \cdot COCH_3$	XII-611	169.61
1299	acetic acid	....	$Cl \cdot CH_2 \cdot CO \cdot OH$	II-194	94.50
1300	acetic anhydride	....	$(Cl \cdot CH_2 \cdot CO)_2O$	II-199	170.98
1301	acetone	....	$Cl \cdot CH_2 \cdot CO \cdot CH_3$	I-653	92.53
1302	acetophenone (*p*)	....	$Cl \cdot C_6H_4 \cdot CO \cdot CH_3$	VII-281	154.60
1303	acetophenone (ω)	phenacyl chloride	$C_6H_5 \cdot CO \cdot CH_2Cl$	VII-282	154.60
1304	acetyl chloride	....	$Cl \cdot CH_2 \cdot CO \cdot Cl$	II-199	112.94
1305	acetylene	chloro-ethyne	$CH:C \cdot Cl$ (or $C:CHCl$)	I-244	60.48
1306	acrylic acid (α)	....	$CH_2:CCl \cdot CO_2H$	II-401	106.51
1307	acrylic acid (β)	....	$ClCH:CH \cdot CO_2H$	II-400	106.51
1308	2-amino-thio-phenol hydro-chloride	(4;1,2)	$Cl \cdot C_6H_3(SH) \cdot NH_2 \cdot HCl$	....	196.10
1309	aniline (*o*)	2-chloroaniline	$Cl \cdot C_6H_4 \cdot NH_2$	XII-597	127.57
1310	aniline (*m*)	3-chloroaniline	$Cl \cdot C_6H_4 \cdot NH_2$	XII-602	127.57
1311	aniline (*p*)	4-chloroaniline	$Cl \cdot C_6H_4 \cdot NH_2$	XII-607	127.57
1312	*o*-anisidine	4-Cl-2-NH$_2$-anisole	$Cl \cdot C_6H_3(OCH_3)NH_2$	XIII-383	157.60
1313	*o*-anisidine HCl	....	$C_7H_8ONCl \cdot HCl$	....	194.06
1314	anisole (*p*)	....	$Cl \cdot C_6H_4 \cdot OCH_3$	VI-186	142.59
1315	anthranilic acid	5-Cl-2-NH$_2$-benzoic	$Cl \cdot C_6H_3(NH_2)CO_2H$	XIV-365	171.58
1316	anthraquinone (1)	....	$C_6H_4(CO)_2C_6H_3Cl$	VII-787	242.66
1317	anthraquinone (2)	....	$C_6H_4(CO)_2C_6H_3Cl$	VII-787	242.66
1318	benzaldehyde (*o*)	....	$Cl \cdot C_6H_4 \cdot CH:O$	VII-233	140.57
1319	benzaldehyde (*m*)	....	$Cl \cdot C_6H_4 \cdot CH:O$	VII-234	140.57
1320	benzaldehyde (*p*)	....	$Cl \cdot C_6H_4 \cdot CH:O$	VII-235	140.57
1321	benzamide (*o*)	....	$Cl \cdot C_6H_4 \cdot CO \cdot NH_2$	IX-336	155.58
1322	benzamide (*m*)	....	$Cl \cdot C_6H_4 \cdot CO \cdot NH_2$	IX-338	155.58
1323	benzamide (*p*)	....	$Cl \cdot C_6H_4 \cdot CO \cdot NH_2$	IX-341	155.58
1324	benzene	phenyl chloride	$C_6H_5 \cdot Cl$	V-199	112.56
1325	benzene-3,5-di-sulfonic acid	(1;3,5)	$Cl \cdot C_6H_3(SO_3H)_2$	*XI-49	272.68

Chloral-antipyrine 567
Chloramide 1285
Chloranil 5746
Chloranol, a mixt. of 1386 and 4117
Chlorazene 1290

Chlorazine 1290
Chlorazone 1290
Chloretone 6145
Chlorex 1996
Chlornol 1386

Table 7-4 (*Continued*)
PHYSICAL CONSTANTS OF ORGANIC COMPOUNDS

No.	Crystalline Form and Color	Specific Gravity	Melting Point °C.	Boiling Point °C.	Solubility in 100 Parts Water	Alcohol	Ether
1286	mn. pr.	$1.619^{50}_{4}°$	51.7*	96.3^{764} d.	$474^{17°}$	v. s.	s.; s. h. CS_2
1287	rhb./bz.-al.	....	150-5** d.	....	i.; s. bz.	2; s. chl.	s.; s. oil
1288	lf./al.-et.	....	103 sl. d.	....	i.; d. h.	s.	s.
1289	col. nd./al. or et.	....	187	....	$0.9^{15°}$	$6.6^{21°}$; $0.07^{21°}$ chl.	v. s.
1290	yel. wh. pd.	....	anh. expl. 175-80	....	14c.; 50 h.	d.; i. bz.	i.; i. chl.
1291	nd./al.	$1.195^{70}_{4}°$	69-70	319^{771mm}	....	....	....
1292	lq.	$1.026^{15°}$	....	157	sl. s.	∞	∞
1293	lq.	....	....	85^{748mm}	....	....	....
1294	mn. pr.	....	120-1	224-5 d.	$10^{24°}$	$9.5^{24°}$ abs.	v. sl. s.
1295	lf.	....	110	....	v. s.	....	s.
1296	nd./aq. ac.	....	88	....	i.	s.	s. bz.
1297	nd./aq. ac.	....	72.5	....	s. CS_2	s.	s. bz.
1298	rhb.	$1.385^{22°}$	176-7	....	sl. s.	s.	v. s.
1299	col. cr.	$1.58^{20}_{20}°$	α62.8 β56.3 γ50.7 δ43.8(?)	189.4	v. s.; v. s. bz.	s.; s. chl.	s.
1300	pr./bz.	....	46	163^{116mm}	v. sl. s. lg.	sl. s. bz.	v. s.
1301	col. lq.	$1.162^{16°}$	-44.5	121	∞	∞	∞
1302	cr.	$1.188^{20°}$	20	232	i.	∞	∞
1303	rhb.	$1.324^{15°}$	54.0	245-7	i.; s. bz.	v. s.	v. s.
1304	col. lq.	$1.498^{20}_{20}°$	....	105	d.	d.	....
1305	gas	....	expl.	-30	....	....	....
1306	nd.	....	65	subl.	s.	s.	s.
1307	lf.	....	84-5	....	s.	s.	s.
1308	yel. cr.	....	208 d.	....	s.	s.	i.
1309	lq.	$1.213^{20}_{4}°$	0	210.5	i.; s. a.	....	s.
1310	lq.	$1.216^{20}_{4}°$	-10.4	230^{767mm}	i.; s. a.	....	s.
1311	rhb.	$1.427^{19°}$	70-1	230-1	s. h.	s.; s. act.	s.; s. CS_2
1312	nd.	....	83-4	....	i. pet.	s.	s.
1313	col. nd.	....	....	....	s.	sl. s.	i.
1314	lq.	....	<-18	198	i.	s.	s.; s. chl.
1315	nd./al.	....	211-2	....	i.; v. s.	$10^{25°}$; s. act. bz.	i. c.
1316	yel. nd./al.	....	162	subl.	i.; s. ac.	sl. s. h.	s. h. bz.
1317	nd./al.	....	208-9	....	i.	s. h. bz.	s. ac.
1318	nd.	$1.29^{8°}$	11	208^{748mm}	v. sl. s.	v. s.	v. s.
1319	pr.	$1.2501^{5°}$	17-8	213-4	v. sl. s.	v. s.	v. s.
1320	pr.	$1.196^{61°}$	47.8	213^{748mm}	s. h.	v. s.	v. s.
1321	rhb.	$1.34^{18°}$	142.5	....	s. h.	s.	s.
1322	nd.	....	134.5	....	s. h.	s.	s.
1323	nd./et.	....	179	....	s. h.	v. s.	v. s.
1324	col. lq.	$1.106^{20}_{4}°$	-45.2	132.0	$0.049^{20°}$	∞; ∞ bz.	∞
1325	hyg. nd.	....	d. 100	....	....	....	....

* Variable m. p., depends upon dissociation.
** Mn. pr./al., m. p. 105-6.
Chloro-acetoacetanilide 41
Chloro-acetonitrile 1399

Chloro-allylene 5343
Chloro-aminoanisole 1312
Chloro-aminotoluene 1469-70

Table 7-4 (*Continued*)
PHYSICAL CONSTANTS OF ORGANIC COMPOUNDS

No.	Name	Synonym	Formula	Beil. Ref.	Formula Weight
1326	**Chloro** benzene sulfonic acid (1,4)	....	$Cl \cdot C_6H_4 \cdot SO_3H$	XI-54	192.62
1327	benzoic acid (*o*)	....	$Cl \cdot C_6H_4 \cdot CO_2H$	IX-334	156.57
1328	benzoic acid (*m*)	....	$Cl \cdot C_6H_4 \cdot CO_2H$	IX-337	156.57
1329	benzoic acid (*p*)	chlorodracylic acid	$Cl \cdot C_6H_4 \cdot CO_2H$	IX-340	156.57
1330	benzonitrile (*p*)	....	$Cl \cdot C_6H_4 \cdot CN$	IX-341	137.57
1331	benzophenone (*p*)	....	$Cl \cdot C_6H_4 \cdot CO \cdot C_6H_5$	VII-419	216.67
1332	benzothiazole (2)	....	$C_6H_4 \cdot N : C(Cl) \cdot S$	XXVII-44	169.63
1333	benzoxazole (2)	....	$C_6H_4 \cdot N : C(Cl) \cdot O$	XXVII-43	153.57
1334	benzoyl chloride	(1,2)	$Cl \cdot C_6H_4 \cdot CO \cdot Cl$	IX-336	175.02
1335	benzoyl chloride	(1,3)	$Cl \cdot C_6H_4 \cdot CO \cdot Cl$	IX-338	175.02
1336	benzoyl chloride	(1,4)	$Cl \cdot C_6H_4 \cdot CO \cdot Cl$	IX-341	175.02
1337	benzyl bromide	(1,2)	$Cl \cdot C_6H_4 \cdot CH_2Br$	*V-155	205.49
1338	benzyl bromide	(1,4)	$Cl \cdot C_6H_4 \cdot CH_2Br$	V-307	205.49
1339	benzyl chloride	(1,4)	$Cl \cdot C_6H_4 \cdot CH_2Cl$	V-297	161.03
1340	bromobenzene	(1,3)	$Cl \cdot C_6H_4 \cdot Br$	V-209	191.46
1341	bromobenzene	(1,4)	$Cl \cdot C_6H_4 \cdot Br$	V-209	191.46
1342	bromoethane (1,1)	ethylidene chlorobromide	$CH_3 \cdot CHClBr$	I-89	143.42
1343	bromoethane (1,2)	....	$Cl \cdot CH_2 \cdot CH_2 \cdot Br$	I-89	143.42
1344	bromomethane	....	$Cl \cdot CH_2Br$	I-67	129.39
1345	buta-1,3-diene (2)	chloroprene	$CH_2 : CCl \cdot CH : CH_2$	....	88.54
1346	*tert*-butylphenol	(4;1,2)	$C_4H_9 \cdot C_6H_3(OH)Cl$	....	184.67
1347	butyric acid (*α*)(*dl*)	....	$C_2H_5 \cdot CHCl \cdot CO_2H$	II-276	122.55
1348	butyric acid (*β*)(*dl*)	....	$CH_3 \cdot CHCl \cdot CH_2 \cdot CO_2H$	II-277	122.55
1349	butyric acid (*γ*)	....	$Cl \cdot (CH_2)_3 \cdot CO_2H$	II-278	122.55
1350	*iso*-butyric acid (*α*)	....	$(CH_3)_2CCl \cdot CO_2H$	II-295	122.55
1351	butyronitrile (*γ*)	....	$Cl \cdot (CH_2)_3 \cdot CN$	II-278	103.55
1352	camphor (*α*)†	3-Cl-*d*-camphor	$Cl \cdot C_{10}H_{15}O$	VII-117	186.68
1353	coumarin (6)	....	$Cl \cdot C_6H_3 \cdot (CH)_2 \cdot CO \cdot O$	XVII-331	180.59
1354	*m*-cresol (*o*)	2-Cl-3-OH-toluene	$CH_3 \cdot C_6H_3(OH)Cl$	**VI-355	142.59
1355	*m*-cresol (*p*)	4-Cl-3-OH-toluene	$CH_3 \cdot C_6H_3(OH)Cl$	*VI-187	142.59
1356	*m*-cresol (6)	6-Cl-3-OH-toluene	$CH_3 \cdot C_6H_3(OH)Cl$	VI-381	142.59
1357	*p*-cresol (*o*)	2-Cl-4-OH-toluene	$CH_3 \cdot C_6H_3(OH)Cl$	VI-402	142.59
1358	crotonic acid (*α*)	....	$CH_3 \cdot CH : CCl \cdot CO_2H$	II-414	120.54
1359	dibromoethane	(1,1,1)	$CH_3 \cdot CClBr_2$	I-92	222.32
1360	dibromoethane	(1,1,2)	$BrCH_2 \cdot CHClBr$	I-92	222.32
1361	dibromomethane	....	$ClCHBr_2$	I-68	208.29
1362	diethyl ether (*β*)	....	$C_2H_5O \cdot CH_2 \cdot CH_2Cl$	I-337	108.57
1363	diiodomethane	....	$ClCHI_2$	I-72	302.28
1364	dinitrobenzene	(3;1,2)	$Cl \cdot C_6H_3(NO_2)_2$	*V-137	202.55
1365	dinitrobenzene (*α*)	(4;1,2)	$Cl \cdot C_6H_3(NO_2)_2$	V-262	202.55
1366	dinitrobenzene	(2;1,3)	$Cl \cdot C_6H_3(NO_2)_2$	V-263	202.55
1367	dinitrobenzene (*α*)	(4;1,3)	$Cl \cdot C_6H_3(NO_2)_2$	V-263	202.55

† See also Nos. 1192-4.
Chloro-biphenyl 1373-5
Chloro-bromopropane 6279
Chloro-bromotoluene 913-4

Chloro-butane 1072-5
Chloro-*iso*-butylene 1567, 4060
Chlorocarbonates, cf. chloroformates.
Chloro-cyanogen 1591

Table 7-4 (*Continued*)
PHYSICAL CONSTANTS OF ORGANIC COMPOUNDS

No.	Crystalline Form and Color	Specific Gravity	Melting Point °C.	Boiling Point °C.	Solubility in 100 Parts		
					Water	Alcohol	Ether
1326	nd.	....	68	$147\text{-}8^{0.1mm}$	s.	s.	i.; i. bz.
1327	mn./aq.	$1.544^{20°}$	140.2	....	$0.208^{25°}$	s.	s.
1328	pr.	....	154.3	....	$0.041^{25°}$ s.h.	s.	s.
1329	tri.	$1.541^{24°}$	239.7	subl.	$0.008^{25°}$	s.	s.
1330	nd./al.	....	93-4	223^{750mm}	i.; v. s. bz.	v. s.	v. s.
1331	nd./et.-al.	....	77-8	332^{771mm}	i.; $95^{25°}$ bz.	$325°$ Me al.	$57^{25°}$
1332	lq.	....	24±	248	....	s.	....
1333	cr.	....	7(1)	201-2 d.	i.	s. HCl	....
1334	lq.	....	-4	$229\text{-}30^{773mm}$	d.	d.	....
1335	lq.	....	....	225	d.	d.	....
1336	lq.	1.377	15-6	220-2	d.	d.	....
1337	col. lq.	....	....	$112\text{-}4^{15mm}$	....	....	....
1338	nd./al.	....	49-50	....	v. s. bz.	s.; v. s. ac.	v. s.
1339	nd.	....	29	213-4	i.	v. s. h.	v. s.
1340	col. oil	$1.630^{20°}_{4}$	-21	195-6	i.	v. s.	v. s.
1341	mn. pr.	$1.576^{71°}_{4}$	67.4	196.3	i.	s. h.	s.
1342	lq.	$1.667^{16°}$	....	82.7	....	....	....
1343	lq.	$1.689^{19°}$	-16.6	106.7	$0.69^{30°}$	....	....
1344	col. lq.	$1.991^{19°}$	<-55	68-9	i.	....	....
1345	col. lq.	$0.958^{20°}_{20°}$	....	59.4	sl. s.	∞	∞
1346	lq.	$1.112^{25°}_{25°}$	<-20*	234-51	i.; ∞ chl.	∞ Me al.	∞ ; ∞ bz.
1347	oily lq.	....	....	101.3^{15mm}	s. h.	∞	∞
1348	col. cr.	$1.186^{20°}_{4}$	16.0-16.5	$109\text{-}10^{17mm}$	....	s.	s.
1349	cr.	1.250^{10mm}	16	116^{13mm} sl. d.	....	s.	s.
1350	col. cr.	....	31	118^{50mm}	v. s.	....	s.
1351	lq.	$1.162^{10°}$	....	195-7	i.	s.	s.
1352	mn. cr.	....	93-4	244-7 sl. d.	v. sl. s. h.	s. h.; s. bz.	s.; s. chl.
1353	nd./al.	....	161-2	....	i. aq. NH_3	s. h.; v. s. CS_2	s. h.; v. s. bz.
1354	cr./pet.	....	55-6	196	sl. s.	s.	s.
1355	col. pr.	$1.215^{15°}$	46	196	i.	s.	s.
1356	col. cr./lg.	....	66	235	i.; s. bz.	s.; s. chl.	s.; s. act.
1357	col. nd.	....	55	228	sl. s.	s.	s.; s. bz.
1358	nd.	....	99.2-99.5	212	$2^{13°}$	v. s.	v. s.
1359	lq.	$2.134^{16°}$	....	$123\text{-}4^{753mm}$	....	....	....
1360	lq.	$2.268^{16°}$	....	162.5-3.0	....	....	....
1361	col. lq.	$2.445^{15°}$	-22±	120^{730} sl. d.	....	....	....
1362	lq.	$0.989^{20°}_{4}$	....	107-8	....	....	....
1363	col. lq.	$3.170°$	-4	200±, d.	....	....	....
1364	cr./et.	....	86.8	....	i.	s.	s.
1365	cr./et.	$1.480^{50°}$	α36.3 β37.1 γ38.8 δ28	315 d.	i.	v. s. h.	v. s.
1366	yel. nd./al.	....	87-8	....	i.	s.	s.
1367	rhb./et.	$α1.697^{22°}$ $β1.680^{20°}_{4}$	α53.4 β43 γ27	315 sl. d.	i.; s. bz.	s. h.	s.; s. CS_2

* Crysts. $+H_2O$, m. p. 18°.
Chloro-cyclohexane 1613
Chloro-cymene 1572

Chloro-decane 1646
Chloro-dimethylacetophenone 2524

Table 7-4 (*Continued*)
PHYSICAL CONSTANTS OF ORGANIC COMPOUNDS

No.	Name	Synonym	Formula	Beil. Ref.	Formula Weight
	Chloro				
1368	dinitrobenzene	(5;1,3)	$Cl \cdot C_6H_3(NO_2)_2$	V-264	202.55
1369	dinitrobenzene	(2;1,4)	$Cl \cdot C_6H_3(NO_2)_2$	V-264	202.55
1370	dinitrobenzoic acid	(5;2,3,1)	$Cl \cdot C_6H_2(NO_2)_2CO_2H$	IX-415	246.56
1371	dinitrobenzoic acid	(5;3,4,1)	$Cl \cdot C_6H_2(NO_2)_2CO_2H$	IX-416	246.56
1372	dinitrophenol (α)	(4;2,6,1,)	$Cl \cdot C_6H_2(NO_2)_2OH$	VI-260	218.55
1373	diphenyl (*o*)	chloro-biphenyl	$Cl \cdot C_6H_4 \cdot C_6H_5$	V-579	188.66
1374	diphenyl (*m*)	xenyl chloride	$Cl \cdot C_6H_4 \cdot C_6H_5$	V-579	188.66
1375	diphenyl (*p*)	. . . .	$Cl \cdot C_6H_4 \cdot C_6H_5$	V-579	188.66
1376	ethoxy diphenyl	(2)(β)	$C_6H_5 \cdot C_6H_4 \cdot O(CH_2)_2Cl$	. . . .	232.71
1377	ethyl acetate (β)	. . . .	$CH_3CO_2 \cdot CH_2 \cdot CH_2Cl$	II-128	122.55
1378	ethylbenzene (*o*)	. . . .	$Cl \cdot C_6H_4 \cdot C_2H_5$	. . . .	140.61
1379	ethylbenzene (*m*)	alkazene 20	$Cl \cdot C_6H_4 \cdot C_2H_5$	. . . .	140.61
1380	ethylbenzene (*p*)	. . . .	$Cl \cdot C_6H_4 \cdot C_2H_5$	V-354	140.61
1381	ethyl 2-chloro-4-*tert*-butyl-phenyl ether (β)	(4;2,1)	$(CH_3)_3C \cdot C_6H_3(Cl) \cdot O \cdot CH_2CH_2Cl$	. . . .	247.17
1382	ethyl chloroformate	β-chloroethyl chlorocarbonate	$Cl \cdot CO \cdot O \cdot CH_2CH_2Cl$	III-11	142.97
1383	ethyl *p*-toluene-sulfonate (β)	. . . .	$CH_3 \cdot C_6H_4 \cdot SO_2 \cdot OCH_2CH_2Cl$	**XI-45	234.70
1384	fumaric acid	. . . .	$CH:CCl(CO_2H)_2$	II-744	150.52
1385	furoic acid (5)	5-chloro-pyromucic	$Cl \cdot C_4H_2O \cdot CO_2H$	XVIII-282	146.53
1386	hydroquinone	(2;1,4)	$Cl \cdot C_6H_3(OH)_2$	VI-849	144.56
1387	iodobenzene (*o*)	. . . .	$Cl \cdot C_6H_4 \cdot I$	V-220	238.46
1388	iodobenzene (*m*)	. . . .	$Cl \cdot C_6H_4 \cdot I$	V-220	238.46
1389	iodobenzene (*p*)	. . . .	$Cl \cdot C_6H_4 \cdot I$	V-221	238.46
1390	iodoethane (1,1)	. . . .	$CH_3 \cdot CHClI$	I-98	190.41
1391	iodoethane (2,1)	. . . .	$ClCH_2 \cdot CH_2I$	I-98	190.41
1392	iodo-hydroxy-quinoline	(5,7,8); nioform	$Cl \cdot I \cdot OH \cdot C_6H:C_3H_3N$	XXI-98	305.50
1393	iodomethane	. . . .	$Cl \cdot CH_2 \cdot I$	I-71	176.38
1394	4-ketodihydro-quinazoline (2)	. . . .	$C_6H_4 \cdot N:C(Cl) \cdot NH \cdot CO$	. . . .	180.59
1395	lepidine (α)	2-Cl-4-Me-quinoline	$C_6H_4:C_3HN(CH_3)Cl$	XX-396	177.64
1396	maleic acid	. . . .	$CH:CCl(CO_2H)_2$	II-752	150.52
1397	malonic acid	. . . .	$ClCH(CO_2H)_2$	II-592	138.51
1398	methyl acetate	. . . .	$CH_3CO_2 \cdot CH_2Cl$	II-152	108.53
1399	methyl cyanide	chloro-acetonitrile	$ClCH_2 \cdot CN$	II-201	75.50
1400	methyl ether	. . . .	$ClCH_2 \cdot O \cdot CH_3$	I-580	80.51
1401	methyl-ethyl ketone	1-Cl-2-butanone	$ClCH_2 \cdot CO \cdot C_2H_5$	I-669	106.55
1402	naphthalene (α)	α-naphthyl chloride	$C_{10}H_7Cl$	V-541	162.62
1403	naphthalene (β)	β-naphthyl chloride	$C_{10}H_7Cl$	V-541	162.62
1404	α-naphthol (2,1)	. . . .	$Cl \cdot C_{10}H_6 \cdot OH$	VI-611	178.62
1405	α-naphthol (4,1)	. . . .	$Cl \cdot C_{10}H_6 \cdot OH$	VI-611	178.62
1406	β-naphthol (1,2)	. . . .	$Cl \cdot C_{10}H_6 \cdot OH$	VI-648	178.62
1407	nitroaniline	(4;2,1)	$Cl \cdot C_6H_3(NO_2)NH_2$	XII-729	172.57
1408	nitroaniline HCl	(4;2,1)	$C_6H_5O_2N_2Cl \cdot HCl$	. . . .	209.03
1409	nitroaniline	(4;3,1)	$Cl \cdot C_6H_3(NO_2)NH_2$	XII-731	172.57

Table 7-4 (*Continued*)
PHYSICAL CONSTANTS OF ORGANIC COMPOUNDS

No.	Crystalline Form and Color	Specific Gravity	Melting Point °C.	Boiling Point °C.	Solubility in 100 Parts		
					Water	Alcohol	Ether
1368	col. nd./al.	....	59	vol. in steam	....	v. s.	v. s.
1369	yel. cr./lg.	....	60	....	i.	s.	s.
1370	nd./aq.	....	199-200	expl.	$0.26^{11°}$	....	....
1371	pr./bz.	....	159	....	v. s. act.	s.	v. s.
1372	yel. mn./al.	$1.742^{22°}$	81-2	subl.	v. sl. s. h.	s.	s.; s. chl.
1373	cr.	....	34	267-8	i.	v. s. lg.	....
1374	cr.	....	89	284-5	....	....	....
1375	lf.	....	77.5	282	i.	s. lig.	....
1376	col. cr.	$1.142^{60°}_{25°}$	54-7	323	v. sl. s.	$12^{25°}$ Me al.	v. s. bz., act.
1377	col. lq.	$1.152^{25°}_{25°}$	<~-20	145	$3^{25°}$	∞	∞
1378	lq.	$1.055^{25°}_{25°}$	-81	179.2	i.	∞	∞
1379	col. lq.	$1.045^{25°}_{25°}$	<-70	183	i.; ∞ bz.	∞ ; ∞ act.	∞ ; ∞ CCl_4
1380	lq.	$1.044^{25°}_{25°}$	-62	184.6	i.	∞	∞
1381	col. lq.	$1.149^{25°}_{25°}$	<-20	$176\text{-}80^{18mm}$	i.; ∞ bz.	∞ ; ∞ act.	∞ ; ∞ CCl_4
1382	lq.	$1.383^{20°}_{4}$	....	152.5^{752mm}	i.; d.	v. s.	v. s.
1383	col. lq.	....	19-20	210^{21mm}	i.	....	....
1384	pl./ac.	....	191.5-2.5	....	v. s.	v. s.	v. s.
1385	lf.	....	176-7	....	$0.3^{20°}$	s.	s.
1386	mn.	....	106	263 sl. d.	v. s.	v. s.	v. s.
1387	col. oil	$1.952^{25°}_{4}$	0.7	234-5	....	....	....
1388	yel. lq.	....	....	230	....	....	....
1389	col. lf./al.	$1.886^{57°}_{4}$	56-7	227.6^{751mm}	i.	s.	....
1390	lq.	$2.054^{19°}$	....	117-9	.,......	....	....
1391	lq.	$2.134^{15.3°}_{0}$	-15.6	140.1	....	....	....
1392	yel. nd./ac.	....	177-8	....	i.; s. h. ac.	2.5	s.; s. chl.
1393	oil	$2.49^{20°}$	....	109	....	....	....
1394	nd.	....	212	....	i.; s. NaOH	s.	s. act.
1395	nd./aq. al.	....	59	296	i.	s.	s.; s. chl.
1396	col. cr.	....	114-5	....	....	v. s.	v. s.
1397	pr.	....	133	....	v. s.	v. s.	v. s.
1398	col. lq.	$1.195^{14°}$	....	$115\text{-}6^{757mm}$	d.	....	....
1399	lq.	$1.193^{20°}$	....	124	....	....	....
1400	lq.	$1.070^{20°}_{4}$	-103.5	59.5^{759mm}	d.	d. h.	....
1401	lq.	$1.081^{3°}$	....	139^{755mm}	i.	....	....
1402	col. lq.	$1.194^{20°}_{4}$	-20	259.3	i.	s.; ∞ CS_2	∞ ; ∞ bz.
1403	lf./al.	$1.266^{16°}$	56-7	$264\text{-}6^{751mm}$	i.	v. s.	v. s.
1404	nd./lg.	....	65-70	....	....	v. s.	v. s.
1405	nd./aq. al.	....	117-20	....	....	v. s.	v. s.
1406	cr./aq.	....	70-1	....	....	v. s.	v. s. bz.
1407	or. nd./aq.	....	116-7	....	v. sl. s. lg.	v. s.	v. s.
1408	or. pr.	....	....	....	sl. s.	s.	i.
1409	yel. nd./aq.	....	102-3	....	sl. s. h.	s.; s. chl.	s.

Table 7-4 (*Continued*)
PHYSICAL CONSTANTS OF ORGANIC COMPOUNDS

No.	Name	Synonym	Formula	Beil. Ref.	Formula Weight
1410	**Chloro** nitroaniline	(2;4,1)	$Cl \cdot C_6H_3(NO_2)NH_2$	XII-733	172.57
1411	nitroaniline HCl	(2;4,1)	$C_6H_5O_2N_2Cl \cdot HCl$	XII-733	209.03
1412	nitrobenzene (*o*)	nitrochlorobenzene	$Cl \cdot C_6H_4 \cdot NO_2$	V-241	157.56
1413	nitrobenzene (*m*)	nitrochlorobenzene	$Cl \cdot C_6H_4 \cdot NO_2$	V-243	157.56
1414	nitrobenzene (*p*)	nitrochlorobenzene	$Cl \cdot C_6H_4 \cdot NO_2$	V-243	157.56
1415	nitrobenzene sulfonic acid	(6;3,1)	$Cl \cdot C_6H_3(NO_2)SO_3H \cdot 2H_2O$	XI-73	273.65
1416	nitroethane (1,1)	....	$CH_3 \cdot CHClNO_2$	I-101	109.51
1417	nitroethane (1,2)	....	$ClCH_2 \cdot CH_2NO_2$	I-101	109.51
1418	nitronaphthalene	(1,4)	$Cl \cdot C_{10}H_6 \cdot NO_2$	V-555	207.62
1419	nitronaphthalene	(7,1)	$Cl \cdot C_{10}H_6 \cdot NO_2$	V-556	207.62
1420	nitropropane	(1,1)	$C_2H_5 \cdot CH(Cl)NO_2$	I-116	123.54
1421	nitropropane	(2,2)	$(CH_3)_2C(Cl)NO_2$	**I-79	123.54
1422	nitrosoethane (1,1)	....	$CH_3 \cdot CH(Cl)NO$	I-99	93.51
1423	pentabromoethane	....	$ClCBr_2 \cdot CBr_3$	I-95	459.02
1424	phenetole (*o*)	....	$Cl \cdot C_6H_4 \cdot O \cdot C_2H_5$	VI-184	156.61
1425	phenetole (*p*)	....	$Cl \cdot C_6H_4 \cdot O \cdot C_2H_5$	VI-187	156.61
1426	phenetole (*β*)	....	$C_6H_5 \cdot O \cdot CH_2CH_2Cl$	VI-142	156.61
1427	phenol (*o*)	....	$Cl \cdot C_6H_4 \cdot OH$	VI-183	128.56
1428	phenol (*m*)	....	$Cl \cdot C_6H_4 \cdot OH$	VI-185	128.56
1429	phenol (*p*)	....	$Cl \cdot C_6H_4 \cdot OH$	VI-186	128.56
1430	phenol sulfonic acid	(4;1,2)	$Cl \cdot C_6H_3(OH) \cdot SO_3H \cdot H_2O$	XI-234	226.64
1431	phenol sulfonate Na	(4;1,2)	$C_6H_4OCl(SO_3Na)$	XI-234	230.60
1432	phenoxyacetic acid	(1,2)	$Cl \cdot C_6H_4 \cdot O \cdot CH_2CO_2H$	**VI-172	186.60
1433	phenylenediamine	(2;1,4)	$Cl \cdot C_6H_3(NH_2)_2$	XIII-117	142.59
1434	phenylenediamine	(2;1,4) HCl	$C_6H_7N_2Cl \cdot 2HCl$	XIII-117	215.51
1435	phenylphenacyl bromide	(4;1',4')	$Cl \cdot C_6H_4 \cdot C_6H_4 \cdot CO \cdot CH_2Br$	....	309.60
1436	*o*-phenylphenol (2)	(2;6,1)	$C_6H_5 \cdot C_6H_3(Cl)OH$	....	204.66
1437	*o*-phenylphenol (4)	(2;4,1)	$C_6H_5 \cdot C_6H_3(Cl)OH$	....	204.66
1438	*p*-phenylphenol (2)	(4;2,1)	$C_6H_5 \cdot C_6H_3(Cl)OH$	....	204.66
1439	phenyl-4-toluene-sulfonate (2)	(4,1;1',2')	$CH_3 \cdot C_6H_4 \cdot SO_3 \cdot C_6H_4 \cdot Cl$	....	282.75
1440	phenyl-4-toluene-sulfonate (4)	(4,1;1',4')	$CH_3 \cdot C_6H_4 \cdot SO_3 \cdot C_6H_4 \cdot Cl$	....	282.75
1441	phthalic acid	(4;1,2)	$Cl \cdot C_6H_3(CO_2H)_2$	IX-816	200.58
1442	propionic acid (*α*)	(*dl*)	$CH_3 \cdot CHCl \cdot CO_2H$	II-248	108.53
1443	propionic acid (*β*)	....	$ClCH_2 \cdot CH_2 \cdot CO_2H$	II-249	108.53
1444	propionyl chloride	(*β*)	$ClCH_2 \cdot CH_2 \cdot CO \cdot Cl$	II-250	126.97
1445	propyl acetate (*β*)	....	$CH_3CO_2 \cdot CH_2 \cdot CHCl \cdot CH_3$	II-129	136.58
1446	propyl acetate (*γ*)	....	$CH_3CO_2 \cdot (CH_2)_2CH_2Cl$	*II-58	136.58
1447	*iso*-propyl acetate	(*β*)	$CH_3CO_2 \cdot CH(CH_3) \cdot CH_2Cl$	II-130	136.58
1448	propyl chloroformate (*γ*)	*γ*-chloropropyl chlorocarbonate	$Cl \cdot CO_2 \cdot (CH_2)_2CH_2Cl$	**III-10	157.00
1449	propyl *p*-toluene-sulfonate (*γ*)	....	$CH_3 \cdot C_6H_4 \cdot SO_3 \cdot (CH_2)_2CH_2Cl$	**XI-45	248.73

Table 7-4 (*Continued*)
PHYSICAL CONSTANTS OF ORGANIC COMPOUNDS

No.	Crystalline Form and Color	Specific Gravity	Melting Point °C.	Boiling Point °C.	Solubility in 100 Parts		
					Water	Alcohol	Ether
1410	yel. nd./aq.	....	107–8	....	sl. s. 50% ac.	s.; i. lg.	s.; s. CS_2
1411	pl.	....	....	....	d.	....	....
1412	mn. nd.	$1.305^{80°}_{4}$	32.5	245.5^{753mm}	i.; s. bz.	s. h.	s.
1413	yel.rhb./al.	$1.343^{50°}_{4}$	44.5(24.0)	235.6	i.; s. chl.	v. s. h.	v. s.; s. ac.
1414	mn. pr.	$1.298^{91°}$	83–4	242^{761mm}	i.	v. s. h.	v. s.
1415	tri./aq.	....	$-2H_2O$, 110	d. 165–70	v. s.	v. sl. s.	sl. s.
1416	col. lq.	$1.258^{20°}_{20}$	....	$124-5^{758mm}$	<0.4	s. alk.	....
1417	col. lq.	$1.405^{7°}$	....	173–4	i.	d. HCl	....
1418	yel. nd./al.	....	85(60–1)	....	i.	s.	s.
1419	yel. nd./al.	....	116	....	i.	s.	s.
1420	lq.	$1.209^{20°}_{20}$	....	140–3	$<0.8^{20°}$	....	....
1421	lq.	$1.193^{20°}_{20}$	....	132	$<0.5^{20°}$	s.; i. alk.	s.
1422	col. lf. or b. lq.	....	65	....	i. c. NaOH	v. s. Me al.	v. s.; v. s. pet.
1423	cr.	....	170 d.	....	....	....	....
1424	col. lq.	....	....	208	s. bz.	s.	s.
1425	lq.	....	21	212	....	s.	s.
1426	col. lq.	$1.147^{25°}_{25}$	27–8	221^{754mm}	v. sl. s.	v. s.; s. bz.	v. s.; s. act.
1427	col. lq.	$1.241^{18.2°}_{15}$	$\alpha7; \beta0; \gamma4.1$	175–6	$2.85^{20°}$	s.	s. alk.
1428	nd.	$1.268^{25°}$	32–3	214	$2.6^{20°}$	s.	s.
1429	nd.	$1.306^{20°}_{4}$	41–3	217	$2.71^{20°}$	v. s.	v. s.
1430	hyg. pl./aq.	....	75–6	....	v. s.; i. bz.	v. sl. s.; i. chl.	v. sl. s.
1431	nd./aq.	....	....	....	s.	sl. s.	....
1432	col. nd./aq.	....	146–7	....	$0.5^{80°}$	$21^{25°}$	$9^{25°}$; i. bz.
1433	nd./bz. lg.	....	64	....	s.	....	....
1434	nd.	....	....	....	....	....	....
1435	col. nd.	....	126–7	....	i.	s. h.	i.
1436	col. lq.	$1.234^{25°}_{25}$	6	317–8	i.; s. alk.	∞	∞
1437	yel. nd.	....	67–71	312	i.; s. CCl_4	$85^{25°}$	$250^{25°}$
1438	col. cr.	....	78–80	323	$0.032^{25°}$; s. al.	$>100^{25°}$ act.	$>100^{25°}$
1439	col. rhb.	....	68–70	....	....	....	....
1440	col. rhb.	....	78–80	....	i.	s. h.	v. s.
1441	nd./al.	....	150.5	$-H_2O, >150$	s.	s.	....
1442	col. lq.	$1.306^{9°}$	<−20	186	∞	∞	∞
1443	col. lf.	....	40–2.5	$203-5^{764mm}$	v. s.	v. s.	v. s.
1444	col. lq.	$1.331^{13°}$	....	$143-5^{763mm}$	i. c.; d. h.	s.; d. h.	v. s.
1445	col. lq.	$1.098^{20°}$	....	$152-3^{750mm}$	i.	s.	s.
1446	lq.	$1.250^{19°}$	....	$163-5^{747mm}$	....	....	....
1447	col. lq.	....	....	149–50	....	....	....
1448	lt. yel. lq.	$1.295^{25°}_{20}$	....	177	d.	d.	....
1449	col. lq.	$1.267^{20°}_{4}$	....	$150-1^{2.5mm}$	....	....	....

Table 7-4 (*Continued*)
PHYSICAL CONSTANTS OF ORGANIC COMPOUNDS

No.	Name	Synonym	Formula	Beil. Ref.	Formula Weight
	Chloro				
1450	pyridine (2)(α)	....	$Cl \cdot C_5H_4N$	XX-230	113.55
1451	pyridine (3)(β)	....	$Cl \cdot C_5H_4N$	XX-230	113.55
1452	pyridine (4)(γ)	....	$Cl \cdot C_5H_4N$	XX-231	113.55
1453	quinoline (2)	....	$Cl \cdot C_9H_6N$	XX-359	163.61
1454	quinoline (3)	....	$Cl \cdot C_9H_6N$	XX-359	163.61
1455	quinoline (4)	....	$Cl \cdot C_9H_6N$	XX-360	163.61
1456	quinoline (5)	....	$Cl \cdot C_9H_6N$	XX-360	163.61
1457	quinoline (6)	....	$Cl \cdot C_9H_6N$	XX-360	163.61
1458	quinoline (7)	....	$Cl \cdot C_9H_6N$	XX-361	163.61
1459	quinoline (8)	....	$Cl \cdot C_9H_6N$	XX-361	163.61
1460	resorcinol	(4;1,3)	$Cl \cdot C_6H_3(OH)_2$	**VI-818	144.56
1461	salicylaldehyde	(5;2,1)	$Cl \cdot C_6H_3(OH)CHO$	VIII-53	156.57
1462	styrene (α)	....	$C_6H_5 \cdot CCl:CH_2$	V-476	138.60
1463	styrene (ω)	....	$C_6H_5 \cdot CH:CHCl$	V-476	138.60
1464	succinic acid (*dl*)	....	$(CHCl \cdot CH_2):(CO_2H)_2$	II-619	152.54
1465	tetrabromoethane	(1,1,2,2)	$Br_2CH \cdot CBr_2Cl$	I-95	380.12
1466	toluene (*o*)	tolyl chloride	$Cl \cdot C_6H_4 \cdot CH_3$	V-290	126.59
1467	toluene (*m*)	tolyl chloride	$Cl \cdot C_6H_4 \cdot CH_3$	V-291	126.59
1468	toluene (*p*)	tolyl chloride	$Cl \cdot C_6H_4 \cdot CH_3$	V-292	126.59
1469	*o*-toluidine (*m*)	(5;1,2)	$Cl \cdot C_6H_3(CH_3)NH_2$	XII-835	141.60
1470	*p*-toluidine (*m*)	(3;1,4)	$Cl \cdot C_6H_3(CH_3)NH_2$	XII-989	141.60
1471	tribromoethane	(1,1,2)	$BrCH_2 \cdot CClBr_2$	I-94	301.22
1472	tribromomethane	....	$Cl \cdot CBr_3$	I-68	287.19
1473	vinylarsine di-chloride	Lewisite	$ClCH:CH \cdot AsCl_2$	**IV-985	207.32
1474	*o*-xylene (3)	xylyl chloride (*o*)	$Cl \cdot C_6H_3(CH_3)_2$	V-363	140.61
1475	*o*-xylene (4)	(4;1,2)	$Cl \cdot C_6H_3(CH_3)_2$	V-363	140.61
1476	*p*-xylene (2)	(2;1,4)	$Cl \cdot C_6H_3(CH_3)_2$	V-384	140.61
1477	**Chloroform**	trichloromethane	$CHCl_3$	I-61	119.38
1478	**Chlorogenin** (pseudo)	....	$C_{27}H_{44}O_4$	....	432.65
1479	**Choleic acid**	deoxycholic acid	$C_{24}H_{40}O_4$	....	392.58
1480	**Cholesterol**	cholesterin	$C_{27}H_{45}OH \cdot H_2O$	....	404.68
1481	**Cholesterol**(eppiallo)	. . ..	$C_{27}H_{45}OH$	....	386.67
1482	**Cholesterol** (*iso*)	....	$C_{27}H_{45}OH$	....	386.67
1483	**Cholesteryl** acetate	....	$CH_3CO_2 \cdot C_{27}H_{45}$	....	428.70
1484	benzoate	....	$C_6H_5CO_2 \cdot C_{27}H_{45}$	....	490.78
1485	benzoate (*iso*)	....	$C_6H_5CO_2 \cdot C_{27}H_{45}$	....	490.78
1486	propionate	....	$C_2H_5CO_2 \cdot C_{27}H_{45}$	....	442.73
1487	**Cholic acid**†	cholalic acid	$C_{23}H_{36}(OH)_3CO_2H \cdot H_2O$	....	426.60
1488	**Choline**	bilineurine	$(CH_3)_3N(OH)CH_2 \cdot CH_2OH$	IV-277	121.18
1489	chloride	....	$C_5H_{14}ON \cdot Cl$	IV-280	139.63
1490	**Chromone**	benzo-γ-pyrone	$C_6H_4 \cdot CO \cdot CH:CH \cdot O$	XVII-327	**146.15**
1491	**Chrysammic acid** (2,4,5,7;1,8)	tetranitro-chrysazine	$(NO_2)_4C_{14}H_2(OH)_2O_2$	VIII-461	**420.21**
1492	**Chrysarobine**	dihydroxy-methyl-anthranol	$C_{15}H_{12}O_3$	VIII-335	240.26
1493	**Chrysene**	....	$C_{18}H_{12}$	V-718	228.30

† See also No. 3440.
Chrysoidine 1698
Chrysoidine orange 1699
Chrysaphanol 1495
Cicutine, cf. alkd.
Cignolin 2296

Cincholepidine 4370
Cinchomeronic acid 5499, 5501
Cinchonamine, cf. alkd.
Cinchonidine, cf. alkd.
Cinchonine, cf. alkd.

Cinchophen 5234
Cinchotine, cf. alkd.
Cinchovatine, cf. alkd.
Cineole 3217
Cinnamein 798

Table 7-4 (*Continued*)
PHYSICAL CONSTANTS OF ORGANIC COMPOUNDS

No.	Crystalline Form and Color	Specific Gravity	Melting Point °C.	Boiling Point °C.	Solubility in 100 Parts		
					Water	Alcohol	Ether
1450	oil	$1.205^{15°}$	....	1667^{14mm}	s.	....	....
1451	lq.	....	....	1487^{44mm}	s.	....	....
1452	lq.	....	....	147-8	s.	....	....
1453	nd./aq. al.	....	37-8	275^{751mm}	i.; v. s. bz.	s.; v. s. lg.	v. s.
1454	hyg. lq.	....	....	255^{743mm}	....	....	....
1455	cr.	$1.251^{\frac{20°}{4}}$	34	$260\text{-}1^{744mm}$	s. HCl	v. s.	v. s.
1456	nd.	....	45	256	....	....	....
1457	nd.	....	40-1	$261\text{-}2^{740mm}$	....	....	....
1458	nd./al.	....	31-2	267-8	....	....	....
1459	yel. oil	....	<-20	288	....	v. s.	v. s.
1460	wh. amor. pd.	....	105-7	255-6	s.	s.	s.
1461	pl./al.	....	99	....	i.; s. alk.	s.	s.
1462	lq.	$1.102^{\frac{18°}{4}}$	-23	199	i.	s.	s.
1463	lq.	$1.110^{\frac{18°}{4}}$	....	199	i.	s.	s.
1464	col. cr.	1.679	153-4	....	v. s.	v. s. ac.	v. sl. s. chl.
1465	cr.	$3.366^{16°}$	32-4	$200\text{-}5^{285mm}$	v. s. chl.	v. s.	v. s.
1466	col. lq.	$1.082^{\frac{20°}{4}}$	-36.5	159.2	i.; ∞ bz.	s.; ∞ chl.	∞
1467	col. lq.	$1.072^{\frac{20°}{4}}$	-47.8	161.6	i.; ∞ bz.	s.; ∞ chl.	∞
1468	col. lq.	$1.070^{\frac{20°}{4}}$	7.5	162.4	i.; ∞ bz.	s.; ∞ chl.	∞
1469	lf./al.	....	29-30	$236\text{-}8^{730mm}$	....	....	....
1470	lq.	$1.151^{20°}$	7	222-3	i.	....	....
1471	lq.	$2.602^{16°}$	....	$165\text{-}7^{285mm}$	i.	....	....
1472	lf.	$2.711^{5°}$	55	160	i.	....	....
1473	col. lq.	$1.888^{\frac{20°}{4}}$	-13	190	i.	v. s.	s.
1474	col. lq.	....	<-20	189.5	i.	∞	∞
1475	col. lq.	$1.069^{\frac{15°}{15}}$	<-20	191.5	....	....	....
1476	col. lq.	....	....	186.8	....	....	....
1477	col. lq.	$1.489^{20°}$	-63.5	61.2	$0.82^{20°}$	∞	∞
1478	cr.	....	268-70	....	i.	sl. s.	v. sl. s.
1479	col. cr./al.	....	172	....	sl. s. c.	s.; s. alk.	s. ac.; s.act.
1480	rhb./al.	1.067	149-51 (anh.)	$subl.^{1.0mm}$	$0.26^{20°}$	$1.1^{17°}$; $1^{78°}$	18; s. bz.
1481	nd./act.	....	84	....	i.; s. bz.	s.; s. chl.	v. s.
1482	nd./et.	....	137-8	....	s. h. ac.	s. h.	v. s.
1483	col. nd.	....	113-4	....	i.	sl. s. h.	v. s.
1484	tet. cr.	....	146.6	....	....	v. sl. s. h.	s.
1485	wh. pd.	....	191-5	....	....	sl. s. h.	s.
1486	col. fl.	....	95-7	....	i.	sl. s.	s.
1487	rhb. cr./aq.	....	200-1 (anh.)	....	0.03; s. ac.	5; s. act.	1.4; s. alk.
1488	syrup	....	....	....	v. s.	v. s. abs.	i.
1489	syrup or hyg. cr.	....	247 d.	subl. in vac. 200	v. s.	v. s.	....
1490	nd./pet.	....	58-9	....	i.; s. bz.	s.	s.; s. chl.
1491	yel. mn.	....	d.	expl.	v. sl. s. h.	s.	s.
1492	yel. lf.	....	202-6	....	i. aq. Na_2CO_3	sl. s. c.; s. H_2SO_4	sl. s. c. bz.
1493	col. rhb.	....	253-4	448	i.	$0.1^{16°}$ abs.	v. sl. s.

Table 7-4 (*Continued*)
PHYSICAL CONSTANTS OF ORGANIC COMPOUNDS

No.	Name	Synonym	Formula	Beil. Ref.	Formula Weight
1494	**Chrysin**	5,7-diOH-flavone	$C_{15}H_{10}O_4$	XVIII-124	254.24
1495	**Chrysophanic acid**	2,4-diOH-2-Me-anthraquinone	$C_{14}H_5(OH)_2(CH_3)O_2$	VIII-470	254.24
1496	**Chrysoquinone**	....	$C_{10}H_6(CO)_2C_6H_4$	VII-827	258.28
1497	**Cinnamic** acid (*cis*)	allocinnamic acid	$C_6H_5{\cdot}CH{:}CH{\cdot}CO_2H$	IX-591	148.16
1498	acid (*trans*)	β-phenyl-acrylic acid	$C_6H_5{\cdot}CH{:}CH{\cdot}CO_2H$	IX-572	148.16
1499	Na salt	....	$C_6H_5{\cdot}CH{:}CH{\cdot}CO_2Na$	IX-580	170.14
1500	acid (*iso*)	....	$C_6H_5{\cdot}CH{:}CH{\cdot}CO_2H$	IX-592	148.16
1501	aldehyde	phenyl acrolein	$C_6H_5{\cdot}CH{:}CH{\cdot}CHO$	VII-348	132.16
1502	amide	cinnamide	$C_6H_5{\cdot}C_2H_2{\cdot}CO{\cdot}NH_2$	IX-587	147.18
1503	anhydride	....	$(C_6H_5CH{:}CH{\cdot}CO)_2O$	IX-586	278.31
1504	carboxylic acid (*o*)	....	$({\cdot}C_6H_4C_2H_2{\cdot})(CO_2H)_2$	IX-898	192.17
1505	chloride	....	$C_6H_5{\cdot}CH{:}CH{\cdot}CO{\cdot}Cl$	IX-587	166.61
1506	**Cinnamal acetophenone**	cinnamylidene acetophenone	$C_6H_5{\cdot}(CH{:}CH)_2{\cdot}CO{\cdot}$ C_6H_5	VII-499	234.30
1507	**Cinnamyl** alcohol	styryl carbinol	$C_6H_5{\cdot}CH{:}CH{\cdot}CH_2OH$	VI-570	134.18
1508	cinnamate	styracin	$C_8H_7CO_2{\cdot}C_9H_9$	IX-585	264.33
1509	**Cinnamoyl-*p*-hydroxyphenyl urea**	elbon	$C_6H_5{\cdot}CH{:}CH{\cdot}CO_2{\cdot}$ $C_6H_4{\cdot}NHCONH_2$	*XIII-170	282.30
1510	**Citraconic** acid (*cis*)	methyl-maleic acid	$CH_3{\cdot}C(CO_2H){:}CH{\cdot}$ CO_2H	II-768	130.10
1511	anhydride	....	$CH_3{\cdot}C{:}CH{\cdot}CO{\cdot}O{\cdot}CO$	XVII-440	112.09
1512	**Citral** (α)	geranial	$C_9H_{15}{\cdot}CHO$	I-753	152.24
1513	semicarbazone	....	$C_{10}H_{16}{:}N{\cdot}NHCONH_2$	III-109	209.29
1514	**Citramalic acid** (*dl*)	methyl-malic acid	$CH_3{\cdot}C(OH)(CO_2H){\cdot}$ $CH_2{\cdot}CO_2H$	III-444	148.12
1515	**Citric** acid	....	$HO_2C{\cdot}CH_2{\cdot}C(OH){\cdot}$ $(CO_2H){\cdot}CH_2{\cdot}CO_2H$	III-556	192.13
1516	acid (*iso*)	....	$HO_2C{\cdot}CHOH{\cdot}CH{\cdot}$ $(CO_2H){\cdot}CH_2CO_2H$	III-555	192.13
1517	amide	citramide	$C_3H_5O(CONH_2)_3$	III-569	189.17
1518	**Citronellal** (*d*)	rhodinal	$C_9H_{17}{\cdot}CHO$	I-745	154.25
1519	**Citronellol** (*d*)	2,6-dimethyl-octene-1-ol-8	$C_{10}H_{20}O$	I-451	156.27
1520	**Citronellyl acetate**	....	$C_{12}H_{22}O_2$	II-139	198.31
1521	**Citrulline** (*dl*)	α-NH$_2$-δ-carbami-dovaleric acid	$NH_2CONH(CH_2)_3{\cdot}$ $CH{\cdot}(NH_2){\cdot}CO_2H$	....	175.19
1522	Cu salt	....	$(C_6H_{12}O_3N_3)_2Cu$	....	411.90
1523	**Cofebrin** hydro-chloride	2,4-diOH-phenyl-propanolamine HCl	$(HO)_2C_6H_3{\cdot}CH(OH){\cdot}$ $CH(NH_2){\cdot}CH_3{\cdot}HCl$	....	219.67
1524	**Collidine** (α)	2-Me-4-Et-pyridine	$CH_3{\cdot}C_5H_3N{\cdot}C_2H_5$	XX-248	121.18
1525	**Collidine** (β)	4-Me-3-Et-pyridine	$CH_3{\cdot}C_5H_3N{\cdot}C_2H_5$	XX-250	121.18
1526	**Collidine** (γ)	2,4,6-triMe-pyridine	$(CH_3)_3C_5H_2N$	XX-250	121.18
1527	**Congo red**	diPh-bis-azonaph-thionic acid	$[C_{10}H_5(NH_2)(SO_3Na){\cdot}$ $N{:}N{\cdot}C_6H_4]_2$	XVI-410	696.68

Table 7-4 (*Continued*)
PHYSICAL CONSTANTS OF ORGANIC COMPOUNDS

No.	Crystalline Form and Color	Specific Gravity	Melting Point °C.	Boiling Point °C.	Solubility in 100 Parts		
					Water	Alcohol	Ether
1494	yel. pl.	....	275	subl.	i.; s. alk.	0.6 c.; 2.5 h.	sl. s.; i. bz.
1495	yel. lf./al.	....	195	subl. sl. d.	i. c.; s. bz.	s. h.; s. act.	sl. s.; sl. s. pet.
1496	or. nd.	....	235-40	subl.	i.	s. h.	v. sl. s.
1497	mn. pr.	$1.2844°$	68	125^{19mm}	....	....	s. pet.
1498	mn. pr.	1.245	133	300	$0.04^{18°}$	$24^{20°}$ abs.	v. s.
1499	wh. cr. pd.	....	d. >115	....	9 c.; 5 h.	0.6	s. gly.
1500	mn. pr.	....	58-9(42)	256 d.	16.6 c. pet.	s.	s.
1501	lq.	$1.110^{20°}_{20}$	−7.5	252±, sl. d.	v. sl. s.	4, 50% al.	∞
1502	nd./bz.	....	145-6	....	sl. s. h.	s.	s.
1503	nd./al.	....	135-6	....	i.	v. sl. s. c.	v. s. h. bz.
1504	nd./et.	....	173-5 d.	$-H_2O$, d.	v. sl. s.	s.; i. bz.	v. sl. s.
1505	cr.	....	35-6	251-3 sl. d.	d.	s. CCl_4	s. pet.
1506	yel. nd./al.	....	$α102$; $β235$	....	s. H_2SO_4	sl. s.	sl. s.
1507	nd.	$1.040^{35°}_{35}$	33	257.5	sl. s.	v. s.	v. s.
1508	nd. or pr.	$1.085^{16.5°}$	44	....	i.	4 c.; 33 h.	33; s. bz.
1509	wh. cr. pd.	....	204	....	i.; i. alk.	s.; s. act.	s. oil
1510	nd.	1.617	92-3	....	$360^{25°}$	s.; v. sl. s. chl.	s.; i. bz.; i. CS_2
1511	lq.	$1.25^{15°}_{4}$	7-8	213-4	d.	v. s.	v. s.
1512	col. oil	$0.890^{17°}_{4}$	....	228-9 sl. d.	i.	∞	∞
1513	cr./ac.	....	$α164$ $β171*$	....	....	....	....
1514	mn. pr.	....	117-9	....	s.; i. bz.	v. s. abs.	v. s. act.
1515	cr.**	$1.542^{20°}_{4}$	153	d.	133 c.	$76^{15°}$, abs.	$2.2^{15°}$, abs.
1516	cr.	....	d. 100	....	s.	....	....
1517	cr./aq.	....	210-5 d.	....	$2.7^{18°}$; $33.3^{100°}$	i.	i.
1518	col. oil	$0.855^{17.5°}$	....	204-8	v. sl. s.	∞	∞
1519	col. oil	$0.848^{20°}_{4}$	....	224-5	v. sl. s.	∞	∞
1520	col. lq.	$0.893^{17.5°}$	....	$172-3^{34mm}$	....	....	....
1521	col. pr./aq. Me al.	....	202-26±	....	s.	i. abs.	i.
1522	cr./aq.	....	257-8 d.	....	i. c.	....	....
1523	wh. cr. pd.	....	178-9 d.	....	66	7	i.
1524	lq.	$0.935^{0°}$	....	177.8^{758mm}	sl. s. c.; v. sl. s. h.	s.	s.; s. bz.
1525	lq.	$0.966^{0°}$	....	$195-6^{753mm}$	i.	s.	v. s.
1526	lq.	$0.917^{15°}$	....	171-2	sl. s. h.	v. s.	....
1527	red	....	....	....	sl. s. h.	sl. s. h. alk.	i.

* Mixture of $α + β$, m. p. 135°.
** Orysts./aq. $+1H_2O$; $-H_2O$, 100-32°.
Coniferin, cf. glcde.
Coniine, cf. alkd.
Conquinine, cf. alkd.
Convallamarin, cf. glcde.
Convallarin, cf. glcde.
Convolvulin, cf. glcde.

Conydrine, cf. alkd.
Conyrine 5443
Coprosterol 2276
Corallin(e) 593
Cordianine 181
Cordol 6110
Coriamyrtin, cf. glcde.
Coriandrol 3974

Corn sugar 3618
Cornutine, cf. alkd.
Corporin 5334
Corulignone 1264
Corybulbine, cf. alkd.
Corycavine, cf. alkd.
Corydaline, cf. alkd.
Corynine, cf. alkd.

Table 7-4 (*Continued*)
PHYSICAL CONSTANTS OF ORGANIC COMPOUNDS

No.	Name	Synonym	Formula	Beil. Ref.	Formula Weight
1528	**Coniferin**	....	$C_{16}H_{22}O_8 \cdot 2H_2O$	....	378.38
1529	**Coniferyl alcohol**	....	$C_{10}H_{12}O_3$	VI-1131	180.21
1530	**Conylene**	octadiene-x,x	C_8H_{14}	I-258	110.20
1531	**Coumaric** acid (*o*) (*trans*)	*o*-hydroxy-cin-namic acid	$HOC_6H_4 \cdot C_2H_2 \cdot CO_2H$	X-288	164.16
1532	acid (*m*)	*m*-hydroxy-cinnamic	$HOC_6H_4 \cdot C_2H_2 \cdot CO_2H$	X-294	164.16
1533	acid (*p*)	*p*-hydroxy-cinnamic	$HOC_6H_4 \cdot C_2H_2 \cdot CO_2H$	X-297	164.16
1534	**Coumarin**	....	$C_9H_6O_2$	XVII-328	146.15
1535	**Coumarin** (*iso*)	....	$C_6H_4 \cdot CH{:}CH \cdot O \cdot CO$ (ring)	XVII-333	146.15
1536	3-carboxylic acid	....	$C_6H_4 \cdot CH{:}C(CO_2H) \cdot$ CO·O (ring)	XVIII-429	190.16
1537	**Coumarinic acid** (*o*)(*cis*)	*o*-hydroxy-cin-namic acid	$HO \cdot C_6H_4 \cdot CH{:}CH \cdot CO_2H$	X-291	164.16
1538	**Coumarone**	....	$C_6H_4 \cdot CH{:}CH \cdot O$ (ring)	XVII-54	118.14
1539	**Creatine**	methyl-guanyl-glycine	$HN{:}C(NH_2)N(CH_3) \cdot CH_2 \cdot CO_2H \cdot H_2O$	IV-363	149.15
1540	**Creatinine**	methyl-glyco-cyamidine	$HN{:}C.N(CH_3) \cdot CH_2 \cdot$ CO·NH (ring)	XXIV-245	113.12
1541	picrate	....	$C_4H_7ON_3 \cdot C_6H_3O_7N_3$	XXIV-246	342.23
1542	picrate	....	$C_4H_7ON_3 \cdot 2C_6H_3O_7N_3$	XXIV-246	571.33
1543	**Creosol**	(3;1,4)	$CH_3O \cdot C_6H_3(CH_3)OH$	VI-878	138.17
1544	**Cresidine**	2-NH_2-4Me-anisole	$CH_3 \cdot C_6H_3(OCH_3)NH_2$	XIII-602	137.18
1545	**Cresol** (*o*)	*o*-methyl-phenol	$CH_3 \cdot C_6H_4 \cdot OH$	VI-349	108.14
1546	**Cresol** (*m*)	*m*-methyl-phenol	$CH_3 \cdot C_6H_4 \cdot OH$	VI-373	108.14
1547	**Cresol** (*p*)	*p*-methyl-phenol	$CH_3 \cdot C_6H_4 \cdot OH$	VI-389	108.14
1548	phthalein (*o*)	....	$C_6H_4 \cdot CO \cdot O \cdot C{:}$ (ring) $(C_6H_3 \cdot CH_3 \cdot OH)_2$	XVIII-153	346.39
1549	phthalin (*o*)	(4′,4″-diOH, 2-CO_2H, 3′,3″-diMe)	$HO_2C \cdot C_6H_4 \cdot CH{:}$ $(C_6H_3 \cdot CH_3 \cdot OH)_2$	X-456	348.40
1550	sulfonphthalein (*o*)	cresol red	$C_6H_4 \cdot SO_2 \cdot O \cdot C{:}$ (ring) $[C_6H_3 \cdot CH_3 \cdot OH]_2$	XIX-91	382.44
1551	sulfonphthalein (*m*)	metacresol purple	$C_{21}H_{18}O_5S$	....	382.44
1552	**Cresyl**† acetate (*o*)	*o*-tolyl acetate	$CH_3 \cdot C_6H_4 \cdot O_2CCH_3$	VI-355	150.18
1553	acetate (*m*)	cresatin	$CH_3 \cdot C_6H_4 \cdot O_2CCH_3$	VI-379	150.18
1554	*p*-toluenesul-fonate (*o*)	(4;1′,2′)	$CH_3 \cdot C_6H_4 \cdot SO_3C_6H_4 \cdot CH_3$	XI-100	262.33
1555	**Crocetin** (*α*)	gardenin	$C_{20}H_{24}O_4$	XXX-106	328.41
1556	**Croconic acid**	crocic acid	$C_5O_3(OH)_2 \cdot 3H_2O$	VIII-488	196.11
1557	**Crotonic** acid (*α*)	butenoic acid	$CH_3 \cdot CH{:}CH \cdot CO_2H$	II-408	86.09
1558	acid (*β*) (*cis*)	*iso*-crotonic acid	$CH_3 \cdot CH{:}CH \cdot CO_2H$	II-412	86.09
1559	aldehyde (*α*)	2-butene-1-al	$CH_3 \cdot CH{:}CH \cdot CHO$	I-728	70.09
1560	amide	crotonamide	$CH_3 \cdot CH{:}CH \cdot CONH_2$	II-412	85.11
1561	anhydride	....	$(CH_3 \cdot CH{:}CH \cdot CO)_2O$	II-411	154.17

† See also under tolyl.
Corytuberine, cf. alkd.
Cotarnine, cf. alkd.
Cotoin 2347
Cotton red-4B 733

Cresatin 1553
Cresol purple 1551
Cresol red 1550
Cresorcinol 2371
Cresot(in)ic acid 3836, 3839, 3841

Cresyl-, cf. also tolyl.
Cresyl benzoate 6013-5
Cresyl ethyl ether 3173-5
Cresyl ethyl sulfide 3176
Cresyl methyl ether 4399-4401

Table 7-4 (*Continued*)
PHYSICAL CONSTANTS OF ORGANIC COMPOUNDS

No.	Crystalline Form and Color	Specific Gravity	Melting Point °C.	Boiling Point °C.	Solubility in 100 Parts		
					Water	Alcohol	Ether
1528	nd.	....	185 (anh.)	d.	0.5 c.; v. s. h.	sl. s. abs.	i.
1529	pr.	....	73–4	....	sl. s. h.	s.	s.
1530	lq.	$0.761^{15°}$	....	126^{738mm}	....	s.	....
1531	nd./aq.	....	207–8	subl.	sl. s. c.	s.	v. sl. s.
1532	pr./aq.	....	191	....	s. h.	s.	s.; s. bz.
1533	cr./aq.	....	206–7 d.	....	s. h.	v. s. h.	v. s.; i. lg.
1534	rhb./et.	$0.935\frac{20}{4}°$	70	290–1	0.3 c.; 2 h.	v. s.	s.
1535	pl./bz.	....	47	$285-6^{719mm}$ sl. d.	i.; s. CS_2	v. s.; v. s. bz.	v. s.
1536	col. nd./ aq.	....	189–90 d.	>290, $-CO^2$	v. sl. s.; $0.7^{25°}$ bz.	$0.9^{25°}$; $0.1^{25°}$ CCl_4	v. sl. s.; v.sl.s.act.
1537	unstable	changes to coumarin	....	....	....	....	....
1538	oil	$1.078\frac{15°}{15}$	<−18	173–4	i.	i. alk.	s.
1539	mn./aq.	....	295 (anh.)	$-H_2O$, 100	$1.4^{18°}$	$0.01^{17°}$	i.
1540	mn.	....	260 d.	....	$8.7^{16°}$	$1^{16°}$ abs.	s. h. al.
1541	yel. nd./aq.	....	220	....	....	....	....
1542	yel. pl./al.	....	161–6	....	....	....	....
1543	pr.	$1.092\frac{20}{20}°$	5.5	$221-2^{765mm}$	v. sl. s.	∞ ; ∞ chl.	∞ ; ∞ bz.
1544	nd./pet.	....	93–4	235	v. sl. s.	s.; s. bz.	s.
1545	cr.	$1.027^{41°}$	30.9	191.0	2.5	$∞^{30°}$	$∞^{30°}$
1546	lq.	$1.034\frac{20}{4}°$	11.9	202.7	0.5	∞	∞
1547	pr.	$1.018^{41°}$	34.8	201.9	1.8	$∞^{36°}$	$∞^{36°}$
1548	red yel. cr./al.	....	223–5	....	sl. s. h.; s. al. KOH	s.; s. al. NH_3	s.; v. sl. s. bz.
1549	nd.	....	217–8	....	....	....	....
1550	red cr./ac.	....	....	....	sl. s.	s.	s. alk.
1551	gn. fl.	....	....	....	sl. s.	sl. s.	i.
1552	col. lq.	....	....	208	v. sl. s.	v. s.	v. s.
1553	col. lq.	....	....	212	s. h.	v. s.	v. s.
1554	nd.	....	....	53–4	....	....	....
1555	red rhb.	....	283–5 d.	....	s. Ac_2O	s. dil. alk.	i. 10% alk.
1556	yel. lf.	....	$-3H_2O$, 100	....	s.	s.	....
1557	col. mn.	$0.964^{79.7°}$	72	189	$8.3^{15°}$	s. h. lg.	....
1558	nd.	$1.031\frac{15°}{4}$	15.5	170–1 d.	$∞^{25°}$	s.	....
1559	col. lq.	$0.853\frac{20}{20}°$	−69	102.2	18	∞	∞
1560	nd./act.	....	159–60	....	2.8 c.	v. s. 0.08 c. bz.	0.09 c.
1561	col. lq.	$1.040^{20°}$	....	$246-8^{766mm}$	....	....	....

Cresyl methyl sulfide 4402 Cresyl salicylate 6038–40 Crocein(e) acid 4512
Cresyl phosphate 6351–3 Cresyl thiophosphate 6356–8 Crocic acid 1556
Cresyl phosphite 6354–5 Cresylic acid, mixt. of 1545–7 Crocin, cf. glcde.
Cresyl phthalate 2841 Croceic acid 4512 Crotonaldehyde 1559

Table 7-4 (*Continued*)
PHYSICAL CONSTANTS OF ORGANIC COMPOUNDS

No.	Name	Synonym	Formula	Beil. Ref.	Formula Weight
1562	**Crotonic** chloride	....	$CH_3 \cdot CH:CH \cdot CO \cdot Cl$	II-411	104.54
1563	nitrile (**trans**)	propenyl cyanide	$CH_3 \cdot CH:CH \cdot CN$	II-412	67.09
1564	**Crotonyl** acetate	crotyl acetate	$CH_3CO_2 \cdot C_4H_7$	II-137	114.15
1565	alcohol	2-butene-1-ol	$CH_3 \cdot CH:CH \cdot CH_2OH$	I-442	72.11
1566	chloride	crotyl chloride	$CH_3 \cdot CH:CH \cdot CH_2Cl$	I-205	90.55
1567	chloride (*iso*)	*iso*-crotyl chloride	$(CH_3)_2C:CH \cdot Cl$	I-209	90.55
1568	iodide	1-iodo-butene-2	$CH_3 \cdot CH:CH \cdot CH_2I$	I-206	182.00
1569	**Cryogenine**	3-semicarbazido-benzamide	$H_2N \cdot CO \cdot NH \cdot NH \cdot C_6H_4 \cdot CO \cdot NH_2$	XV-629	194.19
1570	**Cryptoxanthin**	provitamin A	$C_{40}H_{56}O$	XXX-93	552.89
1571	**Cuminyl** alcohol	cumin alcohol (*p*)	$(CH_3)_2CH \cdot C_6H_4 \cdot CH_2OH$	VI-543	150.22
1572	chloride	1'-chlorocymene	$C_3H_7 \cdot C_6H_4 \cdot CH_2Cl$	V-423	168.67
1573	**Cumyl** phenol (*p*)	(1,4)	$HO \cdot C_6H_4 \cdot C(CH_3)_2 \cdot C_6H_5$	....	212.29
1574	**Cupferon**	....	$C_6H_5 \cdot N(NO) \cdot ONH_4$	XVI-669	155.16
1575	**Curcumin**	....	$C_{21}H_{20}O_6$	VIII-554	368.39
1576	**Cyamelide**	....	$(CNOH)_x$ or $(CNOH)_3$	III-35	$(43.03)_x$
1577	**Cyanamide**	carbamic nitrile	$H_2N \cdot CN$	III-75	42.04
1578	**Cyananilide**	phenyl cyanamide	$C_6H_5 \cdot NH \cdot CN \cdot \frac{1}{2}H_2O$	XII-368	127.15
1579	**Cyanic acid**	....	$HOCN\dagger$	III-33	43.03
1580	**Cyanine**	1,1'-di-*iso*-amyl-4,4'-quinocyanine iodide	$C_{29}H_{35}N_2I$	XXIII-299	538.52
1581	**Cyano**-acetamide	....	$NC \cdot CH_2 \cdot CO \cdot NH_2$	II-589	84.08
1582	acetanilide	....	$C_6H_5 \cdot NH \cdot CO \cdot CH_2CN$	XII-294	160.18
1583	acetic acid	nitrilomalonic acid	$HO_2C \cdot CH_2 \cdot CN$	II-583	85.06
1584	acetoacetic ester	....	$CH_3CO \cdot CH(CN) \cdot CO_2 \cdot C_2H_5$	III-796	155.15
1585	benzoic acid (*p*)	....	$HO_2C \cdot C_6H_4 \cdot CN$	IX-845	147.13
1586	benzyl chloride (*o*)	....	$ClCH_2 \cdot C_6H_4 \cdot CN$	IX-468	151.60
1587	propionic acid (*α*)	....	$CH_3 \cdot CH(CN) \cdot CO_2H$	II-630	99.09
1588	**Cyanogen**	oxalic nitrile	$NC \cdot CN$	II-549	52.04
1589	**Cyanogen** (para)	....	$(CN)_x$	II-553	$(26.02)_x$
1590	bromide	....	$Br \cdot CN$	III-39	105.93
1591	chloride	....	$Cl \cdot CN$	III-38	61.47
1592	iodide	....	$I \cdot CN$	III-41	152.92
1593	sulfide	....	$NC \cdot S \cdot CN$	III-180	84.10
1594	**Cyanuric acid**	....	$(HNCO)_3 \cdot 2H_2O$	XXVI-239	165.11
1595	**Cyclo**-butane	tetramethylene	$CH_2 \cdot CH_2 \cdot CH_2 \cdot CH_2$	V-17	56.11
1596	heptane	suberane	$CH_2 \cdot (CH_2)_5 \cdot CH_2$	V-29	98.19
1597	heptanone	suberone	$CH_2 \cdot (CH_2)_5 \cdot CO$	VII-13	112.17

$\dagger$ HOCN in aq.; NH:C:O (isocyanic acid) as vapor or in ether soln.

Table 7-4 (*Continued*)
PHYSICAL CONSTANTS OF ORGANIC COMPOUNDS

No.	Crystalline Form and Color	Specific Gravity	Melting Point °C.	Boiling Point °C.	Solubility in 100 Parts		
					Water	Alcohol	Ether
1562	col. lq.	$1.091^{20°}$	....	$124\text{-}5^{759mm}$	....	....	....
1563	col. lq.	$0.826^{15°}_{4}$	-51.5	122	d.	d.	....
1564	col. lq.	$0.919^{20°}_{4}$	....	$128.5\text{-}9.5$	2.3	s.	s.
1565	col. lq.	$0.853^{20°}_{4}$	$<\!-30$	$121\text{-}2$	16.6	∞	∞
1566	col. lq.	$0.934^{20°}_{4}$	....	84	i.	∞	∞
1567	lq.	$0.919^{20°}_{4}$	....	68	<0.1; d. h.	....	....
1568	lq.	$1.682^{0°}$	....	$132\text{-}3$ d.	....	....	
1569	col. cr.	....	$172\pm$	....	1	s.; s. act.	s.; s. chl.
1570	pr./bz.-al.	....	$168\text{-}9$	....	i.	s. CS_2	s.; s. chl.
1571	lq.	$0.975^{25°}_{25}$	....	$247\text{-}8$	i.	∞	∞
1572	lq.	$1.020^{22°}_{4}$	....	$225\text{-}9$	....	....	....
1573	col. nd.	....	$74\text{-}5$	187^{10mm}	i.; s. bz.	v. s.	s.; s. act.
1574	col. nd./al.	....	$163\text{-}4$	subl.	v. s.	s. h.	....
1575	red yel. pr.	....	183	....	i.; 0.05 bz.; s. alk.	sl. s. c.; s. ac.	v. sl. s.; i. lg.
1576	wh. amor.	....	....	....	$0.011^{15°}$	i.; sl. s. NH_4OH	i.
1577	col. nd.	$1.073^{48°}_{4}$	$44\text{-}5$	140^{19mm}	v. s.	v. s.	v. s.
1578	cr.	....	47	....	sl. s.	v. s.	v. s.
1579	col. gas	$1.140^{0°}$ (lq.)	-80	-64^{0mm}	sl. s.	s. ac.	s.
1580	hyg., mn., gn.	....	$100\pm$	d. >150	i.; s. chl.	s. h.	i.; s. act.
1581	nd./al.	....	$118\text{-}9$	d.	15 c.	2 c.	....
1582	cr./al.	....	$199\text{-}200$	....	$0.03^{25°}$	....	....
1583	col. hyg. cr.	....	$65\text{-}6$	$108^{0.15mm}$ d. 165	s.	s.	s.
1584	col. nd.	$1.111^{20°}_{4}$ (lq.)	26	$195\text{-}7$	v. sl. s.	v. s.	v. s.
1585	lf./al.	....	219	....	s. h.	s.	s.; s. h. ac.
1586	mn. pr.	....	$60\text{-}1.5$	252^{759mm}	s. h.	s.	....
1587	oil	$1.14^{20°}$	....	$142\text{-}5^{11mm}$	s.	s.	....
1588	col. gas	$0.866^{17.2°}$ 1.804 (A)	-27	-21.2	$450^{20°}$ cc.	$2300^{20°}$ cc.	$500^{20°}$ cc.
1589	brn. bl.	....	$(CN)_2$ at 860	....	i.	i.	sl. s. alk.
1590	nd.	$2.015^{20°}_{4}$	52	61.3^{750mm}	s.	s.	s.
1591	poison. gas	$1.222^{0°}$	-6.9	13.0	$2500^{20°}$ cc.	$10,000^{20°}$ cc.	$5000^{20°}$ cc.
1592	col. nd.	....	$146.5*$	subl. $>$m. p.	s. h.	v. s.	v. s.
1593	pl. or lf.	....	60	d.	v. s. d.	s.	s.
1594	mn./aq.	2.501^{90} anh.	>360	d. to HNCO	$0.27^{17°}$	$0.12^{20°}$	....
1595	col. gas	$0.703^{0°}_{4}$	-50	$11\text{-}12^{726mm}$	i.	v. s.	v. s. act.
1596	oil	$0.810^{20°}_{4}$	-12	$118\text{-}20$	i.	....	....
1597	lq.	$0.950^{21°}$	....	$179\text{-}81$	i.	s.	s.

* Sealed tube.
Cupreine, cf. alkd.
Cupron 720
Curangin, cf. glcde.
Cuscohygrine, cf. alkd.
Cuskhygrine, cf. alkd.
Cuspidatin, cf. glcde.
Cyanacetic ester 3030

Cyanethyl carbonate 3032
Cyano-acetophenone (ω) 742
Cyano-benzene 716
Cyano-diallylamine 1682
Cyano-diethylamine 2143
Cyano-furan 3345
Cyano-guanidine 2080
Cyano-phenyl chloride (*iso*) 5168

Cyano-propyl alcohol 3758
Cyano-propylene oxide 2880
Cyanuramide 4019
Cyanuric acid (*iso*) 3313
Cyanuric chloride 6172
Cyclamin, cf. glcde.
Cyclo-form 1045

Table 7-4 (*Continued*)
PHYSICAL CONSTANTS OF ORGANIC COMPOUNDS

No.	Name	Synonym	Formula	Beil. Ref.	Formula Weight
1598	**Cyclo** heptene	suberene	$CH_2(CH_2)_4CH:CH$ (ring)	V-65	96.17
1599	hexadiene-1,3	*o*-dihydrobenzene	C_6H_8	V-113	80.13
1600	hexadiene-1,4	*p*-dihydrobenzene	C_6H_8	V-113	80.13
1601	hexane	benzene hexahydride	$CH_2 \cdot (CH_2)_4 \cdot CH_2$ (ring)	V-20	84.16
1602	hexanhexol	*d*-inositol	$(CHOH)_6$	VI-1192	180.16
1603	hexanhexol	*i*-inositol; dambose	$(CHOH)_6$	VI-1192	180.16
1604	hexanhexone	triquinoyl	$(CO)_6 \cdot 8H_2O$	VII-907	312.19
1605	hexanol	hexahydro-phenol	$CH_2 \cdot (CH_2)_4 \cdot CHOH$ (ring)	VI-5	100.16
1606	hexanone	pimelin ketone	$CH_2(CH_2)_4 \cdot CO$ (ring)	VII-8	98.15
1607	hexanone oxime	....	$CH_2(CH_2)_4C:NOH$ (ring)	VII-10	113.16
1608	hexanpentol	*d*-quercitol	$CH_2(CHOH)_4 \cdot CHOH$ (ring)	VI-1186	164.16
1609	hexene	benzene tetrahydride	$CH_2(CH_2)_3CH:CH$ (ring)	V-63	82.15
1610	hexyl acetate	....	$CH_3CO_2 \cdot C_6H_{11}$	VI-7	142.20
1611	hexyl anisole (*p*)	....	$CH_3O \cdot C_6H_4 \cdot C_6H_{11}$	**VI-549	190.29
1612	hexyl bromide	....	$C_6H_{11}Br$	V-24	163.06
1613	hexyl chloride	chloro-cyclohexane	$C_6H_{11}Cl$	V-21	118.61
1614	hexyl iodide	....	$C_6H_{11}I$	V-25	210.06
1615	hexyl phenol (*o*)	....	$HO \cdot C_6H_4 \cdot C_6H_{11}$	....	176.26
1616	hexyl phenol (*p*)	....	$HO \cdot C_6H_4 \cdot C_6H_{11}$	VI-583	176.26
1617	octane	octamethylene	$CH_2 \cdot (CH_2)_6 \cdot CH_2$ (ring)	V-35	112.22
1618	pentadiene-1,3	....	$CH_2 \cdot CH:CH \cdot CH:CH$ (ring)	V-112	66.10
1619	pentane	pentamethylene	$CH_2 \cdot (CH_2)_3 \cdot CH_2$ (ring)	V-19	70.14
1620	pentanone	keto-pentamethylene	$CH_2 \cdot (CH_2)_3 \cdot CO$ (ring)	VII-5	84.12
1621	propane	trimethylene	$CH_2 \cdot CH_2 \cdot CH_2$ (ring)	V-15	42.08
1622	**Cymene** (*o*)	isopropyl-toluene	$CH_3 \cdot C_6H_4 \cdot CH(CH_3)_2$	V-419	134.22
1623	**Cymene** (*m*)	*iso*-cymene	$CH_3 \cdot C_6H_4 \cdot CH(CH_3)_2$	V-419	134.22
1624	**Cymene** (*p*)	cymene	$CH_3 \cdot C_6H_4 \cdot CH(CH_3)_2$	V-420	134.22
1625	**Cysteine** (*l*)	....	$HS \cdot CH_2 \cdot CH(NH_2) \cdot CO_2H$	IV-506	121.16
1626	hydrochloride	....	$C_3H_7O_2NS \cdot HCl$	IV-506	157.62
1627	**Cystine** (*l*)	....	$[HO_2C \cdot CH(NH_2) \cdot CH_2S \cdot]_2$	IV-507	240.30
1628	**Daphnetin**	7,8-diOH-coumarin	$O \cdot CO \cdot CH:CH \cdot C_6H_2:$ (ring) $(OH)_2$	XVIII-100	178.15
1629	**Deca**-chlorodiphenyl	....	$Cl_5C_6 \cdot C_6Cl_5$	V-580	498.66
1630	diene-1,3	....	$CH_3 \cdot (CH_2)_5 \cdot CH:CH \cdot CH:CH_2$	I-260	138.25
1631	hydronaphthalene	"Decalin" (*cis*)	$C_{10}H_{18}$	V-92	138.25
1632	hydronaphthalene	"Decalin" (*trans*)	$C_{10}H_{18}$	V-92	138.25

Cyclo-hexane carboxylic acid 3577
Cyclo-hexylamine 326
Cyclo-hexyl benzene 5179
Cystamin 3599
Cysteine thioformacetal 2854

Cystogen 3599
Cytisine, cf. alkd.
D. A. gas 2712
Dagenan 5668
Dahl's acid 4550-1

Table 7-4 (*Continued*)
PHYSICAL CONSTANTS OF ORGANIC COMPOUNDS

No.	Crystalline Form and Color	Specific Gravity	Melting Point °C.	Boiling Point °C.	Solubility in 100 Parts		
					Water	Alcohol	Ether
1598	oil	$0.823^{20°}_{4}$	....	114–5	i.	s.	s.
1599	lq.*	$0.842^{20°}_{20}$	−98	79–80	i.	s.	s.
1600	lq.*	$0.848^{20°}_{20}$	−90±	85.5	i.	s.	s.
1601	col. lq.	$0.779^{20°}_{4}$	6.5	80.7	i.; ∞ act.	∞ ; ∞ bz.	∞ ; $57^{25°}$ Me al.
1602	mn.	....	253	319 in vac.	$4.5^{15°}$	v. sl. s.	i.
1603	mn./aq.	1.752	253	319^{15mm}	$2^{12°}$	i. abs.	i.
1604	nd./aq. HNO^3	....	95–100 d.	....	v. sl. s.; s. alk.	v. sl. s.	v. sl. s.
1605	col. hyg. nd.	$0.962^{20°}_{4}$	25.5	161.1	$3.6^{20°}$	s.	s.
1606	col. oil	$0.947^{19°}_{4}$	−31.2	155.7	s.	s.	s.
1607	pr./lg.	....	89–90	204 sl. d.	s.; sl. s. lg.	v. s.	v. s.
1608	mn./aq.	$1.585^{13°}$	234–5	....	s.	s. h.	i.
1609	lq.	$0.810^{20°}_{4}$	−103.7	83.3	v. sl. s.	v. s.	v. s.
1610	oil	$0.985^{0°}_{4}$	....	174^{750mm}	i.	∞	∞
1611	yel. rhb.	....	58–9	276	i.	s.	v. s.
1612	col. lq.	$1.324^{20°}_{20}$	....	165^{714mm} sl. d.	i.	s.	s.
1613	col. lq.	$0.977^{18°}_{4}$	−43.9	142	i.	∞ bz.	∞
1614	lq.	$1.626^{15°}_{15}$	....	192^{742mm}	....	....	....
1615	cr.	....	50–5	148^{10mm}	v. sl. s.	v. s.	v. s.; s. CCl_4
1616	col. nd./bz.	....	132–3	$155–8^{13mm}$	i.; $42^{5°}$ bz.	$70^{25°}$ act.	$40^{25°}$
1617	cr.	$0.839^{20°}_{4}$	14.4	$150–1^{709mm}$	....	....	....
1618	col. lq.	$0.805^{18°}_{4}$	−85	41–2	i.	∞	∞
1619	col. oil	$0.745^{20°}_{4}$	−93.8	49.3	i.	....	....
1620	oil	$0.948^{20°}$	−51.3	130.7	43.3	....	....
1621	col. gas	$0.720^{-79°}$	−127.4	−32.9	i.	s.	s.
1622	col. lq.	$0.877^{20°}_{4}$	−71.6	178.3	i.	s.	s.
1623	col. lq.	$0.861^{20°}_{4}$	−63.8	175.2	i.	s.	s.
1624	col. lq.	$0.857^{20°}_{4}$	−68.0	177.1	i.	s.	s.
1625	cr. pd.	....	....	....	s.	s. NH_3	s. ac.
1626	col. cr.	....	175–8 d.	....	v. s.	s.	s. act.
1627	pl./aq. HCl	....	d. 258–61	....	$0.01^{19°}$	i.	s. alk.
1628	yel. nd./ aq. al.	....	255–6	subl.	s. h.	v. s. h.; i. chl.	v. sl. s.; i. bz.
1629	rhb.	....	340	....	i.	v. sl. s.	v. sl. s.
1630	lq.	$0.750^{20°}$	....	168–70	....	....	....
1631	lq.	$0.895^{18°}_{4}$	−51	193.3	i.	s.	s.
1632	lq.	$0.872^{20°}_{4}$	−32	185.3	i.	s.	s.

* Probably mixt. of isomers.
Dahlin 3872
Dambose 1603
Daphnin, cf. glcde.

Datiscin, cf. glcde.
Datiscosid, cf. glcde.
Daturine, cf. alkd.

Table 7-4 (*Continued*)
PHYSICAL CONSTANTS OF ORGANIC COMPOUNDS

No.	Name	Synonym	Formula	Beil. Ref.	Formula Weight
1633	**Deca** hydroquinoline	. . . .	$C_9H_{17}N$	XX-156	139.24
1634	methylene glycol	decandiol-1,10	$HOCH_2(CH_2)_8CH_2OH$	I-494	174.29
1635	**Decane** (*n*)	. . . .	$CH_3 \cdot (CH_2)_8 \cdot CH_3$	I-168	142.29
1636	**Decane** (*iso*)	2-methyl-nonane	$(CH_3)_2CH(CH_2)_6CH_3$	I-168	142.29
1637	**Decane**	2,6-diMe-octane	$(CH_3)_2CH(CH_2)_3 \cdot$ $CH(CH_3) \cdot C_2H_5$	I-168	142.29
1638	**Decane** (di-*iso*-amyl)	2,7-diMe-octane	$[(CH_3)_2CH \cdot CH_2CH_2]_2$	I-169	142.29
1639	**Decyl** acetate (*n*)	. . . .	$CH_3CO_2 \cdot C_{10}H_{21}$	II-135	200.32
1640	alcohol (*n*)(*prim*)	decanol-1	$CH_3(CH_2)_8CH_2OH$	I-425	158.29
1641	alcohol (*sec*)	decanol-4	$C_2H_5 \cdot CH_2 \cdot CHOH \cdot$ $(CH_2)_5 \cdot CH_3$	I-426	158.29
1642	alcohol (*tert*)	2-Et-octanol-3	$(C_2H_5)_2COH(CH_2)_4 \cdot$ CH_3	I-426	158.29
1643	alcohol (*tert*)	4-Pr-heptanol-4	$(C_2H_5 \cdot CH_2)_3COH$	I-426	158.29
1644	alcohol(*prim*)	2,6-diMe-3-meth-ylolheptane	$(CH_3)_2CH \cdot CH(CH_2 \cdot$ $OH)(CH_2)_2 \cdot CH:$ $(CH_3)_2$	I-427	158.29
1645	amine (*n*)	. . . .	$CH_3(CH_2)_8CH_2NH_2$	IV-199	157.30
1646	chloride (*n*)(*prim*)	1-chlorodecane	$CH_3(CH_2)_8CH_2Cl$	I-168	176.73
1647	iodide (*n*)(*prim*)	1-iododecane	$CH_3(CH_2)_8CH_2I$	I-168	268.18
1648	nitrate (*n*)	. . . .	$C_9H_{19}CH_2 \cdot O \cdot NO_2$	I-425	203.28
1649	nitrite (*n*)	. . . .	$C_9H_{19}CH_2 \cdot O \cdot NO$	I-425	187.28
1650	**Decylene** (*α*)	decene-1	$CH_2 : CH(CH_2)_7CH_3$	I-223	140.27
1651	**Decylene**	2,7-diMe-octene-2	$(CH_3)_2CH(CH_2)_3 \cdot$ $CH:C(CH_3)_2$	I-224	140.27
1652	**Decylene**	2-Me-5-Et-heptene-5	$(CH_3)_2CH(CH_2)_2 \cdot$ $C(C_2H_5):CHCH_3$	I-224	140.27
1653	**Decylene**	3,3,5-trimethyl-heptene-4	$C_2H_5 \cdot C(CH_3)_2CH:$ $C(CH_3) \cdot C_2H_5$	I-224	140.27
1654	glycol	. . . .	$CH_3 \cdot CHOH \cdot C(C_2H_5) \cdot$ $(C_4H_9)CH_2OH$	. . . .	174.29
1655	**Deguelin**	. . . .	$C_{23}H_{22}O_6$	. . . .	394.43
1656	**Dehydracetic acid**	6-Me-3-aceto-2,4-pyrandione	$C_8H_8O_4$	XVII-559	168.15
1657	**Dehydro**-andro-sterone	dehydro-*iso*-andro-sterone	$C_{19}H_{28}O_2$	. . . .	288.43
1658	benzoylacetic acid	. . . .	$C_6H_5CO \cdot CH \cdot CO \cdot CH:$ $C(C_6H_5) \cdot O \cdot CO$	XVII-575	292.29
1659	**Desoxalic acid**	. . . .	$\cdot CHOH \cdot C(OH):$ $(CO_2H)_3$	III-586	194.10
1660	**Desoxybenzoin**	phenyl-benzyl-ketone	$C_6H_5 \cdot CO \cdot CH_2 \cdot C_6H_5$	VII-431	196.25
1661	**Dextrin**	starch gum	$(C_6H_{10}O_5)x$	. . . .	(162.14)
1662	**Diacetamide**	. . . .	$(CH_3CO)_2NH$	II-181	101.11
1663	**Diacetone** alcohol	diacetone	$(CH_3)_2COH \cdot CH_2 \cdot$ $CO \cdot CH_3$	I-836	116.16
1664	amine	2-NH_2-2-Me-pent-anone-4	$(CH_3)_2C(NH_2)CH_2 \cdot$ $CO \cdot CH_3$	IV-322	115.18
1665	amine acid oxalate	. . . .	$C_6H_{13}ON \cdot C_2H_2O_4 \cdot$ H_2O	IV-322	223.23
1666	glucose	glucose bis-diMe-ketal	$C_{12}H_{20}O_6$	XXXI-155	260.29

Table 7-4 (*Continued*)
PHYSICAL CONSTANTS OF ORGANIC COMPOUNDS

No.	Crystalline Form and Color	Specific Gravity	Melting Point °C.	Boiling Point °C.	Solubility in 100 Parts		
					Water	Alcohol	Ether
1633	pr./lg.	....	48.2-8.5	204^{714mm}	s. h.	v. s.	v. s.
1634	nd./aq. al.	....	72-3	179^{15mm}	i. c.	s.; i. pet.	s. h.
1635	col. lq.	$0.730^{20°}$	−29.7	174.1	i.	∞	∞
1636	col. lq.	$0.726^{20°}_4$	−74.7	167.0	....	....	....
1637	col. lq.	$0.728^{20°}_4$	....	158.5	....	....	....
1638	col. lq.	$0.724^{20°}_4$	−54	159.9	....	....	....
1639	col. lq.	....	....	244	i.	s.	s.; s. bz.
1640	col. oil	$0.830^{20°}_4$	7	232.9	....	s.	....
1641	col. oil	$0.826^{20°}_0$	....	210-1	....	....	....
1642	col. oil	....	....	199	....	....	....
1643	col. oil	$0.834^{21°}_4$	....	190-2	....	....	....
1644	col. oil	$0.849^{0°}$	....	210-2	....	....	....
1645	cr.	....	17	216-8	i.	....	....
1646	lq.	$0.887^{20°}$	....	$180-90^{720mm}$	i.	....	....
1647	lq.	$1.260^{16°}_4$	....	132^{15mm}	....	....	....
1648	lq.	$0.951^{0°}_4$	....	$127-8^{11mm}$	....	....	....
1649	lq.	....	....	$105-8^{12mm}$	....	....	....
1650	col. lq.	$0.740^{30°}$	....	172	i.	∞	∞
1651	col. lq.	$0.748^{20°}_0$	....	$159-62^{650mm}$	i.	....	....
1652	col. lq.	$0.752^{11°}_4$	....	$157-8^{750mm}$	i.	....	....
1653	col. lq.	$0.773^{21°}_{21}$	....	157.5^{759mm}	....	....	....
1654	lq.	$0.945^{20°}_{20}$	....	103^{3mm}	sl. s.	....	....
1655	pa. gn./al.	....	171	....	i.	s.	....
1656	rhb.	....	108.5-9.0	269.9	$0.1^{20°}$; s. h.	v. s. h.	s.
1657	col. cr.	....	148(138)	....	i.	s.	s.
1658	yel. nd./al.	....	171-2	....	i.; s. chl.; s. NH_4OH	sl. s.; s. bz.; s. alk.	s.; sl. s. lg.; s. CS_2
1659	hyg. cr.	....	d. 45	....	v. s.	v. s.	....
1660	pl./al.	....	60	320-2	sl. s. h.	v. s.	v. s.
1661	amor.	1.038	....	....	s.	i. abs.	i.
1662	col. nd.	....	78-9	222.5-3.5	v. s.	sl. s. lg.	sl. s.
1663	lq.	$0.931^{25°}$	−47	167.9	∞	∞	∞
1664	col. lq.	....	....	$25^{0.14mm}$ d. h.	s.	∞	∞
1665	pr./aq.	....	125-6	....	v. s. h.	s. h.	....
1666	col. mn. nd./lg.	....	110-1	subl.	14 h.	s.; s. chl.	s.; s. act.

Table 7-4 (*Continued*)
PHYSICAL CONSTANTS OF ORGANIC COMPOUNDS

No.	Name	Synonym	Formula	Beil. Ref.	Formula Weight
1667	**Diacetyl**	butandione	$CH_3 \cdot CO \cdot CO \cdot CH_3$	I-769	86.09
1668	acetone	....	$(CH_3CO \cdot CH_2)_2CO$	I-808	142.16
1669	*p*-aminophenol	....	$CH_3CO_2 \cdot C_6H_4 \cdot$ NHCOCH$_3$	XIII-464	193.20
1670	aniline (*N*)	N-Ph-diacetamide	$C_6H_5N:(COCH_3)_2$	XII-250	177.20
1671	benzidine	....	$(CH_3CO \cdot NH \cdot C_6H_4)_2$	XIII-227	268.32
1672	carbinol	....	$(CH_3CO)_2:CHOH$	I-852	116.12
1673	glucose	....	$(CH_3CO)_2:C_6H_{10}O_6$	....	264.23
1674	monomethoxime	....	$CH_3CO \cdot C(:NOCH_3) \cdot$ CH$_3$	I-772	115.13
1675	oxyphenyl-isatin	isacen	$C_6H_4 \cdot N:COH \cdot C:$ |_____| $(C_6H_4O_2C \cdot CH_3)_2$	....	401.42
1676	phenylenediamine	(*o*)	$C_6H_4(NHCOCH_3)_2$	XIII-20	192.22
1677	phenylenediamine	(*p*)	$C_6H_4(NHCOCH_3)_2$	XIII-97	192.22
1678	urea (*sym*)	....	$(CH_3CO \cdot NH)_2CO$	*III-29	144.13
1679	**Diacetylene**	butadiyne	$HC:C \cdot C:CH$	I-266	50.06
1680	**Diallyl** amine	....	$(CH_2:CH \cdot CH_2)_2:NH$	IV-208	97.16
1681	barbituric acid (5,5)	dial	$CO \cdot NH \cdot CO \cdot NH \cdot$ | $CO \cdot C:(C_3H_5)_2$ ___|	*XXIV-422	208.22
1682	cyanamide	....	$(C_3H_5)_2:N \cdot CN$	**IV-666	122.17
1683	oxalate	....	$(CO \cdot O \cdot C_3H_5)_2$	II-540	170.17
1684	sulfide	thioallyl ether	$(CH_2:CH \cdot CH_2)_2S$	I-440	114.21
1685	trisulfide	....	$(C_3H_5)_2S_3$	I-441	178.34
1686	**Dialuric acid**	tartronyl urea	$C_4H_4O_4N_2$	XXV-85	144.09
1687	**Diamino-**anisole	(1;2,4)	$CH_3O \cdot C_6H_3(NH_2)_2$	*XIII-204	138.17
1688	anisole HCl	(1;2,4)	$C_7H_{10}ON_2 \cdot 2HCl$	*XIII-204	211.09
1689	anisole HCl	(1;2,5)	$C_7H_{10}ON_2 \cdot 2HCl$	....	211.09
1690	anthraquinone (1,4)	....	$C_6H_4(CO)_2C_6H_2:$ (NH$_2$)$_2$	XIV-197	238.25
1691	anthraquinone (1,5)	....	$(H_2N \cdot C_6H_3)_2(CO)_2$	XIV-203	238.25
1692	anthraquinone (1,8)	....	$(H_2N \cdot C_6H_3)_2(CO)_2$	XIV-212	238.25
1693	anthraquinone (2,3)	....	$C_6H_4(CO)_2C_6H_2:$ (NH$_2$)$_2$	XIV-215	238.25
1694	anthraquinone (2,6)	....	$H_2NC_6H_3(CO)_2C_6H_3 \cdot$ NH$_2$	XIV-215	238.25
1695	anthraquinone (2,7)	....	$H_2NC_6H_3(CO)_2C_6H_3 \cdot$ NH$_2$	XIV-216	238.25
1696	azobenzene (2,2')	....	$(H_2N \cdot C_6H_4 \cdot N:)_2$	XVI-303	212.26
1697	azobenzene (4,4')	azoaniline	$(H_2N \cdot C_6H_4 \cdot N:)_2$	XVI-334	212.26
1698	azobenzene (2,4)	chrysoidine	$(H_2N)_2C_6H_3 \cdot N_2 \cdot C_6H_5$	XVI-383	212.26
1699	azobenzene HCl	chrysoidine orange	$C_{12}H_{12}N_4 \cdot HCl$	XVI-383	248.72
1700	benzoic acid (2,3)	....	$(H_2N)_2C_6H_3 \cdot CO_2H$	XIV-447	152.15
1701	benzoic acid (2,4)	....	$(H_2N)_2C_6H_3 \cdot CO_2H$	XIV-448	152.15
1702	benzoic acid (2,5)	....	$(H_2N)_2C_6H_3 \cdot CO_2H$	XIV-448	152.15

Diacetoxy-ethane 3214
Diacetyl-, cf. also acetyl.
Diacetyl-dioxime 2490
Diacetyl-disulfide 85
Diacetyl-fluorescein 3253

Diacetyl-hydroquinone 3716
Diacetyl-methane 68
Diacetyl-monoxime 4888
Diacetyl-monoxime methyl ether 1674
Diacetyl-morphine, cf. alkd.

Table 7-4 (*Continued*)
PHYSICAL CONSTANTS OF ORGANIC COMPOUNDS

No.	Crystalline Form and Color	Specific Gravity	Melting Point °C.	Boiling Point °C.	Solubility in 100 Parts		
					Water	Alcohol	Ether
1667	yel. gn. lq.	$0.981^{18.5°}_{4}$	Ca. −4	87-8	$25^{15°}$	∞	∞
1668	col. lf.	$1.068^{40°}_{40}$	49	121^{10mm}	s. aq. Na_2CO_3	v. s.	v. s.
1669	lf./aq.	....	150-1	....	....	....	....
1670	pl./lg.	....	37-8	$199-200^{100mm}$	sl. s. c.	s. bz.	s. lg.
1671	nd./ac.	....	330-1	subl. d.	i.	v. sl. s.	v. sl. s.
1672	rhb. nd.	....	67-72	....	v. s. alk.	s. lg.	....
1673	lt. yel. amor.	....	....	....	v. s.	s.	i. bz.
1674	yel. lq.	....	....	125	....	....	....
1675	col. cr.	....	241-2	....	i.; i. dil. HCl	sl. s.	i.
1676	nd./aq.	....	185-6	....	s. h.; s. chl.	s.; s. a.	v. sl. s.
1677	cr./ac.	....	310-2	....	v. sl. s.	v. sl. s.	v. sl. s.
1678	nd./50% ac.	....	153.5	d. 170-80	v. sl. s.	s.	....
1679	gas	$0.736^{0°}_{4}$ ..	−36	9-10	$466^{25°}cc.$	....	....
1680	lq.	$0.789^{20°}_{20}$	−100	111-2	....	....	....
1681	col. lf./aq.	....	171-3	....	0.33 c.; 2 h.	s.; i. pet.	s.; s. act.
1682	col. lq.	....	....	$140-5^{90mm}$	i.	s.; s. bz.	s.
1683	oil	$1.055^{15.5°}$	....	217^{759mm}	i.	s.	....
1684	col. oil	$0.888^{27°}_{4}$	−83	138.6^{758mm}	sl. s.	∞	∞
1685	lq.	$1.085^{15°}$	....	$112-22^{16mm}$	....	....	....
1686	pr.	....	214 d.	....	sl. s. c.	....	....
1687	nd./et.	....	67-8	....	....	....	....
1688	tan, rhb.	....	....	....	v. s.	i.	i.
1689	wh. pd.	....	....	....	v. s.	s.	i.
1690	cr./al.	....	268	....	s. h.	s.	v. s. bz.
1691	red nd.	....	319	subl.	v. sl. s.	sl. s.	sl. s.
1692	cr./al.	....	262	....	i.	s.	sl. s.
1693	cr./ $PhNO_2$	....	>320	....	s. H_2SO_4	sl. s. chl.	s. C_5H_5N
1694	red pr./ C_5H_5N	....	310-20 d.	....	s. H_2SO_4	sl. s. h.	i. chl.
1695	or. nd./al.	....	>330	subl.	i.	sl. s.	s. conc. a.
1696	red lf./al.	....	134	....	v. sl. s.	s. h.	v. s.; v. s. act.
1697	yel. nd./ aq. al.	....	241-3	....	v. sl. s.	s.	sl. s. bz., sl. s. lg.
1698	yel. cr.	....	117.5	....	sl. s. h.	s.	s.; v. s. chl.
1699	bl. cr. or red pd.	....	....	....	s.	s.	....
1700	nd.	....	190-1 d.	....	v. sl. s.	....	....
1701	cr.	....	140±	....	v. sl. s.	....	....
1702	pr./aq.	....	d. 200	....	v. sl. s. h.	v. sl. s.	v. sl. s.

Table 7-4 (*Continued*)
PHYSICAL CONSTANTS OF ORGANIC COMPOUNDS

No.	Name	Synonym	Formula	Beil. Ref.	Formula Weight
	Diamino				
1703	benzoic acid (3,4)	....	$(H_2N)_2C_6H_3 \cdot CO_2H$	XIV-450	152.15
1704	benzoic acid (3,5)	....	$(H_2N)_2C_6H_3 \cdot CO_2H \cdot H_2O$	XIV-453	170.17
1705	benzoic HCl (3,5)	....	$(H_2N)_2C_6H_3 \cdot CO_2H \cdot 2HCl$	XIV-453	225.08
1706	benzophenone (2,2')	....	$(NH_2 \cdot C_6H_4)_2CO$	XIV-87	212.25
1707	benzophenone (3,3')	....	$(NH_2 \cdot C_6H_4)_2CO$	XIV-88	212.25
1708	benzophenone (4,4')	....	$(NH_2 \cdot C_6H_4)_2CO$	XIV-88	212.25
1709	2-butyloxy-5,5'-azopyridine (2',6')	niazo; neotropin	$C_4H_9O \cdot C_5H_3(N) \cdot N_2 \cdot C_5H_2(N)(NH_2)_2$	....	286.34
1710	diphenyl (2,2')	....	$H_2N \cdot C_6H_4 \cdot C_6H_4 \cdot NH_2$	XIII-210	184.24
1711	diphenyl (2,4')	diphenyline	$H_2N \cdot C_6H_4 \cdot C_6H_4 \cdot NH_2$	XIII-211	184.24
1712	diphenyl (3,3')	....	$H_2N \cdot C_6H_4 \cdot C_6H_4 \cdot NH_2$	XIII-213	184.24
1713	diphenyl (4,4')	benzidine	$(H_2N \cdot C_6H_4)_2 \cdot H_2O$ †	XIII-214	202.26
1714	diphenyl (4,4')	benzidine	$(H_2N \cdot C_6H_4)_2$ ‡	XIII-214	184.24
1715	diphenyl (4,4') HCl	benzidine HCl	$(H_2N \cdot C_6H_4)_2 \cdot 2HCl$	XIII-219	257.16
1716	diphenyl H_2SO_4	benzidine sulfate	$(H_2N \cdot C_6H_4)_2 \cdot H_2SO_4$	XIII-219	282.32
1717	diphenylamine (2,4)	....	$C_6H_5 \cdot NH \cdot C_6H_3(NH_2)_2$	XIII-295	199.26
1718	diphenylamine (4,4')	....	$(H_2N \cdot C_6H_4)_2NH$	XIII-110	199.26
1719	diphenylethane (4,4')	diamino-dibenzyl	$(H_2N \cdot C_6H_4 \cdot CH_2)_2$	XIII-248	212.30
1720	diphenyl ether	(4,4')	$H_2N \cdot C_6H_4 \cdot O \cdot C_6H_4 \cdot NH_2$	XIII-441	200.24
1721	diphenylmethane	4,4'-diaminoditane	$(H_2N \cdot C_6H_4)_2CH_2$	XIII-238	198.27
1722	hydrazobenzene (*p*)	biphenin	$(H_2N \cdot C_6H_4 \cdot NH)_2$	XV-653	214.27
1723	phenol (2,4)	....	$(NH_2)_2C_6H_3 \cdot OH$	XIII-549	124.14
1724	phenol (2,4) HCl	amidol	$(NH_2)_2C_6H_3OH \cdot 2HCl$	XIII-550	197.07
1725	phenol (2,5)	....	$(NH_2)_2C_6H_3 \cdot OH$	XIII-553	124.14
1726	phenol (3,4)	....	$(NH_2)_2C_6H_3 \cdot OH$	XIII-564	124.14
1727	phenol (3,5)	....	$(NH_2)_2C_6H_3 \cdot OH$	XIII-567	124.14
1728	phenylacetic acid HCl	....	$(NH_2)_2C_6H_3CH_2 \cdot CO_2H \cdot 2HCl$	XIV-476	239.10
1729	stilbene (2,2')(α)	....	$(H_2N \cdot C_6H_4 \cdot CH:)_2$	XIII-267	210.28
1730	stilbene (2,2')(β)	....	$(H_2N \cdot C_6H_4 \cdot CH:)_2$	XIII-267	210.28
1731	stilbene (4,4')	....	$(H_2N \cdot C_6H_4 \cdot CH:)_2$	XIII-267	210.28
1732	stilbene-disulfonic acid	(4,4'-diNH$_2$-2,2'-diS)	$[H_2N \cdot C_6H_3(SO_3H) \cdot CH]_2$	XIV-798	370.41
1733	triphenylmethane	(4,4'); diamino-tritane	$(H_2N \cdot C_6H_4)_2CH \cdot C_6H_5$	XIII-274	274.37
1734	**Diamyl** amine (*n*)	....	$(C_5H_{11})_2NH$	*IV-378	157.30
1735	amine (*iso*)	....	$(C_5H_{11})_2NH$	IV-182	157.30
1736	amine (*d*)	....	$[C_2H_5CH(CH_3) \cdot CH_2]_2NH$	IV-179	157.30
1737	carbonate (*n*)	amyl carbonate	$(C_5H_{11}O)_2CO$	....	202.30
1738	carbonate (*iso*)	*iso*-amyl carbonate	$(C_5H_{11}O)_2CO$	III-7	202.30
1739	benzene§	....	$(C_5H_{11})_2C_6H_4$	V-470	218.39

† Cryst. from aq. < 60°; ‡ cryst. from aq. > 80°.
§ Mixt. of isomers.
Diamino-butane 5474
Diamino-caproic acid 3986
Diamino-chlorobenzene 1433

Diamino-dibenzyl 1719
Diamino-diphenylsulfide 5886
Diamino-ditane 1721
Diamino-naphthalene 4571-9
Diamino-propane 5464, 6282

Table 7-4 (*Continued*)
PHYSICAL CONSTANTS OF ORGANIC COMPOUNDS

No.	Crystalline Form and Color	Specific Gravity	Melting Point °C.	Boiling Point °C.	Solubility in 100 Parts		
					Water	Alcohol	Ether
1703	lf.	....	210 d.	....	s. h.	....	....
1704	nd./aq.	....	228-36 (anh.)	$-H_2O$, 110	$1.1^{8°}$	s.	s.
1705	nd.	....	253?	....	s.	s.	....
1706	yel. lf./aq. al.	....	134-5	....	i.	s.	....
1707	lt. yel. nd./al.	....	173-4	....	v. sl. s.	s.	s.
1708	yel. nd./ aq. al.	....	237-9	....	v. sl. s. h.	s.	s.
1709	red cr.	....	129	....	sl. s.	s.	s.
1710	nd./al.	....	81	....	....	....	....
1711	nd./aq. al.	....	45	363	v. sl. s.	s.	s.
1712	oil	....	....	....	v. sl. s.	....	s.
1713	cr./aq.	....	105-20	....	....	....	....
1714	cr./aq.	....	127.5-8.7	$400-1^{740mm}$	1 h.	1 h.	2.2abs.
1715	col. lf.	....	....	....	s., d.	s.; i. HCl	i.
1716	col. pl.	....	....	....	$0.0008^{100°}$	v. sl. s. h.	i.
1717	nd.	....	130	....			
1718	lf./aq.	....	158	d.	sl. s.	s.	s.
1719	pl./aq.	....	134-5	subl. sl. d.	sl. s. h.	v. s.	....
1720	cr./al.	....	186-7 d.	....	i.	$i.^{25°}$; $15^{25°}$ act.	i. bz.; i. CCl_4
1721	nd./aq.	....	93-4	$249-53^{15mm}$	sl. s. c.	s.	s.
1722	yel. cr.	....	145	....	s. h.	s.	s.
1723	lf.	....	78-80 d.	....	s. ac.	s.; s. alk.	sl. s.
1724	nd.	....	....	....	s.	sl. s.	....
1725	cr.	....	68	....			
1726	cr.	....	167-8 d.	....			
1727	pr.	....	168-70	....	s.	....	sl. s.
1728	tan nd.	....	....	....	v. s.	sl. s.	i.
1729	yel. pr./al.	....	176	....	s. bz.	s.	s.
1730	red nd./aq.	....	123	....	....	....	....
1731	yel. lf./al.	....	227-8	sl. d.	sl. s. h.	s. Me al.	sl. s. bz.
1732	yel. nd.	....	....	....	v. sl. s.	....	....
1733	cr./et.	....	139*	....	v. sl. s.	s.	s.
1734	col. lq.	....	....	$202-3^{745mm}$	v. sl. s.	v. s.	∞
1735	col. lq.	$0.767\frac{21°}{4}$	-44	188-90	sl. s.	s.; s. chl.	∞
1736	col. lq.	$0.788^{0°}$	....	182-4	sl. s.	s.	s.
1737	col. lq.	....	....	$130-2^{20mm}$	....	....	....
1738	col. lq.	$0.912^{15°}$	....	233	....	....	....
1739	col. lq.	....	....	265±	i.	s.	s.

* Crysts. $+1C_6H_6$/bz., m. p. 104-6°.
Diamino-propanol 3384
Diamino-toluene 6003, 6005, 6007
Diamino-tritane 1733
Diamino-valeric acid 5000

Diamol 1724 or 5260
Diamond green-G base 5672
Diamorphine, cf. alkd.
Di-*iso*-amyl 1638

Table 7-4 (*Continued*)
PHYSICAL CONSTANTS OF ORGANIC COMPOUNDS

No.	Name	Synonym	Formula	Beil. Ref.	Formula Weight
1740	**Diamyl** ether (*n*)	amyl ether	$(C_5H_{11})_2O$	*I-193	158.29
1741	ether (*iso*)	*iso*-amyl ether	$[(CH_3)_2CH(CH_2)_2]_2O$	I-401	158.29
1742	hydroquinone (*tert*)	. . . .	$[C_2H_5C(CH_3)_2]_2:$ $C_6H_2(OH)_2$	VI-952	250.38
1743	ketone (*n*)	caprone	$(C_5H_{11})_2CO$	I-714	170.30
1744	ketone (*iso*)	. . . .	$(C_5H_{11})_2CO$	I-714	170.30
1745	oxalate (*iso*)	. . . .	$(CO_2 \cdot C_5H_{11})_2$	II-540	230.31
1746	phthalate (*n*)	amyl phthalate	$C_6H_4(CO_2 \cdot C_5H_{11})_2$	. . . .	306.41
1747	phthalate (*iso*)	. . . .	$C_6H_4(CO_2 \cdot C_5H_{11})_2$	**IX-587	306.41
1748	succinate (*n*)	amyl succinate	$(CH_2CO_2 \cdot C_5H_{11})_2$	**II-551	258.36
1749	succinate (*iso*)	. . . .	$(CH_2CO_2 \cdot C_5H_{11})_2$	II-611	258.36
1750	succinate (*act*)	. . . .	$(CH_2CO_2 \cdot C_5H_{11})_2$	II-611	258.36
1751	sulfide (*n*)	amyl sulfide	$(C_5H_{11})_2S$	. . . .	174.35
1752	sulfide (*iso*)	. . . .	$(C_5H_{11})_2S$	I-405	174.35
1753	sulfide, di- (*n*)	amyl disulfide	$(C_5H_{11}S)_2$	. . . .	206.41
1754	sulfide, di- (*iso*)	. . . .	$(C_5H_{11}S)_2$	I-406	206.41
1755	sulfite (*n*)	amyl sulfite	$(C_5H_{11}O)_2SO$	. . . .	222.35
1756	sulfone (*iso*)	amyl sulfone	$(C_5H_{11})_2SO_2$	I-406	206.35
1757	tartrate (*iso*)	amyl tartrate	$(HO \cdot CH \cdot CO_2 \cdot C_5H_{11})_2$	*III-179	290.36
1758	**Dianisalacetone**	. . . .	$(CH_3OC_6H_4CH:$ $CH)_2CO$	VIII-354	294.35
1759	**Dianisidine** (*o*)	(3,3'-MeO; 4,4'-NH₂)	$(CH_3O \cdot C_6H_3 \cdot NH_2)_2$	XIII-807	244.30
1760	**Dianisyl**-phenetyl-guanidine HCl	guanicaine	$(C_{23}H_{25}O_3N_3 \cdot HCl$	XIII-487	427.93
1761	**Diazidoethane** (1,2)	1,2-bistriazo-ethane	$(CH_2N_3)_2$	I-103	112.09
1762	**Diazo**-aminoben-zene	benzene-di-azoanilide	$C_6H_5 \cdot N:N \cdot NHC_6H_5$	XVI-687	197.24
1763	aminonaphthalene	(*β*)	$C_{10}H_7 \cdot N_2 \cdot NH \cdot C_{10}H_7$	. . . .	297.36
1764	aminotoluene	(2,2')	$C_7H_7 \cdot N_2NHC_7H_7$	XVI-703	225.30
1765	benzene chloride	. . . .	$C_6H_5 \cdot N_2Cl$	XVI-431	140.57
1766	benzene cyanide	. . . .	$C_6H_5 \cdot N_2CN$	XVI-432	131.14
1767	benzene nitrate	. . . .	$C_6H_5 \cdot N_2NO_3$	XVI-432	167.13
1768	benzene per-bromide	. . . .	$C_6H_5N(Br_3):N$	XVI-431	344.85
1769	benzene sulfonic acid (*o*)	. . . .	$C_6H_4 \cdot N_2 \cdot O \cdot SO_2$	XVI-557	184.18
1770	benzene sulfonic acid (*m*)	. . . .	$C_6H_4 \cdot N_2 \cdot O \cdot SO_2$	XVI-559	184.18
1771	benzene sulfonic acid (*p*)	sulfanilic acid diazide	$C_6H_4 \cdot N_2 \cdot O \cdot SO_2$	XVI-561	184.18
1772	methane	. . . .	CH_2N_2	XXIII-25	42.04
1773	salicylic acid (5)	. . . .	$HO_2C \cdot C_6H_2(OH)N_2$	XVI-553	164.12
1774	uracil (5)	. . . .	$CO \cdot NH \cdot CO \cdot NH \cdot$ $CH:C \cdot N:N \cdot OH$	XXV-565	156.10
1775	**Dibenzalacetone**	cinnamone	$(C_6H_5 \cdot CH:CH)_2CO$	VII-500	234.30
1776	**Dibenzamide**	. . . .	$(C_6H_5 \cdot CO)_2NH$	IX-213	225.25
1777	**Dibenzanthracene**	(1,2,5,6)	$C_{22}H_{14}$	*V-369	278.36

Dianilino-methane 4431
Diathesin 3750
Diatol 2138

Diazine 5477, 5482, 5513
Diazo-acetic ester 3035
Diazo-benzeneimide 6079

Table 7-4 (*Continued*)
PHYSICAL CONSTANTS OF ORGANIC COMPOUNDS

No.	Crystalline Form and Color	Specific Gravity	Melting Point °C.	Boiling Point °C.	Solubility in 100 Parts		
					Water	Alcohol	Ether
1740	lq.	$0.787^{15°}_{4}$	−69.3	**187.5**	i.	∞	∞
1741	col. lq.	$0.777^{20°}_{4}$	....	173.4	i.	∞ ; ∞ chl.	∞
1742	cr./bz.	....	185	....	i. aq. NaOH	s.; s. chl.	s.; v. sl. s. lg.
1743	lf.	$0.829^{15°}_{4}$	14.6	228	i.	v. s.	v. s.
1744	yel. oil	....	....	226	i.	s.	s.
1745	col. lq.	$0.968^{11°}_{11}$	....	265-7	i.	s.	s.
1746	col. lq.	$0.821^{25°}_{4}$	....	$204-6^{11mm}$	....	....	....
1747	col. lq.	$1.022^{15.6°}_{15.6}$	....	225^{40mm}	i.	s.	s.
1748	col. lq.	$0.961^{20°}_{4}$	−9	172^{16mm}	....	....	....
1749	col. lq.	$0.961^{13°}$	....	289.97^{28mm}	i.	s.	s.
1750	col. lq.	$0.959^{20°}_{4}$	....	$178-80^{25mm}$	i.	s.	s.
1751	col. lq.	....	....	$103-5^{12mm}$	....	....	....
1752	lq.	$0.843^{20°}_{4}$	....	216	i.	∞	∞
1753	yel. lq.	....	....	$128-30^{12mm}$	....	....	....
1754	lq.	$0.918^{18°}$	....	250 sl. d.	....	....	....
1755	col. lq.	....	....	$136-8^{15mm}$	....	....	....
1756	nd.	....	31	295	v. sl. s.	v. s.	v. s.
1757	lq.	$1.063^{14.8°}_{4}$	....	195^{16mm}	i.	....	....
1758	yel. lf./EtOAc	....	129-30	....	s. chl., s. bz.	sl. s.	sl. s.
1759	col. lf.	....	131.5	....	i.; s. bz.	s.	s.
1760	col. cr.	....	176	....	6	s.	i. oil
1761	oil	$1.170^{24.9°}$	....	53^{9mm}	....	....	....
1762	yel. lf./al.	....	96-8	d. 150; expl.	i.; v. s. bz.	s. h.	v. s.
1763	red nd./xylene	....	156	....	s. H₂SO₄	....	....
1764	or. cr.	....	51	....	0.05	....	....
1765	hyg. cr./al. et.	....	expl.	....	v. s.	s. abs.	i.; s. act.
1766	yel.	....	69	....	sl. s.	....	....
1767	nd./al. et.	1.37	expl.	....	v. s.	sl. s.	i.
1768	yel. lf.	....	63.5 d.	....	i. d.	sl. s. d.	i. d.
1769	col.	....	expl.	....	d. h.	d. h.	....
1770	col. nd./aq.	....	expl.	....	s.; d. 60°	d. h.	....
1771	col. nd./aq.	....	expl.	....	s. h.	i.; d. h.	s. dil. alk.
1772	gas	expl. 200°	−145	−23	d.	....	s.
1773	yel. cr.	....	expl. 155	....	s. h.	s. Na₂CO₃	i.
1774	red or yel. pl./aq.	....	expl.	....	d. h. a.	....	....
1775	lt. yel. mn.	....	112	d.	v. s. act.	s. h.	sl. s.
1776	rhb./bz.	....	148	....	$0.12^{15°}$	s.	s.
1777	lf./ac.	....	262	subl.	sl. s. ac.	v. s. bz.	i.

Table 7-4 (*Continued*)
PHYSICAL CONSTANTS OF ORGANIC COMPOUNDS

No.	Name	Synonym	Formula	Beil. Ref.	Formula Weight
1778	**Dibenzeneazore-** sorcinol	....	$(C_6H_5 \cdot N_2)_2$: $C_6H_2(OH)_2$	XVI-185	318.34
1779	**Dibenzoyl** disulfide	....	$(C_6H_5 \cdot CO)_2S_2$	IX-424	274.36
1780	ethylenediamine	....	$(C_6H_5CO \cdot NHCH_2)_2$	IX-262	268.32
1781	methane	β-hydroxy chalkone	$(C_6H_5 \cdot CO)_2CH_2$	VII-769	224.26
1782	**Dibenzyl**-acetic acid	....	$(C_6H_5CH_2)_2$: $CHCO_2H$	IX-682	240.30
1783	amine	....	$(C_6H_5 \cdot CH_2)_2NH$	XII-1035	197.28
1784	aniline	....	$(C_6H_5 \cdot CH_2)_2N \cdot C_6H_5$	XII-1037	273.38
1785	fumarate	benzyl fumarate	$(:CH \cdot CO_2 \cdot CH_2C_6H_5)_2$	VI-437	296.33
1786	hydroxylamine (β,β)	....	$(C_6H_5CH_2)_2 : NOH$	XV-19	213.28
1787	ketone	α,α'-diphenyl acetone	$(C_6H_5CH_2)_2 : CO$	VII-445	210.28
1788	maleate	benzyl maleate	$(:CH \cdot CO_2 \cdot CH_2C_6H_5)_2$	VI-437	296.33
1789	phthalate (*o*)	benzyl phthalate	$C_6H_4(CO_2 \cdot C_7H_7)_2$	IX-802	346.39
1790	succinate	....	$(\cdot CH_2CO_2 \cdot C_7H_7)_2$	VI-436	298.34
1791	sulfide	....	$(C_6H_5CH_2)_2S$	VI-455	214.33
1792	tartrate (*d*)	....	$(CHOHCO_2C_7H_7)_2$	*VI-221	330.34
1793	**Dibromo**-acetamide	....	$Br_2CH \cdot CO \cdot NH_2$	II-219	216.87
1794	acetamide (*N*)	acetdibromamide	$CH_3 \cdot CO \cdot NBr_2$	II-182	216.87
1795	acetic acid	....	$Br_2CH \cdot CO_2H$	II-218	217.86
1796	acetylene	dibromoethyne	$BrC \vdots CBr$	I-246	183.84
1797	aminophenol	(4;2,6,1)	$NH_2 \cdot C_6H_2Br_2OH$	XIII-517	266.93
1798	aniline (2,3)	....	$Br_2C_6H_3 \cdot NH_2$	XII-655	250.93
1799	aniline (2,4)	....	$Br_2C_6H_3 \cdot NH_2$	XII-655	250.93
1800	aniline (2,6)	....	$Br_2C_6H_3 \cdot NH_2$	XII-659	250.93
1801	aniline (3,4)	....	$Br_2C_6H_3 \cdot NH_2$	XII-660	250.93
1802	aniline (3,5)	....	$Br_2C_6H_3 \cdot NH_2$	XII-660	250.93
1803	anthracene (9,10)	....	$C_{14}H_8Br_2$	V-665	336.04
1804	anthranilic acid	(3,5)	$Br_2C_6H_2(NH_2)CO_2H$	XIV-371	294.94
1805	anthraquinone (1,3)	....	$Br_2C_{14}H_6O_2$	*VII-414	366.02
1806	anthraquinone (1,5)	....	$Br_2C_{14}H_6O_2$	VII-789	366.02
1807	anthraquinone (2,6)	....	$Br_2C_{14}H_6O_2$	VII-790	366.02
1808	anthraquinone (β)	(2,3)	$Br_2C_6H_2(CO)_2C_6H_4$	VII-790	366.02
1809	anthraquinone (α)	(2,7)	$BrC_6H_3(CO)_2C_6H_3Br$	VII-790	366.02
1810	barbituric acid	(5,5); dibromin	$Br_2C(CONH)_2CO$	XXIV-472	285.89
1811	benzene (*o*)	....	$C_6H_4Br_2$	V-210	235.92
1812	benzene (*m*)	....	$C_6H_4Br_2$	V-211	235.92
1813	benzene (*p*)	....	$C_6H_4Br_2$	V-211	235.92
1814	benzoic acid (2,3)	....	$Br_2C_6H_3 \cdot CO_2H$	IX-357	279.93
1815	benzoic acid (2,4)	....	$Br_2C_6H_3 \cdot CO_2H$	IX-358	279.93
1816	benzoic acid (2,5)	....	$Br_2C_6H_3 \cdot CO_2H$	IX-358	279.93
1817	benzoic acid (2,6)	....	$Br_2C_6H_3 \cdot CO_2H$	IX-358	279.93
1818	benzoic acid (3,4)	....	$Br_2C_6H_3 \cdot CO_2H$	IX-359	279.93
1819	benzoic acid (3,5)	....	$Br_2C_6H_3 \cdot CO_2H$	IX-359	279.93
1820	benzoylene urea (6,8)	....	$Br_2C_6H_2 \cdot NH \cdot CO \cdot$ $NH \cdot CO$	....	319.95

Dibenzo-furan 2763
Dibenzo-pyranol 6453
Dibenzo-pyrone 6451
Dibenzo-pyrrole 1231
Dibenzo-thioxine 5129

Dibenzoyl 697
Dibenzoyl-resorcinol 5553
Dibenzyl-, cf. also benzyl.
Dibenzyl 2717
Dibenzyl disulfide 805

Table 7-4 (*Continued*)
PHYSICAL CONSTANTS OF ORGANIC COMPOUNDS

No.	Crystalline Form and Color	Specific Gravity	Melting Point °C.	Boiling Point °C.	Solubility in 100 Parts		
					Water	Alcohol	Ether
1778	red nd./chl.-al.	....., ...	223-4	....	s. h. chl.	i.	v. sl. s. dil. alk.
1779	pr./CS_2	....	133-5	d.	i.; i. NH_4OH	sl. s. h.	sl. s. h.; s. CS_2
1780	nd./al.	....	250-1	d.	i.	$0.08^{22°}$	....
1781	rhb./al.	....	78(72-3)*	$219\text{-}21^{18mm}$	i.	$4.4^{19.5°}$	s.
1782	nd./aq.	....	88-9	....	v. sl. s. h.	s.; s. bz.	s.; s. ac.
1783	col. oil	$1.028\frac{25°}{25}$	-26	$268\text{-}71^{250mm}$	i.	s.	s.
1784	pr./al.	....	70-1	>300 sl. d.	i.	v. s. h.	v. s.
1785	col. pr./lg.	....	64	239^{14mm}	i.	s.	v. sl. s. c.
1786	nd.	....	124	....	sl. s. h.	s.	s.
1787	cr./aq. al.	....	34-5	330.6	....	....	....
1788	oil	....	....	241^{14mm}	....	....	....
1789	pr./al.	....	42-3	274^{12mm}	v. sl. s.	s.	s.
1790	col. lf./al.	....	45-6	238^{14mm}	i.	s.	s.; s. bz.
1791	rhb./et.	$1.071\frac{50°}{50}$	49	....	i.	s.	s.
1792	cr.	$1.204^{72°}$	50±	$250\text{-}70^{4mm}$	....	....	....
1793	nd.	....	156	....	....	....	....
1794	yel. nd.	....	100	....	s. h.	s.	s.
1795	col. cr.	....	48-50	232-4 d.	v. s.	v. s.	v. s.
1796	poison. lq.	2±	expl. with trace O_2	76.5 in CO_2	i.	s.	s.
1797	nd./al.	....	192-3	....	i.; s. bz.	s.	sl. s.
1798	pl./aq. al.	....	43	....	v. sl. s.	v. s.	v. s.
1799	rhb.	$2.260^{20°}$	79.5	....	....	s.	s. ac.
1800	nd./al.	....	83-4	262-4	....	v. s. abs.	v. s.
1801	lf./aq. al.	....	80-1	subl. 100	i.	s.	....
1802	nd.	....	56.5	....	....	s.	....
1803	yel. nd.	....	221	subl.	s. h. bz.	v. sl. s.	v. sl. s.
1804	nd./al.	....	235-6	....	i.; s. alk.	s.	s.
1805	yel. nd./ac.	....	210	....	s. H_2SO_4	v. sl. s.	v. sl. s.
1806	nd.	....	....	....	....	....	....
1807	yel. cr./Am. al.	....	289-90	....	....	v. sl. s.	s. h. bz.
1808	yel. nd.	....	269-70	subl.	s. bz.	v. sl. s.	s. chl.
1809	yel. nd./ac.	....	245	....	....	v. sl. s. h.	sl. s. h. ac.
1810	cr./aq. HNO_3	....	240 d.	....	3	s.	s.
1811	col. lq.	$1.956\frac{20°}{4}$	1.8	221-2	i.	s.	s.
1812	col. lq.	$1.952\frac{20°}{4}$	-6.9	219^{755mm}	i.	s.	s.
1813	col. pl./al.	$2.261^{18°}$	87.3	220.3	i.; s. chl.	1.6; s. bz.	$71^{25°}$
1814	nd./aq.	....	147-9	....	s. h. lg.	....	....
1815	lf./aq.	....	171-2	subl.	sl. s. h.	....	....
1816	nd./aq.	....	153	....	v. sl. s. c.	s.	s.
1817	nd./aq.	....	146-7	....	s. h.	v. s.	v. s.; v. s. chl.
1818	nd./aq.	....	232-3	subl.	v. sl. s. c.	s.	s.
1819	nd./aq.	....	219-20	subl.	v. sl. s. c.	v. s.	sl. s. c. bz.
1820	yel. nd./glycol	....	305-6	....	i.; i. aq. NaOH	i.; s. h. glycol	i.; s. h. $PhNO_2$

* Enol form, two m. p's.; keto, m. p. 81°.
Dibenzyl ether 806
Dibenzyl sulfone 835

Dibromide of cinnamic acid 1830
Dibromin 1810
Dibromo-acetophenone 967

Table 7-4 *(Continued)*
PHYSICAL CONSTANTS OF ORGANIC COMPOUNDS

No.	Name	Synonym	Formula	Beil. Ref.	Formula Weight
	Dibromo				
1821	butane (1,2)	α-butylene bromide	$C_2H_5 \cdot CHBr \cdot CH_2Br$	I-120	215.93
1822	butane (1,3)	. . . .	$CH_3CHBr \cdot CH_2CH_2Br$	I-120	215.93
1823	butane (1,4)	. . . .	$Br(CH_2)_4Br$	I-120	215.93
1824	butane (2,3)(*dl*)	β-butylene bromide	$(CH_3 \cdot CHBr \cdot)_2$	**II-84	215.93
1825	butane (*iso*)	1,2-diBr-2-Me-propane	$(CH_3)_2CBr \cdot CH_2Br$	I-127	215.93
1826	butyric acid (α,β)	crotonic acid diBr	$CH_3(CHBr)_2CO_2H$	II-284	245.91
1827	butyric acid	*iso*-crotonic diBr	$CH_3(CHBr)_2CO_2H$	II-285	245.91
1828	camphor (α,α')	(3,3-diBr-*d*)	$Br_2C_9H_{14}CO$	VII-125	310.04
1829	cinnamic acid (α,β)	"α-acid"	$C_6H_5CBr:CBrCO_2H$	IX-601	305.96
1830	cinnamic acid (*dl*)	†	$C_6H_5(CHBr)_2CO_2H$	IX-518	307.98
1831	o-cresol (4,6)	(4,6;2,1)	$Br_2C_6H_2(CH_3)OH$	VI-360	265.94
1832	o-cresolsulfon-phthalein	bromcresol purple	$C_{21}H_{16}O_5Br_2S$	. . . .	540.24
1833	diphenyl (4,4')	. . . .	$BrC_6H_4 \cdot C_6H_4Br$	V-580	312.02
1834	diphenyl ether (4,4')	. . . .	$(Br \cdot C_6H_4)_2O$	VI-200	328.01
1835	diphenylsulfone	(4,4')	$(Br \cdot C_6H_4)_2SO_2$	VI-331	376.08
1836	ethane (1,1)	ethylidene dibromide	$CH_3 \cdot CHBr_2$	I-90	187.87
1837	ethane (1,2)	ethylene bromide	$BrCH_2 \cdot CH_2Br$	I-90	187.87
1838	fluorescein	(2,4)	$C_{20}H_{10}O_5Br_2$	XIX-228	490.12
1839	fluorescein	(4,5)	$C_{20}H_{10}O_5Br_2$	XIX-228	490.12
1840	fumaric acid	. . . .	$(:CBr \cdot CO_2H)_2$	II-747	273.88
1841	gallic acid	(2,6;3,4,5;1)	$Br_2C_6(OH)_3CO_2H \cdot H_2O$	X-490	345.94
1842	hexane (1,2)	α-hexylene di-bromide	$CH_3(CH_2)_3CHBr \cdot CH_2Br$	I-144	243.98
1843	hexane (1,5)	. . . .	$CH_3CHBr \cdot (CH_2)_3 \cdot CH_2Br$	I-145	243.98
1844	hexane (1,6)	hexamethylene diBr	$(\cdot CH_2 \cdot CH_2 \cdot CH_2Br)_2$	I-145	243.98
1845	hexane	3,3-dibromo-2,2-dimethyl-butane	$(CH_3)_3C \cdot CBr_2 \cdot CH_3$	I-151	243.98
1846	hexane	3,4-dibromo-2,2-dimethyl-butane	$(CH_3)_3C \cdot CHBr \cdot CH_2Br$	I-151	243.98
1847	hexane	2,3-dibromo-2,3-dimethyl-butane	$(CH_3)_2CBr \cdot CBr(CH_3)_2$	I-152	243.98
1848	8-hydroxy-quinoline (5,7)	. . . .	$Br_2C_9H_4(OH)N$	XXI-97	302.96
1849	iodomethane	. . . .	$ICHBr_2$	I-71	299.74
1850	maleic acid	. . . .	$(:CBr \cdot CO_2H)_2$	II-756	273.88
1851	malonic acid	. . . .	$Br_2C(CO_2H)_2$	II-595	261.86
1852	naphthalene (1,4)	β-dibromonaph-thalene	$C_{10}H_6Br_2$	V-549	285.98
1853	α-naphthol (2,4)	. . . .	$Br_2C_{10}H_5 \cdot OH$	VI-614	301.98
1854	nitromethane	. . . .	$Br_2CH \cdot NO_2$	I-77	218.84
1855	nitrophenol (2,6;4)	. . . .	$Br_2:C_6H_2(NO_2)OH$	VI-247	296.91
1856	pentane (1,4)	. . . .	$CH_3 \cdot CHBr \cdot CH_2 \cdot CH_2 \cdot CH_2Br$	I-131	229.95
1857	pentane (1,5)	pentamethylene diBr	$Br(CH_2)_5Br$	I-131	229.95

†Dibromide of cinnamic acid.
Dibromo-diphenyl 1833
Dibromo-ethene 131-3

Dibromo-ethylbenzene 5647
Dibromo-ethylene 131-3
Dibromo-ethyne 1796

Table 7-4 (*Continued*)
PHYSICAL CONSTANTS OF ORGANIC COMPOUNDS

No.	Crystalline Form and Color	Specific Gravity	Melting Point °C.	Boiling Point °C.	Solubility in 100 Parts		
					Water	Alcohol	Ether
1821	lq.	$1.820\frac{20°}{0}$	−65	166	i.	∞	....
1822	lq.	$1.807^{18.5°}$	....	174-5	....	....	....
1823	col. lq.	$1.819\frac{8.5°}{0}$	−20	197-8	....	....	....
1824	col. lq.	$1.783\frac{20°}{4}$	−34.5	157-8	i.	....	....
1825	col. lq.	$1.759^{20°}$	−70.3	$148-9^{737mm}$	i.	....	....
1826	mn. nd.	....	87	....	sl. s. c.	v. s.	v. s.
1827	nd.	....	58-9	....	sl. s. c.	v. s.	v. s.
1828	col. rhb./lg.	$1.854\frac{21.6°}{4}$	64	subl. >64	i.; s. lg.	$22^{20°}$ abs.	s.; s. bz.
1829	pl./chl. pet.	....	138-9 *	....	i.	s.; s. chl.	s.; sl. s. pet.
1830	mn./chl.	....	199-201	d.	i.; d. h.	s.	s.
1831	nd.	....	56-7	....	v. sl. s.	s.; s. alk.	s.; s. bz.
1832	pink pd.	....	241-2	....	v. sl. s.	s.; s. alk.	i.
1833	mn. pr.	1.897	165-7	355-60	i.	v. sl. s. h.	v. s. bz.
1834	cr.	....	58-60	338-40	v. s. bz.	s.	s.
1835	mn. nd./al.	....	172	....	i.	sl. s. h.	....
1836	lq.	$2.055\frac{20°}{4}$	....	108-10	i.	v. s.	v. s.
1837	col. lq.	$2.180\frac{20°}{4}$	10	131.7	$0.433^{30°}$	∞	∞
1838	red cr. + EtOH/al.	....	300± (anh.)	....	i.; s. ac.	s. h.	v. sl. s.
1839	red cr./al.	....	285	....	sl. s.; s. ac.	s. h.	v. sl. s.
1840	cr./aq.	....	227-9 d.	....	....	v. s.	v. s.
1841	cr./aq.	....	150	−H_2O, 120	$12.4^{15°}$; $200^{100°}$	s.; i. chl.	s.
1842	lq.	$1.596^{13.5°}$	....	87^{16mm}	....	....	....
1843	lq.	$1.599\frac{20°}{4}$	....	$153-4^{100mm}$	....	....	....
1844	lq.	$1.595^{15°}$	....	243	....	....	....
1845	cr.	....	187 **	subl.	....	....	....
1846	col. lq.	$1.616^{0°}$	....	$91-2^{14mm}$	i.; v. s. bz.	v. s.	v. s.; v. s. chl.
1847	long nd.	$1.811^{15°}$	173**	....	d.	s.	v. s.; s. bz.
1848	nd./al.	....	196	subl.	i. c.	sl. s.; s. a.	s.; s. bz.
1849	tablets	....	22.5	$101-4^{50mm}$	soln. d. by light	sl. s. pet.	....
1850	nd./et.	....	123.5	....	v. s.	v. s.	v. s.; i. chl.
1851	pr. or nd.	....	136-7 d.	....	v. s.	v. s.	v. s.
1852	nd./al.	....	82-3	310 d.	....	$1.3^{11°}$	v. s.
1853	nd./al.	....	111	....	i.	s.	s.; s. ac.
1854	lq.	....	....	$58.5-60^{13mm}$	....	....	....
1855	col. pr./al.	....	143-4	d. >145	v. sl. s.; v. s. bz.	s. h.; sl. s. ac.	s. h.; s. h. CS_2
1856	lq.	$1.622\frac{20°}{4}$	....	$196-8^{746}$ d.	....	....	....
1857	lq.	$1.706^{18°}$	−35	221^{763} sl. d.	....	....	....

* β or allo-acid, m. 100°.
** In sealed tube.
Dibromo-hydrin 3388-9

Dibromo-methane 4428
Dibromo-pene 1869

Table 7-4 (*Continued*)
PHYSICAL CONSTANTS OF ORGANIC COMPOUNDS

No.	Name	Synonym	Formula	Beil. Ref.	Formula Weight
1858	**Dibromo** pentane	2,3-dibromo-2-methyl-butane	$(CH_3)_2CBr \cdot CHBr \cdot CH_3$	I-137	229.95
1859	phenol (2,4)	....	$Br_2C_6H_3 \cdot OH$	VI-202	251.92
1860	phenol (2,6)	....	$Br_2C_6H_3 \cdot OH$	VI-202	251.92
1861	phenol (3,4)	....	$Br_2C_6H_3 \cdot OH$	VI-203	251.92
1862	phenol (3,5)	....	$Br_2C_6H_3 \cdot OH$	VI-203	251.92
1863	propane (1,1)	propylidene diBr	$C_2H_5 \cdot CHBr_2$	I-109	201.90
1864	propane (1,2)	propylene bromide	$CH_3 \cdot CHBr \cdot CH_2Br$	I-109	201.90
1865	propane (1,3)	trimethylene diBr	$BrCH_2 \cdot CH_2 \cdot CH_2Br$	I-110	201.90
1866	propane (2,2)	bromacetol	$(CH_3)_2CBr_2$	I-111	201.90
1867	propionic acid (α,α)	....	$CH_3 \cdot CBr_2 \cdot CO_2H$	II-257	231.88
1868	propionic acid (α,β)	....	$CH_2Br \cdot CHBr \cdot CO_2H$	II-258	231.88
1869	propylene (β,γ)	epidibromohydrin (α)	$CH_2 : CBr \cdot CH_2Br$	I-201	199.89
1870	propylene (α,γ)	epidibromohydrin (β)	$CH_2Br \cdot CH : CHBr$	I-201	199.88
1871	propyne-1 (1,3)	....	$CH_2Br \cdot C \vdots CBr$	I-248	197.87
1872	pyridine (2,5)	....	$Br_2C_5H_3N$	XX-233	236.90
1873	pyridine (2,6)	....	$Br_2C_5H_3N$	....	236.90
1874	pyridine (3,5)(α)	....	$Br_2C_5H_3N$	XX-233	236.90
1875	pyruvic acid	....	$CHBr_2 \cdot CO \cdot CO_2H$	III-624	245.87
1876	quinonechloroimide	(2,6)	$Cl \cdot N : C_6H_2 : OBr_2$	VII-640	299.36
1877	succinic acid (*allo*)	....	$Br_2C_2H_2(CO_2H)_2$	II-625	275.89
1878	succinic acid (*meso*)	....	$(CHBr \cdot CO_2H)_2$	II-623	275.89
1879	thymol-sulfon-phthalein	bromthymol blue	$C_{27}H_{28}O_5Br_2S$	*XIX-650	624.40
1880	toluene (2,5)	....	$Br_2C_6H_3 \cdot CH_3$	V-308	249.94
1881	tyrosine (3,5)	l (–)	$C_9H_9O_3NBr_2 \cdot 2H_2O$	XIV-619	375.03
1882	*m*-xylene (4,6)	....	$Br_2C_6H_2(CH_3)_2$	V-374	263.97
1883	**Dibutyl** adipate (*n*)	butyl adipate	$(CH_2CH_2CO_2 \cdot C_4H_9)_2$	**II-575	258.36
1884	amine (*n*)	....	$(C_4H_9)_2NH$	IV-157	129.25
1885	amine (*sec*)	....	$(C_4H_9)_2NH$	IV-162	129.25
1886	amine (*iso*)	....	$(C_4H_9)_2NH$	IV-166	129.25
1887	aminoethyl alcohol (β)(*n*)	....	$(C_4H_9)_2N \cdot CH_2 \cdot CH_2OH$	....	173.30
1888	aminopropyl alcohol (β)(*n*)	....	$(C_4H_9)_2N \cdot CH(CH_3) \cdot CH_2OH$	....	187.33
1889	aminopropyl alcohol (γ)(*n*)	....	$(C_4H_9)_2N \cdot CH_2CH_2 \cdot CH_2OH$	....	187.33
1890	aminopropyl-*p*-aminobenzoate H_2SO_4	butyn	$H_2N \cdot C_6H_4 \cdot CO_2 \cdot (CH_2)_3 \cdot N(C_4H_9)_2 \cdot \frac{1}{2}H_2SO_4$	....	355.49
1891	aniline (*n*)(*N*)	....	$C_6H_5N(C_4H_9)_2$	*XII-160	205.35
1892	carbonate (*n*)	butyl carbonate	$CO(OC_4H_9)_2$	III-6	174.24
1893	carbonate (*iso*)	....	$CO(OC_4H_9)_2$	III-6	174.24
1894	carbonate (*sec*)	....	$CO(OC_4H_9)_2$	III-6	174.24
1895	cyanamide (*n*)	....	$(C_4H_9)_2N \cdot CN$	**IV-635	154.26
1896	α,α'-dibromo-succinate (*n*)	....	$(CHBrCO_2C_4H_9)_2$	....	388.11
1897	disulfide (*n*)	....	$C_4H_9 \cdot S \cdot S \cdot C_4H_9$	**I-400	178.36
1898	ether (*n*)	butyl ether	$(C_2H_5 \cdot CH_2 \cdot CH_2)_2O$	I-369	130.23

Dibromo-propanol 3388-9
Dibromo-propyl alcohol 3388-9

Dibromo-sulfobenzide 1835
Dibromo-xylene (ω) 6491-3

Table 7-4 (*Continued*)
PHYSICAL CONSTANTS OF ORGANIC COMPOUNDS

No.	Crystalline Form and Color	Specific Gravity	Melting Point °C.	Boiling Point °C.	Solubility in 100 Parts		
					Water	Alcohol	Ether
1858	lq.	$1.573^{25°}$	7	170-3 sl. d.	....	....	....
1859	cr.	....	40	238-9	$0.2^{15°}$	v. s.	v. s.
1860	nd./aq.	....	55-6	162^{21mm}	....	v. s.	v. s.
1861	nd./aq.	....	79-80	....	....	v. s.	v. s.
1862	cr./lg.	....	81	....	v. sl. s.	v. s.	v. s.
1863	lq.	....	....	130±	....	....	....
1864	col. lq.	$1.933^{20°}_{4}$	−55.5	141.6	$0.25^{20°}$	s.	v. s.
1865	lq.	$1.987^{15°}_{4}$	−34.2	167.3	$0.168^{30°}$	s.	s.
1866	lq.	$1.783^{20°}$	....	114.5^{740mm}	....	....	....
1867	rhb.	....	61	200-21 sl. d.	....	....	....
1868	pl.	....	64 (51)	220-40 d.	194^{511mm}	v. s.	$304^{10°}$
1869	lq.	$1.934^{20°}_{4}$	....	140-2	....	....	....
1870	lq.	$1.995^{25°}_{4}$	−52	155-6	....	....	....
1871	lq.	$2.137^{0°}$	....	$73-4^{30mm}$	....	....	....
1872	nd./al.	....	94-5	subl.	i.	s.	s.
1873	cr.	....	118-9	....	....	....	....
1874	cr./al.	....	110-2; subl. >100	222	sl. s. h.	s. h.	s.; s. H_2SO_4
1875	nd./et.	....	93	....	s.	....	....
1876	yel. pr./al.	....	83	d. 121	0.0006	sl. s. h.	sl. s. h. ac.
1877	cr.	....	170	d. 180	$>2^{17°}$	v. s.	v. s.
1878	cr.	....	255-6 * d.	subl. >250	$2.0^{17°}$	v. s.; v. sl. s. chl.	v. s.
1879	yel. cr.;	....	....	....	v. sl. s.; s. a.	s.	i.; s. alk.
1880	lq.	$1.813^{19°}$	<−20	236	i.	....	....
1881	rhb./aq.	....	245±, d.	−2H_2O, 120	$5^{16°}$; $3.9^{100°}$	v. sl. s.; s. a.	i.; s. alk.
1882	cr.	....	72	255-6	i.	v. s. h.	....
1883	col. lq.	$0.965^{20°}_{4}$	−38	183^{14mm}	i.	∞	∞
1884	col. lq.	$0.768^{20°}_{20°}$	−62	159^{761mm}	v. s.	∞	∞
1885	col. lq.	$0.783^{0°}_{0°}$	−104.5	132^{758mm}	v. s.	....	....
1886	col. lq.	$0.741^{25°}_{4}$	−70	139-40	v. sl. s.	s.	s.
1887	col. lq.	....	....	$91-6^{7mm}$	....	....	....
1888	col. lq.	....	....	$80-6^{6mm}$	....	....	....
1889	col. lq.	....	....	$110-6^{8mm}$	....	....	....
1890	wh. pd.	....	98-100	....	100; sl. s. chl.	s.; s. act.	i.
1891	lq.	....	....	262.8	i.	∞	∞
1892	col. lq.	$0.924^{20°}$	....	207^{740mm}	i.	s.	....
1893	col. lq.	$0.919^{15°}$	....	190	i.	....	....
1894	col. lq.	....	....	178-80	....	....	....
1895	lq.	....	....	$128-30^{20mm}$	i.	s.	s.
1896	col. lq.	....	....	$171-4^{8mm}$	....	....	....
1897	col. lq.	$0.930^{20°}_{4}$	....	$117-8^{20mm}$	i.	∞	∞
1898	lq.	$0.773^{15°}_{4}$	−97.9	142.4	<0.05	∞	∞

* In sealed tube.
Dibutyl carbinol (*n*) 4921

Dibutyl carbinol (*iso*) 4925

Table 7-4 (*Continued*)
PHYSICAL CONSTANTS OF ORGANIC COMPOUNDS

No.	Name	Synonym	Formula	Beil. Ref.	Formula Weight
1899	**Dibutyl** ether (*iso*)	*iso*-butyl ether	[(CH₃)₂CH·CH₂]₂O	I-376	130.23
1900	ketone (*n*)	nonanone-5	(C₂H₅·CH₂·CH₂)₂CO	I-709	142.24
1901	ketone (*iso*)	valerone	(C₄H₉)₂CO	I-710	142.24
1902	*l*-malate (−)(*n*)	butyl malate	C₄H₄O₅(C₄H₉)₂	III-433	246.31
1903	maleate (*n*)	butyl maleate	(CHCO₂C₄H₉)₂	....	228.29
1904	malonate (*n*)	butyl malonate	CH₂(CO₂C₄H₉)₂	II-581	216.28
1905	oxalate (*n*)	....	(CO₂C₄H₉)₂	II-540	202.25
1906	oxalate (*iso*)	....	(CO₂C₄H₉)₂	II-540	202.25
1907	phenyl-phenyl- phosphate (*tert*) (*p*)	phosphen-2	(C₄H₉·C₆H₄O)₂PO· O·C₆H₅	....	438.51
1908	*o*-phthalate (*n*)	butyl phthalate	C₆H₄(CO₂C₄H₉)₂	**IX-586	278.35
1909	sebacate (*n*)	butyl sebacate	[(CH₂)₄CO₂C₄H₉]₂	II-719	314.47
1910	succinate (*n*)	butyl succinate	(CH₂CO₂C₄H₉)₂	**II-551	230.31
1911	succinate (*iso*)	....	(CH₂CO₂C₄H₉)₂	II-611	230.31
1912	succinate (*sec*)	....	(CH₂CO₂C₄H₉)₂	II-611	230.31
1913	sulfate (*n*)	butyl sulfate	(C₄H₉O)₂SO₂	....	210.29
1914	sulfide (*n*)	butyl sulfide	(C₂H₅·CH₂CH₂)₂S	I-370	146.30
1915	sulfide (*sec*)	....	(C₄H₉)₂S	I-373	146.30
1916	sulfide (*iso*)	....	[(CH₃)₂CH·CH₂]₂S	I-379	146.30
1917	sulfite (*n*)	butyl sulfite	(C₄H₉O)₂SO	**I-397	194.29
1918	sulfone (*n*)	butyl sulfone	(C₄H₉)₂SO₂	I-371	178.30
1919	tartrate (*d*)(*n*)	butyl tartrate	(CHOH·CO₂C₄H₉)₂	III-518	262.31
1920	tartrate (*d*)(*iso*)	....	(CHOH·CO₂C₄H₉)₂	III-518	262.31
1921	thiourea (*n*)	....	(C₄H₉·NH)₂CS	....	188.34
1922	urea (*n*)(*N,N*)	....	(C₄H₉)₂N·CO·NH₂	*IV-372	172.27
1923	**Di-*iso*-butylene** †	....	(CH₃)₂C:CH·C(CH₃)₃	I-222	112.22
1924	**Dicapryl adipate**	(*n*)(*sec*)	[·(CH₂)₂CO₂CH· (CH₃)·(CH₂)₅·CH₃]₂	**II-575	370.58
1925	**Dichloro-**acetal	....	Cl₂CH·CH(OC₂H₅)₂	I-614	187.07
1926	acetaldehyde	2,2-dichloro-ethanal	Cl₂CH·CHO	I-613	112.94
1927	acetamide	....	Cl₂CH·CO·NH₂	II-205	127.96
1928	acetic acid	....	Cl₂CH·CO₂H	II-202	128.94
1929	acetone (*α*)(*uns*)	....	Cl₂CH·CO·CH₃	I-654	126.97
1930	acetone (*β*)(*sym*)	....	(ClCH₂)₂CO	I-655	126.97
1931	acetyl chloride	....	Cl₂CH·CO·Cl	II-204	147.39
1932	aniline (2,3)	....	Cl₂C₆H₃·NH₂	XII-621	162.02
1933	aniline (2,4)	....	Cl₂C₆H₃·NH₂	XII-621	162.02
1934	aniline (2,5)	....	Cl₂C₆H₃·NH₂	XII-625	162.02
1935	aniline (2,6)	....	Cl₂C₆H₃·NH₂	XII-626	162.02
1936	aniline (3,4)	....	Cl₂C₆H₃·NH₂	XII-626	162.02
1937	aniline (3,5)	....	Cl₂C₆H₃·NH₂	XII-626	162.02
1938	aniline sulfonic	3,6-diCl-sulfanilic	Cl₂C₆H₂(NH₂)SO₃H	XIV-707	242.08
1939	anthracene (2,3)	....	C₁₄H₈Cl₂	V-664	247.13
1940	anthracene (9,10)	....	C₁₄H₈Cl₂	V-664	247.13
1941	anthraquinone (1,3)	....	C₆H₄(CO)₂C₆H₂Cl₂	VII-787	277.11
1942	anthraquinone (1,4)	....	C₆H₄(CO)₂C₆H₂Cl₂	VII-787	277.11
1943	anthraquinone (1,5)	....	(ClC₆H₃)₂(CO)₂	VII-787	277.11
1944	anthraquinone (1,6)	....	(ClC₆H₃)₂(CO)₂	*VII-412	277.11
1945	anthraquinone (1,8)	....	(ClC₆H₃)₂(CO)₂	VII-788	277.11
1946	anthraquinone (2,3)	....	C₆H₄(CO)₂C₆H₂Cl₂	VII-788	277.11
1947	anthraquinone (2,6)	....	(ClC₆H₃)₂(CO)₂	VII-788	277.11
1948	anthraquinone (2,7)	....	(ClC₆H₃)₂(CO)₂	VII-788	277.11

† See also octylene.
Dibutyl mercury 4042

Dibutyl urethane 3038

Table 7-4 (*Continued*)
PHYSICAL CONSTANTS OF ORGANIC COMPOUNDS

No.	Crystalline Form and Color	Specific Gravity	Melting Point °C.	Boiling Point °C.	Solubility in 100 Parts		
					Water	Alcohol	Ether
1899	lq.	$0.762^{15°}$	....	122-2.5	i.	∞	∞
1900	lq.	$0.827\frac{13°}{4}$	−5.9	187.7	i.; v. s. CS_2	v. s. chl.	v. s.
1901	oil	$0.806\frac{20°}{4}$	−46.0	168.2	<0.06	∞	∞
1902	lq.	$1.038\frac{20°}{4}$	....	170-1^{13mm}	v. sl. s.	....	....
1903	col. lq.	....	....	141-2^{9mm}	....	s.	....
1904	lq.	$0.981\frac{20°}{4}$	....	251.5	i.	s.	s.
1905	col. lq.	$0.986\frac{20°}{4}$	−29.6	245.5	i.	s.	s.
1906	col. lq.	$1.002^{14°}$	....	228-9	i.	s.	s.
1907	col. lq.	$1.11\frac{25°}{25}$	<0	260-75^{5mm}	i.; ∞ CCl_4	v. s.	∞ bz.
1908	col. lq.	$1.045^{21°}$	....	340	$0.04^{25°}$	∞ ; ∞ bz.	∞ ; ∞ act.
1909	lq.	$0.933^{15°}$	−11	344-5	....		
1910	lq.	$0.965^{20°}$	−29.3	274.5			
1911	col. lq.	$0.974^{15°}$	....	265-6			
1912	col. lq.	$0.974\frac{20°}{4}$	....	256-77^{50mm}			
1913	col. lq.	$1.059\frac{25°}{25}$	....	130-2^{11mm}	i.		
1914	lq.	$0.839\frac{16°}{0}$	−79.7	182	i.		
1915	lq.	$0.832^{23°}$	....	165			
1916	lq.	$0.836^{10°}$	−105.5	172-37^{47mm}	i.	∞	∞
1917	col. lq.	$1.001^{14°}$	....	108-10^{15mm}			
1918	pl./aq.	....	44	....	v. sl. s.		
1919	pr.	$1.098^{15°}$	22-2.5	200-3^{18mm}			
1920	cr.	$1.031\frac{75°}{4}$	73-4	323-5			
1921	col. nd./al.	....	66-7	....	i.	s.	sl. s.
1922	col. hyg. cr.	....	149-50	118-9^{2mm}	....	s.	s.
1923	col. lq.	$0.721\frac{20°}{4}$	−106.5	104.9	....	s.	
1924	lt. yel. lq.	$0.914\frac{20°}{4}$	....	192-4^{3mm}	....		
1925	lq.	$1.138^{14°}$	....	185	....	....	
1926	col. lq.*	....	....	88-90	i.	....	
1927	mn. pr.	....	98	233-4^{745mm}	v. s. h.	v. s.	v. s.
1928	lq.	$1.560\frac{25°}{25}$	9.7 (−4)	194.4	∞	∞	∞
1929	lq.	$1.234^{15°}$	....	120	v. sl. s.	s.	s.
1930	pl. or nd.	$1.383\frac{46°}{4}$	45	173.0-3.4	s.	v. s.	v. s.
1931	lq.	....	....	107-8	d.	d.	∞
1932	nd./lg.	....	23-4	252	sl. s. bz.	s.	v. sl. s.
1933	rhb.	$1.567\frac{20°}{4}$	62-3	245^{759mm}	sl. s.	s.	s.
1934	nd./lg.	....	50	251	v. sl. s.	s.	s.
1935	nd./al.	....	39	....			
1936	nd./lg.	....	71.5	272	i.; sl. s. bz.	$70^{25°}$	v. s.
1937	nd.	....	50.5	259-60	i.	s.	....
1938	nd./aq.	....	....	....	s. h.	....	
1939	yel. lf.	....	261	subl. sl. d.	s. ac.	s. h.	
1940	yel. nd.	....	209-10	....	s. bz.	sl. s.	sl. s.
1941	yel. nd./ac.	....	208-9	....	i.; s. ac.	i.	s. $PhNO_2$
1942	yel. nd./ac.	....	187.5	....	i.; s. h. bz.	v. sl. s.	v. sl. s.
1943	yel. nd./ac.	....	251	....	i.; s. H_2SO_4	sl. s.	s. PhMe
1944	yel. nd./ac.	....	203-4	....	i.	....	....
1945	pa. yel. nd.	....	202-3	....	s. PhMe	sl. s.	s. $PhNO_2$
1946	yel. nd./ac.	....	268-70	....	i.	sl. s.	s. h. bz.
1947	yel. nd./ac.	....	282	....	i.	....	....
1948	yel. nd.	....	210-1	....	i.	s. PhOMe	....

* Polymerizes on standing.
Dicetyl 2866

Dichloro-acetoacetanilide 42

Table 7-4 (*Continued*)
PHYSICAL CONSTANTS OF ORGANIC COMPOUNDS

No.	Name	Synonym	Formula	Beil. Ref.	Formula Weight
	Dichloro				
1949	azo-dicarbonamide	azochloramide	$[Cl \cdot N:C(NH_2) \cdot N:]_2$	....	183.00
1950	azoxybenzene (4,4')		$(ClC_6H_4)_2N_2O$	XVI-625	267.12
1951	barbituric acid (5,5)	dichloromalonyl-urea	$Cl_2C(CONH)_2CO$ ⎿_____⏌	XXIV-472	196.98
1952	benzene (*o*)	....	$C_6H_4Cl_2$	V-201	147.00
1953	benzene (*m*)	....	$C_6H_4Cl_2$	V-202	147.00
1954	benzene (*p*)	....	$C_6H_4Cl_2$	V-203	147.00
1955	benzene sulfonic acid (2,5)	....	$Cl_2C_6H_3 \cdot SO_3H$	XI-55	227.08
1956	benzene sulfonyl chloride (3,4)	....	$Cl_2C_6H_3 \cdot SO_2Cl$	*XI-16	245.51
1957	benzidine (3,3')	....	$(H_2N \cdot C_6H_3Cl)_2$	XIII-234	253.13
1958	benzidine HCl	(3,3')	$C_{12}H_{10}N_2Cl_2 \cdot 2HCl$	XIII-234	326.05
1959	benzoic acid (2,3)	....	$Cl_2C_6H_3 \cdot CO_2H$	IX-342	191.01
1960	benzoic acid (2,4)	....	$Cl_2C_6H_3 \cdot CO_2H$	IX-342	191.01
1961	benzoic acid (2,5)	....	$Cl_2C_6H_3 \cdot CO_2H$	IX-342	191.01
1962	benzoic acid (2,6)	....	$Cl_2C_6H_3 \cdot CO_2H$	IX-343	191.01
1963	benzoic acid (3,4)	....	$Cl_2C_6H_3 \cdot CO_2H$	IX-343	191.01
1964	benzoic acid (3,5)	....	$Cl_2C_6H_3 \cdot CO_2H$	IX-344	191.01
1965	benzophenone (2,4')	....	$(Cl \cdot C_6H_4)_2CO$	VII-420	251.11
1966	benzophenone (4,4')	....	$(Cl \cdot C_6H_4)_2CO$	VII-420	251.11
1967	benzoylene urea (6,8)	Sheibley's reagent	$Cl_2C_6H_2 \cdot NH \cdot CO \cdot$ ⎿_____ NH·CO _____⏌	....	231.04
1968	1-bromoethane (1,1)	....	$CH_3 \cdot CCl_2Br$	I-90	177.86
1969	1-bromoethane (2,2)	....	$Cl_2CH \cdot CH_2Br$	I-90	177.86
1970	bromomethane	....	Cl_2CHBr	I-67	163.83
1971	butane (1,1)(*n*)	butylidene di-chloride	$C_2H_5CH_2CHCl_2$	I-119	127.02
1972	butane (1,2)(*n*)	butylene chloride	$C_2H_5 \cdot CHCl \cdot CH_2Cl$	*I-38	127.02
1973	butane (1,4)(*n*)	tetramethylene diCl	$ClCH_2(CH_2)_2CH_2Cl$	I-119	127.02
1974	butane (*iso*)	1,1-dichloro-2-methyl-propane	$(CH_3)_2CH \cdot CHCl_2$	I-126	127.02
1975	butane (*iso*)	1,3-dichloro-2-methyl-propane	$(ClCH_2)_2CH \cdot CH_3$	**I-88	127.02
1976	butane (*iso*)	1,2-dichloro-2-methyl-propane	$(CH_3)_2CCl \cdot CH_2Cl$	I-126	127.02
1977	butyl alcohol (*tert*)	....	$(CH_2Cl)_2COH \cdot CH_3$	I-382	143.01
1978	butyric acid (α,β)	crotonic acid diCl	$CH_3(CHCl)_2CO_2H$	II-279	157.00
1979	butyric acid	*iso*-crotonic acid diCl	$CH_3(CHCl)_2CO_2H$	II-279	157.00
1980	1,2-dibromoethane (1,1)		$BrCH_2 \cdot CBrCl_2$	I-93	256.76
1981	1,2-dibromoethane (1,2)		$ClBrCH \cdot CHBrCl$	I-93	256.76
1982	1,1-dibromoethane (2,2)		$Br_2CH \cdot CHCl_2$	I-93	256.76
1983	dibromomethane	....	Cl_2CBr_2	I-68	242.74
1984	diethyl carbonate (β,β')		$(ClCH_2CH_2O)_2CO$	**III-5	187.02
1985	diethyl sulfide (β,β')	mustard gas	$(Cl \cdot CH_2CH_2)_2S$	I-349	159.08

Dichloro-*iso*-butane 1975-6

Table 7-4 (*Continued*)
PHYSICAL CONSTANTS OF ORGANIC COMPOUNDS

No.	Crystalline Form and Color	Specific Gravity	Melting Point °C.	Boiling Point °C.	Solubility in 100 Parts		
					Water	Alcohol	Ether
1949	yel. nd.	....	expl. 155	....	i.	s.	sl. s.
1950	yel. nd./al.	....	155–6	subl.	i.	sl. s.	s.
1951	rhb. pr./aq.	....	219–20 d.	....	sl. s.; v. sl. s. bz.	v. s.; v. sl. s. chl.	v. s.; v. s. ac.
1952	col. lq.	$1.306\frac{20°}{4}$	−17.2	180.4	i.; ∞ bz.	∞	∞
1953	col. lq.	$1.288\frac{20°}{4}$	−24.8	172^{766mm}	i.	s.	s.
1954	col. mn.	$1.458^{21°}$	53.1	174.4	i.; s. bz.	∞ h. abs.	v. s.
1955	nd./aq.	....	>100	....	v. s.	....	v. sl. s.
1956	mn./bz.	....	22.4	$170–4^{20mm}$	d.	d.	....
1957	nd./al.	....	132–3	....	i.	s.; s. bz.	s. ac.
1958	nd.	....	....	....	v. sl. s.	s.	....
1959	nd.	....	166	....	sl. s. h.	s.	s.
1960	nd./aq.	....	164	subl.	s. h.	s.	s.
1961	nd./aq.	....	154	301	$0.09^{14°}$	s.	s. alk.
1962	nd./al.	....	140–3	subl.	s. h.	s. bz.	s. alk.
1963	nd./aq.	....	203–4	....	s. h.	v. s.	s. alk.
1964	nd./al.	....	182–3	subl.	....	v. s.	....
1965	col. mn./al.	$1.393^{14°}$	66–7	$214–5^{22mm}$	i.; s. act.	$190^{25°}$ bz.	$48^{25°}$
1966	col. lf./al.	....	145	353^{757mm}	i.; s. act.	$123^{35°}$ bz.	$22^{5°}$
1967	col. nd./al.	....	296	....	i.; i. aq. NaOH	1 h.; s. aq. KOH	i.; v. sl. s. h. gly.
1968	lq.	$1.752^{16°}$	....	$98–9^{758mm}$	....	....	....
1969	lq.	....	....	138	....	....	....
1970	col. lq.	$2.006\frac{15°}{4}$	−56.9	90.1	....	....	....
1971	oil	....	....	113–5	i.	s.	s.; s. chl.
1972	lq.	....	....	124	....	....	....
1973	lq.	....	−38.7	161–3	....	....	....
1974	lq.	$1.011^{12°}$	....	103–5 d.	....	....	....
1975	col. lq.	$1.138^{20°}$	....	135–9	....	....	....
1976	col. lq.	$1.097^{20°}$	....	107–8	....	....	....
1977	col. lq.	$1.277\frac{20°}{4}$	....	174–5	8.2	s.	s.
1978	pr.	....	62.5–3.0	$124–5^{20mm}$	sl. s.	v. s.	$328^{10.5°}$
1979	long pr.	....	78 (73)	131.5^{20mm}	sl. s.	v. s.	v. s.
1980	lq.	$2.270^{16°}$	−66.9	175	i.	....	....
1981	lq.	....	−26	194–5	i.	....	....
1982	lq.	$2.391^{19°}$	....	195–200	i.	....	....
1983	nd.	$2.42\frac{25°}{0}$	22	135±	i.	....	....
1984	pa. yel. lq.	$1.351\frac{20°}{4}$	8.5	240–1	....	....	....
1985	col. oil	$1.275\frac{20°}{4}$	14.5	217 sl. d.	$0.072^{5°}$	s.; s. bz.; s. chl.	s.; s. ac.

Table 7-4 *(Continued)*
PHYSICAL CONSTANTS OF ORGANIC COMPOUNDS

No.	Name	Synonym	Formula	Beil. Ref.	Formula Weight
	Dichloro				
1986	diiodomethane	....	Cl_2CI_2	I-72	336.73
1987	2,2'-dinitrodi-phenyl disulfide	(4,4')	$(O_2N \cdot C_6H_3Cl \cdot S)_2$	VI-341	377.23
1988	dioxane (2,3)	....	$O \cdot (CHCl)_2O(CH_2)_2$	....	157.00
1989	diphenyl (3,3')	....	$Cl \cdot C_6H_4 \cdot C_6H_4 \cdot Cl$	V-579	223.10
1990	diphenyl (4,4')	....	$Cl \cdot C_6H_4 \cdot C_6H_4 \cdot Cl$	V-579	223.10
1991	dipropyl ether (*n*)	(γ,γ')	$(ClCH_2CH_2CH_2)_2O$	**I-370	171.07
1992	dipropyl ether (*iso*)	(β,β')	$[ClCH_2(CH_3)CH]_2O$	**I-370	171.07
1993	ethane (1,1)	ethylidene dichloride	$CH_3 \cdot CHCl_2$	I-83	98.96
1994	ethane (1,2)	ethylene chloride	$ClCH_2 \cdot CH_2Cl$	I-84	98.96
1995	ether (α,β)	....	$ClCH_2 \cdot CHCl \cdot O \cdot C_2H_5$	I-612	143.01
1996	ether (β,β')(*sym*)	diCl-Et ether	$(ClCH_2 \cdot CH_2)_2O$	**I-335	143.01
1997	ethyl alcohol (β,β)	2,2-dichloro-ethanol-1	$Cl_2CH \cdot CH_2OH$	I-338	114.96
1998	ethylene (α,α)	1,1-dichloro-ethene	$CH_2 : CCl_2$	I-186	96.94
1999	fluorescein (2,7)	....	$C_{20}H_{10}O_5Cl_2$	*XIX-722	401.21
2000	fluorescein (3',6')	....	$C_{20}H_{10}O_5Cl_2$	XIX-227	401.21
2001	hexane	2,3-dichloro-2,3-dimethyl-butane	$[(CH_3)_2CCl]_2$	I-152	155.07
2002	hexane	3,3-dichloro-2,2-dimethyl-butane	$(CH_3)_3C \cdot CCl_2 \cdot CH_3$	I-150	155.07
2003	hydroquinone (2,3)	....	$(HO)_2C_6H_2Cl_2$	VI-849	179.00
2004	hydroquinone (2,5)	....	$(HO)_2C_6H_2Cl_2$	VI-850	179.00
2005	1-iodoethane (2,2)	....	$ICH_2 \cdot CHCl_2$	I-98	224.86
2006	iodomethane	....	Cl_2CHI	I-71	210.83
2007	maleic acid	....	$(:CCl \cdot CO_2H)_2$	II-753	184.96
2008	methyl ether (*sym*)	....	$ClCH_2 \cdot O \cdot CH_2Cl$	I-582	114.96
2009	naphthalene (1,2)	....	$C_{10}H_6Cl_2$	V-542	197.07
2010	naphthalene (1,3)	θ-dichloronaphtha-lene	$C_{10}H_6Cl_2$	V-542	197.07
2011	naphthalene (1,4)	β-dichloronaphtha-lene	$C_{10}H_6Cl_2$	V-542	197.07
2012	naphthalene (1,5)	γ-dichloronaphtha-lene	$C_{10}H_6Cl_2$	V-543	197.07
2013	naphthalene (1,6)	η-dichloronaphtha-lene	$C_{10}H_6Cl_2$	V-543	197.07
2014	naphthalene (1,7)	θ'-dichloronaphtha-lene	$C_{10}H_6Cl_2$	V-543	197.07
2015	naphthalene (1,8)	ζ-dichloronaphtha-lene	$C_{10}H_6Cl_2$	V-544	197.07
2016	naphthalene (2,3)	ι-dichloronaphtha-lene	$C_{10}H_6Cl_2$	V-544	197.07
2017	naphthalene (2,6)	ε-dichloronaphtha-lene	$C_{10}H_6Cl_2$	V-544	197.07
2018	naphthalene (2,7)	δ-dichloronaphtha-lene	$C_{10}H_6Cl_2$	V-544	197.07
2019	α-naphthol (2,4)	....	$Cl_2C_{10}H_5OH$	VI-612	213.06
2020	α-naphthoquinone	(2,3)	$O : C_{10}H_4Cl_2 : O$	VII-729	227.05

Dichloro-ethanal 1926
Dichloro-ethanol 1997

Dichloro-ethene 135-6, 1998
Dichloro-ethyl ether 1996
Dichloro-ethylene 135-6

Table 7-4 (*Continued*)
PHYSICAL CONSTANTS OF ORGANIC COMPOUNDS

No.	Crystalline Form and Color	Specific Gravity	Melting Point °C.	Boiling Point °C.	Solubility in 100 Parts		
					Water	Alcohol	Ether
1986	scales	....	85 d.	....	i.	s. d.	....
1987	yel. cr./ act.-al.	....	213-4	....	v. sl. s. CS_2	v. sl. s.	v. sl. s. lg.; sl. s. bz.
1988	col. lq.	....	....	$88\text{-}9^{19mm}$	....	....	....
1989	nd./al.	....	23	322-4	i.	v. s.	v. s.
1990	pr. or nd.	$1.442^{0°}_{4}$	148	315-9	i.; $14^{25°}$ bz.	v. sl. s.	$4^{25°}$
1991	oil	$1.140^{20°}_{20}$	....	215^{745mm}	....	....	....
1992	col. lq.	$1.112^{20°}_{20}$	....	187.3	0.17^{20}	....	....
1993	col. lq.	$1.176^{20°}_{4}$	−97.4	57.3	0.7^{0}; $0.5^{30°}$	∞	∞
1994	col. lq.	$1.253^{20°}_{4}$	−35.3	83.5	0.9^{0}; $0.9^{30°}$	∞ ; ∞ chl.	∞
1995	lq.	$1.174^{23°}$	....	140-5	....	....	....
1996	lq.	$1.222^{20°}_{20}$	....	178.5	1.07^{20}	s.	s.
1997	lq.	$1.145^{15°}$	−46.1	146	sl. s.	s.	s.
1998	lq.	$1.250^{15°}$	....	37	i.	....	....
1999	or. pd.	....	....	....	sl. s.	sl. s.	sl. s.
2000	or. pd.	....	....	....	v. s. alk.	....	....
2001	cr.	....	159-60	....	i.	....	....
2002	cr.	....	151	subl.	i.	....	....
2003	nd./aq.	....	144-5	subl.	i. c. lg.	s.	s.
2004	mn. nd./ act.	$1.815^{24°}$	166-70	subl.	s. h.; s. ac.	v. s.	v. s.
2005	lq.	2.219	....	171-2	i.	....	....
2006	lq.	$2.403^{21.5°}$	....	132	i.	....	....
2007	nd.	....	d. 119-20	....	v. s.; i. chl.	v. s.	v. s.
2008	col. lq.	$1.328^{15°}_{4}$	....	104-5	d.	....	....
2009	nd./al.	$1.315^{48.5°}_{4}$	37	282	....	....	....
2010	nd./al.	....	61.5	291^{775mm}	....	....	....
2011	nd./al.	$1.300^{76°}_{4}$	67-8	$286\text{-}7^{740mm}$	v. s. act.	v. sl. s.	....
2012	lf./al.	....	107	....	i.	s.	s.
2013	nd./al.	....	48-9	....	....	....	....
2014	lf./aq. al.	$1.261^{100°}_{4}$	63-4	285-6	....	s.	s.
2015	cr./al.	$1.292^{100°}_{4}$	88	d.	....	....	....
2016	lf.	....	120	....	....	s. h.	s.
2017	nd./al.	....	135-6	285	....	sl. s.	v. s.
2018	pr.	....	114	....	....	v. s. h.	....
2019	nd./aq. al.	....	106-7	....	s. bz.	s.; s. ac.	s.
2020	yel. nd./al.	....	195-6	....	i.	s. h.	sl. s.

Dichloro-fluoran 3252
Dichloro-hydrin 3393-4

Dichloro-malonyl urea 1951
Dichloro-methane 4429

Table 7-4 *(Continued)*
PHYSICAL CONSTANTS OF ORGANIC COMPOUNDS

No.	Name	Synonym	Formula	Beil. Ref.	Formula Weight
	Dichloro				
2021	4-nitroaniline	(2,6;4,1)	$Cl_2C_6H_2(NO_2)NH_2$	XII-735	207.03
2022	nitrobenzene (2,4)	NO_2-Cl-benzene	$Cl_2C_6H_3 \cdot NO_2$	V-245	192.00
2023	nitrobenzene (2,5)	. . . .	$Cl_2C_6H_3 \cdot NO_2$	V-245	192.00
2024	nitrobenzene (2,6)	. . . .	$Cl_2C_6H_3 \cdot NO_2$	V-246	192.00
2025	nitrobenzene (3,4)	. . . .	$Cl_2C_6H_3 \cdot NO_2$	V-246	192.00
2026	nitrobenzene (3,5)	. . . .	$Cl_2C_6H_3 \cdot NO_2$	V-246	192.00
2027	1-nitroethane (1,1)	. . . .	$Cl_2C(NO_2) \cdot CH_3$	. . . .	143.96
2028	nitrohydrin	diCl-Pr-nitrate	$ClCH_2 \cdot CHCl \cdot CH_2 \cdot NO_3$	I-356	173.98
2029	nitrophenol	(4,6;2,1)	$Cl_2C_6H_2(NO_2)OH$	VI-241	208.00
2030	nitrophenol	(2,6;4,1)	$Cl_2H_6H_2(NO_2)OH$	VI-241	208.00
2031	nitropropane	(1,1,1)	$C_2H_5 \cdot C(NO_2)Cl_2$	. . . .	157.98
2032	1-nitrosoethane (1,1)	. . . .	$CH_3 \cdot CCl_2 \cdot NO$	I-99	127.96
2033	pentane (1,4)	. . . .	$CH_3CHCl(CH_2)_3Cl$	I-131	141.04
2034	pentane (1,5)	amylene chloride	$ClCH_2(CH_2)_3CH_2Cl$	I-131	141.04
2035	pentane (2,3)	. . . .	$C_2H_5(CHCl)_2CH_3$	I-131	141.04
2036	pentane (2,4)	. . . .	$(CH_3CHCl)_2 : CH_2$	*I-43	141.04
2037	pentane	1,4-dichloro-2-methyl-butane	$ClCH_2 \cdot CH(CH_3) \cdot CH_2 \cdot CH_2Cl$	*I-47	141.04
2038	pentane	2,3-dichloro-2-methyl-butane	$(CH_3)_2CCl \cdot CHCl \cdot CH_3$	I-135	141.04
2039	pentane	2,4-dichloro-2-methyl-butane	$(CH_3)_2CCl \cdot CH_2 \cdot CH_2Cl$	I-135	141.04
2040	pentane	3,4-dichloro-2-methyl-butane	$(CH_3)_2CH \cdot CHCl \cdot CH_2Cl$	I-135	141.04
2041	pentane	4,4-dichloro-2-methyl-butane	$(CH_3)_2CH \cdot CH_2 \cdot CHCl_2$	I-135	141.04
2042	phenol (2,4)	. . . .	$Cl_2C_6H_3 \cdot OH$	VI-189	163.00
2043	phenyl-3,4-dichlorobenzene sulfonate (2,4)	. . . .	$Cl_2C_6H_3SO_3 \cdot C_6H_3Cl_2$	. . . .	372.06
2044	phenylhydrazine	(2,5)	$Cl_2C_6H_3 \cdot NH \cdot NH_2$	XV-431	177.03
2045	phenylhydrazine-4-sulfonic acid	(2,5)	$Cl_2C_6H_2(SO_3H) \cdot NH \cdot NH_2$	XV-643	257.10
2046	phenyl-phenyl phosphate (*o,o'*)	phosphen 4	$C_6H_5 \cdot O \cdot PO : (ClC_6H_4 \cdot O)_2$	. . . .	395.18
2047	phenyl-4-toluene sulfonate	. . . .	$CH_3 \cdot C_6H_4 \cdot SO_3 \cdot C_6H_3Cl_2$	. . . .	317.19
2048	*p*-phenylenediamine	(2,6)	$Cl_2C_6H_2(NH_2)_2$	XIII-118	177.03
2049	*p*-phenylenediamine HCl	(2,6)	$Cl_2C_6H_2(NH_2)_2 \cdot HCl$	. . . .	213.50
2050	propane (1,1)	propylidene diCl	$C_2H_5 \cdot CHCl_2$	I-105	112.99
2051	propane (1,2)	propylene chloride	$CH_3 \cdot CHCl \cdot CH_2Cl$	I-105	112.99
2052	propane (1,3)	trimethylene diCl	$ClCH_2 \cdot CH_2 \cdot CH_2Cl$	I-105	112.99
2053	propane (2,2)	acetone chloride	$(CH_3)_2CCl_2$	I-105	112.99
2054	propionic acid (α,α)	. . . .	$CH_3 \cdot CCl_2 \cdot CO_2H$	II-250	142.97
2055	propionic acid (α,β)	. . . .	$ClCH_2 \cdot CHCl \cdot CO_2H$	II-252	142.97
2056	propionitrile (α,α)	. . . .	$CH_3 \cdot CCl_2 \cdot CN$	II-251	123.97
2057	propyl carbonate	(γ,γ')	$[Cl(CH_2)_3O]_2CO$	**III-5	215.08
2058	propylene (α,α)	1,1-diCl-propene-1	$CH_3 \cdot CH : CCl_2$	I-199	110.97
2059	propylene (α,β)	allylene dichloride	$CH_3 \cdot CCl : CHCl$	I-199	110.97

Dichloro-propanol 3393-4
Dichloro-propene 2058-61

Dichloro-propyl acetate 45
Dichloro-propyl carbamate 171

Table 7-4 (*Continued*)
PHYSICAL CONSTANTS OF ORGANIC COMPOUNDS

No.	Crystalline Form and Color	Specific Gravity	Melting Point °C.	Boiling Point °C.	Solubility in 100 Parts		
					Water	Alcohol	Ether
2021	yel. nd./ac.	....	194-5	....	i. h. HCl	s.	....
2022	nd./al.	$1.4398^{80°}$	33	258.5	i.	v. s. h.	∞
2023	tri./al.	$1.6692^{22°}$	54.6	266	i.	v. s. h.	v. s. bz.
2024	mn. pr.	$1.6034^{17°}$	72.5	130^{8mm}	i.	s.	s. CS_2
2025	nd./al.	$1.456^{7.5°}_{4}$	42-3(α)*	255-6	....	....	....
2026	yel. mn.	$1.692^{14°}$	65.4	....	i.	s.	s. ac.
2027	lq.	$1.405^{20°}_{20}$	....	122-5	$<0.5^{20°}$	....	....
2028	lq.	$1.37°$	....	180	i.	....	....
2029	yel. mn./al.	$1.822^{20°}_{4}$	122-3	subl. <100	sl. s.	sl. s.; s. bz.	s; s. chl.
2030	yel. pl./et.	....	125 d.	....	i.; sl. s. bz.	s. h.	v. s.
2031	lq.	$1.314^{20°}_{20}$	....	141-4	$<0.5^{20°}$	....	....
2032	b. oil	$1.252^{19°}$	....	68			
2033	col. lq.	....	....	$58-60^{15mm}$	i.	∞	∞
2034	col. lq.	$1.094^{25°}_{4}$	....	180-1	i.; s. CS_2	s.; s. chl.	s.
2035	col. lq.	....	....	138-9	i.	∞	∞
2036	col. lq.	$1.063^{18°}$	....	147-50	i.	∞	∞
2037	col. lq.	$1.103^{21°}_{4}$	....	170-2	i.	∞	∞
2038	lq.	$1.068^{15°}_{4}$	....	130-5	i.	∞	∞
2039	col. lq.	$1.065^{20°}_{4}$	....	152-4 sl. d.	i.	∞	∞
2040	col. lq.	$1.092^{17.5°}$	....	143-5	i.	∞	∞
2041	col. lq.	$1.05^{24°}$	....	130	i.	∞	∞
2042	nd./bz.	$1.383^{60°}_{25}$	45	209-10	$0.45^{20°}$ v. s. chl.	v. s.; $160^{25°}$ CCl_4	v. s.; v. s. bz.
2043	col. cr.	....	81-2	....	i.	s. h.	s.
2044	nd./aq.	....	105	....	sl. s. h.	s.	s.; s. ac.
2045	nd./aq. HCl	....	....	....	....	....	....
2046	col. lq.	$1.34^{25°}_{25}$	<0	$255-7^{5mm}$	i.; ∞ bz.	v. s.	∞ CCl_4
2047	col. nd.	....	118-9	....	i.	s. h.	i.
2048	nd./aq. al.	....	123.5	....	....	s.	s.
2049	col. pd.	....	....	....	s.	sl. s.	i.
2050	lq.	$1.143^{10°}$	....	87	v. sl. s.	s.	....
2051	col. lq.	$1.155^{20°}_{4}$	<-70	96.4	$0.27^{20°}$	v. s.	v. s.
2052	lq.	$1.186^{20°}_{4}$	....	123-5	$0.27^{25°}$	s.	s.
2053	lq.	$1.091^{20°}_{4}$	-33.8	70.5	i.	s.	∞ CS_2
2054	col. lq.	$1.389^{22°}_{4}$	....	185-90	v. s.	v. s.	....
2055	nd.	....	50 (36)	210 d.	s.	s.	....
2056	col. lq.	$1.431^{15°}$	....	105	i.	∞	∞
2057	col. lq.	....	....	265-70	....	....	....
2058	lq.	$1.176^{19.5°}_{0}$	....	78	i.	....	....
2059	lq.	....	....	75	i.	....	....

*β modification, liquid, changes to α at 15°.
Dichloro-propyl nitrate 2028

Table 7-4 (*Continued*)
PHYSICAL CONSTANTS OF ORGANIC COMPOUNDS

No.	Name	Synonym	Formula	Beil. Ref.	Formula Weight
	Dichloro				
2060	propylene (β,γ)	epidichlorohydrin (α)	$CH_2:CCl\cdot CH_2Cl$	I-199	110.97
2061	propylene (α,γ)	epidichlorohydrin (β)	$CH_2Cl\cdot CH:CHCl$	I-199	110.97
2062	quinazoline (2,4)	....	$C_6H_4\cdot N:CCl\cdot N:CCl$	XXIII-176	199.04
2063	quinoline (2,3)	....	$C_9H_5NCl_2$	XX-361	198.05
2064	quinoline (2,4)	....	$C_9H_5NCl_2$	XX-361	198.05
2065	quinoline (2,7)	....	$C_9H_5NCl_2$	XX-361	198.05
2066	quinoline (5,6)	....	$C_9H_5NCl_2$	XX-361	198.05
2067	quinoline (5,8)	....	$C_9H_5NCl_2$	XX-362	198.05
2068	quinoline (6,8)	....	$C_9H_5NCl_2$	XX-362	198.05
2069	quinoline (7,8)	....	$C_9H_5NCl_2$	....	198.05
2070	quinone (2,5)	....	$Cl_2C_6H_2O_2$	VII-633	176.99
2071	quinone (2,6)	....	$Cl_2C_6H_2O_2$	VII-633	176.99
2072	quinone-chloro-imide	....	$O:C_6H_2Cl_2:NCl$	VII-634	210.45
2073	tetrabromoethane (2,2;1,1,1,2)	(2,2;1,1,1,2)	$Br_3C\cdot CCl_2Br$	I-95	414.56
2074	*p*-toluenesulfon-amide (*N,N*)	dichloramine T	$CH_3\cdot C_6H_4\cdot SO_2NCl_2$	XI-107	240.11
2075	tribromoethane	....	$Br_2CCl\cdot CHBrCl$	I-94	335.66
2076	tribromoethane	....	$Br_2CH\cdot CCl_2Br$	I-94	335.66
2077	tyrosine (3,5)(*dl*)	....	$Cl_2C_6H_2(OH)\cdot CH_2\cdot$ $CHNH_2CO_2H\cdot 2H_2O$	*XIV-670	286.11
2078	**Dicinnamalacetone**	....	$[C_6H_5(CH:CH)_2]_2CO$	VII-524	286.38
2079	**Dicresol** (*o*)	4,4'-diOH;3,3'-diMe	$(CH_3\cdot C_6H_3OH)_2$	VI-1009	214.27
2080	**Dicyandiamide**	param	$H_2N\cdot C(:NH)\cdot NH\cdot CN$	III-91	84.08
2081	**Dicyano**-diamidine	guanyl urea; biuret-amidine	$H_2N\cdot C(:NH)\cdot NH\cdot$ $CONH_2$	III-89	102.10
2082	diamidine H_2SO_4	....	$C_2H_6ON_4\cdot\frac{1}{2}H_2SO_4\cdot$ H_2O	III-90	169.15
2083	**Dicyclohexyl**	decahydro-diPh	$(CH_2(CH_2)_4CH\cdot)_2$	V-108	166.31
2084	adipate	....	$(CH_2CH_2CO_2C_6H_{11})_2$	**VI-11	310.44
2085	amine	....	$(C_6H_{11})_2NH$	XII-6	181.32
2086	maleate	....	$(:CHCO_2C_6H_{11})_2$	....	280.37
2087	oxalate	....	$(CO_2C_6H_{11})_2$	*VI-6	254.33
2088	phthalate	....	$C_6H_4(CO_2C_6H_{11})_2$	IX-799	330.43
2089	**Dicyclopentadiene**	....	$(\cdot CH:CH\cdot CH_2\cdot$ $CH:CH)_2$	V-495	132.21
2090	**Diethanolamine**	iminoethyl alcohol	$HN(CH_2\cdot CH_2OH)_2$	IV-283	105.14
2091	**Diethoxy**-benzene (*o*)	....	$(C_2H_5O)_2C_6H_4$	VI-771	166.22
2092	benzene (*m*)	....	$(C_2H_5O)_2C_6H_4$	VI-814	166.22
2093	benzene (*p*)	....	$(C_2H_5O)_2C_6H_4$	VI-844	166.22
2094	ethyl carbonate	....	$(C_2H_5O\cdot C_2H_4)_2CO_3$	....	206.24
2095	ethyl maleate	....	$(:CHCO_2C_2H_4OC_2H_5)_2$	....	260.29
2096	ethyl phthalate	....	$C_6H_4(CO_2C_2H_4OC_2H_5)_2$	....	310.35
2097	ethyl sebacate	....	$[C_2H_5OC_2H_4CO_2\cdot$ $CH_2(CH_2)_2CH_2\cdot]_2$	....	346.47
2098	quinazoline (2,4)	....	$C_{12}H_{14}O_2N_2$	XXIII-486	218.26

Dichloro-sulfanilic acid 1938
Dichloro-xylene (ω) 6494-6

Dicresyl carbonate 2835
Dicresyl phthalate 2841

Table 7-4 (*Continued*)
PHYSICAL CONSTANTS OF ORGANIC COMPOUNDS

No.	Crystalline Form and Color	Specific Gravity	Melting Point °C.	Boiling Point °C.	Solubility in 100 Parts		
					Water	Alcohol	Ether
2060	col. lq.	$1.204^{25°}$	....	94	i.	∞	∞
2061	col. lq.	$1.233^{17.5°}$	....	106-9	....	....	....
2062	nd.	....	120	....	i.	s.; i. lg.	s.; s. bz.
2063	cr./aq. al.	....	104-5	....	i.; i. alk.	s.; s. bz.	s.; sl. s. lg.
2064	nd./aq. al.	....	67	280-2	v. sl. s. h.	s.; s. chl.	s.; s. bz.
2065	nd./al.	....	98	....	sl. s.	s.	....
2066	nd.	....	85	....	sl. s.	s.	s. pet.
2067	nd./al.	....	92-3	....	....	s.	s.
2068	nd./al.	....	103-4	....	....	s.	....
2069	nd.	....	85.5	....	....	....	....
2070	lt. yel./bz.	....	155-60	....	sl. s. h.	v. s.	v. s.; sl. s. bz.
2071	rhb./al.	....	120-1	subl. <120	sl. s. h.	v. s. h.	s. chl.
2072	yel. nd./al.	....	67-8	d. 170	v. sl. s.	s. h.	v. s.; s. chl.
2073	cr.	....	180 d.	....	....	....	....
2074	pr./chl.-pet.	....	83	....	v. sl. s.	d. h.; s. chl.	120 bz.
2075	col. lq.	$2.626^{21.5°}$	−5	133^{35mm}	....	....	....
2076	col. lq.	$2.632^{\frac{15}{4}°}$	16.8	210	....	....	....
2077	pr./aq.	....	252±, d.	....	4; sl. s. bz.	v. sl. s.	v. sl. s.
2078	yel. nd./al.	....	146	....	i.; s. ac.	s. h.	sl. s.
2079	cr./PhMe	....	161	....	s. h.; s. ac.	s.; s. h. bz.	s.
2080	mn. pl.	$1.401^{4°}$	209-11	d.	$2.3^{13°}$	$1.3^{13°}$	$0.01^{13°}$
2081	cr./al.	....	105	d. 160	s. h.	sl. s. c.	i.
2082	col. nd.	....	193-5	−H_2O, 110	5 c.; 33 h.	sl. s.; s. dil. a.	i.
2083	lq.	$0.886^{\frac{21}{4}°}$	3.6	236^{758mm}	i.; ∞ act.	$7^{25°}$ MeOH	∞ ; ∞ bz.
2084	col. nd.	....	38-9	$208-12^{9mm}$	i.	s.	s.
2085	col. lq.	$0.925^{18°}$	20±	$254-6^{745mm}$	$0.16^{28°}$	v. s.; s. bz.	∞
2086	col. pl.	....	82-3	....	i.	sl. s.	v. s.
2087	cr./MeOH	....	42-3	$190-17^{3mm}$	....	v. s.	v. s.
2088	pr./al.	....	66	....	i.	s.	....
2089	col. cr.	$0.976^{35°}$	32.9 (19)	170 sl. d.	....	v. s.	v. s.
2090	pr.	$1.097^{\frac{20}{4}°}$	28	270^{748mm}	∞ ; i. bz.	∞	v. sl. s.
2091	cr./pet.	....	43-5	....	....	....	....
2092	pr.	....	12.4	$234-5^{756mm}$	i.	s.	s.
2093	mn. lf.	....	71-2	246	i.	s.; s. bz.	s.; s. chl.
2094	col. lq.	....	....	$124-7^{15mm}$	....	....	....
2095	col. lq.	....	....	$174-7^{11mm}$	....	....	....
2096	col. lq.	....	31-3	....	i.	s.	s.
2097	col. lq.	....	−1±	$224-6^{8mm}$	....	....	....
2098	nd.	....	55	....	i.	s.	s.

Diethoxy-ethane 2139
Diethoxy-methane 2172

Table 7-4 (*Continued*)
PHYSICAL CONSTANTS OF ORGANIC COMPOUNDS

No.	Name	Synonym	Formula	Beil. Ref.	Formula Weight
2099	**Diethyl-**acetalde- hyde	2-Et-butyraldehyde	$(C_2H_5)_2CH \cdot CHO$	I-693	100.16
2100	acetamide (*N,N*)	acetyl diEt-amine	$CH_3CO \cdot N(C_2H_5)_2$	IV-110	115.18
2101	acetic acid	. . . .	$(C_2H_5)_2CH \cdot CO_2H$	II-333	116.16
2102	acetoacetic ester	. . . .	$CH_3CO \cdot C(C_2H_5)_2 \cdot CO_2C_2H_5$	III-710	186.25
2103	acetonedicar- boxylate	. . . .	$CO(CH_2CO_2C_2H_5)_2$	III-791	202.21
2104	acetonitrile	3-cyano-pentane	$(C_2H_5)_2CH \cdot CN$	II-334	97.16
2105	acetyl succinate	. . . .	$CH_3CO \cdot C_2H_3 : (CO_2C_2H_5)_2$	III-801	216.24
2106	adipate	. . . .	$[\cdot(CH_2)_2CO_2C_2H_5]_2$	II-652	202.25
2107	allyl-acetamide	novonal	$(C_2H_5)_2C_3H_5 : C \cdot CONH_2$	**II-418	155.24
2108	allyl-malonate	. . . .	$C_3H_5 \cdot CH(CO_2C_2H_5)_2$	II-776	200.24
2109	amine	. . . .	$(C_2H_5)_2NH$	IV-95	73.14
2110	amine hydro- chloride	. . . .	$(C_2H_5)_2NH \cdot HCl$	IV-97	109.60
2111	aminobenzalde- hyde	(*p*)	$(C_2H_5)_2N \cdot C_6H_4 \cdot CHO$	XIV-36	177.25
2112	aminoethanol (*β*)	. . . .	$(C_2H_5)_2N \cdot CH_2CH_2 \cdot OH$	IV-282	117.19
2113	aminoglycerol (*α*)	1-diethylamino- 2,3-propandiol	$(C_2H_5)_2N \cdot CH_2 \cdot CHOH \cdot CH_2OH$	IV-302	147.22
2114	aminophenetole (*m*)	diEt-*m*-pheneti- dine	$C_2H_5O \cdot C_6H_4 \cdot N(C_2H_5)_2$	XIII-410	193.29
2115	aminophenol (*m*)	. . . .	$(C_2H_5)_2N \cdot C_6H_4 \cdot OH$	XIII-408	165.24
2116	aminophenol oxalate (*m*)	hydroxy-diEt- aniline oxalate	$[(C_2H_5)_2N \cdot C_6H_4 \cdot OH]_2 : H_2C_2O_4$	XIII-410	420.51
2117	aminopropyl alco- hol (*β*)	. . . .	$(C_2H_5)_2N \cdot CH(CH_3) \cdot CH_2OH$	**IV-734	131.22
2118	aminopropyl alco- hol (*γ*)	. . . .	$(C_2H_5)_2N \cdot CH_2 \cdot CH_2 \cdot CH_2OH$	IV-288	131.22
2119	aminopropyl cin- namate HCl	apothesine	$(C_2H_5)_2N(CH_2)_3O \cdot CO \cdot C_8H_7 \cdot HCl$	**IX-390	297.83
2120	ammonium diethyl- dithiocarbamate	. . . .	$(C_2H_5)_2N \cdot CS \cdot S \cdot NH_2(C_2H_5)_2$	IV-121	222.42
2121	aniline (*N*)	. . . .	$(C_2H_5)_2N \cdot C_6H_5$	XII-164	149.24
2122	aniline sulfonic acid (*m*)	diethylaminoben- zene sulfonic	$(C_2H_5)_2N \cdot C_6H_4 \cdot SO_3H$	XIV-690	229.30
2123	azelate	ethyl azelate	$(CH_2)_7(CO_2C_2H_5)_2$	II-709	244.33
2124	barbituric acid (5,5)	veronal; barbital	$C_8H_{12}O_3N_2$	XXIV-485	184.20
2125	barbituric Na	medinal	$C_8H_{11}O_3N_2Na$	XXIV-487	206.18
2126	benzene (*o*)	. . . .	$(C_2H_5)_2C_6H_4$	V-426	134.22
2127	benzene (*m*)	. . . .	$(C_2H_5)_2C_6H_4$	V-426	134.22
2128	benzene (*p*)	. . . .	$(C_2H_5)_2C_6H_4$	V-426	134.22
2129	benzoylene urea (1,3)	. . . .	$C_6H_4 \cdot N(C_2H_5) \cdot CO \cdot \underline{\qquad} \ N(C_2H_5) \cdot CO \ \underline{\qquad}$	XXIV-376	218.26
2130	benzylmalonate	. . . .	$C_7H_7 \cdot CH(CO_2C_2H_5)_2$	IX-869	250.30
2131	bromoacetamide (*α*)	neuronal	$(C_2H_5)_2CBr \cdot CO \cdot NH_2$	II-334	194.08
2132	bromomaleate	. . . .	$CH : CBr(CO_2C_2H_5)_2$	II-755	251.08

Diethyl acetal 4
Diethyl-amino–propylene glycol 2113

Diethyl beryllium 845

Table 7-4 (*Continued*)
PHYSICAL CONSTANTS OF ORGANIC COMPOUNDS

No.	Crystalline Form and Color	Specific Gravity	Melting Point °C.	Boiling Point °C.	Solubility in 100 Parts		
					Water	Alcohol	Ether
2099	lq.	$0.816\frac{20°}{20}$	−89	117-8	$0.3^{20°}$	∞	∞
2100	lq.	$0.925^{8.5°}$	....	185-6	....	....	....
2101	col. lq.	$0.920\frac{18°}{0}$	<−15	195-7	sl. s.	s.	s.
2102	col. lq.	$0.965\frac{25°}{4}$	....	212	....	....	....
2103	oil	$1.113\frac{20°}{4}$	....	250	v. sl. s.	∞	∞
2104	oil	....	....	144-6	....	∞	∞
2105	lq.	$1.080\frac{25°}{5}$	....	254-6 d.	i.	s.	....
2106	col. lq.	$1.007\frac{20°}{4}$	−19.8	133.8^{15mm}	$0.43^{30°}$	s.	s.
2107	wh. pd.	....	80	155^{16mm}	0.9	s.	s.
2108	col. lq.	$1.006\frac{25°}{25}$	....	222-3	i.	s.	s.
2109	col. lq.	$0.709\frac{15°}{15}$	−50	55.5^{759mm}	v. s.*	∞	∞
2110	lf./al. et.	$1.048\frac{21°}{4}$	228-9	320-30	$232^{25°}$	sl. s. c.	i.
2111	yel. nd./ aq.	....	41	$1747mm$	....	s.	s.
2112	lq.	$0.885\frac{20°}{20}$	....	162.1	∞	∞	s.
2113	syrup	....	....	233-5	s.	s.	s.; s. chl.
2114	oil	....	....	286	i.	s.	s. ac.
2115	rhb./CS₂- lg.	....	78	276-80	s.; s. chl.	s. CS$_2$	i. lg.
2116	cr.	....	....	155-6	....	....	....
2117	lq.	....	....	$61-3^{25mm}$	....	....	....
2118	lq.	....	....	189.5	v. s.	....	....
2119	col. cr.	....	136	....	s.	s.; sl. s. act.	sl. s.
2120	pa. yel. pl.	....	82-3	....	v. s.	v. s.	v. sl. s.
2121	oil	$0.935\frac{20°}{4}$	−21.3(−34)	217.5	$1.4^{12°}$	s.	s.
2122	cr.	....	270 d.	....	s.	....	....
2123	lq.	$0.973\frac{20°}{4}$	−16	291-2	i.	s.	s.
2124	col. cr.	....	191	subl. vac.	0.7 c.; 8 h.	s. h.	s.; s. alk.
2125	col. cr.	....	....	....	20 c.; 40 h.	0.25	i.
2126	col. lq.	$0.881\frac{20°}{4}$	−31.4	183.5	i.	s.	s.
2127	col. lq.	$0.864\frac{20°}{4}$	−83.9	181.1	i.	s.	s.
2128	col. lq.	$0.862\frac{20°}{4}$	−43.2	183.8	i.	s.	s.
2129	nd.	....	110-11	....	i.	s.	....
2130	lq.	$1.077\frac{15°}{15}$	....	298-300	i.	....	....
2131	col. cr.	....	66-7	d. 160-70	0.8 c.; s. h.	v. s.; s. bz.	v. s.
2132	col. lq.	$1.410^{17.5°}$	....	256	....	....	....

* Forms a hydrate $+1H_2O$, m. p. −19°.

Table 7-4 (*Continued*)
PHYSICAL CONSTANTS OF ORGANIC COMPOUNDS

No.	Name	Synonym	Formula	Beil. Ref.	Formula Weight
	Diethyl				
2133	bromomalonate	Et bromomalonate	$CHBr(CO_2C_2H_5)_2$	II-594	239.07
2134	*n*-butylmalonate	. . . .	$C_4H_9 \cdot CH(CO_2C_2H_5)_2$	*II-282	216.28
2135	butylmalonate	(*sec*)	$C_5H_{10}(CO_2C_2H_5)_2$	II-679	216.28
2136	butylmalonate	(*iso*)	$C_5H_{10}(CO_2C_2H_5)_2$	II-683	216.28
2137	carbanilide	. . . .	$[(C_2H_5)(C_6H_5)N]_2CO$	XII-422	268.36
2138	carbonate	. . . .	$CO(OC_2H_5)_2$	III-5	118.13
2139	cellosolve	diEtO-ethane	$(C_2H_5O \cdot CH_2)_2$	I-468	118.18
2140	chlorofumarate	. . . .	$CH:CCl(CO_2C_2H_5)_2$	II-745	206.63
2141	chloromaleate	. . . .	$CH:CCl(CO_2C_2H_5)_2$	II-753	206.63
2142	citraconate	. . . .	$CH_3 \cdot C_2H(CO_2C_2H_5)_2$	II-771	186.21
2143	cyanamide	cyano-diEt-amine	$(C_2H_5)_2N \cdot CN$	IV-121	98.15
2144	*iso*-cyanine iodide	ethyl red	$C_{23}H_{23}N_2 \cdot I \cdot C_2H_5OH$	XXIII-298	500.43
2145	cyclohexylamine (*N,N*)	hexahydro-diethyl-aniline	$CH_2(CH_2)_4CH \cdot$ $\|_____\|$ $N(C_2H_5)_2$	XII-6	155.29
2146	diacetosuccinate	. . . .	$(CH_3CO)_2(CHCO_2 \cdot C_2H_5)_2$	III-840	258.27
2147	diacetosuccinate	. . . .	$(CH_3CO)_2(CHCO_2 \cdot C_2H_5)_2$	III-840	258.27
2148	diacetosuccinate	. . . . , ...	$(CH_3CO)_2(CHCO_2 \cdot C_2H_5)_2$	III-840	258.27
2149	diacetosuccinate	. . . . , ...	$(CH_3CO)_2(CHCO_2 \cdot C_2H_5)_2$	III-840	258.27
2150	diacetosuccinate	. . . .	$(CH_3CO)_2(CHCO_2 \cdot C_2H_5)_2$	III-840	258.27
2151	diacetosuccinate	. . . .	$(CH_3CO)_2(CHCO_2 \cdot C_2H_5)_2$	III-840	258.27
2152	diacetyl tartrate	. . . .	$(CH_3CO)_2(O \cdot CH \cdot CO_2C_2H_5)_2$	III-515	290.27
2153	dibenzyl malonate	. . . .	$(C_7H_7)_2C(CO_2C_2H_5)_2$	IX-937	340.42
2154	dibromomaleate	. . . .	$(:CBrCO_2C_2H_5)_2$	II-757	329.97
2155	dibromomalonate	. . . .	$Br_2C(CO_2C_2H_5)_2$	II-595	317.97
2156	dibromosuccinate	(*α,α'*)	$(CHBr \cdot CO_2C_2H_5)_2$	II-624	332.00
2157	dibutylmalonate	(*n*)	$(C_4H_9)_2C(CO_2C_2H_5)_2$	*II-295	272.39
2158	diethylmalonate	. . . .	$(C_2H_5)_2C(CO_2C_2H_5)_2$	II-686	216.28
2159	1,4-dihydrocollidine dicarboxylate	(3,5)	$(CH_3)_3C_5H_2N: (CO_2C_2H_5)_2$	XXII-147	267.33
2160	dihydroxymalonate	diEt-mesoxalate	$(HO)_2C(CO_2C_2H_5)_2$	III-769	192.17
2161	dimethylmalonate	. . . .	$(CH_3)_2C(CO_2C_2H_5)_2$	II-648	188.23
2162	dithiocarbamate †	. . . .	$(C_2H_5)_2N \cdot CS \cdot SH$	IV-121	149.28
2163	dithiocarbonate	carbonyl-disulfethyl	$CO(SC_2H_5)_2$	III-211	150.26
2164	dithiocarbonate	Et-xanthate	$C_2H_5O \cdot CS \cdot SC_2H_5$	III-210	150.26
2165	dithioloxalate	. . . .	$(CO \cdot SC_2H_5)_2$	II-565	178.27
2166	disulfide	. . . .	$C_2H_5S \cdot SC_2H_5$	I-347	122.25
2167	ethyl-*iso*-amyl-malonate	. . . .	$C_5H_{11} \cdot C(C_2H_5): (CO_2C_2H_5)_2$	**II-611	258.36
2168	ethyl-*n*-butyl-malonate	. . . .	$C_4H_9 \cdot C(C_2H_5): (CO_2C_2H_5)_2$	II-712	244.33
2169	ethylmalonate	. . . .	$C_2H_5 \cdot CH(CO_2C_2H_5)_2$	II-644	188.23

† Free acid unstable; Na and K salts s. in aq.   Diethyl *iso*-butyl carbinol 4924
Diethyl cadmium 1185   Diethyl carbinol 408

Table 7-4 (*Continued*)
PHYSICAL CONSTANTS OF ORGANIC COMPOUNDS

No.	Crystalline Form and Color	Specific Gravity	Melting Point °C.	Boiling Point °C.	Solubility in 100 Parts		
					Water	Alcohol	Ether
2133	lq.	$1.402^{25}_{4}°$	<-54	233-5 d.	i.	∞	∞
2134	col. lq.	$0.975^{20}_{4}°$	....	235-40^{760mm}	i.	v. s.	v. s.
2135	col. lq.	$0.988^{15°}$	....	233-4^{774mm}	v. sl. s.	v. s.	v. s.
2136	col. lq.	$0.983^{17°}$	....	225	v. sl. s.	v. s.	v. s.
2137	cr./al.	....	79	....	i.	....	....
2138	col. lq.	$0.975^{20}_{4}°$	-43	126.8	i.	∞	∞
2139	col. lq.	$0.842^{20}_{20}°$	....	124^{759mm}	21$^{20°}$	....	....
2140	col. lq.	$1.189^{20}_{4}°$	....	250 sl. d.	....	....	....
2141	col. lq.	$1.191^{25}_{4}°$	....	235 sl. d.	....	....	....
2142	col. lq.	$1.042^{20}_{4}°$	....	230.3	....	....	....
2143	lq.	$0.854^{20}_{4}°$	....	188-9^{748mm}	i.; d. HCl	s.	s.
2144	hyg. gn. met./al.	....	150-2 d.	....	v. sl. s; v. s. aq. a.	s.; s. act.; s. ac.	i.; i. CS$_2$
2145	lq.	$0.872^{0}_{0}°$	....	193	....	....	....
2146	α_1-oil	....	....	....	v. sl. s.	s.	s.; 10 lg.
2147	α_2-cr.	....	20-2	....	v. sl. s.	v. s.	v. s.
2148	α_3-pr./lg.	....	31-2	....	v. sl. s.	v. s.	v. s.; 35 lg.
2149	α_4-oil	....	....	....	....	....	....
2150	β-stable, mn./al.	$1.209^{20}_{4}°$	89-90	....	....	6$^{20°}$ abs.	2 abs.
2151	meta-stable	$1.176^{20}_{4}°$	....	....	....	....	....
2152	mn.	$1.081^{99}_{4}°$	67-8	288.5^{727mm}	s. h.	v. s. h.	v. s.
2153	oil	$1.093^{20}_{4}°$	13-14	243-6$^{18°}$	....	....	....
2154	col. lq.	$1.698^{25}_{25}°$	....	170-5^{15mm}	....	....	....
2155	lq.	....	....	250-6 sl. d.	....	....	....
2156	rhb. nd.	....	68	d. >130	....	....	....
2157	col. lq.	....	....	153-4^{14mm}	....	....	....
2158	col. lq.	$0.985^{20}_{4}°$	....	230	i.	∞	∞
2159	mn. pr./al.	....	131	>315 d.	v. sl. s.; s. chl.	s. h.; s. bz.	sl. s.; sl. s. CS$_2$
2160	col. pl./bz	....	57	200	130$^{22°}$	v. s. abs.	v. s.
2161	col. lq.	$0.994^{25}_{25}°$	....	196.2-6.7	i.	∞	∞
2162	....	....	....	....	i.	....	....
2163	yel. lq.	$1.085^{19°}$	....	198.6-200.1	i.	s.	s.
2164	yel. lq.	$1.085^{19°}$	....	200	i.	s.	s.
2165	yel. nd./et.	....	27-8	238-40^{757mm}	....	....	....
2166	oil	$0.993^{20}_{4}°$	....	152.8-3.4	v. sl. s.	....	....
2167	col. lq.	$0.954^{25}_{25}°$	....	150^{20mm}	....	....	....
2168	lq.	$0.965^{25}_{4}°$	....	245-50^{747mm}	....	....	....
2169	col. lq.	$1.004^{20}_{20}°$	....	211^{748mm}	....	....	....

Diethyl carbinol acetate 401
Diethyl carbitol 2233

Diethyl ether 2911

Table 7-4 (*Continued*)
PHYSICAL CONSTANTS OF ORGANIC COMPOUNDS

No.	Name	Synonym	Formula	Beil. Ref.	Formula Weight
2170	**Diethyl** ethyl-phe-nylmalonate	....	$C_6H_5 \cdot C(C_2H_5):$ $(CO_2C_2H_5)_2$	*IX-384	264.32
2171	ethyl-*iso*-propyl-malonate	....	$C_3H_7 \cdot C(C_2H_5):$ $(CO_2C_2H_5)_2$	II-706	230.31
2172	formal	diEtO-methane	$CH_2(OC_2H_5)_2$†	I-574	104.15
2173	formamide	formyl-diethylamine	$HCON(C_2H_5)_2$	IV-109	101.15
2174	fumarate	....	$(:CH \cdot CO_2C_2H_5)_2$	II-742	172.18
2175	glutaconate	....	$C_3H_4(CO_2C_2H_5)_2$	II-759	186.21
2176	glutarate	....	$(CH_2)_3(CO_2C_2H_5)_2$	II-633	188.23
2177	*n*-heptylmalonate	....	$C_7H_{15}CH(CO_2C_2H_5)_2$	**II-610	258.36
2178	hydrazine (*uns*)	....	$(C_2H_5)_2N \cdot NH_2$	IV-550	88.15
2179	itaconate	....	$C_3H_4(CO_2C_2H_5)_2$	II-762	186.21
2180	ketone	pentanone-3	$(C_2H_5)_2CO$	I-679	86.13
2181	ketone semicar-bazone	....	$(C_2H_5)_2C:N \cdot NH \cdot CO \cdot NH_2$	III-103	143.19
2182	malate (*dl*)	....	$(CHOH \cdot CH_2):$ $(CO_2C_2H_5)_2$	III-437	190.20
2183	maleate	....	$(:CH \cdot CO_2C_2H_5)_2$	II-751	172.18
2184	malonate	malonic ester	$CH_2(CO_2C_2H_5)_2$	II-573	160.17
2185	malonic acid	....	$(C_2H_5)_2C(CO_2H)_2$	II-686	160.17
2186	mesaconate	....	$CH_3 \cdot C_2H(CO_2C_2H_5)_2$	II-766	186.21
2187	methyl-ethyl-mal-onate	....	$C_2H_5 \cdot C(CH_3):$ $(CO_2C_2H_5)_2$	II-664	202.25
2188	muconate	....	$(CH:CH \cdot CO_2C_2H_5)_2$	II-804	198.22
2189	α-naphthylamine (*N*)	....	$C_{10}H_7 \cdot N(C_2H_5)_2$	XII-1223	199.30
2190	4-nitrophthalate	....	$NO_2 \cdot C_6H_3(CO_2C_2H_5)_2$	IX-831	267.24
2191	nitrosamine	nitroso-diethyl-amine	$(C_2H_5)_2N \cdot NO$	IV-129	102.14
2192	oxalacetate	oxalacetic ester	$(CH_2CO)(CO_2C_2H_5)_2$	III-782	188.18
2193	oxalate	ethyl oxalate	$(CO_2C_2H_5)_2$	II-535	146.14
2194	oxamide (*sym*)	....	$(CO \cdot NHC_2H_5)_2$	IV-112	144.17
2195	oxomalonate	....	$CO(CO_2C_2H_5)_2$	III-769	174.15
2196	peroxide	....	$C_2H_5O \cdot OC_2H_5$	I-324	90.12
2197	phosphine	....	$(C_2H_5)_2PH$	IV-582	90.11
2198	phosphoric acid	....	$(C_2H_5O)_2PO \cdot OH$	I-332	154.10
2199	phthalate (*o*)	ethyl phthalate	$C_6H_4(CO_2C_2H_5)_2$	IX-798	222.24
2200	phthalate (*m*)	ethyl isophthalate	$C_6H_4(CO_2C_2H_5)_2$	IX-834	222.24
2201	phthalate (*p*)	ethyl terephthalate	$C_6H_4(CO_2C_2H_5)_2$	IX-844	222.24
2202	pimelate	....	$(CH_2)_5(CO_2C_2H_5)_2$	II-671	216.28
2203	propylmalonate	....	$C_3H_7 \cdot CH(CO_2C_2H_5)_2$	II-657	202.25
2204	*iso*-propylmalonate	....	$C_3H_7 \cdot CH(CO_2C_2H_5)_2$	II-669	202.25
2205	quinolinate (2,3)	....	$C_5H_3N(CO_2C_2H_5)_2$	XXII-151	223.23
2206	sebacate	ethyl sebacate	$(CH_2)_8(CO_2C_2H_5)_2$	II-719	258.36
2207	selenide, mono-	selenium ethyl	$(C_2H_5)_2Se$	I-349	137.08
2208	selenide, di-	ethyl diselenide	$(C_2H_5Se)_2$	I-349	216.04
2209	stilboestrol (*trans*)	stilboestrol	$[HOC_6H_4(C_2H_5)C:]_2 \cdot$ C_2H_5OH	....	314.43
2210	suberate	....	$(CH_2)_6(CO_2C_2H_5)_2$	II-693	230.31
2211	succinate	ethyl succinate	$(CH_2 \cdot CO_2C_2H_5)_2$	II-609	174.20
2212	*iso*-succinate	....	$CH_3 \cdot CH(CO_2C_2H_5)_2$	II-629	174.20
2213	sulfate	ethyl sulfate	$(C_2H_5O)_2SO_2$	I-327	154.19
2214	sulfide	....	$(C_2H_5)_2S$	I-344	90.19
2215	sulfite	....	$(C_2H_5O)_2SO$	I-325	138.19

† Hydrate $+1H_2O$, s. aq., al., et., chl., bz.
Diethyl mercury 4043

Diethyl mesoxalate 2160, 3084
Diethyl phenetidine 2114

Table 7-4 (Continued)
PHYSICAL CONSTANTS OF ORGANIC COMPOUNDS

No.	Crystalline Form and Color	Specific Gravity	Melting Point °C.	Boiling Point °C.	Solubility in 100 Parts		
					Water	Alcohol	Ether
2170	col. lq.	....	....	166^{12mm}	....	....	....
2171	lq.	....	....	232-3	....	....	....
2172	lq.	$0.824\frac{25°}{4}$	−66.5	88.0	$9^{18°}$; $7^{30°}$	∞	∞
2173	col. lq.	$0.908^{19°}$	....	177-8	∞	v. s.	v. s.
2174	col. lq.	$1.052\frac{20°}{4}$	0.6	217.9	....	....	....
2175	col. lq.	$1.050\frac{20°}{4}$	....	236-8	i.	s.	s.
2176	syrup	$1.027\frac{15°}{4}$	−23.8	233.7	$0.882^{20°}$	v. s.	s.
2177	pa. yel. lq.	$0.951^{20°}$	....	$144-6^{8mm}$	....	....	....
2178	hyg. lq.	....	....	96-9	v. s.	v. s.	v. s.
2179	col. lq.	$1.050\frac{15°}{15}$	....	228-9	....	....	....
2180	col. lq.	$0.810\frac{25°}{4}$	−39.9	102.0	$4.7^{20°}$	∞	∞
2181	nd./bz.	....	139	....	....	....	s. bz.
2182	col. lq.	$1.124\frac{21°}{4}$	....	253-5	s.	∞	∞
2183	col. lq.	$1.070\frac{20°}{20}$	−10.5	225	i.	s.	s.
2184	col. lq.	$1.055\frac{20°}{4}$	−51.5	199.3	$2.08^{20°}$	∞	∞
2185	pr./aq.	....	125	d. 170-80	$65^{16°}$	v. s.	v. s.
2186	col. lq.	$1.047\frac{20°}{4}$	....	229	....	....	....
2187	col. lq.	$0.994\frac{15°}{15}$	....	207-8	....	....	....
2188	pr./al.	$0.983\frac{99°}{4}$	63 (13)	200^{12mm}	....	v. s.	....
2189	col. oil	1.005	....	285-90	∞ bz.	∞	∞
2190	pl./al.	....	33-4	$208-10^{15mm}$	i.	s.	s.
2191	yel. oil	$0.943\frac{20°}{4}$	....	176.9	s.	∞	∞
2192	col. oil	$1.131\frac{20°}{4}$	....	$131-2^{24mm}$	i.	∞; ∞ bz.	∞
2193	col. lq.	$1.079\frac{20°}{4}$	−40.6	185.4	v. sl. s.	∞	∞
2194	nd./al.	$1.169^{4°}$	....	179	....	....	....
2195	yel. gn. oil	$1.119\frac{20°}{20}$	−30±	220±	....	s.	s.
2196	lq.	$0.827\frac{15°}{4}$	....	65	v. sl. s.	∞	∞
2197	col. lq.	<1	....	85	....	....	....
2198	lq.	$1.175^{0°}$	....	203.3	....	....	....
2199	col. lq.	$1.121\frac{25°}{25}$	....	289.5	i.	∞	∞
2200	col. lq.	$1.123\frac{25°}{25}$	11.5	302	i.	∞	∞
2201	pr./al.	$1.110\frac{45°}{4}$	43-4	302	i.	s.	s.
2202	col. oil	$0.999\frac{15°}{15}$	−23.8	$252-5^{748mm}$	i.	s.	s.
2203	col. lq.	$0.993\frac{15°}{15}$	....	$225-6^{771mm}$	i.	s.	s.
2204	col. lq.	$0.993\frac{15°}{15}$	....	$215-7^{748mm}$	i.	s.	s.
2205	yel. oil	....	....	280-5 sl. d.	s.	s.	s.; s. bz.
2206	col. lq.	$0.965\frac{20°}{4}$	1.3	306^{773mm}	0.008	s.	s.
2207	lq.	$1.230\frac{28°}{4}$	....	110	i.	....	....
2208	red yel. lq.	$1.696\frac{18°}{4}$	−95	137	....	....	....
2209	col. cr./al.	....	169-71	....	i.; s. alk.	s.	s.
2210	col. lq.	$0.982\frac{20°}{4}$	5.9	282^{763mm}	i.	s.	s.
2211	col. lq.	$1.040\frac{20°}{4}$	−21.3	217.8	i.	∞	∞
2212	col. lq.	$1.021\frac{15°}{15}$	....	201.2-1.4	i.	∞	∞
2213	col. lq.	$1.172\frac{25°}{4}$	−25	210.2 sl. d.	i.; sl. d.	s.; d. h.	∞
2214	col. lq.	$0.837\frac{20°}{4}$	−103.3	92.1	$0.313^{20°}$	∞	∞
2215	col. lq.	$1.077\frac{25°}{4}$	....	157.7	s. d.	s.	....

Diethyl phenylenediamine 303
Diethyl n-propyl carbinol 4966

Diethyl iso-propyl carbinol 4963
Diethyl pyridine 5040-1

Table 7-4 (*Continued*)
PHYSICAL CONSTANTS OF ORGANIC COMPOUNDS

No.	Name	Synonym	Formula	Beil. Ref.	Formula Weight
2216	**Diethyl** sulfone	....	$(C_2H_5)_2SO_2$	I-346	122.19
2217	sulfoxide	....	$(C_2H_5)_2SO$	I-346	106.19
2218	tartrate (*d*)	ethyl tartrate	$(CHOH \cdot CO_2C_2H_5)_2$	III-512	206.20
2219	telluride	tellurium ethyl	$(C_2H_5)_2Te$	I-350	185.72
2220	thiourea (*sym*)	....	$(C_2H_5NH)_2CS$	IV-118	132.23
2221	toluene (3,5)	....	$(C_2H_5)_2C_6H_3 \cdot CH_3$	V-441	148.25
2222	urea (*sym*)	....	$(C_2H_5NH)_2CO$	IV-115	116.16
2223	urea (*uns*)	....	$(C_2H_5)_2N \cdot CO \cdot NH_2$	IV-120	116.16
2224	**Diethylene**-di-acetate	....	$(CH_2CH_2O_2C \cdot CH_3)_2$	II-143	174.20
2225	diamine	piperazine	$NH \cdot C_2H_4 \cdot NH \cdot C_2H_4$	XXIII-4	86.14
2226	diamine hydrate	piperazine hydrate	$C_4H_{10}N_2 \cdot 6H_2O$	XXIII-4	194.23
2227	dilaurate	....	$(C_{11}H_{23}CO_2C_2H_4)_2$	....	454.74
2228	disulfide	dithian-1,4	$S \cdot C_2H_4 \cdot S \cdot C_2H_4$	XIX-3	120.24
2229	glycol diacetate	....	$(CH_3CO_2C_2H_4)_2O$	II-141	190.20
2230	glycol dinitrate	....	$(O_2N \cdot O \cdot C_2H_4)_2O$	**I-521	196.12
2231	glycol monoben-zyl ether	....	$HOCH_2CH_2 \cdot O \cdot CH_2 \cdot CH_2 \cdot O \cdot CH_2C_6H_5$	....	196.25
2232	glycol *n*-butyl ether	butyl carbitol	$HOCH_2CH_2 \cdot O \cdot CH_2 \cdot CH_2 \cdot O \cdot C_4H_9$	**I-521	162.23
2233	glycol diethyl ether	diethyl carbitol	$(C_2H_5O \cdot CH_2CH_2)_2O$	**I-520	162.23
2234	glycol monolaur-ate	glaurin	$C_{11}H_{23}CO_2C_2H_4 \cdot O \cdot C_2H_4OH$	....	288.43
2235	imide oxide	morpholine	$NH \cdot (CH_2)_2 \cdot O \cdot (CH_2)_2$	XXVII-5	87.12
2236	imide oxide ethanol	morpholine ethanol	$C_4H_8O:N \cdot C_2H_4OH$	....	131.18
2237	oxide	dioxan-1,4	$O:(CH_2)_4:O$	XIX-3	88.11
2238	triamine	....	$(NH_2 \cdot C_2H_4)_2NH$	IV-255	103.17
2239	**Difluoro**-acetic acid	....	$F_2CH \cdot CO_2H$	II-193	96.03
2240	1-bromoethane (2,2)	....	$F_2CH \cdot CH_2Br$	I-89	144.95
2241	1-chloroethane (2,2)	....	$F_2CH \cdot CH_2Cl$	I-83	100.50
2242	1,2-dibromoethane	(2,2)	$F_2CBr \cdot CH_2Br$	I-92	223.85
2243	1,1-dibromoethane	(2,2)	$Br_2CH \cdot CHF_2$	I-92	223.85
2244	1,1-dichloroethane	(2,2)	$F_2CH \cdot CHCl_2$	I-85	134.94
2245	1,2-dichloroethyl-ene (1,2)	....	$FClC:CClF$	....	132.93
2246	dichloromethane	Freon-12	F_2CCl_2	I-61	120.91
2247	diphenyl (2,2')	....	$F \cdot C_6H_4 \cdot C_6H_4 \cdot F$	....	190.19
2248	diphenyl (4,4')	....	$F \cdot C_6H_4 \cdot C_6H_4 \cdot F$	V-579	190.19
2249	ethane (1,2)	ethylene fluoride	$FCH_2 \cdot CH_2F$	I-82	66.05
2250	ethylene (α,α)	1,1-difluoro-ethene	$CH_2:CF_2$	I-186	64.04
2251	1-iodoethane (2,2)	....	$F_2CH \cdot CH_2I$	I-98	191.95
2252	tetrabromoethane	(1,1,2,2;1,2)	$FBr_2C \cdot CFBr_2$	I-95	381.66
2253	tetrachloroethane	(1,1,1,2;2,2)	$Cl_3C \cdot CF_2Cl$	I-86	203.83
2254	tetrachloroethane	(1,1,2,2;1,2)	$FCl_2C \cdot CFCl_2$	....	203.83
2255	tribromoethane	(1,1,2;1,2)	$Br_2FC \cdot CHBrF$	I-94	302.75
2256	**Diformyl hydrazine**	(*sym*)	$(\cdot NH \cdot CHO)_2$	II-93	88.07
2257	**Difurfural**-cyclo-hexanone	....	$(C_4H_3O \cdot CH:)_2:C_6H_6O$	....	254.29

Diethyl tin 5949
Diethyl toluidine 4216-8
Diethyl zinc 6503
Diethylene glycol 3453

Table 7-4 (*Continued*)
PHYSICAL CONSTANTS OF ORGANIC COMPOUNDS

No.	Crystalline Form and Color	Specific Gravity	Melting Point °C.	Boiling Point °C.	Solubility in 100 Parts		
					Water	Alcohol	Ether
2216	rhb. pl.	$1.357^{20°}_{4}$	73-4	248	$15.6^{16°}$	v. s. bz.	s. h.
2217	syrup	....	15	88-90^{15mm}	v. s.	s.	s.
2218	lq.	$1.204^{20°}_{4}$	17	280	sl. s.	∞	∞
2219	red-yel. lq.	$1.599^{15°}_{4}$	....	137-8	i.	s.	....
2220	cr.	....	77	....	s.	s.	....
2221	lq.	$0.879^{20°}_{4}$	....	198-200	i.	∞	∞
2222	nd./al.	1.042	112.5	263	v. s.	v. s.	v. s.
2223	nd./et.	....	75	....	v. s.	v. s.	$2.6^{22°}$
2224	lq.	$1.048^{15°}$	12	230^{751mm}	....	....	
2225	rhb./al.	$1.110^{24°}_{20}$	110	145-6	$15^{20°}$	v. s.	i.
2226	col. cr.	....	44	125-30	v. s.	s.	i.
2227	....	....	28-30	....	....	....	....
2228	mn./et.	....	111-2	199-200	i.; v. s. CS$_2$	s.	s.
2229	lq.	$1.116^{20°}_{4}$	17-9	250	∞	....	....
2230	lq.	$1.377^{25°}_{4}$	−11.3	161±	$0.4^{24°}$	v. sl. s.	v. s.
2231	col. lq.	....	....	176-9^{18mm}	....	....	....
2232	col. lq.	$0.956^{20°}_{20}$	−68.1	231	∞ ; ∞ oils	v. s.	v. s.
2233	col. lq.	$0.909^{20°}_{20}$	−44.3	188	∞	v. s.	v. s.
2234	straw colrd. oil	0.960	17-8	>270	i.	s.	s.
2235	col. oil	$1.002^{20°}_{20}$	−3.1	128.9	∞	∞	s.
2236	col. lq.	$1.072^{20°}_{20}$	....	226	∞	....	....
2237	col. lq.	$1.034^{20°}_{4}$	11.8	101.4	∞	s.	s.
2238	lq.	$0.954^{20°}_{20}$	−39	208 sl. d.	∞	∞	i.
2239	col. lq.	$1.525^{20°}_{4}$	−0.35	134.2^{766mm}	∞	∞	∞
2240	col. lq.	$1.817^{20°}_{4}$	−74.5	57.3	v. sl. s.	∞	∞
2241	col. lq.	>1	....	36	....	....	....
2242	col. lq.	$2.242^{12.2°}$	−56.5	93	i.	....	....
2243	col. lq.	$2.312^{20°}$	....	107.5^{760mm}	....	....	....
2244	col. lq.	$1.494^{17°}$	....	60	i.	∞	∞
2245	gas	....	−112	20.9	....	....	....
2246	gas	$1.486^{-30°}$	−155	−29.2	$5.7^{26°}$ cc.	s.	s.
2247	cr.	....	117	....	....	....	....
2248	col. mn./al.	$1.336^{25°}_{4}$	94-5	254-5	i.; s. oil	s.; s. chl.	s.
2249	gas	....	....	....	....	....	....
2250	col. gas	....	....	....	i.	$150^{18°}$ cc.	$150^{18°}$ cc., chl.
2251	col. lq.	$2.243^{12.2°}$	....	89.5	i.	....	....
2252	col. cr.	....	62.5	186.5^{758mm}	i.	s.	s. ac.
2253	col. cr.	....	52	91	i.	sl. s. c.	v. s.
2254	col. cr.	$1.645^{25°}_{4}$	24.7	92.8	i.; v. s. bz.	v. s.	v. s.
2255	col. lq.	$2.603^{20°}$	....	146	i.	....	....
2256	pr.	....	159-60	....	v. s.	sl. s.	i.
2257	yel. nd	....	144-5	....	i.	s. h.	sl. s.

Diethylene glycol ethyl ether 1232
Diethylene glycol methyl ether 4177

Difluoro-ethene 2250
Difluoro-methane 4433

Table 7-4 *(Continued)*
PHYSICAL CONSTANTS OF ORGANIC COMPOUNDS

No.	Name	Synonym	Formula	Beil. Ref.	Formula Weight
2258	**Difurfural** cyclopentanone (1,3;5)	pyroxanthin	$(C_4H_3O \cdot CH:)_2:$ C_5H_4O	XIX-140	240.26
2259	**Diglycerol**	. . . .	$[(HO)_2C_3H_5]_2O$	I-513	166.18
2260	**Diglycol chlorohydrin**	Cl-OH-diEt ether	$Cl \cdot CH_2CH_2 \cdot O \cdot$ CH_2CH_2OH	I-467	124.57
2261	**Diglycolic acid**	. . . .	$(HO_2C \cdot CH_2)_2O \cdot H_2O$	III-234	152.10
2262	**Diglycyl glycine**	. . . .	$H_2N(CH_2CONH)_2 \cdot$ CH_2CO_2H	IV-374	189.17
2263	**Diheptyl-**acetic acid (*n*)	. . . .	$(C_7H_{15})_2CH \cdot CO_2H$	II-376	256.43
2264	ether (*n*)	heptyl ether	$(C_7H_{15})_2O$	I-414	214.39
2265	ketone (*n*)	caprylone	$[CH_3(CH_2)_6]_2CO$	I-717	226.41
2266	sulfide (*n*)	heptyl sulfide	$(C_7H_{15})_2S$	I-415	230.46
2267	**Dihexyl ketone** (*n*)	oenanthone	$[CH_3(CH_2)_5]_2CO$	I-715	198.35
2268	**Dihydracrylic acid**	Na salt	$O(CH_2CH_2CO_2Na)_2$	III-297	206.11
2269	**Dihydrazino-**diphenyl	(2,2')	$(H_2N \cdot NH \cdot C_6H_4)_2$	XV-584	214.27
2270	diphenyl	(4,4')	$(H_2N \cdot NH \cdot C_6H_4)_2$	XV-585	214.27
2271	**Dihydro-**acridine (9,10)	. . . .	$C_6H_4 \cdot NH \cdot C_6H_4 \cdot CH_2$	XX-443	181.24
2272	anthracene (9,10)	. . . .	$C_6H_4:(CH_2)_2:C_6H_4$	V-641	180.25
2273	anthranol (9,10;9)	. . . .	$C_6H_4 \cdot CH_2 \cdot C_6H_4 \cdot CHOH$	VI-697	196.25
2274	carveol	. . . .	$C_{10}H_{18}O$	VI-63	154.25
2275	carvone (*l*)	. . . .	$C_{10}H_{16}O$	VII-83	152.24
2276	cholesterol	coprosterol	$C_{27}H_{47}OH$	. . . .	388.68
2277	ethylanthracene (9,10;9)	. . . .	$C_6H_4:C_2H_3(C_2H_5):$ C_6H_4	V-649	208.31
2278	naphthalene (1,4)	. . . .	$C_{10}H_{10}$	V-519	130.19
2279	phenazine (9,10)	hydrazophenylene	$NH \cdot C_6H_4 \cdot NH \cdot C_6H_4$	XXIII-209	182.23
2280	*o*-phthalic acid (1,4)	. . . .	$C_6H_6(CO_2H)_2$	IX-781	168.15
2281	*o*-phthalic acid (2,4)	. . . .	$C_6H_6(CO_2H)_2$	IX-781	168.15
2282	*o*-phthalic acid (2,6)	. . . .	$C_6H_6(CO_2H)_2$	IX-782	168.15
2283	*p*-phthalic acid (1,4)	dihydro-terephthalic	$C_6H_6(CO_2H)_2$	IX-785	168.15
2284	quinoline	. . . .	C_9H_9N	. . . .	131.18
2285	resorcinol (*m*)	hydroresorcin	$CO \cdot (CH_2)_3 \cdot COCH_2$	VII-554	112.13
2286	toluene (1,2)	Me-cyclohexadiene	$CH_3 \cdot C_6H_7$	V-115	94.16
2287	toluene (1,3)	. . . .	$CH_3 \cdot C_6H_7$	V-115	94.16
2288	toluene (2,4)	. . . .	$CH_3 \cdot C_6H_7$	V-115	94.16
2289	*o*-xylene (*x,x*)	diMe-cyclohexadiene	C_8H_{12}	V-118	108.18
2290	*m*-xylene (1,5)	. . . .	C_8H_{12}	V-119	108.18
2291	*p*-xylene (1,3)	. . . .	C_8H_{12}	V-120	108.18
2292	**Dihydroxy-**acetone	. . . .	$(HO \cdot CH_2)_2CO$	I-846	90.08
2293	anthracene (β)(1,5)	rufol	$C_{14}H_8(OH)_2$	VI-1032	210.23
2294	anthracene (α)(1,8)	chrysazol	$C_{14}H_8(OH)_2$	VI-1033	210.23

Difurfuroyl 3328
Difuryl-diketone 3328
Digallic acid 5687
Digitalin, cf. glcde.
Digitin, cf. glcde.

Digitonin, cf. glcde.
Digitoxin, cf. glcde.
Dihexyl 2856
Dihydro-benzene 1599-1600

Table 7-4 (*Continued*)
PHYSICAL CONSTANTS OF ORGANIC COMPOUNDS

No.	Crystalline Form and Color	Specific Gravity	Melting Point °C.	Boiling Point °C.	Solubility in 100 Parts		
					Water	Alcohol	Ether
2258	or. nd./al.; red yel./ bz.	....	162-3	d.	i.	s. h.	v. sl. s.; v. sl. s. CS_2
2259	lq.	....	....	220-30^{10mm}	s. h.	....	i.
2260	lq.	$1.170\frac{20}{20}°$	....	196.8	∞	....	....
2261	cr.	....	148	d.	v. s.	v. s.	sl. s.
2262	nd.	....	246 d.	....	v. s. h.	i.	i.
2263	cr.	$0.877\frac{25}{4}°$	26-7	240-50^{85mm}	v. sl. s.	v. s.	v. s.; v. s. bz.
2264	lq.	$0.806^{20°}$	....	261.9	i.	s.	s.
2265	cr./al.	....	39-40	278	....	s.	....
2266	lq.	....	....	298	....	....	....
2267	lf./al.	$0.825^{30°}$	33	264	v. s. chl.	v. s.; s. lg.	v. s.
2268	cr.	....	....	....	i. c. 95% al.	s. h. 90% al.	....
2269	lf./bz.	....	110	....	s. h.; v. s. chl.	v. s.; v. s. h. bz.	i. pet.
2270	lf.	....	165-7 d.	....	sl. s. h.	v. sl. s.	v. sl. s.
2271	col. cr./al.	....	169	subl.; d. 300	i.	s. h.	s.
2272	mn. pr.	$0.897\frac{11}{4}°$	108.5	305	i.; v. s. bz.	v. s.	v. s.
2273	nd./pet.	....	76	....	s. h.	s.	s.; s. bz.
2274	lq.	$0.927\frac{20}{4}°$	....	224-5	....	....	....
2275	oil	$0.925\frac{20}{4}°$	....	221-2	....	....	....
2276	nd.	....	104-5	210-20^{1mm}	s. chl., pet.	s. abs.	s.; s. bz.
2277	oil	$1.049\frac{18}{18}°$	....	320-3 sl. d.	i.; ∞ bz.	∞	∞
2278	col. pl.	$0.997\frac{12}{4}°$	25-8	211-2	i.	s.	s.
2279	rhb. lf.	....	212	....	i.	v. sl. s.	i. bz.
2280	mn.	....	153	....	$1.6^{6°}$	....	....
2281	mn. pr.	....	179-80	....	sl. s. h.	s.	....
2282	tri.	....	215	....	$0.3^{25°}$; 6 h.	s.	s. act.
2283	nd./aq.	....	subl. d.	....	0.01 c.	....	....
2284	lq.	....	....	220-6	....	....	....
2285	pr./bz.	....	105-6 sl. d.	....	v. s.; s. act.	v. s.; s. h. bz.	v. sl. s. abs.
2286	lq.	....	....	108	....	....	....
2287	lq.	0.835	....	110	....	....	....
2288	lq.	$0.827\frac{20}{4}°$	....	106	....	....	....
2289	col. lq.	....	....	134-5	....	....	s.
2290	col. lq.	$0.823\frac{20}{4}°$	....	129-30^{745mm}	....	....	....
2291	col. lq.	$0.830\frac{20}{4}°$	....	135-8	....	....	....
2292	cr.	....	68-75	....	v. s.; i. lg.	sl. s.	sl. s.
2293	yel./aq. al.	....	265±, d.	....	s. alk.	s.	s.; s. bz.
2294	yel./aq. al.	....	225 d.	....	s. alk.	s.	s.; s. bz.

Dihydro-estrone 2900
Dihydro-ketoacridine 145
Dihydro-morphinone hydrochloride, cf. alkd.
Dihydro-myrcene 2514

Dihydro-naringenin 5264
Dihydro-phytol 5295
Dihydro-pyrrole 5524
Dihydroxy-acetophenone 5550

Table 7-4 (*Continued*)
PHYSICAL CONSTANTS OF ORGANIC COMPOUNDS

No.	Name	Synonym	Formula	Beil. Ref.	Formula Weight
2295	**Dihydroxy** 9-anthranol (3,4)	anthrarobin	$C_{14}H_7(OH)_3$	VIII-330	226.23
2296	*x*-anthranol (1,8)	anthralin	$C_{14}H_7(OH)_3$	VIII-332	226.23
2297	anthraquinone (1,2)	alizarin	$C_6H_4(CO)_2C_6H_2:$ $(OH)_2$	VIII-439	240.22
2298	anthraquinone (1,3)	purpuroxanthin	$C_6H_4(CO)_2C_6H_2:$ $(OH)_2$	VIII-448	240.22
2299	anthraquinone (1,4)	quinizarin	$C_6H_4(CO)_2C_6H_2:$ $(OH)_2$	VIII-450	240.22
2300	anthraquinone (1,5)	anthrarufin	$(HO \cdot C_6H_3)_2(CO)_2$	VIII-453	240.22
2301	anthraquinone (1,8)	chrysazin	$(HO \cdot C_6H_3)_2(CO)_2$	VIII-458	240.22
2302	anthraquinone (2,3)	hystazin; hystazarin	$C_6H_4(CO)_2C_6H_2:$ $(OH)_2$	VIII-462	240.22
2303	anthraquinone (2,6)	anthraflavic acid	$(HO \cdot C_6H_3)_2(CO)_2$	VIII-463	240.22
2304	anthraquinone (2,7)	*iso*-anthraflavic acid	$C_{14}H_8O_4 \cdot H_2O$	VIII-466	258.23
2305	anthraquinone carboxylic acid †	(2,4;1 or 1,3;2) munjistin	$C_6H_4(CO)_2C_6H:$ $(OH)_2(CO_2H)$	X-1036	284.23
2306	anthraquinone carboxylic acid	(1;8,3) rheic acid; rhein	$HO \cdot C_6H_3(CO)_2C_6H_2:$ $(OH)(CO_2H)$	X-1033	284.23
2307	azobenzene-4'- sulfonic acid (2,4)	. . . .	$(HO)_2C_6H_3 \cdot N:N \cdot$ $C_6H_4 \cdot SO_3H$	XVI-275	294.29
2308	azobenzene-4'- sulfonate Na (2,4)	tropeolin O; resorcin yellow	$C_{12}H_9O_5N_2SNa \cdot$ $2\frac{1}{2}H_2O$	XVI-275	361.30
2309	behenic acid	. . . .	$C_{22}H_{44}O_4$	III-410	372.59
2310	benzaldehyde (2,4)	*β*-resorcyl aldehyde	$(HO)_2C_6H_3 \cdot CHO$	VIII-241	138.12
2311	benzaldehyde (3,4)	protocatechuic ald.	$(HO)_2C_6H_3 \cdot CHO$	VIII-246	138.12
2312	benzene (*o*)	pyrocatechin	$C_6H_4(OH)_2$	VI-759	110.11
2313	benzene (*m*)	resorcinol	$C_6H_4(OH)_2$	VI-796	110.11
2314	benzene (*p*)	hydroquinone	$C_6H_4(OH)_2$	VI-836	110.11
2315	benzoic acid (2,3)	. . . .	$(HO)_2C_6H_3CO_2H \cdot aq.$	X-375	154.12
2316	benzoic acid (2,4)	resorcylic acid (*β*)	$(HO)_2C_6H_3CO_2H \cdot aq.$	X-377	154.12
2317	benzoic acid (2,5)	gentisic acid	$(HO)_2C_6H_3 \cdot CO_2H$	X-384	154.12
2318	benzoic acid (2,6)	resorcylic acid (*γ*)	$(HO)_2C_6H_3 \cdot CO_2H \cdot$ H_2O	X-388	172.14
2319	benzoic acid (3,4)	protocatechuic acid	$(HO)_2C_6H_3 \cdot CO_2H \cdot$ H_2O	X-389	172.14
2320	benzoic acid (3,5)	resorcylic acid (*α*)	$(HO)_2C_6H_3 \cdot CO_2H \cdot$ $1\frac{1}{2}H_2O$	X-404	181.15
2321	benzophenone (2,4)	benzoyl resorcin (4)	$(HO)_2C_6H_3 \cdot CO \cdot C_6H_5$	VIII-312	214.22
2322	benzophenone (2,5)	benzoyl hydroqui- none (2)	$(HO)_2C_6H_3 \cdot CO \cdot C_6H_5$	VIII-312	214.22
2323	benzophenone (2,2')	salicyl-phenol (2)	$(HO \cdot C_6H_4)_2CO$	VIII-313	214.22
2324	benzophenone (2,3')	salicyl-phenol (3)	$(HO \cdot C_6H_4)_2CO$	VIII-315	214.22
2325	benzophenone (2,4')	salicyl-phenol (4)	$(HO \cdot C_6H_4)_2CO$	VIII-315	214.22

† See also No. 176.
Dihydroxy-anthraquinone carboxylic acid 176   Dihydroxy-anthraquinone quinoline 174

Table 7-4 (*Continued*)
PHYSICAL CONSTANTS OF ORGANIC COMPOUNDS

No.	Crystalline Form and Color	Specific Gravity	Melting Point °C.	Boiling Point °C.	Solubility in 100 Parts		
					Water	Alcohol	Ether
2295	yel. nd./ aq. al.	....	208	....	sl. s.; s. ac.	s.; s. alk.	s.; s. act.
2296	yel. pd.	....	178-80	....	i.; s. fat	s. h.; s. bz.	s. dil. alk.
2297	red rhb.	....	289-90	430	$0.03^{100°}$	v. s.	v. s.; s. alk.
2298	yel. lf./bz.	....	263-4	....	i.; s. act.	sl. s.	s. h. ac.
2299	red nd./al.	....	200-2	subl. sl. d.	s. alk.	s. H_2SO_4	s.
2300	yel. lf.	....	280	subl.	i.	sl. s.	s.
2301	red-yel./ al.	....	191	....	s. H_2SO_4	s.; s. alk.	s.; s. ac.
2302	yel. brn. nd.	....	>280	....	s. H_2SO_4	v. sl. s. h.	v. sl. s.
2303	yel. nd./ al.	....	>330	....	s. H_2SO_4	$1.4^{17°}$	i.
2304	yel. nd./ aq. al.	....	>330; subl.	d. $-H_2O$, 100	s. H_2SO_4	sl. s.	v. sl. s.
2305	yel. lf./ac.	....	230-1	subl. d. > m. p.	s. h.; s. alk.	s. h.; s. H_2SO_4	s.; s. chl.
2306	yel. nd./ Me al.	....	321-2	subl.	i.; s. alk.; s. pyr.	sl. s.; sl. s. bz.	sl. s.; sl. s. chl.
2307	red lf.	....	....	....	$0.2^{19°}$; i. aq. HCl	v. sl. s.	i.
2308	brn. lf.	....	....	....	$0.4^{23°}$; $20^{100°}$	....	....
2309	cr./al.	....	132-3 (99)	....	....	$0.1^{18°}$ abs.	i.
2310	yel. nd./aq.	....	135-6	$220-8^{22mm}$	v. s.	v. s.	v. s.
2311	cr./aq.	....	153-4	....	5 c.; 33 h.	100 h.	v. s.
2312	nd./aq.	$1.344^{4°}$	105	245.6	$45.1^{20°}$	v. s.; s. chl.	v. s.; s. bz.
2313	col. rhb.	$1.272^{15°}$	110(108)	275.9	$147.3^{12.5°}$	v. s.	v. s.
2314	cr.	$1.332^{15°}$	172.3	285^{730mm}	$6^{15°}$	v. s.	v. s.
2315	cr./aq.	....	204(anh.)	$-aq.$ 100	....	....	....
2316	cr.	....	204-6 (anh.) d.	....	$0.26^{17°}$; s. h.	s.	s.
2317	nd./aq.	....	199-200	....	s.; i. CS_2	s.; i. chl.	s.; i. bz.
2318	nd./aq.	....	148-67 (anh.) d.	....	v. s. h.	....	....
2319	nd./aq.	1.542	199 d.	....	$2^{14°}$	v. s.	s.
2320	pr. or nd.	....	237 (anh.)	....	v. s. h.	v. s.	v. s.
2321	nd./h. aq.	....	143-4	....	i. c.	s.	s.; sl. s. bz.
2322	yel. nd./ aq. al.	....	122-4	....	s. bz.	s.	s.
2323	yel. pr./lg.	....	59-60	330-40 sl. d.	i.; v. s. chl.	v. s.	v. s.
2324	yel. cr./et.	....	126	....	....	....	....
2325	lt. yel./aq.	....	147-8	....	sl. s. h.	s. h.; s. bz.	$44^{25°}$

Dihydroxy-azobenzene 609-11, 662   Dihydroxy-benzaldehyde dimethyl ether 2405

Table 7-4 (*Continued*)
PHYSICAL CONSTANTS OF ORGANIC COMPOUNDS

No.	Name	Synonym	Formula	Beil. Ref.	Formula Weight
2326	**Dihydroxy** benzophenone (3,3')	....	$(HO \cdot C_6H_4)_2CO$	VIII-316	214.22
2327	benzophenone (3,4')	....	$(HO \cdot C_6H_4)_2CO$	VIII-316	214.22
2328	benzophenone (4,4')	....	$(HO \cdot C_6H_4)_2CO$	VIII-316	214.22
2329	butane (1,2)	α-butylene glycol	$C_2H_5 \cdot CHOH \cdot CH_2OH$	I-477	90.12
2330	butane (1,3)	β-butylene glycol	$CH_3 \cdot CHOH \cdot CH_2 \cdot CH_2OH$	I-477	90.12
2331	butane (1,4)	butandiol-1,4	$HO(CH_2)_4OH$	I-478	90.12
2332	butane (2,3)	butylene glycol (pseudo) (meso)	$CH_3 \cdot (CHOH)_2 \cdot CH_3$	I-479	90.12
2333	cinnamic acid (2,4)	umbellic acid	$(HO)_2C_6H_3 \cdot C_2H_2 \cdot CO_2H$	X-434	180.16
2334	cinnamic acid (2,5)	....	$(HO)_2C_6H_3 \cdot C_2H_2 \cdot CO_2H$	X-435	180.16
2335	cinnamic acid (3,4)	caffeic acid	$C_9H_8O_4$	X-436	180.16
2336	dinaphthyl (α)	α-dinaphthol	$(HO \cdot C_{10}H_6 \cdot)_2$	VI-1053	286.33
2337	dinaphthyl (2,2', 1,1')	β-dinaphthol	$(HO \cdot C_{10}H_6 \cdot)_2$	VI-1051	286.33
2338	diphenyl (2,5)	Ph-hydroquinone	$(HO)_2C_6H_3 \cdot C_6H_5$	VI-989	186.21
2339	diphenyl (3,4)	Ph-catechol (4)	$(HO)_2C_6H_3 \cdot C_6H_5$	VI-990	186.21
2340	diphenyl (2,2')	o,o'-diphenol	$(HO \cdot C_6H_4 \cdot)_2$	VI-989	186.21
2341	diphenyl (2,4')	δ-diphenol	$(HO \cdot C_6H_4 \cdot)_2$	VI-990	186.21
2342	diphenyl (3,3')	....	$(HO \cdot C_6H_4 \cdot)_2$	VI-991	186.21
2343	diphenyl (4,4')	γ-diphenol	$(HO \cdot C_6H_4 \cdot)_2$	VI-991	186.21
2344	diphenylmethane (4,4')		$(HO \cdot C_6H_4 \cdot)_2CH_2$	VI-995	200.24
2345	ethylamino-1-hydroxy-benzene (4)		$HO \cdot C_6H_4 \cdot N(CH_2 \cdot CH_2OH)_2$	....	197.24
2346	hexane (2,3)	2,3-hexandiol	$CH_3(CH_2)_2(CHOH)_2 \cdot CH_3$	I-484	118.18
2347	4-methoxy-benzophenone (2,6)	cotoin	$C_6H_5 \cdot CO \cdot C_6H_2 \vdots (OH)_2(OCH_3)$	VIII-419	244.25
2348	naphthalene (1,2)	β-naphthohydroquinone	$C_{10}H_6(OH)_2$	VI-975	160.17
2349	naphthalene (1,3)	naphthoresorcinol	$C_{10}H_6(OH)_2$	VI-978	160.17
2350	naphthalene (1,4)	α-naphthohydroquinone	$C_{10}H_6(OH)_2$	VI-979	160.17
2351	naphthalene (1,5)	....	$C_{10}H_6(OH)_2$	VI-980	160.17
2352	naphthalene (1,6)	....	$C_{10}H_6(OH)_2$	VI-981	160.17
2353	naphthalene (1,7)	....	$C_{10}H_6(OH)_2$	VI-981	160.17
2354	naphthalene (1,8)	....	$C_{10}H_6(OH)_2$	VI-981	160.17
2355	naphthalene (2,3)	....	$C_{10}H_6(OH)_2$	VI-982	160.17
2356	naphthalene (2,6)	....	$C_{10}H_6(OH)_2$	VI-984	160.17
2357	naphthalene (2,7)	....	$C_{10}H_6(OH)_2$	VI-985	160.17
2358	naphthalene di-acetate (1,5)	....	$(CH_3CO_2)_2C_{10}H_6$	VI-981	244.25
2359	naphthalene-3,6-disulfonic acid	(1,8); chromotropic acid	$(HO)_2C_{10}H_4 \vdots (SO_3H)_2 \cdot 2H_2O$	XI-307	356.33
2360	phenanthrene (3,4)	morphol	$C_{14}H_8(OH)_2$	VI-1034	210.23
2361	phenyl-α-alanine (3,4)(l-)	hydroxytyrosine; dopa	$(HO)_2C_6H_3 \cdot CH_2 \cdot CHNH_2 \cdot CO_2H$	*XIV-681	197.19

Dihydroxy–chlorobenzene 1460
Dihydroxy–coumarin 1628, 2898
Dihydroxy–diethylsulfide 5900
Dihydroxy–ethylamine 2090

Dihydroxy–ethyl ether 3453
Dihydroxy–flavone 1494
Dihydroxy–fluoran 3253
Dihydroxy–fluorane 3720

Table 7-4 (*Continued*)
PHYSICAL CONSTANTS OF ORGANIC COMPOUNDS

No.	Crystalline Form and Color	Specific Gravity	Melting Point °C.	Boiling Point °C.	Solubility in 100 Parts		
					Water	Alcohol	Ether
2326	nd./aq.	....	163-4	....	sl. s.	s. alk.	....
2327	nd./aq.	....	197-200	....	....	....	....
2328	cr./aq.	....	207-10	....	s. h.	s.; i. CS_2	s.; i. chl.
2329	lq.	$1.006\frac{17.5°}{0}$	....	192-4	v. s.	v. s.	....
2330	oil	$1.006\frac{20°}{20}$	....	206.5	∞	v. s.	i.
2331	oil	$1.020^{20°}$	19	230^{759mm}	∞	....	v. sl. s.
2332	lq.	$1.048^{0°}$	34	181.7^{742mm}	∞	∞	∞
2333	yel. pd.	....	d. 240-60	....	s. h.	s.; i. bz.	i.; i. lg.
2334	cr./aq.	....	207 d.	....	....	....	....
2335	yel. pr./aq.	....	195-213 d.	....	s. h.	s.	sl. s.
2336	pl./al.	....	300	....	i.; s. alk.	s.; sl. s. bz.	v. s.; sl. s. chl.
2337	nd./al.	....	218	subl.	i.; s. alk.	s.	v. s.
2338	nd./aq. al.	....	102-3	....	....	....	....
2339	cr./pet.	....	136-7	>360	s. h.	v. s.; s. CS_2	v. s.; s. chl
2340	pr./toluene	....	103-9	$325-6^{755mm}$	s.	v. s.	v. s.
2341	mn. pr.	....	160-1	324	sl. s. h.	v. s.	v. s.
2342	nd./aq.	....	123	247^{18mm}	s. h.	v. s.	v. s.
2343	rhb./al.	1.25	278-81	....	sl. s.	v. s.	v. s.
2344	lf./aq.	....	158 (148)	....	i. CS_2	s.	v. s.
2345	col. fl.	....	140-1	....	i.	sl. s.	sl. s.
2346	cr.	$0.967^{0°}$	60	206-7	∞	s.	....
2347	yel. pr./al.	....	130-1	....	sl. s. h.; s. alk.	s.; s. bz.; s. CS_2	s.; s. chl.; s. act.
2348	lf.	....	60±	....	s. alk.	....	....
2349	lf.	....	124-5	....	v. s.	v. s.	v. s.
2350	long nd.	....	183-8	....	s. h.	v. s.	v. s.
2351	pr./aq.	....	258-60 d.	....	sl. s.	s.	v. s.
2352	pr./bz.	....	137-8	....	....	sl. s. c.	v. s.
2353	nd./aq.	....	178	....	v. s. h.	v. s.	v. s.
2354	nd. or lf.	....	140	....	sl. s. h.	v. s. bz.	v. s.
2355	mn.	....	159-60	....	sl. s. h.	v. s.	v. s.
2356	rhb./aq.	....	216-8	....	v. s. h.	v. s.	v. s.
2357	nd./aq.	....	190	subl. sl. d.	s. h.	v. s.	v. s.
2358	cr./bz.	....	159-60	....	....	....	....
2359	nd. or lf.	....	....	....	v. s.; i. aq. NaCl	i.	i.
2360	nd.	....	143	....	s.	s. alk.	....
2361	pr. or nd./aq.; lf./aq. al.	....	280 d.	....	$0.5^{20°}$; $2.5^{100°}$	i.; i. chl.; v. sl. s. bz.	i.; i. pet.; v. sl. s. CS_2

Table 7-4 (*Continued*)
PHYSICAL CONSTANTS OF ORGANIC COMPOUNDS

No.	Name	Synonym	Formula	Beil. Ref.	Formula Weight
	Dihydroxy				
2362	pyridine (2,4)	....	$(HO)_2C_5H_3N$	XXI-160	111.10
2363	pyridine (2,6)	....	$(HO)_2C_5H_3N\cdot\frac{1}{2}H_2O$	XXI-161	120.11
2364	quinone (2,5)	....	$(HO)_2C_6H_2O_2$	VIII-377	140.10
2365	stearic acid (4,9) (*dl*)	....	$C_{18}H_{36}O_4$	III-407	316.49
2366	stearic acid (*dl*)	(9,10)	$C_{18}H_{36}O_4$	III-408	316.49
2367	tartaric acid	....	$[(HO)_2C\cdot CO_2H]_2$	III-830	182.09
2368	tartaric Na salt	....	$[(HO)_2C\cdot CO_2Na]_2\cdot 3H_2O$	III-832	280.10
2369	thiazole	mustard oil acetic acid	$HO\cdot C{:}CH\cdot S\cdot C(OH){:}N$ ⌐_____⌐	XXVII-233	117.13
2370	toluene (2,3)	....	$CH_3\cdot C_6H_3(OH)_2$	VI-872	124.14
2371	toluene (2,4)	4-Me-resorcinol	$CH_3\cdot C_6H_3(OH)_2$	VI-872	124.14
2372	toluene (2,5)	toluhydroquinone	$CH_3\cdot C_6H_3(OH)_2$	VI-874	124.14
2373	toluene (2,6)	2-Me-resorcinol	$CH_3\cdot C_6H_3(OH)_2$	VI-878	124.14
2374	toluene (3,4)	homocatechol	$CH_3\cdot C_6H_3(OH)_2$	VI-878	124.14
2375	*o*-xylene (3,5)	4,5-diMe-resorcinol	$(CH_3)_2C_6H_2(OH)_2$	VI-908	138.17
2376	*o*-xylene (3,6)	2,3-diMe-hydro-quinone	$(CH_3)_2C_6H_2(OH)_2$	VI-908	138.17
2377	*m*-xylene (2,4)	2,4-diMe-resorcinol	$(CH_3)_2C_6H_2(OH)_2$	VI-911	138.17
2378	*m*-xylene (2,5)	2,6-diMe-hydro-quinone	$(CH_3)_2C_6H_2(OH)_2$	VI-911	138.17
2379	*p*-xylene (2,5)	hydrophloron	$(CH_3)_2C_6H_2(OH)_2$	VI-915	138.17
2380	*p*-xylene (2,6)	*β*-orcin	$(CH_3)_2C_6H_2(OH)_2$	VI-918	138.17
2381	**Diiodo**-acetic acid	....	$I_2CH\cdot CO_2H$	II-224	311.85
2382	acetylene	diiodo-ethyne	$IC{:}CI$	I-246	277.83
2383	anthranilic acid	(3,5;2,1)	$I_2C_6H_2(NH_2)CO_2H$	*XIV-554	388.93
2384	atophan	*p*-iodophenyl-6-iodoquinoline-4-carboxylic acid	$C_{16}H_9O_2NI_2$	....	501.06
2385	aniline (2,4)	....	$I_2C_6H_3\cdot NH_2$	XII-675	344.92
2386	benzene (*o*)	*o*-phenylene iodide	$C_6H_4I_2$	V-225	329.91
2387	benzene (*m*)	....	$C_6H_4I_2$	V-225	329.91
2388	benzene (*p*)	....	$C_6H_4I_2$	V-227	329.91
2389	butane (1,4)	tetramethylene di-iodide	$I(CH_2)_4I$	I-123	309.92
2390	ethane (1,1)	ethylidene diiodide	$CH_3\cdot CHI_2$	I-99	281.86
2391	ethane (1,2)	ethylene iodide	$ICH_2\cdot CH_2I$	I-99	281.86
2392	fluorescein	....	$C_{20}H_{10}O_5I_2$	....	584.11
2393	hexane (1,6)	hexamethylene di-iodide	$I(CH_2)_6I$	I-147	337.97
2394	2-hydroxybenzoic acid (3,5)	3,5-diI-salicylic acid	$I_2C_6H_2(OH)CO_2H$	X-113	389.92
2395	4-hydroxybenzoic	(3,5;4,1)	$I_2C_6H_2(OH)CO_2H$	X-180	389.92
2396	pentane (1,5)	pentamethylene di-iodide	$I(CH_2)_5I$	I-133	323.94
2397	propane (1,2)	propylene iodide	$CH_3\cdot CHI\cdot CH_2I$	I-115	295.89
2398	propane (2,2)	iodoacetol	$(CH_3)_2CI_2$	I-115	295.89
2399	propane (1,3)	trimethylene diiodide	$I(CH_2)_3I$	I-115	295.89
2400	thiophene (2,5)	....	$I_2C_4H_2S$	XVII-35	335.93
2401	tyrosine (3,5)(*l*)	iodogorgonic acid	$C_9H_9O_3NI_2$	XIV-619	432.99

Dihydroxy-propane 5465, 6286
Dihydroxy-propionic acid 3379
Dihydroxy- purine, cf. alkd.

Dihydroxy-xanthone 3227
Diiodo-ethene 137-8
Diiodo-ethyne 2382

Table 7-4 (Continued)
PHYSICAL CONSTANTS OF ORGANIC COMPOUNDS

No.	Crystalline Form and Color	Specific Gravity	Melting Point °C.	Boiling Point °C.	Solubility in 100 Parts		
					Water	Alcohol	Ether
2362	yel. nd./aq.	....	255-65 d.	....	s. h.	s.	v. sl. s.
2363	nd./aq.	....	195 d.	....	sl. s.	s. h.	i.
2364	yel. nd.	....	subl. 215- 20 sl. d.	....	i. c.	s.	i. c.
2365	lf./al.	....	136.5	....	v. s. h. al.	$0.6^{19°}$	$0.2^{18°}$
2366	lf.	....	99-100	....	v. sl. s. c.	$3.6^{18°}$	sl. s.
2367	wh. pd.	....	114-5 d.	....	v. s.; d. h.	....	....
2368	cr.	....	d., $-H_2O$, $-CO_2$	....	$0.04^{0°}$	....	....
2369	rhb.	....	126-8	179^{19mm}; subl.	v. s. h.	s.	s.
2370	lf./bz.	....	68	238-40 sl. d. chl.	s.; v. s.	v. s.	s.; v. s. bz.
2371	cr./bz. pet.	....	104-5	267-70	s.; i. pet.	s.	s.; sl. s. bz.
2372	lf./bz.	....	127-9	....	s.	s.	s.; sl. s. bz.
2373	cr.	....	116	264	v. s.	v. s.	v. s.
2374	pr./bz.	$1.129\frac{7.4°}{4}$	65	252	v. s.	v. s.	v. s.
2375	nd./bz.*	....	136-7	subl.	s.; sl. s. lg.	s.	s.
2376	cr./aq.	....	221 sl. d.	....	....	....	....
2377	nd.	....	149-50	....	s.	v. s.	v. s.
2378	nd./ xylene	....	150-1	....	....	....	....
2379	lf./aq.	....	213	subl.	s. h.	v. s.	v. s.; i. bz.
2380	cr./aq. al.	....	163	277-80	s. h.	s.	s.
2381	lt. yel. nd.	....	110 (95-6)	....	s.	s.	s.
2382	col. rhb.	....	78-82	d. 80-100	sl. s. c. lg.	....	....
2383	pr./al.	....	234-5	....	i. h.	sl. s. h. bz.	v. s.
2384	yel. pd.	....	291-2	....	i.	s. h.	v. sl. s.
2385	rhb.	2.75	95-6	....	s. h.	v. s. h.	v. s.
2386	cr./lg.	$2.54^{20°}$	27	286.5	i.	sl. s. c.	....
2387	cr./al. et.	$2.47^{25°}$	40.4	284.7^{757mm}	i.	sl. s.	....
2388	lf./al.	....	129.4	285	i.	sl. s.	....
2389	lq.	$2.307^{18°}$	5.8	$120-5^{12mm}$	....	....	....
2390	lq.	$2.84^{0°}$	....	177-9	i.	v. s.	v. s.
2391	yel. mn.	$2.132^{10°}$	81-2	d.	sl. s.	s.	s.
2392	or.-red pd.	....	....	....	sl. s.	s.	s. alk.
2393	col. nd.	$2.05^{18°}$	9.5	$163^{17.5mm}$	....	....	....
2394	yel.-wh. nd./al.	....	230±, d.	....	0.07 c.; 0.15 h.	s.	s.
2395	nd./aq. al.	....	278-9 d.	....	i. h.	v. s.	v. s.
2396	oil	$2.194^{18°}$	9	149^{20mm}	....	....	....
2397	lq.	$2.490^{18.5°}$	....	d.	....	....	....
2398	lq.	$2.15^{0°}$	....	147-8 d.	....	....	....
2399	lq.	$2.576^{15°}$	-13	224	v. sl. s.	s.	s.
2400	col. lf./al.	....	40-1	....	i.	s.	s.; s. chl.
2401	nd./aq. al.	....	213 d.	....	$0.288^{15°}$	d. h. aq.	s. dil. alk.

* Cryst./aq. + $1H_2O$, m. p. 115-7°.
Diiodoform 5812
Diiodo-methane 4434

Diiodo-phenol sulfonic acid 5626
Diiodo-propanol 3401
Diiodo-salicylic acid 2394

Table 7-4 (*Continued*)
PHYSICAL CONSTANTS OF ORGANIC COMPOUNDS

No.	Name	Synonym	Formula	Beil. Ref.	Formula Weight
	Diiodo				
2402	tyrosine (3,5)(*dl*)	....	$C_9H_9O_3NI_2$	XIV-622	432.99
2403	**Dimethallyl ether** (*β*)	....	$[CH_2:C(CH_3)\cdot CH_2]_2O$	....	126.20
2404	**Dimethoxy-**aniline	(2,5;1)	$(CH_3O)_2C_6H_3\cdot NH_2$	XIII-738	153.18
2405	benzaldehyde (2,4)	2-MeO-anisaldehyde	$(CH_3O)_2C_6H_3\cdot CHO$	VIII-242	166.18
2406	benzene (*o*)	veratrole	$(CH_3O)_2C_6H_4$	VI-771	138.17
2407	benzene (*m*)	resorcinol di Me ether	$(CH_3O)_2C_6H_4$	VI-813	138.17
2408	benzene (*p*)	hydroquinone di Me ether	$(CH_3O)_2C_6H_4$	VI-843	138.17
2409	benzophenone (4,4′)	*p*-anisoyl-anisole	$(CH_3O\cdot C_6H_4)_2CO$	VIII-317	242.28
2410	ethyl adipate	....	$[(CH_2)_2\cdot CO_2(CH_2)_2\cdot OCH_3]_2$	....	262.31
2411	ethyl carbonate	....	$[CH_3O(CH_2)_2O]_2CO$	....	178.19
2412	ethyl phthalate	....	$[CH_3O(CH_2)_2O_2C]_2\cdot C_6H_4$	**IX-597	282.30
2413	2-methyl-3,4-dihydroisoquinoline chloride (6,7)	Iodal	$C_{12}H_{16}O_2N\cdot Cl\cdot 3\frac{1}{2}H_2O$	XXI-170	304.77
2414	quinazoline (2,4)	....	$(CH_3O)_2C_8H_4N_2$	XXIII-486	190.20
2415	**Dimethyl-**acetal	dimethyl aldehyde	$CH_3\cdot CH(OCH_3)_2$	I-603	90.12
2416	acetamide (*N,N*)	acet-dimethylamide	$CH_3CO\cdot N(CH_3)_2$	IV-59	87.12
2417	acetanilide (2,3)	aceto-*o*-xylidide (*vic*)	$(CH_3)_2C_6H_3NH\cdot COCH_3$	XII-1101	163.22
2418	acetanilide (2,4)	aceto-*m*-xylidide (*uns*)	$(CH_3)_2C_6H_3NH\cdot COCH_3$	XII-1118	163.22
2419	acetanilide (2,5)	aceto-*p*-xylidide	$C_8H_9NH\cdot COCH_3$	XII-1137	163.22
2420	acetanilide (2,6)	aceto-*m*-xylidide (*vic*)	$C_8H_9NH\cdot COCH_3$	XII-1109	163.22
2421	acetanilide (3,4)	aceto-*o*-xylidide (*uns*)	$C_8H_9NH\cdot COCH_3$	XII-1104	163.22
2422	acetanilide (3,5)	aceto-*m*-xylidide (*sym*)	$C_8H_9NH\cdot COCH_3$	XII-1131	163.22
2423	acetoacetic ester	....	$CH_3CO\cdot C(CH_3)_2CO_2\cdot C_2H_5$	III-695	158.20
2424	acrylic acid (*β,β*)	....	$(CH_3)_2C:CH\cdot CO_2H$	II-432	100.12
2425	acrylic acid (*α,β*) (*cis*)	angelic acid	$CH_3\cdot CH:C(CH_3)\cdot CO_2H$	II-428	100.12
2426	acrylic acid (*α,β*) (*trans*)	tiglic acid	$CH_3\cdot CH:C(CH_3)\cdot CO_2H$	II-430	100.12
2427	acrylic aldehyde (*trans*)	tiglic aldehyde	$CH_3\cdot CH:C(CH_3)\cdot CHO$	I-733	84.12
2428	adipate	methyl adipate	$[(CH_2)_2CO_2CH_3]_2$	II-652	174.20
2429	amine	....	$(CH_3)_2NH$	IV-39	45.08
2430	amine hydrochloride	....	$(CH_3)_2NH\cdot HCl$	IV-41	81.55
2431	amino-azobenzene (*p*)	benzene-azo-di Me-aniline	$(CH_3)_2N\cdot C_6H_4\cdot N: N\cdot C_6H_5$	XVI-312	225.30
2432	amino-azobenzene (*p*) HCl	....	$C_{14}H_{15}N_3\cdot HCl$	XVI-312	261.76

Table 7-4 (*Continued*)
PHYSICAL CONSTANTS OF ORGANIC COMPOUNDS

No.	Crystalline Form and Color	Specific Gravity	Melting Point °C.	Boiling Point °C.	Solubility in 100 Parts		
					Water	Alcohol	Ether
2402	pl./aq.	....	>200 d.	....	$0.567^{75°}$	....	....
2403	lq.	$0.816\frac{20}{4}°$	....	134	<0.1	....	....
2404	sc./pet.	....	80–1	270 sl. d.	s.	v. s. h.	v. s. h. pet.
2405	nd./aq. al.	....	69–70	165^{10mm}	i.; s. bz.	s.; s. lg.	s.
2406	cr.	$1.091\frac{15}{15}°$	22.5(20.9)	206.3	v. sl. s.	s.	s.
2407	lq.	$1.062\frac{15}{15}°$	−52	216.5–7.5	v. sl. s.	v. s.	v. s.
2408	lf.	$1.053\frac{55}{55}°$	56	212.6	v. sl. s.	v. s.	v. s.
2409	nd./al.	....	144–5	....	s. bz.	s. h.	s. chl.
2410	col. lq.	....	....	$194–6^{20mm}$	....	....	....
2411	col. lq.	....	....	$115–7^{13mm}$	....	....	....
2412	col. lq.	$1.171^{15°}$	....	$178–80^{1mm}$	....	....	....
2413	yel. nd./ aq. act.	....	61–2; 186 d. (anh.)	....	v. s.	v. s.	sl. s. chl.
2414	col. nd.	....	75	....	i.	s.	s.
2415	lq.	$0.850\frac{20}{4}°$	−113.2	64^{748mm}	∞	∞	∞
2416	col. lq.	$0.943\frac{20}{4}°$	−20	165.5^{754mm}	....	....	....
2417	nd./al.	....	132–3	....	s. h.	s.	s.
2418	nd./aq. al.	....	128–9	....	v. sl. s. c.	s.	....
2419	nd./aq.	....	141–2	....	s. h.	s. h. toluene	....
2420	nd.	....	174–6	....	....	....	....
2421	pr./aq. al.	....	98–9	....	i.	v. s.	....
2422	nd./al.	....	144.5	....	i.	s.	....
2423	oil	$0.977\frac{20}{20}°$	....	184–5	v. sl. s.	s.	s.
2424	mn./aq.	$1.006^{24°}$	69–70	194–5	s.	s.	s.
2425	mn. nd.	$0.983^{47°}$	45	185	s. h.	s.	s.
2426	tri. pl.	$0.964\frac{76}{4}°$	64.5	198.5	s. h.	s.	s.
2427	lq.	$0.870\frac{18}{4}°$	....	115.8^{739mm}	2.5	∞	∞
2428	col. lq.	$1.063\frac{20}{0}°$	10–1	115^{13mm}	i.	....	....
2429	col. lq.	$0.680\frac{0}{0}°$	−92.2	6.9	v. s.	s.	s.
2430	nd./al.	....	170–1	....	$369^{25°}$	v. s.	i.; $16.9^{25°}$ chl.
2431	yel. lf./al.	....	116–7	d.	i.; s. oil	s.; s. bz.	s.; s. chl.
2432	red-vl. cr.	....	168–74	....	....	....	....

Table 7-4 (*Continued*)
PHYSICAL CONSTANTS OF ORGANIC COMPOUNDS

No.	Name	Synonym	Formula	Beil. Ref.	Formula Weight
	Dimethyl				
2433	amino-azobenzene-arsinic acid (*p*)	....	$(CH_3)_2N \cdot C_6H_4 \cdot N{:}N \cdot$ $C_6H_4 \cdot As(OH)_2$	....	333.22
2434	amino-azo-ben-zene-arsonic acid (*p*)	....	$(CH_3)_2N \cdot C_6H_4 \cdot N{:}N \cdot$ $C_6H_4 \cdot AsO(OH)_2$	XVI-885	349.22
2435	amino-benzalde-hyde	(1,2)	$(CH_3)_2N \cdot C_6H_4 \cdot CHO$	XIV-25	149.19
2436	amino-benzalde-hyde (1,4)	Ehrlich's reagent	$(CH_3)_2N \cdot C_6H_4 \cdot CHO$	XIV-31	149.19
2437	amino-benzal-rhodanine-5 (*p*)	....	$(CH_3)_2N \cdot C_6H_4 \cdot CH{:}$ C_3HONS_2	XXVII-433	264.37
2438	amino-benzene-1-azo-1-naphtha-lene (4)	butter yellow	$(CH_3)_2N \cdot C_6H_4 \cdot N{:}N \cdot$ $C_{10}H_7$	XVI-321	275.36
2439	amino-benzene-1-azo-2-naphtha-lene (4)	....	$(CH_3)_2N \cdot C_6H_4 \cdot N{:}N \cdot$ $C_{10}H_7$	XVI-321	275.36
2440	amino-benzoic acid (*o*)	*N,N*-diMe-an-thranilic acid	$(CH_3)_2N \cdot C_6H_4 \cdot CO_2H$	XIV-325	165.19
2441	amino-benzoic acid	(*p*)	$(CH_3)_2N \cdot C_6H_4 \cdot CO_2H$	XIV-426	165.19
2442	amino-benzo-phenone (*p*)	....	$(CH_3)_2N \cdot C_6H_4 \cdot CO \cdot$ C_6H_5	XIV-82	225.29
2443	amino-ethanol	....	$(CH_3)_2N \cdot CH_2 \cdot CH_2OH$	IV-276	89.14
2444	amino-glycerol (*α*)	γ-dimethylamino-propylene glycol	$(CH_3)_2N \cdot CH_2 \cdot CHOH \cdot$ CH_2OH	IV-302	119.16
2445	amino-phenol (*m*)	....	$(CH_3)_2N \cdot C_6H_4 \cdot OH$	XIII-405	137.18
2446	amino-phenol (*p*)	....	$(CH_3)_2N \cdot C_6H_4 \cdot OH$	XIII-442	137.18
2447	amino-phenol H_2SO_4	(*p*)	$C_8H_{11}ON \cdot \frac{1}{2}H_2SO_4$	XIII-442	186.22
2448	amino-phenyl acetate (*p*)	....	$CH_3CO_2 \cdot C_6H_4 \cdot$ $N(CH_3)_2$	XIII-443	179.22
2449	ammonium-di-methyl-dithiocar-bamate	....	$(CH_3)_2N \cdot CS \cdot S \cdot$ $NH_2(CH_3)_2$	**IV-577	166.31
2450	aniline †	....	$(CH_3)_2N \cdot C_6H_5$	XII-141	121.18
2451	aniline sulfonic acid (*m*)	....	$(CH_3)_2N \cdot C_6H_4 \cdot SO_3H$	XIV-690	201.25
2452	aniline sulfonic acid (*p*)	....	$(CH_3)_2N \cdot C_6H_4 \cdot SO_3H \cdot$ H_2O	XIV-699	219.26
2453	anthracene (1,3)	....	$(CH_3)_2C_{14}H_8$	V-678	206.29
2454	anthracene (2,3)	....	$(CH_3)_2C_{14}H_8$	V-678	206.29
2455	anthracene (2,6)	....	$(CH_3)_2C_{14}H_8$	V-678	206.29
2456	arsine	cacodyl hydride	$(CH_3)_2AsH$	IV-599	106.00
2457	benzamide (*N*)	....	$C_6H_5CO \cdot N(CH_3)_2$	IX-201	149.19
2458	benzoic acid (2,3)	hemellitic acid	$(CH_3)_2C_6H_3 \cdot CO_2H$	IX-531	150.18
2459	benzoic acid (2,4)	xylic acid	$(CH_3)_2C_6H_3 \cdot CO_2H$	IX-531	150.18
2460	benzoic acid (2,5)	....	$(CH_3)_2C_6H_3 \cdot CO_2H$	IX-534	150.18
2461	benzoic acid (2,6)	....	$(CH_3)_2C_6H_3 \cdot CO_2H$	IX-531	150.18
2462	benzoic acid (3,4)	....	$(CH_3)_2C_6H_3 \cdot CO_2H$	IX-535	150.18

† See also xylidine.
Dimethyl-amino-azobenzene carboxylic acid 4376
Dimethyl-amino-benzophenonimide 592

Dimethyl-amino-propylene glycol 2444
Dimethyl-*n*-amyl carbinol 4967
Dimethyl-aniline 6471-5, 6477

Table 7-4 (*Continued*)
PHYSICAL CONSTANTS OF ORGANIC COMPOUNDS

No.	Crystalline Form and Color	Specific Gravity	Melting Point °C.	Boiling Point °C.	Solubility in 100 Parts		
					Water	Alcohol	Ether
2433	vl. cr.	....	....	....	i.	sl. s.	i.
2434	red pd.	....	....	....	s. a.	s. alk.	....
2435	yel. oil	....	....	244	s. dil. ac.	s.	s.
2436	lf./aq.	....	74-5	176-7^{17mm}	i.; s. ac.	s.	s.
2437	red nd./al.	....	240-70 d.	......,....	i.; s. conc. a.	sl. s. h.; v. sl. s. bz.	v. sl. s.; v. sl. s. chl.
2438	red-bl. pr./act.	....	132-4	....	s. conc. H_2SO_4	sl. s.; s. chl.	sl. s.; s. act.
2439	yel. brn. cr./bz.- lq.	....	175	....	....	....	....
2440	nd./et.	....	70	....	s.	s.	sl. s. h.
2441	nd./al.	....	238-9	....	s.	s.	sl. s.
2442	lf./al.	....	90-2	....	i.	v. s. h.	v. s.
2443	col. lq.	0.887$^{20°}_{4}$	−59	135^{758mm}	∞	....	....
2444	syrup	....	....	220^{749mm}	s.	s.	s.; s. chl.
2445	nd./lg.	....	85	265-8	sl. s. h.	s.	s.
2446	cr./et.-lg.	....	78	....	v. sl. s. lg.	s.	s.
2447	cr.	....	209-10	....	v. s.	sl. s.	....
2448	nd. or pl./ aq. al.	....	78-9	....	v. sl. s.; v. sl. s₀ ac.	s.; v. s. bz.	s.; s. lg.; v. s. chl.
2449	pa. yel. pl.	....	131-3	....	v. s.	v. s.	v. sl. s.
2450	yel. lq.	0.956$^{20°}_{4}$	2.5	194.2	i.	s.; s. chl.	s.
2451	....	....	d. 266	....	s.	....	....
2452	pr. or lf.	....	270 (anh.) −H₂O, 135°	....	s. h.; s. ac.	v. sl. s.	v. sl. s.
2453	lf./et.	....	83	....	....	....	....
2454	lf.	....	252	....	v. s. bz.	....	....
2455	red-yel.; gn.-yel. or silvery	....	243-4	subl.	....	sl. s.	s. bz.
2456	col. lq.	1.213$^{29°}$	....	35.6^{747mm}	....	∞	∞
2457	col. cr.	....	41-2	272-3	v. s.	....	....
2458	pr./al.	....	144	....	v. sl. s. h.	s.	....
2459	mn. or tri.	....	126-7	267^{727mm}	sl. s. h.	s. h.	s. bz., chl.
2460	nd al.	1.069$^{20°}_{4}$	132	268	v. sl. s. h.	v. s. c.	....
2461	col. nd./lg.	....	116	....	....	sl. s. c. lg.	v. s.
2462	pr./al.	....	165-6	subl.	v. sl. s. h.	s.	....

Dimethyl-anthranilic acid 2440
Dimethyl-azobenzene 612-4
Dimethyl-benzene 6454, 6457, 6462

Dimethyl-benzene sulfonyl chloride 6461
Dimethyl-benzidine 5955, 5959
Dimethyl-benzophenone 2838

Table 7-4 (*Continued*)
PHYSICAL CONSTANTS OF ORGANIC COMPOUNDS

No.	Name	Synonym	Formula	Beil. Ref.	Formula Weight
2463	**Dimethyl** benzoic acid (3,5)	mesitylenic acid	$(CH_3)_2C_6H_3 \cdot CO_2H$	IX-536	150.18
2464	benzoylene urea	....	$C_6H_4 \cdot N(CH_3) \cdot CO \cdot$ $\overline{\quad N(CH_3) \cdot CO \quad}$	XXIV-375	190.20
2465	benzylamine (*N*)	....	$C_6H_5 \cdot CH_2 \cdot N(CH_3)_2$	XII-1019	135.21
2466	benzyl carbinol	....	$C_6H_5CH_2 \cdot C(CH_3)_2OH$	VI-523	150.22
2467	carbanilide	*N,N'*-diMe-*N,N'*-diPh urea	$[C_6H_5(CH_3)N]_2CO$	XII-418	240.31
2468	carbonate	methyl carbonate	$CO(OCH_3)_2$	III-4	90.08
2469	citraconate	....	$CH_3 \cdot C_2H(CO_2CH_3)_2$	II-770	158.16
2470	cyclohexan-3,5-diol (1,1)	....	$(CH_2 \cdot CHOH)_2CH_2 \cdot C:$ $\overline{\quad (CH_3)_2 \quad}$	*VI-371	144.22
2471	cyclohexane (*o*) *	hexahydro-*o*-xylene	$(CH_3)_2C_6H_{10}$	V-36	112.22
2472	cyclohexane (*m*) *	hexahydro-*m*-xylene	$(CH_3)_2C_6H_{10}$	V-36	112.22
2473	cyclohexane (*p*) *	hexahydro-*p*-xylene	$(CH_3)_2C_6H_{10}$	V-39	112.22
2474	2,4-dicarbethoxy-pyrrole (3,5)	....	$C_{12}H_{17}O_4N$	XXII-133	239.27
2475	dihydroresorcinol (5,5)	1,1-diMe-cyclo-hexan-3,5-dione	$(CH_3)_2C_6H_6O_2$	VII-559	140.18
2476	1,3-diphenylcyclo-butane (1,3)	α-Me-styrene dimer	$[C_6H_5(CH_3)C \cdot CH_2]_2$	V-652	236.36
2477	dithiocarbamate Na	....	$(CH_3)_2N \cdot CS_2Na \cdot$ $2\frac{1}{2}H_2O$	IV-75	188.24
2478	ether	methyl ether	$CH_3 \cdot O \cdot CH_3$	I-281	46.07
2479	6-ethoxyquinoline (2,4)		$C_2H_5O \cdot C_9H_4N(CH_3)_2$	....	201.27
2480	6-ethoxyquinoline ethiodide (2,4)	....	$C_{13}H_{15}ON \cdot C_2H_5I$	....	357.24
2481	ethylacetic acid	....	$(CH_3)_2(C_2H_5) \vdots C \cdot$ CO_2H	II-335	116.16
2482	ethylamine	....	$(CH_3)_2N \cdot C_2H_5$	IV-94	73.14
2483	ethylbenzene (3,4)	4-Et-*o*-xylene	$C_2H_5 \cdot C_6H_3(CH_3)_2$	V-427	134.22
2484	ethylbenzene (3,5)	5-Et-*m*-xylene	$C_2H_5 \cdot C_6H_3(CH_3)_2$	V-429	134.22
2485	ethylbenzene (2,4)	6-Et-*m*-xylene	$C_2H_5 \cdot C_6H_3(CH_3)_2$	V-428	134.22
2486	ethylbenzene (2,5)	2-Et-*p*-xylene	$C_2H_5 \cdot C_6H_3(CH_3)_2$	V-428	134.22
2487	fumarate	methyl fumarate	$(:CH \cdot CO_2 \cdot CH_3)_2$	II-741	144.13
2488	furan (2,5)	α,α'-diMe-furan	$(CH_3)_2C_4H_2O$	XVII-41	96.13
2489	glutaric acid (α,α)	....	$(CH_3)_2C_3H_4(CO_2H)_2$	II-676	160.17
2490	glyoxime	diacetyl-dioxime	$(CH_3 \cdot C:NOH)_2$	I-772	116.12
2491	guanidine H_2SO_4 (*uns*)	....	$(CH_3)_2N \cdot C(NH) \cdot$ $NH_2 \cdot \frac{1}{2}H_2SO_4$	**IV-574	136.16
2492	1,3-heptadiene-(2,6)	*iso*-geraniolene	C_9H_{16}	I-260	124.23
2493	1,5-hexadiene (2,5)	....	$[CH_2:C(CH_3) \cdot CH_2 \cdot]_2$	I-259	110.20
2494	hydrazine (*sym*)	hydrazomethane	$CH_3NH \cdot NHCH_3$	IV-547	60.10
2495	hydrazine (*uns*)	....	$(CH_3)_2N \cdot NH_2$	IV-547	60.10
2496	hydrazine HCl	(*uns*)	$(CH_3)_2N \cdot NH_2 \cdot HCl$	IV-547	96.56
2497	2-(2-hydroxy-phenyl)-imida-zole (4,5)	....	$CH_3 \cdot C:C(CH_3) \cdot N:C \cdot$ $\overline{\quad (C_6H_4OH) \cdot NH \quad}$	XXIII-391	188.23

Dimethyl-benzyl phenol 1573   * cis
Dimethyl-butadiene 2769
Dimethyl-butandiol 5312
Dimethyl-butyl carbinol (*n*) 3535
Dimethyl-butyl carbinol (*iso*) 3530

Dimethyl-butyl carbinol (*tert*) 3528
Dimethyl cadmium 1186
Dimethyl-chloroarsine 1179
Dimethyl-cyclohexadiene 2289-91
Dimethyl-diazine 2533

Table 7-4 (*Continued*)
PHYSICAL CONSTANTS OF ORGANIC COMPOUNDS

No.	Crystalline Form and Color	Specific Gravity	Melting Point °C.	Boiling Point °C.	Solubility in 100 Parts		
					Water	Alcohol	Ether
2463	mn./aq. al.	....	166	subl.	v. sl. s. h.	v. s. c.	....
2464	col. nd.	....	167-8	....	i.	s.	....
2465	lq.	0.915^0	....	$183\text{-}4^{765mm}$	sl. c.; v. sl. h.	∞	∞
2466	nd.	$0.979\frac{16}{4}^{\circ}$	24	228	....	....	....
2467	mn. pr./al.	....	121	350	i.	s.	s.; s. bz.
2468	col. lq.	$1.070\frac{20}{4}^{\circ}$	0.5	89-90	i.	∞	∞
2469	col. oil	$1.121\frac{15}{15}^{\circ}$	....	210.5	3^{15^0}	....	....
2470	pl./act.	....	146-7	....	s.; s. bz.	s.; s. chl.	v. sl. s.
2471	lq.	$0.796\frac{20}{4}^{\circ}$	-50.0	129.7	i.	∞	∞
2472	lq.	$0.766\frac{20}{4}^{\circ}$	-75.6	120.1	i.	∞	∞
2473	lq.	$0.783\frac{20}{4}^{\circ}$	-87.4	124.3	....	....	....
2474	cr./aq. al. or ac.	....	134-5	....	i.; s. chl.; s. H_2SO_4	s.; s. bz.; i. HCl	v. sl. s.; s. ac.
2475	yel. mn. nd./aq.	....	148-9	....	0.416^{25^0}; s. chl.	6.6 h. bz.; s. EtOAc	v. sl. s.
2476	cr./et.	....	52.6	307.5^{760mm}	i.; 125^{25^0} act.	5^{25^0}; 190^{25^0} bz.	147^{25^0}; 104^{25^0} CCl_4
2477	cr.	....	$-2H_2O$, 115	$-2\frac{1}{2}H_2O$, 130	....	....	....
2478	gas	1.617(A)	-138.5	-23.7	3700^{18^0} cc.	s.	s.
2479	col. nd.	....	86-8	....	i.	v. s.	v. s.
2480	yel. rhb.	....	163-4	....	i.	s.	i.
2481	lq.	....	-14	187	v. sl. s.	s.	s.
2482	col. lq.	....	....	37.5	....	....	....
2483	col. lq.	$0.875\frac{20}{4}^{\circ}$	-67.0	189.8	i.	s.	s.
2484	col. lq.	$0.866\frac{20}{4}^{\circ}$	-84.2	183.8	i.	s.	s.
2485	lq.	$0.876\frac{20}{4}^{\circ}$	-63.0	188.4	i.	s.	s.
2486	lq.	$0.877\frac{20}{4}^{\circ}$	-53.7	186.9	i.	s.	s.
2487	col. tri.	1.045^{106^0}	102	193.3	s. c. chl.	sl. s.	sl. s.
2488	col. lq.	$0.888\frac{20}{4}^{\circ}$	....	93-4	i.; i. alk.	∞	∞
2489	nd./HCl	....	90	....	v. s.	v. s.	v. s.
2490	col. cr.	....	240-6	....	0.062^{20^0}	v. s.	v. s.
2491	col. cr./aq.	....	285-7 d.	....	v. s.	i.	i.
2492	lq.	$0.765\frac{10}{4}^{\circ}$	....	$143\text{-}5^{755mm}$	i.	....	....
2493	lq.	$0.749\frac{21}{21}^{\circ}$	<-80	$113\text{-}4^{755mm}$	i.	....	....
2494	hyg. lq.	$0.827\frac{20}{4}^{\circ}$	....	81^{747mm}	∞	∞	∞
2495	hyg. lq.	0.791^{22^0}	....	62.5^{717mm}	v. s.	v. s.	v. s.
2496	col. cr.	....	82-3	....	s.	s.	i.
2497	nd./aq. al.	....	218	....	....	....	....

Dimethyl-diphenyl urea 2467
Dimethyl-disulfide 4222
Dimethyl-ethyl carbinol 409
Dimethyl-ethylene 1148-9
Dimethyl-ethylene glycol 1154

Dimethyl-ethylene oxide 1155
Dimethyl-furfurane carboxylic acid 6396
Dimethyl-hexyl carbinol (*n*) 4922
Dimethyl-hydantoin 66
Dimethyl-hydroquinone 2376

Table 7-4 (*Continued*)
PHYSICAL CONSTANTS OF ORGANIC COMPOUNDS

No.	Name	Synonym	Formula	Beil. Ref.	Formula Weight
2498	**Dimethyl** itaconate	....	$C_3H_4(CO_2CH_3)_2$	II-762	158.16
2499	ketene	....	$(CH_3)_2C:CO$	I-731	70.09
2500	malate (*l*)	methyl malate	$C_2H_4O(CO_2CH_3)_2$	III-429	162.14
2501	maleate	methyl maleate	$(:CH\cdot CO_2\cdot CH_3)_2$	II-751	144.13
2502	malonate	methyl malonate	$CH_2(CO_2\cdot CH_3)_2$	II-572	132.12
2503	malonic acid	....	$(CH_3)_2C(CO_2H)_2$	II-647	132.12
2504	mesaconate	....	$CH_3\cdot C_2H(CO_2CH_3)_2$	II-765	158.16
2505	naphthylamine (α)	....	$(CH_3)_2N\cdot C_{10}H_7$	XII-1221	171.24
2506	naphthylamine (β)	....	$(CH_3)_2N\cdot C_{10}H_7$	XII-1273	171.24
2507	naphthalene (1,4)	....	$(CH_3)_2C_{10}H_6$	V-570	156.23
2508	naphthalene (2,3)	guajen	$(CH_3)_2C_{10}H_6$	*V-268	156.23
2509	naphthalene (2,6)	....	$(CH_3)_2C_{10}H_6$	V-570	156.23
2510	α-naphthoquinoline	(2,4)	$(CH_3)_2C_{13}H_7N$	XX-475	207.28
2511	β-naphthoquinoline	(2,4)	$(CH_3)_2C_{13}H_7N$	XX-476	207.28
2512	4-nitrophthalate	....	$NO_2C_6H_3(CO_2CH_3)_2$	IX-830	239.19
2513	nitrosamine	nitroso-dimethyl-amine	$(CH_3)_2N\cdot NO$	IV-84	74.08
2514	2,6-octadiene (2,6)	dihydromyrcene	$C_{10}H_{18}$	I-260	138.25
2515	2,7-octadiene (2,6)	linaloolene	$C_{10}H_{18}$	I-261	138.25
2516	2,6-octadiene-8-ol (2,6)	nerol	$C_{10}H_{18}O$	I-459	154.25
2517	2,6-octadiene-8-ol (2,6)	geraniol	$C_{10}H_{18}O$	I-457	154.25
2518	2,7-octadiene-6-ol (2,6)	*l*-linalool	$C_{10}H_{18}O$	I-460	154.25
2519	octanol-6 (2,6)	linalool tetrahydride	$C_{10}H_{22}O$	I-426	158.29
2520	octanol-8 (2,6)	geraniol tetrahydride	$C_{10}H_{22}O$	I-426	158.29
2521	oxalate	methyl oxalate	$(CO_2\cdot CH_3)_2$	II-534	118.09
2522	oxamide (*sym*)	....	$(CO\cdot NHCH_3)_2$	IV-61	116.12
2523	oxamide (*uns*)	....	$(CH_3)_2N\cdot (CO)_2NH_2$	IV-61	116.12
2524	phenacyl chloride (2,4)	ω-Cl-2,4-di Me-acetophenone	$(CH_3)_2C_6H_3\cdot CO\cdot CH_2Cl$	VII-324	182.65
2525	phenanthrene (9,10)	....	$(CH_3)_2C_{14}H_8$	V-680	206.29
2526	phenyl carbinol	....	$C_6H_5\cdot C(CH_3)_2\cdot OH$	VI-506	136.20
2527	*p*-phenylenedi-amine	(*sym*)	$(CH_3NH)_2C_6H_4$	XIII-71	136.20
2528	*p*-phenylenedi-amine oxalate	(*sym*)	$(CH_3NH)_2C_6H_4\cdot 2H_2C_2O_4$	....	316.27
2529	phosphine	....	$(CH_3)_2PH$	IV-580	62.05
2530	*o*-phthalate	methyl phthalate	$C_6H_4(CO_2CH_3)_2$	IX-797	194.19
2531	*m*-phthalate	Me *iso*-phthalate	$C_6H_4(CO_2CH_3)_2$	IX-834	194.19
2532	*p*-phthalate	Me-terephthalate	$C_6H_4(CO_2CH_3)_2$	IX-843	194.19
2533	pyrazine (2,5)	ketine; glycoline	$CH_3C:CHN:C(CH_3)\cdot$ ─── CH:N ───|	XXIII-96	108.14
2534	pyrazole (3,5)	....	$(CH_3)_2N:C\cdot CH:C\cdot NH$ |─── ───|	XXIII-74	96.13
2535	pyrrol (1,2)	....	$CH_3\cdot C(CH)_3\cdot N\cdot CH_3$ |─── ───|	....	95.15
2536	pyrrol (2,3)	....	$(CH_3)_2C_4H_2NH$	XX-172	95.15
2537	pyrrol (2,4)	....	$(CH_3)_2C_4H_2NH$	XX-172	95.15

Dimethyl-itaconic acid 5700
Dimethyl-ketazine 3932
Dimethy-ketol 43
Dimethyl ketone 56

Dimethyl mercury 4044
Dimethyl-octane 1637-8
Dimethyl-octenol 1519
Dimethyl-pentanol 3529-31

Table 7-4 (*Continued*)
PHYSICAL CONSTANTS OF ORGANIC COMPOUNDS

No.	Crystalline Form and Color	Specific Gravity	Melting Point °C.	Boiling Point °C.	Solubility in 100 Parts		
					Water	Alcohol	Ether
2498	mn./Me al.	$1.124^{18°}_{4}$	38	208	....	....	....
2499	yel. lq.	....	−97.5	34^{750mm}	d.	d.	s.
2500	lq.	$1.233^{20°}_{4}$	....	242	s.	∞	∞
2501	col. lq.	$1.151^{21°}$	7.6	205	....	....	....
2502	col. lq.	$1.154^{20°}_{4}$	−62	180-1	i.	∞	∞
2503	col. pr.	$1.357^{18°}$	192-3 d.	d. > 130	$10^{13°}$	v. s. abs.	v. s.
2504	col. lq.	$1.121^{20°}_{20}$	....	205.5-6.5	....	....	....
2505	col. oil	$1.042^{20°}$	....	274.5^{711mm}	i.	s.	s.
2506	col. cr.	$1.039^{70°}_{70}$	46-7	304-5	i.	s.	s.
2507	lq.	$1.016^{20°}_{4}$	<−18	264-6	i.	....	....
2508	lf./al.	....	104.0-4.5	$265-6^{767mm}$	i.	sl. s.	s. bz.
2509	lf./al.	$1.142^{0°}_{4}$	110-1	$261-2^{762mm}$	i.	sl. s.	
2510	nd./pet.	....	43-4	....	v. s. h. pet.	i. 90% al.	v. s.
2511	nd./et.	....	126-7	>300 d.	v. sl. s. h.	s.; s. ac.	s.; s. act.
2512	cr./aq. al.	....	66-7.5	....	....	....	....
2513	yel. oil	$1.006^{20°}_{4}$	....	153^{774mm}	s.	s.	s.
2514	lq.	$0.775^{21°}_{4}$	....	171.5-3.5	....	....	....
2515	lq.	$0.788^{20°}$	....	165-8	....	....	....
2516	oil	$0.881^{15°}$	....	$224-5^{745mm}$	....	....	....
2517	col. lq.	$0.883^{15°}$	<−15	230	i.	∞	∞
2518	oil	$0.862^{20°}$	....	$197-200^{756mm}$	v. sl. s.	∞	∞
2519	lq.	$0.836^{15°}_{4}$	....	196-7	....	....	....
2520	lq.	$0.849^{0°}_{4}$	....	212-3	....	....	....
2521	col. mn.	$1.148^{54°}$	54	163.3	6	s.	s.
2522	nd./aq.	$1.34°$	212-7	subl.	$2.5^{9.4°}$; v. s. h.	s.; d. h. alk.	v. sl. s.; s. chl.
2523	pl./bz.	....	104	....	v. s.	v. s.	v. sl.s.
2524	lf.	....	62-3	....	....	s.	....
2525	pr./dil. ac.	....	139	subl.	s. ac.	sl. s.; s. bz.	s. chl.
2526	pr.	$0.972^{19°}_{4}$	23	215-20 sl. d.	....	....	....
2527	cr./pet.	....	53	150^{17mm}	v. sl. s.	s.	s.
2528	col. cr.	....	194-6 d.	....	s.	sl. s.	i.
2529	col. lq.	<1	....	25	i.	....	....
2530	col. lq.	$1.189^{25°}_{25}$	....	280^{734mm}	0.43	....	....
2531	nd./aq. al.	....	67-8	....	i.	....	....
2532	rhb./al.	....	140	>300; subl.	0.3 h.	s. h.	s.
2533	pl.	$0.990^{18°}_{4}$	15	155	∞	∞	∞
2534	pl.	$0.884^{26°}_{4}$	106-7	218^{759mm}	s.; s. bz.	s.; s. chl.	s.
2535	lq.	....	....	65^{14mm}	....	....	....
2536	lq.	....	....	165	....	....	....
2537	lq.	$0.927^{14°}_{4}$	....	171	sl. s.	s.	s.; s. bz.

Dimethyl-phenol 6465-70
Dimethyl-phenyl hydrazine 6500-2
Dimethyl-phenylenediamine 305-7

Dimethyl-propyl carbinol (*n*) 3631
Dimethyl-propyl carbinol (*iso*) 3629
Dimethyl-pyridine 3981-4

Table 7-4 (*Continued*)
PHYSICAL CONSTANTS OF ORGANIC COMPOUNDS

No.	Name	Synonym	Formula	Beil. Ref.	Formula Weight
2538	**Dimethyl** pyrrol (2,5)	....	$(CH_3)_2C_4H_2NH$	XX-172	95.15
2539	quinolinate (2,3)	....	$C_5H_3N(CO_2CH_3)_2$	XXII-151	195.18
2540	quinoline (2,3)	....	$(CH_3)_2C_9H_5N$	XX-406	157.22
2541	quinoline (2,4)	4-Me-quinaldine	$(CH_3)_2C_9H_5N$	XX-407	157.22
2542	quinoline H_2SO_4	(2,4)	$C_{11}H_{11}N \cdot \frac{1}{2}H_2SO_4$	XX-408	206.26
2543	quinoline EtI	(2,4)	$C_{11}H_{11}N \cdot C_2H_5I$	XX-408	313.18
2544	quinoline (2,6)	6-Me-quinaldine	$(CH_3)_2C_9H_5N$	XX-408	157.22
2545	quinoline EtI	(2,6)	$C_{11}H_{11}N \cdot C_2H_5I$	XX-409	313.18
2546	quinoline (3,4)	....	$(CH_3)_2C_9H_5N$	XX-410	157.22
2547	quinoline (5,8)	....	$(CH_3)_2C_9H_5N$	XX-411	157.22
2548	quinoline (6,8)	....	$(CH_3)_2C_9H_5N$	XX-411	157.22
2549	quinone (2,3)	*o*-xyloquinone	$(CH_3)_2C_6H_2O_2$	VII-656	136.15
2550	quinone (2,5)	*p*-xyloquinone	$(CH_3)_2C_6H_2O_2$	VII-658	136.15
2551	quinone (2,6)	*m*-xylo-*p*-quinone	$(CH_3)_2C_6H_2O_2$	VII-657	136.15
2552	sebacate	methyl sebacate	$(CH_2)_8(CO_2CH_3)_2$	*II-293	230.31
2553	selenide	selenium methyl	$(CH_3)_2Se$	I-291	109.03
2554	succinate	methyl succinate	$(CH_2CO_2 \cdot CH_3)_2$	II-609	146.14
2555	*iso*-succinate	....	$CH_3 \cdot CH(CO_2 \cdot CH_3)_2$	II-628	146.14
2556	succinic acid (*meso*)	....	$(CH_3CH)_2(CO_2H)_2$	II-665	146.14
2557	succinic acid (α,α)	....	$(CH_3)_2C \cdot CH_2(CO_2H)_2$	II-661	146.14
2558	succinic acid (*dl*)	....	$(CH_3CH)_2(CO_2H)_2$	II-667	146.14
2559	sulfate	methyl sulfate	$(CH_3O)_2SO_2$	I-283	126.13
2560	sulfide	methyl sulfide	$(CH_3)_2S$	I-288	62.13
2561	sulfite	....	$(CH_3O)_2SO$	I-282	110.13
2562	sulfone	....	$(CH_3)_2SO_2$	I-289	94.13
2563	sulfoxide	....	$(CH_3)_2SO$	I-289	78.13
2564	tartrate (*d*)	....	$(CHOH \cdot CO_2CH_3)_2$	III-510	178.14
2565	tartrate (racemic)	diMe-racemate	$(CH_3)_2C_4H_4O_6$	III-527	178.14
2566	telluride	tellurium methyl	$(CH_3)_2Te$	I-291	157.67
2567	thetin	....	$HO \cdot S(CH_3)_2 \cdot CH_2 \cdot CO_2H$	III-247	138.19
2568	thiourea (*sym*)	....	$(CH_3NH)_2C:S$	IV-70	104.17
2569	toluidine (*o*)(*N*)	....	$CH_3 \cdot C_6H_4 \cdot N(CH_3)_2$	XII-785	135.21
2570	toluidine (*m*)(*N*)	....	$CH_3 \cdot C_6H_4 \cdot N(CH_3)_2$	XII-857	135.21
2571	toluidine (*p*)(*N*)	....	$CH_3 \cdot C_6H_4 \cdot N(CH_3)_2$	XII-902	135.21
2572	urea (*sym*)	....	$(CH_3NH)_2C:O$	IV-65	88.11
2573	urea (*uns*)	....	$(CH_3)_2N \cdot CO \cdot NH_2$	IV-73	88.11
2574	**Dinaphthyl** (1,1')	α,α'-binaphthyl	$C_{10}H_7 \cdot C_{10}H_7$	V-726	254.33
2575	**Dinaphthyl** (2,2')	....	$C_{10}H_7 \cdot C_{10}H_7$	V-727	254.33
2576	amine (β,β')	....	$C_{10}H_7 \cdot NH \ C_{10}H_7$	XII-1278	269.35
2577	ketone (α,α')	....	$(C_{10}H_7)_2CO$	VII-539	282.35
2578	ketone (α,β')	....	$(C_{10}H_7)_2CO$	VII-539	282.35
2579	ketone (β,β')	....	$(C_{10}H_7)_2CO$	VII-539	282.35
2580	methane (α,α')	....	$(C_{10}H_7)_2CH_2$	V-728	268.36
2581	methane (α,β')	....	$(C_{10}H_7)_2CH_2$	*V-360	268.36
2582	methane (β,β')	....	$(C_{10}H_7)_2CH_2$	V-729	268.36
2583	sulfide (α,α')	....	$(C_{10}H_7)_2S$	VI-623	286.40
2584	sulfide (β,β')	....	$(C_{10}H_7)_2S$	VI-659	286.40
2585	urea (α)(*sym*)	α,α dinaphthyl carbamide	$(C_{10}H_7NH)_2CO$	XII-1238	312.38

Dimethyl-resorcinol 2375, 2377, 2380, 6479   Dimethyl zinc 6504
Dimethyl-thiophene 5928-31   Dinaphthol 2336-7
Dimethyl-xanthine, cf. alkd.   Dinaphthyl carbamide 2585

Table 7-4 (*Continued*)
PHYSICAL CONSTANTS OF ORGANIC COMPOUNDS

No.	Crystalline Form and Color	Specific Gravity	Melting Point °C.	Boiling Point °C.	Solubility in 100 Parts		
					Water	Alcohol	Ether
2538	oil	$0.935\frac{20°}{4}$	....	169	v. sl. s.	v. s.	v. s.; v. sl. s. alk.
2539	col. pl./bz.	....	53-4	....	s.; s. CS_2	s.; i. lg.	s.; s. bz.
2540	pl.		68-9	261^{729mm}	sl. s.	v. s.	s.; s. lg.
2541	lq.	$1.061^{15°}$	....	264-5	i.	s.	s.
2542	nd./al.		225-8 d.		v. s. h.	v. sl. s.	....
2543	yel. nd./al.		226-7		....	s. h.	....
2544	cr./et.		60	266-7	v. sl. s. h.	s.	s.
2545	cr.		225-7	....			
2546	cr.		65	290^{737mm}	i.	....	....
2547	lq.	$1.070^{21°}$	4-5	265^{736mm}	i.	....	....
2548	lq.	$1.067^{4°}$	....	268-9	i.	....	....
2549	yel. nd.		55	subl.	sl. s.	s.	s.
2550	yel. tri.		124-5	subl.	sl. s. h.	v. s. h.	s.; s. bz.
2551	yel. nd.		72-3	subl.	....		
2552	nd. or pl./et.	$0.988\frac{2.8°}{4}$	38 (26)	293^{754mm}	....		
2553	lq.	$1.408\frac{16°}{4}$	....	58.2	i.		
2554	col. cr.	$1.126\frac{15°}{15}$	18.2	195.9	0.9	3	....
2555	lq.	$1.028\frac{25°}{25}$	....	179	i.	∞	∞
2556	tri.	1.314	209	d.	$<3^{14°}$	v. s.	v. s.
2557	tri.	1.323	142	d. 165	$7.5^{14°}$	v. s.	v. sl. s.
2558	rhb. pr.	1.339	129	d. to anhyd.	$3^{14°}$	v. s.	v. s.
2559	poison. oil	$1.352\frac{0°}{4}$	−26.8	188.3-8.6	v. sl. s.	∞	∞
2560	oil	$0.846\frac{21°}{4}$	−98.3	37.3	i.	s.	s.
2561	lq.	$1.046\frac{16.2°}{4.1}$	....	126.5^{756mm}	d.	s.	s.
2562	pr.		109	238	....	....	....
2563	oil		....	volt. 100 d.	v. s.	v. s.	v. s.
2564	cr.	$1.328\frac{20°}{4}$	61.5 (48)	280	s.*	$200^{15°}$	v. s. bz.*
2565	mn./al.	$1.260\frac{8.9-6°}{4}$	89	282	....	$21^{15°}$	
2566	lt. yel. oil	<1	....	82	s.	s.	s.
2567	delq. cr.		d.−H_2O	....	s.	sl. s.	....
2568	hyg., delq.		61-2	....	v. s.	v. s.	sl. s. et.
2569	lq.	$0.929\frac{20°}{4}$	−61.3	185-6	i.	s.	s.
2570	yel. oil	$0.941\frac{20°}{4}$	....	213-5	i.	s.	s.
2571	lq.	$0.937\frac{20°}{4}$	....	210-1	i.	s.	s.
2572	pr./chl. et.	1.142	106	268-70	v. s.	v. s.	v. sl. s.
2573	mn./al.	1.255	182-5	....	s.	sl. s. c.	v. sl. s.
2574	lf./al.		160(144-6)	$240-4^{12mm}$	i.; s. bz.	s. h.	s.
2575	pl.		187.8	452^{753mm}	i.	v. sl. s. c.	v. sl. s. c.
2576	lf./bz.		171-2	471	i.	sl. s. h.	v. s. h. bz.
2577	nd./et.		104	....	s. H_2SO_4	s. h.	v. s. bz.
2578	nd./al.		135-6	....	s. bz.	$1.3^{14°}$	sl. s. h.
2579	lf. or nd.		α125.5 β164.5	....	....	$α0.4^{19°}$ $β0.08^{19°}$	v. sl. s.; v. s. chl.
2580	pr. or nd./ al.		109	>360	v. s. bz.	0.8 c.; 6.6 h.	v. s.
2581	pr./al.		95-6	....	s. bz.	s. EtOAc	....
2582	nd./al.		92	....	....	s.	s. bz.
2583	nd./al.		110	$289-90^{15mm}$		v. sl. s.	v. s. CS_2
2584	lf./al.		151	$295-6^{15mm}$	i.	v. sl. s.	v. s. CS_2
2585	nd./ac.		297-8	subl.	....	sl. s. h.	....

* Modification m. p. 61.5° less soluble.
Dinaphthyl ether 4560-2
Dinaphthyl hydrazine 3690-1

Dinaphthyl mercury 4045
Dinicotinic acid 5502

Table 7-4 (*Continued*)
PHYSICAL CONSTANTS OF ORGANIC COMPOUNDS

No.	Name	Synonym	Formula	Beil. Ref.	Formula Weight
2586	**Dinitro**-acetanilide (2,4)	....	$(NO_2)_2C_6H_3 \cdot NH \cdot COCH_3$	XII-754	225.16
2587	aniline (2,3)	....	$(NO_2)_2C_6H_3 \cdot NH_2$	XII-747	183.12
2588	aniline (2,4)	....	$(NO_2)_2C_6H_3 \cdot NH_2$	XII-747	183.12
2589	aniline (2,5)	....	$(NO_2)_2C_6H_3 \cdot NH_2$	XII-757	183.12
2590	aniline (2,6)	....	$(NO_2)_2C_6H_3 \cdot NH_2$	XII-758	183.12
2591	aniline (3,4)	....	$(NO_2)_2C_6H_3 \cdot NH_2$	XII-758	183.12
2592	aniline (3,5)	....	$(NO_2)_2C_6H_3 \cdot NH_2$	XII-759	183.12
2593	anisole (2,4)	α-dinitroanisole	$CH_3O \cdot C_6H_3(NO_2)_2$	VI-254	198.14
2594	anthranilic acid (3,5)	....	$C_6H_2(CO_2H)(NO_2)_2 \cdot NH_2$	XIV-379	227.13
2595	anthraquinone (1,5)	....	$(NO_2C_6H_3)_2(CO)_2$	VII-793	298.21
2596	anthraquinone (1,8)	....	$(NO_2 \cdot C_6H_3)_2(CO)_2$	VII-795	298.21
2597	anthraquinone (2,7)	β-diNO$_2$-anthraquinone	$(NO_2 \cdot C_6H_3)_2(CO)_2$	VII-795	298.21
2598	azoxybenzene (4,4')	....	$(NO_2 \cdot C_6H_4)_2N_2O$	XVI-628	288.22
2599	benzalacetone (3,3')	....	$(NO_2C_6H_4CH:CH)_2:CO$	VII-506	324.30
2600	benzaldehyde (2,4)	....	$(NO_2)_2C_6H_3 \cdot CHO$	VII-264	196.12
2601	benzaldehyde (2,6)	....	$(NO_2)_2C_6H_3 \cdot CHO$	*VII-144	196.12
2602	benzene (*o*)	....	$(NO_2)_2C_6H_4$	V-257	168.11
2603	benzene (*m*)	....	$(NO_2)_2C_6H_4$	V-258	168.11
2604	benzene (*p*)	....	$(NO_2)_2C_6H_4$	V-261	168.11
2605	benzene sulfonic acid	(2,4;1)	$(NO_2)_2C_6H_3 \cdot SO_3H \cdot 3H_2O$	XI-78	302.23
2606	benzene sulfonic	Na salt (2,4;1)	$C_6H_3O_7N_2SNa \cdot H_2O$	XI-78	288.17
2607	benzoic acid (2,4)	....	$(NO_2)_2C_6H_3 \cdot CO_2H$	IX-411	212.12
2608	benzoic acid (2,5)	....	$(NO_2)_2C_6H_3 \cdot CO_2H$	IX-412	212.12
2609	benzoic acid (2,6)	....	$(NO_2)_2C_6H_3 \cdot CO_2H$	IX-412	212.12
2610	benzoic acid (3,4)	....	$(NO_2)_2C_6H_3 \cdot CO_2H$	IX-413	212.12
2611	benzoic acid (3,5)	....	$(NO_2)_2C_6H_3 \cdot CO_2H$	IX-413	212.12
2612	benzophenone (4,4')	....	$(NO_2 \cdot C_6H_4)_2CO$	VII-428	272.22
2613	benzoyl chloride	(3,5)	$(NO_2)_2C_6H_3 \cdot COCl$	IX-414	230.57
2614	benzoylene urea	(6,8)	$C_8H_4O_6N_4$	*XXIV-344	252.14
2615	benzyl chloride (2,4)	....	$(NO_2)_2C_6H_3 \cdot CH_2Cl$	V-344	216.58
2616	*o*-cresol (4,6)	3,5-diNO$_2$-2-OH-toluene	$(NO_2)_2C_6H_2(CH_3) \cdot OH$	VI-368	198.14
2617	*m*-cresol (2,6)	2,4-diNO$_2$-3-OH-toluene	$(NO_2)_2C_6H_2(CH_3) \cdot OH$	VI-387	198.14
2618	diazo-aminobenzene (4,4')	....	$NO_2C_6H_4 \cdot N:N \cdot NH \cdot C_6H_4NO_2$	XVI-700	287.24
2619	diethylaniline	(2,4	$(NO_2)_2C_6H_3N(C_2H_5)_2$	XII-750	239.23
2620	diphenyl (2,2')	....	$(NO_2 \cdot C_6H_4 \cdot)_2$	V-583	244.21
2621	diphenyl (3,3')	....	$(NO_2 \cdot C_6H_4 \cdot)_2$	V-583	244.21
2622	diphenyl (4,4')	....	$(NO_2 \cdot C_6H_4 \cdot)_2$	V-584	244.21
2623	diphenyl (2,4')	....	$(NO_2 \cdot C_6H_4 \cdot)_2$	V-584	244.21

Dinitro-aminophenol 5303
Dinitro-bromobenzene 930-1

Dinitro-bromobenzoic acid 932
Dinitro-chlorobenzene 1364-9

Table 7-4 (*Continued*)
PHYSICAL CONSTANTS OF ORGANIC COMPOUNDS

No.	Crystalline Form and Color	Specific Gravity	Melting Point °C.	Boiling Point °C.	Solubility in 100 Parts		
					Water	Alcohol	Ether
2586	nd./al.	....	120	....	i. c.	v. s. h.	s.
2587	or. yel./al.	....	127	....	....	v. s.	s.
2588	yel. or gn. mn.	$1.615^{14°}$	187-8	....	v. sl. s. h.	$0.76^{21°}$	sl. s. h. HCl
2589	or. yel./al.	....	137	....	....	v. s.	....
2590	yel. nd./al.	....	138-40	....	i.	s. h. bz.	i. lg.
2591	yel. nd./aq.	....	154	....	....	v. s.	s.
2592	yel. nd./aq.	....	159-60	....	sl. s. bz.	s.	s.
2593	col. mn.	$1.341^{20°}$	94-5	....	sl. s. h.	$1.5^{20°}$	....
2594	yel. pl./al.	....	268	....	$0.02^{15°}$	sl. s.	....
2595	yel. nd.	....	>330	subl.	i.	v. sl. s.	v. sl. s.
2596	yel. pr.	....	312	....	s. Ac_2O	....	....
2597	nd./ac.	....	262	subl.	i.	sl. s.	sl. s.
2598	yel. nd./ bz.	....	192-3	....	....	....	....
2599	cr./Ac_2O	....	237	....	i.	i.	i
2600	yel. pr.	....	69-70	$190-210^{10mm}$	i.	v. s.	v. s.
2601	lf./ac.	....	123	....	s. h.	s.; sl. s. CS_2	s.; v. sl. s. lg.
2602	col. mn.	$1.59^{18°}$	117-8	319^{774mm}	0.01 c.	$1.9^{21°}$	$5.7^{18°}$ bz.
2603	col. rhb.	$1.575\frac{20°}{4}$	89.8	300-2	$0.39^{9°}$	$3.3^{20°}$ abs.	$39.5^{18°}$ bz.
2604	col. mn.	$1.625^{18°}$	173-4	299^{777mm}	$0.18^{100°}$	$0.18^{21°}$	$2.6^{18°}$ bz.
2605	pa. yel., hyg. pr.	....	106-8	$-3H_2O$, >130	s.; s. ac.	s.; i. bz.; i. lg.	v. sl. s.
2606	cr./aq.	....	....	....	v. s.	....	....
2607	cr./aq.	....	179-80	....	$1.85^{25°}$	s.	$0.71^{30°}$ bz.
2608	mn. pr.	....	177-9	....	s. h.	s.	s.
2609	nd./aq.	....	202	....	v. s. h.	....	....
2610	nd.	....	163-4	....	$0.67^{25°}$	s.	s.
2611	mn. pr.	....	204-5	subl.	2 h.	v. s.	sl. s.
2612	col. nd./ac.	....	189	....	i.	....	....
2613	nd./bz.	....	68-9	196^{12mm}	d.	d.	....
2614	gn. yel. pr.	....	274-5	....	$0.016^{23°}$	sl. s. h. s. h. ac.	s. NaOH; sl. s. act.
2615	pl./et.	....	34	....	i.	s.	s.
2616	yel. pr./al.	....	86-7	....	v. sl. s.; v. sl. s. lg.	$10^{15°}$; s. alk.	s.; s. act.
2617	or. red nd.	....	99	....	....	....	....
2618	yel. nd./al.	....	224-6 d.	....	i.; v. sl. s. chl.	sl. s. h.; v. sl. s. bz.	s.
2619	yel. rhb./ act.	$1.374^{15°}$	80	....	v. sl. s. lg.	v. s. h.	v. s. h.
2620	yel. nd.	1.45	124	....	i.	s. h.	s.
2621	cr.	....	198	....	i.	$0.6^{20°}$	....
2622	nd./al.	1.445	233	....	i.	$1.5^{20°}$	sl. s. bz.
2623	mn.	1.474	93.5	....	....	v. s. h.	....

Dinitro-chlorobenzoic acid 1370-1
Dinitro-chlorophenol 1372

Dinitro-dihydroxyquinone 4586

Table 7-4 (*Continued*)
PHYSICAL CONSTANTS OF ORGANIC COMPOUNDS

No.	Name	Synonym	Formula	Beil. Ref.	Formula Weight
2624	**Dinitro** diphenyl-amine (2,4)	....	$(NO_2)_2C_6H_3\cdot NH\cdot C_6H_5$	XII-751	259.22
2625	diphenylamine (2,4')	....	$(NO_2\cdot C_6H_4)_2NH$	XII-715	259.22
2626	diphenylamine (4,4')	....	$(NO_2\cdot C_6H_4)_2NH$	XII-716	259.22
2627	diphenyl ether (2,2')	....	$(NO_2\cdot C_6H_4)_2O$	VI-219	260.21
2628	diphenyl ether (4,4')	....	$(NO_2\cdot C_6H_4)_2O$	VI-232	260.21
2629	diphenylmethane	(4,4')	$(NO_2\cdot C_6H_4)_2CH_2$	V-595	258.24
2630	ethane (1,1)	....	$CH_3\cdot CH(NO_2)_2$	I-102	120.07
2631	hexane (1,1)	....	$C_5H_{11}\cdot CH(NO_2)_2$	I-147	176.17
2632	hydroquinone acetate (2,6)	(2,6;1,4)	$(NO_2)_2C_6H_2(OH)\cdot O_2C\cdot CH_3$	VI-858 †	242.15
2633	hydroquinone di-acetate (2,6)	....	$(NO_2)_2C_6H_2:$ $(O_2C\cdot CH_3)_2$	*VI-419	284.18
2634	4'-hydroxydiphen-ylamine (2,4)	....	$(NO_2)_2C_6H_3\cdot NH\cdot$ $C_6H_4\cdot OH$	XIII-444	275.22
2635	mesitylene (2,4)	*eso*-dinitro-mesitylene	$(NO_2)_2C_6H(CH_3)_3$	V-411	210.19
2636	methane	....	$(NO_2)_2CH_2$	I-77	106.04
2637	naphthalene (1,3)	γ-dinitronaphthalene	$(NO_2)_2C_{10}H_6$	V-557	218.17
2638	naphthalene (1,5)	α-dinitronaphthalene	$(NO_2)_2C_{10}H_6$	V-558	218.17
2639	naphthalene (1,6)	δ-dinitronaphthalene	$(NO_2)_2C_{10}H_6$	V-559	218.17
2640	naphthalene (1,8)	β-dinitronaphthalene	$(NO_2)_2C_{10}H_6$	V-559	218.17
2641	α-naphthol (2,4)	....	$(NO_2)_2C_{10}H_5\cdot OH$	VI-617	234.17
2642	α-naphthol (2,4) Na	Martius yellow	$C_{10}H_5O_5N_2Na\cdot H_2O$	VI-618	274.17
2643	β-naphthol (1,6)	....	$(NO_2)_2C_{10}H_5\cdot OH$	VI-655	234.17
2644	phenanthraquinone	(2,7)	$O:C_{14}H_6(NO_2)_2:O$	VII-807	298.21
2645	phenetole (2,4)	α-dinitrophenetole	$(NO_2)_2C_6H_3\cdot OC_2H_5$	VI-254	212.16
2646	phenol (2,3)	ϵ-dinitrophenol	$(NO_2)_2C_6H_3\cdot OH$	VI-251	184.11
2647	phenol (2,4)	α-dinitrophenol	$(NO_2)_2C_6H_3\cdot OH$	VI-251	184.11
2648	phenol (2,4) Na	....	$C_6H_3O_5N_2Na\cdot H_2O$	*VI-126	224.11
2649	phenol (2,5)	γ-dinitrophenol	$(NO_2)_2C_6H_3\cdot OH$	VI-256	184.11
2650	phenol (2,6)	β-dinitrophenol	$(NO_2)_2C_6H_3\cdot OH$	VI-257	184.11
2651	phenol (3,4)	δ-dinitrophenol	$(NO_2)_2C_6H_3\cdot OH$	VI-257	184.11
2652	phenol (3,5)	θ-dinitrophenol	$(NO_2)_2C_6H_3\cdot OH$	VI-258	184.11
2653	phenylacetic acid (2,4)	....	$(NO_2)_2C_6H_3\cdot CH_2\cdot CO_2H$	IX-459	226.15
2654	phenylcarbazide (4,4')	....	$(NO_2\cdot C_6H_4\cdot NH\cdot NH)_2:CO$	....	332.28
2655	phenyl disulfide	(2,2')	$(NO_2\cdot C_6H_4\cdot S\cdot)_2$	VI-338	308.34
2656	phenyl disulfide	(4,4')	$(NO_2\cdot C_6H_4\cdot S\cdot)_2$	VI-340	308.34
2657	phenylhydrazine	(2,4)	$(NO_2)_2C_6H_3\cdot NHNH_2$	XV-489	198.14
2658	propane (1,1)	....	$C_2H_5\cdot CH(NO_2)_2$	I-117	134.09
2659	propane (2,2)	....	$(CH_3)_2C(NO_2)_2$	I-117	134.09
2660	resorcinol (2,4)	....	$(NO_2)_2C_6H_2(OH)_2$	VI-827	200.11
2661	resorcinol (4,6)	....	$(NO_2)_2C_6H_2(OH)_2$	VI-828	200.11
2662	salicylic acid (3,5)	3,5-diNO$_2$-2-OH-benzoic acid	$(NO_2)_2C_6H_2(OH)\cdot CO_2H\cdot H_2O$	X-122	246.13

† See also correction in *VI-419
Dinitro-diphenyl disulfide 2655-6

Dinitro-glycol 3462
Dinitro-hydroxybenzoic acid 2662

Table 7-4 (*Continued*)
PHYSICAL CONSTANTS OF ORGANIC COMPOUNDS

No.	Crystalline Form and Color	Specific Gravity	Melting Point °C.	Boiling Point °C.	Solubility in 100 Parts		
					Water	Alcohol	Ether
2624	red nd./bz.	....	157	....	s. act.	s. h.	s. chl.
2625	red nd.	....	220-2	....	i.	sl. s.	$0.4^{22°}$ act.
2626	yel. nd.	....	214-5	....	i.	sl. s.	$5.66^{20°}$ act.
2627	nd./al.	....	114.5	....	i.	$0.7^{20°}$	....
2628	nd./al.	....	142-3	....	i.	3 h.	sl. s.
2629	nd./bz.	....	183-5	....	s. h. ac.	i. c.	sl. s.
2630	lq.	$1.350_{23.5}^{23.5°}$	....	185-6	v. sl. s.	s.	s.
2631	yel. oil	>1	....	d.	v. sl. s.	v. s.	v. s.
2632	yel. nd./al.	....	95-6	....	s. chl.	s.	s.
2633	nd./al.	....	135-6	....	i.	s. h.	s. ac.
2634	red nd./ aq. al.	....	195-6	....	s. alk.	s.	v. sl. s. bz.
2635	rhb./al.	....	86	....	i.	s. h.	....
2636	lq.	....	<−15	100 d.	s.	....	....
2637	lt. yel. nd.	....	144-5	subl.	i.	s.	....
2638	nd./ac.	....	216	subl.	i.	s. h. bz.	v. sl. s. CS_2
2639	cr./ac.	....	166	....	i.	....	....
2640	rhb. pl.	....	170-2	d.	i.	$0.19^{19°}$, 88% al.	$0.72^{19°}$ bz.
2641	yel. nd./al.	....	138	....	v. sl. s. h.	sl. s.	sl. s.; s. ac.
2642	yel. red nd.	....	....	....	v. s.	....	....
2643	yel. nd.	....	195	....	v. sl. s. h.	s.	v. s.
2644	yel. cr./ac.	....	303-5	....	....	sl. s.	sl. s. ac.
2645	col. nd./al.	....	86	d.	v. sl. s.	s.	....
2646	yel. mn./al.	$1.681^{20°}$	144-5	....	sl. s.	v. s. h.	v. s.
2647	yel. rhb.	$1.683^{24°}$	112.9	subl.	0.5 c.; 5 h.	$4^{20°}$; s. bz.	v. s. h.
2648	cr.	....	....	....	$425°$	$2.7^{25°}$ abs.	$12^{5°}$ act.
2649	yel. mn./ aq.	....	104	....	sl. s.	v. s. h.	s.; s. alk.
2650	yel. rhb./ aq.	....	63-4	....	s. h.	s. h.	s.; s. chl.
2651	col. tri./aq.	1.672	134	....	....	v. s.	v. s.
2652	mn./HCl	1.702	123	....	....	v. s.	v. s.
2653	col. nd./ aq.	....	160	d. 179-80	....	....	....
2654	brn. pd.	....	261-3	....	i.	s.	i.
2655	yel. nd./bz.	....	195	d.	v. sl. s. ac.	v. sl. s.	v. sl. s. act.
2656	pl./al.	....	181	....	....	....	....
2657	vl. pr./al.	....	197-8 d.	....	i.	s. dil. a.	i.
2658	acidic oil	$1.258^{22°}$	....	189	s. alk.	....	....
2659	cr.	....	53	185.5	v. sl. s.	i. alk.	....
2660	yel. lf.	....	165 d.	expl.	v. sl. s.	s.	s. alk.
2661	yel. pr.	....	215	subl.	s. h.	s.	s.
2662	pl./aq.	....	173 d.	....	s. c.	v. s.	v. s.

Table 7-4 (*Continued*)
PHYSICAL CONSTANTS OF ORGANIC COMPOUNDS

No.	Name	Synonym	Formula	Beil. Ref.	Formula Weight
2663	**Dinitro** stilbene (4,4')	low m.p. form	$(NO_2 \cdot C_6H_4 \cdot CH:)_2$	V-637	270.25
2664	stilbene (4,4')	high m.p. form	$(NO_2 \cdot C_6H_4 \cdot CH:)_2$	V-637	270.25
2665	thymol (2,4)	(2,4;3,1,6)	$(NO_2)_2C_6H(CH_3) \cdot (OH)(C_3H_7)$	VI-543	240.22
2666	toluene (2,3)	....	$(NO_2)_2C_6H_3 \cdot CH_3$	V-339	182.14
2667	toluene (2,4)	....	$(NO_2)_2C_6H_3 \cdot CH_3$	V-339	182.14
2668	toluene (2,5)	....	$(NO_2)_2C_6H_3 \cdot CH_3$	V-341	182.14
2669	toluene (2,6)	....	$(NO_2)_2C_6H_3 \cdot CH_3$	V-341	182.14
2670	toluene (3,4)	....	$(NO_2)_2C_6H_3 \cdot CH_3$	V-341	182.14
2671	toluene (3,5)	....	$(NO_2)_2C_6H_3 \cdot CH_3$	V-341	182.14
2672	tyrosine (3,5)(*l*)	....	$C_9H_9O_7N_3 \cdot H_2O$	*XIV-668	289.20
2673	*o*-xylene (3,4)	....	$(NO_2)_2C_6H_2(CH_3)_2$	V-369	196.16
2674	*o*-xylene (3,6)	....	$(NO_2)_2C_6H_2(CH_3)_2$	V-369	196.16
2675	*o*-xylene (4,5)	....	$(NO_2)_2C_6H_2(CH_3)_2$	V-369	196.16
2676	*o*-xylene (3,5)	....	$(NO_2)_2C_6H_2(CH_3)_2$	V-369	196.16
2677	*m*-xylene (2,4)	....	$(NO_2)_2C_6H_2(CH_3)_2$	V-379	196.16
2678	*m*-xylene (4,6)	....	$(NO_2)_2C_6H_2(CH_3)_2$	V-380	196.16
2679	*p*-xylene (2,3)	β-dinitro-*p*-xylene	$(NO_2)_2C_6H_2(CH_3)_2$	V-387	196.16
2680	*p*-xylene (2,5)	γ-dinitro-*p*-xylene	$(NO_2)_2C_6H_2(CH_3)_2$	V-388	196.16
2681	*p*-xylene (2,6)	α-dinitro-*p*-xylene	$(NO_2)_2C_6H_2(CH_3)_2$	V-388	196.16
2682	**Dinonyl ketone** (*n*)	caprinone	$(C_9H_{19})_2CO$	I-718	282.51
2683	**Dioctyl** acetic acid (*n*)	....	$[CH_3(CH_2)_7]_2CH \cdot CO_2H$	II-388	284.49
2684	ether (*n*)	....	$[CH_3(CH_2)_7]_2O$	I-419	242.45
2685	ketone (*n*)	pelargone	$[CH_3(CH_2)_7]_2CO$	I-718	254.46
2686	**Dioxindole**	oxindole; hydrindic acid	$C_6H_4 \cdot CHOH \cdot CO \cdot NH$ (joined)	XXI-578	149.15
2687	**Diphenetyl urea** (4,4')	di-(*p*-EtO-phenyl)-urea	$(C_2H_5O \cdot C_6H_4 \cdot NH)_2 : CO$	XIII-481	300.36
2688	**Diphenic** acid (2,2')	diphenic acid (1,10)	$(\cdot C_6H_4 \cdot CO_2H)_2$	IX-922	242.23
2689	acid (3,3')	diphenic acid (2,9)	$(\cdot C_6H_4 \cdot CO_2H)_2$	IX-927	242.23
2690	acid (2,3')	diphenic acid (1,9)	$(\cdot C_6H_4 \cdot CO_2H)_2$	IX-926	242.23
2691	acid (2,4')	diphenic acid (1,8)	$(\cdot C_6H_4 \cdot CO_2H)_2$	IX-926	242.23
2692	**Diphenoxy**-diethyl ether (β,β')	....	$(C_6H_5OCH_2CH_2)_2O$	*VI-84	258.32
2693	ethane (1,2)	....	$(C_6H_5O \cdot CH_2 \cdot)_2$	VI-146	214.27
2694	propane (1,3)	....	$(C_6H_5O \cdot CH_2)_2CH_2$	VI-147	228.29
2695	**Diphenyl**	biphenyl; xenene	$C_6H_5 \cdot C_6H_5$	V-576	154.21
2696	acetamidine (*N,N'*)	ethenyl-diPh-amidine	$CH_3C(NHC_6H_5): NC_6H_5$	XII-248	210.28
2697	acetic acid	....	$(C_6H_5)_2CH \cdot CO_2H$	IX-673	212.25
2698	amine	....	$(C_6H_5)_2NH$	XII-174	169.23
2699	amine HCl	....	$(C_6H_5)_2NH \cdot HCl$	XII-180	205.69
2700	amine sulfate	....	$(C_6H_5)_2NH \cdot H_2SO_4$	XII-180	267.31
2701	amine sulfonic acid	*N*-Ph-sulfanilic	$C_6H_5 \cdot NH \cdot C_6H_4 \cdot SO_3H$	XIV-699	249.29
2702	amine sulfonic	Ba salt	$(C_{12}H_{10}O_3NS)_2Ba$	XIV-699	633.90
2703	benzene (*o*)	diphenyl-phenylene	$C_6H_5 \cdot C_6H_4 \cdot C_6H_5$	**V-611	230.31
2704	benzene (*m*)	*iso*-diphenyl-benzene	$C_6H_5 \cdot C_6H_4 \cdot C_6H_5$	V-695	230.31
2705	benzene (*p*)	terphenyl	$C_6H_5 \cdot C_6H_4 \cdot C_6H_5$	V-695	230.31

Table 7-4 (*Continued*)
PHYSICAL CONSTANTS OF ORGANIC COMPOUNDS

No.	Crystalline Form and Color	Specific Gravity	Melting Point °C.	Boiling Point °C.	Solubility in 100 Parts		
					Water	Alcohol	Ether
2663	red yel./chl.	....	210-6	....	s. bz., act.	v. sl. s.	v. sl. s.
2664	yel. lf./ac.	....	288-92	....	s. act.	sl. s.	sl. s.
2665	yel. pr./pet.	....	55	....	v. sl. s.	v. s.	v. s.
2666	nd./pet.	$1.263^{111°}$	61-3	....	....	....	....
2667	nd./CS_2	$1.321^{71°}$	70	300 sl. d.	$0.03^{22°}$	$1.21^{5°}$	$9^{15°}$
2668	nd./al.	$1.282^{111°}$	52.5	....	v. s. bz.	v. s.	v. s. CS_2
2669	rhb.	$1.283^{111°}$	60.5 *	....	....	s.	....
2670	nd./CS_2	$1.259^{111°}$	60-1	....	i.	$2.2^{17°}$ CS_2	....
2671	mn. pr.	$1.277^{111°}$	92-3	subl.	sl. s.	s. h.	s.
2672	yel. pl.	....	$-H_2O$, 140	d. >220	s. alk.	s. dil. a.	....
2673	nd./al.	....	82	expl. 413	sl. s. pet.	sl. s.	v. s.
2674	col. cr./al.	....	56	....	s. pet.	s.	s.
2675	col. nd./al.	....	115-6	....	sl. s. h.	sl. s. c.	sl. s. c. pet.
2676	yel. nd./al.	....	75-6	expl. 438	v. s. bz.	v. s. act.	v. s. chl.
2677	cr./al.	....	83-4	....	....	....	....
2678	col. pr./al.	....	93-4	....	i.	s. h.	....
2679	mn. pr./al.	....	90-3	....	i.	v. s. h.	....
2680	yel. nd./al.	....	147-8	....	i.	s. h.	s. h.
2681	nd./al.	....	123.5	....	....	....	....
2682	lf./al.	....	58	>350 sl. d.	....	s.	s.
2683	lf./al.	....	39	$270-5^{100mm}$	i.	s. abs.	....
2684	lq.	$0.805^{\frac{17}{17}°}$	....	291.7	sl. s.	s.	s.
2685	pl./Me al.	....	52-3	....	....	sl. s. c.	s. Me al.
2686	rhb./al.	....	180 (violet)	195 d.	8.3 c.; 16.6 h.	6.6 c.; 10 h.	s. alk.
2687	nd./ac.; pr./al.	....	225-6	....	....	....	....
2688	mn. pr.	....	228-9	subl.	sl. s.	s.	s.
2689	lf./al.	....	356-7	....	v. sl. s. h.	s. h.	s.
2690	nd./aq.	....	216	....	sl. s. h.	s.	....
2691	lf./al.	....	251-2	....	....	....	....
2692	col. nd./aq. al.	....	66-7	....	i.	s. h.	s.
2693	lf./abs. al.	....	97-8	....	i.; $27^{25°}$ bz.	s. h.; $23^{25°}$ act.	$9^{25°}$
2694	lf./al.	....	61	$338-40^{762mm}$	i.	s.	s.
2695	col. mn.	$0.992^{\frac{73}{4}°}$	70.5	256.1	i.	$9.98^{19.5°}$	$6.57^{19.5°}$ Me al.
2696	nd./al.	....	132-3	....	s. a.	s. h.	s.
2697	nd./aq.	....	148	....	s. h.	s.	s.
2698	col. mn.	$1.160^{\frac{20}{20}°}$; $1.054^{61°}$	52.9	302	i.; $58^{19.5°}$ Me al.	$56^{19.5°}$; s. bz.	s.; s. CS_2; s. ac.
2699	col. nd./al.	....	....	....	s.	s.	....
2700	col. cr.	....	123-5	....	i.	s.	s. H_2SO_4
2701	col. lf.	....	....	....	s.	s.	i.
2702	col. lf.	....	....	....	v. sl. s.	....	....
2703	pr./Me al.	....	56-7	332	i.	s.; s. chl.	s.; s. act.
2704	nd./aq. al.	....	86-7	363	i.; s. bz.	s. h.	s.; s. ac.
2705	lf. or nd.	$1.234^{\frac{0}{4}°}$	212-3	376	s. h. bz.	v. sl. s. h.	sl. s.

Dioxy-pyrimidine 6383   *β, m. 65.5°
Diphenol 2340-3   γ, m. 48°
Diphenyl-acetamide 84.1
Diphenyl-acetone 1787
Diphenyl-acetylene 5954
Diphenyl-amine orange 4996

Table 7-4 (*Continued*)
PHYSICAL CONSTANTS OF ORGANIC COMPOUNDS

No.	Name	Synonym	Formula	Beil. Ref.	Formula Weight
	Diphenyl				
2706	benzidine (*N*)	....	$(C_6H_5 \cdot NH \cdot C_6H_4)_2$	XIII-223	336.44
2707	butyl-phenyl phosphate (*p-tert*)	phosphen I	$C_4H_9 \cdot C_6H_4 \cdot O \cdot PO:$ $(O \cdot C_6H_5)_2$	....	382.40
2708	carbamine chloride	diPh-carbamyl chloride	$(C_6H_5)_2N \cdot CO \cdot Cl$	XII-428	231.68
2709	"carbazone" (*sym*)	....	$C_6H_5 \cdot N:N \cdot CO \cdot NH \cdot$ $NH \cdot C_6H_5$	XVI-24	240.27
2710	carbohydrazide (1,5)	diPh-carbazide	$(C_6H_5NH \cdot NH)_2CO$	XV-292	242.28
2711	carbonate	phenyl carbonate	$(C_6H_5O)_2CO$	VI-158	214.22
2712	chloroarsine	sneezing gas	$(C_6H_5)_2AsCl$	XVI-845	264.59
2713	*o*-chlorophenyl phosphate	phosphen 3	$(C_6H_5O)_2PO \cdot$ OC_6H_4Cl	....	360.74
2714	*iso*-cyanate	*p*-xenylcarbimide	$C_6H_5 \cdot C_6H_4N:CO$	XII-1319	195.22
2715	disulfide	phenyl disulfide	$(C_6H_5S \cdot)_2$	VI-323	218.34
2716	disulfoxide	phenyl oxydisulfide	$(C_6H_5 \cdot SO)_2$	VI-324	250.34
2717	ethane (*sym*)	dibenzyl	$(C_6H_5 \cdot CH_2 \cdot)_2$	V-598	182.27
2718	ethane (*uns*)	....	$(C_6H_5)_2CH \cdot CH_3$	V-605	182.27
2719	ether	phenyl ether	$(C_6H_5)_2O$	VI-146	170.21
2720	ethylenediamine (*N*)(*sym*)	ethylene-diphenyl-diamine	$(C_6H_5 \cdot NH \cdot CH_2 \cdot)_2$	XII-543	212.30
2721	formamidine (*N,N'*)	methenyl-diphenyl-amine	$HC(:NC_6H_5)NHC_6H_5$	XII-236	196.25
2722	glyoxalone (4,5)	4,5-diPh-imidazolone	$C_6H_5 \cdot C:C(C_6H_5) \cdot$ $\underset{\text{NH} \cdot \text{CO} \cdot \text{NH}}{\vert\underline{\qquad\qquad}\vert}$	XXIV-211	236.28
2723	guanidine	melaniline	$(C_6H_5NH)_2C:NH$	XII-369	211.27
2724	hydantoin (5,5)	dilantin	$\underset{(C_6H_5)_2}{CO \cdot NH \cdot CO \cdot NH \cdot C:}$ $\vert\underline{\qquad\qquad}\vert$	XXIV-410	252.28
2725	hydrazine (*α,α*)†	....	$(C_6H_5)_2N \cdot NH_2$	XV-122	184.24
2726	hydrazine HCl	(*α,α*), *uns*, or *N,N*	$(C_6H_5)_2N \cdot NH_2 \cdot HCl$	XV-123	220.70
2727	hydrazine (*p*)	4-hydrazino-diPh	$C_6H_5 \cdot C_6H_4 \cdot NH \cdot NH_2$	XV-576	184.24
2728	iodonium hydroxide	known only in soln.	$(C_6H_5)_2I \cdot OH$	V-219	298.13
2729	iodonium iodide	....	$(C_6H_5)_2I \cdot I$	V-219	408.02
2730	ketene	....	$(C_6H_5)_2C:CO$	VII-471	194.24
2731	methane	ditane	$(C_6H_5)_2CH_2$	V-588	168.24
2732	*α*-naphthyl carbinol	....	$(C_6H_5)_2C(OH) \cdot C_{10}H_7$	VI-729	310.40
2733	nitrosamine	nitroso-diPh-amine	$(C_6H_5)_2N \cdot NO$	XII-580	198.23
2734	phenanthracene	(9,10)	$C_{14}H_8(C_6H_5)_2$	V-747	330.43
2735	phosphate	....	$(C_6H_5O)_2PO \cdot OH$	VI-178	250.19
2736	phosphate	....	$(C_6H_5)_2HPO_4 \cdot 2H_2O$	*VI-95	286.22
2737	phthalate	phenyl phthalate	$C_6H_4(CO_2C_6H_5)_2$	IX-801	318.33
2738	piperazine (*N,N'*)	....	$(C_6H_5N)_2(CH_2)_4$	XXIII-8	238.34
2739	propionic acid (*β,β*)	*β*-Ph-hydrocin-namic acid	$(C_6H_5)_2CH \cdot CH_2 \cdot$ CO_2H	IX-680	226.28
2740	propiophenone	(*β,β*)	$(C_6H_5)_2CH \cdot CH_2 \cdot$ $CO \cdot C_6H_5$	VII-524	286.38

† See also No. 3685
Diphenyl-benzylamine 804
Diphenyl-black base-P 314
Diphenyl-carbamyl chloride 2708

Diphenyl-carbazide 2710
Diphenyl carbinol 691
Diphenyl-carbonimide 2714
Diphenyl-carboxylic acid 5160-2

Diphenyl-dichloromethane 726
Diphenyl-diimide 599
Diphenyl-ethyl carbamate 2761
Diphenyl-ethylene 5642-3

Table 7-4 (*Continued*)
PHYSICAL CONSTANTS OF ORGANIC COMPOUNDS

No.	Crystalline Form and Color	Specific Gravity	Melting Point °C.	Boiling Point °C.	Solubility in 100 Parts		
					Water	Alcohol	Ether
2706	lf./PhMe	....	242	....	s. ac.	sl. s.	sl. s. bz.
2707	col. lq.	1.16$\frac{25}{25}$°	<0	245-60^{5mm}	i.	v. s.; ∞ CCl$_4$	∞ bz.
2708	lf./al.	....	85-6	....	d. h.	d. alk.	....
2709	or. red nd.	....	157±, d.	....	i.	s.; s. bz.	s. chl.
2710	cr./al.	....	175	....	v. sl. s. h.	s. h.	i.; s. act.
2711	nd./al.	1.272$^{14°}$	80	302-6	i.	v. s.	s.; s. CCl$_4$
2712	rhb.	1.583$^{40°}$	43-4	333, in CO$_2$	0.2, d.	20	s.; s. bz.
2713	col. lq.	1.30$\frac{25}{25}$°	<0	256-60^{15mm}	i.	v. s.	∞ bz.; ∞ CCl$_4$
2714	nd./et.	....	58-60	283 d.	....	....	v. s.
2715	nd./al.	....	61	310	i.	s.	v. s.
2716	mn./al.	....	45	d.	i.; i. alk.	s. h.	s.
2717	col. pr.	0.978$\frac{50}{50}$°	52.0	284	i.; i. NH$_3$	s.; s. SO$_2$	v. s.
2718	col. oil	1.004$^{20°}$	−21.5	272	i.	∞	∞
2719	col. rhb.	1.073$^{20°}$	26.9	258.3	v. sl. s.; s. ac.	5$^{-10°}$, 87% al.	∞ ; s. bz.
2720	lt./aq. al.	....	65-7	....	....	v. s.	v. s.
2721	nd./bz.	....	139-43	sl. d.	v. s. chl.	0.05$^{9°}$ pet.	s.
2722	nd./al.	....	324-5	....	i.; s. h. ac.	s.; v. sl. s. bz.	v. sl. s.; v. sl. s. lg.
2723	mn./al.	....	147-8	d. >170	v. sl. s. c.; s. dil. a.	9.1$^{20°}$ 90% al.	sl. s.; s. h. bz.
2724	cr./al.	....	295-8	....	i.; s. alk.; sl. s. bz.	2; s. ac.; sl. s. chl.	sl. s.; 3 act.
2725	tri.	1.190$^{16°}$	44 (36)	220^{50mm}	sl. s.	s.	s.
2726	nd./al.-HCl	....	....	....	i.	v. s.	i. HCl
2727	lf./al.	....	135-6 d.	....	v. sl. s.	v. sl. s. lg.	....
2728	....	....	....	....	....	....	....
2729	yel. nd./al.	....	182	....	....	v. sl. s. h.	....
2730	red-yel. lq.	1.104$\frac{20}{4}$°	....	265-70	....	....	....
2731	col. pr.	1.001$\frac{26}{4}$°	26.6	265	i.	s.	v. s.
2732	cr./lg.	....	137-8	d.	i.; s. bz.	s. h.	v. s.
2733	yel. mn.	....	66-7	....	v. s. h. bz.	s. h.	....
2734	col. nd./al.	....	249-50	subl.	s. bz.	sl. s.	s.
2735	nd./chl. lg.	....	70	....	3	s.; s. bz.	s.; s. chl.
2736	col. cr./aq.	1.242$\frac{60}{25}$°	51	−2H$_2$O, 100	3$^{25°}$	4$^{25°}$ CCl$_4$	100$^{25°}$
2737	col. pr./al.	1.572$^{74°}$	73	* 405^{759mm}	i.; s. act.	sl. s.	sl. s.
2738	nd./Me al.	....	164	230-5^{12mm}	i.; 7$^{25°}$ bz.	4$^{25°}$ act.	2$^{23°}$
2739	nd./aq. al.	....	154-5	....	v. sl. s.	s.	....
2740	nd./al.	....	96	....	v. sl. s. lg.; s. bz.	s. h.; s. act.	sl. s.; s. chl.

* Heated rapidly; decomposed when heated slowly.

Diphenyl-glycolic acid 699
Diphenyl-imidazolone 2722
Diphenyl-ketone 724
Diphenyl-ketoxime 728
Diphenyl mercury 4047
Diphenyl-methylenediamine 4431
Diphenyl-oxide 2719
Diphenyl-phenoxy ethane 2853
Diphenyl-phenylene 2703-5
Diphenyl-phthalide 5292

Table 7-4 (Continued)
PHYSICAL CONSTANTS OF ORGANIC COMPOUNDS

No.	Name	Synonym	Formula	Beil. Ref.	Formula Weight
2741	**Diphenyl** quino-methane	fuchsone	$(C_6H_5)_2C{:}C_6H_4{:}O$	VII-520	258.32
2742	selenium	selenium diphenyl	$(C_6H_5)_2Se$	VI-345	233.17
2743	selenium dichloride	....	$(C_6H_5)_2SeCl_2$	VI-346	304.08
2744	semicarbazide(1,1)	....	$(C_6H_5)_2N{\cdot}NHCONH_2$	XV-304	227.27
2745	semicarbazide(1,4)	....	$(C_6H_5NH)_2({\cdot}NHCO{\cdot})$	XV-288	227.27
2746	semicarbazide(2,4)	....	$C_6H_5N(NH_2){\cdot}CO{\cdot}NH{\cdot}C_6H_5$	XV-277	227.27
2747	semicarbazide(4,4)	....	$(C_6H_5)_2N{\cdot}CONHNH_2$	*XII-257	227.27
2748	succinate	phenyl succinate	$(CH_2{\cdot}CO_2C_6H_5)_2$	VI-155	270.29
2749	sulfide	phenyl sulfide	$(C_6H_5)_2S$	VI-299	186.28
2750	sulfone	sulfobenzide	$(C_6H_5)_2SO_2$	VI-300	218.28
2751	4-sulfonic acid	Na salt	$C_6H_5{\cdot}C_6H_4{\cdot}SO_3Na$	XI-192	256.26
2752	sulfoxide	....	$(C_6H_5)_2SO$	VI-300	202.28
2753	thiocarbazone	dithizone	$C_6H_5N{:}N{\cdot}CS{\cdot}NH{\cdot}NH{\cdot}C_6H_5$	XVI-26	256.33
2754	thiocarbo-hydrazide (1,5)	diPh-thiocarbazide	$(C_6H_5NH{\cdot}NH)_2CS$	XV-299	258.35
2755	*iso*-thiocyanate	*p*-xenyl mustard oil	$C_{12}H_9N{:}CS$	XII-1319	211.29
2756	tolyl methane (*m*)	3-methyl tritane	$CH_3{\cdot}C_6H_4{\cdot}CH(C_6H_5)_2$	V-710	258.37
2757	tolyl methane (*p*)	4-methyl tritane	$CH_3{\cdot}C_6H_4{\cdot}CH(C_6H_5)_2$	V-710	258.37
2758	triketone	....	$(C_6H_5{\cdot}CO)_2CO$	VII-871	238.25
2759	urea (*sym*)	carbanilide	$(C_6H_5{\cdot}NH)_2CO$	XII-352	212.25
2760	urea (*uns*)	....	$(C_6H_5)_2N{\cdot}CO{\cdot}NH_2$	XII-429	212.25
2761	urethane	*N*-diPh-ethyl carbamate	$(C_6H_5)_2N{\cdot}CO_2C_2H_5$	XII-427	241.29
2762	*o*-xenyl phosphate	phosphen 5	$C_6H_5{\cdot}C_6H_4O{\cdot}PO{:}(O{\cdot}C_6H_5)_2$	....	402.39
2763	**Diphenylene** oxide	dibenzofurane	$(C_6H_4)_2O$	XVII-70	168.20
2764	**Dipicryl**-amine (2,4,6,2',4',6')	hexaNO$_2$-diPh-amine	$[(NO_2)_3C_6H_2]_2NH$	XII-766	439.21
2765	sulfide	(2,4,6,2',4',6')	$[(NO_2)_3C_6H_2]_2S$	VI-344	456.26
2766	**Dipiperonalace-tone**	dipiperonylidene-acetone	$(CH_2O_2{:}C_6H_3{\cdot}CH{:}CH)_2CO$	XIX-446	322.32
2767	**Di-*iso*-propanol-amine**	....	$(C_3H_6{\cdot}OH)_2NH$	**IV-737	133.19
2768	**Dipropargyl**	hexadiyne-1,5	$(CH{:}C{\cdot}CH_2{\cdot})_2$	I-266	78.11
2769	**Di-*iso*-propenyl**	2,3-dimethyl-buta-diene-1,3	$[CH_2{:}C(CH_3)]_2$	I-256	82.15
2770	**Dipropyl** acetic acid (*n*)	....	$(C_3H_7)_2CH{\cdot}CO_2H$	II-350	144.22
2771	adipate (*n*)	*n*-propyl adipate	$[(CH_2)_2CO_2C_3H_7]_2$	**II-574	230.31
2772	amine (*n*)	....	$(C_2H_5{\cdot}CH_2)_2NH$	IV-138	101.19
2773	amine (*iso*)	....	$(C_3H_7)_2NH$	IV-154	101.19
2774	aminobenzalde-hyde	(*n*)(*p*)	$(C_3H_7)_2N{\cdot}C_6H_4{\cdot}CHO$	....	205.30
2775	aniline (*n*)(*N*)	....	$C_6H_5N(C_3H_7)_2$	XII-167	177.29
2776	barbituric acid (5,5)	proponal	$CO{\cdot}NH{\cdot}CO{\cdot}NH{\cdot}\overline{}CO{\cdot}C{:}(CH_2{\cdot}C_2H_5)_2$	XXIV-492	212.25
2777	benzene (*iso*)(*o*)	....	$[(CH_3)_2CH]_2C_6H_4$	V-447	162.28
2778	benzene (*iso*)(*m*)	alkazene 12	$[(CH_3)_2CH]_2C_6H_4$	V-447	162.28

Diphenyl selenide 2742
Diphenyl-thiocarbazide 2754
Diphenyl-thiourea 5892
Diphenylene carbinol 3247

Diphenylene disulfide 5877
Diphenylene-imine 1231
Diphenylene ketone 3248
Diphenylene ketone oxide 6451

Table 7-4 (*Continued*)
PHYSICAL CONSTANTS OF ORGANIC COMPOUNDS

No.	Crystalline Form and Color	Specific Gravity	Melting Point °C.	Boiling Point °C.	Solubility in 100 Parts		
					Water	Alcohol	Ether
2741	brn. yel. nd.	....	168-9	....	i.	v. s. act.	sl. s. h.
2742	oil	$1.338\frac{16}{4}°$	2.5	301-2	....	∞	∞
2743	pa. yel. pr.	....	182-5	....	i.	i.	i.
2744	nd./al.	....	195	....	....	s.	s. bz.
2745	nd./al.	....	179-80	....	v. sl. s. h.	s.; s. bz.	i.
2746	col. lf./al.	....	165.5 *	....	v. s. ac.	s.; s. bz.	s.; s. chl.
2747	pr./al.	....	155-6	....	s. h.; s. bz.	s.; s. chl.	v. sl. s.
2748	lf./al.	....	122-3	330	i.; s. bz.	s. CS_2	s.
2749	col. lq.	$1.119\frac{15}{15}°$	<-40	296-7	i.; ∞ bz.	s. h.	∞ ; ∞ CS_2
2750	nd./aq.	$1.248\frac{25}{4}°$	128-9	379	sl. s. h.	s. h.	s. bz.
2751	col. cr.	....	....	....	$1.5^{25°}$	$0.03^{25°}$ CCl_4	$0.03^{25°}$
2752	pr./lg.	....	70.5	340 sl. d.	sl. s. pet.	s.	s.; s. bz.
2753	b.-bl. cr.	....	....	....	i.; s. alk.	v. sl. s.; s. H_2SO_4	v. sl. s.; sl. s. chl.
2754	pr./al.	....	150±	d.	v. sl. s. bz.	v. sl. s.	v. sl. s. ac.
2755	nd./et.	....	58	....	....	....	v. s.
2756	pr./al.	$1.07^{16°}$	60.5-1.5	$353-477^{4mm}$	v. s. bz.	v. s.	v. s.
2757	pr./Me al.	....	72	>360	sl. s. pet.	v. s. h.	v. s. bz.
2758	yel. nd./lg.	....	69-70	289^{175mm}	i.	sl. s.	s.
2759	rhb.	1.239	241-2	260-2	v. sl. s.	sl. s. h.	s.
2760	rhb.	1.276	189	....	v. sl. s.	s.	s.; s. chl.
2761	pr./lg.	....	72	>360	s.; s. pet.	v. s.; s. bz.	v. s.
2762	col. lq.	$1.20\frac{60}{4}°$	<0	$250-85^{5mm}$	i.; ∞ bz.	v. s.	∞ CCl_4
2763	lf./al.	....	86-7	287-8	i.; s. bz.	s. h.	v. s.
2764	yel. pr./ac.	....	244-6 d.	....	i.; s. 93% HNO_3	v. sl. s. act.	i.
2765	yel. lf./ac.	....	226-30	expl. 290	i.	v. sl. s.	v. sl. s.
2766	yel. nd./bz.	....	185	....	i.; i. lg.	sl. s.	s. chl.; s. act.
2767	....	$0.989\frac{45}{20}°$	42	248.7	....	....	....
2768	lq.	$0.805\frac{20}{4}°$	-6	85.4	i.	s.	v. s.
2769	lq.	$0.727\frac{20}{4}°$	-76	68.5	....	....	....
2770	col. lq.	$0.922\frac{0}{4}°$	....	221-2	v. sl. s.	....	....
2771	col. lq.	$0.979\frac{20}{4}°$	-20.3	$143-5^{10mm}$	i.	s.	s.
2772	col. lq.	$0.739\frac{20}{4}°$	-39.6	110-1	s.	∞	∞
2773	col. lq.	$0.722^{22°}$	-61	83.57^{43mm}	s.	s.	....
2774	lt. yel. lq.	....	....	$204-6^{20mm}$	....	....	....
2775	yel. oil	$0.910^{20.4°}$	....	245.4	i.	s.	s.
2776	cr.	....	145	....	v. sl. s. c.; $1.4^{100°}$	v. s.	v. s.
2777	col. lq.	$0.858\frac{25}{25}°$	....	210^{760mm}	i.	∞	∞
2778	col. lq.	$0.860\frac{25}{25}°$	-105	202^{760mm}	i.	∞	∞

* Decomposes to form 1,4-derivative
Diphenyline 1711
Diphosgene 6159
Diphthalimido-ethane 3205

Dipicolinic acid 5500
Diplosal 5585
Dipropasin 2779
Dipropenyl 3573

Table 7-4 (*Continued*)
PHYSICAL CONSTANTS OF ORGANIC COMPOUNDS

No.	Name	Synonym	Formula	Beil. Ref.	Formula Weight
2779	**Dipropyl**				
	carbanilid-4,4'-dicarboxylate (*n*)	dipropasin	$CO(NH \cdot C_6H_4 \cdot CO_2 \cdot CH_2 \cdot C_2H_5)_2$	XIV-434	384.44
2780	carbonate (*n*)	*n*-propyl carbonate	$(C_2H_5 \cdot CH_2 \cdot O)_2CO$	III-6	146.19
2781	α,α'-dibromo-succinate (*n*)	....	$(CHBr \cdot CO_2C_3H_7)_2$	....	360.05
2782	disulfide (*n*)	....	$(C_2H_5 \cdot CH_2 \cdot S \cdot)_2$	I-360	150.31
2783	ether (*n*)	*n*-propyl ether	$(C_2H_5 \cdot CH_2)_2O$	I-354	102.18
2784	ether (*iso*)	*iso*-propyl ether	$[(CH_3)_2CH]_2O$	I-362	102.18
2785	ether (*n-iso*)	....	$C_2H_5CH_2 \cdot O \cdot CH(CH_3)_2$	I-362	102.18
2786	ketone (*n*)	heptanone-4	$(C_2H_5 \cdot CH_2)_2CO$	I-699	114.19
2787	ketone (*iso*)	2,4-di Me-pen-tanone-3	$[(CH_3)_2CH]_2CO$	I-703	114.19
2788	maleate (*n*)	propyl maleate	$(CH \cdot CO_2C_3H_7)_2$	II-752	200.24
2789	malonate (*n*)	....	$CH_2(CO_2C_3H_7)_2$	II-581	188.23
2790	nitrosamine (*n*)	nitroso-diPr-amine	$(C_3H_7)_2N \cdot NO$	IV-146	130.19
2791	nitrosamine (*iso*)	....	$[(CH_3)_2CH]_2N \cdot NO$	IV-156	130.19
2792	oxalate (*n*)	propyl oxalate	$(CO_2CH_2 \cdot C_2H_5)_2$	II-539	174.20
2793	oxalate (*iso*)	....	$(CO_2 \cdot C_3H_7)_2$	II-539	174.20
2794	phthalate (*n*)	propyl phthalate	$(C_6H_4(CO_2 \cdot C_3H_7)_2$	**IX-586	250.30
2795	phthalate (*iso*)	....	$C_6H_4(CO_2 \cdot C_3H_7)_2$	IX-798	250.30
2796	succinate (*n*)	propyl succinate	$(CH_2 \cdot CO_2 \cdot CH_2 \cdot C_2H_5)_2$	II-611	202.25
2797	succinate (*iso*)	....	$(CH_2 \cdot CO_2 \cdot C_3H_7)_2$	II-611	202.25
2798	sulfate (*n*)	*n*-propyl sulfate	$(C_2H_5 \cdot CH_2O)_2SO_2$	I-354	182.24
2799	sulfide (*n*)	*n*-propyl sulfide	$(C_2H_5 \cdot CH_2)_2S$	I-359	118.24
2800	sulfide (*iso*)	....	$[(CH_3)_2CH]_2S$	I-367	118.24
2801	sulfite (*n*)	....	$(C_2H_5 \cdot CH_2 \cdot O)_2SO$	I-354	166.24
2802	sulfone (*n*)	....	$(C_2H_5 \cdot CH_2)_2SO_2$	I-359	150.24
2803	sulfone (*iso*)	....	$[(CH_3)_2CH]_2SO_2$	I-367	150.24
2804	tartrate (*n*)(*d*)	propyl tartrate	$(CHOH \cdot CO_2 \cdot C_3H_7)_2$	III-516	234.25
2805	tartrate (*iso*)(*d*)	....	$(CHOH \cdot CO_2 \cdot C_3H_7)_2$	III-517	234.25
2806	urea (*n*)(*sym*)	....	$(C_2H_5 \cdot CH_2 \cdot NH)_2CO$	IV-142	144.22
2807	urea (*n*)(*uns*)	....	$(C_3H_7)_2N \cdot CO \cdot NH_2$	IV-143	144.22
2808	**Dipropylene glycol**	....	$(CH_3CHOHCH_2)_2O$	**I-537	134.18
2809	"**Dipyridine**"(?)	nicotyrine	$C_{10}H_{10}N_2$	XXIII-185	158.20
2810	**Dipyridyl** (2,2')	(α,α')	$(C_5H_4N)_2$	XXIII-199	156.19
2811	**Dipyridyl** (4,4')	....	$(C_5H_4N)_2$	XXIII-200	156.19
2812	**Dipyridyl** (2,3')	....	$C_{10}H_8N_2$	XXIII-200	156.19
2813	**Dipyridyl** (3,3')	....	$C_{10}H_8N_2$	XXIII-200	156.19
2814	**Diquinoyl** (2,3')	....	$(C_9H_6N)_2$	XXIII-293	256.31
2815	**Diquinoyl** (3,7')(β)	....	$(C_9H_6N)_2$	XXIII-295	256.31
2816	**Diquinoyl** (6,6')(γ)	....	$(C_9H_6N)_2$	XXIII-295	256.31
2817	**Diresorcinol** (5,5')	....	$[C_6H_3(OH)_2]_2 \cdot 2H_2O$	VI-1164	254.24
2818	**Dithio-**acetic acid	....	$CH_3 \cdot CS \cdot SH$	II-233	92.18
2819	carbamic acid	....	$NH_2 \cdot CS \cdot SH$	III-216	93.17
2820	hydroquinone (*p*)	....	$HS \cdot C_6H_4 \cdot SH$	VI-867	142.24
2821	oxamide	rubeanic acid	$(\cdot C:S \cdot NH_2)_2$	II-565	120.20
2822	resorcinol (*m*)	....	$HS \cdot C_6H_4 \cdot SH$	VI-834	142.24
2823	salicylic acid	(1,2)	$HO \cdot C_6H_4 \cdot CS \cdot SH$	X-134	170.25
2824	**Ditolualacetone** (*p*)	dixylidene acetone	$(CH_3 \cdot C_6H_4 \cdot CH: CH)_2CO$	VII-508	262.35
2825	**Ditolyl** (2,2')	....	$(CH_3 \cdot C_6H_4 \cdot)_2$	V-608	182.27
2826	**Ditolyl** (2,3')	....	$(CH_3 \cdot C_6H_4 \cdot)_2$	V-609	182.27

Dipropyl carbinol (*n*) 3543
Dipropyl carbinol (*iso*) 3531
Dipropyl mercury 4046
Dipropyl zinc 6505-6

Di-*iso*-propylidene acetone 5268
Disalicylic acid 5585
Disalicylide 3747
Dispermin 2226

Table 7-4 (*Continued*)
PHYSICAL CONSTANTS OF ORGANIC COMPOUNDS

No.	Crystalline Form and Color	Specific Gravity	Melting Point °C.	Boiling Point °C.	Solubility in 100 Parts		
					Water	Alcohol	Ether
2779	wh. pd.	....	171-2	....	i.	s.	....
2780	col. lq.	$0.968^{22°}$	....	168.2	....	....	....
2781	lt. yel. lq.	....	....	$147\text{-}50^{7mm}$	i.	s.	s.
2782	col. lq.	$0.814^{17°}$	$-102\pm$	193-5	i.	s.	s.
2783	col. lq.	$0.752^{\frac{15}{4}°}$	-122	90.1	sl. s.	∞	∞
2784	col. lq.	$0.725^{\frac{20.8°}{0}}$	-60	68.5-9.0	0.2	∞	∞
2785	col. lq.	$0.747^{\frac{12.5°}{0}}$	....	82-3	$0.5^{20°}$	....	....
2786	col. lq.	$0.816^{\frac{22}{4}°}$	-32.5	143.5	0.43	∞	∞
2787	col. lq.	$0.806^{\frac{20}{4}°}$	....	123.7	v. sl. s.	∞	∞ ; s. bz.
2788	col. lq.	$1.030^{\frac{18.4°}{4}}$	....	$114\text{-}7^{6mm}$	....	....	....
2789	col lq.	$1.009^{\frac{20°}{4}}$	....	228.3	....	....	....
2790	yel. oil	$0.916^{\frac{20°}{4}}$	....	205.9	v. sl. s.	....	....
2791	cr./et.	....	46	194.5	v. sl. s.	s.	s.; s. bz.
2792	col. lq.	$1.038^{\frac{0}{0}}$	-51.7	213.5	d. h.	....	....
2793	col. lq.	....	....	$190\pm$	....	....	....
2794	col. lq.	....	....	$129\text{-}32^{1mm}$	i.	s.	s.
2795	col. lq.	....	....	$160\text{-}19^{mm}$	i.	s.	s.
2796	col. lq.	$1.001^{\frac{20°}{4}}$	-10.4	250.8	....	....	....
2797	col. lq.	$1.019^{\frac{0}{0}}$	....	247.1	....	....	....
2798	oil	$1.106^{\frac{20°}{4}}$	d. 140-70	120^{20mm}	....	....	v. s. pet.
2799	lq.	$0.839^{\frac{20°}{4}}$	....	$142\text{-}377^{2mm}$	i.	s.	s.
2800	lq.	....	....	120.5^{763mm}	....	....	....
2801	col. lq.	$1.030^{\frac{20°}{4}}$	....	194	i.	s.	s.
2802	scales	$1.028^{\frac{50°}{4}}$	29-30	....	....	....	....
2803	cr.	....	36	....	v. s.	....	....
2804	lq.	$1.139^{\frac{20°}{4}}$	....	303	sl. s.	s.	s.
2805	col. lq.	$1.130^{20°}$	....	275	....	....	....
2806	nd./aq.	....	105	255	sl. s. c.	v. s.	v. s.
2807	nd./pet.	....	76	....	v. s.	....	....
2808	col. lq.	$1.025^{\frac{20°}{20}}$	....	231.8	∞	....	....
2809	lq.	$1.124^{13°}$	....	280-1	v. sl. s. h.	s.	s.
2810	pr./pet.	....	69.5	272.5	0.5	s.; s. bz.	s.; s. chl.
2811	pl.	....	111-2	304.8	s. h.	v. s.	v. s.
2812	lq.	....	....	296	v. sl. s.	....	....
2813	hyg. nd.	$1.164^{20°}$	68	295.5-6.5	∞	∞	sl. s.
2814	yel. nd.	....	175.5-6.0	>400 sl. d.	i. h.	s. h.	s.
2815	pl./al.	....	192.5	subl.	i.	s. h.	sl. s.
2816	mn./al.	....	178	....	v. sl. s. h.	sl. s.	sl. s.; s. bz.
2817	pl. or nd.	....	310 (anh.)	$-2H_2O, 100$	s. h.	sl. s.	i. ac.
2818	red-yel. oil	$1.24^{20°}$	....	371^{5mm}	i. c.	v. s.	v. s.; s. bz.
2819	col. nd.	....	....	....	v. s. d.	v. s.; d. h.	v. s.
2820	lf./aq. al.	....	98	....	s. bz.	s.; s. lg.	s. ac.
2821	or. red cr.	....	d. 190-200	subl.	v. sl. s. c.	s.; s. alk.	i.
2822	cr.	....	27	243	....	....	....
2823	yel. nd./ pet.	....	48-50	....	sl. s.	s.	s.; s. bz.
2824	yel. nd./al.	....	175	....	....	....	....
2825	cr./al.	$0.955^{10°}$	17.8	258^{738mm}	v. s. bz.	v. s.	v. s.
2826	lq.	$0.998^{22°}$	....	273-4	....	v. s.	v. s.

Distyryl ketone 1775
Disulfo acid-C 4534
Disulfo acid-S 4533

Ditane 2731
Dithizone 2753
Dithymol-diiodide 5940

Table 7-4 (*Continued*)
PHYSICAL CONSTANTS OF ORGANIC COMPOUNDS

No.	Name	Synonym	Formula	Beil. Ref.	Formula Weight
2827	**Ditolyl** (3,3′)	....	$(CH_3 \cdot C_6H_4 \cdot)_2$	V-609	182.27
2828	**Ditolyl** (4,4′)	....	$(CH_3 \cdot C_6H_4 \cdot)_2$	V-610	182.27
2829	**Ditolyl** (2,4′)	....	$(CH_3 \cdot C_6H_4 \cdot)_2$	V-609	182.27
2830	amine (2,2′)	....	$(CH_3 \cdot C_6H_4)_2NH$	XII-787	197.28
2831	amine (3,3′)	....	$(CH_3 \cdot C_6H_4)_2NH$	XII-858	197.28
2832	amine (4,4′)	....	$(CH_3 \cdot C_6H_4)_2NH$	XII-907	197.28
2833	carbonate (*o*)	dicresyl carbonate	$(CH_3 \cdot C_6H_4 \cdot O)_2CO$	VI-356	242.28
2834	carbonate (*m*)	....	$(CH_3 \cdot C_6H_4 \cdot O)_2CO$	VI-379	242.28
2835	carbonate (*p*)	....	$(CH_3 \cdot C_6H_4 \cdot O)_2CO$	VI-398	242.28
2836	disulfide (*p*)	....	$(CH_3 \cdot C_6H_4 \cdot S)_2$	VI-425	246.40
2837	guanidine (*o*)	....	$(C_7H_7 \cdot NH)_2C{:}NH$	XII-803	239.32
2838	ketone (*p*)	diMe-benzophenone	$(CH_3 \cdot C_6H_4)_2CO$	VII-451	210.28
2839	*m*-phenylene- diamine (*p*)	....	$(CH_3 \cdot C_6H_4NH)_2C_6H_4$	XIII-42	288.41
2840	*p*-phenylene- diamine (*p*)	....	$(CH_3 \cdot C_6H_4NH)_2C_6H_4$	XIII-81	288.41
2841	phthalate (*o*)	*o*-cresyl phthalate	$C_6H_4(CO_2 \cdot C_7H_7)_2$	....	346.39
2842	sulfide (*p*)	diMe-diPh sulfide	$(CH_3 \cdot C_6H_4)_2S$	VI-419	214.33
2843	sulfone (*p*)	....	$(CH_3 \cdot C_6H_4)_2SO_2$	VI-419	246.33
2844	thiourea (*o*)(*sym*)	....	$(CH_3 \cdot C_6H_4 \cdot NH)_2CS$	XII-807	256.37
2845	thiourea (*m*)(*sym*)	....	$(CH_3 \cdot C_6H_4 \cdot NH)_2CS$	XII-864	256.37
2846	thiourea (*p*)(*sym*)	....	$(CH_3 \cdot C_6H_4 \cdot NH)_2CS$	XII-948	256.37
2847	urea (*o*)(*sym*)	....	$(CH_3 \cdot C_6H_4 \cdot NH)_2CO$	XII-801	240.31
2848	urea (*m*)(*sym*)	....	$(CH_3 \cdot C_6H_4 \cdot NH)_2CO$	XII-863	240.31
2849	urea (*p*)(*sym*)	....	$(CH_3 \cdot C_6H_4 \cdot NH)_2CO$	XII-948	240.31
2850	**Diundecyl ketone**	laurone	$(C_{11}H_{23})_2CO$	I-719	338.62
2851	**Diurea**	*p*-urazine	$CO{:}(NH \cdot NH)_2{:}CO$	XXVI-204	116.08
2852	**Divinyl sulfide**	....	$(CH_2{:}CH)_2S$	I-434	86.16
2853	**Dixenoxy ethane** (*o*)	diPh-phenoxy ethane	$(C_6H_5 \cdot C_6H_4 \cdot O \cdot CH_2)_2$	....	366.46
2854	**Djenkolic acid** (*l*)(−)	cysteine thioform- acetal	$[HO_2C \cdot CH(NH_2) \cdot CH_2S]_2CH_2$	....	254.33
2855	**Docosane** (*n*)	....	$CH_3(CH_2)_{20}CH_3$	I-174	310.61
2856	**Dodecane** (*n*)	dihexyl	$CH_3(CH_2)_{10}CH_3$	I-171	170.34
2857	**Dodecane**	2,4,5,7-tetra- methyl-octane	$C_{12}H_{26}$	I-171	170.34
2858	**Dodecyl** acetate (*n*)	....	$CH_3 \cdot CO_2 \cdot C_{12}H_{25}$	II-136	228.38
2859	alcohol (*n*)	dodecanol-1	$CH_3(CH_2)_{10}CH_2OH$	I-428	186.34
2860	bromide (*n*)	lauryl bromide	$CH_3(CH_2)_{10}CH_2Br$	**I-133	249.24
2861	chloride (*n*)	lauryl chloride	$CH_3(CH_2)_{10}CH_2Cl$	**I-133	204.79
2862	cyanide (*n*)	....	$CH_3(CH_2)_{10}CH_2CN$	II-364	195.35
2863	**Dodecylene** (*α*)	dodecene-1	$CH_3(CH_2)_9CH{:}CH_2$	I-225	168.33
2864	**Dodecyne-2**	....	$CH_3(CH_2)_8C{:}C \cdot CH_3$	I-261	166.31
2865	**Doryl**	carbamylcholine chloride	$NH_2 \cdot CO_2 \cdot CH_2 \cdot CH_2 \cdot N(CH_3)_3Cl$	....	182.65
2866	**Dotriacontane** (*n*)	dicetyl	$CH_3(CH_2)_{30}CH_3$	I-177	450.88
2867	**Echitin**	....	$C_{32}H_{52}O_2$	....	468.77
2868	**Eicosane** (*n*)	....	$CH_3(CH_2)_{18}CH_3$	I-174	282.56
2869	**Eicosyl alcohol**	eicosanol-1	$CH_3(CH_2)_{18}CH_2OH$	I-431	298.56
2870	**Elaidic acid**	....	$C_{17}H_{33} \cdot CO_2H$	II-469	282.47
2871	**Eleostearic acid**	....	$C_{18}H_{32}O_2$	II-497	280.45

Table 7-4 (*Continued*)
PHYSICAL CONSTANTS OF ORGANIC COMPOUNDS

No.	Crystalline Form and Color	Specific Gravity	Melting Point °C.	Boiling Point °C.	Solubility in 100 Parts		
					Water	Alcohol	Ether
2827	oil	$0.999^{16°}_{4}$	5-7	287^{713mm}	i.; s. bz.	s.	s.
2828	mn./et.	....	121-2	295	i.; s. bz.	v. sl. s. c.	s.
2829	lq.	....	....	273-6	i.; s. bz.	s.	s.
2830	lq.	....	....	$312^{727.5mm}$	v. sl. s.	....	....
2831	lq.	....	<-12	320-4	v. sl. s.	s.	s.
2832	nd.	....	79	330.5	v. sl. s.	....	....
2833	nd./al.	....	60	....	s. ac.	....	....
2834	col. cr.	....	47-9	....	i.	s. h.	s.
2835	nd./al.	....	111-3	....	v. sl. s.	sl. s. c.	....
2836	nd./al.	....	46	$210-5^{20mm}$	....	s.	v. s.
2837	cr./aq. al.	$1.10^{20°}_{4}$	178-9	....	v. sl. s.	s. h.	s.
2838	rhb./al.	....	95	333^{725mm}	i.; s. CS_2	v. s. abs.	v. s.
2839	nd./al.	....	138-9	d.	sl. s. bz.; sl. s. ac.	sl. s. c.	sl. s.
2840	lf.	....	182	....	sl. s. bz.; sl. s. ac.	v. sl. s. c.	v. sl. s. pet.
2841	col. cr.	....	110-2	....	i.	s. h.	i.
2842	nd./al.	....	56-7	>300	i.; s. bz.	v. s. h.	v. s.
2843	pr./bz.	....	158-9	405^{714mm}	s. CS_2	s. h.	s. chl.
2844	nd./al.	....	158	216-8	i.	v. s. h.	i.
2845	nd.	....	122	....	v. sl. s. h.	s.	s. bz.
2846	rhb.	....	178	....	i.	v. sl. s. h.	i.
2847	nd./ac.	....	255-6	....	i.	sl. s. h.	s. h. ac.
2848	nd./al.	....	223-5	....	i.	s. h.	....
2849	cr./al.	....	266-8	....	i.	sl. s. c.	sl. s.
2850	pl.	$0.809^{69°}_{4}$	69-70	....	....	i. c.	....
2851	mn./aq.	....	267-70	....	sl. s. c.	sl. s.	sl. s. h. ac.
2852	oil	$0.917^{15°}_{4}$	....	85-6	v. sl. s.	∞	∞
2853	col. cr.	....	101-2	....	$0.02^{25°}$	$19^{25°}$ act.	$55^{25°}$ bz.
2854	nd.	....	300-50	....	....	....	....
2855	cr.	$0.778^{44°}_{4}$	44.5	224.5^{15mm}	i.	4 h.	v. s.
2856	lq.	$0.751^{20°}_{4}$	-9.6	214.5	i.	v. s.	v. s.
2857	oily lq.	....	....	208-10	s. lg.	....	....
2858	col. lq.	....	....	$151-2^{15mm}$	....	....	....
2859	lf.	$0.831^{24°}_{4}$	24	255-9	i.	s.	s.
2860	col. lq.	....	ca -10	$175-80^{45mm}$	i.	s.	s.
2861	col. lq.	....	....	$128-30^{11mm}$	i.	....	s.
2862	col. lq.	....	8-9	275	....	v. s.	v. s.
2863	col. lq.	$0.762^{15°}_{4}$	-31.5	96^{15mm}	i.	v. s.	v. s.
2864	lq.	$0.792^{15°}_{4}$	-9	105^{15mm}	....	....	....
2865	col. cr.	....	203-4 d.	....	100	2	i.
2866	cr./chl.	$0.780^{70.1°}$	70-1	310^{15mm}	s. h. ac.	v. sl. s. c.	s. h.
2867	lf.	....	170	....	v. s. chl.	$0.07^{15°}$, 80% al.	sl. s.
2868	cr.	$0.778^{86.7°}_{4}$	36.5	205^{15mm}	i.	....	∞
2869	col. wax	....	65-6	220^{3mm}	s. h. pet.	v. s. c.	s. h. bz.
2870	lf./al.	$0.851^{79.4°}_{4}$	51-2	288^{100mm}	i.	v. s.	v. s.
2871	α lf./al. / β nd./al.	....	α48-9 / β72	$235^{12mm}±$, sl. d.	αv. s. et. / βv. s. CS_2	β s. h. ac.	....

Table 7-4 (*Continued*)
PHYSICAL CONSTANTS OF ORGANIC COMPOUNDS

No.	Name	Synonym	Formula	Beil. Ref.	Formula Weight
2872	**Ellagic acid**	dilactone	$C_{14}H_2O_4(OH)_4 \cdot 2H_2O$	XIX-261	338.23
2873	**Embelin**	embelic acid	$C_{18}H_{26}O_2(OH)_2$	. . . .	308.42
2874	**Emodin**	4,5,7-triOH-2-Me-anthraquinone	$C_{15}H_7O_2(OH)_3$	VIII-520	270.24
2875	**Eosine**	tetraBr-fluorescein	$C_{20}H_8O_5Br_4$	XIX-228	647.92
2876	salt	soluble eosine	$C_{20}H_6O_5Br_4Na_2$	XIX-230	691.88
2877	**Ephedrine** (*l*) †	. . . .	$C_6H_5 \cdot CHOH \cdot CH \cdot (CH_3) \cdot NHCH_3$	XIII-636	165.24
2878	*pseudo-* **Ephedrine** (*d*) †	*d*-isoephedrine	$C_6H_5 \cdot CHOH \cdot CH \cdot (CH_3) \cdot NHCH_3$	XIII-637	165.24
2879	**Epichlorohydrin** (α)	chloropropylene oxide	$O \cdot CH_2 \cdot CH \cdot CH_2 \cdot Cl$	XVII-6	92.53
2880	**Epicyanohydrin**	cyanopropylene oxide	$O \cdot CH_2 \cdot CH \cdot CH_2CN$	XVIII-261	83.09
2881	**Epiiodohydrin** (α)	iodopropylene oxide	$O \cdot CH_2 \cdot CH \cdot CH_2I$	XVII-10	183.98
2882	**Epinine**	3,4-diOH-Ph-EtMe-amine	$(HO)_2C_6H_3 \cdot CH_2 \cdot CH_2 \cdot NHCH_3$	*XIII-325	167.21
2883	**Equilenin** (*dl*)	(synthetic)	$C_{18}H_{18}O_2$	. . . .	266.34
2884	**Equilenin** (*d*)	(from mares)	$C_{18}H_{18}O_2$	. . . .	266.34
2885	benzoate	. . . .	$C_{18}H_{17}O \cdot O_2C \cdot C_6H_5$	. . . .	370.45
2886	**Equilenin** (*l*)	(synthetic)	$C_{18}H_{18}O_2$	. . . .	266.34
2887	*iso-***Equilenin** (*dl*)	. . . .	$C_{18}H_{18}O_2$	. . . .	266.34
2888	*iso-***Equilenin** (*d*)	14-epi-equilenin	$C_{18}H_{18}O_2$	. . . .	266.34
2889	**Ergosterol**	ergosterin	$C_{28}H_{44}O$	. . . .	396.66
2890	**Ergothioneine** (*d*)	thiohistidine betaine	$C_9H_{15}O_2N_3S \cdot 2H_2O$	XXV-521	265.33
2891	**Eriodictyol** §	5,7,3′,4′-tetraOH-flavanone	$(HO)_2C_7H_2O_2 \cdot C_6H_3(OH)$	VIII-543	244.21
2892	**Erucic acid**	docosenoic acid	$CH_3(CH_2)_7CH : CH \cdot (CH_2)_{11}CO_2H$	II-472	338.58
2893	**Erucyl alcohol**	docosene-9-ol-22	$C_{22}H_{44}O$	I-453	324.60
2894	**Erythritol** (*dl*)	butantetrol-1,2,3,4	$(CHOH \cdot CH_2OH)_2$	I-525	122.12
2895	tetranitrate	nitro-erythrite	$C_4H_6(ONO_2)_4$	I-527	302.11
2896	**Erythrosin**	tetraiodo-fluorescein	$C_{20}H_8O_5I_4$	XIX-231	835.90
2897	salt	. . . .	$C_{20}H_6O_5I_4Na_2$	. . . .	879.87
2898	**Esculetin**	6,7-diOH-coumarin	$C_9H_6O_4 \cdot H_2O$	XVIII-98	196.16
2899	**Esculin**	. . . .	$C_{15}H_{16}O_9 \cdot 1\frac{1}{2}H_2O$	. . . .	367.31
2900	**Estradiol** (β)	dihydroestrone	$CH_3 \cdot C_{17}H_{19}(OH)_2$	. . . .	272.39
2901	**Estragole**	*p*-methoxy-allyl-phenol	$CH_2 : CH \cdot CH_2 \cdot C_6H_4 \cdot OCH_3$	VI-571	148.21
2902	**Estriol**	estrone hydrate	$C_{18}H_{24}O_3$	. . . .	288.39
2903	**Estrone**	ketohydroxy-estrin	$CH_3 \cdot C_{17}H_{18}O(OH)$	. . . .	270.37
2904	**Ethane**	. . . .	$CH_3 \cdot CH_3$	I-80	30.07
2905	sulfinic acid	ethyl sulfinic acid	$C_2H_5 \cdot SO \cdot OH$	IV-1	94.13
2906	sulfonic acid	ethyl sulfonic acid	$C_2H_5 \cdot SO_2 \cdot OH$	IV-5	110.13
2907	**Ethanolamine**	amino-ethyl alcohol	$NH_2 \cdot CH_2 \cdot CH_2OH$	IV-274	61.08

† See also Alkaloid Table.
§ Beilstein shows structure as a chalcone.
Elon 4117
Embelic acid 2873
Embutal 4583

Emerald green base 5762
Emetine, cf. alkd.
Empirin 115
Enanthaldehyde 3521
Enanthaldoxime 3522

Enanthic acid 3520
Endoiodin 3591
Ephedrine (iso) 2878
Epiallocholesterol 1481
Epicarin 3790

Table 7-4 (*Continued*)
PHYSICAL CONSTANTS OF ORGANIC COMPOUNDS

No.	Crystalline Form and Color	Specific Gravity	Melting Point °C.	Boiling Point °C.	Solubility in 100 Parts		
					Water	Alcohol	Ether
2872	yel. pd.	$1.667^{18°}$	d.	$-2H_2O >$ 120	v. sl. s. h.	sl. s.	i.
2873	or. lf.	....	143	subl.	i.; s. alk.	s.	sl. s. pet.
2874	red nd./ac.	....	255-7	....	i.; s. alk.	s.; s. bz.; s. chl.	s.; s. ac.
2875	col. cr./ac.	....	....	....	i.	s.	sl. s. h. ac.
2876	red brn. pd.	....	....	....	s.	s.	....
2877	cr./et.	....	40	255	5	500	s.; s. chl.
2878	rhb. pl.	....	117	....	sl. s. c.	s.	s.
2879	lq.	$1.183\frac{25}{25}$	-57.2	117^{756mm}	<5	∞	∞
2880	pr.	....	162	....	s. h.	s.	d. a.
2881	lq.	$2.03^{13°}$	....	160-80	i.	....	....
2882	col. cr./al.	....	188-9	....	sl. s.	sl. s. h.	....
2883	lf./act. al.	....	288 (278)	....	v. sl. s.	....	....
2884	nd./aq. al.	....	251 (vac.)	....	v. sl. s.	s.	s.
2885	cr.	....	223 (vac.)	....	....	s.	s.
2886	col. cr.	....	251 (vac.)	....	v. sl. s.	s.	s.
2887	col. cr.	....	223 (206)	....	v. sl. s.	s.	s.
2888	col. cr.	....	273 (258)	....	v. sl. s.	s.	s.
2889	cr.	1.04	160-3	....	i.; s. chl.	s.; s. bz.	s.
2890	col. mn./aq.	....	$290 \pm$, d.	....	$11.6^{20°}$	v. sl. s.	i.; i. chl.
2891	col. pl./al.	....	267	....	sl. s. h.	sl. s. h.	s. alk.
2892	nd./al.	$0.860\frac{5.5}{4}$	33-4	281^{30mm}	i.	v. s.	v. s.
2893	cr.	....	34.5	$241-2^{10mm}$	v. s. ac.	v. s.	v. s. bz.
2894	tet. pr.	$1.451\frac{20}{4}$	121.5	329-31	60	sl. s. c.	i.
2895	lf./al.	....	61	expl.	i. c.	s.; s. glyc.	s.
2896	or. cr./et.	....	....	....	i.; i. bz.	sl. s.; i. chl.	i. abs.
2897	brn. pd.	....	....	....	s.	s.	....
2898	nd.	....	>270 d. (anh.)	....	s. h.	s.	v. sl. s.
2899	pr.	....	160 d. (anh.)	$-H_2O, 120-$ 30	$0.15^{11°}$	4 h.; s. ac.	i. abs.
2900	nd./bz. act.	....	223 (175)	d.	s. dioxane	s.	s.
2901	lq.	$0.965^{21°}$	....	$214-6^{764mm}$	....	....	....
2902	cr./al. EtOAc	....	283 (275)	....	0.003	s.	sl. s.
2903	col. cr./al.	....	259-61 *	....	i.; s. alk.	s.; s. bz.	s.; s. chl.
2904	col. gas	$0.546^{-88°}$ 1.049(A)	-183.2	-88.6	$4.7^{20°}$ cc.; $1.8^{80°}$cc.	150 cc., abs.	....
2905	syrup	....	....	....	s. alk.	....	....
2906	hyg. cr.	$1.334^{25°}$	-17	....	s.	s.	s. alk.
2907	col. oil	$1.022^{20°}$	10.5	171^{757mm}	∞ ; sl. s. bz.	∞ ; s. chl.	1

* Two other forms, m. p. 254° and 256°.

Epidibromohydrin 1869-70	Equisetic acid 141	Eriodictyol (homo) 3672
Epidichlorohydrin 2060-1	Ergosterin 2889	Eriodictyonone 3672
Epihydrin alcohol 3436	Ergotinine, cf. alkd.	Erythrene 1019
Epinephrine 157	Ergotoxine, cf. alkd.	Esculetin methyl ether 5604
	Ericin 4078	Esculin, cf. glcde.

Table 7-4 (*Continued*)
PHYSICAL CONSTANTS OF ORGANIC COMPOUNDS

No.	Name	Synonym	Formula	Beil. Ref.	Formula Weight
2908	**Ethenyl-** aminophenol	2-methyl- benzoxazole	$C_6H_4N:C(CH_3)O$ |_____|	XXVII-46	133.15
2909	aminothiophenol	2-methyl- benzothiazole	$C_6H_4N:C(CH_3)S$ |_____|	XXVII-46	149.22
2910	**p,p'**-diethoxy- diphenylamidine HCl	holocaine-HCl; phenacaine-HCl	$C_2H_5O\cdot C_6H_4\cdot NH\cdot C:$ $(CH_3)N\cdot C_6H_4\cdot$ $OC_2H_5\cdot HCl\cdot H_2O$	XIII-468	352.86
2911	**Ether**	(di)ethyl ether	$(C_2H_5)_2O$	I-314	74.12
2912	**Ethoxy-**acetic acid	glycolic ethyl ether	$C_2H_5O\cdot CH_2\cdot CO_2H$	III-233	104.11
2913	acetoanil (**p**)	*N*-ethylidene-*p*- phenetidine	$C_2H_5O\cdot C_6H_4\cdot N:$ $CH\cdot CH_3$	. . . .	163.22
2914	benzaldehyde (**o**)	. . . .	$C_2H_5O\cdot C_6H_4\cdot CHO$	VIII-43	150.18
2915	benzoic acid (**o**)	. . . .	$C_2H_5O\cdot C_6H_4\cdot CO_2H$	X-64	166.18
2916	benzoic acid (**m**)	. . . .	$C_2H_5O\cdot C_6H_4\cdot CO_2H$	X-138	166.18
2917	benzoic acid (**p**)	. . . .	$C_2H_5O\cdot C_6H_4\cdot CO_2H$	X-156	166.18
2918	diphenyl (**p**)	. . . .	$C_2H_5O\cdot C_6H_4\cdot C_6H_5$	. . . .	198.27
2919	diphenyl sulfide	(**p**)	$C_2H_5O\cdot C_6H_4\cdot S\cdot C_6H_5$	. . . .	230.33
2920	ethyl adipate (**β**)	. . . .	$(CH_2)_4(CO_2\cdot CH_2\cdot$ $CH_2\cdot O\cdot C_2H_5)_2$	. . . .	290.36
2921	ethyl *o*-benzoyl- benzoate (**β**)	. . . .	$C_6H_5\cdot CO\cdot C_6H_4\cdot CO_2\cdot$ $CH_2CH_2\cdot OC_2H_5$	. . . .	298.34
2922	ethyl carbamate	(**β**)	$NH_2\cdot CO_2CH_2CH_2$ OC_2H_5	. . . .	133.15
2923	ethyl chloroform- ate (**β**)	**β**-EtO-ethyl chloro- carbonate	$Cl\cdot CO_2CH_2\cdot CH_2\cdot$ OC_2H_5	. . . .	152.58
2924	ethyl glycolate	(**β**)	$HOCH_2CO_2\cdot$ $(CH_2)_2OC_2H_5$	. . . .	148.16
2925	ethyl α-hydroxy- *iso*-butyrate	. . . .	$(CH_3)_2C(OH)CO_2\cdot$ $CH_2CH_2\cdot OC_2H_5$	. . . .	176.21
2926	ethyl lactate (**β**)	. . . .	$CH_3\cdot CHOH\cdot CO_2CH_2\cdot$ $CH_2\cdot OC_2H_5$	. . . .	162.19
2927	phenol (**o**)	guaethol	$C_2H_5O\cdot C_6H_4\cdot OH$	VI-771	138.17
2928	phenol (**m**)	resorcinol ethyl ether	$C_2H_5O\cdot C_6H_4\cdot OH$	VI-814	138.17
2929	phenol (**p**)	hydroquinone ethyl ether	$C_2H_5O\cdot C_6H_4\cdot OH$	VI-843	138.17
2930	**Ethoxyl-**aniline	**β**-anilino-ethanol	$C_6H_5\cdot NH(C_2H_4OH)$	XII-182	137.18
2931	piperidine	**γ**-pipecolyl carbinol	$C_5H_{10}N\cdot C_2H_4OH$	XXI-4	129.20
2932	**Ethyl** abietate	. . . .	$C_{19}H_{29}CO_2\cdot C_2H_5$	. . . .	330.52
2933	acetamide (**N**)	. . . .	$CH_3CO\cdot NH\cdot C_2H_5$	IV-109	87.12
2934	acetanilide	acetethylanilide	$CH_3CO\cdot N(C_2H_5)C_6H_5$	XII-246	163.22
2935	acetate	acetic ether	$CH_3CO_2\cdot C_2H_5$	II-125	88.11
2936	acetate, ortho	ethenyl triEt-ether	$CH_3C(OC_2H_5)_3$	II-129	162.23
2937	acetoacetate	acetoacetic ester	$CH_3CO\cdot CH_2\cdot CO_2\cdot$ C_2H_5	III-632	130.14
2938	acetoacetic ester	. . . .	$CH_3CO\cdot CH(C_2H_5)\cdot$ $CO_2\cdot C_2H_5$	III-691	158.20
2939	acetopyruvate	oxalacetone	$CH_3CO\cdot CH_2\cdot CO\cdot$ $CO_2\cdot C_2H_5$	III-747	158.16
2940	acetylglycolate	. . . .	$CH_3CO_2\cdot CH_2\cdot CO_2\cdot$ C_2H_5	III-237	146.14
2941	acetylsalicylate	. . . .	$CH_3CO_2\cdot C_6H_4\cdot CO_2\cdot$ C_2H_5	X-75	208.22

Table 7-4 (*Continued*)
PHYSICAL CONSTANTS OF ORGANIC COMPOUNDS

No.	Crystalline Form and Color	Specific Gravity	Melting Point °C.	Boiling Point °C.	Solubility in 100 Parts		
					Water	Alcohol	Ether
2908	col. lq.	1.137^{0}	9-10	200-1	i.	s.	∞
2909	lq.	....	12-4	238	i.	s.	s. HCl
2910	col. cr./aq.	....	189 (anh.)	.......,	2	s.	i.; s. chl.
2911	col. lq.	$0.708^{\frac{25}{4}0}$	$\alpha -116.3$ $\beta -123.3$	34.6	$7.5^{20 0}$	∞	∞ chl.
2912	col. lq.	$1.102^{\frac{20}{4}0}$	....	206-7 sl. d.	s.	s.	s.
2913	yel. lq.	....	....	$160-2^{7mm}$	....	....	....
2914	col.	....	20-2(6)	247-9	....	∞	∞
2915	oil	....	21-3	d. 300±	v. sl. s.	....	....
2916	nd./aq.	....	135-7	subl.	v. sl. s. h.	s.	s.
2917	nd.	....	196-8	....	v. sl. s. h.	....	....
2918	col. fl.	....	71-2	....	i.	s. h.	s.
2919	yel. lq.	....	....	$186-9^{10mm}$	....	....	....
2920	col. lq.	....	....	$164-6^{4mm}$	....	....	....
2921	col. pl.	....	51-3	....	i.	s.	v. s.
2922	col. cr.	....	61-2	....	s.	s.	sl. s.
2923	lt. yel. lq.	....	....	$55-65^{15mm}$	....	....	....
2924	col. lq.	....	....	$98-100^{7mn}$	....	....	....
2925	col. lq.	....	....	204-6	....	....	....
2926	col. lq.	....	....	$99-101^{10mm}$	....	....	....
2927	oil	....	28	216-7	sl. s.	∞	∞
2928	yel. lq.	....	....	246-7	v. sl. s.	s.	s.; s. bz.
2929	lf./aq.	....	66-7	246-7	v. s. h.	v. s.	v. s.
2930	lq.	$1.097^{\frac{20}{20}0}$	35±	286	$4.6^{20 0}$	s.	s.; s. chl.
2931	lq.	$1.006^{\frac{15}{4}0}$	....	227-8	∞	∞	....
2932	lq.	$1.020^{\frac{20}{20}0}$	d. >280	200^{4mm}	i.	....	....
2933	oily lq.	$0.942^{4.5 0}$	....	205	∞	∞	i. aq. alk.
2934	rhb./aq.	$0.994^{\frac{60}{4}0}$	53-4	258	i.	s.	v. s.
2935	col. lq.	$0.901^{\frac{20}{4}0}$	−83.6	77.2	$8.5^{15 0}$	∞	∞
2936	col. lq.	$0.885^{\frac{25}{4}0}$	....	144-6	i.c.	∞ ; ∞ chl.	∞ act.
2937	col. lq.	$1.025^{\frac{20}{4}0}$	−45	180^{755mm}	$131^{7 0}$; ∞	∞ ; ∞ chl.	∞
2938	oil	$0.986^{\frac{20}{4}0}$	....	198	sl. s.	∞	∞
2939	cr.	$1.125^{\frac{20}{4}0}$	18	213-5	....	....	....
2940	lq.	$1.099^{17 0}$	....	180	v. sl. s.	....	....
2941	col. lq.	$1.157^{15 0}$	....	272 sl. d.	i.	s.	s.

Table 7-4 (*Continued*)
PHYSICAL CONSTANTS OF ORGANIC COMPOUNDS

No.	Name	Synonym	Formula	Beil. Ref.	Formula Weight
2942	**Ethyl** acrylate	(polymerizes readily)	$CH_2{:}CH{\cdot}CO_2{\cdot}C_2H_5$	II-399	100.12
2943	acrylic acid (β)	pentenoic acid	$C_2H_5{\cdot}CH{:}CH{\cdot}CO_2H$	II-426	100.12
2944	adipate (mono)	Et hydrogen adipate	$C_2H_5O_2C{\cdot}(CH_2)_4{\cdot}$ CO_2H	* II-277	174.20
2945	alaninate HCl (*dl*)	alanine ethyl ester HCl	$CH_3CH(NH_2)CO_2{\cdot}$ $C_2H_5{\cdot}HCl$	IV-390	153.61
2946	alcohol	ethanol; alcohol	$CH_3{\cdot}CH_2OH$	I-292	46.07
2947	allophanate	....	$NH_2CO{\cdot}NHCO_2C_2H_5$	III-69	132.12
2948	allylacetoacetate	....	$CH_3CO{\cdot}CH(C_3H_5){\cdot}$ $CO_2C_2H_5$	III-738	170.21
2949	allyl ether	....	$CH_2{:}CH{\cdot}CH_2{\cdot}O{\cdot}C_2H_5$	I-438	86.13
2950	amine	amino-ethane	$C_2H_5{\cdot}NH_2$	IV-87	45.08
2951	amine hydro-bromide	....	$C_2H_5NH_2{\cdot}HBr$	IV-91	126.00
2952	amine hydro-chloride	....	$C_2H_5NH_2{\cdot}HCl$	IV-91	81.55
2953	aminobenzoate (*o*)	ethyl anthranilate	$NH_2{\cdot}C_6H_4{\cdot}CO_2C_2H_5$	XIV-319	165.19
2954	aminobenzoate (*m*)	....	$NH_2{\cdot}C_6H_4{\cdot}CO_2C_2H_5$	XIV-389	165.19
2955	aminobenzoate (*p*)	anesthesine	$NH_2{\cdot}C_6H_4{\cdot}CO_2C_2H_5$	XIV-422	165.19
2956	aminocrotonate (β)	....	$CH_3C(NH_2){:}CH{\cdot}$ $CO_2C_2H_5$	III-654	129.16
2957	amino ethanol (β)	....	$C_2H_5NH{\cdot}CH_2{\cdot}CH_2OH$	IV-282	89.14
2958	amino phenol (*o*)	....	$C_2H_5NH{\cdot}C_6H_4{\cdot}OH$	XIII-364	137.18
2959	amino phenol (*m*)	....	$C_2H_5NH{\cdot}C_6H_4{\cdot}OH$	XIII-408	137.18
2960	amino phenol (*p*)	....	$C_2H_5NH{\cdot}C_6H_4{\cdot}OH$	XIII-443	137.18
2961	amyl ether (*act.*)	....	$C_2H_5{\cdot}O{\cdot}C_5H_{11}$	I-387	116.20
2962	amyl ether (*iso*)	....	$C_2H_5{\cdot}O{\cdot}C_5H_{11}$	I-401	116.20
2963	*n*-amyl ketone	octanone-3	$C_2H_5{\cdot}CO{\cdot}C_5H_{11}$	I-706	128.22
2964	*iso*-amyl ketone	....	$C_2H_5{\cdot}CO{\cdot}C_5H_{11}$	I-706	128.22
2965	aniline †	Et-Ph-amine	$C_6H_5{\cdot}NH{\cdot}C_2H_5$	XII-159	121.18
2966	aniline sulfonic acid (*m*)	*N*-ethyl-metanilic acid	$C_2H_5NH{\cdot}C_6H_4{\cdot}SO_3H$	XIV-690	201.25
2967	aniline sulfonate	Na salt	$C_8H_{10}O_3NSNa{\cdot}2H_2O$	XIV-690	259.26
2968	anisate (*p*)	....	$CH_3O{\cdot}C_6H_4{\cdot}CO_2{\cdot}C_2H_5$	X-159	180.21
2969	anthracene (*9*)	....	$(C_6H_4)_2C_2H{\cdot}C_2H_5$	V-678	206.29
2970	arsine	....	$C_2H_5{\cdot}AsH_2$	IV-601	106.00
2971	arsonic acid	....	$C_2H_5{\cdot}AsO(OH)_2$	IV-614	154.00
2972	benzamide (*N*)	....	$C_6H_5CO{\cdot}NH{\cdot}C_2H_5$	IX-202	149.19
2973	benzene	phenylethane	$C_6H_5{\cdot}C_2H_5$	V-351	106.17
2974	benzene sulfonate	....	$C_6H_5{\cdot}SO_3{\cdot}C_2H_5$	XI-30	186.23
2975	benzoate	....	$C_6H_5{\cdot}CO_2{\cdot}C_2H_5$	IX-110	150.18
2976	benzoic acid (*o*)	....	$C_2H_5{\cdot}C_6H_4{\cdot}CO_2H$	IX-526	150.18
2977	benzoic acid (*m*)	....	$C_2H_5{\cdot}C_6H_4{\cdot}CO_2H$	IX-528	150.18
2978	benzoic acid (*p*)	....	$C_2H_5{\cdot}C_6H_4{\cdot}CO_2H$	IX-529	150.18
2979	*o*-benzoylbenzoate	....	$C_6H_5CO{\cdot}C_6H_4{\cdot}CO_2{\cdot}$ C_2H_5	X-749	254.29
2980	benzoylformate	....	$C_6H_5CO{\cdot}CO_2C_2H_5$	X-657	178.19
2981	benzoylene urea (1)	....	$C_{10}H_{10}O_2N_2$	....	190.20
2982	benzoylene urea (3)	....	$C_{10}H_{10}O_2N_2$	XXIV-375	190.20
2983	benzylacetoacetate	....	$CH_3CO{\cdot}CH(C_7H_7){\cdot}$ $CO_2C_2H_5$	X-710	220.27
2984	benzylaniline	benzyl-Et-aniline	$C_6H_5N(C_2H_5)C_7H_7$	XII-1026	211.31

† See also Nos. 319-21.
Ethyl adipate 2944
Ethyl aminobenzene sulfonic acid 2966

Ethyl anilinocrotonate 39
Ethyl anthranilate 2953

Table 7-4 (*Continued*)
PHYSICAL CONSTANTS OF ORGANIC COMPOUNDS

No.	Crystalline Form and Color	Specific Gravity	Melting Point °C.	Boiling Point °C.	Solubility in 100 Parts		
					Water	Alcohol	Ether
2942	col. lq.	$0.925^{15°}$	−72	100-1	sl. s.	....	....
2943	col. lq.	$0.992^{15°}$	9-10	197-200	$6.3^{20°}$	....	....
2944	hyg. cr./et.-pet.	....	29	180^{19mm}	....	....	....
2945	col. hyg. cr.	....	85-7	....	v. s.	s.	i.
2946	col. lq.	$*0.789^{20°}_{4}$	−114.5	78.4	∞	∞ chl.	∞
2947	cr./bz.	....	197-8	d.	s. h.	$0.5^{21°}$	$0.1^{20°}$
2948	lq.	$0.992^{17.6°}_{4}$	....	206 sl. d.	....	....	....
2949	lq.	$0.765^{20°}_{4}$	....	$66-7^{42.9mm}$	i.	∞	∞
2950	col. lq.	0.689^{15}_{15}	−80.6	16.6	∞	∞	∞
2951	mn. nd./al.	1.741	....	....	s. act.	v. s.	i. chl.
2952	mn.	1.216	108-9	....	$240^{17°}$	v. s.	i.
2953	cr.	$1.117^{20°}_{4}$	13	266-8	....	s.	s.
2954	oil	....	....	294	sl. s. h.	∞	∞
2955	cr./al.	....	91-2	....	i.	s.	s.
2956	mn. pr.	$1.021^{20°}_{4}$	33.9(20)	210-5 d.	i.	s.	s.; s. bz.
2957	oil	$0.914^{20°}_{4}$	....	$167-9^{751mm}$	v. s.	v. s.	v. s.
2958	pl.	....	108-9	....	i.; sl. s. CS₂	v. s.; s. h. bz.	sl. s.
2959	cr./bz. lg.	....	62	176^{12mm}	s. h.	s.; sl. s. lg.	s.; v. s. chl.
2960	nd./aq.	....	100	....	s. h.	s.	s.
2961	lq.	$0.759^{18°}_{4}$	....	$108-9^{736mm}$	i.	∞	∞
2962	lq.	$0.764^{18°}$	....	112	i.	∞	∞
2963	lq.	$0.850^{0°}$	....	$169-70^{738mm}$	i.	∞	∞
2964	lq.	$0.830^{20°}$	....	163.5	i.	∞	∞
2965	lq.	$0.963^{20°}_{4}$	−65.8	205.5	i.	∞	∞
2966	nd./aq.	....	d. 294	....	$2.15^{15°}$	....	....
2967	lf./aq. al.	....	....	....	....	....	....
2968	lq.	$1.103^{25°}_{25}$	7-8	269-70	i.	s.	s.
2969	lf./al.	$1.041^{99°}$	59-60	....	i.	s.	....
2970	col. lq.	$1.217^{22°}$	....	36	$0.01^{19°}$	....	....
2971	cr./al.	....	99.5	....	$70^{27°}$	$39.4^{25°}$	....
2972	nd./aq.	....	70-1	298-300	sl. s. h.	....	....
2973	col. lq.	$0.867^{20°}_{4}$	−95.0	136.2	$0.01^{15°}$	∞; i. NH₃	∞; s. SO₂
2974	col. lq.	$1.219^{17°}_{4}$	....	156^{15mm}	d. h.	∞; ∞ bz.	∞; ∞ chl.
2975	col. lq.	$1.052^{15°}_{15}$	−34.7	212.4	i.; ∞ pet.	∞; ∞ chl.	∞
2976	nd./h. aq.	....	68	259	v. sl. s.	s.	s.
2977	nd./aq. al.	$1.042^{100°}_{4}$	47	....	i. c.	s.	....
2978	pr./al.	....	112-3	....	s. h.	s.	s.
2979	col. rhb. pr.	$1.122^{64.4°}_{4}$	58	....	....	v. s.	v. s.
2980	lq.	$1.122^{25.1°}_{4}$	....	265^{766mm}	....	....	....
2981	cr.	....	215-7	....	i.; s. NaOH	sl. s.	s. H₂SO₄
2982	nd.	....	197.5-8.5	....	i.	s.	s. NaOH
2983	col. lq.	$1.036^{15.5°}_{16.5}$	....	283-4	i.	∞	∞
2984	lt. yel. oil	$1.034^{18.5°}$	....	$165-7^{9mm}$	i.; ∞ chl.	18	∞

* See also special table of specific gravities.   "Ethyl" base 5762
Ethyl azelate 2123   Ethyl benzoylacetate 739

Table 7-4 (*Continued*)
PHYSICAL CONSTANTS OF ORGANIC COMPOUNDS

No.	Name	Synonym	Formula	Beil. Ref.	Formula Weight
	Ethyl				
2985	benzylbenzene (*p*)	....	$C_2H_5 \cdot C_6H_4 \cdot CH_2 \cdot C_6H_5$	V-614	196.29
2986	benzyl ether	....	$C_2H_5O \cdot CH_2 \cdot C_6H_5$	VI-431	136.20
2987	benzyl ketone	....	$C_2H_5 \cdot CO \cdot CH_2 \cdot C_6H_5$	VII-314	148.21
2988	benzyl-*o*-toluidine	Et *o*-tolyl-benzyl-amine	$CH_3 \cdot C_6H_4 \cdot N(C_2H_5) \cdot CH_2 \cdot C_6H_5$	XII-1033	225.34
2989	bromide	bromoethane	$C_2H_5 \cdot Br$	I-88	108.97
2990	bromoacetate	....	$Br \cdot CH_2 \cdot CO_2 \cdot C_2H_5$	II-214	167.01
2991	α-bromoaceto-acetate	....	$CH_3CO \cdot CHBr \cdot CO_2 \cdot C_2H_5$	III-664	209.05
2992	bromobutyrate	(α)(*n*)	$C_2H_5 \cdot CHBr \cdot CO_2C_2H_5$	II-282	195.06
2993	bromo-*iso*-butyrate	(α)	$(CH_3)_2CBr \cdot CO_2C_2H_5$	II-296	195.06
2994	α-bromo-*n*-caproate	....	$CH_3(CH_2)_3CHBr \cdot CO_2C_2H_5$	II-325	223.12
2995	α-bromopropionate	....	$CH_3 \cdot CHBr \cdot CO_2C_2H_5$	II-255	181.04
2996	β-bromopropionate	....	$Br(CH_2)_2CO_2 \cdot C_2H_5$	II-256	181.04
2997	α-bromovalerate (*n*)	....	$CH_3(CH_2)_2CHBr \cdot CO_2C_2H_5$	II-302	209.09
2998	α-bromo-*iso*-val-erate	....	$(CH_3)_2CH \cdot CHBr \cdot CO_2C_2H_5$	II-317	209.09
2999	*n*-butyl-barbituric acid (5,5)	sonneryl; neonal	$(CONH)_2CO \cdot C:$ │_____│ $(C_2H_5)(C_4H_9)$	....	212.25
3000	*n*-butyl ether	....	$C_2H_5 \cdot O \cdot C_2H_4 \cdot C_2H_5$	I-369	102.18
3001	*iso*-butyl ether	....	$C_2H_5 \cdot O \cdot CH_2 \cdot CH:$ $(CH_3)_2$	I-376	102.18
3002	*tert*-butyl ether	....	$C_2H_5 \cdot O \cdot C(CH_3)_3$	I-381	102.18
3003	*n*-butyl ketone	heptanone-3	$C_2H_5 \cdot CO \cdot C_4H_9$	I-699	114.19
3004	*iso*-butyl ketone	....	$C_2H_5 \cdot CO \cdot C_4H_9$	I-700	114.19
3005	*n*-butyl sulfide	....	$C_2H_5 \cdot S \cdot C_4H_9$	**I-399	118.24
3006	*n*-butyrate	....	$C_2H_5 \cdot CH_2 \cdot CO_2 \cdot C_2H_5$	II-270	116.16
3007	*iso*-butyrate	....	$(CH_3)_2CH \cdot CO_2 \cdot C_2H_5$	II-291	116.16
3008	*n*-caprate	....	$CH_3(CH_2)_8 \cdot CO_2C_2H_5$	II-356	200.32
3009	*n*-caproate	....	$CH_3(CH_2)_4 \cdot CO_2C_2H_5$	II-323	144.22
3010	*n*-caprylate	....	$CH_3(CH_2)_6 \cdot CO_2C_2H_5$	II-348	172.27
3011	carbazole (*N*)	....	$C_{12}H_8N \cdot C_2H_5$	XX-436	195.27
3012	carbonate, ortho †	....	$C(OC_2H_5)_4$	III-5	192.26
3013	carbostyril	....	$C_6H_4CH:C(C_2H_5) \cdot$ │_____ $NH \cdot CO$ ___│	XXI-115	173.22
3014	chaulmoograte	chaulmestrol; moogrol	$C_5H_7(CH_2)_{12}CO_2C_2H_5$	IX-80	308.51
3015	chloride	chloroethane	$C_2H_5 \cdot Cl$	I-82	64.52
3016	chloroacetate	....	$Cl \cdot CH_2 \cdot CO_2 \cdot C_2H_5$	II-197	122.55
3017	chloroacetoacetate	....	$Cl \cdot CH_2 \cdot CO \cdot CH_2 \cdot CO_2 \cdot C_2H_5$	III-663	164.59
3018	α-chloroaceto-acetate	....	$CH_3CO \cdot CHCl \cdot CO_2 \cdot C_2H_5$	III-662	164.59
3019	chloroformate	ethyl chlorocar-bonate	$Cl \cdot CO_2 \cdot C_2H_5$	III-10	108.53
3020	α-chloropropionate	....	$CH_3 \cdot CHCl \cdot CO_2C_2H_5$	II-249	136.58

† See also No. 2138.
Ethyl borate 6186
Ethyl bromomalonate 2133
Ethyl-butyl-acetaldehyde 3062
Ethyl-butyl alcohol 3638

Ethyl-butylamine 3640
Ethyl-butyl carbamate 1140-1
Ethyl-butyl carbinol (*iso*) 3537
Ethyl-butyl carbinol (*sec*) 3541
Ethyl butylmalonate 2134-6

Table 7-4 (Continued)
PHYSICAL CONSTANTS OF ORGANIC COMPOUNDS

No.	Crystalline Form and Color	Specific Gravity	Melting Point °C.	Boiling Point °C.	Solubility in 100 Parts		
					Water	Alcohol	Ether
2985	lq.	$0.985^{19°}$	....	294-5	s. chl.	s.	s.
2986	oil	$0.949^{20°}_{4}$	....	$187\text{-}9^{732mm}$	i.	∞	∞
2987	lq.	$0.998^{17°}$	....	230^{755mm}	i.	s.	∞
2988	yel. oil	....	....	$153\text{-}7^{10mm}$	....	....	....
2989	col. lq.	$1.460^{20°}_{4}$	−118.9	38.4	$1.06^{0°}$; $0.93^{0°}$	∞	∞
2990	lq.	$1.506^{20°}_{20}$	....	168	i.	∞	∞
2991	lq.	$1.429^{14°}_{4}$	....	210-5 d.	....	....	....
2992	col. lq.	$1.330^{20°}_{20}$	....	177.5^{765mm} sl. d.	i.	∞	∞
2993	col. lq.	$1.329^{20°}_{20}$	....	163.6^{762mm}	i.	∞	∞
2994	lq.	....	....	205-10	....	....	....
2995	col. lq.	$1.394^{20°}_{4}$	....	160-5 sl. d.	i.	∞	∞
2996	lq.	$1.412^{18°}_{4}$	....	$72\text{-}5^{15mm}$	....	....	....
2997	lq.	$1.226^{18°}_{4}$	....	190-2	i.	∞	∞
2998	lq.	$1.278^{12°}_{12}$	....	186	....	....	....
2999	col. cr./ aq. al.	....	127-8	....	sl. s.; s. dil. alk.	20	11
3000	col. lq.	$0.752^{20°}_{20}$	....	91.4	i.	∞	∞
3001	col. lq.	0.751	....	78-80	i.	∞	∞
3002	col. lq.	$0.752^{20°}$	....	70^{758mm}	i.	∞	∞
3003	lq.	$0.816^{20°}_{20}$	−39	149-50	i.	∞	∞
3004	lq.	$0.815^{17°}_{4}$	....	135^{735mm}	i.	∞	∞
3005	yel. lq.	$0.857^{25°}_{25}$	....	143-5	....	....	....
3006	col. lq.	$0.874^{25°}_{4}$	−100.8	121.6	$0.68^{25°}$	∞	∞
3007	col. lq.	$0.866^{20°}_{4}$	−88.2	110.0	sl. s.	∞	∞
3008	lq.	$0.856^{25°}_{4}$	−18.0(−20)	244.9	i.; ∞ chl.	∞	∞
3009	col. lq.	$0.859^{20°}_{20}$	−67.5	167.9	i.	∞	∞
3010	col. lq.	$0.871^{15°}_{4}$	−43.2(−59)	208.5	i.	∞	∞
3011	lf./et.	....	67-8	....	i.	v. s. h.	v. s.
3012	col. lq.	$0.919^{19°}_{4}$	....	158-9	....	....	....
3013	cr./aq. HCl	....	168	....	....	....	....
3014	pa. yel. lq.	$0.906^{15°}_{4}$	....	230^{20mm}	i.	∞	∞ chl.
3015	col. lq.	$0.903^{10°}$	−138	12.3	$0.45^{0°}$	∞	∞
3016	col. lq.	$1.159^{20°}_{4}$	−26	144	i.	∞	∞
3017	lq.	$1.218^{17°}_{4}$	−8±	220^{756mm}	v. sl. s.	∞	∞
3018	lq.	$1.19^{14°}_{18}$	....	193 d.	v. sl. s.	s.	s.
3019	col. lq.	$1.138^{20°}_{4}$	−80.6	94-5	d.	∞; ∞ bz.	∞; ∞ chl.
3020	col. lq.	$1.087^{20°}_{4}$	....	147-8	i.	∞	∞

Ethyl-butyraldehyde 2099
Ethyl-butyric acid 2101
Ethyl-cacodyl 579
Ethyl-caproic acid 1221

Ethyl carbamate 6392
Ethyl carbonate 2138
Ethyl-carbylamine 3029
Ethyl chlorocarbonate 3019

Table 7-4 (Continued)
PHYSICAL CONSTANTS OF ORGANIC COMPOUNDS

No.	Name	Synonym	Formula	Beil. Ref.	Formula Weight
	Ethyl				
3021	β-chloropropionate	. . . .	$Cl\cdot CH_2\cdot CH_2\cdot CO_2C_2H_5$	II-250	136.58
3022	chlorosulfate	ethyl chlorosulfonate	$C_2H_5O\cdot SO_2Cl$	I-327	144.58
3023	cinnamate (*trans*)	. . . .	$C_6H_5\cdot C_2H_2\cdot CO_2C_2H_5$	IX-581	176.22
3024	crotonate (α)	. . . .	$C_3H_5\cdot CO_2\cdot C_2H_5$	II-411	114.15
3025	crotonic acid (α)	. . . .	$CH_3\cdot CH:C(C_2H_5)\cdot CO_2H$	II-440	114.15
3026	crotonic acid (α)	. . . .	$CH_3\cdot CH:C(C_2H_5)\cdot CO_2H$	II-440 / *II-408	114.15 / 114.15
3027	cyanate	. . . .	$NCO(C_2H_5)$	. . . .	71.08
3028	*iso*-cyanate	. . . .	$C_2H_5\cdot N:CO$	IV-122	71.08
3029	*iso*-cyanide	ethyl carbylamine	$C_2H_5\cdot NC$	IV-107	55.08
3030	cyanoacetate	cyanacetic ester	$NC\cdot CH_2\cdot CO_2\cdot C_2H_5$	II-585	113.12
3031	α-cyanocinnamate	Et β-Ph-α-cyano-acrylate	$C_6H_5\cdot CH:C(CN)\cdot CO_2\cdot C_2H_5$	IX-894	201.23
3032	cyanoformate	cyanethyl carbonate	$NC\cdot CO_2\cdot C_2H_5$	II-547	99.09
3033	cyclopentan-1-one-2-carboxylate	. . . .	$CO(CH_2)_3CH\cdot CO_2\cdot C_2H_5$	X-597	156.18
3034	diacetoacetate	diacetoethyl acetate	$(C_2H_3O)_2CH\cdot CO_2C_2H_5$	III-751	172.18
3035	diazoacetate	diazoethyl acetate	$N_2CH\cdot CO_2\cdot C_2H_5$	* III-211	114.10
3036	dibromoacetate	. . . .	$Br_2CH\cdot CO_2\cdot C_2H_5$	II-219	245.91
3037	α,β-dibromobutyrate	. . . .	$CH_3(CHBr)_2CO_2C_2H_5$	II-284	273.96
3038	di-*n*-butylcarbamate	(*N,N*)	$(C_4H_9)_2N\cdot CO_2C_2H_5$	. . . .	201.31
3039	dichloroacetate	. . . .	$Cl_2CH\cdot CO_2\cdot C_2H_5$	II-203	157.00
3040	dichloroarsine	. . . .	$C_2H_5\cdot AsCl_2$	IV-603	174.89
3041	3,5-diiodosalicylate	(3,5;2,1)	$I_2C_6H_2(OH)CO_2C_2H_5$	X-114	417.97
3042	β,β-dimethylacrylate	. . . .	$(CH_3)_2C:CHCO_2C_2H_5$	II-433	128.17
3043	3,5-dinitrobenzoate	. . . .	$(NO_2)_2C_6H_3\cdot CO_2C_2H_5$	IX-414	240.17
3044	diphenylamine	. . . .	$C_2H_5\cdot N:(C_6H_5)_2$	XII-181	197.28
3045	diphenylphosphine	. . . .	$C_2H_5\cdot P:(C_6H_5)_2$	XVI-759	214.25
3046	fluoride	fluoroethane	$CH_3\cdot CH_2F$	I-82	48.06
3047	formamide (*N*)	. . . .	$H\cdot CO\cdot NHC_2H_5$	IV-108	73.10
3048	formate	. . . .	$H\cdot CO_2\cdot C_2H_5$	II-19	74.08
3049	formate, ortho	aethon	$H\cdot C(OC_2H_5)_3$	II-20	148.20
3050	furoate (α)	ethyl pyromucate	$C_4H_3O\cdot CO_2\cdot C_2H_5$	XVIII-275	140.14
3051	furoate (β)	. . . .	$C_4H_3O\cdot CO_2\cdot C_2H_5$	. . . .	140.14
3052	furoylacetate (α)	. . . .	$C_4H_3O\cdot CO\cdot CH_2\cdot CO_2\cdot C_2H_5$	XVIII-408	182.18
3053	β-furylacrylate	. . . .	$C_4H_3O\cdot CH:CH\cdot CO_2\cdot C_2H_5$	XVIII-300	166.18
3054	glutaric acid (α)	. . . .	$C_2H_5\cdot C_3H_5(CO_2H)_2$	II-676	160.17
3055	glutaric acid (β)	. . . .	$C_2H_5\cdot CH(CH_2CO_2H)_2$	II-676	160.17
3056	glycerate	. . . .	$(HO)_2C_3H_3\cdot CO_2C_2H_5$	III-397	134.13
3057	glycinate	. . . .	$H_2N\cdot CH_2\cdot CO_2\cdot C_2H_5$	IV-340	103.12
3058	glycinate HCl	. . . .	$C_4H_9O_2N\cdot HCl$	IV-342	139.58
3059	glycine (*N*)	. . . .	$C_2H_5NH\cdot CH_2\cdot CO_2H$	IV-349	103.12
3060	glycolate	. . . .	$HO\cdot CH_2\cdot CO_2\cdot C_2H_5$	III-236	104.11

Table 7-4 (*Continued*)
PHYSICAL CONSTANTS OF ORGANIC COMPOUNDS

No.	Crystalline Form and Color	Specific Gravity	Melting Point °C.	Boiling Point °C.	Solubility in 100 Parts		
					Water	Alcohol	Ether
3021	col. lq.	$1.109\frac{20°}{4}$	....	$162\text{-}3^{765mm}$	....	....	....
3022	lq.	$1.263^{18°}$	....	58^{20mm}	s. chl.	s. bz.	s.
3023	col. lq.	$1.049\frac{20°}{4}$	12(7.5)	271	i.	∞	∞
3024	col. lq.	$0.924\frac{15°}{4}$	....	138^{748mm}	i.	s.	s.
3025	col. pr.	....	41.5	209	v. sl. s.	v. s.	v. s.
3026	oily lq.*; nd./aq.	.... $0.958\frac{50°}{4}$	−35 41-2	199.5^{750mm} $204\text{-}7^{755mm}$	i.	∞	∞
3027	lq.	$1.127^{15°}$	....	d.	i.	∞	∞
3028	lq.	$0.907\frac{16°}{4}$	....	60	d.	....	....
3029	col. lq.	$0.738^{25°}$	<−66	78-9	sl. s.	....	s.
3030	col. lq.	$1.062\frac{20°}{4}$	−22.5	208^{753mm}	$22^{50°}$; 980°	∞	∞
3031	nd./al.**	....	50-1	360 sl. d.	s. chl.; s. ac.	12 c.	s.; s. bz.
3032	lq.	$1\ 003\frac{20°}{4}$	....	115-6	....	....	....
3033	col. lq.	$1.098^{0°}$	....	218 sl. d.	....	....	....
3034	col. lq.	$1.089\frac{25°}{25}$	....	209-11 sl. d.	sl. s.	v. s.	v. s.
3035	yel. oil	$1.085\frac{18°}{4}$	−22	$140\text{-}17^{20mm}$	sl. s.	∞	∞
3036	oil	$1.903\frac{20°}{20}$	....	192-4	i.	∞	∞
3037	lq.	....	....	$123\text{-}4^{30mm}$	....	....	....
3038	col. lq.	....	....	$101\text{-}3^{6mm}$	....	....	....
3039	col. lq.	$1.282\frac{20°}{4}$	....	158	i.	∞	∞
3040	col. lq.	$1.742\frac{15°}{4}$	....	156	sl. s.	∞	∞
3041	col. lf./al.	....	133	d. >200°	i. c.	s. h.	sl. s.
3042	lq.	$0.922\frac{21°}{21}$	....	154-5	v. s. CS_2; v. s. lg.	v. s.; v. s. chl.	v. s.; v. s. bz.
3043	nd./al.	$1.295^{111°}$	91-2	....	....	$0.6^{13°}$	s. h. al.
3044	lq.	....	....	295-7	i.	s.	....
3045	oil	....	....	293	....	s.	s. bz.
3046	gas	1.7(A)	....	−32	$198^{14°}$cc.	v. s.	....
3047	lq.	$0.952^{21°}$	<−30	197-9	∞	∞	∞
3048	col. lq.	$0.923\frac{20°}{4}$	−79.4	54.2	$11^{18°}$	∞	∞
3049	lq.	$0.939\frac{25°}{25}$	−76	145-6	v. sl. s.	∞	∞
3050	lf.	$1.117\frac{20.8°}{20}$	34	195^{766mm}	i.	∞	∞
3051	lq.	$1.038\frac{20°}{20}$	....	$65\text{-}7^{14mm}$	....	....	....
3052	lt. yel. oil	$1.165^{17°}$	....	$143\text{-}5^{10mm}$	i.; v. s. NH_4OH	s.	s.
3053	yel. cr.	$1.09\frac{20°}{4}$	24.5	232-3	i.	∞	∞
3054	col. cr.	....	60.5	250-60 sl. d.	v. s.	v. s.	v. s.
3055	pr./chl.	....	73	....	v. s.	v. s.	v. s.
3056	lq.	$1.191\frac{15°}{15}$	....	121^{14mm}	s.	s.	....
3057	col. oil	$1.028\frac{20°}{4}$	<−20	148^{748mm} sl. d.	∞	∞	∞
3058	nd.	....	144	subl.	v. s.	v. s.	d. alk.
3059	lf./al.	....	>160 d.	....	s.	s.	....
3060	col. lq.	$1.087\frac{15°}{4}$	....	160	....	v. s.	v. s.

* Easily converted to solid modification.
** Also a liquid form, d. on boiling & forms some solid modification.
Ethyl dithiooxalate 2165
Ethyl ether 2911

Ethyl ethyl-acetoacetate 2938
Ethyl ethyl-amyl-malonate 2167
Ethyl ethyl-butyl-malonate 2168
Ethyl N-ethyl-carbamate 3184
Ethyl-ethylene 1147

Table 7-4 *(Continued)*
PHYSICAL CONSTANTS OF ORGANIC COMPOUNDS

No.	Name	Synonym	Formula	Beil. Ref.	Formula Weight
3061	**Ethyl** glycol ether	....	$HO \cdot CH_2 \cdot CH_2 \cdot O \cdot C_2H_5$	I-467	90.12
3062	hexaldehyde (2)	Et-Bu-acetaldehyde	$C_4H_9 \cdot CH(C_2H_5) \cdot CHO$	I-707	128.22
3063	hydracrylate	....	$HO \cdot CH_2 \cdot CH_2 \cdot CO_2C_2H_5$	III-297	118.13
3064	hydrazine	....	$C_2H_5 \cdot NH \cdot NH_2$	IV-550	60.10
3065	hydrocinnamate	....	$C_6H_5(CH_2)_2CO_2C_2H_5$	IX-511	178.23
3066	hydrogen oxalate	ethyl oxalic acid	$HO_2C \cdot CO_2 \cdot C_2H_5$	II-535	118.09
3067	hydrogen phthalate (*o*)	....	$C_6H_4(CO_2C_2H_5) \cdot CO_2H$	IX-797	194.19
3068	hydrogen tetra-chlorophthalate	....	$HO_2C \cdot C_6Cl_4 \cdot CO_2 \cdot C_2H_5$	IX-820	331.97
3069	hydrogen sulfate	ethyl sulfuric acid	$C_2H_5O \cdot SO_2 \cdot OH$	I-325	126.13
3070	hydrogen tartrate (*d*)	....	$HO_2C \cdot (CHOH)_2 \cdot CO_2C_2H_5$	III-512	178.14
3071	hydroselenide	....	$C_2H_5 \cdot SeH$	I-349	109.03
3072	*p*-hydroxybenzo-ate †	....	$HO \cdot C_6H_4 \cdot CO_2C_2H_5$	X-159	166.18
3073	α-hydroxy-*iso*-butyrate	....	$(CH_3)_2C(OH)CO_2 \cdot C_2H_5$	III-315	132.16
3074	hydroxylamine (α)	....	$C_2H_5O \cdot NH_2$	I-336	61.08
3075	hydroxylamine (β)	....	$C_2H_5 \cdot NH \cdot OH$	IV-535	61.08
3076	hypochlorite	....	$C_2H_5 \cdot O \cdot Cl$	I-324	80.51
3077	iodide	iodoethane	$CH_3 \cdot CH_2I$	I-96	155.97
3078	iodoacetate	....	$ICH_2 \cdot CO_2 \cdot C_2H_5$	II-222	214.00
3079	lactate	....	$CH_3 \cdot CHOH \cdot CO_2C_2H_5$	III-280	118.13
3080	laurate	....	$CH_3(CH_2)_{10}CO_2C_2H_5$	II-361	228.38
3081	levulinate	....	$CH_3 \cdot CO(CH_2)_2 \cdot CO_2 \cdot C_2H_5$	III-675	144.17
3082	malonic acid	....	$C_2H_5 \cdot CH(CO_2H)_2$	II-643	132.12
3083	mercaptan	ethanthiol	$C_2H_5 \cdot SH$	I-340	62.13
3084	mesoxalate	....	$CO(CO_2C_2H_5)_2$	III-769	174.15
3085	methylaceto-acetate	Me-acetoacetic ester	$CH_3CO \cdot CH(CH_3) \cdot CO_2 \cdot C_2H_5$	III-679	144.17
3086	methyl acrylate	....	$C_3H_5 \cdot CO_2 \cdot C_2H_5$	II-423	114.15
3087	S-methylxanthate	....	$C_2H_5O \cdot CS \cdot SCH_3$	III-210	136.24
3088	myristate	....	$C_{13}H_{27}CO_2 \cdot C_2H_5$	II-365	256.43
3089	naphthalene (α)	....	$C_2H_5 \cdot C_{10}H_7$	V-569	156.23
3090	naphthalene (β)	....	$C_2H_5 \cdot C_{10}H_7$	V-569	156.23
3091	α-naphthylamine	....	$C_2H_5 \cdot NH \cdot C_{10}H_7$	XII-1222	171.24
3092	β-naphthylamine	....	$C_2H_5 \cdot NH \cdot C_{10}H_7$	XII-1274	171.24
3093	α-naphthyl ether	....	$C_{10}H_7 \cdot O \cdot C_2H_5$	VI-606	172.23
3094	β-naphthyl ether	neroline; bromelia	$C_{10}H_7 \cdot O \cdot C_2H_5$	VI-641	172.23
3095	nicotinate	....	$C_5H_4N \cdot CO_2C_2H_5$	XXII-39	151.17
3096	nitrate	nitric ether	$C_2H_5 \cdot O \cdot NO_2$	I-329	91.07
3097	nitrite	nitrous ether	$C_2H_5 \cdot O \cdot NO$	I-329	75.07
3098	nitroacetate	....	$NO_2 \cdot CH_2 \cdot CO_2 \cdot C_2H_5$	II-225	133.10
3099	nitrobenzoate (*o*)	....	$NO_2 \cdot C_6H_4 \cdot CO_2C_2H_5$	IX-372	195.18
3100	nitrobenzoate (*m*)	....	$NO_2 \cdot C_6H_4 \cdot CO_2C_2H_5$	IX-378	195.18
3101	nitrobenzoate (*p*)	....	$NO_2 \cdot C_6H_4 \cdot CO_2C_2H_5$	IX-390	195.18
3102	nitrocinnamate (*o*)	....	$NO_2 \cdot C_6H_4 \cdot CH:CH \cdot CO_2 \cdot C_2H_5$	IX-605	221.21
3103	nitrocinnamate (*m*) (*trans*)	....	$NO_2 \cdot C_6H_4 \cdot CH:CH \cdot CO_2 \cdot C_2H_5$	IX-606	221.21
3104	nitrocinnamate (*p*)	....	$NO_2 \cdot C_6H_4 \cdot CH:CH \cdot CO_2 \cdot C_2H_5$	IX-607	221.21

† See also ethyl salicylate.
Ethyl ethyl-malonate 2169
Ethyl ethyl-phenyl-malonate 2170
Ethyl ethyl-propyl-malonate 2171
Ethyl S-ethylxanthate 2164

Ethyl fumarate 2174
Ethyl glutaconate 2175
Ethyl glutarate 2176
Ethyl glycolic acid 2912
Ethyl *n*-heptylate 3108

Ethyl heptyl-malonate 2177
Ethyl hexenal 3141
Ethyl-hexyr carbinol 4920
Ethyl itaconate 2179
Ethyl malate 2182

Table 7-4 (*Continued*)
PHYSICAL CONSTANTS OF ORGANIC COMPOUNDS

No.	Crystalline Form and Color	Specific Gravity	Melting Point °C.	Boiling Point °C.	Solubility in 100 Parts		
					Water	Alcohol	Ether
3061	lq.	$0.935\frac{15}{15}°$	....	134-57^{48mm}	....	....	s. lq. NH$_3$
3062	col. lq.	$0.820\frac{20}{20}°$	<-100	163-4	0.072$^{20°}$	....	....
3063	col. lq.	$1.064^{25°}$	....	185-90	∞	∞	∞
3064	hyg. lq.	....	....	99.5^{709mm}	v. s.	v. s.	v. s.
3065	lq.	$1.015\frac{20}{4}°$	....	249	i.	s.	s.
3066	col. lq.	$1.218\frac{20}{4}°$	....	1171^{5mm}	....	s.	....
3067	oil	....	2	d.	sl. s.	s.	s.
3068	cr.	....	94-5	d. 150	i.; s. aq. alk. carb.	s.	s.
3069	syrup	$1.316^{17°}$	....	d.	∞ ; d. h.	∞ ; d. h.	∞
3070	col. hyg. pr.	....	90±	....	s.; d. h.	s.	i.
3071	lq.	$1.395\frac{24}{4}°$	....	53.5	i.	....	....
3072	col. cr.	....	116-8	297-8	180°	72$^{25°}$ act.	45$^{25°}$
3073	col. lq.	....	....	149-50	d. h.	....	....
3074	col. lq.	$0.883^{7.5°}$	....	68	∞	∞	∞
3075	nd./lg.	$0.908\frac{20}{4}°$	59 d.	....	v. s.	v. s.	sl. s.
3076	yel. lq.	$1.013\frac{-6}{4}°$	expl.	36^{752mm}	∞ bz.	∞ chl.	∞
3077	col. lq.	$1.933\frac{20}{4}°$	-110.9	72.4	0.42$^{20°}$	∞	∞
3078	col. oil	$1.817\frac{12.7°}{4}$	....	178-80	....	....	....
3079	oil	$1.030\frac{25}{4}°$	....	155	∞	∞	∞
3080	oil	$0.868\frac{13}{4}°$	-10.7	269	i.	s.	∞
3081	col. lq.	$1.016\frac{20}{20}°$	....	205.2^{756mm}	v. s.	∞	....
3082	col. pr.	....	111.5	d. 160	v. s.	v. s.	v. s.
3083	lq.	$0.839\frac{20}{4}°$	-147	35.1	1.5; s. alk.	s.	s.
3084	yel. gn. lq.	$1.119\frac{20}{20}°$	-30±	220±	....	s.	s.
3085	col. lq.	$1.019\frac{20}{4}°$	....	186.8	i.	s.	s.
3086	col. lq.	$0.913^{15.6}$	....	118	i.	s.	s.
3087	lq.	$1.119\frac{25}{4}°$	....	183-4	i.	s.	s.
3088	col. cr.	$0.856\frac{20}{4}°$	10.5-1.5	295	i.	sl. s.	sl. s.
3089	lq.	$0.990\frac{25}{25}°$	-27	258^{758mm} sl. d.	i.	∞	∞
3090	lq.	$1.002\frac{25}{0}°$	-19	251	i.	∞	∞
3091	oil	$1.060\frac{20}{4}°$	....	303^{723mm}	i.	s.	s.
3092	oil	$1.057\frac{20}{4}°$	....	316-7	i.	s.	s.
3093	cr.	$1.061\frac{20}{20}°$	5.5	276.4	i.	s.	s.
3094	pl.	$1.064\frac{20}{20}°$	37.5	282	i.; s. chl.	s.; s. bz.	s.; s. pet.
3095	col. oil	....	8-9	225	sl. s.	s.	s.; s. bz.
3096	col. lq.	$1.100\frac{25}{4}°$	-102	87-8	1.35$^{5°}$	∞	∞
3097	lq.	$0.900^{15.5°}$	....	17	v. sl. s.	∞	∞
3098	col. lq.	$1.199\frac{20}{4}°$	....	105-7^{25mm}	v. sl. s.	∞	....
3099	tri.	....	30	149^{10mm}	i.	s.	s.
3100	mn. pr.	....	41	298±	i.	s.	s.
3101	tri./al.	....	57	....	i.	s.	s.
3102	yel. rhb.	....	44	....	v. s. bz.	v. s. h.	v. s.
3103	col. mn./ al.	....	78-9	....	i.	sl. s.	sl. s.
3104	wh. tri.	....	141-2	....	i.	v. sl. s. c.	s. ac.

Ethyl maleate 2183
Ethyl malonate 2184
Ethyl mesaconate 2186
Ethyl-metanilic acid 2966

Ethyl N-methylcarbamate 4416
Ethyl-morphine HCl, cf. alkd.
Ethyl muconate 2188
Ethyl mustard oil 3159

Table 7-4 (*Continued*)
PHYSICAL CONSTANTS OF ORGANIC COMPOUNDS

No.	Name	Synonym	Formula	Beil. Ref.	Formula Weight
3105	**Ethyl** nitrolic acid	....	$CH_3 \cdot C(NOH) \cdot NO_2$	II-189	104.07
3106	octyl ether (*n*)	....	$C_2H_5 \cdot O \cdot C_8H_{17}$	I-419	158.29
3107	octyl ketone (*n*)	undecanone-3	$C_2H_5 \cdot CO \cdot C_8H_{17}$	I-713	170.30
3108	oenanthylate	ethyl *n*-heptylate	$C_6H_{13}CO_2 \cdot C_2H_5$	II-340	158.24
3109	oleate	....	$C_{17}H_{33}CO_2 \cdot C_2H_5$	II-467	310.52
3110	oxanilate	....	$C_6H_5 \cdot NH \cdot CO \cdot CO_2 \cdot C_2H_5$	XII-282	193.20
3111	oxindole (1)	*N*-Et-oxindole	$C_8H_6ON \cdot C_2H_5$	XXI-283	161.21
3112	palmitate	....	$C_{15}H_{31}CO_2 \cdot C_2H_5$	II-372	284.49
3113	pelargonate	....	$C_8H_{17}CO_2 \cdot C_2H_5$	II-353	186.30
3114	phenol (*o*)	phlorol	$C_2H_5 \cdot C_6H_4 \cdot OH$	VI-470	122.17
3115	phenol (*m*)	....	$C_2H_5 \cdot C_6H_4 \cdot OH$	VI-471	122.17
3116	phenol (*p*)	....	$C_2H_5 \cdot C_6H_4 \cdot OH$	VI-472	122.17
3117	γ-phenoxybutyrate	....	$C_6H_5O \cdot CH_2(CH_2)_2 \cdot CO_2 \cdot C_2H_5$	**VI-159	208.26
3118	phenylacetate	....	$C_6H_5 \cdot CH_2 \cdot CO_2 \cdot C_2H_5$	IX-434	164.21
3119	phenylacetic acid	....	$C_6H_5 \cdot CH(C_2H_5)CO_2H$	IX-541	164.21
3120	phenylacetylene	1-phenyl-butyne-1	$C_6H_5 \cdot C \vdots C \cdot C_2H_5$	V-517	130.19
3121	phenylbromo-acetate	(*dl*)	$C_6H_5 \cdot CHBrCO_2C_2H_5$	IX-452	243.11
3122	phenyl carbinol (*dl*)	sec-Ph-Pr-alcohol	$C_6H_5 \cdot CHOH \cdot C_2H_5$	VI-502	136.20
3123	phenyl carbinol (*l*)	....	$C_6H_5 \cdot CHOH \cdot C_2H_5$	VI-502	136.20
3124	phenyl-ethanol-amine	β-ethylanilino-ethyl alcohol	$C_6H_5N(C_2H_5)CH_2 \cdot CH_2OH$	XII-183	165.24
3125	phenylglycinate	*N*-Ph-glycine Et ester	$C_6H_5NH \cdot CH_2CO_2 \cdot C_2H_5$	XII-470	179.22
3126	phenylhydantoin (*dl*) (5,5)	nirvanol	$CO \cdot NH \cdot CO \cdot NH \cdot C \vert \underline{} \vert (C_2H_5)(C_6H_5)$	*XXIV-348	204.23
3127	phenylhydrazine	(α,α)	$C_6H_5N(C_2H_5) \cdot NH_2$	XV-119	136.20
3128	phenylhydrazine	(α,β)	$C_6H_5 \cdot NH \cdot NH \cdot C_2H_5$	XV-120	136.20
3129	phenyl ketone	propiophenone	$C_6H_5 \cdot CO \cdot C_2H_5$	VII-300	134.18
3130	2-phenyl-6-methyl-cinchoninate	neocincophen; neo-quinophan; novatophan	$CH_3 \cdot C_9H_4N(C_6H_5) \cdot CO_2 \cdot C_2H_5$	*XXII-520	291.35
3131	phenylpropiolate	....	$C_6H_5 \cdot C \vdots C\ CO_2 \cdot C_2H_5$	IX-634	174.20
3132	phenyl sulfide	thiophenetole	$C_6H_5 \cdot S \cdot C_2H_5$	VI-297	138.23
3133	phenylsulfone	....	$C_6H_5 \cdot SO_2 \cdot C_2H_5$	VI-297	170.23
3134	phenyl urea (*N,N'*)	....	$C_2H_5NH \cdot CO \cdot NHC_6H_5$	XII-348	164.21
3135	phosphine	....	$C_2H_5 \cdot PH_2$	IV-581	62.05
3136	propargyl ether	....	$CH \vdots C \cdot CH_2 \cdot O \cdot C_2H_5$	I-454	84.12
3137	propiolate	....	$CH \vdots C \cdot CO_2 \cdot C_2H_5$	II-477	98.10
3138	propionate	....	$C_2H_5 \cdot CO_2 \cdot C_2H_5$	II-240	102.13
3139	*n*-propylaceto-acetate	....	$CH_3CO \cdot CH(C_3H_7) \cdot CO_2 \cdot C_2H_5$	III-700	172.23
3140	*iso*-propylaceto-acetate	....	$CH_3CO \cdot CH(C_3H_7) \cdot CO_2 \cdot C_2H_5$	III-702	172.23
3141	β-*n*-propyl-acrolein (α)	2-Et-hexenal	$C_3H_7 \cdot CH \vdots C(C_2H_5) \cdot CHO$	I-744	126.20
3142	*n*-propyl ether	....	$C_2H_5 \cdot O \cdot CH_2 \cdot C_2H_5$	I-354	88.15
3143	*iso*-propyl ether	....	$C_2H_5 \cdot O \cdot CH(CH_3)_2$	I-362	88.15
3144	*n*-propyl ketone	hexanone-3	$C_2H_5 \cdot CO \cdot CH_2 \cdot C_2H_5$	I-690	100.16
3145	*iso*-propyl ketone	2-Me-pentanone-3	$C_2H_5 \cdot CO \cdot CH(CH_3)_2$	I-691	100.16
3146	pyridine (2)	....	$C_2H_5 \cdot C_5H_4N$	XX-241	107.16

Ethyl nitrophthalate 2190
Ethyl nitroso-ethylcarbamate 4886
Ethyl nitroso-methylcarbamate 4891
Ethyl orthoformate 3049
Ethyl orthopropionate 6193

Ethyl oxalate 2193
Ethyl oxalic acid 3066
Ethyl oxamate 5015
Ethyl-phenylamine 2965
Ethyl-phenyl-barbituric acid 5113

Table 7-4 (*Continued*)
PHYSICAL CONSTANTS OF ORGANIC COMPOUNDS

No.	Crystalline Form and Color	Specific Gravity	Melting Point °C.	Boiling Point °C.	Solubility in 100 Parts		
					Water	Alcohol	Ether
3105	rhb.	....	88 d.	....	s.	s.	s.
3106	lq.	$0.801^{0°}$	....	189	i.	s.	s.
3107	lq.	$0.827\frac{20°}{4}$	12.5	227	i.	s.	s.
3108	col. lq.	$0.872\frac{20°}{4}$	−66.1	187-8	$0.029^{20°}$	∞	∞; ∞ chl.
3109	oil	$0.869\frac{20°}{4}$	<−15	$216-8^{15mm}$	i.	∞	∞
3110	pl. or pr./ al.	....	66-7	260-300 (sl. d.)	sl. s. h.	s.	s.
3111	nd./act.	....	97	....	sl. s. h.	....	....
3112	col. nd.	$0.858\frac{25°}{4}$	24-5	191^{10mm}	i.	s.	s.
3113	col. lq.	$0.866^{17.5°}$	−36.8(−44)	227.0	i.	∞	∞
3114	col. lq.	$1.018\frac{25°}{25}$	−45	$207-8^{756mm}$	v. sl. s.	∞; s. bz.	∞
3115	col. lq.	$1.001\frac{25°}{25}$	−4	214^{752mm}	v. sl. s.	∞	∞
3116	nd.	....	46-7	218.5-9.5	v. sl. s.	$1200^{25°}$	$710^{25°}$
3117	lq.	$1.048\frac{23°}{25}$	....	$156-62^{20mm}$	....	....	....
3118	col. lq.	$1.033\frac{20°}{4}$	....	227	i.	∞	∞
3119	pl./et.	....	42	270	....	....	....
3120	lq.	$0.923^{21°}$	....	201-3	i.	s.	s.
3121	oil	$1.415\frac{20°}{4}$	....	145^{15mm}	....	....	....
3122	col. lq.	$0.994\frac{23°}{0}$	....	219-20 sl. d.	i.	s.	s.
3123	col. lq.	$0.998\frac{13.8°}{4}$	....	....	i.	s.	s.
3124	lq.	$1.04\frac{20°}{20}$	37.2	268^{740mm}	$0.5^{20°}$	....	....
3125	lf.	....	57-8	273 sl. d.	v. sl. s. h.	v. s. h.	v. s.
3126	col. nd./ al.	....	199-200	....	0.06 c.; 0.9 h.	6; i. bz.; s. alk.	0.6; s. ac.
3127	oil	$1.018^{15°}$	....	237	....	....	....
3128	oil	$1.004\frac{15°}{15}$	....	*237-40	sl. s.	v. s.	v. s.
3129	pl.	$1.012\frac{20°}{20}$	21	218	i.	s.	s.
3130	yel. cr./al.	....	75-6	....	i.; s. a.	s. h.	s.; s. chl.
3131	oil	$1.063\frac{13°}{4}$	....	260-70 sl. d.	....	....	....
3132	lq.	$1.024\frac{15°}{4}$	....	205-6	....	....	....
3133	mn.	$1.010^{22°}$	42	>300	s. h.	s.	s.; s. a.
3134	nd./aq. al.	....	99	....	....	s.	....
3135	col. lq.	<1	....	25	....	....	....
3136	lq.	$0.833\frac{20°}{4}$	....	80-2	sl. s.	∞	....
3137	col. lq.	$0.968\frac{15°}{4}$	....	119^{745mm}	i.; v. s. chl.	v. s.	v. s.
3138	col. lq.	$0.896\frac{15°}{4}$	−73.9	99.1	$2.4^{20°}$	∞	∞
3139	lq.	$0.948\frac{25°}{4}$	....	223.6	....	....	....
3140	lq.	$0.960\frac{25°}{4}$	....	205 d.	v. sl. s.	∞	∞
3141	col. lq.	$0.848\frac{20°}{4}$	....	174.5^{748mm}	$0.07^{20°}$	s.	∞; ∞ bz.
3142	col. lq.	$0.739\frac{20°}{4}$	<−79	62-3	sl. s.	∞	∞
3143	col. lq.	$0.745^{0°}$	....	54	sl. s.	∞	∞
3144	col. lq.	$0.813\frac{21.8°}{4}$	....	123-4	v. sl. s.	∞	∞
3145	col. lq.	$0.814\frac{18°}{0}$	....	114^{745mm}	v. sl. s.	v. s.	∞
3146	lq.	$0.950^{0°}$	....	148.5^{753mm}	sl. s.	∞	v. s.

*In nitrogen.

Ethyl phenyl-benzyl-carbamate 5165	Ethyl-phenylnitrosoamine 4885	Ethyl-propyl carbinol (*n*) 3626
Ethyl pnenyl-carbamate 5254	Ethyl phosphate 6188	Ethyl-propyl carbinol (*iso*) 3632
Ethyl phenyl-cyanoacrylate 3031	Ethyl phosphite 6192	Ethyl propyl-malonate 2203-4
Ethyl-phenyl ether 5111	Ethyl phthalate 2199, 3067	
	Ethyl pimelate 2202	

Table 7-4 (*Continued*)
PHYSICAL CONSTANTS OF ORGANIC COMPOUNDS

No.	Name	Synonym	Formula	Beil. Ref.	Formula Weight
3147	**Ethyl** pyridine (3)	β-lutidine	$C_2H_5 \cdot C_5H_4N$	XX-242	107.16
3148	pyridine (4)	....	$C_2H_5 \cdot C_5H_4N$	XX-243	107.16
3149	pyrrole (*N*)	Et-pyrrolylamine	$C_4H_4N \cdot C_2H_5$	XX-163	95.15
3150	pyruvate	....	$CH_3CO \cdot CO_2 \cdot C_2H_5$	III-616	116.12
3151	salicylate (*o*)	Et hydroxybenzoic †	$HO \cdot C_6H_4 \cdot CO_2 \cdot C_2H_5$	X-73	166.18
3152	silicate, ortho	....	$Si(OC_2H_5)_4$	I-334	208.33
3153	stearate	....	$C_{17}H_{35}CO_2 \cdot C_2H_5$	II-379	312.54
3154	succinic acid	....	$HO_2C \cdot CH(C_2H_5) \cdot$ $CH_2 \cdot CO_2H$	II-660	146.14
3155	sulfone chloride	Et-sulfonyl chloride	$C_2H_5 \cdot SO_2 \cdot Cl$	IV-6	128.58
3156	thioacetate (*S*)	....	$CH_3CO \cdot S \cdot C_2H_5$	II-232	104.17
3157	thiocarbamate	dithiourethane	$NH_2 \cdot CS \cdot S \cdot C_2H_5$	III-218	121.22
3158	thiocyanate	....	$C_2H_5 \cdot S \cdot CN$	III-175	87.14
3159	*iso*-thiocyanate	ethyl mustard oil	$C_2H_5 \cdot N:C:S$	IV-123	87.14
3160	thioglycolate	....	$HS \cdot CH_2CO_2 \cdot C_2H_5$	III-255	120.17
3161	thioglycolic acid	(*S*)	$C_2H_5 \cdot S \cdot CH_2 \cdot CO_2H$	III-248	120.17
3162	thiourea	....	$C_2H_5 \cdot NH \cdot CS \cdot NH_2$	IV-117	104.17
3163	toluate (*o*)	....	$CH_3 \cdot C_6H_4 \cdot CO_2 \cdot C_2H_5$	IX-463	164.21
3164	toluate (*m*)	....	$CH_3 \cdot C_6H_4 \cdot CO_2 \cdot C_2H_5$	IX-476	164.21
3165	toluate (*p*)	....	$CH_3 \cdot C_6H_4 \cdot CO_2 \cdot C_2H_5$	IX-484	164.21
3166	toluene (*o*)	methyl-ethyl-benzene	$C_2H_5 \cdot C_6H_4 \cdot CH_3$	V-396	120.20
3167	toluene (*m*)	....	$C_2H_5 \cdot C_6H_4 \cdot CH_3$	V-396	120.20
3168	toluene (*p*)	....	$C_2H_5 \cdot C_6H_4 \cdot CH_3$	V-397	120.20
3169	*p*-toluene sulfonate	....	$CH_3 \cdot C_6H_4 \cdot SO_3 \cdot C_2H_5$	XI-99	200.26
3170	*o*-toluidine	....	$CH_3 \cdot C_6H_4 \cdot NHC_2H_5$	XII-786	135.21
3171	*m*-toluidine	....	$CH_3 \cdot C_6H_4 \cdot NHC_2H_5$	XII-857	135.21
3172	*p*-toluidine	....	$CH_3 \cdot C_6H_4 \cdot NHC_2H_5$	XII-904	135.21
3173	*o*-tolyl ether	Et *o*-cresyl ether	$CH_3 \cdot C_6H_4 \cdot O \cdot C_2H_5$	VI-352	136.20
3174	*m*-tolyl ether	....	$CH_3 \cdot C_6H_4 \cdot O \cdot C_2H_5$	VI-376	136.20
3175	*p*-tolyl ether	....	$CH_3 \cdot C_6H_4 \cdot O \cdot C_2H_5$	VI-393	136.20
3176	*p*-tolyl sulfide	Et *p*-cresyl sulfide	$CH_3 \cdot C_6H_4 \cdot S \cdot C_2H_5$	VI-417	152.26
3177	triazoacetate	....	$N_3CH_2 \cdot CO_2 \cdot C_2H_5$	II-229	129.12
3178	tribromoacetate	....	$Br_3C \cdot CO_2 \cdot C_2H_5$	II-221	324.81
3179	trichloroacetate	....	$Cl_3C \cdot CO_2 \cdot C_2H_5$	II-209	191.44
3180	tripropylbenzene	(*iso*)	$C_2H_5 \cdot C_6H_2(C_3H_7)_3$	....	232.41
3181	undecylate (*n*)	....	$C_{10}H_{21}CO_2 \cdot C_2H_5$	II-358	214.35
3182	*n*-undecyl ketone	tetradecanone-3	$C_2H_5 \cdot CO \cdot C_{11}H_{23}$	I-716	212.38
3183	urea	....	$C_2H_5 \cdot NH \cdot CO \cdot NH_2$	IV-115	88.11
3184	urethane	Et *N*-Et carbamate	$C_2H_5 \cdot NHCO_2C_2H_5$	IV-114	117.15
3185	valerate (*n*)	....	$CH_3(CH_2)_3CO_2C_2H_5$	II-301	130.19
3186	*iso*-valerate	....	$(CH_3)_2CH \cdot CH_2 \cdot$ $CO_2 \cdot C_2H_5$	II-312	130.19
3187	vanillate (4,3;1)	....	$HO(CH_3O)C_6H_3 \cdot$ $CO_2 \cdot C_2H_5$	X-397	196.20
3188	vanillin	3-EtO-4-OH-ben-zaldehyde; ethovan	$HO(C_2H_5O)C_6H_3 \cdot$ CHO	VIII-256	166.18
3189	vinyl carbinol	pentene-1-ol-3	$CH_2:CH \cdot CHOH \cdot C_2H_5$	I-443	86.13
3190	vinyl sulfide	....	$CH_2:CH \cdot S \cdot C_2H_5$	I-434	88.17
3191	violet base	hexaEt-4,4′,4″-tri-NH_2-triPh-carbinol	$[(C_2H_5)_2N \cdot C_6H_4]_3$ $C \cdot OH$	XIII-759	473.71

† See also No. 3072
Ethyl pyromucate 3050
Ethyl-pyrrolylamine 3149
Ethyl red 2144
Ethyl sebacate 2206

Ethyl selenomercaptan 3071
Ethyl suberate 2210
Ethyl succinate 2211-2
Ethyl sulfate 2213
Ethyl sulfinic acid 2905

Ethyl sulfonic acid 2906
Ethyl sulfonyl chloride 3155
Ethyl sulfuric acid 3069
Ethyl tartrate 2218, 3070
Ethyl thioncarbamate 6450

Table 7-4 (*Continued*)
PHYSICAL CONSTANTS OF ORGANIC COMPOUNDS

No.	Crystalline Form and Color	Specific Gravity	Melting Point °C.	Boiling Point °C.	Solubility in 100 Parts		
					Water	Alcohol	Ether
3147	lq.	$0.959^{9°}_{4}$	....	165.3	sl. s. c.; v. sl. s. h.	s.	s.
3148	lq.	$0.936^{20°}$	....	166	s. aq. a.	....	....
3149	lq.	$0.888^{16°}$	....	130-1	i. c.; s. a.	∞	∞
3150	col. lq.	$1.060^{16°}_{4}$	....	155	i.	∞	∞
3151	col. lq.	$1.136^{15°}_{4}$	1.3	233-4	i.	∞	∞
3152	lq.	$0.936^{20°}_{20}$	−82.5	168.6	sl. d.	∞	....
3153	col. cr.	$0.848^{36.3°}$	33.8(31.1)	199-201^{10mm}	i.	s.	s.
3154	col. pr.	....	98	....	v. s.	v. s.	v. s.
3155	lq.	$1.357^{22.5°}$	....	177.5	sl. d.	sl. d.	v. s.
3156	lq.	$0.976^{28°}_{4}$	....	116-7	i.	v. s.	v. s.
3157	lf./et.	....	41-2	d.	i.	v. s.	v. s.
3158	lq.	$0.996^{25°}_{4}$	−85.5	145^{765mm}	i.	∞	∞
3159	col. lq.	$1.004^{15°}_{4}$	−5.9	131-2	i.	∞	∞
3160	lq.	$1.096^{15°}$	....	156-8	....	....	....
3161	oil	$1.150^{20°}_{4}$	−8.7	123-4^{15mm}	∞	s.	s.
3162	nd.	....	114 (113)	....	v. s.	v. s.	....
3163	lq.	$1.032^{25°}_{25}$	<−10	227	i.	∞	∞
3164	lq.	$1.030^{20°}_{4}$	....	231^{750mm}	i.	∞	∞
3165	lq.	$1.024^{25°}_{25}$	....	235.5	i.	∞	∞
3166	lq.	$0.881^{20°}_{4}$	−80.8	165.2	i.	∞	∞
3167	lq.	$0.865^{20°}_{4}$	−95.6	161.3	i.	∞	∞
3168	lq.	$0.861^{20°}_{4}$	−62.4	162.0	i.	∞	∞
3169	mn. pr./al.	$1.166^{48°}_{4}$	33-4	221.3	i.	s.	s.
3170	lq.	$0.948^{25°}_{4}$	<−15	215-6	i.	....	....
3171	lq.	....	....	221-2	....	....	....
3172	lq.	$0.942^{25°}_{4}$	....	217	i.	....	....
3173	lq.	$0.959^{13.3°}_{4}$	....	184	....	....	....
3174	lq.	$0.956^{0°}_{0}$	....	192	....	....	....
3175	lq.	$0.966^{0°}_{0}$	....	188-9	....	....	....
3176	lq.	$1.002^{17.5°}$	....	220-1	....	....	....
3177	col. oil	$1.127^{20°}_{20}$	....	70^{20mm}	....	....	....
3178	col. lq.	$2.230^{20°}_{20}$	....	225	i.	∞	∞
3179	col. lq.	$1.383^{20°}_{4}$	....	167-8	i.	∞	∞
3180	col. cr.	....	106.9	260	i.	$1.5^{25°}$	$105^{25°}$
3181	lq.	....	....	140^{20mm}	i.	....	....
3182	cr./Me al.	....	34	152^{16mm}	....	....	....
3183	nd.	$1.213^{18°}$	92	....	v. s.	80	i.
3184	col. lq.	$0.981^{20°}_{4}$	....	174-6	$63^{15°}$	d. h. alk.	....
3185	col. lq.	$0.877^{20°}$	−91.2	145.5	$0.24^{25°}$	∞	∞
3186	col. lq.	$0.867^{20°}_{4}$	−99.3	135	$0.17^{20°}$	∞ ; ∞ bz.	∞
3187	nd.	....	44	291-3	i.; s. alk.	v. s.	v. s.
3188	col. pl./aq.	....	77-8	....	v. sl. s.	s.	s.
3189	lq.	$0.840^{19.5°}_{0}$	....	114.5-4.7	....	....	....
3190	col. lq.	$0.887^{14°}_{0}$	....	90.5-1.5	....	....	....
3191	cr./lg.	....	....	....	....	sl. d.	....

Ethyl thionurethane 6450
Ethyl thiourethane 5921
Ethyl-tolyl-benzylamine 2988
Ethyl vinyl ether 6438
Ethyl xylene 2483-6

Ethylal 2172
Ethylene acetate 3445
Ethylene benzoate 3447
Ethylene bromide 1837
Ethylene chloride 1994

Ethylene chlorobromide 1343
Ethylene diacetate 3445
Ethylene dibenzoate 3447
Ethylene dibutyrate 3450
Ethylene dicaprate 3451

Table 7-4 (*Continued*)
PHYSICAL CONSTANTS OF ORGANIC COMPOUNDS

No.	Name	Synonym	Formula	Beil. Ref.	Formula Weight
3192	**Ethylene**	ethene	$CH_2:CH_2$	I-180	28.05
3193	bromohydrin	glycol bromohydrin	$Br \cdot CH_2 \cdot CH_2OH$	I-338	124.97
3194	chlorohydrin	2-chloro-ethanol-1	$ClCH_2 \cdot CH_2OH$	I-337	80.51
3195	cyanohydrin	β-OH-propionitrile	$HO \cdot CH_2 \cdot CH_2 \cdot CN$	III-298	71.08
3196	diamine	....	$NH_2 \cdot CH_2 \cdot CH_2 \cdot NH_2$	IV-230	60.10
3197	diamine HBr	....	$(CH_2NH_2)_2 \cdot 2HBr$	IV-232	221.93
3198	diamine HCl	....	$(CH_2NH_2)_2 \cdot 2HCl$	IV-232	133.02
3199	diamine hydrate	....	$(CH_2NH_2)_2 \cdot H_2O$	IV-230	78.11
3200	diamine *iso-*valerate	*iso-*valeryl-ethylene-diamine	$C_{12}H_{28}O_4N_2$	....	264.37
3201	dicaprate	....	$(C_9H_{19}CO_2)_2C_2H_4$	....	370.58
3202	diphenyl ether	....	$(C_6H_5 \cdot O \cdot CH_2)_2$	VI-146	214.27
3203	diphenylsulfone	....	$(C_6H_5 \cdot SO_2 \cdot CH_2)_2$	VI-302	310.39
3204	diphenylthioether	....	$(C_6H_5 \cdot S \cdot CH_2)_2$	VI-301	246.40
3205	diphthalimide (*N,N'*)	....	$[C_6H_4(CO)_2:N \cdot CH_2]_2$	XXI-492	320.31
3206	disulfonic acid	....	$(HO_3S \cdot CH_2)_2$	IV-11	190.19
3207	ethylidene oxide	glycol-ethylidene-diacetal	$CH_3 \cdot CH:(OCH_2 \cdot)_2$	XIX-8	88.11
3208	iodohydrin	2-iodo-ethanol-1	$ICH_2 \cdot CH_2OH$	I-339	171.97
3209	mercaptan	ethandithiol-1,2	$HS \cdot CH_2 \cdot CH_2 \cdot SH$	I-471	94.20
3210	oxide	....	$CH_2 \cdot CH_2 \cdot O$	XVII-4	44.05
3211	thiocyanate	....	$(CH_2 \cdot SCN)_2$	III-178	144.22
3212	urea	....	$(CH_2NH)_2CO$	XXIV-2	86.09
3213	**Ethylidene** acetone	Me-propenyl ketone	$CH_3 \cdot CH:CH \cdot CO \cdot CH_3$	I-732	84.12
3214	diacetate	diacetoxyethane (1,1)	$CH_3 \cdot CH(O_2C \cdot CH_3)_2$	II-152	146.14
3215	diurethane	....	$CH_3 \cdot CH:(NH \cdot CO_2 \cdot C_2H_5)_2$	III-24	204.23
3216	urea	....	$CH_3 \cdot CH:(NHCONH)$	III-60	86.09
3217	**Eucalyptole**	cineole	$C_{10}H_{18}O$	XVII-24	154.25
3218	**Eugenol** (1,3,4)	allyl guaiacol	$C_6H_3(C_3H_5)(OCH_3) \cdot OH$	VI-961	164.21
3219	acetate	....	$CH_3 \cdot CO_2 \cdot C_{10}H_{11}O$	VI-965	206.24
3220	benzoate	benzoyl eugenol	$C_6H_5CO_2C_{10}H_{11}O$	IX-135	268.32
3221	cinnamate	....	$C_8H_7CO_2C_{10}H_{11}O$	IX-586	294.35
3222	methyl ether	....	$C_3H_5 \cdot C_6H_3:(OCH_3)_2$	VI-963	178.23
3223	*iso-***Eugenol** (1,3,4)	propenyl guaiacol	$C_6H_3(C_3H_5)(OCH_3) \cdot OH$	VI-955	164.21
3224	methyl ether	Me isoeugenol	$C_3H_5 \cdot C_6H_3:(OCH_3)_2$	VI-956	178.23
3225	**Eugetinic acid**	(5;6,3,1)	$CH_3O \cdot C_6H_2(OH) \cdot (C_3H_5)CO_2H$	X-441	208.22
3226	**Eupittonic acid**	....	$C_{19}H_8(OCH_3)_6O_3$	VIII-574	470.48
3227	**Euxanthone**	1,7-diOH-xanthone	$CO:(C_6H_3OH)_2O$	XVIII-113	228.21
3228	**Evans blue**	....	$C_{34}H_{24}O_{14}N_6S_4Na_4$	....	960.82
3229	**Evernic acid**	....	$C_{16}H_{13}O_6 \cdot OCH_3$	X-416	332.31
3230	**Everninic acid**	(6,4;2,1)	$HO(CH_3O)C_6H_2 \cdot (CH_3)CO_2H$	X-413	182.18
3231	**Evipan** (1,5;5)	*N*-Me-cyclohexenyl-methylbarbituric	$C_4HO_3N_2(CH_3)_2 \cdot C_6H_9$	....	236.27
3232	**Fenchene** (*dl*)	....	$C_{10}H_{16}$	V-163	136.24
3233	**Fenchone** (*d*)	....	$C_{10}H_{16}O$	VII-96	152.24

Table 7-4 (*Continued*)
PHYSICAL CONSTANTS OF ORGANIC COMPOUNDS

No.	Crystalline Form and Color	Specific Gravity	Melting Point °C	Boiling Point °C	Solubility in 100 Parts		
					Water	Alcohol	Ether
3192	col. gas	$0.975(A)$ $0.566^{-192°}_{4}$	-169.2	-103.7	$25.6°°$ cc.	360 cc.	s.
3193	col. lq.	$1.772^{20°}_{4}$	....	150 sl. d.	s.	s.	i. pet.
3194	col. lq.	$1.202^{20°}_{4}$	-67.5	$128.6*$	∞	∞	∞
3195	lq.	$1.059^{0°}$	-46.2	$220\text{-}275^{4mm}$	∞	∞	$2.3^{15°}$
3196	col. lq.	$0.900^{20°}_{20}$	8.5	117.2	∞	∞; i. bz.	0.3
3197	col. pr.	....	....	....	s.	i.	i.
3198	mn. pr.	....	subl.	....	s.	i.	i.
3199	col. lq.	$0.963^{21°}_{4}$	10	118	∞	....	....
3200	wh. pd.	....	129	....	s.	s.	....
3201	col. pl.	....	37-8	....	i.	s.	v. s.
3202	lf./abs. al.	....	97-8	....	i.	s. h.	s.; s. chl.
3203	nd./al.	....	179-80	....	v. sl. s. h.	s. h.; s. bz.	v. s. h. ac.
3204	nd.	....	70.0-0.5	....	i.	....	....
3205	nd./ac.	....	233-4	....	i.	s. h.	i.; s. bz.
3206	nd.	....	104	....	v. s.	v. s.	s.
3207	lq.	$0.987^{15°}_{4}$	....	82.5^{766mm}	66.6; i. aq. CaCl$_2$	∞	∞
3208	col. lq.	$2.197^{20°}_{4}$	....	85^{25mm}	s.	s.	....
3209	lq.	$1.12^{23.5°}$	....	146	v. s. alk.	v. s.	s. NH$_4$OH
3210	lq.	$0.887^{7°}_{4}$	-111.7	10.7	∞	∞	v. s.
3211	pl. or nd.	....	90	d.	sl. s.	s.	s.
3212	nd.	....	131	....	s.	s. h.	v. sl. s.
3213	col. lq.	$0.856^{20°}$	....	$122\text{-}4^{745mm}$	s.	....	....
3214	col. lq.	$1.061^{12°}$	18.85	168^{740mm}	sl. s.	∞	d. alk.
3215	nd.	....	125-6	$170\text{-}80^{20mm}$	s. h.	s.	s.
3216	nd.	....	154	d. 160	i.	sl. s.	i.
3217	col. oil	$0.927^{20°}$	1.5	176-7	$1.9^{15°}$	∞; s. oils	∞; s. ac.
3218	oil	$1.066^{20°}_{4}$	-9.2	254.8	v. sl. s.	∞; ∞ chl.	∞
3219	pl./al.	$1.087^{15°}_{15}$	29-30	$281\text{-}2^{752mm}$	i.	s.	s.
3220	col. cr.	....	70.5	360	i.; s. act.	s. h.	s.; s. chl.
3221	col. nd.	....	90-1	....	i.; s. act.	s. h.	s.; s. chl.
3222	lq.	$1.055^{15°}$	....	248-9	i.	25, 60% al.	∞
3223	oil	$1.091^{15°}_{15}$	16-9	267.5	v. sl. s.	∞	∞
3224	col. lq.	$1.055^{22°}$	....	263-4	i.	s.	s.
3225	pr./aq.	....	124	d.	v. sl. s. c.	s.; s. (NH$_4$)$_2$CO$_3$	s.
3226	or. nd./al.	....	200 d.	....	s. alk.	sl. s. h. abs.	s. ac.
3227	yel. nd.	....	240	subl. sl. d.	i.; s. alk.	s. h.	sl. s.
3228	b. pd.	....	....	....	s.	i.	i.
3229	pr./al.	....	168-9 d.	....	v. sl. s. h.	s. h.	v. sl. s.
3230	cr./aq.	....	170-1 d.	....	v. sl. s. h.; s. EtOAc	s. h.; s. dil. NaOH	v. sl. s.
3231	col. cr. pd.	....	143-5	....	sl. s.	s. h.	sl. s.
3232	lq.	$0.868^{19°}$	....	155-6	....	....	....
3233	oil	$0.948^{18°}$	5-6	193-5	i.	v. s.	v. s.

* Constant b. p. mixt., 42.5% aq. 95.8°.

Ethylene tribromide 6105	Ethylidene phenetidine 2913	Evipal 3231
Ethylene trichloride 6155	Ethyne 130	Evodiamine, cf. alkd.
Ethylidene chlorobromide 1342	Eubornyl 863	Ewer and Pick's acid 4462
Ethylidene dibromide 1836	Eudermol, cf. alkd.	Exalgin 4086
Ethylidene dichloride 1993	Eupyrin 6420	F acid 4510
Ethylidene diiodide 2390	Euresol 114	Fantan 5171
Ethylidene dimethyl ether 2415	Eurisol 114	Fast green-J base 5672
	Euxanthic acid, cf. glcde.	Fast red-D 3732

Table 7-4 (*Continued*)
PHYSICAL CONSTANTS OF ORGANIC COMPOUNDS

No.	Name	Synonym	Formula	Beil. Ref.	Formula Weight
3234	**Fenchyl** alcohol (*dl*)	....	$C_{10}H_{17}OH$	VI-71	154.25
3235	alcohol (*d*)(*α*)	fenchol	$C_{10}H_{17}OH$	VI-70	154.25
3236	**Ferulic acid** (3;4,1)	MeO–OH–cinnamic acid	$CH_3O \cdot C_6H_3(OH) \cdot$ $CH:CHCO_2H$	X-436	194.19
3237	*iso*-**Ferulic acid** (4;3,1)	hesperetinic acid	$CH_3O \cdot C_6H_3(OH) \cdot$ $CH:CHCO_2H$	X-437	194.19
3238	**Filicic acid**	filicin (1,1;2,4,6)	$(CH_3)_2C_6H_4(:O)_3$	VII-856	154.17
3239	**Filixic acid**	....	$C_{35}H_{40}O_{12}$	VIII-576	652.70
3240	**Flavianic** acid	2,4–diNO$_2$–1–naph-thol-7-sulfonic	$(NO_2)_2C_{10}H_4(OH) \cdot$ $SO_3H \cdot 3H_2O$	XI-275	368.28
3241	sodium salt	naphthol yellow S	$C_{10}H_4O_8N_2SNa_2 \cdot 3H_2O$	XI-275	412.24
3242	**Flavaniline** (*α*)	*p*-aminophenyl-lepidine	$NH_2 \cdot C_6H_4 \cdot C_9H_5N \cdot$ CH_3	XXII-469	234.30
3243	**Flavone**	3-phenyl-benzo-4-pyrone	$C_6H_4 \cdot OC(C_6H_5):$ $\quad\mid$ $\quad CH \cdot CO$ $\underline{\quad\quad}\mid$	XVII-373	222.25
3244	**Fluoran**	....	$C_{20}H_{12}O_3$	XIX-146	300.32
3245	**Fluoranthene**	idryl	$C_{16}H_{10}$	V-685	202.26
3246	**Fluorene**	....	$C_6H_4 \cdot CH_2 \cdot C_6H_4$ $\mid\quad\quad\quad\mid$	V-625	166.22
3247	**Fluorenol**	diphenylene carbinol	$C_6H_4 \cdot C_6H_4 \cdot CHOH$ $\mid\quad\quad\quad\mid$	VI-691	182.22
3248	**Fluorenone**	diphenylene ketone	$C_6H_4 \cdot CO \cdot C_6H_4$ $\mid\quad\quad\quad\mid$	VII-465	180.21
3249	oxime	....	$C_{12}H_8:C:NOH$	VII-467	195.22
3250	**Fluorescein**	resorcinol phthalein	$C_{20}H_{12}O_5$	XIX-222	332.32
3251	sodium salt	uranin	$C_{20}H_{10}O_5Na_2$	XIX-225	376.28
3252	chloride	3,6–diCl–fluoran	$C_{20}H_{10}O_3Cl_2$	XIX-147	369.21
3253	diacetate	3,6–diAcO–fluoran	$C_{24}H_{16}O_7$	XIX-227	416.39
3254	**Fluorescin**	resorcinol phthalin	$C_{20}H_{14}O_5$	XVIII-358	334.33
3255	**Fluoro**-acetic acid	....	$FCH_2 \cdot CO_2H$	II-193	78.04
3256	aniline (*p*)	....	$F \cdot C_6H_4 \cdot NH_2$	XII-597	111.12
3257	anisole (*p*)	....	$F \cdot C_6H_4 \cdot OCH_3$	*VI-98	126.13
3258	benzene	phenyl fluoride	$C_6H_5 \cdot F$	V-198	96.11
3259	benzoic acid (*o*)	....	$F \cdot C_6H_4 \cdot CO_2H$	IX-333	140.12
3260	benzoic acid (*m*)	....	$F \cdot C_6H_4 \cdot CO_2H$	IX-333	140.12
3261	benzoic acid (*p*)	....	$F \cdot C_6H_4 \cdot CO_2H$	IX-333	140.12
3262	bromobenzene (*p*)	....	$F \cdot C_6H_4 \cdot Br$	V-209	175.01
3263	chlorobenzene (*o*)	....	$F \cdot C_6H_4 \cdot Cl$	*V-110	130.55
3264	chlorobenzene (*m*)	....	$F \cdot C_6H_4 \cdot Cl$	....	130.55
3265	chlorobenzene (*p*)	....	$F \cdot C_6H_4 \cdot Cl$	V-201	130.55
3266	chlorobromo-methane	....	$CHClBrF$	I-67	147.38
3267	chloromethane	....	CH_2ClF	I-60	68.48
3268	dibromoethane	(2;1,1)	$FCH_2 \cdot CHBr_2$	I-92	205.86
3269	dibromoethane	(2;1,2)	$FCHBr \cdot CH_2Br$	I-92	205.86
3270	1,1-dichloro-1,2-di-bromoethane (2)	....	$Cl_2CBr \cdot CHFBr$	I-93	274.75
3271	dichloromethane	Freon-21	$FCHCl_2$	I-61	102.92
3272	iodobenzene (*p*)	....	$F \cdot C_6H_4 \cdot I$	V-220	222.00
3273	naphthalene (*α*)	naphthyl fluoride	$C_{10}H_7F$	V-540	146.17
3274	nitrobenzene (*p*)	....	$NO_2 \cdot C_6H_4 \cdot F$	V-241	141.10

Table 7-4 (*Continued*)
PHYSICAL CONSTANTS OF ORGANIC COMPOUNDS

No.	Crystalline Form and Color	Specific Gravity	Melting Point °C.	Boiling Point °C.	Solubility in 100 Parts		
					Water	Alcohol	Ether
3234	col. cr.	$0.935^{40°}$	33-5*	201	sl. s.	....	....
3235	col. pr.	$0.964^{20°}_4$	42	201-2	sl. s.	s.	s.; s. pet.
3236	pr./aq.	....	169-70	d.	s. h.	s.; sl. s. bz.	sl. s.
3237	nd. or pr.	....	225-8	....	sl. s. h.	s.	s.; i. lg.
3238	col. cr./al.	....	213-5 d.	subl. sl. d.	1.4 h.	10 h.	sl. s.
3239	cr.	....	184 d.	....	i.; s. chl.	i. abs.	sl. s.; s. bz.
3240	pa. yel. nd./HCl	....	100; 140-50(anh.)	d. >175	v. s.	v. s.	....
3241	yel. pd.	....	....	....	s.	....	....
3242	yel. nd./al.	....	83-4	....	i.	s.	s.; s. bz.
3243	col./lg.	....	97	....	i.	s.	s.
3244	nd.	....	184	....	s. H_2SO_4	s. HNO_3	....
3245	nd./al.	$1.252^{0°}_4$	109-10	250-160mm	i.	v. s. h.	v. s.
3246	col. cr./al.	$1.203^{0°}_4$	115-6	293-5	i.; i.NH_3; s. SO_2	s. h.	s.
3247	nd./aq.	....	153	....	....	s.	s.
3248	yel. rhb.	....	83-4	341.5	i.	v. s.	v. s.
3249	nd.	....	192-3	....	s. chl.	i. pet.	....
3250	yel. red pd.	....	d. >290	....	v. sl. s. h.	s. h.; s. alk.	s. h. ac.
3251	or. red pd.	....	....	....	s.	sl. s.	....
3252	col. cr.	....	258-60	....	i.; s. chl.	v. sl. s.	s. h. bz.
3253	col. cr./et. bz.	....	201-5	....	i. c. alk.	v. sl. s.	s. ac.
3254	col. nd./ac.	....	125-7**	....	i.; s. alk.	s.	s.
3255	cr.	....	33	165	....	....	....
3256	oil	$1.152^{25°}_4$	-0.8	187.4	v. sl. s.	....	....
3257	col. lq.	....	-43.5	156-7	....	∞	∞
3258	col. lq.	$1.024^{20°}_4$	-41.9	84.8	i.	∞	∞
3259	nd./aq.	....	120-2	....	s. h.	s.	s.
3260	lf./aq.	....	124	....	sl. s. h.	....	....
3261	mn./aq.	....	184-6	....	sl. s. h.	s.	s.
3262	col. lq.	$1.597^{20°}_4$	-8	152^{755mm}	i.	v. s.	v. s.
3263	lq.	....	-42.5	138^{774mm}	....	....	....
3264	col. lq.	....	....	123-5	....	....	....
3265	col. lq.	$1.226^{20.5°}_4$	-26.9	130^{756mm}	....	....	....
3266	col. lq.	$1.906^{16°}$	<-65	38	....	....	....
3267	gas	....	....	....	d.	....	....
3268	lq.	....	....	117.5	....	....	....
3269	col. lq.	$2.257^{17°}$	-54	122.5^{761mm}	....	....	....
3270	col. lq.	$2.130^{23°}$	....	163.5	....	....	....
3271	gas	$1.426^{0°}$	-135	8.9	i.	s.	s.
3272	col. lq.	$1.925^{15°}$	-18 (-27)	183.2^{760mm}	i.	s.	s.
3273	lq.	$1.133^{19.5°}_4$	-8	212-6	s. bz.	s.; s. ac.	s. chl.
3274	pa. yel. cr.	$1.320^{0°}_4$	27(21.6)	206.7	....	....	....

* Rapid heating, m. p. 37-8°.
** Solvent free crysts. m. p. 253-4°.
Fluoro-ethane 3046

Fluoro-ethylene 6439
Fluoro-methane 4240
Fluoro-octane 4978

Table 7-4 (*Continued*)
PHYSICAL CONSTANTS OF ORGANIC COMPOUNDS

No.	Name	Synonym	Formula	Beil. Ref.	Formula Weight
	Fluoro				
3275	pentabromoethane	....	$Br_3C \cdot CBr_2F$	I-95	442.57
3276	pentachloroethane	....	$Cl_3C \cdot CCl_2F$	....	220.29
3277	tetrabromoethane	....	$Br_3C \cdot CHBrF$	I-95	363.66
3278	tetrachloroethane	....	$Cl_3C \cdot CHClF$	I-86	185.84
3279	toluene (*o*)	....	$CH_3 \cdot C_6H_4 \cdot F$	V-290	110.13
3280	toluene (*m*)	....	$CH_3 \cdot C_6H_4 \cdot F$	V-290	110.13
3281	toluene (*p*)	....	$CH_3 \cdot C_6H_4 \cdot F$	V-290	110.13
3282	tribromoethane	....	$BrCH_2 \cdot CFBr_2$	I-93	284.76
3283	tribromoethane	....	$Br_2CH \cdot CHBrF$	I-93	284.76
3284	trichloroethane	....	$Cl_2CH \cdot CHClF$	I-85	151.40
3285	trichloroethylene	....	$FClC:CCl_2$	....	149.38
3286	trichloromethane	Freon-11	Cl_3CF	I-64	137.37
3287	**Fluoroform**	trifluoromethane	CHF_3	I-59	70.01
3288	**Formaldehyde**	methanal	$H \cdot CH:O$	I-558	30.03
3289	**Formaldehyde** (meta)	*α*-trioxymethylene	$(CH_2O)_3$	XIX-381	90.08
3290	**Formaldehyde** (para)	....	$(CH_2O)_x \cdot xH_2O$	I-566	(30.03)
3291	acetamide	formicin †	$CH_3CONH \cdot CH_2OH$	II-178	89.09
3292	bisulfite Na	....	$CH_2OH \cdot SO_3Na \cdot H_2O$	I-578	152.10
3293	hydrosulfite	hydrosulfite N. F.	$CH_2O \cdot Na_2S_2O_4 \cdot H_2O$	I-578	222.15
3294	oxime	formaldoxime	$H_2C:NOH$	I-590	45.04
3295	sulfoxalate Na	formosul	$CH_2OH \cdot SO_2Na \cdot 2H_2O$	I-577	154.12
3296	**Formamide**	....	$H \cdot CO \cdot NH_2$	II-26	45.04
3297	**Formamidine**	salts only known	$HN:CH \cdot NH_2$	II-90	44.06
3298	**Formamidoxime**	isuretin	$H_2N \cdot CH:NOH$	II-91	60.06
3299	**Formanilide**	formylaniline	$C_6H_5 \cdot NH \cdot CHO$	XII-230	121.14
3300	sodium salt	....	$C_6H_5 \cdot N(Na) \cdot CHO$	XII-233	143.13
3301	**Formic acid**	methanoic acid	$H \cdot CO_2H$	II-8	46.03
3302	sodium salt	sodium formate	$H \cdot CO_2Na$	II-14	68.00
3303	thallium salt	thallous formate	$H \cdot CO_2Tl$	II-16	249.39
3304	**Formyl** acetone	....	$HO \cdot CH:CH \cdot CO \cdot CH_3$	I-767	86.09
3305	amino-1,3-di-methylbenzene (2)	form-*m*-xylidide (*vis*)	$(CH_3)_2C_6H_3NH \cdot OCH$	XII-1109	149.19
3306	diphenylamine	*N*-Ph-formanilide	$H \cdot CO \cdot N(C_6H_5)_2$	XII-235	197.24
3307	hydrazine	formhydrazide	$H_2N \cdot NH \cdot CHO$	II-93	60.06
3308	phenylhydrazine (*β*)	....	$C_6H_5NH \cdot NH \cdot CHO$	XV-233	136.15
3309	thioaldehyde	....	$CH_2 \cdot (S \cdot CH_2)_2 \cdot S$	XIX-382	138.27
3310	**Fraxin** ‡	....	$C_{16}H_{18}O_{10}$ + aq.	XXXI-249	370.32
3311	**Fructosamine** (*d*)	*iso*-glucosamine	$C_6H_{11}O_5(NH_2)$	IV-332	179.17
3312	**Fucose**	....	$C_5H_{11}O_4 \cdot CHO$	I-876	164.16
3313	**Fulminuric acid**	*iso*-cyanuric acid	$NC \cdot CH(NO_2) \cdot CONH_2$	II-598	129.08
3314	**Fumaric** acid (*trans*)	butendioic acid	$(:CH \cdot CO_2H)_2$	II-737	116.07
3315	amide	....	$(:CH \cdot CO \cdot NH_2)_2$	II-743	114.10
3316	chloride	fumaryl chloride	$(:CH \cdot CO \cdot Cl)_2$	II-743	152.97
3317	**Furfural** (3)	....	$C_4H_3O \cdot CHO$	....	96.09
3318	**Furfural** (2)	furfurol	$C_4H_3O \cdot CHO$	XVII-272	96.09

† Formicin is a commercial syrupy product.   ‡ See also Glucoside table.

Fluoro-propane 5406-7	Formaldoxime 3294	Formosul 3295
Folliculin 2903	Formamine 3599	Formyl-aniline 3299
Formaldehyde diacetate 4430	Formhydrazide 3307	Formyl-diethylamine 2173
Formaldehyde diethylacetal 2172	Formicin 3291	Formyl-salicylic acid 168
Formaldehyde dimethylacetal 4424	Formin 3599	Formyl-xylidine 3305
Formaldehyde dipropylacetal 5458	Formonitrile 3714	Form-xylidide 3305
Formaldehyde trimethyleneacetal 6285	Formopan 3295	Forsling's acid-I 4551

Table 7-4 (*Continued*)
PHYSICAL CONSTANTS OF ORGANIC COMPOUNDS

No.	Crystalline Form and Color	Specific Gravity	Melting Point °C.	Boiling Point °C.	Solubility in 100 Parts		
					Water	Alcohol	Ether
3275	col. cr.	....	176 d.	subl. 120	s. bz., chl.	sl. s. c.	s.
3276	col. cr.	$1.74^{25°}$	100	136.8	i.	....	....
3277	lq.	$1.939^{16°}$	....	103.5^{523mm}	....	....	....
3278	col. lq.	$1.631^{16.7°}$	....	116.5	....	....	....
3279	col. lq.	$1.004^{13.2°}$	−80±	113-4	i.	s.	s.
3280	col. lq.	$0.997^{13.4°}$	−110.8	115-6	i.	s.	s.
3281	col. lq.	$1.001^{16°}_{4}$	....	116-7	i.	s.	s.
3282	col. lq.	$2.605^{17.5°}$	....	162.7^{757mm}	....	....	....
3283	col. lq.	$2.674^{18°}$	....	178	....	....	....
3284	col. lq.	$1.539^{20°}_{4}$	....	103	i.	∞	∞
3285	col. lq.	$1.530^{25°}$	−82	71.0	d.		
3286	col. lq.	$1.494^{17.2°}$	....	24.9	i.	∞	∞
3287	gas	lq. $1.47^{-84°}$	....	$20^{40atm.}$	75 cc.	500 cc.	sl. s. chl.
3288	gas	$0.815^{-20°}$	−92±	−21	v. s.	v. s.	v. s.
3289	wh. solid	$1.176^{5°}$	64; subl. 46	114.5^{759mm}	$21^{25°}$; ∞ h.	s.	s.; sl. s. pet.
3290	wh. amor.	....	150-60	subl. 120±	$20-30^{18°}$	i.	i.
3291	hyg. cr.	1.2	50-2±	d.	v. s.	v. s.	i.
3292	nd./aq.	....	....	....	s.; d. alk.	sl. s.; d. a.	s. Me al.
3293	nd.	....	....	....	s.	....	....
3294	col. lq.	....	....	84	10-20	....	....
3295	hyg. nd./aq.	....	63-4	d. >125	50-60	i. abs.	i.; i. bz.
3296	hyg. lq.	$1.133^{20°}_{4}$	2.5	$109^{15 mm}$	∞	∞	v. sl. s.
3297	....	....	....	....	....	....	....
3298	rhb.	....	104-5	....	v. s.; i. bz.	v. sl. s.	v. sl. s.
3299	mn.	$1.147^{15°}_{15}$	47	216^{120mm}	sl. s.	v. s.	s.
3300	....	....	....	....	d.	v. sl. s.	....
3301	col. lq.	$1.220^{20°}_{4}$	8.40	100.8	∞	∞	....
3302	mn.	1.919	255	d.	70	sl. s.	i.; s. gly.
3303	nd./al.	$4.967^{104°}_{4}$	104(101)	....	$500^{10°}$	sl. s.	v.sl.s.chl.
3304	exists only as salts	....	....	....	....	....	....
3305	nd./ai.	....	164-5*	....	....	s.	....
3306	rhb./al.	$1.230^{20°}_{4}$	73-4	$189-90^{13mm}$	i.	s.	s.; s. bz.
3307	pl. or nd.	....	54	....	v. s. chl.	v. s.; s. bz.	v. s.
3308	lf./al.	....	145-6	....	s. h.	s.	sl. s. bz.
3309	pr./chl.	....	247(218)	subl. 150	sl. s. h.	sl. s.	sl. s.; s. bz.
3310	nd./al.	....	$-H_2O$, 110	d. 201-5	s. h.	s. h.	i.
3311	syrup	....	....	....	....	s.	i.
3312	nd.	....	145	....	v. s.	v. sl. s.	....
3313	pr./ai.	....	145 d.	....	v. s.; i. bz.	v. s.	v. sl. s.
3314	col. pr.	$1.635^{20°}_{4}$	286-7**; subl. 7^{200}	290	$0.71^{7°}$; $9.8^{100°}$	$5.75^{29.7°}$	$0.72^{5°}$
3315	pr./aq.	....	d. 265-70	....	s. h.	sl. s. h.	i. ac.; i. chl.
3316	col. lq.	$1.415^{20°}_{20}$	....	161-4	d.	d.	....
3317	lq.	$1.111^{20°}_{20}$	....	144^{732mm}	....	....	....
3318	lq.	$1.159^{20°}_{20}$	−38.7	161.7^{760mm}	$9.1^{13°}$	∞	∞

* When heated rapidly, m. p. 176-7°.   ** In sealed tube.

Forsling's acid-II 4550	Fructose 3616	Fumarine, cf. alkd.
Frangulin, cf. glcde.	Fructosone 3368	Fumaryl chloride 3316
Frangulinic acid 2874	Fruit sugar 3616	Furacrolein 3347
Franguloside, cf. glcde.	Fuchsine 5571	Furan 3327
Fraxin, cf. glcde.	Fuchsine (para) 5039	Furan carboxylic acid **3339-40**
Fraxinin, cf. glcde.	Fuchsine carbinol base 5570	Furan tetrahydride 5773
Freon 2246, 3271, 3286, 6206	Fuchsone 2741	Furfuracrolein 3347
		Furfuraldehyde 3318

Table 7-4 (Continued)
PHYSICAL CONSTANTS OF ORGANIC COMPOUNDS

No.	Name	Synonym	Formula	Beil. Ref.	Formula Weight
3319	**Furfural** acetone	....	$C_4H_3O \cdot CH:CH \cdot CO \cdot CH_3$	XVII-306	136.15
3320	acetophenone (ω)	....	$C_4H_3O \cdot CH:CH \cdot CO \cdot C_6H_5$	XVII-353	198.22
3321	diacetate	....	$C_4H_3O \cdot CH(O_2C \cdot CH_3)_2$	XVII-278	198.18
3322	oxime (α)(*syn*)	furfuraldoxime	$C_4H_3O \cdot CH:NOH$	XVII-281	111.10
3323	oxime (β)(*anti*)	....	$C_4H_3O \cdot CH:NOH$	XVII-281	111.10
3324	phenylhydrazone	....	$C_4H_3O \cdot CH:N \cdot NH \cdot C_6H_5$	XVII-282	186.22
3325	2-thiohydantoin (5)	....	$CO \cdot NH \cdot CS \cdot NH \cdot C:$ $\overline{}$ $CH \cdot C_4H_3O$	....	194.21
3326	**Furfuramide**	....	$(C_5H_4O)_3N_2$	XVII-281	268.27
3327	**Furfuran**	furan	$CH:CH \cdot CH:CH \cdot O$ $\overline{}$	XVII-27	68.08
3328	**Furil** (α,α)	difurfuroyl	$(C_4H_3O \cdot CO)_2$	XIX-166	190.16
3329	dioxime (*syn*)(α)	....	$(C_4H_3O \cdot C:NOH)_2 \cdot H_2O$	XIX-166	238.20
3330	dioxime (β)	....	$(C_4H_3O \cdot C:NOH)_2$	XIX-166	220.19
3331	**Furfuryl** acetate	....	$C_4H_3O \cdot CH_2 \cdot O_2C \cdot CH_3$	XVII-112	140.14
3332	alcohol	α-furyl carbinol	$C_4H_3O \cdot CH_2OH$	XVII-112	98.10
3333	amine	....	$C_4H_3O \cdot CH_2NH_2$	XVIII-584	97.12
3334	butyrate (α)	....	$C_3H_7CO_2 \cdot CH_2 \cdot C_4H_3O$	....	168.19
3335	chloride (α)	....	$C_4H_3O \cdot CH_2Cl$	....	116.55
3336	furoate (α)	furfuryl pyromucate	$C_4H_3O \cdot CO_2 \cdot CH_2 \cdot C_4H_3O$	....	192.17
3337	mercaptan (α)	....	$C_4H_3O \cdot CH_2 \cdot SH$	....	114.17
3338	propionate (α)	....	$C_2H_5CO_2 \cdot CH_2 \cdot C_4H_3O$	....	154.17
3339	**Furoic** acid (β)(3)	....	$C_4H_3O \cdot CO_2H$	*XVIII-439	112.09
3340	acid (α)(2)	pyromucic acid	$C_4H_3O \cdot CO_2H$	XVIII-272	112.09
3341	amide (α)	pyromucamide	$C_4H_3O \cdot CO \cdot NH_2$	XVIII-276	111.10
3342	amide (β)	....	$C_4H_3O \cdot CO \cdot NH_2$	....	111.10
3343	anhydride (α)	pyromucic anhydride	$(C_4H_3O \cdot CO)_2O$	XVIII-276	206.16
3344	chloride (α)	furoyl chloride	$C_4H_3O \cdot COCl$	XVIII-276	130.53
3345	nitrile (α)	2-cyanofuran	$C_4H_3O \cdot CN$	XVIII-278	93.09
3346	**Furoin** (α,α)	....	$C_4H_3O \cdot CHOH \cdot CO \cdot C_4H_3O$	XIX-204	192.17
3347	**Furyl**-acrolein	fur (fur)acrolein	$C_4H_3O \cdot CH:CH \cdot CHO$	XVII-305	122.12
3348	acrolein oxime	....	$C_4H_3O \cdot (CH)_3NOH$	....	137.14
3349	acrylamide	....	$C_4H_3O \cdot CH:CH \cdot CO \cdot NH_2$	XVIII-300	137.14
3350	acrylic acid	(stable form)	$C_4H_3O \cdot CH:CH \cdot CO_2H$	XVIII-300	138.12
3351	acrylic acid	(labile form)	$C_6H_5O \cdot CO_2H$	XVIII-301	138.12
3352	**Gallamide** (3,4,5;1)	gallamic acid	$(HO)_3C_6H_2 \cdot CONH_2 \cdot 1\frac{1}{2}H_2O$	X-487	196.16
3353	**Gallocyanine**	solid violet	$C_{15}H_{12}O_5N_2$	XXVII-438	300.27
3354	**Geranyl** acetate	geraniol acetate	$CH_3CO_2 \cdot C_{10}H_{17}$	II-140	196.29
3355	*n*-butyrate	....	$C_3H_7 \cdot CO_2 \cdot C_{10}H_{17}$	II-272	224.35
3356	formate	....	$H \cdot CO_2 \cdot C_{10}H_{17}$	II-23	182.26
3357	**Germanium tetra ethyl**	tetraEt germanium	$(C_2H_5)_4Ge$	IV-631	188.84

Table 7-4 (*Continued*)
PHYSICAL CONSTANTS OF ORGANIC COMPOUNDS

No.	Crystalline Form and Color	Specific Gravity	Melting Point °C.	Boiling Point °C.	Solubility in 100 Parts		
					Water	Alcohol	Ether
3319	col. nd.	....	39-40	135-7^{33mm}	i.	s.	s.; s. chl.
3320	oil	1.114$^{20°}$	....	317	....	....	....
3321	col. cr./pet.	....	52-3	220	i.; s. bz.	s.	v. s.
3322	nd./lg.	....	90-1	201-8 sl. d.	v. sl. s. c.	s.; s. bz.	s.; s. ac.
3323	nd./lg.	....	74-5	....	sl. s.	v. s.; s. bz.	v. s.
3324	yel. lf./al.	....	97-8	....	i.; i. lg.	s.	s.
3325	gn. cr.	....	250-2	....	i.	sl. s.	i.
3326	nd./al.	....	117-21	250 d.	i.; d. a.	s.	s.
3327	col. lq.	0.937$^{20°}_4$	....	31-2^{756mm}	i.	s.	s.
3328	yel. nd./bz.	....	165-6	....	i.; s. chl.	sl. s.	sl. s.
3329	col. nd./aq.	....	166-8	....	v. sl. s. bz.	v. s.	v. s.
3330	cr./lg.-et.	....	188-90 d.	....	....	....	v. sl. s.
3331	col. oil	1.118$^{20°}_4$	....	175-7	i.	s.	s.
3332	oil	1.129$^{25°}_4$	....	169.5^{752mm}	∞; slow d.	s.	s.
3333	lq.	<1	....	145^{754mm}	∞	s.	s.
3334	col. lq.	1.053$^{20°}_4$	....	212-3	v. sl. s.	s.	∞
3335	col. lq.	1.178$^{20°}_4$	....	49^{26mm}	i.	s.	s.
3336	col. lq.	α1.33 β1.40	α19.5 β27.5	>350^{760} d.	i.; i. pet.	∞; ∞ bz.	∞; ∞ chl.
3337	col. oil	1.132$^{20°}_4$	....	84^{65mm}	i.	....	....
3338	col. lq.	1.109$^{20°}_4$	....	195-6	v. sl. s.	s.	∞
3339	col. nd./aq.	....	121-2	105-10^{12mm}	s. h.	s.	v. s.
3340	mn. pr.	....	133-4 subl.>100	230-2 d. >250	3.6$^{15°}$; 25 h.	s.	s.
3341	cr.	....	142-3	subl. >100	....	....	....
3342	cr.	....	169	....	....	....	....
3343	nd./al.	....	73	325 d.	....	s.	s.
3344	col. lq.	....	−2 to 0	173-6	sl. d. h.	sl. d. h.	s.
3345	col. lq.	....	....	146-8	v. sl. s.	∞	∞
3346	pa. brn. nd./Me al.	....	138-9	d.	v. sl. s. h.	sl. s. h.; s. h. toluene	sl. s.
3347	yel. nd./lg.	....	54	>200 d.	i.	s.	s.
3348	yel. nd.	....	103-5	....	i.	s.	s.
3349	sc./aq.	....	168-9	....	v. sl. s. c.	....	....
3350	nd./aq.	....	141	286 d.; in vac. subl.	0.2 c.; s. h.	s.; v. s. ac.; 1.5$^{19°}$ bz.	v. s.; i. lg.; i. CS$_2$
3351	pr. or pl.	....	103-4	....	s. h.	7.5$^{19°}$ bz.	....
3352	lf./aq.	....	....	244 d. (anh.)	s. h.	s.	....
3353	gn. cr.	....	....	....	v. sl. s. h.	s.; s. act.	s. alk.
3354	col. lq.	0.917$^{15°}$	....	242-5^{764} d.	v. sl. s.	s.	∞
3355	col. lq.	0.901$^{17°}_4$	....	151-3^{18mm}	i.	s.	s.
3356	col. lq.	0.927$^{20°}_4$	....	113-4^{15mm}	i.	s.	s.
3357	col. lq.	0.991$^{24.5°}_{24.5}$	−90	163.5	i.	s.	s.

Table 7-4 (*Continued*)
PHYSICAL CONSTANTS OF ORGANIC COMPOUNDS

No.	Name	Synonym	Formula	Beil. Ref.	Formula Weight
3358	**Gluco-α-heptose** (*d*)	....	$C_7H_{14}O_7$	I-934	210.19
3359	**Gluconic** acid	....	$CH_2OH(CHOH)_4 \cdot CO_2H$	III-542	196.16
3360	Ca salt	Ca gluconate	$C_{12}H_{22}O_{14}Ca \cdot H_2O$	III-544	448.40
3361	δ-lactone	glucono-δ-lactone	$C_6H_{10}O_6$	*XVIII-405	178.14
3362	**Glucose** (α) penta-acetate (*d*)	penta-acetyl glucose	$C_6H_7O_6(COCH_3)_5$	XXXI-119	390.35
3363	(β)pentacetate (*d*)	....	$C_6H_7O_6(COCH_3)_5$	XXXI-120	390.35
3364	pentabutyrate	....	$C_6H_7O_6(COC_3H_7)_5$	....	530.62
3365	pentapropionate	....	$C_6H_7O_6(COC_2H_5)_5$	....	460.48
3366	phenylhydrazone	(*d*)(α)	$C_6H_5NH \cdot N:C_6H_{12}O_5$	XV-221	270.29
3367	phenylhydrazone (β)	....	$C_6H_5NH \cdot N:C_6H_{12}O_5$	XV-221	270.29
3368	**Glucosone** (*d*)	fructosone	$C_4H_9O_4 \cdot CO \cdot CHO$	I-932	178.14
3369	**Glucosoxime** (*d*)	....	$C_6H_{12}O_5:NOH$	I-902	195.17
3370	**Glutaconic acid** (trans)	pentendioic acid	$HO_2C \cdot CH_2 \cdot CH: CH \cdot CO_2H$	II-758	130.10
3371	**Glutamic** acid (*dl*)	glutaminic acid	$HO_2C \cdot CHNH_2 \cdot (CH_2)_2 \cdot CO_2H$	IV-493	147.13
3372	acid (*l*)(+)	α-NH₂-glutaric	$C_5H_9O_4N$	IV-488	147.13
3373	acid HCl (*d*)	acidulin	$C_5H_9O_4N \cdot HCl$	IV-491	183.59
3374	amide (*dl*)	glutamine	$HO_2C \cdot C_3H_5NH_2 \cdot CO \cdot NH_2$	IV-491	146.15
3375	amide (*l*)(+)	*l*-glutamine	$C_5H_{10}O_3N_2$	IV-491	146.15
3376	**Glutaric** acid	pentandioic acid	$CH_2(CH_2 \cdot CO_2H)_2$	II-631	132.12
3377	nitrile	....	$CH_2(CH_2 \cdot CN)_2$	II-635	94.12
3378	**Glutathione**	....	$C_{10}H_{17}O_6N_3S$	**IV-931	307.33
3379	**Glyceric** acid (*dl*)	dihydroxy-propionic	$HOCH_2CHOH \cdot CO_2H$	III-392	106.08
3380	aldehyde (*dl*)	....	$HOCH_2CHOH \cdot CHO$	I-845	90.08
3381	**Glycerol**	glycerin	$CHOH(CH_2OH)_2$	I-502	92.10
3382	acetate, mono	monacetin †	$C_5H_{10}O_4$	II-146	134.13
3383	acetate, di	diacetin	$HO \cdot C_3H_5(O_2C \cdot CH_3)_2$	II-147	176.17
3384	amine, di (αγ)	1,3-diNH₂-2-pro-panol	$CHOH(CH_2NH_2)_2$	IV-290	90.13
3385	amine, mono (α)	1-NH₂-2,3-pro-pandiol	$CH_2OH \cdot CHOH \cdot CH_2 \cdot NH_2$	IV-301	91.11
3386	*iso*-amyl ether	(di)(αγ)	$(C_5H_{11}OCH_2)_2:CHOH$	I-513	232.37
3387	*iso*-amyl ether	(mono)(α)	$CH_2OH \cdot CHOH \cdot CH_2 \cdot O \cdot C_5H_{11}$	I-513	162.23
3388	bromohydrin, di (α)	1,3-dibromo-2-pro-panol	$(CH_2Br)_2:CHOH$	I-365	217.90
3389	bromohydrin, di (β)	2,3-dibromo-1-pro-panol	$Br \cdot CH_2 \cdot CHBr \cdot CH_2OH$	I-357	217.90
3390	*n*-butyl ether	(di)(αγ)	$(C_4H_9OCH_2)_2:CHOH$	....	204.31
3391	*n*-butyl ether	(mono)(α)	$CH_2OH \cdot CHOH \cdot CH_2 \cdot O \cdot C_4H_9$	....	148.20
3392	*n*-butyrate, mono	α-monobutyrin	$C_3H_7CO_2 \cdot CH_2 \cdot CHOH \cdot CH_2OH$	II-273	162.19
3393	chlorohydrin, di (α)	1,3-dichloro-2-pro-panol	$(CH_2Cl)_2:CHOH$	I-364	128.99
3394	chlorohydrin, di (β)	2,3-dichloro-1-pro-panol	$Cl \cdot CH_2 \cdot CHCl \cdot CH_2OH$	I-356	128.99

† Mixture of both isomers.
Gluco-chloral 1289
Gluco-gallin, cf. glcde.
Glucosamine (*iso*) 3311
Glucose 3618

Glucose dimethyl-ketal 60
Glucosido-methyl salicylate, cf. glcde.
Glucosido-salicylaldehyde, cf. glcde.
Gluside 5579
Glutamine 3374-5

Table 7-4 (*Continued*)
PHYSICAL CONSTANTS OF ORGANIC COMPOUNDS

No.	Crystalline Form and Color	Specific Gravity	Melting Point °C.	Boiling Point °C.	Solubility in 100 Parts		
					Water	Alcohol	Ether
3358	rhb.	....	180-90 d.	....	$10^{14°}$; v. s. h.	v. sl. s. abs. al.	....
3359	nd./al. et.	....	softens 110; 131	....	s.	i. abs. al.	i.
3360	col. nd./aq.	....	....	....	$3.8^{16.5°}$	i.	20 h. aq.
3361	nd./al.	....	153	....	d.	....	....
3362	mn. nd./al.	....	112-3	....	$0.15^{18.5°}$; i. pet.	$1.3^{19°}$ abs. al.	$2.8^{15°}$
3363	mn. nd./al.	....	132-4	subl. vac.	$0.1^{18°}$; i. pet.	$0.8^{19°}$ abs. al.	$2.1^{15°}$
3364	pa. yel. lq.	$1.094^{25°}$	....	$228^{1.5mm}$	i.	s.	s.; s. chl.
3365	pa. yel. lq.	$1.151^{25°}$	....	....	i.	s.	s.; s. chl.
3366	col. cr.	....	159-60	....	s.	sl. s. c.	sl. s. c.
3367	col. cr.	....	140-1	....	sl. s.	sl. s. c.	v. sl. s. c.
3368	amor.	....	....	....	s.	s. h.	i.
3369	nd./Me al.	....	137-8	....	v. s.	v. sl. s.	i.
3370	pr./et.	....	137-8	....	v. s.	v. s.	v. s.
3371	cr./aq.	1.460	199 d.	....	$1.5^{20°}$	v. sl. s.	v. sl. s.
3372	rhb.	$1.538^{\frac{20°}{4}}$	224-5 d.	....	$0.7^{20°}$	i. abs.	i.
3373	rhb.	....	202-4*	....	33.3 c.*	i. HCl	....
3374	nd./aq.	....	256	....	$3.6^{18°}$	v. sl. s.	....
3375	nd./aq. al.	....	186	....	$3.6^{18°}$	$0.0005^{25°}$	v. sl. s.
3376	col. cr.	$1.429^{15°}$	97.5	200^{20mm}	$63.9^{20°}$	v. s. abs.	v. s.; s. bz.
3377	col. lq.	$0.991^{\frac{15°}{4}}$	−29.5(−32)	285-7	s.	s.	i.
3378	col. pr./al.	....	190-2 d.	....	$10^{0°}$; d.h.	i. abs.	i.; i. ac.
3379	syrup	....	....	....	∞	∞	i.
3380	nd. or pr.	$1.455^{\frac{18°}{18}}$	142	$140-50^{0.8mm}$	$3^{18°}$	sl. s.	sl. s.; i. bz.
3381	col. lq.	$1.261^{\frac{20°}{4}}$	18.2**	290	∞; i. bz.	∞; i. pet.	i.; i. chl.
3382	col. oil	$1.20^{\frac{20°}{4}}$	....	158^{165mm}	v. s.	v. s.	sl. s.; i. bz.
3383	col. lq.	$1.178^{\frac{15°}{15}}$	40	$175-6^{40mm}$	s.	s.	sl. s.
3384	col. hyg.	$1.096^{\frac{25°}{25}}$	42	$130^{10mm} \pm$	∞; d. act	∞; i. bz.	i.; i. CCl_4
3385	oil	$1.185^{\frac{25°}{25}}$	....	264 sl. d.	s.	s.	i.; i. bz.
3386	col. lq.	$0.903^{\frac{25°}{25}}$	....	270-2	i.	∞	∞
3387	col. lq.	$0.987^{\frac{25°}{25}}$	....	260-2	s.	∞	∞
3388	col. lq.	$2.135^{\frac{21°}{21}}$	....	214.8	i.	....	....
3389	col. lq.	$2.120^{\frac{20°}{4}}$	....	219 sl. d.	i.	∞; ∞ act.	∞; ∞ bz.
3390	col. lq.	....	....	$121-5^{13mm}$	i.	∞	∞
3391	col. lq.	$0.945^{\frac{25°}{25}}$	....	$133-7^{18mm}$	s.	∞	∞
3392	col. lq.	$1.129^{18°}$	....	269-71	∞	....	....
3393	col. lq.	$1.367^{\frac{20°}{4}}$	<−20	174.3^{760mm}	$12^{20°}$, $19^{70°}$	∞	∞
3394	col. lq.	$1.362^{\frac{20°}{4}}$	....	182-3	14.5	∞	∞

* Heated rapidly, m. p. 213°; d. – HCl, in aq.
** Solidifies at much lower temp.
Glutaminic acid 3371

Glycerin(e) 3381
Glycerol, cf. also glyceryl.

Table 7-4 (*Continued*)
PHYSICAL CONSTANTS OF ORGANIC COMPOUNDS

No.	Name	Synonym	Formula	Beil. Ref.	Formula Weight
	Glycerol				
3395	chlorohydrin, mono	α-chlorohydrin	$CH_2Cl\cdot CHOH\cdot CH_2OH$	I-473	110.54
3396	chlorohydrin, mono	β-chlorohydrin	$CHCl(CH_2OH)_2$	I-476	110.54
3397	formal *	....	$H_2C:C_3H_6O_3$	....	104.11
3398	formate, di *	....	$(HCO_2)_2C_3H_6O$	II-24	148.12
3399	furfural *	....	$C_5H_4O:C_3H_6O_3$	....	170.17
3400	guaiacol ether, mono	guaiamar; aresol; oreson	$CH_3O\cdot C_6H_4\cdot O\cdot$ $C_3H_5(OH)_2$	....	198.22
3401	iodohydrin, di (α)	1,3-diiodo-2-propanol; iotone	$(CH_2I)_2:CHOH$	I-366	311.89
3402	iodohydrin, mono (α)	3-iodo-1,2-propandiol; alival	$CH_2I\cdot CHOH\cdot CH_2OH$	I-475	201.99
3403	methyl ether, di	($\alpha\gamma$)	$(CH_3OCH_2)_2CHOH$	I-512	120.15
3404	methyl ether, mono	(α)	$CH_3O\cdot C_3H_5(OH)_2$	I-512	106.12
3405	nitrate, di (1,2)	....	$C_3H_5(OH)(O\cdot NO_2)_2$	I-515	182.09
3406	nitrate, di (1,3)	....	$CHOH(CH_2NO_3)_2$	I-515	182.09
3407	nitrate, mono (α)	....	$(HO)_2C_3H_5\cdot O\cdot NO_2$	I-514	137.09
3408	nitrate, mono (β)	....	$(CH_2OH)_2CHO\cdot NO_2$	I-515	137.09
3409	phenyl ether, di	($\alpha\gamma$)	$(C_6H_5OCH_2)_2CHOH$	VI-149	244.29
3410	phenyl ether, mono (α)	antodyne	$C_6H_5\cdot O\cdot CH_2\cdot CHOH\cdot$ CH_2OH	VI-149	168.19
3411	phosphoric acid	....	$(HO)_2C_3H_5\cdot OPO_3H_2$	I-517	172.08
3412	phosphate, sodium	(β)	$(HOCH_2)_2CH\cdot O\cdot$ $PO(ONa)_2\cdot 5H_2O$	*I-517	306.12
3413	propionate, mono	α-monopropionin	$C_2H_5CO_2\cdot CH_2\cdot$ $CHOH\cdot CH_2OH$	*II-107	148.16
3414	salicylate, mono (α or β)	monosalicylin	$HO\cdot C_6H_4CO_2\cdot$ $C_3H_5(OH)_2$	X-82	212.20
3415	stearate, mono	α-monostearin	$C_{17}H_{35}CO_2\cdot CH_2\cdot$ $CHOH\cdot CH_2OH$	II-380	358.57
3416	stearate, mono	β-monostearin	$C_{17}H_{35}CO_2\cdot CH:$ $(CH_2OH)_2$	II-380	358.57
3417	**Glyceryl** ether	....	$C_3H_5O_3:C_3H_5$	XIX-393	130.14
3418	triacetate	triacetin	$(CH_3CO_2)_3C_3H_5$	II-147	218.21
3419	triarachidate	triarachin	$(C_{19}H_{39}CO_2)_3C_3H_5$	II-390	975.67
3420	tribenzoate	tribenzoin	$(C_6H_5CO_2)_3C_3H_5$	IX-140	404.42
3421	tributyrate	tributyrin	$(C_3H_7CO_2)_3C_3H_5$	II-273	302.37
3422	tricaprate	tricaprin	$(C_9H_{19}CO_2)_3C_3H_5$	II-356	554.86
3423	tricaproate	tricaproin	$(C_5H_{11}CO_2)_3C_3H_5$	II-324	386.53
3424	tricaprylate	tricaprylin	$(C_7H_{15}CO_2)_3C_3H_5$	II-348	470.70
3425	triheptylate	triheptylin	$(C_6H_{13}CO_2)_3C_3H_5$	**II-295	428.61
3426	trilaurate	trilaurin	$(C_{11}H_{23}CO_2)_3C_3H_5$	II-362	639.02
3427	trimyristate	trimyristin	$(C_{13}H_{27}CO_2)_3C_3H_5$	II-367	723.18
3428	trinitrate	nitroglycerin	$(O_2N\cdot O)_3C_3H_5$	I-516	227.09
3429	trinitrite	....	$(ON\cdot O)_3C_3H_5$	I-514	179.09
3430	trioleate	triolein	$(C_{17}H_{33}CO_2)_3C_3H_5$	II-468	885.46
3431	tripalmitate	tripalmitin	$(C_{15}H_{31}CO_2)_3C_3H_5$	II-373	807.35
3432	tripropionate	tripropionin	$(C_2H_5CO_2)_3C_3H_5$	*II-107	260.29
3433	tristearate	tristearin	$(C_{17}H_{35}CO_2)_3C_3H_5$	II-383	891.51
3434	trivalerate	trivalerin	$(C_4H_9CO_2)_3C_3H_5$	....	344.45

* Mixture of both isomers.
Glyceryl, cf. also glycerol.

Glyceryl-amine 3384–5
Glyceryl diacetate 3383

Table 7-4 (*Continued*)
PHYSICAL CONSTANTS OF ORGANIC COMPOUNDS

No.	Crystalline Form and Color	Specific Gravity	Melting Point °C.	Boiling Point °C.	Solubility in 100 Parts		
					Water	Alcohol	Ether
3395	col. lq.	1.318$\frac{25}{25}$°	....	213	∞	∞	s.
3396	col. lq.	1.321$\frac{20}{4}$°	....	146^{18mm}	s.	∞ ; s. bz.	∞ ; ∞ act
3397	col. lq.	1.220$\frac{25}{25}$°	....	192-5	∞	∞	∞
3398	col. lq.	1.304^{15v}	....	163-6^{30mm}	i. CS$_2$	....	
3399	yel. lq.	1.267$\frac{25}{25}$°	....	163-7^{22mm}	s.	∞	sl. s.
3400	col. cr. pd.	....	78-9	....	5	s.	s.; s. chl.
3401	yel. oil	2.41$^{5°}$	freezes −16 to −20	d.	1.3; d. alk.	∞ ; ∞ bz.	∞ ; ∞ chl.
3402	col. cr.	2.03$^{13°}$	49 ±	....	v. s.	v. s.	....
3403	col. lq.	1.004$\frac{25}{4}$°	....	169	∞	∞	∞
3404	col. lq.	1.111$\frac{25}{4}$°	....	220	∞	∞	s.
3405	oil	....	expl.	....	d. by a.	d. by alk.	....
3406	oil	1.47$^{15°}$	<−30; expl.	146-8^{15mm} sl. d.	d. by a., alk.	v. s.	v. s.
3407	col. pr.	1.40	58-9	155-60*	70$^{15°}$	v. s.	v. sl. s.
3408	lf.	1.40	54	155-60	....	v. s.	sl. s.
3409	col. hex.	1.179$\frac{24}{4}$°	80-1	200-10^{10mm}	i.; s. bz.	s. h.	s.; s. chl.
3410	col. nd./ abs. et.	1.225$\frac{30}{4}$°	69-70 (53-4)	200^{22mm}	∞ ; v. sl. s. lg.	∞ ; v. s. bz.	∞
3411	syrup	1.59$^{14°}$	−20	....	∞	∞ abs.	....
3412	pl./aq.	....	98-100	d. >130	66 c.	....	....
3413	col. lq.	1.154$\frac{20}{4}$°	....	141-4^{6mm}	....	....	....
3414	col. nd./ et.	....	76	....	1 c.; v. s. h.	s.; ∞ gly.	sl. s.; s. h. bz.
3415	col. nd./ Me al.	0.984$\frac{20}{4}$°	82(57)	....	i.	s. h.	s.
3416	nd. or wax	....	....	....	i.	s. h.	s.
3417	lq.	1.091$^{18°}$	....	171-2	∞	∞	∞
3418	col. lq.	1.161$\frac{17}{4}$°	−78	258-9	7.17$^{15°}$	∞ ; ∞ chl.	∞ ; ∞ bz.
3419	col. cr.	....	72.2	....	i.	s. h.	s. h.
3420	nd./Me al.	1.228$^{12°}$	75-6	d.	i.; s. bz.	s. h.	s.; s. chl.
3421	col. lq.	1.032$\frac{20}{4}$°	<−75	305-9	i.	s.	s.
3422	col. cr.	0.921$\frac{40}{4}$°	31(25)	....	i.; s. bz.	s. h.	v. s.
3423	col. lq.	0.987$\frac{20}{4}$°	−25	....	i.	s. 85% al.	s.
3424	col. lq.	0.954$\frac{20}{4}$°	8.3(−21)	....	i.	s. 85% al.	s.
3425	col. lq.	0.969$^{20°}$	....	223-5^{3mm}	i.	s.	s.
3426	col. nd.	0.894$\frac{60}{4}$°	46.4(36)**	....	v. s. bz.	sl. s. c.	v. s.
3427	lf.	0.885$\frac{60}{4}$°	56.5(49)**	....	i.; v. s. bz.	v. s.	v. s. chl.
3428	col. or yel. oil	1.594$\frac{20}{4}$°	13.3(2.0)	160^{15mm}; expl. 270	0.18$^{20°}$	54$^{20°}$ abs.	∞
3429	yel. lq.	1.291$\frac{10}{18}$°	....	150 sl. d.	d.; i. CS$_2$	d.; s. chl.	s.; s. bz.
3430	col. oil	0.915$^{15o°}$	−4	240^{18mm} sl. d.	i.	sl. s.	v. s.
3431	col. nd.	0.866$\frac{8}{4}$°	65.1 (55; 45)	310-20$^{0.1mm}$	i.	0.004$^{21°}$ abs. al.	v. s.
3432	lq.	1.100$\frac{29}{18}$°	....	177-82^{20mm}	....	....	....
3433	col pr.	0.862$\frac{8}{4}$°	70.8(54.5; 64.5)	....	i.; s. bz.	s. h.	s. h.; sl. s. pet.
3434	pa. yel. lq	1.030$^{20°}$	....	152-5^{1mm}	i.	s.	s.

* Not explosive.
** Probably several other modifications.

Glyceryl tribromohydrin 6112
Glyceryl trichlorohydrin 6166

Table 7-4 (*Continued*)
PHYSICAL CONSTANTS OF ORGANIC COMPOUNDS

No.	Name	Synonym	Formula	Beil. Ref.	Formula Weight
	Glyceryl				
3435	trivalerate (*iso*)	tri-*iso*-valerin	$(C_4H_9CO_2)_3C_3H_5$	II-314	344.45
3436	**Glycide**	glycidol; epihydrin alcohol	$CH_2 \cdot O \cdot CH \cdot CH_2OH$	XVII-104	74.08
3437	**Glycine**	amino-acetic acid	$NH_2 \cdot CH_2 \cdot CO_2H$	IV-333	75.07
3438	amide	....	$NH_2 \cdot CH_2 \cdot CO \cdot NH_2$	IV-343	74.08
3439	anhydride	diketopiperazine	$C_4H_6O_2N_2$	XXIV-264	114.10
3440	**Glycocholic** acid	cholylglycine; "cholic acid"	$C_{24}H_{39}O_4 \cdot NH \cdot CH_2 \cdot CO_2H$	....	465.64
3441	sodium salt	Na glycocholate	$C_{26}H_{42}O_6NNa$	....	487.62
3442	**Glycocyamine**	guanyl glycine	$HN:C(NH_2) \cdot NH \cdot CH_2CO_2H$	IV-359	117.11
3443	**Glycogen**	animal starch	$(C_6H_{10}O_5)x$	....	(162.14)
3444	**Glycol**	ethandiol-1,2	$HOCH_2 \cdot CH_2OH$	I-465	62.07
3445	acetate, di	ethylene acetate	$(CH_3CO_2 \cdot CH_2 \cdot)_2$	II-142	146.14
3446	acetate, mono	....	$CH_3CO_2 \cdot CH_2CH_2OH$	II-141	104.11
3447	benzoate, di	ethylene benzoate	$(C_6H_5CO_2 \cdot CH_2)_2$	IX-129	270.29
3448	benzyl ether, di	....	$(C_6H_5CH_2 \cdot O \cdot CH_2)_2$	....	242.32
3449	benzyl ether, mono	benzyl cellosolve	$C_7H_7O \cdot C_2H_4OH$	....	152.19
3450	butyrate, di	ethylene dibutyrate	$(C_3H_7CO_2 \cdot CH_2)_2$	II-272	202.25
3451	capr(o)ate, di	ethylene dicaprate	$(C_5H_{11}CO_2 \cdot CH_2)_2$	....	258.36
3452	caprylate, di	ethylene dicaprylate	$(C_7H_{15}CO_2 \cdot CH_2)_2$	**II-303	314.47
3453	ether	diethylene glycol	$(HO \cdot CH_2 \cdot CH_2)_2O$	I-468	106.12
3454	formal	dioxolane; glycol-methylene ether	$O \cdot CH_2 \cdot CH_2 \cdot O \cdot CH_2$	XIX-2	74.08
3455	formate, di	ethylene formate	$(HCO_2 \cdot CH_2)_2$	II-23	118.09
3456	formate, mono	....	$HCO_2 \cdot CH_2 \cdot CH_2OH$	II-23	90.08
3457	guaiacyl ether, di	....	$(CH_3OC_6H_4O \cdot CH_2)_2$	....	274.32
3458	laurate, di	ethylene dilaurate	$(C_{11}H_{23}CO_2 \cdot CH_2)_2$	II-361	426.69
3459	methyl ether, di	di MeO-ethane	$(CH_3 \cdot O \cdot CH_2)_2$	I-467	90.12
3460	methyl ether, mono	methyl cellosolve	$CH_3O(CH_2)_2OH$	I-467	76.10
3461	myristate, di	ethylene dimyristate	$(C_{13}H_{27}CO_2 \cdot CH_2)_2$	II-366	482.79
3462	nitrate, di	dinitroglycol	$(O_2N \cdot O \cdot CH_2)_2$	I-469	152.06
3463	nitrite, di	ethylene nitrite	$(ON \cdot O \cdot CH_2)_2$	I-469	120.07
3464	palmitate, di	ethylene dipalmitate	$(C_{15}H_{31}CO_2 \cdot CH_2)_2$	II-373	538.90
3465	propionate, di	ethylene dipropionate	$(C_2H_5CO_2 \cdot CH_2)_2$	II-242	174.20
3466	salicylate, mono	glysal; spirosal	$C_7H_5O_3 \cdot C_2H_4OH$	X-81	182.18
3467	stearate, di	ethylene distearate	$(C_{17}H_{35}CO_2 \cdot CH_2)_2$	II-380	595.01
3468	**Glycolic** acid	hydroxy-acetic acid	$HO \cdot CH_2 \cdot CO_2H$	III-228	76.05
3469	Ca salt	calcium glycolate	$C_4H_6O_6Ca \cdot 4H_2O$	III-232	262.23
3470	aldehyde	hydroxy-acetaldehyde	$HO \cdot CH_2 \cdot CHO$	I-817	60.05
3471	aldehyde-diethyl-acetal	....	$HO \cdot CH_2 \cdot CH: (OC_2H_5)_2$	I-818	134.18
3472	amide	hydroxy-acetamide	$HO \cdot CH_2 \cdot CO \cdot NH_2$	III-240	75.07
3473	anhydride	....	$(HO \cdot CH_2 \cdot CO)_2O$	III-239	134.09
3474	nitrile	....	$HO \cdot CH_2 \cdot CN$	III-242	57.05
3475	**Glycolide**	glycolid	$CH_2 \cdot CO_2 \cdot CH_2 \cdot CO \cdot O$	XIX-153	116.07
3476	**Glycyl** glycine	....	$H_2N \cdot CH_2 \cdot CO \cdot NH \cdot CH_2 \cdot CO_2H$	IV-371	132.12
3477	glycine ester	....	$C_3H_7O_2N_2 \cdot CO_2C_2H_5$	IV-373	160.17
3478	**Glyoxal**	ethandial	$O:CH \cdot CH:O$	I-759	58.04

Table 7-4 (*Continued*)
PHYSICAL CONSTANTS OF ORGANIC COMPOUNDS

No.	Crystalline Form and Color	Specific Gravity	Melting Point °C.	Boiling Point °C.	Solubility in 100 Parts		
					Water	Alcohol	Ether
3435	oil	$0.998^{20°}$	....	$153\text{-}6^{2mm}$	....	....	....
3436	col. lq.	$1.114^{16}_{16}°$	....	166-7 sl. d.	∞ ; sl. s. pet.	∞ ; sl. s. xylene	∞ ; s. bz.
3137	mn.	$1.575^{50°}$	232-6 d.	....	23 c.	0.1 c.	l.
3438	hyg. nd.	....	65-7	....	v. s.	v. s.	v. sl. s.
3439	pl./aq.	....	275 d.	subl. 260	s. h.	s.; d. a.	d. alk.
3440	nd.	....	154-5 d.	....	$3.3^{20°}$; $8.5^{100°}$	v. s.; v. sl. s. bz.	$0.09^{20°}$; v. sl. s. chl.
3441	hyg. pd.	....	....	....	s.	s.	....
3442	nd./aq.	....	270-80	....	$0.45^{15°}$	v. sl. s.	v. sl. s.
3443	wh. pd.	....	240	....	s.	i.	i.
3444	col. lq.	$1.113^{19}_{4}°$	-15.6	197.9	∞	∞	1.0
3445	col. lq.	$1.109^{14}_{4}°$	-31	190.5	$14.3^{22°}$	∞	∞
3446	col. lq.	1.108	....	182	∞	∞	....
3447	rhb./et.	....	73-4	>360	i.	....	s.
3448	col. lq.	....	....	$131\text{-}6^{2mm}$	i.	....	....
3449	col. lq.	1.07	<-75	256	0.4	....	....
3450	col. lq.	$1.024^{0°}$	....	240	i.	v. s.	v. s.
3451	....	....	37-8	....	i.	s.	....
3452	pa. yel. lq.	....	22	....	i.	....	....
3453	lq.	$1.118^{20}_{20}°$	-10.5	244.8	∞ ; i. bz.	∞ ; i. chl.	i.
3454	lq.	$1.060^{20}_{4}°$	....	75-6	∞		
3455	lq.			174	....	s.	s.
3456	lq.	$1.199^{15}_{4}°$		180	∞		
3457	col. nd.	....	138-9	....	v. sl. s.	s. h.	....
3458	col. amor.	....	52-4	188^{20mm}	i.	v. s.	v. s.
3459	lq.	$0.863^{20}_{4}°$	-58	84-5	∞	....	
3460	col. lq.	$0.965^{20}_{4}°$	-85.1	124-5	∞	∞	∞ ; ∞ bz.
3461	....	....	63-4	$208^{0.1mm}$	....	....	
3462	yel. lq.	$1.488^{20}_{4}°$	-22.3	expl. 114	i.; d. alk.	s.; ∞bz.	∞ ; ∞act.
3463	lq.	$1.216^{0°}$	<-15	96-8	i.; d. alk.	s. d.	s.
3464	nd./al. chl.	....	71-2	$226^{0.1mm}$	i.	s.	s.
3465	lq.	$1.045^{25°}$	....	211-2	sl. s.	∞	∞
3466	col. cr.	....	37	$169\text{-}70^{12mm}$	$0.97^{22°}$	v. s.; s. bz.	v. s.
3467	lf.	....	76-7	$241^{0.1mm}$	i.	$0.12^{40°}$	v. s.
3468	nd./aq.	....	79(63)	d.	v. s.	$90^{25°}$	v. s.
3469	nd.	....	-4H₂O,110°	....	$1.3^{15°}$	$5^{100°}$ aq.	....
3470	col. pl.	$1.366^{100°}$	95-7	....	v. s.	v. s. h.	v. sl. s.
3471	lq.	$0.888^{24°}$	....	167	d. h.	....	....
3472	rhb.	....	120	....	v. s.	sl. s.	....
3473	cr. pd.	....	128-30	....	i. c.; d. h.	i.	i.
3474	col. oil	$1.104^{19°}$	<-72	183 sl. d.	v. s.	v. s.	v. s.; i. bz.
3475	lf./al.	....	86-7	....	s. h.; s. ac.	s. h.	sl. s. c.; s. h. chl.
3476	lf./aq.	....	d. 215-20	....	v. s. h.	sl. s.	i.
3477	nd.	....	88-9	....	v. s.	v. s.	sl. s.
3478	yel. cr.	$1.14^{20°}$	15	51^{776mm}	....	v. s. abs.	v. s. abs.

Glycol iodohydrin 3208
Glycol methylene ether 3454
Glycol phenyl ether 5136
Glycolic ethyl ether 2912
Glycolic methyl ether 4063

Glycoline 2533
Glycoluril 139
Glycolyl thiourea 5907
Glycolyl urea 3679

Glycovanillin, cf. glcde.
Glycophillin, cf. glcde.
Glyoxalic acid 3480
Glyoxalin 3851

Table 7-4 (*Continued*)
PHYSICAL CONSTANTS OF ORGANIC COMPOUNDS

No.	Name	Synonym	Formula	Beil. Ref.	Formula Weight
3479	**Glyoxime**	....	$HON:CH \cdot CH:NOH$	I-761	88.07
3480	**Glyoxylic acid**	glyoxalic acid	$HO_2C \cdot CH:O \cdot H_2O$	III-594	92.05
3481	**Guaiacol** acetate	....	$CH_3OC_6H_4 \cdot O_2CCH_3$	VI-774	166.18
3482	benzoate	benzosol	$CH_3OC_6H_4O_2C_7H_5$	IX-130	228.25
3483	*n*-caproate	....	$CH_3OC_6H_4O_2C_6H_{11}$	....	222.29
3484	carbonate	duotal	$(CH_3O \cdot C_6H_4O)_2CO$	VI-776	274.28
3485	cinnamate	styracol	$CH_3OC_6H_4O_2C_9H_7$	IX-585	254.29
3486	salicylate	guaiacol salol	$C_{14}H_{12}O_4$	X-81	244.25
3487	valerate	....	$CH_3OC_6H_4O_2C_5H_9$	....	208.26
3488	**Guanidine**	imino-urea	$(H_2N)_2C:NH$	III-82	59.07
3489	acetate	....	$CH_5N_3 \cdot HC_2H_3O_2$	III-86	119.12
3490	carbonate	....	$(CH_5N_3)_2 \cdot H_2CO_3$	III-86	180.17
3491	hydrochloride	....	$CH_5N_3 \cdot HCl$	III-86	95.54
3492	nitrate	....	$CH_5N_3 \cdot HNO_3$	III-86	122.08
3493	sulfate	....	$(CH_5N_3)_2 \cdot H_2SO_4 \cdot$ $\frac{1}{2}H_2O$	III-86	225.23
3494	thiocyanate	....	$CH_5N_3 \cdot HCNS$	III-169	118.16
3495	**Gynocardic acid**	....	$C_{17}H_{33}CO_2H$	*IX-45	282.47
3496	**Hedonal**	Me-*n*-Pr-carbinol-urethane	$H_2N \cdot CO_2 \cdot CH(CH_3) \cdot$ $(CH_2CH_2CH_3)$	III-29	131.18
3497	**Helenin**	alantolactone	$C_{15}H_{20}O_2$	XVII-327	232.33
3498	**Hematein**	....	$C_{16}H_{12}O_6$	XVIII-227	300.27
3499	**Hematin**	phenodin	$C_{34}H_{33}O_5N_4Fe$	....	633.51
3500	**Hematoxylin**	hydroxybrasilin	$C_{16}H_{14}O_6 \cdot 3H_2O$	XVII-219	356.33
3501	**Hemin**	....	$C_{34}H_{32}O_4N_4ClFe$	....	651.96
3502	**Hemipinic acid** (3,4;1,2)	3,4-diMeO-phthalic acid	$(CH_3O)_2C_6H_2:$ $(CO_2H)_2 +$ aq.	X-543	226.19
3503	**Heneicosane** (*n*)	....	$CH_3(CH_2)_{19}CH_3$	I-174	296.58
3504	**Heneicosene**-9	....	$CH_3(CH_2)_7CH:$ $CH(CH_2)_{10}CH_3$	I-227	294.57
3505	**Hentriacontane** (*n*)	....	$CH_3(CH_2)_{29}CH_3$	I-177	436.86
3506	**Hepta**-chloropropane	(1,1,1,2,2,3,3)	$Cl_3C \cdot CCl_2 \cdot CHCl_2$	I-108	285.21
3507	cosane (*n*)	....	$CH_3(CH_2)_{25}CH_3$	I-176	380.75
3508	decane (*n*)	....	$CH_3(CH_2)_{15}CH_3$	I-173	240.48
3509	decyl alcohol	heptadecanol-9	$[CH_3(CH_2)_7]_2CHOH$	I-430	256.48
3510	diene-2,4	....	$CH_3 \cdot (CH:CH)_2 \cdot C_2H_5$	I-257	96.17
3511	**Heptandiol**-1,7	....	$HO \cdot (CH_2)_7 \cdot OH$	I-489	132.20
3512	**Heptane** (*n*)	....	$CH_3 \cdot (CH_2)_5 \cdot CH_3$	I-154	100.21
3513	**Heptane** (*iso*)	2-methyl-hexane	$(CH_3)_2CH \cdot C_4H_9$	I-156	100.21
3514	**Heptane**	3-methyl-hexane	$C_3H_7 \cdot CH(CH_3) \cdot C_2H_5$	I-157	100.21
3515	**Heptane**	2,2-dimethyl-pentane	$(CH_3)_3C \cdot CH_2 \cdot C_2H_5$	I-157	100.21
3516	**Heptane**	2,4-dimethyl-pentane	$[(CH_3)_2CH]_2CH_2$	I-158	100.21
3517	**Heptane**	3,3-dimethyl-pentane	$(CH_3)_2C(C_2H_5)_2$	I-158	100.21
3518	**Heptane**	3-ethyl-pentane	$(C_2H_5)_3CH$	I-157	100.21
3519	**Heptane**	2,2,3-triMe-butane	$(CH_3)_3C \cdot CH(CH_3)_2$	**I-121	100.21
3520	**Heptanoic acid**	oenanthylic acid	$CH_3(CH_2)_5CO_2H$	II-338	130.19
3521	aldehyde	oenanthole	$CH_3(CH_2)_5 \cdot CHO$	I-695	114.19
3522	aldehyde oxime	enanthaldoxime	$CH_3(CH_2)_5 \cdot CH:NOH$	I-698	129.20
3523	amide	*n*-heptamide	$CH_3(CH_2)_5CO \cdot NH_2$	II-340	129.20

Table 7-4 (*Continued*)
PHYSICAL CONSTANTS OF ORGANIC COMPOUNDS

No.	Crystalline Form and Color	Specific Gravity	Melting Point °C.	Boiling Point °C.	Solubility in 100 Parts		
					Water	Alcohol	Ether
3479	pr./aq.	....	178	....	v. s. h.	v. s.	v. s.
3480	mn./aq.	....	98	....	v. s.	v. sl. s.	v. sl. s.
3481	col. lq.	1.156°	....	240-1	....	∞	∞
3482	col. cr. pd.	....	57-8	....	v. sl. s.	s. h.	s.; s. chl.
3483	col. lq.	....	....	168-70^{15mm}	i.	s.	s.
3484	cr.	....	86-8	....	i.	2; s. h.	7; s. bz.
3485	nd./al.	....	130	....	i.	s.; s. bz.	s.; s. chl.
3486	col. cr./al.	....	65	....	i.	s.	s.; s. chl.
3487	yel. lq.	1.05	....	265	i.	s.; s. bz.	s.; s. chl.
3488	hyg. col. cr.	....	50±	....	v. s.	s.	....
3489	nd.	....	229-30	....	v. s.	v. s.	i.
3490	col. cr.	1.25	197 d.	....	50$^{24°}$	0.2$^{24°}$	i. NH_3
3491	col. cr.	....	....	....	v. s.	s.	....
3492	col. cr.	....	214	....	14$^{25°}$	1.6$^{25°}$	0.6$^{25°}$ act.
3493	col. cr.	....	....	....	v. s.	i.	....
3494	col. lf.	....	118	....	73^{00}	135$^{15°}$ aq.	....
3495	col. lf./al.	....	67.5	....	i.	s.	s.
3496	col. nd.	....	74-6	215±	0.8$^{35°}$	s.	s.; s. chl.
3497	nd./aq. al.	....	76	275	v. sl. s. h.	s.; s. bz.	s.; s. chl.
3498	silv. nd.	....	250 d.	....	0.06$^{20°}$; s. NH_4OH	sl. s.; i. chl.	0.013$^{20°}$; i. bz.
3499	brn. pd.	....	>200	....	i.; s. alk.	i.; i. chl.	i.
3500	cr./aq. $(NH_4)_2SO_3$	....	140(anh.)	−H₂O, 100-20	s. h.	s.; s. aq. borax	s.
3501	b. bl. rhb.		....	....	i.; s. ac.	i.	i.; i. chl.
3502	mn.	....	182-6 d.	....	sl. s. c.	s.	0.09
3503	col. cr.	$0.778^{40.4°}_{5}$	40.4	215^{15mm}	i.	....	....
3504	col. lq.	$0.802^{20°}_{4}$	3	201-211^{11mm}	i.	....	....
3505	col. cr.	$0.781^{68.1°}_{4}$	68.1	302^{15mm}	i.	0.7$^{15°}$ chl.	sl. s.
3506	col. cr.	$1.805^{34°}_{4}$	30	247-8	....	....	....
3507	col. cr.	$0.780^{59.5°}_{4}$	59.5	270^{15mm}	i.	....	....
3508	col. cr.	$0.775^{20°}_{4}$	22.5	303	i.	sl. s.	s.
3509	col. pl.	....	61	....	....	sl. s. aq. al.	....
3510	col. lq.	$0.733^{21.5°}_{4}$	....	107	....	....	....
3511	col. cr.	....	22.5	262	v. s.	v. s.	i.
3512	col. lq.	$0.684^{20°}_{4}$	−90.6	98.4^{760mm}	0.0052$^{18°}$	sl. s.	∞; ∞ chl.
3513	col. lq.	$0.679^{20°}_{4}$	−118.3	90.1	i.	s.	∞
3514	col. lq.	$0.687^{20°}_{4}$	−119.4	92.0	i.	s.	∞
3515	col. lq.	$0.674^{20°}_{4}$	−123.8	79.2	i.	s.	∞
3516	col. lq.	$0.673^{20°}_{4}$	−119.2	80.5	i.	s.	∞
3517	col. lq.	$0.693^{20°}_{4}$	−134.5	86.1	i.	s.	∞
3518	col. lq.	$0.698^{20°}_{4}$	−118.6	93.5	i.	s.	∞
3519	col. lq.	$0.690^{20°}_{4}$	−25.0	80.9	i.	s.	∞
3520	col. lq.	$0.922^{15°}_{4}$	−7.5	223.0	0.25$^{15°}$	s.	s.
3521	col. lq.	$0.822^{15°}_{4}$	−43.3	152.8	0.02$^{20°}$	∞	∞
3522	pl./al.	$0.858^{54.7°}_{4}$	56	195	sl. s. c.	s.	s.
3523	nd./al.	$0.849^{11.2°}_{4}$	95-6	250-8	v. s.	v. s.	v. s.

Table 7-4 (*Continued*)
PHYSICAL CONSTANTS OF ORGANIC COMPOUNDS

No.	Name	Synonym	Formula	Beil. Ref.	Formula Weight
	Heptanoic				
3524	anhydride (*n*)	heptylic anhydride	$(C_6H_{13}CO)_2O$	II-340	242.36
3525	**Heptanthiol**-2	heptyl mercaptan	$CH_3 \cdot CH(SH) \cdot C_5H_{11}$	I-415	132.27
3526	**Heptyl** acetate (*n*)	....	$CH_3CO_2 \cdot C_7H_{15}$	II-134	158.24
3527	alcohol (*n*)	heptanol-1	$CH_3(CH_2)_5CH_2OH$	I-414	116.20
3528	alcohol	2,2,3-trimethyl-butanol-3	$(CH_3)_3C \cdot C(CH_3)_2 \cdot OH$	I-418	116.20
3529	alcohol	2,3-dimethyl-pentanol-3	$(CH_3) \cdot C_2H_5$	I-417	116.20
3530	alcohol	2,4-dimethyl-pentanol-2	$(CH_3)_2C(OH) \cdot CH_2 \cdot CH(CH_3)_2$	I-417	116.20
3531	alcohol	2,4-di Me-pentanol-3	$[(CH_3)_2CH]_2CHOH$	I-417	116.20
3532	alcohol	3-ethyl-pentanol-2	$(C_2H_5)_2CH \cdot CHOH \cdot CH_3$	I-416	116.20
3533	alcohol	3-ethyl-pentanol-3	$(C_2H_5)_3COH$	I-417	116.20
3534	alcohol	2-methyl-hexanol-1	$CH_3(CH_2)_3CH(CH_3) \cdot CH_2OH$	I-415	116.20
3535	alcohol	2-methyl-hexanol-2	$(CH_3)_2C(OH)(CH_2)_3 \cdot CH_3$	I-415	116.20
3536	alcohol	2-methyl-hexanol-3	$(CH_3)_2CH \cdot CHOH \cdot (CH_2)_2CH_3$	I-416	116.20
3537	alcohol	2-methyl-hexanol-4	$(CH_3)_2CH \cdot CH_2 \cdot CHOH \cdot C_2H_5$	I-416	116.20
3538	alcohol	2-methyl-hexanol-5	$(CH_3)_2CH(CH_2)_2 \cdot CHOH \cdot CH_3$	I-416	116.20
3539	alcohol (*iso*)	2-methyl-hexanol-6	$(CH_3)_2CH(CH_2)_3 \cdot CH_2OH$	I-416	116.20
3540	alcohol	3-methyl-hexanol-3	$C_2H_5 \cdot CH_2 \cdot C(CH_3) \cdot (OH) \cdot C_2H_5$	I-416	116.20
3541	alcohol	3-methyl-hexanol-4	$C_2H_5 \cdot CH(CH_3) \cdot CHOH \cdot C_2H_5$	I-416	116.20
3542	alcohol	heptanol-2	$CH_3(CH_2)_4 \cdot CHOH \cdot CH_3$	I-415	116.20
3543	alcohol	heptanol-4	$(CH_3 \cdot CH_2 \cdot CH_2)_2 : CHOH$	I-415	116.20
3544	amine (*n*)	....	$C_7H_{15} \cdot NH_2$	IV-193	115.22
3545	bromide (*n*)	1-bromoheptane	$CH_3(CH_2)_5 \cdot CH_2Br$	I-155	179.11
3546	chloride (*n*)	1-chloroheptane	$CH_3(CH_2)_5 \cdot CH_2Cl$	I-154	134.65
3547	chloride	3-chloro-2,3-di-methyl-petane	$(CH_3)_2CH \cdot CCl(CH_3) \cdot C_2H_5$	I-157	134.65
3548	formate (*n*)	....	$HCO_2 \cdot C_7H_{15}$	II-22	144.22
3549	iodide	1-iodo-2-methyl-hexane	$ICH_2 \cdot CH(CH_3) \cdot (CH_2)_3 \cdot CH_3$	I-157	226.10
3550	malonic acid (*n*)	....	$C_7H_{15}CH(CO_2H)_2$	II-721	202.25
3551	nitrite (*n*)	....	$CH_3(CH_2)_6 \cdot O \cdot NO$	I-415	145.20
3552	oenanthylate	*n*-heptyl *n*-heptylate	$C_6H_{13}CO_2 \cdot C_7H_{15}$	II-340	228.38
3553	propionate (*n*)	....	$C_2H_5CO_2 \cdot C_7H_{15}$	II-241	172.27
3554	**Heptylene** (*α*)	heptene-1	$CH_3(CH_2)_4 \cdot CH:CH_2$	I-219	98.19
3555	**Heptylene**	2,4-dimethyl-pentene-2	$(CH_3)_2C:CH \cdot CH: (CH_3)_2$	I-220	98.19
3556	**Heptylene**	2,3,3-triMe-butene-1	$(CH_3)_3C \cdot C(CH_3):CH_2$	**I-199	98.19
3557	alcohol	2-Me-penten-4-ol-2	$C_3H_5 \cdot CH_2 \cdot C(CH_3)_2 \cdot OH$	I-445	114.19
3558	**Heptyne**-1	oenanthylidene	$CH_3(CH_2)_4 \cdot C : CH$	I-256	96.17

Heptanol 3527-43
Heptanone 2786, 3003, 4122
Heptene 3554-6
Heptyl, cf. also diheptyl.
Heptyl aldehyde 3521

Heptyl carbinol 4961
Heptyl cyanide 1226
Heptyl ether 2264
Heptyl heptylate 3552
Heptyl mercaptan 3525

Table 7-4 (*Continued*)
PHYSICAL CONSTANTS OF ORGANIC COMPOUNDS

No.	Crystalline Form and Color	Specific Gravity	Melting Point °C.	Boiling Point °C.	Solubility in 100 Parts		
					Water	Alcohol	Ether
3524	lq.	$0.932^{21°}$	17	258-68	....	....	....
3525	lq.	$0.835^{20°}$	....	$174\text{-}5^{765mm}$	i.	....	....
3526	col. lq.	$0.875^{\frac{15°}{4}}$	-50.2	192.5	i.	s.	s.
3527	col. lq.	$0.824^{\frac{20°}{4}}$	.34.6	175^{756mm}	$0.18^{25°}$	∞	∞
3528	col. lq.	....	15-7	131-2	hydrate $+1H_2O$	∞	∞
3529	lq.	$0.840^{\frac{20°}{4}}$	<-30	$138\text{-}40^{750mm}$	....	∞	∞
3530	lq.	$0.816^{\frac{20°}{4}}$	<-20	$132\text{-}3^{760mm}$	i.	∞	∞
3531	col. lq.	$0.829^{\frac{20°}{4}}$	....	140	v. sl. s.	∞	∞
3532	lq.	$0.853^{0°}$	....	151^{743mm}	....	∞	∞
3533	col. oil	$0.839^{\frac{20°}{4}}$	....	142^{764mm}	sl. s.	∞	∞
3534	lq.	$0.831^{\frac{13°}{4}}$	....	$162\text{-}4^{750mm}$	....	∞	∞
3535	col. lq.	$0.815^{\frac{20°}{20}}$	....	$141\text{-}27^{55mm}$	i.	∞	∞
3536	col. oil	$0.827^{\frac{16.7°}{4}}$	....	145-6	i.	∞	∞
3537	lq.	....	....	$147\text{-}8^{756mm}$	....	∞	∞
3538	lq.	$0.819^{17.5°}$	....	148-50	....	∞	∞
3539	lq.	$0.825^{\frac{11.5°}{4}}$	....	170.5^{755mm}	v. sl. s.	∞	∞
3540	lq.	$0.828^{\frac{18°}{4}}$	....	140.3^{745mm}	....	∞	∞
3541	lq.	$0.852^{0°}$	....	149-50	i.	s.	s.
3542	lq.	$0.819^{\frac{20°}{4}}$	....	160.4	$0.4^{20°}$	s.	s.
3543	lq.	$0.820^{\frac{20°}{4}}$	-37	156	i.	s.	s.
3544	col. lq.	$0.777^{20°}$	-23	155	v. sl. s.	∞	∞
3545	lq.	$1.145^{\frac{15°}{4}}$	-58.1	180.0	i.	∞	∞
3546	lq.	$0.881^{16°}$	....	159.2^{750mm}	i.	∞	∞
3547	lq.	$0.884^{22°}$	<-15	$135\text{-}8^{757mm}$	i.	∞	∞
3548	lq.	$0.883^{\frac{15°}{4}}$	-46.4	178.1	i.	s.	s.
3549	lq.	$1.366^{\frac{21°}{4}}$	....	$78\text{-}9^{19mm}$	i.	s.	s.
3550	cr./bz.	....	95 d.	....	i.	s.	s.; v. s. act.
3551	lq.	$0.894^{0°}$	....	155	i.	....	s.
3552	col. lq.	$0.864^{\frac{15°}{4}}$	-33.3	277.2	i.	s.	s.
3553	lq.	$0.872^{\frac{15°}{4}}$	-50.9	210.0	i.	s.	s.
3554	col. lq.	$0.697^{\frac{20°}{4}}$	-119.2	93.3	i.	s.	s.
3555	col. lq.	$0.696^{\frac{20°}{4}}$	....	82.4	i.	s.	s.
3556	lq.	$0.705^{20°}$	-111.4	77.9	....	....	....
3557	lq.	$0.831^{\frac{18°}{0}}$	-73	119.5	....	....	....
3558	lq.	$0.732^{\frac{20°}{4}}$	-81	99	....	....	....

Heptyl sulfide 2266
Heptylic acid 3520
Heptylic anhydride 3524
Herapathite, cf. alkd.
Heroin, cf. alkd.

Hesperetinic acid 3237
Hesperetole 6441
Hesperidin, cf. glcde.
Heteroauxin 3865
Hexabromo-cyclohexane 666

Hexachloro-cyclohexane 668-9
Hexadecanoic acid 5022
Hexadecanol 1275
Hexadecene 3570
Hexadecyl, cf. cetyl.

Table 7-4 (*Continued*)
PHYSICAL CONSTANTS OF ORGANIC COMPOUNDS

No.	Name	Synonym	Formula	Beil. Ref.	Formula Weight
3559	**Hesperetin**†	5,7,3'-triOH-4'-MeO-flavanone	$C_{16}H_{14}O_6$	VIII-544	302.29
3560	**Hesperidin**	....	$C_{28}H_{34}O_{15}$	....	610.57
3561	**Hexabromo-**benzene	perbromobenzene	C_6Br_6	V-215	551.52
3562	ethane	perbromoethane	$Br_3C\cdot CBr_3$	I-96	503.48
3563	**Hexachloro-**benzene	perchlorobenzene	C_6Cl_6	V-205	284.78
3564	ethane	carbon hexachloride	$Cl_3C\cdot CCl_3$	I-87	236.74
3565	chlorophenol	(low melting)	C_6OCl_6	VI-194	300.78
3566	chlorophenol	(high melting)	C_6OCl_6	VII-144	300.78
3567	**Hexacosane** (*n*)	....	$CH_3(CH_2)_{24}\cdot CH_3$	I-175	366.72
3568	**Hexacosane** (*iso*) (impure?)	cerane	$(CH_3)_2CH\cdot (CH_2)_{22}\cdot CH_3$	*I-70	366.72
3569	**Hexadecane** (*n*)	cetane	$CH_3(CH_2)_{14}\cdot CH_3$	I-172	226.45
3570	**Hexadecylene** (*α*)	hexadecene-1	$CH_3(CH_2)_{13}\cdot CH:CH_2$	I-226	224.43
3571	**Hexadecyne-**1	....	$CH_3(CH_2)_{13}\cdot C:CH$	I-262	222.42
3572	**Hexadecyne-**2	cetylene	$CH_3(CH_2)_{12}\cdot C:C\cdot CH_3$	I-262	222.42
3573	**Hexadiene-**2,4	dipropenyl	$(CH_3\cdot CH:CH\cdot)_2$	I-254	82.15
3574	**Hexadiene-**1,5	diallyl	$(CH_2:CH\cdot CH_2\cdot)_2$	I-253	82.15
3575	**Hexaethylbenzene**	"alkazene 6"	$(C_2H_5)_6C_6$	V-471	246.44
3576	**Hexahydro-**anthracene	anthracene hexahydride	$C_{14}H_{16}$	V-573	184.28
3577	benzoic acid	cyclohexane-carboxylic acid	$CH_2(CH_2)_4\cdot CH\cdot CO_2H$	IX-7	128.17
3578	cumene	....	$C_3H_7\cdot C_6H_{11}$	V-41	126.24
3579	cymene (*p*)	menthane; terpane	$CH_3\cdot C_6H_{10}\cdot C_3H_7$	V-47	140.27
3580	mellitic acid	....	$C_6H_6(CO_2H)_6$	IX-1007	348.22
3581	mesitylene	(1,3,5)	$C_6H_9(CH_3)_3$	V-45	126.24
3582	naphthalene		$C_{10}H_{14}$	V-433	134.22
3583	*o*-phthalic acid (*cis*)	....	$C_6H_{10}(CO_2H)_2$	IX-730	172.18
3584	*o*-phthalic acid (*dl*)	(*trans*)	$C_6H_{10}(CO_2H)_2$	IX-730	172.18
3585	*p*-phthalic acid	(*cis*)	$C_6H_{10}(CO_2H)_2$	IX-733	172.18
3586	*p*-phthalic acid	(*trans*)	$C_6H_{10}(CO_2H)_2$	IX-733	172.18
3587	salicylic acid (*o*)	....	$HO\cdot C_6H_{10}\cdot CO_2H$	X-5	144.17
3588	**Hexahydroxy-**benzene	....	$C_6(OH)_6$	VI-1198	174.11
3589	**Hexaiodobenzene**	periodobenzene	C_6I_6	V-230	833.49
3590	**Hexamethyl-**benzene	....	$C_6(CH_3)_6$	V-450	162.28
3591	diamino-*iso*-propanol-diiodide	endoiodin	$[I(CH_3)_3N\cdot CH_2]_2:CHOH$	....	430.11
3592	pararosaniline base	4,4',4''-tris-diMe-amino-triPh-carbinol	$[(CH_3)_2N\cdot C_6H_4]_3C\cdot OH$	XIII-755	389.55
3593	pararosaniline chloride	(dye salt) ‡	$C_{25}H_{30}N_3Cl\cdot 9H_2O$	XIII-756	570.13
3594	pararosaniline-hydroxymethylate	(dye salt) §	$C_{26}H_{33}N_3Cl_2$	XIII-758	458.48
3595	4,4',4''-triamino-triphenyl-methane	(leuco base)	$[(CH_3)_2N\cdot C_6H_4]_3CH$	XIII-315	373.55

† Beilstein lists this compound as a chalkone.
‡ Commercial product usually mixed with pentamethyl deriv.
§ Base not isolated; comes into commerce as $ZnCl_2$ double salt.
Hexadecyl acetylene 4942
Hexadecyl bromide 1276

Hexadienoic acid 5625
Hexadiyne 2768

Hexaethyl-pararosaniline 3191
Hexahydro-aniline 326
Hexahydro-benzene 1601
Hexahydro-cresol 4200-1
Hexahydro-diethylaniline 2145

Table 7-4 (*Continued*)
PHYSICAL CONSTANTS OF ORGANIC COMPOUNDS

No.	Crystalline Form and Color	Specific Gravity	Melting Point °C.	Boiling Point °C.	Solubility in 100 Parts		
					Water	Alcohol	Ether
3559	yel. wh. cr./ EtOAc	....	226-7 d.	....	i.; v. sl. s. alk.	s.; sl. s. bz.	s.; sl. s. chl.
3560	hyg. nd.	....	251 $\pm$, d.	....	0.02 h.	sl. s.	i.; i. bz.
3561	nd./bz.	....	316	....	i. h.	$0.01^{20°}$	v. sl. s.
3562	rhb.	3.823	170 d.	....	v. s. CS_2	v. sl. s. h.	v. sl. s.
3563	mn.	$2.044^{24°}$	228-31	309^{742mm}	i.; s. bz.	v. sl. s. h.	s. h.
3564	rhb.	$2.091^{\frac{20°}{4}}$	186.9-7.4*	184.4	$0.005^{22°}$	v. s.	v. s.
3565	yel. cr.	....	46	....	....	....	....
3566	yel. cr.	....	106	d. 210	i.	s.; s. chl.	s. pet.
3567	cr.	$0.779^{\frac{5.7°}{4}}$	56.6	262^{15mm}	s. bz.	v. sl. s.	$3.5^{15°}$chl.
3568	pl./et.	....	61	$207^{0.7mm}$	i.	s.	s.
3569	lf.	$0.774^{\frac{20°}{4}}$	18.5(16.2)	287.5	i.	∞	∞
3570	lq.	$0.784^{\frac{15°}{4}}$	4	274	i.	....	....
3571	cr.	$0.797^{20°}$	15	155^{15mm}	i.	....	....
3572	pl.	$0.804^{\frac{20°}{4}}$	20	160^{15mm}	i.	....	....
3573	oil	$0.720^{20°}$	....	80	i.	....	....
3574	lq.	$0.691^{20°}$	-141	59.6	i.	....	....
3575	pr./al.	$0.831^{\frac{13.0°}{4}}$	130	298.3	i.; v. s. bz.	$0.75^{25°}$	$8^{25°}$
3576	lf.	....	63	290	i.; v. s. bz.	v. s.	v. s.
3577	pr.	$1.034^{\frac{22°}{4}}$	30-1	232-4	sl. s.	v. s.; s. chl.	v. s.; s. pet.
3578	lq.	$0.790^{\frac{20°}{4}}$	-90.6	154.7	i.	v. s.	v. s.
3579	lq.	$0.793^{\frac{20°}{0}}$	....	169-70	i.	v. s.	v. s.
3580	syrup	....	d., $-H_2O$	....	v. s.	v. s.	sl. s.
3581	lq.	$0.787^{4°}$	....	135-8	....	....	....
3582	lq.	$0.934^{\frac{23°}{0}}$	....	200	....	....	....
3583	tri./aq.	....	192	d., $-H_2O$ >192	>$0.2^{20°}$	s.	....
3584	mn./aq.	....	215-21	....	$0.2^{20°}$	s. act.	....
3585	lf./aq.	....	....	....	>$0.08^{17°}$	s.	s.; s. chl.
3586	mn./aq.	....	300 $\pm$	subl.	$0.08^{17°}$; 1.3 h.	s.; s. act.	sl. s.; i. chl.
3587	cr./aq.	....	111	....	s.	s.	s.; sl. s. bz.
3588	nd./aq. HCl	....	d. 200	....	sl. s. c.	sl. s.	sl. s.; sl. s. bz.
3589	red nd./bz.	....	340-50 d.	....	i.	i.	i.
3590	pl./al.	....	165.5	265	i.	$0.2^{0°}$; 4.6 h.	v. s.; v. s. bz.
3591	col. cr. pd.	....	275 d.	....	s.	sl. s.	i.; i. act.
3592	col. cr./bz.	....	219	....	i.; s. bz.; s. pet.	sl. s.; s. CS_2	s.; s. chl.
3593	gn. brn. met. cr./aq.	....	$-8H_2O$, 70-80°	....	s.	s.	v. s. chl.
3594	gn. pd.	....	....	....	s.	sl. s.	....
3595	lf./al.	....	178-9	....	i. c.; s. bz.	s. h.; s. ac.	s.; s. chl.

* Sealed tube.
Hexahydro-phenol 1605
Hexahydro-pyridine 5322
Hexahydro-toluene 4199

Hexahydro-triphenyltriazine 512
Hexahydro-xylene 2471-3
Hexahydroxy-anthraquinone 5575
Hexahydroxy-hexahydrobenzene 1602-3

Hexalin 1605
Hexamethyl-ethane 4945

Table 7-4 (*Continued*)
PHYSICAL CONSTANTS OF ORGANIC COMPOUNDS

No.	Name	Synonym	Formula	Beil. Ref.	Formula Weight
3596	**Hexamethylene-** diamine	....	$NH_2 \cdot (CH_2)_6 \cdot NH_2$	IV-269	116.21
3597	dicyanide	....	$NC \cdot (CH_2)_6 \cdot CN$	II-694	136.20
3598	glycol	hexandiol-1,6	$HO \cdot (CH_2)_6 \cdot OH$	I-484	118.18
3599	tetramine	urotropine; formin; methenamine	$(CH_2)_6N_4$	I-583	140.19
3600	tetramine-alliodide	allyliodourotropine	$C_6H_{12}N_4 \cdot C_3H_5I$	I-588	308.17
3601	camphorate	amphotropin	$C_{12}H_{24}N_8 \cdot C_{10}H_{16}O_4$	*I-314	480.62
3602	salicylate	saliformin	$C_6H_{12}N_4 \cdot C_7H_6O_3$	....	278.31
3603	**Hexane** (*n*)	....	$CH_3 \cdot (CH_2)_4 \cdot CH_3$	I-142	86.18
3604	**Hexane** (*iso*)	2-methyl-pentane	$(CH_3)_2CH \cdot (CH_2)_2 \cdot CH_3$	I-148	86.18
3605	**Hexane**	2,2-dimethyl-butane	$(CH_3)_3C \cdot C_2H_5$	I-150	86.18
3606	**Hexane**	2,3-dimethyl-butane	$[(CH_3)_2CH]_2$	I-151	86.18
3607	**Hexane**	3-methyl-pentane	$(C_2H_5)_2CH \cdot CH_3$	I-149	86.18
3608	**Hexanhexol**	dulcitol	$HOCH_2(CHOH)_4 \cdot CH_2OH$	I-544	182.17
3609	**Hexanhexol**	*d*-mannitol; mannite	$HOCH_2(CHOH)_4 \cdot CH_2OH$	I-534	182.17
3610	hexaacetate	mannitol hexa-acetate	$C_6H_8O_6(COCH_3)_6$	II-150	434.40
3611	**Hexanhexol**	*d*-sorbitol; sorbite	$HOCH_2(CHOH)_4 \cdot CH_2OH$	I-533	182.17
3612	hexaacetate	sorbitol hexaacetate	$C_6H_8O_6(COCH_3)_6$	II-150	434.40
3613	**Hexatriacontane**	....	$C_{36}H_{74}$	I-178	506.99
3614	**Hexenyl** alcohol	....	$C_6H_{11}OH$	I-446	100.16
3615	alcohol	1-hexen-5-ol	$CH_2:CH(CH_2)_2 \cdot CHOH \cdot CH_3$	I-444	100.16
3616	**Hexose**	*d*-fructose; levulose	$C_5H_{12}O_5:CO$	I-918	180.16
3617	**Hexose**	*d*-galactose	$C_5H_{11}O_5 \cdot CHO$	I-911	180.16
3618	**Hexose**	*d*-glucose; dextrose	$C_5H_{11}O_5 \cdot CHO$	I-879	180.16
3619	**Hexose**	*d*-mannose; semi-nose	$C_5H_{11}O_5 \cdot CHO$	I-905	180.16
3620	**Hexose**	*d*-sorbose; sorbinose	$C_6H_{12}O_6$	I-927	180.16
3621	**Hexyl** acetate (*n*)	....	$CH_3CO_2 \cdot C_6H_{13}$	II-132	144.22
3622	acetate	Me-*iso*-Bu carbinol acetate	$CH_3CO_2 \cdot CH(CH_3) \cdot CH_2CH(CH_3)_2$	II-133	144.22
3623	acetate (*iso*)	....	$CH_3CO_2 \cdot C_4H_7(CH_3)_2$	II-133	144.22
3624	alcohol (*n*)	hexanol-1	$CH_3(CH_2)_4CH_2OH$	I-407	102.18
3625	alcohol (*n*)	hexanol-2	$CH_3 \cdot CHOH \cdot C_4H_9$	I-408	102.18
3626	alcohol (*n*)	hexanol-3	$C_2H_5 \cdot CHOH \cdot C_3H_7$	I-408	102.18
3627	alcohol	2,2-diMe-butanol-3	$(CH_3)_3C \cdot CHOH \cdot CH_3$	I-412	102.18
3628	alcohol	2,2-diMe-butanol-4	$(CH_3)_3CCH_2CH_2OH$	I-412	102.18
3629	alcohol	2,3-diMe-butanol-2	$(CH_3)_2CH \cdot C(CH_3)_2 \cdot OH$	I-413	102.18
3630	alcohol	2-Me-pentanol-1	$C_2H_5 \cdot CH_2 \cdot CH(CH_3) \cdot CH_2OH$	I-409	102.18
3631	alcohol	2-Me-pentanol-2	$(CH_3)_2COH \cdot CH_2 \cdot C_2H_5$	I-409	102.18
3632	alcohol	2-Me-pentanol-3	$(CH_3)_2CH \cdot CHOH \cdot C_2H_5$	I-410	102.18
3633	alcohol	2-Me-pentanol-4	$CH_3 \cdot CHOH \cdot CH_2 \cdot CH(CH_3)_2$	I-410	102.18

Table 7-4 (*Continued*)
PHYSICAL CONSTANTS OF ORGANIC COMPOUNDS

No.	Crystalline Form and Color	Specific Gravity	Melting Point °C.	Boiling Point °C.	Solubility in 100 Parts		
					Water	Alcohol	Ether
3596	lf.	....	42	204-5	v. s.	s.	....
3597	col. lq.	$0.954^{18°}$	-3.5	185^{15mm}	....	....	....
3598	nd./aq.	....	42	250	s.	s.	sl. s. h.
3599	col. rhb.	....	....	subl. vac. 230-70	$81^{12°}$	$31^{2°}$ abs.	v. sl. s.
3600	col. cr.	....	148±, d.	....	v. s.	i. chl.	i.
3601	col. cr.	....	....	....	10	s.; s. chl.	i.; i. bz.
3602	col. cr.	....	....	....	s.	s.	....
3603	col. lq.	$0.659^{\frac{20°}{4}}$	-95.3	68.7	i.; ∞ chl.	$50^{33°}$	∞
3604	lq.	$0.654^{\frac{20°}{4}}$	-153.7	60.3	i.	....	s.
3605	lq.	$0.649^{\frac{20°}{20}}$	-99.7	49.7	i.	....	s.
3606	lq.	$0.662^{\frac{20°}{4}}$	-128.4	58.0	i.	....	s.
3607	lq.	$0.664^{\frac{20°}{4}}$	-118	63.3	i.	....	s.
3608	mn.	$1.466^{15°}$	189	$290\text{-}5^{3mm}$	$3.2^{15°}$; v. s. h.	v. sl. s.	i.
3609	col. rhb.	$1.489^{\frac{20°}{4}}$	167-9	$290\text{-}5^{3mm}$	$13^{14°}$	$0.01^{14°}$ abs. al.	i.
3610	col. cr.	....	122-4	....	i.	sl. s. h.	i.; s. ac.
3611	cr. $+\frac{1}{2}$ or $1H_2O$	....	110-2 (anh.)*	....	v. s.	v. s. h.	....
3612	col. cr./al.	....	99-100	....	sl. s.	sl. s.	i.
3613	cr.	$0.782^{76°}$	75.8	$265^{1.0mm}$	sl. s. chl.	v. sl. s.	sl. s.
3614	lq.	$0.891^{10°}$	....	137^{765mm}	$10^{10°}$	∞	∞
3615	lq.	$0.842^{\frac{16.2°}{17.5}}$	....	140^{759mm}	v. sl. s.	....	....
3616	nd./aq.	$1.669^{17.5°}$	95-105	....	v. s.	$8.5^{18°}$ abs. al.	s. act.
3617	pr.	....	165.5 (anh.)**	....	$10.3^{0°}$ $68.3^{25°}$	$0.6^{38.5°}$ 85% al.	s. pyr.
3618	rhb.***	$1.544^{25°}$	146(anh.)	....	$82^{17.5°}$	sl. s.	i.
3619	rhb.	$1.539^{\frac{20°}{4}}$	132	....	$248^{17°}$	v. sl. s. abs.	i.
3620	rhb.	$1.654^{15°}$	165	....	$55^{17°}$	v. sl. s. abs.	sl. s. Me al.
3621	col. lq.	$0.878^{\frac{15°}{4}}$	-60.9	171.5	i.	v. s.	v. s.
3622	col. lq.	$0.860^{\frac{20°}{20}}$	-63.8	146-7	$0.13^{20°}$	....	....
3623	col. lq.	....	....	159^{755mm}	i.	v. s.	v. s.
3624	col. lq.	$0.822^{\frac{15°}{4}}$	-51.6	157.5	$0.6^{20°}$	∞	∞
3625	col. lq.	$0.818^{\frac{16.8°}{4}}$	....	139-40	sl. s.	∞	∞
3626	col. lq.	$0.818^{\frac{20°}{4}}$	....	135	sl. s.	∞	∞
3627	lq. or nd.	$0.812^{25°}$	5.5	120-1	v. sl. s.	∞	∞
3628	oily lq.	....	-60	....	v. sl. s.	∞	∞
3629	lq.	$0.821^{\frac{20°}{0}}$	-14	120-1	v. sl. s.	∞	∞
3630	lq.	$0.826^{\frac{20°}{4}}$	....	148^{762mm}	v. sl. s.	∞	∞
3631	lq.	$0.809^{\frac{20°}{4}}$	-107	122.5-3.5^{762mm}	v. sl. s.	∞	∞
3632	lq.	$0.826^{\frac{18°}{4}}$	....	127.5^{721mm}	v. sl. s.	∞	∞
3633	lq.	$0.813^{\frac{20°}{4}}$	....	131.8	$1.6^{20°}$	∞	∞

* Crysts./al., m. 87-95°.
** Crysts. $+ 1H_2O$, m. 118-20°.
*** Crysts. $+ 1H_2O$, mn. pl.
Hexene 3655-61

Hexophan 3801
Hexone 4164
Hexyl, cf. also dihexyl.

Table 7-4 (*Continued*)
PHYSICAL CONSTANTS OF ORGANIC COMPOUNDS

No.	Name	Synonym	Formula	Beil. Ref.	Formula Weight
3634	**Hexyl** alcohol	2-Me-pentanol-5	$(CH_3)_2CH\cdot CH_2\cdot CH_2\cdot CH_2OH$	I-411	102.18
3635	alcohol (*act.*)	3-Me-pentanol-1	$C_2H_5\cdot CH(CH_3)\cdot CH_2\cdot CH_2OH$	I-411	102.18
3636	alcohol (*act.*)	3-Me-pentanol-2	$CH_3\cdot CHOH\cdot CH\cdot (CH_3)\cdot C_2H_5$	I-411	102.18
3637	alcohol	3-Me-pentanol-3	$C_2H_5\cdot C(CH_3)\cdot (OH)\cdot C_2H_5$	I-411	102.18
3638	alcohol	3-methylol-pentane	$(C_2H_5)_2CH\cdot CH_2OH$	I-412	102.18
3639	amine (*n*)	....	$CH_3(CH_2)_5\cdot NH_2$	IV-188	101.19
3640	amine	2-Et Bu-amine	$(C_2H_5)_2CHCH_2\cdot NH_2$	IV-192	101.19
3641	bromide (*n*)	1-bromohexane	$CH_3(CH_2)_4CH_2Br$	I-144	165.08
3642	bromide (*iso*)	5-bromo-2-methyl-pentane	$(CH_3)_2CH\cdot (CH_2)_2\cdot CH_2Br$	I-148	165.08
3643	bromide	2-bromo-2,3-diMe-butane	$(CH_3)_2CH\cdot CBr\!:(CH_3)_2$	I-152	165.08
3644	chloride (*n*)	1-chlorohexane	$CH_3\cdot (CH_2)_4\cdot CH_2Cl$	I-143	120.62
3645	chloride	2-chlorohexane	$CH_3(CH_2)_3CHCl\cdot CH_3$	I-144	120.62
3646	chloride	1-chloro-2,3-diMe-butane	$(CH_3)_2CH\cdot CH\!:(CH_2Cl)CH_3$	I-151	120.62
3647	chloride	2-chloro-2,3-diMe-butane	$(CH_3)_2CH\cdot CCl(CH_3)_2$	I-151	120.62
3648	cyanide (*n*)	oenanthylic nitrile	$CH_3(CH_2)_5\cdot CN$	II-341	111.19
3649	formate (*n*)	....	$HCO_2\cdot C_6H_{13}$	II-22	130.19
3650	iodide (*n*)	1-iodohexane	$CH_3\cdot (CH_2)_4\cdot CH_2I$	I-146	212.07
3651	iodide	2-iodo-2,3-dimethyl-butane	$(CH_3)_2CI\cdot CH(CH_3)_2$	I-153	212.07
3652	mercaptan (*n*)	hexanthiol-1	$CH_3(CH_2)_4\cdot CH_2\cdot SH$	I-408	118.24
3653	nitrite (*n*)	....	$C_6H_{13}\cdot O\cdot NO$	I-407	131.18
3654	resorcinol (*n*)	(1;2,4)	$C_6H_{13}\cdot C_6H_3(OH)_2$	**VI-904	194.28
3655	**Hexylene** (*α*)	1-hexene	$CH_3(CH_2)_3CH\!:\!CH_2$	I-215	84.16
3656	**Hexylene**	2,3-diMe-butene-1	$(CH_3)_2CH\cdot C(CH_3)\!:\!CH_2$	I-218	84.16
3657	**Hexylene**	3,3-diMe-butene-1	$(CH_3)_3C\cdot CH\!:\!CH_2$	....	84.16
3658	**Hexylene**	2,3-diMe-butene-2	$(CH_3)_2C\!:\!C(CH_3)_2$	I-218	84.16
3659	**Hexylene**	2-methyl-pentene-2	$(CH_3)_2C\!:\!CH\cdot C_2H_5$	I-217	84.16
3660	**Hexylene** (*cis?*)	3-methyl-pentene-2	$(C_2H_5)(CH_3)C\!:\!CH\cdot CH_3$	I-217	84.16
3661	*trans?*	3-methyl-pentene-2	C_6H_{12}	I-217	84.16
3662	glycol (2,3)	hexandiol-2,3	$C_3H_7\cdot (CHOH)_2\cdot CH_3$	I-484	118.18
3663	**Hexylphenyl-carbinol** (*n*)	α-Ph-*n*-heptyl alcohol	$C_6H_{13}\cdot CHOH\cdot C_6H_5$	*VI-272	192.30
3664	**Hippuric** acid	benzoyl glycine	$C_6H_5CO\cdot NHCH_2\cdot CO_2H$	IX-225	179.18
3665	calcium salt	Ca hippurate	$(C_9H_8O_3N)_2Ca\cdot 3H_2O$	IX-229	450.46
3666	sodium salt	Na hippurate	$C_9H_8O_3NNa$	IX-229	201.16
3667	**Histamine**	β-imidazolyl-4-ethylamine	$C_3H_3N_2\cdot C_2H_4\cdot NH_2$	*XXV-629	111.15
3668	hydrochloride, di	....	$C_5H_9N_3\cdot 2HCl$	*XXV-630	184.08
3669	**Histidine** (*l*)(−)	β-imidazolyl-α-alanine	$C_3H_3N_2\cdot CH_2\cdot CH(NH_2)\cdot CO_2H$	XXV-513	155.16
3670	hydrochloride, di	....	$C_6H_9O_2N_3\cdot 2HCl$	XXV-515	228.08

Table 7-4 (*Continued*)
PHYSICAL CONSTANTS OF ORGANIC COMPOUNDS

No.	Crystalline Form and Color	Specific Gravity	Melting Point °C.	Boiling Point °C.	Solubility in 100 Parts		
					Water	Alcohol	Ether
3634	lq.	$0.816_4^{20°}$	....	$151\text{-}27^{47mm}$	v. sl. s.	∞	∞
3635	lq.	$0.826^{20°}$	....	152-3	i.	s.	s.
3636	oil	$0.831_4^{18°}$	....	135^{762mm}	....	....	....
3637	lq.	$0.824_0^{20°}$	<−38	122.5^{758mm}	i.	....	....
3638	lq.	$0.833_{20}^{20°}$	<−15	148.9	0.43	....	....
3639	col. lq.	$0.763_4^{25°}$	−19	$129\text{-}30^{742mm}$	v. sl. s.	∞	∞
3640	col. lq.	....	....	125.3	s.	∞	∞
3641	lq.	$1.173^{20°}$	....	155.5^{744mm}	....	....	....
3642	col. lq.	$1.168^{20°}$	....	146-7	....	....	....
3643	col. cr.	$1.177_0^{20°}$	24-5	$132\text{-}3^{742mm}$	....	....	....
3644	lq.	$0.876_4^{20°}$	....	132.9^{765mm}	i.	....	....
3645	lq.	$0.869_4^{21°}$	....	122.5^{754mm}	....	....	....
3646	lq.	$0.887^{22°}$	....	122	....	....	....
3647	lq.	$0.878^{19°}$	−10.4	112	....	....	....
3648	col. lq.	$0.813_4^{15°}$	−64	184.6	i.	....	....
3649	lq.	$0.886_4^{20°}$	−62.7	155.5	....	∞	∞
3650	lq.	$1.439_4^{20°}$	....	180^{763mm}	....	....	....
3651	nd.	$1.446_0^{20°}$	....	140^{749mm}	....	....	....
3652	col. lq.	$0.849^{20°}$	....	$149\text{-}50^{768mm}$	i.	....	....
3653	yel. lq.	$0.885^{20°}$	....	$129\text{-}30^{774mm}$	i.	s.	s.
3654	col. nd.	....	68-70	179^{7mm}	0.05	v. s.; s. bz.	s.; s. act.
3655	col. lq.	$0.673_4^{20°}$	−139	63.6	i.	∞	∞
3656	col. lq.	$0.678^{20°}$	−140.0	55.6	i.; ∞ CS_2	∞	∞
3657	col. lq.	$0.653^{20°}$	−115.5	41.2	i.	....	....
3658	lq.	$0.708^{20°}$	−75.4	73.2^{760mm}	....	s. act.	s.
3659	lq.	$0.686^{20°}$	....	67.2	s. 2 vol. H_2SO_4 + 1 aq.		
3660	lq.	$0.699^{20°}$	−138.4	70.5	....	....	....
3661	lq.	$0.694^{20°}$	−135.2	67.8	....	....	....
3662	cr.	$0.967^{0°}$	60	206-7	∞	s.	....
3663	lq.	0.946	....	275	....	....	....
3664	rhb.	$1.371_4^{20°}$	189-90	d.	$0.4^{20°}$; i. pet.	s. h.; i. bz.	$0.25^{18°}$
3665	col. pr.	1.32	....	....	5 c.; 16 h.	....	....
3666	wh. pd.	....	....	....	s.	s.	....
3667	col. hyg. cr./aq.	....	86*	$209\text{-}10^{18mm}$	v. s.; s. h. chl.	s.	i.
3668	pr./aq. al.	....	239-46 (d.)	....	s.	sl. s.	i.
3669	lf./aq.	....	d. 287-8	....	s.	v. sl. s.	i.
3670	rhb.	....	245	....	s. d.	i.	i.

* Sealed tube.
Homohydroquinone 2372
Homopyrocatechin 2374

Homosalicylaldehyde 4379
Homosalicylic acid 3833, 3836, 3839, 3841
Hot stuff gas 1985

Table 7-4 (*Continued*)
PHYSICAL CONSTANTS OF ORGANIC COMPOUNDS

No.	Name	Synonym	Formula	Beil. Ref.	Formula Weight
3671	**Histidine** hydrochloride, mono	....	$C_6H_9O_2N_3 \cdot HCl \cdot H_2O$	XXV-515	209.63
3672	**Homo-**eriodictyol	5,7,4'-triOH-3'-methoxy-flavanone	$C_{16}H_{14}O_6$†	....	302.29
3673	gentisinic acid	2,5-diOH-phenyl-acetic acid	$(HO)_2C_6H_3CH_2 \cdot CO_2H \cdot H_2O$	X-407	186.17
3674	phthalic acid (*o*)	....	$HO_2C \cdot C_6H_4 \cdot CH_2 \cdot CO_2H$	IX-857	180.16
3675	**Hordenine** (*p*)‡	....	$HO \cdot C_6H_4 \cdot CH_2 \cdot CH_2 \cdot N(CH_3)_2$	XIII-626	165.24
3676	sulfate	....	$(C_{10}H_{15}ON)_2 \cdot H_2SO_4 \cdot H_2O$	XIII-626	446.57
3677	sulfate	....	$(C_{10}H_{15}ON)_2 \cdot H_2SO_4 \cdot 2H_2O$	*XIII-236	464.58
3678	**Hydantoic acid**	....	$H_2N \cdot CO \cdot NH \cdot CH_2 \cdot CO_2H$	IV-359	118.09
3679	**Hydantoin**	glycolyl urea	$HN \cdot CH_2 \cdot CO \cdot NH \cdot CO$ ‖————‖	XXIV-242	100.08
3680	**Hydnocarpic acid**	....	$C_5H_7(CH_2)_{10}CO_2H$	IX-79	252.40
3681	**Hydracetamide**	....	$(CH_3 \cdot CH)_3N_2$	I-608	112.18
3682	**Hydracrylic acid**	ethylene lactic acid	$HO \cdot CH_2CH_2 \cdot CO_2H$	III-295	90.08
3683	**Hydratropic acid**	α-Ph-propionic acid	$C_6H_5CH(CH_3)CO_2H$	IX-524	150.18
3684	**Hydrazinobenzoic acid** (*p*)	....	$H_2N \cdot NH \cdot C_6H_4 \cdot CO_2H$	XV-631	152.15
3685	**Hydrazo-**benzene	*N,N'*-diphenyl-hydrazine	$C_6H_5NH \cdot NHC_6H_5$	XV-123	184.24
3686	benzoic acid (*o*)	....	$(HO_2C \cdot C_6H_4 \cdot NH \cdot)_2$	XV-626	272.26
3687	benzoic acid (*m*)	....	$(HO_2C \cdot C_6H_4 \cdot NH \cdot)_2$	XV-629	272.26
3688	benzoic acid (*p*)	....	$(HO_2C \cdot C_6H_4 \cdot NH \cdot)_2$	XV-632	272.26
3689	dicarbonamide	....	$(H_2N \cdot CO \cdot NH \cdot)_2$	III-116	118.10
3690	naphthalene (α,α')	....	$C_{10}H_7NH \cdot NHC_{10}H_7$	XV-562	284.36
3691	naphthalene (β,β')	....	$C_{10}H_7NH \cdot NHC_{10}H_7$	XV-569	284.36
3692	toluene (*o*)	....	$(CH_3 \cdot C_6H_4 \cdot NH \cdot)_2$	XV-497	212.30
3693	toluene (*m*)	....	$(CH_3 \cdot C_6H_4 \cdot NH \cdot)_2$	XV-506	212.30
3694	toluene (*p*)	....	$(CH_3 \cdot C_6H_4 \cdot NH \cdot)_2$	XV-511	212.30
3695	**Hydrindene** (1,2)	indane	$C_6H_4 \cdot CH_2 \cdot CH_2 \cdot CH_2$ ‖————————‖	V-486	118.16
3696	dicarboxylic acid	(β,β)	$C_9H_8(CO_2H)_2$	IX-904	206.20
3697	dicarboxylic acid	(α,β)	$C_9H_8(CO_2H)_2$	*IX-391	206.20
3698	**Hydrindone** (α)	indanone-1	$C_6H_4 \cdot CH_2 \cdot CH_2 \cdot CO$ ‖————————‖	VII-360	132.16
3699	**Hydrindone** (β)	indanone-2	$C_6H_4 \cdot CH_2 \cdot CO \cdot CH_2$ ‖————————‖	VII-363	132.16
3700	**Hydro-**benzamide	tribenzaldiamine	$C_6H_5CH:(N:CHC_6H_5)_2$	VII-215	298.39
3701	benzoin	....	$(C_6H_5 \cdot CHOH \cdot)_2$	VI-1003	214.27
3702	benzoin (*iso*)	....	$C_{14}H_{12}(OH)_2$	VI-1004	214.27
3703	carbostyril	....	C_9H_9ON	XXI-288	147.18
3704	cinnamic acid	β-Ph-propionic acid	$C_6H_5CH_2CH_2CO_2H$	IX-508	150.18
3705	cinnamic alcohol	γ-Ph-Pr-alcohol	$C_6H_5(CH_2)_3OH$	VI-503	136.20
3706	cinnamic aldehyde	....	$C_6H_5CH_2CH_2 \cdot CHO$	VII-304	134.18

† Beilstein gives structure as a chalcone.
‡ See also Alkaloid table.
HS (hot stuff) gas 1985
Hydracetin 111

Hydraergotocin, cf. alkd.
Hydramine, mixt. of 2314 and 5260
Hydrastine, cf. alkd.
Hydrastinine, cf. alkd.

Table 7-4 (*Continued*)
PHYSICAL CONSTANTS OF ORGANIC COMPOUNDS

No.	Crystalline Form and Color	Specific Gravity	Melting Point °C.	Boiling Point °C.	Solubility in 100 Parts		
					Water	Alcohol	Ether
3671	col. rhb./ aq.	....	250-5	$-H_2O$, 100	s.	i.	i.
3672	col. nd./ ac.	....	224-5	....	i.; i. chl.	s.; i. bz.	sl. s. EtOAc
3673	pr./aq.	....	152(anh.)	....	v. s.	v. s.	v. s.; i. bz.
3674	cr./aq.	....	175-80	....	s. h.	v. s.	sl. s.
3675	col. rhb.	....	117.8 subl. > 140	$173-4^{11mm}$	s.; i. pet.; s. chl.	v. s.; s. dil. a.	v. s.; s. alk.
3676	pr. nd.	....	205(anh.)	....	s.	v. sl. s.	....
3677	col. cr.	....	197	....	s.	sl. s.	i.
3678	mn.	....	160-1	....	$3^{20°}$; s. h.	$0.5^{20°}$; s. h.	....
3679	nd.	....	220-1	....	s. h.; s. alk.	1.7 h.	i.
3680	lf./al.	....	59-60	....	i.	v. sl. s.	v. sl. s.
3681	yel. pd.	....	....	....	v. s. d.	v. s.	....
3682	syrup	....	....	d.	....	....	....
3683	col. lq.	$1.1^{0°}$	<−20	266-7	v. sl. s.	....	....
3684	nd. or pl./ aq.	....	220-5 d.	....	sl. s. h.	....	....
3685	lt. yel./al.	$1.158^{16°}$	131	d.	v. sl. s.	$5^{16°}$	i. ac.
3686	lf./al.	....	205	....	i.	s. h.	s. alk.
3687	lt. yel./al.	....	....	....	i.	sl. s. h.	....
3688	nd./al.	....	....	....	v. sl. s.	sl. s. h.	....
3689	pl./aq.	....	254-9	....	$0.02^{16°}$; sl. s. h.	i.; s. in 30%NaOH	i.
3690	lf./bz.	....	274(271)	....	s. bz.	s.	s.
3691	lt. red lf.	....	140-1	....	i.	v. s.	v. s.
3692	lf./al.	....	165	....	s. bz.	v. sl. s. aq. al.	s.
3693	col. oil	....	....	....	....	v. s.	....
3694	mn.	0.957	133-4(126)	d.	v. s. bz.	v. s.	v. s.
3695	col. lq.	$0.963\frac{16°}{4}$	....	177	i.	∞	∞
3696	lf./aq.	....	199	$-CO_2 > 200$	....	....	....
3697	cr.	....	222	....	v. sl. s. bz.	s.; s. act.	sl. s.
3698	nd./aq.	$1.0994^{2°}$	41-2	243-5	sl. s.	v. s.	v. s.
3699	nd./al.	$1.071^{67°}$	58-61	250-5 sl. d.	i.	v. s.	v. s.
3700	cr./al.	....	101-2	....	i.	v. s.	v. s.
3701	mn.	$0.927^{134°}$	138-9	>300	$0.25^{15°}$	v. s. h.	....
3702	mn./al.	....	121-2	$133^{0.02mm}$	$0.19^{15°}$; 1.25 h.	s.	s.
3703	rhb./al.	....	163	....	i.; s. h. HCl	s.	s.
3704	mn.	$1.071\frac{48.7°}{4}$	48.5	279.8	$0.6^{20°}$	s.; s. chl.	s.; s. bz.
3705	oil	$1.008\frac{20°}{4}$	<−18	235-7	s.	∞	∞
3706	lq.	....	....	$221-4^{744mm}$	i.	16.7	....

Table 7-4 (*Continued*)
PHYSICAL CONSTANTS OF ORGANIC COMPOUNDS

No.	Name	Synonym	Formula	Beil. Ref.	Formula Weight
	Hydro				
3707	cinnamic amide	β-Ph-propionamide	$C_6H_5CH_2CH_2CONH_2$	IX-511	149.19
3708	cinnamic chloride	. . . .	$C_6H_5CH_2CH_2COCl$	IX-511	168.62
3709	cinnamic nitrile	β-Ph-propionitrile	$C_6H_5CH_2CH_2CN$	IX-512	131.18
3710	corulignone (4;3,5)	. . . .	$[HOC_6H_2(OCH_3)_2]_2$	VI-1200	306.32
3711	coumaric acid (*o*)	melilotic acid	$HOC_6H_4(CH_2)_2CO_2H$	X-241	166.18
3712	coumaric acid (*m*)	phenol propionic acid	$HOC_6H_4(CH_2)_2CO_2H$	X-244	166.18
3713	coumaric acid (*p*)	phlorentinic acid	$HOC_6H_4(CH_2)_2CO_2H$	X-244	166.18
3714	cyanic acid	formonitrile	HCN	II-29	27.03
3715	cyanic acid (dimolecular)	prussic acid	HN:CH·NC	II-28	54.05
3716	**Hydroquinone** acetate, di	diacetyl-hydro- quinone (*p*)	$(CH_3CO_2)_2:C_6H_4$	VI-846	194.19
3717	benzyl ether, di	. . . .	$(C_6H_5CH_2O)_2C_6H_4$	VI-845	290.37
3718	benzyl ether, mono	. . . .	$C_6H_5CH_2O·C_6H_4OH$	VI-845	200.24
3719	dicarboxylic acid (2,5;1,4)	(2,5;1,4)	$(HO)_2C_6H_2(CO_2H)_2$	X-554	198.13
3720	**Hydroquino- phthalein**	2,7-dihydroxy- fluorane	$C_{20}H_{12}O_5$	XIX-219	332.32
3721	**Hydroxy**-aceto- phenone (*o*)†	. . . .	$HO·C_6H_4·CO·CH_3$	VIII-85	136.15
3722	acetophenone (*m*)	. . . .	$HO·C_6H_4·CO·CH_3$	VIII-86	136.15
3723	acetophenone (*p*)	. . . .	$HO·C_6H_4·CO·CH_3$	VIII-87	136.15
3724	anthraquinone (1)	. . . .	$C_6H_4(CO)_2C_6H_3·OH$	VIII-338	224.22
3725	anthraquinone (2)	. . . .	$C_6H_4(CO)_2C_6H_3·OH$	VIII-342	224.22
3726	anthraquinone (2)	2-hydroxy- anthraquinone-1,4	$C_{10}H_6(CO)_2CH:COH$	VIII-337	224.22
3727	azobenzene (*o*)	. . . .	$HO·C_6H_4·N:N·C_6H_5$	XVI-90	198.23
3728	azobenzene (*m*)	. . . .	$HO·C_6H_4·N:N·C_6H_5$	XVI-94	198.23
3729	azobenzene (*p*)	. . . .	$HO·C_6H_4·N:N·C_6H_5$	XVI-96	198.23
3730	azobenzene sul- fonic acid (4,4')	. . . .	$HO·C_6H_4·N:N·C_6H_4·SO_3H$	XVI-272	278.29
3731	azobenzene sul- fonic Na	(4,4')	$HO·C_6H_4·N:N·C_6H_4·SO_3Na·2H_2O$	XVI-272	336.29
3732	1,1'-azonaphtha- lene-3,6,4'-tri- sulfonic acid (2) Na	bordeaux S; fast red D	$NaO_3S·C_{10}H_6N:N·C_{10}H_4(OH)(SO_3Na)_2$	*XVI-305	604.48
3733	benzaldehyde (*o*)	salicylaldehyde	$HO·C_6H_4·CHO$	VIII-31	122.12
3734	benzaldehyde (*m*)	. . . .	$HO·C_6H_4·CHO$	VIII-58	122.12
3735	benzaldehyde (*p*)	. . . .	$HO·C_6H_4·CHO$	VIII-64	122.12
3736	benzaldehyde oxime (*o*)	salicylaldoxime	$HO·C_6H_4·CH:NOH$	VIII-49	137.14
3737	benzaldehyde phenylhydrazone	(*o*)	$HO·C_6H_4·CH:N·NH·C_6H_5$	XV-188	212.25
3738	benzamide (*o*)	salicylamide	$HO·C_6H_4·CO·NH_2$	X-87	137.14
3739	benzamide (*m*)	. . . .	$HO·C_6H_4·CO·NH_2$	X-140	137.14
3740	benzamide (*p*)	. . . .	$HOC_6H_4CONH_2·H_2O$	X-164	155.15
3741	benzanilide (*o*)	salicylanilide	$HO·C_6H_4CO·NHC_6H_5$	XII-500	213.24
3742	benzoic acid (*o*)	salicylic acid	$HO·C_6H_4·CO_2H$	X-43	138.12

† See also No. 754.
Hydro-cotarnine, cf. alkd.
Hydro-naphthoquinone 2348, 2350
Hydro-phloron 2379
Hydro-quinine, cf. alkd.

Hydro-quinone 2314
Hydro-quinone diethyl ether 2093
Hydro-quinone dimethyl ether 2408
Hydro-quinone ethyl ether 2929
Hydro-quinone methyl ether 4081

Hydro-resorcin 2285
Hydro-sulfite N.F. 3293
Hydrol 4435
Hydrolit 3295
Hydropyrin 119

Table 7-4 (*Continued*)
PHYSICAL CONSTANTS OF ORGANIC COMPOUNDS

No.	Crystalline Form and Color	Specific Gravity	Melting Point °C.	Boiling Point °C.	Solubility in 100 Parts		
					Water	Alcohol	Ether
3707	nd./aq.	....	104-5	....	....	s.	s.
3708	lq.	$1.135^{21°}$	<-60	225 d.	....	....	....
3709	lq.	$1.001^{18°}$	....	261	....	....	....
3710	mn./al.	....	190±	d.	v. sl. s.	i. CS_2	v. sl. s.
3711	cr./aq.	....	82-3	....	$5^{18°}$; $108^{40°}$	v. s.	v. s.
3712	mn./bz. lg.	....	111	....	i. lg.	s.; s. bz.	s.
3713	mn./et.	....	128-30	....	v. s. h.	v. s.	v. s.; i. CS_2
3714	poison. lq.	$0.688\frac{20}{4}°$	-13.3	25.7	∞	∞	∞
3715	rhb.	....	87	120-5	v. s.	sl. s.	sl. s.
3716	pl./al.	....	123-4	....	sl. s. h.	s.	s.; s. chl.
3717	pl./al.	....	128	....	v. sl. s. ac.	2.5 h.	v. sl. s.
3718	pl./aq.	....	122-3	....	s. h.	s.; s. alk.	s.; s. bz.
3719	yel. cr./al.	....	d.	....	sl. s. h.	sl. s. h.	sl. s.
3720	nd./et.	....	227-32	....	v. sl. s. h.	s.; s. alk.	s.; i. lg.
3721	oil	$1.131\frac{21}{4}°$	4-6	213^{717mm}	sl. s.	∞	∞ ; ∞ ac.
3722	nd. or lf.	$1.099^{109°}$	95-6	296^{756mm}	s. h.	s.	s.; s. bz.
3723	nd./aq. al.	1.109	109	$147-8^{3mm}$	$12^{2°}$; $7.1^{100°}$	s.	s.
3724	or. red/al.	....	194-5	subl.	....	s.	v. s.
3725	yel. nd./al.	....	306-8	subl.	v. sl. s. c.	s.	s.
3726	yel. nd./al.	....	d. 235	subl.	s. alk.	s. alk. carb.	....
3727	or. nd./et.	....	82.5-3	....	sl. s.	s.	s.
3728	yel. pr./bz.	....	114-6	....	0.08 h.	s.	s.
3729	or. rhb./al.	....	155-6	$220-30^{20mm}$ sl. d.	$0.002^{25°}$; 0.08 h.	$31^{25°}$; $0.6^{25°}CCl_4$	v. s.; $2^{25°}$ bz.
3730	yel. red pr./aq.	....	d. > 200	....	sl. s.; i. HCl	s.	i.
3731	red yel. pl.	....	....	....	$0.7^{15°}$; v. s. h.	....	....
3732	red brn. pd.	....	....	....	s.	sl. s.	....
3733	col. oil	$1.153\frac{25}{4}°$	1-2	196.5	sl. s.	∞	∞
3734	nd./aq.	....	106-8	240±	s. h.	s.	s.
3735	nd./aq.	$1.129^{130°}$	116-7	subl.	$1.38^{31°}$	$70^{25°}$ act.	$4^{25°}$ bz.
3736	col. pr./bz.-pet.	....	57-9	d.	v. sl. s. c.	s.; s. dil. HCl	s.; i. lg.; s. bz.
3737	col. nd.	....	142-3	234^{28mm}	sl. s. h.	s. h.	s.
3738	lf./aq.	....	140	d. 270	sl. s.	s.; s. chl.	s.
3739	lf./aq.	....	170.5	....	s. h.	s.; i. chl.	s.; i. CS_2
3740	nd./aq.	....	162(anh.)	$-H_2O$, 100	s. h.	s.; i. chl.	s.; i. CS_2
3741	pr./al.	....	135	d.	v. sl. s. h.	s.; s. bz.; s. chl.	s.; sl. s. CS_2
3742	mn.	$1.443\frac{20}{4}°$	158.3; subl. 76	211^{20mm} ±	$0.16^{4°}$; $2.6^{75°}$	$49.6^{15°}$ abs. al.	$50.5^{15°}$

Hydroxy-acetaldehyde 3470
Hydroxy-acetamide 3472
Hydroxy-acetanilide 33-5
Hydroxy-acetic acid 3468
Hydroxy-acetonaphthone 48

Hydroxy-aniline 345-52
Hydroxy-anthracene 545, 561-2
Hydroxy-benzalacetophenone 3763
Hydroxy-benzene 5118

Table 7-4 (*Continued*)
PHYSICAL CONSTANTS OF ORGANIC COMPOUNDS

No.	Name	Synonym	Formula	Beil. Ref.	Formula Weight
	Hydroxy				
3743	benzoic Ca (*o*)	calcium salicylate	$(C_7H_5O_3)_2Ca \cdot H_2O$†	X-60	332.33
3744	benzoic Na (*o*)	sodium salicylate	$HO \cdot C_6H_4 \cdot CO_2Na$	X-59	160.11
3745	benzoic acid (*m*)	....	$HO \cdot C_6H_4 \cdot CO_2H$	X-134	138.12
3746	benzoic acid (*p*)	....	$HO \cdot C_6H_4 \cdot CO_2H$	X-149	138.12
3747	benzoic anhydride (*o*)	salicylide; β-disalicylide	$(C_6H_4)_2O_2(CO)_2$	XIX-171	240.22
3748	benzonitrile (*o*)	salicylic nitrile	$HO \cdot C_6H_4 \cdot CN$	X-96	119.12
3749	benzophenone (*p*)	....	$C_6H_5 \cdot CO \cdot C_6H_4 \cdot OH$	VIII-158	198.22
3750	benzyl alcohol (*o*)	salicyl alcohol	$HO \cdot C_6H_4 \cdot CH_2OH$	VI-891	124.14
3751	benzyl alcohol (*m*)	....	$HO \cdot C_6H_4 \cdot CH_2OH$	VI-896	124.14
3752	benzyl alcohol (*p*)	....	$HO \cdot C_6H_4 \cdot CH_2OH$	VI-897	124.14
3753	butyric acid (α) (*dl*)	....	$C_2H_5 \cdot CHOH \cdot CO_2H$	III-303	104.11
3754	butyric acid (β)(*l*)	....	$CH_3 \cdot CHOH \cdot CH_2 \cdot CO_2H$	III-307	104.11
3755	butyric acid (β)(*dl*)	....	$HO \cdot C_3H_6 \cdot CO_2H$	III-308	104.11
3756	butyric acid (γ)	....	$HO(CH_2)_3CO_2H$	III-311	104.11
3757	butyric acid (*iso*)(α)	acetonic acid	$(CH_3)_2COH \cdot CO_2H$	III-313	104.11
3758	butyric nitrile (γ)	(*dl*)	$HO \cdot (CH_2)_3 \cdot CN$	III-311	85.11
3759	caproic acid (α)(*dl*)	oxycaproic acid	$CH_3(CH_2)_3CHOH \cdot CO_2H$	III-332	132.16
3760	caproic acid(*iso*)(α)	leucic acid (*dl*)	$HO \cdot C_5H_{10} \cdot CO_2H$	III-336	132.16
3761	caprylic acid (α)	....	$C_6H_{13} \cdot CHOH \cdot CO_2H$	III-348	160.21
3762	caprylic acid (*iso*)(α)	tetramethyl-hydra-crylic acid	$[(CH_3)_2CH]_2 : C(OH) \cdot CO_2H$	III-348	160.21
3763	chalcone (2)‡	*o*-OH-benzalaceto-phenone	$C_6H_5 \cdot CO \cdot CH:CH \cdot C_6H_4 \cdot OH$	VIII-191	224.26
3764	γ-chloro-*n*-butyro-nitrile (β)(*dl*)	....	$ClCH_2 \cdot CHOH \cdot CH_2 \cdot CN$	III-310	119.55
3765	β,β-dimethyl-γ-butyrolactone (α)	(*dl*)§	CHOH·CO·O·|____|CH_2·C:(CH_3)_2	XVIII-3	130.14
3766	β,β-dimethyl-γ-butyrolactone (α)	(*l*)§	$C_6H_{10}O_3$	....	130.14
3767	diphenylamine (*o*)	anilino-phenol	$HO \cdot C_6H_4 \cdot NH \cdot C_6H_5$	XIII-365	185.23
3768	diphenylamine (*m*)	anilino-phenol	$HO \cdot C_6H_4 \cdot NH \cdot C_6H_5$	XIII-410	185.23
3769	diphenylamine (*p*)	anilino-phenol	$HO \cdot C_6H_4 \cdot NH \cdot C_6H_5$	XIII-444	185.23
3770	diphenyl ether (*o*)	phenoxy-phenol	$HO \cdot C_6H_4 \cdot O \cdot C_6H_5$	VI-772	186.21
3771	ethyl ethylenedi-amine	....	$HO \cdot CH_2 \cdot CH_2 \cdot NH \cdot CH_2 \cdot CH_2 \cdot NH_2$	IV-286	104.15
3772	glutamic acid (β)	(*l+*)	$HO_2C \cdot CH_2 \cdot CHOH \cdot CH(NH_2) \cdot CO_2H$	*IV-550	163.13
3773	glutaric acid (α) (*dl*)	....	$HO_2C \cdot CHOH \cdot CH_2 \cdot CH_2 \cdot CO_2H$	III-442	148.12
3774	glutaric acid (β)	....	$HOCH:(CH_2CO_2H)_2$	III-443	148.12

† Crysts. $+ 2H_2O$, $2.8^{16°}$ aq., $-H_2O$ $100°$; $+ 3H_2O$, $2.3^{15°}$, $35.8^{100°}$ aq.
‡ See also No. 1781.
§ Lactone portion of pantothenic acid is the *l*-form.
Hydroxy-benzoquinone-oxime 4900
Hydroxy-brasilin 3500
Hydroxy-butene 4422
Hydroxy-butyraldehyde 169
Hydroxy-chlorobenzaldehyde 1461
Hydroxy-cinchonine, cf. alkd.
Hydroxy-cinnamic acid 1531-3, **1537**
Hydroxy-coniine, cf. alkd.
Hydroxy-coumarin 6369
Hydroxy-cumene 5433-5
Hydroxy-deguelin 5699

Table 7-4 (*Continued*)
PHYSICAL CONSTANTS OF ORGANIC COMPOUNDS

No.	Crystalline Form and Color	Specific Gravity	Melting Point °C.	Boiling Point °C.	Solubility in 100 Parts		
					Water	Alcohol	Ether
3743	col. nd.	....	....	....	1.4 c.; 4 h.	s.	s.
3744	col. pl./al.	....	....	....	$125^{25°}$	$17^{15°}$	25 gly.
3745	rhb./aq.	1.473	201	....	$0.8^{19°}$; s. h.	s. h.	$10^{10°}$
3746	mn./aq. al.	$1.468^{4°}$	214.5-5.5	....	$0.17^{0°}$; $2.7^{55°}$	v. s.	$23^{25°}$; i. CS_2
3747	nd./chl.	....	200-1	d.	i.	s. h.	....
3748	pr./bz.	....	97-8	149^{14mm}	v. sl. s.	v. s.	v. s.
3749	rhb.	....	134-5	....	s. h.	$25^{25°}$ act.	$8^{25°}$
3750	rhb./aq.	$1.161^{25°}$	86-7	subl.	$6.6^{15°}$	v. s.	v. s.; s. bz.
3751	cr./bz.	....	67	300±, d.	v. s. h.	v. s.	v. s.
3752	pr./aq.	....	125	252	v. s.	v. s.	v. s.
3753	hyg. cr.	$1.125^{20°}$	43-4	subl. > 60; 225-60 d.	s.	s.	s.
3754	mn.	....	48-50	....	v. s.	v. s.	v. s.; i. bz.
3755	hyg. syrup	....	....	130^{12mm} ±	....	....	....
3756	lq.	....	<-17	d. slowly room temp.	....	....	....
3757	hyg. pr.	....	79; subl. 50	212	v. s.	v. s.	v. s.; v. sl. s. bz.
3758	lq.	$1.029^{8°}$	....	238-40	s.	s.; i. CS_2	s.; s. chl.
3759	nd.	....	60-2	subl. 100 sl. d.	v. s.	v. s.	v. s.
3760	pl./et. + pet.	....	76-7	....	s.	s.	s.
3761	pl.	....	69.5	....	v. sl. s.	v. s.	v. s.
3762	cr./et.	....	152-3 d.	192-3 d.	v. s.	v. s.	v. s.
3763	yel. lf./aq. al.	....	154-5 d.	....	v. sl. s. CS_2	v. s.	sl. s. chl.
3764	yel. lq.	....	....	250±, d.	s.	s.	s.
3765	col. hyg. nd.	....	56-8	$119-21^{15mm}$	s.; s. bz.	s.; s. chl.; s, CS_2	s.; sl. s. pet.
3766	cr./bz. pet.	....	91-2	....	s.; sl. s. pet.	s.	s.
3767	pr./aq.	....	69-70	$180-9^{20mm}$	sl. s. h.	s.	s.; sl. s. bz.
3768	lf./aq.	....	81.5-2.0	340	sl. s. h.; s. alk.	s.; s. dil. a.	s.; sl. s. lg.
3769	lf.	....	70	330	v. sl. s. c.; s. alk.	s.; s. dil. a.	s.; s. chl.
3770	nd./aq.	....	106-7	$151-5^{11mm}$	sl. s. h.	v. s. h.	v. s. h.
3771	lq.	....	....	238-40	v. s.	v. s.	i.
3772	pr./aq.	....	softens 100	d. > 100	v. s.	i.	i.; v. s. ac.
3773	col. cr.	....	98-100 d.	....	s.	....	....
3774	nd./aq.	....	95	....	v. s.	v. s.	sl. s.

Table 7-4 (*Continued*)
PHYSICAL CONSTANTS OF ORGANIC COMPOUNDS

No.	Name	Synonym	Formula	Beil. Ref.	Formula Weight
	Hydroxy				
3775	hydroquinone tri-acetate (1,2,4)	triacetyl-hydroxy-hydroquinone	$(CH_3CO_2)_3C_6H_3$	VI-1089	252.23
3776	3-methoxyaceto-phenone (4)	acetovanillone; apocynin	$CH_3CO \cdot C_6H_3(OH) \cdot OCH_3$	VIII-272	166.18
3777	methyl benzoic acid	(*o*)	$HO \cdot CH_2 \cdot C_6H_4 \cdot CO_2H$	X-218	152.15
3778	α-naphthoic acid	(2,1)	$HO \cdot C_{10}H_6 \cdot CO_2H$	X-328	188.18
3779	α-naphthoic acid	(5,1)	$HO \cdot C_{10}H_6 \cdot CO_2H$	X-330	188.18
3780	α-naphthoic acid	(6,1)	$HO \cdot C_{10}H_6 \cdot CO_2H$	X-330	188.18
3781	α-naphthoic acid	(7,1)	$HO \cdot C_{10}H_6 \cdot CO_2H$	X-330	188.18
3782	β-naphthoic acid	(1,2)	$HO \cdot C_{10}H_6 \cdot CO_2H$	X-331	188.18
3783	β-naphthoic acid	(3,2)	$HO \cdot C_{10}H_6 \cdot CO_2H$	X-333	188.18
3784	β-naphthoic acid	(5,2)	$HO \cdot C_{10}H_6 \cdot CO_2H$	X-337	188.18
3785	β-naphthoic acid	(7,2)	$HO \cdot C_{10}H_6 \cdot CO_2H$	X-337	188.18
3786	α-naphthoquinone	(2)	$C_{10}H_5O_2 \cdot OH$	VIII-300	174.16
3787	α-naphthoquinone	(5); juglon; nucin	$C_{10}H_5O_2 \cdot OH$	VIII-308	174.16
3788	β-naphthoquinone	(7)	$C_{10}H_5O_2 \cdot OH$	VIII-299	174.16
3789	β-naphthoquinone	(6)	$C_{10}H_5O_2 \cdot OH$	*VIII-638	174.16
3790	naphthyl-*o*-hy-droxy-*m*-toluic acid (β)	epicarin	$HO \cdot C_{10}H_6 \cdot CH_2 \cdot C_6H_3(OH) \cdot CO_2H$	....	294.31
3791	nicotinic acid (α)	....	$HO \cdot C_5H_3N \cdot CO_2H$	XXII-214	139.11
3792	nicotinic acid (*p*)	....	$HO \cdot C_5H_3N \cdot CO_2H$	XXII-215	139.11
3793	oenanthylic acid (α)	....	$C_5H_{11} \cdot CHOH \cdot CO_2H$	III-342	146.19
3794	phenylacetic acid (*o*)	....	$HO \cdot C_6H_4 \cdot CH_2 \cdot CO_2H$	X-187	152.15
3795	phenylacetic acid	(*m*)	$HO \cdot C_6H_4 \cdot CH_2 \cdot CO_2H$	X-189	152.15
3796	phenylacetic acid (*p*)	....	$HO \cdot C_6H_4 \cdot CH_2 \cdot CO_2H$	X-190	152.15
3797	phenylarsonic acid	(*p*)	$HO \cdot C_6H_4 \cdot AsO(OH)_2$	XVI-874	218.04
3798	phenylarsonate Na	(*p*)	$HO \cdot C_6H_4 \cdot As:O (OH)ONa \cdot 2\frac{1}{2}H_2O$	XVI-874	285.06
3799	phenylethylamine (*p*)	tyramine	$HO \cdot C_6H_4 \cdot CH_2 \cdot CH_2 \cdot NH_2$	XIII-625	137.18
3800	phenylglycine (*p*)	photo-glycin	$HO \cdot C_6H_4 \cdot NH \cdot CH_2 \cdot CO_2H$	XIII-488	167.17
3801	phenyl-quinoline-dicarboxylic acid	2-(4-OH-3-CO₂H-Ph)cinchoninic acid	$HO_2C \cdot C_9H_5N \cdot C_6H_3(OH)CO_2H$	*XXII-567	309.28
3802	*o*-phthalic acid	(3;1,2)	$HO \cdot C_6H_3(CO_2H)_2$	X-498	182.13
3803	*o*-phthalic acid	(4;1,2)	$HO \cdot C_6H_3(CO_2H)_2$	X-499	182.13
3804	*m*-phthalic acid	(2;1,3)	$HO \cdot C_6H_3(CO_2H)_2 \cdot H_2O$	X-501	200.15
3805	*m*-phthalic acid	(4;1,3)	$HO \cdot C_6H_3(CO_2H)_2$	X-502	182.13
3806	*m*-phthalic acid	(5;1,3)	$HO \cdot C_6H_3(CO_2H)_2 \cdot 2H_2O$	X-504	218.16
3807	*p*-phthalic acid	(2;1,4)	$HO \cdot C_6H_3(CO_2H)_2$	X-505	182.13
3808	proline (4)(*dl*)	4-OH-pyrrolidine-2-carboxylic acid	$HN:C_4H_5(OH)CO_2H$	XXII-190	131.13
3809	proline (4)(*l*−)	....	$C_5H_9O_3N$	XXII-191	131.13
3810	propiophenone (*o*)	*o*-propionyl-phenol	$HOC_6H_4COC_2H_5$	VIII-102	150.18

Table 7-4 (*Continued*)
PHYSICAL CONSTANTS OF ORGANIC COMPOUNDS

No.	Crystalline Form and Color	Specific Gravity	Melting Point °C.	Boiling Point °C.	Solubility in 100 Parts		
					Water	Alcohol	Ether
3775	nd./abs. al.	....	96-7	>300 sl. d.	d. a.	d. alk.	....
3776	col. pr./ aq.	....	115	295-300	v. s. h.; i. pet.	7.79°	s.; s. bz.
3777	nd.	....	128 d.	....	0.4²⁰°	s.	s.
3778	nd./aq. al.	....	156-7 d.	....	v. sl. s.	v. s. abs.	s.; s. bz.
3779	nd./h. aq.	....	234-7	subl.	s. h.	v. s.	s.; s. ac.
3780	nd./aq.	....	208-9	....	sl. s. h.	v. s.	....
3781	nd./aq.	....	253-4	....	s. h.	s.	....
3782	nd./al.	....	188-91	....	v. sl. s. h.	s.; s. alk.	s.; s. bz.
3783	yel. lf./aq.	....	216	....	sl. s. h.	s.; s. bz.	s.; s. chl.
3784	nd./al.	....	211-2	....	sl. s. h.	s.	....
3785	lf.	....	262	....	s.	s.	s.
3786	red cr./ac.	....	190d.	subl.	sl. s. h.	s.	s.
3787	yel. red/ bz.	....	153-4	d.	i.; s. h. ac.	sl. s. c.	sl. s.
3788	brn. nd.	....	194	....	i. bz.	s.; s. ac.	i.
3789	lt. yel./aq. al.	....	>200	d. > 220	s. h.	s.	....
3790	wh. pd.	....	90±, d.	....	i.	s.; i. chl.	s.; s. act.
3791	nd./aq.	....	256	....	v. sl. s.	....	....
3792	nd./aq.	....	301-2 d.	subl.	sl. s. h.	v. sl. s.	v. sl. s.
3793	pr.	....	65	d.	v. sl. s. c.	....	....
3794	nd./et.	....	145-7	240-3 d.	s.	sl. s. c. chl.	s.
3795	nd./bz. lg.	....	129	190¹¹mm	v. s.	v. s.	v. s.
3796	nd./aq.	....	148	....	v. s. h.	v. s.	v. s.
3797	wh. pd.	....	175-80 d.	....	v. s.	s.	....
3798	nd./aq. al.	....	-2½H₂O, 100	....	s.	....	....
3799	lf./bz.	....	161	175-81⁸mm	<10 h.	10± h.; s. bz.	sl. s. h. xylene
3800	lf./aq.	....	240-1 d.	....	sl. s.; s. alk.	sl. s.	i.
3801	yel. pd.	....	283-4 d.	....	i.; s. alk.	sl. s.; sl. s. EtOAc	i. lg.
3802	nd./aq.	....	150±, d.	....	s.	s.	s.
3803	cr./aq.	....	204-5 d.	....	v. sl. s. bz.	s.	s.; sl. s. pet.
3804	nd./aq.	....	243-4 (anh.)	-H₂O, 100	0.14²⁴°; 2.6¹⁰⁰°	s.	s.; sl. s. chl.
3805	nd./aq.	....	310	....	0.02¹⁰°; 0.71⁰⁰°	s.; s. h. ac.	s.; i. chl.
3806	nd./aq.	....	288(anh.)	-2H₂O, 100	0.035°; 18⁹⁹°	s.	s.; s. bz.
3807	pd./aq.	....	>330	subl. sl. d.	v. sl. s.	s.; s. Me al.	sl. s.
3808	col. pl./ Me al.	....	261-2 d.	....	v. s.	v. sl. s.	....
3809	lf./aq. al.	....	270	....	v. s.	v. sl. s.	....
3810	lq.	....	....	115¹⁵mm	v. sl. s.	s.; s. alk.	s.

Table 7-4 (*Continued*)
PHYSICAL CONSTANTS OF ORGANIC COMPOUNDS

No.	Name	Synonym	Formula	Beil. Ref.	Formula Weight
	Hydroxy				
3811	propiophenone (*p*)	*p*-propionyl-phenol	$HOC_6H_4COC_2H_5$	VIII-102	150.18
3812	pyridine (2)(*α*)	*α*-pyridone	$HO \cdot C_5H_4N$	XXI-43	95.10
3813	pyridine (3)	*β*-pyridone	$HO \cdot C_5H_4N$	XXI-46	95.10
3814	pyridine (*γ*)	*γ*-pyridone	$HO \cdot C_5H_4N \cdot H_2O$	XXI-48	113.12
3815	pyrotartaric acid (*α*)	....	$HO_2C \cdot CH(CH_3) \cdot CHOH \cdot CO_2H$	III-445	148.12
3816	quinaldine (3)	....	$C_{10}H_9ON$	XXI-103	159.19
3817	quinaldine (4)	....	$C_{10}H_9ON$	XXI-104	159.19
3818	quinaldine (5)	....	$C_{10}H_9ON$	XXI-106	159.19
3819	quinaldine (6)	....	$C_{10}H_9ON$	XXI-106	159.19
3820	quinoline (2)(*α*)	carbostyril	$HO \cdot C_9H_6N$	XXI-77	145.16
3821	quinoline (4)(*γ*)	kyanurin	$HO \cdot C_9H_6N \cdot 3H_2O$	XXI-83	199.21
3822	quinoline (5)(*ana*)	....	$HO \cdot C_9H_6N$	XXI-84	145.16
3823	quinoline (6)(*p*)	....	$HO \cdot C_9H_6N$	XXI-85	145.16
3824	quinoline (7)(*m*)	....	$HO \cdot C_9H_6N$	XXI-91	145.16
3825	quinoline (8)(*o*)	oxine	$HO \cdot C_9H_6N$	XXI-91	145.16
3826	*iso*-quinoline (1)	*iso*-carbostyril	$C_6H_4CO \cdot NH \cdot CH:CH$	XXI-100	145.16
3827	quinoline (8) sulfate	chinosol; quinosol	$(HO \cdot C_9H_6N)_2 \cdot H_2SO_4$	XXI-92	388.40
3828	quinoline-5-sulfonic acid (8)	....	$HOC_9H_5N \cdot SO_3H \cdot 2H_2O$	XXII-407	261.26
3829	stearic acid (*α*)	....	$C_{16}H_{33}CHOH \cdot CO_2H$	III-364	300.49
3830	stearic acid (10)	....	$CH_3(CH_2)_7 \cdot CHOH \cdot (CH_2)_8CO_2H$	III-365	300.49
3831	stearic acid (*λ*)	....	$CH_3(CH_2)_5 \cdot CHOH \cdot (CH_2)_{10}CO_2H$	III-366	300.49
3832	*o*-toluic acid (1;2,6)	....	$CH_3C_6H_3(OH)CO_2H$	X-214	152.15
3833	*o*-toluic acid (1;3,2)	Me-salicylic acid (6)	$CH_3C_6H_3(OH)CO_2H$	X-217	152.15
3834	*o*-toluic acid (1;3,6)	....	$CH_3C_6H_3(OH)CO_2H \cdot \frac{1}{2}H_2O$	X-214	161.16
3835	*o*-toluic acid (1;4,2)	....	$CH_3C_6H_3(OH)CO_2H$	X-215	152.15
3836	*m*-toluic acid (1;2,3)	*β*-cresotinic acid	$CH_3C_6H_3(OH)CO_2H$	X-220	152.15
3837	*m*-toluic acid (1;2,5)	....	$CH_3C_6H_3(OH)CO_2H \cdot \frac{1}{2}H_2O$	X-225	161.16
3838	*m*-toluic acid (1;3,5)	....	$CH_3C_6H_3(OH)CO_2H$	X-227	152.15
3839	*m*-toluic acid (1;4,3)	*α*-cresotinic acid	$CH_3C_6H_3(OH)CO_2H$	X-227	152.15
3840	*p*-toluic acid (1;2,4)	....	$CH_3C_6H_3(OH)CO_2H$	X-237	152.15
3841	*p*-toluic acid (1;3,4)	*γ*-cresotinic acid	$CH_3C_6H_3(OH)CO_2H$	X-233	152.15
3842	urea	....	$NH_2 \cdot CO \cdot NHOH$	III-95	76.06
3843	valeric acid (*α*)(*n*)	....	$C_3H_7 \cdot CHOH \cdot CO_2H$	III-320	118.13
3844	*iso*-valeric acid (*α*)	(*dl*)	$(CH_3)_2CH \cdot CHOH \cdot CO_2H$	III-328	118.13
3845	*iso*-valeric acid (*β*)	....	$(CH_3)_2COH \cdot CH_2 \cdot CO_2H$	III-327	118.13

Table 7-4 (*Continued*)
PHYSICAL CONSTANTS OF ORGANIC COMPOUNDS

No.	Crystalline Form and Color	Specific Gravity	Melting Point °C.	Boiling Point °C.	Solubility in 100 Parts		
					Water	Alcohol	Ether
3811	nd./aq.	....	149–50	....	$0.04^{15°}$	$3.3^{100°}$ aq.	$4^{25°}$
3812	nd./bz.	....	106–7	280–1	v. s.	v. s.	s.; sl. s. lg.
3813	nd./bz.	....	129	subl.	s.	s.	sl. s. bz.
3814	mn.	....	92; 148.5 (anh.)	>350	$100^{15°}$	s.; v. sl. s. chl.	i.; i. bz.
3815	mn.	....	123	d.	....	....	....
3816	nd./al.	....	265 d.	....	sl. s.	s.	....
3817	cr.	....	230–1	>360 d.	1 c.; 10 h.	s.; v. sl. s. bz.	v. sl. s.
3818	lf./al.	....	232–4	sl. d.	i.	sl. s. c.	s.
3819	cr./aq.	....	213	sl. d.	sl. s.	s.	s.
3820	pr./al.	....	199–200	subl.	s. h.	v. s.	v. s.
3821	nd./aq.	....	52*	>300 d.	$0.47^{15°}$	v. s. h.	v. sl. s.
3822	nd./al.	....	224 d.	subl.	v. s. h. Na_2CO_3	s.	sl. s.; i. lg.
3823	pr./abs. al.	....	193	>360	v. sl. s. c.	sl. s.; s. alk.	v.sl. s. ;s. a.
3824	pr./abs. al.	....	235–8 d.	....	v. sl. s.	v. s.	s. alk.
3825	pr./aq. al.	....	75–6	266.6^{752mm}	v. sl. s. c.	s.; s. dil. alk.	sl. s.
3826	mn./bz.	....	208–9	subl.	sl. s.	s.; s. chl.	sl. s.; sl. s. bz.
3827	yel. cr. pd.	....	177–8	....	v. s.	sl. s.	i.; s. gly.
3828	pa. yel. cr./HCl	....	275±, d.	....	sl. s. h.	sl. s.	v. sl. s.
3829	nd./chl.	....	92–3	....	v. s. h. bz.	$0.6^{20°}$	s.
3830	tab./al.	....	83–5	....	....	$9.7^{20°}$	$2.4^{20°}$
3831	cr./al.	....	78	....	s. chl.; i. lg.	$13^{20°}$	$5.4^{19°}$
3832	cr./aq.	....	142–5	....	....	....	....
3833	nd./aq.	....	167–8	....	$0.14^{25°}$	v. s.	v. s.
3834	cr./aq.	....	178(anh.)	$-H_2O, 100$	s. h.	v. s.	v. s.; i. chl.
3835	nd./aq.	....	180–2	subl.	s.	v. s.	v. s.
3836	nd./h. aq.	....	163–4	....	s. h.	s.	s.; s. chl.
3837	nd./aq.	....	173–4 (anh.)	$-H_2O, 100$	s. h.	s.	s.; i. CS_2
3838	nd./aq.	....	208–10	subl.	s.	....	....
3839	nd./aq.	....	152–3	subl. sl. d.	s. h.	s.	s.; s. chl.
3840	nd./aq.	....	206–7	subl.	s. h.	s.	s.; i. chl.
3841	nd./aq.	....	177–8	subl.	s.	s.	s. chl.
3842	nd./al.	....	139–40	d.	v. s.	s. h.	....
3843	hyg. pl.	....	34	subl.	v. s.	v. s.	v. s.
3844	rhb.	....	86	....	v. s.	v. s.	v. s.
3845	syrup	....	....	....	v. s.	v. s.	v. s.

* Losses $3H_2O$, 110°; m. p. anh. 201°.

Iodo-gorgonic acid 2401–2	Iodo-nonane 4929	Iodo-propylene 208
Iodo-heptane 3549	Iodo-octadecane 4940	Iodo-propylene glycol 3402
Iodo-hexadecane 1277	Iodo-octane 4980	Iodo-propylene oxide 2881
Iodo-hexahydrobenzene 1614	Iodo-pentane 459–63	Iodo-quinine sulfate, cf. alkd.
Iodo-hexane 3650–1	Iodo-propene 208	Iodo-thymol 5940
Iodo-hydrin 3208, 6287	Iodo-phthalein 5813	Iodo-toluene (ω) 813
Iodo-methane 4287	Iodo-propane 5417–8	Iodol 5817
	Iodo-propyl alcohol 6287	Iodophen 5813

Table 7-4 (*Continued*)
PHYSICAL CONSTANTS OF ORGANIC COMPOUNDS

No.	Name	Synonym	Formula	Beil. Ref.	Formula Weight
3846	**Hydroxy** valeric lactone (γ)	....	$CH_3 \cdot CHCH_2CH_2CO \cdot O$	XVII-235	100.12
3847	**Hyenic acid**	....	$CH_3(CH_2)_{23} \cdot CO_2H$	II-394	382.68
3848	**Hypnoacetin**	....	$C_{16}H_{15}O_3N$	XIII-464	269.30
3849	**Hypogaeic acid**	....	$C_{15}H_{29} \cdot CO_2H$	II-461	254.42
3850	**Imesatin**	....	$C_6H_4 \cdot C(NH)CO \cdot NH$	XXI-440	146.15
3851	**Imidazol**	glyoxalin	$C_3H_4N_2$	XXIII-45	68.08
3852	**Imino**-diacetic acid	....	$HN:(CH_2 \cdot CO_2H)_2$	IV-365	133.10
3853	diaceto-dinitrile	....	$HN:(CH_2 \cdot CN)_2$	IV-367	95.10
3854	**Indandione** (1,3)	α,γ–diketo-hydrindene	$C_6H_4 \cdot CO \cdot CH_2 \cdot CO$	VII-694	146.15
3855	**Indene**	....	$C_6H_4 \cdot CH_2 \cdot CH:CH$	V-515	116.16
3856	**Indican** (β)	indoxyl-β-glucoside	$C_{14}H_{17}O_6N \cdot 3H_2O$	....	349.34
3857	**Indigo**	indigotin	$C_{16}H_{10}O_2N_2$	XXIV-417	262.27
3858	carmine	soluble indigo	$C_{16}H_8O_2N_2(SO_3Na)_2$	XXV-304	466.36
3859	dicarboxylic acid	....	$C_{18}H_{10}O_6N_2$	XXV-273	350.29
3860	disulfonic acid	....	$C_{16}H_8O_2N_2(SO_3H)_2$	XXV-304	422.39
3861	monosulfonic acid	....	$C_{16}H_9O_2N_2(SO_3H)$	XXV-303	342.33
3862	purpurin	indirubin	$(NH \cdot C_6H_4CO \cdot C:)_2$	XXIV-430	262.27
3863	white	....	$(C_6H_4C(OH):C \cdot NH)_2$	XXIII-538	264.29
3864	**Indole**	benzopyrrole	C_8H_7N	XX-304	117.15
3865	**Indolyl**-acetic acid (3)	skatole carboxylic acid (ω)	$C_8H_6N \cdot CH_2CO_2H$	XXII-66	175.19
3866	butyric acid (3)(γ)	....	$C_8H_6N \cdot (CH_2)_3CO_2H$	....	203.24
3867	propionic acid (3)(β)	skatole-ω-acetic	$C_8H_6N \cdot (CH_2)_2CO_2H$	XXII-69	189.22
3868	**Indophenin**	....	$(C_{12}H_7NOS)_2$	XXI-438	426.52
3869	**Indophenol**	....	$HO \cdot C_6H_4 \cdot N:C_6H_4:O$	....	199.21
3870	**Indoxyl**	....	$HN \cdot CH:C(OH) \cdot C_6H_4$	XXI-69	133.15
3871	**Indoxylic acid**	....	$C_6H_4 \cdot COH \cdot C(CO_2H): NH$	XXII-226	177.16
3872	**Inulin**	dahlin; alantin; alant starch	$(C_6H_{10}O_5)_6 \cdot H_2O$	....	990.88
3873	**Iodo**-acetanilide (*p*)	....	$I \cdot C_6H_4 \cdot NH \cdot COCH_3$	XII-671	261.06
3874	acetic acid	....	$I \cdot CH_2 \cdot CO_2H$	II-222	185.95
3875	aniline (*o*)	....	$I \cdot C_6H_4 \cdot NH_2$	XII-669	219.03
3876	aniline (*m*)	....	$I \cdot C_6H_4 \cdot NH_2$	XII-670	219.03
3877	aniline (*p*)	....	$I \cdot C_6H_4 \cdot NH_2$	XII-670	219.03
3878	anisole (*o*)	....	$CH_3O \cdot C_6H_4 \cdot I$	VI-207	234.04
3879	benzene	phenyl iodide	$C_6H_5 \cdot I$	V-215	204.01
3880	benzoic acid (*o*)	....	$I \cdot C_6H_4 \cdot CO_2H$	IX-363	248.02
3881	benzoic acid (*m*)	....	$I \cdot C_6H_4 \cdot CO_2H$	IX-365	248.02
3882	benzoic acid (*p*)	....	$I \cdot C_6H_4 \cdot CO_2H$	IX-366	248.02
3883	benzoyl chloride (*o*)	....	$I \cdot C_6H_4 \cdot CO \cdot Cl$	IX-364	266.47

Table 7-4 (*Continued*)
PHYSICAL CONSTANTS OF ORGANIC COMPOUNDS

No.	Crystalline Form and Color	Specific Gravity	Melting Point °C.	Boiling Point °C.	Solubility in 100 Parts		
					Water	Alcohol	Ether
3846	lq.	$1.050\frac{22}{4}°$	<−18	207-8	∞; i. aq. K_2CO_3	....	....
3847	cr./et.	....	77-8	....	....	v. sl. s. c.	v. s.
3848	lf./al.	....	160±	....	i.	$0.323°$	i.
3849	col. nd.	....	33	236^{15mm}	i.	v. s.	....
3850	yel. pr.	....	175-6	....	i.	s. h.	sl. s.
3851	pr.	....	89-90	255-6	s.	v. s.	sl. s.
3852	rhb.	....	d. 225-36	....	$2.45°$	i.	i.
3853	lf./et.	....	78	....	s.	s.	sl. s.
3854	cr./lg.	....	129-31 d.	....	v. sl. s. c.	s. h.	s. bz.
3855	col. lq.	$0.991\frac{25}{25}°$	−2	181-2	i.	s.	∞
3856	brn. rhb.	....	57; 176-8 (anh.)	....	v. s.	v. s.; sl. s. bz.	sl. s.
3857	b. cr./ aniline	1.35	390-2	subl.	i.	i.	i.; s. h. act.
3858	b. amor.	....	....	....	sl. s. c.	i.	i. aq. NaCl
3859	b. bl. ppt.	....	....	....	s. H_2SO_4	i.	i.; i. chl.
3860	b. amor.	....	....	....	v. s.	v. s.	....
3861	b. amor.	....	d. 200	....	v. s.	v. s.	....
3862	brn. nd.	....	....	subl.	i.	sl. s.	s. ac.
3863	col.-gray	....	....	....	i.; s. alk.	s.	s.
3864	lf./aq.	....	52	253-4	s. h.	s. h.	s.; s. bz.
3865	col. lf./bz.	....	165-8	....	v. sl. s. c.	s.; s. act.	s.; i. chl.
3866	rhb./bz. pet.	....	124-5	....	i.	s.; s. act.	s.; i. chl.
3867	cr./aq.	....	133-4	....	v. sl. s. c.	s.; s. ac.	s.
3868	b. pd.	....	d.	....	i.; s. H_2SO_4	v. sl. s.	v. sl. s.; i. bz.
3869	pl./act. pet.	....	160	....	s.; i. pet.	s.; s. bz.	s.; s. chl.
3870	yel. pr.	....	85	110	s.; s. act.	s.	s.; v. sl. s. pet.
3871	cr.	....	subl. 122-3	d.	v. sl. s; d. h.	....	....
3872	hyg. pd.	1.4(anh.)	d. 160	....	$0.01^{0°}$; $37^{100°}$	$0.02^{16°}$	....
3873	mn.	$1.989^{15-20°}$	183-4	....	s. h.	$6.4^{21°}$	v. s. ac.
3874	col. pl.	....	82-3	....	s.	s.	sl. s.
3875	nd.	....	60-1	....	v. sl. s.	v. s.	v. s.
3876	lf. or nd.	....	33	....	i.	s.	....
3877	nd./aq.	....	67-8	....	sl. s.	s.; s. chl.	s.
3878	yel. lq.	$1.8^{20°}$	....	240-1	i.; s. bz.	∞; ∞ chl.	∞
3879	col. lq.	$1.824\frac{25}{4}°$	−31.3	188.5	i.; ∞ chl.	s.	∞
3880	nd./aq.	2.25	162	....	sl. s. h.	v. s.	v. s.
3881	cr./act.	....	187-8	subi.	si. s.	s.	....
3882	lf.	....	269-70	subl.	sl. s. h.	s.	....
3883	cr.	....	30-1	159^{27mm}	d.	d.	....

Table 7-4 (*Continued*)
PHYSICAL CONSTANTS OF ORGANIC COMPOUNDS

No.	Name	Synonym	Formula	Beil. Ref.	Formula Weight
3884	**Iodo** diphenyl (*o*)	....	I·C_6H_4·C_6H_5	**V-486	280.11
3885	diphenyl (*p*)	....	I·C_6H_4·C_6H_5	V-581	280.11
3886	ethylacetate (*β*)	....	CH_3CO_2·CH_2·CH_2I	II-129	214.00
3887	form	triiodomethane	HCI_3	I-73	393.73
3888	8-hydroxyquino-line-5 sulfonic acid (7)	loretin; ferron	IC_9H_4N(OH)·SO_3H	XXII-408	351.12
3889	lepidine (2)	2-I-4-Me-quinoline	$C_{10}H_8$NI	XX-397	269.09
3890	methyl cyanide	iodoacetonitrile	I·CH_2·CN	II-223	166.95
3891	naphthalene (*α*)	*α*-naphthyl iodide	$C_{10}H_7$I	V-550	254.07
3892	naphthalene (*β*)	*β*-naphthyl iodide	$C_{10}H_7$I	V-552	254.07
3893	nitrobenzene (*o*)	....	I·C_6H_4·NO_2	V-252	249.01
3894	nitrobenzene (*m*)	....	I·C_6H_4·NO_2	V-253	249.01
3895	nitrobenzene (*p*)	....	I·C_6H_4·NO_2	V-253	249.01
3896	phenetole (*o*)	....	I·C_6H_4·O·C_2H_5	VI-207	248.07
3897	phenetole (*p*)	....	I·C_6H_4·O·C_2H_5	VI-208	248.07
3898	phenol (*o*)	....	I·C_6H_4·OH	VI-207	220.01
3899	phenol (*m*)	....	I·C_6H_4·OH	VI-207	220.01
3900	phenol (*p*)	....	I·C_6H_4·OH	VI-209	220.01
3901	propionic acid (*α*)	....	CH_3·CHI·CO_2H	II-261	199.98
3902	propionic acid (*β*)	....	ICH_2·CH_2·CO_2H	II-261	199.98
3903	propyne-1 (1)†	....	CH_3·C⫶CI	I-248	165.96
3904	salicylic acid (3)	*m*-I-salicylic acid	I·C_6H_3(OH)CO_2H	X-112	264.02
3905	toluene (*o*)	*o*-tolyl iodide	I·C_6H_4·CH_3	V-310	218.04
3906	toluene (*m*)	*m*-tolyl iodide	I·C_6H_4·CH_3	V-311	218.04
3907	toluene (*p*)	*p*-tolyl iodide	I·C_6H_4·CH_3	V-312	218.04
3908	trinitromethane	....	IC(NO_2)$_3$	I-79	276.93
3909	*m*-xylene (2)	2-I-1,3-diMe-benzene	I·C_6H_3(CH_3)$_2$	V-375	232.07
3910	**Iodoso**-benzene	....	C_6H_5IO	V-217	220.01
3911	benzoic acid (*o*)	....	OI·C_6H_4·CO_2H	IX-363	264.02
3912	**Iodoxybenzene**	....	C_6H_5IO_2	V-218	236.01
3913	**Ionone** (*α*)	....	$C_{10}H_{16}$:CH·CO·CH_3	VII-168	192.30
3914	**Ionone** (*β*)	....	$C_{10}H_{16}$:CH·CO·CH_3	VII-167	192.30
3915	**Ionone** (*pseudo*)	citrylidene acetone	$C_{13}H_{20}$O	I-757	192.30
3916	semicarbazone (*α*)	....	$C_{13}H_{20}$:NNHCONH_2	VII-169	249.36
3917	semicarbazone (*β*)	....	$C_{13}H_{20}$:NNHCONH_2	VII-168	249.36
3918	**Irone** (*β*)	irone	$C_{14}H_{22}$O	VII-169	206.33
3919	**Isatin**	....	C_6H_4·CO·COH:N⎤ (ring)	XXI-432	147.13
3920	chloride	....	C_6H_4·CO·CCl:N⎤ (ring)	XXI-302	165.58
3921	**Isatinic acid**	anilino-glyoxylic acid	NH_2·C_6H_4·CO·CO_2H	XIV-648	165.15
3922	**Isatoic anhydride**	....	C_6H_4·CO_2·CO·NH⎤ (ring)	XXVII-264	163.13
3923	**Isatoxime** (*β*)	nitroso-oxindol	C_6H_4·N:C(OH)C:NOH⎤ (ring)	XXI-443	162.15
3924	**Isatropic acid** (*α*)	1-Ph-1,2,3,4-tetra-hydronaphthalene-1,4-dicarboxylic acid	($C_9H_8O_2$)$_2$	IX-957	296.33

† See also No. 5344.
Kyanol 513
Kyanurin 3821
L acid 4503
Laburnine, cf. alkd.

Lactic anhydride 3948
Lactophenin 3949
Lactyl urea 4268
Lambda acid 4544
Lantanuric acid 182

Table 7-4 (*Continued*)
PHYSICAL CONSTANTS OF ORGANIC COMPOUNDS

No.	Crystalline Form and Color	Specific Gravity	Melting Point °C.	Boiling Point °C.	Solubility in 100 Parts		
					Water	Alcohol	Ether
3884	lq.	$1.604^{25°}_{25}$	....	$189\text{-}92^{36mm}$	....	....	....
3885	col. cr./al.	....	112-3	320 sl. d.	i.; s. bz.	s. h.	s.; s. ac.
3886	col. lq.	$2.441^{20°}$	....	184^{743mm}	....	....	....
3887	yel. hex.	$4.008^{17°}$	119	subl.	$0.01^{25°}$	$1.5^{17°}$; 11 h.	$13.6^{25°}$
3888	yel. lf.	....	d. 260±	....	0.2 c.; 0.6 h.	sl. s.; s. H_2SO_4	i.; i. bz.; i. chl.
3889	nd./lg.	....	90	....	....	s.	s.; s. lg.
3890	oil	2.307	....	$182\text{-}4^{720}$ d.	s. d.	....	....
3891	oil	$1.734^{15°}$	....	305	i.	∞	∞
3892	lf.	$1.632^{9.9°}_{4}$	54-5	308-10	i.	v. s.	v. s.
3893	yel. nd.	$1.883^{100°}_{4}$	52-4	$288\text{-}9^{729mm}$	i.	s. h.	v. s.
3894	mn.	$1.878^{100°}_{4}$	37-8	280±	....	....	....
3895	yel. nd./ al.	2.273 (solid)	171.5	287^{726mm}	....	....	....
3896	lq.	....	....	245^{736mm}	i.	s.	s.
3897	cr.	....	29	249^{729mm}	i.	s.	s.; s. chl.
3898	nd. or pl.	$1.876^{80°}$	40.4	$186\text{-}7^{160mm}$	s. h.	v. s.	v. s.
3899	nd./lg.	....	40	d.	....	....	s.
3900	nd./aq.	$1.857^{112°}$	93-4	d.	sl. s.	v. s.	v. s.
3901	nd.	....	44.5-5.5	$105^{0.3mm}$	v. sl. s.	s.	s.
3902	lf.	....	82	....	s. h.	v. s.	v. s.
3903	col. oil	$2.08^{22°}$	....	110.2^{757mm}	....	....	....
3904	col. nd./aq.	....	199	....	s. h.	s.	s. alk.
3905	lq.	$1.698^{20°}$	....	211-2	i.	∞	∞
3906	lq.	$1.698^{20°}$	....	213	i.	∞	∞
3907	lf.	$1.678^{40°}$	35-6	211.5	i.	v. s.	v. s.
3908	unstable, yel. pr.	....	55-6	48^{13mm}	i.	s. h.	s. bz., lg.
3909	lq.	....	....	228-30	....	....	....
3910	yel. amor.	....	expl. 210	....	s. h.	s. h.	v. sl. s.
3911	col. lf./aq.	....	>200 d.	....	s. h.	s. h.	v. sl. s.
3912	nd./aq.	....	expl. 236-7	....	s. h.	i.; i. bz.	s. h. ac.
3913	col. oil	$0.930^{20°}$	....	136.1^{17mm}	sl. s.	∞	∞
3914	col. oil	$0.944^{20°}$	....	140^{18mm}	sl. s.	∞	∞
3915	oil	$0.898^{20°}$	....	$143\text{-}5^{12mm}$	....	....	....
3916	cr./bz. lg.	....	107-8 (137-8)	....	....	s.	
3917	nd./al.	....	148-9	....	i.	s.	s.; s. bz.
3918	col. oil	$0.939^{20°}$	....	144^{16mm}	v. sl. s.	v. s.	v. s.
3919	yel. red mn.	....	200-1	....	s. h.	v. s. h.	sl. s.; s. alk.
3920	brn. nd.	....	180±, d.	....	i.	s.	s.; blue
3921	wh. pd.	....	d.	....	s.	....	....
3922	mn./act.	....	240 d.	....	d. h.	$3^{8°}$	sl. s.; 1.3 h. act.
3923	yel. nd.	....	225 d.	....	v. sl. s; s. a.	sl. s.	s. KOH; i. bz.
3924	cr.	....	237	....	sl. s. h.	sl. s.; i. CS_2	i.; i. bz.

Table 7-4 (*Continued*)
PHYSICAL CONSTANTS OF ORGANIC COMPOUNDS

No.	Name	Synonym	Formula	Beil. Ref.	Formula Weight
3925	**Isatropic acid** (β)	....	$(C_9H_8O_2)_2$	IX-957	296.33
3926	**Isomannide**	....	$C_6H_{10}O_4$	I-540	146.14
3927	**Isonitrosoacetone**	....	$CH_3 \cdot CO \cdot CH:NOH$	I-763	87.08
3928	**Itaconic acid**	methylene succinic acid	$CH_2{:}C(CO_2H) \cdot CH_2 \cdot CO_2H$	II-760	130.10
3929	**Itamalic acid**	free acid non-existant	$CH_2OH \cdot CH(CO_2H) \cdot CH_2 \cdot CO_2H$	III-446	148.12
3930	**Kairoline**	*N*-methyl-tetrahy-droquinoline	$C_9H_{10}N \cdot CH_3$	XX-264	147.22
3931	**Kawain**	....	$CH_3O \cdot C_{13}H_{11}O_2$	*XIX-418	230.27
3932	**Ketazine**	acetone azine	$[(CH_3)_2C{:}N \cdot]_2$	I-651	112.18
3933	**Ketene**	....	$H_2C{:}C{:}O$	I-724	42.04
3934	**Keto-**butyric acid (α)	....	$CH_3 \cdot CH_2 \cdot CO \cdot CO_2H$	III-629	102.09
3935	dihydroquinazo-line (4)	....	$C_6H_4 \cdot N{:}CH \cdot NH \cdot CO$	XXIV-143	146.15
3936	pyrrolidine	α,γ-butyrolactam	$NH \cdot (CH_2)_3 \cdot CO \cdot H_2O$	XXI-236	103.12
3937	**Kynurenic acid**	4-hydroxyquinoline-3-carboxylic acid	$C_9H_5N(OH)CO_2H \cdot H_2O$	XXII-230	207.19
3938	**Lacmoid**	resorcinol blue	†	*VI-399	......
3939	**Lactic** acid (*l*) (+)	paralactic acid	$CH_3 \cdot CHOH \cdot CO_2H$	III-261	90.08
3940	acid (*dl*)	....	$CH_3 \cdot CHOH \cdot CO_2H$	III-268	90.08
3941	Ca salt	calcium lactate	$(C_3H_5O_3)_2Ca \cdot 5H_2O$	III-277	308.30
3942	aldehyde	....	$CH_3 \cdot CHOH \cdot CHO$	I-819	74.08
3943	amide	....	$CH_3 \cdot CHOH \cdot CO \cdot NH_2$	III-283	89.09
3944	nitrile	aldehyde-cyanohy-drin	$CH_3 \cdot CHOH \cdot CN$	III-284	71.08
3945	**Lactide** (*dl*)	dilactide	$C_6H_8O_4$	XIX-154	144.13
3946	**Lactoic acid** (*d*)	*d*-galactonic acid	$CH_2OH \cdot (CHOH)_4 \cdot CO_2H$	III-549	196.16
3947	**Lactose**	milk sugar	$C_{12}H_{22}O_{11} \cdot H_2O$	XXXI-407	360.32
3948	**Lactyl** lactic acid	"lactic anhydride"	$C_2H_5O \cdot CO_2 \cdot C_2H_4 \cdot CO_2H$	III-282	162.14
3949	phenetidine (*p*)	lactophenin	$C_2H_5O \cdot C_6H_4 \cdot NH \cdot CO \cdot CHOH \cdot CH_3$	XIII-491	209.25
3950	**Lauric** acid	dodecanoic acid	$CH_3(CH_2)_{10} \cdot CO_2H$	II-359	200.32
3951	ammonium salt	ammonium laurate	$C_{11}H_{23}CO_2NH_4$	*II-156	217.35
3952	aldehyde	dodecanal	$CH_3(CH_2)_{10} \cdot CHO$	I-714	184.32
3953	amide	....	$C_{11}H_{23} \cdot CO \cdot NH_2$	II-363	199.34
3954	anhydride	....	$(C_{11}H_{23}CO)_2O$	II-362	382.63
3955	chloride	lauroyl chloride	$CH_3(CH_2)_{10} \cdot CO \cdot Cl$	II-363	218.77
3956	nitrile	undecyl cyanide	$C_{11}H_{23} \cdot CN$	II-363	181.32
3957	**Lead** tetraethyl	tetraethyl lead	$(C_2H_5)_4Pb$	IV-639	323.44
3958	tetramethyl	tetramethyl lead	$(CH_3)_4Pb$	IV-639	267.33
3959	tetraphenyl	tetraphenyl lead	$(C_6H_5)_4Pb$	XVI-917	515.62
3960	**Lecithin**	protagon	$C_{42}H_{84}O_9PN$	....	778.11
3961	**Lepamine**	....	$C_{20}H_{32}N_2$	....	300.49
3962	**Leuco-**aniline	3-Me; $NH_2,4,4',4''$	$CH_3 \cdot C_6H_3(NH_2)CH{:}(C_6H_4 \cdot NH_2)_2$	XIII-321	303.41
3963	aurine (4,4',4'')	....	$(HO \cdot C_6H_4)_3CH$	VI-1143	292.34
3964	malachite green	$Me_2N, 4,4'$	$[(CH_3)_2N \cdot C_6H_4]_2{:}CH \cdot C_6H_5$	XIII-275	330.45

† Mixed dye; no formula.
Lauryl chloride 2861
Lauth's violet 5922
Lemonflavin, of. quercitrin, glcde.

Lenigallol 5516
Lepidine 4370
Lepidine ethiodide 4371
Lepidone 4179

Table 7-4 (*Continued*)
PHYSICAL CONSTANTS OF ORGANIC COMPOUNDS

No.	Crystalline Form and Color	Specific Gravity	Melting Point °C.	Boiling Point °C.	Solubility in 100 Parts		
					Water	Alcohol	Ether
3925	pl./aq.	....	206	d. to α, 220	....	....	....
3926	hyg., mn.	....	87	176^{30mm}	v. s.	s.; sl. s. chl.	i.; i. bz.
3927	lf./et.	$1.074^{67.5°}$	69	subl.	v. s.	v. sl. s. pet.	v. s.
3928	rhb.	1.63	165-6 sl. d.	....	$8.3^{20°}$; v. sl. s. bz.	$25^{15°}$, 88% al.	v. sl. s.; v. sl. s. chl.
3929	....	....	....	....	....	....	....
3930	lq.	$1.022^{20°}_{4}$	....	$247\text{-}50^{758mm}$	....	v. s.	sl. s.
3931	col. cr.	....	105-6	$195\text{-}7^{0.1mm}$	i.	s.	s.; s. act.
3932	lq.	$0.843^{20°}_{4}$	....	131	∞	∞	∞
3933	col. gas	....	-151	-56	d.	d.	s.; s. act.
3934	hyg. pl.	....	31.5-2.0	85^{21mm}	v. s.	v. s.	v. sl. s.
3935	col. nd.	....	213-4	....	s. h.	s.	s. NaOH
3936	cr.	$1.120^{20°}_{4}$	35	251*	v. s.	....	....
3937	nd.	....	290(anh.)	$-H_2O$, 140-5	$0.09^{100°}$	s. h.	i.
3938	vl. pd.	....	....	....	sl. s.	s.	i.; s. act.
3939	delq. pr.	....	52.8	....	∞	∞	∞
3940	hyg.	$1.249^{15°}_{4}$	16.8	122^{14mm}	∞	∞	∞
3941	col. cr.	....	$-5H_2O$, 100	....	10 c.	3	∞ h. aq.
3942	nd.	....	101-5	....	s.; v. s. ac.	s.; s. act.	i.; i. bz.
3943	cr.	$1.138^{80°}_{4}$	74	....	v. s.	v. s.	....
3944	col. lq.	$0.992^{18°}_{4}$	-40	182-4 sl. d.	∞	∞	i. pet.
3945	tri./al.	$0.862^{10°}_{4}$	124.5	255^{757mm}	v. sl. s. c.; d. h.	v. sl. s. c. abs. al.	....
3946	nd./aq.	....	147.5	....	s.	....	....
3947	col. rhb.	$1.525^{20°}$	202(anh.)	d.	17 c.; 40 h.	i.	i.
3948	lt. yel. oil.	....	....	d. 250-60	v. sl. s.	s.	s.
3949	col. nd./aq.	....	118	....	0.3 c.; 1.8 h.	13; sl. s. lg.	sl. s.; s. h. bz.
3950	col. nd.	$0.871^{50°}_{4}$	44.1	225^{100mm}	i.; s. bz.	s.	s.; s.pet.
3951	waxy	0.88	48-55	$-NH_3 > 50$	s. h.	s.; i. chl.	i.; i. bz.
3952	col. lf.	....	44.5	$184\text{-}5^{100mm}$	i.	s.	s.
3953	col. nd.	....	102	$200^{12.5mm}$	i.	v. s.	....
3954	col. cr.	$0.855^{70°}_{4}$	41	$166^{0.1mm}$	....	....	....
3955	col. lq.	....	-17	145^{18mm}	d.	d.	s.
3956	oil	$0.827^{15°}$	4	198^{100mm}	....	....	....
3957	col. lq.	$1.659^{18°}_{4}$	-136	152^{291mm}	i.; s. bz.	sl. s.	∞
3958	col. lq.	$1.995^{20°}_{4}$	-27.5	110^{760mm}	i.	∞	∞
3959	tet./bz.	$1.530^{20°}$	228-9	d. 270	i.	$0.1^{30°}$	$1.7^{30°}$ bz.
3960	waxy	....	150-200 d.	....	i.; s. chl.	s. h.	s. h.
3961	lq.	....	....	275	....	....	....
3962	cr./aq.	....	100	....	sl. s. h.	v. s.	sl. s.
3963	col. pr./ac.	....	....	....	sl. s.	s.; s. alk.	s. ac.
3964	nd./bz.	....	102(94)	d.	i.	s.; s. bz.	s.; sl. s. lg.

* B. p. of anh. compound.
Leucic acid 3760
Leucine 283, 285-7
Leuco-aniline 6071-3

Leuco-alizarin 2295
Leuco-crystal violet 3595
Leuco-dimethylphenylene green 5833
Leucoline 5538

Table 7-4 (*Continued*)
PHYSICAL CONSTANTS OF ORGANIC COMPOUNDS

No.	Name	Synonym	Formula	Beil. Ref.	Formula Weight
3965	**Leucyl glycine** (*dl*)	....	$C_5H_{12}NCO \cdot NHCH_2 \cdot$ CO_2H	IV-448	188.23
3966	**Levulin**	....	$C_6H_{10}O_5$	I-925	162.14
3967	**Levulinic** acid	acetopropionic acid	$CH_3CO(CH_2)_2CO_2H$	III-671	116.12
3968	aldehyde	....	$CH_3CO(CH_2)_2CHO$	I-774	100.12
3969	**Lichenin**	moss starch	$(C_6H_{10}O_5)_x$	....	(162.14)
3970	**Lignoceric acid**	....	$C_{24}H_{48}O_2$	II-393	368.65
3971	**Limettin**	5,7-diMeO-coumarin	$(CH_3O)_2C_9H_4O_2$	XVIII-97	206.20
3972	**Limonene** (*dl*)	dipentene	$C_{10}H_{16}$	V-137	136.24
3973	**Limonene** (*d* or *l*)	*p*-menthadiene-1,8(9)	$C_{10}H_{16}$	V-133	136.24
3974	**Linalool** (*d*)†	coriandrol	$C_{10}H_{18}O$	I-461	154.25
3975	**Linalyl acetate**	bergamol	$CH_3CO_2 \cdot C_{10}H_{17}$	II-141	196.29
3976	**Linoleic acid**	octadecadienoic acid	$C_{18}H_{32}O_2$	II-496	280.45
3977	**Linolenic acid**	octadecatrienoic acid	$C_{18}H_{30}O_2$	II-499	278.44
3978	**Lophine**	triphenyl-imidazole	$(C_6H_5)_3C{:}C \cdot N{:}C \cdot NH$	XXIII-318	296.38
3979	**Lumisterol**	....	$C_{28}H_{43}OH$	....	396.66
3980	**Luteol**	oxychloro-diphenyl-quinoxaline	$C_{19}H_{14}ONCl$	....	307.79
3981	**Lutidine** (2,5)‡	dimethyl-pyridine	$(CH_3)_2C_5H_3N$	XX-244	107.16
3982	**Lutidine** (2,4) (α,γ)	dimethyl-pyridine	$(CH_3)_2C_5H_3N$	XX-244	107.16
3983	**Lutidine** (2,6)(α,α')	dimethyl-pyridine	$(CH_3)_2C_5H_3N$	XX-244	107.16
3984	**Lutidine** (3,4)(β,γ)	dimethyl-pyridine	$(CH_3)_2C_5H_3N$	XX-246	107.16
3985	**Lycopene**	lycopin	$C_{40}H_{56}$	XXX-81	536.89
3986	**Lysine** (*dl*)	α,ε-diaminocaproic acid	$NH_2 \cdot (CH_2)_4 \cdot CHNH_2 \cdot$ CO_2H	IV-436	146.19
3987	dihydrochloride	....	$C_6H_{14}O_2N_2 \cdot 2HCl$	IV-437	219.11
3988	**Lysine** (*l*+)	....	$(NH_2)_2C_5H_9 \cdot CO_2H$	IV-435	146.19
3989	hydrochloride (*d*)	....	$C_6H_{14}O_2N_2 \cdot 2HCl$	IV-436	219.11
3990	monohydrochloride	(*d*)	$C_6H_{14}O_2N_2 \cdot HCl$	....	182.65
3991	**Lyxose** (α)(*d*)	....	$CH_2(CHOH)_4 \cdot O$	XXXI-56	150.13
3992	**Lyxose** (β)(*d*)	....	$CH_2(CHOH)_4 \cdot O$	XXXI-56	150.13
3993	**Malachite** green§	benzaldehyde green; (zinc salt)	$3C_{23}H_{25}N_2Cl \cdot 2ZnCl_2 \cdot$ $2H_2O$	XIII-745	1403.35
3994	green	(oxalate salt)	$2C_{23}H_{25}N_2 \cdot C_2HO_4 \cdot$ $H_2C_2O_4$	XIII-745	927.03
3995	**Maleic** acid (*cis*)	butendioic acid; toxilic acid	$({:}CH \cdot CO_2H)_2$	II-748	116.07
3996	anhydride	....	$({:}CH \cdot CO)_2O$	XVII-432	98.06
3997	**Malic** acid (*dl*)	hydroxysuccinic acid	$HO_2C \cdot CHOH \cdot CH_2 \cdot$ CO_2H	III-435	134.09
3998	acid (*d* or *l*)	....	$HO \cdot C_2H_3(CO_2H)_2$	III-417	134.09
3999	Ca acid salt (*l*)	Ca bimalate	$Ca(HC_4H_4O_5)_2 \cdot 6H_2O$	....	414.33
4000	acid (α), **iso-**	methyl tartronic acid	$CH_3 \cdot C(OH)(CO_2H)_2$	III-440	134.09
4001	amide (*l*)	*l*-malamide	$HO \cdot C_2H_3(CO \cdot NH_2)_2$	III-435	132.12
4002	**Malonic** acid	propandioic acid	$CH_2{:}(CO_2H)_2$	II-566	104.06
4003	Ca salt	calcium malonate	$CaC_3H_2O_4 \cdot 4H_2O$	II-570	214.19
4004	amide	malonamide	$CH_2{:}(CO \cdot NH_2)_2$	II-582	102.09

† See also No. 2518.
‡ See also No. 3147.
§ In medicinal use without ZnCl₂
Levulose 3616
Lewisite 1473

Light green 3594
Lilacin, cf. glcde.
Linalool tetrahydride 2519
Linaloolene 2515
Linamarin, cf. glcde.
Lindol 6351

Litmopyrin 117
Lobeline, cf. alkd.
Lodal 2413
Loretin 3888
Luminal 5113
Luminol 359

Lupanine, cf. alkd.
Lupinidine, cf. alkd.
Lupinine, cf. alkd.
Lutein 6452
Luteosterone 5334
Lutidinic acid 5498

Table 7-4 (*Continued*)
PHYSICAL CONSTANTS OF ORGANIC COMPOUNDS

No.	Crystalline Form and Color	Specific Gravity	Melting Point °C.	Boiling Point °C.	Solubility in 100 Parts		
					Water	Alcohol	Ether
3965	cr./aq.	....	243 d.	....	6.6 h.	v. sl. s.	v. sl. s.
3966	delq. amor.	....	d. 140-5	....	v. s.	v. sl. s.	i.
3967	lf.	$1.140\frac{20}{20}°$	33.5(18-9)	245-6 sl. d.	v. s.	v. s.	v. s.
3968	col. lq.	$1.018\frac{20}{4}°$	<-21	186-8 sl. d.	∞	∞	s.
3969	col. amor.	....	....	....	s. h.	s. HCl	i.
3970	col. nd.	....	81-2	....	s. bz.	s.; s. ac.	s.; s. CS_2
3971	col. cr./al.	....	147.5	200 sl. d.	i.; s. chl.	s. h.	v. sl. s.
3972	col. lq.	$0.844^{20°}$	....	178	i.	∞	∞
3973	lq.	$0.842\frac{20}{4}°$	-96.9	177	i.	∞	∞
3974	col. oil	$0.868^{20°}$	....	198-200	v. sl. s.	10, 50% al.	∞
3975	col. lq.	$0.895^{20°}$	....	$220^{762}±$, d.	v. sl. s.	∞	∞
3976	lt. yel. oil	$0.903\frac{18}{4}°$	-9.5	$229-30^{16mm}$	i.	∞	∞
3977	oil	$0.914\frac{18}{4}°$	....	$230-2^{17mm}$	i.	s.	v. s.
3978	nd.	....	275	....	i.	2.8 h., abs.	$0.3^{20°}$
3979	col. nd.	....	118	....	i.; s. act.	s.; s. chl.	v. s.
3980	yel. nd.	....	246	....	i.	s.	s.
3981	lq.	$0.938^{0°}$	....	156.5	25 c.; sl. s. h.	∞	∞
3982	lq.	$0.949\frac{0}{4}°$	....	157-9	20 c.; sl. s. h.	s.	s.
3983	lq.	$0.923^{25°}$	-6.6	142-3	∞ c.	sl. s. h. aq.	....
3984	lq.	....	....	163.5-4.5			
3985	red pr./pet.	....	174-5	....	i.; s. h. bz.	v. sl. s. h.	0.03 h.
3986	syrup	....	....	....	s.	....	i.
3987	cr.	....	192-3	....	v. s.	....	....
3988	nd./aq.	....	224-5 d.	....	v. s.	i.	....
3989	cr./HCl	....	193	....	....	....	....
3990	wh. pd.	....	235-6	....	s.	i.	i.
3991	cr./al. et.	....	106-7	....	v. s.	$2.5^{17°}$	....
3992	nd./al.	....	117-8	....	v. s.	sl. s.	....
3993	gn. pr./ aq. al.	....	130±	....	v. s.	v. s.	....
3994	gn. pr.	....	....	....	v. s. h.	s.	....
3995	mn.	1.609	130.5	138±; d., anhyd.	$79^{25°}$; $393^{98°}$	$70^{30°}$	$8^{25°}$
3996	cr./chl.	1.5	52.8	202; subl.	$16.3^{30°}$	v. sl. s. CCl_4	....
3997	col. cr.	$1.601\frac{20}{4}°$	128-9	150 d.	$144^{26°}$; $411^{79°}$	v. s.	v. s.
3998	col. cr.	$1.595\frac{20}{4}°$	99-100	140±, d.	v. s.	v. s.	$8.4^{15°}$
3999	col. rhb.	....	....	....	sl. s. c.	sl. s. ac.	s. a.
4000	col. cr.	....	142 d.	d. 170 ±	v. s.	v. s.	v. s.
4001	pr./aq.	....	156-8 d.	....	s.	....	....
4002	col. tri.	$1.631^{15°}$	130-5 d.	....	$138^{16°}$	$42^{25°}$	$8^{15°}$ abs.
4003	col. nd.	....	$-4H_2O,180$	$-3H_2O$, 100	$0.4^{0°}$	$0.71^{00°}$ aq.	....
4004	tet. or mn.	....	170	....	$8.3^{8°}$	i. abs.	i.

Table 7-4 (*Continued*)
PHYSICAL CONSTANTS OF ORGANIC COMPOUNDS

No.	Name	Synonym	Formula	Beil. Ref.	Formula Weight
4005	**Malonic** anilide	malonanilide	$CH_2(CONH \cdot C_6H_5)_2$	XII-293	254.29
4006	nitrile	methylene dicyanide	$CH_2(CN)_2$	II-589	66.06
4007	**Maltose**	malt sugar	$C_{12}H_{22}O_{11} \cdot H_2O$	XXXI-386	360.32
4008	**Mandelic** acid (*dl*)	....	$C_6H_5 \cdot CHOH \cdot CO_2H$	X-197	152.15
4009	sodium salt	sodium mandelate	$C_6H_5 \cdot CHOH \cdot CO_2Na$	....	174.13
4010	nitrile (*dl*)	....	$C_6H_5 \cdot CHOH \cdot CN$	X-206	133.15
4011	**Mannitol hexani-** trate	....	$C_6H_8(O \cdot NO_2)_6$	I-543	452.17
4012	**Mannoheptose** (*d*)	....	$C_7H_{14}O_7$	I-935	210.19
4013	**Mapharsen**	3-NH_2-4-OH- phenyl-arsinoxide HCl	$O:As \cdot C_6H_3(OH)NH_2 \cdot HCl \cdot \frac{1}{2}C_2H_5OH$	*XVI-447	258.54
4014	**Margaric** acid	heptadecanoic acid	$CH_3 \cdot (CH_2)_{15} \cdot CO_2H$	II-376	270.46
4015	aldehyde	heptadecanal	$CH_3 \cdot (CH_2)_{15} \cdot CHO$	I-717	254.46
4016	**Meconic** acid	....	$C_7H_4O_7 \cdot 3H_2O$	XVIII-503	254.15
4017	lactone	6,7-diMeO-phthal-ide; meconine; opianyl	$CO \cdot O \cdot CH_2 \cdot C_6H_2 : \underset{(OCH_3)_2}{\vert_____\vert}$	XVIII-89	194.19
4018	**Melam**	....	$C_6H_9N_{11}$	III-169	235.21
4019	**Melamine**	cyanuramide	$C_3N_3(NH_2)_3$	XXVI-245	126.12
4020	hydrochloride	....	$C_3H_6N_6 \cdot HCl \cdot \frac{1}{2}H_2O$	XXVI-246	171.59
4021	**Melene**	....	$C_{30}H_{60}$ (or $C_{30}H_{62}$)?	I-227	420.81 422.83
4022	**Melissic acid**	....	$C_{30}H_{61} \cdot CO_2H$	II-396	466.84
4023	**Menthene** (*d*)	....	$C_{10}H_{18}$	V-87	138.25
4024	**Menthol** (α)(*l*)	hexahydrothymol	$C_{10}H_{19}OH$	VI-28	156.27
4025	**Menthol** (*d*)(*neo*)	*p*-menthanol-3	$C_{10}H_{19}OH$	VI-28	156.27
4026	**Menthone** (*l*)	....	$C_{10}H_{18}O$	VII-38	154.25
4027	**Menthyl** acetate (*l*)	....	$CH_3 \cdot CO_2 \cdot C_{10}H_{19}$	VI-32	198.31
4028	acetate (*d*)(*neo*)	....	$CH_3 \cdot CO_2 \cdot C_{10}H_{19}$	....	198.31
4029	amine (*l*)	aminomenthane	$C_{10}H_{19} \cdot NH_2$	XII-26	155.29
4030	benzoate (α)	....	$C_6H_5 \cdot CO_2 \cdot C_{10}H_{19}$	IX-115	260.38
4031	salicylate	salimenthol	$HO \cdot C_6H_4 \cdot CO_2 \cdot C_{10}H_{19}$	X-76	276.38
4032	*iso*-valerate (*l*)	validol	$C_4H_9 \cdot CO_2 \cdot C_{10}H_{19}$	VI-33	240.39
4033	**Mercaptobenzo-** thiazole (2)	....	$C_6H_4 \cdot N:C(SH) \cdot S \overset{}{\underset{\vert_____\vert}{}}$	XXVII-185	167.25
4034	**Mercuri**-hydroxy-phenyl chloride (*o*)	*o*-Cl-mercuriphenol	$HO \cdot C_6H_4 \cdot HgCl$	XVI-959	329.15
4035	hydroxyphenyl chloride (*p*)	*p*-Cl-mercuriphenol	$HO \cdot C_6H_4 \cdot HgCl$	XVI-961	329.15
4036	phenyl acetate	Ph-mercuric acetate	$C_6H_5 \cdot Hg \cdot O_2CCH_3$	XVI-954	336.74
4037	phenyl chloride	Ph-mercuric chloride	$C_6H_5 \cdot HgCl$	XVI-953	313.15
4038	phenyl nitrate	Ph-mercuric nitrate	$C_6H_5 \cdot Hg \cdot O \cdot NO_2$	XVI-953	339.70
4039	tolyl chloride (*p*)	....	$CH_3 \cdot C_6H_4 \cdot HgCl$	XVI-956	327.18
4040	**Mercuric fulminate**	....	$Hg(ONC)_2 \cdot \frac{1}{2}H_2O$	....	293.63
4041	**Mercurochrome**	diNa-diBr-hydroxy-mercury-fluorescein	$C_{20}H_7O_5Br_2Na_2HgOH \cdot 3H_2O$	....	804.72
4042	**Mercury** dibutyl	di-*n*-butyl mercury	$(C_4H_9)_2Hg$	**IV-1049	314.82
4043	diethyl	diethyl mercury	$(C_2H_5)_2Hg$	IV-679	258.71
4044	dimethyl	dimethyl mercury	$(CH_3)_2Hg$	IV-678	230.66

Malonyl thiourea 5887
Malonyl urea 621
Malt sugar 4007
Mandelonitrile glucoside, cf. glcde.
Mannite 3609
Mannitol 3609
Mannitol hexaacetate 3610

Mannoheptitol 5090
Mannose 3619
Marsh gas 4061
Martius yellow 2642
Matecite 4286
Matezite 4286
Mecholin 100

Mecholyl 100
Mecholyl bromide 99
Mecholyl iodide 101
Meconine 4017
Medinal 2125
Melaniline 2723
Meletin 5807

Table 7-4 (*Continued*)
PHYSICAL CONSTANTS OF ORGANIC COMPOUNDS

No.	Crystalline Form and Color	Specific Gravity	Melting Point °C.	Boiling Point °C.	Solubility in 100 Parts		
					Water	Alcohol	Ether
4005	nd./al.	....	229-31	....	i.; s. ac.	v. s. h.	i.
4006	col. cr.	$1.049\frac{3.4°}{4}$	31.7	223-4	13	40	20
4007	col. nd.	$1.540\frac{17°}{0°}$	d.	....	v. s.	v. sl. s. c.	i.
4008	rhb./aq.	$1.300\frac{20°}{4}$	118.1	d.	$16^{20°}$	s.	s.
4009	wh. cr. pd.	....	....	....	99	2	
4010	oil	1.124	-10	d. 170	i.	s.	s.; s. chl.
4011	nd.	$1.604^{0°}$	112-3	expl.	i.	$2.9^{13°}$	$4^{9°}$
4012	nd.	....	134-5	....	v. s.	sl. s. abs.	....
4013	wh. pd.	....	....	....	v. s.; s. alk.	v. s.; v. sl. s. act.	v. sl. s.
4014	col. pl.	$0.853^{60°}$	60.9	227^{100mm}	i.	$32^{28°}$ abs.	v. s.
4015	cr.	....	36	$203\text{-}4^{26mm}$	sl. s. c. al.	v. s. h.	v. s.
4016	rhb.	....	$-3H_2O$, 100	d.	d. h.	s.; d. h. a.	sl. s.
4017	col. nd./aq.	....	102-2.5	subl.	$0.14^{15°}$; $4.5^{100°}$	s.; s. chl.	s.; s. bz.
4018	pd.	....	d.	....	i.	sl. s. a.	s. h. KOH
4019	mn.	$1.573^{250°}$	360 d.	subl.	s. h.	sl. s. h.	i.
4020	nd.	....	....	....	....	i.	....
4021	cr.	$0.913^{25°}$	62-3	$218^{0.5mm}$	i.; v. sl. s. bz.	3.6 h., abs.	v. sl. s.
4022	nd./al.	....	90-1	....	i.	v. s. h.	v. sl. s.
4023	lq.	$0.814\frac{16°}{4}$	....	168-9	....	....	....
4024	col. cr.	$0.890\frac{15°}{15}$	42.5*	216.3	0.04 c.; s. pet.	v. s.; v. s. chl.	v. s.; v. s. ac.
4025	col. lq.	$0.9\pm$	-22	98^{16mm}	....	....	....
4026	oil	$0.896\frac{20°}{20}$	-6.6	207	sl. s.	∞	∞
4027	col. lq.	$0.919\frac{20°}{4}$	....	227	sl. s.	∞	∞
4028	col. cr.	....	37-8	....	i.	s.	v. s.
4029	oil	$0.861\frac{20°}{4}$	....	$209\pm$	s. c.	v. s.	....
4030	rhb./al.	$**1.002\frac{20°}{4}$	54-5	301^{748mm}	....	....	....
4031	col. oil	$1.045\frac{25°}{25}$	....	$203\text{-}5^{20mm}$	i.	s.	s.
4032	col. lq.	$0.907\frac{15°}{15}$	....	129^{9mm}	i.	s.	s.; s. chi.
4033	col. nd./aq. al.	$1.42\frac{20°}{4}$	179	d.	i.; s. alk.; s. ac.	s.; s. alk. carb.	sl. s.
4034	col. cr./aq. al.	....	152.5	....	sl. s.	s.; v. sl. s. chl.	s. h. bz.
4035	lf./act.	....	225-6	....	....	sl. s. h.	....
4036	col. pr./bz.	....	149	....	s. h.	s.; s. ac.	s. bz.
4037	col. lf./bz.	....	251	subl.	i.; s. pyr.	sl. s. h.	s.; s. bz.
4038	pl./al.	....	188-9 d.	....	v. sl. s. h.	s. h.	s. bz.
4039	col. lf./bz.	....	232-3	....	i.	v. sl. s. h.	i.
4040	cr./aq.	4.42(anhr.)	expl.	....	$0.071^{2°}$; $0.174^{9°}$	s.; d. h. aq. KOH	s. NH_4OH
4041	gn. scales	....	....	....	s.	0.02	i.; i. chl.
4042	col. lq.	$1.778\frac{20°}{4}$	....	$120\text{-}3^{23mm}$	i.	sl. s.	v. s.
4043	col. lq.	$2.423\frac{23°}{4}$	....	159	i.	sl. s.	v. s.
4044	col. lq.	$2.954\frac{22°}{4}$	....	95-6	v. sl. s.	v. s.	v. s.

* M. p. β 35.5°; γ 33.5°; δ 31.5°.
** Supercooled liquid.
Melilotic acid 3711
Melin, cf. glcde.
Melissyl alcohol 4449
Melitose 5549

Mellitic acid 667
Mollophanic acid 681
Menthadiene 3973, 5092-3, 5704-5, 5709
Menthane 3579
Menthanol 4025

Menthene 1258
Mercapto-, cf. also thio- and sulfo-.
Mercapto-ethanol 5904
Mercapto-succinic acid 5911

Table 7-4 (*Continued*)
PHYSICAL CONSTANTS OF ORGANIC COMPOUNDS

No.	Name	Synonym	Formula	Beil. Ref.	Formula Weight
4045	**Mercury** dinaphthyl (α)	....	$(C_{10}H_7)_2Hg$	XVI-949	454.92
4046	dipropyl	dipropyl mercury	$(C_2H_5 \cdot CH_2)_2Hg$	IV-679	286.77
4047	diphenyl	diphenyl mercury	$(C_6H_5)_2Hg$	XVI-946	354.80
4048	ditolyl (*o*)	ditolyl mercury	$(CH_3 \cdot C_6H_4)_2Hg$	XVI-947	382.86
4049	ditolyl (*m*)	ditolyl mercury	$(CH_3 \cdot C_6H_4)_2Hg$	XVI-947	382.86
4050	ditolyl (*p*)	ditolyl mercury	$(CH_3 \cdot C_6H_4)_2Hg$	XVI-947	382.86
4051	mercaptide	....	$(C_2H_5 \cdot S)_2Hg$	I-342	322.84
4052	**Merthiolate**	Na ethylmercuri-thiosalicylate	$C_2H_5 \cdot Hg \cdot S \cdot C_6H_4 \cdot CO_2Na$	....	404.81
4053	**Mesityl oxide**	....	$(CH_3)_2C{:}CH \cdot CO \cdot CH_3$	I-736	98.15
4054	**Mesorcin** (1,3,5;2,4)	....	$(CH_3)_3C_6H(OH)_2$	VI-939	152.19
4055	**Mesoxalic acid**	dihydroxy malonic	$(HO)_2C(CO_2H)_2$	III-766	136.06
4056	**Metanil yellow**	Na *m*-sulfonate-azo-diphenylamine	$C_6H_5 \cdot NH \cdot C_6H_4 \cdot N{:}N \cdot C_6H_4 \cdot SO_3Na$	XVI-330	375.38
4057	**Metaphen**®	4-NO_2-3-OH-mercuri-*o*-cresol	$Hg \cdot C_7H_5O_3N$	....	351.71
4058	**Methacrolein**	α-Me-acrolein	$CH_2{:}C(CH_3) \cdot CHO$	I-731	70.09
4059	**Methallyl** alcohol	isopropenyl carbinol	$CH_2{:}C(CH_3) \cdot CH_2OH$	I-443	72.11
4060	chloride (β)	γ-Cl-isobutylene	$CH_2{:}C(CH_3) \cdot CH_2Cl$	I-209	90.55
4061	**Methane**	marsh gas	CH_4	I-56	16.04
4062	**Methionine** (*l*–)	α-NH_2-γ-methyl-thiolbutyric acid	$CH_3 \cdot S \cdot (CH_2)_2 \cdot CHNH_2 \cdot CO_2H$	**IV-938	149.21
4063	**Methoxy**-acetic acid	glycolic methyl ether	$CH_3O \cdot CH_2 \cdot CO_2H$	III-232	90.08
4064	acetophenetidine	kryofine	$C_2H_5O \cdot C_6H_4 \cdot NH \cdot CO \cdot CH_2 \cdot OCH_3$	XIII-489	209.25
4065	acetophenone (*p*)	*p*-acetyl-anisole	$CH_3O \cdot C_6H_4 \cdot COCH_3$	VIII-87	150.18
4066	*tert*-amylbenzene	(*p*)	$CH_3O \cdot C_6H_4 \cdot C_5H_{11}$	VI-549	178.28
4067	benzoic acid (*o*)	salicylic methyl ether	$CH_3O \cdot C_6H_4 \cdot CO_2H$	X-64	152.15
4068	benzoic acid (*m*)	....	$CH_3O \cdot C_6H_4 \cdot CO_2H$	X-137	152.15
4069	benzoic acid (*p*)	*p*-anisic acid	$CH_3O \cdot C_6H_4 \cdot CO_2H$	X-154	152.15
4070	benzophenone (*p*)	*p*-benzoyl anisole	$CH_3O \cdot C_6H_4COC_6H_5$	VIII-159	212.25
4071	benzoyl chloride (*p*)	anisoyl chloride	$CH_3O \cdot C_6H_4 \cdot COCl$	X-163	170.60
4072	benzyl alcohol (*o*)	saligenin-2-Me-ether	$CH_3O \cdot C_6H_4 \cdot CH_2OH$	VI-893	138.17
4073	diphenyl (*o*)	Ph-phenol-Me-ether	$CH_3O \cdot C_6H_4 \cdot C_6H_5$	VI-672	184.24
4074	diphenyl (*p*)	Me Ph-phenol ether	$CH_3O \cdot C_6H_4 \cdot C_6H_5$	VI-674	184.24
4075	ethyl chloro-formate (β)	β-methoxyethyl chlorocarbonate	$Cl \cdot CO_2 \cdot CH_2 \cdot CH_2 \cdot OCH_3$	....	138.55
4076	ethyl formate (β)	....	$HCO_2 \cdot CH_2 \cdot CH_2 \cdot OCH_3$	*II-19	104.11
4077	methylal	....	$(CH_3O \cdot CH_2)_2O$	I-576	106.12
4078	methyl salicylate	mesotan; salmester	$HO \cdot C_6H_4 \cdot CO_2CH_2 \cdot OCH_3$	X-83	182.18
4079	phenol (*o*)	guaiacol	$CH_3O \cdot C_6H_4 \cdot OH$	VI-768	124.14
4080	phenol (*m*)	resorcinol Me ether	$CH_3O \cdot C_6H_4 \cdot OH$	VI-813	124.14
4081	phenol (*p*)	hydroquinone methyl ether	$CH_3O \cdot C_6H_4 \cdot OH$	VI-843	124.14
4082	pyridine (*p*)(γ)	....	$CH_3O \cdot C_5H_4N$	XXI-49	109.13
4083	quinoline (*p*)	quinanisole	$CH_3O \cdot C_9H_6N$	XXI-85	159.19
4084	**Methyl** abietate	abalyn	$C_{19}H_{29}CO_2 \cdot CH_3$	....	316.49
4085	acetamide (*N*)	acet-methylamide	$CH_3 \cdot CO \cdot NH \cdot CH_3$	IV-58	73.10
4086	acetanilide (*N*)	exalgin	$CH_3CO \cdot N(CH_3)C_6H_5$	XII-245	149.19
4087	acetate	....	$CH_3CO_2 \cdot CH_3$	II-124	74.08
4088	acetoacetate	....	$CH_3CO \cdot CH_2 \cdot CO_2 \cdot CH_3$	III-632	116.12

Table 7-4 (*Continued*)
PHYSICAL CONSTANTS OF ORGANIC COMPOUNDS

No.	Crystalline Form and Color	Specific Gravity	Melting Point °C.	Boiling Point °C.	Solubility in 100 Parts		
					Water	Alcohol	Ether
4045	cr./bz.	1.929	243	d.	i.; s. h. chl.	sl. s. h.	s. CS_2
4046	col. lq.	$2.124^{16°}$	....	189-91	i.	sl. s.	v. s.
4047	nd./bz.	$2.318^{4°}$	124-5	>300 d.	i.; s. chl.	sl. s. h.	sl. s.
4048	tri./bz.	....	141(107)	219^{14mm}	....	....	....
4049	nd.	....	102(89)	....	sl. s. chl.	sl. s.	sl. s.
4050	nd./bz.	....	244-6	....	s. h. bz.	sl. s. h.	s. CS_2
4051	lf./al.	....	76-7	....	....	7 h.	....
4052	pd.	....	....	....	100	14	i.; i. bz.
4053	lq.	$0.858^{20°}_{4}$	−59	$129\text{-}30^{750mm}$	$3^{20°}$	∞	∞
4054	lf.	....	149-50	274.5-5.5	v. sl. s.	s.	s.
4055	hyg. cr.	....	119-20 sl. d.	....	v. s.	s.	s.
4056	brn. yel. pd.	....	....	....	s.	....	....
4057	yel. pd.	....	....	....	i.; s. h. ac.	i.; s. dil. $NaOH$	i.; s. NH_4OH
4058	lq.	$0.837^{20°}_{4}$	−81	68.4	6.4	....	....
4059	lq.	$0.852^{20°}_{4}$	....	114.5	25	....	....
4060	col. lq.	$0.926^{20°}_{4}$	....	72-3	<0.1	....	....
4061	col. gas	$(A)0.554^{0°}_{760}$	−182.5	−161.5	$3.3^{20°}$ cc.	$47.1^{20°}$ cc.	$104^{10°}$ cc.
4062	hex. pl./65% al.	....	283 d.	....	s. c.	s.; i. abs.	i.; i. bz.; i. act.
4063	hyg. lq.	$1.177^{20°}_{4}$	8	203-4	∞	∞	∞
4064	col. nd.	....	98-9	....	0.17 c.; 2 h.	s.	s.
4065	pl./et.	$1.082^{41°}_{4}$	38-9	258	v. sl. s.	v. s.	v. s.
4066	lq.	....	....	$113\text{-}4^{13mm}$	....	v. s.	v. s.
4067	pl./aq.	1.180	98.5-9.0	200	$0.5^{30°}$	v. s.	v. s.
4068	nd./aq.	....	107-9	$170\text{-}2^{10mm}$	s. h.	s.	s.
4069	mn./aq.	$1.385^{4°}$	184.2	275-80	$0.03^{19°}$	v. s.	v. s.
4070	pr./et.	....	61-2	$354\text{-}5^{729mm}$	....	v. s.	v. s.
4071	col. nd.	....	22-3(18)	263 sl. d.	i. d.	s. d.	s. bz.
4072	col. lq.	$1.043^{25°}_{25}$	....	248-50	v. sl. s.	s.	∞
4073	pr./pet.	....	29-30	274	....	....	....
4074	lf./al.	....	90	....	....	s. h.	....
4075	col. lq.	....	....	$54\text{-}60^{13mm}$	....	....	....
4076	col. lq.	$1.048^{15°}_{4}$	....	131-1.5	....	....	....
4077	col. lq.	$0.959^{20°}_{4}$	....	106-8	....	....	....
4078	col. oil	$1.2^{15°}$	....	162^{42mm}	v. sl. s.	∞ ; ∞ bz.	∞ ; ∞ chl.
4079	pr.	$1.140^{15°}_{15}$	28.3	205	$1.7^{15°}$	v. s.	v. s.
4080	lq.	>1	<−17.5	243.3-4.3	sl. s.	∞ ; s. alk.	∞
4081	lf./aq.	....	53	243	....	....	v. s. bz.
4082	lq.	....	....	$190\text{-}17^{38mm}$	∞	....	....
4083	lq.	$1.154^{20°}$	26-8	254^{310mm}	....	s.	....
4084	pa. yel. lq.	$1.040^{20°}_{20}$	....	360-5 d.	i.	∞	∞
4085	col. nd.	....	28	206	v. s.	v. s.	v. s.; i. lg.
4086	rhb./al.	$1.004^{105°}_{4}$	102-4	2537^{12mm}	1.7 c.	2	14
4087	col. lq.	$0.933^{20°}_{4}$	−98.7	57.3	$33^{22°}$	∞	∞
4088	col. lq.	$1.077^{20°}_{4}$	−80	169-70 sl. d.	38	∞	∞

Table 7-4 (*Continued*)
PHYSICAL CONSTANTS OF ORGANIC COMPOUNDS

No.	Name	Synonym	Formula	Beil. Ref.	Formula Weight
4089	**Methyl** acetophenone (*p*)	methyl-*p*-tolyl ketone	$CH_3 \cdot C_6H_4 \cdot CO \cdot CH_3$	VII-307	134.18
4090	acetylsalicylate	methyl aspirin; methyl rhodin	$CH_3CO_2 \cdot C_6H_4 \cdot CO_2 \cdot CH_3$	X-73	194.19
4091	acetylurea	....	$CH_3 \cdot NH \cdot CO \cdot NH \cdot CO \cdot CH_3$	IV-66	116.12
4092	acridine (4)	....	$CH_3 \cdot C_{13}H_8N$	XX-470	193.25
4093	acridine (2)	....	$CH_3 \cdot C_{13}H_8N$	XX-470	193.25
4094	acridine (9)	....	$CH_3 \cdot C_{13}H_8N$	XX-470	193.25
4095	acrylate	....	$CH_2{:}CH \cdot CO_2 \cdot CH_3$	II-399	86.09
4096	acrylic acid (α)	....	$CH_2{:}C(CH_3) \cdot CO_2H$	II-421	86.09
4097	hydrogen adipate	Me adipate (mono)	$CH_3O_2C(CH_2)_4 \cdot CO_2H$	II-652	160.17
4098	adipic acid (α)	....	$HO_2C \cdot CH(CH_3) \cdot (CH_2)_3 \cdot CO_2H$	II-672	160.17
4099	adipic acid (β)(*d*)	....	$HO_2C \cdot (CH_2)_2 \cdot CH \cdot (CH_3) \cdot CH_2 \cdot CO_2H$	II-673	160.17
4100	alcohol	wood alcohol	$CH_3 \cdot OH$	I-273	32.04
4101	alizarin (2;3,4)	3-methyl-alizarin	$CH_3(OH)_2C_6H(CO)_2{:}C_6H_4$	VIII-469	254.24
4102	allylamine	....	$CH_3 \cdot NH \cdot C_3H_5$	IV-206	71.12
4103	allyl carbinol	amylene alcohol	$(CH_3)(C_3H_5){:}CHOH$	I-443	86.13
4104	allyl ether	....	$CH_3 \cdot O \cdot CH_2 \cdot CH{:}CH_2$	I-437	72.11
4105	amine	amino-methane	$CH_3 \cdot NH_2$	IV-32	31.06
4106	amine hydrochloride	....	$CH_3NH_2 \cdot HCl$	IV-36	**67.52**
4107	aminoacetate	methyl glycinate	$NH_2 \cdot CH_2 \cdot CO_2 \cdot CH_3$	IV-340	89.09
4108	aminobenzoate (*o*)	methyl anthranilate	$NH_2 \cdot C_6H_4 \cdot CO_2 \cdot CH_3$	XIV-317	151.17
4109	aminobenzoate (*p*)	....	$NH_2 \cdot C_6H_4 \cdot CO_2 \cdot CH_3$	XIV-422	151.17
4110	aminobenzoic acid (*o*)	*N*-methyl anthranilic acid	$CH_3NH \cdot C_6H_4 \cdot CO_2H$	XIV-323	151.17
4111	amino ethanol (β)	....	$CH_3NH \cdot CH_2 \cdot CH_2OH$	IV-276	75.11
4112	3-amino-4-hydroxy benzoate	orthoform new	$NH_2(OH)C_6H_3 \cdot CO_2 \cdot CH_3$	XIV-593	167.17
4113	amino-*p*-hydroxy-benzoic acid (3)	....	$CH_3NH(OH)C_6H_3 \cdot CO_2H$	XIV-593	167.17
4114	amino-2-methyl-heptene (6)	octin	$C_8H_{15} \cdot NHCH_3$	....	141.26
4115	amino-phenol (*o*)	....	$CH_3NH \cdot C_6H_4 \cdot OH$	XIII-362	123.16
4116	amino-phenol H_2SO_4	(*o*)	$(C_7H_9ON)_2 \cdot H_2SO_4$	XIII-362	344.39
4117	amino-phenol H_2SO_4	(*p*); metol; elon	$(C_7H_9ON)_2 \cdot H_2SO_4$	XIII-442	344.39
4118	amino-propionic acid (α)(*dl*)	....	$CH_3 \cdot CH(NHCH_3) \cdot CO_2H$	IV-391	103.12
4119	amylbenzene (*m*)	amyl toluene	$CH_3 \cdot C_6H_4 \cdot C_5H_{11}$	V-446	162.28
4120	*n*-amyl ether	....	$CH_3 \cdot O \cdot C_5H_{11}$	*I-193	102.18
4121	*iso*-amyl ether	....	$CH_3 \cdot O \cdot C_5H_{11}$	I-400	102.18
4122	*n*-amyl ketone	heptanone-2	$CH_3(CH_2)_4 \cdot CO \cdot CH_3$	I-699	114.19
4123	*iso*-amyl ketone	2-Me-hexanone-5	$CH_3 \cdot CO \cdot C_5H_{11}$	I-701	114.19
4124	*iso*-amyl ketoxime	....	$CH_3 \cdot C({:}NOH) \cdot C_5H_{11}$	I-701	129.20
4125	aniline†	Me-Ph-amine	$C_6H_5 \cdot NH \cdot CH_3$	XII-135	107.16

† See also Nos. 5995-6000.
Methoxy-benzaldehyde 531-2
Methoxy-benzene 537
Methoxy-benzyl alcohol 530
Methoxy-cinchonine, cf. alkd.
Methoxy-cinchoninic acid 5537
Methoxy-coniferin, cf. glcde.
Methoxy-hydroxycinnamic acid 3236-7

Methoxy-styrene 6431-3
Methoxy-styryl ketone 528
Methoxy-tetrahydroquinoline 5874
Methoxylamine 4280
Methyl, cf. also dimethyl.
Methyl-acetanilide 26-8
Methyl-acetoacetic ester 3085
Methyl-acetopyrandione 1656

Table 7-4 (*Continued*)
PHYSICAL CONSTANTS OF ORGANIC COMPOUNDS

No.	Crystalline Form and Color	Specific Gravity	Melting Point °C.	Boiling Point °C.	Solubility in 100 Parts		
					Water	Alcohol	Ether
4089	nd.	$0.989^{22°}$	−23	228^{759mm}	i.	v. s.	v. s.
4090	col. cr./al.	....	50-1	$134-6^{9mm}$	i.	s.; s. gly.	s.; s. chl.
4091	mn./aq.	....	180	....	v. s. h.	s. h.	v. sl. s. h.
4092	nd./al.	....	88	....	v. sl. s. h.	s.	....
4093	yel. nd./ aq. al.	....	134	....	s. bz.	s.	s.
4094	pl./lg.	....	114-7	360^{740mm}	....	....	....
4095	col. lq.	$0.974^{0°}$	....	80.3*	....	s.	s.
4096	pr.	$1.015^{20°}_{4}$	15-16	161-3	s. h.	∞	∞
4097	lq.	....	8-9	162^{10mm}	....	....	....
4098	col. cr.	....	64	$216-20^{28mm}$	v. s.	v. s.	v. s.
4099	col. cr.	....	93-4.5	230^{30mm}	v. s.	v. s.	v. s.
4100	col. lq.	$**0.792^{20°}_{4}$	−97.8	64.7	∞	∞	∞
4101	yel. nd.	....	250-2	....	s. alk.	s.	s.
4102	lq.	....	....	64-6	s.	....	....
4103	lq.	$0.834^{20°}_{0}$	....	$115-6^{750mm}$	12.5	∞	∞
4104	lq.	$0.771^{1°}$	....	$42-3^{757mm}$	v. sl. s.	∞	∞
4105	col. gas	$0.699^{-11°}$	−92.5	-6.7^{758mm}	$959^{25°}$ cc. /cc.	s.	....
4106	pl./al.	1.23	226-8	$225-30^{15mm}$	v. s.	23 h. abs.	i.; i. chl.
4107	lq.	....	....	130 d.	....	....	....
4108	col. lq.	$1.168^{18.6°}_{4}$	24(8.2)	135.5^{15mm}	sl. s.	s.	s.
4109	lf./al.	....	112	....	....	....	....
4110	lf./al.	....	179-82	subl. d.	0.2 c.; 0.4 h.	s.	s.
4111	col. oil	$0.937^{20°}$	....	159^{747mm}	∞	∞	∞
4112	nd./bz. cr./chl.	....	142§ 110-1	....	v. sl. s.; i. pet.	18; s. act.	2; s. alk ; s. bz.
4113	cr.	....	142(110)	....	s. h.	s.	s.
4114	col. oil	0.795	....	176-8	i.	s.	s.; s. a.
4115	pl./bz. pet.	....	86-7	....	i.	s.	s. bz.
4116	col. cr.	....	....	....	s.	i.	i.
4117	col. nd./aq.	....	250-60 d.	....	$4^{25°}$	sl. s.	i.
4118	rhb./al.	....	307-17 d.	subl. 292 sl. d.	s.	10 h.; i. c. abs.	....
4119	lq.	$0.868^{22°}$	....	207-8	....	....	....
4120	col. lq.	0.75	....	99-100	....	....	....
4121	lq.	$0.687^{91°}_{4}$	....	91^{765mm}	....	....	....
4122	col. lq.	$0.820^{15°}_{4}$	−35.5	151.5	$0.4^{20°}$	s.	s.
4123	col. lq.	$0.813^{20°}$	−73.9	144	v. sl. s.	∞	∞
4124	oil	$0.888^{20°}_{4}$	....	$195-6^{761mm}$	....	....	....
4125	lq.	$0.986^{20°}_{4}$	−57	196.3	$0.01^{25°}$	s.	∞

* Polymerizes readily.
** See also special table of specific gravities.
§ HCl salt, m. p. 225°; s. 10/100 aq.
Methyl acet-toluidide 102-4
Methyl-acetyl carbinol 43
Methyl-acetylene 225
Methyl-acrolein 4058
Methyl adipate 2428, 4097

Methyl-allene 1018
Methyl-amyl alcohol 3630-7
Methyl-amyl carbinol (*n*) 3542
Methyl-amyl carbinol (*iso*) 3538
Methyl-aniline violet 5076
Methyl anisate 4301

Table 7-4 (*Continued*)
PHYSICAL CONSTANTS OF ORGANIC COMPOUNDS

No.	Name	Synonym	Formula	Beil. Ref.	Formula Weight
4126	**Methyl** *p*-anisyl ketone	*p*-acetonyl-anisole	$CH_3O \cdot C_6H_4 \cdot CH_2 \cdot CO \cdot CH_3$	VIII-106	164.21
4127	anthracene (α)	. . . .	$CH_3 \cdot C_{14}H_9$	V-674	192.26
4128	anthracene (β)	. . . .	$CH_3 \cdot C_{14}H_9$	V-674	192.26
4129	anthraquinone (2)	. . . .	$CH_3 \cdot C_6H_3(CO)_2C_6H_4$	VII-809	222.25
4130	arsenic acid	methyl arsonic acid	$CH_3 \cdot AsO(OH)_2$	IV-613	139.97
4131	arsenious oxide	. . . .	$CH_3 \cdot As{:}O$	IV-610	105.96
4132	arsine	. . . .	$CH_3 \cdot AsH_2$	IV-599	91.97
4133	auramine	. . . .	$[(CH_3)_2N \cdot C_6H_4]_2C{:}$ $N \cdot CH_3$	XIV-93	281.40
4134	azide	. . . .	$CH_3 \cdot N(N)_2$	I-80	57.06
4135	benzamide (*N*)	. . . .	$C_6H_5 \cdot CO \cdot NH \cdot CH_3$	IX-201	135.17
4136	benzimidazole (2)	. . . .	$C_6H_4 \cdot N{:}C(CH_3)NH$	XXIII-145	132.17
4137	benzimidazole (5)	*m*-toliminazole	$CH_3 \cdot C_6H_3 \cdot N{:}CHNH$	XXIII-151	132.17
4138	benzoate	oil of niobe	$C_6H_5 \cdot CO_2 \cdot CH_3$	IX-109	136.15
4139	benzoylacetate	. . . .	$C_6H_5CO \cdot CH_2 \cdot CO_2CH_3$	X-673	178.19
4140	*o*-benzoylbenzoate	. . . .	$C_6H_5CO \cdot C_6H_4 \cdot CO_2 \cdot CH_3$	X-748	240.26
4141	benzoylene urea (1)	. . . .	$C_6H_4N(CH_3)CONHCO$	XXIV-375	176.18
4142	benzoylene urea (3)	. . . .	$C_6H_4NHCON(CH_3)CO$	XXIV-375	176.18
4143	benzoylsalicylate	benzosalin	$C_6H_5CO_2 \cdot C_6H_4 \cdot CO_2 \cdot CH_3$	X-73	256.26
4144	benzylaniline (*N*)	. . . .	$C_6H_5 \cdot CH_2 \cdot N(CH_3) \cdot C_6H_5$	XII-1024	197.28
4145	benzyl ether	. . . .	$CH_3 \cdot O \cdot CH_2 \cdot C_6H_5$	VI-431	122.17
4146	benzyl ketone	phenyl acetone	$CH_3 \cdot CO \cdot CH_2 \cdot C_6H_5$	VII-303	134.18
4147	bromide	bromomethane	$CH_3 \cdot Br$	I-67	94.94
4148	bromoacetate	. . . .	$Br \cdot CH_2 \cdot CO_2 \cdot CH_3$	II-213	152.98
4149	*o*-bromobenzoate	. . . .	$Br \cdot C_6H_4 \cdot CO_2 \cdot CH_3$	IX-348	215.05
4150	*m*-bromobenzoate	. . . .	$Br \cdot C_6H_4 \cdot CO_2 \cdot CH_3$	IX-350	215.05
4151	*p*-bromobenzoate	. . . .	$Br \cdot C_6H_4 \cdot CO_2 \cdot CH_3$	IX-352	215.05
4152	4-bromohydrindene (7)	. . . .	$CH_3(Br)C_6H_2(CH_2)_3$	. . . .	211.11
4153	α-bromolaurate	. . . .	$C_{10}H_{21}CHBrCO_2CH_3$	. . . .	293.25
4154	butadiene-1,3 (2)	isoprene	$CH_2{:}CH \cdot C(CH_3){:}CH_2$	I-252	68.12
4155	butadiene-2,3 (2)	dimethyl-allene (*uns*)	$(CH_3)_2C{:}C{:}CH_2$	I-252	68.12
4156	butylamine (*n*)	. . . .	$CH_3 \cdot NH \cdot C_4H_9$	IV-157	87.17
4157	*iso*-butylamine	. . . .	$CH_3 \cdot NH \cdot C_4H_9$	IV-164	87.17
4158	*n*-butylbenzene (*o*)	butyl-toluene	$CH_3 \cdot C_6H_4 \cdot C_4H_9$	V-437	148.25
4159	*n*-butylbenzene (*m*)	. . . .	$CH_3 \cdot C_6H_4 \cdot C_4H_9$	V-437	148.25
4160	*n*-butylbenzene (*p*)	. . . .	$CH_3 \cdot C_6H_4 \cdot C_4H_9$	V-437	148.25
4161	*n*-butyl ether	. . . .	$CH_3 \cdot O \cdot C_4H_9$	I-369	88.15
4162	*iso*-butyl ether	. . . .	$CH_3 \cdot O \cdot C_4H_9$	I-376	88.15
4163	*n*-butyl ketone	hexanone-2	$CH_3 \cdot CO \cdot C_4H_9$	I-689	100.16
4164	*iso*-butyl ketone	2-Me-pentanone-4	$CH_3 \cdot CO \cdot C_4H_9$	I-691	100.16
4165	butyl ketone (*sec*)	3-Me-pentanone-2	$CH_3 \cdot CO \cdot C_4H_9$	I-693	100.16
4166	butyl ketone (*tert*)	pinacolin	$CH_3 \cdot CO \cdot C(CH_3)_3$	I-694	100.16
4167	*n*-butyl sulfide	. . . .	$CH_3 \cdot S \cdot C_4H_9$	. . . .	104.22

Table 7-4 (*Continued*)
PHYSICAL CONSTANTS OF ORGANIC COMPOUNDS

No.	Crystalline Form and Color	Specific Gravity	Melting Point °C.	Boiling Point °C.	Solubility in 100 Parts		
					Water	Alcohol	Ether
4126	oil	$1.071\frac{17}{17}°$	<−15	267–9	i.	s.	s.
4127	lf./al.	$1.047^{99.4°}$	86	....	s. bz.	....	....
4128	col. lf.	$1.181\frac{0°}{4}$	207	....	i.; s. bz.	v. sl. s.	v. sl. s.
4129	col. nd./al.	....	176–7	subl.	s. H_2SO_4	s.	s.; v. s. bz.
4130	mn.	....	161	....	v. s.	s.	....
4131	cr./CS_2	....	95	275±, d.	....	....	....
4132	col. lq.	....	....	2	0.01	∞	∞
4133	yel. cr./al.	....	133	....	v. sl. s.	s.	v. s. ac.
4134	col. lq.	$0.869\frac{8}{15}°$	expl. > 500	20–1	....	....	....
4135	pl./al.	....	82	291^{765mm}	s.	sl. s. lg.	i. alk.
4136	nd./aq.	....	175–6	....	s. h.; s. NaOH	sl. s.	sl. s.
4137	cr./aq.	....	114	....	....	....	....
4138	col. lq.	$1.087\frac{25}{25}°$	−12.4(−14)	199.5	$0.016^{30°}$	∞	∞
4139	lt. yel. lq.	$1.173\frac{0°}{4}$	....	152^{15mm}	i.	∞	∞
4140	mn. pr.	$1.190\frac{19.4°}{4}$	52	350–2	....	....	....
4141	col. nd.	....	264–5	....	s. h.	sl. s.	s. aq. NaOH
4142	col. nd.	....	242	....	i.	s.	....
4143	pr./al.	....	84–5	$270{-}80^{120mm}$	i. c.	3; s. bz.	s.; s. chi.
4144	lq.	....	9.2	305–6	i.	s.	s.
4145	lq.	$0.971\frac{15}{15}°$	....	174	i.	s.	s.
4146	cr.	$1.003^{20°}$	27	210–2	i.	s.	s.
4147	col. gas	$1.732\frac{0°}{0}$	−93.7	3.5	v. sl. s.	s.; s. CS_2	s.; s. chi.
4148	lq.	....	....	$51{-}2^{15mm}$	....	....	....
4149	lq.	....	....	244–6	....	....	....
4150	pl.	....	31–2	122.5^{15mm}	....	....	....
4151	rhb./aq. al.	1.689	79–80	....	....	s.	s.
4152	lt. yel. lq.	....	....	$120{-}2^{10mm}$	i.	sl. s.	s.
4153	col. lq.	....	....	$145{-}8^{8mm}$	....	....	....
4154	col. lq.	$0.681\frac{20°}{4}$	−146.0	34.1	i.	∞	∞
4155	lq.	$0.680^{20°}$	−120	40	....	....	....
4156	lq.	$0.736\frac{18°}{4}$	−75	91–2	....	....	....
4157	lq.	$0.722^{18°}$	....	76–8	....	....	....
4158	oil	$0.870\frac{18°}{4}$	....	200–1	i.	sl. s.	s.
4159	oil	$0.861\frac{20°}{4}$	....	197.5	i.	sl. s.	s.
4160	oil	$0.864\frac{20°}{4}$	....	197–8	i.	sl. s.	s.
4161	lq.	$0.744\frac{20°}{4}$	−115.5	71	i.	∞	∞
4162	lq.	$0.731\frac{20°}{4}$	....	59^{741mm}	i.	∞	∞
4163	col. lq.	$0.816\frac{15°}{4}$	−56.9	127.2	v. sl. s.	∞	∞
4164	col. lq.	$0.801\frac{20°}{4}$	−84.7	117–9	$2^{20°}$	∞	∞; ∞ bz.
4165	col. lq.	$0.815\frac{18°}{4}$	....	117.8	....	....	....
4166	col. lq.	$0.800^{16°}$	−52.5	106.2	$2.5^{15°}$	s.; v. s. act.	s.
4167	col. lq.	....	....	122–4	....	....	....

Table 7-4 (*Continued*)
PHYSICAL CONSTANTS OF ORGANIC COMPOUNDS

No.	Name	Synonym	Formula	Beil. Ref.	Formula Weight
	Methyl				
4168	butyne-3 (2)	*iso*-propyl acetylene	$(CH_3)_2CH \cdot C \vdots CH$	I-251	68.12
4169	butyrate (*n*)	. . . .	$C_2H_5 \cdot CH_2 \cdot CO_2 \cdot CH_3$	II-270	102.13
4170	*iso*-butyrate	. . . .	$(CH_3)_2CH \cdot CO_2 \cdot CH_3$	II-290	102.13
4171	butyric acid†(*α*)(*dl*)	Me-Et-acetic acid	$C_2H_5 \cdot CH(CH_3)CO_2H$	II-304	102.13
4172	caprate	. . . .	$C_9H_{19} \cdot CO_2 \cdot CH_3$	II-356	186.30
4173	caproate (*n*)	. . . .	$CH_3(CH_2)_4CO_2 \cdot CH_3$	II-323	130.19
4174	caprylate	. . . .	$CH_3(CH_2)_6CO_2 \cdot CH_3$	II-348	158.24
4175	carbamate	"methyl urethane"	$NH_2 \cdot CO_2 \cdot CH_3$	III-21	75.07
4176	carbanilide	. . . .	$C_6H_5 \cdot N(CH_3) \cdot CO \cdot$ $NH \cdot C_6H_5$	XII-418	226.27
4177	carbitol	diethylene-glycol (mono)methyl ether	$CH_3O \cdot CH_2 \cdot CH_2 \cdot O \cdot$ $CH_2 \cdot CH_2 \cdot OH$	. . . .	120.15
4178	carbitol acetate	. . . .	$CH_3CO_2 \cdot C_5H_{11}O_2$	. . . .	162.19
4179	carbostyril (*γ*)	lepidone	$CH_3 \cdot C_9H_6ON$	XXI-107	159.19
4180	carbostyril (*N*)	*N*-Me-*α*-quinolone	$C_6H_4CH \vdots CH \cdot CO \cdot N \cdot$ $\overline{}$ CH_3	XXI-304	159.19
4181	carvacryl ketone	2-aceto-*p*-cymene	$CH_3 \cdot CO \cdot C_{10}H_{13}$	VII-336	176.26
4182	cellosolve acetate	. . . .	$CH_3CO_2 \cdot CH_2 \cdot CH_2 \cdot$ OCH_3	II-141	118.13
4183	chloride	chloromethane	$CH_3 \cdot Cl$	I-59	50.49
4184	chloroacetate	. . . .	$ClCH_2 \cdot CO_2 \cdot CH_3$	II-197	108.53
4185	*o*-chlorobenzoate	. . . .	$Cl \cdot C_6H_4 \cdot CO_2 \cdot CH_3$	IX-336	170.60
4186	*β*-chloroethyl ether	. . . .	$CH_3 \cdot O \cdot CH_2 \cdot CH_2Cl$	I-337	94.54
4187	chloroformate	Me-chlorocarbonate	$Cl \cdot CO_2 \cdot CH_3$	III-9	94.50
4188	4-chloro- hydrindene (7)	. . . .	$CH_3(Cl)C_6H_2(CH_2)_3$	. . . .	166.65
4189	4-chlorophenyl-4- toluene sulfonate	(3,4'-diMe;4-Cl)	$CH_3 \cdot C_6H_4 \cdot SO_3 \cdot$ $C_6H_3(Cl) \cdot CH_3$	. . . .	296.77
4190	cholanthrene	. . . .	$C_{20}H_{13} \cdot CH_3$	. . . .	268.36
4191	cinnamate	. . . .	$C_6H_5 \cdot C_2H_2 \cdot CO_2 \cdot CH_3$	IX-581	162.19
4192	coumarin (*β*)	. . . .	$C_6H_4 \cdot C(CH_3) \vdots CH \cdot CO_2$ $\overline{}$	XVII-336	160.17
4193	crotonate (*α*)	. . . .	$C_3H_5 \cdot CO_2 \cdot CH_3$	II-410	100.12
4194	*iso*-cyanate	methyl-carbonimide	$CH_3 \cdot N \vdots CO$	IV-77	57.05
4195	*iso*-cyanide	methyl-carbylamine	$CH_3 \cdot NC$	IV-56	41.05
4196	cyanoacetate	. . . .	$CH_3O \cdot CO \cdot CH_2 \cdot CN$	II-584	99.09
4197	*iso*-cyanurate	. . . .	$(CH_3)_3(CON)_3$	XXVI-249	171.16
4198	cyclobutane	Me-tetramethylene	$CH_3 \cdot C_4H_7$	V-20	70.14
4199	cyclohexane	toluene hexahydride	$CH_3(CHC_5H_{10})$	V-29	98.19
4200	cyclohexanol (*o*)	hexahydro-*o*-cresol (**cis**) (**dl**)	$CH_3 \cdot C_6H_{10} \cdot OH$	VI-11	114.19
4201	cyclohexanol (*o*)	(*trans*)	$CH_3 \cdot C_6H_{10} \cdot OH$	**VI-18	114.19
4202	cyclohexanol (*p*)	4-methylhexalin (**cis**) (**dl**)	$CH_3 \cdot C_6H_{10} \cdot OH$	VI-14	114.19
4203	cyclohexanol (*p*)	(*trans*)	$CH_3 \cdot C_6H_{10} \cdot OH$	. . . .	114.19
4204	cyclohexanone (*o*)	. . . .	$CH_3 \cdot C_6H_9 \vdots O$	VII-14	112.17
4205	cyclohexanone (*m*)	(*dl*)	$CH_3 \cdot C_6H_9 \vdots O$	VII-17	112.17
4206	cyclohexanone (*p*)	. . . .	$CH_3 \cdot C_6H_9 \vdots O$	VII-18	112.17
4207	cyclohexene (Δ¹)	toluene tetrahydride	$CH_3 \cdot C_6H_9$	V-66	96.17
4208	cyclohexene (Δ³)	. . . .	$CH_2CH \vdots CH(CH_2)_2CH \cdot$ $\overline{}$ CH_3	V-67	96.17

† See also No. 6406.
Methyl carbonate 2468
Methyl-carbonimide 4194
Methyl-carbylamine 4195

Methyl cellosolve 3460
Methyl chlorocarbonate 4187
Methyl-chloroform 6151
Methyl-cinnamyl ketone 626

Table 7-4 (*Continued*)
PHYSICAL CONSTANTS OF ORGANIC COMPOUNDS

No.	Crystalline Form and Color	Specific Gravity	Melting Point °C.	Boiling Point °C.	Solubility in 100 Parts		
					Water	Alcohol	Ether
4168	col. lq.	$0.665^{20°}$	....	28	i.	∞	∞
4169	col. lq.	$0.903^{16°}_4$	−84.8	102.8	1.7	∞	∞
4170	col. lq.	$0.889^{20°}_4$	−87.7	92.3	v. sl. s.	∞	∞
4171	col. lq.	$0.938^{20°}_{20}$	<−80	174-7	sl. s.	∞	∞
4172	lq.	....	−18	223-4	i.	∞	∞
4173	col. lq.	$0.889^{15°}_4$	−71.0	151.3	i.	∞	∞
4174	col. lq.	$0.887^{18°}$	−40	192-4	i.	∞	∞
4175	col. pl.	$1.136^{56°}_4$	54.2	177	$217^{11°}$	$73^{15°}$	s.
4176	nd./al.	....	104	203-5	sl. s. h.	sl. s. c.	v. s.; v. s. bz.
4177	col. lq.	$1.035^{20°}_{20}$	....	193.2	∞	....	....
4178	col. lq.	$1.040^{20°}_{20}$	....	209.1	∞	....	....
4179	nd./aq.	....	223.7	270^{17mm}	sl. s. h.; s. a.	s. h.; i. alk.	sl. s.; sl. s. bz.
4180	col. nd./lg.	....	74	324^{728mm} sl. d.	1.4 c.	s.; s. chl.; sl. s. lg.	s. act.; v. s. bz.
4181	lq.	$0.956^{20°}_4$	<−10	240-4	....	....	....
4182	col. lq.	$1.007^{20°}_{20}$	−65.1	144.5	∞	....	....
4183	col. gas	(A)1.785	−97.7	−24	$280^{16°}$ cc.	$350^{20°}$ cc.	4000 cc., ac.
4184	col. lq.	$1.236^{20°}_4$	−32.7	130^{740mm}	v. sl. s.	∞	∞
4185	col. lq.	....	....	234-5			
4186	col. lq.	$1.031^{20°}_4$	....	90.5	8 c.	∞	∞
4187	col. lq.	$1.236^{15°}$	....	71-2	d.	∞; ∞ chl.	∞; ∞ bz.
4188	yel. lq.	....	....	$111-4^{10mm}$	....		
4189	col. cr.	....	94-5	....	i.	sl. s.	v. s.
4190	yel. nd./bz.	....	176.5-7.5	....	....	....	....
4191	cr.	$1.042^{36°}_0$	33.4	263	i.	v. s.	v. s.
4192	nd./bz.	....	82	....	....	s.	s. bz.
4193	lq.	$0.981^{4°}$	....	128	i.	∞	∞
4194	col. lq.	$0.967^{16°}_4$	....	43-5	....	....	....
4195	col. lq.	$0.734^{18°}_4$	−45	59.6	$10^{15°}$	∞	∞
4196	lq.	$1.123^{15°}_4$	−22.5	203	i.	∞	∞
4197	mn.	....	175-6	274	sl. s. h.	s.	....
4198	lq.	$0.694^{20°}_4$	....	39-42	i.	....	s.
4199	col. lq.	$0.769^{20°}_4$	−126.6	100.9	i.	s.	s.
4200	col. lq.	$0.934^{20°}_4$	−9.5	165	v. sl. s.	∞	∞
4201	col. lq.	$0.924^{20°}_4$	−21	166.5	....	....	....
4202	lq.	$0.913^{21°}_4$	....	$173-4^{750mm}$	v. sl. s.	∞	∞
4203	lq.	$0.912^{21°}_4$	....	$173-4^{745mm}$	....	....	....
4204	lq.	$0.925^{20°}_4$	−14	165.1	i.	s.	s.
4205	lq.	$0.915^{20°}_4$	−73.5	169.6	i.	s.	s.
4206	lq.	$0.916^{20°}_4$	−40.6	171.3	i.	s.	s.
4207	lq.	$0.809^{20°}_4$	....	110-1	i.	s.	s.
4208	lq.	$0.800^{20°}_{20}$	....	103	i.	s.	s.

Methyl citraconate 2469
Methyl citrate 6263
Methyl-cresyl ether 4399-4401

Methyl-cresyl sulfide 4402
Methyl cyanide 49
Methyl-cyclohexadiene 2286-8

Table 7-4 (*Continued*)
PHYSICAL CONSTANTS OF ORGANIC COMPOUNDS

No.	Name	Synonym	Formula	Beil. Ref.	Formula Weight
4209	**Methyl** cyclopentane	....	$CH_3(CH \cdot C_4H_8)$	V-27	84.16
4210	cyclopropane	methyl-trimethylene	$CH_3 \cdot CH \cdot CH_2 \cdot CH_2$	V-18	56.11
4211	diazoaminobenzene (4)	....	$CH_3 \cdot C_6H_4 \cdot N:N \cdot$ $NH \cdot C_6H_5$	XVI-705	211.27
4212	dibromoanthranilate	(3,5)	$Br_2C_6H_2(NH_2) \cdot$ CO_2CH_3	*XIV-553	308.97
4213	dibromocinnamate (α,β)(*dl*)	Me β-Ph-α,β-diBr-propionate	$C_6H_5 \cdot CHBr \cdot CHBr \cdot$ CO_2CH_3	IX-518	322.01
4214	dichloroarsine	....	$CH_3 \cdot AsCl_2$	IV-601	160.86
4215	diethylamine	....	$CH_3 \cdot N(C_2H_5)_2$	IV-99	87.17
4216	diethylaminobenzene (*o*)	*o*-diethyltoluidine	$CH_3 \cdot C_6H_4 \cdot N(C_2H_5)_2$	XII-786	163.26
4217	diethylaminobenzene (*m*)	*m*-diethyltoluidine	$CH_3 \cdot C_6H_4 \cdot N(C_2H_5)_2$	XII-857	163.26
4218	diethylaminobenzene (*p*)	*p*-diethyltoluidine	$CH_3 \cdot C_6H_4 \cdot N(C_2H_5)_2$	XII-904	163.26
4219	diphenylamine (*N*)	....	$CH_3 \cdot N(C_6H_5)_2$	XII-180	183.26
4220	diphenylamine (*p*)	Ph-*p*-tolylamine	$CH_3 \cdot C_6H_4 \cdot NH \cdot C_6H_5$	XII-905	183.26
4221	diphenyl carbinol	....	$(C_6H_5)_2C(OH) \cdot CH_3$	VI-685	198.27
4222	disulfide	dimethyl-disulfide	$CH_3 \cdot S \cdot S \cdot CH_3$	I-291	94.20
4223	epichlorohydrin(β)	....	$O \cdot CH_2 \cdot C(CH_3) \cdot CH_2Cl$	....	106.55
4224	ethyl-acetaldehyde	2-methyl-butanal-1	$C_2H_5 \cdot CH(CH_3) \cdot CHO$	I-682	86.13
4225	β-ethylacrolein(α)	2-Me-pentene-2-al-1	$C_2H_5 \cdot CH \cdot C(CH_3)CHO$	I-735	98.15
4226	ethylamine	....	$CH_3 \cdot NH \cdot C_2H_5$	IV-94	59.11
4227	ethylamine HCl	....	$C_2H_5(CH_3)NH \cdot HCl$	IV-94	95.57
4228	ethylaniline	....	$C_6H_5 \cdot N(CH_3)C_2H_5$	XII-162	135.21
4229	ethyl carbonate	....	$CH_3O \cdot CO \cdot OC_2H_5$	III-4	104.11
4230	ethyl ether	....	$CH_3 \cdot O \cdot C_2H_5$	I-314	60.10
4231	ethyl ethyl-phenyl-malonate	....	$CH_3O_2C \cdot C(C_6H_5):$ $(C_2H_5)CO_2C_2H_5$	....	250.30
4232	ethyl ketone	butanone	$CH_3 \cdot CO \cdot C_2H_5$	I-666	72.11
4233	ethyl ketoxime	butanoxime	$CH_3 \cdot C(:NOH) \cdot C_2H_5$	I-668	87.12
4234	ethyl ketone semi-carbazone	....	$C_3H_8:C:N \cdot NHCONH_2$	III-102	129.16
4235	ethyl malonic acid	....	$C_2H_5(CH_3)C:(CO_2H)_2$	II-664	146.14
4236	ethyl oxalate	....	$CH_3O \cdot (CO)_2 \cdot OC_2H_5$	II-535	132.12
4237	ethyl succinate	....	$C_7H_{12}O_4$	II-609	160.17
4238	ethyl sulfide	....	$CH_3 \cdot S \cdot C_2H_5$	I-343	76.16
4239	ethyl sulfone	....	$CH_3 \cdot SO_2 \cdot C_2H_5$	I-343	108.16
4240	fluoride	fluoromethane	$CH_3 \cdot F$	I-59	34.03
4241	formanilide	....	$C_6H_5(CH_3)N \cdot CHO$	XII-234	135.17
4242	formate	....	$HCO_2 \cdot CH_3$	II-18	60.05
4243	fumaric acid (*trans*)	mesaconic acid	$CH_3 \cdot C(CO_2H):CH \cdot$ CO_2H	II-763	130.10
4244	furfural (2,5)	....	$CH_3 \cdot C_4H_2O \cdot CHO$	XVII-289	110.11
4245	furfurane (2)	sylvane	$CH_3 \cdot C_4H_3O$	XVII-36	82.10
4246	furoate	....	$C_4H_3O \cdot CO_2CH_3$	XVIII-274	126.11
4247	furoylacetate (α)	....	$C_4H_3O \cdot CO \cdot CH_2 \cdot$ CO_2CH_3	....	168.15
4248	furylacrylate (β)	....	$C_4H_3O \cdot CH:CH \cdot$ CO_2CH_3	XVIII-300	152.15
4249	gallate	gallicin	$(HO)_3C_6H_2 \cdot CO_2CH_3$	X-483	184.15
4250	glucoside (α)(*d*)	....	$CH_3O \cdot C_6H_{11}O_5$	XXXI-179	194.19

Table 7-4 (*Continued*)
PHYSICAL CONSTANTS OF ORGANIC COMPOUNDS

No.	Crystalline Form and Color	Specific Gravity	Melting Point °C.	Boiling Point °C.	Solubility in 100 Parts		
					Water	Alcohol	Ether
4209	col. lq.	$0.749^{20°}_{4}$	−142.4	**71.8**	i.	....	s.
4210	col. gas	$0.691^{-20°}_{0}$	....	4-5	i.	....	s.
4211	yel. lf.	....	90-1	d.	i.	....	....
4212	nd.	....	87-8	....	....	s.	....
4213	mn. pl./ CS$_2$	....	115-6	....	....	....	....
4214	col. lq.	$1.838^{20°}_{4}$	−59	133-6	sl. s.	v. s.	s.
4215	col. lq.	....	....	65-7	v. s.	s.	s.
4216	lq.	....	....	208-9^{755mm}	i.	s.	s.
4217	lq.	....	....	231-2	i.	s.	s.
4218	lq.	$0.924^{15.5°}$	....	228-9	....	....	....
4219	col. lq.	$1.048^{20°}_{4}$	−7.6	295-6	i.	s.	s.
4220	cr.	....	87-9	317-8^{728mm}	i.	s.	s.
4221	pr./et.	....	80-1	175-80^{20mm}	v. s. bz.	....	v. s.
4222	lq.	$1.057^{16°}_{4}$	....	116-8	....	....	....
4223	lq.	$1.103^{20°}_{4}$	....	122.3	3.1	....	....
4224	lq.	$0.807^{20°}$	....	90-2	i.	....	....
4225	lq.	$0.854^{25°}_{4}$	....	137.3^{759mm}	i.	....	....
4226	col. lq.	....	....	34-5	....	....	....
4227	lf./al. et.	....	130-3	....	v. s.	v. s.	i.; s. chl.
4228	lq.	....	....	201	i.	∞	∞
4229	lq.	$1.002^{27°}$	−14.5	109.2	i.	∞	∞
4230	col. lq.	$0.697^{21.1°}_{4}$	....	7.6	s.	∞	∞
4231	yel. lq.	....	....	165-70^{20mm}	i.	s.	s.
4232	col. lq.	$0.805^{20°}_{4}$	−86.9	79.6	37	∞	∞ ; ∞ bz.
4233	col. oil	$0.923^{20°}_{4}$	−29.5	152-3	10	∞	∞
4234	lf./aq.	....	148; slowly 143	....	v. s. h.	v. s.	....
4235	pr. or nd.	....	122	d. 180	v. s.	v. s.	v. s.
4236	lq.	$1.156^{0°}_{0}$	....	173.7	i.	v. s.	v. s.
4237	lq.	$1.093^{0°}$	<−20	208.2	i.	v. s.	v. s.
4238	oil	$0.837^{20°}$	−104.8	66.9	i.	∞	∞
4239	nd.	....	36	....	v. s.	v. s.	v. sl. s. c.
4240	gas	....	....	−78.4	1661^{5c} cc.	....	....
4241	lq.	$1.095^{20°}_{4}$	14-5	253^{716mm}	v. sl. s.	v. s.	....
4242	lq.	$0.974^{20°}_{4}$	−99.0	31.8	30$^{20°}$	∞	s. Me al.
4243	cr. pd.	1.466	202	250 d.	2.7$^{18°}$; 118$^{100°}$	30.6$^{17°}$, 90% al.	v. sl. s. chl.
4244	oil	$1.109^{18°}$	....	186-7	3.3	v. s.	....
4245	col. lq.	0.916	....	65^{759mm}	....	....	....
4246	col. lq.	$1.179^{21.4°}_{4}$	....	181.3	i.	∞	∞
4247	yel. lq.	....	....	148-52^{13mm}	i.	s.	s.
4248	col. lq.	....	27.5	227-8^{774mm}	i.; s. lg.	s.; s. bz.	s.
4249	mn./Me al.	....	194-8	....	1.07$^{23°}$	v. s.	s.
4250	col. nd./al.	....	168-9	200$^{0.2mm}$	63	1.6	i.

Methyl-ethylacetic acid 4171
Methyl-ethyl acetone 4165
Methyl-ethyl-acetylene 5087
Methyl-ethylbenzene 3166-8
Methyl-ethyl-*n*-butyl carbinol 4968

Methyl-ethyl carbinol 1035
Methyl-ethyl ketone diethylsulfone 6326
Methyl-ethyl-*n*-propyl carbinol 3540
Methyl-ethyl-*iso*-propyl carbinol 3529
Methyl-ethyl-protocatechuic aldehyde 6419

Table 7-4 (*Continued*)
PHYSICAL CONSTANTS OF ORGANIC COMPOUNDS

No.	Name	Synonym	Formula	Beil. Ref.	Formula Weight
4251	**Methyl** glutaric acid (α)	2-methyl-pentandioic acid	$CH_3 \cdot CH(CO_2H) \cdot CH_2 \cdot CH_2 \cdot CO_2H$	II-655	146.14
4252	glutaric acid (β)	3-Me-pentandioic	$CH_3 \cdot CH(CH_2CO_2H)_2$	II-659	146.14
4253	glycerinate (d)	....	$CH_2OH \cdot CHOH \cdot CO_2 \cdot CH_3$	III-397	120.11
4254	glycerine chlorohydrin (β),mono	....	$ClCH_2 \cdot C(CH_3)OH \cdot CH_2OH$	....	124.57
4255	glycolate	....	$HO \cdot CH_2 \cdot CO_2 \cdot CH_3$	III-236	90.08
4256	glyoxalidine	2-Me-Δ^2-imidazoline	$CH_3 \cdot C \colon N(CH_2)_2NH$	XXIII-31	84.12
4257	glyoxaline (1)	1-methyl-imidazole	$CH_3 \cdot N \cdot CH \colon N \cdot CH \colon CH$	XXIII-46	82.11
4258	guanidine	....	$CH_3NH \cdot CH_3N_2$	IV-68	73.10
4259	guanidine nitrate	....	$C_2H_7N_3 \cdot HNO_3$	IV-69	138.11
4260	guanidine sulfate	....	$(C_2H_7N_3)_2 \cdot H_2SO_4$	....	244.27
4261	heptadecyl ketone	nonadecanone-2	$CH_3(CH_2)_{16} \cdot CO \cdot CH_3$	I-718	282.51
4262	heptenone	....	$(CH_3)_2C \colon CH \cdot CH(CH_2)_2 \cdot CO \cdot CH_3$	I-741	126.20
4263	n-heptylate	Me oenanthylate	$CH_3(CH_2)_5CO_2CH_3$	II-339	144.22
4264	n-heptyl ether	....	$CH_3 \cdot O \cdot (CH_2)_6CH_3$	I-414	130.23
4265	heptyl ketone	nonanone-2	$CH_3 \cdot CO \cdot (CH_2)_6CH_3$	I-709	142.24
4266	hexyl ketone	octanone-2	$CH_3 \cdot CO \cdot (CH_2)_5CH_3$	I-704	128.22
4267	hydantoin (β)	....	$C_4H_6O_2N_2$	XXIV-244	114.10
4268	hydantoin (γ)	lactyl urea	$C_4H_6O_2N_2 \cdot H_2O$	XXIV-279	132.12
4269	hydrazine	....	$CH_3 \cdot NH \cdot NH_2$	IV-546	46.07
4270	hydrazine sulfate	....	$CH_3 \cdot NH \cdot NH_2 \cdot H_2SO_4$	IV-547	144.15
4271	hydrazobenzene (o)	β-Ph-o-tolylhydrazine	$CH_3 \cdot C_6H_4 \cdot NH \cdot NH \cdot C_6H_5$	XV-497	198.27
4272	hydrazobenzene (m)	....	$CH_3 \cdot C_6H_4(NH)_2C_6H_5$	XV-506	198.27
4273	hydrazobenzene (p)	....	$CH_3 \cdot C_6H_4(NH)_2C_6H_5$	XV-511	198.27
4274	hydroxybenzoate (p)	....	$HO \cdot C_6H_4 \cdot CO_2CH_3$	X-158	152.15
4275	3-hydroxy-1,4-naphthoquinone (2)	phthiocol	$CH_3(OH)C_{10}H_4O_2$	....	188.18
4276	hydroxynaphthoquinone acetate	phthiocol monoacetate	$C_{13}H_{10}O_4$	....	230.22
4277	hydroxynaphthoquinone acetate	phthiocol triacetate	$C_{17}H_{16}O_6$	....	316.31
4278	hydroxynaphthoquinone oxime	phthiocol monoxime	$C_{11}H_8O_2 \colon NOH$	....	203.20
4279	hydroxylamine (β)	....	$CH_3 \cdot NH \cdot OH$	IV-534	47.06
4280	hydroxylamine HCl	(α)	$CH_3O \cdot NH_2 \cdot HCl$	I-288	83.52
4281	hypochlorite	....	$CH_3 \cdot O \cdot Cl$	I-282	66.49
4282	indole (1)	....	C_9H_9N	XX-308	131.18
4283	indole (2)	methyl ketole	$C_6H_4 \cdot CH \colon C(CH_3) \cdot NH$	XX-311	131.18
4284	indole (3)	skatole	$C_6H_4C(CH_3) \colon CH \cdot NH$	XX-316	131.18
4285	indole (5)	....	$CH_3 \cdot C_6H_3 \cdot CH \colon CH \cdot NH$	XX-317	131.18
4286	inositol	d-inositol Me ether	$(HO)_5C_6H_6 \cdot OCH_3$	VI-1193	194.19

Table 7-4 (*Continued*)
PHYSICAL CONSTANTS OF ORGANIC COMPOUNDS

No.	Crystalline Form and Color	Specific Gravity	Melting Point °C.	Boiling Point °C.	Solubility in 100 Parts		
					Water	Alcohol	Ether
4251	pr./aq.	....	78-80	214-5^{22mm}	v. s.	v. s.	v. s.
4252	pr. or pl.	....	86-7	....	v. s.	v. s.	v. s.
4253	lq.	1.281$\frac{15}{15}$°	....	119-20^{14mm}	∞	∞	v. sl. s.
4254	lq.	1.237$\frac{20}{4}$°	....	95-6^{7mm}	∞	....	....
4255	lq.	1.168$^{18°}$	....	151.2	....	....	....
4256	col. hyg. cr.	....	105	195-8	s.	s.	i.; s. chl.
4257	lq.	1.036$^{10°}$	−6	197-9	∞	....	
4258	col. delq.	....	d.	....	v. s.	s.	....
4259	col. cr./al.	....	149-50	....	s.	....	....
4260	cr./aq.	....	239-40	....	....	....	....
4261	lf.	0.811$^{55.5°}$	55-6	266.5^{110mm}	v. s. chl.	v. s.	v. s.
4262	lq.	0.860$^{20°}$	−67.1	173-4	i.	∞	∞
4263	lq.	0.885$\frac{15}{4}$°	−55.8	173.8	i.	....	....
4264	lq.	0.795$\frac{0}{0}$°	....	149.8	i.	∞	∞
4265	lq.	0.825$\frac{15}{4}$°	−7.8	194-6	i.	s.	s.
4266	col. lq.	0.819$\frac{20}{4}$°	−21	172.9	i.	∞	∞
4267	pr.	....	156-7	subl.	s.	s.	3
4268	rhb.	....	145(anh.)	....	s.	s.	sl. s.
4269	hyg. lq.	....	<−80	877^{45mm}	s.	∞	∞; i. lg.
4270	cr./Me al.	....	142	....	v. s.	v. sl. s.	....
4271	lf./al.	....	101-2	....	i.	sl. s. c.	s.
4272	lt. yel. cr.	....	59-61	....	i.; s. bz.	v. s.	sl. s.
4273	pl./lg.	....	86-7	....	v. s. bz.	v. s.	v. s.
4274	nd./aq. al.	....	131	270-80 d.	0.25	s.; s. act.	s.
4275	yel. pr./aq. Me al.	....	172-3	vol. in steam	s.; s. dil. alk.	s.	s.
4276	yel. nd.	....	106-7	....	i.; s. alk.	s. Me al.	s. pyr.
4277	col. pr.	....	158-9	....	i.	s. Me al.	....
4278	yel. nd.	....	199-200 d.	....	i.	s.	s. pyr.
4279	hyg. pr.	1.000$\frac{20}{4}$°	42	62.5^{15mm}	v. s.	v. s.	sl. s.
4280	pr./al. et.	....	149	....	s.	s.	i.
4281	gas	....	....	12^{726mm}	....	....	....
4282	lq.	1.071$^{0°}$	<−20	242.4	i.	v. s.	v. s.
4283	nd.	1.07	59-60	272^{750mm}	s. h.; s. HCl	v. s.	v. s.
4284	lf./lg.	....	95	265-6^{755mm}	0.05 c.	s.; s. chl.	s.; s. bz.
4285	nd./aq.	....	58.5	....	s. h.	s.	s.; s. bz.
4286	cr./aq.	1.52	186-7	....	v. s.	i. abs.	i.; i. chl.

Table 7-4 (*Continued*)
PHYSICAL CONSTANTS OF ORGANIC COMPOUNDS

No.	Name	Synonym	Formula	Beil. Ref.	Formula Weight
4287	**Methyl** iodide	iodomethane	CH_3I	I-70	141.94
4288	iodobenzoate (*o*)	....	$I \cdot C_6H_4 \cdot CO_2CH_3$	IX-364	262.05
4289	iodobenzoate (*m*)	....	$I \cdot C_6H_4 \cdot CO_2CH_3$	IX-365	262.05
4290	iodobenzoate (*p*)	....	$I \cdot C_6H_4 \cdot CO_2CH_3$	IX-367	262.05
4291	isatin (*O*)	....	$C_6H_4 \cdot N:C(OCH_3)CO$	XXI-583	161.16
4292	isatin (*N*)	....	$C_6H_4N(CH_3) \cdot CO \cdot CO$	XXI-446	161.16
4293	isatin (5)	*p*-methyl isatin	$CH_3 \cdot C_6H_3 \cdot CO \cdot CO \cdot NH$	XXI-509	161.16
4294	lactate	....	$CH_3 \cdot CHOH \cdot CO_2CH_3$	III-280	104.11
4295	laurate	....	$CH_3(CH_2)_{10}CO_2CH_3$	II-361	214.35
4296	levulinate	....	$CH_3CO(CH_2)_2CO_2CH_3$	III-675	130.14
4297	*d*-mannoside (*α*)	....	$CH_3O \cdot C_6H_{11}O_5$	I-907	194.19
4298	mercaptan	methanthiol	$CH_3 \cdot SH$	I-288	48.11
4299	mercaptobenzo-thiazole (2)	....	$C_6H_4 \cdot N:C(SCH_3) \cdot S$	XXVII-109	181.28
4300	methoxybenzoate (*o*)	....	$CH_3O \cdot C_6H_4CO_2CH_3$	X-71	166.18
4301	methoxybenzoate (*p*)	methyl anisate	$CH_3O \cdot C_6H_4CO_2CH_3$	X-159	166.18
4302	methylanthranilate (*N*)	methylamino-methyl benzoate	$CH_3NH \cdot C_6H_4 \cdot CO_2 \cdot CH_3$	XIV-324	165.19
4303	myristate	....	$CH_3(CH_2)_{12}CO_2CH_3$	II-365	242.41
4304	naphthalene (*α*)	....	$C_{10}H_7 \cdot CH_3$	V-566	142.20
4305	naphthalene (*β*)	....	$C_{10}H_7 \cdot CH_3$	V-567	142.20
4306	1,4-naphthoquin-one (2)	....	$CH_3 \cdot C_{10}H_5O_2$	**VII-656	172.19
4307	naphthohydro-quinonediacetate	(2;1,4)	$CH_3 \cdot C_{10}H_5: (O_2C \cdot CH_3)_2$	**VI-958	258.28
4308	naphthoquinone monoxime	(2;4,1)	$CH_3 \cdot C_{10}H_5O:NOH$	....	187.20
4309	naphthylamine (*α*)	....	$C_{10}H_7 \cdot NHCH_3$	XII-1221	157.22
4310	naphthylamine (*β*)	....	$C_{10}H_7 \cdot NHCH_3$	XII-1273	157.22
4311	naphthyl ether (*α*)	....	$CH_3 \cdot O \cdot C_{10}H_7$	VI-606	158.20
4312	naphthyl ether (*β*)	nerolin; yara yara	$CH_3 \cdot O \cdot C_{10}H_7$	VI-640	158.20
4313	naphthyl ketone (*β*)	*β*-acetonaphthone	$CH_3 \cdot CO \cdot C_{10}H_7$	VII-402	170.21
4314	nitramine	nitraminomethane	$CH_3 \cdot NH \cdot NO_2$	IV-567	76.06
4315	nitrate	....	$CH_3O \cdot NO_2$	I-284	77.04
4316	nitrite	....	$CH_3O \cdot NO$	I-284	61.04
4317	nitrobenzoate (*o*)	....	$NO_2 \cdot C_6H_4 \cdot CO_2CH_3$	IX-372	181.15
4318	nitrobenzoate (*m*)	....	$NO_2 \cdot C_6H_4 \cdot CO_2CH_3$	IX-378	181.15
4319	nitrobenzoate (*p*)	....	$NO_2 \cdot C_6H_4 \cdot CO_2CH_3$	IX-390	181.15
4320	nitrocinnamate (*o*)	....	$NO_2 \cdot C_6H_4 \cdot CH:CH \cdot CO_2CH_3$	IX-605	207.19
4321	nitrocinnamate (*m*)(*trans*)	....	$NO_2 \cdot C_6H_4 \cdot CH:CH \cdot CO_2CH_3$	IX-606	207.19
4322	nitrocinnamate (*p*)(*trans*)	....	$NO_2 \cdot C_6H_4 \cdot CH:CH \cdot CO_2CH_3$	IX-607	207.19
4323	nitrolic acid	....	$HC(:NOH)NO_2$	II-92	90.04
4324	*n*-nonyl ketone	undecanone-2	$CH_3(CH_2)_8 \cdot CO \cdot CH_3$	I-713	170.30
4325	octadiene-4,6 (2)	....	$CH_3 \cdot CH:(CH)_2:CH \cdot CH_2CH(CH_3)_2$	I-260	124.23
4326	*n*-octyl ketone	decanone-2	$CH_3 \cdot CO \cdot C_8H_{17}$	I-711	156.27
4327	oleate	....	$C_{17}H_{33}CO_2CH_3$	II-467	296.50

Table 7-4 (*Continued*)
PHYSICAL CONSTANTS OF ORGANIC COMPOUNDS

No.	Crystalline Form and Color	Specific Gravity	Melting Point °C.	Boiling Point °C.	Solubility in 100 Parts		
					Water	Alcohol	Ether
4287	col. lq.	2.279^{20}_{4}	−66.5	42.4	$1.8^{15°}$	∞	∞
4288	lq.	....	....	$277\text{-}8^{729mm}$	....	....	....
4289	nd.	....	50	$276\text{-}77^{39mm}$	i.; i. lg.	s. h.	s. h.
4290	rhb./et. al.	$2.020^{10°}$	114	subl.	....	....	....
4291	red lf./aq.	....	101-2	....	sl. s. c.	o. alk.	s. h. HCl
4292	red nd./ aq.	....	134	....	s. aq. NaOH	....	....
4293	red nd./ aq. al.	....	187	....	v. sl. s. h.;s.alk.	s.; s. h. HCl	sl. s.
4294	lq.	$1.090^{19°}$	....	144.8	∞; d.	s.	s.
4295	lq.	....	5	148^{18mm}	i.	....	....
4296	lq.	1.047^{20}_{4}	....	196	....	....	....
4297	col. nd./al.	1.473^{7}_{4}	193-4	....	$30.7^{15°}$	3	i.
4298	lq. or gas	$0.896^{0°}$	−121	5.96	s.	v. s.	v. s.
4299	pr./aq. al.	....	52	....	s. conc. a.	....	....
4300	lq.	$1.157^{19}_{4}{}^{°}$	....	245-6	....	s.	....
4301	pl./al.	....	48-9	255-6	i.	s.	s.
4302	cr./pet.	$1.120^{15°}$	18-9	256	i.	s.	s.
4303	cr./al.	....	18-9	295^{751mm}	i.	....	....
4304	oil	$1.025^{14}_{4}{}^{°}$	−30.8	244.8	i.	v. s.	v. s.
4305	mn.	$0.990^{40}_{4}{}^{°}$	34.4	241.1	i.	v. s.	v. s.
4306	yel. nd./ al.	....	106	volt. in steam	i.; sl. s. pet.	s.; s. bz.	s.
4307	col. pr./aq. Me al.	....	114	....	i.	s.	v. s. act.
4308	hex. pl./aq. Me al.	....	166-8	....	i.	s.	s. pyr.
4309	oil	....	....	293-5	s. CS_2	s.	s.
4310	oil	....	....	$308\text{-}10^{761mm}$	....	s.	....
4311	col. oil	$1.096^{14}_{4}{}^{°}$	<−10	265-9	i.	s.	s.; s. bz.
4312	lf./et.	....	73-5	274	sl. s. CS_2	sl. s.	s.; s. bz.
4313	nd./lg.	....	55-6	301-3	....	....	....
4314	nd./et.	$1.243^{49}_{4}{}^{°}$	38	....	v. s.	v. s.	sl. s.
4315	lq.	$1.203^{25°}$	expl.	65	sl. s.	s.	s.
4316	gas	lq. $0.991^{115°}$	....	−12	....	s.	s.
4317	yel. lq.	$1.286^{20°}$	−8	275	i. pet.	s.	s.
4318	col. nd.	....	78.5	279	....	sl. s.	sl. s.
4319	mn.	....	96	....	i.	s.	s.
4320	col. nd./ aq.	....	72-3	$187\text{-}9^{15mm}$	v. sl. s. h.	s. h.	....
4321	yel. pr./ Me al.	....	123-4	d.	i.; v. sl. s. CS_2	v. sl. s.; s. chl.	v. sl. s.; s. bz.
4322	col. nd./al.	....	160-1	281-6	....	sl. s.	i.
4323	nd./et.	....	64-8 d.	....	s.	s.	s.
4324	col. oil	$0.829^{15}_{4}{}^{°}$	12.8	223	i.	s.	s.
4325	lq.	$0.752^{18}_{4}{}^{°}$	....	149	i.	....	....
4326	lq.	$0.828^{15}_{4}{}^{°}$	13.1	209^{750mm}	i.	s.	s.
4327	oil	$0.874^{20}_{4}{}^{°}$	....	$190\text{-}1^{10mm}$	i	∞	∞

Methyl oxalate 4329
Methyl pentenal 4225
Methyl phenol 1545-7
Methyl-phenylamine 4125
Methyl-phenyl carbinol 5186

Methyl phenyl-dibromopropionate 4213
Methyl-phenyl diketone 72
Methyl-phenyl ether 537
Methyl-phenyl glyoxal 72
Methyl-phenyl ketone 51

Table 7-4 (*Continued*)
PHYSICAL CONSTANTS OF ORGANIC COMPOUNDS

No.	Name	Synonym	Formula	Beil. Ref.	Formula Weight
4328	**Methyl** orange	tropeolin D; orange III	$(CH_3)_2N \cdot C_6H_4 \cdot N:N \cdot C_6H_4 \cdot SO_3Na$	XVI-331	327.34
4329	oxalic acid	Me acid oxalate	$CH_3O_2C \cdot CO_2H$	II-534	104.06
4330	palmitate	....	$C_{15}H_{31}CO_2CH_3$	II-372	270.46
4331	pelargonate	methyl nonylate	$CH_3(CH_2)_7CO_2CH_3$	II-353	172.27
4332	2,4-pentanediol (2)	....	$(CH_3)_2C(OH) \cdot CH_2 \cdot CHOH \cdot CH_3$	I-486	118.18
4333	phenacyl bromide (*p*)	ω-Br-*p*-methyl-acetophenone	$CH_3 \cdot C_6H_4 \cdot CO \cdot CH_2Br$	VII-309	213.08
4334	phenazine (2)	tolazine	$CH_3 \cdot C_6H_3 : N_2 : C_6H_4$	XXIII-237	194.24
4335	phenylacetate	....	$C_6H_5 \cdot CH_2 \cdot CO_2CH_3$	IX-434	150.18
4336	phenylenediamine (*N*)(*p*)		$CH_3NH \cdot C_6H_4 \cdot NH_2$	XIII-71	122.17
4337	phenylenediamine HCl (*N*)(*p*)	*p*-amino-methyl-aniline di-HCl	$C_7H_{10}N_2 \cdot 2HCl$	....	195.09
4338	phenylhydrazine (α,α)		$C_6H_5 \cdot N(CH_3) \cdot NH_2$	XV-117	122.17
4339	phenylhydrazine sulfate (α,α)	....	$(C_7H_{10}N_2)_2 \cdot H_2SO_4$	XV-118	342.42
4340	phenylhydrazine (α,β)		$C_6H_5 \cdot NH \cdot NH \cdot CH_3$	XV-118	122.17
4341	phosphine	....	$CH_3 \cdot PH_2$	IV-580	48.02
4342	3-phytyl-1,4-naph-thoquinone (2)	vitamin K-1	$C_{31}H_{46}O_2$	....	450.71
4343	3-phytyl-1,4-naph-thoquinone di-acetate (2)	dihydrovitamin K-1 diacetate	$C_{35}H_{52}O_4$	....	536.80
4344	piperidine (*N*)	....	$C_5H_{10}N \cdot CH_3$	XX-16	99.19
4345	piperidine (2)	α-pipecoline	$CH_3 \cdot CH(CH_2)_4 \cdot NH$	XX-95	99.19
4346	piperidine (3)	β-pipecoline	$CH_2 \cdot NH \cdot (CH_2)_3 \cdot CH \cdot CH_3$	XX-100	99.19
4347	piperidine (4)	γ-pipecoline	$(CH_2)_2 \cdot NH \cdot (CH_2)_2 \cdot CH \cdot CH_3$	XX-101	99.19
4348	propargyl ether	....	$CH_3 \cdot O \cdot CH_2 \cdot C:CH$	I-454	70.09
4349	propionate	....	$C_2H_5 \cdot CO_2CH_3$	II-239	88.11
4350	*n*-propylamine	....	$CH_3 \cdot NH \cdot CH_2 \cdot C_2H_5$	IV-137	73.14
4351	*n*-propylbenzene (*m*)	*m*-propyl toluene	$CH_3 \cdot C_6H_4 \cdot CH_2 \cdot C_2H_5$	V-418	134.22
4352	*n*-propylbenzene (*p*)	*p*-propyl toluene	$CH_3 \cdot C_6H_4 \cdot CH_2 \cdot C_2H_5$	V-419	134.22
4353	*n*-propyl ether	....	$CH_3 \cdot O \cdot CH_2 \cdot C_2H_5$	I-354	74.12
4354	*iso*-propyl ether	....	$CH_3 \cdot O \cdot CH(CH_3)_2$	I-362	74.12
4355	*n*-propyl ketone	pentanone-2	$CH_3 \cdot CO \cdot CH_2 \cdot C_2H_5$	I-676	86.13
4356	propyl ketone oxime	(*n*)	$CH_3 \cdot C(:NOH) \cdot C_3H_7$	I-677	101.15
4357	*iso*-propyl ketone	2-methyl-butanone-3	$CH_3 \cdot CO \cdot CH(CH_3)_2$	I-682	86.13
4358	*iso*-propyl ketoxime	....	$CH_3 \cdot C(:NOH) \cdot C_3H_7$	I-683	101.15
4359	*iso*-propyl ketone semicarbazone	....	$C_3H_7(CH_3)C:N \cdot NH \cdot CO \cdot NH_2$	III-103	143.19
4360	5-pyrazolone (3)	....	$CH_2 \cdot CO \cdot NH \cdot N:C \cdot CH_3$	XXIV-19	98.11

Methyl-phenyl-nitrosoamine 4887
Methyl-phenylphenol ether 4073-4
Methyl phosphate 6268
Methyl phthalate 2530-2
Methyl *iso*-phthalic acid 6397

Methyl picrate 6290
Methyl-picrylnitramine 5873
Methyl pivalate 4404
Methyl-proline-methylpetaine, cf. alkd.

Table 7-4 (*Continued*)
PHYSICAL CONSTANTS OF ORGANIC COMPOUNDS

No.	Crystalline Form and Color	Specific Gravity	Melting Point °C.	Boiling Point °C.	Solubility in 100 Parts		
					Water	Alcohol	Ether
4328	red pd.	....	....	....	0.2 c.	i.	....
4329	cr.	....	37±	108-9^{12mm}	....	....	....
4330	col. cr.	....	29.5-30.5	196^{15mm}	i.	s.	s.
4331	lq.	0.877$^{17.5°}$	....	213-4^{757mm}	i.	s.	s.
4332	lq.	0.924$^{17°}_{4}$	....	196	s.	s.	s.
4333	lf./al.	....	50-1	....	i. d.	v. s. d.	v. s.
4334	nd.	....	117	350±, sl. d.	sl. s. h.	s.; s. H_2SO_4	s.; s. chl.
4335	lq.	1.044$^{16°}$	d. 360	220	i.	∞	∞
4336	col. lf.	....	35.5	257-9.5	s.	s.	s.
4337	yel. pd.	....	....	....	v. s.	sl. s.	i.
4338	yel. lq.	1.038$^{21.6°}_{4}$	....	131^{35mm}	sl. s. h.	∞; ∞ chl.	∞; ∞ bz.
4339	lf.	....	....	....	v. s.	v. sl. s. c.	....
4340	oil	1.04$^{15°}_{15}$	....	110-2$^{12-15mm}$	....	s.	s.
4341	col. gas	....	....	-14^{759mm}	i.	sl. s.	7000$^{0°}$ cc.
4342	pa. yel. oil	....	....	....	i.	s. abs.	s. act.; s. pet.
4343	col. nd./al.	....	62-3	....	i.	v. s.; v. s. bz.	v. s. act.
4344	lq.	0.821$^{15°}$	....	107	....	....	....
4345	lq.	0.862$^{0°}$	....	118-9^{753mm}	s.	i. aq. KOH	....
4346	lq.	0.864$^{0°}_{4}$	....	125-6	s.	....	....
4347	lq.	0.867$^{0°}$	....	127-9	s.	....	....
4348	lq.	0.83$^{12.5°}$	....	63	v. sl. s.	∞	∞
4349	col. lq.	0.915$^{20°}_{4}$	-87.5	79.7	0.5$^{20°}$	∞	∞
4350	lq.	0.720$^{17°}$	....	62-4	s.	s.	....
4351	lq.	0.863$^{16°}$	....	182	i.	s. abs.	∞
4352	lq.	0.864$^{15.4°}_{4}$	-63	183.5	i.	s. abs.	∞
4353	col. lq.	0.738$^{20°}$	....	39.1	sl. s.	∞	∞
4354	col. lq.	0.735$^{20°}_{20}$	....	32.5^{777mm}	v. sl. s.	∞	∞
4355	col. lq.	0.812$^{15°}_{4}$	-77.8	102.4	v. sl. s.	∞	∞
4356	oil	0.910$^{20°}_{4}$	....	168^{748mm}	s.	∞	∞
4357	col. lq.	0.803$^{20°}_{0}$	-92	95	v. sl. s.	∞	∞
4358	lq.	....	....	157-8	s.	∞	∞
4359	cr./et.	....	110-4	....	s.	v. s.	v. s.
4360	pr./aq.; nd./al.	....	216-7	subl.	v. s. h.; sl. s. Na_2CO_3	sl. s. h.	sl. s. aq. a.

Methyl-propenyl ketone 3213
Methyl-*iso*-propyl-acetophenone 4181
Methyl-*iso*-propylbenzene 1622-4
Methyl-propyl carbinol (*n*) 405

Methyl-propyl carbinol (*iso*) 407
Methyl-*iso*-propyl-phenanthrene 5555
Methyl-*iso*-propyl-phenol 1255, 5938
Methyl-pyridine 5300-2

Table 7-4 (*Continued*)
PHYSICAL CONSTANTS OF ORGANIC COMPOUNDS

No.	Name	Synonym	Formula	Beil. Ref.	Formula Weight
4361	**Methyl** pyrogallol (1;3,4,5)	trihydroxy-toluene	$CH_3 \cdot C_6H_2(OH)_3$	VI-1112	140.14
4362	pyrrole (1)	*N*-methyl pyrrole	$C_4H_4N \cdot CH_3$	XX-163	81.12
4363	pyrrole (2)(α)	....	$CH_3 \cdot C_4H_4N$	XX-170	81.12
4364	pyrrole (3)(β)	....	$CH_3 \cdot C_4H_4N$	XX-171	81.12
4365	pyrroline (*N*)	1-Me-3-pyrroline	$CH_3 \cdot N \cdot CH_2(CH)_2CH_2$	XX-133	83.13
4366	pyruvate	....	$C_3H_3O_3 \cdot CH_3$	III-616	102.09
4367	quinoline (2)	quinaldine	$CH_3 \cdot C_9H_6N$	XX-388	143.19
4368	quinoline ethiodide	quinaldine ethiodide	$CH_3 \cdot C_9H_6N:(I)C_2H_5$	XX-392	299.16
4369	quinoline (3)	*Py*-3 (β)	$CH_3 \cdot C_9H_6N$	XX-394	143.19
4370	quinoline (4)	cincholepidine; lepidine (4)	$C_6H_4N:CH \cdot CH:C \cdot CH_3$	XX-395	143.19
4371	quinoline ethiodide	lepidine ethiodide	$CH_3 \cdot C_9H_6N:(I)C_2H_5$	XX-396	299.16
4372	quinoline (6)	*p*-toluquinoline	$CH_3 \cdot C_9H_6N$	XX-398	143.19
4373	quinoline ethiodide	*p*-toluquinoline EtI	$CH_3 \cdot C_9H_6N:(I)C_2H_5$	XX-398	299.16
4374	quinoline (7)	*m*-toluquinoline	$CH_3 \cdot C_9H_6N$	XX-400	143.19
4375	quinoline (8)	*o*-toluquinoline	$CH_3 \cdot C_9H_6N$	XX-401	143.19
4376	red	*p*-diMe-amino- azobenzene-*o'*- carboxylic	$HO_2C \cdot C_6H_4 \cdot N:N \cdot C_6H_4 \cdot N(CH_3)_2$	XVI-329	269.31
4377	red, Na salt	....	$C_{15}H_{14}O_2N_3Na$	*XVI-316	291.29
4378	red, para	(4',4)	$NaO_2C \cdot C_6H_4 \cdot N:N \cdot C_6H_4 \cdot N(CH_3)_2$	XVI-329	291.29
4379	salicylaldehyde (3)	(3;6,1)	$CH_3 \cdot C_6H_3(OH)CHO$	VIII-100	136.15
4380	salicylate (*o*)	oil of wintergreen	$HO \cdot C_6H_4 \cdot CO_2CH_3$	X-70	152.15
4381	stearate	....	$C_{17}H_{35}CO_2CH_3$	II-379	298.51
4382	sulfonic acid	methane sulfonic acid	$CH_3 \cdot SO_3H$	IV-4	96.11
4383	sulfonic chloride	Me-sulfone chloride	$CH_3 \cdot SO_2Cl$	IV-5	114.55
4384	sulfuric acid	Me-hydrogen sulfate	$CH_3O \cdot SO_2 \cdot OH$	I-283	112.10
4385	tetraacetyl-*d*- glucoside (α)	....	$CH_3O \cdot C_{14}H_{19}O_9$	XXXI-180	362.34
4386	tetramethyl-*d*- glucoside (α)	....	$(CH_3O)_5C_6H_7O$	XXXI-180	250.29
4387	thiocyanate	....	$CH_3 \cdot S \cdot CN$	III-175	73.12
4388	*iso*-thiocyanate	methyl mustard oil	$CH_3 \cdot N:C:S$	IV-77	73.12
4389	thiophene (β)	β-thiotolene	$CH_3 \cdot C_4H_3S$	XVII-38	98.17
4390	thiourea	Me-thiocarbamide	$NH_2 \cdot CS \cdot NH \cdot CH_3$	IV-70	90.15
4391	*iso*-thiourea sulfate	....	$(C_2H_6N_2S)_2 \cdot H_2SO_4$	*III-78	278.37
4392	toluate (*o*)	....	$CH_3 \cdot C_6H_4 \cdot CO_2CH_3$	IX-463	150.18
4393	toluate (*m*)	....	$CH_3 \cdot C_6H_4 \cdot CO_2CH_3$	IX-475	150.18
4394	toluate (*p*)	....	$CH_3 \cdot C_6H_4 \cdot CO_2CH_3$	IX-484	150.18
4395	toluene sulfonate	(1,4)	$CH_3 \cdot C_6H_4 \cdot SO_3CH_3$	XI-99	186.23
4396	toluidine (*o*)	....	$CH_3 \cdot C_6H_4 \cdot NHCH_3$	XII-784	121.18
4397	toluidine (*m*)	....	$CH_3 \cdot C_6H_4 \cdot NHCH_3$	XII-856	121.18
4398	toluidine (*p*)	....	$CH_3 \cdot C_6H_4 \cdot NHCH_3$	XII-902	121.18
4399	*o*-tolyl ether	Me *o*-cresyl ether	$CH_3 \cdot O \cdot C_6H_4 \cdot CH_3$	VI-352	122.17
4400	*m*-tolyl ether	*m*-cresyl Me ether	$CH_3 \cdot O \cdot C_6H_4 \cdot CH_3$	VI-376	122.17
4401	*p*-tolyl ether	....	$CH_3 \cdot O \cdot C_6H_4 \cdot CH_3$	VI-392	122.17
4402	*p*-tolyl sulfide	thiocresol Me ether	$CH_3 \cdot S \cdot C_6H_4 \cdot CH_3$	VI-417	138.23
4403	trichloroacetate	....	$Cl_3C \cdot CO_2CH_3$	II-208	177.42
4404	trimethylacetate	methyl pivalate	$(CH_3)_3C \cdot CO_2CH_3$	II-320	116.16

Methyl-quercetin 5556
Methyl-quinaldine 2541, 2544
Methyl quinolinato 2539
Methyl-quinoline carboxylic acid 5528-32
Methyl-quinolone 4180
Methyl-resorcinol 2371, 2373, 4998
Methyl-rhodin 4090
Methyl rosaniline 3593

Methyl-salicylic acid 3833, 4067
Methyl sebacate 2551
Methyl-styrene (α) 5224
Methyl-styrene (dimer) 2476
Methyl-styryl ketone 626
Methyl succinate 2554-5
Methyl-succinic acid 5520
Methyl sulfate 2559

Table 7-4 (*Continued*)
PHYSICAL CONSTANTS OF ORGANIC COMPOUNDS

No.	Crystalline Form and Color	Specific Gravity	Melting Point °C.	Boiling Point °C.	Solubility in 100 Parts		
					Water	Alcohol	Ether
4361	nd./bz.	....	129	....	d. fused alk.	....	....
4362	lq.	$0.920^{10°}$	....	$114\text{-}5^{748mm}$	i.	∞	∞
4363	lq.	0.945	....	$147\text{-}8^{750mm}$	sl. d. a.	....	....
4364	lq.	....	....	$142\text{-}3^{743mm}$	sl. d. a.	....	....
4365	col. lq.	....	....	79-80	∞	s.	s.; s. chl.
4366	lq.	$1.154^{0°}$	....	134-7	....	....	....
4367	lq.	$1.059^{\frac{20°}{4}}$	-1	$244\text{-}5^{750mm}$	v. sl. s.	s. chl.	s.
4368	yel. nd./al.	....	233-4	....	s.	v. sl. s.	i.
4369	cr.	$1.067^{\frac{20°}{4}}$	16	250^{710mm}	i.	s.	s.
4370	lq.	$1.086^{20°}$	9-10	261-3	sl. s.	∞; ∞ bz.	∞; ∞ lg.
4371	yel. cr./al.	....	142-3	....	s.	s.	i.
4372	lq.	$1.068^{20°}$	-22	258-9	v. sl. s.	s.	s.
4373	cr.	....	171-3	....	....	....	....
4374	yel. oil	$1.061^{\frac{20°}{4}}$	39	257.6^{750mm}	v. sl. s.	s.	s.
4375	lq.	$1.073^{20°}$	....	247.8^{760mm}	v. sl. s.	∞	∞
4376	vl. nd. or red pd.	....	179-83	....	v. sl. s.	s.	s. ac.
4377	red pl.	....	....	....	s.	....	....
4378	or. pd.	....	....	....	....	....	....
4379	lf./aq.al.	$1.091^{\frac{5.9°}{4}}$	56	217-8	v. sl. s.	s.	s.; s. chl.
4380	col. lq.	$1.182^{\frac{25°}{25}}$	-8.6	222.9	$0.07^{30°}$	∞; s. chl.	∞
4381	col. cr.	....	38-9	$214\text{-}5^{15mm}$	i.	s.	s.
4382	col. lq.	$1.481^{\frac{18°}{4}}$	20	167^{10mm}	∞	....	....
4383	yel. lq.	$1.481^{\frac{18°}{4}}$	-32	161.5^{730mm}	i. c.; d. h.	....	s.
4384	oil	....	<-30	d.	v. s.	s.	∞ abs.
4385	col. rhb./al.	....	101	....	i.; s. chl.	s.; s. act.	s.; s. pet.
4386	col. lq.	$1.108^{20°}$	....	$145\text{-}50^{13mm}$	sl. s.	s.; s. chl.	sl. s.
4387	lq.	$1.069^{\frac{24°}{4}}$	-51	130-3	v. sl. s.	∞	∞
4388	col. cr.	$1.069^{\frac{37°}{4}}$	35-6	119^{759mm}	v. sl. s.	∞	∞
4389	oil	$1.022^{\frac{20°}{4}}$	-68.9(-74)	115.4	....	....	....
4390	pr.	....	118-9	....	v. s.	v. s.	sl. s.
4391	cr. pd.	....	241-2 d.	....	s. h.	i.	....
4392	col. lq.	$1.073^{15°}$	<-50	213	i.	∞	∞
4393	col. lq.	$1.066^{15°}$	....	215	i.	....	....
4394	cr./pet.	....	33-4	217	i.	v. s.	v. s.
4395	cr./et. lg.	....	28	....	i.; s. bz.	v. s.	v. s.
4396	lq.	$0.973^{15°}$	....	206-7	i.	∞	∞
4397	lq.	....	....	206-7	i.	∞	∞
4398	lq.	$0.935^{\frac{5.5°}{4}}$	....	$209\text{-}11^{761mm}$	i.	∞	∞
4399	lq.	$0.985^{\frac{15°}{15}}$	....	171-2	i.	v. s.	v. s.
4400	lq.	$0.977^{\frac{15°}{15}}$	....	177	....	....	....
4401	lq.	$0.976^{\frac{15°}{15}}$	....	176	....	....	....
4402	lq.	$1.030^{\frac{16°}{4}}$	....	209^{747mm}	....	....	....
4403	lq.	$1.489^{\frac{19.2°}{19.2}}$	-17.5	$152\text{-}3^{765mm}$	d.	d.	s.
4404	lq.	$0.891^{\frac{0°}{4}}$	....	100-2	....	....	....

Table 7-4 (*Continued*)
PHYSICAL CONSTANTS OF ORGANIC COMPOUNDS

No.	Name	Synonym	Formula	Beil. Ref.	Formula Weight
4405	**Methyl** (tyrosine N)($l+$)	ratanhin; andirine	$CH_3NH \cdot C_9H_9O_3$	XIV-612	195.22
4406	umbelliferone (β)	7-OH-4-Me coumarin	$C_{10}H_8O_3$	XVIII-31	176.17
4407	undecylenate	. . . .	$CH_2{:}CH(CH_2)_8CO_2 \cdot CH_3$	II-459	198.31
4408	n-undecyl ketone	tridecanone-2	$CH_3 \cdot CO(CH_2)_{10}CH_3$	I-715	198.35
4409	n-undecyl ketone semicarbazone	. . . .	$C_{11}H_{23}(CH_3)C{:}N \cdot NH \cdot CO \cdot NH_2$	*I-371	255.41
4410	uracil (1)	. . . .	$CO \cdot (CH)_2NH \cdot CO \cdot N \cdot$ $\underset{CH_3}{\lfloor \qquad \qquad \rfloor}$	XXIV-316	126.12
4411	uracil (3)	. . . .	$CO \cdot NH \cdot CO(CH)_2 \cdot N \cdot$ $\underset{CH_3}{\lfloor \qquad \qquad \rfloor}$	XXIV-316	126.12
4412	uracil (4)	. . . .	$(NH \cdot CO)_2CH{:}C \cdot CH_3$ $\lfloor \qquad \qquad \rfloor$	XXIV-342	126.12
4413	uracil (5)	thymine	$(CO \cdot NH)_2CH{:}C \cdot CH_3$ $_7\lfloor \qquad \qquad \rfloor$	XXIV-353	126.12
4414	urea (N)	. . . .	$CH_3 \cdot NH \cdot CO \cdot NH_2$	IV-64	74.08
4415	urea nitrate	. . . .	$C_2H_6ON_2 \cdot HNO_3$	IV-65	137.10
4416	urethane	*cf. also* Me-carbamate	$CH_3 \cdot NH \cdot CO_2C_2H_5$	IV-64	103.12
4417	uric acid (1)	. . . .	$C_6H_6O_3N_4$	XXVI-524	182.14
4418	uric acid (3)	. . . .	$C_6H_6O_3N_4 \cdot \frac{1}{2}H_2O$	XXVI-524	191.15
4419	uric acid (7)	. . . .	$C_6H_6O_3N_4 \cdot H_2O$	XXVI-525	200.16
4420	n-valerate	. . . .	$CH_3 \cdot (CH_2)_3 \cdot CO_2CH_3$	II-301	116.16
4421	*iso*-valerate	. . . .	$(CH_3)_2C_2H_3 \cdot CO_2CH_3$	II-311	116.16
4422	vinyl carbinol	buten-1-ol-3	$CH_2{:}CH \cdot CHOH \cdot CH_3$	I-441	72.11
4423	d-xyloside (β)	. . . .	$CH_3O \cdot C_5H_9O_4$	XXXI-54	164.16
4424	**Methylal**	dimethoxy-methane	$CH_2(OCH_3)_2$	I-574	76.10
4425	**Methylene**-amino-acetonitrile	(dimolecular)	$(CH_2{:}N \cdot CH_2 \cdot CN)_2$	II-89	136.16
4426	blue (base)	N,N,N',N',-tetra-Me-thionine	$[(CH_3)_2N]_2C_{12}H_6 \cdot NS(OH)$	XXVII-393	301.41
4427	blue chloride	Me-thionine chloride	$C_{16}H_{18}N_3S(Cl) \cdot 3H_2O$	XXVII-395	373.90
4428	bromide	dibromomethane	$CH_2{:}Br_2$	I-67	173.85
4429	chloride	dichloromethane	$CH_2{:}Cl_2$	I-60	84.93
4430	diacetate	. . . .	$(CH_3CO_2)_2CH_2$	II-152	132.12
4431	dianiline	. . . .	$(C_6H_5 \cdot NH)_2CH_2$	XII-184	198.27
4432	disulfonic acid	methionic acid	$CH_2{:}(SO_2OH)_2$	I-579	176.17
4433	fluoride	difluoromethane	$CH_2{:}F_2$	I-59	52.02
4434	iodide	diiodomethane	$CH_2{:}I_2$	I-71	267.84
4435	**Michler's** hydrol (p,p')	. . . .	$[(CH_3)_2N \cdot C_6H_4]_2{:}$ $CHOH$	XIII-698	270.38
4436	ketone (p,p')	. . . .	$[(CH_3)_2N \cdot C_6H_4]_2CO$	XIV-89	268.36
4437	**Morpholine-ethan-ol ethyl ether**	. . . .	$O{:}(CH_2CH_2)_2{:}N \cdot C_2H_4 \cdot O \cdot C_2H_5$	. . . .	159.23
4438	**Mucic** acid	. . . .	$(CHOH)_4(CO_2H)_2$	III-581	210.14
4439	acid, allo-	. . . .	$(CHOH)_4(CO_2H)_2$	III-576	210.14

Table 7-4 (Continued)
PHYSICAL CONSTANTS OF ORGANIC COMPOUNDS

No.	Crystalline Form and Color	Specific Gravity	Melting Point °C.	Boiling Point °C.	Solubility in 100 Parts		
					Water	Alcohol	Ether
4405	col. nd.	....	257	d. 280	$0.14^{20°}$; $0.5^{100°}$	0.007 abs.	i.; s. NH$_4$OH
4406	nd./al.*	....	188-9	d.	v. sl. s. h.; s. alk.	s.; s. ac.; i. Na$_2$CO$_3$	v. sl. s.; sl. s. chl.
4407	lq.	$0.889^{15°}$	−27.5	248-9	....	....	....
4408	cr.	$0.823^{28°}$	28-9	260-5	i.	v. s.	v. s.
4409	cr.	....	122-6	....	....	....	
4410	pr./aq. al.	....	174-5	....	v. s.	v. s.	i. bz.
4411	pr./al.	....	232	....	s.; i. dil. HCl	s. h.	s. aq. NaOH
4412	nd./al.	....	d. 270-80	....	$0.7^{22°}$; s. alk.	sl. s.; s. H$_2$SO$_4$	v. sl. s.; s. NH$_4$OH
4413	pl./aq.	....	326 d.	....	$0.4^{25°}$; s. alk.	sl. s.; s. H$_2$SO$_4$	v. sl. s.
4414	pr./aq.	1.204	101-2	d.	v. s.	v. s.	i.
4415	col. cr.	....	128-32	....	....	....	....
4416	col. lq.	$1.009\frac{19}{4}°$	....	170	$69^{16°}$	s.	....
4417	cr.	....	>360 d.	....	0.05 h.	v. sl. s.	....
4418	cr.	....	>360 d.	....	0.2 h.	....	s. alk.
4419	cr.	....	>370 d.	....	$1.3^{100°}$	....	s. alk.
4420	lq.	$0.895\frac{15}{4}°$	−91	127.3	v. sl. s.	∞	∞
4421	col. lq.	$0.881\frac{20}{4}°$	....	116-7^{764mm}	v. sl. s.	∞	∞
4422	lq.	$0.831\frac{20}{4}°$	<−80	97	∞	....	....
4423	cr./EtOAc	....	155-6	....	v. s.	s. h.	5 h. act.
4424	col. lq.	$0.866\frac{15}{4}°$	−105	42.3	33	∞	∞
4425	col. pr./ aq.	....	129.5	....	s. h.	s. h.	i.; i. bz.
4426	dark, amor.	....	....	....	v. s.	v. s.	i.
4427	gn. b. met. /aq. HCl	....	−2H$_2$O, 100	−3H$_2$O, 150	4	2	i.; s. chl.
4428	col. lq.	$2.495\frac{20}{4}°$	−52.7	97.0	$1.173^{0°}$; $1.15^{20°}$	∞; ∞ act.	∞
4429	col. lq.	$1.336\frac{20}{4}°$	−96.7	40.2	$2^{20°}$	∞	∞
4430	lq.	$1.132\frac{20}{20}°$	−23	164-5	sl. s.	v. s.	∞
4431	pl./et. pet.	....	65	208-9 d.	i.; i. pet.	s.	s.
4432	hyg. nd.	....	....	....	$245.8^{25°}$	s.	....
4433	gas	....	....	....	....	....	....
4434	col. lq.	$3.325\frac{20}{4}°$	6.1(5.6)	180 d.	$1.6^{0°}$; $1.4^{20°}$	∞	∞
4435	gn. lf./bz.	....	96-7	....	i.; s. bz.	s. h.	s.; s. ac.
4436	lf./al.	....	174	>360 d.	i.; v. s. bz.	sl. s.	v. sl. s.
4437	col. lq.	$0.965\frac{20}{20}°$	....	206.2	∞	....	
4438	pd.	....	206-14 d. (108)	....	$0.33^{14°}$; $1.6^{100°}$	i.	i.; s. alk.
4439	nd./aq.	....	166-71 d.	....	9 h.	sl. s.	....

Table 7-4 (*Continued*)
PHYSICAL CONSTANTS OF ORGANIC COMPOUNDS

No.	Name	Synonym	Formula	Beil. Ref.	Formula Weight
4440	**Mucic** acid, talo-	(*d* or *l*)	$(CHOH)_4(CO_2H)_2$	III-577	210.14
4441	**Mucobromic acid**	....	$HO_2C \cdot CBr : CBr \cdot CHO$	III-728	257.88
4442	**Mucochloric acid**	....	$HO_2C \cdot CCl : CCl \cdot CHO$	III-727	168.96
4443	**Muconic acid**	....	$(\cdot CH : CH \cdot CO_2H)_2$	II-803	142.11
4444	**Mudarol**	mudarin	$C_{30}H_{47}O(OH)$	....	440.72
4445	**Murexide**	ammonium salt purpuric acid	$C_8H_4O_6N_5 \cdot NH_4 \cdot H_2O$	XXV-499	302.20
4446	**Muscone** (*l*)	3-Me-cyclopenta-decanone	$CH_3 \cdot C_{15}H_{27} : O$	**VII-51	238.42
4447	**Myochrysine**	Na aurothiomalate	$NaO_2C \cdot CH_2 \cdot CH : (SAu)CO_2Na$	....	390.08
4448	**Myricyl** acetate	....	$CH_3CO_2 \cdot C_{31}H_{63}$	*II-63	494.89
4449	alcohol	melissyl alcohol	$C_{31}H_{63}OH$ $C_{30}H_{61}OH(?)$	*I-222	452.86 438.83
4450	**Myristic** acid	tetradecanoic acid	$CH_3(CH_2)_{12} \cdot CO_2H$	II-365	228.38
4451	aldehyde	tetradecanal	$CH_3(CH_2)_{12} \cdot CHO$	I-716	212.38
4452	amide	myristamide	$C_{13}H_{27} \cdot CO_2NH_2$	II-368	243.39
4453	anhydride	....	$(C_{13}H_{27}CO)_2O$	II-367	438.74
4454	anilide	myristanilide	$C_{13}H_{27}CO \cdot NHC_6H_5$	XII-257	303.49
4455	chloride	myristoyl chloride	$C_{13}H_{27} \cdot COCl$	II-368	246.82
4456	nitrile	tridecyl cyanide	$C_{13}H_{27} \cdot CN$	II-368	209.38
4457	**Myristyl bromide**	....	$CH_3(CH_2)_{12}CH_2Br$	**I-136	277.30
4458	**Naphthalene**	....	$C_{10}H_8$	V-531	128.18
4459	dicarboxylic acid	naphthalic acid (1,4)	$C_{10}H_6(CO_2H)_2$	IX-917	216.20
4460	dicarboxylic acid	naphthalic acid (1,8)	$C_{10}H_6(CO_2H)_2$	IX-918	216.20
4461	disulfonic acid (1,5)	Armstrong's acid	$C_{10}H_6(SO_3H)_2$	XI-212	288.30
4462	disulfonic acid (1,6)	(δ)	$C_{10}H_6(SO_3H)_2$	XI-213	288.30
4463	disulfonic acid (2,6)	....	$C_{10}H_6(SO_3H)_2$	XI-215	288.30
4464	disulfonic acid (2,7)	(α)	$C_{10}H_6(SO_3H)_2$	XI-216	288.30
4465	sulfinic acid (α)	....	$C_{10}H_7 \cdot SO_2H$	XI-15	192.24
4466	sulfinic acid (β)	....	$C_{10}H_7 \cdot SO_2H$	XI-16	192.24
4467	sulfone amide (α)	....	$C_{10}H_7 \cdot SO_2NH_2$	XI-157	207.25
4468	sulfone chloride (α)	....	$C_{10}H_7 \cdot SO_2Cl$	XI-157	226.68
4469	sulfone chloride (β)	....	$C_{10}H_7 \cdot SO_2Cl$	XI-173	226.68
4470	sulfonic acid (α)	....	$C_{10}H_7 \cdot SO_3H \cdot 2H_2O$	XI-155	244.27
4471	sulfonic acid (β)	....	$C_{10}H_7 \cdot SO_3H \cdot H_2O$	XI-171	226.25
4472	sulfonic Na salt	(β)	$C_{10}H_7O_3SNa$	XI-171	230.22
4473	tetrachloride	(1,2,3,4)	$C_{10}H_8Cl_4$	V-492	269.99
4474	**Naphthalic anhydride**	(1,8)	$C_{10}H_6 : (CO)_2O$	XVII-521	198.18
4475	**Naphthasultam**	(1,8)	$C_{10}H_7O_2NS$	XXVII-59	205.24
4476	disulfonate Na	(2,4)	$C_{10}H_4O_8NS_3Na_3 \cdot 8\frac{1}{2}H_2O$	XXVII-356	584.44
4477	**Naphthazarin** (5,8;1,4)	5,8-dihydroxy-α-naphthoquinone	$(HO)_2C_{10}H_4O_2$	VIII-412	190.16
4478	*iso*-**Naphthazarin**	(2,3;1,4)	$(HO)_2C_{10}H_4O_2$	VIII-411	190.16
4479	**Naphthoflavone** (α)	7,8-benzoflavone	$C_{19}H_{12}O_2$	XVII-390	272.31
4480	**Naphthoic** acid (α)	....	$C_{10}H_7 \cdot CO_2H$	IX-647	172.19
4481	acid (β)	isonaphthoic acid	$C_{10}H_7 \cdot CO_2H$	IX-656	172.19
4482	aldehyde (α)	....	$C_{10}H_7 \cdot CHO$	VII-400	156.19
4483	aldehyde (β)	....	$C_{10}H_7 \cdot CHO$	VII-401	156.19
4484	amide (α)	....	$C_{10}H_7 \cdot CONH_2$	IX-648	171.20

Mudarln 4444
Munjistin 2305
Murexan 6384
Murrayin, cf. glcde.
Muscarine, cf. alkd.

Muscle sugar 1602-3
Musk, artificial 6296
Musk xylene 6297
Mustard gas 1985
Mustard imitator gas 1473

Mustard oils, cf. also *iso*-thiocyanates.
Mustard oil 220
Mustard oil acetic acid 2369
Mycose 6054
Myrbane oil 4646

Table 7-4 (*Continued*)
PHYSICAL CONSTANTS OF ORGANIC COMPOUNDS

No.	Crystalline Form and Color	Specific Gravity	Melting Point °C.	Boiling Point °C.	Solubility in 100 Parts		
					Water	Alcohol	Ether
4440	lf./act.	....	158± d.	....	v. s.	v. s. h.	i.
4441	pl./et. lg.	....	122-5	....	v. s. h.	v. s.	v. s.
4442	pl./aq.	....	127	....	s. h.	s.	s.; i. lg.
4443	nd./aq.	....	298 d.	....	0.002 c.	s. h.	s. h. ac.
4444	hex./al. et.	....	176	....	i.	s.	s.
4445	gn. red pd.	....	....	....	v. sl. s. c.	i.	i.
4446	col. oil	$0.922^{17°}_{4}$	*	328	v. sl. s.	∞	....
4447	lt. yel. pd.	....	....	....	v. s.	....	....
4448	nd.	....	73-5	311-3^{8mm}	....	....	....
4449	nd. or lf.	$0.777^{95°}$	88	....	i.	v. sl. s. c.; v. sl. s. pet.	v. s.; v. s. bz.
4450	col. lf.	$0.853^{70°}_{4}$	54.2	250.5^{100mm}	i.; v. s. bz.	v. s. abs.	v. s.
4451	cr.	....	23.5	166^{24mm}	....	....	....
4452	nd.	....	102	217^{12mm}	i.	s.	sl. s.
4453	col. cr.	$0.850^{20°}_{4}$	53.5	198$^{0.1mm}$	....	....	....
4454	nd./al.	....	84	113^{10mm}	i.; s. chl.	v. s. bz.	v. s.
4455	col. lq.	....	1-3	168^{15mm}	d.	d.	s.
4456	cr.	$0.828^{19°}_{4}$	19	226.5^{100mm}	s.	s.	....
4457	col. lq.	....	3-5	....	i.	....	s.
4458	col. pl./al.	$1.145^{20°}_{4}$	80.2	218.0	$0.003^{25°}$; v. s. CCl$_4$	$9.5^{19.5°}$; v. s. CS$_2$	v. s.; 46$^{16°}$ bz.
4459	nd.	....	>240	....	i. h.	s.	....
4460	nd./al.	....	d. >140	....	v. sl. s.	s. h.	sl. s.
4461	lf.	....	d.	....	102$^{20°}$	s.	i.
4462	cr.	....	....	d. 125	164$^{20°}$	s.	i.
4463	hyg. lf.	....	....	....	s.	....	....
4464	hyg. nd.	....	....	....	s.	sl. c. HCl	....
4465	nd./aq.	....	98-9	....	s.	s.	sl. s.
4466	nd.	....	105	....	s.	s.	s.
4467	cr./al.	....	150	....	....	v. s.	v. s.
4468	lf./et.	....	68	195^{13mm}	i.	v. s.	v. s.
4469	lf.	....	79	201^{13mm}	i.	sl. s. pet.	s. bz.
4470	cr.	....	90	....	v. s.	v. s.	sl. s.
4471	hyg. cr.	....	125**	....	77$^{30°}$	0.2 h. bz.	....
4472	cr./aq.	....	....	d.	6$^{23.9°}$	....	....
4473	mn./chl.	....	187-9	....	i.	v. sl. s. h.	sl. s. h.
4474	nd./al.	....	273-4	....	sl. s. ac.	v. sl. s.	v. sl. s.
4475	nd./bz.	....	177-8	....	s. h.	sl. s.	s.
4476	yel. lf./aq. al.	....	−8H$_2$O, 160	....	v. s.	sl. s.	....
4477	brn. nd., al.	....	subl. vac.	....	sl. s. h.	sl. s.; s. alk.	sl. s.
4478	brn. lf./ac.	....	276-80	subl.	sl. s. h.	sl. s.	v. sl. s.
4479	cr./aq. al.	....	155-6	....	....	....	....
4480	nd./aq. al.	....	161-3	300	v. sl. s. h.	s. h.	s.
4481	mn.	$1.077^{100°}_{4}$	184	>300	$0.007^{25°}$	s.	s.
4482	lq.	$1.148^{20°}_{4}$	....	291.6	....	....	....
4483	lf./h. aq.	$1.078^{99.4°}$	60.5-1.0	....	s. h.	v. s.	v. s.
4484	nd./al.	....	202	....	....	v. sl. s.	....

* Oxime, m. p. 45°; semicarbazone, m. p. 134°.
** Crysts. + 3H$_2$O, m. p. 83°; anh., m. p. 91°.

Myristamide 4452
Myristanilide 4454
Myristoyl chloride 4455

Myristyl alcohol 5750
Nandinine, cf. alkd.
Napelline, cf. alkd.

Naphthacetol 31
Naphthalene acetic acid 4528
Naphthalene decahydride 1631-2
Naphthalic acid 4459-60
Naphthalidine 4529

Table 7-4 (*Continued*)
PHYSICAL CONSTANTS OF ORGANIC COMPOUNDS

No.	Name	Synonym	Formula	Beil. Ref.	Formula Weight
	Naphthoic				
4485	amide (β)	....	$C_{10}H_7 \cdot CONH_2$	IX-657	171.20
4486	chloride (β)	β-naphthoyl chloride	$C_{10}H_7 \cdot COCl$	IX-657	190.63
4487	nitrile (α)	α-naphthyl cyanide	$C_{10}H_7 \cdot CN$	IX-649	153.19
4488	nitrile (β)	β-naphthyl cyanide	$C_{10}H_7 \cdot CN$	IX-659	153.19
4489	**Naphthol** (α)	α-hydroxy naphthalene	$C_{10}H_7 \cdot OH$	VI-596	144.17
4490	**Naphthol** (β)	β-hydroxy naphthalene	$C_{10}H_7 \cdot OH$	VI-627	144.17
4491	disulfonic acid (β)	(2;3,6); R-acid	$HO \cdot C_{10}H_5(SO_3H)_2$	XI-288	304.30
4492	disulfonic Ca	R-acid Ca salt	$HO \cdot C_{10}H_5(SO_3)_2Ca$	....	342.36
4493	disulfonic K	R-acid K salt	$HO \cdot C_{10}H_5(SO_3K)_2$	....	380.49
4494	disulfonic Na	R-acid Na salt	$HO \cdot C_{10}H_5(SO_3Na)_2$	XI-289	348.26
4495	disulfonic K (β)	(2;6,8); G-acid K	$HO \cdot C_{10}H_5(SO_3K)_2$	....	380.49
4496	disulfonic Na (β)	G-acid Na salt	$HO \cdot C_{10}H_5(SO_3Na)_2$	XI-290	348.26
4497	phthalein (α)	....	$C_8H_4O_2(C_{10}H_6OH)_2$	XVIII-157	418.45
4498	sulfonic acid (α)	(1,2); Schaeffer's (α)	$HO \cdot C_{10}H_6 \cdot SO_3H$	XI-269	224.24
4499	sulfonic Ca (α)	Schaeffer's Ca (α)	$(C_{10}H_7O_4S)_2Ca \cdot H_2O$	XI-270	504.55
4500	sulfonic K (α)	Schaeffer's K	$C_{10}H_7O_4SK \cdot \frac{1}{2}H_2O$	XI-270	271.34
4501	sulfonic Na (α)	Schaeffer's Na	$C_{10}H_7O_4SNa$	XI-269	246.22
4502	sulfonic acid (α)	(1,4); Neville-Winther	$HO \cdot C_{10}H_6 \cdot SO_3H$	XI-271	224.24
4503	sulfonic acid (α)	(1-OH; 5-SO$_3$H)	$HO \cdot C_{10}H_6 \cdot SO_3H$	XI-273	224.24
4504	sulfonic acid (α)	(1,8)	$HO \cdot C_{10}H_6 \cdot SO_3H \cdot H_2O$	XI-275	242.25
4505	sulfonic acid (β)	(2-OH; 1-SO$_3$H)	$HO \cdot C_{10}H_6 \cdot SO_3H$	XI-281	224.24
4506	sulfonic Ca (β)	(2,1); asaprol	$(C_{10}H_7O_4S)_2Ca \cdot 3H_2O$	....	540.58
4507	sulfonic acid (β)	(2,6); Schaeffer's (β)	$HO \cdot C_{10}H_6 \cdot SO_3H$	XI-282	224.24
4508	sulfonic Ca (β)	Schaeffer's Ca (β)	$(C_{10}H_7O_4S)_2Ca \cdot 5H_2O$	XI-282	576.61
4509	sulfonic K (β)	Schaeffer's K (β)	$C_{10}H_7O_4SK \cdot xH_2O$	XI-283	(262.33)
4510	sulfonic acid (β)	(2-OH; 7-SO$_3$H)	$HO \cdot C_{10}H_6 \cdot SO_3H$	XI-285	224.24
4511	sulfonic Na (β)	(2,7); F-acid Na	$C_{10}H_7O_4SNa \cdot 2\frac{1}{2}H_2O$	XI-286	291.26
4512	sulfonic Na (β)	(2,8) Bayer-acid Na	$C_{10}H_7O_4SNa$	XI-286	246.22
4513	**Naphtho**-phenazine	(α,β)	$C_{10}H_6 : N_2 : C_6H_4$	XXIII-276	230.27
4514	quinaldine (α)	....	$C_{13}H_8NCH_3$	XX-471	193.25
4515	quinaldine (β)	....	$C_{13}H_8NCH_3$	XX-471	193.25
4516	quinaldine (γ)	....	$C_{13}H_8NCH_3$	....	193.25
4517	quinoline (α)	....	$C_{13}H_9N$	XX-463	179.22
4518	quinoline (β)	5,6-benzoquinoline	$C_{13}H_9N$	XX-464	179.22
4519	quinoline HCl (β)	....	$C_{13}H_9N \cdot HCl \cdot 2H_2O$	XX-465	251.72
4520	quinone (α)(1,4)	....	$C_{10}H_6O_2$	VII-724	158.16
4521	quinone (β)(1,2)	....	$C_{10}H_6O_2$	VII-709	158.16
4522	quinone-4-sulfonate Na (β)	....	$C_{10}H_5O_2 \cdot SO_3Na$	XI-330	260.20
4523	quinone (amphi)	(2,6)	$C_{10}H_6O_2$	VII-733	158.16
4524	sulfone (1,8)	naphthosultone	$C_{10}H_6O \cdot SO_2$ $\underset{|_____|}{}$	XIX-43	206.22
4525	**Naphthoylaceto-nitrile** (β)	....	$C_{10}H_7 \cdot CO \cdot CH_2 \cdot CN$	....	195.22
4526	**Naphthyl** acetate (α)	....	$CH_3CO_2 \cdot C_{10}H_7$	VI-608	186.21
4527	acetate (β)	....	$CH_3CO_2 \cdot C_{10}H_7$	VI-644	186.21
4528	acetic acid (α)	....	$C_{10}H_7 \cdot CH_2CO_2H$	IX-666	186.21
4529	amine (α)	1-NH$_2$-naphthalene	$C_{10}H_7 \cdot NH_2$	XII-1212	143.19

Naphthane 1631-2
Naphthanthracene 650
Naphthenic acid 3577
Naphthionic acid 4541, 4544

Naphthol yellow 2642
Naphthol yellow-S 3241
Naphtho-hydroquinone 2348, 2350

Naphtho-nitrile 4487-8
Naphtho-picric acid 6305
Naphtho-quinone oxime 4892

Table 7-4 (*Continued*)
PHYSICAL CONSTANTS OF ORGANIC COMPOUNDS

No.	Crystalline Form and Color	Specific Gravity	Melting Point °C.	Boiling Point °C.	Solubility in 100 Parts		
					Water	Alcohol	Ether
4485	pl./al.	....	192	....	s. bz.	s. h.	s.; s. chl.
4486	cr.	....	43	304-6	d.	s. h. bz.	s. h.
4487	nd./lg.	$1.111\frac{25°}{25}$	36-7	299	....	v. s.	....
4488	lf./lg.	$1.094\frac{60°}{60}$	66	305-6	v. sl. s.	s.	s.
4489	mn.	$1.224^{4°}$	96; subl.	278-80	sl. s. h.; s. alk.	v. s.; s. bz.	v. s.; s. chl.
4490	mn.	$1.217^{4°}$	122-3	285-6	0.1c.; 1.25 h.	v. s.	v. s.; s. chl.
4491	delq. nd.	....	....	....	v. s.	v. s.	i.
4492	cr.	....	....	....	$30.6^{25°}$	$61^{90°}$ aq.	....
4493	cr.	....	....	....	$29.5^{25°}$	$58^{100°}$ aq.	....
4494	col. nd.	....	....	....	$25.2^{25°}$	v. sl. s.	$41^{90°}$ aq.
4495	col. cr.	....	....	....	$8.01^{25°}$	....	$32^{90°}$ aq.
4496	col. cr.	....	....	....	$34.2^{20°}$	....	$63^{90°}$ aq.
4497	cr./bz.*	....	253-5*	....	i.	s.	s. alk.
4498	pl./aq.	....	>250	....	v. s. h.	....	i.
4499	cr.	....	....	....	v. sl. s.	....	....
4500	pr./aq.	....	....	....	$2.8^{18°}$	v. sl. s.	i. KCl
4501	cr./al.	....	....	....	$6.29^{25°}$	$23^{90°}$ aq.	....
4502	pl./aq.	....	170 d.	....	v. s.	....	....
4503	hyg. cr.	....	110-20	....	v. s.	....	....
4504	cr.	....	106-7	$-H_2O$, 180	v. s.	....	....
4505	cr.	....	....	....	v. s.	....	....
4506	wh. pd.	....	50±	....	6.6	4	....
4507	lf.	....	125	....	v. s.	v. s.	....
4508	col. lf.	....	....	....	$4.76^{20°}$	s.	$16^{90°}$ aq.
4509	nd. or lf.	....	....	....	v. s. h.	i.	i.
4510	nd./HCl	....	115-6**	d. 150	s.	s.	i.; i. bz.
4511	lf.	....	....	....	$8^{15°}$	....	....
4512	lf.	....	....	....	v. s.	v. sl. s.	....
4513	yel. nd./bz.	....	142.5	>360	sl. s. bz.	v. sl. s.	v. sl. s.
4514	lq.	....	....	>300	....	....	....
4515	nd./aq. al.	....	82	>300	sl. s.	s.	s.
4516	cr.	....	91-2	....	....	....	....
4517	mn./et.	....	52	22^{347mm}	v. sl. s.	s.	s.; s. bz.
4518	pl./h. aq.	....	93.5	350^{721mm}	sl. s. h.; s. aq. a.	v. s.	v. s.; v. s. bz.
4519	nd.	....	....	....	v. s.	i.	....
4520	yel. tri.	....	125-6	subl. 100	v. sl. s. c.	v. s. h.	s.
4521	red nd./et.	....	d. 115-20	....	s.; s. H_2SO_4	s. bz.	s.
4522	cr./50% al.	....	....	....	v. s.	i.	....
4523	red pr.	....	135	....	i. pet.	s. d.	v. sl. s.
4524	pr./bz.	....	154	>360 sl. d.	sl. s. CS_2	sl. s.; v. s. chl.	s. h. bz.
4525	yel. nd.	....	118-20	....	i.	s.	sl. s.
4526	nd./al.	....	46-9	....	....	s.	s.
4527	nd./al.	....	69-70	....	s. chl.	s.	s.
4528	col. nd./aq.	....	133	....	s. h.	3.3	s.; s. bz.
4529	rhb./aq. al.	$1.123\frac{25°}{25}$	50; subl.	300.8	0.17 c.	v. s.	v. s.

* Crysts. $+ C_6H_6$; loses $C_6H_6 > 100°$; m. anh. 253-5°.
** Crysts. $+ 1H_2O$, m. p. 108-9; $+ 2H_2O$, m. p. 95°; $+ 4H_2O$, m. p. 67°.
Naphtho-resorcinol 2349
Naphtho-salol 4569
Naphtho-sultone 4524
Naphthoyl chloride 4486

Table 7-4 (*Continued*)
PHYSICAL CONSTANTS OF ORGANIC COMPOUNDS

No.	Name	Synonym	Formula	Beil. Ref.	Formula Weight
4530	**Naphthyl** amine (β)	amino-naphthalene	$C_{10}H_7 \cdot NH_2$	XII-1265	143.19
4531	amine HCl (α)	....	$C_{10}H_7 \cdot NH_2 \cdot HCl$	XII-1220	179.65
4532	amine HCl (β)	....	$C_{10}H_7 \cdot NH_2 \cdot HCl$	XII-1272	179.65
4533	amine-4,8-disul-fonic acid (α)	α,S-acid; δ-acid	$NH_2 \cdot C_{10}H_5(SO_3H)_2$	XIV-787	303.31
4534	amine-4,8-disul-fonic acid (β)	β,C-acid	$NH_2 \cdot C_{10}H_5(SO_3H)_2$	XIV-786	303.31
4535	amine-5,7-disul-fonic acid (β)	....	$C_{10}H_9O_6NS_2 \cdot 5H_2O$	XIV-783	393.39
4536	amine-5,7-disul-fonic K salt (β)	....	$C_{10}H_8O_6NS_2K$	....	341.41
4537	amine-6,8-disul-fonic acid (β)	β,γ-acid; amido-G acid	$NH_2 \cdot C_{10}H_5(SO_3H)_2 \cdot 4H_2O$	XIV-784	375.38
4538	amine-6,8-disul-fonic K salt (β)	....	$C_{10}H_8O_6NS_2K$	....	341.41
4539	amine-6,8-disul-fonic Na salt (β)	....	$C_{10}H_8O_6NS_2Na$	....	325.30
4540	amine-6,8-disul-fonic Na salt (β)	....	$C_{10}H_7O_6NS_2Na_2$	....	348.28
4541	amine-p-sulfonic acid (α)	naphthionic acid; (1,4)	$NH_2 \cdot C_{10}H_6 \cdot SO_3H$	XIV-739	223.25
4542	amine-p-sulfonic Na (α)	naphthionic Na salt	$C_{10}H_8O_3NSNa \cdot 4H_2O$	XIV-739	317.30
4543	amine sulfonic acid (α)	Schollkopf's acid; peri acid; (1,8)	$NH_2 \cdot C_{10}H_6 \cdot SO_3H$	XIV-752	223.25
4544	amine sulfonic acid	(α)(1,2)	$NH_2 \cdot C_{10}H_6 \cdot SO_3H$	XIV-757	223.25
4545	amine sulfonic acid (α)(4,2)	Cleve's γ-acid	$NH_2 \cdot C_{10}H_6 \cdot SO_3H$	XIV-757	223.25
4546	amine sulfonic (α)	(5,2); Cleve's β-acid	$NH_2 \cdot C_{10}H_6 \cdot SO_3H$	XIV-758	223.25
4547	amine sulfonic (α)	(8,2); Cleve's θ-acid	$NH_2 \cdot C_{10}H_6 \cdot SO_3H$	XIV-765	223.25
4548	amine sulfonic (α)	(5,1); Laurent's acid	$NH_2 \cdot C_{10}H_6 \cdot SO_3H$	XIV-744	223.25
4549	amine sulfonic (β)	(2,1); Tobias' acid	$NH_2 \cdot C_{10}H_6 \cdot SO_3H$	XIV-738	223.25
4550	amine sulfonic (β)	(6,1); Dahl's acid	$NH_2 \cdot C_{10}H_6 \cdot SO_3H$	XIV-748	223.25
4551	amine sulfonic (β)	(7,1); Dahl's acid	$NH_2 \cdot C_{10}H_6 \cdot SO_3H$	XIV-750	223.25
4552	amine sulfonic (β)	(6,2); Bronner's acid	$NH_2 \cdot C_{10}H_6 \cdot SO_3H$	XIV-760	223.25
4553	amine sulfonic (β)	(7,2); δ-acid	$NH_2 \cdot C_{10}H_6 \cdot SO_3H$	XIV-763	223.25
4554	amine-3,6,8-tri-sulfonic acid (α)	Koch acid; amino-H acid	$NH_2 \cdot C_{10}H_4(SO_3H)_3 \cdot 6H_2O$	XIV-801	491.47
4555	amine-3,6,8-tri-sulfonic Na (β)	Koch acid Na; amino-H acid Na	$C_{10}H_6O_9NS_3Na_3$	....	449.32
4556	benzoate (α)	....	$C_6H_5CO_2 \cdot C_{10}H_7$	IX-125	248.28
4557	benzoate (β)	....	$C_6H_5CO_2 \cdot C_{10}H_7$	IX-125	248.28
4558	*iso*-cyanate (α)	....	$C_{10}H_7 \cdot N:CO$	XII-1244	169.18
4559	*iso*-cyanate (β)	....	$C_{10}H_7 \cdot N:CO$	XII-1297	169.18
4560	ether (α)	dinaphthyl ether	$C_{10}H_7 \cdot O \cdot C_{10}H_7$	VI-607	270.33
4561	ether (α,β')	dinaphthyl ether	$C_{10}H_7 \cdot O \cdot C_{10}H_7$	VI-642	270.33
4562	ether (β)	dinaphthyl ether	$C_{10}H_7 \cdot O \cdot C_{10}H_7$	VI-642	270.33
4563	hydrazine (α)	....	$C_{10}H_7 \cdot NH \cdot NH_2$	XV-561	158.20

Naphthylamine red-G 3732
Naphthyl-benzamidine 689
Naphthyl bromide 956-7
Naphthyl chloride 1402-3
Naphthyl cyanide 4487-8

Naphthyl-ethyl ether 3093-4
Naphthyl fluoride 3273
Naphthyl iodide 3891-2
Naphthyl-mercaptan 5913-4
Naphthyl-nitrosohydroxylamine 4581

Table 7-4 (*Continued*)
PHYSICAL CONSTANTS OF ORGANIC COMPOUNDS

No.	Crystalline Form and Color	Specific Gravity	Melting Point °C.	Boiling Point °C.	Solubility in 100 Parts		
					Water	Alcohol	Ether
4530	lf./aq.	$1.061\frac{9.8}{4}°$	111–2	306.1	v. s. h.	s.	s.
4531	nd.	....	....	subl.	3.8^{200}	s.	s.
4532	lf.	....	254	....	v. s.	v. s.	....
4533	cr.	....	....	....	s. alk.	....	....
4534	pr.	....	....	....	s. alk.	v. sl. s. H_2SO_4	....
4535	rhb./aq.	....	....	....	23^{200}	$10^{20°}$ aq.*	$24^{54°}$ aq.*
4536	cr.	....	....	....	$3.4^{18°}$	$67^{1°}$ aq.	....
4537	nd./aq.	....	....	....	9^{200}	sl. s.	....
4538	cr.	....	....	....	12.8^{200}	$347^{8°}$ aq.	....
4539	cr.	....	....	....	$2.7^{18°}$	$127^{8°}$ aq.	....
4540	cr.	....	....	....	v. sl. s.	....	....
4541	nd. + $\frac{1}{2}H_2O$	....	d.	....	$0.03^{10°}$; $0.2^{100°}$	i.; s. alk.	i.
4542	mn. or rhb.	....	$-3\frac{1}{2}H_2O$, 80	$-4H_2O$, 130	s.; v. sl. s. aq. alk.	sl. s.	i.
4543	nd. + $1H_2O$	....	d.	....	$0.02^{21°}$; $0.4^{100°}$	s. ac.	....
4544	nd./aq.	....	262–5 d.	....	0.24^{00}; $3.1^{100°}$	i.	i. bz.
4545	nd.	....	....	....	sl. s.	....	....
4546	cr./aq.	....	....	....	$0.11^{6°}$	i.	i.
4547	nd. + $1H_2O$	....	....	....	$0.46^{25°}$	v. sl. s.	v. sl. s.
4548	cr. + $1H_2O$	....	d.	$-H_2O$, 110	0.1 c.	i.	i.
4549	cr./HCl	....	....	....	d. h.	....	....
4550	pl./aq.	....	....	....	0.03^{200}	i.	i.
4551	pr./aq.	....	....	....	0.06^{200}	i.	....
4552	cr. + $1H_2O$	....	....	....	$0.01^{20°}$; $0.16^{100°}$	....	....
4553	cr. + $1H_2O$	....	....	....	$0.02^{20°}$; $0.31^{100°}$	....	....
4554	cr.	....	....	....	$200^{18°}$	$12.5^{18°}$	....
4555	cr.	....	....	....	7.21^{200}	$16^{54°}$ aq.	....
4556	cr./et. al.	....	56	....	....	....	v. s.
4557	nd./al.	....	107–8	....	i.; s. chl.	s. h.	sl. s.
4558	col. lq.	1.18	....	269–70	d.	s.; s. pet.	s.; s. chl.
4559	lf.	....	55–6	....	d.	v. s.	s.; s. bz.
4560	lf./al.	....	110	>360	i.; s. bz.	sl. s. c.	s.
4561	nd./et. al.	....	81	264^{15mm}	s. bz.	....	s.
4562	nd./al.	....	105; d. 380	250^{19mm} sl. d.	v. s. bz.	s. h.	v. s.; sl. s. c. ac.
4563	col. cr.	....	116–7	203^{20mm}	v. sl. s. c.	v. s. h.	sl. s.

* Solubility of anhy. compd.
Naphthylene ethylene 2
Narceine, cf. alkd.
Narcissine, cf. alkd.
Narcosine, cf. alkd.

Narcotine, cf. alkd.
Naringin, cf. glcde.
Neoarsphenamine 5593
Neobornyval 868
Neocincophen 3130

Neol 376
Neomenthol 4025
Neomenthyl acetate 4028
Neonal 2999
Neoquinophan 3130

Table 7-4 (*Continued*)
PHYSICAL CONSTANTS OF ORGANIC COMPOUNDS

No.	Name	Synonym	Formula	Beil. Ref.	Formula Weight
	Naphthyl				
4564	hydrazine HCl (α)	....	$C_{10}H_7 \cdot NH \cdot NH_2 \cdot HCl$	XV-562	194.67
4565	hydrazine (β)	....	$C_{10}H_7 \cdot NH \cdot NH_2$	XV-568	158.20
4566	hydrazine HCl (β)	....	$C_{10}H_7 \cdot NH \cdot NH_2 \cdot HCl$	XV-568	194.67
4567	hydroxylamine (α)	....	$C_{10}H_7 \cdot NHOH$	XV-32	159.19
4568	salicylate (α)	alphol	$HOC_6H_4CO_2C_{10}H_7$	X-80	264.28
4569	salicylate (β)	betol	$HOC_6H_4CO_2C_{10}H_7$	X-80	264.28
4570	*iso*-thiocyanate (α)	....	$C_{10}H_7 \cdot N:CS$	XII-1244	185.25
4571	**Naphthylene** di-amine (1,2)	diamino-naphthalene	$C_{10}H_6(NH_2)_2$	XIII-196	158.20
4572	diamine (1,4)	....	$C_{10}H_6(NH_2)_2$	XIII-201	158.20
4573	diamine (1,5)	....	$C_{10}H_6(NH_2)_2$	XIII-203	158.20
4574	diamine (1,6)	....	$C_{10}H_6(NH_2)_2$	XIII-204	158.20
4575	diamine (1,7)	....	$C_{10}H_6(NH_2)_2$	XIII-205	158.20
4576	diamine (1,8)	....	$C_{10}H_6(NH_2)_2$	XIII-205	158.20
4577	diamine (2,3)	....	$C_{10}H_6(NH_2)_2$	XIII-207	158.20
4578	diamine (2,6)	....	$C_{10}H_6(NH_2)_2$	XIII-208	158.20
4579	diamine (2,7)	....	$C_{10}H_6(NH_2)_2$	XIII-208	158.20
4580	**Neo**-dorm(e)	α-*iso*-Pr-α-Br-butyramide	$(CH_3)_2CH \cdot C(C_2H_5) \cdot (Br)CONH_2$	**II-299	208.10
4581	cupfer(r)on	α-naphthyl-nitroso-hydroxylamine NH_4	$C_{10}H_7N(NO)ONH_4$	*XVI-396	205.22
4582	synephrine HCl	α-OH-β-methyl-aminoethyl-3-OH benzene HCl	$HO \cdot C_6H_4 \cdot CHOH \cdot CH_2 \cdot NHCH_3 \cdot HCl$	....	203.67
4583	**Nembutal**®	pentobarbital Na	$C_2H_5 \cdot C_4HN_2O_3Na \cdot CH(CH_3) \cdot C_3H_7$	....	248.26
4584	**Neurine**	trimethyl vinyl ammonium hydroxide	$CH_2:CH \cdot N(CH_3)_3 \cdot OH$	IV-203	103.17
4585	**Neutral red**	toluylene red	$C_{15}H_{16}N_4 \cdot HCl$	XXV-401	288.78
4586	**Nitranilic acid**	3,6-diNO$_2$-2,5-diOH-quinone	$(NO_2)_2C_6(OH)_2O_2 \cdot$ aq.	VIII-384	230.09
4587	**Nitro**-acenaphthene	(5-nitro)	$(NO_2)C_{10}H_5:(CH_2)_2$	V-588	199.21
4588	acetanilide (*o*)	....	$NO_2 \cdot C_6H_4 \cdot NHCOCH_3$	XII-691	180.16
4589	acetanilide (*m*)	....	$NO_2 \cdot C_6H_4 \cdot NHCOCH_3$	XII-703	180.16
4590	acetanilide (*p*)	....	$NO_2 \cdot C_6H_4 \cdot NHCOCH_3$	XII-719	180.16
4591	*p*-acetanisidide (3)	(3;1,4)	$NO_2 \cdot C_6H_3(OCH_3) \cdot NHCOCH_3$	XIII-522	210.19
4592	acetic acid	....	$NO_2 \cdot CH_2 \cdot CO_2H$	II-225	105.05
4593	2-acetnaphthalide (1)	1-NO$_2$-2-acetyl-naphthylamine	$NO_2 \cdot C_{10}H_6 \cdot NHCOCH_3$	XII-1313	230.23
4594	acetophenone (ω)	benzoyl-nitro-methane	$C_6H_5 \cdot CO \cdot CH_2 \cdot NO_2$	VII-289	165.15
4595	acetophenone (*m*)	....	$NO_2 \cdot C_6H_4 \cdot CO \cdot CH_3$	VII-288	165.15
4596	*p*-acetphenetidide	(3;1,4)	$NO_2 \cdot C_6H_3(OC_2H_5) \cdot NHCOCH_3$	XIII-522	224.22
4597	acet-*p*-toluidide	3-NO$_2$-4-acetyl-aminotoluene	$NO_2 \cdot C_6H_3(CH_3) \cdot NHCOCH_3$	XII-1002	194.19
4598	4-acetylamino-phenylacetate	(3;4,1)	$NO_2 \cdot C_6H_3(NHCO \cdot CH_3) \cdot O_2CCH_3$	....	238.20
4599	alizarin (3)(β)	alizarin orange	$(HO)_2C_{14}H_5O_2 \cdot NO_2$	VIII-447	285.21
4600	alizarin (4)(α)	....	$(HO)_2C_{14}H_5O_2 \cdot NO_2$	VIII-447	285.21

Table 7-4 (*Continued*)
PHYSICAL CONSTANTS OF ORGANIC COMPOUNDS

No.	Crystalline Form and Color	Specific Gravity	Melting Point °C.	Boiling Point °C.	Solubility in 100 Parts		
					Water	Alcohol	Ether
4564	pl./HCl	....	....	....	s. c.	....	....
4565	col. lf./aq.	....	124-5	....	sl. s. h.	s. h.	sl. s.
4566	nd./aq.	....	233 d.	....	s. h.	....	....
4567	cr./aq.	....	79	....	s.	s.	s.
4568	col. cr.	....	83	....	i.; s. oil	s.	s.
4569	col. cr./al.	....	95	....	i.	s. h.	s.; s. bz.
4570	nd./al.	....	58	....	i.	s. h.	s.
4571	lf./aq.	....	96-8	150-10.5mm	sl. s. h.	s.	s.
4572	pr./h. aq.	....	120	....	sl. s. h.	v. s.	v. s.
4573	pr./et.	....	189.5	....	s. h.	s. h.	s.
4574	nd./aq.	$1.147\frac{99-4°}{4}$	77.5	....	s. h.	s. h.	s. h.
4575	nd./aq.	....	117.5	....	sl. s.	s.	v. sl. s.
4576	cr./aq. al.	$1.127\frac{99-4°}{4}$	66.5	205 12mm	sl. s. h.	∞	∞
4577	lf./et.	....	191-3	....	....	s.	s.
4578	nd./aq.	....	216-8	....	v. sl. s. h.	sl. s.	sl. s.
4579	lf./aq.	....	159	....	....	....	....
4580	col. cr.	....	50-1	subl.	0.7 c.; d. h.	s.; s. pet.	s.; s. bz.
4581	lf./al. NH$_3$	....	125-6	....	s.	s. Me al.	i.
4582	col. cr.	....	139-41	....	s.	s.	....
4583	wh. cr. pd.	....	....	....	s.	s.	i.
4584	syrup, poison.	....	....	....	s.	d. aq. NH$_4$Cl	....
4585	gn. pd.	....	....	....	d.,-HCl	s.	....
4586	yel. pl.	....	-H$_2$O, 100	expl. 170	v. s.	v. s.	i.
4587	yel. nd./lg.	....	101-2	....	s. h.	s.	s.
4588	yel. mn.	1.419 15°	93-4	....	s. h.; s. chl.	s.; v. s. 10% KOH	s.
4589	col. lf.	....	154	....	sl. s. h.	s.	i.; s. chl.
4590	rhb.	....	215-6	....	s. h.	s.; s. KOH	s.
4591	yel. nd./al.	....	117-8	....	s. h.	s.; s. bz.	s.; s. ac.
4592	nd./chl.	....	87-9 d.	....	d.	v. s.	v. s.; i. pet.
4593	yel. rhb. nd./al.	....	123.5	....	sl. s. h.; v. sl. s. lg.	s.; s. bz.; s. ac.	v. sl. s.
4594	lf./aq. al.	....	106-8	....	i. c.	v. s.	v. s.
4595	nd.	....	80-1	202	i.	s.	....
4596	yel. nd./aq.	....	103-4	....	....	s. abs.	s.; s. chl.
4597	yel. nd./pet.	....	96; softens 94	....	sl. s. h.	s.	....
4598	yel. pr./aq. al.	....	147-8	....	....	....	....
4599	yel. lf./al.	....	244 d.	subl. sl. d.	sl. s.	s. chl., bz.	s. aq. alk.
4600	yel. nd./al.	....	289 d.	....	v. sl. s.; s. H$_2$SO$_4$	s. chl., bz	s. aq. alk.

Table 7-4 (*Continued*)
PHYSICAL CONSTANTS OF ORGANIC COMPOUNDS

No.	Name	Synonym	Formula	Beil. Ref.	Formula Weight
	Nitro				
4601	2-aminoanisole (4)	(4;1,2)	$NO_2 \cdot C_6H_3(OCH_3)NH_2$	XIII-389	168.15
4602	2-aminoanisole (5)	(5;1,2)	$NO_2 \cdot C_6H_3(OCH_3)NH_2$	XIII-390	168.15
4603	4-aminoanisole (3)	(3;1,4)	$NO_2 \cdot C_6H_3(OCH_3)NH_2$	XIII-521	168.15
4604	o-aminobenzoic acid (3)	(3;1,2) nitro-anthranilic acid	$NO_2 \cdot C_6H_3(CO_2H) \cdot NH_2$	XIV-373	182.14
4605	o-aminobenzoic acid (4)	(4;1,2)	$NO_2 \cdot C_6H_3(CO_2H) \cdot NH_2$	XIV-374	182.14
4606	o-aminobenzoic acid (5)	(5;1,2)	$NO_2 \cdot C_6H_3(CO_2H) \cdot NH_2$	XIV-375	182.14
4607	o-aminobenzoic acid (6)	(6;1,2)	$NO_2 \cdot C_6H_3(CO_2H) \cdot NH_2$	XIV-378	182.14
4608	m-aminobenzoic acid (5)	(5;1,3)	$NO_2 \cdot C_6H_3(CO_2H) \cdot NH_2$	XIV-415	182.14
4609	p-aminobenzoic acid (2)	(2;1,4)	$NO_2 \cdot C_6H_3(CO_2H) \cdot NH_2$	XIV-439	182.14
4610	o-aminophenol (3)	(3;2,1)	$NO_2 \cdot C_6H_3(NH_2)OH$	....	154.13
4611	o-aminophenol (4)	(4;2,1)	$NO_2 \cdot C_6H_3(NH_2)OH$	XIII-388	154.13
4612	o-aminophenol (5)	(5;2,1)	$NO_2 \cdot C_6H_3(NH_2)OH$	XIII-390	154.13
4613	o-aminophenol (6)	(6;2,1)	$NO_2 \cdot C_6H_3(NH_2)OH$	XIII-391	154.13
4614	m-aminophenol (5)	(5;3,1)	$NO_2 \cdot C_6H_3(NH_2)OH$	XIII-422	154.13
4615	p-aminophenol (2)	(2;4,1)	$NO_2 \cdot C_6H_3(NH_2)OH$	XIII-520	154.13
4616	p-aminophenol (3)	(3;4,1)	$NO_2 \cdot C_6H_3(NH_2)OH$	XIII-521	154.13
4618	aniline (o)	o-nitraniline	$NO_2 \cdot C_6H_4 \cdot NH_2$	XII-687	138.13
4619	aniline (m)	m-nitraniline	$NO_2 \cdot C_6H_4 \cdot NH_2$	XII-698	138.13
4620	aniline (p)	p-nitraniline	$NO_2 \cdot C_6H_4 \cdot NH_2$	XII-711	138.13
4621	aniline-4-sulfonic acid (2)	nitrosulfanilic acid	$NO_2 \cdot C_6H_3(NH_2)SO_3H$	XIV-708	218.19
4622	aniline-2-sulfonic acid (4)	(4;1,2)	$NO_2 \cdot C_6H_3(NH_2)SO_3H$	XIV-686	218.19
4623	anisole (o)	....	$CH_3O \cdot C_6H_4 \cdot NO_2$	VI-217	153.14
4624	anisole (m)	....	$CH_3O \cdot C_6H_4 \cdot NO_2$	VI-224	153.14
4625	anisole (p)	....	$CH_3O \cdot C_6H_4 \cdot NO_2$	VI-230	153.14
4626	anthracene (9)	nitrosoanthron	$C_{14}H_9 \cdot NO_2$	V-666	223.23
4627	anthraquinone (1)	....	$C_6H_4(CO)_2C_6H_3 \cdot NO_2$	VII-791	253.22
4628	anthraquinone (2)	....	$O_2C_{14}H_7 \cdot NO_2$	VII-792	253.22
4629	anthraquinone-2-carbonyl chloride	(1-NO_2)	$O_2C_{14}H_6(NO_2)COCl$	....	315.67
4630	anthraquinone-5-sulfonic acid (1)	(1-NO_2)	$O_2C_{14}H_6(NO_2)SO_3H$	XI-336	333.28
4631	anthraquinone-8-sulfonic acid (1)	(1-NO_2)	$O_2C_{14}H_6(NO_2)SO_3H$	XI-337	333.28
4632	barbituric acid (5)	dilituric acid	$NO_2 \cdot C_4H_3O_3N_2 \cdot 3H_2O$	XXIV-474	227.13
4633	benzaceto-phenone	(1,3)	$NO_2 \cdot C_6H_4 \cdot CH:CH \cdot CO \cdot C_6H_5$	VII-482	253.26
4634	benzal bromide (p)	....	$NO_2 \cdot C_6H_4 \cdot CH:Br_2$	V-336	294.94
4635	benzal chloride (m)	....	$NO_2 \cdot C_6H_4 \cdot CH \cdot Cl_2$	V-332	206.03
4636	benzal chloride (p)	....	$NO_2 \cdot C_6H_4 \cdot CH:Cl_2$	V-332	206.03

Nitro-acetylamino-toluene 4597
Nitro-acetylnaphthylamine 4593
Nitro-allyl 4829
Nitro-aminophenetole 4795
Nitro-aminotoluene 4849–58

Nitro-amylene glycol 4789
Nitro-anisidine 4602
Nitro-anthranilic acid 4604–7
Nitro base 4883

Table 7-4 (*Continued*)
PHYSICAL CONSTANTS OF ORGANIC COMPOUNDS

No.	Crystalline Form and Color	Specific Gravity	Melting Point °C.	Boiling Point °C.	Solubility in 100 Parts		
					Water	Alcohol	Ether
4601	red nd./al.	$1.207^{156°}$	118	....	s. h. bz.	s.; s. ac.	v. sl. s. lg.
4602	pa. yel. nd.	$1.211^{156°}$	139–40	....	....	....	....
4603	red/aq. al.	....	123	....	sl. s.	s.	s.
4604	yel. mn.	$1.558^{15°}$	204	....	i.	v. s.	v. s.
4605	yel. red nd.	....	269.5	....	v. sl. s. h.	s. xylene	....
4606	lt. yel. nd.	....	270–80 d.	....	s. h.	s.	s.
4607	yel. lf./aq.	....	183–4 d.	....	s. h.	v. s.	v. s.
4608	yel. pr./aq.	....	208	....	sl. s.	s. h.	sl. s.; v. s. h. ac.
4609	red nd./ aq.	....	240	....	s. h.	v. s.	v. s. ac.
4610	cr.	....	136	....	....	....	....
4611	or. pr.	....	142–3	....	sl. s. c.	v. s.	v. s.
4612	brn.nd./aq.	....	201–2	....	....	s. h.	....
4613	red nd./ aq. al.	....	110–1	....	v. sl. s. c.	s.; v. s. chl.	v. s.; v. s. bz.
4614	yel. cr.	....	165	....	v. sl. s. bz., chl.	v. s.	v. s.
4615	red nd./al.	....	128–31	....	....	....	....
4616	red pr./et. al.	....	154	....	s.	s.	s.
4618	yel. rhb.	$1.442^{15°}$	71.5	284.1	s. h.	v. s.	v. s.
4619	yel. rhb.	1.43	114	306.4	$0.11^{20°}$	$7.1^{20°}$	$7.9^{20°}$
4620	yel. mn.	$1.437^{14°}$	**147.5**	331.7	$0.08^{18.5°};$ 2.2 h.	$5.8^{20°}$	$6.1^{20°}$
4621	yel. nd.	....	....	....	v. s.; s. aq. H_2SO_4	sl. s.	s. conc. HCl
4622	yel. cr.	....	....	....	....	....	....
4623	col. cr.	$1.254\frac{20°}{4}$	9.5–10.5	272–3	$0.17^{30°}$	∞	∞
4624	nd./al.	$1.373^{18°}$	38	258	i.	s.	....
4625	pr./al.	$1.233^{20°}$	54	274	$0.06^{30°}$	v. s.	v. s.
4626	yel. nd./al.	....	146	>360	i. aq. alk.	sl. s.	v. s. bz.
4627	nd./ac.	....	230	270^{7mm}	i.	sl. s.	v. sl. s.
4628	yel. nd./al.	....	184.5–5.0	$270-17^{mm}$	s. H_2SO_4	v. sl. s. c.	sl. s.; v. s. chl.
4629	tan flakes	....	230 d.	....	i.	i.	i.
4630	yel. cr./aq.	....	....	....	s.	i.	i.
4631	yel. cr./aq.	....	....	....	sl. s.	i.	i.
4632	pr./aq.	....	181–3 (anh.)	....	$0.09^{25°}$	s.; s. alk.	i.
4633	yel. nd./ al. or bz.	....	144–5	....	i. lg.; v. s. bz.	s.; s. chl.	i.; s. ac.
4634	nd./al.	....	82.0–2.5	....	i.	s.	s.
4635	mn.	....	65	....	i.	v. s. h.	v. s.
4636	pr./al.	....	46	....	i.	v. s.	v. s.

Table 7-4 (*Continued*)
PHYSICAL CONSTANTS OF ORGANIC COMPOUNDS

No.	Name	Synonym	Formula	Beil. Ref.	Formula Weight
4637	**Nitro** benzaldehyde (*o*)	....	$NO_2 \cdot C_6H_4 \cdot CH:O$	VII-243	151.12
4638	benzaldehyde phenylhydrazone (*o*)	....	$NO_2 \cdot C_6H_4 \cdot CH:N \cdot NH \cdot C_6H_5$	XV-136	241.25
4639	benzaldehyde (*m*)	....	$NO_2 \cdot C_6H_4 \cdot CH:O$	VII-250	151.12
4640	benzaldehyde (*p*)	....	$NO_2 \cdot C_6H_4 \cdot CH:O$	VII-256	151.12
4641	benzalfurfuralacetone (*m*)	....	$NO_2 \cdot C_6H_4 \cdot CH:CH \cdot CO \cdot CH:CH \cdot C_4H_3O$	....	269.26
4642	benzamide (*o*)	....	$NO_2 \cdot C_6H_4 \cdot CO \cdot NH_2$	IX-373	166.14
4643	benzamide (*m*)	....	$NO_2 \cdot C_6H_4 \cdot CO \cdot NH_2$	IX-381	166.14
4644	benzamide (*p*)	....	$NO_2 \cdot C_6H_4 \cdot CO \cdot NH_2$	IX-394	166.14
4645	benzanilide (*m*)	....	$NO_2C_6H_4CONHC_6H_5$	XII-267	242.24
4646	benzene	oil of mirbane	$C_6H_5 \cdot NO_2$	V-233	123.11
4647	benzeneazochromotropic Na salt	*p*-nitrobenzeneazo-1,8-diOH-naphthalene-3,6-disulfonic Na	$NO_2 \cdot C_6H_4 \cdot N:N \cdot C_{10}H_3(OH)_2(SO_3Na)_2$	....	513.37
4648	benzeneazo-α-naphthol (*p*)	....	$NO_2 \cdot C_6H_4 \cdot N:N \cdot C_{10}H_6(OH)$	XVI-151	293.28
4649	benzeneazoorcinol (*p*)	(4′;4,6,2)	$NO_2 \cdot C_6H_4 \cdot N:N \cdot C_6H_2(OH)_2CH_3$	....	273.25
4650	benzeneazoresorcinol (*m*)	(3′;2,4)	$NO_2 \cdot C_6H_4 \cdot N:N \cdot C_6H_3(OH)_2$	....	259.22
4651	benzeneazoresorcinol (*p*)	(4′;2,4)	$NO_2 \cdot C_6H_4 \cdot N:N \cdot C_6H_3(OH)_2$	XVI-181	259.22
4652	benzeneazosalicylic acid (*p*)	(4′;4,3)	$NO_2 \cdot C_6H_4 \cdot N:N \cdot C_6H_3(OH)CO_2H$	XVI-247	287.23
4653	benzeneazosalicylic Na (*p*)	alizarin yellow R	$C_{13}H_8O_5N_3Na$	....	309.22
4654	benzene sulfonic acid (*o*)	....	$NO_2 \cdot C_6H_4 \cdot SO_3H$	XI-67	203.17
4655	benzene sulfonic acid (*m*)	....	$NO_2 \cdot C_6H_4 \cdot SO_3H$	XI-68	203.17
4656	benzene sulfonic Na	(*m*)	$NO_2 \cdot C_6H_4 \cdot SO_3Na$	XI-68	225.16
4657	benzene sulfonic chloride (*m*)	....	$NO_2 \cdot C_6H_4 \cdot SO_2Cl$	XI-69	221.62
4658	benzhydrazide (*o*)	....	$NO_2 \cdot C_6H_4CO \cdot NH \cdot NH_2$	IX-375	181.15
4659	benzhydrazide (*m*)	....	$NO_2 \cdot C_6H_4CO \cdot NH \cdot NH_2$	IX-388	181.15
4660	benzhydrazide (*p*)	....	$NO_2 \cdot C_6H_4CO \cdot NH \cdot NH_2$	IX-399	181.15
4661	benzidine (2)	....	$NO_2(NH_2)C_6H_3 \cdot C_6H_4 \cdot NH_2$	XIII-235	229.24
4662	benzimidazole (6)	....	$HN \cdot CH:N \cdot C_6H_3 \cdot NO_2$	XXIII-135	163.14
4663	benzoic acid (*o*)	....	$NO_2 \cdot C_6H_4 \cdot CO_2H$	IX-370	167.12
4664	benzoic acid (*m*)	....	$NO_2 \cdot C_6H_4 \cdot CO_2H$	IX-376	167.12
4665	benzoic Na (*m*)	....	$C_7H_4O_4NNa \cdot 3H_2O$	IX-377	243.15
4666	benzoic acid (*p*)	....	$NO_2 \cdot C_6H_4 \cdot CO_2H$	IX-389	167.12
4667	benzonitrile (*o*)	....	$NO_2 \cdot C_6H_4 \cdot CN$	IX-374	148.12
4668	benzonitrile (*m*)	....	$NO_2 \cdot C_6H_4 \cdot CN$	IX-385	148.12

Table 7-4 (*Continued*)
PHYSICAL CONSTANTS OF ORGANIC COMPOUNDS

No.	Crystalline Form and Color	Specific Gravity	Melting Point °C.	Boiling Point °C.	Solubility in 100 Parts		
					Water	Alcohol	Ether
4637	yel. nd./ aq.	....	42-3.5 (37.9)	153^{23mm}	v. sl. s.	v. s.	v. s.; s. bz.
4638	red nd.	....	154-5	....	v. s. act.; i. lg.	sl. s.	sl. s.
4639	nd./aq.	....	58	164^{23mm}	$0.16^{25°}$	v. s. h.	s.; s. chl.
4640	pr./aq.	....	106.5	....	v. sl. s.	v. s.	sl. s.
4641	yel. pd.	....	120-4	....	i.	sl. s. h.	i.
4642	nd./aq. al.	$1.462^{32°}_{4}$	176.6	317	s. h.	s. h.	s.
4643	yel. mn./ aq.	....	142-3	310-5	v. sl. s.	s.	s.
4644	nd./aq.	....	200-1.4	....	sl. s.	s.	s.
4645	lf./al.	....	153-4	subl.	sl. s.	s.	s.; s. bz.
4646	lt. yel. lq.	$1.203^{20°}_{4}$	5.7	210.9	$0.19^{20°}$	v. s.	∞; ∞ bz.
4647	red brn. pd.	....	....	....	s.	i.	....
4648	red gn. nd.	....	234-5	d. 255-60	s. am. al.; v. sl. s. chl.	v. sl. s.; s. xylene	v. sl. s.; v. sl. s. bz.
4649	dark red pd.	....	210-2 d.	....	i.	s.	sl. s.
4650	brn. pd.	....	174-5 d.	....	i.	s.	sl. s.
4651	red pd./ Me al.	....	199-200	....	s. alk.; v. sl. s. ac.	v. sl. s. h.	v. sl. s. toluene
4652	or. brn. nd./aq. ac.	....	254-7 d.	....	s. ac.	s.	sl. s. h. toluene
4653	brn. yel. pd.	....	....	....	s.	....	....
4654	hyg. lf.	....	70	d.	v. s.	s.; s. alk.	i.
4655	hyg. lf.	....	....	....	....	s. h.	....
4656	pl./aq.	....	....	....	v. s.	....	....
4657	mn. pr./ et.; nd./lg.	....	63-4	....	i.; d. h.	s. h.	....
4658	yel. brn./ aq.	....	120-1	....	s.	s.; i. chl.	i.; i. bz.
4659	nd./aq.	....	152	....	sl. s.	sl. s.	i.; i. bz.
4660	yel. nd./aq.	....	210	....	v. sl. s.	v. sl. s.	i.; i. bz.
4661	red nd./aq.	....	143(117)	....	sl. s. h.	....	....
4662	nd./aq.	....	204	....	sl. s.; s. alk. carb.	s.; s. a.; sl. s. chl.	sl. s.; sl. s. bz.
4663	tri./aq.	1.575	147.5	....	$0.65^{20°}$	$28^{11°}$, 90%	$22^{11°}$
4664	mn.	1.494	141-2	....	$0.24^{16.5°}$	$31^{11.7°}$	$25^{10.2°}$
4665	mn.	....	....	....	s. h.	....	....
4666	pa. yel. mn.	$1.550^{32°}_{4}$	240-2	subl.	$0.02^{15°}$	$0.9^{10°}$, 90%	$2.2^{12.5°}$
4667	nd./aq.	....	109-10	....	s. h.	s.	s. ac.
4668	nd./aq.	....	117-8	subl.	s. h.	s.	v. s.

Table 7-4 (*Continued*)
PHYSICAL CONSTANTS OF ORGANIC COMPOUNDS

No.	Name	Synonym	Formula	Beil. Ref.	Formula Weight
4669	**Nitro** benzonitrile (*p*)	. . . .	$NO_2 \cdot C_6H_4 \cdot CN$	IX-397	148.12
4670	benzophenone (*o*)	. . . .	$NO_2 \cdot C_6H_4 \cdot CO \cdot C_6H_5$	VII-425	227.22
4671	benzophenone (*m*)	. . . .	$NO_2 \cdot C_6H_4 \cdot CO \cdot C_6H_5$	VII-425	227.22
4672	benzophenone (*p*)	. . . .	$NO_2 \cdot C_6H_4 \cdot CO \cdot C_6H_5$	VII-425	227.22
4673	benzoyl chloride (*o*)	. . . .	$NO_2 \cdot C_6H_4 \cdot COCl$	IX-373	185.57
4674	benzoylchloride(*m*)	. . . .	$NO_2 \cdot C_6H_4 \cdot COCl$	IX-381	185.57
4675	benzoyl chloride (*p*)	. . . .	$NO_2 \cdot C_6H_4 \cdot COCl$	IX-394	185.57
4676	benzoyl formic acid (*o*)	. . . .	$NO_2 \cdot C_6H_4 \cdot CO \cdot CO_2H \cdot$ aq.	X-664	195.13
4677	benzyl acetate (*p*)	. . . .	$NO_2 \cdot C_6H_4 \cdot CH_2 \cdot O_2C \cdot CH_3$	VI-451	195.18
4678	benzyl alcohol (*o*)	. . . .	$NO_2 \cdot C_6H_4 \cdot CH_2OH$	VI-447	153.14
4679	benzyl alcohol (*m*)	. . . .	$NO_2 \cdot C_6H_4 \cdot CH_2OH$	VI-449	153.14
4680	benzyl alcohol (*p*)	. . . .	$NO_2 \cdot C_6H_4 \cdot CH_2OH$	VI-450	153.14
4681	benzyl bromide (*p*)	ω-Br-*p*-nitrotoluene	$NO_2 \cdot C_6H_4 \cdot CH_2Br$	V-334	216.04
4682	benzyl chloride (*o*)	ω-Cl-*o*-nitrotoluene	$NO_2 \cdot C_6H_4 \cdot CH_2Cl$	V-327	171.58
4683	benzyl chloride (*m*)	. . . .	$NO_2 \cdot C_6H_4 \cdot CH_2Cl$	V-329	171.58
4684	benzyl chloride (*p*)	. . . .	$NO_2 \cdot C_6H_4 \cdot CH_2Cl$	V-329	171.58
4685	benzyl cyanide (*o*)	nitro-α-toluic nitrile	$NO_2 \cdot C_6H_4 \cdot CH_2 \cdot CN$	IX-455	162.15
4686	benzyl cyanide (*m*)	. . . .	$NO_2 \cdot C_6H_4 \cdot CH_2 \cdot CN$	IX-455	162.15
4687	benzyl cyanide (*p*)	. . . .	$NO_2 \cdot C_6H_4 \cdot CH_2 \cdot CN$	IX-456	162.15
4688	bromoform	bromopicrin	$NO_2 \cdot CBr_3$	I-77	297.74
4689	butandiol	2-nitro-2-methyl-1, 3-propandiol	$CH_3 \cdot C \cdot NO_2 : (CH_2OH)_2$	I-480	135.12
4690	butane (1)	1-nitrobutane	$C_2H_5 \cdot CH_2 \cdot CH_2 \cdot NO_2$	I-123	103.12
4691	butane (2)	2-nitrobutane	$C_2H_5 \cdot CHNO_2 \cdot CH_3$	I-123	103.12
4692	butane (α)(*iso*)	1-nitro-2-Me-propane	$(CH_3)_2CH \cdot CH_2 \cdot NO_2$	I-129	103.12
4693	butane (*tert*)	2-nitro-2-Me-propane	$(CH_3)_3C \cdot NO_2$	I-129	103.12
4694	1-butanol (2)	β-NO₂-butyl alcohol	$C_2H_5 \cdot CHNO_2 \cdot CH_2OH$	I-370	119.12
4695	butanol	2NO₂-2-Me-propanol	$(CH_3)_2CNO_2 \cdot CH_2OH$	I-378	119.12
4696	camphor (α)(3)	. . . .	$\overset{\mid}{C_8H_{14}} \cdot CO \cdot CH \cdot NO_2$	VII-129	197.24
4697	chloroform	chloropicrin	$NO_2 \cdot CCl_3$	I-76	164.38
4698	chlorophenol	(1;2,4)	$HO \cdot C_6H_3(Cl) \cdot NO_2$	VI-240	173.57
4699	chlorophenol	(1;2,3)	$HO \cdot C_6H_3(Cl) \cdot NO_2$	VI-239	173.57
4700	chlorophenol	(1;3,4)	$HO \cdot C_6H_3(Cl) \cdot NO_2$	VI-240	173.57
4701	chlorophenol	(1;4,2)	$HO \cdot C_6H_3(Cl) \cdot NO_2$	VI-238	173.57
4702	chlorophenol	(1;4,3)	$HO \cdot C_6H_3(Cl) \cdot NO_2$	VI-239	173.57
4703	chlorophenol	(1;5,2)	$HO \cdot C_6H_3(Cl) \cdot NO_2$	VI-238	173.57
4704	chlorophenol	(1;5,3)	$HO \cdot C_6H_3(Cl) \cdot NO_2$	VI-239	173.57
4705	chlorophenol	(1;6,2)	$HO \cdot C_6H_3(Cl) \cdot NO_2$	VI-239	173.57
4706	chlorophenol	(1;6,3)	$HO \cdot C_6H_3(Cl) \cdot NO_2$	VI-240	173.57
4707	cinnamic acid (*o*)	. . . .	$NO_2 \cdot C_6H_4 \cdot CH:CH \cdot CO_2H$	IX-604	193.16
4708	cinnamic acid (*m*)	. . . .	$NO_2 \cdot C_6H_4 \cdot CH:CH \cdot CO_2H$	IX-605	193.16
4709	cinnamic acid (*p*)	. . . .	$NO_2 \cdot C_6H_4 \cdot CH:CH \cdot CO_2H$	IX-606	193.16
4710	*p*-cresol (2)	3-NO₂-4-OH-toluene	$NO_2 \cdot C_6H_3(CH_3)OH$	VI-412	153.14

Table 7-4 (*Continued*)
PHYSICAL CONSTANTS OF ORGANIC COMPOUNDS

No.	Crystalline Form and Color	Specific Gravity	Melting Point °C.	Boiling Point °C.	Solubility in 100 Parts		
					Water	Alcohol	Ether
4669	yel. lf./al.	....	147-9	....	sl. s.	s. h.	s. ac.
4670	mn./al.	....	105	....	....	sl. s. abs.	....
4671	nd./al.	....	94-5	234^{18mm}	....	s. h.	....
4672	lf./abs. al.	....	138	....	v. sl. s.	s. h.	sl. s. CS_2
4673	ool. or.	....	75(20)	205^{105mm}	d.	d.	s.
4674	cr.	....	34-5	275-8 sl. d.	d.	d.	v. s.
4675	nd./lg.	....	72	154^{15mm}	d.	d.	s.
4676	pr./aq.	....	46-7; 122 d. (anh.)	....	∞ h.	....	....
4677	yel. nd./al.	....	78	....	....	s. h.	....
4678	nd./aq.	....	74	270 sl. d.	sl. s.	v.s.	v. s.
4679	cr.	....	27(13-5)	$175-80^{3mm}$	....	....	....
4680	nd./aq.	....	93(87-9)	185^{12mm}	s. h.	v. s.	v. s.
4681	nd./al.	....	99-100	....	i.	$2^{19°}$	v. s.
4682	cr.	....	48-9	....	i.	$26.3^{30°}$	v. s. h.
4683	yel. nd.	....	45-7	$173-83^{30mm}$	i.	$30.4^{30°}$	v. s.
4684	nd./al.	....	71	....	i.	$8.2^{30°}$	s.
4685	nd./aq.	....	115-6	....	v. s. h.	s.	s.
4686	cr.	....	61-2	....	v. sl. s.	s.	s.
4687	lf./al.	....	116-7	....	i.	s.	....
4688	pr.	$2.811^{12.5°}$	10.3	$*127^{118mm}$	i.	s.	s.
4689	mn. cr.	....	147-9	d.	$80^{20°}$	$45^{20°}$	$4^{20°}$
4690	lq.	....	....	151-2	....	....	....
4691	lq.	$0.988^{0°}$	....	$138-97^{47mm}$	....	....	....
4692	col. lq.	$0.9877°$	....	$158-97^{55mm}$	i.	s.	s.
4693	cr.	....	24	126^{748mm}	i. alk.	∞	∞ ; ∞ bz.
4694	lq.	$1.134\frac{20}{20}°$	−47	105^{10mm}	$20^{20°}$	$∞^{20°}$	$∞^{20°}$
4695	cr./Me al.	....	90.5	95.5^{10mm}	$350^{20°}$	$490^{20°}$	$120^{20°}$
4696	mn./bz.	....	102-3	....	i.	s.	s.; v. s. bz.
4697	lq.	$1.651\frac{22\text{-}8°}{4}$	−64	112.3^{766mm}	$0.17^{18°}$	37 cc., 80% al.	s.
4698	col. nd./aq.	....	111	....	v. s. chl.	v. s.	v. s.
4699	cr./aq.	....	120	....	....	....	....
4700	nd./bz.	....	133	....	....	....	....
4701	yel. mn.	....	87	....	v. sl. s.	s.	v. s.
4702	nd./aq.	....	126-7	....	....	....	....
4703	yel. pr./aq.	....	38.9 (32.7)	....	sl. s.	s.	s.; s. ac.
4704	cr.	....	147	....	....	....	....
4705	yel. nd./aq.	....	70-1	....	v. sl. s.	v. s. chl.	....
4706	nd./aq.	....	118-9	....	....	....	....
4707	nd./al.	....	243-5	subl.	i.	$0.2^{25°}$ abs.	....
4708	col./al.	....	203-5	....	....	$1^{25°}$ abs.	....
4709	lt. yel./al.	....	286-8	....	v. sl. s. h.	$0.01^{25°}$	v. sl. s.; i. CS_2
4710	yel./aq. al.	$1.240\frac{38\text{-}6°}{4}$	32	125^{22mm}	v. sl. s.	v. s.	v. s.

* Explosive.
Nitro-chalcone 4633
Nitro-chloroaniline 1407-11
Nitro-chlorobenzene 1412-4

Nitro-chlorobenzene sulfonic acid 1415
Nitro-chloroethane 1416-7
Nitro-chloronaphthalene 1418-9
Nitro-chloropropane 1420-1

Table 7-4 (*Continued*)
PHYSICAL CONSTANTS OF ORGANIC COMPOUNDS

No.	Name	Synonym	Formula	Beil. Ref.	Formula Weight
	Nitro				
4711	cumene (*o* and *p*)	....	$NO_2 \cdot C_6H_4 \cdot CH(CH_3)_2$	**V-307	165.19
4712	cumene, pseudo-	(5;1,2,4)	$NO_2 \cdot C_6H_2(CH_3)_3$	V-404	165.19
4713	cumene, pseudo-	(6;1,2,4)	$NO_2 \cdot C_6H_2(CH_3)_3$	V-404	165.19
4714	*p*-cymene (2;1,4)	2-nitrocymene	$NO_2 \cdot C_6H_3(CH_3) \cdot C_3H_7$	V-424	179.22
4715	decane	1-nitro-2,7-dimethyl-octane	$(CH_3)_2CH(CH_2)_4 \cdot CH(CH_3)CH_2NO_2$	I-169	187.28
4716	decane	2-nitro-2,7-dimethyl-octane	$(CH_3)_2C(NO_2)(CH_2)_4 \cdot CH(CH_3)_2$	I-169	187.28
4717	diethylaniline (*o*)	....	$NO_2 \cdot C_6H_4 \cdot N(C_2H_5)_2$	*XII-341	194.24
4718	diethylaniline (*m*)	....	$NO_2 \cdot C_6H_4 \cdot N(C_2H_5)_2$	XII-702	194.24
4719	diethylaniline (*p*)	....	$NO_2 \cdot C_6H_4 \cdot N(C_2H_5)_2$	XII-715	194.24
4720	dimethylaniline (*o*)	....	$NO_2 \cdot C_6H_4 \cdot N(CH_3)_2$	XII-690	166.18
4721	dimethylaniline (*m*)	....	$NO_2 \cdot C_6H_4 \cdot N(CH_3)_2$	XII-701	166.18
4722	dimethylaniline (*p*)	....	$NO_2 \cdot C_6H_4 \cdot N(CH_3)_2$	XII-714	166.18
4723	diphenyl (2)	....	$C_6H_5 \cdot C_6H_4 \cdot NO_2$	V-582	199.21
4724	diphenyl (3)	....	$C_6H_5 \cdot C_6H_4 \cdot NO_2$	V-582	199.21
4725	diphenyl (4)	....	$C_6H_5 \cdot C_6H_4 \cdot NO_2$	V-583	199.21
4726	diphenylamine (*o*)	....	$C_6H_5 \cdot NH \cdot C_6H_4 \cdot NO_2$	XII-690	214.23
4727	diphenylamine (*p*)	....	$C_6H_5 \cdot NH \cdot C_6H_4 \cdot NO_2$	XII-715	214.23
4728	diphenylamine-2-sulfonic acid (4)	....	$C_6H_5 \cdot NH \cdot C_6H_3(NO_2) \cdot SO_3H$	XIV-686	294.29
4729	diphenyl ether (2)	....	$C_6H_5 \cdot O \cdot C_6H_4 \cdot NO_2$	VI-218	215.21
4730	diphenyl ether (3)	....	$C_6H_5 \cdot O \cdot C_6H_4 \cdot NO_2$	VI-224	215.21
4731	diphenyl ether (4)	....	$C_6H_5 \cdot O \cdot C_6H_4 \cdot NO_2$	VI-232	215.21
4732	diphenylene oxide	2-NO_2-dibenzfuran	$NO_2 \cdot C_{12}H_7:O$	XVII-72	213.19
4733	ethane	....	$CH_3 \cdot CH_2 \cdot NO_2$	I-99	75.07
4734	ethylacetanilide	(*p*)	$NO_2 \cdot C_6H_4 \cdot N(C_2H_5) \cdot COCH_3$	XII-720	208.22
4735	ethyl alcohol	2-nitro-ethanol-1	$HO \cdot CH_2 \cdot CH_2 \cdot NO_2$	I-339	91.07
4736	ethylbenzene (*o*)	....	$NO_2 \cdot C_6H_4 \cdot C_2H_5$	V-358	151.17
4737	ethylbenzene (*p*)	....	$NO_2 \cdot C_6H_4 \cdot C_2H_5$	V-358	151.17
4738	fluorene (2)	....	$NO_2 \cdot C_{12}H_7:CH_2$	V-628	211.22
4739	formaldehyde-phenylhydrazone	benzene-azo-nitro-methane	$C_6H_5NH \cdot N:CH \cdot NO_2$	XV-235	165.15
4740	furan (2)	....	$NO_2 \cdot C_4H_3O$	XVII-28	113.07
4741	furoic acid (5)	nitro-pyromucic acid	$NO_2 \cdot C_4H_2O \cdot CO_2H$	XVIII-287	157.08
4742	guaiacol (4)	(4;2,1)	$NO_2 \cdot C_6H_3(OCH_3)OH$	VI-788	169.14
4743	guanidine	....	$NO_2 \cdot NH \cdot C(NH) \cdot NH_2$	III-126	104.07
4744	heptane (1)	....	$C_6H_{13} \cdot CH_2 \cdot NO_2$	I-155	145.20
4745	heptane (2)	....	$C_5H_{11} \cdot CHNO_2 \cdot CH_3$	I-156	145.20
4746	heptane	2-NO_2-2,4-di-methylpentane	$(CH_3)_2CNO_2 \cdot CH_2 \cdot CH(CH_3)_2$	I-158	145.20
4747	heptane	3-NO_2-2,2-di-methylpentane	$(CH_3)_3C \cdot CH(NO_2) \cdot C_2H_5$	I-157	145.20
4748	heptane	3-NO_2-3-Et-pentane	$(C_2H_5)_3C \cdot NO_2$	I-157	145.20
4749	hexane (1)	....	$CH_3(CH_2)_4CH_2NO_2$	I-147	131.18

Nitro-dibenzfuran 4732
Nitro-dichloroaniline 2021
Nitro-dichlorobenzene 2022-6

Nitro-dichlorophenol 2029-30
Nitro-dimethylbenzene 4865-70

Table 7-4 (*Continued*)
PHYSICAL CONSTANTS OF ORGANIC COMPOUNDS

No.	Crystalline Form and Color	Specific Gravity	Melting Point °C.	Boiling Point °C.	Solubility in 100 Parts		
					Water	Alcohol	Ether
4711	yel. oil	$1.112°$	−35	224 d.	....	....	....
4712	col. or gn.	....	65(45-6)	265	col.: v. sl. s. pet.	s. h.	gn.: v. s. pet.
4713	pr.	....	20	....	....	....	....
4714	oil	1.067^{20}_{4}	....	152^{15}mm	i.	....	....
4715	lq.	0.925^{21}_{0}	....	235-7 d.	....	....	....
4716	lq.	0.909^{20}_{4}	....	235-7749^{mm} d.	....	....	....
4717	yel. oil	....	....	$153-5^{20}$mm	sl. s.	s.	s.
4718	yel. oil	....	....	288-90	....	....	....
4719	yel. mn./al.	1.225	77-8	....	....	v. s. h.	sl. s. lg.
4720	yel. oil	1.179^{20}_{4}	....	$151-3^{30}$mm	sl. s.	v. s.	v. s.
4721	red mn.	$1.313^{17°}$	60-1	280-5	i.	s.	s.
4722	yel. nd.	....	163-4	....	i.	s. h.	s. h. ac.
4723	rhb.	1.44	37	320	i.	s.	v. s.
4724	yel. lf.	....	61	....	i.	v. s.	v. s. ac.
4725	nd./al.	....	113-4	340	i.	sl. s. c.	v. s.
4726	or./aq. al.	....	75-6	....	....	....	....
4727	yel. nd.	....	132-3	....	i. dil. a.	s.	s. ac.
4728	gn. lf./HCl	....	....	....	v. s.	v. s.	....
4729	yel. oil	$1.258^{15°}$	<−20	235^{60}mm	i.	s. abs.	s. bz.
4730	yel. oil	$1.245^{15°}$	....	$202-41^{4}$mm	....	....	....
4731	tab.	....	56-7	320±	i.	sl. s. c.	v. s.
4732	yel. nd./ac.	....	181-2	....	s. h. ac.	sl. s. h.	sl. s.
4733	lq.	1.052^{20}_{4}	−90	114.876^{1}mm	$4.5^{20°}$; s. a., alk.	∞; ∞ chl.	∞
4734	lf.	....	118-9	....	v. sl. s.; s. bz.	s.; i. lg.	v. sl. s.; s. CS_2
4735	col. lq.	$1.270^{15°}$	<−80	194^{765}mm	v. s.	v. s.	v. s.
4736	col. oil	$1.126^{24.5°}$	−23	227-8	i.	v. s.	s.
4737	col. oil	$1.124^{25°}$	−32	245-6	i.	v. s.	v. s.
4738	nd./50% ac.	....	157-8	....	....	....	....
4739	or. pr.	....	85-6 (75-6)	....	s. lg.	s. chl.	s. bz.
4740	lt. yel./pet.	....	28	....	v. sl. s.	s. alk.	s.
4741	lt. yel./aq.	....	185	subl.	v. sl. s.	s.	s.; i. chl.
4742	yel. nd./aq.	....	104-5	....	....	....	....
4743	nd./aq.	....	246-7	....	$9^{100°}$	sl. s.	v. sl. s.
4744	lt. yel. oil	$0.948^{17°}$	....	193-5	i.	v. s.	v. s.
4745	lq.	$0.947^{0°}$	....	194-8 sl. d.	s. conc. alk.	....	....
4746	lq.	0.956^{0}_{0}	....	$181-2^{742}$mm	....	....	....
4747	lq.	0.940^{20}_{0}	....	$89-90^{40}$mm	....	....	....
4748	lq.	$0.955^{0°}$	....	185-90	....	....	....
4749	col. lo	$0.949^{20°}$	....	$193-4^{765}$mm	i.	v. s.	v. s. al. alk.

Nitro-erythrite 2895
Nitro-fluorobenzene 3274

Nitro-form 6300
Nitro-glycerin(e) 3428

Table 7-4 *(Continued)*
PHYSICAL CONSTANTS OF ORGANIC COMPOUNDS

No.	Name	Synonym	Formula	Beil. Ref.	Formula Weight
4750	**Nitro** hexane	3-NO$_2$-2,2-di-methylbutane	(CH$_3$)$_3$C·CH(NO$_2$)·CH$_3$	I-151	131.18
4751	hexane	2-NO$_2$-2-Me-pentane	(CH$_3$)$_2$C(NO$_2$)·C$_3$H$_7$	I-149	131.18
4752	hippuric acid (*m*)	....	NO$_2$·C$_6$H$_4$·CO·NH·CH$_2$·CO$_2$H	IX-383	224.17
4753	*o*-hydroxybenzoic acid (3;2,1)	3-nitrosalicylic acid	NO$_2$·C$_6$H$_3$(OH)·CO$_2$H·H$_2$O	X-114	201.14
4754	*o*-hydroxybenzoic acid (4;2,1)	4-nitrosalicylic acid	NO$_2$·C$_6$H$_3$(OH)CO$_2$H	X-116	183.12
4755	*o*-hydroxybenzoic acid (5;2,1)	5-nitrosalicylic acid	NO$_2$·C$_6$H$_3$(OH)CO$_2$H	X-116	183.12
4756	*o*-hydroxybenzoic acid (6;2,1)	6-nitrosalicylic acid	NO$_2$·C$_6$H$_3$(OH)CO$_2$H	....	183.12
4757	*m*-hydroxybenzoic acid (2;3,1)	....	NO$_2$C$_6$H$_3$(OH)CO$_2$H·H$_2$O	X-146	201.14
4758	*m*-hydroxybenzoic acid (4;3,1)	....	NO$_2$·C$_6$H$_3$(OH)CO$_2$H	X-146	183.12
4759	*m*-hydroxybenzoic acid (6;3,1)	....	NO$_2$C$_6$H$_3$(OH)CO$_2$H·H$_2$O	X-147	201.14
4760	*p*-hydroxybenzoic acid (3;4,1)	....	NO$_2$·C$_6$H$_3$(OH)CO$_2$H	X-181	183.12
4761	isatin (5)	....	NO$_2$·C$_6$H$_3$·NH·CO·CO $\overline{}$	XXI-456	192.13
4762	malonic dialdehyde	2-nitro-propandial	NO$_2$·CH(CHO)$_2$	I-766	117.06
4763	malonic ester	....	NO$_2$·CH(CO$_2$C$_2$H$_5$)$_2$	II-596	205.17
4764	mesitylene (1,3,5;2)	....	(CH$_3$)$_3$C$_6$H$_2$·NO$_2$	V-410	165.19
4765	methane	....	CH$_3$·NO$_2$	I-74	61.04
4766	*N*-methylacetanilide (*p*)	....	NO$_2$·C$_6$H$_4$·N(CH$_3$)·COCH$_3$	XII-719	194.19
4767	*N*-methylaniline (*o*)	....	NO$_2$·C$_6$H$_4$·NHCH$_3$	XII-689	152.15
4768	*N*-methylaniline (*m*)	....	NO$_2$·C$_6$H$_4$·NHCH$_3$	XII-700	152.15
4769	*N*-methylaniline (*p*)	....	NO$_2$·C$_6$H$_4$·NHCH$_3$	XII-714	152.15
4770	2-methylanthraquinone (1)	....	NO$_2$(CH$_3$)C$_{14}$H$_6$O$_2$	VII-811	267.24
4771	4'-methyldiphenylamine (4)	....	NO$_2$·C$_6$H$_4$·NH·C$_7$H$_7$	XII-906	228.25
4772	naphthalene (α)	....	NO$_2$·C$_{10}$H$_7$	V-553	173.17
4773	naphthalene (β)	....	NO$_2$·C$_{10}$H$_7$	V-555	173.17
4774	naphthoic acid (8,1)	....	NO$_2$·C$_{10}$H$_6$·CO$_2$H	IX-653	217.18
4775	α-naphthol (2,1)	....	NO$_2$·C$_{10}$H$_6$·OH	VI-615	189.17
4776	α-naphthol (4,1)	....	NO$_2$·C$_{10}$H$_6$·OH	VI-615	189.17
4777	β-naphthol (1,2)	....	NO$_2$·C$_{10}$H$_6$·OH	VI-653	189.17
4778	β-naphthol (5,2)	....	NO$_2$·C$_{10}$H$_6$·OH	VI-654	189.17
4779	β-naphthol (8,2)	....	NO$_2$·C$_{10}$H$_6$·OH	VI-655	189.17
4780	α-naphthylamine	(2,1)	NO$_2$·C$_{10}$H$_6$·NH$_2$	XII-1258	188.19
4781	β-naphthylamine	(1,2)	NO$_2$·C$_{10}$H$_6$·NH$_2$	XII-1313	188.19
4782	β-naphthylamine	(5,2)	NO$_2$·C$_{10}$H$_6$·NH$_2$	XII-1314	188.19
4783	β-naphthylamine	(8,2)	NO$_2$·C$_{10}$H$_6$·NH$_2$	XII-1315	188.19

Table 7-4 (*Continued*)
PHYSICAL CONSTANTS OF ORGANIC COMPOUNDS

No.	Crystalline Form and Color	Specific Gravity	Melting Point °C.	Boiling Point °C.	Solubility in 100 Parts		
					Water	Alcohol	Ether
4750	col. pr.	....	40	$167.5-8^{748mm}$	v. s. pet.	v. s.	v. s.
4751	lq.	0.9499^{0}_{0}	....	$172-6^{756mm}$	i. alk.	....	....
4752	nd./aq.	....	165-7	....	$0.4^{23°}$	s.	s.
4753	rhb./aq.	....	123-5; 148-9(anh.)	....	$0.13^{15.5°}$	v. s.; s. bz.	v. s.; s. chl.
4754	col. nd./aq.	....	226-35	....	s. h.; s. chl.	s.; i. lg.	sl. s. bz.
4755	nd./aq.	$1.650^{20°}$	229-32	....	$0.18^{22°}$	v. s.; s. act.	v. s.
4756	cr.	....	130	....	s. act.	sl. s.	v. s.
4757	pl./aq.	....	178	....	v. sl. s.	s.	s.
4758	yel. lf./aq.	....	230	....	v. sl. s.	....	....
4759	yel. nd./aq.	....	169	....	v. s.	v. s.	v. s.
4760	nd./aq.	....	185-6	....	v. s. h.	v. s.	s.
4761	nd./al.	....	248-50	....	sl. s.	s.	s. alk.
4762	pr.	....	50-1	....	d.; v.s.chl.	v. s.	v. s.
4763	col. oil	1.199^{20}_{4}	....	$152-3^{38mm}$	i.	....	....
4764	rhb./al.	....	44	255	....	v. s. h.	....
4765	oil	1.131^{25}_{4}	−28.6	101.3	$9.5^{20°}$	s.; s. alk.	s.
4766	lf./aq.	....	152-3	....	....	s.	s.
4767	red nd./pet.	....	36-7	....	v. sl. s. c.	s.	s.
4768	red yel./al.	....	67-8	....	s. h.	s.	s.
4769	br. yel./al.	....	151-2	....	s. bz.	s.	v. sl. s. lg.
4770	pa. yel. nd./ac.	....	269-70	subl.; d. 330-2	sl. s. bz.; s. $C_6H_5NO_2$	v. sl. s.; sl. s. chl.	v. sl. s.
4771	yel. nd./al. or bz.	....	138-9	....	v. s. ac.; v.sl.s.bz.	v. s. h.	v. sl. s.
4772	yel. nd./al.	$1.223^{61.5°}$	59-60	304	i.; s. CS_2	s.; s. chl.	s.
4773	col./al.	....	79	165^{15mm}	i.	v. s.	v. s.
4774	pr./al.	....	215	....	0.04	4.7	sl. s.
4775	yel. nd./al.	....	128	....	v. sl. s.	s.	....
4776	col. cr.	....	164	....	s. h.	v. s.	v. s. ac.
4777	yel. nd./al.	....	103	....	i.; s. alk.	s.; s. ac.	v. s.
4778	lt. yel./aq.	....	147	....	v. s. h.	v. s.	v. s.
4779	yel. nd./aq.	....	144-5	....	s. bz.	s.; s. chl.	s.
4780	red yel. mn.	....	144	....	....	s.	....
4781	red yel./al.	....	126-7	....	s. h.	s.	s. ac.
4782	red nd./al.	....	143.5	....	s. bz.	s.	i. lg.
4783	red nd.	....	104-5	....	i. lg.	s.	s.

Table 7-4 (*Continued*)
PHYSICAL CONSTANTS OF ORGANIC COMPOUNDS

No.	Name	Synonym	Formula	Beil. Ref.	Formula Weight
4784	**Nitro** nonane (1)	....	$CH_3(CH_2)_7CH_2NO_2$	I-166	173.26
4785	nonane	2-NO$_2$-2,6-di-methylheptane	$(CH_3)_2C(NO_2)\cdot$ $(CH_2)_3CH(CH_3)_2$	I-167	173.26
4786	octane (1)	....	$CH_3(CH_2)_6CH_2NO_2$	I-161	159.23
4787	octane	1-NO$_2$-2,5-di-methylhexane	$(CH_3)_2CH(CH_2)_2\cdot$ $CH(CH_3)CH_2NO_2$	I-163	159.23
4788	octane	2-NO$_2$-2,5-di-methylhexane	$(CH_3)_2C(NO_2)\cdot$ $(CH_2)_2CH(CH_3)_2$	I-163	159.23
4789	pentandiol	2-NO$_2$-2-Et-1,3-propandiol	$C_2H_5\cdot C(NO_2)\cdot$ $(CH_2OH)_2$	I-483	149.15
4790	pentane (1)	....	$CH_3(CH_2)_3CH_2NO_2$	I-133	117.15
4791	pentane (3)	....	$(C_2H_5)_2CH\cdot NO_2$	I-133	117.15
4792	pentane	2-NO$_2$-2-Me-butane	$(CH_3)_2C(NO_2)C_2H_5$	I-140	117.15
4793	pentane	4-NO$_2$-2-Me-butane	$(CH_3)_2CH(CH_2)_2NO_2$	I-140	117.15
4794	phenanthraquinone	(2)	$NO_2\cdot C_6H_3(CO)_2C_6H_4$	VII-806	253.22
4795	*p*-phenetidine (3)	3-NO$_2$-4-NH$_2$-phenetole	$C_6H_3(NH_2)OC_2H_5\cdot$ NO_2	XIII-521	182.18
4796	phenetole (*o*)	....	$C_2H_5O\cdot C_6H_4\cdot NO_2$	VI-218	167.17
4797	phenetole (*m*)	....	$C_2H_5O\cdot C_6H_4\cdot NO_2$	VI-224	167.17
4798	phenetole (*p*)	....	$C_2H_5O\cdot C_6H_4\cdot NO_2$	VI-231	167.17
4799	phenol (*o*)	....	$HO\cdot C_6H_4\cdot NO_2$	VI-213	139.11
4800	phenol Na (*o*)	Na nitrophenoxide	$NaO\cdot C_6H_4\cdot NO_2$	VI-217	161.10
4801	phenol K (*o*)	K nitrophenoxide	$KO\cdot C_6H_4\cdot NO_2\cdot\frac{1}{2}H_2O$	VI-217	186.22
4802	phenol (*m*)	....	$HO\cdot C_6H_4\cdot NO_2$	VI-222	139.11
4803	phenol (*p*)	....	$HO\cdot C_6H_4\cdot NO_2$	VI-226	139.11
4804	phenol Na (*p*)	....	$NaC_6H_4O_3N\cdot 2H_2O$	VI-230	197.13
4805	phenol Na (*p*)	....	$NaC_6H_4O_3N\cdot 4H_2O$	VI-230	233.16
4806	*o*-phenol sulfonic acid	(1;4,2)	$HOC_6H_3(NO_2)SO_3H\cdot$ $3H_2O$	XI-237	273.22
4807	*p*-phenol sulfonic acid	(1;2,4)	$HOC_6H_3(NO_2)SO_3H\cdot$ $3H_2O$	XI-246	273.22
4808	phenylacetate (*o*)	....	$NO_2\cdot C_6H_4\cdot O_2CCH_3$	VI-219	181.15
4809	phenylacetic acid (*p*)	*p*-NO$_2$-α-toluic acid	$NO_2\cdot C_6H_4\cdot CH_2CO_2H$	IX-455	181.15
4810	phenylarsonic acid	(*m*)	$NO_2\cdot C_6H_4\cdot AsO(OH)_2$	XVI-869	247.04
4811	phenyl *iso*-cyanate	(*p*)	$NO_2\cdot C_6H_4\cdot N:CO$	XII-725	164.12
4812	phenylglycine (*p*)	....	$NO_2\cdot C_6H_4\cdot NH\cdot CH_2\cdot$ CO_2H	XII-725	196.16
4813	phenylhydrazine (*m*)	....	$NO_2\cdot C_6H_4\cdot NH\cdot NH_2$	XV-460	153.14
4814	phenylhydrazine (*m*)	HCl salt	$NO_2\cdot C_6H_4\cdot N_2H_3\cdot HCl$	XV-460	189.60
4815	phenylhydrazine (*p*)	....	$NO_2\cdot C_6H_4\cdot NH\cdot NH_2$	XV-468	153.14
4816	phenylhydrazine (*p*)	HCl salt	$NO_2\cdot C_6H_4\cdot N_2H_3\cdot HCl$	XV-468	189.60
4817	phenylpropiolic acid	(*o*)	$NO_2\cdot C_6H_4\cdot C:C\cdot CO_2H$	IX-636	191.14
4818	phenylpropiolic acid	(*p*)	$NO_2\cdot C_6H_4\cdot C:C\cdot CO_2H$	IX-637	191.14
4819	*o*-phthalic acid (3)	....	$NO_2\cdot C_6H_3(CO_2H)_2$	IX-823	211.13
4820	*o*-phthalic acid (4)	....	$NO_2\cdot C_6H_3(CO_2H)_2$	IX-828	211.13

Nitro-orthanilic acid 4622   Nitro-phenylacetonitrile 4685-7

Table 7-4 (*Continued*)
PHYSICAL CONSTANTS OF ORGANIC COMPOUNDS

No.	Crystalline Form and Color	Specific Gravity	Melting Point °C.	Boiling Point °C.	Solubility in 100 Parts		
					Water	Alcohol	Ether
4784	lt. yel. lq.	$0.9231^{7°}$	....	215-8 d.	....	....	....
4785	lq.	$0.915^{18}_{0}°$	....	$113-4^{25mm}$	....	....	....
4786	lt. yel. lq.	$0.935^{20°}$	....	206-10 sl. d.	....	....	....
4787	lq.	....	....	$100-5^{20mm}$	....	....	....
4788	lq.	$0.921^{20}_{0}°$	-18	$201-2^{755mm}$	s. alk.	....	....
4789	nd./aq.	....	57-8	d.	v. s.	s.	s.
4790	col. lq.	$0.948^{20°}$	....	172-3	....	....	....
4791	lq.	$0.958^{0°}$	....	$152-5^{746mm}$	i.	s.	s.
4792	lq.	$0.97^{0°}$	....	$149-51^{748mm}$	....	....	....
4793	lq.	$0.960^{20.6}_{4}°$	....	164^{756mm}	....	....	....
4794	yel. lf./ac.	....	257-8	....	....	v. sl. s.	sl. s. ac.
4795	red pr./al.	....	113	....	....	s. h.	s.; s. chl.
4796	oil	$1.190^{15°}$	5-6	275	v. sl. s.	v. s.	v. s.
4797	yel. cr.	....	34	169^{70mm}	v. sl. s.	v. s. h.	v. s.
4798	col. mn.	$1.18^{15°}$	59-60	283^{758mm}	v. sl. s.	v. s. h.	v. s.
4799	yel. mn.	$1.295^{45°}$	45.0	217.2	$0.21^{20°}$	v. s.	v. s.; s. bz.
4800	red lf./al.	....	....	....	v. s.	v. sl. s. NaOH	....
4801	or. nd./al.	$1.682^{20°}$	....	....	$16^{6°}$	$21^{15°}$ aq.	....
4802	col. mn.	$1.485^{20°}$	96-7	194^{70mm}	$1.35^{20°}$	v. s.	s.; i. pet.
4803	yel. pr.	$1.48^{20°}$	114.0	subl.	$1.6^{25°}$	v. s.	v. s.
4804	or. cr./aq.	....	$-2H_2O$, 110	....	6.5 c.	....	....
4805	yel. mn./aq.	....	....	....	s.	....	....
4806	nd.	....	d. 110	....	v. s.	v. s.	sl. s.
4807	nd./aq.	....	51.5; 141-2 (anh.)	....	v. s.	v. s.	v. s. chl.
4808	col. nd./lg.	....	39-40	253 d.	v. s. bz.	v. s.	v. s.
4809	nd./aq.	....	152-3	....	v. sl. s. c.	s.	s.; s. chl.
4810	lf./aq.	....	d.	....	$2^{18°}$	sl. s. chl.	i. lg.
4811	nd.	....	56-7	....	s. h. lg.	s. chl.	s.; s. bz.
4812	yel. cr./aq.	....	225-30 d.	....	v. sl. s.	v. s. h.	sl. s.
4813	yel. nd./al.	....	93	....	v. sl. s. h.	v. sl. s. bz.	s. ac.
4814	yel. pl.	....	....	....	v. sl. s. c.	v. sl. s. c.	sl. s. HCl
4815	or. red nd.	....	157	....	s. h.	v. s.	s.
4816	or. red pl.	....	214-5 d.	....	s.	i.	i.
4817	nd./aq.	....	157 d.	expl. 155-6	s. h.	s.; i. CS_2	v. sl. s. chl.
4818	nd./al.	....	181-93 d.	....	sl. s.	s. h.	s.; i. pet.
4819	lt. yel./aq.	....	222*	....	$2^{25°}$	v. s. h.	sl. s.
4820	lt. yel. cr.	....	164-5	....	v. s.	v. s.	s.

* Sealed tube.

Table 7-4 (*Continued*)
PHYSICAL CONSTANTS OF ORGANIC COMPOUNDS

No.	Name	Synonym	Formula	Beil. Ref.	Formula Weight
4821	**Nitro** *m*-phthalic acid (5)	....	$NO_2 \cdot C_6H_3(CO_2H)_2$	IX-840	211.13
4822	*p*-phthalic acid (2)	....	$NO_2 \cdot C_6H_3(CO_2H)_2$	IX-851	211.13
4823	phthalic anhydride	(3)	$NO_2 \cdot C_6H_3(CO)_2O$	XVII-486	193.12
4824	phthalide (6)	....	$NO_2 \cdot C_6H_3 \cdot CH_2 \cdot O \cdot CO$	XVII-313	179.13
4825	phthalimide (3)	....	$NO_2 \cdot C_6H_3(CO)_2NH$	XXI-505	192.13
4826	phthalimide (4)	....	$NO_2 \cdot C_6H_3(CO)_2NH$	XXI-506	192.13
4827	propane (1)	....	$CH_3 \cdot CH_2 \cdot CH_2 \cdot NO_2$	I-115	89.09
4828	propane (2)	*β*-nitropropane	$(CH_3)_2CH \cdot NO_2$	I-116	89.09
4829	propene (3,1)	nitroallyl	$CH_2:CH \cdot CH_2 \cdot NO_2$	I-203	87.08
4830	quinaldine (6)	6-NO₂-2-Me-quinoline	$NO_2 \cdot C_9H_5N \cdot CH_3$	XX-394	188.19
4831	quinoline (5)(*ana*)	....	$NO_2 \cdot C_9H_6N$	XX-371	174.16
4832	quinoline (6)(*p*)	....	$NO_2 \cdot C_9H_6N$	XX-372	174.16
4833	quinoline (7)(*m*)	....	$NO_2 \cdot C_9H_6N$	XX-372	174.16
4834	quinoline (8)(*o*)	....	$NO_2 \cdot C_9H_6N$	XX-373	174.16
4835	*iso*-quinoline (5 or 8)	....	$NO_2 \cdot C_6H_3 : C_3H_3N$	XX-386	174.16
4836	resorcinol (2)	(2;1,3)	$NO_2 \cdot C_6H_3(OH)_2$	VI-823	155.11
4837	salol (*α*)	5-nitrosalol	$NO_2 \cdot C_6H_3(OH) \cdot CO_2 \cdot C_6H_5$	X-118	259.22
4838	styrene (*o*)	....	$NO_2 \cdot C_6H_4 \cdot CH:CH_2$	V-478	149.15
4839	styrene (*m*)	....	$NO_2 \cdot C_6H_4 \cdot CH:CH_2$	V-478	149.15
4840	styrene (*p*)	....	$NO_2 \cdot C_6H_4 \cdot CH:CH_2$	V-478	149.15
4841	styrene (*ω*)(*β*)	....	$C_6H_5 \cdot CH:CH \cdot NO_2$	V-478	149.15
4842	tartaric acid	....	$(NO_2 \cdot O \cdot CH \cdot CO_2H)_2$	III-509	240.08
4843	thiophene (2)	....	$NO_2 \cdot C_4H_3S$	XVII-35	129.14
4844	toluene† (*o*)	....	$CH_3 \cdot C_6H_4 \cdot NO_2$	V-318	137.14
4845	toluene (*m*)	....	$CH_3 \cdot C_6H_4 \cdot NO_2$	V-321	137.14
4846	toluene (*p*)	....	$CH_3 \cdot C_6H_4 \cdot NO_2$	V-323	137.14
4847	*o*-toluene sulfonic acid	(1;4,2)	$CH_3 \cdot C_6H_3(NO_2) \cdot SO_3H \cdot 2H_2O$	XI-90	253.23
4848	*p*-toluene sulfonyl chloride	(1;2,4)	$CH_3 \cdot C_6H_3(NO_2) \cdot SO_2Cl$	XI-111	235.65
4849	*o*-toluidine (3;1,2)	....	$NO_2 \cdot C_6H_3(CH_3)NH_2$	XII-843	152.15
4850	*o*-toluidine (4;1,2)	....	$NO_2 \cdot C_6H_3(CH_3)NH_2$	XII-844	152.15
4851	*o*-toluidine (5;1,2)	....	$NO_2 \cdot C_6H_3(CH_3)NH_2$	XII-846	152.15
4852	*o*-toluidine (6;1,2)	....	$NO_2 \cdot C_6H_3(CH_3)NH_2$	XII-848	152.15
4853	*m*-toluidine (2;1,3)	....	$NO_2 \cdot C_6H_3(CH_3)NH_2$	XII-876	152.15
4854	*m*-toluidine (4;1,3)	....	$NO_2 \cdot C_6H_3(CH_3)NH_2$	XII-876	152.15
4855	*m*-toluidine (5;1,3)	....	$NO_2 \cdot C_6H_3(CH_3)NH_2$	XII-877	152.15
4856	*m*-toluidine (6;1,3)	....	$NO_2 \cdot C_6H_3(CH_3)NH_2$	XII-877	152.15
4857	*p*-toluidine (2;1,4)	....	$NO_2 \cdot C_6H_3(CH_3)NH_2$	XII-996	152.15
4858	*p*-toluidine (3;1,4)	....	$NO_2 \cdot C_6H_3(CH_3)NH_2$	XII-1000	152.15
4859	trimethylol-methane	2-NO₂-2-methylol-1,3-propandiol	$NO_2 \cdot C(CH_2OH)_3$	I-520	151.12
4860	tyrosine (3)(*l*)	....	$NO_2 \cdot C_6H_3(OH) \cdot CH_2 \cdot CH(NH_2) \cdot CO_2H$	XIV-620	226.19
4861	undecane (1)	....	$CH_3(CH_2)_9CH_2 \cdot NO_2$	I-170	201.31
4862	uracil (5)	....	$NO_2 \cdot C:CH \cdot (NHCO)_2$	XXIV-320	157.09

† See also No. 5213.
Nitro-propandial 4762
Nitro-*iso*-propyl-methyl-benzene 4714
Nitro-pyromucic acid 4741

Nitro-salicylic acid 4753-6
Nitro-sulfanilic acid 4621
Nitro-toluic acid (*α*) 4809
Nitro-toluic nitrile (*α*) 4685-7

Table 7-4 (*Continued*)
PHYSICAL CONSTANTS OF ORGANIC COMPOUNDS

No.	Crystalline Form and Color	Specific Gravity	Melting Point °C.	Boiling Point °C.	Solubility in 100 Parts		
					Water	Alcohol	Ether
4821	col. cr.	....	255 sl. d.	....	$0.15^{15°}$; $81^{99°}$	v. s.	v. s.
4822	nd./h. aq.	....	263-70	....	s. h.	s. h.	....
4823	col. nd./ac.	....	162-3	....	s. act.	s. h.	v. sl. s. bz.
4824	nd.	....	141	....	i. c.; v. s. h. chl.	sl. s. c.	sl. s.; i. alk. carb.
4825	yel. lf./al.	....	215-6	subl.	i.; s. ac.	s. h.	i.; i. lg.
4826	yel. lf.	....	199-201	subl.	v. sl. s. h.	s.; s. ac.	s. act.
4827	oil	$1.003\frac{20°}{20}$	-108	131.6	$1.4^{20°}$	∞	∞
4828	lq.	$1.024^{0°}$	-93	120.3	$1.7^{20°}$	$∞^{20°}$	$∞^{20°}$
4829	col. lq.	$1.051^{21°}$	....	125-30	i.	s.	s.
4830	yel. nd./aq.	....	163-4	....	v. sl. s.	s.	i.
4831	nd./aq.	....	72	subl.	v. sl. s. h.	s. h.	s. bz.
4832	nd.	....	149-50	subl.	sl. s. c.	sl. s. c.	sl. s.; s. bz.
4833	nd./al.	....	132-3	....	....	v. sl. s. c.	s.
4834	mn./al.	....	88-9	....	s. h.	s.	s.; s. bz.
4835	nd./aq.	....	110	subl.	s. h.	s.	s.; s. bz.
4836	or./aq. al.	....	83-5	d. > 180°	....	....	....
4837	nd./al.	....	151-2	....	....	s.	s. ac.
4838	oil	....	12-13.5	....	s. H_2SO_4	....	....
4839	yel. oil	....	-5	....	v. s. lg.	v. s. abs.	v. s.
4840	pr./lg.	....	29	d.	v. sl. s. c. lg.	s. h.	v. s. h.
4841	yel. pr./al.	....	58	250-60 d.	v. sl. s. h.	s.	v. s.
4842	silky nd.	....	d.	....	d. 0°	v. s. c.; d. h.	v. s.; i. bz.
4843	mn./al.	....	46	224-5	i. alk.	s.	s.
4844	yel. lq.	$1.163\frac{20°}{4}$	(α) -9.3 (β) -3.2	221.7	$0.065^{30°}$; s. pet.	∞; ∞ bz.; s. SO_2	∞; sl. s. NH_3
4845	lq.	$1.157\frac{20°}{4}$	16.1	232.6	$0.050^{30°}$	$8.6^{15°}$	∞
4846	rhb.	$1.123^{55°}$	51.7	238.5	$0.004^{15°}$	s.; s. bz.	$80.8^{15°}$
4847	pl./aq.	....	130 (anh.)	....	$47.7^{23°}$	v. s.	v. s.; s. chl.
4848	pl./et.	....	33-4	....	d.	d.	....
4849	or./45% al.	....	97	....	sl. s.	s.; s. chl.	s.; s. bz.
4850	yel. mn.	$1.365^{15°}$	105-7	....	v. sl. s.	s.	s.
4851	yel. mn.	$1.366^{15°}$	129-30	....	v. sl. s. h.	s.	....
4852	yel. rhb.	$1.378^{15°}$	91-2	305 d.	1.3 h.	v. s.	v. s.; v. s. bz.
4853	yel. nd.	....	53	....	sl. s. c.	s.	....
4854	yel. lf./aq.	....	109-10	....	s. bz.	s.	s.; s. chl.
4855	yel. brn. nd.	....	98	....	sl. s. c.	s.	v. s.; s. bz.
4856	yel. nd./aq.	....	135-8	....	s. h.	s.	s.
4857	yel. nd./aq.	....	81.5	....	sl. s. h.	v. s. h.	s.
4858	red mn.	$1.312^{17°}$	116-7	....	sl. s. h.	s.	....
4859	nd. or pr.	....	165-70 d.	....	$220^{20°}$	$45^{20°}$	$1^{20°}$
4860	pa. yel. nd./aq.	....	237	....	v. sl. s. h.	i.; s. alk.	i.; s. aq. a.
4861	yel. lq.	$0.900^{15°}$	....	d.	....	v. s.	v. s.
4862	col. nd.	....	expl.	....	sl. s. c.	s.	....

Nitroso-amyl ketone 79
Nitroso-anthron 4626
Nitroso-barbituric acid 6445
Nitroso-benzylaniline 828

Nitroso-diethylamine 2191
Nitroso-dimethylamine 2513
Nitroso-diphenylamine 2733
Nitroso-dipropylamine 2790-1

Table 7-4 (*Continued*)
PHYSICAL CONSTANTS OF ORGANIC COMPOUNDS

No.	Name	Synonym	Formula	Beil. Ref.	Formula Weight
4863	**Nitro** urea	....	$NO_2 \cdot NH \cdot CO \cdot NH_2$	III-125	105.05
4864	urethane	....	$NO_2 \cdot NH \cdot CO_2C_2H_5$	III-125	134.09
4865	*o*-xylene (3;1,2)	....	$NO_2 \cdot C_6H_3(CH_3)_2$	V-367	151.17
4866	*o*-xylene (4;1,2)	....	$NO_2 \cdot C_6H_3(CH_3)_2$	V-368	151.17
4867	*m*-xylene (2;1,3)	....	$NO_2 \cdot C_6H_3(CH_3)_2$	V-378	151.17
4868	*m*-xylene (4;1,3)	....	$NO_2 \cdot C_6H_3(CH_3)_2$	V-378	151.17
4869	*m*-xylene (5;1,3)	....	$NO_2 \cdot C_6H_3(CH_3)_2$	V-378	151.17
4870	*p*-xylene (2;1,4)	....	$NO_2 \cdot C_6H_3(CH_3)_2$	V-387	151.17
4871	*m*-xylidine (5), aceto-	(5;1,3,4)	$NO_2 \cdot C_6H_2(CH_3)_2 \cdot NH \cdot COCH_3$	XII-1128	208.22
4872	**Nitron**	4,5-dihydro-1,4-diPh-3,5-phenyl-imino-1,2,4-tri-azole	$C_{20}H_{16}N_4$	XXVI-349	312.38
4873	**Nitroso**-acetone (*iso*)	pyroracemic aldoxime	$CH_3 \cdot CO \cdot CH:NOH$	I-763	87.08
4874	acetophenone (*iso*)	Ph-glyoxaloxime	$C_6H_5 \cdot CO \cdot CH:NOH$	VII-671	149.15
4875	aniline (*p*)	*p*-quinone-imide-oxime tautomer	$ON \cdot C_6H_4 \cdot NH_2$ or $HN:C_6H_4:NOH$	VII-625	122.13
4876	benzene	....	$C_6H_5 \cdot NO$	V-230	107.11
4877	benzoic acid (*o*)	....	$ON \cdot C_6H_4 \cdot CO_2H$	IX-368	151.12
4878	benzoic acid (*m*)	....	$ON \cdot C_6H_4 \cdot CO_2H$	IX-369	151.12
4879	benzoic acid (*p*)	....	$ON \cdot C_6H_4 \cdot CO_2H$	IX-369	151.12
4880	*n*-butyric acid (*iso*)	(α)	$C_2H_5 \cdot C(:NOH) \cdot CO_2H$	III-629	117.11
4881	*m*-cresol (4)	(4;3,1)	$ON \cdot C_6H_3(CH_3)OH$	VII-648	137.14
4882	diethylaniline (*p*)	....	$ON \cdot C_6H_4 \cdot N(C_2H_5)_2$	XII-684	178.24
4883	dimethylaniline (*p*)	....	$ON \cdot C_6H_4 \cdot N(CH_3)_2$	XII-677	150.18
4884	diphenylamine (*p*)	*p*-NO-Ph aniline	$ON \cdot C_6H_4 \cdot NH \cdot C_6H_5$	XII-207	198.23
4885	ethylaniline (*N*)	Et-Ph-nitrosoamine	$C_6H_5 \cdot N(NO)C_2H_5$	XII-580	150.18
4886	*N*-ethyl-urethane	*N*-NO-*N*-Et car-bamate	$C_2H_5N(NO) \cdot CO_2C_2H_5$	IV-129	146.15
4887	methylaniline (*N*)	....	$C_6H_5 \cdot N(NO)CH_3$	XII-579	136.15
4888	methyl-ethyl ketone (*iso*)	diacetyl monoxime	$CH_3COC(:NOH) \cdot CH_3$	I-772	101.11
4889	methyl-*n*-hexyl ketone	(*iso*)	$CH_3 \cdot CO \cdot C(:NOH) \cdot (CH_2)_4CH_3$	I-795	157.21
4890	methyl-propyl ketone (*iso*)	....	$CH_3 \cdot CO \cdot C(:NOH) \cdot C_2H_5$	I-776	115.13
4891	*N*-methyl-urethane	....	$CH_3N(NO) \cdot CO_2C_2H_5$	IV-85	132.12
4892	α-naphthol (1,4)	naphthoquinone oxime tautomer	$HO \cdot C_{10}H_6 \cdot NO$ or $O:]C_{10}H_6:NOH$	VII-727	173.17
4893	α-naphthol (1,2)	2-nitroso-naphthol-1	$HO \cdot C_{10}H_6 \cdot NO$	VII-715	173.17
4894	β-naphthol (2,1)	1-nitroso-naphthol-2	$HO \cdot C_{10}H_6 \cdot NO$	VII-712	173.17
4895	β-naphthylamine	(2,1)	$ON \cdot C_{10}H_6 \cdot NH_2$	VII-717	172.19
4896	nitrobenzene (*o*)	....	$ON \cdot C_6H_4 \cdot NO_2$	V-256	152.11
4897	phenylhydrazine (α)	....	$C_6H_5 \cdot N(NO) \cdot NH_2$	XV-416	137.14
4898	piperidine (*N*)	....	$C_5H_{10}N \cdot NO$	XX-83	114.15
4899	propiophenone (α)	(*iso*); methyl-benzoyl ketoxime (β)	$C_6H_5 \cdot CO \cdot C(:NOH) \cdot CH_3$	VII-677	163.18
4900	resorcinol (4)	2-OH-*p*-benzoqui-none-1-oxime tautomer	$ON \cdot C_6H_3(OH)_2$ or $O:C_6H_3(:NOH)OH$	VIII-235	139.11

Table 7-4 (Continued)
PHYSICAL CONSTANTS OF ORGANIC COMPOUNDS

No.	Crystalline Form and Color	Specific Gravity	Melting Point °C.	Boiling Point °C.	Solubility in 100 Parts		
					Water	Alcohol	Ether
4863	wh. pd./aq.	....	158-9 d.	....	s. h.	s.; sl. s. bz	v. sl. s.
4864	lf./lg.	....	64	....	s.	v. s.	v. s.
4865	yel. oil	$1.147^{15°}$	15	240-5	....	....	....
4866	yel. pr./al.	$1.139\frac{30°}{30}$	29-30	258 sl. d.	i.	$\infty^{30°}$	v. s.
4867	lq.	$1.112^{15°}$	13-5	225^{744mm}	....	....	....
4868	yel. lq.	$1.135^{15°}$	2	244	i.	s.	s.
4869	col. nd./al.	....	74-5	273^{739mm}	....	....	....
4870	yel. lq.	$1.132^{15°}$	....	238.9^{739mm}	....	....	....
4871	yel. nd./aq.	....	172-3	....	....	....	....
4872	yel. lf./al.	....	189-90 d.	....	i.; s. act.	s. h.; s. bz.	v. sl. s.; s. chl.
4873	lf./et.	$1.074^{67.5°}$	69	subl.	v. s.	v. sl. s. pet.	v. s.
4874	mn./aq. al.	....	126-8	....	s. h.	s. alk.	....
4875	b. nd./bz.	....	173-4	....	s.	s. bz.	....
4876	col. rhb.*	....	67.5-8.0	$57-9^{18mm}$	i.; i.NH_3	s.	sl. s. lg.
4877	cr./al.	....	210 d.	....	s. ac.	s. h.	v. sl. s.
4878	col. cr.	....	d. 230	....	....	....	....
4879	yel. pd.	....	d. 250	....	sl. s. bz.	sl. s. h.	sl. s. ac.
4880	nd./aq.	....	169-70	....	sl. s.	s.	sl. s.
4881	nd./aq.	....	158-60	....	sl. s. h.	s.; s. bz.	v. sl. s.
4882	gn. mn.	$1.24^{15°}$	84	....	v. sl. s.	s.	s.
4883	gn. tri.	....	86-7	....	i.	s.	s.
4884	gn. pl./bz.	....	144.6	....	sl. s.; v. s. chl.	v. s.; sl. s. lg.	v. s.; s. bz.
4885	yel. oil	$1.087\frac{20°}{4}$	....	$119-20^{15mm}$	i.	....	....
4886	red oil	$1.071\frac{20°}{4}$	....	86^{36mm}	....	....	....
4887	yel. oil	$1.124\frac{20°}{4}$	14-5	128^{19mm}	....	s.	s.
4888	pr./chl.	....	76	185-6	sl. s.	v. s.; s. alk.	v. s.; s. chl.
4889	cr./lg.	....	58-9	133^{11mm}	....	....	v. s.
4890	lf./lg.	....	58-9	183-7 sl. d.	sl. s. c.	v. s.; v. s. chl.	v. s.
4891	yel. red lq.	$1.122\frac{20°}{4}$	<-20	65^{13mm}	sl. s. h.	∞	∞; ∞ bz.
4892	nd./aq. al.	....	193-4	....	i.	v. s.	v. s.
4893	yel. nd./aq.	....	162-4 d.	....	s. h.	v. s.	sl. s.
4894	brn. pr./al.	....	109.5	....	$0.1^{20°}$	$2.4^{13°}$	v. sl. s. pet.
4895	gn./aq. al.	....	150-2	....	sl. s. h.	v. s. h.	s. h.
4896	yel. cr./act.	....	126**	....	i.; i. pet.	s. h.	v. s. chl.
4897	lt. yel. lf.	....	51	....	s.	poisonous	s.
4898	lt. yel. oil	$1.063\frac{18.5°}{4}$	....	217-8	s.	v. s. aq. a.	....
4899	nd./aq.	....	113-4	....	s. alk.	....	....
4900	yel. nd. + 1H_2O/ aq.	....	$-H_2O$, 105; d. 148	....	s.; s. chl.	v. s.; s. ac.	s.; i. bz.; i. CS_2

* Green as liquid or in solution.
** Green when liquid.
Nonene 4930
Nonyl cyanide 1207
Nonylic acid 5042

Nor-arecaidine, cf. alkd.
Nosophen 5813
Nostal 5390
Novacetyl 118
Novarsenobenzol 5593

Novatophan 3130
Novonal 2107
Nucin 3787
Numal 5370
Nupercaine 1025

7-311

Table 7-4 (*Continued*)
PHYSICAL CONSTANTS OF ORGANIC COMPOUNDS

No.	Name	Synonym	Formula	Beil. Ref.	Formula Weight
4901	**Nitroso** R salt	(1;3,6;2)	$ON \cdot C_{10}H_4(SO_3Na)_2 \cdot OH$	....	377.26
4902	thymol (*p*)	(quinoneoxime tautomer)	$ON \cdot C_{10}H_{12} \cdot OH$ or $O:C_{10}H_{12}:NOH$	VII-664	179.22
4903	toluene (*o*)	....	$CH_3 \cdot C_6H_4 \cdot NO$	V-317	121.14
4904	toluene (*m*)	....	$CH_3 \cdot C_6H_4 \cdot NO$	V-318	121.14
4905	toluene (*p*)	....	$CH_3 \cdot C_6H_4 \cdot NO$	V-318	121.14
4906	triacetonamine (*N*)	....	$C_9H_{16}ON \cdot NO$	XXI-251	184.24
4907	**Nonacosane**	....	$C_{29}H_{60}$	I-176	408.80
4908	**Nonadecane** (*n*)	....	$CH_3 \cdot (CH_2)_{17} \cdot CH_3$	I-174	268.53
4909	**Nonadecylic acid**	....	$CH_3 \cdot (CH_2)_{17} \cdot CO_2H$	II-389	298.51
4910	**Nonane** (*n*)	....	$CH_3 \cdot (CH_2)_7 \cdot CH_3$	I-165	128.26
4911	**Nonane** (*iso*)	2-methyl-octane	$(CH_3)_2CH \cdot C_6H_{13}$	I-166	128.26
4912	**Nonane**	2,4-diMe-heptane	C_9H_{20}	*I-64	128.26
4913	**Nonane**	2,5-diMe-heptane	C_9H_{20}	I-167	128.26
4914	**Nonane**	2,6-diMe-heptane	$[(CH_3)_2CH \cdot CH_2]_2CH_2$	I-167	128.26
4915	**Nonane**	4-ethyl-heptane	$C_2H_5 \cdot CH(CH_2C_2H_5)_2$	I-167	128.26
4916	**Nonane**	3-methyl-octane	$(C_2H_5)(CH_3)CH \cdot (CH_2)_4CH_3$	I-166	128.26
4917	**Nonane**	4-methyl-octane	C_9H_{20}	*I-63	128.26
4918	**Nonyl** alcohol (*n*)	nonanol-1	$CH_3(CH_2)_7CH_2OH$	I-423	144.26
4919	alcohol (*n*)	nonanol-2	$C_7H_{15} \cdot CHOH \cdot CH_3$	I-423	144.26
4920	alcohol (*n*)	nonanol-3	$C_6H_{13} \cdot CHOH \cdot C_2H_5$	I-424	144.26
4921	alcohol (*n*)	nonanol-5	$(C_4H_9)_2CHOH$	I-424	144.26
4922	alcohol	2-methyl-octanol-2	$(CH_3)_2C(OH)C_6H_{13}$	I-424	144.26
4923	alcohol	4-ethyl-heptanol-4	$C_2H_5 \cdot C(OH):(CH_2 \cdot CH_2 \cdot CH_3)_2$	I-424	144.26
4924	alcohol	2-Me-4-Et-hexanol-4	$(C_2H_5)_2C(OH) \cdot CH_2 \cdot CH(CH_3)_2$	I-425	144.26
4925	alcohol	2,6-diMe-heptanol-4	$[(CH_3)_2CH \cdot CH_2]_2:CHOH$	I-425	144.26
4926	amine (*n*)	1-amino-nonane	$CH_3(CH_2)_7CH_2NH_2$	IV-198	143.27
4927	bromide (2)	2-bromo-nonane	$C_7H_{15} \cdot CHBr \cdot CH_3$	I-166	207.16
4928	chloride (2)	2-chloro-nonane	$C_7H_{15} \cdot CHCl \cdot CH_3$	I-166	162.70
4929	iodide (*n*)	1-iodo-nonane	$CH_3(CH_2)_7 \cdot CH_2I$	I-166	254.16
4930	**Nonylene** (*β*)	nonene-2	$C_6H_{13} \cdot CH:CH \cdot CH_3$	I-223	126.24
4931	**Norbixin** (*trans*)	(stable)(*β*)	$C_{24}H_{28}O_4$	XXX-109	380.49
4932	**Norbixin** (*cis*)	(labile)	$C_{24}H_{28}O_4$	XXX-110	380.49
4933	methyl ester (*cis*)	(labile)-bixin	$C_{25}H_{30}O_4$	XXX-112	394.52
4934	**Novocain**	ethocaine; procaine HCl	$C_{13}H_{20}O_2N_2 \cdot HCl$	XIV-424	272.78
4935	base	procaine	$NH_2 \cdot C_6H_4CO_2(CH_2)_2 \cdot N(C_2H_5)_2$	XIV-424	236.32
4936	**Octa**-chloropropane	perchloro-propane	C_3Cl_8	I-108	319.66
4937	cosane	....	$C_{28}H_{58}$	I-176	394.77
4938	decane	*n*-octadecane	$CH_3 \cdot (CH_2)_{16} \cdot CH_3$	I-173	254.50
4939	decyl alcohol (*n*)	octadecanol-1	$CH_3(CH_2)_{16}CH_2OH$	I-431	270.50
4940	decyl iodide (*n*)	1-iodo-octadecane	$CH_3(CH_2)_{16}CH_2I$	I-173	380.40
4941	decylene (*α*)	octadecene-1	$CH_3(CH_2)_{15}CH:CH_2$	I-226	252.49
4942	decyne-1	hexadecyl acetylene	$CH_3(CH_2)_{15}C:CH$	I-262	250.47
4943	**Octandiol** (1,8)	....	$HO \cdot (CH_2)_8 \cdot OH$	I-490	146.23
4944	**Octane** (*n*)	....	$CH_3 \cdot (CH_2)_6 \cdot CH_3$	I-159	114.23
4945	**Octane**	2,2,3,3-tetramethyl-butane	$[(CH_3)_3C]_2$	I-165	114.23

NW acid 4502
Nyctanthin 1555
Octa-decadienoic acid 3976
Octa-decanal 5633
Octa-decanoic acid 5631

Octa-decanol 4939
Octa-decanoyl chloride 5637
Octa-decatrienoic acid 3977
Octa-decene 4941
Octa-decynoic acid 5639

Octa-diene 1530
Octa-methylene 1617
Octa-methylene dicyanide **5606**
Octaldehyde 1222
Octandial 5650

Table 7-4 (*Continued*)
PHYSICAL CONSTANTS OF ORGANIC COMPOUNDS

No.	Crystalline Form and Color	Specific Gravity	Melting Point °C.	Boiling Point °C.	Solubility in 100 Parts		
					Water	Alcohol	Ether
4901	yel. cr.	....	....	....	2.5	sl. s.	....
4902	lt. yel. nd./chl.	....	160-4 sl. d.	....	v. sl. s. h.; s. alk.	s.	s.; s. chl.
4903	nd. or pr.	....	72.5	....	v. s. chl.	v. s.	v. s.
4904	nd.	....	53.5	....	i.	....	s.
4905	col.nd./lg.*	....	48.5	....	v. sl. s.	v.s Me al.	v. s. bz.
4906	nd./aq. al.	1.14	72-3	subl.	....	v. s.	v. s.
4907	cr.	lq. $0.780^{63.8°}$	63.8	346-8^{40mm}	$1.1^{15°}$ chl.	v. s. h.	v. s.
4908	cr.	$0.777^{32°}_{4}$	32	330	i.	sl. s.	s.
4909	lf./al.	....	66.5	297-8^{100mm}	l.	....	....
4910	col. lq.	$0.718^{20°}_{4}$	-53.6	150.8	i.	s. abs.	s.
4911	lq.	$0.713^{20°}_{4}$	-80.4	143.3	....	....	....
4912	col. lq.	$0.715^{20°}_{4}$	....	132.9	i.	∞	∞
4913	col. lq.	$0.717^{20°}_{4}$	....	136	....	....	....
4914	lq.	$0.709^{20°}_{4}$	-102.9	135.2	i.	....	s.
4915	lq.	$0.728^{20°}_{4}$	....	141.2	i.	....	s.
4916	lq.	$0.725^{20°}_{4}$	-107.6	144.2	i.	....	s.
4917	col. lq.	$0.720^{20°}_{4}$	-113.2	142.4	i.	....	....
4918	lq.	$0.828^{20°}_{4}$	-5	213.5	i.	∞	∞
4919	lq.	$0.823^{20°}_{4}$	-35	193-4	i.	s.	s.
4920	lq.	$0.825^{20°}_{4}$	-22	194.5^{750mm}	i.	v. s.	v. s.
4921	oil	$0.823^{20°}$	....	193^{766mm}	i.	∞	∞
4922	lq.	$0.823^{19°}_{19}$	....	178	i.	s.	s.
4923	lq.	$0.835^{20°}$	....	179.5	i.	s.	s.
4924	lq.	$0.840^{22°}$	....	172	i.	s.	s.
4925	lq.	$0.816^{12°}_{4}$	....	172-4^{750mm}	i.	s.	s.
4926	lq.	....	....	201	....	....	....
4927	lq.	$1.081^{20°}$	....	208-9^{767} d.	....	....	....
4928	col. lq.	$0.856^{20°}$	....	190^{764mm}	....	....	....
4929	lq.	$1.287^{16°}_{4}$	....	117^{15mm}	....	....	....
4930	lq.	$0.754^{15°}_{15}$	....	148-9	i.	....	....
4931	b. red cr.	....	>300	....	i.	i.	s. h. pyr.
4932	red nd./ac.	....	254-5	....	i.; s. pyr.	s. alk.	i.
4933	vl. cr./ac.	....	198	....	i.; s. pyr.	sl. s. h.	$0.3^{20°}$ chl.
4934	nd./abs. al.	....	156	....	v. s.	....	i.
4935	cr./lg.	....	59-60**	....	sl. s.	s.	s. chl.
4936	pl.	....	160	268-9^{734mm}	s. lg.	s.	s.
4937	cr.	lq. $0.779^{61.6°}$	61.6	316-8^{40mm}	$2^{15°}$ chl.	....	....
4938	cr.	$0.775^{28°}_{4}$	28.0(27.4)	317	i.	sl. s.	s.; s. act.
4939	lf.	$0.812^{59°}_{4}$	58.5	210.5^{15mm}	i.	s.	....
4940	cr. lf.	....	33.5-4.0	170$^{0.5mm}$	....	v. sl. s. c.	....
4941	cr.	$0.791^{18°}_{4}$	18	179^{15mm}	....	....	....
4942	cr.	$0.796^{30°}$	26	180^{15mm}	....	....	....
4943	nd.	....	63	172^{20mm}	sl. s.	v. s.	sl. s.
4944	col. lq.	$0.703^{20°}_{4}$	-56.8	125.7	$0.002^{16°}$	sl. s.	s.
4945	lf.	....	100.7	106.3	i.	sl. s.	s.

* Green as liquid or in solution.
** Needles $+ 2H_2O$/aq. al., m. p. 51°.
Octandioic acid 5649
Octandione 78
Octanoic acid 1220

Octanol 4961-9
Octanone 2963, 4266
Octanoyl chloride 1225
Octene 4984-90
Octin 4114

Table 7-4 (*Continued*)
PHYSICAL CONSTANTS OF ORGANIC COMPOUNDS

No.	Name	Synonym	Formula	Beil. Ref.	Formula Weight
4946	**Octane**	2,2,3-triMe-pentane	$(CH_3)_3C \cdot CH(CH_3) \cdot C_2H_5$	*I-62	114.23
4947	**Octane** ("iso")	2,2,4-triMe-pentane	$(CH_3)_3C \cdot CH_2 \cdot CH:(CH_3)_2$	**I-127	114.23
4948	**Octane**	2,3,3-triMe-pentane	$C_2H_5 \cdot C(CH_3)_2 \cdot CH:(CH_3)_2$	....	114.23
4949	**Octane**	2,3,4-triMe-pentane	$[(CH_3)_2CH]_2CH \cdot CH_3$	....	114.23
4950	**Octane**	2-Me-3-Et-pentane	$(C_2H_5)_2CH \cdot CH:(CH_3)_2$	I-164	114.23
4951	**Octane**	2,3-dimethyl-hexane	$(CH_3)_2CH \cdot CH(CH_3) \cdot CH_2 \cdot C_2H_5$	*I-62	114.23
4952	**Octane**	2,4-dimethyl-hexane	$C_2H_5 \cdot CH(CH_3) \cdot CH_2 \cdot CH(CH_3)_2$	I-162	114.23
4953	**Octane**	2,5-dimethyl-hexane	$[(CH_3)_2CH \cdot CH_2 \cdot]_2$	I-162	114.23
4954	**Octane**	3,4-dimethyl-hexane	$[C_2H_5 \cdot CH(CH_3) \cdot]_2$	I-163	114.23
4955	**Octane**	3-ethyl-hexane	$(C_2H_5)_2CH \cdot CH_2 \cdot C_2H_5$	*I-62	114.23
4956	**Octane** (*iso*)	2-methyl-heptane	$(CH_3)_2CH(CH_2)_4CH_3$	I-161	114.23
4957	**Octane**	3-methyl-heptane	$C_2H_5 \cdot CH(CH_3) \cdot (CH_2)_3 \cdot CH_3$	I-162	114.23
4958	**Octane**	4-methyl-heptane	$(C_2H_5CH_2)_2CH \cdot CH_3$	I-162	114.23
4959	**Octyl** acetate (*n*)	(*See also* No. 1219)	$CH_3 \cdot CO_2C_8H_{17}$	II-134	172.27
4960	acetate	2-Et-hexyl acetate	$CH_3 \cdot CO_2CH_2 \cdot CH:(C_2H_5)(C_4H_9)$	....	172.27
4961	alcohol (*n*)	octanol-1	$CH_3(CH_2)_6CH_2OH$	I-418	130.23
4962	alcohol (capryl alcohol)	octanol-2	$CH_3(CH_2)_5CHOH \cdot CH_3$	I-419	130.23
4963	alcohol	2-Me-3-Et-pentanol-3	$(C_2H_5)_2C(OH) \cdot CH(CH_3)_2$	I-423	130.23
4964	alcohol	2,2,4-trimethyl-pentanol-4	$(CH_3)_3CCH_2C(OH):(CH_3)_2$	I-423	130.23
4965	alcohol	2-Et-hexanol-1	$C_4H_9(C_2H_5)CH \cdot CH_2OH$	**I-453	130.23
4966	alcohol	3-Et-hexanol-3	$(C_2H_5)_2C(OH) \cdot C_3H_7$	I-421	130.23
4967	alcohol	2-methyl-heptanol-2	$(CH_3)_2C(OH) \cdot CH_2 \cdot (CH_2)_3CH_3$	I-420	130.23
4968	alcohol (*dl*)	3-methyl-heptanol-3	$C_2H_5(CH_3)C(OH) \cdot CH_2 \cdot CH_2 \cdot C_2H_5$	I-421	130.23
4969	alcohol (*dl*)	4-methyl-heptanol-4	$(C_2H_5CH_2)_2C(OH) \cdot CH_3$	I-421	130.23
4970	amine (*n*)	1-amino-octane	$CH_3(CH_2)_7NH_2$	IV-196	129.25
4971	amine (*n*)(*sec*)	2-amino-octane	$C_6H_{13} \cdot CH(NH_2) \cdot CH_3$	IV-196	129.25
4972	amine	2-Et-1-amino-hexane	$C_4H_9(C_2H_5)CH \cdot CH_2 \cdot NH_2$	....	129.25
4973	bromide (*n*)	1-bromo-octane	$CH_3(CH_2)_6CH_2Br$	I-160	193.13
4974	bromide (*n*)(*sec*)	2-bromo-octane	$C_6H_{13} \cdot CHBr \cdot CH_3$	I-160	193.13
4975	chloride (*n*)	1-chloro-octane	$CH_3(CH_2)_6CH_2Cl$	I-159	148.68
4976	chloride (*n*)(*sec*)	2-chloro-octane	$C_6H_{13} \cdot CHCl \cdot CH_3$	I-160	148.68
4977	cinnamate (*n*)(*sec*)	....	$C_6H_5 \cdot CH:CH \cdot CO_2 \cdot CH(CH_3)C_6H_{13}$	*IX-230	260.38
4978	fluoride (*n*)	1-fluoro-octane	$CH_3(CH_2)_6CH_2F$	I-159	132.22
4979	formate (*d*)	β-octyl formate	$HCO_2 \cdot C_8H_{17}$	II-22	158.24
4980	iodide (*n*)	1-iodo-octane	$CH_3(CH_2)_6CH_2I$	I-160	240.13
4981	nitrate (*n*)	....	$CH_3(CH_2)_7O \cdot NO_2$	I-419	175.23
4982	nitrite (*n*)	....	$CH_3(CH_2)_7O \cdot NO$	I-419	159.23

Table 7-4 (*Continued*)
PHYSICAL CONSTANTS OF ORGANIC COMPOUNDS

No.	Crystalline Form and Color	Specific Gravity	Melting Point °C.	Boiling Point °C.	Solubility in 100 Parts		
					Water	Alcohol	Ether
4946	col. lq.	$0.716_4^{20°}$	−112.3	109.8^{760mm}	i.	sl. s.	s.
4947	col. lq.	$0.692_4^{20°}$	−107.4	99.2	i.	sl. s.	s.
4948	col. lq.	$0.726_4^{20°}$	−100.7	114.8	i.	sl. s.	s.
4949	col. lq.	$0.719_4^{20°}$	−109.2	113.5	i.	sl. s.	s.
4950	lq.	$0.719_4^{20°}$	−115.0	115.7	i.	sl. s.	s.
4951	lq.	$0.712_4^{20°}$	....	115.6	i.	sl. s.	s.
4952	lq.	$0.700_4^{20°}$	....	109.4	i.	sl. s.	s.
4953	col. lq.	$0.694_4^{20°}$	−91.2	109.1	i.	sl. s.	s.
4954	lq.	$0.719_4^{20°}$	....	117.7	i.	sl. s.	s.
4955	lq.	$0.714_4^{20°}$	....	118.5	i.	sl. s.	s.
4956	col. lq.	$0.698_4^{20°}$	−109.0	117.7	i.	sl. s.	s.
4957	lq.	$0.706_4^{20°}$	−120.5	118.9	i.	sl. s.	s.
4958	lq.	$0.705_4^{20°}$	−121.0	117.7	i.	sl. s.	s.
4959	col. lq.	$0.885_4^{0°}$	−38.5	210	i.	s.	s.
4960	col. lq.	$0.873_{20}^{20°}$	−93	199	<0.03$^{20°}$	∞	∞
4961	col. lq.	$0.829_4^{20°}$	−16.7	194.5	0.054$^{20°}$	∞	∞ ; ∞ chl.
4962	col. lq.	$0.822_4^{20°}$	−38.6	179–80	0.096$^{25°}$	∞	∞
4963	lq.	$0.830^{20°}$	....	160–1^{750mm}	....	s.	s.
4964	lq.	$0.842^{0°}$	−20	146.5–7.5	i.	sl. s.	s.
4965	col. lq.	$0.834_{20}^{20°}$	< −76	184	0.12$^{0°}$	s.	s.
4966	lq.	$0.838_4^{20°}$	....	160.5	....	∞	∞
4967	lq.	$0.879^{20°}$	....	162	....	....	....
4968	lq.	$0.820^{21°}$	....	162	i.	....	....
4969	lq.	$0.825^{20°}$	....	161.5	i.	....	....
4970	col. lq.	$0.777^{27°}$	....	175–7^{745mm}	v. sl. s.	s.	s.
4971	lq.	$0.771_4^{25°}$	....	164–5	....	....	....
4972	lq.	$0.792_{20}^{20°}$	....	167–8	....	....	....
4973	col. lq.	$1.118^{15°}$	−55	201.5	i.	∞	∞
4974	lq.	$1.099^{22°}$	....	188–9	....	....	....
4975	lq.	$0.892_4^{0°}$	....	182.5–3.5	i.	....	....
4976	lq.	$0.871^{15°}$	....	171–3	i.	....	s.
4977	lq.	$0.972_4^{17°}$	....	240^{60mm}	....	....	....
4978	col. lq.	$0.804^{21°}$	....	142.8	i.	....	....
4979	col. lq.	$0.872^{12.5°}$	−39.1	198.8	i.	....	....
4980	lq.	$1.336_4^{15°}$	−45.7	225.5	....	....	....
4981	lq.	$0.975_4^{9°}$	....	110–2^{20mm}	....	....	....
4982	gn. lq.	$0.862^{17°}$	....	175–7	....	....	....

Table 7-4 (*Continued*)
PHYSICAL CONSTANTS OF ORGANIC COMPOUNDS

No.	Name	Synonym	Formula	Beil. Ref.	Formula Weight
4983	**Octyl** stearate (*n*) (*sec*)	....	$C_{17}H_{35}CO_2 \cdot CH(CH_3) \cdot$ C_6H_{13}	*II-173	396.70
4984	**Octylene** (α)†	octene-1	$CH_3(CH_2)_5CH:CH_2$	I-221	112.22
4985	**Octylene**	2,3,3-trimethyl-pentene-1	$C_2H_5 \cdot C(CH_3)_2 \cdot$ $C(CH_3):CH_2$	....	112.22
4986	**Octylene**	2,4,4-trimethyl-pentene-1	$(CH_3)_3C \cdot CH_2 \cdot$ $C(CH_3):CH_2$	I-222	112.22
4987	**Octylene**	2,3,4-trimethyl-pentene-2	$(CH_3)_2CH \cdot C(CH_3):$ $C(CH_3)_2$	....	112.22
4988	**Octylene**	3,4,4-trimethyl-pentene-2	$(CH_3)_3C \cdot C(CH_3):$ $CH \cdot CH_3$	....	112.22
4989	**Octylene**	3-ethyl-hexene-2	$CH_3 \cdot CH:C(C_2H_5) \cdot$ $CH_2 \cdot C_2H_5$	I-222	112.22
4990	**Octylene**	2-methyl-heptene-2	$(CH_3)_2C:CH \cdot CH_2 \cdot$ $CH_2 \cdot C_2H_5$	I-222	112.22
4991	**Octyne**-1	caprylidene	$CH_3 \cdot (CH_2)_5 \cdot C:CH$	I-258	110.20
4992	**Oleic acid**	....	$CH_3(CH_2)_7CH:CH \cdot$ $(CH_2)_7CO_2H$	II-463	282.47
4993	**Opianic acid**	(5,6;2,1)	$(CH_3O)_2C_6H_2:$ $(CHO)CO_2H$	X-990	210.19
4994	**Orange I**	*p*-sulfobenzene-azo-α-naphthol Na	$NaO_3S \cdot C_6H_4 \cdot N:$ $N \cdot C_{10}H_6 \cdot OH$	*XVI-296	350.33
4995	**Orange II**	*p*-sulfobenzene-azo-β-naphthol Na	$NaO_3S \cdot C_6H_4 \cdot N:$ $N \cdot C_{10}H_6 \cdot OH$	*XVI-296	350.33
4996	**Orange IV**	4'-anilino-azo-ben-zene-4-sulfonic Na	$NaO_3S \cdot C_6H_4 \cdot N:N \cdot$ $C_6H_4 \cdot NH \cdot C_6H_5$	*XVI-319	375.38
4997	**Orcein**	....	$C_{28}H_{24}O_7N_2$	I-886	500.51
4998	**Orcinol**	5-methyl resorcinol	$CH_3 \cdot C_6H_3(OH)_2$	VI-882	124.14
4999	**Orcin-phthalein**	....	$C_{22}H_{16}O_5$	XIX-236	360.37
5000	**Ornithine**	α,δ-diaminovaleric acid	$H_2N \cdot (CH_2)_3 \cdot CH \cdot$ $(NH_2)CO_2H$	IV-420	132.16
5001	**Orsellinic acid** (4,6;2,1)	....	$(HO)_2C_6H_2(CH_3) \cdot$ $CO_2H \cdot aq.$	X-412	168.15
5002	**Ortal sodium**	*n*-hexyl-ethyl-barbituric Na	$C_6H_{13}(C_2H_5):$ $C_4HO_2N_2ONa$	....	262.29
5003	**Orthochrom T**	....	$C_{23}H_{22}N_2 \cdot (C_2H_5I)$	XXIII-310	482.41
5004	**Ostruthin**	....	$C_{19}H_{22}O_3$	....	298.39
5005	**Oxalenediur-amidoxime**	oxaldiureide-dioxime	$[H_2N \cdot CO \cdot NH \cdot C \cdot$ $(:NCH)]_2$	III-65	204.15
5006	**Oxalic** acid	ethandioic acid	$HO_2C \cdot CO_2H$	II-502	90.04
5007	acid	ethandioic acid	$(CO_2H)_2 \cdot 2H_2O$	II-502	126.07
5008	Na salt	sodium oxalate	$(CO_2Na)_2$	II-513	134.00
5009	amide	oxamide	$(CONH_2)_2$	II-545	88.07
5010	anilide	oxanilide	$(CO \cdot NH \cdot C_6H_5)_2$	XII-284	240.26
5011	chloride	oxalyl chloride	$Cl \cdot CO \cdot CO \cdot Cl$	II-542	126.93
5012	imide	oximide	$CO \cdot CO \cdot NH$ $\lfloor_____\rfloor$	XXI-368	71.04
5013	**Oxaluric** acid	....	$NH_2 \cdot CO \cdot NH \cdot CO \cdot$ CO_2H	III-64	132.08
5014	amide	oxalan	$C_3H_5O_3N_3$	III-65	131.09
5015	**Oxamaethane**	ethyl oxamate	$NH_2 \cdot CO \cdot CO_2C_2H_5$	II-544	117.11

† See also di-iso-butylene.
Orange-III 4328
Orange-GS 4996
Orange-N 4996
Orcin (β) 2380

Oreson 3400
Orexin 5182
Orizabin, cf. glcde.
Orthamine 5256
Orthanilic acid 523

Orthoform new 4112
Ortol, mixt. of 2314 and 4116
Ouabain, cf. glcde.
Oxalacetic ester 2192
Oxalacetone 2939

Table 7-4 (*Continued*)
PHYSICAL CONSTANTS OF ORGANIC COMPOUNDS

No.	Crystalline Form and Color	Specific Gravity	Melting Point °C.	Boiling Point °C.	Solubility in 100 Parts		
					Water	Alcohol	Ether
4983	col. cr.	$0.841^{3.8°}_{4}$	34	235^{6mm}	....	....	....
4984	lq.	$0.716^{20°}$	−102.4	121.3	i.	∞	∞
4985	col. lq.	$0.737^{20°}$	−69	108.2	....	....	....
4986	col. lq.	$0.715^{20°}$	−93.5	101.4	....	....	....
4987	col. lq.	$0.743^{20°}$	....	116.3	....	....	....
4988	col. lq.	$0.739^{20°}$	....	112	....	....	....
4989	lq.	$0.737^{20°}$	....	121	i.	....	....
4990	col. lq.	$0.725^{20°}_{0}$	....	122	i.	....	....
4991	lq.	$0.743^{25°}_{4}$	−79	$131-2^{762mm}$	....	....	....
4992	col. nd.	$0.891^{20°}_{4}$	16	$285-6^{100mm}$	i.	∞	∞
4993	nd./aq.	....	150	....	0.25 c.; 1.7 h.	s.	s.
4994	or. red pd.	....	....	....	s.	....	....
4995	or. red pd.	....	....	....	s.	....	....
4996	or. yel. pd.	....	....	....	s.	....	....
4997	brn. cr.	....	....	....	i. chl., CS_2	s.; s. ac.	i.; i. bz.
4998	pr./bz.	$1.290^{4°}$	107-8*	287-90	v. s.	v. s.	v. s.
4999	pr./act.	....	d. >230	....	i.; s. h. ac.	s.; s. alk.	i.; i. bz.
5000	syrup	....	....	....	v. s.	v. s.	sl. s.
5001	cr.	....	176 d. (anh.)	$-H_2O, 100$	s. gly.	v. s.	$22^{20°}$
5002	wh. pd.	....	....	....	v. s.	s.	i.
5003	gn. pr./al.	....	....	....	sl. s. h.	s. h.	....
5004	cr./aq. al.	....	117-9	....	i.; i. pet.	s. h.	s. chl.
5005	nd./aq. al.	....	191-2 d.	....	i. c.; i. bz.; s. a.	s.; i. lg.; s. alk.	i.; i. chl.
5006	col. rhb.	1.90	186-7 d.	subl. > 100	$10^{20°}$; $120^{100°}$	$24^{15°}$ abs.	$1.3^{15°}$ abs.
5007	col. mn.	$1.653^{19°}_{4}$	101.5	$-2H_2O, 100$	....	....	....
5008	cr. pd.	....	....	....	$3.2^{16°}$	i.	6.6 h. aq.
5009	mn.	1.667	417-9 d.	....	$0.04^{7.3°}$; $0.6^{100°}$	i.	i.
5010	pl./bz.	....	251-3	>320	i. h.	sl. s. h.	sl. s. h.
5011	col. lq.	$1.488^{13°}_{4}$	−12	$63-4^{763mm}$	d.	d.	s.
5012	pr.	....	....	....	v. sl. s.; d. h.	sl. s. NH_4OH	....
5013	cr./aq.	....	d. 208-10	....	v. sl. s.; d. h.	i.	i.
5014	cr. ppt.	....	d.	....	i. c.	s. H_2SO_4	s. d. KOH
5015	rhb.	....	114-5	....	s.	v. sl. s. bz.	s.

* Crysts. $+ 1H_2O$/aq., m. p. 56°±.

Oxalan 5014
Oxaldiureide dioxime 5005
Oxalic nitrile 1588
Oxalyl chloride 5011

Oxalyl urea 5036
Oxamide 5009
Oxanilide 5010
Oximide 5012
Oximido-mesoxalyl urea 6445

Oxindole 2686
Oxine 3825
Oxy-, cf. also hydroxy-.
Oxyacanthine, cf. alkd.
Oxycaproic acid 3759

Table 7-4 (*Continued*)
PHYSICAL CONSTANTS OF ORGANIC COMPOUNDS

No.	Name	Synonym	Formula	Beil. Ref.	Formula Weight
5016	**Oxamic** acid	. . . .	$NH_2 \cdot CO \cdot CO_2H$	II-543	89.05
5017	hydrazide	amino-oxamide	$NH_2 \cdot CO \cdot CO \cdot NH \cdot NH_2$	II-559	103.08
5018	**Oxanilic acid**	. . . .	$C_6H_5NH \cdot CO \cdot CO_2H$	XII-281	165.15
5019	**Oxanthranol**	anthrahydroquinone	$C_6H_4(COCHOH)C_6H_4$	VIII-190	210.23
5020	diacetate	. . . .	$C_{14}H_8(O_2C \cdot CH_3)_2$	VI-1034	294.31
5021	**Oxindol** †	. . . .	$C_6H_4 \cdot CH_2 \cdot CO \cdot NH$ |_____|	XXI-282	133.15
5022	**Palmitic** acid	hexadecanoic acid	$CH_3(CH_2)_{14} \cdot CO_2H$	II-370	256.43
5023	aldehyde	hexadecanal	$CH_3(CH_2)_{14} \cdot CHO$	I-717	240.43
5024	aldehyde oxime	. . . .	$C_{15}H_{31} \cdot CH:NOH$	I-717	255.45
5025	amide	. . . .	$C_{15}H_{31} \cdot CO \cdot NH_2$	II-374	255.45
5026	anhydride	. . . .	$(C_{15}H_{31}CO)_2O$	II-374	494.85
5027	chloride	palmityl chloride	$C_{15}H_{31} \cdot CO \cdot Cl$	II-374	274.88
5028	nitrile	pentadecyl cyanide	$C_{15}H_{31} \cdot CN$	II-375	237.43
5029	**Palmitolic acid**	. . . .	$CH_3(CH_2)_7C\vdots C \cdot (CH_2)_5 \cdot CO_2H$	II-494	252.40
5030	**Palmitone**	. . . .	$(C_{15}H_{31})_2CO$	I-719	450.84
5031	**Panthesin**	. . . .	$C_{18}H_{32}O_5N_2S$	. . . .	388.53
5032	**Pantocaine**	*p*-butylaminobenzoyl-dimethyl-aminoethanol	$C_4H_9NH \cdot C_6H_4CO_2 \cdot (CH_2)_2N(CH_3)_2HCl$	. . . .	300.83
5033	**Pantothenic** acid (*dl*)	. . . .	$HOCH_2C(CH_3)_2CH \cdot (OH)CONH \cdot (CH_2)_2 \cdot CO_2H$	. . . .	219.24
5034	acid (*d*)	. . . .	$C_9H_{17}O_5N$	. . . .	219.24
5035	Ca salt (*d*)	. . . .	$(C_9H_{16}O_5N)_2Ca$	. . . .	476.54
5036	**Parabanic acid**	oxalyl urea	$C_3H_2O_3N_2$	XXIV-449	114.06
5037	**Paracotoin**	. . . .	$CH_2O_2 \colon C_{11}H_6O_2$	. . . .	216.20
5038	**Pararosaniline** base (4,4′,4″)	triamino-triphenyl-carbinol	$HO \cdot C(C_6H_4 \cdot NH_2)_3$	XIII-750	305.38
5039	hydrochloride	parafuchsin	$C_{19}H_{18}N_3Cl \cdot H_2O$	XIII-752	341.84
5040	**Parvoline** (α)	2,4-diEt-pyridine	$C_9H_{13}N$	XX-253	135.21
5041	**Parvoline** (β)	3,4-diEt-pyridine	$C_9H_{13}N$	XX-253	135.21
5042	**Pelargonic** acid	nonanoic acid	$CH_3(CH_2)_7 \cdot CO_2H$	II-352	158.24
5043	aldehyde	nonanal	$CH_3(CH_2)_7 \cdot CHO$	I-708	142.24
5044	aldehyde oxime	. . . .	$C_8H_{17} \cdot CH:NOH$	I-708	157.26
5045	amide	nonanamide	$CH_3(CH_2)_7CO \cdot NH_2$	II-353	157.26
5046	chloride	pelargonyl chloride	$CH_3(CH_2)_7 \cdot COCl$	II-353	176.69
5047	nitrile	octyl cyanide	$CH_3(CH_2)_7 \cdot CN$	II-354	139.24
5048	**Penta**-aminobenzene	. . . .	$(NH_2)_5C_6H$	XIII-346	153.19
5048.1	bromoacetone	. . . .	$Br_3C \cdot CO \cdot CHBr_2$	I-659	452.59
5049	bromobenzene	. . . .	Br_5C_6H	V-215	472.62
5050	bromoethane	. . . .	$CHBr_2 \cdot CBr_3$	I-95	424.58
5051	bromophenol	. . . .	$Br_5C_6 \cdot OH$	VI-206	488.62
5052	chloroaniline	. . . .	$Cl_5C_6 \cdot NH_2$	XII-631	265.35
5053	chlorobenzene	. . . .	Cl_5C_6H	V-205	250.34
5054	chloroethane	pentalin	$CHCl_2 \cdot CCl_3$	I-87	202.30
5055	chloroethylbenzene	alkazene-32	$Cl_5C_6 \cdot C_2H_5$	V-355	278.39
5056	chloro-1-keto-1,2, 3,4-tetrahydro-naphthalene	2,2,3,4,4-penta-Cl	$C_{10}H_5OCl_5$	VII-370	318.42
5057	chlorophenol	. . . .	$Cl_5C_6 \cdot OH$	VI-194	266.34
5058	cosane	. . . .	$C_{25}H_{52}$	I-175	352.69

† See also No. 2686.

Table 7-4 (*Continued*)
PHYSICAL CONSTANTS OF ORGANIC COMPOUNDS

No.	Crystalline Form and Color	Specific Gravity	Melting Point °C.	Boiling Point °C.	Solubility in 100 Parts		
					Water	Alcohol	Ether
5016	cr. pd.	....	214 d.	....	v. sl. s.	i. abs.	i.
5017	col. lf./aq.	....	221-3 d.	....	$0.3^{20°}$	i.; s. a.	i.; s. alk.
5018	nd./bz.	....	150	....	sl. s. h.	v. s.	v. s.
5019	yel. nd.	....	....	....	....	s.	s. alk.
5020	nd./ac.	....	108-9	....	....	s.	s. ac.
5021	col. nd./aq.	....	120	227^{73mm}	s. h.	s.	s.; s. alk.
5022	col. pl.	$0.849^{70°}_{4}$	62.8	271.5^{100mm}	i.; v. sl. s. pet.	$9.3^{20°}$ abs.	s.; s. chl.
5023	wh. wax	....	34	$200\text{-}2^{29mm}$	i.	s.	s.
5024	nd./aq. al.	....	88	....	sl. s. bz.	sl. s. pet.	s.; s. chl.
5025	cr. lf.	....	106-7	$235\text{-}6^{12mm}$	i.	v. s.	v. sl. s.
5026	col. cr.	$0.847^{70°}_{4}$	64	....	....	$0.2^{20°}$	s.
5027	col. cr.	....	12	$194\text{-}5^{17mm}$	d.	d.	sl. s.
5028	pl.	$0.822^{31°}_{4}$	31	251.5^{100mm}	i.	s.	s.
5029	nd./aq.	....	42	240^{15mm}	i.	v. s.	v. s.
5030	lf./al.	$0.800^{83°}_{4}$	82.8	....	i.	....	....
5031	wh. pd.	....	157-9	....	33	s.	....
5032	wh. pd.	....	148-50*	....	14	s.	i.
5033	col. syrup	....	....	....	s.	s.	s. act.
5034	syrup	....	....	....	s.	s.	s. act.
5035	col. cr.	....	....	....	s.	s. Me al.	sl. s. act.
5036	mn.	....	243 d.	subl. 100	$4.7^{8°}$	s.	i.
5037	yel. cr.	....	162	....	....	s.	s.; s. chl.
5038	col. lf.	....	205 d.	....	v. sl. s.	s.	i.
5039	gn. met.	....	d. 250	....	$0.31^{22°}$	....	....
5040	lq.	$0.934^{0°}$	....	188	sl. s.	s.	....
5041	lq.	0.916	....	209^{710mm}	....	....	....
5042	col. oil	$0.906^{20°}_{4}$	12.5	253-4	v. sl. s.	s.; s. chl.	s.
5043	lq.	$0.827^{19°}_{19}$	....	185	....	....	....
5044	lf./aq. al.	....	63-4	....	i.	s.	s.
5045	col. cr.	....	99-100	....	i. c.	sl. s.	sl. s.
5046	col. lq.	$0.946^{15°}_{4}$	−60.5	215.4^{760mm}	d.	d.	s.
5047	col. lq.	$0.821^{15°}_{4}$	−34.2	224.0	i.	s.	s.
5048	salts only	....	....	....	....	....	....
5048.1	rhb. nd./al.	....	79-80(72)	subl.	i. c.	v. s.	v. s.
5049	nd./ac.	....	293	....	i.; s. bz.	sl. s.	sl. s.
5050	mn. pr.	3.312	56-7	210^{300mm} d.	i.	s.	v. s.
5051	mn./al.	....	225-6	subl.	i.	sl. s.	sl. s.
5052	nd./al.	....	232	....	sl. s. lg.	s.	s.
5053	nd./al.	$1.834^{17°}$	85-6	275-7	i.	s. h.	v. s.
5054	col. lq.	$1.671^{25°}_{4}$	−29	162	$0.05^{20°}$	∞	∞
5055	cr./al. bz.	$1.552^{60°}_{25}$	53	305±	i.; v. s. bz.	$2^{25°}$ Me al.	$166^{25°}$
5056	mn./bz.	....	156-7	d. >200, −HCl	s. h. bz.	v. s.	i.
5057	mn.	$1.978^{22°}$	188-9	309^{754mm} d.	$0.003^{50°}$	v. s.	$148^{25°}$
5058	cr.	$0.791^{60°}$	53.3	282-440mm	....	s. abs.	$5.4^{15°}$ chl.

* Free base, m. p. 153-4°.
Param 2080
Paramorphine, cf. alkd.
Pararosolic acid 593
Parietic acid 2306
Parillin, cf. glcde.

Paris green 3594
Parodyne 565
Paviin, cf. glcde.
P.D.H. 5261
Pear oil 398
Peganine, cf. alkd.

Pelargonamide 5045
Pelargone 2685
Pelargonyl chloride 5046
Pelletierine, cf. alkd.
Pellotine, cf. alkd.
Penta-acetyl glucose 3362-3

Table 7-4 (*Continued*)
PHYSICAL CONSTANTS OF ORGANIC COMPOUNDS

No.	Name	Synonym	Formula	Beil. Ref.	Formula Weight
5059	**Penta** decanaldoxime	....	$C_{14}H_{29} \cdot CH:NOH$	I-716	241.42
5060	decane (*n*)	....	$CH_3(CH_2)_{13} \cdot CH_3$	I-172	212.42
5061	decyl acetate (*n*)	....	$CH_3CO_2 \cdot C_{15}H_{31}$	II-136	270.46
5062	decyl alcohol (*n*)	pentadecanol-1	$CH_3(CH_2)_{13} \cdot CH_2OH$	I-429	228.42
5063	decyl bromide	1-bromo-pentadecane	$CH_3(CH_2)_{13} \cdot CH_2Br$	I-172	291.32
5064	diene-1,3	piperylene	$CH_3CH:CH \cdot CH:CH_2$	I-251	68.12
5065	diene-2,3		$CH_3CH:C:CHCH_3$	I-251	68.12
5066	erythritol	pentaerythrite	$C(CH_2OH)_4$	I-528	136.15
5067	erythritol tetra-acetate	....	$C(CH_2O_2CCH_3)_4$	II-150	304.30
5068	ethylbenzene		$(C_2H_5)_5C_6H$	V-471	218.39
5069	glycerol		$CH_3 \cdot C(CH_2OH)_3$	I-520	120.15
5070	hydroxy benzo-phenone	mori(n)tannic acid; maclurin	$(HO)_2C_6H_3 \cdot CO \cdot C_6H_2(OH)_3 \cdot H_2O$	VIII-538	280.24
5071	iodobenzene	....	I_5C_6H	V-229	707.60
5072	iodoethane	....	$CHI_2 \cdot CI_3$	*I-31	659.55
5073	methyl-amino-benzene	....	$(CH_3)_5C_6 \cdot NH_2$	XII-1182	163.26
5074	methyl benzene	....	$(CH_3)_5C_6H$	V-443	148.25
5075	methyl benzoic acid	....	$(CH_3)_5C_6 \cdot CO_2H$	IX-569	192.26
5076	methyl para-rosaniline	methyl violet base	$C_{24}H_{29}ON_3$	XIII-755	375.52
5077	methyl pararosani-line chloride	methyl violet dye salt †	$C_{24}H_{28}N_3 \cdot Cl$	XIII-755	393.96
5078	methyl phenol	....	$(CH_3)_5C_6 \cdot OH$	VI-551	164.25
5079	methylene-tetrazol	metrazol; cardiazol	$\cdot N(CH_2)_5C:(N_3)$	....	138.17
5080	triacontane (*n*)	....	$CH_3(CH_2)_{33} \cdot CH_3$	I-177	492.96
5081	**Pentaldol**	2,2-diMe-propan-3-ol-1-al	$(CH_3)_2C(CHO) \cdot CH_2OH$	I-833	102.13
5083	**Pentane**	*n*-pentane	$(C_2H_5)_2CH_2$	I-130	72.15
5084	**Pentane** (*iso*)	2-methyl-butane	$(CH_3)_2CH \cdot C_2H_5$	I-134	72.15
5085	**Pentane** (*neo*)	2,2-dimethyl-propane	$(CH_3)_4C$	I-141	72.15
5086	**Pentyne**-1	propyl acetylene	$C_2H_5 \cdot CH_2 \cdot C:CH$	I-250	68.12
5087	**Pentyne**-2	valerylene	$C_2H_5 \cdot C:C \cdot CH_3$	I-250	68.12
5088	**Perchloro**-ether	....	$Cl_5C_2 \cdot O \cdot C_2Cl_5$	II-210	418.57
5089	methyl mercaptan	....	$Cl_3C \cdot S \cdot Cl$	III-135	185.89
5090	**Perseite** (*d*)	mannoheptitol	$C_7H_{16}O_7$	I-548	212.20
5091	**Phanodorn**	Et-cyclohexenyl-barbituric acid	$(C_2H_5)(C_6H_9): C_4H_2O_3N_2$	....	236.27
5092	**Phellandrene** (*α*)(*d*)	*p*-menthadiene-1,5	$C_{10}H_{16}$	V-129	136.24
5093	**Phellandrene** (*β*)	*p*-menthadiene-2,1(7)	$C_{10}H_{16}$	V-132	136.24
5094	**Phenacyl** acetate	ω-acetoxy-acetophenone	$C_6H_5 \cdot CO \cdot CH_2 \cdot O \cdot CO \cdot CH_3$	VIII-92	178.19
5095	**Phenanthrene**	....	$(C_6H_4CH)_2$	V-667	178.24
5096	hydroquinone	(9,10)	$C_{14}H_8(OH)_2$	VI-1035	210.23
5097	**Phenanthraqui-none**	phenanthrene quinone	$C_6H_4(CO)_2C_6H_4$	VII-796	208.22
5098	**Phenanthrol** (2)	....	$C_{14}H_9OH$	VI-704	194.24
5099	**Phenanthrol** (3)	....	$C_{14}H_9OH$	VI-705	194.24

† Commercial dye mostly penta- with some hexa-methyl derivative.

Penta-diazenone 5480	Penta-methylene-dibromide 1857	Pentane diethylsulfone **5872**
Penta-decanol 5062	Penta-methylene-dicyanide 5311	Pentanoic acid 6399
Penta-decyl cyanide 5028	Penta-methylene-diiodide 2396	Pentanol 404-5, 408
Penta-erythrite 5066	Pentalin 5054	Pentanone 2180, 4355
Penta-hydroxy-flavone 5806-7	Pentanal 6400	Pentanpentol 572, 6478
Penta-methylene 1619	Pentandioic acid 3376	Pentanthiol 467-8
Penta-methylene-diamine 1184	Pentandiolamine 243	Pentaphen 476

Table 7-4 (*Continued*)
PHYSICAL CONSTANTS OF ORGANIC COMPOUNDS

No.	Crystalline Form and Color	Specific Gravity	Melting Point °C.	Boiling Point °C.	Solubility in 100 Parts		
					Water	Alcohol	Ether
5059	nd./aq. al.	....	86	....	sl. s. c. pet.	sl. s. c.	v. s.; sl. s. bz.
5060	col. lq.	$0.770\frac{20^\circ}{4}$	10	270.5	i.	v. s.	v. s.
5061	wax	....	10-1	230^{70mm}	....	....	....
5062	cr.	....	45-6	....	....	....	....
5063	cr.	....	14-5	....	....	....	....
5064	lq.	0.676^{20°	....	42.3	....	....	....
5065	lq.	0.66^{20°	....	40	....	....	....
5066	cr.	....	262	276^{30mm}	5.6^{15°	v. sl.s.	i.
5067	nd./aq. or bz.	$1.273\frac{18^\circ}{4}$	83-4	....	....	....	....
5068	lq.	$0.896\frac{20^\circ}{4}$	<-20	277	i.	....	....
5069	nd./abs. al.	....	199	subl.	v. s.	v. s.	i.
5070	yel. pr./aq.	....	200(anh.)	$-H_2O$, 130-40	0.5^{15°	s.	s.
5071	cr./al.	....	172	....	s. h. ac.	sl. s. c.	sl. s.
5072	col./ac.	....	182-4	....	s. ac.	s.	s. bz.
5073	mn./al.	....	151-2	278-80	i. h.	s.	s.
5074	pr./aq. al.	$0.847\frac{107^\circ}{4}$	54.3	230-1	i.	v. s.	v. s. bz.
5075	nd./aq. al.	....	210.5	subl.	v. sl. s.	v. s. h.	....
5076	salts only	....	....	....	....	....	....
5077	red or b. vl.	....	....	....	s.	....	....
5078	nd./al.	....	125	267	0.15 h.	s. h. aq. NaOH	....
5079	wh. cr. pd.	....	57-8	....	s.	s.	s.
5080	cr.	$0.782\frac{74.7^\circ}{4}$	α75; β72	33^{15mm}	....	....	.
5081	col. nd./al.	....	96-7	$172\text{-}3^{747mm}$ d.	5 c.	2.5 c.	s.
5083	col. lq.	0.626^{20°	-129.7	36.1	0.0361^{6°	∞	∞
5084	col. lq.	0.621^{19°	-159.9	27.9	i.	∞	∞
5085	lq.	$0.613\frac{20^\circ}{4}$	-16.6	9.5	i.	s.	s.
5086	col. lq.	0.691^{20°	-106	40.2	i.	....	s.
5087	lq.	0.711^{20°	-109.3	56.1	i.	....	....
5088	tet.	$1.900^{14.5^\circ}$	69	d.	....	....	....
5089	yel. lq.	$1.695^{17.5^\circ}$	....	149 sl. d.	....	....	....
5090	nd.	1.485	188	....	6.9^{18°	s. h.	447^{4° aq.
5091	wh. cr. pd.	....	173±	....	sl. s.	22	6
5092	lq.	0.845^{20°	....	175	....	i.	s.
5093	lq.	$0.852\frac{20^\circ}{4}$	....	171-2	i.	i.	s.
5094	rhb. pl./et. or lg.	....	48-9	270	i.; s. chl.	s.; sl. s. bz.	s.; sl. s. lg.
5095	pl./al.	1.179^{25°	100.5	340	i.; s. CS_2	2^{14°; 10 h.	v. s.; s. bz.
5096	col. nd.	....	147-8	....	s. h.	v. s.	v. s.; v.s.bz.
5097	or. nd.	1.405^{4°	206.5-7.5	>300; subl.	v. sl. s.	s. h.; s. bz.	sl. s.
5098	lf./aq. al.	....	168-9	....	....	v. s.	v. s.
5099	nd./aq. al.	....	119-22	....	sl. s. h.	v. s.	v. s.

Table 7-4 (*Continued*)
PHYSICAL CONSTANTS OF ORGANIC COMPOUNDS

No.	Name	Synonym	Formula	Beil. Ref.	Formula Weight
5100	**Phenanthrol** (9)	phenanthrone	$C_{14}H_9OH$	VI-706	194.24
5101	**Phenanthroline** (*o*)	(1,10)	$C_{12}H_8N_2$	XXIII-227	180.21
5102	**Phenanthroline** (*m*)	(1,5)	$C_{12}H_8N_2 \cdot 2H_2O$	XXIII-227	216.24
5103	**Phenanthroline** (*p*)	(1,8)	$C_{12}H_8N_2 \cdot 4H_2O$	XXIII-228	252.27
5104	**Phenazine**	azophenylene	$C_6H_4:N_2:C_6H_4$	XXIII-223	180.21
5105	**Phenetidine** (*o*)	*o*-aminophenetole	$C_2H_5O \cdot C_6H_4 \cdot NH_2$	XIII-359	137.18
5106	**Phenetidine** (*m*)	....	$C_2H_5O \cdot C_6H_4 \cdot NH_2$	XIII-404	137.18
5107	**Phenetidine** (*p*)	....	$C_2H_5O \cdot C_6H_4 \cdot NH_2$	XIII-436	137.18
5108	citrate	citrophen	$C_8H_{11}ON \cdot C_6H_8O_7$	XIII-437	329.31
5109	hydrochloride	....	$C_8H_{11}ON \cdot HCl$	XIII-437	173.64
5110	tartrate	vinopyrin	$C_8H_{11}ON \cdot C_4H_6O_6$	XIII-437	287.27
5111	**Phenetole**	ethyl-phenyl ether	$C_2H_5 \cdot O \cdot C_6H_5$	VI-140	122.17
5112	**Phenetyl urea** (*p*)	dulcin; sucrol; valzin	$C_2H_5O \cdot C_6H_4 \cdot NH \cdot CO \cdot NH_2$	XIII-480	180.21
5113	**Phenobarbital**	Et-Ph-barbituric acid; luminal; gardenal	$(C_2H_5)(C_6H_5):C_4H_2O_3N_2$	*XXIV-423	232.24
5114	Na salt	"soluble"	$C_{12}H_{11}O_3N_2Na$	....	254.22
5115	**Phenocoll**	glycine-*p*-phenetidide	$C_2H_5O \cdot C_6H_4 \cdot NH \cdot CO \cdot CH_2 \cdot NH_2 \cdot H_2O$	XIII-506	212.25
5116	hydrochloride	phenamine-HCl	$C_{10}H_{14}O_2N_2 \cdot HCl$	XIII-506	230.70
5117	salicylate	salocoll	$C_{10}H_{14}O_2N_2 \cdot C_7H_6O_3$	....	332.36
5118	**Phenol**	carbolic acid	$C_6H_5 \cdot OH$	VI-113	94.11
5119	Na salt	sodium phenoxide	$C_6H_5ONa \cdot 3H_2O$	VI-136	170.14
5120	Na salt	....	$C_6H_5ONa \cdot 2C_6H_5OH$	*VI-78	304.32
5121	phthalein	....	$C_{20}H_{14}O_4$	XVIII-143	318.33
5122	phthalin	phthalin	$C_{20}H_{16}O_4$	X-455	320.35
5123	sulfonic acid (*o*)	....	$HO \cdot C_6H_4 \cdot SO_3H \cdot \frac{3}{4}H_2O$	XI-234	187.69
5124	sulfonic acid (*m*)	....	$HO \cdot C_6H_4 \cdot SO_3H \cdot 2H_2O$	XI-239	210.21
5125	sulfonic acid (*p*)	....	$HO \cdot C_6H_4 \cdot SO_3H$	XI-241	174.18
5126	sulfonic Na (*p*)	Na sulfocarbolate	$C_6H_5O_4SNa \cdot 2H_2O$	XI-241	232.19
5127	sulfonphthalein	phenol red	$C_{19}H_{14}O_5S$	XIX-91	354.38
5128	tricarboxylic acid	(2;1,3,5)	$HOC_6H_2(CO_2H)_3 \cdot aq.$	X-580	226.14
5129	**Phenoxthine**	phenothioxine; dibenzothioxine (1,4)	$C_6H_4 \cdot O \cdot C_6H_4 \cdot S \cdot$ |_____|	XIX-45	200.26
5130	**Phenoxy**-acetic acid	....	$C_6H_5O \cdot CH_2 \cdot CO_2H$	VI-161	152.15
5131	acetic amide	phenoxyacetamide	$C_6H_5O \cdot CH_2 \cdot CONH_2$	VI-162	151.17
5132	acetone	ω-acetoanisole	$C_6H_5O \cdot CH_2 \cdot COCH_3$	VI-151	150.18
5133	benzoic acid (*o*)	salicylic phenyl ether	$C_6H_5O \cdot C_6H_4 \cdot CO_2H$	X-65	214.22
5134	butyronitrile (γ)	....	$C_6H_5O \cdot (CH_2)_3 \cdot CN$	VI-164	161.21
5135	β'-chloroethyl ether	(β)	$C_6H_5O \cdot (CH_2)_2 \cdot O \cdot CH_2 \cdot CH_2Cl$	**VI-150	200.67
5136	ethyl alcohol	phenyl cellosolve	$C_6H_5O \cdot (CH_2)_2 \cdot OH$	VI-146	138.17
5137	ethyl aniline	(*N*)(β)	$C_6H_5O \cdot (CH_2)_2 \cdot NH \cdot C_6H_5$	**XII-107	213.28
5138	**Phenthiazine**	thiodiphenylamine	$C_6H_4 \cdot NH \cdot C_6H_4 \cdot S \cdot$ |_____|	XXVII-63	199.27
5139	**Phenyl** acetaldehyde	α-toluylaldehyde	$C_6H_5 \cdot CH_2 \cdot CHO$	VII-292	120.15
5140	acetamide	....	$C_6H_5 \cdot CH_2 \cdot CO \cdot NH_2$	IX-437	135.17

Table 7-4 (*Continued*)
PHYSICAL CONSTANTS OF ORGANIC COMPOUNDS

No.	Crystalline Form and Color	Specific Gravity	Melting Point °C.	Boiling Point °C.	Solubility in 100 Parts		
					Water	Alcohol	Ether
5100	col. nd.	....	152-3	....	v. sl. s.	v. s.	v. s.
5101	cr./bz.	....	98-100*	>300	0.3	s.	s.
5102	pl./aq.	....	66; anh. 78	>360	s. h.; i. bz.	v. s.	v. sl. s.
5103	nd./aq.	....	173(anh.)	subl. > 100	s. h.	s.	sl. s.
5104	yel. nd.	....	170-1	>360	v. sl. s.	2 c.; s. h.	ε.
5105	oil	....	<-21	228-9	i.	s.	s.
5106	lq.	....	....	180-205^{100mm}	i.	s.	s.
5107	lq.	1.061$^{15°}$	3-4	254-5	i.	s.	s.
5108	wh. cr. pd.	....	186-8	....	2.5	sl. s.	....
5109	col. pl.	....	233-4	subl.	v. s.	....	....
5110	col. lf.	....	168±	d. 150-5	v. s.	sl. s.	i.
5111	col. lq.	0.965$^{20°}_{4}$	-29.5	170	i.	∞	∞
5112	lf./aq. al.	....	173-4	..:......	0.12$^{15°}$; 0.5$^{100°}$	4	s.
5113	lf./aq.	....	174 (156; 166)	....	0.1 c.; s. h.	14; s. alk.	10
5114	wh. cr. pd.	....	....	....	100	11	i.; i. chl.
5115	nd.	....	95±;100.5 (anh.)	....	sl. s.	s.	....
5116	wh. cr. pd.	....	....	....	5	s.	sl. s.
5117	wh. nd.	....	....	....	0.5 c.	5 h. aq.	
5118	col. nd.	1.071$^{25°}_{4}$ 1.054$^{45°}$	40.9	181.8	8.2$^{15°}$; ∞$^{65.3°}$	∞	∞
5119	col. cr.	....	....	d. by CO_2	23.8 c.	7.4$^{100°}$	0.5$^{25°}$
5120	cr.	....	....	....	80$^{25°}$±	....	
5121	col. rhb.	1.299$^{25°}_{4}$	261-2	....	0.2$^{20°}$; s. h.	10$^{25°}$; s. alk.	5.9 c.; v. sl. s. chl.
5122	nd./aq. al.	....	237-8	....	0.02$^{20°}$	....	....
5123	cr.	....	50 d.	....	v. s.	v. s.	
5124	nd.	....	-1.5H_2O, 100	-2H_2O, 140	....	....	....
5125	delq. nd.	....	....	....	s.	s.	....
5126	col. mn.	....	....	....	24 c.; 125 h.	0.8 c.; 8 h.	....
5127	red nd./ac.	....	....	....	0.08	0.3	i.; s. alk.
5128	cr./aq.	....	-H_2O, 120	d. 180	0.5$^{100°}$	s. h.	sl. s.; i. chl
5129	nd./al.	1.226$^{60°}_{25}$	56-7	185-7^{23mm}	i.; 165$^{25°}$ bz.	7$^{25°}$ Me al. 200$^{25°}$act.	165$^{25°}$
5130	nd./aq.	....	98-9	285 sl. d.	1.2$^{10°}$; v. s. h.	s.; s. ac.	29$^{25°}$; 3$^{25°}$ bz.
5131	nd./aq.	....	101-2	....	v. sl. s. h.	s. h.	....
5132	col. oil	....	....	229-30	v. s.	v. s.	v. s.
5133	lf./aq. al.	....	113-14.5	355 sl. d.	sl. s. h.	v. s.	v. s.
5134	nd.	....	45-6	287-9^{765mm}	....	....	....
5135	pa. yel. lq.	1.149$^{15°}_{15}$	....	138-43^{8mm}	i.	s.	s.
5136	oil	1.109$^{20°}_{20}$	14.0	244.7	2$^{25°}$	∞; s. KOH	∞; ∞ bz.
5137	cr.	1.070$^{60°}_{25}$	44-9	202^{10mm}	i.; v. s. bz.	v. s.; v. s. act.	v. s.; v. s. CCl₄
5138	yel. lf./al.	....	184-5	subl. 371	i.; 3$^{25°}$ bz.	2$^{25°}$; 27$^{25°}$ act.	7$^{25°}$
5139	lq.	1.025$^{20°}$	....	193-4	v. sl. s.	∞	∞
5140	lf.	....	156-7	d. 280-90	v. sl. s. c.	s.	v. sl. s.

* Crysts. + 1H_2O/aq., m. p. 92-3°; anh., m. p. 117°.
Phenol red 5127
Phenol-tetrachlorophthalein 5744
Phenoquin 5234
Phenothiazine 5138
Phenothioxine 5129

Phenoval 1008
Phenoxy-acetamide 5131
Phenoxy-ethyl bromide 969
Phenoxy-phenol 3770
Phenyl-, cf. also diphenyl-.

Table 7-4 (*Continued*)
PHYSICAL CONSTANTS OF ORGANIC COMPOUNDS

No.	Name	Synonym	Formula	Beil. Ref.	Formula Weight
5141	**Phenyl** acetate	acetyl phenol	$C_6H_5 \cdot O \cdot CO \cdot CH_3$	VI-152	136.15
5142	acetic acid	α-toluic acid	$C_6H_5 \cdot CH_2 \cdot CO_2H$	IX-431	136.15
5143	acetanilide	phenacetyl-anilide	$C_6H_5CH_2CO \cdot NHC_6H_5$	XII-275	211.27
5144	acetophenone (*p*)	Me-diphenylyl ketone	$C_{12}H_9 \cdot CO \cdot CH_3$	VII-443	196.25
5145	acetyl chloride	. . . .	$C_6H_5 \cdot CH_2 \cdot CO \cdot Cl$	IX-436	154.60
5146	acetylene	. . . .	$C_6H_5 \cdot C \vdots CH$	V-511	102.13
5147	acridine (9)	. . . .	$C_6H_5 \cdot C_{13}H_8N$	XX-514	255.32
5148	alanine (*dl*)	α-amino-β-phenyl-propionic acid	$C_6H_5CH_2 \cdot CH(NH_2) \cdot CO_2H$	XIV-498	165.19
5149	alanine (β)(*l*)	α-amino-hydrocin-namic acid	$C_6H_5CH_2 \cdot CH(NH_2) \cdot CO_2H$	XIV-495	165.19
5150	alanine (β,β)	β-amino-hydrocin-namic acid	$C_6H_5 \cdot CH(NH_2) \cdot CH_2 \cdot CO_2H$	XIV-493	165.19
5151	angelic acid	α-ethyl-cinnamic acid	$C_6H_5 \cdot CH{:}C(C_2H_5) \cdot CO_2H$	IX-623	176.22
5152	angelic acid (*stereo isomer*)	. . . .	$C_6H_5 \cdot CH{:}C(C_2H_5) \cdot CO_2H$	IX-623	176.22
5153	aniline (*o*)	2-amino-diphenyl	$C_6H_5 \cdot C_6H_4 \cdot NH_2$	XII-1317	169.23
5154	aniline (*m*)	3-amino-diphenyl	$C_6H_5 \cdot C_6H_4 \cdot NH_2$	XII-1318	169.23
5155	aniline (*p*)	4-amino-diphenyl	$C_6H_5 \cdot C_6H_4 \cdot NH_2$	XII-1318	169.23
5156	anthracene (9)	. . . .	$C_6H_5 \cdot C_{14}H_9$	V-725	254.33
5157	anthranilic acid (*N*)(*o*)	*o*-anilino-benzoic acid	$C_6H_5NH \cdot C_6H_4 \cdot CO_2H$	XIV-327	213.24
5158	arsonic acid	Ph-arsinic acid	$C_6H_5 \cdot AsO(OH)_2$	XVI-868	202.04
5159	benzoate	. . . .	$C_6H_5 \cdot CO_2 \cdot C_6H_5$	IX-116	198.22
5160	benzoic acid (*o*)	diPh-carboxylic acid	$C_6H_5 \cdot C_6H_4 \cdot CO_2H$	IX-669	198.22
5161	benzoic acid (*m*)	. . . .	$C_6H_5 \cdot C_6H_4 \cdot CO_2H$	IX-671	198.22
5162	benzoic acid (*p*)	. . . .	$C_6H_5 \cdot C_6H_4 \cdot CO_2H$	IX-671	198.22
5163	benzoylene urea (3)	. . . .	$C_6H_4 \cdot NH \cdot CO \cdot N \cdot$ $\overline{}\vert$ $(C_6H_5) \cdot CO$ $\overline{}\vert$	XXIV-376	238.25
5164	benzylcarbinol	. . . .	$C_6H_5 \cdot CHOH \cdot CH_2 \cdot C_6H_5$	VI-683	198.27
5165	benzylurethane	ethyl *N*-Ph-*N*-benzyl-carbamate	$C_6H_5CH_2(C_6H_5)N \cdot CO_2C_2H_5$	**XII-565	255.32
5166	bromoacetate	. . . .	$Br \cdot CH_2 \cdot CO_2C_6H_5$	VI-154	215.05
5167	carbamate	"phenyl urethane"	$NH_2 \cdot CO_2 \cdot C_6H_5$	VI-159	137.14
5168	carbylamine chloride	*iso*-cyanophenyl chloride	$C_6H_5 \cdot N{:}CCl_2$	XII-447	174.03
5169	chloroacetate	. . . .	$Cl \cdot CH_2 \cdot CO_2C_6H_5$	VI-153	170.60
5170	4-chlorobenzene-sulfonate	. . . .	$Cl \cdot C_6H_4 \cdot SO_3 \cdot C_6H_5$	. . . .	268.72
5171	cinchonoyl-urethane	fantan	$C_6H_5 \cdot C_9H_5N \cdot NH \cdot CO_2C_2H_5$	. . . .	292.34
5172	cinnamic acid (α)(*trans*)	. . . .	$C_6H_5 \cdot CH{:}C(C_6H_5) \cdot CO_2H$	IX-691	224.26
5173	cinnamic acid (α)(*cis*)	. . . .	$C_6H_5 \cdot CH{:}C(C_6H_5) \cdot CO_2H$	IX-693	224.26
5174	cinnamic acid (β)	. . . .	$(C_6H_5)_2C{:}CH \cdot CO_2H$	IX-699	224.26
5175	crotonic acid (γ)(3)	. . . .	$C_{10}H_{10}O_2$	IX-614	162.19

Table 7-4 (*Continued*)
PHYSICAL CONSTANTS OF ORGANIC COMPOUNDS

No.	Crystalline Form and Color	Specific Gravity	Melting Point °C.	Boiling Point °C.	Solubility in 100 Parts		
					Water	Alcohol	Ether
5141	col. lq.	$1.073\frac{25°}{25}$	....	195.8^{756mm}	v. sl. s.	∞ ; ∞ chl.	∞
5142	lf.	$1.081\frac{80°}{4}$	76–7	265.5	s. h.	v. s.	v. s.
5143	pr./al.	....	117–8	....	i. aq. KOH	s.; i. H_2SO_4	s.
5144	pr./act.	....	121	325–7	....	s.	s. act.
5145	col. lq.	$1.168\frac{20°}{4}$	....	170^{250mm}	d.	d.	....
5146	col. lq.	$0.930\frac{20°}{4}$	–43	142–3	i.	∞	∞
5147	yel. pr.	....	181	403–4	i.; s. bz.	s. h.	s.
5148	lf./al.	....	271–3 d.	....	sl. s. c.	v. sl. s. h.	i.
5149	cr./aq.	....	d. 275–83	....	$3.1^{25°}$	v. sl. s.	v. sl. s.
5150	cr./aq.	....	231 d.	....	s. h.; s. alk.	s. h.; s. aq. a.	sl. s.
5151	nd./aq.	....	104–5	....	$0.01^{25°}$; s. h.	v. s.	sl. s. pet.
5152	nd./aq.	....	82	....	sl. s. h.	s.	s.; s. bz.
5153	cr./aq. al.	....	50–2	299^{760mm}	sl. s.	s.	....
5154	nd.	....	30	254^{135mm}	sl. s.	s.	s.
5155	lf./aq. al.	....	53–4	302	s. h.	s.	s.
5156	lf./al.	....	152–3	417	s. h. bz.	s. h.	s. h.
5157	nd./al.	....	185–7	d. > 184	v. sl. s. h.	s. h.	v. sl. s.
5158	cr./aq.	1.760	157–8	$-H_2O > 158$	$3.3^{28°}$	$15.5^{26°}$	i. chl.
5159	mn.	$1.235^{31°}$	70–1	314	i.	v. s. h.	v. s. h.
5160	cr./aq. al.	....	113–4	343–4	i. c.	s.	s. bz.
5161	lf./al.	....	160–1	....	sl. s.	s.	s.
5162	nd./al.	....	225–6.5	subl.	sl. s. h.	s.	s.
5163	nd.	....	282	....	i.	sl. s.	s. NaOH
5164	nd./aq. al.	....	67–8	$167–70^{10mm}$	0.06 h.	$420^{7°}$	v. s.
5165	col. lq.	$1.076\frac{5.5°}{4}$	....	$180–2^{12mm}$	....	....	....
5166	lf./al.	....	32	140^{20mm}	....	....	....
5167	nd./aq.	$1.078^{63.8°}$	141–3	....	sl. s.	s.	....
5168	lq.	....	....	209–10	sl. d. h.	d.	....
5169	nd./al.	....	44–5	230–5	i.	s.	s.
5170	col. cr.	....	90–1	....	i.	s. h.	s.
5171	yel. wh. cr.	....	173–4	....	i.; s. chl.	sl. s.; sl. s. bz.	sl. s.
5172	wh. nd./ aq. al.	....	172	subl.	s. h.	s.	s.
5173	nd.	....	137–8	....	s. h.	....	....
5174	lf./al.	....	162	....	sl. s. h.	s.	s.
5175	pl./bz.	....	65	....	s. bz.	s.; s. CS_2	s.

Table 7-4 (*Continued*)
PHYSICAL CONSTANTS OF ORGANIC COMPOUNDS

No.	Name	Synonym	Formula	Beil. Ref.	Formula Weight
5176	**Phenyl** crotonic acid (*iso*)	....	$C_6H_5 \cdot CH:CH \cdot CH_2 \cdot CO_2H$	IX-612	162.19
5177	*iso*-cyanate	phenyl carbonimide	$C_6H_5 \cdot N:CO$	XII-437	119.12
5178	*iso*-cyanide	phenyl carbylamine	$C_6H_5 \cdot N:C$	XII-191	103.12
5179	cyclohexane	cyclohexylbenzene	$C_6H_5 \cdot CH(CH_2)_4 \cdot CH_2 \cdot$	V-503	160.26
5180	3,4-dichloroben- zene sulfonate	....	$Cl_2C_6H_3 \cdot SO_3 \cdot C_6H_5$	....	303.17
5181	diethanolamine	....	$(HO \cdot CH_2 \cdot CH_2)_2N \cdot C_6H_5$	XII-183	181.24
5182	dihydroquinazo- line	orexin	$C_6H_4 \cdot CH_2 \cdot N(C_6H_5) \cdot$ CH:N	XXIII-137	208.27
5183	ditolylmethane (4,4′)	....	$C_6H_5 \cdot CH(C_6H_4 \cdot CH_3)_2$	V-712	272.39
5184	di-*o*-xenyl phosphate	phosphen 6	$(C_{12}H_9O)_2:PO \cdot OC_6H_5$	....	478.49
5185	ethyl acetate (β)	....	$C_6H_5 \cdot C_2H_4 \cdot O_2CCH_3$	VI-479	164.21
5186	ethyl alcohol (α)	Me-Ph-carbinol	$C_6H_5 \cdot CHOH \cdot CH_3$	VI-475	122.17
5187	ethyl alcohol (β)	benzyl carbinol	$C_6H_5 \cdot CH_2 \cdot CH_2OH$	VI-478	122.17
5188	ethylamine (α)(*dl*)	....	$C_6H_5 \cdot CH(NH_2) \cdot CH_3$	XII-1094	121.18
5189	ethylamine (β)	ω-phenyl-ethylamine	$C_6H_5 \cdot CH_2 \cdot CH_2 \cdot NH_2$	XII-1096	121.18
5190	ethyl benzoate (β)	....	$C_6H_5CO_2 \cdot C_2H_4 \cdot C_6H_5$	....	226.28
5191	ethyl chloride (β)	β-Cl-ethyl benzene	$C_6H_5 \cdot CH_2CH_2Cl$	V-354	140.61
5192	ethyl cinnamate (β)	....	$C_6H_5 \cdot CH:CH \cdot CO_2 \cdot CH_2 \cdot CH_2 \cdot C_6H_5$	....	252.32
5193	*d*-glucosazone	Ph-*d*-fructosazone	$(C_6H_5NHN)_2C_6H_{10}O_4$	XV-225	358.40
5194	glycine (*N*)	anilino-acetic acid	$C_6H_5 \cdot NH \cdot CH_2 \cdot CO_2H$	XII-468	151.17
5195	glycine *o*-car- boxylic acid	....	$HO_2C \cdot C_6H_4 \cdot NHCH_2 \cdot CO_2H$	XIV-348	195.18
5196	hydrazine	....	$C_6H_5 \cdot NH \cdot NH_2$	XV-67	108.14
5197	hydrazine HCl	....	$C_6H_5N_2H_3 \cdot HCl$	XV-108	144.61
5198	hydrazine-*p*-sul- fonic acid	....	$H_2N \cdot NH \cdot C_6H_4 \cdot SO_3H$	XV-639	188.21
5199	hydrazoquinoline	(α)	$C_6H_5(NH)_2C_9H_6N$	XXII-564	235.29
5200	hydroxylamine (β)	....	$C_6H_5 \cdot NH \cdot OH$	XV-2	109.13
5201	hydroxylamine (β)	(oxalate salt)	$2C_6H_5NHOH \cdot H_2C_2O_4$	....	308.29
5202	3-methylpyrazolone	(*N*)	$C_4H_5ON_2 \cdot C_6H_5$	XXIV-20	174.20
5203	morpholine (4)	....	$(CH_2)_2 \cdot O \cdot (CH_2)_2 \cdot N \cdot C_6H_5$	XXVII-6	163.22
5204	naphthalene (α)	....	$C_6H_5 \cdot C_{10}H_7$	V-687	204.27
5205	naphthalene (β)	....	$C_6H_5 \cdot C_{10}H_7$	V-687	204.27
5206	α-naphthylamine	(*N*)	$C_6H_5 \cdot NH \cdot C_{10}H_7$	XII-1224	219.29
5207	β-naphthylamine	(*N*)	$C_6H_5 \cdot NH \cdot C_{10}H_7$	XII-1275	219.29
5208	1-naphthylamine- 8-sulfonic acid	(*N*); phenyl (peri) acid	$C_6H_5 \cdot NH \cdot C_{10}H_6 \cdot SO_3H$	XIV-753	299.35
5209	α-naphthyl carbinol	....	$C_6H_5 \cdot CHOH \cdot C_{10}H_7$	VI-710	234.30
5210	α-naphthyl ketone	benzoyl-naphthalene	$C_6H_5 \cdot CO \cdot C_{10}H_7$	VII-510	232.28
5211	β-naphthyl ketone	....	$C_6H_5 \cdot CO \cdot C_{10}H_7$	VII-511	232.28
5212	nitroamine	....	$C_6H_5 \cdot NH \cdot NO_2$	....	138.13
5213	nitromethane	ω-nitrotoluene	$C_6H_5 \cdot CH_2 \cdot NO_2$	V-325	137.14

Table 7-4 (*Continued*)
PHYSICAL CONSTANTS OF ORGANIC COMPOUNDS

No.	Crystalline Form and Color	Specific Gravity	Melting Point °C.	Boiling Point °C.	Solubility in 100 Parts		
					Water	Alcohol	Ether
5176	nd./aq.	....	88	302 sl. d.	sl. s. h.	v. s.	v. s.
5177	lq.	$1.096\frac{20}{4}$	....	166^{769mm}	d.	d.	v. s.
5178	col. lq.	$0.978^{15°}$	....	78^{40mm}	d.	d.	s.
5179	oil	$0.944\frac{20}{0}$	7-8	239^{745mm}	....	....	....
5180	col. nd.	....	82-5	....	i.	s. h.	v. s.
5181	cr.	$1.120\frac{60}{20}$	58-9	>350 sl. d.	$5^{25°}; 23^{25°}$ bz.	v. s.; v. s. act.	$29^{25°}$
5182	pl./et. lg.	$1.290^{4°}$	95	d.	i.	s.	s.
5183	nd./Me al.	....	55-6	....	v. s. bz.	sl. s. c.	v. s.
5184	col. lq.	$1.20\frac{60}{4}$	....	$285-330^{5mm}$	i.	v. s.	∞ bz.; ∞ CCl$_4$
5185	lq.	$1.05^{122.5°}$	....	232	....	....	....
5186	lq.	$1.019\frac{13}{4}°$	21.4	203.6^{745mm}	i.	∞	∞
5187	col. oil	$1.023\frac{13}{4}°$	−27	$219-21^{750mm}$	$1.6^{20°}$	s. aq. al.	∞
5188	oil	$0.940^{15°}$	ca. −65	187.5^{741mm}	$4.2^{20°}$	∞	∞
5189	lq.	$0.958\frac{24}{4}°$	....	198	s.	∞	∞
5190	yel. lq.	....	....	$204-6^{25mm}$	i.	s.	s.
5191	oil	$1.069\frac{25}{4}°$	....	190-200 sl. d.	....	....	....
5192	col. cr.	....	57-8	....	i.	sl. s.	v. s.
5193	yel. nd.	....	205 d.	....	i.	0.66 abs.	25 c. pyr.
5194	cr.	....	127	....	s.	s.	sl. s.
5195	nd./Me al.	....	218-20	....	sl. s.	s.	s.; i. bz.
5196	lt. yel. oil	$1.097\frac{22.7}{4}°$	19.6*	243.5	sl. s. h.	∞; ∞ chl.	∞; ∞ bz.
5197	lf./al.	....	249 d.	....	v. s.	s.	i.
5198	cr./al.	....	286	....	$0.61^{2°}$; 3 h.	sl. s.	....
5199	nd./al.	....	191	....	a. ohl.	sl. s.	i.; s. ac.
5200	col. nd.	....	81-2	....	2 c.; 10 h.	v. s.	v. s.; v. sl. s. lg.
5201	col. nd.	....	120-2 d.	....	sl. s.	s.	i.
5202	pr./aq.	....	128	191^{17mm}	$12^{0°}$	v. s. h.	v. sl. s.
5203	cr./al. et.	$1.058^{270°}$	57	269.9	$1.0^{20°}$	s.	s.
5204	waxy	....	45±	336-7	v. s. bz.	v. s.	v. s.
5205	lf./al.	....	108-9	345-6	v. s. bz.	sl. s.	sl. s.
5206	pr./al.	....	62	335^{528mm}	s. bz.	s.; s. ac.	s.; s. chl.
5207	rhb./Me al.	....	107-8	395.5	i.; s. h. bz.	v. s. h.	v. s. h.
5208	lf.	....	....	....	v. sl. s.	....	....
5209	cr./al.	....	88	>360	i.	v. s.	v. s.
5210	rhb./al.	....	75.5	385	i.	$2.4^{12°}$ abs.	....
5211	rhb./al.	....	82	398^{754mm}	i.	$2^{12°}$ abs.	....
5212	lf./lg.	....	46	expl. 98	sl. s. c.	v. s.	sl. s. lg.
5213	yel. lq.	$1.160\frac{20}{0}$	....	$141-2^{35}$ sl. d.	....	....	....

* Crysts. $+ \frac{1}{2}H_2O$, m. 24° ±.
Phenyl mercaptan 5915
Phenyl-mercuric acetate 4036
Phenyl-mercuric chloride 4037
Phenyl-mercuric nitrate 4038

Phenyl-methyl-, cf. also methyl-phenyl-.
Phenyl-methylamino-propanol, cf. alkd.
Phenyl-methyl carbinol 5186
Phenyl-methyl-hydrazine 4338-40

Phenyl-methyl ketone 51
Phenyl mustard oil 5241
Phenyl-naphthyl-methane
817-8

Table 7-4 (*Continued*)
PHYSICAL CONSTANTS OF ORGANIC COMPOUNDS

No.	Name	Synonym	Formula	Beil. Ref.	Formula Weight
5214	**Phenyl** phenacyl bromide	ω-Br-*p*-phenyl-acetophenone	$C_6H_5 \cdot C_6H_4 \cdot CO \cdot CH_2 \cdot Br$	....	275.15
5215	phenacyl chloride	ω-Cl-*p*-phenyl-acetophenone	$C_6H_5 \cdot C_6H_4 \cdot CO \cdot CH_2 \cdot Cl$	VII-443	230.70
5216	phenol (*o*)	2-hydroxy-diphenyl	$C_6H_5 \cdot C_6H_4 \cdot OH$	VI-672	170.21
5217	phenol (*m*)	3-hydroxy-diphenyl	$C_6H_5 \cdot C_6H_4 \cdot OH$	VI-673	170.21
5218	phenol (*p*)	4-hydroxy-diphenyl	$C_6H_5 \cdot C_6H_4 \cdot OH$	VI-674	170.21
5219	phosphine	....	$C_6H_5 \cdot PH_2$	XVI-757	110.10
5220	propiolic acid	....	$C_6H_5 \cdot C \vdots C \cdot CO_2H$	IX-633	146.15
5221	propionate	....	$C_2H_5 \cdot CO_2 \cdot C_6H_5$	VI-154	150.18
5222	*n*-propyl bromide	(γ)	$C_6H_5 \cdot (CH_2)_3 \cdot Br$	V-391	199.10
5223	propylene (α,α)	*iso*-allylbenzene	$C_6H_5 \cdot CH:CH \cdot CH_3$	V-481	118.18
5224	propylene (β)	α-methyl-styrene	$C_6H_5 \cdot C(CH_3):CH_2$	V-484	118.18
5225	propylene (γ)	....	$C_6H_5 \cdot CH_2 \cdot CH:CH_2$	V-484	118.18
5226	pyridine (2)(α)	....	$C_6H_5 \cdot C_5H_4N$	XX-424	155.20
5227	pyridine (3)(β)	....	$C_6H_5 \cdot C_5H_4N$	XX-424	155.20
5228	pyridine (4)(γ)	....	$C_6H_5 \cdot C_5H_4N$	XX-424	155.20
5229	pyrrole (*N*)	....	$C_4H_4N \cdot C_6H_5$	XX-164	143.19
5230	quinoline (2)(α)	....	$C_6H_5 \cdot C_9H_6N$	XX-481	205.26
5231	quinoline (4)	....	$C_6H_5 \cdot C_9H_6N$	XX-483	205.26
5232	quinoline (6)(*p*)	....	$C_6H_5 \cdot C_9H_6N$	XX-483	205.26
5233	quinoline (8)(*o*)	....	$C_6H_5 \cdot C_9H_6N$	XX-484	205.26
5234	quinoline-4-carboxylic acid (2)	2-Ph-cinchonic acid; atophan	$C_6H_5 \cdot C_9H_5N \cdot CO_2H$	XXII-103	249.27
5235	salicylate	salol	$HO \cdot C_6H_4 \cdot CO_2 \cdot C_6H_5$	X-76	214.22
5236	semicarbazide (1)	....	$C_6H_5NH \cdot NH \cdot CO \cdot NH_2$	XV-287	151.17
5237	semicarbazide (4)	....	$C_6H_5 \cdot NH \cdot CO \cdot N_2H_3$	XII-378	151.17
5238	stearate	....	$C_{17}H_{35}CO_2 \cdot C_6H_5$	VI-155	360.59
5239	succinic acid (α)	(*dl*)	$C_6H_5 \cdot C_2H_3(CO_2H)_2$	IX-865	194.19
5240	thioacetamide	thio-Ph-acetamide	$C_6H_5 \cdot CH_2 \cdot CS \cdot NH_2$	IX-460	151.23
5241	*iso*-thiocyanate	phenyl mustard oil	$C_6H_5 \cdot N:C:S$	XII-453	135.19
5242	thiohydantoic acid	phenyl-pseudothio-hydantoic acid	$C_6H_5N:C(NH_2) \cdot S \cdot CH_2 \cdot CO_2H$	XII-411	210.26
5243	thiosemicarbazide	(1)	$C_6H_5NH \cdot NH \cdot CS \cdot NH_2$	XV-294	167.23
5244	thiourea	Ph thiocarbamide	$C_6H_5 \cdot NH \cdot CS \cdot NH_2$	XII-388	152.22
5245	toluene (*o*)	....	$C_6H_5 \cdot C_6H_4 \cdot CH_3$	V-596	168.24
5246	toluene (*m*)	....	$C_6H_5 \cdot C_6H_4 \cdot CH_3$	V-596	168.24
5247	toluene (*p*)	....	$C_6H_5 \cdot C_6H_4 \cdot CH_3$	V-597	168.24
5248	*p*-toluene sulfonate	....	$CH_3 \cdot C_6H_4 \cdot SO_3 \cdot C_6H_5$	XI-99	248.30
5249	*o*-tolyl ketone	Me benzophenone	$C_6H_5 \cdot CO \cdot C_6H_4 \cdot CH_3$	VII-439	196.25
5250	*m*-tolyl ketone	....	$C_6H_5 \cdot CO \cdot C_6H_4 \cdot CH_3$	VII-440	196.25
5251	*p*-tolyl ketone	....	$C_6H_5 \cdot CO \cdot C_6H_4 \cdot CH_3$	VII-440	196.25
5252	*p*-tolyl sulfone	Me diphenyl sulfone	$C_6H_5 \cdot SO_2 \cdot C_6H_4 \cdot CH_3$	VI-418	232.30
5253	urea	phenyl carbamide	$C_6H_5 \cdot NH \cdot CO \cdot NH_2$	XII-346	136.15
5254	urethane†(*N*)	Et-Ph-carbamate	$C_6H_5 \cdot NH \cdot CO_2C_2H_5$	XII-320	165.19
5255	xanthydrol (9)	....	$C_6H_5 \cdot C_{13}H_8O(OH)$	XVII-138	274.32
5256	**Phenylene** diamine	1,2-diaminobenzene	$NH_2 \cdot C_6H_4 \cdot NH_2$	XIII-6	108.14
5257	diamine HCl (*o*)	....	$C_6H_4(NH_2)_2 \cdot 2HCl$	XIII-14	181.07
5258	diamine (*m*)	1,3-diaminobenzene	$NH_2 \cdot C_6H_4 \cdot NH_2$	XIII-33	108.14
5259	diamine HCl (*m*)	....	$C_6H_4(NH_2)_2 \cdot 2HCl$	XIII-38	181.07
5260	diamine (*p*)	1,4-diaminobenzene	$NH_2 \cdot C_6H_4 \cdot NH_2$	XIII-61	108.14
5261	diamine HCl (*p*)	....	$C_6H_4(NH_2)_2 \cdot 2HCl$	....	181.07
5262	diamine sulfonic acid (*p*)	2,5-diaminobenzene 1-sulfonic acid	$(NH_2)_2C_6H_3 \cdot SO_3H \cdot 2H_2O$	XIV-712	224.24

† See also No. 5167.

Phenyl-nitrosohydroxylamine 1574
Phenyl-oxydisulfide 2716
Phenyl-pentane 421-4
Phenyl peri acid 5208
Phenyl-phenol methyl ether 4073-4
Phenyl-phenylenediamine 313-4
Phenyl phosphate 6341
Phenyl phosphite 6343
Phenyl phthalate 2737
Phenyl-phthalimide 5273
Phenyl-propionaldehyde 3706
Phenyl-propionamide 3707
Phenyl-propionic acid 3683. 3704
Phenyl-propionitrile 3709
Phenyl-propionyl chloride 3708
Phenyl-propyl alcohol 3122, **3705**
Phenyl-propyl carbinol 5436-7
Phenyl-propyl ketone 5439-40
Phenyl-pyrocatechol 2339
Phenyl succinate 2748
Phenyl-sulfanilic acid 2701

Table 7-4 (*Continued*)
PHYSICAL CONSTANTS OF ORGANIC COMPOUNDS

No.	Crystalline Form and Color	Specific Gravity	Melting Point °C.	Boiling Point °C.	Solubility in 100 Parts		
					Water	Alcohol	Ether
5214	col. nd.	....	124-5	....	....	$1.3^{25°}$	$6.778°$ al.
5215	yel. cr. pd.	....	126-7	....	....	v. s. h.	....
5216	nd./pet.	....	56-7	275	i.; s. lg.	s.; s. alk.	s.
5217	nd./aq.	....	75-8	>300	sl. s.	v. s.	s. KOH
5218	nd./aq. al.	....	164-5	305-8	i.; s. alk.	v. s.	v. s.
5219	lq.	$1.001^{15°}$	....	160-1	....	....	....
5220	nd./aq.	....	136-7	subl.	v. sl. s.	v. s.	v. s.
5221	pr.	$1.047^{\frac{25}{25}°}$	20	211^{760mm}	....	....	....
5222	col. lq.	....	....	$121-2^{20mm}$	....	....	s.
5223	col. lq.	$0.914^{\frac{20}{4}°}$	....	176-7	i.	s.	s.
5224	col. lq.	$0.911^{20°}$	-23.2	165.4	i.	∞	∞
5225	col. lq.	$0.893^{\frac{20}{0}°}$	....	156-7	i.	s.	s.
5226	lq.	>1	....	$269-70^{749mm}$	i.	s.	s.
5227	oil	>1	....	$269-70^{749mm}$	i.	s.	s.
5228	lf./aq.	....	77-8	274-5	s. h.	s.	s.
5229	pl.	....	62	234	i.; s. pet.	s.; s. chl.	s.; s. bz.
5230	nd./aq. al.	....	86	363	sl. s.	s. h.	s.
5231	nd./et.	....	61-2	....	i.	v. s.	v. s.
5232	cr./al.	$1.195^{20°}$	110-1	260^{77mm}	v. sl. s.	s.	s.
5233	lq.	....	....	283^{187mm}	s. bz.	s.	s.
5234	nd./Me al.	....	217-9	....	i.; s. alk.; 4h.Me al.	5 h. abs.; 3 h. act.	s.; s. a.; 0.4 h. bz.
5235	rhb./al.	$1.250^{\frac{20}{4}°}$	**41.4**	$172-3^{12mm}$	0.015	v. s. h.	s.; 80 bz.
5236	lf./aq.	....	174-6	....	s. h.	s.; s. act.	sl. s.
5237	rhb./aq.	....	122	d. 140-70	sl. s. h.	s.	i.
5238	cr.	....	52	267^{15mm}	i.	....	....
5239	cr./aq.	....	168	$-H_2O$ > 168	s. h.; s. ac.	s.; i. lg.	s.; i. bz.
5240	rhb./al.	....	97-8	d.	v. sl. s. h.	s.	s.
5241	col. lq.	$1.138^{\frac{15}{15}°}$	-21	219-20	i.	s.	s.
5242	col. nd.	....	157-8	....	v. sl. s.; s. a.; s. alk.	v. sl. s.; sl. s. CS_2	v. sl. s.; v. sl. s. bz.
5243	pr./al.	....	200-1 d.	....	sl. s.	s. h.	sl. s.
5244	nd./aq.	1.3	154	....	$0.26^{18°}$	s.	$5.9^{100°}$ aq.
5245	lq.	$1.010^{\frac{20}{4}°}$	....	261-4	i.	s. abs.	s.
5246	lq.	$1.031^{0°}$	....	272-7	i.	s. abs.	s.
5247	lf./lg.	$1.015^{27°}$	47-8	267	i.	s. abs.	s.
5248	rhb. nd./al.	....	94-5	....	i.; v. s. bz.	v. s.	v. s.
5249	col. lq.	....	<-18	$312-57^{35mm}$	i.	v. s. 80% al.	v. s.
5250	lq.	$1.08^{17.5°}$	....	321^{761mm}	∞ chl.	∞	∞
5251	mn.	....	59-60(55)	326.5	i.; s. bz.	s.	s.
5252	pl./al.	....	124.5	....	sl. s. ac.	$1.62^{20°}$	sl. s. bz.
5253	mn.	1.302	147	d. 160	s. h.	s.	sl. s.
5254	pl./al.	$1.106^{\frac{30}{4}°}$	52-3	237 sl. d.	i. c.	s.	s.
5255	pr./bz. lg.	....	159-60	subl.	sl. s. ac.	s. h.; s. bz.	s.; i. lg.
5256	lf. /aq.	....	103-4	256-8	$4.2^{35°}$	v. s.	v. s.
5257	nd.	....	....	....	v. s.	....	....
5258	rhb.	$1.139^{\frac{15}{15}°}$	62.8	284-7	v. s.	v. s.	s.
5259	col. nd.	....	....	....	v. s.	s.	....
5260	mn.	....	140	267	$3.8^{24°}$	s.	s.; s. chl.
5261	col. tri.	....	....	....	v. s.	sl. s.	i. HCl
5262	pl./aq.	....	....	....	s.	i.	i.; i. bz.

Table 7-4 *(Continued)*
PHYSICAL CONSTANTS OF ORGANIC COMPOUNDS

No.	Name	Synonym	Formula	Beil. Ref.	Formula Weight
	Phenylene				
5263	diamine sulfonic acid (*p*) HCl	....	$C_6H_8O_3N_2S \cdot 2HCl$	....	261.13
5264	**Phloretin**	dihydronaringenin	$C_{15}H_{14}O_5$	VIII-498	274.28
5265	**Phloroglucinol** tri-ethyl ether	(1,3,5)	$C_6H_3(OC_2H_5)_3$	VI-1103	210.28
5266	trimethyl ether	(1,3,5)	$C_6H_3(OCH_3)_3$	VI-1101	168.19
5267	trioxime (1,3,5)	....	$C_6H_6(:NOH)_3$	XV-34	171.16
5268	**Phorone**	*sym*-diisopropylidene acetone	$[(CH_3)_2C:CH]_2CO$	I-751	138.21
5269	*iso*-**Phorone**	1,1,3-triMe-cyclohexene-3-one-5	$(CH_3)_3C_6H_5O$	VII-65	138.21
5270	**Phosphenyl chloride**	Ph-diCl-phosphine	$C_6H_5 \cdot PCl_2$	XVI-763	178.99
5271	**Phosphobenzene**	....	$C_6H_5 \cdot P:P \cdot C_6H_5$	XVI-824	216.16
5272	**Phthalamidic acid** (*o*)	*o*-phthalamic acid	$NH_2 \cdot CO \cdot C_6H_4 \cdot CO_2H$	IX-809	165.15
5273	**Phthalanil**	*N*-Ph-phthalimide	$C_6H_5 \cdot N:C_8H_4O_2$	XXI-464	223.23
5274	**Phthalic acid** (*o*)	phthalic acid	$C_6H_4(CO_2H)_2$	IX-791	166.13
5275	acid Na	(mono Na salt)	$C_8H_5O_4Na \cdot 2H_2O$	IX-796	224.15
5276	acid (*m*)	isophthalic acid	$C_6H_4(CO_2H)_2$	IX-832	166.13
5277	acid (*p*)	terephthalic acid	$C_6H_4(CO_2H)_2$	IX-841	166.13
5278	aldehyde†(*o*)	....	$C_6H_4(CHO)_2$	VII-674	134.14
5279	aldehyde (*m*)	....	$C_6H_4(CHO)_2$	VII-675	134.14
5280	aldehyde (*p*)	....	$C_6H_4(CHO)_2$	VII-675	134.14
5281	amide (*o*)	phthaldiamide	$C_6H_4(CO \cdot NH_2)_2$	IX-814	164.17
5282	anhydride (*o*)	....	$C_6H_4(CO)_2O$	XVII-469	148.12
5283	dichloride (*o*) (*sym*)	phthalyl dichloride	$C_6H_4(COCl)_2$	IX-805	203.03
5284	dichloride (*o*)(*uns*)	phthalyl dichloride	$C_6H_4 \cdot CO \cdot O \cdot C:Cl_2$	IX-805	203.03
5285	dichloride (*m*)	isophthalyldichloride	$C_6H_4(COCl)_2$	IX-834	203.03
5286	dichloride (*p*)	terephthalyl diCl	$C_6H_4(COCl)_2$	IX-844	203.03
5287	imide (*o*)	*o*-phthalimide	$C_6H_4(CO)_2NH$	XXI-458	147.13
5288	nitrile (*m*)	isophthalic nitrile	$C_6H_4(CN)_2$	IX-836	128.13
5289	nitrile (*p*)	terephthalic nitrile	$C_6H_4(CN)_2$	IX-846	128.13
5290	**Phthalide**	....	$C_6H_4 \cdot CH_2 \cdot O \cdot CO$	XVII-310	134.14
5291	**Phthalonic acid**	(1,2)	$C_6H_4(CO_2H) \cdot CO \cdot CO_2H$	X-857	194.15
5292	**Phthalophenone**	triPh-carbinol-*o*-carboxylic anhyd.	$(C_6H_5)_2C \cdot C_6H_4 \cdot CO \cdot O$	XVII-391	286.33
5293	**Phytadiene**	....	$C_{20}H_{38}$	I-263	278.53
5294	**Phytane**	....	$C_{20}H_{42}$	I-174	282.56
5295	**Phytanol**	dihydrophytol	$C_{20}H_{41}OH$	I-431	298.56
5296	sodium	....	$C_{20}H_{41}ONa$	....	320.54
5297	**Phytene**	....	$C_{20}H_{40}$	I-227	280.54
5298	**Phytol**	....	$C_{20}H_{39}OH$	I-453	296.54
5299	**Picene**	....	$C_{22}H_{14}$	V-735	278.36
5300	**Picoline** (α)	2-methyl-pyridine	$CH_3 \cdot C_5H_4N$	XX-234	93.13
5301	**Picoline** (β)	3-methyl-pyridine	$CH_3 \cdot C_5H_4N$	XX-239	93.13

† See also Nos. 162-4.
Phosgene 1247
Phosphen-1 2707
Phosphen-2 1907
Phosphen-3 2713
Phosphen-4 2046
Phosphen-5 2762
Phosphen-7 6120
Phosphen-8 6165
Phosphen-9 6360
Phosphen-11 1146
Photo-glycin 3800
Photol 4117
Phthal alcohol 6497-9
Phthalamic acid 5272

Table 7-4 (*Continued*)
PHYSICAL CONSTANTS OF ORGANIC COMPOUNDS

No.	Crystalline Form and Color	Specific Gravity	Melting Point °C.	Boiling Point °C.	Solubility in 100 Parts		
					Water	Alcohol	Ether
5263	tan pd.	....	....	....	s.	sl. s.	i.
5264	lf.	....	253-5 d.	....	v. sl. s. h.	∞	$0.35^{16°}$ abs.
5265	cr.	....	43	175^{24mm}	i.	v. s.	v. o.
5266	pr./al.	....	52.5	255.5	i.	v. s.	v. s.
5267	pd.	....	expl. 155	....	v. sl. s.	v. sl. s.	sl. s. chl.
5268	yel. gn. pr.	$0.885^{20°}_{4}$	28	197.2^{743mm}	$0.1^{50°}$	s.	s.
5269	lq.	$0.923^{20°}_{20}$	−8.1	215-6	i.	....	....
5270	fum. lq.	$1.319^{20°}$	....	224.6	d.	∞ bz.	∞ CS_2
5271	lt. yel. pd.	....	149-50	....	i. h.	i.	i.; s. h. bz.
5272	pr.	....	148-9	155 d.	s.	s.	sl. s.
5273	nd./al.	....	208-10	subl.	i.	s.	....
5274	mn./aq.	$1.593^{20°}_{4}$	191* / 231†	....	$0.54^{14°}$; $18^{99°}$	$11.7^{18°}$ abs.	$0.68^{15°}$
5275	pr./aq.	....	$-H_2O$, 100	....	$10^{25°}$	....	i. bz.; i. pet.
5276	nd./h. aq.	....	347-8	subl.	$0.01^{25°}$; 0.2 h.	s.; s. ac.	i. bz.; i. pet.
5277	cr. or amor.	....	425*	subl. >300	0.001 c.	sl. s. h.	i.; s. alk.
5278	lt. yel. nd.	....	56	....	1.4 h.	v. s.	v. s.
5279	nd.	....	89-90	....	sl. s.	v. s.	i. pet.
5280	nd./aq.	....	115-6	$245-8^{771mm}$	1.7 h.	v. s.	s.
5281	cr.	....	221-3 d.	....	v. sl. s. c.	v. sl. s.	i.
5282	rhb.	$1.5274°$	131.5-2.0	284.5	v. sl. s.	s.	sl. s.
5283	col. lq.	$1.414^{25°}_{25}$	16	281.1	d.	d.	s.
5284	cr.	....	88-9	275.2^{720mm}	....	....	....
5285	cr.	....	41	276	d.	d.	....
5286	nd.	....	78-9	259	....	....	....
5287	cr./et.	....	238	subl.	$0.04^{25°}$	5; s. alk.	s. h.; i. bz.
5288	nd.	....	161.5	subl.	s. h.; i. lg.	v. s. h.	v. s. h.
5289	nd./bz.	....	224-6	....	i.; s. h. ac.	sl. s. h.	v. sl. s. h.
5290	nd./aq.	$1.164^{99°}_{4}$	73(65)	290	v. sl. s.	s.	....
5291	pr./bz. al.	....	144.5	....	$115^{15°}$	s.	s.; sl. s. chl.
5292	lf./al.	....	115	419-28 sl. d.	d. h.	s.; s. H_2SO_4	i. c. alk.
5293	lq.	$0.826^{0°}_{4}$	....	$185-8^{22mm}$	∞ pet.	∞ Me al.	∞ ac.
5294	lq.	$0.803^{0°}_{4}$	....	$169.5^{9.5mm}$	v. sl. s. ac.	v. sl. s.	....
5295	oil	$0.840^{20°}_{4}$	....	202^{10mm}	i.	s.	s.
5296	oil	....	....	....	....	s. pet.	s.
5297	col. oil	$0.817^{0°}_{4}$	....	$177-8^{10.5mm}$	v. s. pet.	sl. s. h.	sl. s.
5298	col. oil	$0.852^{20°}_{4}$	....	204^{10mm}	i.	∞ Me al.	∞
5299	col. lf.	....	364	518-20	i.	v. sl. s.	v. sl. s.
5300	col. lq.	$0.944^{20°}_{4}$	−66.6	129.4	v. s.	∞	∞
5301	col. lq.	$0.961^{15°}_{4}$	....	143.5	∞	∞	∞

* Sealed tube.
† Heated rapidly.
Phthaldiamide 5281
Phthalimide 5287
Phthalimino-ethyl bromide 940

Phthalin 5122
Phthalyl dichloride 5283-4
Phthiocol 4275
Phthiocol acetate 4276-7
Phthiocol oxime 4278

Phyllyrin, cf. glcde.
Physostigmine, cf. alkd.
Picein, cf. glcde.
Piceoside, cf. glcde.
Picoline dicarboxylic acid 6398

Table 7-4 (*Continued*)
PHYSICAL CONSTANTS OF ORGANIC COMPOUNDS

No.	Name	Synonym	Formula	Beil. Ref.	Formula Weight
5302	**Picoline** (γ)	4-methyl-pyridine	$CH_3 \cdot C_5H_4N$	XX-240	93.13
5303	**Picramic** acid	4,6-diNO_2-2-NH_2-phenol	$HO \cdot C_6H_2(NH_2)$: $(NO_2)_2$	XIII-394	199.12
5304	sodium salt	sodium picramate	$C_6H_4O_5N_3Na \cdot H_2O$	XIII-395	239.13
5305	**Picrolonic acid**	4-NO_2-3-Me-1-*p*-NO_2-phenylpyra-zolone-5	$C_{10}H_8O_5N_4$	XXIV-51	264.20
5306	**Picropodophyllin**	....	$C_{22}H_{22}O_8$	XIX-424	414.42
5307	**Picrorocellin**	....	$C_{20}H_{22}O_4N_2$	XXV-93	354.41
5308	**Picryl chloride** (1;2,4,6)	....	$Cl \cdot C_6H_2(NO_2)_3$	V-273	247.55
5309	**Pimelic** acid	heptandioic acid	$(CH_2)_5(CO_2H)_2$	II-670	160.17
5310	acid (*iso*)	....	$(C_2H_5)(CH_3)C \cdot CH_2(CO_2H)_2$	II-685	160.17
5311	nitrile	....	$(CH_2)_5(CN)_2$	II-671	122.17
5312	**Pinacol**	2,3-diMe-bu-tandiol-2,3	$[(CH_3)_2C \cdot OH]_2$	I-487	118.18
5313	hydrate	pinacone hydrate	$C_6H_{14}O_2 \cdot 6H_2O$	I-488	226.27
5314	**Pinacoline semi-carbazone**	....	$C_5H_{12}C:N \cdot NH \cdot CO \cdot NH_2$	III-104	157.22
5315	**Pinacolyl acetate**	....	$CH_3CO_2 \cdot C_6H_{13}$	II-133	144.22
5316	**Pinacyanole**	sensitol red	$C_{23}H_{20}N_2 \cdot C_2H_5I$	XXIII-320	480.40
5317	**Pinane** (*d*)	....	$C_{10}H_{18}$	V-93	138.25
5318	**Pinene** (α)(*dl*)	....	$C_{10}H_{16}$	V-144	136.24
5319	hydrochloride	artificial camphor	$C_{10}H_{16} \cdot HCl$	V-94	172.70
5320	**Pinol** (*dl*)	sobrerone	$C_{10}H_{16}O$	XVII-45	152.24
5321	**Piperic acid**	....	$CH_2O_2:C_6H_3 \cdot C_5H_5O_2$	XIX-281	218.21
5322	**Piperidine**	hexazane	$CH_2(CH_2)_4NH$ ‾‾‾‾‾‾‾‾‾‾	XX-6	85.15
5323	**Piperonal**	heliotropine	$CH_2O_2:C_6H_3 \cdot CHO$	XIX-115	150.14
5324	acetophenone	....	$C_8H_6O_2:CH \cdot COC_6H_5$	XIX-141	252.27
5325	**Piperonyl** alcohol	....	$CH_2O_2:C_6H_3 \cdot CH_2OH$	XIX-67	152.15
5326	phloroglucinol-dimethyl ether	protocotoin; (2,4-diMeO)	$(CH_3O)_2C_{14}H_8O_4$	XIX-242	302.29
5327	**Piperonylic acid**	....	$CH_2O_2:C_6H_3 \cdot CO_2H$	XIX-269	166.13
5328	**Proflavine**	3,6-diNH_2-acridine sulfate; trypa-flavine	$C_{13}H_{11}N_3 \cdot H_2SO_4 \cdot H_2O$	*XXII-650	325.35
5329	**Pregnan(e)diol**	23(α),20(α)	$C_{21}H_{36}O_2$	....	320.52
5330	**Pregnan(e)diol**	3(α),20(β)	$C_{21}H_{36}O_2$	....	320.52
5331	**Pregnan(e)diol**	3(β),20(α)	$C_{21}H_{36}O_2$	....	320.52
5332	**Pregnan(e)diol**	3(β),20(β)	$C_{21}H_{36}O_2$	....	320.52
5333	**Progesterone**(α)	....	$C_{21}H_{30}O_2$	....	314.47
5334	**Progesterone**(β]	progestin; corporin	$C_{21}H_{30}O_2$	....	314.47
5335	**Proline** (*l*–)	pyrrolidine-2-carboxylic acid	$HN \cdot (CH_2)_3 \cdot CH \cdot CO_2H$ ‾‾‾‾‾‾‾‾‾‾	XXII-2	115.13
5336	**Prominal**	1-Me-5-Et-5-Ph-barbituric acid	$(C_6H_5)(C_2H_5)(CH_3) \cdot C_4HO_3N_2$	....	246.27
5337	**Prontosil**† (4,1;1',2',4')	prontosil red; rubi-azol; streptozon	$H_2NSO_2 \cdot C_6H_4 \cdot N:N \cdot C_6H_3(NH_2)_2 \cdot HCl$	....	327.79
5338	**Propane**	....	$CH_3 \cdot CH_2 \cdot CH_3$	I-104	44.10
5339	*iso*-**Propanolamine**	α-NH_2-*iso*Pr alcohol	$NH_2 \cdot CH_2CHOH \cdot CH_3$	IV-289	75.11

† *p*-Aminobenzene-sulfonamide also sometimes called "Prontosil."

Table 7-4 (*Continued*)
PHYSICAL CONSTANTS OF ORGANIC COMPOUNDS

No.	Crystalline Form and Color	Specific Gravity	Melting Point °C.	Boiling Point °C.	Solubility in 100 Parts		
					Water	Alcohol	Ether
5302	lq.	$0.957_4^{15°}$	3.7	143.1	∞	∞	∞
5303	red nd./al.	....	169	....	$0.14^{22°}$; s. ac.	s.; sl. s. chl.	sl. s.; s. bz.
5304	red crust	....	....	....	216°	....	....
5305	yel. nd./al.	....	121-2	d. 125	$0.91^{17°}$; 0.93 h.	$4.0^{17°}$ 9 hot	$0.51^{17°}$; $0.6^{17°}$ Me al.
5306	col. nd./bz.	....	228	....	i.; s. alk.	s.	s. chl.
5307	col. pr./al.	....	190-220	....	i.; s. chl.	s. h.	sl. s.
5308	yel. mn.	$1.797^{20°}$	83	....	i.; d. aq. alk.; s.bz.	$4.8^{17°}$; s. h. chl.	$7^{17°}$; sl. s. pet.
5309	mn./aq.	$1.291_4^{25°}$	103-5	272^{100mm}	$2.5^{14°}$	v. s.	v. s.
5310	rhb./aq.	....	103-4	d. 135	$15.4^{15°}$	v. s.	v. s. h.
5311	col. lq.	$0.949^{18°}$	....	$175\text{-}6^{14mm}$	i.	∞	∞
5312	col. nd.	lq. $0.967^{15°}$	41-3(38)*	$171\text{-}2^{739mm}$	sl. s. c.; s. h.	v. s.	v. s.; sl. s. CS_2
5313	cr./aq.	....	46-7**	d. $-H_2O$	sl. s. c.	s.	s.
5314	nd./aq.	....	157	....	sl. s. c.	s.	v. s.
5315	col. lq.	....	....	143^{757mm}	....	....	....
5316	b. gn./al.	....	276-8	....	sl. s.	s.	s. pyr.
5317	col. lq.	$0.839_4^{20°}$	−45	169.4	....	....	....
5318	col. lq.	$0.878_4^{20°}$	−55	154-6	v. sl. s.	∞ abs.	∞
5319	cr. lf.	....	131-2	207-8	i.	33	s.
5320	lq.	$0.953_{20}^{20°}$	....	183-4	s. bz.	s.	s.
5321	yel. nd./al.	....	216-7	subl. sl. d.	i.; 2 h.	3.6 c.	s.; sl. s. bz
5322	lq.	$0.860_4^{20°}$	−10.5	106.4	∞	∞	....
5323	cr./aq.	....	37	263	$0.1^{0°}$; $0.667^{8°}$	$15^{0°}$; $550^{78°}$	∞
5324	yel. nd./al.	....	122	....	s. h.	....	....
5325	nd.	....	52-3	d.	s. h.	∞	∞
5326	lt. yel. mn./al.	....	141-2	....	i.; s. alk.; s. ac.	s.; i. alk. carb.	s.; s. bz.; s. chl.
5327	nd./al.	....	228	subl.	v. sl. s. h.	sl. s. h.	sl. s.
5328	red nd.	....	....	....	1 c.	s.	i.; i. chl.
5329	col. cr./act.	....	237-9	....	sl. s.	s.	s.
5330	col. cr./al.	....	232-4	....	sl. s.	s.	s.
5331	col. cr./et.	....	182-4	....	sl. s.	s.	s.
5332	cr./aq. al.	....	174-6	....	sl. s.	s.	s.
5333	pr.	....	128.5	....	i.	s.	s.
5334	nd./pet.	....	121	....	i.	s.	s.
5335	hyg. nd./ al. et.	....	220-2 d.	....	v. s.	$1.6^{19°}$	i.
5336	col. cr.	....	176	....	s. h.	s.	....
5337	red cr. pd.	....	247-51	....	0.25	....	....
5338	col. gas	(A)1.562 $0.585_4^{-44.5°}$	−187.7	−42.1	$6.5^{17.8°}$ cc.	$92^{616.6°}$ cc.	$129^{916.6°}$ cc.
5339	lq.	$0.973^{18°}$	0	160-1	v. s.	s.	i.

* Crysts. $+ 6H_2O$/aq., m. p. 47°.
** Anhydrous crysts./et., m. p. 38°; b. p. 172-5°.
Planocaine 4935
Poirrier's blue 6344
Polygonin, cf. glcde.
Populin, cf. glcde.
Porphyroxine, cf. alkd.
Potassium myronate, cf. glcde.

Pragmoline 82
Pregnitic acid 680
Prehnitol 5827
Prehnitylic acid 6257
Primulin, cf. glcde.
Procaine 4935

Progestin(e) 5334
Proline-betaine, cf. alkd.
Prontosil album 527
Propadiene 183
Propanal 5352
Propandioic acid 4002
Propandiol 5465, 6286
Propandithiol 6284

Table 7-4 (*Continued*)
PHYSICAL CONSTANTS OF ORGANIC COMPOUNDS

No.	Name	Synonym	Formula	Beil. Ref.	Formula Weight
5340	**Propargyl** acetate	. . . .	$CH_3CO_2 \cdot CH_2 \cdot C \vdots CH$	II-140	98.10
5341	alcohol	propyne-1-ol-3	$HC \vdots C \cdot CH_2OH$	I-454	56.06
5342	bromide	3-bromo-propyne-1	$HC \vdots C \cdot CH_2Br$	I-248	118.97
5343	chloride	chloro-allylene	$HC \vdots C \cdot CH_2Cl$	I-248	74.51
5344	iodide	3-iodo-propyne-1	$HC \vdots C \cdot CH_2I$	I-248	165.96
5345	**Propenyl** chloride†	1-chloro-propene-1	$CH_3 \cdot CH \vdots CHCl$	I-198	76.53
5346	chloride (*iso*)	2-chloro-propene-1	$CH_3 \cdot CCl \vdots CH_2$	I-198	76.53
5347	guaethol	(4;1,2)	$CH_3 \cdot CH \vdots CH \cdot C_6H_3 \vdots$ $(OH)(OC_2H_5)$	. . . .	178.23
5348	**Propiolic** acid	propynoic acid	$HC \vdots C \cdot CO_2H$	II-477	70.05
5349	aldehyde	propargyl aldehyde	$HC \vdots C \cdot CHO$	I-750	54.05
5350	**Propionaldol**	. . . .	$C_2H_5 \cdot CHOH \cdot CH \vdots$ $(CH_3)CHO$	I-836	116.16
5351	**Propionic** acid	propanoic acid	$CH_3 \cdot CH_2 \cdot CO_2H$	II-234	74.08
5352	aldehyde	oropanal	$CH_3 \cdot CH_2 \cdot CHO$	I-629	58.08
5353	aldehyde oxime	propionaldoxime	$C_2H_5 \cdot CH \vdots NOH$	I-631	73.10
5354	aldehyde semi-carbazone (*α*)	. . . .	$C_2H_5 \cdot CH \vdots N \cdot NH \cdot$ $CO \cdot NH_2$	III-101	115.14
5355	aldehyde semi-carbazone	. . . .	$C_2H_5 \cdot CH \vdots N \cdot NH \cdot$ $CO \cdot NH_2$	III-101	115.14
5356	amide	propionamide	$C_2H_5 \cdot CO \cdot NH_2$	II-243	73.10
5357	anhydride	. . . .	$(C_2H_5 \cdot CO)_2O$	II-242	130.14
5358	anilide	propionanilide	$C_2H_5 \cdot CO \cdot NH \cdot C_6H_5$	XII-250	149.19
5359	bromide	propionyl bromide	$C_2H_5 \cdot CO \cdot Br$	II-243	136.98
5360	chloride	propionyl chloride	$C_2H_5 \cdot CO \cdot Cl$	II-243	92.53
5361	fluoride	propionyl fluoride	$C_2H_5 \cdot CO \cdot F$	II-243	76.07
5362	iodide	propionyl iodide	$C_2H_5 \cdot CO \cdot I$	II-243	183.98
5363	nitrile	ethyl cyanide	$CH_3 \cdot CH_2 \cdot CN$	II-245	55.08
5364	**Propyl** acetanilide (*N*)	acetyl *n*-propyl-aniline	$C_6H_5 \cdot N(COCH_3) \cdot$ $CH_2 \cdot C_2H_5$	XII-246	177.25
5365	acetate (*n*)	. . . .	$CH_3CO_2 \cdot CH_2 \cdot C_2H_5$	II-129	102.13
5366	acetate (*iso*)	. . . .	$CH_3CO_2 \cdot CH(CH_3)_2$	II-130	102.13
5367	acetoacetate (*iso*)	. . . .	$CH_3COCH_2CO_2C_3H_7$	III-659	144.17
5368	alcohol (*n*)	propanol-1	$CH_3 \cdot CH_2 \cdot CH_2OH$	I-350	60.10
5369	alcohol (*iso*)	propanol-2	$(CH_3)_2CHOH$	I-360	60.10
5370	allyl barbituric acid (*iso*)	numal; alurate	$(C_3H_5)(C_3H_7) \vdots$ $C_4H_2O_3N_2$	. . . .	210.23
5371	amine (*n*)	1-aminopropane	$CH_3 \cdot CH_2 \cdot CH_2 \cdot NH_2$	IV-136	59.11
5372	amine (*iso*)	2-aminopropane	$(CH_3)_2CH \cdot NH_2$	IV-152	59.11
5373	*p*-aminobenzoate	propesine	$NH_2 \cdot C_6H_4 \cdot CO_2C_3H_7$	XIV-423	179.22
5374	aniline (*n*)(*N*)	. . . .	$C_6H_5 \cdot NH \cdot CH_2 \cdot C_2H_5$	XII-166	135.21
5375	aniline (*iso*)(*p*)	cumidine	$(CH_3)_2CH \cdot C_6H_4 \cdot NH_2$	XII-1147	135.21
5376	arsonic acid (*n*)	. . . .	$C_3H_7 \cdot AsO(OH)_2$	IV-615	168.02
5377	benzaldehyde (*iso*) (*p*)	*p*-cuminic aldehyde	$(CH_3)_2CH \cdot C_6H_4 \cdot CHO$	VII-318	148.21
5378	benzene (*n*)	. . . .	$C_6H_5 \cdot CH_2 \cdot C_2H_5$	V-390	120.20
5379	benzene (*iso*)	cumene	$C_6H_5 \cdot CH(CH_3)_2$	V-393	120.20
5380	benzoate (*n*)	. . . .	$C_6H_5 \cdot CO_2 \cdot CH_2 \cdot C_2H_5$	IX-112	164.21
5381	benzoate (*iso*)	. . . .	$C_6H_5 \cdot CO_2 \cdot CH(CH_3)_2$	IX-112	164.21
5382	benzoic acid (*n*)(*o*)	. . . .	$C_3H_7 \cdot C_6H_4 \cdot CO_2H$	IX-544	164.21
5383	benzoic acid (*iso*) (*o*)	*o*-cuminic acid	$(CH_3)_2CH \cdot C_6H_4 \cdot$ CO_2H	IX-546	164.21
5384	benzoic acid (*n*)(*p*)	. . . .	$C_3H_7 \cdot C_6H_4 \cdot CO_2H$	IX-545	164.21

† Properties for trans-; cis, m. p. −134.8°, b. p. 32.8°.

Propanoic acid 5351
Propanol 5368-9
Propanone 56
Propanselenol 5448
Propanthiol 5424-5
Propargyl aldehyde 5349

Propene 5459
Propenoic acid 149
Propenol 193
Propenyl anisole 511
Propenyl bromide (*iso*) 990
Propenyl carbinol (*iso*) 4059

Propenyl cyanide 1563
Propenyl guaiacol 1280, 3223
Propenyl trimethoxybenzene 584
Propesine 5373
Propine 225
Propionamide 5356

Table 7-4 (*Continued*)
PHYSICAL CONSTANTS OF ORGANIC COMPOUNDS

No.	Crystalline Form and Color	Specific Gravity	Melting Point °C.	Boiling Point °C.	Solubility in 100 Parts		
					Water	Alcohol	Ether
5340	col. lq.	1.005_{4}^{20}°	....	124-5	....	s.	s.
5341	col. lq.	0.972_{4}^{20}°	-17	114-5	s.	∞	∞
5342	lq.	1.58^{20}°	....	82	....	....	....
5343	lq.	1.045^{5}°	....	56	....	....	....
5344	lq.	2.018^{0}°	....	115	....	....	o.
5345	col. lq.	0.935_{5}^{15}°	-99	37	i.	∞ ; ∞ act.	∞ ; ∞ bz.
5346	lq.	0.918^{9}°	-137.4	22.65	i.	....	....
5347	col. pd.	....	85-7	....	i.	s.	s.
5348	col. lq.	1.139_{15}^{15}°	18	102^{200mm}	s.	s.	s.
5349	oil	....	....	59-61	v. s.	....	....
5350	col. lq.	0.986_{4}^{25}°	....	$84-6^{11mm}$	s.	....	....
5351	col. lq.	0.993_{4}^{20}°	-20.8	141.4	∞	∞	∞ ; ∞ chl.
5352	col. lq.	0.807_{4}^{20}°	-81	49.57^{40mm}	20^{20}°	∞	∞
5353	lq.	0.926_{4}^{20}°	21(40?)	130-2	....	....	....
5354	nd./bz. + lg.	....	88-90	....	v. s.	v. s. h. bz.	....
5355	pl./aq.	....	154	less s. bz. than α	....	....	....
5356	rhb. pl.	1.042_{4}^{20}°	79-80	222.2	v. s.	v. s.	v. s.
5357	col. lq.	1.012_{4}^{20}°	-45	167	d.	d.	....
5358	lf./al.	1.175	105-7	....	0.42^{4}°	s.; s. h. aq.	s.
5359	lq.	$1.521_{4}^{16.4}$°	....	103-4	d.	d.	s.
5360	col. lq.	1.065_{4}^{20}°	-94	80	d.	d.	s.
5361	col. lq.	0.972^{15}°	....	44	d.	d.	....
5362	lq.	....	....	127-8	d.	d.	....
5363	col. lq.	0.782_{4}^{20}°	-91.9	97.2	s.	∞	∞
5364	mn. pl./et. or lg.	....	47-8	266^{712mm}	i.	s.	s.
5365	col. lq.	0.886_{4}^{20}°	-95	101.6	1.6^{16}°	∞	∞
5366	col. lq.	0.874_{20}^{20}°	-73.4	88.4	3^{20}°	∞	∞
5367	col. lq.	....	....	185-7	v. sl. s.	s.	s.
5368	col. lq.	0.804_{4}^{20}°	-127	97.2	∞	∞	∞
5369	col. lq.	0.785_{4}^{20}°	-89.5	82.4	∞	∞	∞
5370	col. cr./aq.	....	137-8	....	sl. s.	s.	s.,
5371	col. lq.	0.718_{20}^{20}°	-83	$49-50^{761mm}$	∞ '	∞	∞
5372	col. lq.	0.694_{4}^{15}°	-101	33-4	∞	∞	∞
5373	col. pr.	....	74-6	....	v. sl. s.	s.; s. chl.	s.; s. bz.
5374	lq.	0.949^{18}°	....	222	i.	v. s.	v. s.
5375	lq.	0.953	<-20	225^{761mm}	i.	....	....
5376	col. nd./al.	....	126-7	....	v. s.	v. s.	i.
5377	lq.	0.978_{4}^{20}°	....	236-7	i.	s.	....
5378	col. lq.	0.862_{4}^{20}°	-99.5	159.2	i.	∞ ; s. SO_2	∞ ; i. NH_3
5379	col. lq.	0.862_{4}^{20}°	-96.0	152.4	i.	∞	∞
5380	col. lq.	1.021_{25}^{25}°	-51.6	231	i.	s.	s.
5381	col. lq.	1.010_{25}^{25}°	....	218.5	i.	s.	s.
5382	lf./aq. al.	....	58	272^{739mm}	sl. s. h.	s.	....
5383	pr./aq.	....	51	....	sl. s. h.	s.	s.; s. bz.
5384	pr./h. aq.	....	140-1	....	sl. s. h.	s.	s.; s. bz.

Table 7-4 (*Continued*)
PHYSICAL CONSTANTS OF ORGANIC COMPOUNDS

No.	Name	Synonym	Formula	Beil. Ref.	Formula Weight
5385	**Propyl** benzoic acid (*iso*) (*p*)	cuminic acid	$(CH_3)_2CH \cdot C_6H_4 \cdot CO_2H$	IX-546	164.21
5386	benzoic amide (*iso*)(*p*)	cuminic amide	$(CH_3)_2CH \cdot C_6H_4 \cdot CO \cdot NH_2$	IX-547	163.22
5387	*o*-benzoylbenzoate (*n*)	....	$C_6H_5 \cdot CO \cdot C_6H_4 \cdot CO_2 \cdot CH_2 \cdot C_2H_5$	....	268.32
5388	bromide (*n*)	1-bromopropane	$CH_3 \cdot CH_2 \cdot CH_2Br$	I-108	123.00
5389	bromide (*iso*)	2-bromopropane	$CH_3 \cdot CHBr \cdot CH_3$	I-108	123.00
5390	bromopropenyl barbituric acid	nostal; noctal	$(C_3H_7)(CH_3 \cdot CBr: CH):C_4H_2O_3N_2$	....	289.14
5391	*iso*-butyl ketone (*n*)	....	$C_2H_5 \cdot CH_2 \cdot CO \cdot C_4H_9$	I-706	128.22
5392	*n*-butyrate (*n*)	....	$C_3H_7 \cdot CO_2 \cdot C_3H_7$	II-271	130.19
5393	*iso*-butyrate (*n*)	....	$(CH_3)_2CHCO_2C_3H_7$	II-291	130.19
5394	*n*-butyrate (*iso*)	....	$C_2H_5 \cdot CH_2 \cdot CO_2 \cdot CH: (CH_3)_2$	II-271	130.19
5395	*iso*-butyrate (*iso*)	....	$C_3H_7 \cdot CO_2 \cdot C_3H_7$	II-291	130.19
5396	caproate (*n*)	....	$C_5H_{11} \cdot CO_2 \cdot CH_2 \cdot C_2H_5$	II-323	158.24
5397	caprylate (*n*)	....	$C_7H_{15} \cdot CO_2 \cdot CH_2 \cdot C_2H_5$	II-348	186.30
5398	carbamate (*n*)	....	$NH_2 \cdot CO_2 \cdot CH_2 \cdot C_2H_5$	III-28	103.12
5399	carbonate, ortho-	....	$C(OCH_2 \cdot C_2H_5)_4$	III-6	248.37
5400	chloride (*n*)	1-chloropropane	$CH_3 \cdot CH_2 \cdot CH_2Cl$	I-104	78.54
5401	chloride (*iso*)	2-chloropropane	$CH_3 \cdot CHCl \cdot CH_3$	I-105	78.54
5402	chloroformate (*n*)	*n*-Pr chloro-carbonate	$Cl \cdot CO_2 \cdot CH_2 \cdot C_2H_5$	III-11	122.55
5403	chloroformate (*iso*)	*iso*-Pr chlorocarb.	$Cl \cdot CO_2 \cdot CH(CH_3)_2$	III-12	122.55
5404	*iso*-cyanide (*n*)	*n*-propyl carbylamine	$C_2H_5 \cdot CH_2 \cdot N:C$	IV-141	69.11
5405	*iso*-cyanide (*iso*)	*iso*-Pr carbylamine	$(CH_3)_2CH \cdot N:C$	IV-154	69.11
5406	fluoride (*n*)	1-fluoropropane	$CH_3 \cdot CH_2 \cdot CH_2F$	I-104	62.09
5407	fluoride (*iso*)	2-fluoropropane	$CH_3 \cdot CHF \cdot CH_3$	I-104	62.09
5408	formate (*n*)	....	$H \cdot CO_2 \cdot CH_2 \cdot C_2H_5$	II-21	88.11
5409	formate (*iso*)	....	$H \cdot CO_2 \cdot CH(CH_3)_2$	II-21	88.11
5410	formate, ortho-(*n*)	....	$H \cdot C(O \cdot CH_2 \cdot C_2H_5)_3$	II-21	190.29
5411	furoate (*n*)	....	$C_4H_3O \cdot CO_2 \cdot C_3H_7$	XVIII-275	154.17
5412	β-furylacrylate (*n*)	....	$C_4H_3O \cdot CH:CH \cdot CO_2 \cdot CH_2 \cdot C_2H_5$	....	180.21
5413	glycolate (*iso*)	....	$HOCH_2 \cdot CO_2 \cdot C_3H_7$	**III-172	118.13
5414	*n*-hexyl ketone (*n*)	decanone-4	$C_3H_7 \cdot CO \cdot C_6H_{13}$	I-711	156.27
5415	*p*-hydroxybenzoate	(*n*); nipasol	$HOC_6H_4CO_2 \cdot C_3H_7$	X-160	180.21
5416	hydroxylamine (β)	....	$C_2H_5 \cdot CH_2 \cdot NHOH$	IV-537	75.11
5417	iodide (*n*)	1-iodopropane	$CH_3 \cdot CH_2 \cdot CH_2I$	I-113	169.99
5418	iodide (*iso*)	2-iodopropane	$CH_3 \cdot CHI \cdot CH_3$	I-114	169.99
5419	lactate (*n*)	....	$C_2H_4(OH)CO_2C_3H_7$	III-265	132.16
5420	lactate (*iso*)	....	$C_2H_4(OH)CO_2C_3H_7$	III-282	132.16
5421	levulinate (*n*)	....	$C_4H_7O \cdot CO_2C_3H_7$	III-675	158.20
5422	malonic acid (*n*)	....	$C_3H_7 \cdot CH(CO_2H)_2$	II-657	146.14
5423	malonic acid (*iso*)	....	$C_3H_7 \cdot CH(CO_2H)_2$	II-669	146.14
5424	mercaptan (*n*)	propanthiol-1	$CH_3 \cdot CH_2 \cdot CH_2 \cdot SH$	I-359	76.16
5425	mercaptan (*iso*)	propanthiol-2	$(CH_3)_2CH \cdot SH$	I-367	76.16
5426	α-naphthylamine (*n*)	....	$C_{10}H_7 \cdot NH \cdot CH_2 \cdot C_2H_5$	XII-1224	185.27
5427	nitrate (*n*)	....	$CH_3 \cdot CH_2 \cdot CH_2 \cdot O \cdot NO_2$	I-355	105.09
5428	nitrite (*n*)	....	$CH_3 \cdot CH_2 \cdot CH_2 \cdot O \cdot NO$	I-355	89.09

Table 7-4 (*Continued*)
PHYSICAL CONSTANTS OF ORGANIC COMPOUNDS

No.	Crystalline Form and Color	Specific Gravity	Melting Point °C.	Boiling Point °C.	Solubility in 100 Parts		
					Water	Alcohol	Ether
5385	tri.	1.162	116-7	subl.	$0.015^{25°}$	s.; s. conc. H_2SO_4	s.
5386	nd. or pl.	....	155	....	sl. s. h.	∞	sl. s.
5387	col. lq.	....	....	$223-5^{15mm}$	i.	s.	s.
5388	col. lq.	$1.353^{20°}_{4}$	-109.9	71.0	$0.25^{20°}$	∞	∞
5389	col. lq.	$1.310^{20°}_{4}$	-90.0	59.4	$0.32^{20°}$	∞; ∞ chl.	∞; ∞ bz.
5390	col. cr.	....	178	....	s^{l}. s.	s.	sl. s.
5391	lq.	$0.813^{22°}_{2}$	....	155^{750mm}	i.; s. alk.	s.; s. act.	s.; sl. s. chl.
5392	col. lq.	$0.879^{15°}$	-95.2	142.7	$0.17^{17°}$	∞	∞
5393	col. lq.	$0.884^{°}_{4}$	....	134-5	....	....	....
5394	col. lq.	$0.865^{13°}$	....	128	....	....	....
5395	col. lq.	$0.847^{21°}_{4}$	....	120.8	....	....	....
5396	col. lq.	$0.863^{25°}_{4}$	-68.7	187.2	....	....	....
5397	col. lq.	$0.880^{0°}_{0}$	-43.0	226.4	....	....	....
5398	pr.	....	60-1	200	v. s.	v. s.	v. s.
5399	col. lq.	$0.911^{8°}$	....	224.2	....	....	....
5400	col. lq.	$0.890^{20°}_{4}$	-122.8	46.7	$0.27^{20°}$	∞	∞
5401	col. lq.	$0.859^{20°}$	-117	34.8	$0.31^{20°}$	∞	∞
5402	col. lq.	$1.090^{20°}_{4}$	....	$114-5^{768mm}$	i.; sl. d.	i.; sl. d.	∞; ∞ bz.
5403	col. lq.	....	....	$103-5^{721mm}$	i.	v. s.	v. s.
5404	lq.	$0.753^{25°}$	....	99.5	i.	∞	∞
5405	lq.	$0.760^{0°}$	....	87	i.	∞	∞
5406	gas	....	....	-3	....	....	....
5407	gas	....	....	-11	....	....	....
5408	col. lq.	$0.901^{20°}_{4}$	-92.9	80.9	$2.2^{22°}$	∞	∞
5409	lq.	$0.873^{20°}_{4}$	....	$68-71^{751mm}$	$2.1^{22°}$	....	....
5410	col. lq.	$0.881^{20°}_{4}$	....	196-8	$2.1^{22°}$	....	....
5411	col. lq.	$1.075^{26°}_{4}$	....	211	v. sl. s.	s.	∞
5412	col. lq.	$1.074^{20°}_{4}$	....	119^{7mm}	i.	s.	....
5413	col. lq.	$1.043^{18°}_{4}$	....	164	s.	s.	s.
5414	lq.	$0.824^{21°}_{0}$	-9	206-7	v. sl. s.	∞	∞
5415	col. pr./et.	....	95-6	....	$0.28^{0°}$	$100^{25°}$ Me al.	$50^{25°}$
5416	nd./et.	....	46±	....	v. s.	v. s.	sl. s. lg.
5417	lq.	$1.743^{20°}_{4}$	-98.7	102.5	$0.11^{20°}$	∞	∞
5418	lq.	$1.714^{15°}_{4}$	-90.0	89.5	$0.14^{20°}$	∞; ∞ chl.	∞; ∞ bz.
5419	col. lq.	....	....	$122-3^{150mm}$	s.	s.	s.
5420	col. lq.	....	....	167.5	s.	s.	s.
5421	lq.	$0.990^{20°}_{4}$	....	221	....	....	....
5422	pl./bz.	....	96	d.	v. s.	v. s.	....
5423	col. pr.	....	87	d. 180	....	v. s.	v. s.
5424	lq.	$0.836^{25°}_{4}$	-112	67-8	v. sl. s.	s.	s.
5425	lq.	$0.809^{25°}_{4}$	-130.7	58-60	sl. s.	∞	∞
5426	lt. yel. oil	....	....	$316-8^{771mm}$	i.	....	....
5427	lq.	$1.058^{20°}_{4}$	....	110.5	....	s.	s.
5428	lq.	$0.935^{20°}_{4}$	....	57	....	s.	s.

Table 7-4 (*Continued*)
PHYSICAL CONSTANTS OF ORGANIC COMPOUNDS

No.	Name	Synonym	Formula	Beil. Ref.	Formula Weight
	Propyl				
5429	nitroamine (*n*)	....	$C_2H_5 \cdot CH_2 \cdot NH \cdot NO_2$	IV-570	104.11
5430	phenol (*o*)(*n*)	....	$C_2H_5 \cdot CH_2 \cdot C_6H_4 \cdot OH$	VI-499	136.20
5431	phenol (*m*)(*n*)	....	$C_2H_5 \cdot CH_2 \cdot C_6H_4 \cdot OH$	VI-499	136.20
5432	phenol (*p*)(*n*)	....	$C_2H_5 \cdot CH_2 \cdot C_6H_4 \cdot OH$	VI-500	136.20
5433	phenol (*o*)(*iso*)	*o*-cumenol	$(CH_3)_2CH \cdot C_6H_4 \cdot OH$	VI-504	136.20
5434	phenol (*m*)(*iso*)	*m*-hydroxy-cumene	$(CH_3)_2CH \cdot C_6H_4 \cdot OH$	VI-505	136.20
5435	phenol (*p*)(*iso*)	*p*-hydroxy-cumene	$(CH_3)_2CH \cdot C_6H_4 \cdot OH$	VI-505	136.20
5436	phenyl carbinol (*n*)	....	$C_3H_7 \cdot CHOH \cdot C_6H_5$	VI-522	150.22
5437	phenyl carbinol (*iso*)	....	$C_3H_7 \cdot CHOH \cdot C_6H_5$	VI-523	150.22
5438	phenyl ether (*n*)	....	$C_2H_5 \cdot CH_2 \cdot O \cdot C_6H_5$	VI-142	136.20
5439	phenyl ketone (*n*)	butyrophenone	$C_2H_5 \cdot CH_2 \cdot CO \cdot C_6H_5$	VII-313	148.21
5440	phenyl ketone (*iso*)	*iso*-butyrophenone	$(CH_3)_2CH \cdot CO \cdot C_6H_5$	VII-316	148.21
5441	propionate (*n*)	....	$C_2H_5 \cdot CO_2 \cdot CH_2 \cdot C_2H_5$	II-240	116.16
5442	propionate (*iso*)	....	$C_2H_5 \cdot CO_2 \cdot CH(CH_3)_2$	II-241	116.16
5443	pyridine (2)(α)(*n*)	conyrine	$C_2H_5 \cdot CH_2 \cdot C_5H_4N$	XX-247	121.18
5444	pyridine (2)(α)(*iso*)	....	$(CH_3)_2CH \cdot C_5H_4N$	XX-247	121.18
5445	pyridine (4)(γ)(*iso*)	....	$(CH_3)_2CH \cdot C_5H_4N$	XX-248	121.18
5446	salicylate (*o*)(*n*)	....	$HO \cdot C_6H_4 \cdot CO_2 \cdot C_3H_7$	X-75	180.21
5447	salicylate (*o*)(*iso*)	....	$HO \cdot C_6H_4 \cdot CO_2 \cdot C_3H_7$	....	180.21
5448	selenomercaptan	propanselenol-1	$CH_3 \cdot CH_2 \cdot CH_2 \cdot SeH$	I-360	123.06
5449	succinic acid (*n*)	....	$C_3H_7 \cdot C_2H_3(CO_2H)_2$	II-675	160.17
5450	*iso*-thiocyanate (*n*)	*n*-propyl mustard oil	$CH_3 \cdot CH_2 \cdot CH_2 \cdot N:CS$	IV-145	101.17
5451	thiocyanate (*iso*)	....	$(CH_3)_2CH \cdot S \cdot CN$	III-177	101.17
5452	*p*-toluene sulfonate	(*n*)	$CH_3C_6H_4SO_3C_3H_7$	**XI-45	214.29
5453	urea (*n*)	....	$C_3H_7 \cdot NH \cdot CO \cdot NH_2$	IV-142	102.14
5454	urethane (*n*)	....	$C_3H_7 \cdot NH \cdot CO_2 \cdot C_2H_5$	IV-143	131.18
5455	*n*-valerate (*n*)	....	$C_4H_9CO_2 \cdot C_3H_7$	II-301	144.22
5456	*iso*-valerate (*n*)	....	$C_4H_9CO_2 \cdot CH_2 \cdot C_2H_5$	II-312	144.22
5457	*iso*-valerate (*iso*)	....	$C_4H_9CO_2 \cdot CH(CH_3)_2$	II-312	144.22
5458	**Propylal** (*n*)	methylene diPr ether	$CH_2(OCH_2 \cdot C_2H_5)_2$	I-575	132.20
5459	**Propylene**	propene	$CH_3 \cdot CH:CH_2$	I-196	42.08
5460	bromohydrin (*prim*)	β-Br-propyl alcohol	$CH_3 \cdot CHBr \cdot CH_2OH$	*I-181	139.00
5461	chlorohydrin (*prim*)	2-chloro-propanol-1	$CH_3 \cdot CHCl \cdot CH_2OH$	I-356	94.54
5462	chlorohydrin (*sec*)	1-chloro-propanol-2	$CH_3 \cdot CHOH \cdot CH_2Cl$	I-363	94.54
5463	cyanide	....	$CH_3 \cdot CH(CN) \cdot CH_2 \cdot CN$	II-640	94.12
5464	diamine (*dl*)	1,2-diamino-propane	$CH_3 \cdot CH(NH_2) \cdot CH_2 \cdot NH_2$	IV-257	74.13
5465	glycol (α)	propandiol-1,2	$CH_3 \cdot CHOH \cdot CH_2OH$	I-472	76.10
5466	glycol acetate, di-	....	$(CH_3CO_2)_2CH_2 \cdot CH(CH_3)$	II-142	160.17
5467	glycol acetate, mono-(α)	....	$CH_3CO_2 \cdot CH_2 \cdot CHOH \cdot CH_3$	II-142	118.13
5468	oxide (1,2)	....	$CH_3 \cdot CH \cdot CH_2 \cdot O$ (ring)	XVII-6	58.08
5469	oxide (1,3)	trimethylene oxide	$CH_2 \cdot CH_2 \cdot CH_2 \cdot O$ (ring)	XVII-6	58.08
5470	**Prostigmine bromide**	(1,3)	$Br(CH_3)_3N \cdot C_6H_4 \cdot O_2C \cdot N(CH_3)_2$	....	303.21

Table 7-4 (*Continued*)
PHYSICAL CONSTANTS OF ORGANIC COMPOUNDS

No.	Crystalline Form and Color	Specific Gravity	Melting Point °C.	Boiling Point °C.	Solubility in 100 Parts		
					Water	Alcohol	Ether
5429	col. lq.	$1.104^{15°}$	-22	$128\text{-}9^{40mm}$	sl. s.	∞	∞
5430	lq.	$1.015^{0°}$	....	221-6	v. sl. s.	s.	s.
5431	lq.	....	26	228	v. sl. s.	s.	....
5432	cr.	$1.009^{0°}$	21-2	230-2	v. sl. s.	s.	....
5433	lq.	$1.012^{20°}$	15-6	214-5	v. sl. s.	∞	∞
5434	cr.	....	26	228	v. sl. s.	....	....
5435	nd.	$0.990^{20°}$	61	228.2-9.2	v. sl. s.	$316^{25°}$	$350^{25°}$
5436	oil	$1.021^{\frac{18}{4}°}$	....	$168\text{-}70^{100mm}$	....	....	....
5437	oil	$0.987^{\frac{13.7°}{4}}$	....	218-21	....	....	....
5438	col. lq.	$0.953^{\frac{15}{15}°}$	....	189-90	....	....	....
5439	lq.	$0.990^{\frac{18}{4}°}$	11	231^{727mm}	i.	∞	∞
5440	lq.	$0.985^{\frac{20}{20}°}$	....	220^{746mm}	i.	s.	s.
5441	col. lq.	$0.883^{\frac{20}{4}°}$	-75	122.6	$0.56^{25°}$	∞	∞
5442	col. lq.	$0.893^{0°}$	....	$109\text{-}11^{750mm}$	$0.6^{25°}$	....	....
5443	lq.	<1	....	165-8	....	....	....
5444	lq.	$0.934^{0°}$	....	158-9	sl. s.	....	....
5445	lq.	$0.944^{0°}$	....	177-8	....	sl. s.	....
5446	col. lq.	$1.099^{15°}$	....	238-40	v. sl. s.	∞	∞
5447	col. lq.	$1.010^{25°}$	....	$120\text{-}2^{18mm}$	....	....	....
5448	oil	$1.302^{\frac{20}{4}°}$	....	84	....	....	....
5449	col. cr.	....	92-3	....	s.	2.8 c. chl.	....
5450	lq.	$0.978^{\frac{16}{4}°}$	....	152.7^{743mm}	....	....	....
5451	lq.	$0.963^{20°}$	....	$152\text{-}3^{754mm}$	i.	∞	∞
5452	col. lq.	$1.144^{\frac{20}{4}°}$	<-20	$164\text{-}6^{10mm}$	i.	s.	s.
5453	cr.	....	107	....	s.	....	....
5454	lq.	....	....	192-3	....	....	....
5455	lq.	$0.874^{15°}$	-70.7	167.5	i.	∞	∞
5456	col. lq.	$0.863^{\frac{20}{4}°}$	....	155.9	i.	∞ ; ∞ chl.	∞
5457	col. lq.	$0.854^{17°}$	....	142^{756mm}	....	....	....
5458	lq.	$0.834^{20°}$	-97	137-8	....	....	....
5459	col. gas	(A)1.498 $0.609^{\frac{-47}{4}°}$	-185.3	-47.7	44.6 cc.	1200 cc.	500 cc. ac.
5460	lq.	....	....	$52\text{-}3^{16mm}$	....	....	....
5461	col. lq.	$1.103^{20°}$	....	133-4	s.	s.	s.
5462	col. lq.	$1.115^{\frac{20}{20}°}$	....	$126\text{-}7^{762mm}$	∞	∞	....
5463	col. lq.	....	12	252-4	....	....	....
5464	col. lq.	$0.878^{15°}$	....	119-20	∞	....	....
5465	col. oil	$1.040^{19.4°}$	....	188-9	∞	∞	8
5466	col. lq.	....	-31	190.2	10	....	....
5467	col. lq.	$1.055^{20°}$	....	182-3	s.	....	....
5468	col. lq.	$0.831^{\frac{20}{20}°}$	-104.4	35	$33^{30°}$	∞	∞
5469	lq.	....	....	50	∞	....	....
5470	wh. cr. pd.	....	176 d.	....	v. s.	....	....

Table 7-4 (*Continued*)
PHYSICAL CONSTANTS OF ORGANIC COMPOUNDS

No.	Name	Synonym	Formula	Beil. Ref.	Formula Weight
5471	**Pulegol** (*iso*)(*d*)	....	$C_{10}H_{17}OH$	VI-65	154.25
5472	**Pulegone**	....	$C_{10}H_{16}O$	VII-81	152.24
5473	**Purine**	....	$C_5H_4N_4$	XXVI-354	120.11
5474	**Putrescine**	....	$NH_2 \cdot (CH_2)_4 \cdot NH_2$	IV-264	88.15
5475	hydrochloride	....	$C_4H_{12}N_2 \cdot 2HCl$	IV-264	161.08
5476	**Pyrantin**	*p*-ethoxyphenyl-succinimide	$C_2H_5O \cdot C_6H_4 \cdot N(COCH_2)_2$	XXI-377	219.24
5477	**Pyrazine**	1,4-diazine	$N{:}CH{\cdot}CH{:}N{\cdot}CH{:}CH$	XXIII-91	80.09
5478	**Pyrazole**	....	$N{:}CH{\cdot}CH{:}CH{\cdot}NH$	XXIII-39	68.08
5479	**Pyrazoline**	....	$N{:}CH{\cdot}CH_2{\cdot}CH_2{\cdot}NH$	XXIII-28	70.09
5480	**Pyrazolone** (5)	1,2-pentadiazenone	$CO{\cdot}CH_2{\cdot}CH{:}N{\cdot}NH$	XXIV-13	84.08
5481	**Pyrene**	....	$C_{16}H_{10}$	V-693	202.26
5482	**Pyridazine**	1,2-diazine	$N{:}CH{\cdot}CH{:}CH{\cdot}CH{:}N$	XXIII-89	80.09
5483	**Pyridine**	....	C_5H_5N	XX-181	79.10
5484	hydrochloride	....	$C_5H_5N \cdot HCl$	XX-189	115.57
5485	hydrochloride	....	$C_5H_5N \cdot 2HCl$	XX-189	152.02
5486	carboxylic acid (2)	picolinic acid (2)(α)	$C_5H_4N \cdot CO_2H$	XXII-33	123.11
5487	carboxylic HCl (2)	α-picolinic HCl	$C_5H_4N \cdot CO_2H \cdot HCl$	XXII-34	159.57
5488	carboxylic acid (3)	nicotinic acid (3)(β)	$C_5H_4N \cdot CO_2H$	XXII-38	123.11
5489	carboxylic HCl (3)	nicotinic HCl	$C_5H_4N \cdot CO_2H \cdot HCl$	XXII-39	159.57
5490	carboxylic HNO₃	nicotinic nitrate	$C_5H_4N \cdot CO_2H \cdot HNO_3 \cdot H_2O$	XXII-39	204.14
5491	3–carboxylic acid flavianate	nicotinic flavianate	$C_6H_5O_2N \cdot C_{10}H_6O_8N_2S$	....	437.34
5492	3–carboxylic acid picrate	nicotinic acid picrate	$C_6H_5O_2N \cdot C_6H_3O_7N_3$	*XXII-503	352.22
5493	3–carboxylic amide	nicotin(ic)amide	$C_5H_4N \cdot CONH_2$	XXII-40	122.13
5494	3–carboxylic amide flavianate	nicotinamide flavianate	$C_6H_6ON_2 \cdot C_{10}H_6O_8N_2S$	....	436.36
5495	3–carboxylic amide picrate	nicotinamide picrate	$C_6H_6ON_2 \cdot C_6H_3O_7N_3$	....	351.23
5496	carboxylic acid (4)	*iso*-nicotinic acid (4)	$C_5H_4N \cdot CO_2H$	XXII-45	123.11
5497	dicarboxylic acid (2,3)	quinolinic acid	$C_5H_3N(CO_2H)_2$	XXII-150	167.12
5498	dicarboxylic acid (2,4)	lutidinic acid (2,4)	$C_5H_3N(CO_2H)_2 \cdot H_2O$	XXII-153	185.14
5499	dicarboxylic acid (2,5)	*iso*-cinchomeronic acid	$C_5H_3N(CO_2H)_2 \cdot H_2O$	XXII-153	185.14
5500	dicarboxylic acid (2,6)	dipicolinic acid	$C_5H_3N(CO_2H)_2 \cdot 1\frac{1}{2}H_2O$	XXII-154	194.15
5501	dicarboxylic acid (3,4)	cinchomeronic acid	$C_5H_3N(CO_2H)_2$	XXII-155	167.12
5502	dicarboxylic acid (3,5)	dinicotinic acid	$C_5H_3N(CO_2H)_2$	XXII-160	167.12.
5503	pentacarboxylic acid	....	$C_5N(CO_2H)_5 \cdot 2H_2O$	XXII-190	335.18
5504	pentacarboxylic acid	....	$C_5N(CO_2H)_5 \cdot 3H_2O$	XXII-190	353.20
5505	sulfonic acid (3)	....	$C_5H_4N \cdot SO_3H$	XXII-387	159.16

Puking stuff gas 4697
Pulegomenthol 4025
Pulsatilla camphor 510
Punicine, cf. alkd.
Purpurin 6213-4

Purpuroxanthin 2298
Purpuroxanthin carboxylic acid 2305
Pyracetosalyl 566
Pyramidon 363

Pyrantone-A 1663
Pyridone 3812-4
Pyridyl-methyl-pyrrolidine, cf. alkd.
Pyridyl-piperidine, cf. alkd.
Pyro alcohol 4100

Table 7-4 (*Continued*)
PHYSICAL CONSTANTS OF ORGANIC COMPOUNDS

No.	Crystalline Form and Color	Specific Gravity	Melting Point °C.	Boiling Point °C.	Solubility in 100 Parts		
					Water	Alcohol	Ether
5471	col. lq.	$0.911\frac{20°}{4}$	....	$86\text{-}9^{10mm}$	v. sl. s.	....	....
5472	col. lq.	$0.932\frac{20°}{20}$	....	224^{754mm}	i.	∞	∞
5473	cr./al.	....	217	d.	s.	s. h..	s. toluene
5474	cr.	....	27-8	158-60	v. s.	....	....
5475	nd./aq.	$0.877\frac{25°}{4}$	>290	....	v. s.	i. Me al.	....
5476	pr./al.	....	155-8	....	$0.08^{17°}$; $1.2^{100°}$	s. h.	i.
5477	pr./aq.	$1.031\frac{61°}{4}$	52-3	118^{768mm}	∞	s.	s.; s. HCl
5478	nd./et.	....	70	186-8	s.	s.	s.; s. bz.
5479	lq.	....	....	144	∞	∞	sl. s.
5480	nd./toluene	....	165	subl. d.	s.	v. s.	v. sl. s.
5481	lt. yel. pr.	$1.277\frac{0°}{4}$	150.4	>360	i.	3.1 h. abs.	v. s.
5482	lq.	$1.107\frac{20°}{4}$	-8	208	∞; s. HCl	s.; i. lg.	s.; s. bz.
5483	col. lq.	$0.983\frac{20°}{4}$	-41.5	115.5*	∞	∞	s.
5484	hyg. pl./al.	....	82	218-9	s.	s.	i.; s. chl.
5485	pr.	....	46-7	d. > 55	....	....	i.
5486	nd./al.	....	137-9	d. $-CO_2$	v. s.	v. s.	v. sl. s.
5487	rhb.	....	210-25 d.	....	....	....	....
5488	nd./al.	....	235.2	subl.	s. h.	s. h.	v. sl. s.
5489	pr./aq.	....	274-5	....	....	s.	....
5490	lf. or pr./aq.	....	192-4	....	....	....	....
5491	yel. pl./al.	....	249-50	....	s. h.	sl. s. abs.	i.
5492	pa. yel. pl. or nd./aq.	....	225-7 d.	....	s. h.	sl. s. abs.	i.
5493	col. nd./bz.	....	133	....	100	66.6	v. sl. s.
5494	pa. yel. cr./75% al.	....	269-70 d.	....	s.	sl. s. abs.	i.
5495	yel. pl. or nd.	....	193	....	s. h.	sl. s. abs.	i.
5496	nd./aq.	....	317**	d.	s. h.	sl. s. h.	v. sl. s.
5497	col. mn./aq.	....	d. 110 (slow)	d. > 190 (rapid)	0.67°; s. h.	sl. s.; s. alk.	0.02; i. bz.
5498	pl./aq.	0.942	248-50	....	v. s. h.	s. h.	i.
5499	cr./aq.	....	236-7 (anh.)	subl. d.	sl. s. h.; s. h. HCl	v. sl. s.	v. sl. s.
5500	nd./aq.	....	226 d. (anh.)	....	sl. s. c.	v. sl. s.	v. sl. s.
5501	cr./HCl	....	258-9 d.	subl. d.	sl. s. h.	sl. s.	i.; i. chl.
5502	cr.	....	323 d.	subl. sl. d.	v. sl. s.	v. sl. s.	v. sl. s.; s. HCl
5503	cr./et.	....	$-H_2O$, 100	d.	s.	....	v. sl. s.
5504	cr./aq.	....	d. 220 (anh.)	....	s.	....	v. sl. s.
5505	nd. or lf.	....	d.	....	v. s.	v. sl. s.	i.

* Liquid + $3H_2O$, b. p. 92-3°.
** Sealed tube.

Pyrocatechin 2312	Pyrocatechuic acid 2312	Pyroligneous spirit 4100
Pyrocatechol 2312	Pyrodin 111	Pyromellitic acid 682
Pyrocatechol ethyl ether 2927	Pyrogallic acid 6216	Pyromucic acid 3340
	Pyrogallol 6216	Pyromuc(ic)amide 3341
	Pyrogallol carboxylic acid 6220, 6223	Pyromucic anhydride 3343

Table 7-4 (*Continued*)
PHYSICAL CONSTANTS OF ORGANIC COMPOUNDS

No.	Name	Synonym	Formula	Beil. Ref.	Formula Weight
5506	**Pyridine** tricarboxy- lic acid (2,3,4)	α-carbocinchomer- onic acid	$C_5H_2N(CO_2H)_3 \cdot$ $1\frac{1}{2}H_2O$	XXII-182	238.15
5507	tricarboxylic acid (2,4,5)	berberonic acid	$C_5H_2N(CO_2H)_3 \cdot$ $2H_2O$	XXII-185	247.16
5508	tricarboxylic acid (2,4,6)	trimesitinic acid	$C_5H_2N(CO_2H)_3 \cdot$ $2H_2O$	XXII-185	247.16
5509	tricarboxylic acid (3,4,5)	β-carbocinchomer- onic acid	$C_5H_2N(CO_2H)_3 \cdot$ $3H_2O$	XXII-186	265.17
5510	**Pyridoxin**	2-Me-3-OH-4,5-di- (hydroxymethyl)- pyridine	$C_8H_{11}O_3N$	....	169.18
5511	hydrochloride	....	$C_8H_{11}O_3N \cdot HCl$	....	205.64
5512	3-methyl ether	....	$CH_3O \cdot C_8H_{10}O_2N$	....	183.21
5513	**Pyrimidine**	*m*-diazine (1,3)	CH:CH·CH:N·CH:N	XXIII-89	80.09
5514	**Pyrocoll**	....	$C_4H_3 \cdot N:(CO)_2:N \cdot$ C_4H_3	XXIV-403	186.17
5515	**Pyrogallol** dimethyl ether (1,3;2)	2,6-diMeO-phenol	$(CH_3O)_2C_6H_3 \cdot OH$	VI-1081	154.17
5516	triacetate (1,2,3)	lenigallol	$C_6H_3(O_2CCH_3)_3$	VI-1083	252.23
5517	trimethyl ether (1,2,3)		$C_6H_3(OCH_3)_3$	VI-1081	168.19
5518	**Pyrone** (1,4)	....	$C_5H_4O_2$	XVII-271	96.09
5519	**Pyronine**	....	$C_{17}H_{19}ON_2 \cdot Cl$	XVIII-596	302.81
5520	**Pyrotartaric acid** (*dl*)	methyl succinic acid	$HO_2C(CH_3)CH \cdot$ $CH_2 \cdot CO_2H$	II-636	132.12
5521	**Pyrrole**	....	$(\cdot CH:CH)_2:NH$	XX-159	67.09
5522	carboxylic acid (α)	(2)	$C_4H_3NH(CO_2H)$	XXII-22	111.10
5523	**Pyrrolidine**	tetrahydropyrrole	$(CH_2)_4:NH$	XX-4	71.12
5524	**Pyrroline**	dihydropyrrole	$CH_2 \cdot CH_2 \cdot NH \cdot CH:CH$	XX-133	69.11
5525	**Pyruvic** acid	pyro-racemic acid	$CH_3 \cdot CO \cdot CO_2H$	III-608	88.06
5526	amide	....	$CH_3 \cdot CO \cdot CO \cdot NH_2$	III-620	87.08
5527	nitrile	acetyl cyanide	$CH_3 \cdot CO \cdot CN$	III-620	69.06
5528	**Quinaldine** car- boxylic acid (3)	2-Me-quinoline-3- carboxylic acid	$CH_3 \cdot C_9H_5N \cdot CO_2H$	XXII-83	187.20
5529	4-carboxylic acid	aniluvitonic acid	$CH_3 \cdot C_9H_5N \cdot CO_2H \cdot$ aq.	XXII-85	187.20
5530	5-carboxylic acid	(2;5)	$CH_3 \cdot C_9H_5N \cdot CO_2H$	XXII-86	187.20
5531	6-carboxylic acid	(2;6)	$CH_3 \cdot C_9H_5N \cdot CO_2H$	XXII-87	187.20
5532	8-carboxylic acid	(2;8	$CH_3 \cdot C_9H_5N \cdot CO_2H \cdot$ $1\frac{1}{2}H_2O$	XXII-87	214.22
5533	**Quinaldinic acid**	quinoline-2-car- boxylic acid	$C_9H_6N \cdot CO_2H \cdot 2H_2O$	XXII-71	209.20
5534	**Quinazoline**	....	$C_6H_4 \cdot CH:N \cdot CH:N$	XXIII-175	130.15
5535	**Quinhydrone**	....	$C_6H_4O_2 \cdot C_6H_4(OH)_2$	VII-617	218.21
5536	**Quinic acid** (*l*)	....	$(HO)_4C_6H_7 \cdot CO_2H$	X-535	192.17
5537	**Quininic acid** (6;4)	6-MeO-cinchoninic acid	$CH_3O \cdot C_9H_5N \cdot CO_2H$	XXII-234	203.20
5538	**Quinoline**	leucoline	C_9H_7N	XX-339	129.16
5539	ethiodide	....	$C_9H_7N \cdot (C_2H_5I)$	XX-353	285.13
5540	**Quinoline** (*iso*)	....	C_9H_7N	XX-380	129.16

Table 7-4 (*Continued*)
PHYSICAL CONSTANTS OF ORGANIC COMPOUNDS

No.	Crystalline Form and Color	Specific Gravity	Melting Point °C.	Boiling Point °C.	Solubility in 100 Parts		
					Water	Alcohol	Ether
5506	rhb./aq.	....	249-50 d. (anh.)	$-H_2O$, 115-20	$18°$; s. h.	sl. s.	i.; i. bz.
5507	tri.	....	243	....	s. aq. a.	sl. s. h.	i.; i. bz.
5508	pl./aq. H_2SO_4	....	227 d. (anh.)	subl. d.	v. sl. s.	sl. s.	sl. s.
5509	pl. or lf.	....	d. 261 (anh.)	$-H_2O$, 115	s. h.	....	....
5510	col. nd.	....	160 d.	subl. in vac.	s.; s. act.	s.; sl. s. $CHCl_3$	sl. s.
5511	col. pr.	....	209-10 d.	subl. in vac.	22	1.1	sl. s.; sl. s. act.
5512	cr./chl. lg.	....	103-5	....	s.	s.	s.
5513	cr.	....	20-2	123-4	∞	s.	s.
5514	yel. mn.	....	268-9*	subl.	i.; s. c. H_2SO_4	v. sl. s. c.	v. sl. s.; s. ac.
5515	mn./aq.	....	55-6	262.7	$1.8^{13°}$	s.	s.
5516	wh. cr. pd.	....	164-5	....	i.	s.	d. aq. alk.
5517	rhb.	$1.099\frac{75°}{75}$	47	241	s. bz.	v. s.	v. s.
5518	cr.	$1.190^{40.3°}$	32.5	215-7	v. sl. s.	s.	v. s.
5519	gr. pd.	....	....	....	s.	sl. s.	i.
5520	tri.	1.411	111-2	....	$66^{20°}$	v. s.	v. s.
5521	lq.	$0.970\frac{20°}{4}$	-24	131	$8^{25°}$	s.; s. bz.	s.; i. alk.
5522	mn.	....	192 sl. d.	d. 208.5	s.	s.	s.
5523	lq.	$0.852^{22.5°}$	....	87.5-8.5	∞	∞; ∞ chl.	∞
5524	lq.	$0.910\frac{20°}{4}$	....	90-1	v. s.	∞	∞
5525	col. lq.	$1.267\frac{20°}{4}$	13.6±	165 sl. d.	∞	∞	∞
5526	pl./al.	....	124-5	subl. 100	v. s.	s.	sl. s. c. bz.
5527	lq.	$0.975\frac{20°}{4}$	....	93	....	....	....
5528	nd./al. or bz.	....	235-8 d.	....	i.	v. sl. s.	v. sl. s.
5529	yel. nd./aq.	....	244-6 d.	subl. d.	sl. s.; s. h. ac.	sl. s.; i. h. chl.; s. a.	sl. s.; i. pet.
5530	nd./al.	....	285 d.	subl. d.	v. sl. s.	s.; s. h. a.	i.; i. bz.
5531	nd./al.	....	259-61	subl. d.	v. sl. s. h.	s. h.	....
5532	nd./aq.	....	151	....	v. s. h.	s.; s. a.	s. alk.
5533	nd./aq.	....	156 (anh.)	$-H_2O$, 100	s. h.; s. alk.	s.	s. h. bz.
5534	pl./pet.	....	48.0-8.5	243^{773mm}	v. s.	s.	s.
5535	red brn. rhb.	$1.401^{20°}$	171	subl. sl. d.	s. h.; s. NH_4OH	s.	s.; d. chl.
5536	mn./aq.	1.637	162-3	d.	$40^{9°}$	s.; s. ac.	v. sl. s.
5537	yel. pr./aq. HCl	....	280 d.	....	v. sl. s. c.; s. alk.	1.2 h. abs.; s. a.	v. sl. s.; v. sl. s. bz.
5538	hyg. lq.	$1.095^{20°}$	-15.6	237.1^{747mm}	0.6	∞	∞; s. CS_2
5539	yel. mn./al.	....	159-60	....	$301^{25°}$	$1.8^{25°}$ chl.	i.; s. al.
5540	pl.	$1.091\frac{30°}{4}$	26.5	243.3	sl. s.	....	s. a.

* Sealed tube.

Quercitrinic acid, cf. glcde.	Quinalizarin 5801	Quinine, cf. alkd.	Quinoline blue 1580
Quinaldic acid 5533	Quinanisole 4083	Quinizarin 2299	Quinoline carboxylic
Quinaldine 4367	Quinazine 5548	Quinizarin amide 327	acid 5533
Quinaldine ethiodide 4368	Quinicine, cf. alkd.	Quinoform, cf. alkd.	Quinolinic acid 5497
	Quinidine, cf. alkd.	Quinol 2314	

Table 7-4 (*Continued*)
PHYSICAL CONSTANTS OF ORGANIC COMPOUNDS

No.	Name	Synonym	Formula	Beil. Ref.	Formula Weight
5541	**Quinone** (*o*)	*o*-benzoquinone	$CH \cdot (CH)_3 \cdot CO \cdot CO$	VII-600	108.10
5542	**Quinone** (*p*)	quinone	$CO:(CH:CH)_2:CO$	VII-609	108.10
5543	chlorimide (1,4)	....	$O:C_6H_4:N \cdot Cl$	VII-619	141.56
5544	dichlordiimide (*p*)	....	$Cl \cdot N:C_6H_4:N \cdot Cl$	VII-621	175.02
5545	dioxime (*p*)	....	$HO \cdot N:C_6H_4:N \cdot OH$	VII-627	138.13
5546	monoxime (*p*)	*p*-nitrosophenol tautomer	$O:C_6H_4:N \cdot OH$ or $HO \cdot C_6H_4 \cdot NO$	VII-622	123.11
5547	monoxime Na	(sodium salt)	$NaO \cdot C_6H_4 \cdot NO \cdot 2H_2O$	VII-624	181.12
5548	**Quinoxaline**	quinazine	$C_6H_4 \cdot N:CH \cdot CH:N$	XXIII-176	130.15
5549	**Raffinose**	melitose	$C_{18}H_{32}O_{16} \cdot 5H_2O$	XXXI-462	594.52
5550	**Resacetophenone**	2,4-diOH-aceto-phenone	$(HO)_2C_6H_3 \cdot COCH_3$	VIII-266	152.15
5551	**Resazurin**	"azoresorcin"	$C_{12}H_7O_4N$	XXVII-128	229.19
5552	**Resorcinol** diacetate	(1,3)	$C_6H_4(O_2C \cdot CH_3)_2$	VI-816	194.19
5553	dibenzoate	(1,3)	$C_6H_4(O_2C \cdot C_6H_5)_2$	IX-131	318.33
5554	**Resorufin**	9-OH-iso-phenoxazone	$HC \cdot C_{12}H_6ON:O$	XXVII-128	213.19
5555	**Retene**	1-Me-7-isoPr-phenanthrene	$C_{18}H_{18}$	V-683	234.34
5556	**Rhamnetin**	quercetin-7-Me-ether	$C_{16}H_{12}O_7$	XVIII-245	316.27
5557	**Rhamnitol**	rhamnite	$CH_3 \cdot C_5H_{11}O_5$	I-532	166.18
5558	**Rhamnose** (*β*)	isodulcite	$C_6H_{12}O_5 \cdot H_2O$	I-870	182.17
5559	phenylhydrazone	....	$C_6H_{12}O_4:N \cdot NHC_6H_5$	XV-216	254.29
5560	**Rhodanine**	rhodanic acid	$S \cdot CH_2 \cdot CO \cdot NH \cdot CS$	XXVII-242	133.19
5561	**Rhodeose**	....	$C_5H_{11}O_4 \cdot CHO$	I-876	164.16
5562	**Rhodoxanthin**	....	$C_{40}H_{50}O_2$	XXX-101	562.84
5563	**Riboflavin**	vitamin B_2 or G	$C_{17}H_{20}O_6N_4$	....	376.37
5564	**Ribose** (*d*) (−)	....	$C_4H_9O_4 \cdot CHO$	I-859	150.13
5565	phenylhydrazone (*l*)	....	$C_5H_{10}O_4:N \cdot NHC_6H_5$	XV-215	240.26
5566	**Ricinelaidic** acid	....	$C_{18}H_{34}O_3$	III-388	298.47
5567	sodium salt†	....	$C_{17}H_{33}O \cdot CO_2Na$	....	320.45
5568	**Ricinoleic acid**	....	$C_{18}H_{34}O_3$	III-385	298.47
5569	**Ricinstearolic acid**	....	$C_6H_{13}CHOHCH_2C \vdots C \cdot C_7H_{14}CO_2H$	III-391	296.45
5570	**Rosaniline** (3-Me; 4,4′,4″-NH₂)	fuchsine carbinol base	$(CH_3 \cdot C_6H_3NH_2) \cdot COH(C_6H_4NH_2)_2$	XIII-763	319.41
5571	hydrochloride	fuchsine	$C_{20}H_{20}N_3Cl \cdot aq.$	XIII-765	337.86
5572	**Rosinduline**	....	$HN:C_{10}H_5:NC_6H_4N \cdot C_6H_5$	XXV-348	321.38
5573	**Rosolic acid** ‡	4,4′-diOH-3-Me-fuchsone	$C_{20}H_{16}O_3$	VIII-365	304.35

† Usually a mixture of salts of fatty acids from castor oil.   ‡ See also No. 593.

Quinone-imide-oxime 4875
Quinophan 5234
Quinophenol 3825
Quinosol 3827
Quinotol 1386
Quinotoxine, cf. alkd.
R acid 4491
Racemic acid 5690

Ratanhin 4405
Redoxon 586
Resazoin 5551
Resorcin yellow 2308
Resorcine 2313
Resorcinol 2313
Resorcinol blue 3938
Resorcinol diethyl ether 2092

Resorcinol dimethyl ether 2407
Resorcinol ethyl ether 2928
Resorcinol methyl ether 4080
Resorcinol monacetate 114
Resorcinol phthalein 3250
Resorcinol phthalin 3254
Resorcyl aldehyde (*β*) 2310
Resorcylic acid 2316, 2318, 2320

Table 7-4 (*Continued*)
PHYSICAL CONSTANTS OF ORGANIC COMPOUNDS

No.	Crystalline Form and Color	Specific Gravity	Melting Point °C.	Boiling Point °C.	Solubility in 100 Parts		
					Water	Alcohol	Ether
5541	red pr.	....	d. 60-70	....	v. s. act.	i. pet.	sl. s.
5542	yel. mn./aq.	$1.318\frac{20^\circ}{4}$	112.9	subl.	v. sl. s. c.	s.; s. h. lg.	s.
5543	yel. lq.	....	85-6	expl.	s. h.	s.	s.; s. a.
5544	nd./aq.	....	126 d.	....	v. sl. s. h.	v. s. h.	v. s.
5545	col. or yel.	....	d. 240±	....	v. sl. s. aq. NH_4OH	s. conc. NH_4OH	....
5546	lt. yel. rhb.	....	124-6	d. 144	s. h.	v. s.	v. s.
5547	red nd./al.	....	$-2H_2O$, 100	....	v. s.	s.; s. act.	i.
5548	cr.	$1.133\frac{48^\circ}{4}$	29-30	225-6	s. c.; sl. s. h.	∞	∞; ∞ bz.
5549	cr./aq.	1.465^{0°	80 partly; 118-9 (anh.)	$-H_2O$, 110	14.3^{20°; ∞ h.	0.1^{20°	....
5550	col. cr.	1.18^{141°	144-6	d.	i.; s. pyr.	s. h.; s. ac.	i.; i. bz.
5551	gn. red/ac.	....	d.	....	i.	sl. s.	i.; i. alk.
5552	lq.	....	....	278 sl. d.	....	....	....
5553	pl./al.	....	117	....	....	5^{18° abs.	....
5554	brn./HCl	....	....	....	i.; s. alk.	v. sl. s.	i.
5555	lf./al.	1.13^{16°	98-9	390-4	i.; s. bz.	69 h.; s. CS_2	v. s. h.
5556	yel. n.d/al.	....	>300	....	v. sl. s. h.	s. h.	s. alk.
5557	tri.	....	121	....	v. s.; sl. s. act.	v. s.; sl. s. chl.	v. sl. s.
5558	col. mn.	$1.471\frac{20^\circ}{4}$	126	....	60.8^{21°	54 Me al.	i.
5559	col. lf.	....	159	....	1.25^{15°	....	i.
5560	pa. yel. pr./al.	....	168-70 d.	....	0.2^{25°; s. h.; s.alk.	v. s.; s. NH_4OH	v. s.
5561	nd.	....	144	....	v. s.	v. sl. s.	....
5562	b. bl. lf./bz. Me al.	....	219	....	i.; sl. s. bz.	v. sl. s.; sl. s. chl.	s. pyr.; i. pet.
5563	or. yel. nd.	....	290 d.	....	0.3^{25°	i. abs.	i.; s. pyr.
5564	pl./abs. al.	....	87	....	s.	sl. s.	....
5565	col. cr.	....	154-5 sl. d.	....	v. s.	....	....
5566	nd./al.	....	53	$240\text{-}2^{10mm}$	....	v. s.	v. sl. s. pet.
5567	wh. yel. pd.	....	....	....	s.	s.	....
5568	lq.	0.954^{16°	4-5	$226\text{-}8^{10mm}$	i.	∞; ∞ chl.	∞
5569	nd./al.	....	53	260^{10mm}	....	s.	s.
5570	col. nd./aq.	....	186 d.	....	v. sl. s.; s. a.	sl. s.	i.
5571	gn. red	1.22	d. > 200	....	0.3	s.	i.; s. HCl
5572	brn. nd./et.	....	198-9	....	i.	s.	s.; s. bz.
5573	red lf.	....	308-10 d.	d.	0.12^{25°	v. s. h.	sl. s.; s. alk.

Table 7-4 (*Continued*)
PHYSICAL CONSTANTS OF ORGANIC COMPOUNDS

No.	Name	Synonym	Formula	Beil. Ref.	Formula Weight
5574	**Rotenone**	....	$C_{23}H_{22}O_6$	....	394.43
5575	**Rufigallic acid**	1,2,3,5,6,7-hexa-OH-anthraquinone	$C_{14}H_2O_2(OH)_6$	VIII-567	304.22
5576	**Sabinene**	....	$C_{10}H_{16}$	V-143	136.24
5577	**Saccharic acid** (*d* or *l*)	....	$(CHOH)_4(CO_2H)_2$	III-577	210.14
5578	*iso*-**Saccharic acid**	....	$(CHOH·CH·CO_2H)_2O$	XVIII-364	192.13
5579	**Saccharin**	*o*-benzoyl sulfimide	$C_6H_4·CO·NH·SO_2$ ‖——————‖	XXVII-168	183.19
5580	soluble	(sodium salt)†	$C_7H_4O_3NSNa·2H_2O$	XXVII-170	241.20
5581	**Safrole**	3,4-methylenedioxy-allylbenzene	$CH_2O_2:C_6H_3·$ $CH_2·CH:CH_2$	XIX-39	162.19
5582	*iso*-**Safrole**	3,4-methylenedioxy-propenylbenzene	$CH_2O_2:C_6H_3·$ $CH:CH·CH_3$	XIX-35	162.19
5583	**Salicyl**-acetic acid	....	$C_6H_4(OCH_2CO_2H)·$ CO_2H	X-69	196.16
5584	aldehyde triacetate	acetylsalicylalde-hyde diacetate	$C_6H_4CH:(O_2CCH_3)_2·$ $O_2C·CH_3$	VIII-45	266.25
5585	salicylic acid	disalicylic acid; salysal; diplosal	$HO·C_6H_4·CO_2·C_6H_4·$ CO_2H	X-84	258.23
5586	**Salicylidene**-acetamide	....	$HO·C_6H_4·CH:N·$ $CO·CH_3$	VIII-47	163.18
5587	benzamide	....	$HO·C_6H_4·CH:N·$ $CO·C_6H_5$	IX-212	225.25
5588	*p*-phenetidine	malakin	$HO·C_6H_4·CH:N·$ $C_6H_4·OC_2H_5$	XIII-458	241.29
5589	2-thiohydantoin	5-(2-hydroxyben-zal)-2-thio-hydantoin	$HO·C_6H_4·CH:$ $C_3H_2ON_2S$	*XXV-502	220.25
5590	**Salipyrine**	antipyrene salicylate	$C_{18}H_{18}O_4N_2$	XXIV-32	326.36
5591	**Salyrgan**	mersalyl	$C_{13}H_{16}O_6NHgNa$	....	505.86
5592	**Salvarsan** (1-As;3-NH₂; 4-OH)	arsphenamine; "606"	$(:As·C_6H_3(OH)NH_2·$ $HCl)_2·2H_2O$	*XVI-507	475.04
5593	**Salvarsan** (neo)	neoarsphenamine; novarsenobenzol	$NH_2(OH)C_6H_3As:As·$ $C_6H_3(OH)NH·$ $CH_2OSONa‡$	*XVI-508	466.16
5594	**Santalic acid**	....	$C_{15}H_{14}O_5$	....	274.28
5595	**Santalol** (α)	arheol	$C_{15}H_{24}O$	VI-558	220.36
5596	**Santylyl salicylate**	santyl	$HO·C_6H_4·CO_2·C_{15}H_{23}$	X-80	340.47
5597	**Santonic acid**	....	$C_{15}H_{20}O_4$	X-804	264.32
5598	**Santonin**	santoninic anhydride	$C_{15}H_{18}O_3$	XVII-499	246.31
5599	**Santoninic acid**	....	$C_{15}H_{20}O_4$	X-962	264.32
5600	**Sarcosine**	*N*-methyl glycine	$CH_3NH·CH_2·CO_2H$	IV-345	89.09
5601	hydrochloride	....	$C_3H_7O_2N·HCl$	IV-345	125.56
5602	**Sarsasapogenin,** pseudo	pseudo-smilagenin	$C_{27}H_{44}O_3$	....	416.65
5603	**Schonberg's** reagent	benzylimido-di-(4-methoxyphenyl) methane	$(CH_3O·C_6H_4)_2C:$ $N·CH_2·C_6H_5$	....	331.42

† Usually with 5% H_2O.   ‡ Contains also solvent and inorganic salts.

Rubiazol 5337	S acid 339, 344	Salicin, cf. glcde.
Rufiopin 5799	Sα acid 4533	Salicoyl-phenol 2323-5
Rufol 2293	Sacchar(in)ol 5579	Salicoyl-resorcinol 6225
Rumpff acid 4512	Saccharinose 5579	Salicyl-acetophenone 3763
Rutaecarpine, cf. alkd.	Safranone 670	Salicyl alcohol 3750
Rutin, cf. glcde.	Salamid 3738	Salicyl aldehyde 3733
Rutylidene 6382	Salazolon 5590	Salicyl-aldehyde ethyl ether 2914

Table 7-4 (*Continued*)
PHYSICAL CONSTANTS OF ORGANIC COMPOUNDS

No.	Crystalline Form and Color	Specific Gravity	Melting Point °C.	Boiling Point °C.	Solubility in 100 Parts		
					Water	Alcohol	Ether
5574	pl./al.	....	163	....	i.; s. chl.	sl. s.; s. act.	sl. s.; v. s. pet.
5575	red cr.	....	subl. sl. d.	....	i.; s. H_2SO_4	v. sl. s.; s. alk.	v. sl. s.
5576	lq.	$0.848\frac{15}{15}°$	....	164-5	i.	∞	∞
5577	nd./al.	....	125-6	d. lactone	v. s.	v. s.	i.
5578	rhb.	....	185	d.	s.	s.	v. sl. s.
5579	mn./act.	....	225-8 sl. d.	subl. 300 in vac.	$0.4^{25°}$; sl. s. h.	3.1 c.; sl. s. chl.	1.05 c.; s. alk. carb.
5580	col. pd.	....	....	....	83	2.2	....
5581	col. mn.	$1.100\frac{20}{4}°$	11.2	233-4	i.	s.	∞; ∞ chl.
5582	col. lq.	$1.122\frac{20}{4}°$	6-7	252-3	i.	∞	∞; ∞ bz.
5583	nd./aq.	....	191-2	....	s. h.	s.; s. act.	s.; s. ac.
5584	nd. or pl./al.	....	103-4	....	i.	s. h.; s. CCl_4	s.; s. bz.
5585	col. cr./bz. or chl.	....	148-9	d.	i.; $12^{5°}$ bz.	$45^{25°}$; $66^{25°}$ act.	$28^{25°}$; v. s. CCl_4
5586	yel. pd.	....	d. > 150	....	i.; i. alk. carb.	i. alk.	s. conc. H_2SO_4
5587	yel. pd.	....	d. 190	....	v. s. act.*	s.; s. alk.	i.; v. sl. s. bz.
5588	pa. yel. cr./al.	....	91-2	....	i.	s. h.	s.; s. bz.
5589	nd./ac.	....	251-3	....	....	....	....
5590	cr. pd.	....	91-2	....	$0.5^{15°}$; $4^{100°}$	s.; v. s. chl.	sl. s.
5591	delq. wh.	....	....	....	100	35	i.
5592	yel. pd.	....	....	....	s.	sl. s.	v. sl. s.
5593	yel. pd.	....	....	....	s.; d. by heat, air, light.	sl. s.; v. s. gly.	i.; i. chl.; sl. s. act.
5594	red pr.	....	104	....	i.; s. alk.	∞ abs.	s.
5595	col. oil	$0.977\frac{25}{25}°$	....	300±	i.	s.	....
5596	yel. oil	1.07	....	126.6^{20mm}	i.	s.	s.
5597	col. rhb./aq.	$1.251\frac{26}{4}°$	170-2	285^{15mm}	$0.56^{17°}$	s.	s.; s. chl.
5598	col. pr.	1.187	169-70	....	$0.02^{17.5°}$; $0.4^{100°}$	$2.3^{22.5°}$; $37^{80°}$	$1.3^{17.5°}$; $2.4^{40°}$
5599	rhb. pl./al.	....	$-H_2O$, 120	....	s. h.	s.; s. chl.	sl. s.
5600	rhb.	....	210-5 d.	....	v. s.	sl. s.	....
5601	nd./al.	....	170-2	....	v. s.	v. sl. s.	v. sl. s.
5602	nd./act.	....	171-3	....	sl. s.	s.	s.; s. ac.
5603	pa. yel. cr./abs. al.	....	93	....	i.; sl. s. pet.	s. h.	sl. s.

* Also a modification insoluble in act.
Salicyl-aldehyde glucose, cf. glcde.
Salicyl-aldoxime 3736
Salicyl-amide 3738
Salicyl-anilide 3741
Salicyl-phenol 2323-5
Salicylic acid 3742
Salicylic methyl ether 4067

Salicylic nitrile 3748
Salicylic phenyl ether 5133
Salicylic sulfonic acid 5681
Salicylide 3747
Salicyl(oyl)-resorcinol 6225
Saliformin 3602
Saligenin 3750
Saligenin methyl ether 4072

Saligenol 3750
Salimenthol 4031
Salinigrin, cf. glcde.
Salipyrazolon 5590
Salit 866
Salmester 4078
Salocoll 5117
Salol 5235

Table 7-4 (*Continued*)
PHYSICAL CONSTANTS OF ORGANIC COMPOUNDS

No.	Name	Synonym	Formula	Beil. Ref.	Formula Weight
5604	**Scopoletin**	7-OH-6-MeO-coumarin	$C_{10}H_8O_4$	XVIII-99	192.17
5605	**Sebacic** acid	decandioic acid	$(CH_2)_8(CO_2H)_2$	II-718	202.25
5606	nitrile	....	$(CH_2)_8(CN)_2$	II-720	164.25
5607	**Seleno**-phenol	....	$C_6H_5 \cdot SeH$	VI-345	157.07
5608	urea	....	$NH_2 \cdot CSe \cdot NH_2$	III-227	123.02
5609	**Semicarbazide**	....	$NH_2 \cdot NH \cdot CO \cdot NH_2$	III-98	75.07
5610	hydrochloride	....	$NH_2CON_2H_3 \cdot HCl$	III-100	111.53
5611	**Serine** (*dl*)	....	$HO \cdot CH_2 \cdot CHNH_2 \cdot CO_2H$	IV-512	105.09
5612	**Serine** (*d* or *l*)	....	$C_3H_7O_3N$	IV-505	105.09
5613	*iso*-**Serine** (*dl*)	....	$NH_2 \cdot CH_2 \cdot CHOH \cdot CO_2H$	IV-503	105.09
5614	*iso*-**Serine** (*d* or *l*)	....	$C_3H_7O_3N$	IV-503	105.09
5615	**Silico**-acetic acid	....	$CH_3 \cdot SiO \cdot OH$	IV-629	76.13
5616	benzoic acid	....	$C_6H_5 \cdot SiO \cdot OH$	XVI-911	138.20
5617	**Silicon** methyl	....	$(CH_3)_4Si$	IV-625	88.23
5618	phenyl trichloride	....	$C_6H_5 \cdot SiCl_3$	XVI-911	211.55
5619	tetraethyl	silicononane	$(C_2H_5)_4Si$	IV-625	144.33
5620	**Silver** † fulminate	....	$Ag_2C_2O_2N_2$	I-722	299.77
5621	**Sodium** † acetanilide	....	$CH_3CON(Na) \cdot C_6H_5$	XII-237	157.15
5622	ethyl	....	$C_2H_5 \cdot Na$	*IV-618	52.05
5623	glycerolate	....	$NaC_3H_7O_3$	I-511	114.08
5624	mercaptide	....	$C_2H_5 \cdot S \cdot Na$	I-341	84.12
5625	**Sorbic acid**	hexadienoic acid	$CH_3(CH:CH)_2CO_2H$	II-483	112.13
5626	**Sozoiodolic** acid	2,6-diiodophenol-4-sulfonic acid	$I_2(OH)C_6H_2 \cdot SO_3H \cdot 3H_2O$	XI-245	480.03
5627	Na salt	sozoiodol-sodium	$C_6H_3O_4I_2SNa \cdot 2H_2O$	XI-245	483.98
5628	Hg salt	....	$C_6H_2O_4I_2SHg$	*XI-56	624.54
5629	Zn salt	....	$(C_6H_3O_4I_2S)_2Zn \cdot 6H_2O$	XI-245	1023.38
5630	**Starch**‡	....	$(C_6H_{10}O_5)_x$	....	(162.14)
5631	**Stearic** acid	octadecanoic acid	$CH_3(CH_2)_{16}CO_2H$	II-377	284.47
5632	ammonium salt	amm. stearate	$C_{17}H_{35}CO_2NH_4$	*II-171	301.52
5633	aldehyde	octadecanal	$CH_3(CH_2)_{16}CHO$	I-718	268.49
5634	amide	stearamide	$C_{17}H_{35} \cdot CO \cdot NH_2$	II-384	283.50
5635	anhydride	....	$(C_{17}H_{35}CO)_2O$	II-384	550.96
5636	anilide	stearanilide	$C_{17}H_{35}CO \cdot NHC_6H_5$	XII-257	359.60
5637	chloride	stearyl chloride	$C_{17}H_{35} \cdot CO \cdot Cl$	II-384	302.93
5638	nitrile	heptadecyl cyanide	$C_{17}H_{35} \cdot CN$	II-384	265.49
5639	**Stearolic acid**	octadecynoic acid	$C_8H_{17}C \vdots C(CH_2)_7 \cdot CO_2H$	II-495	280.45
5640	**Stearone**	....	$C_{35}H_{70}O$	I-720	506.95
5641	**Stearoxylic acid**	....	$CH_3(CH_2)_7(CO) \cdot (CH_2)_7CO_2H$	III-761	312.45
5642	**Stilbene** (*trans*)	diPh-ethylene	$C_6H_5 \cdot CH:CH \cdot C_6H_5$	V-630	180.25
5643	**Stilbene** (*cis*)	diPh-ethylene	$C_6H_5 \cdot CH:CH \cdot C_6H_5$	V-633	180.25
5644	**Stovaine**§	benzoyl-EtdiMe-amino-*iso*-propanol HCl	$C_6H_5CO_2C(C_2H_5) \cdot (CH_3)CH_2N(CH_3)_2 \cdot HCl$	IX-175	271.79
5645	**Strophanthidin**	....	$C_{23}H_{32}O_6 \cdot \frac{1}{2}H_2O$	....	413.52
5646	**Styrene**	phenyl-ethylene	$C_6H_5 \cdot CH:CH_2$	V-474	104.15

† See also inorganic compounds.
‡ See also No. 3443.
§ Free base, liquid, b. p. 149^{25mm}, s. organic solvents.
Salt of amber 5653
Salysal 5585
Sandoptal 1038
Santoninic anhydride 5598
Santyl 5596
Saponin. cf. glcde.

Saporubrin, cf. glcde.
Sapotoxin, cf. glcde.
Sarcine, cf. alkd.
Sarcolactic acid 3939
Sarsasaponin, cf. glcde.
Satrapol 4117

Saxin 5579
Scammonin, cf. glcde.
Schaeffer acid 4498, 4507
Schollkopf's acid 4543
Scopolamine, cf. alkd.
Secaline, cf. alkd.
Sedatine 565
Sedormid 217
Selenium diphenyl 2742

Table 7-4 (*Continued*)
PHYSICAL CONSTANTS OF ORGANIC COMPOUNDS

No.	Crystalline Form and Color	Specific Gravity	Melting Point °C.	Boiling Point °C.	Solubility in 100 Parts		
					Water	Alcohol	Ether
5604	nd. or pr.	....	204	subl.	v. sl. s.	s. h.; i. bz.	s. h. ac.
5605	pr./HNO₃	$1.207_4^{25°}$	134.5	294.5^{100mm}	0.1 c.; 2 h.	v. s.	v. s.
5606	col. lq.	....	7-9	$199\text{-}200^{15mm}$	....	....	....
5607	oil	$1.487^{15°}$	....	183.6	v. sl. s.	v. s, CCl_4	v. s.
5608	pr./aq.	....	200± d.	....	$10^{19°}$	$3^{18°}$	$0.6^{18°}$
5609	pr./al.	....	96	....	v. s.	v. s.	i.
5610	pr./aq. al.	....	173 d.	....	v. s.	i. abs.	i.
5611	pr./aq.	....	246 d.	....	$4^{20°}$	i.	i.
5612	col. pr.	....	228 d.	....	$33^{25°}$	i.	....
5613	mn.	....	248	....	$1.5^{20°}$ s. h. aq.	v. sl. s.	v. sl. s.
5614	col. cr./aq.	....	199-200 d.	....	$25^{0°}$	....	....
5615	amor. pd.	....	....	....	i.	i. h. aq. Na_2CO_3	s. conc. KOH
5616	glass/et.	....	−99(−102)	....	i.	s. KOH	s.
5617	lq.	$0.648_4^{19°}$	ign. in air	26.6	i. H_2SO_4	....	....
5618	lq.	$1.326_4^{18.8°}$	....	201.5	d.	d.; s. chl.	s.
5619	col. lq.	$0.768_4^{22°}$	....	154.7	i.	....	....
5620	wh. nd./aq.	....	....	expl.	$0.02^{30°}$	s.	i. HNO_3
5621	cr. pd.	....	....	....	d. h.	....	....
5622	wh. pd.	....	ign. in air	....	d.	i. bz.	....
5623	wh. pd.	....	d. 245	....	d.	s.	i.; i. CS_2
5624	cr./al.	....	....	....	v. s.; d. h.	v. s.; d. h.	....
5625	nd./aq.	....	134.5	228 d.	s. h.	v. s.	v. s.
5626	col. mn. pr.	....	120 (anh.); d. 190	$-3H_2O, 100$	s.	s.	s.
5627	col. cr.	....	....	....	16	6	i.
5628	or. pd.	....	....	....	$0.05^{20°}$	i.; 13.3 aq. 15% NaCl	i.; s. aq. KI
5629	col. nd.	....	....	....	5	30	i.
5630	wh. amor.	$1.50^{21°}$	d.	....	i.	i.	i.
5631	mn./CS_2	$0.847^{69.3°}$	69.6	291^{110mm}	$0.03^{25°}$; s. bz.; s.act.	220°; $100^{50°}$	$6^{15°}$; s. CCl_4
5632	waxy	0.89	73-5	d.	v. sl. s.	s. h.	i.; i. bz.
5633	lf.	....	38	$251\text{-}2^{100mm}$	i.	....	s.
5634	col. cr.	....	108-9	$250\text{-}1^{12}$ sl. d.	i.	s. h.	s. h.
5635	col. cr.	$0.855_4^{8.0°}$	72	....	i.; s. chl.	$0.02^{20°}$abs.	$0.2^{15°}$
5636	nd./al.	....	94-5	153.5^{10mm}	i.; s. act.	s.; s. chl.	v. s.; s. bz.
5637	col. cr.	....	23	215^{15} sl. d.	....	....	....
5638	col. cr.	$0.818_4^{4.1°}$	41	274.5^{100mm}	i.	....	....
5639	pr./al.	....	48	260	i.	s. h.	s.
5640	lf./lg.	$0.798_4^{8.9°}$	88.4	345^{12mm}	i.	sl. s. h.	sl. s. h.
5641	yel. pl.	....	86	....	sl. s. lg.	s. h.	s.
5642	mn./al.	$0.970_{13}^{1.25°}$	124	306-7	i.	$0.9^{17°}$ abs.	$7.9^{14°}$
5643	yel. oil	....	1	145^{13mm}	....	....	....
5644	nd./al.	....	175	....	50; 2 chl.	22 abs.	i.; i. act.
5645	lf./aq. al.	....	170-5*	....	sl. s.; s. bz.	s.; s. chl.	s.; i. pet.
5646	col. lq.	$0.906^{20°}$	−30.6	145.2	v. sl. s.	∞	∞

Table 7-4 (*Continued*)
PHYSICAL CONSTANTS OF ORGANIC COMPOUNDS

No.	Name	Synonym	Formula	Beil. Ref.	Formula Weight
5647	**Styrene** dibromide	α,β-diBr-ethyl-benzene	$C_6H_5CHBr\cdot CH_2Br$	V-356	263.97
5648	oxide	phenylethylene oxide	$C_6H_5\cdot CH\cdot O\cdot CH_2$ (with bracket)	XVII-49	120.15
5649	**Suberic** acid	octandioic acid	$(CH_2)_6(CO_2H)_2$	II-691	174.20
5650	aldehyde	octandial	$(CH_2)_6(CHO)_2$	I-795	142.20
5651	**Succinamic acid**	....	$NH_2\cdot CO\cdot C_2H_4\cdot CO_2H$	II-613	117.11
5652	*iso*-**Succinic acid**	methyl-malonic acid	$CH_3\cdot CH(CO_2H)_2$	II-627	118.09
5653	**Succinic** acid	butandioic acid	$(CH_2\cdot CO_2H)_2$	II-601	118.09
5654	sodium salt	sodium succinate	$C_4H_4O_4Na_2\cdot 6H_2O$	*II-262	270.15
5655	aldehyde	succinic dialdehyde	$(CH_2\cdot CHO)_2$	I-767	86.09
5656	amide	succinamide	$(CH_2\cdot CO\cdot NH_2)_2$	II-614	116.12
5657	anhydride	....	$(CH_2\cdot CO)_2O$	XVII-407	100.07
5658	chloride	succinyl chloride	$(CH_2\cdot COCl)_2$	II-613	154.98
5659	*N*-chloroimide	succinchlorimide	$(CH_2\cdot CO)_2N\cdot Cl$	XXI-380	133.53
5660	imide	succinimide	$(CH_2\cdot CO)_2NH$	XXI-369	99.09
5661	nitrile	ethylene dicyanide	$(CH_2\cdot CN)_2$	II-615	80.09
5662	peroxide	succinyl peroxide	$(HO_2C\cdot C_2H_4\cdot CO)_2O_2$	II-613	234.16
5663	**Sucrose**	cane sugar; beet sugar	$C_{12}H_{22}O_{11}$	XXXI-424	342.30
5664	**Sudan III**	benzene-azo-*p*-benzene-azo-β-naphthol	$C_6H_5\cdot N:N\cdot C_6H_4\cdot N:N\cdot C_{10}H_6OH$	XVI-171	352.40
5665	**Sulfa**-diazine	2-sulfanilamido-pyrimidine	$H_2N\cdot C_6H_4\cdot SO_2\cdot NH\cdot C_4H_3N_2$	....	250.28
5666	guanidine	sulfanilyl-guanidine	$H_2N\cdot C_6H_4\cdot SO_2\cdot NH\cdot C(:NH)\cdot NH_2$	....	214.25
5667	pyrazine	2-sulfanilamido-pyrazine	$H_2N\cdot C_6H_4\cdot SO_2\cdot NH\cdot C_4H_3N_2$	....	250.28
5668	pyridine	2-sulfanilamido-pyridine	$H_2N\cdot C_6H_4\cdot SO_2\cdot NH\cdot C_5H_4N$	....	249.29
5669	pyridine sodium	....	$C_{11}H_{10}O_2N_3SNa$	....	271.28
5670	thiazole	2-sulfanilamido-thiazole	$H_2N\cdot C_6H_4\cdot SO_2\cdot NH\cdot C_3H_2NS$	....	255.32
5671	**Sulfamino**-benzoic acid (*o*)	benzoic acid-(*o*)-sulfamide	$NH_2\cdot SO_2\cdot C_6H_4\cdot CO_2H$	XI-376	201.20
5672	benzoic acid (*m*)	....	$NH_2\cdot SO_2\cdot C_6H_4\cdot CO_2H$	XI-386	201.20
5673	benzoic acid (*p*)	....	$NH_2\cdot SO_2\cdot C_6H_4\cdot CO_2H$	XI-390	201.20
5674	**Sulfarsphenamine**	diNa-3,3'-diNH₂-4,4'-diOH-arseno-benzene-*N*-di-methylenesulfonate	$[NaSO_3\cdot CH_2\cdot HN(OH)\cdot C_6H_3\cdot As:]_2$	*XVI-509	598.23
5675	**Sulfo**-acetic acid	....	$HO_3S\cdot CH_2CO_2H\cdot H_2O$	IV-21	158.13
5676	benzoic acid (*o*)	....	$HO_3S\cdot C_6H_4\cdot CO_2H\cdot 3H_2O$	XI-369	256.23
5677	benzoic acid (*m*)	....	$HO_3S\cdot C_6H_4\cdot CO_2H\cdot 2H_2O$	XI-384	238.22
5678	benzoic acid (*p*)	....	$HO_3S\cdot C_6H_4\cdot CO_2H\cdot 3H_2O$	XI-389	256.23
5679	benzoic anhydride	(1,2)	$C_6H_4(CO)(SO_2):O$	XIX-110	184.17
5680	phenyl-3-methyl-5-pyrazolone	(1-*p*-sulfophenyl)	$HO_3S\cdot C_6H_4\cdot N\cdot C_4H_5ON\cdot H_2O$	XXIV-44	272.28

Table 7-4 (*Continued*)
PHYSICAL CONSTANTS OF ORGANIC COMPOUNDS

No.	Crystalline Form and Color	Specific Gravity	Melting Point °C.	Boiling Point °C.	Solubility in 100 Parts		
					Water	Alcohol	Ether
5647	cr./al.	....	74-5	139-41$^{1.5mm}$	i.; s. lg.	s.; s. bz.	v. s.
5648	lq.	1.052$^{18°}_4$	....	191-2	....	....	s.
5649	nd./aq.	1.266$^{25°}_4$	140-4	279^{100mm}	0.141$^{6°}$	s.; i. chl.	0.815°
5650	oil	....	....	230-40 sl. d.	v. s.	....	....
5651	col. nd.	....	157	....	sl. s.	v. sl. s. abs.	i. bz.
5652	col. nd.	1.455	d. 120-35	....	66$^{20°}$	v. s.	v. s.
5653	col. mn.	1.572$^{25°}_4$	189-90	235(−H_2O)	6.8$^{20°}$; 121$^{100°}$	9.9$^{15°}$	1.2$^{15°}$
5654	wh. pd.	....	−6H_2O, 120	....	2 c.	v. sl. s.	87$^{75°}$ aq.
5655	lq.	1.069$^{18°}_4$	....	169-70 sl. d.	s.	s.	s.
5656	col. nd.	....	242-3	....	0.5$^{15°}$; 11$^{100°}$	i. abs.	i.
5657	col. cr.	1.503	119.6	261	v. sl. s.	v. sl. s. pet.	sl. s.
5658	col. cr.	1.377$^{20°}_4$	16.7	192-3	d.	d.; s. bz.	i. pet.
5659	rhb./bz.	1.65	148	....	s. d.	0.9	sl. s.
5660	cr./act.	1.412$^{16°}$	125-6	287-8	v. s.	s.; i. chl.	v. sl. s.
5661	col.	0.985$^{63°}_4$	57.2	265-7	v. s.	v. s.	sl. s.
5662	col. pl.	....	128 d.	....	33	s.; s. act.	sl. s.; i. bz.
5663	col. mn.	1.588$^{15°}$	170-86 d.	....	*179$^{0°}$	0.9	i.
5664	brn. lf./ac.	....	195	....	i.; i. alk.; s. xylene	sl. s.; s. H_2SO_4	s.; s. oil; s. chl.
5665	col. mn./aq.	....	255-6	....	0.012$^{37°}$	sl. s.	i.
5666	col. mn./aq.	....	189-90	....	0.19$^{37°}$	sl. s.	i.
5667	col. cr.	....	255-7 d.	....	†0.005$^{17°}$	....	....
5668	col. pr./aq.	....	191-2	....	<0.03 c.; 0.05$^{37°}$	25; v. s. aq. HCl	sl. s.; v. s. alk.
5669	col. cr./al.	....	317 d.	....	63$^{25°}$	11	....
5670	brn./75% al.	....	201-2	....	0.09$^{37°}$	....	....
5671	rhb./al.	....	165-7	....	s.	s.	s.
5672	pl./aq.	....	237-8	....	v. sl. s. c.	s.	sl. s.
5673	pr./aq.	....	d. 280	....	sl. s. h.	v. sl. s. bz.	....
5674	yel. pd.	....	....	....	v. s.	v. sl. s.	....
5675	hyg. cr.	....	84-6	245 d.	s.	s.; i. chl.	i. abs.
5676	cr./aq.	....	68-9‡	−3H_2O, 105	v. s.	v. s.	i.
5677	delq. cr.	....	98‡	....	s.	s.; i. bz.	anh. s.
5678	nd./aq.	....	94	m. anh. 260	s.	s.	anh. s.
5679	col. cr./bz.	....	128-9	184-6^{18mm}	s. h.	s. chl.	s.; s. bz.
5680	nd./aq.	....	d. 320	−H_2O, 120	0.52$^{0°}$	v. sl. s. abs.	i.; v. s. alk.

* See also special table of solubility in water.
† Sodium salt very soluble in aq.
‡ Anhydrous crysts., m. p. 141°.
Strophanthin, cf. glcde.
Strychnine, cf. alkd.
Styphnic acid 6312
Stypticin, cf. alkd.
Styptol, cf. alkd.

Styracin 1508
Styracol 3485
Styrolene 5646
Styron 5646
Styrone 1507
Styryl carbinol 1507
Suberane 1596

Suberene 1598
Suberone 1597
Succinamide 5656
Succinimide 5660
Succin-chlorimide 5659
Succinyl chloride 5658
Succinyl peroxide 5662
Sucrol 5112

Table 7-4 (*Continued*)
PHYSICAL CONSTANTS OF ORGANIC COMPOUNDS

No.	Name	Synonym	Formula	Beil. Ref.	Formula Weight
	Sulfo				
5681	salicylic acid (5)	....	$HO_3S \cdot C_6H_3(OH)CO_2H$	XI-411	218.19
5682	salicylic Na	(Na acid salt)	$C_7H_5O_6SNa \cdot 2H_2O$	XI-412	276.20
5683	**Sulfonal**	acetone diethyl-sulfone	$(CH_3)_2C(SO_2 \cdot C_2H_5)_2$	I-662	228.33
5684	**Sylvestrine** (*d* or *l*)	....	$C_{10}H_{16}$	V-125	136.24
5685	**Synthalin**	decamethylene-di-guanidine di-HCl	$[H_2N \cdot C(:NH)NH \cdot (CH_2)_5]_2 \cdot 2HCl$	**IV-712	329.32
5686	**Syringic acid**	gallic acid-3,5-dimethyl ether	$HO(CH_3O)_2 \cdot C_6H_2 \cdot CO_2H$	X-480	198.18
5687	**Tannin**	digallic acid	$(HO)_3C_6H_2 \cdot CO_2 \cdot C_6H_2(OH)_2CO_2H$	....	322.23
5688	**Tartaric** acid (meso)	*i*-tartaric acid	$(CHOH \cdot CO_2H)_2$	III-528	150.09
5689	Ca salt (meso)	Ca meso-tartrate	$C_4H_4O_6Ca \cdot 3H_2O$	III-529	242.20
5690	acid (*racemic*) (*dl*)	"traubensaure"	$(CHOH \cdot CO_2H)_2 \cdot H_2O$	III-522	168.10
5691	acid (*d* or *l*)	"weinsaure"	$(CHOH \cdot CO_2H)_2$	III-481	150.09
5692	amide (*d*)	tartramide	$(CHOH \cdot CONH_2)_2$	III-520	148.12
5693	**Tartramidic acid**	....	$NH_2 \cdot CO \cdot (CHOH)_2 \cdot CO_2H$	III-520	149.10
5694	**Tartrazine**	hydrazine yellow	$C_{16}H_9O_9N_4S_2Na_3$	XXV-252	534.37
5695	**Tartronic acid**	hydroxymalonic acid	$HO \cdot CH(CO_2H)_2 \cdot \frac{1}{2}H_2O$	III-415	129.07
5696	**Taurine**	....	$NH_2 \cdot CH_2 \cdot CH_2 \cdot SO_3H$	IV-528	125.15
5697	**Taurocholic acid**	cholaic acid	$C_{26}H_{45}O_7NS \cdot H_2O$	....	533.75
5698	Na salt		$C_{26}H_{44}O_7NSNa$	....	537.70
5699	**Tephrosin**	hydroxy-deguelin	$C_{23}H_{22}O_7$	....	410.43
5700	**Teraconic acid**	γ-di Me-itaconic acid	$(CH_3)_2C:C(CO_2H) \cdot CH_2 \cdot CO_2H$	II-786	158.16
5701	**Terebic acid**	terebinic acid	$(CH_3)_2C \cdot CH(CO_2H) \cdot \underset{\underline{\quad\quad\quad}}{\quad} CH_2 \cdot CO_2 \underset{\underline{\quad\quad}}{\quad}$	XVIII-377	158.16
5702	**Terpenylic acid** (*dl*)	terpenolic acid	$C_8H_{12}O_4 \cdot H_2O$	XVIII-385	190.20
5703	**Terpin hydrate** (*cis*)	....	$C_{10}H_{20}O_2 \cdot H_2O$	VI-745	190.29
5704	**Terpinene** (α)	*p*-menthadiene (1,3)	$C_{10}H_{16}$	V-126	136.24
5705	**Terpinene** (β)	*p*-menthadiene (3,1)(7)	$C_{10}H_{16}$	V-132	136.24
5706	**Terpinenol-4** (*d*)	terpenol	$C_{10}H_{18}O$	VI-55	154.25
5707	**Terpineol** (α)(*d* or *l*)	....	$C_{10}H_{18}O$	VI-56	154.25
5708	**Terpineol** (α)(*dl*)	....	$C_{10}H_{18}O$	VI-58	154.25
5709	**Terpinolene**	*p*-menthadiene (1,4)(8)	$C_{10}H_{16}$	V-133	136.24
5710	**Terpinyl acetate** (α)	(*dl*)	$C_{10}H_{17} \cdot O_2C \cdot CH_3$	VI-60	196.29
5711	**Testerone** (*trans*)	....	$C_{19}H_{28}O_2$	....	288.43
5712	**Testerone** (*cis*)	....	$C_{19}H_{28}O_2$	....	288.43
5713	**Tetraamino-3,3'-dimethyl-di-phenylmethane**	(4,6,4',6'-tetraNH_2)	$[CH_3(NH_2)_2C_6H_2]_2: CH_2$	XIII-342	256.35
5714	**Tetrabromo-**aniline	(2,3,4,6)	$Br_4C_6H \cdot NH_2$	XII-668	408.73
5715	benzene (1,2,3,5)	....	$Br_4C_6H_2$	V-214	393.72

Table 7-4 (*Continued*)
PHYSICAL CONSTANTS OF ORGANIC COMPOUNDS

No.	Crystalline Form and Color	Specific Gravity	Melting Point °C.	Boiling Point °C.	Solubility in 100 Parts		
					Water	Alcohol	Ether
5681	hyg.nd./aq.	....	120d.(anh.)	....	s.	s.	s.
5682	col. cr.	....	....	....	3.3	i.	....
5683	pr./al.	$1.183^{18.2°}_{4}$	127-8	300 sl. d.	$0.2^{15°}$	$1.5^{15°}$; s. bz.	$0.7^{15°}$; s. chl.
5684	lq.	$0.863^{20°}_{4}$	....	176-7	....	....	....
5685	cr./al. et.	....	193	....	s.	....	....
5686	nd./aq. or et.	....	205-7	....	v. sl. s.	s.; s. chl.	s.
5687	amor. pd.	....	200 d.	....	s.	sl. s. abs.	i. abs.
5688	cr.	**1.737**	159-60	....	$120^{15°}$	....	....
5689	cr./aq.	....	$-3H_2O$, 170	....	0.17 h.	0.03^{18} ac.	$0.09^{100°}$ ac.
5690	tri.	$1.697^{20°}_{4}$	205-6	$-H_2O$, 100	$20.6^{20°}$; $185^{100°}$	$20°$	0.09
5691	mn.	$1.760^{20°}_{4}$	168-70	d.	$139^{20°}$; $343^{100°}$	$25^{15°}$ abs.	$0.4^{15°}$
5692	rhb.	....	195 d.	....	i. bz.	$0.04^{12°}$	i.
5693	rhb./aq.	....	171-2	....	....	....	....
5694	or. yel. pd.	....	....	....	v. s.	i.	....
5695	col. pr./aq.	....	d. 155-8; $-\frac{1}{2}H_2O$,60	subl. 110	v. s.	v. s.	i.; anh. s.
5696	....	....	>240 d.*	....	$6.4^{12°}$	$0.004^{17°}$	i.
5697	delq. nd.	....	125±, d.	....	d. h.	s.	v. sl. s.
5698	yel. pd.	....	....	....	v. s.	v. s. h.	sl. s.
5699	col. pr.	....	198	....	i.	s. act.	s.; s. chl.
5700	tri./et.	....	160-1 d.	....	v. s. h.	v. s.	sl. s.; v. sl. s. bz.
5701	mn./al.	$0.815^{24°}_{4}$	174-5	....	s. h.	s.	$1.7^{10°}$
5702	col. cr./aq.	....	56; 90 (anh.)	subl. 130	s. h.	....	....
5703	rhb.	....	$-H_2O$ > 117	....	$0.4^{15°}$; 3.3 h.	$10^{15°}$; 50 h.	$1^{15°}$; $0.5^{15°}$ chl.
5704	lq.	$0.834^{20°}_{4}$	....	181.5	i.	∞	∞
5705	lq.	$0.838^{22°}$	....	173-4	i.	∞	∞
5706	cr.	$0.936^{15°}$	38-40	$219\text{-}21^{760mm}$	i.	v. s.	v. s.
5707	col. cr.	$0.935^{15°}$	38-40	219-21	i.	v. s.	v. s.
5708	col. cr.	$0.935^{20°}_{20}$	35	$218\text{-}9^{752mm}$	i.	v. s.	v. s.
5709	lq.	$0.862^{20°}_{4}$	....	186-7	i.	∞	∞
5710	lq.	$0.966^{20°}_{4}$	<-50	220 d.	i.	20	....
5711	col.nd./act.	....	154-5	....	i.	s.	s.
5712	cr.	....	220-1	....	i.	s.	s.
5713	lf./aq.	....	203-4	....	v. sl. s.	v. sl. s.	....
5714	nd.	....	116-7	....	....	v. s.	v. s.
5715	nd./al.	....	98.5	329	i.	s. h.	v.s.; v. s. bz.

* The sulfonimide, m. p. 88°.

Sulfanilyl-guanidine 5666	Sulfhydryl-benzoic acid 5918	Sulfon-methane 5683
Sulfarsenol 5674	Sulfo-anthraquinone 559	Sulfur yellow-S 3241
Sulfenazoxine 679	Sulfo-benzide 2750	Sulfuric ether 2911
Sulfenthal 5127	Sulfo-cyanic acid 5897	Superpalite 6159
	Sulfon-ethylmethane 6326	Suprarenine 157

Table 7-4 *(Continued)*
PHYSICAL CONSTANTS OF ORGANIC COMPOUNDS

No.	Name	Synonym	Formula	Beil. Ref.	Formula Weight
	Tetrabromo				
5716	benzene (1,2,4,5)	....	$Br_4C_6H_2$	V-214	393.72
5717	o-cresol	....	$Br_4C_6(OH)\cdot CH_3$	VI-362	423.75
5718	m-cresol-sulfon-phthalein	brom cresol green	$C_{21}H_{14}O_5Br_4S$	....	698.04
5719	ethane (1,1,2,2) (*sym*)	acetylene tetra-bromide	$Br_2CH\cdot CHBr_2$	I-94	345.67
5720	ethane (1,1,1,2)	(*uns*)	$Br_3C\cdot CH_2Br$	I-94	345.67
5721	ethylene	ethylene tetraBr	$Br_2C:CBr_2$	I-192	343.66
5722	phenolphthalein	(3′,5′,3″,5″)	$C_{20}H_{10}O_4Br_4$	XVIII-149	633.94
5723	phenolphthalein	(sodium salt)	$C_{20}H_8O_4Br_4Na_2$	....	677.90
5724	phenolsulfon-phthalein	brom phenol blue	$C_{19}H_{10}O_5Br_4S$	*XIX-649	669.99
5725	phthalic anhydride	....	$C_6Br_4\cdot CO\cdot O\cdot CO$ (with bond line below)	XVII-485	463.72
5726	quinone	bromanil	$O:C_6Br_4:O$	VII-642	423.70
5727	**Tetrabutylammonium iodide** (*n*)	....	$(C_4H_9)_4N\cdot I$	IV-157	369.37
5728	**Tetrachloro**-acetone	(*sym*)	$(Cl_2CH)_2CO\cdot 4H_2O$	I-656	267.92
5729	acetone (*sym*)	....	$(Cl_2CH)_2CO$	I-656	195.86
5730	aniline (2,3,4,5)	....	$Cl_4C_6H\cdot NH_2$	XII-630	230.91
5731	aniline (2,3,5,6)	....	$Cl_4C_6H\cdot NH_2$	**XII-340	230.91
5732	aniline (2,3,4,6)	....	$Cl_4C_6H\cdot NH_?$	XII-630	230.91
5733	benzene (1,2,3,4)	....	$Cl_4C_6H_2$	V-204	215.89
5734	benzene (1,2,3,5)	....	$Cl_4C_6H_2$	V-204	215.89
5735	benzene (1,2,4,5)	....	$Cl_4C_6H_2$	V-205	215.89
5736	1,2-dibromo-ethane	(1,1,2,2)	$Cl_2BrC\cdot CBrCl_2$	I-93	325.65
5737	1,1-dibromo-ethane	(1,2,2,2)	$Cl_3C\cdot CClBr_2$	I-93	325.65
5738	diphenyl (2,4,2′,4′)	....	$(C_6H_3Cl_2)_2$	V-579	291.99
5739	ethane (*sym*)	acetylene tetraCl	$Cl_2CH\cdot CHCl_2$	I-86	167.85
5740	ethane (*uns*)	(1,1,1,2-tetraCl)	$Cl_3C\cdot CH_2Cl$	I-86	167.85
5741	ethylene	ethylene tetraCl	$Cl_2C:CCl_2$	I-187	165.83
5742	hydroquinone	....	$(HO)_2C_6Cl_4$	VI-851	247.89
5743	phenol (2,3,4,6)	....	$HO\cdot C_6HCl_4$	VI-193	231.89
5744	phenolphthalein	(4,5,6,7)	$C_{20}H_{10}O_4Cl_4$	XVIII-148	456.11
5745	o-phthalic acid	....	$Cl_4C_6(CO_2H)_2\cdot\frac{1}{2}H_2O$	IX-819	312.92
5746	quinone	chloranil	$O:C_6Cl_4:O$	VII-636	245.88
5747	**Tetracosane** (*n*)	....	$CH_3\cdot(CH_2)_{22}\cdot CH_3$	I-175	338.67
5748	**Tetradecane** (*n*)	....	$CH_3\cdot(CH_2)_{12}\cdot CH_3$	I-171	198.40
5749	**Tetradecyl** acetate	(*n*)	$CH_3CO_2\cdot C_{14}H_{29}$	II-136	256.43
5750	alcohol (*n*)	tetradecanol-1	$CH_3(CH_2)_{12}CH_2OH$	I-428	214.39
5751	**Tetradecylene** (α)	tetradecene-1	$CH_3(CH_2)_{11}CH:CH_2$	I-226	196.38
5752	**Tetradecyne-2**	....	$CH_3(CH_2)_{10}C:C\cdot CH_3$	I-262	194.36
5753	**Tetraethanolammonium hydroxide**	....	$(HO\cdot CH_2CH_2)_4NOH$	IV-285	211.26
5754	**Tetraethylammonium** bromide	....	$(C_2H_5)_4NBr$	IV-104	210.16
5755	chloride	....	$(C_2H_5)_4NCl\cdot 4H_2O$	IV-104	237.77

Table 7-4 (Continued)
PHYSICAL CONSTANTS OF ORGANIC COMPOUNDS

No.	Crystalline Form and Color	Specific Gravity	Melting Point °C.	Boiling Point °C.	Solubility in 100 Parts		
					Water	Alcohol	Ether
5716	mn./CS$_2$	3.027$^{20°}$	178-80	....	i.	....	....
5717	nd./ac.	....	207-8	....	sl. s. ac.	s.; s. alk.	s.
5718	lt. yel. pd.	....	....	....	sl. s.	s.; s. alk.	i.
5719	col. lq.	2.964$^{20°}_{4}$	0-2; d. >190	151^{54mm}	i.; ∞ ac.	∞; ∞ chl.	∞; ∞ aniline
5720	col. lq.	2.875$^{20°}_{4}$	0; d. 175	103.5$^{13.5mm}$	....	s.	....
5721	cr. pl.	....	56-7	226-7	....	....	....
5722	col. nd./al.	....	294	....	i.; s. alk.	v. sl. s.	s.
5723	b. pd.	....	....	....	v. s.	sl. s.	i.
5724	cr./ac. act.	....	270-1 d.	....	0.07	sl. s.	i.; s. alk.
5725	cr./ac. + xylene	....	275-80	....	v. sl. s. ac.	v. sl. s. bz.	s. PhNO$_2$
5726	yel. mn./bz.	....	300	subl.	i.	s. h.	sl. s.
5727	lf./bz.	....	144-5	....	sl. s.	s.	s.
5728	tri.	....	48-9	....	....	....	....
5729	lq.	....	....	180-2 sl. d.	v. s. bz.	v. s.	v. s.
5730	nd./al.	....	118-20	....	s. bz.	s.; s. ac.	s.
5731	cr.	....	110	....	i.	s.	....
5732	nd./lg.	....	88	....	s. CS$_2$	s.	s. lg.
5733	nd.	....	46-7	254^{761mm}	i.	sl. s.	v. s.
5734	nd./al.	....	54-5	246	i.	sl. s. c.	v. s. CS$_2$
5735	nd./et.	1.858$^{22°}$	138-40	240-6	i.	s. c.	s. bz.
5736	rhb.	2.713	200-5 d.	....	....	....	....
5737	rhb.	2.794	subl.	....	....	s. h.	s.
5738	cr.	....	83	....	0.29$^{20°}$	v. s. h.	sl. s. lg.
5739	col. lq.	1.600$^{20°}_{4}$	-36	146.2	0.29$^{20°}$	∞	∞
5740	lq.	1.588$^{20°}_{4}$	....	129-30	0.02$^{20°}$	∞	∞
5741	col. lq.	1.631$^{15°}_{4}$	-22.4.	121.2	i.	∞; ∞ chl.	∞; ∞ bz.
5742	mn.	....	238-40	subl.	i.; s. act.	20$^{25°}$	20$^{25°}$
5743	nd./lg.	1.6$^{0°}_{4}$	69-70	164^{23mm}	v. sl. s.	v. s.	v. s.
5744	pl./Me al.	....	>300	....	i.; s. alk.	s.; i. chl.	s.; i. bz.
5745	cr./aq.	....	d. -1½H$_2$O*	....	0.61$^{4°}$; 3.0$^{99°}$	s.	s.; v. s. act.
5746	yel.mn./bz.	....	290**	subl. 80°	i. bz.$^{25°}$	i. c.; sl. s. h.	i. c.; sl. s. CCl$_4$
5747	cr.	0.779$^{51°}_{4}$	51.1	324	....	8.4$^{15°}$ chl.	s.
5748	col. lq.	0.765$^{20°}_{4}$	5.5	252.5	i.	v. s.	v. s.
5749	col. cr.	....	12-3	176-7^{15mm}	....	....	....
5750	cr.	0.824$^{38°}_{4}$	38	167-70^{15mm}	<0.02	sl. s.	s.
5751	lq.	0.775$^{15°}_{4}$	-12	127^{15mm}	....	....	....
5752	cr.	0.800$^{15°}_{4}$	6.5	134^{15mm}	....	....	....
5753	....	....	....	....	∞	....	....
5754	cr./abs. al.	....	....	....	....	v. s.	s. chl.
5755	mn.	1.112$^{25°}_{4}$	37.5	....	141$^{25°}$	8.2$^{25°}$ chl.	....

* Anhydride, m. p. 255°.
** Sealed tube.
Terpenolic acid 5702
Terphenyl 2705
Tethrothalein 5722

Tetraacetyl-glucosyl chloride 44
Tetraanhydro-berberine, cf. alkd.
Tetrabase 5834
Tetrabromo-fluorescein 2875-6
Tetrabromo-methane 1242

Tetracaine 5032
Tetrachloro-methane 1243
Tetradecanal 4451
Tetradecanoic acid 4450
Tetradecanol 5750

Table 7-4 (*Continued*)
PHYSICAL CONSTANTS OF ORGANIC COMPOUNDS

No.	Name	Synonym	Formula	Beil. Ref.	Formula Weight
5756	**Tetraethylammo- nium** hydroxide	....	$(C_2H_5)_4NOH$	IV-103	147.26
5757	iodide	....	$(C_2H_5)_4NI$	IV-104	257.16
5758	**Tetraethyl-**benzene	(1,2,4,5)	$(C_2H_5)_4C_6H_2$	V-455	190.33
5759	benzene (1,2,3,4)	....	$(C_2H_5)_4C_6H_2$	V-455	190.33
5760	diaminoben- zophenone	(4,4′)	$[(C_2H_5)_2N \cdot C_6H_4]_2CO$	XIV-98	324.47
5761	diaminodiphenyl- methane	(4,4′)	$[(C_2H_5)_2N \cdot C_6H_4]_2CH_2$	XIII-242	310.49
5762	diaminotriphenyl carbinol	brilliant green base	$[(C_2H_5)_2N \cdot C_6H_4]_2 \cdot$ $C(OH) \cdot C_6H_5$	XIII-746	402.58
5763	diaminotriphenyl carbinol sulfate	brilliant green dye salt	$(C_{27}H_{33}N_2) \cdot HSO_4$	XIII-746	482.65
5764	ethanetetracarb- oxylate (*sym*)	....	$[(C_2H_5O_2C)_2CH]_2$	II-858	318.33
5765	propanetetra- carboxylate	($\alpha,\alpha,\beta,\gamma$)	$(CH_2 \cdot CH \cdot CH) \cdot$ $(CO_2 \cdot C_2H_5)_4$	II-859	332.35
5766	rhodamine	rhodamine B	$C_{28}H_{31}O_3N_2Cl$	XIX-346	479.02
5767	urea	....	$[(C_2H_5)_2N]_2CO$	IV-120	172.27
5768	**Tetraethylene** glycol	....	$(\cdot CH_2OCH_2 \cdot)_3:$ $(\cdot CH_2OH)_2$	I-468	194.23
5769	glycol dimethyl ether	diMeO-tetraglycol	$(CH_3 \cdot CH_2 \cdot CH_2 \cdot$ $O \cdot CH_2 \cdot CH_2)_2O$	....	222.28
5770	pentamine	....	$NH_2 \cdot (CH_2CH_2NH)_3 \cdot$ $CH_2 \cdot CH_2 \cdot NH_2$	....	189.31
5771	**Tetraglycol dichloride**	....	$(Cl \cdot CH_2 \cdot CH_2 \cdot O \cdot$ $CH_2 \cdot CH_2)_2O$	....	231.12
5772	**Tetrahydro-**ben- zoic acid (Δ_1)	....	$CH_2(CH_2)_3CH:C \cdot$ $\vert_____\vert$ CO_2H	IX-41	126.16
5773	furan	tetramethylene oxide	$CH_2(CH_2)_2CH_2 \cdot O$ $\vert_____\vert$	XVII-10	72.11
5774	furfuryl acetate	....	$CH_3CO_2 \cdot CH_2 \cdot C_4H_7O$	**XVII-107	144.17
5775	furfuryl alcohol	....	$C_4H_7O \cdot CH_2OH$	**XVII-106	102.13
5776	furfuryl benzoate	....	$C_6H_5CO_2 \cdot CH_2 \cdot C_4H_7O$	....	206.24
5777	furfuryl butyrate	....	$C_3H_7CO_2 \cdot CH_2 \cdot C_4H_7O$	....	172.23
5778	furfuryl *n*-cap- roate	....	$C_5H_{11}CO_2 \cdot CH_2 \cdot$ C_4H_7O	....	200.28
5779	furfuryl lactate	....	$CH_3 \cdot CHOH \cdot CO_2 \cdot$ $CH_2 \cdot C_4H_7O$	....	174.20
5780	furfuryl laurate	....	$C_{11}H_{23}CO_2 \cdot CH_2 \cdot$ C_4H_7O	....	284.44
5781	furfuryl maleate	....	$(:CH \cdot CO_2 \cdot CH_2 \cdot$ $C_4H_7O)_2$	....	284.31
5782	furfuryl palmitate	....	$C_{15}H_{31}CO_2 \cdot CH_2 \cdot$ C_4H_7O	....	340.55
5783	furfuryl propionate	....	$C_2H_5CO_2 \cdot CH_2 \cdot C_4H_7O$	....	158.20
5784	furfuryl salicylate	....	$HO \cdot C_6H_4CO_2 \cdot$ $CH_2 \cdot C_4H_7O$	....	222.24
5785	furfuryl succinate	....	$(\cdot CH_2 \cdot CO_2 \cdot CH_2 \cdot$ $C_4H_7O)_2$	....	286.33
5786	naphthalene (*trans*) (1,2,3,4)	"Tetralin"	$C_6H_4CH_2(CH_2)_2CH_2$ $\vert_____\vert$	V-491	132.21

Tetradecanone 3182
Tetradecene 5751
Tetraethyl germanium 3357
Tetraethyl lead 3957

Tetraethyl silicon 5619
Tetraethyl tin 5950
Tetrafluoro-methane 1244
Tetrahydro-benzene 1609

Table 7-4 (*Continued*)
PHYSICAL CONSTANTS OF ORGANIC COMPOUNDS

No.	Crystalline Form and Color	Specific Gravity	Melting Point °C.	Boiling Point °C.	Solubility in 100 Parts		
					Water	Alcohol	Ether
5756	only in solns.*	....	*	d.	s.	....	....
5757	col. cr./aq.	1.5594°	>200	....	$45^{25°}$	$1.6^{25°}$ chl.	i.; s. al.
5758	lq.	$0.888\frac{16°}{4}$	13	250	i.	s. abs.	v. s.
5759	lq.	$0.887\frac{20°}{4}$	11.0	248	i.	s. abs.	v. s.
5760	lf./al.	....	95-6	....	....	....	....
5761	cr./al.	....	41-2	253^{10mm}	....	....	....
5762	red brn.	....	....	....	v. sl. s.	s.	s. aq. a.
5763	rhb. gold nd.	....	....	....	v. s.	v. s.	....
5764	tet. pr.	$1.064^{79.5°}$	76	305 d.	i. pet. ..	s.	s.
5765	oil	$1.118\frac{20°}{4}$	....	$200\text{-}1^{14mm}$	....	....	....
5766	lf./HCl	....	....	....	v. s.	v. s.	sl. s. alk.
5767	lq.	$0.886\frac{20°}{4}$	....	210-5	i.	s. a.	i. alk.
5768	lq.	$1.125\frac{20°}{20}$	....	327-8	∞	....	....
5769	lq.	$1.013\frac{20°}{20}$	....	275.8	∞	....	....
5770	lq.	$0.999\frac{20°}{20}$	....	333	∞	....	....
5771	lq.	$1.186\frac{20°}{20}$	....	114^{2mm}	sl. s.	....	....
5772	cr.	$1.072\frac{47°}{4}$	29	240-3	$0.7^{20°}$	....	....
5773	col. lq.	$0.888\frac{21°}{4}$	−108.5	65-6	s.	s.	s.
5774	col. lq.	$1.062\frac{25°}{4}$	....	$192\text{-}4^{740mm}$	∞	∞	∞; ∞ chl.
5775	col. lq.	$1.050\frac{20°}{4}$	....	$177\text{-}8^{743mm}$	∞	∞	∞
5776	lq.	$1.137\frac{20°}{0}$	....	$300\text{-}2^{750mm}$	i.	∞	∞; ∞ chl.
5777	lq.	$1.012\frac{20°}{0}$	....	225-7	i.	∞	∞; ∞ chl.
5778	col. lq.	....	....	$141\text{-}3^{19mm}$	i.	s.	s.
5779	yel. lq.	....	....	$146\text{-}9^{18mm}$	s.	s.	s.
5780	pa. yel. lq.	....	....	$184\text{-}6^{6mm}$	i.	s.	s.
5781	yel. vis-cous lq.	....	....	$190\text{-}3^{2mm}$	sl. s.	s.	s.
5782	pa. yel. lq.	....	20-2	$195\text{-}81.5mm$	i.	s.	s.
5783	lq.	$1.044\frac{20°}{4}$	....	$204\text{-}7^{756mm}$	....	....	....
5784	pa. yel. lq.	....	....	$131\text{-}3^{2mm}$	i.	s.	s.
5785	pa. yel. lq.	....	....	$219\text{-}2^{19mm}$	i.	s.	s.
5786	col. lq.	$0.970\frac{20°}{4}$	−31.5	194	i.; ∞ bz.	s.; ∞ act.	s.

* Crysts. + $4H_2O$, m. p. 49-50°; + $6H_2O$, m. p. 55°.
Tetrahydro-nicotinic acid, cf. alkd.
Tetrahydro-pyrrole 5523
Tetrahydro-toluene 4207-8

Tetrahydroxy-flavanone 2891
Tetraiodo-fluorescein 2896
Tetraiodo-methane 1245

Table 7-4 (*Continued*)
PHYSICAL CONSTANTS OF ORGANIC COMPOUNDS

No.	Name	Synonym	Formula	Beil. Ref.	Formula Weight
	Tetrahydro				
5787	α-naphthol (*ac.*)	....	$C_6H_4 \cdot CHOH \cdot$ $\quad\mid$ $\quad (CH_2)_2 \cdot CH_2$	**VI-541	148.21
5788	α-naphthol (*ar.*)	....	$(CH_2)_4:C_6H_3OH$	VI-578	148.21
5789	β-naphthol (*ac.*)	....	$C_6H_4(CH_2)_2CHOHCH_2$	VI-579	148.21
5790	β-naphthol (*ar.*)	....	$(CH_2)_4:C_6H_3OH$	VI-579	148.21
5791	β-naphthoylaceto-nitrile (*ar.*)	....	$(CH_2)_4:C_6H_3 \cdot CO \cdot$ $CH_2 \cdot CN$	....	199.25
5792	α-naphthylamine (*ac.*)	....	$C_6H_4 \cdot CH(NH_2) \cdot$ $\quad\mid$ $\quad (CH_2)_2 \cdot CH_2$	XII-1200	147.22
5793	α-naphthylamine	(*ar.*)	$(CH_2)_4:C_6H_3 \cdot NH_2$	XII-1197	147.22
5794	β-naphthylamine (*ac.*)(*dl*)	....	$C_6H_4CH_2CH(NH_2) \cdot$ $\quad CH_2 \cdot CH_2$	XII-1200	147.22
5795	β-naphthylamine	(*ar.*)	$(CH_2)_4:C_6H_3 \cdot NH_2$	XII-1198	147.22
5796	o-phthalic acid (Δ₁)	....	$C_8H_{10}O_4$	IX-770	170.17
5797	quinoline (1,2,3,4)	....	$C_9H_{10}NH$	XX-262	133.19
5798	quinone (*p*)	....	$OC:(CH_2)_4:CO$	VII-556	112.13
5799	**Tetrahydroxy-**anthraquinone	rufiopin (1,2,5,6)	$C_{14}H_4O_2(OH)_4$	VIII-549	272.22
5800	anthraquinone (1,3,5,7)	anthrachryson	$C_{14}H_4O_2(OH)_4 \cdot 2H_2O$	VIII-551	308.25
5801	anthraquinone (1,2,5,8)	quinalizarin; alizarin bordeaux	$C_{14}H_4O_2(OH)_4$	VIII-549	272.22
5802	benzene (1,2,4,5)	....	$(HO)_4C_6H_2$	VI-1155	142.11
5803	benzene (1,2,3,5)	....	$(HO)_4C_6H_2$	VI-1154	142.11
5804	3,6-diaminoben-zene HCl	(1,2,4,5)	$(HO)_4C_6(NH_2)_2 \cdot 2HCl$	XIII-842	245.06
5805	flavone (3,7,3',4')	fisetin	$(HO)_4C_9H_3O_2 \cdot C_6H_3:$ $(OH)_2$	XVIII-221	286.24
5806	flavanol (5,7,2',4')	morin; 3,5,7,2',4'-pentaOH-flavone	$C_{15}H_{10}O_7 \cdot 2H_2O$	XVIII-239	338.27
5807	flavanol (5,7,3',4')	quercetin	$C_{15}H_{10}O_7 \cdot 2H_2O$	XVIII-242	338.27
5808	quinone	....	$(HO)_4C_6O_2$	VIII-534	172.10
5809	**Tetraiodo-**benzene	(1,2,3,4)	$I_4C_6H_2$	V-229	581.70
5810	benzene (1,2,3,5)	....	$I_4C_6H_2$	V-229	581.70
5811	benzene (1,2,4,5)	....	$I_4C_6H_2$	V-229	581.70
5812	ethylene	periodo-ethene	$I_2C:CI_2$	I-195	531.64
5813	phenolphthalein	nosophen; iodophen	$C_{20}H_{10}O_4I_4$	XVIII-151	821.92
5814	phenolphthalein	(Na salt); iodeikon	$C_{20}H_8O_4I_4Na_2 \cdot 3H_2O$	XVIII-151	919.93
5815	phenolsulfon-phthalein	....	$C_{19}H_{10}O_5I_4S$	....	857.97
5816	phthalic anhydride	....	$I_4C_6:(CO)_2O$	XVII-486	651.71
5817	pyrrole	iodol	I_4C_4NH	XX-168	570.68

Tetralin 5786
Tetralite 5873
Tetramethyl-alloxantin 232
Tetramethyl-aniline 5818

Tetramethyl-diaminobenzophenone 4436
Tetramethyl-diaminodiphenyl carbinol 4435
Tetramethyl-diaminotriphenyl methane 3964
Tetramethyl-ethylene 3658

Table 7-4 (*Continued*)
PHYSICAL CONSTANTS OF ORGANIC COMPOUNDS

No.	Crystalline Form and Color	Specific Gravity	Melting Point °C.	Boiling Point °C.	Solubility in 100 Parts		
					Water	Alcohol	Ether
5787	col. lq.	$1.090\frac{17}{4}°$	....	140^{17mm}	....	....	....
5788	mn.	....	68.5-9.0	$264-5^{705mm}$	sl. s. h.	v. s.	v. s.
5789	oil	$1.071\frac{20}{4}°$	....	264^{716mm}	v. sl. s.	v. s.	v. s.
5790	nd./lg.	....	58-9	275-6	v. sl. s.	v. s.	v. s.
5791	pa. yel. cr.	....	98-100	....	i.	sl. s.	s.
5792	oil	....	....	246.5^{714mm}	s. h.	s.	s.
5793	oil	$1.067\frac{15}{15}°$	....	275^{12mm}	v. sl. s.	s.	s.
5794	lq.	$1.034\frac{15}{15}°$	....	250^{710} sl. d.	sl. s. h.	s.	s.
5795	nd./lg.	$1.029\frac{22}{4}°$	38	$275-77^{13mm}$	....	s.	s.
5796	lf.	....	120(-H_2O)	....	v. s.	....	....
5797	cr.	$1.070^{4°}$	15-6	251	s.	∞	∞
5798	mn./aq.	....	78	subl. 100	s.	s.	s.
5799	yel. red	....	subl. d.	....	sl. s. h.; s.H_2SO_4	s.; s. h. ac.	v. sl. s.
5800	yel. nd.	....	>360	-H_2O, 150	i.; i. CS_2	sl. s.; s. ac.	v. sl. s.
5801	red nd./ PhNO$_2$	....	>275	subl.	s. alk.; s. H_2SO_4	....	....
5802	lf./ac.	....	215-20	....	v. s.; sl. s. HCl	v. s.	v. s.
5803	nd./aq.	....	165	....	v. s.	v. s.	i. chl., bz.
5804	nd.	....	....	....	v. s.	i. HCl	
5805	yel. nd.	....	>360 d.	....	i. c.; s. alk.	s.; s. act.; sl. s. pet.	sl. s.; sl. s. bz., chl.
5806	yel. nd./al.	....	285 (anh.)	....	$0.03^{20°}$; $0.1^{100°}$	s.; i. CS_2	sl. s.; s. alk.
5807	yel. nd.	....	313-4 (anh.)	....	i. c.; v. sl. s. h.	0.4 c.; 5.5 h.	v. sl. s.; s. alk.
5808	b. bl.	....	....	....	v. s. h.	v. s.	sl. s.
5809	pr./CS_2	....	136	subl.	s. chl.	s.	s.
5810	pr./et.	....	148	subl.	v. s. h. ac.	sl. s.	sl. s.
5811	nd./et.	....	254	subl. in vac.	v. s. CS_2	v. sl. s.	v. sl. s.
5812	yel. mn.	$2.983^{20°}$	187-92	subl. in vac.	v. s. CS_2	v. sl. s. c.	s. ac., bz.
5813	amor. pd.	....	d. 220±	....	i.; s. alk.	sl. s.	s.; s. chl.
5814	pa. b. hyg.	....	d. by CO_2	....	14	sl. s.	....
5815	col. amor.	....	....	....	i.	i.	i.
5816	yel. cr./ac.	....	329-31	subl.	s. PhOH	s. PhNO$_2$	i.
5817	yel./aq. al.	....	d. 140-50	....	0.02	$5.8^{15°}$; s. h.	50

Tetramethyl-ethylene glycol 5312
Tetramethyl-hydracrylic acid 3762
Tetramethyl lead 3958
Tetramethyl-*p*-leucaniline 5844

Tetramethyl-methane 5085
Tetramethyl-piperidone 6056
Tetramethyl-*iso*-propyl alcohol 3531
Tetramethyl silicon 5617

Table 7-4 (*Continued*)
PHYSICAL CONSTANTS OF ORGANIC COMPOUNDS

No.	Name	Synonym	Formula	Beil. Ref.	Formula Weight
5818	**Tetramethyl-**aminobenzene	2,3,4,5-tetramethyl-aniline	$(CH_3)_4C_6H \cdot NH_2$	XII-1175	149.24
5819	aminobenzene	(2,3,5,6); duridine	$(CH_3)_4C_6H \cdot NH_2$	XII-1177	149.24
5820	aminobenzene	(2,3,4,6); *iso*-duridine	$(CH_3)_4C_6H \cdot NH_2$	XII-1175	149.24
5821	ammonium bromide	....	$(CH_3)_4N \cdot Br$	IV-51	154.06
5822	ammonium chloride	....	$(CH_3)_4N \cdot Cl$	IV-51	109.60
5823	ammonium hydroxide	....	$(CH_3)_4NOH \cdot 5H_2O$	IV-50	181.23
5824	amm. hydroxide	....	$(CH_3)_4NOH \cdot 3H_2O$	IV-50	145.19
5825	amm. hydroxide	....	$(CH_3)_4NOH \cdot H_2O$	IV-50	109.17
5826	ammonium iodide	....	$(CH_3)_4N \cdot I$	IV-51	201.05
5827	benzene (1,2,3,4)	prehnitol	$(CH_3)_4C_6H_2$	V-430	134.22
5828	benzene (1,2,3,5)	isodurene	$(CH_3)_4C_6H_2$	V-430	134.22
5829	benzene (1,2,4,5)	durene	$(CH_3)_4C_6H_2$	V-431	134.22
5830	benzidine	....	$[(CH_3)_2N \cdot C_6H_4 \cdot]_2$	XIII-221	240.35
5831	benzoquinone	duroquinone	$(CH_3)_4C_6O_2$	VII-669	164.21
5832	diaminobutane(α,δ)	....	$[(CH_3)_2N \cdot CH_2CH_2 \cdot]_2$	IV-265	144.26
5833	diaminodiphenyl-amine (*p*)	Bindschedler's green leuco base	$[(CH_3)_2N \cdot C_6H_4]_2NH$	XIII-112	255.37
5834	diaminodiphenyl methane (*p,p'*)	Michler's hydride	$[(CH_3)_2N \cdot C_6H_4]_2CH_2$	XIII-239	254.38
5835	diamino-4''-hy-droxy-triphenyl-methane	(4,4')	$[(CH_3)_2N \cdot C_6H_4]_2CH \cdot C_6H_4 \cdot OH$	XIII-737	346.48
5836	diamino-4''-di-methoxy-tri-phenylmethane	(4,4')	$[(CH_3)_2N \cdot C_6H_4]_2CH \cdot C_6H_4 \cdot OCH_3$	XIII-737	360.50
5837	diamino-4'''-methyl-tri-phenylmethane	(4,4')	$[(CH_3)_2N \cdot C_6H_4]_2CH \cdot C_6H_4 \cdot CH_3$	XIII-282	344.50
5838	diamino-3''-nitro-triphenylmethane	(4,4')	$[(CH_3)_2N \cdot C_6H_4]_2CH \cdot C_6H_4 \cdot NO_2$	XIII-279	375.47
5839	*p*-phenylene-diamine	....	$(CH_3)_2N \cdot C_6H_4 \cdot N(CH_3)_2$	XIII-74	164.25
5840	*p*-phenylene-diamine HCl	Wurster's reagent	$C_{10}H_{16}N_2 \cdot 2HCl$	XIII-74	237.17
5841	pyromellitate	(1,2,4,5)	$C_6H_2(CO_2CH_3)_4$	IX-998	310.26
5842	succinic acid	....	$[(CH_3)_2C \cdot CO_2H]_2$	II-706	174.20
5843	thiouram disulfide	....	$[(CH_3)_2N \cdot CS \cdot S \cdot]_2$	IV-76	240.43
5844	triaminotriphenyl-methane	tetramethyl-*p*-leucaniline	$[(CH_3)_2N \cdot C_6H_4]_2CH \cdot C_6H_4 \cdot NH_2$	XIII-314	345.49
5845	urea	....	$[(CH_3)_2N]_2CO$	IV-74	116.16
5846	**Tetranitro-**diphenol	(3,x,3',x;4,4')	$[(NO_2)_2C_6H_2 \cdot OH]_2$	VI-992	366.20
5847	diphenyl (2,4,2',4')	....	$C_{12}H_6(NO_2)_4$	V-585	334.20
5848	diphenyl (3,4,3',4')	....	$C_{12}H_6(NO_2)_4$	V-585	334.20
5849	diphenyl (2,6,2',6')	....	$C_{12}H_6(NO_2)_4$	*V-274	334.20
5850	diphenyldisulfide	(2,2',4,4')	$[(NO_2)_2C_6H_3 \cdot S \cdot]_2$	VI-344	398.33
5851	diphenyl ether	(2,4,2',4')	$[C_6H_3(NO_2)_2]_2O$	VI-255	350.20
5852	diphenylmethane	(2,4,2',4')	$C_{13}H_8(NO_2)_4$	V-596	348.23
5853	methane	....	$C(NO_2)_4$	I-80	196.03

Table 7-4 (*Continued*)
PHYSICAL CONSTANTS OF ORGANIC COMPOUNDS

No.	Crystalline Form and Color	Specific Gravity	Melting Point °C.	Boiling Point °C.	Solubility in 100 Parts		
					Water	Alcohol	Ether
5818	lf./bz.	....	70	259-60	sl. s. h.	s.	s.; s. pet.
5819	pr./h. aq.	....	75	261-2	v. sl. s. c.	s.	s.
5820	cr.	$0.978^{24°}$	23-4	253-5	....	....	....
5821	cr.	1.56	d. > 230	subl. > 360	$55^{15°}$	sl. s. abs.	i.; s. SO_2
5822	col. cr.	1.169	425±*	subl. >300	s.	s. h.	i.; i. chl.
5823	hyg. nd.	....	62-3	d.	$15^{10°}$	$\infty^{63°}$ aq.	....
5824	....	....	59-60	....	....	....	....
5825	....	....	d. 130-5	....	....	....	....
5826	col. pr.	1.84	d. > 230		sl. s. c.	0.1 h.	i.; s. SO_2
5827	lq.	$0.905^{20°}_{4}$	-6.3	205.0	....	....	....
5828	lq.	$0.890^{20°}_{4}$	-24	197.9	i.	s.	....
5829	mn.	$0.838^{81°}_{4}$	79.3	196	i.; s. bz.	s.	s.
5830	nd./bz. pet.	....	193-4	>360	s. h. bz.	v. sl. s.	sl. s.
5831	yel. nd./lg.	....	111	subl. 100	i.; s. bz.	v. s.	v. s.
5832	col. lq.	$0.804^{19°}_{4}$	....	169	∞	s.	s.
5833	pl.	....	119	....	....	s.	....
5834	lf./al.	....	90-1; subl.	300	i.; s. bz.	s. h.; s. ac.	s.; s. chl.
5835	cr./al.	....	163-5	....	i.; s. dil. alk.	v. sl. s. lg.	s. bz.
5836	nd./al.	....	105-6	....	v. s. lg.; v. s. chl.	s.; v. s. CS_2	v. s.
5837	nd./al.	....	94-5	....	i.	s.	s.
5838	yel. pr./al.	....	152	....	i.; s. bz.	sl. s.; sl. s. lg.	sl. s.
5839	lf./aq. al.	....	51	260	sl. s. h.	v. s.; v. s. lg.	v. s.; v. s. chl.
5840	cr.	....	....	....	s.	....	....
5841	lf./Me al.	....	141.5	....	....	sl. s. h.	....
5842	cr.	....	190-200 d.	....	$0.48^{13.5°}$	s.	s.; i. lg.
5843	cr./al. chl.	$1.29^{20°}_{4}$	146	....	i.; s. chl.	v. sl. s.	v. sl. s.
5844	cr./al.	....	151-2	....	....	sl. s.	....
5845	lq.	$0.972^{15°}$	....	177.5	....	v. s.	v. s.
5846	yel. nd.	....	220-5	....	i.	s.	....
5847	yel. pr./bz.	....	165-6	d.	s. bz.; s. ac.	sl. s.	sl. s.
5848	yel. pr.	....	186	....	s. bz.	v. sl. s. lg.	s. ac.
5849	yel. nd./ac.	....	217-8	....	....	....	....
5850	yel. nd.	....	expl. > 280	....	i.; s. pyr.	i.; s. $PhNO_2$	i.; s. $PhNH_2$
5851	cr.	....	195±	....	i.; s. h. bz.	v. sl. s.	sl. s.
5852	lt. yel./ac.	....	172	....	sl. s. bz.	i.	i.
5853	col. lq.	$1.639^{20°}_{4}$	13.8	125.7 sl. d.	i.	v. s.	v. s.

Tetramethylene-glycol diacetate 2224
Tetramethylene oxide 5773
Tetranitro-chrysazine 1491
Tetraphenyl lead 3959

Tetraphenyl-pyrazine 234
Tetraphenyl tin 5952
Thebaine, cf. alkd.
Theelin 2903

*Sealed tube.

Table 7-4 (*Continued*)
PHYSICAL CONSTANTS OF ORGANIC COMPOUNDS

No.	Name	Synonym	Formula	Beil. Ref.	Formula Weight
	Tetranitro				
5854	naphthalene (γ)	(1,3,5,8)	$C_{10}H_4(NO_2)_4$	V-564	308.17
5855	naphthalene (β)	(1,3,6,8)	$C_{10}H_4(NO_2)_4$	V-564	308.17
5856	naphthalene (α)	(1,5,x,x)	$C_{10}H_4(NO_2)_4$	V-564	308.17
5857	phenol (2,3,4,6)	....	$HO \cdot C_6H(NO_2)_4$	VI-292	274.10
5858	phenolsulfon- phthalein	....	$C_{19}H_{10}O_{13}N_4S$	*XIX-650	534.37
5859	**Tetraphenyl-** arsonium bromide	....	$(C_6H_5)_4AsBr \cdot 2H_2O$	....	499.29
5860	ethane (*sym*)	....	$[(C_6H_5)_2CH \cdot]_2$	V-738	334.47
5861	ethylene	....	$(C_6H_5)_2C:C(C_6H_5)_2$	V-743	332.45
5862	guanidine	....	$HN:C[N(C_6H_5)_2]_2$	XII-430	363.47
5863	hydrazine	....	$(C_6H_5)_2N \cdot N(C_6H_5)_2$	XV-125	336.44
5864	methane	....	$(C_6H_5)_4C$	V-739	320.44
5865	urea	....	$[(C_6H_5)_2N]_2CO$	XII-429	364.45
5866	**Tetrapropyl-**ammo- nium iodide (*n*)	....	$(C_2H_5 \cdot CH_2)_4NI$	IV-140	313.27
5867	benzene (*iso*)	(1,2,4,5)	$[(CH_3)_2CH]_4C_6H_2$	**V-358	246.44
5868	**Tetratriacontane** (*n*)	....	$CH_3(CH_2)_{32}CH_3$	I-177	478.94
5869	**Tetrazine** (1,2,4,5)	....	$(:N \cdot CH \cdot N)_2$	XXVI-353	82.07
5870	**Tetrazole**	....	$CH:N \cdot NH \cdot N:N$	XXVI-346	70.05
5871	**Tetrolic acid**	butynoic acid	$CH_3 \cdot C:C \cdot CO_2H$	II-479	84.08
5872	**Tetronal**	pentane-3,3-di- ethyl-sulfone	$(C_2H_5)_2C(SO_2C_2H_5)_2$	I-681	256.39
5873	**Tetryl** (2,4,6)	tri-NO$_2$-phenyl- methylnitramine	$(NO_2)_3C_6H_2 \cdot N(CH_3)NO_2$	XII-770	287.15
5874	**Thallin**	6-methoxy-tetra- hydroquinoline	$CH_3O \cdot C_9H_{10}N$	XXI-61	163.22
5875	**Thialdine**	....	$(CH_3)_3C_3H_4S_2N$	XXVII-461	163.31
5876	**Thiamine chloride**	vitamin B$_1$	$C_{12}H_{17}ON_4ClS \cdot HCl$	....	337.27
5877	**Thianthrene**	diphenylene disul- fide	$(C_6H_4)_2S_2$	XIX-45	216.33
5878	**Thiazole**	....	C_3H_3NS	XXVII-15	85.13
5879	**Thio-**acetaldehyde (α)	sulfaldehyde	$(CH_3 \cdot CHS)_3$	XIX-387	180.35
5880	acetaldehyde (β)	sulfaldehyde	$(CH_3 \cdot CHS)_3$	XIX-387	180.35
5881	acetaldehyde (γ)	sulfaldehyde	$(CH_3 \cdot CHS)_3$	....	180.35
5882	acetamide	aceto-thioamide	$CH_3 \cdot CS \cdot NH_2$	II-232	75.13
5883	acetanilide	....	$C_6H_5 \cdot NH \cdot CS \cdot CH_3$	XII-245	151.23
5884	acetdimethylamide	....	$CH_3 \cdot CS \cdot N(CH_3)_2$	*IV-329	103.19
5885	acetic acid	....	$CH_3 \cdot CO \cdot SH$	II-230	76.12
5886	aniline (4,4')	diNH$_2$-diPH-sulfide	$(NH_2 \cdot C_6H_4)_2S$	XIII-535	216.31
5887	barbituric acid	malonyl thiourea	$C_4H_4O_2N_2S$	XXIV-476	144.15
5888	benzaldehyde (α)	....	$(C_6H_5 \cdot CHS)_3$	XIX-396	366.57
5889	benzaldehyde (β)	....	$(C_6H_5 \cdot CHS)_3$	XIX-397	366.57
5890	benzanilide	....	$C_6H_5 \cdot NH \cdot CS \cdot C_6H_5$	XII-269	213.30
5891	benzoic acid	....	$C_6H_5 \cdot CO \cdot SH$	IX-419	138.19
5892	carbanilide	diphenyl thiourea	$(C_6H_5 \cdot NH)_2CS$	XII-394	228.32
5893	chrome	....	$C_{12}H_{14}ON_4S$	....	262.34
5894	cresol (*o*)	*o*-tolyl mercaptan	$CH_3 \cdot C_6H_4 \cdot SH$	VI-370	124.21
5895	cresol (*m*)	*m*-tolyl mercaptan	$CH_3 \cdot C_6H_4 \cdot SH$	VI-388	124.21
5896	cresol (*p*)	*p*-tolyl mercaptan	$CH_3 \cdot C_6H_4 \cdot SH$	VI-416	124.21

Theelol 2902
Theine, cf. alkd.
Theobromine, cf. alkd.
Theocine, cf. alkd.

Theophylline, cf. alkd.
Thevetin, cf. glcde.
Thienyl alcohol 5924
Thio-, cf. also sulfo-

Table 7-4 (*Continued*)
PHYSICAL CONSTANTS OF ORGANIC COMPOUNDS

No.	Crystalline Form and Color	Specific Gravity	Melting Point °C.	Boiling Point °C.	Solubility in 100 Parts		
					Water	Alcohol	Ether
5854	yel./act.	....	194-5	....	s. HNO$_3$	sl. s.	s. act.
5855	nd./al.	....	203	expl.	i.	....	....
5856	lt. yel./chl.	....	259	expl.	i.	v. sl. s.	....
5857	lt. yel./chl.	....	140	expl.	d. h.	v. sl. s. bz.	v. sl. s. lg.
5858	yel. nd./ ac. act.	....	>200	....	s.	s.	i.
5859	col. cr.	....	281-4 (anh.)	$-2H_2O >$ 100	1.6	s.	sl. s. act.
5860	rhb.	1.182	209-11	379-83	14 h. bz.	0.8 h.	5 h. ac.
5861	tri. or mn.	$1.155_4^{0°}$ ...	221-3	415-25	s. h. bz.	sl. s.	sl. s.
5862	rhb./lg.	....	130-1	....	i.; v. s. bz.	v. s.	v. s.
5863	rhb./al. chl.	....	147-9	....	s. bz.	v. sl. s. h.	s. act.
5864	rhb./bz.	....	285	431	i. lg.	i.; s. h. bz.	i.; i. ac.
5865	rhb.	1.222	183	....	i.	s. h.	....
5866	rhb.	$1.314_4^{25°}$	d. 280±	....	18.6$^{25°}$	55$^{25°}$ chl.	....
5867	cr./al.	....	117	260^{775mm}	i.	1$^{25°}$	85$^{25°}$
5868	cr.	lq. 0.781$^{73°}$	α73;β77(?)	255$^{1.0mm}$	....	....	....
5869	red pr.	....	99	....	s.	s.	s.
5870	lf./al.	....	155	subl.	s.; s. ac.	s.; sl. s. bz.	sl. s.
5871	pl./et.	....	77-8	203	v. s.	v. s.	v. s.
5872	pl./aq.	....	85	....	0.22 c.	5.4$^{15°}$ abs.	10$^{15°}$
5873	yel. mn./al.	1.57$^{19°}$	129	expl. 187	i.; s. bz.	s. h.	s.; s. ac.
5874	pr.	....	42-3	283^{735mm}	sl. s. c.	v. s.; v. s. bz.	v. s.
5875	mn.	1.191$^{18°}$	43	d.	v. sl. s.		v. s.
5876	hyg. pr.	....	252 sl. d.	....	100	0.1	i.; i. bz.
5877	mn./al.	$1.706_4^{19°}$	158-60	364-6	i.	0.25 c.	s. h.
5878	lq.	$1.200_4^{17°}$	....	116.8	sl. s.	s.	s.
5879	cr./al.	....	101	246-7	i.; s. bz.	3.86$^{25°}$	15.6$^{25°}$
5880	cr./al.	....	125-6	246-7	i.; s. chl.	3.97$^{25°}$	13.7$^{25°}$
5881	cr.	....	81	100	....	....	....
5882	mn.	....	115-6	....	v. s.	sl. s.	sl. s.
5883	nd./aq.	....	75-6	d.	i.	i. a.	s. alk.
5884	col. cr.	....	114-6	....	sl. s.	s.	v. s.
5885	yel. lq.	1.074$^{10°}$	<-17	93	s.	∞	∞
5886	nd./aq.	....	108	....	sl. s. h.	s.	s.; s. h. bz.
5887	lf./aq.	....	235 d.	....	sl. s.	s.; s. alk.	s. alk. carb.
5888	nd./bz. al.	....	166-7	....	i.	0.2$^{25°}$	1.1$^{25°}$
5889	nd./bz.	....	225-6 d.	....	s. h. ac.	0.04$^{25°}$	0.4$^{25°}$
5890	yel. pr./al.	....	101-2	....	i.	s.	v. s.
5891	yel. oil	....	24	d.	i.	v. s.	∞
5892	rhb./al.	1.3$^{24°}$	154	d.	i.	v. s.	v. s.
5893	yel. pr./chl.	....	227-8*	....	s.	sl s.	sl. s.
5894	lf.	....	15	194.3	i.	s.	s.
5895	lq.	$1.052_4^{12°}$	<-20	195.4	i.	s.	s.
5896	lf./al.	....	43-4	195	i.	sl. s.	v. s.

* HCl salt, m. p. 217-21°.
Thio-allyl ether 1684
Thio-benzyl alcohol 816

Thio-carbamide 5920
Thio-carbonyl chloride 5916
Thio-cresol methyl ether 4402

Table 7-4 (*Continued*)
PHYSICAL CONSTANTS OF ORGANIC COMPOUNDS

No.	Name	Synonym	Formula	Beil. Ref.	Formula Weight
5897	**Thio** cyanic acid	sulfocyanic acid	$HS \cdot CN$	III-143	59.09
5898	cyanic acid, per-	3,5-dimercapto-1,2, 4-thiodiazol	$CS \cdot NH \cdot CS \cdot NH \cdot S$ $\vert_____\vert$	XXVII-665	150.24
5899	cyanuric acid	....	$C_3H_3N_3S_3$	XXVI-259	177.27
5900	diglycol (β)	β,β'diOH-diEt sulfide	$(HO \cdot CH_2CH_2)_2S$	I-470	122.19
5901	formamide	....	$H \cdot CS \cdot NH_2$	II-95	61.11
5902	formanilide	....	$C_6H_5 \cdot NH \cdot CH:S$	XII-233	137.20
5903	glycerol (1)	....	$HS \cdot CH_2 \cdot C_2H_3(OH)_2$	I-519	108.16
5904	glycol, mono-	ethanol-1-thiol-2	$HO \cdot CH_2 \cdot CH_2 \cdot SH$	I-470	78.13
5905	glycolic acid	....	$HS \cdot CH_2 \cdot CO_2H$	III-245	92.12
5906	glycolic-β-amino-naphthalide	thionalide	$C_{10}H_7 \cdot NH \cdot CO \cdot CH_2 \cdot SH$	....	217.29
5907	hydantoin	glycolyl thiourea	$C_3H_4ON_2S$	XXIV-260	116.14
5908	hydracrylic acid	....	$HS \cdot CH_2CH_2CO_2H$	III-299	106.14
5909	hydroquinone, mono-	(1,4)	$HS \cdot C_6H_4 \cdot OH$	VI-859	126.18
5910	lactic acid	....	$CH_3 \cdot CH(SH) \cdot CO_2H$	III-289	106.14
5911	malic acid (*dl*)	mercaptosuccinic	$(HS \cdot C_2H_3)(CO_2H)_2$	III-439	150.15
5912	naphthene	benzothiophene	C_8H_6S	XVII-59	134.20
5913	naphthol (α)	α-naphthyl mercaptan	$C_{10}H_7 \cdot SH$	VI-621	160.24
5914	naphthol (β)	β-naphthyl mercaptan	$C_{10}H_7 \cdot SH$	VI-657	160.24
5915	phenol	phenyl mercaptan	$C_6H_5 \cdot SH$	VI-294	110.18
5916	phosgene	thiocarbonyl chloride	$Cl_2C:S$	III-134	114.98
5917	propionamide	....	$C_2H_5 \cdot CS \cdot NH_2$	II-264	89.16
5918	salicylic acid (*o*)	sulfhydryl benzoic	$HS \cdot C_6H_4 \cdot CO_2H$	X-125	154.19
5919	semicarbazide	....	$NH_2 \cdot CS \cdot NH \cdot NH_2$	III-195	91.14
5920	urea	thiocarbamide	$NH_2 \cdot CS \cdot NH_2$	III-180	76.12
5921	urethane	ethyl thiourethane	$NH_2 \cdot CO \cdot SC_2H_5$	III-138	105.16
5922	**Thionin**(e)	Lauth's violet	$C_{12}H_9N_3S$	XXVII-391	227.28
5923	**Thiophene**	....	C_4H_4S	XVII-29	84.14
5924	alcohol (α)	thienyl carbinol	$C_4H_3S \cdot CH_2OH$	XVII-113	114.17
5925	aldehyde (α)(2)	....	$C_4H_3S \cdot CHO$	XVII-285	112.15
5926	carboxylic acid (α)	(1,2)	$C_4H_3S \cdot CO_2H$	XVIII-289	128.15
5927	carboxylic acid (β)	(1,3)	$C_4H_3S \cdot CO_2H$	XVIII-292	128.15
5928	**Thioxene** (*o*)	2,5-diMe-thiophene	$(CH_3)_2C_4H_2S$	XVII-41	112.19
5929	**Thioxene** (*o*)	2,3-diMe-thiophene	$(CH_3)_2C_4H_2S$	XVII-40	112.19
5930	**Thioxene** (*m*)	2,4-diMe-thiophene	$(CH_3)_2C_4H_2S$	XVII-41	112.19
5931	**Thioxene** (*m*)	3,4-diMe-thiophene	$(CH_3)_2C_4H_2S$	XVII-42	112.19
5932	**Threonine** (*dl*)	α-amino-β-hydroxy-butyric acid	$CH_3 \cdot CHOH \cdot CH: (NH_2)CO_2H$	IV-514	119.12
5933	**Threonine** (*d*-)	....	$C_4H_9O_3N$	IV-514	119.12
5934	**Threonine**, allo-	(*dl*)	$C_4H_9O_3N$	....	119.12
5935	**Thujone** (α)	....	$C_{10}H_{16}O$	VII-93	152.24
5936	**Thujone** (β)	tanacetone	$C_{10}H_{16}O$	VII-93	152.24
5937	**Thymohydro-quinone**	....	$C_{10}H_{14}O_2$	VI-945	166.22
5938	**Thymol**	5-Me-2-isoPr-phenol	$(CH_3)(C_3H_7)C_6H_3OH$	VI-532	150.22
5939	chloride	(3,6;1,4)	$(CH_3)(C_3H_7)C_6H_2: (OH)Cl$	VI-539	184.67
5940	iodide	4-iodothymol	$C_{10}H_{13}OI$	VI-541	276.12
5941	phthalein	....	$C_{28}H_{30}O_4$	*XVIII-381	430.55

Thio-diphenylamine 5138
Thio-histidine betaine 2890
Thio-phenetole 3132
Thio-phenine 378

Thio-phenylacetamide 5240
Thio-sinamine 221
Thio-trimethylacetamide 6275
Thiols, cf. mercaptans.

Table 7-4 (*Continued*)
PHYSICAL CONSTANTS OF ORGANIC COMPOUNDS

No.	Crystalline Form and Color	Specific Gravity	Melting Point °C.	Boiling Point °C.	Solubility in 100 Parts		
					Water	Alcohol	Ether
5897	col. lq.	....	5±	d.	∞, d.	v. s.	v. s.
5898	not known; salts only	....	....	....	....	....	....
5899	cr.	....	d. 200	....	sl. s. h.	sl. s.	sl. s.
5900	syrup	$1.221^{20°}_{20}$	−10	$164\text{-}6^{20mm}$	∞	s. chl.	....
5901	col. cr.	....	28-9	....	s.; i. bz.	s.; s. act.	s.; s. CS$_2$
5902	nd./aq.	....	138	....	i. aq. HCl	s.	v. s.
5903	yel. lq.	$1.295^{14.4°}$	....	d.	v. sl. s.	∞	i.
5904	lq.	$1.114^{20°}_{4}$	....	153-7	v. s.	v. s.	v.s.; ∞bz.
5905	col. lq.	$1.325^{20°}$	−16.5	123^{29mm}	....	....	....
5906	col. nd.	....	111-2	....	....	s.	s.
5907	nd./aq.	....	d. 200±	....	sl. s.	i.	i.
5908	col. cr.	$1.218^{21°}$	16.8	$111\text{-}2^{15mm}$	∞	∞	∞
5909	cr.	....	32-4	$166\text{-}8^{45mm}$	s.	s. H$_2$SO$_4$	....
5910	oil	....	10±	$98\text{-}9^{14mm}$	∞	∞	∞
5911	col. cr.	....	149-50	....	s.; s. act.	s.; sl. s. bz.	sl. s.
5912	lf.	$1.165^{20°}_{4}$	31-2	220-1	....	....	....
5913	lq.	$1.155^{23°}_{4}$	....	208.5^{200mm}	sl. s.	v. s.	v. s.
5914	cr./al.	....	81	286-8	v. sl. s.	v. s.	v. s.
5915	col. lq.	$1.074^{23°}_{4}$	−14.9	169.5	v. sl. s.	v. s.; ∞ bz.	∞; s. CS$_2$
5916	red lq.	$1.509^{15°}$	....	73.5	d.	d.	s.
5917	lf./bz.	....	41-3	....	v. sl. s.	v. sl. s.	s.; s. bz.
5918	lt. yel. nd.	....	164	subl.	sl. s. h.	sl. s.	s. ac.
5919	nd./aq.	....	181-3 d.	....	s.	s.	....
5920	rhb./al.	$1.405^{20°}_{4}$	180-2	d.	$9.2^{13°}$	s.	sl. s.
5921	lf.	....	108-9	subl. sl. d.	s. h.	s.	s.
5922	brn. bl. lf.	....	....	....	i. c.; i. lg.	sl. s.; s. chl.	sl. s.
5923	col. lq.	$1.070^{15°}_{4}$	−38.3	84	i.; s. bz.	s.	s. H$_2$SO$_4$
5924	lq.	....	....	207	....	....	....
5925	oil	$1.215^{21°}$	....	198	....	....	s.
5926	nd./aq.	....	126.5	260 sl. d.	v. s. h.	v. s.	s.; sl. s. lg.
5927	mn. nd.	....	136	subl.	$0.43^{25°}$	....	....
5928	lq.	$0.976^{17.5°}$	....	136.5-7.5	i.	s.	s.
5929	lq.	$0.994^{21°}$	....	136-7	....	....	....
5930	lq.	$0.996^{20°}$	....	137-8	i.	s.	s.
5931	lq.	$1.008^{23°}_{22}$	....	144-6	....	....	....
5932	col. cr.	....	235-8 d.	....	$20^{25°}$; v. s. h.	i; i. chl.	i.
5933	hex. pl.	....	225-7 d.	....	s.	....	....
5934	col. cr.	....	250-2	....	s. h.	....	....
5935	col. lq.	0.912	....	200-1	i.	s.	....
5936	col. lq.	$0.916^{20°}_{4}$	....	201-3	i.	s.	....
5937	pr.	....	140-3	290	s. h.	v. s.	v. s.
5938	cr./act.	$0.972^{25°}_{25}$	49.6	232.9	$0.09^{19°}$; $0.11^{100°}$	v. s.; v. s. chl.	v. s.; s. alk.
5939	col. pl./lg.	....	59-61	....	0.1; s. dil. alk.	222; 60 bz.	83
5940	nd.	....	68-9	....	sl. s. h.	sl. s.	s.; s. chl.
5941	col. nd./al.	....	252-3	....	i.; s. H$_2$SO$_4$	s.; s. act.	s. dil. alk.

Table 7-4 (*Continued*)
PHYSICAL CONSTANTS OF ORGANIC COMPOUNDS

No.	Name	Synonym	Formula	Beil. Ref.	Formula Weight
5942	**Thymol** sulfonic acid (α)	(3,6;1,4)	$(CH_3)(C_3H_7)C_6H_2:$ $(OH)SO_3H \cdot H_2O$	XI-267	248.30
5943	**Thymoquinone**	....	$O:C_6H_2(CH_3)\cdot$ $(C_3H_7):O$	VII-662	164.21
5944	**Thymotic acid** (*o*)	(1,4;3,2)	$(CH_3)(C_3H_7)C_6H_2:$ $(OH)CO_2H$	X-280	194.23
5945	**Thymyl**-acetate	acetyl thymol	$C_{10}H_{13}\cdot O_2C\cdot CH_3$	VI-537	192.26
5946	amine (1,3;4)	....	$CH_3(NH_2)C_6H_3\cdot C_3H_7$	XII-1171	149.24
5947	**Thyroxine** (*l*)	....	$C_{15}H_{11}O_4NI_4$	*XIV-671	776.88
5948	**Tin** chloride triphenyl	triphenyltin chloride	$(C_6H_5)_3SnCl$	*XVI-540	385.46
5949	diethyl	diethyl tin	$(C_2H_5)_2Sn$	IV-631	176.81
5950	tetraethyl	tetraethyl tin	$(C_2H_5)_4Sn$	IV-632	234.94
5951	tetramethyl	tetramethyl tin	$(CH_3)_4Sn$	IV-632	178.83
5952	tetraphenyl	tetraphenyl tin	$(C_6H_5)_4Sn$	XVI-914	427.12
5953	triethyl	triethyl tin	$[(C_2H_5)_3Sn\cdot]_2$	IV-638	411.75
5954	**Tolane**	diphenyl-acetylene	$C_6H_5\cdot C:C\cdot C_6H_5$	V-656	178.24
5955	**Tolidine** (*o*)	3,3′-di Me-benzidine	$(CH_3\cdot C_6H_3\cdot NH_2)_2$	XIII-256	212.30
5956	hydrochloride	....	$C_{14}H_{16}N_2\cdot 2HCl$	XIII-257	285.22
5957	sulfate	....	$C_{14}H_{16}N_2\cdot H_2SO_4$	XIII-257	310.37
5958	disulfonic acid	(3,4;6)	$[(CH_3)(NH_2)C_6H_2\cdot$ $SO_3H]_2\cdot 1\frac{1}{2}H_2O$	XIV-796	399.44
5959	**Tolidine** (*m*)	2,2′-di Me-benzidine	$(CH_3\cdot C_6H_3\cdot NH_2)_2$	XIII-255	212.30
5960	**Toluene**	methyl-benzene	$CH_3\cdot C_6H_5$	V-280	92.14
5961	azo-1-naphthyl-amine-2 (*o*)	(2,1;1′,2′); yellow OB	$CH_3\cdot C_6H_4\cdot N:N\cdot$ $C_{10}H_6\cdot NH_2$	XVI-373	261.33
5962	disulfonic acid	(2,4)	$CH_3\cdot C_6H_3(SO_3H)_2$	XI-204	252.27
5963	disulfonic Na	(2,4); (Na salt)	$C_7H_6O_6S_2Na_2\cdot 7H_2O$	XI-204	422.34
5964	sulfinic acid (*p*)	....	$CH_3\cdot C_6H_4\cdot SO_2H$	XI-9	156.20
5965	sulfinic Na (*p*)	(sodium salt)	$C_7H_7O_2SNa\cdot 2H_2O$	XI-9	214.22
5966	sulfonic acid (*o*)	....	$CH_3\cdot C_6H_4\cdot SO_3H\cdot 2H_2O$	XI-83	208.23
5967	sulfonic acid (*m*)	....	$CH_3\cdot C_6H_4\cdot SO_3H\cdot aq.$	XI-94	172.20
5968	sulfonic acid (*p*)	....	$CH_3\cdot C_6H_4\cdot SO_3H\cdot H_2O$	XI-97	190.22
5969	sulfonic Na (*p*)	(sodium salt)	$C_7H_7O_3SNa\cdot 2H_2O$	XI-97	230.22
5970	**Toluenesulfonyl-**amide (*o*)	toluenesulfonamide	$CH_3\cdot C_6H_4\cdot SO_2\cdot NH_2$	XI-86	171.22
5971	amide (*p*)	....	$CH_3\cdot C_6H_4\cdot SO_2\cdot NH_2$	XI-104	171.22
5972	anilide (*p*)	toluene-sul-fonanilide	$CH_3\cdot C_6H_4\cdot SO_2\cdot NH\cdot$ C_6H_5	XII-567	247.32
5973	*n*-butylamide (*p*)	....	$CH_3\cdot C_6H_4\cdot SO_2\cdot NH\cdot$ C_4H_9	....	227.33
5974	chloride (*o*)	toluene sulfon-chloride	$CH_3\cdot C_6H_4\cdot SO_2Cl$	XI-86	190.65
5975	chloride (*p*)	....	$CH_3\cdot C_6H_4\cdot SO_2Cl$	XI-103	190.65
5976	di-*n*-butylamide	(*p*)	$CH_3\cdot C_6H_4\cdot SO_2\cdot N:$ $(C_4H_9)_2$	....	283.44
5377	dimethylamide (*p*)	....	$CH_3\cdot C_6H_4\cdot SO_2\cdot N:$ $(CH_3)_2$	**XI-56	199.27
5978	ethylamide (*p*)	....	$CH_3\cdot C_6H_4\cdot SO_2\cdot NH\cdot$ C_2H_5	XI-105	199.27
5979	methylamide	....	$CH_3\cdot C_6H_4\cdot SO_2\cdot NH\cdot$ CH_3	XI-105	185.25
5980	methylanilide	....	$CH_3\cdot C_6H_4\cdot SO_2\cdot N:$ $(CH_3)(C_6H_5)$	XII-575	261.35

Tiglic acid 2426
Tiglic aldehyde 2427
T M A 231
T N T 6316
Tobias' acid 4505, 4549

Tolamine 1290
Tolazine 4334
Toliminazole 4137
Tolualdehyde 5985-7
Toluamide 5988-90

Table 7-4 (*Continued*)
PHYSICAL CONSTANTS OF ORGANIC COMPOUNDS

No.	Crystalline Form and Color	Specific Gravity	Melting Point °C.	Boiling Point °C.	Solubility in 100 Parts		
					Water	Alcohol	Ether
5942	col. pl.	....	115-6	....	v. s.	....	....
5943	yel. tri.	....	46-7	232	v. sl. s.	s.	s.
5944	mn./aq.	....	127	subl.	0.01 c.	s.	s.; s. bz.
5945	lq.	1.009^{0°	....	245^{757mm}	i.; ∞ chl.	∞ ; ∞ bz.	∞
5946	oil	....	....	230	v. sl. s.	s.	s.
5947	col. nd.	....	232 d.	....	i.; s. alk.	i.	i.
5948	col. cr./al.	....	112-3	$240^{13.5mm}$	i.	s.	s.
5949	yel. oil	1.558^{15°	....	d.	i.	s.	s.
5950	col. lq.	$1.199^{\frac{20^\circ}{4}}$	−112	181	i.	....	s.
5951	lq.	$1.291^{\frac{26^\circ}{4}}$	....	78	i.		
5952	tet. pr./chl.	$1.490^{\frac{0^\circ}{4}}$	226	>420	i.; s. pyr.	v. sl. s.	v. sl. s.
5953	lq.	1.412^{0°	....	265-70	i.	i. aq. al.	s. bz.
5954	mn. pr.	$0.966^{\frac{100^\circ}{4}}$	60-2	300	i.	v. s. h.	v. s.
5955	lf.	....	129-31	....	v. sl. s.	s.	s.; s. ac.
5956	scales	....	d. > 340		0.9^{12°	....	
5957	gray	....	d.		0.1c.; d.h.	v. sl. s.	s. dil. a.
5958	nd.	....	−H$_2$O, 150	....	0.2^{18° (anh.)	i.; i. ac.	i.
5959	pr./aq.	....	107-8	....	sl. s. h.	v. s.	v. s.
5960	col. lq.	$0.866^{\frac{20^\circ}{4}}$	−95	110.6	i.; s. act.	∞ abs.	∞
5961	yel. pd.	....	122-5	....	i.; s. oil	s.; s. CCl$_4$	s.; s. bz.
5962	syrup	....	....	....	s.	....	....
5963	pr.	....	....	....	s.	....	....
5964	cr./aq.	....	85-90	....	sl. s.	v. s.; s. h. bz.	v. s.
5965	cr.	....	....	....	s.	....	....
5966	delq. cr.	....	*	$128.8^{0.1mm}$	v. s.	s.	....
5967	syrup	....	....	....	v. s.	s.	....
5968	mn.	....	104-5	$146-7^{0.1mm}$	v. s.	s.	....
5969	lf.	....	....	....	s.	....	....
5970	tet. pr.	....	156	....	0.11^{9°	3.57^{5°	....
5971	mn.	....	137	....	0.2^{9°	7.4^{5°	....
5972	tri./aq. al. or bz.	....	103	....	....	v. s.	....
5973	col. rhb.	....	41-2.5	....	i.	s.	v. s.
5974	oil	1.344^{17°	10	126^{10mm}	i.	....	....
5975	tri.	....	69	134.5^{10mm}	i.	s.	s.; s. bz.
5976	pa. yel. lq.	....	....	$233-4^{20mm}$	i.	s.	s.
5977	nd./pet.	....	86-7	....	i.	sl. s.; v. s. act.	v. s.; v. s. bz.
5978	cr./lg.; pl./aq. al.	....	63-4.5	....	....	....	....
5979	pl./aq. al.	....	77-8	....	v. sl. s.	v. s.	....
5980	mn. pl./ EtOAc	....	93-4	....	....	v. s.	v. s.

* Stable < 100°; forms *p*-isomer 140-50°.
Tolubenzyl alcohol 6016-8
Tolubenzyl-amine 6482-4
Toluene-azo-toluidine 309-12

Tolubenzyl acetate 6481
Toluene hexahydride 4199
Toluene tetrahydride 4207-8
Toluhydroquinone 2372

Table 7-4 (*Continued*)
PHYSICAL CONSTANTS OF ORGANIC COMPOUNDS

No.	Name	Synonym	Formula	Beil. Ref.	Formula Weight
5981	**Toluenesulfonyl** *p*-toluidide (*p*)	....	$CH_3 \cdot C_6H_4 \cdot SO_2 \cdot NH \cdot C_6H_4 \cdot CH_3$	XII-981	261.35
5982	**Toluic** acid†(*o*)	2-Me-benzoic acid	$CH_3 \cdot C_6H_4 \cdot CO_2H$	IX-462	136.15
5983	acid (*m*)	3-Me-benzoic acid	$CH_3 \cdot C_6H_4 \cdot CO_2H$	IX-475	136.15
5984	acid (*p*)	4-Me-benzoic acid	$CH_3 \cdot C_6H_4 \cdot CO_2H$	IX-483	136.15
5985	aldehyde (*o*)	*o*-toluylaldehyde	$CH_3 \cdot C_6H_4 \cdot CHO$	VII-295	120.15
5986	aldehyde (*m*)	*m*-Me-benzaldehyde	$CH_3 \cdot C_6H_4 \cdot CHO$	VII-296	120.15
5987	aldehyde (*p*)	4-Me-benzaldehyde	$CH_3 \cdot C_6H_4 \cdot CHO$	VII-297	120.15
5988	amide (*o*)	*o*-toluamide	$CH_3 \cdot C_6H_4 \cdot CO \cdot NH_2$	IX-465	135.17
5989	amide (*m*)	*m*-toluamide	$CH_3 \cdot C_6H_4 \cdot CO \cdot NH_2$	IX-477	135.17
5990	amide (*p*)	4-toluamide	$CH_3 \cdot C_6H_4 \cdot CO \cdot NH_2$	IX-486	135.17
5991	anhydride (*o*)	....	$(CH_3 \cdot C_6H_4 \cdot CO)_2O$	IX-464	254.29
5992	nitrile (*o*)	*o*-tolunitrile	$CH_3 \cdot C_6H_4 \cdot CN$	IX-466	117.15
5993	nitrile (*m*)	3-tolyl cyanide	$CH_3 \cdot C_6H_4 \cdot CN$	IX-477	117.15
5994	nitrile (*p*)	4-tolunitrile	$CH_3 \cdot C_6H_4 \cdot CN$	IX-489	117.15
5995	**Toluidine** (*o*)	2-methyl aniline	$CH_3 \cdot C_6H_4 \cdot NH_2$	XII-772	107.16
5996	hydrochloride	....	$CH_3 \cdot C_6H_4 \cdot NH_2 \cdot HCl$	XII-782	143.62
5997	**Toluidine** (*m*)	3-methyl aniline	$CH_3 \cdot C_6H_4 \cdot NH_2$	XII-853	107.16
5998	hydrochloride	....	$CH_3 \cdot C_6H_4 \cdot NH_2 \cdot HCl$	XII-856	143.62
5999	**Toluidine** (*p*)	4-methyl aniline	$CH_3 \cdot C_6H_4 \cdot NH_2$	XII-880	107.16
6000	hydrochloride	....	$CH_3 \cdot C_6H_4 \cdot NH_2 \cdot HCl$	XII-896	143.62
6001	**Toluquinone**	*o*-Me-*p*-benzo-quinone	$O:C_6H_3(CH_3):O$	VII-645	122.12
6002	**Toluyl-*o*-benzoic** acid (*p*)	(1,4;1′,2′)	$CH_3 \cdot C_6H_4 \cdot CO \cdot C_6H_4 \cdot CO_2H$	X-759	240.26
6003	**Toluylene** diamine	(1;3,4)(*o*)(*uns*)	$CH_3 \cdot C_6H_3(NH_2)_2$	XIII-148	122.17
6004	diamine HCl	(1;3,4)	$C_7H_{10}N_2 \cdot 2HCl$	XIII-148	195.09
6005	diamine (1;2,4)	(*m*)(*uns*)	$CH_3 \cdot C_6H_3(NH_2)_2$	XIII-124	122.17
6006	diamine HCl (1;2,4)		$C_7H_{10}N_2 \cdot 2HCl$	XIII-129	195.09
6007	diamine (1;2,5)	(*p*)	$CH_3 \cdot C_6H_3(NH_2)_2$	XIII-144	122.17
6008	diamine HCl	(1;2,5)	$C_7H_{10}N_2 \cdot 2HCl$	XIII-144	195.09
6009	diamine H_2SO_4	(1;2,5)	$C_7H_{10}N_2 \cdot H_2SO_4$	XIII-144	220.25
6010	**Tolyl** acetic acid (*o*)	....	$CH_3 \cdot C_6H_4 \cdot CH_2 \cdot CO_2H$	IX-527	150.18
6011	acetic acid (*m*)	....	$CH_3 \cdot C_6H_4 \cdot CH_2 \cdot CO_2H$	IX-528	150.18
6012	acetic acid (*p*)	....	$CH_3 \cdot C_6H_4 \cdot CH_2 \cdot CO_2H$	IX-530	150.18
6013	benzoate (*o*)	*o*-cresyl benzoate	$C_6H_5CO_2 \cdot C_6H_4 \cdot CH_3$	IX-119	212.25
6014	benzoate (*m*)	*m*-cresyl benzoate	$C_6H_5CO_2 \cdot C_6H_4 \cdot CH_3$	IX-120	212.25
6015	benzoate (*p*)	*p*-cresyl benzoate	$C_6H_5CO_2 \cdot C_6H_4 \cdot CH_3$	IX-120	212.25
6016	carbinol (*o*)	*o*-tolubenzyl alcohol	$CH_3 \cdot C_6H_4 \cdot CH_2OH$	VI-484	122.17
6017	carbinol (*m*)	*m*-xylyl alcohol	$CH_3 \cdot C_6H_4 \cdot CH_2OH$	VI-494	122.17
6018	carbinol (*p*)	....	$CH_3 \cdot C_6H_4 \cdot CH_2OH$	VI-498	122.17
6019	*iso*-cyanate (*o*)	*o*-tolylcarbonimide	$CH_3 \cdot C_6H_4 \cdot N:CO$	XII-812	133.15
6020	*iso*-cyanate (*m*)	*m*-tolylcarbonimide	$CH_3 \cdot C_6H_4 \cdot N:CO$	XII-864	133.15
6021	*iso*-cyanate (*p*)	*p*-tolylcarbonimide	$CH_3 \cdot C_6H_4 \cdot N:CO$	XII-955	133.15
6022	hydrazine (*o*)	....	$CH_3 \cdot C_6H_4 \cdot NH \cdot NH_2$	XV-496	122.17
6023	hydrazine HCl (*o*)	....	$C_7H_{10}N_2 \cdot HCl \cdot H_2O$	XV-496	176.65
6024	hydrazine (*m*)	....	$CH_3 \cdot C_6H_4 \cdot NH \cdot NH_2$	XV-506	122.17
6025	hydrazine HCl (*m*)	....	$C_7H_{10}N_2 \cdot HCl$	XV-506	158.63
6026	hydrazine (*p*)	....	$CH_3 \cdot C_6H_4 \cdot NH \cdot NH_2$	XV-510	122.17
6027	hydrazine HCl (*p*)	....	$C_7H_{10}N_2 \cdot HCl$	....	158.63
6028	hydroxylamine	(*o*)(*β*)	$CH_3 \cdot C_6H_4 \cdot NHOH$	XV-13	123.16

† See also No. 5142.
Toluidine sulfonic acid 379-83
Tolunitrile 5992-4
Toluol 5960
Toluphenazine 4334

Toluquinoline 4372, 4374-5
Toluquinone oxime 4881
Toluylaldehyde 5139, 5985-7
Toluylene red 4585

Table 7-4 (*Continued*)
PHYSICAL CONSTANTS OF ORGANIC COMPOUNDS

No.	Crystalline Form and Color	Specific Gravity	Melting Point °C.	Boiling Point °C.	Solubility in 100 Parts		
					Water	Alcohol	Ether
5981	col. tri./ac.	....	117-8	....	....	v. s. h.	....
5982	cr./aq.	$1.062\frac{11.5°}{4}$	104-5	259^{751mm}	$2.17^{100°}$	v. s.	s. chl.
5983	pr./aq.	$1.054\frac{112°}{4}$	110-1	263	$0.09^{15°}$; $1.6^{100°}$	v. s.	v. s.
5984	cr./aq.	....	179-80	274-5	$1.26^{100°}$	v. s.	v. s.
5985	lq.	$1.039\frac{20°}{4}$	....	196-9	sl. s.	∞	∞
5986	lq.	$1.019\frac{20°}{4}$	....	199	sl. s.	∞	∞
5987	lq.	$1.019\frac{16.7°}{4}$	....	204-5	sl. s.	∞	∞
5988	nd./aq.	....	147	....	v. s. h.	v. s.	s.
5989	nd./et.	....	97	....	v. sl. s. bz.	s.	v. sl. s.
5990	nd./aq.	....	159-60	....	s. h.	s.	s.
5991	cr./et.	....	39	>325	....	....	....
5992	col. lq.	$0.998\frac{15°}{15}$	-13	205.2	i.	∞	∞
5993	col. lq.	$0.976^{15°}$	-23	210^{773mm}	0.09 c.	1.7 h. aq.	....
5994	nd./al.	$0.981\frac{30°}{30}$	29.5	217.6	i.	v. s.	v. s.
5995	col. lq.	$0.999\frac{20°}{4}$	-16.4 (-24.4)	200.4	$1.5^{25°}$; s. dil. a.	∞	∞
5996	mn. pr.	....	218-20	242	s.	sl. s.	....
5997	col. lq.	$0.989\frac{20°}{4}$	-31.3	203.4	sl. s.	∞	∞
5998	lf./aq.	....	228	250	$96.3^{12°}$	$61.9^{90°}$	....
5999	cr.	$0.962^{50°}$ $1.046\frac{20°}{4}$	43.8	200.6	$0.74^{21°}$; $1.13^{2°}$	v. s.; s. dil. a.	v. s.; s. CS_2
6000	nd./ac. et.	....	243	257.5	$22.9^{11°}$	$25^{17°}$	i.; i. bz.
6001	yel. nd.	....	68-9	subl.	sl. s. c.	v. s.	v. s.
6002	nd./toluene	....	139-40	d.	v. sl. s. h.; s. act.	v. s.	v. s.
6003	lf./lg.	....	89-90	265	s. c.	....	....
6004	nd.	....	....	....	v. s.	....	....
6005	rhb.	....	99	283-5	s. h.	s.	s.
6006	nd.	....	....	....	s.	....	....
6007	pl./bz.	....	64	273-4	s.; s. h. bz.	s.	s.
6008	lf.	....	....	....	s.	....	....
6009	pd.	....	....	....	$0.8^{11°}$	....	....
6010	nd./aq.	....	88-9	....	s. h.	....	....
6011	nd.	....	60.5-1.5	....	s. h.	....	....
6012	nd./aq.	....	92-4	265-7	s. h.	s.	s.; s. bz.
6013	lq.	....	....	307	i.	....	s.
6014	cr.	....	54-5	313-4	i.	....	....
6015	pl./et. al.	....	71-2	315-6	i.	....	....
6016	nd.	$1.023^{40°}$	34-6	223^{750mm}	1 c.	v. s. abs.	v. s.
6017	lq.	$0.916^{17°}$	<-20	217	5 c.	s.	s.
6018	nd.	....	60°	217	sl. s.	s.	s.
6019	lq.	....	....	184-7	i.; d. h.	d. h.	s.
6020	pa. yel. lq.	....	....	195-8	i.	s.	s.
6021	lq.	....	....	187^{751mm}	....	....	....
6022	nd.	....	56-9	....	sl. s. lg.	v. s.	v. s.
6023	pl.	....	....	....	s.	s.	....
6024	oil	....	....	240-4	i.	s.	s.; s. chl.
6025	nd.	....	....	....	s.	s.	....
6026	rhb.	....	65-6	240-4 sl. d.	sl. s.	v. s.; s. bz.	v. s.
6027	tan pl.	....	....	....	s.	sl.	i.
6028	col./bz. et.	....	44	....	sl. s. lg.	s.	s.

Table 7-4 (*Continued*)
PHYSICAL CONSTANTS OF ORGANIC COMPOUNDS

No.	Name	Synonym	Formula	Beil. Ref.	Formula Weight
6029	**Tolyl** hydroxylamine	$(m)(\beta)$	$CH_3 \cdot C_6H_4 \cdot NHOH$	XV-14	123.16
6030	hydroxylamine	$(p)(\beta)$	$CH_3 \cdot C_6H_4 \cdot NHOH$	XV-15	123.16
6031	α-naphthylamine	$(o)(N)$	$C_{10}H_7 \cdot NH \cdot C_6H_4 \cdot CH_3$	XII-1225	233.32
6032	α-naphthylamine	$(p)(N)$	$C_{10}H_7 \cdot NH \cdot C_6H_4 \cdot CH_3$	XII-1225	233.32
6033	β-naphthylamine	$(o)(N)$	$C_{10}H_7 \cdot NH \cdot C_6H_4 \cdot CH_3$	XII-1277	233.32
6034	β-naphthylamine	$(p)(N)$	$C_{10}H_7 \cdot NH \cdot C_6H_4 \cdot CH_3$	XII-1277	233.32
6035	pyrrole $(o)(N)$	. . . .	$C_4H_4N \cdot C_6H_4 \cdot CH_3$	XX-164	157.22
6036	pyrrole $(m)(N)$	. . . .	$C_4H_4N \cdot C_6H_4 \cdot CH_3$	. . . .	157.22
6037	pyrrole $(p)(N)$	. . . .	$C_4H_4N \cdot C_6H_4 \cdot CH_3$	XX-164	157.22
6038	salicylate (o)	$(1,2;1',2')$	$HO \cdot C_6H_4 \cdot CO_2 \cdot$ $C_6H_4 \cdot CH_3$	X-80	228.25
6039	salicylate (m)	$(1,2;1',3')$	$HO \cdot C_6H_4 \cdot CO_2 \cdot$ $C_6H_4 \cdot CH_3$	X-80	228.25
6040	salicylate (p)	. . . .	$HO \cdot C_6H_4 \cdot CO_2 \cdot$ $C_6H_4 \cdot CH_3$	X-80	228.25
6041	*iso*-thiocyanate (o)	*o*-tolyl mustard oil	$CH_3 \cdot C_6H_4 \cdot N{:}CS$	XII-813	149.22
6042	*iso*-thiocyanate (m)	*m*-tolyl mustard oil	$CH_3 \cdot C_6H_4 \cdot N{:}CS$	XII-865	149.22
6043	*iso*-thiocyanate (p)	*p*-tolyl mustard oil	$CH_3 \cdot C_6H_4 \cdot N{:}CS$	XII-956	149.22
6044	thiosemicarbazide (o)		$CH_3 \cdot C_6H_4 \cdot NH \cdot CS \cdot$ $NH \cdot NH_2$	XII-952	181.26
6045	thiourea (o)	. . . .	$CH_3 \cdot C_6H_4 \cdot NH \cdot CS \cdot$ NH_2	XII-806	166.25
6046	thiourea (m)	. . . .	$CH_3 \cdot C_6H_4 \cdot NH \cdot CS \cdot$ NH_2	XII-863	166.25
6047	thiourea (p)	. . . .	$CH_3 \cdot C_6H_4 \cdot NH \cdot CS \cdot$ NH_2	XII-947	166.25
6048	urea (o)	. . . .	$CH_3 \cdot C_6H_4 \cdot NH \cdot CO \cdot$ NH_2	XII-801	150.18
6049	urea (m)	. . . .	$CH_3 \cdot C_6H_4 \cdot NH \cdot CO \cdot$ NH_2	XII-862	150.18
6050	urea (p)	. . . .	$CH_3 \cdot C_6H_4 \cdot NH \cdot CO \cdot$ NH_2	XII-940	150.18
6051	**Tolylene** cyanide (o)	xylylene dicyanide	$C_6H_4(CH_2 \cdot CN)_2$	IX-874	156.19
6052	cyanide (m)	xylylene dicyanide	$C_6H_4(CH_2 \cdot CN)_2$	IX-875	156.19
6053	cyanide (p)	xylylene dicyanide	$C_6H_4(CH_2 \cdot CN)_2$	IX-875	156.19
6054	**Trehalose**	mycose	$C_{12}H_{22}O_{11} \cdot 2H_2O$	XXXI-378	378.33
6055	**Triacetamide**	. . . .	$(CH_3CO)_3N$	II-181	143.14
6056	**Triacetonamine hydrate**	tetramethyl-piperidone (γ)	$C_9H_{17}ON \cdot H_2O$	XXI-249	173.26
6057	**Triacontane** (n)	. . . .	$CH_3(CH_2)_{28}CH_3$	I-176	422.83
6058	**Triamino-azobenzene**	$(2,4,3')$; Bismarck brown	$NH_2 \cdot C_6H_4 \cdot N{:}N$ $C_6H_3(NH_2)_2$	XVI-386	227.27
6059	benzene $(1,2,3)$	. . . .	$(NH_2)_3C_6H_3$	XIII-294	123.16
6060	benzene $(1,2,4)$	. . . .	$(NH_2)_3C_6H_3$	XIII-294	123.16
6061	benzene HCl	$(1,2,4)$	$(NH_2)_3C_6H_3 \cdot 2HCl$	XIII-295	196.08
6062	benzene $(1,3,5)$	(free base unknown)	$(NH_2)_3C_6H_3$	XIII-299	123.16
6063	benzene HCl	$(1,3,5)$	$(NH_2)_3C_6H_3 \cdot 3HCl$	XIII-299	232.54
6064	benzoic acid $(2,3,5)$	. . . .	$(NH_2)_3C_6H_3 \cdot CO_2H$	XIV-455	167.17

Tolyl iodide 3905-7
Tolyl-mercaptan 5894-6
Tolyl-mercuric chloride 4039
Tolyl-mustard oil 6041-3

Tolyl phosphate 6351-3
Tolyl phosphite 6354-5
Tolyl thiophosphate 6356-8
Tolylene alcohol 6497-9

Table 7-4 (*Continued*)
PHYSICAL CONSTANTS OF ORGANIC COMPOUNDS

No.	Crystalline Form and Color	Specific Gravity	Melting Point °C.	Boiling Point °C.	Solubility in 100 Parts		
					Water	Alcohol	Ether
6029	lf./bz. pet.	....	68.5	....	sl. s. h.	s.	s.; sl. s. lg.
6030	lf./bz.	....	93-4	....	1 c.; 5 h.	v. s.	s.; sl. s. bz.
6031	nd./lg.	....	94-5	....	i.; v. s. bz.	v. s.	v. s.
6032	pr./al.	....	78-9	236^{15mm}	s. bz.	s. h.	s.; sl. s. h. pet.
6033	cr./lg.	....	95-6	400-5	v. s. bz.; v. s. lg.	v. s.; v. s. chl.	v. s.; v. s. act.
6034	lf./al.	....	102-3	....	sl. s. lg.	sl. s. c.	s. bz.
6035	col. oil	....	....	246	v. sl. s. h.	v. s.	v. s. bz.
6036	yel. lq.	....	....	$149-514^{1mm}$	i.	s.	s.
6037	lf./aq. al.	....	82	252^{729mm}	v. sl. s. h.	v. s.	v. s. pet.
6038	col. cr.	....	35	....	i.	s.	s.
6039	col. cr.	....	74	....	i.	s.	s.
6040	col. cr.	....	39	....	i.	s.	s.
6041	col. lq.	$1.104\frac{25°}{25}$	....	239	i.	v. s.	∞
6042	lq.	....	<-20	244^{732mm}	....	....	....
6043	nd./et.	$1.087\frac{25°}{25}$	26	237	d. h.	d. h.	s.
6044	lf./bz.	....	145-6	....	i.	s.	s.; i. lg.
6045	cr./aq.	....	151-2	....	v. s. h.	v. s.	v. sl. s.
6046	pr./al.	....	110-1	....	s. h.	s.	s.
6047	pl./al.	....	188	....	v. sl. s. c.	s. h.	....
6048	lf./al.	....	190-2	....	$0.25^{45°}$; s. h.	s.	s.
6049	lf./aq.	....	142-3	....	....	....	....
6050	nd./aq.	....	180-1	....	$0.34^{5°}$	s.	$0.06^{23°}$
6051	cr./et.	....	59-60	....	....	s.	s.
6052	cr.	....	28-9	$305-10^{300mm}$ sl. d.	i.	s.	s.; s. chl.
6053	nd./aq.	....	98	....	sl. s. h.	s.	s.; s. chl.
6054	rhb./al.	....	97*	$-2H_2O > 130°$	s. h.	sl. s. h.	i.
6055	nd./et.	....	78-9	....	....	....	s.
6056	pl./aq.	....	59; 35 (anh.)	205 sl. d.	s.	s.	s.
6057	cr.	lq. $0.780^{65.9°}$	65.9	$235^{1.0mm}$	s. bz.	sl. s.	s.
6058	or. mn./aq.	....	143.5	....	i.	s.	s.
6059	cr.	....	$103\pm$	336	v. s.	v. s.	v. s.
6060	lf./chl.	....	<100	$340\pm$	v. s.	v. s.	sl. s.; sl. s. chl.
6061	nd.	....	225	....	s.	v. sl. s.	sl. s. HCl
6062	....	....	....	....	....	....	....
6063	col. cr.	....	....	....	s.; d. h.	sl. s.	i.
6064	cr./aq.	....	....	....	s. h.	v. sl. s. h.	i.

* Anhydrous, m. p. 210°.
Toxilic acid 3995
"Traubensaure" 5690

Triacetin 3418
Triacetyl-hydroquinone 3775
Triacetyl-pyrogallol 5516

Table 7-4 *(Continued)*

PHYSICAL CONSTANTS OF ORGANIC COMPOUNDS

No.	Name	Synonym	Formula	Beil. Ref.	Formula Weight
6065	**Triamino** benzoic acid (3,4,5)	. . . .	$(NH_2)_3C_6H_2 \cdot CO_2H \cdot$ $\frac{1}{2}H_2O$	XIV–455	176.18
6066	benzoic acid HCl	(2,4,6)	$(NH_2)_3C_6H_2 \cdot CO_2H \cdot$ 3HCl	XIV–455	276.55
6067	chlorobenzene HCl	(2,4,6)	$(NH_2)_3C_6H_2Cl \cdot 3HCl$	. . . .	266.99
6068	phenol (2,4,6)	. . . .	$(NH_2)_3C_6H_2 \cdot OH$	XIII–569	139.16
6069	toluene (2,4,6)	. . . .	$(NH_2)_3C_6H_2 \cdot CH_3$	XIII–303	137.19
6070	toluene HCl	(2,4,6)	$(NH_2)_3C_6H_2CH_3 \cdot 3HCl$	XIII–303	246.57
6071	triphenylmethane (*o,p′,p″*)	*o*-leucoaniline	$(NH_2 \cdot C_6H_4)_3CH$	XIII–311	289.38
6072	triphenylmethane (*m,p′,p″*)	pseudo-leucoaniline	$(NH_2 \cdot C_6H_4)_3CH$	XIII–312	289.38
6073	triphenylmethane (*p,p′,p″*)	*p*-leucoaniline	$(NH_2 \cdot C_6H_4)_3CH$	XIII–313	289.38
6074	**Triamyl**-amine (*n*)	. . . .	$(C_5H_{11})_3N$	*IV–378	227.44
6075	amine (*iso*)	. . . .	$(C_5H_{11})_3N$	IV–183	227.44
6076	amine (*d*)	. . . .	$[C_2H_5 \cdot CH(CH_3) \cdot$ $CH_2]_3N$	IV–179	227.44
6077	borate (*n*)	*n*-amyl borate	$(C_5H_{11})_3BO_3$	. . . .	272.24
6078	**Triazo**-acetic acid	. . . .	$(N:N):N \cdot CH_2 \cdot CO_2H$	II–229	101.07
6079	benzene	phenyl azide	$C_6H_5 \cdot N_3$	V–276	119.13
6080	**Triazole**	pyrrodiazole	CH:N·NH·CH:N	XXVI–13	69.07
6081	**Tribenzamide**	. . . .	$(C_6H_5CO)_3N$	IX–214	329.36
6082	**Tribenzoyl** methane (α)	. . . .	$(C_6H_5CO)_2C:$ $C(OH) \cdot C_6H_5$	VII–877	328.37
6083	methane (β)	. . . .	$(C_6H_5CO)_3CH$	VII–877	328.37
6084	**Tribenzyl**-amine	. . . .	$(C_6H_5 \cdot CH_2)_3N$	XII–1038	287.41
6085	citrate	benzyl citrate	$(C_6H_5CH_2)_3C_6H_5O_7$	XXVI–421	462.50
6086	**Tribromo**-acetamide	. . . .	$Br_3C \cdot CO \cdot NH_2$	II–221	295.77
6087	acetanilide (2,4,6)	. . . .	$Br_3C_6H_2 \cdot NH \cdot COCH_3$	XII–665	371.87
6088	acetic acid	. . . .	$Br_3C \cdot CO_2H$	II–220	296.76
6089	aniline (2,3,4)	. . . .	$Br_3C_6H_2 \cdot NH_2$	XII–662	329.83
6090	aniline (2,3,5)	. . . .	$Br_3C_6H_2 \cdot NH_2$	XII–662	329.83
6091	aniline (2,4,5)	. . . .	$Br_3C_6H_2 \cdot NH_2$	XII–662	329.83
6092	aniline (2,4,6)	. . . .	$Br_3C_6H_2 \cdot NH_2$	XII–663	329.83
6093	aniline (3,4,5)	. . . .	$Br_3C_6H_2 \cdot NH_2$	XII–668	329.83
6094	anisole (2,4,6)	. . . .	$Br_3C_6H_2 \cdot OCH_3$	VI–205	344.84
6095	benzene (1,2,4)	(*uns*)	$Br_3C_6H_3$	V–213	314.82
6096	benzene (1,2,3)	(*vic*)	$Br_3C_6H_3$	V–213	314.82
6097	benzene (1,3,5)	(*sym*)	$Br_3C_6H_3$	V–213	314.82
6098	benzoic acid (2,4,6)	. . . .	$Br_3C_6H_2 \cdot CO_2H$	IX–360	358.83
6099	benzoic acid (3,4,5)	. . . .	$Br_3C_6H_2 \cdot CO_2H$	IX–361	358.83
6100	butane (1,2,3)	. . . .	$CH_3(CHBr)_2CH_2Br$	I–121	294.83
6101	butane	1,2,3-triBr-2-Me-propane	$(CH_2Br)_2:CBr \cdot CH_3$	I–128	294.83
6102	*tert*-butyl alcohol	acetone-bromoform	$(CH_3)_2C(OH) \cdot CBr_3$	*I–193	310.83
6103	ethane (1,1,2)	vinyl tribromide	$Br \cdot CH_2 \cdot CHBr_2$	I–93	266.77
6104	ethyl alcohol	avertin	$Br_3C \cdot CH_2OH$	**I–338	282.77
6105	ethylene	ethylene tribromide	$Br \cdot CH:CBr_2$	I–191	264.76
6106	mesitylene	(2,4,6;1,3,5)	$Br_3C_6(CH_3)_3$	V–409	356.90
6107	β-naphthol	(1,3,6;2)	$Br_3C_{10}H_4 \cdot OH$	VI–652	380.88
6108	phenol (2,4,6)	(*sym*)	$Br_3C_6H_2 \cdot OH$	VI–203	330.82

Triamino-diphenyl 262
Triamino-triazine 4020
Triamino-triphenyl carbinol 5038

Triarachin 3419
Tribenzal-diamine 3700
Tribenzoin 3420

Table 7-4 (*Continued*)
PHYSICAL CONSTANTS OF ORGANIC COMPOUNDS

No.	Crystalline Form and Color	Specific Gravity	Melting Point °C.	Boiling Point °C.	Solubility in 100 Parts		
					Water	Alcohol	Ether
6065	nd./aq.	....	$-H_2O$, > 100	....	s. h.	i.	i.
6066	pr./aq.	....	....	....	s.	sl. s.	i.
6067	brn. pd.	....	....	....	s.	sl. s.	i.
6068	unstable	....	....	257	....	....	....
6069	oil	....	....	....	....	....	....
6070	nd.	....	....	....	s.; d. h.	....	....
6071	cr./al.	....	165	....	....	....	....
6072	nd./et. lg.*	....	150*	....	v. sl. s. lg.	s.	sl. s.
6073	lf./aq.	....	148	....	sl. s. c.	s. abs.	s. bz.
6074	lq.	....	....	240-5	....	....	....
6075	col. lq.	$0.786\frac{20}{4}°$	....	235	....	....	....
6076	col. lq.	$0.796^{13°}$	....	230-7	i.	v. s.	∞
6077	col. lq.	....	....	$130\text{-}2^{6mm}$	....	....	....
6078	col. hyg.	$1.354^{33°}$	$16\pm$; expl.	93^{3mm}	....	....	....
6079	yel. oil	$1.088\frac{20}{20}°$	expl.	73.5^{24mm}	i.	sl. s.	sl. s.
6080	nd./et.	....	120-1	260	v. s.	v. s.	sl. s.
6081	nd./al.	....	207-8	subl.	s. h. bz.	i. c.	v. sl. s.
6082	cr. ppt.	....	210-20	subl.	0.5/3 cc. chl.	s.	s. act.
6083	nd./al.	....	226-31	....	0.5 c. act. 0.2 chl.	0.01 c.	i. 10% Na_2CO_3
6084	mn./et.	$0.991\frac{9.5}{4}°$	92-3	380-90	v. sl. s.	s. h.	s.
6085	col. pl.	....	51	....	i.	sl. s.	i.
6086	mn.	....	121-2	....	v. sl. s. bz.	s. h.	s. h.
6087	nd./al.	....	232	....	....	....	....
6088	cr.	....	135	245 d.	v. s.; d. h.	v. s.	v. s.
6089	lf./aq. al.	....	100.6	....	sl. s.	v. s.	v. s. bz.
6090	nd./al.	....	91	....	....	....	....
6091	nd./al.	....	85-6	....	v. s. bz.	v. s.	v. s.
6092	rhb./bz.	$2.35\frac{20}{20}°$	119-20	300	i.; s. chl.	v. s. h.	v. s.
6093	nd.	....	118-9	....	i.	s.	s.
6094	nd./al.	....	87-8	....	....	$1^{15°}$	....
6095	nd./al.	....	44-5	275-6	i.; v. s. bz.	v. s. h.	v. s.
6096	col./al.	2.658	87-8	....	....	....	....
6097	nd./al.	....	119-20	271^{765mm}	i.	sl. s. h.	s.
6098	pr./aq.	....	187	....	$0.35^{15°}$	$0.55^{100°}$ aq.	....
6099	nd./aq. al.	....	234-5	....	v. sl. s. h.	s.	s. h.
6100	col. lq.	$2.190\frac{16}{4}°$	$-19\pm$	$110\text{-}3^{19mm}$	....	....	....
6101	col. lq.	$2.211\frac{14}{4}°$	....	222-5	i.	....	....
6102	cr./aq. al.	....	167-76	....	sl. s.	s.	s.
6103	lq.	$2.579\frac{20}{4}°$	−26	$187\text{-}8^{752mm}$	....	....	....
6104	col. cr.	....	79-80	$92\text{-}3^{10mm}$	$2.540°$	s.	s.; s. bz.
6105	lq.	$2.708^{20.5°}$	....	163-4	....	....	....
6106	tri./al.	....	223-6	....	i.	v. sl. s. h.	s. bz.
6107	nd./ac.	....	155	....	....	....	....
6108	nd./aq.	$2.55\frac{20}{20}°$	96	subl.	$0.01^{15°}$; $50^{25°}$ bz.	v. s.; v. s. act.	s.; $12^{25°}$ CCl_4

* Crysts./bz. + $1C_6H_6$, m. p. 145°.
Tribenzylene-benzene 6362
Tribromo-acetaldehyde 876

Tribromo-hydrin 6112
Tribromo-methane 1016
Tribromo-nitromethane 4688

Table 7-4 (*Continued*)
PHYSICAL CONSTANTS OF ORGANIC COMPOUNDS

No.	Name	Synonym	Formula	Beil. Ref.	Formula Weight
	Tribromo				
6109	phenyl acetate	(1;2,4,6)	$CH_3CO_2 \cdot C_6H_2Br_3$	VI-205	372.85
6110	phenyl salicylate	(2,1;2',4',6'); tribromosalol	$HO \cdot C_6H_4 \cdot CO_2 \cdot C_6H_2Br_3$	X-78	450.93
6111	phenyl-*p*-toluene-sulfonate (2,4,6)	....	$CH_3 \cdot C_6H_4 \cdot SO_3 \cdot C_6H_2Br_3$	....	485.01
6112	propane (1,2,3)	glyceryl tribromo-hydrin	$(CH_2Br)_2 {:} CHBr$	I-112	280.80
6113	quinaldine (ω)	....	$C_9H_6N \cdot CBr_3$	....	379.89
6114	resorcinol (2,4,6)	....	$Br_3C_6H(OH)_2$	VI-822	346.82
6115	**Tributyl-**amine (*n*)	....	$(C_4H_9)_3N$	IV-157	185.36
6116	amine (*iso*)	....	$(C_4H_9)_3N$	IV-166	185.36
6117	borate (*n*)	*n*-butyl borate	$B(OC_4H_9)_3$	**I-398	230.16
6118	carbinol (*n*)	....	$(C_4H_9)_3C \cdot OH$	**I-464	200.37
6119	citrate (*n*)	*n*-butyl citrate	$(C_4H_9)_3C_6H_5O_7$	**III-371	360.45
6120	phenyl phosphate (*p; tert*)	phosphen 7	$[(CH_3)_3C \cdot C_6H_4 \cdot O]_3PO$	....	494.62
6121	phosphate (*n*)	*n*-butyl phosphate	$(C_4H_9O)_3PO$	**I-397	266.32
6122	**Tri-*iso*-butylene**	....	$(CH_3)_2C{:}C[C(CH_3)_3]_2$	I-225	168.33
6123	**Tricarballylic acid**	....	$(HO_2C \cdot CH_2)_2 {:} CH \cdot CO_2H$	II-815	176.13
6124	**Trichloro-**acetal	....	$Cl_3C \cdot CH(OC_2H_5)_2$	I-621	221.51
6125	acetal	....	$Cl_3C \cdot CH(OC_2H_5)_2$	I-621	221.51
6126	acetamide	....	$Cl_3C \cdot CO \cdot NH_2$	II-211	162.40
6127	acetanilide (2,4,6)	....	$CH_3 \cdot CO \cdot NH \cdot C_6H_2Cl_3$	XII-628	238.50
6128	acetic acid	....	$Cl_3C \cdot CO_2H$	II-206	163.39
6129	acetyl chloride	....	$Cl_3C \cdot CO \cdot Cl$	II-210	181.83
6130	aniline (2,3,4)	....	$Cl_3C_6H_2 \cdot NH_2$	XII-626	196.46
6131	aniline (2,4,5)	....	$Cl_3C_6H_2 \cdot NH_2$	XII-627	196.46
6132	aniline (2,4,6)	....	$Cl_3C_6H_2 \cdot NH_2$	XII-627	196.46
6133	aniline (3,4,5)	....	$Cl_3C_6H_2 \cdot NH_2$	XII-630	196.46
6134	benzene (1,2,3)	(*vic*)	$Cl_3C_6H_3$	V-203	181.45
6135	benzene (1,2,4)	(*uns*)	$Cl_3C_6H_3$	V-203	181.45
6136	benzene (1,3,5)	(*sym*)	$Cl_3C_6H_3$	V-203	181.45
6137	benzoic acid (2,3,4)	....	$Cl_3C_6H_2 \cdot CO_2H$	IX-345	225.46
6138	benzoic acid (2,3,5)	....	$Cl_3C_6H_2 \cdot CO_2H$	IX-345	225.46
6139	benzoic acid (2,4,5)	....	$Cl_3C_6H_2 \cdot CO_2H$	IX-345	225.46
6140	benzoic acid (2,4,6)	....	$Cl_3C_6H_2 \cdot CO_2H$	IX-345	225.46
6141	benzoic acid (3,4,5)	....	$Cl_3C_6H_2 \cdot CO_2H$	IX-346	225.46
6142	bromoethane	(1;2,2,2)	$BrCH_2 \cdot CCl_3$	I-90	212.31
6143	bromomethane	....	$Br \cdot CCl_3$	I-67	198.28
6144	*tert*-butyl acetate	(β,β,β)	$CH_3CO_2 \cdot C \cdot (CH_3)_2CCl_3$	II-131	219.50
6145	*tert*-butyl alcohol	acetone chloroform	$Cl_3C \cdot COH{:}(CH_3)_2$	I-382	177.46
6146	*o*-cresol (4,5,6)	(4,5,6;2,1)	$Cl_3C_6H(OH) \cdot CH_3$	*VI-175	211.48
6147	*o*-cresol (3,5,6)	(3,5,6;2,1)	$Cl_3C_6H(OH) \cdot CH_3$	*VI-175	211.48
6148	*m*-cresol (2,4,6)	(2,4,6;3,1)	$Cl_3C_6H(OH) \cdot CH_3$	*VI-189	211.48
6149	*p*-cresol (2,3,6)	(2,3,6;4,1)	$Cl_3C_6H(OH) \cdot CH_3$	VI-404	211.48
6150	dibromoethane	(1,1;2,2,2)	$Br_2CH \cdot CCl_3$	I-93	291.21
6151	ethane (1,1,1)	methyl chloroform	$CH_3 \cdot CCl_3$	I-85	133.41
6152	ethane (1,1,2)	vinyl trichloride	$ClCH_2 \cdot CHCl_2$	I-85	133.41
6153	ethyl acetate	....	$CH_3CO_2 \cdot CH_2 \cdot CCl_3$	II-128	191.44
6154	ethyl alcohol	....	$Cl_3C \cdot CH_2OH$	I-338	149.40

Tribromo-salol 6110
Tributyrin 3421
Tricaprin 3422

Tricaproin 3423
Tricaprylin 3424
Trichloro-acetaldehyde 1281

Table 7-4 (*Continued*)
PHYSICAL CONSTANTS OF ORGANIC COMPOUNDS

No.	Crystalline Form and Color	Specific Gravity	Melting Point °C.	Boiling Point °C.	Solubility in 100 Parts		
					Water	Alcohol	Ether
6109	nd./al.	....	82-3	290-300	....	....	....
6110	wh. pd.	....	189	....	i.; s. chl.	sl. s.; s. act.	s. bz.; s. ac.
6111	col. or.	....	113-4	....	i.	s.	s.
6112	cr.	$2.436^{23°}$	16-7	219-21	i.	s.	s.
6113	tan pl.	....	127-8	....	i.	i.	i.
6114	nd./aq.	....	111	....	sl. s. c.	s. h.	s.
6115	col. lq.	$0.778^{20°}_{20}$	....	216.5^{761mm}	i.	s.	∞
6116	col. lq.	$0.764^{25°}_{4}$	-22	191.5	i.	s.	∞
6117	col. lq.	....	....	$103-6^{8mm}$	....	....	....
6118	col. lq.	$0.844^{18°}_{4}$	20	$117-20^{10mm}$	i.	s.	s.
6119	col. lq.	$1.045^{20°}_{20}$	-20	233^{17mm}	i.	∞	∞
6120	col. cr.	....	102-5	300^{5mm}	i.; ∞ bz.	$2^{25°}$	∞ CCl₄
6121	col. lq.	$0.976^{25°}_{25}$	<-80	289 d.	0.6	∞	∞
6122	col. liq.**	$0.759^{20°}_{4}$	<-30**	$178-80^{752mm}$	i.	....	....
6123	pr./aq.	....	165-6	d.	$41^{15°}$	v. s.	$1^{18°}$
6124	lq.	$1.266^{15°}$	....	197	0.5	∞	∞
6125	solid	....	83	230 d.	....	....	....
6126	mn.	....	141	$238-97^{46mm}$	v. sl. s.	v. s.	v. s.
6127	nd.	....	203-4	....	s. 50% ac.	s.; v. sl. s.	v. sl. s.; CS₂ v. sl. s. lg.
6128	cr.	$1.617^{46°}_{15}$	58(50±)	195.5^{754mm}	$120^{25°}$	s.	s.
6129	col. lq.	$1.629^{16.2°}$	....	118	d.	d.	....
6130	nd./lg.	....	67.5	292^{774mm}	s. lg.	v. s.	....
6131	nd./lg.	....	96	270±	sl. s. lg.	s.; s. CS₂	s. 50% ac.
6132	nd./lg.	....	77.5-8.5	262^{746mm}	i. H₃PO₄	s.	s.
6133	nd./aq. al.	....	100	....	....	....	....
6134	cr./al.	....	52-3	218-9	i.	sl. s.	....
6135	col. cr.	$1.446^{26°}$	17	213	i.	....	....
6136	nd.	....	63.5	208.5^{764mm}	i.	sl. s.	s.
6137	nd.	....	186(129)	....	sl. s.	s.	s.
6138	nd./aq.	....	163	....	v. sl. s. c.	s.	s.
6139	nd./aq.	....	163-4	subl.	v. sl. s. c.	s. c. abs.	....
6140	cr./aq.	....	164	....	sl. s.	s.	s.
6141	nd./aq. al.	....	203	subl.	v. sl. s. c.	s.	s.
6142	lq.	1.884	....	151-3	....	....	....
6143	col. lq.	$2.055^{0°}_{4}$	-21	104.1	....	....	....
6144	lq.	....	....	191	i.; v. s. act.	v. s.; v. s. chl.	v. s.; v. s. bz.
6145	col. cr.	....	97*	167	0.8 c.	111	s.; s. chl.
6146	nd./pet.	....	77	....	sl. s.	s.; s. alk.	s.
6147	nd./ac.	....	62	....	i. pet.	s.; s. alk.	s.
6148	col. nd./aq.	....	47	265	sl. s.	s.; s. alk.	s.; s. lg.
6149	nd./pet.	....	66-7	....	sl. s.	s.; s. alk.	s.
6150	lq.	$2.317^{0°}_{4}$	-4.5	$93-5^{14mm}$	....	....	....
6151	lq.	$1.346^{15°}_{4}$	-32.7	74.0	i.	∞	∞
6152	col. lq.	$1.441^{25.5°}_{4}$	-36.7	113.5	$0.44^{20°}$	∞	∞
6153	col. oil	$1.189^{15°}$	....	170^{747mm}	....	....	....
6154	rhb. pl.	$1.550^{23.3°}$	17.8	151^{737mm}	v. sl. s.	∞	∞

* Forms solid solution with H₂O, m. p. 75-97°. **High boiling form, **I-180, d. 0.776, b. p. 195-6°
Trichloro-butaldehyde 1158

Table 7-4 (*Continued*)
PHYSICAL CONSTANTS OF ORGANIC COMPOUNDS

No.	Name	Synonym	Formula	Beil. Ref.	Formula Weight
6155	**Trichloro** ethylene	ethylene trichloride	$ClCH:CCl_2$	I-187	131.39
6156	hydroquinone	(2,3,5)	$(HO)_2C_6HCl_3$	VI-850	213.45
6157	iodomethane	....	$Cl_3C\cdot I$	I-71	245.27
6158	lactic acid	....	$Cl_3C\cdot CHOH\cdot CO_2H$	III-286	193.41
6159	methylchloro-formate	diphosgene	$Cl\cdot CO_2\cdot CCl_3$	III-18	197.83
6160	naphthalene (1,4,5)	(δ)	$C_{10}H_5Cl_3$	V-545	231.51
6161	naphthalene (1,4,6)	(ϵ)	$C_{10}H_5Cl_3$	V-546	231.51
6162	phenol (2,3,5)	....	$Cl_3C_6H_2\cdot OH$	VI-190	197.45
6163	phenol (2,4,5)	....	$Cl_3C_6H_2\cdot OH$	**VI-180	197.45
6164	phenol (2,4,6)	omal	$Cl_3C_6H_2\cdot OH$	VI-190	197.45
6165	phenyl phosphate	(*o*); phosphen 8	$(Cl\cdot C_6H_4\cdot O)_3PO$	....	429.63
6166	propane (1,2,3)	glyceryl trichloro-hydrin	$(ClCH_2)_2:CHCl$	I-106	147.43
6167	propane (1,1,2)	....	$CH_3\cdot CHCl\cdot CHCl_2$	I-106	147.43
6168	*iso*-propyl alcohol	(1,1,1); isopral	$Cl_3C\cdot CHOH\cdot CH_3$	I-365	163.43
6169	quinone	....	$O:C_6HCl_3:O$	VII-634	211.43
6170	tribromoethane	(1,2,2;1,1,2)	$Cl_2CBr\cdot CClBr_2$	I-94	370.11
6171	**Tricosane** (*n*)		$CH_3(CH_2)_{21}CH_3$	I-175	324.64
6172	**Tricyanogen chloride**	cyanuric chloride	$C_3N_3Cl_3$	XXVI-35	184.41
6173	**Tridecane** (*n*)	....	$CH_3(CH_2)_{11}CH_3$	I-171	184.37
6174	**Tridecyl alcohol** (*n*)	tridecanol-1	$CH_3(CH_2)_{11}CH_2OH$	I-428	200.37
6175	**Tridecylene**	....	$C_{13}H_{26}$	I-225	182.35
6176	**Tridecylic** acid	tridecanoic acid	$CH_3(CH_2)_{11}\cdot CO_2H$	II-364	214.35
6177	aldoxime	....	$C_{12}H_{25}\cdot CH:NOH$	I-715	213.37
6178	**Triethanol**-amine	....	$N(CH_2\cdot CH_2\cdot OH)_3$	IV-285	149.19
6179	amine HCl	....	$C_6H_{15}O_3N\cdot HCl$	IV-285	185.65
6180	**Triethylamine**	....	$(C_2H_5)_3N$	IV-99	101.19
6181	hydrobromide	....	$(C_2H_5)_3N\cdot HBr$	IV-101	182.11
6182	hydrochloride	....	$(C_2H_5)_3N\cdot HCl$	IV-101	137.65
6183	**Triethyl** arsine	arsenic triethyl	$(C_2H_5)_3As$	IV-602	162.11
6184	benzene (1,3,5)	(*sym*)	$(C_2H_5)_3C_6H_3$	V-449	162.28
6185	benzene (1,2,4)	(*uns*)	$(C_2H_5)_3C_6H_3$	V-448	162.28
6186	borate	ethyl borate	$(C_2H_5O)_3B$	I-335	146.00
6187	citrate	ethyl citrate	$C_3H_5O(CO_2C_2H_5)_3$	III-568	276.29
6188	phosphate	ethyl phosphate	$(C_2H_5O)_3PO$	I-332	182.16
6189	phosphine	....	$(C_2H_5)_3P$	IV-582	118.16
6190	phosphine oxide	....	$(C_2H_5)_3PO$	IV-592	134.16
6191	phosphine sulfide	....	$(C_2H_5)_3PS$	IV-592	150.22
6192	phosphite	ethyl phosphite	$(C_2H_5O)_3P$	I-330	166.16
6193	propionate, ortho-	Et orthopropionate	$C_2H_5\cdot C(OC_2H_5)_3$	II-240	176.26
6194	silicol	....	$(C_2H_5)_3Si\cdot OH$	IV-627	132.28
6195	silicol ethyl ether	....	$(C_2H_5)_3Si\cdot OC_2H_5$	IV-627	160.33
6196	silicon hydride	....	$(C_2H_5)_3Si\cdot H$	IV-625	116.28
6197	silicon oxide	....	$[(C_2H_5)_3Si]_2O$	IV-627	246.54
6198	stibine	antimony triethyl	$(C_2H_5)_3Sb$	IV-618	208.94
6199	**Triethylene** glycol	....	$(HOCH_2CH_2OCH_2)_2$	I-468	150.18
6200	glycol diacetate	....	$(CH_3CO_2\cdot CH_2\cdot CH_2\cdot O\cdot CH_2)_2$	II-141	234.25
6201	tetramine	....	$(H_2N\cdot CH_2\cdot CH_2\cdot NH\cdot CH_2)_2$	IV-255	146.24
6202	**Trifluoro**-1-chloro-ethane	(1,2,2)	$F_2CH\cdot CHFCl$	I-83	118.49

Trichloro-hydroxy toluene 6146-9
Trichloro-lactic nitrile 1284
Trichloro-methane 1477
Trichloro-nitromethane 4697

Tricresol, mixt. of 1545-7
Tridecanoic acid 6176
Tridecanol 6174
Tridecanone 4408

Table 7-4 (*Continued*)
PHYSICAL CONSTANTS OF ORGANIC COMPOUNDS

No.	Crystalline Form and Color	Specific Gravity	Melting Point °C.	Boiling Point °C.	Solubility in 100 Parts		
					Water	Alcohol	Ether
6155	col. lq.	$1.466^{\frac{20}{4}°}$	−73	87.2	$0.1^{25°}$	∞	∞
6156	pr./aq.	....	134	subl.	$0.6^{15°}$	v. s.	v. s.
6157	lq.	$2.36^{17°}$	−19	42 d.	....	....	....
6158	cr./et.	....	124	$140\text{-}70^{45mm}$	v. s.	v. s.	s.; s. chl.
6159	lq.	$1.653^{14°}$	−57	127.5	**d. ?**	....	....
6160	nd./al.	....	133	....	....	v. s. h.	....
6161	nd./al.	....	65-6	....	....	sl. s. h.	....
6162	nd./aq. al.	....	55	249-50	sl. s. h.	s.	s.; s. lg.
6163	col./pet.	....	61-3	252	i.; s. CCl_4	s.; s. bz.	s.
6164	nd.	$1.490^{\frac{7.8}{0}°}$	68-9	246	$0.09^{25°}$	v. s.	v. s.
6165	col. cr.	$1.38^{\frac{60}{4}°}$	35	$255\text{-}65^{5mm}$	i.; ∞ bz.	v. s.	∞ CCl_4
6166	lq.	$1.391^{\frac{20}{4}°}$	−14.7	156.6	<0.1	∞	∞
6167	oil	$1.372^{25°}$	<−20	140	i.	∞	∞
6168	col. mn.	....	50-1	161.8^{773mm}	3	s.; s. KOH	s.
6169	yel. lf./aq.	....	168-9	subl.	i. c.	v. s. h.	v. s.
6170	pr.	$2.44^{18°}$	178-80 d.		....	....	....
6171	lf.	$0.779^{\frac{47.7}{4}°}$	47.7	234^{15mm}	i.	....	....
6172	mn./et.	1.32	145	190	d. h.; s. ac.	v. s. chl.	s. h. abs.
6173	col. lq.	$0.757^{\frac{20}{4}°}$	−6.2	234	i.	v. s.	v. s.
6174	cr.	$0.822^{\frac{31}{4}°}$	30.5	$155\text{-}6^{15mm}$	....	v. s.	v. s.
6175	lq.	$0.845^{0°}$	....	232.7	i.	s.	v. s.
6176	cr./al.	....	41	$199\text{-}200^{24mm}$	i.	v. s.	v. s.
6177	nd./aq. al.	....	80.5	....	sl. s. bz.	sl. s.	v. s.; s. chl.
6178	col. lq.	$1.126^{\frac{20}{20}°}$	20-1	$277\text{-}9^{150mm}$	∞	∞; s. chl.	sl. s.
6179	cr./al.	....	177	....	v. s.	v. sl. s.	....
6180	col. oil	$0.729^{\frac{20}{20}°}$	−114.7	89.4	∞ < 19°	∞	∞
6181	hex./chl.	1.322	248	subl. > 225	$15^{125°}$	$23^{25°}$ chl.	i.; s. al.
6182	cr./al.	$1.069^{\frac{21}{4}°}$	253-4	subl. > 245	$150^{28°}$	s.; s. chl.	i.
6183	col. lq.	$1.150^{\frac{20}{4}°}$	....	140^{736} sl. d.	i.	∞	∞
6184	lq.	$0.861^{\frac{20}{4}°}$	....	215	i.	s. abs.	s.
6185	lq.	$0.882^{\frac{17}{4}°}$	....	$217\text{-}8^{755mm}$	i.	s. abs.	s.
6186	lq.	$0.864^{\frac{20}{20}°}$	....	120	d.	....	....
6187	oil	$1.137^{\frac{20}{4}°}$	....	294 ±	i.	∞	∞
6188	col. lq.	$1.068^{\frac{20}{4}°}$	....	215-6	$100^{25°}$	s.	s.
6189	col. lq.	$0.800^{\frac{15}{4}°}$	....	127.5^{744mm}	i.	∞	∞
6190	col. nd.	....	52-3(44)	242.8-3.0	∞; i. KOH	∞	sl. s.
6191	hex. pr.	....	94	ign. 70	....	....	....
6192	lq.	$0.969^{\frac{20}{4}°}$	....	$155\text{-}77^{60mm}$	i.	v. s.	v. s.
6193	col. lq.	$0.887^{20°}$	....	161^{766mm}	....	....	....
6194	lq.	$0.871^{0°}$	....	154	i.	....	....
6195	lq.	$0.840^{0°}$	....	153	i.	∞	∞
6196	lq.	$0.721^{\frac{1.5}{4}°}$	....	95-6	i.	i. H_2SO_4	....
6197	lq.	$0.859^{0°}$	....	231	i.	s. H_2SO_4	....
6198	col. lq.	$1.324^{16°}$	<−29	158-9	....	v. s.	v. s.
6199	col. hyg.	$1.125^{\frac{20}{20}°}$	−5	290	∞; i. pet.	∞; ∞ bz.	v. sl. s.
6200	col. lq.	....	....	300	∞	∞	∞
6201	lq.	$0.982^{15°}$	12	266-7	v. s.	v. s.	....
6202	col. lq.	$1.365^{0°}$	....	17	....	v. s.	....

Tridecyl cyanide 4456
Triethyl aluminum 230
Triethyl bismuthine 855

Triethyl boron 871
Triethyl carbinol 3533
Triethyl tin 5953

Table 7-4 (*Continued*)
PHYSICAL CONSTANTS OF ORGANIC COMPOUNDS

No.	Name	Synonym	Formula	Beil. Ref.	Formula Weight
	Trifluoro				
6203	2-chloroethylene	(1,1,2)	$F_2C:CFCl$	....	116.47
6204	1,2-dibromoethane	(1,1,2)	$F_2BrC \cdot CHBrF$	I-92	241.84
6205	1,1,2-tribromo-ethane	(1,2,2)	$Br_2CF \cdot CBrF_2$	I-94	320.74
6206	1,1,2-trichloro-ethane	(1,2,2) Freon-113	$Cl_2CF \cdot CClF_2$	....	187.38
6207	**Triglycol dichloride**	....	$Cl(CH_2 \cdot CH_2 \cdot O)_2 \cdot CH_2 \cdot CH_2 \cdot Cl$	....	187.07
6208	**Triglycyl** glycine	....	$NH_2(CH_2CONH)_3 \cdot CH_2 \cdot CO_2H$	IV-377	246.22
6209	glycine ester	biuret base	$C_7H_{13}O_3N_4 \cdot CO_2C_2H_5$	IV-377	274.28
6210	**Trihexylamine** (*n*)	....	$(C_6H_{13})_3N$	IV-188	269.52
6211	**Trihydroxy**-aceto-phenone (2,3,4)	gallacetophenone; alizarin yellow C	$(HO)_3C_6H_2 \cdot COCH_3$	VIII-393	168.15
6212	anthraquinone	anthragallol (1,2,3)	$C_6H_4(CO)_2C_6H(OH)_3$	VIII-505	256.22
6213	anthraquinone	(1,2,4); purpurin	$C_6H_4(CO)_2C_6H(OH)_3$	VIII-509	256.22
6214	anthraquinone (1,2,6)	flavopurpurin	$HO \cdot C_6H_3(CO)_2C_6H_2: (OH)_2$	VIII-513	256.22
6215	anthraquinone (1,2,7)	anthrapurpurin	$HO \cdot C_6H_3(CO)_2C_6H_2: (OH)_2$	VIII-516	256.22
6216	benzene (1,2,3)(*v*)	pyrogallol	$(HO)_3C_6H_3$	VI-1071	126.11
6217	benzene (1,2,4) (*uns*)	hydroxy-hydro-quinone	$(HO)_3C_6H_3$	VI-1087	126.11
6218	benzene (1,3,5) (*sym*)	phloroglucinol	$(HO)_3C_6H_3 \cdot 2H_2O$	VI-1092	162.14
6219	benzene (1,3,5)	phloroglucinol	$(HO)_3C_6H_3$	VI-1092	126.11
6220	benzoic acid (2,3,4)	pyrogallol carboxylic acid	$(HO)_3C_6H_2 \cdot CO_2H \cdot xH_2O$	X-464	170.12
6221	benzoic acid (2,4,5)	....	$(HO)_3C_6H_2 \cdot CO_2H \cdot \frac{1}{2}H_2O$	X-468	179.13
6222	benzoic acid (2,4,6)	phloroglucinol car-boxylic acid	$(HO)_3C_6H_2 \cdot CO_2H \cdot H_2O$	X-468	188.14
6223	benzoic acid (3,4,5)	gallic acid	$(HO)_3C_6H_2 \cdot CO_2H \cdot H_2O$	X-470	188.14
6224	benzophenone (2,3,4)	alizarin yellow A	$C_6H_5 \cdot CO \cdot C_6H_2(OH)_3 \cdot H_2O$	VIII-417	248.24
6225	benzophenone (2,6,2')	salicyloyl-resorcinol	$(HO)_2C_6H_3 \cdot CO \cdot C_6H_4OH$	VIII-422	230.22
6226	butyrophenone (2,3,4)	*n*-butyropyrogallol	$(HO)_3C_6H_2 \cdot COC_3H_7 \cdot H_2O$	VIII-399	214.22
6227	butyrophenone (*n*)	(2,4,6)	$(HO)_3C_6H_2 \cdot COC_3H_7 \cdot H_2O$	*VIII-691	214.22
6228	glutaric acid	(*d* or *l*)	$(CHOH)_3(CO_2H)_2$	III-553	180.12
6229	glutaric acid (*dl*)	....	$(CHOH)_3(CO_2H)_2$	III-553	180.12
6230	methyl-amino-methane	....	$H_2N \cdot C(CH_2OH)_3$	IV-303	121.14
6231	pyridine (*sym*)	(2,4,6)	$(HO)_3C_5H_2N$	XXI-197	127.10
6232	**Triiodo**-acetic acid	....	$I_3C \cdot CO_2H$	II-225	437.74
6233	benzene (1,2,3)	(*vic*)	$I_3C_6H_3$	V-228	455.80
6234	benzene (1,2,4)	(*uns*)	$I_3C_6H_3$	V-228	455.80
6235	benzene (1,3,5)	(*sym*)	$I_3C_6H_3$	V-228	455.80
6236	benzoic acid	(2,3,5)	$I_3C_6H_2 \cdot CO_2H$	*IX-150	499.81

Trifluoro-methane 3287
Trigemin 364
Trigenolline, cf. alkd.

Trigonelline, cf. alkd.
Triheptylin 3425
Trihydroxy-ethylamine 6178

Table 7-4 (*Continued*)
PHYSICAL CONSTANTS OF ORGANIC COMPOUNDS

No.	Crystalline Form and Color	Specific Gravity	Melting Point °C.	Boiling Point °C.	Solubility in 100 Parts		
					Water	Alcohol	Ether
6203	gas	....	−157.5	−27.9	d.	....	....
6204	col. lq.	$2.254^{14°}$	....	76.5	....	....	....
6205	lq.	$2.567^{7°}$	d. 330	117	....	....	....
6206	lq.	$1.576\frac{20}{4}°$	−35	47.6	i.	∞	∞; ∞ bz.
6207	lq.	$1.197\frac{20}{20}°$	−31.5	241	$1.9^{20°}$	....	....
6208	col. pd.	....	d. > 220	....	$2^{15°}$; $4^{100°}$	v. sl. s.	....
6209	pl./aq.	....	d. 270	....	v. s. c.	v. sl. s.	v. sl. s.
6210	col. lq.	....	....	263–5	v. sl. s.	v. s.	v. s.
6211	col. cr./aq.	....	173	....	0.2 c.; s. h.	s.	v. sl. s. bz.
6212	cr./al. ac.	....	310 d.	subl. 290 ±	v. sl. s.	s.; s. H_2SO_4	s.
6213	red nd./al.	....	256–7	....	sl. s. h.	s.	s.
6214	yel. nd./al.	....	>360	459 sl. d.	v. sl. s. h.	s.	sl. s.
6215	or. nd./al.	....	369	462 sl. d.	sl. s. h.	v. s. h.	sl. s.
6216	nd.	$1.453^{4°}$	133–4	309	$40^{13°}$	s.; sl. s. bz.	s.
6217	mn./aq.	....	140.5	....	v. s.	v. s.	v. s.; sl. s. bz.
6218	rhb.	....	117; −2H_2O, 110	subl. sl. d.	$1.13^{25°}$	v. s.	v. s.
6219	cr.	....	209–19	subl.	....	....	....
6220	nd./aq.	....	206 d. (anh.)	subl. in CO_2	$0.13^{12.5°}$	s.	sl. s.
6221	nd./aq.	....	217–8 d. (anh.)	−$\frac{1}{2}H_2O$, 105	s. h.	s.	....
6222	....	....	....	....	sl. s. c.	s.	v. s.
6223	mn./aq.	$1.694\frac{4}{4}°$	235 d. (anh.)	−H_2O, 100	$11^{3°}$; $3.3^{100°}$	$28^{15°}$ abs.	$2.5^{15°}$; 20 act.
6224	yel. nd./ aq. al.	....	140–1 (anh.)	....	v. sl. s. c.	s.; s. alk.; sl. s. bz.	s.; s. H_2SO_4
6225	yel. pl./al.	....	133–4	....	v. sl. s. h.	s.; s. alk.	s. bz.
6226	yel. nd./aq.	....	76–80; 100 (anh.)	....	....	....	....
6227	nd./aq.	....	179–80 (anh.)	−H_2O, 110	sl. s.	v. s.	v. s.
6228	lf./al.	....	128	....	v. s.	s.	s. act.
6229	pl./act.	....	154–5 d.	....	v. s.	v. s.	s. act.
6230	nd./al.	....	171–2	219–20^{10mm}	v. s.	$0.4^{20°}$	i.; v. sl. s. act.(d.)
6231	nd. or pd.	....	220–30 d.	....	sl. s.; d. h.	i.	i.
6232	yel. lf.	....	150 d.	....	s.	....	....
6233	nd./al.	....	116	....	i.	v. s.	v. s.
6234	nd./al.	....	91.4	....	i.	s.	s. chl.
6235	nd./ac.	....	182–4	....	i.	sl. s.	sl. s.
6236	pr./al.	....	223–4	....	i.; v. sl. s. bz.	s. h.	i.

Trihydroxy-methylanthraquinone 2874
Trihydroxy-toluene 4361
Trihydroxy-triphenylmethane 3963

Table 7-4 (*Continued*)
PHYSICAL CONSTANTS OF ORGANIC COMPOUNDS

No.	Name	Synonym	Formula	Beil. Ref.	Formula Weight
6237	**Triiodo** ethane (1,1,1)	methyl iodoform	$CH_3 \cdot CI_3$	I-99	407.76
6238	phenol (2,4,6)	....	$I_3C_6H_2 \cdot OH$	VI-211	471.80
6239	**Triketohydrin- dene hydrate**	ninhydrin	$C_6H_4(CO)_2 : C(OH)_2$	*VII-475	178.15
6240	**Trimethoxy** benzoic acid (2,3,4)	....	$(CH_3O)_3C_6H_2 \cdot CO_2H$	X-465	212.20
6241	benzoic acid (2,4,5)	asaronic acid	$(CH_3O)_3C_6H_2 \cdot CO_2H$	X-468	212.20
6242	benzoic acid (3,4,5)	gallic acid tri- methyl ether	$(CH_3O)_3C_6H_2 \cdot CO_2H$	X-481	212.20
6243	**Trimethyl-** acetaldehyde	pivalic aldehyde	$(CH_3)_3C \cdot CHO$	I-688	86.13
6244	acetaldehyde oxime	....	$(CH_3)_3C \cdot CH : NOH$	*I-354	101.15
6245	acetic acid	pivalic acid	$(CH_3)_3C \cdot CO_2H$	II-319	102.13
6246	acetonitrile	*tert*-butyl cyanide	$(CH_3)_3C \cdot CN$	II-320	83.13
6247	acetophenone (2,4,6)	acetomesitylene	$(CH_3)_3C_6H_2 \cdot COCH_3$	VII-332	162.23
6248	amine	....	$(CH_3)_3N$	IV-43	59.11
6249	amine HCl	....	$(CH_3)_3N \cdot HCl$	IV-46	95.57
6250	aniline (2,4,5)	pseudocumidine	$(CH_3)_3C_6H_2 \cdot NH_2$	XII-1150	135.21
6251	aniline (2,4,6)	mesidine	$(CH_3)_3C_6H_2 \cdot NH_2$	XII-1160	135.21
6252	arsine	arsenic trimethyl	$(CH_3)_3As$	IV-600	120.03
6253	benzene (1,2,3)	hemimellitene	$(CH_3)_3C_6H_3$	V-399	120.20
6254	benzene (1,2,4)	pseudocumene	$(CH_3)_3C_6H_3$	V-400	120.20
6255	benzene (1,3,5)	mesitylene	$(CH_3)_3C_6H_3$	V-406	120.20
6256	benzoic acid (2,3,5)	γ-*iso*-durylic acid	$(CH_3)_3C_6H_2 \cdot CO_2H$	IX-552	164.21
6257	benzoic acid (2,3,4)	prehnitylic acid	$(CH_3)_3C_6H_2 \cdot CO_2H$	IX-552	164.21
6258	benzoic acid (2,3,6)	....	$(CH_3)_3C_6H_2 \cdot CO_2H$	IX-552	164.21
6259	benzoic acid (2,4,5)	durylic acid	$(CH_3)_3C_6H_2 \cdot CO_2H$	IX-554	164.21
6260	benzoic acid (3,4,5)	α-*iso*-durylic acid	$(CH_3)_3C_6H_2 \cdot CO_2H$	IX-554	164.21
6261	benzoic acid (2,4,6)	β-*iso*-durylic acid	$(CH_3)_3C_6H_2 \cdot CO_2H$	IX-553	164.21
6262	borate	methyl borate	$(CH_3O)_3B$	I-287	103.91
6263	citrate	....	$C_3H_5O(CO_2CH_3)_3$	III-567	234.21
6264	methoxy-propenyl ammonium bromide	esmodil	$(CH_3)_3N(Br) \cdot CH_2 \cdot C(:CH_2) \cdot OCH_3$	....	210.12
6265	phenol (2,4,5)	pseudocumenol	$(CH_3)_3C_6H_2 \cdot OH$	VI-509	136.20
6266	phenol (2,4,6)	mesitol	$(CH_3)_3C_6H_2 \cdot OH$	VI-518	136.20
6267	phenyl ammonium iodide	....	$(CH_3)_3C_6H_5N \cdot I$	XII-159	263.12
6268	phosphate	methyl phosphate	$(CH_3O)_3PO$	I-286	140.08
6269	phosphine	....	$(CH_3)_3P$	IV-580	76.08
6270	quinoline (2,3,4)	....	$(CH_3)_3C_9H_4N$	XX-414	171.24
6271	quinoline (2,3,6)	....	$(CH_3)_3C_9H_4N$	XX-414	171.24
6272	quinoline (2,4,6)	....	$(CH_3)_3C_9H_4N \cdot aq.$	XX-414	171.24
6273	quinoline (2,6,8)	....	$(CH_3)_3C_9H_4N$	XX-415	171.24
6274	stibine	antimony trimethyl	$(CH_3)_3Sb$	IV-617	166.86
6275	thioacetamide	....	$(CH_3)_3C \cdot CS \cdot NH_2$	....	117.21
6276	urea	....	$(CH_3)_2N \cdot CO \cdot NHCH_3$	IV-74	102.14
6277	**Trimethylene-** acetal	....	$CH_3 \cdot CH \cdot O(CH_2)_3 \cdot O$	....	102.13
6278	bromohydrin	α-bromohydrin	$Br \cdot (CH_2)_3 \cdot OH$	I-356	139.00

Table 7-4 (*Continued*)
PHYSICAL CONSTANTS OF ORGANIC COMPOUNDS

No.	Crystalline Form and Color	Specific Gravity	Melting Point °C.	Boiling Point °C.	Solubility in 100 Parts		
					Water	Alcohol	Ether
6237	yel. octahed.	....	95 d.	....	v. s. CS_2; v. s. bz.	v. sl. s.	v. s.; sl. s. lg.
6238	nd./aq. al.	....	157-8	d.	s. act.	2	s.
6239	pr./aq.	....	239-40 d.; sl. d. 139	....	s. h.	s. alk.	v. sl. s.
6240	cr./et.	....	97-9	....	s.	....	....
6241	nd./al.	....	144	300±	s. h.	s.; s. lg.	s. bz.
6242	mn./aq.	....	169-70	225-7^{10mm}	v. sl. s.	v. s.	v. s.; s. chl.
6243	lq.	$0.793^{17°}$	3	74-5	....	....	....
6244	cr.	....	41	65^{20mm}	....	....	....
6245	nd.	$0.9055^{0°}$	35.5	163.8	2.1$^{20°}$	v. s.	v. s.
6246	cr.	....	15-6	105-6	....	....	....
6247	lq.	$0.975^{20°}_{4}$	....	240.5^{735mm}	....	....	....
6248	col. gas	$0.662^{-5°}$	−117.1	2.9	41$^{19°}$	s.	s.
6249	cr./al.	....	271-8 d.	....	s.	s.	i.
6250	nd./aq.	....	66-8	234-5	0.12$^{19°}$	....	....
6251	lq.	0.963	....	229-30	....	....	....
6252	col. lq.	$1.124^{22°}$	....	52.8	sl. s.	....	....
6253	col. lq.	$0.894^{20°}_{4}$	−25.4	176.1	i.	s.	s.
6254	col. lq.	$0.876^{20°}_{4}$	−43.8	169.4	i.	s.; s. SO_2	s.; s. bz.
6255	col. lq.	$0.865^{20°}_{4}$	−44.7(−52)	164.7	i.	s.; ∞ bz.	∞; s. SO_2
6256	pl./lg.	....	127	....	....	....	....
6257	pr./al.	....	167.5	....	....	....	....
6258	nd./aq.	....	105-6	....	....	....	....
6259	nd./bz.	....	149-50	....	v. sl. s. h.	v. s.	v. s.; s. bz.
6260	nd./aq.	....	215-6	...........∘∘	v. sl. s. h.	s.	s.
6261	cr./al.	....	152-5	...........∘∘	s. chl.	s.	s.
6262	lq. burns with	$0.920^{23°}_{4}$	−29 green flame	68.7	d.	∞	∞
6263	tri.	....	78-9	283-7 sl. d.	....	....	....
6264	wh. cr. pd.	....	169	....	s.	s.	....
6265	nd.	....	71-2	231-4	v. sl. s. c.	v. s.	v. s.
6266	nd.	....	72	221	v. sl. s.	v. s.	v. s.
6267	lf./al.	....	228-30 d.	subl.	s.	2$^{8°}$	i. chl.
6268	lq.	$1.197^{19.5°}_{0}$	....	197.2	100$^{25°}$	s.	s.
6269	col. lq.	<1	....	40-2	i.	....	s.
6270	cr.	....	65±	285	....	....	s.
6271	cr./lg.	....	86-7	285	sl. s. bz.	s.	v. s.
6272	hyg. cr.	....	63-4	277-8	sl. s.	sl. s.	v. s.
6273	mn./lg.	....	46	260^{719mm}	i.	v. s.	v. s.
6274	col. lq.	$1.523^{15°}$	....	80.6	v. sl. s.	s.	s.
6275	cr.	....	114-6	....	....	....	....
6276	mn.	1.19	75.5	232.5^{765mm}	v. s.	v. s.	s.
6277	col. lq.	....	....	109-11	s.	s.	s.
6278	lq.	$1.571^{20°}_{4}$	....	98-112^{185mm}	16.6 c.	....	....

Trimethyl-glycocoll, cf. alkd.
Trimethyl-methane 1022
Trimethyl-pentane 4946
Trimethyl-phenylmethane 1052
Trimethyl-pyridine 1526

Trimethyl-trimethylene glycol 4332
Trimethyl-trithiane 5879-81
Trimethyl-vinyl-ammonium hydroxide 4584
Trimethyl-xanthine, cf. alkd.
Trimethylene 1621

Table 7-4 (*Continued*)
PHYSICAL CONSTANTS OF ORGANIC COMPOUNDS

No.	Name	Synonym	Formula	Beil. Ref.	Formula Weight
	Trimethylene				
6279	chlorobromide	3-Cl-1-Br-propane	$Cl\cdot(CH_2)_3\cdot Br$	I-109	157.44
6280	chlorohydrin	α-chlorohydrin	$Cl\cdot(CH_2)_3\cdot OH$	I-356	94.54
6281	diacetate	. . . .	$CH_2(CH_2O_2C\cdot CH_3)_2$	II-143	160.17
6282	diamine	. . . .	$NH_2(CH_2)_3NH_2$	IV-261	74.13
6283	dibutyrate	. . . .	$CH_2(CH_2\cdot O_2C\cdot C_3H_7)_2$	. . . .	216.28
6284	dimercaptan	propandithiol-1,3	$HS\cdot(CH_2)_3\cdot SH$	I-476	108.23
6285	formal	1,3-dioxane	$CH_2\cdot O\cdot(CH_2)_3\cdot O$	XIX-2	88.11
6286	glycol	propandiol-1,3	$HO\cdot(CH_2)_3\cdot OH$	I-475	76.10
6287	iodohydrin	α-iodohydrin	$I\cdot(CH_2)_3\cdot OH$	I-358	185.99
6288	**Trinitro**-acetonitrile	. . . .	$(NO_2)_3C\cdot CN$	II-229	176.05
6289	aniline (2,4,6)	picramide	$(NO_2)_3C_6H_2\cdot NH_2$	XII-763	228.12
6290	anisole (2,4,6)	methyl picrate	$(NO_2)_3C_6H_2\cdot OCH_3$	VI-288	243.13
6291	benzaldehyde (2,4,6)	. . . .	$(NO_2)_3C_6H_2\cdot CHO$	VII-265	241.12
6292	benzene (1,2,3)	(*vic*)	$(NO_2)_3C_6H_3$	*V-140	213.11
6293	benzene (1,3,5)	(*sym*)	$(NO_2)_3C_6H_3$	V-271	213.11
6294	benzene (1,2,4)	(*uns*)	$(NO_2)_3C_6H_3$	V-271	213.11
6295	benzoic acid (2,4,6)	. . . .	$(NO_2)_3C_6H_2\cdot CO_2H$	IX-417	257.12
6296	*tert*-butyltoluene (2,4,6;1,3)	artificial musk	$(NO_2)_3C_6H:(CH_3)\cdot C(CH_3)_3$	V-439	283.24
6297	*tert*-butylxylene (2,4,6;1,3;5)	musk xylene	$(NO_2)_3C_6(CH_3)_2\cdot C(CH_3)_3$	V-448	297.27
6298	*m*-cresol	(2,4,6;1,3)	$(NO_2)_3C_6H(CH_3)\cdot OH$	VI-387	243.13
6299	ethane (1,1,1)	. . . .	$CH_3\cdot C(NO_2)_3$	I-103	165.06
6300	methane	nitroform	$(NO_2)_3CH$	I-79	151.04
6301	naphthalene (1,2,5)	δ-trinitronaphthalene	$(NO_2)_3C_{10}H_5$	V-563	263.17
6302	naphthalene (1,3,5)	α-trinitronaphthalene	$(NO_2)_3C_{10}H_5$	V-563	263.17
6303	naphthalene (1,3,8)	β-trinitronaphthalene	$(NO_2)_3C_{10}H_5$	V-563	263.17
6304	naphthalene (1,4,5)	γ-trinitronaphthalene	$(NO_2)_3C_{10}H_5$	V-563	263.17
6305	α-naphthol (2,4,5)	naphthopicric acid	$(NO_2)_3C_{10}H_4\cdot OH$	VI-619	279.17
6306	orcinol	(2,4,6;1.3,5)	$(NO_2)_3C_6(OH)_2CH_3$	VI-890	259.13
6307	phenol (γ)	(2,3,6)	$(NO_2)_3C_6H_2\cdot OH$	VI-265	229.11
6308	phenol (β)	(2,4,5)	$(NO_2)_3C_6H_2\cdot OH$	VI-265	229.11
6309	phenol (2,3,5)	. . . .	$(NO_2)_3C_6H_2\cdot OH$	VI-264	229.11
6310	phenol (2,4,6)	picric acid	$(NO_2)_3C_6H_2\cdot OH$	VI-265	229.11
6311	phenylhydrazine	(2,4,6)	$(NO_2)_3C_6H_2\cdot N_2H_3$	XV-493	243.14
6312	resorcinol (2,4,6)	styphnic acid	$(NO_2)_3C_6H(OH)_2$	VI-830	245.11
6313	toluene (β)	(2,3,4)	$(NO_2)_3C_6H_2\cdot CH_3$	*V-172	227.13
6314	toluene (ϵ)	(2,3,5)	$(NO_2)_3C_6H_2\cdot CH_3$	*V-172	227.13
6315	toluene (γ)	(2,4,5)	$(NO_2)_3C_6H_2\cdot CH_3$	V-347	227.13
6316	toluene (α)	(2,4,6); TNT	$(NO_2)_3C_6H_2\cdot CH_3$	V-347	227.13
6317	toluene (δ)	(3,4,5)	$(NO_2)_3C_6H_2\cdot CH_3$	*V-173	227.13
6318	trimethyl benzene (2,4,6; 1,3,5)	*eso*-trinitro-mesitylene	$(NO_2)_3C_6(CH_3)_3$	V-412	255.19
6319	trimethyl benzene (4,5,6;1,2,3)	trinitro-hemi-mellitene	$(NO_2)_3C_6(CH_3)_3$	V-400	255.19
6320	trimethyl benzene (3,4,6;1,2,4)	*eso*-trinitro-pseudocumene	$(NO_2)_3C_6(CH_3)_3$	V-405	255.19

Table 7-4 (*Continued*)
PHYSICAL CONSTANTS OF ORGANIC COMPOUNDS

No.	Crystalline Form and Color	Specific Gravity	Melting Point °C.	Boiling Point °C.	Solubility in 100 Parts		
					Water	Alcohol	Ether
6279	col. lq.	$1.638^{8°}$	....	142-3	i.	....	....
6280	lq.	$1.131\frac{20}{4}$	....	160-2	50 c.	s.	s.
6281	col. lq.	$1.070^{19°}$	....	209-10	10	....	....
6282	col. lq.	$0.884\frac{25}{4}$	-23.5	135-67^{38mm}	....	∞	∞
6283	col. lq.	....	....	125-30^{8mm}	i.	s.	s.
6284	oil	....	....	169-70	v. sl. s.	∞ ; ∞ chl.	∞ ; ∞ bz.
6285	col. lq.	$1.034\frac{20}{4}°$	-42	105-6	∞	∞	∞
6286	oil	$1.060\frac{20}{4}°$	....	214	∞	∞ ; i. chl.	i. bz.
6287	lq.	$1.998\frac{20}{4}°$	....	225^{748mm}	v. sl. s.	s.	s.
6288	waxy	....	41.5	expl. 220	d.	d.	s.
6289	yel. mn.	....	188-90	expl.	i.	sl. s.	s. h. act.
6290	col. mn.	$1.408^{20°}$	68.4	....	s. bz.	s. ac.	....
6291	pl./bz.	....	119	....			
6292	lt. gn./al.	....	127.5	....	i.	10 h.	....
6293	col. rhb.	$1.688\frac{20}{4}°$	121(61)	d.	0.04 c.	$1.9^{17.5°}$	$1.5^{17.5°}$
6294	col. cr.	$1.73^{16°}$	61-2	....	....	$5.5^{15.5°}$	$7.1^{15.5°}$
6295	rhb./aq.	....	210-20 d.	....	$2.05^{24°}$	....	....
6296	lt. yel./al.	....	112-3; (105-6)	....	i.	s.	s.; s. bz.
6297	nd./al.	....	110	....	i.	sl. s.	s.
6298	yel. nd./aq.	....	109.5	expl. 150	$0.22^{20°}$; $0.81^{100°}$	v. s.	v. s.
6299	cr.	....	56	....	v. sl. s.	s.; sl. s. lg.	s.
6300	col. cr.	$1.597\frac{24}{4}°$	23; expl.	45-7^{22mm}	s.	....	....
6301	nd./al.	....	112-3	....	....	s.	....
6302	rhb./chl.	....	122-3	....	s. ac.	s.	s. chl.
6303	cr./al.	....	218-9	....	$0.02^{100°}$	$*0.05^{23°}$	$0.13^{15°}$
6304	yel./chl.	....	148-9	....	$1.118°$ bz.	$*0.11^{19°}$	$0.41^{9°}$
6305	yel. nd./aq.	....	190	expl.	sl. s. h.	sl. s.	0.3 c. ac.
6306	yel. nd.	....	162-3	expl.	s. h.	i. aq. a.	v. s. h. bz.
6307	wh. nd.	....	117-8	....	s. h.	v. s.	v. s.
6308	wh. nd./aq.	....	96	....	s. h.	v. s.	v. s.
6309	yel. nd./aq.	....	119-20	....	s. ac.	s.	s. bz.
6310	yel. rhb.	$1.763\frac{20}{4}°$	121.8	expl. > 300	$1.23^{20°}$	$6.23^{20°}$ abs.	$1.08^{13°}$ abs.
6311	yel. nd./al.	....	186	d.	i.; s. ac.	s. h.; i. bz.	i.; i. chl.
6312	yel./act.	1.829	180	....	$0.6^{14°}$	v. s.	v. s.
6313	cr.	$1.620\frac{20}{4}°$	112	expl. 290-310	i.	sl. s. c.	s.
6314	yel. rhb.	....	97.2	d. 335	....	s.	i.
6315	yel. pl/act.	$1.620\frac{20}{4}°$	104	expl. 290	i.	s. h.	v. s.
6316	cr./al.	1.654	80.1	expl. 280	0.15 h.	$1.5^{22°}$	$5^{33°}$
6317	cr.	....	137.5	d. 313	....	$1^{15°}$	....
6318	tri./al.	1.48	232	expl. 415	sl. s. act.	sl. s. h.	sl. s. h.
6319	pl./al.	....	209	....	....	....	....
6320	pr.	....	185	....	....	v. sl. s. h.	v. s. h. bz.

* From 85% alcohol.
Trimyristin 3427
Trinitro-chlorobenzene 5308
Trinitro-cyanomethane 6288

Trinitro-hemimellitene 6319
Trinitro-mesitylene 6318
Trinitro-phenyl-methylnitramine 5873
Trinitro-pseudocumene 6320

Table 7-4 (*Continued*)
PHYSICAL CONSTANTS OF ORGANIC COMPOUNDS

No.	Name	Synonym	Formula	Beil. Ref.	Formula Weight
6321	**Trinitro** triphenyl carbinol	(4,4′,4″)	$(NO_2 \cdot C_6H_4)_3C \cdot OH$	VI-720	395.33
6322	triphenyl methane	(4,4′,4″)	$(NO_2 \cdot C_6H_4)_3CH$	V-707	379.33
6323	*m*-xylene (4,5,6)	. . . .	$(NO_2)_3C_6H(CH_3)_2$	V-381	241.16
6324	*m*-xylene (2,4,6)	. . . .	$(NO_2)_3C_6H(CH_3)_2$	V-381	241.16
6325	*p*-xylene (2,3,6)	*eso*-tri-NO$_2$-*p*-xylene	$(NO_2)_3C_6H(CH_3)_2$	V-389	241.16
6326	**Trional**	Me-Et ketone di-Et sulfone	$(C_2H_5)(CH_3)C$: $(SO_2C_2H_5)_2$	I-671	242.36
6327	**Triphenyl** acetic acid	. . . .	$(C_6H_5)_3C \cdot CO_2H$	IX-712	288.35
6328	amine	. . . .	$(C_6H_5)_3N$	XII-181	245.33
6329	benzene (1,3,5)	. . . .	$(C_6H_5)_3C_6H_3$	V-737	306.41
6330	bromomethane	. . . .	$(C_6H_5)_3C \cdot Br$	V-704	323.24
6331	carbinol	tritanol	$(C_6H_5)_3C \cdot OH$	VI-713	260.34
6332	carbinol methyl ether	methyl-trityl ether	$(C_6H_5)_3C \cdot OCH_3$	VI-716	274.37
6333	chloromethane	. . . .	$(C_6H_5)_3C \cdot Cl$	V-700	278.78
6334	guanidine (α)	. . . .	C_6H_5N:$C(NHC_6H_5)_2$	XII-451	287.37
6335	guanidine (β)	. . . .	HN:$C(NHC_6H_5)N$: $(C_6H_5)_2$	XII-430	287.37
6336	hydrazine	. . . .	$(C_6H_5)_2N \cdot NHC_6H_5$	XV-125	260.34
6337	methane	tritane	$(C_6H_5)_3C \cdot H$	V-698	244.34
6338	methane *o*-carboxylic acid	. . . .	$(C_6H_5)_2CH \cdot C_6H_4 \cdot CO_2H$	IX-714	288.35
6339	methyl	trityl	$(C_6H_5)_3C \cdots$	V-715	243.33
6340	methyl peroxide	. . . .	$[(C_6H_5)_3C \cdot O \cdot]_2$	VI-716	518.66
6341	phosphate	phenyl phosphate	$(C_6H_5O)_3PO$	VI-179	326.29
6342	phosphine	. . . .	$(C_6H_5)_3P$	XVI-759	262.29
6343	phosphite	phenyl phosphite	$(C_6H_5O)_3P$	VI-177	310.29
6344	rosaniline sulfate	Poirrier's blue	$(C_{38}H_{32}N_3)_2SO_4$	XIII-768	1157.46
6345	*p*-rosaniline trisulfonic Na	methyl blue; brilliant cotton blue	$C_{37}H_{27}O_9N_3S_3Na_2$	. . . .	799.81
6346	thiophosphate	. . . .	$(C_6H_5O)_3PS$	VI-181	342.36
6347	**Tri-*iso*-propanol-amine**	. . . .	$N(CH_2 \cdot CHOH \cdot CH_3)_3$	. . . .	191.27
6348	**Tripropylamine** (*n*)	. . . .	$(C_2H_5 \cdot CH_2)_3N$	IV-139	143.27
6349	**Trithio**-carbonic acid	. . . .	$HS \cdot CS \cdot SH$	III-221	110.22
6350	phenyl phosphate	. . . .	$(C_6H_5S)_3PO$	VI-182	374.49
6351	**Tritolyl** phosphate (*o*)	tricresyl phosphate	$(CH_3 \cdot C_6H_4 \cdot O)_3PO$	VI-358	368.37
6352	phosphate (*m*)	tricresyl phosphate	$(CH_3 \cdot C_6H_4 \cdot O)_3PO$	. . . .	368.37
6353	phosphate (*p*)	tricresyl phosphate	$(CH_3 \cdot C_6H_4 \cdot O)_3PO$	VI-401	368.37
6354	phosphite (*m*)	tricresyl phosphite	$(CH_3 \cdot C_6H_4 \cdot O)_3P$	VI-381	352.37
6355	phosphite (*p*)	tricresyl phosphite	$(CH_3 \cdot C_6H_4 \cdot O)_3P$	VI-401	352.37
6356	thiophosphate (*o*)	tricresyl thiophosphate	$(CH_3 \cdot C_6H_4 \cdot O)_3PS$	*VI-173	384.44
6357	thiophosphate (*m*)	. . . .	$(CH_3 \cdot C_6H_4 \cdot O)_3PS$	*VI-203	384.44
6358	thiophosphate (*p*)	. . . .	$(CH_3 \cdot C_6H_4 \cdot O)_3PS$	*VI-203	384.44
6359	**Tritriacontane**	. . . .	$C_{33}H_{68}$	I-177	464.91
6360	**Tri-*o*-xenyl phosphate**	tri-*o*-phenylphenyl phosphate	$(C_6H_5 \cdot C_6H_4 \cdot O)_3PO$	. . . .	554.59
6361	**Tropic acid** (*dl*)	. . . .	$C_6H_5 \cdot CH(CH_2OH) \cdot CO_2H$	X-261	166.18
6362	**Truxene** (α)	tribenzylene benzene	$C_{27}H_{18}$	V-752	342.44

Triolein 3430
Trioxane 3289
Trioxy-methylene 3289
Trioxy-purine 6393
Tripalmitin 3431

Triphenyl antimony 563
Triphenyl-arsine 580
Triphenyl-bismuthine 857
Triphenyl-bismuthine dichloride 858
Triphenyl-carbinol-*o*-carboxylic anhyd. 5292

Table 7-4 (*Continued*)
PHYSICAL CONSTANTS OF ORGANIC COMPOUNDS

No.	Crystalline Form and Color	Specific Gravity	Melting Point °C.	Boiling Point °C.	Solubility in 100 Parts		
					Water	Alcohol	Ether
6321	mn. or rhb.	....	193(167)	....	s. bz.; s. ac.	sl. s. h.	sl. s.
6322	cr./bz.	....	212.5	....	v. sl. s. bz.	v. sl. s. ac.	....
6323	pr./al.	$1.494^{19°}$	125	....	i.	$1.2^{20°}$	sl. s.
6324	yel./al. bz.	$1.604^{19°}$	182	....	i.	$0.04^{20°}$	sl. s.
6325	col. mn./al.	$1.59^{19°}$	139-40	expl. 410	....	....	....
6326	pl./al.	$1.199^{8.5°}_4$	76	....	$0.3^{15°}$	$5.8^{5°}$ abs.	$6.6^{15°}$
6327	mn.	....	264-5 sl. d.	....	sl. s.	s.	v. sl. s. bz.
6328	mn.	$0.774^{0°}_0$	126.5	365	i.; s. act.	sl. s.	s.
6329	rhb.	1.205	170-1	....	v. s. bz.	s. abs.	s.
6330	lt. yel. cr.	1.55	152	230^{15mm}	d.	v. s. CS_2	v. s. bz.
6331	cr./bz.	$1.188^{20°}_4$	162.5	>360	v. s. bz.	v. s.	v. s.
6332	tri./Me al.	....	87-8	....	....	....	....
6333	col. cr.	....	112-3	$230-5^{20mm}$	d.	v. s. CS_2	v. s. bz.
6334	rhb./al.	1.13	145-7	d.	i.	$4.6^{0°}$ abs.	....
6335	pl.	....	131	....	sl. s. bz.	s.	s.
6336	nd./bz. pet.	....	142	....	i.	s.	v. s. bz.
6337	cr.	$1.014^{9.9°}_4$	93.4(81)	$358-9^{754mm}$	i.; 44° bz.	v. s. h.	v. s.
6338	nd./al.	....	162	subl.	i.	s.	s.
6339	col. cr.	....	145-7	d.	i.	sl. s. h.	v. s. chl.
6340	cr./CS_2	....	185-6	....	i.	i.	i.
6341	pr./al.	$1.206^{5.8°}_4$	49-50	245^{11mm}	i.	$155^{25°}$	v. s.
6342	mn./et.	1.194	79	>360*	i.; s. HCl	sl. s.	v. s.; s. bz.
6343	lq.	$1.184^{18°}_{18}$	22-4	360	i.; s. chl.	s.; s. bz.	s.
6344	b. pd.	....	....	....	....	....	....
6345	b. pd.	....	....	....	s.	....	....
6346	pr./al.	$**1.23^{20°}_4$	52-3	>360 d.	i.; s. chl.	s.; s. bz.	s.; s. act.
6347	col. pl.	$0.991^{60°}_{20}$	58	306.5	s.	s.	s.
6348	col. lq.	$0.757^{20°}_4$	-93.5	156.5	v. sl. s.	∞	∞
6349	red oil	$1.47^{17°}_4$	-30	d. 20-30	i. d.	sl. s.	sl. s.
6350	mn./et.	....	114-5	....	i.	s.	s.; s. chl.
6351	lq.	....	....	410 sl. d.	i.; s. bz.	v. s.	v. s.
6352	col. lq.	....	....	$273-5^{17mm}$	i.	s.	s.
6353	nd./aq.	....	77-8	....	v. s. bz.	v. s.	v. s.
6354	col. lq.	....	....	$235-8^{7mm}$	....	....	....
6355	pa. yel. lq.	....	....	$236-9^{7mm}$	....	....	....
6356	col. nd./al.	....	45-6	....	i.	sl. s.	v. s.
6357	rhb. cr.	....	33-4	....	i.	sl. s.	v. s.
6358	nd./al.	....	93-4	....	i.; v. s. bz.	v. s. chl.	sl. s. lg.
6359	cr.	lq. $0.780^{71.8°}$	71.8	328^{15mm}	....	....	....
6360	col. rhb.	....	112-3	....	i.	$i.^{25°}$	$8^{25°}$ bz.; $12^{5°}$ CCl_4
6361	nd./aq.	....	117-8	d.	$2^{15°}$; s. h.	s.; sl. s. bz.	s.; i. CS_2
6362	pl./xylene	....	365-8	....	i.; s. aniline	i.; s. $PhNO_2$	i.

* In hydrogen atm.
** Supercooled liquid.
Triphenyl-dihydroglyoxalin 233
Triphenyl-imidazole 3978
Triphenyl-imidazoline 233

Triphenyl-methyl chloride 6333
Triphenyl-oxazole 698
Triphenyl-stibine 563
Triphenyl-stibine dichloride 564
Triphenyl-tin chloride 5948

Tripropin 3432
Tripropionin 3432
Triptane 3519
Triquinoyl 1604
Tristearin 3433

Table 7-4 (*Continued*)
PHYSICAL CONSTANTS OF ORGANIC COMPOUNDS

No.	Name	Synonym	Formula	Beil. Ref.	Formula Weight
6363	**Tryptamine**	aminoethyl-indole $(\omega)(2)$	$NH \cdot C_6H_4 \cdot CH{:}C \cdot$ $\mid$ _____ $\mid$ $CH_2 \cdot CH_2 \cdot NH_2$	....	160.22
6364	hydrochloride	....	$C_{10}H_{12}N_2 \cdot HCl$	....	196.68
6365	**Tryptophan** (*l*)	β-indolyl-α-alanine	$C_6H_4 \cdot NH \cdot CH{:}C \cdot$ $\mid$ _____ $\mid$ $C_2H_3(NH_2)CO_2H$	XXII-546	204.23
6366	**Tutocaine**	*p*-aminobenzoyl-di-methylamino-1,2-dimethyl propanol HCl	$NH_2 \cdot C_6H_4 \cdot CO_2CH{:}$ $(CH_3) \cdot CH(CH_3) \cdot$ $CH_2 \cdot N(CH_3)_2 \cdot HCl$	....	286.80
6367	**Tyrosine** (*l*) (−)	β-(*p*-hydroxy-phenyl)-alanine	$HO \cdot C_6H_4 \cdot C_2H_3(NH_2) \cdot$ CO_2H	XIV-605	181.19
6368	**Uliron**	4-(4'-aminophenyl-sulfonamido)-phenylsulfondi-methylamide	$NH_2 \cdot C_6H_4 \cdot SO_2 \cdot NH \cdot$ $C_6H_4 \cdot SO_2N(CH_3)_2$	....	355.44
6369	**Umbelliferone**	7-hydroxy-coumarin	$HO \cdot C_6H_3 \cdot CH{:}CH \cdot CO \cdot O$ $\mid$ _____ $\mid$	XVIII-27	162.15
6370	**Undecane** (*n*)	....	$CH_3 \cdot (CH_2)_9 \cdot CH_3$	I-170	156.31
6371	**Undecyl** alcohol (*n*)	undecanol-1	$CH_3(CH_2)_9CH_2OH$	I-427	172.31
6372	alcohol (*n*)(*sec*)	undecanol-2	$C_9H_{19} \cdot CHOH \cdot CH_3$	I-427	172.31
6373	**Undecylene** (α)	undecene-1	$CH_3(CH_2)_8CH{:}CH_2$	I-225	154.30
6374	**Undecylene** (β)	undecene-2	$C_8H_{17} \cdot CH{:}CH \cdot CH_3$	I-225	154.30
6375	alcohol	undecene-1-ol-11	$CH_2{:}CH(CH_2)_9 \cdot OH$	I-452	170.30
6376	**Undecylenic acid**	10-hendecenoic acid	$CH_2{:}CH(CH_2)_8CO_2H$	II-458	184.28
6377	**Undecylic** acid	undecanoic acid	$CH_3 \cdot (CH_2)_9 \cdot CO_2H$	II-358	186.30
6378	aldehyde	undecanal	$CH_3 \cdot (CH_2)_9 \cdot CHO$	I-712	170.30
6379	aldehyde oxime	....	$CH_3(CH_2)_9CH{:}NOH$	I-713	185.31
6380	amide	....	$CH_3(CH_2)_9CO \cdot NH_2$	II-358	185.31
6381	nitrile	decyl cyanide	$CH_3 \cdot (CH_2)_9 \cdot CN$	II-358	167.30
6382	**Undecyne**-1	rutylidene	$CH_3 \cdot (CH_2)_8 \cdot C{:}CH$	I-261	152.28
6383	**Uracil** (2,4)	2,6-dioxy-pyrimi-dine	$CH{:}CH \cdot CO \cdot NH \cdot$ $\mid$ _____ $CO \cdot NH$ ____ $\mid$	XXIV-312	112.09
6384	**Uramil**	5-amino-barbituric acid	$(CONH)_2COCH \cdot NH_2$ $\mid$ _____ $\mid$	XXV-492	143.10
6385	**Uraminobenzoic acid** (*o*)	....	$NH_2 \cdot CO \cdot NH \cdot C_6H_4 \cdot$ CO_2H	....	180.16
6386	**Urea**	carbamide	$NH_2 \cdot CO \cdot NH_2$	III-42	60.06
6387	calcium chloride	afenil	$CaCl_2 \cdot 4CH_4ON_2$	*III-26	351.21
6388	hydrochloride	....	$CO(NH_2)_2 \cdot HCl$	III-54	96.52
6389	nitrate	acidogen nitrate	$CO(NH_2)_2 \cdot HNO_3$	III-54	123.07
6390	oxalate	....	$2CH_4ON_2 \cdot C_2H_2O_4$	III-55	210.15
6391	oxalate	....	$2CH_4ON_2 \cdot C_2H_2O_4 \cdot$ $2H_2O$	III-55	246.18
6392	**Urethane**	ethyl carbamate	$NH_2 \cdot CO_2 \cdot C_2H_5$	III-22	89.09
6393	**Uric acid**	2,6,8-trioxy-purine	$C_5H_4O_3N_4$	XXVI-513	168.11
6394	**Usnic acid** (*d*)	....	$C_{18}H_{16}O_7$	XIX-316	344.32
6395	**Usnic acid** (*dl*)	....	$C_{18}H_{16}O_7$	XIX-316	344.32

Table 7-4 (*Continued*)
PHYSICAL CONSTANTS OF ORGANIC COMPOUNDS

No.	Crystalline Form and Color	Specific Gravity	Melting Point °C.	Boiling Point °C.	Solubility in 100 Parts		
					Water	Alcohol	Ether
6363	cr.	....	120	....	....	....	....
6364	col. cr.	....	255-8	....	s.	sl. s.	i.
6365	hex., rhb. nd.	....	289	....	sl. s. c.; v. s. h.	i.; i. alk.	i.; i. chl.
6366	col. cr. pd.	....	212-5	....	25	2.5	....
6367	nd./aq.	$1.456^{20°}$	>290 d. *314 d.	....	$0.04^{17°}$; $0.65^{100°}$	$0.01^{17°}$; s. alk.	i.; s. a.
6368	col. cr. pd.	....	193-5	....	sl. s.; v. s. alk.	s.	s. act.
6369	nd./aq.	....	224-7	subl.	1 h.	s.; s. HCl	sl. s.; s. ac.
6370	col. lq.	$0.741^{20°}_{4}$	−25.6	194.5	i.	∞	∞
6371	lq.	$0.822^{15°}_{4}$	19	131^{15mm}	<0.02	s.	....
6372	lq.	$0.827^{20°}_{4}$	12	228-9	i.	s.	....
6373	col. lq.	$0.763^{20°}_{4}$	....	188-90	i.	∞	∞
6374	lq.	$0.774^{15°}_{15}$	....	192-3	i.	∞	∞
6375	lq.	$0.850^{15°}$	−2	245^{756mm}	....	....	....
6376	cr.	$0.907^{24°}_{4}$	24.5	295 d.	i.	s.	s.; s. chl.
6377	col. cr.	$0.891^{30°}$	29-30	228^{160mm}	i.	s.	s.
6378	lq.	$0.862^{15°}_{15}$	−4	$116\text{-}7^{18mm}$	....	....	....
6379	nd./Me al.	....	72(61)	....	....	v. s.	v. s.
6380	col. cr.	....	103	....	....	....	....
6381	col. lq.	....	....	253-4	....	....	....
6382	lq.	$0.867^{25°}_{4}$	−33	210-5	i.	s.	s.
6383	nd./aq.	....	338 d.	....	s. h.; s. NH_4OH	i.	i.
6384	nd.	....	>400	....	sl. s. h.; s. alk.	s. c. H_2SO_4	i.; s. NH_3
6385	nd.	....	152	d. 171-2	d. h.	s.; s. Me al.	s. act.
6386	col. pr.	$1.335^{20°}_{4}$	132.7	d.	$100^{17°}$; ∞ h.	$20^{20°}$	sl. s.
6387	wh. hyg. pd.	....	158-60	....	v. s.	sl. s.	i. Me al.
6388	wh. hyg. lf.	....	d. 145	....	s.	....	....
6389	col. mn. pr.	....	152 d.	....	v. s. h.	s.	i. HNO_3
6390	mn. pr.	....	170-1	d.	$4.4^{16°}$	$1.6^{16°}$	i.
6391	....	....	d. > 120	$-H_2O$, 120	....	....	....
6392	col. lf.	$1.11^{20°}_{20}$	49-50	184	v. s.	v. s.	v. s.
6393	cr.	$1.893^{20°}$	d.	....	0.06 h.	i.; s. aq. LiOH	i.
6394	cr.	....	203	d.	i.	v. sl. s.	sl. s.
6395	yel. pr.	....	195-6	....	i.	v. sl. s. c.	$0.3^{20°}$

* Heated rapidly.

Trypaflavine 5328	Tylcalsin 116	Ultraquinine, cf. alkd.
Tryparsamide 583	Tyllithin 117	Umbellic acid 2333
Turicine, cf. alkd.	Tylnatrin 119	Undecanal 6378
Tussol 568	Tyramine 3799	Undecanoic acid 6377
	Ulexine, cf. alkd.	Undecanol 6371-2

Undecanone 3107, 4324
Undecene 6373-4
Undecenol 6375
Undecyl cyanide 3956
Ural 1288

Table 7-4 (*Continued*)
PHYSICAL CONSTANTS OF ORGANIC COMPOUNDS

No.	Name	Synonym	Formula	Beil. Ref.	Formula Weight
6396	**Uvinic acid**	2,5-diMe-furfurane-3-carboxylic acid	$(CH_3)_2C_4HO \cdot CO_2H$	XVIII-297	140.14
6397	**Uvitic acid** (5;1,3)	5-methyl-isophthalic acid	$CH_3 \cdot C_6H_3(CO_2H)_2$	IX-864	180.16
6398	**Uvitonic acid** (2;4,6)	o-picoline-o,p-dicarboxylic acid	$CH_3 \cdot C_5H_2N(CO_2H)_2$	XXII-161	181.15
6399	**Valeric** acid (*n*)	pentanoic acid	$C_2H_5 \cdot CH_2 \cdot CH_2 \cdot CO_2H$	II-299	102.13
6400	aldehyde (*n*)	pentanal	$C_2H_5 \cdot CH_2 \cdot CH_2 \cdot CHO$	I-676	86.13
6401	aldehyde oxime	. . . .	$C_4H_9 \cdot CH{:}NOH$	I-676	101.15
6402	amide (*n*)	. . . .	$CH_3(CH_2)_3CONH_2$	II-301	101.15
6403	anhydride (*n*)	. . . .	$(C_4H_9CO)_2O$	II-301	186.25
6404	chloride (*n*)	n-valeryl chloride	$CH_3 \cdot (CH_2)_3 \cdot COCl$	II-301	120.58
6405	nitrile (*n*)	n-butyl cyanide	$CH_3(CH_2)_3 \cdot CN$	II-301	83.13
6406	*iso*-**Valeric** acid	β-Me-butyric acid	$(CH_3)_2CH \cdot CH_2 \cdot CO_2H$	II-309	102.13
6407	aldehyde	2-methyl-butanal-4	$(CH_3)_2CH \cdot CH_2 \cdot CHO$	I-684	86.13
6408	aldehyde oxime	. . . .	$C_4H_9 \cdot CH{:}NOH$	I-686	101.15
6409	amide	. . . .	$C_4H_9 \cdot CO \cdot NH_2$	II-315	101.15
6410	anhydride	. . . .	$(C_4H_9CO)_2O$	II-314	186.25
6411	anilide	. . . .	$C_4H_9 \cdot CO \cdot NHC_6H_5$	XII-254	177.25
6412	chloride	*iso*-valeryl chloride	$(CH_3)_2CH \cdot CH_2COCl$	II-315	120.58
6413	nitrile	*iso*-butyl cyanide	$(CH_3)_2CH \cdot CH_2 \cdot CN$	II-315	83.13
6414	**Valeryl diethylamide** (*iso*)	valyl	$(CH_3)_2CH \cdot CH_2 \cdot CO \cdot N(C_2H_5)_2$	. . . .	157.26
6415	**Valylene**	. . . .	$CH_2{:}C(CH_3) \cdot C{:}CH$	I-263	66.10
6416	**Vanillalacetone** (1;3,4)	ferulic methyl ketone	$CH_3 \cdot CO \cdot CH{:}CH \cdot C_6H_3(OCH_3)OH$	VIII-291	192.22
6417	**Vanillic** acid (3;4,1)	4-OH-3-MeO-benzoic acid	$(CH_3OC_6H_3(OH) \cdot CO_2H$	X-392	168.15
6418	**Vanillin** (3;4,1)	4-OH-3-MeO-benzaldehyde	$(CH_3O)C_6H_3(OH) \cdot CHO$	VIII-247	152.15
6419	ethyl ether	4-EtO-3-MeO-benzaldehyde	$(CH_3O)C_6H_3{:}(OC_2H_5)CHO$	VIII-256	180.21
6420	p-phenetidine ethyl carboxylate	(4,1;1′,3′,4′); eupyrin	$C_2H_5O \cdot C_6H_4 \cdot N{:}CH \cdot C_6H_3(OCH_3)O \cdot CO_2 \cdot C_2H_5$	. . . .	343.38
6421	*iso*-**Vanillin**	3-OH-4-MeO-benzaldehyde	$(CH_3O)C_6H_3(OH) \cdot CHO$	VIII-254	152.15
6422	**Vanillyl alcohol**	4-OH-3-MeO-benzyl alcohol	$(CH_3O)C_6H_3(OH) \cdot CH_2OH$	VI-1113	154.17
6423	**Veratric** acid	3,4-dimethoxy-benzoic acid	$(CH_3O)_2C_6H_3 \cdot CO_2H$	X-393	182.18
6424	aldehyde (3,4;1)	vanillin methyl ether	$(CH_3O)_2C_6H_3 \cdot CHO$	VIII-255	166.18
6425	**Vinaconic acid**	ethylene malonic acid	$CH_2 \cdot CH_2 \cdot C(CO_2H)_2$ ‾‾‾‾‾‾	IX-722	130.10
6426	**Vinyl** acetate	. . . .	$CH_3CO_2 \cdot CH{:}CH_2$	*II-63	86.09
6427	acetic acid	butenoic acid	$CH_2{:}CH \cdot CH_2 \cdot CO_2H$	II-407	86.09
6428	acrylic acid (β)	. . . .	$CH_2{:}(CH)_2{:}CH \cdot CO_2H$	II-481	98.10
6429	alcohol	ethenol	$CH_2{:}CH \cdot OH$	I-601	44.05
6430	amine†	. . . .	$CH_2{:}CH \cdot NH_2$	IV-203	43.07
6431	anisole (*o*)	o-methoxy styrene	$CH_2{:}CH \cdot C_6H_4 \cdot OCH_3$	VI-560	134.18
6432	anisole (*m*)	m-methoxy styrene	$CH_2{:}CH \cdot C_6H_4 \cdot OCH_3$	VI-561	134.18
6433	anisole (*p*)	p-methoxy styrene	$CH_2{:}CH \cdot C_6H_4 \cdot OCH_3$	VI-561	134.18
6434	bromide	bromo-ethylene	$CH_2{:}CH \cdot Br$	I-188	106.96

† This product is actually ethylene imine, $CH_2 \cdot CH_2 \cdot NH$.

Uraline 1288	Urea chloride 1229	Uroxin 188
Uralium 1288	Ureido-hydantoin, cf. alkd.	Ursin, cf. glcde.
Uranin 3251	Ureous acid, cf. alkd.	Ursol-D 5260
Urazine 2851	Urotropine 3599	Ursol-P 349
		Vacciniin, cf. glcde.

Table 7-4 (*Continued*)
PHYSICAL CONSTANTS OF ORGANIC COMPOUNDS

No.	Crystalline Form and Color	Specific Gravity	Melting Point °C.	Boiling Point °C.	Solubility in 100 Parts		
					Water	Alcohol	Ether
6396	cr./aq.	....	135	subl.	0.25 h.	s.	v. s.
6397	nd./aq.	....	290-1	subl.	sl. s. h.	s.	s.
6398	cr. pd.	....	274-82 d.	....	i. c.; s. a.	s. h. aniline	v. sl. s. h. bz.
6399	col. lq.	$0.940\frac{20°}{4}$	-34.5(-59)	186.4	$3.3^{16°}$	∞	∞
6400	lq.	$0.819^{11°}$	-92	103.4	v. sl. s.	s.	s.
6401	cr.	....	52	....	....	....	....
6402	mn. pl.	1.023	106	....	v. s.	v. s.	v. s.
6403	lq.	$0.922\frac{17°}{4}$	-56.1	227.5	d. h.	....	....
6404	col. lq.	$1.016^{15°}$	-110	127-8	d.	d.	....
6405	col. lq.	$0.804\frac{15°}{4}$	-96	140.8	i.	s.	s.
6406	col. lq.	$0.925\frac{20°}{4}$	-29.3	176.5	$4.2^{20°}$	∞; ∞ chl.	∞
6407	col. lq.	$0.803^{17°}$	-51	92.5	sl. s.	s.	s.
6408	oil	$0.893\frac{20°}{4}$	....	161.3^{759mm}	....	....	....
6409	mn.	$0.965\frac{20°}{4}$	135-7	232	s.	s.	s.
6410	col. lq.	$0.929\frac{27°}{4}$	113-4	215	....	....	....
6411	nd./lg.	1.078	113-4	....	sl. s. h.	s.	s.
6412	col. lq.	$0.989\frac{20°}{4}$	....	116-7.5	d.	d.	s.
6413	col. lq.	$0.795\frac{15°}{4}$	-100.9	130.3	....	....	....
6414	col. lq.	....	....	210	4	s.	s.
6415	col. lq.	....	....	50	....	....	....
6416	yel. nd./aq. al.	....	129-30	....	v. sl. s.	s.; s. conc. H$_2$SO$_4$	s.; s. bz.
6417	nd./aq.	....	207	subl.	$0.12^{14°}$; $2.5^{100°}$	v. s.	v. s.
6418	mn.	1.056	82-3.5	285(in CO$_2$)	$1^{14°}$; $57^{5°}$	v. s.; v. s. chl.	v. s.; v. s. CS$_2$
6419	mn.	....	64-5	subl.	v. sl. s. h.	s.	s.
6420	yel. cr.	....	87-8	....	sl. s.	s. h.	s.; s. chl.
6421	mn.	$1.196\frac{20°}{4}$	115-7	subl. sl. d.	s. h.	s.; sl. s. CS$_2$	s.
6422	mn./aq.	....	115	d.	v. s. h.	v. s.	v. s.
6423	cr./aq.*	....	180-1	subl.	$0.05^{14°}$; $0.6^{100°}$	v. s.	v. s.
6424	nd./et.	....	44-7	280-5	sl. s. h.	s.	s.
6425	tri./et.	....	175	210^{30mm}	v. s.	s.	....
6426	col. lq.**	$0.932\frac{20°}{4}$	-92.8	72-3	$2^{20°}$	∞	∞
6427	col. lq.	$1.013\frac{15°}{15}$	-39	163	s.	∞	∞
6428	hyg. pr./et.	....	80	d. 110-5	s. h.	s.	s.; sl. s. pet.
6429	not known;	acetalde-	hyde	des-	motrope	....	....
6430	lq.	0.832	....	56	....	s.	....
6431	lq.	$1.005\frac{17.2°}{4}$	....	195-200	i.	s.	s.
6432	lq.	....	....	$89-90^{14mm}$	i.	s.	s.
6433	lq.	$1.000\frac{13°}{4}$	....	$204-5^{756mm}$	i.	s.	s.
6434	lq.	$1.529\frac{11°}{4}$	-137.8	15.8	i.	∞	∞

* Crysts. $+ 1H_2O < 50°$.
** Polymerizes in light.
Valamin 497
Valdivin, cf. glcde.
Valerone 1901

Valerophenone 1117-8
Valeryl chloride 6404, 6412
Valerylene 5087
Validol 4032
Valine 391-5

Valisan 863
Valyl 6414
Valzin 5112
Vanillal 3188
Vanillin methyl ether 6424

Table 7-4 (*Continued*)
PHYSICAL CONSTANTS OF ORGANIC COMPOUNDS

No.	Name	Synonym	Formula	Beil. Ref.	Formula Weight
6435	**Vinyl** chloride	chloro-ethylene	$CH_2:CH\cdot Cl$	I-186	62.50
6436	ether	....	$(CH_2:CH)_2O$	I-433	70.09
6437	ethyl alcohol	allyl carbinol	$C_3H_5\cdot CH_2OH$	I-441	72.11
6438	ethyl ether	ethyl-vinyl ether	$CH_2:CH\cdot O\cdot C_2H_5$	I-433	72.11
6439	fluoride	fluoro ethylene	$CH_2:CH\cdot F$	I-186	46.04
6440	glycolic acid	....	$CH_2:CH\cdot CHOH\cdot CO_2H$	III-370	102.09
6441	guaiacol (1;3,4)	hesperetole	$CH_2:CH\cdot C_6H_3(OH)\cdot$ (OCH_3)	VI-954	150.18
6442	iodide	iodoethylene	$CH_2:CH\cdot I$	I-192	153.95
6443	phenol (*o*)	*o*-hydroxy styrene	$CH_2:CH\cdot C_6H_4\cdot OH$	VI-560	120.15
6444	phenol (*m*)	*m*-hydroxy styrene	$CH_2:CH\cdot C_6H_4\cdot OH$	VI-561	120.15
6445	**Violuric** acid	nitroso-barbituric acid	$CO\cdot(NHCO)_2\cdot C:NOH$	XXIV-506	157.09
6446	sodium salt	....	$C_4H_2O_4N_3Na$	XXIV-507	179.07
6447	**Xanthene**	2,2'-methylene-diphenyl ether	$C_6H_4\cdot CH_2\cdot C_6H_4\cdot O$	XVII-73	182.22
6448	**Xanthic acid***	xanthogenic acid	$C_2H_5O\cdot CS\cdot SH$	III-209	122.21
6449	**Xanthine**†	2,6-dioxy-purine	$C_5H_4O_2N_4$	XXVI-447	152.11
6450	**Xanthogenamide**	Et-thioncarbamate	$C_2H_5O\cdot CS\cdot NH_2$	III-137	105.16
6451	**Xanthone**	benzophenone oxide	$OC:(C_6H_4)_2O$	XVII-354	196.21
6452	**Xanthophyll**	lutein	$C_{40}H_{56}O_2$	XXX-95	568.89
6453	**Xanthydrol**	9-OH-xanthene	$HOCH:(C_6H_4)_2O$	XVII-129	198.22
6454	**Xylene** (*o*)	1,2-dimethylbenzene	$C_6H_4(CH_3)_2$	V-362	106.17
6455	sulfonic acid	(1,2;4)	$(CH_3)_2C_6H_3\cdot SO_3H\cdot$ $2H_2O$	XI-121	222.26
6456	sulfonic Na	(1,2;4)	$(CH_3)_2C_6H_3\cdot SO_3Na\cdot$ $5H_2O$	XI-121	298.29
6457	**Xylene** (*m*)	1,3-dimethylbenzene	$C_6H_4(CH_3)_2$	V-370	106.17
6458	sulfonic acid	(1,3;4)	$(CH_3)_2C_6H_3\cdot SO_3H\cdot$ $2H_2O$	XI-123	222.26
6459	sulfonic Na	(1,3;4)	$(CH_3)_2C_6H_3\cdot SO_3Na\cdot$ H_2O	XI-123	226.23
6460	sulfonic Ca	(1,3;4)	$[(CH_3)_2C_6H_3\cdot SO_3]_2Ca$	....	410.53
6461	sulfonyl chloride	(1,3;4)	$(CH_3)_2C_6H_3\cdot SO_2Cl$	XI-123	204.68
6462	**Xylene** (*p*)	1,4-dimethylbenzene	$C_6H_4(CH_3)_2$	V-382	106.17
6463	sulfonic acid	(1,4;2)	$(CH_3)_2C_6H_3\cdot SO_3H\cdot$ $2H_2O$	XI-127	222.26
6464	sulfonic Na	(1,4;2)	$(CH_3)_2C_6H_3\cdot SO_3Na\cdot$ H_2O	XI-127	226.23
6465	**Xylenol** (*vic*)(*o*)	2,3-dimethyl phenol	$(CH_3)_2C_6H_3\cdot OH$	VI-480	122.17
6466	**Xylenol** (*uns*)(*o*)	3,4-dimethyl phenol	$(CH_3)_2C_6H_3\cdot OH$	VI-480	122.17
6467	**Xylenol** (*sym*)(*m*)	3,5-dimethyl phenol	$(CH_3)_2C_6H_3\cdot OH$	VI-492	122.17
6468	**Xylenol** (*vic*)(*m*)	2,6-dimethyl phenol	$(CH_3)_2C_6H_3\cdot OH$	VI-485	122.17
6469	**Xylenol** (*uns*)(*m*)	2,4-dimethyl phenol	$(CH_3)_2C_6H_3\cdot OH$	VI-486	122.17
6470	**Xylenol** (*p*)	2,5-dimethyl phenol	$(CH_3)_2C_6H_3\cdot OH$	VI-494	122.17
6471	**Xylidine** (*vic*)(*o*)	2,3-dimethyl aniline	$(CH_3)_2C_6H_3\cdot NH_2$	XII-1101	121.18
6472	**Xylidine** (*uns*)(*o*)	3,4-dimethyl aniline	$(CH_3)_2C_6H_3\cdot NH_2$	XII-1103	121.18
6473	**Xylidine** (*sym*)(*m*)	3,5-dimethyl aniline	$(CH_3)_2C_6H_3\cdot NH_2$	XII-1131	121.18
6474	**Xylidine** (*vic*)(*m*)	2,6-dimethyl aniline	$(CH_3)_2C_6H_3\cdot NH_2$	XII-1107	121.18
6475	**Xylidine** (*uns*)(*m*)	2,4-dimethyl aniline	$(CH_3)_2C_6H_3\cdot NH_2$	XII-1111	121.18
6476	acetate	....	$C_8H_{11}N\cdot HC_2H_3O_2$	....	181.24
6477	**Xylidine** (*p*)	2,5-dimethyl aniline	$(CH_3)_2C_6H_3\cdot NH_2$	XII-1135	121.18

† See also Alkaloid table.
Vanirome 3188
Vasicine, cf. alkd.
Veratridine, cf. alkd.
Veratrine, cf. alkd.

Veratrole 2406
Veratroylaconine, cf. alkd.
Veritol 4117
Veronal 2124
Vesipyrin 120

Vicianin, cf. glcde.
Vicine, cf. alkd.
Victoria yellow 4056
Vinetine, cf. alkd.
Vinopyrin 5110

* Really (new nomenclature) ethyl xanthic acid. Xanthic acid (hypothetical) is HOCSSH.

Table 7-4 (Continued)
PHYSICAL CONSTANTS OF ORGANIC COMPOUNDS

No.	Crystalline Form and Color	Specific Gravity	Melting Point °C.	Boiling Point °C.	Solubility in 100 Parts		
					Water	Alcohol	Ether
6435	gas	$0.908^{25°}_{25}$	−160	−13.9	sl. s.	s.	v. s.
6436	col. lq.	$0.773^{20°}_{20}$	....	28.3	v. sl. s.	∞	∞
6437	lq.	$0.838^{17.5°}_{4}$	....	113.5^{748mm}	s.	....	....
6438	lq.	$0.763^{14.5°}_{17.5}$	−115.3	35.5	v. sl. s.	s.	....
6439	col. gas	$0.853^{-26°}$	−160.5	−72.2	i.	$100^{20°}$ cc.	$550^{20°}$ cc. act.
6440	hyg. nd.	....	33–40	$129-30^{12mm}$	v. s.	s.	s.; i. CS_2
6441	cr.	....	57	....	sl. s.	s.	s.
6442	lq.	$2.08^{0°}$	....	56	....	....	....
6443	nd.	$1.061^{19.2}_{4}$	29	108^{15mm}	s. alk.	s.	s.
6444	oil	....	....	$114-6^{17mm}$	....	....	....
6445	rhb.	....	224 d.	−H_2O, 100	s. h.	s.	....
6446	lf./al.	....	100.5	315	sl. s.	....	....
6447	lf./al.	....	100.5	315	sl. s.; s. H_2SO_4	sl. s. c.	s.; s. bz.
6448	oil	>1	−53	d. 24	v. sl. s.	....	....
6449	pd.	....	d. > 150	....	$0.26^{17°}$	$0.03^{17°}$	s. KOH
6450	mn.	....	40–1	....	$2.3^{20°}$	∞	∞
6451	nd./al.	....	173–4	$349-50^{730mm}$	sl. s. h.; s. chl.	0.7 c.; 8.5 h.	sl. s.; sl. s. lg.; s. bz.
6452	brn. red	....	190–3	....	i.; s. chl.	sl. s. h.	sl. s.
6453	col./aq. al.	....	122–3 d.	....	v. sl. s.	s.	s. chl.
6454	col. lq.	$0.880^{20°}_{4}$	−25.2	144.4	i.	∞ abs.	∞
6455	pl./aq. H_2SO_4	....	d.	....	s.	....	....
6456	pr.	....	....	....	s.	....	....
6457	col. lq.	$0.864^{20°}_{4}$	−47.9	139.1	i.	∞ abs.	∞
6458	cr./aq. H_2SO_4	....	63–4	....	s.	....	....
6459	pl.	....	....	....	s.	....	....
6460	col. cr.	....	....	....	s.	i.	i.
6461	cr.	....	34	$135-6^{11mm}$	d. h.	d. h.	....
6462	pl.	$0.861^{20°}_{4}$	13.3	138.4	i.	s.	v. s.
6463	col. lf./aq.	....	86	$149^{0.1mm}$	s.	s. chl.	....
6464	mn.pr./aq.	$1.522^{15°}$	....	....	s.	....	....
6465	nd./aq.	....	75	218	s.	s.	....
6466	cr./10% al.	$1.023^{17°}_{15}$	64.5–6.5	225^{757mm}	sl. s.	s.	∞
6467	nd./aq.	....	68	219.5	sl. s.	s.	....
6468	lf.	....	48–9	212	s. h.	s.	....
6469	nd.	$1.036^{20°}_{4}$	25–6	211.5^{766mm}	....	....	....
6470	mn.	....	74.5	211.5–3.5	....	....	....
6471	lq.	$0.991^{50°}$	<−15	223	v. sl. s.	s.	s.
6472	pr. lg.	$1.076^{17.5°}$	49–50	224–6	v. sl. s. c.	s. pet.	....
6473	oil	$0.972^{20°}_{4}$	....	221–2	....	....	....
6474	lq.	$0.980^{15°}$	10–2	216–7	....	....	....
6475	lq.	$0.978^{19.6°}_{4}$	....	213–4	v. sl. s.	....	....
6476	cr.	....	70	....	v. sl. s.	....	....
6477	oil	$0.979^{21°}_{4}$	15.5	215^{739mm}	v. sl. s.	....	....

Table 7-4 (*Continued*)
PHYSICAL CONSTANTS OF ORGANIC COMPOUNDS

No.	Name	Synonym	Formula	Beil. Ref.	Formula Weight
6478	**Xylitol**	pentanpentol	$C_5H_{12}O_5$	I-531	152.15
6479	**Xylorcinol** (*m*)	4,6-diMe-resorcinol	$(CH_3)_2C_6H_2(OH)_2$	VI-912	138.17
6480	*l*-**Xylose** (+)	wood sugar	$C_4H_9O_4 \cdot CHO$	I-865	150.13
6481	**Xylyl** acetate (*p*)	*p*-tolubenzyl acetate	$CH_3C_6H_4CH_2 \cdot O_2C \cdot CH_3$	VI-498	164.21
6482	amine (*o*)	*o*-tolubenzylamine	$CH_3 \cdot C_6H_4 \cdot CH_2 \cdot NH_2$	XII-1106	121.18
6483	amine (*m*)	....	$CH_3 \cdot C_6H_4 \cdot CH_2 \cdot NH_2$	XII-1134	121.18
6484	amine (*p*)	....	$CH_3 \cdot C_6H_4 \cdot CH_2 \cdot NH_2$	XII-1141	121.18
6485	bromide (*o*)	ω-bromo-*o*-xylene	$CH_3 \cdot C_6H_4 \cdot CH_2 \cdot Br$	V-365	185.07
6486	bromide (*m*)	ω-bromo-*m*-xylene	$CH_3 \cdot C_6H_4 \cdot CH_2 \cdot Br$	V-374	185.07
6487	bromide (*p*)	ω-bromo-*p*-xylene	$CH_3 \cdot C_6H_4 \cdot CH_2 \cdot Br$	V-385	185.07
6488	chloride† (*o*)	ω-chloro-*o*-xylene	$CH_3 \cdot C_6H_4 \cdot CH_2 \cdot Cl$	V-364	140.61
6489	chloride (*m*)	....	$CH_3 \cdot C_6H_4 \cdot CH_2 \cdot Cl$	V-373	140.61
6490	chloride (*p*)	....	$CH_3 \cdot C_6H_4 \cdot CH_2 \cdot Cl$	V-384	140.61
6491	**Xylylene** dibromide (*o*)	ω,ω'-diBr-*o*-xylene	$C_6H_4(CH_2Br)_2$	V-366	263.97
6492	dibromide (*m*)	....	$C_6H_4(CH_2Br)_2$	V-374	263.97
6493	dibromide (*p*)	....	$C_6H_4(CH_2Br)_2$	V-385	263.97
6494	dichloride (*o*)	ω,ω' diCl-*o*-xylene	$C_6H_4(CH_2Cl)_2$	V-364	175.06
6495	dichloride (*m*)	....	$C_6H_4(CH_2Cl)_2$	V-373	175.06
6496	dichloride (*p*)	....	$C_6H_4(CH_2Cl)_2$	V-384	175.06
6497	glycol (*o*)	phthal alcohol	$C_6H_4(CH_2OH)_2$	VI-910	138.17
6498	glycol (*m*)	....	$C_6H_4(CH_2OH)_2$	VI-914	138.17
6499	glycol (*p*)	....	$C_6H_4(CH_2OH)_2$	VI-919	138.17
6500	**Xylyl** hydrazine	(2,4;1)	$(CH_3)_2C_6H_3 \cdot NHNH_2$	XV-549	136.20
6501	hydrazine (2,5;1)	diMe-Ph-hydrazine	$(CH_3)_2C_6H_3 \cdot NHNH_2$	XV-552	136.20
6502	hydrazine (2,6;1)	diMe-Ph-hydrazine	$(CH_3)_2C_6H_3 \cdot NHNH_2$	XV-548	136.20
6503	**Zinc** diethyl	zinc ethide	$(C_2H_5)_2Zn$	IV-672	123.49
6504	dimethyl	dimethyl zinc	$(CH_3)_2Zn$	IV-671	95.44
6505	dipropyl (*n*)	*n*-dipropyl zinc	$(C_2H_5 \cdot CH_2)_2Zn$	IV-675	151.55
6506	di-*iso*-propyl	....	$[(CH_3)_2CH]_2Zn$	IV-675	151.55
6507	**Zingerone**	3-MeO-4-OH-benzylacetone	$HO(CH_3O)C_6H_3 \cdot CH_2 \cdot CH_2 \cdot COCH_3$	*VIII-623	194.23

† See also Nos. 1474-6.
Vitamin-B6 5510
Vitamin-C 586
Vitamin-D2 1187
Vitamin-G 5563
Vitamin-H 853
Vitamin-K 4342
Vomicine, cf. alkd.
Waldivin, cf. glcde.

War gases 1247; 1303; 1473; 1985; 4697
"Weinsaure" 5691
Westrosol 6155
White damp 1239
White tar 4458
Wintergreen oil 4380
Wood alcohol 4100
Wood naphtha 4100

Wood spirit 4100
Wood sugar 6480
Wrightine, cf. alkd.
Wurster's reagent 5840
Xanthaline, cf. alkd.
Xanthenol 6453
Xanthogenic acid 6448
Xanthopuccine, cf. alkd.
Xanthopurpurin 2298

Table 7-4 (*Continued*)
PHYSICAL CONSTANTS OF ORGANIC COMPOUNDS

No.	Crystalline Form and Color	Specific Gravity	Melting Point °C.	Boiling Point °C.	Solubility in 100 Parts		
					Water	Alcohol	Ether
6478	syrup	....	....	....	s.	....	....
6479	lf./chl.	....	124-5	276-9	v. s.	v. s.	v. s.
6480	nd.	$1.535°$	153-4	....	$117^{20°}$	v. sl. s. c.	i.
6481	lq.	....	....	227	....	....	....
6482	oil	$0.977^{1.0°}_{0}$	0	205.6^{745mm}	....	....	....
6483	oil	$0.965^{2.0°}_{0}$	....	205^{750mm}	i.	s.	s.
6484	lq.	$0.952^{2.0°}_{0}$	12.6-13.2	204^{739mm}	v. sl. s.	....	....
6485	pr.	$1.381^{23°}$	21	223-4	i.	s.	s.
6486	col. lq.	$1.371^{23°}$	....	212-5 sl. d.	i.	s.	s.
6487	nd./al.	1.324	38	$218-20^{740mm}$	i.	v. s. chl.	v. s. h.
6488	col. lq.	....	d. 170±	195-203	i.	∞ abs.	∞
6489	lq.	$1.064^{20°}$	d. 170±	195-6	i.	∞ abs.	∞
6490	oil	....	d. 170±	200-2	i.	∞ abs.	∞
6491	rhb.	$1.988°$	94.5	d.	16.6 pet.	s.	20
6492	mn./chl.	$1.959°$	76-7	$135-40^{20mm}$	33 pet.	v. s. chl.	v. s.
6493	mn./bz.	$2.012°$	145-7	245	d. >80°	v. s. h. chl.	$2.7^{20°}$
6494	mn.?	$1.393°$	55	239-41	v. s. chl.	v. s.	v. s.
6495	cr.	$1.302^{20°}$	34.2	250-5	i.		
6496	mn.	$1.417°$	100.5	240-5 d.	i.; s. act.	s.; s. chl.	v. sl. s.
6497	pl./et.	....	64.2-4.8	....	$>25^{18°}$	$>25^{18°}$	$25^{18°}$
6498	cr./bz.	lq. $1.161^{18°}$	46-7	$154-9^{13mm}$	v. s.	....	s.
6499	nd.	....	115-6	....	sl. s.	v. s.	v. s.
6500	nd./et.	....	85	d.	v. sl. s.	v. s.	s.
6501	col. nd.	....	78	....	i.	s.	s.
6502	nd./pet.	....	46±	....	s. lg.	....	....
6503	col. lq.	$1.182^{18°}$	−28	118	d.	d.	....
6504	col. lq.	$1.386^{11°}$	−40	46	d.	d.	....
6505	col. lq.	$1.072^{\frac{21°}{21}}$	....	158-60	d.	d.	....
6506	col. lq.	....	....	$94-8^{40mm}$	d.	d.	....
6507	col. cr./ et. pet.	....	40-1	....	sl. s.	s. dil. alk.	s.; sl. s. pet.

Table 7-5
PHYSICAL CONSTANTS OF ALKALOIDS
Compiled by F. E. SHEIBLEY, Ph.D.

Names of the compounds in the table below are arranged alphabetically. No compound is listed more than once in the table and each compound is given a number.

Synonyms. At the bottom of each page is an alphabetical listing of names for compounds which are to be found in the main body of the table but under a different name; the number following the name refers to the numerical place of this compound in the table; thus, *acetyl-benzoyl-aconine 3* indicates that this compound is the 3d compound in the table where it will be found listed under the name *aconitine*.

Appearance. In addition to the crystalline form and color, the solvent used in purification is often given; thus, "rhomb./al." indicates that rhombic crystals were obtained when the compound was crystallized from alcohol.

Optical Properties are given in the column headed "[a]"; only the directions of the specific rotations have been indicated, since the factors affecting the determination of this property are too variable to come within the scope of the table.

Color Reactions with Sulfuric Acid, where available, have been included as an

---

Abbreviations used in the table

a., acid	d, dextro-rotatory
abs., absolute	d., decomposes or decomposition
ac. a., acetic acid	deliq., deliquescent
act., acetone	dil., dilute
al., alcohol	dk., dark
alk., alkali (i. e. aqueous NaOH or KOH)	diss., dissociates
am. al., amyl alcohol	effl., efflorescent
amor., amorphous	et., ether $(C_2H_5)_2O$
anh., anhydrous	Et, ethyl (C_2H_5)
aq., aqueous; water	EtOAc, ethyl acetate
bl., blue	gly., glycerol
B. P., boiling point	grn., green
br., brown	h., hot
bz., benzene, C_6H_6	hex., hexagonal
c., cold	hyg., hygroscopic
chl., chloroform, $CHCl_3$	i., insoluble
colorl., colorless	*in.*, inactive
cryst., crystals or crystalline	*l*, levo-rotatory
	leaf., leaflets or leaves

No.	Name	[a]	Formula	Appearance	Melting Point °C.
1	**Aconine**	d	$C_{25}H_{41}O_9N$	hyg., amor.	132
2	salts	l	....	hyg.	....
3	**Aconitine**	d	$C_{34}H_{47}O_{11}N$	rhomb. pr./chl.	204
4	hydrobromide	l	$C_{34}H_{47}O_{11}N \cdot HBr \cdot 2\frac{1}{2}H_2O$	hex. tab./aq.	sint. 160
5	hydrobromide	...	$C_{34}H_{47}O_{11}N \cdot HBr \cdot \frac{1}{2}H_2O$	need./al.	206–7
6	hydrochloride	l	$C_{34}H_{47}O_{11}N \cdot HCl \cdot 3H_2O$	cryst.	149; 170*
7	**Adenine**	...	$C_5H_5N_5$	need. + $3H_2O$/aq.	360–5 d. subl. 220
8	**Agmatine**	...	$HN:C(NH_2) \cdot NH \cdot (CH_2)_4NH_2$	....	....

* Melting point of the anhydrous compound.

Acetyl-benzoyl-aconine 3   Acidol 34

Table 7-5 (*Continued*)
PHYSICAL CONSTANTS OF ALKALOIDS

aid in the making of rapid preliminary examinations.

Solubilities expressed by numbers are given in parts by weight of solvent required to dissolve one part of the alkaloid at a temperature of approximately 25°C. Because of the wide discrepancies existing between many of these figures as found in the literature no claim to accuracy can be made, and the values stated are perhaps best considered as upper limits.

References. The information given in the table has been collected mainly from the following sources: Henry: **Plant Alkaloids,** 4th edition, published by The Blakiston Co., Philadelphia-Toronto (1949); Manske and Holmes, **The Alkaloids,** published by Academic Press, New York (1950–1960); **Merck's Index,** 7th edition, published by Merck and Co., Inc., Rahway, N. J. (1960); Heilbron: **Dictionary of Organic Compounds,** published by Oxford University Press, New York (1934); Beilstein, **Handbuch der Organischen Chemie,** 3d edition.

lig., ligroin	r., red
liq., liquid	rhomb., rhombic
lt., light	s., soluble
lustr., lustrous	sint., sinters
Me, methyl (CH₃)	sl., slight or slightly
MeOH, methyl alcohol	subl., sublimes
met., metallic	tab., tabular
mon., monoclinic	tricl., triclinic
need., needles	trim., trimetric
octahedrl., octahedral	v., very
org., orange	v. s., very soluble
orthorhomb., orthorhombic	v. sl. s., very slightly soluble
pa., pale	wh., white
pet., petroleum ether	yel., yellow
powd., powder	∞, soluble in all proportions;
pr., prisms	i.e., miscible
pyr., pyridine	>, greater than

No.	Reaction with H₂SO₄	Solubility Expressed in Parts of Solvent Required to Dissolve 1 Part Alkaloid					
		Water	Alcohol	Ether	Chloroform	Benzene	Others
1	....	v. s.	v. s.	i.	s.	....	i. pet.
2	....	....	....	....	....	....	....
3	colorl. when pure	3300	23	47	3	6.2	....
4	....	s.	s.	....	....	....	....
5	....	....	....	....	....	....	....
6	....	s.	s.	....	....	....	....
7	....	1086 c.; 40 h.	sl. s. h.	i.	i.	s. a.	s. h. NH₄OH
8	....	....	....	....	....	....	....

ψ-Aconitine 185   Acraconitine 185

Table 7-5 (*Continued*)
PHYSICAL CONSTANTS OF ALKALOIDS

No.	Name	[α]	Formula	Appearance	Melting Point °C.
9	sulfate	...	$C_5H_{14}N_4 \cdot H_2SO_4$	colorl. cryst./aq. MeOH	226–9
10	**Allantoin†**	*in*	$C_4H_6O_3N_4$	need./h. aq.	235–6
11	**Anabasine**	*l*	$C_{10}H_{14}N_2$	colorl. liq.	B.P. 276
12	**Anhalonine**	*l*	$C_{12}H_{15}O_3N$	wh. need.	85
13	**Apomorphine**	...	$C_{17}H_{17}O_2N$	pr. + $1Et_2O$/et.	170 d.
14	hydrochloride	*l*	$C_{17}H_{17}O_2N \cdot HCl \cdot \frac{1}{2}H_2O$	pr./aq.	....
15	**Apoquinine**	*l*	$C_{19}H_{22}O_2N_2$	need./et.	180–90 d.
16	**Arecoline**	*in*	$C_8H_{13}O_2N$	very alkaline oil	B.P. 209
17	hydrobromide	*in*	$C_8H_{13}O_2N \cdot HBr$	pr./al.	169–71
18	hydrochloride	...	$C_8H_{13}O_2N \cdot HCl$	cryst.	158
19	**Aspidospermine**	*l*	$C_{22}H_{30}O_2N_2$	need./al. or pet.	208
20	**Atisine**	*l*	$C_{22}H_{33}O_2N$	wh., amor.	indefinite
21	hydrochloride	*d*	$C_{22}H_{33}O_2N \cdot HCl$	prisms	296
22	**Atropine**	*in*	$C_{17}H_{23}O_3N$	colorl. pr.	118, subl.
23	sulfate	*in*	$(C_{17}H_{23}O_3N)_2 \cdot H_2SO_4 \cdot H_2O$	need.	194*
24	**Bebeerine, α**	*l*	$C_{18}H_{19}O_3N$	pr./MeOH	214
25	**Bebeerine, β**	*d*	$C_{18}H_{19}O_3N$	yel., amor.	142–50
26	hydrochloride	...	$C_{18}H_{19}O_3N \cdot HCl$	need. or scales	259–60
27	**Benzoylecgonine**	*l*	$C_{16}H_{19}O_4N \cdot 4H_2O$	lustr. need./aq.	90–2; 193–5*
28	**Berberine**	*in*	$C_{20}H_{19}O_5N \cdot 6H_2O$	red-yel. need./aq.	145d.
29	bisulfate	...	$C_{20}H_{17}O_4N \cdot H_2SO_4$	yel. need.	....
30	chloroform	...	$C_{20}H_{19}O_5N \cdot CHCl_3$	tricl. tab./chl.	179
31	hydrochloride	...	$C_{20}H_{17}O_4N \cdot HCl \cdot 2H_2O$	org. need. or yel. powd.	....
32	nitrate	...	$C_{20}H_{17}O_4N \cdot HNO_3$	yel. need.	....
33	**Betaine†**	*in*	$C_5H_{11}O_2N \cdot H_2O$	sweet deliq. cryst.; anh. at 100°	293*
34	hydrochloride	...	$C_5H_{11}O_2N \cdot HCl$	mon. cryst.	227–8 d.
35	**Brucine**	*l*	$C_{23}H_{26}O_4N_2 \cdot 4H_2O$	mon. pr./al.	105; 178*
36	hydrochloride	...	$C_{23}H_{26}O_4N_2 \cdot HCl$	wh. need.	....
37	nitrate	...	$C_{23}H_{26}O_4N_2 \cdot HNO_3 \cdot 2H_2O$	wh. pr.	230 d.*
38	sulfate	...	$(C_{23}H_{26}O_4N_2)_2 \cdot H_2SO_4 \cdot 7H_2O$	long need.	....
39	**Caffeine**	*in*	$C_8H_{10}O_2N_4 \cdot H_2O$	need./al.; anh. 100°	235*; subl. 178
40	citrate (true)	...	$C_8H_{10}O_2N_4 \cdot C_6H_8O_7$	mon.	
41	hydrochloride	...	$C_8H_{10}O_2N_4 \cdot HCl \cdot 2H_2O$	mon.	d. 80–100
42	mercurichloride	...	$C_8H_{10}O_2N_4 \cdot HgCl_2$	colorl. need.	246
43	sulfate	...	$C_8H_{10}O_2N_4 \cdot H_2SO_4$	wh. need.	....
44	triiodide	...	$C_8H_{10}O_2N_4I_2 \cdot HI \cdot 1\frac{1}{2}H_2O$	long grn. met. pr.	171
45	**Canadine**	*l*	$C_{20}H_{21}O_4N$	silky need./al.	133–4
46	**Carpaine**	*d*	$C_{14}H_{25}O_2N$	pr./al.	121
47	**Carpiline**	*d*	$C_{16}H_{18}O_3N_2$	wh. cryst.	187
48	**Cephaeline**	*l*	$C_{28}H_{38}O_4N_2$	fine need./et.	107–8; 120–30*

* Melting point of the anhydrous compound.
† See also listing in the table Physical Constants of Organic Compounds.
Amino-butyl-guanidine 8   6-Amino-purine 7
2-Amino-6-hydroxy-purine 115   Anhaline 121

Table 7-5 (*Continued*)
PHYSICAL CONSTANTS OF ALKALOIDS

No.	Reaction with H_2SO_4	Solubility Expressed in Parts of Solvent Required to Dissolve 1 Part Alkaloid					
		Water	Alcohol	Ether	Chloroform	Benzene	Others
9	....	s.	v. sl. s.	....	....	....	....
10	....	132 c.; 30 h.	5000 abs.	i.	....	....	s. NaOH
11	....	s.	s.	s.	....	s.	....
12	....	s.	s.	s.	s.	....	....
13	colorl.	sl. s.	s.	sl. s.	v. s.	sl. s.	s. alk.
14	....	39.5	38.2	1864	sl. s.	....	....
15	fluorescent dil.	s. h.	s.	sl. s.	v. s.	v. s.	s. KOH
16	....	s.	s.	s.	s.	....	....
17	....	1	8 c.; 2 h.	sl. s.	sl. s.	....	....
18	....	s.	s.	....	....	....	....
19	colorl.	6000	48	106	s.	s.	s. dil. a.
20	....	sl. s.	s.	s.	s.	s.	s. dil. a.
21	....	v. s.	v. s.	i.	....	....	....
22	colorl.	300	1.46	16.6	1.56	s.	s. dil. a.
23	colorl.	0.38	3.7	2140	620	....	....
24	br.→r. h.	i.	sl. s.	sl. s.	s.	....	s. a., act.
25	....	i.	sl. s.	....	s.	s.	s. dil. a.
26	....	s.	s.	....	....	....	....
27	....	s. h.	s.	i.	....	s. dil. a.	s. alk.
28	grn.→yel.	22	100	v. sl. s.	sl. s.	sl. s.	unstable
29	....	100; s. h.	sl. s.	....	....	....	....
30	....	....	....	....	....	....	....
31	....	400; s. h.	s. h.	i.	i.	....	....
32	....	sl. s.	....	....	....	....	....
33	....	s.	s.	sl. s.	....	....	....
34	....	1.7	15	i.	i.	....	....
35	dk. yel. h.; HNO_3→r.	320 c.; 150 h.	1.1	133	7.5	88	i. alk.
36	....	s.	s.	....	....	....	....
37	....	s.	s.	....	....	....	....
38	....	75 c.; 10 h.	84	....	254	....	....
39	+$K_2Cr_2O_7$ →grn.	45.6	53.2	375 c.; 339 h.	8 c.; 6.4 h.	88 c.; 18.9 h.	s. EtOAc sl. s. pet.
40	....	s. d.	s. d.	....	....	....	....
41	....	s. d.	s. d.	....	....	....	....
42	....	260	....	....	....	....	....
43	....	s. d.	s. d.	....	....	....	....
44	....	i.	s.	....	sl. s.	....	....
45	....	i.	s.	v. s.	v. s.	v. s.	....
46	....	sl. s.	9	33	s.	5.5	s. dil. a.
47	....	s. h.	....	sl. s.	s.	s.	....
48	....	i.	s.	sl. s.	s.	s.	s. alk.; s. dil. a.

Aphrodine 244
Apocupreine 15
Aticine 20
Atroscine 129

Banisterino 118
Baptitoxine 92
Barytine 138
Benzoyl-pseudotropine 235

Table 7-5 (*Continued*)
PHYSICAL CONSTANTS OF ALKALOIDS

No.	Name	[α]	Formula	Appearance	Melting Point °C.
49	**Cevadine**	d	$C_{32}H_{49}O_9N \cdot 2C_2H_5OH$	rhomb.; becomes anh. 130–40	205*
50	**Chelerythrine**	in	$C_{21}H_{19}O_5N \cdot C_2H_5OH$	pr. leaf./al.	207
51	**Chelidonine**	d	$C_{20}H_{19}O_5N \cdot H_2O$	mon. tab./dil. HCl	135–6*
52	**Cinchonamine**	d	$C_{19}H_{24}ON_2$	orthorhomb.need./al.	185
53	**Cinchonidine**	l	$C_{19}H_{22}ON_2$	trim. pr./al.	210.5
54	hydrochloride	l	$C_{19}H_{22}ON_2 \cdot HCl \cdot 2H_2O$	pyramids or pr.	242*
55	sulfate	l	$(C_{19}H_{22}ON_2)_2 \cdot H_2SO_4 \cdot 3H_2O$	mon. pr.	240 d.*
56	**Cinchonine**	d	$C_{19}H_{22}ON_2$	rhomb. pr./al.	264
57	bisulfate	...	$C_{19}H_{22}ON_2 \cdot H_2SO_4 \cdot 4H_2O$	octahedrl.	....
58	hydrochloride	d	$C_{19}H_{22}ON_2 \cdot HCl \cdot 2H_2O$	mon.	217–8 d.*
59	sulfate	d	$(C_{19}H_{22}ON_2)_2 \cdot H_2SO_4 \cdot 2H_2O$	rhomb.	198.5*
60	**Cinchotine**	d	$C_{19}H_{24}ON_2$	pr. or scales	268–9
61	**Cinnamylcocaine**	l	$C_{19}H_{23}O_4N$	need./bz.	121
62	**Cocaine**	l	$C_{17}H_{21}O_4N$	mon.pr./al.;need. /aq.	98
63	chromate	...	$C_{17}H_{21}O_4N \cdot H_2CrO_4 \cdot H_2O$	org. yel. leaf.	127
64	hydrochloride	l	$C_{17}H_{21}O_4N \cdot HCl$	short pr./al.	195
65	**Coclaurine**	l	$C_{17}H_{19}O_3N$	bitter need.	221
66	**Codamine**	...	$C_{20}H_{25}O_4N$	pr./al.	121
67	**Codeine**	l	$C_{18}H_{21}O_3N \cdot H_2O$	rhomb. pr./aq.	155*
68	hydrochloride	l	$C_{18}H_{21}O_3N \cdot HCl \cdot 2H_2O$	need., pr./aq.	280 d.
69	phosphate	l	$C_{18}H_{21}O_3N \cdot H_3PO_4 \cdot 2H_2O$	need. or pr.	235 d.
70	sulfate	l	$(C_{18}H_{21}O_3N)_2 \cdot H_2SO_4 \cdot 5H_2O$	rhomb. pr.	278 d.
71	**Colchicine**	l	$C_{22}H_{25}O_6N$	yel. varnish; yel. need./EtOAc	143–7*; 155–7
72	chloroform	...	$C_{22}H_{25}O_6N \cdot CHCl_3$	need./chl.	d. 60–70
73	**Columbamine**	...	$C_{20}H_{21}O_5N$	free base unknown	....
74	chloride	...	$C_{20}H_{20}O_4NCl \cdot 2\frac{1}{2}H_2O$	yel. need.	194
75	chloride	...	$C_{20}H_{20}O_4NCl \cdot 4H_2O$	br. pr.	184
76	**Conessine**	d	$C_{24}H_{40}N_2$	leaf. or need./act.	123–5
77	**Conhydrine**	d	$C_8H_{17}ON$	wh. cryst./et.	121; B.P. 226
78	**Coniine**	d	$C_3H_7 \cdot C_5H_{10}N$	colorl. liq.	−2; B.P. 166–7
79	hydrochloride	d	$C_8H_{17}N \cdot HCl$	rhombs/aq.	220
80	picrate	...	$C_8H_{17}N \cdot C_6H_3O_7N_3$	yel. need./h. aq.	75
81	**Corybulbine**	l	$C_{21}H_{25}O_4N$	light-sensitive crysts.	238
82	**Corycavine**	in	$C_{21}H_{21}O_5N$	rhomb. tab./al.	218–9

* Melting point of the anhydrous compound.

Chinicine 193
Chinidine 195
Chinotine 195
Choline sinapate 215

α-Chondodendrine 24
Cicutine 78
Cinchovatine 53
Cinnamoylcocaine 61

Table 7-5 (*Continued*)
PHYSICAL CONSTANTS OF ALKALOIDS

No.	Reaction with H_2SO_4	Solubility Expressed in Parts of Solvent Required to Dissolve 1 Part Alkaloid					
		Water	Alcohol	Ether	Chloroform	Benzene	Others
49	yel.→r.	sl. s.	10	12	s.	....	s. CS_2
50	grn.→yel.	i.	sl. s.	sl. s.	v. s.	s. dil. a.	sl. s. act.
51	crimson with guaiacum	i.	v. s.	v. s.	s.	....	s. am. al.
52	....	v. sl. s.	30	100	s. h.	s. h.	s. dil. a.
53	no fluorescence	5000	20	200	s.	....	s. dil. a.
54	....	20	s.	300	v. s.	....	....
55	....	63 c.; 20 h.	72 c.; 32 h.	v. sl. s.	923	....	....
56	....	3670	48 c.; 20 h.	370	165	....	s. am. al.
57	....	0.4	0.8	....	....	....	....
58	....	22 c.; 3.5 h.	1 c.	275	22	....	....
59	....	60 c.; 30 h.	10 c.; 6 h.	3230	70	....	....
60	....	1300	sl. s.	534	v. sl. s.	....	....
61	....	i.	s.	s.	s.	s.	....
62	colorl.	600	5	2.5	1.1	s.	s. act.
63	....	sl. s.	....	....	....	....	....
64	....	0.4	2.6	i.	19	....	s. act.
65	....	sl. s.	v. s. h.	sl. s.	sl. s.	i.	s. alk.; s. dil. a.
66	....	s. h.	s.	s.	s.	s.	s. dil. a.
67	h.→bl.	120	1.6 c.; 1 h.	12.5	0.66	10.4	68 NH_4OH
68	....	20 c.; 1 h.	145	....	....	....	....
69	....	2.25	261	1310	6700	....	....
70	....	30 c.; 6.3 h.	1200	i.	i.	....	....
71	yel.→r. h.	22	s.	157	s.	88	i. pet.
72	....	d. h.	....	....	....	....	....
73	....	....	....	....	....	....	....
74	....	s.	s.	....	....	....	....
75	....	s.	s.	....	....	....	....
76	....	v. sl. s.	s.	s.	s.	....	....
77	....	sl. s.	s.	s.	s.	....	....
78	colorl.	100	v. s.	v. s.	sl. s.	s.	sl. s. CS_2
79	....	2	s.	....	s.	....	....
80	....	....	s.	s.	....	....	....
81	....	v. sl. s.	s. h.	v. sl. s.	s.	s.	....
82	....	i.	v. sl. s.	....	s.	s. dil. a.	i. alk.

Cinnamoylecgonine methyl ester 61
Coffearin 234
Conchinine 195
Conicine 78

Conquinine 195
Conydrine 77
Cordianine 10
Cornutine 107

Table 7-5 (*Continued*)
PHYSICAL CONSTANTS OF ALKALOIDS

No.	Name	[a]	Formula	Appearance	Melting Point °C.
83	**Corydaline**	*d*	$C_{22}H_{27}O_4N$	colorl. pr./al.	135
84	**Corytuberine**	*d*	$C_{19}H_{21}O_4N$	silky need./et.	240
85	**Cotarnine**	...	$C_{12}H_{15}O_4N$	need./bz.	132–3 d.
86	hydrochloride	...	$C_{12}H_{14}O_3NCl \cdot 2H_2O$	pa. yel. silky need.	197 d.
87	phthalate	...	$(C_{12}H_{14}O_3N)_2 \cdot$ $C_6H_4(CO_2)_2$	yel. cryst. or powd.	103
88	**Cryptopine**	*in*	$C_{21}H_{23}O_5N$	pr./al. or bz.	220–1
89	**Cupreine**	*l*	$C_{19}H_{22}O_2N_2 \cdot 2H_2O$	pr./et.	198*
90	**Cuscohygrine**	*in*	$C_{13}H_{24}ON_2$	oil	B.P. 215⁵⁰
91	hydrate	...	$C_{13}H_{24}ON_2 \cdot 3\frac{1}{2}H_2O$	need.	40–1; 120–30**
92	**Cytisine**	*l*	$C_{11}H_{14}ON_2$	large rhomb. cryst.	152–3
93	**Delphinine**	*d*	$C_{33}H_{45}O_9N$	plates/al.	198–200
94	hydrochloride	...	$C_{33}H_{45}O_9N \cdot HCl$	need./MeOH + et.	208–10
95	**Diacetylmorphine**	...	$C_{21}H_{23}O_5N$	bitter cryst./MeOH	172
96	hydrochloride	*l*	$C_{21}H_{23}O_5N \cdot HCl \cdot H_2O$	cryst. powd.	230 d.
97	**Dilaudid**	...	$C_{17}H_{19}O_3N \cdot HCl$	cryst. powd.	. . . .
98	**Dionin**		$C_{19}H_{23}O_3N \cdot HCl \cdot H_2O$	wh. cryst. powd.	123 d.; 170 d.*
99	**Ecgonine**	*l*	$C_9H_{15}O_3N \cdot H_2O$	mon. pr./al.	198; 205*
100	hydrochloride	*l*	$C_9H_{15}O_3N \cdot HCl$	rhomb. or tricl. tab.	246
101	**Emetine**	*l*	$C_{29}H_{40}O_4N_2$	plates/al. or et.	74
102	hydrochloride	*d*	$C_{29}H_{40}O_4N_2 \cdot 2HCl \cdot 7H_2O$	woolly need./h. aq.; thick pr./c. satd. soln.	235–55; dry, d.
103	**Ephedrine ‡**	*l*	$C_{10}H_{15}ON$	unctuous, colorl. cryst.	40; B.P. 255
104	hydrochloride	*l*	$C_{10}H_{15}ON \cdot HCl$	need.	216 d.
105	sulfate	*l*	$(C_{10}H_{15}ON)_2 \cdot H_2SO_4$	wh. odorless cryst.	245 d.
106	**Ergotinine**	*d*	$C_{35}H_{39}O_5N_5$	long need./al.	239 d.
107	**Ergotoxine**	*l*	$C_{35}H_{41}O_6N_5$	pr./bz.	190–200
108	**Evodiamine**	*d*	$C_{19}H_{17}ON_3$	yel. leaf./al.	278
109	hydrate	*in*	$C_{19}H_{19}O_2N_3$	rhomb. leaf.	146–7
110	**Gelsemine**	*d*	$C_{20}H_{22}O_2N_2$	wh. cryst.	178
111	acetone	...	$C_{20}H_{22}O_2N_2 \cdot (CH_3)_2CO$	pr./act.	−act. at 120
112	hydrochloride	*d*	$C_{20}H_{22}O_2N_2 \cdot HCl$	pr./aq. al. or aq.	300
113	**Glaucine**	*d*	$C_{21}H_{25}O_4N$	yel. rhomb. pr.	119–20
114	**Gnoscopine**	*in*	$C_{22}H_{23}O_7N$	long need./MeOH	232 d.
115	**Guanine**	*in*	$C_5H_5ON_5$	wh. cryst. powd.	>360 d.
116	**Guvacine**	...	$C_6H_9O_2N \cdot H_2O$	short rods/dil. al.	285 d.

* Melting point of the anhydrous compound.
** Becomes anhydrous at 120–30°.
‡ See also listing in table Physical Properties of Organic Compounds.

Corynine 244	Diamorphine 95	
Cuskhygrine 90	Dihydromorphinone hydrochloride 97	
Daturine 22	2,6-Dihydroxy-purine 243	1,3-Dimethyl-xanthine 233
Dehydromorphine 188	Dimethoxy-strychnine 35	3,7-Dimethyl-xanthine 232

Table 7-5 (*Continued*)
PHYSICAL CONSTANTS OF ALKALOIDS

No.	Reaction with H_2SO_4	Solubility Expressed in Parts of Solvent Required to Dissolve 1 Part Alkaloid					
		Water	Alcohol	Ether	Chloroform	Benzene	Others
83	....	i.	s. h.	v. s.	v. s.	s.	i. alk.
84	....	s. h.	s.	i.	i.	i.	....
85	....	s. h.	s.	s.	....	s. dil. a.	s. NH₄OH
86	....	1	4	....	....	....	....
87	....	v. s.	....	....	....	....	....
88	violet→ grn.→yel.	v. sl. s.	100 h.	v. sl. s.	sl. s.	v. sl. s.	s. h. pyr.
89	dil.→no fluores- cence	i.	s.	sl. s.	sl. s.	sl. s.	s. alk.; i. NH₄OH
90	....	∞	s.	s.	....	s.	....
91	....	s.	s.	s. †	....	s. †	
92	....	s.	s.	i.	s.	s.	i. pet.
93	+malic a. →org.→bl.	50,000	20	10	15	....	....
94	....	....	....	....			
95	....	1700	24	70	2.2	s.	s. dil. a.; s. alk.
96	....	2	s.	i.	i.	....	....
97	....	s.	s.	i.	....	....	....
98	....	7	1.4	i.	i.	....	....
99	....	4	50	i.	i.	i.	s. EtOAc
100	....	s.	sl. s.	....	....	....	....
101	....	1000	s.	s.	s.	sl. s.	....
102	....	4	s.	s.			
103	....	20	0.2	s.	s.	....	s. oils
104	....	3	12	i.	....	....	....
105	....	1.2	76	i.	....	....	....
106	with et.→ org.→bl.	sl. s.	200 c.; 52 h.	1020	s.	77 h.	26 act.
107	ditto	i.	s.	sl. s.	s.	s. h.	s. NaOH
108	....	i.	sl. s.	sl. s.	sl. s.	i.	i. dil. a.
109	....	....	....	....	....	....	....
110	colorl.→yel. br.→yel. grn.	sl. s.	s.	s.	s.	s.	s. dil. a.
111	....	....	....	....	....	....	
112	....	s.	sl. s.	....	....	....	....
113	colorl.→bl. in time	s. h.	v. s.	s.	v. s.	sl. s.	sl. s. pet.
114	....	i.	1500	....	s. h.	sl. s.	i. alk.
115	....	i.	v. sl. s.	v. sl. s.	s. a.	v. sl. s. NH₄OH	s. KOH
116	....	s.	....	i.	....	....	....

† With separation of droplets of water.

Table 7-5 (*Continued*)
PHYSICAL CONSTANTS OF ALKALOIDS

No.	Name	[α]	Formula	Appearance	Melting Point °C.
117	**Harmaline**	*in*	$C_{13}H_{14}ON_2$	pr./al. + bz.	250 d.
118	**Harmine**	*in*	$C_{13}H_{12}ON_2$	rhomb. pr./al.	257–9 d.
119	**Homoatropine**	...	$C_{16}H_{21}O_3N$	deliq. pr./et.	99–100
120	hydrobromide	...	$C_{16}H_{21}O_3N \cdot HBr$	rhomb.	217–8 d.
121	**Hordenine**	*in*	$C_{10}H_{15}ON$	orthorhomb. pr.	117.8; subl. 140–50
122	sulfate	...	$(C_{10}H_{15}ON)_2 \cdot H_2SO_4 \cdot 2H_2O$	colorl. cryst.	208–10*
123	**Hydrastine**	*l*	$C_{21}H_{21}O_6N$	colorl. rhomb. pr./al.	132
124	hydrochloride	*d*	$C_{21}H_{21}O_6N \cdot HCl$	hyg. powd.	116
125	**Hydrastinine**	*in*	$C_{11}H_{13}O_3N$	need./lig.	116–7
126	hydrochloride	*in*	$C_{11}H_{12}O_2NCl$	yel. need.	212 d.
127	**Hydrocotarnine**	*in*	$C_{12}H_{15}O_3N \cdot \frac{1}{2}H_2O$	mon. pr./al.	55–6
128	**Hydroquinine**	*l*	$C_{20}H_{26}O_2N_2 \cdot 2H_2O$	need./chl. or et.	172*
129	**Hyoscine**	*l*	$C_{17}H_{21}O_4N$	syrup or cryst./et.	59
130	hydrobromide	*l*	$C_{17}H_{21}O_4N \cdot HBr \cdot 3H_2O$	rhomb. tab. or need./aq.	194–7*
131	**Hyoscyamine**	*l*	$C_{17}H_{23}O_3N$	silky need./aq. al.	106–8
132	hydrobromide	*l*	$C_{17}H_{23}O_3N \cdot HBr$	deliq. pr.	152
133	hydrochloride	...	$C_{17}H_{23}O_3N \cdot HCl$	wh. cryst.	149–51
134	sulfate	*l*	$(C_{17}H_{23}O_3N)_2 \cdot H_2SO_4 \cdot 2H_2O$	need./al.	206*
135	**Hypaphorine**	*d*	$C_{14}H_{18}O_2N_2 \cdot 2H_2O$	large mon. cryst./aq.	255*
136	**Hypoxanthine**	*in*	$C_5H_4ON_4$	minute need.	d. 150
137	**Japaconitine**	*d*	$C_{34}H_{47}O_{11}N$	need./al., et. or chl.	202–9
138	**Jervine**	*l*	$C_{26}H_{37}O_3N \cdot 2H_2O$	long grouped pr.	238–42
139	**Laudanine**	*in*	$C_{20}H_{25}O_4N$	pr./aq. al.	166–7
140	**Laudanosine**	*d*	$C_{21}H_{27}O_4N$	need./bz.	90
141	**Lobeline**	*l*	$C_{22}H_{27}O_2N$	broad colorl. need.	130–1
142	hydrochloride	*l*	$C_{22}H_{27}O_2N \cdot HCl$	wh. granular powd.	180
143	**Lupanine**	*d*	$C_{15}H_{24}ON_2$	need./pet.; very alkaline	40
144	hydrochloride	*d*	$C_{15}H_{24}ON_2 \cdot HCl \cdot 2H_2O$	....	127–8; 250–2*
145	**Lupinine**	*l*	$C_{10}H_{19}ON$	rhomb. cryst./pet.	69–71
146	hydrochloride	*l*	$C_{10}H_{19}ON \cdot HCl$	rhomb. pr./aq. al.	212–3
147	**Lycorine**	*l*	$C_{16}H_{17}O_4N$	colorl. pr./al.	280 d.
148	**Mezcaline**	*in*	$C_{11}H_{17}O_3N$	colorl. alk. oil or cryst.	B.P. 180^{12} 35–6
149	**Morphine**	*l*	$C_{17}H_{19}O_3N \cdot H_2O$	trim. pr./al.	254 d.*
150	acetate	*l*	$C_{17}H_{19}O_3N \cdot CH_3CO_2H \cdot 3H_2O$	cryst. powd./al.	200 d.

* Melting point of the anhydrous compound.

Herapathite 207	Hydro-berberine 45	*dl*-Hyoscyamine 22
Heroin 95	Hydro-cinchonine 60	Iodoquinine sulfate 207
Hydra-ergotocin 244	Hydroxy-cinchonine 89	Laburnine 92
	Hydroxy-coniine 77	
	6-Hydroxy-purine 136	
	Hydroxy-stachydrine 238	

Table 7-5 (*Continued*)
PHYSICAL CONSTANTS OF ALKALOIDS

No.	Reaction with H₂SO₄	Solubility Expressed in Parts of Solvent Required to Dissolve 1 Part Alkaloid					
		Water	Alcohol	Ether	Chloroform	Benzene	Others
117	....	v. sl. s.	s. h.	sl. s.	....	....	s. dil. a.
118	yel. with grn. fluor- escence	v. sl. s.	sl. s.	sl. s.	s.	....	s. dil. a.
119	....	sl. s.	s.	s.	s.	s.	s. dil. a.
120	....	6	32	i.	625	....	....
121	....	s.	v. s.	v. s.	s.	sl. s.	s. dil. a.; s. alk.
122	....	s.	sl. s.	i.	....	....	....
123	olive grn. with (NH₄)₂MoO₄	i.	170	175	1.4	15	....
124	....	s.	s.	v. sl. s.	sl. s.	....	....
125	....	s. h.	s.	s.	s.	d.	s. a.
126	....	v. s.	v. s.	300	286	....	....
127	yel.→r.h.	i.	v. s.	v. s.	v. s.	v. s.	i. alk.
128	dil.→fluor- escence	v. sl. s.	s.	s.	s.	s. act.; i. pet.	s. NH₄OH
129	h.→bl.	sl. s.	s.	s.	s.	sl. s.	sl. s. pet.
130	....	1.5	16	i.	750	....	....
131	colorl.	281	s.	49	1.5	132	s. dil. a.
132	....	v. s.	2.5	1610	1.7	....	....
133	....	s.	s.	....	....	....	....
134	....	0.5	4.5	v. sl. s.	v. sl. s.	....	....
135	....	v. s.	v. s.	....	....	....	i.
136	....	1370 c.; 69.5¹⁰⁰●	900 h.	i.	....	s. a.	s. alk.
137	....	i.	s.	s.	s.	....	i. pet.
138	yel.→grn. h.	i.	s.	sl. s.	s.	sl. s.	s. act.
139	red	v. sl. s.	s.	600	v. s.	v. s.	s. alk. carb.
140	rose red → red violet 150°	i.	s.	19	s.	s. h.	i. alk.
141	red-br.	v. sl. s.; d. h.	s. h.	s.	s.	s.	v. sl. s. pet.
142	....	40	10	....	v. s.	....	....
143	....	s.	s.	s.	s.	....	....
144	....	....	....	....	....	....	....
145	....	s.	s.	s.	s.	s.	sl. s. pet.
146	....	s.	....	....	....	....	....
147	+MoO₃→ grn.→bl.	i.	sl. s.	sl. s.	sl. s.	sl. s. EtOAc	s. a.
148	yel.→violet	s.	s.	i.	s.	s.	i. pet.
149	pink→grn. h.→br.	3533 c.; 1075 h.	170 c.; 80 h.	4450	1525	9000	475 EtOAc
150	....	2.25	17.3	i.	710	....	6.5 gly.

Lupinidine 220
Lycine 34
Macleyine 183
Mescaline 148

Methoxy-cinchonine 199
Methyl-benzoyl- ecgonine 62
Methyl-granatonine 190

Methyl-hydrocupreine 128
Methyl-6-methoxy-7,8- methylenedioxy-tetra- hydro-isoquinoline 12

O-Methyl-morphine 67
N-Methyl-proline- methylbetaine 222
Methyl-theobromine 39

Table 7-5 (*Continued*)
PHYSICAL CONSTANTS OF ALKALOIDS

No.	Name	[α]	Formula	Appearance	Melting Point °C.
151	hydrochloride	*l*	$C_{17}H_{19}ON \cdot HCl \cdot 3H_2O$	silky need./aq.	200 d.
152	sulfate	*l*	$(C_{17}H_{19}O_3N)_2 \cdot H_2SO_4 \cdot 5H_2O$	silky or cubic cryst./aq.	250 d.*
153	**Muscarine**	*d*	$C_8H_{19}O_3N$	deliq. cryst.	....
154	**Nandinine**	*d*	$C_{19}H_{19}O_4N$	leaf.	145–6
155	**Narceine**	*in*	$C_{23}H_{27}O_8N \cdot 3H_2O$	need. or pr./aq.	145*
156	bisulfate	...	$C_{23}H_{27}O_8N \cdot H_2SO_4 \cdot 10H_2O$	cryst. powd.	d.→yel.
157	hydrochloride	...	$C_{23}H_{27}O_8N \cdot HCl \cdot 3H_2O$	cryst./HCl	192*
158	**Narcotine**	*l*	$C_{22}H_{23}O_7N$	long need./h. al.	176
159	hydrochloride	...	$C_{22}H_{23}O_7N \cdot HCl \cdot H_2O$	lustrous cryst.	197–8
160	**Nicotine**	*l*	$C_{10}H_{14}N_2$	colorl. oil	B.P. 246^{730}
161	hydrochloride	*d*	$C_{10}H_{14}N_2 \cdot 2HCl$	deliq. cryst.	....
162	picrate	...	$C_{10}H_{14}N_2 \cdot 2C_6H_3O_7N_3$	yel. need. or pr./al.	218
163	salicylate	*d*	$C_{10}H_{14}N_2 \cdot C_7H_6O_3$	wh. plates	117–8
164	tartrate	*d*	$C_{10}H_{14}N_2 \cdot 2C_4H_6O_6 \cdot 2H_2O$	reddish-wh. cryst.	88–90
165	**Oxyacanthine**	*d*	$C_{38}H_{38}O_6N_2$	need./al. or et.	216–7
166	nitrate	...	$C_{38}H_{38}O_6N_2 \cdot 2HNO_3 \cdot 4H_2O$	need.	195–200
167	**Papaveraldine**	...	$C_{20}H_{19}O_5N$	cryst./bz. or pet.	210
168	**Papaverine**	*in*	$C_{20}H_{21}O_4N$	rhomb. pr. or need./al.-et.	147–8
169	hydrochloride	...	$C_{20}H_{21}O_4N \cdot HCl$	mon. pl./aq.	220–1 d.
170	**Paraconiine**	*in*	$C_8H_{15}N$	yel. liq.	B.P. 168–70
171	**Pelletierine**	*in*	$C_8H_{15}ON$	colorl. oil	B.P. 106^{21}; 195^{760}
172	**Pellotine**	...	$C_{13}H_{19}O_3N$	plates/al.	110–2
173	hydrochloride	...	$C_{13}H_{19}O_3N \cdot HCl$	wh. cryst.	....
174	**Physostigmine**	*l*	$C_{15}H_{21}O_2N_3$	hyg. cryst. (2 forms)	86–7; 105–6
175	**Pilocarpidine**	*d*	$C_{10}H_{14}O_2N_2$	viscid oil	....
176	nitrate	*d*	$C_{10}H_{14}O_2N_2 \cdot HNO_3$	pr./aq.	137
177	**Pilocarpine**	*d*	$C_{11}H_{16}O_2N_2$	colorl. oil or need.	34
178	hydrochloride	*d*	$C_{11}H_{16}O_2N_2 \cdot HCl$	pr. or need.	204–5
179	nitrate	*d*	$C_{11}H_{16}O_2N_2 \cdot HNO_3$	pr./al. or aq.	176–8
180	**Piperine**	*in*	$C_{17}H_{19}O_3N$	mon. need./al.	129–30
181	periodide	...	$(C_{17}H_{19}O_3N)_2 \cdot HI \cdot I_2$	steel bl. need.	145
182	**Porphyroxine**	*l*	$C_{19}H_{23}O_4N$	pr./lig.	134–5
183	**Protopine**	*in*	$C_{20}H_{19}O_5N$	mon. cryst./al.	207–8
184	**Protoveratrine**	...	$C_{32}H_{51}O_{11}N$	rectangular tab.	245–50 d.
185	**Pseudoaconitine**	*d*	$C_{36}H_{51}O_{12}N$	rhombs/chl. + et.	212–4
186	**Pseudoephedrine**	*d*	$C_{10}H_{15}ON$	rhomb. tab./et.	118–9
187	hydrochloride	...	$C_{10}H_{15}ON \cdot HCl$	need.	176
188	**Pseudomorphine**	*l*	$C_{34}H_{36}O_6N_2 \cdot 3H_2O$	crusts or need.	d. 327

* Melting point of the anhydrous compound.
Napelline 3
Narcissine 147
Narcosine 158
dl-Narcotine 114
Nepaline 185

Neriine 76
Nor-arecaidine 116
Opianin 158
Opin 182
Oxy-dimorphine 188

Table 7-5 (*Continued*)
PHYSICAL CONSTANTS OF ALKALOIDS

No.	Reaction with H_2SO_4	Solubility Expressed in Parts of Solvent Required to Dissolve 1 Part Alkaloid					
		Water	Alcohol	Ether	Chloroform	Benzene	Others
151	....	17.2 c.; 0.5 h.	42	i.	i.	....	19 gly.
152	....	15.3	452	i.	i.	....	....
153	....	v. s.	v. s.	sl. s.	sl. s.	....	d.a.; stable alk.
154	....	sl. s.	s.	s.	s.	s.	s. dil. a.
155	br.→r. h.	769 c.; 220 h.	945 c.; s. h.	i.	v. sl. s.	i.; s. alk.	s. NH$_4$OH; s. dil. a.
156	....	s.→basic salt	s. h.	s.	s.	....	....
157	....	s. h.	s. h.	....	....	....	s. MeOH
158	yel. grn.→ r. h.	3300	100	166	3	22	s. h. alk.
159	....	diss.	s.	....	s.	....	....
160	colorl.	s.	∞	∞	∞	....	s. pet.
161	....	s.	s.	....	....	....	....
162	....	....	....	....	....	....	low solubiiity
163	....	s.	s.	....	....	....	....
164	....	v. s.	v. s.	....	....	....	....
165	colorl.	i.	s.	s.	s.	s.	s. dil. a.
166	....	sl. s.	....	....	....	....	....
167	....	i.	sl. s.	sl. s.	s.	s.	s. a.; sl. s. pet.
168	colorl.→ rose r. h.	i.	45 c.; 4 h.	250	s. h.	s. h.	13 pyr.
169	....	37	s.	i.	s.	....	....
170	....	v. sl. s.	∞	∞	....	....	....
171	grn. with $K_2Cr_2O_7$	20	s.	s.	s.	s.	....
172	yel.→r. with HNO$_3$	v. sl. s.	s.	s.	s.	....	sl. s. pet.
173	....	s.	....	....	....	....	....
174	colorl.→yel.	sl. s.	s.	s.	s.	s.	....
175	....	s.	s.	....	s.	....	....
176	....	2	82	....	....	....	....
177	colorl.	v. s.	v. s.	sl. s.	v. s.	v. sl. s.	i. pet.; s. alk.
178	....	0.3	3	i.	545	....	....
179	....	4	60	i.	i.	....	....
180	dk. r.→ br. blk.	sl. s.	12	26	2.6	s.	i. pet.
181	....	....	s.	....	v. s.	....	s. dil. a.
182	red	sl. s.	s.	s.	s.	s.	s. dil. a.
183	yel.→ violet→grn.	i.; sl.s. act.	900	1000	15	v. sl. s.	sl. s. NH$_4$OH
184	grn.→bl.→ violet	i.	s. h.	sl. s.	s.	i.	i. pet.
185	....	v. sl. s.	s.	sl. s.	s.	....	*l* salts
186	....	sl. s.	s.	s.	s.	....	....
187	....	s.	s.	....	....	....	....
188	+sucrose→ dk. grn.→ br.	i.	i.	i.	i.	s. alk.	s. h. NH$_4$OH; s. pyr.

Oxy-neurine HCl 34
6-Oxy-purine 136
Paramorphine 228
Peganine 239
1-Phenyl-2-methylamino-
    propanol 103

Pilosine 47
Pitayine 195
Proline-betaine 222
2-Propyl-piperidine 78
Pseudocinchonine 60

Table 7-5 (*Continued*)
PHYSICAL CONSTANTS OF ALKALOIDS

No.	Name	[α]	Formula	Appearance	Melting Point °C.
189	hydrochloride	*l*	$C_{34}H_{36}O_6N_2 \cdot 2HCl \cdot 2H_2O$	cryst. powd.	....
190	**Pseudopelletierine**	*in*	$C_9H_{15}ON$	anh. prism. tab.	48; B.P. 246
191	**Pseudotropine**	*in*	$C_8H_{15}ON$	tab. or pr./et.	108
192	**Pukateine**	*l*	$C_{18}H_{17}O_3N$	cryst./et.	200
193	**Quinicine**	*d*	$C_{20}H_{24}O_2N_2$	yel. oil; hardens on standing	(60)
194	oxalate	*d*	$(C_{20}H_{24}O_2N_2)_2 \cdot H_2C_2O_4 \cdot 9H_2O$	pr./chl. or need./al.	149
195	**Quinidine**	*d*	$C_{20}H_{24}O_2N_2$	pr.+al./al.; tab.+ et./et.	174–5*
196	bisulfate	...	$C_{20}H_{24}O_2N_2 \cdot H_2SO_4 \cdot 4H_2O$	hair-like need. or pr.	....
197	hydrochloride	*d*	$C_{20}H_{24}O_2N_2 \cdot HCl \cdot H_2O$	asbestos-like pr.	258–9 d.*
198	sulfate	*d*	$(C_{20}H_{24}O_2N_2)_2 \cdot H_2SO_4 \cdot 2H_2O$	pr. or need./h. aq.	....
199	**Quinine**	*l*	$C_{20}H_{24}O_2N_2$	wh. need. or powd.	175
200	arsenate	...	$3(C_{20}H_{24}O_2N_2) \cdot 2H_3AsO_4 \cdot 5H_2O$	wh. effl. cryst.	....
201	bisulfate	*l*	$C_{20}H_{24}O_2N_2 \cdot H_2SO_4 \cdot 7H_2O$	pr./aq. or al.	160 d.*
202	formate	*l*	$C_{20}H_{24}O_2N_2 \cdot HCO_2H$	cryst. powd.; need.	109 d.
203	hydrate	*l*	$C_{20}H_{24}O_2N_2 \cdot 3H_2O$	efflorescent; anh. at 100°	57
204	hydrobromide	...	$C_{20}H_{24}O_2N_2 \cdot HBr \cdot H_2O$	hyg., silky need.	152–200
205	hydrochloride	*l*	$C_{20}H_{24}O_2N_2 \cdot HCl \cdot 2H_2O$	effl., silky need.	158–60*
206	hydrochloride, di-	*l*	$C_{20}H_{24}O_2N_2 \cdot 2HCl$	wh. powd. or need.	180–5
207	iodosulfate	...	$4C_{20}H_{24}O_2N_2 \cdot 3H_2SO_4 \cdot 2HI \cdot I_4 \cdot 6H_2O$	pl./al.; r. or grn. by reflected or transmitted light	$-H_2O$, 100
208	salicylate	...	$C_{20}H_{24}O_2N_2 \cdot C_7H_6O_3 \cdot H_2O$	need./aq.	195
209	sulfate	*l*	$(C_{20}H_{24}O_2N_2)_2 \cdot H_2SO_4 \cdot 7H_2O$	efflorescent need.	235*
210	sulfate	*l*	$(C_{20}H_{24}O_2N_2)_2 \cdot H_2SO_4 \cdot 2H_2O$	by drying in air	205
211	valerate	...	$C_{20}H_{24}O_2N_2 \cdot C_5H_{10}O_2 \cdot H_2O$	cryst. powd.	95
212	**Rhoeadine**	...	$C_{21}H_{21}O_6N$	small pr. or need.	245–7 d.
213	**Ricinine**	...	$C_8H_8O_2N_2$	pr. or tab./al. or aq.	201, subl.
214	**Rutaecarpine**	...	$C_{18}H_{13}ON_3$	yel. pl.; need./EtOAc	260–2
215	**Sinapine**	...	$C_{16}H_{25}O_6N$	free base unknown	....
216	bisulfate	...	$C_{16}H_{24}O_5NHSO_4 \cdot 3H_2O$	leaf./al.	127*
217	thiocyanate	...	$C_{16}H_{24}O_5NSCN \cdot H_2O$	pale yel. need./aq.	178
218	**Solanidine**	*l*	$C_{26}H_{41}ON$	need./et. or al.	219
219	**Solanine**	*l*	$C_{44}H_{71}O_{15}N$	slender need./al.	244–54 d.

* Melting point of the anhydrous compound.
Pseudo-punicine 190
Punicine 171
Pyridyl-*N*-methyl-pyrrolidine 160
2-(3-Pyridyl)-piperidine 11

Quebrachine 244
α-Quinidine 53
β-Quinine 195

Table 7-5 (*Continued*)
PHYSICAL CONSTANTS OF ALKALOIDS

No.	Reaction with H_2SO_4	Solubility Expressed in Parts of Solvent Required to Dissolve 1 Part Alkaloid					
		Water	Alcohol	Ether	Chloroform	Benzene	Others
189	....	70	....	....	....	....	....
190	+CrO₃→ grn.	s.	s.	s.	s.	....	sl. s. pet.
191	....	v. s.	v. s.	sl. s.	s.	....	very alkaline
192	org.→r. and violet h.	i.	s.	161	s.	s. pyr.	s. alk
193	no fluorescence	v. sl. s.	s.	s.	s.	....	....
194	....	s. h.	s.	....	s.	....	....
195	dil.→bl. fluorescence	2000 c.; 800 h.	26	22	2.3	s.	sl. s. lig.
196	....	8; fluorescence	....	....	....	....	....
197	....	60 c.; v. s. h.	s.	sl. s.	s.	....	....
198	....	100 c.; 15 h.	8	v. sl. s.	15	i.	....
199	colorl.→lt. yel.→br. h.	1750	0.6	22.6	1.9	166 c.; 30 h.	s. CS₂; s. NH₄OH
200	....	650 c.; 120 h.	200 c.; 50 h.	....	....	....	s. dil. a.
201	....	fluorescence; 9	19	1770	920	....	18 gly.
202	....	19	s.	v. sl. s.	s.	....	....
203	....	1560 c.; 800 h.	0.6	1.4	1.6	70	212 gly.
204	dil.→fluorescence	40 c.; 3 h.	1	23	1	....	9 gly.
205	ditto	16 c.; 0.5 h.	0.6	340	1	....	9 gly.
206	....	0.6	5	v. sl. s.	7	....	....
207	....	1000 h.	800 c.; 50 h.	....	....	....	60 h. ac. a.
208	....	1500	14	114	38	....	18 gly.
209	dil.→fluorescence	725 c.; 30 h.	60	sl. s.	1000	....	24 gly.
210	....	810 c.; 30 h.	96	sl. s.	sl. s.	....	....
211	....	70 c.; 40 h.	2	10	....	....	....
212	purple r.	1200	700	800	v. sl. s.	i.	s. a., d.
213	....	s. h.; i. pet.	sl. s. c.; s. h.	sl. s.	s.	sl. s.	forms no salts
214	bright yel.	i.	sl. s.	s.	s.	s.	....
215	....	....	....	....	....	....	....
216	....	s.	s. h.	i.	....	....	....
217	....	sl. s.	sl. s.	....	....	....	....
218	....	v. sl. s. h.	s. h.	sl. s.	s.	....	....
219	yel.→rose →r.	i.	s. h.	i.	i.	i.	....

Quinoform 202
Quinotoxine 193
Rheadin 212
Ricidine 213

Sarcine 136
Scopolamine 129
Secaline 106
Sophorine 92

Table 7-5 (*Continued*)
PHYSICAL CONSTANTS OF ALKALOIDS

No.	Name	[a]	Formula	Appearance	Melting Point °C.
220	**Sparteine**	*l*	$C_{15}H_{26}N_2$	colorl. oil	B.P. 325[754] in H_2
221	bisulfate	...	$C_{15}H_{26}N_2 \cdot H_2SO_4 \cdot 5H_2O$	transparent cryst.	150–2*
222	**Stachydrine**	...	$C_7H_{13}O_2N \cdot H_2O$	deliq. cryst.	235 d*
223	oxalate	...	$C_7H_{13}O_2N \cdot H_2C_2O_4$	need.	105–7
224	**Strychnine**	*l*	$C_{21}H_{22}O_2N_2$	colorl. rhombs/al.	286–8; B.P. 270[5]
225	hydrochloride	...	$C_{21}H_{22}O_2N_2 \cdot HCl \cdot 2H_2O$	efflorescent pr.	....
226	nitrate	*l*	$C_{21}H_{22}O_2N_2 \cdot HNO_3$	shining need.	....
227	sulfate	...	$(C_{21}H_{22}O_2N_2)_2 \cdot H_2SO_4 \cdot 5H_2O$	effl., mon. pr.	200 d.*
228	**Thebaine**	*l*	$C_{19}H_{21}O_3N$	leaf. or pr./al.	193
229	hydrochloride	*l*	$C_{19}H_{21}O_3N \cdot HCl \cdot H_2O$	large rhombs or yel. powd.	....
230	*iso*-**Thebaine**	*d*	$C_{19}H_{21}O_3N$	rhomb./al. or et.	203–4
231	sulfate	...	$(C_{19}H_{21}O_3N)_2 \cdot H_2SO_4$	....	120–1 d.
232	**Theobromine**	*in*	$C_7H_8O_2N_4$	minute rhomb. cryst.; mon./h. aq.	330†; subl. 290
233	**Theophylline**	*in*	$C_7H_8O_2N_4 \cdot H_2O$	mon. tab. or need./h. aq.	269–72
234	**Trigonelline**	...	$C_7H_7O_2N \cdot H_2O$	hyg. pr./al.	218 d.*
235	**Tropacocaine**	*in*	$C_{15}H_{19}O_2N$	need. or plates	49
236	hydrochloride	...	$C_{15}H_{19}O_2N \cdot HCl$	need.; pl./aq. al.	283 d.
237	**Tropine**	*in*	$C_8H_{15}ON$	hyg. tab./abs. et.	63
238	**Turicine**	*d*	$C_7H_{13}O_3N \cdot H_2O$	sweet, effl. pr. or need./aq. al.	260 d.*
239	**Vasicine**	*l*	$C_{11}H_{12}ON_2$	need./al.	211–2
240	**Veratridine**	*in*	$C_{36}H_{51}O_{11}N$	amor.; yel.	180
241	**Vicine**	*l*	$C_{10}H_{16}O_7N_4 \cdot 2H_2O$	need.	239–42
242	**Vomicine**	*d*	$C_{22}H_{24}O_4N_2$	need./aq. al.	278–80
243	**Xanthine**	*in*	$C_5H_4O_2N_4 \cdot H_2O$	small pl.; anh. at 125°	>150 d.
244	**Yohimbine**	*d*	$C_{21}H_{26}O_3N_2$	need./aq. al.	247–8
245	hydrochloride	*d*	$C_{21}H_{26}O_3N_2 \cdot HCl$	plates	302
246	nitrate	...	....	colorl. pr.	276
247	thiocyanate	...	....	rectangular pr./h. aq.	233–4
248	**Zygadenine**	*l*	$C_{39}H_{63}O_{10}N$	need./bz.	200–1

* Melting point of the anhydrous compound.
† Sealed tube.

Stypticin 86	Tetrahydro-nicotinic acid 116	Trimethyl-glycocoll 33
Styptol 87	Theine 39	1,3,7-Trimethyl-xanthine 39
Telepathine 118	Theocine 233	Tropine mandelate 119
Tetraanhydro-berberine 45	Trigenolline 234	Ulexine 92

Table 7-5 (*Continued*)
PHYSICAL CONSTANTS OF ALKALOIDS

No.	Reaction with H_2SO_4	Solubility Expressed in Parts of Solvent Required to Dissolve 1 Part Alkaloid					
		Water	Alcohol	Ether	Chloroform	Benzene	Others
220	colorl.	328	s.	s.	s.	i.	....
221	colorl.	1.1	2.4	i.	i.	....	....
222	....	s.	s.	i.	i.	s. dil. a.	d. in air
223	....	....	i. c.	....	....	....	....
224	colorl.	6400 c.; 3100 h.	110 c.; 28 h.	v. sl. s.	6	150	173 PhMe
225	....	35	60	i.	....	....	i. HCl
226	....	42 c.; 10 h.	120	i.	156	....	60 gly.
227	....	31 c.; 7 h.	65	i.	325	....	7 gly.
228	blood r.	i.	10	135	19	18	i. alk.
229	....	12	s.	....	....	....	....
230	....	....	s.	sl. s.	s.	....	....
231	....	....	....	....	....	....	....
232	....	2000 c.; 150 h.	1775 c.; 260 h.	3125 h.	157 c.; 100 h.	100,000	s. alk.; 4700 h. CCl_4
233	....	120 c.; s. h.	64	sl. s.	164	s. alk.	s. NH_4OH
234	....	v. s.	s.	v. sl. s.	v. sl. s.	i.	neutral reaction
235	....	i.	s.	s.	v. s.	v. s.	s. dil. NH_4OH
236	....	s.	sl. s.	....	....	....	....
237	....	v. s.	v. s.	s.	s.	s.	....
238	....	v. s.	sl. s.	....	....	....	....
239	colorl.	sl. s.	s.	sl. s.	s.	sl. s.	i. pet.; s. dil. a.
240	yel.	s.	....	sl. s.	....	....	....
241	yel.	sl. s.	i. abs.	....	....	s. dil. a.	s. MeOH
242	+$CrO_3 \rightarrow$ deep r.	....	s. h.	sl. s.	s.	....	s. act.
243	....	14,400 c.; 1500 h.	2400	3000	....	....	s. alk.; s. a.
244	colorl. + $K_2Cr_2O_7 \rightarrow$ dirty grn.	v. sl. s.	s.	sl. s.	s.	s. h.	....
245	....	120	400	....	....	....	....
246	....	....	....	....	....	....	....
247	....	....	....	....	....	....	....
248	or. $\rightarrow$ cherry r.	....	s.	....	s.	s.	....

Ultraquinine 89
5-Ureido-hydantoin 10
Ureous acid 243
Veratrine, a mixture of 49,240 *et al.*
Veratrine (crystallized) 49

Veratroylaconine 185
Vinetine 165
Viridine 138
Wrightine 76

Xanthaline 167
Xanthopuccine 45
Yageine 118
Yajeine 118

Table 7-6
PHYSICAL CONSTANTS OF GLUCOSIDES
Compiled by F. E. SHEIBLEY, Ph.D.

Names of the compounds in the table below are arranged alphabetically.

Appearance. In addition to the crystalline form and color, the solvent used in purification is often given; thus, "rhomb./al." indicates that rhombic crystals were obtained when the compound was crystallized from alcohol.

Optical Properties. These are indicated by the symbols d, l, or in. following the name. Most of the optically active glucosides are levo-rotatory.

Solubilities. Most glucosides are soluble in cold or hot water. The solubilities given for alcohol refer to the ordinary alcohol of approximately 95% concentration. Solubilities expressed by numbers are given in parts by weight of solvent required to dissolve one part of the glucoside. Because of discrepancies existing between many of these figures as found in the literature no claim to accuracy can be made,

---

Abbreviations used in the table

a., acid
abs., absolute
ac., a., acetic acid
act., acetone
al., alcohol
alk., alkali (i. e. aqueous NaOH or KOH)
amor., amorphous
anh., anhydrous
aq., aqueous; water
br., brown
bz., benzene, C_6H_6
c., cold
Chl., chloroform, $CHCl_3$

colorl., colorless
conc., concentrated
cryst., crystals or crystalline
crystn., crystallization
d., decomposes or decomposition
d, dextro-rotatory
dil., dilute
et., ether $(C_2H_5)_2O$
EtOAc, ethyl acetate
h., hot
hyd., hydrate
hyg., hygroscopic

No.	Name	Formula	Appearance	Melting Point °C.
1	Absinthin	$C_{30}H_{40}O_8$	glossy need. or yel. amor. powd.	68
2	Aesculin (l)	$C_{15}H_{16}O_9 \cdot 2H_2O$	pr./aq. or dil. al.	205 d.
3	Amygdalin (l)	$C_{20}H_{27}O_{11}N \cdot 3H_2O$	orthorhomb. pr./aq.; glossy scales ($+2H_2O$)/ 80% al.	214–6*
4	Antiarin	$C_{27}H_{42}O_{10} \cdot 4H_2O$	plates/aq.	220–5
5	Apiin (l)	$C_{26}H_{28}O_{14} \cdot H_2O$	glossy need. or yel. cryst. powd.	228
6	Arbutin (l)	$C_{12}H_{16}O_7 \cdot H_2O$	long silky need./aq.	195–200*
7	Baptisin (l)	$C_{28}H_{30}O_{14} \cdot 3H_2O$	thin wh. need./al. slowly becomes anh.	sint. 150; m. 249–51
8	Bryonin (d)	$C_{34}H_{50}O_9$	amor. bright yel. powd.	softens 208
9	Carminic acid	$C_{22}H_{20}O_{13}$	purplish-br. mass or bright r. powd.; r. pr.	d. 136

* Melting point of the anhydrous compound.

Abietin 15	Asebotin 55
Acocantherin 51	Aurantiin 49
Arthanitin 22	

Table 7-6 (*Continued*)
PHYSICAL CONSTANTS OF GLUCOSIDES

and the numerical values stated are perhaps best considered as upper limits.

Hydrolytic Products. By hydrolysis glucosides are split into a sugar (generally glucose) or a mixture of sugars, and a principle characteristic of the glucoside. The hydrolysis is usually effected with hot dilute sulfuric acid or hydrochloric acid, and less often with emulsin or baryta water.

References. The information given in the table has been collected mainly from the following sources:—

J. J. L. van Rijn: **Die Glykoside.** Published by Borntraeger, Berlin (1900).

Merck's Index, 7th edition. Published by Merck and Co., Inc., Rahway, N. J.

Beilstein: **Handbuch der Organischen Chemie,** 3d edition.

i., insoluble	pet., petroleum ether
in, inactive	powd., powder
K_2CO_3, aq. soln. of	pr., prisms
KOH, aq. soln. of	pyr., pyridine
l, levo-rotatory	r., red
leaf., leaflets or leaves	rhomb., rhombic
m-, meta	s., soluble
m., melts	sint., sinters
m.p., M.P., melting point	sl., slight or slightly
mon., monoclinic	v., very
need., needles	wh., white
org., orange	yel., yellow
orthorhomb., orthorhombic	>, greater than

No.	Solubility Expressed in Parts of Solvent Required to Dissolve 1 Part Glucoside					Hydrolytic Products		
	Water	Alcohol	Ether	Chloro-form	Others	Principle	M.P. °C.	Sugar
1	sl. s.	s.	s.	s.	s. bz.; s. NaOH	resinous substance	......	glucose
2	576 c.; s. h.†	24 h.	sl. s.	s. h.	v. s. dil. alk.	aesculetin	270 d.	glucose
3	12 c.; v. s. h.	720 c.; 9 h.	i.	......	....	hydrocyanic acid benzaldehyde	−12 −26	glucose
4	s.	s.	sl. s.	......	....	antiarigenin	180	antiarose
5	sl. s. c.; s. h.	s. h.	i.	......	....	apigenin	subl. 292–5	glucose + apiose
6	8 c.; 1 h.	13	i.	i.	i. CS_2	hydroquinone	170.3	glucose
7	sl. s. c.; s. h.	sl. s. dil.	v. sl. s.	v. sl. s.	s. glacia. ac. a.	baptigenin	296–8	glucose + rhamnose
8	s.	s	i.	i.	....	bryogenin (resinous)	......	glucose
9	s.	s.	v. sl. s.	i.	s. alk.; o. oono. H_2SO_4	carmine red	......	sugar

† Aqueous solutions fluoresce faint blue.

Avenein 35

Avornin 30

6-Benzoyl-*d*-glucose 79

6-Benzoyl-salicin 59

Table 7-6 (*Continued*)
PHYSICAL CONSTANTS OF GLUCOSIDES

No.	Name	Formula	Appearance	Melting Point °C.
10	Cerberin (*l*)	$C_{27}H_{40}O_8$	glossy cryst./et.	191–2
11	α–Chinovin (*d*)	$C_{30}H_{48}O_8$	rosettes of small need./al.	....
12	β–Chinovin (*d*)	$C_{30}H_{48}O_8$	scales/dil. al.	235 d.
13	Clavicepsin (*d*)	$C_{18}H_{34}O_{16} \cdot H_2O$	wh. cryst.	91; 198*
14	Colocynthin	$C_{56}H_{84}O_{23}$	microscopic pr.	....
15	Coniferin (*l*)	$C_{16}H_{22}O_8 \cdot 2H_2O$	wh. pointed satiny need.	185
16	Convallamarin (*l*)	$C_{23}H_{44}O_{12}$, mixture	wh., cryst. powd.	....
17	Convallarin (*l*)	$C_{34}H_{62}O_{11}$, mixture	rectangular pr.	....
18	Convolvulin (*l*)	$C_{54}H_{96}O_{27}$	wh. amor. powd.	155–8
19	Coriamyrtin (*d*)	$C_{15}H_{18}O_5$	mon. pr.	228–30
20	Crocin	$C_{44}H_{64}O_{26} \cdot H_2O$	br. red cryst.	186 d.
21	Curangin (*d*)	$C_{48}H_{77}O_{20}$	amor.	172
22	Cyclamin (*l*)	$C_{27}H_{38}O_{13}$	wh. amor. powd.; microscopic cryst.	236
23	Daphnin (*l*)	$C_{15}H_{16}O_9 \cdot 2H_2O$	pr. or need./aq.	215 d.
24	Datiscin (*l*)	$C_{27}H_{30}O_{15} \cdot 4H_2O$	glossy need. or pl./aq.	192
25	Digitalin	$C_{36}H_{56}O_{14}$	wh. cryst. powd.	229
26	Ligitonin (*l*)	$C_{55}H_{90}O_{29}$	wh. cryst. powd.	sint. 225; d. 235
27	Digitoxin**	$C_{41}H_{64}O_{13}$	wh. leaf.	255–6*
28	Digitoxin hydrate	$C_{41}H_{64}O_{13} \cdot 6H_2O$	leaf./al.	145
29	Euxanthic acid	$C_{19}H_{16}O_{10} \cdot H_2O$	straw yel. need.	155–8 d.
30	Frangulin (*l*)	$C_{21}H_{20}O_9 \cdot H_2O$	org. need./aq. pyr.	246–9
31	Fraxin	$C_{16}H_{18}O_{10}$	need./al.	205
32	Fustin	$C_{36}H_{26}O_{14}$	wh. glossy need./aq.	217 d.
33	Gaultherin (*l*)	$C_{14}H_{18}O_8 \cdot H_2O$	need. or pr./al.	179–80
34	Glucogallin (*l*)	$C_{13}H_{16}O_{10}$	wh.-yel. cryst.	193 d.
35	Glycovanillin (*l*)	$C_{14}H_{18}O_8 \cdot 2H_2O$	wh. need./dil. al.	192
36	Glycyphyllin	$C_{21}H_{24}O_9 \cdot 3H_2O$	long, thin, glossy pr./et.	175–80 d.
37	Gratiolin	$C_{43}H_{70}O_{15}$	fine, glossy need.	235–7 d.
38	α–Hederin (*l*)	$C_{41}H_{64}O_{11}$	wh. need.	256–7
39	Helicin ‡ (*l*)	$C_{13}H_{16}O_7 \cdot \frac{3}{4}H_2O$	fine, radiating need./aq.	175 *; —aq. 100
40	Helleborein (*l*)	$C_{37}H_{56}O_{18}$	warts of fine need./al.	270

* Melting point of the anhydrous compound.
** Crystalline digitalin.
‡ Does not occur naturally but is obtained by oxidizing salicin.

Cuspidatin 58
Datiscosid 24
Digitalin, crystalline 27

Digitin 26
Esculin 2
Esculinic acid 2

Table 7-6 (*Continued*)
PHYSICAL CONSTANTS OF GLUCOSIDES

No.	Solubility Expressed in Parts of Solvent Required to Dissolve 1 Part Glucoside					Hydrolytic Products		
	Water	Alcohol	Ether	Chloro-form	Others	Principle	M.P. °C.	Sugar
10	sl. s.	12	sl. s.	9	sl. s. CCl$_4$	cerberetin	85.5	glucose
11	i.	s. dil.	v. sl. s.	v. sl. s.	s. alk.	chinovic acid	d. 295	chinovose
12	....	v. s. ‡	i.	......	i. EtOAc	chinovic acid	d. 295	chinovose
13	s.	sl. s.	i.	i.	i. bz.	d-mannitol	166	glucose
14	s.	s. h.	i.	......	....	colocynthein (resinous)	......	glucose
15	200 c.; v. s. h.	sl. s.	i.	......	....	coniferyl alcohol †	73–4	glucose
16	s.	s. dil.	sl. s.	i.	....	convallamaretin	......	sugar
17	sl. s.	s.	i.	......	....	convallaretin	......	sugar
18	sl. s.	s.	i.	sl. s.	s. EtOAc	methylethyl-acetic and other acids	......	glucose, rhodeose
19	sl. s. c.	s. h.	s.	s.	....	indefinite	......	sugar
20	sl. s. c.; s. h.	sl. s. abs.	i.	......	....	crocetin	285	gentiobiose
21	v. sl. s.	s.	sl. s.	......	s. aq. act.	curangaegenin	132	rhamnose + a little glucose
22	sl. s.	57	i.	i.	i. bz.	cyclamiretin	198	fructose, cyclose
23	sl. s. c.; s. h.	v. s. h.	i.	......	s. alk.	daphnetin	253–6 d.	glucose
24	s.	s.	i.	......	....	datiscetin	276	rhamnose
25	1000	12	sl. s.	sl. s.	s. MeOH	digitaligenin	210–2	glucose, digitalose
26	sl. s.	s.	v. sl. s.	v. sl. s.	s. MeOH	digitogenin	softens 250	glucose, galactose
27	v. sl. s.	s.	v. sl. s.	s.	....	digitoxigenin	230	digitoxose
28	....	....	......	......	....	....	......	....
29	s. h.	s. h.	v. sl. s.	......	s. alk.	euxanthone	240	glycuronic acid
30	i.	sl. s. h.	i.	s. alk.	s. h. bz.	frangula-emodin	256–7	rhamnose
31	sl. s. c.; s. h.	s. h.	i.	......	§	fraxetin	227	glucose
32	v. s. h.	v. s.	sl. s.	......	s. alk.	fisetin	>360	rhamnose
33	slowly s.	s.	v. sl. s.	v. sl. s.	v. sl. s. act.	methyl salicylate	−8.3	glucose
34	s.	s.	sl. s.	i.	s. alk.	gallic acid	235 d.	glucose
35	s.	sl. s.	i.	......	... .	vanillin	80	glucose
36	s. h.	s.	s.	i.	i. bz.	phloretin	180	isodulcite
37	s. h.	s.	i.	......	....	gratiogenin	198	glucose
38	i.	s.	i.	......	s. ac. a.	hederagenin	331	arabinose, rhamnose
39	60 c.; v. s. h.	s.	sl. s.	......	....	salicylaldehyde	−7	glucose
40	v. s.	sl. s.	i.	......	....	helleboretin	>200	glucose, arabinose, ac. a.

‡Heat evolved; forms penta-alcoholate, m.p. 70–80°, which separates.
†By the action of emulsin; heating with dilute acids gives resinous material.
§Solutions fluoresce blue.

Franguloside 30   Glucosido-salicylaldehyde 39
Fraxinin 31   Gratus strophanthin 51
Glucosido-methyl salicylate 33   Helixin 38

Table 7-6 (*Continued*)
PHYSICAL CONSTANTS OF GLUCOSIDES

No.	Name	Formula	Appearance	Melting Point °C.
41	Helleborin	$C_{28}H_{36}O_6$	lustrous need.	>250
42	Hesperidin	$C_{28}H_{34}O_{15}$	wh., micro. need./aq. MeOH	251–2; d. 254
43	Indican	$C_{14}H_{17}O_6N \cdot 3H_2O$	need./aq.	57–8; 176–8 *
44	Iridin	$C_{24}H_{26}O_{13}$	wh. need.→yel. in air	208
45	Jalapin (*l*)	$C_{34}H_{56}O_{16}$	colorl. amor. mass	131–50
46	Linamarin (*l*)	$C_{10}H_{17}O_6N$	need.	142–3
47	Maclayin	$C_{17}H_{32}O_{10}$	deliquescent cryst.	158–65
48	Murrayin	$C_{18}H_{22}O_{10}$	microscopic need.	170
49	Naringin (*l*)	$C_{27}H_{32}O_{14} \cdot 8H_2O$	need./aq.	82
50	Ononin	$C_{25}H_{26}O_{11}$	small pr., need. or plates	210
51	Ouabain (*l*)	$C_{29}H_{44}O_{12} \cdot 7H_2O$	transparent plates	185 *
52	Parillin (*l*)	$C_{26}H_{44}O_{10} \cdot 2\frac{1}{2}H_2O$	fine plates	177
53	Periplocin (*d*)	$C_{30}H_{48}O_{12}$	long, thin need.	205
54	Phillyrin	$C_{27}H_{34}O_{11}$	need. or plates	162
55	Phloridzin (*l*)	$C_{21}H_{24}O_{10} \cdot 2H_2O$	small, wh., silky need.	108 **
56	Picein (*l*)	$C_{14}H_{18}O_7 \cdot H_2O$	need./aq.	194 *
57	Picrocrocin (*l*)	$C_{16}H_{26}O_7$	pr./et.-chl.-MeOH mixt.	154–6
58	Polygonin	$C_{21}H_{20}O_{10}$	glossy, yel. need.	202–3
59	Populin (*l*)	$C_{20}H_{22}O_8 \cdot 2H_2O$	wh. cryst. powd. or very fine need./aq.	180 *
60	Prulaurasin (*l*)	$C_{14}H_{17}O_6N$	bitter need.	120–2
61	Quercitrin (*l*)	$C_{21}H_{20}O_{11} \cdot 2H_2O$	pale yel. need. or plates	182–5; 250–2 *
62	Robinin	$C_{32}H_{40}O_{19} \cdot 7\frac{1}{2}H_2O$	yellowish need./aq.	195 *
63	Ruberythric acid	$C_{25}H_{26}O_{13}$	small, citron yel. need.; yel. pr./aq.	258–60
64	Rubiadin glucoside	$C_{21}H_{20}O_9$	yel. need./glacial ac. a.	270 d.
65	Rutin	$C_{27}H_{30}O_{16} \cdot 2H_2O$	bright yel. need./aq.	188–90
66	Salicin (*l*)	$C_{13}H_{18}O_7$	glossy need., plates or rhomb. pr./aq.	199–201, then re-melts 230–40
67	Saponin	$C_{32}H_{52}O_{17}$	wh. amor. powd.	d. 195
68	Saporubrin (*l*)	$(C_{18}H_{28}O_{10})_4$	wh. amor. powd.	....
69	Sapotoxin	$C_{17}H_{26}O_{10}$	wh. amor. powd.	....
70	Sarsasaponin (*l*)	$C_{44}H_{76}O_{20} \cdot 7H_2O$	long need./al	sint. 200; m. 248

* Melting point of the anhydrous compound.
** Melts at 108°, then solidifies at 130° and remelts at 170–1 d.

iso-Hesperidin 49	*dl*-Mandelonitrile-glucoside 60
Indoxyl-β-glucoside 43	Melin 65
Kalmin 55	Methoxy-coniferin 74
Laricin 15	Monotropitoside 33
Lilacin 74	Orizabin 45
Macleyin 47	Paviin 31

Table 7-6 (*Continued*)
PHYSICAL CONSTANTS OF GLUCOSIDES

No.	Solubility Expressed in Parts of Solvent Required to Dissolve 1 Part Glucoside					Hydrolytic Products		
	Water	Alcohol	Ether	Chloro-form	Others	Principle	M.P. °C.	Sugar
41	i.	sl. s.	sl. s.	s.	....	helleboresin	d. >140	glucose
42	v. sl. s.	sl. s.	i.	i.	v. s. dil. alk.	hesperetin	226	glucose, rhamnose
43	v. s.	v. s.	sl. s.	sl. s.	sl. s. bz.	indigo	390–2 d.	glucose
44	v. sl. s.	s. h.	i.	i.	sl. s. act.	irigenin	186	glucose
45	sl. s	s.	s. h.	s.	d. alk.	jalapinolic acid	67–9	sugars
46	v. s.	sl. s.	sl. s.	sl. s.	s. h. act.	hydrocyanic acid	−12	glucose, act.
47	....	sl. s.	i.	i.	....	maclayetin	209–10	glucose
48	s. h.	s.	i.	......	s. alk.	murrayetin	110	glucose
49	sl. s. c.; s. h.	v. s.	i.	i.	v. s. h. ac. a.	naringenin	251	glucose, rhamnose
50	sl. s. h.	s. h.	i.	......	s. h. KOH	formononetin	265	glucose
51	100 c.; 5 h.	20 c.; 8 h.	i.	i.	i. EtOAc	acocanthic acid lactone	......	rhamnose
52	v. sl. s. c.; 20 h.	s.	i.	s.	i. pet.	parigenin	......	sugars
53	125 c.	s.	v. sl. s.	v. sl. s.	i. bz.	periplogenin	185	glucose
54	s. h.	s.	i.	s. h.	....	phillygenin	......	glucose
55	1000 c.; s. h.	4	sl. s.	i.	....	phloretin	262–4	glucose
56	sl. s.	s.	s.	i.	s. ac. a.	p-hydroxy-acetophenone	109	glucose
57	v. s.	v. s.	sl. s.	sl. s.	i. bz.	safranol	......	glucose
58	sl. s. h.	sl. s. h.	i.	......	....	emodin	254	glucose
59	i. c.; sl. s. h.	sl. s. c.; s. h.	i.	s. dil. a.	s. dil. alk.	saligenin	87	glucose, benzoic a.
60	s.	s.	i.	......	....	dl-mandelonitrile	−10	glucose
61	i. c.; sl. s. h.	s.	sl. s.	......	s. alk.	quercetin	313–4 *	rhamnose
62	s. h.	s. h.	i.	......	s. alk.	caempferol	271	rhamnose
63	sl. s. c.; s. h.	v. sl. s. abs.	v. sl. s.	s. alk. →r.	i. bz.	alizarin	290	glucose; xylose
64	v. sl. s. h.	s.	s.	......	i. K₂CO₃	rubiadin	290	glucose
65	s. h.	s. h.	i.	i.	s. alk.	quercetin	313–4 *	glucose, rhamnose
66	23 c.; 3 h.	72	i.	i.	s. alk.	saligenin	87	glucose
67	s. †	i.	i.	i.	i. bz.	sapogenin	257–60	sugar
68	v. s. †	s. dil.	i.	i.	i. bz.	sapogenin	257–60	glucose
69	s. †	s. dil.	i.	i.	s. alk.	sapotoxin-sapogenin	......	sugar
70	v. s. †	s. h.	v. sl. s.	......	....	sarsasapogenin	197–8	glucose

† Aqueous solutions foam on shaking.
Phaseolunatin 46
Phillyroside 54
Phlorizin 55
Phyllyrin 54
Piceoside 56
Potassium myronate 72
Primulin 22

Quercimelin 61
Quercitrinic acid 61
Rhamnin 83
Rhamnoxanthin 30
Rhodeoretin 18
Rubianic acid 63
Salinigrin 56
Scammonin 45

Table 7-6 (*Continued*)
PHYSICAL CONSTANTS OF GLUCOSIDES

No.	Name	Formula	Appearance	Melting Point °C.
71	Sinalbine (*l*)	$C_{30}H_{42}O_{15}N_2S_2 \cdot 5H_2O$	pale yel. need.	83–4; 139 *
72	Sinigrin (*l*)	$C_{10}H_{16}O_9NS_2K \cdot H_2O$	rhomb. pr./aq.; need./al.	127–9; 179 *
73	Strophanthin (*in*)	$C_{31}H_{48}O_{12}$	microcryst., hyg.	179
74	Syringin (*l*)	$C_{17}H_{24}O_9 \cdot H_2O$	rosettes of long need./aq.	192
75	Tampicin	$C_{34}H_{54}O_{14}$	amor., colorl. to yel.	130
76	Tannic acid (*d*)	$C_{76}H_{52}O_{46}$	yel. to br. amor. bulky powd. or spongy masses; shining scales	d. 210–5
77	Thevetin (*l*)	$C_{42}H_{66}O_{18} \cdot 3H_2O$	need./al.	210
78	Thujin	$C_{20}H_{22}O_{12}$	yel. microscopic tablets	. . . .
79	Vacciniin (*d*)	$C_6H_{11}O_6COC_6H_5 \cdot H_2O$	cryst./aq. act.	104–6
80	Valdivin (*in*)	$C_{36}H_{48}O_{20} \cdot 2H_2O$	hexagonal pr.	230 d.*
81	Vicianin (*l*)	$C_{19}H_{25}O_{10}N$	wh. need.	147–8
82	Violutin (*l*)	$C_{19}H_{26}O_{12}$	wh. cryst.	169–72
83	Xanthorhamnin (*d*)	$C_{34}H_{42}O_{20}$	yel. micro need. + EtOH/al.	−EtOH, at 120

* Melting point of anhydrous compound.
Sinigroside 72
Smilacin 52

Sophorin 65
Strophanthin Thoms 51

Table 7-6 (*Continued*)
PHYSICAL CONSTANTS OF GLUCOSIDES

No.	Solubility Expressed in Parts of Solvent Required to Dissolve 1 Part Glucoside					Hydrolytic Products		
	Water	Alcohol	Ether	Chloro-form	Others	Principle	M.P. °C.	Sugar
71	s.	sl. s.	i.	......	i. CS_2	sinapine sulfate, and higher mustard oils	127 *	glucose
72	s.	sl. s. c.; i. abs.	i.	i.	i. bz.	allyl mustard oil	−80	glucose
73	43	s.	i.	i. CS_2	i. bz.	strophanthidin	170; 235 *	sugars; no glucose
74	s. h.	s. h.	i.	......	....	syringenin	......	glucose
75	i.	s.	s.	......	....	tampicolic acid	......	sugar
76	v. s.	sl. s.	v. sl. s.	v. sl. s.	v. sl. s. act.	gallic acid	235 d.*	glucose
77	s. h.	s.	sl. s.	sl. s.	s. EtOAc	thevetigenin	140	glucose
78	s. h.	s.	......	......	....	thujetin	......	glucose
79	s.	s.	i.	sl. s.	sl. s. bz.	benzoic acid	121.7	glucose
80	s. h.	s. dil.	i.	v. s.	....	?	......	sugar
81	s. h.	sl. s.	......	i.	i. bz.	benzaldehyde hydrocyanic acid	−26 −12	vicianose
82	s.	s.	i.	......	i. act.	methy.salicylate	−8.3	glucose; arabinose
83	v. s.; d.	s.	i.	i.	i. bz.	rhamnetin	>300	rhamnose

Ursin 6

Violutiside 82
Waldivin 80

PROPERTIES OF HORMONES

JOHN B. STANBURY, M.D., and EDWIN D. BRANSOME, JR., M.D.

Unit of Experimental Medicine, Department of Nutrition
and Food Science, Massachusetts Institute of Technology,
Cambridge, Mass.

HORMONAL STEROIDS

Steroids are non-saponifiable lipids (they are not rendered water soluble by treatment with alkali) which are secondary alcohols and have a perhydrocyclopentanophenanthrene nucleus in common. The structure consisting of three hexane (A, B, and C) and one pentane (D) nuclei, has a conventional order of numbering of its 17 carbon atoms: illustrated in the two-dimensional structure below.

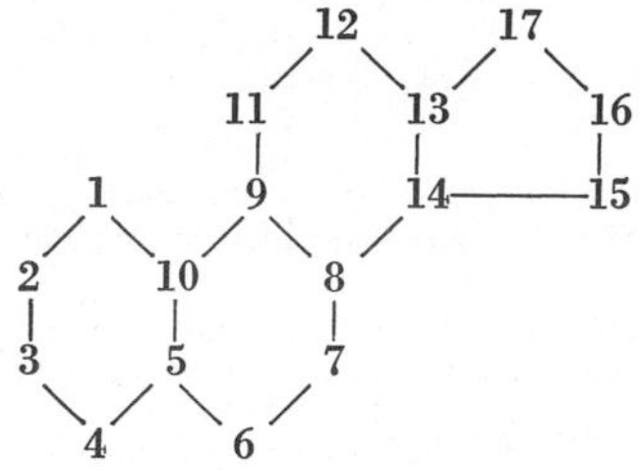

Steroid hormones are synthesized by the adrenal cortex, ovary, testis, and placenta. They have a number of biological activities: effects on intermediary metabolism (glucocorticoids), upon salt retention (mineralocorticoids), upon sexual characteristics and function (male androgens, female estrogens, female reproductive progestogens). Several hundred natural steroids have been isolated and characterized so far; most do not have the biologic activity of hormones and are either biosynthetic precursors of steroid hormones or metabolic products. Only hormonal steroids and some of their synthetic analogs will be briefly mentioned here. Further information may be discovered through perusal of the appropriate references on p. 775.

Nomenclature. Steroid hormones differ in respect to which —H, —OH, or =O radicals, and —CH₃ or carbon side chains are attached to the basic carbon skeleton. They may also differ in the presence and position of double bonds between skeleton carbon atoms. Small differences exert great effects on the physical and biological behavior of the compounds. Four types of names for steroids are used: (1) a trade name if the compound is sold as a drug, (2) a common or trivial name, (3) a modification of a trivial name, (4) a modification of the parent hydrocarbon. An example may be provided by the adrenal cortex hormone, hydrocortisone:

(1) Cortef, Cort-Dome, etc.; (2) Hydrocortisone, cortisol, Kendall's compound F; (3) 17-hydroxycorticosterone; (4) 11β,17,21-trihydroxy pregn-4-ene-3,20-dione.

The parent compounds named according to substitution of the 17 carbon skeleton are: estrane (one —CH₃ group, C-18, attached to C-13); androstane (an additional —CH₃, C-19 at C-10), and pregnane (androstane plus a side chain, C-20 and C-21, from C-17). Further modifications and their descriptions follow.

(*a*) *Double bond:* This is indicated in the spelling of the hydrocarbon and the position, e.g., pregnane (0); pregn-4-ene (1); pregn-1,4-diene (2); pregn-1,4,9-triene (3).

(*b*) *Oxygen substitution:* This may be a hydroxyl group (suffix **ol** preceded by carbon atom number or prefix **hydroxy**) or a ketone group (suffix **one**; prefix **keto** or **oxo**).

(*c*) *Isomerism:* Hydroxyl groups or hydrogens at locations of steroisomerism may be designated as α if they extend below the relatively flat plane of the ring and are represented by a dotted line; if above the plane of the ring, they are β and are indicated by a solid line.

(*d*) *Other changes:* **nor** means shortening of a side chain or elimination of a skeleton carbon. **Deoxy:** substitution of —H for —OH; **Dehydro:** loss of an —H; **Dihydro:** substitution of 2H for a double bond. (Please see References 1 and 2 for further information.)

PROPERTIES OF COMMON STEROIDS

In their free alcohol forms, these compounds (except for the estrogens) have little or no solubility in water. Identification is accomplished by chromatographic (paper, column, thin layer, gas-liquid) separation and subsequent purification to constant specific radioactivity after addition of labeled standard. Physicochemical characteristics which are useful include infrared absorption spectra (cf. standards), O.R.D. (optical rotatory dispersion), fluorescence and precipitability with digitonin.

Abbreviations used in the tables:

a. acid	aq., water	i., insoluble	s., soluble
alk., alkali	But., butanol	IEP, isoelectric point	solv., solvent
al., alcohol	dil., dilute	ORD, optical rotatory dispersion	vol., volatile

Table 7-7
STEROID HORMONES

No.	Name		Biology	
	Trivial	Chemical	Main source	Action
1	**Hydrocortisone** (cortisol)	11β,17α,21-trihydroxy-pregn-4-ene-3,20-dione	Adrenal cortex	Glucocorticoid. Anti-inflammatory. Rx as ester.
2	**Cortisone** 11-dehydrohydro-cortisone	17α,21-dihydroxy-pregn-4-ene-3,11,20-trione	Metabolism of (1) synthetic	Glucocorticoid. Anti-inflammatory. Rx as ester. 0.7 times potency of hydrocortisone in rats and humans.
3	**Corticosterone**	11β,21-dihydroxy-pregn-4-ene-3,20-dione	Adrenal cortex	Replaces (1) in some species (rats, mice, birds).
4	**Aldosterone**	11β,21-dihydroxy-18-formyl-pregn-4-ene-3,20-dione	Adrenal cortex (synthetic)	Mineralocorticoid. Potency: about 25 times deoxycorticosterone re NaCl retention.
5	**Deoxycorticosterone** (DOC)	21-hydroxy-pregn-4-ene-3,20-dione	Adrenal cortex (synthetic)	Mineralocorticoid. Rx as esters.
6	**9α fluorohydrocortisone** (fluorinef)	9α-fluoro-11β,17α,21-trihydroxy-pregn-4-ene-3,20-dione	Synthetic only	Mineralocorticoid (and glucocorticoid). NaCl retaining potency equal to aldosterone.
7	**Prednisolone** (Δ-1 hydrocortisone)	11β,17α,21-trihydroxy-pregn-1,4-diene-3,20-dione	Synthetic only	Glucocorticoid. Anti-inflammatory potency: 4 times hydrocortisone in man; 3.5 times in rats.
8	**Methyl prednisolone** (6α methyl Δ-1 hydrocortisone)	11β,17α,21-trihydroxy-6α methyl-pregn-1,4-diene-3,20-dione	Synthetic only	Same. Rx as alcohol or ester. Anti-inflammatory potency: 5 times hydrocortisone in man and rats.
9	**Dexamethasone** (16α methyl-9α-fluoro Δ-1 hydrocortisone)	9α fluoro-11β,17α,21-trihydroxy-16α-methyl-pregn-1,4-diene-3,20-diene	Synthetic only	Same. Anti-inflammatory potency: 30 times hydrocortisone in man; 150 times in rats.
10	**Triamcinolone** (16α-hydroxy-9α fluoro Δ-1 hydrocortisone	9α fluoro-11β, 16α-17α 21-tetrahydroxy-pregn-1,4-diene-3,20-dione	Synthetic only	Same. Anti-inflammatory potency: 4 times hydrocortisone in man; 3 times in rats.
11	**Estradiol 17β**	3β,17β dihydroxy-estr-1,3,5-triene	Ovary	Estrogenic: the principal biological estrogen. Rx as esters.

Table 7-7 (*Continued*)
STEROID HORMONES

| No. | Chemistry | | | | |
	Formula	Mol. wt.	Melting point, °C	Identification (O.R.D., etc.)	Abs. max.
1	$C_{21}H_{30}O_5$	362.47	217–220	$[\alpha]^{25}D +150–156°$ (dioxane) green fluorescence in H_2SO_4	242 mμ (methanol)
2	$C_{21}H_{28}O_5$	360.44	220–224	$[\alpha]^{25}D +209°$ (ethanol) green fluorescence in H_2SO_4	237 mμ
3	$C_{21}H_{30}O_4$	346.45	180–182	$[\alpha]^{55}D +223°$ (ethanol) green fluorescence in H_2SO_4	240 mμ
4	$C_{21}H_{28}O_5$	360.45	108–112 (hydrate)	$[\alpha]^{25}D +161°$ (chloroform)	240 mμ
5	$C_{21}H_{30}O_3$	330.45	141–142	$[\alpha]^{22}D +178°$ (ethanol) no H_2SO_4 fluorescence	240 mμ
6	$C_{21}H_{29}FO_5$	380.46	260–262*	$[\alpha]^{23}D +139°$ (ethanol)	239 mμ
7	$C_{21}H_{28}O_5$	360.44	240–241	$[\alpha]^{25}D +102°$ (dioxane)	242 mμ
8	$C_{22}H_{30}O_5$	374.46	228–237	$[\alpha]^{20}D +83°$ (dioxane)	243 mμ
9	$C_{22}H_{29}FO_5$	392.45	262–264	$[\alpha]^{25}D +73°$ (chloroform) for 21-acetate	239 mμ
10	$C_{25}H_{31}FO_8$	478.52	186–188	$[\alpha]^{25}D +22°$ (chloroform)	239 mμ
11	$C_{18}H_{24}O_2$	272.37	173–179	$[\alpha]^{25}D +76–83°$ (dioxane)	225, 280 mμ

Table 7-7 (*Continued*)
STEROID HORMONES

No.	Name		Biology	
	Trivial	Chemical	Main source	Action
12	**Estrone** (theelin)	3β hydroxy-estr-1,3,5-triene-17-one	Ovary, placenta	Estrogenic (weak).
13	**Estriol**	3β,16α,17β,tri-hydroxy-estr-1,3,5-triene	Ovary	Estrogenic (weak).
14	**Progesterone**	Pregn-4-ene-3,20-dione	Ovary, placenta	Progestogen. Maintains secretory endometrium.
15	**Pregnanediol**	3α,20α-dihydroxy-pregnane	Metabolism of (14)	Metabolite of progesterone; presence in urine used in pregnancy test.
16	**Testosterone**	17β-hydroxy-4-androsten-3-one	Testis, synthetic	Androgen; the principal biologic androgen. Rx as esters.

* Decomposition point.

Table 7-7 (*Continued*)
STEROID HORMONES

No.	Chemistry				
	Formula	Mol. wt.	Melting point, °C	Identification (O.R.D., etc.)	Abs. max.
12	$C_{18}H_{22}O_2$	270.36	258–262	$[\alpha]^{25}D +158–168°$ (dioxane); precipitated by digitonin	283, 285 mμ
13	$C_{18}H_{24}O_3$	288.37	282	$[\alpha]^{25}D +58°$ (dioxane); precipitated by digitonin	280 mμ
14	$C_{21}H_{30}O_2$	314.45	121	$[\alpha]^{20}D +172–182°$ (dioxane)	240 mμ
15	$C_{21}H_{36}O_2$	320.50	238	$[\alpha]^{20}D +27.4°$ (ethanol) water soluble not precipitated by digitonin	
16	$C_{19}H_{28}O_2$	288.41	155	$[\alpha]^{24}D +109°$ (ethanol)	238 mμ

Table 7-8

HORMONAL AMINO ACIDS AND POLYPEPTIDES

These include water-soluble hormones from the anterior and posterior pituitary, thyroid, and parathyroid glands; the adrenal medulla; the pancreas; and the gastrointestinal tract.

Name	Action	Source	Formula; mol. wt.	Method of testing	Use and method of administration
Adrenocorticotropin (ACTH, corticotropin).	Maintenance and stimulation of adrenal corticosteroid biosynthesis.	Anterior pituitary.	Polypeptide, 39 residues. Sol.: aq.-s.	Bioassay—adrenal ascorbic acid depletion; adrenal vein corticosterone concentration; immunoassay.	Parenterally, supplanted by corticosteroids in rheumatoid arthritis, asthma, and other allergic conditions. Used to maintain the integrity of the adrenal cortex.
Chorionic gonadotropin (HCG, HPL).	Stimulates rupture of ovarian follicles and formation of corpora lutea; stimulates testosterone production in males and seminiferous tubule growth in males.	Urine of pregnant women, placenta.	Glyco-polypeptide, about 30,000. Sol.: aq.-s., glycerol-s.	Immunoassay; mouse ovarian weight increase; ovulation in mouse or rabbit; spermatozoa discharge in frog.	Cryptorchism; male hyogonadism; some cases of male and female sterility.
Calcitonin (Thyrocalcitonin).	Regulation of plasma calcium levels; hypocalcemic.	Bovine and porcine thyroid glands.	Polypeptide.	Plasma calcium lowering in rats.	Commercial preparations for human use not available.
Follicle stimulating hormone (FSH, perganol (commercial preparation).	Stimulates graafian follicle maturation in the ovary; stimulates spermatogenesis and seminiferous tubule growth in the ovary.	Anterior pituitary; postmenopausal urine.	Polypeptide; 30,000 (human). Sol.: aq.-s.; 50% al.-s.	Bioassay, mouse ovarian weight increase; radioimmunoassay.	Induction of ovulation (in combination with chorionic gonadotropin) parenterally.
Glucagon.	Transient elevation of blood sugar; inhibition of intestinal motility.	Pancreas—α cells.	Polypeptide 29 residues, 3,485. Sol.: dil. a.-s.; dil. alk.-s.	Bioassay (glycogenolysis); immunoassay	For diagnosis in glycogen deposition disease parenterally; rarely for hypoglycemia.
Growth hormone (Somatotropin, GH, STH).	Species specific promotion of somatic growth; effects on intermediary metabolism.	Anterior pituitary, bovine, porcine.	Polypeptide, about 45,000. Sol.: aq.-s.	Radioimmunoassay; bioassay for growth in hypophysectomized rats.	Occasionally in hypopituitary dwarfism, intramuscularly.
Insulin.	Promotes uptake of blood glucose by liver, muscle; other metabolic effects.	Pancreas—β cells.	Polypeptide; α chain 21 residues; β chain 30 residues, 5,700. Sol. aq.-i.; dil. a. + alk.-s.	Bioassay, hypoglycemic effect. Radioimmunoassay.	In diabetes mellitus subcutaneously; occasionally intramuscularly; by subcutaneous injection in combination with zinc, protamine or globin for prolonged effect.

Name	Action	Source	Formula: mol. wt.	Method of testing	Use and method of administration
L-triiodothyronine β-[4-(4-hydroxy-3-iodophenoxy)-3,5-di-iodophenyl] alanine (liothyronine).	More rapid but the same as thyroxine; present in very low concentration.	Thyroid gland; synthetic.	$C_{15}H_{12}I_3NO_4$, 651.01. Sol. aq.-i.; al.-i.; dil. alk.-s.	Iodometric; bioassay; chromatography.	In hypothyroid states orally and occasionally parenterally.
Leutenizing hormone (L.H.; ICSH).	Stimulates graafian follicle rupture and corpus luteum formation; with FSH stimulates estrogen synthesis; with prolactin, progesterone synthesis.	Pituitary, ovine, porcine.	Polypeptide 26,000 (human). Sol.: aq.-s.	Rat ventral prostate hypertrophy radioimmunoassay.	Pure preparation not commercially available.
L-thyroxine—β[(β,5-diiodo-4-hydroxy-phenoxy)-3,5-diiodophenyl] alanine; 3,5,3'5'-tetraiodo-thyronine.	Control of metabolic rate; also promotion of growth.	Thyroid gland; synthetic.	$C_{15}H_{11}O_4NI_4$; 776.93. Sol.: aq.-i.; al.-i.; alk.-s.; al. + a.-s.; vol. solv.-i.	Iodometric; bioassay; chromatography.	In hypothyroid states orally and occasionally parenterally.
Vasopressin (antidiuretic hormone; ADH; Petressin).	Antidiuretic action on kidneys.	Bovine neurohypophysis; synthetic.	Octapeptide—lysine vasopressin (swine): mol. wt. 1056; arginine vasopressin mol. wt. 1084. IEP: pH 10.9. Sol.: aq.-s.	Blood pressure rise in dogs. Radioimmunoassay.	In diabetes insipidus as tannate; in oil intramuscularly crude posterior pituitary powder; by nasal insufflation.
Melanocyte stimulating hormone (α-MSH; β-MSH). (Intermedian, MSH, melanotropic hormone).	Increase of skin pigmentation.	Neurohypophysis.	Polypeptides; α-MSH: 14 residues; β-MSH: 22 residues (human). Shares same amino acid sequence with ACTH.	Darkening of frog skin.	Preparations for human use not available.
Melatonin, N-acetyl-5-methoxy-tryptamine.	Decreases skin pigmentation.	Pineal gland; synthetic.	$C_{13}H_{15}O_2N_2$, 231.	Blanching of pigmented frog skin.	Not used in man.
Norepinephrine (noradrenaline)	Neuro-humoral transmitter and regulates blood pressure by stimulating vasoconstriction.	Sympathetic nervous system postganglionic fibers, adrenal medulla.	$C_8H_{11}O_3N$, 169.18. Benzene ring, ethylamine side chain with +OH groups at ring C_3, C_4 and the side chain βC. Sol.: aq.-s.	Bioassay, (blood pressure rise); colorimetry, fluorescence on oxidation, chromatography.	Support of blood pressure in shock (intravenous).

Table 7-8 (*Continued*)

HORMONAL AMINO ACIDS AND POLYPEPTIDES

Name	Action	Source	Formula: mol. wt.	Method of testing	Use and method of administration
Oxytocin (pitocin).	Promotes uterine contraction, milk ejection.	Bovine neurohypophysis; synthetic.	Octapeptide, 1007. IEP: pH 7.7. Sol.: aq.-s.; But.-s.	Bioassay—isolated uterus contraction. Avian vasodepression.	To induce labor and to promote involution of postpartum uterus; promotion of milk ejection; by intranasal spray.
Parathyroid hormone (parathormone).	Regulation of plasma calcium; hypercalcemic; stimulation of calcium absorption by intestine, breakdown of bone structure.	Bovine parathyroid glands.	Polypeptide, about 9,000. Sol.: dil. a.-s.	Increase in plasma calcium in dog. Radioimmunoassay.	Occasionally in initial control of parathyroprivic tetany.
Prolactin (lactogenic hormone; luteotropin).	Stimulates lactation; with L.H., promotes progesterone synthesis by ovary.	Ovine pituitary.	Polypeptide, 23,500. Sol.: aq.-i.; saline-s.; al. + a.-s.	Pigeon crop sac hypertrophy.	No preparation available for human use.
Thyrotropin (TSH).	Maintenance of thyroid function.	Anterior pituitary; bovine, porcine.	Glyco-polypeptide, about 10,000. Sol.: aq.-s.	Thyroid iodine depletion—murine or chick.	Diagnostic for distinguishing primary and secondary myxedema.
Epinephrine (adrenaline).	Support of blood pressure by increasing cardiac output; other metabolic effects including increased oxygen consumption.	Adrenal medulla.	$C_9H_{13}O_3N$, 183.21. Structure of norepinephrine with methyl substitution of the amine. Sol.: aq.-s.	Cf. norepinephrine.	Intravenously or subcutaneously in allergic reactions.

HORMONES OF THE GASTROINTESTINAL TRACT

A number of polypeptides are secreted by mucosal cells of the gastrointestinal tract in response to chemical or mechanical stimuli. Their mode of action is not yet very well understood.

Hormone	Origin	Target
Secretin	Duodenum	Pancreatic secretion; bile flow
Gastrin	Pylorus	Gastric secretion
Cholecystokinin	Small intestine	Contraction of gall bladder
Enterogastrone	Small intestine	Inhibits gastric secretion, contraction
Pancreozymin	Upper intestine	Pancreatic secretion

REFERENCES

1. L. F. Fieser and M. Fieser, "Steroids," Reinhold Publ. Co., New York (1959).

2. R. I. Dorfman and F. Ungar, "Metabolism of Steroid Hormones," Academic Press, New York (1965).

3. A. B. Eisenstein (ed.), "The Adrenal Cortex," Little, Brown & Co., Boston (1967).

4. P. H. Katzman and W. H. Elliott, Chemistry of Estrogens (review), in M. Florkin and E. H. Statz (eds.), "Comprehensive Biochemistry," p. 47, Elsevier, Amsterdam (1963).

5. G. I. Fujimoto and R. W. Ledeen, Chemistry of Androgens and Other C^{19} Steroids (review), in M. Florkin and E. H. Statz (eds.), "Comprehensive Biochemistry," p. 33, Elsevier, Amsterdam (1963).

6. E. Diczfalusy and P. Troen, Placental Hormones (review), *Vitamins, Hormones* 19, *229 (1961)*.

7. P. G. Stecher, M. J. Finkel, O. H. Siegmund, and B. M. Szatranski (eds.), "The Merck Index of Chemicals and Drugs," Rahway, N.J. (1960).

8. "The United States Pharmacopeia," 17th revision, Mack Co., Easton, Pa., 1965.

COMMERCIAL ORGANIC MATERIALS

Table 7-9
CONSTANTS OF FATS, OILS, AND WAXES

Classification: Class SV, semi-drying vegetable oil; NVO, non-drying vegetable oil of the olive oil type; AF, animal fat; AW, animal wax; IW, insect wax; NVR, non-drying vegetable oil of the rape oil

No.	Name	Class	Specific Gravity $\frac{15°}{15}$C.
1	Acorn, *Quercus agrifolia*	SV	0.916
2	Almond, *Prunus amygdalus*	NVO	0.914-0.921
3	Apricot kernel, *Prunus Armeniaca*	NVO	0.915-0.926
4	Beef marrow, *Bos taurus*	AF	0.931-0.938
5	Beef tallow, *Bos taurus*	AF	0.895
6	Beechnut, *Fagus sylvatica F. Americana*	SV	0.922
7	Beeswax (ordinary), *Apis mellifera*	IW	0.953-0.970
	Indian	IW	0.953-0.970
8	Black mustard, *Sinapis nigra*	NVR	0.915-0.919
9	Black walnut, *Juglans nigra; (cf. Walnut)*	DV	0.918-0.921
10	Bone fat, *Sevum ossis*	AF	0.914-0.916
11	Brazil nut, *Bertholletia excelsis*	SV	0.917-0.918
12	Butter fat, *Vaccae lactis adeps*	AF	$0.907\text{-}0.912\frac{40°}{15}$
13	Candelilla, *Euphorbia cerifera*	VW	0.981-0.994
14	Candlenut, *Aleurites moluccana*	DV	0.925
15	Candlenut, *Aleurites triloba*	DV	0.927
16	Carnauba wax, *Corypha cerifera*		
	No. 1 Yellow	VW	0.990-0.996
	No. 3 Crude	VW	0.994-1.010
	No. 3 Refined	VW	0.990-0.996
17	Castor, *Ricinus communis*	NVC	0.960-0.967
18	Ceresine	MW	0.900-0.920
19	Chaulmoogra, U. S. P. X Revision	SV	$0.950^{25°}$
20	Chaulmoogra, *Taraktogenos Kurzii*	SV	0.943-0.954
21	Cherry kernel, *Prunus cerasus*	NVO	0.918-0.929
22	Chicken fat, *Gallus domesticus*	AF	0.924
23	Chinese insect wax, *Coccus cerifera*	IW	0.950-0.970
24	Chinese vegetable tallow, *Stillingia sebifera*	VF	0.918-0.922
25	Coconut, *Cocos butyracea; C. nucifera*	VF	0.926
26	Cocoa (Cacao) butter, *Theobroma cacao*	VF	0.964-0.974
27	Cod liver, *Gadus morrhua*	MA	0.922-0.931
28	Corn (Maize), *Zea Mays*	SV	0.921-0.928
29	Cottonseed, *Gossypium* Species	SV	$0.917\text{-}0.918\frac{25°}{15}$
30	Cottonseed stearin, *Gossypium*	VF	$0.867\text{-}0.868^{100°}$
31	Croton, *Croton tiglium*	SV	0.942-0.944
32	Date kernel, *Phoenix dactylifera*	NVO	—
33	Deer fat, *Cervus elephus*	AF	0.962-0.967
34	Dolphin, *Delphinus globiceps*	MA	0.908-0.930
35	Esparto, *Stipa tenacissima*	VW	0.985-0.995
36	Goat's butter, *Capellae lactis adeps*	AF	$0.917\text{-}0.935\frac{37\cdot7°}{37\cdot7}$
37	Goose fat, *Anser cinereus*	AF	$0.923\text{-}0.930^{37\cdot7}$
38	Grape seed, *Vitis vinefera*	NVC	0.917-0.933
39	Hazelnut, *Corylus avellana*	NVO	0.917
40	Hemp seed, *Cannibis sativa*	DV	0.928-0.934
41	Herring, *Clupea harengus*	MA	0.920-0.939
42	Horse fat, *Equus caballus*	AF	0.919-0.933
43	Human fat	AF	0.9033
44	Japan wax, *Rhus succedaneum*	VW	0.970-0.998
			$0.875^{100°}$

Table 7-9 (*Continued*)
CONSTANTS OF FATS, OILS, AND WAXES

type; DV, drying vegetable oil; VW, vegetable wax; NVC, non-drying vegetable oil of the castor oil type; VF, vegetable fat; MA, marine animal or fish oil; MW, mineral wax; NA, non-drying animal oil; SMW, semi-mineral wax; Sp., sperm oil.

No.	Solidification point °C.	Acid value	Saponification value	Iodine value	Reichert-Meissl value
1	−10	....	199.3	100.0	....
2	−15 to −20	0.5–3.5	183.3–207.6	93–103.4	0.5
3	−17	3.5	191.4–198.2	100–108.7	0.2
4	29 to 31	1.6	196–199	39–55.4	2
5	31 to 38	0.25	196–200	35.4–42.3	0.25
6	−17	....	191–196	97–111	....
7	62 to 66	17.0–21.0	88–100	8–11	....
	61 to 67	5.0–10.5	87–117	4–10.5	....
8	16	5.7–7.3	173–175	99–110	....
9	turbid −12	8.6–9.0	190.1–191.5	141–142.7	....
10	15 to 17	29.6–53	185–198	46–55.8	0.2–1.7
11	0 to 3	1.4	193	90–106	....
12	20 to 23	0.45–35.4	210–230	26–38	17.0–34.5
13	73 to 77	18.6–23.9	55.0–64.2	....	....
14	< −18	2	189–195	163–164	1.2
15	....	....	202–204	139–143.8	....
16					
	86 to 88	1.5–2.5	75–86	....	....
	86 to 90	3.0–8.5	75–89	....	....
	86 to 89	3.0–5.0	76–85	7.0–14.5	....
17	turbid −12	0.12–0.8	175–183	84	1.4
	solid −17 to −18				
18	56 to 82	0	0	4.0–8.0	....
19	<25	....	196–213	98–104	....
20	20 to 25	0.79–21.5	196–213	97.6–110.4	....
21	−19 to −20	1.1	193.3–195	110–114.3	....
22	21 to 27	1.2	193–204.6	66–71.5	1.8
23	80 to 85	1.9–8.9	78–93	1.0–2.5	....
24	24 to 34	2.4	179–206	23–40.5	0.2–0.9
25	14 to 22	2.5–10	253.4–262	6.2–10	6.6–7.5
26	21.5–23	1.1–1.9	192.8–195	32.8–41.7	0.3–1
27	−3	5.6	171–189	137–166	0.2
28	−10 to −20	1.37–2.02	187–193	111–128	4.3
29	+12 to −13	0.6–0.9	194–196	103–111.3	0.95
30	16 to 22	4–10	195	88.7–93.6	0.22
31	−8 to −18	27–30.9	193–215	108–109	12–13.6
32	18.1	....	211	52.3	0.88
33	....	0.8–5.3	194.5–200	26–36	0.68
34	+5 to −3	2–12	body 203.4 jaw 290	126.9 body 32.8 jaw	body 46.9 jaw 65.9
35	75 to 79	22.0–27.0	58.0–72.5	7.0–15.0	....
36	....	....	233–236	25–37	20.8–27.7
37	22 to 24	0.59	191–193	58–67	0.2–0.98
38	−10 to −17	0.75	171–191	94.3–135	0.46
39	−17 to −18	....	191–197	87	0.99
40	−15 to −28	0.45	190–195	145–161.7	....
41	....	1.8–44	170–194	102–149	....
42	20 to 45	0–2.4	195–200	75–86	1.6–2.1
43	15	....	193–200	57–73	....
44	49 to 56	4.0–15.0	210–235	4.0–15.0	....

Table 7-9 (Continued)
CONSTANTS OF FATS, OILS, AND WAXES

No.	Name	Class	Specific Gravity $\frac{15°}{15}$C.
45	Lard oil, *Sus scrofa*	NA	0.913-0.915
46	Lard oil (fatty tissue), *Sus scrofa*	AF	0.934-0.938
			$0.861\frac{100}{15.5}°$
47	Laurel (bayberry), *Laurus nobilis*	VF	$0.880^{100°}$
48	Linseed, *Linum usitatissimum*	DV	0.930-0.938
49	Menhaden, *Alosa manhaden—Brevortia tyrannus*	MA	0.923-0.933
50	Montan:		
	Reibeck	SMW	0.995-1.040
	Domestic	SMW	1.020-1.040
51	Mutton tallow, *Ovis aries*	AF	0.937-0.953
			$0.858\frac{100}{15.5}°$
52	Myrtle wax, *Myrica cerifera—M. Carolinensis*	VF	$0.995-0.875^{100°}$
53	Neat's foot, *Bos taurus*	NA	0.913-0.918
54	Nutmeg (mace) butter, *Myristica officinalis*	VF	0.945-0.966
55	Olive, *Olea Europaea sativa*	NVO	0.9140-0.918
56	Ouricury, *Syagrus coronata* (Mart.)	VW	0.990-1.010
57	Ozokerite	MW	0.900-0.996
58	Palm, *Elaeis guineensis* (W. Africa)	VF	$0.924; 0.858^{100°}$
59	Palm kernel, *Elaeis guineensis* (S. America)	VF	....
60	Palm kernel, *Elaeis guineensis* (W. Africa)	VF	$0.866-0.873^{100°}$
61	Peach kernel, *Amygdalus Persica*	NVO	0.918-0.925
62	Peanut, *Arachis hypogaea*	NVO	0.917-0.926
63	Perilla, *Perilla ocimoides*	DV	0.930-0.937
64	Pistachio nut, *Pistacia vera*	NVO	0.913-0.919
65	Plum kernel, *Prunus domestica* and *P. damascena*	NVO	0.912-0.913
66	Poppy seed, *Papaver somniferum*	DV	0.924-0.926
67	Porpoise—body oil, *Delphinus phocaena*	MA	0.926
68	Pumpkin seed, *Cucurbita pepo*	SV	0.923-0.925
69	Rabbit fat, *Lepus cuniculus*	AF	0.934-0.936
			$0.861\frac{100}{15}°$
70	Rape seed, *Brassica campestris*	NVR	0.913-0.917
71	Safflower, *Carthamus tinctorius*	DV	0.925-0.928
72	Sardine, *Clupea pilchardus* and *C. scombrinus*	MA	0.920-0.934
73	Seal, *Phoca* species	MA	0.915-0.926
74	Sesame, *Sesamum indicum*	SV	$0.919\frac{25}{25}°$
75	Shark, *Selache (cetorhinus) maxima et al.*	MA	0.916-0.919
76	Soya, soy or soja bean, *Soja hispida; Dolichos hispida*	SV	0.924-0.927
77	Sperm, *Physeter macrocephalus*	Sp	0.878-0.884
78	Spermaceti, *Cetaceum*	AW	0.905-0.960
79	Spermaceti, *Physeter macrocephalus*	AW	0.905-0.945
79.1	Sugar cane wax, *Saccharum officinarum*	VW	0.970-0.980
80	Sunflower, *Helianthus annuus*	DV	0.924-0.926
81	Tallow, *Oleum adipis bovis*	NA	0.914-0.919
82	Tallow, *Sevum*	AF	0.925-0.950
83	Tung,—China wood, *Aleurites fordii*	DV	0.939-0.949
84	Tung,—China wood, *Aleurites montana*	DV	0.939-0.949
85	Walnut, *Juglans regia* (cf. Black walnut)	DV	0.925-0.927
86	Whale, *Balaena mysticetus*	MA	0.917-0.924
87	White mustard seed, *Sinapis alba*	NVR	0.912-0.916
88	Wool fat, *Ovis aries*	AW	0.970-0.973

Table 7-9 (*Continued*)
CONSTANTS OF FATS, OILS, AND WAXES

No.	Solidification point °C.	Acid value	Saponification value	Iodine value	Reichert-Meissl value
45	+4 to −2	0.1-2.5	193-198	62.5-79	0-0.2
46	27.1 to 29.9	0.5-0.8	195-203	47-66.5	0.5-0.8
47	25	26.3	198-199	68-80	1.6
48	−19 to −27	1-3.5	188-195	175-202	0.95
49	−5	3-11.6	189-192.9	148-185	1.2
50					
	83 to 89	35.0-45.0	80-99	....	....
	80 to 86	35.0-45.0	100-115	....	....
51	32 to 41	1.7-14	195-196	48-61	....
52	39 to 43	3-4.4	205.5-211.7	3.9-9.5	0.5
53	+10 to −2	0.1-0.6	193-199	57.5-75	0.9-1.2
54	41 to 42	17.2	154-178	40-81	1.1-4.2
55	turbid +2, ppt. −6	0.3-1.0	185-196	79-88	0.6-1.5
56	86 to 89	12.0-18.8	88.0-95.8	....	....
57	56 to 82	0	0	4.0-8.0	....
58	35 to 42	10	200-205	49.2-58.9	0.9-1.9
59	27.4	0.33-0.55	220.2-231.4	25.5-31.6	....
60	....	....	243-255	10.5-17.5	5-6.8
61	−20	1-1.5	191-193	92-99.7	....
62	3	0.8	186-194	88-98	0.4
63	....	....	188-194	185-206	....
64	−5 to −10	....	191	83-87	....
65	−5 to −8	0.55	191-193	100-103.6	....
66	−16 to −18	2.5	193-195	128-141	0.6
67	−16	body 1.2 jaw 5.0	body 203.4 jaw 253-272	body 126.9 jaw 30.9-49.6	body 46.9 jaw 132
68	−15	....	188-193	121-130	4.45
69	17 to 23	1.4-7.2	199-203	70-99.8	0.7-2.8
70	−10	0.36-1.0	168-179	94-105	0.0-7.9
71	−13 to −18	0.6	188-203	122-141	0.0-0.2
72	20 to 22	4-25	187.7-196	150-193	0.5-1
73	3	1.9-40	187.5-196.2	130-152	0.2
74	−4 to −6	9.8	188-193	103-117	1.1-1.2
75	....	....	157-164	115-139	....
76	−10 to −16	0.3-1.8	189-193.5	122-134	0.5-2.8
77	15.5	13.2	120-137	80-84	0.6
78	41 to 49	0.5-3.0	121-135	2.5-8.5	
79	42 to 47	0.5-2.8	126-135	3.8-9.5	....
79.1	76-82	20-30	55-70	....	....
80	−17	11.2	188-193	129-136	0.5
81	2 to 7.5	0.2-0.25	193.5-199	56-60.5	0.3
82	....	....	193-198	35-45	0.5-1.0
83	<17	2	190-197	163-171	1.10
84	<17	2	190-197	163-171	0.35
85	−15 to −27	2.5	190.1-197	139-150	0.92
86	−2 to 0	1.9	160-202	90-146	14
87	−8 to −16	5.4	171-174	94-98.4	....
88	38 to 40	59.8	82-130	17-29	5-8

Table 7-9 (*Continued*)
CONSTANTS OF FATS, OILS, AND WAXES

No.	Name	Index of Refraction at 25° C.	Hehner value	Acetyl value	Unsaponifiable matter
1	Acorn	....	....	....	....
2	Almond	1.4593-1.4646*	96.0	9.6	0.75
3	Apricot kernel	1.4636-1.4705	....	12.2	....
4	Beef marrow	....	....	....	....
5	Beef tallow	1.4552-1.4587*	96-96.5	2.7-8.6	....
6	Beechnut	1.4698	95-96	....	....
7	Beeswax (ordinary)	1.4445-1.4473†	....	13.0-16.0	52-55
	Indian	1.4380-1.4420‡	....	....	52-55
8	Black mustard	1.4718	96	....	3.3
9	Black walnut	....	95.8	....	....
10	Bone fat	....	91-95	11.3	0.5-1.8
11	Brazil nut	1.4671	....	....	....
12	Butter fat	1.4555-1.4578*	87.6-89.6	1.9-8.6	0.3-0.6
13	Candelilla	1.4565-1.4610‡	....	....	....
14	Candlenut *Al. mol.*	1.4760-1.4790	95-96	9.8	0.5-9
15	Candlenut *Al. tri.*	1.4760-1.4790	....	....	....
16	Carnauba wax, yellow	1.4490-1.4525§	....	....	50-55
	crude	1.4460-1.4520§	....	....	50-55
	refined	1.4465-1.4500§	....	54.0-56.0	50-55
17	Castor	1.4771	....	146-150.5	0.6
18	Ceresine	1.4320-1.4370‡	....	....	100
19	Chaulmoogra, U.S.P. X Revision	....	....	....	....
20	Chaulmoogra, *Taraktogenos Kurzii*	1.4777-1.4779	....	....	2.4-2.6
21	Cherry kernel	1.4635	....	....	....
22	Chicken fat	1.4580*	94.6	45	....
23	Chinese insect wax	1.4566*	....	....	49-55
24	Chinese vegetable tallow	1.4470-1.4579*	95.3	....	....
25	Coconut	1.453	82.3-90.5	2.3-6.9	....
26	Cocoa (Cacao) butter	1.4537-1.4580*	94-95	1.97	....
27	Cod liver	1.4758-1.4783	95.3	1.15	0.54-2.68
28	Corn	1.4733	93-95	7.5-11.5	1.5-2.8
29	Cottonseed	1.4743-1.4752	95.7	21-25	1.1
30	Cottonseed stearin	1.4700-1.4725	96.5	....	....
31	Croton	1.4710*	89.0	19.8-38.6	0.6
32	Date kernel	1.4535-1.4633	95.2	....	....
33	Deer fat	....	95.8	....	0.52
34	Dolphin	1.4665	93.1	....	2
35	Esparto	1.4550-1.4590‡	....	....	15-20
36	Goat's butter	1.4499-1.4551*	....	....	....
37	Goose fat	1.4583-1.4626	94.5-95.3	....	....
38	Grape seed	1.4713-1.4725	92	13.5-14.5	1.6
39	Hazelnut	1.4667	95.5	3.2	0.5
40	Hemp seed	1.4740-1.4745*	....	....	1.08
41	Herring	1.4665-1.4729	95-96	....	1-2
42	Horse fat	1.4658-1.4702	95-98	....	....
43	Human fat	1.4593-1.4607*	94-96	....	....
44	Japan wax	1.4520-1.4585‡	89-91	17.0-25.0	1.0-3
45	Lard oil	1.4607*	97	2.6	0.6
46	Lard oil (fatty tissue)	1.4609-1.4620	93-95	2.6	....
47	Laurel (bayberry)	1.4783	....	....	....
48	Linseed	1.4797-1.4802	94.5-95.5	....	0.4-1.2
49	Menhaden	1.4787	....	....	0.6-1.43
50	Montan, Reibeck	....	....	....	30-45
	Domestic	....	....	....	....
51	Mutton tallow	1.4545-1.4585*	95.5	....	....
52	Myrtle wax	1.4511*	92-94	....	....

* At 40° C; † at 65° C; ‡ at 80° C; § at 95° C.

Table 7-9 (*Continued*)
CONSTANTS OF FATS, OILS, AND WAXES

No.	Maumené number	Insoluble fatty acids			
		Melting point °C.	Solidification point or Titer	Acid value	Iodine value
1	....	25	....	....	....
2	51-54	13-14	9.5-11.5	196-207	93-96.5
3	42.5	2.3-4.5	....	197	99.4-108
4	....	44-46	....	204.5	44-56
5	....	42.5-44	37.9-46.2	197-202	41.3
6	64	23-24	17	....	114
7	....	....	....	....	....
8	43	16-17	13.4-13.7	187.1	109.6
9	....	0	....	....	....
10	....	41-43	39-43	200	48-57
11	....	28-30	31.1-32.2	....	....
12	....	38-41	33-39	210-233	36-52
13	....	....	....	....	....
14	....	20-21	....	....	....
15	....	....	17.8	....	....
16	....	....	....	....	....
17	46-7	13	3	192	87-93
18	....	....	....	....	....
19	....	....	....	....	....
20	....	44-45	39.6	215	103
21	45	19-21	13-15	....	104-114
22	....	38-40	32-34	200.8	64.6
23	....	92	....	....	....
24	....	39-57	45.2-47.2	202-208	30-55
25	21	24-27	21.2-25.2	258-273	8.4-9
26	....	48-53	47.2-49.2	190-198	33-39
27	102-113	21.8-38	17.5-24.3	204-207	130-170
28	74-86	17-20	14-16	198.4	113-126
29	75-81	34.5	....	202-208	111-115
30	....	27-45	39.9-51	....	94
31	....	17-19	17-19	201	112
32	....	....	....	....	....
33	....	50-64	46-50	....	....
34	....	....	....	....	....
35	....	....	....	....	....
36	....	....	....	....	....
37	....	36.6-40	31-34	202	65.3
38	53	23-25	18-20	187.4	99
39	36	22-25	19-20	201	91-98
40	97	17-21	15.6-16.6	....	141
41	....	30-32	....	179	....
42	46-54	31.3-53.4	33.7-45	203	72-87
43	....	35.5	....	....	64
44	....	53-56.5	54-55	214	....
45	40-47	33-38.4	27-33	....	....
46	24-28	37-46.6	36-42.4	202	64.2
47	116	....	15.1	....	82
48	103-126	20-24	16-20.6	197	179-192
49	123-128	....	....	....	....
50	....	....	....	....	....
51	....	33.5-49	40-48.5	210	34.8
52	....	47-48	....	231	....

Table 7-9 (*Continued*)
CONSTANTS OF FATS, OILS, AND WAXES

No.	Name	Index of Refraction at 25° C.	Hehner value	Acetyl value	Unsaponifiable matter
53	Neat's foot	1.4643-1.4685	94.8-95.9	7.7-9.3	0.12-0.65
54	Nutmeg	1.4700-1.4812*	....	....	....
55	Olive	1.4657-1.4667	95	10.5	0.4-1.0
56	Ouricury	1.4530-1.4555§	....	....	....
57	Ozokerite	1.4320-1.4370‡	....	....	100
58	Palm	1.4603-1.4639*	94.5-97	15.7	....
59	Palm kernel (S. America)	....	....	....	....
60	Palm kernel (W. Africa)	1.4492-1.4543*	91-91.5	7.6	....
61	Peach kernel	1.4682-1.4701	94-96	6.5	....
62	Peanut	1.4620-1.4653*	95	3.5	0.5-0.9
63	Perilla	1.4753*	95.8	....	....
64	Pistachio nut	1.4672	96	....	....
65	Plum kernel	1.4679-1.4702	95.2	....	....
66	Poppy seed	1.4739-1.4742	95.4	....	0.43
67	Porpoise—body oil	1.4622-1.4625	body 85.5 jaw 68-72	....	body 3.7 jaw 16-17
68	Pumpkin seed	1.4724-1.4739	96	....	....
69	Rabbit fat	1.4586*	99.5	....	....
70	Rape seed	1.4649-1.4659	94.5-96.3	14.75	1.48
71	Safflower	1.4769	95	16.1	....
72	Sardine	1.4763-1.4852	93.3-96	21-22	0.98
73	Seal	1.4742-1.4762	93-96	33-34	0.3-1.0
74	Sesame	1.4704-1.4717	95	....	0.9-1.3
75	Shark	1.4825	87-97	11.9	2.8-15.2
76	Soya	1.4723-1.4756	93-94.5	4.9	1.27-1.54
77	Sperm	1.4573	....	4.5-6.4	37-41
78	Spermaceti (*Cet.*)	....	....	2.0-3.0	50-55
79	Spermaceti (*Phys.*)	....	....	2.6	51.5
80	Sunflower	1.4659-1.4721	95	....	0.31
81	Tallow (oil)	....	....	....	....
82	Tallow (sevum)	....	95-96	....	....
83	Tung (*Al. ford.*)	1.515-1.520	96	....	0.4-0.8
84	Tung (*Al. mont.*)	1.515-1.520	....	....	0.4-0.8
85	Walnut	1.4770	93.4-95.4	....	....
86	Whale	1.4679-1.4724	93-95	11-23	1-4
87	White mustard seed	1.4649*	96-97	....	....
88	Wool fat	1.4784-1.4822	91	23	39-44

* At 40° C;   † at 65° C;   ‡ at 80° C;   § at 95° C.

Synthetic Waxes

No.	Name	Melting point °C.	Color
1	Acrawax (Glyco)	95-97	tan
2	Acrawax B (Glyco)	86-90	light brown
3	Acrawax C (Glyco)	140-142	light brown
4	Armowax (Armour Chemical Div.)	132	light tan
5	Carbowax 1500 (Carbide & Carbon)	30-40	white
6	Carbowax 4000	54-57	white
7	Castor Wax (Bakers Castor Oil Co	84-87	white
8	Chlorowax (Diamond Alkali Co.)	100	cream
9	Gersthofen Wax (Formerly IG Wax)	79-82	yellow
10	Opal Wax (Du Pont)	77-81	white
11	Santowax (Monsanto)	56-104	yellow

Table 7-9 (*Continued*)
CONSTANTS OF FATS, OILS, AND WAXES

No.	Maumené number	Insoluble fatty acids			
		Melting point °C.	Solidification point or Titer	Acid value	Iodine value
53	47-58.5	29-41	16-26.5	200.6	62-76
54	....	42.5	36	....	....
55	35-52	26-30	16.9-26.4	193-198	86-90
56	....	....	....	....	....
57	....	....	....	....	....
58	....	50	42.5-45.5	204-207	53
59	....	....	....	....	....
60	....	25-28.5	20-25.5	258-264	12
61	42.5	10-18	13-13.5	201-205	94-102
62	44-67	26-36	30.5-39	202	96-103
63	124	−5	....	....	200-211
64	44.5-45	17-20	13-14	....	89-96
65	45	12.4-18	....	....	95.7-104
66	71-88	20.5	17-19	199	139
67	body 50	....	....	207	126
68	....	26	26-28	197	134
69	....	39-50	35-41	218	64
70	50-67	18.5-20	11.7-13.6	185	99-106
71	....	11-17	7-12	199	148
72	....	30-34.8	28.2	177-185	....
73	....	22-23	13-17	193	186-201
74	61-68.5	25-35	24	197-201	110-116
75	....	21-22	....	....	....
76	59-61	26.2-27.5	24	198	115-140
77	51	13.4	11.9	23.6	83-86
78	....	....	....	....	....
79	....	....	....	....	....
80	60-75	22-24	18-19.8	202	124-134
81	....	43-46	35-37.5	197-202	25-40
82	....	....	40-50	....	....
83	....	31-44	....	189-198	144-159
84	....	....	....	....	....
85	103	15-20	14.3	200	150
86	85-92	14-27	10-24	....	131
87	44-49	15-16	9-10	185.8	95
88	....	41.8	....	....	17

Synthetic Waxes (*Continued*)

No.	Specific gravity $\frac{15°}{15°}$ C.	Flash point °C.	Acid value	Saponification value	Iodine value
1	1.04	230	....	....	....
2	0.97	235	2.0	....	....
3	0.97	285	10.0	....	....
4	....	250	12.0	17.0	....
5	1.151	430	....	....	....
6	1.204	535	....	....	....
7	0.98-1.00	....	2.0	175-185	3-6
8	1.62-1.70	....	....	....	....
9	1.01-1.02	....	17-25	158-178	....
10	0.98-1.00	....	2.0	175-185	2-5
11	1.097	191	....	....	....

Table 7-10
PHYSICAL AND CHEMICAL PROPERTIES OF NATURAL RESINS

For a review of modified natural resins and synthetic resins (ester gums, maleated and limed rosin, phenolated resins, etc.) see (1) for resins and of synthetic resins of interest to the paint, printing ink and allied trades: *Bull. 738* (May, 1950) of Scientific Section of Natl. Paint, Varnish & Lacquer Assoc., Inc.—Resin Index of 1950; (2) for resins of interest to the plastics trade, the latest *Modern Plastics Encyclopedia* published by Plastics Catalogue Corp.

Classification. The 3d column headed "Class," places each resin into one of the following groups:

I. Natural gum-resins of vegetable origin, which contain some resinous constituents in admixture with carbohydrate bodies, so that the resulting complex will yield some water-soluble constituents.

II. Natural resins of animal origin.

III. Natural resins of vegetable origin—largely mixtures of resin acids, alcohols and esters, with some resenes or hydrocarbon bodies. This class is divided into the following groups:

    A. Oleo-resins and balsams—very fresh exudations containing volatile solvents which act as carriers.

    B. New resins—same as IIIA—with the solvents gone completely.

Name	Source	Class	Sp. Gr.
Gum Accroides Used to color spirit varnishes and nitrocellulose lacquers, and in sealing wax. Composition is 85% p-coumaric ester of xanthoresino-tannol.	A species of *Xanthorroea* in Australia Yellow variety Red variety	III. B III. B	....
Accra Copal	W. Africa	III. C	1.033
Amber Used in jewelry; it is the hardest and most highly fossilized resin. Composition is 28% succino-abietic acid, 70% esters of succinic acid and succino-resinol.	Baltic coast and Burma. Fossil resin from extinct conifers	III. D	1.052 (1.05–1.11)
Gum Ammoniacum Composition:—20-26% gum, 65-75% resin, balance water, ash, etc.	*Dorema Ammoniacum,* from Persia and Africa	I.	1.19–1.21
Red Angola Copal Commonly used in Europe; less in in U. S. A.	Semi-fossil from Angola, W. Africa	III. C	1.066–1.068
White Angola Copal	W. Africa	III. C	1.055
Gum Arabic (Acacia) Used in adhesives and emulsions. Varieties:—gum senegal, kordofan.	Dried juice from bark of *Acacia senegal*	True gum	....
Asafoetida Composition:—25% gum, 60% resin ester, 7% essential oil. U. S. P. requires ash to be under 10%. Used in treatment of heaves.	Species of *Ferula*	I.	....
Batu An important variety of East India resin (*q.v.*).			

Table 7-10 (*Continued*)
PHYSICAL AND CHEMICAL PROPERTIES OF NATURAL RESINS

C. Semi-fossil resins—the products of the action of the elements on new resins over a relatively short time. These resins are not entirely soluble unless melted and partially depolymerized.

D. Fossil resins—those exposed for long periods of time. They are substantially insoluble.

IV. Natural hydrocarbon resins.

The 4th column gives the **specific gravity**, the 5th gives the **softening point** (S. P.), and the 6th gives the **melting point** (M. P.). The next three columns give the **acid, saponification, and iodine numbers** in that order. Figures enclosed in parentheses indicate the extreme values reported by different investigators. The 10th and last column gives the solubility of the resins in various solvents. The figures 1 to 12 refer to the following solvents: 1-ethyl alcohol; 2-methyl alcohol; 3-ether; 4-benzene; 5-acetone; 6-amyl alcohol; 7-chloroform; 8-aniline; 9-benzaldehyde; 10-carbon tetrachloride; 11-turpentine; 12-amyl acetate. The letter "s." (soluble), indicates that over 90% of the resin is soluble in the solvents numbered there-after; "p. s." (partly soluble), indicates that from 41-90% of the resin will dissolve, while "i." (insoluble), indicates that 40% or less dissolves. The **ester number** is indicated by **E**, and the methoxyl number by **M**.

S. P., °C.	M. P., °C.	Acid No.	Sap. No.	Iod. No.	Solubility
....	....	64–88	98–176	156–192	s. 1
....	....	64–106	18–25	....	....
75	120 (106–156)	98 (46–129)	140 (133–168)	58 (58–62)	s. 6, 8, 12; p. s. 1, 3, 5, 7, 9; i. 2, 4, 10, 11.
175	300 (280–315)	15 (15–34)	115 (86–145)	62 E.=71–91	v. sl. s. or i. 1, 2, 3, 4, 5, 6, 7, 8, 9, 10, 11, 12.
....	....	Resin:- 57–135	Resin:- E.=19–98	M.=8.6–11.0	....
90	>300 (305)	128 (128–143)	132 (132–162)	63–137 E.=58–62	s. 5, 6, 8, 9, 12; p. s. 1, 3; i. 2, 4, 7, 10, 11.
45	95 (95–125)	127 (57–127)	160 (132–160)	130	s. 5, 6, 8, 9, 12; p. s. 1, 2, 3, 4, 7; i. 10, 12.
....	....	2–8	56–90	0.5	s. water.
....	....	Resin.- 11–82	Resin:- E.=82–214	M.=7–18	s. in 90% alcohol over 50%

Table 7-10 (*Continued*)
PHYSICAL AND CHEMICAL PROPERTIES OF NATURAL RESINS

Name	Source	Class	Sp. Gr.
Benguela Copal An uncommon resin.	W. Africa	III. C	1.058 (1.035-1.062)
Gum Benzoin Composition:—69% cinnamic acid esters, 30% cinnamic acid, 1% or less of vanillin.	Sumatra and Siam *Styrax Benzoin*	III. A, B	1.063-1.092
Brazil Copal	S. America *Hymenae courbaril*	III. C	1.053
Cameron Copal	W. Africa	III. C	1.052
Canada Balsam Used in optometric work.	*Abies balsamea*	III. A	0.90
Coal Resin A dark colored natural hydro-carbon resin	Utah coal	IV.	1.03
Colombia Copal	S. America	III. C	1.054
Congo Copal Composition:—89% resin acids, 8.5% resenes. It is the commonest fossil varnish resin, widely used because of its low price. Loango copal is a variety of Congo.	Belgian Congo	III. C	1.061
Copaiba Balsam Oleo resin, named, as are many others, from port of shipment. Used in medicine.	S. American species of *Copaiba*	III. A	0.916-0.995
Dammar A common resin, used in varnish and lacquer. It is dissolved in coal tar solvents and alcohol, which precipitate incompatible ingredients.	East Indian trees of *Shorea* variety	III. B	1.031 (1.031-1.123)
Demerara Copal Semi fossil gum. Not in common use.	British Guiana	III. D	1.047
Dragon's Blood Used as red coloring agent.	E. Indies (Sumatra) *Calamus draco*	III. A, B	1.25
East India A very common resin, widely used in the varnish industry. Composition:—12% resin acids, 78% resenes.	Species of *Dipterocarpus* in East Indies	III. B, C	....
Gum Elemi Contains some essential oils, mixed with resins. Used in lacquers and spirit varnishes.	Mainly from the Philippines; *Canarium commune*	III. A, B	1.018-1.083

Table 7-10 (*Continued*)
PHYSICAL AND CHEMICAL PROPERTIES OF NATURAL RESINS

S. P., °C.	M. P., °C.	Acid No.	Sap. No.	Iod. No.	Solubility
65 (65-95)	165 (140-215)	123 (123-137)	157 (135-168)	61-85 E. = 50-64	s. 6, 8, 12; p. s. 1, 2, 3, 5, 7, 9; i. 4, 10, 11.
....	< 100	127-142 (98-142)	190-207 (148-207)	57-76 M. = 13-44	s. completely in 1.
50	100	123 (123-130)	133 (133-143)	123-134 E. = 1-5	s. 6, 8, 12; p. s. 1, 2, 3, 4, 5, 7, 9, 10, 11.
100 (96-110)	150 (110-150)	160 (129-160)	70 (70-168)	65-70	s. 8; p. s. 3, 6, 9, 12; i, 1, 2, 4, 5, 7, 10, 11.
....	Liquid	88-106	105-116 E. = 4-10	Refr. Index = 1.532; M. = 0	....
....	160-165	very low	0	....	s. 3, 4, 7, 10, 11; i. 1, 2, 5, 6, 12
90	>300	119	156	....	s. 6, 8, 12; p. s. 1, 3, 5, 7, 9; i. 2, 4, 10, 11.
90 (90-95)	195 (115-195)	132 (132-151)	132-179	58-59	s. 6, 8, 12; p. s. 1, 2, 3, 5, 9; i. 4, 7, 10, 11.
....	....	34-98	E. = 2-33	....	....
75	100 E. = 4-20 M. = 0	35 (21-35)	39 (31-47)	64-142	s. 3, 4, 7, 8, 9, 10, 11, 12; p. s. 1, 2, 5, 6.
90	180	98	102	....	p. s. 3, 7, 9, 12; i. 1, 2, 4, 5, 8, 10, 11.
....	100 E. = 142	11 M. = 25-34	153	54-98	s. 1, 3; p. s. 7, CS_2, petr. ether, ethyl acetate.
....	....	21	35	....	s. coal tar hydrocarbons; i. 1, esters.
....	soft	15-22 E. = 3-24	24-28 M. = 0-2.5	81-175	....

Table 7-10 (*Continued*)
PHYSICAL AND CHEMICAL PROPERTIES OF NATURAL RESINS

Name	Source	Class	Sp. Gr.
Galbanum Composition:—64% alcohol-soluble resin, 27% gum, 9% essential oil. Used in medicine.	*Ferula galbaniflua*	I.	1.109-1.133
Gamboge Composition:—65-75% resin, 20-25% gum. Used as a yellow coloring agent and as a purgative.	Siam. *Garcinia Hanburii*	I.	....
Gilsonite A natural, dark, asphaltic resin.	Mineral deposits	IV.	1.015
Guaiacum Used in medicine.	From wood of *Guajacum officinale* and *G. sanctum*	III. B	....
Gurjun Balsam	Java and Cochin, China *Dipterocarpus*	III. A	0.960-0.966
Jalap, Resin of Used in medicine; purgative.	*Exogonium jalapa*	III. B	....
Karaya Gum (Indian gum) Resembles tragacanth. Used in hair "wave set" and in foods.	India; Africa *Sterculia urens*	I.	1.461-1.480
Kauri, Bush Composition:—64.5% resin acids, 12% resenes.	A recent to semi-fossil variety of Kauri. It is the softest grade of Kauri on the market.	III. B, C	1.030-1.038
Kauri, Fossil Composition:—69% resin acids, 9.5% resenes. A common resin used for high grade varnishes.	*Agathis australis* (the New Zealand Kauri Pine)	III. C, D	1.053
Kissel Copal Said to make poor varnishes.	Africa	III. C	1.066
Gum Lac (Shellac) (Data given is for orange shellac.)	Secretion of *Coccus lacca*, the lac insect	II.	1.182 (1.113-1.214)
Madagascar Copal A hard fossil resin, used little because of its high price.	Fossil gum from Madagascar	III. D	1.056
Manila, Soft Used as a cheap shellac substitute in spirit varnishes.	E. Indies. Trees of *Hymenea* group	III. B	1.060
Manila, Hard The hardest grade used in the varnishes. A very common resin.	E. Indies	III. C	1.065

Table 7-10 (*Continued*)
PHYSICAL AND CHEMICAL PROPERTIES OF NATURAL RESINS

S. P., °C.	M. P., °C.	Acid No.	Sap. No.	Iod. No.	Solubility
....	....	Resin:- 19–40	Resin:- E.=55– 91	....	....
....	....	79–100 E.=57– 67	148 M.=0–2.4	71 (116)	s. 1; emulsifies in water.
115–120	130–140	very low	0	....	s. 3, 4, 7, 10, 11; i. 1, 2, 5, 6, 12.
....	....	23–44	M.=74– 84	....	54–74% soluble in ether.
....	Sticky liquid	....	....	....	s. 4, 7, CS_2; p. s. 1.
....	....	12–13	E.=120	M.=0	....
....	....	13–23	....	....	....
50	125	82 (74–82)	87 (79–102)	74–170	s. 1, 6, 8, 9, 12; p. s. 2, 3, 4, 5, 7; i. 10, 11.
90	185	79 (63–79)	90 E.=26– 36	120 (120–164)	s. 6, 8, 9, 12; p. s. 1, 5, 7; i. 2, 3, 4, 10, 11.
65	110	70	118	....	s. 8; p. s. 1, 3, 5, 7, 9, 12; i. 2, 4, 6, 10, 11.
95	150	61 (55–65) E.=150	201 (200–212) Unsap.= 3.5	18 (15–18)	s. alcohols, borax solutions; p. s. 3, 5, 7, CS_2, ethyl acetate; i. 4, toluene, petr. ether.
130	300	66	78.5	126	p. s. 6, 8, 9, 12; i. 1, 2, 3, 4, 5, 7, 10, 11.
45	120	145 (136–150)	185 (185–196)	91–111	s. 1, 2, 5, 6, 8, 9, 12; p. s. 3, 4, 7; i. 10, 11.
80	190	73 E.=44–50	87	90 (86–91)	s. 6, 8, 9; p. s. 1, 3, 5, 7; i. 2, 4, 10, 11.

Table 7-10 (*Continued*)
PHYSICAL AND CHEMICAL PROPERTIES OF NATURAL RESINS

Name	Source	Class	Sp. Gr.
Mastic Formerly much used in lacquers and spirit varnishes. Not so popular now.	Evergreen shrubs in Mediterranean regions. *Pistacia lentiscus*	III. B	1.057 (1.04-1.07)
Mecca Balsam (Balm of Gilead) Used as a tonic in the Orient.	*Balsamodendron gileadense* in Arabia	III. A	....
Myrrh (Bdellium is a variety.) Has 50-60% gum. Used in medicine.	Turkey, India, Arabia, Somaliland. Species of *Comniphora*	I.	....
Peru Balsam Has 55-66% cinnamein. Used in medicine and perfumery.	Central America. Black liquid from *Toluifera pereirae*	III. A	1.14-1.151
Olibanum or **Frankincense** Used in perfumery.	*Boswellia Carterii*, etc. Somaliland and S. Australia	I.	....
Pontianak Composition:—84.5% resin acids, 4% resenes. A recent to semi-fossil resin. Similar to, but harder and less odorous than soft manila. Used in spirit varnishes.	Borneo	III. B, C	1.037
Rosin or **Colophony** Largely abietic acid, $C_{20}H_{30}O_2$. Commonest and cheapest resin. Graded according to color. Used in paints, soap, paper, etc.	*Pinus* species. Mostly *Pinus palustris*. Gum obtained on distillation	III. B	1.045-1.086
Gum Sandarac Composition:—85% sandaracolic acid, 10% callitrolic acid. Common resin for use in spirit varnishes.	N. E. Africa. Small trees: *Callitris quadrivalvis*	III. B	1.073
Sierra Leone Copal The hardest of the recent resins.	W. Africa. *Guibourtia Copalifera*	III. B	1.072
Balsam Storax Used in medicine.	Levant and Greece. *Styrax officinalis*	III. A	....
Balsam Tolu	S. America. *Toluifera balsamum*	III. A	....
Gum Thus A soft, fresh oleoresin from slashed trees, yielding turpentine and rosin on distillation.	*Pinus* species	III. A	....
Gum Tragacanth Swells in water—used in adhesives.	Asia Minor and Persia. *Astragalus gummifer*	Gum	....

Table 7-10 (*Continued*)
PHYSICAL AND CHEMICAL PROPERTIES OF NATURAL RESINS

S. P., °C.	M. P., °C.	Acid No.	Sap. No.	Iod. No.	Solubility
80-93	95	63 (50-71)	79 (70-194)	64-159 E. = 23-29	s. 3, 4, 6, 7, 8, 9, 10, 11, 12; p. s. 1, 2, 5.
....	....	....	....	....	....
....	M. = ca. 13	Resin:- 42-70	Resin:- 159-216	Resin:- E. = 95-145	....
....	....	....	....	....	s. 1, 3.
....	E. = 7-131	42-50	M. = 5-7	....	....
55	135	134	186	119-142	s. 1, 5, 6, 8, 9, 12; p. s. 2, 3, 7; i. 4, 10, 11.
70-80	120-135	155-175 E. = 8-23	167-194 M. = 0	80-220	s. nearly all organic solvents; p. s. 3, petr. ether.
....	145	140 (154)	154 (142-174)	66-160 E. = 1-11 M. = 0	s. 1, 3, 6, 8, 12; p. s. 2, 5, 7, 9; i. 4, 10, 11.
60	130 (130-200)	110 (73-130)	123 (123-158)	63-133	s. 6, 8, 9, 12; p. s. 2, 3, 4, 5, 7; i. 1, 10, 11.
....	....	128-131	191-206	65	s. 3, 1 (hot).
....	....	....	....	....	....
....	....	108-145	140-161	E. = 3-60	....
....	....	....	....	....	....

Table 7-10 (*Continued*)
PHYSICAL AND CHEMICAL PROPERTIES OF NATURAL RESINS

Name	Source	Class	Sp. Gr.
Venice Turpentine	European larch. *Larix europaea*	III. A	1.094– 1.190
Zanzibar Copal Composition:—85% resin acids, 10% essential oils, 5% resenes. A common resin, little used only because of its high price. Soluble in oils after depolymerization.	Zanzibar and E. Africa *Trachylobium verrucosum*	III. D	1.054– 1.063

GLYCERIDE CONTENT OF DRYING OILS

The values in the table give the percentage content of various fatty acid esters present in the glyceride structure of the common drying oils.

The composition of different samples of the same drying oil may vary considerably. These variations are often reflected in changes in the constants of the oil, particularly the iodine value.

Oil	Saturated	Oleate	Linoleate	Linolenate	Eleostearate
Cottonseed	25	40	35	..	..
Soybean	14	26	52	8	..
Dehydrated Castor	5	10	85	..	..
Linseed	10	18	17	55	..
Perilla	7	14	16	63	..
Tung	5	7	3	..	85
Oiticica	10	6	10	..	74*

* Licanate, $(C_{17}H_{27}OCO_2)_3C_3H_5$

Table 7-10 (*Continued*)
PHYSICAL AND CHEMICAL PROPERTIES OF NATURAL RESINS

S. P., °C.	M. P. ,°C.	Acid No.	Sap. No.	Iod. No.	Solubility
E. =30-56	M. =0	67-101	81-127	144	s. 1, 3, 4, 7, petr. ether.
150	300	93 (87-93)	93 (70-93)	115-123	p. s. 8, 12; i. 1, 2, 3, 4, 5, 6, 7, 9, 10, 11.

Table 7-11
THE VITAMINS

In compiling data for this revision certain facts became evident in making additions to the previous table. Some of the new factors which have now been added have not as yet been shown to be indispensable to man. The importance of some are only now being recognized and clinical experimentation is practically just beginning. However, it is known that the animal body does not function properly where the intake of these factors is low. Some of these factors have been shown to be synthesized in the intestines of some animals by the intestinal flora and consequently it is very difficult to produce a recognizable deficiency. Almost invariably a deficiency of any factor in man is accompanied by a deficiency of other factors and this gives the chief impetus to preparation of poly-vitamin products. Thus far no factor essential for lower animals, bacteria, etc., has been found which is not also essential for man. Some signs of deficiencies are apparently attributable to more than one vitamin. Some deficiencies are only partially cured by one or another vitamin while complete cure is obtained only by administration of all vitamins concerned. Most of the vitamins are interdependent upon the presence of each other for proper functioning.

INTERNATIONAL VITAMIN STANDARDS

1 Unit vitamin A ⇌ 0.344 microgram (μg) pure vitamin A acetate ⇌ 0.3 microgram pure vitamin A alcohol; 1 unit provitamin A = 0.6 microgram β-carotene.
1 Unit vitamin B_1 ⇌ 3.3 micrograms thiamine chloride.
1 Unit vitamin C ⇌ 50 micrograms of l-ascorbic acid.
1 Unit vitamin D ⇌ 1 milligram of standard solution of irradiated ergosterol (corresponding to 0.1 microgram of the ergosterol used in its preparation or 0.025 microgram crystalline vitamin D_2 (viosterol or calciferol).

Name and Synonym	Physiological Effect	Concentrates for Medicinal Use	Sources* Plant	Animal
Vitamin A Axerophthol Provitamins A (α-, β-, and γ-carotenes, cryptoxanthin, aphanin, aphanicin, echinenone, myxoxanthin, leprotene (and probably other carotinoids) Anti-infective Anti-ophthalmic (Fat soluble)	Increases resistance to infection of eyes, sinuses, ears, kidneys, skin, and mucous membranes. Prevents xerophthalmia, hyperkeratosis of skin and mucous membranes. Promotes appetite and growth. Essential for reproduction. The vitamin is stored in the body and is required at all ages. Deficiency results in night blindness and injury to the nervous system. Adult human requirements approximately 3000 to 5000 units daily.	Carotene, obtained from carrots or vegetables, is converted by the body into vitamin A. Unsaponifiable fractions of halibut liver oil, cod liver oil and other fish liver oils.	All yellow and green vegetables. Raw carrots, yellow turnips, yellow corn, spinach, squash, canned and fresh tomatoes, sweet potatoes, oranges, prunes, apricots and pimento peppers.	Halibut liver oil, cod liver oil and other fish liver oils, fresh milk, cream, butter, cheddar and cream cheese, egg yolk, liver and kidney.
Thiamine chloride Vitamin B_1 Anti-neuritic	Promotes appetite and digestion by stimulating metabolism. Vital to health at all ages. Required by the mother for reproduction and	Synthetic thiamine chloride. From yeast and wheat germ.	Whole grain cereals, green leafy vegetables, water cress, carrots,	Milk, cheese, raw oysters, brain, kidney and liver.

* See also table: Composition and Use of Foods.

	Functions and requirements	Synthetic / Preparation	Dietary source	Occurrence
Anti-beriberi (Soluble in water and in 90% alcohol; heat labile; acid stable; adsorbed from acid solution)	lactation. Prevents the diseases beriberi and polyneuritis. Promotes normal oxidation of carbohydrates in the body and formation of fat from carbohydrates. Adult human requirements approximately 1.5 to 2.0 milligrams daily.		canned tomatoes, green corn, baked potatoes, dried lima beans, turnips and nuts.	Widely distributed over entire plant and animal kingdoms.
p-Aminobenzoic acid PAB Chromotrichia factor Anti-gray-hair factor	Influences enzyme activity. In experimental rats it apparently prevents achromotrichia. Inhibits oxidation of adrenaline. Nutrilite for certain bacteria. Antagonizes sulfanilamide and sulfapyridine.	Synthetic. Extracts of yeast.	Yeast.	
Choline (Choline is replaceable by methionine or betaine with estimated efficacy of 30%; also replaceable by dimethylthetin or dimethyl-β-propiothetin)	Maintains integrity of tissues. In chickens it is necessary for normal nutrition and egg production. Deficiency results in marked deposition of liver fat. Acts as transmethylating agent. Deficiency in young rats produces severe renal lesions. Important in lipid metabolism. Deficiency causes a type of cirrhosis of the liver, hemorrhagic degeneration of kidneys, necrosis of kidney cortex, enlargement of spleen. Deficiency results in anemia in the rat. Necessary for normal growth and lactation of rat.	Synthetic choline chloride. As lecithin in phospholipids.	Rice bran, yeast.	Liver. Present in phospholipids of practically all animal and plant cells.
Ascorbic acid Vitamin C Cevitamic acid Anti-scorbutic (Soluble in water; heat labile at alkaline reaction)	Prevents and cures scurvy. Important in formation and maintenance of connective tissue in the body; essential for normal tooth and bone formation and strength of capillary walls. Inhibits dental caries, pyorrhea, certain gum infections, anorexia, anemia, failure of wound healing and susceptibility to infection. Important in carbohydrate metabolism and in controlling infective processes. Body stores readily depleted. Adult human requirements 75 to 100 milligrams daily.	Synthetic l-ascorbic acid.	Oranges, lemons, grapefruit, raw and leafy vegetables, raw apples, tomatoes, carrots and paprika.	Adrenal cortex, pituitary, ovary, thymus, intestinal wall, and milk.
Vitamin D Anti-rachitic (Fat soluble)	Prevents and cures rickets in young and osteoporosis in adults. Promotes absorption of calcium and phosphorus, and highly important for normal metabolism of these mineral elements in the young. Prevents spasmophilia and osteomalacia. Limited storage in the body. Approximate human requirements: for premature babies 600 to 800 units with milk; for full term infants 300 to 400 units with milk; for children 300 to 400 units	The effect of this vitamin is produced by the exposure of the body to direct sunlight or ultraviolet light. Ergosterol and foods exposed to (or irradiated with) ultraviolet. Unsaponifiable fractions of liver oils of halibut, cod, and many	Practically none.	Cod, halibut and other fish liver oils, butter, egg yolk, and fish.

Table 7-11 (*Continued*)
THE VITAMINS

Name and Synonym	Physiological Effect	Sources*		
		Concentrates for Medicinal Use	Plant	Animal
	and for pregnant and nursing women 800 units daily.	other fish. Cholesterol and several of its derivatives after exposure to ultraviolet light.		
Pantothenic acid Pantothen Filtrate factor Factor II Chick anti-dermatitis factor	Prevents dermatitis in chicks. Deficiency may produce lesions in nervous system or endocrine systems, or may involve skin or hair. May play a part in maintaining integrity of nervous tissue. Necessary in treatment of peripheral neuritis along with other B-vitamins. Prevents hemorrhagic adrenals and achromotrichia in rats. Prevents duodenal ulcers and atrophy of duodenal mucosa. Essential for growth in various experimental animals and plants. Needed for reproduction in chicks and increases egg hatchability. Is necessary for acetylations in the rat. It is a constituent of coenzyme A. As coenzyme A it yields citrate from acetate and oxaloacetate.	Synthetic calcium *d*-pantothenate. Liver and yeast extracts.	Rice bran, wheat bran, brewer's yeast, Irish potato, taro root, alfalfa, wheat germ, molasses, peanuts, cabbage, cucumbers, and rolled oats.	Mammalian liver, egg yolk, eggs, canned salmon, whole milk, kidney, heart, spleen, brain, pancreas, tongue, lung, muscle, beef.
Pyridoxin Pyridoxal Pyridoxamine Vitamin B$_6$ Adermin Factor Y Anti-dermatitis Anti-acrodynia factor (Soluble in water and in dilute alcohol; heat and acid stable)	Deficiency of this vitamin causes dermatitis in rats; anemia in dogs and pigs, and is presumably essential for hemoglobin formation. Controls both iron and copper absorption in rats. In experimental animals deficiency produces convulsions resembling human epileptic fits. Requirements proportional to metabolism. Apparently concerned with metabolism of the amino acids to carbohydrates. Deficiency impairs liver function, produces erythrodermic dermatosis, impairs growth.	Synthetic pyridoxin. From the germ and bran of wheat and rice; yeast, and liver extracts.	Grains, chiefly in the germ, yeast, maize, and molasses, seeds, legumes, spinach, and lettuce.	Liver, heart muscles, milk, and egg yolk.
Riboflavin Vitamin B$_2$ Vitamin G (Soluble in water and in dilute alcohol; heat and acid stable)	Plays an important role in vision mechanism. In fowls important in production and hatchability of eggs. Deficiency produces cheilosis, glossitis, seborrheic lesions about eyes and nose, vascularizing keratitis, and mild photophobia. Is a component of several enzymes. Plays part in metabolism of amino acids and carbohydrates. Deficiency produces loss of	Synthetic *d*-riboflavin. From milk whey, liver and yeast. Fermentation residues.	Fruits, leafy vegetables, soy beans, wheat germ, yeast.	Kidney, liver, milk, eggs.

* See also table: Composition and Use of Foods.

Name / Factor	Properties and Functions	Synthetic / Other Sources	Plant Sources	Animal Sources
Vitamin E α-, β-, and γ-Tocopherols Anti-sterility (Fat soluble)	hair, injury to nervous system, failure to grow, yellow liver, anemia and death. Adult human requirements approximately 2 to 3 milligrams daily. Absence of this vitamin causes sterility in either sex. It is stored in the body to a certain degree. Exerts favorable influence on growth. Limited value in certain cases of muscular dystrophy and amyotropic lateral sclerosis. Aids in resisting protein deficiency and prevents nutritional liver damage.	Synthetic dl-α-tocopherol. Unsaponifiable fraction of wheat germ oil.	Vegetable oils, lettuce, beans, whole wheat, rice, barley, corn, and water cress.	Meat, liver, milk, and eggs.
Nicotinic acid Niacin Nicotinic acid amide Niacin amide Anti-pellagra Pellagra preventive (P-P) factor	Prevents and cures black tongue in dogs and pellagra in man. Is a component of several enzymes. Influences carbohydrate, protein, water, and heavy metals metabolism. Deficiency produces a certain type of anemia. Adult human requirement approximately 10 to 20 milligrams daily.	Synthetic nicotinic acid, its amide, and all hydrolyzable nicotinoyl derivatives. From liver and yeast.	Yeast, wheat germ, peas, beans, green leafy vegetables.	Liver, lean meat, milk, egg yolk.
Vitamin K Anti-hemorrhagic Coagulation factor Prothrombin factor Phylloquinones Menadione (2-methyl-1,4-naphthoquinone) (Fat soluble, heat stable, alkali labile)	Promotes normal blood coagulation time. In chicks it prevents intestinal, subcutaneous and intramuscular hemorrhage. Deficiency reduces prothrombin formation, hemorrhagic diathesis in newborn, certain types of hepatic and biliary diseases.	Synthetic K-vitamers; i.e., many related 1,4-naphthoquinones and compounds readily converted to them, most important is 2-methyl-1,4-naphthoquinone, which replaces commercially the vitamin, and is known commercially as menadione.	All green vegetables and green leaves, tomatoes, oat sprouts, dried carrot tops, hemp-seed oil, cotton-seed oil, soy-bean oil, and bacterial action on rice bran.	Hog-liver fat, egg yolk, bacterial action on casein and fish meal. K_2 from putrefied fish meal.
Biotin Vitamin H Coenzyme R Anti-egg-white factor Curative factor of egg-white injury Bios II$_a$ Bios II$_b$ Factor X Factor W	Essential in vital economy of animals and in normal pigment metabolism. Takes part in carbon dioxide fixation; deaminase activity stimulates lipid utilization. Important in intermediary metabolism. Prevents dermatitis. Prevents perosis in chicks and fatty livers. Deficiency results in seborrheic desquamative dermatitis, emaciation and death. Nutrilite for certain bacteria and for yeast. It is antagonized by avidin which occurs in egg white.	Synthetic. Extracts of egg yolk, liver, milk-sugar residues, fermentation residues.	Widely distributed in nature. Yeast, seeds, nuts, tomatoes, carrots, corn and other vegetables.	Liver, kidney, pancreas, egg yolk, pork, and milk. Occurs in small amounts in all mammals and birds; tissue of eye only exception found so far.
Inositol (Meso-inositol) Bios I Anti-alopecia factor	Promotes growth, cures baldness in mice, important in fat metabolism, prevents fatty liver, favors growth of yeast and other microorganisms.	Liver extracts. Wheat bran extracts in the form of the hexa-phosphate ester (phytic acid) which on hydrolysis yields inositol.	Citrus fruits, nuts, yeast, various whole cereals and grains.	Muscle, kidney, liver, brain, milk, and eggs.

Table 7-11 (*Continued*)
THE VITAMINS

Name and Synonym	Physiological Effect	Sources*		
		Concentrates for Medicinal Use	Plant	Animal
Folic acid Pteroylglutamic acid (L. casei factor)	Nutrilite for yeast and certain bacteria. Anti-anemia factor. Appears to be a growth factor for rats. Deficiency in mice is accompanied by reduction in cellular elements of blood and arrest of bone marrow maturation.	Extracts of yeast, milk sugar residue, spinach. Synthetic pteroylglutamic acid.	Yeast, mushrooms. Leaves of many plants.	Milk. Present in all animal tissues.
Citrin Vitamin P Permeability vitamin	Promotes tissue and capillary permeability. Prolongs life of scorbutic guinea pigs. Deficiency causes nutritional purpura.	Extracts of citrus fruits.	Citrus fruit juices and the leaves of several plants.	None.
Vitamin B_{12} Antipernicious factor Animal protein factor	Prevents: pernicious anemia; megaloblastic anemia of infancy; nutritional macrocytic anemia; sprue; nutritional glossitis. Required for gestation and lactation.	Liver extracts.		Liver, milk, fish meal.

* See also table: Composition and Use of Foods.

Table 7-12
FORMULAS OF THERMOPLASTIC AND THERMOSETTING MATERIALS AND SYNTHETIC RUBBERS

A. Thermoplastic Materials

(1) Polyethylene

$CH_2:CH_2 \longrightarrow [—CH_2 \cdot CH_2—]n$

Ethylene   Polyethylene
                Polythene

(2) Polystyrene

$CH(C_6H_5):CH_2 \longrightarrow [—CH(C_6H_5) \cdot CH_2—]n$

Styrene   Styron,
               Lustron

(3) Polyvinylchloride

$CH_2:CHCl \longrightarrow [—CH_2 \cdot CHCl—]n$

Vinyl   Geon,
chloride   Koroseal

(4) Polyvinylchloride Acetate

$CH_2:CHCl + CH_2:CH(OAc) \longrightarrow [—CH_2 \cdot CHCl \cdot CH_2 \cdot CH(OAc) \cdot CH_2 \cdot CHCl—]n$

Vinyl   Vinyl   Vinylite V
chloride   acetate

(5) Polyvinylidene Chloride

$CH_2:CHCl + CH_2:CCl_2 \longrightarrow [—CH_2 \cdot CCl_2 \cdot CH_2 \cdot CHCl—]n$

Vinyl   Vinylidene   Saran,
chloride   chloride   Geon, Velon

(6) Polytetrafluoroethylene

$CF_2:CF_2 \longrightarrow [—CF_2 \cdot CF_2—]n$

Tetra-   Teflon
fluoro-
ethylene

(7) Polyvinyl Butyral

$[—CH_2 \cdot CH(OAc)—]n \longrightarrow [—CH_2 \cdot CH(OH)—]n + C_3H_7 \cdot CHO \longrightarrow$

Polyvinyl   Polyvinyl   Butyr-
acetate   a,cohol   aldehyde

$$\left[\begin{array}{c} —CH_2 \cdot CH \cdot CH_2 \cdot CH— \\ | \quad\quad\quad\quad | \\ O—CH—O \\ | \\ C_3H_7 \end{array} \right]n$$

Saflex, Butacite,
Vinylite X

(8) Polymethyl Methacrylate

$CH_2:C(CH_3) \cdot CO \cdot OCH_3 \longrightarrow [—CH_2 \cdot C(CH_3)(CO \cdot OCH_3)—]n$

Methyl methacrylate   Plexiglas, Lucite

(9) Nylon

$C_6H_{11}OH \quad N:C(CH_2)_4C:N \longleftarrow [(CH_2)_2CO_2H]_2 + H_2N(CH_2)_6NH_2$
Cyclohexanol

           Adiponitrile   Adipic acid   Hexamethylene
                                                 diamine

$\longrightarrow [—CO \cdot (CH_2)_4 \cdot CO \cdot NH \cdot (CH_2)_6 \cdot NH—]n$

                           Nylon

(10) Cellulose Nitrate

$C_6H_{10}O_5 + HNO_3 \longrightarrow [—C_6H_7O_2(OH)(ONO_2)_2—]n$

Cellulose   Nitric   Celluloid
          acid

(11) Cellulose Acetate

$C_6H_{10}O_5 + Ac_2O \longrightarrow [—C_6H_7O_2(OH)(OAc)_2—]n$

Cellulose   Acetic   Lumarith,
          anhydride   Tenite

(12) Cellulose Acetate Butyrate

$C_6H_{10}O_5 + Ac \cdot O \cdot COC_3H_7 \longrightarrow [—C_6H_7O_2(OH)(OAc)(O_2CC_3H_7)—]n$

Cellulose   Acetic-butyric   Tenite II
          anhydride

(13) Ethyl Cellulose

$C_6H_{10}O_5 + C_2H_5Cl \longrightarrow [—C_6H_7O_2(OH)(OC_2H_5)_2—]n$

Cellulose   Ethyl   Ethocel
          chloride

Table 7-12 (*Continued*)
FORMULAS OF THERMOPLASTIC AND THERMOSETTING MATERIALS AND SYNTHETIC RUBBERS

B. Thermosetting Materials

(14) Phenol-Formaldehyde

C_6H_5OH + HCHO $\longrightarrow$ [$-C_6H_2(OH)CH_2-$]n

  Phenol   Formal-   Durez,
           dehyde   Resinox

(15) Phenol-Furfural

C_6H_5OH + $C_4H_3O(CHO)$ $\longrightarrow$ [$-C_6H_3(OH)\cdot\overset{|}{C}H\cdot C_4HO=$]$n$

  Phenol   Furfural   Durite

(16) Urea-Formaldehyde

$CO(NH_2)_2$ + HCHO $\longrightarrow$ [$-CH_2\cdot\overset{|}{N}\cdot CO\cdot NH-$]$n$

  Urea   Formal-   Plaskon,
          dehyde   Beetle

(17) Melamine-Formaldehyde

$H_2N\cdot CN$ $\longrightarrow$ $C_3H_3(NH_2)_3$ + HCHO $\longrightarrow$ [$-NH\cdot C_3N_3:(NHCH_2-)_2$]$n$

  Cyan-   Melamine   Formal-   Melmac,
  amide   dehyde   Resimene

(18) Polyester

$(CHCO)_2O$ + $(HOCH_2CH_2)_2O$ + $C_6H_5CH{:}CH_2$

  Maleic   Diglycol   Styrene
  anhydride

$$\longrightarrow \begin{bmatrix} -CH-CH\cdot CH_2\cdot CH(C_6H_5)- \\ \overset{|}{O}{:}C \ \ O{:}\overset{|}{C}\cdot O\cdot(CH_2)_2\cdot O\cdot(CH_2)_2\cdot O- \end{bmatrix} n$$

(19) Silicone Rubber

$Cl_2Si(CH_3)_2$ $\longrightarrow$ $(HO)_2Si(CH_3)_2$ $\longrightarrow$ [$-Si(CH_3)_2\cdot O-$]n

  Dimethylchloro-   Silastic
  silane

C. Synthetic Rubbers

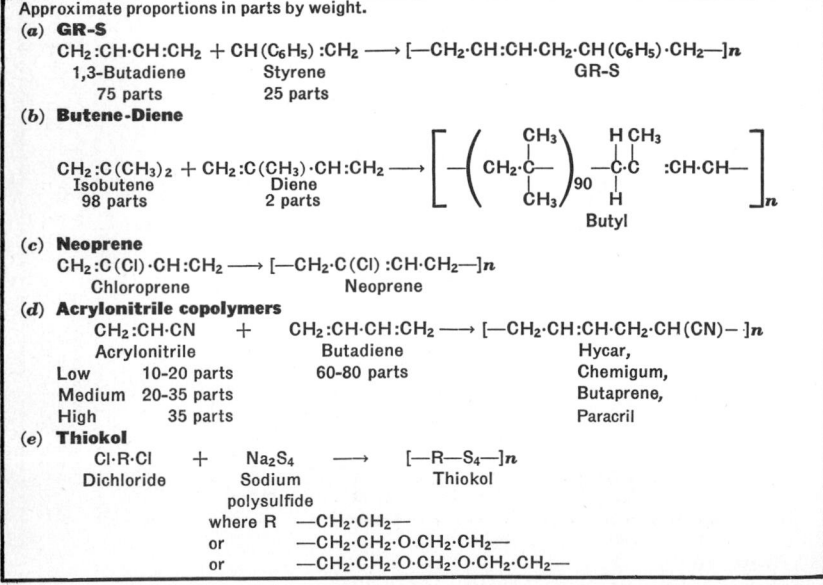

Approximate proportions in parts by weight.

(a) GR-S

$CH_2{:}CH\cdot CH{:}CH_2$ + $CH(C_6H_5){:}CH_2$ $\longrightarrow$ [$-CH_2\cdot CH{:}CH\cdot CH_2\cdot CH(C_6H_5)\cdot CH_2-$]$n$

  1,3-Butadiene   Styrene   GR-S
   75 parts   25 parts

(b) Butene-Diene

$CH_2{:}C(CH_3)_2$ + $CH_2{:}C(CH_3)\cdot CH{:}CH_2$ $\longrightarrow$ $\left[-\left(CH_2\cdot\overset{CH_3}{\underset{CH_3}{\overset{|}{\underset{|}{C}}}}\right)_{90}\overset{H\ CH_3}{\underset{H}{-\overset{|}{C}\cdot\overset{|}{C}}}{:}CH\cdot CH-\right]n$

  Isobutene   Diene
  98 parts   2 parts   Butyl

(c) Neoprene

$CH_2{:}C(Cl)\cdot CH{:}CH_2$ $\longrightarrow$ [$-CH_2\cdot C(Cl){:}CH\cdot CH_2-$]$n$

  Chloroprene   Neoprene

(d) Acrylonitrile copolymers

$CH_2{:}CH\cdot CN$ + $CH_2{:}CH\cdot CH{:}CH_2$ $\longrightarrow$ [$-CH_2\cdot CH{:}CH\cdot CH_2\cdot CH(CN)-$]$n$

  Acrylonitrile   Butadiene   Hycar,
  Low   10-20 parts   60-80 parts   Chemigum,
  Medium 20-35 parts   Butaprene,
  High   35 parts   Paracril

(e) Thiokol

$Cl\cdot R\cdot Cl$ + Na_2S_4 $\longrightarrow$ [$-R-S_4-$]n

  Dichloride   Sodium   Thiokol
              polysulfide

        where R $-CH_2\cdot CH_2-$
        or $-CH_2\cdot CH_2\cdot O\cdot CH_2\cdot CH_2-$
        or $-CH_2\cdot CH_2\cdot O\cdot CH_2\cdot O\cdot CH_2\cdot CH_2-$

Table 7-13

PROPERTIES OF THERMOPLASTIC AND THERMOSETTING MATERIALS

A. Thermoplastic Materials*

Property	(1) Poly-ethylene	(2) Poly-styrene	(3) Polyvinyl Chloride Non-Rigid	(4) Polyvinylchloride Acetate		(5) Poly-vinylidene Chloride	(6) Poly-tetrafluoro-ethylene
				Rigid	Non-Rigid		
Fabrication							
Bulk factor	2.2-3.6	2.0-2.3	......	2	2	2	......
Injectn. moldg. temp., °F	325-375	325-500	330-375	280-300	300-330	300-400	......
Injectn. moldg. press., psi $\times 10^{-3}$	4.5-30	10-30	15-25	18-30	7-20	10-30	......
Mold shrinkage, mils/in.	20-50	2-8	17	1	20-100	5-15	......
Physical							
Specific gravity	0.92	1.06	1.18-1.65	1.35-1.45	1.15-1.45	1.65-1.72	2.1-2.3
Specific volume, in.^{3}/lb	30	26	16-20	19-21	17-23	16-17	12-13
Coefficient thermal expansion, linear $\times 10^5$	18	6-8	7	6.9	7-25	19	9
Specific heat, cal/g	0.52	0.32	0.36	0.24	0.3-0.5	0.32	0.25
Thermal conductivity $\times 10^4$	8	1.9	3.9	4	3.9	2.2	6
Heat distortion temperature, °F	115-122	165-190	......	125-135	......	150-180	266
Heat resistance—continuous, °F	......	150-170	125	130	150	160-200	550
Flammability, in./min	1.1-1.4	2.0	0-0.3	0	0-0.7	None	None
Water absorption, %	0.01-0.03	0.04-0.06	0.4-1.3	0.15	0.1-1.0	<0.1	0
Mechanical							
Impact strength, Izod ft-lbs/in.	No break	0.26-0.6	......	0.25-0.50	......	0.3-1.0	4
Tensile strength, psi $\times 10^{-3}$	1.4-2.4	3-8.5	1-2.6	4.5-8	1-3.2	4-7	2-4.5
Elongation at break, %	200-600	1.5-3.5	200-390	......	225-370	15-17	200-300
Flexural strength, psi $\times 10^{-3}$	1.5-1.7	4.8-19	......	7.5-13	......	4.5-5.5	2.0
Compressive strength, psi $\times 10^{-3}$	......	11.5-17	......	9-10	......	......	1.7
Electrical							
Dielec. strength, short, v/mil	400-475	500-700	325-425	400-425	150-400	350-400	450
Vol. resistivity, ohms-cm	10^7-10^9	10^7-10^{19}	10^{11}-10^{15}	>10^{14}	10^7	10^{14}-10^{16}	10^{16}
Dielec. constant, 60 cycles	2.3	2.5-2.7	6.2-6.4	3.2-3.3	8.1-8.3	3-5	2
" , 1,000 cycles	2.3	2.5-2.7	4.2-4.9	3.1-3.2	6.9-8.4	3-5	2
" , 10^6 cycles	2.3	2.5-2.7	......	3.0-3.1	......	3-5	2
Power factor, 60 cycles	0.3-0.5	0.06-0.5	97-150	7-10	50-100	30-80	0.2
" , 1,000 cycles	0.3-0.5	0.05-0.5	109-115	11-13	85-108	30-150	0.2
" , 10^6 cycles	0.3-0.5	0.10-0.5	......	18-19	......	30-50	0.2

* The values were chosen as the most representative for each type. Considerable variation may occur since the properties of a molded article depend not only on the plastic used but on many other factors including conditions of forming and design of the molded part itself. Improvement of specific properties may often be accomplished by changes in formulation. Often improvement in one property may be accomplished at the expense of other properties.

Table 7-13 (Continued)
PROPERTIES OF THERMOPLASTIC AND THERMOSETTING MATERIALS

A. Thermoplastic Materials

Property	(7) Polyvinyl Butyral Plasticized	(8) Polymethyl Methacrylate	(9) Nylon	(10) Cellulose Nitrate	(11) Cellulose Acetate	(12) Cellulose Acetate Butyrate	(13) Ethyl Cellulose
Fabrication							
Bulk factor	......	1.7-2.5	2.3	......	2.0-2.6	2.0-2.4	2.0-2.5
Injectn. moldg. temp., °F	250-340	325-480	510-600	......	300-500	330-430	350-500
Injectn. moldg. press., psi $\times 10^{-3}$	15-30	10-30	Low	......	8-32	8-32	3-30
Mold shrinkage, mils/in.	4-20	3-6	12-15	......	1-7	1-5	1-10
Physical							
Specific gravity	1.05-1.50	1.18-1.20	1.14-1.16	1.35-1.57	1.26-1.5	1.1-1.23	1.07-1.18
Specific volume, in.^{3}/lb	18-26	18-20	24	20	20-22	22	23-26
Coefficient thermal expansion, linear $\times 10^5$	......	7-9	10.3	9-16	8-16	11-17	9-16
Specific heat, cal/g	0.4	0.35	0.43	0.34-0.38	0.4	0.4	0.36
Thermal conductivity $\times 10^4$	......	4-6	6	3.1-5.1	4-8	4-8	3-5
Heat distortion point, °F	......	135-200	165-170	110-150	100-235	103-182	100-180
Heat resistance—continuous, °F	......	120-140	......	Poor	140-220	140-220	140-220
Flammability, in./min	......	0.5-1.0	Self ext.	V. rapid	2-self ext.	1.5-self ext.	2-self ext.
Water absorption, %	1-2	0.3-0.6	1.5	0.7-4.0	2-6	1.3-2.4	0.7-4.0
Mechanical							
Impact strength, Izod ft-lbs/in.	0.5-3.5	0.2-0.4	0.6-0.9	0.2-8.0	0.5-5.7	0.4-9.4	3-11
Tensile strength, psi $\times 10^{-3}$	150-450	4-6	9-10.5	3-10	1.5-8	1.4-6.5	2.5-8
Elongation at yield, %	......	1-10	45-55	25-50	10-75	35-100	2-100
Flexural strength, psi $\times 10^{-3}$	......	10-20	11-13	6-15	2-16	1.6-11	......
Compressive strength, psi $\times 10^{-3}$	......	9-15	14-16	20-30	10-37	......	8-20
Electrical							
Dielec. strength, short, v/mil	175-375	500	350-400	250-365	250-265	250-400	470-550
Vol. resistivity, ohms-cm	10^{10}	10^{15}-10^{19}	10^{11}-10^{13}	10^{11}	10^{10}-10^{12}	10^{10}-10^{12}	10
Dielec. constant, 60 cycles	5.6	3.4-3.6	4-5	6.7-7.3	3.7-7.5	3.5-6.4	3.2-4
" , 1,000 cycles	4-13	3.3-3.5	4.5-5	......	3.5-7.0	......	3.3-8
" , 10^6 cycles	4	2.8-3.3	3.4-4.0	......	3.2-7.0	......	3.3-7
Power factor, 60 cycles	115	50-60	14-50	60-150	10-60	10-40	8-30
" , 1,000 cycles	10-100	40-70	20-50	......	10-60	......	8-15
" , 10^6 cycles	20-100	28-33	40-70	......	10-100	......	17-36

B. Thermosetting Materials*
(14) Phenol-Formaldehyde

Property	No Filler	Wood Flour	Mica	Asbestos	Fabric	Sisal Felt	No Filler Cast
Fabrication							
Bulk factor	2.3	2.2	2.5	2-12	4-18	2-5	......
Molding temp., °F	300	300-350	350	350	350	300	200
Molding press., psi $\times 10^{-3}$	3	4	5	2-6	8	3	None
Colors	Dark			Limited			
Mold shrinkage, mils/in.	10	7	3	3	6	20	......
Molding qualities	Fair	Excellent	Fair	Good-Fair	Fair	Fair	......
Physical							
Specific gravity	1.27	1.37	1.6-2.0	1.80	1.40	0.7-1.4	1.30
Specific volume, in.^{3}/lb	22	20	14-17	1.6	19-21	20-40	22
Coefficient thermal expansion, linear $\times 10^5$	2.5-6	3-4.4	2	2	1-3	7-20	5-15
Specific heat, cal/g	0.40	0.4	0.28-0.32	0.28-0.32	0.30-0.35	0.3-0.4	0.4
Thermal conductivity $\times 10^4$	3-6	4-7	10-14	8-16	4-7	2-8	3-5
Heat distortion point, °F	240-260	260-300	210-320	285-350	250-300	320	100-170
Heat resistance—continuous, °F	250	300	250-300	350-400	250	250	160
Water absorption, %	0.1-0.2	0.8	0.02-0.1	0.1-0.3	0.5-1.8	0.5-15	0.2-2.0
Mechanical							
Impact strength, Izod ft-lbs/in.	0.4-0.5	0.2-0.4	......	......	1-8	3-16	0.3-0.4
Tensile strength, psi $\times 10^{-3}$	7-10	6.5-9.5	4.5-7	4-8	4-8	7-12	2-9
Elongation at yield, %	1-1.5	0.4-0.8	......	......	0.7	......	......
Flexural strength, psi $\times 10^{-3}$	12-17	9-12	7-13	7-13	9-13	16-20	6-9
Compressive strength, psi $\times 10^{-3}$	10-30	24-32	15-25	15-25	15-30	10-35	9-25
Electrical							
Dielec. strength, short, v/mil	300-400	300-375	325-500	200-350	200-325	25-400	75-450
Vol. resistivity, ohms-cm	10^{12}	10^{11}	10^{11}-10^{14}	10^9-10^{12}	10^{10}-10^{12}	10^{11}-10^{12}	10^{11}-10^{12}
Dielec. constant, 60 cycles	5-6	5.5-7	5-6	12-50	6-10	5-10	6.5-30
" , 1,000 cycles	4-5	4.8-6	4.5-5.5	18-50	5.5-9	......	5-18
" , 10^6 cycles	4.5-5.5	4.5-5.5	4.5-5.2	5.5-8.0	4.5-6	3-5	5-11
Power factor, 60 cycles	0.3-0.65	0.16-0.84	0.01-0.06	0.15-0.4	0.08-0.3	0.1-0.3	0.4-4.0
" , 1,000 cycles	0.14-0.45	0.14-0.36	0.01-0.04	0.1-0.4	0.04-0.18	......	1.2-3.6
" , 10^6 cycles	0.07-0.15	0.13-0.28	0.05-0.03	0.05-0.2	0.03-0.06	0.3-0.5	0.05-1.1
Arc resistance, secs	......						120-250

* The values were chosen as the most representative for each type. Considerable variation may occur since the properties of a molded article depend not only on the plastic used but on many other factors including conditions of forming and design of the molded part itself. Improvement of specific properties may often be accomplished by changes in formulation. Often improvement in one property may be accomplished at the expense of other properties.

Table 7-13 (Continued)

PROPERTIES OF THERMOPLASTIC AND THERMOSETTING MATERIALS

B. Thermosetting Materials

Property	(15) Phenol-Furfural		(16) Urea-Formaldehyde	(17) Melamine-Formaldehyde			(18) Poly-ester	(19) Silicone Rubber
	Wood Flour	Fabric	Cellulose	Cellulose	Asbestos	Fabric	None	Mineral
Fabrication								
Bulk factor	2.2-3	2.5	2.2-3	2.2-2.7	2.1-2.5	5-10	Cast	1-1.5
Molding temp., °F	280-400	270-360	275-370	275-370	275-370	275-340		300
Molding press., psi ×10⁻³	1-5	1-8	1.56	1-6	1-7	3-5		0.1-0.8
Colors	Limited		Unltd.	Unltd.	Gray			Unltd.
Mold shrinkage, mils/in.	5-10	3-70	5-11	6-12	3-5	3-5		4-6
Molding qualities	Excellent	Good	Excellent	Excellent	Good	Good	Excellent	Good
Physical								
Specific gravity	1.32-1.42	1.3-1.45	1.45-1.55	1.40-1.55	1.7-2.00	1.5	1.32-1.40	1.4-2.0
Specific volume, in.³/lb	20	20	18-19	18-19	14-16	18	20-21	14-20
Coefficient thermal expansion, linear ×10⁵	3-7	2-16	2.5-3	2.5-4.5	2-4.5		5.5-10	.....
Specific heat, cal/g	0.3-0.4	0.3-0.4	0.4				0.3-0.5	.....
Thermal conductivity ×10⁴	3-12	3-8	7-10	10	13.7	11	5	.....
Heat distortion point, °F	240-300	240-285	260-280	285-300	265	310	140-200	.....
Heat resistance—continuous, °F	300	240-300	180	210	250-400	250	220	500
Water absorption, %	0.2-0.6	0.5-2.5	0.75-2.0	0.08-1.7	0.08-1.4	0.3-0.6	0.2	0.25-1.0
Mechanical								
Impact strength, Izod ft-lbs/in.	0.2-0.4	0.6-4.8	0.24-0.36	0.24-0.35	0.3-0.4	1.0	0.3-0.4	0.200-0.60
Tensile strength, psi×10⁻³	5-8.5	5.5-8	6-13	7-13	6-7	7-8	5-6	100-225
Elongation at yield, %								
Flexural strength, psi×10⁻³	8-15	8-13	10-18	9-16	9-11	10-15	8-14	.....
Compressive strength, psi×10⁻³	16-36	20-32	27-43	30	21-32	30	19-23	.....
Electrical								
Dielec. strength, short, v/mil	300-550	150-450	300-400	300-400	350-400	250-350	570-1275	290-450
Vol. resistivity, ohms-cm	10⁹-10¹²	10⁹-10¹¹	10¹²-10¹³	10¹⁰-10¹³	10¹¹		10¹⁴	10¹¹
Dielec. constant, 60 cycles	5-12	5-10	7-9.5	8.0-9.5	6.4-9.9	7.7-8.3	3.45	.....
" , 1,000 cycles	4-10	4-9	6.4-9	7.6-8.7		7.2-7.6	3.50	.....
" , 10⁶ cycles	4-9	4-8	6.3-7.5	7.2-8.0	6.7	6.7	3.6	3.2-7.4
Power factor, 60 cycles	0.04-6.3	0.06-0.3	0.035-0.045	0.025-0.075	0.07-0.17	0.075-0.12	0.2	.....
" , 1,000 cycles	0.04-0.8	0.06-0.2	0.027-0.055	0.015-0.035		0.035-0.05	0.014	.....
" , 10⁶ cycles	0.03-0.1	0.03-0.1	0.027-0.04	0.025-0.45	0.041	0.035-0.036	0.06	0.003
Arc resistance, secs			100-140	125	120-140		126	

Table 7-14.
PROPERTIES OF NATURAL AND SYNTHETIC RUBBERS*

	Natural Rubber	(a) Styrene-Butadiene Copolymer, GR-S	(b) Butene-Diene Copolymer	(c) Neoprene Poly-chloroprene	(d) Butadiene-Acrylonitrile Copolymers — Acrylonitrile Content			(e) Poly-sulfide Thiokol
					Low	Medium	High	
Density	0.92	0.94	0.915	1.23	0.96	0.98	1.00	1.35
Refractive index	1.52	1.53	1.51	1.56	1.54	1.52	1.54	1.65
Specific heat, cal/g	0.452	0.454	0.464	....	....	....	....	....
Gum Stocks:								
Tensile strength, psi	3,100	300	2,000	2,800	500	700	900	1,000
Elongation, %	775	380	800	600	400	400	500	200
Black-loaded Stocks:								
Tensile strength, psi	3,900	3,000	2,200	3,600	3,000	3,500	4,000	1,000
Elongation, %	780	650	600	350	400	450	400	200
Stress 300%, psi	1,400	1,200	8,000	2,000	1,500	1,500	1,500	....
Swelling, % by volume, in:								
Kerosene at 25°C	200	100	300	60	10	8	4	4
Benzene at 25°C	200	200	300	150	....	120	....	50
Acetone at 25°C	25	30	30	20	60	100	150	25
Mineral oil at 70°C	120	150	130	10	10	5	2	1
Brittle point, °C	−56	−60	−45	−40	−40	−30	−1	−35
Rel. permeability to H_2	50	....	....	12	....	20	....	4
Rel. permeability to air	11	....	1	....	....	....	....	....
Insulation resist, ohms/cm	10^{17}	10^{15}	10^{16}	10^{10}	10^{10}	10^{10}	10^{10}	10^{15}
Resilience, %	90	75	50	75	....	74	63	62
Tear resistance, psi	1,640	550	1,000	1,100	....	....	....	....
Creep, 70°C	26	14.6	....	62	....	17	....	....

* The values were chosen as the most representative for each type. Considerable variation may occur since properties of a rubber compound can be varied greatly by choice of fillers, softeners, resins, accelerators, condition of cure, and numerous other factors.

Section 8

SPECTROSCOPY

X-RAY METHODS

An X-ray tube operating at a voltage V (in keV) emits a continuous X-ray spectrum, the minimum wavelength of which is given by $\lambda_{min} = 12.398/V$ with the wavelength expressed in angstroms. For expressing the wavelength in kX units, divide by the factor 1.00202. Tables 8-1 and 8-2 are based on the K and L wavelength values as published by Y. Cauchois and H. Hulubei (Tables de Constantes et Données Numériques, I. Longueurs d'Onde des Émissions X et des Discontinuités d'Absorption X, Hermann, Paris, 1947) and by the International Union of Crystallography [International Tables for X-ray Crystallography, Kynoch Press, Birmingham, England, 1962]. Wavelength accuracy is only to about 1 in 25,000 except for the lines employed in X-ray diffraction work.

Use of energy-proportional detectors for X-rays creates a need for energy values of K and L absorption edges (Table 8-3) and emission series (Table 8-4). These values were obtained by a conversion to keV of tabulated experimental wavelength values and smoothed by a fit to Moseley's law. Although values are listed to 1 eV, chemical form may shift absorption edges and emission lines as much as 10 to 20 eV. S. Fine and C. F. Hendee [*Nucleonics*, **13**(3), 36 (1955)] also give values for $K\beta_2$, $L\gamma_1$, and $L\beta_2$ lines.

The relative intensities of X-ray emission lines from targets varies for different elements. However, one can assume a ratio of $K\alpha_1/K\alpha_2 = 2$ for the commonly used targets. The ratio of $K\alpha_2/K\beta_1$ from these targets varies from 6 to 3.5. The intensities of $K\beta_2$ radiations amount to about one per cent of that of the corresponding $K\alpha_1$ radiation. In practical applications these ratios have to be corrected for differential absorption in the window of the tube and air path, the ratio of scattering factors for, and differential absorption in the crystal, and for sensitivity characteristics of the detector. Generalizing, the intensities of radiations from the K and L series are as follows:

Emission line	$K\alpha_1$	$K\alpha_2$	$K\beta_1$	$K\beta_2$	$L\alpha_1$	$L\alpha_2$	$L\beta_1$	$L\beta_2$	$L\gamma_1$
Relative intensity	500	250	80–150	5	100	10	30	60	40

For angles at which the $K\alpha_1$, $K\alpha_2$ doublet is not resolved, a mean wavelength [$K\bar{\alpha} = (2K\alpha_1 + K\alpha_2)/3$] can be used.

Table 8-1
WAVELENGTHS OF X-RAY EMISSION SPECTRA
IN ANGSTROMS

Atomic No.	Element	$K\alpha_2$	$K\alpha_1$	$K\beta_1$	$L\alpha_1$	$L\beta_1$
3	Li	240				
4	Be	113				
5	B	67				
6	C	44				
7	N	31.60				
8	O	23.71				
9	F	18.31				
10	Ne	14.616		14.464		
11	Na	11.909		11.617	407.6	
12	Mg	9.889		9.558	251.0	
13	Al	8.3392	8.3367	7.981	169.8	
14	Si	7.1277	7.1253	6.7681	123	
15	P	6.1549		5.8038		
16	S	5.3747	5.3720	5.0317		
17	Cl	4.7305	4.7276	4.4031		
18	Ar	4.1946	4.1916	3.8848		
19	K	3.7446	3.7412	3.4538	42.7	
20	Ca	3.3616	3.3583	3.0896	36.32	35.95
21	Sc	3.0345	3.0311	2.7795	31.33	31.01
22	Ti	2.75207	2.7484	2.5138	27.39	27.02
23	V	2.5073	2.5035	2.2843	24.26	23.85
24	Cr	2.29351	2.28962	2.08480	21.67	21.28
25	Mn	2.1057	2.1018	1.9102	19.45	19.12
26	Fe	1.93991	1.93597	1.75653	17.567	17.255
27	Co	1.79278	1.78892	1.62075	15.968	15.667
28	Ni	1.66169	1.65784	1.50010	14.566	14.279
29	Cu	1.54433	1.54051	1.39217	13.330	13.053
30	Zn	1.4389	1.4351	1.2952	12.257	11.985
31	Ga	1.3439	1.3400	1.20784	11.290	11.023
32	Ge	1.2580	1.2540	1.1289	10.435	10.174
33	As	1.1798	1.1758	1.0573	9.671	9.414
34	Se	1.1088	1.1047	0.9921	8.990	8.736
35	Br	1.0438	1.0397	0.9327	8.375	8.125
36	Kr	0.9841	0.9801	0.8785	7.822	7.574
37	Rb	0.9296	0.9255	0.8286	7.3181	7.076
38	Sr	0.8794	0.8752	0.7829	6.8625	6.6237
39	Y	0.8330	0.8279	0.7407	6.4485	6.2117
40	Zr	0.7901	0.7859	0.7017	6.0702	5.8358
41	Nb	0.7504	0.7462	0.6657	5.7240	5.4921
42	Mo	0.713543	0.70926	0.632253	5.4063	5.1768
43	Tc	0.6793	0.6749	0.6014	5.1126	4.8782
44	Ru	0.6474	0.6430	0.5725	4.8455	4.6204
45	Rh	0.6176	0.6132	0.5456	4.5973	4.3739
46	Pd	0.5898	0.5854	0.5205	4.3676	4.1460
47	Ag	0.563775	0.559363	0.49701	4.1541	3.9344

Table 8-1 (*Continued*)
WAVELENGTHS OF X-RAY EMISSION SPECTRA
IN ANGSTROMS

Atomic No.	Element	$K\alpha_2$	$K\alpha_1$	$K\beta_1$	$L\alpha_1$	$L\beta_1$
48	Cd	0.5394	0.5350	0.4751	3.9563	3.7381
49	In	0.5165	0.5121	0.4545	3.7719	3.5552
50	Sn	0.4950	0.4906	0.4352	3.5999	3.3848
51	Sb	0.4748	0.4703	0.4171	3.4392	3.2256
52	Te	0.4558	0.4513	0.4000	3.2891	3.0767
53	I	0.4378	0.4333	0.3839	3.1485	2.9373
54	Xe	0.4204	0.4160	0.3685	3.016	2.807
55	Cs	0.4048	0.4003	0.3543	2.9016	2.8920
56	Ba	0.3896	0.3851	0.3408	2.7752	2.5674
57	La	0.3753	0.3707	0.3280	2.6651	2.4583
58	Ce	0.3617	0.3571	0.3158	2.5612	2.3558
59	Pr	0.3487	0.3441	0.3042	2.4627	2.2584
60	Nd	0.3565	0.3318	0.2933	2.3701	2.1666
61	Pm	0.3249	0.3207	0.2821	2.282	2.0796
62	Sm	0.3137	0.3190	0.2731	2.1994	1.9976
63	Eu	0.3133	0.2985	0.2636	2.1206	1.9202
64	Gd	0.2932	0.2884	0.2544	2.0460	1.8462
65	Tb	0.2834	0.2788	0.2460	1.9755	1.7763
66	Dy	0.2743	0.2696	0.2376	1.9088	1.7100
67	Ho	0.2655	0.2608	0.2302	1.8447	1.6468
68	Er	0.2572	0.2525	0.2226	1.7843	1.5873
69	Tm	0.2491	0.2444	0.2153	1.7263	1.5299
70	Yb	0.2415	0.2368	0.2088	1.6719	1.4756
71	Lu	0.2341	0.2293	0.2021	1.6194	1.4235
72	Hf	0.2270	0.2222	0.1955	1.5696	1.3740
73	Ta	0.2203	0.2155	0.1901	1.5219	1.3270
74	W	0.213813	0.208992	0.184363	1.4764	1.2818
75	Re	0.2076	0.2028	0.1789	1.4329	1.2385
76	Os	0.2016	0.1968	0.1736	1.3911	1.1972
77	Ir	0.1959	0.1910	0.1685	1.3513	1.1578
78	Pt	0.1904	0.1855	0.1637	1.3130	1.1198
79	Au	0.1851	0.1802	0.1590	1.2764	1.0836
80	Hg	0.1799	0.1750	0.1544	1.2411	1.0486
81	Tl	0.1750	0.1701	0.1501	1.2074	1.0152
82	Pb	0.1703	0.1654	0.1460	1.1750	0.9822
83	Bi	0.1657	0.1608	0.1419	1.1439	0.9520
84	Po	0.1608	0.1559	0.1382	1.1138	0.9222
85	At	0.1570	0.1521	0.1343	1.0850	0.8936
86	Rn	0.1529	0.1479	0.1307	1.0572	0.8659
87	Fr	0.1489	0.1440	0.1272	1.030	0.840
88	Ra	0.1450	0.1401	0.1237	1.0047	0.8137
89	Ac	0.1414	0.1364	0.1205	0.9799	0.7890
90	Th	0.1378	0.1328	0.1174	0.9560	0.7652
91	Pa	0.1344	0.1294	0.1143	0.9328	0.7422
92	U	0.1310	0.1259	0.1114	0.9105	0.7200

Table 8-1 (*Continued*)
WAVELENGTHS OF X-RAY EMISSION SPECTRA
IN ANGSTROMS

Atomic No.	Element	$K\alpha_2$	$K\alpha_1$	$K\beta_1$	$L\alpha_1$	$L\beta_1$
93	Np	0.1278	0.1226	0.1085	0.8893	0.6984
94	Pu	0.1246	0.1195	0.1058	0.8682	0.6777
95	Am	0.1215	0.1165	0.1031	0.8481	0.6576
96	Cm	0.1186	0.1135	0.1005	0.8287	0.6388
97	Bk	0.1157	0.1107	0.0980	0.8098	0.6203
98	Cf	0.1130	0.1079	0.0956	0.7917	0.6023
99	Es	0.1103	0.1052	0.0933	0.7740	0.5850
100	Fm	0.1077	0.1026	0.0910	0.7570	0.5682

Table 8-2
WAVELENGTHS OF ABSORPTION EDGES IN ANGSTROMS

Atomic No.	Element	K	L_I	L_{II}	L_{III}
3	Li	226.5			
4	Be	110.68			
5	B	66.289			
6	C	43.68			
7	N	30.99			
8	O	23.32			
9	F	17.913			
10	Ne	14.183			
11	Na	11.478		400	
12	Mg	9.512	197.4	247.92	
13	Al	7.951	142.5	170	
14	Si	6.745	105.1	126.48	
15	P	5.787	81.0	96.84	
16	S	5.018	64.23	76.05	
17	Cl	4.397	52.08	61.37	62.93
18	Ar	3.871	43.19	50.39	50.60
19	K	3.436	36.35	42.02	42.17
20	Ca	3.070	31.07	35.20	35.49
21	Sc	2.757	26.83	30.16	30.53
22	Ti	2.497	23.39	26.83	27.37
23	V	2.269	20.52	23.70	24.26
24	Cr	2.07012	16.7	17.9	20.7
25	Mn	1.896	16.27	18.90	19.40
26	Fe	1.74334	14.60	17.17	17.53
27	Co	1.60811	13.34	15.53	15.93
28	Ni	1.48802	12.27	14.13	14.58
29	Cu	1.38043	11.27	13.01	13.29
30	Zn	1.283	10.33	11.86	12.13
31	Ga	1.195	9.54	10.61	11.15
32	Ge	1.116	8.73	9.97	10.23

Table 8-2 (*Continued*)
WAVELENGTHS OF ABSORPTION EDGES IN ANGSTROMS

Atomic No.	Element	K	L_I	L_{II}	L_{III}
33	As	1.044	8.108	9.124	9.367
34	Se	0.9800	7.505	8.417	8.646
35	Br	0.9199	6.925	7.752	7.989
36	Kr	0.8655	6.456	7.165	7.395
37	Rb	0.8155	5.997	6.643	6.863
38	Sr	0.7697	5.582	6.172	6.387
39	Y	0.7276	5.233	5.756	5.962
40	Zr	0.6888	4.867	5.378	5.583
41	Nb	0.6529	4.581	5.025	5.223
42	Mo	0.61977	4.299	4.719	4.912
43	Tc	0.5888	4.064	4.427	4.629
44	Ru	0.5605	3.841	4.179	4.369
45	Rh	0.5338	3.626	3.942	4.130
46	Pd	0.5092	3.428	3.724	3.908
47	Ag	0.48582	3.254	3.514	3.698
48	Cd	0.4641	3.084	3.326	3.504
49	In	0.4439	2.926	3.147	3.324
50	Sn	0.4247	2.778	2.982	3.156
51	Sb	0.4066	2.639	2.830	3.000
52	Te	0.3897	2.510	2.687	2.855
53	I	0.3738	2.390	2.553	2.719
54	Xe	0.3585	2.274	2.429	2.592
55	Cs	0.3447	2.167	2.314	2.474
56	Ba	0.3314	2.068	2.204	2.363
57	La	0.3184	1.973	2.103	2.258
58	Ce	0.3065	1.891	2.009	2.164
59	Pr	0.2952	1.811	1.924	2.077
60	Nd	0.2845	1.735	1.843	1.995
61	Pm	0.2743	1.668	1.766	1.918
62	Sm	0.2646	1.598	1.702	1.845
63	Eu	0.2555	1.536	1.626	1.775
64	Gd	0.2468	1.477	1.561	1.709
65	Tb	0.2384	1.421	1.501	1.649
66	Dy	0.2305	1.365	1.438	1.579
67	Ho	0.2229	1.319	1.390	1.535
68	Er	0.2157	1.269	1.339	1.483
69	Tm	0.2089	1.222	1.288	1.433
70	Yb	0.2022	1.181	1.243	1.386
71	Lu	0.1958	1.140	1.198	1.341
72	Hf	0.1898	1.099	1.154	1.297
73	Ta	0.1839	1.061	1.113	1.255
74	W	0.17837	1.025	1.074	1.215
75	Re	0.1731	0.9901	1.036	1.177
76	Os	0.1678	0.9557	1.001	1.140
77	Ir	0.1629	0.9243	0.9670	1.106

Table 8-2 (*Continued*)
WAVELENGTHS OF ABSORPTION EDGES IN ANGSTROMS

Atomic No.	Element	K	L_I	L_{II}	L_{III}
78	Pt	0.1582	0.8914	0.9348	1.072
79	Au	0.1534	0.8638	0.9028	1.040
80	Hg	0.1492	0.8353	0.8779	1.009
81	Tl	0.1447	0.8079	0.8436	0.9793
82	Pb	0.1408	0.7815	0.8155	0.9503
83	Bi	0.1371	0.7565	0.7891	0.9234
84	Po	0.1332	0.7322	0.7638	0.8970
85	At	0.1295	0.7092	0.7387	0.8720
86	Rn	0.1260	0.6868	0.7153	0.8479
87	Fr	0.1225	0.6654	0.6929	0.8248
88	Ra	0.1192	0.6446	0.6711	0.8027
89	Ac	0.1161	0.6248	0.6500	0.7813
90	Th	0.1129	0.6061	0.6301	0.7606
91	Pa	0.1101	0.5875	0.6106	0.7411
92	U	0.1068	0.5697	0.5919	0.7233
93	Np	0.1045	0.5531	0.5742	0.7042
94	Pu	0.1018	0.5366	0.5571	0.6867
95	Am	0.0992	0.5208	0.5404	0.6700
96	Cm	0.0967	0.5060	0.5246	0.6532
97	Bk	0.0943	0.4913	0.5093	0.6375
98	Cf	0.0920	0.4771	0.4945	0.6223
99	Es	0.0897	0.4636	0.4801	0.6076
100	Fm	0.0875	0.4506	0.4665	0.5935

Table 8-3
CRITICAL X-RAY ABSORPTION ENERGIES IN keV

Atomic No.	Element	K	L_I	L_{II}	L_{III}
1	H	0.0136			
2	He	0.0246			
3	Li	0.0547			
4	Be	0.112			
5	B	0.187			
6	C	0.284			
7	N	0.400			
8	O	0.532			
9	F	0.692			
10	Ne	0.874	0.048		0.022
11	Na	1.08	0.055		0.034
12	Mg	1.30	0.0628		0.0502
13	Al	1.559	0.0870		0.0720
14	Si	1.838	0.118		0.0977
15	P	2.142	0.153		0.128

Table 8-3 (*Continued*)
CRITICAL X-RAY ABSORPTION ENERGIES IN keV

Atomic No.	Element	K	L_I	L_{II}	L_{III}
16	S	2.469	0.193	0.163	0.162
17	Cl	2.822	0.238	0.202	0.201
18	Ar	3.200	0.287	0.246	0.244
19	K	3.606	0.341	0.295	0.292
20	Ca	4.038	0.399	0.350	0.346
21	Sc	4.496	0.462	0.411	0.407
22	Ti	4.966	0.530	0.462	0.456
23	V	5.467	0.604	0.523	0.515
24	Cr	5.988	0.679	0.584	0.574
25	Mn	6.542	0.762	0.656	0.644
26	Fe	7.113	0.849	0.722	0.709
27	Co	7.713	0.929	0.798	0.783
28	Ni	8.337	1.02	0.877	0.858
29	Cu	8.982	1.10	0.954	0.935
30	Zn	9.662	1.20	1.05	1.02
31	Ga	10.39	1.30	1.17	1.14
32	Ge	11.10	1.42	1.24	1.21
33	As	11.87	1.529	1.358	1.32
34	Se	12.65	1.66	1.472	1.431
35	Br	13.48	1.791	1.599	1.552
36	Kr	14.32	1.92	1.729	1.674
37	Rb	15.197	2.064	1.863	1.803
38	Sr	16.101	2.212	2.004	1.937
39	Y	17.053	2.387	2.171	2.096
40	Zr	17.998	2.533	2.308	2.224
41	Nb	18.986	2.700	2.467	2.372
42	Mo	20.003	2.869	2.630	2.525
43	Tc	21.050	3.045	2.796	2.680
44	Ru	22.117	3.227	2.968	2.839
45	Rh	23.210	3.404	3.139	2.995
46	Pd	24.356	3.614	3.338	3.181
47	Ag	25.535	3.828	3.547	3.375
48	Cd	26.712	4.019	3.731	3.541
49	In	27.929	4.226	3.929	3.732
50	Sn	29.182	4.445	4.139	3.911
51	Sb	30.497	4.708	4.391	4.137
52	Te	31.817	4.953	4.621	4.347
53	I	33.164	5.187	4.855	4.559
54	Xe	34.551	5.448	5.103	4.783
55	Cs	35.974	5.706	5.360	5.014
56	Ba	37.432	5.995	5.629	5.250
57	La	38.923	6.264	5.902	5.490
58	Ce	40.43	6.556	6.169	5.728
59	Pr	41.99	6.837	6.446	5.968
60	Nd	43.57	7.134	6.728	6.215

Table 8-3 (*Continued*)
CRITICAL X-RAY ABSORPTION ENERGIES IN keV

Atomic No.	Element	K	L_I	L_{II}	L_{III}
61	Pm	45.19	7.431	7.022	6.462
62	Sm	46.85	7.742	7.316	6.720
63	Eu	48.51	8.059	7.624	6.984
64	Gd	50.23	8.383	7.942	7.251
65	Tb	52.00	8.713	8.258	7.520
66	Dy	53.77	9.053	8.587	7.795
67	Ho	55.61	9.395	8.918	8.074
68	Er	57.47	9.754	9.270	8.362
69	Tm	59.38	10.12	9.622	8.656
70	Yb	61.31	10.49	9.985	8.949
71	Lu	63.32	10.87	10.35	9.248
72	Hf	65.37	11.28	10.75	9.567
73	Ta	67.46	11.68	11.14	9.883
74	W	69.51	12.09	11.54	10.20
75	Re	71.67	12.52	11.96	10.53
76	Os	73.87	12.97	12.38	10.86
77	Ir	76.11	13.41	12.82	11.21
78	Pt	78.35	13.865	13.26	11.55
79	Au	80.67	14.351	13.731	11.92
80	Hg	83.08	14.838	14.205	12.278
81	Tl	85.52	15.344	14.695	12.65
82	Pb	87.95	15.861	15.200	13.03
83	Bi	90.54	16.386	15.709	13.42
84	Po	93.16	16.925	16.233	13.81
85	At	95.73	17.481	16.777	14.21
86	Rn	98.45	18.054	17.331	14.61
87	Fa	101.1	18.628	17.893	15.02
88	Ra	103.9	19.228	18.473	15.44
89	Ac	107.7	19.829	19.071	15.86
90	Th	109.8	20.452	19.673	16.278
91	Pa	112.4	21.096	20.295	16.720
92	U	115.0	21.757	20.944	17.163
93	Np	118.2	22.411	21.585	17.606
94	Pu	121.2	23.117	22.250	18.062
95	Am	124.3	23.795	22.935	18.524
96	Cm	127.2	24.502	23.629	18.992
97	Bk	131.3	25.231	24.344	19.466
98	Cf	133.6	26.010	25.070	19.954
99	Es	138.1	26.729	25.824	20.422
100	Fm	141.5	27.503	26.584	20.912

Table 8-4
X-RAY EMISSION ENERGIES IN keV

Atomic No.	Element	$K\beta_1$	$K\alpha_1$	$L\beta_1$	$L\alpha_1$
3	Li		0.052		
4	Be		0.110		
5	B		0.185		
6	C		0.282		
7	N		0.392		
8	O		0.523		
9	F		0.677		
10	Ne		0.851		
11	Na	1.067	1.041		
12	Mg	1.297	1.254		
13	Al	1.553	1.487		
14	Si	1.832	1.740		
15	P	2.136	2.015		
16	S	2.464	2.308		
17	Cl	2.815	2.622		
18	Ar	3.192	2.957		
19	K	3.589	3.313		
20	Ca	4.012	3.691	0.344	0.341
21	Sc	4.460	4.090	0.399	0.395
22	Ti	4.931	4.510	0.458	0.452
23	V	5.427	4.952	0.519	0.512
24	Cr	5.946	5.414	0.581	0.571
25	Mn	6.490	5.898	0.647	0.636
26	Fe	7.057	6.403	0.717	0.704
27	Co	7.649	6.930	0.790	0.775
28	Ni	8.264	7.477	0.866	0.849
29	Cu	8.904	8.047	0.948	0.928
30	Zn	9.571	8.638	1.032	1.009
31	Ga	10.263	9.251	1.122	1.096
32	Ge	10.981	9.885	1.216	1.186
33	As	11.725	10.543	1.317	1.282
34	Se	12.495	11.221	1.419	1.379
35	Br	13.290	11.923	1.526	1.480
36	Kr	14.112	12.649	1.638	1.587
37	Rb	14.960	13.394	1.752	1.694
38	Sr	15.834	14.164	1.872	1.806
39	Y	16.736	14.957	1.996	1.922
40	Zr	17.666	15.774	2.124	2.042
41	Nb	18.621	16.614	2.257	2.166
42	Mo	19.607	17.478	2.395	2.293
43	Tc	20.612	18.370	2.538	2.424
44	Ru	21.655	19.278	2.683	2.558
45	Rh	22.721	20.214	2.834	2.696
46	Pd	23.816	21.175	2.990	2.838
47	Ag	24.942	22.162	3.151	2.984

Table 8-4 (*Continued*)
X-RAY EMISSION ENERGIES IN keV

Atomic No.	Element	$K\beta_1$	$K\alpha_1$	$L\beta_1$	$L\alpha_1$
48	Cd	26.093	23.172	3.316	3.133
49	In	27.274	24.207	3.487	3.287
50	Sn	28.483	25.270	3.662	3.444
51	Sb	29.723	26.357	3.843	3.605
52	Te	30.993	27.471	4.029	3.769
53	I	32.292	28.610	4.220	3.937
54	Xe	33.644	29.779	4.422	4.111
55	Cs	34.984	30.970	4.620	4.286
56	Ba	36.376	32.191	4.828	4:467
57	La	37.799	33.440	5.043	4.651
58	Ce	39.255	34.717	5.262	4.840
59	Pr	40.746	36.023	5.489	5.034
60	Nd	42.269	37.359	5.722	5.230
61	Pm	43.811	38.726	5.956	5.431
62	Sm	45.400	40.124	6.206	5.636
63	Eu	47.027	41.529	6.456	5.846
64	Gd	48.718	42.983	6.714	6.059
65	Tb	50.391	44.470	6.979	6.275
66	Dy	52.178	45.985	7.249	6.495
67	Ho	53.934	47.528	7.528	6.720
68	Er	55.690	49.099	7.810	6.948
69	Tm	57.487	50.730	8.103	7.181
70	Yb	59.352	52.360	8.401	7.414
71	Lu	61.282	54.063	8.708	7.654
72	Hf	63.209	55.757	9.021	7.898
73	Ta	65.210	57.524	9.341	8.145
74	W	67.233	59.310	9.670	8.396
75	Re	69.298	61.131	10.008	8.651
76	Os	71.404	62.991	10.354	8.910
77	Ir	73.549	64.886	10.706	9.173
78	Pt	75.736	66.820	11.069	9.441
79	Au	77.968	68.794	11.439	9.711
80	Hg	80.258	70.821	11.823	9.987
81	Tl	82.558	72.860	12.210	10.266
82	Pb	84.922	74.957	12.611	10.549
83	Bi	87.335	77.097	13.021	10.836
84	Po	89.809	79.296	13.441	11.128
85	At	92.319	81.525	13.873	11.424
86	Rn	94.877	83.800	14.316	11.724
87	Fr	97.483	86.119	14.770	12.029
88	Ra	100.136	88.485	15.233	12.338
89	Ac	102.846	90.894	15.712	12.650
90	Th	105.592	93.334	16.200	12.966
91	Pa	108.408	95.851	16.700	13.291
92	U	111.289	98.428	17.218	13.613

Table 8-4 (*Continued*)
X-RAY EMISSION ENERGIES IN keV

Atomic No.	Element	$K\beta_1$	$K\alpha_1$	$L\beta_1$	$L\alpha_1$
93	Np	114.181	101.005	17.740	13.945
94	Pu	117.146	103.653	18.278	14.279
95	Am	120.163	106.351	18.829	14.618
96	Cm	123.235	109.098	19.393	14.961
97	Bk	126.362	111.896	19.971	15.309
98	Cf	129.544	114.745	20.562	15.661
99	Es	132.781	117.646	21.166	16.018
100	Fm	136.075	120.598	21.785	16.379

Filters. The K spectra of the light metals, often used as target material in the production of X-rays for diffraction studies, contain three strong lines, α_1, α_2 and β_1, of which the α lines form a doublet with a narrow wavelength separation. The $K\beta$ radiation can be eliminated by using a thin foil filter, usually of the element of next lower atomic number to that of the target element; the $K\alpha$ lines are transmitted with a relatively small loss of intensity. Table 8-5, restricted to the K wavelengths of target elements in common use, lists the calculated thicknesses of β filters required to reduce the $K\beta_1/K\alpha_1$ integrated intensity ratio to $1/100$.

Table 8-5
β FILTERS FOR COMMON TARGET ELEMENTS

Target Element	$K\bar{\alpha}$, Å	Excitation Voltage, keV	Absorber	$K\beta_1/K\alpha_1 = 1/100$ Thickness, mm	g/cm^2	% Loss $K\alpha_1$
Ag	0.560834	25.52	Pd	0.062	0.074	60
Mo	0.71069	20.00	Zr	0.081	0.053	57
Cu	1.54178	8.981	Ni	0.015	0.013	45
Ni	1.65912	8.331	Co	0.013	0.011	42
Co	1.79021	7.709	Fe	0.012	0.009	39
Fe	1.93728	7.111	Mn	0.011	0.008	38
			MnO_2	0.026	0.013	45
Cr	2.29092	5.989	V	0.011	0.007	37
			V_2O_5	0.036	0.012	48
	$L\alpha_1$			$L\beta_1/L\alpha_1 = 1/100$		% Loss $L\alpha_1$
W	1.4763	10.200	Cu	0.035		77

Interplanar Spacings. Diffractometer alignment procedures require the use of a well-prepared polycrystalline specimen. Two standard samples found to be suitable are silicon and α-quartz (including Novaculite). The 2θ values of several of the most intense reflections for these materials are listed in Table 8-6 (Tables of Interplanar Spacings *d vs* Diffraction Angle 2θ for Selected Targets, Picker Nuclear, White Plains, N.Y., 1966). To convert to *d* for $K\bar{\alpha}$ or to *d* for $K\alpha_2$, multiply the tabulated *d* value (Table 8-6) for $K\alpha_1$ by the factor given below:

Element	$K\bar{\alpha}$	$K\alpha_2$
W	1.00769	1.02307
Ag	1.00263	1.00789
Mo	1.00202	1.00604
Cu	1.00082	1.00248
Ni	1.00077	1.00232
Co	1.00072	1.00216
Fe	1.00067	1.00204
Cr	1.00057	1.00170

Table 8-6
INTERPLANAR SPACINGS FOR $K\alpha_1$ RADIATION, *d vs* 2θ

α-quartz (Including Novaculite)

hkl d(Å)	100 4.260	101 3.343	110 2.458	102 2.282	200 2.128	112 1.817	202 1.672	211 1.541	203 1.375	301 1.372
W $K\alpha_1$: 2θ	2.81	3.58	4.87	5.25	5.63	6.59	7.17	7.78	8.72	8.74
Ag $K\alpha_1$: 2θ	7.53	9.60	13.07	14.08	15.10	17.71	19.26	20.91	23.47	23.52
Mo $K\alpha_1$: 2θ	9.55	12.18	16.59	17.88	19.19	22.51	24.49	26.61	29.89	29.96
Cu $K\alpha_1$: 2θ	20.83	26.64	36.52	39.45	42.44	50.16	54.86	59.98	68.14	68.31
Ni $K\alpha_1$: 2θ	22.44	28.71	39.42	42.60	45.85	54.28	59.44	65.08	74.15	74.34
Co $K\alpha_1$: 2θ	24.24	31.04	42.68	46.15	49.71	58.98	64.68	70.96	81.16	81.38
Fe $K\alpha_1$: 2θ	26.27	33.66	46.38	50.20	54.11	64.38	70.75	77.83	89.50	89.74
Cr $K\alpha_1$: 2θ	31.18	40.05	55.52	60.22	65.09	78.11	86.42	95.96	112.73	113.11

Silicon

hkl d(Å)	111 3.1353	220 1.91997	311 1.63736	400 1.357630	331 1.24584	422 1.1085	511,333 1.0451	440 0.959986	531 0.917922	620 0.858637
W $K\alpha_1$: 2θ	3.82	6.24	7.32	8.83	9.62	10.82	11.48	12.50	13.07	13.98
Ag $K\alpha_1$: 2θ	10.24	16.75	19.67	23.78	25.95	29.23	31.04	33.88	35.48	38.02
Mo $K\alpha_1$: 2θ	12.99	21.29	25.02	30.28	33.08	37.32	39.67	43.36	45.45	48.79
Cu $K\alpha_1$: 2θ	28.44	47.30	56.12	69.13	76.38	88.03	94.96	106.71	114.10	127.55
Ni $K\alpha_1$: 2θ	30.66	51.16	60.83	75.26	83.42	96.80	104.96	119.42	129.12	149.76
Co $K\alpha_1$: 2θ	33.15	55.53	66.22	82.42	91.77	107.59	117.71	137.42	154.04	
Fe $K\alpha_1$: 2θ	35.97	60.55	72.48	90.96	101.97	121.67	135.70			
Cr $K\alpha_1$: 2θ	42.83	73.21	88.72	114.97	133.53					

Analyzing Crystals. The range of wavelengths usable with various analyzing crystals are governed by the d spacings of the crystal planes and by the geometric limits to which the goniometer can be rotated. The d value should be small enough to make the angle 2θ greater than approximately 10 or 15 deg, even at the shortest wavelength used; otherwise excessively long analyzing crystals would be needed to prevent the direct fluorescent beam from entering the detector. A small d value is also favorable for producing a large dispersion of the spectrum to give good separation of adjacent lines. On the other hand, a small d value imposes an upper limit to the range of wavelengths that can be analyzed. Actually the goniometer is limited mechanically to about 150 deg for a 2θ value. A final requirement is the reflection efficiency and minimization of higher-order reflections. Table 8-7 gives a list of crystals commonly used for X-ray spectroscopy.

The long-wavelength analyzers are prepared by dipping an optical flat into the film of the metal fatty acid about 50 times to produce a layer 180 molecules in thickness.

Lithium fluoride is the optimum crystal for all wavelengths less than 3 Å. Pentaerythritol (PET) and potassium hydrogen phthalate (KAP) are usually the crystals of choice for wavelengths from 3 to 20 Å. Two crystals suppress even-ordered reflections: silicon (111) and calcium fluoride (111).

Table 8-7
ANALYZING CRYSTALS FOR X-RAY SPECTROSCOPY

Crystal	Reflecting Plane	2d Spacing, Å	Reflectivity
Quartz	$50\bar{5}2$	1.624	Low
Aluminum	111	2.338	High
Topaz	303	2.712	Medium
Quartz	$20\bar{2}3$	2.750	Low
Lithium fluoride	220	2.848	High
Silicon	111	3.135	High
Quartz	112	3.636	Medium
Lithium fluoride	200	4.028	High
Sodium chloride	200	5.639	High
Calcium fluoride	111	6.32	High
Quartz	$10\bar{1}1$	6.686	High
Quartz	$10\bar{1}0$	8.50	Medium
Pentaerythritol (PET)	002	8.742	High
Ethylenediamine tartrate (EDT)	020	8.808	Medium
Ammonium dihydrogen phosphate (ADP)	110	10.648	Low
Gypsum	020	15.185	Medium
Mica	002	19.92	Low
Potassium hydrogen phthalate (KAP)	$10\bar{1}1$	26.4	Medium
Lead palmitate		45.6	
Strontium behenate		61.3	
Lead stearate		100.4	Medium

Mass Absorption Coefficients. Radiation traversing a layer of substance is diminished in intensity by a constant fraction per centimeter thickness x of material. The emergent radiant power P, in terms of incident radiant power P_0, is given by

$$P = P_0 \exp\left(-\mu x\right)$$

which defines the total linear absorption coefficient μ. Since the reduction of intensity is determined by the quantity of matter traversed by the primary beam, the absorber thickness is best expressed on a mass basis, in g/cm^2. The mass absorption coefficient μ/ρ, expressed in units cm^2/g, where ρ is the density of the material, is approximately independent of the physical state of the material and, to a good approximation, is additive with respect to the elements composing a substance.

Table 8-8 contains values of μ/ρ for the common target elements employed in X-ray work. A more extensive set of mass absorption coefficients for K, L, and M emission lines within the wavelength range from 0.7 to 12 Å is contained in Heinrich's paper in T. D. McKinley, K. F. J. Heinrich, and D. B. Wittry (eds.), "The Electron Microprobe," pp. 351–377, Wiley, New York, 1964. This article should be consulted to ascertain the probable accuracy of the values and for a compilation of coefficients and exponents employed in the computations.

Table 8-8
MASS ABSORPTION COEFFICIENTS FOR $K\alpha_1$ LINES AND W $L\alpha_1$ LINE

Emitter	Ag $K\alpha_1$	Mo $K\alpha_1$	Cu $K\alpha_1$	Ni $K\alpha_1$	Co $K\alpha_1$	Fe $K\alpha_1$	Cr $K\alpha_1$	W $L\alpha_1$
Wavelength, Å	0.559	0.709	1.541	1.658	1.789	1.936	2.290	1.476
Absorber								
1 H	0.37	0.38	0.43	0.4	0.4	0.5	0.5	0.4
2 He	0.16	0.18	0.37	0.4	0.4	0.5	0.7	0.3
3 Li	0.18	0.22	0.50	0.6	0.7	0.9	1.5	0.4
4 Be	0.22	0.30	1.2	1.5	1.9	2.3	3.7	1.1
5 B	0.30	0.45	2.5	3.1	3.9	4.9	7.9	2.2
6 C	0.42	0.50	4.6	5.7	7.1	8.8	14.2	4.1
7 N	0.60	0.83	7.5	9.3	11.5	14.4	23.1	6.7
8 O	0.80	1.45	12.9	15.8	19.5	24.5	39.4	11.4
9 F	1.00	1.9	16.5	20.3	25.2	31.4	50.3	14.6
10 Ne	1.41	2.6	22.8	27.9	34.6	43.1	69.0	20.1
11 Na	1.75	3.5	30.3	37.2	45.9	57.2	91.4	26.8
12 Mg	2.27	4.6	39.5	48.4	59.8	74.6	119.1	34.9
13 Al	2.74	5.8	49.6	60.7	75.0	93.4	149.0	43.9
14 Si	3.44	7.3	61.4	75.2	92.8	115.5	183.8	54.4
15 P	4.20	8.8	74.7	91.4	112.9	140.5	223.6	66.2
16 S	5.15	10.6	89.2	109.2	134.7	167.4	266.1	79.1
17 Cl	5.86	12.4	104.8	128.2	158.1	196.6	312.4	92.8

Table 8-8 (*Continued*)
MASS ABSORPTION COEFFICIENTS FOR $K\alpha_1$ LINES
AND W $L\alpha_1$ LINE

Emitter Wavelength, Å Absorber	Ag $K\alpha_1$ 0.559	Mo $K\alpha_1$ 0.709	Cu $K\alpha_1$ 1.541	Ni $K\alpha_1$ 1.658	Co $K\alpha_1$ 1.789	Fe $K\alpha_1$ 1.936	Cr $K\alpha_1$ 2.290	W $L\alpha_1$ 1.476
18 Ar	6.40	14.5	121.4	148.5	183.0	227.3	360.7	107.6
19 K	8.0	16.7	139.8	171	211	262	415	124
20 Ca	9.7	18.9	158.6	194	239	296	469	141
21 Sc	10.5	21.8	180.5	221	272	337	534	160
22 Ti	11.8	25.3	203	247	304	378	597	180
23 V	13.3	27.7	228	278	342	424	77	202
24 Cr	15.7	31.0	254	311	382	474	88	226
25 Mn	17.4	34.5	282	344	423	63.5	101	250
26 Fe	19.9	38.1	311	380	57.6	71.4	113	276
27 Co	21.8	42.1	341	52.8	64.9	80.6	127	303
28 Ni	25.0	46.4	48.3	58.9	72.5	90.0	142	333
29 Cu	26.4	50.7	53.7	65.5	80.6	100.0	158	47.6
30 Zn	28.2	55.4	59.5	72.7	89.4	110.9	175	52.8
31 Ga	30.8	60.1	65.9	80.5	99.0	122.8	194	58.5
32 Ge	33.5	65.2	72.3	88.2	108.6	134.7	213	64.1
33 As	36.5	70.5	79.1	96.6	118.9	147	233	70.2
34 Se	38.5	76.0	86.1	105.1	129.4	161	254	76.4
35 Br	42.3	82.5	93.9	114.7	141.2	175	277	83.4
36 Kr	45.0	88.3	101.9	124.5	153.2	190	300	90.5
37 Rb	48	95	84	103	127	158	252	98
38 Sr	52	102	90	110	137	170	271	106
39 Y	56	109	97	119	147	183	292	114
40 Zr	61	17	104	128	158	197	314	122
41 Nb	66	18	112	138	170	212	338	132
42 Mo	71	19	119	146	180	225	358	140
43 Tc	76	20	128	157	194	241	384	150
44 Ru	12	22	137	168	207	258	410	160
45 Rh	13	23	146	179	221	275	438	171
46 Pd	14	24	155	190	235	292	466	182
47 Ag	15	26	165	202	249	310	493	193
48 Cd	15	28	174	213	263	327	520	204
49 In	16	30	185	227	280	347	553	217
50 Sn	17	32	195	239	295	367	583	229
51 Sb	19	34	206	252	310	386	612	241
52 Te	19	36	216	265	326	405	644	253
53 I	21	37	230	281	346	431	684	269
54 Xe	22	39	239	293	361	448	710	280
55 Cs	24	42	332	404	495	612	822	295
56 Ba	25	44	349	425	522	645	622	311
57 La	26	46	365	444	545	673	647	325
58 Ce	28	48	383	466	571	603	216	341
59 Pr	29	51	401	487	597	453	229	356
60 Nd	31	54	420	510	534	473	241	373

Table 8-8 (*Continued*)
MASS ABSORPTION COEFFICIENTS FOR $K\alpha_1$ LINES
AND W $L\alpha_1$ LINE

Emitter Wavelength, Å	Ag $K\alpha_1$ 0.559	Mo $K\alpha_1$ 0.709	Cu $K\alpha_1$ 1.541	Ni $K\alpha_1$ 1.658	Co $K\alpha_1$ 1.789	Fe $K\alpha_1$ 1.936	Cr $K\alpha_1$ 2.290	$W\ L\alpha_1$ 1.476
Absorber								
61 Pm	32	56	440	535		164	254	392
62 Sm	33	59	L_I 456	473	417	173	268	406
63 Eu	35	61	405	354	148	182	282	423
64 Gd	36	64	L_{II} 424	370	156	191	296	
65 Tb	38	67	316	135	164	201	311	393 L_I
66 Dy	39	70	L_{III} 329	141	172	211	327	293 L_{II}
67 Ho	41	72	123	148	181	222	343	304
68 Er	43	75	129	156	189	233	360	316 L_{III}
69 Tm	45	79	135	163	199	244	377	120
70 Yb	46	82	141	171	208	256	395	126
71 Lu	48	84	148	179	218	267	414	132
72 Hf	51	88	155	187	228	280	433	138
73 Ta	52	91	162	196	238	293	453	144
74 W	55	95	169	204	249	306	473	151
75 Re	57	98	176	213	260	319	494	157
76 Os	59	102	184	223	271	333	515	164
77 Ir	61	106	192	232	283	347	538	171
78 Pt	64	109	200	242	295	362	560	179
79 Au	67	113	209	252	307	377	584	186
80 Hg	69	117	218	263	321	394	609	194
81 Tl	72	121	227	275	334	411	635	203
82 Pb	74	125	236	286	348	428	662	211
83 Bi	78	129	247	298	363	446	690	220
84 Po		131	258	311	380	466	721	230
85 At			269	325	397	487	753	240
86 Rn	85		281	340	414	509	787	251
87 Fr		89	294	356	433	532	823	262
88 Ra	91		307	372	453	556	861	274
89 Ac			322	389	474	582	900	287
90 Th	97		337	408	497	610	944	301
91 Pa			353	427	520	639	988	315
92 U	104		372	450	548	673	898	332
93 Np			392	474	578	709	945	350
94 Pu		54	418	505	615	755	835	373

ELECTRONIC EMISSION AND ABSORPTION SPECTROSCOPY

The tables of emission and absorption lines are presented in two parts. In Table 8-9 the data are arranged by element in alphabetical order of chemical symbol, whereas in Table 8-10 the sensitive lines of the elements are arranged in order of decreasing wavelengths.

The wavelengths in column 2 of Table 8-9 are all normal air wavelengths and are given to the nearest 0.01 Å, except for band systems. A Roman numeral II following the wavelength signifies a singly ionized atom, the letter *d* an unresolved double line, and *t* a triplet line.

The relative intensity numbers in column 3 of Table 8-9 are taken from W. F. Meggers, C. H. Corliss, and B. F. Scribner, "Tables of Spectral-line Intensities, Part I," National Bureau of Standards Monograph 32, U.S. Government Printing Office, Washington, D.C., 1961. All emission lines are assigned relative intensities proportional to their limiting detectabilities. In a fully exposed spectrogram of copper containing 0.1 atomic per cent of another element any faint but unmistakable line at a given wavelength is assigned unit intensity. For example, a line of intensity 10 should show plainly at 0.01 atomic per cent, while one of intensity 1000 should be easily seen at 0.0001 atomic per cent (one in a million). Actual values of detection limits reported in the literature for a dc arc and a spark-porous cup are included in columns 4 and 5, respectively.

The flame emission detection limit is quite dependent on instrument and operating variables, particularly the detector, the fuel and oxidant gases, and the slit width. Many of the data are for a Beckman model DU spectrophotometer or a 0.5-meter Jarrell-Ash monochromator, equipped with a 1P28 multiplier phototube and a sprayer-burner combination. For example, a value of 0.6 μg/ml per 0.1 mV for silver implies that the balancing potentiometer moved through one division out of 100 total divisions (0.1 mV for a 10-mV recorder) when a solution containing 0.6 μg/ml of silver was sprayed into the flame.

The atomic absorption sensitivity values are also dependent on instrument and operating variables, particularly lamp current, slit width, burner type, path length, and type of flame. Sensitivity is defined as the concentration of test element required to cause an absorption of 1 per cent (0.004 absorbance unit). A table for conversion of per cent absorption values into absorbance units is in Section 2. The concentration range in which the stated sensitivity is valid is given in parentheses for many wavelengths. I am indebted to Mr. T. C. Rains, Analytical Chemistry Division, National Bureau of Standards, Washington, D.C., for many of the atomic absorption data.

Abbreviations in the table

AA, air-acetylene flame
AH, air-hydrogen flame
AP, air-propane flame
NA, nitrous oxide-acetylene flame
(N-A)A, (nitrous oxide-air)-acetylene flame
OA, oxygen-acetylene flame
OH, oxygen-hydrogen flame
n, organic solvent aspirated directly into flame
r, fuel-rich flame
w, lean flame
z, emission from innerconal gases

Table 8-9
EMISSION AND ABSORPTION LINES 1900 to 9000 Å

Element	Wavelength, Å	Relative Intensity	DC Arc, $\mu g/g$	Spark-porous Cup, $\mu g/ml$	Flame Emission, $\mu g/ml/0.1\ mV$	Atomic Absorption: Sensitivity, $\mu g/ml/1\%$ Abs (conc'n range for which valid)
Ag	3280.68	5500	0.1	0.02	1.0 OH	0.13 AA (1–10)
	3382.89	2800			0.6 OH	0.22 AA (1–50)
Al	2269.09	2				9 NAr (10–1000)
	2367.06	18				8 NAr (10–1000)
	2373.13 d	36				5 NAr (100–1000)
	2567.99	24				16 NAr (100–1000)
	2575.10 d	48				11 NAr (100–1000)
	3082.16	320				4 NAr (10–500)
	3092.71 d	650	1.0			2.2 NAr (5–100)
	3944.03	450			0.3 OAn	5 NAr (10–200)
	3961.53	900	1.0	0.3	0.5 OAn	3 NAr (10–200)
AlO	4842				0.5 OAn	
As	1890					1.7 AA; 2.8 AH
	1936.96	17				1.3 AAw (3–50)
						1.2 AH (3–100)
	1971.97	28				1.8 AAw (3–100)
						2.3 AH (3–100)
	2288.12	44		3	4.0 OAnz	
	2349.84	85	30	3	2.2 OAnz	
	2780.22	140	50	10	13.0 OAnz	
Au	2427.95	200	2	30		0.3 AA and NA
	2675.95	340	2	20	5 OA	0.6 AA
B	2088.93	7				50 NAr (50–1000)
	2089.59	11				45 NAr (50–1000)
	2496.78	240	2	2	7 OAnrz	63 NAr (200–1000)
	2497.73	480	2	1		35 NAr (200–1000)
BO$_2$	4530				30 OAn (50%)	
	4715				10 OAn (50%)	
	4920				5 OAn (50%)	
	5180				3 OAn (50%)	
	5476				3 OAn (50%)	
	5790				6 OAn (50%)	

Table 8-9 (*Continued*)
EMISSION AND ABSORPTION LINES 1900 to 9000 Å

Element	Wavelength, Å	Relative Intensity	DC Arc, μg/g	Spark-porous Cup, μg/ml	Flame Emission, μg/ml/0.1 mV	Atomic Absorption: Sensitivity, μg/ml/1% Abs (conc'n range for which valid)
Ba	2304.24 II	28		0.5		
	2335.27 II	55		0.5		
	3071.58	18	100			45 (N–A) Ar (100–1000)
	3501.11	50				92 (N–A) Ar (200–1000)
	4130.66 II	150		4		
	4554.03 II	6500	0.1	0.1	0.06 OA	
	4934.09 II	2000			0.08 OA	
	5535.48	650	0.1		0.03 OA	3 (N–A) Ar (10–500)
Be	2348.61	300	0.1	0.02	1.0 OAnr	0.3 NAr
	3130.42 II	480	0.1	0.02		
	3131.07 II	320		0.003		
	3321.34 t	100	1			
Bi	2061.70	55				5 AA (5–500)
	2110.26	10				17 AA (50–500)
	2228.25	3			11.5 OAn	1.6 AA (5–100)
	2230.61	14				0.7 AA (5–100)
	2276.58	5			6.4 OAn	9 AA (10–1000)
	2897.98	400	10			
	3067.72	3600	1	1		2 AA (5–500)
Br(InBr)	3758				1.6 AH	
C	2478.57	10				
Ca	2398.56	4			0.001 NA	
	3933.67 II	4200	0.1	0.01	0.005 OAz	
	3968.47 II	2200			0.01 OAz	
	4226.73	1100	0.1		0.07 OA	0.08 AA (1–50)
	4454.78	140	1.0			
CaOH	5540				0.25 OH	
	6220				1.6 OH	
Cd	2265.02 II	110		0.2		
	2288.02	1500	10	0.2	10 OH; 4 OAn	0.07 AA (1–30)
	3261.06	32	10		5 OH; 2 OAn	38 AA (50–1000)
Ce	3801.53 II	200	40			
	3942.75 II	190		25		
	3999.24 II	200	60			
	4012.38 II	190		10		
	4040.76 II	150	10			
	4186.60 II	250		3		
	5200.12 d	6				30 NA
	5223.49	28				30 NA
	5697.00	32				39 NA
	5699.23	40			16 NAn	
Cl (CuCl)	4354				10 OH	
(InCl)	3599				0.7 AH	
CN	3883				20 OHn	
	3851				60 OHn	
Co	2174.60	2				3.5 AA (5–100)
	2286.16 II	26		0.5		

Table 8-9 (*Continued*)
EMISSION AND ABSORPTION LINES 1900 to 9000 Å

| Element | Wavelength, Å | Relative Intensity | Detection limit | | | Atomic Absorption: Sensitivity, μg/ml/1% Abs (conc'n range for which valid) |
			DC Arc, μg/g	Spark-porous Cup, μg/ml	Flame Emission, μg/ml/0.1 mV	
	2309.02	24				7 AA (10–500)
	2363.79 II	30		0.5		
	2407.25	140				0.02 AA (1–50)
	2424.93	130			1.7 OAr	0.2 AA (1–50)
	2435.83	25				1.4 AA (5–100)
	2521.36	180	3			0.8 AA (3–100)
	2987.16	36				12 AA (50–1000)
	2989.59	36				12 AA (50–1000)
	3044.00	160				2.3 AA (5–500)
	3405.12	700			6.2 OA	
	3412.63	140			5.2 OA	0.7 AA (3–100)
	3431.58	160			10 OA	
	3443.64	550			14 OA	
	3453.50	1300	1	2.0	3.4 OA; 0.8 OAn	
	3465.80	320			6.4 OA	6 AA (10–1000)
	3474.02	500			8 OA	10 AA (50–1000)
	3506.32	440			6 OA	
	3512.48	240			10 OA	
	3526.85	400			4 OA	4 AA (10–1000)
	3873.12	240			6 OA	
	4121.32	190			15 OA	
Cr	2364.71	3				40 AAr (50–500)
	2677.16 II	200		0.1		
	2835.63 II	280	10	0.3		
	2843.25 II	190		0.3		
	3578.69	2400			0.2 OAn	0.22 AAr (1–50)
	3593.49	2100			0.25 OAn	0.29 AAr (1–50)
	3605.33	1600			0.33 OAn	0.35 AAr (1–30)
	3615.64	11				23 AAr
	4254.35	1700	1		0.10 OAn	0.6 AAr (1–50)
	4274.80	1300			0.13 OAn	0.8 AAr (1–100)
	4289.72	850			0.17 OAn	1.1 AAr (1–100)
	5204.52	440				49 A
	5208.44	900				19 AA
Cs	4555.36	40	10		2 OH	65 AA (100–500)
	4593.18	20			8 OH	200 AA (500–1000)
	8521.10	1500	10		0.5 OH	0.8 AA (10–100)
	8943.50	800			0.5 OH	13 AA (50–100)
Cu	2024.34	2				0.6 AA (5–50)
	2178.94	8				0.5 AA (10–50)
	2181.72	6				0.7 AA (10–50)
	2225.70	4				1.2 AA (5–500)
	2441.64	4				34 AA (100–500)
	2492.15	36				7 AA (10–100)
	3247.54	5000	0.2	0.2	0.6 OA	0.1 AA (1–10)
	3273.96	2500		0.05	0.8 OA	0.2 AA (1–10)
Dy	3407.79 II	480		2		
	3531.70 II	7000	1			
	4000.48 II	650	10			
	4045 99	1000			0.07 NA	0.75 NA
	4186.78	950				0.9 NA
	4191.60	180				16 NA

Table 8-9 (*Continued*)
EMISSION AND ABSORPTION LINES 1900 to 9000 Å

Element	Wavelength, Å	Relative Intensity	DC Arc, μg/g	Spark-porous Cup, μg/ml	Flame Emission, μg/ml/0.1 mV	Atomic Absorption: Sensitivity, μg/ml/1% Abs (conc'n range for which valid)
	4194.85	550				1.4 NA
	4211.72	1300			0.5 OA	0.7 NA
	4225.14	220				26 NA
DyO	5280				0.11 OAn; 1 OH	
	5400				0.20 OAn	
	5490				0.14 OAn	
	5730				0.08 OAn	
	5830				0.1 OAn; 5 OH	
Er	3372.76 II	750	5	2		
	3499.11 II	650	10			
	3692.64 II	700	10			
	3892.69	340				3.7 NA
	3896.25 II	420	10			
	3906.34 II	850	4			
	4007.97	1100	60		2 OA	0.9 NA
	4087.65	280				0.9 NA; 6.4 AA
	4151.10	550	300			1.3 NA
ErO	5040				0.07 OAn	
	5520				0.1 OAn; 2 OH	
	5650				3 OH	
Eu	3819.67 II	3400	10			
	4205.05 II	4000		0.5		
	4435.56 II	900	10			
	4594.03	750			0.05 OAn	0.8 NAr
	4627.22	650			0.06 OAn	0.9 NA
	4661.88	550			0.2 OAn	1.1 NA
F(CaF)	5291		100		300 OA	
(MgF)	3594		325			
(SrF)	6633		225			
Fe	2166.77	15				0.7 AAr (1–100)
	2382.04	60	0.5			
	2395.62	60	0.5			
	2483.27	280				0.15 AAr (1–30)
	2484.19	90				0.18 AAr (1–50)
	2487.97	4				0.3 AAr (1–30)
	2489.75 2490.64 } d	180				0.6 AAr (3–50)
	2510.83	90				1.6 AAr (5–500)
	2524.39	50				4 AAr (10–100)
	2527.43	140				0.8 AAr (5–100)
	2599.40 II	200	0.2			
	2719.02	260				0.5 AAr (1–100)
	2744.07	30				4 AAr (30–400)
	3020.64	280			0.2 OAn	0.5 AAr (5–50)
	3440.61	400			0.44 OAn	2.4 AAr (10–500)
	3581.20	600			0.8 OAn; 15 OA	
	3719.95	600	0.2		0.12 OAn; 3 OA	1 AAr (5–100)
	3734.87	700			0.16 OAn	
	3748.26	140			0.23 OAn	4 AAr (30–1000)
	3859.91	420			0.14 OAn; 3 OA	2.1 AAr (30–500)
	3886.28	180				6 AAr (50–1000)
	3920.26	36			3.5 OAn	40 AAr (300–1000)
	3927.92	70			10 OA	24 AAr (100–1000)

Table 8-9 (*Continued*)
EMISSION AND ABSORPTION LINES 1900 to 9000 Å

Element	Wavelength, Å	Relative Intensity	Detection limit			Atomic Absorption: Sensitivity, $\mu g/ml/1\%$ Abs (conc'n range for which valid)
			DC Arc, $\mu g/g$	Spark-porous Cup, $\mu g/ml$	Flame Emission, $\mu g/ml/0.1$ mV	
Ga	2874.24	500				2.3 AA
	2943.64	950	10	0.5		2.4 AA
	4032.98	1000		10	1.2 OA	6.2 AA
	4172.06	2000	1		0.5 OA	3.7 AA
Gd	3422.47 II	700		4		
	3684.13	200				17 NA
	3768.39 II	850	4			
	3783.05	280				17 NA
	3850.97	500		0.5		
	4058.22	240				19 NA
	4078.70	260				17 NA
	4190.78	200				40 NA
	4251.73 II	160	10			
	4346.46 *d*	200			25 OA	27 NA
	4401.86	130			2 NA	
GdO	4640				0.1 OAn; 1 OA	
	5680				0.2 OAn; 2 OA	
	6010				0.05 OAn	
	6220				0.03 OAn	
Ge	2592.54	500				4 NA
	2651.18 *d*	1200	1.0	0.5	2 OAn	2 NA
	2691.34	500				10 NA
	2709.63	850				5 NA
	2754.59	650				5 NA
	3039.06	750	1.0	10		40 NA
Hf	2641.41 II	120		5		
	2773.36 II	110	100			
	2820.22 II	140		4		
	2866.37	240				18 NA
	2898.26	200				40 NA
	2916.48	220	100			
	3072.88	240				14 NA
	3682.24	220			75 OAnz	45 NA
Hg	2536.52	1500	5	10	2.5 OAn	2 AA
	3650.15	280	100			
	4046.56	180				
	4358.35	400				
	5460.74	320				
Ho	3398.98 II	900		0.5		
	3891.02 II	1500	30			
	4053.93	900				1.9 NA
	4103.84	1000			0.5 OA	1.4 NA
	4163.03	900				2.4 NA
HoO	5160				0.1 OAn	
	5320				0.2 OAn	
	5660				0.05 OAn; 2 OA	
I(InI)	4099				2.0 AH	
In	2710.26	160				8.8 AA
	3039.36	800			8.0 OH	0.9 AA
	3256.09	1300	20	10	2.2 OH	1.0 AA
	3258.56	300				11 AA
	4101.76	1700		3	0.14 OH	2.6 AA
	4511.31	1800	2		0.07 OH; 0.3 OA	2.8 AA

Table 8-9 (*Continued*)
EMISSION AND ABSORPTION LINES 1900 to 9000 Å

Element	Wavelength, Å	Relative Intensity	Detection limit			Atomic Absorption: Sensitivity, μg/ml/1% Abs (conc'n range for which valid)
			DC Arc, μg/g	Spark-porous Cup, μg/ml	Flame Emission, μg/ml/0.1 mV	
Ir	2088.82	13				8 AA
	2543.97	380				20 AA
	2639.71	170				13 AA
	2664.79	200				15 AA
	2849.72	280	10			18 AA
	3220.78	500	10	10		
K	4044.14	32	100	200	1.7 OH	6 AA (10–1000)
	4047.20	16				13 AA (50–1000)
	7664.91	1800	1		0.2 OH	0.06 AA (2–10)
	7698.98	900				0.1 AA (2–10)
La	3337.49 II	200		0.3		
	3949.10 II	900	10	1.0		
	4086.72 II	550	5			
	4187.32	28			40 OA	49 NA
	5501.34	36			8 NA	34 NA
	5791.34	34			2 NA	
LaO	4371/6				0.06 OAn	
	4418.24				0.06 OAn	
	5406/8				1.4 OAn	
	5430				1.4 OAn	
	5600				0.18 OAn	
	7410				0.005 OAn	
	7910				0.005 OAn	
Li	2741.20	5				15 AA (100–1000)
	3232.61	17	10	10	46 OA	18 AA (50–1000)
	4602.86	13			13 OA	
	6103.64	320			4 OA	192 AP
	6707.84	3600	0.1	0.1	0.07 OA	0.04 AA (0.1–10)
Lu	2615.42 II	1200	0.5			
	2911.39 II	600	10	0.5		
	3312.11	360			6 OA	21 NA
	3359.56	440				12 NA
	3567.84	280				27 NA
LuO	4680				0.05 OAn; 3 OH	
	5170				0.05 OAn; 4 OH	
Mg	2025.82					2.0 AA
	2795.53 II	1000	0.2	0.01		
	2802.70 II	600		0.003		
	2852.13	6000	0.2		0.2 OAr	0.008 AAw
MgOH	3702				1.4 OH	
	3810/30				1.6 OH	
Mn	2576.10 II	1200	1	0.02		
	2593.73 II	800		0.05		
	2213.85	4				1.3 AA (2–500)
	2794.82	800				0.06 AA (1–20)
	2798.27	650				0.08 AA (1–10)
	2801.06	480	1			0.12 AA (1–50)
	4030.76	2000	1		0.1 OA	0.6 AA (2–100)
	4033.07	1400				0.8 AA (2–200)
	4034.49	800				1.0 AA (2–200)
Mo	2816.15 II	220		1.0		

Table 8-9 (*Continued*)
EMISSION AND ABSORPTION LINES 1900 to 9000 Å

Element	Wavelength, Å	Relative Intensity	DC Arc, μg/g	Spark-porous Cup, μg/ml	Flame Emission, μg/ml/0.1 mV	Atomic Absorption: Sensitivity, μg/ml/1% Abs (conc'n range for which valid)
					Detection limit	
	3112.12	170			48 OAn	20 (N–A) Ar (100–1000)
	3132.59	1800	10	0.3	2 OAn	0.8 (N–A) Ar (2–200)
	3158.16	750			10 OAn	4 (N–A) Ar (10–1000)
	3170.35	1100	1		4 OAn	2 (N–A) Ar (5–500)
	3193.97	950			4 OAn	3.5 (N–A) Ar (5–200)
	3208.83	380			19 OAn	14 (N–A) Ar (100–1000)
	3798.25	3200	1		0.5 OAn	2 (N–A) Ar (5–200)
	3864.11	2800			0.6 OAn	4 (N–A) Ar (5–500)
	3902.96	1800			0.7 OAn	4 (N–A) Ar (10–500)
	5506.49	480			9 OAn	
Na	3302.32/.99	30	10	35	12.5 OH	2.8 AA (10–500)
	5889.95	2000	0.1		0.001 OH	0.016 AA (0.1–5)
	5895.92	1000				0.03 AA (0.1–10)
Nb	2950.88 II	180		2		
	3349.06	200			72 OAn	27 NA
	3580.27	800			43 OAn	27 NA
	3713.01	340			56 OAn	
	3742.39	180			31 OAn	
	4058.94	1700			13 OAn	36 NA
	4079.73	1200			17 OAn	32 NA
	4100.92	700			19 OAn	42 NA
	4123.81	550			28 OAn	40 NA
Nd	4012.25 II	220		5		
	4061.09 II	280	20			
	4303.58	320	10			
	4634.24	30				10 NA
	4896.93	24				14 NA
	4924.53	40			5 OAn	8 NA
NdO	5550				0.2 OAn	
	6630				0.4 OAn; 1 OH	
	7020				0.2 OAn; 1 OH	
	7120				0.4 OAn; 1 OH	
Ni	2289.98	18				0.6 AA (5–50)
	2310.96	30				0.22 AA (3–50)
	2320.03	44				0.15 AA (1–30)
	2345.54	26				0.5 AA (5–100)
	3002.49	320	3			0.7 AA (3–300)
	3037.94	140				2 AA (10–300)
	3050.82	280				0.7 AA (3–300)
	3232.96	100				5 AA (10–500)
	3369.57	260				3 AA (3–300)
	3391.05	120				6 AA (10–500)
	3392.99	300			10 OA	2.4 AA (3–100)
	3414.76	750	3	0.8	4 OA	
	3417.8					0.6 AA (1–100)
	3433.56	240			10 OA	2.5 AA (5–500)
	3446.26	440			10 OA	
	3458.47	460			5 OA	
	3461.65	460			11 OA	1.1 AA (1–500)
	3492.96	500			6 OA	

Table 8-9 (*Continued*)
EMISSION AND ABSORPTION LINES 1900 to 9000 Å

Element	Wavelength, Å	Relative Intensity	Detection limit DC Arc, μg/g	Detection limit Spark-porous Cup, μg/ml	Flame Emission, μg/ml/0.1 mV	Atomic Absorption: Sensitivity, μg/ml/1% Abs (conc'n range for which valid)
	3515.05	600			5 OA	
	3524.54	750			2 OA; 0.2 OAn	0.6 AA (1–50)
	3619.39	600			7 OA	
NO	2363				10 OH; 14 AH (sheathed)	
Os	2637.13	360				1.8 NA
	2644.11	180				4.8 NA
	2714.64	280				4.2 NA
	2806.91	260				4.6 NA
	2909.06	900	10			1.0 NA
	3018.04	460				3.2 NA
	3058.66	900		15		1.6 NA
	3301.56	800				3.6 NA
	4260.85	440				30 NA; 21 AA
	4420.47	440			10 OAnz	19 NA
P	2135.47 ⎤					240 AA or NA
	2136.20 ⎦					
	2534.01	70		5	230 OAnr	
	2535.65	60	30		50 OAnr	
	2553.28	38	100		120 OAnr	
HPO	5100				13 AH reversed	
	5249				6 AH reversed	
	5600				8 AH reversed	
PO	2464.2				3 OAnr; 19 OHn	
Pb	2022.02	5				7 AA (20–1000)
	2053.27	8				6 AA (20–1000)
	2169.99	22			550 OHn	0.23 AA (1–50)
	2614.18	700			13 OHn	
	2801.99	1000			10 OHn	
	2833.06	950	1	4	6 OHn	0.6 AA (1–100)
	3639.58	550			4 OHn	
	3683.48	1400			2 OHn; 21 OA	
	4057.83	3400	1		2 OHn; 14 OA	
Pd	2447.91	65				0.3 AA
	2476.42	100		2		0.3 AA
	2763.09	160				1.0 AA
	3404.58	2600	1	2	0.1 OHn; 1 OAn	1.2 AA
	3421.24	1400			1.0 OHn; 5 OAn	
	3516.94	1300			0.3 OHn; 3 OAn	
	3609.55	2200			0.2 OHn; 2 OAn	
	3634.70	2200			0.1 OHn; 1 OAn	
Pr	3908.41 II	320	25			
	4100.75 II	260		2		
	4225.33 II	340	10			
	4914.03	12				19 NA
	4951.36	34			0.4 OAn; 15 OA	13 NA
	5133.42	24			0.4 OAn	23 NA
Pt	2144.23	6				7.3 AA
	2174.67	7				3.3 AA
	2487.17	100			200 OAn	
	2628.03	110			100 OAn	5.3 AA
	2659.45	280		1	15 OAn; 13 OHn	2.2 AA

Table 8-9 (*Continued*)
EMISSION AND ABSORPTION LINES 1900 to 9000 Å

Element	Wavelength, Å	Relative Intensity	Detection limit			Atomic Absorption: Sensitivity, μg/ml/1% Abs (conc'n range for which valid)
			DC Arc, μg/g	Spark-porous Cup, μg/ml	Flame Emission, μg/ml/0.1 mV	
	2830.30	140				7.4 AA
	3064.71	320	2		15 OAn; 10 OHn	4.6 AA
Rb	4201.85	32	100		4 OH	12 AA (25–100)
	4215.56	16			15 OH	24 AA (50–100)
	7800.23	3000	1		0.6 OH	0.2 AA (0.5–10)
	7947.60	1500			0.7 OH	0.35 AA (1–10)
Re	2274.62	24				29 NA
	2287.51	40				20 NA
	2294.49	160				25 NA
	3451.88	1600	10		11 OAn	29 NA
	3460.46	5500	1	10	3 OAn	11 NA
	3464.73	4000		5	5 OAn	19 NA
	4889.14	220			8 OAn	
	5275.56	160			12 OAn	
Rh	3396.85	480				0.8 AA
	3434.89	700	1	0.7	3 OAn; 2 OHn	0.3 AA
	3502.52	500			6 OAn; 2 OHn	1.3 AA
	3507.32	240				1.3 AA
	3528.02	750			4 OAn; 5 OHn	
	3657.99	700			4 OAn; 3 OHn	1.7 AA
	3692.36	800			1 OAn; 1 OHn	0.6 AA
Ru	3436.74	650	10			
	3498.94	850		2	2 OHn; 3 OAn	1.3 AA
	3728.03	1000			0.5 OHn	
	3799.35	700			0.4 OAn	
	3925.92	300				14 AA
S_2	3645				5 AH (reversed shielded)	
	3740				4 AH	
	3837				3 AH	
	3940				3 AH	
	4050				4 AH	
	4150				4 AH	
SO_2	2070				10 AH	
Sb	2068.33	55				0.8 AA (2–100)
	2127.39	5				20 AA (100–1000)
	2175.81	38				0.6 AA (1–100)
	2179.19	7				1.5 AA
	2311.47	45			0.5 AH	1.5 AA (5–500)
	2528.52				1 OAn	
	2598.05	600	7	2	0.6 OAn	
	2877.92	140	10			
Sc	3269.91	400				1.6 NA
	3273.63	500				5 NA
	3353.73 II	900	2			
	3613.84 II	2500	0.5	0.05		
	3630.75 II	1800	1			
	3642.79 II	1200	1			
	3907.49	1800	10		1 OAn	0.5 NA
	3911.81	2100			0.7 OAnr	0.5 Na
	4020.40	1800			0.1 OAnr	0.9 NA
	4023.69	1800				0.7 NA

Table 8-9 (*Continued*)
EMISSION AND ABSORPTION LINES 1900 to 9000 Å

Element	Wavelength, Å	Relative Intensity	DC Arc, μg/g	Spark-porous Cup, μg/ml	Flame Emission, μg/ml/0.1 mV	Atomic Absorption: Sensitivity, μg/ml/1% Abs (conc'n range for which valid)
	4054.55	500				1.4 NA
	4246.83 II	1400	1			
ScO	6700				0.01 OAn	
Se	1960.26	34		70		0.7 AAw (1–100)
	2039.85	40			3 OHn	11 AA
	2062.79	15				43 AA
	2074.79	3				50 AAw (100–500)
Si	2216.67	3				9 NA
	2506.90	170			12 OAnr	6 NA
	2514.32	160			14 OAnr	7 NA
	2516.11	360	1	1	4 OAnr; 10 OAr	2 NA
	2519.21	120			16 OAnr	11 NA
	2524.11	240			14 OAnr	8 NA
	2528.51	400			12 OAnr	7 NA
Sm	3609.49 II	280	20			
	3634.29 II	280	40	3		
	3885.29 II	280	20			
	4296.74	110				8.5 NA
	4424.34 II	200	10			24 NA
	4760.27	75				
	4783.10	60			5 OA	
	5200.59	34				13 NA
SmO	6140				0.1 OAn	
	6240				0.1 OAn	
	6400				0.1 OAn	
	6520				0.1 OAn; 3 OHn	
Sn	2246.02	45		2		
	2334.80	38			17 OAn	5.5 AH
	2429.49	420			2 OAnr	
	2661.24	140				19 AH
	2839.99	1400	3		0.3 NA	
	2863.33	1000				10 AH (50–1000)
	3034.12	850			9 OAn	
	3175.02	550	1			
Sr	3464.46 II	65	10		26 OAn	
	4077.71 II	4600	0.1	0.006		
	4607.33	650	0.1		0.09 OH; 0.06 OA	0.06 AA or NA
SrOH	6060				0.3 OH; 0.6 OA	
Ta	2608.63	160				21 NA
	2685.15 II	180		2		
	2714.67	300				11 NA
	2775.88	90				21 NA
	3012.54 II	240	100			
	3311.16	140	100			
	4812.75	20			18 OAnz	
Tb	3509.17 II	600	15	3		
	3901.35	150			4 OAnrz	12 NA
	4061.59	120				13 NA
	4278.52 II	70	10			
	4318.85	200			10 OAn	9 NA
	4326.47	280			0.5 NA	8 NA
	4338.45	160				14 NA

Table 8-9 (*Continued*)
EMISSION AND ABSORPTION LINES 1900 to 9000 Å

Element	Wavelength, Å	Relative Intensity	Detection limit			Atomic Absorption: Sensitivity, μg/ml/1% Abs (conc'n range for which valid)
			DC Arc, μg/g	Spark-porous Cup, μg/ml	Flame Emission, μg/ml/0.1 mV	
TbO	5340				2 OHn; 0.1 OAn	
	5730				0.1 OAn	
	6080				0.2 OAn	
	6120				0.3 OAn	
Te	2142.75	55			7 OAn	0.5 AA (10–100)
	2259.04	6			7 OAn	4 AA (10–500)
	2383.25	55	100		4 OAn	67 AA
	2385.76	70	70	10	2 OAn; 380 OA	43 AA
Th	3244.46	20				850 NA
	3392.03 II	90	10			
	3539.59	48	35			
Ti	3186.51	200				3 NA
	3234.52 II	550		0.1		
	3241.99 II	220	10			
	3341.88	480			16 OAn	
	3349.04 II	1000		3		
	3354.64	340				2.9 NA
	3371.45	360			16 OAn	2.0 NA
	3372.80 II	480	1			
	3635.46	400			8 OAn	4.4 NA
	3642.68	550			7 OAn	2.2 NA
	3653.50	600	1		6 OAn	2.5 NA
	3741.06	280				2.6 NA
	3752.86	440			6 OAn	2.5 NA
	3948.67	380			13 OAn	5.0 NA
	3958.21	440			6 OAn	5.0 NA
	3989.76	480			6 OAn	5.2 NA
	3998.64	650			5 OAn	16 OAn
Tl	2767.87	440	20	3		0.3 AA
	2580.14	70				30 AA
	3519.24	2000	10			
	3775.72	1200			0.6 OH	1 AA
	5350.46	1800	1		1.2 OH	
Tm	3131.26 II	700		2		
	3462.20 II	800	5			
	3717.92	650				0.5 NA
	4094.19	750				1.2 NA
	4105.84	700				1.0 NA
	4203.73	440				1.2 NA
	4359.93	200				3.6 NA
TmO	4850				0.12 OAn	
	4900				0.16 OAn	
	5350				0.16 OAn; 3 OHn	
	5570				0.10 OAn	
U	3566.60	95				180 NA
	3584.88	130				120 NA
	3670.07 II	160		100		
	4241.67 II	75	100			
	5915.40	20			10 OAnz	
V	3033.82 II	38		0.3		
	3066.38	320			9 OAn	7 NA (20–200)
	3093.11 II	500		1		2 NA
	3183.98	700	1		5 OAn	

Table 8-9 (*Continued*)
EMISSION AND ABSORPTION LINES 1900 to 9000 Å

Element	Wavelength, Å	Relative Intensity	DC Arc, μg/g	Spark-porous Cup, μg/ml	Flame Emission, μg/ml/0.1 mV	Atomic Absorption: Sensitivity, μg/ml/1% Abs (conc'n range for which valid)
	3185.40	500	10		6 OAn	1.9 NA (5–100)
	3271.12 II	500		1		
	3840.75	280			6 OAn	
	3855.84	320			5 OAn	
	4379.24	950	1		3 OAn	9 NA
	4384.72	550			5 OAn	
	4389.97	380			6 OAn	9 NA
W	2397.09 II	34		3		
	2551.35	280				5 NA
	2658.04 II	50		10		
	2946.98	300	100			
	4008.75	950		30	4 OAnz	20 NA
	4294.61	450	100			
Y	3242.28 II	800	10	0.2		
	3710.30 II	1500		0.1		
	4077.38	950				5.7 NA
	4102.38	1000			3 OAn	5 NA
	4128.31	900				5.4 NA
	4142.85	750				11 NA
	4374.94	1200	1			
YO	4280				0.1 OAn	
	5980				0.02 OAn	
	6140				0.02 OAn	
Yb	2464.49	65				1.5 NA
	2671.98	55				14 NA
	3289.37 II	2600	10	0.04		
	3464.36	340				0.8 NA
	3694.19 II	3200	0.5			
	3987.98	1900			0.02 OAn	0.3 NA
	5556.48	140			0.2 OH; 0.06 OAn	
Zn	2138.56	1000	10		77 OAn	0.02 AA (0.1–10)
	3075.90	26	20			60 AA (300–1000)
	3345.02	140	10	4		
	4810.53	110	100			
Zr	3273.05 II	160		2		
	3391.98 II	900	1	0.2		
	3438.23 II	750	10			
	3519.60	320			52 OAnz	20 NA
	3601.19	550			75 OAnz	15 NA

Table 8-10
SENSITIVE LINES OF THE ELEMENTS

In this table the sensitive lines of the elements are arranged in order of decreasing wavelengths. In the column headed Sensitivity the most sensitive line of the neutral or un-ionized atom is indicated by U1, and other lines by U2, U3, etc., in order of decreasing sensitivity. For the singly ionized atom the corresponding designations are V1, V2, V3, etc. Where U1 is not given, the most sensitive lines lie outside the range of 10,000 to 2,000 Å.

SPECTROSCOPY

The table is taken by permission of George R. Harrison and the Technology Press from *M.I.T. Wavelength Tables*, 1939 ed., John Wiley & Sons, New York.

The *abbreviations* used in this table are those employed in the *M.I.T. Tables*, and are as follows:

bh, band head
d, double line
h, hazy, diffuse
l, shaded, or displaced to longer wavelengths
r, narrow self-reversal
R, wide self-reversal
s, shaded, or displaced to shorter wavelengths

w, wide or complex
W, very wide or complex
I, line classified as being emitted by the un-ionized atom
II, line classified as being emitted by the singly ionized atom
?, line listed in M.I.T. Tables, but not found on a plate made from pure sample of the element in question

Wavelength	Element		Intensity Arc	Intensity Spk [Dis]	Sensitivity	Wavelength	Element		Intensity Arc	Intensity Spk [Dis]	Sensitivity
9237.49	S	I	....	[200]	U6	5777.665	Ba	I	500R	100R	U2
9228.11	S	I	....	[200]	U5	5688.224	Na	I	300	....	..
9212.91	S	I	....	[200]	U4	5682.657	Na	I	80	....	..
8943.50	Cs	I	2000R	....	U2	5679.56	N	II	....	[500]	V2
8521.10	Cs	I	5000R	....	U1	5676.02	N	II	....	[100]	V4
8115.311	Ar	I	....	[5000]	U2	5666.64	N	II	....	[300]	V3
7947.60	Rb	I	5000R	....	U2	5608.8	Pb	II	....	[40]	V2
7800.227	Rb	I	9000R	....	U1	5570.2895	Kr	I	....	[2000]	U3
7775.433	O	I	....	[100]	U4	5535.551	Ba	I	1000R	200R	U1
7774.138	O	I	....	[300]	U3	5519.115	Ba	I	200R	60R	U3
7771.928	O	I	....	[1000]	U2	5465.487	Ag	I	1000R	500R	U4
7698.979	K	I	5000R	....	U2	5464.61	I	II	....	[900]	..
7664.907	K	I	9000R	....	U1	5460.740	Hg	I	....	[2000]	..
7503.867	Ar	I	....	[700]	U4	5455.146	La	I	200	1	U3
7450.00	Rn	I	....	[600]	U2	5424.616	Ba	I	100R	30R	U4
7067.217	Ar	I	....	[400]	U3	5400.562	Ne	I	....	2000	..
7055.42	Rn	I	....	[400]	U3	5350.46	Tl	I	5000R	2000R	U1
6965.430	Ar	I	....	[400]	U3	5291.0	bhCaF		200	....	..
6902.46	F	I	....	[500]	U3	5218.202	Cu	I	700	....	U3
6856.02	F	I	....	[1000]	U2	5209.067	Ag	I	1500R	1000R	U3
6707.844	Li	I	3000R	200	U1	5208.436	Cr	I	500R	100	U4
6562.79	H	I	....	[3000]	U2	5206.039	Cr	I	500R	200	U5
6438.4696	Cd	I	2000	1000	..	5204.518	Cr	I	400R	100	U6
6402.246	Ne	I	....	[2000]	..	5183.618	Mg	I	500wh	300	..
6362.347	Zn	I	1000Wh	500	..	5172.699	Mg	I	200wh	100wh	..
6249.929	La	I	300	....	U1	5167.343	Mg	I	100wh	50	..
6243.36	Al	II	....	100	V3	5161.188	I	II	....	[300]	..
6231.76	Al	II	....	30	..	5153.235	Cu	I	600	....	U4
6103.642	Li	I	2000R	300	U3	5105.541	Cu	I	500	....	U5
5930.648	La	I	250	....	U2	5007.213	Ti	I	200	40	..
5895.923	Na	I	5000R	500R	U2	4999.510	Ti	I	200	80	..
5889.953	Na	I	9000R	1000R	U1	4991.066	Ti	I	200	100	..
5875.618	He	I	....	[1000]	U3	4981.733	Ti	I	300	125	U1
5870.9158	Kr	I	....	[3000]	U2	4962.263	Sr	I	40	....	U4
5852.488	Ne	I	....	[2000]	..	4934.086	Ba	II	400h	400h	V2

Table 8-10 (*Continued*)
SENSITIVE LINES OF THE ELEMENTS

Wave-length	Element		Intensity		Sensi-tivity	Wave-length	Element		Intensity		Sensi-tivity
			Arc	Spk [Dis]					Arc	Spk [Dis]	
4889.17	Re	I	2000w	....	U2	4420.468	Os	I	400R	100	..
4872.493	Sr	I	25	....	U3	4390.865	Sm	II	150	150	..
4861.327	H	I	....	[500]	U3	4389.974	V	I	80R	60R	..
4832.075	Sr	I	200	8	U2	4384.722	V	I	125R	125R	..
4825.91	Ra	I	....	[800]	U1	4379.238	V	I	200R	200R	U1
4819.46	Cl	II	....	[200]	V4	4358.35	Hg	I	3000w	500	..
4816.71	Br	II	....	[300]	V3	4305.447	Sr	II	40	....	..
4810.534	Zn	I	400w	300h	..	4303.573	Nd		100	40	..
4810.06	Cl	II	....	[200]	V3	4302.108	W	I	60	60	U1
4794.54	Cl	II	....	[250]	V2	4294.614	W	I	50	50	U2
4785.50	Br	II	....	[400]	V2	4289.721	Cr	I	3000R	8000r	U3
4772.312	Zr	I	100	....	..	4274.803	Cr	I	4000R	800r	U2
4742.25	Se	I	....	[500]	U6	4267.27	C	II	....	500	V2
4739.478	Zr	I	100	....	..	4267.02	C	II	....	350	V3
4739.03	Se	I	....	[800]	U5	4254.346	Cr	I	5000R	1000	U1
4730.78	Se	I	....	[1000]	U4	4241.669	U		40	50	..
4722.552	Bi	I	1000	100	..	4226.728	Ca	I	500R	50W	U1
4722.159	Zn	I	400w	300h	..	4226.570	Ge	I	200	50	..
4710.075	Zr	I	60	....	..	4225.327	Pr		50	40	..
4704.86	Br	II	....	[250]	V1	4215.556	Rb	I	1000R	300	U4
4696.25	S	I	....	[15]	U9	4215.524	Sr	II	300r	400W	V2
4695.45	S	I	....	[30]	U8	4211.719	Dy		200	15	..
4694.13	S	I	....	[500]	U7	4205.046	Eu	II	200R	50	..
4687.803	Zr	I	125	....	U4	4201.851	Rb	I	2000R	500	U3
4685.75	He	II	....	[300]	..	4189.518	Pr		100	50	..
4682.28	Ra	II	....	[800]	V2	4186.599	Ce	II	80	25	..
4680.138	Zn	I	300w	200h	..	4179 422	Pr		200	40	..
4674.848	Y	I	80	100	U1	4177.321	Nd		15	25	..
4671.226	Xe	I	....	[2000]	U2	4172.056	Ga	I	2000R	1000R	U1
4643.695	Y	I	50	100	U2	4167.966	Dy		50	12	..
4624.276	Xe	I	....	[1000]	U3	4165.606	Ce	II	40	6	..
4607.331	Sr	I	1000R	50R	U1	4137.095	Cb	I	100	60	U5
4603.00	Li	I	800	....	U4	4130.664	Ba	II	50r	60Wh	V3
4593.177	Cs	I	1000R	50	U4	4129.737	Eu	II	150R	50R	..
4555.355	Cs	I	2000R	100	U3	4123.810	Cb	I	200	125	U4
4554.042	Ba	II	1000R	200	V1	4123.228	La	II	500	500	V4
4524.741	Sn		500wh	50	..	4109.98	N	I	....	[1000]	U2
4518.57	Lu		300	40	..	4103.37	N	III	....	[80]	..
4511.323	In	I	5000R	4000R	U1	4101.773	In	I	2000R	1000R	U2
4500.977	Xe	I	....	[500]	U4	4100.923	Cb	I	300w	200w	U3
4454.781	Ca	I	200	....	U2	4099.94	N	I	....	[150]	U3
4434.960	Ca	I	150	....	U3	4097.31	N	III	....	[100]	..
4434.321	Sm	II	200	200	V2	4093.161	Hf	II	25	20	..
4425.441	Ca	I	100	....	U4	4079.729	Cb	I	500w	200w	U2
4424.342	Sm	II	300	300	V1	4077.974	Dy		150r	100	..

Table 8-10 (*Continued*)
SENSITIVE LINES OF THE ELEMENTS

Wave-length	Element		Intensity		Sensi-tivity	Wave-length	Element		Intensity		Sensi-tivity
			Arc	Spk [Dis]					Arc	Spk [Dis]	
4077.714	Sr	II	400r	500W	V1	3774.332	Y	II	12	100	..
4077.340	La	II	600	400	V3	3768.405	Gd		20	20	..
4062.817	Pr		150	50	..	3761.917	Tm		200	120	..
4058.938	Cb	I	1000w	400w	U1	3761.333	Tm		250	150	..
4057.820	Pb	I	2000R	300R	U1	3748.264	Fe	I	500	200	U4
4047.201	K	I	400	200	U4	3748.17	Ho		60	40	..
4046.561	Hg	I	200	300	..	3745.903	Fe	I	150	100	U5
4045.983	Dy		150	12	..	3745.564	Fe	I	500	500	U3
4044.140	K	I	800	400	U3	3737.133	Fe	I	1000r	600	U2
4040.762	Ce	II	70	5	..	3719.935	Fe	I	1000R	700	U1
4034.490	Mn	I	250r	20	U3	3710.290	Y	II	80	150	V1
4033.073	Mn	I	400r	20	U2	3694.203	Yb		500R	1000R	..
4032.982	Ga	I	1000R	500R	U2	3692.652	Er		20	12	..
4030.755	Mn	I	500r	20	U1	3692.357	Rh	I	500hd	150wd	..
4023.688	Sc	I	100	25	U3	3683.471	Pb	I	300	50	U2
4020.399	Sc	I	50	20	U4	3672.579	U		8	15	..
4019.137	Th		8	8	..	3663.276	Hg	I	500	400	U5
4012.388	Ce	I, II	60	20	..	3657.987	Rh	I	500W	200W	..
4008.753	W	I	45	45	U3	3654.833	Hg	I	....	[200]	U4
4000.454	Dy		400	300	..	3653.496	Ti	I	500	200	U2
3987.994	Yb		1000R	500R	..	3650.146	Hg	I	200	500	U3
3968.468	Ca	II	500R	500R	V2	3646.196	Gd		200w	150	..
3961.527	Al	I	3000	2000	U1	3642.785	Sc	II	60	50	V3
3951.154	Nd		40	30	..	3642.675	Ti	I	300	125	..
3949.106	La	II	1000	800	V2	3639.580	Pb	I	300	50h	..
3944.032	Al	I	2000	1000	U2	3635.463	Ti	I	200	100	..
3933.666	Ca	II	600R	600R	V1	3634.695	Pd		2000R	1000R	U3
3911.810	Sc	I	150	30	U1	3633.123	Y	II	50	100	..
3907.476	Sc	I	125	25	U2	3630.740	Sc	II	50	70	V2
3906.316	Er		25	12	..	3613.836	Sc	II	40	70	V1
3905.528	Si	I	20	15W	..	3613.790	W	II	10	30	..
3902.963	Mo	I	1000R	500R	U3	3610.510	Cd	I	1000	500	..
3891.785	Ba	II	18	25	V4	3609.548	Pd	I	1000R	700R	..
3891.02	Ho		200	40	..	3601.193	Zr	I	400	15	U1
3888.646	He	I	....	[1000]	U2	3601.040	Th		8	10	..
3874.18	Tb		200	200	..	3600.734	Y	II	100	300	..
3864.110	Mo	I	1000R	500R	U2	3596.179	Ru	I	30	100	U3
3848.75	Tb		100	200	..	3572.473	Zr	II	60	80	V4
3838.258	Mg	I	300	200	U2	3561.74	Tb		200	200	..
3832.306	Mg	I	250	200	U3	3554.43	Lu		50	150	..
3829.350	Mg	I	100w	150	U4	3552.172	U		8	12	..
3814.42	Ra	II	....	[2000]	V1	3547.682	Zr	I	200	12	U2
3798.252	Mo	I	1000R	1000R	U1	3538.75	Th		....	50	..
3788.697	Y	II	30	30	..	3529.813	Co	I	1000R	30	U3
3775.72	Tl	I	3000R	1000R	U2	3524.541	Ni	I	1000R	100wh	..

Table 8-10 (*Continued*)
SENSITIVE LINES OF THE ELEMENTS

Wave-length	Element		Intensity		Sensi-tivity	Wave-length	Element		Intensity		Sensi-tivity
			Arc	Spk [Dis]					Arc	Spk [Dis]	
3519.605	Zr	I	100	10	U3	3302.988	Na	I	300R	150R	U4
3519.24	Tl	I	2000R	1000R	U3	3302.588	Zn	I	800	300	U3
3516.943	Pd	I	1000R	500R	..	3302.323	Na	I	600R	300R	U3
3515.054	Ni	I	1000R	50h	..	3290.59	Th		....	40h	..
3513.645	Ir	I	100h	100	U2	3289.37	Yb		500R	1000R	..
3509.17	Tb		200	200	..	3282.333	Zn	I	500R	300	U4
3499.104	Er		18	15	..	3280.683	Ag	I	2000R	1000R	U1
3498.942	Ru	I	500R	200	U1	3273.962	Cu	I	3000R	1500R	U2
3496.210	Zr	II	100	100	V3	3269.494	Ge	I	300	300	U3
3492.956	Ni	I	1000R	100h	U2	3267.945	Os	I	400R	30	..
3474.887	Sr	II	80	50	..	3267.502	Sb	I	150	150Wh	..
3472.48	Lu		50	150	..	3262.328	Sn	I	400h	300h	U3
3466.201	Cd	I	1000	500	..	3262.290	Os	I	500R	50	..
3465.800	Co	I	2000R	25	U2	3261.057	Cd	I	300	300	..
3464.57	Sr	II	200	200	..	3258.564	In	I	500R	300R	U5
3462.21	Tm		200	100	..	3256.090	In	I	1500R	600R	U3
3460.47	Re	I	1000W	....	U1	3247.540	Cu	I	5000R	2000R	U1
3453.505	Co	I	3000R	200	U1	3242.280	Y	II	60	100	..
3451.41	B	II	5	30	V2	3232.61	Li	I	1000R	500	U2
3438.230	Zr	II	250	200	V2	3232.499	Sb	I	150	250wh	..
3437.015	Ir	I	20	15	..	3229.75	Tl	I	2000	800	..
3436.737	Ru	I	300R	150	U2	3225.479	Cb	II	150w	800wr	..
3434.893	Rh		1000R	200r	U1	3220.780	Ir	I	100	30	U1
3421.24	Pd	I	2000R	1000R	U2	3215.560	W	I	10	9	..
3414.765	Ni	I	1000R	50wh	U1	3194.977	Cb	II	30	300	..
3406.664	Ta		70w	18s	..	3185.396	V	I	500R	400R	U2
3405.120	Co	I	2000R	150	..	3183.982	V	I	500R	400R	..
3404.580	Pd	I	2000R	1000R	U1	3183.406	V	I	200R	100R	..
3403.653	Cd	I	800	500h	..	3179.332	Ca	II	100	400w	V3
3397.07	Lu		50	20r	..	3175.019	Sn	I	500h	400hr	..
3396.85	Rh	I	1000w	500	..	3163.402	Cb	II	15	8	..
3391.975	Zr	II	300	400	V1	3158.869	Ca	II	100	300w	V4
3383.761	Ti	II	70	300R	..	3134.718	Hf	II	80	125	..
3382.891	Ag	I	1000R	700R	U2	3131.072	Be	II	200	150	V2
3380.711	Sr	II	150	200	..	3130.786	Cb	II	100	100	..
3372.800	Ti	II	80	400R	V3	3130.416	Be	II	200	200	V1
3361.213	Ti	II	100	600R	V2	3125.284	V	II	80	200R	..
3349.035	Ti	II	125	800R	V1	3118.383	V	II	70	200R	V4
3345.020	Zn	I	800	300	U2	3110.706	V	II	70	300R	V3
3323.092	Rh	I	1000	200	..	3102.299	V	II	70	300R	V2
3321.343	Be	I	1000r	30	U2	3094.183	Cb	II	100	1000	V1
3321.086	Be	I	100	....	U3	3093.108	V	II	100R	400R	V1
3321.013	Be	I	50	....	U4	3092.713	Al	I	1000	1000	U3
3318.840	Ta		125	35	..	3082.155	Al	I	800	800	U4
3311.162	Ta		300w	70w	U1	3072.877	Hf	I	80	18	..

Table 8-10 (*Continued*)
SENSITIVE LINES OF THE ELEMENTS

Wave-length	Element		Intensity		Sensi-tivity	Wave-length	Element		Intensity		Sensi-tivity
			Arc	Spk [Dis]					Arc	Spk [Dis]	
3071.591	Ba	I	100R	50R	U5	2837.602	C	II	....	40	V5
3067.716	Bi	I	3000hR	2000wh	U1	2836.710	C	II	....	200	V4
3064.712	Pt	I	2000R	300R	U1	2835.633	Cr	II	100	400r	V1
3058.66	Os	I	500R	500	..	2833.069	Pb	I	500R	80R	..
3039.356	In	I	1000R	500R	U4	2830.295	Pt	I	1000R	600r	..
3039.064	Ge	I	1000	1000	U2	2820.224	Hf	II	40	100	..
3034.121	Sn	I	200wh	150wh	..	2816.179	Al	II	10	100	V2
3009.147	Sn	I	300h	200h	..	2816.154	Mo	II	200	300h	V1
2997.967	Pt	I	1000R	200r	..	2809.625	Bi	I	200w	100	..
2989.029	Bi	I	250wh	100wh	..	2802.695	Mg	II	150	300	V2
2976.586	Ru		60	200	..	2802.19	Au		....	200	..
2965.546	Ru		60	200	..	2795.53	Mg	II	150	300	V1
2945.668	Ru		60	300	..	2780.521	Bi	I	200w	100	..
2943.637	Ga	I	10	20r	U3	2780.197	As	I	75R	75	U5
2940.772	Hf	I	60	12	..	2773.357	Hf	II	25	60	..
2938.298	Bi	I	300w	300w	..	2769.67	Te	I	....	[30]	..
2936.77	Ho		....	1000R	..	2767.87	Tl	I	400R	300R	..
2929.794	Pt	I	800R	200w	..	2748.58	Cd	II	5	200	..
2924.792	Ir	I	25wh	15	..	2712.410	Ru		80	300	..
2918.32	Tl	I	400R	200R	..	2709.626	Ge	I	30	20	..
2916.481	Hf	I	50	15	..	2692.065	Ru		8	200	..
2911.39	Lu		100	300	..	2678.758	Ru		100	300	..
2909.116	Mo	II	25	40h	V5	2675.95	Au	I	250R	100	U2
2909.061	Os	I	500R	400	U1	2669.166	Al	II	3	100	V1
2904.408	Hf	I	30	6	..	2659.454	Pt	I	2000R	500R	U2
2898.71	As	I	25r	40	..	2658.722	Pd	II	20	300	..
2898.259	Hf	I	50	12	..	2651.575	Ge	I	30	20	..
2897.975	Bi	I	500WR	500WR	U2	2651.178	Ge	I	40	20	..
2894.84	Lu		60	200	..	2650.781	Be	I	25	....	U5
2890.994	Mo	II	30	50h	V4	2641.406	Hf	II	40	125	..
2881.578	Si	I	500	400	U1	2631.553	Al	II	....	40	..
2877.915	Sb	I	250W	150	..	2614.178	Pb		200r	80	..
2874.244	Ga	I	10	15r	U4	2605.688	Mn	II	100R	500R	V3
2871.508	Mo	II	100	100h	V3	2598.062	Sb	I	200	100	..
2863.327	Sn	I	300R	300R	U2	2593.729	Mn	II	200R	1000R	V2
2860.934	Cr	II	60	100	V5	2589.167	W	II	15d	25	..
2860.452	As	I	50r	50	..	2576.104	Mn	II	300R	2000R	V1
2855.676	Cr	II	60	200Wh	V4	2573.09	Cd	II	3	150	..
2854.581	Pd	II	4	500h	..	2557.958	Zn	II	10	300	V3
2852.129	Mg	I	300R	100R	U1	2554.93	P	I	60	[20]	..
2849.838	Cr	II	80	150r	V3	2553.28	P	I	80	[20]	U3
2849.725	Ir	I	40h	20h	..	2536.519	Hg	I	2000R	1000R	U2
2848.232	Mo	II	125	200h	V2	2535.65	P	I	100	[30]	U2
2843.252	Cr	II	125	400r	V2	2534.01	P	I	50	[20]	..
2839.989	Sn	I	300R	300R	U1	2530.70	Te	I	....	[30]	..

Table 8-10 (*Continued*)
SENSITIVE LINES OF THE ELEMENTS

Wave-length	Element		Intensity		Sensi-tivity	Wave-length	Element		Intensity		Sensi-tivity
			Arc	Spk [Dis]					Arc	Spk [Dis]	
2528.535	Sb	I	300R	200	..	2307.857	Co	II	25	50w	..
2528.516	Si	I	400	500	U2	2304.235	Ba	II	60R	80R	..
2519.822	Co	II	40	200	..	2296.89	C	III	....	200	..
2516.881	Hf	II	35	100	..	2288.12	As	I	250R	5	U3
2516.123	Si	I	500	500	U3	2288.018	Cd	I	1500R	300R	U1
2513.028	Hf	II	25	70	..	2287.084	Ni	II	100	500	V1
2506.899	Si	I	300	200	U4	2286.156	Co	II	40	300*l*	V1
2505.739	Pd	II	3	30	..	2276.578	Bi	I	100R	40	..
2502.001	Zn	II	20	400w	V4	2270.213	Ni	II	100	400	V2
2498.784	Pd	II	4	150h	..	2265.017	Cd	II	25d	300	V2
2497.733	B	I	500	400	U1	2264.457	Ni	II	150	400	V3
2496.778	B	I	300	300	U2	2253.86	Ni	II	100	300	V4
2488.921	Pd	II	10	30	..	2246.995	Cu	II	30	500	V3
2478.573	C	I	400	[400]	U2	2246.412	Ag	II	25	300hs	V3
2456.53	As	I	100r	8	U4	2203.505	Pb	II	50W	5000R	V1
2437.791	Ag	II	60	500wh	V2	2192.260	Cu	II	25	500h	V2
2427.95	Au	I	400R	100	U1	2175.890	Sb	I	300	40	U2
2413.309	Fe	II	60	100h	V5	2169.994	Pb	I	1000R	1000R	..
2410.517	Fe	II	50	70h	V4	2144.382	Cd	II	50	200R	V1
2404.882	Fe	II	50	100wh	V3	2142.75	Te	I	60R	....	..
2397.091	W	II	18	30	..	2138.56	Zn	I	800R	500	U1
2395.625	Fe	II	50	100wh	V2	2135.976	Cu	II	25	500w	V1
2388.918	Co	II	10	35	..	2068.38	Sb	I	300R	3	U1
2385.76	Te	I	600	[300]	U2	2062.788	Se	I	....	[800]	U3
2383.25	Te	I	500	[300]	U3	2062.38	I		....	[900]	..
2382.039	Fe	II	40r	100R	V1	2061.91	Zn	II	100	100	V2
2378.622	Co	II	25	50w	..	2061.70	Bi	I	300R	100	..
2370.77	As	I	50r	3	..	2039.851	Se	I	....	[1000]	U2
2369.67	As	I	40r	....	..	2025.51	Zn	II	200	200	V1
2363.787	Co	II	25	50	..						
2349.84	As	I	250R	18	U3						
2348.610	Be	I	2000R	50	U1						
2335.269	Ba	II	60R	100R	..						
2312.84	Cd	II	1	200	..						
2311.469	Sb	I	150R	50	..						

Section 9

THERMODYNAMIC PROPERTIES

HEATS AND FREE ENERGIES OF FORMATION, ENTROPIES AND HEAT CAPACITIES OF ELEMENTS AND COMPOUNDS

The tables contain values of the enthalpy and Gibbs (formerly free) energy of formation, entropy, and heat capacity at 298.15°K (25°C). No values are given in these tables for metal alloys or other solid solutions, fused salts, or for substances of undefined chemical composition.

For a more complete listing of compounds see the tables of "Selected Values of Chemical Thermodynamic Properties," *National Bureau of Standards Technical Notes* 270-3, 270-4, 270-5 by D. D. Wagman et al., Washington, D.C.

The physical state of each substance is indicated in the column headed State as crystalline solid (c), liquid (liq), gaseous (g), or amorphous (amorp). Solutions in water are listed as aqueous (aq).

The values of the thermodynamic properties of the pure substances given in these tables are, for the substances in their standard states, defined as follows: For a pure solid or liquid, the standard state is the substance in the condensed phase under a pressure of one atmosphere. For a gas the standard state is the hypothetical ideal gas at unit fugacity, in which state the enthalpy is that of the real gas at the same temperature and at zero pressure.

The values of $\Delta Hf°$ and $\Delta Gf°$ given in the tables represent the change in the appropriate thermodynamic quantity when one gram-formula weight of the substance in its standard state is formed, isothermally at the indicated temperature, from the elements, each in its appropriate standard reference state. The standard reference state at 25°C for each element (except phosphorus) has been chosen to be the standard state that is thermodynamically stable at 25°C and one atmosphere pressure. For phosphorus the standard reference state is the crystalline white form. The standard reference states are indicated in the tables by the fact that the values of $\Delta Hf°$ and $\Delta Gf°$ are exactly zero.

The values of $S°$ represent the virtual or "thermal" entropy of the substance in the standard state at 298.15°K, omitting contributions from nuclear spins. Isotope mixing effects are also excluded except in the case of the (^{1}H-^{2}H) system.

Solutions in water are designated as aqueous, and the concentration of the solution is expressed in terms of the number of moles of solvent associated with one mole of the solute. If no concentration is indicated, the solution is assumed to be dilute. The standard state for a solute in aqueous solution is taken as the hypothetical ideal solution of unit molality (indicated as std. state, $m = 1$). In this state the partial molal enthalpy and the heat capacity of the solute are the same as in the infinitely dilute real solution (as, ∞).

The value of $\Delta Hf°$ given for a solute in its standard state is the apparent molal enthalpy of formation of the substance in the infinitely dilute real solution. The experimental value for a heat of dilution is obtained directly as the difference between the two values of $\Delta Hf°$ at the corresponding concentrations.

Values of $\Delta Hf°$ and $\Delta Gf°$ (or $\Delta Ff°$) in the tables are expressed in kilocalories per mole; values of $S°$ and $C_p°$ are expressed in calories per degree per mole.

Table 9-1
ELEMENTS AND INORGANIC COMPOUNDS

Formula and Description	State	$\Delta Hf°$	$\Delta Gf°$	$S°$	$C_p°$
Actinium					
Ac_2O_3	c	-444			
Aluminum					
Al	c	0	0	6.77	5.82
	g	78.0	68.3	39.30	5.11
Al^{3+} std. state, $m = 1$	aq	-127	-116	-76.9	
$Al(BH_4)_3$	liq	-3.9	34.6	69.1	46.5
	g	3	35	90.6	
AlBr	g	-1	-10	57.22	8.50
$AlBr_3$	c	-126.0	-120.7	44	24.3
std. state, $m = 1$	aq	-214	-191	-17.8	
Al_4C_3	c	-49.9	-46.9	21.26	27.91
$Al(CH_3)_3$	liq	-32.6	-2.4	50.05	37.19
	g	-17.7			
$Al_2(CH_3)_6$	g	-55.19	-2.34	125.4	
$Al(OAc)_3$	c	-452.3			
AlCl	g	-11.4	-17.7	54.50	8.36
$AlCl_3$	c	-168.3	-150.3	26.45	21.95
std. state, $m = 1$	aq	-247	-210	-36.4	
$AlCl_3 \cdot 6H_2O$	c	-643.3	-542.4	90	
Al_2Cl_6	g	-308.5	-291.7	117	
AlF	g	-61.7	-67.8	51.36	7.63
AlF_3	c	-359.5	-340.6	15.88	17.95
	g	-287.9	-284.0	66.2	14.97
$AlF_3 \cdot 3H_2O$	c	-549.1	-490.4	50	
AlH	g	61.96	55.25	44.88	7.02
AlH_3	c	-11			
AlI	g	15.66			8.60
AlI_3	c	-75.0	-71.9	38	23.6
std. state, $m = 1$	aq	-167	-153	2.9	
AlN	c	-76.0	-68.6	4.82	7.20
$Al(NO_3)_3$ std. state, $m = 1$	aq	-276	-196	28.1	
$Al(NO_3)_3 \cdot 6H_2O$	c	-681.28	-526.74	111.8	103.5
$Al(NO_3)_3 \cdot 9H_2O$	c	-897.96	(-700.2)	$(136.)$	
AlP	c	-39.8			
$AlPO_4$ berlinite	c	-404.4	-382.7	21.70	22.27
AlO	g	21.8	15.6	52.17	7.38
AlO_2^- std. state, $m = 1$	aq	-219.6	-196.8	-5	

Table 9-1 (*Continued*)
ELEMENTS AND INORGANIC COMPOUNDS

Formula and Description	State	$\Delta Hf°$	$\Delta Gf°$	$S°$	$C_p°$
Al_2O_3 α, corundum	c	−400.5	−378.2	12.17	18.89
δ	c	−398			
ρ	c	−391			
κ	c	−397			
γ	c	−395			
$Al_2O_3 \cdot H_2O$ boehmite	c	−472.0	−436.3	23.15	31.37
diaspore	c	−478	−440	16.86	25.22
$Al_2O_3 \cdot 3H_2O$ gibbsite	c	−612.5	−546.7	33.51	44.49
bayerite	c	−610.1			
$Al(OH)^{2+}$ std. state, $m = 1$	aq		−165.9		
$Al(OH)_3$	amorp	−305			
$Al(OH)_4^-$ std. state, $m = 1$	aq	−356.2	−310.2	28	
AlS	g	48.02	35.88	55.09	7.98
Al_2S_3	c	−173		(23)	
$Al_2(SO_4)_3$	c	−822.38	−740.95	57.2	62.00
std. state, $m = 1$	aq	−906	−766	−139.4	
$Al_2(SO_4)_3 \cdot 6H_2O$	c	−1269.53	−1104.82	112.1	117.8
Al_2Se_3	c	−135			
Al_2SiO_5 andalusite	c	−655.9	−620.8	22.28	29.33
kyanite	c	−656.4	−620.5	20.03	29.09
sillimanite	c	−662.6	−627.6	22.99	29.30
$Al_2Si_2O_7 \cdot 2H_2O$ kaolinite	c	−979.6	−903.0	48.5	58.62
halloysite	c	−975.1	−898.5	48.6	58.86
Al_2Te_3	c	−78			
Ammonium					
NH_3	g	−11.02	−3.94	45.97	8.38
undissoc; std. state, $m = 1$	aq	−19.19	−6.35	26.6	
	aq, 1	−18.011	−18.011		
	aq, 10	−19.074	−19.074		
	aq, 100	−19.167	−19.167		
NH_4^+ std. state, $m = 1$	aq	−31.67	−18.97	27.1	19.1
NH_4OH	liq	−86.33	−60.74	39.57	37.02
undissoc; std. state, $m = 1$	aq	−87.505	−63.04	43.3	
ionized; std. state, $m = 1$	aq	−86.64	−56.56	24.5	−16.4
	aq, 1	−86.875			
	aq, 2	−87.078			
	aq, 10	−87.396			
	aq, 100	−87.483			
$NH_4Al(SO_4)_2$	c	−562.2	−487.2	51.7	54.12
std. state, $m = 1$	aq	−593	−491	−40.2	
NH_4AsO_2 std. state, $m = 1$	aq	−134.21	−102.63	37.0	
$NH_4H_2AsO_3$ std. state, $m = 1$	aq	−202.51	−159.32	53.5	
$NH_4H_2AsO_4$	c	−523.3	−199.1	41.12	36.13
std. state, $m = 1$	aq	−249.06	−199.01	55.1	
$(NH_4)_2HAsO_4$	c	−282.4			
std. state, $m = 1$	aq	−279.9	−208.7	53.8	
$(NH_4)_3AsO_4$	c	−307.4			
std. state, $m = 1$	aq	−307.28	−211.91	42.4	
NH_4BO_2 std. state, $m = 1$	aq	−216.27	−181.24	18.2	
NH_4Br	c	−64.73	−41.9	27	23
std. state, $m = 1$	aq	−60.72	−43.82	46.8	−14.8
	aq, 100	−60.614			
	aq, 1000	−60.650			
NH_4BrO	aq	−54.2	−27.0	37	
NH_4BrO_3	aq	−51.7	−18.6	66.1	
$(NH_4)_2CO_3$ std. state, $m = 1$	aq	−225.18	−164.11	40.6	
NH_4HCO_3	c	−203.0	−159.2	28.9	
std. state, $m = 1$	aq	−197.06	−159.23	48.9	
NH_4 carbamate	c	−154.17	−107.09	31.9	

Table 9-1 (*Continued*)
ELEMENTS AND INORGANIC COMPOUNDS

Formula and Description	State	$\Delta Hf°$	$\Delta Gf°$	$S°$	$C_p°$
NH$_4$CN	c	0.10			32
std. state, $m = 1$	aq	4.3	22.2	49.6	
NH$_4$CNO cyanate	c	−72.75			
std. state, $m = 1$	aq	−66.6	−42.3	52.6	
NH$_4$CNS thiocyanate	c	−18.8			
std. state, $m = 1$	aq	−13.40	3.18	61.6	9.5
NH$_4$ formate	c	−135.63			
std. state, $m = 1$	aq	−133.38	−102.9	49	−1.9
NH$_4$ acetate	c	−147.26			
std. state, $m = 1$	aq	−147.83	−107.26	47.8	17.6
NH$_4$ chloroacetate	c	−153.7			
NH$_4$ trichloroacetate	c	−156.7			
(NH$_4$)$_2$C$_2$O$_4$	c	−268.72			
	aq, 2100	−260.6	−196.2		
NH$_4$ dithiocarbamate	c	−30.3			
NH$_4$Cl	c	−75.15	−48.51	22.6	20.1
std. state, $m = 1$	aq	−71.62	−50.34	40.6	−13.5
	aq, 10	−71.567			
	aq, 100	−71.487			
NH$_4$ClO std. state, $m = 1$	aq	−57.3	−27.8	37	
NH$_4$ClO$_2$ std. state, $m = 1$	aq	−47.6	−14.9	51.3	
NH$_4$ClO$_3$ std. state, $m = 1$	aq	−55.4	−19.8	65.9	
NH$_4$ClO$_4$	c	−70.58	−21.25	44.5	
std. state, $m = 1$	aq	−62.58	−21.03	70.6	
NH$_4$HCrO$_4$ std. state, $m = 1$	aq	−241.6	−201.8	71.1	
(NH$_4$)$_2$CrO$_4$	c	−279.0			
std. state, $m = 1$	aq	−273.5	−211.90	66.2	
(NH$_4$)$_2$Cr$_2$O$_7$	c	−431.8			
std. state, $m = 1$	aq	−419.5	−348.9	116.8	
NH$_4$Cr(SO$_4$)$_2$ · 12H$_2$O	c			170.9	168.5
NH$_4$F	c	−110.89	−83.36	17.20	15.60
std. state, $m = 1$	aq	−111.17	−85.61	23.8	−6.4
NH$_4$HF$_2$	c	−191.9	−155.6	27.61	25.50
std. state, $m = 1$	aq	−187.01	−157.15	49.2	
NH$_4$I	c	−48.14	−26.9	28	
std. state, $m = 1$	aq	−44.86	−31.30	53.7	−14.9
	aq, 100	−44.784			
NH$_4$IO std. state, $m = 1$	aq	−57.4	−28.2	25.8	
NH$_4$IO$_3$	c	−92.2			
std. state, $m = 1$	aq	−84.6	−49.6	55.4	
NH$_4$IO$_4$	aq	−66.9			
NH$_4$N$_3$ azide	c	27.6	65.5	26.9	
	aq	34.1	64.3	52.7	
NH$_4$NO$_2$	c	−61.3			
std. state, $m = 1$	aq	−56.7	−27.9	60.6	−4.2
NH$_4$NO$_3$	c	−87.37	−43.98	36.11	33.3
std. state, $m = 1$	aq	−81.23	−45.58	62.1	−1.6
	aq, 10	−82.470			
	aq, 100	−81.340			
(NH$_4$)$_2$O	liq	−102.94	−63.84	63.94	59.08
NH$_4$PO$_3$	aq	−265.2			
NH$_4$H$_2$PO$_2$ hypophosphite	c	−180.0			
NH$_4$H$_2$PO$_4$	c	−345.38	−289.33	36.32	34.00
std. state, $m = 1$	aq	−341.49	−289.14	48.7	
(NH$_4$)$_2$HPO$_4$	c	−374.50			45
std. state, $m = 1$	aq	−372.17	−298.28	46.2	
(NH$_4$)$_3$PO$_4$	c	−399.6			
std. state, $m = 1$	aq	−400.3	−300.4	28	
(NH$_4$)$_4$P$_2$O$_7$ std. state, $m = 1$	aq	−669.5	−534.6	80	

Table 9-1 (*Continued*)
ELEMENTS AND INORGANIC COMPOUNDS

Formula and Description	State	$\Delta Hf°$	$\Delta Gf°$	$S°$	$C_p°$
$(NH_4)_2PoCl_6$ std. state, $m = 1$	aq		−176		
$(NH_4)_2PtCl_6$	c	−192.0			56.8
NH_4ReO_4	c	−226.0	−185.2	55.6	
NH_4HS	c	−37.5	−12.1	23.3	
std. state, $m = 1$	aq	−35.9	−16.09	42.1	
$(NH_4)_2S$ std. state, $m = 1$	aq	−55.4	−17.4	50.7	
$(NH_4)_2S_2$ std. state, $m = 1$	aq	−56.1	−18.9	61.0	
NH_4HSO_3	c	−183.7			
std. state, $m = 1$	aq	−181.34	−145.12	60.5	
NH_4HSO_4	c	−245.45			
std. state, $m = 1$	aq	−243.75	−199.66	58.6	−0.9
	aq, 200	−245.65			
$(NH_4)_2SO_3$	c	−211.6			
std. state, $m = 1$	aq	−215.2	−154.2	47.2	
$(NH_4)_2SO_4$	c	−282.23	−215.56	52.6	44.81
std. state, $m = 1$	aq	−280.66	−215.77	58.6	−31.8
	aq, 100	−280.407			
$(NH_4)_2S_2O_3$	aq	−219.2			
$(NH_4)_2S_2O_4$ std. state, $m = 1$	aq	−243.4	−181.4	76	
$(NH_4)_2S_2O_6$	aq	−349.7			
$(NH_4)_2S_2O_7$	aq	−398.2			
$(NH_4)_2S_2O_8$	c	−392.5			
std. state, $m = 1$	aq	−383.3	−303.3	113.5	
$(NH_4)_2S_4O_6$	aq	−355.92			
$(NH_4)_2Sb_2S_4$ std. state, $m = 1$	aq	−115.7	−61.7	41.7	
NH_4HSe	c	−31.8	−5.6	23.1	
std. state, $m = 1$	aq	−27.9	−8.5	46	
$(NH_4)_2Se$ std. state, $m = 1$	aq		−7.0		
NH_4HSeO_3 std. state, $m = 1$	aq	−154.65	−117.33	60.2	
NH_4HSeO_4 std. state, $m = 1$	aq	−170.7	−127.1	62.8	
$(NH_4)_2SeO_3$ std. state, $m = 1$	aq	−185.0	−126.3	57.3	
$(NH_4)_2SeO_4$	c	−209.0			
std. state, $m = 1$	aq	−206.5	−143.4	67.1	
$(NH_4)_2SiF_6$ hexagonal	c	−640.94	−565.38	66.98	54.52
$(NH_4)_2SnCl_6$	c	−295.6			
NH_4HTe	c	0.3			
$NH_4H_5TeO_6$	aq	−333.2			
$(NH_4)_2TeO_3$	aq	−205.9			
NH_4VO_3	c	−251.7	−212.3	33.6	30.91
Antimony					
Sb III	c	0	0	10.92	6.03
IV explosive	amorp	2.54			
	g	62.7	53.1	43.06	4.97
Sb_2	g	56.3	44.7	60.90	8.70
Sb_4	g	49.0	33.8	84	
$SbBr_3$	c	−62.0	−57.2	49.5	
$SbCl$	g	−6.22			8.49
$SbCl_2$	g	−18.5			
$SbCl_3$	c	−91.34	−77.37	44.0	25.8
	g	−75.0	−72.0	80.71	18.33
$SbCl_5$	liq	−105.2	−83.7	72	
	g	−94.25	−79.91	96.04	28.95
SbF	g	−11.29			7.97
SbF_3	c	−218.8			
SbH_3	g	34.681	35.31	55.61	9.81
Sb_2H_4	g	57.2			
SbI_3	c	−24.0			
SbN	g	63.66			7.41
SbO	g	47.67			
SbO^+ std. state, $m = 1$	aq		−42.33		

Table 9-1 (*Continued*)
ELEMENTS AND INORGANIC COMPOUNDS

Formula and Description	State	$\Delta Hf°$	$\Delta Gf°$	$S°$	$C_p°$
SbO_2^- std. state, $m = 1$	aq		-81.32		
Sb_2O_3	c	-164.9			
Sb_2O_4	c	-216.9	-190.2	30.4	27.39
Sb_2O_5	c	-232.3	-198.2	29.9	
Sb_4O_6 II, cubic	c	-344.3	-303.1	52.8	
I, orthorhombic	c	338.7	-299.5	50.0	40.40
$HSbO_2$ undissoc; std. state, $m = 1$	aq	-116.6	-97.4	11.1	
$Sb(OH)_3$	c		-163.8		
undissoc; std. state, $m = 1$	aq	-184.9	-154.1	27.8	
H_3SbO_4	aq	-216.8			
$HSb(OH)_6$	aq	-353.4			
$SbOCl$	c	-89.4			
$SbOF$ undissoc; std. state, $m = 1$	aq		-116.5		
Sb_2S_3 black	c	-41.8	-41.5	43.5	28.65
orange	amorp	-35.2			
$Sb_2S_4^{2-}$ std. state, $m = 1$	aq	-52.4	-23.8	-12.5	
$Sb_2(SO_4)_3$	c	-574.2			
Sb_2Te_3	c	-13.5	-13.2	56	
Argon					
Ar	g	0	0	36.982	4.968
std. state, $m = 1$	aq	-2.9	3.9	14.2	
Arsenic					
As α, gray	c	0	0	8.4	5.89
γ, yellow, cubic	c	3.5			
β	amorp	1.0			
As_2	g	53.1	41.1	57.2	8.366
As_4	g	34.4	22.1	75	
$AsBr_3$	c	-47.2		(53)	
	liq	-43.1			
	g	-31	-38	86.94	18.92
$AsCl_3$	liq	-72.9	-62.0	51.7	
	g	-62.5	-59.5	78.17	18.10
AsF_3	liq	(-226)	-198	(43)	
AsH_3	g	15.88	16.47	53.22	9.10
As_2H_4	g	35.2			
AsI_3	c	-13.9	-14.2	50.92	25.28
	g			92.79	19.27
AsN	g	46.91	40.15	53.9	7.27
AsO	g	16.72			
AsO_2^- std. state, $m = 1$	aq	-102.54	-83.66	9.9	
AsO_4^{3-} std. state, $m = 1$	aq	-212.27	-55.00	-38.9	
As_2O_4	c	-189.72		(36)	
As_2O_5	c	-221.05	-187.0	25.2	27.85
As_4O_6 octahedral	c	-314.04	-275.46	51.2	45.72
monoclinic	c	-313.0	-275.82	56	
	g	-289.0	-262.4	91	
$HAsO_2$ undissoc; std. state, $m = 1$	aq	-109.1	-96.25	30.1	
H_2AsO_3 undissoc; std. state, $m = 1$	aq	-170.84	-140.35	26.4	
H_3AsO_3 undissoc; std. state, $m = 1$	aq	-177.4	-152.94	46.6	
$HAsO_4^{2-}$ undissoc; std. state, $m = 1$	aq	-216.62	-170.82	-0.4	
$H_2AsO_4^-$ undissoc; std. state, $m = 1$	aq	-217.39	-180.04	28	
H_3AsO_4	c	-216.6			
undissoc; std. state, $m = 1$	aq	-215.7	-183.1	44	
As_2S_2	c	-34.1			
As_2S_3	c	-40.4	-40.3	39.1	27.8
Astatine					
At	c	0	0	29.0	
Barium					
Ba	c	0	0	16	
	g	42	34.60	40.70	5.10

Table 9-1 (*Continued*)
ELEMENTS AND INORGANIC COMPOUNDS

Formula and Description	State	$\Delta Hf°$	$\Delta Gf°$	$S°$	$C_p°$
Ba^{2+}	aq	−128.67	−134.0	3	
$Ba_3(AsO_4)_2$	c	−817.8			
$BaHAsO_4 \cdot H_2O$	c	−411.5			
$Ba(H_2AsO_4)_2 \cdot 2H_2O$	c	−694.7			
$Ba(acetate)_2$	c	−355.1			
	aq, 400	−361.5			
$Ba(acetate)_2 \cdot 3H_2O$	c	−567.3			
$BaBr_2$	c	−169		(35)	
std. state, $m = 1$	aq	−186.47	−183.1	42	
$BaBr_2 \cdot 2H_2O$	c	−326.3			
$Ba(HCO_2)_2$ formate	c	−326.5			
	aq, 400	−324.5			
$Ba(CN)_2$	c	−47.9			
	aq	−50.7			
$Ba(CN)_2 \cdot 2H_2O$	c	−191.1			
$Ba(CNO)_2$ cyanate	c	−209.9			
$BaCO_3$	c	−291.8	−272.2	26.8	
std. state, $m = 1$	aq	−290.30	−260.2	−10	
$Ba(HCO_3)_2$ std. state, $m = 1$	aq	−459.0	−414.6	48	
$BaC_2O_4 \cdot 2H_2O$	c	−470.1			
$BaCl_2$	c	−205.5	−193.8	30	
	aq, 50	−208.09			
std. state, $m = 1$	aq	−208.72	−196.7	29	
$BaCl_2 \cdot 2H_2O$	c	−349.35	−309.7	48.5	
$Ba(OCl)_2$	aq	−176.1			
$Ba(ClO_2)_2$	c	−158.2			
$Ba(ClO_3)_2$	c	−181.7			
	aq	−175.6			
$Ba(ClO_4)_2$	c	−194.3			
$Ba(ClO_4)_2 \cdot 3H_2O$	c	−405.4			
$BaCrO_4$	c	−341.3			
BaF_2	c	−286.9	−274.5	23.03	
	aq	−286.0	−265.3		
BaH	g	52	46	−33.7	52.97
BaH_2	c	−40.9	−31.5		
BaI_2	c	−144.6	−143	39	
std. state, $m = 1$	aq	−155.41	−158.7	55	
$BaI_2 \cdot 2H_2O$	c	−290.9			
$Ba(IO_3)_2$	c	−264.5			
	aq	−238.4			
$Ba(IO_3)_2 \cdot H_2O$	c	−319.6			
$BaMoO_4$	c	−373.8			
$Ba(N_3)_2$	c	−8.0			
	aq	0.2			
Ba_3N_2	c	−90.6	(−73.4)	36.4	
$Ba(NH_2)_2$	c	−78.9			
$Ba(NO_2)_2$	c	−174.0			
	aq, 800	−178.6	−150.75		
$Ba(NO_2)_2 \cdot H_2O$	c	−254.5			
$Ba(NO_3)_2$	c	−237.06	−190.0	51.1	
std. state, $m = 1$	aq	−227.41	−186.8	73	
BaO	c	−133.5	−126.3	16.8	
BaO_2	c	−151.9	−139.5	22.62	
$BaO_2 \cdot 8H_2O$	c	−719.3			
$Ba(OH)_2$	c	−226.2			
std. state, $m = 1$	aq	−238.58	−209.2	−2	
$Ba(OH)_2 \cdot H_2O$	c	−299.0			
$Ba_3(PO_4)_2$	c	−998.0			
$BaHPO_4$	c	−465.8			

Table 9-1 (*Continued*)
ELEMENTS AND INORGANIC COMPOUNDS

Formula and Description	State	$\Delta Hf°$	$\Delta Gf°$	$S°$	$C_p°$
Ba(H$_2$PO$_2$)$_2$ hypophosphite	aq, 400	−423.6			
Ba(H$_2$PO$_2$)$_2$ · H$_2$O	c	−429.0			
Ba(H$_2$PO$_4$)$_2$	c	−749.6			
BaPtCl$_6$	c	−286.8			
	aq	−296.1			
BaS	c	106.0			
	g	41			
	aq	−118.4			
Ba(HS)$_2$	aq	−134.8			
Ba(HSO$_3$)$_2$	aq	−430.7			
BaSO$_3$	c	−282.6			
BaSO$_4$	c	−350.2	−323.4	31.6	
std. state, $m = 1$	aq	−345.57	−311.3	7	
BaS$_2$O$_6$	aq	−409.3			
BaS$_2$O$_6$ · 2H$_2$O	c	−552.5			
BaS$_2$O$_8$	aq	−454.0			
BaS$_2$O$_8$ · 4H$_2$O	c	−738.7			
BaS$_4$O$_6$	aq	−401.3			
BaSeO$_4$	c	−280.0			
BaSiF$_6$	c	−691.6			
BaSiO$_3$	c	−359.5			
Ba$_2$SiO$_4$	c	−496.8			
BaWO$_4$	c	−407.7			
Beryllium					
Be	c	0	0	2.28	
	g	76.6	67.6	32.55	
BeBr$_2$	c	−79.4	(−76.5)	29	
	aq	−142	−127.9		
BeCl$_2$	c	−118	(−102.9)	(23)	
BeCl$_2$ · 4H$_2$O	c	−436.8			
BeF$_2$	aq	−251.4			
	c	−241.2	(−216)	(17)	
BeH	g	78.1	71.3	40.84	6.96
BeI$_2$	c	−50.6	(−39.4)	(31)	
	aq	−112	−103.4		
BeMoO$_4$	c	−330			
Be$_3$N$_2$	c	−133.5	−121.4	12.0	
Be(NO$_3$)$_2$	aq	−188.3			
BeO	c	−143.1	−136.1	3.37	
	g	11.8	5.7	47.18	
Be(OH)$_2$	c	−216.8			
BeS	c	−55.9			
BeSO$_4$	c	−286.0			
	aq, 400	−304.1	−254.8		
BeSO$_4$ · 4H$_2$O	c	−576.3			
Bismuth					
Bi	c	0	0	13.56	6.10
	g	49.5	40.2	44.669	4.968
Bi$_2$	g	52.5			8.83
BiAsO$_4$	c		−148		
BiBr$_3$	c	63	(−56)	(54)	26
BiCl	c	−31.2	−25.9	22.6	
BiCl$_3$	c	−90.6	−75.3	42.3	25
	g	−63.5	−61.2	85.74	19.04
BiF$_3$	c	(−216)	(−200)	(34)	
BiH$_3$	g	66.4			
BiI$_3$	c	−24.0	−41.9		
BiO$^+$ std. state, $m = 1$	aq		−35.0		
Bi$_2$O$_3$	c	−137.16	−118.0	36.2	27.13

Table 9-1 (*Continued*)
ELEMENTS AND INORGANIC COMPOUNDS

Formula and Description	State	$\Delta Hf°$	$\Delta Gf°$	$S°$	$C_p°$
BiO(OH)	c		−88.0		
Bi(OH)$_3$	c	−170.0			
BiOBr	c		−71.0		
BiOCl	c	−87.7	−77.0	28.8	
BiONO$_3$	c		−67.0		
BiS	g	43	29	68	
Bi$_2$S$_3$	c	−34.2	−33.6	47.9	29.2
Bi$_2$(SO$_4$)$_3$	c	−608.1			
BiSe	g	42.0			
BiTe	g	42.8			
Bi$_2$Te$_3$	c	−18.5	−18.4	62.36	28.8
Boron					
B	c	0	0	1.40	2.65
	amorp	0.9		1.56	2.86
	g	134.5	124.0	36.65	4.971
B$_2$	g	198.5	185.0	48.23	7.30
BBr	g	56.9	46.7	53.75	7.87
BBr$_3$	liq	−57.3	−57.0	54.9	
	g	−49.15	−55.56	77.47	16.20
B$_4$C	c	−17	−17	6.48	12.62
B(CH$_3$)$_3$	liq	−34.2	−7.7	57.1	
	g	−29.7	−8.6	75.2	21.15
BCl	g	35.73	28.90	50.94	7.57
BCl$_3$	liq	−102.1	−92.6	49.3	25.5
	g	−96.50	−92.91	69.31	14.99
B$_2$Cl$_4$	liq	−125.0	−111.1	62.7	32.9
BOCl	g	−75			
(BOCl)$_3$	g	−390.4	−370.5	91	
BClF$_2$	g	−212.8	−209.4	65	
BCl$_2$F	g	−154.2	−150.9	68	
BF	g	−29.2	−35.8	47.89	7.07
BF$_3$	g	−271.75	−267.77	60.71	12.06
BF$_4^-$ std. state, $m = 1$	aq	−376.4	−355.4	43	
B$_2$F$_4$	g	−145			
BOF	g	−145			
HBF$_4$	aq	−375.5			
BH	g	107.46	100.29	41.05	6.97
BH$_3$	g	24			
BH$_4^-$ std. state, $m = 1$	aq	11.51	27.31	26.4	
B$_2$H$_6$	g	8.5	20.7	55.45	13.60
B$_4$H$_{10}$	g	15.8			
B$_5$H$_9$	liq	10.20	41.03	44.03	36.12
BI$_3$	g	17.00	4.96	83.43	16.92
BN	c	−60.8	−54.6	3.54	4.71
	g	154.75	146.87	50.71	7.04
B$_3$N$_3$H$_6$	liq	−129.3	−93.88	47.7	
BO	g	6	−1	48.62	6.98
BO$_2$	g	−71.8	−73.1	54.84	10.28
BO$_2^-$ std. state, $m = 1$	aq	−184.60	−162.27	−8.9	
B$_2$O$_2$	g	−108.7	−110.5	57.93	13.69
B$_2$O$_3$	c	−304.20	−285.30	12.90	15.04
	amorp	−299.84	−282.6	18.6	14.6
	g	−201.67	−198.85	66.85	15.98
B$_4$O$_7^{2-}$ std. state, $m = 1$	aq		−622.6		
HBO$_2$ cubic	c	−192.17			
monoclinic	c	−189.83	−172.9	9	
orthorhombic	c	−188.52	−172.5	12	
	g	−134.3	−131.7	57.35	10.09
H$_3$BO$_3$	c	−261.55	−231.60	21.23	19.45
	g	−237.6			

Table 9-1 (*Continued*)
ELEMENTS AND INORGANIC COMPOUNDS

Formula and Description	State	$\Delta Hf°$	$\Delta Gf°$	$S°$	$C_p°$
unionized, std. state, $m = 1$	aq	−256.29	−231.56	38.8	
$B(OH)_4^-$ std. state, $m = 1$	aq	−321.23	−275.65	24.5	
BP cubic	c	−19			
BS	b	81.74	69.02	51.65	7.18
B_2S_3	c	−57.5			
Bromine					
Br	g	26.741	19.701	41.805	4.968
Br^-	g	−55.9			
Br^- std. state, $m = 1$	aq	−29.05	−24.85	19.7	−33.9
Br_2	liq	0	0	36.384	18.090
	g	7.387	0.751	58.641	8.61
std. state, $m = 1$	aq	−0.62	0.94	31.2	
	CCl_4	0.71	0.36	37.6	
Br_3^- std. state, $m = 1$	aq	−31.17	−25.59	51.5	
BrCl	g	3.50	−0.23	57.36	8.36
Br_2Cl^- std. state, $m = 1$	aq	−40.7	−30.7	45.1	
BrF	g	−22.43	−26.09	54.70	7.88
BrF_3	liq	−71.9	−57.5	42.6	29.78
	g	−61.09	−54.84	69.89	15.92
BrF_5	liq	−109.6	−84.1	53.8	
BrO	g	30.06	25.87	56.75	7.67
BrO^- std. state, $m = 1$	aq	−22.5	−8.0	10	
BrO_2	c	11.6			
BrO_3^- std. state, $m = 1$	aq	−20.0	0.4	39.0	
HBrO undissoc; std. state, $m = 1$	aq	−27.0	−19.7	34	
$HBrO_3$ std. state, $m = 1$	aq	−20.0	0.4	39.0	
HBr std. state, $m = 1$	aq	−29.05	−24.85	19.7	−33.9
	aq, 1	−17.38			
	aq, 2	−22.40			
	aq, 5	−26.706			
	aq, 10	−27.953			
	aq, 100	−28.815			
	g	−8.70	−12.77	47.463	6.965
Cadmium					
Cd γ	c	0	0	12.37	6.21
α	c	−0.14	−0.14	12.37	
	g	26.77	18.51	40.066	4.968
$CdAs_2$	c	−4.2			
Cd_3As_2	c	−10.0			
$Cd_3(AsO_4)_2$	c		−410.2		
$Cd(BO_2)_2$	c		−354.87		
$CdBr_2$	c	−75.57	−70.82	32.8	18.32
std. state, $m = 1$	aq	−76.24	−68.24	21.9	
$CdBr_2 \cdot 4H_2O$	c	−356.73	−298.287	75.6	
$CdCl_2$	c	−93.57	−82.21	27.55	17.85
std. state, $m = 1$	aq	−98.04	−81.286	9.5	
$CdCl_2 \cdot \frac{5}{2}H_2O$	c	−270.54	−225.644	54.3	
$CdCl_3^-$ std. state, $m = 1$	aq	−134.1	−116.4	48.5	
$Cd(ClO_4)_2$ std. state, $m = 1$	aq	−79.96	−22.66	69.5	
$Cd(ClO_4)_2 \cdot 6H_2O$	c	−490.6			
$Cd(CN)_2$	c	38.8			
std. state, $m = 1$	aq	53.9	63.9	27.5	
$Cd(CN)_4^{2-}$ std. state, $m = 1$	aq	102.3	121.3	77	
$Cd(CNS)_2$ thiocyanate	c	12.43			
std. state, $m = 1$	aq	18.40	25.76	51.4	
$CdCO_3$	c	−179.4	−160.0	22.1	
CdC_2O_4	c	−218.1			
std. state, $m = 1$	aq	−215.3	−179.6	−6.6	
$Cd(acetate)_2$ std. state, $m = 1$	aq	−250.46	−195.12	23.9	
$Cd(formate)_2$ std. state, $m = 1$	aq	−221.56	−186.23	26	

Table 9-1 (*Continued*)
ELEMENTS AND INORGANIC COMPOUNDS

Formula and Description	State	$\Delta Hf°$	$\Delta Gf°$	$S°$	$C_p°$
Cd fulminate	c	90			
CdF_2	c	−167.4	−154.8	18.5	
std. state, $m = 1$	aq	−172.14	−151.82	−24.1	
CdI_2	c	−48.6	−48.13	38.5	19.11
std. state, $m = 1$	aq	−44.52	−43.20	35.7	
CdI_3^- std. state, $m = 1$	aq		−62.0		
CdI_4^{2-} std. state, $m = 1$	aq	−81.7	−75.5	78	
$Cd(IO_3)_2$	c		−90.13		
std. state, $m = 1$	aq	−123.9	−79.7	39.1	
$Cd(N_3)_2$	c	108			
std. state, $m = 1$	aq	113.38	147.9	34.1	
Cd_3N_2	c	38.7			
$Cd(NH_3)_4^{2+}$ std. state, $m = 1$	aq	−107.6	−54.1	80.4	
$Cd(NO_3)_2$	c	−109.06			
std. state, $m = 1$	aq	−117.26	−71.76	52.5	
$Cd(NO_3)_2 \cdot 4H_2O$	c	−394.11			
Cd_3P_2	c	−27.4			
$Cd_3(PO_4)_2$	c		−587.1		
CdO	c	−61.7	−54.6	13.1	10.38
CdO_2^{2-} std. state, $m = 1$	aq		−68.0		
$CdOH^+$ std. state, $m = 1$	aq		−62.4		
$HCdO_2^-$ std. state, $m = 1$	aq		−86.9		
$Cd(OH)_2$ precipitated	c	−134.0	−113.2	23	
std. state, $m = 1$	aq	−128.08	−93.73	−22.6	
$Cd(OH)_3^-$ std. state, $m = 1$	aq		−143.6		
$Cd(OH)_4^{2-}$ std. state, $m = 1$	aq		−181.3		
CdS	c	−38.7	−37.4	15.5	
$CdSO_4$	c	−223.06	−196.65	29.407	23.80
std. state, $m = 1$	aq	−235.46	−196.51	−12.7	
$CdSO_4 \cdot H_2O$	c	−296.26	−255.46	36.814	32.16
$CdSO_4 \cdot {}^8/_3 H_2O$	c	−413.33	−350.224	54.883	50.97
CdSb	c	−3.44	−3.11	22.2	
Cd_3Sb_2	c	−13.9			
$CdSeO_3$	c	−137.5	−119.0	34.0	
std. state, $m = 1$	aq	−139.8	−106.9	−14.4	
$CdSeO_4$	c	−151.3	−127.1	39.3	
std. state, $m = 1$	aq	−161.3	−124.0	−4.6	
$CdSiO_3$	c	−284.20	−264.20	23.3	21.17
CdTe	c	−22.1	−22.0	24	
Calcium					
Ca	c	0	0	9.95	6.30
	g	42.2	34.14	36.99	4.97
$CaHAsO_4 \cdot H_2O$	c	−410	−363	35	
$CaBr_2$	c	−161.3	−157.5	(31)	
std. state, $m = 1$	aq	−187.57	−181.33	25.4	
$CaBr_2 \cdot 6H_2O$	c	−597.2			
$CaO \cdot B_2O_3$	c	−483.3	−457.7	25.1	
$CaO \cdot 2B_2O_3$	c	−798.8	−752.4	32.2	
CaC_2	c	−15.0	−16.2	16.8	
$CaCO_3$ calcite	c	−288.45	−269.78	22.2	
aragonite	c	−288.49	−269.53	21.2	
$Ca(HCO_3)_2$ std. state, $m = 1$	aq	−460.13	−412.80	32.2	
$Ca(formate)_2$	c	−323.5			
CaC_2O_4	c	−332.2			
$CaC_2O_4 \cdot 2H_2O$	c	−469.1	−416.9	47	
$Ca(acetate)_2$	c	−355.0			
$Ca(CN)_2$	c	−44.2			
	aq		−54		
$CaCN_2$ cyanamide	c	−84.0			

Table 9-1 (*Continued*)
ELEMENTS AND INORGANIC COMPOUNDS

Formula and Description	State	$\Delta Hf°$	$\Delta Gf°$	$S°$	$C_p°$
$CaCl_2$	c	−190.4	−179.5	27.2	
std. state, $m = 1$	aq	−209.82	−194.88	13.1	
	aq, 25	−208.51			
$CaCl_2 \cdot H_2O$	c	−265.1			
$CaCl_2 \cdot 2H_2O$	c	−335.5			
$CaCl_2 \cdot 4H_2O$	c	−480.2			
$CaCl_2 \cdot 6H_2O$	c	−623.15			
$CaOCl_2$	c	−178.6			
	aq	−189.1			
$Ca(OCl)_2$	aq	−180.0			
$CaCrO_4$	c	−329.6	−305.3	32	
	aq	−336.0			
CaF_2	c	−290.3	−277.7	16.46	
std. state, $m = 1$	aq	−287.09	−264.34	−17.8	
CaH	g	58.7			
CaH_2	c	−45.1	−35.8	10	
CaI_2	c	−127.5	(−126.4)	(34)	
std. state, $m = 1$	aq	−156.51	−156.88	39.1	
$CaI_2 \cdot 8H_2O$	c	−700.7			
Ca_3N_2	c	−108.2	−93.2	25.4	
$Ca(NO_2)_2$	c	−178.3			
	aq	−180.5			
$Ca(NO_3)_2$	c	−224.0	−177.34	46.2	
std. state, $m = 1$	aq	−228.51	−185.00	56.8	
	aq, 10	−229.48			
$Ca(NO_3)_2 \cdot 2H_2O$	c	−368.00	−293.51	64.3	
$Ca(NO_3)_2 \cdot 4H_2O$	c	−509.37	−406.5	81	
CaO	c	−151.79	−144.4	9.5	
CaO_2	c	(−156.5)	(−143.5)	(15.4)	
$Ca(OH)_2$	c	−235.80	−214.33	18.2	
std. state, $m = 1$	aq	−239.68	−207.37	−13.2	
Ca_3P_2	c	−120.5			
$Ca_3(PO_4)_2 \quad \alpha$	c	−986.2	−929.7	57.6	
β	c	−988.2	−932.0	56.4	
$CaHPO_4$	c	−435.2	−401.5	21	
$CaHPO_4 \cdot 2H_2O$	c	−576.0	−514.6	40	
$Ca(H_2PO_4)_2$	ppt	−744.4			
CaS	c	−115.3	−114.1	13.5	
$CaSO_3 \cdot 2H_2O$	c	−421.2	−374.1	44	
$CaSO_4$ anhydrite	c	−342.42	−315.56	25.5	
α	c	−340.27	−313.52	25.9	
β	c	−339.21	−312.46	25.9	
$CaSO_4 \cdot \frac{1}{2}H_2O \quad \alpha$	c	−376.47	−343.02	31.2	
β	c	−375.97	−342.78	32.1	
$CaSO_4 \cdot 2H_2O$	c	−483.06	−429.19	46.36	
CaS_2O_3	aq, 1000	−283.4			
$CaS_2O_3 \cdot 6H_2O$	c		−602.2		
CaSe	c	−50.8		16	
$CaSi_2$	c	−36			
Ca_2Si	c	−50			
$CaSiO_3 \quad \alpha$	c	−377.4	−357.4	20.9	
β, wollastonite	c	−378.6	−358.2	19.6	
$Ca_2SiO_4 \quad \beta$	c	−538.0			
γ	c	−539.0			
Ca_3SiO_5	c	−688.4			
$CaWO_4$	c	−392.5			
Carbon *					
C graphite	c	0	0	1.372	2.038

*For the values of organic compounds see the table following zirconium compounds.

Table 9-1 (*Continued*)
ELEMENTS AND INORGANIC COMPOUNDS

Formula and Description	State	$\Delta Hf°$	$\Delta Gf°$	$S°$	$C_p°$
diamond	c	0.4533	0.6930	0.568	1.4617
	g	171.291	160.442	37.7597	4.9805
CBr_4 monoclinic	c	4.5	11.4	50.8	34.5
	g	19	16	85.55	21.79
$CHBr_3$	liq	−6.8	−1.2	52.8	31
	g	4	2	79.07	17.02
CH_3Br	g	−8.4	−6.2	58.86	10.14
CCl_4	liq	−32.37	−15.60	51.72	31.49
	g	−24.6	−14.49	74.03	19.91
$CHCl_3$	liq	−32.14	−17.62	48.2	27.2
	g	−24.65	−16.82	70.65	15.70
CH_2Cl_2	liq	−29.03	−16.09	42.5	23.9
	g	−22.10	−15.75	64.56	12.18
CH_3Cl	g	−19.32	−13.72	56.04	9.74
CCl_3Br	g	−11.0	−5.1	79.55	20.38
CF_3	g	−114			
CF_4	g	−221	−210	62.50	14.60
CF_3Br	g	−153.6	−147.3	71.14	16.57
CF_3Cl	g	−166	−156	68.16	15.98
CF_2Cl_2	g	−114	−105	71.86	17.27
$CFCl_3$	liq	−72.02	−56.61	53.86	29.05
CI_4	g			93.65	22.91
CHI_3	c	33.7			
	g			85.1	17.92
CH_2I_2	liq	16.0	21.6	41.6	32
	g	27.0	22.9	74.0	13.79
CH_3I	liq	−3.7	3.2	39.0	30
	g	3.1	3.5	60.71	10.54
CN	g	109	102	48.4	6.97
CN^- std. state, $m = 1$	aq	36.0	41.2	22.5	
$CN \cdot N_3$ cyanogen azide	c	92.6			
HCN	liq	26.02	29.86	26.97	16.88
	g	32.3	29.8	48.20	8.57
std. state, $m = 1$	aq	36.0	41.2	22.5	
nonionized, std. state, $m = 1$	aq	25.6	28.6	29.8	
CNBr	c	33.58			
	g	44.5	39.5	59.32	11.22
CNCl	g	32.97	31.32	56.42	10.75
CNI	c	39.71	44.22	23.0	
	g	53.9	47.0	61.35	11.54
CO	g	−26.416	−32.780	47.219	6.959
CO_2	g	−94.051	−94.254	51.06	8.87
undissoc; std. state, $m = 1$	aq	−98.90	−92.26	28.1	
CO_3^{2-} std. state, $m = 1$	aq	−161.84	−126.17	−13.6	
C_3O_2	liq	−28.03	−25.10	43.28	25.8
	g	−22.28	−26.08	65.852	16.029
HCO	g	−4.12	−7.76	53.68	8.26
$HCOO^-$ std. state, $m = 1$	aq	−101.71	−83.9	22	−21.0
HCO_3^- std. state, $m = 1$	aq	−165.39	−140.26	21.8	
$HCHO^+$	g	224.2			
H_2CO_3 std. state, $m = 1$	aq	−167.22	−148.94	44.8	
$COBr_2$	g	−23.0	−26.5	73.85	14.78
$COCl_2$	g	−52.3	−48.9	67.74	13.78
COF_2	g	−151.7	−148.0	61.78	11.19
HCNO cyanic acid, ionized	aq	−34.90	−23.3	25.5	
nonionized	aq	−36.90	−28.0	34.6	
HNCO isocyanic acid	g			56.85	10.72
CNO^- std. state, $m = 1$	aq	−34.9	−23.3	25.5	
CS	g	56	44	50.30	7.12

Table 9-1 (*Continued*)
ELEMENTS AND INORGANIC COMPOUNDS

Formula and Description	State	$\Delta Hf°$	$\Delta Gf°$	$S°$	$C_p°$
CS_2	liq	21.44	15.60	36.17	18.1
	g	28.05	16.05	56.82	19.85
COS	g	−33.96	−40.47	55.32	9.92
$CSCl_2$	g			71.51	15.18
HCNS undissoc; std. state, $m = 1$	aq		23.31		
CNS^-	aq	18.27	22.15	34.5	−9.6
HNCS isothiocyanic acid	g	30.5	27.0	59.2	11.2
NOSCN std. state, $m = 1$	aq	55.2	63.5	51.2	
$SC(SH)_2$ trithiocarbonic acid	liq	6.0	7.0	52	35.8
Cerium					
Ce	c	0	0	13.64	
$CeBr_3$	c	−192	(−185)	(45)	
$CeCl_3$	c	−260.3			
std. state, $m = 1$	aq	−293.9	−264.5	−5	
CeF_3	c	−391.0	(−372.1)	(24)	
CeF_4	c	−442.0	(−420)	(37)	
Ce_3H_8	c	−170			
CeI_3	c	−163.0	(−161)	(50)	
std. state, $m = 1$	aq	−213.9	−207.5	34	
CeN	c	−78.3	−70.8		
CeO_2	c	(−260)	−245.9	14.88	
Ce_2O_3	c	(435)	(−411.5)	(21.8)	
$CeO_3 \cdot 2H_2O$	c	−389			
CeS_2	c	−153.9			
Ce_2S_3	c	−298.7			
$Ce(SO_4)_2$	c	−560			
$Ce_2(SO_4)_3$ std. state, $m = 1$	aq	−998.3	−873.0	−76	
$Ce_2(SO_4)_3 \cdot 5H_2O$	c	−1308			
$Ce_2(SO_4)_3 \cdot 8H_2O$	c		−1340.2		
Cesium					
Cs	c	0	0	19.8	
	g	18.83	12.24	41.94	
$CsAl(SO_4)_2 \cdot 12H_2O$	c	−1449.5	−1218.5	164	
CsBr	c	−94.3	−91.6	29	
std. state, $m = 1$	aq	−88.1	−91.98	51.1	
$CsBrO_3$	c			38.8	
CsCN	c	−27			
CsCNO cyanate	c	−96			
CsCNS thiocyanate	c	−50			
Cs_2CO_3	c	−267.4			
	aq, 200	−280.0			
$CsHCO_3$	c	−228			
	aq	−224.4	−210.6		
CsCl	c	−103.5			
std. state, $m = 1$	aq	−99.2	−88.76	45.0	
$CsClO_4$	c	−103.86	−73.28	41.89	
std. state, $m = 1$	aq	−90.6	−69.98	75.3	
CsF	c	−126.9			
std. state, $m = 1$	aq	−135.9	−133.49	29.5	
$CsHF_2$	c	−216.1			
	aq	−212.8			
CsH	g	29.0	24.3	51.25	
CsI	c	−80.5	−79.7	31	
std. state, $m = 1$	aq	−72.6	−79.76	57.9	
$CsNH_2$	c	−25.4			
$CsNO_2$	c	−85			
$CsNO_3$	c	−118.11			
std. state, $m = 1$	aq	−108.6	−93.82	86.8	
Cs_2O	c	−75.9			

Table 9-1 (Continued)
ELEMENTS AND INORGANIC COMPOUNDS

Formula and Description	State	$\Delta Hf°$	$\Delta Gf°$	$S°$	$C_p°$
Cs_2O_2	c	−96.2			
CsOH	c	−97.2			
std. state, $m = 1$	aq	−114.2	−105.00	29.3	
	aq, 200	−114.0			
$CsReO_4$	c	−257.2			
	aq	−249.5			
Cs_2S	c	−81.1			
CsHS	c	−62.9			
Cs_2SO_4	c	−349.8			
std. state, $m = 1$	aq	−335.3	−312.16	67.7	
$CsHSO_4$	c	−274.0			
CsHSe	c	−36.7			
Cs_2SiF_6	c	−669.5			
Chlorine					
Cl	g	29.082	25.262	39.457	5.220
Cl^-	g	−58.8			
Cl^- std. state, $m = 1$	aq	−39.952	−31.372	13.5	−32.6
Cl_2	g	0	0	53.288	8.104
ClF	g	−13.02	−13.37	52.05	7.66
ClF_3	liq	−45.3			
	g	−39.0	−29.4	67.28	15.26
Cl_2F_6	g	−81.1	−56.7	117	
HCl	g	−22.062	−22.777	44.646	6.96
std. state, $m = 1$	aq	−39.952	−31.372	13.5	−32.6
ClO	g	24.34	23.45	54.14	7.52
ClO^- std. state, $m = 1$	aq	−25.6	−8.8	10	
ClO_2	g	24.5	28.8	61.36	10.03
ClO_2^- std. state, $m = 1$	aq	−15.9	4.1	24.2	
ClO_3	g	37			
ClO_3^- std. state, $m = 1$	aq	−23.7	−0.8	38.8	
ClO_4^- std. state, $m = 1$	aq	−30.91	−2.06	43.5	
Cl_2O	g	19.2	23.4	63.60	10.85
Cl_2O_7	liq	56.9			
ClO_3F	g	−5.7	11.5	66.65	15.52
HClO	g			56.54	8.88
undissoc; std. state, $m = 1$	aq	−28.9	−19.1	34	
$HClO_2$ undissoc; std. state, $m = 1$	aq	−12.4	1.4	45.0	
$HClO_3$ std. state, $m = 1$	aq	−23.7	−0.8	38.8	
$HClO_4$	liq	−9.70			
std. state, $m = 1$	aq	−30.91	−2.06	43.5	
$HClO_4 \cdot H_2O$	c	−91.35			
$HClO_4 \cdot 2H_2O$	liq	−162.04			
$ClF_3 \cdot HF$	g	−107.7	−91.8	86	
Chromium					
Cr	c	0	0	5.68	5.58
	g	94.8	84.1	41.68	4.97
Cr^{2+} std. state, $m = 1$	aq	−34.3			
$CrBr_2$	c	−72.2			
	g	−17			
$[Cr(H_2O)_6]Br_3$ purple	c	−550.4			
Cr_3C_2	c	−19.3	−19.5	20.42	23.53
Cr_7C_3	c	−38.7	−39.9	48.0	49.92
$Cr_{23}C_6$	c	−87.2	−89.3	145.8	149.2
$CrCl_2$	c	−94.5	−85.1	27.56	17.01
	g	−30.7			
	aq	−114.2			
$CrCl_2 \cdot 2H_2O$	c	−237.1			
$CrCl_2 \cdot 3H_2O$	c	−308.9			
$CrCl_2 \cdot 4H_2O$	c	−384.4			

Table 9-1 (*Continued*)
ELEMENTS AND INORGANIC COMPOUNDS

Formula and Description	State	$\Delta Hf°$	$\Delta Gf°$	$S°$	$C_p°$
$CrCl_3$	c	−133.0	−116.2	29.4	21.94
$CrCl_4$	g	−102	(−91.6)		
CrO_2Cl_2	liq	−138.5	−122.1	53.0	
	g	−128.6	−119.9	78.8	20.2
$[Cr(H_2O)_6]Cl_3$ violet	c	−586.2			
CrI_2	c	186	(−172)		
	g	−99			
CrF_3	c	−277	−260	22.44	18.82
CrF_4	c	−298			
Cr_7H_2	c	−3.8			
CrI_2	c	−37.5			
	g	24			
	aq	−60.1			
CrI_3	c	−49.0			
CrN	c	−29.8			11.0
Cr_2N	c	−30.5			17.3
$Cr(NO_3)_3 \cdot 9H_2O$	c				109.2
CrO	g	53⁰ᐟᴷ			
CrO_2	c	−143			
	g	−14⁰ᐟᴷ			
CrO_3	c	−140.9			
	g	−92.2			
Cr_2O_3	c	−272.4	−252.9	19.4	28.38
$Cr_2O_3 \cdot H_2O$	c	−360			
$Cr_2O_3 \cdot 2H_2O$	c	−441			
$Cr_2O_3 \cdot 3H_2O$	c	−519			
Cr_3O_4	c	−366			
CrO_4^{2-} std. state, $m = 1$	aq	−210.60	−173.96	12.00	
$Cr_2O_7^{2-}$ std. state, $m = 1$	aq	−356.2	−311.0	62.6	
$HCrO_4^-$ std. state, $m = 1$	aq	−209.9	−182.8	44.0	
$CrO_2(OH)_2$	g	−174			
$Cr(OH)_3$	c	−254.3			
$[Cr(H_2O)_6]^{3+}$	aq	−477.8			
$Cr_2(SO_4)_3$	c				67.5
$Cr_2(SO_4)_3 \cdot 18H_2O$	c				223
CrSb	c				12.7
$CrSb_2$	c				19.7
CrSi	c	−15			9.2
$CrSi_2$	c	−23			12.7
Cr_3Si	c	−30			19.3
Cr_5Si_3	c	−64			34.9
Cobalt					
Co hexagonal	c	0	0	7.18	5.93
face-centered cubic	c	0.11	0.06	7.34	
Co^{2+} std. state, $m = 1$	aq	−13.9	−13.0	−27	
Co^{3+} std. state, $m = 1$	aq	22	32	−73	
$Co_3(AsO_4)_2$	c		−387.4		
$Co(BO_2)_2$	c		−325.8		
$CoBr_2$	c	−52.8			
std. state, $m = 1$	aq	−72.0	−62.7	12	
$CoBr_2 \cdot 6H_2O$	c	−482.8			
$CoCO_3$	c	−170.4			
$CoCl_2$	c	−74.7	−64.5	26.09	18.76
std. state, $m = 1$	aq	−93.8	−75.7	0	
$CoCl_2 \cdot H_2O$	c	−147			
$CoCl_2 \cdot 2H_2O$	c	−220.6	−182.8	45	
$CoCl_2 \cdot 6H_2O$	c	−505.6	−412.4	82	
$Co(ClO_4)_2$ std. state, $m = 1$	aq	−75.7	−17.1	60	
$Co(ClO_4)_2 \cdot 6H_2O$	c	−487.2			

Table 9-1 (*Continued*)
ELEMENTS AND INORGANIC COMPOUNDS

Formula and Description	State	$\Delta Hf°$	$\Delta Gf°$	$S°$	$C_p°$
Co(CNO)$_2$ cyanate	c	−51.8			
Co(CNS)$_2$ thiocyanate	c	24.2			
Co(formate)$_2$	c	−208.7			
CoC$_2$O$_4$	c	−203.5			
std. state, $m = 1$	aq	−211.1	−174.1	−16	
CoF$_2$	c	−165.4	−154.7	19.59	16.44
CoF$_3$	c	−193.8			
CoI$_2$	c	−21.2			
std. state, $m = 1$	aq	−40.3	−37.7	26	
Co(IO$_3$)$_2$ std. state, $m = 1$	aq	−119.7	−74.2	30	
Co(IO$_3$)$_2$ · 2H$_2$O	c	−258.6	−190.2	64	
Co(NH$_3$)$_6^{3+}$ std. state, $m = 1$	aq	−139.8	−38.9	40	
Co(NH$_3$)$_6^{2+}$ std. state, $m = 1$	aq		−45.3		
[Co(NH$_3$)$_6$]Br$_2$	c	−216.4			
[Co(NH$_3$)$_6$]Br$_3$	c	−239.7	−119.8	77.7	78.1
std. state, $m = 1$	aq	−227.0	−113.4	99	
[Co(NH$_3$)$_6$]Cl$_2$	c	−238.0			
[Co(NH$_3$)$_6$]Cl$_3$	c	−268.7			76.6
std. state, $m = 1$	aq	−259.7	−133.0	80	
[Co(NH$_3$)$_5$Cl]Cl$_2$	c	−243.1	−139.3	87.5	57.2
[Co(NH$_3$)$_6$](ClO$_4$)$_3$	c	−247.3	−54.3	152	
std. state, $m = 1$	aq	−232.5	−45.1	170	
[Co(NH$_3$)$_6$]I$_2$	c	−189.9			69.1
[Co(NH$_3$)$_6$]I$_3$	c	−104.8			74.3
[Co(NH$_3$)$_5$NO$_2$]$^{2+}$ std. state, $m = 1$	aq	−146.6	−41.3	43	
[Co(NH$_3$)$_5$NO$_2$](NO$_3$)$_2$	c	−260.2	−100.0	83	
	aq	−245.7	−94.5	113	
[Co(NH$_3$)$_6$](NO$_3$)$_3$ std. state, $m = 1$	aq	−288.5	−118.7	145	
	c	−306.4	−126.8	112	
Co(NO$_3$)$_2$	c	−100.5			
std. state, $m = 1$	aq	−113.0	−66.2	43	
Co(NO$_3$)$_2$ · 2H$_2$O	c	−244.2			
Co(NO$_3$)$_2$ · 6H$_2$O	c	−528.49			
CoO	c	−56.87	−51.20	12.66	13.20
Co$_3$O$_4$	c	−213	−185	24.5	29.5
Co(OH)$_2$ blue	c		−107.6		
pink	c	−129.0	−108.6	19	
std. state, $m = 1$	aq	−123.8	−88.2	−32	
Co(OH)$_3$	c	−171.3			
Co$_2$P	c	−45			
Co$_3$(PO$_4$)$_2$	c		−573.3		
CoHPO$_4$	c		−282.5		
CoS	c	−19.8			
Co$_2$S$_3$ pptd	c	−35.2			
CoSO$_4$	c	−212.3	−187.0	28.2	
std. state, $m = 1$	aq	−231.2	−191.0	−22	
CoSO$_4$ · 6H$_2$O	c	−641.4	−534.35	87.86	84.46
CoSO$_4$ · 7H$_2$O	c	−712.22	−591.26	97.05	93.33
CoSe	c	−14.6			
CoSeO$_3$ · 2H$_2$O	c	−266.5			
CoSi	c	−24.0	−23.6	10.3	10.6
Co$_2$SiO$_4$	c	−353			
CoTe$_2$	c	−31			
Co$_3$Te$_4$	c	−77			
Copper					
Cu	c	0	0	7.923	5.840
	g	80.86	71.37	39.74	4.968
Cu$^+$ std. state, $m = 1$	aq	17.13	11.95	9.7	
Cu^{2+} std. state, $m = 1$	aq	15.48	15.66	−23.8	
Cu$_2$	g	115.72	103.24	57.71	8.75

Table 9-1 (*Continued*)
ELEMENTS AND INORGANIC COMPOUNDS

Formula and Description	State	$\Delta Hf°$	$\Delta Gf°$	$S°$	$C_p°$
Cu_3As	c	−2.8			
$Cu_3(AsO_4)_2$	c		−310.9		
std. state, $m = 1$	aq	−378.10	−263.02	−149.2	
CuBr	c	−25.0	−24.1	22.97	13.08
$CuBr_2$	c	−33.9			
$CuBr_2 \cdot 4H_2O$	c	−317.0			
$CuCO_3 \cdot Cu(OH)_2$ malachite	c	−251.3	−213.6	44.5	
$2CuCO_3 \cdot Cu(OH)_2$ azurite	c	−390.1			
CuCl	c	−32.8	−28.65	20.6	11.6
$CuCl_2^-$ std. state, $m = 1$	aq		−57.4		
$CuCl_3^{2-}$ std. state, $m = 1$	aq		−90		
$2CuCl \cdot CO \cdot 2H_2O$	c	−225			
$2CuCl \cdot C_2H_2$	c	−23.3	−7.63	50.7	
$3CuCl \cdot C_2H_2$	c	−56.4	−36.52	71.0	
$CuCl_2$	c	−52.6	−42.0	25.83	13.82
undissoc, std. state, $m = 1$	aq		−47.3		
$CuCl_2 \cdot 2H_2O$	c	−196.3	−156.8	40	
$Cu(ClO_4)_2$ std. state, $m = 1$	aq	−46.34	11.54	63.2	
$Cu(ClO_4)_2 \cdot 6H_2O$	c	−460.9			
CuCN	c	23.0	26.6	20.2	
$Cu(CN)_2^-$ std. state, $m = 1$	aq		61.6		
$Cu(CN)_3^{2-}$ std. state, $m = 1$	aq		96.5		
$Cu(CN)_4^{3-}$ std. state, $m = 1$	aq		135.4		
CuCNS thiocyanate	c		16.7		
std. state, $m = 1$	aq	35.40	34.10	44.2	
$Cu(CNS)_2$ std. state, $m = 1$	aq	52.02	59.96	45.2	
$Cu(acetate)_2$	c	−213.5			
std. state, $m = 1$	aq	−216.84	−160.92	17.6	
$Cu(formate)_2$	c	−186.7			
std. state, $m = 1$	aq	−187.94	−152.1	20	
Cu(I) fulminate	c	26.3			
CuC_2O_4	c		−158.2		
std. state, $m = 1$	aq	−181.7	−145.5	−12.9	
$Cu(C_2O_4)_2^{2-}$ std. state, $m = 1$	aq	−380.5	−319.3	35	
CuF^+ std. state, $m = 1$	aq	−62.4	−52.7	−16	
CuF_2	c	−129.7			
$CuF_2 \cdot 2H_2O$	c		−234.6		
$CuFeO_2$	c	−127.3	−114.7	21.2	19.13
$CuFe_2O_4$	c	−230.69	−205.26	33.7	35.52
CuH	c	5.1			
	g	70			
CuI	c	−16.2	−16.6	23.1	12.92
$Cu(IO_3)_2$ std. state, $m = 1$	aq	−90.3	−45.5	32.8	
$Cu(IO_3)_2 \cdot H_2O$	c	−165.4	−112.0	59.1	
$CuMoO_4$	c	−225			
CuN_3	c	66.7	82.4	24	
$Cu(N_3)_2$	c	143.0			
Cu_3N	c	17.8			22
$Cu(NH_3)^{2+}$ std. state, $m = 1$	aq	−9.3	3.72	2.9	
$Cu(NH_3)_2^{2+}$ std. state, $m = 1$	aq	−34.0	−7.28	26.6	
$Cu(NH_3)_3^{2+}$ std. state, $m = 1$	aq	−58.7	−17.48	47.7	
$Cu(NH_3)_4^{2+}$ std. state, $m = 1$	aq	−83.3	−26.60	65.4	
$Cu(NO_3)_2$	c	−72.4			
std. state, $m = 1$	aq	−83.64	−37.56	46.2	
$Cu(NO_3)_2 \cdot 6H_2O$	c	−504.5			
CuO	c	−37.6	−31.0	10.19	10.11
Cu_2O	c	40.3	34.9	22.26	15.21
$Cu(OH)_2$	c	−107.5			
std. state, $m = 1$	aq	−94.46	−59.53	−28.9	
CuP_2	c	−29			

Table 9-1 (*Continued*)
ELEMENTS AND INORGANIC COMPOUNDS

Formula and Description	State	$\Delta Hf°$	$\Delta Gf°$	$S°$	$C_p°$
Cu_3P	c	−36.2			
$Cu_2P_2O_7$	c		−448.0		
std. state, $m = 1$	aq	−511.8	−427.4	−76	
$Cu_3(PO_4)_2$	c		−490.3		
CuS	c	−12.7	−12.8	15.9	11.43
Cu_2S α	c	−19.0	−20.6	28.9	18.24
$CuSO_4$	c	−184.36	−158.2	26	23.9
std. state, $m = 1$	aq	−201.84	−162.31	−19.0	
	aq, 100	−200.374			
$CuSO_4 \cdot H_2O$	c	−259.52	−219.46	34.9	32
$CuSO_4 \cdot 3H_2O$	c	−402.56	−334.65	52.9	49
$CuSO_4 \cdot 5H_2O$	c	−544.85	−449.344	71.8	67
Cu_2SO_3 std. state, $m = 1$	aq	−117.6	−92.4	12	
Cu_2SO_4	c	−179.6			
CuSe	c	−9.45			
$CuSe_2$	c	−10.3			
Cu_2Se	c	−14.2			
$CuSeO_3$	c		−83.2		
std. state, $m = 1$	aq	−106.2	−72.7	−20.7	
$CuSeO_4$	c	−114.36			
$CuSeO_4 \cdot 5H_2O$	c	−466.96			
$CuWO_4$	c	−250.0			
Dysprosium					
Dy	c	0	0		
$DyBr_3$	c	−209			
$DyCl_3$ β	c	−237.8			
γ	c	−234.8			
std. state, $m = 1$	aq	−286.1	−255.2		
DyI_3	c	−144.5			
std. state, $m = 1$	aq	−206.1	−198.2		
$Dy(OH)_3$	c		−305.8		
$Dy_2(SO_4)_3$ std. state, $m = 1$	aq	−982.7	−854.4		
$Dy_2(SO_4)_3 \cdot 8H_2O$	c	−1322.0			
Erbium					
Er	c	0	0		
$ErBr_3$	c	−205			
$Er(acetate)_2$	aq	−512.8			
$Er(acetate)_2 \cdot 4H_2O$	c	−785.6			
$ErCl_3$	c	−213.8			
std. state, $m = 1$	aq	−282.4	−251.5		
ErI_3	c	−140.0			
std. state, $m = 1$	aq	−202.4	−194.5		
$Er(OH)_3$	c	−326.8			
$Er_2(SO_4)_3$ std. state, $m = 1$	aq	−975.3	−847.0		
$Er_2(SO_4)_3 \cdot 8H_2O$	c		−1313.9		
Europium					
Eu	c	0	0		
	g	87		45.10	
$EuBr_3$	c	−202			
$EuCl_3$	c	−247.1			
std. state, $m = 1$	aq	−289.4	−259.2		
$Eu_2(SO_4)_3$ std. state, $m = 1$	aq	−989.3	−862.2		
$Eu_2(SO_4)_3 \cdot 8H_2O$	c		−1331.0		
Fluorine					
F	g	18.88	14.80	37.917	5.436
F^-	g	−64.7			
F_2	g	0	0	48.44	7.48
HF	liq	−71.65		18.02	12.35
	g	−64.8	−65.3	41.508	6.963

Table 9-1 (*Continued*)
ELEMENTS AND INORGANIC COMPOUNDS

Formula and Description	State	$\Delta Hf°$	$\Delta Gf°$	$S°$	$C_p°$
HF undissoc; std. state, $m = 1$	aq	−76.50	−70.95	21.2	
ionized; std. state, $m = 1$	aq	−79.50			
	aq, 2	−75.79			
	aq, 10	−76.235			
	aq, 100	−76.340			
HF_2^- std. state, $m = 1$	aq	−155.34	−138.18	22.1	
FO	g	41			
Gadolinium					
Gd	c	0	0	14	
	g	87	77	46.41	
$GdBr_3$	c	−214			
$GdCl_3$	c	−245.5			
std. state, $m = 1$	aq	−288.9	−256.6	−7.6	
GdI_3	c	−147.6			
std. state, $m = 1$	aq	−208.9	−201.6	−31.3	
$Gd(OH)_3$	c		−308.1		
$Gd_2(SO_4)_3$ std. state, $m = 1$	aq	−988.3	−861.2	−81.9	
$Gd_2(SO_4)_3 \cdot 8H_2O$	c	−1518.9	−1329.8	155.8	
Gallium					
Ga	c	0	0	9.77	6.18
	liq	1.33			
	g	66.2	57.1	40.38	6.06
Ga_2	g	104.8			
GaAs	c	−17	−16.2	15.34	11.05
Ga_2C_2	g	134			
$Ga(CH_3)_3$	liq	−18.7			
	g	−10.8			
GaBr	g	−11.9	−21.5	60.2	8.70
$GaBr_3$	c	−92.4	−86.0	43	
	g	−70			
$GaBr_4^-$ std. state, $m = 1$	aq	−158.2	−131.5	8.6	
GaCl	g	−19.1	−25.4	57.4	8.50
$GaCl_3$	c	−125.4	−108.7	34	
	g	−107.0			
GaF	g	−60.2			7.95
GaF_3	c	−278	−259.4	20	
GaH	g	52.7	46.3	46.69	7.00
GaI	g	6.9			8.76
GaI_3	c	−57.1			
GaN	c	−26.4			
	g	42	36	54	
GaO	g	66.8	60.6	55.2	7.66
Ga_2O_3 rhombic	c	−260.3	−238.6	20.31	22.00
GaOH	g	−27.4			
$Ga(OH)^{2+}$ std. state, $m = 1$	aq		−90.9		
$Ga(OH)_2^+$ std. state, $m = 1$	aq		−142.8		
GaO_3^{3-} std. state, $m = 1$	aq		−148		
$H_2GaO_3^-$ std. state, $m = 1$	aq		−178		
$Ga(OH)_3$	c	−230.5	−198.7	24	
GaP	c	−21			
$GaPO_4$	c		−310.1		
$Ga_2(SO_4)_3$	c				62.4
GaSb	c	−10.0	−9.3	18.18	11.60
Ga_2Te_3	c				41.2
Germanium					
Ge	c	0	0	7.43	5.580
	g	90.0	80.3	40.103	7.345
Ge_2	g	113.08	99.5	60.4	8.5
GeBr	g	56.32			8.87

Table 9-1 (*Continued*)
ELEMENTS AND INORGANIC COMPOUNDS

Formula and Description	State	$\Delta Hf°$	$\Delta Gf°$	$S°$	$C_p°$
$GeBr_2$	g	−15.0	−25.5	79.1	
$GeBr_4$	liq	−83.1	−79.2	67.1	
	g	−71.7	−76.0	94.66	24.34
GeH_3Br	g			65.66	13.47
GeC	g	151			
$Ge(C_2H_5)_4$	liq	−49.5			
GeCl	g	37	30	58.8	8.81
$GeCl_4$	liq	−127.1	−110.6	58.7	
	g	−118.5	−109.3	83.08	22.97
GeH_3Cl	g			63.00	13.08
$GeHCl_3$	liq			53.6	
GeF	g	−7.97			8.30
GeF_4	g	−284.4		72.36	19.56
GeH_4	g	21.7	27.1	51.87	10.76
Ge_2H_6	liq	32.82			
	g	38.8			
Ge_3H_8	liq	46.3			
	g	54.2			
GeI_2	c	−21	−20	32	
	g	11.2	−1.0	76	
GeI_4	c	−33.9	−34.5	64.8	
	g	−13.6	−25.4	102.49	24.89
GeH_3I	g			67.65	13.75
Ge_3N_4	c	−15.1			
GeO brown	c	−50.7	−56.7	12	
yellow	c		−49.5		
	g	−11.04	−17.49	53.58	7.39
GeO_2 hexagonal	c	−131.7	−118.8	13.21	12.45
tetragonal	c	−138.7			
Ge_2O_2	g	−112			
Ge_2O_3	g	−212			
H_2GeO_3	aq	−195.73			
GeP	c	−5	−4	15	
GeS	c	−16.5	−17.1	17	
	g	22	10	56	8.05
GeS_2	c	−45.3			
GeSe	c	−22.0			
	g	22.84			8.42
GeSi	g	127			
GeTe	c	−6			
	g	42			8.59
$GeTe_2$	g	44			
Gold					
Au	c	0	0	11.33	6.075
	g	87.5	78.0	43.115	4.968
Au_2	g	123.1			8.808
AuBr	c	−3.34			
$AuBr_3$	c	−12.73			
$AuBr_4^-$ std. state, $m = 1$	aq	−45.8	−40.0	80.3	
$Au(CN)_2^-$ std. state, $m = 1$	aq	57.9	68.3	41	
AuCl	c	−8.6			
$AuCl_3$	c	−28.1			
$AuCl_4^-$ std. state, $m = 1$	aq	−77.0	−56.72	63.8	
$HAuCl_4$ std. state, $m = 1$	aq	−77.0	−56.22	63.8	
AuF_3	c	−86.9			
AuH	g	70.5	63.5	50.441	6.968
AuI	c	0			
AuO_3^{3-} std. state, $m = 1$	aq		−12.4		
$Au(OH)_3$ precipitated	c	−101.5	−75.77	45.3	
Au_2P_3	g	−23.8			

Table 9-1 (*Continued*)
ELEMENTS AND INORGANIC COMPOUNDS

Formula and Description	State	$\Delta Hf°$	$\Delta Gf°$	$S°$	$C_p°$
Hafnium					
Hf hexagonal	c	0	0	10.41	6.15
	g	148.0	137.8	44.642	4.972
HfB	c	−47			
HfB$_2$	c	−80.3	−79.4	10.2	11.89
HfC	c	−60.1			
HfCl$_4$	c	−236.70	−215.42	45.6	28.80
	g	−211.4			
	aq	−302.7			
HfF$_4$ monoclinic	c	−461.4	−437.5	27	
	g	−399.1			
HfN	c	−88.3			
HfO	g	12			
HfO$_2$	c	−273.6	−245.5	14.18	14.40
hydrous ppt		−269.0			
HfOOH$^+$	aq	−279.5			
Helium					
He	g	0	0	30.124	4.9679
Holmium					
Ho	c	0	0		
HoBr$_3$	c	−225			
HoCl$_3$	c	−232.8			
std. state, $m = 1$	aq	−283.8	−253.2		
HoI$_3$	c	−141.7			
std. state, $m = 1$	aq	−203.8	−196.2		
Ho$_2$(SO$_4$)$_3$ std. state, $m = 1$	aq	−978.1	−850.4		
Ho$_2$(SO$_4$)$_3 \cdot 8H_2O$	c		−1318.0		
Hydrogen					
H	g	52.095	48.581	27.391	4.9679
^{1}H	g	52.095	48.580	27.391	4.9679
^{2}H	g	52.981	49.360	29.455	4.9679
H$^+$ std. state, $m = 1$	aq	0	0	0	0
^{1}H$_2$	g	0	0	31.208	6.889
^{2}H$_2$	g	0	0	34.620	6.978
^{1}H^{2}H	g	0.076	−0.350	34.343	6.978
H$_2$ std. state, $m = 1$	aq	−1.0	4.2	13.8	
OH	g	9.31	8.18	43.890	7.143
OH$^+$	g	317.5			
OH$^-$	g	−33.67			
std. state, $m = 1$	aq	−54.970	−37.594	−2.57	−35.5
HO$_2$	g	5			
HO$_2^-$ std. state, $m = 1$	aq	−38.32	−16.1	5.7	
H$_2$O	liq	−68.315	−56.687	16.71	17.995
^{2}H$_2$O	liq	−70.411	−58.195	18.15	20.16
^{1}H^{2}HO	liq	−69.285	−57.817	18.95	
H$_2$O	g	−57.796	−54.634	45.104	8.025
^{2}H$_2$O deuterium oxide	g	−59.560	−56.059	47.378	8.19
H$_2$O$^+$	g	234.3			
H$_2$O$_2$	liq	−44.88	−28.78	26.2	21.3
	g	−32.58	−25.24	55.6	10.3
undissoc; std. state, $m = 1$	aq	−45.69	−32.05	34.4	
	aq, 1	−45.365			
	aq, 10	−45.670			
H$_2$O$_2^+$	g	220.7			
Indium					
In	c	0	0	13.82	6.39
	g	58.15	49.89	41.51	4.98
In$_2$	g	91.04			
InAs	c	−14.0	−12.8	18.1	11.42

Table 9-1 (*Continued*)
ELEMENTS AND INORGANIC COMPOUNDS

Formula and Description	State	$\Delta Hf°$	$\Delta Gf°$	$S°$	$C_p°$
InBr	c	−41.9	−40.4	27	
	g	−13.6	−22.54	61.99	8.76
InBr$_3$	c	−102.5			
	g	−67.4			
InCl II	c	−44.5			
	g	−18			
InCl$_3$	c	−128.4			
	g	−89.4			
In$_2$Cl$_3$	g	−103.6			
InF	g	−48.61			
InH	g	51.5	45.49	49.60	7.07
InI	g	1.8	−9.0	63.87	8.80
	c	−27.8	−28.8	31	
InI$_3$	c	−57			
InN	c	−4.2			
InO	g	92.5	87.1	56.5	7.78
In$_2$O$_3$	c	−221.27	−198.55	24.9	22
InOH	g	−19			
In(OH)$^{2+}$ std. state, $m = 1$	aq	−88.5	−74.8	−21	
In(OH)$_2^+$ std. state, $m = 1$	aq	−148	−125.5	6	
InP	c	−21.2	−18.4	14.3	10.86
InS	c	−33.0	−31.5	16	
	g	90			
In$_2$S	g	15	3.1	76	
In$_2$S$_3$	c	−102	−98.6	39.1	28.20
In$_2$(SO$_4$)$_3$	c	−666	−583	65	67
InSe	c	−28			
In$_2$Se$_3$	c	−82			
InSb	c	−7.3	−6.1	20.6	11.82
	g	82.3			
InTe	c	−23			
In$_2$Te$_3$	c	−47			
Iodine					
I	g	25.535	16.798	43.184	4.968
I$^-$	g	−47.0			
std. state, $m = 1$	aq	−13.19	−12.33	26.6	−34.0
I$_2$	c	0	0	27.757	13.011
	g	14.923	4.627	62.28	8.82
std. state, $m = 1$	aq	5.4	3.92	32.8	
std. state, $m = 1$	CCl$_4$	6.0	2.66	39.0	
I$_3^-$ std. state, $m = 1$	aq	−12.3	−12.3	57.2	
IBr	g	9.76	0.89	61.822	8.71
IBr$_2^-$ std. state, $m = 1$	aq		−29.4		
BrI$_2^-$ std. state, $m = 1$	aq	−30.6	−26.3	47.2	
IBrCl$^-$ std. state, $m = 1$	aq		−35.0		
ICl	liq	−5.71	−3.25	32.3	
	g	4.25	−1.30	59.140	8.50
ICl$_2^-$ std. state, $m = 1$	aq		−38.5		
ICl$_3$	c	−21.4	−5.34	40.0	
I$_2$Cl$^-$ std. state, $m = 1$	aq		−31.7		
IF	g	−22.86	−28.32	56.42	7.99
IF$_5$	liq	−206.7			
	g	−196.58	−179.68	78.3	23.7
IF$_7$	g	−225.6	−195.6	82.8	32.6
HI	g	6.33	0.41	49.351	6.969
std. state, $m = 1$	aq	−13.19	−12.33	26.6	−34.0
HIO undissoc; std. state, $m = 1$	aq	−33.0	−23.7	22.8	
HIO$_3$	c	−55.0			
	aq, ∞	−52.2			

Table 9-1 (*Continued*)
ELEMENTS AND INORGANIC COMPOUNDS

Formula and Description	State	$\Delta Hf°$	$\Delta Gf°$	$S°$	$C_p°$
H_5IO_6	aq	−180.4			
IO	g	41.84	35.80	58.65	7.86
IO^- std. state, $m = 1$	aq	−25.7	−9.2	−1.3	
IO_3^- std. state, $m = 1$	aq	−52.9	−30.6	28.3	
IO_4^-	aq	−35.2			
I_2O_5	c	−37.78			
Iridium					
Ir	c	0	0	8.48	6.00
	g	159.0	147.7	46.240	4.968
$IrCl_3$	c	−58.7	−43	27	
	g	25	24	90	
$IrCl_6^{2-}$	aq	−148.1	(−129)	(50)	
$IrCl_6^{3-}$	aq	−179.5	(−109)	(70)	
IrF_6	c	−138.54	−110.34	59.2	
	g	−130	−110	85.5	28.94
IrO_2	c	−65.5	(−42)	(15)	
IrO_3	g	1.9	(7)	(69)	
IrS_2	c	−33	(−32)	(15)	
Ir_2S_3	c	−56	(−53)	(23)	
Iron					
Fe α	c	0	0	6.52	6.00
	g	99.5	88.6	43.112	6.137
Fe^{2+} std. state, $m = 1$	aq	−21.3	−18.85	−32.9	
Fe^{3+} std. state, $m = 1$	aq	−11.6	−1.1	−75.5	
$FeAl_2O_4$	c	−470	−442	25.4	29.53
FeAsS	c	−10	−12	29	
$FeBr_2$	c	−59.7			
	g	−11			
std. state, $m = 1$	aq	−79.4	−68.55	6.5	
$FeBr_3$	c	−64.1			
	g	−29.6			
std. state, $m = 1$	aq	−98.8	−75.7	−16.4	
Fe_3C α-cementite	c	6.0	4.8	25.0	25.3
$Fe(CN)_6^{3-}$ std. state, $m = 1$	aq	134.3	174.3	64.6	
$Fe(CN)_6^{4-}$ std. state, $m = 1$	aq	108.9	166.09	22.7	
$H_2Fe(CN)_6^{2-}$ std. state, $m = 1$	aq	108.9	157.37	52	
$FeCO_3$ siderite	c	−177.00	−159.35	22.2	19.63
$Fe(CO)_5$	liq	−185.0	−168.6	80.8	57.5
	g	−175.4	−166.65	106.4	
$Fe_2(C_2O_4)_3$ std. state, $m = 1$	aq	−614.8	−485.5	−118.3	
$Fe(acetate)_3$ std. state, $m = 1$	aq	−360.1	−266.0	−13.4	
$FeCNS^{2+}$ std. state, $m = 1$	aq	5.6	17.0	−31	
$FeCl^{2+}$ std. state, $m = 1$	aq	−43.1	−34.4	−27	
$FeCl_2$	c	−81.69	−72.26	28.19	18.32
std. state, $m = 1$	aq	−101.2	−81.59	−5.9	
	aq, 100	−99.88			
$FeCl_2 \cdot 2H_2O$	c	−227.8			
$FeCl_2 \cdot 4H_2O$	c	−370.3			
$FeCl_3$	c	−95.48	−79.84	34.0	23.10
std. state, $m = 1$	aq	−131.5	−95.2	−35.0	
$FeCl_3 \cdot 6H_2O$	c	−531.5			
Fe_2Cl_6	g	−156.5			
$Fe(ClO_4)_2$ std. state, $m = 1$	aq	−83.1	−22.97	54.1	
$Fe(ClO_4)_2 \cdot 6H_2O$	c	−494.4			
$FeCr_2O_4$	c	−345.3	−321.2	34.9	31.94
FeF_2	c			20.79	16.28
std. state, $m = 1$	aq	−180.3	−152.13	−39.5	
FeF_3 std. state, $m = 1$	aq	−250.1	−201.0	−85.4	
FeI_2	c	−27.0			
std. state, $m = 1$	aq	−47.7	−43.51	20.3	

Table 9-1 (*Continued*)
ELEMENTS AND INORGANIC COMPOUNDS

Formula and Description	State	$\Delta Hf°$	$\Delta Gf°$	$S°$	$C_p°$
FeI_3	g	17			
std. state, $m = 1$	aq	−51.2	−38.1	4.3	
$FeMoO_4$	c	−257	−233	30.9	28.31
$Fe_2(MoO_4)_3$	c	−702			
Fe_4N	c	−2.5	0.9	37	
$Fe(NO_3)_3$ std. state, $m = 1$	aq	−160.3	−80.9	29.5	
$Fe(NO_3)_3 \cdot 9H_2O$	c	−785.2			
$Fe_{0.947}O$ wistite	c	−63.64	−58.59	13.74	11.50
FeO	c	−65.0			
Fe_2O_3 hematite	c	−197.0	−177.4	20.89	24.82
Fe_3O_4 magnetite	c	−267.3	−242.7	35.0	34.28
$FeOH^+$ std. state, $m = 1$	aq	−77.6	−66.3	−7	
$Fe(OH)^{2+}$ std. state, $m = 1$	aq	−69.5	−54.83	−34	
$Fe(OH)_2$ pptd	c	−136.0	−116.3	21	
$Fe(OH)_2^-$ std. state, $m = 1$	aq		−104.7		
$Fe(OH)_3$ pptd	c	−196.7	−166.5	25.5	
FeP	c	−30			
FeP_2	c	−46			
Fe_2P	c	−39			
Fe_3P	c	−39			
$FePO_4$	c	−310.1			
$FePO_4 \cdot 2H_2O$ strengite	c	−451.3	−396.2	40.93	43.15
FeS iron-rich pyrrhotite	c	−23.9	−24.0	14.41	12.08
FeS_2 pyrite	c	−42.6	−39.9	12.65	14.86
Fe_7S_8 sulfur-rich pyrrhotite	c	−176.0	−178.9	116.1	95.26
$FeSO_4$	c	−221.9	−196.2	25.7	24.04
std. state, $m = 1$	aq	−238.6	−196.82	−28.1	
$FeSO_4 \cdot 7H_2O$	c	−720.50	−599.97	97.8	94.28
$Fe_2(SO_4)_3$	c	−617.0			
std. state, $m = 1$	aq	−675.2	−536.1	−136.4	
FeSe	c	−18.0			
$FeSe_2$	c			20.75	17.42
FeSi	c	−17.6	−17.6	11.0	11.4
$FeSi_2$ β-lebanite	c	−19.4	−18.7	13.3	15.79
Fe_3Si	c	−22.4	−22.6	24.8	23.50
Fe_2SiO_4 fayalite	c	−353.7	−329.6	34.7	31.76
FeTe	c	−15.0			
$FeWO_4$	c	−276	−252	31.5	27.39
$Fe_2(WO_4)_3 \cdot 8H_2O$	c	−1355			
Krypton					
Kr	g	0	0	39.191	4.968
KrF_2	c	14.4			
Lanthanum					
La	c	0	0	13.7	
	g	88	79	43.67	
$LaBr_3$	c	−233			
$La_2(CN_2)_3$	c	−229			
$LaCl_3$	c	−263.6			
std. state, $m = 1$	aq	−296.3	−266.9	−5	
La_3H_8	c	−382			
LaI_3	c	−167.4			
std. state, $m = 1$	aq	−216.3	−209.9	34	
LaN	c	−72.1	−64.6		
La_2O_3	c	−428.5			
$La(OH)_3$	c		−312.8		
LaS_2	c	−156.7			
La_2S_3	c	−306.8			
$La_2(SO_4)_3$ std. state, $m = 1$	aq	−1003.1	−877.8	−76	
$La_2(SO_4)_3 \cdot 8H_2O$	c		−14.031		

Table 9-1 (*Continued*)
ELEMENTS AND INORGANIC COMPOUNDS

Formula and Description	State	$\Delta Hf°$	$\Delta Gf°$	$S°$	$C_p°$
Lead					
Pb	c	0	0	15.49	6.32
	g	46.6	38.7	41.889	4.968
$PbBr_2$	c	−66.6	−62.60	38.6	19.15
$Pb(BrO_3)_2$	c		−11.95		
$Pb(CH_3)_4$	liq	23.4			
	g	32.48			
$Pb(C_2H_5)_4$	liq	12.6			
	g	26.19			
$Pb(acetate)_2$	c	−230.5			
$PbCl_2$	c	−85.90	−75.08	32.5	
$PbCl_4$	liq	−78.7			
$Pb(ClO_3)_2$ undissoc; std. state	aq		−6.6		
PbClF	c	−127.8	−116.7	29.1	
$PbCO_3$	c	−167.1	−149.5	31.3	20.89
PbC_2O_4	c	−203.5	−179.3	34.9	25.2
$PbO \cdot PbCO_3$	c	−219.5	−195.2	48.8	
$PbCrO_4$	c	−222.5			
PbF_2	c	−158.7	−147.5	26.4	
PbF_4	c	−225.1			
PbI_2	c	−41.94	−41.50	41.79	18.49
$Pb(IO_3)_2$	c	−118.4	−84.0	74.8	
$PbMoO_4$	c	−251.4	−227.4	39.7	28.61
$Pb(N_3)_2$ monoclinic	c	114.3	149.3	35.4	
orthorhombic	c	113.8	148.7	35.7	
$Pb(NO_3)_2$	c	−108.0			
PbO yellow	c	−51.94	−44.91	16.42	10.94
red	c	−52.34	−45.16	15.9	10.95
PbO_2	c	−66.3	−51.95	16.4	15.45
Pb_2O_3	c			36.3	25.74
Pb_3O_4	c	−171.7	−143.7	50.5	35.1
$HPbO_2^-$ std. state, $m = 1$	aq		−80.90		
$Pb(OH)_2$	c		−108.1	21	
precipitated	c	−123.3			
H_2PbO_2 undissoc; std. state, $m = 1$	aq		−95.8		
$Pb(OH)_3^-$ std. state, $m = 1$	aq		−137.6		
$Pb_3(PO_4)_2$	c	−620.3	−581.4	84.4	61.25
$Pb(ReO_4)_2 \cdot 2H_2O$	c	−534	−455	74	
PbS	c	−24.0	−23.6	21.8	11.83
$Pb(SCN)_2$	c		32.1		
$PbSO_3$	c	−160.1			
$PbSO_4$	c	−219.87	−194.36	35.51	24.667
PbSe	c	−24.6	−24.3	24.5	12.0
$PbSeO_3$	c	−128.5			
$PbSeO_4$	c	−145.6	−120.7	40.1	
$PbSiO_3$	c	−273.83	−253.86	26.2	21.52
Pb_2SiO_4	c	−325.8	−299.4	44.6	32.78
PbTe	c	−16.9	−16.6	26.3	12.08
$PbWO_4$	c				28.63
Lithium					
Li	c	0	0	6.70	
	g	38	29.19	33.14	
$LiAlH_4$	c	−24.1			18.2
LiBr	c	−83.72			
	b	−41	−50	53.78	
std. state, $m = 1$	aq	−95.45	−94.79	22.7	
$LiBr \cdot 2H_2O$	c	−229.94			
$LiBrO_3$	aq	−77.9	−65.70		
Li_2C_2	c	−14.2			

Table 9-1 (*Continued*)
ELEMENTS AND INORGANIC COMPOUNDS

Formula and Description	State	$\Delta Hf°$	$\Delta Gf°$	$S°$	$C_p°$
LiCN	aq, 200	−30.3	−31.35		
LiCNO cyanate	aq	−101.2	−94.12		
Li_2CO_3	c	−290.54	−270.66	21.60	
	aq, ∞	−294.74	−266.66	−5.9	
$LiHCO_3$	aq, ∞	−231.73	−210.53	29.5	
$LiC_2H_3O_2$	aq	−183.9	−160.0		
LiCl	g	−53	−58	51.01	
	c	−97.70			
	aq, 3	−102.62			
	aq, ∞	−106.58	−101.57	16.6	
$LiCl \cdot H_2O$	c	−170.31	−151.2	24.8	
$LiCl \cdot 2H_2O$	c	−242.1			
$LiCl \cdot 3H_2O$	c	−313.5			
$LiClO_3$	aq	−87.5	−70.95		
$LiClO_4$	aq	−106.3	−81.4		
LiF	c	−144.7	−139.5	8.57	
	aq, ∞	−145.21	−136.30	1.1	
$LiHF_2$	aq, 400	−220.4			
Li_2SiF_6	c	−688.9			
LiH	g	30.7	25.2	40.77	
	c	−21.61	−16.72	5.9	8.3
LiI	g	−16	−26	55.68	
	c	−71.2			
	aq, ∞	−79.92	−82.57	29.5	
$LiI \cdot \frac{1}{2}H_2O$	c	−103.8			
$LiI \cdot H_2O$	c	−141.16			
$LiI \cdot 2H_2O$	c	−213.03			
$LiI \cdot 3H_2O$	c	−285.02			
$LiIO_3$	aq	−121.3	−102.95		
Li_3N	c	−47.2	−37.33		
$LiNH_2$	c	−43.50			
$LiNO_2$	c	−96.6			
$LiNO_3$	c	−115.28			
	aq, 3	−115.1			
	aq, ∞	−115.93	−96.63	38.4	
$LiNO_3 \cdot 3H_2O$	c	−328.6			
Li_2O	c	−142.4			
Li_2O_2	c	−151.7	−138.0		
	aq	−159.0			
LiOH	c	−116.45	−106.1	12	
	aq, 11	−120.51			
	aq, ∞	−121.51	−107.82	0.9	
$LiOH \cdot H_2O$	c	−188.77	−164.8	22	
$LiReO_4$	c	−253.4			
$LiReO_4 \cdot H_2O$	c	−324.9			
$LiReO_4 \cdot 2H_2O$	c	−395.1			
Li_2SO_4	c	−342.83	−314.66		
	aq, 18	−348.40			
	aq, ∞	−350.01	−317.78	10.9	
$Li_2SO_4 \cdot H_2O$	c	−414.20	−375.07		
Li_2Se	c	−91.1			
Li_2SiO_3	gls	−376.7			
Lutetium					
Lu	c	0	0		
	g	87		44.14	
$LuBr_3$	c	−200			
$LuCl_3$ γ	c	−227.9			
	aq, ∞	−280.2	−249.0		
LuI_3 β	c	−133.2			
	aq, ∞	−200.2	−192.0		

Table 9-1 (*Continued*)
ELEMENTS AND INORGANIC COMPOUNDS

Formula and Description	State	$\Delta Hf°$	$\Delta Gf°$	$S°$	$C_p°$
$Lu_2(SO_4)_3$	aq, ∞	-970.9	-842.0		
$Lu_2(SO_4)_3 \cdot 8H_2O$	c		-1308.1		
Magnesium					
Mg	c	0	0	7.77	
	g	35.9	27.6	35.50	
$Mg(AlH_4)_2$	c	-23.1			
$Mg_3(AsO_4)_2$	c	-731.3			
	aq	-749	-630.14		
$MgHAsO_4$	aq	-349.2			
$Mg(H_2AsO_4)_2$	aq	-541.0			
$Mg(NH_4)AsO_4 \cdot 6H_2O$	c	-800.7			
Mg_3Bi_2	c	-36.5			
$MgBr_2$	c	-123.7			
	aq, 25	-166.66			
	aq, ∞	-168.21	-158.14	10.4	
$MgBr_2 \cdot 6H_2O$	c	-575.4	-491.0	95	
$Mg(CN)_2$	aq	-31.9	-29.08		
$MgCN_2$	c	-60.3			
$MgCO_3$	c	-266	-246	15.7	
$Mg(C_2H_3O_2)_2$	aq	-344.6	-286.38		
$MgCl_2$	c	-153.40	-141.57	21.4	
	aq, 10	-185.65			
	aq, ∞	-190.46	-171.69	-1.9	
$MgCl_2 \cdot H_2O$	c	-231.15	-206.11	32.8	
$MgCl_2 \cdot 2H_2O$	c	-305.99	-267.32	43.0	
$MgCl_2 \cdot 4H_2O$	c	-454.00	-390.49	63.1	
$MgCl_2 \cdot 6H_2O$	c	-597.42	-305.65	87.5	
$MgCl_2 \cdot KCl \cdot 6H_2O$	c	-702.0			
$MgCl_2 \cdot MgO$	c	-312.6			
$MgCl_2 \cdot MgO \cdot 6H_2O$	c	-742.1			
$MgCl_2 \cdot MgO \cdot 16H_2O$	c	-1440.8			
$Mg(OH)Cl$	c	-191.3	-175.0	19.8	
$Mg(ClO_4)_2$	c	-140.6			
	aq	-172.5			
$Mg(ClO_4)_2 \cdot 2H_2O$	c	-290.7			
$Mg(ClO_4)_2 \cdot 4H_2O$	c	-438.6			
$Mg(ClO_4)_2 \cdot 6H_2O$	c	-583.2			
$MgCrO_4$	c	-318.3			
	aq	-321.2			
MgF_2	c	-263.5	-250.8	13.68	
MgH	g	41	34	47.61	
MgI_2	c	-86.0			
	aq, ∞	-137.15	-133.69	24.1	
$MgMoO_4$	c	-329.9			
Mg_3N_2	c	-110.24	-100.8		
$Mg(NO_3)_2$	c	-188.72	-140.63	39.2	
	aq, 15	-208.410			
	aq, ∞	-209.15	-161.81	41.8	
$Mg(NO_3)_2 \cdot 2H_2O$	c	-336.63			
$Mg(NO_3)_2 \cdot 6H_2O$	c	-624.36			
MgO	c	-143.84	-136.13	6.4	
MgO_2	c	-148.9			
$Mg(OH)_2$	c	-221.00	-199.27	15.09	
$Mg_3(PO_4)_2$	c	-961.5			
$MgHPO_4$	aq	-422.2			
$Mg(NH_4)PO_4 \cdot 6H_2O$	c	-881.0			
MgS	c	-83.0			
	aq	-108			
$MgSO_3$	c	-241.0			

Table 9-1 (*Continued*)
ELEMENTS AND INORGANIC COMPOUNDS

Formula and Description	State	$\Delta Hf°$	$\Delta Gf°$	$S°$	$C_p°$
MgSO$_3$ · 3H$_2$O	c	−461.6			
MgSO$_3$ · 6H$_2$O	c	−673.4			
MgSO$_4$	c	−305.5	−280.5	21.9	
	aq, 200	−326.23			
	aq, ∞	−327.31	−286.33	−24.0	
MgSO$_4$ · 2H$_2$O	c	−381.9			
MgSO$_4$ · 4H$_2$O	c	−595.5			
MgSO$_4$ · 6H$_2$O	c	−736.6			
MgSO$_4$ · 7H$_2$O	c	−808.7			
MgS$_2$O$_3$	aq	−256.1			
MgS$_2$O$_3$ · 3H$_2$O	c	−454.2			
MgS$_2$O$_3$ · 6H$_2$O	c	−669.4			
Mg$_3$Sb$_2$	c	−68.1			
Mg$_2$Si	c	−18.6			
MgSiO$_3$	c	−357.9	−337.2	16.2	
MgSiO$_4$	c	−488.2	−459.8	22.7	
Mg$_2$Sn	c	−17.0			
MgTe	c	−50			
MgWO$_4$	c	−345.2			
Manganese					
Mn α	c	0	0	7.65	6.29
β	c			8.22	6.34
γ	c	0.37	0.34	7.75	9.59
	g	67.1	57.0	41.49	4.97
MnAs	c	−14			
Mn$_3$(AsO$_4$)$_2$	c	−512.8			
MnBr$_2$	c	−92.0	−89	(33)	
std. state, $m = 1$	aq	−110.9	−97.8		
MnBr$_2$ · H$_2$O	c	−168.5			
MnBr$_2$ · 4H$_2$O	c	−380.1			
Mn$_3$C	c	1.1	1.3	23.6	22.33
Mn$_7$C$_3$	c	−10			
MnCO$_3$ natural	c	−213.7	−195.2	20.5	19.48
pptd	c	−211	−194	27.0	
MnC$_2$O$_4$	c	−246.2			
undissoc; std. state, $m = 1$	aq	−248.5	−221.0	16.1	
MnC$_2$O$_4$ · 2H$_2$O	c	−389.2	−338.2	48	
MnC$_2$O$_4$ · 3H$_2$O	c	−459.1			
Mn(acetate)$_2$	c	−274.4			
Mn(acetate)$_2$ · 4H$_2$O	c	−558.8			
Mn$_2$(CO)$_{10}$	c	−401.0			
	g	−386.0			
MnCl	g	10.1			8.05
MnCl$_2$	c	−115.03	−105.29	28.26	17.43
std. state, $m = 1$	aq	−132.66	−117.3	9.3	−53
MnCl$_2$ · H$_2$O	c	−188.8	−166.4	41.6	
MnCl$_2$ · 2H$_2$O	c	−261.0	−225.2	52.3	
MnCl$_2$ · 4H$_2$O	c	−403.3	−340.3	72.5	
Mn^{2+} std. state, $m = 1$	aq	−52.76	−54.5	−17.6	12
MnF$_2$	c	−189	−179	22.05	15.96
MnI$_2$	c	−59.3	(−65)	(37)	
2H$_2$O	c	−201.4			
4H$_2$O	c	−343.9			
Mn(IO$_3$)$_2$	c	−160	−124.4	63	
MnMoO$_4$	c	−284.8	−261	(32)	
MnN$_6$ azide	c	92.2			
Mn$_5$N$_2$	c	−48.8			
Mn(NO$_3$)$_2$	c	−137.73			
std. state, $m = 1$	aq	−151.9	−107.8	52	−29

Table 9-1 (*Continued*)
ELEMENTS AND INORGANIC COMPOUNDS

Formula and Description	State	$\Delta Hf°$	$\Delta Gf°$	$S°$	$C_p°$
Mn(NO$_3$)$_2$·6H$_2$O glassy	amorp	−566.9			
	liq	−557.27			
MnO	c	−92.07	−86.74	14.27	10.86
	g	29.6			
MnO$_2$	c	−124.29	−111.18	12.68	12.94
pptd	amorp	−120.1			
Mn$_2$O$_3$	c	−229.2	−210.6	26.4	25.73
Mn$_3$O$_4$	c	−331.7	−306.7	37.2	33.38
MnO$_4^-$ std. state, $m = 1$	aq	−129.4	−106.9	45.7	
MnO$_4^{2-}$ std. state, $m = 1$	aq	−156	−119.7	14	
Mn(OH)$_2$ pptd	amorp	−166.2	−147.0	23.7	
MnP	c	−27			
MnP$_3$	c	−51			
Mn$_3$(PO$_4$)$_2$	c	−744.9			
MnHPO$_4$	c		−332.5		
MnS green	c	−51.2	−52.2	18.7	11.94
pink pptd	amorp	−51.1			
MnSO$_4$	c	−254.60	−228.83	26.8	24.02
std. state, $m = 1$	aq	−270.1	−232.5		
MnSO$_4$·H$_2$O α	c	−329.0	−289.9	(37)	
β	c	−322.2			
MnSO$_4$·4H$_2$O	c	−539.7			
MnSO$_4$·5H$_2$O	c	−610.2			78
MnSO$_4$·7H$_2$O	c	−750.3			
Mn$_2$(SO$_4$)$_3$	c	−666.9			
MnS$_2$O$_6$·6H$_2$O	c	−751.0			
MnSb	c	−12			
MnSe	c	−25.5	−26.7	21.7	12.20
MnSiO$_3$	c	−315.7	−296.5	21.3	20.66
Mn$_2$SiO$_4$	c	−413.6	−390.1	39.0	31.04
MnTe	c			22.4	17.49
MnWO$_4$	c	−311.9			29.7
Mercury					
Hg	liq	0	0	18.17	6.688
	g	14.655	7.613	41.79	4.968
HgBr$_2$	c	−40.8	−36.6	41	
undissoc; std. state, $m = 1$	aq	−38.4	−34.2	41	
HgBr$_3^-$ std. state, $m = 1$	aq	−70.1	−62.0	62	
HgBr$_4^{2-}$ std. state, $m = 1$	aq	−103.0	−88.7	74	
Hg$_2$Br$_2$	c	−49.45	−43.278	52	
HgBrCl std. state, $m = 1$	aq	−45.5	−38.7	40	
HgBrI undissoc; std. state, $m = 1$	aq	−28.1	−26.7	46	
Hg(CN)$_2$	c	63.0			
	g	91			
undissoc; std. state, $m = 1$	aq	66.5	74.6	37.3	
Hg(CN)$_3^-$ std. state, $m = 1$	aq	94.9	110.7	51.4	
Hg(CN)$_4^{2-}$ std. state, $m = 1$	aq	125.8	147.8	71	
Hg(II) fulminate	c	64			
Hg(I) acetate	c	−199.6	−153.3	78	
Hg$_2$CO$_3$	c	−132.3	−111.9	43	
HgC$_2$O$_4$	c	−162.1			
Hg$_2$C$_2$O$_4$	c		−141.8		
Hg(CNS)$_2$ undissoc; std. state	aq	46.9	60.1	37.6	
Hg(CNS)$_4^{2-}$ std. state, $m = 1$	aq	78.0	98.3	109	
Hg$_2$(CNS)$_2$ thiocyanate	c		54.4		
HgCl	g	20.1	15.0	62.09	8.68
HgCl$_2$	c	−53.6	−42.7	34.9	
undissoc; std. state, $m = 1$	aq	−51.7	−41.4	37	
HgCl$_3^-$ std. state, $m = 1$	aq	−92.9	−73.9	50	

Table 9-1 (*Continued*)
ELEMENTS AND INORGANIC COMPOUNDS

Formula and Description	State	$\Delta Hf°$	$\Delta Gf°$	$S°$	$C_p°$
$HgCl_4^{2-}$ std. state, $m = 1$	aq	−132.4	−106.8	70	
Hg_2Cl_2	c	−63.39	−50.377	46.0	
HgF	g	1.0	−2.4	53.98	8.26
Hg_2F_2	c		−104.1		
HgH	g	57.36	51.63	52.46	7.16
HgI	g	31.64	21.14	67.26	8.99
HgI_2 red	c	−25.2	−24.3	43	
yellow	c	−24.6			
	g	−4.1	−14.3	80.31	14.60
undissoc; std. state, $m = 1$	aq	−19.0	−18.0	42	
HgI_3^- std. state, $m = 1$	aq	−36.5	−35.5	72	
HgI_4^{2-} std. state, $m = 1$	aq	−56.2	−50.6	86	
Hg_2I_2	c	−29.00	−26.53	55.8	
$Hg_2(N_3)_2$	c	142.0	178.4	49	
$Hg(NO_3)_2 \cdot \frac{1}{2}H_2O$	c	−93.8			
$Hg_2(NO_3)_2 \cdot 2H_2O$	c	−207.5			
HgO red, orthorhombic	c	−21.71	−13.995	16.80	10.53
yellow	c	−21.62	−13.964	17.0	
hexagonal	c	−21.4	−13.92	17.6	
$Hg(OH)_2$ undissoc; std. state	aq	−84.9	−65.7	34	
HgS red	c	−13.9	−12.1	19.7	11.57
black	c	−12.8	−11.4	21.1	
$HgSO_4$	c	−169.1			
Hg_2SO_4	c	−177.61	−149.589	47.96	31.54
$Hg(HS)_2$ undissoc; std. state, $m = 1$	aq		−6.4		
HgSe	c	−11			
	g	18.1	7.5	63.82	8.8
$HgSeO_3$	c		−68.0		
Hg_2SeO_3	c		−71.1		
HgTe	c	−10			
	g			65.73	
Molybdenum					
Mo	c	0	0	6.85	5.75
	g	157.3	146.4	43.461	4.968
$MoBr_2$	c	−62.4	−53		18.3
$MoBr_3$	c	(−34)			
$MoBr_4$	c	−76.8			
MoO_2Br_2	c	−150.4			
MoB	c	−21			
MoC	c	−2.4			
Mo_2C	c	−10.9			
$Mo(CO)_6$	c	−234.9	−209.8	77.9	57.90
	g	−218.0	−204.6	117	49
$MoCl_2$	c	−67.4	(−35)		
$MoCl_3$	c	−92.5		(32.6)	
$MoCl_4$	c	−114.8	−58.5	44.7	
	g	−90			
$MoCl_5$	c	−126.0	(−68.5)	53	
	g	−103			
$MoOCl_2$	c	−126			
MoO_2Cl_2	c	−171.4			
	g	−151.6			
	aq	−190.4			
$MoO_2Cl_2 \cdot H_2O$	c	−245.4			
$MoOCl_4$	c	−153.0			
MoF_6	liq	−378.95	−352.08	62.06	40.58
	g	−372.29	−351.88	83.75	28.82
MoI_2	g	32			

Table 9-1 (*Continued*)
ELEMENTS AND INORGANIC COMPOUNDS

Formula and Description	State	$\Delta Hf°$	$\Delta Gf°$	$S°$	$C_p°$
MoI$_3$		(−15)	(−15)		
Mo$_2$N	c	−19.50			
MoO	g	101			
MoO$_2$	c	−140.76	−127.40	11.06	13.38
MoO$_3$	c	−178.08	−159.66	18.58	17.92
	g	−78			
	aq	−172.5			
MoO$_4$	aq	−158.0			
MoO$_5$	aq	−139.6			
MoO$_4^{2-}$ std. state	aq	−238.5	−199.9	6.5	
H$_2$MoO$_4$ white	c	−250.0			
	aq	−240.8			
H$_2$MoO$_4$ · H$_2$O yellow	c	−325			
MoS$_2$	c	−56.2	−54.0	14.96	15.19
Mo$_2$S$_3$	c	−87			
MoSi$_2$	c	−28			
Mo$_3$Si	c	−23	−23	25.4	22.23
Mo$_5$Si$_3$	c	−68			
Neodymium					
Nd	g	87			
	c	0	0		
NdCl$_3$ α	c	−254.3			
	aq, ∞	−291.3	−261.6		
NdCl$_3$ · 6H$_2$O	c	−692.3			
Nd(OH)$_3$	c		−309.6		
NdI$_3$ α	c	−158.9			
	aq, ∞	−211.3	−204.6		
Nd$_2$O$_3$	c	−442.0			
Nd$_2$S$_3$	c	−281.8			
Nd$_2$(SO$_4$)$_3$	c	−948.1			
	aq, ∞	−993.1	−867.2		
Nd$_2$(SO$_4$)$_3$ · 5H$_2$O	c	−1318.4			
Nd$_2$(SO$_4$)$_3$ · 8H$_2$O	c	−1524.7	−1334.5		
Neon					
Ne	g	0	0	34.95	4.968
Neptunium					
Np	c	0	0		
NpBr$_3$	c	−174			
NpBr$_4$	c	−183			
NpCl$_3$	c	−216			
NpCl$_4$	c	−237			
NpCl$_5$	c	−246			
NpF$_3$	c	−360			
NpF$_4$	c	−428			
NpI$_3$	c	−120			
NpO$_2$	c	−246			
Nickel					
Ni	c	0	0	7.14	6.23
	g	102.7	91.9	43.519	5.583
Ni^{2+} std. state, $m = 1$	aq	−12.9	−10.9	−30.8	
Ni$_3$(AsO$_4$)$_2$	c		−377.5		
Ni(BO$_2$)$_2$	c		−347.3		
NiBr$_2$	c	−50.7			
std. state, $m = 1$	aq	−71.0	−60.6	8.6	
Ni$_3$C	c	16.1			
Ni(CN)$_2$ pptd	c	30.5			
Ni(CN)$_4^{2-}$ std. state, $m = 1$	aq	87.9	112.8	52	
Ni(CNO)$_2$ cyanate	c	−54.4			
Ni(CNS)$_2$ thiocyanate	c	22.8			

Table 9-1 (*Continued*)
ELEMENTS AND INORGANIC COMPOUNDS

Formula and Description	State	$\Delta Hf°$	$\Delta Gf°$	$S°$	$C_p°$
Ni(CO)$_4$	liq	−151.3	−140.6	74.9	48.9
	g	−144.10	−140.36	98.1	34.70
Ni(acetate)$_2$ std. state, $m = 1$	aq	−245.2	−187.5	10.6	
NiCO$_3$	c		−146.4		
NiC$_2$O$_4$	c	−204.8			
std. state, $m = 1$	aq	−210.1	−172.0	−19.9	
NiCl$_2$	c	−72.976	−61.918	23.34	17.13
std. state, $m = 1$	aq	−92.8	−73.6	−3.6	
NiCl$_2 \cdot 2$H$_2$O	c	−220.4	−181.7	42	
NiCl$_2 \cdot 4$H$_2$O	c	−362.5	−295.2	58	
NiCl$_2 \cdot 6$H$_2$O	c	−502.67	−409.54	82.3	
Ni(ClO$_4$)$_2$ std. state, $m = 1$	aq	−74.7	−15.0	56.2	
Ni(ClO$_4$)$_2 \cdot 6$H$_2$O	c	−486.6			
NiF$_2$	c	−155.7	−144.4	17.59	15.31
NiI$_2$	c	−18.7			
std. state, $m = 1$	aq	−39.3	−35.6	22.4	
Ni(IO$_3$)$_2$	c	−116.9	−78.0	51	
Ni(NO$_3$)$_2$	c	−99.2			
std. state, $m = 1$	aq	−112.0	−64.2	39.2	
Ni(NO$_3$)$_2 \cdot 6$H$_2$O	c	−528.6			111
NiO	c	−57.3	−50.6	9.08	10.59
Ni$_2$O$_3$	c	−117.0			
Ni(OH)$_2$	c	−126.6	−106.9	21	
Ni(OH)$_3$ pptd	c	−160			
Ni$_3$P	c	−50.2			
Ni$_5$P$_2$	c	−97.7			
Ni$_2$P$_2$O$_7$	c		−497.9		
Ni$_3$(PO$_4$)$_2$	c		−562.4		
NiS	c	−19.6	−19.0	12.66	11.26
pptd	c	−18.5			
Ni$_3$S$_2$	c	−48.5	−47.1	32.0	28.12
NiSO$_4$	c	−208.63	−181.6	22	33
std. state, $m = 1$	aq	−230.2	−188.9	−26.0	
NiSO$_4 \cdot 6$H$_2$O tetrahedral green	c	−641.21	−531.78	79.391	78.36
NiSO$_4 \cdot 7$H$_2$O	c	−711.36	−588.49	90.57	87.14
NiSe	c	−14.1			
NiSeO$_3 \cdot 2$H$_2$O	c	−271.11			
NiSi	c	−20.6			10.9
Ni$_2$Si	c	−33.6			16.8
NiTe	c	−12.8			12.5
Ni$_4$W	c	−43			29.0
NiWO$_4$	c	−270.9			29.1
Niobium					
Nb	c	0	0	8.70	5.88
	g	173.5	162.8	44.490	7.208
NbBr$_5$	c	−132.9			
	g	−104.8			
NbC	c	−33.2	−32.7	8.46	8.81
Nb$_2$C	c	−45.4	−44.4	15.3	14.48
NbCl$_3$	g	−86			
NbCl$_4$	c	−166.0			
	g	−134			
NbCl$_5$	c	−190.6	−163.3	50.3	35.4
	g	−168.2	−154.4	95.71	28.88
NbCo$_2$	c	−13.7	−13.2	22	
NbCo$_3$	c	−14.1	−13.7	29	
NbCr$_2$	c	−5.0	−5.0	19.97	17.45
NbF$_5$	c	−433.5	−406.1	38.3	32.2
	g	−415.8	−401.1	76.9	23.2

Table 9-1 (*Continued*)
ELEMENTS AND INORGANIC COMPOUNDS

Formula and Description	State	$\Delta Hf°$	$\Delta Gf°$	$S°$	$C_p°$
$NbFe_2$	c	−11.1	−11.8	24	13.0
$NbGe_2$	c	−20.8			
NbI_5	c	−64.2			
NbN	c	−56.2	−49.2	8.25	9.32
Nb_2N	c	−59.9			
NbO	c	−97.0	−90.5	11.5	9.86
	g	51	44	57.09	7.36
NbO_2	c	−190.3	−177.0	13.03	13.74
	g	−51.3	−52.3	61.0	
NbO_3^- ionic strength = 1	aq		−222.8		
Nb_2O_5 high-temperature form	c	−454.0	−422.1	32.80	31.57
$Nb(OH)_5$ undissoc; ionic strength = 1	aq		−346.2		
$NbOCl_2$	c	−185.1			
$NbOCl_3$	c	−210.2	−187	34	
	g	−179.8	−171.6	85.6	22.0
Nitrogen					
N	g	112.979	108.886	36.613	4.968
N_2	g	0	0	45.77	6.961
N_3^- std. state, $m = 1$	aq	65.76	83.2	25.8	
HN_3	g	70.3	78.4	57.09	10.44
undissoc; std. state, $m = 1$	aq	62.16	76.9	34.9	
NCl_3	liq	55			
NH_3	g	−11.02	−3.94	45.97	8.38
undissoc; std. state, $m = 1$	aq	−19.19	−6.35	26.6	
	aq, 1	−18.011			
	aq, 10	−19.074			
NH_4^+ std. state, $m = 1$	aq	−31.67	−18.97	27.1	19.1
N_2H_4	liq	12.10	35.67	28.97	23.63
	g	22.80	38.07	56.97	11.85
undissoc; std. state, $m = 1$	aq	8.20	30.6	33	
$N_2H_5^+$ std. state, $m = 1$	aq	−1.8	19.7	36	16.8
N_2H_5Br	c	−37.2			
std. state, $m = 1$	aq	−30.8	−5.2	55.7	−17.1
$N_2H_5Br \cdot HBr$	c	−64.8			
N_2H_5Cl	c	−47.0			
std. state, $m = 1$	aq	−41.8	−11.7	49.5	−15.8
$N_2H_5Cl \cdot HCl$	c	−87.8			
$N_2H_5ClO_4$	c	−42.2			
std. state, $m = 1$	aq	−32.7	17.6	79.7	
N_2H_5OH	liq	−58.01			
	g	−49.0	−18.9	63	
undissoc; std. state, $m = 1$	aq	−60.11	−26.1	49.7	17.5
$N_2H_5NO_3$	c	−60.13			
std. state, $m = 1$	aq	−51.41	−6.91	71	
$(N_2H_5)_2SO_4$	c	−229.2			
std. state, $m = 1$	aq	−221.0	−138.6	77	−36
HNF_2	g			60.40	10.37
NO	g	21.57	20.69	50.347	7.133
NO_2	g	7.93	12.26	57.35	8.89
NO_2^- std. state, $m = 1$	aq	−25.0	−8.9	33.5	−23.3
NO_3	g	16.95	27.36	60.36	11.22
NO_3^- nitrate, std. state, $m = 1$	aq	−49.56	−26.61	35.0	−20.7
peroxynitrite	aq	−10.7			
N_2O	g	19.61	24.90	52.52	9.19
$HN_2O_2^-$	aq	−9.4	18.2	34	
$N_2O_2^{2-}$ hyponitrite	aq	−4.1	33.2	6.6	
N_2O_3	g	20.01	33.32	74.61	15.68
N_2O_4	g	2.19	23.38	72.70	18.47
N_2O_5	g	2.7	27.5	85.0	20.2

Table 9-1 (*Continued*)
ELEMENTS AND INORGANIC COMPOUNDS

Formula and Description	State	$\Delta Hf°$	$\Delta Gf°$	$S°$	$C_p°$
NOBr	g	19.64	19.70	65.38	10.87
NOCl	g	12.36	15.79	62.52	10.68
NO_2Cl	g	3.0	13.0	65.02	12.71
$NOClO_4$	c	−36.9			
NO_2ClO_4	c	8.7			
NOF	g	−15.9	−12.2	59.27	9.88
NO_2F	g	26		62.2	11.9
$N_2O_3(SO_3)_2$	c	−253			
HNO_2 *cis*	g	−18.64	−10.27	59.43	10.70
trans	g	−19.15	−10.82	59.54	11.01
undissoc; std. state, $m = 1$	aq	−28.5	−13.3	36.5	
HNO_3	g	−32.28	−17.87	63.64	12.75
	liq	−41.40	−19.10	37.19	
std. state, $m = 1$	aq	−49.56	−26.61	35.0	−20.7
	aq, 1	−44.845			
	aq, 2	−46.500			
	aq, 10	−49.192			
	aq, 100	−49.440			
NH_2OH	c	−25.5			
	aq	−21.7	−5.60	40	
$NH_2OH_2^+$	aq	−32.8			
$NH_2OH \cdot HCl$	c	−75.9			
$(NH_2OH)_2 \cdot H_2SO_4$	aq	−281.3			
$NH_2OH \cdot H_2SO_4$	aq	−246.7			
$H_2N_2O_2$ hyponitrous acid	aq	−13.7	8.6	52	
NH_2NO_2 nitramide	c	−21.4			
N_4S_4	c	128.0			
NSF	g			62.07	10.55
NSF_3	g			68.48	17.18
NSe	c	42.3			
Osmium					
Os	c	0	0	7.8	5.9
	g	189	178	46.000	4.968
$OsCl_3$	c	−45.5	−29	31	
$OsCl_4$	c	−60.9	−38	37	
	g	−19			
OsF_6 cubic	c			58.8	
	g			85.56	28.88
OsO_2	c		−46		
OsO_3	g	−67.8			
OsO_4 yellow	c	−94.2	−72.9	34.4	
white	c	−92.2	−72.6	40.1	
	g	−80.6	−70.0	70.2	17.7
$Os(OH)_4$	amorp		−161.0		
OsS_2	c	−34.9	−32	13	
Oxygen					
O	g	59.553	55.389	38.467	5.237
O_2	g	0	0	49.003	7.016
O_3	g	34.1	39.0	57.08	9.37
OF_2	g	−5.2	−1.1	59.11	10.35
O_2F_2	g	4.3			
O_2F_3	g	3.8			
Palladium					
Pd	c	0	0	8.98	6.21
	g	90.4	81.2	39.90	4.968
Pd^{2+}	aq, ∞	40.5	42.2	−28	
$PdBr_2$	c	−24.9			
$PdBr_4^{2-}$	aq, ∞	−88.8	−76.0	70	
$Pd(CN)_2$	c	56.9			

Table 9-1 (*Continued*)
ELEMENTS AND INORGANIC COMPOUNDS

Formula and Description	State	$\Delta Hf°$	$\Delta Gf°$	$S°$	$C_p°$
PdCl$_2$	c	−41.0	−29.9	25	
PdCl$_4^{2-}$ 1M HCl	aq, ∞	−124.8	−99.6	62	
PdCl$_6^{2-}$ 1M HCl	aq, ∞	−143	−102.8	65	
Pd(CNS)$_2$	c			56.0	
Pd$_2$H	c	−4.7	−1.2	21.9	
PdI$_2$	c	−15.2	−15.0	36	
PdO	c	−20.4			7.5
Pd(OH)$_2$ pptd	c	−89.0	−72		
Pd(OH)$_4$ pptd	c	−156.0	−115	(35)	
PdS	c	−18	−16	11	
PdS$_2$	c	−19.4	−17.8	19	
PdTe	c			21.42	12.23
PdTe$_2$	c			30.25	18.31
Phosphorus					
P α, white	c	0	0	9.82	5.698
red, triclinic	c	−4.2	−2.9	5.45	5.07
black	c	−9.4			
red	amorp	−1.8			
P$_2$	g	75.20	66.51	38.978	4.968
P$_4$	g	14.08	5.85	66.89	16.05
PBr$_3$	liq	−44.1	−42.0	57.4	
	g	−33.3	−38.9	83.17	18.16
PBr$_5$	c	−64.5			
PCl$_3$	liq	−76.4	−65.1	51.9	
	g	−68.6	−64.0	74.49	17.17
PCl$_5$	c	−106.0			
	g	−89.6	−72.9	87.11	26.96
PF	g			53.74	7.56
PF$_3$	g	−219.6	−214.5	65.28	14.03
PF$_5$	g	−381.4			
PH$_3$	g	1.3	3.2	50.22	8.87
std. state, $m = 1$	aq	−2.16	0.35	48.2	
PH$_4^+$ std. state, $m = 1$	aq		16.2		
P$_2$H$_4$	liq	−1.2	16.0	40	
	g	5.0			
PH$_4$Br	c	−30.5	−11.4	26.3	
PH$_4$Cl	c	−34.7			
PH$_4$I	c	−16.7	0.2	29.4	26.2
PH$_4$OH std. state, $m = 1$	aq	−70.48	−56.34	65	
PI$_3$	c	−10.9			
	g			89.45	18.73
PN	g	32.76	27.47	50.45	7.10
P$_3$N$_5$	c	−71.4			
P$_3$N$_3$Cl$_6$	c	−194.1			
	g	−175.9			
P$_4$N$_4$Cl$_8$	c	−259.2			
	g	−236.1			
PO	g			53.221	
PO$_3^-$	aq	−233.5			
PO$_4^{3-}$ std. state, $m = 1$	aq	−305.3	−243.5	−53	
P$_2$O$_7^{4-}$ std. state, $m = 1$	aq	−542.8	−458.7	−28	
P$_4$O$_6$	c	−392.0			
P$_4$O$_{10}$ hexagonal	c	−713.2	−644.8	54.70	50.60
	amorp	−727			
POBr$_3$	c	−109.6			
	g			85.97	21.48
POCl$_3$	liq	−142.7	−124.5	53.17	33.17
	g	−133.48	−122.60	77.76	20.30
POF$_3$	g	−289.5	−277.9	68.11	16.41

Table 9-1 (*Continued*)
ELEMENTS AND INORGANIC COMPOUNDS

Formula and Description	State	$\Delta Hf°$	$\Delta Gf°$	$S°$	$C_p°$
HPO_3	c	−226.7			
	aq	−233.5			
HPO_3^{2-}	aq	−231.6			
HPO_4^{2-} std. state, $m = 1$	aq	−308.83	−260.34	−8.0	
$H_2PO_4^-$ std. state, $m = 1$	aq	−309.82	−270.17	21.6	
H_3PO_4	c	−305.7	−267.5	26.41	25.35
ionized, std. state, $m = 1$	aq	−305.3	−243.5	−53	
undissoc; std. state, $m = 1$	aq	−307.92	−273.10	37.8	
	aq, 1	−304.69			
	aq, 2	−305.60			
	aq, 5	−306.87			
$H_2PO_3^-$	aq	−231.7			
H_3PO_3	c	−230.5			
	aq	−230.6			
$H_2PO_2^-$	aq	−146.7			
H_3PO_2	c	−144.5			
$H_4P_2O_7$	c	−535.6			
undissoc; std. state, $m = 1$	aq	−542.2	−485.7	64	
$HP_2O_7^{3-}$ std. state, $m = 1$	aq	−543.7	−471.6	11	
$H_2P_2O_7^{2-}$ std. state, $m = 1$	aq	−544.6	−480.5	39	
$H_3P_2O_7^-$ std. state, $m = 1$	aq	−544.1	−483.6	51	
$H_4P_2O_5$	aq	−393.6			
H_2PO_3F undissoc; std. state, $m = 1$	aq		−287.5		
P_2S_3	c	−19.2			
$PSBr_3$	c			55.2	
	g			89.07	22.69
$PSCl_3$	g			80.47	21.39
Platinum					
Pt	c	0	0	9.95	6.18
	g	135.1	124.4	45.960	6.102
$PtBr$	c	−11	(−7)	(28)	
$PtBr_2$	c	−23.1		(44)	
$PtBr_3$	c	−30.9		(56)	
$PtBr_4$	c	−38.0		(68)	
$PtBr_4^{2-}$	aq	−89	(−69)	(47)	
$PtBr_6^{2-}$	aq	−114	(−89)	(67)	
$PtCl$	c	−9		(27)	
$PtCl_2$	c	−26.5		(28)	
std. state, $m = 1$	aq		−18.3		
$PtCl_3$	c	−41.6	(−32)	(36)	
$PtCl_4$	c	−56.5	(−41)	(42)	
	aq	−76.2			
$PtCl_4^{2-}$ std. state, $m = 1$	aq	−120.3	−88.1	40	
$PtCl_6^{2-}$ std. state, $m = 1$	aq	−161	−117	52.6	
PtF_6 cubic	c			56.3	
	g			83.23	29.35
PtI_4	c	−17.4			
PtO_2	g	41.0	40.1	62	
Pt_3O_4	c	−39			
$Pt(OH)_2$	c	−84.1	(−66)	(30)	
PtS	c	−19.5	−18.2	13.16	10.37
PtS_2	c	−26.0	−23.8	17.85	15.75
$PtTe$	c			19.41	11.93
$PtTe_2$	c			30.25	18.31
Polonium					
Po	c	0	0		
Po^{2+} std. state, $m = 1$	aq		17		
Po^{4+} std. state, $m = 1$	aq		70		
$PoCl_6^{2-}$ std. state, $m = 1$	aq		−138		

Table 9-1 (*Continued*)
ELEMENTS AND INORGANIC COMPOUNDS

Formula and Description	State	$\Delta Hf°$	$\Delta Gf°$	$S°$	$C_p°$
Po(OH)$_2^{4+}$ std. state, $m = 1$	aq		−113		
Po(OH)$_4$	c		−130		
PoS	c		52		
Potassium					
K	g	21.51	14.62	30.30	
	c	0	0	15.2	
K$_2$O · Al$_2$O$_3$ · 4SiO$_2$ leucite	c	−1406.4			
K$_2$O · Al$_2$O$_3$ · 6SiO$_2$ microcline	c	−1816			
(adularia)	c	−1842			
K$_2$SiF$_6$	c	−671			
KAl(SO$_4$)$_2$	c	−589.24	−534.29	48.9	
	aq	−616.9			
KAl(SO$_4$)$_2$ · H$_2$O	c	−667.5			
KAl(SO$_4$)$_2$ · 2H$_2$O	c	−742			
KAl(SO$_4$)$_2$ · 3H$_2$O	c	−814			
KAl(SO$_4$)$_2$ · 12H$_2$O	c	−1447.74	−1227.8	164.3	
K$_3$AsO$_3$	aq	−323.0			
K$_3$AsO$_4$	aq	−390.3	−355.7		
KH$_2$AsO$_4$	c	−271.5	−237.0	37.08	
	aq	−276.2			
KBr	c	−93.73	−90.63	23.05	
	aq, 10	−89.77			
	aq, 50	−89.07			
	aq, 500	−88.87	−89		
	aq, 1000	−88.88			
	aq, ∞	−88.94	−92.04	43.8	
KBrO	aq	−82.0			
KBrO$_3$	c	−79.4	−58.2	35.65	
	aq, ∞	−69.6	−56.6	63.4	
K$_2$CO$_3$	c	−273.93			
	aq, 50	−281.02			
	aq, 1000	−281.56	−264		
K$_2$CO$_3$ · ½H$_2$O	c	−210.43			
K$_2$CO$_3$ · 1½H$_2$O	c	−283.40			
KHCO$_3$	c	−229.3			
	aq, 1500	−224.5	−207.7		
K(HCO$_2$) formate	c	−158.0			
	aq	−157.7			
KC$_2$H$_3$O$_2$	c	−173.2			
	aq, 10	−175.6			
	aq, ∞	−176.88	−156.7		
K$_2$C$_2$O$_4$	c	−320.8			
	aq, 100	−316.88			
	aq, 5000	−316.90	−393.1		
	aq, ∞	−317.1			
K$_2$C$_2$O$_4$ · H$_2$O	c	−392.17			
KCN	c	−26.90			
	aq	−24.1	−28		
KCNO cyanate	c	−98.5			
	aq	−93.5	−90.85		
KCNS thiocyanate	c	−48.62			
	aq, 2	−45.46			
	aq, 8	−44.32			
	aq, 50	−43.11			
	aq, ∞	−42.8	−44		
KCl	g	−51.6	−56.2	57.24	
	c	−104.18	−97.59	19.76	
	aq, 50	−100.11			
	aq, ∞	−100.06	−98.82	37.7	
KClO	aq	−85.4			

Table 9-1 (*Continued*)
ELEMENTS AND INORGANIC COMPOUNDS

Formula and Description	State	$\Delta Hf°$	$\Delta Gf°$	$S°$	$C_p°$
$KClO_3$	c	−93.50	−69.29	34.17	
	aq, 100	−84.09			
	aq, ∞	−83.54	−68.09	63.5	
$KClO_4$	c	−103.6	−72.7	36.1	
	aq, 500	−91.58			
	aq, ∞	−91.45	−70.04	68.0	
K_2CrO_4	c	−330.49			
	aq, 17.2	−328.2			
	aq, 54	−327.0			
	aq, ∞	−326.0	−306		
$K_2Cr_2O_7$	c	−485.90			
	aq, 135	−470.41			
	aq, 800	−469.50	−441		
	aq, 2000	−468.7			
KF	c	−134.46	−127.42	15.91	
	aq, 4	−137.35			
	aq, 50	−138.51			
	aq, ∞	−138.70	−133		
$KF \cdot 2H_2O$	c	−277.0	−242.7	36	
$KF \cdot 4H_2O$	c	−418.0			
KHF_2	c	−219.98	−203.73	24.92	
	aq, 25	−213.73			
	aq, 200	−213.4			
$KF \cdot 2HF$	c	−296.7			
$KF \cdot 3HF$	c	−373.0			
KI	c	−78.31	−77.03	24.94	
	aq, 6	−74.94			
	aq, ∞	−73.41	−79.82	50.6	
KIO_3	c	−121.5	−101.7	36.20	
	aq, 200	−115.16			
	aq, ∞	−115.0	−99.9	52.2	
KIO_4	aq, 1200	−97.6			
$KMnO_4$	c	−194.4	−170.6	41.04	
	aq, 140	−184.44			
	aq, 400	−183.9	−168.0		
K_2MoO_4	aq, 880	−374.1			
KNH_2	c	−28.3			
KNO_2	c	−88.5			
	aq, 400	−85.2	−76		
KNO_3	c	−117.76	−93.96	31.77	
	aq, 400	−109.48	−93.68		
	aq, ∞	−109.41	−93.88	69.5	
K_2O	c	−86.4			
K_2O_2	c	−118			
K_2O_4	c	−134			
KOH	c	−101.78			
	aq, 3	−111.77			
	aq, 5	−113.31			
	aq, 25	−114.70			
	aq, 500	−114.86	−105		
	aq, ∞	−115.00	−105.06	22.0	
$KOH \cdot \frac{1}{2}H_2O$	c	−161.7			
$KOH \cdot H_2O$	c	−179.6			
$KOH \cdot 2H_2O$	c	−251.2			
K_2OsCl_6	c	−280			
K_3PO_3	aq	−397.5			
K_3PO_4	aq	−478.7	−443.3		
KH_2PO_4	c	−374.9	−326		
	aq, 35	−370.70			
	aq, 755	−370.4			

Table 9-1 (*Continued*)
ELEMENTS AND INORGANIC COMPOUNDS

Formula and Description	State	$\Delta Hf°$	$\Delta Gf°$	$S°$	$C_p°$
KReO$_4$	c	−264.02			
K$_3$RhCl$_6$	c	−343			
K$_2$S	c	−100			
	aq, 7	−107.04			
	aq, 10	−108.95			
	aq, 20	−110.17			
	aq, 25	−110.27			
	aq, 50	−110.17			
	aq, 200	−109.94	−111.4		
K$_2$S · 2H$_2$O	c	−243.0			
K$_2$S · 5H$_2$O	c	−456.7			
KHS	c	−63.2			
	aq, 10	−64.35			
	aq, 200	−64.1			
KHS · ¼H$_2$O	c	−80.1			
K$_2$S$_4$	c	−113.0			
	aq	−114.6			
K$_2$S$_4$ · ½H$_2$O	c	−150.5			
K$_2$S$_4$ · 2H$_2$O	c	−258.7			
K$_2$SO$_3$	c	−266.9			
	aq	−269.1	−251.3		
KHSO$_3$	aq, 385	−209.7			
K$_2$SO$_4$	c	−342.66	−314.62	42.0	
	aq, 1000	−336.75	−310		
	aq, 5000	−336.81			
KHSO$_4$	c	−276.8			
	aq, 20	−273.38			
	aq, 100	−273.44			
	aq, 800	−274.3			
K$_2$S$_2$O$_5$	c	−362.6			
	aq	−351.9			
K$_2$S$_2$O$_5$ · 1½H$_2$O	c	−397.0			
K$_2$S$_2$O$_6$	c	−413.6			
	aq, 400	−401.0			
K$_2$S$_2$O$_8$	c	−458.3			
	aq	−445.3			
K$_2$S$_4$O$_6$	c	−422			
	aq	−410			
K$_2$S$_5$O$_6$ · 1½H$_2$O	c	−515.4			
K$_2$Se	c	−79.3			
	aq	−88.5	−99		
K$_2$Se · 9H$_2$O	c	−722.0			
K$_2$Se · 14H$_2$O	c	−1066			
K$_2$Se · 19H$_2$O	c	−1417			
KHSe	c	−35.9			
	aq	−35.4			
K$_2$SeO$_4$	aq, 440	−265.3	−240		
KHSeO$_4$	aq, 220	−202.8			
K$_2$TeO$_3$	aq	−261.6			
K$_2$TeO$_4$	aq	−290.3			
KVO$_3$	aq	−284.0			
KVO$_4$	c	−273.9			
	aq	−270.4			
KVO$_5$	aq	−254.9			
Praseodymium					
Pr	g	87			
	c	0	0		
PrBr$_3$	c	−225			
PrCl$_3$ α	c	−257.8			
	aq, ∞	−292.8	−263.1		

Table 9-1 (*Continued*)
ELEMENTS AND INORGANIC COMPOUNDS

Formula and Description	State	$\Delta Hf°$	$\Delta Gf°$	$S°$	$C_p°$
$PrCl_3 \cdot H_2O$	c	−330.8			
$PrCl_3 \cdot 7H_2O$	c	−764.5			
PrI_3 α	c	−162.0			
	aq, ∞	−212.8	−206.1		
$Pr(NO_3)_3$	aq	−318.8			
PrO_2	c	−234.0			
Pr_2O_3	c	−444.5			
Pr_6O_{11}	c	−1391			
$Pr(OH)_3$	c		−310.7		
$Pr_2(SO_4)_3$	aq, ∞	−996.1	−870.2		
$Pr_2(SO_4)_3 \cdot 8H_2O$	c		−1337.0		
Promethium					
Pm	c	0	0		
$PmCl_3$	c	−251.9			
	aq, ∞	−290.5			
Radium					
Ra	g	31	23	42.15	
	c	0	0	17	
$RaBr_2$	c	−195			
$RaCl_2 \cdot 2H_2O$	c	−351	−311.7	50	
$Ra(NO_3)_2$	c	−237	−190.3	52	
RaO	c	−125			
$RaSO_4$	c	−352	−326.0	34	
Radon					
Rn	g	0	0	42.09	4.968
Rhenium					
Re	c	0	0	8.81	6.09
	g	184.0	173.2	45.131	4.968
Re^- std. state, $m = 1$	aq	11	2.4	55	
$ReAs_2$	c	1.3			
$ReBr_3$	c	−40			
Re_3Br_9	g	−69			
$ReCl_3$	c	−63	−45	29.6	22.08
$ReCl_5$	c	−89			
Re_3Cl_9	g	−137			
$ReCl_6^{2-}$ std. state, $m = 1$	aq	−182	−141	60	
H_2ReCl_4	c	−152			
ReF_6	g	−273			
ReO_2	c	−101	−88	41	
$ReO_2 \cdot 2H_2O$ pptd	c	−236			
ReO_3	c	−144.6	−127	61.5	
Re_2O_7	c	−296.4	−254.8	49.5	39.7
	g	−263	−237.6	108	
$HReO_4$	c	−182.2	−156.9	37.8	
	g	−159			
std. state, $m = 1$	aq	−188.2	−166.0	48.1	−3.2
ReS_2	c	−43			
Re_2S_7	c	−107			
Re_3Si	c	0			
Rhodium					
Rh	c	0	0	7.53	5.97
	g	133.1	122.1	44.383	5.022
$RhCl_2$	g	30.3			
$RhCl_3$	c	−71.5	(−54)	(21)	
	g	16			
Rh_2O_3	c	−82	(−65)	(27)	24.8
$Rh(OH)_3$	c		(−112)		
Rubidium					
Rb	g	20.51	13.35	40.63	
	c	0	0	16.6	

Table 9-1 (*Continued*)
ELEMENTS AND INORGANIC COMPOUNDS

Formula and Description	State	$\Delta Hf°$	$\Delta Gf°$	$S°$	$C_p°$
RbAl(SO$_4$)$_2$	c	−569			
RbAl(SO$_4$)$_2$ · H$_2$O	c	−647			
RbAl(SO$_4$)$_2$ · 2H$_2$O	c	−722			
RbAl(SO$_4$)$_2$ · 3H$_2$O	c	−797			
RbAl(SO$_4$)$_2$ · 12H$_2$O	c	−1448.0			
RbBr	c	−93.03	−90.38	25.88	
	aq, 200	−87.77			
	aq, ∞	−87.8	−92.02	49.0	
Rb$_2$CO$_3$	c	−269.6			
	aq, 200	−279.4	−263.8		
	aq, 2000	−278.0			
Rb$_2$CO$_3$ · H$_2$O	c	−344.2			
Rb$_2$CO$_3$ · 1½H$_2$O	c	−381.5			
Rb$_2$CO$_3$ · 3½H$_2$O	c	−521.7			
RbHCO$_3$	c	−228.5			
	aq	−224.1	−209.1		
RbCN	aq	−25.9			
RbCNS thiocyanate	c	−54			
	aq	−41.7			
RbCl	c	−102.9	−98.48		
	aq, 100	−98.90			
	aq, 1000	−98.84			
	aq, ∞	−98.9	−98.80	42.9	
RbClO$_3$	c	−93.8	−69.8	36.3	
	aq, ∞	−82.4	−68.07	68.7	
RbClO$_4$	c	−103.87	−73.19	38.4	
	aq, ∞	−90.3	−70.02	73.2	
RbF	c	−131.28			
	aq, 100	−137.45			
	aq, ∞	−137.6	−133.53	27.4	
RbF · ⅓H$_2$O	c	−156.12			
RbF · 1½H$_2$O	c	−240.33			
RbHF$_2$	c	−217.3			
	aq	−212.5			
RbI	c	−78.5	−77.8	28.21	
	aq, ∞	−72.3	−79.80	55.8	
Rb$_2$IrCl$_6$	c	−290.9			
RbNO$_3$	c	−117.04			
	aq, 60	−109.00			
	aq, 200	−108.54			
	aq, ∞	−108.5	−93.86	64.7	
RbNH$_2$	c	−25.7			
Rb$_2$O	c	−78.9			
Rb$_2$O$_2$	c	−101.7			
RbOH	c	−98.9			
	aq, 3.2	−110.5			
	aq, 200	−113.7			
	aq, ∞	−113.9	−105.05	27.2	
RbOH · H$_2$O	c	−177.8			
RbOH · 2H$_2$O	c	−250.8			
RbReO$_4$	c	−256.9			
	aq	−249.2			
Rb$_2$S	c	−83.2			
	aq, 500	−107.8			
RbHS	c	−62.4			
	aq	−63.1			
RbHSO$_4$	c	−273.7			
	aq, 300	−270.4			
Rb$_2$SO$_4$	c	−340.50			
	aq, 400	−334.51			
	aq, ∞	−334.7	−312.24	33.8	

Table 9-1 (*Continued*)
ELEMENTS AND INORGANIC COMPOUNDS

Formula and Description	State	$\Delta Hf°$	$\Delta Gf°$	$S°$	$C_p°$
RbHSe	c	−35.5			
	aq	−34.3			
Rb_2SiF_6	c	−678.4			
Ruthenium					
Ru	c	0	0	6.82	5.75
	g	153.6	142.4	44.550	5.144
$RuBr_3$	c	−33			
$RuCl_3$ black	c	−49			
	g	−0.3			
$RuCl_4$	g	−12.4			
$RuCl_5(OH)^{2-}$	aq, ∞		−168.7		
RuF_5	c	−213.4			
	g	−189			
RuI_3	c	−15.7			
RuO_2	c	−72.9			
hydrated	amorp		−51.3		
RuO_3	g	−18.7			
RuO_4	c	−57.2	−36.4	35.0	
	liq	−54.6	−36.4	43.8	
	g	−44.0	−33.4	69.3	18.14
RuO_4^-	aq, ∞		−58.7		
RuO_4^{2-}	aq, ∞		−72.6		
RuS_2	c	−47	−44		
Samarium					
Sm	g	87		43.74	
	c	0	0		
$SmBr_3$	c	−216			
$SmCl_3$ α	c	−249.8			
	aq, ∞	−289.9	−259.9		
SmI_3 β	c	−153.4			
	aq, ∞	−209.9	−202.9		
$Sm(OH)_3$	c		−308.8		
$Sm_2(SO_4)_3$	aq, ∞	−990.3	−863.8		
$Sm_2(SO_4)_3 \cdot 8H_2O$	c		−1332.6		
Scandium					
Sc	c	0	0	8.28	6.10
	g	90.3	80.32	41.75	5.28
Sc_2	g	154.9	141.6	61	8.7
Sc^{3+} std. state, $m = 1$	aq	−146.8	−140.2	−61	
$ScBr_2$	g			77.6	13.0
$ScBr_3$	c	−177.6			
ScCl	g	26.9	20.6	56.00	8.40
$ScCl_2$	g			72.5	12.6
$ScCl_3$	c	−221.1			
	aq, 5500	−268.8			
$ScCl_3 \cdot 6H_2O$	c	−671.6			
$Sc(CNS)^{2+}$ std. state, $m = 1$	aq		−119.6		
ScF	g	−33.2	−39.3	53.11	7.74
ScF_2	g	−153.5	−156.6	67.0	11.5
ScF_3	c	−389.4	−371.8	22	
	g	−298	−295	71.8	16.2
ScI_2	g			81.6	13.3
ScO	g	−13.68	−19.90	53.65	7.38
Sc_2O	g	−6.9			11.2
Sc_2O_3	c	−456.22	−434.85	18.4	22.52
$Sc(OH)^{2+}$ std. state, $m = 1$	aq	−205.9	−191.5	−32	
$Sc(OH)_3$	c	−325.9	−294.8	24	
$Sc(OH)_2Cl$	c	−303	−276.3	26	
ScS	g	41.8	29.7	56.3	8.0

Table 9-1 (*Continued*)
ELEMENTS AND INORGANIC COMPOUNDS

Formula and Description	State	$\Delta Hf°$	$\Delta Gf°$	$S°$	$C_p°$
$Sc(SO_4)^+$ std. state, $m = 1$	aq		−321.7		
$Sc(SO_4)_2^-$ std. state, $m = 1$	aq		−501.5		
$Sc_2(SeO_3)_3 \cdot 10H_2O$	c	−1326.5			
Selenium					
Se hexagonal, black	c	0	0	10.144	6.062
monoclinic, red	c	1.6			
	g	54.27	44.71	42.21	4.978
glassy	amorp	1.2			
Se_2	g	34.9	23.0	60.2	8.46
Se^{2-} std. state, $m = 1$	aq		30.9		
Se_6	g	39.2			
$SeBr_2$	g	−5			
Se_2Br_2	g	7			
$SeCl_2$	g	−7.6			
$SeCl_4$	c	−43.8			
Se_2Cl_2	liq	−19.7			
	g	4			
SeF_6	g	−267	−243	74.99	26.4
HSe^- std. state, $m = 1$	aq	3.8	10.5	19	
H_2Se	g	7.1	3.8	52.32	8.30
std. state, $m = 1$	aq	4.6	5.3	39.1	
SeO	g	12.75	6.41	55.9	7.47
SeO_2	c	−53.86			
SeO_3	c	−39.9			
SeO_3^{2-} std. state, $m = 1$	aq	−121.7	−88.4	3	
SeO_4^{2-} std. state, $m = 1$	aq	−143.2	−105.5	12.9	
Se_2O_5	c	−97.6			
$SeOCl_2$	g	−6			
$HSeO_3^-$ std. state, $m = 1$	aq	−122.98	−98.36	32.3	
$HSeO_4^-$ std. state, $m = 1$	aq	−139.0	−108.1	35.7	
H_2SeO_3	c	−125.35			
undissoc; std. state, $m = 1$	aq	−121.29	−101.87	49.7	
H_2SeO_4	c	−126.7			
	aq, ∞	−140.3			
Silicon					
Si	c	0	0	4.50	4.78
	amorp	1.0			
	g	108.9	98.3	40.12	5.318
Si_2	g	142	128	54.92	8.22
SiBr	g	50			9.23
$SiBr_4$	liq	−109.3	−106.1	66.4	
	g	−99.3	−103.2	90.29	23.21
SiC β, cubic	c	−15.6	−15.0	3.97	6.42
α, hexagonal	c	−15.0	−14.4	3.94	6.38
	g	177			
SiCl	g	45.39			8.81
$SiCl_2$	g	−39.59	−42.35	67.0	12.16
$SiCl_4$	liq	−164.2	−148.16	57.3	34.73
	g	−157.03	−147.47	79.02	21.57
SiH_3Cl	g			59.88	12.20
SiH_2Cl_2	g			68.26	14.45
$SiHCl_3$	liq	−128.9	−115.34	54.4	
SiF	g	1.7	−5.8	53.94	7.80
SiF_2	g	−148	−150	60.38	10.49
SiF_4	g	−385.98	−375.88	67.49	17.60
SiF_6^{2-} std. state, $m = 1$	aq	−571.0	−525.7	29.2	
SiH_3F	g			56.95	11.33
$SiHF_3$	g			64.96	14.47
SiH	g			47.42	6.98

Table 9-1 (*Continued*)
ELEMENTS AND INORGANIC COMPOUNDS

Formula and Description	State	$\Delta Hf°$	$\Delta Gf°$	$S°$	$C_p°$
SiH_4	g	8.2	13.6	48.88	10.24
Si_2H_6	g	19.2	30.4	65.14	19.31
Si_3H_8	liq	22.1			
	g	28.9			
SiI_4	c	−45.3			
SiN	g	116.28	109.01	51.78	7.21
Si_3N_4	c	−177.7		24.2	
SiO_2 quartz	c	−217.72	−204.75	10.00	10.62
cristobalite	c	−217.37	−204.56	10.20	10.56
tridymite	c	−217.27	−204.42	10.4	10.66
	amorp	−215.94	−203.33	11.2	10.6
H_2SiO_3	c	−284.1	−261.1	32	
undissoc; std. state, $m = 1$	aq	−282.7	−258.0	26	
H_4SiO_4	c	−354.0	−318.6	46	
undissoc; std. state, $m = 1$	aq	−351.0	−314.7	43	
SiS	g	26.88	14.56	53.43	7.71
SiS_2	c	−49.5			
SiSe	g	23.78			8.04
$SiSe_2$	c	−7			
SiTe	g	30.99			8.31
Silver					
Ag	c	0	0	10.17	6.059
	g	68.01	58.72	41.321	4.968
Ag_2	g	97.99	85.75	61.43	8.84
Ag^+ std. state, $m = 1$	aq	25.234	18.433	17.37	5.2
Ag^{2+} in $4M$ $HClO_4$, std. state	aq	64.2	64.3	−21	
Ag_3AsO_4	c		−129.7		
AgBr	c	−23.99	−23.16	25.6	12.52
std. state, $m = 1$	aq	−3.82	−6.42	37.1	−28.7
$AgBrO_3$	c	−6.5	13.0	36.5	
AgCl	c	−30.370	−26.244	23.0	12.14
	g			58.75	8.57
std. state, $m = 1$	aq	−14.718	−12.939	30.9	−27.4
$AgClO_2$	c	2.10	18.1	32.16	20.87
std. state, $m = 1$	aq	9.3	22.5	41.6	
$AgClO_3$	c	−6.1			
std. state, $m = 1$	aq	1.5	17.6	56.2	
$AgClO_4$	c	−7.44			
std. state, $m = 1$	aq	−5.68	16.37	60.9	
Ag_2CrO_4	c	−174.89	−153.40	52.0	34.00
AgCN	c	34.9	37.5	25.62	15.95
$Ag(CN)_2^-$ std. state, $m = 1$	aq	64.6	73.0	46	
$AgCN_2$ cyanamide	c	56.2			
AgONC fulminate	c	43			
AgOCN cyanate	c	−22.8	−13.9	29	
AgSCN thiocyanate	c	21.0	24.23	31.3	15
Ag_2CO_3	c	−120.9	−104.4	40.0	26.83
$Ag_2C_2O_4$	c	−160.9	−139.6	50	
Ag acetate	c	−95.3	−73.56	35.8	
AgF	c	−48.9			
std. state, $m = 1$	aq	−54.27	−48.21	14.1	−20.3
$AgF · 2H_2O$	c	−191.4	−160.4	41.8	31
AgF_2	c	−87.3			
AgI	c	−14.78	−15.82	27.6	13.58
std. state, $m = 1$	aq	12.04	6.10	44.0	−28.8
$AgIO_3$	c	−40.9	−22.4	35.7	24.60
std. state, $m = 1$	aq	−27.7	−12.2	45.7	
Ag_2MoO_4	c	−200.9	−178.8	51	
AgN_3	c	73.8	89.9	24.9	

Table 9-1 (*Continued*)
ELEMENTS AND INORGANIC COMPOUNDS

Formula and Description	State	$\Delta Hf°$	$\Delta Gf°$	$S°$	$C_p°$
Ag_3N	c	47.6			
$Ag(NH_3)_2^+$ std. state, $m = 1$	aq	−26.60	−4.12	58.6	
$AgNO_2$	c	−10.77	4.56	30.64	19.17
$AgNO_3$	c	−29.73	−8.00	33.68	22.24
Ag_2O	c	−7.42	−2.68	29.0	15.74
AgO	c	−2.73	3.40	13.810	10.76
Ag_2O_2	c	−5.8	6.6	28	21
Ag_2O_3	c	8.1	29.0	24	
$AgOH$ std. state, $m = 1$	aq	−29.736	−19.161	14.80	−30.3
AgP_2	c	−11.0			
AgP_3	c	−16.6			
Ag_3PO_4	c		−210		
$Ag_4P_2O_7$	c	−453			
$AgReO_4$	c	−176	−151.9	36.6	
Ag_2S α, orthorhombic	c	−7.79	−9.72	34.42	18.29
β	c	−7.03	−9.43	36.0	
Ag_2SO_3	c	−117.3	−98.3	37.8	
std. state, $m = 1$	aq	−101.4	−79.4	27.8	
Ag_2SO_4	c	−171.10	−147.82	47.9	31.40
std. state, $m = 1$	aq	−166.85	−141.10	39.6	−60
Ag_2Se	c	−9	−10.6	36.02	19.54
Ag_2SeO_3	c	−87.3	−72.7	55.0	
Ag_2SeO_4	c	−100.5	−79.9	59.4	
Ag_2Te	c	−8.9	10.3	37.0	20.9
Ag_2WO_4	c	−221.2			
Sodium					
Na	g	25.98	18.67	36.72	
	c	0	0	12.2	
$NaAlO_2$	c	−273			
NaH_2AsO_3	aq, 400	−227.7			
NaH_2AsO_4	aq, 300	−273.4			
Na_2HAsO_4	aq, 400	−329.5			
Na_3AsO_3	aq, 500	−314.6			
Na_3AsO_4	c	−365			
	aq, 500	−381.5	−341.17		
$Na_3AsO_4 \cdot 12H_2O$	c	−1213.9			
$NaBO_2$	c	−253			
	aq, 300	−241.1			
$NaBO_3$ peroxyborate	aq	−220.0			
$NaBO_3 \cdot 4H_2O$	c	−504.8			
$Na_2B_4O_7$	c	−777.7			
	aq, 900	−787.9			
$Na_2B_4O_7 \cdot 4H_2O$	c	−1072.9			
$Na_2B_4O_7 \cdot 5H_2O$	c	−1143.5			
$Na_2B_4O_7 \cdot 10H_2O$	c	−1497.2			
Na_3BiO_4	c	−288			
NaBr	g	−36.33			
	c	−86.03			
	aq, 8	−86.89			
	aq, ∞	−86.18	−87.16	33.7	
$NaBr \cdot 2H_2O$	c	−227.25			
NaBrO	aq	−79.1			
$NaBrO_3$	aq, 400	−68.89	−57.59		
NaCN	c	−21.46			
$NaCN \cdot \frac{1}{2}H_2O$	c	−56.19			
$NaCN \cdot 2H_2O$	c	−162.25			
NaCNO cyanate	c	−95.6			
	aq, 2000	−90.9			
NaNHCN	aq	−34.5			

Table 9-1 (*Continued*)
ELEMENTS AND INORGANIC COMPOUNDS

Formula and Description	State	$\Delta Hf°$	$\Delta Gf°$	$S°$	$C_p°$
NaCNS thiocyanate	c	−41.73			
	aq, 3	−41.13			
	aq, 100	−40.16			
	aq, ∞	−40.1			
Na_2CO_3	c	−270.3	−250.4	32.5	
	aq, 15	−278.13			
	aq, 40	−277.30			
	aq, 200	−276.17			
	aq, 400	−275.9	−251.4		
$Na_2CO_3 \cdot H_2O$	c	−341.8			
$Na_2CO_3 \cdot 7H_2O$	c	−765.1			
$Na_2CO_3 \cdot 10H_2O$	c	−975.6			
$NaHCO_3$	c	−226.5	−203.6	24.4	
	aq	−222.5	−203.9		
$Na_2CO_3 \cdot NaHCO_3 \cdot 2H_2O$	c	−641.2			
$NaHCO_2$ formate	c	−155.03			
	aq, 400	−154.92			
$NaHCO_2 \cdot 2H_2O$	c	−296.6			
$NaHCO_2 \cdot 3H_2O$	c	−364.2			
$NaC_2H_3O_2$	c	−169.8			
	aq, 3	−171.20			
	aq, 6	−172.57			
	aq, 25	−173.63			
	aq, 6400	−174.07	−152.3		
	aq, ∞	−174.12			
$NaC_2H_3O_2 \cdot 3H_2O$	c	−383.50			
$Na_2C_2O_4$	c	−314.3			
	aq, 600	−310.5	−283.4		
$NaHC_2O_4$	c	−257.8			
	aq, 400	−252.7			
$NaHC_2O_4 \cdot H_2O$	c	−330.2			
NaCl	g	−43.50			
	c	−98.23	−91.79	17.30	
	aq, 8	−97.78			
	aq, 100	−97.25			
	aq, 400	−97.21	−93.92		
	aq, 1000	−97.23			
	aq, ∞	−97.30	−93.94	27.6	
NaClO	aq	−82.7			
$NaClO_2$	c	−72.65			
	aq, 1000	−73.80			
$NaClO_3$	c	−85.73			
	aq, 400	−80.74	−62.8		
	aq, 1000	−80.73			
	aq, 5000	−80.74			
	aq, ∞	−80.78	−63.21	53.4	
$NaClO_4$	c	−92.18			
	aq, 400	−88.76			
	aq, 5000	−88.67			
	aq, ∞	−88.69	−65.16	57.9	
Na_2CrO_4	c	−317.6			
	aq, 600	−320.6	−296		
$Na_2CrO_4 \cdot 4H_2O$	c	−601.3			
$Na_2Cr_2O_7$	aq, 600	−463.4	−431		
NaF	c	−136.0	−129.3	14.0	
	aq, ∞	−135.94	−128.67	12.1	
$NaHF_2$	c	−216.6			
	aq	−211.2			
NaH	g	29.88	24.78	44.93	
	c	−13.7	−9.3		

Table 9-1 (*Continued*)
ELEMENTS AND INORGANIC COMPOUNDS

Formula and Description	State	$\Delta Hf°$	$\Delta Gf°$	$S°$	$C_p°$
NaI	g	−20.94			
	c	−68.84			
	aq, 8	−71.50			
	aq, 15	−71.19			
	aq, 30	−70.82			
	aq, 100	−70.60			
	aq, 1000	−70.58			
	aq, 5000	−70.61			
	aq, ∞	−70.65	−70.94	40.5	
NaI · 2H$_2$O	c	−211.05			
NaIO$_3$	aq, 100	−111.93			
	aq, 500	−112.13	−94.8		
	aq, ∞	−112.2			
Na$_2$IrCl$_6$	c	−233.4			
Na$_2$MoO$_4$	c	−368			
	aq, 800	−368.6	−333		
NaNH$_2$	c	−28.4			
NaNO$_2$	c	−85.9			
	aq	−82.6	−71.0		
NaNO$_3$	c	−111.54	−87.45	27.8	
	aq, 6	−108.47			
	aq, 15	−107.81			
	aq, 100	−106.83			
	aq, 1000	−106.60			
	aq, 5000	−106.61			
	aq, ∞	−106.65	−89.00	49.4	
Na$_2$O	c	−99.4	−90.0	17.4	
Na$_2$O$_2$	c	−120.6	−105.0		
NaOH	c	−101.99	−90.60		
	aq, 3	−108.89			
	aq, 6	−111.52			
	aq, 25	−112.22			
	aq, 100	−112.11			
	aq, 300	−112.11			
	aq, 1000	−112.14			
	aq, ∞	−112.24	−100.18	11.9	
NaOH · H$_2$O	c	−175.17	−149.00	20.2	
NaPO$_3$	c	−288.6			
	aq, 600	−292.8			
Na$_3$PO$_3$	aq, 1000	−389.1			
NaH$_2$PO$_3$	c	−289.4			
	aq, 600	−290.5			
NaH$_2$PO$_3$ · 2½H$_2$O	c	−454.8			
Na$_2$HPO$_3$	c	−338.0			
	aq, 800	−347.5			
Na$_2$HPO$_3$ · 5H$_2$O	c	−684.2			
NaH$_2$PO$_4$	aq, 300	−367.7			
Na$_2$HPO$_4$	c	−417.4			
	aq, 200	−423.89			
	aq, 1000	−423.44			
Na$_2$HPO$_4$ · 2H$_2$O	c	−560.2			
Na$_2$HPO$_4$ · 7H$_2$O	c	−913.3			
Na$_2$HPO$_4$ · 12H$_2$O	c	−1266.4			
Na$_3$PO$_4$	c	−460			
	aq, 300	−475.0	−428.7		
	aq, 1000	−473.9			
Na$_3$PO$_4$ · 12H$_2$O	c	−1309.0			
NaNH$_4$HPO$_4$	aq, 500	−398.8			
NaNH$_4$HPO$_4$ · 4H$_2$O	c	−682.7			

Table 9-1 (*Continued*)
ELEMENTS AND INORGANIC COMPOUNDS

Formula and Description	State	$\Delta Hf°$	$\Delta Gf°$	$S°$	$C_p°$
$NaH_3P_2O_7$	c	−602.7			
	aq, 120	−603.9			
$NaH_3P_2O_7 \cdot H_2O$	c	−670.6			
$Na_2H_2P_2O_7$	c	−663.4			
	aq, 1500	−661.6			
$Na_2H_2P_2O_7 \cdot 6H_2O$	c	−1085.5			
$Na_3HP_2O_7$	c	−711.4			
	aq, 1500	−718.5			
$Na_3HP_2O_7 \cdot H_2O$	c	−788.2			
$Na_3HP_2O_7 \cdot 6H_2O$	c	−1135.7			
$Na_4P_2O_7$	c	−760.8			
	aq, 1500	−772.9			
$Na_4P_2O_7 \cdot 10H_2O$	c	−1468.2			
Na_2PbO_3	c	−205			
Na_2PtBr_6	c	−221.8			
	aq	−231.7			
$Na_2PtBr_6 \cdot 6H_2O$	c	−650.1			
Na_2PtCl_4	aq	−240.0	−216.8		
Na_2PtCl_6	c	−273.6			
	aq	−282.0			
$Na_2PtCl_6 \cdot 2H_2O$	c	−418.5			
$Na_2PtCl_6 \cdot 6H_2O$	c	−702.5			
Na_2PtI_6	aq	−170.1			
$NaReO_4$	c	−249.4			
	aq	−247.6			
Na_2S	c	−89.2			
	aq, 20	−105.42			
	aq, 100	−104.55			
	aq, 800	−104.36	−101.8		
$Na_2S \cdot 4\frac{1}{2}H_2O$	c	−416.9			
$Na_2S \cdot 5H_2O$	c	−452.7			
$Na_2S \cdot 9H_2O$	c	−736.7			
$NaHS$	c	−56.5			
	aq, 4	−61.60			
	aq, 8	−62.26			
	aq, 20	−61.77			
	aq, 100	−61.36			
	aq, 800	−61.25			
$NaHS \cdot 2H_2O$	c	−199.27			
Na_2S_2	aq	−104.6			
Na_2S_3	aq	−106.5			
Na_2S_4	c	−98.4			
Na_2SO_3	c	−260.6	−239.5	34.9	
	aq	−263.8			
$Na_2SO_3 \cdot 7H_2O$	c	−753.4			
$NaHSO_3$	aq	−206.6			
Na_2SO_4	c	−330.90	−302.78	35.73	
	aq, 100	−331.64			
	aq, 1000	−331.22			
	aq, 3000	−331.25			
	aq, ∞	−331.46	−302.52	32.9	
$Na_2SO_4 \cdot 10H_2O$	c	−1033.48	−870.93	141.7	
$NaHSO_4$	c	−269.2			
	aq, 200	−270.6			
$NaHSO_4 \cdot H_2O$	c	−339.2			
$Na_2SO_4 \cdot (NH_4)_2SO_4 \cdot H_2O$	c	−691.5			
$Na_2S_2O_3$	c	−267.0			
	aq	−269			
$Na_2S_2O_3 \cdot 5H_2O$ I	c	−621.89			
II	c	−620.60			

Table 9-1 (*Continued*)
ELEMENTS AND INORGANIC COMPOUNDS

Formula and Description	State	$\Delta Hf°$	$\Delta Gf°$	$S°$	$C_p°$
$Na_2S_2O_5$	c	-349.1			
	aq	-345			
$Na_2S_2O_6$	c	-399.9			
	aq	-394.6			
$Na_2S_2O_6 \cdot 2H_2O$	c	-542.5			
$Na_2S_3O_6$	aq	-409			
$Na_2S_3O_6 \cdot 3H_2O$	c	-623.0			
$Na_2S_4O_6$	aq	-405			
$Na_2S_4O_6 \cdot 2H_2O$	c	-550.0			
Na_3SbO_4	c	-352			
Na_2Se	c	-63.0			
	aq	-82.9	-89.4		
$Na_2Se \cdot 4\frac{1}{2}H_2O$	c	-398.2			
$Na_2Se \cdot 9H_2O$	c	-709.1			
$Na_2Se \cdot 16H_2O$	c	-1199.4			
NaHSe	c	-27.8			
	aq	-32.7			
Na_2SeO_3	aq	-236.6			
$NaHSeO_3$	aq	-180.7			
Na_2SeO_4	c	-258			
	aq	-259.7	-230.3		
$NaHSeO_4$	aq	-201.1			
Na_2SiO_3	c	-363	-341	27.2	
$Na_2SiO_3 \cdot 5H_2O$	c	-720.0			
$Na_2SiO_3 \cdot 9H_2O$	c	-1002.0			
Na_2SiF_6	c	-677			
	aq, 600	-671.2			
$NaHSiF_6$	aq, 400	-614.1			
Na_2SnO_3	c	-276			
Na_4SnO_4	aq, 1200	-455.5			
Na_2Te	c	-84.0			
Na_2Te_2	c	-101.5			
Na_2TeO_4	c	-313			
Na_2UO_4	c	-501			
$Na_2U_2O_7 \cdot 1\frac{1}{2}H_2O$	c	-880			
$(Na_2O_2)_2 \cdot UO_4$	aq	-596			
$(Na_2O_2)_2 \cdot UO_4 \cdot 9H_2O$	c	-1225			
Na_3VO_4	c	-420			
Na_2WO_4	c	-395			
	aq, 220	-380.9	-345		
Na_2ZnO_2	c	-188			
Strontium					
Sr	g	39.2	26.3	39.33	
	c	0	0	13.0	
$Sr_3(AsO_4)_2$	c	-800.7			
$SrHAsO_4$	aq	-344.8			
$Sr(H_2AsO_4)_2$	aq	-561.4			
$SrBr_2$	c	-171.7			
	aq, 100	-187.56			
	aq, ∞	-188.2	-182.1	29.2	
$SrBr_2 \cdot H_2O$	c	-246.2			
$SrBr_2 \cdot 6H_2O$	c	-604.4			
$SrCO_3$	c	-291.2	-271.9	23.2	
$Sr(HCO_3)_2$	aq, ∞	-460.7	-413.8	36.0	
$Sr(HCO_2)_2$ formate	c	-325.5			
	aq	-326.4			
$Sr(HCO_2)_2 \cdot 2H_2O$	c	-468.3			
$Sr(C_2H_3O_2)_2$	c	-356.7			
	aq	-362.7	-311.8		

Table 9-1 (*Continued*)
ELEMENTS AND INORGANIC COMPOUNDS

Formula and Description	State	$\Delta Hf°$	$\Delta Gf°$	$S°$	$C_p°$
$Sr(C_2H_3O_2)_2 \cdot \frac{1}{2}H_2O$	c	−391.2			
SrC_2O_4	aq, ∞	−326.1	−293.1	3	
$SrC_2O_4 \cdot H_2O$	c		−359.6		
$SrC_2O_4 \cdot 2\frac{1}{2}H_2O$	c	−502.9			
$Sr(CN)_2$	aq	−57.8	−54.5		
$Sr(CN)_2 \cdot 4H_2O$	c	−335.2			
$SrCl_2$	c	−198.0	−186.7	28	
	aq, ∞	−210.43	−195.7	16.9	
$SrCl_2 \cdot H_2O$	c	−271.7			
$SrCl_2 \cdot 2H_2O$	c	−343.7			
$SrCl_2 \cdot 6H_2O$	c	−627.1			
SrF_2	c	−290.3			
SrH	g	52.4	45.8	49.43	
SrH_2	c	−42.3			
SrI_2	c	−135.5			
	aq, ∞	−157.1	−157.7	42.9	
$SrI_2 \cdot H_2O$	c	−212.2			
$SrI_2 \cdot 2H_2O$	c	−282.8			
$SrI_2 \cdot 6H_2O$	c	−571.2			
$Sr(N_3)_2$	c	48.9			
Sr_3N_2	c	−93.4	−76.5		
$Sr(NO_2)_2$	c	−179.3			
	aq	−180.4			
$Sr(NO_3)_2$	c	−233.25			
	aq, 20	−231.95			
	aq, 200	−229.18			
	aq, 1000	−228.84			
	aq, 2000	−228.84			
	aq, ∞	−229.02	−185.8	60.06	
$Sr(NO_3)_2 \cdot 4H_2O$	c	−514.5			
SrO	c	−141.1	−133.8	13.0	
SrO_2	c	−153.6	−139.0		
$SrO_2 \cdot 8H_2O$	c	−722.6			
$Sr(OH)_2$	c	−229.3			
	aq, ∞	−240.29	−208.2	−14.4	
$Sr(OH)_2 \cdot H_2O$	c	−302.3			
$Sr(OH)_2 \cdot 8H_2O$	c	−801.2			
$Sr_3(PO_4)_2$	c	−987.3			
$SrHPO_4$	c	−431.3			
$Sr(H_2PO_4)_2 \cdot H_2O$	c	−819.4			
SrS	g	19			
	c	−108.1			
$SrSO_4$	c	−345.3	−318.9	29.1	
	aq, ∞	−347.38	−310.3	−5.3	
SrS_2O_6	aq	−410.8			
$SrS_2O_6 \cdot 4H_2O$	c	−693.1			
SrSe	c	−78.7			
$SrSi_2$	c	−150			
$SrSiO_3$	c	−371.2			
Sr_2SiO_4	c	−520.6			
$SrWO_4$	c	−398.3			
Sulfur					
S rhombic	c	0	0	7.60	5.41
monoclinic	c	0.08			
	g	66.636	56.951	40.084	5.658
S_2	g	30.68	18.96	54.51	7.76
S_8	g	24.45	11.87	102.98	37.39
S^{2-} std. state, $m = 1$	aq	7.9	20.5	−3.5	
S_2Br_2	liq	−3			

Table 9-1 (*Continued*)
ELEMENTS AND INORGANIC COMPOUNDS

Formula and Description	State	$\Delta Hf°$	$\Delta Gf°$	$S°$	$C_p°$
SCl_2	liq	−12			
	g	−4.7			
S_2Cl_2	g	−4.4	−7.6	79.2	17.6
S_3Cl_2	liq	−12.4			
S_4Cl_2	liq	−10.2			
S_5Cl_2	liq	−8.8			
SCl_4	liq	−13.7			
SF_4	g	−185.2	−174.8	69.77	17.45
SF_6	g	−289	−264.2	69.72	23.25
std. state, $m = 1$	aq	−293.0	−259.3	39.8	
S_2F_{10}	liq	(−485)			
SF_2Cl	g	−250.5	−226.9	76.26	24.9
HS	g	34.10	27.08	46.74	7.72
HS^- std. state, $m = 1$	aq	−4.2	2.88	15.0	
H_2S	g	−4.93	−8.02	49.16	8.18
std. state, $m = 1$	aq	−9.5	−6.66	29	
H_2S_2	liq	−4.33			20.1
	g	3.71			
H_2S_3	liq	−3.57			
	g	7.29			
H_2S_4	liq	−2.99			
	g	10.57			
H_2S_5	liq	−2.49			
	g	13.84			
H_2S_6	liq	−1.99			
SO	g	1.496	−4.741	53.02	7.21
SO_2	liq	−76.6			
	g	−70.944	−71.748	59.30	9.53
undissoc; std. state, $m = 1$	aq	−77.194	−71.871	38.7	
	aq	−80.584			
SO_3 β	c	−108.63	−88.19	12.5	
	liq	−105.41	−88.04	22.85	
	g	−94.58	−88.69	61.34	12.11
SO_3^{2-} std. state, $m = 1$	aq	−151.9	−116.3	−7	
SO_4^{2-} std. state, $m = 1$	aq	−217.32	−177.97	4.8	−70
$S_2O_3^{2-}$	aq	−155.9			
$S_2O_4^{2-}$ std. state, $m = 1$	aq	−180.1	−143.5	22	
$S_2O_6^{2-}$	aq	−286.4			
$S_2O_7^{2-}$	aq	−334.9			
$S_2O_8^{2-}$ std. state, $m = 1$	aq	−320.0	−265.4	59.3	
$S_3O_6^{2-}$	aq	−286.7			
$S_4O_6^{2-}$	aq	−292.58			
$SOBr_2$	g	−21.8			
$SOCl_2$	liq	−58.7			
	g	−50.8	−47.4	74.01	15.9
SO_2Cl_2	liq	−94.2			
	g	−87.0	−76.5	74.53	18.4
$S_2O_5Cl_2$	liq	−168.7			56
SOF_2	g			66.58	13.58
SO_2F_2	g			67.86	15.78
SO_3F^-	aq	−193.0			
HSO_3^- std. state, $m = 1$	aq	−149.67	−126.15	33.4	
HSO_4^- std. state, $m = 1$	aq	−212.08	−180.69	31.5	−20
H_2SO_3 undissoc; std. state, $m = 1$	aq	−145.51	−128.56	55.5	
	aq	−148.899			
H_2SO_4	liq	−194.548	−164.938	37.501	33.20
std. state, $m = 1$	aq	−217.32	−177.97	4.8	−70
	aq, 1	−201.193			
	aq, 2	−204.455			

Table 9-1 (*Continued*)
ELEMENTS AND INORGANIC COMPOUNDS

Formula and Description	State	$\Delta Hf°$	$\Delta Gf°$	$S°$	$C_p°$
	aq, 5	−208.288			
	aq, 10	−210.451			
	aq, 100	−212.150			
	aq, 1000	−213.275			
$H_2SO_4 \cdot H_2O$	liq	−269.508	−227.182	50.56	51.35
$H_2SO_4 \cdot 2H_2O$	liq	−341.085	−286.770	66.06	62.34
$HS_2O_4^-$ std. state, $m = 1$	aq		−146.9		
$H_2S_2O_4$ std. state, $m = 1$	aq		−147.4		
$H_2S_2O_6$	aq	−286.4			
$H_2S_2O_7$	c	−304.4			27
$H_2S_2O_8$ std. state, $m = 1$	aq	−320.0	−265.4	59.3	
HSO_3Cl	liq	−143.7			
HSO_3F	liq	−186			28.00
$SO_2(NH_2)_2$ sulfamide	c	−129.3			
Tantalum					
Ta	c	0	0	9.92	6.06
	g	186.9	176.7	44.241	4.985
TaB_2	c	−46			
TaC	c	−35.0	−34.6	10.11	8.79
Ta_2C	c	−51.0	−50.8	20.7	
$TaBr_5$	c	−143.0			
	g	−115.6			
$TaCl_3$	c	−132.2			
$TaCl_4$	c	−167.7			
	g	−134.0			
$TaCl_5$	c	−205.3			
	g	−181.3			
TaF_5	c	−454.97			
std. state, undissoc	aq		−137.6		
TaF_6^- std. state	aq		−209.2		
TaF_7^{2-} std. state	aq		−280.2		
Ta_2H	c	−7.8	−16.5	18.9	21.7
TaN	c	−60.1			9.7
Ta_2N	c	−65			
TaO	g	60	53	57.6	7.31
TaO_2	g	−41	−43	64	12.5
Ta_2O_5 β	c	−489.0	−456.8	34.2	32.30
	aq	−496.7			
$TaOCl_3$	g	−186.6			
TaS_2	c	−111			
$TaSi_2$	c	−28			
Ta_5Si_3	c	−76			
Technetium					
Tc	c	0	0		
	g	162		43.25	4.97
Tc_2O_7	c	−266			
$HTcO_4$	c	−167			
std. state, $m = 1$	aq	−173			
Tellurium					
Te	c	0	0	11.88	6.15
	g	47.02	37.55	43.65	4.968
	amorp	2.7			
Te_2	g	40.2	28.2	64.06	8.78
$TeBr_4$	c	−45.5			
$TeCl_4$	c	−78.0			33.1
TeF_6	g	−315			
H_2Te	g	23.8			
TeO	g	15.6	9.2	57.7	7.19
TeO_2	c	−77.1	−64.6	19.0	

Table 9-1 (*Continued*)
ELEMENTS AND INORGANIC COMPOUNDS

Formula and Description	State	$\Delta Hf°$	$\Delta Gf°$	$S°$	$C_p°$
TeO_3^{2-}	aq	−142.6			
H_2TeO_3 std. state, $m = 1$	aq		−76.2		
$Te(OH)_3^+$ std. state, $m = 1$	aq	−145.4	−118.6	26.7	
H_6TeO_6	c	−310.4			
TeSe	q	38.0	26.0	63.5	
Terbium					
Tb	g	87			
	c	0	0		
$TbCl_3$ β	c	−241.6			
	aq, ∞	−288.5	−257.9		
$Tb_2(SO_4)_3$	aq, ∞	−987.5	−859.8		
$Tb_2(SO_4)_3 \cdot 8H_2O$	c		−1328.2		
Thallium					
Tl	c	0	0	15.34	6.29
	g	43.55	35.24	43.225	4.968
TlBr	c	−41.4	−40.00	28.8	
	g	−9.0			
std. state, $m = 1$	aq	−27.77	−32.59	49.7	
$TlBr^{2+}$ std. state, $m = 1$	aq	9.0	13.5	−13	
$TlBr_2^+$ std. state, $m = 1$	aq	−26.1	−21.4	20	
$TlBr_2^-$ std. state, $m = 1$	aq	−53.8	−58.8	84	
$TlBr_3$ undissoc; std. state	aq	−59.7	−53.7	49	
std. state, $m = 1$	aq	−40.2	−23.2	13	
$TlBr_4^-$ std. state, $m = 1$	aq	−90.9	−84.2	80	
$TlBrO_3$	c	−32.6	−12.70	40.3	
std. state, $m = 1$	aq	−18.7	−7.3	69.0	
TlCl	c	−48.79	−44.20	26.59	12.17
	g	−16.2			
std. state, $m = 1$	aq	−36.67	−39.11	43.5	
undissoc; std. state, $m = 1$	aq	−41.10	−39.91	41.3	
$TlCl^{2+}$ std. state, $m = 1$	aq	1.0	9.7	−19	
$TlCl_2^+$ std. state, $m = 1$	aq	−43.0	−29.6	7	
$TlCl_2^-$ std. state, $m = 1$	aq		−70.7		
$TlCl_3$	c	−75.3			
std. state, $m = 1$	aq	−72.9	−42.8	−5.5	
undissoc; std. state, $m = 1$	aq	−84.0	−65.6	32	
$TlCl_4^-$ std. state, $m = 1$	aq	−124.1	−100.8	58	
$TlClO_3$	aq	−22.4	−8.5	68.8	
Tl_2CO_3	c	−167.3	−146.9	37.1	
Tl(I) acetate	c	−126.1			
Tl(I) fulminate	c	27.6			
TlCNS thiocyanate	c	6.8	9.21	39	
std. state, $m = 1$	aq	19.55	14.41	64.5	
Tl_2CrO_4	c	−225.8	−205.9	67.5	
TlF	c	−77.6			
	g	−43.6			
std. state, $m = 1$	aq	−78.22	−74.38	26.7	
$TlHF_2$	c			34.92	21.35
TlI	c	−29.6	−29.97	30.5	
	g	1.7			
std. state, $m = 1$	aq	−11.91	−20.07	56.6	
TlI_2^- std. state, $m = 1$	aq		−35.1		
TlI_4^- std. state, $m = 1$	aq		−39.3		
$TlIO_3$	c	−63.9	−45.86	42.2	
std. state, $m = 1$	aq	−51.6	−38.3	58.3	
TlN_3	c	55.8	70.38	35.1	
$TlNO_3$	c	−58.30	−36.44	38.4	23.78
std. state, $m = 1$	aq	−48.28	−34.35	65.0	
Tl_2O	c	−42.7	−35.2	30	

Table 9-1 (*Continued*)
ELEMENTS AND INORGANIC COMPOUNDS

Formula and Description	State	$\Delta Hf°$	$\Delta Gf°$	$S°$	$C_p°$
Tl$_2$O$_3$	c	−74.5			
Tl$_2$O$_4$	c	−83.0			
TlOH	c	−57.1	−46.8	21	
std. state, $m = 1$	aq	−53.69	−45.33	27.4	
Tl(OH)$_3$	c		−121.2		
Tl$_2$S	c	−23.2	−22.4	36	
Tl$_2$SO$_4$	c	−222.7	−198.49	55.1	
Tl$_2$Se	c	−14	−14.1	41	
Tl$_2$SeO$_4$	c	−151	−126.4	56	
Tl$_2$Te	c	−22			
Thorium					
Th	c	0	0	13.6	
ThBr$_4$	c	−227.1			
	aq	−298.6			
ThBr$_4 \cdot$ 7H$_2$O	c	−753.7			
ThBr$_4 \cdot$ 10H$_2$O	c	−971.6			
ThBr$_4 \cdot$ 12H$_2$O	c	−1116.0			
ThOBr$_2$	c	−252.8			
ThC$_2$	c	−45			
ThCl$_4$	c	−285			
	aq	−343.0			
ThCl$_4 \cdot$ 2H$_2$O	c	−437.4			
ThCl$_4 \cdot$ 4H$_2$O	c	−589.2			
ThCl$_4 \cdot$ 7H$_2$O	c	−805.9			
ThCl$_4 \cdot$ 8H$_2$O	c	−877.6			
ThCl$_4 \cdot$ NH$_4$Cl	c	−373.6			
ThCl$_4 \cdot$ 2NH$_4$Cl $\cdot$ 10H$_2$O	c	−1172.6			
ThOCl$_2$	c	−274.8			
Th(OH)Cl$_3 \cdot$ H$_2$O	c	−377.4			
ThF$_4$	c	−477			
ThH$_4$	c	−43			
ThI$_4$	c	−131			
ThOI$_2$	c	−228.2			
ThOI$_2 \cdot$ 3½H$_2$O	c	−479.3			
Th(OH)I$_3 \cdot$ 10H$_2$O	c	−952.4			
Th$_3$N$_4$	c	−308	−282		
Th(NO$_3$)$_4$	aq, 20	−379.10			
	aq, 50	−380.36			
	aq, 100	−380.48			
	aq, 500	−380.5			
ThO$_2$	c	−292	−280.1		
Th(OH)$_4$	c*	−421.5			
Th$_2$S$_3$	c	−262.0			
Th(SO$_4$)$_2$	c	−602			
	aq	−616.8			
Th(SO$_4$)$_2 \cdot$ 4H$_2$O	c	−882.7			
Th(SO$_4$)$_2 \cdot$ 8H$_2$O	c	−1168.6			
ThOSO$_4$	c	−487			
Thulium					
Tm	c	0	0		
TmCl$_3$ γ	c	−229.5			
	aq, ∞	−281.4	−250.5		
TmI$_3$ β	c	−137.8			
	aq, ∞	−201.4	−193.5		
Tin					
Sn I, white	c	0	0	12.32	6.45
II, gray	c	−0.50	0.03	10.55	6.16
	g	72.2	63.9	40.243	5.081
SnBr$_2$	c	−58.2			

Table 9-1 (*Continued*)
ELEMENTS AND INORGANIC COMPOUNDS

Formula and Description	State	$\Delta Hf°$	$\Delta Gf°$	$S°$	$C_p°$
SnBr$_4$	c	−90.2	−83.7	63.2	
	g	−75.2	−79.2	98.43	24.71
SnCl$_2$	c	−79.3			
std. state, $m = 1$	aq	−78.8	−71.6	41	
SnCl$_2 \cdot 2H_2O$	c	−220.2			
SnCl$_3^-$ in aq HCl, std. state	aq	−116.4	−102.8	62	
SnCl$_4$	liq	−122.2	−105.2	61.8	39.5
SnH$_4$	g	38.9	45.0	54.39	11.70
Sn$_2$H$_6$	g	65.6			
SnI$_2$	c	−34.3			
SnI$_4$	c				20.3
SnO	c	−68.3	−61.4	13.5	10.59
	g			55.45	7.55
SnO$_2$	c	−138.8	−124.2	12.5	12.57
Sn(OH)$_2$ pptd	c	−134.1	−117.5	37	
Sn(OH)$_4$ pptd	c	−265.3			
	g			106.6	25.2
SnS	c	−24	−23.5	18.4	11.77
	g	28.5			
SnS$_2$	c			20.9	16.76
Sn(SO$_4$)$_2$	c	−389.4			
std. state, $m = 1$	aq		−354.2		
SnSe	c	−21.7			
	g	30.8			
SnTe	c	−14.6			
	g	38.4			
Titanium					
Ti	c	0	0	7.32	5.98
TiAs	c	−35.8			
TiBr$_2$	c	−96			
TiBr$_3$	c	−131.1	−125.2	42.2	24.31
TiBr$_4$	c	−147.4	−140.9	58.2	31.43
	g	−131.3	−135.8	95.2	24.1
TiB$_2$	c	−77.4	−76.4	6.81	10.58
TiC	c	−44.1	−43.2	5.79	8.04
TiCl$_2$	c	−122.8	−111.0	20.9	16.69
TiCl$_3$	c	−172.3	−156.2	33.4	23.22
TiCl$_4$	liq	−192.2	−176.2	60.31	34.70
	g	−182.4	−173.7	84.8	22.8
	aq, 1600	−250.3			
TiF$_4$	amorp	−394.2	−372.7	32.02	27.31
	g	−371.0			
H$_2$TiF$_6$	aq	−573.7			
TiH$_2$	c	−28.6	−19.2	7.1	7.2
TiI$_2$	c	−63			
	g	−13			
TiI$_4$	c	−89.8	−88.8	59.6	30.03
	g	−66.4			
TiN	c	−80.8	−74.0	7.23	8.86
TiO α	c	−124.2	−118.3	8.31	9.55
	g	4	−3	56.0	7.81
TiO^{2+} in HClO$_4$ medium	aq	−164.9			
TiO$_2$ anatase	c	−224.6	−211.4	11.93	13.26
Brookite	c	−225.1			
rutile	c	−225.8	−212.6	12.03	13.15
	amorp	−210			
hydrated ppt		−219.8			
Ti$_2$O$_3$	c	−363.5	−342.8	18.83	23.27
Ti$_3$O$_5$	c	−587.8	−553.9	30.9	37.00

Table 9-1 (*Continued*)
ELEMENTS AND INORGANIC COMPOUNDS

Formula and Description	State	$\Delta Hf°$	$\Delta Gf°$	$S°$	$C_p°$
TiOCl	c	−180			
TiP	c	−67.6			
TiS	c	−57			
TiS$_2$	c			18.73	16.23
TiSi	c	−31			
TiSi$_2$	c	−32			
Ti$_5$Si$_3$	c	−138			
Tungsten					
W	c	0	0	7.80	5.80
	g	203.0	192.9	41.549	5.093
WBr$_5$	c	−75.6			
WBr$_6$	c	−83.3			
WO$_2$Br$_2$	c	−170.3			
WOBr$_4$	c	−130.1			
WC	c	−9.69			
W$_2$C	c	−6.3			
W(CO)$_6$	c	−227.9			
WCl$_2$	c	−61			
WCl$_4$	c	−112			
	g	−73			
WCl$_5$	c	−118.6			
	g	−100.8			
WCl$_6$	c	−144.0			
	g	−122.8			
W$_2$Cl$_{10}$	g	−210			
WO$_2$Cl$_2$	c	−187.2			
	g	−164			
WOCl$_4$	c	−161.7			
	g	−140			
WF$_6$	c	−418.2			
	liq	−417.7	−389.93	60.1	
	g	−411.5	−390.1	81.49	28.45
WO$_2$	c	−140.94	−127.61	12.08	13.41
WO$_3$	c	−201.45	−182.62	18.14	17.63
WO$_4^{2-}$ std. state	aq	−257.1			
H$_2$WO$_4$	c	−270.5			
	g	−229			
WS$_2$	c	−50			
WSi$_2$	c	−22			
Uranium					
U	g	125			
	c	0	0	12.03	
UBr$_3$	c	−170.1	−164.7	49	
UBr$_4$	c	−196.6	−188.5	58	
UC$_2$	c	−42	−42	14	
UCl$_3$	c	−213.0	−196.9	37.99	
UCl$_4$	c	−251.2	−230.0	47.4	
UCl$_5$	c	−262.1	−237.4	62	
UCl$_6$	c	−272.4	−241.5	68.3	
UF$_3$	c	−357	−339	26	
UF$_4$	c	−443	−421	36.1	
UF$_5$	c	−488	−461	43	
UF$_6$	g	−505	−485	90.76	
UH$_3$	c	−30.4			
UI$_3$	c	−114.7	−115.3	56	
UI$_4$	c	−127.0	−126.1	65	
UIBr$_3$	c	−177.1			
UICl$_3$	c	−219.9	−204.4	54	
UN	c	−80	−75	18	

Table 9-1 (*Continued*)
ELEMENTS AND INORGANIC COMPOUNDS

Formula and Description	State	$\Delta Hf°$	$\Delta Gf°$	$S°$	$C_p°$
U_2N_3	c	−213	−194	29	
UO_2	c	−270	−257	18.6	
UO_3	c	−302	−283	23.57	
$UO_3 \cdot H_2O$	c	−375.4			
$UO_3 \cdot 2H_2O$	c	−446.2			
$UO_4 \cdot 2H_2O$	c	−436			
U_3O_8	c	−898			
$U(SO_4)_2$	c	−563			
UO_2Br_2	aq	−308.2			
$UO_2(C_2H_3O_2)_2$	aq	−484.0			
$UO_2(C_2H_3O_2)_2 \cdot 2H_2O$	c	−624.9			
$UO_2(C_2H_3O_2)_2 \cdot$ $NH_4C_2H_3O_2 \cdot 6H_2O$	c	−1045.8			
UO_2Cl_2	aq	−331			
UO_2CrO_4	aq	−456.7			
$UO_2CrO_4 \cdot 5\frac{1}{2}H_2O$	c	−838.8			
$UO_2(NO_3)_2$	c	−329.2	−273.1	66	
	aq, ∞	−349.1	−289.2	53	
$UO_2(NO_3)_2 \cdot H_2O$	c	−404.8	−335.3	76	
$UO_2(NO_3)_2 \cdot 2H_2O$	c	−480.0	−396.6	85	
$UO_2(NO_3)_2 \cdot 3H_2O$	c	−552.2	−454.7	94	
$UO_2(NO_3)_2 \cdot 6H_2O$	c	−764.3	−625.0	120.85	
UO_2SO_4	aq, ∞	−467.3	−413.7	−13	
$UO_2SO_4 \cdot 3H_2O$	c	−666.8	−586.0	63	
Vanadium					
V	c	0	0	6.91	5.95
	g	122.90	108.32	43.544	6.217
VBr_2	c	−87.3			
	g	−37.1			
VBr_3	c	−103.6			
	g	−56.8			
VBr_4	g	−80.5			
VCl_2	c	−108	−97	23.2	17.26
	g	−61.3			
VCl_3	c	−138.8	−122.2	31.3	22.27
VCl_4	liq	−136.1	−120.4	61	
	g	−125.6	−117.6	86.6	23.0
$V(CO)_6$	g	−236			
VF_3	c			23.18	21.62
VF_4	c	−335.4			
VF_5	liq	−353.8	−328.2	42.0	
	g	−342.7	−327.4	76.67	23.56
VI_2	c	−60.1			
	g	−1			
VI_3	c	−64.7			
VI_4	g	−29.3			
VN	c	−51.9	−45.7	8.91	9.08
VO	c	−103.2	−96.6	9.3	10.86
	g	25	18	55.8	7.3
VO^{2+} std. state	aq	−116.3	−106.7	−32.0	
VO_2	g	−57			
VO_2^+ std. state	aq	−155.3	−140.3	−10.1	
VO_3^- std. state	aq	−212.3	−187.3	12	
V_2O_3	c	−293.5	−272.3	23.5	24.67
V_2O_4 α	c	−341.1	−315.1	24.5	27.96
V_2O_5	c	−370.6	−339.3	31.3	30.51
V_3O_5	c	−465	−434	39	
V_4O_7	c	−635	−591	52	
V_6O_{13}	c	−1062			

Table 9-1 (*Continued*)
ELEMENTS AND INORGANIC COMPOUNDS

Formula and Description	State	$\Delta Hf°$	$\Delta Gf°$	$S°$	$C_p°$
HVO$_4^{2-}$ std. state	aq	−277.0	−233.0	4	
H$_2$VO$_4^-$ std. state	aq	−280.6	−244.0	29	
[VO$_2$H$_2$O$_2$]$^+$ std. state	aq		−178.4		
VOCl	c	−138.4			
VO$_2$Cl	c	−185.6			
VOCl$_2$	c	−165.0			
VOCl$_3$	liq	−175.6	−159.8	58.4	
	g	−166.25	−157.58	82.26	21.49
VOSO$_4$	c	−312.9	−279.6	26.0	
std. state, undissoc	aq		−282.6		
VOSCN$^+$ std. state	aq	−98	−86	8	
V$_2$S$_3$	c	−227			
VSi$_2$	c	−73			
V$_2$Si	c	−37			
V$_3$Si	c	−26			
V$_5$Si$_3$	c	−94			
Xenon					
Xe	g	0	0	40.529	4.968
XeF$_4$	c	−62.5	−29.4		
	g	−51.5	−50		28.334
XeF$_2$	c	−39.2			
	g	−25.9	−37		
XeF$_6$	s	−86			
	g	−71			
XeO$_3$	c	96			
XeOF$_4$	liq	35			
XeO$_2$F$_2$	c	35			
Ytterbium					
Yb	c	0	0		
	g	87		41.30	
YbBr$_3$	c	−185			
YbCl$_3$	c	−228.7			
	aq	−280.7	−249.5		
Yb$_2$(SO$_4$)$_3$	aq	−971.9	−843.0		
Yb$_2$(SO$_4$)$_3 \cdot$ 8H$_2$O	c		−1308.8		
Yttrium					
Y	c	0	0	10.62	6.34
	g	100.7	91.1	42.87	6.18
Y$_2$	g	163.5	150.7	64	8.7
Y^{3+} std. state, $m = 1$	aq	−172.9	−156.8	−60	
YBr^{2+} std. state, $m = 1$	aq	−203.8	−191.6	−43	
YC$_2$	c	−26	−26	13	
	g	142.6	128.4	61	10.7
Y$_2$(CO$_3$)$_3$			−752.4		
Y$_2$(C$_2$O$_4$)$_3 \cdot$ 9H$_2$O	c		−1363.8		
Y(acetate)$_3$	aq	−516.1			
Y(SCN)$^{2+}$ std. state, $m = 1$	aq		−144.7		
YCl	g	47.8	41.5	58.33	8.56
YCl^{2+} std. state, $m = 1$	aq	−214.0	−198.7	−46	
YCl$_3$	c	−239.0			
	g	−179.3			18
	aq, 4000	−291.96			
YCl$_3 \cdot$ 6H$_2$O	c	−691.3	−592.1	92	
YF	g	−33	−39	55.38	7.92
YF$_3$	c	−410.8	−393.1	24	
	g	−308.0	−305.4	74.5	16.8
YH$_2$	c	−37.5	−27.8	9.17	8.24
YH$_3$	c	−47.3	−33.2	10.02	10.36
YI$_3$	c	−147.4			
Y(IO$_3$)$_3$	c		−271.2		

Table 9-1 (*Continued*)
ELEMENTS AND INORGANIC COMPOUNDS

Formula and Description	State	$\Delta Hf°$	$\Delta Gf°$	$S°$	$C_p°$
YO	g	−9.3	−15.5	55.88	7.53
Y_2O_2	g	−127.4			15.8
Y_2O_3	c	−455.38	−434.19	23.68	24.50
$Y(OH)^{2+}$ std. state, $m = 1$	aq		−210.1		
$Y(OH)_3$	c		−308.6		
$Y(OH)_2Cl$	c		−297.9		
$Y(ReO_4)_3$	c	−701.9	−629.4	88	
YS	g	41.7	29.7	58	8.2
Zinc					
Zn	c	0	0	9.95	6.07
	g	31.245	22.748	38.450	4.968
$ZnAs_2$	c	−7.6			
Zn_3As_2	c	−0.8			
$Zn(BO_2)_2$	c		−373.58		
$ZnBr_2$	c	−78.55	−74.60	33.1	
std. state, $m = 1$	aq	−94.88	−84.84	12.6	−57
$ZnBr_2 \cdot 2H_2O$	c	−224.0	−191.1	47.5	
$ZnCO_3$	c	−94.26	−174.85	19.7	19.05
ZnC_2O_4 std. state, $m = 1$	aq	−234.0	−196.2	−15.9	
$ZnC_2O_4 \cdot 2H_2O$	c	−374.0	−321.7	46.7	
$Zn(C_2O_4)_2^{2-}$ std. state, $m = 1$	aq	−430.7	−367.5	31	
$Zn(formate)_2$	c	−235.8			
std. state, $m = 1$	aq	−240.20	−202.9	17	
$Zn(acetate)_2$	c	−257.8			
std. state, $m = 1$	aq	−269.10	−211.72	14.6	
$Zn(CN)_2$	c	22.9			
$Zn(CN)_4^{2-}$ std. state, $m = 1$	aq	81.8	106.8	54	
$ZnCl_2$	c	−99.20	−88.296	26.64	17.05
	g	−63.6			
std. state, $m = 1$	aq	−116.68	−97.88	0.2	−54
$ZnCl_3^-$ std. state, $m = 1$	aq		−129.2		
$ZnCl_4^{2-}$ std. state, $m = 1$	aq		−159.2		
$Zn(ClO_4)_2$ std. state, $m = 1$	aq	−98.60	−39.26	60.2	
$Zn(ClO_4)_2 \cdot 6H_2O$	c	−509.89	−371.8	130.4	
ZnF_2	c	−182.7	−170.5	17.61	15.69
std. state, $m = 1$	aq	−195.78	−168.42	−33.4	−40
ZnI_2	c	−49.72	−49.94	38.5	
std. state, $m = 1$	aq	−63.16	−59.80	25.2	−57
$Zn(IO_3)_2$	c		−103.68		
std. state, $m = 1$	aq	−142.6	−96.3	29.8	
$Zn(N_3)_2$	c	52			
Zn_3N_2	c	−5.4			26
$Zn(NO_3)_2$	c	−115.6			
ionized, std. state	aq	−135.90	−88.36	43.2	−30
$Zn(NO_3)_2 \cdot 6H_2O$	c	−551.30	−423.79	109.2	77.2
$Zn(NH_3)_4^{2+}$ std. state, $m = 1$	aq	−127.5	−72.2	72	
ZnO	c	−83.24	−76.08	10.43	9.62
ZnO_2^{2-} std. state, $m = 1$	aq		−91.85		
$ZnO_2 \cdot 2H_2O$	c	−207.6			
$Zn(OH)^+$ std. state, $m = 1$	aq		−78.9		
$HZnO_2^-$ std. state, $m = 1$	aq		−109.26		
$Zn(OH)_2$ β	c	−153.42	−132.31	19.4	
ϵ	c	−152.74	−132.68	19.5	17.3
std. state, $m = 1$	aq	−146.72	−110.33	−31.9	−60
$Zn(OH)_3^-$ std. state, $m = 1$	aq		−165.95		
$Zn(OH)_4^{2-}$ std. state, $m = 1$	aq		−205.23		
Zn_3P_2	c	−113			
$Zn(PO_3)_2$	c	−497.9			
$Zn_2P_2O_7$	c	−600.0			
$Zn_3(PO_4)_2$	c	−691.3			

Table 9-1 (*Continued*)
ELEMENTS AND INORGANIC COMPOUNDS

Formula and Description	State	$\Delta Hf°$	$\Delta Gf°$	$S°$	$C_p°$
ZnS wurtzite	c	−46.04			
sphalerite	c	−49.23	−48.11	13.8	11.0
$ZnSO_4$	c	−234.9	−209.0	28.6	
std. state, $m = 1$	aq	−254.10	−213.11	−22.0	−59
$ZnSO_4 \cdot H_2O$	c	−311.78	−270.58	33.1	
$ZnSO_4 \cdot 6H_2O$	c	−663.83	−555.64	86.9	85.49
$ZnSO_4 \cdot 7H_2O$	c	−735.60	−612.59	92.9	91.64
ZnS_2O_6	aq	−323.2			
$ZnS_2O_6 \cdot 6H_2O$	c	−735.2			
ZnSb	c	−3.5			
ZnSe	c	39	39	20	
$ZnSeO_3$ std. state, $m = 1$	aq	−158.5	−123.5	−23.7	
$ZnSeO_3 \cdot H_2O$	c	−222.5	−189.5	39	
$ZnSeO_4$	c	−158.8			
std. state, $m = 1$	aq	−180.0	−140.6	−13.9	
$ZnSeO_4 \cdot 6H_2O$	c	−587.5			
$ZnSiO_3$	c	−301.2			
Zn_2SiO_4	c	−391.19	−364.06	31.4	29.48
ZnTe	c	−28.1			
$ZnWO_4$	c	−293			30.0
Zirconium					
Zr α, hexagonal	c	0	0	9.32	6.06
	g	145.5	135.4	43.32	6.37
ZrB_2	c	−78.0	−77.0	8.59	11.53
$ZrBr_4$	c	−181.8			
	g	−153.8			
ZrC	c	−48.5	−47.7	7.96	9.06
ZrCl	c	−63			
	g	60			
$ZrCl_2$	c	−120			
$ZrCl_3$	c	−179			
$ZrCl_4$	c	−234.35	−212.7	43.4	28.63
	g	−208.0	−199.7	88.0	23.49
	aq	−293.3			
ZrF_2	g	−135			
ZrF_3	g	−265			
ZrF_4 β, monoclinic	c	−456.8	−432.6	25.00	24.79
	g	−400.0	−391.1	76.3	20.9
ZrH_2	c	−40.4	−30.8	8.37	7.40
ZrI_4	c	−115.1			
	g	−84.1			
ZrN	c	−87.2	−80.4	9.29	9.66
	g	134			
ZrO	g	15			
ZrO_2 α, monoclinic	c	−263.04	−249.24	12.04	13.43
hydrated precipitate			−260.4		
ZrO_3 precipitate		−241			
$ZrO(OH)^+$	aq	−270.1			
$ZrOBr_2$	aq	−259.9			
$ZrOBr_2 \cdot 8H_2O$	c	−808.4			
$ZrOCl_2$	aq	−280.3			
$ZrOCl_2 \cdot 8H_2O$	c	−829.8			
ZrS_2	c	−135.3			
$Zr(SO_4)_2$	c	−529.9			
$Zr(SO_4)_2 \cdot 4H_2O$	c	−825.6			
$ZrO(SO_4)_2^{2-}$	aq	−630.1			
ZrSi	c	−37			
$ZrSi_2$	c	−38			
$ZrSiO_4$	c	−486.0	−458.7	20.1	23.58

Table 9-2
ORGANIC COMPOUNDS (CARBON)

Substance	State	$\Delta Hf°$	$\Delta Gf°$	$S°$	$C_p°$
Acetaldehyde	liq	−45.96	−30.64	38.3	
	g	−39.72	−30.81	59.8	13.7
	aq, 1000	−50.35			
Acetaldoxime (high-melting)	c	−18.6			
(low-melting)	liq	−19.5			
Acetamide	c	−76.0			16
Acetic acid	liq	−115.8	−93.2	38.2	29.7
	g	−103.31	−89.4	67.5	15.9
ionized, std. state, $m = 1$	aq	−116.16	−88.29	20.7	−1.5
nonionized; std. state, $m = 1$	aq	−116.10	−94.78	42.7	
	aq, 10	−115.807			
Acetic anhydride	liq	−155.16	−121.75		
	g	−148.82	−119.29		
Acetone	liq	−59.32	−37.22		
	g	−51.79	−36.50		
Acetonitrile	liq	12.8	23.7	35.76	21.86
	g	20.9	25.0	58.67	12.48
Acetyl bromide	liq	−53.39			
Acetyl chloride	liq	−65.44	−49.73	48.0	28
	g	−58.20	−49.20	70.5	16.2
Acetyl fluoride	liq	−110.83			
	g	−104.9			
Acetyl iodide	liq	−38.84			
Acetylene	g	54.19	50.00	48.00	10.50
std. state, $m = 1$	aq	50.54	51.88	29.5	
Acrolein	liq	−27.97	−16.17		
	g	−20.50	−15.45		
Adipic acid	liq	−235.51	−177.17		
	g	−216.19	−163.96		
5-Aminotetrazole	c	49.7			
n-Amyl acetate	liq		−77.90		
n-Amyl alcohol (1-pentanol)	liq	−85.35	−38.63	61.9	49.8
	g	−71.74	−35.22	96.1	31.8
sec-Amyl alcohol (2-methyl-2-butanol)	liq	−90.66	−41.83	54.8	59.1
	g	−78.7	−39.6	87.5	
tert-Amyl alcohol	liq	−96.46	−47.60		
Aniline	liq	7.34*	35.40		
Anthracene	c	27.60*	64.80		
l-Asparagine	c	−187.10*	−125.20		
l-Aspartic acid	c	−231.34*	−173.50		
Benzaldehyde	liq	−21.23	2.24		
	g	−9.57	5.85		
Benzene	liq	11.63	29.40		
	g	19.82	30.99	64.34	
Benzoic acid	c	−93.21*	−60.10		
Benzophenone	c	−10.00*	31.40		
Benzoquinone	c		−21.41		
Bromal CBr_3CHO	liq	−31.13			
Bromal hydrate $CBr_3CH(OH)_2$	c	−112			
Butadiene-1,3	liq	21.63*			
	g	26.75			
n-Butane	g	−29.81	−3.75	74.10	
Butene-1	g	0.280	17.22	73.48	
Butene-2 cis	g	−1.36	16.05	71.90	
trans	g	−2.41	15.32	70.86	
n-Butyl alcohol	liq	−78.18	−39.00	54.5	42.2
	g	−65.65	−36.11	86.90	26.29
iso-Butyl alcohol	liq	−79.85	(−39.4)		43.7
	g	−67.62	(−38.3)		27

*At 18°C.

Table 9-2 (*Continued*)
ORGANIC COMPOUNDS (CARBON)

Substance	State	$\Delta Hf°$	$\Delta Gf°$	$S°$	$C_p°$
sec-Butyl alcohol	liq	−81.06	−42.33	53.4	47.5
	g	−70.00	−40.12	85.8	27.1
tert-Butyl alcohol	c	−87.45	−44.18	40.89	34.92
(undercooled)	liq	−85.86	−44.18	46.20	52.61
	g	−74.68	−42.45	77.92	27.10
n-Butylbenzene	liq	−18.67*	27.50		
	g	−3.30	34.62	104.91	
tert-Butylbenzene	liq	−20.17*	29.00		
iso-Butyl bromide	liq		−4.60		
tert-Butyl bromide	liq		−5.6		
Butyne-1	g	39.70	48.52		
Butyne-2	g	35.37	44.73		
n-Butyraldehyde (butanal)	g	−52.40	−73.24		
Chloral CCl_3CHO	liq	−56.45			36
	g	−47.0			
	aq, 500	−68.4			
Chloral hydrate	c	−137.7			34
Chloroacetaldehyde	liq	−61.5			
Chloroacetamide	c	−81.1			
Chloroacetic acid	c, I	−122.3			
	c, II	−122.12			
	c, III	−121.46			
ionized	aq	−119.813			
nonionized; std. state, $m = 1$	aq	−118.92			
Chloroacetyl chloride	liq	−67.85			
Chloroacetyl nitrile	g			80.42	22.96
Creatinine	c	−52.80*	−2.9		
Cyanamide $CNNH_2$	c	14.1			
Cyanoguanidine $C{=}NH(NH_2)(NHC{\equiv}N)$	c	5.4	42.9	30.90	28.40
Cyclohexane	liq	−36.70*	6.80		
	g	−29.43	8.03		
Cyclohexanol	liq	−85.60*	−34.30		
Cyclohexene	liq	−15.29*	18.20		
Cyclopentane	liq	−25.31	8.70		
	g	−18.46	9.23		
p-Cymene	liq	−22.57*	24.60		
n-Decane	liq	−78.61*	−2.90		
	g	−59.7	8.23	129.19	
Decene-1	g	−29.67	28.99	128.98	
Diazirine CH_2N_2	g			56.87	10.19
Diazomethane CH_2N_2	g			58.02	12.55
Dibenzoylethane	c	−65.30*	−2.50		
Dibenzoylethylene	c	−32.40*	21.50		
Dibenzyl	c	9.47*	60.80		
1,2-Dibromoethane	liq	−19.4	−5.0	53.37	32.51
	g	−9.16	−2.47	79.1	20.7
Dichloroacetic acid	liq	−119.0			
ionized	aq	−122.4			
nonionized	aq	−120.4			
1,1-Dichloroethane	liq	−38.3	−18.1	50.61	30.18
	g	−30.93	−17.35	72.90	18.22
1,2-Dichloroethane	liq	−39.49	−19.03	49.84	30.9
	g	−31.02	−17.67	73.68	18.8
1,1-Dichloroethylene	liq	−5.8	5.85	48.17	26.60
	g	0.58	6.01	69.04	16.03
1,2-Dichloroethylene cis	liq	−6.6	5.27	47.42	27
	g	0.90	6.28	69.19	15.55
trans	liq	−5.53	6.52	46.81	27
	g	1.47	6.82	69.29	15.93

* At 18°C.

Table 9-2 (*Continued*)
ORGANIC COMPOUNDS (CARBON)

Substance	State	$\Delta Hf°$	$\Delta Gf°$	$S°$	$C_p°$
Diethylamine	liq	−24.7			
	g	−17.07			
	aq, 30	−31.9			
Diethylammonium nitrate	c	−100.1			
	aq	−94.7			
Diethyl ether ($C_2H_5)_2O$	liq	−66.82			
	g	−60.26			
Diethyl ketone	liq	−73.8			
Diethyl sulfide	liq	−28.43	2.81	64.36	40.97
	g	−19.86	4.34	87.96	27.97
Difluoroacetamide	c	−167.3			
Difluoroacetic acid	liq	−207.8			
1,1-Difluoroethane	g	−114.3	−100.6	67.5	16.2
2,2-Difluoroethanol	liq	−164.8			
2,2-Difluorochloroethylene					
CF_2=CHCl	g	−75.4	−69.1	72.39	17.23
2,2-Difluorotetrachloroethane	g	−117.1	−97.3	91.5	29.5
1,1-Difluoroethene	g	−78.6	−73.0	63.6	14.36
1,2-Diiodoethane	c	0.1	13.8	47	
	g	15.9	18.8	83.2	19.2
Dimethylamine	liq	−10.5	16.7	43.58	32.9
	g	−4.41	16.35	65.24	16.9
std. state, $m = 1$	aq	−16.88	13.85	31.8	
$(CH_3)_2NH_2^+$ std. state, $m = 1$	aq	−28.74	−0.80	41.2	
Dimethylammonium chloride	aq, ∞	−68.69	−32.17	54.7	
Dimethyl ether	g	−43.99	−26.93	63.64	15.39
1,1-Dimethylhydrazine	liq	11.8	49.4	47.32	39.21
	g	20.18	50.31	72.28	
1,2-Dimethylhydrazine	liq	13.3	50.8	47.60	40.88
	g	22.70	52.20	74.39	
Dimethylnitrosamine	liq	0.7			
Dimethyl sulfide	liq	−15.55	1.45	46.94	28.23
	g	−8.90	1.73	68.32	17.71
Dimethyl sulfite	liq	−126.7			
Dimethyl sulfone	c	−107.8	−72.3	34	
	g	−88.7	−65.2	74.2	23.9
Dimethyl sulfoxide	liq	−48.6	−23.7	45.0	35.2
	g	−35.96	−19.48	73.20	21.26
	aq, 2	−50.67			
	aq, 10	−52.21			
o-Dinitrobenzene	c	1.00*	49.40		
m-Dinitrobenzene	c	−5.20*	42.90		
Diphenyl	c	20.87*	57.40		
Diphenylcarbinol	c	−20.38*	30.70		
Diphenylmethane	c	19.72*	62.60		
2,3-Dithiabutane CH_3SSCH_3	liq	−14.82	1.67	56.26	34.92
	g	−5.64	3.64	80.46	22.54
Dithiocarbamic acid NH_2CSSH	aq	7.2			
n-Dodecane	liq	−90.45	3.90		
	g	−69.52	12.33	147.55	
Dulcitol	c	−317.71*	−223.10		
Durene	c	−32.57	19.00		
iso-Durene	liq	−31.07*	15.90		
n-Eicosane	g	−108.93	28.71	221.02	
dl-Erythritol	c	−214.81*	−149.40		
Ethane	g	−20.24	−7.86	54.85	12.58
Ethane-1,2-dithiol	liq	−12.83			
Ethanethiol	liq	−17.53	−1.28	49.48	28.17
	g	−10.95	−1.05	70.77	17.37

* At 18°C.

Table 9-2 (*Continued*)
ORGANIC COMPOUNDS (CARBON)

Substance	State	$\Delta Hf°$	$\Delta Gf°$	$S°$	$C_p°$
Ethoxide ion std. state, $m = 1$	aq		−24.5		
Ethyl C_2H_5 radical	g	25	31	59.2	
Ethyl acetate	liq	−110.72	−76.11		
	g	−102.02	−74.93		
Ethyl alcohol	liq	−66.37	−41.80	38.4	26.64
	g	−56.19	−40.29	67.54	15.64
std. state, $m = 1$	aq	−68.9	−43.44	35.5	
	aq, 1	−66.587			
	aq, 10	−68.063			
Ethylamine	liq	−17.7			31
	g	−11.27			16.7
	aq, 400	−24.2			
Ethylammonium ion	aq	−37.3			
Ethylbenzene	liq	−2.98	28.61		
	g	7.12	31.21	86.15	
Ethyl bromide	liq	−21.99	−6.64	47.5	24.1
	g	−15.42	−6.34	68.50	15.42
Ethyl chloride	liq	−32.63	−14.20	45.60	24.94
	g	−26.81	−14.45	65.94	15.01
Ethyl hydrogen peroxide	liq	−58			
Ethyl hydrogen sulfate	aq	−209.3			
Ethyl iodide	liq	−9.6	3.5	50.6	27.5
	g	−1.84	4.57	73.1	16.0
Ethylnitramine	liq	−22.4			
Ethyl nitrate	liq	−45.49	−10.29	59.08	40.7
	g	−36.82	−8.82	83.23	
Ethyl nitrite	liq	−30.8			
	g	−24.9			
Ethyl propanoate	liq	−122.16	−79.16		
	g	−112.36	−77.37		
Ethylene	g	12.49	16.28	52.45	10.41
Ethylene chlorohydrin	liq	−70.6			
Ethylenediamine	liq	−5.82		50	
	aq, 200	−13.32			
Ethylenediammonium chloride	c	−122.7			
	aq. 5000	−115.92			
Ethylene glycol	liq	−108.70	−77.25	39.9	35.8
	aq, 1	−109.01			
	aq, 10	−109.89			
	g	−92.53	−71.26		
Ethyleneimine	liq	21.97			
	g	30.2			
Ethylene oxide	liq	−18.60	−2.83	36.77	21.02
	g	−12.58	−3.12	57.94	11.45
Ethylene sulfite	liq	−119.3			
Fluoroacetamide	c	−118.7			
Fluoroacetic acid	c	−164.5			
2-Fluoroethanol	liq	−111.3			
Formaldehyde	g	−28	−27	52.26	8.46
unhydrolyzed	aq	−35.9	−31.02		
Formamide	c	−61.6			
	liq	−60.7			
	g			59.41	10.84
	aq, 200	−59.2			
Formic acid	liq	−101.51	−86.38	30.82	23.67
	g	−90.48			
nonionized; std. state, $m = 1$	aq	−101.68	−89.0	39	
ionized; std. state, $m = 1$	aq	−101.71	−83.9	22	−21.0

Table 9-2 (*Continued*)
ORGANIC COMPOUNDS (CARBON)

Substance	State	$\Delta Hf°$	$\Delta Gf°$	$S°$	$C_p°$
Formic acid (dimer)	g	−195.08			
Formylurea	c	−117.8			
	aq, 2000	−110.3			
Fumaric acid	c	−194.28	−156.70		
d-Galactose	c		−219.60		
d-Glucose	aq		−217.02		
	c		−215.8		
d-Glutamic acid	c	−236.60	−170.40		
Glycerol	liq	−159.16	−113.65		
Glycine	c	−126.22	−88.09	24.74	23.71
ionized; std. state, $m = 1$	aq	−112.280	−75.278	28.54	
nonionized; std. state, $m = 1$	aq	−122.846	−88.618	37.84	
$NH_3^+CH_2COOH$ std. state, $m = 1$	aq	−123.784	−91.824	45.46	
Glycolic acid	c	−158.7			
	aq, 400	−155.6			
Glycolic nitrile	liq	−33.7			
Glyoxal	c	−84.2			
Glyoxime	c	−21.2			
Glyoxylic acid	c	−199.7			
Guanidine	c	−18.1			
Guanidine carbonate	c	−232.10	−133.23	70.6	61.87
Guanidine nitrate	c	−92.5			
Guanidine perchlorate	c	−74.8			
	aq	−64.4			
Guanidine sulfate	c	−288.0			
	aq	−281.2			
n-Heptadecane	g	−94.15	22.56	193.47	
n-Heptane	liq	−53.63	0.42	77.92	
	g	−44.89	2.09	101.64	
1-Heptanol	liq	−95.3	−33.9	77.9	66.5
	g	−79.4	−29.0	114.8	42.7
Heptene-1	g	−14.85	22.84	101.43	
Hexachloroethane I, cubic	c	−46.0			
	g	−33.9	−13.8	95.3	32.7
Hexadecane	g	−89.23	20.52	184.28	
Hexamethylbenzene	c	−41.17*	25.20		
Hexamethylenediamine	g	−30.57	28.91		
n-Hexane	liq	−47.52	−0.91	70.76	
	g	−39.96	0.05	92.45	
2-Methylpentane	liq	−48.82	−1.88	69.21	
	g	−41.66	−1.11	90.65	
3-Methylpentane	liq	−48.28	−1.34	69.22	
	g	−41.02	−0.51	90.77	
2,2-Dimethylbutane	liq	−51.00	−2.66	65.18	
	g	−44.35	−2.33	85.72	
2,3-Dimethylbutane	liq	−49.48	−1.66	66.27	
	g	−42.49	−0.95	87.33	
Hexene-1	g	−9.96	20.80	92.25	
Hexene-2 *cis*	g	−11.56	19.18	92.35	
trans	g	−12.56	18.46	91.40	
n-Hexyl alcohol	liq	−90.7	−36.4	69.2	56.6
	g	−75.9	−32.4	105.2	37.2
Hydroquinone	c	−87.57*	−52.70		
Ketene	liq	−18.78	−13.32		
	g	−14.6	−14.8	59.16	12.37
dl-Lactic acid	liq	−161.65*	−124.40		
Lactose α	c		−418.20		
β			−373.70		

*At 18°C.

Table 9-2 (*Continued*)
ORGANIC COMPOUNDS (CARBON)

Substance	State	$\Delta Hf°$	$\Delta Gf°$	$S°$	$C_p°$
Maleic acid	c	−187.48*	−149.40		
l-Malic acid	c		−211.45		
Maltose β	c		−412.60		
Mannitol	c	−316.51	−222.20		
Metaldehyde $\frac{1}{4}(CH_3CHO)_4$	c	−56.2			
Methane	g	−17.88	−12.13	44.92	8.439
Methyl acrylate	liq	−82.76	−58.13		
	g	−70.10	−56.78		
Methyl alcohol	liq	−57.04	−39.76	30.3	19.5
	g	−47.96	−38.72	57.29	10.49
std. state, $m = 1$	aq	−58.779			
	aq, 1	−57.425			
	aq, 10	−58.439			
Methylamine	liq	−11.3	8.5	35.90	
	g	−5.49	7.67	58.15	12.7
std. state, $m = 1$	aq	−16.77	4.94	29.5	
Methylammonium ion std. state, $m = 1$	aq	−29.86	−9.55	34.1	
Methylammonium chloride	c	−71.20	−37.99	33.13	21.27
std. state, $m = 1$	aq	−69.82	−40.92	47.6	
Methylammonium hydroxide					
CH_3NH_3OH std. state, $m = 1$	aq	−84.83	−47.14	31.5	
nonionized; std. state, $m = 1$	aq	−85.08	−51.75	46.2	
Methylammonium nitrate	c	−84.7			
std. state, $m = 1$	aq	−79.43	−36.16	69.1	
Methyl cyanate *iso*	liq	−22.0			
Methyl cyanide *iso*	liq	28.0	38.1	38	
	g	35.6	39.6	58.99	12.65
Methylcyclohexane	liq	−46.55*	3.6		
	g	−36.99	6.52		
Methylcyclopentane	g	−25.50	8.55		
Methyl formate	g	−83.7	−71.37		
	liq	−90.60	−71.53	29	
Methylhydrazine	liq	12.9	43.0	39.66	32.25
	g	22.55	44.66	66.61	17.0
Methyl mercaptan	liq	−11.08	−1.85	40.44	21.64
	g	−5.34	−2.23	60.96	12.01
Methyl nitrate	liq	−38.0	−10.4	51.9	37.6
	g	−29.8	−9.4	76.1	
Methyl nitrite	g	−16.5	0.2		
2-Methylpropane	g	0.280	17.22	73.48	
2-Methylpropene	g	−3.34	14.58	70.17	
Methyl thiocyanate CH_3SCN	liq	28.4			
	g	38.3			
Methyl isothiocyanate CN_3NCS	c	19.0			
	g	31.3	34.5	69.29	15.65
Methylene sulfate CH_2SO_4	c	−164.6			
Naphthalene	c	15.96*	45.20		
o-Nitroaniline	c	−4.60*	41.40		
m-Nitroaniline	c	−5.60*	40.40		
p-Nitroaniline	c	−9.60*	36.40		
Nitrobenzene	liq		36.40		
o-Nitrobenzoic acid	c	−95.60*	−48.40		
m-Nitrobenzoic acid	c	−101.60*	−54.20		
p-Nitrobenzoic acid	c	−102.60*	−55.50		
Nitroethane	liq	−33.5			33
	g	−23.56			
aci form	aq	−30.7			
nitro form	aq	−32			

* At 18°C.

Table 9-2 (*Continued*)
ORGANIC COMPOUNDS (CARBON)

Substance	State	$\Delta Hf°$	$\Delta Gf°$	$S°$	$C_p°$
Nitroguanidine	c	−22.13			
Nitromethane	liq	−27.03	−3.47	41.05	25.33
	g	−17.86	−1.65	65.69	13.70
Nitrourea	c	−67.5			
Nonadecane	g	−104.00	26.66	211.83	
Nonane	g	−54.74	6.18	120.00	
1-Nonanol	liq	−109.7	−33.1	92.0	
Nonene-1	g	−24.74	26.94	119.80	
Octadecane	g	−99.08	24.61	202.65	
Octane	g	−49.82	4.14	110.82	
	liq	−59.74	1.77	85.50	
1-Octanol	liq	−101.6	−32.6	84.7	
	g	−84.4	−27.1	124.1	48.2
Oxalic acid	c	−197.7	−166.8	28.7	
std. state, $m = 1$	aq	−197.2	−161.1	10.9	
Oxalic acid dihydrate	c	−340.9			
Oxamic acid	c	−160.4			
	aq, 800	−153.1			
Oxamide	c	−123.0			
Palmitic acid	c	−215.80*	−80.00		
Paraldehyde $\frac{1}{3}(CH_3CHO)_3$	liq	−54.73			
Pentacosane	c	−190.56	13.40		
Pentadecane	g	−84.31	18.47	175.10	
Pentadiene-1,4	g		40.22		
Pentamethylbenzene	c	−35.92*	21.90		
Pentane	g	−35.00	−1.96	83.27	
	liq	−41.36	−2.21	62.79	
2-Methylbutane	liq	−42.85	−3.59	62.38	
	g	−36.92	−3.50	81.98	
2,2-Dimethylpropane	liq	−44.98*			
	g	−39.67	−3.64	73.23	
Pentene-1	g	−5.00	18.79	83.08	
Pentene-2 *cis*	g	−6.71	17.17	82.76	
trans	g	−7.59	16.58	81.81	
2-Methyl-1-butene	g	−8.68	15.51	81.73	
3-Methyl-1-butene	g	−6.92	17.87	79.70	
2-Methyl-2-butene	g	−10.17	14.27	80.90	
Pentyne-1	g	34.50	50.17		
Pentyne-2	g	30.80	46.41		
3-Methyl-1-butyne	g	32.60	49.12		
Phenanthrene	c	23.10	60.00		
Phenol	liq	−37.80	−11.02		
	g	−21.71	−6.26		
Picric acid	c		10.00		
Prehnitene	liq	−28.17*	20.20		
Propane	liq	−28.64*			
	g	−24.82	−5.61	64.51	
Propene	g	4.88	14.99	63.80	
Propionaldehyde	g	−49.15	−33.96		
Propionic acid	g	−108.75	−88.27		
	liq	−121.7	−91.65		
Propionic anhydride	g	−147.32	−109.78		
	liq	−161.53	−113.66		
n-Propyl alcohol	g	−61.17	−38.83		
	liq	−71.87	−39.84		
iso-Propyl alcohol	g	−62.41	−38.20		
	liq	−74.32	−38.83		
n-Propylbenzene	g	1.87	32.81		
	liq	−9.18	29.60		

* At 18°C.

Table 9-2 (*Continued*)
ORGANIC COMPOUNDS (CARBON)

Substance	State	$\Delta Hf°$	$\Delta Gf°$	$S°$	$C_p°$
iso-Propylbenzene	g	0.94	32.74		
	liq	−9.85	29.71		
Propyne	g	44.32	46.31		
Pyrocatechol	c	−85.57*	−51.40		
Quinhydrone	c		−77.19		
Resorcinol	c	−87.57*	−53.20		
Semicarbazide std. state, $m = 1$	aq	−39.9	−9.7	71.2	
l-Sorbose	c		−217.10		
Stilbene	c	32.38*	75.80		
Styrene	liq	24.72*			
	g	35.11*			
Succinic acid	c	−224.99*	−178.80		
Sucrose	c		−371.60		
Taurine $NH_2CH_2CH_2SO_3H$	c	−187.7	−134.3	36.8	33.6
ionized; std. state, $m = 1$	aq	−171.92	−121.76	47.8	
nonionized; std. state, $m = 1$	aq	−181.92	−134.12	55.7	
1,1,1,2-Tetrachloroethane	g			85.07	24.55
1,1,2,2-Tetrachloroethane	liq	−47.0	−22.7	59.0	39.6
	g	−35.7	−19.6	86.69	24.09
1,1,1,2-Tetrachlorodifluoroethane	g	−117.1	−97.3	91.5	29.5
Tetradecane	g	−79.38	16.42	165.92	
Tetrafluoroethylene C_2F_4	g	−155.5	−147.2	71.69	19.23
Teflon $1/n(C_2F_4)_n$	c	−196.1			
Tetraiodoethylene	c	73			
Tetraphenylmethane	c	62.20*	137.20		
Tetrazole	c	56.7			
Thiacyclopropane	liq	12.41	22.52	38.84	
	g	19.69	23.19	61.01	12.83
Thiolacetic acid	liq	−52.1			
Thiophene	liq		26.30		
Thiourea	c	−21.1			
	aq, 100	−15.6			
Thiocyanogen $(SCN)_2$	liq	74.3			
Thiourea nitrate	c	−73.1			
Toluene	g	11.95	29.23	76.42	
Trichloroacetamide	c	−87			
Trichloroacetic acid	c	−120.7			
ionized	aq	−123.4			
	aq, 100	−122.56			
1,1,2-Trichloroethane	liq	−43.5	−21.5	55.60	
	g	−33.94	−19.35	80.57	21.27
Trichloroethylene	liq	−10.1	2.9	54.6	28.8
	g	−1.86	4.31	77.6	19.18
Tridecane	g	−74.45	14.37	156.74	
Triethylamine	liq	−32.1			
	g	−22.9			
Triethylammonium ion	aq	−51.8			
Triethylammonium chloride	c	−92.2			
	aq, 25	−91.68			
Triethylammonium nitrate	c	−107.0			
	aq	−101.4			
Trimethylamine	liq	−11.0	24.1	49.82	32.31
	g	−5.81	23.65	68.6	
std. state, $m = 1$	aq	−18.17	22.22	31.9	
Trimethylammonium ion					
std. state, $m = 1$	aq	−26.99	8.90	47.0	
Trimethylammonium chloride	c	−67.29			
std. state, $m = 1$	aq	−66.94			

* At 18°C.

Table 9-2 (*Continued*)
ORGANIC COMPOUNDS (CARBON)

Substance	State	$\Delta Hf°$	$\Delta Gf°$	$S°$	$C_p°$
Trimethylammonium nitrate	c	−82.2			
std. state, $m = 1$	aq	−76.55	−17.71	82.0	
1,2,3-Trimethylbenzene	liq	−14.01	25.68	66.40	
	g	−2.29	29.32	93.50	
1,2,4-Trimethylbenzene	liq	−14.79	24.46	67.93	
	g	−3.33	27.91	94.71	
1,3,5-Trimethylbenzene	liq	−15.18	24.83	65.35	
	g	−3.84	28.17	92.15	
Triphenylamine	c	58.70*	120.50		
Triphenylcarbinol	c	3.96*	69.70		
Triphenylethylene	c	64.22*	123.00		
Triphenylmethane	c	41.76*	101.40		
Undecane	g	−65.60	10.28	138.37	
Urea	c	−79.56	−47.04	25.00	22.26
	aq, 5	−76.466			
std. state, $m = 1$	aq	−75.954			
Urea nitrate	c	−134.8			
	aq, 400	−124.3			
Urea oxalate	c	−365.3			
	aq, 500	−348.1			
Vinyl bromide	g	18.7	19.3	65.90	13.27
Vinyl chloride	liq	3.5			
	g	8.5	12.4	63.07	12.84
$1/n(CH_2{=}CHCl)_n$	c	−22.5			
o-Xylene	liq	−5.84	26.37	58.91	
	g	4.54	29.18	84.31	
m-Xylene	liq	−6.08	25.73	60.27	
	g	4.12	28.41	85.49	
p-Xylene	liq	−5.84	26.31	59.12	
	g	4.29	28.95	84.23	

* At 18°C.

LATENT HEATS

Table 9-3

HEATS OF FUSION AND VAPORIZATION OF THE ELEMENTS AND INORGANIC COMPOUNDS

For the freezing point or the normal boiling point unless otherwise stated. Values in parentheses are uncertain.

Abbreviations Used in the Table

subl, sublimation   tr, transition point ΔS_v, entropy of vaporization, in electrostatic units or cal/(mole)(deg)

Substance	Heat of Fusion, kcal/mole	Heat of Vaporization, kcal/mole
Actinium	(3.4)	(95.0)
Aluminum		
Al	2.57	67.95
AlBr₃	2.71	16.08
AlCl₃*	8.50	15.61
AlF	5.00	38.00
AlF₃*	16.38	77.00 subl
Na₃AlF₆	3.98	18.50
AlI₃	(26.0)	
Al₂O₃		
Antimony		
Sb	4.740	46.230
SbBr₃	3.51	(12.0)
SbCl₃	3.03	10.30
SbCl₅	2.40	11.57
SbF₃		10.9¹²⁵°
SbH₃	5.08	
SbI₃ orthorhombic		23.2²⁵°
Sb₄O₆	(27.0)	17.82
$\Delta H_{tr} = 1.39$ (c → rb)		
Sb₄S₆	11.20	
Argon	0.2808	1.558
Arsenic		
As	5.100	7.630 subl
AsBr₃	4.10	10.0
AsCl₃	2.42	7.57
AsF₃	12.47	7.10
AsF₅	2.80	4.98
AsH₃	4.34	
AsI₃	2.20	19.2
As₄O₆ monoclinic	8.00	14.30
orthorhombic	15.86	(8.22 rh → mn)
Astatine		
Barium		
Ba	(2.85)	(10.8)
BaBr₂	1.83	36.07
BaCl₂	(6)	(50)
BaF₂	5.37	(50)
BaI₂	3.00	70.00
Ba(NO₃)₂	(6.8)	(45)
BaO	(5.9)	
Ba(OH)₂	13.80	
	3.39	

Table 9-3 (Continued)
HEATS OF FUSION AND VAPORIZATION OF THE ELEMENTS AND INORGANIC COMPOUNDS

Substance	Heat of Fusion, kcal/mole	Heat of Vaporization, kcal/mole	Substance	Heat of Fusion, kcal/mole	Heat of Vaporization, kcal/mole
$Ba_3(PO_4)_2$	18.60		$B_{10}H_{14}$	5.25	18.15
$BaSO_4$	9.70		B_2H_5Br		6.23
Beryllium			BI_3	(1.0)	(10.0)
Be	2.80	70.40	$B_3H_6N_3$		7.67
BeO	17.00	117.0	B_2O_3	5.50	(70.)
$BeBr_2$	(4.5)	(22.0)	**Bromine**		
$BeCl_2$	(3.0)	(25.0)	Br_2	2.520	7.170
BeF_2	(6.0)	(40.0)	$BrCl$	2.4	8.3
BeI_2	(4.5)	(19.0)	$BrCN$		43.3 subl
Bismuth			BrF		6.0
Bi	2.60	36.20	BrF_3 ($\Delta S_v = 25.3$ e.u.)	2.875	10.24
$BiBr_3$	5.19	18.02	BrF_5 ($\Delta S_v = 23.2$ e.u.)	1.355	7.310
$BiCl_3$	2.60	17.35	**Cadmium**		
BiF_3	(6.2)	(28.0)	Cd	1.45	23.87
BiF_5		14.9	$CdBr_2$	5.00	27.00
BiI_3		5.00	$CdCl_2$	5.30	29.86
Bi_2O_3	6.80		CdF_2	5.40	52.0
Bi_2S_5	8.90		CdI_2	3.66	25.40
Boron			$Cd(NO_3)_2 \cdot 4H_2O$	7.80	
B	5.30		CdO		53.82 subl
BBr_3	(0.70)	7.298	$CdSO_4$	4.79	
BCl_3	(0.50)	5.70	**Calcium**		
B_2Cl_4 ($\Delta S_v = 23.7$ e.u.)	2.579	8.029	Ca	2.23	35.84
BF_3	0.480	4.62	$CaBr_2$	4.18	(50.0)
B_2F_4 ($\Delta S_v = 28$ e.u.)		6.700	$CaCO_3$	(12.7)	
B_2H_6		3.45	CaC_2	$(1.33 \text{ tr})^{720°K}$	
B_4H_{10}	1.06	6.47	$CaCl_2$	6.78	
B_5H_9		6.8	CaF_2 $(1.140 \text{ tr})^{1424°K}$	6.78	(55.0)
B_5H_{11}		7.7	CaI_2	(5.0)	83.00
B_6H_{10}		9.16	$Ca(NO_3)_2$	5.12	(35)

Table 9-3 (Continued)
HEATS OF FUSION AND VAPORIZATION OF THE ELEMENTS AND INORGANIC COMPOUNDS

Substance	Heat of Fusion, kcal/mole	Heat of Vaporization, kcal/mole
CaO	12.00	
$CaSiO_3$	13.40	
$CaSO_4$	6.70	
$CaTiO_3$	0.54	
Carbon		
C graphite	25.000 (ΔS_m = 5.8 e.u.)	78 subl (ΔS_v = 18.6 e.u.)
CBr_4 (1.43 tr)$^{47°C}$	0.950	(9.7)
CCl_4	0.644	7.283
CF_4	0.167	3.010
CH_4	0.224	2.040
CI_4	(1.15)	(12.0)
C_2N_2	1.938	5.576
$CNBr$	11.30 subl	11.010
$CNCl$	2.720	6.290
CNF		5.780
CNI	14.20 subl	13.980
C_3O_2 (ΔS_m = 8.020 e.u.)	1.291	5.92 (ΔS_v = 21.17 e.u.)
CO (ΔS_m = 2.949 e.u.)	0.201	1.444 (ΔS_v = 17.67 e.u.)
CO_2 (ΔS_m = 8.77 e.u.)	1.90	3.88 (ΔS_v = 17.91 e.u.)
(ΔS_{subl} = 30.98 e.u.)		6.030 subl
COS	1.129	4.423
$COCl_2$	1.371	5.825
CS_2	1.049	
Cerium		
Ce (0.70 tr)$^{730°C}$	1.238	18.67 subl
$CeBr_3$	(8.0)	(44.0)
$CeCl_3$	(8.0)	40.8
CeF_3	(9.0)	(62.0)
CeF_4	(10.0)	dec

Substance	Heat of Fusion, kcal/mole	Heat of Vaporization, kcal/mole
CeI_3 (ΔS_m = 12.0 e.u.)	12.4	(40.0)
Cesium		
Cs	0.50	16.32
$CsBr$		35.99
$CsCl$	3.60	35.69
CsF	2.45	34.33
CsI		35.93
$CsNO_3$	3.25	
$CsOH$	1.60	
Chlorine		
Cl_2	1.531	4.878
ClF_3	1.819	6.580
ClF_5		5.477
Cl_2O		6.28
ClO_2		6.52
ClO_3		9.5
Cl_2O_7	12.3 subl	8.29
Chromium		
Cr	3.500	83.360
$CrBr_2$	(6.5)	(35.0)
$CrBr_3$		(54) subl
$CrCl_2$ (ΔS_m = 7.1 e.u.)	7.72	47.50
$CrCl_3$ (ΔS_{subl} = 48.4 e.u.)		60 subl
$CrCl_4$	(2.0)	(9.0)
CrF_2		86.8 subl
CrF_3	(5.5)	60 subl
CrF_4	(6.0)	(14.0)
CrI_2		(24)
Cr_2O_3	4.20	
CrO_3	3.77	

Table 9-3 (Continued)

HEATS OF FUSION AND VAPORIZATION OF THE ELEMENTS AND INORGANIC COMPOUNDS

Substance	Heat of Fusion, kcal/mole	Heat of Vaporization, kcal/mole
CrO$_2$Cl$_2$		8.250
CrO$_2$F$_2$		8.2
Cobalt		
Co	5.6	91.400
CoBr$_2$	3.640	52 subl
CoCl$_2$ (ΔS_v = 26.11 e.u.)	54 subl	34.62
CoF$_2$	10.72	78 subl
CoI$_2$		46 subl
Co(NO$_3$)$_2 \cdot$ 6H$_2$O	8.79	
Copper		
Cu	3.120	72.800
CuBr		35.1 subl
CuCl		37.5 subl
CuCl$_2$		46.3 subl
CuF$_2$		63 subl
CuI		43.3 subl
Cu(NO$_3$)$_2 \cdot$ 6H$_2$O	8.70	
Cu$_2$O	13.40	
CuO	2.82	
Cu$_2$S	5.50	
Dysprosium	(4.1)	60.0
Erbium	(4.1)	(70.0)
Europium	(2.5)	(42)
Fluorine, F$_2$		1.64
Gadolinium	(3.7)	(74.5)
Gallium		
Ga	1.336	
GaCl$_3$	2.55	
Germanium		
Ge	8.1	79.9
GeBr$_4$		8.560
GeCl$_4$		7.030
GeF$_2$		27.00 subl
GeH$_4$		3.608
Ge$_2$H$_6$		5.99
Ge$_3$H$_8$		7.55
GeHCl$_3$		8.000
Ge(CH$_3$)$_4$		6.460
Gold	2.955	77.540
Hafnium		
Hf	5.2	158
Hf(BH$_4$)$_4$	2.85	24.0 subl
HfBr$_4$		23.8
HfCl$_4$	18	63.0 subl
HfF$_4$		48.9 subl
HfI$_4$		
Helium	0.0033	0.0194
Holmium	(4.1)	(60)
Hydrogen		
H$_2$	0.028	0.216
D$_2$	0.057	0.293
T$_2$		0.322
HBr	0.575	4.210
HCl	0.505	3.860
HCN	0.040	6.027
HF	1.094	1.8
HI	0.686	4.724
HN$_3$		7.29
H$_2$O	1.436	9.717
D$_2$O	1.501	9.945

Table 9-3 (Continued)
HEATS OF FUSION AND VAPORIZATION OF THE ELEMENTS AND INORGANIC COMPOUNDS

Substance	Heat of Fusion, kcal/mole	Heat of Vaporization, kcal/mole	Substance	Heat of Fusion, kcal/mole	Heat of Vaporization, kcal/mole
H_2O_2	2.920	10.530	$FeBr_2$	15.5	31.6
HNO_3	0.600		Fe_3C (0.180 tr)[190°]	12.33	
H_3PO_2	2.310		$FeCl_2$	45.3 subl	29.99
H_3PO_3	3.070		$FeCl_3$	10.30	6.02
H_3PO_4	2.520		$Fe(CO)_5$	3.250	9.000
$H_4P_2O_7$	8.30		FeF_3		60.8 subl
H_2S	0.5676	4.463	FeI_2	53.0 subl	36.76
H_2S_2	1.805	7.497	FeO	7.70	
H_2S_3	*	9.327	Fe_2O_3 (0.160 tr)[677°]		
H_2S_4		11.261	(0.0 tr)[777°]	dec	
H_2S_5		13.340	Fe_3O_4 (0.0 tr)[627°]	33.00	
H_2Se		4.75	FeS	5.000	
H_2SeO_4	3.450		**Krypton**	0.3907	2.158
H_2SO_4	2.360		**Lanthanum**	(2.7)	95.5
$HOSO_2F$		8.4	**Lead**		
H_2Te		5.7	Pb	1.141	42.88
Indium			$PbBr_2$	4.43	27.694
In	0.781		$PbCl_2$	5.80	29.604
$In(CH_3)_3$	3.77		PbF_2	1.86	38.34
Iodine			PbI_2	6.01	24.846
I_2	3.74	9.970	$PbMoO_4$	(25.8)	
IBr		12.433 subl	PbO yellow	2.80	51.00
ICl	1.85	9.95	(0.250 tr red → yel)[489°]		
IF_5	3.84	7.46 subl	PbS	4.15	
IF_7	(6.3)	(134.7)	$PbSO_4$	9.60	
Iridium			$PbWO_4$	(15.2)	(50.0)
Iron			**Lithium**		
Fe (0.410 tr)[760°]	3.670	83.900	Li	0.723	32.19
(0.210 tr)[906°]			$LiBO_2$	(5.57)	36.94 subl
(0.110 tr)[1301°]					

Table 9-3 (Continued)
HEATS OF FUSION AND VAPORIZATION OF THE ELEMENTS AND INORGANIC COMPOUNDS

Substance	Heat of Fusion, kcal/mole	Heat of Vaporization, kcal/mole	Substance	Heat of Fusion, kcal/mole	Heat of Vaporization, kcal/mole
LiBr	2.900	35.420	MnF_2		76.1 subl
LiCl	3.200	35.960	MnF_3		68 subl
LiF	(2.36)	50.970	MnO	13.00	
Li_2HfCl_6	8.80	10.9	Mn_3O_4 (4.97 tr)$^{172°}$	(39.0)	
LiH	5.010	38.44 subl	MnO_3F		8.1
LiI	(1.42)		$MnSiO_3$	(8.1)	
Li_2MoO_4	4.20		$MnTiO_3$	(8.0)	
LiOH	2.48		**Mercury**		
Li_2SiO_3	7.21		Hg	0.549	14.137
Li_4SiO_4	7.43		$HgBr_2$	3.960	14.080
Li_2SO_4	3.04		$HgCl_2$	4.150	14.080
Li_2WO_4	(6.7)		HgF_2	(5.5)	(22.0)
Li_2ZrCl_6	9.30		HgI_2 (0.650 tr)$^{130°}$	4.50	14.263
Lutetium			$HgSO_4$	(1.44)	
Lu	(4.6)	(59)	**Molybdenum**		
Magnesium			Mo	6.600	142.000
Mg	2.140	30.750	$MoCl_2$	6.00	
$MgBr_2$	8.30	(35.0)	$MoCl_3$		52.00 subl
$MgCl_2$	10.30	32.70	$MoCl_4$		25.00 subl
MgF_2	13.90	65.00	$MoCl_5$	22 subl	14
MgI_2	(5.3)	(25.0)	MoF_4	1.020	12.090
Mg_3N_2 (0.110 tr)$^{550°}$			MoF_5		12.37
MgO	18.50		MoF_6 (1.96 tr o-rh → cub)	0.92	6.94
$Mg_3(PO_4)_2$	(11.3)		MoO_3	12.54	33.00
$MgSiO_3$	14.70		$MoOCl_3$		25 subl
$MgSO_4$	3.50		$MoOCl_4$		16.19
Manganese			**Neodymium**	(2.6)	67.8
Mn (0.535 tr)$^{727°}$	3.50	52.52	**Neon**	0.0801	0.422
(0.545 tr)$^{1101°}$			**Neptunium**	2.6	
Mn_3C (3.57 tr)$^{1037°}$			**Nickel**		
$MnCl_2$	8.97	35.6			

9-77

Table 9-3 (Continued)
HEATS OF FUSION AND VAPORIZATION OF THE ELEMENTS AND INORGANIC COMPOUNDS

Substance	Heat of Fusion, kcal/mole	Heat of Vaporization, kcal/mole
Ni	4.210	88.870
NiBr$_2$		54.84
NiCl$_2$	18.47	53.81
Ni(CO)$_4$		7.00
NiF$_2$		79.4 subl
NiO (0.0 tr) 252° and 292°		
Ni$_2$S	2.98	
Ni$_3$S$_2$	5.800	
Niobium		
Nb		166.5
NbBr$_5$	6.40	19.9
NbCl$_4$	26.5 subl	31.4 subl
NbCl$_5$	22.8 subl	13.1
NbF$_5$		12.9
NbN	(14.5)	
Nb$_2$O$_5$	24.2	
NbOCl$_3$		28.1 subl
Nitrogen		
N$_2$	0.1723	1.333
NF$_3$		2.769
N$_2$F$_2$ cis	3.670	21.9
trans	3.400	21.0
N$_2$F$_4$	3.17	15.9
ClNF$_2$		4.350
NH$_3$	1.377	5.559
N$_2$H$_4$		9.760
NH$_4$SCN	(4.7)	
HNF$_2$	1.46	5.940
NH$_4$NO$_3$	1.563	
N$_2$O		3.956
NO		3.293
N$_2$O$_3$	0.550	9.40
N$_2$O$_4$ equil. mixture		14
N$_2$O$_5$ equil. mixture	3.501	9.110
NOCl		14.9 subl
NOF		6.14
NO$_2$F		4.607
N$_3$P$_3$Cl$_6$		4.30
N$_4$P$_4$Cl$_8$	5.0	13.2
N$_3$P$_3$F$_6$	5.3	15.6
NH$_4$HSO$_4$	3.42	7.6
Osmium		
OsF$_6$	4.06	6.84
OsO$_4$ yellow	2.34	9.45
white		
Oxygen		
O$_2$	0.1063	1.630
O$_3$		2.88
OF$_2$		2.650
O$_2$F$_2$		4.583
O$_3$F$_2$		4.58
Palladium		
Pd	4.12	
PdCl$_2$	9.58	
Phosphorus		
P white	0.157	2.97
red triclinic		7.69 subl 25°
black		8.21 subl
PBr$_3$	4.5	11.6

Table 9-3 (Continued)

HEATS OF FUSION AND VAPORIZATION OF THE ELEMENTS AND INORGANIC COMPOUNDS

Substance	Heat of Fusion, kcal/mole	Heat of Vaporization, kcal/mole	Substance	Heat of Fusion, kcal/mole	Heat of Vaporization, kcal/mole
PBr_5		4.2	**Plutonium**		
$PBrF_2$		5.45	Pu		79.71
PBr_2F		7.34	$PuBr_3$	13.4	56.5
PCl_3		7.8	$PuCl_3$	15.2	72.8 subl
PCl_5		15.5 subl	PuF_3	(13)	89.6 subl
$PClF_2$		5.95	PuF_4	(10.2)	71.6 subl
PCl_2F		5.45	PuF_6	4.456	11.6 subl
P_2ClF_2		4.20	PuI_3	(12)	
$PCl(NCO)_2$		11.2	PuO_2		133.8
$PCl_2(NCS)$		12.6	**Polonium**		
PF_3		3.49	PoO_2	(3.0)	14.4
PF_5	0.270	4.11	**Potassium**		
PH_3		3.489	K	0.554	18.53
P_2H_4		6.89	KBr	5.00	37.06
PI_3		(10.5)	KCl	6.10	38.84
$P(NCO)_3$		11.9	KSCN	2.25	
P_4O_6	3.360	10.38	K_2CO_3	7.80	
P_4O_{10}	6.5	16.2	K_2CrO_4	6.92	
$POBr_3$		10.9	$K_2Cr_2O_7$	8.77	
$POCl_3$	3.11	8.06	KF	6.75	41.275
$POClF_2$		6.08	KI	4.10	34.691
$POCl_2F$		7.40	KNO_3	2.84	
POF_3	3.60	5.55	K_2O_2	6.10	
$PQ(NCO)_3$		13.41	K_2O_3	7.03	
$PQ(NCS)_3$		14.82	KO_2	3.92	
$PSClF_2$		5.70	KOH		30.85
$PSCl_2F$		6.89	KPO_3	2.11	
PSF_3		4.68	K_3PO_4	8.90	
$PS(NCS)_3$		14.82	$K_4P_2O_7$	14.00	
Platinum	4.70	(107.0)	$KReO_4$	20.40	
			K_2SO_4	8.10	

Table 9-3 (Continued)
HEATS OF FUSION AND VAPORIZATION OF THE ELEMENTS AND INORGANIC COMPOUNDS

Substance	Heat of Fusion, kcal/mole	Heat of Vaporization, kcal/mole
K_2WO_4	4.65	
K_2ZrCl_6	5.50	
Praseodymium	2.70	79.5
Promethium	(3.0)	(70)
Protactinium	(3.5)	(110)
Radium	(2.0)	(32.7)
Radon	(0.693)	4.01
Rhenium		
Re	7.90	169.0
ReF_5	1.107	13.88
ReF_6	1.799	6.867
ReF_7		9.154
ReO_2		65.64 subl
ReO_3	5.20	49.8 subl
Re_2O_7	15.8	17.7
Re_2O_8	3.80	
$ReOCl_4$		10.9
$ReOF_4$	3.23	14.590
$ReOF_5$		7.720
ReO_2F_3	8.94 subl$^{>30°}$	15.7
ReS_2		22.66 subl
Rhodium	(5.2)	(118.4)
Rubidium		
Rb	0.525	18.11
RbBr	3.70	37.12
RbCl	4.40	36.92
RbF	4.13	39.51
RbI	2.99	35.96
$RbNO_3$	1.34	dec
RbOH	1.62	

Substance	Heat of Fusion, kcal/mole	Heat of Vaporization, kcal/mole
Ruthenium	(6.1)	(135.7)
Samarium	(2.65)	(45.8)
Scandium	(3.85)	(72.85)
Selenium		
Se	1.30	25.49
Se_2		20.60
Se_6		11.2
SeF_4		6.6 subl
SeF_6		4.75
H_2Se		22.583
SeO_2	1.01	10.2
$SeOCl_2$		6.77
SeO_2F_2		
Silicon		
Si	11.10	105
$SiBr_4$		9.05
$SiBrCl_2F$		5.954
$SiBr_2ClF$		7.095
$SiBrF_3$		4.405
$SiBr_2F_2$		5.929
$SiBr_3F$		7.833
$SiCl_4$	1.845	6.860
Si_3C_8		12.34
$(SiCl_3)_2O$		8.820
SiF_4	(0.80)	6.130 subl
Si_2F_6	3.900	10.400
$SiClF_3$		4.460
$SiCl_2F_2$		5.080
SiH_4		2.960
Si_2H_6		5.110

Table 9-3 (Continued)
HEATS OF FUSION AND VAPORIZATION OF THE ELEMENTS AND INORGANIC COMPOUNDS

Substance	Heat of Fusion, kcal/mole	Heat of Vaporization, kcal/mole	Substance	Heat of Fusion, kcal/mole	Heat of Vaporization, kcal/mole
Si_3H_8		6.780	$NaClO_3$	5.29	
Si_4H_{10}		8.890	$NaCN$	(4.40)	37.28
$SiHBr_3$		8.322	$NaSCN$	4.45	
SiH_2Br_2		6.838	Na_2CO_3	7.000	
SiH_3Br		5.648	NaF	7.78	53.26
Si_2H_5Br		7.000	$NaFeCl_4$	4.30	
$SiHCl_3$		6.360	NaI	5.24	
$(SiH_3)_3N$		6.850	Na_2MoO_4	3.60	
$(SiH_3)_2O$		5.350	$NaNO_3$	3.76	
SiO_2 cristobalite	2.100		Na_2O_2	5.86	
(0.20 tr)$^{260°}$			$NaOH$	2.00	
quartz	3.400		$NaPO_3$	(5.0)	
(0.29 tr)$^{575°}$			$Na_4P_2O_7$	(13.7)	
tridymite	2.150		Na_2S	(1.2)	
(0.040 tr)$^{117°}$			Na_2SiO_3	10.3	
Silver			$Na_2Si_2O_5$	8.46	
Ag	2.700	60.960	Na_2SO_4	5.83	
$AgBr$	2.18	47.4	Na_2TiO_3	16.8	
$AgCl$	3.155	47.5	Na_2WO_4	5.80	
$AgCN$	2.75		Na_2ZrCl_6	4.0	
AgF	4.0	42.8	**Strontium**		
AgI (1.47 tr)$^{150°}$	2.25	34.45	Sr	2.20	33.61
$AgNO_3$	2.755		$SrBr_2$	4.78	
Ag_2S	3.360		$SrCl_2$	4.10	
Ag_2SO_4	(4.3)		SrF_2	4.26	71.0
Sodium			$Sr_3(PO_4)_2$	18.50	
Na	0.622	21.28	**Sulfur**		
$NaBO_2$	8.660		S rhombic		2.20
$NaBr$	6.140	37.95	S_2Cl_2		8.61
$NaCl$	6.85	40.81	SF_4		6.32

Table 9-3 (Continued)

HEATS OF FUSION AND VAPORIZATION OF THE ELEMENTS AND INORGANIC COMPOUNDS

Substance	Heat of Fusion, kcal/mole	Heat of Vaporization, kcal/mole	Substance	Heat of Fusion, kcal/mole	Heat of Vaporization, kcal/mole
SF_6	5.60 subl	4.08	TcO_3F	5.377	9.453
S_2F_{10}		6.15	**Tellurium**		
SO_2	1.769	5.960	Te	4.180	12.100
SO_3 (mp 17°)	2.27	13.45 subl	$TeCl_4$		18.4
(mp 30.5°)	3.21	13.91 subl	TeF_4		8.2
(mp 62.2°)	6.09	15.91 subl	TeF_6		6.7 subl
$SOBr_2$		10.40	Te_2F_{10}		9.44
$SOCl_2$		7.48	H_2Te		5.7
SOF_2		5.18	TeO_2	59 subl	51.7
$SOClF$		5.90	**Terbium**		
SOF_4		5.1	Tb	(3.9)	(70)
SOF_6 (F_5SOF)		5.21	**Thallium**		
SO_2Cl_2		6.70	Tl	1.030	38.81
SO_2F_2		4.79	TlBr	5.99	23.8
SO_3BrF		8.67	TlCl	4.26	24.42
SO_3F_2 (FSO_2OF)		5.35	Tl_2CO_3	4.40	
$S_2O_5Cl_2$		13.2	TlI	3.125	25.03
S_2O_5ClF		8.1	$TlNO_3$	2.29	
$S_2O_5F_2$		7.6	Tl_2S	3.00	
Tantalum			Tl_2SO_4	5.50	
Ta	7.500	180.00	**Thorium**		
$TaBr_5$	10.9	14.8	Th	(3.74)	(130)
$TaCl_4$		30.2 subl	$ThBr_4$	9.50	34.5
$TaCl_5$	9.2	14.0	$ThCl_4$	22.50	36.5
TaF_5		13.0	ThI_4	8.00	31.5
TaI_5	1.6	18.1	ThO_2	291.1	
Ta_2O_5	48.00		**Thullium**		
Technetium			Tm	(4.4)	(51.0)
Tc	5.50	138.0	**Tin**		
TcF_6 cubic	1.128	7.427	Sn	1.72	69.40

Table 9-3 (Continued)
HEATS OF FUSION AND VAPORIZATION OF THE ELEMENTS AND INORGANIC COMPOUNDS

Substance	Heat of Fusion, kcal/mole	Heat of Vaporization, kcal/mole
$SnBr_2$	1.72	32.50
$SnBr_4$	3.00	(10.5)
$SnCl_2$	3.05	19.50
$SnCl_4$	2.19	8.325
$Sn(CH_3)_4$		7.32
SnH_4		4.42
SnI_2	(3.0)	24.00
Titanium		
Ti	4.50	102.5
$TiBr_4$	3.08	13.2
$TiCl_2$		46.8 subl
$TiCl_3$		42 subl
$TiCl_4$	2.33	9.5
TiF_4		22.87 subl
TiI_2		48.4 subl
TiI_4	(3.0)	13.97
TiO (0.82 tr)[991°]	14.00	
TiO_2 rutile	15.5	
Ti_2O_3 (0.215 tr)[200°]	38.40	
Tungsten		
W	8.42	191.000
WBr_5	5.0	14.5
WCl_4		39 subl
WCl_5		15.7
WCl_6	16.7 subl	14.5
WF_6 cubic	(16) subl / 0.42	6.33
WO_3	13.94	13
$WOBr_4$	14.7	15.16
$WOCl_4$	4.98	15.16
WOF_4	2.26	14.23
WO_2Cl_2		26.3 subl
Uranium		
U (0.700 tr)[672°] (1.145 tr)[772°]		
UBr_3	3.20	110.00
UCl_3	11.00	45.00
UCl_4	9.00	41.00
UCl_5	10.30	33.00
UF_4	8.50	57.50
UF_6	5.70	11.43 subl
UI_4	15.0	30.7
Vanadium		
V	(4.2)	109.6
VBr_2		45 subl
VBr_3		43.3 subl
VCl_2		44.8 subl
VCl_3		44.7 subl
VCl_4	2.3	9.9
VF_5	11.94	10.62
VI_2		44 subl
VO		
V_2O_3	15.0	
V_2O_4	(28)	
V_2O_5	27.21	63.00
VO_2Cl	15.56	9.6
Xenon		
Xe	0.5495	3.020
XeF_2		12.3 subl
XeF_4		15.3 subl
XeF_6		15.3 subl

Table 9-3 (Continued)
HEATS OF FUSION AND VAPORIZATION OF THE ELEMENTS AND INORGANIC COMPOUNDS

Substance	Heat of Fusion, kcal/mole	Heat of Vaporization, kcal/mole	Substance	Heat of Fusion, kcal/mole	Heat of Vaporization, kcal/mole
Ytterbium			ZnO	4.47	
Yb	(2.2)	(37.1)	ZnS	(9.0)	
Yttrium			**Zirconium**		
Y	(4.1)	(94.0)	Zr (0.920 tr)$^{862°}$	4.0	139.0
Zinc			ZrBr$_4$		25.8 subl
Zn	1.765	27.56	ZrCl$_2$	7.3	
ZnBr$_2$	4.00	24.25	ZrCl$_4$	25.3 subl	16.3
ZnCl$_2$	5.54	28.70	ZrF$_2$	14.5	
ZnF$_2$	(7.0)	44.00	ZrF$_4$	15.35	40.75
Zn(C$_2$H$_5$)$_2$		8.96	ZrI$_4$		29.0 subl
ZnI$_2$	4.50	23.00	ZrO$_2$ (1.42 tr)$^{1205°}$	20.80	

Table 9-4
HEAT OF VAPORIZATION OF COMPOUNDS

For elements see special table.

The heat of vaporization (ΔHv) is given in gram-calories (15°) per gram. To convert to: (a) joules (abs) per gram multiply by 4.185; (b) joules (int) per gram multiply by 4.183; (c) btu per pound multiply by 1.8.

Name	Formula	t, °C	ΔHv
Acetal	$C_6H_{14}O_2$	102.9	66.2
Acetaldehyde	CH_3CHO	21	136.2
Acetic acid	CH_3CO_2H	118.3	96.8
....	....	140	94.4
....	....	220	81.2
....	....	321.6	0
anhydride	$(CH_3CO)_2O$	137	66.2
Acetone	$(CH_3)_2CO$	0	134.7
....	....	20	131.9
....	....	40	128.1
....	....	56.1	124.5
....	....	80	118.3
....	....	100	112.8
....	....	235	0
Acetonitrile	CH_3CN	80	173.7
Acetophenone	$C_6H_5COCH_3$	203.7	77.2
Acetyl chloride	CH_3COCl	51	78.9
Air	....	......	50.97
Allyl alcohol	C_3H_5OH	96	163
Aluminum bromide	$AlBr_3$	256.4	20.5
chloride	$AlCl_3$	180.2	100.3*
iodide	AlI_3	385.5	18.83
Ammonia†	NH_3	−33.4	327.1
....	....	−20	317.6
....	....	−10	309.7
....	....	0	301.6
....	....	17	296.5
....	....	40	263.1
....	....	55	245.1
Ammonium chloride	NH_4Cl	350	70.9
Amyl alcohol (n)	$C_5H_{11}OH$	131	120.2
alcohol (t)	$C_5H_{11}OH$	102	105.8
amine (n)	$C_5H_{11}NH_2$	95	98.7
bromide (n)	$C_5H_{11}Br$	129	48.3
ether (n)	$(C_5H_{11})_2O$	170	69.5
iodide (n)	$C_5H_{11}I$	155	47.6
Anethole (p)	$C_3H_5C_6H_4OCH_3$	232	71.4
Aniline	$C_6H_5NH_2$	183	103.7
Antimony trichloride	$SbCl_3$	219	45.4
trioxide	Sb_2O_3	1425	30.6
Arsenic trichloride	$AsCl_3$	122	41.8
trioxide	As_2O_3	457.2	36.1
Benzaldehyde	C_6H_5CHO	179	86.5
Benzene	C_6H_6	0	107.0
....	....	25	103.6
....	....	40	100.7
....	....	60	97.5
....	....	80.1	94.14
....	....	100	90.6
....	....	120	86.5
....	....	140	82.6
....	....	160	78.5

* Heat of sublimation for solid to gas.

Table 9-4 (*Continued*)
HEAT OF VAPORIZATION OF COMPOUNDS

Name	Formula	t, °C	ΔHv
Benzene	C_6H_6	220	62.1
....	....	260	43.8
....	....	288.5	0
Benzonitrile	C_6H_5CN	189	87.7
Benzyl alcohol	$C_6H_5CH_2OH$	204.3	112.3
Bismuth bromide	$BiBr_3$	461	40.2
chloride	$BiCl_3$	441	55.0
Boron bromide	BBr_3	91.3	29.1
chloride	BCl_3	10	38.2
fluoride	BF_3	−100.9	68.1
hydride	B_2H_6	−92.4	133.1
hydride	B_3H_{10}	16	152.1
hydride	B_5H_9	58	121.9
hydride	B_5H_{11}	67	130.4
Bromobenzene	C_6H_5Br	155.9	57.6
Butanes:			
n-butane	C_4H_{10}	−0.50	92.09
....	....	10	89.8
....	....	25	86.6
iso-butane	C_4H_{10}	−11.72	87.56
....	....	0	85.1
....	....	10	82.4
....	....	25	78.6
Butenes:			
1-butene	C_4H_8	−6.25	93.36
....	....	25	86.8
cis-2-butene	C_4H_8	3.72	99.46
....	....	25	86.8
trans-2-butene	C_4H_8	0.88	96.94
....	....	25	91.8
iso-butene	C_4H_8	−6.90	94.22
....	....	25	87.7
n-Butyl acetate	$CH_3CO_2C_4H_9$	124.0	73.8
alcohol	C_4H_9OH	116.8	141.3
formate	$HCO_2C_4H_9$	105.1	86.8
iodide	C_4H_9I	129.5	45.9
sec-Butyl alcohol	C_4H_9OH	98.1	134
tert-Butyl alcohol	C_4H_9OH	83	130.5
iso-Butyl acetate	$CH_3CO_2C_4H_9$	115.5	73.8
alcohol	C_4H_9OH	106.9	138.1
n-butyrate	$C_3H_7CO_2C_4H_9$	157	64.5
iso-butyrate	$C_3H_7CO_2C_4H_9$	148	63.3
formate	$HCO_2C_4H_9$	97	78.5
propionate	$C_2H_5CO_2C_4H_9$	137	65.9
n-valerate	$C_4H_9CO_2C_4H_9$	169	57.8
iso-valerate	$C_4H_9CO_2C_4H_9$	169	60.5
n-Butyric acid	$C_3H_7CO_2H$	163.5	114.0
nitrile	C_3H_7CN	117.4	114.9
iso-Butyric acid	$C_3H_7CO_2H$	154	111.6
Cadmium chloride	$CdCl_2$	967	162.9
iodide	CdI_2	796	69.3
oxide	CdO	1559	41.9*
Capronitrile	$C_5H_{11}CN$	156	88.2
Carbon dioxide	CO_2	−60	87.2
....	....	−50	83.4
....	....	−40	79.6

* Heat of sublimation for solid to gas.

Table 9-4 (*Continued*)
HEAT OF VAPORIZATION OF COMPOUNDS

Name	Formula	t, °C	ΔHv
Carbon dioxide	CO_2	−30	71.4
....	....	−20	66.9
....	....	−10	61.4
....	....	0	55.0
....	....	10	46.6
....	....	20	35.1
....	....	30	11.9
disulfide	CS_2	0	89.4
....	....	46.3	84.1
....	....	100	75.5
....	....	140	67.4
monoxide	CO	−192	50.4
oxychloride	$COCl_2$	8.0	60.6
oxysulfide	COS	−50.2	73.6
tetrachloride	CCl_4	0	52.1
tetrafluoride	CF_4	−127.9	35.3
Carvacrol	$C_{10}H_{13}OH$	237	68.1
Cesium bromide	CsBr	1300	172.7
chloride	CsCl	1303	222.1
fluoride	CsF	1251	228.4
iodide	CsI	1280	140.9
Chloral	CCl_3CHO	......	54
hydrate	$CCl_3CH(OH)_2$	96	131.9
Chlorine dioxide	ClO_2	10.9	105.2
fluoride	ClF_3	11.3	63.7
heptoxide	Cl_2O_7	79	46.4
monoxide	Cl_2O	2.0	72.3
Chlorobenzene	C_6H_5Cl	130.6	77.6
β-Chloroethyl acetate	$CH_3CO_2C_2H_4Cl$	141.5	80.8
alcohol	ClC_2H_4OH	126.5	122.9
Chloroform	$CHCl_3$	0	64.7
....	....	40	60.9
....	....	61.5	59.0
....	....	100	55.2
....	....	263	0
Chlorosulfonic acid	$ClSO_3H$	151	110.2
Chlorotoluene (*o*)	$ClC_6H_4CH_3$	158.1	72.6
(*p*)	$ClC_6H_4CH_3$	160.4	73.1
Chromyl chloride	CrO_2Cl_2	117	53.3
Cobalt chloride	$CoCl_2$	1050	209.2
Cresol (*m*)	$CH_3C_6H_4OH$	202	100.6
Cyanogen	$(CN)_2$	0	103
chloride	CNCl	13	135
fluoride	CNF	−72.8	128.4*
iodide	CNI	141	91.4*
Cyclopentane	C_5H_{10}	25	97.1
Cyclohexane	C_6H_{12}	25	93.8
....	....	80.7	85.6
Cyclohexanol	$C_6H_{11}OH$	161.6	108.2
Cyclohexene	C_6H_{10}	81.6	88.7
Cyclohexyl chloride	$C_6H_{11}Cl$	142	74.8
Cymene (*p*)	$C_3H_7C_6H_4CH_3$	176	67.6
n-Decane	$C_{10}H_{22}$	160	60.2
Dichloroacetic acid	Cl_2CHCO_2H	194.4	77.2
Dichlorodifluoromethane	CCl_2F_2	−29.8	40.4
Diethyl amine	$(C_2H_5)_2NH$	58	91.0

* Heat of sublimation for solid to gas.

Table 9-4 (*Continued*)
HEAT OF VAPORIZATION OF COMPOUNDS

Name	Formula	t, °C	ΔHv
Diethyl carbonate	$(C_2H_5)_2CO_3$	126	73.1
ketone	$(C_2H_5)_2CO$	101	90.8
oxalate	$(CO_2C_2H_5)_2$	185	67.6
Dibutylamine (*iso*)	$(C_4H_9)_2NH$	134	65.7
Dimethyl aniline	$C_6H_5N(CH_3)_2$	193	80.8
carbonate	$(CH_3)_2CO_3$	90	88.2
Dimethylcyclohexane (1,1)	C_8H_{16}	25	80.9
....	....	119.5	70.7
(1,2) (cis)	C_8H_{16}	25	84.6
....	....	129.7	72.9
(1,2) (trans)	C_8H_{16}	25	81.7
....	....	123.4	71.1
(1,3) (cis)	C_8H_{16}	25	83.5
....	....	124.5	72.1
(1,3) (trans)	C_8H_{16}	25	81.4
....	....	120.1	70.9
(1,4) (cis)	C_8H_{16}	25	83.1
....	....	124.3	71.9
(1,4) (trans)	C_8H_{16}	25	80.7
....	....	119.4	70.4
Dimethylcyclopentane (1,1)	C_7H_{14}	25	82.5
....	....	87.5	74.6
(1,2) (cis)	C_7H_{14}	25	86.4
....	....	99.3	77.0
(1,2) (trans)	C_7H_{14}	25	83.9
....	....	91.9	75.5
(1,3) (cis)	C_7H_{14}	......	......
....	....	......	......
(1,3) (trans)	C_7H_{14}	25	83.6
....	....	90.8	75.3
Dipropylamine (*n*)	$(C_3H_7)_2NH$	108	75.7
Ethane	C_2H_6	−88.9	116.9
....	....	−40	97.5
....	....	−20	87.0
....	....	0	75.0
Ethyl acetate	$CH_3CO_2C_2H_5$	0	102.0
alcohol	C_2H_5OH	78.3	204.3
amine	$C_2H_5NH_2$	15	14.6
benzene	$C_6H_5C_2H_5$	25	95.1
....	....	136.2	81.0
benzoate	$C_6H_5CO_2C_2H_5$	213	64.5
bromide	C_2H_5Br	38.4	59.9
n-butyrate	$C_3H_7CO_2C_2H_5$	118.4	74.7
caprylate	$C_7H_{15}CO_2C_2H_5$	207	60.5
chloride	C_2H_5Cl	4.7	92.9
....	....	15	92.5
....	....	20	92.2
....	....	25	92.0
cyclohexane	C_8H_{16}	25	86.2
....	....	131.8	73.9
cyclopentane	C_7H_{14}	25	88.6
....	....	103.5	78.3
ether	$(C_2H_5)_2O$	34.6	83.9
formate	$HCO_2C_2H_5$	53.3	97.2
iodide	C_2H_5I	71.2	45.6
nonylate	$C_8H_{17}CO_2C_2H_5$	227	58.1

Table 9-4 (*Continued*)
HEAT OF VAPORIZATION OF COMPOUNDS

Name	Formula	t, °C	ΔHv
Ethyl propionate	$C_2H_5CO_2C_2H_5$	97.6	80.1
n-propyl ether	$C_2H_5OC_3H_7$	60.0	82.7
n-valerate	$C_4H_9CO_2C_2H_5$	98	77.2
iso-valerate	$C_4H_9CO_2C_2H_5$	144	67.9
Ethylene	C_2H_4	−103.7	115.4
bromide	$(CH_2Br)_2$	130.8	46.2
chloride	$(CH_2Cl)_2$	0	85.3
....	....	82.3	77.3
chlorohydrin	$ClCH_2CH_2OH$	126.5	123.9
glycol	$(CH_2OH)_2$	197	191.1
oxide	$(CH_2)_2O$	13	138.6
Ethylidene chloride	CH_3CHCl_2	0	76.7
....	....	60	67.1
Fluorine oxide	F_2O	−144.8	49.1
Formic acid	HCO_2H	101	119.9
Furane	$(CH)_4O$	31	95.3
Furfural	C_4H_3OCHO	160.5	107.5
Germanium bromide	$GeBr_4$	189	21.8
chloride	$GeCl_4$	84	32.8
hydride	GeH_4	−89.1	46.7
....	Ge_2H_6	31.4	39.0
....	Ge_3H_8	110.6	33.4
tetramethyl	$Ge(CH_3)_4$	44	48.7
Heptanes:			
n-heptane	C_7H_{16}	25	87.2
....	....	98.4	76.5
2-Me-hexane	C_7H_{16}	25	83.0
....	....	90.1	73.4
3-Me-hexane	C_7H_{16}	25	83.7
....	....	92	74.1
3-Et-pentane	C_7H_{16}	25	84.0
....	....	93.5	74.3
2,2-DiMe-pentane	C_7H_{16}	25	77.4
....	....	79.2	69.7
2,3-DiMe-pentane	C_7H_{16}	25	81.7
....	....	89.8	72.9
2,4-DiMe-pentane	C_7H_{16}	25	78.4
....	....	80.5	70.9
3,3-DiMe-pentane	C_7H_{16}	25	78.8
....	....	86.1	70.6
2,2,3-TriMe-butane	C_7H_{16}	25	76.4
....	....	80.9	69.3
Heptyl alcohol (*n*)	$C_7H_{15}OH$	176	104.9
Hexanes:			
n-hexane	C_6H_{14}	0	89.1
....	....	25	87.5
....	....	68.7	80.5
2-Me-pentane	C_6H_{14}	25	82.8
....	....	60.3	76.9
3-Me-pentane	C_6H_{14}	25	84.0
....	....	63.3	78.4
2,2-DiMe-butane	C_6H_{14}	25	76.8
....	....	49.7	73.8
2,3-DiMe-butane	C_6H_{14}	25	80.8
....	....	58.0	76.5
Hexene	C_6H_{12}	0	92.8

Table 9-4 (*Continued*)
HEAT OF VAPORIZATION OF COMPOUNDS

Name	Formula	t, °C	ΔHv
Hydrogen bromide	HBr	−66.7	52.0
chloride	HCl	−85	105.9
cyanide	HCN	20	210.2
fluoride	HF	19.4	360
iodide	HI	−37.2	33.9
peroxide	H_2O_2	158	302.1
selenide	H_2Se	−41.3	60.1
sulfide	H_2S	−61.4	131.9
telluride	H_2Te	−2.2	43.6
Iodine fluoride	IF_7	4	28.7*
Iron carbonyl	$Fe(CO)_5$	105	45.9
chloride (ous)	$FeCl_2$	1026	236.7
Lead bromide	$PbBr_2$	916	76.8
chloride	$PbCl_2$	954	106.4
fluoride	PbF_2	1292	160.7
iodide	PbI_2	872	53.9
oxide	PbO	1472	229.9
Limonene	$C_{10}H_{16}$	165	69.5
Lithium bromide	LiBr	1310	410.1
chloride	LiCl	1382	876.4
fluoride	LiF	1681	1967.9
iodide	LiI	1171	304.5
Magnesium chloride	$MgCl_2$	1418	343.4
Manganese chloride	$MnCl_2$	1190	235.5
Mercuric bromide	$HgBr_2$	319	39.1
chloride	$HgCl_2$	304	51.9
iodide	HgI_2	354	31.4
Mesityl oxide	$C_6H_{10}O$	128	85.8
Methane	CH_4	−161.6	121.9
Methyl acetate	$CH_3CO_2CH_3$	0	114.0
....	....	56.3	98.1
alcohol	CH_3OH	0	284.3
....	....	64.7	262.8
n-amyl ketone	$CH_3COC_5H_{11}$	149.2	82.7
aniline	$C_6H_5NHCH_3$	194	95.6
n-butyl ketone	$CH_3COC_4H_9$	127	82.4
n-butyrate	$C_3H_7CO_2CH_3$	102.6	79.8
iso-butyrate	$C_3H_7CO_2CH_3$	91.1	78.1
chloride	CH_3Cl	−23.8	102.3
....	....	15	96.0
....	....	25	94.6
cyclohexane	C_7H_{14}	25	86.1
....	....	100.9	76.9
cyclopentane	C_6H_{12}	25	89.8
....	....	71.8	83.2
2-Et-benzene (1)	C_9H_{12}	25	94.9
....	....	165.2	77.3
3-Et-benzene(1)	C_9H_{12}	25	93.3
....	....	161.3	76.6
4-Et-benzene (1)	C_9H_{12}	25	92.7
....	....	162.1	76.4
ethyl ketone	$CH_3COC_2H_5$	78.2	106.0
ethyl ketoxime	C_4H_8NOH	182	115.9
formate	HCO_2CH_3	31.3	112.4
hexyl ketone	$CH_3COC_6H_{13}$	173	74.1
iodide	CH_3I	42	45.9

Hydrazine, N_2H_4, 10.0 kg cal/mole at 113.5°C.

* Heat of sublimation for solid to gas.

Table 9-4 (*Continued*)
HEAT OF VAPORIZATION OF COMPOUNDS

Name	Formula	t, °C	ΔHv
Methyl phenyl ether	$CH_3OC_6H_5$	153	81.5
propionate	$C_2H_5CO_2CH_3$	79.0	87.6
iso-propyl ketone	$CH_3COC_3H_7$	92	89.8
n-valerate	$C_4H_9CO_2CH_3$	116	70.0
iso-valerate	$C_4H_9CO_2CH_3$	116	72.4
Methylal	$C_3H_8O_2$	42	89.8
Methylene chloride	CH_2Cl_2	40.5	78.6
Molybdenum fluoride	MoF_6	36	28.6
Naphthalene	$C_{10}H_8$	218	75.5
Nickel carbonyl	$Ni(CO)_4$	42.5	41.0
chloride	$NiCl_2$	987	373.2*
Nitric acid	HNO_3	86.0	114.9
oxide	NO	−151.7	110.2
Nitrobenzene	$C_6H_5NO_2$	210	79.1
Nitromethane	CH_3NO_2	99.9	135.0
Nitrogen fluoride	NF_3	−129.0	42.3
oxychloride	$NOCl$	−6.4	93.7
pentoxide	N_2O_5	50	44.8
Nitrous oxide	N_2O	−88.5	89.8
....	....	−20	66.9
....	....	0	59.7
....	....	20	43.7
....	....	35	13.5
Nonanes:			
n-nonane	C_9H_{20}	25	86.5
....	....	150.8	70.4
2-Me-octane	C_9H_{20}	25	83.2
....	....	143.3	68.3
3-Me-octane	C_9H_{20}	25	83.4
....	....	144.8	68.5
4-Me-octane	C_9H_{20}	25	83.4
....	....	142.5	68.2
3-Et-heptane	C_9H_{20}	25	83.5
....	....	143	68.5
4-Et-heptane	C_9H_{20}	25	83.5
....	....	142	68.3
2,2-DiMe-heptane	C_9H_{20}	25	78.8
....	....	130.5	64.8
2,3-DiMe-heptane	C_9H_{20}	25	81.6
....	....	140.7	67.3
2,4-DiMe-heptane	C_9H_{20}	25	79.9
....	....	133	65.9
2,5-DiMe-heptane	C_9H_{20}	25	79.9
....	....	136	66.4
2,6-DiMe-heptane	C_9H_{20}	25	79.8
....	....	135.2	66.2
3,3-DiMe-heptane	C_9H_{20}	25	79.5
....	....	137.3	65.8
3,4-DiMe-heptane	C_9H_{20}	25	81.7
....	....	143	67.8
3,5-DiMe-heptane	C_9H_{20}	25	80.1
....	....	136	66.4
4,4-DiMe-heptane	C_9H_{20}	25	79.5
....	....	138	65.9
2-Me-3-Et-hexane	C_9H_{20}	25	81.7
....	....	139	67.1

* Heat of sublimation for solid to gas.

Table 9-4 (*Continued*)
HEAT OF VAPORIZATION OF COMPOUNDS

Name	Formula	t, °C	ΔHv
2-Me-4-Et-hexane	C_9H_{20}	25	80.1
....	....	136	66.4
3-Me-3-Et-hexane	C_9H_{20}	25	80.2
....	....	143	66.6
3-Me-4-Et-hexane	C_9H_{20}	25	81.9
....	....	143	67.8
2,3,3-TriMe-hexane	C_9H_{20}	25	78.1
....	....	134	64.8
2,2,4-Tri-Me-hexane	C_9H_{20}	25	75.6
....	....	126.5	63.4
2,2,5-TriMe-hexane	C_9H_{20}	25	74.9
....	....	124.1	62.9
2,3,3-TriMe-hexane	C_9H_{20}	25	78.7
....	....	138	65.2
2,3,4-TriMe-hexane	C_9H_{20}	25	80.0
....	....	140	66.5
2,3,5-TriMe-hexane	C_9H_{20}	25	77.2
....	....	131.4	64.9
2,2,4-TriMe-hexane	C_9H_{20}	25	76.1
....	....	131	63.9
3,3,4-TriMe-hexane	C_9H_{20}	25	78.8
....	....	139	65.5
3,3-DiEt-pentane	C_9H_{20}	25	80.8
....	....	146.5	67.1
2,2-DiMe-3-Et-pentane	C_9H_{20}	25	78.3
....	....	133.8	64.9
2,3-DiMe-3-Et-pentane	C_9H_{20}	25	79.3
....	....	142	65.8
2,4-DiMe-3-Et-pentane	C_9H_{20}	25	80.0
....	....	136.7	66.0
2,2,3,3-TetraMe-pentane	C_9H_{20}	25	76.4
....	....	140.2	65.7
2,2,3,4-TetraMe-pentane	C_9H_{20}	25	76.4
....	....	133.0	63.9
2,2,4,4-TetraMe-pentane	C_9H_{20}	25	71.0
....	....	122.3	61.2
2,3,3,4-TetraMe-pentane	C_9H_{20}	25	77.8
....	....	141.5	65.1
Octanes:			
n-octane	C_8H_{18}	25	86.8
....	....	125.7	73.2
2-Me-heptane	C_8H_{18}	25	83.0
....	....	117.6	70.3
3-Me-heptane	C_8H_{18}	25	83.4
....	....	118.9	71.3
4-Me-heptane	C_8H_{18}	25	83.0
....	....	117.7	70.9
3-Et-hexane	C_8H_{18}	25	83.0
....	....	118.5	71.7
2,2-DiMe-hexane	C_8H_{18}	25	78.0
....	....	106.8	67.7
2,3-DiMe-hexane	C_8H_{18}	25	81.2
....	....	115.6	70.2
2,4-DiMe-hexane	C_8H_{18}	25	79.0
....	....	109.4	68.5
2,5-DiMe-hexane	C_8H_{18}	25	79.2

Table 9-4 (*Continued*)
HEAT OF VAPORIZATION OF COMPOUNDS

Name	Formula	t, °C	ΔHv
2,5-DiMe-hexane	....	109.1	68.6
3,3-DiMe-hexane	C_8H_{18}	25	78.5
....	....	112.0	68.5
3,4-DiMe-hexane	C_8H_{18}	25	81.6
....	....	117.7	70.2
2-Me-3-Et-pentane	C_8H_{18}	25	80.6
....	....	115.7	69.7
3-Me-3-Et-pentane	C_8H_{18}	25	79.5
....	....	118.3	69.3
2,2,3-TriMe-pentane	C_8H_{18}	25	77.2
....	....	109.8	67.3
2,2,4-TriMe-pentane	C_8H_{18}	25	73.5
....	....	99.2	64.9
2,3,3-TriMe-pentane	C_8H_{18}	25	77.9
....	....	114.8	68.1
2,3,4-TriMe-pentane	C_8H_{18}	25	78.9
....	....	113.5	68.4
2,2,3,3-TetraMe-butane	C_8H_{18}	25	89.6*
....	....	106.3	66.2
Octyl alcohol (*n*)	$C_8H_{17}OH$	196	97.5
alcohol (*sec*) (*dl*)	$C_8H_{17}OH$	180	94.4
Osmium tetroxide	OsO_4	130	37.2
Pentanes:			
n-pentane	C_5H_{12}	25	87.5
....	....	36.1	85.4
iso-pentane	C_5H_{12}	25	81.5
....	....	27.9	81.0
2,2-DiMe-propane	C_5H_{12}	9.5	75.4
....	....	25	72.2
Phosphine	PH_3	−87.7	102.6*
Phosphorus oxychloride	$POCl_3$	105.1	54.6
pentoxide	P_2O_5	591	72.8
trichloride	PCl_3	78	51.4
trioxide	P_2O_3	174	47.2
Picoline (α)	$C_5H_4NCH_3$	129	90.8
Piperidine	$C_5H_{11}N$	106	89.4
Potassium bromide	KBr	1376	321.0
chloride	KCl	1417	542
fluoride	KF	1505	721.2
hydroxide	KOH	1324	579.3
iodide	KI	1330	224.1
Propane	C_3H_8	−42.1	101.8
....	....	−20	95.3
....	....	0	89.6
....	....	20	83.4
Propionic acid	$C_2H_5CO_2H$	139.4	98.8
Propionitrile	C_2H_5CN	97	134.3
n-Propyl acetate	$CH_3CO_2C_3H_7$	100.4	80.3
alcohol	C_3H_7OH	97.2	164.4
benzene	$C_6H_5C_3H_7$	25	91.9
....	....	159.2	76.0
n-butyrate	$C_3H_7CO_2C_3H_7$	143.6	68.3
iso-butyrate	$C_3H_7CO_2C_3H_7$	134	63.8
formate	$HCO_2C_3H_7$	80	88.1
propionate	$C_2H_5CO_2C_3H_7$	120.6	74.9
iso-valerate	$C_4H_9CO_2C_3H_7$	156	64.5

* Heat of sublimation for solid to gas.

Table 9-4 (Continued)
HEAT OF VAPORIZATION OF COMPOUNDS

Name	Formula	t, °C	ΔHv
iso-Propyl alcohol	C_3H_7OH	82.3	159.4
benzene (cumene)	$C_6H_5C_3H_7$	25	89.8
....	....	152.4	74.6
Propylene	C_3H_6	−47.7	104.6
Pyridine	C_5H_5N	114.1	107.4
Rhenium heptoxide	Re_2O_7	362.4	372.7
Rubidium bromide	RbBr	1352	224.4
chloride	RbCl	1381	305.3
fluoride	RbF	1408	378.1
iodide	RbI	1304	169.3
Salicylaldehyde	C_6H_5OCHO	196	74.8
Selenium oxide	SeO_2	317	188.3*
Silicochloroform	$SiHCl_3$	31.8	47.0
Silicon bromohydride	SiH_3Br	2.4	50.9
chlorotrifluoride	SiF_3Cl	−70.1	37.0
dibromohydride	SiH_2Br_2	70.5	36.0
dichlorodifluoride	SiF_2Cl_2	−31.5	37.1
hydride (silane)	SiH_4	−111.6	92.2
hydride (silicoethane)	Si_2H_6	−14.3	82.2
hydride (trisilane)	Si_3H_8	53.1	73.5
hydride (tetrasilane)	Si_4H_{10}	100	72.7
oxychloride	Si_2OCl_6	135.6	31.0
tetrachloride	$SiCl_4$	57	36.1
tetrafluoride	SiF_4	−94.8	58.9*
trifluoride	Si_2F_6	−18.9	61.1*
Silver chloride	AgCl	1564	296.7
iodide	AgI	1506	146.7
Sodium bromide	NaBr	1392	368.8
chloride	NaCl	1465	697.6
cyanide	NaCN	1500	760.8
fluoride	NaF	1704	1268.1
hydroxide	NaOH	1388	787.5
iodide	NaI	1300	262.8
Sulfur chloride	S_2Cl_2	138	63.9
dioxide	SO_2	−10.8	94.9
....	....	0	91.3
....	....	10	87.7
....	....	20	84.1
....	....	30	80.8
....	....	40	71.2
....	....	50	73.8
....	....	60	70.3
hexafluoride	SF_6	−63.5	38.3*
pentoxy dichloride	$SO_3 \cdot SO_2Cl_2$	140	61.2
trioxide	SO_3	53	118.5
Sulfuric acid	H_2SO_4	326	122.1
Sulfuryl chloride	SO_2Cl_2	69.1	49.4
Tetrachloroethane	$(CHCl_2)_2$	145.0	55.1
Tetrachloroethylene	$(CCl_2)_2$	120.7	50.1
Tellurium hexafluoride	TeF_6	−38.6	27.7*
tetrachloride	$TeCl_4$	392	62.5
Thallium bromide	TlBr	819	83.7
chloride	TlCl	807	101.8
iodide	TlI	823	75.6
Thionyl bromide	$SOBr_2$	139.5	47.7
chloride	$SOCl_2$	82	54.5

* Heat of sublimation for solid to gas.

Table 9-4 (*Continued*)
HEAT OF VAPORIZATION OF COMPOUNDS

Name	Formula	t, °C	ΔHv
Tin chloride (ous)	$SnCl_2$	623	109.4
chloride (ic)	$SnCl_4$	100	31.8
....	....	110	31.2
....	....	120	30.5
....	....	200	24.6
....	....	280	15.6
hydride	SnH_4	−52.3	36.0
tetramethyl	$Sn(CH_3)_4$	78.3	40.9
Titanium tetrachloride	$TiCl_4$	136	44.0
Toluene	$C_6H_5CH_3$	25	98.6
....	....	110.6	86.8
Toluidine (*o*)	$C_7H_7NH_2$	198	95.1
Trichloroethylene	C_2HCl_3	85.7	57.3
TriMe-benzene (1,2,3)	$C_6H_3(CH_3)_3$	25	97.6
....	....	176.2	79.6
(1,2,4) (pseudocumene)	$C_6H_3(CH_3)_3$	25	95.3
....	....	169.3	78.0
(1,3,5) (mesitylene)	$C_6H_3(CH_3)_3$	25	94.4
....	....	164.7	77.6
Tungsten fluoride	WF_6	17.3	21.3
Turpentine	$C_{10}H_{16}$	156	68.6
Uranium hexafluoride	UF_6	55.1	28.4*
Valeric acid (*n*)	$C_4H_9CO_2H$	184.6	103.2
acid (*iso*)	$C_4H_9CO_2H$	176.3	101.1
Valeronitrile (*n*)	C_4H_9CN	129	96.3
Water†	H_2O	0	595.9
....	....	10	590.4
....	....	20	584.9
....	....	30	579.5
....	....	40	574.0
....	....	50	568.5
....	....	60	563.2
....	....	70	557.5
....	....	80	551.5
....	....	90	545.8
....	....	100	539.55
....	....	110	532.9
....	....	120	525.7
....	....	130	518.5
....	....	140	511.1
....	....	150	503.5
....	....	160	495.6
....	....	170	487.2
....	....	180	478.6
Water, heavy	D_2O	101.4	496.5
Xylene (*o*)	$C_6H_4(CH_3)_2$	25	97.8
....	....	144.4	82.9
(*m*)	$C_6H_4(CH_3)_2$	25	96.0
....	....	139.1	82.0
(*p*)	$C_6H_4(CH_3)_2$	25	95.4
....	....	138.4	81.2
Zinc chloride	$ZnCl_2$	732	210.6
diethyl	$Zn(C_2H_5)_2$	118	72.6
Zirconium bromide	$ZrBr_4$	357	62.8*
chloride	$ZrCl_4$	311	108.5*
iodide	ZrI_4	431	48.5*

* Heat of sublimation for solid to gas.

HEAT OF COMBUSTION OF VARIOUS SUBSTANCES

In the table below, gaseous products of combustion are to be assumed unless otherwise stated. To convert gram-calories per gram to B.T.U. per pound, multiply the values below by 1800.

Substance	kg.-cal. per gram	Substance	kg.-cal. per gram
Asphalt	9.532	Petroleum	10.30 to 10.33
Castor oil	8.850		
Cod liver oil	9.40	Pitch	8.4
Cottonseed oil	9.40	Rape-seed oil	9.45
Dynamite (75%)	1.29	Sawdust, pressed	4.065
Graphite	α 7.840; β 7.856	Spermaceti	9.95
Guncotton	1.0563	Sulfur, monoclinic (to SO_2)	2.241
Hemoglobin	5.89	Sulfur, rhombic (to SO_2)	2.221
Hydrogen (to H_2O, liq.)	34.18	Whale oil	9.47
Hydrogen (to H_2O, steam)	29.15	Wood, beech (13% H_2O)	4.168
Linseed oil	9.41	Wood, birch (11.8% H_2O)	4.207
Mining powder	0.5088	Wood, oak (13.3% H_2O)	3.990
Paraffin	10.34	Wood, pine (12.2% H_2O)	4.422
Paraffin oil	9.80		

Table 9-5
HEAT OF COMBUSTION OF ORGANIC COMPOUNDS
(Kharasch, *Bu. Stand. Jour. Research*, Vol. 2, p. 359 (1929))

The heat of combustion is expressed in kilogram calories per gram formula weight of the compound as given in the table; the values are based upon a combustion at atmospheric pressure and at 20°C, the final combustion products consisting of liquid water, gaseous carbon dioxide and gaseous nitrogen. To convert to British Thermal Units (B.T.U.) multiply the value in kg-cal. by 3.9685.

Name	Formula	Physical state	Heat of combustion in kg. cal.
Acetaldehyde	CH_3CHO	liq.	279.0
Acetamide	CH_3CONH_2	solid	282.6
Acetanilide	$C_6H_5NHCOCH_3$	solid	1010.4
Acetic acid	CH_3COOH	liq.	209.4
Acetic anhydride	$(CH_3CO)_2O$	liq.	431.9
Acetone	CH_3COCH_3	liq.	426.8
Acetonitrile	CH_3CN	liq.	302.4
Acetophenone	$CH_3COC_6H_5$	solid	988.9
Acetylacetone	$CH_3COCH_2COCH_3$	liq.	615.9
Acetylene	C_2H_2	gas	312.0
Acrolein	$CH_2:CHCHO$	liq.	389.6
Acrylic acid	$CH_2:CHCOOH$	liq.	327.5
Adipic acid	$HOOC.(CH_2)_4.COOH$	solid	669.0
Alanine	$CH_3CH(NH_2)COOH$	solid	387.7
Aldol	$CH_3CHOHCH_2CHO$	liq.	546.6
Alizarin	$C_{14}H_6O_2(OH)_2-(1,2)$	solid	1448.9
Allylacetic acid	$C_3H_5.CH_2COOH$	liq.	641.6
Allyl alcohol	$CH_2:CHCH_2OH$	liq.	442.4
Allylamine	$CH_2:CHCH_2NH_2$	gas	528.1
Allyldiethyl carbinol	$C_3H_5(C_2H_5)_2COH$	liq.	1207
Allyldimethyl carbinol	$C_3H_5(CH_3)_2COH$	liq.	886.5
Allylene	$CH_3C:CH$	gas	465.1
p-Aminoazobenzene	$C_6H_5N_2C_6H_4NH_2$	solid	1574.0
p-Aminophenol	$HOC_6H_4NH_2$	solid	760.0
Amygdalin	$C_{20}H_{27}O_{11}N$	solid	2348.4

Table 9-5 (*Continued*)
HEAT OF COMBUSTION OF ORGANIC COMPOUNDS

Name	Formula	Physical state	Heat of combustion in kg. cal.
Amylacetate	$C_4H_9COOC_2H_5$	liq.	1042.5
Amyl alcohol (ferm.)	$C_4H_9.CH_2OH$	liq.	793.7
Amylpropargyl alconol	$C_8H_{13}OH$	liq.	1192
Amylene	C_5H_{10}	liq.	803.4
Anethole	$C_{10}H_{12}O$	solid	1324.4
Angelic acid	$C_4H_7.COOH$	solid	634.8
Aniline	$C_6H_5NH_2$	liq.	811.7
p-Anisidine	$CH_3OC_6H_4NH_2$	solid	924.0
Anisole	$CH_3OC_6H_5$	liq.	905.1
Anthracene	$C_6H_4:(CH)_2:C_6H_4$	solid	1700.4
Anthraquinone	$C_6H_4:(CO)_2:C_6H_4$	solid	1544.5
Arabinose	$C_5H_{10}O_5$	solid	559.9
Arabitol	$C_5H_{12}O_5$	solid	661.2
Arachidic acid	$C_{20}H_{40}O_2$	solid	3025.9
Asparagine	$C_4H_8N_2O_3$	solid	463.3
Atropic acid	$CH_2:C(C_6H_5).COOH$	solid	1044
Azelaic acid	$(CH_2)_7(COOH)_2$	solid	1141.7
Azobenzene	$C_6H_5N:NC_6H_5$	solid	1555.2
Azoxybenzene	$(C_6H_5)_2O$	solid	1534.5
Behenic acid	$C_{22}H_{44}O_2$	solid	3338.4
Benzalacetone	$CH_3COCH:CHC_6H_5$	solid	1257.4
Benzaldehyde	C_6H_5CHO	liq.	841.3
Benzamide	$C_6H_5CONH_2$	solid	847.6
Benzanilide	$C_6H_5CONHC_6H_5$	solid	1575.5
Benzene	C_6H_6	liq.	782.3
Benzenediazonium nitrate	$C_6H_5N_2NO_3$	solid	782.6
Benzidine	$NH_2C_6H_4.C_6H_4NH_2$	solid	1560.9
Benzil	$C_6H_5CO.COC_6H_5$	solid	1624.6
Benzilic acid	$C_{14}H_{12}O_3$	solid	1618
*Benzoic acid	C_6H_5COOH	solid	771.2
Benzoic anhydride	$(C_6H_5CO)_2O$	solid	1555.1
Benzoin	$C_6H_5COCHOHC_6H_5$	solid	1671.4
Benzonitrile	C_6H_5CN	liq.	865.5
Benzophenone	$C_6H_5COC_6H_5$	solid	1556.5
Benzoyl chloride	C_6H_5COCl	liq.	782.8
Benzoyl peroxide	$(C_6H_5CO)_2O_2$	solid	1551.7
Benzyl alcohol	$C_6H_5CH_2OH$	liq.	894.3
Benzyl amine	$C_6H_5CH_2NH_2$	liq.	969.4
Benzyl carbylamine	$C_6H_5CH_2NC$	liq.	1046.5
Benzyl chloride	$C_6H_5CH_2Cl$	liq.	886.4
Benzyl cyanide	$C_6H_5CH_2CN$	liq.	1023.5
Benzylethylamine	$C_6H_5CH_2NHC_2H_5$	liq.	1290
Bicyclohexane (0, 1, 3)	C_6H_{10}	liq.	912.5
Borneo camphene	$C_{10}H_{16}$	liq.	1470.2
Borneol	$C_{10}H_{18}O$	liq.	1469.6
Brucine	$C_{23}H_{26}O_4N_2$	solid	2933.0
n-Butyl alcohol	C_4H_9OH	liq.	638.6
tert.-Butyl alcohol	$(CH_3)_3COH$	liq.	629.3
n-Butylamine	$C_4H_9NH_2$	liq.	710.6
sec.-Butylamine	$(C_2H_5)(CH_3):CHNH_2$	liq.	713.0
tert.-Butylamine	$(CH_3)_3CNH_2$	liq.	716.0
tert.-Butyl benzene	$(CH_3)_3C.C_6H_5$	liq.	1400.4
n-Butyramide	$C_2H_5.CH_2CONH_2$	solid	596.0
n-Butyric acid	$C_2H_5.CH_2COOH$	liq.	524.3
n-Butyronitrile	$C_2H_5.CH_2CN$	liq.	613.3

*Accepted value by International Union of Pure and Applied Chemistry, Lyons, 1923.

Table 9-5 (*Continued*)
HEAT OF COMBUSTION OF ORGANIC COMPOUNDS

Name	Formula	Physical state	Heat of combustion in kg. cal.
Caffeine	$C_8H_{10}O_2N_4$	solid	1014.2
Camphene	$C_{10}H_{16}$	solid	1468.8
Camphor	$C_{10}H_{16}O$	solid	1411.0
Cane sugar	$C_{12}H_{22}O_{11}$	solid	1349.6
Capric acid	$C_{10}H_{20}O_2$	solid	1458.1
Caproic acid	$CH_3(CH_2)_4COOH$	liq.	831.0
Carbon disulfide	CS_2	liq.	246.6
Carbon tetrachloride	CCl_4	liq.	37.3
Carbonyl sulfide	COS	gas	130.5
Carvacrol	$C_{10}H_{13}OH$	liq.	1354.5
Cetyl alcohol	$C_{16}H_{33}OH$	solid	2504.5
Cetyl palmitate	$C_{15}H_{31}CO_2C_{16}H_{33}$	solid	4872.8
Chloroacetic acid	$CH_2Cl.COOH$	solid	171.0
o-Chlorobenzoic acid	$ClC_6H_4\cdot COOH$	solid	734.5
Chloroform	$CHCl_3$	liq.	89.2
Chrysene	$C_{18}H_{12}$	solid	2139.1
Cinnamic acid (*trans*)	$C_6H_5CH{:}CHCOOH$	solid	1040.2
Cinnamic aldehyde	$C_6H_5CH{:}CHCHO$	liq.	1112.3
Cinnamic anhydride	$(C_6H_5CH{:}CHCO)_2O$	solid	2091.3
d-Citrene	$C_{10}H_{16}$	liq.	1473.0
Citric acid (anhydrous)	$H_3C_6H_5O_7$	solid	474.5
Codeine	$C_{18}H_{21}O_3N\cdot H_2O$	solid	2327.6
Coniine	$C_8H_{17}N$	liq.	1275.5
Creatine (anhydrous)	$C_4H_9O_2N_3$	solid	559.8
Creatinine	$C_4H_7ON_3$	solid	563.4
o-Cresol	$CH_3.C_6H_4OH$	solid	879.5
o-Cresol	$CH_3.C_6H_4OH$	liq.	882.6
m-Cresol	$CH_3.C_6H_4OH$	liq.	880.5
p-Cresol	$CH_3.C_6H_4OH$	solid	880.0
p-Cresol	$CH_3.C_6H_4OH$	liq.	882.5
m-Cresol methylether	$CH_3.C_6H_4OCH_3$	liq.	1057.0
Crotonaldehyde	$CH_3.CH{:}CHCHO$	liq.	542.1
Crotonic acid	$CH_3.CH{:}CHCOOH$	solid	477.7
Cyanacetic acid	$CH_2(CN).COOH$	solid	298.8
Cyanogen	$(CN)_2$	gas	258.3
Cyclobutyl carbinol	$C_4H_7.CH_2OH$	liq.	748
Cycloheptanol	$CH_2(CH_2)_5CHOH$	liq.	1050.2
Cyclohexanol	$CH_2(CH_2)_4CHOH$	liq.	890.7
Cycloheptene	C_7H_{12}	liq.	1049.9
Cycloheptane	$(CH_2)_7$	liq.	1087.3
Cyclohexane	$(CH_2)_6$	liq.	937.8
Cyclohexene	C_6H_{10}	liq.	891.9
Cyclopentane	$(CH_2)_5$	liq.	783.6
Cyclopropane	$(CH_2)_3$	gas	496.8
Cymene (p)	$CH_3.C_6H_4.C_3H_7$	liq.	1402.8
Decahydronaphthalene (*cis*)	$C_{10}H_{18}$	liq.	1502.5
Decahydronaphthalene (*trans*)	$C_{10}H_{18}$	liq.	1499.5
Decane	$C_{10}H_{22}$	liq.	1610.2
Dextrose	$C_6H_{12}O_6$	solid	673.0
Diallyl	$(CH_3CH{:}CH_2)_2$	vapor	903.4
Diamyl ether	$(C_5H_{11})_2O$	liq.	1609.3
Diamylene	$C_{10}H_{20}$	liq.	1582.2
Dibenzyl	$(C_6H_5CH_2)_2$	solid	1810.6

Table 9-5 (*Continued*)
HEAT OF COMBUSTION OF ORGANIC COMPOUNDS

Name	Formula	Physical state	Heat of combustion in kg. cal.
Dibenzylamine	$(C_6H_5CH_2)_2NH$	solid	1853.0
o-Dichlorobenzene	$C_6H_4Cl_2$	liq.	671.8
Diethylacetic acid	$(C_2H_5)_2CH.COOH$	liq.	830.8
Diethylamine	$(C_2H_5)_2NH$	liq.	716.9
Diethyl aniline	$C_6H_5N(C_2H_5)_2$	liq.	1451.6
Diethyl carbonate	$(C_2H_5O)_2CO$	liq.	647.9
Diethyl ether	$(C_2H_5)_2O$	liq.	651.7
Diethyl ketone	$(C_2H_5)_2CO$	liq.	735.6
Diethyl malonate	$CH_2(COOC_2H_5)_2$	liq.	860.4
Diethyl oxalate	$(COOC_2H_5)_2$	liq.	716.0
Diethyl succinate	$(CH_2COOC_2H_5)_2$	liq.	1007.3
Dihydrobenzene	C_6H_8	liq.	847.8
Δ_1-Dihydronaphthalene	$C_{10}H_{10}$	liq.	1296.3
Δ_1-Dihydronaphthalene	$C_{10}H_{10}$	solid	1298.3
Dihydroxyanthraquinone (1, 2)	$(HO)_2C_{14}H_6O_2$	solid	1448.9
Diisoamyl	$(C_5H_{11})_2$	liq.	1615.8
Diisobutylene	$(C_4H_8)_2$	liq.	1252.4
Diisopropyl	$(C_3H_7)_2$	vapor	993.9
Diisopropyl ketone	$(C_3H_7)_2CO$	liq.	1045.5
Dimethylamine	$(CH_3)_2NH$	liq.	416.7
Dimethyl aniline	$C_6H_5N(CH_3)_2$	liq.	1142.7
Dimethyl carbonate	$(CH_3O)_2CO$	liq.	340.8
Dimethyl ether	$(CH_3)_2O$	gas	347.6
Dimethylethyl carbinol	$(CH_3)_2C_2H_5COH$	liq.	784.6
Dimethyl fumarate	$(CHCOOCH_3)_2$	solid	664.3
Dimethyl hexane (2, 5)	C_8H_{18}	liq.	1303.3
Dimethyl hexane (3, 4)	C_8H_{18}	liq.	1303.7
Dimethyl maleate	$(CHCOOCH_3)_2$	solid	669.2
Dimethyl malonic acid	$(CH_3)_2C(COOH)_2$	solid	515.1
Dimethyl oxalate	$(COOCH_3)_2$	solid	401.9
Dimethyl pentane (2, 2)	C_7H_{16}	liq.	1148.9
Dimethyl pentane (2, 3)	C_7H_{16}	liq.	1148.9
Dimethyl pentane (2, 4)	C_7H_{16}	liq.	1148.9
Dimethyl pentane (3, 3)	C_7H_{16}	liq.	1147.9
Dimethyl phthalate	$C_6H_4(COOCH_3)_2$	liq.	1119.7
α, β-Dimethyl styrene	$C_{20}H_{16}$	solid	1357.2
Dimethyl succinate	$(CH_2COOCH_3)_2$	solid	703.3
m-Dinitrobenzene	$C_6H_4(NO_2)_2$	solid	696.8
Dinitrophenol (2, 4)	$(HO)C_6H_3(NO_2)_2$	solid	648.0
Dinitrotoluene (2, 4)	$CH_3.C_6H_3(NO_2)_2$	solid	852.8
Diphenyl	$C_6H_5.C_6H_5$	solid	1493.6
Diphenylamine	$(C_6H_5)_2NH$	solid	1536.2
Diphenyl carbinol	$(C_6H_5)_2CHOH$	solid	1615.4
Diphenyl methane	$(C_6H_5)_2CH_2$	solid	1655.0
Diphenylnitrosamine	$(C_6H_5)_2N.NO$	solid	1532.6
Diphenylstyrene	$C_{20}H_{16}$	solid	2496
Dipropargyl	$(CH{:}C.CH_2)_2$	vapor	882.9
Dipropyl ketone	$(C_3H_7)_2CO$	liq.	1050.5
Dulcitol	$C_6H_{14}O_6$	solid	729.1
Durene (1, 2, 4, 5)	$(CH_3)_4C_6H_2$	solid	1393.6
Eicosane	$C_{20}H_{42}$	solid	3183.1
Erythritol	$C_4H_{10}O_4$	solid	504.1
Ethane	$CH_3.CH_3$	gas	368.4
Ethyl acetate	$CH_3COOC_2H_5$	liq.	536.9
Ethyl acetoacetate	$CH_3COCH_2COOC_2H_5$	liq.	690.8

Table 9-5 (*Continued*)
HEAT OF COMBUSTION OF ORGANIC COMPOUNDS

Name	Formula	Physical state	Heat of combustion in kg. cal.
Ethyl alcohol	C_2H_5OH	liq.	327.6
Ethylamine	$C_2H_5NH_2$	liq.	408.5
Ethyl aniline	$C_6H_5NHC_2H_5$	liq.	1121.5
Ethylbenzene	$C_2H_5.C_6H_5$	liq.	1091.2
Ethyl benzoate	$C_6H_5COOC_2H_5$	liq.	1098.7
Ethyl bromide	C_2H_5Br	vapor	340.5
Ethyl *n*-butyrate	$C_2H_5.CH_2COOC_2H_5$	liq.	851.2
Ethyl carbylamine	C_2H_5NC	liq.	477.1
Ethyl chloride	C_2H_5Cl	vapor	316.7
Ethyl cycloheptane	$C_7H_{13}.C_2H_5$	liq.	1406.8
Ethyl formate	$HCOOC_2H_5$	liq.	391.7
Ethyl hexane (3)	$C_3H_7.CH(C_2H_5)_2$	liq.	1302.3
Ethyl iodide	C_2H_5I	liq.	356.0
Ethyl isobutyrate	$(CH_3)_2CH.COOC_2H_5$	liq.	845.7
Ethyl isocyanate	C_2H_5NCO	liq.	424.5
Ethyl mercaptan	C_2H_5SH	gas	452.0
Ethyl mercaptan	C_2H_5SH	liq.	517.2
Ethyl nitrate	$C_2H_5ONO_2$	vapor	322.4
Ethyl nitrite	C_2H_5ONO	vapor	332.6
Ethylpentane (3)	$(C_2H_5)_3CH$	liq.	1149.9
Ethyl propionate	$C_2H_5COOC_2H_5$	liq.	690.8
Ethyl salicylate	$HOC_6H_4COOC_2H_5$	liq.	1051.2
Ethylsuccinic anhydride	$C_6H_8O_3$	liq.	684.8
Ethyl thiocyanate	C_2H_5CNS	liq.	613.8
Ethylurea	$C_2H_5NHCONH_2$	solid	472
Ethyl valerate	$C_4H_9COOC_2H_5$	liq.	1017.5
Ethylene	C_2H_4	gas	331.6
Ethylene chloride	$ClCH_2.CH_2Cl$	vapor	271.0
Ethylene diamine	$(CH_2NH_2)_2$	liq.	452.6
Ethylene glycol	$HOCH_2.CH_2OH$	liq.	281.9
Ethylene iodide	$ICH_2.CH_2I$	solid	324.8
Ethylene oxide	$O.CH_2.CH_2$	liq.	302.1
Ethylidene chloride	$CH_3.CHCl_2$	liq.	267.1
Eugenol	$C_{10}H_{12}O_2$	liq.	1286.6
Fenchane	$C_{10}H_{18}$	liq.	1502.6
Fluorene	$CH_2:(C_6H_4)_2$	solid	1584.9
Fluorobenzene	C_6H_5F	liq.	747.2
Formaldehyde	CH_2O	gas	134.1
Formamide	$HCONH_2$	solid	134.9
Formanilide	$HCONHC_6H_5$	solid	861.0
Formic acid	$HCOOH$	liq.	62.8
l-Fructose	$C_6H_{12}O_6$	solid	675.6
Fucose	$C_6H_{12}O_5$	solid	712.0
Fumaric acid (*trans*)	$(CHCOOH)_2$	solid	320.0
Furfural	$C_4H_3O.CHO$	liq.	559.5
Galactose	$C_6H_{12}O_6$	solid	670.7
Gallic acid	$(HO)_3C_6H_2COOH$	solid	633.7
Geranic acid	$C_{10}H_{16}O_2$	liq.	1379
l-Glucosan	$C_6H_{10}O_5$	solid	678
d-Glucose	$C_6H_{12}O_6$	solid	673.0
Glutamic acid	$C_5H_9O_4N$	solid	542.4
Glutaric acid	$(CH_2)_3(COOH)_2$	solid	514.9
Glutaric anhydride	$C_5H_6O_3$	solid	528.0
Glycerol	$C_3H_5(OH)_3$	liq.	397.0

Table 9-5 (*Continued*)
HEAT OF COMBUSTION OF ORGANIC COMPOUNDS

Name	Formula	Physical state	Heat of combustion in kg. cal.
Glyceryl tributyrate	$C_{15}H_{26}O_6$	liq.	1941.1
Glycine	$H_2N.CH_2COOH$	solid	234.5
Glycogen	$1/x(C_6H_{10}O_5)_x$	solid	4186.8
Glycol	$(CH_2OH)_2$	liq.	281.9
Glycolic acid	$HOCH_2COOH$	solid	166.6
Glycolic nitrile	$HOCH_2CN$	liq.	256.7
Glycylglycine	$C_4H_8O_3N_2$	solid	470.7
Guanine	$C_5H_5ON_5$	solid	586
n-Heptaldehyde	$C_6H_{13}CHO$	liq.	1062.4
n-Heptane	C_7H_{16}	liq.	1149.9
Heptine (1)	$CH:C(CH_2)_4CH_3$	liq.	1091.2
n-Heptyl alcohol	$C_6H_{13}CH_2OH$	liq.	1104.9
Heptylamine	$C_7H_{15}NH_2$	liq.	1178.9
Heptylic acid	$C_6H_{13}COOH$	liq.	986.1
n-Hexane	C_6H_{14}	liq.	989.8
Hexachlorobenzene	C_6Cl_6	solid	509.0
Hexachloroethane	C_2Cl_6	solid	110.0
Hexadecane	$C_{16}H_{34}$	solid	2559.1
Hexahydrobenzoic acid	$C_6H_{11}COOH$	liq.	934
Hexahydronaphthalene	$C_{10}H_{14}$	liq.	1419.3
Hexamethylbenzene	$C_6(CH_3)_6$	solid	1711.9
Hexamethylenetetramine	$(CH_2)_6N_4$	solid	1006.7
Hexamethylethane	$(CH_3)_3C.C(CH_3)_3$	solid	1301.8
Hexylamine	$C_6H_{13}NH_2$	liq.	1022.2
Hexylene	C_6H_{12}	liq.	952.6
Hippuric acid	$C_6H_5CONHCH_2COOH$	solid	1012.4
Hydantoic acid	$C_3H_6O_3N_2$	solid	308.6
Hydrazobenzene	$C_6H_5NH.NHC_6H_5$	solid	1597.3
Hydrobenzoin	$C_{14}H_{14}O_2$	solid	1723
α, β-Hydromuconic acid	$C_6H_8O_4$	solid	628.8
Hydrosorbic acid	$C_6H_{10}O_2$	liq.	795
Hydroquinol	$C_6H_4(OH)_2$ (1, 4)	solid	683.7
Hydroquinoldimethyl ether	$C_6H_4(OCH_3)_2$	solid	1014.7
p-Hydroxyazobenzene	$C_6H_5N_2C_6H_4OH$	solid	1502.0
o-Hydroxybenzaldehyde	$(HO)C_6H_4.CHO$	liq.	796.0
m-Hydroxybenzaldehyde	$(HO)C_6H_4.CHO$	solid	788.7
p-Hydroxybenzaldehyde	$(HO)C_6H_4.CHO$	solid	792.7
m-Hydroxybenzoic acid	$(HO)C_6H_4.COOH$	solid	726.1
p-Hydroxybenzoic acid	$(HO)C_6H_4.COOH$	solid	725.4
β-Hydroxybutyraldehyde	$CH_3CHOHCH_2CHO$	liq.	546.6
Indigo	$C_{16}H_{10}O_2N_2$	solid	1815.0
Indole	C_8H_7N	solid	1022.2
Inositol	$C_6H_6(OH)_6$	solid	662.1
Iodoform	CHI_3	solid	161.9
Isatin	$C_8H_5O_2N$	solid	868
Isoamylamine	$C_5H_{11}NH_2$	liq.	866.8
Isobutane	$(CH_3)_3CH$	gas	683.4
Isobutyl alcohol	$C_3H_7.CH_2OH$	liq.	638.2
Isobutylamine	$C_3H_7.CH_2NH_2$	liq.	713.6
Isobutyl chloride	$C_3H_7.CH_2Cl$	solid	635.5
Isobutylene	$CH_2:C(CH_3)_2$	gas	647.2
Isobutyraldehyde	$(CH_3)_2CH.CHO$	vapor	596.8
Isobutyramide	$(CH_3)_2CHCONH_2$	solid	595.9
Isobutyric acid	$(CH_3)_2CH.COOH$	liq.	517.4
Isoeugenol	$C_{10}H_{12}O_2$	liq.	1277.6

Hydrazine, N_2H_4, liquid, 148.6 kg cal/mole.

Table 9-5 (*Continued*)
HEAT OF COMBUSTION OF ORGANIC COMPOUNDS

Name	Formula	Physical state	Heat of combustion in kg. cal.
Isopentane	C_5H_{12}	gas	843.5(?)
Isopentane	C_5H_{12}	liq.	838.3(?)
Isophthalic acid	$C_6H_4(COOH)_2$ (1, 3)	solid	768.3
Isopropyl alcohol	$(CH_3)_2CHOH$	liq.	474.8
Isopropylbenzene	$C_6H_5.CH(CH_3)_2$	liq.	1247.3
Isopropyl iodide	$(CH_3)_2CHI$	liq.	509.1
Isopropyltoluene	$CH_3.C_6H_4.CH(CH_3)_2$ (1, 3)	liq.	1409.5
Isopropyltoluene	$CH_3.C_6H_4.CH(CH_3)_2$ (1, 4)	liq.	1402.8
Isosafrole	$C_{10}H_{10}O_2$	liq.	1233.9
Isostilbene	$C_{14}H_{12}$	liq.	1770.9
Lactic acid	$CH_3CHOH.COOH$	liq.	326.0
Lactose (anhydrous)	$C_{12}H_{22}O_{11}$	solid	1350.8
Lactose octoacetate	$C_{28}H_{38}O_{19}$	solid	3029.3
Lauric acid	$C_{12}H_{24}O_2$	solid	1771.7
Laurolene	C_8H_{14}	liq.	1192.7
Leucine	$C_6H_{13}O_2N$	solid	855.6
d-Limonene	$C_{10}H_{16}$	liq.	1471.2
l-Limonene	$C_{10}H_{16}$	liq.	1457.2
Maleic acid (*cis*)	$(CHCOOH)_2$	solid	326.1
Maleic anhydride	$(CHCO)_2O$	solid	333.9
l-Malic acid	$C_2H_3OH:(COOH)_2$	solid	320.1
Malonamide	$CH_2(CONH_2)_2$	solid	359
Malonic acid	$CH_2(COOH)_2$	solid	207.2
Malononitrile	$CH_2(CN)_2$	solid	394.8
Maltose	$C_{12}H_{22}O_{11}$	solid	1350.2
Mandelic acid	$C_6H_5CHOHCOOH$	solid	890.3
d-Mannitol	$C_6H_{14}O_6$	solid	727.6
Menthene	$C_{10}H_{18}$	liq.	1523.2
Menthol	$C_{10}H_{20}O$	solid	1508.8
Mesitylene	$C_6H_3(CH_3)_3$ (1, 3, 5)	liq.	1243.6
Mesityl oxide	$CH_3COCH:C(CH_3)_2$	liq.	846.7
Mesotartaric acid	$(CHOH)_2(COOH)_2$	solid	276.0
Methane	CH_4	gas	210.8
Methyl acetate	CH_3COOCH_3	liq.	381.2
Methylal	$C_3H_8O_2$	liq.	462.8
Methyl alcohol	CH_3OH	liq.	170.9
Methylamine	CH_3NH_2	liq.	256.1
Methyl aniline	$C_6H_5NHCH_3$	liq.	973.5
Methyl benzoate	$C_6H_5COOCH_3$	liq.	943.5
Methyl bromide	CH_3Br	vapor	184.0
Methylbutyl ketone	$CH_3COC_4H_9$	liq.	895.2
Methyl *tert*-butyl ketone	$CH_3COC(CH_3)_3$	solid	891.8
Methyl butyrate	$C_3H_7COOCH_3$	liq.	692.8
Methyl carbylamine	CH_3NC	liq.	320.1
Methyl chloride	CH_3Cl	gas	164.2
Methyl cinnamate	$C_6H_5CH:CHCOOCH_3$	solid	1213.0
Methylcyclobutane	$CH_3CHCH_2CH_2CH_2$ $\overline{}$	liq.	784.2
Methylcycloheptane	$C_7H_{13}.CH_3$	liq.	1244.5
Methylcyclohexane	$C_6H_{11}.CH_3$	liq.	1091.8
Methylcyclopentane	$CH_3CHCH_2CH_2CH_2CH_2$ $\overline{}$	liq.	937.9
Methyldiethyl carbinol	$(CH_3)(C_2H_5)_2COH$	liq.	927.0
Methylene chloride	CH_2Cl_2	vapor	106.8
Methylene iodide	CH_2I_2	liq.	178.4

Table 9-5 (*Continued*)
HEAT OF COMBUSTION OF ORGANIC COMPOUNDS

Name	Formula	Physical state	Heat of combustion in kg. cal.
Methylethyl ether	$CH_3OC_2H_5$	vapor	503.4
Methylethyl ketone	$CH_3COC_2H_5$	liq.	582.3
Methyl formate	$HCOOCH_3$	liq.	233.1
Methylheptane (2)	$CH_3.C_7H_{18}$	liq.	1306.1
Methylhexane (2)	$CH_3.C_6H_{13}$	liq.	1148.9
Methylhexane (3)	$CH_3.C_6H_{13}$	liq.	1148.9
Methylhexyl ketone	$CH_3COC_6H_{13}$	liq.	1205.1
Methyl iodide	CH_3I	liq.	194.7
Methyl isobutyrate	$(CH_3)_2CHCOOCH_3$	liq.	694.2
Methyl isocyanate	CH_3NCO	liq.	269.4
Methylisopropyl ketone	$CH_3COCH(CH_3)_2$	liq.	733.9
Methyl lactate	$CH_3CHOHCOOCH_3$	liq.	497.2
Methyl mercaptan	CH_3SH	gas	297.6
Methyl propionate	$C_2H_5COOCH_3$	vapor	552.3
Methylpropyl ketone	$CH_3COC_3H_7$	liq.	735.6
Methyl salicylate	$HOC_6H_4COOCH_3$	liq.	898.3
Milk sugar	$C_{12}H_{22}O_{11}$	solid	1350.8
Methyl tetryl	$C_8H_7O_8N_5$	solid	1009
Methyl thiocyanate	CH_3CNS	liq.	453.1
Morphine	$C_{17}H_{19}O_3N.H_2O$	solid	2146.3
Mucic acid	$C_6H_{10}O_8$	solid	483.6
Myristic acid	$C_{14}H_{28}O_2$	solid	2085.8
Naphthalene	$C_{10}H_8$	solid	1232.5
Naphthalic acid	$C_{10}H_6(COOH)_2$	solid	1244
α-Naphthoic acid	$C_{10}H_7COOH$	solid	1231.8
β-Naphthoic acid	$C_{10}H_7COOH$	solid	1227.6
α-Naphthol	$C_{10}H_7OH$	solid	1185.4
β-Naphthol	$C_{10}H_7OH$	solid	1187.2
α-Naphthonitrile	$C_{10}H_7CN$	solid	1326.2
β-Naphthonitrile	$C_{10}H_7CN$	solid	1321.0
α-Naphthoquinone	$C_{10}H_6O_2$	solid	1100.8
β-Naphthoquinone	$C_{10}H_6O_2$	solid	1106.4
α-Naphthylamine	$C_{10}H_7NH_2$	solid	1263.5
β-Naphthylamine	$C_{10}H_7NH_2$	solid	1261.0
Narceine	$C_{23}H_{27}O_8N.2H_2O$	solid	2802.9
Narcotine	$C_{22}H_{23}O_7N$	solid	2644.5
Nicotine	$C_{10}H_{14}N_2$	liq.	1427.7
m-Nitroacetanilide	$NO_2.C_6H_4NHCOCH_3$	solid	970
o-Nitroaniline	$NO_2.C_6H_4NH_2$	solid	765.8
p-Nitroaniline	$NO_2.C_6H_4NH_2$	solid	761.0
m-Nitroaniline	$NO_2.C_6H_4NH_2$	solid	765.2
m-Nitrobenzaldehyde	$NO_2.C_6H_4CHO$	solid	800.4
Nitrobenzene	$C_6H_5NO_2$	liq.	739.2
m-Nitrobenzoic acid	$NO_2.C_6H_4COOH$	solid	729.1
Nitroethane	$CH_3.CH_2NO_2$	liq.	322.2
Nitroglycerine	$C_3H_5(NO_3)_3$	liq.	432.4
Nitromethane	CH_3NO_2	liq.	169.4
p-Nitromethylaniline	$NO_2.C_6H_4NHCH_3$	solid	924
p-Nitrophenetole	$NO_2.C_6H_4OC_2H_5$	liq.	1006
o-Nitrophenol	$NO_2.C_6H_4OH$	solid	689.1
m-Nitrophenol	$NO_2.C_6H_4OH$	solid	684.4
p-Nitrophenol	$NO_2.C_6H_4OH$	solid	688.8
Nitropropane	$C_2H_5.CH_2NO_2$	liq.	477.9
o-Nitrotoluene	$NO_2.C_6H_4CH_3$	liq.	897.0
p-Nitrotoluene	$NO_2C_6H_4CH_3$	solid	888.6

Table 9-5 (*Continued*)
HEAT OF COMBUSTION OF ORGANIC COMPOUNDS

Name	Formula	Physical state	Heat of combustion in kg. cal.
Octahydronaphthalene	$C_{10}H_{16}$	liq.	1461.7
n-Octane	C_8H_{18}	liq.	1302.7
Octyl alcohol	$C_8H_{17}OH$	liq.	1262.0
Oleic acid	$C_{17}H_{33}COOH$	liq.	2657.0
Opianic acid	$C_{10}H_{10}O_5$	solid	1090
Oxamic acid	$H_2NCO.COOH$	solid	130
Oxalic acid	$(COOH)_2$	solid	60.2
Oxamide	$(CONH_2)_2$	solid	203.2
Palmitic acid	$C_{15}H_{31}COOH$	solid	2398.4
Papaverine	$C_{20}H_{21}O_4N$	solid	2478.1
Paraformaldehyde	$1/n(CH_2O)n$	solid	122
Pentamethylbenzene	$(CH_3)_5C_6H$	solid	1554.0
n-Pentane	C_5H_{12}	gas	838.3
n-Pentane	C_5H_{12}	liq.	833.4
Phenacetin	$C_2H_5OC_6H_4NHCOCH_3$	solid	1285.2
Phenanthraquinone	$C_{14}H_8O_2$	solid	1544.0
Phenanthrene	$C_{14}H_{10}$	solid	1692.5
Phenetole	$C_2H_5OC_6H_5$	liq.	1060.3
Phenol	C_6H_5OH	solid	732.2
Phenylacetic acid	$C_6H_5CH_2COOH$	solid	930.2
Phenylacetylene	$C_6H_5C:CH$	liq.	1024.2
Phenylalanine	$C_9H_{11}O_2N$	solid	1111.3
Phenylbutene	$C_6H_5CH_2CH:CHCH_3$	liq.	1361.2
p-Phenylenediamine	$C_6H_4(NH_2)_2$	solid	843.4
Phenylglycine	$C_6H_5NHCH_2COOH$	solid	955.1
Phenylhydrazine	$C_6H_5NH.NH_2$	solid	875.4
Phenylhydroxylamine	C_6H_5NHOH	liq.	803.7
Phenyl iodide	C_6H_5I	liq.	770.7
Phloroglucinol	$C_6H_3(OH)_3$	solid	635.7
Phthalic acid	$C_6H_4(COOH)_2$	solid	771.0
Phthalide	$C_8H_6O_2$	solid	884
Phthalic anhydride	$C_6H_4(CO)_2O$	solid	783.4
Phthalimide	$C_8H_5O_2N$	solid	849.5
Picric acid	$HOC_6H_2(NO_2)_3$ (1, 2, 4, 6)	solid	611.8
Pinacol	$C_6H_{14}O_2$	solid	897.6
Pinacoline	$(CH_3)_3CCOCH_3$	solid	891.8
Piperidine	$C_5H_{11}N$	liq.	826.6
Piperonal	$C_8H_6O_3$	solid	870.7
Piperonylic acid	$C_8H_6O_4$	solid	803.5
Propane	C_3H_8	gas	526.3
Propine	$HC:CCH_3$	gas	465.1
Propionaldehyde	C_2H_5CHO	liq.	434.2
Propionamide	$C_2H_5CONH_2$	solid	439.9
Propionic acid	C_2H_5COOH	liq.	367.2
Propionic anhydride	$(C_2H_5CO)_2O$	liq.	746.6
Propionitrile	C_2H_5CN	liq.	456.4
n-Propyl alcohol	$C_2H_5CH_2OH$	liq.	480.5
Propylamine	$C_3H_7NH_2$	liq.	558.3
n-Propylbenzene	$C_6H_5.C_3H_7$	liq.	1246.4
Propyl bromide	C_3H_7Br	vapor	497.3
Propyl carbylamine	C_3H_7NC	liq.	639.6
Propyl chloride	C_3H_7Cl	vapor	478.3
Propylene	$CH_2:CHCH_3$	gas	490.2
Propylene glycol	$CH_3CHOHCH_2OH$	liq.	431.0
n-Propyl iodide	C_3H_7I	liq.	514.3

Table 9-5 (*Continued*)
HEAT OF COMBUSTION OF ORGANIC COMPOUNDS

Name	Formula	Physical state	Heat of combustion in kg. cal.
n-Propyltoluene (1, 3)	$CH_3.C_6H_4.C_3H_7$	liq.	1405.4
Pseudocumene (1, 2, 4)	$(CH_3)_3C_6H_3$	liq.	1241.7
Pyridine	C_5H_5N	liq.	658.5
Pyrocatechol (1, 2)	$C_6H_4(OH)_2$	solid	684.8
Pyrogallol (1, 2, 3)	$C_6H_3(OH)_3$	solid	638.7
Pyromellitic acid	$C_{10}H_6O_8$	solid	776.8
Pyrrole	C_4H_5N	liq.	567.7
Quercitol	$C_6H_7(OH)_5$	solid	704.2
Quinoline	C_9H_7N	liq.	1123.5
Quinone	$O:C_6H_4:O$	solid	656.6
Raffinose	$C_{18}H_{32}O_6$	solid	2025.5
Retene	$C_{18}H_{18}$	solid	2306.8
Resorcinol (1, 3)	$C_6H_4(OH)_2$	solid	683.0
Resorcinol dimethylether	$C_6H_4(OCH_3)_2$	liq.	1022.6
Rhamnose	$C_6H_{12}O_5$	solid	718.3
Rhamnose triacetate	$C_{12}H_{18}O_8$	solid	1351
Saccharic acid lactone	$C_6H_{10}O_5$	solid	656.6
Safrole	$C_{10}H_{10}O_2$	liq.	1244.1
Salicylaldehyde	HOC_6H_4CHO	liq.	796.0
*Salicylic acid	HOC_6H_4COOH	solid	723.1
Sarcosine	CH_3NHCH_2COOH	solid	401.1
Sebacic acid	$HOOC.(CH_2)_8COOH$	solid	1297.3
Skatole	$CH_3.C_8H_6N$	liq.	1170.5
d-Sorbose	$C_6H_{12}O_6$	solid	668.3
Starch	$1/x(C_6H_{10}O_5)_x$	solid	677.5
Stearic acid	$C_{17}H_{35}COOH$	solid	2711.8
Stearolic acid	$C_{18}H_{32}O_2$	solid	2629
Stilbene	$C_{14}H_{12}$	solid	1765
Strychnine	$C_{21}H_{22}O_2N_2$	solid	2685.7
Styrene	$C_6H_5CH:CH_2$	liq.	1047.1
Suberic acid	$HOOC.(CH_2)_6COOH$	solid	985.2
Succinamide	$(CH_2CONH_2)_2$	solid	509.2
Succinic acid	$(CH_2COOH)_2$	solid	357.1
Succinic acid nitrile	$(CH_2CN)_2$	liq.	545.7
Succinic anhydride	$(CH_2CO)_2O$	solid	369.6
Succinimide	$(CH_2CO)_2NH$	solid	437.9
Sucrose	$C_{12}H_{22}O_{11}$	solid	1349.6
Sylvestrene	$C_{10}H_{16}$	liq.	1464.7
d-Tartaric acid	$(CHOH)_2(COOH)_2$	solid	275.1
dl-Tartaric acid (anhydrous)	$(CHOH)_2(COOH)_2$	solid	278.4
Taurine	$C_2H_7O_3NS$	solid	382.9
Terephthalic acid (1, 4)	$C_6H_4(COOH)_2$	solid	770.4
Terpin hydrate	$C_{10}H_{22}O_3$	solid	1451.0
Terpineol	$C_{10}H_{18}O_2$	solid	1469.5
Tetrachloroquinone	$O:C_6Cl_4:O$	solid	517.7
Tetrahydrobenzene	C_6H_{10}	liq.	891.9
Tetrahydronaphthalene	$C_{10}H_{12}$	liq.	1352.4
Tetrahydroquinaldine	$C_{10}H_{13}N$	liq.	1382
Tetramethylmethane	$(CH_3)_4C$	gas	842.6
Tetrolic acid	$C_4H_4O_2$	solid	452.4
Tetryl	$C_7H_5N_5O_8$	solid	842.3
Thebaine	$C_{19}H_{21}O_3N$	solid	2441.3
Thialdine	$C_6H_{13}NS_2$	solid	1264
Thiophene	C_4H_4S	liq.	670.5
Thiourea	$CS(NH_2)_2$	solid	342.8

*Recommended as a secondary thermochemical standard.

Table 9-5 (*Continued*)
HEAT OF COMBUSTION OF ORGANIC COMPOUNDS

Name	Formula	Physical state	Heat of combustion in kg. cal.
Thujane	$C_{10}H_{18}$	liq.	1506.4
Thymol	$C_{10}H_{14}O$	liq.	1353.4
Thymol	$C_{10}H_{14}O$	solid	1349.7
Thymoquinone	$C_{10}H_{12}O_2$	solid	1271.3
Toluene	$C_6H_5.CH_3$	liq.	934.2
o-Toluic acid	$CH_3.C_6H_4.COOH$	solid	928.9
m-Toluic acid	$CH_3.C_6H_4.COOH$	solid	928.6
p-Toluic acid	$CH_3.C_6H_4.COOH$	solid	926.9
o-Toluidine	$CH_3.C_6H_4.NH_2$	liq.	964.3
m-Toluidine	$CH_3.C_6H_4.NH_2$	liq.	965.3
p-Toluidine	$CH_3.C_6H_4.NH_2$	solid	958.4
o-Tolunitrile	$CH_3.C_6H_4CN$	liq.	1030.3
Toluquinone	$C_7H_6O_2$	solid	803.2
Triaminotriphenyl carbinol	$(H_2NC_6H_4)_3COH$	solid	2483.5
Tribenzylamine	$(C_6H_5CH_2)_3N$	solid	2762.1
Trichloroacetic acid	$Cl_3C.COOH$	solid	92.8
Triethylamine	$(C_2H_5)_3N$	liq.	1036.8
Triethyl carbinol	$(C_2H_5)_3COH$	liq.	1080.0
Triisoamylamine	$(C_5H_{11})_3N$	liq.	2459.3
Triisobutylamine	$(C_4H_9)_3N$	liq.	1973.6
Trimesic acid	$C_9H_6O_6$	solid	767
Trimethylamine	$(CH_3)_3N$	liq.	578.6
Trimethylbutane (2, 2, 3)	$(CH_3)_3C_4H_7$	liq.	1147.9
Trimethyl carbinol	$(CH_3)_3COH$	liq.	629.3
Trimethylene	$\overline{CH_2CH_2CH_2}$	gas	496.8
Trimethylene nitrile	C_4H_5N	liq.	581
Trimethylethylene	$(CH_3)_2C{:}CH.CH_3$	liq.	796.0
Trimethylethylene	$(CH_3)_2C{:}CH.CH_3$	vapor	803.6
Trimethylpentane (2, 2, 4)	$(CH_3)_3C_5H_9$	liq.	1303.9
Trimethyl succinic acid	$C_7H_{12}O_4$	solid	829.9
Trinitrobenzene (1, 3, 5)	$C_6H_3(NO_2)_3$	solid	663.7
Trinitroglycerol	$C_3H_5(NO_3)_3$	liq.	432.4
Trinitrotoluene (1, 2, 4, 6)	$CH_3.C_6H_2(NO_2)_3$	solid	820.7
Triphenylamine	$(C_6H_5)_3N$	solid	2267.8
Triphenylbenzene (1, 3, 5)	$(C_6H_5)_3C_6H_3$	solid	2936.7
Triphenyl carbinol	$(C_6H_5)_3COH$	solid	2340.8
Triphenylmethane	$(C_6H_5)_3CH$	solid	2388.7
Triphenyl methyl	$(C_6H_5)_3C$	solid	2378.5
Tyrosine	$C_9H_{11}O_3N$	solid	1070.2
Undecolic acid	$C_{11}H_{18}O_2$	solid	1538
Undecylenic acid	$C_{11}H_{20}O_2$	solid	1580
Undecylic acid	$C_{11}H_{22}O_2$	solid	1615.9
Urea	$CO(NH_2)_2$	solid	151.6
Urethane	$NH_2COOC_2H_5$	solid	397.2
Uric acid	$C_5H_4O_3N_4$	solid	460.2
Uvitic acid	$C_9H_8O_4$	solid	928.3
n-Valeric acid	C_4H_9COOH	liq.	681.6
Vanillin (1, 2, 4)	$HOC_6H_3(OCH_3).CHO$	solid	914.1
Veronal	$C_8H_{12}N_2O_3$	solid	983
o-Xylene	$C_6H_4(CH_3)_2$	liq.	1091.7
m-Xylene	$C_6H_4(CH_3)_2$	liq.	1088.4
p-Xylene	$C_6H_4(CH_3)_2$	liq.	1089.1
Xylose	$C_5H_{10}O_5$	solid	561.5

Table 9-6
HEATS OF SOLUTION

The values given in this table are expressed in kilo-calories per gram mole of compound as shown by the formula and for a dilution as indicated (where this is not indicated the dilution is such that further dilution is without an appreciable, thermal effect); a positive value indicates an evolution of heat during solution. To convert from kilogram calories to B. T. U. multiply by the factor 3.9685.

Name	Formula	Dilution in moles of water	Heat of solution in kg-cal.	Temp. °C.
Acetic acid	$CH_3 \cdot COOH$	200	+0.375	18
Aluminum bromide	$AlBr_3$	2970	+85.3	9
Aluminum chloride	$AlCl_3$	1250	+76.845	18
Aluminum chloride	$AlCl_3$	960	+76.3	9
Aluminum iodide	AlI_3	2260	+89.0	9
Aluminum sulfate	$Al_2(SO_4)_3 \cdot 6H_2O$	....	+56.0	15
Aluminum sulfate	$Al_2(SO_4)_3 \cdot 18H_2O$	....	+8.02	15
Ammonium acetate	$NH_4C_2H_3O_2$	200	+0.25	24
Ammonium bicarbonate	NH_4HCO_3	220–440	−6.3	15
Ammonium bisulfide	NH_4HS	890	−3.25	12.5
Ammonium bisulfate	NH_4HSO_4	200	−0.02	18
Ammonium bromide	NH_4Br	200	−4.38	18
Ammonium chloride	NH_4Cl	200	−3.88	18
Ammonium chloroplatinate	$(NH_4)_2PtCl_4$	660	−8.48	15
Ammonium cyanide	NH_4CN	820	−4.4	15
Ammonium ferrocyanide	$(NH_4)_4Fe(CN)_6 \cdot 3H_2O$	....	−6.8	14
Ammonium fluoride	NH_4F	....	−1.5	15
Ammonium fluosilicate	$(NH_4)_2SiF_6$	2400	−8.4	7
Ammonium iodide	NH_4I	200	−3.55	15
Ammonium nitrate	NH_4NO_3	200	−6.32	18
Ammonium nitrate	NH_4NO_3	220–440	−6.2	10–15
Ammonium nitrite	NH_4NO_2	400	−4.75	12.5
Ammonium oxalate	$(NH_4)_2C_2O_4$	345–690	−8.0	15
Ammonium oxalate	$(NH_4)_2C_2O_4 \cdot H_2O$	395–790	−11.5	15
Ammonium sulfate	$(NH_4)_2SO_4$	400	−2.37	18
Ammonium sulfite	$(NH_4)_2SO_3$	220	−1.54	8
Ammonium persulfate	$(NH_4)_2S_2O_8$	1100	−9.7	9.4
Ammonium thiocyanate	NH_4CNS	....	−5.67	12
Antimony pentachloride	$SbCl_5$	1100	+35.2	15
Antimony trichloride	$SbCl_3$	....	+8.91	15
Arsenic acid	H_3AsO_4	....	−0.4	15
Arsenic trioxide	As_2O_3	....	−7.55	15
Barium acetate	$Ba(C_2H_3O_2)_2$	600	+5.2	10.8
Barium acetate	$Ba(C_2H_3O_2)_2 \cdot 3H_2O$	800	−1.15	18
Barium bromide	$BaBr_2$	400	+4.98	18
Barium bromide	$BaBr_2 \cdot 2H_2O$	400	−4.13	18
Barium chlorate	$Ba(ClO_3)_2$	500–1000	−6.7	10
Barium chlorate	$Ba(ClO_3)_2 \cdot H_2O$	600	−11.24	18
Barium chloride	$BaCl_2$	400	+2.07	18
Barium chloride	$BaCl_2 \cdot 2H_2O$	400	−4.93	18
Barium chloride	$BaCl_2 \cdot 2H_2O$	560	−5.2	10
Barium cyanide	$Ba(CN)_2$	....	+0.89	9
Barium cyanide	$Ba(CN)_2 \cdot H_2O$	....	−2.1	5
Barium cyanide	$Ba(CN)_2 \cdot 2H_2O$	....	−2.56	7
Barium ferrocyanide	$Ba_2Fe(CN)_6 \cdot 6H_2O$	....	−11.4	13.5
Barium fluoride	BaF_2	....	−1.90	15
Barium hydroxide	$Ba(OH)_2$	....	+12.26	18
Barium hydroxide	$Ba(OH)_2 \cdot H_2O$	....	+8.7	12
Barium hydroxide	$Ba(OH)_2 \cdot 8H_2O$	400	−15.21	18
Barium hypophosphate	$Ba(H_2PO_2)_2 \cdot H_2O$	800	+0.29	18

Table 9-6 (*Continued*)
HEATS OF SOLUTION

Name	Formula	Dilution in moles of water	Heat of solution in kg-cal.	Temp. °C.
Barium iodide	BaI_2	....	+10.3	16
Barium iodide	$BaI_2 \cdot 7H_2O$	500	+6.85	18
Barium nitrate	$Ba(NO_3)_2$	800	−5.7	12
Barium nitrate	$Ba(NO_3)_2 \cdot H_2O$	800	−8.6	12
Barium nitride	BaN_6	700	−7.8	19.8
Barium oxide	BaO	666	+35.64	15
Barium oxide	$BaO \cdot H_2O$	666	+11.40	15
Barium oxide	$BaO \cdot 2H_2O$	666	+7.06	15
Barium oxide	$BaO \cdot 9H_2O$	666	−14.50	15
Barium perchlorate	$Ba(ClO_4)_2$	550–1100	−1.8	10
Barium perchlorate	$Ba(ClO_4)_2 \cdot 3H_2O$	650–1300	−9.4	15
Barium persulfate	$BaS_2O_8 \cdot 4H_2O$	1600	−11.8	12
Barium sulfate	$BaSO_4$	....	−5.58	18
Barium sulfide	BaS	....	+7.0	15
Beryllium chloride	$BeCl_2$	....	+44.5	15
Beryllium sulfate	$BeSO_4 \cdot 4H_2O$	400	+1.1	18
Bismuth chloride	$BiCl_3$	1600	+7.83	18
Boric acid	$B_2O_3 \cdot 3H_2O$	800	−10.79	16
Boric oxide	B_2O_3	....	+7.3	15
Cadmium bromide	$CdBr_2$	400	+0.44	18
Cadmium bromide	$CdBr_2 \cdot 4H_2O$	600	−7.29	18
Cadmium chloride	$CdCl_2$	400	+3.01	18
Cadmium chloride	$CdCl_2 \cdot H_2O$	400	+0.625	15
Cadmium chloride	$CdCl_2 \cdot 2H_2O$	400	+0.76	18
Cadmium iodide	CdI_2	400	−0.96	18
Cadmium nitrate	$Cd(NO_3)_2 \cdot H_2O$	400	+4.18	18
Cadmium nitrate	$Cd(NO_3)_2 \cdot 4H_2O$	400	−5.04	18
Cadmium sulfate	$CdSO_4$	400	+10.690	15
Cadmium sulfate	$CdSO_4 \cdot H_2O$	400	+6.05	18
Cadmium sulfate	$CdSO_4 \cdot \frac{8}{3}H_2O$	400	+2.539	15
Calcium acetate	$Ca(C_2H_3O_2)_2$	440	+7.0	15.5
Calcium acetate	$Ca(C_2H_3O_2)_2 \cdot H_2O$	600	+5.4	17
Calcium bromide	$CaBr_2$	400	+24.51	18
Calcium bromide	$CaBr_2 \cdot 6H_2O$	400	−1.09	18
Calcium chloride	$CaCl_2$	300	+17.41	18
Calcium chloride	$CaCl_2 \cdot 6H_2O$	400	−4.31	19.3
Calcium ferrocyanide	$Ca_2Fe(CN)_6 \cdot 12H_2O$	....	+4.6	10
Calcium fluoride	CaF_2	....	+2.70	15
Calcium hydroxide	$Ca(OH)_2$	2500	+2.79	18
Calcium iodide	CaI_2	....	+28.12	15
Calcium nitrate	$Ca(NO_3)_2$	400	+3.95	18
Calcium nitrate	$Ca(NO_3)_2 \cdot 4H_2O$	400	−7.25	18
Calcium oxide	CaO	2500	+18.33	18
Calcium sulfate	$CaSO_4$	....	+4.44	18
Calcium sulfate	$CaSO_4 \cdot 2H_2O$	....	−0.30	18
Calcium sulfide	CaS	....	+6.1	15
Cesium bisulfate	$CsHSO_4$	....	−3.73	15
Cesium bromide	$CsBr$	....	−6.73	15
Cesium chloride	$CsCl$	....	−4.75	15
Cesium fluoride	CsF	....	+8.37	15
Cesium hydroxide	$CsOH$	....	+16.423	15
Cesium hydroxide	$CsOH \cdot H_2O$	....	+4.317	15
Cesium iodide	CsI	....	−8.25	15
Cesium oxide	Cs_2O	....	+83.2	15
Cesium sulfate	Cs_2SO_4	....	−4.97	15

Table 9-6 (*Continued*)
HEATS OF SOLUTION

Name	Formula	Dilution in moles of water	Heat of solution in kg-cal.	Temp. °C.
Chromium bromide (ic)	$CrBr_3 \cdot 6H_2O$-green	500	+0.7	15
Chromium bromide (ic)	$CrBr_3 \cdot 6H_2O$-blue	....	+14.35	15
Chromium chloride (ic)	$CrCl_3$	....	+35.9	15
Chromium chloride (ic)	$2CrCl_3 \cdot 13H_2O$-gray	....	+24.04	15
Chromium chloride (ic)	$2CrCl_3 \cdot 13H_2O$-green	....	−0.1	15
Chromium chloride (ous)	$CrCl_2$	....	+18.6	15
Chromium chloride (ous)	$CrCl_2 \cdot 4H_2O$	....	+2.0	15
Chromium sulfate (ic)	$Cr_2(SO_4)_3 \cdot 8H_2O$	....	+13.6	15
Chromium sulfate (ic)	$Cr_2(SO_4)_3 \cdot 16H_2O$	....	+6.2	15
Chromium trioxide	CrO_3	220	+1.9	19
Cobalt chloride	$CoCl_2$	400	+18.34	18
Cobalt chloride	$CoCl_2 \cdot 6H_2O$	400	−2.85	18
Cobalt nitrate	$Co(NO_3)_2$	280	+11.82	8
Cobalt nitrate	$Co(NO_3)_2 \cdot 6H_2O$	400	−4.96	18
Cobalt sulfate	$CoSO_4 \cdot 7H_2O$	800	−3.57	18
Copper acetate	$Cu(C_2H_3O_2)_2$	320	+2.4	16
Copper acetate	$Cu(C_2H_3O_2)_2 \cdot H_2O$	400	+0.2	18
Copper bromide	$CuBr_2$	400	+8.25	18
Copper bromide	$CuBr_2 \cdot 4H_2O$	....	−1.5	7.5
Copper chloride	$CuCl_2$	600	+11.08	18
Copper chloride	$CuCl_2 \cdot 2H_2O$	400	+4.21	18
Copper nitrate	$Cu(NO_3)_2$	280	+10.47	8
Copper nitrate	$Cu(NO_3)_2 \cdot 6H_2O$	400	−10.71	18
Copper sulfate	$CuSO_4$	400	+15.80	18
Copper sulfate	$CuSO_4 \cdot H_2O$	400	+9.34	18
Copper sulfate	$CuSO_4 \cdot 5H_2O$	400	−2.75	18
Cyanogen	C_2N_2	....	−73.0	15
Dysprosium sulfate	$Dy_2(SO_4)_3 \cdot 8H_2O$	1200	+6.3	18
Erbium acetate	$Er(C_2H_3O_2)_3 \cdot 4H_2O$	1500	+0.7	18
Gold bromic acid	$HAuBr_4 \cdot 5H_2O$	1000	−11.40	18
Gold bromide	$AuBr_3$	2000	−3.71	18
Gold chloride	$AuCl_3$	900	+4.45	18
Gold chloride	$AuCl_3 \cdot 2H_2O$	600	−1.69	18
Gold chloric acid	$HAuCl_4 \cdot 3H_2O$	400	−3.55	18
Gold chloric acid	$HAuCl_4 \cdot 4H_2O$	400	−5.83	18
Hydrobromic acid	HBr	....	+20.0	15
Hydrochloric acid	HCl	....	+17.4	15
Hydroferrocyanic acid	$H_4Fe(CN)_6$	200	+0.4	10
Hydriodic acid	HI	....	+19.6	15
Hydrogen selenide	H_2Se	....	+9.3	15
Hydrogen sulfide	H_2S	....	+4.56	15
Hydroxylamine	NH_2OH	....	−2.8	15
Iodine pentoxide	I_2O_5	....	−1.79	15
Iron chloride (ous)	$FeCl_2$	350	+17.85	18
Iron chloride (ous)	$FeCl_2 \cdot 2H_2O$	1100	+8.7	20
Iron chloride (ous)	$FeCl_2 \cdot 4H_2O$	400	+2.75	18
Iron chloride (ic)	$FeCl_3$	1000	+32.68	18
Iron chloride (ic)	$FeCl_3 \cdot 2\frac{1}{2}H_2O$	1200	+21.0	18
Iron chloride (ic)	$FeCl_3 \cdot 6H_2O$	1200	+5.65	20.8
Iron nitrate (ic)	$Fe(NO_3)_3 \cdot 9H_2O$	480	−9.0	15
Iron sulfate (ous)	$FeSO_4 \cdot 7H_2O$	400	−4.51	18
Lead acetate	$Pb(C_2H_3O_2)_2$	440	+1.4	16
Lead acetate	$Pb(C_2H_3O_2)_2 \cdot 3H_2O$	800	−6.14	18
Lead bromide	$PbBr_2$	2500	−10.04	18
Lead chloride	$PbCl_2$	1800	−6.8	18
Lead dithionate	$PbS_2O_6 \cdot 4H_2O$	400	−8.54	18

Table 9-6 (*Continued*)
HEATS OF SOLUTION

Name	Formula	Dilution in moles of water	Heat of solution in kg-cal.	Temp. °C.
Lead nitrate	$Pb(NO_3)_2$	400	-7.61	18
Lithium bromide	$LiBr$	....	$+11.25$	15
Lithium chloride	$LiCl$	230	$+8.37$	15
Lithium chloride	$LiCl \cdot H_2O$	....	$+4.211$	15
Lithium chloride	$LiCl \cdot 2H_2O$	....	$+0.981$	15
Lithium fluoride	LiF	....	-1.04	15
Lithium fluosilicate	Li_2SiF_6	800	$+1.8$	15
Lithium hydroxide	$LiOH$	400	$+5.82$	15
Lithium hydroxide	$LiOH \cdot H_2O$	....	$+0.720$	18
Lithium iodide	LiI	....	$+14.76$	15
Lithium nitrate	$LiNO_3$	100	$+0.30$	18
Lithium oxide	Li_2O	222	$+30.76$	15
Lithium oxide	$4Li_2O \cdot 5H_2O$	888	$+8.182$	15
Lithium oxide	$4Li_2O \cdot 3H_2O$	888	$+16.026$	15
Lithium peroxide	Li_2O_2	....	$+7.19$	15
Lithium selenide	Li_2Se	....	$+10.7$	20
Lithium selenide	$Li_2Se \cdot 9H_2O$	1146–6426	-12.2	15
Lithuim sulfate	Li_2SO_4	200	$+6.05$	18
Lithium sulfate	$Li_2SO_4 \cdot H_2O$	400	$+3.41$	18
Magnesium ammonium sulfate	$MgSO_4 \cdot (NH_4)_2SO_4 \cdot 6H_2O$	....	-9.7	15
Magnesium bromide	$MgBr_2$	....	$+43.3$	15
Magnesium chloride	$MgCl_2$	800	$+35.92$	18
Magnesium chloride	$MgCl_2 \cdot 6H_2O$	400	$+2.95$	18
Magnesium dithionate	$MgS_2O_6 \cdot 6H_2O$	400	-2.96	18
Magnesium fluoride	MgF_2	....	$+2.778$	15
Magnesium hydroxide	$Mg(OH)_2$	....	-0.0	18
Magnesium iodide	MgI_2	....	$+49.8$	15
Magnesium nitrate	$Mg(NO_3)_2 \cdot 6H_2O$	400	-4.22	18
Magnesium potassium sulfate	$MgSO_4 \cdot K_2SO_4$	600	$+10.60$	18
Magnesium potassium sulfate	$MgSO_4 \cdot K_2SO_4 \cdot 6H_2O$	600	-10.02	18
Magnesium sodium sulfate	$MgSO_4 \cdot Na_2SO_4$	....	$+16.7$	19
Magnesium sulfate	$MgSO_4$	400	$+20.28$	18
Magnesium sulfate	$MgSO_4 \cdot H_2O$	400	$+13.30$	18
Magnesium sulfate	$MgSO_4 \cdot 7H_2O$	400	-3.80	18
Manganese chloride	$MnCl_2$	350	$+16.01$	18
Manganese chloride	$MnCl_2 \cdot 4H_2O$	400	$+1.54$	18
Manganese dithionate	$MnS_2O_6 \cdot 6H_2O$	400	-1.93	18
Manganese nitrate	$Mn(NO_3)_2$	280	$+12.39$	14
Manganese nitrate	$Mn(NO_3)_2 \cdot 6H_2O$	400	-6.15	18
Manganese potassium sulfate	$MnSO_4 \cdot K_2SO_4$	600	$+6.38$	18
Manganese potassium sulfate	$MnSO_4 \ K_2SO_4 \cdot 4H_2O$	600	-6.44	18
Manganese sodium sulfate	$MnSO_4 \cdot Na_2SO_4$	....	$+13.0$	15
Manganese sodium sulfate	$MnSO_4 \cdot Na_2SO_4 \cdot 2H_2O$	....	$+3.2$	15
Manganese sodium sulfate	$MnSO_4 \cdot Na_2SO_4 \cdot 6H_2O$	....	-9.7	15
Manganese sulfate	$MnSO_4$	400	$+13.79$	18
Manganese sulfate	$MnSO_4 \cdot H_2O$	400	$+7.82$	18
Manganese sulfate	$MnSO_4 \cdot 5H_2O$	400	$+0.04$	18
Mercurous acetate	$HgC_2H_3O_2$	222	-3.8	13.7
Mercuric bromide	$HgBr_2$	....	-3.4	12
Mercuric chloride	$HgCl_2$	300	-3.30	18

Table 9-6 (*Continued*)
HEATS OF SOLUTION

Name	Formula	Dilution in moles of water	Heat of solution in kg-cal.	Temp. °C.
Mercuric cyanide	$Hg(CN)_2$	....	-2.97	18
Mercuric potassium bromide	HgK_2Br_4	600	-9.75	18
Mercuric potassium chloride	$HgKCl_3$	770	-9.5	14
Mercuric potassium chloride	$HgKCl_3 \cdot H_2O$	800	-11.3	14
Mercuric potassium chloride	HgK_2Cl_4	930	-15.0	14
Mercuric potassium chloride	$HgK_2Cl_4 \cdot H_2O$	970	-16.66	14
Mercuric potassium iodide	HgK_2I_4	800	-9.81	18
Neodymium chloride	$NdCl_3$	....	$+35.4$	17
Neodymium chloride	$NdCl_3 \cdot 6H_2O$	....	$+7.6$	15
Neodymium iodide	NdI_3	....	$+48.9$	19
Neodymium sulfate	$Nd_2(SO_4)_3$	....	$+36.5$	14
Nickel chloride	$NiCl_2$	400	$+19.17$	18
Nickel chloride	$NiCl_2 \cdot 6H_2O$	400	-1.16	18
Nickel dithionate	$NiS_2O_6 \cdot 6H_2O$	400	-2.42	18
Nickel nitrate	$Ni(NO_3)_2$	280	$+11.88$	18
Nickel nitrate	$Ni(NO_3)_2 \cdot 6H_2O$	400	-7.47	18
Nickel sulfate	$NiSO_4 \cdot 7H_2O$	800	-4.25	18
Oxalic acid	$H_2C_2O_4 \cdot 2H_2O$	530	-8.59	15
Perchloric acid	$HClO_4$	....	$+20.3$	15
Phosphoric acid, meta	HPO_3	....	$+10.1$	15
Phosphoric acid, ortho	H_3PO_4	120	$+2.69$	15
Phosphoric acid, pyro	$H_4P_2O_7$	....	$+7.9$	15
Phosphorous acid, hypo	H_3PO_2	....	-0.17	15
Phosphorous acid, ortho	H_3PO_3 (solid)	120	-0.13	15
Phosphorous acid, pyro	$H_4P_2O_5$	550	$+35.6$	15
Phosphorus trichloride	PCl_3	1000	$+65.14$	15
Phosphorus pentoxide	P_2O_5	550	$+35.6$	15
Platinum brom acid	$H_2PtBr_6 \cdot 9H_2O$	....	-2.9	15
Platinum bromide	$PtBr_4$	....	$+9.86$	15
Platinum chlor acid	$H_2PtCl_6 \cdot 6H_2O$	450	$+4.34$	15
Platinum chloride	$PtCl_4$	....	$+19.6$	17
Potassium acetate	$KC_2H_3O_2$	200	$+3.34$	18
Potassium acid arsenate	KH_2AsO_4	....	-4.9	18
Potassium acid carbonate	$KHCO_3$	....	-5.3	10–15
Potassium acid fluoride	$KF \cdot HF$	400	-6.0	15
Potassium acid flouride	$KF \cdot 2HF$	....	-8.0	15
Potassium acid fluoride	$KF \cdot 3HF$	....	-8.6	15
Potassium acid iodate	$KH(IO_3)_2$	865	-11.8	15
Potassium acid oxalate	KHC_2O_4	....	-9.6	15
Potassium acid sulfate	$KHSO_4$	200	-3.80	18
Potassium aluminum sulfate	$KAl(SO_4)_2 \cdot 12H_2O$	1200	-10.1	18
Potassium bromate	$KBrO_3$	200	-9.76	18
Potassium bromide	KBr	200	-5.08	18
Potassium bromoplatinate	K_2PtBr_6	2000	-12.26	18
Potassium carbonate	K_2CO_3	400	$+6.49$	18
Potassium carbonate	$K_2CO_3 \cdot \frac{1}{2}H_2O$	400	$+4.28$	18
Potassium carbonate	$K_2CO_3 \cdot 1\frac{1}{2}H_2O$	400	-0.38	18
Potassium chlorate	$KClO_3$	400	-10.04	18
Potassium chloride	KCl	100	-4.19	21
Potassium chloropalladate	K_2PdCl_6	....	-15.00	18
Potassium chloroplatinate	K_2PtCl_4	600	-12.22	18
Potassium chloroplatinate	K_2PtCl_6	....	-13.76	18
Potassium chromate	K_2CrO_4		-5.25	15
Potassium cyanate	$KCNO$		-5.2	20

Table 9-6 (*Continued*)
HEATS OF SOLUTION

Name	Formula	Dilution in moles of water	Heat of solution in kg-cal.	Temp. °C.
Potassium cyanide	KCN	175	−3.01	18
Potassium dichromate	$K_2Cr_2O_7$	400	−16.70	18
Potassium dithionate	$K_2S_2O_6$	500	−13.01	18
Potassium ferricyanide	$K_3Fe(CN)_6$	400	−14.4	12
Potassium ferrisulfate	$KFe(SO_4)_2 \cdot 12H_2O$	1000	−16.0	15
Potassium ferrocyanide	$K_4Fe(CN)_6$	820	−12.0	12
Potassium ferrocyanide	$K_4Fe(CN)_6 \cdot 3H_2O$	940	−16.9	11
Potassium fluoride	KF	....	+3.6	20
Potassium fluoride	$KF \cdot 2H_2O$	....	−1.0	20
Potassium hydroxide	KOH	250	+13.29	18
Potassium hydroxide	$KOH \cdot 2H_2O$	170	−0.03	15
Potassium iodate	KIO_3	500	−6.78	18
Potassium iodide	KI	200	−5.11	18
Potassium nitrate	KNO_3	200	−8.52	18
Potassium oxalate	$K_2C_2O_4$	465–930	−4.74	10–15
Potassium oxalate	$K_2C_2O_4 \cdot H_2O$	800	−7.41	18
Potassium oxide	K_2O	....	+75.0	15
Potassium pentathionate	$K_2S_5O_6 \cdot 1\frac{1}{2}H_2O$	2030	−13.1	9.5
Potassium perchlorate	$KClO_4$	460	−12.13	20
Potassium permanganate	$KMnO_4$	700	−10.2	16
Potassium persulfate	$K_2S_2O_8$	3300	−14.55	9
Potassium phosphate dihydrogen	KH_2PO_4	....	−4.85	15
Potassium selenide	KSe	1762–1965	+8.5	13
Potassium selenide	$KSe \cdot 9H_2O$	921–4844	−19.2	14
Potassium selenide	$KSe \cdot 14H_2O$	2145–5914	−20.4	13
Potassium selenide	$KSe \cdot 19H_2O$	....	−29.3	14
Potassium silver cyanide	$KAg(CN)_2$	440	−8.55	11
Potassium stannichloride	K_2SnCl_6	800	−3.38	18
Potassium stannochloride	$K_2SnCl_4 \cdot H_2O$	600	−13.42	18
Potassium sulfate	K_2SO_4	400	−6.38	18
Potassium sulfide	K_2S	....	+22.7	
Potassium tetrasulfide	K_2S_4	600	+1.4	10
Potassium tetrasulfide	$K_2S_4 \cdot \frac{1}{2}H_2O$	....	−1.212	15.7
Potassium tetroxalate	$KHC_2O_4 \cdot H_2C_2O_4$	....	−15.7	15
Potassium tetrathionate	$K_2S_4O_6$	500	−13.15	18
Potassium thiosulfate	$K_2S_2O_3$	950	−5.0	10
Potassium thiosulfate	$K_2S_2O_3 \cdot H_2O$	....	−6.2	14
Potassium thrithionate	$K_2S_3O_6$	500	−12.46	18
Rubidium bicarbonate	$RbHCO_3$	....	+4.731	15
Rubidium bisulfate	$RbHSO_4$	....	−3.73	15
Rubidium bromide	RbBr	....	−5.96	15
Rubidium chloride	RbCl	....	−4.50	15
Rubidium fluoride	RbF	....	+5.80	15
Rubidium hydroxide	RbOH	....	+14.264	15
Rubidium hydroxide	$RbOH \cdot H_2O$	....	+3.70	15
Rubidium hydroxide	$RbOH \cdot 2H_2O$	....	−0.65	15
Rubidium iodide	RbI	....	−6.50	15
Rubidium oxide	Rb_2O	....	+80.0	15
Rubidium sulfate	Rb_2SO_4	....	− 6.66	15
Rubidium sulfide	Rb_2S	....	+24.6	15
Selenic acid	H_2SeO_4	....	+16.8	15
Selenium dioxide	SeO_2	....	−0.74	15
Silver acetate	$AgC_2H_3O_2$	120	−4.3	10
Silver chloride	AgCl	....	−15.75	15

Table 9-6 (*Continued*)
HEATS OF SOLUTION

Name	Formula	Dilution in moles of water	Heat of solution in kg-cal.	Temp. °C.
Silver dithionate	$Ag_2S_2O_6 \cdot 2H_2O$	400	−10.36	18
Silver fluoride	AgF	....	+3.4	10
Silver fluoride	$AgF \cdot 2H_2O$	....	−1.5	10
Silver nitrate	$AgNO_3$	200	−5.44	18
Silver nitrite	$AgNO_2$	....	−8.8	15
Silver sulfate	Ag_2SO_4	1400	−4.48	18
Sodium acetate	$NaC_2H_3O_2$	250	+4.1	5.7
Sodium acetate	$NaC_2H_3O_2$	200	+3.87	18
Sodium acetate	$NaC_2H_3O_2 \cdot 3H_2O$	400	−4.81	18
Sodium ammonium phosphate	$NaNH_4HPO_4 \cdot 4H_2O$	800	−10.75	18
Sodium arsenate	$Na_3AsO_4 \cdot 12H_2O$	670	−12.6	18–20
Sodium bicarbonate	$NaHCO_3$	....	−4.3	15
Sodium bifluoride	$NaF \cdot HF$	400	−6.2	12
Sodium bisulfate	$NaHSO_4$	200	+1.19	18
Sodium bisulfide	$NaHS$	....	+4.4	10–16
Sodium bisulfide	$NaHS \cdot 2H_2O$	....	−1.5	17.5
Sodium borate (tetra)	$Na_2B_4O_7$	....	+10.2	15
Sodium borate (tetra)	$Na_2B_4O_7 \cdot 10H_2O$	1600	−25.86	18
Sodium bromide	$NaBr$	200	−0.19	18
Sodium bromide	$NaBr \cdot 2H_2O$	300	−4.71	18
Sodium bromoplatinate	Na_2PtBr_6	600	+9.99	18
Sodium bromoplatinate	$Na_2PtBr_6 \cdot 6H_2O$	800	−8.55	18
Sodium carbonate	Na_2CO_3	400	+5.64	18
Sodium chlorate	$NaClO_3$	180–360	−5.6	10
Sodium chloride	$NaCl$	100	−1.18	18
Sodium chloroplatinate	Na_2PtCl_6	800	+8.54	18
Sodium chloroplatinate	$Na_2PtCl_6 \cdot 6H_2O$	900	−10.63	18
Sodium chromate	Na_2CrO_4	360–720	+2.2	10.5
Sodium chromate	$Na_2CrO_4 \cdot 4H_2O$	650	−7.6	11
Sodium chromate	$Na_2CrO_4 \cdot 10H_2O$	760	−15.8	10.5
Sodium cyanate	$NaCNO$	....	−4.8	12.8
Sodium cyanide	$NaCN$	100	−0.5	9
Sodium cyanide	$NaCN \cdot \frac{1}{2}H_2O$	100	−1.0	6
Sodium cyanide	$NaCN \cdot 2H_2O$	....	−4.4	9
Sodium dithionate	$Na_2S_2O_6$	400	−5.37	18
Sodium dithionate	$Na_2S_2O_6 \cdot 2H_2O$	400	−11.65	18
Sodium fluoride	NaF	400	−0.6	12
Sodium formate	$NaCHO_2$	150	−0.52	11.5
Sodium hydroxide	$NaOH$	200	+9.94	18
Sodium iodide	NaI	200	+1.22	18
Sodium iodide	$NaI \cdot 2H_2O$	300	−4.01	18
Sodium nitrate	$NaNO_3$	200	−5.03	18
Sodium oxide	Na_2O	....	+55.0	15
Sodium perchlorate	$NaClO_4$	200–400	−3.5	10
Sodium phosphate (tri)	$Na_3PO_4 \cdot 12H_2O$	670	−14.5	18-20
Sodium phosphate (di)	Na_2HPO_4	400	+5.64	18
Sodium phosphate (di)	$Na_2HPO_4 \cdot 2H_2O$	400	−0.39	18
Sodium phosphate (di)	$Na_2HPO_4 \cdot 7H_2O$	....	−11.3	15
Sodium phosphate (di)	$Na_2HPO_4 \cdot 12H_2O$	400	−22.83	18
Sodium phosphate (pyro)	$Na_4P_2O_7$	800	+11.85	18
Sodium phosphate (pyro)	$Na_4P_2O_7 \cdot 10H_2O$	800	−11.67	18
Sodium phosphite (di)	Na_2HPO_3	555	+9.15	13.5
Sodium phosphite (di)	$Na_2HPO_3 \cdot 5H_2O$	555	−4.6	13.5
Sodium phosphite (mono)	NaH_2PO_3	555	+0.75	13

Table 9-6 (*Continued*)
HEATS OF SOLUTION

Name	Formula	Dilution in moles of water	Heat of solution in kg-cal.	Temp. °C.
Sodium phosphite (mono)	$NaH_2PO_3 \cdot 2\frac{1}{2}H_2O$	555	−5.3	15
Sodium selenide	Na_2Se	789–2587	+18.6	14
Sodium selenide	$Na_2Se \cdot 4\frac{1}{2}H_2O$	1030–2125	−7.9	13
Sodium selenide	$Na_2Se \cdot 9H_2O$	723–1352	−10.6	12
Sodium selenide	$Na_2Se \cdot 16H_2O$	1476–3572	−22.0	14
Sodium sulfate	Na_2SO_4	400	+0.46	18
Sodium sulfate	$Na_2SO_4 \cdot H_2O$	400	−1.90	18
Sodium sulfate	$Na_2SO_4 \cdot 10H_2O$	400	−18.76	18
Sodium sulfide	Na_2S	584–1027	+15.0	14.5
Sodium sulfide	$Na_2S \cdot 4\frac{1}{2}H_2O$	589–1059	−5.0	17
Sodium sulfide	$Na_2S \cdot 5H_2O$	513–1167	−6.6	17
Sodium sulfide	$Na_2S \cdot 9H_2O$	774–1495	−16.72	13
Sodium sulfite	Na_2SO_3	....	+2.5	10
Sodium sulfite	$Na_2SO_3 \cdot 7H_2O$	490	−11.2	10
Sodium tetra sulfide	Na_2S_4	....	+9.8	16.5
Sodium tetrathionate	$Na_2S_4O_6 \cdot 2H_2O$	620	−9.7	9.6
Sodium thiosulfate	$Na_2S_2O_3$	440	+1.7	15
Sodium thiosulfate	$Na_2S_2O_3 \cdot 5H_2O$	400	−11.37	18
Sodium trithionate	$Na_2S_3O_6 \cdot 3H_2O$	675	−10.1	10
Strontium acetate	$Sr(C_2H_3O_2)_2$	300	+5.6	11.5
Strontium acetate	$Sr(C_2H_3O_2)_2 \cdot \frac{1}{2}H_2O$	440	+5.3	12
Strontium bromide	$SrBr_2$	400	+16.11	18
Strontium bromide	$SrBr_2 \cdot 6H_2O$	400	−7.22	18
Strontium chloride	$SrCl_2$	400	+11.14	18
Strontium chloride	$SrCl_2 \cdot 6H_2O$	400	−7.50	18
Strontium cyanide	$Sr(CN)_2 \cdot 4H_2O$	100	−4.15	8
Strontium dithionate	$SrS_2O_6 \cdot 4H_2O$	400	−9.25	18
Strontium fluoride	SrF_2	....	−2.10	15
Strontium hydroxide	$Sr(OH)_2$	....	+11.64	18
Strontium hydroxide	$Sr(OH)_2 \cdot 8H_2O$	....	−14.64	18
Strontium iodide	SrI_2	....	+20.5	12
Strontium iodide	$SrI_2 \cdot 7H_2O$	....	−4.47	15
Strontium nitrate	$Sr(NO_3)_2$	400	−4.62	18
Strontium nitrate	$Sr(NO_3)_2 \cdot 4H_2O$	400	−12.30	18
Strontium oxide	SrO	....	+29.34	18
Strontium oxide	$SrO \cdot 0.14H_2O$	1111	+26.10	15
Strontium oxide	$SrO \cdot H_2O$	1111	+10.33	15
Strontium oxide	$SrO \cdot 2H_2O$	1111	+5.26	15
Strontium oxide	$SrO \cdot 9H_2O$	1111	−14.27	15
Strontium selenide	$SrSe$	....	+7.4	15
Sulfur dioxide	SO_2	300	+1.50	15
Sulfur trioxide	SO_3	1600	+39.17	15
Sulfur heptoxide	S_4O_7	....	+37.29	15
Sulfuric acid	H_2SO_4	1600	+17.85	15
Thallium chloride	$TlCl$	4500	−10.10	18
Thallium hydroxide	$TlOH$	235	−3.15	18
Thallium nitrate	$TlNO_3$	300	−9.97	18
Thallium oxide	Tl_2O	570	−3.08	18
Thallium sulfate	Tl_2SO_4	1600	−8.28	18
Tin bromide (ic)	$SnBr_4$	970	+16.6	10.5
Tin bromide (ous)	$SnBr_2$	1080	−1.6	15
Tin chloride (ic)	$SnCl_4$	720	+28.5	10.5
Tin chloride (ous)	$SnCl_2$	300	+0.35	18
Tin chloride (ous)	$SnCl_2 \cdot 2H_2O$	200	−5.37	18
Uranyl acetate	$UO_2(C_2H_3O_2)_2 \cdot 2H_2O$	1000–2500	−4.3	18–20

Table 9-6 (*Continued*)
HEATS OF SOLUTION

Name	Formula	Dilution in moles of water	Heat of solution in kg-cal.	Temp. °C.
Uranyl chloride	$UO_2Cl_2 \cdot H_2O$	1000–2500	+2.0	18–20
Uranyl chromate	$UO_2CrO_4 \cdot 5\frac{1}{2}H_2O$	1000–2500	−6.3	18–20
Uranyl nitrate	$UO_2(NO_3)_2 \cdot 3H_2O$	1000–2500	−3.7	18–20
Uranyl potassium chloride	$UO_2Cl_2 \cdot 2KCl \cdot 2H_2O$	1000–2500	+2.0	18–20
Uranyl sulfate	$UO_2SO_4 \cdot 3H_2O$	1000–2500	+5.1	18–20
Ytterbium sulfate	$Y_2(SO_4)_3 \cdot 8H_2O$	1200	+10.7	18
Zinc acetate	$Zn(C_2H_3O_2)_2$	720	+9.8	22.5
Zinc acetate	$Zn(C_2H_3O_2)_2 \cdot H_2O$	800	+7.0	22.5
Zinc acetate	$Zn(C_2H_3O_2)_2 \cdot 2H_2O$	500	+4.2	10.2
Zinc ammonium chloride	$3ZnCl_2 \cdot 6NH_4Cl \cdot H_2O$	....	+3.23	13
Zinc bromide	$ZnBr_2$	400	+15.03	18
Zinc chloride	$ZnCl_2$	300	+15.63	18
Zinc dithionate	$ZnS_2O_6 \cdot 6H_2O$	400	−2.42	1
Zinc formate	$Zn(CHO_2)_2$	500	+4.0	15
Zinc iodide	ZnI_2	400	+11.31	18
Zinc nitrate	$Zn(NO_3)_2 \cdot 6H_2O$	400	−5.84	18
Zinc potassium sulfate	$ZnSO_4 \cdot K_2SO_4$	600	+7.91	18
Zinc potassium sulfate	$ZnSO_4 \cdot K_2SO_4 \cdot 6H_2O$	600	−11.90	18
Zinc sulfate	$ZnSO_4$	400	+18.43	18
Zinc sulfate	$ZnSO_4 \cdot H_2O$	400	+9.95	18
Zinc sulfate	$ZnSO_4 \cdot 7H_2O$	400	−4.26	18
Zirconyl nitrate	$ZrO(NO_3)_2 \cdot 2H_2O$	....	+2.17	15
Zirconyl nitrate	$ZrO(NO_3)_2 \cdot 3\frac{1}{2}H_2O$	....	−1.92	15

Table 9-7
SPECIFIC HEAT OF GASES AND VAPORS

The values of γ in the table below are the ratios of the specific heat at constant pressure (Cp) to that at constant volume (Cv). To convert the values below into joules per gram per degree C., multiply by 4.185.*

Gas or Vapor	Temp. °C.	Press. atm.	Specific Heat in cal. 15° per gram per °C.		$\gamma = Cp/Cv$
			Const. press.	Const. vol.	
Acetaldehyde, C_2H_4O	30	1	......	......	1.14
Acetic acid, $C_2H_4O_2$	118–140	1	1.50	......	1.15[136°]
" "	140–180	1	1.27	......	....
" "	180–220	1	0.951	......	....
Acetone, C_3H_6O	26–110	1	0.347	......	....
"	130–230	1	0.412	......	....
Acetylene, C_2H_2	15	1	0.3830	0.3039	1.26
"	−71	1	0.3513	0.2676	1.31
Air	0	1	0.2399	0.1710	1.403
"	100	1	0.2404	0.1715	1.401
"	200	1	0.2413	0.1726	1.398
"	400	1	0.2430	0.1744	1.393
"	600	1	0.2471	0.1785	1.385
"	800	1	0.2521	0.1835	1.376
"	1000	1	0.2571	0.1885	1.365
"	1400	1	0.2700	0.2014	1.341
"	2000	1	0.2949	0.2263	1.088
"	0	20	0.2490	......	....
"	0	100	0.2800	......	....
"	50	220	0.2961	......	....
"	−50	10	0.2440	......	....
"	−100	10	0.2581	......	....
"	−140	10	0.4079	......	....
Allyl chloride, C_3H_5Cl	14	0.2	......	......	1.137
Ammonia, NH_3	15	1	0.5232	0.3995	1.310
"	15	3.5	......	......	1.41
Argon, Ar	−180	1	1.33	0.0754	1.76
"	15	1	0.1252	0.07531	1.668
Arsenic trichloride, $AsCl_3$	160–270	1	0.112	......	....
Benzene, C_6H_6	90	1	0.3254	0.2958	1.10
"	34–115	1	0.301	......	....
"	120–220	1	0.370	......	....
Bromine, Br_2	20–350	0.3–1.5	......	......	1.32
"	19–388	1	0.0550	......	....
iso-Butane, C_4H_{10}	15	1	......	......	1.11
Carbon dioxide, CO_2	0	1	0.1973	0.1506	1.310
" "	15	1	0.1988	0.1525	1.304
" "	100	1	0.2064	0.1626	1.281
" "	400	1	0.2379	0.1927	1.235
" "	1000	1	0.2763	0.2312	1.195

* Cal. 15°C/g/°C = btu 60°F./lb./°C

Table 9-7 (*Continued*)
SPECIFIC HEAT OF GASES AND VAPORS

Gas or Vapor	Temp. °C.	Press. atm.	Specific Heat in cal. 15° per gram per °C. Const. press.	Const. vol.	$\gamma = C_p/C_v$
Carbon dioxide, CO_2	2000	1	0.3091	0.2640	1.171
,, ,,	38	24.25	0.2879	......	....
,, ,,	38	54.1	0.3259	......	....
,, ,,	38	85.4	0.9950	......	....
,, ,,	−75	1	0.184	0.144	1.37
Carbon disulfide, CS_2	99.7	sat. vap.	......	......	1.63
,, ,,	80–190	1	0.157	......	.....
,, ,,	17	0.3	0.157	......	....
Carbon monoxide, CO	−180	1	0.259	0.176	1.41
,, ,,	15	1	0.2478	0.1766	1.404
Carbon tetrachloride, CCl_4	0	1	0.140	......	....
,, ,,	30	1	0.132	......	....
,, ,,	70	1	0.115	......	.....
,, ,,	20	0.1	......	......	1.13
Chlorine, Cl_2	15	1	0.1150	0.08486	1.355
,,	16	0.5	......	......	1.34
Chloroform, $CHCl_3$	27–118	1	0.145	......	....
,,	120–230	1	0.157	......	....
,,	20	0.15	......	......	1.15
,,	100	1	......	......	1.15
Cyanogen, $(CN)_2$	15	1	0.4095	0.3260	1.256
Cyclohexane, C_6H_{12}	80	1	......	......	1.08
,,	100	1	0.413	......	.....
Ethane, C_2H_6	−82	1	0.347	0.271	1.28
,,	15	1	0.3860	0.3136	1.22
,,	50	1	......	......	1.21
,,	100	1	......	......	1.19
Ethyl acetate, $C_4H_8O_2$	35–113	1	0.2369	......	....
,, ,,	110–220	1	0.4010	......	....
Ethyl alcohol, C_2H_6O	90	1	0.406	0.359	1.13
,, ,,	100–223	1	0.454	......	......
Ethyl bromide, C_2H_5Br	14	0.3	......	......	1.19
,, ,,	28–116	1	0.161	......	....
,, ,,	80–200	1	0.190	......	....
Ethyl chloride, C_2H_5Cl	−30	0.137	0.21	......	....
,, ,,	0	0.137	0.22	......	....
,, ,,	40	0.137	0.239	......	....
,, ,,	40	1.37	0.244	......	....
,, ,,	16	0.3–0.5	......	......	1.19
Ethyl cyanide, C_3H_5N	114–223	1	0.4260	......	....
Ethyl ether, $C_4H_{10}O$	35	1	0.4449	0.4119	1.08
,, ,,	27–189	1	0.4619	......	1.086800
,, ,,	200–300	1	0.5331	......	....
,, ,,	16	0.28	0.459	......	....
Ethyl sulfide, $C_4H_{10}S$	120–223	1	0.399	......	....
Ethylene, C_2H_4	−91	1	0.3086	0.2285	1.35
,,	15	1	0.3592	0.2858	1.255
,,	15–100	1	0.399	......	1.181000
,,	25–200	1	0.430	......	....
Ethylene chloride, $C_2H_4Cl_2$	19	0.06	......	......	1.137
,, ,,	23	0.2	......	......	1.134
,, ,,	111–221	1	0.229	......	....
Ethylidene chloride, $C_2H_4Cl_2$	110–220	1	0.230	......	....
Helium, He	−180	1	1.248	0.752	1.660
n-Hexane, C_6H_{14}	80	1	0.3646	......	1.08

Table 9-7 (*Continued*)
SPECIFIC HEAT OF GASES AND VAPORS

Gas or Vapor	Temp. °C.	Press. atm.	Specific Heat in cal. 15° per gram per °C.		$\gamma = Cp/Cv$
			Const. press.	Const. vol.	
Hydrogen, H$_2$	−181	1	2.64	1.66	1.597
"	−76	1	3.15	2.17	1.453
"	0	....	3.39	2.40	1.410
"	15	1	3.388	2.402	1.410
"	100	....	3.428	2.442	1.404
"	400	....	3.533	2.547	1.387
"	1000	....	3.740	2.754	1.358
"	2000	....	4.088	3.102	1.318
Hydrogen bromide, HBr	30	0.3–1.5	......	......	1.42
" "	11–100	1	0.0820	......	......
Hydrogen chloride, HCl	15	1	0.1939	0.1375	1.41
" "	100	1	......	......	1.40
" "	10–190	1	0.185	......	......
Hydrogen cyanide, HCN	65	1	......	......	1.31
" "	140	1	......	......	1.28
" "	210	1	......	......	1.24
Hydrogen iodide, HI	20–100	1	......	......	1.40
Hydrogen sulfide, H$_2$S	−57	1	0.292	0.227	1.29
" "	−45	1	0.279	0.215	1.30
" "	15	1	0.2532	0.1918	1.32
" "	18	0.5	......	......	1.32
" "	10–190	1	0.2430	......	......
Iodine, I$_2$	185	1	......	......	1.30
"	206–377	1	0.0337	......	......
Krypton, Kr	19	1	......	......	1.68
Mercury, Hg	360	0.5–1	......	......	1.67
Methane, CH$_4$	−115	1	0.450	0.319	1.41
"	−74	1	0.498	0.370	1.35
"	15	1	0.5282	0.4032	1.31
"	10–200	1	0.5931	......	......
Methyl acetate, C$_3$H$_6$O$_2$	15	1	......	......	1.14
Methyl alcohol, CH$_4$O	77	1	0.390	0.324	1.203
" "	100–223	1	0.4581	......	......
Methyl bromide, CH$_3$Br	18	0.3–0.6	......	......	1.27
Methyl chloride, CH$_3$Cl	16	0.8	......	......	1.28
Methyl ether, C$_2$H$_6$O	6–30	1	......	......	1.11
Methyl iodide, CH$_3$I	20	0.3	......	......	1.286
Neon, Ne	19	1	......	......	1.64
Nitric oxide, NO	−80	1	0.244	0.177	1.38
" "	−45	1	0.239	0.172	1.39
" "	15	1	0.2328	0.1664	1.400
" "	10–180	1	0.232	......	......
Nitrogen, N$_2$	−181	1	0.256	0.174	1.47
"	15	1	0.2477	0.1765	1.404
Nitrogen dioxide, NO$_2$	27–67	1	1.62	......	......
" "	27–100	1	1.70	......	......
" "	27–200	1	0.851	......	......
" "	27–300	1	0.630	......	......
Nitrous oxide, N$_2$O	−70	1	0.190	0.142	1.34
" "	−30	1	0.200	0.153	1.31
" "	15	1	0.2004	0.1538	1.303
" "	100	1	......	......	1.28
" "	25–200	1	0.224	......	......
Oxygen, O$_2$	−181	1	0.2284	0.1575	1.45
"	−76	1	0.214	0.152	1.415

Table 9-7 (*Continued*)
SPECIFIC HEAT OF GASES AND VAPORS

Gas or Vapor	Temp. °C.	Press. atm.	Specific Heat in cal. 15° per gram per°C.		$\gamma = Cp/Cv$
			Const. press.	Const. vol.	
Oxygen O_2	15	1	0.2177	0.1554	1.401
"	100	....	0.2187	0.1563	1.399
"	200	....	0.2194	0.1570	1.397
"	400	....	0.2209	0.1585	1.394
n-Pentane, C_5H_{12}	86	1	0.4093	0.3769	1.086
iso-Pentane, C_5H_{12}	58	1	0.449	......	....
"	100	1	0.471	......	....
Phosphorus, P	300	1	......	......	1.17
Phosphorus trichloride, PCl_3	110–250	1	0.135	......	....
Potassium, K	850	1	......	......	1.77
Propane, C_3H_8	16	0.5	......	......	1.13
n- or iso-Propyl chloride, C_3H_7Cl	21	0.1	......	......	1.13
Silicon tetrachloride, $SiCl_4$	14	0.15	......	......	1.13
" "	90–230	1	0.132	......	....
Sodium, Na	750–920	1	......	......	1.68
Stannic chloride, $SnCl_4$	149–273	1	0.0939	......	....
Sulfonyl chloride, SO_2Cl_2	19–98	1	0.114	......	....
Sulfur dioxide, SO_2	15	1	0.1516	0.1175	1.29
" "	20	0.5	......	......	1.27
" "	20	2.5	......	......	1.35
" "	10–190	1	0.134	......	....
Titanium tetrachloride, $TiCl_4$	160–270	1	0.129	......	....
Water, H_2O	100	1	0.4836	0.3652	1.324
"	200	1	0.4791	0.3657	1.310
"	400	1	0.4801	0.3690	1.301
"	700	1	0.5024	0.3918	1.282
"	1000	1	0.5488	0.4384	1.252
"	1400	1	0.6470	0.5366	1.206
"	2000	1	0.8290	0.7187	1.155
"	2300	1	0.8634	0.7530	1.146
"	200	2	0.4769	......	....
"	200	4	0.4939	......	....
"	200	6	0.5150	......	....
"	200	10	0.5689	......	....
"	200	14	0.6710	......	....
Xenon, Xe	19	1	......	......	1.66

KOPP'S RULE

This rule, which should be used only where experimental values are lacking, states that the specific heat of a compound is approximately equal to the sum of the heat capacities of the constituent elements and that an *approximate* value expressed in gram calories per gram formula weight can be calculated by assigning the following atomic heat capacities to the elements:

For solids: C, 1.8; H, 2.3; O, 4.0; S, 5.4; P, 5.4; F, 5.0; Si, 3.8; B, 2.7; all other elements, 6.2.

For liquids: C, 2.8; H, 4.8; O, 6.0; S, 7.4; P, 7.4; F, 7.0; Si, 5.8; B, 4.7; all other elements, 8.0.

Example. For $BaCO_3$: $6.2 + 1.8 + (3 \times 4.0) = 20.0$ g cal/g formula weight; or, $20.0 \div 197.37 = 0.1013$ g cal/g. Found by experiment, 0.0999.

Table 9-8
SPECIFIC HEAT OF ELEMENTS, COMPOUNDS, AND VARIOUS SUBSTANCES*

The values are given in gram-calories (15°) per gram per degree Centigrade. To convert to joules per gram per degree Centigrade multiply by 4.185.

Name	Formula	State	Specific Heat	Temp. °C.
Acetal	$CH_3CH(OC_2H_5)_2$	L	0.4669	0
....	....	L	0.5198	19 to 99
Acetic acid	CH_3COOH	S	0.487	0
....	....	L	0.468	0
Acetone	$(CH_3)_2CO$	S	0.540	−210
....	....	L	0.506	0
....	....	L	0.528	20
Acetonitrile	CH_3CN	L	0.541	21 to 76
Acetophenone	$C_6H_5COCH_3$	L	0.474	20 to 196
Acetyl chloride	CH_3COCl	L	0.339	0
Albite	$Na_2O \cdot Al_2O_3 \cdot 6SiO_2$	S	0.212	100
Allyl acetate	$CH_3COOC_3H_5$	L	0.431	0
alcohol	CH_2CHCH_2OH	L	0.665	21 to 96
benzoate	$C_6H_5COOC_3H_5$	L	0.388	20
butyrate	$C_3H_7COOC_3H_5$	L	0.451	20
chloride	CH_2CHCH_2Cl	L	0.313	0
chloroacetate	$ClCH_2COOC_3H_5$	L	0.3959	20
dichloroacetate	$Cl_2CHCOOC_3H_5$	L	0.3319	20
isobutyrate	$C_3H_7COOC_3H_5$	L	0.448	20
propionate	$C_2H_5COOC_3H_5$	L	0.451	20
trichloroacetate	$Cl_3CCOOC_3H_5$	L	0.2879	20
valerate	$C_4H_9COOC_3H_5$	L	0.451	20
Aluminum	Al	S	0.0092	−240.6
....	....	S	0.0165	−233
....	....	S	0.0889	−190
....	....	S	0.1466	−192 to −82
....	....	S	0.1962	−76 to −1
....	....	S	0.217	17 to 100
....	....	S	0.236	15 to 435
....	....	S	0.274	500
Aluminum chloride	$AlCl_3$	S	0.188	−22 to 15
....	$AlCl_3 \cdot 6H_2O$	S	0.313	35
fluoride	AlF_3	S	0.229	35
....	$2AlF_3 \cdot 7H_2O$	S	0.342	35
hydroxide	$Al(OH)_3$	S	0.103	−100
....	$Al(OH)_3$	S	0.177	0
....	$Al(OH)_3$	S	0.202	50
....	$Al(OH)_3$	S	0.215	100
oxide	Al_2O_3	S	0.0992	−100
....	Al_2O_3	S	0.174	0
....	Al_2O_3	S	0.198	50
....	Al_2O_3	S	0.21	100
sulfate	$Al_2(SO_4)_3$	S	0.184	50
....	$Al_2(SO_4)_3 \cdot 18H_2O$	S	0.354	34
o-Aminobenzoic acid	$NH_2C_6H_4COOH$	S	0.254	85
....	$NH_2C_6H_4COOH$	L	0.435	145
m-Aminobenzoic acid	$NH_2C_6H_4COOH$	S	0.253	120
....	$NH_2C_6H_4COOH$	L	0.435	174
p-Aminobenzoic acid	$NH_2C_6H_4COOH$	S	0.287	128
....	$NH_2C_6H_4COOH$	L	0.444	186

*See also the table: Specific Heat of Gases and Vapors.

Table 9-8 (*Continued*)
SPECIFIC HEAT OF ELEMENTS, COMPOUNDS, AND VARIOUS SUBSTANCES

Name	Formula	State	Specific Heat	Temp.°C.
Ammonia	NH_3	S	0.502	−103 to −188
....	....	L	1.047	−60
....	....	L	1.098	0
....	....	L	1.125	20
....	....	L	1.48	100
Ammonium bromide	NH_4Br	S	0.210	20
chloride	NH_4Cl	S	0.121	−200
....	....	S	0.263	−100
....	....	S	0.357	0
....	....	S	0.389	50
iodide	NH_4I	S	0.111	0
....	....	S	0.118	50
nitrate	NH_4NO_3	S	0.306	−100
....	....	S	0.397	0
....	....	S	0.4146	50
....	....	S	0.428	100
sulfate	$(NH_4)_2SO_4$	S	0.283	−100
....	....	S	0.337	0
....	....	S	0.345	50
prim-Amyl alcohol (d)	$C_5H_{11}OH$	L	0.712	22 to 125
tert-Amyl alcohol	$C_5H_{11}OH$	L	0.753	20 to 99
Amylene	C_5H_{10}	L	0.282	0
Andalusite	Al_2SiO_5	S	0.228	100
Anethole	$C_9H_9OCH_3$	L	0.551	22.5
Aniline	$C_6H_5NH_2$	S	0.741	?
....	....	L	0.478	0
....	....	L	0.521	50
....	....	L	0.547	100
Anisole	$CH_3OC_6H_5$	L	0.483	20 to 152
Anorthite	$CaO \cdot Al_2O_3 \cdot (SiO_2)_2$	S	0.205	100
Anthracene	$C_{14}H_{10}$	S	0.308	50
....	....	S	0.350	100
Anthraquinone	$C_6H_4(CO)_2C_6H_4$	S	0.258	0
Antimony	Sb	S	0.0462	−186 to −79
....	....	S	0.0468	−188 to +20
....	....	S	0.0503	20
....	....	S	0.0513	100
....	....	S	0.0520	200
....	....	S	0.0537	300
Antimony trioxide	Sb_2O_3	S	0.934	60
trisulfide	Sb_2S_3	S	0.0829	0
....	....	S	0.884	100
Apiol	$C_{12}H_{14}O_4$	S	0.299	10
Arsenic	As	S	0.0704	−188 to +20
Arsenic, gray cryst.	As	S	0.0822	0 to 100
Arsenic, black amorph.	As	S	0.0861	0 to 100
Arsenic trichloride	$AsCl_3$	L	0.177	56
Arsenous oxide	As_2O_3	S	0.117	0
Asbestos	....	S	0.195	20 to 98
Azobenzene	$(C_6H_5N)_2$	S	0.330	28
Barium	Ba	S	0.068	−185 to +20
Barium carbonate	$BaCO_3$	S	0.0999	0
....	....	S	0.110	100
chlorate	$Ba(ClO_3)_2 \cdot H_2O$	S	0.158	32

Table 9-8 (*Continued*)
SPECIFIC HEAT OF ELEMENTS, COMPOUNDS, AND VARIOUS SUBSTANCES

Name	Formula	State	Specific Heat	Temp.°C.
Barium chloride	$BaCl_2$	S	0.08523	0
....	$BaCl_2 \cdot 2H_2O$	S	0.140	0
formate	$Ba(CHO_2)_2$	S	0.137	0
molybdate	$BaMoO_4$	S	0.11	15
nitrate	$Ba(NO_3)_2$	S	0.148	47
sulfate	$BaSO_4$	S	0.111	0
thiosulfate	BaS_2O_3	S	0.162	58
Basalt	....	S	0.20	20 to 100
Benzaldehyde	C_6H_5CHO	L	0.428	22 to 172
Benzene	C_6H_6	S	0.0399	−250
....	....	S	0.124	−200
....	....	S	0.227	−100
....	....	S	0.299	−50
....	....	L	0.389	5
....	....	L	0.406	20
....	....	L	0.444	60
....	....	L	0.473	90
Benzoic acid	C_6H_5COOH	S	0.287	20
Benzonitrile	C_6H_5CN	L	0.441	22 to 186
Benzophenone	$(C_6H_5)_2CO$	S	0.115	−150
....	....	S	0.220	−50
....	....	S	0.275	0
....	....	S	0.303	20
....	....	L	0.383	3 to 40
Benzyl alcohol	$C_6H_5CH_2OH$	L	0.511	20 to 100
chloride	$C_6H_5CH_2Cl$	L	0.323	0
Beryl	$3BeO \cdot Al_2O_3 \cdot 6SiO_2$	S	0.20	57
Beryllium	Be	S	0.397	0 to 46
....	....	S	0.425	0 to 100
....	....	S	0.505	0 to 300
Beryllium oxide	BeO	S	0.260	50
sulfate	$BeSO_4$	S	0.198	50
Betol	$HOC_6H_4COOC_{10}H_7$	S	0.129	−150
....	....	S	0.167	−100
....	....	L	0.356	19 to 63
Bismuth	Bi	S	0.0218	−253 to −196
....	....	S	0.0284	−188 to +20
....	....	S	0.0285	−79 to +17
....	....	S	0.0303	17 to 100
....	....	S	0.0303	18
....	....	S	0.0338	271
....	....	L	0.0363	280 to 360
Bismuth oxide	Bi_2O_3	S	0.0569	50
....	....	S	0.0593	100
sulfide	Bi_2S_3	S	0.0600	50
Borax	$Na_2B_4O_7 \cdot 10H_2O$	S	0.385	35
Boron, amorph.	B	S	0.071	−191 to −78
Boron, cryst.	B	S	0.2518	0 to 100
....	....	S	0.165	−78 to 0
....	....	S	0.307	0 to 100
....	....	S	0.357	0 to 234
Boron nitride	BN	S	0.349	200
Bromine	Br	S	0.070	−191 to −81
....	....	S	0.084	−78 to −20

Table 9-8 (*Continued*)
SPECIFIC HEAT OF ELEMENTS, COMPOUNDS, AND VARIOUS SUBSTANCES

Name	Formula	State	Specific Heat	Temp.°C.
Bromine	Br	L	0.107	1 to 32
Bromobenzene	C_6H_5Br	L	0.231	20
o-Bromochlorobenzene	BrC_6H_4Cl	S	0.192	−34
....	....	L	0.215	0
m-Bromochlorobenzene	BrC_6H_4Cl	S	0.150	−52
....	....	L	0.212	0
p-Bromochlorobenzene	BrC_6H_4Cl	S	0.150	−40
....	....	S	0.170	0
o-Bromoiodobenzene	BrC_6H_4I	S	0.143	−50
....	....	L	0.160	5 to 100
m-Bromoiodobenzene	BrC_6H_4I	S	0.143	−75 to −15
....	....	L	0.158	5 to 100
p-Bromoiodobenzene	BrC_6H_4I	S	0.150	−40
α-Bromonaphthalene	$C_{10}H_7Br$	S	0.260	41
Bromophenol	HOC_6H_4Br	L	0.316	18 to 77
Brucite	$MgO \cdot H_2O$	S	0.311	35
n-Butane	C_4H_{10}	L	0.550	0
n-Butyl alcohol	C_4H_9OH	L	0.526	2.3
....	....	L	0.563	19.2
butyrate	$C_3H_7COOC_4H_9$	L	0.459	20
chloride	C_4H_9Cl	L	0.451	20
formate	$HCOOC_4H_9$	L	0.459	20
propionate	$C_2H_5COOC_4H_9$	L	0.459	20
valerate	$C_4H_9COOC_4H_9$	L	0.459	20
n-Butyric acid	C_3H_7COOH	L	0.515	20 to 100
n-Butyronitrile	C_3H_7CN	L	0.547	21 to 113
Cadmium	Cd	S	0.0308	−253 to −196
....	....	S	0.0498	−186 to −79
....	....	S	0.0537	−79 to +18
....	....	S	0.0549	20
....	....	S	0.0566	100
....	....	S	0.0594	200
....	....	S	0.0617	300
Cadmium chloride	$2CdCl_2 \cdot 5H_2O$	S	0.243	0
nitrate	$Cd(NO_3)_2 \cdot 4H_2O$	S	0.260	40
sulfate	$CdSO_4 \cdot 8H_2O$	S	0.195	0
....	....	S	0.200	20
sulfide	CdS	S	0.0882	0
....	....	S	0.0922	50
Calcium	Ca	S	0.0714	−253 to −196
....	....	S	0.157	−185 to +20
....	....	S	0.145	0 to 20
....	....	S	0.149	0 to 100
....	....	S	0.152	0 to 157
....	....	S	0.168	20.4
Calcium carbide	CaC_2	S	0.239	20 to 500
carbonate	$CaCO_3$	S	0.203	0
....	....	S	0.214	100
chloride	$CaCl_2$	S	0.164	61
....	$CaCl_2 \cdot 6H_2O$	S	0.320	0
....	....	S	0.552	33 to 99
fluoride	CaF_2	S	0.204	0
....	....	S	0.212	40
formate	$Ca(HCO_2)_2$	S	0.238	0

Table 9-8 (*Continued*)
SPECIFIC HEAT OF ELEMENTS, COMPOUNDS, AND VARIOUS SUBSTANCES

Name	Formula	State	Specific Heat	Temp. °C.
Calcium hydride	CaH_2	S	0.0466	−200
hydroxide	$Ca(OH)_2$	S	0.260	0
....	....	S	0.288	50
molybdate	$CaMoO_4$	S	0.165	15
oxide	CaO	S	0.177	0
....	....	S	0.197	100
silicate	$CaSiO_3$	S	0.195	100
sulfate	$CaSO_4 \cdot 2H_2O$	S	0.265	36
sulfite	$CaSO_3 \cdot 2H_2O$	S	0.272	9
tungstate	$CaWO_4$	S	0.104	15
Calcspar	....	S	0.2005	0 to 100
Camphene	$C_{10}H_{16}$	S	0.380	35
Capric acid	$CH_3(CH_2)_8COOH$	S	0.695	8
Caproic acid	$CH_3(CH_2)_4COOH$	L	0.533	29 to 105
Capronitrile	$C_5H_{11}CN$	L	0.542	18 to 156
Caprylic acid	$CH_3(CH_2)_6COOH$	S	0.628	−2
Carbon, charcoal	C	S	0.165	0 to 24
gas carbon	C	S	0.204	24 to 68
graphite	C	S	0.005	−243
....	....	S	0.0175	−203
....	....	S	0.060	−188 to −78
....	....	S	0.160	11
....	....	S	0.254	138
....	....	S	0.445	642
diamond	C	S	0.0005	−233
....	....	S	0.0025	−185
....	....	S	0.019	−188 to −78
....	....	S	0.079	−78 to +18
....	....	S	0.113	11
....	....	S	0.222	140
....	....	S	0.303	247
....	....	S	0.441	606
Carbon dioxide	CO_2	S	0.124	−225
monoxide	CO	S	0.417	−220
....	....	S	0.457	206
tetrachloride	CCl_4	S	0.0812	−200
....	....	S	0.182	−80
....	....	S	0.201	−40
....	....	L	0.198	0
....	....	L	0.201	20
Carborundum	....	S	0.162	3 to 44
Carvacrol	$C_{10}H_{13}OH$	L	0.577	24 to 233
Catechol	$C_6H_4(OH)_2$	L	0.462	0
Cellulose, dry	....	S	0.37	......
Cement, powder	....	S	0.20	20 to 100
Cerium	Ce	S	0.033	−253 to −196
....	....	S	0.0448	0 to 100
....	....	S	0.0511	20 to 100
Ceric oxide	CeO_2	S	0.0870	0
....	....	S	0.0946	50
Cerous sulfate	$CeSO_4$	S	0.117	50
....	$CeSO_4 \cdot 5H_2O$	S	0.201	50
Cesium	Cs	S	0.048	0 to 26
....	....		0.0587	56

Table 9-8 (*Continued*)
SPECIFIC HEAT OF ELEMENTS, COMPOUNDS, AND VARIOUS SUBSTANCES

Name	Formula	State	Specific Heat	Temp.°C.
Cesium bromide	$CsBr$	S	0.0581	10
chloride	$CsCl$	S	0.0746	10
fluoride	CsF	S	0.0796	10
iodide	CsI	S	0.0478	10
Chalcopyrite	$FeCuS_2$	S	0.129	48
Chalk	....	S	0.214	20 to 99
Charcoal	....	S	0.16	10
Chloral	CCl_3CHO	L	0.250	17 to 53
Chloral alcoholate	$CCl_3CH(OH)OC_2H_5$	S	0.509	78
hydrate	$CCl_3CH(OH)_2$	S	0.213	32
....	....	L	0.470	55 to 88
Chlorine	Cl	L	0.226	0 to 24
Chloro-acetic acid	$ClCH_2COOH$	S	0.363	60
benzene	C_6H_5Cl	L	0.309	20
benzoic acid (*o*)	ClC_6H_4COOH	L	0.392	0
benzoic acid (*m*)	ClC_6H_4COOH	S	0.232	94
benzoic acid (*p*)	ClC_6H_4COOH	S	0.228	80
....	....	S	0.242	180
....	....	L	0.547	226
phenol (*o*)	ClC_6H_4OH	L	0.401	0 to 20
toluene	$ClC_6H_4CH_3$	L	0.316	0
Chloroform	$CHCl_3$	L	0.232	0
....	....	L	0.226	15
....	....	L	0.234	20
Chromic oxide	Cr_2O_3	S	0.168	0
....	....	S	0.189	50
sulfate	$Cr_2(SO_4)_3$	S	0.172	50
....	$Cr_2(SO_4)_3 \cdot 5H_2O$	S	0.200	50
Chromium	Cr	S	0.0142	−253 to −196
....	....	S	0.0793	−188 to +20
....	....	S	0.098	−79 to +17
....	....	S	0.1039	0
....	....	S	0.110	17 to 100
....	....	S	0.1202	18 to 500
....	....	S	0.112	100
....	....	S	0.133	400
Chrysoberyl	Al_2BeO_4	S	0.20	50
Clay, dry	....	S	0.22	20 to 100
Cobalt	Co	S	0.0207	−253 to −196
....	....	S	0.0827	−188 to +20
....	....	S	0.1041	0 to 100
....	....	S	0.1035	15 to 100
....	....	S	0.1047	15 to 185
....	....	S	0.121	300
....	(at transformation temp.)	S	0.145 ⎫ 0.125 ⎭	508
....	....	S	0.160	800
....	....	S	0.184	1000
....	(at transformation temp.)	S	0.270 ⎫ 0.170 ⎭	1112
Cobaltite	$CoAsS$	S	0.098	58
Cobaltous nitrate	$Co(NO_3)_2 \cdot 6H_2O$	S	0.373	32
sulfate	$CoSO_4 \cdot 7H_2O$	S	0.342	48
Columbium pentoxide	Cb_2O_5	S	0.101	50

Table 9-8 (*Continued*)
SPECIFIC HEAT OF ELEMENTS, COMPOUNDS, AND VARIOUS SUBSTANCES

Name	Formula	State	Specific Heat	Temp.°C.
Copper	Cu	S	0.0031	−253
....	....	S	0.0245	−253 to −196
....	....	S	0.029	−213
....	....	S	0.047	−193
....	....	S	0.0788	−188 to +20
....	....	S	0.0883	−79 to +18
....	....	S	0.0909	0
....	....	S	0.0912	20
....	....	S	0.09305	15 to 100
....	....	S	0.0928	50
....	....	S	0.0942	100
....	....	S	0.0963	200
....	....	S	0.1259	900
Copper ammonium sulfate	$CuSO_4(NH_4)_2SO_4 \cdot 6H_2O$	S	0.256	0
Copper carbonate	$2CuO \cdot CO_2 \cdot H_2O$	S	0.177	57
sulfate	$CuSO_4$	S	0.166	50
sulfate	$CuSO_4 \cdot H_2O$	S	0.172	0
....	....	S	0.191	50
....	$CuSO_4 \cdot 3H_2O$	S	0.228	9
....	$CuSO_4 \cdot 5H_2O$	S	0.253	0
....	....	S	0.287	50
o-Cresol	$CH_3C_6H_4OH$	L	0.499	0 to 20
m-Cresol	$CH_3C_6H_4OH$	L	0.479	0 to 20
p-Cresyl methyl ether	$CH_3OC_6H_4CH_3$	L	0.405	0
Crotonic acid	$CH_3CHCHCOOH$	S	0.520	38
....	....	L	0.500	71.4
Cryolite	$3NaF \cdot AlF_3$	S	0.2519	43
Cupric chloride	$CuCl_2$	S	0.139	58
oxide	CuO	S	0.125	0
....	....	S	0.144	100
sulfide	CuS	S	0.129	0
....	....	S	0.151	100
Cuprous iodide	CuI	S	0.0658	0
....	....	S	0.0671	50
oxide	Cu_2O	S	0.110	0
....	....	S	0.116	100
selenide	Cu_2Se	S	0.104	60
sulfide	Cu_2S	S	0.148	0
....	....	S	0.166	50
Cyamelide	$(CNOH)_3$	S	0.263	40
Cyanamide	NH_2CN	S	0.547	20
Cyanogen	$(CN)_2$	S	0.167	−188 to +78
Cyanuric acid	$(HNCO)_3$	S	0.318	40
Cyclo-hexanol	$C_6H_{11}OH$	L	0.417	15 to 18
hexanone	$C_6H_{10}O$	L	0.433	15 to 18
Decylene-2	$C_{10}H_{20}$	L	0.469	0 to 50
Dextrose	$C_6H_{12}O_6$	S	0.0155	−250
....	....	S	0.277	0
....	....	S	0.275	20
Dextrin	$(C_6H_{10}O_5)_x$	S	0.292	0 to 90
Diallyl oxalate	$(COOC_3H_5)_2$	L	0.426	20
succinate	$(CH_2COOC_3H_5)_2$	L	0.425	20
Diamylene	$C_{10}H_{20}$	L	0.545	20 to 130
Dibenzyl	$(C_6H_5CH_2)_2$	S	0.363	28

Table 9-8 (*Continued*)
SPECIFIC HEAT OF ELEMENTS, COMPOUNDS, AND VARIOUS SUBSTANCES

Name	Formula	State	Specific Heat	Temp.°C.
o-Dibromobenzene	$C_6H_4Br_2$	S	0.249	−36
....	....	L	0.180	0
m-Dibromobenzene	$C_6H_4Br_2$	S	0.134	−25
....	....	L	0.175	0
p-Dibromobenzene	$C_6H_4Br_2$	S	0.139	−50
Dibutyl oxalate	$(COOC_4H_9)_2$	L	0.441	20
Dichloroacetic acid	$Cl_2CHCOOH$	S	0.406	......
....	....	L	0.350	21 to 106
o-Dichlorobenzene	$C_6H_4Cl_2$	S	0.185	−48.5
....	....	L	0.270	0
m-Dichlorobenzene	$C_6H_4Cl_2$	S	0.186	−52
....	....	L	0.270	0
p-Dichlorobenzene	$C_6H_4Cl_2$	S	0.219	−50
....	....	L	0.298	53 to 99
Dicyandiamide	$C_2H_4N_4$	S	0.456	0 to 204
Diethylamine	$(C_2H_5)_2NH$	L	0.518	22.5
Diethylaniline	$C_6H_5N(C_2H_5)_2$	L	0.452	20
Diethyl carbonate	$CO(OC_2H_5)_2$	L	0.464	20 to 100
ketone	$(C_2H_5)_2CO$	L	0.557	20 to 98.5
malate	$C_2H_4O(COOC_2H_5)_2$	L	0.475	24 to 186
malonate	$CH_2(COOC_2H_5)_2$	L	0.433	20
oxalate	$(COOC_2H_5)_2$	L	0.433	20
succinate	$C_2H_4(COOC_2H_5)_2$	L	0.452	20
Dihydronaphthalene	$C_{10}H_{10}$	L	0.346	18 to 28
o-Diiodobenzene	$C_6H_4I_2$	S	0.109	0
....	....	L	0.136	0
m-Diiodobenzene	$C_6H_4I_2$	S	0.100	−52
....	....	L	0.140	34.2 to 99.6
p-Diiodobenzene	$C_6H_4I_2$	S	0.101	−50
Diisoamyl	$C_{10}H_{22}$	L	0.590	21.5 to 155
Diisoamyl oxalate	$(COOC_5H_{11})_2$	L	0.449	20
Diisobutylamine	$(C_4H_9)_2NH$	L	0.571	22 to 130
Dimethylaniline	$C_6H_5N(CH_3)_2$	L	0.418	0 to 20
Dimethyl carbonate	$CO(OCH_3)_2$	L	0.452	19.8 to 88
Dimethyl oxalate	$(COOCH_3)_2$	S	0.212	10
Dimethylpyrone	$(CH_3)_2C_5H_2O_2$	S	0.368	50
....	....	L	0.55	166
o-Dinitrobenzene	$C_6H_4(NO_2)_2$	S	0.252	−160
....	....	L	0.349	0
m-Dinitrobenzene	$C_6H_4(NO_2)_2$	S	0.248	−160
....	....	L	0.405	90
p-Dinitrobenzene	$C_6H_4(NO_2)_2$	S	0.259	119
Diopside	$CaMg(SiO_3)_2$	S	0.193	50
Diphenyl	$(C_6H_5)_2$	S	0.385	40
Diphenylamine	$(C_6H_5)_2NH$	S	0.337	26
....	....	L	0.464	53
Diphenyl oxide	$(C_6H_5)_2O$	L	0.399	30
Dipropylamine	$(C_3H_7)_2NH$	L	0.464	53
Dipropyl ketone	$(C_3H_7)_2CO$	L	0.552	20 to 140
malonate	$CH_2(COOC_3H_7)_2$	L	0.433	20
oxalate	$(COOC_3H_7)_2$	L	0.433	20
succinate	$(CH_2COOC_3H_7)_2$	L	0.452	20
Dodecane	$C_{12}H_{26}$	L	0.500	0 to 50
Dodecylene	$C_{12}H_{24}$	L	0.457	0 to 50

Table 9-8 (*Continued*)
SPECIFIC HEAT OF ELEMENTS, COMPOUNDS, AND VARIOUS SUBSTANCES

Name	Formula	State	Specific Heat	Temp. °C.
Dulcitol	$C_6H_8(OH)_6$	S	0.282	20
Ebonite	. . . .	S	0.40	20 to 100
Erbium oxide	Er_2O_3	S	0.065	50
Erythritol	$C_4H_6(OH)_4$	S	0.351	60
Ethyl acetate	$CH_3COOC_2H_5$	L	0.459	20
acetoacetate	$CH_3COCH_2COOC_2H_5$	L	0.477	20 to 100
alcohol (cryst.)	C_2H_5OH	S	0.232	−190
alcohol (vitreous)	C_2H_5OH	S	0.260	−190
. . . .	. . . .	L	0.456	−100
. . . .	. . . .	L	0.535	0
. . . .	. . . .	L	0.581	25
. . . .	. . . .	L	0.824	100
benzene	$C_6H_5C_2H_5$	L	0.409	30
benzoate	$C_6H_5COOC_2H_5$	L	0.389	20
bromide	C_2H_5Br	L	0.216	5 to 10
. . . .	. . . .	L	0.215	15 to 20
butyrate	$C_3H_7COOC_2H_5$	L	0.459	20
chloride	C_2H_5Cl	L	0.368	0
chloroacetate	$ClCH_2COOC_2H_5$	L	0.418	9 to 138
dichloroacetate	$Cl_2CHCOOC_2H_5$	L	0.329	20
ether	$(C_2H_5)_2O$	L	0.517	−50
. . . .	. . . .	L	0.529	0
. . . .	. . . .	L	0.547	30
. . . .	. . . .	L	0.803	120
. . . .	. . . .	L	1.041	180
formate	$HCOOC_2H_5$	L	0.510	14 to 49
iodide	C_2H_5I	L	0.162	0
. . . .	. . . .	L	0.172	60
isobutyrate	$C_3H_7COOC_2H_5$	L	0.459	20
propionate	$C_2H_5COOC_2H_5$	L	0.459	20
sulfide	$(C_2H_5)_2S$	L	0.470	0
. . . .	. . . .	L	0.477	15 to 20
trichloroacetate	$Cl_3CCOOC_2H_5$	L	0.295	10 to 81
valerate	$C_4H_9COOC_2H_5$	L	0.459	20
Ethylene bromide	$C_2H_4Br_2$	L	0.174	20
chloride	$C_2H_4Cl_2$	L	0.2791	−30
. . . .	. . . .	L	0.301	20
. . . .	. . . .	L	0.319	60
Ferric oxide	Fe_2O_3	S	0.148	0
. . . .	. . . .	S	0.182	100
. . . .	. . . .	S	0.263	400
Ferrosoferric oxide (magnetite)	Fe_3O_4	S	0.151	0
. . . .	. . . .	S	0.179	100
Ferrous carbonate	$FeCO_3$	S	0.194	54
sulfate	$FeSO_4$	S	0.167	45
. . . .	$FeSO_4 \cdot 4H_2O$	S	0.284	9
. . . .	$FeSO_4 \cdot 7H_2O$	S	0.325	0
. . . .	. . . .	S	0.337	10
sulfide	FeS	S	0.135	0
Formamide	$HCONH_2$	L	0.551	19
Formic acid	$HCOOH$	S	0.387	−22
. . . .	. . . .	S	0.430	0
. . . .	. . . .	L	0.437	0
. . . .	. . . .	L	0.511	15.5

Table 9-8 (*Continued*)
SPECIFIC HEAT OF ELEMENTS, COMPOUNDS, AND VARIOUS SUBSTANCES

Name	Formula	State	Specific Heat	Temp. °C.
Formic acid	HCOOH	L	0.526	20 to 100
Furfural	$(C_4H_3O)CHO$	L	0.418	20 to 100
Gallium	Ga	L	0.080	13 to 110
....		S	0.079	12 to 23
Gallium sesqui-oxide	Ga_2O_3	S	0.105	50
Germanium	Ge	S	0.074	0 to 100
....	....	S	0.0773	0 to 211
Germanium oxide	GeO_2	S	0.129	50
Glass, normal thermometer	....	S	0.1988	19 to 100
Glass, crown	....	S	0.161	10 to 50
Glass, flint	....	S	0.117	10 to 50
Glutaric acid	$(CH_2)_3(COOH)_2$	S	0.299	20
Glycerol	$C_3H_5(OH)_3$	S	0.0471	−250
....	....	S	0.115	−200
....	....	S	0.217	−100
....	....	S	0.330	0
....	....	L	0.540	0
....	....	L	0.600	50
....	....	L	0.669	100
Glycol	$(CH_2OH)_2$	S	0.323	−40
....	....	L	0.544	0
....	....	L	0.571	14.9
Gold	Au	S	0.016	−253 to −196
....	....	S	0.0297	−188 to +20
....	....	S	0.0297	−79 to +17
....	....	S	0.0316	0 to 100
....	....	S	0.0312	18
....	....	S	0.031	17 to 100
....	....	S	0.0345	0 to 900
Gold iodide	AuI	S	0.0404	0
....		S	0.0432	50
Granite	....	S	0.192	12 to 100
Heptaldehyde	$C_6H_{13}CHO$	L	0.365	0
n-Heptane	C_7H_{16}	L	0.365	0
Heptylene (B. P. 98°)	C_7H_{14}	L	0.488	0 to 50
Heptylic acid	$C_6H_{13}COOH$	L	0.558	9
Hexachloroethane	C_2Cl_6	S	0.174	25
Hexadecane	$C_{16}H_{34}$	S	0.495	19
n-Hexadecane (B. P. 275°)	$C_{16}H_{34}$	L	0.496	0 to 50
Hexadiene (1, 5)	C_6H_{10}	L	0.407	0
o-Hexahydrocresol	$CH_3C_6H_{10}OH$	L	0.418	15 to 18
m-Hexahydrocresol	$CH_3C_6H_{10}OH$	L	0.422	15 to 18
p-Hexahydrocresol	$CH_3C_6H_{10}OH$	L	0.423	15 to 18
n-Hexane	C_6H_{14}	L	0.600	20 to 100
Hexylene	C_6H_{12}	L	0.506	0 to 50
Hydrazine	N_2H_4	L	0.74	25
Hydrogen	H	L	0.231	−253
Hydrogen peroxide	H_2O_2	S	0.471	−25
....	....	L	0.578	0
Ice see water	....	....	......	....
India rubber (Para)	....	S	0.481	? to 100
Indium	In	S	0.0263	−186 to −79
....	....	S	0.0303	−79 to +18
....	....	S	0.0323	18 to 100
Indium sesqui-oxide	In_2O_3	S	0.0808	50

Table 9-8 (*Continued*)
SPECIFIC HEAT OF ELEMENTS, COMPOUNDS, AND VARIOUS SUBSTANCES

Name	Formula	State	Specific Heat	Temp. °C.
Iodine	I	S	0.0361	−253 to −196
....	....	S	0.031	−243
....	....	S	0.043	−193
....	....	S	0.0467	−189 to −76
....	....	S	0.0516	−76 to 0
....	....	S	0.054	9 to 98
....	....	L	0.108	107 to 180
Iodine chloride	ICl	L	0.158	15 to 77
Iodobenzene	C_6H_5I	S	0.191	40
Iridium	Ir	S	0.0099	−253 to −196
....	....	S	0.0263	−186 to −79
....	....	S	0.0302	−79 to +18
....	....	S	0.0323	18 to 100
....	....	S	0.0371	0 to 900
....	....	S	0.0401	0 to 1400
Iron	Fe	S	0.0175	−253 to −196
....	....	S	0.0721	−186 to −79
....	....	S	0.1000	−79 to +18
....	....	S	0.1045	0
....	....	S	0.113	18 to 100
....	....	S	0.1137	97
....	....	S	0.138	300
....	....	S	0.138	0 to 650
....	....	S	0.195	650
....	....	S	0.23	850
Iron, cast	Fe	S	0.1189	20 to 100
Iron, electrolytic	Fe	S	0.1939	0 to 1600
Iron, hard drawn	Fe	S	0.1146	20 to 100
Iron, wrought	Fe	S	0.1152	15 to 100
Iron, diarsenide	$FeAs_2$	S	0.0860	50
Iron disulfide	FeS_2	S	0.128	50
Iron, see also ferrous and ferric	....	....	......	......
Isoamyl acetate	$CH_3COOC_5H_{11}$	L	0.459	20
alcohol	$C_5H_{11}OH$	L	0.502	0
....	....	L	0.535	20
....	....	L	0.688	75.5
amine	$C_5H_{11}NH_2$	L	0.614	22 to 91
butyrate	$C_3H_7COOC_5H_{11}$	L	0.459	20
formate	$HCOOC_5H_{11}$	L	0.459	20
isobutyrate	$C_3H_7COOC_5H_{11}$	L	0.459	20
propionate	$C_2H_5COOC_5H_{11}$	L	0.459	20
succinate	$C_2H_4(COOC_5H_{11})_2$	L	0.449	0
valerate	$C_4H_9COOC_5H_{11}$	L	0.459	20
Isobutane	C_4H_{10}	L	0.550	0
Isobutyl acetate	$CH_3COOC_4H_9$	L	0.459	20
alcohol	C_4H_9OH	L	0.716	21 to 109
butyrate	$C_3H_7COOC_4H_9$	L	0.459	20
succinate	$C_2H_4(COOC_4H_9)_2$	L	0.442	0
Isobutyric acid	C_3H_7COOH	L	0.450	20
Isoheptane	C_7H_{16}	L	0.501	0 to 50
Isopentane	C_5H_{12}	L	0.527	8
Isopropyl alcohol	C_3H_7OH	S	0.0507	−200
Isovaleric acid	C_4H_9COOH	L	0.590	23 to 93
Lactose	$C_{12}H_{22}O_{11}$	S	0.287	20

Table 9-8 (*Continued*)
SPECIFIC HEAT OF ELEMENTS, COMPOUNDS, AND VARIOUS SUBSTANCES

Name	Formula	State	Specific Heat	Temp. °C.
Lactose	$C_{12}H_{22}O_{11} \cdot H_2O$	S	0.299	20
Lanthanum	La	S	0.0322	−253 to −196
....	....	S	0.0448	0 to 100
Lanthanum molybdate	$La_2(MoO_4)_3$	S	0.115	15
Lanthanum sesqui-oxide	La_2O_3	S	0.0750	50
Lauric acid	$C_{11}H_{23}COOH$	S	0.430	−30
....	....	L	0.515	57
Lead	Pb	S	0.0120	−253
....	....	S	0.0220	−233
....	....	S	0.0275	−173
....	....	S	0.0293	−192 to +20
....	....	S	0.0291	−186 to −79
....	....	S	0.0292	−100
....	....	S	0.0300	−79 to +18
....	....	S	0.0305	0
....	....	S	0.0305	20 to 100
....	....	S	0.0313	100
....	....	S	0.0338	300
....	....	L	0.0410	360
Lead ammonium chloride	$2PbCl_2 \cdot NH_4Cl$	S	0.0865	10
Lead arsenate	$Pb_3As_2O_8$	S	0.0729	55
borate	PbB_2O_4	S	0.0903	57
bromide	$PbBr_2$	S	0.0502	0
....	....	S	0.0530	50
....	....	L	0.0779	550
carbonate	$PbCO_3$	S	0.0800	32
chloride	$PbCl_2$	S	0.0649	0
....	....	S	0.0681	100
....	....	S	0.0741	300
....	....	L	0.121	540
chromate	$PbCrO_4$	S	0.0908	35
dioxide	PbO_2	S	0.0619	0
....	....	S	0.0650	50
fluoride	PbF_2	S	0.0719	9
iodide	PbI_2	S	0.0417	0
....	....	S	0.0437	100
molybdate	$PbMoO_4$	S	0.100	15
monoxide	PbO	S	0.0483	0
....	....	S	0.0509	50
....	....	S	0.0521	100
nitrate	$Pb(NO_3)_2$	S	0.115	45
pyrophosphate	$Pb_2P_2O_7$	S	0.0820	55
silicate	$PbSiO_3$	S	0.0779	60
sulfate	$PbSO_4$	S	0.0839	45
sulfide	PbS	S	0.0502	0
....	....	S	0.0511	100
tetraborate	PbB_4O_7	S	0.10	57
thiosulfate	PbS_2O_3	S	0.0918	58
tungstate	$PbWO_4$	S	0.0769	15
Leather, dry	....	S	0.36	....
Levulose	$C_6H_{12}O_6$	S	0.275	20
Limonite	$2Fe_2O_3 \cdot 3H_2O$	S	0.22	60
Lithium	Li	S	0.1924	−253 to −196
....	....	S	0.52	−190 to −80

Table 9-8 (*Continued*)
SPECIFIC HEAT OF ELEMENTS, COMPOUNDS, AND
VARIOUS SUBSTANCES

Name	Formula	State	Specific Heat	Temp. °C.
Lithium	Li	S	0.5997	−100
....	....	S	0.7951	0
....	....	S	0.9063	50
....	....	S	1.09	0 to 100
....	....	S	1.0407	100
....	....	L	1.3745	190
Lithium chloride	LiCl	S	0.282	55
fluoride	LiF	S	0.373	10
hydride	LiH	S	0.980	0
....	....	S	1.07	50
hydroxide	LiOH	S	0.327	0
....	....	S	0.356	50
nitrate	$LiNO_3$	S	0.387	210
....	....	L	0.390	280
thiosulfate	$Li_2S_2O_3$	S	0.0920	58
Magnesium	Mg	S	0.0713	−253 to −196
....	....	S	0.189	−186 to −79
....	....	S	0.222	−185 to +20
....	....	S	0.233	−79 to +18
....	....	S	0.248	17 to 100
....	....	S	0.2421	28
....	....	S	0.3235	325
....	....	L	0.4352	625
Magnesium carbonate	$MgCO_3$	S	0.200	25
chloride	$MgCl_2$	S	0.194	48
....	$MgCl_2 \cdot 6H_2O$	S	0.378	44
nitrate	$Mg(NO_3)_2 \cdot 6H_2O$	S	0.887	55
oxide	MgO	S	0.209	0
....	....	S	0.232	50
sulfate	$MgSO_4$	S	0.222	61
....	$MgSO_4 \cdot H_2O$	S	0.239	9
....	$MgSO_4 \cdot 6H_2O$	S	0.349	9
....	$MgSO_4 \cdot 7H_2O$	S	0.361	12
Malachite	$2CuO \cdot CO_2 \cdot H_2O$	S	0.177	57
Malonic acid	$CH_2(COOH)_2$	S	0.275	20
Maltose	$C_{12}H_{22}O_{11}$	S	0.320	20
Manganese	Mn	S	0.0229	−253 to −196
....	....	S	0.093	−188 to +20
....	....	S	0.0979	−100
....	....	S	0.1072	0
....	....	S	0.1143	100
....	....	S	0.1652	500
Manganese dioxide	MnO_2	S	0.152	0
....	....	S	0.163	50
Manganic oxide	Mn_2O_3	S	0.162	58
....	$Mn_2O_3 \cdot 3H_2O$	S	0.177	38
Manganous nitrate	$Mn(NO_3)_2 \cdot 6H_2O$	S	0.373	47
oxide	MnO	S	0.158	58
sulfate	$MnSO_4$	S	0.182	61
sulfide	MnS	S	0.14	60
sulfate	$MnSO_4 \cdot 5H_2O$	S	0.323	32
Mannitol	$C_6H_8(OH)_6$	S	0.313	0
Marble	....	S	0.21	0 to 100
Melamine	$C_3H_6N_6$	S	0.351	40

Table 9-8 (*Continued*)
SPECIFIC HEAT OF ELEMENTS, COMPOUNDS, AND VARIOUS SUBSTANCES

Name	Formula	State	Specific Heat	Temp. °C.
Mercuric bromide	$HgBr_2$	S	0.0519	100
chloride	$HgCl_2$	S	0.0640	0
....	....	S	0.0669	100
cyanide	$Hg(CN)_2$	S	0.100	29
Iodide (red)	HgI_2	S	0.0404	0
....	....	S	0.0413	50
oxide	HgO	S	0.0485	0
....	....	S	0.0521	50
sulfide	HgS	S	0.0506	0
....	....	S	0.0520	50
Mercurous chloride	$HgCl$	S	0.0499	0
....	....	S	0.0512	50
sulfate	Hg_2SO_4	S	0.0616	0
....	....	S	0.0680	50
Mercury	Hg	S	0.0232	−253 to −196
See also special table				
....	....	S	0.0266	−213
....	....	S	0.0285	−183
....	....	S	0.0315	−78 to −40
....	....	SL	0.032	−185 to +20
....	....	L	0.0334	−35.6 to −3.4
....	....	L	0.03346	0
....	....	L	0.03325	20
....	....	L	0.03308	40
....	....	L	0.03294	60
....	....	L	0.03267	100
....	....	L	0.03196	200
....	....	L	0.0322	250
Mesitylene	$(CH_3)_3C_6H_3$	L	0.393	0
Mesityl oxide	$C_6H_{10}O$	L	0.521	21 to 121
Methyl acetate	CH_3COOCH_3	L	0.468	15
Methylal	$C_3H_8O_2$	L	0.521	15 to 41
Methyl alcohol	CH_3OH	L	0.566	0
....	....	L	0.600	20
aniline	$C_6H_5NHCH_3$	L	0.513	20 to 197
benzoate	$C_6H_5COOCH_3$	L	0.363	0
butyl ketone	$CH_3COC_4H_9$	L	0.553	21 to 127
n-butyrate	$C_3H_7COOCH_3$	L	0.459	20
chloroacetate	$ClCH_2COOCH_3$	L	0.382	20
cyclohexanone (o)	$C_7H_{12}O$	L	0.463	15 to 18
cyclohexanone (m)	$C_7H_{12}O$	L	0.441	15 to 18
cyclohexanone (p)	$C_7H_{12}O$	L	0.441	15 to 18
dichloroacetate	$Cl_2CHCOOCH_3$	L	0.311	20
ethyl ketone	$CH_3COC_2H_5$	L	0.549	20 to 78
ethyl ketoxime	$(CH_3)(C_2H_5)NOH$	L	0.650	22 to 152
formate	$HCOOCH_3$	L	0.516	13 to 29
hexyl ketone	$CH_3COC_6H_{13}$	L	0.552	22 to 168
isobutyl ketone	$CH_3COC_4H_9$	L	0.459	20
isopropyl ketone	$CH_3COC_3H_7$	L	0.525	20 to 91
trichloroacetate	$Cl_3CCOOCH_3$	L	0.267	20
valerate	$C_4H_9COOCH_3$	L	0.459	20
Methylene chloride	CH_2Cl_2	L	0.288	15 to 40
Mica (Mg)	....	S	0.2061	20 to 98
Molybdenum	Mo	S	0.0141	−253 to −196

Table 9-8 (*Continued*)
SPECIFIC HEAT OF ELEMENTS, COMPOUNDS, AND VARIOUS SUBSTANCES

Name	Formula	State	Specific Heat	Temp. °C.
Molybdenum	Mo	S	0.062	−185 to +20
....	....	S	0.072	15 to 93
....	....	S	0.0722	20 to 550
....	....	S	0.0647	60
....	....	S	0.0750	475
Molybdenum trioxide	MoO_3	S	0.134	54
Myristic acid	$C_{13}H_{27}COOH$	S	0.381	0
....	....	L	0.539	56 to 100
Naphthalene	$C_{10}H_8$	S	0.281	−130
....	....	L	0.402	87.5
α-Naphthol	$C_{10}H_7OH$	S	0.240	50
β-Naphthol	$C_{10}H_7OH$	S	0.252	61
α-Naphthylamine	$C_{10}H_7NH_2$	S	0.270	0
....	....	L	0.475	53.2
Nickel	Ni	S	0.0208	−253 to −196
....	....	S	0.092	−185 to +20
....	....	S	0.0743	−186 to −79
....	....	S	0.0983	−79 to +18
....	....	S	0.1034	0 to 20
....	....	S	0.1089	15 to 100
....	....	S	0.1128	100
....	....	S	0.1140	0 to 200
....	....	S	0.1256	0 to 400
....	....	S	0.131	0 to 800
....	....	S	0.1607	1000
Nickel carbonyl	$Ni(CO)_4$	S	0.165	−188 to −78
nitrate	$Ni(NO_3)_2 \cdot 6H_2O$	S	0.473	80
sulfate	$NiSO_4$	S	0.225	58
....	$NiSO_4 \cdot 6H_2O$	S	0.313	35
sulfide	NiS	S	0.116	0
....	....	S	0.128	100
Nitro-aniline (*o*)	$NO_2C_6H_4NH_2$	S	0.269	−160
....	....	L	0.400	0
aniline (*m*)	$NO_2C_6H_4NH_2$	S	0.275	−160
....	....	L	0.392	0
aniline (*p*)	$NO_2C_6H_4NH_2$	S	0.276	−160
....	....	L	0.427	0
benzene	$C_6H_5NO_2$	L	0.339	30
....	....	L	0.3429	90
....	....	L	0.394	120
benzoic acid (*o*)	$NO_2C_6H_4COOH$	S	0.256	−163
....	....	L	0.314	0
benzoic acid (*m*)	$NO_2C_6H_4COOH$	S	0.247	−160
....	....	L	0.405	0
benzoic acid (*p*)	$NO_2C_6H_4COOH$	S	0.247	−160
....	....	L	0.449	238
methane	CH_3NO_2	L	0.412	17
naphthalene (α)	$C_{10}H_7NO_2$	S	0.236	0
....	....	L	0.365	58.6
Nitrogen	N	L	0.474	−208 to −196
Nitrogen pentoxide	N_2O_5	S	0.239	−80 to −5
Nonane	C_9H_{20}	L	0.503	0 to 50
Nonylene	C_9H_{18}	L	0.485	0 to 50
n-Octane	C_8H_{18}	L	0.578	20 to 123

Table 9-8 (*Continued*)
SPECIFIC HEAT OF ELEMENTS, COMPOUNDS, AND VARIOUS SUBSTANCES

Name	Formula	State	Specific Heat	Temp. °C.
Octylene	C_8H_{16}	L	0.486	0 to 50
Olive oil	....	L	0.471	6.6
Orthoclase	$K_2O \cdot Al_2O_3 \cdot 6SiO_2$	S	0.205	100
Osmium	Os	S	0.0078	−253 to −196
....	....	S	**0.0311**	19 to 98
Oxalic acid	$(COOH)_2$	S	0.259	−200 to +50
....	$(COOH)_2 \cdot 2H_2O$	S	0.338	0
....	....	S	0.385	50
Oxygen	O	L	0.35	−200 to −183
Palladium	Pd	S	0.0190	−253 to −196
....	....	S	0.0528	−186 to +18
....	....	S	0.0567	−79 to +18
....	....	S	0.0586	18
....	....	S	0.0592	0 to 100
....	....	S	0.0632	0 to 500
....	....	S	0.0672	0 to 900
Palmitic acid	$C_{15}H_{31}COOH$	S	0.167	−180
....	....	S	0.251	−100
....	....	S	0.306	−50
....	....	S	0.382	0
....	....	S	0.430	20
....	....	L	0.653	65 to 104
Paraffin	....	S	0.694	0 to 20
Paraldehyde	$(CH_3CHO)_3$	L	0.436	0
Pentadecane	$C_{15}H_{32}$	L	0.497	0 to 50
Pentadecylene	$C_{15}H_{30}$	L	0.471	0 to 50
Petroleum	....	L	0.511	21 to 58
Phenetole	$C_2H_5OC_6H_5$	L	0.446	20
Phenol	C_6H_5OH	L	0.561	14 to 26
Phosphorus (yellow)	P	S	0.0774	−253 to −196
....	....	S	0.169	−188 to +20
....	....	S	0.178	−186 to +20
....	....	S	0.190	7 to 30
Phosphorus (red)	P	S	0.0431	−253 to −196
....	....	S	0.1829	0 to 51
Phosphorus (liquid)	P	L	0.2045	49 to 98
Phosphorus trichloride	PCl_3	S	0.203	33
Phthalic acid	$C_6H_4(COOH)_2$	S	0.232	20
α-Picoline	C_6H_7N	L	0.4339	22 to 124
Picric acid	$HOC_6H_2(NO_2)_3$	S	0.165	−100
....	....	S	0.240	0
....	....	S	0.263	50
....	....	S	0.297	100
Piperidine	$C_5H_{11}N$	L	0.523	20 to 98
Platinum	Pt	S	0.0135	−253 to −196
....	....	S	0.0293	−180 to +18
....	....	S	0.0307	0 to 10
....	....	S	0.03224	15 to 100
....	....	S	0.0347	0 to 500
....	....	S	0.0275	100
....	....	S	0.0356	500
....	....	S	0.0344	600
....	....	S	0.0369	800
....	....	S	0.0382	1000

Table 9-8 (*Continued*)
SPECIFIC HEAT OF ELEMENTS, COMPOUNDS, AND VARIOUS SUBSTANCES

Name	Formula	State	Specific Heat	Temp. °C.
Platinum	Pt	S	0.0398	1200
....	....	S	0.0368	1500
Porcelain	....	S	0.26	19 to 950
Potassium	K	S	0.1280	−253 to −196
....	....	S	0.170	−185 to +20
....	....	S	0.1728	0
....	....	S	0.188	0 to 22
....	....	S	0.192	22 to 56
....	....	L	0.217	78 to 100
....	....	L	0.224	100 to 157
Potassium acetate	$KC_2H_3O_2$	S	0.272	20
acid arsenate	KH_2AsO_4	S	0.174	31
acid sulfate	$KHSO_4$	S	0.244	35
aluminum sulfate (alum)	$K_2SO_4 \cdot Al_2(SO_4)_3 \cdot 24H_2O$	S	0.324	0
bromide	KBr	S	0.104	0
....	....	S	0.108	100
carbonate	K_2CO_3	S	0.210	47
chlorate	$KClO_3$	S	0.191	0
....	....	S	0.205	50
choride	KCl	S	0.162	0
....	....	S	0.168	100
chloroplatinate	K_2PtCl_6	S	0.112	30
chromate	K_2CrO_4	S	0.186	46
dichromate	$K_2Cr_2O_7$	S	0.178	0
....	....	L	0.0335	297
ferricyanide	$K_3Fe(CN)_6$	S	0.232	26
ferrocyanide	$K_4Fe(CN)_6$	S	0.210	0
....	....	S	0.225	50
....	$K_4Fe(CN)_6 \cdot 3H_2O$	S	0.267	0
....	....	S	0.285	50
fluoride	KF	S	0.199	0
....	....	S	0.204	50
metaborate	$K_2B_2O_4$	S	0.225	57
nitrate	KNO_3	S	0.214	0
....	....	S	0.240	100
....	....	L	0.0332	380
perchlorate	$KClO_4$	S	0.189	30
phosphate	KH_2PO_4	S	0.208	33
pyrophosphate	$K_4P_2O_7$	S	0.191	58
sulfate	K_2SO_4	S	0.176	0
....	....	S	0.191	100
thiosulfate	$K_2S_2O_3$	S	0.196	60
tetraborate	$K_2B_4O_7$	S	0.22	57
Propane	C_3H_8	L	0.576	0
Propionaldehyde	C_2H_5CHO	L	0.522	0
Propionic acid	C_2H_5COOH	S	0.726	−33
....	....	L	0.560	20 to 137
Propionitrile	C_2H_5CN	L	0.538	19 to 95
n-Propyl-acetate	$CH_3COOC_3H_7$	L	0.459	20
alcohol	C_3H_7OH	S	0.170	−200
....	....	S	0.497	−130
....	....	L	0.435	−100
....	....	L	0.526	0
....	....	L	0.586	25

Table 9-8 (*Continued*)
SPECIFIC HEAT OF ELEMENTS, COMPOUNDS, AND VARIOUS SUBSTANCES

Name	Formula	State	Specific Heat	Temp. °C.
n-Propyl benzene	$C_3H_7C_6H_5$	L	0.400	0
benzoate	$C_6H_5COOC_3H_7$	L	0.398	20
butyrate	$C_3H_7COOC_3H_7$	L	0.459	20
chloroacetate	$ClCH_2COOC_3H_7$	L	0.414	20
formate	$HCOOC_3H_7$	L	0.459	20
isobutyrate	$C_3H_7COOC_3H_7$	L	0.459	20
phenyl ether	$C_3H_7OC_6H_5$	L	0.429	0
propionate	$C_2H_5COOC_3H_7$	L	0.459	20
valerate	$C_4H_9COOC_3H_7$	L	0.459	20
Proustite	Ag_3AsS_3	S	0.081	50
Pseudocumene	$(CH_3)_3C_6H_3$	L	0.414	20
Pyrargyrite	Ag_3SbS_3	S	0.076	50
Pyridine	C_5H_5N	L	0.431	21 to 108
Pyrosulfuric acid	$H_2S_2O_7$	L	0.334	35
Pyrotartaric acid	$C_5H_8O_4$	S	0.301	20
Quartz	....	S	0.188	12 to 100
Quinhydrone	$C_{12}H_{10}O_4$	S	0.0165	−250
....	....	S	0.0980	−200
....	....	S	0.256	0
Quinol	$C_6H_4(OH)_2$	S	0.0246	−250
....	....	S	0.268	−150
Quinoline	C_9H_7N	L	0.352	0 to 20
Quinone	$C_6H_4O_2$	S	0.0311	−250
....	....	S	0.113	−200
....	....	S	0.282	−150
Resorcinol	$C_6H_4(OH)_2$	S	0.269	−160
....	....	L	0.452	0
Rhodium	Rh	S	0.0134	−253 to −196
....	....	S	0.0580	10 to 97
Rock salt	....	S	0.219	13 to 45
Rubidium	Rb	S	0.0711	−253 to −196
....	....	S	0.0802	0
....	....	L	0.0908	50
Rubidium bromide	RbBr	S	0.0743	10
carbonate	Rb_2CO_3	S	0.122	33
chloride	RbCl	S	0.101	10
fluoride	RbF	S	0.115	10
iodide	RbI	S	0.0581	10
Ruthenium	Ru	S	0.0109	−253 to −196
....	....	S	0.061	0 to 100
Salicylaldehyde	HOC_6H_4CHO	L	0.382	18
Salol	$HOC_6H_4COOC_6H_5$	S	0.289	32
....	....	L	0.391	44.1
Scandium oxide	Sc_2O_3	S	0.168	−150 to +40
Selenium	Se	S	0.0361	−253 to −196
....	....	S	0.068	−188 to +18
(amorph.)	Se	S	0.095	18 to 38
(cryst.)	Se	S	0.084	22 to 63
Silicon (melted)	Si	S	0.0303	−253 to −196
(cryst.)	Si	S	0.0271	−253 to −196
....	....	S	0.123	−185 to +20
(amorph.)	Si	S	0.091	−190 to −80
....	....	S	0.147	−79 to +17
....	....	S	0.179	3 to 50

Table 9-8 (*Continued*)
SPECIFIC HEAT OF ELEMENTS, COMPOUNDS, AND
VARIOUS SUBSTANCES

Name	Formula	State	Specific Heat	Temp. °C.
Silicon (cryst.)	Si	S	0.136	−40
....	....	S	0.170	21
....	....	S	0.196	129
Silicon carbide	SiC	S	0.143	0
....	....	S	0.194	100
....	....	S	0.280	600
chloride	$SiCl_4$	S	0.237	−200 to −80
Silver	Ag	S	0.0242	−253 to −196
....	....	S	0.0175	−233
....	....	S	0.040	−193
....	....	S	0.0496	−186 to −70
....	....	S	0.0544	−79 to +18
....	....	S	0.0556	0
....	....	S	0.05625	15 to 100
....	....	S	0.0557	58
....	....	S	0.0581	500
....	....	S	0.076	800
....	....	L	0.0748	907 to 1100
Silver bromide	AgBr	S	0.0695	0
....	....	S	0.0734	100
....	....	L	0.0760	500
chloride	AgCl	S	0.0848	0
....	....	S	0.0906	50
....	....	L	0.129	490
cyanate	AgCNO	S	0.124	40
iodide	AgI	S	0.0548	0
....	....	S	0.0593	100
nitrate	$AgNO_3$	S	0.146	50
....	....	L	0.187	250
selenide	Ag_2Se	S	0.0693	37 to 187
sulfide	Ag_2S	S	0.0719	0
....	....	S	0.0748	50
Sodium	Na	S	0.1519	−253 to −196
....	....	S	0.229	−186
....	....	S	0.253	−185 to +20
....	....	S	0.266	−80
....	....	S	0.279	−40
....	....	S	0.293	0
....	....	S	0.290	27
....	....	L	0.323	100
....	....	L	0.319	140
Sodium acetate	$NaC_2H_3O_2$	S	0.333	38
....	$NaC_2H_3O_2 \cdot 3H_2O$	S	0.344	0
....	$NaC_2H_3O_2 \cdot 3H_2O$	L	0.846	61.8
....	....	S	0.602	40
borate	$Na_2B_2O_4$	S	0.253	57
bromide	NaBr	S	0.118	0
....	....	S	0.124	100
carbonate	Na_2CO_3	S	0.256	45
chlorate	$NaClO_3$	L	0.325	280
chloride	NaCl	S	0.204	0
....	....	S	0.217	100
fluoride	NaF	S	0.258	0
....	....	S	0.279	100

Table 9-8 (*Continued*)
SPECIFIC HEAT OF ELEMENTS, COMPOUNDS, AND VARIOUS SUBSTANCES

Name	Formula	State	Specific Heat	Temp. °C.
Sodium formate	$NaCHO_2$	S	0.306	46
iodide	NaI	S	0.0829	0
....	....	S	0.0848	50
nitrate	$NaNO_3$	S	0.247	0
....	....	S	0.270	50
....	....	L	0.430	350
phosphate	$Na_2HPO_4 \cdot 7H_2O$	S	0.351	0
....	....	S	0.406	50
....	$Na_2HPO_4 \cdot 12H_2O$	S	0.404	0
....	....	S	0.464	50
phosphite	$NaPO_3$	S	0.217	30
pyrophosphate	$Na_4P_2O_7$	S	0.227	50
sulfate	Na_2SO_4	S	0.202	0
....	....	S	0.220	100
tetraborate	$Na_2B_4O_7$	S	0.234	45
(borax)	$Na_2B_4O_7 \cdot 10H_2O$	S	0.385	35
thiosulfate	$Na_2S_2O_3$	S	0.220	9
....	$Na_2S_2O_3 \cdot 5H_2O$	S	0.346	21
....	....	L	0.570	13 to 98
Stannic chloride	$SnCl_4$	S	0.115	−100
....	....	L	0.148	14 to 98
oxide	SnO_2	S	0.0898	45
sulfide	SnS_2	S	0.119	54
Stannous chloride	$SnCl_2$	S	0.102	60
sulfide	SnS	S	0.0839	56
Stearic acid	$C_{17}H_{35}COOH$	S	0.399	15
....	....	L	0.550	74 to 137
Strontium (impure)	Sr	S	0.0550	−253 to −196
Strontium carbonate	$SrCO_3$	S	0.146	54
chloride	$SrCl_2$	S	0.119	58
molybdate	$SrMoO_4$	S	0.148	15
nitrate	$Sr(NO_3)_2$	S	0.182	32
sulfate	$SrSO_4$	S	0.143	48
Succinic acid	$(CH_2COOH)_2$	S	0.248	0
Sucrose	$C_{12}H_{22}O_{11}$	S	0.299	20
Sulfur	S	S	0.0546	−253 to −196
(rhombic)	S	S	0.1520	−71
....	....	S	0.1728	0 to 54
(monoclinic)	S	S	0.1498	−72
....	....	S	0.1809	0 to 52
....	....	L	0.235	119 to 147
....	....	L	0.331	201 to 233
Sulfur chloride	S_2Cl_2	L	0.220	12 to 70
dioxide	SO_2	S	0.229	−185 to −103
....	....	L	0.313	−20
....	....	L	0.318	0
....	....	L	0.327	20
....	....	L	0.418	100
Sulfuric acid	H_2SO_4	S	0.239	−30
....	....	S	0.270	0
....	....	L	0.339	10
Talc	$Mg_3Si_4O_{11} \cdot H_2O$	S	0.208	57
Tantalum	Ta	S	0.033	−185 to +20
....	....	S	0.036	58

Table 9-8 (*Continued*)
SPECIFIC HEAT OF ELEMENTS, COMPOUNDS, AND VARIOUS SUBSTANCES

Name	Formula	State	Specific Heat	Temp. °C.
Tantalum	Ta	S	0.043	1400
Tartaric acid	$(CHOHCOOH)_2$	S	0.287	36
....	$(CHOHCOOH)_2 \cdot H_2O$	S	0.308	0
....	....	S	0.366	50
Tellurium	Te	S	0.0288	−253 to −196
....	....	S	0.047	−188 to +18
(cryst.)	Te	S	0.0483	15 to 100
....	....	S	0.0490	15 to 300
Tetrachloroethylene	C_2Cl_4	S	0.198	0
....	....	L	0.211	20
Tetradecane	$C_{14}H_{30}$	L	0.497	0 to 50
Tetradecylene	$C_{14}H_{28}$	L	0.453	0 to 50
Tetryl	$C_7H_5N_5O_8$	S	0.212	0
....	....	S	0.236	100
Thallium	Tl	S	0.0235	−253 to −196
....	....	S	0.038	−185 to +20
....	....	S	0.0326	20 to 100
....	....	S	0.0311	28
Thallium bromide	TlBr	L	0.080	500
bromide	TlBr	S	0.0526	390
chloride	TlCl	L	0.059	480
....	....	S	0.0520	0
....	....	S	0.0542	100
Thorium	Th	S	0.0197	−253 to −196
....	....	S	0.0276	0 to 100
Thorium chloride	ThCl	S	0.406	30
dioxide	ThO_2	S	0.0571	0
....	....	S	0.0589	50
sulfate	$Th(SO_4)_2$	S	0.0980	50
Thymol	$C_{10}H_{14}O$	S	0.315	0
m-Thymol	$C_{10}H_{14}O$	S	0.0567	50
Tin	Sn	S	0.0286	−253 to −196
....	....	S	0.0486	−186 to −79
....	....	S	0.0518	−79 to +18
....	....	S	0.0536	0
....	....	S	0.0541	20
....	....	S	0.0565	100
....	....	L	0.0608	250 to 350
....	....	L	0.0758	1100
(gray)	Sn	S	0.0589	0 to 18
Titanium	Ti	S	0.0205	−253 to −196
....	....	S	0.036	−185 to +20
....	....	S	0.142	20
....	....	S	0.1125	0 to 100
....	....	S	0.1563	0 to 333
Titanium chloride	$TiCl_4$	S	0.177	−25
....	....	L	0.192	13 to 99
oxide	TiO_2	S	0.168	0
Toluene	$C_6H_5CH_3$	L	0.386	0
....	....	L	0.421	50
....	....	L	0.470	100
o-Toluic acid	$CH_3C_6H_4COOH$	S	0.277	54
m-Toluic acid	$CH_3C_6H_4COOH$	S	0.239	54
p-Toluic acid	$CH_3C_6H_4COOH$	S	0.271	130

Table 9-8 (*Continued*)
SPECIFIC HEAT OF ELEMENTS, COMPOUNDS, AND VARIOUS SUBSTANCES

Name	Formula	State	Specific Heat	Temp. °C.
p-Toluidine	$CH_3C_6H_4NH_2$	S	0.337	0
....	....	S	0.387	20
Topaz	$2(AlF)O \cdot SiO_2$	S	0.205	52
Trichloroacetic acid	Cl_3CCOOH	S	0.459	......
Trichloroethane	$C_2H_3Cl_3$	L	0.2660	20
Trichloroethylene	C_2HCl_3	L	0.223	20
Tridecane	$C_{13}H_{28}$	L	0.499	0 to 50
Tridecylene	$C_{13}H_{26}$	L	0.457	0 to 50
Trimethyl carbinol	$(CH_3)_3COH$	S	0.559	−4
Trinitrotoluene	$CH_3C_6H_2(NO_2)_3$	S	0.170	−100
....	....	S	0.311	0
....	....	S	0.385	100
....	....	L	0.335	?
Trinitroxylene	$(CH_3)_2C_6H(NO_2)_3$	S	0.423	20 to 50
Triphenylmethane	$(C_6H_5)_3CH$	S	0.189	0
Tungsten	W	S	0.0095	−253 to −196
....	....	S	0.036	−185 to +20
....	....	S	0.034	15 to 93
....	....	S	0.034	20 to 100
....	....	S	0.041	1350
Tungsten trioxide	WO_3	S	0.0743	0
....	....	S	0.0832	50
Turpentine, oil	....	L	0.411	0
Undecane	$C_{11}H_{24}$	L	0.501	0 to 50
Undecylene	$C_{11}H_{22}$	L	0.482	0 to 50
Uranium	U	S	0.0138	−253 to −196
....	....	S	0.028	0 to 98
....	....	S	0.026	11 to 98
Uranium oxide	U_3O_8	S	0.0671	0
....	....	S	0.0750	50
Urea	$(NH_2)_2CO$	S	0.320	20
Valeronitrile	C_4H_9CN	L	0.520	23 to 121
Vanadium	V	S	0.1153	0 to 100
Vulcanite	....	S	0.3312	20 to 100
Water	H_2O	S	0.0361	−250
See also special table				
....	....	S	0.156	−200
....	....	S	0.246	−150
....	....	S	0.332	−100
....	....	S	0.435	−40
....	....	S	0.492	0
....	....	L	1.0000	15
Wood	....	S	0.42	......
o- Xylene	$(CH_3)_2C_6H_4$	L	0.411	30
m- Xylene	$(CH_3)_2C_6H_4$	L	0.387	16 to 35
p- Xylene	$(CH_3)_2C_6H_4$	L	0.397	30
o- Xylene dibromide	$C_8H_8Br_2$	L	0.183	15 to 40
dichloride	$C_8H_8Cl_2$	L	0.2829	15 to 40
tetrachloride	$C_8H_6Cl_4$	L	0.2399	15 to 40
Xylyl ethyl ether (2, 4)	$C_{10}H_{14}O$	L	0.4170	0
Ytterbium oxide	Yb_2O_3	S	0.0650	50
Yttrium oxide	Y_2O_3	S	0.112	57
Zinc	Zn	S	0.0384	−253 to −196
....	....	S	0.0268	−233

Table 9-8 (*Continued*)
SPECIFIC HEAT OF ELEMENTS, COMPOUNDS, AND VARIOUS SUBSTANCES

Name	Formula	State	Specific Heat	Temp. °C.
Zinc	Zn	S	0.063	−193
....	....	S	0.0836	−192 to +20
....	....	S	0.080	−186 to −79
....	....	S	0.0895	−79 to +18
....	....	S	0.0903	1
....	....	S	0.0924	20
....	....	S	0.0951	100
....	....	S	0.1040	300
....	....	S	0.110	419
....	....	L	0.121	419
Zinc acetate	$Zn(C_2H_3O_2)_2$	S	0.270	45
....	$Zn(C_2H_3O_2)_2 \cdot 3H_2O$	S	0.410	85
carbonate	$ZnCO_3$	S	0.142	0
chloride	$ZnCl_2$	S	0.136	60
nitrate	$Zn(NO_3)_2 \cdot 6H_2O$	S	0.318	30
oxide	ZnO	S	0.114	0
....	....	S	0.129	100
....	....	S	0.15	600
sulfate	$ZnSO_4$	S	0.174	50
....	$ZnSO_4 \cdot H_2O$	S	0.194	9
....	$ZnSO_4 \cdot 6H_2O$	S	0.299	9
....	$ZnSO_4 \cdot 7H_2O$	S	0.322	0
sulfide	ZnS	S	0.116	0
....	....	S	0.118	100
Zircon	$ZrO_2 \cdot SiO_2$	S	0.131	36
Zirconium	Zr	S	0.0262	−253 to −196
....	....	S	0.068	0 to 100
Zirconium dioxide	ZrO_2	S	0.103	0

Table 9-9
SPECIFIC HEAT OF MERCURY AND WATER
In gram-calories (15° C.) per gram per °C.

Temp. °C.	Mercury	Water	Temp. °C.	Mercury	Water
−10	....	0.48*	80	0.03283	1.00239
−6	....	1.0119	85	....	1.00329
−4	....	1.0105	90	0.03274	1.00433
−2	....	1.0097	95	....	1.00534
0	0.03346	1.00874	100	0.03267	1.00645
5	0.03340	1.00477	110	0.03260	1.0116
10	0.03335	1.00184	120	0.03253	1.0144
15	0.03330	1.00000	130	0.03246	1.0174
20	0.03325	0.99859	140	0.03239	1.0206
25	0.03320	0.99765	150	0.03232	1.0240
30	0.03316	0.99745	160	0.03225	1.0275
35	0.03312	0.99743	170	0.03218	1.0313
40	0.03308	0.99761	180	0.03211	1.0353
45	....	0.99790	190	0.03203	1.0395
50	0.03300	0.99829	200	0.03196	1.0439
55	....	0.99873	220	0.03185	1.0769
60	0.03294	0.99934	240	0.03217	1.0939
65	....	1.00001	260	0.03215	1.1126
70	0.03290	1.00077	280	....	1.1329
75	....	1.00158	300	....	1.1549

Value for solid state at −10°C.

Table 9-10
HEAT CAPACITY STANDARDS

In the following tables for water, mercury, and aluminum oxide the units of temperature are based on the International Temperature Scale of 1948 using 0°C = 273.16°K. Values expressed in calories are in units of the "defined" calorie. This calorie which is independent of the properties of water is by definition equal to 4.1833 international joules. Redefined in January 1948 this calorie is equal to 4.1840 absolute joules. One absolute joule = 0.999835 international joule = 0.239006 calorie.

A. Water

The values in the following table were determined by Ginnings and Furukawa of the National Bureau of Standards. Published in the *J. Am. Chem. Soc.*, **75**, 522 (1953), they are reproduced here by permission of the American Chemical Society. For international use in calorimetry, the International Committee on Weights and Measures has approved a heat capacity table (Procès-Verbaux of the International Committee on Weights and Measures, Session of 1950, p. 92) which differs from this table by less than 0.005 joule per degree per gram mole.

Values in the columns headed *C-satd.* and *C-p* give the heat capacities (specific heats) at saturation pressure and at 1 atmosphere (760 mm Hg) pressure, respectively. The units in the columns headed *J* are in absolute joules per degree per gram mole (mol. wt. = 18.016); the units in the columns headed *cal.* are in calories per degree per gram.

	C-satd.		C-p			C-satd.		C-p	
°C	J	cal.	J	cal.	°C	J	cal.	J	cal.
0	75.993	1.00814	75.985	1.00804	55	75.350	0.99962	75.348	0.99959
5	75.714	1.00445	75.706	1.00434	60	75.385	1.00008	75.385	1.00008
10	75.532	1.00203	75.525	1.00194	65	75.428	1.00065	75.428	1.00065
15	75.417	1.00050	75.410	1.00040	70	75.426	1.00129	75.478	1.00132
20	75.345	0.99955	75.339	0.99947	75	75.532	1.00203	75.536	1.00208
25	75.303	0.99899	75.298	0.99893	80	75.594	1.00285	75.601	1.00295
30	75.282	0.99871	75.278	0.99866	85	75.667	1.00382	75.675	1.00391
35	75.277	0.99865	75.273	0.99860	90	75.746	1.00489	75.757	1.00502
40	75.283	0.99873	75.280	0.99869	95	75.835	1.00605	75.850	1.00625
45	75.298	0.99893	75.295	0.99889	100	75.934	1.00736	75.954	1.00763
50	75.320	0.99922	75.318	0.99919					

B. Mercury

The values in the following table were determined by Douglas, Ball and Ginnings, and published in *J. Research Natl. Bur. Standards*, **46**, 334 (1951). Values in the columns headed *C-satd.*, *C-p*, and *C-v* are those for liquid-vapor equilibrium, maintenance of constant pressure, and maintenance of constant volume, respectively. The units are in "defined calories." (1 calorie = 4.1840 absolute joules = 4.1833 international joules.)

	Liquid			Vapor
°C	C-satd.	C-p	C-v	C-p
−38.88*	0.033686	0.033686	0.02975	0.02476
−20	0.033534	0.033534	0.02941	0.02476
0	0.033817	0.033817	0.02920	0.02476
20	0.033240	0.033240	0.02876	0.02476
25	0.033206	0.033206	0.02867	0.02476
40	0.033110	0.033110	0.02845	0.02476
60	0.032987	0.032987	0.02816	0.02476
80	0.032877	0.032877	0.02789	0.02476
100	0.032776	0.032776	0.02764	0.02476
120	0.032686	0.032686	0.02739	0.02476
140	0.032606	0.032606	0.02716	0.02476

* Triple point.

Table 9-10 (*Continued*)
HEAT CAPACITY STANDARDS

B. Mercury (*Continued*)

	Liquid			Vapor
°C	C-satd.	C-p	C-v	C-p
160	0.032536	0.032536	0.02693	0.02476
180	0.032476	0.032476	0.02674	0.02477
200	0.032426	0.032426	0.02659	0.02477
220	0.032386	0.032386	....	0.02477
240	0.032356	0.032356	....	0.02477
260	0.032335	0.032336	....	0.02478
280	0.032324	0.032325	....	0.02479
300	0.032321	0.032323	....	0.02480
320	0.032328	0.032330	....	0.02481
340	0.032343	0.032346	....	0.02482
356.58†	0.032362	0.032366	....	0.02484
360	0.032367	0.032371	....	0.02484
380	0.032398	0.032404	....	0.02486
400	0.032437	0.032445	....	0.02489
420	0.032483	0.032494	....	0.02492
440	0.032536	0.032550	....	0.02495
460	0.032596	0.032614	....	0.02499
480	0.032661	0.032684	....	0.02503
500	0.032733	0.032762	....	0.02507

C. Aluminum Oxide

The values in the following table were determined by Ginnings and Furukawa of the National Bureau of Standards on aluminum oxide in the form of synthetic sapphire (corundum) in pieces that passed a #10 and were retained by a #40 sieve and with impurities between 0.01 and 0.02% by weight. Heat capacity values below the experimental range were obtained by extrapolating a Debye equation fitted to the experimental values at the lowest temperature. The units in the columns headed J are in absolute joules per degree per gram mole (mol. wt. = 101.96) at constant pressure of 1 atmosphere (760 mm Hg); the units in the columns headed *cal.* are in calories per degree per gram at constant pressure of 1 atmosphere. Published in the *J. Am. Chem. Soc.*, **75**, 522 (1953), they are reproduced here by permission of the American Chemical Society.

	C-p			C-p	
°K	J	cal.	°K	J	cal.
0	0.0	0.0	65	3.620	0.008486
5	0.0012	0.000003	70	4.582	0.01074
10	0.0094	0.000022	75	5.668	0.01329
15	0.0316	0.000074	80	6.895	0.01616
20	0.0759	0.000178	85	8.247	0.01933
25	0.1417	0.000332	90	9.692	0.02272
30	0.2627	0.000616	95	11.223	0.02631
35	0.4377	0.001026	100	12.84	0.03010
40	0.6907	0.001619	110	16.31	0.03823
45	1.039	0.002436	120	20.05	0.04700
50	1.492	0.003497	130	23.96	0.05617
55	2.069	0.004850	140	27.96	0.06554
60	2.780	0.006517	150	31.99	0.07499

† Boiling point.

Table 9-10 (*Continued*)
HEAT CAPACITY STANDARDS

C. Aluminum Oxide (*Continued*)

	C-p			C-p	
°K	J	cal.	°K	J	cal.
160	35.99	0.08436	560	110.21	0.25835
170	39.94	0.09362	570	110.82	0.25978
180	43.79	0.10263	580	111.40	0.26114
190	47.53	0.11147	590	111.97	0.26247
200	51.14	0.11988	600	112.51	0.26374
210	54.60	0.12799	610	113.03	0.26495
220	57.92	0.13577	620	113.54	0.26615
230	61.09	0.14320	630	114.03	0.26730
240	64.12	0.15030	640	114.50	0.26841
250	67.01	0.15708	650	114.95	0.26946
260	69.75	0.16350	660	115.40	0.27051
270	72.36	0.16962	670	115.82	0.27150
280	74.84	0.17543	680	116.24	0.27248
290	77.19	0.18094	690	116.64	0.27342
298.16	79.01	0.18521	700	117.03	0.27433
300	79.41	0.18615	720	117.77	0.27607
310	81.52	0.19109	740	118.46	0.27768
320	83.50	0.19574	760	119.12	0.27923
330	85.39	0.20017	780	119.74	0.28069
340	87.18	0.20436	800	120.33	0.28207
350	88.88	0.20834	820	120.88	0.28336
360	90.52	0.21219	840	121.40	0.28458
370	92.06	0.21580	860	121.90	0.28575
380	93.51	0.21920	880	122.37	0.28685
390	94.88	0.22241	900	122.82	0.28791
400	96.18	0.22545	920	123.24	0.28889
410	97.40	0.22832	940	123.65	0.28985
420	98.55	0.23102	960	124.03	0.29074
430	99.64	0.23357	980	124.40	0.29161
440	100.69	0.23603	1000	124.74	0.29241
450	101.68	0.23835	1020	125.08	0.29321
460	102.64	0.24061	1040	125.39	0.29393
470	103.54	0.24271	1060	125.69	0.29464
480	104.42	0.24477	1080	125.98	0.29531
490	105.25	0.24672	1100	126.26	0.29597
500	106.05	0.24860	1120	126.52	0.29658
510	106.81	0.25038	1140	126.77	0.29718
520	107.55	0.25211	1160	127.01	0.29773
530	108.26	0.25378	1180	127.24	0.29827
540	108.93	0.25535	1200	127.47	0.29881
550	109.59	0.25690			

Table 9-11
CRITICAL PROPERTIES

Substance	t_c, °C	P_c, atm	ρ_c, g/cm³
Acetaldehyde (ethanal)	188		
Acetic acid	321.3	57.1	0.351
Acetic anhydride	296	46.2	
Acetone	236.5	47.2	0.278
Acetonitrile	274.7	47.7	0.237
Acetylene (ethyne)	35.18	60.59	0.231
Air	−140.6	37.2	0.313
Allene	120		
Allyl alcohol	272		
Allyl sulfide	380		
Ammonia	132.4	111.3	0.235
Aniline	426	52.4	0.34
Anisole	368	41.2	
Antimony tribromide	904.5	56	
Argon	−122.44	48.00	0.5307
Benzaldehyde	352	21.5	
Benzene	288.94	48.34	0.302
Benzonitrile	426.2	41.6	
Biphenyl	516	38	0.307
Boron pentafluoride	197		
Boron tribromide	300		
Boron trichloride	178.8	38.2	
Boron trifluoride	−12.3	49.2	
Bromine	311	102	1.18
Bromobenzene	397	44.6	0.485
Bromoethane (ethyl bromide)	230.7	61.5	0.507
Bromomethane (methyl bromide)	191		
Bromopentafluorobenzene	397	44.6	
Bromotrifluoromethane	67.0	39.2	0.76
1,3-Butadiene	152	42.7	0.245
n-Butane	152.01	37.7	0.228
Butane nitrile (butyronitrile)	309.1	37.4	
Butanoic acid (n-butyric acid)	355	52	0.304
1-Butanol (n-butyl alcohol)	289.78	43.55	0.270
2-Butanol (sec-butyl alcohol)	262.80	41.39	0.276
2-Butanone (ethyl methyl ketone)	262.4	41.0	0.270
1-Butene	146.4	39.7	0.234
2-Butene cis	162.40	41.5	0.240
trans	155.46	40.5	0.236
n-Butyl acetate	306		
n-Butylamine	251	41	
n-Butylbenzene	387.3	28.49	0.270
1-Butyne (ethylacetylene)	190.5		
2-Butyne (dimethylacetylene)	215.5		
Carbon dioxide	31.04	72.85	0.468
Carbon disulfide	279	78	0.44
Carbon monoxide	−140.23	34.53	0.301
Carbon tetrachloride	283.15	44.97	0.558
Carbonyl chloride (phosgene)	182	56	0.52
Carbonyl sulfide	105	61	
Chlorine	144.0	76.1	0.573

Table 9-11 (*Continued*)
CRITICAL PROPERTIES

Substance	t_c, °C	P_c, atm	ρ_c, g/cm³
Chlorine pentafluoride	142.6	51.9	0.565
Chlorine trifluoride	153.5		
Chlorobenzene	359.2	44.6	0.365
1-Chloro-1,1-difluoroethane	137.1	40.7	0.435
2-Chloro-1,1-difluoroethylene	127.4	44.0	0.499
Chlorodifluoromethane (freon 22)	96.0	49.12	0.525
Chloroethane (ethyl chloride)	187.2	52	
Chloroform	263.4	54	0.50
Chloromethane (methyl chloride)	143.1	65.92	0.353
Chloropentafluoroethane	80.0	31.16	0.613
1-Chloropropane (*n*-propyl chloride)	230	45.2	
3-Chloropropene (allyl chloride)	241		
Chlorotrifluoromethane (freon 13)	28.9	38.7	0.579
o-Cresol	424.4	49.4	0.384
m-Cresol	432.6	45.0	0.346
p-Cresol	431.4	50.8	0.391
Cyanogen	128	59	
Cyclohexane	280.3	40.2	0.273
Cyclohexene	287.26		
Cyclopentane	238.5	44.49	0.27
Cyclopentene	232.9		
Cyclopropane	124.65	54.23	
Cymene	385		
Decalin *cis*	429.0		
trans	413.8		
1-Decanol	427		0.264
Decanenitrile (caprylonitrile)	348.8	32.1	
Deuterium (equilibrium)	−234.90	16.28	0.0668
(normal)	−234.81	16.43	
Deuterium bromide (DBr)	88.8		
Deuterium chloride	50.3		
Deuterium hydride (DH)	−237.25	14.64	0.0481
Deuterium iodide	148.6		
Diborane (B_2H_6)	16.7	39.53	
Dibromomethane (methylene bromide)	310	71	
1,2-Dibromoethane	309.8	70.6	
Dichlorodifluoromethane (freon 12)	111.80	40.71	0.558
1,1-Dichloroethane	250	50	0.42
1,2-Dichloroethane	288	53	0.44
1,1-Dichloroethylene	271.0		
1,2-Dichloroethylene	243.3	54.4	
Dichlorofluoromethane (freon 21)	178.5	51.0	0.522
Dichloromethane (methylene chloride)	237	60	
1,1-Dichloro-1,2,2,2-tetrafluoroethane	145.5	32.6	0.582
1,2-Dichloro-1,1,2,2-tetrafluoroethane (freon 114)	145.7	32.2	0.582
Dideuterium oxide (D_2O)	371.0	215.7	0.363
Diethylamine	223.5	36.6	0.243
1,4-Diethylbenzene	384.73	27.66	
Diethyl ether	193.55	35.9	0.265
Difluoroamine (HNF_2)	130	93	
Difluorochloromethane	96.4	48.5	0.525
Difluorodiazine *cis* (N_2F_2)	−1	70	
trans	−13	55	

Table 9-11 (*Continued*)
CRITICAL PROPERTIES

Substance	t_c, °C	P_c, atm	ρ_c, g/cm³
Difluorodichloromethane (freon 12)	111.8	40.71	0.558
1,1-Difluoroethane	113.5	44.37	0.365
1,1-Difluoroethylene	30.1	43.75	0.417
Dihydrogen disulfide	299	58.3	
Dihydrogen trisulfide	465	50.6	
Dihydrogen tetrasulfide	582	43.1	
Dihydrogen pentasulfide	657	38.4	
Dihydrogen hexasulfide	707	36	
Dihydrogen heptasulfide	742	33	
Dihydrogen octasulfide	767	32	
Dimethylamine	164.5	52.4	
N,N-Dimethylaniline	414	35.8	
2,2-Dimethylbutane	215.58	30.40	0.240
2,3-Dimethylbutane	226.78	30.86	0.241
Dimethyl ether	126.9	53	0.242
2,2-Dimethylhexane	276.65	24.96	0.239
2,3-Dimethylhexane	290.27	25.94	0.244
2,4-Dimethylhexane	280.30	25.23	0.242
2,5-Dimethylhexane	276.84	24.54	0.237
3,3-Dimethylhexane	288.80	26.19	0.258
3,4-Dimethylhexane	295.63	26.57	0.245
Dimethyl oxalate	355	39.3	
2,2-Dimethylpentane	247.29	27.37	0.241
2,3-Dimethylpentane	264.14	28.70	0.255
2,4-Dimethylpentane	246.58	27.01	0.240
3,3-Dimethylpentane	263.19	29.07	0.242
2,2-Dimethylpropane (neopentane)	160.60	31.57	0.238
2,3-Dimethylpyridine (2,3-lutidine)	382.3		
2,4-Dimethylpyridine	374		
2,5-Dimethylpyridine	371.0		
2,6-Dimethylpyridine	350.6		
3,4-Dimethylpyridine	410.6		
3,5-Dimethylpyridine	394.1		
N,N-Dimethyl-o-toluidine	395	30.8	
Dioxane	314	51.4	0.370
Di-n-propylamine	277	31	
3-4-Dithiahexane (ethyl disulfide)	369		
Ethane	32.28	48.16	0.203
Ethanethiol (ethyl mercaptan)	226	54.2	0.300
Ethanol (ethyl alcohol)	243.1	62.96	0.276
Ethyl acetate	250.1	37.99	0.308
Ethylacetylene	190.5		
Ethylal (propionaldehyde)	254		
Ethyl allyl ether	245		
Ethylamine	183	55.5	
Ethylbenzene	343.94	35.62	0.284
Ethyl bromide	230.7	61.5	0.507
Ethyl n-butanoate (ethyl n-butyrate)	293	30	0.28
Ethyl chloride	187.2	52	
Ethyl crotonate	326		
Ethylcyclopentane	296.3	33.5	0.262
Ethyl disulfide	369		

Table 9-11 (*Continued*)
CRITICAL PROPERTIES

Substance	t_c, °C	P_c, atm	ρ_c, g/cm^3
Ethyl ether, *see* Diethyl ether			
Ethylene (ethene)	9.21	49.66	0.218
Ethylene oxide	196	71.0	0.314
Ethyl fluoride	102.16	49.62	
Ethyl formate	235.3	46.76	0.323
3-Ethylhexane	292.27	25.74	0.251
Ethyl mercaptan	226	54.2	0.300
Ethyl-2-methylpropanoate (ethyl isobutyrate)	280	30	0.28
3-Ethylpentane	267.42	28.53	0.241
Ethyl propanoate (ethyl propionate)	272.9	33.18	0.296
Ethyl *n*-propyl ether	226.4	32.1	0.361
Ethyl sulfide	284	39.1	0.279
Ethyltoluene ortho	380	31	0.28
meta	363	31	0.28
para	363	31	0.28
Fluorine	−129.0	55	0.63
Fluorobenzene	286.6	44.6	0.354
Fluorodichloromethane	178.5	51.0	0.522
Fluoroethane	102.16	49.62	
Fluorotrichloromethane (freon 11)	198.0	43.2	0.554
Fluoromethane	44.55	58.0	0.300
Germanium tetrachloride	276.9	38	
Helium	−267.96	2.261	0.06930
Helium-3	−269.81	1.167	0.041
n-Heptane	267.0	27.00	0.232
1-Heptanol	360		0.267
1-Heptene	264.08		
1,5-Hexadiene	234.4		
Hexamethylbenzene	494		
n-Hexane	234.2	29.3	0.233
1-Hexanol	337		0.268
1-Hexene	230.83		
Hydrazine	380	145	
Hydrogen (equilibrium)	−240.17	12.77	0.0308
(normal)	−239.91	12.80	0.0310
Hydrogen bromide	89.8	84	
Hydrogen chloride	51.40	81.5	0.42
Hydrogen cyanide	183.5	53.2	0.195
Hydrogen deuteride, *see* Deuterium hydride			
Hydrogen fluoride	188	64	0.29
Hydrogen iodide	150.7	81	
Hydrogen selenide	137	88	
Hydrogen sulfide	100.4	88.9	0.31
Iodine	535		
Iodobenzene	448	44.6	0.581
Iodomethane (methyl iodide)	255		
Isobutyl acetate	288		
Isobutylbenzene	377	31	
Isobutyl butanoate	338		

Table 9-11 (*Continued*)
CRITICAL PROPERTIES

Substance	t_c, °C	P_c, atm	ρ_c, g/cm³
Isobutyl formate	278	38.3	0.29
Isobutyl-3-methyl butanoate	348		
Isobutyl propanoate	319		
Isopropylbenzene	357.9	31.67	0.28
Isopropyl ether	226.9	28.4	0.265
Isoquinoline	530		
Isoxazole	278.9		
Krypton	−63.75	54.20	0.9085
Mercury	900	180	
Methane	−82.60	45.44	0.162
Methanethiol (methyl mercaptan)	196.8	71.4	0.332
Methanol (methyl alcohol)	239.43	79.9	0.272
Methoxybenzene (anisole)	368	41.2	
Methyl acetate	233.7	46.33	0.325
Methylamine	156.9	73.6	
N-Methylaniline	428	51.3	
2-Methylbutane (isopentane)	187.24	33.37	0.236
3-Methylbutanoic acid (isovaleric acid)	361		
Methyl butanoate	281.3	34.28	0.300
3-Methyl-1-butanol	306.25		
2-Methyl-2-butanol	272		
3-Methyl-2-butanone (methyl isopropyl ketone)	280.2	38.0	0.278
2-Methyl-1-butene	192	34	
2-Methyl-2-butene	197	34	
Methylcyclohexane	299.1	34.32	0.285
Methylcyclopentane	259.6	37.4	0.264
Methyl ether, *see* Dimethyl ether			
Methylethylamine	223.5	36.6	0.243
Methyl ethyl ether	164.7	43.4	0.272
2-Methyl-3-ethylpentane	293.87	26.65	0.258
3-Methyl-3-ethylpentane	303.36	27.71	0.251
Methyl ethyl sulfide	260	42	
Methyl formate	214.0	59.25	0.349
2-Methylheptane	286.42	24.52	0.234
3-Methylheptane	290.45	25.13	0.246
4-Methylheptane	288.52	25.09	0.240
2-Methylhexane	257.16	26.98	0.238
3-Methylhexane	262.04	27.77	0.248
Methylhydrazine	292	79.3	0.170
Methyl-2-methyl propanoate	267.55	33.87	0.301
1-Methylnaphthalene	499		
2-Methylnaphthalene	488		
Methyl oxalate	260	9.48	
2-Methylpentane	224.30	29.71	0.235
3-Methylpentane	231.20	30.83	0.235
4-Methyl-2-pentanone (methyl isobutyl ketone)	298.3	32.3	
2-Methylpropane (isobutane)	134.98	36.00	0.221
Methyl propanoate	257.4	39.52	0.312
2-Methylpropanoic acid (isobutyric acid)	336	40	0.302
2-Methyl-1-propanol (isobutyl alcohol)	274.58	42.39	0.272
2-Methyl-2-propanol (*t*-butyl alcohol)	233.0	39.20	0.270

Table 9-11 (*Continued*)
CRITICAL PROPERTIES

Substance	t_c, °C	P_c, atm	ρ_c, g/cm^3
2-Methylpropene	144.73	39.48	0.235
2-Methylpyridine (α-picoline)	348		
3-Methylpyridine (β-picoline)	372		
4-Methylpyridine (γ-picoline)	373		
Methyl sulfide	229.9	54.6	0.309
Naphthalene	475.2	39.98	0.31
Neon	−228.71	26.86	0.4835
Niobium pentachloride	534	0.46	0.68
Nitric oxide	−92.9	64.6	0.52
Nitrogen-14	−146.89	33.54	0.3110
Nitrogen-15	−146.8	33.5	0.332
Nitrogen dioxide (equilibrium)	158.2	100	0.557
Nitrogen trifluoride	−39.3	44.7	
Nitromethane	315	62.3	0.352
Nitrous oxide	36.434	71.596	0.4525
Nitryl fluoride	76.3		
n-Nonane	321.41	22.8	
1-Nonanol	404		0.264
n-Octane	295.61	24.54	0.232
1-Octanol	385		0.266
2-Octanol	364		
1-Octene	293.4		
Oxygen	−118.38	50.14	0.419
Oxygen difluoride (fluorine oxide)	−58.0	48.9	0.553
Ozone	−12.10	54.6	0.436
Paraldehyde	290		
Pentafluorochloroacetone	137.5	28.4	
Pentafluorobenzene	258.8	34.7	
1,1,2-H-Pentafluoropropane (refrigerant 245)	106.96	30.96	0.491
n-Pentane	196.5	33.35	0.237
Pentanoic acid (*n*-valeric acid)	378		
1-Pentanol	313		0.270
2-Pentanone (methyl *n*-propyl ketone)	290.8	38.4	0.286
3-Pentanone (diethyl ketone)	287.8	36.9	0.256
1-Pentene	191.59	40	
2-Pentene *cis*	203	36	
trans	202	36	
n-Pentyl formate	303		
1-Pentyne (propylacetylene)	220.3		
Perchloryl fluoride	95.2	53.0	0.64
Perfluoroacetone (hexafluoroacetone)	84.1	28.0	
Perfluorobenzene	243.57	32.61	
Perfluoro-*n*-butane	113.2	22.93	0.629
Perfluoro-(2-butyltetrahydrofuran)	227.1	15.86	0.707
Perfluorocyclobutane	115.22	27.41	0.616
Perfluorocyclohexane	184.0	24	
Perfluorocyclohexene	188.6		
Perfluoro-*n*-decane	269.2	14.3	
Perfluoroethane	19.7		0.617
Perfluoroethene (tetrafluoroethylene)	33.3	38.92	0.58

Table 9-11 (*Continued*)
CRITICAL PROPERTIES

Substance	t_c, °C	P_c, atm	ρ_c, g/cm³
Perfluoro-*n*-heptane	201.6	16.0	0.584
Perfluoro-1-heptene	205.0		
Perfluoro-*n*-hexane	174.5	18.8	
Perfluoro-1-hexene	181.2		
Perfluoromethane (tetrafluoromethane)	−45.6	36.9	0.630
Perfluoro-(methylcyclohexane)	213.6	23	
Perfluoronaphthalene	399.9		
Perfluoro-*n*-nonane	250.8	15.4	
Perfluoro-*n*-octane	229	16.4	
Perfluoro-*n*-pentane	149	20.1	
Perfluoro-*n*-propane	71.9	26.45	0.628
Phenetole	374	33.8	
Phenol	421.1	60.5	0.41
Phosphine	51.3	64.5	
Phosphonium chloride	49.1	72.7	
Phosphorus bromide difluoride ($PBrF_2$)	113		
Phosphorus chloride difluoride	89.17	44.61	
Phosphorus dibromide fluoride	254		
Phosphorus dichloride fluoride	189.84	49.3	
Phosphorus pentachloride	372		
Phosphorus trichloride	285.5		
Phosphorus trifluoride	−2.05	42.69	
Phosphoryl chloride difluoride ($POClF_2$)	150.6	43.4	
Phosphoryl trichloride	329		
Phosphoryl trifluoride	73.3	41.8	
Piperidine	320.9		
Propane	96.67	41.94	0.217
Propanenitrile (propionitrile)	291.2	41.3	0.240
Propanoic acid (propionic acid)	339	53	0.32
1-Propanol (*n*-propyl alcohol)	263.56	51.02	0.275
2-Propanol (isopropyl alcohol)	235.16	47.02	0.273
Propene (propylene)	91.8	45.6	0.233
n-Propyl acetate	276.2	33.19	0.269
n-Propylamine	233.8	46.8	
n-Propylbenzene	365.15	31.58	0.273
n-Propyl formate	264.9	40.08	0.309
n-Propyl propanoate	305		
Propyne (methylacetylene)	129.23	55.54	0.245
Pyridine	346.8	55.6	0.312
Pyrrole	366.6		
Pyrrolidine	295.4	55.4	0.286
Quinoline	509		
Radon	103.84	62	1.6
Silicon chloride trifluoride ($SiClF_3$)	34.5	34.2	
Silicon hydride (silane)	−3.5	47.8	
Silicon tetrachloride	233		
Silicon tetrafluoride	−14.1	36.7	
Silicon trichlorofluoride	165.3	35.3	
Sulfane di- (H_2S_2)	572	58.3	
tri- (H_2S_3)	738	50.6	

Table 9-11 (*Continued*)
CRITICAL PROPERTIES

Substance	t_c, °C	P_c, atm	ρ_c, g/cm³
tetra- (H_2S_4)	855	43.1	
penta- (H_2S_5)	930	38.4	
Sulfur	1040	116	
Sulfur dioxide	157.50	77.9	0.5240
Sulfur hexafluoride	45.55	37.11	0.734
Sulfur tetrafluoride	91		
Sulfur trioxide	218.2	83.8	0.633
Tantalum pentachloride	494	0.43	0.89
1,1,2,2-Tetrachloro-1,2-difluoroethane	278		
1,1,2,2-Tetrachloroethane	388.0		
Tetrachloroethene	347.1		
Tetrafluorohydrazine (F_2NNF_2)	36	77	
Tetrahydrothiophene	358.8		
1,2,4,5-Tetramethylbenzene	402	29	
2,2,3,3-Tetramethylbutane	294.7	28.3	0.248
o-Terphenyl	617.8	38.5	0.306
m-Terphenyl	651.7	34.6	0.300
p-Terphenyl	652.8	32.8	0.302
2-Thiabutane (methyl ethyl sulfide)	260	42	
4-Thia-1,5-heptadiene (diallyl sulfide)	380		
3-Thiapentane (diethyl sulfide)	284	39.1	0.284
2-Thiapropane (dimethyl sulfide)	229.9	54.6	0.309
Thiophene	307	56.2	0.385
Thymol	425		
Tin(IV) chloride	318.7	37.0	0.742
Titanium tetrachloride	355		
Toluene	318.57	40.55	0.292
Toluonitrile	450		
Trichloroethene	271.0	49.5	
Trichlorofluoromethane	198.0	43.5	0.554
1,2,2-Trichloro-1,1,2-trifluoroethane (freon 113)	214.1	33.7	0.576
1H-Tridecafluorohexane	198.6		
Triethylamine	262	30	0.26
Trifluoroacetic acid	218.1	32.15	0.559
1,1,1-Trifluoroethane	73.1	37.09	0.434
Trifluoromethane	25.74	47.73	0.525
Trimethylamine	160.1	40.2	0.233
1,2,3-Trimethylbenzene	391.3	34.09	
1,2,4-Trimethylbenzene	375.90	31.90	
1,3,5-Trimethylbenzene	364.13	30.86	
2,2,3-Trimethylbutane	257.96	29.15	0.252
2,2,3-Trimethylpentane	290.28	26.94	0.262
2,2,4-Trimethylpentane	270.74	25.34	0.244
2,3,3-Trimethylpentane	300.34	27.83	0.251
2,3,4-Trimethylpentane	293.19	26.94	0.248
2,4,6-Trimethyl-*s*-trioxane (paraldehyde)	290		
1H-Undecafluoropentane	170.8		
Uranium hexafluoride	230.2	45.5	
Vinyl fluoride (fluoroethene)	54.7	51.7	0.320

Table 9-11 (*Continued*)
CRITICAL PROPERTIES

Substance	t_c, °C	P_c, atm	ρ_c, g/cm³
Water	374.2	218.3	0.325
Water, heavy, *see* Dideuterium oxide			
Xenon	16.59	57.62	1.105
o-Xylene	357.1	36.84	0.288
m-Xylene	343.82	34.95	0.282
p-Xylene	343.0	34.65	0.280
2,3-Xylenol	449.7	48	0.26
2,4-Xylenol	434.4	43	0.24
2,5-Xylenol	449.9	48	0.26
2,6-Xylenol	427.8	42	0.24
3,4-Xylenol	456.7	49	0.27
3,5-Xylenol	442.4	36	0.20

Table 9.12
LOGARITHM OF EQUILIBRIUM CONSTANT, $\log_{10} K$

Source: Table Oy, American Petroleum Institute Research Project 44, at Natl. Bu. Stand., Washington, D. C.

1. K_f for formation from the elements; e.g., in the formation of water: $H_2 + \frac{1}{2}O_2 \rightarrow H_2O$; $K_f = (H_2O)/((H_2)/(\sqrt{O_2}))$.
2. K_p for the combustion gas reactions where CO or CO_2 are formed:

(a) $C + H_2O \rightarrow CO + H_2$
(b) $CO_2 + C \rightarrow 2CO$
(c) $CO + H_2O \rightarrow CO_2 + H_2$

(d) $C + 2H_2O \rightarrow CO_2 + 2H_2$
(e) $CO + \frac{1}{2}O_2 \rightarrow CO_2$

For example, in equation a, $K_f = (CO)(H_2)/(C)(H_2O)$; thus, at 900°K, $K_p = \log K_f(CO) + \log K_f(H_2) - \log K_f(C) - \log K_f(H_2O)$
$= 11.1254 + 0 - 0 - 11.4978 = -0.3724$

$\log_{10} K_f$ represents the logarithm (to the base 10) of the equilibrium constant for the reaction of forming the given compound in the gaseous state from the elements with all the reactants and products in the gaseous state and at the temperature indicated.
$\log_{10} K_f = -\Delta F_f^\circ/2.302585\, RT = -\Delta F_f^\circ\, 0.004575651\, T$; ΔF_f° in kcal/mole, T in °K.

The value of $\log_{10} K_f$ at all temperatures for the elements hydrogen, nitrogen, oxygen (all in gaseous state) and for carbon (as solid graphite) is zero. For all substances listed in the table the value of $\log_{10} K_f$ at 0°K is infinite. Interpolation to other temperatures in the interval 298.16° to 1500°K may be made by appropriate graphical or analytical methods. For temperatures below 298.16°K, values may be estimated by extrapolating to lower temperatures the values for 298.16°, 500°, and 700°K.

The values in this table are given to more significant figures than are warranted by the absolute accuracy of the individual values in order to retain the internal consistency of the several thermodynamic functions of a single substance, and also to retain the significance of the increments with temperature of a given thermodynamic function.

Compound	log K						
				Temperature in °K			
	298.16°	500°	700°	900°	1100°	1300°	1500°
Water, H₂O	40.0470	22.8857	15.5834	11.4978	8.8830	7.0637	5.7256
Carbon monoxide, CO	24.0477	16.2526	12.9646	11.1254	9.9443	9.1174	8.5043
" " (a)	-15.9993	-6.6331	-2.6188	-0.3724	1.0613	2.0537	2.7787
" " (b)	-20.9961	-8.7528	-3.5737	-0.7155	1.0860	2.3185	3.2100
Carbon dioxide, CO₂	69.0915	41.2580	29.5029	22.9663	18.8026	15.9163	13.7986
" " (c)	4.4968	2.1197	0.9549	0.3431	-0.0247	-0.2648	-0.4313
" " (d)	-11.0025	-4.5134	-1.6639	-0.0293	1.0366	1.7889	2.3474
" " (e)	45.0438	25.0054	16.5383	11.8409	8.8583	6.7989	5.2943
Methane, CH₄	8.8985	3.4273	0.9529	-0.4881	-1.4345	-2.1006	-2.5923
Ethane, C₂H₆	5.7613	-0.5105	-3.4019	-5.1004	-6.2147	-6.9972	-7.5744
Propane, C₃H₈	4.1150	-3.5973	-7.1584	-9.2395	-10.5966	-11.5457	-12.2499

Table 9-12 (*Continued*)
LOGARITHM OF EQUILIBRIUM CONSTANT, $\log_{10} K$

Compound	298.16°	500°	700°	900°	1100°	1300°	1500°
Butanes, C₄H₁₀:							
n-Butane	2.7516	−6.4987	−10.7697	−13.2552	−14.8746	−16.0130	−16.8445
iso-Butane	3.1489	−6.5949	−11.0660	−13.6680	−15.3610	−16.5404	−17.4026
Pentanes, C₅H₁₂:							
n-Pentane	1.4366	−9.4282	−14.4304	−17.3406	−19.2342	−20.5603	−21.5270
iso-Pentane	2.5655	−8.8687	−14.1088	−17.1487	−19.1190	−20.4914	−21.4949
2,2-DiMe-propane	2.6681	−9.5680	−15.1298	−18.3313	−20.4005	−21.8346	−22.8747
Hexanes, C₆H₁₄:							
n-Hexane	−0.037	−12.440	−18.149	−21.466	−23.623	−25.130	−26.232
2-Me-pentane	0.704	−12.147	−18.071	−21.493	....	....	....
3-Me-pentane	0.213	−12.457	−18.302	−21.668	....	....	....
2,2-DiMe-butane	1.723	−11.933	−18.199	−21.784	....	....	....
2,3-DiMe-butane	0.535	−12.532	−18.539	−21.993	....	....	....
Heptanes, C₇H₁₆:							
n-Heptane	−1.532	−15.469	−21.880	−25.602	−28.018	−29.704	−30.944
2-Me-hexane	−0.718	−15.124	−21.752	−25.577	....	....	....
3-Me-hexane	−0.806	−15.036	−21.608	−25.398	....	....	....
3-Et-pentane	−1.898	−15.963	−22.498	−26.272	....	....	....
2,2-DiMe-pentane	−0.066	−15.277	−22.270	−26.269	....	....	....
2,3-DiMe-pentane	−0.117	−14.818	−21.618	−25.541	....	....	....
2,4-DiMe-pentane	−0.528	−15.421	−22.283	−26.223	....	....	....
3,3-DiMe-pentane	−0.462	−15.307	−22.142	−26.066	....	....	....
2,2,3-TriMe-butane	−0.557	−15.626	−22.554	−26.524	....	....	....
Octanes, C₈H₁₈:							
n-Octane	−3.035	−18.498	−25.611	−29.735	−32.413	−34.278	−35.654
2-Me-heptane	−2.243	−18.174	−25.486	−29.691	....	....	....
3-Me-heptane	−2.412	−18.139	−25.386	−29.548	....	....	....
4-Me-heptane	−2.932	−18.629	−25.876	−30.033	....	....	....
3-Et-hexane	−2.895	−18.529	−25.801	−29.970	....	....	....
2,2-DiMe-hexane	−1.876	−18.445	−26.063	−30.405	....	....	....
2,3-DiMe-hexane	−3.101	−18.865	−26.113	−30.264	....	....	....
2,4-DiMe-hexane	−2.052	−18.236	−25.701	−29.973	....	....	....

log K — Temperature in °K

Table 9-12 (*Continued*)
LOGARITHM OF EQUILIBRIUM CONSTANT, $\log_{10} K$

Compound	298.16°	500°	700°	log K Temperature in °K 900°	1100°	1300°	1500°
Octanes (cont.):							
2,5-DiMe-hexane	−1.832	−18.249	−25.817	−30.160	. . . .	. . . .	. . . .
3,3-DiMe-hexane	−2.324	−18.577	−26.060	−30.337	. . . .	. . . .	. . . .
3,4-DiMe-hexane	−3.643	−19.350	−26.575	−30.716	. . . .	. . . .	. . . .
2-Me-3-Et-pentane	−3.724	−19.337	−26.526	−30.648	. . . .	. . . .	. . . .
3-Me-3-Et-pentane	−3.489	−19.363	−26.688	−30.852	. . . .	. . . .	. . . .
2,2,3-TriMe-pentane	−2.998	−19.223	−26.707	−30.978	. . . .	. . . .	. . . .
2,2,4-TriMe-pentane	−2.294	−18.804	−26.407	−30.745	. . . .	. . . .	. . . .
2,3,3-TriMe-pentane	−3.313	−19.267	−26.632	−30.835	. . . .	. . . .	. . . .
2,3,4-TriMe-pentane	−3.167	−19.184	−26.572	−30.801	. . . .	. . . .	. . . .
2,2,3,3-TetraMe-butane	−3.577	−20.189	−27.777	−32.100	. . . .	. . . .	. . . .
Ethene, C_2H_4	−11.9345	−8.4119	−7.0797	−6.3996	−5.9918	−5.7206	−5.5249
Propene, C_3H_6	−10.9685	−9.7870	−9.5156	−9.4490	−9.4391	−9.4426	−9.4455
Butenes, C_4H_8:							
1-Butene	−12.6199	−12.8637	−13.2403	−13.5442	−13.7707	−13.9358	−14.0537
cis-2-Butene	−11.7330	−12.5224	−13.1803	−13.6636	−14.0119	−14.2670	−14.4516
trans-2-Butene	−11.2316	−12.2872	−13.0314	−13.5553	−13.9273	−14.1950	−14.3898
2-Me-propene	−10.6826	−11.9996	−12.8344	−13.3989	−13.7940	−14.0782	−14.2836
Pentenes, C_5H_{10}:							
1-Pentene	−13.7704	−15.6377	−16.7617	−17.4898	−17.9911	−18.3556	−18.6070
cis-2-Pentene	−12.7574	−15.1876	−16.5825	−17.4833	−18.0971	−18.5415	−18.8518
trans-2-Pentene	−12.1495	−14.8045	−16.2717	−17.2066	−17.8393	−18.2874	−18.6119
2-Me-1-butene	−11.3680	−14.3451	−15.9408	−16.9409	−17.6130	−18.0943	−18.4355
3-Me-1-butene	−13.1017	−15.5156	−16.8408	−17.6825	−18.2545	−18.6621	−18.9408
2-Me-2-butene	−10.4572	−13.9086	−15.7414	−16.8936	−17.6701	−18.2218	−18.6222
Hexenes, C_6H_{12}:							
1-Hexene	−15.2491	−18.6647	−20.4918	−21.6299	−22.3959	−22.9410	−23.3230
cis-2-Hexene	−14.2249	−18.1511	−20.2382	−21.5337	. . . .	. . . .	. . . .
trans-2-Hexene	−13.5291	−17.7155	−19.8900	−21.2278	. . . .	. . . .	. . . .
cis-3-Hexene	−14.7494	−18.6730	−20.7418	−22.0259	. . . .	. . . .	. . . .
trans-3-Hexene	−13.8262	−18.0179	−20.1803	−21.5110	. . . .	. . . .	. . . .
2-Me-1-pentene	−12.8135	−17.2955	−19.5871	−20.9840	. . . .	. . . .	. . . .

Table 9-12 (*Continued*)
LOGARITHM OF EQUILIBRIUM CONSTANT, $\log_{10} K$

Compound	log K — Temperature in °K						
	298.16°	500°	700°	900°	1100°	1300°	1500°
Hexenes (cont.):							
3-Me-1-pentene	−14.8655	−18.5657	−20.4888	−21.6720	......	......	......
4-Me-1-pentene	−14.5865	−18.5063	−20.5432	−21.7932	......	......	......
2-Me-2-pentene	−12.1480	−17.0859	−19.5911	−21.1236	......	......	......
cis-3-Me-2-pentene	−12.6171	−17.3656	−19.7909	−21.2790	......	......	......
trans-3-Me-2-pentene	−12.6171	−17.3656	−19.7909	−21.2790	......	......	......
cis-4-Me-2-pentene	−13.8377	−18.2302	−20.4724	−21.8462	......	......	......
trans-4-Me-2-pentene	−13.0261	−17.6895	−20.0378	−21.4740	......	......	......
2-Et-1-butene	−13.5690	−17.8819	−20.1003	−21.4611	......	......	......
2,3-DiMe-1-butene	−12.7782	−17.5844	−19.9712	−21.4175	......	......	......
3,3-DiMe-1-butene	−13.9578	−18.5922	−20.9860	−22.3805	......	......	......
2,3-DiMe-2-butene	−12.1073	−17.3503	−20.0122	−21.6491	......	......	......
Acetylene, C_2H_2	−36.6490	−20.6290	−13.8925	−10.1702	−7.8158	−6.1956	−5.0134
Propyne, C_3H_4	−33.9469	−21.0781	−15.7228	−12.8007	−10.9660	−9.7072	−8.7879
1-Butyne, C_4H_6	−35.5616	−24.0341	−19.3627	−16.8427	−15.2680	−14.1910	−13.4030
2-Butyne, C_4H_6	−32.7823	−22.5629	−18.4752	−16.2932	−14.9409	−14.0186	−13.3462
1-Pentyne, C_5H_8	−36.7712	−26.8440	−22.9082	−20.8084	−19.5046	−18.6227	−17.9662
2-Pentyne, C_5H_8	−34.0177	−25.2377	−21.8297	−20.0457	−18.9519	−18.2095	−17.6668
3-Me-1-butyne, C_5H_8	−36.0061	−26.6495	−22.9486	−20.9810	−19.7622	−18.9315	−18.3115

Section 10

PHYSICAL PROPERTIES

SOLUBILITIES

Table 10-1
SOLUBILITY OF GASES IN WATER

The column headed "α" gives the volume of gas reduced to standard conditions (0°C and 760 mm) dissolved in one volume of water when the pressure of the gas (without the aqueous tension) is 760 mm; this value "α" is the "absorption coefficient." The column headed "l" gives the volume of the gas expressed in cubic centimeters dissolved in one volume of water at a total pressure (partial pressure of the gas plus the aqueous tension at the stated temperature) of 760 mm. The column headed "q" gives the weight of gas in grams dissolved in 100 grams of water at a total pressure (partial pressure of the gas plus the aqueous tension at the stated temperature) of 760 mm.

Temp. Deg. C.	ACETYLENE		AIR*		AMMONIA		BROMINE	
	α	q	cc/1000cc	%Oxygen in dissd. air.	α	q	α	q
0	1.73	0.200	29.18	34.91	1176	89.5	60.5	42.9
1	1.68	0.194	28.42	34.87	....	....	....	....
2	1.63	0.188	27.69	34.82	....	....	54.1	38.3
3	1.58	0.182	26.99	34.78	....	....	....	....
4	1.53	0.176	26.32	34.74	1047	79.6	48.3	34.2
5	1.49	0.171	25.68	34.69	....	....	....	....
6	1.45	0.167	25.06	34.65	....	....	43.3	30.6
7	1.41	0.162	24.47	34.60	....	....	....	....
8	1.37	0.157	23.90	34.56	947	72.0	38.9	27.5
9	1.34	0.154	23.36	34.52	....	....	....	....
10	1.31	0.150	22.84	34.47	....	....	35.1	24.8
11	1.27	0.146	22.34	34.43	....	....	....	....
12	1.24	0.142	21.87	34.38	857	65.1	31.5	22.2
13	1.21	0.138	21.41	34.34	837	63.6	....	....
14	1.18	0.135	20.97	34.30	....	....	28.4	20.0
15	1.15	0.131	20.55	34.25	....	....	....	....
16	1.13	0.129	20.14	34.21	775	58.7	25.7	18.0
17	1.10	0.125	19.75	34.17	....	....	....	....
18	1.08	0.123	19.38	34.12	....	....	23.4	16.4
19	1.05	0.119	19.02	34.08	....	....	....	....
20	1.03	0.117	18.68	34.03	702	53.1	21.3	14.9
21	1.01	0.115	18.34	33.99	....	....	....	....
22	0.99	0.112	18.01	33.95	....	....	19.4	13.5
23	0.97	0.110	17.69	33.90	....	....	....	....
24	0.95	0.107	17.38	33.86	639	48.2	17.7	12.3
25	0.93	0.105	17.08	33.82	....	....	....	....
26	0.91	0.102	16.79	33.77	....	....	16.3	11.3
27	0.89	0.100	16.50	33.73	....	....	....	....
28	0.87	0.098	16.21	33.68	586	44.0	15.0	10.3
29	0.85	0.095	15.92	33.64	....	....	....	....
30	0.84	0.094	15.64	33.60	....	....	13.8	9.5
35	....	.....	.....	.....	....	....	....	....
40	....	.....	.....	.....	....	....	9.4	6.3
45	....	.....	.....	.....	....	....	....	....
50	....	.....	.....	.....	....	....	6.5	4.1
60	....	.....	.....	.....	....	....	4.9	2.9
70	....	.....	.....	.....	....	....	3.8	1.9
80	....	.....	.....	.....	....	....	3.0	1.2
90	....	.....	.....	.....	....	....	....	....
100	....	.....	.....	.....	....	....	....	....

* Free from NH_3 and CO_2; total pressure of air + aq. tension is 760 mm.

Table 10-1 (*Continued*)
SOLUBILITY OF GASES IN WATER

Temp. °C.	CARBON DIOXIDE α	CARBON DIOXIDE q	CARBON MONOXIDE α	CARBON MONOXIDE q	CHLORINE l	CHLORINE q	ETHANE α	ETHANE q	ETHYLENE α	ETHYLENE q	HYDROGEN α	HYDROGEN q
0	1.713	0.3346	0.03537	0.004397	....	....	0.09874	0.01317	0.226	0.0281	0.02148	0.0001922
1	1.646	0.3213	0.03455	0.004293	....	....	0.09476	0.01263	0.219	0.0272	0.02126	0.0001901
2	1.584	0.3091	0.03375	0.004191	....	....	0.09093	0.01212	0.211	0.0262	0.02105	0.0001881
3	1.527	0.2978	0.03297	0.004092	....	....	0.08725	0.01162	0.204	0.0253	0.02084	0.0001862
4	1.473	0.2871	0.03222	0.003996	....	....	0.08372	0.01114	0.197	0.0244	0.02064	0.0001843
5	1.424	0.2774	0.03149	0.003903	....	....	0.08033	0.01069	0.191	0.0237	0.02044	0.0001824
6	1.377	0.2681	0.03078	0.003813	....	....	0.07709	0.01025	0.184	0.0228	0.02025	0.0001806
7	1.331	0.2589	0.03009	0.003725	....	....	0.07400	0.00983	0.178	0.0220	0.02007	0.0001789
8	1.282	0.2492	0.02942	0.003640	....	....	0.07106	0.00943	0.173	0.0214	0.01989	0.0001772
9	1.237	0.2403	0.02878	0.003559	....	....	0.06826	0.00906	0.167	0.0207	0.01972	0.0001756
10	1.194	0.2318	0.02816	0.003479	3.148	0.9972	0.06561	0.00870	0.162	0.0200	0.01955	0.0001740
11	1.154	0.2239	0.02757	0.003405	3.047	0.9654	0.06328	0.00838	0.157	0.0194	0.01940	0.0001725
12	1.117	0.2165	0.02701	0.003332	2.950	0.9346	0.06106	0.00808	0.152	0.0188	0.01925	0.0001710
13	1.083	0.2098	0.02646	0.003261	2.856	0.9050	0.05894	0.00780	0.148	0.0183	0.01911	0.0001696
14	1.050	0.2032	0.02593	0.003194	2.767	0.8768	0.05694	0.00753	0.143	0.0176	0.01897	0.0001682
15	1.019	0.1970	0.02543	0.003130	2.680	0.8495	0.05504	0.00727	0.139	0.0171	0.01883	0.0001668
16	0.985	0.1903	0.02494	0.003066	2.597	0.8232	0.05326	0.00703	0.136	0.0167	0.01869	0.0001654
17	0.956	0.1845	0.02448	0.003007	2.517	0.7979	0.05159	0.00680	0.132	0.0162	0.01856	0.0001641
18	0.928	0.1789	0.02402	0.002947	2.440	0.7738	0.05003	0.00659	0.129	0.0158	0.01844	0.0001628
19	0.902	0.1737	0.02360	0.002891	2.368	0.7510	0.04858	0.00639	0.125	0.0153	0.01831	0.0001616
20	0.878	0.1688	0.02319	0.002838	2.299	0.7293	0.04724	0.00620	0.122	0.0149	0.01819	0.0001603
21	0.854	0.1640	0.02281	0.002789	2.238	0.7100	0.04589	0.00602	0.119	0.0146	0.01805	0.0001588
22	0.829	0.1590	0.02244	0.002739	2.180	0.6918	0.04459	0.00584	0.116	0.0142	0.01792	0.0001575
23	0.804	0.1540	0.02208	0.002691	2.123	0.6739	0.04335	0.00567	0.114	0.0139	0.01779	0.0001561
24	0.781	0.1493	0.02174	0.002646	2.070	0.6572	0.04217	0.00551	0.111	0.0135	0.01766	0.0001548
25	0.759	0.1449	0.02142	0.002603	2.019	0.6413	0.04104	0.00535	0.108	0.0131	0.01754	0.0001535
26	0.738	0.1406	0.02110	0.002560	1.970	0.6259	0.03997	0.00520	0.106	0.0129	0.01742	0.0001522
27	0.718	0.1366	0.02080	0.002519	1.923	0.6112	0.03895	0.00506	0.104	0.0126	0.01731	0.0001509
28	0.699	0.1327	0.02051	0.002479	1.880	0.5975	0.03799	0.00493	0.102	0.0123	0.01720	0.0001496
29	0.682	0.1292	0.02024	0.002442	1.839	0.5847	0.03709	0.00480	0.100	0.0121	0.01709	0.0001484
30	0.665	0.1257	0.01998	0.002405	1.799	0.5723	0.03624	0.00468	0.098	0.0118	0.01699	0.0001474
35	0.592	0.1105	0.01877	0.002231	1.602	0.5104	0.03230	0.00412	....	....	0.01666	0.0001425
40	0.530	0.0973	0.01775	0.002075	1.438	0.4590	0.02915	0.00366	....	....	0.01644	0.0001384
45	0.479	0.0860	0.01690	0.001933	1.322	0.4228	0.02660	0.00327	....	....	0.01624	0.0001341
50	0.436	0.0761	0.01615	0.001797	1.225	0.3925	0.02459	0.00294	....	....	0.01608	0.0001287
60	0.359	0.0576	0.01488	0.001522	1.023	0.3295	0.02177	0.00239	....	....	0.01600	0.0001178
70	....	....	0.01440	0.001276	0.862	0.2793	0.01948	0.00185	....	....	0.0160	0.000102
80	....	....	0.01430	0.000980	0.683	0.2227	0.01826	0.00134	....	....	0.0160	0.000079
90	....	....	0.0142	0.00057	0.39	0.127	0.0176	0.0008	....	....	0.0160	0.000046
100	....	....	0.0141	0.00000	0.00	0.000	0.0172	0.0000	....	....	0.0160	0.000000

Table 10-1 (Continued)
SOLUBILITY OF GASES IN WATER

Temp. °C.	HYDROGEN SULFIDE		METHANE α	METHANE q	NITRIC OXIDE α	NITRIC OXIDE q	NITROGEN* α	NITROGEN* q	OXYGEN α	OXYGEN q	SULFUR DIOXIDE l	SULFUR DIOXIDE q
0	4.670	0.7066	0.05563	0.003959	0.07381	0.009833	0.02354	0.002942	0.04889	0.006945	79.789	22.83
1	4.522	0.6839	0.05401	0.003842	0.07184	0.009564	0.02297	0.002869	0.04758	0.006756	77.210	22.09
2	4.379	0.6619	0.05244	0.003728	0.06993	0.009305	0.02241	0.002798	0.04633	0.006574	74.691	21.37
3	4.241	0.6407	0.05093	0.003619	0.06809	0.009057	0.02187	0.002730	0.04512	0.006400	72.230	20.66
4	4.107	0.6201	0.04946	0.003515	0.06632	0.008816	0.02135	0.002663	0.04397	0.006232	69.828	19.98
5	3.977	0.6001	0.04805	0.003410	0.06461	0.008584	0.02086	0.002600	0.04287	0.006072	67.485	19.31
6	3.852	0.5809	0.04669	0.003312	0.06298	0.008361	0.02037	0.002537	0.04180	0.005918	65.200	18.65
7	3.732	0.5624	0.04539	0.003217	0.06140	0.008147	0.01990	0.002477	0.04080	0.005773	62.973	18.02
8	3.616	0.5446	0.04413	0.003127	0.05990	0.007943	0.01945	0.002419	0.03983	0.005632	60.805	17.40
9	3.505	0.5276	0.04292	0.003039	0.05846	0.007747	0.01902	0.002365	0.03891	0.005498	58.697	16.80
10	3.399	0.5112	0.04177	0.002955	0.05709	0.007560	0.01861	0.002312	0.03802	0.005368	56.647	16.21
11	3.300	0.4960	0.04072	0.002879	0.05587	0.007393	0.01823	0.002263	0.03718	0.005246	54.655	15.64
12	3.206	0.4814	0.03970	0.002805	0.05470	0.007233	0.01786	0.002216	0.03637	0.005128	52.723	15.09
13	3.115	0.4674	0.03872	0.002733	0.05357	0.007078	0.01750	0.002170	0.03559	0.005014	50.849	14.56
14	3.028	0.4540	0.03779	0.002665	0.05250	0.006930	0.01717	0.002126	0.03486	0.004906	49.033	14.04
15	2.945	0.4411	0.03690	0.002599	0.05147	0.006788	0.01685	0.002085	0.03415	0.004802	47.276	13.54
16	2.865	0.4287	0.03606	0.002538	0.05049	0.006652	0.01654	0.002045	0.03348	0.004703	45.578	13.05
17	2.789	0.4169	0.03525	0.002478	0.04956	0.006524	0.01625	0.002006	0.03283	0.004606	43.939	12.59
18	2.717	0.4056	0.03448	0.002422	0.04868	0.006400	0.01597	0.001970	0.03220	0.004514	42.360	12.14
19	2.647	0.3948	0.03376	0.002369	0.04785	0.006283	0.01570	0.001935	0.03161	0.004426	40.838	11.70
20	2.582	0.3846	0.03308	0.002319	0.04706	0.006173	0.01545	0.001901	0.03102	0.004339	39.374	11.28
21	2.517	0.3745	0.03243	0.002270	0.04625	0.006059	0.01522	0.001869	0.03044	0.004252	37.970	10.88
22	2.456	0.3648	0.03180	0.002222	0.04545	0.005947	0.01498	0.001838	0.02988	0.004169	36.617	10.50
23	2.396	0.3554	0.03119	0.002177	0.04469	0.005838	0.01475	0.001809	0.02934	0.004087	35.302	10.12
24	2.338	0.3463	0.03061	0.002133	0.04395	0.005733	0.01454	0.001780	0.02881	0.004007	34.026	9.76
25	2.282	0.3375	0.03006	0.002091	0.04323	0.005630	0.01434	0.001751	0.02831	0.003931	32.786	9.41
26	2.229	0.3290	0.02952	0.002050	0.04254	0.005530	0.01413	0.001724	0.02783	0.003857	31.584	9.06
27	2.177	0.3208	0.02901	0.002011	0.04188	0.005435	0.01394	0.001698	0.02736	0.003787	30.422	8.73
28	2.128	0.3130	0.02852	0.001974	0.04124	0.005342	0.01376	0.001672	0.02691	0.003718	29.314	8.42
29	2.081	0.3055	0.02806	0.001938	0.04063	0.005252	0.01358	0.001647	0.02649	0.003651	28.210	8.10
30	2.037	0.2983	0.02762	0.001904	0.04004	0.005165	0.01342	0.001624	0.02608	0.003588	27.161	7.80
35	1.831	0.2648	0.02546	0.001733	0.03734	0.004757	0.01256	0.001501	0.02440	0.003315	22.489	6.47
40	1.660	0.2361	0.02369	0.001586	0.03507	0.004394	0.01184	0.001391	0.02306	0.003082	18.766	5.41
45	1.516	0.2110	0.02238	0.001466	0.03311	0.004059	0.01130	0.001300	0.02187	0.002858	.....	.....
50	1.392	0.1883	0.02134	0.001359	0.03152	0.003758	0.01088	0.001216	0.02090	0.002657	.....	.....
60	1.190	0.1480	0.01954	0.001144	0.02954	0.003237	0.01023	0.001052	0.01946	0.002274	.....	.....
70	1.022	0.1101	0.01825	0.000926	0.02810	0.002668	0.00977	0.000851	0.01833	0.001856	.....	.....
80	0.917	0.0765	0.01770	0.000695	0.02700	0.001984	0.00958	0.000660	0.01761	0.001381	.....	.....
90	0.84	0.041	0.01735	0.00040	0.0265	0.00113	0.0095	0.00038	0.0172	0.00079	.....	.....
100	0.81	0.000	0.0170	0.00000	0.0263	0.00000	0.0095	0.00000	0.0170	0.00000	.....	.....

*Atmospheric Nitrogen containing 98.815% N_2 by volume + 1.185% inert gases.

Table 10-2
SOLUBILITIES OF INORGANIC COMPOUNDS IN WATER
AT VARIOUS TEMPERATURES

This table shows the amount of anhydrous substance which is soluble in 100 grams of water at the temperature in degrees Centigrade as indicated; where the formula is preceded by † the value is expressed in grams of substance in 100 ml of saturated solution. Solid Phase gives the hydrated form in equilibrium with the saturated solution.

	Substance	Formula	Solid Phase	0°C.	10°C.
1	**Aluminum** chloride	AlCl$_3$	6H$_2$O	....	....
2	sulfate	Al$_2$(SO$_4$)$_3$	18H$_2$O	31.2	33.5
3	**Ammonium** aluminum sulfate	(NH$_4$)$_2$Al$_2$(SO$_4$)$_4$	24H$_2$O	2.1	4.99
4	bicarbonate	NH$_4$HCO$_3$	....	11.9	15.8
5	bromide	NH$_4$Br	....	60.6	68
6	cadmium sulfate	(NH$_4$)$_2$Cd(SO$_4$)$_2$	6H$_2$O	....	....
7	chloride	NH$_4$Cl	....	29.4	33.3
8	chloroplatinate	(NH$_4$)$_2$PtCl$_6$	....	....	0.7
9	chromate	(NH$_4$)$_2$CrO$_4$	....	....	......
10	chromium sulfate	(NH$_4$)$_2$Cr$_2$(SO$_4$)$_4$	24H$_2$O	....	......
11	cobaltous sulfate	(NH$_4$)$_2$Co(SO$_4$)$_2$	6H$_2$O	6.0	9.5
12	dichromate	(NH$_4$)$_2$Cr$_2$O$_7$	....	....	......
13	dihydrogen phosphite	NH$_4$H$_2$PO$_3$	....	171	....
14	ferric sulfate	(NH$_4$)$_2$Fe$_2$(SO$_4$)$_4$	24H$_2$O	....	......
15	ferrous sulfate	(NH$_4$)$_2$Fe(SO$_4$)$_2$	6H$_2$O	12.5	17.2
16	hydrogen phosphate	(NH$_4$)$_2$HPO$_4$	....	....	....
17	iodide	NH$_4$I	....	154.2	163.2
18	lithium sulfate	NH$_4$LiSO$_4$	....	....	55.23
19	magnesium phosphate	NH$_4$MgPO$_4$	6H$_2$O	0.023	....
20	manganese phosphate	NH$_4$MnPO$_4$	7H$_2$O	....	....
21	nitrate	NH$_4$NO$_3$	....	118.3	....
22	oxalate	(NH$_4$)$_2$C$_2$O$_4$	1H$_2$O	2.2	3.1
23	†perchlorate	NH$_4$ClO$_4$†	....	11.56	....
24	persulfate	(NH$_4$)$_2$S$_2$O$_8$	....	58.2	....
25	selenate	(NH$_4$)$_2$SeO$_4$	....	....	1.22¹²°
26	sulfate	(NH$_4$)$_2$SO$_4$	....	70.6	73.0
27	thioantimonate	(NH$_4$)$_3$SbS$_4$	4H$_2$O	71.2	....
28	thiocyanate	NH$_4$CNS	....	119.8	144
29	vanadate (meta)	NH$_4$VO$_3$	....	....	....
30	**Antimonous** chloride	SbCl$_3$	....	601.6	....
31	fluoride	SbF$_3$	....	384.7	... ...
32	sulfide	Sb$_2$S$_3$	....	....	....
33	**Arsenic oxide**	As$_2$O$_5$	....	59.5	62.1
34	**Arsenious sulfide**	As$_2$S$_3$	....	5.17× 10^{-5} at 18°	
35	**Barium** acetate	Ba(C$_2$H$_3$O$_2$)$_2$	3H$_2$O	59	63
36	acetate	Ba(C$_2$H$_3$O$_2$)$_2$	1H$_2$O	....	....
37	bromate	Ba(BrO$_3$)$_2$	1H$_2$O	0.287	0.441
38	bromide	BaBr$_2$	2H$_2$O	98	101
39	carbonate	BaCO$_3$	....	....	0.0016⁸°
40	chlorate	Ba(ClO$_3$)$_2$	1H$_2$O	20.34	26.95
41	chloride	BaCl$_2$	2H$_2$O	31.6	33.3
42	chromate	BaCrO$_4$	....	0.0002	0.00028
43	hydroxide	Ba(OH)$_2$	8H$_2$O	1.67	2.48
44	iodate	Ba(IO$_3$)$_2$	1H$_2$O	0.008	0.014
45	iodide	BaI$_2$	6H$_2$O	170.2	185.7
46	iodide	BaI$_2$	2H$_2$O	....	....
47	molybdate	BaMoO$_4$	....	....	....
48	nitrate	Ba(NO$_3$)$_2$	....	5.0	7.0
49	nitrite	Ba(NO$_2$)$_2$	1H$_2$O	....	....

Table 10-2 (*Continued*)
SOLUBILITIES OF INORGANIC COMPOUNDS IN WATER
AT VARIOUS TEMPERATURES

20°C.	30°C.	40°C.	50°C.	60°C.	70°C.	80°C.	90°C.	100°C.	
$69.86^{15°}$	.....	.....	.....	....	....	....	....	.....	1
36.4	40.4	46.1	52.2	59.2	66.1	73.0	80.8	89.0	2
7.74	10.94	14.88	20.10	26.70	....	....	....	$109.7^{95°}$	3
21	27	.....	.....	....	....	....	....	.....	4
75.5	83.2	91.1	99.2	107.8	116.8	126	135.6	145.6	5
$72.3^{25°}$	.....	.....	.....	....	....	....	....	.....	6
37.2	41.4	45.8	50.4	55.2	60.2	65.6	71.3	77.3	7
....	.....	.....	.....	....	....	....	....	1.25	8
....	40.4	.....	.....	....	....	....	....	.....	9
$10.78^{25°}$	.....	.....	.....	....	....	....	....	.....	10
13.0	17.0	22.0	27.0	33.5	40.0	49.0	....	.....	11
....	47.17	.....	.....	....	....	....	....	.....	12
$190^{14.5°}$	$260^{31°}$	.....	.....	....	....	....	....	.....	13
....	$44.15^{25°}$	.....	.....	....	....	....	....	.....	14
....	.....	33	40	....	52	....	....	.....	15
131^{15}	.....	.....	.....	....	....	....	....	.....	16
172.3	181.4	190.5	199.6	208.9	218.7	228.8	....	250.3	17
....	55.93	.....	56.25	....	56.69	....	....	.....	18
0.052	.....	0.036	0.030	0.040	0.016	0.019	....	.....	19
0	.....	0	.....	0	0.005	0.007	....	.....	20
192	241.8	297.0	344.0	421.0	499.0	580.0	740.0	871.0	21
4.4	5.9	8.0	10.3	....	....	....	....	.....	22
20.85	.....	30.58	.....	39.05	....	48.19	....	57.01	23
....	.....	.....	.....	....	....	....	....	.....	24
....	.....	.....	.....	....	....	....	....	.....	25
75.4	78.0	81.0	.....	88.0	....	95.3	....	103.3	26
91.2	119.8	.....	.....	....	....	....	....	.....	27
170	207.7	.....	.....	....	....	....	....	.....	28
0.48	0.84	1.32	1.78	....	3.05	....	....	.....	29
931.5	1068.0	1368.0	1917.0	4531.0	....	∞	....	.....	30
444.7	563.6	.....	.....	....	....	....	....	.....	31
$0.000175^{18°}$	.....	.....	.....	....	....	....	....	.....	32
65.8	69.5	71.2	.....	73.0	....	75.1	....	76.7	33
....	.....	.....	.....	....	....	....	....	.....	34
71	.....	.....	.....	....	....	....	....	.....	35
....	75	79	77	74	74	....	....	75	36
0.656	0.959	1.33	1.75	2.32	3.01	3.65	4.45	5.71	37
104	109	114	118	123	128	135	....	149	38
$0.0022^{18°}$	0.0024 at 24.2°	.....	.....	....	....	....	....	.....	39
33.80	41.70	49.61	.....	66.81	....	84.84	....	104.9	40
35.7	38.2	40.7	43.6	46.4	49.4	52.4	....	58.8	41
0.00037	0.00046	.....	.....	....	....	....	....	.....	42
3.89	5.59	8.22	13.12	20.94	....	101.4	....	.....	43
0.022	0.031	0.041	0.056	0.074	0.093	0.115	0.141	0.197	44
203.1	219.6	.....	.....	....	....	....	....	.....	45
....	.....	231.9	.....	247.3	....	261.0	....	271.7	46
$0.0058^{23°}$	.....	.....	.....	....	....	....	....	.....	47
9.2	11.6	14.2	17.1	20.3	....	27.0	....	34.2	48
67.5	.....	.....	.....	....	....	205.8	....	300	49

Table 10-2 (*Continued*)
SOLUBILITIES OF INORGANIC COMPOUNDS IN WATER
AT VARIOUS TEMPERATURES

	Substance	Formula	Solid Phase	0°C.	10°C.
1	**Barium** oxalate	BaC_2O_4	....	....	0.00168°
2	perchlorate	$Ba(ClO_4)_2$	$3H_2O$	205.8	....
3	sulfate	$BaSO_4$	....	1.15×10^{-4}	2.0×10^{-4}
4	**Beryllium** potassium fluoride	$BeF_2.2KF$	....	....	....
5	sodium fluoride	$BeF_2.2NaF$	....	....	....
6	sulfate	$BeSO_4$	$6H_2O$	....	....
7	sulfate	$BeSO_4$	$4H_2O$	....	....
8	sulfate	$BeSO_4$	$2H_2O$	....	....
9	**Bismuth** sulfide	Bi_2S_3	....	....	....
10	**Boric acid**	H_3BO_3	....	2.66	3.57
11	**Boron oxide**	B_2O_3	....	1.1	1.5
12	**Bromine**	Br_2	....	4.22	3.4
13	**Cadmium** bromide	$CdBr_2$	....	61.08	....
14	chloride	$CdCl_2$	$4H_2O$	97.59	125.1
15	chloride	$CdCl_2$	$2\frac{1}{2}H_2O$	90.01	....
16	chloride	$CdCl_2$	$1H_2O$	....	135.1
17	cyanide	$Cd(CN)_2$	....	....	....
18	hydroxide	$Cd(OH)_2$	....	....	....
19	iodide	CdI_2	....	79.8	83.2
20	sulfate	$CdSO_4$	....	76.48	76.00
21	sulfide	CdS	....	....	....
22	**Calcium** acetate	$Ca(C_2H_3O_2)_2$	$2H_2O$	37.4	36.0
23	acetate	$Ca(C_2H_3O_2)_2$	$1H_2O$	....	....
24	bicarbonate	$Ca(HCO_3)_2$	....	16.15	....
25	bromide	$CaBr_2$	$6H_2O$	125	132
26	bromide	$CaBr_2$	$4H_2O$	....	....
27	chloride	$CaCl_2$	$6H_2O$	59.5	65.0
28	chloride	$CaCl_2$	$2H_2O$	....	....
29	fluoride	CaF_2	....	....	....
30	hydroxide	$Ca(OH)_2$	....	0.185	0.176
31	iodate	$Ca(IO_3)_2$	$6H_2O$	0.10	0.17
32	iodate	$Ca(IO_3)_2$	$1H_2O$	....	....
33	iodide	CaI_2	....	181.9	194.1
34	nitrate	$Ca(NO_3)_2$	$4H_2O$	102.0	115.3
35	nitrate	$Ca(NO_3)_2$	$3H_2O$	....	....
36	nitrate	$Ca(NO_3)_2$	....	....	....
37	nitrite	$Ca(NO_2)_2$	$4H_2O$	62.07	....
38	nitrite	$Ca(NO_2)_2$	$2H_2O$	....	....
39	oxalate	CaC_2O_4	....	....	6.7×10^{-4} at 13°
40	sulfate	$CaSO_4$	$2H_2O$	0.1759	0.1928
41	**Carbon** dioxide, 760 mm	CO_2	....	0.3346	0.2318
42	monoxide, 760 mm	CO	....	0.0044	0.0035
43	**Cerous** sulfate	$Ce_2(SO_4)_3$	$9H_2O$	20.75	....
44	sulfate	$Ce_2(SO_4)_3$	$8H_2O$	18.98	....
45	sulfate	$Ce_2(SO_4)_3$	$5H_2O$	....	....
46	sulfate	$Ce_2(SO_4)_3$	$4H_2O$	....	....
47	**Cesium** chlorate	$CsClO_3$	....	2.46	3.8
48	chloride	$CsCl$	....	161.4	174.7
49	fluoride	CsF	$1\frac{1}{2}H_2O$	....	....

Table 10-2 (*Continued*)
SOLUBILITIES OF INORGANIC COMPOUNDS IN WATER
AT VARIOUS TEMPERATURES

20°C.	30°C.	40°C.	50°C.	60°C.	70°C.	80°C.	90°C.	100°C.	
$0.0022^{18°}$	0.0024 at 24.2°	.....	.....	....	....	....	....	.....	1
289.1	.....	358.7	426.3	....	495.2	....	562.3	.....	2
2.4×10^{-4}	2.85×10^{-4}	.....	3.36×10^{-4}	....	....	....	....	4.13×10^{-4}	3
2	.....	.....	.....	....	....	....	....	5.2	4
1.4	.....	.....	.....	....	....	....	....	2.9	5
....	52	.....	60.67	....	....	....	....	.....	6
....	43.78	46.74	.....	....	62	....	83	100	7
....	.....	.....	.....	....	....	84.76	98	110	8
1.8×10^{-5} at 18°	.....	.....	.....	....	....	....	....	.....	9
5.04	6.60	8.72	11.54	14.81	18.62	23.75	30.38	40.25	10
2.2	.....	4.0	.....	6.2	....	9.5	....	15.7	11
3.20	3.13	.....	.....	....	....	....	....	.....	12
....	132.0	154.1	.....	157.1	....	165.2	....	.....	13
....	.....	.....	.....	....	....	....	....	.....	14
....	132.1	.....	.....	....	....	....	....	.....	15
134.5	.....	135.3	.....	136.5	....	140.4	....	147.0	16
$1.71^{5°}$	.....	.....	.....	....	....	....	....	.....	17
....	2.6×10^{-4} at 25°	.....	.....	....	....	....	....	.....	18
86.2	89.7	93.8	97.4	....	....	....	....	127.6	19
76.60	.....	78.54	.....	83.68	....	....	63.13	60.77	20
9×10^{-7} at 18°	.....	.....	.....	....	....	....	....	.....	21
34.7	33.8	33.2	.....	32.7	....	33.5	....	.....	22
....	.....	.....	.....	....	....	....	31.1	29.7	23
16.60	.....	17.05	.....	17.50	....	17.95	....	18.40	24
143	.....	.....	.....	....	....	....	....	.....	25
....	.....	213	.....	278	....	295	....	.....	26
74.5	102	.....	.....	....	....	....	....	.....	27
....	.....	.....	.....	136.8	141.7	147.0	152.7	159	28
$0.00161^{18°}$	$0.00172^{26°}$	.....	.....	....	....	....	....	.....	29
0.165	0.153	0.141	0.128	0.116	0.106	0.094	0.085	0.077	30
....	0.42	0.61	0.90	1.38	....	....	....	.....	31
....	.....	0.52	0.59	0.65	....	0.80	....	0.95	32
208.8	222.6	242.5	.....	284.6	....	354.5	....	426.3	33
129.3	152.6	195.9	.....	....	....	....	....	.....	34
....	.....	237.5	281.5	....	....	....	....	.....	35
....	.....	.....	.....	....	....	358.7	....	363.6	36
76.68	.....	.....	.....	....	....	....	....	.....	37
....	.....	.....	.....	132.6	151.9	....	244.8	.....	38
6.8×10^{-4} at 25°	9.5×10^{-4} at 50°	14×10^{-4} at 95°	.....	....	....	....	....	.....	39
....	0.2090	0.2097	0.2038	....	0.1966	....	.....	0.1619	40
0.1688	0.1257	0.0973	0.0761	0.0576	....	....	....	0	41
0.0028	0.0024	0.0021	0.0018	0.0015	0.0013	0.0010	0.0006	0	42
10.08	6.79	.....	4.674	3.874	....	....	....	.....	43
9.52	.....	5.947	.....	4.04	....	....	....	.....	44
....	.....	.....	.....	3.247	....	1.204	....	0.46	45
....	.....	6.056	3.42	2.35	....	1.01	....	0.42	46
6.2	9.5	13.8	19.4	26.2	34.7	45.0	58.0	79.0	47
186.5	197.3	208.0	218.5	229.7	239.5	250.0	260.1	270.5	48
$366.5^{18°}$	.....	.....	.....	....	....	....	....	.....	49

Table 10-2 (*Continued*)
SOLUBILITIES OF INORGANIC COMPOUNDS IN WATER
AT VARIOUS TEMPERATURES

	Substance	Formula	Solid Phase	0°C.	10°C.
1	**Cesium** hydroxide	CsOH	....	....	....
2	iodate	$CsIO_3$	....	....	....
3	nitrate	$CsNO_3$	....	9.33	14.9
4	perchlorate	$CsClO_4$	....	0.8	1.0
5	periodate	$CsIO_4$	....	....	2.15[15]
6	sulfate	Cs_2SO_4	....	167.1	173.1
7	**Chlorine, 760 mm**	Cl_2	....	1.46	0.980
8	**Chromic anhydride**	CrO_3	....	164.9	....
9	**Cobaltous** chloride	$CoCl_2$	$6H_2O$	41.6	46.0
10	chloride	$CoCl_2$	$1H_2O$	....	....
11	iodide	CoI_2	$1H_2O$	138.1	159.7
12	nitrate	$Co(NO_3)_2$	$6H_2O$	84.03	....
13	nitrate	$Co(NO_3)_2$	$3H_2O$	....	....
14	sulfate	$CoSO_4$	$7H_2O$	25.55	30.55
15	sulfide	CoS	....	....	....
16	**Cupric** chloride	$CuCl_2$	$2H_2O$	70.7	73.76
17	iodide	CuI_2	....	....	....
18	nitrate	$Cu(NO_3)_2$	$6H_2O$	81.8	95.28
19	nitrate	$Cu(NO_3)_2$	$3H_2O$	....	....
20	sulfate	$CuSO_4$	$5H_2O$	14.3	17.4
21	sulfide	CuS	....	....	....
22	**Cuprous chloride**	CuCl	....	....	....
23	**Ferric chloride**	$FeCl_3$	....	74.4	81.9
24	**Ferrous** bromide	$FeBr_2$	$6H_2O$	102.0	....
25	chloride	$FeCl_2$	$4H_2O$	....	64.5
26	chloride	$FeCl_2$	....	....	....
27	nitrate	$Fe(NO_3)_2$	$6H_2O$	71.02	....
28	sulfate	$FeSO_4$	$7H_2O$	15.65	20.51
29	sulfate	$FeSO_4$	$1H_2O$	....	....
30	sulfide	FeS	....	....	....
31	**Hydrobromic acid, 760 mm**	HBr	....	221.2	210.3
32	**Hydrochloric acid, 760 mm**	HCl	....	82.3	....
33	**Iodine**	I_2	....	....	....
34	**Lanthanum sulfate**	$La_2(SO_4)_3$	$9H_2O$	3	....
35	**Lead** acetate	$Pb(C_2H_3O_2)_2$	$3H_2O$	....	....
36	bromide	$PbBr_2$	....	0.4554	....
37	carbonate	$PbCO_3$	....	....	....
38	chloride	$PbCl_2$	....	0.6728	....
39	chromate	$PbCrO_4$	....	....	....
40	fluoride	PbF_2	....	....	0.060
41	iodide	PbI_2	....	0.0442	....
42	nitrate	$Pb(NO_3)_2$	....	38.8	48.3
43	sulfate	$PbSO_4$	....	0.0028	0.0035
44	sulfide	PbS	....	....	....
45	**Lithium** bromide	LiBr	$2H_2O$	143	166
46	bromide	LiBr	$1H_2O$	....	....
47	carbonate	Li_2CO_3	....	1.54	1.43
48	chloride	LiCl	....	67	72
49	hydroxide	LiOH	$1H_2O$	12.7	12.7
50	iodide	LiI	$3H_2O$	151	157
51	iodide	LiI	$1H_2O$	....	....
52	nitrate	$LiNO_3$	$3H_2O$	53.4	61.0
53	nitrate	$LiNO_3$	$1½H_2O$	....	....

Table 10-2 (*Continued*)
SOLUBILITIES OF INORGANIC COMPOUNDS IN WATER AT VARIOUS TEMPERATURES

20°C.	30°C.	40°C.	50°C.	60°C.	70°C.	80°C.	90°C.	100°C.	
$385.7^{15°}$	.....	.....	.....	....	....	....	...	.....	1
$2.6^{24°}$	.....	.....	.....	.....	....	....	...	.....	2
23.0	33.9	47.2	64.4	83.8	107.0	134.0	163.0	197.0	3
1.6	2.6	4.0	5.4	7.3	9.8	14.4	20.5	30.0	4
......	....	....	....	....	....	....	...	.....	5
178.7	184.1	189.9	194.9	199.9	205.0	210.3	214.9	220.3	6
0.716	0.562	0.451	0.386	0.324	0.274	0.219	0.125	0	7
......	.....	174.0	182.1	....	....	....	217.5	206.8	8
50.4	53.5	.....	.....	....	....	....	.....	.....	9
......	.....	69.5	88.7	90.5	....	98.0	....	104.1	10
187.4	233.3	300.0	376.2	....	....	400.0	....	.....	11
100.0	.....	126.8	.....	....	....	....	....	.....	12
......	.....	.....	.....	163.2	184.8	212.5	334.9	.....	13
36.21	42.26	48.85	55.2	60.4	65.7	70	....	83	14
3.79×10^{-4} at 18°	.....	.....	.....	....	....	....	...	.....	15
77.0	80.34	83.8	87.44	91.2	....	99.2	....	107.9	16
1.107	.....	.....	.....	....	....	....	....	.....	17
125.1	.....	.....	.....	....	....	....	....	.....	18
......	.....	159.8	.....	178.8	....	207.8	....	.....	19
20.7	25	28.5	33.3	40	....	55	....	75.4	20
3.3×10^{-5} at 18°	.....	.....	....	....	....	....	....	.....	21
$1.52^{25°}$	.....	.....	.....	....	....	....	....	.....	22
91.8	.....	.....	315.1	....	....	525.8	....	535.7	23
115.1	122.2	128.3	.....	144	....	159.7	....	177.8	24
......	73.0	77.3	82.5	88.7	....	100	....	.....	25
......	.....	.....	.....	....	....	....	105.3	105.8	26
83.8	.....	.....	.....	165.6	....	....	....	.....	27
26.5	32.9	40.2	48.6	....	....	....	....	.....	28
......	.....	.....	.....	....	50.9	43.6	37.3	.....	29
$0.000616^{18°}$	.....	.....	.....	....	....	....	....	.....	30
198	.....	.....	171.5	....	....	....	....	130	31
......	67.3	63.3	59.6	56.1	....	....	....	.....	32
0.029	0.04	0.056	0.078	....	....	....	....	.....	33
......	1.9	.....	1.5	....	....	....	....	0.69	34
......	$55.04^{25°}$	.....	.....	....	....	....	....	.....	35
0.85	1.15	1.53	1.94	2.36	....	3.34	....	4.75	36
0.00011	.....	.....	.....	....	....	....	....	.....	37
0.99	1.20	1.45	1.70	1.98	....	2.62	....	3.34	38
7×10^{-6}	.....	.....	.....	....	....	....	....	.....	39
0.064	0.068	.....	.....	....	....	....	....	.....	40
0.068	0.090	0.125	0.164	0.197	....	0.302	....	0.436	41
56.5	66	75	85	95	....	115	....	138.8	42
0.0041	0.0049	0.0056	.....	....	....	....	....	.....	43
8.6×10^{-5} at 18°	.....	.....	....	....	....	....	....	.....	44
177	191	205	.....	....	....	....	....	.....	45
......	.....	.....	214	224	....	245	....	266	46
1.33	1.25	1.17	1.08	1.01	....	0.85	....	0.72	47
78.5	84.5	90.5	97	103	....	115	....	127.5	48
12.8	12.9	13	13.3	13.8	....	15.3	....	17.5	49
165	171	179	187	202	230	....	....	.....	50
......	.....	.....	.....	....	....	435	....	481	51
......	132.6	.....	.....	....	....	....	....	.....	52
......	.....	.....	156.4	175.3	....	....	....	.....	53

Table 10-2 (*Continued*)
SOLUBILITIES OF INORGANIC COMPOUNDS IN WATER
AT VARIOUS TEMPERATURES

	Substance	Formula	Solid Phase	0°C.	10°C.
1	**Lithium** nitrate	$LiNO_3$	....	....	....
2	sulfate	Li_2SO_4	$1H_2O$	35.3	35.0
3	**Magnesium** bromide	$MgBr_2$	$6H_2O$	91.0	94.5
4	chloride	$MgCl_2$	$6H_2O$	52.8	53.5
5	hydroxide	$Mg(OH)_2$	....	....	....
6	iodide	MgI_2	$8H_2O$	120.8	....
7	nitrate	$Mg(NO_3)_2$	$6H_2O$	66.55	....
8	sulfate	$MgSO_4$	$7H_2O$	....	30.9
9	sulfate	$MgSO_4$	$6H_2O$	40.8	42.2
10	sulfate	$MgSO_4$	$1H_2O$	....	....
11	**Manganous** bromide	$MnBr_2$	$4H_2O$	127.3	135.9
12	bromide	$MnBr_2$	$2H_2O$	....	....
13	chloride	$MnCl_2$	$4H_2O$	63.4	68.1
14	chloride	$MnCl_2$	$2H_2O$	....	....
15	nitrate	$Mn(NO_3)_2$	$6H_2O$	101.98	118.0
16	nitrate	$Mn(NO_3)_2$	$3H_2O$	....	....
17	sulfate	$MnSO_4$	$7H_2O$	53.23	60.0î
18	sulfate	$MnSO_4$	$5H_2O$	....	59.5
19	sulfate	$MnSO_4$	$4H_2O$	....	....
20	sulfate	$MnSO_4$	$1H_2O$	....	....
21	sulfide	MnS	....	....	....
22	**Mercuric** bromide	$HgBr_2$	....	....	....
23	chloride	$HgCl_2$	....	3.6	4.8
24	cyanide	$Hg(CN)_2$	....	....	9.3î$3.5°$
25	iodide	HgI_2	....	....	....
26	**Mercurous** bromide	$HgBr$	....	....	....
27	chloride	$HgCl$	....	0.00014	....
28	iodide	HgI	....	....	....
29	**Molybdic oxide**	MoO_3	$2H_2O$	....	....
30	**Neodymium sulfate**	$Nd_2(SO_4)_3$	....	9.5	....
31	**Nickel** bromide	$NiBr_2$	$6H_2O$	112.8	122.2
32	carbonate	$NiCO_3$	....	....	....
33	chloride	$NiCl_2$	$6H_2O$	53.9	59.5
34	iodide	NiI_2	$6H_2O$	124.2	135.3
35	nitrate	$Ni(NO_3)_2$	$6H_2O$	79.58	....
36	nitrate	$Ni(NO_3)_2$	$3H_2O$	....	....
37	sulfate	$NiSO_4$	$7H_2O$	27.22	32
38	sulfate	$NiSO_4$	$6H_2O$	....	....
39	sulfide	NiS	....	....	....
40	**Nitric oxide, 760 mm**	NO	....	0.00984	0.00757
41	**Nitrous oxide**	N_2O	....	....	0.1705
42	**Ozone**	O_3	....	0.0039	0.0029
43	**Potassium** acetate	$KC_2H_3O_2$	$1½H_2O$	216.7	233.9
44	acetate	$KC_2H_3O_2$	$½H_2O$	....	....
45	alum	$K_2SO_4 \cdot Al_2(SO_4)_3$	$24H_2O$	3.0	4.0
46	bicarbonate	$KHCO_3$	....	22.4	27.7
47	bisulfate	$KHSO_4$	....	36.3	....
48	bitartrate	$KHC_4H_4O_6$	....	0.32	0.40
49	bromate	$KBrO_3$	....	3.1	4.8
50	bromide	KBr	....	53.5	59.5
51	carbonate	K_2CO_3	$2H_2O$	105.5	108
52	chlorate	$KClO_3$	....	3.3	5
53	chloride	KCl	....	27.6	31.0
54	chloroplatinate	K_2PtCl_6	....	0.74	0.90

Table 10-2 (*Continued*)
SOLUBILITIES OF INORGANIC COMPOUNDS IN WATER AT VARIOUS TEMPERATURES

20°C.	30°C.	40°C.	50°C.	60°C.	70°C.	80°C.	90°C.	100°C.	
....	....	....	....	....	194.1	....	....	....	1
34.2	33.5	32.7	32.5	31.9	....	30.7	....	29.9	2
96.5	99.2	101.6	104.1	107.5	....	113.7	....	120.2	3
54.5	....	57.5	....	61.0	....	66.0	....	73.0	4
$0.00009^{18°}$	....	....	....	....	....	....	....	....	5
139.8	....	173.2	....	....	....	....	....	....	6
....	....	84.74	....	....	....	....	137.0	....	7
35.5	40.8	45.6	....	....	....	....	....	....	8
44.5	45.3	....	50.4	53.5	59.5	64.2	69.0	74.0	9
....	....	....	....	....	....	62.9	....	68.3	10
146.9	157.1	169.0	181.7	197.0	212.5	....	....	....	11
....	....	....	....	....	....	224.7	225.7	228.0	12
73.9	80.71	88.59	98.15	....	....	....	....	....	13
....	....	....	....	108.6	110.6	112.7	114.1	115.3	14
142.7	....	....	....	....	....	....	....	....	15
....	206.6	....	....	....	....	....	....	....	16
....	....	....	....	....	....	....	....	....	17
62.9	67.76	....	....	....	....	....	....	....	18
64.5	66.44	68.8	72.6	....	....	....	....	....	19
....	....	....	58.17	55.0	52.0	48.0	42.5	34.0	20
$0.00062^{18°}$	....	....	....	....	....	....	....	....	21
0.5	....	....	....	....	....	....	25		22
6.5	8.3	10.2	....	16.2	....	30.0	....	61.3	23
....	....	....	....	....	....	....	....	$53.85^{101.1°}$	24
....	$0.00591^{25°}$	....	....	....	....	....	....	....	25
3.9×10^{-6} at 25°	....	....	....	....	....	....	....	....	26
0.0002	....	0.0007	....	....	....	....	....	....	27
2×10^{-8} at 25°	....	....	....	....	....	....	....	....	28
0.138	0.264	0.476	0.687	1.206	2.055	2.106	....	....	29
....	5	....	3.7	....	....	2.7	....	....	30
130.9	138.1	144.5	150	152.5	....	153.8	....	155.1	31
$0.0092^{25°}$	....	....	....	....	....	....	....	....	32
64.2	68.9	73.3	78.3	82.2	85.2	....	....	87.6	33
148.1	161.1	173.9	183.0	184.1	185.7	187.4	188.2	....	34
96.31	....	122.2	....	....	....	....	....	....	35
....	....	....	....	163.1	169.1	....	235.1	....	36
....	42.46	....	....	....	....	....	....	....	37
....	....	....	50.15	54.80	59.44	63.17	....	76.7	38
$0.00036^{18°}$	....	....	....	....	....	....	....	....	39
0.00618	0.00517	0.00440	0.00376	0.00324	0.00267	0.00199	0.00114	0	40
0.1211	....	....	....	....	....	....	....	....	41
0.0021	0.0007	0.0004	0.0001	0	....	....	....	....	42
255.6	283.8	323.3	....	....	....	....	....	....	43
....	....	....	337.3	350	364.8	380.1	396.3	....	44
5.9	8.39	11.70	17.00	24.75	40.0	71.0	109.0	....	45
33.2	39.1	45.4	....	60.0	....	....	....	....	46
51.4	....	67.3	....	....	....	....	....	121.6	47
0.53	0.90	1.32	1.83	2.46	....	4.6	....	6.95	48
6.9	9.5	13.2	17.5	22.7	....	34.0	....	50.0	49
65.2	70.6	75.5	80.2	85.5	90.0	95.0	99.2	104.0	50
110.5	113.7	116.9	121.2	126.8	133.1	139.8	147.5	155.7	51
7.4	10.5	14	19.3	24.5	....	38.5	....	57	52
34.0	37.0	40.0	42.6	45.5	48.3	51.1	54.0	56.7	53
1.12	1.41	1.76	2.17	2.64	3.19	3.79	4.45	5.18	54

Table 10-2 (*Continued*)
SOLUBILITIES OF INORGANIC COMPOUNDS IN WATER
AT VARIOUS TEMPERATURES

	Substance	Formula	Solid Phase	0°C.	10°C,
1	**Potassium** chromate	K_2CrO_4	....	58.2	60.0
2	dichromate	$K_2Cr_2O_7$	....	5	7
3	ferricyanide	$K_3Fe(CN)_6$	....	31	36
4	hydroxide	KOH	$2H_2O$	97	103
5	hydroxide	KOH	$1H_2O$	....	....
6	iodate	KIO_3	....	4.73	....
7	iodide	KI	....	127.5	136
8	nitrate	KNO_3	....	13.3	20.9
9	nitrite	KNO_2	....	278.8	....
10	perchlorate	$KClO_4$	....	0.75	1.05
11	permanganate	$KMnO_4$	....	2.83	4.4
12	†persulfate	$K_2S_2O_8$†	†	1.62	2.60
13	sulfate	K_2SO_4	....	7.35	9.22
14	thiocyanate	KCNS	....	177.0	....
15	**Rubidium** chlorate	$RbClO_3$	....	2.14	......
16	chloride	RbCl	....	77	84.4
17	nitrate	$RbNO_3$	....	19.5	33
18	perchlorate	$RbClO_4$	....	0.5	0.6
19	sulfate	Rb_2SO_4	....	36.4	42.6
20	**Selenic** acid	H_2SeO_4	$1H_2O$	426	....
21	acid	H_2SeO_4	....	....	....
22	**Selenious acid**	H_2SeO_3	....	90.1	122.2
23	**Silver** acetate	$AgC_2H_3O_2$	....	0.72	0.88
24	arsenate	Ag_3AsO_4	....	....	....
25	arsenite	Ag_3AsO_3	....	....	....
26	bromide	AgBr	....	....	....
27	carbonate	Ag_2CO_3	....	....	....
28	chloride	AgCl	....	....	8.9×10^{-5}
29	chromate	Ag_2CrO_4	....	0.0014	....
30	cyanide	AgCN	....	....	....
31	ferricyanide	$Ag_3Fe(CN)_6$	....	....	....
32	iodate	$AgIO_3$	....	....	0.003
33	iodide	AgI	....	....	....
34	nitrate	$AgNO_3$	....	122	170
35	nitrite	$AgNO_2$	....	0.155	0.220
36	sulfate	Ag_2SO_4	....	0.573	0.695
37	sulfide	Ag_2S	....	....	....
38	**Sodium** acetate	$NaC_2H_3O_2$	$3H_2O$	36.3	40.8
39	acetate	$NaC_2H_3O_2$	....	119	121
40	bicarbonate	$NaHCO_3$	....	6.9	8.15
41	bromate	$NaBrO_3$	....	27.5	....
42	bromide	NaBr	$2H_2O$	79.5	....
43	bromide	NaBr	....	....	....
44	carbonate	Na_2CO_3	$10H_2O$	7	12.5
45	carbonate	Na_2CO_3	$1H_2O$	....	....
46	chlorate	$NaClO_3$	....	79	89
47	chloride	NaCl	....	35.7	35.8
48	chromate	Na_2CrO_4	$10H_2O$	31.70	50.17
49	chromate	Na_2CrO_4	$4H_2O$	....	....
50	chromate	Na_2CrO_4	....	....	....
51	dichromate	$Na_2Cr_2O_7$	$2H_2O$	163.0	....
52	dichromate	$Na_2Cr_2O_7$	....	....	....
53	dihydrogen phosphate	NaH_2PO_4	$2H_2O$	57.9	69.9
54	dihydrogen phosphate	NaH_2PO_4	$1H_2O$	....	....
55	dihydrogen phosphate	NaH_2PO_4	....	....	....
56	ferrocyanide	$Na_4Fe(CN)_6$	....	....	....

Table 10-2 (*Continued*)
SOLUBILITIES OF INORGANIC COMPOUNDS IN WATER
AT VARIOUS TEMPERATURES

20°C.	30°C.	40°C.	50°C.	60°C.	70°C.	80°C.	90°C.	100°C.	
61.7	63.4	65.2	66.8	68.6	70.4	72.1	73.9	75.6	1
12	20	26	34	43	52	61	70	80	2
43	50	60	.....	66	....	....	....	82.6[104]	3
112	126	.....	.....	....	....	....	....	.....	4
....	.....	.....	140	....	....	....	....	178	5
8.13	11.73	12.8	.....	18.5	....	24.8	....	32.2	6
144	152	160	168	176	184	192	200	208	7
31.6	45.8	63.9	85.5	110.0	138	169	202	246	8
298.4	.....	334.9	.....	....	....	....	....	412.8	9
1.80	2.6	4.4	6.5	9	11.8	14.8	18	21.8	10
6.4	9.0	12.56	16.89	22.2	....	....	....	.....	11
4.49	7.19	9.89	.....	....	....	....	....	.....	12
11.11	12.97	14.76	16.50	18.17	19.75	21.4	22.8	24.1	13
217.5	.....	.....	.....	....	....	....	....	.....	14
5.4	8	.....	15.98	....	....	....	....	62.8	15
91.1	97.6	103.5	109.3	115.5	121.4	127.2	133.1	138.9	16
53.3	81.3	116.7	155.6	200	251	309	375	452	17
1.0	1.5	2.3	3.5	4.85	6.72	9.2	12.7	18	18
48.2	53.5	58.5	63.1	67.4	71.4	75.0	78.7	81.8	19
567	.....	.....	.....	....	....	....	....	.....	20
1329	1718	2757	ω	....	....	....	....	.....	21
166.7	235.6	344.4	380.8	383.9	383.9	383.9	385.4	.....	22
1.04	1.21	1.41	1.64	1.89	2.18	2.52	....	.....	23
0.00085	.....	.....	....	....	....	....	....	.....	24
0.00115	.....	.....	....	....	....	....	....	.....	25
8.4×10^{-6}	.....	.....	....	....	....	....	....	.....	26
0.0032	.....	.....	....	....	....	....	....	0.05	27
1.5×10^{-4}	.....	.....	0.0005	....	....	....	....	0.002	28
....	0.0036	.....	0.0053	....	0.008	....	....	0.011	29
2.2×10^{-5}	.....	.....	....	....	....	....	....	.....	30
6.6×10^{-5}	.....	.....	....	....	....	....	....	.....	31
0.004	.....	.....	....	0.018	....	....	....	.....	32
....	3×10^{-7}	.....	....	3×10^{-6}	....	....	....	.....	33
222	300	376	455	525	....	669	....	952	34
0.340	0.5	0.715	0.995	1.363	....	....	....	.....	35
0.796	0.888	0.979	1.08	1.15	1.22	1.30	1.36	1.41	36
1.3×10^{-16}	.....	.....	....	....	....	....	....	.....	37
46.5	54.5	65.5	83	139	....	....	....	.....	38
123.5	126	129.5	134	139.5	146	153	161	170	39
9.6	11.1	12.7	14.45	16.4	....	....	....	.....	40
34.5	.....	50.2	.....	62.5	....	75.7	....	90.9	41
90.5	97.6	105.8	116.0	....	....	....	....	.....	42
....	.....	.....	....	....	....	118.3	....	121.2	43
21.5	38.8	.....	....	....	....	....	....	.....	44
....	50.5	48.5	.....	46.4	....	45.8	....	45.5	45
101	113	126	140	155	172	189	....	230	46
36.0	36.3	36.6	37.0	37.3	37.8	38.4	39.0	39.8	47
88.7	.....	.....	....	....	....	....	....	.....	48
....	88.7	95.96	104	114.6	....	....	....	.....	49
....	.....	.....	....	....	123.0	124.8	....	125.9	50
177.8	.....	.....	244.8	....	316.7	376.2	....	.....	51
....	.....	.....	....	....	....	....	....	426.3	52
85.2	106.5	138.2	.....	....	....	....	....	.....	53
....	.....	.....	158.6	....	....	....	....	.....	54
....	.....	.....	....	179.3	190.3	207.3	225.3	246.6	55
17.9	.....	30	.....	....	....	59	....	63	56

Table 10-2 (*Continued*)
SOLUBILITIES OF INORGANIC COMPOUNDS IN WATER
AT VARIOUS TEMPERATURES

	Substance	Formula	Solid Phase	0°C.	10°C.
1	**Sodium** hydrogen arsenate	Na_2HAsO_4	$12H_2O$	7.3	15.5
2	hydrogen phosphate	Na_2HPO_4	$12H_2O$	1.67	3.6
3	hydrogen phosphate	Na_2HPO_4	$7H_2O$	....	....
4	hydrogen phosphate	Na_2HPO_4	$2H_2O$	....	....
5	hydrogen phosphate	Na_2HPO_4	....	....	....
6	hydroxide	NaOH	$4H_2O$	42	....
7	hydroxide	NaOH	$3½H_2O$	....	51.5
8	hydroxide	NaOH	$1H_2O$	....	....
9	hydroxide	NaOH	....	....	....
10	iodate	$NaIO_3$	....	2.5	....
11	iodide	NaI	$2H_2O$	158.7	168.6
12	iodide	NaI	....	....	....
13	nitrate	$NaNO_3$	....	73	80
14	nitrite	$NaNO_2$	....	72.1	78.0
15	oxalate	$Na_2C_2O_4$	....	....	....
16	phosphate, tri-	Na_3PO_4	$12H_2O$	1.5	4.1
17	pyrophosphate	$Na_4P_2O_7$	$10H_2O$	3.16	3.95
18	selenate	Na_2SeO_4	$10H_2O$	13.30	....
19	selenate	Na_2SeO_4	....	....	....
20	sulfate	Na_2SO_4	$10H_2O$	5.0	9.0
21	sulfate	Na_2SO_4	$7H_2O$	19.5	30
22	sulfate	Na_2SO_4	....	....	....
23	sulfide	Na_2S	$9H_2O$	....	15.42
24	sulfide	Na_2S	$5½H_2O$	....	....
25	sulfide	Na_2S	$6H_2O$	....	....
26	sulfite	Na_2SO_3	$7H_2O$	13.9	20
27	sulfite	Na_2SO_3	....	....	....
28	tetraborate	$Na_2B_4O_7$	$10H_2O$	1.3	1.6
29	tetraborate	$Na_2B_4O_7$	$5H_2O$	....	....
30	tungstate	Na_2WO_4	$10H_2O$	57.58	....
31	tungstate	Na_2WO_4	$2H_2O$	71.61	....
32	vanadate (meta)	$NaVO_3$	$2H_2O$	....	....
33	vanadate (meta)	$NaVO_3$	....	....	....
34	**Stannous** chloride	$SnCl_2$	....	83.9	....
35	iodide	SnI_2	....	....	....
36	sulfate	$SnSO_4$	....	....	....
37	**Strontium** acetate	$Sr(C_2H_3O_2)_2$	$4H_2O$	36.9	43.61
38	acetate	$Sr(C_2H_3O_2)_2$	$½H_2O$	....	42.95
39	bromide	$SrBr_2$	$6H_2O$	85.2	93
40	chloride	$SrCl_2$	$6H_2O$	43.5	47.7
41	chloride	$SrCl_2$	$2H_2O$	....	....
42	iodide	SrI_2	$6H_2O$	165.3	....
43	iodide	SrI_2	$2H_2O$	....	....
44	nitrate	$Sr(NO_3)_2$	$1H_2O$	52.7	....
45	nitrate	$Sr(NO_3)_2$	$4H_2O$	40.1	....
46	nitrate	$Sr(NO_3)_2$	....	....	....
47	oxalate	SrC_2O_4	$1H_2O$	0.0033	0.0044
48	sulfate	$SrSO_4$	....	0.0113	....
49	**Sulfur dioxide, 760 mm**	SO_2	....	22.83	16.21
50	**Telluric** acid	H_2TeO_4	$6H_2O$	16.17	35.52
51	acid	H_2TeO_4	$2H_2O$	....	33.85
52	**Thallium** bromate	$TlBrO_3$	....	....	....
53	bromide	TlBr	....	0.024	0.029
54	chlorate	$TlClO_3$	....	2.8	....
55	chloride	TlCl	....	0.21	0.25

Table 10-2 (*Continued*)
SOLUBILITIES OF INORGANIC COMPOUNDS IN WATER
AT VARIOUS TEMPERATURES

20°C.	30°C.	40°C.	50°C.	60°C.	70°C	80°C.	90°C.	100°C.	
26.5	37	47	.....	65	....	85	....	.....	1
7.7	20.8	.....	.....	....	....	....	....	.....	2
....	.....	51.8	.....	....	....	....	....	.....	3
....	.....	.....	80.2	82.9	88.1	92.4	102.9	.....	4
....	.....	.....	.....	....	....	....	....	102.2	5
....	.....	.....	.....	....	....	....	....	.....	6
....	.....	.....	.....	....	....	....	....	.....	7
109	119	129	145	174	....	....	....	.....	8
....	.....	.....	.....	....	....	....	313	347	9
9	.....	15	.....	21	....	27	....	34	10
178.7	190.3	205.0	227.8	256.8	....	....	....	.....	11
....	.....	.....	.....	....	294	296	....	302	12
88	96	104	114	124	....	148	....	180	13
84.5	91.6	98.4	104.1	....	....	132.6	....	163.2	14
3.7	.....	.....	.....	....	....	....	....	6.33	15
11	20	31	43	55	....	81	....	108	16
6.23	9.95	13.50	17.45	21.83	....	30.04	....	40.26	17
....	78.73	.....	.....	....	....	....	....	.....	18
....	.....	.....	80.15	....	....	....	....	72.83	19
19.4	40.8	.....	.....	....	....	....	....	.....	20
44	.....	.....	.....	....	....	....	....	.....	21
....	.....	48.8	46.7	45.3	....	43.7	....	42.5	22
18.8	22.5	28.5	.....	....	....	....	....	.....	23
....	.....	.....	39.82	42.69	45.73	51.40	59.23	.....	24
....	.....	.....	36.4	39.1	43.31	49.14	57.28	.....	25
26.9	36	.....	.....	....	....	....	....	.....	26
....	.....	28	28.2	28.8	....	28.3	....	.....	27
2.7	3.9	.....	10.5	20.3	....	....	....	.....	28
....	.....	.....	.....	....	24.4	31.5	41	52.5	29
....	.....	.....	.....	....	....	....	....	.....	30
72.4	.....	.....	.....	....	....	91.2	....	97.2	31
$15.3^{25°}$	.....	30.2	.....	68.4	....	....	....	.....	32
$21.10^{25°}$	.....	26.23	.....	32.97	36.9	$38.8^{75°}$	....	.....	33
$269.8^{15°}$	.....	.....	.....	....	....	....	....	.....	34
1.0	1.2	1.4	1.7	2.1	2.5	3.0	3.4	4.0	35
19	.....	.....	.....	....	....	....	....	18	36
....	.....	.....	.....	....	....	....	....	.....	37
41.6	39.5	.....	37.35	....	36.24	36.10	....	36.4	38
102.4	111.9	123.2	135.8	150	....	181.8	....	222.5	39
52.9	58.7	65.3	72.4	81.8	....	....	....	.....	40
....	.....	.....	.....	....	85.9	90.5	....	100.8	41
177.8	.....	191.5	.....	217.5	....	270.4	....	.....	42
....	.....	.....	.....	....	....	....	365.2	383.1	43
64.0	.....	.....	83.8	97.2	....	....	130.4	139	44
70.5	.....	.....	.....	....	....	....	....	.....	45
....	88.6	90.1	.....	93.8	96	98	100	.....	46
0.0046	0.0057	.....	.....	....	....	....	....	.....	47
0.0114	0.0114	.....	.....	....	....	....	....	.....	48
11.29	7.81	5.41	4.5	....	....	....	....	.....	49
....	.....	.....	.....	....	....	....	....	.....	50
....	50.06	57.19	.....	77.52	....	106.4	....	155.3	51
0.346	.....	0.736	.....	....	....	....	....	.....	52
0.042	.....	.....	.....	....	....	....	....	.....	53
3.92	.....	.....	12.67	....	....	36.65	....	57.31	54
0.33	0.42	0.52	0.63	0.8	....	1.2	....	1.8	55

Table 10-2 (*Continued*)
SOLUBILITIES OF INORGANIC COMPOUNDS IN WATER
AT VARIOUS TEMPERATURES

	Substance	Formula	Solid Phase	0°C.	10°C.
1	**Thallium** hydroxide	TlOH	....	25.44	....
2	iodide	TlI	....	....	0.0036
3	nitrate	TlNO$_3$	....	3.91	6.22
4	perchlorate	TlClO$_4$	....	6	8.04
5	selenate	Tl$_2$SeO$_4$	....	....	2.13
6	sulfate	Tl$_2$SO$_4$	....	2.70	3.70
7	sulfide	Tl$_2$S	....	....	....
8	**Thorium** selenate	Th(SeO$_4$)$_2$	....	0.498	....
9	sulfate	Th(SO$_4$)$_2$	9H$_2$O	0.74	0.98
10	sulfate	Th(SO$_4$)$_2$	8H$_2$O	1.0	1.25
11	sulfate	Th(SO$_4$)$_2$	6H$_2$O	1.50	....
12	sulfate	Th(SO$_4$)$_2$	4H$_2$O	....	....
13	**Uranyl nitrate**	UO$_2$(NO$_3$)$_2$	6H$_2$O	98.0	108.3
14	**Ytterbium sulfate**	Yb$_2$(SO$_4$)$_3$	8H$_2$O	44.2	38.4
15	**Zinc** bromide	ZnBr$_2$	2H$_2$O	389.0	....
16	bromide	ZnBr$_2$	....	....	....
17	chlorate	ZnClO$_3$	6H$_2$O	145.0	152.5
18	chlorate	ZnClO$_3$	4H$_2$O	....	....
19	iodide	ZnI$_2$	2H$_2$O	430.9	457.4
20	iodide	ZnI$_2$	....	429.4	....
21	nitrate	Zn(NO$_3$)$_2$	6H$_2$O	94.78	....
22	nitrate	Zn(NO$_3$)$_2$	3H$_2$O	....	....
23	sulfate	ZnSO$_4$	7H$_2$O	41.9	47
24	sulfate	ZnSO$_4$	6H$_2$O	....	....
25	sulfate	ZnSO$_4$	1H$_2$O	....	....

Table 10-3
SOLUBILITY OF CANE (OR BEET) SUGAR IN WATER

Temp. °C.	Per cent sugar	Grams of sugar in 100g H$_2$O	Temp. °C.	Per cent sugar	Grams of sugar in 100g H$_2$O
0	64.18	179.2	50	72.25	260.4
5	64.87	184.7	55	73.20	273.1
10	65.58	190.5	60	74.18	287.3
15	66.33	197.0	65	75.88	315.0
20	67.09	203.9	70	76.22	320.4
25	67.89	211.4	75	77.27	339.9
30	68.80	219.5	80	78.36	362.1
35	69.55	228.4	85	79.46	386.9
40	70.42	238.1	90	80.61	415.7
45	71.32	248.7	95	81.77	448.5
			100	82.97	487.2

Table 10-2 (*Continued*)
SOLUBILITIES OF INORGANIC COMPOUNDS IN WATER
AT VARIOUS TEMPERATURES

20°C.	30°C.	40°C.	50°C.	60°C.	70°C.	80°C.	90°C.	100°C.	
......	39.9	49.5	.....	73.8	....	106	126.1	148.3	1
0.006	0.008	0.015	.....	0.035	.....	0.070	....	0.120	2
9.55	14.3	20.9	30.4	46.2	69.5	111.0	200.0	414.0	3
......	19.72	.....	39.62	....	65.32	81.49	....	166.6	4
2.8	.....	.....	.....	.....	....	8.5	....	10.9	5
4.87	6.16	.....	9.21	10.92	12.74	14.61	16.53	18.45	6
0.0022	.....	.....	.....	....	....	....	....	.....	7
......	.....	.....	.....	....	....	....	....	1.97	8
1.38	1.995	2.998	5.22	....	....	....	....	.....	9
1.62	.....	.....	.....	....	....	....	....	.....	10
1.90	2.45	.....	.....	6.64	....	....	....	.....	11
......	.....	4.04	2.54	1.63	1.09	....	....	.....	12
125.5	.....	.....	203	....	....	....	....	.....	13
......	21.0	.....	....	10.4	7.22	6.92	5.83	4.67	14
446.4	528.1	.....	.....	....	....	....	....	.....	15
......	.....	591.1	.....	618.4	....	644.6	....	672.2	16
......	.....	.....	.....	....	....	....	....	.....	17
200.3	209.2	223.2	273.1	....	....	....	....	.....	18
484.8	.....	.....	.....	....	....	....	....	.....	19
......	.....	445.2	.....	467.2	....	490	....	510.5	20
118.3	.....	.....	.....	....	....	....	....	.....	21
......	.....	206.9	.....	....	....	....	....	.....	22
54.4	.....	.....	.....	.....	....	....	....	.....	23
......	.....	70.1	76.8	....	....	....	....	.....	24
......	.....	.....	.....	....	....	86.6	83.7	80.8	25

Table 10-4
COMPOSITION OF SEA WATER

Reprinted from *The Oceans*, by Sverdrup, Johnson, and Fleming, with permission from the publishers, Prentice-Hall, Inc., New York.

The table below lists all of the elements known to occur in sea water as dissolved solids, except hydrogen, oxygen and dissolved gases. They are not given as ions but as the amounts of the individual elements which occur in water of 19.00% chlorinity. Cadmium, chromium, cobalt and tin are merely mentioned because they have been found in the ash of marine organisms, and hence it is implied that they occur in sea water although so far they have not been obtained directly.

Element	parts per million, mg/kg	Element	parts per million, mg/kg
Chlorine	18980	Copper	0.001–0.01
Sodium	10561	Zinc	0.005
Magnesium	1272	Lead	0.004
Sulfur	884	Selenium	0.004
Calcium	400	Cesium	0.002
Potassium	380	Uranium	0.0015
Bromine	65	Molybdenum	0.0005
Carbon	28	Thorium	<0.0005
Strontium	13	Cerium	0.0004
Boron	4.6	Silver	0.0003
Silicon	0.02–4.0	Vanadium	0.0003
Fluorine	1.4	Lanthanum	0.0003
Nitrogen*	0.01–0.7	Yttrium	0.0003
Aluminum	0.5	Nickel	0.0001
Rubidium	0.2	Scandium	0.00004
Lithium	0.1	Mercury	0.00003
Phosphorus	0.001–0.10	Gold	0.000006
Barium	0.05	Radium	$0.2–3 \times 10^{-10}$
Iodine	0.05	Cadmium	
Arsenic	0.01–0.02	Chromium	
Iron	0.002–0.02	Cobalt	
Manganese	0.001–0.01	Tin	

* In dissolved compounds and not as dissolved atmospheric nitrogen.

VAPOR PRESSURES

Table 10-5
VAPOR PRESSURE OF MERCURY

Temp. °C.	mm of Hg	Temp. °C.	mm of Hg	Temp. °C.	mm of Hg
−38	0.0_5145	88	0.1413	214	26.826
−36	0.0_5197	90	0.1582	216	28.504
−34	0.0_5266	92	0.1769	218	30.271
−32	0.0_5359	94	0.1976	220	32.133
−30	0.0_5478	96	0.2202	222	34.092
−28	0.0_5630	98	0.2453	224	36.153
−26	0.0_5828	100	0.2729	226	38.318
−24	0.0_4108	102	0.3032	228	40.595
−22	0.0_4140	104	0.3366	230	42.989
−20	0.0_4181	106	0.3731	232	45.503
−18	0.0_4232	108	0.4132	234	48.141
−16	0.0_4298	110	0.4572	236	50.909
−14	0.0_4380	112	0.5052	238	53.812
−12	0.0_4481	114	0.5576	240	56.855
−10	0.0_4606	116	0.6150	242	60.044
− 8	0.0_4762	118	0.6776	244	63.384
− 6	0.0_4954	120	0.7457	246	66.882
− 4	0.0_3119	122	0.8198	248	70.543
− 2	0.0_3149	124	0.9004	250	74.375
0	0.0_3185	126	0.9882	252	78.381
+ 2	0.0_3228	128	1.084	254	82.568
4	0.0_3276	130	1.186	256	86.944
6	0.0_3335	132	1.298	258	91.518
8	0.0_3406	134	1.419	260	96.296
10	0.0_3490	136	1.551	262	101.28
12	0.0_3588	138	1.692	264	106.48
14	0.0_3706	140	1.845	266	111.91
16	0.0_3846	142	2.010	268	117.57
18	0.001009	144	2.188	270	123.47
20	0.001201	146	2.379	272	129.62
22	0.001426	148	2.585	274	136.02
24	0.001691	150	2.807	276	142.69
26	0.002000	152	3.046	278	149.64
28	0.002359	154	3.303	280	156.87
30	0.002777	156	3.578	282	164.39
32	0.003261	158	3.873	284	172.21
34	0.003823	160	4.189	286	180.34
36	0.004471	162	4.528	288	188.79
38	0.005219	164	4.890	290	197.57
40	0.006079	166	5.277	292	206.70
42	0.007067	168	5.689	294	216.17
44	0.008200	170	6.128	296	226.00
46	0.009497	172	6.596	298	236.21
48	0.01098	174	7.095	300	246.80
50	0.01267	176	7.626	302	257.78
52	0.01459	178	8.193	304	269.17
54	0.01677	180	8.796	306	280.98
56	0.01925	182	9.436	308	293.21
58	0.02206	184	10.116	310	305.89
60	0.02524	186	10.839	312	319.02
62	0.02883	188	11.607	314	332.62
64	0.03287	190	12.423	316	346.70
66	0.03740	192	13.287	318	361.26
68	0.04251	194	14.203	320	376.33
70	0.04825	196	15.173	322	391.92
72	0.05469	198	16.200	324	408.04
74	0.06189	200	17.287	326	424.71
76	0.06993	202	18.437	328	441.94
78	0.07889	204	19.652	330	459.74
80	0.08880	206	20.936	332	478.13
82	0.1000	208	22.292	334	497.12
84	0.1124	210	23.723	336	516.74
86	0.1261	212	25.233	338	537.00

Table 10-5 (*Continued*)
VAPOR PRESSURE OF MERCURY

Temp. °C.	mm of Hg	Temp. °C.	mm of Hg	Temp. °C.	mm of Hg
340	557.90	374	1028.9	520	7691
342	579.45	376	1064.4	550	10650
344	601.69	378	1100.9	600	22.87atm.
346	624.64	380	1138.4	650	35.49atm.
348	648.30	382	1177.0	700	52.51atm.
350	672.69	384	1216.6	750	74.86atm.
352	697.83	386	1257.3	800	103.31atm.
354	723.73	388	1299.1	850	138.42atm.
350	750.43	390	1341.9	900*	180.92atm.
358	777.92	392	1386.1	950	226.58atm.
360	806.23	394	1431.3	1000	290.5atm.
362	835.38	396	1477.7	1050	358.1atm.
364	865.36	398	1525.2	1100	437.3atm.
366	896.23	400	1574.1	1150	521.3atm.
368	928.02	430	2464	1200	616.8atm.
370	960.66	460	3715	1250	721.4atm.
372	994.34	490	5420	1300	835.9atm.

* Critical point.

Table 10-6
VAPOR PRESSURE OF LIQUID AMMONIA, NH_3

t°C.	p in atm	t°C.	p in atm	t°C.	p in atm
−78	0.0582	−6	3.3677	66	29.784
−76	0.0683	−4	3.6405	68	31.211
−74	0.0797	−2	3.9303	70	32.687
−72	0.0929	0	4.2380	72	34.227
−70	0.1078	+2	4.5640	74	35.813
−68	0.1246	4	4.9090	76	37.453
−66	0.1437	6	5.2750	78	39.149
−64	0.1651	8	5.6610	80	40.902
−62	0.1891	10	6.0685	82	42.712
−60	0.2161	12	6.4985	84	44.582
−58	0.2461	14	6.9520	86	46.511
−56	0.2796	16	7.4290	88	48.503
−54	0.3167	18	7.9310	90	50.558
−52	0.3578	20	8.4585	92	52.677
−50	0.4034	22	9.0125	94	54.860
−48	0.4536	24	9.5940	96	57.111
−46	0.5087	26	10.2040	98	59.429
−44	0.5693	28	10.8430	100	61.816
−42	0.6357	30	11.512	102	64.274
−40	0.7083	32	12.212	104	66.804
−38	0.7875	34	12.943	106	69.406
−36	0.8738	36	13.708	108	72.084
−34	0.9676	38	14.507	110	74.837
−32	1.0695	40	15.339	112	77.668
−30	1.1799	42	16.209	114	80.578
−28	1.2992	44	17.113	116	83.570
−26	1.4281	46	18.056	118	86.644
−24	1.5671	48	19.038	120	89.802
−22	1.7166	50	20.059	122	93.045
−20	1.8774	52	21.121	124	96.376
−18	2.0499	54	22.224	126	99.796
−16	2.2349	56	23.372	128	103.309
−14	2.4328	58	24.562	130	106.913
−12	2.6443	60	25.797	132	110.613
−10	2.8703	62	27.079	132.3	111.3(c.p.)
− 8	3.1112	64	28.407		

Table 10-7
PARTIAL PRESSURES OF AQUEOUS AMMONIA SOLUTIONS
Perman, *Jour. Chem. Soc. 83, 1168 (1903)*.

°C	% Wt. NH₃	Partial Pressure mm Hg		°C	% Wt. NH₃	Partial Pressure mm Hg	
		NH₃	H₂O			NH₃	H₂O
0	4.72	11.4	5.1	+30.09	9.75	120.0	28.5
	9.15	24.8	5.3		12.77	175.0	26.6
	14.73	51.3	4.1		17.76	290.2	24.8
	19.62	82.5	3.0		17.84	291.1	24.3
	22.90	116.6	2.8		21.47	404.6	22.1
+10	4.16	16.5	9.1	+40	3.79	61.1	53.5
	8.26	37.2	8.8		7.36	133.0	50.7
	12.32	64.2	7.6		11.06	218.5	49.1
	15.88	95.1	7.0		15.55	353.6	44.1
	20.54	149.2	6.2		17.33	427.7	.....
	21.83	169.8	5.5		20.85	576.1	37.8
+19.9	4.18	27.4	16.4	+50	3.29	79.1	89.6
	6.50	45.8	16.1		5.90	151.3	87.1
	6.55	46.0	16.0		8.91	246.6	83.0
	7.72	56.2	15.6		11.57	341.7	80.6
	10.15	80.6	15.1		14.15	451.4	77.0
	10.75	86.3	14.7		14.94	487.1	75.2
	16.64	166.1	12.9				
	19.40	215.6	12.3	+60	3.86	136.9	144.1
	23.37	302.4	10.3		5.77	215.9	.....
					7.78	300.4	138.5
+30.09	3.93	41.2	31.1		9.37	375.7	135.5
	7.43	86.3	29.2		11.31	475.8	130.4

Table 10-8
VAPOR PRESSURE OF WATER IN MILLIMETERS OF MERCURY
For Temperatures from $-10°$ to $118°C$

The values in the table below are for water in contact with its own vapor. Where the water is in contact with air at a temperature $t°C$, the following correction must be added: Correction for temperatures up to $40°C = p(0.0775 - 3.13 \times 10^{-4}t)/100$; for temperatures above $50°C = p(0.0652 - 8.75 \times 10^{-5}t)/100$.

$t°C.$	$\dfrac{p}{mm\ Hg}$	$t°C.$	$\dfrac{p}{mm\ Hg}$	$t°C.$	$\dfrac{p}{mm\ Hg}$
−10.0	2.149	−4.5	3.284	1.0	4.926
−9.9	2.167	−4.4	3.309	1.1	4.962
−9.8	2.184	−4.3	3.334	1.2	4.998
−9.7	2.201	−4.2	3.359	1.3	5.034
−9.6	2.219	−4.1	3.384	1.4	5.070
−9.5	2.236	−4.0	3.410	1.5	5.107
−9.4	2.254	−3.9	3.436	1.6	5.144
−9.3	2.271	−3.8	3.461	1.7	5.181
−9.2	2.289	−3.7	3.487	1.8	5.219
−9.1	2.307	−3.6	3.514	1.9	5.256
−9.0	2.326	−3.5	3.540	2.0	5.294
−8.9	2.343	−3.4	3.567	2.1	5.332
−8.8	2.362	−3.3	3.593	2.2	5.370
−8.7	2.380	−3.2	3.620	2.3	5.408
−8.6	2.399	−3.1	3.647	2.4	5.447
−8.5	2.418	−3.0	3.673	2.5	5.486
−8.4	2.437	−2.9	3.702	2.6	5.525
−8.3	2.456	−2.8	3.730	2.7	5.565
−8.2	2.475	−2.7	3.757	2.8	5.605
−8.1	2.495	−2.6	3.785	2.9	5.645
−8.0	2.514	−2.5	3.813	3.0	5.685
−7.9	2.533	−2.4	3.841	3.1	5.725
−7.8	2.553	−2.3	3.871	3.2	5.766
−7.7	2.572	−2.2	3.898	3.3	5.807
−7.6	2.593	−2.1	3.927	3.4	5.848
−7.5	2.613	−2.0	3.956	3.5	5.889
−7.4	2.633	−1.9	3.986	3.6	5.931
−7.3	2.654	−1.8	4.016	3.7	5.973
−7.2	2.674	−1.7	4.045	3.8	6.015
−7.1	2.695	−1.6	4.075	3.9	6.058
−7.0	2.715	−1.5	4.105	4.0	6.101
−6.9	2.736	−1.4	4.135	4.1	6.144
−6.8	2.757	−1.3	4.165	4.2	6.187
−6.7	2.778	−1.2	4.196	4.3	6.230
−6.6	2.800	−1.1	4.227	4.4	6.274
−6.5	2.822	−1.0	4.258	4.5	6.318
−6.4	2.843	−0.9	4.289	4.6	6.363
−6.3	2.866	−0.8	4.320	4.7	6.408
−6.2	2.887	−0.7	4.353	4.8	6.453
−6.1	2.909	−0.6	4.385	4.9	6.498
−6.0	2.931	−0.5	4.416	5.0	6.543
−5.9	2.955	−0.4	4.448	5.1	6.589
−5.8	2.976	−0.3	4.480	5.2	6.635
−5.7	3.000	−0.2	4.513	5.3	6.681
−5.6	3.022	−0.1	4.546	5.4	6.728
−5.5	3.046	0.0	4.579	5.5	6.775
−5.4	3.069	+0.1	4.613	5.6	6.822
−5.3	3.092	0.2	4.647	5.7	6.869
−5.2	3.115	0.3	4.681	5.8	6.917
−5.1	3.139	0.4	4.715	5.9	6.965
−5.0	3.163	0.5	4.750	6.0	7.013
−4.9	3.187	0.6	4.785	6.1	7.062
−4.8	3.211	0.7	4.820	6.2	7.111
−4.7	3.235	0.8	4.855	6.3	7.160
−4.6	3.259	0.9	4.890	6.4	7.209

Table 10-8 (*Continued*)
VAPOR PRESSURE OF WATER IN MILLIMETERS OF MERCURY

t°C.	p mm Hg	t°C.	p mm Hg	t°C.	p mm Hg
6.5	7.259	12.8	11.085	19.1	16.581
6.6	7.309	12.9	11.158	19.2	16.685
6.7	7.360	13.0	11.231	19.3	16.789
6.8	7.411	13.1	11.305	19.4	16.894
6.9	7.462	13.2	11.379	19.5	16.999
7.0	7.513	13.3	11.453	19.6	17.105
7.1	7.565	13.4	11.528	19.7	17.212
7.2	7.617	13.5	11.604	19.8	17.319
7.3	7.669	13.6	11.680	19.9	17.427
7.4	7.722	13.7	11.756	20.0	17.535
7.5	7.775	13.8	11.833	20.1	17.644
7.6	7.828	13.9	11.910	20.2	17.753
7.7	7.882	14.0	11.987	20.3	17.863
7.8	7.936	14.1	12.065	20.4	17.974
7.9	7.990	14.2	12.144	20.5	18.085
8.0	8.045	14.3	12.223	20.6	18.197
8.1	8.100	14.4	12.302	20.7	18.309
8.2	8.155	14.5	12.382	20.8	18.422
8.3	8.211	14.6	12.462	20.9	18.536
8.4	8.267	14.7	12.543	21.0	18.650
8.5	8.323	14.8	12.624	21.1	18.765
8.6	8.380	14.9	12.706	21.2	18.880
8.7	8.437	15.0	12.788	21.3	18.996
8.8	8.494	15.1	12.870	21.4	19.113
8.9	8.551	15.2	12.953	21.5	19.231
9.0	8.609	15.3	13.037	21.6	19.349
9.1	8.668	15.4	13.121	21.7	19.468
9.2	8.727	15.5	13.205	21.8	19.587
9.3	8.786	15.6	13.290	21.9	19.707
9.4	8.845	15.7	13.375	22.0	19.827
9.5	8.905	15.8	13.461	22.1	19.948
9.6	8.965	15.9	13.547	22.2	20.070
9.7	9.025	16.0	13.634	22.3	20.193
9.8	9.086	16.1	13.721	22.4	20.316
9.9	9.147	16.2	13.809	22.5	20.440
10.0	9.209	16.3	13.898	22.6	20.565
10.1	9.271	16.4	13.987	22.7	20.690
10.2	9.333	16.5	14.076	22.8	20.815
10.3	9.395	16.6	14.166	22.9	20.941
10.4	9.458	16.7	14.256	23.0	21.068
10.5	9.521	16.8	14.347	23.1	21.196
10.6	9.585	16.9	14.438	23.2	21.324
10.7	9.649	17.0	14.530	23.3	21.453
10.8	9.714	17.1	14.622	23.4	21.583
10.9	9.779	17.2	14.715	23.5	21.714
11.0	9.844	17.3	14.809	23.6	21.845
11.1	9.910	17.4	14.903	23.7	21.977
11.2	9.976	17.5	14.997	23.8	22.110
11.3	10.042	17.6	15.092	23.9	22.243
11.4	10.109	17.7	15.188	24.0	22.377
11.5	10.176	17.8	15.284	24.1	22.512
11.6	10.244	17.9	15.380	24.2	22.648
11.7	10.312	18.0	15.477	24.3	22.785
11.8	10.380	18.1	15.575	24.4	22.922
11.9	10.449	18.2	15.673	24.5	23.060
12.0	10.518	18.3	15.772	24.6	23.198
12.1	10.588	18.4	15.871	24.7	23.337
12.2	10.658	18.5	15.971	24.8	23.476
12.3	10.728	18.6	16.071	24.9	23.616
12.4	10.799	18.7	16.171	25.0	23.756
12.5	10.870	18.8	16.272	25.1	23.897
12.6	10.941	18.9	16.374	25.2	24.039
12.7	11.013	19.0	16.477	25.3	24.182

Table 10-8 (*Continued*)
VAPOR PRESSURE OF WATER IN MILLIMETERS OF MERCURY

t°C.	$\frac{p}{\text{mm Hg}}$	t°C.	$\frac{p}{\text{mm Hg}}$	t°C.	$\frac{p}{\text{mm Hg}}$
25.4	24.326	31.7	35.062	38.0	49.692
25.5	24.471	31.8	35.261	38.1	49.961
25.6	24.617	31.9	35.462	38.2	50.231
25.7	24.764	32.0	35.663	38.3	50.502
25.8	24.912	32.1	35.865	38.4	50.774
25.9	25.060	32.2	36.068	38.5	51.048
26.0	25.209	32.3	36.272	38.6	51.323
26.1	25.359	32.4	36.477	38.7	51.600
26.2	25.509	32.5	36.683	38.8	51.879
26.3	25.660	32.6	36.891	38.9	52.160
26.4	25.812	32.7	37.099	39.0	52.442
26.5	25.964	32.8	37.308	39.1	52.725
26.6	26.117	32.9	37.518	39.2	53.009
26.7	26.271	33.0	37.729	39.3	53.294
26.8	26.426	33.1	37.942	39.4	53.580
26.9	26.582	33.2	38.155	39.5	53.867
27.0	26.739	33.3	38.369	39.6	54.156
27.1	26.897	33.4	38.584	39.7	54.446
27.2	27.055	33.5	38.801	39.8	54.737
27.3	27.214	33.6	39.018	39.9	55.030
27.4	27.374	33.7	39.237	40.0	55.324
27.5	27.535	33.8	39.457	40.2	55.91
27.6	27.696	33.9	39.677	40.4	56.51
27.7	27.858	34.0	39.898	40.6	57.11
27.8	28.021	34.1	40.121	40.8	57.72
27.9	28.185	34.2	40.344	41.0	58.34
28.0	28.349	34.3	40.569	41.2	58.96
28.1	28.514	34.4	40.796	41.4	59.58
28.2	28.680	34.5	41.023	41.6	60.22
28.3	28.847	34.6	41.251	41.8	60.86
28.4	29.015	34.7	41.480	42.0	61.50
28.5	29.184	34.8	41.710	42.2	62.14
28.6	29.354	34.9	41.942	42.4	62.80
28.7	29.525	35.0	42.175	42.6	63.46
28.8	29.697	35.1	42.409	42.8	64.12
28.9	29.870	35.2	42.644	43.0	64.80
29.0	30.043	35.3	42.880	43.2	65.48
29.1	30.217	35.4	43.117	43.4	66.16
29.2	30.392	35.5	43.355	43.6	66.86
29.3	30.568	35.6	43.595	43.8	67.56
29.4	30.745	35.7	43.836	44.0	68.26
29.5	30.923	35.8	44.078	44.2	68.97
29.6	31.102	35.9	44.320	44.4	69.69
29.7	31.281	36.0	44.563	44.6	70.41
29.8	31.461	36.1	44.808	44.8	71.14
29.9	31.642	36.2	45.054	45.0	71.88
30.0	31.824	36.3	45.301	45.2	72.62
30.1	32.007	36.4	45.549	45.4	73.36
30.2	32.191	36.5	45.799	45.6	74.12
30.3	32.376	36.6	46.050	45.8	74.88
30.4	32.561	36.7	46.302	46.0	75.65
30.5	32.747	36.8	46.556	46.2	76.43
30.6	32.934	36.9	46.811	46.4	77.21
30.7	33.122	37.0	47.067	46.6	78.00
30.8	33.312	37.1	47.324	46.8	78.80
30.9	33.503	37.2	47.582	47.0	79.60
31.0	33.695	37.3	47.841	47.2	80.41
31.1	33.888	37.4	48.102	47.4	81.23
31.2	34.082	37.5	48.364	47.6	82.05
31.3	34.276	37.6	48.627	47.8	82.87
31.4	34.471	37.7	48.891	48.0	83.71
31.5	34.667	37.8	49.157	48.2	84.56
31.6	34.864	37.9	49.424	48.4	85.42

Table 10-8 (*Continued*)
VAPOR PRESSURE OF WATER IN MILLIMETERS OF MERCURY

t°C.	p mm Hg	t°C.	p mm Hg	t°C.	p mm Hg
48.6	86.28	71.0	243.9	92.2	571.26
48.8	87.14	71.5	249.3	92.4	575.55
49.0	88.02	72.0	254.6	92.6	579.87
49.2	88.90	72.5	260.2	92.8	584.22
49.4	89.79	73.0	265.7	93.0	588.60
49.6	90.69	73.5	271.5	93.2	593.00
49.8	91.59	74.0	277.2	93.4	597.43
50.0	92.51	74.5	283.2	93.6	601.89
50.5	94.86	75.0	289.1	93.8	606.38
51.0	97.20	75.5	295.3	94.0	610.90
51.5	99.65	76.0	301.4	94.2	615.44
52.0	102.09	76.5	307.7	94.4	620.01
52.5	104.65	77.0	314.1	94.6	624.61
53.0	107.20	77.5	320.7	94.8	629.24
53.5	109.86	78.0	327.3	95.0	633.90
54.0	112.51	78.5	334.2	95.2	638.59
54.5	115.28	79.0	341.0	95.4	643.30
55.0	118.04	79.5	348.1	95.6	648.05
55.5	120.92	80.0	355.1	95.8	652.82
56.0	123.80	80.5	362.4	96.0	657.62
56.5	126.81	81.0	369.7	96.2	662.45
57.0	129.82	81.5	377.3	96.4	667.31
57.5	132.95	82.0	384.9	96.6	672.20
58.0	136.08	82.5	392.8	96.8	677.12
58.5	139.34	83.0	400.6	97.0	682.07
59.0	142.60	83.5	408.7	97.2	687.04
59.5	145.99	84.0	416.8	97.4	692.05
60.0	149.38	84.5	425.2	97.6	697.10
60.5	152.91	85.0	433.6	97.8	702.17
61.0	156.43	85.5	442.3	98.0	707.27
61.5	160.10	86.0	450.9	98.2	712.40
62.0	163.77	86.5	459.8	98.4	717.56
62.5	167.58	87.0	468.7	98.6	722.75
63.0	171.38	87.5	477.9	98.8	727.98
63.5	175.35	88.0	487.1	99.0	733.24
64.0	179.31	88.5	496.6	99.2	738.53
64.5	183.43	89.0	506.1	99.4	743.85
65.0	187.54	89.5	515.9	99.6	749.20
65.5	191.82	90.0	525.76	99.8	754.58
66.0	196.09	90.2	529.77	100.0	760.00
66.5	200.53	90.4	533.80	102	815.86
67.0	204.96	90.6	537.86	104	875.06
67.5	209.57	90.8	541.95	106	937.92
68.0	214.17	91.0	546.05	108	1004.4
68.5	218.95	91.2	550.18	110	1074.6
69.0	223.73	91.4	554.35	112	1148.7
69.5	228.72	91.6	558.53	114	1227.2
70.0	233.7	91.8	562.75	116	1309.9
70.5	238.8	92.0	566.99	118	1397.2

Table 10-9
VAPOR PRESSURE OF ICE IN MILLIMETERS OF MERCURY
For Temperatures from −99° to 0°C

The values in the table below are for ice in contact with its own vapor. Where the ice is in contact with air at a temperature $t°C$, the following correction must be added: Correction $= 20p/100\,(t+273)$.

$t°C.$	p mm Hg	$t°C.$	p mm Hg	$t°C.$	p mm Hg
−99	0.000012	−41	0.0862	−25.3	0.462
−98	0.000015	−40	0.0966	25.2	0.467
−97	0.000018	−39	0.1081	−25.1	0.471
−96	0.000022	−38	0.1209	−25.0	0.476
−95	0.000027	−37	0.1351	−24.9	0.481
−94	0.000033	−36	0.1507	−24.8	0.486
−93	0.000040	−35	0.1681	−24.7	0.490
−92	0.000048	−34	0.1873	−24.6	0.495
−91	0.000058	−33	0.2084	−24.5	0.500
−90	0.000070	−32	0.2318	−24.4	0.505
−89	0.000084	−31	0.2575	−24.3	0.510
−88	0.00010	−30	0.2859	−24.2	0.515
−87	0.00012	−29.9	0.289	−24.1	0.520
−86	0.00014	−29.8	0.292	−24.0	0.526
−85	0.00017	−29.7	0.295	−23.9	0.531
−84	0.00020	−29.6	0.298	−23.8	0.536
−83	0.00024	−29.5	0.301	−23.7	0.541
−82	0.00029	−29.4	0.304	−23.6	0.547
−81	0.00034	−29.3	0.307	−23.5	0.552
−80	0.00040	−29.2	0.311	−23.4	0.558
−79	0.00047	−29.1	0.314	−23.3	0.563
−78	0.00056	−29.0	0.317	−23.2	0.569
−77	0.00066	−28.9	0.320	−23.1	0.574
−76	0.00077	−28.8	0.324	−23.0	0.580
−75	0.00090	−28.7	0.327	−22.9	0.586
−74	0.00105	−28.6	0.330	−22.8	0.592
−73	0.00123	−28.5	0.334	−22.7	0.597
−72	0.00143	−28.4	0.337	−22.6	0.603
−71	0.00167	−28.3	0.341	−22.5	0.609
−70	0.00194	−28.2	0.344	−22.4	0.615
−69	0.00225	−28.1	0.348	−22.3	0.621
−68	0.00261	−28.0	0.351	−22.2	0.627
−67	0.00302	−27.9	0.355	−22.1	0.633
−66	0.00349	−27.8	0.359	−22.0	0.640
−65	0.00403	−27.7	0.362	−21.9	0.646
−64	0.00464	−27.6	0.366	−21.8	0.652
−63	0.00534	−27.5	0.370	−21.7	0.659
−62	0.00614	−27.4	0.374	−21.6	0.665
−61	0.00703	−27.3	0.377	−21.5	0.672
−60	0.00808	−27.2	0.381	−21.4	0.678
−59	0.00925	−27.1	0.385	−21.3	0.685
−58	0.0106	−27.0	0.389	−21.2	0.691
−57	0.0121	−26.9	0.393	−21.1	0.698
−56	0.0138	−26.8	0.397	−21.0	0.705
−55	0.0157	−26.7	0.401	−20.9	0.712
−54	0.0178	−26.6	0.405	−20.8	0.719
−53	0.0203	−26.5	0.409	−20.7	0.726
−52	0.0230	−26.4	0.414	−20.6	0.733
−51	0.0261	−26.3	0.418	−20.5	0.740
−50	0.0296	−26.2	0.422	−20.4	0.747
−49	0.0334	−26.1	0.426	−20.3	0.754
−48	0.0378	−26.0	0.430	−20.2	0.761
−47	0.0426	−25.9	0.435	−20.1	0.769
−46	0.0481	−25.8	0.439	−20.0	0.776
−45	0.0541	−25.7	0.444	−19.9	0.783
−44	0.0609	−25.6	0.448	−19.8	0.791
−43	0.0684	−25.5	0.453	−19.7	0.799
−42	0.0768	−25.4	0.457	−19.6	0.806

Table 10-9 (*Continued*)
VAPOR PRESSURE OF ICE IN MILLIMETERS OF MERCURY

t°C.	$\frac{p}{\text{mm Hg}}$	t°C.	$\frac{p}{\text{mm Hg}}$	t°C.	$\frac{p}{\text{mm Hg}}$
−19.5	0.814	−13.0	1.490	−6.5	2.649
−19.4	0.822	−12.9	1.504	−6.4	2.672
−19.3	0.830	−12.8	1.518	−6.3	2.695
−19.2	0.838	−12.7	1.532	−6.2	2.718
−19.1	0.846	−12.6	1.546	−6.1	2.742
−19.0	0.854	−12.5	1.559	−6.0	2.765
−18.9	0.862	−12.4	1.574	−5.9	2.790
−18.8	0.870	−12.3	1.588	−5.8	2.813
−18.7	0.879	−12.2	1.602	−5.7	2.838
−18.6	0.887	−12.1	1.617	−5.6	2.862
−18.5	0.895	−12.0	1.632	−5.5	2.887
−18.4	0.904	−11.9	1.646	−5.4	2.912
−18.3	0.912	−11.8	1.661	−5.3	2.937
−18.2	0.921	−11.7	1.676	−5.2	2.962
−18.1	0.930	−11.6	1.691	−5.1	2.987
−18.0	0.939	−11.5	1.707	−5.0	3.013
−17.9	0.947	−11.4	1.722	−4.9	3.039
−17.8	0.956	−11.3	1.737	−4.8	3.065
−17.7	0.966	−11.2	1.753	−4.7	3.091
−17.6	0.975	−11.1	1.769	−4.6	3.117
−17.5	0.984	−11.0	1.785	−4.5	3.144
−17.4	0.993	−10.9	1.800	−4.4	3.171
−17.3	1.002	−10.8	1.817	−4.3	3.198
−17.2	1.012	−10.7	1.833	−4.2	3.225
−17.1	1.021	−10.6	1.849	−4.1	3.252
−17.0	1.031	−10.5	1.866	−4.0	3.280
−16.9	1.041	−10.4	1.883	−3.9	3.308
−16.8	1.051	−10.3	1.899	−3.8	3.336
−16.7	1.060	−10.2	1.916	−3.7	3.364
−16.6	1.070	−10.1	1.934	−3.6	3.393
−16.5	1.080	−10.0	1.950	−3.5	3.422
−16.4	1.091	−9.9	1.968	−3.4	3.451
−16.3	1.101	−9.8	1.985	−3.3	3.480
−16.2	1.111	−9.7	2.003	−3.2	3.509
−16.1	1.121	−9.6	2.021	−3.1	3.539
−16.0	1.132	−9.5	2.039	−3.0	3.568
−15.9	1.142	−9.4	2.057	−2.9	3.599
−15.8	1.153	9.3	2.075	−2.8	3.630
−15.7	1.164	9.2	2.093	−2.7	3.660
−15.6	1.175	9.1	2.112	−2.6	3.691
−15.5	1.186	9.0	2.131	−2.5	3.722
−15.4	1.196	−8.9	2.149	−2.4	3.753
−15.3	1.208	−8.8	2.168	−2.3	3.785
−15.2	1.219	−8.7	2.187	−2.2	3.816
−15.1	1.230	−8.6	2.207	−2.1	3.848
−15.0	1.241	−8.5	2.226	−2.0	3.880
−14.9	1.253	−8.4	2.246	−1.9	3.913
−14.8	1.264	−8.3	2.266	−1.8	3.946
−14.7	1.276	−8.2	2.285	−1.7	3.979
−14.6	1.288	−8.1	2.306	−1.6	4.012
−14.5	1.300	−8.0	2.326	−1.5	4.045
−14.4	1.312	−7.9	2.346	−1.4	4.079
−14.3	1.324	−7.8	2.367	−1.3	4.113
−14.2	1.336	−7.7	2.387	−1.2	4.147
−14.1	1.348	−7.6	2.408	−1.1	4.182
−14.0	1.361	−7.5	2.429	−1.0	4.217
−13.9	1.373	−7.4	2.450	−0.9	4.252
−13.8	1.386	−7.3	2.472	−0.8	4.287
−13.7	1.399	−7.2	2.493	−0.7	4.323
−13.6	1.411	−7.1	2.515	−0.6	4.359
−13.5	1.424	−7.0	2.537	−0.5	4.395
−13.4	1.437	−6.9	2.559	−0.4	4.431
−13.3	1.450	−6.8	2.581	−0.3	4.467
−13.2	1.464	−6.7	2.603	−0.2	4.504
−13.1	1.477	−6.6	2.626	−0.1	4.542
				0.0	4.579

Table 10-10
VAPOR PRESSURES OF VARIOUS SUBSTANCES

The values in the following table were selected from the following sources from which additional data on some compounds not listed here may be found: *International Critical Tables*, vol. 3, pp 199–246; Landolt-Bornstein *Physikalish-Chemische Tabellen*, 5th ed., vol. 2, pp 1332–1381; Stull, *Ind. Eng. Chem.*, **39**, 517 (1947); Rossini et al., *Selected Values of Properties of Hydrocarbons and Related Compounds*, Am. Petroleum Inst. Research Project 44; Dreisbach, *Dow Physical Property Sheets*.

For a set of tables listing the numerical values of the constants A, B, and C in the Antoine equation for innumerable compounds belonging to 23 classes of organic compounds (hydrocarbons, alcohols, aldehydes, ketones, acids, esters, amines, etc.) see Dreisbach's *P-V-T Relationships of Organic Compounds*, published by Handbook Publishers, Inc., Sandusky, Ohio, U. S. A. (1952). This compilation can be used also to estimate vapor pressure data for those organic compounds for which there are no published data. This book of tables, which has been in constant use in the laboratories and plants of The Dow Chemical Co., will be of value wherever vapor pressure-temperature data are needed.

For most of the compounds in the following table, values are given for the constants A, B, and C for use in the Antoine equation which is:

$$\text{(Equation 1)} \qquad \log_{10}P = A - B/(C + t)$$

where P is the vapor pressure of the compound in mm of mercury and where t is the temperature in degrees centigrade.

For the other compounds in the following table, values for the constants B and C are given for use in the equation:

$$\text{(Equation 2)} \qquad \log_{10}P = \frac{-52.23\,B}{T} + C$$

where P is the vapor pressure of the compound in mm of mercury and where T is the absolute temperature ($t°\text{C} + 273.1$).

Name	Formula	Range, °C	A	B	C
Acenaphthene	$C_{12}H_{10}$	147 to 288	Equation 2	54.279	8.033
Acetaldehyde	C_2H_4O	−75 to −45	7.3839	1216.8	250
		−45 to +70	6.81089	992.0	230
Acetic acid	$C_2H_4O_2$	0 to 36	7.80307	1651.2	225
		36 to 170	7.18807	1416.7	211
anhydride	$C_4H_6O_3$	100 to 140	Equation 2	45.585	8.688
Acetone	C_3H_6O	....	7.02447	1161.0	224
Acetonitrile	C_2H_3N	....	7.11988	1314.4	230
Acetophenone	C_8H_8O	30 to 100	Equation 2	55.117	9.1352
Acetylene	C_2H_2	−140 to −82	Equation 2	21.914	8.933
Acrylonitrile	C_3H_3N	−20 to +140	7.03855	1232.53	222.47
Aluminum chloride	$AlCl_3$	70 to 190	Equation 2	115	16.24
oxide	Al_2O_3	1840 to 2200	Equation 2	540	14.22
Aminobenzotrifluoride (*m*)	$C_7H_6NF_3$	0 to 96	7.65186	1940.6	218.0
		96 to 300	7.17030	1650.21	193.58
Ammonia†	NH_3	−83 to +60	7.55466	1002.711	247.885
Ammonium bromide	NH_4Br	250 to 400	Equation 2	90.208	9.9404
chloride	NH_4Cl	100 to 400	Equation 2	83.486	10.0164
cyanide	NH_4CN	7 to 17	Equation 2	41.484	9.978
iodide	NH_4I	300 to 400	Equation 2	95.730	10.2700
Amyl-benzene (*n*)	$C_{11}H_{16}$	15 to 104	7.35171	1858.37	212.0
		104 to 270	7.04709	1670.68	195.6
cyclopentane (*n*)	$C_{10}H_{20}$	....	6.929	1526	197
mercaptan (*n*)	$C_5H_{12}S$	0 to 39	7.33940	1581.0	230
		39 to 180	6.93311	1369.479	211.314

† See special table.

Table 10-10 (*Continued*)
VAPOR PRESSURES OF VARIOUS SUBSTANCES

Name	Formula	Range, °C	A	B	C
Amyl-naphthalene (β, n)	$C_{15}H_{18}$	25 to 190	7.4117	2266.6	192.6
		190 to 370	7.0600	2005.9	170
Aniline	C_6H_7N	....	7.24179	1675.3	200
Anisole	C_7H_8O	....	6.98926	1453.8	200
Anthracene	$C_{14}H_{10}$	100 to 160	Equation 2	72	8.91
		223 to 342	Equation 2	59.219	7.910
Anthraquinone	$C_{14}H_8O_2$	224 to 286	Equation 2	110.05	12.305
		285 to 370	Equation 2	63.985	8.002
Antimony	Sb	1070 to 1325	Equation 2	189	9.051
tribromide	$SbBr_3$	235 to 324	Equation 2	55	8.005
trichloride	$SbCl_3$	170 to 253	Equation 2	49.44	8.090
triiodide	SbI_3	330 to 445	Equation 2	64.15	7.831
Argon	Ar	−207.62 to −189.19	Equation 2	7.8145	7.5741
Arsenic	As	440 to 815	Equation 2	133	10.800
		800 to 860	Equation 2	47.1	6.692
trichloride	$AsCl_3$	50 to 100	Equation 2	39.11	7.953
trioxide	As_2O_3	100 to 310	Equation 2	111.35	12.127
		315 to 490	Equation 2	52.12	6.513
Barium	Ba	930 to 1130	Equation 2	350	15.765
Benzene*	C_6H_6	....	6.90565	1211.033	220.790
Benzoic acid	$C_7H_6O_2$	60 to 110	Equation 2	63.82	9.033
Benzonitrile	C_7H_5N	....	6.74631	1436.72	181.0
Benzophenone	$C_{13}H_{10}O$	48 to 202	7.34966	2331.4	195.0
		202 to 330	7.13450	2025.1	170
Benzotrifluoride	$C_7H_5F_3$	−20 to +180	7.00708	1331.30	220.58
Benzoyl chloride	C_7H_5OCl	140 to 200	Equation 2	45.416	7.9245
Benzyl alcohol	C_7H_8O	20 to 113	7.81844	1950.3	194.36
		113 to 300	6.95916	1461.64	153.0
Bismuth	Bi	1210 to 1420	Equation 2	200	8.876
trichloride	$BiCl_3$	91 to 213	Equation 2	13.125	2.681
Boron tribromide	BBr_3	−40 to +90	Equation 2	33.32	7.655
trichloride	BCl_3	....	6.18811	756.89	214.0
trimethyl	$B(CH_3)_3$	−118 to −20	Equation 2	22.171	7.4595
Bromine	Br_2	....	6.83298	1133.0	228.0
Bromo-benzene	C_6H_5Br	....	6.88339	1440.1	204
cumene (o)	$C_9H_{11}Br$	....	6.99354	1666.7	195
cumene (p)	$C_9H_{11}Br$	20 to 113	7.5131	2062.5	228.2
		113 to 262	7.15534	1825.32	208.0
cyclohexane	$C_6H_{11}Br$	0 to 68	7.34139	1778.81	235
		68 to 260	6.97980	1572.19	217.38
diphenyl oxide (p)	$C_{12}H_9OBr$	25 to 190	7.0093	1902.7	153.3
		190 to 400	6.68143	1683.84	132.90
ethylbenzene (o)	C_8H_9Br	20 to 97	7.30701	1831.9	216.62
		97 to 255	6.96150	1621.24	198.0
ethylbenzene (p)	C_8H_9Br	....	6.98209	1632.60	193
naphthalene (α)	$C_{10}H_7Br$	....	7.00350	1927.05	186.0
styrene (o)	C_8H_7Br	....	6.91038	1631.2	195
styrene (p)	C_8H_7Br	....	7.01490	1682.5	195
toluene (p)	C_7H_7Br	10 to 85	7.22838	1743.67	218.0
		85 to 280	7.00762	1612.35	206.36
Butadienes:					
1,2-butadiene	C_4H_6	−60 to +80	7.1619	1121.0	251.00
1,3-butadiene	C_4H_6	−80 to +65	6.85941	935.531	239.554
Butanes:					
n-butane	C_4H_{10}	....	6.83029	945.90	240.00
iso-butane	C_4H_{10}	....	6.74808	882.80	240.00

* At triple point, 5.525°C; 35.856mm.

Table 10-10 (*Continued*)
VAPOR PRESSURES OF VARIOUS SUBSTANCES

Name	Formula	Range, °C	A	B	C
Butenes:					
1-butene	C_4H_8	....	6.84290	926.10	240.00
cis-2-butene	C_4H_8	....	6.86926	960.100	237.00
trans-2-butene	C_4H_8	....	6.86952	960.80	240.00
2-Me-1-propene	C_4H_8	....	6.84134	923.200	240.00
Butyl bromide (*sec*)	C_4H_9Br	−10 to +150	6.82724	1229.08	220
chloride (*n*)	C_4H_9Cl	....	6.75197	1125.8	212
cyclopentane (*n*)	C_9H_{18}	0 to 63	7.2840	1662.5	223.0
		63 to 220	6.9189	1460.0	205.0
mercaptan (*tert*)	$C_4H_{10}S$	−20 to +110	6.78781	1115.565	221.314
naphthalene (*α, n*)	$C_{14}H_{16}$	25 to 170	7.43447	2227.7	202.2
		170 to 345	7.0814	1971.5	180
naphthalene (*β, n*)	$C_{14}H_{16}$	25 to 170	7.43808	2242.2	202.3
		170 to 345	7.0848	1984.3	180
n-valerate (*iso*)	$C_9H_{18}O_2$	90 to 170	Equation 2	44.482	8.143
Butyric acid (*n*)	$C_4H_8O_2$	0 to 82	7.85941	1800.7	200
		82 to 210	7.38423	1542.6	179
acid (*iso*)	$C_4H_8O_2$	0 to 73	7.86161	1775.4	205.0
		73 to 190	7.40246	1529.2	185.0
acetanilide (*n*)	$C_{12}H_{17}ON$	60 to 170	7.6952	2356.3	210.5
		170 to 380	7.32668	2085.31	188.08
alcohol (*n*)	$C_4H_{10}O$	75 to 117.5	Equation 2	46.774	9.1362
alcohol (*tert*)	$C_4H_{10}O$	....	8.13596	1582.4	218.9
aniline (*n*) (*N*)	$C_{10}H_{15}N$	....	7.07888	1748.91	175.0
benzene (*n*)	$C_{10}H_{14}$	....	6.98317	1577.965	201.378
benzene (*iso*)	$C_{10}H_{14}$	....	6.93033	1526.384	204.171
benzene (*sec*)	$C_{10}H_{14}$	0 to 75	7.09350	1621.00	212.352
		75 to 240	6.95097	1540.174	205.101
benzene (*tert*)	$C_{10}H_{14}$	....	6.92050	1504.572	203.328
Cadmium	Cd	150 to 320.9	Equation 2	109	8.564
		500 to 840	Equation 2	99.9	7.897
iodide	CdI_2	385 to 450	Equation 2	122.2	9.269
Calcium	Ca	500 to 700	Equation 2	195	9.697
		960 to 1100	Equation 2	370	16.240
Camphor	$C_{10}H_{16}O$	0 to 180	Equation 2	53.559	8.799
Carbazole	$C_{12}H_9N$	244 to 352	Equation 2	64.715	8.280
Carbon	C	3880 to 4430	Equation 2	540	9.596
dioxide	CO_2	....	9.64177	1284.07	268.432
disulfide	CS_2	−10 to +160	6.85145	1122.50	236.46
monoxide	CO	−210 to −165	6.24020	230.274	260.0
oxysulfide	COS	−80 to −50	Equation 2	19.22	7.383
suboxide	C_3O_2	−100 to +6	Equation 2	25.46	7.640
tetrachloride	CCl_4	....	6.93390	1242.43	230.0
Cesium	Cs	200 to 350	Equation 2	73.4	6.949
bromide	CsBr	978 to 1305	Equation 2	153.6	7.990
chloride	CsCl	986 to 1295	Equation 2	163.2	8.340
fluoride	CsF	1033 to 1255	Equation 2	140.9	7.703
iodide	CsI	1052 to 1280	Equation 2	185.7	9.124
Chlorine	Cl_2	....	6.86773	821.107	240
dioxide	ClO_2	−59 to +11	Equation 2	27.26	7.893
Chloro-aniline (*o*)	C_6H_6NCl	20 to 108	7.56265	1998.6	220.0
		108 to 300	7.19240	1762.74	200.0
aniline (*m*)	C_6H_6NCl	15 to 125	7.55939	2073.75	215
		125 to 310	7.23603	1857.75	196.64
benzene	C_6H_5Cl	0 to 42	7.10690	1500.0	224.0
		42 to 230	6.94504	1413.12	216.0

Table 10-10 (*Continued*)
VAPOR PRESSURES OF VARIOUS SUBSTANCES

Name	Formula	Range, °C	A	B	C
Chloro-benzotrichloride (*o*)	$C_7H_4Cl_4$	30 to 150	7.50430	2228.07	220.0
		150 to 350	7.11794	1951.37	196.27
bromobenzene (*p*)	C_6H_4ClBr	23 to 63	Equation 2	69.755	11.629
cumene (*o*)	$C_9H_{11}Cl$	10 to 90	7.14087	1687.0	205.81
		90 to 270	6.99207	1599.61	198.0
cumene (*p*)	$C_9H_{11}Cl$	....	6.98784	1623.51	197
ethylbenzene (*o*)	C_8H_9Cl	....	6.98169	1556.0	201.0
ethylbenzene (*m*)	C_8H_9Cl	....	6.99082	1577.3	200
ethylbenzene (*p*)	C_8H_9Cl	....	6.98309	1577.0	200
phenol (*o*)	C_6H_5OCl	15 to 80	7.24196	1668.0	210.0
		80 to 200	6.87731	1471.61	193.17
propionitrile (*β*)	C_3H_4NCl	0 to 84	7.32973	1732.55	211.79
		84 to 240	7.20085	1657.25	205.3
styrene (*o*)	C_8H_7Cl	....	6.86644	1541.1	198
styrene (*p*)	C_8H_7Cl	....	6.84248	1545.0	198
toluene (*o*)	C_7H_7Cl	0 to 65	7.36797	1735.8	230.0
		65 to 220	6.94763	1497.2	209.0
Chloroform	$CHCl_3$	−30 to +150	6.90328	1163.03	227.4
Cobalt	Co	2375	Equation 2	309	7.571
Copper	Cu	2100 to 2310	Equation 2	468	12.344
bromide (ous)	Cu_2Br_2	997 to 1351	Equation 2	79.9	5.460
chloride (ous)	Cu_2Cl_2	878 to 1369	Equation 2	80.7	5.454
iodide(ous)	Cu_2I_2	991 to 1154	Equation 2	80.7	5.570
Cresol (*o*)	C_7H_8O	....	6.97943	1479.4	170.0
Cresol (*m*)	C_7H_8O	....	7.62336	1907.24	201.0
Cresol (*p*)	C_7H_8O	....	7.00592	1493.0	160.0
Cyanogen	C_2N_2	−72 to −28	Equation 2	32.437	9.6539
		−32 to −6	Equation 2	23.75	7.808
bromide	CNBr	−17 to +35	Equation 2	47.051	10.328
chloride	CNCl	−5 to +40	Equation 2	27.1	7.840
Cyclo-hexane*	C_6H_{12}	−50 to +200	6.84498	1203.526	222.863
hexene	C_6H_{10}	....	6.88617	1229.973	224.104
octatetraene	C_8H_8	0 to 50	7.30765	1635.0	230
		50 to 210	7.06926	1504.036	218.534
pentane	C_5H_{10}	....	6.88676	1124.162	231.361
pentene	C_5H_8	....	6.92066	1121.818	233.446
Decahydro-naphthalene (*cis*)	$C_{10}H_{18}$	15 to 95	7.41527	1898.13	225.0
		95 to 270	7.04387	1668.10	205.0
naphthalene (*trans*)	$C_{10}H_{18}$	10 to 85	7.25013	1807.04	228.0
		85 to 250	6.88657	1583.28	208
Decane (*n*)	$C_{10}H_{22}$	10 to 80	7.31509	1705.60	212.59
		70 to 260	6.95367	1501.268	194.480
Decene-1	$C_{10}H_{20}$	....	6.96034	1501.872	197.578
Decylcyclopentane (*n*)	$C_{15}H_{30}$	....	6.971	1798	160.4
Dibenzyl ketone	$C_{15}H_{14}O$	285 to 325	Equation 2	62.118	8.257
Dibromo-benzene (*o*)	$C_6H_4Br_2$	20 to 117	7.50128	2093.7	230
		117 to 300	7.10265	1825.77	207.0
propane(1,2)	$C_3H_6Br_2$	0 to 50	7.30398	1644.4	232.0
		50 to 250	6.89105	1419.60	212.0
propane (1,3)	$C_3H_6Br_2$	0 to 71	7.54984	1890.56	240.0
		71 to 275	7.19874	1678.26	222.0
Dichloro-benzene (*o*)	$C_6H_4Cl_2$	....	6.92400	1538.3	200
benzene (*m*)	$C_6H_4Cl_2$	....	6.88045	1496.2	201
benzene (*p*)	$C_6H_4Cl_2$	....	6.89797	1507.3	201

At triple point, 6.67°C; 39.96 mm.

Table 10-10 (*Continued*)
VAPOR PRESSURES OF VARIOUS SUBSTANCES

Name	Formula	Range, °C	A	B	C
Dichloro–benzotrichloride (3,4)	$C_7H_3Cl_5$	20 to 167	7.43954	2190.0	200
		167 to 340	6.98524	1868.905	172.00
benzyl chloride (2,4)	$C_7H_5Cl_3$	20 to 138	7.50457	2125.9	213.8
		138 to 350	7.14735	1881.38	192.93
toluene (3,4)	$C_7H_6Cl_2$	0 to 105	7.34394	1882.5	215.0
		105 to 330	6.97925	1655.44	195.0
p-xylene (2,5)	$C_8H_8Cl_2$	68 to 119	7.83214	2314.5	245.0
		119 to 280	7.45697	2049.3	223.5
Diethyl-amine	$C_4H_{11}N$	−30 to +100	6.83188	1057.2	212.0
aniline (*N*)	$C_{10}H_{15}N$	....	7.25396	1810.51	198.0
benzene(*o*)	$C_{10}H_{14}$	....	6.99016	1577.894	200.554
benzene (*m*)	$C_{10}H_{14}$	....	7.00600	1576.261	201.004
benzene (*p*)	$C_{10}H_{14}$	....	7.00054	1589.273	202.019
disulfide	$C_4H_{10}S_2$	15 to 61	7.34989	1695.00	227.29
		61 to 230	6.97507	1485.970	208.958
ether	$C_4H_{10}O$	....	6.78574	994.195	220.0
ketone	$C_5H_{10}O$	....	6.85791	1216.3	204
sulfide	$C_4H_{10}S$	0 to 150	6.92836	1257.833	218.662
Dimethyl-amine	C_2H_7N	−80 to −30	7.42061	1085.7	233.0
		−30 to +65	7.18553	1008.4	227.353
aniline	$C_8H_{11}N$	....	7.12954	1675.84	201.0
cyclohexane (1,1)	C_8H_{16}	....	6.80225	1323.861	218.053
cyclohexane (1,*cis*-2)	C_8H_{16}	....	6.84164	1369.525	216.040
cyclohexane (1,*cis*-3)	C_8H_{16}	....	6.84293	1340.658	218.281
cyclohexane (1,*cis*-4)	C_8H_{16}	....	6.83699	1347.794	216.360
cyclohexane (1,*trans*-4)	C_8H_{16}	....	6.82180	1332.613	218.791
cyclopentane (1,1)	C_7H_{14}	....	6.81724	1219.474	221.946
cyclopentane (1,*cis*-2)	C_7H_{14}	....	6.85008	1269.140	220.209
cyclopentane (1,*trans*-2)	C_7H_{14}	....	6.84422	1242.748	221.686
cyclopentane (1,*cis*-3)	C_7H_{14}	....	6.83817	1240.023	221.621
cyclopentane (1,*trans*-3)	C_7H_{14}	....	6.83715	1237.456	222.005
disulfide	$C_2H_6S_2$	15 to 180	6.97792	1346.342	218.863
ether	C_2H_6O	....	6.73669	791.184	230.0
3-ethylbenzene (1,2)	$C_{10}H_{14}$	....	7.0488	1646.00	201.00
4-ethylbenzene (1,2)	$C_{10}H_{14}$	....	7.0493	1633.0	202.00
2-ethylbenzene (1,3)	$C_{10}H_{14}$	....	7.0440	1632.00	202.00
4-ethylbenzene (1,3)	$C_{10}H_{14}$	....	7.0427	1629.00	203.00
5-ethylbenzene (1,3)	$C_{10}H_{14}$	....	7.0459	1615.00	204.00
2-ethylbenzene (1,4)	$C_{10}H_{14}$	....	7.0301	1622.00	204.00
formamide	C_3H_7ON	15 to 60	7.3438	1624.7	216.2
		60 to 350	6.99608	1437.84	199.83
hexene (1,*trans*-2)	C_8H_{16}	....	6.83722	1356.100	219.342
hexene (1,*trans*-3)	C_8H_{16}	....	6.83866	1345.859	215.598
naphthalene (1,3)	$C_{12}H_{12}$	20 to 148	7.6347	2295.4	232.6
		148 to 310	7.2698	2076.0	210
naphthalene (1,4)	$C_{12}H_{12}$	20 to 148	7.6347	2345.8	232.6
		148 to 310	7.2698	2076.0	210
naphthalene (1,6)	$C_{12}H_{12}$	20 to 148	7.6347	2345.8	232.6
		148 to 310	7.2698	2076.0	210.0
naphthalene (1,7)	$C_{12}H_{12}$	20 to 148	7.6347	2345.8	232.6
		148 to 310	7.2698	2076.0	210.0
naphthalene (1,8)	$C_{12}H_{12}$	25 to 150	7.40789	2123.2	201.2
		150 to 320	7.0564	1879	180
naphthalene (2,3)	$C_{12}H_{12}$	20 to 155	7.40396	2111.9	201.1
		155 to 315	7.0527	1869	180

Table 10-10 (*Continued*)
VAPOR PRESSURES OF VARIOUS SUBSTANCES

Name	Formula	Range, °C	A	B	C
Dimethyl-naphthalene (2,6)	$C_{12}H_{12}$	20 to 150	7.3968	2080.3	200.8
		150 to 310	7.0460	1841	180
naphthalene (2,7)	$C_{12}H_{12}$	25 to 150	7.39875	2085.9	200.9
		150 to 310	7.0478	1846	180
sulfide	C_2H_6S	−50 to +130	6.93138	1081.587	229.746
thiophene (2,3)	C_6H_8S	−10 to +50	7.26811	1615.8	228.5
		50 to 205	6.9249	1430.0	212
thiophene (2,4)	C_6H_8S	−10 to +50	7.34146	1639.2	228.6
		50 to 205	6.9939	1450.7	212.0
thiophene (2,5)	C_6H_8S	−10 to +47	7.30659	1613.2	229.6
		47 to 200	6.9611	1427.7	213.2
thiophene (3,4)	C_6H_8S	−5 to +54	7.3438	1657.8	228.2
		54 to 205	6.9961	1467.1	211.5
Diphenyl-amine	$C_{12}H_{11}N$	278 to 284	Equation 2	57.35	8.088
methane	$C_{13}H_{12}$	217 to 283	Equation 2	52.36	7.967
oxide	$C_{12}H_{10}O$	25 to 147*	7.4531	2115.2	206.8
		147 to 325	7.09894	1871.92	185.84
Dipropyl-benzene (*iso*) (*o*)	$C_{12}H_{18}$	0 to 100	7.14679	1736.45	203.69
		100 to 260	7.07875	1694.92	200.0
benzene (*iso*) (*m*)	$C_{12}H_{18}$	15 to 100	7.14587	1732.91	203.5
		100 to 260	7.08134	1693.57	200.0
benzene (*iso*) (*p*)	$C_{12}H_{18}$	0 to 105	7.25702	1828.3	208.45
		105 to 270	7.08043	1718.36	198.8
ether (*n*)	$C_6H_{14}O$	8 to 90	Equation 2	34.295	7.821
sulfide (*n*)	$C_6H_{14}S$	0 to 53	7.28307	1598.88	222.17
		53 to 195	6.93897	1414.975	205.846
sulfide (*iso*)	$C_6H_{14}S$	−20 to +33	7.21420	1501.30	228.19
		33 to 180	6.87419	1328.624	212.684
Dodecane (*n*)	$C_{12}H_{26}$	5 to 120	7.35518	1867.55	202.59
		115 to 320	6.98059	1625.928	180.311
Dodecene-1	$C_{12}H_{24}$	....	6.97522	1619.862	182.271
Dodecylcyclopentane (*n*)	$C_{17}H_{34}$	....	6.985	1900	151
Eicosane (*n*)	$C_{20}H_{42}$	25 to 223	8.7603	3113.0	204.07
		223 to 420	7.0225	1948.7	127.8
Eicosene-1	$C_{20}H_{40}$	25 to 222	8.0714	2716.5	189.2
		222 to 335	6.859	1807.9	113.3
Ethane	C_2H_6	....	6.80266	656.40	256.00
Ethyl acetate	$C_4H_8O_2$	−20 to +150	7.09808	1238.71	217.0
alcohol	C_2H_6O	....	8.04494	1554.3	222.65
amine	C_2H_7N	−70 to −20	7.09137	1019.7	225.0
		−20 to +90	7.05413	987.31	220.0
aniline (*N*)	$C_8H_{11}N$	....	7.20621	1751.43	200.0
benzene	C_8H_{10}	....	6.95719	1424.255	213.206
bromide	C_2H_5Br	−50 to +130	6.89285	1083.8	231.7
n-butyrate	$C_6H_{12}O_2$	45 to 121	Equation 2	39.318	8.093
chloride	C_2H_5Cl	−65 to +70	6.80270	949.62	230
cyclohexane	C_8H_{16}	....	6.87041	1384.036	215.128
cyclopentane	C_7H_{14}	....	6.88709	1298.599	220.675
formate	$C_3H_6O_2$	−30 to +235	7.11700	1176.6	223.4
mercaptan	C_2H_6S	−40 to +100	6.95206	1084.531	231.385
naphthalene (*α*)	$C_{12}H_{12}$	....	6.9599	1791.4	180.5
naphthalene (*β*)	$C_{12}H_{12}$	....	7.0819	1886.0	191.0
propionate	$C_5H_{10}O_2$	−20 to +160	7.07293	1298.30	210.7
iso-propyl sulfide	$C_5H_{12}S$	−30 to +23	7.23240	1461.12	230.09
		23 to 180	6.89130	1293.058	215.041

* Supercooled.

Table 10-10 (*Continued*)
VAPOR PRESSURES OF VARIOUS SUBSTANCES

Name	Formula	Range, °C	A	B	C
Ethyl-styrene (*m*)	$C_{10}H_{12}$	....	7.03928	1614.0	198
styrene (*p*)	$C_{10}H_{12}$	....	6.90071	1570.9	198
thiophene-3	C_6H_8S	−10 to +46	7.29798	1606.8	229.6
		46 to 200	6.9530	1422.0	213.2
toluene (*o*)	C_9H_{12}	....	7.00314	1535.374	207.300
toluene (*m*)	C_9H_{12}	....	7.01582	1529.184	208.509
toluene (*p*)	C_9H_{12}	....	6.99802	1527.113	208.921
Ethylene	C_2H_4	....	6.74756	585.00	255.00
bromide	$C_2H_4Br_2$	....	7.06245	1469.70	220.1
chloride	$C_2H_4Cl_2$	....	7.18431	1358.46	232.2
glycol	$C_2H_6O_2$	25 to 112	8.2621	2197.0	212.0
		112 to 340	7.8808	1957.0	193.8
oxide	C_2H_4O	−70 to +100	7.40783	1181.31	250.60
Ethylidene chloride	$C_2H_4Cl_2$	0 to 30	Equation 2	31.706	7.909
Fluorene	$C_{13}H_{10}$	161 to 300	Equation 2	56.615	8.059
Fluorobenzene	C_6H_5F	−40 to +180	6.93667	1736.35	220.0
Formic acid	CH_2O_2	....	6.94459	1295.26	218.0
Furan	C_4H_4O	−35 to +90	6.97523	1060.851	227.740
Germanium tetrachloride	$GeCl_4$	10.4 to 86	Equation 2	38.5	7.340
Gold	Au	2315 to 2500	Equation 2	385	9.853
Helium*	He	....	16.1313	282.126	290
Heptadecane (*n*)	$C_{17}H_{36}$	20 to 190	7.8369	2440.20	194.59
		190 to 320	7.0115	1847.82	145.52
Heptadecene-1	$C_{17}H_{34}$	25 to 186	7.8223	2411.8	193.76
		186 to 310	6.920	1774.6	139.7
Heptanes:					
n-heptane	C_7H_{16}	....	6.90240	1268.115	216.900
2-methylhexane	C_7H_{16}	....	6.87318	1236.026	219.545
3-methylhexane	C_7H_{16}	....	6.86764	1240.196	219.223
3-ethylpentane	C_7H_{16}	....	6.87564	1251.827	219.887
2,2-dimethylpentane	C_7H_{16}	....	6.81480	1190.033	223.303
2,3-dimethylpentane	C_7H_{16}	....	6.85382	1238.017	221.823
2,4-dimethylpentane	C_7H_{16}	....	6.82621	1192.041	221.634
3,3-dimethylpentane	C_7H_{16}	....	6.82667	1228.663	225.316
2,2,3-trimethylbutane	C_7H_{16}	....	6.79230	1200.563	226.050
Heptene-1	C_7H_{14}	....	6.90069	1257.505	219.179
Heptyl-benzene (*n*)	$C_{13}H_{20}$	15 to 138	7.64140	2194.61	218.0
		138 to 295	7.19114	1885.77	192.0
cyclopentane	$C_{12}H_{24}$	....	6.942	1649	182
naphthalene (α) (*n*)	$C_{17}H_{22}$	25 to 215	7.5213	2467.9	194.2
		215 to 405	7.1631	2184	170
naphthalene (β) (*n*)	$C_{17}H_{22}$	25 to 215	7.5249	2474.6	194.3
		215 to 405	7.1665	2190	170
Hexa-chloropropene	C_3Cl_6	20 to 109	7.2664	1863.7	212.95
		109 to 267	6.92329	1649.33	193.87
decane (*n*)	$C_{16}H_{34}$	....	7.03044	1831.317	154.528
decene-1	$C_{16}H_{32}$	25 to 173	7.7237	2298.3	194.8
		173 to 300	6.936	1755.2	148.4
decylcyclopentane (*n*)	$C_{21}H_{42}$	....	7.021	2070	132
fluorohexane (*n*)	C_6F_{14}	−30 to +150	7.12338	1205.37	227.0
Hexanes:					
n-hexane	C_6H_{14}	....	6.87776	1171.530	224.366
2-methylpentane	C_6H_{14}	....	6.83910	1135.410	226.572
3-methylpentane	C_6H_{14}	....	6.84887	1152.368	227.129
2,2-dimethylbutane	C_6H_{14}	....	6.75483	1081.176	229.343

* These values are doubtful.

Table 10-10 (*Continued*)
VAPOR PRESSURES OF VARIOUS SUBSTANCES

Name	Formula	Range, °C	A	B	C
Hexanes (cont.):					
2,3-dimethylbutane	C_6H_{14}	....	6.80983	1127.187	228.900
Hexene-1	C_6H_{12}	....	6.86572	1152.971	225.849
Hexyl-benzene (*n*)	$C_{12}H_{18}$	15 to 122	7.62263	2104.10	220.0
		122 to 290	7.18284	1813.74	195.5
cyclopentane (*n*)	$C_{11}H_{22}$	....	6.934	1589	189
naphthalene (α) (*n*)	$C_{16}H_{20}$	25 to 200	7.4546	2345.8	193.3
		200 to 375	7.1003	2076	170
naphthalene (β) (*n*)	$C_{16}H_{20}$	25 to 200	7.46221	2359.38	193.4
		200 to 380	7.1075	2088	170
Hydrazine	N_2H_4	−10 to +39	8.26230	1881.6	238.0
		39 to 250	7.77306	1620.0	218.0
Hydrogen	H_2	−259.2 to −248	5.92088	71.615	276.337
bromide	HBr	−120 to −87*	8.4622	1112.4	270
		−120 to −60	6.88059	732.68	250
chloride	HCl	−127 to −60	7.06145	710.584	255.0
cyanide	HCN	−85 to −40	7.80196	1425.0	265.0
		−40 to +70	7.29761	1206.79	247.532
fluoride	HF	−55 to +105	8.38036	1952.55	335.52
iodide	HI	−97 to −51	Equation 2	24.16	8.259
		−50 to −34	Equation 2	21.58	7.630
peroxide	H_2O_2	10 to 90	Equation 2	48.53	8.853
selenide	H_2Se	−66 to −26	Equation 2	20.21	7.431
sulfide	H_2S	−110 to −83	Equation 2	20.69	7.880
telluride	H_2Te	−46 to 0	Equation 2	22.76	7.260
Iodine	I_2	....	7.26304	1697.87	204.0
Iodobenzene	C_6H_5I	....	6.89506	1562.87	201.0
Iron	Fe	2220 to 2450	Equation 2	309	7.482
chloride (ous)	$FeCl_2$	700 to 930	Equation 2	135.2	8.33
Krypton	Kr	−188.7 to −169	Equation 2	10.065	7.1770
Lauric acid	$C_{12}H_{24}O_2$	164 to 205	Equation 2	74.386	9.768
Lead	Pb	525 to 1325	Equation 2	188.5	7.827
bromide	$PbBr_2$	735 to 918	Equation 2	118	8.064
chloride	$PbCl_2$	500 to 950	Equation 2	141.9	8.961
fluoride	PbF_2	1078 to 1289	Equation 2	165.1	8.391
Lithium bromide	LiBr	1010 to 1265	Equation 2	152.7	8.068
chloride	LiCl	1045 to 1325	Equation 2	155.9	7.939
fluoride	LiF	1398 to 1666	Equation 2	218.4	8.753
iodide	LiI	940 to 1140	Equation 2	143.6	8.011
Magnesium	Mg	900 to 1070	Equation 2	260	12.993
Maleic anhydride	$C_4H_2O_3$	60 to 160	Equation 2	46.34	7.825
Manganese	Mn	1510 to 1900	Equation 2	267	9.300
Mercury†	Hg	100 to 200	7.46905	2771.898	244.831
		200 to 300	7.7324	3003.68	262.482
		300 to 400	7.69059	2958.841	258.460
		400 to 800	7.7531	3068.195	273.438
bromide (ic)	$HgBr_2$	130 to 270	Equation 2	79.8	10.094
chloride (ic)	$HgCl_2$	60 to 130	Equation 2	85.03	10.888
		130 to 270	Equation 2	78.85	10.094
		275 to 309	Equation 2	61.02	8.409
chloride (ous)	Hg_2Cl_2	....	8.52151	3110.96	168.0
iodide (ic)	HgI_2	100 to 250	Equation 2	82.34	10.057
		266 to 360	Equation 2	62.77	8.115
Methane	CH_4	solid‡	7.69540	532.20	275.00
		liquid	6.61184	389.93	266.00

* Solid.   † See special table.   ‡ At triple point, −182.48°C; 87.7 mm.

Table 10-10 (Continued)
VAPOR PRESSURES OF VARIOUS SUBSTANCES

Name	Formula	Range, °C	A	B	C
Methyl acetate	$C_3H_6O_2$	....	7.20211	1232.83	228.0
alcohol	CH_4O	−20 to +140	7.87863	1473.11	230.0
amine	CH_5N	−93 to −45	6.91831	883.054	223.112
		−45 to +50	6.91205	838.116	214.237
aniline	C_7H_9N	....	7.22584	1728.17	202.0
anthranilate	$C_8H_9O_2N$	20 to 150	7.57558	2196.85	210.0
		150 to 260	7.24299	1963.72	190.35
benzoate	$C_8H_8O_2$	25 to 100	7.4312	1871.5	213.9
		100 to 260	7.07832	1656.25	195.23
n-butyrate	$C_5H_{10}O_2$	....	6.97211	1272.73	208.5
iso-butyrate	$C_5H_{10}O_2$	....	7.02835	1265.00	212.7
chloride	CH_3Cl	−47 to −10	Equation 2	21.988	7.481
cyclohexane	C_7H_{14}	0 to 210	6.82689	1272.864	221.630
cyclopentane	C_6H_{12}	....	6.86283	1186.059	226.042
dichloroarsine	CH_3Cl_2As	−17 to +35	Equation 2	43.686	8.6944
1-ethylcyclopentane (1)	C_8H_{16}	0 to 33	7.19158	1520.9	233.0
		33 to 185	6.87149	1355.287	218.092
2-ethylcyclopentane (1,cis)	C_8H_{16}	−20 to +38	7.28998	1591.91	235.0
		38 to 200	6.90561	1388.307	216.888
2-ethylcyclopentane (1,trans)	C_8H_{16}	−20 to +33	7.2565	1533.4	232.0
		33 to 190	6.8844	1356.0	217.5
3-ethylcyclopentane (1,cis)	C_8H_{16}	−25 to +33	7.1526	1493.5	230.0
		33 to 185	6.8838	1355.0	217.5
3-ethylcyclopentane (1,trans)	C_8H_{16}	−20 to +32	7.1441	1489.4	230.0
		32 to 185	6.8743	1351.0	217.5
ethyl ether	C_3H_8O	0 to 25	Equation 2	26.262	7.769
ethyl ketone	C_4H_8O	....	6.97421	1209.6	216
ethyl sulfide	C_3H_8S	−20 to +130	6.93849	1182.562	224.784
fluoride	CH_3F	−102 to −76	Equation 2	17.053	7.445
formate	$C_2H_4O_2$	....	7.13623	1111.0	229.2
naphthalene (α)	$C_{11}H_{10}$	....	7.06899	1852.674	197.716
naphthalene (β)	$C_{11}H_{10}$	....	7.06850	1840.268	198.395
propionate	$C_4H_8O_2$	−2.5 to +257	7.12841	1257.14	216.4
propyl ether	$C_4H_{10}O$	−0.5 to +40	Equation 2	28.952	7.729
n-propyl sulfide	$C_4H_{10}S$	−40 to +14	7.30058	1451.26	234.43
		>14	6.95545	1284.334	219.662
salicylate	$C_8H_8O_3$	175 to 215	Equation 2	48.67	8.008
styrene (o)	C_9H_{10}	10 to 75	7.27793	1709.5	220.0
		75 to 255	6.88461	1485.41	200.0
styrene (m)	C_9H_{10}	10 to 72	7.27534	1695.4	220.0
		72 to 250	6.87928	1471.44	200.0
styrene (p)	C_9H_{10}	10 to 73	7.27787	1700.2	220.0
		73 to 250	6.88108	1476.10	200.0
styrene (α)	C_9H_{10}	....	6.92366	1486.88	202.4
styrene (β)	C_9H_{10}	....	6.92339	1499.80	201.0
thiophene (2)	C_5H_6S	−23 to +29	7.28307	1498.9	229.6
		25 to 200	6.93897	1326.474	214.309
thiophene (3)	C_5H_6S	−20 to +30	7.33318	1541.1	232.4
		30 to 200	6.98611	1363.862	216.784
Methylene chlorobromide	CH_2ClBr	−10 to +155	6.92776	1165.95	220.0
Molybdenum	Mo	1800 to 2240	Equation 2	680	10.844
Morpholine	C_4H_9ON	0 to 44	7.71813	1745.8	235.0
		44 to 170	7.16030	1447.70	210.0
Myristic acid	$C_{14}H_{28}O_2$	190 to 224	Equation 2	75.783	9.541
Naphthalene	$C_{10}H_8$	....	6.84577	1606.529	187.227

Table 10-10 (*Continued*)
VAPOR PRESSURES OF VARIOUS SUBSTANCES

Name	Formula	Range, °C	A	B	C
Naphthol (α)	$C_{10}H_8O$	....	7.28421	2077.56	184.0
Naphthol (β)	$C_{10}H_8O$	....	7.34714	2135.00	183.0
Neon	Ne	....	7.57352	183.34	285.0
Nickel	Ni	2360	Equation 2	309	7.600
carbonyl	$Ni(CO)_4$	2 to 40	Equation 2	29.8	7.780
Nitric oxide	NO	−200 to −161	Equation 2	16.423	10.048
		−163.7 to −148	Equation 2	13.04	8.440
Nitro-aniline (*o*)	$C_6H_6O_2N_2$	150 to 260	Equation 2	63.881	8.8684
aniline (*m*)	$C_6H_6O_2N_2$	170 to 260	Equation 2	65.88	8.8188
aniline (*p*)	$C_6H_6O_2N_2$	190 to 260	Equation 2	77.345	9.5595
benzene	$C_6H_5O_2N$	112 to 209	Equation 2	48.955	8.192
benzotrifluoride (*m*)	$C_7H_4O_2NF_3$	10 to 105	7.65315	2006.1	220.0
		104 to 280	7.18025	1710.60	195.12
ethylbenzene (*o*)	$C_8H_9O_2N$	15 to 127	7.15618	1774.65	185
		127 to 260	6.76008	1535.70	163.2
methane	CH_3O_2N	47 to 100	Equation 2	36.914	8.033
toluene (*o*)	$C_7H_7O_2N$	50 to 225	Equation 2	48.114	7.9728
toluene (*m*)	$C_7H_7O_2N$	55 to 235	Equation 2	50.128	8.0655
toluene (*p*)	$C_7H_7O_2N$	80 to 240	Equation 2	49.95	7.9815
Nitrogen*	N_2	−210 to −180	6.86606	308.365	273.2
pentoxide	N_2O_5	−30 to +30	Equation 2	57.18	12.647
tetroxide	N_2O_4	−100 to −40	Equation 2	55.16	13.400
		−40 to −10	Equation 2	45.44	11.214
		−8 to +43.2	Equation 2	33.43	8.814
trioxide	N_2O_3	−25 to 0	Equation 2	39.4	10.30
Nitrosyl chloride	NOCl	−61.5 to −5.4	Equation 2	25.5	7.870
Nitrous oxide	N_2O	....	7.00467	661.88	250.0
Nona-decane (*n*)	$C_{19}H_{40}$	20 to 160	8.7262	3041.10	207.30
		160 to 410	7.0192	1916.96	131.66
decene-1	$C_{19}H_{38}$	25 to 211	7.9894	2615.0	190.5
		211 to 330	6.881	1800.3	122.1
Nonanes:					
n-nonane	C_9H_{20}	−10 to +60	7.26430	1607.12	217.54
		60 to 230	6.93513	1428.811	201.619
2-methyloctane	C_9H_{20}	0 to 53	7.0564	1482.7	212.6
		53 to 190	6.9179	1410.0	206.0
3-methyloctane	C_9H_{20}	0 to 53	7.2073	1569.5	220.2
		53 to 170	6.9102	1411.0	206.0
4-methyloctane	C_9H_{20}	0 to 52	7.25812	1588.7	222.3
		52 to 165	6.9155	1406.0	206.0
3-ethylheptane	C_9H_{20}	20 to 190	6.901	1403	206
4-ethylheptane	C_9H_{20}	25 to 190	6.905	1397	206.0
2,2-dimethylheptane	C_9H_{20}	−20 to +43	7.1783	1521.1	223.0
		43 to 180	6.8580	1355.0	208.00
2,3-dimethylheptane	C_9H_{20}	−10 to +50	7.2655	1593.6	225
		50 to 190	6.887	1392.0	207.0
2,4-dimethylheptane	C_9H_{20}	20 to 180	6.869	1360	208
2,5-dimethylheptane	C_9H_{20}	0 to 47	7.238	1560.0	223.9
		47 to 185	6.881	1372.0	207
2,6-dimethylheptane	C_9H_{20}	−10 to +46	7.2559	1567.1	225.0
		46 to 180	6.8725	1366.0	207.0
3,3-dimethylheptane	C_9H_{20}	0 to 47	6.989	1447.5	215.7
		47 to 190	6.869	1385	210.0
3,4-dimethylheptane	C_9H_{20}	−10 to +50	7.2525	1590	225.0
		50 to 190	6.897	1400	208.0

* For nitrogen, J. Res., Bu. Standards, 53, 263 (1954), gives log P (in mm) equal to 6.49594 − 255.821/(T − 6.600); b. p. 77.364°K.

Table 10-10 (*Continued*)
VAPOR PRESSURES OF VARIOUS SUBSTANCES

Name	Formula	Range, °C	A	B	C
Nonanes (cont.):					
3,5-dimethylheptane	C_9H_{20}	0 to 46	7.1962	1541.80	223
		46 to 180	6.878	1375	208.0
4,4-dimethylheptane	C_9H_{20}	−10 to +45	7.174	1539.15	225
		45 to 180	6.858	1373	210.0
2-methyl-3-ethylhexane	C_9H_{20}	0 to 49	6.954	1438.7	213.3
		49 to 190	6.872	1381.0	208.0
2-methyl-4-ethylhexane	C_9H_{20}	20 to 185	6.854	1362	209.0
3-methyl-3-ethylhexane	C_9H_{20}	20 to 190	6.863	1404	212.0
3-methyl-4-ethylhexane	C_9H_{20}	20 to 49	7.1985	1565.93	224.0
		49 to 190	6.885	1399	209.0
2,2,3-trimethylhexane	C_9H_{20}	0 to 43	7.2457	1577.6	230.0
		43 to 185	6.8448	1366.0	211.0
2,2,4-trimethylhexane	C_9H_{20}	25 to 175	6.8391	1344.0	213.00
2,2,5-trimethylhexane	C_9H_{20}	20 to 170	6.83531	1324.049	210.737
2,3,3-trimethylhexane	C_9H_{20}	0 to 46	7.1584	1556.77	228
		46 to 190	6.8474	1391.0	213.00
2,3,4-trimethylhexane	C_9H_{20}	0 to 47	7.170	1566.4	227.1
		47 to 190	6.867	1395	211.0
2,3,5-trimethylhexane	C_9H_{20}	−20 to +41	7.2544	1570.9	230.0
		41 to 180	6.8505	1359.0	211.00
2,4,4-trimethylhexane	C_9H_{20}	−15 to +40	7.27274	1591.6	234.0
		40 to 180	6.85163	1368.723	214.047
3,3,4-trimethylhexane	C_9H_{20}	0 to 49	7.1507	1566.6	227
		49 to 190	6.8557	1401.0	212.00
3,3-diethylpentane	C_9H_{20}	−10 to +52	7.28499	1669.2	235.0
		52 to 205	6.89262	1451.245	215.575
2,2-dimethyl-3-ethylpentane	C_9H_{20}	−10 to +43	7.2047	1564.7	230
		43 to 180	6.8482	1376.0	213.00
2,3-dimethyl-3-ethylpentane	C_9H_{20}	0 to 49	7.160	1580	229
		49 to 200	6.853	1414	214
2,4-dimethyl-3-ethylpentane	C_9H_{20}	0 to 45	7.2018	1577.6	230
		45 to 190	6.8524	1389.0	213.00
2,2,3,3-tetramethylpentane	C_9H_{20}	−10 to +47	7.16257	1577.3	230
		47 to 200	6.82876	1397.483	213.703
2,2,3,4-tetramethylpentane	C_9H_{20}	25 to 180	6.83173	1374.042	214.762
2,2,4,4-tetramethylpentane	C_9H_{20}	20 to 170	6.79710	1325.183	216.093
2,3,3,4-tetramethylpentane	C_9H_{20}	0 to 48	7.27444	1644.37	235.0
		48 to 200	6.85961	1417.473	214.705
Nonene-1	C_9H_{18}	. . . .	6.95387	1435.359	205.535
Nonylcyclopentane (*n*)	$C_{14}H_{28}$	. . . .	6.967	1757	168
Octadecane (*n*)	$C_{18}H_{38}$	20 to 200	7.9117	2542.00	193.4
		200 to 350	7.0156	1883.73	139.46
Octadecene-1	$C_{18}H_{36}$	25 to 199	7.9033	2512.0	191.9
		199 to 320	6.901	1789.4	130.9
Octanes:					
n-octane	C_8H_{18}	−20 to +40	7.37200	1587.81	230.07
		20 to 200	6.92374	1355.126	209.517
2-methylheptane	C_8H_{18}	. . . .	6.91735	1337.468	213.693
3-methylheptane	C_8H_{18}	. . . .	6.89944	1331.530	212.414
4-methylheptane	C_8H_{18}	. . . .	6.90065	1327.661	212.568
3-ethylhexane	C_8H_{18}	. . . .	6.89098	1327.884	212.595
2,2-dimethylhexane	C_8H_{18}	. . . .	6.83715	1273.594	215.072
2,3-dimethylhexane	C_8H_{18}	. . . .	6.87004	1315.503	214.157
2,4-dimethylhexane	C_8H_{18}	. . . .	6.85305	1287.876	214.790
2,5-dimethylhexane	C_8H_{18}	. . . .	6.85984	1287.274	214.412

Table 10-10 (*Continued*)
VAPOR PRESSURES OF VARIOUS SUBSTANCES

Name	Formula	Range, °C	A	B	C
Octanes (cont.):					
3,3-dimethylhexane	C_8H_{18}	....	6.85121	1307.882	217.439
3,4-dimethylhexane	C_8H_{18}	....	6.87986	1330.035	214.863
2-methyl-3-ethylpentane	C_8H_{18}	....	6.86358	1318.120	215.306
3-methyl-3-ethylpentane	C_8H_{18}	....	6.86731	1347.209	219.684
2,2,3-trimethylpentane	C_8H_{18}	....	6.82546	1294.875	218.420
2,2,4-trimethylpentane	C_8H_{18}	....	6.81189	1257.840	220.735
2,3,3-trimethylpentane	C_8H_{18}	....	6.84353	1328.046	220.375
2,3,4-trimethylpentane	C_8H_{18}	....	6.85396	1315.084	217.526
2,2,3,3-tetramethylbutane	C_8H_{18}	solid	7.73092	1601.54	224.76
		liquid	6.87665	1329.93	226.36
2,2,3,3-tetramethylbutane	C_8H_{18}	solid	7.92864	1709.428	233.634
		100.8 to 160	6.87665	1327.8	226.0
Octene-1	C_8H_{16}	....	6.93263	1353.486	212.764
Octylcyclopentane (*n*)	$C_{13}H_{26}$	....	6.957	1704	175
Osmium fluoride	OsF_8	38 to 47.3	Equation 2	29.2	7.650
oxide	OsO_4	−38 to +40.1	Equation 2	56.5	10.7100
Oxalic acid	$C_2H_2O_4$	55 to 105	Equation 2	90.5026	12.2229
Oxygen	O_2	−210 to −160	6.98983	370.757	273.2
Ozone	O_3	....	6.72602	566.95	260.0
Penta-decane (*n*)	$C_{15}H_{32}$	15 to 160	7.6991	2242.42	198.72
		160 to 350	7.0017	1768.82	158.49
decene-1	$C_{15}H_{30}$	20 to 159	7.6282	2185.6	196.2
		159 to 300	6.9503	1730.30	157.02
decylcyclopentane (*n*)	$C_{20}H_{40}$	....	7.013	2029	136
Pentadienes:					
pentadiene (1,2)	C_5H_8	−40 to +120	7.01100	1154.420	234.652
pentadiene (1,*cis*-3)	C_5H_8	−45 to +120	6.94178	1118.371	231.327
pentadiene (1,*trans*-3)	C_5H_8	−45 to +120	6.92257	1108.937	232.338
pentadiene (1,4)	C_5H_8	−60 to +95	6.84880	1025.016	232.354
pentadiene (2,3)	C_5H_8	−40 to +130	6.88603	1086.636	223.040
3-methyl-1,2-butadiene	C_5H_8	−50 to +110	7.005	1130	234
2-methyl-1,3-butadiene	C_5H_8	−50 to +95	6.90334	1080.996	234.668
Pentanes:					
n-pentane	C_5H_{12}	....	6.85221	1064.63	232.000
iso-pentane	C_5H_{12}	....	6.78967	1020.012	233.097
2,2-dimethylpropane	C_5H_{12}	....	6.73812	950.84	237.00
Pentenes:					
pentene-1	C_5H_{10}	....	6.84650	1044.895	233.516
pentene (*cis*-2)	C_5H_{10}	−50 to +105	6.87274	1067.951	230.585
pentene (*trans*-2)	C_5H_{10}	−50 to +105	6.90575	1083.987	232.965
2-methyl-1-butene	C_5H_{10}	−50 to +100	6.87314	1053.780	232.788
3-methyl-1-butene	C_5H_{10}	−60 to +80	6.82618	1013.474	236.816
2-methyl-2-butene	C_5H_{10}	−50 to +110	6.91562	1095.088	232.842
Phenanthrene	$C_{14}H_{10}$	203 to 347	Equation 2	57.247	7.771
Phenetidine (*p*)	$C_8H_{11}ON$	20 to 148	7.87653	2216.5	198.6
		148 to 300	7.16534	1750.62	160.0
Phenetole	$C_8H_{10}O$	....	7.03013	1518.4	196
Phenol	C_6H_6O	....	7.13617	1518.1	175.0
Phosgene	$COCl_2$	−68 to +68	6.84297	941.25	230
Phosphine	PH_3	....	6.70101	643.72	256.0
Phosphonium bromide	PH_4Br	−80 to +40	Equation 2	48.115	10.9561
iodide	PH_4I	10 to 60	Equation 2	51.854	10.9500
Phosphorus (white)	P	20 to 44.1	Equation 2	63.123	9.6511
(violet)	P	380 to 590	Equation 2	108.51	11.0842
pentachloride	PCl_5	....	9.42740	2422.17	208.0

Table 10-10 (*Continued*)
VAPOR PRESSURES OF VARIOUS SUBSTANCES

Name	Formula	Range, °C	A	B	C
Phosphorus (cont.):					
trichloride	PCl_3	0 to 70	Equation 2	31.86	7.681
Phthalic anhydride	$C_8H_4O_3$	160 to 285	Equation 2	54.92	8.022
Platinum	Pt	1425 to 1765	Equation 2	486	7.786
Potassium	K	260 to 760	Equation 2	84.9	7.183
bromide	KBr	906 to 1063	Equation 2	168.1	8.2470
		1095 to 1375	Equation 2	163.8	7.936
chloride	KCl	906 to 1105	Equation 2	174.5	8.3526
		1116 to 1418	Equation 2	169.7	8.130
fluoride	KF	1278 to 1500	Equation 2	207.5	9.000
hydroxide	KOH	1170 to 1327	Equation 2	136	7.330
iodide	KI	843 to 1028	Equation 2	157.6	8.0957
		1063 to 1333	Equation 2	155.7	7.949
Propadiene	C_3H_4	−100 to +40	5.6457	441.0	194.0
Propane	C_3H_8	....	6.82973	813.20	248.00
Propionic acid	$C_3H_6O_2$	0 to 60	7.71558	1690	210
		60 to 185	7.35027	1497.775	194.12
Propionitrile	C_3H_5N	....	6.92886	1285.78	220
Propyl acetate	$C_5H_{10}O_2$	0 to 170	7.06665	1304.10	210.0
alcohol (*n*)	C_3H_8O	....	7.99733	1569.70	209.5
alcohol (*iso*)	C_3H_8O	0 to 113	6.66040	813.055	132.93
benzene (*n*)	C_9H_{12}	....	6.95142	1491.297	207.140
benzene (*iso*)	C_9H_{12}	....	6.93666	1460.793	207.777
bromide (*iso*)	C_3H_7Br	0 to 30	Equation 2	30.76	7.722
chloride (*n*)	C_3H_7Cl	0 to 50	Equation 2	28.894	7.593
chloride (*iso*)	C_3H_7Cl	0 to 30	Equation 2	27.242	7.493
cyclohexane (*n*)	C_9H_{18}	....	6.90610	1472.05	209.0
cyclohexane (*iso*)	C_9H_{18}	....	6.91487	1468.40	209.5
cyclopentane (*n*)	C_8H_{16}	....	6.90392	1384.386	213.159
cyclopentane (*iso*)	C_8H_{16}	....	6.88622	1379.415	217.969
formate	$C_4H_8O_2$	....	7.04006	1235.00	216.1
iodide (*n*)	C_3H_7I	0 to 30	Equation 2	35.334	7.826
iodide (*iso*)	C_3H_7I	0 to 30	Equation 2	32.978	7.629
α-methylstyrene (*p, iso*)	$C_{12}H_{16}$	20 to 120	7.69688	2099.17	218.0
		120 to 130	7.22972	1799.7	193.0
naphthalene (α) (*n*)	$C_{13}H_{14}$	20 to 155	7.41108	2136.5	201.3
		155 to 336	7.0594	1890.8	180
naphthalene (β) (*n*)	$C_{13}H_{14}$	20 to 160	7.41227	2141.9	201.4
		160 to 335	7.0605	1895.5	180
propionate (*n*)	$C_6H_{12}O_2$	45 to 125	Equation 2	39.221	8.0525
styrene (*p, iso*)	$C_{11}H_{14}$	....	7.09845	1683.5	195.0
toluene (*o, n*)	$C_{10}H_{14}$	....	7.0023	1594.00	201.95
toluene (*m, n*)	$C_{10}H_{14}$	....	7.0160	1591.00	202.95
toluene (*p, n*)	$C_{10}H_{14}$	....	6.9926	1589.00	203.15
toluene (*o, iso*)	$C_{10}H_{14}$	....	6.9427	1549.00	203.20
toluene (*m, iso*)	$C_{10}H_{14}$	....	6.9428	1540.00	203.98
toluene (*p, iso*)	$C_{10}H_{14}$	....	6.9260	1538.00	203.10
Propylene	C_3H_6	....	6.81960	785.00	247.00
oxide (1,2)	C_3H_6O	−35 to +130	7.06492	1113.6	232
Propyne	C_3H_4	−73 to −13	Equation 2	23.893	7.877
Quinoline	C_9H_7N	180 to 240	Equation 2	49.72	7.969
Radon	Rn	....	6.6964	717.986	250
Rubidium	Rb	250 to 370	Equation 2	76	6.976
bromide	RbBr	1050 to 1365	Equation 2	165	8.223
chloride	RbCl	1142 to 1395	Equation 2	198.6	9.111

Table 10-10 (*Continued*)
VAPOR PRESSURES OF VARIOUS SUBSTANCES

Name	Formula	Range, °C	A	B	C
Rubidium (cont.):					
fluoride	RbF	1142 to 1400	Equation 2	183.2	8.570
iodide	RbI	1075 to 1325	Equation 2	156.6	8.067
Selenic acid	H_2SeO_4	25 to 56	Equation 2	82.4	14.130
Selenium	Se	....	6.96158	3256.55	110.0
dioxide	SeO_2	....	6.57781	1879.81	179.0
Silica	SiO_2	1860 to 2230	Equation 2	506	13.43
Silicon	Si	1200 to 1320	Equation 2	170	5.950
hexahydride	Si_2H_6	−115 to −14.6	Equation 2	21.7	7.258
octahydride	Si_3H_8	−70 to +52	Equation 2	29.85	7.676
tetrachloride	$SiCl_4$	−70 to +5	Equation 2	30.1	7.644
tetrahydride	SiH_4	−160 to −112	Equation 2	12.69	6.996
Silver	Ag	1650 to 1950	Equation 2	250	8.762
chloride	AgCl	1255 to 1442	Equation 2	185.5	8.179
Sodium	Na	180 to 883	Equation 2	103.3	7.553
bromide	NaBr	1138 to 1394	Equation 2	161.6	7.948
chloride	NaCl	976 to 1155	Equation 2	180.3	8.3297
		1156 to 1430	Equation 2	185.8	8.548
cyanide	NaCN	800 to 1360	Equation 2	155.52	7.472
fluoride	NaF	1562 to 1701	Equation 2	218.2	8.640
hydroxide	NaOH	1010 to 1402	Equation 2	132	7.030
iodide	NaI	1063 to 1307	Equation 2	165.1	8.371
Strontium	Sr	940 to 1140	Equation 2	360	16.056
Styrene	C_8H_8	....	6.92409	1420.0	206
Sulfur	S	....	6.69535	2285.37	155.0
chloride	S_2Cl_2	0 to 138	Equation 2	35.99	7.455
dioxide	SO_2	....	7.32776	1022.80	240.0
trioxide	SO_3	24 to 48	Equation 2	43.45	10.022
Tetra-bromobutane (1,2,3,4)	$C_4H_6Br_4$	to 170	7.39417	2104.4	185.0
		170 to 325	6.98938	1819.00	160.0
chloroethylene	C_2Cl_4	....	7.02003	1415.49	221.0
decane (*n*)	$C_{14}H_{30}$	15 to 145	7.6133	2133.75	200.8
		145 to 340	6.9957	1725.46	165.75
decene-1	$C_{14}H_{28}$	20 to 144	7.5289	2069.7	197.6
		144 to 280	6.9615	1699.76	165.53
decylcyclopentane	$C_{19}H_{38}$	....	7.003	1987	141
methylbenzene (1,2,3,4)	$C_{10}H_{14}$	....	7.0584	1689.10	199.28
methylbenzene (1,2,3,5)	$C_{10}H_{14}$	....	7.0769	1674.00	200.94
methylbenzene (1,2,4,5)	$C_{10}H_{14}$	....	7.0790	1671.00	201.23
Thallium	Tl	950 to 1200	Equation 2	120	6.140
bromide	TlBr	634 to 817	Equation 2	105.4	7.940
chloride	TlCl	665 to 807	Equation 2	105.2	7.974
fluoride	TlF	282 to 298	Equation 2	105	12.52
iodide	TlI	693 to 822	Equation 2	105.4	7.902
Thiophene	C_4H_4S	−10 to 180	6.95926	1246.038	221.354
Tin	Sn	1950 to 2270	Equation 2	328	9.643
chloride (ic)	$SnCl_4$	−52 to −38	Equation 2	46.74	9.824
hydride (ic)	SnH_4	−148 to −49	Equation 2	19.14	7.400
Titanium tetrachloride	$TiCl_4$	−10 to +45	7.09106	1491.4	220.0
		45 to 225	6.69428	1287.91	201.2
Toluene	C_7H_8	....	6.95464	1344.800	219.482
Toluidine (*o*)	C_7H_9N	....	7.12032	1654.7	190.0
Toluidine (*m*)	C_7H_9N	20 to 105	7.3903	1810.8	200.0
		105 to 280	7.03984	1602.51	181.85
Toluidine (*p*)	C_7H_9N	....	7.26022	1758.55	201.0

Table 10-10 (*Continued*)
VAPOR PRESSURES OF VARIOUS SUBSTANCES

Name	Formula	Range, °C	A	B	C
Tribromo-ethane (1,1,2)	$C_2H_3Br_3$	20 to 90	7.33723	1788.6	215
		90 to 300	6.94373	1562.12	195.55
propane (1,2,3)	$C_3H_5Br_3$	15 to 115	7.27269	1893.3	210.0
		115 to 330	7.09534	1779.19	200
Trichloro-benzene (1,2,4)	$C_6H_3Cl_3$	20 to 109	7.5553	2064.4	230.1
		109 to 278	7.19508	1827.00	210.0
ethane (1,1,2)	$C_2H_3Cl_3$	....	6.85189	1262.57	205.17
ethene (1,1,2)	C_2HCl_3	....	7.02808	1315.04	230.0
Tridecane (*n*)	$C_{13}H_{28}$	15 to 132	7.5360	2016.19	203.02
		132 to 330	6.9887	1677.43	172.90
Tridecene-1	$C_{13}H_{26}$	15 to 128	7.4196	1947.0	198.8
		128 to 265	6.9692	1662.68	173.90
Tridecylcyclopentane (*n*)	$C_{18}H_{36}$	....	6.993	1945	146
Triethylamine	$C_6H_{15}N$	0 to 130	6.8264	1161.4	205.0
Trimethyl-amine	C_3H_9N	−90 to −40	7.01174	1014.2	243.1
		−60 to +50	6.81628	937.49	235.35
benzene (1,2,3)	C_9H_{12}	....	7.04082	1593.958	207.078
benzene (1,2,4)	C_9H_{12}	....	7.04383	1573.267	208.564
benzene (1,3,5)	C_9H_{12}	....	7.07436	1569.622	209.578
cyclopentane (1,1,2)	C_8H_{16}	10 to 175	6.82205	1309.618	218.557
cyclopentane (1,1,3)	C_8H_{16}	0 to 165	6.80947	1275.998	219.899
cyclopentane (1,*cis*-2,*cis*-3)	C_8H_{16}	−20 to +34	7.12618	1492.2	230.0
		34 to 190	6.84846	1349.0	217.0
cyclopentane (1,*cis*-2,*trans*-3)	C_8H_{16}	−10 to +29	7.10790	1463.0	230.0
		29 to 180	6.84802	1331.0	218.0
cyclopentane (1,*trans*-2,*cis*-3)	C_8H_{16}	20 to 170	6.82682	1301.0	219.5
cyclopentane (1,*cis*-2-*trans*-4)	C_8H_{16}	−25 to +29	7.09414	1455.4	230.0
		29 to 180	6.85448	1333.894	218.952
cyclopentane (1,*trans*-2,*cis*-4)	C_8H_{16}	0 to 170	6.84970	1306.153	219.808
cyclopentane (1,*cis*-2,*cis*-4)	C_8H_{16}	....	6.842	1335	219
Trinitrotoluene	$C_7H_5O_6N_3$	....	3.8673	1259.406	160
Tungsten	W	2230 to 2770	Equation 2	897	9.920
Undecane (*n*)	$C_{11}H_{24}$	15 to 100	7.3685	1803.90	208.32
		100 to 310	6.97674	1566.65	187.48
Undecene-1	$C_{11}H_{22}$	....	6.96662	1562.469	189.743
Undecylcyclopentane (*n*)	$C_{16}H_{32}$	....	6.974	1854	157
Uranium fluoride	UF_6	0 to 69	Equation 2	41.73	9.521
Urethane	$C_3H_7O_2N$	....	7.42164	1758.21	205.0
Valeric acid (*n*)	$C_5H_{10}O_2$	20 to 102	7.90344	1882.4	190.0
		102 to 250	7.57366	1694.37	175.0
acid (*iso*)	$C_5H_{10}O_2$	....	8.34173	2065.37	208.0
Valeronitrile (*n*)	C_5H_9N	−10 to +50	7.34296	1623.90	225.0
		50 to 200	7.03576	1458.25	210.5
Vinyl chloride	C_2H_3Cl	−100 to +50	6.49712	783.4	230.0
Water†	H_2O	0 to 60	8.10765	1750.286	235.0
		60 to 150	7.96681	1668.21	228.0
Xenon	Xe	....	6.6788	573.480	260
Xylene (*o*)	C_8H_{10}	....	6.99891	1474.679	213.686
Xylene (*m*)	C_8H_{10}	....	7.00908	1462.266	215.105
Xylene (*p*)	C_8H_{10}	....	6.99052	1453.430	215.307
Zinc	Zn	250 to 419.4	Equation 2	133	9.200

† See special table.

BOILING POINTS

Table 10-11
BOILING POINT OF WATER

A. At Various Barometric Pressures

Temp. °C.	0.0° mm of Hg	0.2° mm of Hg	0.4° mm of Hg	0.6° mm of Hg	0.8° mm of Hg
80	355.40	358.28	361.19	364.11	367.06
81	370.03	373.01	376.02	379.05	382.09
82	385.16	388.25	391.36	394.49	397.64
83	400.81	404.00	407.22	410.45	413.71
84	416.99	420.29	423.61	426.95	430.32
85	433.71	437.12	440.55	444.01	447.49
86	450.99	454.51	458.06	461.63	465.22
87	468.84	472.48	476.14	479.83	483.54
88	487.28	491.04	494.82	498.63	502.46
89	506.32	510.20	514.11	518.04	521.99
90	525.97	529.98	534.01	538.07	542.15
91	546.26	550.40	554.56	558.75	562.96
92	567.20	571.47	575.76	580.08	584.43
93	588.80	593.20	597.63	602.09	606.57
94	611.08	615.62	620.19	624.79	629.41
95	634.06	638.74	643.45	648.19	652.96
96	657.75	662.58	667.43	672.32	677.23
97	682.18	687.15	692.15	697.19	702.25
98	707.35	712.47	717.63	722.81	728.03
99	733.28	738.56	743.87	749.22	754.59
100	760.00	765.44	770.91	776.42	781.95

B. At Various Pressures

The table below gives the boiling point of water in degrees Centigrade at various pressures in millimeters of mercury at 0°C.

Pressure mm of Hg	Tenths of a millimeter of Hg									
	0.0	0.1	0.2	0.3	0.4	0.5	0.6	0.7	0.8	0.9
690	97.317°	321	325	329	333	337	341	345	349	353
691	97.357°	361	365	369	373	377	381	385	389	393
692	97.397°	401	405	409	413	417	421	425	429	433
693	97.437°	441	445	449	453	457	461	465	469	473
694	97.477°	480	484	488	492	496	500	504	508	512
695	97.516°	520	524	528	532	536	540	544	548	552
696	97.556°	560	564	568	572	576	580	584	588	592
697	97.596°	599	603	607	611	615	619	623	627	631
698	97.635°	639	643	647	651	655	659	663	667	671
699	97.675°	678	682	686	690	694	698	702	706	710
700	97.714°	718	722	726	730	734	738	742	746	750
701	97.754°	757	761	765	769	773	777	781	785	789
702	97.793°	797	801	805	809	813	816	820	824	828
703	97.832°	836	840	844	848	852	856	860	864	868
704	97.872°	876	879	883	887	891	895	899	903	907
705	97.911°	915	919	923	927	931	935	939	943	947
706	97.950°	954	958	962	966	970	974	977	981	985
707	97.989°	993	996	*000	*004	*008	*012	*016	*020	*024
708	98.028°	032	036	040	043	047	051	055	059	063
709	98.067	071	075	079	082	086	090	094	098	102
710	98.106°	110	114	118	121	125	129	133	137	141
711	98.145°	149	153	157	160	164	168	172	176	180
712	98.184°	188	192	195	199	203	207	211	215	219
713	98.223°	227	230	234	238	242	246	250	254	258
714	98.262°	266	270	274	278	282	286	290	293	297
715	98.301°	304	308	312	316	320	323	327	331	335
716	98.339°	343	347	351	355	358	362	366	370	374
717	98.378°	382	385	389	393	397	401	405	409	413
718	98.417°	420	424	428	432	436	440	443	447	451
719	98.455°	459	463	467	470	474	478	482	486	490

Table 10-11 (*Continued*)
BOILING POINT OF WATER

B. At Various Pressures

Pressure mm of Hg	Tenths of a millimeter of Hg									
	0.0	0.1	0.2	0.3	0.4	0.5	0.6	0.7	0.8	0.9
720	98.494°	497	501	505	509	513	517	520	524	528
721	98.532°	536	540	544	547	551	555	559	563	567
722	98.571°	574	578	582	586	590	593	597	601	605
723	98.609°	613	617	620	624	628	632	636	640	644
724	98.648°	652	655	059	663	667	671	675	678	682
725	98.686°	689	693	697	701	705	709	712	716	720
726	98.724°	728	732	735	739	743	747	751	755	758
727	98.762°	766	770	774	777	781	785	789	793	797
728	98.801°	804	808	812	816	819	823	827	831	835
729	98.839°	843	846	850	854	858	861	865	869	873
730	98.877°	880	884	888	892	896	899	903	907	911
731	98.915°	918	922	926	930	934	937	941	945	949
732	98.953°	956	960	964	968	972	975	979	983	987
733	98.991°	994	998	*002	*006	*010	*013	*017	*021	*025
734	99.029°	032	036	040	044	048	051	055	059	063
735	99.067°	070	074	078	082	085	089	093	097	101
736	99.105°	109	112	116	119	123	127	131	135	138
737	99.142°	146	150	153	157	161	165	169	172	176
738	99.180°	184	187	191	195	199	203	206	210	214
739	99.218°	221	225	229	233	236	240	244	248	252
740	99.255°	259	263	267	270	274	278	282	285	289
741	99.293°	297	300	304	308	312	316	319	323	327
742	99.331°	334	338	342	346	349	353	357	361	364
743	99.368°	372	376	379	383	387	391	394	398	402
744	99.406°	409	413	417	421	424	428	432	436	439
745	99.443°	447	451	454	458	462	466	469	473	477
746	99.481°	484	488	492	495	499	503	507	510	514
747	99.518°	522	525	529	533	537	540	544	548	551
748	99.555°	559	563	566	570	574	578	581	585	589
749	99.593°	596	600	604	607	611	615	619	622	626
750	99.630°	633	637	641	645	648	652	656	659	663
751	99.667°	671	674	678	682	686	689	693	697	700
752	99.704°	708	712	715	719	723	726	730	734	738
753	99.741°	745	749	752	756	760	764	767	771	775
754	99.778°	782	786	790	793	797	801	804	808	812
755	99.815°	819	823	827	830	834	838	841	845	849
756	99.852°	856	860	863	867	871	875	878	882	886
757	99.889°	893	897	900	904	908	911	915	919	923
758	99.926°	930	934	937	941	945	948	952	956	959
759	99.963°	967	970	974	978	982	985	989	993	996
760	100.000°	004	007	011	015	018	022	026	029	033
761	100.037°	040	044	048	052	055	059	063	066	070
762	100.073°	077	081	085	088	092	096	099	103	107
763	100.110°	114	118	121	125	129	132	136	140	143
764	100.147°	151	154	158	162	165	169	173	176	180
765	100.184°	187	191	195	198	202	206	209	213	216
766	100.220°	224	227	231	235	238	242	246	249	253
767	100.257°	260	264	268	271	275	279	283	286	290
768	100.293°	297	300	304	308	311	315	319	322	326
769	100.330°	333	337	341	344	348	352	355	359	363
770	100.366°	370	373	377	381	384	388	392	395	399
771	100.402°	406	410	414	417	421	424	428	432	435
772	100.439°	442	446	450	453	457	461	464	468	472
773	100.475°	479	483	486	490	493	497	501	504	508
774	100.511°	515	519	522	526	530	533	537	540	544
775	100.547°	551	555	559	562	566	569	573	577	580
776	100.584°	588	591	595	598	602	606	609	613	616
777	100.620°	624	627	631	634	638	642	645	649	653
778	100.656°	660	663	667	671	674	678	681	685	689
779	100.692°	696	699	703	707	710	714	718	721	725

Table 10-11 (*Continued*)
BOILING POINT OF WATER

B. At Various Pressures

Pressure	Tenths of a millimeter of Hg									
mm of Hg	0.0	0.1	0.2	0.3	0.4	0.5	0.6	0.7	0.8	0.9
780	100.728°	732	735	739	743	746	750	753	757	761
781	100.764°	768	772	775	779	782	786	789	793	797
782	100.800°	804	807	811	815	818	822	825	829	833
783	100.836°	840	843	847	851	854	858	861	865	869
784	100.872°	876	879	883	886	890	894	897	901	904
785	100.908°	912	915	919	922	926	929	933	937	940
786	100.944°	947	951	954	958	962	965	969	972	976
787	100.980°	984	988	991	995	998	*002	*006	*009	*013
788	101.016°	020	023	027	030	034	037	040	044	047
789	101.051°	054	058	062	065	069	072	076	079	083
790	101.087°	090	094	097	101	104	108	112	115	119
791	101.122°	126	129	133	136	140	144	147	151	154
792	101.158°	161	165	168	172	176	179	183	186	190
793	101.194°	197	201	204	208	211	215	218	222	225
794	101.229°	232	236	239	243	246	250	254	257	261
795	101.265°	269	272	276	279	283	287	290	294	297
796	101.300°	303	307	310	314	317	321	324	328	332
797	101.335°	339	342	346	349	353	356	360	363	367
798	101.371°	375	378	382	385	388	392	395	399	402
799	101.406°	409	413	416	420	423	427	430	434	437
800	101.441°									

Pressure, atm.	Boiling Point, °C.	Pressure, atm.	Boiling Point, °C.	Pressure, atm.	Boiling Point, °C.	Pressure, atm.	Boiling Point, °C.
0.5	80.9	7	164.2	14	194.1	21	213.9
1	100.0	8	169.6	15	197.4	22	216.2
2	119.6	9	174.5	16	200.4	23	218.5
3	132.9	10	179.0	17	203.4	24	220.8
4	142.9	11	183.2	18	206.1	25	222.9
5	151.1	12	187.1	19	208.8	26	225.0
6	158.1	13	190.7	20	211.4	27	227.0

Table 10-12
CORRECTION OF BOILING POINT TEMPERATURE AT VARIOUS PRESSURES TO NORMAL PRESSURE (760 mm. Hg)

To correct for small differences in barometric pressure the following formula is employed:

$$T_c = T_o + C(760 - P_{mm})*$$

where T_c = corrected boiling point.
T_o = observed boiling point.
P = atmospheric pressure in millimeters of mercury.
C = a constant having the value of 0.037 at $25 - 40°C$; 0.043 at $41 - 75°C$; 0.044 at $76 - 100°C$; 0.046 at $101 - 120°C$; 0.048 at $121 - 140°C$; 0.051 at $141 - 155°C$; 0.055 at $156 - 220°C$; 0.057 at $221 - 300°C$; 0.064 at $310 - 325°C$.

A. S. T. M. METHOD

In the *A. S. T. M. Standards on Petroleum Products and Lubricants*, September 1937, page 96, the Sydney Young equation is used for calculating the proper correction to the observed boiling point. In this equation the temperature range as well as the pressure is considered. The Sydney Young equation is as follows:

For Centigrade scale: $C_c = 0.00012 (760 - P) (273 + t_c)$,
For Fahrenheit scale: $C_f = 0.00012 (760 - P) (460 + t_f)$

in which C_c and C_f are, respectively, corrections to be made on the observed temperatures t_c or t_f, and P is the actual barometric pressure in millimeters of mercury.

The table which follows is a convenient approximation of the corrections as calculated from the equations above. These approximations will suffice for ordinary work. For more precise work, the correction should be calculated using the Sydney Young equation and the temperature when 50% of the material has boiled. A new correction should be made for each variation of more than 10°C. or 18°F.

As pointed out by Hoyt, *Jour. Chem. Ed.* **11**, *405 (1934)*, the Sydney Young equation is applicable only to non-polar substances with low dielectric constants such as the hydrocarbons for which it was proposed by the *A. S. T. M.* For other types of compounds a more exact value can be obtained by use of the following equations and constants:

For Centigrade scale: $C_c = K (760 - P) (273 + t_c)$,
For Fahrenheit scale: $C_f = K (760 - P) (460 + t_f)$,

where the value of the constant K is for:

Hydrocarbons	0.000125	Esters	0.000121
Halogen derivatives	0.000125	Ketones	0.000121
Ethers	0.000125	Amines	0.000118
Aldehydes	0.000125	Alcohols	0.000100

The variation in the case of acids is too wide to permit of a single value for K.

* This formula is applicable only to liquids of particular types such as a hydrocarbon, an ether, an aldehyde, a ketone, an ester, or a halogen derivative.

Table 10-12 *(Continued)*
CORRECTION OF BOILING POINT TEMPERATURES

Boiling Point		Correction for Each One mm. Difference in Pressure*	
t_c in °C.	t_f in °F.	C_c in °C.	C_f in °F.
10–30	50–86	0.035	0.063
30–50	86–122	.038	.068
50–70	122–158	.040	.072
70–90	158–194	.042	.076
90–110	194–230	.045	.081
110–130	230–266	.047	.085
130–150	266–302	.050	.089
150–170	302–338	.052	.094
170–190	338–374	.054	.098
190–210	374–410	.057	.102
210–230	410–446	.059	.106
230–250	446–482	.062	.111
250–270	482–518	.064	.115
270–290	518–554	.066	.119
290–310	554–590	.069	.124
310–330	590–626	.071	.128
330–350	626–662	.074	.132
350–370	662–698	.076	.137
370–390	698–734	.078	.141
390–410	734–770	.081	.145

* To be added in case barometric pressure is below 760 mm., to be subtracted in case the pressure is above 760 mm.

DREISBACH METHOD

A formula (2) has been developed from the Antoine equation (1) which gives the rate of change of boiling point with pressure:

$$(1) \quad \log p = A - [B \div (t^0 C + 230)]$$
$$(2) \quad dt/dp = B \div [2.3026p(A - \log p)^2]$$

This equation will give an accurate value for dt/dp at any pressure but cannot be used to correct boiling points at various barometric readings to boiling points at 760 mm pressure since the dt/dp value varies with pressure. For equations to correct for these variations and for the numerical values of the constants A and B for innumerable compounds belonging to 23 classes of organic compounds (hydrocarbons, alcohols, aldehydes, ketones, acids, esters, amines, etc.) see Dreisbach's *P-V-T RELATIONSHIPS OF ORGANIC COMPOUNDS*, published by Handbook Publishers, Inc., Sandusky, Ohio, U. S. A. (1952).

The use of this book of tables is simple and does not involve complicated mathematical operations. All of the computations have been done by the author. Its use involves merely the reading of temperature values in the body of the table corresponding to pressure values, ranging from 760 mm to 0.1 mm, which are shown across the top of the page. Values for latent heats of vaporization, vapor and liquid densities, index of refraction and flash points are also given.

This book of tables, which has been in constant use in the laboratories and plants of The Dow Chemical Co., will be of value wherever such data are required for not only compounds for which data are recorded in the literature but also for those compounds for which experimental values have not as yet been determined.

Table 10-13
CALCULATION OF BOILING POINTS OF ORGANIC COMPOUNDS

Cf. Kinney, *Jour. Am. Chem. Soc. 60, 3032 (1938); Ind. Eng. Chem. 32, 559 (1940).*

Boiling points of organic compounds may be calculated from their structures as follows: Obtain the molecular *B.P.N.* by adding together the appropriate boiling point numbers (*b.p.n.*) for the various atoms and structural groupings given in Table A. From the *B.P.N.* obtain the B.P. (boiling point in degrees Celsius) using Table B.

Certain rules must be followed in applying the data:

1. The *b.p.n.'s* of 0.8 and 1.0 are applied to the carbon and hydrogen atoms, respectively, of the longest aliphatic chain in the molecule. It is essential that the longest aliphatic chain be used as the base because the boiling point is dependent, in part, upon this factor.
2. All groups or atoms attached to this chain are assigned *b.p.n.'s* as indicated in Table A.
3. With many atoms and groups the position of the group, or the presence of other groups, alters the *b.p.n.* These, as far as are known, are given in Table A.

Example: To calculate the boiling point of 1-methyl-4-(1-methylethyl)-1-cyclopentene, $(CH_3)_2$: $CH \cdot CH \cdot CH_2 \cdot C(CH_3){:}CH \cdot CH_2$.

Carbon in the longest aliphatic chain (2 × 0.8)	1.6
Hydrogen in the longest aliphatic chain (4 × 1.0)	4.0
Methyl radical attached to the aliphatic chain	3.05
Carbon in the cyclopentene ring (5 × 0.8)	4.0
Hydrogen in the cyclopentene ring (6 × 1.0)	6.0
B.P.N. for the 5-membered ring	2.5
B.P.N. for the ethylenic linkage, type $R_2C{:}CHR$	2.3
Methyl attached to the cyclopentene ring	3.05
Calculated *B.P.N.*, sum of the items above	26.50
Calculated B.P., from Table B	142.5°C.
Observed B.P.	143.1°C.

Example: To calculate the boiling point of 2,8-dimethyl-5-nonanone, $[(CH_3)_2CH \cdot CH_2 \cdot CH_2]_2CO$.

Carbon in the longest aliphatic chain (9 × 0.8)	7.2
Hydrogen in the longest aliphatic chain (16 × 1.0)	16.0
Two methyl radicals (2 × 3.05)	6.1
One carbonyl oxygen, type RCH_2COCH_2R	7.5
Calculated *B.P.N.*, sum of the items above	36.8
Calculated B.P., from Table B	222.5°C.
Observed B.P.	226.0°C.

A. Atomic and Group Boiling Point Numbers

Carbon, in the main chain	0.8		Type of olefinic linkage	
Hydrogen, attached to the main chain	1.0		$CH_2{=}CH_2$	1.2
			$RCH{=}CH_2$	1.5
			$RCH{=}CHR$	1.9
Radicals, saturated, attached to the main chain or to cyclic rings			$R_2C{=}CHR$	2.3
			$R_2C{=}CR_2$	2.8
Methyl	3.05			
Ethyl	5.5		Radicals, unsaturated, attached to main chain	
Propyl	7.0			
Butyl	9.7		Methylene	4.4
2,2-Dimethyl grouping	−0.4		Ethylidene	7.0
			Vinyl	5.4
Two or three alkyls attached to adjacent carbons of saturated main chains of 6 carbons or less	+0.5		Propylidene	9.0
			Butylidene	10.4
			Type of acetylenic linkage	
			$HC{\equiv}CH$	4.0
Four or more alkyls attached to adjacent carbons of saturated main chains of 6 carbons or less	+1.0		$RC{\equiv}CH$	4.4
			$RC{\equiv}CCH_3$	5.4
			$RC{\equiv}CR$	4.8

Table 10-13 (*Continued*)
CALCULATION OF BOILING POINTS OF ORGANIC COMPOUNDS

A. Atomic and Group Boiling Point Numbers

Type of diolefin		$R_2CHOCHR_2$, R_3COCHR_2	1.1
Allenes	4.8	R_3COCR_3	(0.2?)
Conjugated, normal values of double bonds plus	0.8	**Aldehyde** $=O$	
Not conjugated, normal values of double bonds only		RCH_2CHO	8.2
		R_2CHCHO	7.6
		R_3CCHO	7.0
Type of triolefin		**Ketone** $=O$	
All bonds conjugated, normal values of double bonds plus	2.4	RCH_2COCH_3	8.0
Two bonds conjugated, normal values of double bonds plus	0.8	$R_2CHCOCH_3$, RCH_2COCH_2R	7.5
No conjugation, normal values of double bonds only		R_3CCOCH_3, $R_2CHCOCH_2R$	7.0
		R_3CCOCH_2R, $R_2CHCOCHR_2$	6.5
		R_3CCOCH_2R	(6.0?)
		R_3CCOCR_3	(5.5?)
Type of diacetylene		**Ester** $-OO-$	
1,3-Diacetylenes, normal values of triple bonds only		RCH_2COOCH_3, CH_3COOCH_2R	8.5
All other conjugated, normal values of triple bonds plus	3.0	$R_2CHCOOCH_3$, RCH_2COOCH_2R, $HCOOCHR_2$, $CH_3COOCHR_2$	7.6
No conjugation, normal values of triple bonds only		$R_3CCOOCH_3$, $R_2CHCOOCH_2R$, $RCH_2COOCHR_2$, $HCOOCR_3$, CH_3COOCR_3	6.7
Type of enyne		$R_3CCOOCH_2R$, $R_2CHCOOCHR_2$, RCH_2COOCR_3	5.8
Conjugated, normal values of bonds plus	0.8	$R_3CCOOCHR_2$, $R_2CHCOOCR_3$	4.9
No conjugation, normal values of bonds only		$R_3CCOOCR_3$	4.0
Type of dienyne		**Acid** $-COOH$	
Conjugated, normal values of bonds plus	2.4	RCH_2COOH	19.3
No conjugation, normal values of bonds only		$R_2CHCOOH$	18.6
		R_3CCOOH	17.9
Cyclic radicals: Add 0.8 for each carbon, 1.0 for each hydrogen, the normal values of any unsaturated linkages, and the following values for the ring:		**Amine, primary** $-NH_2$	
		RCH_2NH_2	7.3
		R_2CHNH_2	6.2
		R_3CNH_2	5.1
Cyclo–propyl, -butyl, -pentyl, -hexyl, respectively 2.1, 2.3, 2.5, 2.7		**Amine, secondary** $-NH-$	
Cyclo–heptyl, -octyl, etc., 3.4, 3.9, etc. (add 0.5 for each additional CH_2 in the ring)		RCH_2NHCH_3	5.0
		$R_2CHNHCH_3$, RCH_2NHCH_2R	4.0
		R_3CNHCH_3, $R_2CHNHCH_2R$	3.5
		R_3CNHCH_2R, $R_2CHNHCHR_2$	3.0
Alcohol $-OH$		**Amine, tertiary** $=N-$	
RCH_2OH	10.8	$RCH_2N(CH_3)_2$	2.0
R_2CHOH	8.8	$R_2CHN(CH_3)_2$, $(RCH_2)_2NCH_3$	1.5
R_3COH	6.8	$R_3CN(CH_3)_2$, $(R_2CH)_2NCH_3$, $(RCH_2)_3N$	1.25
Ether $-O-$		**Cyanide** $-CN$	
RCH_2OCH_3, R_2CHOCH_3, R_3COCH_3	2.9	RCH_2CN	14.0
		R_2CHCN	12.8
RCH_2OCH_2R, R_2CHOCH_2R, R_3COCH_2R	2.0	R_3CCN	11.6
		Isocyanide $-NC$	
		RCH_2NC	12.2
		R_2CHNC	11.1
		R_3CNC	10.0

Table 10-13 (*Continued*)
CALCULATION OF BOILING POINTS OF ORGANIC COMPOUNDS

B. Molecular Boiling Point Numbers and Their Boiling Points

Calculated from the formula: $B.P. = 230.14 \sqrt[3]{B.P.N.} - 543$

B. P. N.	B. P., °C.	Av. Increase per 0.1 Unit, °C.	B. P. N.	B. P., °C.	Av. Increase per 0.1 Unit, °C
5	−149.5	2.50	39	237.5	0.66
6	−124.5	2.18	40	244.1	0.65
7	−102.7	2.00	41	250.6	0.64
8	−82.7	1.84	42	257.0	0.63
9	−64.3	1.71	43	263.3	0.62
10	−47.2	1.60	44	269.5	0.61
11	−31.2	1.51	45	275.6	0.60
12	−16.1	1.43	46	281.6	0.59
13	−1.8	1.35	47	287.5	0.59
14	+11.7	1.29	48	293.4	0.58
15	24.6	1.23	49	299.2	0.57
16	36.9	1.19	50	304.9	0.56
17	48.8	1.14	51	310.5	0.55
18	60.2	1.10	52	316.0	0.54
19	71.1	1.06	53	321.5	0.55
20	81.7	1.03	54	326.9	0.53
21	92.0	0.99	55	332.2	0.53
22	101.9	0.96	56	337.5	0.52
23	111.5	0.94	57	342.7	0.52
24	120.9	0.90	58	347.9	0.50
25	129.9	0.89	59	352.9	0.51
26	138.8	0.86	60	358.0	0.50
27	147.4	0.85	61	363.0	0.49
28	155.9	0.82	62	367.9	0.49
29	164.1	0.80	63	372.8	0.48
30	172.1	0.79	64	377.6	0.47
31	180.0	0.76	65	382.3	0.47
32	187.6	0.76	66	387.0	0.48
33	195.2	0.74	67	391.8	0.45
34	202.6	0.73	68	396.3	0.47
35	209.9	0.70	69	401.0	0.45
36	216.9	0.70	70	405.5	0.45
37	223.9	0.68	71	410.0	
38	230.7	0.68			

Table 10-14
ORGANIC SOLVENTS

In the table below the compounds are arranged in order of increasing boiling points.
For methods of testing the purity, the purification, and the drying of many of the solvents listed below, the reader is referred to the monograph *ORGANIC SOLVENTS*, by Weissberger & Proskauer, Oxford University Press, London & New York, 1935.

Name	B. P. 760 mm °C	Name	B. P. 760 mm °C
Ethyl chloride	13	Trichloroethylene	87.2
Ethylene oxide	14	iso-Propyl acetate	88.4
Furan	31–2	iso-Butyl bromide	91.5
Methyl formate	32	2,5-Dimethyl-furan	93–4
Diethyl ether	34.6	Ethyl chloroformate	94–5
Propylene oxide	35	Allyl alcohol	96.6
n-Pentane	36.1	1,2-Dichloropropane	96.8
Ethyl bromide	38.4	n-Propyl alcohol	97.8
Methylene chloride	40	n-Heptane	98.4
Methylal	42.3	Ethyl propionate	99.1
Carbon disulfide	46.3	sec-Butyl alcohol	99.5
Ethyl formate	54	iso-Amyl chloride	99.7
Acetone	56.5	Ligroin (high boiling)	100–150
Methyl acetate	57.1	Formic acid	100.8
Ethylidene dichloride	57.3	Methyl-cyclohexane	100.9
Acetylene dichloride*	59–61	Dioxane (1,4)	101.1
Ligroin (low boiling)	60–120	Nitromethane	101.5
Chloroform	61.2	n-Propyl acetate	101.6
Methyl alcohol	64.7	Diethyl ketone	101.7
Tetrahydrofuran	65–6	tert-Amyl alcohol	102
Di-iso-propyl ether	68.5–69.0	Acetal	102.2
n-Hexane	68.7	n-Butyl formate	106.9
iso-Butyl chloride	68.9	iso-Butyl alcohol	107–8
Trichloroethane (1,1,1)	74.1	Acetylene dibromide*	110
Dioxolane	75–6	Toluene	110.6
Carbon tetrachloride	76.8	sec-Butyl acetate	112–3
Ethyl acetate	77.1	Trichloroethane (1,1,2)	113.5
n-Butyl chloride	77.9	Nitroethane	114.8
Ethyl alcohol	78.4	Pyridine	115–6
Methyl-ethyl ketone	79.6	Pentan-3-ol	115.6
2-Methyl-tetrahydrofuran	80.0	Epichlorohydrin	117
Benzene	80.1	n-Butyl alcohol	117
Cyclohexane	80.7	iso-Butyl acetate	118
n-Propyl formate	81.3	Methyl-iso-butyl ketone	118
Acetonitrile	82	Acetic acid	118.1
iso-Propyl alcohol	82.5	Propylene glycol monomethyl ether	119
tert-Butyl alcohol	82.9	Ethyl n-butyrate	120–1
Cyclohexene	83.3	2-Nitropropane	120.3
Ethylene chloride	83.7	iso-Amyl bromide	120.4
Thiophene	84	Tetrachloroethylene	120.8

* Mixture of isomers.

Table 10-14 (*Continued*)
ORGANIC SOLVENTS

Name	B. P. 760 mm °C	Name	B. P. 760 mm °C
Di-*iso*-propyl ketone	123.7	Bromobenzene	156.2
Ethylene glycol diethyl ether		Monoethyl-glycol acetate	156.5
(Diethyl Cellosolve)	124	Hexan-1-ol	157.2
Ethylene glycol monomethyl		Trichloropropane (1,2,3)	158
ether (Methyl Cellosolve)	124–5	Ethylene glycol mono-*iso*-butyl	
n-Octane	125.7	ether	158.8
Diethyl carbonate	126		
		Cyclohexanol	160–1
n-Butyl acetate	126	*iso*-Amyl propionate	160.2
sec-Propylene chlorohydrin	127	Heptan-2-ol	160.4
Ethylene chlorohydrin	128.8	Furfural	161.7
Mesityl oxide	130.1	Pentachloroethane	162
Ethylene bromide	131.5		
		Diacetone alcohol	167.9
1-Nitropropane	131.6	Di-*iso*-butyl ketone	168.1
2-Methyl-pentan-4-ol	131.8	Methyl acetoacetate†	169–70
iso-Butyl carbinol	132.0	Furfuryl alcohol	170
Chlorobenzene	132.1	Methyl *o*-tolyl ether	171–2
Xylene*	*ca* 133		
		Ethylene glycol mono-*n*-butyl-	
Cyclohexylamine	134	ether	171.2
Ethylene glycol monoethyl		Phenetole	172
ether (Cellosolve)	135.1	Di-*iso*-amyl ether	173.4
Ethylbenzene	136.2	*n*-Decane	174.0
n-Amyl alcohol (*prim.*)	138	Glycol diformate	174
Acetic anhydride	139.6		
		Cyclohexyl acetate	174–5
Di-*iso*-propyl carbinol	140	2,6-Dimethylheptan-4-ol	174–5
Acetylacetone	140.5	α,α'-Dichlorohydrin	174.3
iso-Amyl acetate	142	Furfuryl acetate	175–7
Di-*n*-butyl ether	142.4	Methyl *p*-tolyl ether	176
Ethylene glycol mono-*iso*-			
propyl ether	144	Eucalyptole	176–7
		Methyl *m*-tolyl ether	177
		p-Cymene	177.1
Monomethylglycol acetate		Dichloroethyl ether (*sym*)	178.5
(Methyl-cellosolve Acetate)	144.5	*iso*-Amyl *n*-butyrate	178.6
Acetylene tetrachloride	146.3		
2-Methylpentan-1-ol	148	*o*-Dichlorobenzene	179
3-Methylol-pentane	148.9	Octan-2-ol	179–80
n-Amyl acetate	149	Ethyl acetoacetate	180
		Ethylene glycol mono-*iso*-amyl	
Ethyl-*n*-butyl ketone	149–50	ether	181
Bromoform	150.5	Phenol	181.4
n-Nonane	150.8		
Methyl-*n*-amyl ketone	151.5	2-Ethylhexan-2-ol	184
iso-Propylbenzene	152.4	Aniline	184.4
		Diethyl oxalate	186
Anisole	154–5	Diethylene glycol diethyl ether	188
Ethyl lactate	155	α-Propylene glycol	188–9
Heptan-4-ol	156		
Cyclohexanone	155–6	Ethyl benzyl ether	190
Heptan-4-ol	156	Glycol diacetate	190.5
		Benzonitrile	190.7

† Slight decompn.

Table 10-14 (*Continued*)
ORGANIC SOLVENTS

Name	B. P. 760 mm °C	Name	B. P. 760 mm °C
Decalin*	191.7	n-Propyl benzoate	231
Dimethylaniline	193	Tributyl borate	231
		Decan-1-ol	232.9
iso-Amyl iso-valerate	194	Benzyl cyanide	233–4
Octan-1-ol	194–5	Quinoline	238
Acetonyl acetone	194.1		
Diethylene glycol monomethyl ether	194.2	Ethylene glycol monophenyl ether (Phenyl Cellosolve)	244.7
Diethylene glycol monoethyl ether	195	Diethylene glycol	244.8
		n-Dibutyl oxalate	245.5
Glycol	197.4	Diethylene glycol monobutyl ether acetate	246.4
Methyl benzoate	198–9	Triethylene glycol monoethyl ether	248
Diethyl malonate	198.9		
o-Toluidine	199.7		
"Carbitol"	200.1	n-Butyl benzoate	250
		Diethylene glycol mono-n-hexyl ether	252
p-Toluidine	200.3		
Acetophenone	203	Diethylene glycol di-n-butyl ether (Dibutyl Carbitol)	255
Ethylene glycol dibutyl ether	203	Triacetin	258–9
Methyl-phenyl carbinol	203.9	α-Chloronaphthalene	259.3
Benzyl alcohol	204.7		
Tetralin	206–7	iso-Amyl benzoate	262
Butan-1,3-diol	206.5	α-Monobutyrin	269–71
γ-Valerolactone	207–8	Ethyl cinnamate	271
Ethylene glycol monomethyl ether acetal	207.2	o-Nitroanisole	272–3
Camphor	209.1	Tetraethylene glycol dimethyl ether	275.8
Diethylene glycol monomethyl ether acetate	209.1	iso-Amyl salicylate	277–8
o-Chloroaniline	210.5	α-Bromonaphthalene	281.1
Nitrobenzene	210.9	Dimethyl phthalate	282
Ethyl benzoate	211–2	Glycerol	290
iso-Phorone	215–6	Triethylene glycol	290
Diethylene glycol monoethyl ether acetate	217.7	Diethyl phthalate	298–9
Naphthalene	217.9	Benzyl benzoate	323–4
Acetamide	222	Tetraethylene glycol dibutyl ether	330
Methyl salicylate	222.2	n-Dibutyl phthalate	340
Diethyl maleate	225		

* Mixture of isomers.

AZEOTROPIC MIXTURES

These tables give the data for various binary and ternary azeotropic or constant boiling mixtures.

For a very complete listing of azeotropes see Horsley, *Ind. Eng. Chem., Analytical Edition, 19, 508 (1947).*

For a thorough discussion of Azeotropy, see Lecat, *Traite de Chimie Organique,* V. Grignard, Mason et Cie, Paris, 1935, Tome I, pp. 121 to 267; or Young, *Distillation Principles and Processes,* Macmillan, London, 1922.

The boiling points listed are in °C. at 760 mm.

These data are taken from Lecat, *"L'Azeotropisme",* Brussels, 1918; *Ann. soc. sci. Bruxelles 45B, 169–76, 284–94 (1926); 47B, i, 21–7, 63–71, 108–14, 149–58 (1927); 48B, i, 13–22, 54–62, 113–26; ii, 1–18 (1928); 49B, ii, 17–47, 109–43 (1929); 50B, 21–33 (1930); 55B, 43–7, 253–65 (1935); 56B, 41–54, 221–34 (1936); Rec. trav. chim. Pays-Bas 45, 620–7 (1926); 46, 240–7 (1927); 47, 13–18 (1928),* unless otherwise designated as follows:

(1) Young and Fortey, *Trans. Chem. Soc., 81, 717, 739, 752 (1902); 83, 45 (1903).*
(2) Roscoe, *Quart. Jour. Chem. Soc., 13, 146 (1861); 15, 270 (1862).*
(3) Bogin, *U. S. P. Re. 17, 157.*
(4) Young, *"Distillation Principles and Processes",* Macmillan, London, 1922.
(5) Miller and Bliss, *Ind. Eng. Chem., 32, 123 (1940).*
(6) Hannotte, *Bull. Soc. Chim. Belg., 35, 85–109 (1926).*
(7) Wuyts, *ibid., 33, 167–192 (1924).*
(8) Beduwe, *ibid., 34, 41–55 (1925).*
(9) Atkins, *Trans. Chem. Soc., 117, 218–20 (1920).*
(10) Booklet by Shell Chemical Co., "Organic Chemicals Manufactured by Shell Chemical Company", 1939.
(11) Azeotropic Data, Advances in Chemistry Series (6), American Chemical Society (1952).

Table 10-15
BINARY AZEOTROPES CONTAINING WATER

	B.P. 760 mm.		% By weight	
	Other component	Azeotrope	Water	Other component
A—*Alcohols*				
Ethyl alcohol (1)	78.4	78.1	4.5	95.5
n-Propyl alcohol (1)	97.2	87.7	28.3	71.7
iso-Propyl alcohol (1)	82.5	80.4	12.1	87.9
n-Butyl alcohol	117.8	92.4	38	62
iso-Butyl alcohol	108.0	90.0	33.2	66.8
sec-Butyl alcohol	99.5	88.5	32.1	67.9
tert-Butyl alcohol	82.8	79.9	11.7	88.3
n-Amyl alcohol (Pentanol-1)	137.8	96.0	54.0	46.0
Prim-*iso*-amyl alcohol (Methyl-2-butanol-4)	131.4	95.2	49.6	50.4
tert-Amyl alcohol (Methyl-2-butanol-2)	102.3	87.4	27.5	72.5
sec-Amyl alcohol (Pentanol-3)	115.4	91.7	36.0	64.0
Act.-*sec*-amyl alcohol (Pentanol-2)	119.3	92.5	38.5	61.5
n-Hexyl alcohol	157.9	97.8	75	25
n-Heptyl alcohol	176.2	98.7	83	17
n-Octyl alcohol	195.2	99.4	90	10

Table 10-15 (*Continued*)
BINARY AZEOTROPES CONTAINING WATER

	B.P. 760 mm.		% By weight	
	Other component	Azeotrope	Water	Other component
Allyl alcohol	97.0	88.2	27.1	72.9
Benzyl alcohol	205.2	99.9	91	9
Furfuryl alcohol	169.4	98.5	80	20
B—Hydrocarbons				
Benzene	80.2	69.3	8.9	91.1
Toluene	110.8	84.1	19.6	80.4
C—Substituted Hydrocarbons				
Ethylene chloride	83.7	72	8.3	91.7
Propylene chloride	96.8	78	12	88
D—Ethers				
Diethyl ether	34.5	34.2	1.3	98.7
Di-*iso*-prepyl ether	68.4	62.2	4.5	95.5
Ethyl *n*-propyl ether	63.6	59.5	4	96
Di-*iso*-butyl ether	122.2	88.6	23	77
Di-*iso*-amyl ether	172.6	97.4	54	46
Diphenyl ether	259.3	99.3	96.8	3.2
Phenetole	170.4	97.3	59	41
Anisole	153.9	95.5	40.5	59.5
Resorcinol diethyl ether	235.0	99.7	91	9
E—Esters				
n-Propyl formate	80.9	71.9	3.6	96.4
n-Butyl formate	106.8	83.8	15	85
iso-Butyl formate	98.4	80.4	7.8	92.2
n-Amyl formate	132.0	91.6	28.4	71.6
iso-Amyl formate	123.9	89.7	23.5	76.5
Benzyl formate	202.3	99.2	80	20
Ethyl acetate	77.1	70.4	6.1	93.9
n-Propyl acetate	101.6	82.4	14	86
iso-Propyl acetate	91.0	77.4	6.2	93.8
n-Butyl acetate	126.2	90.2	28.7	71.3
iso-Butyl acetate	117.2	87.5	19.5	80.5
n-Amyl acetate	148.8	95.2	41	59
iso-Amyl acetate	142.1	93.8	36.2	63.8
Benzyl acetate	214.9	99.6	87.5	12.5
Phenyl acetate	195.7	98.9	75.1	24.9
Methyl propionate	79.9	71.4	3.9	96.1
Ethyl propionate	99.2	81.2	10	90
n-Propyl propionate	122.1	88.9	23	77
iso-Butyl propionate	136.9	92.8	32.2	67.8
iso-Amyl propionate	160.3	96.6	48.5	51.5
Methyl butyrate	102.7	82.7	11.5	88.5
Ethyl butyrate	120.1	87.9	21.5	78.5
n-Propyl butyrate	142.8	94.1	36.4	63.6
n-Butyl butyrate	165.7	97.2	53	47
iso-Butyl butyrate	156.8	96.3	46	54
iso-Amyl butyrate	178.5	98.1	63.5	36.5
Methyl *iso*-butyrate	92.3	77.7	6.8	93.2
Ethyl *iso*-butyrate	110.1	85.2	15.2	84.8
n-Propyl *iso*-butyrate	133.9	92.2	30.8	69.2

Table 10-15 (*Continued*)
BINARY AZEOTROPES CONTAINING WATER

	B.P. 760 mm.		% By weight	
	Other component	Azeotrope	Water	Other component
iso-Butyl iso-butyrate	147.3	95.5	39.4	60.6
iso-Amyl iso-butyrate	168.9	97.4	56.0	44.0
Methyl iso-valerate	116.3	87.2	19.2	80.8
Ethyl iso-valerate	134.7	92.2	30.2	69.8
n-Propyl iso-valerate	155.8	96.2	45.2	54.8
iso-Butyl iso-valerate	168.7	97.4	55.8	44.2
iso-Amyl iso-valerate	193.5	98.8	74.1	25.9
Ethyl caproate	166.8	97.2	54	46
Methyl cinnamate	261.9	99.9	95.5	4.5
Methyl benzoate	199.5	99.1	79.2	20.8
Ethyl benzoate	212.4	99.4	84.0	16.0
n-Propyl benzoate	230.9	99.7	90.9	9.1
n-Butyl benzoate	249.8	99.9	94	6
iso-Butyl benzoate	242.2	99.8	92.6	7.4
iso-Amyl benzoate	262.3	99.9	95.6	4.4
Ethyl phenylacetate	228.8	99.7	91.3	8.7
Ethyl nitrate	87.7	74.4	22	78
n-Propyl nitrate	110.5	84.8	20	80
iso-Butyl nitrate	122.9	89.0	25	75
F—Organic Acids				
Formic acid (Max.)	100.8	107.3	22.5	77.5
Acetic acid	118.1	No	Azeotrope	
Propionic acid	141.1	99.98	82.3	17.7
Butyric acid	163.5	99.4	81.6	18.4
iso-Butyric acid	154.5	99.3	79	21
G—Inorganic Acids (2)				
Nitric acid (Max.)	86.0	120.5	32	68
Perchloric acid (Max.)	110.0	203	28.4	71.6
Hydrofluoric acid (Max.)	19.4	120	63	37
Hydrochloric acid (Max.)	−84	110	79.76	20.24
Hydrobromic acid (Max.)	−73	126	52.5	47.5
Hydriodic acid (Max.)	−34	127	43	57
H—Ketones				
Methyl ethyl ketone	79.6	73.5	11	89
Methyl n-propyl ketone (10)	102.0	83.3	19.5	80.5
Methyl iso-butyl ketone (10)	115.9	87.9	24.3	75.7
Mesityl oxide (10)	129.5	91.8	34.8	65.2
Diacetone alcohol (10)	166	98.8	87.3	12.7
I—Aldehydes				
Furfural	161.5	97.5	65	35
Butyraldehyde (3)	75.7	68	6	94
J—Amines				
Pyridine	115.5	92.6	43	57

Table 10-16
BINARY AZEOTROPES CONTAINING ALCOHOLS

	B.P. 760 mm.		% By weight	
	Other component	Azeotrope	Alcohol	Other component
A—*Methyl Alcohol* (B.P. 64.7°)				
Methylal	42.3	41.9	8.2	91.8
Dimethyl sulfide	37.3	34.0	15	85
Methyl borate	68.7	54.6	32	68
Dimethyl carbonate	90.4	62.7	70	30
Methyl acetate	57.0	53.8	18.7	81.3
Methyl propionate	79.8	62.5	47.5	52.5
Ethyl formate	54.1	51.0	16	84
Ethyl acetate	77.1	62.3	44	56
iso-Propyl acetate	91.0	64.5	80	20
n-Pentane	36.2	30.8	9	91
n-Hexane	68.9	50.6	28	72
n-Heptane	98.5	59.1	51.5	48.5
Benzene	80.2	58.3	39.6	60.4
Toluene	110.8	63.8	69	31
Cyclohexane	80.8	54.2	37.2	62.8
Nitromethane	101.2	64.6	91	9
Methyl iodide	42.6	38.0	6.5	93.5
Chloroform	61.1	53.5	12.6	87.4
Carbon tetrachloride (4)	76.8	55.7	20.6	79.4
Ethyl bromide	38.4	35.0	4.5	95.5
Ethylene chloride	83.7	61.0	32	68
n-Propyl chloride	46.6	40.5	9.5	90.5
n-Propyl bromide	71.0	54.5	21	79
iso-Propyl chloride	36.3	33.4	6	94
iso-Propyl bromide	59.8	48.6	15.0	85.0
iso-Propyl iodide	89.4	61.0	38	62
n-Butyl chloride	78.1	57.0	27	73
iso-Butyl bromide	91.0	61.3	41.7	58.3
Di-*n*-propyl ether	90.4	63.8	72	28
Acetone	56.5	55.7	12.1	87.9
B—*Ethyl Alcohol* (B.P. 78.3°)				
Methyl acetate	57.0	56.9	3	97
Methyl propionate	79.7	72.0	33	67
Ethyl nitrate	87.7	71.9	44	56
Ethyl acetate	77.1	71.8	30.8	69.2
Ethyl propionate	99.2	78.0	75	25
n-Propyl formate	80.8	71.8	38	62
n-Propyl acetate	101.6	78.2	85	15
iso-Propyl acetate	91.0	76.8	57	43
Benzene	80.2	68.2	32.4	67.6
Toluene	110.8	76.7	68	32
n-Pentane	36.2	34.3	5	95
n-Hexane	68.9	58.7	21	79
n-Heptane	98.5	70.9	49	51
n-Octane	125.6	77.0	78	22
Methyl iodide	42.6	41.2	3.2	96.8
Ethylene chloride	83.7	70.5	37	63
Carbon tetrachloride	76.8	65.1	15.8	84.2
Allyl chloride	45.7	44	5	95

Table 10-16 (*Continued*)
BINARY AZEOTROPES CONTAINING ALCOHOLS

	B.P. 760 mm.		% By weight	
	Other component	Azeotrope	Alcohol	Other component
Chloroform	61.1	59.4	7	93
n-Propyl chloride	46.7	45.0	6	94
n-Propyl bromide	71.0	62.8	20.5	79.5
n-Propyl iodide	102.4	75.4	44	56
iso-Propyl chloride	36.3	35.6	2.8	97.2
iso-Propyl bromide	59.8	55.6	10.5	89.5
iso-Propyl iodide	89.4	71.5	27	73
n-Butyl chloride	78.1	65.7	20.3	79.7
n-Butyl bromide	100.3	75.0	43	57
iso-Butyl bromide	91.0	72.5	31	69
Acetal	103.6	78.0	76	24
Di-n-propyl ether	90.4	74.5	44	56
Methyl ethyl ketone	79.6	74.8	40	60
C—iso-Propyl Alcohol (B.P. 82.5°)				
Methyl propionate	79.8	76.4	37	63
Ethyl acetate	77.1	75.3	25	75
iso-Propyl acetate	91.0	81.3	60	40
Benzene	80.2	71.9	33.3	66.7
Toluene	110.8	81.3	79	21
n-Pentane	36.2	35.5	6	94
n-Hexane	68.9	62.7	23	77
n-Heptane	98.5	76.3	54	46
Carbon tetrachloride	76.8	69.0	18	82
Chloroform	61.1	60.8	4.2	95.8
Allyl bromide	70.8	66.5	20	80
Ethyl iodide	72.3	67.1	15	85
Ethylene chloride	83.7	74.7	43.5	56.5
iso-Propyl bromide	59.8	57.8	12	88
iso-Propyl iodide	89.4	76.0	32	68
n-Propyl chloride	46.7	46.4	2.8	97.2
n-Propyl bromide	71.0	66.8	20.5	79.5
n-Propyl iodide	102.4	79.8	42	58
n-Butyl chloride	78.1	70.8	23	77
Acetal	103.6	81.3	63	37
Ethyl n-propyl ether	63.6	62.0	10	90
Di-iso-propyl ether (5)	82.3	66.2	14.1	85.9
Methyl ethyl ketone	79.0	77.5	32	68
D—n-Propyl Alcohol (B.P. 97.2°)				
Methyl butyrate	102.7	94.4	49	51
Ethyl propionate	99.2	93.4	48	52
n-Propyl formate	80.8	80.65	3	97
n-Propyl acetate	101.6	94.7	51	49
Benzene	80.2	77.1	16.9	83.1
Toluene	110.8	92.4	52.5	47.5
n-Hexane	68.9	65.7	4	96
Carbon tetrachloride	76.8	73.1	11.5	88.5
Chlorobenzene	132.0	96.9	83	17
Ethylene chloride	83.7	80.7	19	81
n-Propyl bromide	71.0	69.7	9	91
n-Butyl chloride	78.1	74.8	18	82
Acetal	103.6	92.4	37	63
Di-n-propyl ether	90.4	85.7	30	70

Table 10-16 (*Continued*)
BINARY AZEOTROPES CONTAINING ALCOHOLS

	B.P. 760 mm.		% By weight	
	Other component	Azeotrope	Alcohol	Other component
E—*iso-Butyl Alcohol* (B.P. 107.9°)				
Methyl butyrate	102.7	101.3	25	75
iso-Butyl formate	97.9	97.4	12	88
iso-Butyl acetate	117.5	107.6	92	8
Benzene	80.2	79.8	9.3	90.7
Toluene	110.8	100.9	44.5	55.5
m-Xylene	139.0	107.7	87	13
Cyclohexane	80.8	78.1	14	86
n-Hexane	68.9	68.3	2.5	97.5
Chlorobenzene	132.0	107.1	63	37
Ethylene chloride	83.7	83.5	6.5	93.5
n-Butyl chloride	78.1	77.7	4	96
n-Butyl bromide	100.3	95.0	21	79
iso-Butyl bromide	91.0	88.8	12	88
iso-Butyl iodide	120.4	104.0	36	64
iso-Amyl chloride	99.8	94.5	22	78
Acetal	103.6	98.2	20	80
F—*n-Butyl Alcohol* (B.P. 117.8°)				
Ethyl butyrate	120.0	115.7	64	36
n-Butyl formate	106.6	105.8	23.7	76.3
n-Butyl acetate	126.2	117.2	47	53
Toluene	110.8	105.7	27	73
m-Xylene	139.0	116.0	80	20
Cyclohexane	80.8	79.8	4	96
n-Heptane	98.5	94.4	18	82
Carbon tetrachloride	76.8	76.6	2.5	97.5
Chlorobenzene	132.0	115.3	56	44
iso-Butyl bromide	91.0	90.2	7	93
iso-Butyl iodide	120.4	110.5	30	70
Acetal	103.6	101	13	87
G—*iso-Amyl Alcohol* (B.P. 131.4°)				
iso-Amyl formate	123.8	123.7	10	90
iso-Amyl acetate	142.1	131.3	98.5	1.5
Toluene	110.8	110.0	14	86
m-Xylene	139.0	127.0	53.3	46.7
Chlorobenzene	132.0	124.3	35	65
iso-Amyl bromide	120.3	116.8	21	79
iso-Amyl iodide	147.7	129.2	54	46
Paraldehyde	124.0	122.9	22	78
H—*n-Amyl Alcohol* (B.P. 137.8°)				
n-Amyl formate	132.0	130.4	43	57
Di-*n*-butyl ether	142.1	134.0	52	48
I—*Cyclohexyl Alcohol* (B.P. 160.7°)				
o-Xylene	143.6	143.0	14	86
iso-Amyl iodide	147.5	147.0	10	90
Furfural	161.4	155.6	45	55
Di-*iso*-amyl ether	172.6	158.8	78	22
Phenetole	170.4	159.2	72	28

Table 10-16 (*Continued*)
BINARY AZEOTROPES CONTAINING ALCOHOLS

	B.P. 760 mm.		% By weight	
	Other component	Azeotrope	Alcohol	Other component
J—*Allyl Alcohol* (B.P. 97.0°)				
Methyl butyrate	102.7	93.8	55	45
n-Propyl acetate	101.6	94.2	53	47
Benzene	80.2	76.8	17.4	82.6
Toluene	110.8	92.4	50	50
Cyclohexane	80.8	74	20	80
Allyl iodide	102.0	89.4	28	72
Chlorobenzene	132.0	96.2	85	15
Ethylene chloride	83.7	79.9	18	82
K—*Benzyl Alcohol* (B.P. 205.2°)				
Nitrobenzene	210.8	204.0	58	42
Iodobenzene	188.6	187.8	12	88
o-Bromotoluene	181.4	181.25	7	93
Naphthalene	218.1	204.1	60	40
m-Cresol (Max.)	202.2	207.1	61	39
Dimethyl aniline	194.1	193.9	6.5	93.5
L—*Ethylene Glycol* (B.P. 197.4°)				
Ethyl benzoate	212.6	186.1	46.5	53.5
iso-Amyl acetate	142.1	141.95	3	97
Diphenyl	254.9	192.0	64	36
Mesitylene	164.6	156.0	13	87
Naphthalene	218.1	183.9	51	49
Toluene	110.8	110.2	6.5	93.5
m-Xylene	139.0	135.6	15	85
Ethylene bromide	131.7	129.8	4	96
Nitrobenzene	210.9	185.9	59	41
Chlorobenzene	132.0	130.1	94.4	5.6
Benzyl chloride	179.3	167.0	30	70
Benzyl alcohol	205.1	193.1	56	44
Di-n-butyl ether	142.1	140.0	10	90
Diphenyl ether	259.3	193.1	60	40
Anisole	153.9	150.5	10.5	89.5
β-Phenylethyl alcohol	219.4	194.4	69	31
Acetophenone	202.1	185.7	52	48
Aniline	184.4	180.6	24	76
Dimethyl aniline	194.1	175.9	33.5	66.5
o-Cresol	191.1	189.6	27	73
M—*Glycerol* (B.P. 291.0°)				
n-Propyl benzoate	230.9	228.8	8	92
n-Butyl benzoate	249.8	243.0	17	83
Diphenyl	254.9	243.8	55	45
Naphthalene	218.1	215.2	10	90
p-Dibromobenzene	220.3	217.1	10	90
Diphenyl ether	257.7	246.3	22	78

Table 10-17
BINARY AZEOTROPES CONTAINING ORGANIC ACIDS

| | B.P. 760 mm. | | % By weight | |
	Other component	Azeotrope	Acid	Other component
A—Formic Acid (B.P. 100.8°)				
Benzene	80.2	71.7	31	69
Toluene	110.8	85.8	50	50
m-Xylene	139.0	94.2	70.2	29.8
n-Pentane	36.2	34.2	10	90
n-Hexane	68.9	60.6	28	72
n-Heptane	98.5	78.2	43.5	56.5
n-Octane	125.8	90.5	63	37
Chloroform	61.2	59.2	15	85
Carbon tetrachloride	76.8	66.7	18.5	81.5
Carbon disulfide	46.3	42.6	17	83
Methyl iodide	42.6	42.1	6	94
Ethyl bromide	38.4	38.2	3	97
Ethylene chloride	83.6	77.4	14	86
Ethylene bromide	131.7	94.7	51.5	48.5
n-Propyl chloride	46.7	45.6	8	92
n-Propyl bromide	71.0	64.7	27	73
iso-Propyl chloride	34.8	34.7	1.5	98.5
iso-Propyl bromide	59.4	56.0	14	86
iso-Butyl chloride	68.9	63.0	19	81
iso-Amyl chloride	99.8	80.0	33.5	66.5
Chlorobenzene	132.0	95.0	55	45
Diethyl ketone (Max.)	102.2	105.4	33	67
Methyl n-propyl ketone (Max.)	102.3	105.3	32	68
B—Acetic Acid (B.P. 118.5°)				
Benzene	80.2	80.05	2	98
Toluene	110.8	105.0	34	66
m-Xylene	139.0	115.4	72.5	27.5
n-Heptane	98.5	92.3	30	70
n-Octane	125.8	109.0	50	50
Carbon tetrachloride	76.8	76.6	3	97
Ethylene bromide	131.7	114.4	55	45
iso-Propyl iodide	89.2	88.3	9	91
n-Butyl bromide	100.4	97.6	18	82
iso-Butyl bromide	91.3	90.2	12	88
iso-Amyl chloride	99.8	97.2	18.5	81.5
Chlorobenzene	132.0	114.7	58.5	41.5
C—Propionic Acid (B.P. 140.9°)				
m-Xylene	139.0	132.7	35.5	64.5
Ethylene bromide	131.7	127.8	17.5	82.5
iso-Butyl iodide	120.4	119.5	9	91
iso-Amyl bromide	120.3	119.2	10	90
Chlorobenzene	132.0	128.9	18	82
Anisole	153.9	140.8	96	4
D—Butyric Acid (B.P. 162.5°)				
m-Xylene	139.0	138.3	6	94
Ethylene bromide	131.7	131.1	3.5	96.5
Chlorobenzene	132.0	131.8	2.8	97.2

Table 10-17 (*Continued*)
BINARY AZEOTROPES CONTAINING ORGANIC ACIDS

	B.P. 760 mm.		% By weight	
	Other component	Azeotrope	Acid	Other component
Benzyl chloride	179.3	160.8	65	35
Anisole	153.9	152.9	12	88
Furfural	161.5	159.4	42.5	57.5
E—Iso-Butyric Acid (B.P. 154.4°)				
m-Xylene	139.0	136.8	14	86
Ethylene bromide	131.7	130.5	6.5	93.5
Chlorobenzene	132.0	131.2	8	92
Benzyl chloride	179.3	153.5	80	20
Anisole	153.9	148.5	42	58
F—iso-Valeric Acid (B.P. 176.5°)				
Mesitylene	164.6	162.8	20	80
Benzyl chloride	179.3	171.2	36	64
Benzaldehyde	179.2	174.5	68	32
iso-Amyl butyrate	178.5	176.1	70	30
Phenetole	170.5	168.5	20	80
G—Caproic Acid (B.P. 205.2°)				
Naphthalene	218.1	202.0	70	30
Benzyl chloride	179.3	179.0	3	97
Nitrobenzene	210.8	202.0	70	30
H—Caprylic Acid (B.P. 237.5°)				
Naphthalene	218.1	216.2	6	94
p-Dibromobenzene	220.3	218.8	10	90
I—Chloroacetic Acid (B.P. 189.4°)				
Mesitylene	164.6	162	17	83
Naphthalene	218.1	187.1	78	22
Benzyl chloride	179.3	173.8	25	75
p-Dichlorobenzene	174.1	167.6	24.5	75.5
o-Cresol	191.1	187.5	54	46
J—Phenylacetic Acid (B.P. 266.5°)				
α-Chloronaphthalene	262.7	255.9	30	70
Diphenyl	255.9	252.2	23.3	76.7
iso-Amyl benzoate	262.0	259.9	26	74
Methyl cinnamate	261.9	261.8	3	97
Diphenyl ether	259.3	255.4	27.8	72.2
K—Benzoic Acid (B.P. 250.5°)				
Diphenyl	255.9	245.9	50.5	49.5
Naphthalene	218.1	217.7	5	95
p-Dibromobenzene	220.3	219.5	3.8	96.2
p-Nitrotoluene	239.0	237.4	11	89
Ethyl salicylate	234.0	233.85	6	94
iso-Butyl benzoate	241.9	241.2	12	88
Diphenyl ether	259.3	247.0	59	41

Table 10-18
TERNARY AZEOTROPES CONTAINING WATER AND ALCOHOLS

	B.P. 760 mm.		% By weight		
	Other component	Azeotrope	Water	Alcohol	Other component
A—Ethyl Alcohol (B.P. 78.3°)					
Ethyl acetate (6)	77.1	70.3	7.8	9.0	83.2
Diethyl formal (7)	87.5	73.2	12.1	18.4	69.5
Diethyl acetal (8)	103.6	77.8	11.4	27.6	61.0
Cyclohexane	80.8	62.1	7	17	76
Benzene (4)	80.2	64.9	7.4	18.5	74.1
Chloroform	61.2	55.5	3.5	4.0	92.5
Carbon tetrachloride	76.8	61.8	4.3	9.7	86.0
Ethyl iodide	72.3	61	5	9	86
Ethylene chloride	83.7	66.7	5	17	78
B—n-Propyl Alcohol (B.P. 97.2°)					
n-Propyl formate (6)	80.9	70.8	13	5	82
n-Propyl acetate (6)	101.6	82.2	21.0	19.5	59.5
Di-n-propyl formal (7)	137.4	86.4	8.0	44.8	47.2
Di-n-propyl acetal (8)	147.7	87.6	27.4	51.6	21.0
Di-n-propyl ether (7)	91.0	74.8	11.7	20.2	68.1
Cyclohexane	80.8	66.6	8.5	10.0	81.5
Benzene (4)	80.2	68.5	8.6	9.0	82.4
Carbon tetrachloride	76.8	65.4	5	11	84
Diethyl ketone	102.2	81.2	20	20	60
C—iso-Propyl Alcohol (B.P. 82.5°)					
Cyclohexane	80.8	64.3	7.5	18.5	74.0
Benzene (4)	80.2	66.5	7.5	18.7	73.8
D—n-Butyl Alcohol (B.P. 117.8°)					
n-Butyl formate (6)	106.6	83.6	21.3	10.0	68.7
n-Butyl acetate (6)	126.2	89.4	37.3	27.4	35.3
Di-n-butyl ether (7)	141.9	91	29.3	42.9	27.7
E—iso-Butyl Alcohol (B.P. 108.0°)					
iso-Butyl formate (6)	94.4	80.2	17.3	6.7	76.0
iso-Butyl acetate	117.2	86.8	30.4	23.1	46.5
F—tert-Butyl Alcohol (B.P. 82.6°)					
Benzene (4)	80.2	67.3	8.1	21.4	70.5
Carbon tetrachloride (9)	76.8	64.7	3.1	11.9	85.0
G—n-Amyl Alcohol (B.P. 137.8°)					
n-Amyl formate (6)	131.0	91.4	37.6	21.2	41.2
n-Amyl acetate (6)	148.8	94.8	56.2	33.3	10.5
H—iso-Amyl Alcohol (B.P. 131.4°)					
iso-Amyl formate (6)	124.2	89.8	32.4	19.6	48.0
iso-Amyl acetate (6)	142.0	93.6	44.8	31.2	24.0
I—Allyl Alcohol (B.P. 97.0°)					
n-Hexane	69.0	59.7	5	5	90
Cyclohexane	80.8	66.2	8	11	81
Benzene	80.2	68.2	8.6	9.2	82.2
Carbon tetrachloride	76.8	65.2	5	11	84

Table 10-19
MOLECULAR DEPRESSION OF THE FREEZING POINT

The constant K_F gives the depression of the freezing point in °C. produced by solution of one gram molecular weight of a substance in 1000 grams of any one of the solvents listed below.

Compound	Freezing Point °C.	K_F	Heat of Fusion gram-cal. per gram	Compound	Freezing Point °C.	K_F	Heat of Fusion gram-cal. per gram
Acetic acid	16.7	3.9	43.2	β-Naphthol	122.5	11.25	31.3
Aluminum bromide	97.5	26.8	10.47	Nitrobenzene	5.7	8.1	22.46
Barium chloride	962	108	27.8	Phenanthrene	99.3	12.0	25.0
Benzene	5.5	5.12	30.1	Phenol	42	7.27	29.0
Benzophenone	48.5	9.8	23.5	Salol	43	12.3	16.13
Bromoform	7.8	14.4	10.9	Stannic bromide	31	28.0	6.49
Camphor*	178.4	37.7	10.74	Stearic acid	69	4.5	47.6
Cyclohexane	6.5	20.0	7.4	Strontium chloride	873	107	25.6
m-Dinitrobenzene	89.8	10.6	24.7	Sulfuric acid	10.5	6.81	24.03
1, 4-Dioxane	10.5	4.9	.....	p-Toluidine	44.5	5.2	39.9
Diphenyl	70.0	8.0	28.8	Tribromophenol	96	20.4	13.4
Diphenylamine	52.9	8.6	25.2	Triphenyl methane	93	12.45	21.1
Ethylene bromide	9.98	12.5	13.5	Triphenyl phosphate	49.9	11.76	17.56
Formamide	2	3.8	.....	Urethane	49.7	5.14	40.9
Formic acid	8.6	2.77	58.9	Water	0	1.86	79.67
Naphthalene	80.2	6.9	36.0	p-Xylene	13.2	4.3	38.1

* A K_F of 40 is better. According to Rast (**Ber.** 55, 1051, 3727 (1922) mol. wt. determinations may be made with an ordinary thermometer.

Table 10-20
MOLECULAR ELEVATION OF THE BOILING POINT

The constant K_B gives the elevation of the boiling point in °C. produced by solution of one gram molecular weight of a substance in 1000 grams of any one of the solvents listed below. In the column headed *Barometric Correction* is given the number of degrees for each mm. of difference between the barometric reading and 760 mm. of mercury to be subtracted from K_B if the pressure is lower, or added if higher, than 760 mm. Cf. Hoyt and Fink, *J. Phys. Chem.* 41, 453 (1937).

Compound	Boiling Point °C.	K_B	Barometric Correction per mm. Hg	Heat of Vaporization gram-cal. per gram.
Acetic Acid	118.5	3.07	0.0008	96.8
Acetone	56.00	1.71	0.0004	124.5
Aniline	184.3	3.52	0.0009	103.7
Benzene	80.15	2.53	0.0007	94.3
Bromobenzene	155.83	6.26	0.0016	57.6
Camphor	208.25	5.95	0.0015	74.23
Carbon disulfide	46.13	2.34	0.0006	84.1
Carbon tetrachloride	76.50	5.03	0.0013	46.4
Chlorobenzene	131.98	4.15	0.0011	77.6
Chloroform	60.19	3.63	0.0009	59.0
Cyclohexane	80.88	2.79	0.0007	85.6
Diphenyl	254.9	7.08	0.0018	78.42
Ethanol	78.26	1.22	0.0003	204
Ethyl acetate	77.13	2.77	0.0007	88.4
Ethyl ether	34.42	2.02	0.0005	83.9
Ethyl iodide	72.5	5.05	0.0013	45.6
Ethylene bromide	131.5	6.44	0.0016	46.2
n-Heptane	98.42	3.43	0.0008	76.3
n-Hexane	68.59	2.75	0.0007	79.3
Iodobenzene	188.47	8.53	0.0021	....
Methanol	64.67	0.83	0.0002	262.8
Methyl acetate	57.1	2.15	0.0005	98.1
Naphthalene	218.0	5.65	0.0014	75.5
Nitrobenzene	210.85	5.24	0.0013	79.1
n-Octane	125.80	4.02	0.0010	70.9
n-Pentane	36.00	2.04	0.0005	....
Phenol	181.2	3.56	0.0009	114.3
Stannic chloride	114.1	9.43	0.0024	30.3
Toluene	110.7	3.33	0.0008	....
Water	100.00	0.512	0.0001	539.6

MELTING POINTS

Table 10-21
FREEZING MIXTURES

The table below gives the percent of the anhydrous material as shown in the column on the left in the eutectic mixture with ice. The eutectic temperature is the lowest temperature which can be obtained from a mixture of the substance with ice. To obtain the maximum cooling effect, the freezing mixture should be prepared with ice rather than with water and the other ingredient should be cooled to 0°C. For most purposes sodium chloride and ice, or calcium chloride and ice, mixtures are most common providing temperatures of -21.2°C. and -55°C. respectively.

Formula of Substance	%	Eutectic Temp. °C.	Formula of Substance	%	Eutectic Temp. °C.
$BaCl_2$	22.5	−7.8	$MnSO_4$	32.2	−10.5
$CaCl_2$	29.8	−55	NH_4Cl	18.6	−15.8
$Ca(NO_3)_2$	35	−16	NH_4NO_3	41.2	−17.35
$CuCl_2$	36	−40	$(NH_4)_2SO_4$	38.3	−19.05
$Cu(NO_3)_2$	36	−24	$NaBr$	40.3	−28
$CuSO_4$	11.9	−1.6	Na_2CO_3	5.9	−2.1
$FeCl_3$	33.1	−55	$NaCl$	23.3	−21.13
$FeSO_4$	13.04	−1.824	$NaOH$	19	−28
HCl	24.8	−86	$NaNO_3$	37	−18.5
HNO_3	32.7	−43	Na_2SO_4	12.7	−3.55
K_2CO_3	39.5	−36.5	$Na_2S_2O_3$	30	−11
KCl	19.75	−11.1	$NiSO_4$	20.6	−4.15
K_2CrO_4	36.6	−11.3	SO_3	32	−75
KOH	31.5	−65	$SrCl_2$	26	−18.7
KNO_3	10.9	−2.9	$Sr(NO_3)_2$	24.5	−5.75
$MgCl_2$	21.6	−33.6	$ZnCl_2$	51	−62
$Mg(NO_3)_2$	34.6	−29	$Zn(NO_3)_2$	39.4	−29
$MgSO_4$	19	−3.9	$ZnSO_4$	27.2	−6.55

Table 10-22
NON-AQUEOUS COOLING BATHS

Low temperatures may be produced by mixtures of various substances with carbon dioxide snow or by evaporation of low boiling liquids.*

Substance	Deg. C.	Substance	Deg. C.
Alcohol-carbon dioxide	−72	Nitrogen, boiling point	−196
Ammonia, boiling point	−33.4	Oxygen, boiling point	−183
Chloroform-carbon dioxide	−77	Sulfur dioxide, boiling point	−10
Ether-carbon dioxide	ca −78		
Ethyl chloride, boiling point	+12.5		
Liquid air, boiling point	−190		

* The assertion that certain baths with CO_2 snow, especially the acetone bath, give a temperature materially lower than pure dry CO_2 snow is erroneous. Cf. Thiel and Caspar, *Z. physik. Chem.* **86**, *257–93 (1914)*; *C. A.* **8**, *1229 (1914)*. Liquid nitrogen (but not liquid air, which is hazardous) poured into ethyl alcohol until a slush is formed gives a cooling bath −115° to −125°C.

According to Kolb, *Chemist-Analyst*, 1-methoxy-2-propanol is a very good liquid for dry ice cooling baths. This solvent is available from the Dow Chemical Co., Midland, Mich., under the trade name DOWANOL 33-B.

Table 10-23
COMPOSITIONS OF AQUEOUS ANTIFREEZE SOLUTIONS
FREEZING POINT OF ETHYL ALCOHOL-WATER MIXTURES †

Specific Gravity 20°/4°C. (68°F.)	% alcohol by weight	% alcohol by volume	Freezing Point	
			°C.	°F.
0.99363	2.5	3.13	−1.0	30.2
0.98971	4.8	6.00	−2.0	28.4
0.98658	6.8	8.47	−3.0	26.6
0.98006	11.3	14.0	−5.0	23.0
0.97670	13.8	17.0	−6.1	21.0
0.97336	16.4	20.2	−7.5	18.5
0.97194	17.5	21.5	−8.7	16.3
0.97024	18.8	23.1	−9.4	15.1
0.96823	20.3	24.8	−10.6	12.9
0.96578	22.1	27.0	−12.2	10.0
0.96283	24.2	29.5	−14.0	6.8
0.95914	26.7	32.4	−16.0	3.2
0.95400	29.9	36.1	−18.9	−2.0
0.94715	33.8	40.5	−23.6	−10.5
0.93720	39.0	46.3	−28.7	−19.7
0.92193	46.3	53.8	−33.9	−29.0
0.90008	56.1	63.6	−41.0	−41.8
0.86311	71.9	78.2	−51.3	−60.3

FREEZING POINT OF METHYL (WOOD) ALCOHOL-WATER MIXTURES †

Specific Gravity 15.6°C. (60°F.)	% alcohol by weight	% alcohol by volume	Freezing Point	
			°C.	°F.
0.993	3.9	5	−2.2	28
0.986	8.1	10	−5.0	23
0.980	12.2	15	−8.3	17
0.974	16.4	20	−11.7	11
0.968	20.6	25	−15.6	4
0.963	24.9	30	−20.0	−4
0.956	29.2	35	−25.0	−13
0.949	33.6	40	−30.0	−22
0.942	38.0	45	−35.6	−32

FREEZING POINT OF PRESTONE—WATER MIXTURES *

% Prestone		Specific Gravity	Freezing Point	
By Weight	By Volume	15°/15C. (59°F.)	°C.	°F.
10	9.2	1.013	−3.6	25.6
15	13.8	1.019	−5.6	22.0
20	18.3	1.026	−7.9	17.8
25	23.0	1.033	−10.7	12.8
30	28.0	1.040	−14.0	6.8
40	37.8	1.053	−22.3	−8.2
50	47.8	1.067	−33.8	−28.8
60	58.1	1.079	−49.3	−56.7

† Values are for pure alcohol. Since some commercial antifreezes contain small amounts of water, slightly higher volume concentrations than those given in the table may be required. Antifreezes also contain corrosion inhibitors and other additives to make them function properly as cooling liquids. These affect freezing point slightly and specific gravity to a greater degree. If a protection table is furnished by the manufacturer it should be used in preference to the values given above for the pure substance.

* Eveready Prestone (manufactured by the National Carbon Co.), marketed for antifreeze purposes, is 97% ethylene glycol containing fractional percentages of soluble and insoluble ingredients to prevent foaming, creepage and water corrosion in automobile cooling systems.

Table 10-23 (*Continued*)
COMPOSITIONS OF AQUEOUS ANTIFREEZE SOLUTIONS

Specific Gravity 15.6°C. (60°F.)	% alcohol by volume	Freezing Point	
		°C.	°F.
0.990	5	−1.7	29
0.984	10	−3.3	26
0.978	15	−6.1	21
0.972	20	−8.3	17
0.964	25	−11.1	12
0.955	30	−14.4	6
0.945	35	−17.8	0
0.933	40	−18.3	−1
0.922	45	−18.9	−2
0.910	50	−20.0	−4
0.899	55	−21.7	−7
0.887	60	−23.3	−10
0.875	65	−24.4	−12
0.864	70	−26.7	−16
0.852	75	−32.2	−26
0.840	80	−41.7	−43

FREEZING POINT OF PROPYLENE GLYCOL-WATER MIXTURES †

Specific Gravity 15.6°C. (60°F.)	% glycol by volume	Freezing Point	
		°C.	°F.
1.004	5	−1.1	30
1.006	10	−2.2	28
1.012	15	−3.9	25
1.017	20	−6.7	20
1.020	25	−8.9	16
1.024	30	−12.8	9
1.028	35	−16.1	3
1.032	40	−20.6	−5
1.037	45	−26.7	−16
1.040	50	−33.3	−28

FREEZING POINT OF GLYCEROL (GLYCERINE)—WATER MIXTURES *

% Glycerol by Weight	Specific Gravity 15°/15°C. (59°F.)	Specific Gravity 20°/20°C. (68°F.)	Freezing Point	
			°C.	°F.
10	1.02415	1.02395	−1.6	29.1
20	1.04935	1.04880	−4.8	23.4
30	1.07560	1.07470	−9.5	14.9
40	1.10255	1.10135	−15.5	4.3
50	1.12985	1.12845	−22.0	−7.4
60	1.15770	1.15605	−33.6	−28.5
70	1.18540	1.18355	−37.8	−36.0
80	1.21290	1.21090	−19.2	−2.3
90	1.23950	1.23755	−1.6	29.1
100	1.26557	1.26362	17.0	62.6

† See footnote on preceding page.
* The values are those reported by Bosart and Snoddy (*Jour. Ind. Eng. Chem.*, **19**, 506 (1927)), and Lane (*Jour. Ind. Eng. Chem.*, **17**, 924 (1925)) but modified by adding 2°F to all temperatures below 0°F in accordance with the suggestion of the Procter and Gamble Co.

Table 10-23 (*Continued*)
COMPOSITIONS OF AQUEOUS ANTIFREEZE SOLUTIONS

% MgCl₂ by weight	Spec. Grav. 15.6°C. (60°F.)	Freezing Point °C.	Freezing Point °F.	% MgCl₂ by weight	Spec. Grav. 15.6°C. (60°F.)	Freezing Point °C.	Freezing Point °F.
5	1.043	−3.11	26.4	18	1.161	−22.1	−7.7
6	1.051	−3.89	25.0	19	1.170	−25.6	−12.2
7	1.060	−4.72	23.5	20	1.180	−27.4	−17.3
8	1.069	−5.67	21.8	21	1.190	−30.6	−23.0
9	1.078	−6.67	20.0	22	1.200	−32.8	−27.0
10	1.086	−7.83	17.9	23	1.210	−28.9	−20.0
11	1.096	−9.05	15.7	24	1.220	−25.6	−14.0
12	1.105	−10.5	13.1	25	1.230	−23.3	−10.0
13	1.114	−12.1	10.3	26	1.241	−21.1	−6.0
14	1.123	−13.7	7.3	27	1.251	−19.4	−3.0
15	1.132	−15.6	4.0	28	1.262	−18.3	−1.0
16	1.142	−17.6	0.4	29	1.273	−17.2	+1.0
17	1.151	−19.7	−3.5	30	1.283	−16.7	2.0

FREEZING POINT OF SODIUM CHLORIDE BRINES
Compiled in collaboration with C. D. Looker, Ph.D., International Salt Co., Inc.

% NaCl by weight	Spec. Grav. 15°C. (59°F.)	Freezing Point °C.	Freezing Point °F.	% NaCl by weight	Spec. Grav. 15°C. (59°F.)	Freezing Point °C.	Freezing Point °F.
0	1.000	0.00	32.0	15	1.112	−10.88	12.4
1	1.007	−0.58	31.0	16	1.119	−11.90	10.6
2	1.014	−1.13	30.0	17	1.127	−12.93	8.7
3	1.021	−1.72	28.9	18	1.135	−14.03	6.7
4	1.028	−2.35	27.8	19	1.143	−15.21	4.6
5	1.036	−2.97	26.7	20	1.152	−16.46	2.4
6	1.043	−3.63	25.5	21	1.159	−17.78	+0.0
7	1.051	−4.32	24.2	22	1.168	−19.19	−2.5
8	1.059	−5.03	22.9	23	1.176	−20.69	−5.2
9	1.067	−5.77	21.6	23.3 (*E*)	1.179	−21.13	−6.0
10	1.074	−6.54	20.2	24	1.184	−17.0*	+1.4*
11	1.082	−7.34	18.8	25	1.193	−10.4*	13.3*
12	1.089	−8.17	17.3	26	1.201	−2.3*	27.9*
13	1.097	−9.03	15.7	26.3	1.203	0.0*	32.0*
14	1.104	−9.94	14.1				

* Saturation temperatures of sodium chloride dihydrate; at these temperatures NaCl·2H₂O separates leaving the brine of the eutectic composition (*E*).

PROPYLENE GLYCOL-GLYCEROL

Propylene glycol, a satisfactory antifreeze with the advantage of being non-toxic, can be combined with glycerol, also an efficient non-toxic antifreeze, to give a mixture that can be tested for freezing point with an ethylene glycol (Prestone) hydrometer. A mixture of 70% propylene glycol and 30% glycerol (% by weight of water-free materials), when diluted, can be tested on the standard instrument used for ethylene glycol solutions.

Table 10-23 (Continued)
COMPOSITIONS OF AQUEOUS ANTIFREEZE SOLUTIONS; FREEZING POINT OF CALCIUM CHLORIDE BRINES

Reproduced from "Dow Calcium Chloride for Refrigeration," copyright 1929 by the Dow Chemical Co., by permission.

Density of Brine @ 60°F. Specific Gravity Scale	Baume Scale	Freezing Point °F.	Freezing Point °C.	Ammonia Suction Pressure lbs. gauge Corresponding to Freezing Point	Calcium Chloride Content of Brine in Per Cent — Anhydrous $CaCl_2$	Calcium Chloride Content of Brine in Per Cent — Dow 73-75% $CaCl_2$	Anhydrous Calcium Chloride Content of Brine† — Pounds Per Gallon	Anhydrous Calcium Chloride Content of Brine† — Pounds Per Cu. Ft.	Weight of Brine @ 60°F. — Pounds Per Gallon	Weight of Brine @ 60°F. — Pounds Per Cu. Ft.	Spec. Heat −20	Spec. Heat −10	Spec. Heat 0	Spec. Heat +10	Spec. Heat +20	Heat Cap. −20	Heat Cap. −10	Heat Cap. 0	Heat Cap. +10	Heat Cap. +20
1.00	0.0	+32.0	0.0	47.6	0.0	0.0	0.0	0.0	8.34	62.4	…	…	…	…	…	…	…	…	…	…
1.01	1.4	+31.1	−0.5	46.4	1.1	1.48	0.093	0.69	8.41	63.1	…	…	…	…	…	…	…	…	…	…
1.02	2.8	+30.2	−1.0	45.3	2.3	3.11	0.196	1.46	8.50	63.7	…	…	…	…	…	…	…	…	…	…
1.03	4.2	+29.1	−1.8	43.9	3.5	4.73	0.301	2.25	8.59	64.3	…	…	…	…	…	…	…	…	…	…
1.04	5.6	+28.0	−2.2	42.6	4.7	6.35	0.407	3.04	8.67	64.9	…	…	…	…	…	…	…	…	…	…
1.05	6.9	+27.0	−2.8	41.4	5.8	7.99	0.508	3.80	8.75	65.5	…	…	…	…	…	…	…	…	…	…
1.06	8.2	+25.9	−3.2	40.1	7.0	9.46	0.619	4.62	8.84	36.2	…	…	…	…	…	…	…	…	…	…
1.07	9.5	+24.6	−4.1	38.6	8.1	10.93	0.723	5.40	8.92	66.8	…	…	…	…	…	…	…	…	…	…
1.08	10.7	+23.4	−4.8	37.2	9.2	12.41	0.829	6.19	9.00	67.4	…	…	…	…	…	…	…	…	…	…
1.09	12.0	+21.7	−5.9	35.4	10.4	14.08	0.945	7.06	9.09	68.1	…	…	…	…	…	…	…	…	…	…
1.10	13.2	+20.3	−6.5	33.8	11.4	15.40	1.05	7.81	9.17	68.7	…	…	…	…	0.829	…	…	…	…	7.60
1.11	14.4	+18.5	−7.5	31.9	12.5	16.90	1.16	8.64	9.25	69.3	…	…	…	…	0.815	…	…	…	…	7.55
1.12	15.5	+16.5	−8.6	29.9	13.5	18.25	1.26	9.42	9.34	69.9	…	…	…	…	0.802	…	…	…	…	7.49
1.13	16.7	+14.4	−9.9	27.9	14.6	19.75	1.38	10.28	9.42	70.6	…	…	…	…	0.789	…	…	…	…	7.44
1.14	17.8	+12.0	−11.1	25.6	15.6	21.05	1.48	11.08	9.50	71.2	…	…	…	…	0.776	…	…	…	…	7.38
1.15	18.9	+9.7	−12.4	23.5	16.6	22.41	1.59	11.89	9.58	71.8	…	…	…	0.759	0.764	…	…	…	7.28	7.33
1.16	20.0	+7.0	−13.9	21.2	17.6	23.80	1.70	12.71	9.67	72.4	…	…	…	0.748	0.753	…	…	…	7.24	7.29
1.17	21.1	+4.1	−15.6	18.9	18.6	25.15	1.81	13.56	9.75	73.1	…	…	…	0.737	0.742	…	…	…	7.19	7.24
1.18	22.1	+1.4	−17.0	16.8	19.5	26.38	1.92	14.34	9.83	73.7	…	…	…	0.727	0.732	…	…	…	7.15	7.20
1.19	23.2	−2.2	−19.0	14.2	20.5	27.75	2.02	15.20	9.92	74.3	…	…	0.712	0.717	0.722	…	…	7.07	7.12	7.17
1.20	24.2	−5.8	−21.0	11.7	21.5	29.06	2.13	16.07	10.01	74.9	…	…	0.703	0.707	0.712	…	…	7.03	7.08	7.13
1.21	25.2	−9.4	−23.0	9.4	22.4	30.30	2.25	16.89	10.08	75.5	…	…	0.694	0.699	0.704	…	…	7.00	7.05	7.10
1.22	26.2	−13.2	−25.1	7.3	23.3	31.50	2.37	17.71	10.17	76.2	…	0.681	0.686	0.691	0.696	…	6.93	6.98	7.03	7.08
1.23	27.1	−17.1	−27.2	5.1	24.2	32.70	2.48	18.54	10.25	76.8	…	0.674	0.679	0.683	0.688	…	6.92	6.96	7.01	7.06
1.24	28.1	−21.3	−29.6	3.1	25.1	33.97	2.60	19.39	10.33	77.4	0.663	0.667	0.672	0.676	0.681	6.84	6.89	6.94	6.99	7.04
1.25	29.0	−25.8	−32.1	1.3	26.0	35.13	2.71	20.25	10.41	78.1	0.657	0.661	0.665	0.669	0.674	6.84	6.89	6.93	6.98	7.02
1.26	29.9	−30.8	−34.9	*1.7	26.9	36.40	2.83	21.12	10.50	78.7	0.651	0.655	0.659	0.663	0.667	6.84	6.88	6.92	6.96	7.00
1.27	30.8	−36.4	−38.0	*6.1	27.8	37.60	2.94	21.93	10.58	79.3	0.646	0.650	0.653	0.657	0.660	6.84	6.88	6.91	6.95	6.98
1.28	31.7	−44.1	−42.3	*11.2	28.7	38.80	3.06	22.89	10.66	79.9	0.642	0.645	0.647	0.650	0.654	6.84	6.87	6.90	6.93	6.97
1.29	32.6	−59.8	−51.0	*18.5	29.6	40.00	3.18	23.79	10.75	80.5	0.637	0.640	0.642	0.645	0.648	6.85	6.90	6.90	6.93	6.97
1.30	33.5	−41.8	−41.0	*9.9	30.5	41.20	3.30	24.70	10.84	81.2	0.633	0.635	0.637	0.640	0.643	6.86	6.88	6.91	6.94	6.97

* Inches of mercury below one standard atmosphere (29.92 in.)

† For Dow 73-75% $CaCl_2$ multiply by 1.25 and for weight of Dow Flake 77-80% Calcium Chloride multiply by 1.28.

Table 10-24
PHYSICAL PROPERTIES OF REFRIGERANTS

Compiled from information contained in a private communication from KINETIC CHEMICALS, INC., Wilmington, Del., U.S.A. and from Cir. No. 2, Am. Soc. Refrigeration Engineers (1926).

Refrigerant	Ammonia	Carbon Dioxide	Ethyl Chloride	Methyl Chloride	Sulfur Dioxide	Freon*	F—11*	F—21*	F—114*
Chemical symbol	NH_3	CO_2	C_2H_5Cl	CH_3Cl	SO_2	CCl_2F_2	CCl_3F	$CHCl_2F$	$C_2Cl_2F_4$
Molecular weight	17.032	44.00	64.50	50.48	64.06	120.914	137.371	102.922	170.914
Color of liquid	colorless	colorless	colorless	colorless	colorless	colorless	colorless	colorless	colorless
Odor	pungent aromatic	odorless	pungent etherial odor; sweetish taste	similar to chloroform, but less sweet	pungent characteristic	etherial odor	etherial odor	etherial odor	etherial odor
Density of liquid (Water =1)	0.6818	See Note X	0.9232	0.998	1.4601	1.445	1.568	1.446	1.570
at, deg. C.	−33.35		0.0	−24.09	−10.0	−15.0	−15.0	−15.0	−15.0
Density of gas, grams 1 liter (See Note Y)	0.7708	1.9768	2.31	2.3045	2.9267	5.44	6.20	4.60	7.75
(Air =1)	0.5962	1.5290		1.7824	2.2636	4.21	4.80	3.56	6.0
Boiling point at 1 atm., deg. C.	−33.35	See Note X	13.1	−24.09	−10.0	−29.8	23.7	8.9	3.5
deg. F.	−28.03		55.6	−11.36	14.0	−21.7	74.67	48.0	38.4
Melting point, deg. C.	−77.70	−78.52	−138.7	−91.5	−75.2	−155.0	−88.0	−127.0	−105.5
deg. F.	−107.86	−109.34	−217.7	−132.7	−103.4	−247.0	−126.4	−196.6	−158.0
Critical temperature, deg. C.	132.9	31.00	182.8	143.12	157.12	111.7	196.6		146.1
deg. F.	271.2	87.80	361.0	289.6	314.82	233.0	386.0		295.0
Critical pressure, atm. abs.	112.3	72.85	53.3	65.93	77.65	39.4	41.7		35.5
lbs. per sq. in. abs.	1651	1071	784	969.2	1141.5	580	612		550
Specific heat of constant pressure (C_p)	0.5202	0.2025	0.273	0.24	0.1511	0.1476	0.147	0.18	
Specific heat of constant volume (C_v)	0.4011	0.1558		0.20		0.1297	0.1296	0.1607	
Ratio of specific heats (C_p/C_v)	1.2969	1.3003	1.1257	1.1991	1.256	1.138	1.135	1.12	1.106
at deg. C.		0.0	73	66—86	16—34				
Latent heat of vaporization at 1 atm., BTU per lb.	589.4	256.3 See Note X	168.6	180.6	172.3	71.95	78.8	101.8	58.3
Suction pressure at 5°F., lb. per sq. in. abs.	34.5	331.8	4.65	20.89	11.82	26.51	2.96	5.5	7.2
Head pressure at 86°F., lb. per sq. in. abs.	168.5	1039.6	27.1	95.53	65.9	107.9	18.3	30.5	35.8

*Trade names for dichlorodifluoromethane, trichloromonofluoromethane, dichloromonofluoromethane, and dichlorotetrafluoroethane.
Note X—Carbon Dioxide not a liquid at atmospheric pressure.
Note Y—Density of gas at 0°C. (32°F.) and 760 mm. (1 atm.), except for Ethyl Chloride for which no temperature has been given.

Table 10-25
MELTING POINTS OF COMPOUNDS USEFUL FOR DETERMINING APPROXIMATE MELTING POINTS WITH THE MICROSCOPE

Reprinted by permission from *Handbook of Chemical Microscopy*, by Chamot and Mason, published by John Wiley & Sons, Inc.

M.P. °C.	Compound	M.P. °C.	Compound
20	Acetophenone	137	Picrolonic acid
22	Anethole	140	p-Phenylenediamine
27	Diphenylmethane	145	Anthranilic acid
30	o-Cresol	148	p-Nitroaniline; 2, 4-Dinitro-
35	p-Cresol		resorcinol
41	Phenol	150	Ammonium thiocyanate
43	Salol	153	Citric acid; Methylglyoxime
45	o-Nitrophenol	156	Benzenesulphonimide
48	Chloral hydrate; Urethane	159	Salicylic acid
50	α-Naphthylamine	164	Cupferron
52	Thymol	169	Hydroquinone
53	p-Dichlorobenzene	170	Santonin
58	Trichloroacetic acid	171	Dimethylamine hydrochloride
63	m-Phenylenediamine	172	p-Dinitrobenzene
67	Coumarin; Azobenzene	173	Potassium thiocyanate
69	Diphenyl	175	Quinine; Narcotine
70	Pyrazole	178	Brucine
72	o-Nitroaniline	184	Quinine citrate
74	Hedonal	185	Succinic acid; Cinchonamine
75	Borax	186	Saccharose
76	Trional	189	Nitron
80	Naphthalene	191	Veronal
81	Vanillin	193	Chrysophanic acid
86	Saligenin	198	Aniline hydrochloride
87	p-Dibromobenzene	202	Salicin; Lactose
90	m-Dinitrobenzene	210	Cinchonidine
93	Triphenylmethane; Salipyrin	212	Silver nitrate
96	α-Naphthol; m-Nitrophenol	218	Anthracene
100	Phenanthrene; Exalgin	219	Phloroglucinol
104	o-Phenylenediamine	228	Saccharin
105	Pyrocatechin	235	Carbanilide
108	Pyramidon	237	Caffeine
110	Resorcinol; β-Naphthylamine	244	Carbazole
112	m-Nitroaniline	246	Dimethylglyoxime
113	p-Nitrophenol	248	Sodium chlorate
114	Acetanilide; Ammonium acetate	261	Phenolphthalein
116	Atropine; Quinone	268	Strychnine
117	o-Dinitrobenzene	280	Lead acetate (anhydrous)
118	Chrysoidine	285	Anthraquinone
119	Iodoform	288	Sulfanilic acid
122	Benzoic acid; Picric acid; β-Naphthol	290	Alizarin
123	Dionine	297	Potassium nitrite
128	Sulfonal	302	Mercurous chloride
129	o-Tolidine	324	Sodium acetate (anhydrous)
131	Maleic acid	337	Theobromine
133	Urea; Hydrastine	368	Potassium chlorate
134	Pyrogallol	398	Potassium dichromate

HUMIDITY

Table 10-26
MASS OF WATER VAPOR IN SATURATED AIR

The values in this table have been selected from those in the 1947 American Society of Refrigeration Engineer's Brochure on Psychrometry and are reprinted by permission.

To convert grains per cubic foot to pounds per cubic foot multiply by 0.0001429.

°F	°C	grains/ft³	grams/m³	°F	°C	grains/ft³	grams/m³
....	......	.....	.....	42	5.6	3.080	7.048
....	......	.....	.....	44	6.7	3.313	7.581
−24	−31.1	0.132	0.302	46	7.8	3.561	8.149
−22	−30.0	0.149	0.341	48	8.9	3.826	8.755
−20	−28.9	0.165	0.378	50	10.0	4.108	9.401
−18	−27.8	0.184	0.421	52	11.1	4.407	10.08
−16	−26.7	0.205	0.469	54	12.2	4.723	10.81
−14	−25.6	0.229	0.524	56	13.3	5.063	11.59
−12	−24.4	0.255	0.584	58	14.4	5.419	12.40
−10	−23.3	0.283	0.648	60	15.6	5.798	13.27
−8	−22.2	0.314	0.719	62	16.7	6.201	14.19
−6	−21.1	0.348	0.796	64	17.8	6.627	15.17
−4	−20.0	0.386	0.883	66	18.9	7.080	16.20
−2	−18.9	0.427	0.977	68	20.0	7.561	17.30
0	−17.8	0.473	1.229	70	21.1	8.064	18.45
2	−16.7	0.522	1.318	72	22.2	8.605	19.69
4	−15.6	0.576	1.439	74	23.3	9.169	20.98
6	−14.4	0.636	1.574	76	24.4	9.763	22.34
8	−13.3	0.702	1.698	78	25.6	10.39	23.78
10	−12.2	0.772	1.860	80	26.7	11.06	25.31
12	−11.1	0.851	2.046	82	27.8	11.76	26.91
14	−10.0	0.935	2.231	84	28.9	12.50	28.60
16	−8.9	1.027	2.439	86	30.0	13.28	30.39
18	−7.8	1.128	2.668	88	31.1	14.10	32.27
20	−6.7	1.236	2.938	90	32.2	14.96	34.23
22	−5.6	1.356	3.224	92	33.3	15.86	36.29
24	−4.4	1.484	3.517	94	34.4	16.82	38.49
26	−3.3	1.624	3.824	96	35.6	17.82	40.78
28	−2.2	1.775	4.147	98	36.7	18.88	43.21
30	−1.1	1.940	4.488	100	37.8	19.99	45.74
32	0.0	2.119	4.849	102	38.9	21.15	48.40
34	1.1	2.287	5.234	104	40.0	22.38	51.21
36	2.2	2.466	5.643	106	41.1	23.65	54.12
38	3.3	2.657	6.080	108	42.2	24.98	57.16
40	4.4	2.861	6.547	110	43.3	26.39	60.39

Table 10-26 (*Continued*)
MASS OF WATER VAPOR IN SATURATED AIR

°F	°C	grains/ft³	grams/m³	°F	°C	grains/ft³	grams/m³
112	44.4	27.86	63.75	162	72.2	94.72	216.8
114	45.6	29.40	67.28	164	73.3	98.97	226.5
116	46.7	31.00	70.94	166	74.4	103.4	236.7
118	47.8	32.70	74.83	168	75.6	108.0	247.2
120	48.9	34.46	78.86	170	76.7	112.8	258.1
122	50.0	36.29	83.05	172	77.8	117.8	269.5
124	51.1	38.21	87.44	174	78.9	122.9	281.2
126	52.2	40.23	92.06	176	80.0	128.2	293.4
128	53.3	42.33	96.87	178	81.1	133.7	305.9
130	54.4	44.50	101.8	180	82.2	139.4	319.0
132	55.6	46.79	107.1	182	83.3	145.3	332.4
134	56.7	49.18	112.5	184	84.4	151.3	346.3
136	57.8	51.66	118.2	186	85.6	157.7	360.8
138	58.9	54.23	124.1	188	86.7	164.2	375.7
140	60.0	56.91	130.2	190	87.8	170.9	391.2
142	61.1	59.75	136.7	192	88.9	177.9	407.1
144	62.2	62.67	143.4	194	90.0	185.1	423.5
146	63.3	65.67	150.3	196	91.1	192.5	440.5
148	64.4	68.83	157.5	198	92.2	200.2	458.0
150	65.6	72.16	165.1	200	93.3	208.1	476.2
152	66.7	75.54	172.9				
154	67.8	79.12	181.1				
156	68.9	82.79	189.5				
158	70.0	86.62	198.2				
160	71.1	90.59	207.3				

Table 10-27
EFFICIENCY OF DRYING AGENTS

Weight of Residual Water per Liter of Gas Dried at 25°C.

Drying agent	Milligrams of water
P_2O_5	2×10^{-5}
BaO	1×10^{-4}
$Mg(ClO_4)_2$	5×10^{-4}
KOH (fused)	2×10^{-3}
H_2SO_4	3×10^{-3}
$CaSO_4$	4×10^{-3}
$CaCl_2$ (granular)	$0.14 - 0.25$

Table 10-28
HUMIDITY AND DEW POINT FROM WET AND DRY BULB READINGS

In the table below $d-w$ is the difference between the dry and the wet bulb thermometer readings, $r.h.$ is the relative humidity, and $a.h.$ is the absolute humidity.

Temp. °C.	$d-w=0°C.$ r. h. %	dew point °C.	a. h. mm of Hg	$d-w=1°C.$ r. h. %	dew point °C.	a. h. mm of Hg	$d-w=2°C.$ r. h. %	dew point °C.	a. h. mm of Hg	$d-w=3°C.$ r. h. %	dew point °C.	a. h. mm of Hg
−20	100	−20	0.8	..	......	....	..	......	....	..	......	....
−10	100	−10	1.9	66	−14.6	1.3	32	22.1	0.6	..	......	....
− 5	100	− 5	3.0	75	− 8.3	2.3	51	−12.7	1.5	27	−19.4	0.8
0	100	0	4.6	81	− 2.5	3.7	63	− 5.5	2.9	45	− 9.3	2.1
+ 5	100	+ 5	6.5	86	+ 2.8	5.6	72	+ 0.3	4.7	58	− 2.3	3.8
10	100	10	9.2	88	8.1	8.1	76	6.1	7.0	65	+ 3.8	6.0
15	100	15	12.8	90	13.4	11.5	80	11.6	10.2	71	9.7	9.0
16	100	16	13.6	90	14.4	12.3	81	12.7	11.0	71	10.8	9.7
17	100	17	14.5	90	15.4	13.1	81	13.7	11.8	72	12.0	10.5
18	100	18	15.5	91	16.5	14.0	82	14.8	12.6	73	13.1	11.3
19	100	19	16.5	91	17.5	15.0	82	15.9	13.5	74	14.2	12.1
20	100	20	17.5	91	18.5	16.0	83	16.9	14.5	74	15.3	13.0
21	100	21	18.7	91	19.5	17.0	83	18.0	15.5	75	16.4	14.0
22	100	22	19.8	92	20.6	18.2	83	19.1	16.5	76	17.5	15.0
23	100	23	21.1	92	21.6	19.3	84	20.1	17.6	76	18.6	16.0
25	100	25	23.8	92	23.6	21.9	84	22.2	20.1	77	20.7	18.3
30	100	30	31.8	93	28.7	29.5	86	27.4	27.3	79	26.0	25.2

Temp. C.	$d-w=4°C.$ r. h. %	dew point °C	a. h. mm of Hg	$d-w=5°C.$ r. h. %	dew point °C.	a. h. mm of Hg	$d-w=6°C.$ r. h. %	dew point °C.	a. h. mm of Hg	$d-w=7°C.$ r. h. %	dew point °C.	a. h. mm of Hg
0	28	−14.6	1.3	11	−24.2	0.5	..	......	....	..	......	....
5	45	− 5.3	2.9	32	− 9.3	2.1	19	−15.2	1.2	6	−27.1	0.4
10	54	+ 1.2	5.0	44	− 1.5	4.0	34	− 4.6	3.1	24	− 8.7	2.2
12	57	3.9	6.0	48	+ 1.2	5.0	38	− 1.6	4.0	29	− 4.9	3.0
14	60	6.4	7.2	51	4.0	6.1	42	+ 1.3	5.0	34	− 1.6	4.0
15	61	7.6	7.8	52	5.4	6.7	44	2.8	5.6	36	− 0.1	4.5
16	62	8.8	8.5	54	6.7	7.3	46	4.3	6.2	37	+ 1.5	5.1
17	64	10.0	9.2	55	8.0	8.0	47	5.6	6.8	39	3.1	5.7
18	65	11.2	10.0	56	9.2	8.7	49	7.0	7.5	41	4.6	6.3
19	65	12.4	10.8	58	10.5	9.5	50	8.3	8.2	43	6.0	7.0
20	66	13.5	11.6	59	11.7	10.3	51	9.6	9.0	44	7.4	7.7
21	67	14.7	12.5	60	12.9	11.1	52	10.9	9.8	46	8.8	8.5
22	68	15.8	13.5	61	14.1	12.0	54	12.2	10.6	47	10.1	9.3
23	69	16.9	14.5	61	15.2	13.0	55	13.4	11.5	48	11.4	10.1
25	70	19.2	16.7	63	17.5	15.0	57	15.8	13.5	50	14.0	12.0
30	73	24.6	23.3	67	23.2	21.3	61	21.6	19.4	55	20.0	17.6

Table 10-28 (*Continued*)
HUMIDITY AND DEW POINT FROM WET AND DRY BULB READINGS

Temp. °C.	d−w=8°C. r.h. %	dew point °C.	a.h. mm of Hg	d−w=9°C. r.h. %	dew point °C.	a.h. mm of Hg	d−w=10°C. r.h. %	dew point °C.	a.h. mm of Hg	d−w=11°C. r.h. %	dew point °C.	a.h. mm of Hg
8	7	−22.9	0.6	..	......	....	..	.....	....	..	......	....
10	14	−14.5	1.3	5	−26.0	0.4	..	.....	....	..	......	....
12	20	− 9.1	2.1	11	−15.5	1.2	..	.....	....	..	......	....
14	25	− 5.0	3.0	17	− 9.5	2.0	9	−16.3	1.1	..	......	....
15	27	− 3.2	3.5	20	− 7.1	2.5	12	−12.6	1.5	5	−22.6	0.6
16	30	− 1.5	4.0	22	− 5.0	3.0	15	− 9.6	2.0	8	−16.8	1.0
17	32	+ 0.1	4.6	24	− 3.1	3.5	17	− 7.1	2.5	10	−12.8	1.5
18	34	1.8	5.2	27	− 1.3	4.1	20	− 4.9	3.0	13	− 9.6	2.0
19	35	3.4	5.8	29	+ 0.4	4.7	22	− 2.9	3.6	15	− 6.9	2.5
20	37	5.0	6.5	30	2.1	5.3	24	− 1.0	4.2	18	− 4.6	3.1
21	39	6.4	7.2	32	3.8	6.0	26	+ 0.8	4.8	20	− 2.5	3.7
22	40	7.9	8.0	34	5.4	6.7	28	2.6	5.5	22	− 0.6	4.3
23	42	9.3	8.8	36	6.9	7.5	30	4.3	6.2	24	+ 1.3	5.0
24	43	10.7	9.6	37	8.4	8.3	31	5.9	7.0	26	3.1	5.7
25	44	12.0	10.5	38	9.9	9.1	33	7.5	7.8	27	4.9	6.5
26	46	13.3	11.5	40	11.3	10.0	34	9.0	8.6	29	6.6	7.3
28	48	15.9	13.5	42	14.0	12.0	37	11.9	10.5	32	9.7	9.0
30	50	18.4	15.8	44	16.6	14.1	39	14.7	12.5	34	12.7	11.0

Table 10-29
RELATIVE HUMIDITIES AND AQUEOUS TENSIONS OF AQUEOUS SOLUTIONS OF H_2SO_4, NaOH AND $CaCl_2$ AT 25°C

Concentrations are expressed in percentage of anhydrous solute by weight. Stokes and Robinson, *Ind. Eng. Chem.*, **41**, 2013 (1949).

% Humidity	Aqueous Tension	% H_2SO_4	% NaOH	% $CaCl_2$	% Humidity	Aqueous Tension	% H_2SO_4	% NaOH	% $CaCl_2$
100	23.756	0.00	0.00	0.00	50	11.88	43.10	28.15	35.64
95	22.57	11.02	5.54	9.33	45	10.69	45.41	29.86	37.61
90	21.38	17.91	9.83	14.95	40	9.50	47.71	31.58	39.62
85	20.19	22.88	13.32	19.03	35	8.31	50.04	33.38	41.83
80	19.00	26.79	16.10	22.25	30	7.13	52.45	35.29	44.36
75	17.82	30.14	18.60	24.95	25	5.94	55.01	37.45	.....
70	16.63	33.09	20.80	27.40	20	4.75	57.76	40.00	.....
65	15.44	35.80	22.80	29.64	15	3.56	60.80	43.32	.....
60	14.25	38.35	24.66	31.73	10	2.38	64.45	47.97	.....
55	13.07	40.75	26.42	33.71	5	1.19	69.44	.....	.....

Table 10-30
SOLUTIONS FOR MAINTAINING CONSTANT HUMIDITY

A saturated aqueous solution in contact with an excess of a definite solid phase at a given temperature will maintain constant humidity in an enclosed space. The table below gives a number of salts suitable for this purpose listed in order of decreasing humidity. The aqueous tension of a solution at a given temperature divided by the vapor pressure of pure water at the same temperature is equal to the relative humidity. In the following table the relative humidity is expressed in per cent and the aqueous tension of the saturated solution in mm of mercury pressure.

Values in the table are at 20°C. For additional solutions and for values at other temperatures, see Carr and Harris, *Ind. Eng. Chem.*, **41**, 2014 (1949), and Stokes and Robinson, *ibid.*, **41**, 2013 (1949).

Solid Phase	% Humidity	Aqueous Tension
$Pb(NO_3)_2$	98	17.2
$Na_2SO_4 \cdot 10H_2O$	93	16.3
KBr	84	14.7
NaCl	75.7	13.85
$NaNO_2$	66	11.6
$NaBr \cdot 2H_2O$	57.9	10.3
$Na_2Cr_2O_7 \cdot 2H_2O$	52	9.12
$Zn(NO_3)_2 \cdot 6H_2O$	42	7.36
$CaCl_2 \cdot 6H_2O$	32.3	5.66
$KC_2H_3O_2$	20	3.51
$ZnCl_2 \cdot 1.5H_2O$	10	1.74

DENSITY AND SPECIFIC GRAVITY

HYDROMETERS

Various hydrometers and the relation between the various scales.

Alcoholometer. This hydrometer is used in determining the density of aqueous ethyl alcohol solutions; the reading in degrees is numerically the same as the percentage of alcohol by volume. The scale known as *Tralle* gives the percentage by volume. (See also formulas for conversion to percent by weight, percent by volume, and percent proof.) *Wine* and *Must* hydrometer relations are given below.

Ammoniameter. This hydrometer, employed in finding the density of aqueous ammonia solutions, has a scale graduated in equal divisions from 0° to 40°. To convert the reading to specific gravity multiply by 3 and subtract the resulting number from 1000.

Balling Hydrometer. See under **Saccharometer.**

Barkometer or **Barktrometer.** This hydrometer which is used in determining the density of tanning liquors has a scale from 0° to 80° Bk; the number to the right of the decimal point of a specific gravity degree is the corresponding Bk degree; thus, a specific gravity of 1.015 is 15°Bk.

Baumé Hydrometers. For liquids heavier than water.—This hydrometer was originally based on the density of a 10% sodium chloride solution which was given the value of 10° and the density of pure water which was given the value of 0°, the interval between these two values was divided into 10 equal parts. Other reference points have been taken with the result that so much confusion exists that there are about 36 different scales in use, many of which are incorrect. In general a Baumé hydrometer should have inscribed on it the temperature at which it was calibrated and also the temperature of the water used in relating the density to a specific gravity. The following expression gives the relation between the specific gravity and several of the Baumé scales:—

Specific gravity $= \dfrac{m}{m - \text{Baumé}}$

$m = 145$ at 60°/60°F (15.56°C) for the American Scale

$m = 144$ for the old scale used in Holland

$m = 146.3$ at 15°C for the Gerlach Scale

$m = 144.3$ at 15°C for the Rational Scale generally used in Germany

See also special table for conversion to density and Twaddell scale.

For liquids lighter than water.— Originally the density of a solution of 1 gram of sodium chloride in 9 grams of water at 12.5°C was given a value of 0°Bé and pure water a value of 10°Bé. The scale between these points was divided into ten equal parts and these divisions were repeated throughout the scale giving a relation which could be expressed by the formula: Specific gravity $= 145.88/(135.88 + \text{Bé})$, which is approximately equal to $146/(136 + \text{Bé}.)$ Other scales have since come into more general use such as that of the Bu. of Standards in which the specific gravity of water at 60°/60°F $= 140/(130 + \text{Bé}.)$ and that of the American Petroleum Institute (A.P.I. Scale) in which the specific gravity at 60°/60°F $= 141.5/(131.5 + \text{API}°.)$

See also special table for conversion to density and Twaddell scale.

Beck's Hydrometer. This hydrometer is graduated to show a reading of 0° in pure water and a reading of 30° in a solution with a specific gravity of 0.850, with equal scale divisions above and below these two points.

Brix Hydrometer. See under Saccharometers.

Cartier's Hydrometer. This hydrometer shows a reading of 22° when immersed in a solution having a density of 22° Baumé but the scale divisions are smaller than on the Baumé hydrometer in the ratio of 16 Cartier to 15 Baumé.

Fatty Oil Hydrometer. The gradu-

HYDROMETERS

ations on this hydrometer are in specific gravity within the range 0.908 to 0.938. The letters on the scale correspond to the specific gravity of the various common oils as follows: *R*, rape; *O*, olive; *A*, almond; *S*, sesame; *HL*, hoof oil; *HP*, hemp; *C*, cotton seed; *L*, linseed. See also Oleometer below.

Lactometers. These hydrometers are used in determining the density of milk. The various scales in common use are the following:

New York Board of Health has a scale graduated into 120 equal parts, 0° being equal to the specific gravity of water and 100° being equal to a specific gravity of 1.029.

Quevenne lactometer is graduated from 15° to 40° corresponding to specific gravities from 1.015 to 1.040.

Soxhlet lactometer has a scale from 25° to 35° corresponding to specific gravities from 1.025 to 1.035 respectively.

Oleometer. A hydrometer for determining the density of vegetable and sperm oils with a scale from 50° to 0° corresponding to specific gravities from 0.870 to 0.970. See also *Fatty Oil* hydrometer above.

Saccharometers. These hydrometers are used in determining the density of sugar solutions. Solutions of the same concentration but of different carbohydrates have very nearly the same specific gravity and in general a concentration of 10 grams of carbohydrate per 100 cc of solution shows a specific gravity of 1.0386. Thus, the wt. of sugar in 1000 ml soln. is, (a) For conc. <12g/100 ml: (wt. of 1000 ml soln. −1000) ÷ 0.386; (b) for conc. >12g/100 ml: (wt. of 1000 ml soln. − 1000) ÷ 0.385.

Brix hydrometer is graduated so that the number of degrees is identical with the percentage by weight of cane sugar and is used at the temperature indicated on the hydrometer.

Balling's saccharometer is used in Europe and is practically identical with the Brix hydrometer.

Bates brewers' saccharometer which is used in determining the density of malt worts is graduated so that the divisions express pounds per barrel (32 gallons). The relation between degrees Bates (=b) and degrees Balling (=B) is shown by the following formula: B = 260b/(360+b).

See also below under *Wine and Must.*

Salinometer. This hydrometer, which is used in the pickling and meat packing plants, is graduated to show percentage of saturation of a sodium chloride solution. An aqueous solution is completely saturated when it contains 26.4% pure sodium chloride. The range from 0% to 26.4% is divided into 100 parts, each division therefore representing 1% of saturation. In another type of salinometer, the degrees correspond to percentages of sodium chloride expressed in grams of sodium chloride per 100 cc of water.

Sprayometer (Parrot and Stewart).— This hydrometer which is used in determining the density of *lime sulfur* solutions has two scales; one scale is graduated from 0° to 38° Baumé and the other scale is from 1.000 to 1.350 specific gravity.

Tralle Hydrometer. See *alcoholometer* above.

Twaddell Hydrometer. This hydrometer which is used only for liquids heavier than water, has a scale such that when the reading is multiplied by 5 and added to 1000 the resulting number is the specific gravity with reference to water as 1000. To convert specific gravity at 60°/60°F to Twaddell degrees, take the decimal portion of the specific gravity value and multiply it by 200; thus a specific gravity of 1.032 = 0.032 × 200 = 6.4° Tw. See also special table for conversion to density and Baumé scale.

Wine and Must Hydrometer—This instrument has three scales. One scale shows readings of 0° to 30° Brix for sugar (see Brix hydrometer above); another scale from 0° to 15° Tralle is used for sweet wines to indicate the percentage of alcohol by volume; and a third scale from 0° to 20° Tralle is used for tart wines to indicate the percentage of alcohol by volume.

SPECIFIC GRAVITY CORRECTIONS FOR THE BUOYANT EFFECT OF THE AIR

Determinations made with a pyknometer.

$$D_{vac} = \frac{W_2}{W_1} d - 0.0012 \left(\frac{W_2 d}{W_1} - 1 \right)$$

$$S_{vac} = \frac{W_2}{W_1} - 0.0012 \left(\frac{W_2}{W_1} - 1 \right)$$

Where

D_{vac} = density of the liquid in grams per milliliter at t°C corrected for the buoyant effect of air.

W_1 = weight in air of the water required to fill the pyknometer at t°C.

W_2 = weight in air of the liquid required to fill the pyknometer at t°C.

d = density of water in grams per milliliter at t°C.

S_{vac} = specific gravity of the liquid at t°C. referred to water at t°C corrected for the buoyant effect of air.

When the weight of the water is determined at a temperature of t°C, and that of the liquid at a different temperature t′, the equations above are modified as follows:

$$D_{vac} = \frac{W_2}{W_1} d - 0.0012 \left(\frac{W_2}{W_1} d - 1 \right) + 0.000026 \ (t' - t°) \left(\frac{W_2}{W_1} d \right)$$

$$S_{vac} = \frac{W_2}{W_1} - 0.0012 \left(\frac{W_2}{W_1} - 1 \right) + 0.000026 \ (t' - t°) \left(\frac{W_2}{W_1} \right)$$

Determinations made with a plummet or sinker.

The equations above may also be used when the density is determined with plummet or sinker, but in this case

W_1 = weight of the plummet in air minus its weight in water.

W_2 = weight of the plummet in air minus its weight in the liquid.

CONVERSION OF SPECIFIC GRAVITY AT 25°/25° C. TO DENSITY AT ANY TEMPERATURE FROM 0° TO 40° C.

Cf. Dreisbach, *Ind. Eng. Chem., Anal. Ed. 12, 160 (1940).*

Liquids change volume with change in temperature but the amount of this change, β (coefficient of cubical expansion), varies widely with different liquids, and to some extent for the same liquid at different temperatures. See special table, *Coefficients of Cubical Expansion for Various Liquids and Aqueous Solutions.*

The table below, which is calculated from the relationship:

$$F\beta_t = \frac{\text{density of water at } 25°C. \; (=0.99705)}{[1-\beta\,(25-t)]} \quad (1)$$

may be used to find d^t, the density (weight of 1 cc.), of a liquid at any temperature (t) between 0° and 40°C. if the specific gravity at 25°/25°C. (S) and the coefficient of cubical expansion (β) are known. Substitutions are made in the equations:

$$d^t = SF\beta_t \quad (2)$$
$$S = \frac{d^t}{F\beta_t} \quad (3)$$

TABLE—FACTORS $(F\beta_t)$

Density $t°C. = $ Sp. Gr. $25°/25° \times F\beta_t$

°C. $*\beta \times 10^3$	0	5	10	15	20	25	30	35	40
1.3	1.0306	1.0237	1.0169	1.0102	1.0036	0.99705	0.99065	0.9843	0.9780
1.2	1.0279	1.0216	1.0154	1.0092	1.0031	0.99705	0.9911	0.9853	0.9794
1.1	1.0253	1.0195	1.0138	1.0082	1.0026	0.99705	0.9916	0.9963	0.9809
1.0	1.0227	1.0174	1.0123	1.0072	1.0021	0.99705	0.9921	0.9872	0.98234
0.9	1.0200	1.0153	1.0107	1.0060	1.0016	0.99705	0.99262	0.9882	0.9838
0.8	1.0174	1.0133	1.0092	1.0051	1.0011	0.99705	0.9931	0.98918	0.9851
0.7	1.0148	1.0113	1.0077	1.0041	1.0006	0.99705	0.9936	0.99015	0.98672
0.6	1.0122	1.0092	1.0061	1.0031	1.0001	0.99705	0.9941	0.9911	0.9882
0.5	1.0097	1.0072	1.0046	1.0021	0.99958	0.99705	0.9944	0.9921	0.9897
0..	1.0071	1.0051	1.0031	1.0011	0.99908	0.99705	0.9951	0.9931	0.9911

$*\beta = $ coefficient of cubical expansion.

EXAMPLES

All examples are based upon an assumed coefficient of cubical expansion, β, of 1.3×10^{-3}.

Example 1. To find the density of a liquid at 20°C., d^{20}, which has a specific gravity (S) of $1.2500\frac{25°}{25°}$.

From the table above $F\beta_t$ at 20°C. $= 1.0036$.

$$d^{20} = d^t = SF\beta_t = 1.2500 \times 1.0036 = 1.2545$$

CONVERSIONS OF DENSITY—SPECIFIC GRAVITY

Example 2. To find the density at 20°C. (d^{20}) of a liquid which has a specific gravity of $1.2500\frac{17°}{4°}$.

Since the density of water at 4°C. is equal to 1, specific gravity at $17°/4° = d^{17} = 1.2500$.

Substitution in equation *3* with $F\beta_t$ at 17°C., by interpolation from the table, equal to 1.00756, gives

$$Sp.\ gr.\ 25°/25° = S = 1.2500 \div 1.00756$$

Substitution of this value for S in equation *2* with $F\beta_t$ at 20°C., from the table, equal to 1.0036, gives

$$d^{20} = d^t = (1.2500 \div 1.00756) \times 1.0036 = 1.2451$$

Example 3. To find the specific gravity at 20°/4°C. of a liquid which has a specific gravity of $1.2500\frac{25°}{4°}$.

Since the density of water at 4°C. is equal to 1, specific gravity $25°/4° = d^{25} = 1.2500$; and, specific gravity $20°/4° = d^{20}$.

Substitution in equation *3*, with $d^t = 1.2500$; and, with $F\beta_t$ at 25°C., from the table, equal to 0.99705, gives

$$Sp.\ gr.\ 25°/25° = S = 1.2500 \div 0.99705$$

Substitution of this value for S in equation *2*, with $F\beta_t$ at 20°C., from the table, equal to 1.0036, gives

$$Sp.\ gr.\ 20°/4° = d^{20} = (1.2500 \div 0.99705) \times 1.0036 = 1.2582$$

Example 4. To find the density at 25°C. of a liquid which has a specific gravity of $1.2500\frac{15°}{15°}$.

Since the density of water (see special table, *Absolute Density of Water*) at 15°C. = 0.99910,

$$d^{15} = sp.\ gr.\ 15°/15° \times 0.99910 = 1.2500 \times 0.99910$$

Substitution in equation *3*, with $F\beta_t$ at 15°C., from the table, equal to 1.0102, gives

$$Sp.\ gr.\ 25°/25° = S = (1.2500 \times 0.99910) \div 1.0102$$

Substitution of this value for S in equation *2*, with $F\beta_t$ at 25°, from the table, equal to 0.99705, gives

$$d^{25} = d^t = (1.2500 \times 0.99910 \div 1.0102) \times 0.99705 = 1.2326$$

DENSITY OF AQUEOUS SOLUTIONS

The values in the following tables have been computed from the values **given under** *specific gravity* and *per cent by weight* by means of the following factors:

$$\text{Degrees Baumé} = 145 - \frac{145}{\text{Sp. gravity}}$$

Degrees Twaddell = (Sp. gravity −1)×200
Grams per Liter = Sp. gravity × per cent by weight × 10
Pounds per U. S. Gallon = Grams per Liter × 0.0083454
Pounds per Cubic Foot = Grams per Liter × 0.06243

See also the tables:—"*Specific Gravity of Aqueous Solutions*", "*Specific Gravity of Alcohol Water Solutions,*" and "*Hydrometer Conversion Tables.*"

Table 10-31
ACETIC ACID

Density of Aqueous Acetic Acid Solutions at $\frac{20°}{4°}$ C. Computed from Values Given in the International Critical Tables.

Specific Gravity	Weight of $HC_2H_3O_2$ in solution expressed in			Per Cent $HC_2H_3O_2$	Degrees	
	Grams per Liter	Pounds per U. S. Gallon	Pounds per Cubic Foot		Baumé	Twaddell
0.9982	....	....	....	0	....	....
0.9996	9.996	0.08342	0.6241	1	....	....
1.0012	20.02	0.1671	1.250	2	0.2	0.24
1.0025	30.08	0.2510	1.878	3	0.4	0.50
1.0040	40.16	0.3352	2.507	4	0.6	0.80
1.0055	50.28	0.4196	3.139	5	0.8	1.10
1.0069	60.41	0.5042	3.772	6	1.0	1.38
1.0083	70.58	0.5890	4.406	7	1.2	1.66
1.0097	80.78	0.6741	5.043	8	1.4	1.94
1.0111	91.00	0.7594	5.681	9	1.6	2.22
1.0125	101.3	0.8450	6.321	10	1.8	2.50
1.0139	111.5	0.9308	6.963	11	2.0	2.78
1.0154	121.8	1.017	7.607	12	2.2	3.08
1.0168	132.2	1.103	8.252	13	2.4	3.36
1.0182	142.5	1.190	8.899	14	2.6	3.64
1.0195	152.9	1.276	9.547	15	2.8	3.90
1.0209	163.3	1.363	10.20	16	3.0	4.18
1.0223	173.8	1.450	10.85	17	3.2	4.46
1.0236	184.2	1.538	11.50	18	3.3	4.72
1.0250	194.8	1.625	12.16	19	3.5	5.00
1.0263	205.3	1.713	12.81	20	3.7	5.26
1.0276	215.8	1.801	13.47	21	3.9	5.52
1.0288	226.3	1.889	14.13	22	4.1	5.76
1.0301	236.9	1.977	14.79	23	4.2	6.02
1.0313	247.5	2.066	15.45	24	4.4	6.26
1.0326	258.2	2.154	16.12	25	4.6	6.52
1.0338	268.8	2.243	16.78	26	4.7	6.76
1.0349	279.4	2.332	17.44	27	4.9	6.98
1.0361	290.1	2.421	18.11	28	5.1	7.22
1.0372	300.8	2.510	18.78	29	5.2	7.44
1.0384	311.5	2.600	19.45	30	5.4	7.68
1.0395	322.2	2.689	20.12	31	5.5	7.90
1.0406	333.0	2.779	20.79	32	5.7	8.12
1.0417	343.8	2.869	21.46	33	5.8	8.34
1.0428	354.6	2.959	22.13	34	6.0	8.56
1.0438	365.3	3.049	22.81	35	6.1	8.76
1.0449	376.2	3.139	23.48	36	6.2	8.98
1.0459	387.0	3.230	24.16	37	6.4	9.18
1.0469	397.8	3.320	24.84	38	6.5	9.38
1.0479	408.7	3.411	25.51	39	6.6	9.58

Table 10-31 (*Continued*)
ACETIC ACID

Specific Gravity	Weight of HC₂H₃O₂ in solution expressed in			Per Cent HC₂H₃O₂	Degrees	
	Grams per Liter	Pounds per U. S. Gallon	Pounds per Cubic Foot		Baumé	Twaddell
1.0488	419.5	3.501	26.19	40	6.8	9.76
1.0498	430.4	3.592	26.87	41	6.9	9.96
1.0507	441.3	3.683	27.55	42	7.0	10.14
1.0516	452.2	3.774	28.23	43	7.1	10.32
1.0525	463.1	3.865	28.91	44	7.2	10.50
1.0534	474.0	3.956	29.59	45	7.4	10.68
1.0542	484.9	4.047	30.27	46	7.5	10.84
1.0551	495.9	4.138	30.96	47	7.6	11.02
1.0559	506.8	4.230	31.64	48	7.7	11.18
1.0567	517.8	4.321	32.33	49	7.8	11.34
1.0575	528.8	4.413	33.01	50	7.9	11.50
1.0582	539.7	4.504	33.69	51	8.0	11.64
1.0590	550.7	4.596	34.38	52	8.1	11.80
1.0597	561.6	4.687	35.06	53	8.2	11.94
1.0604	572.6	4.779	35.75	54	8.3	12.08
1.0611	583.6	4.870	36.43	55	8.4	12.22
1.0618	594.6	4.962	37.12	56	8.4	12.36
1.0624	605.6	5.054	37.81	57	8.5	12.48
1.0631	616.6	5.146	38.49	58	8.6	12.62
1.0637	627.6	5.237	39.18	59	8.7	12.74
1.0642	638.5	5.329	39.86	60	8.8	12.84
1.0648	649.5	5.421	40.55	61	8.8	12.96
1.0653	660.5	5.512	41.23	62	8.9	13.06
1.0658	671.5	5.604	41.92	63	9.0	13.16
1.0662	682.4	5.695	42.60	64	9.0	13.24
1.0666	693.3	5.786	43.28	65	9.1	13.32
1.0671	704.3	5.878	43.97	66	9.1	13.42
1.0675	715.2	5.969	44.65	67	9.2	13.50
1.0678	726.1	6.060	45.33	68	9.2	13.56
1.0682	737.1	6.151	46.01	69	9.3	13.64
1.0685	748.0	6.242	46.69	70	9.3	13.70
1.0687	758.8	6.332	47.37	71	9.3	13.74
1.0690	769.7	6.423	48.05	72	9.4	13.80
1.0693	780.6	6.514	48.73	73	9.4	13.86
1.0694	791.4	6.604	49.40	74	9.4	13.88
1.0696	802.2	6.695	50.08	75	9.4	13.92
1.0698	813.0	6.785	50.76	76	9.5	13.96
1.0699	823.8	6.875	51.43	77	9.5	13.98
1.0700	834.6	6.965	52.10	78	9.5	14.00
1.0700	845.3	7.054	52.77	79	9.5	14.00
1.0700	856.0	7.144	53.44	80	9.5	14.00
1.0699	866.6	7.232	54.10	81	9.5	13.98
1.0698	877.2	7.321	54.77	82	9.5	13.96
1.0696	887.8	7.409	55.42	83	9.4	13.92
1.0693	898.2	7.496	56.08	84	9.4	13.86
1.0689	908.6	7.582	56.72	85	9.4	13.78
1.0685	918.9	7.669	57.37	86	9.3	13.70
1.0680	929.2	7.754	58.01	87	9.2	13.60
1.0675	939.4	7.840	58.65	88	9.2	13.50
1.0668	949.5	7.924	59.27	89	9.1	13.36
1.0661	959.5	8.007	59.90	90	9.0	13.22
1.0652	969.3	8.089	60.52	91	8.9	13.04
1.0643	979.2	8.171	61.13	92	8.8	12.86
1.0632	988.8	8.252	61.73	93	8.6	12.64
1.0619	998.2	8.330	62.32	94	8.5	12.38
1.0605	1007	8.408	62.90	95	8.3	12.10
1.0588	1016	8.483	63.46	96	8.1	11.76
1.0570	1025	8.556	64.01	97	7.8	11.40
1.0549	1034	8.627	64.54	98	7.6	10.98
1.0524	1042	8.695	65.04	99	7.2	10.48
1.0498	1050	8.761	65.54	100	6.9	9.96

Table 10-32
ALBUMEN

Density of Aqueous Albumen Solutions at 15.5°C.

| Specific Gravity | Weight of Albumen in solution expressed in | | | | Degrees | |
	Grams per Liter	Pounds per U. S. Gallon	Pounds per Cubic Foot	Per Cent Albumen	Baumé	Twaddell
1.0026	10.03	0.08367	0.6259	1	0.37	0.52
1.0054	20.11	0.1678	1.255	2	0.77	1.08
1.0078	30.23	0.2523	1.888	3	1.12	1.56
1.0130	50.65	0.4227	3.162	5	1.85	2.60
1.0261	102.6	0.8563	6.406	10	3.66	5.22
1.0384	155.8	1.300	9.724	15	5.32	7.68
1.0515	210.3	1.755	13.13	20	7.06	10.30
1.0644	266.1	2.221	16.61	25	8.72	12.88
1.0780	323.4	2.699	20.19	30	10.42	15.60
1.0919	382.2	3.189	23.86	35	13.12	18.38
1.1058	442.3	3.691	27.61	40	13.78	21.16
1.1204	504.2	4.208	31.48	45	15.48	24.08
1.1352	567.6	4.737	35.44	50	17.16	27.04
1.1511	633.1	5.283	39.52	55	18.90	30.22

Table 10-33
AMMONIA

Specific Gravity of Aqueous Ammonia Solutions.

Adopted 1903 as Standard by the Manufacturing Chemists' Association of the United States.

Authority: W. C. Ferguson.

Specific Gravity determinations were made at 60°F., compared with water at 60°F.

From the Specific Gravities the corresponding degrees Baumé were calculated by the following formula:

$$\text{Baumé} = \frac{140}{\text{Sp. Gr.}} - 130.$$

* Baumé Hydrometers for use with this table must be graduated by the above formula, which formula should always be printed on the scale.

Atomic weights from F. W. Clarke's table of 1901. $O = 16$.

Be°	Sp. Gr.	%NH₃	Be°	Sp. Gr.	%NH₃	Be°	Sp. Gr.	%NH₃
10.00	1.0000	.00	15.00	.9655	8.49	20.00	.9333	17.76
10.25	.9982	.40	15.25	.9639	8.93	20.25	.9318	18.24
10.50	.9964	.80	15.50	.9622	9.38	20.50	.9302	18.72
10.75	.9947	1.21	15.75	.9605	9.83	20.75	.9287	19.20
11.00	.9929	1.62	16.00	.9589	10.28	21.00	.9272	19.68
11.25	.9912	2.04	16.25	.9573	10.73	21.25	.9256	20.16
11.50	.9894	2.46	16.50	.9556	11.18	21.50	.9241	20.64
11.75	.9876	2.88	16.75	.9540	11.64	21.75	.9226	21.12
12.00	.9859	3.30	17.00	.9524	12.10	22.00	.9211	21.60
12.25	.9842	3.73	17.25	.9508	12.56	22.25	.9195	22.08
12.50	.9825	4.16	17.50	.9492	13.02	22.50	.9180	22.56
12.75	.9807	4.59	17.75	.9475	13.49	22.75	.9165	23.04
13.00	.9790	5.02	18.00	.9459	13.96	23.00	.9150	23.52
13.25	.9773	5.45	18.25	.9444	14.43	23.25	.9135	24.01
13.50	.9756	5.88	18.50	.9428	14.90	23.50	.9121	24.50
13.75	.9739	6.31	18.75	.9412	15.37	23.75	.9106	24.99
14.00	.9722	6.74	19.00	.9396	15.84	24.00	.9091	25.48
14.25	.9705	7.17	19.25	.9380	16.32	24.25	.9076	25.97
14.50	.9689	7.61	19.50	.9365	16.80	24.50	.9061	26.46
14.75	.9672	8.05	19.75	.9349	17.28	24.75	.9047	26.95

Table 10-33 (*Continued*)
AMMONIA

Be°	Sp. Gr.	%NH₃	Be°	Sp. Gr.	%NH₃	Be°	Sp. Gr.	%NH₃
25.00	.9032	27.44	26.50	.8946	30.38	28.00	.8861	33.32
25.25	.9018	27.93	26.75	.8931	30.87	28.25	.8847	33.81
25.50	.9003	28.42	27.00	.8917	31.36	28.50	.8833	34.30
25.75	.8989	28.91	27.25	.8903	31.85	28.75	.8819	34.79
26.00	.8974	29.40	27.50	.8889	32.34	29.00	.8805	35.28
26.25	.8960	29.89	27.75	.8875	32.83			

Allowance For Temperature

The coefficient of expansion for Ammonia Solutions, varying with the temperature, correction must be applied according to the following table:

Degrees Baumé	Corrections to be added for each degree below 60° F.		Corrections to be subtracted for each degree above 60° F.			
	40° F.	50° F.	70° F.	80° F.	90° F.	100° F.
14°	.015° Bé	.017° Bé	.020° Bé	.022° Bé	.024° Bé	.026° Bé
16°	.021° Bé	.023° Bé	.026° Bé	.028° Bé	.030° Bé	.032° Bé
18°	.027° Bé	.029° Bé	.031° Bé	.033° Bé	.035° Bé	.037° Bé
20°	.033° Bé	.036° Bé	.037° Bé	.038° Bé	.040° Bé	.042° Bé
22°	.039° Bé	.042° Bé	.043° Bé	.045° Bé	.047° Bé	
26°	.053° Bé	.057° Bé	.057° Bé	.059° Bé		

Density of Aqueous Ammonia Solutions at $\frac{20°}{4°}$ C. Computed from Values Given in the International Critical Tables.

Specific Gravity	Weight of NH₃ in solution expressed in				Degrees	
	Grams per Liter	Pounds per U. S. Gallon	Pounds per Cubic Foot	Per Cent NH₃	Baumé	Twaddell
0.9939	9.939	0.08294	0.6205	1	10.9	.....
0.9895	19.79	0.1652	1.235	2	11.5	.....
0.9811	39.24	0.3275	2.450	4	12.7	.....
0.9730	58.38	0.4872	3.645	6	13.9	.....
0.9651	77.21	0.6443	4.820	8	15.1	.....
0.9575	95.75	0.7991	5.978	10	16.2	.....
0.9501	114.0	0.9515	7.118	12	17.3	.....
0.9430	132.0	1.102	8.242	14	18.5	.....
0.9362	149.8	1.250	9.352	16	19.5	.....
0.9295	167.3	1.396	10.45	18	20.6	.....
0.9229	184.6	1.540	11.52	20	21.7	.....
0.9164	201.6	1.682	12.59	22	22.8	.....
0.9101	218.4	1.823	13.64	24	23.8	.....
0.9040	235.0	1.962	14.67	26	24.9	.....
0.8980	251.4	2.098	15.70	28	25.9	.....
0.8920	267.6	2.233	16.71	30	27.0	.....

Table 10-33 (*Continued*)
AMMONIA
Determined in Sealed Tubes at $\frac{15°}{4}$ C.

| Specific Gravity | Weight of NH₃ in solution expressed in | | | | Degrees | |
	Grams per Liter	Pounds per U. S. Gallon	Pounds per Cubic Foot	Per Cent NH₃	Baumé	Twaddell
0.849	382.1	3.188	23.85	45	34.9	.....
0.832	416.0	3.472	25.97	50	38.3	.....
0.815	448.3	3.741	27.98	55	41.8	.....
0.796	477.6	3.986	29.82	60	45.9	.....
0.776	504.4	4.209	31.49	65	50.4	.....
0.755	528.5	4.411	32.99	70	55.4	.....
0.733	549.8	4.588	34.32	75	61.0	.....
0.711	568.8	4.747	35.51	80	66.9	.....
0.688	584.8	4.880	36.51	85	73.5	.....
0.665	598.5	4.995	37.36	90	80.5	.....
0.642	609.9	5.090	38.08	95	88.1	.....
0.618	618.0	5.157	38.58	100	96.5	.....

Table 10-34
ARSENIC ACID

Density of Aqueous Arsenic Acid Solutions at $\frac{15°}{4}$ C. Computed from Values Given in the International Critical Tables.

| Specific Gravity | Weight of H₃AsO₄ in solution expressed in | | | | Degrees | |
	Grams per Liter	Pounds per U. S. Gallon	Pounds per Cubic Foot	Per Cent H₃AsO₄	Baumé	Twaddell
1.0057	10.06	0.08393	0.6279	1	0.8	1.14
1.0124	20.25	0.1690	1.264	2	1.8	2.48
1.0260	41.04	0.3425	2.562	4	3.7	5.20
1.0398	62.39	0.5207	3.895	6	5.6	7.96
1.0538	84.30	0.7036	5.263	8	7.4	10.76
1.0681	106.8	0.8914	6.668	10	9.3	13.62
1.0826	129.9	1.084	8.110	12	11.1	16.52
1.0975	153.7	1.282	9.592	14	12.9	19.50
1.1128	178.0	1.486	11.12	16	14.7	22.56
1.1285	203.1	1.695	12.68	18	16.5	25.70
1.1447	228.9	1.911	14.29	20	18.3	28.94
1.1614	255.5	2.132	15.95	22	20.2	32.28
1.1785	282.8	2.360	17.66	24	22.0	35.70
1.1961	311.0	2.595	19.41	26	23.8	39.22
1.2143	340.0	2.837	21.23	28	25.6	42.86
1.2331	369.9	3.087	23.09	30	27.4	46.62
1.2829	449.0	3.747	28.03	35	32.0	56.58
1.3370	534.8	4.463	33.39	40	36.6	67.40
1.3959	628.2	5.242	39.22	45	41.1	79.18
1.4602	730.1	6.093	45.58	50	45.7	92.40
1.5304	841.7	7.024	52.55	55	50.3	106.08
1.6070	964.2	8.047	60.20	60	54.8	121.40
1.6904	1099	9.170	68.60	65	59.2	138.08
1.7811	1247	10.40	77.84	70	63.6	156.22

Table 10-34 (*Continued*)
ARSENIC ACID

Specific Gravity	Weight of As$_2$O$_5$ in solution expressed in			Per Cent As$_2$O$_5$	Degrees	
	Grams per Liter	Pounds per U. S. Gallon	Pounds per Cubic Foot		Baumé	Twaddell
1.0057	8.142	0.06795	0.5083	0.8096	0.8	1.14
1.0124	16.39	0.1368	1.023	1.619	1.8	2.48
1.0260	33.23	0.2773	2.075	3.239	3.7	5.20
1.0398	50.51	0.4216	3.154	4.858	5.6	7.96
1.0538	68.25	0.5696	4.261	6.477	7.4	10.76
1.0681	86.47	0.7217	5.399	8.096	9.3	13.62
1.0826	105.2	0.8778	6.567	9.716	11.1	16.52
1.0975	124.3	1.038	7.763	11.33	12.9	19.50
1.1128	144.1	1.203	8.997	12.95	14.7	22.56
1.1285	164.4	1.372	10.26	14.57	16.5	25.70
1.1447	185.3	1.547	11.57	16.19	18.3	28.94
1.1614	206.8	1.726	12.91	17.81	20.2	32.28
1.1785	229.0	1.911	14.30	19.43	22.0	35.70
1.1961	251.8	2.101	15.72	21.05	23.8	39.22
1.2143	275.3	2.297	17.19	22.67	25.6	42.86
1.2331	299.5	2.500	18.70	24.29	27.4	46.62
1.2829	363.6	3.034	22.70	28.34	32.0	56.58
1.3370	433.1	3.614	27.04	32.39	36.6	67.40
1.3959	508.5	4.244	31.75	36.43	41.1	79.18
1.4602	591.1	4.933	36.90	40.48	45.7	92.40
1.5304	681.5	5.687	42.55	44.53	50.3	106.08
1.6070	780.7	6.515	48.74	48.58	54.8	121.40
1.6904	889.7	7.425	55.54	52.63	59.2	138.08
1.7811	1009	8.423	63.01	56.67	63.6	156.22

Table 10-35
CHROMIC ACID

Density of Aqueous Chromic Acid Solutions at $\frac{15°}{4}$ C. Computed from Values Given in the International Critical Tables.

Specific Gravity	Weight of CrO$_3$ in solution expressed in			Per Cent CrO$_3$	Degrees	
	Grams per Liter	Pounds per U. S. Gallon	Pounds per Cubic Foot		Baumé	Twaddell
1.006	10.06	0.08395	0.6280	1	0.9	1.2
1.014	20.28	0.1692	1.266	2	2.0	2.8
1.030	41.20	0.3438	2.572	4	4.2	6.0
1.045	62.70	0.5233	3.914	6	6.2	9.0
1.060	84.80	0.7077	5.294	8	8.2	12.0
1.076	107.6	0.8980	6.717	10	10.2	15.2
1.093	131.2	1.095	8.188	12	12.3	18.6
1.110	155.4	1.297	9.702	14	14.4	22.0
1.127	180.3	1.505	11.26	16	16.3	25.4
1.145	206.1	1.720	12.87	18	18.4	29.0
1.163	232.6	1.941	14.52	20	20.3	32.6
1.181	259.8	2.168	16.22	22	22.2	36.2
1.200	288.0	2.403	17.98	24	24.2	40.0
1.220	317.2	2.647	19.80	26	26.2	44.0
1.240	347.2	2.898	21.68	28	28.1	48.0
1.260	378.0	3.155	23.60	30	29.9	52.0
1.313	459.6	3.835	28.69	35	34.6	62.6
1.371	548.4	4.577	34.24	40	39.2	74.2
1.435	645.8	5.389	40.31	45	44.0	87.0
1.505	752.5	6.280	46.98	50	48.7	101.0
1.581	869.6	7.257	54.29	55	53.3	116.2
1.663	997.8	8.327	62.29	60	57.8	132.6

Table 10-36
CITRIC ACID

Density of Aqueous Citric Acid Solutions at 15° C. Computed from the Values of Gerlach.

| Specific Gravity | Weight of $C_6H_8O_7 \cdot H_2O$ in solution expressed in | | | | Degrees | |
	Grams per Liter	Pounds per U. S. Gallon	Pounds per Cubic Foot	Per Cent $C_6H_8O_7 \cdot H_2O$	Baumé	Twaddell
1.0074	20.15	0.1681	1.258	2	1.1	1.48
1.0149	40.60	0.3388	2.534	4	2.1	2.98
1.0227	61.36	0.5121	3.831	6	3.2	4.54
1.0309	82.47	0.6883	5.149	8	4.3	6.18
1.0392	103.9	0.8673	6.488	10	5.5	7.84
1.0470	125.6	1.049	7.844	12	6.5	9.40
1.0549	147.7	1.232	9.220	14	7.6	10.98
1.0632	170.1	1.420	10.62	16	8.6	12.64
1.0718	192.9	1.610	12.04	18	9.7	14.36
1.0805	216.1	1.803	13.49	20	10.8	16.10
1.0889	239.6	1.999	14.96	22	11.8	17.78
1.0972	263.3	2.198	16.44	24	12.8	19.44
1.1060	287.6	2.400	17.95	26	13.9	21.20
1.1152	312.3	2.606	19.49	28	15.0	23.04
1.1244	337.3	2.815	21.06	30	16.0	24.88
1.1332	362.6	3.026	22.64	32	17.1	26.66
1.1422	388.3	3.241	24.24	34	18.1	28.44
1.1515	414.5	3.460	25.88	36	19.1	30.30
1.1612	441.3	3.682	27.55	38	20.1	32.24
1.1709	468.4	3.909	29.24	40	21.2	34.18
1.1814	496.2	4.141	30.98	42	22.3	36.28
1.1899	523.6	4.369	32.69	44	23.1	37.98
1.1998	551.9	4.606	34.46	46	24.2	39.96
1.2103	580.9	4.848	36.27	48	25.2	42.06
1.2204	610.2	5.092	38.09	50	26.2	44.08
1.2307	640.0	5.341	39.95	52	27.2	46.14
1.2410	670.1	5.593	41.84	54	28.2	48.20
1.2514	700.8	5.848	43.75	56	29.1	50.28
1.2627	732.4	6.112	45.72	58	30.2	52.54
1.2738	764.3	6.378	47.71	60	31.2	54.76
1.2849	796.6	6.648	49.73	62	32.2	56.98
1.2960	829.4	6.922	51.78	64	33.1	59.20
1.3071	862.7	7.199	53.86	66	34.1	61.42

Table 10-37
FERRIC CHLORIDE

Density of Aqueous Ferric Chloride Solutions at $\frac{20°}{4}$ C. Computed from Values Determined by The Dow Chemical Co. and Reprinted by Permission.

| Specific Gravity | Weight of $FeCl_3$ in solution expressed in | | | | Degrees | |
	Grams per Liter	Pounds per U. S. Gallon	Pounds per Cubic Foot	Per Cent $FeCl_3$	Baumé	Twaddell
1.009	10.09	0.08420	0.6299	1	1.3	1.8
1.017	20.34	0.1697	1.270	2	2.4	3.4
1.026	30.78	0.2569	1.922	3	3.7	5.2
1.034	41.36	0.3452	2.582	4	4.8	6.8
1.043	52.15	0.4352	3.256	5	6.0	8.6
1.052	63.12	0.5267	3.940	6	7.2	10.4
1.060	74.20	0.6192	4.632	7	8.2	12.0
1.069	85.52	0.7137	5.339	8	9.4	13.8
1.077	96.93	0.8089	6.051	9	10.4	15.4
1.086	108.6	0.9063	6.780	10	11.5	17.2

Table 10-37 (*Continued*)
FERRIC CHLORIDE

| Specific Gravity | Weight of FeCl₃ in solution expressed in | | | | Degrees | |
	Grams per Liter	Pounds per U. S. Gallon	Pounds per Cubic Foot	Per Cent FeCl₃	Baumé	Twaddell
1.094	120.3	1.004	7.510	11	12.5	18.8
1.105	132.6	1.107	8.278	12	13.8	21.0
1.115	145.0	1.210	9.052	13	15.0	23.0
1.125	157.5	1.314	9.832	14	16.1	25.0
1.135	170.3	1.421	10.63	15	17.2	27.0
1.144	183.0	1.527	11.42	16	18.3	28.8
1.154	196.2	1.637	12.25	17	19.4	30.8
1.164	209.5	1.748	13.08	18	20.4	32.8
1.174	223.1	1.862	13.93	19	21.5	34.8
1.185	237.0	1.978	14.80	20	22.6	37.0
1.195	251.0	2.095	15.67	21	23.7	39.0
1.206	265.3	2.214	16.56	22	24.6	41.2
1.216	279.7	2.334	17.46	23	25.8	43.2
1.226	294.2	2.455	18.37	24	26.7	45.2
1.237	309.3	2.581	19.31	25	27.8	47.4
1.249	324.7	2.710	20.27	26	28.9	49.8
1.260	340.2	2.839	21.24	27	29.9	52.0
1.271	355.9	2.970	22.22	28	30.9	54.2
1.283	372.1	3.105	23.23	29	32.0	56.6
1.295	388.5	3.242	24.25	30	33.0	59.0
1.307	405.2	3.381	25.30	31	34.1	61.4
1.319	422.1	3.522	26.35	32	35.1	63.8
1.331	439.2	3.665	27.42	33	36.1	66.2
1.344	457.0	3.814	28.53	34	37.1	68.8
1.357	475.0	3.964	29.65	35	38.1	71.4
1.370	493.2	4.116	30.79	36	39.2	74.0
1.383	511.7	4.270	31.94	37	40.2	76.6
1.396	530.5	4.427	33.12	38	41.1	79.2
1.410	549.9	4.589	34.33	39	42.2	82.0
1.423	569.2	4.750	35.53	40	43.1	84.6
1.436	588.8	4.914	36.76	41	44.0	87.2
1.449	608.6	5.079	37.99	42	44.9	89.8
1.462	628.7	5.247	39.25	43	45.8	92.4
1.475	649.0	5.416	40.52	44	46.7	95.0
1.488	669.6	5.588	41.80	45	47.6	97.6
1.501	690.5	5.762	43.11	46	48.4	100.2

Table 10-38
FORMIC ACID

Density of Aqueous Formic Acid Solutions at $\frac{20°}{4°}$ C. Computed from Values Given in the International Critical Tables.

| Specific Gravity | Weight of HCO₂H in solution expressed in | | | | Degrees | |
	Grams per Liter	Pounds per U. S. Gallon	Pounds per Cubic Foot	Per Cent HCO₂H	Baumé	Twaddell
0.9982	.....	.....	.....	0	....	.....
1.0019	10.02	0.08361	0.6255	1	0.3	0.38
1.0044	20.09	0.1676	1.254	2	0.6	0.88
1.0070	30.21	0.2521	1.886	3	1.0	1.40
1.0093	40.37	0.3369	2.520	4	1.3	1.86
1.0115	50.58	0.4221	3.157	5	1.6	2.30
1.0141	60.85	0.5078	3.799	6	2.0	2.82
1.0170	71.19	0.5941	4.444	7	2.4	3.40
1.0196	81.57	0.6807	5.092	8	2.8	3.92
1.0221	91.99	0.7677	5.743	9	3.1	4.42

Table 10-38 (*Continued*)
FORMIC ACID

Specific Gravity	Weight of HCO$_2$H in solution expressed in			Per Cent HCO$_2$H	Degrees	
	Grams per Liter	Pounds per U. S. Gallon	Pounds per Cubic Foot		Baumé	Twaddell
1.0246	102.5	0.8551	6.397	10	3.5	4.92
1.0271	113.0	0.9429	7.053	11	3.8	5.42
1.0296	123.6	1.031	7.713	12	4.2	5.92
1.0321	134.2	1.120	8.376	13	4.5	6.42
1.0345	144.8	1.209	9.042	14	4.8	6.90
1.0370	156.6	1.298	9.711	15	5.2	7.40
1.0393	166.3	1.388	10.38	16	5.5	7.86
1.0417	177.1	1.478	11.06	17	5.8	8.34
1.0441	187.9	1.568	11.73	18	6.1	8.82
1.0464	198.8	1.659	12.41	19	6.4	9.28
1.0488	209.8	1.751	13.10	20	6.8	9.76
1.0512	220.8	1.842	13.78	21	7.1	10.24
1.0537	231.8	1.935	14.47	22	7.4	10.74
1.0561	242.9	2.027	15.16	23	7.7	11.22
1.0585	254.0	2.120	15.86	24	8.0	11.70
1.0609	265.2	2.213	16.56	25	8.3	12.18
1.0633	276.5	2.307	17.26	26	8.6	12.66
1.0656	287.7	2.401	17.96	27	8.9	13.12
1.0681	299.1	2.496	18.67	28	9.3	13.62
1.0705	310.4	2.591	19.38	29	9.6	14.10
1.0729	321.9	2.686	20.09	30	9.9	14.58
1.0753	333.3	2.782	20.81	31	10.2	15.06
1.0777	344.9	2.878	21.53	32	10.5	15.54
1.0800	356.4	2.974	22.25	33	10.7	16.00
1.0823	368.0	3.071	22.97	34	11.0	16.46
1.0847	379.6	3.168	23.70	35	11.3	16.94
1.0871	391.4	3.266	24.43	36	11.6	17.42
1.0895	403.1	3.364	25.17	37	11.9	17.90
1.0919	414.9	3.463	25.90	38	12.2	18.38
1.0940	426.7	3.561	26.64	39	12.5	18.80
1.0963	438.5	3.660	27.38	40	12.7	19.26
1.0990	450.6	3.760	28.13	41	13.1	19.80
1.1015	462.6	3.861	28.88	42	13.4	20.30
1.1038	474.6	3.961	29.63	43	13.6	20.76
1.1062	486.7	4.062	30.39	44	13.9	21.24
1.1085	498.8	4.163	31.14	45	14.2	21.70
1.1108	511.0	4.264	31.90	46	14.5	22.16
1.1130	523.1	4.366	32.66	47	14.7	22.60
1.1157	535.5	4.469	33.43	48	15.0	23.14
1.1185	548.1	4.574	34.22	49	15.4	23.70
1.1207	560.4	4.676	34.98	50	15.6	24.14
1.1223	572.4	4.777	35.73	51	15.8	24.46
1.1244	584.7	4.879	36.50	52	16.0	24.88
1.1269	597.3	4.984	37.29	53	16.3	25.38
1.1295	609.9	5.090	38.08	54	16.6	25.90
1.1320	622.6	5.196	38.87	55	16.9	26.40
1.1342	635.2	5.301	39.65	56	17.2	26.84
1.1361	647.6	5.404	40.43	57	17.4	27.22
1.1381	660.1	5.509	41.21	58	17.6	27.62
1.1401	672.7	5.614	41.99	59	17.8	28.02
1.1424	685.4	5.720	42.79	60	18.1	28.48
1.1448	698.3	5.828	43.60	61	18.3	28.96
1.1473	711.3	5.936	44.41	62	18.6	29.46
1.1493	724.1	6.043	45.20	63	18.8	29.86
1.1517	737.1	6.151	46.02	64	19.1	30.34
1.1543	750.3	6.262	46.84	65	19.4	30.86
1.1565	763.3	6.370	47.65	66	19.6	31.30
1.1584	776.1	6.477	48.45	67	19.8	31.68
1.1604	789.1	6.585	49.26	68	20.0	32.08
1.1628	802.3	6.696	50.09	69	20.3	32.56

Table 10-38 (*Continued*)
FORMIC ACID

| Specific Gravity | Weight of HCO₂H in solution expressed in | | | | Degrees | |
	Grams per Liter	Pounds per U. S. Gallon	Pounds per Cubic Foot	Per Cent HCO₂H	Baumé	Twaddell
1.1655	815.9	6.809	50.93	70	20.6	33.10
1.1677	829.1	6.919	51.76	71	20.8	33.54
1.1702	842.5	7.031	52.60	72	21.1	34.04
1.1728	856.1	7.145	53.45	73	21.4	34.56
1.1752	869.6	7.258	54.29	74	21.6	35.04
1.1769	882.7	7.366	55.11	75	21.8	35.38
1.1785	895.7	7.475	55.92	76	22.0	35.70
1.1801	908.7	7.583	56.73	77	22.1	36.02
1.1818	921.8	7.693	57.55	78	22.3	36.36
1.1837	935.1	7.804	58.38	79	22.5	36.74
1.1860	948.8	7.918	59.23	80	22.7	37.20
1.1876	962.0	8.028	60.05	81	22.9	37.52
1.1896	975.5	8.141	60.90	82	23.1	37.92
1.1914	988.9	8.252	61.73	83	23.3	38.28
1.1929	1002	8.362	62.56	84	23.5	38.59
1.1953	1016	8.479	63.43	85	23.7	39.06
1.1976	1030	8.595	64.30	86	23.9	39.52
1.1994	1043	8.708	65.14	87	24.1	39.88
1.2012	1057	8.822	65.99	88	24.3	40.24
1.2028	1070	8.934	66.83	89	24.5	40.56
1.2044	1084	9.046	67.67	90	24.6	40.88
1.2059	1097	9.158	68.51	91	24.8	41.18
1.2078	1111	9.273	69.37	92	25.0	41.56
1.2099	1125	9.390	70.25	93	25.2	41.98
1.2117	1139	9.505	71.11	94	25.3	42.34
1.2140	1153	9.625	72.00	95	25.6	42.80
1.2158	1167	9.741	72.87	96	25.7	43.16
1.2170	1180	9.852	73.70	97	25.9	43.40
1.2183	1194	9.964	74.54	98	26.0	43.66
1.2202	1208	10.08	75.42	99	26.2	44.04
1.2212	1221	10.19	76.24	100	26.3	44.24

Table 10-39
GLYCEROL

Density of Aqueous Glycerol Solutions at $\frac{20°}{4°}$ C. Computed from the Values of Bosart and Snoddy. *Ind. Eng. Chem. 20, 1377 (1928)*.

| Specific Gravity | Weight of C₃H₈O₃ in solution expressed in | | | | Degrees | |
	Grams per Liter	Pounds per U. S. Gallon	Pounds per Cubic Foot	Per Cent C₃H₈O₃	Baumé	Twaddell
0.99823	.....	....	......	0	...	....
1.00060	10.01	0.08350	0.6247	1	0.1	0.12
1.00300	20.06	0.1674	1.252	2	0.4	0.60
1.00540	30.16	0.2517	1.883	3	0.8	1.08
1.00780	40.31	0.3364	2.516	4	1.1	1.56
1.01015	50.51	0.4215	3.153	5	1.5	2.03
1.01255	60.75	0.5070	3.793	6	1.8	2.51
1.01495	71.05	0.5929	4.436	7	2.1	2.99
1.01730	81.38	0.6792	5.081	8	2.5	3.46
1.01970	91.77	0.7659	5.729	9	2.8	3.94
1.02210	102.2	0.8530	6.381	10	3.1	4.42
1.02455	112.7	0.9406	7.036	11	3.5	4.91
1.02705	123.2	1.028	7.691	12	3.8	5.41
1.02955	133.7	1.116	8.347	13	4.1	5.91
1.03200	144.5	1.206	9.020	14	4.5	6.40

Table 10-39 (*Continued*)
GLYCEROL

Specific Gravity	Weight of C3H8O3 in solution expressed in			Per Cent C3H8O3	Degrees	
	Grams per Liter	Pounds per U. S. Gallon	Pounds per Cubic Foot		Baumé	Twaddell
1.03450	155.2	1.295	9.688	15	4.8	6.90
1.03695	165.9	1.385	10.36	16	5.2	7.39
1.03945	176.7	1.475	11.03	17	5.5	7.89
1.04195	187.6	1.565	11.71	18	5.8	8.39
1.04440	198.4	1.656	12.39	19	6.2	8.88
1.04690	209.4	1.748	13.07	20	6.5	9.38
1.04950	220.4	1.839	13.76	21	6.8	9.90
1.05205	231.5	1.932	14.45	22	7.2	10.41
1.05465	242.6	2.025	15.15	23	7.5	10.93
1.05720	253.7	2.117	15.84	24	7.8	11.44
1.05980	265.0	2.212	16.54	25	8.2	11.96
1.06240	276.2	2.305	17.24	26	8.5	12.48
1.06495	287.5	2.399	17.95	27	8.8	12.99
1.06755	298.9	2.494	18.66	28	9.2	13.51
1.07010	310.3	2.590	19.37	29	9.5	14.02
1.07270	321.8	2.686	20.09	30	9.8	14.54
1.07535	333.4	2.782	20.81	31	10.2	15.07
1.07800	345.0	2.879	21.54	32	10.5	15.60
1.08070	356.6	2.976	22.26	33	10.8	16.14
1.08335	368.3	3.074	22.99	34	11.2	16.67
1.08600	380.1	3.172	23.73	35	11.5	17.20
1.08865	391.9	3.271	24.47	36	11.8	17.73
1.09135	403.8	3.370	25.21	37	12.1	18.27
1.09400	415.7	3.469	25.95	38	12.5	18.80
1.09665	427.7	3.569	26.70	39	12.8	19.33
1.09930	439.7	3.669	27.45	40	13.1	19.86
1.10200	451.8	3.770	28.21	41	13.4	20.40
1.10470	464.0	3.872	28.97	42	13.8	20.94
1.10740	476.2	3.974	29.73	43	14.1	21.48
1.11010	488.4	4.076	30.49	44	14.4	22.02
1.11280	500.8	4.179	31.26	45	14.7	22.56
1.11550	513.1	4.282	32.03	46	15.0	23.10
1.11820	525.6	4.386	32.81	47	15.3	23.64
1.12090	538.0	4.490	33.59	48	15.6	24.18
1.12360	550.6	4.595	34.37	49	16.0	24.72
1.12630	563.2	4.700	35.16	50	16.3	25.26
1.12905	575.8	4.805	35.95	51	16.6	25.81
1.13180	588.5	4.911	36.74	52	16.9	26.36
1.13455	601.3	5.018	37.54	53	17.2	26.91
1.13730	614.1	5.125	38.34	54	17.5	27.46
1.14005	627.0	5.233	39.14	55	17.8	28.01
1.14280	640.0	5.341	39.96	56	18.1	28.56
1.14555	653.0	5.450	40.77	57	18.4	29.11
1.14830	665.9	5.557	41.57	58	18.7	29.66
1.15105	679.1	5.667	42.40	59	19.0	30.21
1.15380	692.3	5.778	43.22	60	19.3	30.76
1.15655	705.5	5.887	44.04	61	19.6	31.31
1.15930	718.8	5.999	44.87	62	19.9	31.86
1.16205	732.1	6.110	45.71	63	20.2	32.41
1.16475	745.4	6.221	46.54	64	20.5	32.95
1.16750	758.9	6.333	47.38	65	20.8	33.50
1.17025	772.4	6.446	48.22	66	21.0	34.05
1.17300	785.9	6.559	49.06	67	21.3	34.60
1.17575	800.0	6.676	49.94	68	21.6	35.15
1.17850	813.2	6.786	50.77	69	21.9	35.70
1.18125	826.9	6.901	51.62	70	22.2	36.25
1.18395	840.6	7.015	52.48	71	22.5	36.79
1.18670	854.4	7.130	53.34	72	22.8	37.34
1.18940	868.3	7.246	54.21	73	23.1	37.88
1.19215	882.2	7.362	55.08	74	23.3	38.43

Table 10-39 (*Continued*)
GLYCEROL

Specific Gravity	Weight of $C_3H_8O_3$ in solution expressed in				Degrees	
	Grams per Liter	Pounds per U. S. Gallon	Pounds per Cubic Foot	Per Cent $C_3H_8O_3$	Baumé	Twaddell
1.19485	896.1	7.478	55.93	75	23.6	38.97
1.19760	910.0	7.594	56.81	76	23.9	39.52
1.20030	924.2	7.713	57.70	77	24.2	40.06
1.20305	938.4	7.831	58.58	78	24.4	40.61
1.20575	952.4	7.948	59.46	79	24.7	41.15
1.20850	966.8	8.068	60.36	80	25.0	41.70
1.21115	981.0	8.187	61.24	81	25.2	42.23
1.21380	995.3	8.306	62.14	82	25.5	42.76
1.21650	1010	8.429	63.05	83	25.8	43.30
1.21915	1024	8.543	63.91	84	26.0	43.83
1.22180	1039	8.671	64.86	85	26.3	44.36
1.22445	1053	8.785	65.72	86	26.5	44.89
1.22710	1068	8.913	66.68	87	26.8	45.42
1.22975	1082	9.029	67.54	88	27.1	45.95
1.23245	1097	9.155	68.49	89	27.3	46.49
1.23510	1112	9.280	68.42	90	27.6	47.02
1.23770	1126	9.397	70.30	91	27.8	47.54
1.24035	1141	9.521	71.23	92	28.1	48.07
1.24300	1156	9.646	72.16	93	28.3	48.60
1.24560	1171	9.771	73.09	94	28.6	49.12
1.24825	1186	9.896	74.03	95	28.8	49.65
1.25080	1201	10.02	74.96	96	29.1	50.16
1.25335	1216	10.15	75.90	97	29.3	50.67
1.25590	1231	10.27	76.84	98	29.5	51.18
1.25850	1246	10.40	77.78	99	29.8	51.70
1.26108	1261	10.52	78.72	100	30.0	52.216

Table 10-40
HYDROCHLORIC ACID

Specific Gravity of Aqueous Hydrochloric Acid Solutions. Adopted 1903 as Standard by the Manufacturing Chemists' Association of the United States.

Authority: W. C. Ferguson.

Specific Gravity determinations were made at 60° F., compared with water at 60°F. From the Specific Gravities, the corresponding degrees Baumé were calculated by the following formula:

$$\text{Baumé} = 145 - \frac{145}{\text{Sp. Gr.}}$$

* Baumé Hydrometers for use with this table must be graduated by the above formula which formula should always be printed on the scale.

Atomic weights from F. W. Clarke's table of 1901. O = 16.

ALLOWANCE FOR TEMPERATURE:
10 — 15° Be. — 1/40° Be. or .0002 Sp. Gr. for 1° F.
15 — 22° Be. — 1/30° Be. or .0003 Sp. Gr. for 1° F.
22 — 25° Be. — 1/28° Be. or .00035 Sp. Gr. for 1° F.

Be.°	Sp. Gr.	Tw.°	%HCl.	Be.°	Sp. Gr.	Tw.°	%HCl.	Be.°	Sp. Gr.	Tw.°	%HCl.
1.00	1.0069	1.38	1.40	6.50	1.0469	9.38	9.40	9.00	1.0662	13.24	13.26
2.00	1.0140	2.80	2.82	6.75	1.0488	9.76	9.78	9.25	1.0681	13.62	13.65
3.00	1.0211	4.22	4.25	7.00	1.0507	10.14	10.17	9.50	1.0701	14.02	14.04
4.00	1.0284	5.68	5.69	7.25	1.0526	10.52	10.55	9.75	1.0721	14.42	14.43
5.00	1.0357	7.14	7.15	7.50	1.0545	10.90	10.94	10.00	1.0741	14.82	14.83
5.25	1.0375	7.50	7.52	7.75	1.0564	11.28	11.32	10.25	1.0761	15.22	15.22
5.50	1.0394	7.88	7.89	8.00	1.0584	11.68	11.71	10.50	1.0781	15.62	15.62
5.75	1.0413	8.26	8.26	8.25	1.0603	12.06	12.09	10.75	1.0801	16.02	16.01
6.00	1.0432	8.64	8.64	8.50	1.0623	12.46	12.48	11.00	1.0821	16.42	16.41
6.25	1.0450	9.00	9.02	8.75	1.0642	12.84	12.87	11.25	1.0841	16.82	16.81

Table 10-40 (*Continued*)
HYDROCHLORIC ACID

Be.°	Sp. Gr.	Tw.°	%HCl.	Be.°	Sp. Gr.	Tw.°	%HCl.	Be.°	Sp. Gr.	Tw.°	%HCl.
11.50	1.0861	17.22	17.21	18.0	1.1417	28.34	27.92	21.8	1.1770	35.40	34.83
11.75	1.0881	17.62	17.61	18.1	1.1426	28.52	28.09	21.9	1.1779	35.58	35.02
12.00	1.0902	18.04	18.01	18.2	1.1435	28.70	28.26	22.0	1.1789	35.78	35.21
12.25	1.0922	18.44	18.41	18.3	1.1444	28.88	28.44	22.1	1.1798	35.96	35.40
12.50	1.0943	18.86	18.82	18.4	1.1453	29.06	28.61	22.2	1.1808	36.16	35.59
12.75	1.0964	19.28	19.22	18.5	1.1462	29.24	28.78	22.3	1.1817	36.34	35.78
13.00	1.0985	19.70	19.63	18.6	1.1471	29.42	28.95	22.4	1.1827	36.54	35.97
13.25	1.1006	20.12	20.04	18.7	1.1480	29.60	29.13	22.5	1.1836	36.72	36.16
13.50	1.1027	20.54	20.45	18.8	1.1489	29.78	29.30	22.6	1.1846	36.92	36.35
13.75	1.1048	20.96	20.86	18.9	1.1498	29.96	29.48	22.7	1.1856	37.12	36.54
14.00	1.1069	21.38	21.27	19.0	1.1508	30.16	29.65	22.8	1.1866	37.32	36.73
14.25	1.1090	21.80	21.68	19.1	1.1517	30.34	29.83	22.9	1.1875	37.50	36.93
14.50	1.1111	22.22	22.09	19.2	1.1526	30.52	30.00	23.0	1.1885	37.70	37.14
14.75	1.1132	22.64	22.50	19.3	1.1535	30.70	30.18	23.1	1.1895	37.90	37.36
15.00	1.1154	23.08	22.92	19.4	1.1544	30.88	30.35	23.2	1.1904	38.08	37.58
15.25	1.1176	23.52	23.33	19.5	1.1554	31.08	30.53	23.3	1.1914	38.28	37.80
15.50	1.1197	23.94	23.75	19.6	1.1563	31.26	30.71	23.4	1.1924	38.48	38.03
15.75	1.1219	24.38	24.16	19.7	1.1572	31.44	30.90	23.5	1.1934	38.68	38.26
16.0	1.1240	24.80	24.57	19.8	1.1581	31.62	31.08	23.6	1.1944	38.88	38.49
16.1	1.1248	24.96	24.73	19.9	1.1590	31.80	31.27	23.7	1.1953	39.06	38.72
16.2	1.1256	25.12	24.90	20.0	1.1600	32.00	31.45	23.8	1.1963	39.26	38.95
16.3	1.1265	25.30	25.06	20.1	1.1609	32.18	31.64	23.9	1.1973	39.46	39.18
16.4	1.1274	25.48	25.23	20.2	1.1619	32.38	31.82	24.0	1.1983	39.66	39.41
16.5	1.1283	25.66	25.39	20.3	1.1628	32.56	32.01	24.1	1.1993	39.86	39.64
16.6	1.1292	25.84	25.56	20.4	1.1637	32.74	32.19	24.2	1.2003	40.06	39.86
16.7	1.1301	26.02	25.72	20.5	1.1647	32.94	32.38	24.3	1.2013	40.26	40.09
16.8	1.1310	26.20	25.89	20.6	1.1656	33.12	32.56	24.4	1.2023	40.46	40.32
16.9	1.1319	26.38	26.05	20.7	1.1666	33.32	32.75	24.5	1.2033	40.66	40.55
17.0	1.1328	26.56	26.22	20.8	1.1675	33.50	32.93	24.6	1.2043	40.86	40.78
17.1	1.1336	26.72	26.39	20.9	1.1684	33.68	33.12	24.7	1.2053	41.06	41.01
17.2	1.1345	26.90	26.56	21.0	1.1694	33.88	33.31	24.8	1.2063	41.26	41.24
17.3	1.1354	27.08	26.73	21.1	1.1703	34.06	33.50	24.9	1.2073	41.46	41.48
17.4	1.1363	27.26	26.90	21.2	1.1713	34.26	33.69	25.0	1.2083	41.66	41.72
17.5	1.1372	27.44	27.07	21.3	1.1722	34.44	33.88	25.1	1.2093	41.86	41.99
17.6	1.1381	27.62	27.24	21.4	1.1732	34.64	34.07	25.2	1.2103	42.06	42.30
17.7	1.1390	27.80	27.41	21.5	1.1741	34.82	34.26	25.3	1.2114	42.28	42.64
17.8	1.1399	27.98	27.58	21.6	1.1751	35.02	34.45	25.4	1.2124	42.48	43.01
17.9	1.1408	28.16	27.75	21.7	1.1760	35.20	34.64	25.5	1.2134	42.68	43.40

Density of Aqueous Hydrochloric Acid Solutions at $\frac{20°}{4°}$ C. Computed from Values Given in the International Critical Tables.

Specific Gravity	Weight of HCl in solution expressed in					Degrees	
	Grams per Liter	Pounds per U.S. Gallon	Pounds per Cubic Foot	N	%	Baumé	Twaddell
1.0032	10.03	0.08372	0.6263	0.275	1	0.5	0.64
1.0082	20.16	0.1683	1.259	0.553	2	1.2	1.64
1.0181	40.72	0.3399	2.542	1.12	4	2.6	3.62
1.0279	61.67	0.5147	3.850	1.69	6	3.9	5.58
1.0376	83.01	0.6927	5.182	2.28	8	5.3	7.52
1.0474	104.7	0.8741	6.539	2.87	10	6.6	9.48
1.0574	126.9	1.059	7.922	3.48	12	7.9	11.48
1.0675	149.5	1.247	9.330	4.10	14	9.2	13.50
1.0776	172.4	1.439	10.76	4.73	16	10.4	15.52
1.0878	195.8	1.634	12.22	5.37	18	11.7	17.56
1.0980	219.6	1.833	13.71	6.02	20	12.9	19.60
1.1083	243.8	2.035	15.22	6.69	22	14.2	21.66
1.1187	268.5	2.241	16.76	7.36	24	15.4	23.74
1.1290	293.5	2.450	18.33	8.05	26	16.6	25.80
1.1392	319.0	2.662	19.91	8.75	28	17.7	27.84

Table 10-40 (*Continued*)
HYDROCHLORIC ACID

| Specific Gravity | Weight of HCl in solution expressed in | | | | | Degrees | |
	Grams per Liter	Pounds per U. S. Gallon	Pounds per Cubic Foot	N	%	Baumé	Twaddell
1.1492	344.8	2.877	21.53	9.46	30	18.8	29.84
1.1593	371.0	3.096	23.16	10.2	32	19.9	31.86
1.1691	397.5	3.317	24.82	10.9	34	21.0	33.82
1.1789	424.4	3.542	26.50	11.6	36	22.0	35.78
1.1885	451.6	3.769	28.20	12.4	38	23.0	37.70
1.1980	479.2	3.999	29.92	13.1	40	24.0	39.60

Table 10-41
HYDROCYANIC ACID

Density of Aqueous Hydrocyanic Acid Solutions at $\frac{15°}{4}$ C. Computed from Values Given in the International Critical Tables.

| Specific Gravity | Weight of HCN in solution expressed in | | | Per Cent HCN | Degrees | |
	Grams per Liter	Pounds per U. S. Gallon	Pounds per Cubic Foot		Baumé	Twaddell
0.998	9.980	0.08329	0.6231	1	10.3	.....
0.996	19.92	0.1662	1.244	2	10.6	.....
0.993	39.72	0.3315	2.480	4	11.0	.....
0.989	59.34	0.4952	3.705	6	11.4	.....
0.984	78.72	0.6569	4.914	8	12.3	.....
0.978	97.80	0.8162	6.106	10	13.2	.....
0.971	116.5	0.9724	7.274	12	14.2	.....
0.964	135.0	1.126	8.426	14	15.2	.....
0.956	153.0	1.277	9.549	16	16.4	.....

Density of Aqueous Hydrocyanic Acid Solutions at $\frac{20°}{4}$ C.

| Specific Gravity | Weight of HCN in solution expressed in | | | Per Cent HCN | Degrees | |
	Grams per Liter	Pounds per U. S. Gallon	Pounds per Cubic Foot		Baumé	Twaddell
0.759	607.2	5.067	37.91	80	54.4	.....
0.752	616.6	5.146	38.50	82	56.2	.....
0.745	625.8	5.223	39.07	84	57.9	.....
0.738	634.7	5.297	39.62	86	59.7	.....
0.731	643.3	5.368	40.16	88	61.5	.....
0.724	651.6	5.438	40.68	90	63.4	.....
0.717	659.6	5.505	41.18	92	65.3	.....
0.711	668.3	5.578	41.72	94	66.9	.....
0.704	675.8	5.640	42.19	96	68.9	.....
0.697	683.1	5.700	42.64	98	70.9	.....
0.691	691.0	5.767	43.14	100	72.6	.....

Table 10-42
HYDROFLUOROSILICIC ACID

Density of Aqueous Hydrofluosilicic Acid Solutions at 17.5° C. Computed from the Values of Stolba.

| Specific Gravity | Weight of H_2SiF_6 in solution expressed in | | | Per Cent H_2SiF_6 | Degrees | |
	Grams per Liter	Pounds per U. S. Gallon	Pounds per Cubic Foot		Baumé	Twaddell
1.0040	5.020	0.04189	0.3134	0.5	0.6	0.80
1.0080	10.08	0.08412	0.6293	1.0	1.2	1.60
1.0120	15.18	0.1267	0.9477	1.5	1.7	2.40
1.0161	20.32	0.1696	1.269	2.0	2.3	3.22
1.0201	25.50	0.2128	1.592	2.5	2.9	4.02
1.0242	30.73	0.2564	1.918	3.0	3.4	4.84
1.0283	35.99	0.3004	2.247	3.5	4.0	5.66
1.0324	41.30	0.3446	2.578	4.0	4.6	6.48

Table 10-42 (*Continued*)
HYDROFLUOROSILICIC ACID

| Specific Gravity | Weight of H_2SiF_6 in solution expressed in | | | Per Cent H_2SiF_6 | Degrees | |
	Grams per Liter	Pounds per U. S. Gallon	Pounds per Cubic Foot		Baumé	Twaddell
1.0366	46.65	0.3893	2.912	4.5	5.1	7.32
1.0407	52.04	0.4343	3.249	5.0	5.7	8.14
1.0449	57.47	0.4796	3.588	5.5	6.2	8.98
1.0491	62.95	0.5253	3.930	6.0	6.8	9.82
1.0533	68.46	0.5714	4.274	6.5	7.3	11.06
1.0576	74.03	0.6178	4.622	7.0	7.9	11.52
1.0618	79.64	0.6646	4.972	7.5	8.4	12.36
1.0661	85.29	0.7118	5.325	8.0	9.0	13.22
1.0704	90.98	0.7593	5.680	8.5	9.5	14.08
1.0747	96.72	0.8072	6.038	9.0	10.1	14.94
1.0791	102.5	0.8555	6.400	9.5	10.6	15.82
1.0834	108.3	0.9041	6.764	10.0	11.2	16.68
1.0878	114.2	0.9532	7.131	10.5	11.7	17.56
1.0922	120.1	1.003	7.500	11.0	12.2	18.44
1.0966	126.1	1.052	7.873	11.5	12.8	19.32
1.1011	132.1	1.103	8.249	12.0	13.3	20.22
1.1055	138.2	1.153	8.627	12.5	13.8	21.10
1.1100	144.3	1.204	9.009	13.0	14.4	22.00
1.1145	150.5	1.256	9.393	13.5	14.9	22.90
1.1190	156.7	1.307	9.780	14.0	15.4	23.80
1.1236	162.9	1.360	10.17	14.5	16.0	24.72
1.1281	169.2	1.412	10.56	15.0	16.5	25.62
1.1327	175.6	1.465	10.96	15.5	17.0	26.54
1.1373	182.0	1.519	11.36	16.0	17.5	27.46
1.1419	188.4	1.572	11.76	16.5	18.0	28.38
1.1466	194.9	1.627	12.17	17.0	18.5	29.32
1.1512	201.5	1.681	12.58	17.5	19.0	30.24
1.1559	208.1	1.736	12.99	18.0	19.6	31.18
1.1606	214.7	1.792	13.40	18.5	20.1	32.12
1.1653	221.4	1.848	13.82	19.0	20.6	33.06
1.1701	228.2	1.904	14.24	19.5	21.1	34.02
1.1748	235.0	1.961	14.67	20.0	21.6	34.96
1.1796	241.8	2.018	15.10	20.5	22.1	35.92
1.1844	248.7	2.076	15.53	21.0	22.6	36.88
1.1892	255.7	2.134	15.96	21.5	23.1	37.84
1.1941	262.7	2.192	16.40	22.0	23.6	38.82
1.1989	269.8	2.251	16.84	22.5	24.1	39.78
1.2038	276.9	2.311	17.29	23.0	24.6	40.76
1.2087	284.0	2.370	17.73	23.5	25.0	41.74
1.2136	291.3	2.431	18.18	24.0	25.5	42.72
1.2186	298.6	2.492	18.64	24.5	26.0	43.72
1.2235	305.9	2.553	19.10	25.0	26.5	44.70
1.2285	313.3	2.614	19.56	25.5	27.0	45.70
1.2335	320.7	2.676	20.02	26.0	27.5	46.70
1.2385	328.2	2.739	20.49	26.5	27.9	47.70
1.2436	335.8	2.802	20.96	27.0	28.4	48.72
1.2486	343.4	2.866	21.44	27.5	28.9	49.72
1.2537	351.0	2.930	21.92	28.0	29.3	50.74
1.2588	358.8	2.994	22.40	28.5	29.8	51.76
1.2639	366.5	3.059	22.88	29.0	30.3	52.78
1.2691	374.4	3.124	23.37	29.5	30.7	53.82
1.2742	382.3	3.190	23.86	30.0	31.2	54.84
1.2794	390.2	3.257	24.36	30.5	31.7	55.88
1.2846	398.2	3.323	24.86	31.0	32.1	56.92
1.2898	406.3	3.391	25.36	31.5	32.6	57.96
1.2951	414.4	3.459	25.87	32.0	33.0	59.02
1.3003	422.6	3.527	26.38	32.5	33.5	60.06
1.3065	431.1	3.598	26.92	33.0	34.0	61.30
1.3109	439.2	3.665	27.42	33.5	34.4	62.18
1.3162	447.5	3.735	27.94	34.0	34.8	63.24

Table 10-43
NITRIC ACID

Specific Gravity of Aqueous Nitric Acid Solutions. Adopted 1903 as Standard by the Manufacturing Chemists' Association of the United States. Authority: W. C. Ferguson.

Specific Gravity determinations were made at 60°F., compared with water at 60°F.

From the Specific Gravities, the corresponding degrees Baumé were calculated by the following formula:

$$\text{Baumé} = 145 - \frac{145}{\text{Sp. Gr.}}$$

Baumé Hydrometers for use with this table must be graduated by the above formula, which formula should always be printed on the scale.

Atomic weight from F. W. Clarke's table of 1901. O = 16.

ALLOWANCE FOR TEMPERATURE:

At 10° — 20° Be. — 1/30° Be. or .00029 Sp. Gr. = 1°F.
20° — 30° Be. — 1/23° Be. or .00044 Sp. Gr. = 1°F.
30° — 40° Be. — 1/20° Be. or .00060 Sp. Gr. = 1°F.
40° — 48.5° Be. — 1/17° Be. or .00084 Sp. Gr. = 1°F.

Be.°	Sp. Gr.	Tw.°	% HNO₃	Be.°	Sp. Gr.	Tw.°	% HNO₃	Be.°	Sp. Gr.	Tw.°	% HNO₃
10.00	1.0741	14.82	12.86	21.25	1.1718	34.36	28.02	32.50	1.2889	57.78	45.68
10.25	1.0761	15.22	13.18	21.50	1.1741	34.82	28.36	32.75	1.2918	58.36	46.14
10.50	1.0781	15.62	13.49	21.75	1.1765	35.30	28.72	33.00	1.2946	58.92	46.58
10.75	1.0801	16.02	13.81	22.00	1.1789	35.78	29.07	33.25	1.2975	59.50	47.04
11.00	1.0821	16.42	14.13	22.25	1.1813	36.26	29.43	33.50	1.3004	60.08	47.49
11.25	1.0841	16.82	14.44	22.50	1.1837	36.74	29.78	33.75	1.3034	60.68	47.95
11.50	1.0861	17.22	14.76	22.75	1.1861	37.22	30.14	34.00	1.3063	61.26	48.42
11.75	1.0881	17.62	15.07	23.00	1.1885	37.70	30.49	34.25	1.3093	61.86	48.90
12.00	1.0902	18.04	15.41	23.25	1.1910	38.20	30.86	34.50	1.3122	62.44	49.35
12.25	1.0922	18.44	15.72	23.50	1.1934	38.68	31.21	34.75	1.3152	63.04	49.83
12.50	1.0943	18.86	16.05	23.75	1.1959	39.18	31.58	35.00	1.3182	63.64	50.32
12.75	1.0964	19.28	16.39	24.00	1.1983	39.66	31.94	35.25	1.3212	64.24	50.81
13.00	1.0985	19.70	16.72	24.25	1.2008	40.16	32.31	35.50	1.3242	64.84	51.30
13.25	1.1006	20.12	17.05	24.50	1.2033	40.66	32.68	35.75	1.3273	65.46	51.80
13.50	1.1027	20.54	17.38	24.75	1.2058	41.16	33.05	36.00	1.3303	66.06	52.30
13.75	1.1048	20.96	17.71	25.00	1.2083	41.66	33.42	36.25	1.3334	66.68	52.81
14.00	1.1069	21.38	18.04	25.25	1.2109	42.18	33.80	36.50	1.3364	67.28	53.32
14.25	1.1090	21.80	18.37	25.50	1.2134	42.68	34.17	36.75	1.3395	67.90	53.84
14.50	1.1111	22.22	18.70	25.75	1.2160	43.20	34.56	37.00	1.3426	68.52	54.36
14.75	1.1132	22.64	19.02	26.00	1.2185	43.70	34.94	37.25	1.3457	69.14	54.89
15.00	1.1154	23.08	19.36	26.25	1.2211	44.22	35.33	37.50	1.3488	69.76	55.43
15.25	1.1176	23.52	19.70	26.50	1.2236	44.72	35.70	37.75	1.3520	70.40	55.97
15.50	1.1197	23.94	20.02	26.75	1.2262	45.24	36.09	38.00	1.3551	71.02	56.52
15.75	1.1219	24.38	20.36	27.00	1.2288	45.76	36.48	38.25	1.3583	71.66	57.08
16.00	1.1240	24.80	20.69	27.25	1.2314	46.28	36.87	38.50	1.3615	72.30	57.65
16.25	1.1262	25.24	21.03	27.50	1.2340	46.80	37.26	38.75	1.3647	72.94	58.23
16.50	1.1284	25.68	21.36	27.75	1.2367	47.34	37.67	39.00	1.3679	73.58	58.82
16.75	1.1306	26.12	21.70	28.00	1.2393	47.86	38.06	39.25	1.3712	74.24	59.43
17.00	1.1328	26.56	22.04	28.25	1.2420	48.40	38.46	39.50	1.3744	74.88	60.06
17.25	1.1350	27.00	22.38	28.50	1.2446	48.92	38.85	39.75	1.3777	75.54	60.71
17.50	1.1373	27.46	22.74	28.75	1.2473	49.46	39.25	40.00	1.3810	76.20	61.38
17.75	1.1395	27.90	23.08	29.00	1.2500	50.00	39.66	40.25	1.3843	76.86	62.07
18.00	1.1417	28.34	23.42	29.25	1.2527	50.54	40.06	40.50	1.3876	77.52	62.77
18.25	1.1440	28.80	23.77	29.50	1.2554	51.08	40.47	40.75	1.3909	78.18	63.48
18.50	1.1462	29.24	24.11	29.75	1.2582	51.64	40.89	41.00	1.3942	78.84	64.20
18.75	1.1485	29.70	24.47	30.00	1.2609	52.18	41.30	41.25	1.3976	79.52	64.93
19.00	1.1508	30.16	24.82	30.25	1.2637	52.74	41.72	41.50	1.4010	80.20	65.67
19.25	1.1531	30.62	25.18	30.50	1.2664	53.28	42.14	41.75	1.4044	80.88	66.42
19.50	1.1554	31.08	25.53	30.75	1.2692	53.84	42.58	42.00	1.4078	81.56	67.18
19.75	1.1577	31.54	25.88	31.00	1.2719	54.38	43.00	42.25	1.4112	82.24	67.95
20.00	1.1600	32.00	26.24	31.25	1.2747	54.94	43.44	42.50	1.4146	82.92	68.73
20.25	1.1624	32.48	26.61	31.50	1.2775	55.50	43.89	42.75	1.4181	83.62	69.52
20.50	1.1647	32.94	26.96	31.75	1.2804	56.08	44.34	43.00	1.4216	84.32	70.33
20.75	1.1671	33.42	27.33	32.00	1.2832	56.64	44.78	43.25	1.4251	85.02	71.15
21.00	1.1694	33.88	27.67	32.25	1.2861	57.22	45.24	43.50	1.4286	85.72	71.98

Table 10-43 (*Continued*)
NITRIC ACID

Be.°	Sp. Gr.	Tw.°	% HNO₃.	Be.°	Sp. Gr.	Tw.°	% HNO₃.	Be.°	Sp. Gr.	Tw.°	% HNO₃.
43.75	1.4321	86.42	72.82	45.50	1.4573	91.46	79.03	47.25	1.4834	96.68	86.98
44.00	1.4356	87.12	73.67	45.75	1.4610	92.20	80.04	47.50	1.4872	97.44	88.32
44.25	1.4392	87.84	74.53	46.00	1.4646	92.92	81.08	47.75	1.4910	98.20	89.76
44.50	1.4428	88.56	75.40	46.25	1.4684	93.68	82.18	48.00	1.4948	98.96	91.35
44.75	1.4464	89.28	76.28	46.50	1.4721	94.42	83.33	48.25	1.4987	99.74	93.13
45.00	1.4500	90.00	77.17	46.75	1.4758	95.16	84.48	48.50	1.5026	100.52	95.11
45.25	1.4536	90.72	78.07	47.00	1.4796	95.92	85.70				

Density of Aqueous Nitric Acid Solutions at $\frac{20°}{4}$ C. Computed from Values Given in the International Critical Tables.

Specific Gravity	Grams per Liter	Pounds per U.S. Gallon	Pounds per Cubic Foot	N	%	Baumé	Twaddell
1.0036	10.04	0.08375	0.6265	0.159	1	0.5	0.72
1.0091	20.18	0.1684	1.260	0.320	2	1.3	1.82
1.0146	30.44	0.2540	1.900	0.483	3	2.1	2.92
1.0201	40.80	0.3405	2.547	0.647	4	2.9	4.02
1.0256	51.28	0.4280	3.201	0.814	5	3.6	5.12
1.0312	61.87	0.5163	3.863	0.982	6	4.4	6.24
1.0369	72.58	0.6057	4.531	1.15	7	5.2	7.38
1.0427	83.42	0.6961	5.208	1.32	8	5.9	8.54
1.0485	94.37	0.7875	5.891	1.50	9	6.7	9.70
1.0543	105.4	0.8799	6.582	1.67	10	7.5	10.86
1.0602	116.6	0.9733	7.281	1.85	11	8.2	12.04
1.0661	127.9	1.068	7.987	2.03	12	9.0	13.22
1.0721	139.4	1.163	8.701	2.21	13	9.8	14.42
1.0781	150.9	1.260	9.423	2.39	14	10.5	15.62
1.0842	162.6	1.357	10.15	2.58	15	11.3	16.84
1.0903	174.4	1.456	10.89	2.77	16	12.0	18.06
1.0964	186.4	1.555	11.64	2.96	17	12.8	19.28
1.1026	198.5	1.656	12.39	3.15	18	13.5	20.52
1.1088	210.7	1.758	13.15	3.34	19	14.2	21.76
1.1150	223.0	1.861	13.92	3.54	20	15.0	23.00
1.1213	235.5	1.965	14.70	3.74	21	15.7	24.26
1.1276	248.1	2.070	15.49	3.94	22	16.4	25.52
1.1340	260.8	2.177	16.28	4.13	23	17.1	26.80
1.1404	273.7	2.284	17.09	4.34	24	17.9	28.08
1.1469	286.7	2.393	17.90	4.55	25	18.6	29.38
1.1534	299.9	2.503	18.72	4.76	26	19.4	30.68
1.1600	313.2	2.614	19.55	4.97	27	20.0	32.00
1.1666	326.6	2.726	20.39	5.18	28	20.7	33.32
1.1733	340.3	2.840	21.24	5.40	29	21.4	34.66
1.1800	354.0	2.954	22.10	5.62	30	22.1	36.00
1.1867	367.9	3.070	22.97	5.84	31	22.8	37.34
1.1934	381.9	3.187	23.84	6.06	32	23.5	38.68
1.2002	396.1	3.305	24.73	6.29	33	24.2	40.04
1.2071	410.4	3.425	25.62	6.51	34	24.9	41.42
1.2140	424.9	3.546	26.53	6.74	35	25.6	42.80
1.2205	439.4	3.667	27.43	6.97	36	26.2	44.10
1.2270	454.0	3.789	28.34	7.20	37	26.8	45.40
1.2335	468.7	3.912	29.26	7.44	38	27.5	46.70
1.2399	483.6	4.036	30.19	7.67	39	28.1	47.98
1.2463	498.5	4.160	31.12	7.91	40	28.7	49.26
1.2527	513.6	4.286	32.06	8.15	41	29.3	50.54
1.2591	528.8	4.413	33.01	8.39	42	29.8	51.82
1.2655	544.2	4.541	33.97	8.64	43	30.4	53.10
1.2719	559.6	4.670	34.94	8.88	44	31.0	54.38
1.2783	575.2	4.801	35.91	9.13	45	31.6	55.66

Table 10-43 (*Continued*)
NITRIC ACID

Specific Gravity	Weight of HNO₃ in solution expressed in					Degrees	
	Grams per Liter	Pounds per U.S. Gallon	Pounds per Cubic Foot	N	%	Baumé	Twaddell
1.2847	591.0	4.932	36.89	9.38	46	32.1	56.94
1.2911	606.8	5.064	37.88	9.63	47	32.7	58.22
1.2975	622.8	5.198	38.88	9.88	48	33.2	59.50
1.3040	639.0	5.332	39.89	10.1	49	33.8	60.80
1.3100	655.0	5.466	40.89	10.4	50	34.3	62.00
1.3160	671.2	5.601	41.90	10.7	51	34.8	63.20
1.3219	687.4	5.737	42.91	10.9	52	35.3	64.38
1.3278	703.7	5.873	43.93	11.1	53	35.8	65.56
1.3336	720.1	6.010	44.96	11.4	54	36.3	66.72
1.3393	736.6	6.147	45.99	11.7	55	36.7	67.86
1.3449	753.1	6.285	47.02	12.0	56	37.2	68.98
1.3505	769.8	6.424	48.06	12.2	57	37.6	70.10
1.3560	786.5	6.563	49.10	12.5	58	38.1	71.20
1.3614	803.2	6.703	50.15	12.7	59	38.5	72.28
1.3667	820.0	6.843	51.19	13.0	60	38.9	73.34
1.3719	836.9	6.984	52.25	13.3	61	39.3	74.38
1.3769	853.7	7.124	53.30	13.5	62	39.7	75.38
1.3818	870.5	7.265	54.35	13.8	63	40.1	76.36
1.3866	887.4	7.406	55.40	14.1	64	40.4	77.32
1.3913	904.3	7.547	56.46	14.4	65	40.8	78.26
1.3959	921.3	7.689	57.52	14.6	66	41.1	79.18
1.4004	938.3	7.830	58.58	14.9	67	41.5	80.08
1.4048	955.3	7.972	59.64	15.2	68	41.8	80.96
1.4091	972.3	8.114	60.70	15.4	69	42.1	81.82
1.4134	989.4	8.257	61.77	15.7	70	42.4	82.68
1.4176	1006	8.400	62.84	16.0	71	42.7	83.52
1.4218	1024	8.543	63.91	16.3	72	43.0	84.36
1.4258	1041	8.686	64.98	16.5	73	43.3	85.16
1.4298	1058	8.830	66.05	16.8	74	43.6	85.96
1.4337	1075	8.974	67.13	17.1	75	43.9	86.74
1.4375	1093	9.117	68.20	17.3	76	44.1	87.50
1.4413	1110	9.262	69.28	17.6	77	44.4	88.26
1.4450	1127	9.406	70.36	17.9	78	44.7	89.00
1.4486	1144	9.550	71.44	18.2	79	44.9	89.72
1.4521	1162	9.695	72.52	18.4	80	45.1	90.42
1.4555	1179	9.839	73.60	18.7	81	45.4	91.10
1.4589	1196	9.984	74.69	19.0	82	45.6	91.78
1.4622	1214	10.13	75.77	19.3	83	45.8	92.44
1.4655	1231	10.27	76.85	19.5	84	46.1	93.10
1.4686	1248	10.42	77.93	19.8	85	46.3	93.72
1.4716	1266	10.56	79.01	20.1	86	46.5	94.32
1.4745	1283	10.71	80.09	20.4	87	46.7	94.90
1.4773	1300	10.85	81.16	20.6	88	46.8	95.46
1.4800	1317	10.99	82.23	20.9	89	47.0	96.00
1.4826	1334	11.14	83.30	21.2	90	47.2	96.52
1.4850	1351	11.28	84.36	21.4	91	47.4	97.00
1.4873	1368	11.42	85.42	21.7	92	47.5	97.46
1.4892	1385	11.56	86.46	22.0	93	47.6	97.84
1.4912	1402	11.70	87.51	22.3	94	47.8	98.24
1.4932	1419	11.84	88.56	22.5	95	47.9	98.64
1.4952	1435	11.98	89.61	22.8	96	48.0	99.04
1.4974	1452	12.12	90.68	23.0	97	48.2	99.48
1.5008	1471	12.27	91.82	23.3	98	48.4	100.16
1.5056	1491	12.44	93.05	23.7	99	48.7	101.12
1.5129	1513	12.63	94.45	24.0	100	49.2	102.58

Table 10-44
OXALIC ACID

Density of Aqueous Oxalic Acid Solutions at 17.5° C. Computed from the Values of Gerlach.

| Specific Gravity | Weight of $H_2C_2O_4 \cdot 2H_2O$ in solution expressed in | | | Per Cent $H_2C_2O_4 + 2H_2O$ | Degrees | |
	Grams per Liter	Pounds per U. S. Gallon	Pounds per Cubic Foot		Baumé	Twaddell
1.0035	10.04	0.08375	0.6265	1	0.5	0.70
1.0070	20.14	0.1681	1.257	2	1.0	1.40
1.0105	30.32	0.2530	1.893	3	1.5	2.10
1.0140	40.56	0.3385	2.532	4	2.0	2.80
1.0175	50.88	0.4246	3.176	5	2.5	3.50
1.0210	61.26	0.5112	3.824	6	3.0	4.20
1.0245	71.72	0.5985	4.477	7	3.5	4.90
1.0280	82.24	0.6863	5.134	8	4.0	5.60
1.0315	92.84	0.7747	5.796	9	4.4	6.30
1.0350	103.5	0.8637	6.462	10	4.9	7.00
1.0385	114.2	0.9533	7.132	11	5.4	7.70
1.0420	125.0	1.044	7.806	12	5.8	8.40
1.0455	135.9	1.134	8.485	13	6.3	9.10

| Specific Gravity | Weight of $H_2C_2O_4$ in solution expressed in | | | Per Cent $H_2C_2O_4$ | Degrees | |
	Grams per Liter	Pounds per U. S. Gallon	Pounds per Cubic Foot		Baumé	Twaddell
1.0035	7.166	0.05980	0.4474	0.7141	0.5	0.70
1.0070	14.38	0.1200	0.8977	1.428	1.0	1.40
1.0105	21.64	0.1806	1.351	2.142	1.5	2.10
1.0140	28.97	0.2418	1.809	2.857	2.0	2.80
1.0175	36.33	0.3032	2.268	3.571	2.5	3.50
1.0210	43.75	0.3651	2.731	4.285	3.0	4.20
1.0245	51.21	0.4274	3.197	4.999	3.5	4.90
1.0280	58.73	0.4901	3.666	5.713	4.0	5.60
1.0315	66.29	0.5533	4.139	6.427	4.4	6.30
1.0350	73.91	0.6168	4.614	7.141	4.9	7.00
1.0385	81.58	0.6809	5.093	7.856	5.4	7.70
1.0420	89.30	0.7452	5.575	8.570	5.8	8.40
1.0455	97.06	0.8100	6.060	9.284	6.3	9.10

Table 10-45
PHOSPHORIC ACID

Density of Aqueous Phosphoric Acid Solutions at $\frac{20°}{4°}$ C. Computed from Values Given in the International Critical Tables.

| Specific Gravity | Weight of H_3PO_4 in solution expressed in | | | Per Cent H_3PO_4 | Degrees | |
	Grams per Liter	Pounds per U. S. Gallon	Pounds per Cubic Foot		Baumé	Twaddell
1.0038	10.04	0.08377	0.6267	1	0.6	0.76
1.0092	20.18	0.1684	1.260	2	1.3	1.84
1.0200	40.80	0.3405	2.547	4	2.8	4.00
1.0309	61.85	0.5162	3.862	6	4.3	6.18
1.0420	83.36	0.6957	5.204	8	5.8	8.40
1.0532	105.3	0.8789	6.575	10	7.3	10.64
1.0647	127.8	1.066	7.976	12	8.8	12.94
1.0764	150.7	1.258	9.408	14	10.3	15.28
1.0884	174.1	1.453	10.87	16	11.8	17.68
1.1008	198.1	1.654	12.37	18	13.3	20.16

Table 10-45 (*Continued*)
PHOSPHORIC ACID

| Specific Gravity | Weight of H_3PO_4 in solution expressed in | | | Per Cent H_3PO_4 | Degrees | |
	Grams per Liter	Pounds per U. S. Gallon	Pounds per Cubic Foot		Baumé	Twaddell
1.1134	222.7	1.858	13.90	20	14.8	22.68
1.1263	247.8	2.068	15.47	22	16.3	25.26
1.1395	273.5	2.282	17.07	24	17.8	27.90
1.1529	299.8	2.502	18.71	26	19.2	30.58
1.1665	326.6	2.726	20.39	28	20.7	33.30
1.1805	354.2	2.956	22.11	30	22.2	36.10
1.216	425.6	3.552	26.57	35	25.8	43.2
1.254	501.6	4.186	31.31	40	29.4	50.8
1.293	581.9	4.856	36.32	45	32.9	58.6
1.335	667.5	5.571	41.67	50	36.4	67.0
1.379	758.5	6.330	47.35	55	39.9	75.8
1.426	855.6	7.140	53.42	60	43.3	85.2
1.475	958.8	8.001	59.85	65	46.7	95.0
1.526	1068	8.915	66.69	70	50.0	105.2
1.579	1184	9.883	73.93	75	53.2	115.8
1.633	1306	10.90	81.56	80	56.2	126.6
1.689	1436	11.98	89.63	85	59.2	137.8
1.746	1571	13.11	98.10	90	62.0	149.2
1.770	1628	13.59	101.7	92	63.1	154.0
1.794	1686	14.07	105.3	94	64.2	158.8
1.819	1746	14.57	109.0	96	65.3	163.8
1.844	1807	15.08	112.8	98	66.4	168.8
1.870	1870	15.61	116.7	100	67.5	174.0

Density of Aqueous Phosphoric Acid Solutions at $\frac{20°}{4°}$ C. Computed from Values Given in the International Critical Tables.

| Specific Gravity | Weight of P_2O_5 in solution expressed in | | | Per Cent P_2O_5 | Degrees | |
	Grams per Liter	Pounds per U. S. Gallon	Pounds per Cubic Foot		Baumé	Twaddell
1.0038	7.272	0.06068	0.4540	0.7244	0.6	0.76
1.0092	14.62	0.1220	0.9129	1.449	1.3	1.84
1.0200	29.56	0.2467	1.845	2.898	2.8	4.00
1.0309	44.80	0.3739	2.797	4.346	4.3	6.18
1.0420	60.38	0.5039	3.770	5.795	5.8	8.40
1.0532	76.29	0.6367	4.763	7.244	7.3	10.64
1.0647	92.55	0.7724	5.778	8.693	8.8	12.94
1.0764	109.1	0.9109	6.814	10.14	10.3	15.28
1.0884	126.1	1.053	7.875	11.59	11.8	17.68
1.1008	143.5	1.198	8.961	13.04	13.3	20.16
1.1134	161.3	1.346	10.07	14.49	14.8	22.68
1.1263	179.5	1.498	11.21	15.94	16.3	25.26
1.1395	198.2	1.654	12.37	17.39	17.8	27.90
1.1529	217.1	1.812	13.55	18.83	19.2	30.58
1.1665	236.6	1.974	14.77	20.28	20.7	33.30
1.1805	256.5	2.141	16.01	21.73	22.2	36.10
1.216	308.3	2.573	19.24	25.35	25.8	43.2
1.254	363.4	3.033	22.69	28.98	29.4	50.8
1.293	421.5	3.518	26.32	32.60	32.9	58.6
1.335	483.5	4.035	30.19	36.22	36.4	67.0
1.379	549.4	4.585	34.30	39.84	39.9	75.8
1.426	619.7	5.172	38.69	43.46	43.3	85.2
1.475	694.6	5.797	43.36	47.09	46.7	95.0
1.526	773.8	6.458	48.31	50.71	50.0	105.2
1.579	857.9	7.159	53.56	54.33	53.2	115.8

Table 10-45 (*Continued*)
PHOSPHORIC ACID

| Specific Gravity | Weight of P_2O_5 in solution expressed in | | | | Degrees | |
	Grams per Liter	Pounds per U. S. Gallon	Pounds per Cubic Foot	Per Cent P_2O_5	Baumé	Twaddell
1.633	946.3	7.897	59.08	57.95	56.2	126.6
1.689	1040	8.679	64.92	61.57	59.2	137.8
1.746	1138	9.500	71.07	65.20	62.0	149.2
1.770	1180	9.844	73.64	66.64	63.1	154.0
1.794	1222	10.19	76.26	68.09	64.2	158.8
1.819	1265	10.56	78.97	69.54	65.3	163.8
1.844	1309	10.92	81.72	70.99	66.4	168.8
1.870	1355	11.30	84.57	72.44	67.5	174.0

Table 10-46
POTASSIUM HYDROXIDE

Density of Aqueous Potassium Hydroxide Solutions at $\frac{15°}{4°}$ C. Computed from Values Given in the International Critical Tables.

| Specific Gravity | Weight of KOH in solution expressed in | | | | Degrees | |
	Grams per Liter	Pounds per U. S. Gallon	Pounds per Cubic Foot	Per Cent KOH	Baumé	Twaddell
1.0083	10.08	0.08415	0.6295	1	1.2	1.66
1.0175	20.35	0.1698	1.270	2	2.5	3.50
1.0267	30.80	0.2570	1.923	3	3.8	5.34
1.0359	41.44	0.3458	2.587	4	5.0	7.18
1.0452	52.26	0.4361	3.263	5	6.3	9.04
1.0544	63.26	0.5280	3.950	6	7.5	10.88
1.0637	74.46	0.6214	4.648	7	8.7	12.74
1.0730	85.84	0.7164	5.359	8	9.9	14.60
1.0824	97.42	0.8130	6.082	9	11.0	16.48
1.0918	109.2	0.9112	6.816	10	12.2	18.36
1.1013	121.1	1.011	7.563	11	13.3	20.26
1.1108	133.3	1.112	8.322	12	14.5	22.16
1.1203	145.6	1.215	9.092	13	15.6	24.06
1.1299	158.2	1.320	9.874	14	16.7	25.98
1.1396	170.9	1.427	10.67	15	17.8	27.92
1.1493	183.9	1.535	11.48	16	18.8	29.86
1.1590	197.0	1.644	12.30	17	19.9	31.80
1.1688	210.4	1.756	13.13	18	20.9	33.76
1.1786	223.9	1.869	13.98	19	22.0	35.72
1.1884	237.7	1.984	14.84	20	23.0	37.68
1.1984	251.7	2.100	15.71	21	24.0	39.68
1.2083	265.8	2.218	16.60	22	25.0	41.66
1.2184	280.2	2.339	17.49	23	26.0	43.68
1.2285	294.8	2.461	18.41	24	27.0	45.70
1.2387	309.7	2.584	19.33	25	27.9	47.74
1.2489	324.7	2.710	20.27	26	28.9	49.78
1.2592	340.0	2.837	21.33	27	29.8	51.84
1.2695	355.5	2.966	22.19	28	30.8	53.90
1.2800	371.2	3.098	23.17	29	31.7	56.00
1.2905	387.2	3.231	24.17	30	32.6	58.10
1.3010	403.3	3.366	25.18	31	33.6	60.20
1.3117	419.7	3.503	26.20	32	34.5	62.34
1.3224	436.4	3.642	27.24	33	35.4	64.48
1.3331	453.3	3.783	28.30	34	36.2	66.62
1.3440	470.4	3.926	29.37	35	37.1	68.80
1.3549	487.8	4.071	30.45	36	38.0	70.98
1.3659	505.4	4.218	31.55	37	38.8	73.18
1.3769	523.2	4.366	32.66	38	39.7	75.38
1.3879	541.3	4.517	33.79	39	40.5	77.58
1.3991	559.6	4.670	34.94	40	41.4	79.82

Table 10-46 (*Continued*)
POTASSIUM HYDROXIDE

Specific Gravity	Weight of KOH in solution expressed in			Per Cent KOH	Degrees	
	Grams per Liter	Pounds per U. S. Gallon	Pounds per Cubic Foot		Baumé	Twaddell
1.4103	578.2	4.826	36.10	41	42.2	82.06
1.4215	597.0	4.982	37.27	42	43.0	84.30
1.4329	616.1	5.142	38.47	43	43.8	86.58
1.4443	635.5	5.303	39.67	44	44.6	88.86
1.4558	655.1	5.467	40.90	45	45.4	91.16
1.4673	675.0	5.633	42.14	46	46.2	93.46
1.4790	695.1	5.801	43.40	47	47.0	95.80
1.4907	715.5	5.971	44.67	48	47.7	98.14
1.5025	736.2	6.144	45.96	49	48.5	100.50
1.5143	757.2	6.319	47.27	50	49.2	102.86
1.5262	778.4	6.496	48.59	51	50.0	105.24
1.5382	799.9	6.675	49.94	52	50.7	107.64

Table 10-47
SODIUM HYDROXIDE*

Density of Aqueous Sodium Hydroxide Solutions at $\frac{20°}{4}$ C. Computed from Values Given in the International Critical Tables.

Specific Gravity	Weight of NaOH in solution expressed in			Per Cent NaOH	Degrees	
	Grams per Liter	Pounds per U. S. Gallon	Pounds per Cubic Foot		Baumé	Twaddell
1.0095	10.10	0.08425	0.6302	1	1.4	1.90
1.0207	20.41	0.1704	1.274	2	2.9	4.14
1.0318	30.95	0.2583	1.932	3	4.5	6.36
1.0428	41.71	0.3481	2.604	4	6.0	8.56
1.0538	52.69	0.4397	3.289	5	7.4	10.76
1.0648	63.89	0.5332	3.989	6	8.8	12.96
1.0758	75.31	0.6285	4.701	7	10.2	15.16
1.0869	86.95	0.7256	5.428	8	11.6	17.38
1.0979	98.81	0.8246	6.169	9	12.9	19.58
1.1089	110.9	0.9254	6.923	10	14.2	21.78
1.1309	135.7	1.133	8.472	12	16.8	26.18
1.1530	161.4	1.347	10.08	14	19.2	30.60
1.1751	188.0	1.569	11.74	16	21.6	35.02
1.1972	215.5	1.798	13.45	18	23.9	39.44
1.2191	243.8	2.035	15.22	20	26.1	43.82
1.2411	273.0	2.279	17.05	22	28.2	48.22
1.2629	303.1	2.529	18.92	24	30.2	52.58
1.2848	334.0	2.788	20.85	26	32.1	56.96
1.3064	365.8	3.053	22.84	28	34.0	61.28
1.3279	398.4	3.325	24.87	30	35.8	65.58
1.3490	431.7	3.603	26.95	32	37.5	69.80
1.3696	465.7	3.886	29.07	34	39.1	73.92
1.3900	500.4	4.176	31.24	36	40.7	78.00
1.4101	535.8	4.472	33.45	38	42.2	82.02
1.4300	572.0	4.774	35.71	40	43.6	86.00
1.4494	608.7	5.080	38.00	42	45.0	89.88
1.4685	646.1	5.392	40.34	44	46.3	93.70
1.4873	684.2	5.710	42.71	46	47.5	97.46
1.5065	723.1	6.035	45.14	48	48.8	101.30
1.5253	762.7	6.365	47.61	50	49.9	105.06

* For a table giving corrections to readings between 50° F and 130° F and taken with an hydrometer graduated at 60/60° F, see table by Griswold, Ind. Eng. Chem., Analytical Ed., 9, 388 (1937).

Table 10-48
SODIUM SILICATE

Density of Aqueous Sodium Silicate Solutions at $\frac{20°}{4}$ C. Computed from Values Given in the International Critical Tables.

Specific Gravity	Weight of $Na_2O + 3.9SiO_2$ in solution expressed in			Per Cent $Na_2O + 3.9SiO_2$	Degrees	
	Grams per Liter	Pounds per U. S. Gallon	Pounds per Cubic Foot		Baumé	Twaddell
1.006	10.06	0.08395	0.6280	1	0.9	1.2
1.014	20.28	0.1692	1.266	2	2.0	2.8
1.030	41.20	0.3438	2.572	4	4.2	6.0
1.046	62.76	0.5238	3.918	6	6.4	9.2
1.063	85.04	0.7097	5.309	8	8.6	12.6
1.080	108.0	0.9013	6.742	10	10.7	16.0
1.098	131.8	1.100	8.226	12	12.9	19.6
1.116	156.2	1.304	9.754	14	14.6	23.2
1.134	181.4	1.514	11.33	16	17.1	26.8
1.153	207.5	1.732	12.96	18	19.2	30.6
1.172	234.4	1.956	14.63	20	21.3	34.4
1.191	262.0	2.187	16.36	22	23.3	38.2
1.211	290.6	2.426	18.14	24	25.3	42.2
1.232	320.3	2.673	20.00	26	27.3	46.4
1.253	350.8	2.928	21.90	28	29.3	50.6
1.275	382.5	3.192	23.88	30	31.3	55.0
1.298	415.4	3.466	25.93	32	33.3	59.6

Specific Gravity	Weight of $Na_2O + 3.36SiO_2$ in solution expressed in			Per Cent $Na_2O + 3.36SiO_2$	Degrees	
	Grams per Liter	Pounds per U. S. Gallon	Pounds per Cubic Foot		Baumé	Twaddell
1.006	10.06	0.08395	0.6280	1	0.9	1.2
1.014	20.28	0.1692	1.266	2	2.0	2.8
1.030	41.20	0.3438	2.572	4	4.2	6.0
1.047	62.82	0.5243	3.922	6	6.5	9.4
1.065	85.20	0.7110	5.319	8	8.9	13.0
1.083	108.3	0.9038	6.761	10	11.1	16.6
1.101	132.1	1.103	8.248	12	13.3	20.2
1.120	156.8	1.309	9.789	14	15.5	24.0
1.139	182.2	1.521	11.38	16	17.7	27.8
1.159	208.6	1.741	13.02	18	19.9	31.8
1.179	235.8	1.968	14.72	20	22.0	35.8
1.200	264.0	2.203	16.48	22	24.2	40.0
1.222	293.3	2.448	18.31	24	26.3	44.4
1.244	323.4	2.699	20.19	26	28.4	48.8
1.267	354.8	2.961	22.15	28	30.6	53.4
1.290	387.0	3.230	24.16	30	32.6	58.0
1.314	420.5	3.509	26.25	32	34.7	62.8
1.339	455.3	3.799	28.42	34	36.7	67.8
1.365	491.4	4.101	30.68	36	38.8	73.0
1.393	529.3	4.418	33.05	38	40.9	78.6

Table 10-48 (*Continued*)
SODIUM SILICATE

Specific Gravity	Weight of $Na_2O + 2.40SiO_2$ in solution expressed in			Per Cent $Na_2O + 2.40SiO_2$	Degrees	
	Grams per Liter	Pounds per U. S. Gallon	Pounds per Cubic Foot		Baumé	Twaddell
1.007	10.07	0.08404	0.6287	1	1.0	1.4
1.016	20.32	0.1696	1.269	2	2.3	3.2
1.034	41.36	0.3452	2.582	4	4.8	6.8
1.052	63.12	0.5268	3.941	6	7.2	10.4
1.071	85.68	0.7150	5.349	8	9.6	14.2
1.090	109.0	0.9096	6.805	10	12.0	18.0
1.110	133.2	1.112	8.316	12	14.4	22.0
1.130	158.2	1.320	9.876	14	16.7	26.0
1.151	184.2	1.537	11.50	16	19.0	30.2

Specific Gravity	Weight of $Na_2O + 2.44SiO_2$ in solution expressed in			Per Cent $Na_2O + 2.44SiO_2$	Degrees	
	Grams per Liter	Pounds per U. S. Gallon	Pounds per Cubic Foot		Baumé	Twaddell
1.285	359.8	3.003	22.46	28	32.2	57.0
1.309	392.7	3.277	24.52	30	34.2	61.8
1.334	426.9	3.562	26.65	32	36.3	66.8
1.360	462.4	3.859	28.87	34	38.4	72.0
1.387	499.3	4.167	31.17	36	40.5	77.4
1.415	537.7	4.487	33.57	38	42.5	83.0
1.445	578.0	4.824	36.08	40	44.7	89.0

Specific Gravity	Weight of $Na_2O + 2.06SiO_2$ in solution expressed in			Per Cent $Na_2O + 2.06SiO_2$	Degrees	
	Grams per Liter	Pounds per U. S. Gallon	Pounds per Cubic Foot		Baumé	Twaddell
1.007	10.07	0.08404	0.6287	1	1.0	1.4
1.016	20.32	0.1696	1.269	2	2.3	3.2
1.035	41.40	0.3455	2.585	4	4.9	7.0
1.054	63.24	0.5278	3.948	6	7.4	10.8
1.073	85.84	0.7164	5.359	8	9.9	14.6
1.093	109.3	0.9122	6.824	10	12.3	18.6
1.113	133.6	1.115	8.338	12	14.7	22.6
1.134	158.8	1.325	9.911	14	17.1	26.8
1.156	185.0	1.544	11.55	16	19.6	31.2
1.178	212.0	1.770	13.24	18	21.9	35.6
1.200	240.0	2.003	14.98	20	24.2	40.0
1.223	269.1	2.245	16.80	22	26.4	44.6
1.247	299.3	2.498	18.68	24	28.7	49.4
1.271	330.5	2.758	20.63	26	30.9	54.2
1.296	362.9	3.028	22.65	28	33.1	59.2
1.321	396.3	3.307	24.74	30	35.2	64.2
1.346	430.7	3.595	26.89	32	37.3	69.2
1.371	466.1	3.890	29.10	34	39.2	74.2
1.397	502.9	4.197	31.40	36	41.2	79.4
1.423	540.7	4.513	33.76	38	43.1	84.6
1.450	580.0	4.840	36.21	40	45.0	90.0
1.520	684.0	5.708	42.70	45	49.6	104.0
1.594	797.0	6.651	49.76	50	54.0	118.8
1.673	920.2	7.679	57.44	55	58.3	134.6

Table 10-48 (*Continued*)
SODIUM SILICATE

| Specific Gravity | Weight of $Na_2O + 1.69SiO_2$ in solution expressed in | | | Per Cent $Na_2O + 1.69SiO_2$ | Degrees | |
	Grams per Liter	Pounds per U. S. Gallon	Pounds per Cubic Foot		Baumé	Twaddell
1.007	10.07	0.08404	0.6287	1	1.0	1.4
1.017	20.34	0.1697	1.270	2	2.4	3.4
1.036	41.44	0.3458	2.587	4	5.0	7.2
1.056	63.36	0.5208	3.956	6	7.7	11.2
1.077	86.16	0.7190	5.379	8	10.4	15.4
1.098	109.8	0.9163	6.855	10	12.9	19.6
1.119	134.3	1.121	8.383	12	15.4	23.8
1.141	159.7	1.333	9.973	14	17.9	28.2
1.163	186.1	1.553	11.62	16	20.3	32.6
1.186	213.5	1.782	13.33	18	22.7	37.2
1.210	242.0	2.020	15.11	20	25.2	42.0
1.234	271.5	2.266	16.95	22	27.5	46.8
1.259	302.2	2.522	18.86	24	29.8	51.8
1.284	333.8	2.786	20.84	26	32.1	56.8
1.310	366.8	3.061	22.90	28	34.3	62.0
1.337	401.1	3.347	25.04	30	36.6	67.4
1.365	436.8	3.645	27.27	32	38.8	73.0
1.394	474.0	3.955	29.59	34	41.0	78.8
1.424	512.6	4.278	32.00	36	43.2	84.8
1.456	553.3	4.617	34.54	38	45.4	91.2

Table 10-49
SODIUM HYDROGEN SULFITE

Density or Aqueous Sodium Acid Sulfite Solutions at $\dfrac{15.6°}{15.6°}$ C.

| Specific Gravity | Weight of $NaHSO_3$ in solution expressed in | | | Per Cent $NaHSO_3$ | Degrees | |
	Grams per Liter	Pounds per U. S. Gallon	Pounds per Cubic Foot		Baumé	Twaddell
1.007	10.27	0.0857	0.6412	1.0	1	1.4
1.021	31.25	0.2607	1.951	3.1	3	4.2
1.036	52.92	0.4416	3.304	5.1	5	7.1
1.051	75.55	0.6304	4.716	7.2	7	10.1
1.066	99.16	0.8274	6.190	9.3	9	13.2
1.082	123.6	1.031	7.715	11.4	11	16.4
1.099	149.0	1.243	9.300	13.6	13	19.7
1.115	175.7	1.466	10.97	15.8	15	23.1
1.133	203.5	1.698	12.70	18.0	17	26.6
1.151	232.5	1.940	14.51	20.2	19	30.2
1.169	262.4	2.190	16.38	22.4	21	33.9
1.189	293.7	2.451	18.33	24.7	23	37.7
1.208	326.1	2.721	20.36	27.0	25	41.7
1.229	359.7	3.001	22.45	29.3	27	45.8
1.250	394.6	3.293	24.64	31.6	29	50.0
1.272	430.7	3.594	26.89	33.9	31	54.4
1.295	469.3	3.916	29.30	36.3	33	58.9
1.318	511.2	4.266	31.91	38.8	35	63.6
1.343	554.5	4.627	34.62	41.3	37	68.5
1.368	598.1	4.991	37.34	43.7	39	73.6

Table 10-49 (*Continued*)
SODIUM HYDROGEN SULFITE

| Specific Gravity | Weight of SO₂ in solution expressed in | | | | Degrees | |
	Grams per Liter	Pounds per U. S. Gallon	Pounds per Cubic Foot	Per Cent SO₂	Baumé	Twaddell
1.007	6.323	0.05277	0.3947	0.627	1	1.4
1.021	19.24	0.1605	1.201	1.88	3	4.2
1.036	32.58	0.2719	2.034	3.15	5	7.1
1.051	46.51	0.3881	2.904	4.43	7	10.1
1.066	61.04	0.5094	3.811	5.73	9	13.2
1.082	76.08	0.6349	4.750	7.03	11	16.4
1.099	91.70	0.7653	5.725	8.35	13	19.7
1.115	108.2	0.9026	6.752	9.70	15	23.1
1.133	125.3	1.045	7.819	11.06	17	26.6
1.151	143.1	1.194	8.934	12.44	19	30.2
1.169	161.6	1.348	10.09	13.81	21	33.9
1.189	180.8	1.509	11.29	15.21	23	37.7
1.208	200.8	1.676	12.53	16.62	25	41.7
1.229	221.4	1.848	13.82	18.02	27	45.8
1.250	243.0	2.028	15.17	19.44	29	50.0
1.272	265.1	2.213	16.55	20.85	31	54.4
1.295	289.0	2.411	18.04	22.32	33	58.9
1.318	314.7	2.626	19 65	23.87	35	63.6
1.343	341.4	2.849	21.31	25.43	37	68.5
1.368	368.2	3.073	22.99	26.91	39	73.6

Table 10-50
SUCROSE

Density of Aqueous Sucrose Solutions at $\frac{20°}{4}$ C. Computed from the Values of Plato.

| Specific Gravity | Weight of $C_{12}H_{22}O_{11}$ in solution expressed in | | | | Degrees | |
	Grams per Liter	Pounds per U. S. Gallon	Pounds per Cubic Foot	Per Cent $C_{12}H_{22}O_{11}$	Baumé	Twaddell
0.9982	.....	.....	.....	0	....	.....
1.0021	10.02	0.08363	0.6256	1	0.3	0.42
1.0060	20.12	0.1679	1.256	2	0.9	1.20
1.0099	30.30	0.2528	1.891	3	1.4	1.98
1.0139	40.56	0.3385	2.532	4	2.0	2.78
1.0179	50.90	0.4247	3.177	5	2.5	3.58
1.0219	61.31	0.5117	3.828	6	3.1	4.38
1.0259	71.81	0.5993	4.483	7	3.7	5.18
1.0299	82.39	0.6876	5.144	8	4.2	5.98
1.0340	93.06	0.7766	5.810	9	4.8	6.80
1.0381	103.8	0.8663	6.481	10	5.3	7.62
1.0423	114.7	0.9568	7.158	11	5.9	8.46
1.0465	125.6	1.048	7.840	12	6.4	9.30
1.0507	136.6	1.140	8.527	13	7.0	10.14
1.0549	147.7	1.232	9.220	14	7.6	10.98
1.0592	158.9	1.326	9.919	15	8.1	11.84
1.0635	170.2	1.420	10.62	16	8.7	12.70
1.0678	181.5	1.515	11.33	17	9.2	13.56
1.0721	193.0	1.610	12.05	18	9.7	14.42
1.0765	204.5	1.707	12.77	19	10.3	15.30
1.0810	216.2	1.804	13.50	20	10.9	16.20
1.0854	227.9	1.902	14.23	21	11.4	17.08
1.0899	239.8	2.001	14.97	22	12.0	17.98
1.0944	251.7	2.101	15.71	23	12.5	18.88
1.0990	263.8	2.201	16.47	24	13.1	19.80

Table 10-50 (*Continued*)
SUCROSE

| Specific Gravity | Weight of C12H22O11 in solution expressed in | | | Per Cent C12H22O11 | Degrees | |
	Grams per Liter	Pounds per U. S. Gallon	Pounds per Cubic Foot		Baumé	Twaddell
1.1036	275.9	2.302	17.22	25	13.6	20.72
1.1082	288.1	2.405	17.99	26	14.2	21.64
1.1128	300.5	2.507	18.76	27	14.7	22.56
1.1175	312.9	2.611	19.53	28	15.2	23.50
1.1222	325.4	2.716	20.32	29	15.8	24.44
1.1270	338.1	2.822	21.11	30	16.3	25.40
1.1318	350.9	2.928	21.90	31	16.9	26.36
1.1366	363.7	3.035	22.71	32	17.4	27.32
1.1415	376.7	3.144	23.52	33	18.0	28.30
1.1463	389.7	3.253	24.33	34	18.5	29.26
1.1513	403.0	3.363	25.16	35	19.1	30.26
1.1562	416.2	3.474	25.99	36	19.6	31.24
1.1612	429.6	3.586	26.82	37	20.1	32.24
1.1663	443.2	3.699	27.67	38	20.7	33.26
1.1713	456.8	3.812	28.52	39	21.2	34.26
1.1764	470.6	3.927	29.38	40	21.7	35.28
1.1816	484.5	4.043	30.24	41	22.3	36.32
1.1868	498.5	4.160	31.12	42	22.8	37.36
1.1920	512.6	4.278	32.00	43	23.4	38.40
1.1972	526.8	4.396	32.89	44	23.9	39.44
1.2025	541.1	4.516	33.78	45	24.4	40.50
1.2079	555.6	4.637	34.69	46	25.0	41.58
1.2132	570.2	4.759	35.60	47	25.5	42.64
1.2186	584.9	4.881	36.52	48	26.0	43.72
1.2241	599.8	5.006	37.45	49	26.5	44.82
1.2296	614.8	5.131	38.38	50	27.1	45.92
1.2351	629.9	5.257	39.32	51	27.6	47.02
1.2406	645.1	5.384	40.27	52	28.1	48.12
1.2462	660.5	5.512	41.23	53	28.6	49.24
1.2519	676.0	5.642	42.20	54	29.2	50.38
1.2575	691.6	5.772	43.18	55	29.7	51.50
1.2632	707.4	5.903	44.16	56	30.2	52.64
1.2690	723.3	6.036	45.16	57	30.7	53.80
1.2748	739.4	6.170	46.16	58	31.3	54.96
1.2806	755.6	6.305	47.17	59	31.8	56.12
1.2865	771.9	6.442	48.19	60	32.3	57.30
1.2924	788.4	6.579	49.22	61	32.8	58.48
1.2983	804.9	6.718	50.25	62	33.3	59.66
1.3043	821.7	6.857	51.30	63	33.8	60.86
1.3103	838.6	6.998	52.35	64	34.3	62.06
1.3163	855.6	7.140	53.41	65	34.8	63.26
1.3224	872.8	7.284	54.49	66	35.3	64.48
1.3286	890.2	7.429	55.57	67	35.9	65.72
1.3347	907.6	7.574	56.66	68	36.4	66.94
1.3409	925.2	7.721	57.76	69	36.9	68.18
1.3472	943.0	7.870	58.87	70	37.4	69.44
1.3535	961.0	8.020	59.99	71	37.9	70.70
1.3598	979.1	8.171	61.12	72	38.4	71.96
1.3661	997.3	8.322	62.26	73	38.9	73.22
1.3725	1016	8.476	63.41	74	39.4	74.50
1.3790	1034	8.631	64.57	75	39.9	75.80
1.3854	1053	8.787	65.73	76	40.3	77.08
1.3920	1072	8.945	66.91	77	40.8	78.40
1.3985	1091	9.103	68.10	78	41.3	79.70
1.4051	1110	9.264	69.30	79	41.8	81.02

Table 10-50 (*Continued*)
SUCROSE

Specific Gravity	Weight of $C_{12}H_{22}O_{11}$ in solution expressed in				Degrees	
	Grams per Liter	Pounds per U. S. Gallon	Pounds per Cubic Foot	Per Cent $C_{12}H_{22}O_{11}$	Baumé	Twaddell
1.4117	1129	9.425	70.51	80	42.3	82.34
1.4184	1149	9.588	71.73	81	42.8	83.68
1.4251	1169	9.752	72.95	82	43.2	85.02
1.4318	1188	9.918	74.19	83	43.7	86.36
1.4386	1208	10.08	75.44	84	44.2	87.72
1.4454	1229	10.25	76.70	85	44.7	89.08
1.4522	1249	10.42	77.97	86	45.1	90.44
1.4591	1269	10.59	79.25	87	45.6	91.82
1.4660	1290	10.77	80.54	88	46.1	93.20
1.4730	1311	10.94	81.84	89	46.6	94.60
1.4800	1332	11.12	83.16	90	47.0	96.00
1.4870	1353	11.29	84.48	91	47.5	97.40
1.4941	1375	11.47	85.81	92	48.0	98.82
1.5012	1396	11.65	87.16	93	48.4	100.24
1.5083	1418	11.83	88.51	94	48.9	101.66
1.5155	1440	12.02	89.88	95	49.3	103.10
1.5227	1462	12.20	91.26	96	49.8	104.54
1.5299	1484	12.38	92.65	97	50.2	105.98
1.5372	1506	12.57	94.05	98	50.7	107.44
1.5445	1529	12.76	95.46	99	51.1	108.90
1.5518	1552	12.95	96.88	100	51.6	110.36

Table 10-51
SULFURIC ACID

Specific Gravity of Aqueous Sulfuric Acid Solutions Adopted 1904 as Standard by the Manufacturing Chemists' Association of the United States.
Authorities: W. C. Ferguson; H. P. Talbot.

Specific Gravity determinations were made at 60°F., compared with water at 60°F. From the Specific Gravities, the corresponding degrees Baumé were calculated by the following formula:

$$\text{Baumé} = 145 - \frac{145}{\text{Sp. Gr.}}$$

Baumé Hydrometers for use with this table must be graduated by the above formula, which formula should always be printed on the scale.

66° Baumé = Sp. Gr. 1.8354.

1 cu. ft. water at 60°F. weighs 62.37 lbs. avoirdupois.

Atomic weights from F. W. Clarke's table of 1901. O = 16.

H_2SO_4 = 100 per cent

	H_2SO_4	O. V.	60°
O. V.	93.19	100.00	119.98
60°	77.67	83.35	100.00
50°	62.18	66.72	80.06

Acids stronger than 66° Bé should have their percentage compositions determined by chemical analysis.

* Calculated from Pickering's results, Journal of London Chemical Society, vol. 57, p. 363.

Be°	Sp. Gr.	Tw.°	Per Cent H_2SO_4	Weight of 1 cu. ft. in lbs. Av.	Per Cent O. V.	Pounds O. V. in 1 cu. ft.	*Freezing (Melting) Point
0	1.0000	0.0	0.00	62.37	0.00	0.00	32.0° F.
1	1.0069	1.4	1.02	62.80	1.09	0.68	31.2° F.
2	1.0140	2.8	2.08	63.24	2.23	1.41	30.5° F.
3	1.0211	4.2	3.13	63.69	3.36	2.14	29.8° F.
4	1.0284	5.7	4.21	64.14	4.52	2.90	28.9° F.

Table 10-51 (*Continued*)
SULFURIC ACID

Be°	Sp. Gr.	Tw.°	Per Cent H_2SO_4	Weight of 1 cu. ft. in lbs. Av.	Per Cent O. V.	Pounds O. V. in 1 cu. ft.	*Freezing (Melting) Point
5	1.0357	7.1	5.28	64.60	5.67	3.66	28.1° F.
6	1.0432	8.6	6.37	65.06	6.84	4.45	27.2° F.
7	1.0507	10.1	7.45	65.53	7.99	5.24	26.3° F.
8	1.0584	11.7	8.55	66.01	9.17	6.06	25.1° F.
9	1.0662	13.2	9.66	66.50	10.37	6.89	24.0° F.
10	1.0741	14.8	10.77	66.99	11.56	7.74	22.8° F.
11	1.0821	16.4	11.89	67.49	12.76	8.61	21.5° F.
12	1.0902	18.0	13.01	68.00	13.96	9.49	20.0° F.
13	1.0985	19.7	14.13	68.51	15.16	10.39	18.3° F.
14	1.1069	21.4	15.25	69.04	16.36	11.30	16.6° F.
15	1.1154	23.1	16.38	69.57	17.58	12.23	14.7° F.
16	1.1240	24.8	17.53	70.10	18.81	13.19	12.6° F.
17	1.1328	26.6	18.71	70.65	20.08	14.18	10.2° F.
18	1.1417	28.3	19.89	71.21	21.34	15.20	7.7° F.
19	1.1508	30.2	21.07	71.78	22.61	16.23	4.8° F.
20	1.1600	32.0	22.25	72.35	23.87	17.27	+ 1.6° F.
21	1.1694	33.9	23.43	72.94	25.14	18.34	− 1.8° F.
22	1.1789	35.8	24.61	73.53	26.41	19.42	− 6.0° F.
23	1.1885	37.7	25.81	74.13	27.69	20.53	−11 ° F.
24	1.1983	39.7	27.03	74.74	29.00	21.68	−16 ° F.
25	1.2083	41.7	28.28	75.36	30.34	22.87	−23 ° F.
26	1.2185	43.7	29.53	76.00	31.69	24.08	−30 ° F.
27	1.2288	45.8	30.79	76.64	33.04	25.32	−39 ° F.
28	1.2393	47.9	32.05	77.30	34.39	26.58	−49 ° F.
29	1.2500	50.0	33.33	77.96	35.76	27.88	−61 ° F.
30	1.2609	52.2	34.63	78.64	37.16	29.22	−74 ° F.
31	1.2719	54.4	35.93	79.33	38.55	30.58	−82 ° F.
32	1.2832	56.6	37.26	80.03	39.98	32.00	−96 ° F.
33	1.2946	58.9	38.58	80.74	41.40	33.42	−97 ° F.
34	1.3063	61.3	39.92	81.47	42.83	34.90	−91 ° F.
35	1.3182	63.6	41.27	82.22	44.28	36.41	−81 ° F.
36	1.3303	66.1	42.63	82.97	45.74	37.95	−70 ° F.
37	1.3426	68.5	43.99	83.74	47.20	39.53	−60 ° F.
38	1.3551	71.0	45.35	84.52	48.66	41.13	−53 ° F.
39	1.3679	73.6	46.72	85.32	50.13	42.77	−47 ° F.
40	1.3810	76.2	48.10	86.13	51.61	44.45	−41 ° F.
41	1.3942	78.8	49.47	86.96	53.08	46.16	−35 ° F.
42	1.4078	81.6	50.87	87.80	54.58	47.92	−31 ° F.
43	1.4216	84.3	52.26	88.67	56.07	49.72	−27 ° F.
44	1.4356	87.1	53.66	89.54	57.58	51.56	−23 ° F.
45	1.4500	90.0	55.07	90.44	59.09	53.44	−20 ° F.
46	1.4646	92.9	56.48	91.35	60.60	55.36	−14 ° F.
47	1.4796	95.9	57.90	92.28	62.13	57.33	−15 ° F.
48	1.4948	99.0	59.32	93.23	63.65	59.34	−18 ° F.
49	1.5104	102.1	60.75	94.20	65.18	61.40	−22 ° F.
50	1.5263	105.3	62.18	95.20	66.72	63.52	−27 ° F.
51	1.5426	108.5	63.66	96.21	68.31	65.72	−33 ° F.
52	1.5591	111.8	65.13	97.24	69.89	67.96	−39 ° F.
53	1.5761	115.2	66.63	98.30	71.50	70.28	−49 ° F.
54	1.5934	118.7	68.13	99.38	73.11	72.66	−59 ° F.
55	1.6111	122.2	69.65	100.48	74.74	75.10	··
56	1.6292	125.8	71.17	101.61	76.37	77.60	·· Below
57	1.6477	129.5	72.75	102.77	78.07	80.23	·· −40
58	1.6667	133.3	74.36	103.95	79.79	82.95	··
59	1.6860	137.2	75.99	105.16	81.54	85.75	− 7
60	1.7059	141.2	77.67	106.40	83.35	88.68	+12.6° F.
61	1.7262	145.2	79.43	107.66	85.23	91.76	27.3° F.
62	1.7470	149.4	81.30	108.96	87.24	95.06	39.1° F.
63	1.7683	153.7	83.34	110.29	89.43	98.63	46.1° F.
64	1.7901	158.0	85.66	111.65	91.92	102.63	46.4° F.

Table 10-51 (*Continued*)
SULFURIC ACID

Be°	Sp. Gr.	Tw.°	Per Cent H₂SO₄	Weight of 1 cu. ft. in lbs. Av.	Per Cent O. V.	Pounds O. V. in 1 cu. ft.	*Freezing (Melting) Point
64¼	1.7957	159.1	86.33	112.00	92.64	103.75	43.6° F.
64½	1.8012	160.2	87.04	112.34	93.40	104.93	41.1° F.
64¾	1.8068	161.4	87.81	112.69	94.23	106.19	37.9° F.
65	1.8125	162.5	88.65	113.05	95.13	107.54	33.1° F.
65¼	1.8182	163.6	89.55	113.40	96.10	108.97	24.6° F.
65½	1.8239	164.8	90.60	113.76	97.22	110.60	13.4° F.
65¾	1.8297	165.9	91.80	114.12	98.51	112.42	− 1 ° F.
66	1.8354	167.1	93.19	114.47	100.00	114.47	−29 ° F.

APPROXIMATE BOILING POINTS			Per Cent 60°	Pounds 60° in 1 cu. ft.	Per Cent 50°	Pounds 50° in 1 cu. ft.
50° Bé, 295° F.						
60° Bé, 386° F.						
61° Bé, 400° F.			61.93	53.34	77.36	66.63
62° Bé, 415° F.			63.69	55.39	79.56	69.19
63° Bé, 432° F.			65.50	57.50	81.81	71.83
64° Bé, 451° F.			67.28	59.66	84.05	74.53
65° Bé, 485° F.			69.09	61.86	86.30	77.27
66° Bé, 538° F.						
			70.90	64.12	88.56	80.10
FIXED POINTS			72.72	66.43	90.83	82.98
			74.55	68.79	93.12	85.93

Sp. Gr.	Per Cent H₂SO₄	Sp. Gr.	Per Cent H₂SO₄				
				76.37	71.20	95.40	88.94
				78.22	73.68	97.70	92.03
1.0000	.00	1.5281	62.34				
1.0048	.71	1.5440	63.79	80.06	76.21	100.00	95.20
1.0347	5.14	1.5748	66.51	81.96	78.85	102.38	98.50
1.0649	9.48	1.6272	71.00	83.86	81.54	104.74	101.85
1.0992	14.22	1.6679	74.46	85.79	84.33	107.15	105.33
				87.72	87.17	109.57	108.89
1.1353	19.04	1.7044	77.54				
1.1736	23.94	1.7258	79.40	89.67	90.10	112.01	112.55
1.2105	28.55	1.7472	81.32	91.63	93.11	114.46	116.30
1.2513	33.49	1.7700	83.47	93.67	96.26	117.00	120.24
1.2951	38.64	1.7959	86.36	95.74	99.52	119.59	124.31
				97.84	102.89	122.21	128.52
1.3441	44.15	1.8117	88.53				
1.3947	49.52	1.8194	89.75	100.00	106.40	124.91	132.91
1.4307	53.17	1.8275	91.32	102.27	110.10	127.74	137.52
1.4667	56.68	1.8354	93.19	104.67	114.05	130.75	142.47
1.4822	58.14			107.30	118.34	134.03	147.82
				110.29	123.14	137.76	153.81

ALLOWANCE FOR TEMPERATURE						
At 10° Bé, .029° Bé or .00023 Sp. Gr. =1° F.			111.15	124.49	138.84	155.50
At 20° Bé, .036° Bé or .00034 Sp. Gr. =1° F.			112.06	125.89	139.98	157.25
At 30° Bé, .035° Bé or .00039 Sp. Gr. =1° F.			113.05	127.40	141.22	159.14
At 40° Bé, .031° Bé or .00041 Sp. Gr. =1° F.			114.14	129.03	142.57	161.17
At 50° Bé, .028° Bé or .00045 Sp. Gr. =1° F.			115.30	130.75	144.02	163.32
At 60° Bé, .026° Bé or .00053 Sp. Gr. =1° F.			116.65	132.70	145.71	165.76
At 63° Bé, .026° Bé or .00057 Sp. Gr. =1° F.			118.19	134.88	147.63	168.48
At 66° Bé, .0235°Bé or .00054 Sp. Gr. =1° F.			119.98	137.34	149.87	171.56

Density of Aqueous Sulfuric Acid Solutions at $\frac{20°}{4°}$ C. Computed from Values Given in the International Critical Tables.

Specific Gravity	Grams per Liter	Pounds per U. S. Gallon	Pounds per Cubic Foot	N	%	Baumé	Twaddell
	Weight of H₂SO₄ in solution expressed in					Degrees	
1.0051	10.05	0.08388	0.6275	0.205	1	0.7	1.02
1.0118	20.24	0.1689	1.263	0.413	2	1.7	2.36
1.0184	30.55	0.2550	1.907	0.623	3	2.6	3.68
1.0250	41.00	0.3422	2.560	0.836	4	3.5	5.00
1.0317	51.59	0.4305	3.220	1.05	5	4.5	6.34

Table 10-51 (*Continued*)
SULFURIC ACID

Specific Gravity	Weight of H$_2$SO$_4$ in solution expressed in					Degrees	
	Grams per Liter	Pounds per U.S. Gallon	Pounds per Cubic Foot	N	%	Baumé	Twaddell
1.0385	62.31	0.5200	3.890	1.27	6	5.4	7.70
1.0453	73.17	0.6106	4.568	1.49	7	6.3	9.06
1.0522	84.18	0.7025	5.255	1.72	8	7.2	10.44
1.0591	95.32	0.7955	5.951	1.94	9	8.1	11.82
1.0661	106.6	0.8897	6.656	2.17	10	9.0	13.22
1.0731	118.0	0.9851	7.369	2.41	11	9.9	14.62
1.0802	129.6	1.082	8.092	2.64	12	10.8	16.04
1.0874	141.4	1.180	8.825	2.88	13	11.7	17.48
1.0947	153.3	1.279	9.568	3.13	14	12.5	18.94
1.1020	165.3	1.379	10.32	3.37	15	13.4	20.40
1.1094	177.5	1.481	11.08	3.62	16	14.3	21.88
1.1168	189.9	1.584	11.85	3.87	17	15.2	23.36
1.1243	202.4	1.689	12.63	4.13	18	16.0	24.86
1.1318	215.0	1.795	13.43	4.38	19	16.9	26.36
1.1394	227.9	1.902	14.23	4.65	20	17.7	27.88
1.1471	240.9	2.010	15.04	4.91	21	18.6	29.42
1.1548	254.1	2.120	15.86	5.18	22	19.4	30.96
1.1626	267.4	2.232	16.69	5.45	23	20.3	32.52
1.1704	280.9	2.344	17.54	5.73	24	21.1	34.08
1.1783	294.6	2.458	18.39	6.01	25	21.9	35.66
1.1862	308.4	2.574	19.25	6.29	26	22.8	37.24
1.1942	322.4	2.691	20.13	6.57	27	23.6	38.84
1.2023	336.6	2.809	21.02	6.86	28	24.4	40.46
1.2104	351.0	2.929	21.91	7.16	29	25.2	42.08
1.2185	365.6	3.051	22.82	7.46	30	26.0	43.70
1.2267	380.3	3.174	23.74	7.76	31	26.8	45.34
1.2349	395.2	3.298	24.67	8.06	32	27.6	46.98
1.2432	410.3	3.424	25.61	8.37	33	28.4	48.64
1.2515	425.5	3.551	26.56	8.68	34	29.1	50.30
1.2599	441.0	3.680	27.53	8.99	35	29.9	51.98
1.2684	456.6	3.811	28.51	9.31	36	30.7	53.68
1.2769	472.5	3.943	29.50	9.64	37	31.4	55.38
1.2855	488.5	4.077	30.50	9.96	38	32.2	57.10
1.2941	504.7	4.212	31.51	10.3	39	33.0	58.82
1.3028	521.1	4.349	32.53	10.6	40	33.7	60.56
1.3116	537.8	4.488	33.57	11.0	41	34.5	62.32
1.3205	554.6	4.628	34.62	11.3	42	35.2	64.10
1.3294	571.6	4.771	35.69	11.7	43	35.9	65.88
1.3384	588.9	4.915	36.76	12.0	44	36.7	67.68
1.3476	606.4	5.061	37.86	12.4	45	37.4	69.52
1.3569	624.2	5.209	38.97	12.7	46	38.1	71.38
1.3663	642.2	5.359	40.09	13.1	47	38.9	73.26
1.3758	660.4	5.511	41.23	13.5	48	39.6	75.16
1.3854	678.8	5.665	42.38	13.8	49	40.3	77.08
1.3951	697.6	5.821	43.55	14.2	50	41.1	79.02
1.4049	716.5	5.979	44.73	14.6	51	41.8	80.98
1.4148	735.7	6.140	45.93	15.0	52	42.5	82.96
1.4248	755.1	6.302	47.14	15.4	53	43.2	84.96
1.4350	774.9	6.467	48.38	15.8	54	44.0	87.00
1.4453	794.9	6.634	49.63	16.2	55	44.7	89.06
1.4557	815.2	6.803	50.89	16.6	56	45.4	91.14
1.4662	835.7	6.975	52.17	17.0	57	46.1	93.24
1.4768	856.5	7.148	53.47	17.5	58	46.8	95.36
1.4875	877.6	7.324	54.79	17.9	59	47.5	97.50
1.4983	899.0	7.502	56.12	18.3	60	48.2	99.66
1.5091	920.6	7.682	57.47	18.8	61	48.9	101.82
1.5200	942.4	7.865	58.83	19.2	62	49.6	104.00
1.5310	964.5	8.049	60.22	19.7	63	50.3	106.20
1.5421	986.9	8.236	61.61	20.1	64	51.0	108.42
1.5533	1010	8.426	63.03	20.6	65	51.7	110.66

Table 10-51 (*Continued*)
SULFURIC ACID

Specific Gravity	Weight of H₂SO₄ in solution expressed in					Degrees	
	Grams per Liter	Pounds per U. S. Gallon	Pounds per Cubic Foot	N	%	Baumé	Twaddell
1.5646	1033	8.618	64.47	21.1	66	52.3	112.92
1.5760	1056	8.812	65.92	21.5	67	53.0	115.20
1.5874	1079	9.008	67.39	22.0	68	53.7	117.48
1.5989	1103	9.207	68.88	22.5	69	54.3	119.78
1.6105	1127	9.408	70.38	23.0	70	55.0	122.10
1.6221	1152	9.611	71.90	23.5	71	55.6	124.42
1.6338	1176	9.817	73.44	24.0	72	56.3	126.76
1.6456	1201	10.03	75.00	24.5	73	56.9	129.12
1.6574	1226	10.24	76.57	25.0	74	57.5	131.48
1.6692	1252	10.45	78.16	25.5	75	58.1	133.84
1.6810	1278	10.66	79.76	26.1	76	58.7	136.20
1.6927	1303	10.88	81.37	26.6	77	59.3	138.54
1.7043	1329	11.09	82.99	27.1	78	59.9	140.86
1.7158	1355	11.31	84.62	27.6	79	60.5	143.16
1.7272	1382	11.53	86.26	28.2	80	61.1	145.44
1.7383	1408	11.75	87.90	28.7	81	61.6	147.66
1.7491	1434	11.97	89.54	29.2	82	62.1	149.82
1.7594	1460	12.19	91.17	29.8	83	62.6	151.88
1.7693	1486	12.40	92.78	30.3	84	63.0	153.86
1.7786	1512	12.62	94.38	30.8	85	63.5	155.72
1.7872	1537	12.83	95.95	31.3	86	63.9	157.44
1.7951	1562	13.03	97.50	31.9	87	64.2	159.02
1.8022	1586	13.24	99.01	32.3	88	64.5	160.44
1.8087	1610	13.43	100.5	32.8	89	64.8	161.74
1.8144	1633	13.63	101.9	33.3	90	65.1	162.88
1.8195	1656	13.82	103.4	33.8	91	65.3	163.90
1.8240	1678	14.00	104.8	34.2	92	65.5	164.80
1.8279	1700	14.19	106.1	34.7	93	65.7	165.58
1.8312	1721	14.37	107.5	35.1	94	65.8	166.24
1.8337	1742	14.54	108.8	35.5	95	65.9	166.74
1.8355	1762	14.71	110.0	35.9	96	66.0	167.10
1.8364	1781	14.87	111.2	36.3	97	66.0	167.28
1.8361	1799	15.02	112.3	36.7	98	66.0	167.22
1.8342	1816	15.15	113.4	37.0	99	65.9	166.84
1.8305	1831	15.28	114.3	37.3	100	65.8	166.10

FUMING SULFURIC ACID

This table may be used in calculating the quantity of concentrated sulfuric acid to be added to an oleum to obtain an oil of any desired lower strength in free SO_3. Gerster (*Chem. Zeit.* **11**, 3 (1887)) gives the following formula for this purpose: $x = 100(b - a)/(a - c)$, where x represents the quantity of sulfuric acid to be added to 100 parts of the oleum, a the total SO_3 per 100 parts of the acid desired, b the total SO_3 per 100 parts of the original strong oleum, and c the SO_3 per 100 parts of the acid used for dilution. The values for a and b are taken from the table; c is obtained by multiplying the percentage of H_2SO_4 present in the acid used for dilution by 0.816

Sp Gr 35°/15°	°Be	% Free SO₃	%H₂SO₄	Equivalent to			
				Total SO₃	100% H₂SO₄	98% H₂SO₄	93.19% (66° Be) H₂SO₄
1.8186	65.3	0	100	81.63	100.00	102.04	107.31
1.8228	65.5	1	99	81.82	100.23	102.27	107.55
1.8270	65.6	2	98	82.00	100.45	102.50	107.79
1.8315	65.8	3	97	82.18	100.67	102.73	108.03
1.8360	66.0	4	96	82.37	100.90	102.96	108.28

Table 10-51 (*Continued*)
FUMING SULFURIC ACID

Sp Gr 35°/15°	°Be	% Free SO$_3$	%H$_2$SO$_4$	Equivalent to			
				Total SO$_3$	100% H$_2$SO$_4$	98% H$_2$SO$_4$	93.19% (66° Be) H$_2$SO$_4$
1.8393	66.2	5	95	82.55	101.13	103.19	108.52
1.8425	66.3	6	94	82.73	101.35	103.42	108.76
1.8461	66.5	7	93	82.92	101.58	103.65	109.00
1.8498	66.6	8	92	83.10	101.80	103.88	109.24
1.8531	66.8	9	91	83.29	102.03	104.11	109.48
1.8565	66.9	10	90	83.47	102.25	104.34	109.72
1.8596	67.0	11	89	83.65	102.47	104.57	109.96
1.8627	67.2	12	88	83.84	102.70	104.80	110.21
1.8660	67.3	13	87	84.02	102.92	105.03	110.45
1.8692	67.4	14	86	84.20	103.15	105.26	110.69
1.8724	67.6	15	85	84.39	103.38	105.49	110.93
1.8756	67.7	16	84	84.57	103.60	105.71	111.17
1.8793	67.8	17	83	84.75	103.82	105.94	111.41
1.8830	68.0	18	82	84.94	104.05	106.17	111.65
1.8875	68.2	19	81	85.12	104.27	106.40	111.90
1.8919	68.4	20	80	85.31	104.50	106.63	112.14
1.8969	68.6	21	79	85.49	104.73	106.86	112.38
1.9020	68.7	22	78	85.67	104.95	107.09	112.62
1.9056	68.9	23	77	85.86	105.18	107.32	112.86
1.9092	69.1	24	76	86.04	105.40	107.55	113.10
1.9125	69.2	25	75	86.22	105.62	107.78	113.34
1.9158	69.3	26	74	86.41	105.85	108.01	113.59
1.9189	69.4	27	73	86.59	106.07	108.24	113.83
1.9220	69.6	28	72	86.78	106.30	108.47	114.07
1.9250	69.7	29	71	86.96	106.53	108.70	114.31
1.9280	69.8	30	70	87.14	106.75	108.93	114.55
1.9309	69.9	31	69	87.33	106.98	109.16	114.79
1.9338	70.0	32	68	87.51	107.20	109.39	115.03
1.9372	70.1	33	67	87.69	107.42	109.62	115.28
1.9405	70.3	34	66	87.88	107.65	109.85	115.52
1.9439	70.4	35	65	88.06	107.87	110.08	115.76
1.9474	70.5	36	64	88.24	108.10	110.31	116.00
1.9504	70.7	37	63	88.43	108.33	110.54	116.24
1.9534	70.8	38	62	88.61	108.55	110.76	116.48
1.9559	70.9	39	61	88.80	108.78	110.99	116.73
1.9584	71.0	40	60	88.98	109.00	111.22	116.97
1.9598	71.0	41	59	89.16	109.22	111.45	117.21
1.9612	71.1	42	58	89.35	109.45	111.68	117.45
1.9627	71.1	43	57	89.53	109.67	111.91	117.69
1.9643	71.2	44	56	89.71	109.90	112.14	117.93
1.9658	71.2	45	55	89.90	110.13	112.37	118.17
1.9672	71.3	46	54	90.08	110.35	112.60	118.41
1.9687	71.3	47	53	90.27	110.58	112.83	118.66
1.9702	71.4	48	52	90.45	110.80	113.06	118.90
1.9717	71.5	49	51	90.63	111.02	113.29	119.14
1.9733	71.5	50	50	90.82	111.25	113.52	119.38
1.9741	71.5	51	49	91.00	111.48	113.75	119.62
1.9749	71.6	52	48	91.18	111.70	113.98	119.86
1.9755	71.6	53	47	91.37	111.93	114.21	120.11
1.9760	71.6	54	46	91.55	112.15	114.44	120.35
1.9766	71.6	55	45	91.73	112.37	114.67	120.59
1.9772	71.7	56	44	91.92	112.60	114.90	120.83
1.9763	71.6	57	43	92.10	112.82	115.13	121.07
1.9754	71.6	58	42	92.29	113.05	115.36	121.31
1.9746	71.6	59	41	92.47	113.28	115.59	121.55
1.9738	71.5	60	40	92.65	113.50	115.82	121.79
1.9723	71.5	61	39	92.84	113.73	116.05	122.04
1.9709	71.4	62	38	93.02	113.95	116.28	122.28
1.9691	71.4	63	37	93.20	114.17	116.51	122.52
1.9672	71.3	64	36	93.39	114.40	116.74	122.76

Table 10-51 (*Continued*)
FUMING SULFURIC ACID

Sp Gr 35°/15°	°Be	% Free SO₃	%H₂SO₄	Equivalent to			
				Total SO₃	100% H₂SO₄	98% H₂SO₄	93.19% (66° Be) H₂SO₄
1.9654	71.2	65	35	93.57	114.62	116.96	123.00
1.9636	71.2	66	34	93.76	114.85	117.19	123.24
1.9618	71.1	67	33	93.94	115.08	117.42	123.49
1.9600	71.0	68	32	94.12	115.30	117.65	123.73
1.9582	71.0	69	31	94.31	115.53	117.88	123.97
1.9564	70.9	70	30	94.49	115.75	118.11	124.21
1.9533	70.8	71	29	94.67	115.97	118.34	124.45
1.9502	70.6	72	28	94.86	116.20	118.57	124.69
1.9472	70.6	73	27	95.04	116.42	118.80	124.93
1.9442	70.4	74	26	95.22	116.65	119.03	125.18
1.9411	70.3	75	25	95.41	116.88	119.26	125.42
1.9379	70.2	76	24	95.59	117.10	119.49	125.66
1.9347	70.1	77	23	95.78	117.33	119.72	125.90
1.9315	69.9	78	22	95.96	117.55	119.95	126.14
1.9283	69.8	79	21	96.14	117.77	120.18	126.38
1.9251	69.7	80	20	96.33	118.00	120.41	126.62
1.9217	69.5	81	19	96.51	118.22	120.64	126.86
1.9183	69.4	82	18	96.69	118.45	120.87	127.11
1.9149	69.3	83	17	96.88	118.68	121.10	127.35
1.9115	69.1	84	16	97.06	118.90	121.23	127.59
1.9080	69.0	85	15	97.25	119.13	121.56	127.83
1.9046	68.8	86	14	97.43	119.35	121.79	128.07
1.9013	68.7	87	13	97.61	119.57	122.02	128.31
1.8980	68.6	88	12	97.80	119.80	122.25	128.56
1.8934	68.4	89	11	97.98	120.03	122.48	128.80
1.8888	68.2	90	10	98.16	120.25	122.70	129.04
1.8844	68.1	91	9	98.35	120.48	122.93	129.98
1.8800	68.0	92	8	98.53	120.70	123.16	129.52
1.8756	67.7	93	7	98.71	120.92	123.39	129.76
1.8712	67.5	94	6	98.90	121.15	123.62	130.00
1.8659	67.3	95	5	99.08	121.37	123.85	130.25
1.8605	67.1	96	4	99.27	121.60	124.08	130.49
1.8546	66.8	97	3	99.45	121.83	124.31	130.73
1.8488	66.6	98	2	99.63	122.05	124.54	130.97
1.8429	66.3	99	1	99.82	122.28	124.77	131.21
1.8370	66.1	100	0	100.00	122.50	125.00	131.45

SULFURIC ACID, as SO₃

Specific Gravity	Weight of SO₃ in solution expressed in				Degrees	
	Grams per Liter	Pounds per U. S. Gallon	Pounds per Cubic Foot	Per Cent SO₃	Baumé	Twaddell
1.0250	33.47	0.2793	2.089	3.265	3.5	5.00
1.0317	42.11	0.3514	2.629	4.082	4.5	6.34
1.0385	50.86	0.4245	3.175	4.898	5.4	7.70
1.0453	59.73	0.4985	3.729	5.714	6.3	9.06
1.0522	68.71	0.5734	4.290	6.530	7.2	10.44
1.0591	77.81	0.6494	4.858	7.347	8.1	11.82
1.0661	87.03	0.7263	5.433	8.163	9.0	13.22
1.0731	96.36	0.8041	6.016	8.979	9.9	14.62
1.0802	105.8	0.8831	6.606	9.796	10.8	16.04
1.0874	115.4	0.9630	7.204	10.61	11.7	17.48

Table 10-51 (*Continued*)
SULFURIC ACID, as SO_3

Specific Gravity	Weight of SO_3 in solution expressed in			Per Cent SO_3	Degrees	
	Grams per Liter	Pounds per U. S. Gallon	Pounds per Cubic Foot		Baumé	Twaddell
1.0947	125.1	1.044	7.810	11.43	12.5	18.94
1.1020	134.9	1.126	8.424	12.24	13.4	20.40
1.1094	144.9	1.209	9.046	13.06	14.3	21.88
1.1108	155.0	1.293	9.676	13.88	15.2	23.36
1.1243	165.2	1.379	10.31	14.69	16.0	24.86
1.1318	175.5	1.465	10.96	15.51	16.9	26.36
1.1394	186.0	1.552	11.61	16.33	17.7	27.88
1.1471	196.6	1.641	12.28	17.14	18.6	29.42
1.1548	207.4	1.731	12.95	17.96	19.4	30.96
1.1626	218.3	1.822	13.63	18.78	20.3	32.52
1.1704	229.3	1.914	14.32	19.59	21.1	34.08
1.1783	240.5	2.007	15.01	20.41	21.9	35.66
1.1862	251.8	2.101	15.72	21.22	22.8	37.24
1.1942	263.2	2.197	16.43	22.04	23.6	38.84
1.2023	274.8	2.293	17.16	22.86	24.4	40.46
1.2104	286.5	2.391	17.89	23.67	25.2	42.08
1.2185	298.4	2.490	18.63	24.49	26.0	43.70
1.2267	310.4	2.591	19.38	25.31	26.8	45.34
1.2349	322.6	2.692	20.14	26.12	27.6	46.98
1.2432	334.9	2.795	20.91	26.94	28.4	48.64
1.2515	347.3	2.899	21.68	27.75	29.1	50.30
1.2599	360.0	3.004	22.47	28.57	29.9	51.98
1.2684	372.7	3.111	23.27	29.39	30.7	53.68
1.2769	385.7	3.219	24.08	30.20	31.4	55.38
1.2855	398.8	3.328	24.89	31.02	32.2	57.10
1.2941	412.0	3.438	25.72	31.84	33.0	58.82
1.3028	425.4	3.550	26.56	32.65	33.7	60.56
1.3116	439.0	3.663	27.41	33.47	34.5	62.32
1.3205	452.7	3.778	28.26	34.29	35.2	64.10
1.3294	466.6	3.894	29.13	35.10	35.9	65.88
1.3384	480.7	4.012	30.01	35.92	36.7	67.68
1.3476	495.0	4.131	30.90	36.73	37.4	69.52
1.3569	509.5	4.252	31.81	37.55	38.1	71.38
1.3663	524.2	4.375	32.73	38.37	38.9	73.26
1.3758	539.1	4.499	33.65	39.18	39.6	75.16
1.3854	554.1	4.625	34.60	40.00	40.3	77.08
1.3951	569.4	4.752	35.55	40.82	41.1	79.02
1.4049	584.9	4.881	36.51	41.63	41.8	80.98
1.4148	600.6	5.012	37.49	42.45	42.5	82.96
1.4248	616.4	5.144	38.48	43.26	43.2	84.96
1.4350	632.6	5.279	39.49	44.08	44.0	87.00
1.4453	648.9	5.415	40.51	44.90	44.7	89.06
1.4557	665.4	5.553	41.54	45.71	45.4	91.14
1.4662	682.2	5.693	42.59	46.53	46.1	93.24
1.4768	699.2	5.835	43.65	47.35	46.8	95.36
1.4875	716.4	5.979	44.73	48.16	47.5	97.50
1.4983	733.8	6.124	45.81	48.98	48.2	99.66
1.5091	751.5	6.271	46.91	49.80	48.9	101.82
1.5200	769.3	6.420	48.03	50.61	49.6	104.00
1.5310	787.4	6.571	49.15	51.43	50.3	106.20
1.5421	805.7	6.723	50.30	52.24	51.0	108.42
1.5533	824.2	6.878	51.45	53.06	51.7	110.66
1.5646	843.0	7.035	52.63	53.88	52.3	112.92
1.5760	862.0	7.193	53.81	54.69	53.0	115.20
1.5874	881.2	7.354	55.01	55.51	53.7	117.48
1.5989	900.6	7.516	56.22	56.33	54.3	119.78
1.6105	920.3	7.680	57.45	57.14	55.0	122.10
1.6221	940.1	7.846	58.69	57.96	55.6	124.42
1.6338	960.3	8.014	59.95	58.77	56.3	126.76

Table 10-51 (*Continued*)
SULFURIC ACID, as SO$_3$

Specific Gravity	Weight of SO$_3$ in solution expressed in				Degrees	
	Grams per Liter	Pounds per U. S. Gallon	Pounds per Cubic Foot	Per Cent SO$_3$	Baumé	Twaddell
1.6456	980.6	8.184	61.22	59.59	56.9	129.12
1.6574	1001	8.355	62.50	60.41	57.5	131.48
1.6692	1022	8.528	63.80	61.22	58.1	133.84
1.6810	1043	8.703	65.11	62.04	58.7	136.20
1.6927	1064	8.879	66.42	62.86	59.3	138.54
1.7043	1085	9.056	67.75	63.67	59.9	140.86
1.7158	1106	9.234	69.08	64.49	60.5	143.16
1.7272	1128	9.413	70.42	65.30	61.1	145.44
1.7383	1149	9.592	71.76	66.12	61.6	147.66
1.7491	1171	9.771	73.09	66.94	62.1	149.82
1.7594	1192	9.948	74.42	67.75	62.6	151.88
1.7693	1213	10.12	75.74	68.57	63.0	153.86
1.7786	1234	10.30	77.05	69.39	63.5	155.72
1.7872	1255	10.47	78.33	70.20	63.9	157.44
1.7951	1275	10.64	79.59	71.02	64.2	159.02
1.8022	1295	10.80	80.82	71.84	64.5	160.44
1.8087	1314	10.97	82.04	72.65	64.8	161.74
1.8144	1333	11.12	83.22	73.47	65.1	162.88
1.8195	1352	11.28	84.38	74.28	65.3	163.90
1.8240	1370	11.43	85.52	75.10	65.5	164.80
1.8279	1388	11.58	86.63	75.92	65.7	165.58
1.8312	1405	11.73	87.72	76.73	65.8	166.24
1.8337	1422	11.87	88.78	77.55	65.9	166.74
1.8355	1438	12.00	89.80	78.37	66.0	167.10
1.8364	1454	12.14	90.78	79.18	66.0	167.28
1.8361	1469	12.26	91.70	80.00	66.0	167.22
1.8342	1482	12.37	92.54	80.81	65.9	166.84
1.8305	1494	12.47	93.29	81.63	65.8	166.10

Table 10-52
TANNIC ACID

Density of Aqueous Tannic Acid Solutions at 15° C. Computed from the Values of Trammer.

Specific Gravity	Weight of C$_{14}$H$_{10}$O$_9$ in solution expressed in				Degrees	
	Grams per Liter	Pounds per U. S. Gallon	Pounds per Cubic Foot	Per Cent C$_{14}$H$_{10}$O$_9$	Baumé	Twaddell
1.0040	10.04	0.08379	0.6268	1.0	0.6	0.80
1.0044	11.05	0.09220	0.6898	1.1	0.6	0.88
1.0048	12.06	0.1006	0.7528	1.2	0.7	0.96
1.0052	13.07	0.1091	0.8158	1.3	0.8	1.04
1.0056	14.08	0.1175	0.8789	1.4	0.8	1.12
1.0060	15.09	0.1259	0.9421	1.5	0.9	1.20
1.0064	16.10	0.1344	1.005	1.6	0.9	1.28
1.0068	17.12	0.1428	1.069	1.7	1.0	1.36
1.0072	18.13	0.1513	1.132	1.8	1.0	1.44
1.0076	19.14	0.1598	1.195	1.9	1.1	1.52
1.0080	20.16	0.1682	1.259	2.0	1.2	1.60
1.0084	21.18	0.1767	1.322	2.1	1.2	1.68
1.0088	22.19	0.1852	1.386	2.2	1.3	1.76
1.0092	23.21	0.1937	1.449	2.3	1.3	1.84
1.0096	24.23	0.2022	1.513	2.4	1.4	1.92
1.0100	25.25	0.2107	1.576	2.5	1.4	2.00
1.0104	26.27	0.2192	1.640	2.6	1.5	2.08
1.0108	27.29	0.2278	1.704	2.7	1.5	2.16
1.0112	28.31	0.2363	1.768	2.8	1.6	2.24
1.0116	29.34	0.2448	1.831	2.9	1.7	2.32

Table 10-52 (*Continued*)
TANNIC ACID

| Specific Gravity | Weight of $C_{14}H_{10}O_9$ in solution expressed in | | | Per Cent $C_{14}H_{10}O_9$ | Degrees | |
	Grams per Liter	Pounds per U. S. Gallon	Pounds per Cubic Foot		Baumé	Twaddell
1.0120	30.36	0.2534	1.895	3.0	1.7	2.40
1.0124	31.38	0.2619	1.959	3.1	1.8	2.48
1.0128	32.41	0.2705	2.023	3.2	1.8	2.56
1.0132	33.44	0.2790	2.087	3.3	1.9	2.64
1.0136	34.46	0.2876	2.151	3.4	1.9	2.72
1.0140	35.49	0.2962	2.216	3.5	2.0	2.80
1.0144	36.52	0.3048	2.280	3.6	2.1	2.88
1.0148	37.55	0.3133	2.344	3.7	2.1	2.96
1.0152	38.58	0.3219	2.408	3.8	2.2	3.04
1.0156	39.61	0.3305	2.473	3.9	2.2	3.12
1.0160	40.64	0.3392	2.537	4.0	2.3	3.20
1.0164	41.67	0.3478	2.602	4.1	2.3	3.28
1.0168	42.71	0.3564	2.666	4.2	2.4	3.36
1.0172	43.74	0.3650	2.731	4.3	2.5	3.44
1.0176	44.77	0.3737	2.795	4.4	2.5	3.52
1.0180	45.81	0.3823	2.860	4.5	2.6	3.60
1.0184	46.85	0.3910	2.925	4.6	2.6	3.68
1.0188	47.88	0.3996	2.989	4.7	2.7	3.76
1.0192	48.92	0.4083	3.054	4.8	2.7	3.84
1.0196	49.96	0.4169	3.119	4.9	2.8	3.92
1.0200	51.00	0.4256	3.184	5.0	2.8	4.00
1.0401	104.0	0.8680	6.493	10.0	5.6	8.02

Table 10-53
TARTARIC ACID

Density of Aqueous Tartaric Acid Solutions at 15° C. Computed from the Values of Gerlach.

| Specific Gravity | Weight of $H_2C_4H_4O_6$ in solution expressed in | | | Per Cent $H_2C_4H_4O_6$ | Degrees | |
	Grams per Liter	Pounds per U. S. Gallon	Pounds per Cubic Foot		Baumé	Twaddell
1.0045	10.05	0.08383	0.6271	1	0.6	0.90
1.0090	20.18	0.1684	1.260	2	1.3	1.80
1.0179	40.72	0.3398	2.542	4	2.6	3.58
1.0273	61.64	0.5144	3.848	6	3.9	5.46
1.0371	82.97	0.6924	5.180	8	5.2	7.42
1.0469	104.7	0.8737	6.536	10	6.5	9.38
1.0565	126.8	1.058	7.915	12	7.8	11.30
1.0661	149.3	1.246	9.318	14	9.0	13.22
1.0761	172.2	1.437	10.75	16	10.3	15.22
1.0865	195.6	1.632	12.21	18	11.5	17.30
1.0969	219.4	1.831	13.70	20	12.8	19.38
1.1072	243.6	2.033	15.21	22	14.0	21.44
1.1175	268.2	2.238	16.74	24	15.2	23.50
1.1282	293.3	2.448	18.31	26	16.5	25.64
1.1393	319.0	2.662	19.92	28	17.7	27.86
1.1505	345.2	2.880	21.55	30	19.0	30.10
1.1615	371.7	3.102	23.20	32	20.2	32.30
1.1726	398.7	3.327	24.89	34	21.3	34.52
1.1840	426.2	3.557	26.61	36	22.5	36.80
1.1959	454.4	3.793	28.37	38	23.8	39.18
1.2078	483.1	4.032	30.16	40	25.0	41.56
1.2198	512.3	4.275	31.98	42	26.1	43.96
1.2317	541.9	4.523	33.83	44	27.3	46.34
1.2441	572.3	4.776	35.73	46	28.5	48.82
1.2568	603.3	5.034	37.66	48	29.6	51.36

Table 10-53 (*Continued*)
TARTARIC ACID

| Specific Gravity | Weight of $H_2C_4H_4O_6$ in solution expressed in | | | | Degrees | |
	Grams per Liter	Pounds per U. S. Gallon	Pounds per Cubic Foot	Per Cent $H_2C_4H_4O_6$	Baumé	Twaddell
1.2696	634.8	5.298	39.63	50	30.8	53.92
1.2828	667.1	5.567	41.64	52	32.0	56.56
1.2961	699.9	5.841	43.69	54	33.1	59.22
1.3093	733.2	6.119	45.77	56	34.3	61.86

Table 10-54
ZINC CHLORIDE

Specific Gravity of Aqueous Zinc Chloride Solutions Adopted 1922 as Standard by the Manufacturing Chemists' Association of the United States.

Authorities: W. H. Whitney, E. H. Hartle, C. H. Sakryd.

Checked by: L. C. Drefahl, R. M. Meiklejohn, R. O. Fernandez.

Specific gravity determinations were made at 60° F., compared with water at 60° F., on solutions of Zinc Chloride made from chemically pure zinc and Hydrochloric Acid, these solutions exhibiting a neutral reaction to methyl orange when diluted below 5° Bé with water.

The corresponding degrees Baumé were calculated from the Specific Gravities by the following formula:

$$\text{Baumé} = 145 - \frac{145}{\text{Sp. Gr.}}$$

Baumé Hydrometers for use with this table must be graduated by the above formula, which formula should always be printed on the scale.
1 cubic foot of water weighs 62.37 lbs. avoirdupois.
Atomic weights used: F. W. Clarke's table of 1912. O = 16.

* The percentage composition on all fixed points was determined by actual chemical analysis, both Zinc and Chlorine being determined.

Bé°	Sp. Gr.	% $ZnCl_2$	Weight of 1 cu. ft. in lbs.	Lbs. $ZnCl_2$ in 1 cu. ft.	% 50% $ZnCl_2$	Lbs. 50% $ZnCl_2$ in 1 cu. ft.
0	1.0000	0	62.37	0	0	0
1	1.0069	.76	62.80	.4773	1.52	.9546
2	1.0140	1.53	63.24	.9676	3.06	1.9352
3	1.0211	2.29	63.69	1.4585	4.58	2.9170
4	1.0284	3.05	64.14	1.9563	6.10	3.9126
5	1.0357	3.81	64.60	2.4613	7.62	4.9226
6	1.0432	4.63	65.06	3.0123	9.26	6.0246
7	1.0507	5.45	65.53	3.5714	10.90	7.1428
8	1.0584	6.27	66.01	4.1388	12.54	8.2776
9	1.0662	7.09	66.50	4.7149	14.18	9.4298
10	1.0741	7.91	66.99	5.2989	15.82	10.5978
11	1.0821	8.78	67.49	5.9256	17.56	11.8512
12	1.0902	9.65	68.00	6.5620	19.30	13.1240
13	1.0985	10.52	68.51	7.2073	21.04	14.4146
14	1.1069	11.39	69.04	7.8637	22.78	15.7274

Table 10-54 (*Continued*)
ZINC CHLORIDE

Bé°	Sp. Gr.	% ZnCl₂	Weight of 1 cu. ft. in lbs.	Lbs. ZnCl₂ in 1 cu. ft.	% 50% ZnCl₂	Lbs. 50% ZnCl₂ in 1 cu. ft.
15	1.1154	12.26	69.57	8.5293	24.52	17.0586
16	1.1240	13.21	70.10	9.2602	26.42	18.5204
17	1.1328	14.15	70.65	9.9970	28.30	19.9940
18	1.1417	15.10	71.21	10.7527	30.20	21.5054
19	1.1508	16.04	71.78	11.5135	32.08	23.0270
20	1.1600	16.98	72.35	12.2850	33.96	24.5700
21	1.1694	17.96	72.94	13.1000	35.92	26.2000
22	1.1789	18.94	73.53	13.9266	37.88	27.8532
23	1.1885	19.92	74.13	14.7667	39.84	29.5334
24	1.1983	20.90	74.74	15.6207	41.80	31.2414
25	1.2083	21.88	75.36	16.4888	43.76	32.9776
26	1.2185	22.88	76.00	17.3888	45.76	34.7776
27	1.2288	23.88	76.64	18.3016	47.76	36.6032
28	1.2393	24.89	77.30	19.2400	49.78	38.4800
29	1.2500	25.89	77.96	20.1838	51.78	40.3676
30	1.2609	26.90	78.64	21.1542	53.80	42.3084
31	1.2719	27.91	79.33	22.1410	55.82	44.2820
32	1.2832	28.91	80.03	23.1367	57.82	46.2734
33	1.2946	29.92	80.74	24.1574	59.84	48.3148
34	1.3063	30.93	81.47	25.1987	61.86	50.3974
35	1.3182	31.93	82.22	26.2528	63.86	52.5056
36	1.3303	32.94	82.97	27.3303	65.88	54.6606
37	1.3426	33.95	83.74	28.4297	67.90	56.8594
38	1.3551	34.96	84.52	29.5482	69.92	59.0964
39	1.3679	35.97	85.32	30.6896	71.94	61.3792
40	1.3810	36.98	86.13	31.8509	73.96	63.7018
41	1.3942	38.02	86.96	33.0622	76.04	66.1244
42	1.4078	39.05	87.80	34.2859	78.10	68.5718
43	1.4216	40.09	88.67	35.5478	80.18	71.0956
44	1.4356	41.12	89.54	36.8188	82.24	73.6376
45	1.4500	42.16	90.44	38.1295	84.32	76.2590
46	1.4646	43.21	91.35	39.4723	86.42	78.9446
47	1.4796	44.26	92.28	40.8431	88.52	81.6862
48	1.4948	45.32	93.23	42.2518	90.64	84.5036
49	1.5104	46.37	94.20	43.6805	92.74	87.3610
50	1.5263	47.43	95.20	45.1534	94.86	90.3068
51	1.5426	48.48	96.21	46.6426	96.96	93.2852
52	1.5591	49.54	97.24	48.1727	99.08	96.3454
53	1.5761	50.60	98.30	49.7398	101.20	99.4796
54	1.5934	51.66	99.38	51.3397	103.32	102.6794
55	1.6111	52.72	100.48	52.9731	105.44	105.9462
56	1.6292	53.80	101.61	54.6662	107.60	109.3324
57	1.6477	54.88	102.77	56.4002	109.76	112.8004
58	1.6667	55.97	103.95	58.1808	111.94	116.3616
59	1.6860	57.06	105.16	60.0043	114.12	120.0086
60	1.7059	58.15	106.40	61.8716	116.30	123.7432
61	1.7262	59.23	107.66	63.7670	118.46	127.5340
62	1.7470	60.30	108.96	65.7029	120.60	131.4058
63	1.7683	61.37	110.29	67.6850	122.74	135.3700
64	1.7901	62.44	111.65	69.7143	124.88	139.4286
65	1.8125	63.52	113.05	71.8094	127.04	143.6188
66	1.8354	64.68	114.47	74.0392	129.36	148.0784
67	1.8590	65.85	115.95	76.3531	131.70	152.7062
68	1.8831	67.02	117.45	78.7150	134.04	157.4300
69	1.9079	68.19	119.00	81.1461	136.38	162.2922
70	1.9333	69.36	120.58	83.6343	138.72	167.2686

Table 10-54 (*Continued*)
ZINC CHLORIDE

APPROXIMATE ALLOWANCE FOR TEMPERATURE		*FIXED POINTS		
		Bé°	Sp. Gr.	Per Cent ZnCl$_2$
		5.08	1.0363	3.88
At 5° Bé = 0.024° Bé		10.16	1.0754	8.05
At 10° Bé = 0.029° Bé		15.35	1.1184	12.59
At 15° Bé = 0.029° Bé		20.35	1.1633	17.32
At 20° Bé = 0.033° Bé		25.14	1.1707	22.02
At 25° Bé = 0.033° Bé	For each 1° Fahrenheit			
At 30° Bé = 0.033° Bé		30.00	1.2609	26.90
At 35° Bé = 0.033° Bé		35.07	1.3190	32.00
At 40° Bé = 0.033° Bé		40.15	1.3829	37.14
At 45° Bé = 0.033° Bé		44.99	1.4499	42.15
At 50° Bé = 0.030° Bé		50.14	1.5285	47.58
At 55° Bé = 0.028° Bé				
At 60° Bé = 0.027° Bé		55.05	1.6120	52.77
At 65° Bé = 0.027° Bé		60.13	1.7085	58.29
At 70° Bé = 0.024° Bé		65.11	1.8150	63.65
		70.05	1.9345	69.42

Table 10-55
RELATIVE DENSITY AND VOLUME OF WATER* AND OF MERCURY AT VARIOUS TEMPERATURES

These values are based upon the mass of one milliliter of water at 4°C. as unity. To convert to absolute density in C. G. S. units multiply the value for relative density by 0.999973.

	WATER			MERCURY		
Temp. Degree C.	Relative Density g/ml	Volume ml/g		Relative Density g/ml	Volume ml/g	Temp. Degree C.
−20	....	....		13.6450	0.0732869	−20
−19	....	....		13.6425	0.0733002	−19
−18	....	....		13.6400	0.0733135	−18
−17	....	....		13.6375	0.0733269	−17
−16	....	....		13.6350	0.0733402	−16
−15	....	....		13.6326	0.0733535	−15
−14	....	....		13.6301	0.0733670	−14
−13	....	....		13.6276	0.0733805	−13
−12	....	....		13.6251	0.0733940	−12
−11	....	....		13.6226	0.0734074	−11
−10	0.99815	1.00186		13.6202	0.0734205	−10
−9	0.99843	1.00157		13.6177	0.0734338	−9
−8	0.99869	1.00131		13.6152	0.0734472	−8
−7	0.99892	1.00108		13.6127	0.0734608	−7
−6	0.99912	1.00088		13.6103	0.0734739	−6
−5	0.99930	1.00070		13.6078	0.0734873	−5
−4	0.99945	1.00055		13.6053	0.0735006	−4
−3	0.99958	1.00042		13.6029	0.0735140	−3
−2	0.99970	1.00031		13.6004	0.0735273	−2
−1	0.99979	1.00021		13.5979	0.0735407	−1
0	0.99987	1.00013		13.5955	0.0735540	0
1	0.99993	1.00007		13.5930	0.0735674	1
2	0.99997	1.00003		13.5905	0.0735808	2
3	0.99999	1.00001		13.5880	0.0735943	3
4	1.00000	1.00000		13.5856	0.0736075	4
5	0.99999	1.00001		13.5831	0.0736209	5
6	0.99997	1.00003		13.5806	0.0736342	6
7	0.99993	1.00007		13.5782	0.0736476	7
8	0.99988	1.00012		13.5757	0.0736610	8
9	0.99981	1.00019		13.5732	0.0736746	9
10	0.99973	1.00027		13.5708	0.0736877	10
11	0.99963	1.00037		13.5683	0.0737011	11
12	0.99952	1.00048		13.5658	0.0737145	12
13	0.99940	1.00060		13.5634	0.0737278	13
14	0.99927	1.00073		13.5609	0.0737412	14
15	0.99913	1.00087		13.5584	0.0737550	15
16	0.99897	1.00103		13.5560	0.0737680	16
17	0.99880	1.00120		13.5535	0.0737813	17
18	0.99862	1.00138		13.5511	0.0737947	18
19	0.99843	1.00157		13.5486	0.0738081	19
20	0.99823	1.00177		13.5461	0.0738215	20
21	0.99802	1.00198		13.5437	0.0738348	21
22	0.99780	1.00221		13.5412	0.0738487	22
23	0.99756	1.00244		13.5388	0.0738617	23
24	0.99732	1.00268		13.5363	0.0738754	24
25	0.99707	1.00294		13.5339	0.0738888	25
26	0.99681	1.00320		13.5314	0.0739022	26
27	0.99654	1.00347		13.5290	0.0739153	27
28	0.99626	1.00375		13.5265	0.0739289	28
29	0.99597	1.00405		13.5241	0.0739423	29
30	0.99567	1.00435		13.5216	0.0739558	30
31	0.99537	1.00466		13.5191	0.0739691	31
32	0.99505	1.00497		13.5167	0.0739825	32
33	0.99473	1.00530		13.5142	0.0739960	33
34	0.99440	1.00563		13.5118	0.0740094	34
35	0.99406	1.00598		13.5094	0.0740228	35

*See also the table: Properties of Saturated Steam.

Table 10-55 (*Continued*)
RELATIVE DENSITY AND VOLUME OF WATER AND OF MERCURY AT VARIOUS TEMPERATURES

	WATER			MERCURY	
Temp. Degree C.	Relative Density g/ml	Volume ml/g	Relative Density g/ml	Volume ml/g	Temp. Degree C.
36	0.99371	1.00633	13.5069	0.0740362	36
37	0.99336	1.00669	13.5045	0.0740496	37
38	0.99299	1.00706	13.5020	0.0740630	38
39	0.99262	1.00743	13.4996	0.0740765	39
40	0.99224	1.00782	13.4971	0.0740898	40
41	0.99186	1.00821	13.4947	0.0741033	41
42	0.99147	1.00861	13.4922	0.0741167	42
43	0.99107	1.00901	13.4898	0.0741301	43
44	0.99066	1.00943	13.4873	0.0741435	44
45	0.99025	1.00985	13.4849	0.0741570	45
46	0.98982	1.01028	13.4825	0.0741704	46
47	0.98940	1.01072	13.4800	0.0741838	47
48	0.98896	1.01116	13.4776	0.0741972	48
49	0.98852	1.01162	13.4751	0.0742107	49
50	0.98807	1.01207	13.4727	0.0742241	50
51	0.98762	1.01254	13.4703	0.0742375	51
52	0.98715	1.01301	13.4678	0.0742510	52
53	0.98669	1.01349	13.4654	0.0742644	53
54	0.98621	1.01398	13.4630	0.0742778	54
55	0.98573	1.01448	13.4605	0.0742913	55
60	0.98324	1.01705	13.4484	0.0743585	60
65	0.98059	1.01979	13.4362	0.0744257	65
70	0.97781	1.02270	13.4241	0.0744930	70
75	0.97489	1.02576	13.4120	0.0745602	75
80	0.97183	1.02899	13.3999	0.0746276	80
85	0.96865	1.03237	13.3878	0.0746949	85
90	0.96534	1.03590	13.3757	0.0747623	90
95	0.96192	1.03959	13.3637	0.0748294	95
100	0.95838	1.04343	13.3516	0.0748971	100
110	0.9510	1.0515	13.328	0.075032	110
120	0.9434	1.0601	13.304	0.075167	120
130	0.9352	1.0693	13.280	0.075303	130
140	0.9264	1.0794	13.256	0.075439	140
150	0.9173	1.0902	13.232	0.075575	150
160	0.9075	1.1019	13.208	0.075712	160
170	0.8973	1.1145	13.184	0.075849	170
180	0.8866	1.1279	13.160	0.075986	180
190	0.8750	1.1429	13.137	0.076124	190
200	0.8628	1.1590	13.113	0.076262	200
210	0.850	1.177	13.089	0.076400	210
220	0.837	1.195	13.065	0.076539	220
230	0.823	1.215	13.042	0.076678	230
240	0.809	1.236	13.018	0.076818	240
250	0.794	1.259	12.994	0.076958	250
260	....	....	12.970	0.077099	260
270	....	....	12.947	0.077239	270
280	....	....	12.923	0.077371	280
290	....	....	12.899	0.077513	290
300	....	....	12.876	0.077666	300
			Under 20 ATM.		
310	....	....	12.853	0.077810	310
320	....	....	12.829	0.077954	320
330	....	....	12.805	0.078099	330
340	....	....	12.781	0.078245	340
350	....	....	12.758	0.078391	350
360	....	....	12.734	0.078538	360
370	....	....	12.710	0.078685	370
380	....	....	12.686	0.078834	380
390	....	....	12.662	0.078993	390
400	....	....	12.638	0.079133	400
450	....	....	12.517	0.079897	450
500	....	....	12.395	0.080686	500

Table 10-56
ABSOLUTE DENSITY OF WATER

This table gives the weight in grams of a cubic centimeter of water at temperatures from 0° to 30°C. Water attains its maximum density at 3.98°C. at which temperature the density is 0.999973 (C. G. S.)

Temp. °C.	Density	Temp. °C.	Density	Temp. °C.	Density	Temp. °C.	Density
0.0	0.999841	7.6	0.999872	15.2	0.999069	22.8	0.997585
0.2	9854	7.8	9861	15.4	9038	23.0	7538
0.4	9866	8.0	9849	15.6	9007	23.2	7490
0.6	9878	8.2	9837	15.8	8975	23.4	7442
0.8	9889	8.4	9824	16.0	8943	23.6	7394
1.0	9900	8.6	9810	16.2	8910	23.8	7345
1.2	9909	8.8	9796	16.4	8877	24.0	7296
1.4	9918	9.0	9781	16.6	8843	24.2	7246
1.6	9927	9.2	9766	16.8	8809	24.4	7196
1.8	9934	9.4	9751	17.0	8774	24.6	7146
2.0	9941	9.6	9734	17.2	8739	24.8	7095
2.2	9947	9.8	9717	17.4	8704	25.0	7044
2.4	9953	10.0	9700	17.6	8668	25.2	6992
2.6	9958	10.2	9682	17.8	8632	25.4	6941
2.8	9962	10.4	9664	18.0	8595	25.6	6888
3.0	9965	10.6	9645	18.2	8558	25.8	6836
3.2	9968	10.8	9625	18.4	8520	26.0	6783
3.4	9970	11.0	9605	18.6	8482	26.2	6729
3.6	9972	11.2	9585	18.8	8444	26.4	6676
3.8	9973	11.4	9564	19.0	8405	26.6	6621
4.0	9973	11.6	9542	19.2	8365	26.8	6567
4.2	9973	11.8	9520	19.4	8325	27.0	6512
4.4	9972	12.0	9498	19.6	8285	27.2	6457
4.6	9970	12.2	9475	19.8	8244	27.4	6401
4.8	9968	12.4	9451	20.0	8203	27.6	6345
5.0	9965	12.6	9427	20.2	8162	27.8	6289
5.2	9961	12.8	9402	20.4	8120	28.0	6232
5.4	9957	13.0	9377	20.6	8078	28.2	6175
5.6	9952	13.2	9352	20.8	8035	28.4	6118
5.8	9947	13.4	9326	21.0	7992	28.6	6060
6.0	9941	13.6	9299	21.2	7948	28.8	6002
6.2	9935	13.8	9272	21.4	7904	29.0	5944
6.4	9927	14.0	9244	21.6	7860	29.2	5885
6.6	9920	14.2	9216	21.8	7815	29.4	5826
6.8	9911	14.4	9188	22.0	7770	29.6	5766
7.0	9902	14.6	9159	22.2	7724	29.8	5706
7.2	9893	14.8	9129	22.4	7678	30.0	5646
7.4	9883	15.0	9099	22.6	7632		

Table 10-57
SPECIFIC GRAVITY OF AQUEOUS SOLUTIONS

% by weight	AgF $\frac{18°C}{4°C}$	AgNO₃ $\frac{20°C}{4°C}$	AlCl₃ $\frac{18°C}{4°C}$	Al(NO₃)₃ $\frac{18°C}{4°C}$	Al₂(SO₄)₃ $\frac{15°C}{4°C}$	AuCl₃ $\frac{15°C}{4°C}$	Ba(C₂H₃O₂)₂ $\frac{18°C}{4°C}$	BaBr₂ $\frac{20°C}{4°C}$	BaCl₂ $\frac{20°C}{4°C}$	BaI₂ $\frac{20°C}{4°C}$	Ba(NO₃)₂ $\frac{18°C}{4°C}$
1	1.0088	1.0070	1.0075	1.0065	1.0093	1.0060	1.0059	1.0156	1.0159	1.0154	1.0151
2	1.0191	1.0154	1.0164	1.0144	1.0195	1.0132	1.0133	....	....	....	....
3	1.0399	1.0327	1.0344	1.0305	1.0404	1.0281	1.0282	1.0335	1.0341	1.0331	1.0320
4	....	....	....	....	....	....	....	....	....	....	....
5	....	....	....	....	....	....	....	....	....	....	....
6	1.0613	1.0506	1.0526	1.0469	1.0618	1.0434	1.0433	1.0519	1.0528	1.0513	1.0494
7	1.0838	1.0690	1.0711	1.0638	1.0837	1.0591	1.0587	1.0710	1.0721	1.0701	1.0674
8	....	....	....	....	....	....	....	....	....	....	....
9	....	....	....	....	....	....	....	....	....	....	....
10	1.1071	1.0882	1.0900	1.0811	1.1062	1.0750	1.0745	1.0907	1.0921	1.0896	....
11	....	....	....	....	....	....	....	....	....	....	....
12	1.1316	1.1080	1.1093	1.0989	1.1293	....	1.0908	1.111	1.1128	1.1099	....
13	....	....	....	....	....	....	....	....	....	....	....
14	1.1572	1.1284	1.1290	1.1171	1.1529	....	1.1075	1.1323	1.1342	1.1308	....
15	....	....	....	....	....	....	....	....	....	....	....
16	1.1837	1.1495	1.1491	1.1357	1.1770	....	1.1246	1.1543	1.1564	1.1525	....
17	1.2110	1.1715	....	1.1549	1.2017	....	1.1421	1.1770	1.1793	1.1750	....
18	....	....	....	....	....	....	....	....	....	....	....
19	....	....	....	....	....	....	....	....	....	....	....
20	1.2394	1.1942	....	1.1745	1.2272	....	1.1599	1.2006	1.2031	1.1984	....
21	....	....	....	1.1946	1.2534	....	1.1782	....	....	....	....
22	....	....	....	....	....	....	....	....	1.2277	....	....
23	....	....	....	1.2153	1.2803	....	1.1970	....	....	....	....
24	....	....	....	....	....	....	....	1.2634	1.2531	1.2610	....
25	1.3146	1.2545	....	....	....	....	....	....	....	....	....
30	1.40	1.3205	....	1.2805	1.3079	....	1.2554	1.3325	....	1.3289	....
35	1.48	1.3931	....	....	....	....	1.3069	1.4087	....	1.404	....
40	1.58	1.4743	....	....	....	....	1.3608	1.4926	....	1.490	....
45	1.70	1.565	....	....	....	....	....	....	....	1.587	....
50	1.85	1.668	....	....	....	....	....	....	....	1.698	....
60	2.26	1.916	....	....	....	....	....	....	....	1.970	....
70	....	2.2333	....	....	....	....	....	....	....	....	....
80	....	....	....	....	....	....	....	....	....	....	....

% by weight	BaS 15°C	Be(NO₃)₂ 18°/4°	CaBr₂ 20°/4°	Ca(C₂H₃O₂)₂ 18°/4°	CaCl₂ 20°/4°	CaI₂ 20°/4°	Ca(NO₃)₂ 18°/4°	CdBr₂ 20°/4°	CdCl₂ 20°/4°	CdI₂ 20°/4°	Cd(NO₃)₂ 18°/4°
1	.....	.....	.....	1.0043	.....	.....	.....	.....	.....	.....	.....
2	1.0104	1.0108	1.0152	1.0100	1.0148	1.0150	1.0137	1.0158	1.0159	1.0153	1.0154
3	1.0172	.....	.....	.....	.....	.....	.....	.....	.....	.....	.....
4	1.0243	1.0233	1.0326	1.0215	1.0316	1.0323	1.0291	1.0339	1.0339	1.0328	1.0326
5	1.0346	.....	.....	.....	.....	.....	.....	.....	.....	.....	.....
6	1.0417	1.0361	1.0504	1.0331	1.0486	1.0500	1.0448	1.0524	1.0524	1.0507	1.0502
7	.....	.....	.....	.....	.....	.....	.....	.....	.....	.....	.....
8	1.0561	1.0491	1.0688	1.0447	1.0659	1.0683	1.0608	1.0714	1.0715	1.0690	1.0683
9	.....	.....	.....	.....	.....	.....	.....	.....	.....	.....	.....
10	1.0718	1.0624	1.0877	1.0563	1.0835	1.0873	1.0771	1.0910	1.0912	1.0879	1.0869
11	.....	.....	.....	.....	.....	.....	.....	.....	.....	.....	.....
12	1.0867	1.0761	1.1071	1.0679	1.1015	1.1069	1.0937	1.1112	1.1115	1.1075	1.1061
13	.....	.....	.....	.....	.....	.....	.....	.....	.....	.....	.....
14	1.1020	1.0902	1.1272	1.0795	1.1198	1.1273	1.1106	1.1322	1.1324	1.1278	1.1261
15	.....	.....	.....	.....	.....	.....	.....	.....	.....	.....	.....
16	1.1184	1.1046	1.1480	1.0912	1.1386	1.1485	1.1279	1.1540	1.1540	1.1489	1.1468
17	1.1367	.....	.....	.....	.....	.....	.....	.....	.....	.....	.....
18	.....	1.1193	1.1696	1.1029	1.1578	1.1703	1.1455	1.1766	1.1762	1.1709	1.1682
19	.....	.....	.....	.....	.....	.....	.....	.....	.....	.....	.....
20	1.1492	1.1344	1.1919	1.1146	1.1775	1.1928	1.1636	1.2000	1.1992	1.1937	1.1904
21	1.1547	.....	.....	.....	.....	.....	.....	.....	.....	.....	.....
22	.....	1.1499	.....	1.1263	.....	.....	.....	.....	.....	.....	.....
23	1.1688	.....	.....	.....	.....	.....	.....	.....	.....	.....	.....
24	.....	1.1657	.....	.....	.....	.....	.....	.....	.....	.....	.....
25	1.1835	.....	1.2499	.....	1.2284	1.2530	1.2106	1.2605	1.2604	1.2546	1.2488
30	1.2192	.....	1.3125	.....	1.2816	1.3195	1.260	1.3286	1.3273	1.3219	1.3124
35	.....	.....	1.381	.....	1.3373	1.3928	1.311	1.4049	1.4010	1.3967	1.3822
40	.....	.....	1.457	.....	1.3957	1.4734	1.365	1.4902	1.4833	1.4801	1.4590
45	.....	.....	1.541	.....	.....	.....	1.422	.....	1.5748	1.5726	1.5438
50	.....	.....	1.635	.....	.....	.....	.....	.....	1.6762	.....	1.6356
60	.....	.....	.....	.....	.....	.....	.....	.....	.....	.....	.....
70	.....	.....	.....	.....	.....	.....	.....	.....	.....	.....	.....
80	.....	.....	.....	.....	.....	.....	.....	.....	.....	.....	.....

Table 10-57 (*Continued*)
SPECIFIC GRAVITY OF AQUEOUS SOLUTIONS

% by weight	CdSO₄ 18°C/4°C	Ce₂(SO₄)₃ 15°C/4°C	CoCl₂ 20°C/4°C	Co(NO₃)₂ 20°C/4°C	CrBr₃ 18°C/4°C	CrCl₃-violet 18°C/4°C	CrCl₃-green 18°C/4°C	Cr₂(SO₄)₃-violet 15°C/4°C	Cr₂(SO₄)₃-green 15°C/4°C	CsBr 20°C/4°C	CsCl 20°C/4°C
1	1.0086	1.0090	1.0073	1.0064	1.0074	1.0076	1.0071	1.0091	1.0081	1.00612	1.00593
2	1.0182	1.0190	1.0165	1.0145	1.0162	1.0166	1.0157	1.0191	1.0172	1.01412	1.01374
3	.....	.....	.....	.....	.....	.....	.....	.....	.....	.....	.....
4	1.0383	1.0395	1.0350	1.0315	1.0344	1.0349	1.0332	1.0395	1.0358	1.03048	1.02969
5	.....	.....	.....	.....	.....	.....	.....	.....	.....	.....	.....
6	1.0590	1.0606	1.0538	1.0485	1.0532	1.0535	1.0510	1.0604	1.0551	1.04734	1.04609
7	.....	.....	.....	.....	.....	.....	.....	.....	.....	.....	.....
8	1.0803	1.0823	1.0735	1.0660	1.0726	1.0724	1.0691	1.0817	1.0751	1.06472	1.06297
9	.....	.....	.....	.....	.....	.....	.....	.....	.....	.....	.....
10	1.1023	1.1047	1.0940	1.084	1.0926	1.0917	1.0876	1.1034	1.0958	1.08265	1.08036
11	.....	.....	.....	.....	.....	.....	.....	.....	.....	.....	.....
12	1.1250	1.1279	1.1150	1.103	1.1132	1.1114	1.1065	1.1257	1.1172	1.10116	1.09828
13	.....	.....	.....	.....	.....	.....	.....	.....	.....	.....	.....
14	1.1485	1.1520	1.1365	1.122	1.1345	1.1316	.....	1.1486	1.1392	1.12029	1.11676
15	.....	.....	.....	.....	.....	.....	.....	.....	.....	.....	.....
16	1.1729	1.1770	1.1585	1.142	1.1565	.....	.....	1.1722	1.1618	1.14007	1.13582
17	.....	.....	.....	.....	.....	.....	.....	.....	.....	.....	.....
18	1.1982	1.2030	1.1815	1.163	1.1794	.....	.....	1.1966	1.1851	1.16053	1.15549
19	.....	.....	.....	.....	.....	.....	.....	.....	.....	.....	.....
20	1.2243	1.2300	1.2050	1.184	1.2031	.....	.....	1.2218	1.2091	1.18170	1.17580
21	.....	.....	.....	.....	.....	.....	.....	.....	.....	.....	.....
22	.....	1.2582	.....	.....	1.2277	.....	.....	1.2479	1.2339	1.20362	1.19679
23	.....	.....	.....	.....	.....	.....	.....	.....	.....	.....	.....
24	1.2940	1.2876	.....	1.239	1.2532	.....	.....	1.2750	1.2594	1.22634	1.21849
25	.....	.....	.....	.....	.....	.....	.....	.....	.....	.....	.....
30	1.3714	.....	.....	1.299	1.3355	.....	.....	.....	.....	1.29973	1.28817
35	1.4551	.....	.....	.....	.....	.....	.....	.....	.....	1.36764	1.35218
40	1.5470	.....	.....	.....	.....	.....	.....	.....	.....	1.44275	1.44245
45	.....	.....	.....	.....	.....	.....	.....	.....	.....	1.52626	1.49993
50	.....	.....	.....	.....	.....	.....	.....	.....	.....	1.61970	1.58575
60	.....	.....	.....	.....	.....	.....	.....	.....	.....	.....	1.78859
70	.....	.....	.....	.....	.....	.....	.....	.....	.....	.....	.....
80	.....	.....	.....	.....	.....	.....	.....	.....	.....	.....	.....

% by weight	CsI 20°/4°C	CsNO$_3$ 20°/4°C	Cs$_2$SO$_4$ 20°/4°C	CuCl$_2$ 20°/4°C	Cu(NO$_3$)$_2$ 20°/4°C	CuSO$_4$ 20°/4°C	FeCl$_3$ 20°/4°C	Fe(NO$_3$)$_3$ 18°/4°C	FeSO$_4$ 18°/4°C	Fe$_2$(SO$_4$)$_3$ 17.5°/4°C	FeSO$_4$·(NH$_4$)$_2$SO$_4$ 25°/4°C
1	1.00608	1.00566	1.0061	1.007	1.007	1.009	1.009	1.0065	1.0085	1.007	1.005
2	1.01402	1.01319	1.0144	1.017	1.015	1.019	1.017	1.0144	1.0180	1.016	1.013
3	....	....	....	....	....	....	1.026	....	....	....	....
4	1.03029	1.02859	1.0316	1.036	1.032	1.040	1.034	1.0304	1.0375	1.033	1.029
5	....	....	....	....	....	....	1.043	....	....	....	....
6	1.04707	1.04443	1.0494	1.056	1.050	1.062	1.052	1.0468	1.0575	1.050	1.045
7	1.06438	1.06072	1.0676	1.075	1.069	1.084	1.060	1.0636	1.0785	1.067	1.062
8	1.08225	1.07745	1.0870	1.096	1.088	1.107	1.069	1.0810	1.1000	1.084	1.080
9	....	....	....	....	....	....	1.077	....	....	....	....
10	1.10071	1.09463	1.1071	1.117	1.107	1.131	1.086	1.0989	1.1220	1.103	1.098
11	1.11979	1.11227	1.1275	1.138	1.126	1.154	1.094	1.1172	1.1445	1.122	1.116
12	....	....	....	....	....	....	1.105	....	....	....	....
13	1.13953	....	1.1484	1.160	1.147	1.180	1.115	1.1359	1.1675	1.141	....
14	1.15996	....	1.1696	1.182	1.168	1.206	1.125	1.1551	1.1905	1.161	....
15	1.18112	....	1.1913	1.205	1.189	....	1.135	1.1748	1.2135	1.181	....
16	....	....	....	....	....	....	1.144	....	....	....	....
17	1.20305	....	1.2137	....	....	....	1.154	....	....	....	....
18	1.22580	....	1.2375	....	1.248	....	1.164	1.2281	....	1.241	....
19	....	....	....	....	....	....	1.174	....	....	....	....
20	....	....	....	....	....	....	1.185	....	....	....	....
21	1.29944	....	....	....	....	....	1.195	....	....	1.307	....
22	1.36776	....	....	....	....	....	1.206	....	....	1.376	....
23	1.44354	....	....	....	....	....	1.216	....	....	1.449	....
24	1.52803	....	....	....	....	....	1.226	....	....	1.528	....
25	....	....	....	....	....	....	1.237	....	....	....	....
30	1.62278	....	....	....	....	....	1.295	....	....	1.613	....
35	....	....	....	....	....	....	1.357	....	....	1.798	....
40	....	....	....	....	....	....	1.423	....	....	....	....
45	....	....	....	....	....	....	1.488	....	....	....	....

Table 10-57 (Continued)
SPECIFIC GRAVITY OF AQUEOUS SOLUTIONS

% by weight	Fe$_2$(SO$_4$)$_3$·(NH$_4$)$_2$SO$_4$ 15°/4°C	H$_3$BO$_3$ 15°/15°C	HBr 20°/4°C	HClO$_3$ 18°/4°C	HClO$_4$ 15°/4°C	HF 20°/4°C	HI 20°/4°C	HIO$_3$ 18°/4°C	HIO$_4$ 17°/4°C	H$_2$O$_2$ 18°/4°C	H$_2$SeO$_4$ 20°/4°C
1	1.0072	1.0034	1.0053	1.0044	1.0050	.....	1.0054	1.0071	1.0076	1.0022	1.0059
2	1.0157	1.0069	1.0124	1.0103	1.0109	.....	1.0127	1.0157	1.0165	1.0058	1.0136
3	.....	1.0106	.....	.....	.....	.....	.....	.....	.....	.....	.....
4	1.0324	1.0147	1.0269	1.0222	1.0228	.....	1.0277	1.0334	1.0349	1.0131	1.0291
5	.....	1.0248*	.....	.....	.....	1.017	.....	.....	.....	.....	.....
6	1.0498	.....	1.0417	1.0344	1.0348	.....	1.0431	1.0517	1.0539	1.0204	1.0447
7	.....	.....	.....	.....	.....	.....	.....	.....	.....	.....	.....
8	1.0675	.....	1.0568	1.0468	1.0471	.....	1.0589	1.0706	1.0737	1.0277	1.0605
9	.....	.....	.....	.....	.....	.....	.....	.....	.....	.....	.....
10	1.0856	.....	1.0723	1.0594	1.0597	1.035	1.0751	1.0900	1.0944	1.0351	1.0766
11	.....	.....	.....	.....	.....	.....	.....	.....	.....	.....	.....
12	1.1038	.....	1.0883	1.0723	1.0726	.....	1.0918	1.1100	1.1161	1.0425	1.0931
13	.....	.....	.....	.....	.....	.....	.....	.....	.....	.....	.....
14	1.1224	.....	1.1048	1.0856	1.0859	.....	1.1091	1.1306	1.1388	1.0499	1.1101
15	.....	.....	.....	.....	.....	1.053	.....	.....	.....	.....	.....
16	1.1412	.....	1.1219	1.0991	1.0995	.....	1.1270	1.1519	1.1623	1.0574	1.1276
17	.....	.....	.....	.....	.....	.....	.....	.....	.....	.....	.....
18	1.1609	.....	1.1396	1.1130	1.1135	.....	1.1456	1.1740	1.1865	1.0649	1.1455
19	.....	.....	.....	.....	.....	.....	.....	.....	.....	.....	.....
20	1.1809	.....	1.1579	1.1273	1.1279	1.070	1.1649	1.1969	1.2116	1.0725	1.1639
21	.....	.....	.....	.....	.....	.....	.....	.....	.....	.....	.....
22	.....	.....	1.1767	1.1419	1.1428	.....	1.1850	1.2206	1.2376	1.0802	1.1829
23	.....	.....	.....	.....	.....	.....	.....	.....	.....	.....	.....
24	.....	.....	1.1961	1.1568	1.1581	.....	1.2059	1.2450	1.2647	1.0880	1.2026
25	.....	.....	.....	.....	.....	1.086	.....	.....	.....	.....	.....
30	.....	.....	1.2580	.....	1.2067	1.101	1.2737	1.3218	1.3545	1.1122	1.2653
35	.....	.....	1.3150	.....	1.2991	1.116	1.3357	1.3900	.....	1.1327	1.3819
40	1.3799	.....	1.3772	.....	1.3521	1.130	1.4029	1.4640	.....	1.1536	.....
45	.....	.....	1.4446	.....	.....	1.143	1.4755	.....	.....	1.1749	.....
50	.....	.....	1.5173	.....	1.4103	1.155	.....	.....	.....	1.1966	1.5186
60	.....	.....	1.6787	.....	1.5389	.....	.....	.....	.....	1.2416	1.6850
70	.....	.....	.....	.....	1.6736	.....	.....	.....	.....	1.2897	1.8870
80	.....	.....	.....	.....	.....	.....	.....	.....	.....	1.3406	2.1220

*Saturated solution.

% by weight	$HgCl_2$ $\frac{20°C}{4°C}$	K-alum* $\frac{17.5°}{17.5°C}$	KBr $\frac{20°C}{4°C}$	$KBrO_3$ $\frac{20°C}{4°C}$	K_2CO_3 $\frac{20°C}{4°C}$	$KC_2H_3O_2$ $\frac{18°C}{4°C}$	$K_2C_4H_4O_6$ $\frac{20°C}{4°C}$	$KCNS$ $\frac{18°C}{4°C}$	KCl $\frac{20°C}{4°C}$	$KClO_3$ $\frac{20°C}{4°C}$	K_2CrO_4 $\frac{18°C}{4°C}$
1	1.0065	1.0049	1.0054	1.0056	1.0072	1.0038	1.0048	1.0035	1.0046	1.0045	1.0066
2	1.0150	1.0100	1.0127	1.0131	1.0163	1.0089	1.0114	1.0085	1.0110	1.0109	1.0147
3	1.0236	1.0152	......	1.0206	......	......	......	......	......	1.0174	......
4	1.0323	1.0205	1.0275	1.0282	1.0345	1.0191	1.0248	1.0186	1.0239	1.0241	1.0311
5	1.0411	1.0258	......	1.0359	......	......	......	......	......	......	......
6	......	1.0310	1.0426	......	1.0529	1.0293	1.0383	1.0288	1.0369	......	1.0477
7	......	1.0362	......	......	......	......	......	......	......	......	......
8	......	1.0415	1.0581	......	1.0715	1.0395	1.0519	1.0391	1.0500	......	1.0647
9	......	1.0469	......	......	......	......	......	......	......	......	......
10	......	1.0523	1.0740	......	1.0904	1.0497	1.0657	1.0495	1.0633	......	1.0821
11	......	1.0578	......	......	......	......	......	......	......	......	......
12	......	1.0635	1.0903	......	1.1096	1.0599	1.0798	1.0601	1.0768	......	1.0999
13	......	1.0690	......	......	......	......	......	......	......	......	......
14	......	......	1.1070	......	1.1291	1.0703	1.0941	1.0708	1.0905	......	1.1181
15	......	......	......	......	......	......	......	......	......	......	......
16	......	......	1.1242	......	1.1490	1.0808	1.1087	1.0817	1.1043	......	1.1366
17	......	......	1.1419	......	......	......	......	......	......	......	......
18	......	......	......	......	1.1692	1.0914	1.1236	1.0927	1.1185	......	1.1555
19	......	......	......	......	......	......	......	......	......	......	......
20	......	......	1.1601	......	1.1898	1.1022	1.1387	1.1039	1.1328	......	1.1748
21	......	......	1.1788	......	1.2107	1.1131	1.1540	1.1152	1.1474	......	1.1945
22	......	......	......	......	......	......	......	......	......	......	......
23	......	......	......	......	1.2320	1.1241	1.1696	1.1266	1.1623	......	1.2147
24	......	......	1.1980	......	......	......	......	......	......	......	......
25	......	......	......	......	......	......	......	......	......	......	......
30	......	......	1.2593	......	1.2979	1.1579	1.2181	1.1618	......	......	1.2784
35	......	......	1.3147	......	1.3548	1.1868	1.2606	1.1899	......	......	......
40	......	......	1.3746	......	1.4141	1.2162	1.3051	1.2200	......	......	......
45	......	......	......	......	1.4759	1.2460	1.3516	1.2517	......	......	......
50	......	......	......	......	1.5404	1.2761	1.4001	1.2849	......	......	......
60	......	......	......	......	......	1.3372	......	1.3554	......	......	......
70	......	......	......	......	......	......	......	1.4307	......	......	......
80	......	......	......	......	......	......	......	......	......	......	......

* $K_2SO_4 \cdot Al_2(SO_4)_3 \cdot 24H_2O$.

Table 10-57 (Continued)
SPECIFIC GRAVITY OF AQUEOUS SOLUTIONS

% by weight	$K_2Cr_2O_7$ 20°/4°C	$K_2Cr_2(SO_4)_4$- green 15°/4°C	KF 18°/4°C	$K_3Fe(CN)_6$ 20°/4°C	$K_4Fe(CN)_6$ 20°/4°C	KI 20°/4°C	KIO_3 20°/4°C	KNO_3 20°/4°C	KN_3 20°/4°C	K_2SO_4 20°/4°C	$KHSO_4$ 20°/4°C
1	1.0052	1.007	1.0072	1.0034	1.0051	1.0055	1.0068	1.0045	1.0037	1.0063	1.0051
2	1.0122	1.016	1.0159	1.0090	1.0119	1.0130	1.0155	1.0108	1.0093	1.0145	1.0120
3	......	......	......	......	......	......	......	......	......	......	......
4	1.0264	1.034	1.0334	1.0201	1.0256	1.0281	1.0243	1.0234	1.0206	1.0310	1.0260
5	......	......	......	......	......	......	......	......	......	......	......
6	1.0408	1.052	1.0512	1.0314	1.0395	1.0437	......	1.0363	1.0319	1.0477	1.0403
7	......	......	......	......	......	......	......	......	......	......	......
8	1.0554	1.070	1.0693	1.0427	1.0536	1.0597	......	1.0494	1.0434	1.0646	1.0549
9	......	......	......	......	......	......	......	......	......	......	......
10	1.0703	1.089	1.0877	1.0542	1.0678	1.0761	......	1.0627	1.0550	1.0817	1.0698
11	......	......	......	......	......	......	......	......	......	......	......
12	......	1.109	1.1064	1.0656	1.0823	1.0930	......	1.0762	1.0669	......	1.0850
13	......	......	......	......	......	......	......	......	......	......	......
14	......	1.129	1.1254	1.0774	1.0971	1.1104	......	1.0899	1.0790	......	1.1004
15	......	......	......	......	......	......	......	......	......	......	......
16	......	1.150	1.1448	1.0890	1.1120	1.1284	......	1.1039	1.0913	......	1.1161
17	......	1.171	1.1646	1.1010	......	1.1469	......	1.1181	1.1036	......	......
18	......	......	......	......	......	......	......	......	......	......	......
19	......	......	......	......	......	......	......	......	......	......	......
20	......	1.193	1.1847	1.1130	......	1.1660	......	1.1326	1.1161	......	......
21	......	1.216	1.2052	......	......	1.1857	......	1.1473	1.1288	......	......
22	......	......	......	......	......	......	......	......	......	......	......
23	......	......	......	......	......	......	......	......	......	......	......
24	......	1.239	1.2260	......	......	1.2060	......	1.1623	1.1417	......	......
25	......	......	......	......	......	......	......	......	......	......	......
30	......	1.315	......	......	......	1.2712	......	......	1.1809	......	......
35	......	1.383	......	......	......	1.3308	......	......	......	......	......
40	......	1.456	......	......	......	1.3959	......	......	......	......	......
45	......	1.533	......	......	......	1.4672	......	......	......	......	......
50	......	1.615	......	......	......	1.5458	......	......	......	......	......
60	......	......	......	......	......	......	......	......	......	......	......
70	......	......	......	......	......	......	......	......	......	......	......
80	......	......	......	......	......	......	......	......	......	......	......

% by weight	La(NO₃)₃ 18°C/4°C	LiBr 20°C/4°C	LiCl 20°C/4°C	LiI 20°C/4°C	LiNO₃ 20°C/4°C	LiOH 20°C/4°C	Li₂SO₄ 20°C/4°C	MgBr₂ 20°C/4°C	MgCl₂ 20°C/4°C	MgI₂ 20°C/4°C	Mg(NO₃)₂ 20°C/4°C
1	1.0076	1.0055	1.0041	1.0056	1.00409	1.0102	1.0068	.....	.....	.....	.....
2	1.0167	1.0128	1.0099	1.0131	1.01002	1.0217	1.0155	1.0151	1.0146	1.0149	1.0132
3	.....	.....	.....	.....	.....	.....	.....	.....	.....	.....	.....
4	1.0353	1.0277	1.0215	1.0284	1.02198	1.0437	1.0329	1.0324	1.0311	1.0321	1.0285
5	.....	.....	.....	.....	.....	.....	.....	.....	.....	.....	.....
6	1.0545	1.0429	1.0330	1.0442	1.03413	1.0650	1.0505	1.0501	1.0478	1.0498	1.0441
7	.....	.....	.....	.....	.....	.....	.....	.....	.....	.....	.....
8	1.0742	1.0585	1.0444	1.0604	1.04647	1.0862	1.0684	1.0683	1.0646	1.0680	1.0600
9	.....	.....	.....	.....	.....	.....	.....	.....	.....	.....	.....
10	1.0945	1.0746	1.0559	1.0771	1.05903	1.1074	1.0863	1.0871	1.0816	1.0869	1.0762
11	.....	.....	.....	.....	.....	.....	.....	.....	.....	.....	.....
12	1.1153	1.0910	1.0675	1.0943	1.07181	.....	1.1044	1.1065	1.0989	1.1065	1.0928
13	.....	.....	.....	.....	.....	.....	.....	.....	.....	.....	.....
14	1.1368	1.1079	1.0792	1.1120	1.08482	.....	1.1228	1.1265	1.1164	1.1268	1.1098
15	.....	.....	.....	.....	.....	.....	.....	.....	.....	.....	.....
16	1.1589	1.1253	1.0910	1.1303	1.09807	.....	1.1411	1.1471	1.1342	1.1478	1.1272
17	.....	.....	.....	.....	.....	.....	.....	.....	.....	.....	.....
18	1.1817	1.1432	1.1029	1.1492	1.11159	.....	1.1599	1.1683	1.1523	1.1695	1.1449
19	.....	.....	.....	.....	.....	.....	.....	.....	.....	.....	.....
20	1.2052	1.1616	1.1150	1.1688	1.12539	.....	1.1789	1.1903	1.1706	1.1920	1.1630
21	.....	.....	.....	.....	.....	.....	.....	.....	.....	.....	.....
22	1.2295	1.1806	1.1274	1.1890	1.13948	.....	1.1984	.....	.....	.....	1.1815
23	.....	.....	.....	.....	.....	.....	.....	.....	.....	.....	.....
24	1.2547	1.2002	1.1399	1.2099	1.15387	.....	1.2182	1.2482	1.2184	1.2519	1.2004
25	.....	.....	.....	.....	.....	.....	.....	.....	.....	.....	.....
30	1.3360	1.2629	1.1791	1.2772	1.19879	.....	.....	1.3110	1.2688	1.3180	.....
35	.....	1.3204	.....	1.3393	1.2392	.....	.....	1.379	.....	1.3914	.....
40	.....	1.3836	.....	1.4078	1.2837	.....	.....	1.452	.....	1.4730	.....
45	.....	1.4535	.....	1.4840	.....	.....	.....	1.532	.....	.....	.....
50	.....	.....	.....	1.5692	.....	.....	.....	.....	.....	.....	.....
60	.....	.....	.....	1.7748	.....	.....	.....	.....	.....	.....	.....
70	.....	.....	.....	.....	.....	.....	.....	.....	.....	.....	.....
80	.....	.....	.....	.....	.....	.....	.....	.....	.....	.....	.....

Table 10-57 (Continued)
SPECIFIC GRAVITY OF AQUEOUS SOLUTIONS

% by weight	$MgSO_4$ 20°C/4°C	$MnBr_2$ 18°C/4°C	$MnCl_2$ 18°C/4°C	$Mn(NO_3)_2$ 18°C/4°C	$MnSO_4$ 20°C/4°C	N_2H_4 15°C/4°C	NH_2OH 20°C/4°C	NH_4Br 25°C/4°C	$(NH_4)_2CO_3$ 15°C/4°C	$NH_4C_2H_3O_2$ 25°C/4°C	NH_4Cl 20°C/4°C
1	....	1.0071	1.0069	1.0063	1.0080	1.0002	1.0002	1.0027	1.0026	0.9992	1.0013
2	1.0186	1.0157	1.0153	1.0140	1.0178	1.0013	1.0023	1.0084	1.0061	1.0013	1.0045
3	....	....	....	....	....	....	....	....	....	....	....
4	1.0392	1.0332	1.0324	1.0298	1.0378	1.0034	1.0065	1.0198	1.0130	1.0055	1.0107
5	....	....	....	....	....	....	....	....	....	....	....
6	1.0602	1.0511	1.0498	1.0459	1.0583	1.0056	1.0107	1.0314	1.0199	1.0096	1.0168
7	....	....	....	....	....	....	....	....	....	....	....
8	1.0816	1.0695	1.0676	1.0624	1.0794	1.0077	1.0149	1.0432	1.0267	1.0136	1.0227
9	....	....	....	....	....	....	....	....	....	....	....
10	1.1034	1.0886	1.0859	1.0794	1.1012	1.0099	1.0192	1.0552	1.0335	1.0176	1.0286
11	....	....	....	....	....	....	....	....	....	....	....
12	1.1256	1.1083	1.1046	1.0969	1.1236	1.0121	1.0235	1.0674	1.0403	1.0216	1.0344
13	....	....	....	....	....	....	....	....	....	....	....
14	1.1484	1.1287	1.1238	1.1149	1.1467	1.0143	1.0278	1.0799	1.0471	1.0255	1.0401
15	....	....	....	....	....	....	....	....	....	....	....
16	1.1717	1.1498	1.1435	1.1333	1.1705	1.0164	1.0322	1.0927	1.0539	1.0294	1.0457
17	....	....	....	....	....	....	....	....	....	....	....
18	1.1955	1.1716	1.1638	1.1522	1.1950	1.0186	1.0366	1.1058	1.0607	1.0331	1.0512
19	....	....	....	....	....	....	....	....	....	....	....
20	1.2198	1.1942	1.1846	1.1717	....	1.0207	1.0410	1.1191	1.0675	1.0368	1.0567
21	....	....	....	....	....	....	....	....	....	....	....
22	1.2447	1.2176	1.2061	1.1918	....	1.0228	1.0454	1.1327	1.0742	1.0404	1.0621
23	....	....	....	....	....	....	....	....	....	....	....
24	1.2701	1.2419	1.2283	1.2125	....	1.0248	1.0499	1.1466	1.0808	1.0439	1.0674
25	....	....	....	....	....	....	....	....	....	....	....
30	....	1.3206	1.2988	1.2781	....	1.0305	1.0637	1.1901	1.1006	1.0540	....
35	....	....	....	1.3367	....	1.035	1.0755	....	1.1157	1.0618	....
40	....	....	....	1.3993	....	1.038	1.0875	1.2702	1.1294	1.0691	....
45	....	....	....	1.4662	....	1.042	1.0997	....	1.1417	1.0760	....
50	....	....	....	1.5378	....	1.044	1.1122	....	....	....	....
60	....	....	....	....	....	1.047	....	....	....	....	....
70	....	....	....	....	....	1.046	....	....	....	....	....
80	....	....	....	....	....	1.040	....	....	....	....	....

% by weight	NH_4F 18°/4°	NH_4I 18°/4°	NH_4NO_3 20°/4°	$(NH_4)_2SO_4$ 20°/4°	Na_3AsO_4 17°/4°	Na_2HAsO_4 14°/4°	$Na_2B_4O_7 \cdot 10H_2O$* 15°/15°	$NaBr$ 20°/4°	$NaBrO_3$ 18°/4°	Na_2CO_3 20°/4°	$NaHCO_3$ 18°/4°
1	1.0034	1.0050	1.0023	1.0041	1.0097	1.0083	1.0049	1.0060	1.0064	1.0086	1.0059
2	1.0085	1.0114	1.0064	1.0101	1.0207	1.0175	1.0099	1.0139	1.0143	1.0190	1.0132
3	.....	.....	.....	.....	.....	.....	1.0149	.....	.....	.....	1.0206
4	1.0178	1.0244	1.0147	1.0220	1.0431	1.0365	1.0199	1.0298	1.0305	1.0398	1.0280
5	.....	.....	.....	.....	.....	.....	1.0249	.....	.....	.....	1.0354
6	1.0265	1.0377	1.0230	1.0338	1.0659	1.0563	1.0299	1.0462	1.0471	1.0606	1.0429
7	.....	.....	.....	.....	.....	.....	.....	.....	.....	.....	1.0505
8	1.0346	1.0513	1.0313	1.0456	1.0892	1.0768	.....	1.0631	1.0641	1.0816	1.0581
9	.....	.....	.....	.....	.....	.....	.....	.....	.....	.....	.....
10	1.0420	1.0652	1.0397	1.0574	1.1130	1.0980	.....	1.0803	1.0816	1.1029	.....
11	.....	.....	.....	.....	.....	.....	.....	.....	.....	.....	.....
12	1.0487	1.0795	1.0482	1.0691	1.1373	1.1197	.....	1.0981	1.0996	1.1244	.....
13	.....	.....	.....	.....	.....	.....	.....	.....	.....	.....	.....
14	1.0547	1.0942	1.0567	1.0808	.....	1.1419	.....	1.1164	1.1182	1.1463	.....
15	.....	.....	.....	.....	.....	.....	.....	.....	.....	.....	.....
16	.....	1.1093	1.0653	1.0924	.....	1.1645	.....	1.1352	1.1373	.....	.....
17	.....	.....	.....	.....	.....	.....	.....	.....	.....	.....	.....
18	.....	1.1248	1.0740	1.1039	.....	.....	.....	1.1546	1.1569	.....	.....
19	.....	.....	.....	.....	.....	.....	.....	.....	.....	.....	.....
20	.....	1.1407	1.0828	1.1154	.....	.....	.....	1.1745	1.1771	.....	.....
21	.....	1.1570	1.0916	1.1269	.....	.....	.....	1.1951	1.1979	.....	.....
22	.....	.....	.....	.....	.....	.....	.....	.....	.....	.....	.....
23	.....	.....	.....	.....	.....	.....	.....	.....	.....	.....	.....
24	.....	1.1737	1.1005	1.1383	.....	.....	.....	1.2163	1.2193	.....	.....
25	.....	.....	.....	.....	.....	.....	.....	.....	.....	.....	.....
30	.....	1.2265	1.1277	1.1721	.....	.....	.....	1.2841	1.3565	.....	.....
35	.....	1.2745	1.1512	1.2000	.....	.....	.....	1.3462	.....	.....	.....
40	.....	1.3264	1.1754	1.2277	.....	.....	.....	1.4138	.....	.....	.....
45	.....	1.3823	1.2003	1.2552	.....	.....	.....	.....	.....	.....	.....
50	.....	.....	1.2258	1.2825	.....	.....	.....	.....	.....	.....	.....
60	.....	.....	.....	.....	.....	.....	.....	.....	.....	.....	.....
70	.....	.....	.....	.....	.....	.....	.....	.....	.....	.....	.....
80	.....	.....	.....	.....	.....	.....	.....	.....	.....	.....	.....

Table 10-57 (Continued)
SPECIFIC GRAVITY OF AQUEOUS SOLUTIONS

% by weight	$NaC_2H_3O_2$ $\frac{20°}{4°}$	$Na_2C_4H_4O_6$ $\frac{20°}{4°}$	$NaKC_4H_4O_6$ $\frac{20°}{4°}$	$NaCl$ $\frac{20°}{4°}$	$NaClO_4$ $\frac{18°}{4°}$	Na_2CrO_4 $\frac{18°}{4°}$	$Na_2Cr_2O_7$ $\frac{15°}{15°}$	$NaIO_3$ $\frac{19.5°}{19.5°}$	NaN_3 $\frac{20°}{4°}$	$NaNO_2$ $\frac{15°}{4°}$	$NaNO_3$ $\frac{20°}{4°}$
1	1.0033	1.0052	1.0049	1.0053	1.0051	1.0074	1.008	1.010	1.0040	1.0058	1.0049
2	1.0084	1.0123	1.0116	1.0125	1.0116	1.0163	1.015	1.019	1.010	1.0125	1.0117
3	.....	.....	.....	.....	.....	.....	.....	1.028	.....	.....	.....
4	1.0186	1.0266	1.0252	1.0268	1.0247	1.0344	1.028	1.036	1.022	1.0260	1.0254
5	.....	.....	.....	.....	.....	.....	.....	1.044	.....	.....	.....
6	1.0289	1.0410	1.0390	1.0413	1.0381	1.0529	1.045	1.054	1.034	1.0397	1.0392
7	.....	.....	.....	.....	.....	.....	.....	1.065	.....	.....	.....
8	1.0392	1.0555	1.0530	1.0559	1.0517	1.0718	1.061	1.075	1.047	1.0535	1.0532
9	.....	.....	.....	.....	.....	.....	.....	1.085	.....	.....	.....
10	1.0495	1.0702	1.0673	1.0707	1.0656	1.0912	1.077	1.095	1.059	1.0675	1.0674
11	.....	.....	.....	.....	.....	.....	.....	.....	.....	.....	.....
12	1.0598	1.0851	1.0818	1.0857	1.0798	1.1110	1.095	.....	1.072	1.0816	1.0819
13	.....	.....	.....	.....	.....	.....	.....	.....	.....	.....	.....
14	1.0702	1.1002	1.0965	1.1009	1.0943	1.1312	1.112	.....	1.085	1.0959	1.0967
15	.....	.....	.....	.....	.....	.....	.....	.....	.....	.....	.....
16	1.0807	1.1156	1.1114	1.1162	1.1090	1.1518	1.130	.....	1.099	1.1103	1.1118
17	.....	.....	.....	.....	.....	.....	.....	.....	.....	.....	.....
18	1.0913	1.1313	1.1265	1.1319	1.1241	1.1728	1.147	.....	1.112	1.1248	1.1272
19	.....	.....	.....	.....	.....	.....	.....	.....	.....	.....	.....
20	1.1021	1.1471	1.1419	1.1478	1.1396	1.1942	1.166	.....	1.126	1.1394	1.1429
21	.....	.....	.....	.....	.....	.....	.....	.....	.....	.....	.....
22	1.1130	1.1633	1.1576	1.1640	1.1554	1.2160	1.184	.....	1.140	.....	1.1589
23	.....	.....	.....	.....	.....	.....	.....	.....	.....	.....	.....
24	1.1240	1.1797	1.1735	1.1804	1.1717	1.2383	1.204	.....	1.155	.....	1.1752
25	.....	.....	.....	.....	.....	.....	.....	.....	.....	.....	.....
30	1.2225	.....	1.2225	.....	1.2227	.....	1.266	.....	1.202	.....	1.2256
35	.....	.....	.....	.....	.....	.....	1.323	.....	.....	.....	1.2701
40	.....	.....	.....	.....	.....	.....	1.383	.....	.....	.....	1.3175
45	.....	.....	.....	.....	.....	.....	1.447	.....	.....	.....	1.3683
50	.....	.....	.....	.....	.....	.....	.....	.....	.....	.....	.....
60	.....	.....	.....	.....	.....	.....	.....	.....	.....	.....	.....
70	.....	.....	.....	.....	.....	.....	.....	.....	.....	.....	.....
80	.....	.....	.....	.....	.....	.....	.....	.....	.....	.....	.....

% by weight	$Na_2HPO_4 \cdot 12H_2O$ 19°/19°C	$Na_3PO_4 \cdot 12H_2O$ 15°/15°C	Na_2S 18°/4°C	Na_2SO_3 19°/4°C	Na_2SO_4 20°/4°C	$Na_2S_2O_3$ 20°/4°C	Na_2WO_4 20°/4°C	$NiBr_2$ 18°/4°C	$NiCl_2$ 20°/4°C	$Ni(NO_3)_2$ 18°/4°C	$NiSO_4$ 18°/4°C
1	1.0041		1.0098	1.0078	1.0073	1.0065	1.0074	1.0078	1.008	1.0070	1.009
2	1.0083	1.0086	1.0211	1.0172	1.0164	1.0148	1.0166	1.0170	1.018	1.0155	1.020
3	1.0125										
4	1.0166		1.0440	1.0363	1.0348	1.0315	1.0354	1.0359	1.037	1.0330	1.042
5	1.0208										
6	1.0250	1.0263	1.0672	1.0556	1.0535	1.0483	1.0546	1.0555	1.057	1.0508	1.063
7	1.0292										
8	1.0332		1.0907	1.0751	1.0724	1.0654	1.0742	1.0758	1.078	1.0693	1.085
9	1.0375										
10	1.0418	1.0455	1.1146	1.0948	1.0915	1.0827	1.0944	1.0969	1.099	1.0882	1.109
11	1.0460										
12	1.0503		1.1388	1.1146	1.1109	1.1003	1.1154	1.1188	1.121	1.1076	1.133
13											
14		1.0633	1.1634	1.1346	1.1306	1.1182	1.1372	1.1414	1.143	1.1277	1.158
15											
16			1.1885	1.1549	1.1506	1.1365	1.1598	1.1648	1.167	1.1484	1.183
17											
18		1.0827	1.2140	1.1755	1.1709	1.1551	1.1833	1.1889	1.191	1.1696	1.209
19											
20					1.1915	1.1740	1.2076	1.2140	1.215	1.1914	
21											
22		1.1025			1.2124	1.1932	1.2328				
23											
24					1.2336	1.2128	1.2590	1.2815	1.280	1.2493	
25											
30						1.2739	1.3444		1.353	1.3114	
35						1.3273	1.5156			1.3777	
40						1.3827					
45											
50											
60											
70											
80											

Table 10-57 (Continued)
SPECIFIC GRAVITY OF AQUEOUS SOLUTIONS

% by weight	$Pb(C_2H_3O_2)_2$ 18°/4°	$Pb(NO_3)_2$ 18°/4°	RbBr 20°/4°	RbCl 20°/4°	RbI 20°/4°	$RbNO_3$ 20°/4°	Rb_2SO_4 20°/4°	SO_2 15.5°/4°	$Sa(NO_3)_3$ 18°/4°	$SnCl_2$ 15°/4°	$SnCl_4$ 15°/4°
1	1.0061	1.0074	1.00593	1.00561	1.00591	1.0053	1.0066	1.0040	1.0061	1.0068	1.007
2	1.0137	1.0163	1.01372	1.01307	1.01370	1.0125	1.0150	1.0091	1.0136	1.0146	1.015
3											
4	1.0290	1.0344	1.02965	1.02825	1.02963	1.0272	1.0322	1.0191	1.0289	1.0306	1.031
5											
6	1.0446	1.0529	1.04604	1.04379	1.04604	1.0422	1.0499	1.0292	1.0445	1.0470	1.047
7											
8	1.0605	1.0720	1.06291	1.05971	1.06296	1.0575	1.0680	1.0393	1.0604	1.0638	1.064
9											
10	1.0768	1.0918	1.08028	1.07604	1.08041	1.0731	1.0864	1.0493	1.0766	1.0810	1.081
11											
12	1.0936	1.1123	1.09817	1.09281	1.09842	1.0892	1.1052		1.0931	1.0986	1.099
13											
14	1.1109	1.1336	1.11661	1.11004	1.11701	1.1057	1.1246		1.1099	1.1167	1.117
15											
16	1.1283	1.1557	1.13563	1.12775	1.13621	1.1227	1.1446		1.1271	1.1353	1.135
17											
18	1.1473	1.1789	1.15526	1.14596	1.15605	1.1401	1.1652		1.1447	1.1545	1.154
19											
20	1.1663	1.2030	1.17554	1.16469	1.17657	1.1580	1.1864		1.1628	1.1743	1.173
21											
22	1.1860	1.2277	1.19650	1.18396	1.19781	1.1763	1.2083		1.1814	1.1948	1.192
23											
24	1.2063	1.2529	1.21817	1.20379	1.21980	1.1952	1.2309		1.2005	1.2159	1.212
25											
30	1.2711	1.3289	1.28784	1.26691	1.29061	1.2552	1.3028			1.2837	1.278
35	1.3304		1.35191	1.32407	1.35598					1.3461	1.337
40	1.3994		1.42233	1.38599	1.42806					1.4145	1.403
45			1.50010	1.45330	1.50792					1.4897	1.475
50			1.58639	1.52675	1.59691					1.5729	1.555
60					1.80924					1.7695	1.742
70											1.971
80											

% by weight	$SrBr_2$ 20°/4°	$SrCl_2$ 20°/4°	SrI_2 20°/4°	$Sr(NO_3)_2$ 20°/4°	$UO_2(NO_3)_2$ 17°/4°	WO_3 17.5°/4°	$ZnBr_2$ 20°/4°	$ZnCl_2$ 20°/4°	ZnI_2 20°/4°	$Zn(NO_3)_2$ 18°/4°	$ZnSO_4$ 20°/4°
1	1.0157	1.0161	1.0154	1.015	1.008	1.0078	1.0167	1.0167	1.016	1.0154	1.0190
2	.....	.....	.....	.....	1.017	1.0170	.....	.....	.....	.....	.....
3	1.0337	1.0344	1.0331	1.031	1.036	1.0357	1.0354	1.0350	1.034	1.0322	1.0403
4	.....	.....	.....	.....	.....	.....	.....	.....	.....	.....	.....
5	.....	.....	.....	.....	.....	.....	.....	.....	.....	.....	.....
6	1.0522	1.0532	1.0513	1.048	1.054	1.0550	1.0544	1.0532	1.053	1.0496	1.0620
7	1.0712	1.0726	1.0701	1.065	1.072	1.075	1.0738	1.0715	1.072	1.0675	1.0842
8	.....	.....	.....	.....	.....	.....	.....	.....	.....	.....	.....
9	.....	.....	.....	.....	.....	.....	.....	.....	.....	.....	.....
10	1.0907	1.0925	1.0896	1.083	1.091	1.096	1.0935	1.0819	1.091	1.0859	1.1071
11	1.1109	1.1130	1.1099	1.101	1.110	1.118	1.1135	1.1085	1.111	1.1048	1.1308
12	.....	.....	.....	.....	.....	.....	.....	.....	.....	.....	.....
13	1.1317	1.1341	1.1308	1.119	1.129	1.141	1.1338	1.1275	1.131	1.1244	1.1553
14	.....	.....	.....	.....	.....	.....	.....	.....	.....	.....	.....
15	.....	.....	.....	.....	.....	.....	.....	.....	.....	.....	.....
16	1.1532	1.1558	1.1526	1.138	1.149	1.166	1.1544	1.1468	1.152	1.1445	1.1806
17	1.1757	1.1781	1.1753	1.158	1.171	1.191	1.1753	1.1665	1.174	1.1652	.....
18	.....	.....	.....	.....	.....	.....	.....	.....	.....	.....	.....
19	.....	.....	.....	.....	.....	.....	.....	.....	.....	.....	.....
20	1.1992	1.2010	1.1990	1.179	1.194	1.217	1.1965	1.1866	1.197	1.1865	.....
21	.....	.....	.....	.....	1.218	1.244	.....	.....	.....	.....	.....
22	.....	.....	.....	.....	.....	.....	.....	.....	.....	.....	.....
23	.....	.....	.....	.....	.....	.....	.....	.....	.....	.....	.....
24	.....	.....	.....	.....	.....	.....	.....	.....	.....	.....	.....
25	1.2620	1.2600	1.2608	1.233	1.243	1.272	1.2543	1.2380	1.258	1.2427	.....
30	1.3300	1.3250	1.3295	1.290	1.322	1.363	1.3170	1.2928	1.325	1.3029	.....
35	1.4050	1.3960	1.4058	1.352	.....	1.546	1.3859	1.3522	1.398	1.3678	.....
40	1.4890	.....	1.4904	1.419	1.466	.....	1.4620	1.4173	1.478	1.4378	.....
45	1.5830	.....	1.5844	.....	.....	.....	1.5470	1.4890	1.566	1.5134	.....
50	1.6860	.....	.....	.....	1.649	.....	1.6430	1.5681	1.663	1.5944	.....
60	.....	.....	.....	.....	.....	.....	1.8690	1.749	1.893	.....	.....
70	.....	.....	.....	.....	.....	.....	.....	1.962	2.202	.....	.....
80	.....	.....	.....	.....	.....	.....	.....	.....	.....	.....	.....

Table 10-58
ETHYL ALCOHOL

The table below gives the values for $d_{4°}^{t°}$ = g/ml *in vacuo*.
Osborne-McKelvy-Bearce, *Bull. Bu. Stand.*, Vol 9, p. 424 (1913)

% Alcohol by weight	$d\frac{10°}{4°}C$	$d\frac{20°}{4°}C$	$d\frac{25°}{4°}C$	$d\frac{30°}{4°}C$
0	0.99973	0.99823	0.99708	0.99568
1	0.99785	0.99636	0.99520	0.99379
2	0.99602	0.99453	0.99336	0.99194
3	0.99426	0.99275	0.99157	0.99014
4	0.99258	0.99103	0.98984	0.98839
5	0.99098	0.98938	0.98817	0.98670
6	0.98946	0.98780	0.98656	0.98507
7	0.98801	0.98627	0.98500	0.98347
8	0.98660	0.98478	0.98346	0.98189
9	0.98524	0.98331	0.98193	0.98031
10	0.98393	0.98187	0.98043	0.97875
11	0.98267	0.98047	0.97897	0.97723
12	0.98145	0.97910	0.97753	0.97573
13	0.98026	0.97775	0.97611	0.97424
14	0.97911	0.97643	0.97472	0.97278
15	0.97800	0.97514	0.97334	0.97133
16	0.97692	0.97387	0.97199	0.96990
17	0.97583	0.97259	0.97062	0.96844
18	0.97473	0.97129	0.96923	0.96697
19	0.97363	0.96997	0.96782	0.96547
20	0.97252	0.96864	0.96639	0.96395
21	0.97139	0.96729	0.96495	0.96242
22	0.97024	0.96592	0.96348	0.96087
23	0.96907	0.96453	0.96199	0.95929
24	0.96787	0.96312	0.96048	0.95769
25	0.96665	0.96168	0.95895	0.95607
26	0.96539	0.96020	0.95738	0.95442
27	0.96406	0.95867	0.95576	0.95272
28	0.96268	0.95710	0.95410	0.95098
29	0.96125	0.95548	0.95241	0.94922
30	0.95977	0.95382	0.95067	0.94741
31	0.95823	0.95212	0.94890	0.94557
32	0.95665	0.95038	0.94709	0.94370
33	0.95502	0.94860	0.94525	0.94180
34	0.95334	0.94679	0.94337	0.93986
35	0.95162	0.94494	0.94146	0.93790
36	0.94986	0.94306	0.93952	0.93591
37	0.94805	0.94114	0.93756	0.93390
38	0.94620	0.93919	0.93556	0.93186
39	0.94431	0.93720	0.93353	0.92979
40	0.94238	0.93518	0.93148	0.92770
41	0.94042	0.93314	0.92940	0.92558
42	0.93842	0.93107	0.92729	0.92344
43	0.93639	0.92897	0.92516	0.92128
44	0.93433	0.92685	0.92301	0.91910
45	0.93226	0.92472	0.92085	0.91692
46	0.93017	0.92257	0.91868	0.91472
47	0.92806	0.92041	0.91649	0.91250
48	0.92593	0.91823	0.91429	0.91028
49	0.92379	0.91604	0.91208	0.90805
50	0.92162	0.91384	0.90985	0.90580
51	0.91943	0.91160	0.90760	0.90353
52	0.91723	0.90936	0.90534	0.90125
53	0.91502	0.90711	0.90307	0.89896
54	0.91279	0.90485	0.90079	0.89667
55	0.91055	0.90258	0.89850	0.89437
56	0.90831	0.90031	0.89621	0.89206
57	0.90607	0.89803	0.89392	0.88975
58	0.90381	0.89574	0.89162	0.88744
59	0.90154	0.89344	0.88931	0.88512
60	0.89927	0.89113	0.88699	0.88278
61	0.89698	0.88882	0.88466	0.88044

Table 10-58 (*Continued*)
ETHYL ALCOHOL

% Alcohol by weight	$d\frac{10°}{4°}C$	$d\frac{20°}{4°}C$	$d\frac{25°}{4°}C$	$d\frac{30°}{4°}C$
62	0.89468	0.88650	0.88233	0.87809
63	0.89237	0.88417	0.87998	0.87574
64	0.89006	0.88183	0.87763	0.87337
65	0.88774	0.87948	0.87527	0.87100
66	0.88541	0.87713	0.87291	0.86863
67	0.88308	0.87477	0.87054	0.86625
68	0.88074	0.87241	0.86817	0.86387
69	0.87839	0.87004	0.86579	0.86148
70	0.87602	0.86766	0.86340	0.85908
71	0.87365	0.86527	0.86100	0.85667
72	0.87127	0.86287	0.85859	0.85426
73	0.86888	0.86047	0.85618	0.85184
74	0.86648	0.85806	0.85376	0.84941
75	0.86408	0.85564	0.85134	0.84698
76	0.86168	0.85322	0.84891	0.84455
77	0.85927	0.85079	0.84647	0.84211
78	0.85685	0.84835	0.84403	0.83966
79	0.85442	0.84590	0.84158	0.83720
80	0.85197	0.84344	0.83911	0.83473
81	0.84950	0.84096	0.83664	0.83224
82	0.84702	0.83848	0.83415	0.82974
83	0.84453	0.83599	0.83164	0.82724
84	0.84203	0.83348	0.82913	0.82473
85	0.83951	0.83095	0.82660	0.82220
86	0.83697	0.82840	0.82405	0.81965
87	0.83441	0.82583	0.82148	0.81708
88	0.83181	0.82323	0.81888	0.81448
89	0.82919	0.82062	0.81626	0.81186
90	0.82654	0.81797	0.81362	0.80922
91	0.82386	0.81529	0.81094	0.80655
92	0.82114	0.81257	0.80823	0.80384
93	0.81839	0.80983	0.80549	0.80111
94	0.81561	0.80705	0.80272	0.79835
95	0.81278	0.80424	0.79991	0.79555
96	0.80991	0.80138	0.79706	0.79271
97	0.80698	0.79846	0.79415	0.78981
98	0.80399	0.79547	0.79117	0.78684
99	0.80094	0.79243	0.78814	0.78382
100	0.79784	0.78934	0.78506	0.78075

Table 10-58 (*Continued*)
ETHYL ALCOHOL

Specific gravity of aqueous solutions of ethyl alcohol at $\frac{20°}{20°}$ C and at $\frac{25°}{25°}$ C.

% Alcohol by weight	Specific gravity $\frac{20°}{20°}$C	Specific gravity $\frac{25°}{25°}$C	% Alcohol by weight	Specific gravity $\frac{20°}{20°}$C	Specific gravity $\frac{25°}{25°}$C
0	1.00000	1.00000	52	0.91097	0.90799
1	0.99813	0.99811	53	0.90872	0.90571
2	0.99629	0.99627	54	0.90645	0.90343
3	0.99451	0.99447	55	0.90418	0.90113
4	0.99279	0.99274	56	0.90191	0.89883
			57	0.89962	0.89654
5	0.99113	0.99106			
6	0.98955	0.98945	58	0.89733	0.89423
7	0.98802	0.98788	59	0.89502	0.89191
8	0.98653	0.98634	60	0.89271	0.88959
9	0.98505	0.98481	61	0.89040	0.88725
			62	0.88807	0.88491
10	0.98361	0.98330	63	0.88574	0.88256
11	0.98221	0.98184			
12	0.98084	0.98039	64	0.88339	0.88020
13	0.97948	0.97897	65	0.88104	0.87783
14	0.97816	0.97757	66	0.87869	0.87547
			67	0.87632	0.87309
15	0.97687	0.97619	68	0.87396	0.87071
16	0.97560	0.97484			
17	0.97431	0.97346	69	0.87158	0.86833
18	0.97301	0.97207	70	0.86920	0.86593
19	0.97169	0.97065	71	0.86680	0.86352
			72	0.86440	0.86110
20	0.97036	0.96922	73	0.86200	0.85869
21	0.96901	0.96778			
22	0.96763	0.96630	74	0.85958	0.85626
23	0.96624	0.96481	75	0.85716	0.85383
24	0.96483	0.96329	76	0.85473	0.85140
25	0.96339	0.96176	77	0.85230	0.84895
			78	0.84985	0.84650
26	0.96190	0.96018			
27	0.96037	0.95856	79	0.84740	0.84404
28	0.95880	0.95689	80	0.84494	0.84157
29	0.95717	0.95520	81	0.84245	0.83909
30	0.95551	0.95345	82	0.83997	0.83659
31	0.95381	0.95168			
			83	0.83747	0.83408
32	0.95207	0.94986	84	0.83496	0.83156
33	0.95028	0.94802	85	0.83242	0.82902
34	0.94847	0.94613	86	0.82987	0.82646
35	0.94662	0.94422	87	0.82729	0.82389
36	0.94473	0.94227			
			88	0.82469	0.82128
37	0.94281	0.94031	89	0.82207	0.81865
38	0.94086	0.93830	90	0.81942	0.81600
39	0.93886	0.93626	91	0.81674	0.81331
40	0.93684	0.93421	92	0.81401	0.81060
41	0.93479	0.93212			
			93	0.81127	0.80785
42	0.93272	0.93001	94	0.80848	0.80507
43	0.93062	0.92787	95	0.80567	0.80225
44	0.92849	0.92571	96	0.80280	0.79939
45	0.92636	0.92355	97	0.79988	0.79648
46	0.92421	0.92137			
			98	0.79688	0.79349
47	0.92204	0.91917	99	0.79383	0.79045
48	0.91986	0.91697	100	0.79074	0.78736
49	0.91766	0.91475			
50	0.91546	0.91251			
51	0.91322	0.91026			

Table 10-59
METHYL ALCOHOL

Specific gravity of aqueous solutions of methyl alcohol by weight.

% Alcohol by weight	$d\dfrac{0°}{4°}C$	$d\dfrac{10°}{4°}C$	$d\dfrac{20°}{4°}C$	% Alcohol by weight	$d\dfrac{0°}{4°}C$	$d\dfrac{10°}{4°}C$	$d\dfrac{20°}{4°}C$
0	0.9999	0.9997	0.9982	52	0.9250	0.9182	0.9114
1	0.9981	0.9980	0.9965	53	0.9230	0.9162	0.9094
2	0.9963	0.9962	0.9948	54	0.9211	0.9142	0.9073
3	0.9946	0.9945	0.9931	55	0.9191	0.9122	0.9052
4	0.9930	0.9929	0.9914	56	0.9172	0.9101	0.9032
5	0.9914	0.9912	0.9896	57	0.9151	0.9080	0.9010
6	0.9899	0.9896	0.9880	58	0.9131	0.9060	0.8988
7	0.9884	0.9881	0.9863	59	0.9111	0.9039	0.8968
8	0.9870	0.9865	0.9847	60	0.9090	0.9018	0.8946
9	0.9856	0.9849	0.9831	61	0.9068	0.8998	0.8924
10	0.9842	0.9834	0.9815	62	0.9046	0.8977	0.8902
11	0.9829	0.9820	0.9799	63	0.9024	0.8955	0.8879
12	0.9816	0.9805	0.9784	64	0.9002	0.8933	0.8856
13	0.9804	0.9791	0.9768	65	0.8980	0.8911	0.8834
14	0.9792	0.9778	0.9754	66	0.8958	0.8888	0.8811
15	0.9780	0.9764	0.9740	67	0.8935	0.8865	0.8787
16	0.9769	0.9751	0.9725	68	0.8913	0.8842	0.8763
17	0.9758	0.9739	0.9710	69	0.8891	0.8818	0.8738
18	0.9747	0.9726	0.9696	70	0.8869	0.8794	0.8715
19	0.9736	0.9713	0.9681	71	0.8847	0.8770	0.8690
20	0.9725	0.9700	0.9666	72	0.8824	0.8747	0.8665
21	0.9714	0.9687	0.9651	73	0.8801	0.8724	0.8641
22	0.9702	0.9673	0.9636	74	0.8778	0.8699	0.8616
23	0.9690	0.9660	0.9622	75	0.8754	0.8676	0.8592
24	0.9678	0.9646	0.9607	76	0.8729	0.8651	0.8567
				77	0.8705	0.8626	0.8542
25	0.9666	0.9632	0.9592				
26	0.9654	0.9618	0.9576	78	0.8680	0.8602	0.8518
27	0.9642	0.9604	0.9562	79	0.8657	0.8577	0.8494
28	0.9629	0.9590	0.9546	80	0.8634	0.8551	0.8469
29	0.9616	0.9575	0.9531	81	0.8610	0.8527	0.8446
				82	0.8585	0.8501	0.8420
30	0.9604	0.9560	0.9515	83	0.8560	0.8475	0.8394
31	0.9590	0.9546	0.9499				
32	0.9576	0.9531	0.9483	84	0.8535	0.8449	0.8366
33	0.9563	0.9516	0.9466	85	0.8510	0.8422	0.8340
34	0.9549	0.9500	0.9450	86	0.8483	0.8394	0.8314
35	0.9534	0.9484	0.9433	87	0.8456	0.8367	0.8286
				88	0.8428	0.8340	0.8258
36	0.9520	0.9469	0.9416				
37	0.9505	0.9453	0.9398	89	0.8400	0.8314	0.8230
38	0.9490	0.9437	0.9381	90	0.8374	0.8287	0.8202
39	0.9475	0.9420	0.9363	91	0.8347	0.8261	0.8174
40	0.9459	0.9403	0.9345				
41	0.9443	0.9387	0.9327	92	0.8320	0.8234	0.8146
				93	0.8293	0.8208	0.8118
42	0.9427	0.9370	0.9309	94	0.8266	0.8180	0.8090
43	0.9411	0.9352	0.9290	95	0.8240	0.8152	0.8062
44	0.9395	0.9334	0.9272	96	0.8212	0.8124	0.8034
45	0.9377	0.9316	0.9252				
46	0.9360	0.9298	0.9234	97	0.8186	0.8096	0.8005
				98	0.8158	0.8068	0.7976
47	0.9342	0.9279	0.9214	99	0.8130	0.8040	0.7948
48	0.9324	0.9260	0.9196	100	0.8102	0.8009	0.7917
49	0.9306	0.9240	0.9176				
50	0.9287	0.9221	0.9156				
51	0.9269	0.9202	0.9135				

Table 10-60
SPECIFIC GRAVITY OF AIR AT VARIOUS TEMPERATURES

The table below gives the weight in grams $\times 10^4$ of one milliliter of air at 760 mm of mercury pressure and at the temperature indicated. Density in grams per milliliter is the same as the specific gravity referred to water at 4°C. as unity. To convert to density referred to air at 70°F. as unity, divide the values below by 12.00.

t°C.	Sp.Gr. $\times 10^4$	t°C.	Sp.Gr. $\times 10^4$	t°C.	Sp.Gr. $\times 10^4$	t°C.	Sp.Gr. $\times 10^4$
−25	14.240	15	12.255	60	10.596	140	8.541
−24	14.182	16	12.213	62	10.532	142	8.500
−23	14.125	17	12.170	64	10.470	144	8.459
−22	14.069	18	12.129	66	10.408	146	8.419
−21	14.013	19	12.087	68	10.347	148	8.379
−20	13.957	20	12.046	70	10.286	150	8.339
−19	13.902	21	12.004	72	10.227	155	8.242
−18	13.847	22	11.964	74	10.168	160	8.147
−17	13.793	23	11.923	76	10.109	165	8.054
−16	13.739	24	11.883	78	10.052	170	7.963
−15	13.685	25	11.843	80	9.995	175	7.874
−14	13.632	26	11.803	82	9.938	180	7.787
−13	13.580	27	11.764	84	9.882	185	7.702
−12	13.527	28	11.725	86	9.828	190	7.619
−11	13.476	29	11.686	88	9.773	195	7.537
−10	13.424	30	11.647	90	9.719	200	7.457
−9	13.373	31	11.609	92	9.666	205	7.379
−8	13.322	32	11.570	94	9.613	210	7.303
−7	13.272	33	11.533	96	9.561	215	7.228
−6	13.222	34	11.495	98	9.509	220	7.155
−5	13.173	35	11.458	100	9.458	230	7.013
−4	13.124	36	11.420	102	9.408	240	6.881
−3	13.075	37	11.383	104	9.358	250	6.753
−2	13.026	38	11.347	106	9.308	260	6.624
−1	12.978	39	11.310	108	9.259	270	6.504
0	12.931	40	11.274	110	9.211	280	6.389
+1	12.883	41	11.238	112	9.163	290	6.277
2	12.836	42	11.202	114	9.116	300	6.166
3	12.790	43	11.167	116	9.069	310	6.062
4	12.743	44	11.132	118	9.022	320	5.942
5	12.697	45	11.097	120	8.976	330	5.847
6	12.652	46	11.062	122	8.931	340	5.755
7	12.606	47	11.027	124	8.886	350	5.664
8	12.561	48	10.993	126	8.841	360	5.578
9	12.517	49	10.958	128	8.797	370	5.493
10	12.472	50	10.924	130	8.753	380	5.407
11	12.428	52	10.857	132	8.710	400	5.248
12	12.385	54	10.791	134	8.667	420	5.101
13	12.341	56	10.725	136	8.625	440	4.952
14	12.298	58	10.660	138	8.583	460	4.812

DENSITY OF MOIST AIR

The density of moist air depends upon the temperature, the humidity, and the barometric pressure. It is expressed by the equation:

$$d_t = D_t \times \frac{P - 0.3783e}{760}$$

where d_t is the density of the moist air at the temperature t; D_t is the density of dry air at the temperature t (see the table *Specific Gravity of Air*); P is the height of the barometer after correction and reduction to standard conditions, and is expressed in millimeters of mercury (see the section on *Barometry*); e is the vapor pressure of water at the temperature of the dew point and is expressed in millimeters of mercury (see the table *Vapor Pressure of Water*).

Example. To find the density of moist air at a temperature of 20°C, with a dew point of 10°C, and a corrected barometric pressure of 750 mm.

Reference to the table of *Specific Gravity of Air* shows that D at 20°C is equal to 0.0012046 g/ml. Reference to the table of *Vapor Pressure of Water* shows that at 10°C (the temperature of the dew point) e is equal to 9.209 mm. Therefore,

$$d = 0.0012046 \times \frac{750 - (0.3783 \times 9.209)}{760}$$

$$d = 0.0011832 \text{ g/ml} = 1.1832 \text{ g/l}$$

DIELECTRIC CONSTANTS

The dielectric constant (ϵ) of a material, sometimes called the specific inductive capacity, is a measure of the ratio of electric displacement to electric field intensity. The dielectric constant of a vacuum is unity.

Table 10-61
DIELECTRIC CONSTANTS—GASES

Recommended Values of Reference Gases at 20°C and 1 Atmosphere
Maryott and Buckley, *Natl. Bur. Standards, Cir. 537 (1953)*.

The values of the dielectric constant (ϵ) can be adjusted to somewhat different conditions of temperature and pressure by means of the equation

$$\frac{(\epsilon - 1)_{t,p}}{(\epsilon - 1)_{20°,1atm}} = \frac{p}{760[1 + 0.003411(t - 20)]}$$

where p is the pressure in millimeters of mercury, and t is the temperature in degrees C. The errors associated with this equation probably do not exceed 0.1% for CO_2 and 0.02% for the remaining gases at temperatures between 10° and 30° and pressures between 700 and 800 mm.

Gas	ϵ	Gas	ϵ
He	1.0000650 ± 4	Ar	1.0005172 ± 4
H$_2$	1.0002538 ± 3	N$_2$	1.0005480 ± 5
O$_2$	1.0004947 ± 2	CO$_2$	1.000922 ± 1
	Air (dry, CO$_2$ free)		1.0005364 ± 3

In the following table the values are from various sources. The frequency in cycles per second is given in the column headed f.

Gas	t °C	f	ϵ	Gas	t °C	f	ϵ
Acetaldehyde	20	<3¶	1.0213	n-Heptane	100	<3¶	1.0035
Acetone	100	<3¶	1.0159	Hydrogen bromide	0	<3¶	1.00313
Acetyl chloride	20	<3¶	1.0217	Hydrogen chloride	0	<3¶	1.0046
Acetylene	0	<3¶	1.00134	Hydrogen iodide	0	<3¶	1.00234
Ammonia	0	1¶	1.0072	Hydrogen sulfide	0	1¶	1.0040
β-Amylene	100	<3¶	1.0028	Mercury	400	3*	1.00074
Benzene	100	<3¶	1.0028	Methane	0	<3¶	1.00094
Bromine	180	<3¶	1.0128	Methyl alcohol	100	<3¶	1.0057
Butylene	0	<3¶	1.00319	Methyl bromide	100	<3¶	1.0068
Carbon disulfide	0	<3¶	1.0029	Methyl chloride	100	<3¶	1.0069
Carbon monoxide	0	a	1.00070	Methyl iodide	100	<3¶	1.0063
Carbon tetrachloride	110	<3¶	1.0030	Methylamine	100	<3¶	1.0038
Chloroform	120	<3¶	1.0042	Methylene chloride	100	<3¶	1.0065
Dimethylamine	100	<3¶	1.0033	Neon	0	<3¶	1.000127
Ethane	0	<3¶	1.00150	Nitromethane	100	<3¶	1.0247
Ethyl alcohol	100	<3¶	1.0061	Nitrous oxide (N$_2$O)	0	2¶	1.00113
Ethyl bromide	20	<3¶	1.0139	n-Pentane	100	<3¶	1.0025
Ethyl chloride	20	<3¶	1.0132	n-Propyl chloride	20	<3¶	1.0143
Ethyl ether	100	<3¶	1.0049	iso-Propyl chloride	20	<3¶	1.0152
Ethyl formate	100	<3¶	1.0083	Sulfur dioxide	0	<3¶	1.0093
Ethyl iodide	20	<3¶	1.0140	Toluene	126	<3¶	1.0043
Ethyl iodide	100	<3¶	1.0089	Vinyl bromide	20	<3¶	1.0081
Ethylene	0	<3¶	1.00144	Water (steam)	110	<3¶	1.0126
				Water (steam)	140	<3¶	1.00785

* $\times 10^8$.   ¶ $\times 10^6$.   a Frequency of 50 cycles per second.

Table 10-62
DIELECTRIC CONSTANTS—LIQUIDS
Recommended Values of Reference Liquids at 25°C
Maryott and Smith, *Natl. Bur. Standards, Cir. 514 (1951)*.

The values of a or α may be used for interpolating or for extrapolating over a limited range of temperature by means of the equations

$$\epsilon_t' = \epsilon_t - a(t' - t) \tag{1}$$

$$\text{Log}_{10}\,\epsilon_t' = \text{Log}_{10}\,\epsilon_t - \alpha\,(t' - t) \tag{2}$$

Liquid	ϵ	a or (α)	Liquid	ϵ	a or (α)
Benzene	2.274	0.0020	Hydrogen	1.228*	0.0034
Carbon tetrachloride	2.228	0.0020	Methanol	32.63	0.00260 (α)
Chlorobenzene	5.621	0.00133 (α)	Nitrobenzene	34.82	0.00225 (α)
Cyclohexane	2.015	0.0016	Oxygen	1.507†	0.0024
Ethylene chloride	10.36	0.00240 (α)	Water	78.54‡	0.00200 (α)

* At 20.4°K. † At 80.0°K.
‡ $\epsilon = 78.54[1 - 4.579(10^{-3})(t - 25) + 1.19(10^{-5})(t - 25)^2 - 2.8(10^{-8})(t - 25)^3]$; av. dev. ± 0.03%.

All of the substances and the values given in the following table, with the exception of those entered as "Oils," have been selected from a very comprehensive list in the Natl. Bur. of Standards' Cir. 514. Abbreviations: t, temp. °C; T, temp. °K; $(t_1; t_2)$, the limits (range) of temperature between which a (or α) is considered applicable for use in equations *1* or *2* given above; f, frequency in cycles per second; the temperature of measurement is given in parentheses after the name.

Liquid (°C)	ϵ	a (or α) $\times 10^2$	Range $t_1; t_2$	Liquid (°C)	ϵ	a (or α) $\times 10^2$	Range $t_1; t_2$
Acetaldehyde (21°)	21.1[a]	.....	.....	Arsenic tri-chloride (20°)	12.6[a]	.....	.....
Acetic acid (20°)	6.15	.....	.....	Arsine (−100°)	2.50	0.43	−116; −72
anhydride (19°)	20.7	.....	.....	Benzaldehyde (20°)	17.8	.....	.....
Acetone (25°)	20.7	0.205 (α)	−60; 40	Benzonitrile (25°)	25.20	0.157 (α)	0; 25
Acetonitrile (20°)	37.5	16	15; 25	Benzoyl chlo-ride (20°)	23	.....	.....
Acetyl chloride (22°)	15.8	.....	.....	Benzyl alcohol (20°)	13.1	.....	.....
Allyl chloride (20°)	8.2	.....	.....	chloride (13°)	7.0	.....	.....
Aluminum bro-mide (100°)	3.38	0.33	100; 240	Bromine (20°)	3.09	0.7	0; 50
Ammonia (−33.4°)	22.4	.....	.....	Bromo-benzene (25°)	5.40	0.115 (α)	0; 70
Amyl acetate (20°)	4.75	1.2	at 20	cyclohexane (25°)	7.92	0.140 (α)	1; 55
alcohol (*n*) (25°)	13.9	0.23 (α)	15; 35	naphtha-lene-1 (25°)	4.83	0.87	25; 55
bromide (*n*) (25°)	6.32	0.152 (α)	−45; 55	Bromoform (20°)	4.39	0.105 (α)	10; 70
bromide (*iso*) (20°)	6.05	2.3	−18; 23	Butyl acetate (*n*) (20°)	5.01	1.4	20; 40
Aniline (20°)	6.89	0.148 (α)	0; 50	acetate (*iso*) (20°)	5.29	1.6	at 20
Antimony tri-chloride (75°)	33[b]	.....	.....	alcohol (*n*) (25°)	17.1	0.335 (α)	25; 70
pentachloride (20°)	3.22	0.46	2; 47	alcohol (*iso*) (25°)	17.7	0.377 (α)	20; 90
Argon (−191°)	1.538	0.34	−191; −184	bromide (*n*) (20°)	7.07	0.150 (α)	10; 90

[a] $f = 4 \times 10^8$.   [b] $f = 3.6 \times 10^8$.

Table 10-62 (*Continued*)
DIELECTRIC CONSTANTS—LIQUIDS

Liquid (°C)	ϵ	a (or α) $\times 10^2$	Range $t_1; t_2$	Liquid (°C)	ϵ	a (or α) $\times 10^2$	Range $t_1; t_2$
bromide (*iso*) (25°)	7.18	2.8	1; 55	Dichloro-benzene (*o*) (25°)	9.93	0.194 (α)	0; 50
bromide (*sec*) (25°)	8.64	3.30	1; 55	benzene (*m*) (25°)	5.04	0.120 (α)	0; 50
bromide (*tert*) (50°)	10.15	5.20	−15; 55	benzene (*p*) (50°)	2.41	0.18	50; 80
chloride (*n*) (20°)	7.39	0.173 (α)	−10; 70	methane (20°)	9.08^j	.....	−80; 25
chloride (*iso*) (20°)	6.05	0.160 (α)	−40; 23	propane-2,2 (20°)	10.2	0.247 (α)	−33; 20
chloride (*tert*) (0°)	10.95	0.225 (α)	−23; 30	Diethyl oxalate (21°)	8.1^a	.....	.....
Butyric acid (*n*) (20°)	2.97	−0.23	10; 70	sulfate (20°)	29.2	0.24 (α)	−25; 20
aldehyde (*n*) (26°)	13.4	.....	.....	Dimethyl-amine (25°)	5.26	.....	.....
Carbon dioxide (0°)	1.60^c	.....	.....	aniline (*N,N*) (20°)	4.91	2	at 20
disulfide (20°)	2.641	0.268	−90; 130	sulfate (20°)	42.6	.....	.....
tetrachloride (20°)	2.238	0.200	−10; 60	Dioxane-1,4 (25°)	2.209	0.170	20; 50
"Cellosolve" acetate (30°)	7.57	3.1	30; 50	Diphenyl (75°)	2.53	0.18	75; 155
Chloral (20°)	4.94	0.17 (α)	15; 45	methane (25°)	2.57	0.14	20; 50
Chlorine (−50°)	2.10	0.31	−65; −33	Di-*iso*-propenyl (25°)	2.10	0.17	−50; 50
Chloro-acetic acid (60°)	12.3	2	60; 80	Dodecane (*n*) (20°)	2.014	0.120	10; 150
napthalene-1 (25°)	5.04	1.07	1; 55	Ethyl acetate (25°)	6.02	1.5	at 25
phenol (*o*) (25°)	6.31	2.7	25; 58	alcohol (25°)	24.3	0.270 (α)	−5; 70
phenol (*p*) (55°)	9.47	3.7	55; 65	amine (10°)	6.94*	.....	−20; 10
Chloroform (20°)	4.806	0.160 (α)	0; 50	benzoate (20°)	6.02	2.1	20; 40
Cresol (*o*) (25°)	11.5	11	25; 30	bromide (20°)	9.39	0.196 (α)	−30; 30
Cresol (*m*) (25°)	11.8	0.41 (α)	15; 50	butyrate (18°)	5.10	1.0	at 20
Cresol (*p*) (58°)	9.91	.....	.....	carbamate (50°)	14.2	5.2	50; 70
Cyanogen (23°)	2.52	.....	.....	ether (20°)	4.34	0.217 (α)	−40; 30
Cyclo-hexane (20°)	2.023	0.160	10; 60	" (−116°)	10.4	.....	.....
hexanol (25°)	15.0	0.437 (α)	20; 66	iodide (20°)	7.82	0.150 (α)	−20; 70
hexanone (20°)	18.3	.....	.....	mercaptan (15°)	6.91	.....	.....
pentanol (20°)	18	0.38 (α)	at 20	oleate (28°)	3.17	0.48	28; 122
Cymene (*p*) (20°)	2.24	0.16	4; 60	palmitate (20°)	3.20	0.4	20; 40
Decahydro-naphthalene (*cis*) (20°)	2.197	0.11	20; 100	propionate (19°)	5.65	1.8	at 19
Decahydro-naphthalene (*trans*) (20°)	2.172	0.11	20; 100	salicylate (30°)	7.99	2	30; 40
Decane (*n*) (20°)	1.991	0.130	10; 110	stearate (40°)	2.98	0.6	32; 50
Deuterium oxide (20°K)	1.277	0.4	18.8; 21.2°K	thiocyanate (21°)	29.3	.....	.....
oxide (25°)	78.25^d	.....	0.4; 98	*iso*-thiocyanate (21°)	19.5	.....	.....
Dibutyl phthalate (30°)	6.43	1.98	30; 35	toluene (*p*) (25°)	2.24	0.19	25; 45
				trichloroacetate (20°)	7.8	2.8	2; 60
				Ethylene bromide (25°)	4.78	0.60	10; 55
				diamine (20°)	14.2	10	10; 27
				glycol (25°)	37.7	0.224 (α)	20; 100
				Fluorine (−202°)	1.54	0.19	−216; −190

[a] $f = 4 \times 10^8$. [c] At pressure of 50 atm.
[d] $\epsilon = 78.25[1 - 4.617(10^{-3}) (t - 25) + 1.22(10^{-5}) (t - 25)^2 - 2.7(10^{-8}) (t - 25)^3]$; av. dev. ± 0.04%.
[j] $\epsilon = (3320/T) - 2.24$.
* $\epsilon = 6.94 - 0.036(t - 10) + 0.0004(t - 10)^2$.

Table 10-62 (*Continued*)
DIELECTRIC CONSTANTS—LIQUIDS

Liquid (°C)	ϵ	a (or α) $\times 10^2$	Range t_1; t_2	Liquid (°C)	ϵ	a (or α) $\times 10^2$	Range t_1; t_2
Formic acid				Nonane (*n*)			
(16°)	58.5ᵃ	.....	.....	(20°)	1.972	0.135	−10; 90
amide				Octane (*n*)			
(20°)	109	72	18; 25	(20°)	1.948	0.130	−50; 50
Furfural (20°)	41.9	.....	.....	"Iso-octane"			
Germanium				(20°)	1.940	0.142	−100; 100
tetrachlo-				Octyl alcohol			
ride (25°)	2.43	0.240	0; 55	(*n*) (20°)	10.34	0.410 (α)	20; 60
Glycerol (25°)	42.5	0.208 (α)	0; 100	Oleic acid (20°)	2.46	.....	.....
Helium				Oil, almond			
(2.63°K)	1.055	.....	.....	(20°)	2.83	.....	.....
Heptane (*n*)				castor (11°)	4.67	.....	.....
(20°)	1.924	0.140	−50; 50	cottonseed			
Hexane (*n*)				(14°)	3.10	.....	.....
(20°)	1.890	0.155	−10; 50	lemon (21°)	2.25	.....	.....
Hexyl alcohol				linseed (13°)	3.35	.....	.....
(*n*) (25°)	13.3	0.35 (α)	15; 35	olive (20°)	3.11	.....	.....
bromide (*n*)				peanut (11°)	3.03	.....	.....
(25°)	5.82	1.73	25; 55	petroleum	2.13	.....	.....
Hydrazine				rapeseed			
(20°)	52.9	0.21 (α)	0; 25	(16°)	2.85	.....	.....
bromide				sesame (13°)	3.02	.....	.....
(−85°)	7.00	0.26 (α)	−85; −70	turpentine			
chloride				(20°)	2.23	.....	.....
(−15°)	6.35	0.288(α)	−85; −15	Paraldehyde			
cyanide				(25°)	13.9	.....	.....
(20°)	114.9	0.63 (α)	18; 26	Pentane (*n*)			
fluoride				(20°)	1.844	0.160	−50; 30
(0°)	84	.....	.....	Pentanone-2			
iodide				(20°)	15.4	0.195 (α)	−40; 80
(−50°)	3.39	0.8	−51; −37	Pentanone-3			
peroxide				(20°)	17.0	0.225 (α)	0; 80
(0°)	84.2§	.....	−30; 20	Pentene-1 (20°)	2.10	.....	.....
sulfide				Phenol (60°)	9.78	0.32 (α)	40; 70
(−78.5°)	9.05	.....	.....	Phenyl acetate			
Iodobenzene				(20°)	5.23	0.7	at 20
(20°)	4.63	.....	.....	ether (30°)	3.65	0.7	30; 50
Isoprene (25°)	2.10	0.24	−75; 25	*iso*-thiocyanate			
Lactic acid				(20°)	10.4ᵃ	.....	.....
(*dl*) (17°)	22	.....	.....	Phosphorus			
Methyl acetate				(46°)	4.06	.....	.....
(25°)	6.68	2.2	25; 40	oxychloride			
benzoate				(22°)	13.3	.....	.....
(20°)	6.59	0.14 (α)	20; 50	trichloride			
ether (25°)	5.02	2.38	25; 100	(25°)	3.43	0.84	17; 60
salicylate				Propane (0°)	1.61	0.20	−90; 15
(30°)	9.41	3.1	30; 40	Propyl acetate			
Mesitylene				(*n*) (19°)	5.69	0.8	at 19
(20°)	2.27	.....	.....	alcohol (*n*)			
Mesityl oxide				(25°)	20.1	0.293 (α)	20; 90
(20°)	15.1ᵃ	.....	.....	alcohol (*iso*)			
Naphthalene				(25°)	18.3	0.310 (α)	20; 70
(85°)	2.54	.....	.....	bromide (*n*)			
Nitro-aniline (*o*)				(25°)	8.09	3.35	1; 55
(90°)	34.5	3	90; 110	bromide (*iso*)			
aniline (*p*)				(25°)	9.46	4.40	1; 55
(160°)	56.3	6	160; 180	ether (*iso*)			
benzene				(25°)	3.88	1.8	0; 25
(25°)	34.82	0.225 (α)	10; 80	Pyridine (25°)	12.3	.....	.....
ethane (30°)	28.0	11.4	30; 35	Quinoline (25°)	9.00	.....	.....
methane				Salicylaldehyde			
(30°)	35.87	0.189 (α)	12; 92	(30°)	17.1	7	30; 40
phenol (*o*)				Selenium			
(50°)	17.3	6.4	50; 60	(250°)	5.40	0.25	237; 301
propane-1				Silicon tetra-			
(30°)	23.2	10.1	30; 35	chloride			
propane-2				(16°)	2.40	.....	.....
(30°)	25.5	10.9	30; 35	Sulfur (118°)	3.52	.....	.....
Nitrogen				dioxide			
(−203°)	1.454	0.29	−210; −195	(20°)#	14.1	7.7	14; 140
Nitrous oxide				monochloride			
(0°)#	1.61	0.6	−6; 14	(15°)	4.79	0.146 (α)	−41; 15

ᵃ $f = 4 \times 10^8$.   § $\epsilon = 84.2 - 0.62t + 0.0032t^2$.
\# At pressure = vapor pressure.

Table 10-62 (*Continued*)
DIELECTRIC CONSTANTS—LIQUIDS

Liquid (°C)	ϵ	a (or α) $\times 10^2$	Range $t_1 ; t_2$	Liquid (°C)	ϵ	a (or α) $\times 10^2$	Range $t_1 ; t_2$
trioxide (18°)	3.11	.....	.....	Triethylamine			
Sulfuryl chloride				(25°)	2.42	.....	.....
(22°)	10.0	.....	.....	Trimethylamine			
Tetrabromo-				(25°)	2.44	0.52	0; 25
ethane				Tripalmitin			
(*sym*)				(60°)	2.92	0.32	60; 70
(22°)	7.0	.....	.....	Triphenyl-			
Tetrachloro-				methane			
ethylene				(100°)	2.45	0.14	94; 175
(25°)	2.30	0.20	25; 90	Tristearin (70°)	2.78	0.34	70; 80
Thionyl chloride				Undecane (*n*)			
(20°)	9.25	3.9	at 20	(20°)	2.005	0.125	10; 130
Tin tetrachloride				Vanadyl tri-			
(20°)	2.87	0.30	−30; 20	chloride			
Titanium tetra-				(25°)	3.4[b]	.....	.....
chloride				Water (200°)†	34.59‡	.....	100; 370
(20°)	2.80	0.20	−20; 20	Xylene (*m*)			
Toluene (25°)	2.379	0.243	0; 90	(20°)	2.374	0.195	−40; 180
Tributyl phos-				Xylene (*p*)			
phate (30°)	7.95	2.74	30; 35	(20°)	2.270	0.160	20; 130
Trichloro-							
ethane-1,1,1							
(0°)	7.1	3.6	−33; 2				

[b] $f = 3.6 \times 10^8$. † See also preceding Reference Liquids at 25°.
‡ $\epsilon = 5321/T + 233.76 - 0.9297T + 0.001417T^2$.

Table 10-63
DIELECTRIC CONSTANTS—SOLIDS

The frequency in cycles per second is given in the column headed f. The symbol A stands for "audio frequency" and signifies that the condenser frequency lies between 20 and 20,000 vibrations or cycles per second and is audible to the average person. Values for the dielectric constant (ϵ) in the following table are at ordinary room temperatures (17 to 22° C) unless otherwise given after the name of the solid. The symbols $\perp$ and $\parallel$ signify that the determinations were made perpendicular and parallel respectively to the optical axes in the case of crystals, and to the fibers in the case of woods. The word "mean" in parentheses following the name shows that the mean value of the three different axes in monoclinic and rhombic crystals is given.

Solid	f	ϵ	Solid	f	ϵ
Acetamide	4*	4.0	Paper	<3¶	2.0
Acetanilide	...	2.9	Paraffin	<3¶	2.10
Acetic acid (2° C)	4*	4.1	Phenanthrene	4*	2.80
Aluminum oleate	4*	2.40	Phenol (10° C)	4*	4.3
Ammonium bromide	1¶	7.1	Phosphorus, red	1*	4.1
" chloride	1¶	7.0	" , yellow	1*	3.6
Antimony trichloride	1*	5.34	Porcelain	<3¶	5.7-6.8
Apatite $\perp$	3*	9.50	Potassium aluminum sulfate	1¶	3.8
" $\parallel$	3*	7.41	Potassium carbonate (15° C)	1*	5.6
Asphalt	<3¶	2.68	" chlorate	6§	5.1
Barium chloride (anhyd.)	6§	11.4	" chloride	A	5.03
" " (2 H$_2$O)	6§	9.4	" chromate	6§	7.3
" nitrate	6§	5.9	" iodide	6§	5.6
" sulfate (15° C)	1*	11.4	" nitrate	6§	5.0
Beryl $\perp$	A	7.02	" sulfate	6§	5.9
" $\parallel$	A	6.08	Quartz $\perp$	3§	4.34
Calcite $\perp$	A	8.5	" $\parallel$	3§	4.27
" $\parallel$	A	8.0	Resorcinol	4*	3.2
Calcium carbonate	1¶	6.14	Ruby $\perp$	A	13.27
" fluoride	A	7.36	" $\parallel$	A	11.28
Calcium sulfate (2 H$_2$O)	A	5.66	Rutile $\perp$	1*	86
Cassiterite $\perp$	100†	23.4	" $\parallel$	1*	170
" $\parallel$	100†	24	Selenium	1*	6.6
d-Cocaine	5*	3.10	Shellac	<3¶	2.9-3.7
Cupric oleate	4*	2.80	Silver bromide	1¶	12.2
" oxide (15° C)	1*	18.1	" chloride	1¶	11.2
" sulfate (anhyd.)	6§	10.3	" cyanide	1¶	5.6
" " (5 H$_2$O)	6§	7.8	Smithsonite $\perp$	100†	9.3
Diamond	A	16.5	" $\parallel$	1†	9.4
"	1*	5.5	Sodium carbonate (anhyd.)	6§	8.4
Diphenylmethane	4*	2.7	Sodium carbonate (10 H$_2$O)	6§	5.3
Dolomite $\perp$	1*	8.0	" chloride	A	6.12
" $\parallel$	1*	6.8	" nitrate	...	5.2
Ferrous oxide (15° C)	1*	14.2	" oleate	4*	2.75
Glass, borosilicate	<3¶	6.4-7.7	" perchlorate	6§	5.4
" , flint	<3¶	6.6-9.9	Sucrose (mean)	3*	3.32
" , lead	<3¶	5.4-8.0	Sulfur (mean)	...	4.0
Iodine	1*	4	Thallium chloride	1¶	46.9
Lead acetate	1¶	2.6	p-Toluidine	4*	3.0
" carbonate (15° C)	1*	18.6	Tourmaline $\perp$	A	7.10
Lead chloride	1¶	4.2	" $\parallel$	A	6.3
" monoxide (15° C)	1*	25.9	Urea	4*	3.5
" nitrate	6§	37.7	Water (ice) (−5° C)	40#	4.6
" oleate	4*	3.27	" " (−5° C)	3#	40
" sulfate	1¶	14.3	" " (−5° C)	1#	73
" sulfide (15° C)	1*	17.9	" " (−22° C)	40#	2.2
Malachite (mean)	100†	7.2	" " (−22° C)	500	40
Mercuric chloride	1¶	3.2	Wood, beech $\parallel$	<3¶	2.5-4.8
Mercurous chloride	1¶	9.4	" , " $\perp$	<3¶	3.6-7.7
Mica	<3¶	5.6-6.6	" , oak $\parallel$	<3¶	2.4-4.2
Naphthalene	4*	2.52	" , " $\perp$	<3¶	3.6-6.8
Ozokerite	<3¶	2.21	Zircon $\perp$ and $\parallel$	1*	12

† × 10¹⁰.   * × 10⁸.   § × 10⁷.   ¶ × 10⁶.   ‡ × 10⁵.   # × 10².

DIPOLE MOMENTS

Table 10-64
DIPOLE MOMENTS OF VARIOUS SUBSTANCES

In this table and in the following table of *GROUP MOMENTS* the dipole moments are expressed in Debye units, usually represented by the symbol λ, which are equal to 10^{-18} electrostatic units. A dipole consists of two equal electrical charges, very close together and opposite in sign. Its magnitude is measured by its moment which is the product of either charge by the distance between the two charges.

Substance	λ	Substance	λ
Acetone	2.7	Diphenyl ether	1.15
Ammonia	1.46	Hydrazine	1.19
iso-Amyl cyanide	3.53	Hydrogen	0
n-Amyl fluoride	1.85	Hydrogen bromide	0.78
tert-Amyl fluoride	1.92	Hydrogen chloride	1.03
Benzene	0	Hydrogen cyanide	2.93
Benzophenone	2.95	Hydrogen iodide	0.38
Boron trichloride	0	Hydrogen peroxide	2.1
Bromine	0	Hydrogen sulfide	1.10
Butylamine	1.40	Iodine	0
Carbon dioxide	0	Nitrogen	0
Carbon disulfide	0	Paraffin	0
Carbon tetrachloride	0	*iso*-Propyl cyanide	3.61
Chlorine	0	Rosin	8.8
p-Chloronitrobenzene	2.78	Stannic chloride	0
Cyclohexane	0	Stannic iodide	0
o-Dichlorobenzene	2.51	Sulfur dioxide	1.6
Dioxane	0	Water	1.84

Table 10-65
GROUP MOMENTS

The following data are taken by permission of the Publishers from "Dipole Moments" by C. P. Smyth in *PHYSICAL METHODS OF ORGANIC CHEMISTRY*, A. Weissberger, ed.; copyright, Interscience Publishers, New York (1946). For further information concerning dipole moments and their applications reference can be made to the afore-mentioned work, Volume II.

Each moment value given is that which is to be found when a group listed in the first column is attached to one of the groups listed across the top of the table. For example, the moment listed for C_6H_5-Cl is that of chlorobenzene.

Group	C_6H_5-		CH_3-		C_2H_5-		Degrees Angle
	Gas	Soln.	Gas	Soln.	Gas	Soln.	
-CH$_3$	0.36	0.4	0	....	0	....	180
-OCH$_3$	1.35	1.25	1.30	....	....	....	55
-SCH$_3$	....	1.27	....	1.40	....	....	52
-NH$_2$	1.48	1.53	1.23	....	1.2	1.38	..
-I	1.6	1.30	1.64	1.5	1.87	1.8	0
-Br	1.75	1.52	1.80	1.8	2.01	1.9	0
-Cl	1.72	1.55	1.87	1.7	2.05	1.8	0
-F	1.57	1.43	1.81	....	1.92	....	0
-OH	1.4	1.6	1.69	1.66	1.69	1.7	62
-CO$_2$H	....	1.7	1.73	1.6	1.73	1.7	74
-CO$_2$CH$_3$	....	1.9	1.67	1.75	1.76	1.9	70
-CHO	3.1	2.8	2.72	2.5	2.73	2.5	58
-COCH$_3$	3.00	2.9	2.84	2.74	2.78	....	59
-CN	4.39	4.0	3.94	3.4	4.00	3.57	0
-NO$_2$	4.21	3.98	3.50	3.1	3.68	3.3	0

REFRACTIVE INDEX

Table 10-66
INDEX OF REFRACTION (LIQUIDS)

Arrangement. The substances are arranged in alphabetical order. No substance is listed more than once. To ascertain if a compound is listed in the table it is to be looked for under the alphabetical listing according to its common name or by reference to an accompanying *formula index*. In using the table for the identification of an unknown liquid, reference should be made to the table *Refractive Index Finding Numbers* and in this case, consideration must be given to the effect of temperature. When possible, values for the refractive index were selected at or near 20°C. The index of refraction of a liquid decreases with rise in temperature and this thermal variation differs with various substances but is approximately 0.0004 per 1°C.

Specific Gravity. As an aid in applying formulas for calculation of refractive index (*q. v.*) the specific gravity is given at the same temperature as used in the refractive index determination.

Beil. Ref. In the column so headed will be found the reference to the volume and page numbers of the 4th edition of Beilstein: *Handbuch der Organischen Chemie*. Where the Roman numeral, which indicates the volume number, is preceded by an asterisk, reference is to the first supplementary volumes; two asterisks, to the second supplementary volumes.

It is to be noted that the symbols α, β, γ which are often employed in reporting refractive index of liquids, refer to the wave length of light used and signify the wave length corresponding to the Hα, Hβ and Hγ spectrum lines. In the case of refractive index of biaxial crystals the symbols α, β and γ have a different significance (*q. v.*).

References. The information in the table has been collected mainly from the following sources:

Beilstein: *Handbuch der Organischen Chemie*, 4th edition. Published by Springer, Berlin.

International Critical Tables, Vol. VII. Published by McGraw-Hill Book Co., New York.

U. S. Bureau of Standards: *American Petroleum Institute Research Project 44*.

No.	Substance	Formula	t,°C.	d_4^t	n_D^t	Beil. Ref.
1	**Acenaphthene** (m.p. 95°)	$C_{12}H_{10}$	98.8	1.024	1.6048	V–586
2	**Acetal**	$CH_3CH(OC_2H_5)_2$	20	0.831	1.3819	I–603
3	**Acetaldehyde**	CH_3CHO	18	0.783	1.3392	I–594
4	oxime	$CH_3CH{:}NOH$	20.4	0.965	1.4257	I–608
5	**Acetic** acid	CH_3CO_2H	22.9	1.046	1.3715	II–96
6	anhydride	$(CH_3CO)_2O$	20	1.082	1.3901	II–166
7	**Acetimidoethyl ether**	$HN{:}C(CH_3)OC_2H_5$	18.8	0.873	1.4035	II–182
8	**Acetoin**	$CH_3CHOHCOCH_3$	15	1.002	1.4194	I–827
9	**Acetol**	CH_3COCH_2OH	20	1.082	1.4295	I–821
10	**Acetone**	$(CH_3)_2CO$	20	0.791	1.3591	I–635
11	cyanohydrin	$(CH_3)_2C(OH)CN$	19	0.932	1.4000	III–316
12	oxime (m.p. 60°)	$(CH_3)_2C{:}NOH$	75	.....	1.42	I–649
13	**Aceto-**nitrile	CH_3CN	16.5	0.78	1.3460	II–183
14	phenone	$CH_3COC_6H_5$	18.8	1.03	1.5338	VII–271
15	thienone (α)	$CH_3COC{:}CHCH{:}$ | $CH{\cdot}S$ |__	21.8	1.168	1.5663	XVII–287
16	**Acetonyl acetone**	$(CH_3COCH_2)_2$	20	0.974	1.4232	I–788
17	**Acetoxy-**acetone	$CH_3CO_2CH_2COCH_3$	20	1.075	1.4150	II–155
	propiophenone (α)	$C_6H_5COCH(CH_3)O{\cdot}OCCH_3$	19.7	1.113	1.5153	*VIII–547
18	γ-valero-lactone (γ)	$CH_3CO_2C(CH_3)CH_2{\cdot}CH_2COO$ (m.p. 78°)	78.6	1.123	1.4236	XVIII–2

Table 10-66 (*Continued*)
INDEX OF REFRACTION—LIQUIDS

No.	Substance	Formula	t,°C.	d_4^t	n_D^t	Beil. Ref.
19	**Acetoxymethylene-** cyclohexan-2- one (1)	$CH_3CO_2CH:C_6H_8O$	15.6	1.115	1.4994	*VIII-509
20	diethyl ketone	$C_2H_5COC(CH_3):$ CHO_2CCH_3	18.3	1.034	1.4649	*II-73
21	menthone	$CH_3CO_2C_{11}H_{17}O$	16	1.032	1.4875	*VIII-512
22	pinacoline (m.p. 44°)	$(CH_3)_3CCOCH:$ CHO_2CCH_3	55.3	0.965	1.4443	*II-73
23	**Acetyl-**acetone	$(CH_3CO)_2CH_2$	18.5	0.977	1.4518	I-777
24	acetone-*O*-ethyl ether	$CH_3C(OC_2H_5):CH \cdot$ $COCH_3$	16.3	0.948	1.4665	I-844
25	acetone-*O*-methyl ether	$CH_3COCH:C(OCH_3) \cdot$ CH_3	20	0.971	1.4688	*I-426
26	bromide	CH_3COBr	15.8	1.663	1.4537	II-174
27	camphor (3)	$C_{12}H_{18}O_2$	19	1.031	1.4939	VII-596
28	chloride	CH_3COCl	20	1.105	1.3898	II-173
29	ethylamino- naphthalene (β)	$C_{10}H_7N(C_2H_5)CO \cdot$ CH_3	23.1	1.089	1.6031	XII-1285
30	indazole (2) (*stable*)	$CH_3CON \cdot N:C_6H_4:CH$ ‖	20	1.186	1.588	*XXIII-33
31	methylheptenone	$C_{10}H_{16}O_2$	15	0.945	1.4849	I-804
32	6-methyl-indazole (2) (*stable*)	$OH_3CON \cdot N:C_6H_3 \cdot$ ‖ $(CH_3):CH$ ‖	5	1.162	1.587	....
33	5-methyltetra- hydroindazole (2)	$CH_3CON \cdot CH:C_6H_7 \cdot$ ‖ $(CH_3):N$ ‖	20	1.079	1.521	....
34	salicylaldehyde (*o*) (m.p. 37°)	$CH_3CO_2C_6H_4CHO$	44.5	1.158	1.5225	VIII-44
35	tetrahydroindazole (2)	$CH_3CON \cdot CH:C_6H_8:N$ ‖	20	1.119	1.531	....
36	tropein	$C_{10}H_{17}O_2N$	15.4	1.063	1.4769	XXI-18
37	**Acetylene** dibromide (*cis*)	$BrCH:CHBr$	17.5	2.271	1.5431	I-190
38	" (*trans*)	$BrCH:CHBr$	17.5	2.267	1.5505	I-190
39	dichloride (*cis*)	$ClCH:CHCl$	15	1.299	1.4519	I-187
40	" (*trans*)	$ClCH:CHCl$	15	1.265	1.4490	I-187
41	dicyanide	$NCC:CCN$	25	0.970	1.4647	*II-317
42	diiodide (*cis*)	$ICH:CHI$	11.2	3.1	1.706	I-194
43	**Acrylic** acid	$CH_2:CHCO_2H$	20	1.051	1.4224	II-397
44	aldehyde	$CH_2:CHCHO$	20	0.841	1.3998	I-725
45	**Adiponitrile**	$(CH_2CH_2CN)_2$	20(?)	0.950	1.4597	II-653
46	**Allyl** acetate	$CH_3CO_2CH_2CH:CH_2$	20	0.928	1.4045	II-136
47	acetic acid	$CH_2:CHCH_2CH_2 \cdot$ CO_2H	17.7	0.984	1.43	II-425
48	acetone	$C_3H_5CH_2COCH_3$	15.4	0.848	1.4213	I-734
49	acetylindazole (2) (*stable*)	$C_{12}H_{12}ON_2$	22.6	1.117	1.571	....
50	alcohol	$CH_2:CHCH_2OH$	20	0.854	1.4135	I-436
51	allylphenyl ether (*o*)	$CH_2:CHCH_2C_6H_4 \cdot$ $OCH_2CH:CH_2$	20	0.963	1.524	*VI-282
52	amine	$CH_2:CHCH_2NH_2$	21.8	0.761	1.4194	IV-205
53	anisole (*o*)	$CH_3OC_6H_4CH_2CH:$ CH_2	20	0.972	1.524	*VI-282

Table 10-66 (*Continued*)
INDEX OF REFRACTION—LIQUIDS

No.	Substance	Formula	t,°C.	d_4^t	n_D^t	Beil. Ref.
54	**Allyl** benzene	$C_6H_5CH_2CH:CH_2$	20(?)	0.909	1.5143	V-484
55	bromide	$CH_2:CHCH_2Br$	20	1.398	1.4655	I-201
56	chloride	$CH_2:CHCH_2Cl$	20	0.938	1.4154	I-198
57	o-cresol (*o*)	$CH_2:CHCH_2C_6H_3:$ $(CH_3)\,OH\,(1,2,3)$	20	1.000	1.538	*VI-287
58	o-cresyl ether	$CH_3C_6H_4OCH_2CH:$ CH_2	20	0.965	1.517	*VI-171
59	cyanide	$CH_2:CHCH_2CN$	16	0.837	1.4079	II-408
60	di-*o*, *o*-allyl- phenyl ether	$(C_3H_5)_2C_6H_3OC_3H_5$	20	0.950	1.523	*VI-301
61	guaiacol (*o*)	$CH_2:CHCH_2\cdot$ $C_6H_3(OCH_3)\,OH$	20	1.066	1.539	*VI-461
62	guaiacyl ether	$CH_3OC_6H_4OCH_2\cdot$ $CH:CH_2\,(1,2)$	20	1.054	1.533	VI-772
63	2-hydroxy- diphenyl (3)	$C_6H_5C_6H_3(OH)\cdot$ $(CH_2CH:CH_2)$	20	1.072	1.601	**VI-664
64	1-indazole	$C_3H_5N\cdot C_6H_4\cdot CH:N$ $\mid\underline{\qquad\qquad}\mid$	20	1.067	1.582	....
65	2-indazole	$C_3H_5N\cdot N:C_6H_4:CH$ $\mid\underline{\qquad\qquad}\mid$	20	1.074	1.598	....
66	5-methyl- indazole (1)	$CH_3C_6H_3CH:N\cdot N\cdot$ $\mid\underline{\qquad\qquad}\mid$ C_3H_5	20	1.047	1.576	....
67	5-methyl- indazole (2)	$C_3H_5N\cdot N:C_6H_3\cdot$ $\mid\underline{\qquad\qquad}$ $(CH_3):CH$ $\underline{\qquad\qquad}\mid$	20	1.050	1.590	....
68	phenol (*o*)	$CH_2:CHCH_2C_6H_4OH$	20	1.021	1.545	*VI-282
69	phenol (*p*)	$CH_2:CHCH_2C_6H_4OH$	18	1.02	1.5441	*VI-571
70	tetrahydro- furfuryl ether	$C_3H_5OCH_2OC_4H_7$	20	0.957	1.4498	....
71	thymol (*o*) (1,2,3,4)	$(CH_3)\,(C_3H_5)\,(OH)\cdot$ $(C_3H_7)\,C_6H_2$	20	0.966	1.526	....
72	thymyl ether	$(CH_3)\,(CH_2:CH\cdot$ $CH_2O)\,C_6H_3\,(C_3H_7)$	20	0.935	1.508	**VI-498
73	*iso*-thiocyanate	$CH_2:CHCH_2NCS$	20	1.013	1.5266	IV-214
74	thiourea (m.p. 78°)	$CH_2:CHCH_2NH\cdot$ $CSNH_2$	78.1	1.2	1.60	IV-211
75	**Allyloxy-*m*-xylene** (*sym*)	$(CH_3)_2C_6H_3OCH_2\cdot$ $CH:CH_2$	20	0.948	1.514	*VI-244
76	**Aluminum** triethyl	$(C_2H_5)_3Al$	6.5	.....	1.480	IV-643
77	trimethyl	$(CH_3)_3Al$	12	.....	1.432	IV-643
78	**Amino**-ethylbenzene (*o*)	$NH_2C_6H_4C_2H_5$	22	0.983	1.5584	XII-1089
79	" (*p*)	$NH_2C_6H_4C_2H_5$	22	0.975	1.5529	XII-1090
80	hexahydrobenzene	$C_6H_{11}NH_2$	24	0.865	1.4575	XII-5
81	**Ammonia**	NH_3	16.5	.....	1.325	....
82	**Amyl** acetate (*n*)	$CH_3CO_2C_5H_{11}$	20	0.879	1.4012	II-131
83	" (*iso*) †	$CH_3CO_2C_5H_{11}$	18.1	0.87	1.4014	II-132
84	" ‡	$CH_3CO_2CH_2CH\cdot$ $(CH_3)\,C_2H_5$	20	0.88	1.4012	II-131
85	acetic acid (*iso*)	$(CH_3)_2C_4H_7CO_2H$	19(?)	0.912	1.4209	II-342
86	α-aceto-cyano- acetate (*iso*)	$HC:(COCH_3)\,(CN)\cdot$ $(CO_2C_5H_{11})$	20	1.033	1.4676	III-798
87	acetylcamphor- carboxylate	$C_{18}H_{28}O_4$	20.5	1.022	1.4772	X-38

† Commonly called amyl acetate.
‡ From *l*-amyl alcohol.

Table 10-66 (*Continued*)
INDEX OF REFRACTION—LIQUIDS

No.	Substance	Formula	$t,°C.$	d_4^t	n_D^t	Beil. Ref.
88	**Amyl** alcohol (*n*)	$CH_3(CH_2)_3CH_2OH$	20	0.817	1.4101	I-383
89	" (*sec*) (*n*)	$C_2H_5CH_2CHOHCH_3$	20	0.809	1.4053	I-384
90	" (*prim*) (*iso*)	$C_4H_9CH_2OH$	15	0.813	1.4085	I-392
91	" (*sec*) (*iso*)	$(CH_3)_2CHCHOHCH_3$	20	0.818	1.3973	I-391
92	" (*tert*)	$(CH_3)_2COHC_2H_5$	20	0.809	1.406	I-388
93	amine (*iso*)	$C_4H_9CH_2NH_2$	17.9	0.751	1.4096	IV-180
94	*iso*-amylidene-amine (*iso*) §	$(CH_3)_2CH(CH_2)_2N:$ $CHCH_2CH(CH_3)_2$	18.9	0.772	1.4258	....
95	anisate (*iso*)	$CH_3OC_6H_4CO_2C_5H_{11}$ (*p*)	20	1.040	1.5085	....
96	anthracene (*iso*) (9) (*meso*) (m.p. 58°)	$C_{19}H_{20}$	71.1	1.002	1.6364	I-685
97	benzene (*n*)	$CH_3(CH_2)_4C_6H_5$	20	0.86	1.4943	V-382
98	" (*iso*)	$C_5H_{11}C_6H_5$	15.6	0.863	1.4867	V-434
99	" (*tert*)	$C_5H_{11}C_6H_5$	23	0.8	1.4915	V-436
100	benzoate	$C_6H_5CO_2C_5H_{11}$	20	0.989	1.494	....
101	bromide (*n*)	$CH_3(CH_2)_3CH_2Br$	20	1.218	1.4444	I-131
102	" (*iso*)	$(CH_3)_2CHCH_2CH_2Br$	20	1.215	1.4412	I-136
103	" (*tert*)	$C_2H_5CBr(CH_3)_2$	20(?)	1.216	1.4421	I-136
104	*n*-butyrate	$C_2H_5CH_2CO_2C_5H_{11}$	20	0.87	1.4110	II-271
105	*iso*-butyrate	$(CH_3)_2CHCO_2C_5H_{11}$	20	0.859	1.4076	....
106	camphorcarboxylate (*iso*)	$C_{16}H_{26}O_3$	17.2	1.014	1.4760	X-645
107	carbamate (*iso*) (m.p. 64°)	$NH_2CO_2C_5H_{11}$	70.6	0.944	1.4175	III-30
108	chloride (*n*)	$CH_3(CH_2)_3CH_2Cl$	20	0.878	1.4119	I-130
109	" (*iso*)	$(CH_3)_2CHCH_2CH_2Cl$	18	0.800	1.4112	I-135
110	" (*tert*)	$(CH_3)_2CClC_2H_5$	18	0.869	1.407	I-134
111	chloroacetate	$ClCH_2CO_2C_5H_{11}$	20	1.055	1.434	II-198
112	3-chloro-campho-3-carboxylate	$C_{16}H_{25}O_3Cl$	19	1.091	1.4866	X-647
113	α-crotonate	$C_9H_{16}O_2$	20	0.896	1.4371	II-411
114	cyanide (*n*)	$CH_3(CH_2)_4CN$	14.3	0.80	1.4085	II-324
115	" (*iso*)	$(CH_3)_2CH(CH_2)_2CN$	22.2	0.80	1.4048	II-329
116	dichloroamine (*iso*)	$(CH_3)_2CHCH_2·$ CH_2NCl_2	25.7	1.027	1.4438	....
117	ether (*iso*)	$(C_5H_{11})_2O$	20	0.78	1.408	I-401
118	fluoride (*n*)	$CH_3(CH_2)_4F$	20	0.788	1.3574	**I-94
119	formate (*iso*)	$HCO_2C_5H_{11}$	20	0.871	1.3977	II-22
120	furylacrylate (*n*)	$C_4H_3OCH:CHCO_2·$ C_5H_{11}	20	1.032	1.5083	....
121	1-indazole (*iso*)	$C_6H_4CH:N·N·C_5H_{11}$	20	1.001	1.546	....
122	2-indazole (*iso*)	$N:C_6H_4CH:N·C_5H_{11}$	20	1.007	1.558	....
123	2-indazole-car-boxylate (*iso*)	$C_6H_4CH·N(CO_2·$ $C_5H_{11})·N$	15.1	1.110	1.543	....
124	iodide (*n*)	$CH_3(CH_2)_3CH_2I$	20	1.510	1.4955	I-133
125	mercaptan (*n*)	$CH_3(CH_2)_3CH_2SH$	20(?)	0.857	1.4437	I-384
126	" (*iso*)	$C_5H_{11}SH$	20	0.835	1.4412	I-405

§ Probably not a pure material.

Table 10-66 (*Continued*)
INDEX OF REFRACTION—LIQUIDS

No.	Substance	Formula	t,°C.	d_4^t	n_D^t	Beil. Ref.
127	**Amyl** β-methoxya-crylic acid (β) (m.p. 55°)	$C_5H_{11}C(OCH_3):CH\cdot CO_2H$	60.8	0.969	1.4588	III-382
128	nitrate (*iso*)	$C_5H_{11}ONO_2$	21.7	0.996	1.4122	I-403
129	nitrite (*n*)	$CH_3(CH_2)_3CH_2ONO$	20(?)	0.853	1.3851	I-384
130	" (*iso*)	$C_5H_{11}ONO$	20.7	0.871	1.3871	I-402
131	phenyl ether (*iso*)	$C_5H_{11}OC_6H_5$	22.1	0.920	1.49	VI-143
132	phenyl ketone (*iso*)	$C_5H_{11}COC_6H_5$	15(?)	0.962	1.5125	VII-334
133	propiolic acid (*n*)	$C_5H_{11}C:CCO_2H$	12.6	0.962	1.4634	II-487
134	propiolic diethyl acetal	$C_5H_{11}C:CCH:(OC_2H_5)_2$	9.5	0.886	1.4421	I-751
135	propiolic nitrile	$C_5H_{11}C:CCN$	13	0.851	1.4553	II-488
136	propionate (*iso*)	$C_2H_5CO_2C_5H_{11}$	20	0.870	1.4065	II-241
137	*iso*-propylmalonate (*iso*)	$CH_2(CO_2C_5H_{11})\cdot(CO_2C_3H_7)$	20	0.96	1.4293	....
138	salicylate (*iso*)	$HOC_6H_4CO_2C_5H_{11}$	20	1.05	1.506	X-76
139	thiophene (*iso*) (α)	$C_5H_{11}C:CHCH:CH\cdot S$	20	0.941	1.498	....
140	valerate (*n*)	$C_4H_9CO_2C_5H_{11}$	20	0.87	1.4145	II-301
141	*iso*-valerate (*iso*)	$C_4H_9CO_2C_5H_{11}$	18.7	0.86	1.4130	II-312
142	**Amylene glycol** (1,4)	$C_5H_{10}(OH)_2$	16.8	0.996	1.4439	I-480
143	" " (1,5)	$CH_2(CH_2CH_2OH)_2$	20	0.994	1.4499	I-481
144	**Anethole** (*p*)	$CH_3CH:CHC_6H_4\cdot OCH_3$	21	0.990	1.5614	VI-566
145	**Angelic acid** (*cis*) (m.p. 45°)	$CH_3CH:C(CH_3)\cdot CO_2H$	76	0.954	1.43	II-428
146	**Angelolactone** (α)	$CH_3C:CHCH_2CO\cdot O$	20	1.084	1.448	XVII-252
147	" (β)	$CH_3CHCH:CHCO\cdot O$	20	1.074	1.460	XVII-253
148	**Anhydro-δ-aceto-butyl alcohol**	$CH_3C:CHCH_2CH_2\cdot CH_2O$	20	0.906	1.446	XVII-21
149	**Aniline**	$C_6H_5NH_2$	20	1.022	1.5863	XII-59
150	**Anisaldehyde** (*o*)	$CH_3OC_6H_4CHO$	20.2	1.133	1.5597	VIII-43
151	" (*p*)	$CH_3OC_6H_4CHO$	12.7	1.13	1.5764	VIII-67
152	oxime methyl ether (*o*)	$CH_3OC_6H_4CH:NOCH_3$	20	1.092	1.554	VIII-77
153	**Anisidine** (*o*)	$CH_3OC_6H_4NH_2$	20	1.093	1.5754	XIII-358
154	" (*p*) (m.p. 58°)	$CH_3OC_6H_4NH_2$	67	1.061	1.5559	XIII-435
155	**Anisole**	$C_6H_5OCH_3$	20.6	0.994	1.5173	VI-138
156	**Anthracene octa-hydride** (m.p. 71°)	$C_{14}H_{18}$	88.8	0.965	1.5363	V-526
157	**Anthranil**	$O\cdot N:C_6H_4:CH$	20	1.18	1.586	XXVII-39
158	**Antimony** penta-chloride	$SbCl_5$	14	.....	1.601	....
159	trimethyl	$(CH_3)_3Sb$	15	1.52	1.48	IV-617
160	**Apiol** (m.p. 30°)	$C_6H(O_2CH_2)(O\cdot CH_3)_2C_3H_5$	14	1.176	1.5380	*XIX-642
161	" (*iso*) (m.p. 56°)	$C_{12}H_{14}O_4$	12	1.197	1.5703	XIX-85
162	**Apocyclene** (m.p. 43°)	C_9H_{14}	40	0.871	1.4514	*V-66

Table 10-66 (*Continued*)
INDEX OF REFRACTION—LIQUIDS

No.	Substance	Formula	t,°C.	d_4^t	n_D^t	Beil. Ref.
163	**Arsenic triethyl**	$(C_2H_5)_3As$	20	1.150	1.467	IV-602
164	**Asaron** (1,2,4,5)	$CH_3CH:CHC_6H_2(O\cdot CH_3)_3$	11	1.091	1.5719	VI-1129
165	**Ascaridol**	$C_{10}H_{16}O_2$	20	0.999	1.4769	XIX-17
166	**Atractylene**	$C_{15}H_{24}$	20	0.927	1.5057	V-470
167	**Azelaic acid** (m.p. 107°)	$(CH_2)_7(CO_2H)_2$	107.3	1.0	1.43	II-707
168	**Azoxybenzene** (m.p. 36°)	$C_6H_5\cdot N(:O):N\cdot C_6H_5$	26	1.24	1.6644	XVI-622
169	**Benzal-acetone** (m.p. 41°)	$C_6H_5CH:CHCOCH_3$	45.9	.....	1.5836	VII-364
170	acetophenone (m.p. 62°)	$C_6H_5CH:CHCOC_6H_5$	62.3	1.071	1.6458	VII-478
171	bromide	$C_6H_5CHBr_2$	15(?)	1.51	1.541	V-308
172	chloride	$C_6H_5CHCl_2$	19.4	1.3	1.5515	V-297
173	dipropyl ketone (α)	$C_6H_5CH:C(C_2H_5)\cdot COC_3H_7$	23.9	0.972	1.5476	VII-379
174	pinacoline (m.p. 40°)	$C_{13}H_{16}O$	40	1.2	1.5473	VI-183
175	**Benzaldehyde**	C_6H_5CHO	19.5	1.049	1.5456	VII-174
176	oxime (*anti*) (m.p. 35°)	$C_6H_5CH:NOH$	20	1.111	1.5908	VII-218
177	oxime *O*-ethyl ether (α)	$C_6H_5CH:NOC_2H_5$	21.7	0.992	1.5350	VII-223
178	oxime *O*-methyl ether (α)	$C_6H_5CH:NOCH_3$	20	1.020	1.548	VII-223
179	**Benzene**	C_6H_6	20	0.879	1.5011	V-179
180	**Benzhydrylamine**	$(C_6H_5)_2CHNH_2$	21.5	1.064	1.5963	XII-1323
181	**Benzoic acid** (m.p. 122°)	$C_6H_5CO_2H$	131.9	1.2	1.504	IX-92
182	anhydride (m.p. 42°)	$(C_6H_5CO)_2O$	15	1.199	1.5767	IX-164
183	nitrile	C_6H_5CN	20	1.005	1.5289	IX-275
184	**Benzo**-phenone (*labile*) (m.p. 26.5°)	$(C_6H_5)_2CO$	23.4	1.108	1.6060†	VII-410
185	phenone (*labile*) (m.p. 48°)	$(C_6H_5)_2CO$	19	1.091	1.6077†	VII-410
186	phenone (m.p. 48.5°)	$(C_6H_5)_2CO$	45.2	1.089	1.5975	VII-410
187	trichloride	$C_6H_5CCl_3$	19.2	1.37	1.5584	V-300
188	**Benzoyl**-acetone (m.p. 61°)	$C_6H_5COCH_2COCH_3$	77.8	1.08	1.5678	VII-680
189	acetone *O*-methyl ether	$C_6H_5C(OCH_3):CH\cdot COCH_3$	16	1.068	1.5598	VIII-559
190	chloride	C_6H_5COCl	20	1.212	1.5537	IX-182
191	**Benzyl** acetate	$CH_3CO_2CH_2C_6H_5$	21	1.06	1.5242	VI-435
192	alcohol	$C_6H_5CH_2OH$	20	1.043	1.5396	VI-428
193	amine	$C_6H_5CH_2NH_2$	19.5	0.980	1.5441	XII-1013
194	aniline (m.p. 37°)	$C_6H_5CH_2NHC_6H_5$	24.8	1.065	1.6118	XII-1023
195	azide	$C_6H_5CH_2N_3$	24.9	1.066	1.5341	V-350
196	benzoate	$C_6H_5CO_2CH_2C_6H_5$	21.5	1.12	1.5685	IX-121
197	*iso*-butyrate	$(CH_3)_2CHCO_2C_7H_7$	18(?)	1.016	1.4910	VI-436
198	chloride	$C_6H_5CH_2Cl$	17.4	1.103	1.5391	V-292
199	chloroacetate	$ClCH_2CO_2CH_2C_6H_5$	20	1.2	1.523	VI-435

† Supercooled.

Table 10-66 (*Continued*)
INDEX OF REFRACTION—LIQUIDS

No.	Substance	Formula	t,°C.	$d_4{}^t$	$n_D{}^t$	Beil. Ref.
200	**Benzyl-**crotonolac- tone (γ)	$C_6H_5CH_2CHCH:$ $\quad\quad\vert\_\_\_\_$ $\quad CHCO\cdot O$ $\quad\_\_\_\_\vert$	20	1.149	1.563	XVII-340
201	cyanide	$C_6H_5CH_2CN$	20	1.018	1.5242	IX-441
202	cyclohex-1-ene (1)	$C_6H_5CH_2C:CH\cdot$ $\quad\quad\vert\_\_\_$ $\quad (CH_2)_3CH_2$ $\quad\_\_\_\_\vert$	14.1	0.969	1.5443	V-524
203	cyclohex-1-ene-6- one (1)	$C_6H_5CH_2C:CH\cdot$ $\quad\quad\vert\_\_\_$ $\quad (CH_2)_3CO$ $\quad\_\_\_\_\vert$	21.7	1.061	1.5628	*VII-208
204	*p*-cymene (2)	$(CH_3)(C_7H_7)C_6H_3\cdot$ $CH(CH_3)_2$	20	0.963	1.5564	V-620
205	dichloroacetate	$Cl_2CHCO_2CH_2C_6H_5$	20	1.3	1.526	VI-435
206	3,5-dimethyl- pyrazole (1)	$CH_3C:CHC(CH_3):$ $\quad\vert\_\_\_$ $\quad N\cdot N(CH_2C_6H_5)$ $\quad\_\vert$	20	1.034	1.549	....
207	lactate	$CH_3CHOHCO_2C_7H_7$	25	.....	1.5252	**VI-420
208	laurate	$C_{11}H_{23}CO_2CH_2C_6H_5$	20	0.95	1.483	**VI-417
209	3-methyl-4-chloro- pyrazole (1)	$CH:C(Cl)\cdot C(CH_3):$ $\quad\vert\_\_\_$ $\quad N\cdot N\cdot CH_2C_6H_5$ $\quad\_\vert$	20	1.156	1.556	....
210	5-methyl-4-chloro- pyrazole (1)	$N:CHC(Cl):C\cdot$ $\quad\vert\_\_\_$ $\quad (CH_3)\cdot N(C_7H_7)$ $\quad\_\_\_\_\vert$	20	1.178	1.560	....
211	methylnitroamine	$C_6H_5CH_2N(NO_2)\cdot$ CH_3	25.7	1.171	1.5515	....
212	3-methylpyrazole (1)	$CH:CHC(CH_3):N\cdot N\cdot$ $\quad\vert_____\vert$ $\quad CH_2C_6H_5$	20	1.049	1.554	....
213	myristate (m.p. 20°)	$C_{13}H_{27}CO_2CH_2C_6H_5$	20	0.93	1.482	**VI-417
214	oleate	$C_{17}H_{33}CO_2CH_2C_6H_5$	25	0.933	1.4875	*VI-221
215	palmitate (m.p. 36°)	$C_{15}H_{31}CO_2CH_2C_6H_5$	50	0.9	1.4689	**VI-417
216	stearate (m.p. 46°)	$C_{17}H_{35}CO_2CH_2C_6H_5$	50	0.908	1.4663	*VI-221
217	tetrahydroindazole (1)	$C_6H_8\cdot CH:N\cdot NC_7H_7$ $\quad\vert_____\vert$	20	1.089	1.573	....
218	tetrahydroindazole (2)	$C_7H_7N\cdot N:C_6H_8:CH$ $\quad\vert_____\vert$	20	1.086	1.579	....
219	toluene (*p*)	$C_6H_5CH_2C_6H_4CH_3$	19.3	0.99	1.5710	V-607
220	trichloroacetate	$Cl_3CCO_2CH_2C_6H_5$	18.8	1.38	1.5288	VI-436
221	*d*-valerate	$C_4H_9CO_2CH_2C_6H_5$	20.4	0.98	1.4922	VI-436
222	**Bis-diethylacetal**	$[(C_2H_5O)_2CHC:]_2$	26	0.953	1.4328	I-805
223	**Bismuth trimethyl**	$(CH_3)_3Bi$	15	2.3	1.56	IV-622
224	**Bornyl** acetate (*d*)	$CH_3CO_2C_{10}H_{17}$	22.6	0.99	1.4623	VI-78
225	acetate (*dl*)	$CH_3CO_2C_{10}H_{17}$	20	0.985	1.4630	VI-85
226	acetate (*iso*)	$CH_3CO_2C_{10}H_{17}$	20(?)	0.984	1.4649	VI-89
227	*n*-butyrate (*d*)	$C_2H_5CH_2CO_2C_{10}H_{17}$	15	0.966	1.4638	VI-79
228	formate (*d*)	$HCO_2C_{10}H_{17}$	15	1.017	1.4708	VI-78
229	propionate (*d*)	$C_2H_5CO_2C_{10}H_{17}$	15	0.979	1.4644	VI-79
230	*n*-valerate (*d*)	$C_4H_9CO_2C_{10}H_{17}$	15	0.956	1.4628	VI-79
231	*iso*-valerate	$C_4H_9CO_2C_{10}H_{17}$	22.2	0.95	1.4600	VI-79

Table 10-66 (*Continued*)
INDEX OF REFRACTION—LIQUIDS

No.	Substance	Formula	t,°C.	d_4^t	n_D^t	Beil. Ref.
232	**Brassidic** acid (m.p. 60°)	$C_{21}H_{41}CO_2H$	57	0.859	1.45	II-474
233	anhydride (m.p. 64°)	$C_{44}H_{82}O_3$	100	0.8	1.4366	....
234	**Bromo**-aniline (*m*)	$BrC_6H_4NH_2$	20.4	1.579	1.6260	XII-633
235	anisole (*o*)	$BrC_6H_4OCH_3$	20	1.502	1.5725	VI-197
236	" (*p*)	$BrC_6H_4OCH_3$	20	1.457	1.5605	VI-199
237	benzene	C_6H_5Br	20	1.495	1.5604	V-206
238	2-butene (2)	$CH_3CBr{:}CHCH_3$	25.4	1.322	1.4583	I-205
239	*p-iso*-butyro-*o*-cresol (α)	$(CH_3)(OH)C_6H_3{\cdot}COCBr(CH_3)_2$	22.3	1.367	1.5714	*VIII-556
240	*p-iso*-butyro-*o*-cresyl acetate (α)	$CH_3CO_2C_6H_3(CH_3){\cdot}COCBr(CH_3)_2$	25.7	1.317	1.5288	*VIII-556
241	*n*-butyrophenone (α)	$C_2H_5CHBrCOC_6H_5$	20	1.364	1.562	VII-314
242	*iso*-butyrophenone (α)	$(CH_3)_2CBrCOC_6H_5$	20	1.355	1.5567	VII-316
243	crotonaldehyde (α)	$CH_3CH{:}CBrCHO$	23	1.566	1.5165	*I-380
244	2,3-dimethyl-butane (2)	$(CH_3)_2CHBr(CH_3)_2$	20(?)	1.177	1.4517	I-152
245	3,5-dimethyl-cyclo-hexane (1)	$(CH_3)_2C_6H_9Br$	18.2	1.207	1.4832	V-38
246	2,2-dimethyl-propane (1)	$(CH_3)_3CCH_2Br$	20	1.260	1.4369	I-141
247	ethyl acetate (β)	$CH_3CO_2C_2H_4Br$	20	1.462	1.4433	II-128
248	" " (α)	$CH_3CO_2C_2H_4Br$	20	1.514	1.4550	II-153
249	methyl methyl ether	$BrCH_2OCH_3$	20	1.598	1.4562	I-582
250	naphthalene (α)	$C_{10}H_7Br$	20	1.482	1.6582	V-547
251	*m*-nitrobenzene	$BrC_6H_4NO_2$	20.1	1.704	1.5979	V-248
252	nonane (2)	$CH_3(CH_2)_6CHBrCH_3$	20(?)	1.081	1.4536	I-166
253	propionic acid (α) (*dl*)	$CH_3CHBrCO_2H$	20	1.700	1.4753	II-254
254	propiophenone (α)	$C_6H_5COCHBrCH_3$	19.6	1.430	1.5718	VII-302
255	propylene (α)	$CH_3CH{:}CHBr$†	20	1.428	1.4554	I-200
256	" (α)	$CH_3CH{:}CHBr$‡	15.8	1.417	1.4549	I-200
257	propylene (2)	$CH_3CBr{:}CH_2$	15.8	1.397	1.4467	I-200
258	styrene (α)	$C_6H_5CBr{:}CH_2$	20	1.406	1.588	V-477
259	" (ω) §	$C_6H_5CH{:}CHBr$	16.6	1.42	1.6091	V-477
260	" (ω) ¶	$C_6H_5CH{:}CHBr$	22.5	1.425	1.5990	V-477
261	thiophene (α)	$BrC{:}CH{\cdot}CH{:}CH{\cdot}S$ $\underline{\hspace{4em}}$	20	1.684	1.59	XVII-33
262	toluene (*o*)	$BrC_6H_4CH_3$	20(?)	1.422	1.5608	V-304
263	" (*m*)	$BrC_6H_4CH_3$	20	1.410	1.551	V-305
264	" (*p*) (m.p. 28°)	$BrC_6H_4CH_3$	20	1.390	1.5490	V-305
265	1,1,4-trimethyl-cyclohexane (4)	$(CH_3)_3C_6H_8Br$	17.4	1.170	1.4800	*V-17
266	1,3,5-trimethyl-cyclohexane (1)	$(CH_3)_3C_6H_8Br$	11.1	1.175	1.4828	*V-18
267	*o*-xylene (4)	$BrC_6H_3(CH_3)_2$ (4,1,2)	18.4	1.37	1.5571	V-365
268	*p*-xylene (2)	$BrC_6H_3(CH_3)_2$ (2,1,4)	18.5	1.36	1.5514	V-385
269	**Bromoform**	$HCBr_3$	19	2.892	1.5980	I-68
270	**Butenyl-anisole** (α) (*p*)	$CH_3OC_6H_4CH{:}CH{\cdot}C_2H_5$	19	0.980	1.5559	VI-575
271	**Butyl**-acetate (*n*)	$CH_3CO_2CH_2C_2H_5$	20	0.881	1.3951	II-130
272	acetate (*iso*)	$CH_3CO_2C_4H_9$	18.8	0.870	1.3907	II-131

† α-form. ‡ β-form. § Isomer No. 1. ¶ Isomer No. 2.

Table 10-66 (*Continued*)
INDEX OF REFRACTION—LIQUIDS

No.	Substance	Formula	t,°C.	d_4^t	n_D^t	Beil. Ref.
273	**Butyl** acetate (*sec*)	$CH_3CO_2C_4H_9$	20	0.870	1.3877	II-131
274	acetate (*tert*)	$CH_3CO_2C(CH_3)_3$	20(?)	0.896	1.3947	II-131
275	alcohol (*n*)	$C_2H_5CH_2CH_2OH$	20	0.810	1.3991	I-367
276	" (*iso*)	$(CH_3)_2CHCH_2OH$	17.5	0.805	1.3968	I-373
277	" (*sec*)	$CH_3CHOHC_2H_5$	22	0.806	1.3924	I-371
278	" (*tert*)	$(CH_3)_3COH$	20	0.786	1.3878	I-379
279	amine (*n*)	$C_2H_5CH_2CH_2NH_2$	20	0.740	1.401	IV-156
280	" (*iso*)	$(CH_3)_2CHCH_2NH_2$	17	0.74	1.3988	IV-163
281	" (*sec*)	$C_4H_9NH_2$	20	0.724	1.394	IV-160
282	" (*tert*)	$(CH_3)_3CNH_2$	18	0.698	1.3794	IV-173
283	*d*-amyl ether (*iso*) †	$(CH_3)_2CHCH_2O\cdot$ C_5H_{11}	20.2	0.77	1.4008	I-387
284	anisate (*p*)	$CH_3OC_6H_4CO_2C_4H_9$	20	1.054	1.508	....
285	" (*p*) (*iso*)	$CH_3OC_6H_4CO_2C_4H_9$	20	1.052	1.507	....
286	benzene (*n*)	$C_6H_5C_4H_9$	20	0.860	1.4898	V-413
287	" (*iso*)	$C_6H_5CH_2CH(CH_3)_2$	20	0.853	1.4865	V-414
288	" (*sec*)	$C_6H_5CH(CH_3)C_2H_5$	20	0.862	1.4902	V-414
289	" (*tert*)	$C_6H_5C(CH_3)_3$	20	0.867	1.4926	V-416
290	benzoate (*sec*) (*d*)	$C_6H_5CO_2C_4H_9$	20	1.000	1.4930	*IX-63
291	bromide (*n*)	$C_2H_5CH_2CH_2Br$	20	1.279	1.4398	I-119
292	" (*iso*)	$(CH_3)_2CHCH_2Br$	20	1.264	1.436	I-126
293	" (*sec*)	$C_2H_5CHBrCH_3$	25.3	1.251	1.4344	I-119
294	" (*tert*)	$(CH_3)_3CBr$	20.5	1.211	1.428	I-127
295	*n*-butyrate (*n*)	$C_2H_5CH_2CO_2C_4H_9$	20	0.872	1.4049	II-271
296	" (*iso*)	$C_2H_5CH_2CO_2C_4H_9$	18.4	0.862	1.4030	II-271
297	*iso*-butyrate (*iso*)	$(CH_3)_2CHCO_2CH_2\cdot$ $CH(CH_3)_2$	20	0.87	1.3999	II-291
298	"carbitol"	$C_4H_9O(CH_2)_2O\cdot$ $(CH_2)_2OH$	27	0.95	1.4290	**I-521
299	"cellosolve"	$C_4H_9OCH_2CH_2OH$	20	0.900	1.4191	**I-519
300	chloride (*n*)	$C_2H_5CH_2CH_2Cl$	20	0.884	1.4015	I-118
301	" (*iso*)	$(CH_3)_2CHCH_2Cl$	17.8	0.879	1.3970	I-124
302	" (*sec*)	$C_2H_5CHClCH_3$	25.2	0.87	1.3953	I-119
303	" (*tert*)	$(CH_3)_3CCl$	17.8	0.843	1.3869	I-125
304	chloroformate (*n*)	$ClCO_2CH_2CH_2C_2H_5$	20	1.078	1.41	**III-11
305	decyl ether	$C_4H_9OC_{10}H_{21}$	20	0.801	1.4278	....
306	dichloroamine (*iso*)	$(CH_3)_2CHCH_2NCl_2$	24	1.090	1.4484	IV-171
307	ether (*n*)	$(C_2H_5CH_2CH_2)_2O$	20	0.769	1.3992	I-369
308	2-ethylhexyl phthalate	$C_6H_4(CO_2C_4H_9)\cdot$ $CO_2C_8H_{17}$	20	1.01	1.4888	....
309	formate (*n*)	$HCO_2CH_2CH_2C_2H_5$	20	0.89	1.3891	II-21
310	" (*iso*)	$HCO_2CH_2CH(CH_3)_2$	20	0.885	1.3857	II-21
311	furylacrylate (*n*)	$C_4H_3OCH:CHCO_2\cdot$ C_4H_9	20	1.048	1.5129	....
312	indazole (1) (*n*)	$C_6H_4\overset{\shortmid}{C}H:N\cdot N\cdot C_4H_9$	20	1.019	1.553	....
313	" (2) (*n*)	$C_4H_9N:CH\cdot C_6H_4:N$	20	1.025	1.562	....
314	iodide (*n*)	$C_2H_5CH_2CH_2I$	20	1.617	1.4998	I-123
315	" (*iso*)	$(CH_3)_2CHCH_2I$	20	1.606	1.4960	I-128
316	mercaptan (*iso*)	$(CH_3)_2CHCH_2SH$	20	0.836	1.4386	I-378
317	nitrate (*iso*)	$(CH_3)_2CHCH_2ONO_2$	20	1.015	1.4028	I-377
318	nitrite (*iso*)	$(CH_3)_2CHCH_2ONO$	22.1	0.87	1.3715	I-377
319	nitroamine (*n*)	$C_2H_5CH_2CH_2NHNO_2$	23	1.058	1.4604	IV-571
320	" (*sec*)	$C_4H_9NHNO_2$	21.9	1.057	1.4581	IV-571
321	oleate (*n*)	$C_{17}H_{33}CO_2C_4H_9$	20	0.864	1.4522	**II-439

† Mixture of isomers?

Table 10-66 (Continued)
INDEX OF REFRACTION—LIQUIDS

No.	Substance	Formula	t,°C.	d_4^t	n_D^t	Beil. Ref.
322	**Butyl**-phenyl ketone (*iso*)	$(CH_3)_2CHCH_2CO \cdot$ C_6H_5	15.3	0.99	1.5139	VII-329
323	" " (*tert*)	$(CH_3)_3CCOC_6H_5$	19.2	0.968	1.5086	VII-330
324	piperidine (α) (*iso*)	$C_4H_9CH_2 \cdot$ $\mathcal{L}$ $(CH_2)_3NHCH$ $\mathcal{L}$	21.7	0.851	1.4553	XX-128
325	propionate (*iso*)	$C_2H_5CO_2C_4H_9$	20	0.869	1.3975	II-241
326	" (*sec*)	$C_2H_5CO_2C_4H_9$	20	0.866	1.3952	II-241
327	*iso*-propylmalonate (*n*)	$CH_2(CO_2C_4H_9)(CO_2 \cdot$ $C_3H_7)$	20	0.97	1.4311	....
328	ricinoleate (*n*)	$HOC_{17}H_{32}CO_2C_4H_9$	22	0.906	1.4566	III-388
329	" (*iso*)	$HOC_{17}H_{32}CO_2CH_2 \cdot$ $CH(CH_3)_2$	22	0.903	1.4538	III-388
330	styryl ketone (*n*) (m.p. 40°)	$C_4H_9COCH:CHC_6H_5$	44.7	0.960	1.5518	VII-377
331	*iso*-thiocyanate (*iso*)	$(CH_3)_2CHCH_2NCS$	14	0.964	1.5005	IV-171
332	toluene (*p*) (*tert*)	$CH_3C_6H_4C(CH_3)_3$	13.3	0.867	1.4947	V-439
333	*iso*-valerate (*iso*)	$(CH_3)_2CHCH_2CO_2 \cdot$ $CH_2CH(CH_3)_2$	20	0.854	1.4057	II-312
334	valerate (*sec*)	$C_4H_9CO_2CH(CH_3) \cdot$ (C_2H_5)	20	0.860	1.4070	II-131
335	**Butylene** (α) (b.p. −5°)	$C_2H_5CH:CH_2$	20	.....	1.3962	I-203
336	chlorohydrin (β, γ)	$CH_3CHClCHOHCH_3$	20(?)	1.105	1.4438	I-373
337	phenylacetylene (*iso*)	$(CH_3)_2C:CHC:$ CC_6H_5	13	0.930	1.5828	V-568
338	**Butyne** (1)	$C_2H_5C:CH$	?	0.7	1.3962	I-248
339	" (2)	$CH_3C:CCH_3$	20	0.693	1.392	I-249
340	**Butyraldehyde** (*n*)	$C_2H_5CH_2CHO$	20	0.817	1.3843	I-662
341	" (*iso*)	$(CH_3)_2CHCHO$	20	0.794	1.3730	I-671
342	oxime (*iso*)	$(CH_3)_2CHCH:NOH$	20.5	0.902	1.4302	*I-350
343	**Butyrchloral**	$CH_3CHClCICl_2CHO$	20	1.396	1.4755	I-664
344	**Butyric acid** (*n*)	$C_2H_5CH_2CO_2H$	20	0.964	1.3979	II-264
345	" " (*iso*)	$(CH_3)_2CHCO_2H$	20	0.949	1.3930	II-288
346	**Butyro**-anisole (*p*) (*iso*)	$(CH_3)_2CHCO \cdot$ $C_6H_4OCH_3$	16.6	1.050	1.5393	*VIII-553
347	*p*-chlorophenol (*o*) (*iso*)	$(HO)(Cl)C_6H_3CO \cdot$ $CH(CH_3)_2$	20	1.192	1.552	....
348	*o*-cresol (*o*) (*iso*)	$(CH_3)_2CHCOC_6H_3:$ $(CH_3)OH$	20	1.047	1.537	....
349	*m*-cresol (*o*) (*iso*)	$(CH_3)_2CHCOC_6H_3:$ $(CH_3)(OH)$	20	1.042	1.540	....
350	*p*-cresol (*o*) (*iso*)	$(CH_3)_2CHCOC_6H_3:$ $(CH_3)OH$	21.3	1.043	1.5355	*VIII-556
351	*o*-cresyl acetate (*o, iso*) (1,2,3)	$(CH_3)(CH_3CO_2):$ $C_6H_3(COC_3H_7)$	20	1.074	1.514	....
352	*p*-methoxytoluene (*m*) (*iso*)	$CH_3C_6H_3(OCH_3)CO \cdot$ $CH(CH_3)_2$	13.7	1.024	1.5218	*VIII-556
353	nitrile (*n*)	$C_2H_5CH_2CN$	20	0.79	1.3808	I-335
354	thienone (α)	$C_3H_7COC:CHCH:$ $\mathcal{L}$ $CH \cdot S$ $\mathcal{L}$	20	1.098	1.542	....

Table 10-66 (*Continued*)
INDEX OF REFRACTION—LIQUIDS

No.	Substance	Formula	t,°C.	d_4^t	n_D^t	Beil. Ref.
355	**Butyro-*uns-m-*** xylenol (*o*) (*iso*)	$(CH_3)_2C_6H_2(OH)\cdot$ $COCH(CH_3)_2$	20	1.028	1.573	....
356	**Butyroin**	$C_2H_5CH_2CHOHCO\cdot$ $CH_2C_2H_5$	16.7	0.911	1.4346	I-840
357	**Butyryl**-acetophe- none (*ω*)	$C_6H_5(COCH_2)_2C_2H_5$	20	1.049	1.5712	VII-689
358	camphor (3)	$C_{14}H_{22}O_2$	17.4	1.016	1.4923	VII-598
359	chloride (*n*)	$C_2H_5CH_2COCl$	20	1.028	1.4121	II-274
360	chloride (*iso*)	$(CH_3)_2CHCOCl$	20	1.017	1.4079	II-293
361	indazole (2) (*n*) (*stable*)	$C_2H_5CH_2CON:CH\cdot$ $\underline{\qquad\qquad}$ $C_6H_4:N$ $\underline{\qquad}$	20	1.120	1.565	....
362	indazole (2) (*iso*) (*stable*)	$(CH_3)_2CHCON:CH\cdot$ $\underline{\qquad\qquad}$ $C_6H_4:N$ $\underline{\qquad}$	20	1.112	1.562	....
363	**Cadaverine**	$H_2N(CH_2)_5NH_2$	16	0.88	1.46	IV-266
364	**Cadinene** (*l*)	$C_{15}H_{24}$	20	0.918	1.5073	V-459
365	**Cadmium** diethyl	$(C_2H_5)_2Cd$	18.1	1.65	1.5680	IV-677
366	dimethyl	$(CH_3)_2Cd$	17.9	1.984	1.5849	IV-677
367	**Camphenamine**	$C_{10}H_{17}N$	20	0.940	1.4935	XII-50
368	**Camphene**	$C_{10}H_{16}$	78(?)	0.822	1.44	V-156
369	" $-d(l)$ (m.p. 43°)	$C_{10}H_{16}$	50	.....	1.4564	V-156
370	**Campholene**	C_9H_{16}	20	0.803	1.4441	I-81
371	**Camphoroxime-*d*** ethyl ether	$C_8H_{14}CH_2C:NOC_2H_5$ $\underline{\qquad\qquad}$	17	0.947	1.4785	VII-114
372	methyl ether	$C_8H_{14}CH_2C:NOCH_3$ $\underline{\qquad\qquad}$	22.5	0.958	1.4795	VII-114
373	**Camphoryl** chloride (*trans*) (*l*)	$C_8H_{14}(COCl)_2$	20.7	1.227	1.4988	IX-754
374	dichloride (*d*) (*cis*)	$C_{10}H_{14}O_2Cl_2$	19.9	1.245	1.5013	IX-754
375	**Camphylamine** (*α*)	$C_{10}H_{19}N$	17.8	0.874	1.4728	XII-40
376	**Capric** acid (m.p. 31°)	$CH_3(CH_2)_8CO_2H$	30	0.895	1.4308	II-355
377	aldehyde (*n*)	$CH_3(CH_2)_8CHO$	22	0.82	1.4273	I-711
378	**Caproic** acid (*n*)	$CH_3(CH_2)_4CO_2H$	19.6	0.921	1.4145	II-321
379	acid (*iso*)	$(CH_3)_2CHCH_2H_4CO_2H$	20	0.925	1.4150	II-327
380	aldehyde (*n*)	$CH_3(CH_2)_4CHO$	20	0.834	1.4279	I-688
381	**Caproimido ethyl** ether (*iso*)	$C_5H_{11}C(:NH)O\cdot$ C_2H_5	17.9	0.858	1.4200	II-329
382	**Caproyl chloride** (*n*)	$CH_3(CH_2)_4COCl$	20	.....	1.4867	II-324
383	**Caprylic** acid (*n*)	$CH_3(CH_2)_6CO_2H$	21	0.91	1.4268	II-347
384	aldehyde (*n*)	$CH_3(CH_2)_6CHO$	20	0.821	1.4217	I-704
385	**Carane**	$C_{10}H_{18}$	20	0.838	1.4567	*V-47
386	**Carbethoxy-7-** methyltetrahy- droindazole (2)	$C_2H_5O_2CCH_2N\cdot CH:$ $\underline{\qquad\qquad}$ $C_6H_7(CH_3):N$ $\underline{\qquad}$	20	1.093	1.506	....
387	"**Carbitol**"	$C_2H_5O(CH_2)_2O\cdot$ $(CH_2)_2OH$	26	0.98	1.4244	**I-520
388	**Carbon** disulfide	CS_2	20	1.263	1.6276	III-197
389	suboxide	$OC:C:CO$	0	1.114	1.4538	I-805
390	tetrachloride	CCl_4	15	1.604	1.4630	I-65
391	**Carone** (*d*)	$C_{10}H_{16}O$	18.8	0.958	1.4788	VII-91

Table 10-66 (*Continued*)
INDEX OF REFRACTION—LIQUIDS

No.	Substance	Formula	t,°C.	d_4^t	n_D^t	Beil. Ref.
392	**Carvacrol**	$(CH_3)(C_3H_7)C_6H_3OH$ (1,4,2)	18.6	0.98	1.5245	VI-527
393	**Carvacrylamine** (1,2,4)	$C_6H_3(CH_3)(NH_2)\cdot$ (C_3H_7)	19	0.995	1.543	XII-1171
394	**Carvenone** (*dl*)	$C_{10}H_{16}O$	20	0.926	1.4825	VII-78
395	**Carvo-**monthone (*d*)	$O_{10}H_{18}$	18	0.83	1.4563	V-84
396	menthol	$C_{10}H_{19}OH$	20	0.908	1.4619	VI-26
397	menthone	$C_{10}H_{18}O$	20	0.9	1.4553	VII-34
398	tanacetone (*dl*)	$(CH_3)_2CHCHCH_2\cdot$ $\overline{\;\;\;\;\;\;\;\;\;\;\;\;\;\;\;}\,\rvert$ $COC(CH_3):CHCH_2$ $\overline{\;\;\;\;\;\;\;\;\;\;\;\;\;\;\;}\,\rvert$	20	0.935	1.4806	VII-77
399	**Carvone** (*d*)	$C_{10}H_{14}O$	18.7	0.96	1.4994	VII-153
400	**Caryophyllene** (α)	$C_{15}H_{24}$	20	0.903	1.4998	I-464
401	**Cedrene**	$C_{15}H_{24}$	20	0.929	1.5034	V-461
402	**"Cellosolve acetate"**	$CH_3CO_2CH_2CH_2\cdot$ OC_2H_5	20	0.973	1.4030	II-141
403	**Cerotic acid** (m.p. 78°)	$C_{25}H_{51}CO_2H$	79	0.836	1.44	II-394
404	**Cetyl** acetate	$CH_3CO_2C_{16}H_{33}$	33.9	0.85	1.4358	II-136
405	alcohol (m.p. 50°)	$CH_3(CH_2)_{14}CH_2OH$	79	0.798	1.43	I-429
406	iodido (m.p. 22°)	$CH_3(CH_2)_{14}CH_2I$	20	1.123	1.4806	I-172
407	**Chavibetol** (1,3,4)	$CH_2:CHCH_2\cdot$ $C_6H_3(OH)(OCH_3)$	16	1.065	1.5397	VI-963
408	**Chloral**	Cl_3CCHO	20	1.512	1.4557	I-616
409	**Chloro-**acetic acid (m.p. 63°)	$ClCH_2CO_2H$	65	1.370	1.4297	II-194
410	acetylacetone (3)	$CH_3COCHClCOCH_3$	16.7	1.169	1.4798	I-785
411	aniline (*o*)	$ClC_6H_4NH_2$	20	1.213	1.5895	XII-597
412	" (*m*)	$ClC_6H_4NH_2$	20	1.216	1.5931	XII-602
413	anthracene (1) (m.p. 81°)	$C_{14}H_9Cl$	99.5	1.171	1.6959	*V-324
414	benzaldehyde (*o*)	ClC_6H_4CHO	20	1.252	1.567	VII-233
415	" (*m*)	ClC_6H_4CHO	20	1.241	1.5650	VII-234
416	" (*p*) (m.p. 47°)	ClC_6H_4CHO	61	1.196	1.5553	VII-235
417	benzene	C_6H_5Cl	20	1.107	1.5251	V-199
418	bromobenzene (*o*)	ClC_6H_4Br	20	1.644	1.5814	V-209
419	" (*m*)	ClC_6H_4Br	20	1.630	1.577	V-209
420	1-bromoethylene (2) (*cis*)	$BrCH:CHCl$	15	1.797	1.4982	I-189
421	1-bromoethylene (2) (*trans*)	$BrCH:CHCl$	20	1.77	1.4998	I-189
422	butyric acid (β)	$CH_3CHClCH_2CO_2H$	19.9	1.186	1.4421	II-277
423	butyryl chloride (β)	$CH_3CHClCH_2COCl$	20	1.217	1.4511	II-278
424	*d*-camphoryldi- chloride (3) (m.p. 26°)	$C_{10}H_{13}O_2Cl_3$	31.3	1.322	1.5080	IX-751
425	diethyl ether (β)	$C_2H_5OCH_2CH_2Cl$	20	0.989	1.4113	I-337
426	1,3-dimethyl-4- (β,β dichloro- ethyl) benzene (5)	$(CH_3)_2(Cl)C_6H_2\cdot$ CH_2CHCl_2	16.1	1.262	1.5528	. . . .

Table 10-66 (*Continued*)
INDEX OF REFRACTION—LIQUIDS

No.	Substance	Formula	$t,°C.$	d_4^t	n_D^t	Beil. Ref.
427	**Chloro**-1,3-dimethyl- 1-dichloromethyl- 4-methene-cyclo- hexa-2,5-diene (5)	$(CH_3)_2(Cl)\cdot$ $C_6H_2(:CH_2)\cdot CHCl_2$	17.3	1.269	1.5681	*V-206
428	ethyl acetate (2)	$CH_3CO_2CH_2CH_2Cl$	20	1.17	1.4247	II-128
429	ethyl ether (2)	$(ClCH_2CH_2)_2O$	20	1.220	1.457	**I-335
430	fumaryl chloride	$(ClCOCOCl):CHCOCl$	17.6	1.566	1.5217	*II-303
431	hexane (1)	$CH_3(CH_2)_4CH_2Cl$	20	0.876	1.4194	I-143
432	" (2)	$CH_3CHCl(CH_2)_3CH_3$	21	0.869	1.4142	I-144
433	maleyl chloride (*uns*)	$ClC:CHCCl_2\cdot O\cdot CO$ $\lfloor\underline{\quad\quad\quad}\rfloor$	18.1	1.605	1.5136	....
434	1-methyl-4- (*β,β*-dichloro- ethyl)-benzene (3)	$(CH_3)(Cl)C_6H_3CH_2\cdot$ $CHCl_2$	20.3	1.287	1.5501	*V-194
435	methyl methyl ether	$ClCH_2OCH_3$	20	1.070	1.3974	I-580
436	naphthalene (*α*)	$C_{10}H_7Cl$	20	1.194	1.6332	V-541
437	" (*β*) (m.p. 56°)	$C_{10}H_7Cl$	70.7	1.138	1.6079	V-541
438	pentane (2)	$CH_3(CH_2)_2CHClCH_3$	19.5	0.870	1.4062	I-131
439	phenol (*o*)	HOC_6H_4Cl	40	1.2	1.5473	VI-183
440	" (*m*) (m.p. 28°)	HOC_6H_4Cl	40	1.2	1.5565	VI-185
441	phenol (*p*) (m.p. 41°)	HOC_6H_4Cl	40	1.2	1.5579	VI-186
442	propan-2-ol (1)	$CH_3CHOHCH_2Cl$	20	1.114	1.4392	I-363
443	propan-1-ol (2)	$CH_3CHClCH_2OH$	20	1.103	1.4362	I-356
444	quinoline (2) (m.p. 37°)	$C_6H_4\cdot CH:CH\cdot$ $\lfloor\underline{\quad\quad\quad}$ $C(Cl):N$ $\underline{\quad\quad\quad}\rfloor$	24.6	1.246†	1.6342†	XX-359
445	styrene (*α*)	$C_6H_5CCl:CH_2$	16.6	1.10	1.5623	V-476
446	" (*ω*)	$C_6H_5CH:CHCl$	17.8	1.11	1.5774	V-476
447	sulfonic acid	$HOSO_2Cl$	14	.....	1.437	....
448	thiophene (*α*)	$ClC:CHCH:CH\cdot S$ $\lfloor\underline{\quad\quad\quad\quad}\rfloor$	20	1.278	1.55	XVII-32
449	toluene (*o*)	$ClC_6H_4CH_3$	20.2	1.081	1.5247	V-290
450	" (*m*)	$ClC_6H_4CH_3$	18.7	1.07	1.5225	V-291
451	" (*p*)	$ClC_6H_4CH_3$	24.4	1.06	1.5193	V-292
452	1,2,2-tribromo- ethane (1)	$ClBrCHCHBr_2$	20	2.65	1.603	....
453	**Chloroform**	$CHCl_3$	19	1.490	1.4457	I-61
454	**Chloroprene**	$CH_2:CClCH:CH_2$	20	0.958	1.4583	....
455	**Chromane**	$C_6H_4CH_2CH_2CH_2\cdot O$ $\lfloor\underline{\quad\quad\quad\quad}\rfloor$	20	1.059	1.544	XVII-52
456	**Cineol**	$C_{10}H_{18}O$	20	0.927	1.4584	XVII-24
457	**Cinnamic aldehyde**	$C_6H_5CH:CHCHO$	20	1.110	1.6195	VII-348
458	**Cinnamyl alcohol** (m.p. 33°)	$C_6H_5CH:CHCH_2OH$	36.1	1.04	1.5754	VI-570
459	**Cinnamoyl chloride** (m.p. 36°)	$C_6H_5CH:CHCOCl$	42.5	.....	1.6136	IX-587
460	**Citraconic an- hydride**	$C_5H_4O_3$	20	1.245	1.4717	....
461	**Citral** (*α*)	$C_9H_{15}CHO$	17	0.890	1.4895	I-753
462	" (*β*)	$C_{10}H_{16}O$	19	0.888	1.4900	I-755

† Supercooled.

Table 10-66 (*Continued*)
INDEX OF REFRACTION—LIQUIDS

No.	Substance	Formula	t,°C.	d_4^t	n_D^t	Beil. Ref.
463	**Citronellal** (*d*)	$C_9H_{17}CHO$	14	0.86	1.455	I-745
464	**Citronellol** (*d*)	$C_{10}H_{20}O$	20	0.85	1.459	I-451
465	**Citronellyl** acetate	$C_{12}H_{22}O_2$	17.5	0.893	1.4456	II-139
466	formate	$HCO_2C_{10}H_{19}$	20	0.884	1.4556	II-23
467	**Clovene**	$C_{15}H_{24}$	18	0.930	1.5007	V-468
468	**Cocaine** (*pseudo*) (*d*) (m.p. 41°)	$C_{17}H_{21}O_4N$	99.6	1.102	1.5022	XXII-206
469	**Cocaine** (*dl*) (m.p. 81°)	$C_{17}H_{21}O_4N$	99.5	1.103	1.5021	XXII-211
470	**Coniceine** (γ)	$C_8H_{15}N$	18.4	0.875	1.4607	XX-145
471	**Coniine** (*d*)	$C_8H_{17}N$	21.9	0.84	1.4512	XX-111
472	**Coumalin**	$CH{:}CHCH{:}CH{\cdot}$ $\lfloor\quad\quad\quad$ $O{\cdot}CO$ $\quad\quad\rfloor$	20	1.198	1.5298	XVII-271
473	**Coumaran**	$C_6H_4CH_2CH_2{\cdot}O$ $\lfloor\quad\quad\quad\quad\rfloor$	19	1.057	1.5420	XVII-50
474	**Coumaranone** (*iso*)	$O{\cdot}C_6H_4CH_2CO$ $\lfloor\quad\quad\quad\rfloor$	20	1.217	1.5523	XVII-309
475	**Coumarone**	$C_6H_4CH{:}CH{\cdot}O$ $\lfloor\quad\quad\quad\rfloor$	22.7	1.09	1.5645	XVII-54
476	**Creosol** (3,1,4)	$CH_3OC_6H_3(CH_3)OH$	25	1.09	1.5353	VI-878
477	**Cresol** (*o*) (m.p. 30°)	$CH_3C_6H_4OH$	40	1.04	1.5372	VI-349
478	" (*m*)	$CH_3C_6H_4OH$	18	1.04	1.5425	VI-373
479	" (*p*) (m.p. 35°)	$CH_3C_6H_4OH$	35	1.03	1.5	VI-389
480	**Cresyl acetate** (*p*)	$CH_3CO_2C_6H_4CH_3$	23	1.050	1.4991	VI-397
481	**Crotonic** acid (α) (m.p. 72°)	$CH_3CH{:}CHCO_2H$	72	0.96	1.43	II-408
482	acid (β)	$CH_3CH{:}CHCO_2H$	19.6	1.026	1.446	II-412
483	aldehyde (α)	$CH_3CH{:}CHCHO$	20.5	0.852	1.4362	I-728
484	anhydride	$(CH_3CH{:}CHCO)_2O$	20	1.040	1.4745	II-411
485	nitrile	$CH_3CH{:}CHCN$	20	0.822	1.4156	II-412
486	**Crotonoyl**-anisole	$CH_3CH{:}CHCO{\cdot}$ $C_6H_4OCH_3$	11.7	1.091	1.587	....
487	chloride	C_3H_5COCl	17.9	1.09	1.4600	II-411
488	**Crotonyl** alcohol	$CH_3CH{:}CHCH_2OH$	20	0.854	1.4240	I-442
489	**Cuminic aldehyde** (*p*)	$(CH_3)_2CHC_6H_4CHO$	20(?)	0.978	1.5301	VII-318
490	**Cyanamide** (m.p. 44°)	H_2NCN	48	1.073	1.4418	III-75
491	**Cyano**-acetic ester	$CH_3COCH(CN)CO_2{\cdot}$ C_2H_5	20	1.111	1.4710	III-796
492	cyclohexan-2-one (1)	$CH_2(CH_2)_3COCHCN$ $\lfloor\quad\quad\quad\quad\rfloor$	20	1.070	1.477	....
493	2-ethoxycyclo-hex-1-ene (1)	$C_2H_5OC{:}C(CN)CH_2{\cdot}$ $\lfloor\quad\quad\quad\quad$ $(CH_2)_2CH_2$	20	1.019	1.494	....
494	Δ^1-*iso*-heptenic acid (α) (m.p. 53°)	$(CH_3)_2CHCH_2CH{:}$ $C(CN)CO_2H$	99.6	0.972	1.4469	II-792
495	heptoic acid (α)	$C_2H_5(CH_2)_3{\cdot}$ $CH(CN)CO_2H$	20	1.022	1.448	....
496	1-methyl-3-*iso*-propylcyclo-pentane (2)	$C_{10}H_{17}N$	16.2	0.881	1.45	*IX-17
497	tetrabromide (*iso*) (m.p. 38.5°)	$Br_2C{:}N{\cdot}N{:}CBr_2$	55	2.680	1.6465	III-120

Table 10-66 (*Continued*)
INDEX OF REFRACTION—LIQUIDS

No.	Substance	Formula	t,°C.	d_4^t	n_D^t	Beil. Ref.
498	**Cyclo**-acetobutyl methyl ether	$C_7H_{14}O_2$	20	0.946	1.427	....
499	butane (b.p. 12°)	CH₂CH₂CH₂CH₂	0	0.703	1.3752	V-17
500	butane carboxylic acid	CH₂(CH₂)₂CHCO₂H	19	1.053	1.4434	IX-5
501	butanol	CH₂CH₂CH₂CHOH	20	0.92	1.4335	VI-4
502	citral (β)	$C_{10}H_{16}O$	13.3	0.96	1.4971	VII-87
503	fenchene	$C_{10}H_{16}$	20.4	0.859	1.4521	V-165
504	heptane	CH₂(CH₂)₅CH₂	20	0.810	1.4440	V-29
505	heptanone	CH₂(CH₂)₅CO	21.9	0.95	1.4603	VII-13
506	heptene	CH₂(CH₂)₄CH:CH	20	0.823	1.4530	V-65
507	hexane	CH₂(CH₂)₄CH₂	20	0.779	1.4262	V-20
508	hexanol (m.p. 24°)	CH₂(CH₂)₄CHOH	37	0.9	1.461	VI-5
509	hexanone	CH₂(CH₂)₄CO	19	0.947	1.4503	VII-8
510	hex-1-ene pro-pionic acid (α)	C_6H_9CH(CH₃)CO₂H	20.3	1.010	1.4753	IX-51
511	hexylacetic acid (m.p. 32°)	CH₂(CH₂)₃CH₂CH·CH₂CO₂H	33.5	1.001	1.46	IX-14
512	hexyl bromide	CH₂(CH₂)₄CHBr	14.6	1.3	1.4956	V-24
513	hexyl chloride	CH₂(CH₂)₄CHCl	18	0.977	1.4555	V-21
514	γ-hydroxy valeric aldehyde	CH₃CHCH₂CH₂·CHOH·O	20	1.019	1.435	*I-421
515	octane	CH₂(CH₂)₆CH₂	20	0.839	1.4586	V-35
516	pentadiene	CH₂CH:CHCH:CH	20	0.803	1.45	V-112
517	pentane	CH₂CH₂CH₂CH₂CH₂	20	0.745	1.4065	V-19
518	pentane carboxylic acid	CH₂(CH₂)₃CHCO₂H	20	1.053	1.4532	IX-6
519	pentanol	CH₂(CH₂)₃CHOH	20	0.947	1.4529	VI-5
520	pentanone	CH₂(CH₂)₃CO	20	0.948	1.4366	VII-5
521	pentene	CH:CHCH₂CH₂CH₂	14	0.775	1.4208	V-61
522	pentylacetic acid	CH₂(CH₂)₃CHCH₂·CO₂H	18	1.022	1.45	IX-10
523	propane carboxylic acid	CH₂CH₂CHCO₂H	20.9	1.087	1.4370	IX-4
524	propyl carbinol	(CH₂)₂CHCH₂OH	15.1	0.90	1.4313	VI-4

Table 10-66 (*Continued*)
INDEX OF REFRACTION—LIQUIDS

No.	Substance	Formula	t,°C.	d_4^t	n_D^t	Beil. Ref.
525	**Cymene** (*o*)	$CH_3C_6H_4CH(CH_3)_2$	20	0.877	1.5006	V-419
526	" (*m*)	$CH_3C_6H_4CH(CH_3)_2$	20	0.861	1.4930	V-419
527	" (*p*)	$CH_3C_6H_4CH(CH_3)_2$	20	0.857	1.4909	V-420
528	**Decahydronaph-** thalene (*trans*)	$C_{10}H_{18}$	20	0.877	1.4701	V-92
529	**Decane** (*n*)	$CH_3(CH_2)_8CH_3$	14.9	0.73	1.4108	I-168
530	**Decene**-1	$CH_2:CH(CH_2)_7CH_3$	16	0.75	1.4313	I-223
531	**Decyl** acetate (*n*)	$CH_3CO_2(CH_2)_9CH_3$	20	.....	1.4273	II-135
532	alcohol (*n*)	$CH_3(CH_2)_8CH_2OH$	20	0.830	1.4372	I-425
533	bromide (*n*)	$C_{10}H_{21}Br$	20	1.068	1.4550	**I-130
534	*n*-butyrate (*n*)	$C_2H_5CH_2CO_2CH_2 \cdot$ $(CH_2)_8CH_3$	20	0.862	1.4308	....
535	caprate (*n*)	$C_9H_{19}CO_2C_{10}H_{21}$	20	0.859	1.4423	II-356
536	chloride (*n*)	$CH_3(CH_2)_8CH_2Cl$	20	0.887	1.4380	I-168
537	formate (*n*)	$HCO_2CH_2(CH_2)_8CH_3$	20	0.873	1.4271	....
538	iodide (*n*)	$CH_3(CH_2)_8CH_2I$	20	1.256	1.4859	I-168
539	phenyl ether	$CH_3(CH_2)_9OC_6H_5$	20	0.898	1.4862	....
540	propionate (*n*)	$C_2H_5CO_2C_{10}H_{21}$	20	0.864	1.4291	....
541	**Diacetyl**	$(CH_3CO)_2$	18.5	0.981	1.3933	I-769
542	acetone (m.p. 49°)	$(CH_3CO)_2(CH_2CO \cdot$ $CH_2)$	60	1.06	1.4787	I-808
543	**Diallyl**-acetic acid	$(CH_2:CHCH_2)_2CH \cdot$ CO_2H	21.6	0.947	1.4508	II-489
544	acetone (*sym*)	$(CH_2:CHCH_2CH_2)_2:$ CO	14	0.868	1.4504	I-751
545	" (*uns*)	$(CH_2:CHCH_2)_2CH \cdot$ $COCH_3$	20.9	0.859	1.4462	I-751
546	phenol (*o,o*)	$(CH_2:CHCH_2)_2:$ C_6H_3OH	20	0.988	1.540	*VI-301
547	sulfide	$(CH_2:CHCH_2)_2S$	26.8	0.888	1.4877	I-440
548	**Diamyl**-amine (*iso*)	$(C_5H_{11})_2NH$	21.1	0.767	1.4229	IV-182
549	aniline (*iso*)	$C_6H_5N(C_5H_{11})_2$	24.9	0.889	1.5063	XII-169
550	cyanamide (*iso*)	$[(CH_3)_2CHCH_2 \cdot$ $CH_2]_2NCN$	23.7	0.846	1.4392	IV-186
551	oxalate (*iso*)	$(CO_2C_5H_{11})_2$	11	0.968	1.42	II-540
552	sulfide (*iso*)	$(C_5H_{11})_2S$	20	0.843	1.4524	I-405
553	**Diazidoethane**	$(CH_2N_3)_2$	24.9	1.170	1.4798	I-103
554	**Dibenzyl**-amine	$(C_6H_5CH_2)_2NH$	21.6	1.026	1.5743	XII-1035
555	methane	$(C_6H_5CH_2)_2CH_2$	20	1.007	1.5760	V-613
556	**Dibromo**-benzene	$C_6H_4Br_2$ (*o*)	17.4	1.964	1.6117	V-210
557	" (*m*)	$C_6H_4Br_2$	17.4	1.962	1.6083	V-211
558	" (*p*) (m.p. 88°)	$C_6H_4Br_2$	99.3	1.9	1.5743	V-211
559	butane (1,3)	$CH_3CHBrCH_2CH_2Br$	20	1.806	1.507	I-120
560	ethane (1,1)	CH_3CHBr_2	20	2.055	1.5128	I-90
561	hexane (1,2)	$CH_3(CH_2)_3CHBr \cdot$ CH_2Br	13.5	1.596	1.5060	I-144
562	" (1,6)	$(CH_2CH_2CH_2Br)_2$	15	1.595	1.5111	I-145
563	2-methylpropane (1,2)	$(CH_3)_2CBrCH_2Br$	20	1.759	1.509	I-127
564	pentane (1,5)	$BrCH_2(CH_2)_3CH_2Br$	15	1.70	1.5146	I-131
565	" (2,3)	$C_2H_5(CHBr)_2CH_3$	20	1.678	1.5078	I-132
566	propane (1,2)	$CH_3CHBrCH_2Br$	20	1.933	1.5203	I-109
567	" (1,3)	$BrCH_2CH_2CH_2Br$	15	1.987	1.5249	I-110
568	propan-1-ol (2,3)	$BrCH_2CHBrCH_2OH$	25	2.126	1.5577	I-357

Table 10-66 (*Continued*)
INDEX OF REFRACTION—LIQUIDS

No.	Substance	Formula	t,°C.	d_4^t	n_D^t	Beil. Ref.
569	**Dibromo**-propan-2-ol (1,3)	BrCH$_2$CHOHCH$_2$Br	25	2.120	1.5495	I-365
570	propylene (1,2)	C$_3$H$_4$Br$_2$ (*cis*)	17.4	2.024	1.5337	**I-171
571	propylene (1,2) (*trans*)	C$_3$H$_4$Br$_2$	17.4	2.003	1.5369	**I-171
572	thiophene (α,α')	BrC:CH·CH:CBr·S	20	2.141	1.63	XVII-33
573	1,1,4-trimethyl-cyclohexane (3,4)	(CH$_3$)$_3$C$_6$H$_7$Br$_2$	18.7	1.532	1.5281	*V-17
574	valeronitrile (α,β)	C$_2$H$_5$(CHBr)$_2$CN	20	1.756	1.520	**II-269
575	**Dibutyl**-amine (*n*)	(C$_4$H$_9$)$_2$NH	20	0.76	1.4097	IV-157
576	" (*iso*)	(C$_4$H$_9$)$_2$NH	19.6	0.747	1.4093	IV-166
577	ketone (*iso*)	(C$_4$H$_9$)$_2$CO	21	0.805	1.412	I-710
578	oxalate (*n*)	(CO$_2$C$_4$H$_9$)$_2$	20	1.0	1.4229	II-540
579	phthalate (*n*)	C$_6$H$_4$(CO$_2$C$_4$H$_9$)$_2$	25	1.045	1.4925	....
580	succinate (*sec*)	(CH$_2$CO$_2$C$_4$H$_9$)$_2$	25.3	0.97	1.4238	II-611
581	tartrate (*n*)	(CHOHCO$_2$C$_4$H$_9$)$_2$	20	1.090	1.4463	III-518
582	**Dichloro**-acetic acid	Cl$_2$CHCO$_2$H	20	1.563	1.4659	II-202
583	acetone (β) (*sym*) (m.p. 45°)	(ClCH$_2$)$_2$CO	46	1.383	1.4714	I-655
584	acetyl-acetone (3,3)	CH$_3$COCCl$_2$COCH$_3$	16.8	1.307	1.4589	I-785
585	benzene (*o*)	C$_6$H$_4$Cl$_2$	17	1.30	1.5524	V-201
586	" (*m*)	C$_6$H$_4$Cl$_2$	17.3	1.28	1.5472	V-202
587	" (*p*) (m.p. 53°)	C$_6$H$_4$Cl$_2$	69.9	1.4	1.5266	V-203
588	1-bromoethylene (1,2)	ClBrC:CHCl	16	1.91	1.5190	I-190
589	crotonic acid (γ,γ) (m.p. 43°)	Cl$_2$CHCH:CHCO$_2$H	99.4	1.333	1.4597	**II-397
590	crotononitrile (γ,γ)	Cl$_2$CHCH:CHCN	20	1.305	1.4974	**II-397
591	crotonoyl chloride (γ,γ)	Cl$_2$CHCH:CHCOCl	20	1.442	1.4991	**II-397
592	cyclopropane (1,1)	CH$_2$CH$_2$CCl$_2$	17.7	1.211	1.4402	V-17
593	1,2-dibromo-ethylene (1,2)	BrClC:CClBr	16	2.303	1.579	**I-164
594	ethane (1,1)	CH$_3$CHCl$_2$	20	1.175	1.4166	I-83
595	ethyl sulfide	(CH$_3$CHCl)$_2$S	20	1.27	1.5313	**I-685
596	2-methylbutane (2,3)	(CH$_3$)$_2$CClCHClCH$_3$	18	1.06	1.445	I-135
597	1-methyl-4-(β,β-dichloroethyl)-benzene (3,5)	(CH$_3$)Cl$_2$C$_6$H$_2$CH$_2$·CHCl$_2$	19.2	1.397	1.5678	*V-194
598	1-methyl-1-di-chloromethyl-4-methene-cyclo-hexa-2,5-diene (3,5)	(CH$_3$)Cl$_2$(CHCl$_2$)·C$_6$H$_2$:CH$_2$	16.7	1.408	1.5801	*V-194
599	methyl ether (*sym*)	(ClCH$_2$)$_2$O	20	1.315	1.4346	I-582
600	naphthalene (1,2) (m.p. 37°)	C$_{10}$H$_6$Cl$_2$	48.5	1.315	1.6338	V-542
601	naphthalene (1,4) (m.p. 68°)	C$_{10}$H$_6$Cl$_2$	75.9	1.300	1.6228	V-542
602	naphthalene (1,7) (m.p. 64°)	C$_{10}$H$_6$Cl$_2$	99.5	1.261	1.6092	V-543
603	naphthalene (1,8)	C$_{10}$H$_6$Cl$_2$	99.8	1.292	1.6236	V-544

Table 10-66 (*Continued*)
INDEX OF REFRACTION—LIQUIDS

No.	Substance	Formula	$t,°C.$	d_4^t	n_D^t	Beil. Ref.
604	**Dichloro**-propane (1,2)	$CH_3CHClCH_2Cl$	20	1.157	1.4387	I-105
605	" (2,2)	$(CH_3)_2CCl_2$	20	1.093	1.4093	I-105
606	(2-*p*-tolyl)-butane (1,1)	$C_2H_5CH(C_6H_4CH_3)\cdot CHCl_2$	19.8	1.112	1.5283	*V-209
607	1,1,2-tribromo-ethane (1,2)	$Br_2CCl\cdot CHBrCl$	15.5	2.63	1.5989	I-94
608	γ-valerolactone (α,δ)	$ClCH_2CHCH_2CHCl\cdot$ $\underset{CO\cdot O}{\underline{\qquad}}$	24.6	1.436	1.4962	*XVII-131
609	**Dicyclopentadiene** (m.p. 33°)	$\underline{\qquad}$ $(CH:CHCH_2\cdot$ $\underline{\qquad}$ $CH:CH)_2$ $\underline{\qquad}$	35	0.976	1.5050	V-495
610	**Didecyl ether** (*n*)	$(C_{10}H_{21})_2O$	20	0.819	1.4418	....
611	**Diethanolamine**	$HN(CH_2CH_2OH)_2$	20	1.097	1.4776	IV-283
612	**Diethoxydiphenyl**	$(C_2H_5OC_6H_4)_2$	47.4	1.043	1.5602	VI-989
613	**Diethyl**-acetic acid	$(C_2H_5)_2CHCO_2H$	10.2	0.933	1.4179	II-333
614	acetoacetic ester	$CH_3COC(C_2H_5)_2CO_2\cdot$ C_2H_5	17	0.971	1.4326	III-710
615	acetone di-carboxylate	$CO(CH_2CO_2C_2H_5)_2$	23.6	1.107	1.4378	III-791
616	acetylmalonate	$CH_3COCH(CO_2\cdot$ $C_2H_5)_2$	26	1.083	1.4374	III-796
617	acetylmalonate-*O*-acetate	$CH_3CO_2C(CH_3):C:$ $(CO_2C_2H_5)_2$	17.7	1.117	1.451	*III-163
618	acetylmalonate-*O*-propionate	$C_2H_5CO_2C(CH_3):C:$ $(CO_2C_2H_5)_2$	18.6	1.098	1.4519	*III-163
619	acetylsuccinate	$CH_3COC_2H_3(CO_2\cdot$ $C_2H_5)_2$	16	1.09	1.438	III-801
620	adipate	$(CH_2CH_2CO_2C_2H_5)_2$	20	1.009	1.4281	II-652
621	amine	$(C_2H_5)_2NH$	19	0.712	1.3871	IV-95
622	aminonaphthalene (α)	$C_{10}H_7N(C_2H_5)_2$ (α)	18.1	1.007	1.5933	XII-1223
622.1	aminonaphthalene (β)	$C_{10}H_7N(C_2H_5)_2$	21.6	1.025	1.6321	XII-1275
623	aniline	$C_6H_5N(C_2H_5)_2$	22.3	0.93	1.5411	XII-164
624	benzalmalonate (m.p. 32°)	$C_6H_5CH:C(CO_2\cdot$ $C_2H_5)_2$	20.4	1.105	1.5389	IX-892
625	benzene (*o*)	$(C_2H_5)_2C_6H_4$	20	0.881	1.5031	V-426
626	" (*m*)	$(C_2H_5)_2C_6H_4$	20	0.864	1.4953	V-426
627	" (*p*)	$(C_2H_5)_2C_6H_4$	20	0.862	1.4949	V-426
628	α-bromoacetyl malonate	$BrCH_2COCH(CO_2\cdot$ $C_2H_5)_2$	18.1	1.391	1.4602	*III-279
629	*n*-butyrylmalonate	$C_2H_5CH_2COCH:$ $(CO_2C_2H_5)_2$	20	1.055	1.4451	III-807
630	*iso*-butyryl-malonate	$(CH_3)_2CHCOCH:$ $(CO_2C_2H_5)_2$	20	1.050	1.4403	III-811
631	*d*-camphorate	$C_{14}H_{24}O_4$	19.1	1.032	1.456	IX-752
632	*trans*-camphorate (*l*)	$C_{14}H_{24}O_4$	21.6	1.029	1.4545	*IX-333
633	camphoric carbon-ate	$C_8H_{14}C:C(OCO_2\cdot$ $C_2H_5)(CO_2C_2H_5)$	20	1.076	1.4739	X-38
634	*O*-carbethoxy-β-hydroxyethylidene malonate	$CH_3C(OCO_2C_2H_5):$ $C(CO_2C_2H_5)_2$	20	1.127	1.4484	*III-163

Table 10-66 (*Continued*)
INDEX OF REFRACTION—LIQUIDS

No.	Substance	Formula	t,°C.	d_4^t	n_D^t	Beil. Ref.
635	**Diethyl** carbonate	$CO(OC_2H_5)_2$	20	0.975	1.3846	III-5
636	α-chloroacetyl malonate	$ClCH_2COCH(CO_2 \cdot C_2H_5)_2$	12.3	1.195	1.4446	*III-279
637	chlorofumarate	$CH:CCl(CO_2C_2H_5)_2$	18.3	1.19	1.4578	II-745
638	chloromalonate	$ClCH(CO_2C_2H_5)_2$	19.3	1.205	1.4353	II-593
639	citraconate	$CH_3C_2H(CO_2C_2H_5)_2$	20.3	1.062	1.4468	II-771
640	crotonylidene malonate	$CH_3CH:CHCH:C(CO_2C_2H_5)_2$	20	1.045	1.481	**II-676
641	cyanamide	$(C_2H_5)_2NCN$	48	0.8	1.4126	IV-121
642	cyanomalonate	$NCCH(CO_2C_2H_5)_2$	20	1.093	1.4263	II-811
643	*dl*-dehydrocamphorate	$(CH_3)_3C_6H_3(CO_2 \cdot C_2H_5)_2$	20	1.039	1.4644	*IX-345
644	diacetylmalonate	$(CH_3CO)_2C(CO_2 \cdot C_2H_5)_2$	20	1.115	1.4502	III-838
645	diallylmalonate	$(CH_2:CHCH_2)_2: C(CO_2C_2H_5)_2$	17.2	0.996	1.4475	II-807
646	diethylmalonate	$(C_2H_5)_2C(CO_2C_2H_5)_2$	16.6	0.99	1.4252	II-686
647	*cis*-α,β-dimethylglutaconate	$CH_3C(CO_2C_2H_5): C(CH_3)CH_2CO_2 \cdot C_2H_5$	20	1.024	1.454	*II-314
648	*cis*-α,β-dimethylglutaconate (*trans*)	$C_{11}H_{18}O_4$	20	1.031	1.454	*II-314
649	α,α'-dimethylglutarate (*mal*)	$CH_2[CH(CH_3)CO_2 \cdot C_2H_5]_2$	20	0.980	1.423	....
650	α,α'-dimethylglutarate	$C_{11}H_{20}O_4$	20	0.977	1.422	*II-591
651	dimethylmalonate	$(CH_3)_2C(CO_2C_2H_5)_2$	24.1	0.995	1.4105	II-648
652	dimethylsuccinate (*sym*) (*mal*)	$(CH_3CHCO_2C_2H_5)_2$	20	0.996	1.423	II-668
653	dimethylsuccinate (*sym*) (*fum*)	$(CH_3CHCO_2C_2H_5)_2$	20	0.995	1.421	II-667
654	dimethylsuccinate (*uns*)	$(CH_3)_2CCH_2(CO_2 \cdot C_2H_5)_2$	20	0.993	1.421	II-663
655	disulfide	$(C_2H_5S)_2$	20	0.993	1.5063	I-347
656	dithiocarbonate	$CO(SC_2H_5)_2$	18.2	1.084	1.5370	III-210
657	ethylacetylmalonate	$CH_3COC(C_2H_5): (CO_2C_2H_5)_2$	19.3	1.054	1.4338	III-812
658	ethylcyanomalonate	$NCC(C_2H_5)(CO_2 \cdot C_2H_5)_2$	20	1.052	1.4267	II-818
659	ethylmalonate	$C_2H_5CH(CO_2C_2H_5)_2$	14.8	1.01	1.4180	II-644
660	ethylidenemalonate	$CH_3CH:C(CO_2 \cdot C_2H_5)_2$	16.1	1.043	1.4408	II-773
661	ethoxymethylenemalonate	$C_2H_5OCH:C(CO_2 \cdot C_2H_5)_2$	15.3	1.081	1.4631	III-469
662	fumarate	$(CHCO_2C_2H_5)_2$	20.4	1.052	1.4410	II-742
663	glutaconate	$C_3H_4(CO_2C_2H_5)_2$	19	1.05	1.4466	II-759
664	glutarate	$CH_2(CH_2CO_2C_2H_5)_2$	20	1.025	1.4241	II-633
665	hexahydrohomophthalate (*cis*)	$C_{13}H_{22}O_4$	20	1.040	1.454	....
666	hexahydrohomophthalate (*trans*)	$C_{13}H_{22}O_4$	20	1.038	1.453	....
667	hexahydrophthalate (*m*) (*cis*)	$C_6H_{10}(CO_2C_2H_5)_2$	20	1.054	1.453	....
668	hexahydrophthalate (*trans*)	$C_{12}H_{20}O_4$	20	1.040	1.450	....

Table 10-66 (*Continued*)
INDEX OF REFRACTION—LIQUIDS

No.	Substance	Formula	t,°C.	d_4^t	n_D^t	Beil. Ref.
669	**Diethyl** hexahydro-*iso*-phthalate	$C_6H_{10}(CO_2C_2H_5)_2$ (*m*) (*cis*)	20	1.045	1.452	....
670	hexahydro-*tere*-phthalate (*cis*)	$C_6H_{10}(CO_2C_2H_5)_2$ (*p*)	20.6	1.052	1.452	....
671	hydrazo-*iso*-butyrate	$[HNC(CH_3)_2CO_2\cdot C_2H_5]_2$	20.5	0.999	1.434	IV-560
672	hydroxymethylene-malonate	$HOCH:C(CO_2C_2H_5)_2$	14.3	1.128	1.4561	III-787
673	imidocarbonate	$(C_2H_5O)_2C:NH$	18.2	0.964	1.4170	III-37
674	indane-1,3-dione (2,2)	$C_6H_4COC(C_2H_5)_2CO$ ⌐_____⌐	20	1.064	1.5393	*VII-381
675	indazole (2,3)	$C_2H_5N\cdot N:C_6H_4:C\cdot$ ⌐____⌐ C_2H_5	20	1.039	1.579	....
676	isoprenedi-carboxylate	$C_2H_5O_2CCH:C(CH_3)\cdot CH:CHCO_2C_2H_5$	20	1.056	1.500	*II-318
677	itaconate	$C_3H_4(CO_2C_2H_5)_2$	20.2	1.045	1.4388	II-762
678	ketene	$(C_2H_5)_2C:CO$	14.9	0.834	1.4136	I-740
679	ketone	$(C_2H_5)_2CO$	16.6	0.82	1.3939	I-679
680	ketoxime	$(C_2H_5)_2C:NOH$	20	0.914	1.4454	I-680
681	malate	$(CHOHCH_2)(CO_2\cdot C_2H_5)_2$	20	1.128	1.4362	III-437
682	maleate	$(CHCO_2C_2H_5)_2$	20.1	1.067	1.4416	II-751
683	malonate	$CH_2(CO_2C_2H_5)_2$	20	1.055	1.4143	II-573
684	mesaconate	$CH_3C_2H(CO_2C_2H_5)_2$	20	1.047	1.4494	II-766
685	mesoxalate	$CO(CO_2C_2H_5)_2$	15.6	1.136	1.4187	III-769
686	methylcyano-malonate	$NCC(CH_3)(CO_2\cdot C_2H_5)_2$	20	1.070	1.4232	II-814
687	5-methylcouma-ranone (2,2)	$CH_3C_6H_3CO\cdot$ ⌐_____ $C(C_2H_5)_2\cdot O$ _____⌐	22.6	1.032	1.531	....
688	β-methyl-α-ethyl-glutaconate (*cis*)	$C_2H_5C(CO_2C_2H_5)$: $C(CH_3)CH_2CO_2\cdot C_2H_5$	20	1.019	1.455	*II-315
689	β-methyl-α-ethyl-glutaconate (*trans*)	$C_{12}H_{20}O_4$	20	1.014	1.453	*II-315
690	methylethyl-malonate	$C_2H_5C(CH_3)$: $(CO_2C_2H_5)_2$	18.2	0.99	1.4190	II-664
691	β-methylgluta-conate (*cis*)	$C_2H_5O_2CCH$: $C(CH_3)CH_2CO_2\cdot C_2H_5$	20	1.034	1.452	*II-311
692	β-methylgluta-conate (*trans*)	$C_{10}H_{16}O_4$	20	1.034	1.452	*II-311
693	N-methylpyrrol-idene-2,5-di-acetate	$CH_3NC_4H_6(CH_2CO_2\cdot C_2H_5)_2$	15.5	1.047	1.4597	....
694	muconate (m.p. 64°)	$(CH:CHCO_2C_2H_5)_2$	99.1	0.983	1.468	II-804
695	naphthalate (1,8) (m.p. 58°)	$C_{10}H_6(CO_2C_2H_5)_2$	70	1.140	1.5586	IX-919
696	naphthalene (1,4)	$C_{10}H_6(C_2H_5)_2$	13.1	0.998	1.5970	**V-471
697	nitrosoamine	$(C_2H_5)_2NNO$	19.9	0.943	1.4386	IV-129
698	oxalacetate	$(CH_2CO)(CO_2C_2H_5)_2$	16.6	1.172	1.4561	III-782
699	oxalate	$(CO_2C_2H_5)_2$	20	1.079	1.4101	II-535

Table 10-66 (*Continued*)
INDEX OF REFRACTION—LIQUIDS

No.	Substance	Formula	t,°C.	d_4^t	n_D^t	Beil. Ref.
700	**Diethyl** α-oxalyl-γ-phenylbutyrate	$C_6H_5(CH_2)_2CH \cdot$ $(CO_2C_2H_5)$ $CO \cdot$ $CO_2C_2H_5$	20	1.112	1.493	....
701	pentane (3,3)	$(C_2H_5)_4C$	20	0.752	1.4200	**I-129
702	phosphite	$(C_2H_5O)_2POH$	20	1.074	1.4082	*I-166
703	phthalate (*o*)	$C_6H_4(CO_2C_2H_5)_2$	17.7	1.12	1.5029	IX-798
704	*iso*-phthalate (*m*)	$C_6H_4(CO_2C_2H_5)_2$	17.5	1.13	1.5082	IX-834
705	phthalide (3,3) (m.p. 55°)	$C_6H_4CO \cdot OC(C_2H_5)_2$ $\lfloor \quad \quad \rfloor$	61.5	1.040	1.5087	XVII-325
706	phthalylmalonate (m.p. 75°)	$C_6H_4CO \cdot OC:C(CO_2 \cdot$ $\lfloor \quad \quad \quad \rfloor$ $C_2H_5)_2$	84.4	1.189	1.5411	XVIII-498
707	pimelate	$(CH_2)_5(CO_2C_2H_5)_2$	20	0.99	1.4305	II-671
708	propionylmalonate	$C_2H_5COCH(CO_2 \cdot$ $C_2H_5)_2$	16.4	1.079	1.4436	III-800
709	propionylmalonate-*O*-acetate	$C_2H_5CO_2C(CH_3):$ $C(CO_2C_2H_5)_2$	18.6	1.098	1.452	*III-163
710	propionylmalonate-*O*-propionate	$C_2H_5CO_2C(C_2H_5):$ $C(CO_2C_2H_5)_2$	19.1	1.079	1.4511	*III-163
711	propylcyano-malonate	$NCC(C_3H_7)(CO_2 \cdot$ $C_2H_5)_2$	20	1.033	1.4291	II-820
712	*iso*-propyli-denemalonate	$(CH_3)_2C:C(CO_2 \cdot$ $C_2H_5)_2$	17	1.028	1.4486	II-781
713	pyrotartrate	$CH_3CH(CO_2C_2H_5) \cdot$ $CH_2CO_2C_2H_5$	19.1	1.012	1.4198	II-639
714	styrene (β,β)	$C_6H_5CH:C(C_2H_5)_2$	18.7	0.892	1.5168	V-502
715	suberate	$[(CH_2)_3CO_2C_2H_5]_2$	20	0.982	1.4328	II-693
716	succinate	$(CH_2CO_2C_2H_5)_2$	20	1.040	1.4201	II-609
717	*iso*-succinate	$CH_3CH(CO_2C_2H_5)_2$	20	1.018	1.4131	II-629
718	succinylmalonate (m.p. 68°)	$COCH_2CH_2COC:$ $\lfloor \quad \quad \quad \rfloor$ $(CO_2C_2H_5)_2$	73	1.167	1.4725	XVIII-489
719	sulfate	$(C_2H_5O)_2SO_2$	20	1.178	1.3902	I-327
720	sulfide	$(C_2H_5)_2S$	20.5	0.837	1.4425	I-344
721	sulfite	$(C_2H_5O)_2SO$	11	1.08	1.4198	I-325
722	tartrate	$(CHOHCO_2C_2H_5)_2$	25	1.20	1.4454	III-512
723	tetramethyl-succinate	$[(CH_3)_2CCO_2C_2H_5]_2$	20	0.995	1.436	II-707
724	thiocarbonate	$CS(OC_2H_5)_2$	17.5	1.027	1.4601	III-133
725	trimethylsuccinate	$(CH_3)_2CCHCH_3 \cdot$ $(CO_2C_2H_5)_2$	20	0.993	1.427	*II-595
726	*iso*-valerylmalonate	$(CH_3)_2CHCH_2CO \cdot$ $CH(CO_2C_2H_5)_2$	20	1.034	1.4447	....
727	**Diethylenediamine** (m.p. 106°)	$NHC_2H_4NHC_2H_4$ $\lfloor \quad \quad \quad \rfloor$	113	.....	1.446	XXIII-4
728	**Difluoro**-acetic acid	F_2CHCO_2H	20	1.526	1.3419	II-193
729	benzene (*m*)	$C_6H_4F_2$	20	1.172	1.4417	*V-108
730	" (*p*)	$C_6H_4F_2$	20	1.164	1.4375	V-199
731	1-bromoethane (2,2)	CHF_2CH_2Br	20	1.8	1.39	I-89
732	1,2-dibromoethane (1,1)	CF_2BrCH_2Br	17.5	2.24	1.4482	I-92
733	1,1-dibromoethane (2,2)	Br_2CHCHF_2	20	2.312	1.4622	I-92
734	1-iodoethane (2,2)	CHF_2CH_2I	12.2	2.243	1.4681	I-98

Table 10-66 (*Continued*)
INDEX OF REFRACTION—LIQUIDS

No.	Substance	Formula	t,°C.	d_4^t	n_D^t	Beil. Ref.
735	**Difluoro**-1,1,2-tribro-moethane (1,2)	$CBr_2FCHBrF$	17.5	2.6	1.5079	I-94
736	**Difurfuryl ether**	$(C_4H_3OCH_2)_2O$	20	1.140	1.5088	....
737	**Dehydrocamphoryl chloride** (*d*) (m.p. 50°)	$(CH_3)_3C_5H_3(COCl)_2$	48	1.219	1.5043	*IX-345
738	**Dihydro**-amyl-anthracene (*iso*) (9) (*meso*)	$C_{19}H_{22}$	44.4	0.994	1.5626	V-653
739	benzene (1,2)	$(CH_2)_2(CH:CH)_2$ ⌊_____⌋	20	0.842	1.4744	V-113
740	" (1,4)	C_6H_8	25	0.84	1.4681	V-113
741	carveol	$C_{10}H_{18}O$	20(?)	0.927	1.4817	VI-63
742	carvone (*l*)	$C_{10}H_{16}O$	20	0.925	1.4711	VII-83
743	eucarvone (α)	$C_{10}H_{16}O$	18.9	0.923	1.4674	VII-73
744	methylindole (2?)	$C_9H_{11}N$	23.4	1.020	1.5687	....
745	naphthalene (Δ^1)	$C_{10}H_{10}$	18.3	0.998	1.5832	V-519
746	" (Δ^2)	$C_{10}H_{10}$	32.7	0.993	1.5549	V-519
747	toluene (1,3)	$CH_3C_6H_7$	20	0.835	1.4763	V-115
748	" (2,4)	$CH_3C_6H_7$	20	0.827	1.4680	V-115
749	*o*-xylene ($\Delta^{1,3}$)	$(CH_3)_2C_6H_6$	19.5	0.830	1.4789	V-120
750	" ($\Delta^{1,5}$)	$(CH_3)_2C_6H_6$	20	0.823	1.4675	V-119
751	*m*-xylene	$CH_3CHCH_2C(CH_3):$ ⌊_____ CHCH:CH _____⌋	19.7	0.823	1.4721	V-119
752	*p*-xylene ($\Delta^{1,3}$)	$(CH_3)_2C_6H_6$	19	0.83	1.4797	V-119
753	**Dihydroxynaph-thalene** (1,5) di-*iso*-amyl ether (m.p. 96°)	$C_{10}H_6(OC_5H_{11})_2$	99.9	0.941	1.5160	**VI-950
754	di-*iso*-amyl ether (2,3)	$C_{10}H_6(OC_5H_{11})_2$	20	0.991	1.545	**VI-955
755	di-*iso*-amyl ether (2,7)	$C_{10}H_6(OC_5H_{11})_2$	99.6	0.934	1.5155	**VI-956
756	**Diiodo**-hexane (1,6)	$ICH_2(CH_2)_4CH_2I$	15	2.1	1.5890	I-147
757	pentane (1,5)	$ICH_2(CH_2)_3CH_2I$	15	2.20	1.6046	I-133
758	propane (1,3)	$ICH_2CH_2CH_2I$	15	2.576	1.6363	I-115
759	**Dimethoxy**-α-chromene (4,6)	$CH_3C_6H_3C(CH_3):$ ⌊_____ CHCH_2·O _____⌋	20	1.042	1.5623	*XVII-29
760	**Dimethyl**-acetal	$CH_3CH(OCH_3)_2$	20	0.850	1.37	I-603
761	acetamide (*N,N*)	$CH_3CON(CH_3)_2$	22.5	0.94	1.4371	IV-59
762	acetoacetic ester	$CH_3COC(CH_3)_2CO_2·$ C_2H_5	18.8	0.98	1.4183	III-695
763	acetophenone (2,4)	$(CH_3)_2C_6H_3COCH_3$	18.6	0.996	1.5301	VII-324
764	" (3,4)	$(CH_3)_2C_6H_3COCH_3$	15.5	1.007	1.5408	VII-323
765	acetonylacetone (ω,ω)	$(CH_3)_2CHCOCH_2·$ CH_2COCH_3	23	0.934	1.4317	I-796
766	3-acetoxycouma-rone (2,5) (m.p. 29°)	$CH_3C_6H_3C(O_2C·$ ⌊_____ CH_3):C(CH_3)·O _____⌋	20	1.132	1.5389	*XVII-66
767	acetylacetone	$CH_3COC(CH_3)_2CO·$ CH_3	27.8	0.950	1.4262	I-794

Table 10-66 (*Continued*)
INDEX OF REFRACTION—LIQUIDS

No.	Substance	Formula	t,°C.	$d_4{}^t$	$n_D{}^t$	Beil. Ref.
768	**Dimethyl-**δ**-acetyl-** valeric acid (β,β)	$CH_3CO(CH_2)_2C$: $(CH_3)_2CH_2CO_2H$	20	1.032	1.4564	**III-249
769	acetylene dibromide	CH_3CBr:$CBrCH_3$	25.5	1.318	1.4562	I-206
770	acryl chloride (β,β)	$(CH_3)_2C$:$CHCOCl$	12.4	1.065	1.479	*II-193
771	adipate	$(CH_2CH_2CO_2CH_3)_2$	20	1.063	1.4286	II-652
772	allylanisole	$(CH_3)_2(C_3H_5)C_6H_2\cdot$ OCH_3 (1,3,4,5)	20	0.960	1.524	*VI-293
773	aminoethanol	$(CH_3)_2NCH_2CH_2OH$	20	0.887	1.43	IV-276
774	aminonaphthalene (β) (m.p. 46°)	$C_{10}H_7N(CH_3)_2$	53.2	1.028	1.6443	XII-1273
775	amino-*o*-xylene (4)	$(CH_3)_2C_6H_3N(CH_3)_2$ (1,2,4)	20	0.916	1.5201	XII-1103
776	" (*m*) (2)	$(CH_3)_2C_6H_3N(CH_3)_2$ (1,3,2)	20	0.915	1.5131	XII-1108
777	" (*m*) (*uns*)	$(CH_3)_2C_6H_3N(CH_3)_2$ (1,3,4)	20	0.939	1.5481	XII-1115
778	angelolactone	$(CH_3)C$:$CHC(CH_3)_2\cdot$ ⌐ $CO\cdot O$ ⌐	20	0.976	1.433	....
779	aniline	$C_6H_5N(CH_3)_2$	20	0.956	1.5587	XII-141
780	benzene (*o*)	$(CH_3)_2C_6H_4$	20	0.880	1.5054	V-362
781	" (*m*)	$(CH_3)_2C_6H_4$	20	0.864	1.4972	V-370
782	" (*p*)	$(CH_3)_2C_6H_4$	20	0.861	1.4958	V-382
783	4-bromopyrazole (1,3)	$CH_3N\cdot CH$:$CBr\cdot C\cdot$ ⌐ (CH_3):N ⌐	20	1.490	1.519	....
784	buta-1,3-diene (2,3)	$[CH_2$:$C(CH_3)]_2$	20	0.726	1.4391	I-256
785	butane (2,2)	$(CH_3)_3CC_2H_5$	20	0.649	1.3688	I-150
786	" (2,3)	$[(CH_3)_2CH]_2$	20	0.661	1.3750	I-151
787	butan-3-ol (2,2)	$(CH_3)_3CCHOHCH_3$	20	0.82	1.4146	I-412
788	butan-2-ol (2,3)	$(CH_3)_2CHC$: $(CH_3)_2OH$	20(?)	0.821	1.4140	I-413
789	but-3-ene (2,2)	CH_2:$CHC(CH_3)_3$	20	0.653	1.3760	I-217
790	but-1-ene (2,3)	CH_2:$C(CH_3)\cdot$ $CH(CH_3)_2$	20	0.678	1.3904	I-218
791	but-2-ene (2,3)	$(CH_3)_2C$:$C(CH_3)_2$	20	0.708	1.4122	I-218
792	butenyl carbinol	$(CH_3)_2C(OH)C_4H_7$	18	0.854	1.443	*I-489
793	camphorate	$C_{12}H_{20}O_4$	16.9	1.077	1.4633	IX-750
794	5-chlorocoumaranone (2,2) (m.p. 67°)	$ClC_6H_3COC(CH_3)_2O$ ⌐————⌐	99.6	1.148	1.522	....
795	2-chlorocoumaranone (2,5)	$CH_3C_6H_3COC\cdot$ ⌐ $(ClCH_3)\cdot O$ ⌐	20	1.217	1.555	....
796	5-chlorocyclohexa-3,5-diene (1,3)	$(CH_3)_2C_6H_5Cl$	15.4	1.007	1.5046	V-119
797	3-chlorocyclohex-3-ene-5-one (1,1)	$(CH_3)_2C_6H_5OCl$	19.7	1.084	1.4955	*VII-49
798	chloroformamide	$ClCON(CH_3)_2$	22.1	1.166	1.4520	IV-73
799	4-chloropyrazole (1,3)	$CH_3N\cdot CH$:$CCl\cdot$ ⌐ $C(CH_3)$:N ⌐	20	1.153	1.492	....

Table 10-66 (*Continued*)
INDEX OF REFRACTION—LIQUIDS

No.	Substance	Formula	t,°C.	d_4^t	n_D^t	Beil. Ref.
800	**Dimethyl**-4-chloro-pyrazole (1,5)	N:CHCClC(CH₃)N· ⌐_____⌐ CH₃	20	1.166	1.497	....
801	chromene (α) (2,2)	C₆H₄(CH)₂C· ⌐_____ (CH₃)₂·O _____⌐	20	1.013	1.5512	XVII-64
802	citraconate	CH₃C₂H(CO₂CH₃)₂	20.1	1.12	1.4475	II-770
803	coumarane (2,5)	CH₃C₆H₃CH(CH₃)· ⌐_____ CH₂·O _____⌐	13.7	1.015	1.5279	*XVII-24
804	" (2,7)	CH₃C₆H₃CH₂CH· ⌐_____ (CH₃)·O _____⌐	13.8	1.015	1.5297	*XVII-24
805	coumarone (2,5)	CH₃C₆H₃CH:C· ⌐_____ (CH₃)·O _____⌐	20	1.031	1.5534	XVII-62
806	" (2,7)	CH₃C₆H₃CH:C· ⌐_____ (CH₃)·O _____⌐	20	1.036	1.5546	XVII-62
807	" (3,5)	CH₃C₆H₃C(CH₃): ⌐_____ CH·O ____⌐	22.2	1.035	1.5485	XVII-62
808	" (3,6)	CH₃C₆H₃C(CH₃): ⌐_____ CH·O ____⌐	20	1.046	1.5505	XVII-62
809	" (5,7)	C₁₀H₁₂O	18.9	1.026	1.5358	*XVII-24
810	cyclohexa-1,3-diene-2-carboxylic acid (m.p. 41°)	(CH₃)₂C₆H₅CO₂H	18.5†	1.045	1.4938	IX-82
811	cyclohexane (1,1)	(CH₃)₂(C·C₅H₁₀)	20	0.781	1.4290	V-35
812	cyclohexane (1,2) (*cis*)	(CH₃)₂C₆H₁₀	20	0.796	1.4360	V-36
813	cyclohexane (1,2) (*trans*)	(CH₃)₂C₆H₁₀	20	0.776	1.4270	V-36
814	cyclohexane (1,3) (*cis*) ‡	(CH₃)₂C₆H₁₀	20	0.766	1.4229	V-36
815	cyclohexane (1,3) (*trans*) §	(CH₃)₂C₆H₁₀	20	0.785	1.4309	V-36
816	cyclohexane (1,4) (*cis*)	(CH₃)₂C₆H₁₀	20	0.783	1.4297	V-38
817	cyclohexane (1,4) (*trans*)	(CH₃)₂C₆H₁₀	20	0.763	1.4209	V-38
818	cyclohexan-2-ol (1,1)	(CH₃)₂CCHOH· ⌐_____ (CH₂)₃CH₂ _____⌐	20	0.923	1.4648	VI-12
819	cyclohexan-3-ol (1,1)	(CH₃)₂C₆H₉OH	19.7	0.909	1.4589	VI-17

† Supercooled.
‡ Boiling point 120.09°.
§ Boiling point 124.45°.

Table 10-66 (*Continued*)
INDEX OF REFRACTION—LIQUIDS

No.	Substance	Formula	t,°C.	d_4^t	n_D^t	Beil. Ref.
820	**Dimethyl-**cyclo-hexan-4-ol (1,1)	$(CH_3)_2C_6H_9OH$	24.7	0.922	1.4613	*VI-13
821	cyclohexan-1-ol (1,2)	$CH_3COHCH(CH_3)\cdot$ $\underline{\qquad\qquad}$ $(CH_2)_3CH_2$ $\underline{\qquad\qquad}$	13.7	0.925	1.4625	VI-17
822	cyclohexan-4-ol (1,2)	$(CH_3)_2C_6H_9OH$	16	0.907	1.458	VI-17
823	cyclohexan-1-ol (1,3)	$(CH_3)_2C_6H_9OH$	22.9	0.903	1.4541	VI-17
824	cyclohexan-4-ol (1,3)	$(CH_3)_2C_6H_9OH$	16	0.912	1.458	VI-18
825	cyclohexan-1-ol (1,4)	$(CH_3)_2C_6H_9OH$	23.9	0.906	1.4553	*VI-13
826	cyclohexan-2-ol (1,4)	$(CH_3)_2C_6H_9OH$	16	0.907	1.455	VI-19
827	cyclohexan-1-ol (3,5) (*cis*)	$(CH_3)_2C_6H_9OH$	20	0.911	1.4540	VI-18
828	cyclohexan-1-ol (3,5) (*trans*)	$(CH_3)_2C_6H_9OH$	16	0.902	1.457	VI-18
829	cyclohexan-1-ol (3,5)	$(CH_3)_2C_6H_9OH¶$	18.6	0.900	1.4560	VI-18
830	cyclohexan-2-one (1,1)	$(CH_3)_2C_5H_8CO$	20	0.913	1.4486	*VII-17
831	cyclohexan-3-one (1,1)	$(CH_3)_2C_5H_8CO$	18.6	0.907	1.4476	VII-22
832	cyclohexan-4-one (1,1)	$(CH_3)_2C_5H_8CO$	24.2	0.928	1.4533	*VII-17
833	cyclohexan-4-one (1,2)	$(CH_3)_2C_5H_8CO$	12.6	0.912	1.4509	VII-23
834	cyclohexan-2-one (1,3)	$(CH_3)_2C_5H_8CO$	11.9	0.92	1.4519	VII-23
835	cyclohexan-5-one (1,3)	$(CH_3)_2C_5H_8CO$	18.2	0.896	1.4433	VII-24
836	cyclohexanyl-5 acetate (1,3)	$(CH_3)_2C_6H_9\cdot O_2CCH_3$	20.2	0.928	1.4376	VI-18
837	cyclohex-3-ene (1,1)	$(CH_3)_2C_6H_8$	16.2	0.806	1.4452	V-72
838	cyclohex-4-ene-5-acetic acid (1,3)	$(CH_3)_2C_6H_7CH_2\cdot$ CO_2H	20.3	0.995	1.4773	*IX-34
839	cyclohex-1-ene-3-ol (1,3)	$(CH_3)_2C_6H_7OH$	17.5	0.933	1.4771	VI-50
840	cyclohex-3-ene-5-one (1,3)	$(CH_3)_2C_5H_6CO$	22	0.94	1.4819	VII-60
841	cyclohexyl-1-chloride (1,2)	$CH_3C(Cl)CH(CH_3)\cdot$ $(CH_2)_3CH_2$	13.3	0.971	1.4672	*V-15
842	cyclohexyl-1-chloride (1,4)	$(CH_3)_2C_6H_9Cl$	18.5	0.944	1.4537	*V-16
843	cyclopentane (1,1)	$(CH_3)_2(C\cdot C_4H_8)$	20	0.754	1.4136	V-33
844	cyclopentane (1,2) (*cis*)	$(CH_3)_2C_5H_8$	20	0.773	1.4222	V-34
845	cyclopentane (1,2) (*trans*)	$(CH_3)_2C_5H_8$	20	0.751	1.4120	V-34
846	cyclopentane (1,3) (*cis*)	$(CH_3)_2C_5H_8$	20	0.749	1.4111	V-34
847	cyclopentane (1,3) (*trans*)	$(CH_3)_2C_5H_8$	20	0.745	1.4089	V-34

¶ Mixture of stereoisomers.

Table 10-66 (*Continued*)
INDEX OF REFRACTION—LIQUIDS

No.	Substance	Formula	t,°C.	d_4^t	n_D^t	Beil. Ref.
848	**Dimethyl-***d***-dehydro-**camphorate	$(CH_3)_3C_5H_3(CO_2 \cdot CH_3)_2$	18.1	1.088	1.4736	*IX-344
849	*iso*-dehydro-camphorate	$(CH_3)_3C_5H_3(CO_2 \cdot CH_3)_2$	17.5	1.085	1.4678	*IX-345
850	1-dichloromethyl-cyclohexa-2,5-dlene-4-one (1,3) (m.p. 56°)	$(CH_3)_2C_5H_3COCHCl_2$	20	1.241	1.5440	VII-150
851	1-dichloromethyl-cyclohexa-3,5-diene-2-one	$(CH_3)_2(CHCl_2) \cdot C_5H_3CO$	20	1.225	1.5336	....
852	1-dichloromethyl-cyclohex-3-ene (1,4)	$(CH_3)_2C_6H_7 \cdot CHCl_2$	20	1.125	1.5005	*V-40
853	4-dichloromethyl-cyclohex-1-ene-3-one (1,4) (m.p. 40°)	$(CH_3)_2C_5H_5CO \cdot CHCl_2$	45.8	1.207	1.5151	VII-67
854	4,4-diethylcyclo-hexan-2-ol (1,1)	$(CH_3)_2(C_2H_5)_2 \cdot C_6H_7OH$	13.6	0.922	1.4759	*VI-33
855	4,4-diethylcyclo-hexan-2-one (1,1)	$(CH_3)_2(C_2H_5)_2 \cdot C_5H_6CO$	15.2	0.911	1.4631	*VII-48
856	2,2-diethylIndane-1,3-dione (4,7) (m.p. 51°)	$(CH_3)_2C_6H_2COC \cdot \underline{\qquad} (C_2H_5)_2CO \underline{\qquad}$	62.4	1.017	1.5260	*VII-382
857	diethylmalonate	$(C_2H_5)_2C(CO_2CH_3)_2$	24.5	1.031	1.4253	II-686
858	diethylmethane-tetracarboxylate	$C(CO_2CH_3)_2(CO_2 \cdot C_2H_5)_2$	17.2	1.183	1.4333	*II-331
859	2,6-diethylphenol methyl ether (3,5)	$(CH_3)_2(C_2H_5)_2 \cdot C_6HOCH_3$	20	0.946	1.513	*VI-272
860	dimethylmalonate	$(CH_3)_2C(CO_2CH_3)_2$	24.2	1.059	1.4131	II-648
861	disulfide	$(CH_3S)_2$	16	1.057	1.5219	I-291
862	2-ethoxycouma-ranone (2,5)	$CH_3C_6H_3COC \cdot \underline{\qquad} (CH_3)(OC_2H_5) \cdot O \underline{\qquad}$	14.1	1.103	1.5322	*XVII-304
863	3-ethylbenzene (1,2)	$(CH_3)_2C_6H_3C_2H_5$	20	0.892	1.5117	....
864	4-ethylbenzene	$C_{10}H_{14}$ (1,2)	20	0.875	1.5031	V-427
865	2-ethylbenzene	$C_{10}H_{14}$ (1,3)	20	0.890	1.5107	....
866	4-ethylbenzene	$C_{10}H_{14}$ (1,3)	20	0.876	1.5038	V-428
867	5-ethylbenzene	$C_{10}H_{14}$ (1,3)	20	0.865	1.4981	V-429
868	2-ethylbenzene	$C_{10}H_{14}$ (1,4)	20	0.877	1.5043	V-428
869	3-ethylchromanone (2,6)	$CH_3C_6H_3COCH \cdot \underline{\qquad} (C_2H_5)CH(CH_3) \cdot O \underline{\qquad}$	21	1.063	1.539	....
870	2-ethylcouma-ranone (2,5)	$CH_3C_6H_3COC(CH_3) \cdot \underline{\qquad} (C_2H_5) \cdot O \underline{\qquad}$	19.6	1.056	1.5360	*XVII-168

Table 10-66 (*Continued*)
INDEX OF REFRACTION—LIQUIDS

No.	Substance	Formula	t,°C.	d_4^t	n_D^t	Beil. Ref.
871	**Dimethyl**-2-ethylcou-maranone (4,6)	$(CH_3)_2C_6H_2COCH\cdot$ $\mid_____$ $(C_2H_5)\cdot O$ $\_\_\_\_\_\mid$	20	1.067	1.5426	....
872	3-ethylcoumarone (2,5)	$CH_3C_6H_3C(C_2H_5):$ $\mid_____$ $C(CH_3)\cdot O$ $\_\_\_\_\_\mid$	20	1.003	1.5433	....
873	2-ethyl-7-hydroxy-hydrindone (3,4)	$(CH_3)(OH)C_6H_2CH\cdot$ $\mid_____$ $(CH_3) CH\cdot$ $(C_2H_5) CO$ $\_\_\_\_\_\mid$	20	1.073	1.548	....
874	2-ethyl-3-meth-oxybenzene (1,5)	$(CH_3)_2C_2H_5C_6H_2\cdot$ (CH_3O)	20	0.956	1.515	*VI-268
875	3-ethylpentane (2,2)	$(CH_3)_3CCH(C_2H_5)_2$	20	0.735	1.4123	....
876	" (2,3)	$(CH_3)_2CHC(CH_3):$ $(C_2H_5)_2$	20	0.754	1.419	....
877	" (2,4)	$[(CH_3)_2CH]_2CHC_2H_5$	20	0.738	1.4137	....
878	3-ethylpyrazine (2,5)	$C_8H_{12}N_2$	24	0.966	1.5014	XXIII-99
879	α-ethylstyrene (β,β)	$C_6H_5C(C_2H_5):C:$ $(CH_3)_2$	17.5	0.889	1.5142	V-502
880	4-ethylidene-cyclohexa-2,5-diene (1,1)	$(CH_3)_2C_6H_4:CHCH_3$	20	0.857	1.5135	V-427
881	formamide	$HCON(CH_3)_2$	22.4	0.948	1.4294	IV-58
882	furan (2,5)	$CH_3C:CHCH:C\cdot$ $\mid_____$ $(CH_3)\cdot O$ $\_\_\_\_\_\mid$	21.6	0.887	1.4351	XVII-41
883	furazan	$CH_3C:NON:CCH_3$ $\mid_____\mid$	18.9	1.050	1.4271	XXVII-565
884	hepta-1,3-diene (2,6)	C_9H_{16}	10	0.765	1.4620	I-260
885	heptane (2,2)	$(CH_3)_3C(CH_2)_4\cdot$ CH_3	20	0.711	1.402	....
886	" (2,3)	$(CH_3)_2CHCH(CH_3)$ $(CH_2)_3CH_3$	20	0.726	1.4085	....
887	" (2,4)	$(CH_3)_2CHCH_2CH:$ $(CH_3)(CH_2C_2H_5)$	20	0.716	1.403	*I-64
888	" (2,5)	$(CH_3)_2CH(CH_2)_2\cdot$ $CH(CH_3)(C_2H_5)$	20	0.715	1.4038	I-167
889	" (2,6)	$[(CH_3)_2CHCH_2]_2CH_2$	20	0.709	1.4007	I-167
890	" (3,3)	$C_2H_5C(CH_3)_2CH_2\cdot$ $CH_2C_2H_5$	20	0.725	1.4085	**I-129
891	" (3,4)	$C_2H_5CH(CH_3)CH:$ $(CH_3)(CH_2C_2H_5)$	20	0.731	1.4108	....
892	" (3,5)	$[C_2H_5(CH_3)CH]_2CH_2$	20	0.723	1.407	....
893	" (4,4)	$(C_2H_5CH_2)_2C:$ $(CH_3)_2$	20	0.725	1.408	....
894	heptan-4-ol (2,4)	$(CH_3)_2CHCH_2COH:$ $(CH_3)CH_2C_2H_5$	18	0.827	1.4318	I-425
895	heptan-5-ol (2,5)	$(CH_3)_2CH(CH_2)_2\cdot$ $COH(CH_3)C_2H_5$	20	0.829	1.4310	I-425

Table 10-66 (*Continued*)
INDEX OF REFRACTION—LIQUIDS

No.	Substance	Formula	t,°C.	d_4^t	n_D^t	Beil. Ref.
896	**Dimethyl-**heptan-4-ol (2,6)	$[(CH_3)_2CHCH_2]_2$:CHOH	21	0.81	1.423	I-425
897	heptan-4-ol (3,5)	$[C_2H_5(CH_3)CH]_2$:CHOH	20	0.836	1.4322	**I-458
898	hept-2-ene-6-ol (2,6)	$(CH_3)_2C$:CH$(CH_2)_2$·COH$(CH_3)_2$	20	0.855	1.4506	I-450
899	hexadecylethylene	$(CH_3)_2C$:CHC$_{16}H_{33}$	79.1	.....	1.428	....
900	hexa-1,5-diene (2,5)	$[CH_2$:C$(CH_3)CH_2·]_2$	20	0.75	1.4309	I-259
901	hexane (2,2)	$(CH_3)_3C(CH_2)_3·CH_3$	20	0.695	1.3935	**I-126
902	" (2,3)	$(CH_3)_2CHCH(CH_3)·CH_2C_2H_5$	20	0.712	1.4013	*I-62
903	" (2,4)	$(CH_3)_2CHCH_2·CH(CH_3)C_2H_5$	20	0.700	1.3953	I-162
904	" (2,5)	$[(CH_3)_2CHCH_2]_2$	20	0.694	1.3925	I-162
905	" (3,3)	$C_2H_5C(CH_3)_2CH_2·C_2H_5$	20	0.710	1.4001	*I-62
906	" (3,4)	$[C_2H_5CH(CH_3)]_2$	20	0.719	1.4042	I-163
907	hexan-4-ol (2,4)	$(CH_3)_2CHCH_2COH·(CH_3)(C_2H_5)$	18	0.83	1.4286	I-422
908	hexan-2-ol (2,5)	$(CH_3)_2CH(CH_2)_2·C(CH_3)_2OH$	20	0.823	1.4209	I-422
909	hex-3-ene (2,2) (*cis?*)	$(CH_3)_3CCH$:CHC$_2H_5$	20	0.719	1.4099	....
910	hex-3-ene (2,2) (*trans*)	$(CH_3)_3CCH$:CHC$_2H_5$	....	.....	....	....
911	hex-4-ene (2,2) (*cis?*)	CH_3CH:CHCH$_2$·C$(CH_3)_3$	20	0.716	1.410	....
912	hex-4-ene (2,2) (*trans*)	CH_3CH:CHCH$_2$·C$(CH_3)_3$	....	.....	....	....
913	hex-5-ene (2,2)	CH_2:CHCH$_2$CH$_2$·C$(CH_3)_3$	20	0.709	1.405	....
914	hex-2-ene (2,3)	$(CH_3)_2C$:C(CH_3)·CH$_2C_2H_5$	20	0.740	1.4260	....
915	hex-3-ene (2,3) (*cis?*)	$(CH_3)_2CHC(CH_3)$:CHC$_2H_5$	20	0.728	1.416	....
916	hex-3-ene (2,3) (*trans*)	$(CH_3)_2CHC(CH_3)$:CHC$_2H_5$	......	.....	....	....
917	hex-4-ene (2,3) (*cis?*)	CH_3CH:CHCH(CH_3)·CH$(CH_3)_2$	20	0.725	1.413	....
918	hex-4-ene (2,3) (*trans*)	CH_3CH:CHCH(CH_3)·CH$(CH_3)_2$	....	.....	....	....
919	hex-5-ene (2,3)	CH_2:CHCH$_2$·CH(CH_3)·CH:$(CH_3)_2$	20	0.728	1.414	....
920	hex-2-ene (2,4)	$(CH_3)_2C$:CHCH:$(CH_3)(C_2H_5)$	20	0.725	1.411	....
921	hex-3-ene (2,4) (*cis?*)	$(CH_3)_2CHCH$:C:$(CH_3)(C_2H_5)$	20	0.714	1.409	....
922	hex-3-ene (2,4) (*trans*)	$(CH_3)_2CHCH$:C:$(CH_3)(C_2H_5)$	....	.....	....	....
923	hex-4-ene (2,4) (*cis?*)	CH_3CH:C$(CH_3)CH_2$·CH$(CH_3)_2$	20	0.725	1.416	....
924	hex-4-ene (2,4) (*trans*)	CH_3CH:C$(CH_3)CH_2$·CH$(CH_3)_2$	....	.....	....	....

Table 10-66 (*Continued*)
INDEX OF REFRACTION—LIQUIDS

No.	Substance	Formula	t,°C.	d_4^t	n_D^t	Beil. Ref.
925	**Dimethyl**-hex-5-ene (2,4)	CH_2:$CHCH(CH_3)$· CH_2·$CH(CH_3)_2$	20	0.708	1.404	....
926	hex-2-ene (2,5)	$(CH_3)_2C$:$CHCH_2$· $CH(CH_3)_2$	20	0.720	1.4140	....
927	hex-3-ene (2,5) (*cis?*)	$[(CH_3)_2CH \cdot CH:]_2$	20	0.710	1.403	....
928	hex-3-ene (2,5) (*trans*)	$[(CH_3)_2CH \cdot CH:]_2$	....	.....	......	....
929	hex-1-ene (3,3)	CH_2:$CHC(CH_3)_2$· $CH_2C_2H_5$	20	0.714	1.4070	....
930	hex-4-ene (3,3) (*cis*)	CH_3CH:$CHC(C_2H_5)$· $(CH_3)_2$	20	0.722	1.413	....
931	hex-5-ene (3,3)	CH_2:$CHCH_2C(CH_3)_2$· C_2H_5	20	0.720	1.4102	....
932	hex-1-ene (3,4)	CH_2:$CHCH(CH_3)CH$: $(CH_3)(C_2H_5)$	20	0.724	1.413	....
933	hex-2-ene (3,4) (*cis?*)	CH_3CH:$C(CH_3)CH$: $(CH_3)(C_2H_5)$	20	0.737	1.418	....
934	hex-2-ene (3,4) (*trans*)	CH_3CH:$C(CH_3)CH$: $(CH_3)(C_2H_5)$	....	.....	......	....
935	hex-3-ene (3,4) (*cis?*)	$[(C_2H_5)(CH_3)C:]_2$	20	0.747	1.430	*I-94
936	hex-3-ene (3,4) (*trans*)	$[(C_2H_5)(CH_3)C:]_2$	....	.....	......	*I-94
937	Δ^2-hexenic acid (β,δ)	$(CH_3)_2CHCH_2C$· (CH_3):$CHCO_2H$	15.8	0.936	1.4480	....
938	hydrazine (*sym*)	$CH_3NHNHCH_3$	20	0.827	1.4209	IV-547
939	" (*uns*)	$(CH_3)_2NNH_2$	22.3	0.791	1.4075	IV-547
940	hydrind-1-one (2,2) (m.p. 44°)	$C_6H_4CH_2C(CH_3)_2CO$ $\lfloor_____\rfloor$	13.9	1.031	1.5408	*VII-198
941	hydrind-1-one (3,3)	$C_6H_4C(CH_3)_2$· $\lfloor\_\_\_\_$ CH_2CO $\_\_\rfloor$	20	1.027	1.5427	....
942	7-hydroxyhydrindone (2,5)	$(CH_3)(HO)C_6H_2$· $\lfloor\_\_\_\_$ $CH_2CH(CH_3)CO$ $\_\_\_\rfloor$	20	1.123	1.567	....
943	2-hydroxystyrene (1',5)	$(CH_3)(HO)C_6H_3$· $C(CH_3)$:CH_2	20	1.012	1.5514	VI-577
944	isoprene dicarboxylate (m.p. 39°)	CH_3O_2CCH:$C(CH_3)$· CH:$CHCO_2CH_3$	42.7	1.096	1.505	*II-319
945	isoxazole (α,γ)	CH_3C:CHC· $\lfloor\_\_\_\_$ (CH_3):N·O $\_\_\_\rfloor$	20	0.983	1.442	XXVII-17
946	itaconate	$C_3H_4(CO_2CH_3)_2$	20	1.122	1.4441	II-762
947	1-keto-1,2,3,4-tetrahydronaphthalene (m.p. 49°)	$(CH_3)_2C_6H_2CH_2$· $\lfloor_____$ CH_2CH_2CO $\_\_\rfloor$	15.5	1.065	1.5650	....
948	malate	$C_2H_4O(CO_2CH_3)_2$	20	1.233	1.4425	III-429
949	maleate	$(CHCO_2CH_3)_2$	19.9	1.151	1.4416	II-751
950	malonate	$H_2C(CO_2CH_3)_2$	20	1.154	1.4140	II-572
951	mesaconate	$CH_3C_2H(CO_2CH_3)_2$	19.9	1.121	1.4557	II-765

Table 10-66 (*Continued*)
INDEX OF REFRACTION—LIQUIDS

No.	Substance	Formula	t,°C.	d_4^t	n_D^t	Beil. Ref.
952	**Dimethyl-**2-methoxy-coumaranone (2,5)	$C_7H_6 \cdot CO \cdot C \cdot O \cdot$ $|$_____$|$ $(CH_3)(OCH_3)$	22.6	1.130	1.5383	*XVIII-304
953	3-methoxycoumarone (2,5)	$CH_3C_6H_3C(OCH_3)$: $|$_____ $C(CH_3) \cdot O$ _____$|$	19	1.081	1.5440	*XVII-65
954	3-methylalhept-3-ene (2,6)	$(CH_3)_2CHCH_2CH$: $C(CHO)CH(CH_3)_2$	17.4	0.854	1.45	I-748
955	4-methylene-cyclo-hexa-2,5-diene (1,1)	$(CH_3)_2C_6H_4:CH_2$	15.8	0.836	1.5030	V-399
956	1-methylenecyclohexane (3,5)	$(CH_3)_2(CH_2:)C_6H_8$	14.6	0.792	1.4463	*V-40
957	3-methylenepentane (2,2)	$CH_2:C(C_2H_5)C:$ $(CH_3)_3$	20	0.728	1.416	....
958	3-methylenepentane (2,4)	$CH_2:C[CH(CH_3)_2]_2$	20	0.722	1.409	....
959	naphthalene (1,2)	$(CH_3)_2C_{10}H_6$	20	1.019	1.616	*V-267
960	" (1,4)	$(CH_3)_2C_{10}H_6$	16.4	1.02	1.6157	V-570
961	" (1,6)	$(CH_3)_2C_{10}H_6$	20	1.002	1.607	*V-268
962	naphthylamine (α)	$C_{10}H_7N(CH_3)_2$	20(?)	1.042	1.624	XII-1221
963	nitroamine (m.p. 57°)	$(CH_3)_2NNO_2$	72.3	1.109	1.4462	IV-85
964	nitrosoamine	$(CH_3)_2NNO$	18.4	1.01	1.4374	IV-84
965	nona-2,8-diene-6-ol (2,6)	$(CH_3)_2C_9H_{13}OH$	16.6	0.866	1.4647	I-462
966	octa-2,6-diene (2,6)	$C_{10}H_{18}$	19	0.78	1.4497	I-260
967	octa-2,7-diene (2,6)	$C_{10}H_{18}$	20(?)	0.788	1.455	I-261
968	octa-2,6-diene-8-ol (2,6)	$C_9H_{15}CH_2OH$	15	0.883	1.4773	I-457
969	octa-2,7-diene-6-ol (2,6)	$C_{10}H_{18}O$	20	0.862	1.4624	I-460
970	octane (2,6)	$(CH_3)_2CH(CH_2)_3 \cdot$ $CH(CH_3)C_2H_5$	15	0.74	1.4135	I-168
971	" (2,7)	$[(CH_3)_2CHCH_2CH_2]_2$	18.1	0.72	1.4092	I-169
972	octan-6-ol (2,6)	$C_{10}H_{22}O$	15	0.836	1.4388	I-426
973	octan-8-ol (2,6)	$C_{10}H_{22}O$	18	0.830	1.438	I-426
974	octa-3,5,7-triene (2,6)	$(CH_3)_2CHCH:CH \cdot$ $CH:C(CH_3)CH:$ CH_2	20	0.809	1.5436	....
975	oct-2-ene (2,6)	$(CH_3)_2C:CH(CH_2)_2 \cdot$ $CH(CH_3)C_2H_5$	20	0.75	1.4303	I-224
976	" (2,7)	$(CH_3)_2CH(CH_2)_3 \cdot$ $CH:C(CH_3)_2$	20	0.75	1.4304	I-224
977	oct-2-ene-6-ol (2,6)	$(CH_3)_2C:CH(CH_2)_2 \cdot$ $COH(CH_3)C_2H_5$	15.9	0.858	1.4555	I-452
978	oxalate (m.p. 54°)	$(CO_2CH_3)_2$	56.6	1.14	1.3915	II-534
979	penta-1,3-diene (2,4)	$(CH_3)_2C:CHC:$ $(CH_3)CH_2$	23	0.734	1.4390	I-257
980	penta-2,4-diene (2,4)	$(CH_3)_2C:CHC(CH_3)$: CH_2	12	0.749	1.4468	I-257
981	pentane (2,2)	$(CH_3)_3CCH_2C_2H_5$	20	0.674	1.3822	I-157

Table 10-66 (*Continued*)
INDEX OF REFRACTION—LIQUIDS

No.	Substance	Formula	t,°C.	d_4^t	n_D^t	Beil. Ref.
982	**Dimethyl**-pentane (2,3)	$(CH_3)_2CHCH(CH_3)\cdot$ C_2H_5	20	0.695	1.3920	I-157
983	" (2,4)	$[(CH_3)_2CH]_2CH_2$	20	0.673	1.3815	I-158
984	" (3,3)	$(CH_3)_2C(C_2H_5)_2$	20	0.693	1.3909	I-158
985	pentan-2-ol (2,4)	$(CH_3)_2COHCH_2\cdot$ $CH(CH_3)_2$	20	0.82	1.4172	I-417
986	pentan-3-ol (2,4)	$[(CH_3)_2CH]_2CHOH$	20	0.829	1.4226	I-417
987	pent-3-ene (2,2) (*cis?*)	$(CH_3)_3CCH:CH\cdot$ CH_3	20	0.688	1.399	*I-92
988	pent-3-ene (2,2) (*trans*)	$(CH_3)_3CCH:CH\cdot$ CH_3	....	.....	.....	....
989	pent-4-ene (2,2)	$(CH_3)_3CCH_2CH:$ CH_2	20	0.683	1.3918	....
990	pent-1-ene (2,3)	$CH_2:C(CH_3)CH:$ $(CH_3)(C_2H_5)$	20	0.710	1.403	....
991	pent-2-ene (2,3)	$(CH_3)_2C:C(CH_3)\cdot$ C_2H_5	20	0.728	1.421	*I-92
992	pent-3-ene (2,3) (*cis?*)	$CH_3CH:C(CH_3)\cdot$ $CH(CH_3)_2$	20	0.713	1.407	....
993	pent-3-ene (2,3) (*trans*)	$CH_3CH:C(CH_3)\cdot$ $CH(CH_3)_2$	....	.....	......	....
994	pent-4-ene (2,3)	$CH_2:CHCH(CH_3)\cdot$ $CH(CH_3)_2$	20	0.701	1.399	....
995	pent-1-ene (2,4)	$CH_2:C(CH_3)CH_2\cdot$ $CH(CH_3)_2$	20	0.694	1.397	....
996	pent-2-ene (2,4)	$(CH_3)_2C:CHCH:$ $(CH_3)_2$	20	0.696	1.403	I-220
997	pent-1-ene (3,3)	$CH_2:CHC(CH_3)_2C_2H_5$	20	0.696	1.399	....
998	phenyl carbinol	$C_6H_5COH(CH_3)_2$	18.5	0.972	1.5314	VI-506
999	5-phenylcyclo-hexa-3,5-diene (1,3)	$(CH_3)_2C_6H_5C_6H_5$	17	0.972	1.5749	V-572
1000	5-phenylcyclo-hex-3-ene (1,3)	$(CH_3)_2(C_6H_5)C_6H_7$	20	0.942	1.5341	*V-254
1001	phthalate (*o*)	$C_6H_4(CO_2CH_3)_2$	20.8	1.192	1.5155	IX-797
1002	piperidylamine (γ)	$C_7H_{15}N$	19.2	0.759	1.4220	IV-222
1003	7-*iso*-propyl-coumarone (2,4)	$(CH_3)(C_3H_7):$ $C_6H_2CH:C(CH_3)\cdot O$ $\mid\underline{}\mid$	20	0.994	1.5401	....
1004	4-propylidene-cyclohexa-2,5-diene (1,1)	$(CH_3)_2C_6H_4:CHC_2H_5$	20	0.858	1.5042	**V-334
1005	pyrazine (2,5)	$CH_3C:CHN:C(CH_3)\cdot$ $\mid\underline{}$ $CH:N$ $\underline{}\mid$	23.6	0.986	1.4992	XXIII-96
1006	pyrazole (1,3)	$CH:CH\cdot C(CH_3):N\cdot N\cdot$ $\mid\underline{}\mid$ CH_3	20	0.966	1.473	....
1007	pyrazole (3,4) (m.p. 58°)	$(CH_3)_2C_3H_2N_2$	99.3	0.933	1.4657	XXIII-72
1008	pyrotartrate	$CH_3CH(CO_2CH_3)\cdot$ $CH_2CO_2CH_3$	19.6	1.069	1.4204	II-639
1009	pyrrole (2,4)	$C_4H_2(CH_3)_2NH$	13.7	0.927	1.4993	XX-172
1010	" (2,5)	$CH_3C:CHCH:C:$ $\mid\underline{}$ $(CH_3)NH$ $\underline{}\mid$	20	0.935	1.5036	XX-172

Table 10-66 (*Continued*)
INDEX OF REFRACTION—LIQUIDS

No.	Substance	Formula	t,°C.	d_4^t	n_D^t	Beil. Ref.
1011	**Dimethyl**-styrene (α,β)	$C_6H_5C(CH_3):CH$ CH_3	19.7	0.910	1.5339	V-488
1012	styrene (β,β)	$C_6H_5CH:C(CH_3)_2$	19.6	0.899	1.5273	V-489
1013	succinate	$(CH_2CO_2CH_3)_2$	20	1.121	1.4198	II-609
1014	sulfate	$(CH_3O)_2SO_2$	20	1.33	1.3874	I-283
1015	tetrahydroindazole (2,7)	$CH_3N\cdot CH:C_6H_7\cdot$ $\underset{\rule{1.5cm}{0.4pt}}{(CH_3):N}$	20	1.012	1,512	....
1016	5,6,7,8-tetrahy-dronaphthalene (1,3)	$(CH_3)_2C_6H_2CH_2\cdot$ $\underset{\rule{1.5cm}{0.4pt}}{CH_2CH_2CH_2}$	21	0.959	1.5409	**V-398
1017	thiophene (β,β)	$(CH_3)_2C_4H_2S$	15	0.99	1.522	XVII-40
1018	" (2,5)	$(CH_3)_2C_4H_2S$	19	0.994	1.5142	XVII-41
1019	o-toluidine (N)	$CH_3C_6H_4N(CH_3)_2$	23.3	0.93	1.5244	XII-785
1020	m-toluidine (N)	$CH_3C_6H_4N(CH_3)_2$	20	0.941	1.5492	XII-857
1021	p-toluidine (N)	$CH_3C_6H_4N(CH_3)_2$	20	0.937	1.5471	XII-902
1022	p-tolyl carbinol	$CH_3C_6H_4COH:$ $(CH_3)_2$	20	0.977	1.5162	VI-544
1023	m-tolyl ketone	$CH_3COC_6H_4CH_3$	15.3	1.011	1.5327	VII-307
1024	undecyl carbinol	$(CH_3)_2COC_{11}H_{23}$	81.8	.....	1.417	I-429
1025	**Dinitroethane** (1,1)	$CH_3CH(NO_2)_2$	20	1.35	1.4488	I-102
1026	**Diosphenol** acetate	$CH_3CO_2C_{10}H_{15}O$	20	1.036	1.481	VIII-9
1027	ethyl ether	$C_2H_5OC_{10}H_{15}O$	20	0.977	1.485	VIII-9
1028	methyl ether	$CH_3OC_{10}H_{15}O$	20	0.988	1.485	VIII-9
1029	**Dioxan** (1,4)	$O:(CH_2)_4:O$	20	1.033	1.4221	XIX-3
1030	**Dipentene** (*dl*)	$C_{10}H_{16}$	20.9	0.840	1.4744	V-137
1031	**Diphenyl** (m.p. 69°)	$C_6H_5\cdot C_6H_5$	77.1	0.99	1.5882	V-576
1032	acetaldehyde	$(C_6H_5)_2CHCHO$	20	1.100	1.5921	VII-438
1033	ethane (*uns*)	$(C_6H_5)_2CHCH_3$	20	1.004	1.5761	V-605
1034	ketene	$(C_6H_5)_2C:CO$	14.1	1.11	1.615	VII-471
1035	methane (m.p. 27°)	$(C_6H_5)_2CH_2$	16.8	1.01	1.5788	V-588
1036	sulfide	$(C_6H_5)_2S$	18.5	1.117	1.635	VI-299
1037	**Diphenylene oxide** (m.p. 87°)	$\underset{\rule{1.5cm}{0.4pt}}{C_6H_4\cdot C_6H_4\cdot O}$	99.3	1.089	1.6079	XVII-70
1038	**Dipropargyl**	$(CH:CCH_2\cdot)_2$	20	0.805	1.4414	I-266
1039	**Dipropyl**-amine (n)	$(C_2H_5CH_2)_2NH$	19.5	0.739	1.4046	IV-138
1040	carbimide	$C_3H_7N:C:NC_3H_7$	23	0.838	1.4454	IV-145
1041	cyanamide	$(C_2H_5CH_2)_2NCN$	24.4	0.857	1.4298	IV-144
1042	fumarate (n)	$(CHCO_2C_3H_7)_2$	20	1.022	1.4435	II-742
1043	ketone (n)	$(C_2H_5CH_2)_2CO$	22	0.82	1.4073	I-699
1044	" (*iso*)	$[(CH_3)_2CH]_2CO$	20(?)	0.806	1.417	I-703
1045	ketoxime	$(C_3H_7)_2C:NOH$	24.1	0.888	1.4472	I-700
1046	malate (l)	$(CHOHCH_2):(CO_2\cdot$ $C_3H_7)_2$	20	1.075	1.4380	III-432
1047	maleate (n)	$(CHCO_2C_3H_7)_2$	20	1.029	1.4437	II-752
1048	nitrosoamine	$(C_3H_7)_2NNO$	20	0.916	1.4446	IV-146
1049	**Dodecane** (n)	$CH_3(CH_2)_{10}CH_3$	20(?)	0.751	1.4209	I-171
1050	**Dotriacontane** (n)	$CH_3(CH_2)_{30}CH_3$	79.4	0.775	1.43	I-177
1051	**Eicosane** (n) (m.p. 38°)	$CH_3(CH_2)_{18}CH_3$	42.9	0.77	1.434	I-174
1052	**Elaidic acid** (m.p. 51°)	$C_{17}H_{33}CO_2H$	79.4	0.851	1.44	II-469
1053	**Elemicin** (1,3,4,5)	$CH_2:CHCH_2\cdot$ $C_6H_2(OH)_3$	20(?)	1.063	1.5285	VI-1131

Table 10-66 (*Continued*)
INDEX OF REFRACTION—LIQUIDS

No.	Substance	Formula	t,°C.	d_4^t	n_D^t	Beil. Ref.
1054	**Elemol** (α) (m.p. 46°)	$C_{15}H_{26}O$	20(?)	0.941	1.5030	*VI-66
1055	**Epichlorohydrin** (α)	OCH_2CHCH_2Cl ⌊___⌋	16.1	1.18	1.4397	XVII-6
1056	**Epidibromohydrin** (β)	$BrCH_2CH{:}CHBr$	25	1.995	1.5378	I-201
1057	**Erucic acid** (m.p. 33°)	$CH_3(CH_2)_7CH{:}CH{\cdot}$ $(CH_2)_{11}CO_2H$	55	0.860	1.45	II-472
1058	anhydride (m.p. 48°)	$(C_{21}H_{41}CO)_2O$	100	.....	1.4347	II-474
1059	**Estragole** (1,4)	$CH_2{:}CHCH_2C_6H_4{\cdot}$ OCH_3	17.5	0.971	1.5230	VI-571
1060	**Ethanolamine**	$H_2NCH_2CH_2OH$	20	1.022	1.4539	IV-274
1061	**Ether**	$(C_2H_5)_2O$	24.8	0.708	1.3497	I-314
1062	**Ethoxy** acetic acid	$C_2H_5OCH_2CO_2H$	20	1.102	1.4194	III-233
1063	allylphenol (p)	$C_2H_5OC_6H_4CH_2CH{:}$ CH_2	12.2	0.961	1.5179	VI-572
1064	5-bromotoluene (2)	$CH_3C_6H_3(OC_2H_5)$ Br	11.4	1.359	1.5486	*VI-176
1065	coumarone (3)	$C_6H_4C(OC_2H_5){:}CH{\cdot}O$ ⌊_____⌋	17.9	1.106	1.5477	*XVII-59
1066	1,3-dimethylben-zene (2)	$(CH_3)_2C_6H_3OC_2H_5$	13.9	0.942	1.4969	VI-485
1067	methylene-cyclo-hexan-2-one (1)	$C_9H_{14}O_2$	10.8	1.029	1.5002	*VIII-509
1068	methylene-diethyl ketone	$C_2H_5COC(CH_3){:}$ $CHOC_2H_5$	18.2	0.945	1.4717	*I-427
1069	methylene-menthone	$C_2H_5OC_{11}H_{17}O$	15.2	0.970	1.4895	*VIII-512
1070	methylene-pinacoline	$(CH_3)_3CCOCH{:}$ $CHOC_2H_5$	14.2	0.916	1.4605	*I-427
1071	styrene (α)	$C_6H_5C(OC_2H_5){:}CH_2$	17.5	0.977	1.5304	VI-563
1072	thionaphthene (3)	$C_6H_4C(OC_2H_5){:}$ ⌊_____⌋ CH·S ⌊___⌋	17.9	1.159	1.6050	*XVII-61
1073	toluene (o)	$CH_3C_6H_4OC_2H_5$	13.3	0.959	1.5079	VI-352
1074	" (m)	$CH_3C_6H_4OC_2H_5$	20	0.949	1.513	VI-376
1075	" (p)	$CH_3C_6H_4OC_2H_5$	17.6	0.8	1.5058	VI-393
1076	γ-valerolactone (δ)	$C_2H_5OCH_2CHCH_2{\cdot}$ ⌊___⌋ $CH_2CO{\cdot}O$ ⌊_____⌋	24.9	1.072	1.4408	*XVIII-297
1077	o-xylene (1,2,3)	$(CH_3)_2C_6H_3OC_2H_5$	20	0.952	1.510	VI-480
1078	" (1,2,4)	$(CH_3)_2C_6H_3OC_2H_5$	20	0.950	1.510	VI-481
1079	m-xylene (1,3,5)	$(CH_3)_2C_6H_3OC_2H_5$	20	0.939	1.505	VI-493
1080	" (1,3,4)	$(CH_3)_2C_6H_3OC_2H_5$	14	0.949	1.5069	*VI-241
1081	p-xylene	$(CH_3)_2C_6H_3OC_2H_5$	20	0.938	1.504	VI-495
1082	**Ethyl** acetate	$CH_3CO_2C_2H_5$	19.2	0.901	1.3728	II-125
1083	acetoacetate	$CH_3COCH_2CO_2C_2H_5$	20	1.025	1.4198	III-632
1084	acetoacetic ester	$CH_3COCH(C_2H_5){\cdot}$ $CO_2C_2H_5$	18.7	0.99	1.4226	III-691
1085	α'-acetothienone (α)	$CH_3COC{:}CHCH{:}C{\cdot}$ ⌊___⌋ $(C_2H_5){\cdot}S$ ⌊_____⌋	20	1.079	1.551	XVII-297
1086	acetonoxalate	$CH_3COCH_2COCO_2{\cdot}$ C_2H_5	17.6	1.128	1.4750	III-747

Table 10-66 (*Continued*)
INDEX OF REFRACTION—LIQUIDS

No.	Substance	Formula	t,°C.	d_4^t	n_D^t	Beil. Ref.
1087	**Ethyl**-acetonox- alate *O*-ethyl ether	$CH_3COCH:C(O\cdot C_2H_5)CO_2C_2H_5$	20	1.060	1.4689	III-877
1088	acetoxymethylene acetate	$CH_3CO_2CH:CHCO_2\cdot C_2H_5$	18.1	1.090	1.4474	III-370
1089	acetoxymethylene propionate	$CH_3CO_2C(CH_3):CH\cdot CO_2C_2H_5$	20	1.060	1.4447	III-373
1090	acetylacetone	$CH_3COCH(C_2H_5)CO\cdot CH_3$	16.9	0.956	1.4397	I-794
1091	β-acetylacrylate	$CH_3COCH:CHCO_2\cdot C_2H_5$	20	1.037	1.453	III-731
1092	4-acetylnaphthalene (1)	$C_2H_5C_{10}H_6COCH_3$	15.7	1.092	1.6198	....
1093	acid-*d*-camphorate (α)	$C_8H_{14}(CO_2H)(CO_2\cdot C_2H_5)$	16.8	1.102	1.4737	IX-751
1094	alcohol ‡	C_2H_5OH	20.5	0.788	1.3610	I-292
1095	allocinnamate	$C_6H_5CH:CHCO_2C_2H_5$	20	1.048	1.5451	IX-594
1096	allocinnamyl- idene acetate	$C_6H_5\cdot(CH:CH)_2\cdot CO_2C_2H_5$	20	1.043	1.619	....
1097	allomethylcam- phorate (*o*)	$C_{13}H_{22}O_4$	17.7	1.057	1.4604	....
1098	allylaceto-acetate	$CH_3COCH(CH_2CH: CH_2)CO_2C_2H_5$	17.6	0.992	1.4388	III-738
1099	allyl ether	$C_2H_5OCH_2CH:CH_2$	20	0.765	1.3881	I-438
1100	1-allyl-3-methyl- pyrazole-5-car- boxylate	$C_3H_5N\cdot C(CO_2C_2H_5):$ $CHC(CH_3):N$	20	1.046	1.489	....
1101	1-allyl-5-methyl- pyrazole-5-car- boxylate	$C_3H_5NC(CH_3):CH\cdot$ $C(CO_2C_2H_5):N$	20	1.077	1.502	....
1102	β-aminocrotonate	$CH_3C(NH_2):CH\cdot CO_2C_2H_5$	18.8	1.022	1.5007	III-654
1103	aminoethanol (β)	$C_2H_5NHCH_2CH_2OH$	20	0.914	1.444	IV-282
1104	aminonaphthalene (α)	$C_{10}H_7NHC_2H_5$	15.1	1.06	1.6477	XII-1222
1105	aminonaphthalene (β)	$C_{10}H_7NHC_2H_5$	21.3	1.06	1.6544	XII-1274
1106	*cis-p*-amino-*o,o,o'*- trimethylhexa- hydrobenzoate	$(CH_3)_3C_6H_7(NH_2)\cdot CO_2C_2H_5$	20	0.990	1.475	XIV-307
1107	*trans-p*-amino- *o,o,o'*-trimethyl- hexahydrobenz- oate	$C_{12}H_{23}O_2N$	20	0.980	1.473	XIV-308
1108	*n*-amyl ketone	$C_2H_5COC_5H_{11}$	20	0.8	1.4156	I-706
1109	*iso*-amyl malonate	$CH_2(CO_2C_2H_5)(CO_2\cdot C_5H_{11})$	20	0.95	1.4275	....
1110	amylpropiolate	$C_5H_{11}C:CCO_2C_2H_5$	25	0.910	1.4453	II-487
1111	angelate	$C_4H_7CO_2C_2H_5$	19.5	0.918	1.431	II-429
1112	aniline	$C_6H_5NHC_2H_5$	20	0.963	1.5559	XII-159
1113	anisate (*p*)	$CH_3OC_6H_4CO_2C_2H_5$	20	1.106	1.5249	X-159
1114	anthracene (9)	$C_{16}H_{14}$ (m.p. 59°)	99.2	1.041	1.6762	V-678
1115	anthranilate (*o*)	$NH_2C_6H_4CO_2C_2H_5$	20	1.117	1.5646	XIV-319

‡ See special tables.

Table 10-66 (*Continued*)
INDEX OF REFRACTION—LIQUIDS

No.	Substance	Formula	t,°C.	d_4^t	n_D^t	Beil. Ref.
1116	**Ethyl** atropate	$C_6H_5C(:CH_2)CO_2\cdot C_2H_5$	16.1	1.051	1.5261	*IX-252
1117	benzene	$C_6H_5C_2H_5$	20	0.867	1.4959	V-351
1118	benzoate	$C_6H_5CO_2C_2H_5$	18.7	1.046	1.5057	IX-110
1119	benzoic acid (*o*) (m.p. 68°)	$C_2H_5C_6H_4CO_2H$	99.6	1.042	1.5101	IX-526
1120	benzoic acid (*m*) (m.p. 47°)	$C_2H_5C_6H_4CO_2H$	100	1.042	1.5345	IX-528
1121	benzoylacetate	$C_6H_5COCH_2CO_2\cdot C_2H_5$	16	1.12	1.5312	X-674
1122	benzoylformate	$C_6H_5COCO_2C_2H_5$	25.1	1.122	1.5190	X-657
1123	benzoylformyl- acetate	$C_6H_5COCH(CHO)\cdot CO_2C_2H_5$	24.9	1.140	1.5313	....
1124	β-benzylamino- acetoacetate	$CH_3C(:NC_7H_7)CH_2\cdot CO_2C_2H_5$	20.6	1.060	1.5579	....
1125	α-benzylcinnamate (m.p. 38°)	$C_6H_5CH:C(CH_2\cdot C_6H_5)CO_2C_2H_5$	16.1	1.08	1.5885	*IX-299
1126	benzyl ether	$C_2H_5OCH_2C_6H_5$	20	0.949	1.4955	VI-431
1127	*d*-bornyl ether	$C_2H_5OC_{10}H_{17}$	24.7	0.9	1.4555	VI-78
1128	bornylenecar- boxylate	$C_{13}H_{20}O_2$	19.8	0.983	1.4760	*IX-51
1129	brassidate (m.p. 30°)	$C_{24}H_{46}O_2$	35	.....	1.4587	II-475
1130	bromide	C_2H_5Br	20	1.460	1.4239	I-88
1131	bromoacetate	$BrCH_2CO_2C_2H_5$	13	1.514	1.4542	II-214
1132	α-bromoaceto- acetate	$CH_3COCHBrCO_2\cdot C_2H_5$	14.1	1.429	1.4634	III-664
1133	γ-bromoaceto- acetate	$BrCH_2COCH_2CO_2\cdot C_2H_5$	18.1	1.528	1.4831	III-664
1134	bromobenzoate (*o*)	$BrC_6H_4CO_2C_2H_5$	20	1.439	1.543	IX-348
1135	" (*m*)	$BrC_6H_4CO_2C_2H_5$	20	1.430	1.543	IX-350
1136	" (*p*)	$BrC_6H_4CO_2C_2H_5$	20	1.430	1.547	IX-352
1137	α-bromopropionate	$CH_3CHBrCO_2C_2H_5$	20(?)	1.394	1.4469	II-255
1138	buta-1,3-diene (2)	$CH_2:C(C_2H_5)CH:CH_2$	20	0.730	1.445	....
1139	butan-1-ol (2)	$(C_2H_5)_2CHCH_2OH$	20	0.831	1.4208	I-412
1140	butyl acetate (2)	$CH_3CO_2CH_2\cdot CH(C_2H_5)_2$	20	0.879	1.4103	....
1141	*n*-butylcarbamate	$C_4H_9NHCO_2C_2H_5$	25.6	0.943	1.4278	IV-158
1142	*iso-* "	$C_4H_9NHCO_2C_2H_5$	20	0.943	1.4288	IV-168
1143	*sec-* "	$C_4H_9CH(CH_3)NH\cdot CO_2C_2H_5$	26	0.940	1.4267	IV-162
1144	*tert*-butyl ether	$C_2H_5OC(CH_3)_3$	20	0.752	1.3794	I-381
1145	*n*-butyl malonate	$CH_2(CO_2C_2H_5)(CO_2\cdot C_4H_9)$	20	0.98	1.4242	*II-528
1146	*iso-* "	$CH_2(CO_2C_2H_5)(CO_2\cdot C_4H_9)$	20	0.97	1.4248	....
1147	*sec-* "	$CH_2(CO_2C_2H_5)(CO_2\cdot C_4H_9)$	20	0.99	1.4284	....
1148	*n*-butylnitro- carbamate	$C_4H_9N(NO_2)CO_2\cdot C_2H_5$	20.7	1.099	1.4455	IV-159
1149	*iso*-butylnitro- carbamate	$C_4H_9N(NO_2)CO_2\cdot C_2H_5$	20.5	1.094	1.4433	IV-172
1150	*sec*-butylnitro- carbamate	$C_4H_9N(NO_2)CO_2\cdot C_2H_5$	21.4	1.086	1.4397	IV-163
1151	*n*-butyrate	$C_2H_5CH_2CO_2C_2H_5$	18	0.88	1.3930	II-270

Table 10-66 (*Continued*)
INDEX OF REFRACTION—LIQUIDS

No.	Substance	Formula	t,°C.	d_4^t	n_D^t	Beil. Ref.
1152	**Ethyl-*iso*-butyrate**	$(CH_3)_2CHCO_2C_2H_5$	20	0.871	1.3903	II-291
1153	*α-n*-butyro-*α*-cyanoacetate	$C_2H_5CH_2COCH\cdot(CN)CO_2C_2H_5$	20	1.056	1.4614	III-807
1154	*α-iso*-butyro-*α*-cyanoacetate	$(CH_3)_2CHCO\cdot CH(CN)CO_2C_2H_5$	20	1.054	1.4606	III-811
1155	camphor (*α*)	$C_{12}H_{20}O$	20	0.937	1.4719	VII-142
1156	camphorcarboxylate	$C_{10}H_{15}OCO_2C_2H_5$	24.3	1.053	1.4736	X-644
1157	chavibetol (1,3,4)	$CH_2{:}CHCH_2C_6H_3(O\cdot C_2H_5)(OCH_3)$	11.5	1.013	1.5276	....
1158	*iso*-chavibetol	$CH_3CH{:}CHC_6H_3(O\cdot C_2H_5)(OCH_3)$	11	1.039	1.5602	VI-958
1159	chloroacetate	$ClCH_2CO_2C_2H_5$	20	1.159	1.4216	II-197
1160	o-chlorobenzoate	$ClC_6H_4CO_2C_2H_5$	20	1.189	1.523	IX-336
1161	m- "	$ClC_6H_4CO_2C_2H_5$	20	1.181	1.520	IX-338
1162	p- "	$ClC_6H_4CO_2C_2H_5$	20	1.181	1.524	IX-340
1163	*β*-chlorobutyrate	$CH_3CHClCH_2CO_2\cdot C_2H_5$	20	1.052	1.4246	II-277
1164	o-chlorocinnamate	$ClC_6H_4CH{:}CHCO_2\cdot C_2H_5$	20	1.171	1.568	....
1165	*α*-chlorocrotonate	$CH_3CH{:}CClCO_2C_2H_5$	20.1	1.108	1.453	II-415
1166	*α*-chloro-*iso*-crotonate	$CH_3CH{:}CClCO_2C_2H_5$	20	1.100	1.453	**II-396
1167	*β*-chlorocrotonate†	$CH_3CCl{:}CHCO_2C_2H_5$	19.6	1.106	1.459	II-416
1168	*β*-chloro-*iso*-crotonate	$CH_3CCl{:}CHCO_2C_2H_5$	18.7	1.092	1.4542	II-417
1169	chloroformate	$ClCO_2C_2H_5$	20	1.138	1.3955	III-10
1170	*α*-chloropropionate	$CH_3CHClCO_2C_2H_5$	20	1.087	1.4185	II-249
1171	*β*-chloropropionate	$ClCH_2CH_2CO_2C_2H_5$	20	1.109	1.4254	II-250
1172	*β*-chloro-*α* (*p*-tolyl)-acrylate	$CH_3C_6H_4C({:}CHCl)\cdot CO_2C_2H_5$	19.7	1.136	1.5358	*IX-257
1173	cinnamate (*trans*)	$C_6H_5CH{:}CHCO_2C_2H_5$	20	1.049	1.5598	IX-581
1174	cinnamylidene-acetate (m.p. 25°)	$C_6H_5CH{:}CH\cdot CH{:}CHCO_2C_2H_5$	42.2	1.029	1.613	IX-639
1175	coumalate (m.p. 36°)	$C_2H_5O_2CC{:}CH\cdot O\cdot \underline{\qquad}|$ $COCH{:}CH$ $|\underline{\qquad}|$	39.2	1.209	1.5066	XVIII-406
1176	coumarilate (m.p. 31°)	$C_6H_4CH{:}C(CO_2\cdot \underline{\qquad}|$ $C_2H_5)\cdot O$ $|\underline{\qquad}|$	32.2	1.162	1.5619	XVIII-308
1177	crotonate (*α*)	$C_3H_5CO_2C_2H_5$	20	0.919	1.4246	II-411
1178	*iso*-crotonate	$CH_3CH{:}CHCO_2C_2H_5$	20	0.925	1.4246	II-411
1179	crotonic acid (*α*) §	$CH_3CH{:}C(C_2H_5)\cdot CO_2H$	20	0.976	1.4511	II-440
1180	" (m.p. 45°) #	$CH_3CH{:}C(C_2H_5)\cdot CO_2H$	56.1	0.948	1.4426	II-439
1181	*iso*-cyanide	C_2H_5NC	25	0.744	1.36	IV-107
1182	cyanoacetate	$NCCH_2CO_2C_2H_5$	20.5	1.062	1.4179	II-585
1183	*α*-cyanocrotonate	$CH_3CH{:}C(CN)CO_2\cdot C_2H_5$	20	1.024	1.4525	**II-655
1184	1-cyano-cyclo-hexan-2-one (1)	$C_2H_5(CN){:}$ $C\cdot CO(CH_2)_3CH_2$ $|\underline{\qquad}|$	20	1.008	1.466	....
1185	cyanoformate	$NCCO_2C_2H_5$	20	1.003	1.3821	II-547

† Ester of acid m.p. 94°.
§ For liquid variety.
# For solid variety.

Table 10-66 (*Continued*)
INDEX OF REFRACTION—LIQUIDS

No.	Substance	Formula	$t,°C.$	d_4^t	n_D^t	Beil. Ref.
1186	**Ethyl** α-cyano-Δ^1-*iso*-heptenate	$(CH_3)_2CHCH_2CH:$ $C(CN)CO_2C_2H_5$	20	0.966	1.4549	**II-665
1187	α-cyano-Δ^1-pentenate	$C_2H_5CH:C(CN)\cdot$ $CO_2C_2H_5$	20	1.000	1.4529	**II-657
1188	cyclohexane	$C_2H_5(CHC_5H_{10})$	20	0.788	1.4330	V-35
1189	cyclohex-1-ene (1)	$C_2H_5C:CH\cdot$ $\vert\_\_\_\_$ $(CH_2)_3CH_2$ $\_\_\_\_\_\vert$	19.1	0.824	1.4567	V-71
1190	cyclohex-1-ene acetate	$C_6H_9CH_2CO_2C_2H_5$	16.2	0.989	1.4691	*IX-23
1191	α-cyclohexene-propionate	$CH_2(CH_2)_3CH:C\cdot$ $\vert_____\vert$ $CH(CH_3)CO_2\cdot$ C_2H_5	17.5	0.968	1.4646	IX-51
1192	cyclohexylidene-acetate	$CH_2(CH_2)_4C:CHCO_2\cdot$ $\vert_____\vert$ C_2H_5	16.9	0.983	1.4808	*IX-24
1193	cyclopentane	$C_2H_5(CHC_4H_8)$	20	0.766	1.4198	**V-19
1194	cyclopentan-1-ol-1-carboxylate	$CH_2(CH_2)_3COHCO_2\cdot$ $\vert_____\vert$ C_2H_5	16	1.060	1.4516	*X-3
1195	cymenc (2)	$C_6H_3(CH_3)(C_2H_5)\cdot$ $(C_3H_7)(1,2,4)$	15.6	0.871	1.4988	V-448
1196	*n*-decyl ether	$C_2H_5OC_{10}H_{21}$	20	0.794	1.4223	....
1197	diacetoacetate	$(C_2H_3O)_2CHCO_2\cdot$ C_2H_5	18.3	1.08	1.4695	III-751
1198	diallylacetate	$(CH_2:CHCH_2)_2CH\cdot$ $CO_2C_2H_5$	19.1	0.897	1.4364	II-489
1199	diallylacetoacetate	$CH_3COC(CO_2C_2H_5):$ $(CH_2CH:CH_2)_2$	17.4	0.982	1.4572	III-741
1200	diazoacetate	$N_2CHCO_2C_2H_5$	17.6	1.085	1.4588	*III-211
1201	dibromoacetate	$Br_2CHCO_2C_2H_5$	12.5	1.91	1.5017	II-219
1202	2,7-dibromofluo-rene-9-acetate (m.p. 87°)	$C_{17}H_{14}O_2Br_2$	99.8	1.512	1.6130	....
1203	α,β-dibromo-valerate	$C_2H_5CHBrCHBr\cdot$ $CO_2C_2H_5$	20	1.613	1.496	**II-269
1204	dichloroacetate	$Cl_2CHCO_2C_2H_5$	20	1.282	1.4386	II-203
1205	γ,γ-dichloro-crotonate	$Cl_2CHCH:CHCO_2\cdot$ C_2H_5	20	1.229	1.4619	**II-397
1206	γ,γ-dichloro-α,β-dibromo-butyrate	$Cl_2CH(CHBr)_2CO_2\cdot$ C_2H_5	20	1.830	1.5235	**II-257
1207	α,β-dichloro-propionate	$CH_2ClCHClCO_2C_2H_5$	20	1.246	1.4482	II-252
1208	β,β-dichloro-α (*p*-tolyl)-propionate	$Cl_2CHCH(C_6H_4\cdot$ $CH_3)CO_2C_2H_5$	18.7	1.202	1.5214	*IX-214
1209	α,α-diethoxy-propionate	$CH_3C(OC_2H_5)_2CO_2\cdot$ C_2H_5	17.1	0.979	1.4151	III-617
1210	α,α-diethyl-(*o*-methoxy-benzoyl) acetate	$(C_2H_5)_2C(CO_2C_2H_5)\cdot$ $COC_6H_4OCH_3$	20	1.086	1.5140	....

Table 10-66 (*Continued*)
INDEX OF REFRACTION—LIQUIDS

No.	Substance	Formula	t,°C.	d_4^t	n_D^t	Beil. Ref.
1211	**Ethyl** Δ^1-dihydro-α-naphthoate (*stable*)	$C_6H_4CH_2CH_2CH:C\cdot$ $\lfloor$ _____ $\rfloor$ $CO_2C_2H_5$	20	1.102	1.559	X-642
1212	Δ^2-dihydro-α-naphthoate (*labile*)	$C_{10}H_9CO_2C_2H_5$	20	1.089	1.535	....
1213	Δ^1-dihydro-β-naphthoate	$C_6H_4CH_2CH_2C(CO_2\cdot$ $\rule{2cm}{0.4pt}$ $C_2H_5:CH$ $\rule{1cm}{0.4pt}\rfloor$	20	1.093	1.578	....
1214	Δ^3-dihydro-β-naphthoate	$C_{10}H_9CO_2C_2H_5$	20	1.085	1.551	....
1215	4,6-dimethyl-3-acetoxycoumarone (2)	$(CH_3)_2C_6H_2C(O_2C\cdot$ $\rule{2cm}{0.4pt}$ $CH_3):C(C_2H_5)\cdot O$ $\rule{2.5cm}{0.4pt}\rfloor$	20	1.078	1.5303	....
1216	β,β-dimethyl-acrylate	$(CH_3)_2C:CHCO_2\cdot$ C_2H_5	20	0.913	1.435	II-433
1217	α,β-dimethyl-cinnamate	$C_6H_5C(CH_3):C:$ $(CH_3)CO_2C_2H_5$	15.2	1.020	1.5212	*IX-260
1218	2,5-dimethyl-coumaran-3-one-2-carboxylate	$CH_3C_6H_3COC(CH_3)\cdot$ $\rule{2cm}{0.4pt}$ $(CO_2C_2H_5)\cdot O$ $\rule{2.5cm}{0.4pt}\rfloor$	20	1.161	1.536	*XVIII-492
1219	3,5-dimethyl-coumarilate (m.p. 55°)	$CH_3C_6H_3C(CH_3):C\cdot$ $\rule{2cm}{0.4pt}$ $(CO_2C_2H_5)\cdot O$ $\rule{2.5cm}{0.4pt}\rfloor$	59	1.093	1.5464	XVIII-310
1220	α-(1,3-dimethyl-cyclohexa-3,5-diene)-5-*iso*-butyrate	$(CH_3)_2C_6H_5C:$ $(CH_3)_2(CO_2C_2H_5)$	13.4	0.969	1.4895	*IX-52
1221	1,3-dimethyl-cyclohexan-5-ol-5-acetate	$(CH_3)_2C_6H_8(OH)\cdot$ $CH_2CO_2C_2H_5$	14.9	0.990	1.4566	*X-12
1222	1,3-dimethyl-cyclohex-4-ene-5-acetate	$(CH_3)_2C_6H_7CH_2CO_2\cdot$ C_2H_5	20.3	0.938	1.458	*IX-35
1223	1,3-dimethyl-cyclohex-3-ene-5-hydroxy-5-acetate	$(CH_3)_2C_6H_6(OH)\cdot$ $CH_2CO_2C_2H_5$	17.6	1.012	1.4713	*X-14
1224	3,5-dimethyl-cyclohexylidene-5-acetate	$(CH_3)_2C_6H_8:CHCO_2\cdot$ C_2H_5	20.3	0.937	1.4575	*IX-35
1225	α-(1,3-dimethyl-cyclohexylidene)-5-propionate	$(CH_3)_2C_6H_8:C:$ $(CH_3)CO_2C_2H_5$	20	0.969	1.5069	....
1226	dimethyloxamate	$(CH_3)_2NCOCO_2C_2H_5$	22.5	1.074	1.4414	IV-61
1227	β,δ-dimethyl-sorbate	$(CH_3)_2C:CHC(CH_3)$ $:CHCO_2C_2H_5$	18.5	0.934	1.488	II-490
1228	1,4-diphenyl-but-3-ene (1)	$C_2H_5CH(C_6H_5)CH_2\cdot$ $CH:CHC_6H_5$	20	0.992	1.5880	*V-317
1229	erucate	$C_{21}H_{41}CO_2C_2H_5$	20	0.865	1.4543	II-473

Table 10-66 (*Continued*)
INDEX OF REFRACTION—LIQUIDS

No.	Substance	Formula	t,°C.	d_4^t	n_D^t	Beil. Ref.
1230	**Ethyl**-ethenyl-*o*-phenylene-amidine	$C_6H_4 \cdot N:C(CH_3) \cdot N \cdot$ $\underline{\qquad\qquad}$ C_2H_5	19.8	1.073	1.5802	XXIII–145
1231	α-ethoxyacrylate	$CH_2:C(OC_2H_5) CO_2 \cdot$ C_2H_5	17.2	0.996	1.4324	III–370
1232	β-ethoxyacrylate	$C_2H_5OCH:CHCO_2 \cdot$ C_2H_5	18.1	0.995	1.448	III–370
1233	β-ethoxy-β,n-amylacrylate	$C_5H_{11}C(OC_2H_5):CH \cdot$ $CO_2C_2H_5$	23.2	0.931	1.4518	III–382
1234	β-ethoxy-cinnamate	$C_6H_5C(OC_2H_5):CH \cdot$ $CO_2C_2H_5$	16	1.074	1.5336	X–301
1235	3-ethoxycoumar-ilate	$C_6H_4C(OC_2H_5):C \cdot$ $\underline{\qquad\qquad}$ $(CO_2C_2H_5) \cdot O$ $\underline{\qquad\quad}$	14.3	1.168	1.5592	*XVIII–457
1236	β-ethoxycrotonate (m.p. 31°)	$CH_3C(OC_2H_5):CH \cdot$ $CO_2C_2H_5$	37.6	0.966	1.4468	III–373
1237	β-ethoxy-α-methylacrylate	$C_2H_5OCH:C(CH_3) \cdot$ $CO_2C_2H_5$	21.6	0.979	1.4501	III–377
1238	α-ethoxymethyl-eneacetoacetate	$C_2H_5OCH:C(CO \cdot$ $CH_3) CO_2C_2H_5$	16.9	1.071	1.4762	III–878
1239	ethylbenzoyl-acetate	$C_6H_5COCH(C_2H_5) \cdot$ $CO_2C_2H_5$	20	1.066	1.5082	X–710
1240	α-ethylcinnamate	$C_6H_5CH:C(C_2H_5) \cdot$ $CO_2C_2H_5$	13.8	1.025	1.5425	*IX–259
1241	2-ethylcoumaran-3-one-2-carboxyl-ate	$C_6H_4COC(C_2H_5) \cdot$ $\underline{\qquad\qquad}$ $(CO_2C_2H_5) \cdot O$ $\underline{\qquad\quad}$	18.6	1.155	1.5366	*XVIII–491
1242	-ethylcrotonate†	$CH_3CH:C(C_2H_5) \cdot$ $CO_2C_2H_5$	20	0.899	1.428	II–440
1243	" ‡	$CH_3CH:C(C_2H_5) \cdot$ $CO_2C_2H_5$	20	0.908	1.435	II–440
1244	α-ethyl-β-hydroxycarb-ethoxy-α,β-crotonate	$C_2H_5O_2C \cdot O \cdot C(CH_3):$ $C(C_2H_5) CO_2C_2H_5$	20	1.056	1.4431	III–380
1245	1-ethylindazole-3-carboxylate	$C_2H_5N \cdot C_6H_4 \cdot C(CO_2 \cdot$ $\underline{\qquad\qquad}$ $C_2H_5):N$ $\underline{\qquad\quad}$	17.3	1.143	1.566	....
1246	2-ethylindazole-3-carboxylate	$\underline{\qquad\qquad}$ $N \cdot C_6H_4 \cdot C(CO_2 \cdot$ $\underline{\qquad\qquad}$ $(C_2H_5) \cdot NC_2H_5$ $\underline{\qquad\quad}$	22.8	1.129	1.571	....
1247	ethyloximino-cyanoacetate	$NCC(:NOC_2H_5) \cdot$ $CO_2C_2H_5$	20	1.082	1.4530	III–775
1248	ethyl-*iso*-phorone carboxylate	$(CH_3)_3C_6H_3(OC_2H_5) \cdot$ $CO_2C_2H_5$	19	1.011	1.4829	X–37
1249	α-ethylsorbate	$CH_3CH:CHCH:C:$ $(C_2H_5) CO_2C_2H_5$	20	0.931	1.494	**II–455
1250	ethylsulfonate	$C_2H_5SO_2OC_2H_5$	22	1.145	1.4196	IV–6
1251	ethylidene-acetoacetate	$CH_3COC(CO_2C_2H_5):$ $CHCH_3$	18	1.023	1.4528	III–736
1252	eugenyl ether	$C_2H_5OC_{10}H_{11}O$	18.3	1.014	1.53	VI–964

† For ester of liquid variety.
‡ For ester of the solid acid variety.

Table 10-66 (*Continued*)
INDEX OF REFRACTION—LIQUIDS

No.	Substance	Formula	t,°C.	d_4^t	n_D^t	Beil. Ref.
1253	**Ethyl** fluoroacetate	$FCH_2CO_2C_2H_5$	20	1.093	1.3767	II-193
1254	formate	$HCO_2C_2H_5$	20	0.906	1.3598	II-19
1255	formylcamphor	$C_{13}H_{20}O_2$	17.9	1.004	1.5047	....
1256	furoate (m.p. 34°)	$C_4H_3OCO_2C_2H_5$	40	1.097	1.4699	XVIII-275
1257	furylacrylate	$C_4H_3OCH:CHCO_2\cdot$ C_2H_5	20	1.09	1.5286	XVIII-300
1258	glycinate	$H_2NCH_2CO_2C_2H_5$	20	1.028	1.4242	IV-340
1259	glycol ether	$HOCH_2CH_2OC_2H_5$	20	0.930	1.4080	I-467
1260	heptane (3)	$(C_2H_5)_2CH(CH_2)_3\cdot$ CH_3	20	0.727	1.4092	....
1261	" (4)	$C_2H_5CH(CH_2C_2H_5)_2$	20	0.730	1.4109	I-167
1262	α,β-*iso*-heptenate	$(CH_3)_2CHCHCH_2CH:$ $CHCO_2C_2H_5$	20	0.889	1.436	**II-411
1263	heptylate	$C_6H_{13}CO_2C_2H_5$	15	0.88	1.4144	II-340
1264	hexahydrobenzoate	$C_6H_{11}CO_2C_2H_5$	16	0.962	1.452	IX-8
1265	hexan-1-al (2)	$C_4H_9CH(C_2H_5)CHO$	20	0.823	1.4160	I-707
1266	hexane (3)	$(C_2H_5)_2CHCH_2C_2H_5$	20	0.714	1.4016	*I-62
1267	hexan-1-ol (2)	$C_4H_9CH(C_2H_5)\cdot$ CH_2OH	20	0.833	1.4300	....
1268	hexan-3-ol (3)	$(C_2H_5)_2COHCH_2\cdot$ C_2H_5	13	0.84	1.4322	I-421
1269	hexan-1-ol acetate (2)	$C_4H_9CH(C_2H_5)\cdot$ $CH_2O_2CCH_3$	20	0.871	1.4204	....
1270	hexan-1-ol butyrate (2)	$C_3H_7CO_2C_8H_{17}$	20	0.87	1.4266	....
1271	hexan-1-ol formate (2)	$HCO_2CH_2CH(C_2H_5)\cdot$ C_4H_9	20	0.88	1.4232	....
1272	hexan-1-ol phthalate (2)	$C_6H_4(CO_2C_8H_{17})_2$	20	0.98	1.4867	....
1273	hexan-1-ol pro-pionate (2)	$C_2H_5CO_2C_8H_{17}$	20	0.87	1.4238	....
1274	α,β-hexenate	$CH_3(CH_2)_2CH:CH\cdot$ $CO_2C_2H_5$	20	0.901	1.434	II-434
1275	α,β-*iso*-hexenate	$(CH_3)_2CHCH:CH\cdot$ $CO_2C_2H_5$	20	0.896	1.432	II-439
1276	hex-1-ene (3)	$CH_2:CHCH(C_2H_5)\cdot$ $CH_2C_2H_5$	20	0.715	1.407	....
1277	hex-2-ene (3) (*cis?*)	$CH_3CH:C(C_2H_5)\cdot$ $CH_2C_2H_5$	20	0.737	1.424	I-222
1278	" (*trans*)	$CH_3CH:C(C_2H_5)\cdot$ $CH_2C_2H_5$	20	.....	....	....
1279	hex-3-ene (3)	$(C_2H_5)_2C:CHC_2H_5$	20	0.729	1.418	....
1280	hex-4-ene (3) (*cis*)	$CH_3CH:CHCH:$ $(C_2H_5)_2$	20	0.725	1.412	....
1281	hex-5-ene (3)	$CH_2:CHCH_2CH:$ $(C_2H_5)_2$	20	0.726	1.412	....
1282	hex-2-ene-1-al (2)	$C_2H_5CH_2CH:$ $C(C_2H_5)CHO$	20	0.848	1.4535	I-744
1283	hex-5-ene-3-ol (3)	$CH_2:CHCH_2COH:$ $(C_2H_5)_2$	21.7	0.847	1.4410	I-449
1284	Δ^2-hexenic acid (α)	$CH_3CH_2CH:CHCH:$ $(C_2H_5)CO_2H$	20	0.928	1.4421	**II-414
1285	*n*-hexylcarbamate	$C_6H_{13}NHCO_2C_2H_5$	26.8	0.923	1.4341	IV-188
1286	hexylpropiolate	$C_6H_{13}C:CCO_2C_2H_5$	25	0.903	1.4464	II-490
1287	hydracrylate	$HOCH_2CH_2CO_2C_2H_5$	23	1.07	1.4271	III-297

Table 10-66 (*Continued*)
INDEX OF REFRACTION—LIQUIDS

No.	Substance	Formula	$t,°C.$	d_4^t	n_D^t	Beil. Ref.
1288	**Ethyl** hydrocinnamate	$C_6H_5(CH_2)_2CO_2\cdot$ C_2H_5	20	1.015	1.4954	IX-511
1289	hydrocoumarilate	$C_6H_4CH_2CH(CO_2\cdot$ $\underline{\quad\quad\quad}$ $C_2H_5)\cdot O$ $\underline{\quad\quad}\rfloor$	13.9	1.154	1.5250	XVIII-305
1290	hydrogen fumarate (m.p. 70°)	$HO_2CCH\colon CHCO_2\cdot$ C_2H_5	83	1.115	1.4409	II-741
1291	hydrogen malonate	$CH_2(CO_2H)(CO_2\cdot$ $C_2H_5)$	20	1.176	1.4271	II-573
1292	hydrogen phthalate	$C_6H_4(CO_2C_2H_5)\cdot$ (CO_2H) (*o*)	22	.....	1.5088	IX-797
1293	hydrogen pyrotartrate	$CH_3CH(CO_2C_2H_5)\cdot$ CH_2CO_2H §	20.2	1.098	1.4312	II-639
1294	β-hydroxy-carbethoxy-α, β-crotonate	$C_2H_5O_2C\cdot O\cdot C(CH_3)\colon$ $CHCO_2C_2H_5$	20	1.093	1.4435	III-374
1295	hydroxymethylene-acetoacetate	$CH_3COC(\colon CHOH)\cdot$ $CO_2C_2H_5$	16.9	1.134	1.4738	III-749
1296	α-hydroxy-methylene-propionate	$HOCH\colon C(CH_3)CO_2\cdot$ C_2H_5	20	1.024	1.4270	III-669
1297	hydroxylamine (m.p. 59°)	C_2H_5NHOH	63.9	0.9	1.4152	IV-535
1298	indazole (1)	$C_2H_5N\cdot C_6H_4CH\colon N$ $\underline{\quad\quad\quad}\rfloor$	15.7	1.07	1.5747	....
1299	" (2)	$C_2H_5N\colon N\colon C_6H_4\colon CH$ $\underline{\quad\quad\quad}\rfloor$ $\lceil\underline{\quad}\rfloor$	20	1.073	1.591	XXIII-124
1300	indazole-2-carboxylate	$N\cdot C_6H_4\cdot CH\cdot N\cdot CO_2$ $\lfloor\underline{\quad\quad\quad}\rfloor$ C_2H_5	13.2	1.206	1.569	....
1301	indene-β-carb-oxylate (m.p. 50°)	$C_6H_4CH\colon C(CO_2\cdot$ $\underline{\quad\quad\quad}$ $C_2H_5)CH_2$ $\underline{\quad\quad}\rfloor$	73.9	1.054	1.5481	*IX-268
1302	iodide	CH_3CH_2I	18.5	1.93	1.5133	I-96
1303	iodoacetate	$ICH_2CO_2C_2H_5$	12.7	1.817	1.5079	II-222
1304	1-keto-1,2,3,4-tetra-hydronaph-thalene (7)	$C_2H_5C_6H_3COCH_2\cdot$ $\underline{\quad\quad\quad}$ CH_2CH_2 $\underline{\quad}\rfloor$	17.2	1.056	1.5599	....
1305	lactate	$CH_3CHOHCO_2C_2H_5$	20	1.031	1.4125	III-280
1306	laurate	$CH_3(CH_2)_{10}CO_2\cdot$ C_2H_5	12.9	0.868	1.44	II-361
1307	*iso*-lauronolate	$CH_2C(CH_3)_2CH\cdot$ $\underline{\quad\quad\quad}$ $(CH_3)CH\colon CCO_2\cdot$ $\underline{\quad}\rfloor$ C_2H_5	20	0.953	1.4650	IX-58
1308	levulinate	$CH_3CO(CH_2)_2CO_2\cdot$ C_2H_5	16.2	1.017	1.4242	III-675
1309	2-menthatriene (*p*)	$C_{12}H_{18}$¶	18	0.886	1.5041	V-448
1310	menthyl ether	$C_2H_5OC_{10}H_{19}$	17.1	0.86	1.4435	VI-31
1311	mercaptan	C_2H_5SH	20	0.839	1.4306	I-340

§ Or $CH_3CH(CO_2H)CH_2CO_2C_2H_5$
¶ Probably a mixture of isomers.

Table 10-66 (*Continued*)
INDEX OF REFRACTION—LIQUIDS

No.	Substance	Formula	$t,°C.$	d_4^t	n_D^t	Beil. Ref.
1312	**Ethyl-**mesitylene	$(CH_3)_3C_6H_2C_2H_5$ (1,3,5,2)	16.4	0.889	1.5127	V-442
1313	methoxybenzoate (*o*)	$CH_3OC_6H_4CO_2C_2H_5$	14.6	1.116	1.5238	X-74
1314	methoxybenzoate (*m*)	$CH_3OC_6H_4CO_2C_2H_5$	16.4	1.103	1.5170	X-139
1315	*p*-methoxycin-namate (m.p. 49°)	$CH_3OC_6H_4CH:CH\cdot CO_2C_2H_5$	59.2	1.066	1.5619	X-299
1316	3-methoxycoumar-ilate (m.p. 59°)	$C_6H_4C(OCH_3):C\cdot$ $\overline{\quad\quad}$ $(CO_2C_2H_5)\cdot O$ $\overline{\quad\quad}$	70	1.160	1.5499	*XVIII-456
1317	1-methoxy-2-naphthoate	$CH_3OC_{10}H_6CO_2C_2H_5$	20	1.146	1.588	. . . .
1318	3-methoxy-2-naphthoate	$CH_3OC_{10}H_6CO_2C_2H_5$ (3,2)	20	1.159	1.560	. . . .
1319	3-methoxy-3-phthalide	$C_6H_4C(C_2H_5)(O\cdot$ $\overline{\quad\quad}$ $CH_3)OCO$ $\overline{\quad\quad}$	20.4	1.132	1.5199	*XVIII-305
1320	methylaceto-acetate	$CH_3COCH(CH_3)CO_2\cdot$ C_2H_5	17.8	1.02	1.4207	III-679
1321	5-methyl-3-acetoxycoumar-one (2)	$CH_3C_6H_3C(O_2C\cdot$ $\overline{\quad\quad}$ $CH_3):C(C_2H_5)\cdot O$ $\overline{\quad\quad}$	20	1.107	1.5343	*XVII-68
1322	methylcinnamate (*o*)	$CH_3C_6H_4CH:CH\cdot$ $CO_2C_2H_5$	16.3	1.043	1.5555	*IX-256
1323	methylcinnamate (*p*)	$CH_3C_6H_4CH:CH\cdot$ $CO_2C_2H_5$	15.9	1.034	1.5630	IX-617
1324	methylcinnamate (*α*)	$C_6H_5CH:C(CH_3)\cdot$ $CO_2C_2H_5$	20.6	1.032	1.5475	IX-616
1325	methylcinnamate (*β*)	$C_6H_5C(CH_3):CHCO_2\cdot$ C_2H_5	16.6	1.039	1.5456	IX-614
1326	methylcinnamyl-ideneacetate (*α*)	$C_6H_5CH:CH\cdot CH:$ $C(CH_3)CO_2C_2H_5$	20	1.041	1.615	. . . .
1327	methylcinnamyl-ideneacetate (*β*)	$C_6H_5CH:CH\cdot C(CH_3):$ $CHCO_2C_2H_5$	20	1.043	1.602	. . . .
1328	methylcinnamyl-ideneacetate (*γ*)	$C_6H_5CH:C(CH_3)\cdot$ $CH:CHCO_2C_2H_5$	20	1.037	1.601	. . . .
1329	2-methyl-coumar-ane-5-carboxyl-ate	$C_2H_5O_2CC_6H_3CH_2\cdot$ $\overline{\quad\quad}$ $CH(CH_3)\cdot O$ $\overline{\quad\quad}$	14.3	1.118	1.5363	*XVIII-442
1330	2-methyl-coumar-an-3-one-2-car-boxylate	$C_6H_4COC(CH_3)\cdot$ $\overline{\quad\quad}$ $(CO_2C_2H_5)\cdot O$ $\overline{\quad\quad}$	16.7	1.192	1.5390	*XVIII-491
1331	3-methylcoumar-ilate (m.p. 51°)	$C_6H_4C(CH_3):C(CO_2\cdot$ $\overline{\quad\quad}$ $C_2H_5)\cdot O$ $\overline{\quad\quad}$	56.6	1.116	1.5482	XVIII-309
1332	6-methylcoumar-ilate (m.p. 42°)	$CH_3C_6H_3CH:C(CO_2\cdot$ $\overline{\quad\quad}$ $C_2H_5)\cdot O$ $\overline{\quad\quad}$	43.9	1.119	1.5558	*XVIII-443

Table 10-66 (*Continued*)
INDEX OF REFRACTION—LIQUIDS

No.	Substance	Formula	t,°C.	d_4^t	n_D^t	Beil. Ref.
1333	**Ethyl** 1-methylcyclo-hex-1-ene-3-methenecar-boxylate	$CH_3C_6H_7:CHCO_2 \cdot C_2H_5$	20	0.998	1.5319	IX-82
1334	1-methylcyclo-hex-(1 or 2)-ene-2-acetate	$(CH_3) C_6H_8CH_2CO_2 \cdot C_2H_5$	18.4	0.948	1.4539	*IX-29
1335	1-methylcyclo-hex-(2 or 3)-ene-3-acetate	$(CH_3) C_6H_8CH_2CO_2 \cdot C_2H_5$	15.2	0.951	1.4559	*IX-30
1336	2-methylcyclo-hexylidene-1-acetate	$(CH_3) C_6H_9:CHCO_2 \cdot C_2H_5$	14.8	0.959	1.4791	*IX-29
1337	3-methylcyclo-hexylidene-1-acetate	$CH_3C_6H_9:CHCO_2 \cdot C_2H_5$	15	**0.957**	1.4773	*IX-30
1338	4-methylcyclo-hexylidene-1-acetate	$CH_3C_5H_9C:CHCO_2 \cdot C_2H_5$	17	0.994	1.4738	*IX-31
1339	1-methyl-1-dichloromethyl-cyclohexa-2,5-diene-4-methene-carboxylate	$(CH_3) (CHCl_2) C_6H_4: CHCO_2C_2H_5$	22.1	1.215	1.5691	*IX-214
1340	4-methyl-4-dichloromethyl-cyclohex-1-ene-3-one (1)	$(C_2H_5) (CH_3) C_5H_5 \cdot COCHCl_2$	17	1.196	1.5233	VII-87
1341	3-methyl-4-ethoxybenzoate	$(CH_3) (C_2H_5O) C_6H_3 \cdot CO_2C_2H_5$	20	1.057	1.5169	*X-98
1342	5-methyl-3-ethoxycoumar-ilate (m.p. 47°)	$CH_3C_6H_3C(OC_2H_5): \\ \ \underline{\quad\quad\quad} \\ C(CO_2C_2H_5) \cdot O \\ \underline{\quad\quad\quad}$	52.6	1.104	1.5404	*XVIII-460
1343	methylethyl-acetoacetate	$CH_3COC(CH_3) \cdot (C_2H_5) CO_2C_2H_5$	17.7	0.973	1.426	III-703
1344	methylfurazan	$C_2H_5C:N \cdot O \cdot N:CCH_3 \\ \underline{\quad\quad\quad}$	18.8	1.018	1.4326	XXVII-565
1345	α-methyl-β-hydroxy-car-bethoxy-α,β-cro-tonate	$C_2H_5O_2COC(CH_3): \\ C(CH_3) CO_2C_2H_5$	20	1.078	1.4398	III-378
1346	1-methylindene-2-carboxylate (m.p. 38°)	$CH_3C_9H_6CO_2C_2H_5$	39.1	1.074	1.5609	IX-644
1347	α-methyl-p-methoxycin-namate	$CH_3OC_6H_4CH:C: \\ (CH_3) (CO_2C_2H_5)$	15.6	1.089	1.5701	*X-138
1348	5-methyl-3-methoxycoumar-ilate (m.p. 29°)	$CH_3C_6H_3C(OCH_3): \\ \ \underline{\quad\quad\quad} \\ C(CO_2C_2H_5) \cdot O \\ \underline{\quad\quad\quad}$	23.8	1.170	1.5628	*XVIII-460
1349	α-methyl-β-naphthoate	$CH_3C_{10}H_6CO_2C_2H_5$	20	1.113	1.5950	....

Table 10-66 (*Continued*)
INDEX OF REFRACTION—LIQUIDS

No.	Substance	Formula	t,°C.	d_4^t	n_D^t	Beil. Ref.
1350	**Ethyl** Δ^1-α-methyl-β-naphthoate	$C_6H_4CH_2CH_2C(CO_2\cdot$ $\underline{\qquad}$ $C_2H_5):CCH_3$ $\underline{\qquad}$	20	1.087	1.572	....
1351	methylnitro-carbamate	$CH_3N(NO_2)CO_2\cdot$ C_2H_5	23	1.229	1.4483	IV-86
1352	methylnitroso-carbamate	$CH_3N(NO)CO_2\cdot$ C_2H_5	20	1.122	1.4363	IV-85
1353	methyl-β-phenylglycidate	$C_6H_5C(CH_3)CH\cdot O\cdot$ $\underline{\qquad}$ $CO_2C_2H_5$	20	1.091	1.5045	XVIII-306
1354	methyl-*iso*-phor-onecarboxylate	$(CH_3)_3C_6H_3(OCH_3)\cdot$ $CO_2C_2H_5$	19	1.025	1.4818	*X-18
1355	5-methyl-2-*iso*-propylcyclohexa-1,5-diene-1-carboxylate	$(CH_3)(C_3H_7)C_6H_5\cdot$ $CO_2C_2H_5$	16.9	0.977	1.5001	*IX-49
1356	1-methyl-3-*iso*-propylcyclopen-tane-2-car-boxylate	$C_9H_{17}CO_2C_2H_5$	11.8	0.918	1.44	*IX-17
1357	α-methylsorbate	$CH_3CH:CHCH:$ $C(CH_3)CO_2C_2H_5$	20	0.947	1.498	....
1358	γ-methylsorbate	$CH_3CH:C(CH_3)CH:$ $CHCO_2C_2H_5$	20	0.946	1.499	**II-454
1359	5-methyltetrahy-droindazole (1)	$CH_3C_6H_7:[CH:N\cdot N\cdot$ $C_2H_5]$	20	0.985	1.502	....
1360	5-methyltetrahy-droindazole (2)	$CH_3C_6H_7[:CH\cdot N\cdot$ $(C_2H_5)\cdot N:]$	20	0.987	1.500	....
1361	7-methyltetrahy-droindazole (2)	$C_2H_5N\cdot CH:C_6H_7\cdot$ $\underline{\qquad}$ $(CH_3):N$ $\underline{\qquad}$	20	0.990	1.507	....
1362	naphthoate (α)	$C_{10}H_7CO_2C_2H_5$	20	1.122	1.594	IX-648
1363	" (β) (m.p. 26°)	$C_{10}H_7CO_2C_2H_5$	22.7	1.114	1.5951	IX-657
1364	naphthyl ether (α)	$C_2H_5OC_{10}H_7$	20(?)	1.061	1.5992	VI-606
1365	" " (β) (m.p. 37°)	$C_2H_5OC_{10}H_7$	35.6	1.05	1.5975	VI-641
1366	nitrate	$C_2H_5ONO_2$	21.5	1.104	1.3848	I-329
1367	nona-1,8-diene-5-ol (5)	$[CH_2:CHCH_2CH_2]_2:$ $COHC_2H_5$	20.7	0.866	1.4623	*I-239
1368	α,β-nonenate	$C_6H_{13}CH:CHCO_2\cdot$ C_2H_5	20	0.889	1.442	II-453
1369	orthocarbonate	$C(OC_2H_5)_4$	18.5	0.919	1.3935	III-5
1370	orthoformate	$HC(OC_2H_5)_3$	18.8	0.897	1.3922	II-20
1371	palmitate	$CH_3(CH_2)_{14}CO_2C_2H_5$	34.3	.....	1.4347	II-372
1372	pentane (3)	$(C_2H_5)_3CH$	20	0.698	1.3934	I-157
1373	pentan-3-ol (3)	$(C_2H_5)_3COH$	20	0.839	1.4314	I-417
1374	α,β-pentenate	$CH_3CH_2CH:CHCO_2\cdot$ C_2H_5	20	0.909	1.431	*II-191
1375	pent-1-ene (3)	$CH_2:CHCH(C_2H_5)_2$	20	0.699	1.398	**I-198
1376	pent-2-ene (3)	$(C_2H_5)_2C:CHCH_3$	20	0.722	1.4143	I-200
1377	phenylacetate	$C_6H_5CH_2CO_2C_2H_5$	18.5	1.034	1.4992	IX-434
1378	γ-phenylbutyrate	$C_6H_5(CH_2)_3CO_2\cdot$ C_2H_5	20	1.001	1.494	*IX-211

Table 10-66 (*Continued*)
INDEX OF REFRACTION—LIQUIDS

No.	Substance	Formula	$t,°C.$	d_4^t	n_D^t	Beil. Ref.
1379	**Ethyl** phenylcarbamate (m.p. 52°)	$C_6H_5NHCO_2C_2H_5$	30.4	1.106	1.5376	XII-320
1380	phenyl carbinol	$C_2H_5CHOHC_6H_5$	20	0.996	1.5200	VI-502
1381	α-phenylcinnamate (m.p. 28°)	$C_6H_5CH:C(C_6H_5)\cdot CO_2C_2H_5$	18.6	1.097	1.5972	*IX-294
1382	α-(β-phenylethyl)-acetoacetate	$C_6H_5(CH_2)_2CH(CO\cdot CH_3)CO_2C_2H_5$	20	1.048	1.498	*X-339
1383	phenylhydrazine (α,α)	$C_6H_5N(C_2H_5)NH_2$	20.8	1.018	1.571	XV-119
1384	phenylhydrazine (α,β)	$C_6H_5NHNHC_2H_5$	15	1.004	1.55	XV-120
1385	phenyl ketone	$C_2H_5COC_6H_5$	20	1.011	1.5270	VII-300
1386	phenylpropiolate	$C_6H_5C:CCO_2C_2H_5$	13	1.063	1.5566	IX-634
1387	*iso*-phoronecarboxylate	$O:C_6H_4(CH_3)_3CO_2\cdot C_2H_5$	20.4	1.029	1.4788	*X-303
1388	piperidine (*N*)	$C_5H_{10}NC_2H_5$	18.9	0.826	1.4445	XX-17
1389	" (β)	$C_7H_{15}N$	23.2	0.857	1.4531	XX-106
1390	propargyl ether	$CH:CCH_2OC_2H_5$	20	0.833	1.4039	I-454
1391	α-propiocyanoacetate	$C_2H_5COCH(CN)\cdot CO_2C_2H_5$	20	1.076	1.4603	III-800
1392	propiolate	$CH:CCO_2C_2H_5$	15	0.968	1.42	II-477
1393	propionate	$C_2H_5CO_2C_2H_5$	20	0.891	1.3839	I-240
1394	β-*n*-propyl-acrolein (α)	$C_2H_5CH_2CH:C:(C_2H_5)CHO$	20	0.848	1.4535	I-744
1395	*n*-propyl ether	$C_2H_5OCH_2C_2H_5$	20	0.739	1.3695	I-354
1396	*n*-propyl ketone	$C_2H_5COCH_2C_2H_5$	22	0.813	1.3990	I-690
1397	α-*iso*-propylidene-acetoacetate	$CH_3COC(CO_2C_2H_5):C(CH_3)_2$	19.9	1.001	1.4523	III-738
1398	pseudocumene (*sym*)	$(CH_3)_3C_6H_2C_2H_5$ (1,2,4,5)	15.8	0.887	1.5105	V-442
1399	pseudocumyl ether	$(CH_3)_3C_6H_2OC_2H_5$ (1,2,4,5)	20	0.942	1.509	VI-510
1400	pyridine (2)	$C_2H_5\cdot C_5H_4N$	22.5	0.94	1.5021	XX-241
1401	pyruvate	$CH_3COCO_2C_2H_5$	15.6	1.060	1.4083	III-616
1402	ricinoleate	$HOC_{17}H_{32}CO_2C_2H_5$	22	0.915	1.4618	II-387
1403	salicylate (*o*)	$HOC_6H_4CO_2C_2H_5$	14.4	1.14	1.5251	X-73
1404	sorbate	$CH_3(CH:CH)_2CO_2\cdot C_2H_5$	20	0.956	1.5023	II-484
1405	styrene (β)	$C_6H_5CH:CHC_2H_5$	20	0.910	1.5387	V-487
1406	styryl ether	$C_6H_5CH:CHOC_2H_5$	21.2	0.971	1.5502	VI-564
1407	styryl ketone (m.p. 38°)	$C_2H_5COCH:CHC_6H_5$	50.3	0.987	1.5684	VII-373
1408	Δ¹-tetrahydro-benzoate	$C_6H_9CO_2C_2H_5$	20	0.998	1.469	IX-41
1409	5,6,7,8-tetrahydro-naphthalene (2)	$C_2H_5C_{10}H_{11}$	17.6	0.950	1.5347	**V-397
1410	Δ¹-tetrahydro-*m*-toluate	$CH_3C_6H_8CO_2C_2H_5$	20	0.975	1.470	IX-47
1411	thiocyanate	C_2H_5SCN	22.9	0.994	1.4653	III-175
1412	*iso*-thiocyanate	C_2H_5NCS	18	0.998	1.5142	IV-123
1413	thionaphthene-2-carboxylate (m.p. 37°)	$C_{11}H_{10}O_2S$	55.4	1.172	1.592	....
1414	thiophene (α)	$C_2H_5C:CHCH:CH\cdot S$	20	0.992	1.513	XVII-39

Table 10-66 (*Continued*)
INDEX OF REFRACTION—LIQUIDS

No.	Substance	Formula	t,°C.	d_4^t	n_D^t	Beil. Ref.
1415	**Ethyl** α-thiophene-carboxylate	SCH:CHCH:CCO$_2$· $\overline{\quad\quad}$ C$_2$H$_5$	20	1.159	1.526	XVIII-289
1416	tiglate	CH$_3$CH:C(CH$_3$)CO$_2$· C$_2$H$_5$	19.5	0.924	1.436	II-431
1417	o-toloxyacetate	CH$_3$C$_6$H$_4$OCH$_2$CO$_2$· C$_2$H$_5$	12.9	1.085	1.5061	*VI-172
1418	α,p-toloxy-*iso*-butyrate	(CH$_3$)$_2$C(OC$_6$H$_4$· CH$_3$)CO$_2$C$_2$H$_5$	20	1.027	1.4893	VI-399
1419	toluate (o)	CH$_3$C$_6$H$_4$CO$_2$C$_2$H$_5$	21.6	1.033	1.5070	IX-463
1420	" (m)	CH$_3$C$_6$H$_4$CO$_2$C$_2$H$_5$	21.6	1.03	1.5050	IX-476
1421	" (p)	CH$_3$C$_6$H$_4$CO$_2$C$_2$H$_5$	18.2	1.03	1.5089	IX-484
1422	p-tolyl ketone	C$_2$H$_5$COC$_6$H$_4$CH$_3$	20.7	0.990	1.5271	VII-317
1423	triazoacetate	N$_3$CH$_2$CO$_2$C$_2$H$_5$	24.9	1.13	1.4349	II-229
1424	tribromoacetate	Br$_3$CCO$_2$C$_2$H$_5$	12.5	2.2	1.5438	II-221
1425	trichloroacetate	Cl$_3$CCO$_2$C$_2$H$_5$	20	1.383	1.4507	II-209
1426	γ,γ,γ-trichloro-β-acetoxybutyrate	CH$_3$CO$_2$CH(CCl$_3$)· CH$_2$CO$_2$C$_2$H$_5$	14.1	1.340	1.4646	*III-117
1427	γ,γ,γ-trichloro-crotonate	Cl$_3$CCH:CHCO$_2$· C$_2$H$_5$	14.2	1.338	1.4869	*II-190
1428	trifluoroacetate	F$_3$CCO$_2$C$_2$H$_5$	20	1.19	1.306	....
1429	trimethylacrylate	(CH$_3$)$_2$C:C(CH$_3$)· CO$_2$C$_2$H$_5$	19.3	0.907	1.4299	II-443
1430	p,α,β-trimethyl-cinnamate	CH$_3$C$_6$H$_4$C(CH$_3$):C: (CH$_3$)CO$_2$C$_2$H$_5$	18.4	1.003	1.5192	*IX-264
1431	tropane-2-car-boxylate	C$_{11}$H$_{19}$O$_2$N	20.5	1.041	1.4756	XXII-18
1432	tropidine-2-carboxylate	C$_{11}$H$_{17}$O$_2$N	21.3	1.063	1.4948	....
1433	tropinone-2-carboxylate	C$_{11}$H$_{17}$O$_3$N	20.8	1.121	1.4954	....
1434	urethane	C$_2$H$_5$NHCO$_2$C$_2$H$_5$	20	0.981	1.4219	IV-114
1435	n-valerate	CH$_3$(CH$_2$)$_3$CO$_2$C$_2$H$_5$	20	0.877	1.39	II-301
1436	*iso*-valerate	(CH$_3$)$_2$CHCH$_2$CO$_2$· C$_2$H$_5$	18.4	0.87	1.3974	II-312
1437	*iso*-valerylcam-phor-carboxylate	C$_{18}$H$_{28}$O$_4$	22.7	1.022	1.4764	X-38
1438	β-vinylacrylate	CH$_2$:CHCH:CHCO$_2$· C$_2$H$_5$	20	0.938	1.476	**II-451
1439	**Ethylene** bromide	BrCH$_2$CH$_2$Br	20	2.182	1.5379	I-90
1440	bromohydrin	BrCH$_2$CH$_2$OH	20	1.772	1.4969	I-338
1441	chloride	ClCH$_2$CH$_2$Cl	20	1.253	1.4443	I-84
1442	chlorohydrin	ClCH$_2$CH$_2$OH	20	1.199	1.4419	I-337
1443	diamine	(CH$_2$NH$_2$)$_2$	26.1	0.892	1.4540	IV-230
1444	diamine hydrate	C$_2$H$_8$N$_2$·H$_2$O	20.5	0.963	1.4500	IV-230
1445	dipiperidine (γ)	C$_{12}$H$_{24}$N$_2$	17.8	0.921	1.4887	XX-67
1446	ethylidene oxide	CH$_3$CH:(OCH$_2$)$_2$	20	0.981	1.40	XIX-8
1447	fluorohydrin	FCH$_2$CH$_2$OH	20	1.114	1.3639	*I-170
1448	oxide	(CH$_2$)$_2$O	8.4	0.886	1.3599	XVII-4
1449	**Ethylidene** acetone	CH$_3$CH:CHCOCH$_3$	19.6	0.862	1.4390	I-732
1450	cyclohexane	CH$_3$CH:C(CH$_2$)$_4$CH$_2$ $\overline{\quad\quad}$	20	0.822	1.4626	V-71
1451	**Eucarvone**	C$_{10}$H$_{14}$O	21.2	0.948	1.5087	VII-151
1452	**Eugenol** (*iso*) (1,3,4)	C$_6$H$_3$(C$_3$H$_5$)(O· CH$_3$)OH	19	1.08	1.5739	VI-955

Table 10-66 (*Continued*)
INDEX OF REFRACTION—LIQUIDS

No.	Substance	Formula	t,°C.	d_4^t	n_D^t	Beil. Ref.
1453	**Eugenol** (1,3,4)	$C_6H_3(C_3H_5)(O\cdot CH_3)OH$	19.4	1.07	1.5416	VI-961
1454	acetate (m.p. 31°)	$C_{12}H_{14}O_3$	20	1.084	1.5207	VI-965
1455	*iso*-amyl ether	$C_5H_{11}OC_{10}H_{11}O$	14.8	0.973	1.5128	VI-964
1456	ethyl ether (*iso*)	$CH_3CH:CHC_6H_3(O\cdot CH_3)(OC_2H_5)$	11	1.044	1.5607	VI-957
1457	**Farnesol**	$C_{15}H_{26}O$	20	0.88	1.4899	I-464
1458	**Fenchene**	$C_{10}H_{16}$	20	0.867	1.4705	V-163
1459	**Fencholamide** (m.p. 94°)	$C_{10}H_{19}ON$	108.2	0.907	1.45	IX-32
1460	**Fencholenamide** (α) (m.p. 113°)	$CH_3C_5H_6C(CH_3)_2\cdot CONH_2$	117.9	0.933	1.47	IX-67
1461	**Fenchone** (*d*)	$C_{10}H_{16}O$	18	0.948	1.4636	VII-96
1462	**Fenchyl alcohol** (*l*) (*iso*)	$C_{10}H_{18}O$	15	0.961	1.4801	VI-72
1463	**Fenchylene**	$C_{10}H_{16}$	20	0.838	1.4502	*V-80
1464	**Fluoro**-aniline (*o*)	$FC_6H_4NH_2$	20	1.151	1.5407	*XII-296
1465	" (*m*)	$FC_6H_4NH_2$	20	1.16	1.5455	XII-597
1466	" (*p*)	$FC_6H_4NH_2$	20	1.152	1.536	XII-597
1467	benzene	C_6H_5F	20	1.024	1.4677	V-198
1468	1,2-dibromoethane (2)	$FBrCHCH_2Br$	19	2.26	1.5012	I-92
1469	nitrobenzene (*o*)	$FC_6H_4NO_2$	17.2	1.338	1.5323	V-241
1470	" (*m*)	$FC_6H_4NO_2$	17.2	1.33	1.5207	V-241
1471	" (*p*) (m.p. 26°)	$FC_6H_4NO_2$	56	1.30	1.5150	V-241
1472	phenol (*m*)	FC_6H_4OH	20	1.222	1.514	*VI-97
1473	" (*p*) (m.p. 48°)	FC_6H_4OH	56	1.189	1.5010	VI-183
1474	1,1,1,2-tetrabromo-ethane (2)	$Br_3CCHBrF$	16	1.939	1.5971	I-95
1475	toluene (*o*)	$CH_3C_6H_4F$	20	1.001	1.4704	V-290
1476	" (*m*)	$CH_3C_6H_4F$	20	0.999	1.4691	V-290
1477	" (*p*)	$CH_3C_6H_4F$	20	0.99	1.470	V-290
1478	1,1,2-tribromo-ethane (1)	$BrCH_2CFBr_2$	17.5	2.605	1.5022	I-93
1479	1,1,2-tribromo-ethane (2)	$FBrCHCHBr_2$	18	2.674	1.5638	I-93
1480	trichloromethane	CCl_3F	18.5	1.494	1.3865	I-64
1481	**Formamide**	$HCONH_2$	20	1.139	1.4472	II-26
1482	**Formanilide** (m.p. 47°)	$C_6H_5NHCH:O$	25	1.14	1.5876	XII-230
1483	**Formic acid**	HCO_2H	20	1.220	1.3714	II-8
1484	**Formyl-menthone** (2)	$C_{11}H_{18}O_2$	15.5	1.001	1.5000	VII-568
1485	**Fumaryl chloride**	$(CHCOCl)_2$	18.1	1.41	1.5004	II-743
1486	**Furan**	$(CH:CH)_2O$	20	0.937	1.4216	XVII-27
1487	**Furfural** (2)	C_4H_3OCHO	20	1.159	1.5261	XVII-272
1488	" (3)	$C_5H_4O_2$	20	1.111	1.4945	....
1489	**Furfuryl** alcohol	$C_4H_3OCH_2OH$	20	1.129	1.4852	XVII-112
1490	chloride	$C_4H_3OCH_2Cl$	20	1.178	1.4941	....
1491	**Furylimido ethyl** ether	$OC_4H_3\cdot C(:NH)O\cdot C_2H_5$	17.8	1.078	1.4930	XVIII-278
1492	**Geranic acid**	$C_{10}H_{16}O_2$	20.2	0.951	1.4870	II-491
1493	**Geraniolene**	C_9H_{16}	21.5	0.768	1.4453	I-260
1494	**Geranyl** acetate	$CH_3CO_2C_{10}H_{17}$	15	0.917	1.4628	II-140

Table 10-66 (*Continued*)
INDEX OF REFRACTION—LIQUIDS

No.	Substance	Formula	t,°C.	d_4^t	n_D^t	Beil. Ref.
1495	**Geranyl**-acetic acid	$C_{12}H_{20}O_2$	21 (?)	0.938	1.4739	*II-210
1496	amine	$C_{10}H_{19}N$	20 (?)	0.83	1.4727	*IV-398
1497	chloride	$C_{10}H_{17}Cl$	20 (?)	0.92	1.4741	*I-123
1498	formate	$HCO_2C_{10}H_{17}$	20	0.909	1.4659	II-23
1499	**Germanium tetra-ethyl**	$Ge(C_2H_5)_4$	20 (?)	.....	1.400	IV-631
1500	**Glycerol**	$CHOH(CH_2OH)_2$	20	1.26	1.4729	I-502
1501	dichlorohydrin (1,3) (α)	$CHOH(CH_2Cl)_2$	16.9	1.351	1.4802	I-364
1502	**Glyceryl** triacetate	$(CH_3CO_2)_3C_3H_5$	20	1.16	1.4306	II-147
1503	tributyrate	$(C_3H_7CO_2)_3C_3H_5$	20	1.032	1.4359	II-273
1504	tricaproate	$(C_5H_{11}CO_2)_3C_3H_5$	20	0.988	1.4427	II-324
1505	tricaprylate	$(C_7H_{15}CO_2)_3C_3H_5$	20	0.954	1.4482	II-348
1506	triformate	$(HCO_2)_3C_3H_5$	18	1.320	1.4412	*II-19
1507	trilaurate (m.p. 46°)	$(C_{11}H_{23}CO_2)_3C_3H_5$	60	0.894	1.4404	II-362
1508	trimyristate (m.p. 56°)	$(C_{13}H_{27}CO_2)_3C_3H_5$	60	0.885	1.4429	II-367
1509	trioleate	$(C_{17}H_{33}CO_2)_3C_3H_5$	60	0.9	1.4561	II-468
1510	tripalmitate (m.p. 65°)	$(C_{15}H_{31}CO_2)_3C_3H_5$	80	0.866	1.4381	II-373
1511	tristearate (m.p. 70°)	$(C_{17}H_{35}CO_2)_3C_3H_5$	60	0.9	1.4441	II-383
1512	**Glycol**	$(CH_2OH)_2$	20	1.114	1.4318	I-465
1513	diacetate	$(CH_3CO_2CH_2)_2$	15	1.108	1.4183	II-142
1514	ether	$(HOCH_2CH_2)_2O$	20	1.117	1.4475	I-468
1515	**Glycolic** aldehyde† (m.p. 97°)	CH_2OHCHO	11	1.4	1.4811	I-817
1516	nitrile	$HOCH_2CN$	19	1.104	1.4117	III-242
1517	**Glyoxal**	$(CHO)_2$	20	1.14	1.3828	I-759
1518	**Glutaric** acid (m.p. 97.5°)	$CH_2(CH_2CO_2H)_2$	106.4	1.192	1.4188	II-631
1519	nitrile	$CH_2(CH_2CN)_2$	23.2	0.99	1.4365	II-635
1520	**Glutaryl** chloride	$CH_2(CH_2COCl)_2$	20.2	1.324	1.4728	II-634
1521	**Guaiacol** (o) (m.p. 28°)	$CH_3OC_6H_4OH$	?	1.1	1.54	VI-768
1522	**Guajene**	$C_{15}H_{24}$	20	0.909	1.5005	V-468
1523	**Hemimellitenyl ethyl ether**	$(CH_3)_3C_6H_2OC_2H_5$ (1,2,3,5)	20	0.956	1.516	**VI-480
1524	**Heneicosane** (n) (m.p. 40°)	$CH_3(CH_2)_{19}CH_3$	45.3	0.775	1.4344	I-174
1525	**Hepta**-cosane (n) (m.p. 60°)	$CH_3(CH_2)_{25}CH_3$	65	0.77	1.4345	I-176
1526	decane (n)	$CH_3(CH_2)_{15}CH_3$	20 (?)	0.778	1.4405	I-173
1527	2,4-diene	$CH_3CH:CHCH:CH\cdot C_2H_5$	21.5	0.733	1.4486	I-257
1528	1,5-diene-4-ol	$CH_2:CHCH_2CHOH\cdot CH:CHCH_3$	20	0.861	1.454	*I-235
1529	**Heptaldehyde**	$CH_3(CH_2)_5CHO$	19.9	0.817	1.4125	I-695
1530	oxime (m.p. 58°)	$CH_3(CH_2)_5CH:NOH$	83.9	0.833	1.421	I-698
1531	oxime-O-methyl ether	$CH_3(CH_2)_5CH:NO\cdot CH_3$	19.4	0.838	1.4243	I-698
1532	**Heptane** (n)	$CH_3(CH_2)_5CH_3$	20	0.684	1.3876	I-154
1533	**Heptanoic** acid	$CH_3(CH_2)_5CO_2H$	19.8	0.918	1.4216	II-338
1534	anhydride	$(C_6H_{13}CO)_2O$	15	0.94	1.4335	II-340
1535	**Heptanol** (2)	$C_5H_{11}CHOHCH_3$	20	0.819	1.4209	I-415

† Supercooled.

Table 10-66 (*Continued*)
INDEX OF REFRACTION—LIQUIDS

No.	Substance	Formula	t,°C.	d_4^t	n_D^t	Beil. Ref.
1536	**Heptanol** (3)	$C_4H_9CHOHC_2H_5$	20	0.823	1.4206	*I-205
1537	" (4)	$C_3H_7CHOHC_3H_7$	20	0.820	1.4166	I-415
1538	**Heptan-2-thiol**	$C_5H_{11}CH(SH)CH_3$	20(?)	0.835	1.4460	I-415
1539	**Heptatriene** (1,3,5)	CH_2:CHCH:CHCH: CHCH$_3$	20	0.764	1.516	*I-126
1540	**Heptene** (1)	CH_2:CH(CH$_2$)$_4$CH$_3$	20	0.697	1.3994	I-219
1541	" (2) (*cis*)	CH_3CH:CHC$_4$H$_9$	20	0.708	1.406	I-219
1542	" (2) (*trans*)	CH_3CH:CHC$_4$H$_9$	20	0.704	1.406	I-219
1543	" (3) (*cis?*)	C_2H_5CH:CHC$_3$H$_7$	20	0.701	1.404	I-220
1544	" (3) (*trans*)	C_2H_5CH:CHC$_3$H$_7$	20	.....	......	....
1545	**Hept-2-ene-4-ol**	CH_3CH:CHCHOH· CH$_2$C$_2$H$_5$	20	0.845	1.4373	I-447
1546	**Heptenic acid** (*iso*) (α,β)	$(CH_3)_2$CHCH$_2$CH: CHCO$_2$H	20	0.941	1.4444	II-445
1547	**Heptenonitrile** (Δ^2) (*iso*)	$(CH_3)_2$CHCH$_2$CH: CHCN	20	0.826	1.430	**II-411
1548	**Heptenoyl** chloride (α,β) (*iso*)	$(CH_3)_2$CHCH$_2$CH: CHCOCl	20	0.991	1.461	**II-411
1549	**Heptyl** acetate (*n*)	CH_3CO$_2$C$_7$H$_{15}$	20	0.87	1.4153	II-134
1550	alcohol (*n*)	CH_3(CH$_2$)$_5$CH$_2$OH	22.4	0.817	1.4233	I-414
1551	amine (*n*)	CH_3(CH$_2$)$_5$CH$_2$NH$_2$	26.5	0.77	1.42	IV-193
1552	fluoride (*n*)	CH_3(CH$_2$)$_5$CH$_2$F	20	0.804	1.3861	I-56
1553	*n*-heptylate	CH_3(CH$_2$)$_5$CO$_2$CH$_2$· (CH$_2$)$_5$CH$_3$	18.8	0.865	1.4318	II-340
1554	iodide (*n*)	CH_3(CH$_2$)$_5$CH$_2$I	20	1.4	1.4594	I-155
1555	**Heptyne** (1)	CH:C(CH$_2$)$_4$CH$_3$	12.6	0.738	1.4136	I-256
1556	**Hexa**-cosane (*n*) (m.p. 60°)	CH_3(CH$_2$)$_{24}$CH$_3$	65	0.8	1.4333	I-175
1557	decene (1)	CH_3(CH$_2$)$_{13}$CH:CH$_2$	19	0.782	1.4420	I-226
1558	diene (1,2)	CH_2:C:CHC$_3$H$_7$	20	0.717	1.428	**I-229
1559	" (1,3) (*cis?*)	CH_2:CHCH:CHC$_2$H$_5$	20	0.705	1.438	I-253
1560	" (1,3) (*trans*)	CH_2:CHCH:CHC$_2$H$_5$	20	.....	......	....
1561	" (1,4) (*cis?*)	CH_2:CHCH$_2$CH: CHCH$_3$	20	0.695	1.410	I-253
1562	" (1,4) (*trans*)	CH_2:CHCH$_2$CH: CHCH$_3$	20	.....	......	....
1563	" (1,5)	$(CH_2$:CHCH$_2$)$_2$	20	0.691	1.4042	I-253
1564	" (2,3)	CH_3CH:C:CHC$_2$H$_5$	20	0.680	1.395	....
1565	" (2,4) (*cis, cis?*)	$(CH_3$CH:CH)$_2$	20	0.720	1.450	I-254
1566	diene (2,4) (*cis, trans*)	$(CH_3$CH:CH)$_2$	20	.....	......	....
1567	diene (2,4) (*trans, trans*)	$(CH_3$CH:CH)$_2$	20	.....	......	....
1568	ethylbenzene (m.p. 129°)	$(C_2H_5)_6$C$_6$	130	0.831	1.48	V-471
1569	**Hexahydro-**benzoic acid (m.p. 31°)	CH_2(CH$_2$)$_4$CHCO$_2$H $\lfloor$_____$\rfloor$	33.8	1.03	1.46	IX-7
1570	benzyl alcohol	CH_2(CH$_2$)$_4$CH· $\lfloor$_____$\rfloor$ CH$_2$OH	20	0.928	1.4649	VI-14
1571	cresol (*o*)	CH_3C$_6$H$_{10}$OH	20	0.933	1.461	VI-11
1572	" (*m*) (*l*)	CH_3C$_6$H$_{10}$OH	20	0.914	1.4581	VI-12
1573	" (*m*) (*dl*)	CH_3C$_6$H$_{10}$OH	20	0.923	1.4590	VI-13
1574	" (*p*)	CH_3C$_6$H$_{10}$OH	20	0.918	1.4573	VI-14

Table 10-66 (*Continued*)
INDEX OF REFRACTION—LIQUIDS

No.	Substance	Formula	t,°C.	d_4^t	n_D^t	Beil. Ref.
1575	**Hexahydro-**cymene (*p*)	$CH_3C_6H_{10}C_3H_7$	20	0.793	1.4380	V-47
1576	mesitylene	$(CH_3)_3C_6H_9$ (1,3,5)	13.1	0.778	1.4318	V-45
1577	naphthalene	$C_{10}H_{14}$	18.4	0.94	1.5331	V-433
1578	xylene (*o*)	$(CH_3)_2C_6H_{10}$	17.9	0.78	1.4302	V-36
1579	" (*m*)	$(CH_3)_2C_6H_{10}$	20.8	0.770	1.4250	V-36
1580	" (*p*)	$(CH_3)_2C_6H_{10}$	20	0.769	1.4244	V-38
1581	**Hexane** (*n*)	$CH_3(CH_2)_4CH_3$	20	0.659	1.3749	I-142
1582	**Hexanol** (2)	$CH_3CHOHC_4H_9$	20	0.814	1.4135	I-408
1583	" (3)	$C_2H_5CHOHCH_2C_2H_5$	20	0.818	1.4141	I-408
1584	**Hexan-6-ol-2-one**	$CH_3COCH_2(CH_2)_2\cdot$ CH_2OH	20	0.990	1.450	I-835
1585	**Hexatriene** (1,3,5)	$CH_2{:}CHCH{:}CHCH{:}$ CH_2	16.2	0.741	1.49	I-263
1586	**Hexene** (1)	$CH_2{:}CH(CH_2)_3CH_3$	20	0.673	1.3876	I-215
1587	" (2) (*cis*)	$CH_3CH{:}CHCH_2C_2H_5$	20	0.685	1.3954	I-215
1588	" (2) (*trans*)	$CH_3CH{:}CHCH_2C_2H_5$	20	0.678	1.3935	I-215
1589	" (3) (*cis*)	$C_2H_5CH{:}CHC_2H_5$	20	0.680	1.3934	I-215
1590	" (3) (*trans*)	$C_2H_5CH{:}CHC_2H_5$	20	0.678	1.3938	I-215
1591	**Hex-2-ene-4-ol**	$CH_3CH{:}CHCHOH\cdot$ C_2H_5	25	0.841	1.4312	I-445
1592	**Hexenic** acid (α,β) (m.p. 32°)	$C_3H_7CH{:}CHCO_2H$	40	0.95	1.4466	II-434
1592.1	acid (β,γ)	$C_2H_5CH{:}CHCH_2\cdot$ CO_2H	23	0.964	1.44	II-435
1593	acid (*iso*) (α,β)	$(CH_3)_2CHCH{:}CH\cdot$ CO_2H	16	0.96	1.4506	II-438
1594	**Hexenonitrile** (β,γ)	$C_2H_5CH{:}CHCH_2CN$	20	0.841	1.432	**II-404
1595	" (*iso*) (α,β)	$(CH_3)_2CHCH{:}CHCN$	20	0.855	1.4374	II-439
1596	**Hexenoyl chloride** (β,γ)	$C_2H_5CH{:}CHCH_2\cdot$ $COCl$	20	1.014	1.447	....
1597	**Hexenoyl chloride** (*iso*) (α,β)	$(CH_3)_2CHCH{:}CH\cdot$ $COCl$	20	1.018	1.461	....
1598	**Hexyl** alcohol (*n*)	$CH_3(CH_2)_4CH_2OH$	20	0.820	1.4133	I-407
1599	bromide (*n*)	$CH_3(CH_2)_4CH_2Br$	20	1.173	1.4478	I-144
1600	bromide (*iso*)	$(CH_3)_2CH(CH_2)_2\cdot$ CH_2Br	20(?)	1.168	1.4490	I-148
1601	cyanide (*n*)	$CH_3(CH_2)_5CN$	20(?)	0.81	1.4195	II-341
1602	iodide (*n*)	$CH_3(CH_2)_4CH_2I$	20	1.439	1.4929	I-146
1603	nitrite (*n*)	$CH_3(CH_2)_4CH_2ONO$	20	0.885	1.4018	I-407
1604	nitroamine (*N*)	$C_2H_5(CH_2)_3\cdot$ CH_2NHNO_2	25.2	1.004	1.4593	IV-572
1605	**Hexylpropiolic** acid	$C_6H_{13}C{:}CCO_2H$	25	0.944	1.4596	II-490
1606	diethylacetal	$CH_3(CH_2)_5C{:}$ $CCH(OC_2H_5)_2$	12.1	0.881	1.4424	I-751
1607	nitrile	$C_6H_{13}C{:}CCN$	14.4	0.849	1.4564	II-490
1608	**Homocatechol** (m.p. 65°)	$CH_3C_6H_3(OH)_2$ (1,3,4)	73.6	1.129	1.5425	VI-878
1609	**Hydrazine**	H_2NNH_2	22	.....	1.470	....
1610	**Hydrindene** (1,2)	$C_6H_4CH_2CH_2CH_2$ $\rule{1cm}{0.4pt}$	20.8	0.96	1.5351	V-486
1611	**Hydrindone** (α) (m.p. 41°)	$C_6H_4CH_2CH_2CO$ $\rule{1cm}{0.4pt}$	45	1.09	1.5608	VII-360
1612	**Hydrindone** (β) (m.p. 58°)	$C_6H_4CH_2COCH_2$ $\rule{1cm}{0.4pt}$	66.3	1.07	1.538	VII-363

Table 10-66 (*Continued*)
INDEX OF REFRACTION—LIQUIDS

No.	Substance	Formula	$t,°C.$	d_4^t	n_D^t	Beil. Ref.
1613	**Hydro-**cinnamic acid (m.p. 48°)	$C_6H_5CH_2CH_2CO_2H$	80	1.06	1.50	IX-508
1614	cinnamyl alcohol	$C_6H_5CH_2CH_2CH_2OH$	20	1.008	1.5357	VI-503
1615	cyanic acid	HCN	19	0.698	1.254	II-29
1616	**Hydrogen** bromide	HBr	10	.....	1.325	....
1617	chloride	HCl	?	.....	1.256	....
1618	disulfide	H_2S_2	?	.....	1.885	....
1619	iodide	HI	12	.....	1.466	....
1620	peroxide	H_2O_2	?	.....	1.333	....
1621	"	$H_2O_2.2H_2O$	22	.....	1.414	....
1622	sulfide	H_2S	?	.....	1.374	....
1623	**Hydroxy-aceto-phenone** (*o*)	$HOC_6H_4COCH_3$	21.3	1.131	1.5550	VIII-85
1624	**Hydroxy-aceto-phenone** (*m*) (m.p. 95°)	$HOC_6H_4COCH_3$	109	1.099	1.54	VIII-86
1625	**Hydroxy-aceto-phenone** (*p*) (m.p. 109°)	$HOC_6H_4COCH_3$	109	1.109	1.56	VIII-87
1626	*o*-(*α*-**Hydroxy-iso-butyro**)-*p*-**cresol** (m.p. 55°)	$(CH_3)_2COHCO \cdot C_6H_3(OH)(CH_3)$ (2,1,4)	54	1.094	1.5347	*VIII-624
1627	**Hydroxy-**Δ^1-crotonolactone (*γ*) (m.p. 55°)	$CH_2CH:CHCO \cdot O$ $\lfloor \underline{\hspace{2cm}} \rfloor$	99.3	1.262	1.4563	III-727
1628	diethylaceto-*p*-cresol (*α*)	$(C_2H_5)_2COHCO \cdot C_6H_3(CH_3)OH$ (1,3,6)	21.1	1.080	1.540	....
1629	dihydrocinnamyl alcohol (*o*)	$HOC_6H_4CH_2CH_2 \cdot CH_2OH$	15	1.126	1.5558	VI-928
1630	2,5-dimethoxy-propiophenone (*α*)	$(CH_3O)_2C_6H_3CO \cdot CHOHCH_3$	13.4	1.149	1.5353	*VIII-690
1631	glutaric nitrile (2)	$C_3H_5OH(CN)$	20	1.181	1.4805	*III-157
1632	methene-4,5-dimethylcyclo-hexan-2-one (1)	$(CH_3)_2C_6H_6O:CHOH$	20	1.022	1.4977	....
1633	3-methylaceto-phenone (6) (m.p. 50°)	$CH_3C_6H_3(OH)CO \cdot CH_3$	53	1.080	1.5410	VIII-111
1634	methylene-cyclo-hexan-2-one (1)	$CH_2(CH_2)_3COC:$ $\lfloor \underline{\hspace{2cm}} \rfloor$ CHOH	20.5	1.085	1.5122	VII-558
1635	methylene diethyl ketone	$C_2H_5COCH(CH_3) \cdot CHO$	48.9	0.965	1.4579	*I-406
1636	methylene-3-methylcyclo-hexan-2-one (1)	$CH_3C_6H_7O:CHOH$	15.4	1.056	1.507	*VII-314
1637	methylene-pinacoline	$(CH_3)_3CCOCH:$ CHOH	20	0.935	1.4527	I-794
1638	(*α*-**Hydroxypropi-onyl**)-*p*-**meth-oxytoluene** (*m*)	$(CH_3)(CH_3O)C_6H_3 \cdot COCHOHCH_3$	19.2	1.099	1.5263	*VIII-624
1639	**Hydroxy-**propio-phenone (*α*)	$C_6H_5COCHOHCH_3$	22.8	1.104	1.5356	*VIII-547

Table 10-66 (*Continued*)
INDEX OF REFRACTION—LIQUIDS

No.	Substance	Formula	$t,°C.$	d_4^t	n_D^t	Beil. Ref.
1640	**Hydroxy**-styrene (*o*) (m.p. 29°)	$HOC_6H_4CH:CH_2$	17.4	1.06	1.5572	VI-560
1641	valeric lactone (γ)	$CH_3CHCH_2\cdot$ $\underline{\qquad}$ $CH_2CO\cdot O$ $\underline{\qquad}$	22.4	1.050	1.4331	XVII-235
1642	**Hydroxylamine**	NH_2OH	23.5	.....	1.440	....
1643	**Indene**	$C_6H_4CH_2CH:CH$ $\underline{\qquad}$	16.1	1.01	1.5739	V-515
1644	**Indoxazene**	$C_6H_4CH:N\cdot O$ $\underline{\qquad}$	20	1.170	1.563	XXVII-39
1645	**Iodo**-benzene	C_6H_5I	17.8	1.835	1.6213	V-215
1646	2,3-dimethyl-butane (2)	$(CH_3)_2CICH(CH_3)_2$	20	1.446	1.5047	I-153
1647	2-methylbutane (1)	$C_2H_5CH(CH_3)CH_2I$	15	1.53	1.4981	I-138
1648	2-methylhexane (1)	$CH_3(CH_2)_3CH:$ $(CH_3)CH_2I$	21	1.366	1.4891	I-157
1649	methyl methyl ether	ICH_2OCH_3	20	2.030	1.5472	I-538
1650	naphthalene (α)	$C_{10}H_7I$	14	1.747	1.7054	V-550
1651	" (β) (m.p. 54°)	$C_{10}H_7I$	99.4	1.632	1.6662	V-552
1652	propan-1-ol (3)	$ICH_2CH_2CH_2OH$	20	1.998	1.5585	I-358
1653	pseudocumene (m.p. 37°)	$(CH_3)_3C_6H_2I$ (1,2,4,5)	65.3	1.511	1.5813	V-404
1654	toluene (*o*)	$CH_3C_6H_4I$	20	1.023	1.6085	V-639
1655	*o*-xylene (*vic*)	$(CH_3)_2C_6H_3I$ (1,2,3)	20	1.639	1.6072	V-367
1656	" (*uns*)	$(CH_3)_2C_6H_3I$ (1,2,4)	20	1.631	1.604	V-367
1657	*m*-xylene (*sym*)	$(CH_3)_2C_6H_3I$	20	1.607	1.596	V-376
1658	" (*uns*)	$(CH_3)_2C_6H_3I$ (1,3,4)	20	1.623	1.599	V-376
1659	*p*-xylene	$(CH_3)_2C_6H_3I$	20	1.613	1.598	V-386
1660	**Ionone** (α)	$C_{10}H_{16}:CHCOCH_3$	22.3	0.93	1.4984	VII-168
1661	" (β)	$C_{10}H_{16}:CHCOCH_3$	18.9	0.95	1.5198	VII-167
1662	" (pseudo)	$C_{13}H_{20}O$	18.8	0.90	1.5312	I-757
1663	**Irone** (β)	$C_{10}H_{16}:CHCOCH_3$	20	0.939	1.5017	VII-169
1664	**Isoxazole**	$OCH:CHCH:N$ $\underline{\qquad}$	20	1.078	1.428	XXVII-14
1665	**Kairoline**	$C_9H_{10}N\cdot CH_3$	23.1	1.019	1.5802	XX-264
1666	**Ketazine**	$[(CH_3)_2C:N]_2$	24.5	0.84	1.4510	I-651
1667	**Keto**-octahydro-phenanthrene (m.p. 81°)	$C_{14}H_{16}O$	100	1.064	1.5628	....
1668	1,2,3,4-tetra-hydro-naphthalene (1)	$C_6H_4COCH_2CH_2CH_2$ $\underline{\qquad}$	15.5	1.099	1.5712	VII-370
1669	**Lactic** acid	$CH_3CHOHCO_2H$	20	1.24	1.44	III-268
1670	nitrile	$CH_3CHOHCN$	18.4	0.992	1.4058	III-284
1671	**Lauric** acid (m.p. 48°)	$CH_3(CH_2)_{10}CO_2H$	60	0.8	1.4267	II-359
1672	**Laurolene**	$(CH_3)_3C_5H_5$ (1,2,3)	20(?)	0.797	1.4438	V-75
1673	**Laurone** (m.p. 69°)	$(C_{11}H_{23})_2CO$	79.5	0.8	1.4283	I-719
1674	**Lead** butyl-tri-*iso*-amyl	$(C_5H_{11})_3Pb\cdot CH_2\cdot$ $CH(CH_3)_2$	19.5	1.252	1.4962	*IV-596
1675	diethyl-di-*iso*-butyl	$(C_2H_5)_2Pb(C_4H_9)_2$	20	1.444	1.5086	*IV-594
1676	diethyl-di-*n*-propyl	$(C_2H_5)_2Pb(C_3H_7)_2$	20	1.533	1.5149	*IV-592
1676.1	diethyl-*n*-propyl-*iso*-amyl	$(C_2H_5)_2(C_3H_7)Pb\cdot$ C_5H_{11}	22.1	1.439	1.5066	*IV-595
1677	diethyl-*n*-propyl-*iso*-butyl	$(C_2H_5)_2(C_3H_7)Pb\cdot$ (C_4H_9)	20	1.489	1.5120	*IV-594

Table 10-66 (*Continued*)
INDEX OF REFRACTION—LIQUIDS

No.	Substance	Formula	t,°C.	d_4^t	n_D^t	Beil. Ref.
1678	**Lead** dimethyl-di-*iso*-amyl	$(CH_3)_2Pb(C_5H_{11})_2$	20	1.430	1.5005	*IV-596
1679	dimethyl-di-*iso*-butyl	$(CH_3)_2Pb(C_4H_9)_2$	20	1.505	1.5024	*IV-594
1680	dimethyl-diethyl	$(CH_3)_2Pb(C_2H_5)_2$	20	1.785	1.5177	*IV-591
1681	dimethyl-di-*n*-propyl	$(CH_3)_2Pb(C_3H_7)_2$	20	1.627	1.5086	*IV-592
1682	dimethyl-ethyl-*iso*-amyl	$(CH_3)_2(C_2H_5)Pb\cdot(C_5H_{11})$	21.7	1.558	1.5052	*IV-595
1683	dimethyl-ethyl-*iso*-butyl	$(CH_3)_2(C_2H_5)Pb\cdot C_4H_9$	20.7	1.623	1.5078	*IV-593
1684	dimethyl-*n*-propyl-*iso*-amyl	$(CH_3)_2(C_3H_7)Pb\cdot(C_5H_{11})$	22	1.503	1.5020	. . . .
1685	ethyl-tri-*iso*-amyl	$C_2H_5Pb(C_5H_{11})_3$	19.6	1.292	1.4983	*IV-596
1686	ethyl-tri-*iso*-butyl	$C_2H_5Pb(C_4H_9)_3$	22.1	1.376	1.5055	*IV-594
1687	ethyl-tri-*n*-propyl	$(C_2H_5)Pb(C_3H_7)_3$	21.3	1.485	1.5115	*IV-592
1688	methyl-diethyl-*iso*-amyl	$(CH_3)(C_2H_5)_2Pb\cdot C_5H_{11}$	20.8	1.523	1.5078	*IV-595
1689	methyl-diethyl-*n*-propyl	$(CH_3)(C_2H_5)_2Pb\cdot(C_3H_7)$	22.1	1.640	1.5141	*IV-592
1690	methyl-ethyl-*n*-propyl-*iso*-amyl	$(CH_3)(C_2H_5):Pb(C_3H_7)(C_5H_{11})$	21	1.479	1.5064	*IV-595
1691	methyl-tri-*iso*-amyl	$CH_3Pb(C_5H_{11})_3$	22	1.313	1.4962	*IV-596
1692	methyl-tri-*iso*-butyl	$(C_4H_9)_3PbCH_3$	20	1.397	1.5030	*IV-594
1693	methyl-triethyl	$CH_3Pb(C_2H_5)_3$	20	1.713	1.5183	*IV-591
1694	methyl-tri-*n*-propyl	$(C_3H_7)_3PbCH_3$	22.3	1.522	1.5091	. . . .
1695	propyl-tri-*iso*-amyl	$C_2H_5CH_2Pb(C_5H_{11})_3$	22	1.274	1.4970	*IV-596
1696	propyl-tri-*iso*-butyl	$C_2H_5CH_2Pb(C_4H_9)_3$	19.6	1.351	1.5056	*IV-594
1697	tetra-*iso*-amyl	$(C_5H_{11})_4Pb$	20.5	1.233	1.4946	*IV-596
1698	tetra-*iso*-butyl	$[(CH_3)_2CHCH_2]_4Pb$	20.2	1.324	1.5042	*IV-594
1699	tetraethyl	$(C_2H_5)_4Pb$	20	1.653	1.5198	IV-639
1700	tetramethyl	$(CH_3)_4Pb$	20	1.995	1.5120	IV-639
1701	tri-*iso*-butyl-*iso*-amyl	$(C_4H_9)_3Pb(C_5H_{11})$	20.6	1.298	1.5010	*IV-596
1702	triethyl-*iso*-amyl	$(C_2H_5)_3PbC_5H_{11}$	20	1.484	1.5099	*IV-595
1703	triethyl-*iso*-butyl	$(C_2H_5)_3PbC_4H_9$	20	1.531	1.5127	*IV-594
1704	triethyl-*n*-propyl	$(C_2H_5)_3PbC_3H_7$	20	1.589	1.5168	*IV-592
1705	trimethyl-*iso*-amyl	$(CH_3)_3PbC_5H_{11}$	20.3	1.525	1.4926	*IV-595
1706	trimethyl-*n*-butyl	$(CH_3)_3PbC_4H_9$	20	1.678	1.5046	*IV-593
1707	trimethyl-*iso*-butyl	$(CH_3)_3PbC_4H_9$	20	1.672	1.5026	*IV-593
1708	trimethyl-ethyl	$(CH_3)_3PbC_2H_5$	20	1.882	1.5154	*IV-591
1709	trimethyl-*n*-propyl	$(CH_3)_3PbC_3H_7$	20	1.767	1.5095	*IV-592
1710	tri-*n*-propyl-*iso*-amyl	$(C_2H_5CH_2)_3PbC_5H_{11}$	21	1.381	1.5047	*IV-594
1711	tri-*n*-propyl-*iso*-butyl	$(C_3H_7)_3PbC_4H_9$	22.6	1.403	1.5067	*IV-594
1712	**Levulinic** acid (m.p. 33°)	$CH_3CO(CH_2)_2CO_2H$	17	1.143	1.4429	III-671
1713	aldehyde	$CH_3CO(CH_2)_2CHO$	21.5	1.02	1.4257	I-774
1714	**Limonene** *d* (*l*)	$C_{10}H_{16}$	19.6	0.843	1.4727	V-133
1715	**Linalool** (*d*)	$C_{10}H_{18}O$	20	0.868	1.4623	I-461
1716	**Linalyl** acetate	$CH_3CO_2C_{10}H_{17}$	21	0.89	1.4544	II-141
1717	**Malonic nitrile** (m.p. 30°)	$CH_2(CN)_2$	34.2	1.049	1.4146	II-589
1718	**Malonyl chloride**	$CH_2(COCl)_2$	22.1	1.449	1.4617	II-582
1719	**Menthane** (*o*)	$C_{10}H_{20}$	21	0.814	1.447	V-46

Table 10-66 (*Continued*)
INDEX OF REFRACTION—LIQUIDS

No.	Substance	Formula	t,°C.	d_4^t	n_D^t	Beil. Ref.
1720	**Menthane** (*m*)	$C_{10}H_{20}$	20(?)	0.79	1.4420	V-46
1721	**Menthene** (*d*)	$C_{10}H_{18}$	20	0.81	1.4524	V-87
1722	**Menthol** (*l*) (*α*)	$C_{10}H_{19}OH$	25§	0.88	1.458	VI-28
1723	**Menthone**	$C_{10}H_{18}O$	20	0.896	1.4498	VII-38
1724	" (*dl*)	$C_{10}H_{18}O$	20(?)	0.898	1.4521	*VII-36
1725	**Menthyl** acetate	$C_{12}H_{22}O_2$	20	0.919	1.4467	VI-32
1726	amine (*l*)	$C_{10}H_{21}N$	20	0.860	1.4606	XII-26
1727	chloride (*sec*)	$C_{10}H_{19}Cl$	20	0.941	1.4642	V-49
1728	" (*tert*)	$C_{10}H_{19}Cl$	18.8	0.948	1.4655	V-49
1728.1	*iso*-valerate	$C_{15}H_{28}O_2$	20	0.906	1.4485	VI-33
1729	**Mercury** diethyl	$(C_2H_5)_2Hg$	23.2	2.423	1.5399	IV-679
1730	dimethyl	$(CH_3)_2Hg$	22.2	2.954	1.5327	IV-678
1731	**Mesityl**-amine (*ω*)	$(CH_3)_2C_6H_3CH_2NH_2$	20.5	0.950	1.5305	XII-1163
1732	ethyl ether	$(CH_3)_2C_6H_3CH_2O\cdot$ C_2H_5 (1,3,5)	20	0.930	1.496	....
1733	methyl ether	$CH_3OC_6H_2(CH_3)_3$ (1,2,4,6)	14.4	0.953	1.5055	VI-519
1734	oxide	$(CH_3)_2C:CHCOCH_3$	20	0.858	1.4440	I-736
1735	oxide oxime (*α*)	$C_6H_{10}:NOH$	21	0.942	1.4908†	I-738
1736	oxide oxime (*α*) (m.p. 49°)	$C_6H_{10}:NOH$	40	0.969	1.4900‡	I-739
1737	oxide oxime acetate (*α*)	$(CH_3)_2C:CHC(:NO\cdot$ $OCCH_3)(CH_3)$	20	0.990	1.478	II-186
1738	oxide oxime acetate (*β*)	$(CH_3)_2C:CHC(:NO\cdot$ $OCCH_3)(CH_3)$	20	0.990	1.478	II-186
1739	oxide oxime benzyl ether (*α*)	$(CH_3)_2C:CHC(CH_3):$ $NOCH_2C_6H_5$	20	0.987	1.533	VI-441
1740	**Methoxy**-acetic acid	$CH_3OCH_2CO_2H$	20	1.177	1.4168	III-232
1741	acetophenone (*o*)	$CH_3OC_6H_4COCH_3$	23.6	1.085	1.5379	VIII-85
1742	" (*m*)	$CH_3OC_6H_4COCH_3$	19	1.094	1.5367	VIII-86
1743	" (*p*) (m.p. 38°)	$CH_3OC_6H_4COCH_3$	41.3	1.082	1.5468	VIII-87
1744	benzaldehyde (*m*)	$CH_3OC_6H_4CHO$	20	1.119	1.5530	VIII-59
1745	coumarone (3)	$C_6H_4C(OCH_3):CH\cdot O$ ∣———∣	19.3	1.144	1.5589	*XVII-59
1746	1,3-dimethylben-zene (2)	$CH_3OC_6H_3(CH_3)_2$	20	0.956	1.502	VI-485
1747	3-methylaceto-phenone (6)	$(CH_3O)(CH_3)C_6H_3\cdot$ $COCH_3$	13.8	1.069	1.5377	*VIII-549
1748	methylene-*d*-cam-phor (m.p. 39°)	$C_{12}H_{18}O_2$	47.3	1.002	1.4999	VIII-28
1749	naphth-1-aldehyde (4) (m.p. 34°)	$CH_3OC_{10}H_6CHO$	43.7	1.183	1.6530	VIII-147
1750	phthalide (*α*)	$C_6H_4CO\cdot O\cdot CH(O\cdot$ ∣———∣ $CH_3)$	22.8	1.215	1.5352	XVIII-17
1751	styrene (*α*)	$C_6H_5C(OCH_3):CH_2$	20	1.00	1.5422	VI-563
1752	" (*β*)	$C_6H_5CH:CHOCH_3$	24.3	1.0	1.5620	VI-564
1753	" (*o*)	$CH_3OC_6H_4CH:CH_2$	17.4	1.003	1.5570	VI-560
1754	" (*p*)	$CH_3OC_6H_4CH:CH_2$	20(?)	1.0	1.5642	VI-561
1755	thionaphthene (3)	$C_6H_4C(OCH_3):CH\cdot S$ ∣———∣	13.6	1.209	1.6277	XVII-120
1756	toluene (*m*)	$CH_3C_6H_4OCH_3$	12.9	0.978	1.5164	VI-376
1757	*m*-toluic aldehyde (*o*)	$(CH_3)(CH_3O)C_6H_3\cdot$ CHO	20	1.098	1.5534	....

§ Supercooled.
† Labile; alkali stable form.
‡ Melted.

Table 10-66 (*Continued*)
INDEX OF REFRACTION—LIQUIDS

No.	Substance	Formula	$t,°C.$	d_4^t	n_D^t	Beil. Ref.
1758	**Methoxy**-valero-δ-lactone (γ)	$CH_3OCH_2CHCH_2\cdot$ $\mid\_\_\_\_$ $CH_2CO\cdot O$ $\_\_\_\_\mid$	23.2	1.120	1.4453	*XVIII-297
1759	o-xylene (*vic*) (m.p. 29°)	$(CH_3)_2C_6H_3OCH_3$ (1,2,3)	39.7	0.960	1.5120	VI-480
1760	o-xylene (*uns*)	$(CH_3)_2C_6H_3OCH_3$ (1,2,4)	20	0.969	1.517	VI-481
1761	m-xylene (*sym*)	$(CH_3)_2C_6H_3OCH_3$ (1,3,5)	20	0.958	1.513	VI-493
1762	" (*uns*)	$(CH_3)_2C_6H_3OCH_3$ (1,3,4)	12.9	0.970	1.5173	VI-486
1763	p-xylene	$(CH_3)_2C_6H_3OCH_3$	20	0.963	1.515	VI-494
1764	**Methyl acetate**	$CH_3CO_2CH_3$	20	0.934	1.3594	II-124
1765	acetoacetate	$CH_3COCH_2CO_2CH_3$	20	1.076	1.4196	III-632
1766	α′-acetothienone (α) (m.p. 25°)	$CH_3COC:CHCH:C\cdot$ $\mid\_\_\_\_$ $(CH_3)\cdot S$ $\_\_\_\_\mid$	24.7	1.119	1.560	XVII-296
1767	β-acetoxystyrene (α)	$C_6H_5C(CH_3):CH\cdot$ $O\cdot COCH_3$	15.2	1.057	1.5428	*VI-285
1768	2-acetoxystyrene (1′)	$CH_3CO_2C_6H_4CH:$ $CHCH_3$	13.5	1.048	1.5175	*VI-284
1769	3-acetoxystyrene (1′)	$CH_3CO_2C_6H_4\cdot$ $C(CH_3):CH_2$	14.1	1.061	1.5325	*VI-285
1770	acetylacetone	$CH_3COCH(CH_3)CO\cdot$ CH_3	13.5	0.978	1.4380	I-791
1771	acetylcamphor-carboxylate	$C_{14}H_{20}O_4$	19.8	1.082	1.4844	X-37
1772	2-acetylnaphthalene (1)	$CH_3C_{10}H_6COCH_3$	16.1	1.105	1.6270	. . . .
1773	(5) 8-acetylnaphthalene (2)	$CH_3C_{10}H_6COCH_3$	13.3	1.099	1.6244	. . . .
1774	acrylate¶	$CH_2:CHCO_2CH_3$	20	0.95	1.4600	II-399
1775	acrylic acid (α)	$CH_2:C(CH_3)CO_2H$	20	1.015	1.4314	II-421
1776	alcohol	CH_3OH	20	0.791	1.3288	I-274
1777	allocinnamate	$C_6H_5CH:CHCO_2CH_3$	20	1.078	1.5575	IX-594
1778	alloethyl camphorate (o)	$C_{13}H_{22}O_4$	17.1	1.049	1.4589	. . . .
1779	allo-β-methyl-cinnamate (m.p. 27°)	$C_6H_5C(CH_3):CHCO_2\cdot$ CH_3	38.1	1.037	1.5276	*IX-253
1780	allo-β-methyl-o-methoxycinnamate (m.p. 44°)	$CH_3OC_6H_4C(CH_3):$ $CHCO_2CH_3$	17	1.098	1.5390	*X-138
1781	allylcamphor carboxylate (m.p. 75°)	$C_{15}H_{22}O_3$	20.2	1.067	1.4911	X-653
1782	allyl carbinol	$(CH_3)(CH_2:CH\cdot$ $CH_2)CHOH$	20	0.834	1.425	I-443
1783	allyl ether	$CH_3OCH_2CH:CH_2$	?	0.7	1.38±	I-437
1784	2-allylphenol (4)	$CH_3C_6H_3(OH)CH_2\cdot$ $CH:CH_2$	20	1.002	1.541	*VI-287
1785	aminobenzoate (o)	$NH_2C_6H_4CO_2CH_3$	20	1.16	1.5844	XIV-317
1786	aminoethanol (β)	$CH_3NHCH_2CH_2OH$	20	0.937	1.4385	IV-276
1787	o-aminophenyl-sulfide	$CH_3\cdot S\cdot C_6H_4NH_2$	17	1.133	1.6263	XIII-399

¶ Liquid polymer.

Table 10-66 (*Continued*)
INDEX OF REFRACTION—LIQUIDS

No.	Substance	Formula	t,°C.	d_4^t	n_D^t	Beil. Ref.
1788	**Methyl** *n*-amyl ether	$CH_3OC_5H_{11}$	20	0.754	1.3849	**I-417
1789	*n*-amyl ketone	$CH_3CO(CH_2)_4CH_3$	20	0.817	1.4083	I-699
1790	*iso*-amyl ketoxime	$CH_3C(NOH)C_5H_{11}$	20(?)	0.888	1.4448	I-701
1791	β-amyl-β-meth- oxyacrylate	$C_5H_{11}C(OCH_3):CH\cdot$ CO_2CH_3	23.1	0.961	1.4549	III-382
1792	amylpropiolate	$C_5H_{11}C:CCO_2CH_3$	12.3	0.934	1.4509	II-487
1793	Δ^1-angelolactone (β)	$CH_3C:C(CH_3)CH_2\cdot$ $\begin{array}{c}\text{\textbar}\underline{\hspace{2cm}}\\ CO\cdot O\\ \underline{\hspace{1cm}}\text{\textbar}\end{array}$	20	1.056	1.4545	*XVII-139
1794	aniline	$C_6H_5NHCH_3$	21.5	0.986	1.5702	XII-135
1795	anthracene (α) (m.p. 86°)	$CH_3C_{14}H_9$	99.4	1.047	1.6803	V-674
1796	anthracene (9) (m.p. 80°)	$C_{15}H_{12}$	99.4	1.066	1.6959	**V-586
1797	anthranil	$\begin{array}{c}\underline{\hspace{0.5cm}}\text{\textbar}\quad\text{\textbar}\underline{\hspace{0.5cm}}\\ C_6H_4N\cdot O\cdot CCH_3\\ \text{\textbar}\underline{\hspace{1.5cm}}\text{\textbar}\end{array}$	20	1.133	1.5780	XXVII-45
1798	benzalacetone (α,α)	$C_6H_5CH:C(CH_3)\cdot$ $COCH_3$	45.3	1.002	1.5691	VII-373
1799	benzoate	$C_6H_5CO_2CH_3$	18.7	1.092	1.5144	IX-109
1800	benzoylacetate	$C_6H_5COCH_2CO_2CH_3$	24.7	1.15	1.5365	X-673
1801	α-benzoyl-acetone (α)	$C_6H_5COCH(CH_3)\cdot$ $COCH_3$	20	1.079	1.5382	*VII-369
1802	*o*-benzoylbenzoate (m.p. 52°)	$C_6H_5COC_6H_4CO_2\cdot$ CH_3	19.7	1.190	1.5913	X-748
1803	benzylcarbinol (*d*)	$C_6H_5CH_2CHOHCH_3$	20	0.991	1.5174	VI-503
1804	benzyl ketone	$CH_3COCH_2C_6H_5$	20	1.028	1.5168	VII-303
1805	benzylthiophene (α)	$\begin{array}{c}S\cdot CH:CH\cdot CH:C\cdot\\ \text{\textbar}\underline{\hspace{2cm}}\text{\textbar}\\ CH_2C_6H_4CH_3\end{array}$	20	1.083	1.586	....
1806	*d*-bornyl ether	$CH_3OC_{10}H_{17}$	23.4	0.914	1.4624	I-78
1807	5-bromocoumaran (2)	$\begin{array}{c}BrC_6H_3CH_2CH\cdot\\ \text{\textbar}\underline{\hspace{2cm}}\\ (CH_3)\cdot O\\ \underline{\hspace{1.5cm}}\text{\textbar}\end{array}$	14.4	1.458	1.5727	*XVII-23
1808	5-bromocoumarone (2) (m.p. 20°)	$\begin{array}{c}BrC_6H_3\cdot CH:\\ \text{\textbar}\underline{\hspace{2cm}}\\ C(CH_3)\cdot O\\ \underline{\hspace{1.5cm}}\text{\textbar}\end{array}$	34.8	1.477	1.5915	....
1809	*o*-bromophenyl sulfide	$CH_3\cdot S\cdot C_6H_4Br$	20	1.522	1.6319	**VI-300
1810	α-bromopropionate	$CH_3CHBrCO_2CH_3$	20	1.492	1.4506	II-253
1811	β-bromopropionate	$BrCH_2CH_2CO_2CH_3$	20	1.519	1.4570	II-256
1812	4-bromopyrazole (3) (m.p. 76°)	$\begin{array}{c}CH:CBrC(CH_3):\\ \text{\textbar}\underline{\hspace{2cm}}\\ N\cdot NH\\ \underline{\hspace{1cm}}\text{\textbar}\end{array}$	99.6	1.564	1.52	XXIII-61
1813	buta-1,3-diene (2)	$CH_2:CHC(CH_3):CH_2$	20	0.681	1.4216	I-252
1814	buta-2,3-diene (2)	$CH_2:C:C(CH_3)_2$	20	0.680	1.410	I-252
1815	but-1-ene (2)	$CH_2:C(CH_3)C_2H_5$	20	0.650	1.3778	I-211
1816	but-2-ene (2)	$(CH_3)_2C:CHCH_3$	20	0.662	1.3874	I-211
1817	but-3-ene (2)	$CH_2:CHCH(CH_3)_2$	15	0.627	1.3643	I-213
1818	*n*-butylamine	$CH_3NHC_4H_9$	18.1	0.736	1.4018	IV-157
1819	*n*-butylbenzene (*o*)	$CH_3C_6H_4C_4H_9$	18.3	0.870	1.4966	V-437

Table 10-66 (*Continued*)
INDEX OF REFRACTION—LIQUIDS

No.	Substance	Formula	t,°C.	d_4^t	n_D^t	Beil. Ref.
1820	**Methyl**-n-butyl-benzene (*m*)	$CH_3C_6H_4C_4H_9$	18.4	0.86	1.4932	V-437
1821	n-butylbenzene (*p*)	$CH_3C_6H_4C_4H_9$	14.2	0.861	1.4912	V-437
1822	butylcarbamate (*n*)	$C_4H_9NHCO_2CH_3$	23.3	0.969	1.4289	IV-158
1823	" (*iso*)	$C_4H_9NHCO_2CH_3$	21.6	0.965	1.4275	IV-168
1824	" (*sec*)	$C_4H_9CH(CH_3)NH\cdot CO_2CH_3$	24.9	0.965	1.4263	IV-162
1825	5-*tert*-butyl-coumarone (2)	$(CH_3)_3CC_6H_3CH:$ $\overline{\qquad}$ $C(CH_3)\cdot O$ $\overline{\qquad}$	20	0.988	1.5355	....
1826	n-butyl ether	$CH_3OCH_2CH_2C_2H_5$	20	0.744	1.38	I-369
1827	n-butyl hydrazine (*uns*)	$CH_3(C_4H_9)NNH_2$	21.3	0.804	1.4259	IV-552
1828	*iso*-butyl ketone	$CH_3COCH_2CH:$ $(CH_3)_2$	20	0.800	1.3959	I-691
1829	*sec*-butyl ketone	$CH_3COCH(CH_3)\cdot$ (C_2H_5)	18	0.815	1.4002	I-693
1830	n-butylnitroamine	$CH_3N(NO_2)C_4H_9$	23.6	1.023	1.4563	IV-158
1831	*iso*-butylnitroamine	$(CH_3)_2CH\cdot CH_2N(NO_2)CH_3$	22.5	1.021	1.4561	IV-172
1832	n-butylnitro-carbamate	$C_4H_9N(NO_2)CO_2\cdot CH_3$	22	1.139	1.4486	IV-159
1833	*iso*-butylnitro-carbamate	$(CH_3)_2CH\cdot CH_2N(NO_2)CO_2\cdot CH_3$	21.9	1.135	1.4461	IV-172
1834	*sec*-butylnitro-carbamate	$C_4H_9N(NO_2)CO_2\cdot CH_3$	23.3	1.125	1.4417	IV-163
1835	α,β-butylene-oxide-δ-car-boxylate	$CH_3O_2CCH_2CH_2\cdot CHCH_2O$ $\overline{\qquad}$	24.7	1.073	1.4259	....
1836	but-3-yne (2)	$CH:CCH(CH_3)_2$	20	0.665	1.378	I-251
1837	n-butyrate	$C_2H_5CH_2CO_2CH_3$	20	0.898	1.39	II-270
1838	*iso*-butyrate	$(CH_3)_2CHCO_2CH_3$	20	0.891	1.3840	II-290
1839	butyric acid (α)	$(C_2H_5)(CH_3):CH\cdot CO_2H$	20	0.941	1.4051	II-304
1840	α-butyrocyano-acetate	$C_2H_5CH_2COCH\cdot (CN)CO_2CH_3$	20	1.093	1.4763	III-807
1841	camphidine (*N*)	$C_{11}H_{21}N$	19	0.901	1.4763	....
1842	camphocarboxylate	$C_{12}H_{18}O_3$	18	1.085	1.4833	X-644
1843	carbamate (m.p. 54°)	$NH_2CO_2CH_3$	55.6	1.136	1.4125	III-21
1844	"carbitol"	$CH_3O(CH_2)_2O\cdot (CH_2)_2OH$	27	1.03	1.4264	....
1845	carvacryl ether	$CH_3(OCH_3)C_6H_3\cdot CH(CH_3)_2$	20	0.944	1.506	VI-529
1846	"cellosolve"	$CH_3OCH_2CH_2OH$	20	0.964	1.4028	....
1847	chloroacetate	$ClCH_2CO_2CH_3$	20	1.236	1.4221	II-197
1848	α-chlorocrotonate	$CH_3CH:CClCO_2CH_3$	22.6	1.159	1.4563	*II-189
1849	β-chloroethyl ether	$CH_3OCH_2CH_2Cl$	20	1.046	1.4111	I-337
1850	chloroformate	$ClCO_2CH_3$	20	1.223	1.3868	III-9
1851	chromane (6)	$CH_3C_6H_3(CH_2)_3\cdot O$ $\overline{\qquad}$	14.3	1.037	1.5421	*XVII-24

Table 10-66 (*Continued*)
INDEX OF REFRACTION—LIQUIDS

No.	Substance	Formula	t,°C.	d_4^t	n_D^t	Beil. Ref.
1852	**Methyl**-chromanone (6) (m.p. 35°)	CH₃C₆H₃COCH₂· I_____I CH₂O ___I	57.1	1.125	1.5553	*XVII-163
1853	cinnamate (m.p. 35°)	C₆H₅CH:CHCO₂CH₃	21.4	1.05	1.5766	IX-581
1854	coumarane (2)	C₆H₄CH₂CH(CH₃)·O I_____I	13.7	1.036	1.5338	*XVII-23
1855	" (5)	CH₃C₆H₃CH₂CH₂·O I_____I	18.6	1.046	1.5404	XVII-53
1856	coumaranone (2)	C₆H₄COCH(CH₃)·O I_____I	19.7	1.153	1.5632	XVII-122
1857	" (5) (m.p. 54°)	CH₃C₆H₃COCH₂·O I_____I	53.8	1.151	1.5652	XVII-123
1858	*o*-coumarketone methyl ether (m.p. 48°)	CH₃OC₆H₄CH:CH· COCH₃	61.4	1.053	1.5859	*VIII-559
1859	coumarone (2)	C₆H₄CH:C(CH₃)·O I_____I	20	1.054	1.5589	XVII-60
1860	" (3)	C₆H₄C(CH₃):CH·O I_____I	23.4	1.054	1.5526	XVII-60
1861	" (5)	CH₃C₆H₃CH:CH·O I_____I	19	1.060	1.5573	XVII-61
1862	" (7)	CH₃C₆H₃CH:CH·O I_____I	17	1.05	1.5525	XVII-61
1863	*o*-cresyl diketone ethyl ether (m.p. 64°)	(C₂H₅O)(CH₃)C₆H₃· COCOCH₃	65	1.052	1.5127	*VIII-628
1864	*p*-cresyl diketone methyl ether (m.p. 60°)	CH₃C₆H₃(OCH₃)· COCOCH₃	16†	1.119	1.5378	*VIII-628
1865	cresyl ether (*o*)	CH₃OC₆H₄CH₃	15.3	0.985	1.5189	VI-352
1866	" " (*p*)	CH₃OC₆H₄CH₃	19.3	0.97	1.5124	VI-392
1867	cyclobutane	CH₃CHCH₂CH₂CH₂ I_____I	18	0.69	1.3847	V-20
1868	cyclohexane	CH₃(CHC₅H₁₀)	20	0.769	1.4231	V-29
1869	cyclohexan-1-ol (1)	CH₃COH(CH₂)₄CH₂ I_____I	20	0.930	1.4620	VI-11
1870	cyclohexan-2-one (1)	CH₃C₅H₉CO	20	0.926	1.4493	VII-14
1873	cyclohexan-3-one (1)	CH₃C₅H₉CO	20	0.896	1.4478	VII-17
1874	cyclohexan-4-one (1)	CH₃C₅H₉CO	20	0.914	1.4458	VII-18
1875	cyclohexene-acetate	CH₂(CH₂)₃CH:C· I_____I CH₂CO₂CH₃	18.2	1.003	1.4687	*IX-23
1876	cyclohex-(1 or 6)-ene-2-acetic acid (2)	CH₃C₆H₈CH₂CO₂H	15.8	1.028	1.4817	*IX-29
1877	cyclohex-(2 or 3)-ene-3-acetic acid (1)	(CH₃)C₆H₈CH₂· CO₂H	15.4	1.027	1.4824	*IX-30

† Supercooled.

Table 10-66 (*Continued*)
INDEX OF REFRACTION—LIQUIDS

No.	Substance	Formula	t,°C.	d_4^t	n_D^t	Beil. Ref.
1878	**Methyl**-cyclohex-1-ene-3-one (1)	CH$_3$C:CHCO· | ____ (CH$_2$)$_2$CH$_2$ ____|	18.7	0.971	1.4955	VII-54
1879	α-cyclohexene-propionate	C$_8$H$_{13}$CO$_2$CH$_3$	18.3	0.986	1.4665	*IX-29
1880	cyclohexyli-deneacetate	CH$_2$(CH$_2$)$_4$C:CHCO$_2$· | ____| CH$_3$	19.7	0.999	1.4831	*IX-24
1881	cyclopentane	CH$_3$(CHC$_4$H$_8$)	20	0.749	1.4097	V-27
1882	cyclopentan-1-ol-1-carboxylate	CH$_2$(CH$_2$)$_3$C(OH)· | ____| CO$_2$CH$_3$	17	1.105	1.4567	*X-3
1883	1-cyanohexan-2-one (1)	CH$_3$C$_5$H$_8$(CO)CN	20	1.015	1.461	....
1884	1-cyanohexan-2-one (3)	CH$_3$C$_5$H$_8$(CO)CN	20	1.040	1.471	....
1885	cymene (2)	(CH$_3$)$_2$CHC$_6$H$_3$: (CH$_3$)$_2$ (1,2,4)	15.5	0.874	1.5000	V-440
1886	n-decyl ether	CH$_3$OC$_{10}$H$_{21}$	20	0.797	1.4218	....
1887	diacetone ether	(CH$_3$)$_2$C(OCH$_3$)· CH$_2$COCH$_3$	20	0.909	1.487	**I-876
1888	γ,γ-dichloro-crotonate	Cl$_2$CHCH:CHCO$_2$· CH$_3$	20	1.302	1.4694	**II-397
1889	1-dichloromethyl-cyclohexane (1)	(CH$_3$)(CHCl$_2$): C(CH$_2$)$_4$CH$_2$ | _____|	23.7	1.127	1.4895	*V-14
1890	1-dichloromethyl-cyclohexan-2-one (1)	CH$_3$C$_5$H$_8$COCHCl$_2$	17.5	1.235	1.5008	*VII-17
1891	1-dichloromethyl-cyclohexan-4-one (1) (m.p. 47°)	(CH$_3$)(CHCl$_2$)C$_5$H$_8$: CO	52.8	1.221	1.4978	*VII-18
1892	1-dichloromethyl-4-ethyl-cyclohex-2-ene-6-one (1)	(CH$_3$)(C$_2$H$_5$)C$_5$H$_5$· COCHCl$_2$	17.4	1.167	1.5029	VII-86
1893	dichloromethyl-ketodihydroben-zene (o) (m.p. 33°)	(CH$_3$)(CHCl$_2$)CCO· | ____ CH:CHCH:CH _____|	33.5	1.270	1.5328	VII-149
1894	dichloromethyl-ketodihydroben-zene (p) (m.p. 55°)	(CH$_3$)(CHCl$_2$): CCH:CHCOCH:CH | _____|	56.2	1.26	1.5342	VII-149
1895	1-dichloromethyl-4-iso-propylcyclo-hex-3-ene-2-one (1)	(C$_3$H$_7$)(CH$_3$)C$_5$H$_5$· COCHCl$_2$	15.8	1.165	1.5177	VII-138
1896	2,4-diethylnaph-thalene (1)	(CH$_3$)(C$_2$H$_5$)$_2$C$_{10}$H$_5$	13.3	0.987	1.5923	**V-473
1897	α,α-dimethylaceto-acetate	CH$_3$COC(CH$_3$)$_2$CO$_2$· CH$_3$	16.8	1.012	1.4217	III-695
1898	β,β-dimethyl-acrylate	(CH$_3$)$_2$C:CHCO$_2$CH$_3$	19.8	0.934	1.4321	II-433
1899	α,p-dimethylcin-namate	CH$_3$C$_6$H$_4$CH: C(CH$_3$)CO$_2$CH$_3$	18.3	1.051	1.5640	*IX-261

Table 10-66 (*Continued*)
INDEX OF REFRACTION—LIQUIDS

No.	Substance	Formula	t,°C.	$d_4{}^t$	$n_D{}^t$	Beil. Ref.
1900	**Methyl-**β**,**p**-dimethyl-**cinnamate (m.p. 45°)	$CH_3C_6H_4C(CH_3):$ $CHCO_2CH_3$	60.2	1.014	1.5408	IX-624
1901	1,4-dimethylcyclo-hexa-1,3-diene-2-carboxylate	$(CH_3)_2C_6H_5CO_2CH_3$	18.2	0.997	1.4755	IX-83
1902	1,4-dimethylcyclo-hex-3-ene-2-carboxylate	$(CH_3)_2C_6H_7CO_2CH_3$	20	0.970	1.4588	*IX-31
1903	3,5-dimethylcyclo-hexylideneacetate	$(CH_3)_2C_6H_8:CHCO_2 \cdot$ CH_3	15	0.957	1.4763	*IX-35
1904	α,m-dimethyl-o-methoxycin-namate	$(CH_3)(CH_3O)C_6H_3 \cdot$ $CH:C(CH_3)CO_2 \cdot$ CH_3	14.6	1.099	1.5642	*X-141
1905	α,α-dimethyloxy-propionate	$CH_3C(OCH_3)_2CO_2 \cdot$ CH_3	17.6	1.068	1.4122	*III-219
1906	diphenylamine (N)	$(C_6H_5)_2NCH_3$	20	1.048	1.6193	XII-180
1907	l-ecgoninate	$C_{10}H_{17}O_3N$	20.5	1.147	1.4877	XXII-198
1908	erucate	$C_{21}H_{41}CO_2CH_3$	20	0.870	1.4558	**II-446
1909	3-ethoxycoumarone (5)	$CH_3C_6H_3C(OC_2H_5):$ $\vert$_____ $CH \cdot O$ _____$\vert$	16.6	1.083	1.5440	*XVII-64
1910	4-ethyl-2-acetyl-naphthalene (1)	$C_{10}H_5(CH_3)(C_2H_5) \cdot$ $(COCH_3)$	15.1	1.074	1.6118	. . . .
1911	β-ethylacrolein (α)	$C_3H_6:C(CH_3) \cdot CHO$	20	0.856	1.4454	I-735
1912	2-ethylbenzene (1)	$CH_3C_6H_4C_2H_5$	20	0.881	1.5045	V-396
1913	3-ethylbenzene (1)	$CH_3C_6H_4C_2H_5$	20	0.865	1.4966	V-396
1914	4-ethylbenzene (1)	$CH_3C_6H_4C_2H_5$	20	0.861	1.4950	V-397
1915	α-ethyl-β-chloro-styrene (4)	$CH_3C_6H_4C(C_2H_5):$ $CHCl$	19.9	1.036	1.5418	*V-239
1916	4-ethylcyclohexa-1,3-diene (1)	$CH_3C_6H_6C_2H_5$	19.4	0.837	1.4825	V-121
1917	2-ethylcyclohexa-1,5-diene-1-car-boxylic acid (5)	$(CH_3)(C_2H_5)C_6H_5 \cdot$ CO_2H	14.8	1.028	1.4989	*IX-47
1918	1-ethylcyclo-pentane (1)	$CH_3 \cdot C_5H_8 \cdot C_2H_5$	20	0.781	1.4272	. . . .
1919	2-ethylcyclo-pentane (1) (*cis*)	$CH_3 \cdot C_5H_8 \cdot C_2H_5$	20	0.785	1.4295	V-39
1920	2-ethylcyclopen-tane (1) (*trans*)	$CH_3 \cdot C_5H_8 \cdot C_2H_5$	20	0.769	1.4219	V-39
1921	3-ethylcyclopen-tane (1) (*cis*)	$CH_3 \cdot C_5H_8 \cdot C_2H_5$	20	0.772	1.420	V-39
1922	3-ethylcyclopen-tane (1) (*trans*)	$CH_3 \cdot C_5H_8 \cdot C_2H_5$	20	0.762	1.4186	V-39
1923	5-ethylhept-5-ene (2)	$(CH_3)_2CH(CH_2)_3 \cdot$ $C(C_2H_5):CHCH_3$	20(?)	0.75	1.4271	I-224
1924	3-ethylhexane (2)	$(CH_3)_2CHCH:$ $(C_2H_5)(CH_2C_2H_5)$	20	0.731	1.411	. . . .
1925	4-ethylhexane (2)	$(CH_3)_2CHCH_2CH:$ $(C_2H_5)_2$	20	0.723	1.407	. . . .
1926	3-ethylhexane (3)	$(C_2H_5)_2(CH_3)C \cdot$ $CH_2C_2H_5$	20	0.741	1.415	. . . .
1927	4-ethylhexane (3)	$(C_2H_5)(CH_3)CH \cdot$ $CH(C_2H_5)_2$	20	0.742	1.416	. . . .

Table 10-66 (*Continued*)
INDEX OF REFRACTION—LIQUIDS

No.	Substance	Formula	t,°C.	d_4^t	n_D^t	Beil. Ref.
1928	**Methyl**-4-ethylhexan-4-ol (2)	$(C_2H_5)_2COHCH_2CH$:$(CH_3)_2$	13	0.8	1.4346	I-425
1929	1-ethylindazole (3)	$C_2H_5N \cdot C_6H_4C \cdot$ (CH_3) :N	20	1.035	1.561	XXIII-142
1930	2-ethylindazole (3)	$C_2H_5N \cdot N$:C_6H_4:$C \cdot$ CH_3	20	1.058	1.587	XXIII-142
1931	1-ethylindazole (5)	$CH_3C_6H_3CH$:$N \cdot N \cdot$ C_2H_5	20	1.043	1.568	....
1932	2-ethylindazole (5)	$C_2H_5N \cdot N$:$C_6H_3 \cdot$ (CH_3) :CH	20	1.050	1.581	....
1933	ethylketazine	$[CH_3(C_2H_5)C$:$N]_2$	26.6	0.834	1.4516	I-669
1934	ethyl ketone	$CH_3COC_2H_5$	15.9	0.81	1.3807	I-666
1935	ethyl ketoxime	$CH_3C(NOH)C_2H_5$	20(?)	0.923	1.4428	I-668
1936	2-ethylnaphthalene (1)	$CH_3C_{10}H_6C_2H_5$	15.4	1.001	1.6014	**V-470
1937	4-ethylnaphthalene (1)	$CH_3C_{10}H_6C_2H_5$	13.1	1.009	1.6057	**V-470
1938	ethylnitroamine	$CH_3N(NO_2)C_2H_5$	19.2	1.097	1.4571	IV-130
1939	ethyloximino-cyanoacetate	$NCC(:NOC_2H_5) \cdot$ CO_2CH_3	20	1.124	1.4578	III-775
1940	3-ethylpentane (2)	$(C_2H_5)_2CHCH$:$(CH_3)_2$	20	0.719	1.4040	I-164
1941	" (3)	$(C_2H_5)_3CCH_3$	20	0.727	1.4078	....
1942	3-ethylpent-1-ene (2)	CH_2:$C(CH_3)CH$:$(C_2H_5)_2$	20	0.730	1.415	....
1943	3-ethylpent-4-ene (2)	$(CH_3)_2CHCH(C_2H_5) \cdot$ CH:CH_2	20	0.726	1.410	....
1944	3-ethylpent-1-ene (3)	CH_2:$CHC(CH_3)$:$(C_2H_5)_2$	20	0.731	1.418	....
1945	3-methylpyrazole (1)	CH:$CHC(CH_3)$:$N \cdot N$ C_2H_5	20	0.945	1.472	....
1946	ethyl pyrotartrate	$C_5H_6O_4(CH_3)(C_2H_5)$	21.9	1.029	1.4196	II-639
1947	eugenyl ether	CH_2:$CHCH_2C_6H_3(O \cdot$ $CH_3)_2$ (1,3,4)	17	1.036	1.5383	VI-963
1948	*iso*-eugenyl ether (1,3,4)	CH_2CH:$CH \cdot$ $C_6H_3(OCH_3)_2$	19.1	1.052	1.57	VI-956
1949	fluorene (9) (m.p. 46°)	$C_6H_4 \cdot C_6H_4 \cdot CHCH_3$	66.2	1.026	1.6101	V-642
1950	formate	HCO_2CH_3	20	0.974	1.344	II-18
1951	furan (2)	$CH_3C_4H_3O$	20	0.913	1.4344	XVII-36
1952	furfuryl ether	$CH_3OCH_2C_4H_3O$	20	1.02	1.4570	XVII-112
1953	furoate	$C_4H_3OCO_2CH_3$	20	1.18	1.4860	XVIII-274
1954	geranate	$C_{11}H_{18}O_2$	19.1	0.923	1.4714	*II-210
1955	hemimellitenyl ether	$(CH_3)_3C_6H_2OCH_3$ (1,2,3,5)	20	0.976	1.524	**VI-480
1956	hepta-1,6-diene-4-ol (4)	$(CH_2$:$CHCH_2)_2COH \cdot$ CH_3	20.6	0.844	1.4417	I-456
1957	heptane (2)	$(CH_3)_2CHC_5H_{11}$	20	0.698	1.3950	I-161

Table 10-66 (*Continued*)
INDEX OF REFRACTION—LIQUIDS

No.	Substance	Formula	$t,°C.$	d_4^t	n_D^t	Beil. Ref.
1958	**Methyl**-heptane (3)	$C_2H_5CH(CH_3)\cdot$ $(CH_2)_3CH_3$	20	0.706	1.3985	I-162
1959	" (4)	$(C_2H_5CH_2)_2CHCH_3$	20	0.705	1.3979	I-162
1960	heptan-2-ol (2)	$CH_3(CH_2)_4COH:$ $(CH_3)_2$	20(?)	0.879	1.4303	I-420
1961	heptan-3-ol (2)	$(CH_3)_2CHCHOH\cdot$ C_4H_9	20	0.825	1.4204	I-421
1962	heptan-4-ol (2)	$(CH_3)_2CHCH_2\cdot$ $CHOHC_3H_7$	20	0.821	1.4203	I-421
1963	heptan-5-ol (2)	$(CH_3)_2CH(CH_2)_2\cdot$ $CHOHC_2H_5$	20	0.808	1.4201	I-421
1964	heptan-6-ol (2)	$(CH_3)_2CH(CH_2)_3\cdot$ $CHOHCH_3$	20	0.813	1.4238	I-421
1965	heptan-3-ol (3)	$(CH_3)(C_2H_5)COH\cdot$ (C_4H_9)	19	0.827	1.4274	I-421
1966	hept-1-ene (2)	$CH_2:C(CH_3)C_5H_{11}$	20	0.721	1.4123	....
1967	hept-2-ene (2)	$(CH_3)_2C:CHC_4H_9$	20	0.725	1.4170	I-222
1968	hept-3-ene (2) (*cis?*)	$(CH_3)_2CHCH:CH\cdot$ $CH_2C_2H_5$	20	0.706	1.407	....
1969	hept-3-ene (2) (*trans*)	$(CH_3)_2CHCH:CH\cdot$ $CH_2C_2H_5$	....	.....	......	....
1970	hept-4-ene (2) (*cis?*)	$C_2H_5CH:CHCH_2\cdot$ $CH(CH_3)_2$	20	0.713	1.410	....
1971	hept-4-ene (2) (*trans*)	$C_2H_5CH:CHCH_2\cdot$ $CH(CH_3)_2$	....	.....	......	....
1972	hept-5-ene (2) (*cis?*)	$CH_3CH:CHCH_2CH_2\cdot$ $CH(CH_3)_2$	20	0.718	1.412	....
1973	hept-5-ene (2) (*trans*)	$CH_3CH:CHCH_2CH_2\cdot$ $CH(CH_3)_2$	....	.....	......	....
1974	hept-6-ene (2)	$CH_2:CH(CH_2)_3CH:$ $(CH_3)_2$	20	0.712	1.4070	*I-94
1975	hept-1-ene (3)	$CH_2:CHCH(CH_3)\cdot$ C_4H_9	20	0.711	1.406	....
1976	hept-2-ene (3) (*cis?*)	$CH_3CH:C(CH_3)\cdot$ C_4H_9	20	0.729	1.419	....
1977	hept-2-ene (3) (*trans*)	$CH_3CH:C(CH_3)\cdot$ C_4H_9	....	.....	......	....
1978	hept-3-ene (3) (*cis?*)	$C_2H_5C(CH_3):CH\cdot$ $CH_2C_2H_5$	20	0.728	1.418	....
1979	hept-3-ene (3) (*trans*)	$C_2H_5C(CH_3):CH\cdot$ $CH_2C_2H_5$	....	.....	......	....
1980	hept-4-ene (3) (*cis?*)	$C_2H_5CH:CHCH:$ $(CH_3)(C_2H_5)$	20	0.713	1.410	....
1981	hept-4-ene (3) (*trans*)	$C_2H_5CH:CHCH:$ $(CH_3)(C_2H_5)$	....	.....	......	....
1982	hept-5-ene (3) (*cis?*)	$CH_3CH:CHCH_2CH:$ $(CH_3)(C_2H_5)$	20	0.723	1.414	....
1983	hept-5-ene (3) (*trans*)	$CH_3CH:CHCH_2CH:$ $(CH_3)(C_2H_5)$	....	.....	......	....
1984	hept-6-ene (3)	$CH_2:CHCH_2CH_2\cdot$ $CH(CH_3)(C_2H_5)$	20	0.716	1.4094	....
1985	hept-1-ene (4)	$CH_2:CHCH_2CH:$ $(CH_3)(C_3H_7)$	20	0.717	1.410	....
1986	hept-2-ene (4) (*cis?*)	$CH_3CH:CHCH:$ $(CH_3)(C_3H_7)$	20	0.716	1.410	....

Table 10-66 (*Continued*)
INDEX OF REFRACTION—LIQUIDS

No.	Substance	Formula	t,°C.	d_4^t	n_D^t	Beil. Ref.
1987	**Methyl**-hept-2-ene (4) (*trans*)	$CH_3CH:CHCH:$ $(CH_3)(C_3H_7)$	....	.....	......	....
1988	hept-3-ene (4) (*cis?*)	$C_2H_5CH:C(CH_3)\cdot$ $CH_2C_2H_5$	20	0.725	1.417	I-222
1989	hept-3-ene (4) (*trans*)	$C_2H_5CH:C(CH_3)\cdot$ $CH_2C_2H_5$	....	.....	......	....
1990	hept-2-ene-6-ol (2)	$(CH_3)_2C:CH(CH_2)_2\cdot$ $CHOHCH_3$	20	0.854	1.4505	I-448
1991	hept-1-ene-4-ol (4)	$CH_2:CHCH_2COH\cdot$ $(CH_3)CH_2C_2H_5$	21.3	0.834	1.4376	I-449
1992	hept-2-ene-6-one (2)	$(CH_3)_2C:CH(CH_2)_2\cdot$ $COCH_3$	20	0.860	1.4403	I-741
1993	*n*-heptylate	$CH_3(CH_2)_5CO_2CH_3$	20	0.88	1.4114	II-339
1994	*n*-heptyl ketone	$CH_3(CH_2)_6COCH_3$	20(?)	0.832	1.4279	I-709
1995	hexane (2)	$(CH_3)_2CH(CH_2)_3CH_3$	20	0.679	1.3849	I-156
1996	" (3)	$C_2H_5CH(CH_3)\cdot$ $CH_2C_2H_5$	20	0.687	1.3887	I-157
1997	hexan-1-ol (2)	$CH_3(CH_2)_3CH:$ $(CH_3)CH_2OH$	20	0.82	1.4226	I-415
1998	hexan-2-ol (2)	$(CH_3)_2COH(CH_2)_2\cdot$ CH_3	20(?)	0.816	1.4159	I-415
1999	hexan-3-ol (2)	$(CH_3)_2CHCHOH\cdot$ $CH_2C_2H_5$	20	0.82	1.4215	I-416
2000	hexan-6-ol (2)	$(CH_3)_2CH(CH_2)_3\cdot$ CH_2OH	20(?)	0.82	1.4254	I-416
2001	hexan-3-ol (3)	$C_2H_5CH_2C:$ $(CH_3)(OH)C_2H_5$	20	0.82	1.423	I-416
2002	hex-1-ene (2)	$CH_2:C(CH_3)C_4H_9$	20	0.700	1.404	....
2003	hex-2-ene (2)	$(CH_3)_2C:CHC_3H_7$	20	0.709	1.410	**I-197
2004	hex-3-ene (2) (*cis?*)	$(CH_3)_2CHCH:CH\cdot$ C_2H_5	20	0.694	1.399	....
2005	hex-3-ene (2) (*trans*)	$(CH_3)_2CHCH:CH\cdot$ C_2H_5	....	.....	......	....
2006	hex-4-ene (2) (*cis*)	$CH_3CH:CHCH_2\cdot$ $CH(CH_3)_2$	20	0.700	1.400	....
2007	hex-4-ene (2) (*trans*)	$CH_3CH:CHCH_2\cdot$ $CH(CH_3)_2$	20	0.700	1.400	....
2008	hex-5-ene (2)	$CH_2:CHCH_2CH_2\cdot$ $CH(CH_3)_2$	20	0.692	1.3970	*I-91
2009	hex-1-ene (3)	$CH_2:CHCH(CH_3)\cdot$ C_3H_7	20	0.695	1.397	....
2010	hex-2-ene (3) (*cis?*)	$CH_3CH:C(CH_3)\cdot$ C_3H_7	20	0.712	1.410	*I-91
2011	hex-2-ene (3) (*trans*)	$CH_3CH:C(CH_3)\cdot$ C_3H_7	....	.....	......	....
2012	hex-3-ene (3) (*cis?*)	$C_2H_5C(CH_3):CH\cdot$ C_2H_5	20	0.703	1.407	**I-198
2013	hex-3-ene (3) (*trans*)	$C_2H_5C(CH_3):CH\cdot$ C_2H_5	....	.....	......	....
2014	hex-4-ene (3) (*cis*)	$CH_3CH:CHCH:$ $(CH_3)(C_2H_5)$	20	0.701	1.399	....
2015	hex-4-ene (3) (*trans*)	$CH_3CH:CHCH:$ $(CH_3)(C_2H_5)$	20	0.698	1.399	....
2016	hex-5-ene (3)	$CH_2:CHCH_2CH:$ $(CH_3)(C_2H_5)$	20	0.697	1.399	....

Table 10-66 (*Continued*)
INDEX OF REFRACTION—LIQUIDS

No.	Substance	Formula	t,°C.	d_4^t	n_D^t	Beil. Ref.
2017	**Methyl**-hex-5-ene-2-ol (2)	CH$_2$:CH(CH$_2$)$_2$COH: (CH$_3$)$_2$	16.2	0.838	1.4345	I-447
2018	Δ²-hexenic acid (α)	C$_2$H$_5$CH:CHCH: (CH$_3$)CO$_2$H	20	0.942	1.4395	**II-410
2019	" (γ)	CH$_3$CH$_2$C(CH$_3$):CH· CH$_2$CO$_2$H	20	0.966	1.4503	**II-411
2020	*n*-hexyl ketone	CH$_3$(CH$_2$)$_5$COCH$_3$	20	0.819	1.4161	I-704
2021	*n*-hexyl ketoxime	CH$_3$C(:NOH)· C$_6$H$_{13}$	20	0.886	1.4511	I-705
2022	*n*-hexylpropiolate	C$_6$H$_{13}$C:CCO$_2$CH$_3$	12.5	0.924	1.4518	II-490
2023	hydracrylate	HOCH$_2$CH$_2$CO$_2$CH$_3$	20	1.12	1.432	*III-112
2024	hydrindone (β)	C$_6$H$_4$CH$_2$CH(CH$_3$)· [⎯⎯⎯⎯⎯] CO ⎯|	20	1.066	1.5538	VII-372
2025	hydrogen pyro-tartrate	CH$_3$CH(CO$_2$CH$_3$)· CH$_2$CO$_2$H†	20.7	1.144	1.4323	II-639
2026	2-hydroxystyrene (1′)	HOC$_6$H$_4$C(CH$_3$):CH$_2$	18.5	1.030	1.5490	VI-572
2027	3-hydroxystyrene (α)	HOC$_6$H$_4$C(CH$_3$): CH$_2$	13.0	1.048	1.5734	*VI-284
2028	hydroxylamine (β)	CH$_3$NHOH	20	1.000	1.4164	IV-534
2029	indazole (1)	C$_8$H$_6$N$_2$	99.2	1.032	1.5522	....
2030	indene (1)	C$_6$H$_4$C(CH$_3$):CHCH$_2$ |⎯⎯⎯⎯⎯⎯|	27	0.968	1.5591	V-520
2031	iodide	CH$_3$I	21	2.277	1.5293	I-70
2032	isoxazole (α)	CH$_3$C:CHCH:NO |⎯⎯⎯⎯|	20	1.023	1.44	XXVII-16
2033	" (γ)	CH$_3$CCH:CH·O·N |⎯⎯⎯⎯|	20	1.022	1.44	XXVII-16
2034	1-keto-1,2,3,4-tetrahydronaph-thalene (7) (m.p. 33°)	CH$_3$C$_6$H$_3$COCH$_2$· |⎯⎯⎯ CH$_2$CH$_2$ ⎯|	35	1.057	1.5567	....
2035	1-keto-1,2,3,4-tetrahydronaph-thalene (2)	C$_6$H$_4$COCH(CH$_3$)· |⎯⎯⎯ CH$_2$CH$_2$ ⎯|	20.9	1.060	1.5515	*VII 197
2036	lactate	CH$_3$CHOHCO$_2$CH$_3$	16	1.08	1.4156	III-280
2037	*iso*-lauronolate	(CH$_3$)$_3$C$_5$H$_4$CO$_2$CH$_3$	20	0.970	1.469	IX-58
2038	*p*-menthatriene (2)	C$_{11}$H$_{16}$	19	0.874	1.5012	V-441
2039	methenyl-*o*-phenyleneamidine	C$_6$H$_4$·N(CH$_3$)·CH:N |⎯⎯⎯⎯⎯⎯|	17.1	1.128	1.6013	XXIII-132
2040	α-methoxyacrylate	CH$_2$:C(OCH$_3$)CO$_2$· CH$_3$	20	1.068	1.4316	*III-134
2041	*o*-methoxybenzoate	CH$_3$OC$_6$H$_4$CO$_2$CH$_3$	19.5	1.157	1.5342	X-71
2042	2-methoxyben-zoylacetone	(CH$_3$)(CH$_3$O)C$_6$H$_3$· COCH$_2$COCH$_3$	20	1.118	1.5847	....
2043	β-methoxycin-namate	C$_6$H$_5$C(OCH$_3$):CH· CO$_2$CH$_3$	24.7	1.118	1.5497	*X-133
2044	*o*-methoxycin-namate	CH$_3$OC$_6$H$_4$CH:CH· CO$_2$CH$_3$	16.7	1.137	1.5854	X-291
2045	7-methoxycoumar-one (2)	CH$_3$OC$_6$H$_3$·CH: |⎯⎯⎯ C(CH$_3$)·O ⎯⎯|	20	1.119	1.5636	....

† Or CH$_3$CH(CO$_2$H)CH$_2$CO$_2$CH$_3$

Table 10-66 (*Continued*)
INDEX OF REFRACTION—LIQUIDS

No.	Substance	Formula	$t,°C.$	d_4^t	n_D^t	Beil. Ref.
2046	**Methyl**-3-methoxy-coumarone (5)	$CH_3C_6H_3C(OCH_3)$: $\|$—————— $CH·O$ ——$\|$	24.3	1.107	1.5504	*XVII-64
2047	6-methoxyhydrin-done (2)	$CH_3OC_6H_3COCH·$ $\|$—————— $(CH_3)\,CH_2$ ———$\|$	16.9	1.119	1.5588	*VIII-560
2048	o-methoxyhydro-cinnamate	$CH_3OC_6H_4CH_2CH_2·$ CO_2CH_3	18.4	1.095	1.5126	X-242
2049	2-methoxystyrene (1′)	$CH_3OC_6H_4C(CH_3)$: CH_2	19	0.99	1.5315	VI-572
2050	3-methoxystyrene (1′)	$CH_3OC_6H_4C(CH_3)$: CH_2	20(?)	0.99	1.5417	VI-573
2051	α-methylaceto-acetate	$CH_3COCH(CH_3)\,CO_2·$ CH_3	23.8	1.031	1.4163	III-679
2052	β-methylcinnamate (m.p. 29°)	$C_6H_5C(CH_3):CHCO_2·$ CH_3	37.8	1.054	1.5505	IX-614
2053	p-methylcinnamate (m.p. 57°)	$CH_3C_6H_4CH:CH·$ CO_2CH_3	56.6	1.027	1.5581	*IX-256
2054	1-methylcyclohex-(2 or 3)-ene-3-acetate	$CH_3C_6H_8CH_2CO_2·$ CH_3	15.7	0.969	1.4582	*IX-30
2055	1-methylcyclohex-1-ene-4-acetate	$CH_3C_6H_8CH_2CO_2·$ CH_3	16.9	0.961	1.4540	*IX-30
2056	2-methylcyclo-hexylidene-1-acetate	$CH_2(CH_2)_3CH·$ $\|$—————— $(CH_3)\,C:CHCO_2·$ ———$\|$ CH_3	14.2	0.977	1.4807	*IX-51
2057	3-methylcyclo-hexylidene-1-acetate	$(CH_3)\,C_6H_9:CHCO_2·$ CH_3	15.5	0.975	1.4793	*IX-30
2058	4-methylcyclo-hexylidene-1-acetate	$(CH_3)\,C_6H_9:CHCO_2·$ CH_3	16.9	0.970	1.4758	*IX-31
2059	α-methyl-o-methoxycin-namate	$CH_3OC_6H_4CH:$ $C(CH_3)\,CO_2CH_3$	15.2	1.126	1.5721	X-311
2060	β-methyl-o-meth-oxycinnamate	$CH_3OC_6H_4C(CH_3)$: $CHCO_2CH_3$	16.7	1.104	1.5488	*X-137
2061	β-methyl-2-meth-oxy-5-methyl-cinnamate (m.p. 36°)	$(CH_3)\,(CH_3O)\,C_6H_3·$ $C(CH_3):CHCO_2·$ CH_3	12.7	1.087	1.5472	*X-141
2062	6-methyl-1-naph-thoate	$CH_3C_{10}H_6CO_2CH_3$	20	1.135	1.606	*IX-279
2063	methyloximino-cyanoacetate	$NCC(:NOCH_3)\,CO_2·$ CH_3	20	1.177	1.4591	III-775
2064	β-methylphenyl-glycidate	$C_6H_5C(CH_3)·CH·$ $\|$—————— $(CO_2CH_3)·O$ ————$\|$	20	1.123	1.5131	XVIII-306
2065	α-methylstyrene (p)	$CH_3C_6H_4C(CH_3)$: CH_2	18.7	0.902	1.5345	V-490

Table 10-66 (*Continued*)
INDEX OF REFRACTION—LIQUIDS

No.	Substance	Formula	t,°C.	d_4^t	n_D^t	Beil. Ref.
2066	**Methyl**-3-methylene-cyclohex-1-ene (1)	$CH_3C_6H_7$:CH_2	17	0.839	1.4872	*V-64
2067	3-methylenehexane (2)	$C_2H_5CH_2C(:CH_2)\cdot$ $CH(CH_3)_2$	20	0.725	1.414	....
2068	4-methylenehexane (2)	CH_2:$C(C_2H_5)CH_2\cdot$ $CH(CH_3)_2$	20	0.720	1.4105	....
2069	5-methylenehexane (2)	CH_2:$C(CH_3)CH_2\cdot$ $CH_2\cdot CH(CH_3)_2$	20	0.717	1.4105	....
2070	2-methylenehexane (3)	CH_2:$C(CH_3)CH$: $(CH_3)(C_3H_7)$	20	0.725	1.414	....
2071	4-methylenehexane (3)	CH_2:$C(C_2H_5)CH$: $(CH_3)(C_2H_5)$	20	0.729	1.4142	....
2072	5-methylenehexane (3)	CH_2:$C(CH_3)CH_2CH$: $(CH_3)(C_2H_5)$	20	0.720	1.411	....
2073	3-methylene-pentane (2)	$(CH_3)_2CH\cdot C(C_2H_5)$: CH_2	20	0.715	1.410	....
2074	naphth-1-aldehyde (4) (m.p. 34°)	$CH_3C_{10}H_6CHO$	38.2	1.125	1.6457	....
2075	naphthalene (α)	$C_{10}H_7CH_3$	20	1.00	1.618	V-566
2076	" (β) (m.p. 37°)	$C_{10}H_7CH_3$	39.9	0.994	1.6026	V-567
2077	naphthyl ether (α)	$CH_3OC_{10}H_7$	13.9	1.096	1.6232	VI-606
2078	naphthyl ketone (α)	$CH_3COC_{10}H_7$	21.5	1.117	1.6280	VII-401
2079	nitroamine	CH_3NHNO_2 (m.p. 38°)	49	1.243	1.4616	IV-567
2080	nonane (2)	$(CH_3)_2CHC_7H_{15}$	20	0.72	1.408	I-168
2081	" (3)	$C_2H_5(CH_3)CHC_6H_{13}$	20	0.735	1.4126	....
2082	" (5)	$(C_4H_9)_2CHCH_3$	20	0.732	1.4116	I-168
2083	*n*-nonyl ketone	$CH_3(CH_2)_8COCH_3$	17.3	0.83	1.4300	I-713
2084	octa-4,6-diene (2)	CH_3CH:$(CH)_2$:$CH\cdot$ $CH_2CH(CH_3)_2$	18	0.752	1.4543	I-260
2085	octane (2)	$(CH_3)_2CHC_6H_{13}$	20	0.713	1.4031	I-166
2086	" (3)	$(CH_3)(C_2H_5)\cdot$ CHC_5H_{11}	20	0.721	1.4062	I-166
2087	" (4)	$C_3H_7CH(CH_3)(C_4H_9)$	20	0.720	1.4061	*I-63
2088	octan-4-ol (3)	$C_2H_5(CH_2)_2CHOH\cdot$ $CH(CH_3)C_2H_5$	20	0.834	1.4317	**I-457
2089	palmitate (m.p. 30°)	$C_{15}H_{31}CO_2CH_3$	80.7	.....	1.4175	II-372
2090	penta-1,3-diene (2) (*cis?*)	CH_2:$C(CH_3)CH$:$CH\cdot$ CH_3	20	0.719	1.446	I-255
2091	penta-1,3-diene (2) (*trans*)	CH_2:$C(CH_3)CH$:$CH\cdot$ CH_3	....	.....	......	....
2092	penta-2,3-diene (2)	$(CH_3)_2C$:C:$CHCH_3$	20	0.711	1.425	I-255
2093	penta-2,4-diene (2)	CH_2:$CHCH$:$C(CH_3)_2$	20	0.719	1.451	I-255
2094	penta-3,4-diene (2)	CH_2:C:$CHCH(CH_3)_2$	20	0.708	1.424	**I-232
2095	penta-1,2-diene (3)	CH_2:C:$C(CH_3)C_2H_5$	20	0.715	1.425	I-255
2096	penta-1,3-diene (3) (*cis?*)	CH_2:$CHC(CH_3)$:$CH\cdot$ CH_3	20	0.735	1.452	*I-118
2097	penta-1,3-diene (3) (*trans*)	CH_2:$CHC(CH_3)$:$CH\cdot$ CH_3	....	.....	......	....
2098	penta-1,4-diene (2)	CH_2:$C(CH_3)CH_2\cdot$ CH:CH_2	20	0.694	1.405	....
2099	penta-1,4-diene (3)	$(CH_2$:$CH)_2CHCH_3$	20	0.695	1.405	....
2100	pentane (2)	$(CH_3)_2CHC_3H_7$	20	0.653	1.3715	I-148
2101	" (3)	$(C_2H_5)_2CHCH_3$	20	0.664	1.3765	I-149

Table 10-66 (*Continued*)
INDEX OF REFRACTION—LIQUIDS

No.	Substance	Formula	t,°C.	d_4^t	n_D^t	Beil. Ref.
2102	**Methl**-pentan-1-ol (2)	$C_2H_5CH_2CH(CH_3)\cdot$ CH_2OH	20	0.829	1.4178	I-409
2103	pentan-3-ol (2)	$(CH_3)_2CHCHOH\cdot$ C_2H_5	20	0.824	1.4135	I-410
2104	pentan-4-ol (2)	$CH_3CHOHCH_2\cdot$ $CH(CH_3)_2$	20	0.806	1.4087	I-410
2105	pentan-5-ol (2)	$(CH_3)_2CH(CH_2)_2\cdot$ CH_2OH	20	0.816	1.4142	I-411
2106	pentan-2-ol (3)	$CH_3CHOHCH(CH_3)\cdot$ C_2H_5	18	0.831	1.4205	I-411
2107	pentan-3-ol (3)	$(C_2H_5)_2COHCH_3$	20	0.824	1.4196	I-411
2108	pentan-4-ol acetate (2)	$CH_3CO_2CH(CH_3)\cdot$ $CH_2CH(CH_3)_2$	20	0.858	1.4008	........,....
2109	pent-1-ene (2)	$CH_2:C(CH_3)CH_2C_2H_5$	20	0.682	1.3925	I-217
2110	pent-2-ene (2)	$(CH_3)_2C:CHCH_2H_5$	20	0.686	1.4004	I-217
2111	pent-3-ene (2) (*cis?*)	$(CH_3)_2CHCH:CH\cdot$ CH_3	20	0.672	1.389	*I-90
2112	pent-3-ene (2) (*trans?*)	$(CH_3)_2CHCH:CH\cdot$ CH_3	20	0.670	1.388	*I-90
2113	pent-4-ene (2)	$CH_2:CHCH_2CH:$ $(CH_3)_2$	20	0.665	1.384	....
2114	pent-1-ene (3)	$CH_2:CHCH(CH_3)\cdot$ C_2H_5	20	0.670	1.384	....
2115	pent-2-ene (3) (*cis?*)	$CH_3CH:C(CH_3)C_2H_5$	20	0.699	1.4045	I-217
2116	pent-2-ene (3) (*trans?*)	$CH_3CH:C(CH_3)C_2H_5$	20	0.694	1.4016	I-217
2117	pent-3-ene-2-ol (2)	$CH_3CH:CHCOH:$ $(CH_3)_2$	20	0.835	1.4302	I-445
2118	phenylacetate	$C_6H_5CH_2CO_2CH_3$	16	1.044	1.5091	IX-434
2119	1-phenyl-hepta-1,3-diene (6)	$C_6H_5(CH:CH)_2CH_2\cdot$ $CH(CH_3)_2$	20	0.951	1.5855	V-525
2120	1-phenyl-hexa-1,3-diene (5)	$C_6H_5(CH:CH)_2\cdot$ $CH(CH_3)_2$	20	0.925	1.5873	V-524
2121	phenylhydrazine (α)	$C_6H_5N(CH_3)NH_2$	21.6	1.038	1.5824	XV-117
2122	" (β)	$C_6H_5NHNHCH_3$	23.1	1.03	1.572	XV-118
2123	μ-phenyloxazoline (β)	$N:C(C_6H_5)\cdot O\cdot CH\cdot$ $\vert \qquad\qquad \vert$ $(CH_3)\cdot CH_2$ $\vert_____\vert$	22.9	1.070	1.5460	XXVII-51
2124	phenylpropiolate (m.p. 25°)	$C_6H_5C\vdots CCO_2CH_3$	30.5	1.077	1.5601	IX-634
2125	o-phthalaldehyde-carboxylate	$HCOC_6H_4CO_2CH_3$	20.3	1.195	1.541	*X-316
2126	phthalide	$C_6H_4CO\cdot OCH(CH_3)$ $\vert_____\vert$	20	1.158	1.5447	XVII-318
2127	piperidine (N)	$C_5H_{10}NCH_3$	20	0.818	1.4464	XX-16
2128	" (2)	$CH_2(CH_2)_3NHCH\cdot$ $\vert_____\vert$ CH_3	23.6	0.844	1.4464	XX-95
2129	" (3)	$CH_3CH\cdot$ $\vert\_\_$ $(CH_2)_3NHCH_2$ $_____\vert$	24.3	0.845	1.4463	XX-100

Table 10-66 (*Continued*)
INDEX OF REFRACTION—LIQUIDS

No.	Substance	Formula	t,°C.	d_4^t	n_D^t	Beil. Ref.
2130	**Methyl**-piperidine (4)	$CH_2CH_2NHCH_2\cdot$ $\|$____ CH_2CHCH_3 ____$\|$	21.6	0.845	1.4378	XX-101
2131	propionate	$C_2H_5CO_2CH_3$	18.9	0.917	1.3770	II-239
2132	propiophenone-*o*- carboxylate	$C_2H_5COC_6H_4CO_2\cdot$ CH_3	15.2	1.129	1.5240	*IX-334
2133	2-*iso*-propyl-3- acetoxycouma- rone (5)	$CH_3C_6H_3C(O_2C\cdot$ $\|$____ $CH_3):C(C_3H_7)\cdot O$ _____$\|$	20	1.081	1.5281	*XVII-69
2134	2-*n*-propylcouma- ranone (5)	$CH_3C_6H_3COCH\cdot$ $\|$____ $(C_3H_7)\cdot O$ ____$\|$	20	1.064	1.545	....
2135	2-*iso*-propyl- coumaranone (5) (m.p. 26°)	$CH_3C_6H_3COCH\cdot$ $\|$____ $(C_3H_7)\cdot O$ ____$\|$	13.8	1.071	1.5429	*XVII-69
2136	3-*iso*-propyl- cyclopentane (1)	$(CH_3)CH\cdot$ $CH(CH_2)_2CH\cdot$ $\|$____ $(CH_3)CH_2$ ____$\|$	19	0.773	1.4250	*V-18
2137	4-*iso*-propyl-1- dichloromethyl- cyclohex-2-ene- 6-one (1)	$(C_3H_7)(CH_3)C_5H_5\cdot$ $COCHCl_2$	16.8	1.142	1.5014	VII-138
2138	*n*-propyl ether	$CH_3OCH_2C_2H_5$	14.3	0.74	1.3602	I-354
2139	*iso*-propyl ether	$CH_3OCH(CH_3)_2$	20	0.735	1.3576	I-362
2140	*n*-propyl ketone	$CH_3COCH_2C_2H_5$	20.2	0.81	1.3895	I-676
2141	*iso*-propyl ketone	$CH_3COCH(CH_3)_2$	16.0	0.805	1.3879	I-682
2142	*n*-propyl ketoxime	$CH_3C(NOH)C_3H_7$	20	0.910	1.4450	I-677
2143	6-*iso*-propyl- phenol (2)	$(CH_3)(C_3H_7):$ C_6H_3OH	15.2	0.987	1.5239	VI-526
2144	pseude-cumenyl ether	$(CH_3)_3C_6H_2OCH_3$ (1,2,4,5)	20	0.964	1.519	VI-510
2145	pyran (α)	$CH_3CHCH:CHCH:$ $\|$____ $CH\cdot O$ ____$\|$	20	0.917	1.4532	....
2146	pyrazine	$CH_3CCHNCHCHN$ $\|$_____$\|$	18.7	1.030	1.5067	XXIII-94
2147	pyrazole (1)	$C_4H_6N_2$	13.7	0.993	1.4786	XXIII-40
2148	" (3)	$C_4H_6N_2$	16.3	1.01	1.4971	XXIII-50
2149	" (5)	$N:C(CH_3)CH:$ $\|$____ $CH\cdot NH$ ____$\|$	13.7	1.023	1.4941	XXIII-50
2150	pyrrole (1)	$C_4H_4NCH_3$	16	0.915	1.4888	XX-163
2151	quinoline (α)	$C_6H_4\cdot CH:CH\cdot C\cdot$ $\|$____ $(CH_3):N$ ____$\|$	25.4	1.054	1.6091	XX-388

Table 10-66 (*Continued*)
INDEX OF REFRACTION—LIQUIDS

No.	Substance	Formula	t,°C.	d_4^t	n_D^t	Beil. Ref.
2152	**Methyl**-quinoline (3)	$C_6H_4 \cdot CH:C(CH_3) \cdot$ |_____ $CH:N$ ___|	20	1.067	1.6171	XX-394
2153	" (4)	$C_6H_4 \cdot C(CH_3):CH \cdot$ |_____ $CH:N$ ___|	20	1.087	1.6206	XX-395
2154	" (6)	$CH_3C_6H_3(CH)_3:N$ |_____|	20	1.065	1.6157	XX-397
2155	" (7)	$CH_3C_6H_3(CH)_3:N$ |_____|	20.8	1.067	1.6149	XX-400
2156	" (8)	$CH_3C_6H_3(CH)_3:N$ |_____|	20.8	1.072	1.6162	XX-401
2157	*iso*-quinoline (α)	$C_6H_4 \cdot CH:CH \cdot N:C \cdot$ |_____| CH_3	20.5	1.076	1.6095	XX-403
2158	ricinoleate	$HOC_{17}H_{32}CO_2CH_3$	22	0.924	1.4588	II-387
2159	salicylate	$HOC_6H_4CO_2CH_3$	18.1	1.186	1.5377	X-70
2160	stilbene (*m*) (m.p. 53°)	$CH_3C_6H_4CH:CH \cdot$ C_6H_5	60.5	0.989	1.6363	*V-311
2161	styrene (α)	$C_6H_5C(CH_3):CH_2$	20	0.911	1.5386	V-484
2162	styrene (β) (*cis?*)	$C_6H_5CH:CHCH_3$	20	0.911	1.545	V-481
2163	" (β) (*trans?*)	$C_6H_5CH:CHCH_3$	....	.....	......	....
2164	styrene (*o*)	$CH_3C_6H_4CH:CH_2$	14.1	0.916	1.5477	*V-233
2165	" (*p*)	$CH_3C_6H_4CH:CH_2$	16.4	0.900	1.5447	V-485
2166	(α,β-tetrahydro-benz)-2,3-isox-azole (1)	$CH_3C_6H_7 \cdot CH:N \cdot O$ |_____|	20	1.062	1.492	....
2167	(β,γ-tetrahydro-benz)-2,3-isox-azole (1)	$CH_3C_6H_7(:CH \cdot O \cdot N:)$	20	1.059	1.491	....
2168	tetrahydro-indazole (2)	$CH_3N \cdot CH:C_6H_8:N$ |_____|	20	1.030	1.517	....
2169	5,6,7,8-tetrahydro-naphthalene (2)	$CH_3C_6H_3(CH_2)_3CH_2$ |_____|	15.1	0.954	1.5372	**V-394
2170	tetrahydroquinoline (2)	$C_{10}H_{13}N$	20	1.02	1.5727	XX-283
2171	thiocyanate	CH_3SCN	23.8	1.069	1.4680	III-175
2172	*iso*-thiocyanate (m.p. 35°)	CH_3NCS	40	1.07	1.5245	IV-77
2173	thionaphthene (5)	$CH_3C_6H_3CH:CH \cdot S$ |_____|	21.7	1.111	1.6148	*XVII-27
2174	thiophene (2)	$CH_3C_4H_3S$	15.3	1.02(?)	1.524	XVII-38
2175	" (3)	$CH_3C_4H_3S$	14.8	1.02(?)	1.524	XVII-38
2176	thymyl ether	$(CH_3)(C_3H_7)C_6H_3 \cdot$ OCH_3	28	0.931	1.5024	VI-536
2177	tolyl carbinol (*m*)	$CH_3CHOHC_6H_4CH_3$	20	0.993	1.5243	*VI-254
2178	" " (*p*)	$CH_3CHOHC_6H_4CH_3$	20	1.003	1.5332	VI-508
2179	tolyl ketone (*o*)	$CH_3COC_6H_4CH_3$	16.8	1.001	1.5242	VII-306
2180	" " (*p*)	$CH_3COC_6H_4CH_3$	22	0.989	1.5283	VII-307
2181	γ,γ,γ-trichloro-β-acetoxybutyrate	$Cl_3CCH(O_2CCH_3) \cdot$ $CH_2CO_2CH_3$	14.5	1.394	1.4682	*III-117
2182	γ,γ,γ-trichloro-crotonate	$Cl_3CCH:CHCO_2CH_3$	21.2	1.397	1.4898	*II-190

Table 10-66 (*Continued*)
INDEX OF REFRACTION—LIQUIDS

No.	Substance	Formula	t,°C.	d_4^t	n_D^t	Beil. Ref.
2183	**Methyl**-urethane	$CH_3NHCO_2C_2H_5$	18.9	1.009	1.4200	IV-64
2184	**Methylal**	$CH_2(OCH_3)_2$	20	0.88	1.3534	I-574
2185	**Methylene** chloride	CH_2Cl_2	20	1.336	1.4237	I-60
2186	cyclohexane	$CH_2:C(CH_2)_4CH_2$	20	0.803	1.4528	V-69
		$\underline{\quad\quad\quad\quad}$				
2187	heptane (3)	$CH_2:C(C_2H_5)C_4H_9$	20	0.727	1.4157	**I-201
2188	" (4)	$(C_2H_5CH_2)_2C:CH_2$	20	0.724	1.4136	....
2189	hexane (3)	$CH_2:C(C_2H_5)C_3H_7$	20	0.708	1.405	....
2190	iodide	CH_2I_2	15	3.33	1.7425	I-71
2191	pentane (3)	$CH_2:C(C_2H_5)_2$	20	0.689	1.3969	**I-195
2192	**Myrcene**	$C_{10}H_{16}$	20	0.798	1.4707	I-264
2193	**Myristic acid**	$CH_3(CH_2)_{12}CO_2H$	60	0.858	1.4308	II-365
	(m.p. 58°)					
2194	**Naphthaldehyde** (α)	$C_{10}H_7CHO$	19.3	1.15	1.6546	VII-400
2195	" (β)	$C_{10}H_7CHO$	99.4	1.078	1.6211	VII-401
	(m.p. 61°)					
2196	**Naphthalene**	$C_{10}H_8$	98.4	.....	1.5823	V-531
	(m.p. 80°)					
2197	decahydride	$C_{10}H_{18}$	18	0.895	1.4804	V-92
2198	**Naphthol** (α)	$C_{10}H_7OH$	98.7	1.099	1.6206	VI-596
	(m.p. 96°)					
2199	**Naphthonitrile** (α)	$C_{10}H_7CN$	17.8	1.11	1.6298	IX-649
	(m.p. 36°)					
2200	**Naphthylamine** (β)	$C_{10}H_7NH_2$	98.4	1.061	1.6493	XII-1265
	(m.p. 111°)					
2201	**Naphthylamine** (α)	$C_{10}H_7NH_2$	51	1.1	1.6703	XII-1212
	(m.p. 50°)					
2202	**Naphthylene**-					
	diamine (1,6)	$C_{10}H_6(NH_2)_2$	99.4	1.147	1.7083	XIII-204
	(m.p. 78°)					
2203	diamine (1,8)	$C_{10}H_6(NH_2)_2$	99.4	1.127	1.6828	XIII-205
	(m.p. 66°)					
2204	**Neomenthol** (l)	$C_{10}H_{20}O$	20	0.900	1.4603	*VI-29
2205	**Nerolidol**	$C_{15}H_{26}O$	20	0.87	1.4898	I-464
2206	**Nicoteine** (*iso*)	$C_{10}H_{12}N_2$	20	1.098	1.5749	....
2207	**Nicotine**	$C_{10}H_{14}N_2$	20	1.009	1.5286	XXIII-111
2208	" (*meta*)	$C_{10}H_{14}N_2$	19.7	1.002	1.5551	....
2209	**Nitric acid**	HNO_3	16.4	.....	1.397	....
2210	**Nitro**-anisole (o)	$CH_3OC_6H_4NO_2$	20	1.254	1.5620	VI-217
2211	anisole (p)	$CH_3OC_6H_4NO_2$	60	1.219	1.5707	VI-230
	(m.p. 54°)					
2212	benzene	$C_6H_5NO_2$	20	1.21	1.5524	V-233
2213	benzyl chloride (o)	$NO_2C_6H_4CH_2Cl$	61.5	.....	1.5557	V-327
	(m.p. 49°)					
2214	benzyl chloride (m)	$NO_2C_6H_4CH_2Cl$	61.5	.....	1.5577	V-329
	(m.p. 45°)					
2215	benzyl chloride (p)	$NO_2C_6H_4CH_2Cl$	61.5	.....	1.5647	V-329
	(m.p. 71°)					
2216	bromoform	Br_3CNO_2	12.5	2.811	1.5831	I-77
2217	chloroform	Cl_3CNO_2	22.8	1.651	1.4608	I-76
2218	p-cresol (2)	$(CH_3)(HO)C_6H_3NO_2$	38.6	1.240	1.5763	VI-412
	(m.p. 37°)	(1,4,2)				
2219	2,6-dimethyl-	$(CH_3)_2C(NO_2)\cdot$	18	0.915	1.4326	I-167
	heptane (2)	$(CH_2)_3CH(CH_3)_2$				
2220	2,5-dimethyl-	$(CH_3)_2C(NO_2)\cdot$	20	0.921	1.4316	I-163
	hexane (2)	$(CH_2)_2CH(CH_3)_2$				

Table 10-66 (*Continued*)
INDEX OF REFRACTION—LIQUIDS

No.	Substance	Formula	t,°C.	d_4^t	n_D^t	Beil. Ref.
2221	**Nitro**-2,7-dimethyl-octane (1)	$(CH_3)_2CH(CH_2)_4\cdot CH(CH_3)CH_2NO_2$	21	0.925	1.4426	I-169
2222	2,7-dimethyl-octane (2)	$(CH_3)_2CNO_2\cdot (CH_2)_4CH(CH_3)_2$	20	0.909	1.4357	I-169
2223	2,4-dimethyl-pentane (2)	$(CH_3)_2C(NO_2)CH_2\cdot CH(CH_3)_2$	30	0.94	1.4235	I-158
2224	ethane	$CH_3CH_2NO_2$	24.3	1.047	1.3901	I-100
2225	methane	CH_3NO_2	20	1.139	1.3935	I-74
2226	2-methylbutane (4)	$(CH_3)_2CHCH_2\cdot CH_2NO_2$	20.6	0.960	1.4181	I-140
2227	pentane (1)	$CH_3(CH_2)_3CH_2NO_2$	20(?)	0.948	1.4218	I-133
2228	phenetole (*o*)	$C_2H_5OC_6H_4NO_2$	20	1.19	1.5425	VI-218
2229	piperidine (1)	$C_5H_{10}O_2N_2$	26.4	1.152	1.4954	XX-84
2230	propane (1)	$CH_3CH_2CH_2NO_2$	24.3	1.002	1.4003	I-115
2231	toluene (*o*)	$CH_3C_6H_4NO_2$	20.4	1.163	1.5474	V-318
2232	" (*m*)	$CH_3C_6H_4NO_2$	21	1.16	1.5470	V-321
2233	" (*p*) (m.p. 52°)	$CH_3C_6H_4NO_2$	62.5	1.13	1.5346	V-323
2234	**Nitrogen dioxide**	NO_2	−90	.....	1.330	....
2235	**Nitroso**-di-*iso*-butylamine	$[(CH_3)_2CH\cdot CH_2]_2NNO$	21.2	0.892	1.4439	IV-172
2236	ethylurethane	$C_2H_5N(NO)CO_2\cdot C_2H_5$	20	1.071	1.4322	IV-129
2237	methylaniline (*N*)	$C_6H_5N(CH_3)NO$	20	1.124	1.5769	XII-579
2238	piperidine (*N*)	$C_5H_{10}N\cdot NO$	18.5	1.063	1.4933	XX-83
2239	**Nitrous oxide**	N_2O	16	.....	1.193	....
2240	**Nonacosane** (m.p. 64°)	$C_{29}H_{60}$	65	0.77	1.4364	I-176
2241	**Nonane** (*n*)	$CH_3(CH_2)_7CH_3$	20	0.718	1.4055	I-165
2242	**Nonanol** (1)	$CH_3(CH_2)_7CH_2OH$	20	0.828	1.4311	I-423
2243	" (2)	$CH_3(CH_2)_6CHOH\cdot CH_3$	20(?)	0.823	1.4353	I-423
2244	" (3)	$C_2H_5CHOHC_6H_{13}$	20	0.825	1.4308	I-424
2245	" (5)	$(C_4H_9)_2CHOH$	20	0.823	1.430	I-424
2246	**Nondecane** (*n*) (m.p. 32°)	$CH_3(CH_2)_{17}CH_3$	34.6	0.775	1.436	I-174
2247	**Nonylenic** acid (α,β)	$CH_3(CH_2)_5CH:CH\cdot CO_2H$	20	0.930	1.454	II-453
2248	nitrile (α,β)	$C_6H_{13}CH:CHCN$	20	0.833	1.445	**II-417
2249	**Nonylenyl chloride** (α,β)	$CH_3(CH_2)_5CH:CH\cdot COCl$	20	0.967	1.460	*II-194
2250	**Nonyl formate** (γ)	$HCO_2C_9H_{19}$	20	0.869	1.421	**II-32
2251	**Nopinane**	C_9H_{16}	20(?)	0.86	1.4641	*V-42
2252	**Octacosane** (m.p. 65°)	$C_{28}H_{56}$	65	0.77	1.4354	I-176
2253	**Octadecane** (*n*)	$CH_3(CH_2)_{16}CH_3$	35.2	0.77	1.4349	I-173
2254	**Octahydrophenan-threne**	$C_{14}H_{18}$	20	1.025	1.5668	V-527
2255	**Octane** (*n*)	$CH_3(CH_2)_6CH_3$	20	0.703	1.3975	I-159
2256	**Octanol** (1)	$CH_3(CH_2)_6CH_2OH$	20.5	0.826	1.4304	I-418
2257	" (2)	$CH_3CHOHC_6H_{13}$	20	0.822	1.4256	I-419
2258	" (2) †	$CH_3CHOHC_6H_{13}$	20	0.822	1.4244	I-419
2259	**Octene** (1)	$CH_2:CH(CH_2)_5CH_3$	20	0.716	1.4088	I-221
2260	" (2) (*cis*)	$CH_3CH:CHC_5H_{11}$	20	0.724	1.4150	**I-200
2261	" (2) (*trans*)	$CH_3CH:CHC_5H_{11}$	20	0.720	1.4132	**I-200
2262	" (3) (*cis*)	$C_2H_5CH:CHC_4H_9$	20	0.721	1.4135	....

† Mixture of *l* and *dl*.

Table 10-66 (*Continued*)
INDEX OF REFRACTION—LIQUIDS

No.	Substance	Formula	t,°C.	d_4^t	n_D^t	Beil. Ref.
2263	**Octene** (3) (*trans*)	$C_2H_5CH:CHC_4H_9$	20	0.715	1.4126	....
2264	" (4) (*cis*)	$(C_2H_5CH_2CH:)_2$	20	0.723	1.4144	....
2265	" (4) (*trans*)	$(C_2H_5CH_2CH:)_2$	20	0.714	1.4118	....
2266	**Octracenone**	$C_6H_8 \cdot CH:C_6H_6O:CH$ $\lfloor\rule{1cm}{0pt}\rfloor$	62.4	1.076	1.5707	....
2267	**Octyl** acetate (*n*)	$CH_3CO_2C_8H_{17}$	20	0.88	1.4204	II-134
2268	amine (*n*)	$CH_3(CH_2)_7NH_2$	26.8	0.777	1.43	IV-196
2269	amine (*n*) (*sec*)	$CH_3(CH_2)_5 \cdot$ $CH(NH_2)CH_3$	24.5	0.771	1.4232	IV-196
2270	fluoride (*n*)	$CH_3(CH_2)_6CH_2F$	20	0.803	1.3947	I-159
2271	formate	$HCO_2C_8H_{17}$	20	0.87	1.414	II-22
2272	iodide (*n*)	$CH_3(CH_2)_6CH_2I$	20	1.333	1.489	I-160
2273	**Octyne** (1)	$CH:C(CH_2)_5CH_3$	12.5	0.753	1.4208	I-258
2274	**Oleic** acid	$CH_3(CH_2)_7CH:CH \cdot$ $(CH_2)_7CO_2H$	17.7	0.895	1.46	II-463
2275	acid ozonide	$C_{18}H_{34}O_5$	20.5	1.028	1.4622	II-466
2276	aldehyde	$C_{17}H_{33}CHO$	20	0.85	1.4557	I-749
2277	**Osmium tetroxide**	OsO_4	45	.....	1.56	....
2278	**Oxalyl chloride**	$(COCl)_2$	12.9	1.489	1.4340	II-542
2279	**Palmitic acid** (m.p. 64°)	$CH_3(CH_2)_{14}CO_2H$	70	0.849	1.4304	II-370
2280	**Palmitone** (m.p. 83°)	$(C_{15}H_{31})_2CO$	93.5	0.79	1.4297	I-719
2281	**Paraldehyde**	$(C_2H_4O)_3$	20	0.994	1.4049	XIX-385
2282	**Patschoulene**	$C_{15}H_{24}$	20	0.930	1.4984	V-470
2283	**Pelargonic** acid	$CH_3(CH_2)_7CO_2H$	20(?)	0.906	1.4306	II-352
2284	aldehyde	$CH_3(CH_2)_7CHO$	18.6	0.827	1.4242	I-708
2285	**Pelletierine** (*pseudo*) (m.p. 49°)	$C_9H_{15}ON$	99.5	1.001	1.4760	XXI-261
2286	**Pentachloro**-ethane	C_2HCl_5	24	1.67	1.5025	I-87
2287	propane (1,1,2,3,3)	$HCCl_2CHClCHCl_2$	20	1.6	1.512	*I-34
2288	**Pentacosane** (m.p. 55°)	$C_{25}H_{52}$	80	0.8	1.4262	I-175
2289	**Pentadiene** (1,2)	$CH_2:C:CHC_2H_5$	20	0.692	1.421	**I-226
2290	**Pentadiene** (1,3) (*cis*)	$CH_2:CHCH:CHCH_3$	20	0.691	1.4359	I-251
2291	**Pentadiene** (1,3) (*trans*)	$CH_2:CHCH:CHCH_3$	20	0.676	1.4299	I-251
2292	**Pentadiene** (1,4)	$(CH_2:CH)_2CH_2$	20	0.660	1.388	I-251
2293	**Pentadiene** (2,3)	$(CH_3CH:)_2C$	20	0.66	1.39	I-251
2294	**Penta-2,3-dione**	$CH_3COCOCH_2CH_3$	19	0.957	1.4014	I-776
2295	**Pentaethylbenzene**	$(C_2H_5)_5C_6H$	20.3	0.896	1.52	V-471
2296	**Pentamethyl-benzene** (m.p. 53°)	$(CH_3)_5C_6H$	72.8	0.9	1.5049	V-443
2297	**Pentane** (*n*)	$CH_3(CH_2)_3CH_3$	20	0.626	1.3575	I-130
2298	" (*iso*)	$(CH_3)_2CHC_2H_5$	20	0.620	1.3537	I-134
2299	**Pentan-1,2-diol** (*dl*)	$C_3H_7CHOHCH_2OH$	20	0.980	1.441	**I-548
2300	**Pentanol** (3)	$(C_2H_5)_2CHOH$	14.7	0.808	1.4124	I-385
2301	**Pentene** (1)	$CH_2:CHCH_2C_2H_5$	20	0.641	1.3714	I-210
2302	" (2) (*cis*)	$CH_3CH:CHC_2H_5$	20	0.656	1.3830	I-210
2303	" (2) (*trans*)	$CH_3CH:CHC_2H_5$	20	0.648	1.379	I-210
2304	**Pent-1-ene-3-ol**	$CH_2:CHCHOHC_2H_5$	20	0.837	1.4240	I-443
2305	**Pent-2-ene-4-ol**	$CH_3CH:CHCHOH \cdot$ CH_3	18	0.839	1.4282	I-443
2306	**Pentenic** acid (α,β)	$C_2H_5CH:CHCO_2H$	14.7	0.992	1.454	II-426

Table 10-66 (*Continued*)
INDEX OF REFRACTION—LIQUIDS

No.	Substance	Formula	t,°C.	d_4^t	n_D^t	Beil. Ref.
2307	**Pentenic** acid (β,γ)	$CH_3CH:CHCH_2CO_2H$	18.8	0.99	1.4357	II-426
2308	**Pentenonitrile** (α,β)	$CH_3CH_2CH:CHCN$	20	0.828	1.433	II-426
2309	" (β,γ)	$CH_3CH:CHCH_2CN$	20	0.841	1.423	II-427
2310	**Pentenoyl** chloride (α,β)	$C_2H_5CH:CHCOCl$	20	1.063	1.465	**II-400
2311	chloride (β,γ)	$CH_3CH:CHCH_2COCl$	20	1.064	1.448	....
2312	**Pentyne** (1)	$CH:CCH_2C_2H_5$	20	0.691	1.385	I-250
2313	" (2)	$CH_3C:CC_2H_5$	20	0.711	1.4039	I-250
2314	**Perchloromethyl** mercaptan	Cl_3CSCl	17.5	1.695	1.5484	III-135
2315	**Petroselinic acid** (m.p. 34°)	$C_{18}H_{34}O_2$	40	0.870	1.4533	II-462
2316	**Phellandral**	$C_{10}H_{16}O$	20	0.93	1.4903	VII-77
2317	**Phellandrene** (α)	$C_{10}H_{16}$	20	0.845	1.4835	V-129
2318	" (β)	$C_{10}H_{16}$	20	0.852	1.4788	V-131
2319	**Phenanthrene** (m.p. 99°)	$(C_6H_4CH)_2$	129	.....	1.6567	V-667
2320	**Phenetole**	$C_6H_5OC_2H_5$	18.8	0.97	1.5084	VI-140
2321	**Phenol** (m.p. 42°)	C_6H_5OH	45	1.07	1.54	VI-113
2322	**Phenyl**-acetaldehyde	$C_6H_5CH_2CHO$	19.6	1.025	1.5255	VII-292
2323	acetate	$CH_3CO_2C_6H_5$	20	1.078	1.503	VI-152
2324	acetic acid (m.p. 77°)	$C_6H_5CH_2CO_2H$	79.8	1.081	1.51	IX-431
2325	acetylene	$C_6H_5C:CH$	17.4	0.93	1.5501	V-511
2326	Δ^2-angelolactone (δ)	$C_6H_5CH_2C:CHCH_2\cdot$ $\underset{\lfloor}{\mid}\quad CO\cdot O \quad\underset{\rfloor}{}$	20	1.113	1.529	....
2327	buta-1,3-diene (1)	$C_6H_5CH:CHCH:CH_2$	16	0.931	1.6128	V-517
2328	*iso*-cyanate	$C_6H_5N:CO$	19.6	1.096	1.5368	XII-437
2329	cyclohexane	$C_6H_5CH(CH_2)_4CH_2$	16	0.955	1.589	V-503
2330	cyclohex-1-ene	$C_6H_5C:CH\cdot$ $(CH_2)_3CH_2$	20	0.993	1.5718	V-523
2331	cymene (2)	$(CH_3)(C_6H_5)C_6H_3\cdot$ $CH(CH_3)_2$	15	0.978	1.5680	V-619
2332	2,4-dimethyl-pyrazole (1)	$C_6H_5N\cdot N:C(CH_3)\cdot$ $CH:CCH_3$	19.3	1.057	1.5738	XXIII-75
2333	3,4-dimethyl-pyrazole (1)	$CH:C(CH_3)\cdot$ $C(CH_3):N\cdot NC_6H_5$	20	1.061	1.587	XXIII-72
2334	3,5-dimethyl-pyrazole (1)	$CH_3C:CHC(CH_3):$ $N\cdot NC_6H_5$	20	1.055	1.574	XXIII-75
2335	4,5-dimethyl-pyrazole (1)	$CH_3C:C(CH_3)\cdot CH:$ $N\cdot NC_6H_5$	20	1.065	1.575	....
2336	ether	$(C_6H_5)_2O$	24	1.07	1.5826	VI-146
2337	ethyl alcohol (α)	$C_6H_5CHOHCH_3$	20	1.01	1.5211	VI-475

Table 10-66 (*Continued*)
INDEX OF REFRACTION—LIQUIDS

No.	Substance	Formula	t,°C.	d_4^t	n_D^t	Beil. Ref.
2338	**Phenyl**-ethyl alcohol (β)	$C_6H_5CH_2CH_2OH$	18	1.02	1.5267	VI-478
2339	ethylamine (β)	$C_6H_5CH_2CH_2NH_2$	24(?)	0.958	1.53	XII-1096
2340	hexa-1,3-diene (1)	$C_6H_5(CH:CH)_2\cdot$ C_2H_5	20	0.919	1.5989	V-522
2341	hydrazine	$C_6H_5NHNH_2$	20	1.098	1.6081	XV-67
2342	3-hydroxy-but-1-yne (1)	$C_6H_5C\vdots CCHOHCH_3$	12.6	1.036	1.5731	VI-588
2343	3-hydroxy-pent-1-yne (1)	$C_6H_5C\vdots CCHOHCH_2\cdot$ CH_3	13	1.014	1.5633	VI-589
2344	3-hydroxy-4,4,4-trichlorobut-1-yne (1)	$C_6H_5C\vdots CCHOHCCl_3$	9.5	1.354	1.5917	VI-588
2345	*p*-menthatriene (2)	$C_{16}H_{18}$	20(?)	0.971	1.5674	V-619
2346	3-methyl-5-chloro-pyrazole (1)	$ClC:CH\cdot C(CH_3):$ $\underline{\qquad}$ $N\cdot NC_6H_5$ $\underline{\quad}$	20	1.193	1.580	XXIII-54
2347	5-methyl-3-chloro-pyrazole (1)	$CH_3C:CH\cdot C(Cl):$ $\underline{\qquad}$ $N\cdot NC_6H_5$ $\underline{\quad}$	20	1.205	1.584	XXIII-55
2348	5-methylpyrazole (1)	$CH_3C:CH\cdot CH:N\cdot N\cdot$ $\underline{\qquad}$ C_6H_5	20	1.086	1.582	XXIII-52
2349	β-methylstyrene (α) (m.p. 52°)	$(C_6H_5)_2C:CHCH_3$	63.5	0.981	1.5791	V-644
2350	nitromethane	$C_6H_5CH_2NO_2$	20	1.16	1.5323	V-325
2351	nitromethane methyl ether (*aci*)	$C_6H_5CH:N(OCH_3):O$	15	1.124	1.5871	....
2352	oxazoline (μ)	$N:C(C_6H_5)\cdot O\cdot CH_2$ $\underline{\quad} CH_2 \underline{\qquad}$	23.4	1.122	1.5655	XXVII-47
2353	penta-1,3-diene (1)	$C_6H_5\cdot(CH:CH)_2\cdot CH_3$	15.2	0.933	1.6000	V-521
2354	pent-2-ene (1)	$C_6H_5CH_2CH:CHC_2H_5$	16.2	0.888	1.5089	V-497
2355	propiolic acetal	$C_6H_5C\vdots CCH(O\cdot$ $C_2H_5)_2$	13	0.994	1.5222	VII-383
2356	propiolic aldehyde	$C_6H_5C\vdots CCHO$	12.6	1.068	1.6079	VII-383
2357	propiolic nitrile (m.p. 41°)	$C_6H_5C\vdots CCN$	41.5	1.005	1.5854	IX-636
2358	pyrazole (1)	$CH:CH\cdot CH:N\cdot N\cdot$ $\underline{\qquad}$ C_6H_5	20	1.111	1.596	XXIII-40
2359	styrene (α)	$(C_6H_5)_2C:CH_2$	20	1.023	1.6085	V-639
2360	tetrahydro-indazole (1)	$C_6H_8\cdot CH:N\cdot NC_6H_5$ $\underline{\qquad}$	99.9	1.061	1.568	....
2361	tetrahydro-indazole (2)	$C_6H_5N\cdot N:C_6H_8:CH$ $\underline{\qquad}$	99.6	1.054	1.578	....
2362	*iso*-thiocyanate	$C_6H_5N:CS$	23.4	1.126	1.6492	XII-453
2363	1-*p*-tolyl-ethane (1)	$CH_3C_6H_4CH(C_6H_5)\cdot$ CH_3	17.2	0.985	1.5659	V-614
2364	3,4,5-trimethyl-pyrazole (1)	$CH_3C:C(CH_3)\cdot C\cdot$ $\underline{\qquad}$ $(CH_3):N\cdot NC_6H_5$ $\underline{\qquad}$	20	1.044	1.571	XXIII-82
2365	vinyl acetate	$C_6H_5CH:CHO_2CCH_3$	22.9	1.066	1.5494	VI-564

Table 10-66 (*Continued*)
INDEX OF REFRACTION—LIQUIDS

No.	Substance	Formula	t,°C.	d_4^t	n_D^t	Beil. Ref.
2366	**Phenylenediamine** (*m*) (m.p. 63°)	$C_6H_4(NH_2)_2$	57.7	1.107	1.6339	XIII-33
2367	**Phorone** (m.p. 28°)	$[(CH_3)_2C:CH]_2CO$	29	0.88	1.4950	I-751
2368	**Phosphenyl chloride**	$C_6H_5PCl_2$	7	1.32	1.6053	XVI-763
2369	**Phosphine**	PH_3	17.5	.....	1.317	....
2370	**Phosphorus** oxychloride	$POCl_3$	25.1	.....	1.460	....
2371	tribromide	PBr_3	26.6	.....	1.697	....
2372	trichloride	PCl_3	14	.....	1.516	....
2373	**Phthalide** (m.p. 73°)	$C_6H_4CH_2 \cdot O \cdot CO$ $\mid_____\mid$	99.1	1.164	1.5356	XVII-310
2374	**Phthalyl** dichloride (*o*) (*sym*)	$C_6H_4(COCl)_2$	20	1.41	1.5692	IX-805
2375	dichloride (*m*) (*sym*) (m.p. 41°)	$C_6H_4(COCl)_2$	46.9	.....	1.5700	IX-834
2376	**Phytol**	$C_{20}H_{39}OH$	20	0.852	1.4638	I-453
2377	**Picoline** (*α*)	$CH_3C_5H_4N$	20	0.95	1.501	XX-234
2378	" (*β*)	$CH_3C_5H_4N$	24	0.95	1.5043	XX-239
2379	**Pinane** (*d*)	$C_{10}H_{18}$	20	0.839	1.4624	V-93
2380	**Pinene** (*α*) (*dl*)	$C_{10}H_{16}$	23.5	0.86	1.4653	V-144
2381	**Pinic acid** (m.p. 102°)	$C_9H_{14}O_4$	109.4	1.093	1.4482	IX-743
2382	**Pinocamphane**	$C_{10}H_{18}$	20	0.855	1.4609	*V-48
2383	**Pinol**	$C_{10}H_{16}O$	20	0.942	1.4715	XVII-45
2384	**Pinylamine**	$C_{10}H_{17}N$	20	0.940	1.5036	XII-54
2385	**Piperidine**	$CH_2(CH_2)_4NH$ $\mid_____\mid$	20	0.860	1.4530	XX-6
2386	**Propargyl** acetate	$CH_3CO_2CH_2C:CH$	20	1.005	1.4205	II-140
2387	alcohol	$CH:CCH_2OH$	20	0.972	1.4306	I-454
2388	**Propenyl**-anisole (*o*)	$CH_3OC_6H_4CH:CH \cdot CH_3$	20	0.993	1.5604	VI-565
2389	*p*-cresol (*o*)	$(CH_3)(HO)C_6H_3CH: CHCH_3$	9	1.029	1.5777	*VI-287
2390	4-methoxytoluene (2) (*n*)	$(CH_3)(CH_3O)C_6H_3 \cdot CH:CHCH_3$	12.3	0.986	1.5557	*VI-287
2391	4-methoxytoluene (3) (*iso*)	$CH_3C(:CH_2) \cdot C_6H_3(OCH_3)(CH_3)$	17.6	0.966	1.5308	*VI-288
2392	phenol (*o*) (m.p. 38°)	$CH_3CH:CHC_6H_4OH$	14.3	1.044	1.5837	*VI-279
2393	phenyl ketone	$CH_3CH:CHCOC_6H_5$	20	1.028	1.5601	VII-368
2394	**Propio**-*p*-cresol	$(HO)(CH_3)C_6H_3(CO \cdot C_2H_5)$ (1,4,2)	13.8	1.084	1.5485	VIII-120
2395	*p*-methoxytoluene (*m*)	$(CH_3O)(CH_3)C_6H_3 \cdot COC_2H_5$ (4,1,3)	20	1.047	1.5313	VIII-120
2396	phenone-*o*-carboxylic acid (m.p. 93°)	$C_2H_5COC_6H_4CO_2H$	99.1	1.140	1.5177	X-701
2397	thienone (*α*)	$C_2H_5COC:CHCH:$ $\mid_____$ $CH \cdot S$ $\_\_\_\mid$	20	1.131	1.556	XVII-295
2398	**Propiolic** acid	$CH:CCO_2H$	15	1.139	1.4429	II-477
2399	nitrile	$HC:CCN$	17	0.816	1.3870	*II-208
2400	**Propion**-aldehyde	C_2H_5CHO	20	0.807	1.3636	I-629
2401	aldoxime	$C_2H_5CH:NOH$	20	0.926	1.4287	I-631
2402	amide	$C_2H_5CONH_2$	20	1.042	1.43	II-243
2403	imidoethyl ether	$HN:C(C_2H_5)OC_2H_5$	16.9	0.871	1.4103	II-245

Table 10-66 (*Continued*)
INDEX OF REFRACTION—LIQUIDS

No.	Substance	Formula	t,°C.	d_4^t	n_D^t	Beil. Ref.
2404	**Propionic** acid	$C_2H_5CO_2H$	19.9	0.992	1.3874	II-234
2405	anhydride	$(C_2H_5CO)_2O$	20	1.012	1.4038	II-242
2406	nitrile	C_2H_5CN	19	0.78	1.3681	II-245
2407	**Propionyl**-acetone	$C_2H_5COCH_2COCH_3$	20	0.959	1.4516	I-788
2408	acetophenone (ω)	$C_6H_5COCH_2COC_2H_5$	20	1.078	1.5844	VII-687
2409	bromide	C_2H_5COBr	16.4	1.521	1.4570	II-243
2410	camphor (3)	$C_{13}H_{20}O_2$	18.1	1.019	1.4905	VII-697
2411	chloride	C_2H_5COCl	20	1.065	1.4051	II-243
2412	tropein	$C_{11}H_{19}O_2N$	19.6	1.040	1.4743	....
2413	o-xylene (*uns*)	$(CH_3)_2C_6H_3CO\cdot$ C_2H_5 (1,2,4)	20	0.871	1.5007	*VII-176
2414	**Propyl** acetate	$CH_3CO_2CH_2C_2H_5$	20	0.886	1.3844	II-129
2415	acetate (*iso*)	$CH_3CO_2CH(CH_3)_2$	20	0.872	1.3770	II-130
2416	α'-acetothienone (α)	$C_3H_7C{:}CHCH{:}C\cdot$ $\mid$ $(COCH_3)\cdot S$ $\mid$	20	1.061	1.545	XVII-300
2417	acetylacetone	$CH_3COCH(C_3H_7)\cdot$ $COCH_3$	20	0.935	1.4429	**I-846
2418	alcohol (*n*)	$C_2H_5CH_2OH$	20	0.804	1.3854	I-350
2419	alcohol (*iso*)	$(CH_3)_2CHOH$	20	0.785	1.3776	I-360
2420	amine (*n*)	$C_2H_5CH_2NH_2$	16.6	0.72	1.3901	IV-136
2421	amine (*iso*)	$(CH_3)_2CHNH_2$	15.4	0.694	1.3770	IV-152
2422	anisate (*p*)	$CH_3OC_6H_4CO_2C_3H_7$	20	1.09	1.514	....
2423	benzene (*n*)	$C_6H_5CH_2C_2H_5$	20	0.862	1.4920	V-390
2424	benzene (*iso*)	$C_6H_5CH(CH_3)_2$	20	0.862	1.4915	V-393
2425	benzoate (*n*)	$C_6H_5CO_2CH_2C_2H_5$	20.3	1.027	1.5000	IX-112
2426	bromide (*n*)	$C_2H_5CH_2Br$	20	1.354	1.4341	I-108
2427	" (*iso*)	$(CH_3)_2CHBr$	20	1.310	1.4251	I-108
2428	*n*-butyrate (*n*)	$C_2H_5CH_2CO_2CH_2\cdot$ C_2H_5	20	0.87	1.4005	II-271
2429	*iso*-butyrate (*n*)	$(CH_3)_2CHCO_2CH_2\cdot$ C_2H_5	20	0.87	1.3959	II-291
2430	chloride (*n*)	$C_2H_5CH_2Cl$	20	0.890	1.3884	I-104
2431	" (*iso*)	$(CH_3)_2CHCl$	15	0.868	1.3811	**I-73
2432	chloroformate (*n*)	$ClCO_2CH_2C_2H_5$	20	1.090	1.4035	III-11
2433	1-cyanocyclohexan- 2-one (1)	$C_3H_7(CN)\cdot$ $CCO(CH_2)_3CH_2$ $\mid_____\mid$	20	0.997	1.467	....
2434	cyclohexan-1-ol (*iso*) (1)	$(CH_3)_2CH\cdot$ $COH(CH_2)_4CH_2$ $\mid_____\mid$	15.5	0.914	1.4642	*VI-15
2435	cyclopentane (*n*)	$C_3H_7\cdot C_5H_9$	20	0.776	1.4263	**V-22
2436	" (*iso*)	$C_3H_7\cdot C_5H_9$	20	0.777	1.4259	**V-22
2437	cymene (2) (1,2,4)	$(CH_3)(C_3H_7)C_6H_3\cdot$ $CH(CH_3)_2$	15	0.869	1.4959	V-452
2438	*n*-decyl ether (*n*)	$C_2H_5CH_2OC_{10}H_{21}$	20	0.797	1.4249	....
2439	dichloroamine (*n*)	$C_2H_5CH_2NCl_2$	23	1.145	1.4525	IV-145
2440	4,6-dimethyl- coumaranone (2, *iso*)	$(CH_3)_2C_6H_2COCH\cdot$ $\mid$ $(C_3H_7)\cdot O$ $\mid$	20	1.048	1.5400	....
2441	ether (*n*)	$(C_2H_5CH_2)_2O$	14.5	0.75	1.3832	I-354
2442	" (*iso*)	$[(CH_3)_2CH]_2O$	23	0.72	1.3678	I-362
2443	formate (*n*)	$HCO_2CH_2C_2H_5$	20	0.901	1.3771	II-21
2444	furfuryl ether (*n*)	$C_2H_5CH_2OCH_2C_4H_3O$	20	0.966	1.4523	XXVII-112

Table 10-66 (*Continued*)
INDEX OF REFRACTION—LIQUIDS

No.	Substance	Formula	$t,°C.$	d_4^t	n_D^t	Beil. Ref.
2445	**Propyl**-furylacrylate (*n*)	$C_4H_3OCH:CHCO_2\cdot$ C_3H_7	20	1.074	1.5229	....
2446	furylacrylate (*iso*)	$C_4H_3OCH:CHCO_2\cdot$ $CH(CH_3)_2$	20	1.050	1.5146	....
2447	glyoxalidine	$C_3H_7C:NCH_2CH_2NH$	17	0.964	1.4871	XXIII-35
2448	hepta-1,6-diene-4-ol (4)	$(CH_2:CHCH_2)_2COH\cdot$ $CH_2C_2H_5$	21.1	0.865	1.4568	I-462
2449	heptan-4-ol (4)	$(C_2H_5CH_2)_3COH$	21	0.834	1.4356	I-426
2450	hept-1-ene-4-ol (4)	$(C_3H_7)_2COHCH_2\cdot$ $CH:CH_2$	19	0.845	1.4445	I-452
2451	2-hydroxydiphenyl (3)	$(C_3H_7)(OH)C_6H_3\cdot$ C_6H_5	20	1.047	1.579	....
2452	indazole (1)	$C_2H_5CH_2N\cdot C_6H_4\cdot$ $CH:N$	20	1.039	1.563	....
2453	" (1, *iso*)	$(CH_3)_2CHN\cdot C_6H_4\cdot$ $CH:N$	20	1.037	1.561	....
2454	" (2, *iso*)	$(CH_3)_2CHN\cdot N:$ $C_6H_4:CH$	20	1.051	1.580	....
2455	iodide (*n*)	$C_2H_5CH_2I$	20	1.743	1.5055	I-113
2456	" (*iso*)	$(CH_3)_2CHI$	20	1.703	1.4997	I-114
2457	menthatriene (2) †	$C_{13}H_{20}$	15	0.880	1.5027	V-452
2458	nitrate (*n*)	$C_2H_5CH_2ONO_2$	20	1.058	1.3972	I-355
2459	nitrite (*n*)	$C_2H_5CH_2ONO$	20	0.935	1.3613	I-355
2460	nonane (5)	$(C_4H_9)_2CHC_3H_7$	20	0.756	1.4228	**I-134
2461	phenyl ketone (*n*)	$C_2H_5CH_2COC_6H_5$	18.3	0.990	1.5202	VII-313
2462	" " (*iso*)	$(CH_3)_2CHCOC_6H_5$	16.6	0.99	1.5192	VII-316
2463	propenyl-*o*-phenyleneamidine	$C_3H_7N\cdot C_6H_4\cdot N:C\cdot$ C_2H_5	20.8	1.034	1.5626	....
2464	propionate (*n*)	$C_2H_5CO_2CH_2C_2H_5$	20	0.883	1.3933	II-240
2465	propylideneamine	$C_2H_5CH:NC_3H_7$	17.5	0.760	1.4127	IV-141
2466	*iso*-propylmalonate	$CH_2(CO_2C_3H_7)_2$	20	0.98	1.4259	....
2467	styryl ketone (*n*)	$C_3H_7COCH:CHC_6H_5$	24	0.992	1.5680	VII-376
2468	" " (*iso*)	$(CH_3)_2CHCOCH:$ CHC_6H_5	24.6	0.986	1.5669	VII-376
2469	*iso*-thiocyanate	$C_2H_5CH_2NCS$	16	0.978	1.5085	IV-145
2470	toluene (*o*)	$CH_3C_6H_4CH_2C_2H_5$	20	0.874	1.4993	V-418
2471	" (*m*)	$CH_3C_6H_4CH_2C_2H_5$	20	0.862	1.4951	V-418
2472	" (*p*)	$CH_3C_6H_4CH_2C_2H_5$	20	0.859	1.493	V-419
2473	*p*-tolyl ketone (*n*)	$C_2H_5CH_2COC_6H_4CH_3$	20	0.97	1.5250	VII-330
2474	" " (*iso*)	$(CH_3)_2CHCOC_6H_4\cdot$ CH_3	21.2	0.968	1.5187	VII-331
2475	*iso*-valerate (*n*)	$(CH_3)_2CHCH_2CO_2\cdot$ $CH_2C_2H_5$	17.8	0.87	1.4041	II-312
2476	*iso*-valerate (*iso*)	$(CH_3)_2CHCH_2CO_2\cdot$ $CH(CH_3)_2$	20(?)	0.85	1.397	II-312
2477	*o*-xylene (*uns*) (*n*)	$(CH_3)_2C_6H_3C_3H_7$ (1,2,4)	20	0.866	1.4955	V-440
2478	**Propylene** glycol (*α*)	$CH_3CHOHCH_2OH$	27	1.03	1.4293	I-472

† Probably a mixture of isomers.

Table 10-66 (*Continued*)
INDEX OF REFRACTION—LIQUIDS

No.	Substance	Formula	$t,°C.$	d_4^t	n_D^t	Beil. Ref.
2479	**Propylene** mono-acetate (α)	$CH_3CO_2CH_2CHOH\cdot$ CH_3	20(?)	1.055	1.4197	II-142
2480	**Propylidene** cyclo-hexane (*iso*)	$(CH_3)_2C{:}C(CH_2)_4\cdot$ $\underline{\quad\quad\quad\quad}$ CH_2 $-\vert$	20	0.836	1.4723	V-77
2481	cyclopentanone (*iso*, 1)	$(CH_3)_2C{:}CCO\cdot$ $\underline{\quad\quad}$ $(CH_2)_2CH_2$ $\underline{\quad\quad\quad}$	20	0.957	1.4932	*VII-52
2482	**Pulegene**	C_9H_{16}	22	0.791	1.4380	V-80
2483	**Pulegol** (*iso*) (*l*)	$C_{10}H_{18}O$	20	0.911	1.4723	**VI-70
2484	" (*iso*)	$C_{10}H_{18}O$	20(?)	0.914	1.4729	VI-65
2485	**Pulegone**	$C_{10}H_{16}O$	20	0.932	1.4880	VII-81
2486	" (*iso*)	$C_{10}H_{16}O$	20	0.919	1.4690	VII-85
2487	**Pulenenone** (α,β)	$(CH_3)_2CCOCH{:}C\cdot$ $\underline{\quad\quad\quad\quad}$ $(CH_3)CH_2CH_2$ $\underline{\quad\quad\quad\quad}$	16.5	0.932	1.4796	VII-67
2488	" (β,γ)	$(CH_3)_3C_6H_5CO$	17.7	0.906	1.4558	VII-66
2489	**Pulenol** (*dl*)	$C_9H_{18}O$	16.5	0.91	1.4632	VI-22
2490	" (β,γ)	$(CH_3)_2CCH{:}CHCH\cdot$ $\underline{\quad\quad\quad\quad}$ $(CH_3)CH_2CHOH$ $\underline{\quad\quad\quad\quad}$	18.5	0.921	1.4740	VI-50
2491	**Pyrazine** (m.p. 53°)	$N{:}CHCH{:}NCH{:}CH$ $\underline{\quad\quad\quad\quad}$	60.9	1.031	1.4953	XXIII-91
2492	**Pyrazole** (m.p. 70°)	$CH{:}CHCH{:}N\cdot NH$ $\underline{\quad\quad\quad\quad}$	99.8	1.001	1.4703	XXIII-39
2493	**Pyridazine**	$CH{:}CHCH{:}CHN{:}N$ $\underline{\quad\quad\quad\quad}$	23.5	1.104	1.5231	XXIII-89
2494	**Pyridine**	C_5H_5N	20	0.982	1.509	XX-181
2495	**Pyridyl-α-aldehyde**	C_6H_5ON	18.5	1.13	1.5389	*XXI-287
2496	**Pyrone** (1,4) (m.p. 32.5°)	$C_5H_4O_2$	40.3	1.190	1.5238	XVII-271
2497	**Pyrrole**	$(CH{:}CH)_2NH$	20	0.968	1.5091	XX-159
2498	**Pyruvic acid**	CH_3COCO_2H	15.3	1.3	1.4303	III-608
2499	**Quinoline**	$C_6H_4\cdot CH{:}CH\cdot CH{:}N$ $\underline{\quad\quad\quad\quad}$	18.2	1.09	1.6283	XX-339
2500	" (*iso*)	$C_6H_4\cdot CH{:}CH\cdot N{:}CH$ $\underline{\quad\quad\quad\quad}$	25.1	1.09	1.6223	XX-380
2501	**Quinoxaline** (m.p. 30°)	$C_6H_4N{:}CH\cdot CH{:}N$ $\underline{\quad\quad\quad\quad}$	48	1.133	1.6231	XXIII-176
2502	**Sabinene**	$C_{10}H_{16}$	20	0.842	1.4678	V-143
2503	**Sabinol**	$C_{10}H_{16}O$	20(?)	0.943	1.488	VI-98
2504	**Safrole**	$CH_2{:}O_2{:}C_6H_3C_3H_5$	19	1.097	1.5390	XIX-39
2505	" (*iso*) (*trans*)	$CH_3CH{:}CHC_6H_3{:}O_2{:}$ CH_2	19.3	1.12	1.5759	XIX-35
2506	**Salicylaldehyde** (*o*)	HOC_6H_4CHO	19.7	1.166	1.5736	VIII-31
2507	**Salicylaldehyde** (*homo*) (*p*) (m.p. 56°)	$HOC_6H_3(CH_3)CHO$ (6,3,1)	59.2	1.091	1.5466	VIII-100
2508	**Santalene** (α)	$C_{15}H_{24}$	15	0.913	1.4921	V-462
2509	" (β)	$C_{15}H_{24}$	20	0.894	1.4946	V-463
2510	" (γ)	$C_{15}H_{24}$	20	0.936	1.5042	V-463
2511	**Santalol** (α)	$C_{15}H_{24}O$	20	0.978	1.4992	VI-558

Table 10-66 (*Continued*)
INDEX OF REFRACTION—LIQUIDS

No.	Substance	Formula	t,°C.	d_4^t	n_D^t	Beil. Ref.
2512	**Santalol** (β)	$C_{15}H_{24}O$	20	0.972	1.5091	VI-558
2513	**Santene**	C_9H_{14}	20	0.87	1.4644	V-122
2514	**Sebacic acid**	$(CH_2)_8(CO_2H)_2$	133	.....	1.42	II-718
	(m.p. 134°)					
2515	**Sebacyl chloride**	$[(CH_2)_4COCl]_2$	18.3	.....	1.4684	II-719
2516	**Selenium diethyl**	$(C_2H_5)_2Se$	20	1.24	1.4768	I-349
2517	n-propyl hydride	$C_2H_5CH_2SeH$	20	1.302	1.4756	I-360
2518	**Silicon** methoxide	$Si(OCH_3)_4$	?	.....	1.368	....
2519	methyl-triethyl	$(CH_3)Si(C_2H_5)_3$	20	0.744	1.4160	....
2520	n-propyl trichloride	$(C_3H_7)SiCl_3$	20.3	1.196	1.4312	IV-630
2521	tetra-n-butyl	$(C_4H_9)_4Si$	20	0.801	1.4465	....
2522	tetraethyl	$(C_2H_5)_4Si$	20	0.766	1.4268	IV-625
2523	tetramethyl	$(CH_3)_4Si$	18.7	0.648	1.3591	IV-625
2524	tri-n-amyl fluoride	$(C_5H_{11})_3SiF$	20	0.839	1.4305	....
2525	tri-n-butyl fluoride	$(C_4H_9)_3SiF$	20	0.837	1.4250	....
2526	triethyl-n-amyl	$(C_2H_5)_3Si(C_5H_{11})$	20	0.784	1.4377	....
2527	triethyl-n-butyl	$(C_2H_5)_3Si(C_4H_9)$	20	0.779	1.4348	*IV-580
2528	triethyl-n-decyl	$(C_2H_5)_3Si(C_{10}H_{21})$	20	0.804	1.4472	....
2529	triethyl fluoride	$(C_2H_5)_3SiF$	20	0.835	1.3900	....
2530	triethyl-n-heptyl	$(C_2H_5)_3Si(C_7H_{15})$	20	0.791	1.4422	....
2531	triethyl-n-hexyl	$(C_2H_5)_3Si(C_6H_{13})$	20	0.788	1.4400	....
2532	triethyl-n-octyl	$(C_2H_5)_3Si(C_8H_{17})$	20	0.797	1.4438	....
2533	triethyl-n-propyl	$(C_2H_5)_3Si(C_3H_7)$	20	0.772	1.4308	*IV-580
2534	trimethyl-n-amyl	$(CH_3)_3Si(C_5H_{11})$	20	0.731	1.4096	....
2535	trimethyl-n-butyl	$(CH_3)_3Si(C_4H_9)$	20	0.718	1.4030	*IV-580
2536	trimethyl-n-decyl	$(CH_3)_3Si(C_{10}H_{21})$	20	0.771	1.4310	....
2537	trimethyl-n-dodecyl	$(CH_3)_3Si(C_{12}H_{25})$	20	0.780	1.4358	....
2538	trimethyl-ethyl	$(CH_3)_3Si(C_2H_5)$	20	0.685	1.3820	*IV-579
2539	trimethyl-n-heptyl	$(CH_3)_3Si(C_7H_{15})$	20	0.751	1.4201	....
2540	trimethyl-n-hexyl	$(CH_3)_3Si(C_6H_{13})$	20	0.742	1.4154	....
2541	trimethyl-n-octyl	$(CH_3)_3Si(C_8H_{17})$	20	0.758	1.4242	....
2542	trimethyl-n-propyl	$(CH_3)_3Si(C_3H_7)$	20	0.702	1.3929	*IV-580
2543	trimethyl-n-tetradecyl	$(CH_3)_3Si(C_{14}H_{29})$	20	0.791	1.4410	....
2544	tri-n-propyl fluoride	$(C_3H_7)_3SiF$	20	0.834	1.4107	....
2545	**Sorbyl chloride**	$CH_3(CH:CH)_2COCl$	20	1.065	1.5562	II-484
2546	**Sparteine**	$C_{15}H_{26}N_2$	19	1.024	1.5289	....
2547	" (*iso*)	$C_{15}H_{26}N_2$	17	1.028	1.5332	....
2548	**Spinacene**	$C_{29}H_{48}$	20	0.862	1.4956	*I-129
2549	**Stearic acid**	$CH_3(CH_2)_{16}CO_2H$	70	0.847	1.4335	II-377
	(m.p. 69°)					
2550	**Styrene**	$C_6H_5CH:CH_2$	20	0.906	1.5469	V-474
2551	**Suberyl chloride**	$[(CH_2)_3COCl]_2$	20.6	1.172	1.4685	II-694
2552	**Succinic** aldehyde	$(CH_2CHO)_2$	18	1.069	1.4262	I-767
2553	nitrile (m.p. 55°)	$(CH_2CN)_2$	63.1	0.985	1.4165	II-615
2554	**Succinyl chloride**	$(CH_2COCl)_2$	15.2	1.39	1.4735	II-613
2555	**Sulfur** dichloride	SCl_2	14	.....	1.557	....
2556	dioxide	SO_2	20(?)	.....	1.410	....
2557	monobromide	S_2Br_2	....	.....	1.736	....
2558	monochloride	S_2Cl_2	14	.....	1.666	....
2559	**Sulfuric acid**	H_2SO_4	20(?)	.....	1.429	....
2560	**Sulfuryl chloride**	SO_2Cl_2	20(?)	.....	1.444	....
2561	**Sylvestrine**	$C_{10}H_{16}$	20	0.848	1.4757	V-125
2562	**Tellurium** diethyl	$(C_2H_5)_2Te$	15	1.599	1.5182	I-350
2563	diphenyl	$(C_6H_5)_2Te$	15.8	1.603	1.6930	VI-347
2564	**Terpinene** (α)	$C_{10}H_{16}$	19.4	0.836	1.4781	V-126

Table 10-66 (*Continued*)
INDEX OF REFRACTION—LIQUIDS

No.	Substance	Formula	t,°C.	d_4^t	n_D^t	Beil. Ref.
2565	**Terpinene** (β)	$C_{10}H_{16}$	22	0.838	1.4754	V-132
2566	**Terpinenol** (4) (*d*)	$C_{10}H_{18}O$	19	0.93	1.4785	VI-55
2567	" (α) *d* (*l*)	$C_{10}H_{18}O$	16	0.93	1.4762	VI-56
2568	" (α) (*dl*)	$C_{10}H_{18}O$	20	0.936	1.4827	VI-58
	(m.p. 35°)					
2569	**Terpineol** (β)	$C_{10}H_{18}O$	20	0.919	1.4747	VI-62
	(m.p. 33°)					
2570	**Terpineol** (γ)	$C_{10}H_{18}O$	80	0.895	1.47	VI-61
	(m.p. 69°)					
2571	**Terpinolene**	$C_{10}H_{16}$	19	0.855	1.480	V-133
2572	**Tetrabromo**-butane	$Br_2CHCH_2CH_2CHBr_2$	20	2.529	1.6077	I-121
	(1,1,4,4)					
2573	ethane (1,1,1,2)	Br_3CCH_2Br	20	2.875	1.6277	I-94
2574	" (1,1,2,2)	$Br_2CHCHBr_2$	20	2.967	1.6380	I-94
2575	**Tetrachloro**-ethane	CH_2ClCCl_3	23.2	1.58	1.4816	I-86
	(1,1,1,2)					
2576	ethane (1,1,2,2)	$Cl_2CHCHCl_2$	18.5	1.6	1.4921	I-86
2577	ethylene	$Cl_2C:CCl_2$	20	1.619	1.5055	I-187
2578	**Tetracosane** (*n*)	$CH_3(CH_2)_{22}CH_3$	65	0.77	1.4303	I-175
	(m.p. 54°)					
2579	**Tetradecane** (*n*)	$CH_3(CH_2)_{12}CH_3$	20	0.765	1.4459	I-171
2580	**Tetraethyl**-benzene	$(C_2H_5)_4C_6H_2$	19.6	0.887	1.5085	V-455
	(1,2,3,4)					
2581	**Tetraethyl**-benzene	$(C_2H_5)_4C_6H_2$	16	0.888	1.5041	V-455
	(1,2,4,5)					
2582	citrate	$C_5H_9O(CO_2C_2H_5)_3$	20	1.102	1.4484	III-568
2583	cyclobutane-2,4-	$(C_2H_5)_4C_2(CO)_2$	17.8	0.946	1.4518	*VII-321
	dione	(1,1,3,3)				
2584	diaminobenzene (*o*)	$C_6H_4[N(C_2H_5)_2]_2$	12.6	0.927	1.5213	....
2585	diaminonaphthalene	$C_{10}H_6[N(C_2H_5)_2]_2$	99.6	0.932	1.5394	....
	(1,4) (m.p. 47°)					
2586	diaminonaphthalene	$C_{10}H_6[N(C_2H_5)_2]_2$	26.8	0.989	1.5742	....
	(1,5) (m.p. 41°)					
2587	diaminonaphthalene	$C_{10}H_6[N(C_2H_5)_2]_2$	18	1.012	1.5951	....
	(1,8) (m.p. 31°)					
2588	dioxalylmalonate	$(C_2H_5O_2CCO)_2C:$	14.9	1.193	1.4567	*III-199
		$(CO_2C_2H_5)_2$				
2589	methane-tetra-	$C(CO_2C_2H_5)_4$	17	1.136	1.436	*II-331
	carboxylate					
2590	oxalylmethane-	$C_2H_5O_2CCOC(CO_2\cdot$	15.5	1.167	1.4403	*III-298
	tricarboxylate	$C_2H_5)_3$				
2591	*m*-phenylene-	$C_6H_4[N(C_2H_5)_2]_2$	11.8	0.952	1.5537	....
	diamine					
2592	**Tetrahydro**-	$CH_2CH:CH\cdot$	22.1	0.808	1.4451	V-63
	benzene	$\underline{\qquad\qquad}$				
		$(CH_2)_2CH_2$				
		$\underline{\qquad\qquad}$				
2593	benzisoxazole (β,γ)	$C_6H_8(:CH\cdot O\cdot N:)$	20	1.095	1.498	....
2594	benzoic acid (Δ^1)	$CH_2(CH_2)_3CH:C\cdot$	47.2	1.072	1.4903	IX-41
	(m.p. 29°)	$\underline{\qquad\qquad\qquad}$				
		CO_2H				
2595	cumene	$(CH_3)_2CHC:CH\cdot$	15.2	0.83	1.4615	V-77
		$\underline{\qquad\qquad}$				
		$(CH_2)_3CH_2$				
		$\underline{\qquad\qquad}$				
2596	cumene (*pseudo*)	$(CH_3)_3C_6H_7$ (1,2,4)	16.2	0.808	1.4499	*V-40

Table 10-66 (*Continued*)
INDEX OF REFRACTION—LIQUIDS

No.	Substance	Formula	t,°C.	d_4^t	n_D^t	Beil. Ref.
2597	**Tetrahydro-**fur-furyl alcohol	$C_4H_7OCH_2OH$	19	1.052	1.4502	....
2598	furfuryl benzyl ether	$C_4H_7OCH_2OCH_2 \cdot C_6H_5$	20	1.048	1.5174	....
2599	furfuryl chloride	$C_4H_7OCH_2Cl$	20	1.110	1.4560	....
2600	furoic acid	$C_4H_7OCO_2H$	20	1.193	1.4585	*XVIII-435
2601	mesitylene	$(CH_3)_3C_6H_7$ (1,3,5)	21	0.797	1.4447	*V-40
2602	naphthalene (1,2,3,4)	$C_6H_4CH_2(CH_2)_2CH_2$	20	0.97	1.5402	V-491
2603	α-naphthol (1,2,3,4)	$C_6H_4(CH_2)_3CHOH$	17	1.09	1.5671	....
2604	α-naphthylamine (*ar*)	$NH_2C_6H_3CH_2 \cdot (CH_2)_2CH_2$	23.1	1.054	1.5896	XII-1197
2605	β-naphthylamine (*ac*)	$C_6H_4CH_2CH(NH_2) \cdot CH_2CH_2$	22.2	1.03	1.5604	XII-1200
2606	β-naphthylamine (*ar*) (m.p. 38°)	$NH_2C_6H_3(CH_2)_3CH_2$	22.2	1.029	1.5604	XII-1198
2607	quinoline (1,2,3,4)	$C_6H_4CH_2CH_2CH_2NH$	23.9	1.05	1.5933	XX-262
2608	*iso*-quinoline	$C_9H_{11}N$ (1,2,3,4)	23.1	1.07	1.5798	XX-275
2609	toluene (1,2,5,6)	$CH_3C_6H_9$	16	0.8	1.4443	V-67
2610	" (1,2,3,4)	$CH_3C_6H_9$	20	0.805	1.4454	V-67
2611	" (Δ¹)	$CH_3C_6H_9$	20	0.810	1.4496	V-66
2612	*m*-toluic acid (Δ¹)	$CH_2(CH_2)_2CH \cdot (CH_3)CH:CCO_2H$	69	1.013	1.476	IX-47
2613	*o*-xylene (Δ¹)	$(CH_3)_2C_6H_8$	20	0.823	1.4587	V-72
2614	*m*-xylene (1,2,3,4)	$(CH_3)_2C_6H_8$	20	0.798	1.441	V-76
2615	*p*-xylene (1,2,3,6)	$(CH_3)_2C_6H_8$	20	0.802	1.4459	V-74
2616	**Tetramethyl-**ben-zene (1,2,3,4)	$(CH_3)_4C_6H_2$	20	0.905	1.5201	V-430
2617	benzene (1,2,3,5)	$(CH_3)_4C_6H_2$	20	0.890	1.5125	V-430
2618	" (1,2,4,5)	$(CH_3)_4C_6H_2$	20	0.889	1.512	V-431
2619	butenediol (*cis*) (m.p. 76°)	$[(CH_3)_2C(OH)CH:]_2$	69	0.869	1.422	*I-260
2620	butenediol (*trans*) (m.p. 69°)	$[(CH_3)_2C(OH)CH:]_2$	99.2	0.863	1.422	*I-261
2621	coumaranone (2,2,4,6) (m.p. 41°)	$(CH_3)_2C_6H_2CO \cdot C(CH_3)_2 \cdot O$	20	1.060	1.5428	....
2622	cyclohexa-2,5-diene-4-ol (1,1,2,4)	$(CH_3)_4C_6H_3OH$	20	0.925	1.4824	**VI-102
2623	cyclohexane (1,2,4,5)	$(CH_3)_4C_6H_8$	13.1	0.791	1.4372	*V-23
2624	cyclohexan-1-ol (1,2,4,5)	$(CH_3)_4C_6H_7OH$	18.1	0.898	1.4595	*VI-30
2625	cyclohexan-2-ol (m.p. 53°) (1,1,4,4)	$(CH_3)_4C_6H_7OH$	30.2	0.894	1.4571	*VI-30
2626	cyclohexan-2-one (m.p. 18°) (1,1,4,4)	$(CH_3)_4C_5H_6CO$	14.3	0.893	1.4485	*VII-38

Table 10-66 (*Continued*)
INDEX OF REFRACTION—LIQUIDS

No.	Substance	Formula	t,°C.	d_4^t	n_D^t	Beil. Ref.
2627	**Tetramethyl-**cyclo-hex-1-ene	$(CH_3)_4C_6H_6$ (1,2,4,5)	16.5	0.820	1.4588	*V-45
2628	cyclohexyl-1-chloride	$(CH_3)_4C_6H_7Cl$ (1,2,4,5)	20	0.936	1.4624	*V-24
2629	4-ethylidene-cyclohexa-2,5-diene (1,1,2,5)	$(CH_3)_4C_6H_2$:CHCH$_3$	20	0.880	1.5125	**V-341
2630	4-methene-cyclohexa-2,5-diene (1,1,2,6)	$(CH_3)_4C_6H_2$:CH$_2$	20	0.879	1.5149	**V-336
2631	4-methene-cyclohexa-2,5-diene (1,1,2,5)	$(CH_3)_4C_6H_2$:CH$_2$	20	0.877	1.5147	**V-336
2632	pentane (2,2,3,3)	$(CH_3)_3CC(CH_3)_2 \cdot C_2H_5$	20	0.757	1.4234	....
2633	" (2,2,3,4)	$(CH_3)_3CCH(CH_3) \cdot CH(CH_3)_2$	20	0.739	1.4146	....
2634	" (2,2,4,4)	$[(CH_3)_3C]_2CH_2$	20	0.720	1.4068	....
2635	" (2,3,3,4)	$[(CH_3)_2CH]_2C(CH_3)_2$	20	0.755	1.4220	....
2636	piperidine (*N*,2,4,6)	$(CH_3)_4C_5H_7N$	19.7	0.823	1.4430	XX-129
2637	pyrazole (1,3,4,5)	CH_3C:(C·CH$_3$)$_2$:N·N $\underset{CH_3}{\mid_____\mid}$	20	0.947	1.480	XXIII-82
2638	**Tetranitromethane**	$C(NO_2)_4$	17	1.645	1.4399	I-80
2639	**Tetrolic aldehyde**	CH_3C:CCHO	17	0.927	1.4467	*I-388
2640	**Thionaphthene** (m.p. 32°)	$C_6H_4 \cdot CH$:CH·S $\mid_____\mid$	36.2	1.15	1.6332	XVII-59
2641	**Thionyl chloride**	$SOCl_2$	10	.....	1.527	....
2642	**Thiophene**	CH:CHCH:CHS $\mid_____\mid$	19.7	1.065	1.5287	XVII-29
2643	**Thiophenol**	C_6H_5SH	23.2	1.074	1.5861	VI-294
2644	**Thiophosgene**	$SCCl_2$	15(?)	1.509	1.5442	III-134
2645	**Thujane**	$C_{10}H_{18}$	20	0.814	1.4376	V-93
2646	**Thujene** (*α*)	$C_{10}H_{16}$	20	0.830	1.4515	V-142
2647	" (*β*)	$C_{10}H_{16}$	20	0.821	1.4471	V-143
2648	**Thujone** (*α*)	$C_{10}H_{16}O$	20	0.912	1.4530	VII-93
2649	" (*β*)	$C_{10}H_{16}O$	17.6	0.92	1.4522	VII-93
2650	**Thujyl alcohol**	$C_{10}H_{17}OH$	20	0.921	1.4635	VI-68
2651	**Thymol** (m.p. 52°)	$(CH_3)(C_3H_7)$: C_6H_3OH (1,4,3)	20	0.969	1.5227	VI-532
2652	**Tiglic** acid (*trans*) (m.p. 64°)	CH_3CH:C(CH$_3$)· CO_2H	76	0.964	1.44	II-430
2653	aldehyde	CH_3CH:C(CH$_3$)CHO	9.6	0.88	1.4512	I-733
2654	**Tin** tetraethyl	$(C_2H_5)_4Sn$	19.7	1.199	1.4724	IV-632
2655	tetramethyl	$(CH_3)_4Sn$	20(?)	1.3	1.5201	IV-632
2656	triethyl	$[(C_2H_5)_3Sn]_2$	17.8	1.4	1.5374	IV-638
2657	**Titanium** tetrachloride	$TiCl_4$	10.5	.....	1.61	....
2658	**Tolualdehyde** (*o*)	$CH_3C_6H_4CHO$	19	1.038	1.5485	VII-295
2659	" (*m*)	$CH_3C_6H_4CHO$	21.4	1.02	1.5407	VII-296
2660	" (*p*)	$CH_3C_6H_4CHO$	16.6	1.019	1.5469	VII-297
2661	**Toluene**	$C_6H_5CH_3$	20	0.867	1.4969	V-280
2662	**Toluic** acid (*o*) (m.p. 105°)	$CH_3C_6H_4CO_2H$	114.6	1.062	1.512	IX-462
2663	**Toluic** acid (*m*) (m.p. 110°)	$CH_3C_6H_4CO_2H$	111.6	1.054	1.51	IX-475

Table 10-66 (*Continued*)
INDEX OF REFRACTION—LIQUIDS

No.	Substance	Formula	t,°C.	d_4^t	n_D^t	Beil. Ref.
2664	**Toluidine** (*o*)	$CH_3C_6H_4NH_2$	20	0.999	1.5728	XII-772
2665	" (*m*)	$CH_3C_6H_4NH_2$	22.4	0.99	1.5711	XII-853
2666	" (*p*) (m.p. 45°)	$CH_3C_6H_4NH_2$	59.1	0.97	1.5532	XII-880
2667	**Tolunitrile** (*o*)	$CH_3C_6H_4CN$	23.1	0.99	1.5272	IX-466
2668	**Toluquinoxaline**	$CH_3C_6H_3 \cdot N:(CH)_2:N$	18.4	1.118	1.6211	XXIII-184
2669	**Tolyl** *iso*-butyrate (*o*)	$(CH_3)_2CHCO_2C_6H_4 \cdot CH_3$	20	1.001	1.491	....
2670	α-ethylcrotonate (*p*)	$CH_3CH:C(C_2H_5) \cdot CO_2C_6H_4CH_3$	20	1.009	1.513	....
2671	**Triallyl** citrate	$(CH_2CO_2C_3H_5)_2: C(OH)CO_2C_3H_5$	17.4	1.136	1.4714	*III-197
2672	phenol (*o,o,p*)	$(C_3H_5)_3C_6H_2OH$	20	0.974	1.541	*VI-322
2673	tricarballylate	$(CH_2CO_2C_3H_5)_2: CHCO_2C_3H_5$	15.9	1.097	1.4675	*II-322
2674	**Triamyl**-amine†	$(C_5H_{11})_3N$	22	0.845	1.4128	....
2675	borate†	$(C_5H_{11}O)_3B$	22	0.845	1.4128	I-404
2676	**Triazobenzene**	$C_6H_5N_3$	22.5	1.086	1.5642	V-276
2677	**Triazole** (1,2,5)	$CH:N \cdot NH \cdot N:CH$	25.3	1.186	1.4854	XXVI-11
2678	**Tribromo**-butane (1,2,3)	$CH_3(CHBr)_2CH_2Br$	20	2.19	1.567	I-121
2679	ethane (1,1,2)	$BrCH_2CHBr_2$	20	2.579	1.5890	I-93
2680	ethylene	$BrCH:CBr_2$	20	2.708	1.5992	I-191
2681	pentane (1,2,3)	$C_2H_5(CHBr)_2CH_2Br$	20	2.09	1.559	**I-98
2682	propane (1,2,3)	$BrCH_2CHBrCH_2Br$	23	2.436	1.6	I-112
2683	**Tributyl**-amine (*n*)	$(C_4H_9)_3N$	20	0.778	1.4291	IV-157
2684	amine (*iso*)	$(C_4H_9)_3N$	17.3	0.771	1.4252	IV-666
2685	borate (*iso*)	$(C_4H_9O)_3B$	20	0.858	1.4086	I-377
2686	citrate (*n*)	$C_6H_5O_7(C_4H_9)_3$	20	1.04	1.4460	....
2687	phosphate (*n*)	$(C_4H_9O)_3PO$	20	0.979	1.4245	**I-397
2688	**Tricaprin** (m.p. 31°)	$C_{33}H_{62}O_6$	40	0.921	1.4446	II-356
2689	**Trichloro**-benzene	$C_6H_3Cl_3$ (1,2,4)	20	1.45	1.5671	V-204
2690	bromomethane	Cl_3CBr	20	1.94	1.5300	I-67
2691	crotononitrile	$Cl_3CCH:CHCN$	20	1.420	1.5083	**II-398
2692	crotonyl chloride	$Cl_3CCH:CHCOCl$	20	1.528	1.5176	**II-398
2693	1,1-dibromo-ethane (2,2,2)	Br_2CHCCl_3	25.7	2.3	1.5299	I-93
2694	ethane (1,1,1)	CH_3CCl_3	21	1.32	1.4377	I-85
2695	" (1,1,2)	$CH_2ClCHCl_2$	22	1.446	1.4719	I-85
2696	ethylene	$CHCl:CCl_2$	19.8	1.463	1.4777	I-187
2697	furan (2,3,4)	C_4HOCl_3	20	1.547	1.5057	....
2698	methylchloro-formate	$ClCO_2CCl_3$	22	1.65	1.4566	III-18
2699	non-3-yne-2-ol (1,1,1)	$Cl_3CCHOHC: C(CH_2)_4CH_3$	9.9	1.234	1.5000	I-456
2700	**Tricosane** (*n*) (m.p. 47°)	$CH_3(CH_2)_{21}CH_3$	80.8	0.7	1.4236	I-175
2701	**Tricresyl phosphate** (*o*)	$(CH_3C_6H_4O)_3PO$	20	1.17	1.556	VI-358
2702	**Tridecane** (*n*)	$CH_3(CH_2)_{11}CH_3$	16.8	0.76	1.4419	I-171
2703	**Triethanolamine**	$N(CH_2CH_2OH)_3$	20	1.124	1.4852	IV-285
2704	**Triethyl** aconitate	$C_{12}H_{18}O_6$	20	1.106	1.4556	II-852

† Contains amyl group in its various isomeric forms.

Table 10-66 (*Continued*)
INDEX OF REFRACTION—LIQUIDS

No.	Substance	Formula	t,°C.	d_4^t	n_D^t	Beil. Ref.
2705	**Triethyl**-acetyl-methanetricar-boxylate	$CH_3COC(CO_2C_2H_5)_3$	13.5	1.128	1.4366	*III-294
2706	amine	$(C_2H_5)_3N$	20	0.728	1.4003	IV-99
2707	benzene (1,3,5)	$(C_2H_5)_3C_6H_3$	17	0.86	1.4951	V-449
2708	borate	$(C_2H_5O)_3B$	20	0.88	1.3808	I-335
2709	citrate	$C_3H_5O(CO_2C_2H_5)_3$	20	1.137	1.4455	III-568
2710	methane-tricarboxylate	$HC(CO_2C_2H_5)_3$	16.2	1.108	1.4275	II-810
2711	oxalylmalonate	$C_2H_5O_2CCOCH:$ $(CO_2C_2H_5)_2$	16.1	1.160	1.4496	III-850
2712	phosphate	$(C_2H_5O)_3PO$	17.1	1.07	1.4067	I-332
2713	phosphine	$(C_2H_5)_3P$	15	0.800	1.458	IV-582
2714	phosphite	$(C_2H_5O)_3P$	10	0.97	1.4131	I-330
2715	**Trifluoro**-1,2-di-bromoethane (1,1,2)	$CBrF_2CHBrF$	14	2.254	1.4145	I-92
2716	1,1,2-tribromo-ethane (1,2,2)	$CFBr_2CBrF_2$	20(?)	2.6	1.4666	I-94
2717	**Triformal**-ethylamine	$(HCH:NC_2H_5)_3$	20.2	0.894	1.4604	XXVI-2
2718	methylamine	$(H_2C:NCH_3)_3$	19	0.918	1.4632	XXVI-1
2719	propylamine	$(H_2C:NC_3H_7)_3$	18.8	0.880	1.4597	XXVI-3
2720	**Trimethyl**-acetic acid (m.p. 35.5°)	$(CH_3)_3CCO_2H$	36.5	0.91	1.3931	II-319
2721	acetophenone (2,4,5)	$(CH_3)_3C_6H_2COCH_3$	14.9	1.004	1.5411	VII-333
2722	acetophenone (2,4,6)	$(CH_3)_3C_6H_2COCH_3$	20	0.975	1.5175	VII-332
2723	benzene (1,2,3)	$(CH_3)_3C_6H_3$	20	0.894	1.5139	V-399
2724	" (1,2,4)	$(CH_3)_3C_6H_3$	20	0.876	1.5049	V-400
2725	" (1,3,5)	$(CH_3)_3C_6H_3$	20	0.865	1.4994	V-406
2726	butane (2,2,3)	$(CH_3)_3CCH(CH_3)_2$	20	0.690	1.3895	**I-121
2727	but-3-ene (2,2,3)	$(CH_3)_3CC(CH_3):$ CH_2	20	0.705	1.4029	I-221
2728	5-chloropyrazole (1,3,4)	$CH_3NCCl:C(CH_3)C \cdot$ $\vert \qquad\qquad \vert$ $(CH_3):N$ $\vert\!_____\vert$	20	1.108	1.487	....
2729	α-chromene (2,4,6)	$CH_3C_6H_3C(CH_3):$ $\vert \qquad\qquad$ $CHCH(CH_3) \cdot O$ $\!_____\vert$	20	1.012	1.5513	....
2730	α-chromene (3,4,6)	$CH_3C_6H_3C(CH_3):$ $\vert \qquad\qquad$ $C(CH_3) \cdot CH_2 \cdot O$ $\!_____\vert$	20	1.029	1.5575	....
2731	coumaranone (2,2,5)	$CH_3C_6H_3CO \cdot C \cdot$ $\vert \qquad\qquad \vert$ $(CH_3)_2 \cdot O$ $\!_____\vert$	19.3	1.073	1.5404	*XVII-166
2732	coumaranone (2,2,7) (m.p. 68°)	$CH_3C_6H_3COC \cdot$ $\vert \qquad\qquad \vert$ $(CH_3)_2 \cdot O$ $\!_____\vert$	100.6	1.001	1.507	....

Table 10-66 (*Continued*)
INDEX OF REFRACTION—LIQUIDS

No.	Substance	Formula	t,°C.	d_4^t	n_D^t	Beil. Ref.
2733	**Trimethyl-**coumara-none (2,4,6)	(CH$_3$)$_2$C$_6$H$_2$COCH· $\vert\underline{\quad\quad}$ (CH$_3$)·O $\underline{\quad\quad}\vert$	19.2	1.099	1.5574	*XVII-68
2734	coumarone (2,3,5)	CH$_3$C$_6$H$_3$C(CH$_3$):C· $\vert\underline{\quad\quad\quad}$ (CH$_3$)·O $\underline{\quad\quad}\vert$	20	1.021	1.5506	....
2735	cyclohexane (1,1,3)	(CH$_3$)$_3$C$_6$H$_9$	25.3	0.787	1.4339	V-42
2736	" (1,2,4)	(CH$_3$)$_3$C$_6$H$_9$	16.9	0.780	1.431	V-42
2737	cyclohexan-2-ol (1,1,2)	(CH$_3$)$_3$C$_6$H$_8$OH	19.9	0.926	1.4679	*VI-16
2738	cyclohexan-5-ol (1,2,4)	(CH$_3$)$_3$C$_6$H$_8$OH	19.2	0.899	1.4585	*VI-17
2739	cyclohexan-1-ol (1,3,5)	(CH$_3$)$_3$C$_6$H$_8$OH	16.3	0.888	1.4537	*VI-17
2740	cyclohexan-4-one (1,1,3)	(CH$_3$)$_3$C$_5$H$_7$CO	19	0.902	1.4545	VII-29
2741	cyclohexan-5-one (1,2,4)	(CH$_3$)$_3$C$_5$H$_7$CO	17.4	0.899	1.4501	*VII-25
2742	cyclohex-2-ene (1,1,2)	(CH$_3$)$_3$C$_6$H$_7$	20.4	0.822	1.4560	*V-39
2743	cyclohex-3-ene (1,1,4)	(CH$_3$)$_3$C$_6$H$_7$	23.2	0.802	1.4442	*V-39
2744	cyclohex-1-ene (1,2,3)	(CH$_3$)$_3$C$_6$H$_7$	11.8	0.835	1.4630	*V-40
2745	cyclohex-1-ene-3-ol (1,3,5)	(CH$_3$)$_3$C$_6$H$_6$OH	19.3	0.913	1.4735	*VI-36
2746	cyclohex-2-ene-4-one (1,1,3)	(CH$_3$)$_3$C$_5$H$_5$CO	16.3	0.933	1.4795	VII-65
2747	cyclohexyl-1-chloride (1,3,5)	(CH$_3$)$_3$C$_6$H$_8$Cl	20	0.929	1.4550	*V-18
2748	cyclopentane (1,1,2)	(CH$_3$)$_3$C$_5$H$_7$	20	0.773	1.4230	V-39
2749	cyclopentane (1,1,3)	(CH$_3$)$_3$C$_5$H$_7$	20	0.748	1.4112	*V-16
2750	cyclopentane (1,2,3) (*cis, cis, cis*)	(CH$_3$)$_3$C$_5$H$_7$	20	0.779	1.4263	V-40
2751	cyclopentane (1,2,3) (*cis, cis, trans*)	(CH$_3$)$_3$C$_5$H$_7$	20	0.770	1.4219	V-40
2752	cyclopentane (1,2,3) (*cis, trans, cis*)	(CH$_3$)$_3$C$_5$H$_7$	20	0.754	1.4144	V-40
2753	cyclopentane (1,2,4) (*cis, cis, cis*)	(CH$_3$)$_3$C$_5$H$_7$	20	0.766	1.422	*V-16
2754	cyclopentane (1,2,4) (*cis, cis, trans*)	(CH$_3$)$_3$C$_5$H$_7$	20	0.763	1.4185	*V-16
2755	cyclopentane (1,2,4) (*cis, trans, cis*)	(CH$_3$)$_3$C$_5$H$_7$	20	0.747	1.4106	*V-16

Table 10-66 (*Continued*)
INDEX OF REFRACTION—LIQUIDS

No.	Substance	Formula	t,°C.	d_4^t	n_D^t	Beil. Ref.
2756	**Trimethyl**-5-(β,β-di-chloroethyl)-benzene (1,2,3)	$(CH_3)_3C_6H_2CH_2\cdot$ $CHCl_2$	20	1.144	1.5439	**V-336
2757	1-dichloromethyl-cyclohex-3-ene-2-one (1,4,5)	$(CH_3)_3C_5H_4COCHCl_2$	20	1.197	1.5238	....
2758	4-ethylidenecyclo-hexa-2,5-diene (1,1,3)	$(CH_3)_3C_6H_3:CHCH_3$	20	0.879	1.5160	V-442
2759	hexane (2,2,3)	$(CH_3)_3CCH(CH_3)\cdot$ $CH_2C_2H_5$	20	0.729	1.4105	....
2760	" (2,2,4)	$(CH_3)_3CCH_2CH:$ $(CH_3)(C_2H_5)$	20	0.716	1.4033	....
2761	" (2,2,5)	$(CH_3)_3CCH_2CH_2\cdot$ $CH(CH_3)_2$	20	0.707	1.3996	*I-64
2762	" (2,3,3)	$(CH_3)_2CHC(CH_3)_2\cdot$ $CH_2C_2H_5$	20	0.738	1.4143	....
2763	" (2,3,4)	$(CH_3)_2CHCH(CH_3)\cdot$ $CH(CH_3)C_2H_5$	20	0.741	1.415	....
2764	" (2,3,5)	$(CH_3)_2CHCH(CH_3)\cdot$ $CH_2CH(CH_3)_2$	20	0.722	1.4060	**I-129
2765	" (2,4,4)	$C_2H_5C(CH_3)_2CH_2\cdot$ $CH(CH_3)_2$	20	0.725	1.4075	....
2766	" (3,3,4)	$C_2H_5C(CH_3)_2CH:$ $(CH_3)(C_2H_5)$	20	0.745	1.4178	....
2767	hexan-4-ol (3,3,4)	$(C_2H_5)(CH_3)_2CCOH:$ $(CH_3)(C_2H_5)$	20	0.832	1.4345	I-425
2768	2-hydroxystyrene (2',2',5)	$(CH_3)(OH)C_6H_3CH:$ $C(CH_3)_2$	21.5	0.991	1.5501	*VI-293
2769	isoxazole	$N:C(CH_3)C(CH_3):$ $\vert\underline{}$ $C(CH_3)\cdot O$ $\underline{}\vert$	20	0.981	1.452	XXVII-18
2770	4-methenecyclo-hexa-2,5-diene (1,1,2)	$(CH_3)_3C_6H_3:CH_2$	20	0.866	1.5139	V-429
2771	2-methoxystyrene (2',2',5)	$CH_3(CH_3O)C_6H_3\cdot$ $CH:C(CH_3)_2$	17.1	0.968	1.5425	*VI-293
2772	pentane (2,2,3)	$(CH_3)_3CCH(CH_3)\cdot$ C_2H_5	20	0.716	1.4030	*I-62
2773	" (2,2,4)	$(CH_3)_3CCH_2CH:$ $(CH_3)_2$	20	0.692	1.3915	I-164
2774	" (2,3,3)	$C_2H_5C(CH_3)_2\cdot$ $CH(CH_3)_2$	20	0.726	1.4075	....
2775	" (2,3,4)	$(CH_3)_2CHCH(CH_3)\cdot$ $CH(CH_3)_2$	20	0.719	1.4042	....
2776	pent-1-ene (2,3,3)	$CH_2:C(CH_3)C(CH_3)_2\cdot$ C_2H_5	20	0.737	1.418	....
2777	" (2,3,4)	$CH_2:C(CH_3)CH\cdot$ $(CH_3)CH(CH_3)_2$	20	0.729	1.415	....
2778	pent-4-ene (2,2,3)	$CH_2:CHCH(CH_3)\cdot$ $C(CH_3)_3$	20	0.719	1.412	....
2779	" (2,2,4)	$CH_2:C(CH_3)CH_2\cdot$ $C(CH_3)_3$	20	0.715	1.4086	....
2780	" (2,3,3)	$CH_2:CHC(CH_3)_2\cdot$ $CH(CH_3)_2$	20	0.729	1.414	....
2781	piperidine (2,4,6)	$(CH_3)_3C_5H_8N$	19	0.830	1.4435	XX-126

Table 10-66 (*Continued*)
INDEX OF REFRACTION—LIQUIDS

No.	Substance	Formula	$t,°C.$	d_4^t	n_D^t	Beil. Ref.
2782	**Trimethyl-**pyrazole (1,3,4)	CH:C(CH₃)C·$\underline{\qquad}$ (CH₃):N·NCH₃$\underline{\qquad}$	20	0.951	1.476	. . . .
2783	" (1,3,5)	CH₃C:CHC(CH₃):$\underline{\qquad}$ N·N(CH₃)$\underline{\quad}$	20	1.034	1.549	XXIII-75
2784	pyrrole (2,3,5)	CH₃C:CHC(CH₃):$\underline{\qquad}$ C(CH₃)NH$\underline{\qquad}$	15.8	0.935	1.509	XX-177
2785	styrene (α,β,β)	C₆H₅C(CH₃):C: (CH₃)₂	20	0.892	1.5164	V-498
2786	**Trimethylene** bromohydrin	BrCH₂CH₂CH₂OH	20	1.571	1.4883	I-356
2787	chlorohydrin	ClCH₂CH₂CH₂OH	20	1.131	1.4469	I-356
2788	glycol	HOCH₂CH₂CH₂OH	20	1.060	1.4398	I-475
2789	**Triphenyl-**methane (m.p. 94°)	(C₆H₅)₃CH	99	1.014	1.59	V-698
2790	phosphine (m.p. 79°)	(C₆H₅)₃P	69	1.1	1.5248	XVI-759
2791	**Tripropyl-**amine (*n*)	(C₂H₅CH₂)₃N	19.4	0.757	1.4176	IV-139
2792	methane	(C₃H₇)₃CH	20	0.73	1.414	. . . .
2793	**Trithioacetone**	(CH₃CSCH₃)₃	20	1.068	1.542	XIX-389
2794	**Tropacocaine** (m.p. 49°)	C₁₅H₁₉O₂N	100	1.043	1.5080	XXI-39
2795	**Tropane**	C₈H₁₅N	21.6	0.93	1.4766	XX-141
2796	**Tropidine**	C₈H₁₃N	18.4	0.95	1.4904	XX-177
2797	**Tropine** (m.p. 63°)	C₈H₁₅ON	99.8	1.016	1.4811	XXI-16
2798	**Tropinone** (m.p. 41°)	C₈H₁₃ON	99.6	0.987	1.4621	XXI-258
2799	**Tropylidene**	HC:(CH)₅CH₂ $\underline{\qquad}$	20	0.888	1.5261	V-280
2800	**Undecane** (*n*)	CH₃(CH₂)₉CH₃	20(?)	0.741	1.4158	I-170
2801	**Undecan-2-ol**	C₉H₁₉CHOHCH₃	20	0.827	1.4369	I-427
2802	**Undec-1-ene**	CH₃(CH₂)₈CH:CH₂	20	0.763	1.4284	I-225
2803	**Undec-2-ene**	C₈H₁₇CH:CHCH₃	15(?)	0.774	1.4333	I-225
2804	**Undecyl** alcohol (*n*)	CH₃(CH₂)₉CH₂OH	23	0.833	1.4392	I-427
2805	phenyl ketone (m.p. 47°)	CH₃(CH₂)₁₀COC₆H₅	52.4	0.897	1.4850	VII-345
2806	**Undecylenic acid** (β)	CH₃(CH)₂C₇H₁₄· CO₂H	24	0.907	1.45	II-458
2807	**Undecylic** acid (m.p. 29°)	CH₃(CH₂)₉CO₂H	45.2	. . . . .	1.4294	II-358
2808	aldehyde	CH₃(CH₂)₉CHO	23	0.825	1.4322	I-712
2809	**Urethane** (m.p. 49°)	NH₂CO₂C₂H₅	52	0.9	1.4144	III-22
2810	**Valeric** acid (*n*)	C₂H₅CH₂CH₂CO₂H	19.1	0.94	1.409	II-299
2811	acid (*iso*)	(CH₃)₂CHCH₂CO₂H	20	0.93	1.4043	II-309
2812	aldehyde (*n*)	C₂H₅CH₂CH₂CHO	20	0.82	1.3882	I-676
2813	" (*iso*)	(CH₃)₂CHCH₂CHO	20	0.802	1.3902	I-684
2814	aldoxime (*iso*)	C₄H₉CH:NOH	20	0.893	1.4367	I-686
2815	anhydride (*iso*)	(C₄H₉CO)₂O	26	0.93	1.4147	II-314
2816	**Valero-***p*-cresol (*o*) (*iso*)	(CH₃)₂CHCH₂CO· C₆H₃(OH)(CH₃)	20	1.028	1.5320	*VIII-557
2817	nitrile (*n*)	CH₃(CH₂)₃CN	18	0.801	1.3917	II-301

Table 10-66 (*Continued*)
INDEX OF REFRACTION—LIQUIDS

No.	Substance	Formula	$t,°C.$	d_4^t	n_D^t	Beil. Ref.
2818	**Valero-**thienone (α) (*iso*)	$C_4H_9COC{:}CHCH{:}$ CH·S	20	1.062	1.536	....
2819	**Valeryl-**aceto-phenone (ω) (*iso*)	$C_6H_5COCH_2COCH_2·$ $CH(CH_3)_2$	20	1.031	1.5605	VII-690
2820	chloride (n)	$CH_3(CH_2)_3COCl$	20	1.01	1.42	II-301
2821	" (*iso*)	$(CH_3)_2CHCH_2COCl$	20	0.989	1.4156	II-315
2822	indazole (2) (*iso*) (*stable*)	$C_{12}H_{14}ON_2$	16.3	1.088	1.555	....
2823	**Veratrole** (o)	$C_6H_4(OCH_3)_2$	21	1.08	1.53	VI-771
2824	**Verbenene**	$C_{10}H_{14}$	20	0.882	1.4986	*V-207
2825	**Vinyl-**acetic acid	$CH_2{:}CHCH_2CO_2H$	15	1.013	1.4257	II-407
2826	bromide	$CH_2{:}CHBr$	20(?)	1.5	1.4462	I-188
2827	ethyl alcohol	$C_3H_5CH_2OH$	17.5	0.838	1.4146	I-441
2828	**Water**†	H_2O	20	0.998	1.3330	....
2829	**Xylidine** (o) (*vic*)	$(CH_3)_2C_6H_3NH_2$ (1,2,3)	20	0.99	1.570	XII-1101
2830	" (o) (*sym*)	$(CH_3)_2C_6H_3NH_2$ (1,3,5)	20	0.972	1.558	XII-1131
2831	" (m) (*vic*)	$(CH_3)_2C_6H_3NH_2$ (1,3,2)	20	0.979	1.561	XII-1107
2832	" (m) (*uns*)	$(CH_3)_2C_6H_3NH_2$ (1,3,4)	19.6	0.978	1.5607	XII-1111
2833	" (p)	$(CH_3)_2C_6H_3NH_2$ (1,4,2)	21.3	0.979	1.5591	XII-1135
2834	**Xylylamine** (p)	$CH_3C_6H_4CH_2NH_2$	20	0.952	1.5364	XII-1141
2835	**Zingiberene**	$C_{15}H_{24}$	20	0.873	1.4940	V-461

† See special table.

Table 10-67
REFRACTIVE INDEX OF LIQUIDS FINDING NUMBERS

The compounds below are arranged in the order of ascending values of refractive index, an arrangement intended to serve as an aid in the identification of liquids. The numbers listed under the bold faced refractive index heading refer to the compounds with corresponding numerical listings in the table *Index of Refraction— Liquids*. Thus, under the refractive index 1.360 is listed *No. 1181* which is *ethyl isocyanide, n 1.36*.

1.193	**1.369**
2239	785
1.254	**1.370**
1615	760; 1395
1.256	**1.371**
1617	1483; 2301
1.306	**1.372**
1428	5; 318; 2100
1.317	**1.373**
2369	341; 1082
1.325	**1.374**
81; 1616	1622
1.329	**1.375**
1776	499; 786; 1581
1.330	**1.376**
2234	789
1.333	**1.377**
1620; 2828	1253; 2101; 2131; 2415; 2421; 2443
1.339	**1.378**
3	1815; 1836; 2419
1.342	**1.379**
728	282; 1144
1.344	**1.380**
1950	1783; 1826; 2303
1.346	**1.381**
13	353; 1934; 2431; 2708
1.350	**1.382**
1061	2; 981; 983; 1185; 2538
1.353	**1.383**
2184	1517; 2302; 2441
1.354	**1.384**
2298	340; 1393; 1838; 2113; 2114; 2414
1.357	**1.385**
118	129; 635; 1366; 1788; 1867; 1995; 2312; 2418
1.358	**1.386**
2139; 2297	310; 1552
1.359	**1.387**
10; 1764; 2523	130; 303; 621; 1014; 1480; 1816; 1850; 2399; 2404
1.360	**1.388**
1181; 1254; 1448; 2138	273; 278; 1099; 1532; 1586; 2112; 2141; 2292; 2430; 2812
1.361	**1.389**
1094; 2459	309; 1996; 2111
1.364	**1.390**
1447; 1817; 2400	6; 28; 719; 731; 790; 1152; 1435; 1837; 2140; 2224; 2293; 2420; 2529; 2726; 2813
1.368	
2406; 2442; 2518	

Table 10-67 (*Continued*)
REFRACTIVE INDEX OF LIQUIDS FINDING NUMBERS

1.391

272; 984

1.392

277; 339; 978; 982; 989; 1370; 2773; 2817

1.393

345; 541; 904; 1151; 1372; 1589; 2109; 2464; 2542; 2720

1.394

281; 679; 901; 1369; 1588; 1590; 2225

1.395

271; 274; 302; 326; 903; 1564; 1587; 1957; 2270

1.396

335; 338; 1169; 1828; 2429

1.397

91; 276; 301; 435; 995; 1436; 2008; 2009; 2191; 2209; 2458; 2476

1.398

119; 325; 344; 1375; 1959; 2255

1.399

275; 280; 307; 987; 988; 994; 997; 1396; 1540; 1958; 2004; 2005; 2014; 2015; 2016

1.400

11; 44; 297; 905; 1446; 1499; 1829; 2006; 2007; 2110; 2230; 2706; 2761

1.401

82; 83; 84; 279; 283; 889; 902; 2108; 2294; 2428

1.402

300; 885; 1266; 1603; 1818; 2116

1.403

296; 317; 402; 887; 927; 928; 990; 996; 1846; 2085; 2535; 2727; 2760; 2772

1.404

7; 888; 906; 925; 1390; 1543; 1544; 1563; 1940; 2002; 2313; 2405; 2432; 2475; 2775; 2811

1.405

46; 89; 115; 295; 913; 1039; 1839; 2098; 2099; 2115; 2189; 2281; 2411

1.406

92; 333; 438; 1541; 1542; 1670; 1975; 2086; 2087; 2241; 2764

1.407

110; 136; 334; 517; 892; 929; 992; 993; 1043; 1276; 1925; 1968; 1969; 1974; 2012; 2013; 2634; 2712

1.408

59; 105; 117; 360; 702; 893; 939; 1259; 1401; 1789; 1941; 2080; 2765; 2774

1.409

90; 114; 576; 605; 847; 886; 890; 921; 922; 958; 971; 1260; 1984; 2104; 2259; 2685; 2686; 2779; 2810

1.410

88; 93; 304; 575; 699; 909; 910; 911; 912; 931; 1140; 1561; 1562; 1814; 1881; 1943; 1970; 1971; 1980; 1981; 1985; 1986; 1987; 2003; 2010; 2011; 2073; 2403; 2534; 2556

1.411

104; 109; 425; 529; 651; 846; 891; 920; 1261; 1849; 1924; 1993; 2068; 2069; 2072; 2544; 2749; 2755; 2759

1.412

108; 128; 359; 577; 791; 845; 875; 1280; 1281; 1516; 1905; 1966; 1972; 1973; 2082; 2265; 2300; 2778

1.413

141; 641; 717; 860; 917; 918; 930; 932; 1305; 1529; 1598; 1843; 2081; 2261; 2263; 2465; 2674; 2675; 2714

1.414

50; 432; 678; 683; 788; 843; 877; 919; 926; 950; 970; 1263; 1376; 1555; 1582; 1583; 1621; 1982; 1983; 2067; 2070; 2071; 2103; 2105; 2188; 2262; 2264; 2271; 2752; 2762; 2780; 2792; 2809

1.415

17; 56; 140; 378; 379; 787; 1209; 1297; 1549; 1717; 1926; 1942; 2260; 2540; 2633; 2715; 2763; 2777; 2815; 2827

1.416

485; 915; 916; 923; 924; 957; 1108; 1265; 1927; 1998; 2020; 2028; 2036; 2051; 2187; 2519; 2800; 2821

1.417

594; 673; 985; 1024; 1044; 1537; 1740; 1967; 1988; 1989; 2553

1.418

107; 613; 659; 762; 933; 934; 1182; 1279; 1513; 1944; 1978; 1979; 2089; 2102; 2226; 2766; 2776; 2791

1.419

8; 52; 299; 431; 685; 690; 876; 1062; 1170; 1518; 1922; 1976; 1977; 2754

1.420

12; 381; 551; 701; 713; 716; 721; 1008; 1013; 1083; 1193; 1250; 1269; 1392; 1551; 1601; 1765; 1921; 1946; 1961; 1962; 1963; 2107; 2183; 2267; 2479; 2514; 2539; 2820

1.421

48; 85; 521; 653; 654; 817; 908; 938; 991; 1049; 1139; 1320; 1530; 1535; 1536; 2106; 2250; 2273; 2289; 2386

1.422

43; 384; 650; 844; 1002; 1029; 1159; 1196; 1434; 1486; 1533; 1813; 1847; 1886; 1897; 1920; 1999; 2227; 2619; 2620; 2635; 2751; 2753

1.423

16; 548; 578; 649; 652; 686; 814; 896; 986; 1084; 1271; 1550; 1868; 1997; 2001; 2269; 2309; 2460; 2632; 2748

1.424

18; 387; 488; 580; 664; 1130; 1145; 1258; 1273; 1277; 1278; 1308; 1531; 1580; 1964; 2094; 2185; 2223; 2258; 2284; 2304; 2541; 2700

1.425

428; 646; 857; 1146; 1163; 1171; 1177; 1178; 1579; 1782; 2000; 2092; 2095; 2136; 2427; 2438; 2525; 2684; 2687

Table 10-67 (*Continued*)
REFRACTIVE INDEX OF LIQUIDS FINDING NUMBERS

1.426

4; 94; 507; 642; 767; 914; 1343; 1713; 1824; 1827; 1835; 1844; 2257; 2288; 2435; 2436; 2466; 2552; 2750; 2825

1.427

377; 383; 498; 531; 537; 658; 725; 813; 883; 1143; 1270; 1287; 1291; 1296; 1671; 1918; 1923; 1965; 2522

1.428

294; 305; 380; 620; 899; 1109; 1141; 1147; 1242; 1558; 1664; 1673; 1823; 1994; 2305; 2710; 2802

1.429

137; 298; 540; 711; 771; 811; 881; 907; 1142; 1822; 2401; 2478; 2559; 2683; 2807

1.430

9; 47; 145; 167; 342; 405; 409; 481; 773; 816; 935; 936; 975; 976; 1041; 1050; 1267; 1429; 1547; 1578; 1919; 1960; 2083; 2117; 2245; 2256; 2268; 2279; 2280; 2291; 2402; 2498; 2578

1.431

327; 376; 524; 530; 534; 707; 815; 895; 900; 1111; 1293; 1311; 1373; 1374; 1502; 1591; 1775; 2193; 2242; 2244; 2283; 2387; 2520; 2524; 2533; 2536; 2736

1.432

77; 765; 894; 897; 1231; 1268; 1275; 1512; 1553; 1576; 1594; 1898; 2023; 2025; 2040; 2088; 2220; 2236; 2808

1.433

222; 614; 715; 778; 858; 1188; 1344; 1556; 1641; 2219; 2308; 2803

1.434

111; 293; 501; 657; 671; 1051; 1274; 1285; 1524; 1534; 1951; 2278; 2426; 2549; 2735

1.435

356; 514; 599; 638; 882; 1058; 1216; 1243; 1371; 1423; 1525; 1928; 2017; 2243; 2252; 2253; 2527; 2767

1.436

292; 404; 443; 483; 681; 723; 812; 1198; 1262; 1352; 1416; 1503; 2222; 2240; 2246; 2290; 2307; 2449; 2537; 2589

1.437

113; 233; 246; 447; 520; 523; 532; 616; 761; 964; 1519; 1545; 1595; 2623; 2705; 2801; 2814

1.438

536; 615; 619; 730; 836; 973; 1046; 1510; 1559; 1560; 1575; 1770; 1991; 2130; 2482; 2526; 2645; 2694

1.439

316; 442; 550; 604; 677; 697; 784; 972; 979; 1098; 1204; 1449; 1786; 2804

1.440

291; 368; 403; 592; 630; 1052; 1055; 1090; 1150; 1306; 1345; 1356; 1507; 1592.1; 1642; 1669; 1992; 2018; 2032; 2033; 2531; 2590; 2638; 2652; 2788

1.441

102; 126; 660; 662; 1038; 1076; 1226; 1283; 1290; 1506; 1526; 2299; 2543; 2614

1.442

103; 134; 422; 490; 535; 610; 682; 729; 945; 949; 1284; 1368; 1442; 1557; 1606; 1720; 1834; 1956; 2530; 2702

1.443

247; 500; 720; 792; 835; 948; 1149; 1180; 1244; 1504; 1508; 1712; 1935; 2221; 2398; 2417; 2636

1.444

22; 101; 116; 125; 142; 336; 370; 504; 708; 946; 1042; 1047; 1103; 1294; 1310; 1441; 1511; 1546; 1672; 1734; 2235; 2532; 2560; 2609; 2743; 2781

1.445

596; 629; 636; 680; 722; 726; 837; 1040; 1048; 1089; 1110; 1138; 1388; 1493; 1758; 1790; 1911; 2142; 2248; 2450; 2592; 2601; 2610; 2688

1.446

148; 453; 465; 482; 545; 581; 727; 956; 963; 1148; 1286; 1538; 1833; 1874; 2090; 2091; 2127; 2128; 2129; 2579; 2615; 2709; 2826

1.447

257; 494; 639; 663; 980; 1045; 1088; 1137; 1236; 1481; 1592; 1596; 1719; 1725; 2521; 2528; 2639; 2647; 2787

1.448

146; 306; 495; 634; 645; 732; 802; 831; 937; 1207; 1232; 1351; 1505; 1514; 1599; 1873; 2311; 2381; 2582

1.449

40; 684; 712; 830; 1025; 1527; 1600; 1728.1; 1832; 1870; 2626

1.450

70; 143; 232; 496; 509; 516; 522; 544; 644; 668; 954; 966; 1057; 1237; 1444; 1459; 1463; 1565; 1566; 1567; 1584; 1723; 2019; 2596; 2597; 2611; 2711; 2741; 2806

1.451

162; 423; 471; 543; 617; 710; 833; 898; 1179; 1425; 1593; 1666; 1792; 1810; 1990; 2021; 2093; 2653

1.452

23; 39; 244; 321; 503; 552; 618; 669; 670; 691; 692; 709; 798; 834; 1194; 1233; 1264; 1397; 1721; 1724; 1933; 2022; 2096; 2097; 2407; 2444; 2583; 2646; 2649; 2769

1.453

506; 518; 519; 666; 667; 689; 832; 1091; 1165; 1166; 1183; 1187; 1247; 1251; 1389; 1637; 2145; 2186; 2315; 2385; 2439; 2648

1.454

26; 252; 329; 389; 647; 648; 665; 823; 827; 842; 1060; 1131; 1168; 1229; 1282; 1334; 1394; 1443; 1528; 1716; 2055; 2084; 2247; 2306; 2739

1.455

135; 248; 255; 256; 324; 397; 463; 533; 632; 688; 825; 826; 967; 1186; 1791; 1793; 2740; 2747

1.456

249; 369; 395; 408; 466; 513; 631; 672; 698; 768; 769; 829; 951; 977; 1127; 1335; 1509;

Table 10-67 (*Continued*)
REFRACTIVE INDEX OF LIQUIDS FINDING NUMBERS

1607; 1627; 1830; 1831; 1848; 1908; 2276; 2488; 2599; 2704; 2742

1.457
328; 385; 429; 828; 1189; 1199; 1221; 1574; 1811; 1882; 1938; 1952; 2448; 2588; 2625; 2698

1.458
80; 238; 320; 454; 456; 637; 822; 824; 1222; 1224; 1572; 1635; 1722; 1939; 2054; 2409; 2713

1.459
127; 464; 515; 584; 819; 1129; 1167; 1200; 1554; 1573; 1604; 1778; 1902; 2063; 2158; 2600; 2613; 2627; 2738

1.460
45; 147; 231; 319; 363; 487; 505; 511; 589; 628; 693; 724; 1097; 1391; 1569; 1605; 1774; 2204; 2249; 2274; 2370; 2624; 2717; 2719

1.461
470; 508; 820; 1070; 1153; 1154; 1548; 1571; 1597; 1726; 1883; 2217; 2382

1.462
224; 396; 733; 884; 969; 1205; 1367; 1402; 1715; 1718; 1806; 1869; 2079; 2275; 2379; 2595; 2628; 2798

1.463
133; 225; 230; 390; 661; 793; 821; 855; 1132; 1450; 1494; 2489; 2718; 2744

1.464
227; 229; 643; 1461; 1727; 2251; 2376; 2434; 2513; 2650

1.465
20; 41; 226; 818; 965; 1191; 1307; 1411; 1426; 1570; 2310; 2380

1.466
55; 216; 582; 1007; 1184; 1498; 1619; 1728

1.467
24; 163; 743; 841; 1879; 2433; 2716

1.468
86; 694; 734; 740; 748; 750; 849; 1467; 2171; 2181; 2502; 2515; 2673; 2737

1.469
25; 215; 1087; 1190; 1408; 1476; 1875; 1888; 2037; 2486; 2551

1.470
528; 1197; 1256; 1410; 1460; 1475; 1477; 1609; 2492; 2570

1.471
228; 491; 583; 742; 1223; 1458; 1884; 1954; 2192; 2671

1.472
460; 751; 1068; 1155; 1945; 2383; 2480; 2483; 2654; 2695

1.473
375; 718; 1006; 1107; 1496; 1500; 1520; 1714; 2484

1.474
633; 739; 848; 1030; 1093; 1156; 1295; 1338; 1495; 1497; 2412; 2490; 2554; 2745

1.475
253; 484; 510; 1086; 1106; 2565; 2569

1.476
106; 343; 747; 854; 1128; 1238; 1431; 1437; 1438; 1840; 1841; 1901; 1903; 2058; 2285; 2517; 2561; 2567; 2612; 2782

1.477
36; 87; 105; 492; 830; 839; 968; 1337; 2516; 2795

1.478
611; 1737; 1738; 2564; 2696

1.479
371; 391; 542; 749; 770; 1336; 1387; 2057; 2147; 2318; 2566

1.480
76; 159; 265; 372; 410; 553; 752; 1462; 1501; 1568; 2197; 2487; 2571; 2637; 2746

1.481
398; 406; 640; 1026; 1192; 1515; 1631; 2056; 2797

1.482
213; 741; 840; 1354; 1876; 1877; 2575; 2622

1.483
208; 245; 266; 394; 1133; 1248; 1842; 1880; 1916; 2568

1.484
1771; 2317

1.485
31; 1027; 1028; 1489; 2677; 2703; 2805

1.486
538; 539; 1953

1.487
98; 112; 287; 382; 1272; 1427; 1492; 1887; 2066; 2447; 2728

1.488
21; 214; 547; 1227; 1907; 2485; 2503; 2786

1.489
308; 1100; 1418; 1445; 1648; 2150; 2272

1.490
131; 286; 288; 461; 462; 1069; 1220; 1457; 1585; 1736; 1889; 2182; 2205; 2316; 2594; 2796

1.491
197; 527; 1735; 1781; 1821; 2167; 2410; 2669

1.492
99; 221; 358; 799; 2166; 2423; 2424; 2508; 2576

1.493
289; 290; 526; 579; 700; 1491; 1602; 1705; 1820; 2238; 2472; 2481

1.494
27; 97; 100; 367; 493; 810; 1249; 1378; 1490; 2149; 2835

1.495
332; 626; 627; 1288; 1432; 1433; 1488; 1697; 1914; 2229; 2367; 2471; 2491; 2509; 2707

1.496
124; 315; 512; 608; 782; 797; 1117; 1126; 1203; 1674; 1691; 1732; 1878; 2437; 2477; 2548

Table 10-67 (*Continued*)
REFRACTIVE INDEX OF LIQUIDS FINDING NUMBERS

1.497
502; 590; 781; 800; 1066; 1440; 1695; 1819; 1913; 2148; 2661

1.498
139; 420; 867; 1357; 1382; 1632; 1647; 1660; 1685; 1891; 2282; 2593

1.499
19; 373; 399; 480; 591; 1005; 1009; 1195; 1358; 1377; 1917; 2470; 2511; 2725; 2824

1.500
314; 400; 421; 479; 676; 1067; 1355; 1360; 1484; 1485; 1613; 1748; 1885; 2425; 2456; 2699

1.501
179; 331; 374; 467; 525; 852; 878; 1102; 1468; 1473; 1522; 1678; 1701; 1890; 2038; 2137; 2377; 2413

1.502
468; 469; 1101; 1201; 1359; 1400; 1404; 1478; 1663; 1679; 1684; 1746; 2176

1.503
401; 625; 703; 864; 955; 1054; 1692; 1707; 1892; 2286; 2323; 2457

1.504
181; 737; 866; 868; 1004; 1010; 1081; 1309; 1698; 2378; 2384; 2510; 2581

1.505
609; 780; 796; 944; 1079; 1255; 1353; 1420; 1646; 1682; 1706; 1710; 1912; 2296; 2724

1.506
138; 166; 386; 561; 655; 1075; 1118; 1417; 1686; 1690; 1696; 1733; 1845; 2455; 2577; 2697

1.507
285; 364; 559; 1080; 1175; 1225; 1361; 1419; 1636; 1677; 1711; 2146; 2732

1.508
72; 120; 284; 424; 549; 565; 704; 735; 1073; 1239; 1303; 1683; 1688; 2320; 2691; 2794

1.509
95; 323; 563; 705; 736; 1292; 1399; 1421; 1451; 1675; 1681; 1694; 2118; 2354; 2469; 2494; 2497; 2512; 2580; 2784

1.510
1077; 1078; 1119; 1702; 1709; 2324; 2663

1.511
562; 865; 1398

1.512
863; 1015; 1634; 1677; 1687; 1700; 1759; 1866; 2287; 2618; 2662

1.513
132; 311; 560; 776; 859; 1074; 1302; 1312; 1414; 1455; 1703; 1761; 1863; 2048; 2064; 2617; 2629; 2670

1.514
54; 75; 322; 351; 433; 879; 880; 1018; 1210; 1412; 1472; 1689; 1799; 2422; 2723; 2770

1.515
17; 564; 853; 874; 1471; 1676; 1708; 1763; 2446; 2630; 2631

1.516
753; 755; 1001; 1022; 1523; 1539; 1756; 2372; 2758; 2785

1.517
58; 155; 243; 714; 1314; 1341; 1704; 1760; 1762; 1803; 1804; 2168; 2598

1.518
1063; 1680; 1693; 1768; 1895; 2396; 2562; 2692; 2722

1.519
451; 588; 783; 1122; 1430; 1865; 2144; 2462; 2474

1.520
566; 574; 775; 1161; 1319; 1380; 1661; 1699; 1812; 2295; 2461; 2616; 2655

1.521
33; 1208; 1217; 1454; 1470; 2337; 2584

1.522
352; 430; 794; 861; 1017; 2355

1.523
34; 60; 199; 450; 1059; 1160; 1340; 2445; 2493; 2651

1.524
51; 53; 191; 201; 772; 1019; 1162; 1206; 1313; 1955; 2132; 2143; 2174; 2175; 2177; 2179; 2496; 2757

1.525
207; 392; 417; 449; 567; 1113; 1289; 1403; 2172; 2473; 2790

1.526
71; 205; 856; 1116; 1415; 1487; 1638; 2322; 2799

1.527
73; 587; 1012; 1385; 1422; 2338; 2641; 2667

1.528
573; 606; 803; 1157; 1779; 2133; 2180

1.529
183; 220; 240; 1053; 1257; 2031; 2207; 2326; 2546; 2642

1.530
472; 489; 763; 804; 1071; 1215; 1252; 2339; 2690; 2693; 2823

1.531
35; 595; 687; 998; 1121; 1123; 1662; 1731; 2391; 2395

1.532
862; 1333; 1469; 2049; 2350; 2816

1.533
62; 1023; 1577; 1730; 1739; 1769; 1893; 2178; 2547

1.534
14; 195; 570; 851; 1000; 1011; 1234; 1321; 1854; 1894; 2041

1.535
177; 476; 1120; 1212; 1409; 1610; 1626; 1630; 1750; 2065; 2233

Table 10-67 (*Continued*)
REFRACTIVE INDEX OF LIQUIDS FINDING NUMBERS

1.536
156; 350; 809; 870; 1172; 1218; 1329; 1466; 1614; 1639; 1825; 2373; 2818; 2834

1.537
348; 477; 571; 656; 1241; 1742; 1800; 2169; 2328; 2656

1.538
57; 160; 952; 1056; 1379; 1439; 1612; 1741; 1747; 1801; 1864; 1947; 2159

1.539
61; 198; 346; 624; 674; 766; 869; 1330; 1405; 1780; 2161; 2495; 2504; 2585

1.540
192; 349; 407; 546; 1003; 1342; 1521; 1624; 1628; 1729; 1855; 2321; 2440; 2602; 2731

1.541
171; 623; 706; 764; 940; 1016; 1464; 1633; 1784; 1900; 2125; 2659; 2672; 2721

1.542
354; 473; 1453; 1751; 1851; 1915; 2050; 2793

1.543
37; 123; 393; 478; 871; 872; 941; 1134; 1135; 1240; 1608; 1767; 2135; 2228; 2621; 2771

1.544
69; 193; 202; 455; 850; 953; 974; 1424; 1909; 2644; 2756

1.545
68; 754; 1095; 2126; 2134; 2162; 2163; 2165; 2416

1.546
121; 175; 1219; 1325; 1465; 2123

1.547
174; 439; 586; 1021; 1136; 1649; 1743; 2061; 2231; 2232; 2507; 2550; 2660

1.548
173; 178; 777; 873; 1065; 1301; 1324; 1331; 2164; 2314

1.549
206; 264; 807; 1020; 1064; 2026; 2060; 2365; 2394; 2658; 2783

1.550
434; 448; 569; 1316; 1384; 1406; 2043; 2046; 2325; 2768

1.551
88; 263; 268; 801; 808; 943; 1085; 1214; 2052; 2729; 2734

1.552
172; 211; 330; 347; 474; 585; 2029; 2035; 2212

1.553
79; 312; 426; 805; 1744; 1757; 1860; 1862; 2666

1.554
152; 190; 212; 2024; 2591

1.555
416; 746; 795; 806; 1623; 1852; 2208; 2822

1.556
154; 204; 209; 270; 1112; 1322; 1332; 1629; 2213; 2390; 2397; 2545; 2701

1.557
242; 267; 440; 1386; 1640; 1753; 1861; 2034; 2555; 2733

1.558
78; 122; 187; 441; 568; 1124; 1777; 2053; 2214; 2730; 2830

1.559
090; 779; 1211; 1235; 1052; 1745; 1859; 2030; 2047; 2681; 2833

1.560
150; 189; 210; 223; 237; 612; 1158; 1173; 1304; 1318; 1625; 1766; 2124; 2277; 2388; 2393; 2605; 2606

1.561
144; 236; 262; 1346; 1456; 1611; 1929; 2453; 2819; 2831; 2832

1.562
241; 313; 362; 445; 759; 1176; 1315; 1752; 2210

1.563
200; 203; 738; 1323; 1348; 1644; 1667; 1856; 2343; 2452; 2463

1.564
1479; 1754; 1899; 1904; 2045; 2676

1.565
361; 415; 475; 947; 1115; 1857; 2215

1.566
15; 1245; 2352; 2363

1.567
414; 942; 2254; 2345; 2468; 2603; 2678; 2689

1.568
188; 365; 427; 597; 1164; 1407; 1931; 2331; 2360; 2467

1.569
196; 744; 1300; 1339; 1798; 2374

1.570
161; 1347; 1794; 1948; 2375; 2829

1.571
49; 219; 239; 357; 1246; 1383; 1668; 2211; 2266; 2364; 2665

1.572
164; 254; 1350; 2059; 2122; 2330

1.573
217; 235; 355; 1807; 2027; 2170; 2342; 2664

1.574
554; 558; 1452; 1643; 2332; 2334; 2506; 2586

1.575
153; 458; 999; 1298; 2206; 2335

1.576
66; 151; 555; 1033; 2218; 2505

1.577
182; 419; 446; 1853; 2237

1.578
1213; 1797; 2361; 2389

1.579
218; 593; 675; 1035; 2349; 2451

Table 10-67 (*Continued*)
REFRACTIVE INDEX OF LIQUIDS FINDING NUMBERS

1.580
598; 1230; 1665; 2346; 2454; 2608

1.581
418; 1653; 1932

1.582
64; 2121; 2196; 2348

1.583
337; 745; 2216; 2336

1.584
169; 1785; 2347; 2392; 2408

1.585
366; 2042; 2044; 2357

1.586
149; 157; 1805; 1858; 2119; 2643

1.587
32; 486; 1930; 2120; 2333; 2351

1.588
30; 258; 1031; 1228; 1317; 1482

1.589
756; 1125; 2329; 2679

1.590
67; 261; 411; 2604; 2789

1.591
176; 1299; 1802

1.592
1032; 1413; 1808; 1896; 2344

1.593
412; 622; 2607

1.594
1362

1.595
1349; 1363; 2587

1.596
180; 1657; 2358

1.597
695; 1381; 1474

1.598
65; 186; 251; 269; 1365; 1659

1.599
260; 607; 1364; 1658; 2340; 2680

1.600
74; 2353; 2682

1.601
63; 158; 1328; 1936; 2039

1.602
1327

1.603
29; 452; 2076

1.604
1656

1.605
1; 757; 1072; 2368

1.606
184; 1937; 2062

1.607
961; 1655

1.608
185; 437; 557; 1037; 2341; 2356; 2572

1.609
259; 602; 1654; 2151; 2359

1.610
1949; 2157; 2657

1.612
194; 556; 1910

1.613
1174; 1202; 2327

1.614
459

1.615
1034; 1326; 2155; 2173

1.616
959; 960; 2154; 2156

1.617
2152

1.618
2075

1.619
1096; 1906

1.620
457; 1092

1.621
1645; 2153; 2195; 2198; 2668

1.622
2500

1.623
601; 2077; 2501

1.624
603; 962; 1773

1.626
234; 1787

1.627
1772

1.628
388; 1755; 2078; 2499; 2573

1.630
572; 2199

1.632
622.1; 1809

1.633
436; 2640

1.634
444; 600; 2366

Table 10-67 (*Continued*)
REFRACTIVE INDEX OF LIQUIDS FINDING NUMBERS

	1.635		**1.666**
1036		1651; 2558	
	1.636		**1.670**
96; 758; 2160		2201	
	1.638		**1.676**
2574		1114	
	1.644		**1.680**
774		1795	
	1.646		**1.683**
170; 2074		2203	
	1.647		**1.693**
497		2563	
	1.648		**1.696**
1104		413; 1796	
	1.649		**1.697**
2200; 2362		2371	
	1.653		**1.705**
1749		1650	
	1.654		**1.706**
1105		42	
	1.655		**1.708**
2194		2202	
	1.657		**1.736**
2319		2557	
	1.658		**1.743**
250		2190	
	1.664		**1.885**
168		1618	

Table 10-68
INDEX OF REFRACTION OF WATER AT VARIOUS TEMPERATURES AND FOR VARIOUS WAVE LENGTHS OF LIGHT

From Values Given in the *International Critical Tables, Vol. 7, p. 13.*

t°C.	Wave length of light			
	$n_\alpha(C)$	$n_D(D)$	$n_\beta(F)$	$n_\gamma(G')$
0	....	1.3340	....	....
10	1.3318	1.3337	1.3378	1.3411
20	1.33115	1.33300	1.33714	1.34035
30	1.3302	1.3320	1.3360	1.3392
40	1.3288	1.3306	1.3347	1.3379
50	1.3274	1.3290	1.3332	1.3364
60	1.3257	1.3272	1.3315	1.3346
70	1.3237	1.3252	1.3294	1.3325

Table 10-69
INDEX OF REFRACTION OF AQUEOUS HALOGEN ACID SOLUTIONS AT 18°C.

$$n_D^{18}$$

Acid	Normality of the acid solution				
	0.1	0.5	1.0	2.0	4.0
HCl	....	1.33753	1.34168	1.34977	1.36480
HBr	1.33452	1.33952	1.34571	1.35805	1.38271
HI	1.33525	1.34346	1.35387	1.37497	....
HClO$_3$	....	1.33773	1.34211	1.35078	....
HIO$_3$	1.33568	1.34498	1.35642	1.37910	1.42393
HIO$_4$	....	1.3445*	....	....	....

*At 26°C.

Table 10-70
REFRACTIVE INDEX OF AQUEOUS GLYCEROL SOLUTIONS AT 20.0°C.

L. F. Hoyt, *Ind. Eng. Chem. 26, 329 (1934)*.

% Glycerol by weight	$n_D^{20°}$	% Glycerol by weight	$n_D^{20°}$	% Glycerol by weight	$n_D^{20°}$
0	1.33303	35	1.37740	70	1.42789
1	1.33416	36	1.37874	71	1.42938
2	1.33530	37	1.38008	72	1.43087
3	1.33645	38	1.38143	73	1.43236
4	1.33762	39	1.38278	74	1.43385
5	1.33880	40	1.38413	75	1.43534
6	1.33999	41	1.38548	76	1.43683
7	1.34118	42	1.38683	77	1.43832
8	1.34238	43	1.38818	78	1.43982
9	1.34359	44	1.38953	79	1.44135
10	1.34481	45	1.39089	80	1.44290
11	1.34604	46	1.39227	81	1.44450
12	1.34729	47	1.39368	82	1.44612
13	1.34834	48	1.39513	83	1.44770
14	1.34980	49	1.39660	84	1.44930
15	1.35106	50	1.39809	85	1.45085
16	1.35233	51	1.39958	86	1.45237
17	1.35361	52	1.40107	87	1.45389
18	1.35490	53	1.40256	88	1.45539
19	1.35619	54	1.40405	89	1.45689
20	1.35749	55	1.40554	90	1.45839
21	1.35879	56	1.40703	91	1.45989
22	1.36010	57	1.40852	92	1.46139
23	1.36141	58	1.41001	93	1.46290
24	1.36272	59	1.41150	94	1.46443
25	1.36404	60	1.41299	95	1.46597
26	1.36536	61	1.41448	96	1.46752
27	1.36669	62	1.41597	97	1.46909
28	1.36802	63	1.41746	98	1.47071
29	1.36936	64	1.41895	99	1.47234
30	1.37070	65	1.42044	100	1.47399
31	1.37204	66	1.42193		
32	1.37338	67	1.42342		
33	1.37472	68	1.42491		
34	1.37606	69	1.42640		

Table 10-71
REFRACTIVE INDEX OF AQUEOUS ETHYL ALCOHOL SOLUTIONS

The table below gives the readings of the Zeiss Immersion Refractometer for various concentrations of ethyl alcohol in water. The values were calculated by B. H. St. John from the data of Doroshevskii and Dvorzhanchik. Reprinted by permission from the Official and Tentative Methods of Analysis, 3d Edition (1930), published by The Association of Official Agricultural Chemists at Washington, D. C.

Scale Reading	20°C.		23°C.		25°C.	
	% by volume	% by weight	% by volume	% by weight	% by volume	% by weight
13.2	....	....	....	....	0.00	0.00
13.4	....	....	....	....	0.18	0.14
13.6	....	....	....	....	0.35	0.28
13.8	....	....	0.10	0.08	0.53	0.42
14.0	....	....	0.28	0.22	0.70	0.56
14.2	....	....	0.45	0.36	0.88	0.70
14.4	....	....	0.63	0.50	1.06	0.85
14.6	0.16	0.13	0.80	0.64	1.24	0.99
14.8	0.34	0.27	0.98	0.78	1.40	1.11
15.0	0.52	0.41	1.16	0.92	1.55	1.23
15.2	0.69	0.55	1.32	1.05	1.71	1.36
15.4	0.85	0.68	1.47	1.17	1.86	1.48
15.6	1.03	0.82	1.62	1.29	2.01	1.60
15.8	1.21	0.96	1.77	1.41	2.17	1.72
16.0	1.36	1.08	1.92	1.53	2.33	1.85
16.2	1.51	1.20	2.08	1.65	2.48	1.97
16.4	1.66	1.32	2.24	1.78	2.62	2.09
16.6	1.81	1.44	2.39	1.90	2.77	2.21
16.8	1.96	1.56	2.53	2.02	2.92	2.33
17.0	2.11	1.68	2.69	2.14	3.06	2.44
17.2	2.26	1.80	2.82	2.25	3.21	2.56
17.4	2.41	1.92	2.97	2.37	3.36	2.68
17.6	2.56	2.04	3.12	2.49	3.51	2.80
17.8	2.70	2.15	3.27	2.61	3.66	2.92
18.0	2.85	2.27	3.42	2.73	3.81	3.04
18.2	3.00	2.39	3.57	2.85	3.96	3.16
18.4	3.15	2.51	3.71	2.96	4.11	3.28
18.6	3.30	2.63	3.86	3.08	4.26	3.40
18.8	3.45	2.75	4.01	3.20	4.41	3.52
19.0	3.59	2.86	4.16	3.32	4.56	3.64
19.2	3.73	2.98	4.31	3.44	4.70	3.76
19.4	3.88	3.10	4.46	3.56	4.85	3.88
19.6	4.03	3.22	4.61	3.68	5.00	4.00
19.8	4.17	3.33	4.75	3.80	5.15	4.12
20.0	4.32	3.45	4.90	3.92	5.29	4.23
20.2	4.47	3.57	5.05	4.04	5.44	4.35
20.4	4.61	3.68	5.20	4.16	5.58	4.47
20.6	4.75	3.80	5.34	4.27	5.72	4.58
20.8	4.90	3.92	5.48	4.39	5.87	4.70
21.0	5.04	4.03	5.62	4.50	6.02	4.81
21.2	5.19	4.15	5.77	4.62	6.16	4.93
21.4	5.33	4.26	5.91	4.73	6.30	5.05
21.6	5.47	4.38	6.06	4.85	6.44	5.16
21.8	5.61	4.49	6.20	4.97	6.59	5.28
22.0	5.76	4.61	6.34	5.08	6.73	5.39
22.2	5.90	4.72	6.49	5.20	6.87	5.51
22.4	6.05	4.84	6.63	5.31	7.01	5.62
22.6	6.19	4.95	6.77	5.43	7.16	5.74
22.8	6.33	5.07	6.91	5.54	7.31	5.86
23.0	6.47	5.18	7.06	5.66	7.45	5.97
23.2	6.61	5.30	7.20	5.77	7.59	6.08
23.4	6.75	5.41	7.34	5.89	7.73	6.20
23.6	6.90	5.53	7.48	6.00	7.87	6.31
23.8	7.04	5.64	7.62	6.11	8.00	6.42

Table 10-71 (*Continued*)
REFRACTIVE INDEX—ETHYL ALCOHOL

Scale Reading	20°C.		23°C.		25°C.	
	% by volume	% by weight	% by volume	% by weight	% by volume	% by weight
24.0	7.18	5.76	7.76	6.22	8.14	6.53
24.2	7.32	5.87	7.90	6.34	8.28	6.65
24.4	7.46	5.99	8.04	6.45	8.42	6.76
24.6	7.60	6.10	8.17	6.56	8.55	6.87
24.8	7.74	6.21	8.31	6.67	8.69	6.98
25.0	7.88	6.32	8.45	6.79	8.84	7.10
25.2	8.01	6.43	8.59	6.90	8.98	7.21
25.4	8.14	6.54	8.73	7.01	9.12	7.33
25.6	8.28	6.65	8.86	7.12	9.26	7.44
25.8	8.42	6.76	9.00	7.23	9.39	7.55
26.0	8.55	6.87	9.14	7.35	9.53	7.67
26.2	8.69	6.98	9.28	7.46	9.67	7.78
26.4	8.82	7.09	9.42	7.57	9.81	7.90
26.6	8.96	7.20	9.55	7.68	9.95	8.01
26.8	9.10	7.31	9.69	7.79	10.09	8.12
27.0	9.23	7.42	9.83	7.91	10.23	8.24
27.2	9.37	7.54	9.97	8.02	10.37	8.35
27.4	9.51	7.65	10.10	8.13	10.51	8.46
27.6	9.65	7.76	10.24	8.24	10.65	8.57
27.8	9.79	7.87	10.38	8.35	10.79	8.69
28.0	9.92	7.98	10.51	8.46	10.93	8.80
28.2	10.06	8.09	10.65	8.58	11.06	8.91
28.4	10.19	8.20	10.79	8.69	11.20	9.02
28.6	10.32	8.31	10.93	8.80	11.33	9.13
28.8	10.46	8.42	11.06	8.91	11.47	9.24
29.0	10.59	8.53	11.20	9.02	11.61	9.36
29.2	10.73	8.64	11.33	9.13	11.75	9.47
29.4	10.86	8.74	11.47	9.24	11.88	9.58
29.6	10.99	8.85	11.60	9.35	12.01	9.69
29.8	11.12	8.96	11.74	9.46	12.15	9.80
30.0	11.26	9.07	11.87	9.57	12.29	9.91
30.2	11.38	9.18	12.00	9.68	12.42	10.02
30.4	11.51	9.28	12.13	9.79	12.56	10.13
30.6	11.64	9.39	12.27	9.90	12.70	10.24
30.8	11.78	9.50	12.40	10.01	12.84	10.36
31.0	11.91	9.60	12.54	10.12	12.97	10.47
31.2	12.04	9.71	12.67	10.23	13.11	10.58
31.4	12.17	9.82	12.81	10.34	13.24	10.69
31.6	12.30	9.92	12.94	10.45	13.37	10.80
31.8	12.43	10.03	13.07	10.55	13.51	10.91
32.0	12.57	10.14	13.20	10.66	13.64	11.02
32.2	12.70	10.25	13.34	10.77	13.77	11.13
32.4	12.83	10.35	13.47	10.88	13.91	11.24
32.6	12.96	10.46	13.60	10.99	14.04	11.35
32.8	13.09	10.57	13.73	11.10	14.17	11.46
33.0	13.22	10.68	13.86	11.21	14.31	11.57
33.2	13.35	10.79	13.99	11.31	14.44	11.68
33.4	13.48	10.89	14.13	11.42	14.58	11.79
33.6	13.61	11.00	14.26	11.53	14.71	11.90
33.8	13.74	11.10	14.39	11.64	14.85	12.01
34.0	13.86	11.21	14.52	11.75	14.98	12.12
34.2	13.99	11.31	14.65	11.85	15.11	12.23
34.4	14.12	11.41	14.78	11.96	15.25	12.34
34.6	14.25	11.52	14.91	12.07	15.38	12.45
34.8	14.37	11.62	15.05	12.18	15.51	12.56
35.0	14.50	11.73	15.18	12.28	15.65	12.67
35.2	14.62	11.83	15.31	12.39	15.78	12.78
35.4	14.75	11.93	15.44	12.50	15.91	12.89
35.6	14.87	12.04	15.56	12.60	16.05	13.00
35.8	15.00	12.14	15.69	12.71	16.18	13.11

Table 10-71 (*Continued*)
REFRACTIVE INDEX—ETHYL ALCOHOL

Scale Reading	20°C.		23°C.		25°C.	
	% by volume	% by weight	% by volume	% by weight	% by volume	% by weight
36.0	15.13	12.24	15.82	12.82	16.31	13.21
36.2	15.25	12.35	15.95	12.92	16.44	13.32
36.4	15.38	12.45	16.08	13.03	16.56	13.43
36.6	15.51	12.56	16.21	13.14	16.69	13.53
36.8	15.03	12.66	16.34	13.24	16.82	13.64
37.0	15.76	12.77	16.47	13.35	16.95	13.75
37.2	15.89	12.87	16.60	13.45	17.08	13.86
37.4	16.01	12.97	16.72	13.56	17.21	13.96
37.6	16.14	13.08	16.85	13.66	17.34	14.07
37.8	16.26	13.18	16.98	13.77	17.46	14.17
38.0	16.39	13.28	17.11	13.87	17.59	14.28
38.2	16.51	13.38	17.23	13.98	17.72	14.38
38.4	16.64	13.49	17.36	14.08	17.85	14.49
38.6	16.76	13.59	17.48	14.19	17.97	14.59
38.8	16.89	13.69	17.61	14.29	18.10	14.70
39.0	17.01	13.79	17.74	14.40	18.23	14.80
39.2	17.14	13.90	17.86	14.50	18.35	14.91
39.4	17.26	14.00	17.99	14.60	18.48	15.01
39.6	17.39	14.10	18.11	14.71	18.61	15.12
39.8	17.51	14.21	18.24	14.81	18.73	15.22
40.0	17.63	14.31	18.36	14.92	18.86	15.32
40.2	17.76	14.41	18.49	15.02	18.99	15.43
40.4	17.88	14.51	18.61	15.12	19.11	15.53
40.6	18.01	14.62	18.74	15.23	19.24	15.64
40.8	18.13	14.72	18.86	15.33	19.37	15.74
41.0	18.25	14.82	18.99	15.43	19.49	15.85
41.2	18.37	14.92	19.11	15.53	19.62	15.95
41.4	18.50	15.03	19.24	15.64	19.75	16.06
41.6	18.62	15.13	19.36	15.74	19.87	16.16
41.8	18.74	15.23	19.48	15.84	20.00	16.27
42.0	18.87	15.33	19.61	15.94	20.13	16.37
42.2	18.99	15.43	19.73	16.05	20.25	16.48
42.4	19.11	15.53	19.86	16.15	20.38	16.58
42.6	19.23	15.63	19.98	16.25	20.50	16.69
42.8	19.36	15.74	20.10	16.35	20.63	16.79
43.0	19.48	15.84	20.23	16.46	20.75	16.90
43.2	19.60	15.94	20.35	16.56	20.88	17.00
43.4	19.72	16.04	20.47	16.66	21.01	17.10
43.6	19.85	16.14	20.60	16.76	21.13	17.21
43.8	19.97	16.24	20.72	16.87	21.25	17.31
44.0	20.09	16.34	20.84	16.97	21.38	17.41
44.2	20.21	16.44	20.96	17.07	21.50	17.52
44.4	20.33	16.55	21.09	17.17	21.63	17.62
44.6	20.45	16.65	21.21	17.28	21.75	17.73
44.8	20.58	16.75	21.33	17.38	21.88	17.83
45.0	20.70	16.85	21.45	17.48	22.00	17.93
45.2	20.82	16.95	21.58	17.58	22.13	18.04
45.4	20.94	17.05	21.70	17.68	22.25	18.14
45.6	21.06	17.15	21.82	17.79	22.38	18.25
45.8	21.18	17.25	21.94	17.89	22.51	18.36
46.0	21.30	17.35	22.07	17.99	22.64	18.47
46.2	21.42	17.45	22.19	18.09	22.76	18.57
46.4	21.54	17.55	22.32	18.20	22.89	18.68
46.6	21.66	17.65	22.44	18.30	23.02	18.79
46.8	21.78	17.75	22.57	18.41	23.15	18.90
47.0	21.90	17.85	22.69	18.51	23.28	19.00
47.2	22.02	17.95	22.82	18.62	23.41	19.11
47.4	22.15	18.06	22.94	18.72	23.54	19.22
47.6	22.27	18.16	23.07	18.83	23.67	19.33
47.8	22.39	18.26	23.20	18.93	23.80	19.43

Table 10-71 (*Continued*)
REFRACTIVE INDEX—ETHYL ALCOHOL

Scale Reading	20°C.		23°C.		25°C.	
	% by volume	% by weight	% by volume	% by weight	% by volume	% by weight
48.0	22.51	18.36	23.32	19.04	23.93	19.54
48.2	22.63	18.46	23.45	19.14	24.06	19.65
48.4	22.75	18.56	23.58	19.25	24.19	19.76
48.6	22.87	18.66	23.71	19.36	24.32	19.87
48.8	22.99	18.76	23.83	19.47	24.45	19.98
49.0	23.12	18.86	23.96	19.57	24.59	20.09
49.2	23.24	18.96	24.09	19.68	24.72	20.20
49.4	23.36	19.07	24.22	19.79	24.85	20.31
49.6	23.48	19.17	24.35	19.89	24.98	20.42
49.8	23.61	19.27	24.48	20.00	25.11	20.54
50.0	23.73	19.38	24.61	20.11	25.25	20.65
50.2	23.85	19.48	24.74	20.22	25.38	20.76
50.4	23.98	19.58	24.86	20.33	25.51	20.87
50.6	24.10	19.69	24.99	20.44	25.65	20.98
50.8	24.22	19.79	25.12	20.55	25.78	21.10
51.0	24.35	19.89	25.25	20.66	25.91	21.21
51.2	24.47	20.00	25.38	20.77	26.05	21.32
51.4	24.59	20.10	25.51	20.87	26.18	21.44
51.6	24.72	20.20	25.64	20.98	26.32	21.55
51.8	24.84	20.31	25.77	21.09	26.45	21.66
52.0	24.96	20.41	25.90	21.20	26.59	21.78
52.2	25.09	20.52	26.03	21.31	26.72	21.89
52.4	25.21	20.62	26.16	21.42	26.86	22.01
52.6	25.34	20.72	26.29	21.53	26.99	22.12
52.8	25.46	20.83	26.42	21.64	27.13	22.24
53.0	25.59	20.93	26.56	21.75	27.27	22.35
53.2	25.71	21.04	26.69	21.86	27.40	22.47
53.4	25.84	21.14	26.82	21.97	27.54	22.58
53.6	25.96	21.25	26.95	22.08	27.67	22.70
53.8	26.09	21.36	27.08	22.20	27.81	22.81
54.0	26.22	21.47	27.21	22.31	27.95	22.93
54.2	26.34	21.57	27.35	22.42	28.08	23.04
54.4	26.47	21.68	27.48	22.53	28.22	23.16
54.6	26.59	21.79	27.61	22.65	28.36	23.27
54.8	26.72	21.90	27.75	22.76	28.49	23.39
55.0	26.85	22.00	27.88	22.87	28.63	23.51
55.2	26.97	22.11	28.01	22.98	28.77	23.62
55.4	27.10	22.21	28.15	23.10	28.90	23.74
55.6	27.23	22.32	28.28	23.21	29.04	23.86
55.8	27.35	22.42	28.41	23.32	29.18	23.97
56.0	27.48	22.53	28.54	23.43	29.31	24.09
56.2	27.60	22.64	28.68	23.54	29.45	24.20
56.4	27.73	22.74	28.81	23.66	29.58	24.32
56.6	27.85	22.85	28.94	23.77	29.72	24.43
56.8	27.98	22.95	29.07	23.88	29.86	24.55
57.0	28.10	23.06	29.20	23.99	29.99	24.66
57.2	28.23	23.17	29.34	24.11	30.13	24.78
57.4	28.35	23.27	29.47	24.22	30.27	24.90
57.6	28.48	23.38	29.60	24.33	30.41	25.01
57.8	28.60	23.49	29.73	24.44	30.55	25.13
58.0	28.73	23.59	29.87	24.55	30.69	25.25
58.2	28.86	23.70	29.99	24.66	30.83	25.37
58.4	28.98	23.81	30.13	24.78	30.97	25.49
58.6	29.11	23.91	30.26	24.89	31.11	25.61
58.8	29.23	24.02	30.40	25.00	31.25	25.73
59.0	29.36	24.13	30.53	25.12	31.40	25.86
59.2	29.49	24.24	30.67	25.24	31.54	25.98
59.4	29.61	24.34	30.81	25.36	31.68	26.10
59.6	29.74	24.45	30.94	25.47	31.83	26.23
59.8	29.87	24.56	31.08	25.59	31.97	26.35

Table 10-71 (*Continued*)
REFRACTIVE INDEX—ETHYL ALCOHOL

Scale Reading	20°C.		23°C.		25°C.	
	% by volume	% by weight	% by volume	% by weight	% by volume	% by weight
60.0	29.99	24.67	31.22	25.71	32.12	26.48
60.2	30.12	24.77	31.36	25.83	32.27	26.61
60.4	30.25	24.88	31.50	25.95	32.41	26.73
60.6	30.38	24.99	31.64	26.07	32.56	26.86
60.8	30.51	25.10	31.78	26.19	32.71	26.99
61.0	30.64	25.21	31.92	26.31	32.86	27.12
61.2	30.77	25.32	32.06	26.43	33.01	27.24
61.4	30.90	25.44	32.20	26.55	33.16	27.37
61.6	31.03	25.55	32.34	26.67	33.31	27.50
61.8	31.16	25.66	32.49	26.79	33.46	27.63
62.0	31.29	25.77	32.63	26.92	33.60	27.76
62.2	31.43	25.88	32.77	27.04	33.75	27.88
62.4	31.56	25.99	32.91	27.16	33.90	28.01
62.6	31.69	26.11	33.06	27.29	34.05	28.15
62.8	31.83	26.23	33.20	27.41	34.21	28.28
63.0	31.96	26.35	33.35	27.54	34.36	28.42
63.2	32.10	26.46	33.50	27.66	34.52	28.55
63.4	32.23	26.58	33.64	27.79	34.67	28.69
63.6	32.37	26.70	33.79	27.91	34.83	28.82
63.8	32.51	26.82	33.93	28.04	34.98	28.96
64.0	32.65	26.94	34.08	28.17	35.15	29.10
64.2	32.79	27.05	34.23	28.30	35.31	29.24
64.4	32.92	27.17	34.39	28.43	35.48	29.38
64.6	33.06	27.29	34.54	28.57	35.64	29.52
64.8	33.20	27.41	34.69	28.70	35.80	29.67
65.0	33.34	27.53	34.84	28.83	35.97	29.81
65.2	33.48	27.65	34.99	28.96	36.13	29.95
65.4	33.62	27.77	35.15	29.10	36.30	30.10
65.6	33.76	27.89	35.30	29.23	36.46	30.24
65.8	33.90	28.01	35.46	29.37	36.63	30.39
66.0	34.04	28.13	35.62	29.51	36.79	30.54
66.2	34.18	28.26	35.77	29.64	36.96	30.68
66.4	34.33	28.38	35.93	29.78	37.13	30.83
66.6	34.47	28.51	36.09	29.92	37.30	30.98
66.8	34.62	28.64	36.25	30.06	37.48	31.13
67.0	34.76	28.76	36.41	30.20	37.65	31.28
67.2	34.91	28.89	36.57	30.34	37.83	31.44
67.4	35.05	29.01	36.73	30.49	38.00	31.59
67.6	35.20	29.14	36.90	30.63	38.18	31.74
67.8	35.35	29.27	37.06	30.77	38.35	31.89
68.0	35.50	29.41	37.23	30.91	38.53	32.05
68.2	35.65	29.54	37.39	31.06	38.70	32.21
68.4	35.80	29.67	37.56	31.21	38.88	32.37
68.6	35.95	29.80	37.73	31.35	39.06	32.53
68.8	36.10	29.93	37.90	31.50	39.24	32.69
69.0	36.25	30.06	38.07	31.65	39.43	32.86
69.2	36.41	30.20	38.24	31.79	39.61	33.02
69.4	36.56	30.33	38.41	31.94	39.80	33.18
69.6	36.72	30.47	38.58	32.09	39.98	33.34
69.8	36.87	30.61	38.75	32.25	40.17	33.51
70.0	37.02	30.74	38.92	32.41	40.35	33.67
70.2	37.19	30.88	39.10	32.57	40.53	33.84
70.4	37.35	31.01	39.28	32.72	40.72	34.00
70.6	37.51	31.16	39.46	32.88	40.90	34.17
70.8	37.67	31.30	39.64	33.04	41.08	34.33
71.0	37.83	31.44	39.82	33.20	41.27	34.50
71.2	37.99	31.59	40.00	33.36	41.46	34.67
71.4	38.16	31.73	40.18	33.52	41.64	34.83
71.6	38.32	31.87	40.36	33.68	41.83	35.00
71.8	38.49	32.01	40.54	33.84	42.02	35.17

Table 10-71 (*Continued*)
REFRACTIVE INDEX—ETHYL ALCOHOL

Scale Reading	20°C.		23°C.		25°C.	
	% by volume	% by weight	% by volume	% by weight	% by volume	% by weight
72.0	38.65	32.17	40.72	34.00	42.21	35.34
72.2	38.82	32.32	40.90	34.16	42.40	35.51
72.4	38.98	32.47	41.08	34.33	42.58	35.68
72.6	39.16	32.62	41.26	34.49	42.77	35.85
72.8	39.33	32.77	41.45	34.65	42.96	36.02
73.0	39.50	32.92	41.63	34.81	43.15	36.18
73.2	39.67	33.07	41.81	34.98	43.33	36.35
73.4	39.84	33.22	41.99	35.14	43.52	36.52
73.6	40.02	33.37	42.17	35.31	43.70	36.68
73.8	40.19	33.53	42.36	35.47	43.89	36.85
74.0	40.36	33.68	42.54	35.64	44.08	37.02
74.2	40.53	33.83	42.72	35.80	44.28	37.20
74.4	40.71	33.98	42.91	35.97	44.48	37.38
74.6	40.88	34.14	43.09	36.13	44.67	37.56
74.8	41.05	34.30	43.28	36.30	44.87	37.75
75.0	41.23	34.46	43.46	36.47	45.07	37.93
75.2	41.41	34.61	43.65	36.64	45.29	38.12
75.4	41.58	34.77	43.83	36.81	45.50	38.31
75.6	41.76	34.93	44.02	36.97	45.71	38.50
75.8	41.94	35.09	44.21	37.15	45.92	38.69
76.0	42.12	35.25	44.41	37.33	46.12	38.88
76.2	42.30	35.41	44.60	37.50	46.34	39.08
76.4	42.48	35.58	44.80	37.68	46.56	39.29
76.6	42.66	35.74	44.99	37.86	46.78	39.49
76.8	42.84	35.90	45.19	38.04	47.00	39.69
77.0	43.02	36.07	45.40	38.23	47.23	39.90
77.2	43.20	36.24	45.60	38.42	47.45	40.11
77.4	43.39	36.40	45.81	38.60	47.68	40.32
77.6	43.57	36.57	46.01	38.79	47.91	40.54
77.8	43.76	36.74	46.23	38.98	48.14	40.75
78.0	43.94	36.91	46.45	39.18	48.37	40.97
78.2	44.13	37.08	46.67	39.39	48.60	41.18
78.4	44.32	37.25	46.89	39.59	48.84	41.40
78.6	44.51	37.42	47.11	39.80	49.07	41.62
78.8	44.70	37.59	47.34	40.00	49.31	41.84
79.0	44.89	37.76	47.56	40.21	49.54	42.05
79.2	45.08	37.94	47.79	40.42	49.77	42.27
79.4	45.28	38.11	48.01	40.63	50.01	42.49
79.6	45.48	38.30	48.23	40.84	50.24	42.71
79.8	45.68	38.48	48.46	41.05	50.48	42.93
80.0	45.88	38.67	48.68	41.26	50.71	43.15

Table 10-72
CONVERSION OF SCALE DIVISIONS OF THE IMMERSION REFRACTOMETER TO REFRACTIVE INDEX

Originally the immersion instrument was designed with a single prism (A) of crown glass which allowed only the range from n_D 1.32 to 1.36 to be covered. By a choice of glass and prism angle it has been found possible with a battery of six prisms to extend the upper limit of the range to 1.54. Each prism of the series overlaps the one immediately above so that there is no lost region from 1.32 to 1.54. The ranges covered by these prisms are: A–1.32539 to 1.36674; B–1.36428 to 1.40608; C–1.39860 to 1.43830; D–1.43620 to 1.47562; E–1.47320 to 1.51335; F–1.50969 to 1.54409. Prism A is the standard to which the values in the table below apply. For the other prisms (B to F) the index of refraction is not standardized as between different manufacturers nor for different lots issued by the same manufacturer. For these prisms (B to F) it is necessary to use the table supplied with any particular instrument and impossible to publish any conversion tables which would be universally applicable.

Example: Scale division 25.1 on prism A corresponds to the refractive index $n_D = 1.33705 + 0.000038 = 1.33709$.

Scale Division	n_D	Scale Division	n_D	Scale Division	n_D
-5	1.32539	33	1.34010		
-4	2578	34	4048		
-3	2618			70	1.35388
-2	2657			71	5425
-1	2696	35	4086	72	5461
		36	4124	73	5497
0	2736	37	4162	74	5533
1	2775	38	4199		
2	2814	39	4237	75	5569
3	2854			76	5606
4	2893	40	4275	77	5642
		41	4313	78	5678
5	2932	42	4350	79	5714
6	2971	43	4388		
7	3010	44	4426	80	5750
8	3049			81	5786
9	3087	45	4463	82	5822
		46	4500	83	5858
10	3126	47	4537	84	5894
11	3165	48	4575		
12	3204	49	4612	85	5930
13	3242			86	5966
14	3281	50	4650	87	6002
		51	4687	88	6038
15	3320	52	4724	89	6074
16	3358	53	4761		
17	3397	54	4798	90	6109
18	3435			91	6145
19	3474	55	4836	92	6181
		56	4873	93	6217
20	3513	57	4910	94	6252
21	3551	58	4947		
22	3590	59	4984	95	6287
23	3628			96	6323
24	3667	60	5021	97	6359
		61	5058	98	6394
25	3705	62	5095	99	6429
26	3743	63	5132		
27	3781	64	5169	100	6464
28	3820			101	6500
29	3858	65	5205	102	6535
		66	5242	103	6570
30	3896	67	5279	104	6605
31	3934	68	5316		
32	3972	69	5352	105	6640

Table 10-73
INDEX OF REFRACTION CALCULATIONS
For Organic Liquids.

Reference: Smiles, *The Relation between Chemical Constitution and some Physical Properties* (1910). Published by Longmans, Green and Co., London.

In the Lorentz and Lorenz formula:

$$\frac{n^2-1}{n^2+2}\times\frac{1}{d}=k$$

n is the index of refraction of a liquid at a given temperature, d is the density of the substance at the same temperature, and k, which is nearly independent of the temperature, is a constant known as the *specific refraction*. The product $M\times k$, where M is the molecular weight of the substance, is the *molecular refraction;* this quantity (Mk) is additive and can be calculated for most organic liquid compounds from empirically deduced atomic and structural refractive constants. Such refractive constants (k) for sodium (D) light are given in the following table:

	k		k
C	2.418	N	[1]2.45; [2]3.21; [3]2.65; [4]3.59; [5]3.00;
H	1.100		[6]4.36; [7]2.48; [8]2.47; [9]3.05; [10]3.79;
O	(a) 1.525; (b) 2.211; (c) 1.643;		[11]3.93; [12]2.65; [13]2.27; [14]2.71;
	(d) 1.64		[15]3.776; [16]3.901; [17]4.10; [18]3.46
S	(e) 7.69; (f) 7.97; (g) 7.91; (h)8.11	NO$_2$	[19]7.59; [20]7.44; [21]6.72; [22]7.30;
Se	(i) 11.17		[23]7.51
Hal.	(j) 0.95; (k) 1.1; (m) 5.967;	NO	[24]5.91; [25]5.37
	(n) 8.865; (p) 13.900		
Str.	(q) 1.733; (r) 2.398; (s) 0.71;		
	(t) 0.48		

In the form of: (a) (OH); (b) C:O; (c) ether OR; (d) ester OR; (e) SH; (f) RSR; (g) RCNS; (h) RSSR; (i) in dialkyl selenides; (j) one F on carbon; F (k) in polyfluorides; (m) Cl; (n) Br; (p) I; (q) double bond; (r) triple bond; (s) 3-membd. ring; (t) 4-membd. ring; (1) aliph. prim. amine; (2) arom. prim. amine; (3) aliph. sec. amine; (4) arom. sec. amine; (5) aliph. tert. amine; (6) arom. tert. amine; (7) hydroxylamine; (8) hydrazine; (9) aliph. cyanide; (10) arom. cyanide; (11) aliph. oxime; (12) prim. amide; (13) sec. amide; (14) tert. amide; (15) imidine; (16) oximido; (17) carbimido; (18) hydrazone; (19) alkyl nitrate; (20) alkyl nitrite; (21) nitro paraffin; (22) nitro arom.; (23) nitramine; (24) nitrite; (25) nitrosoamine.

Example: Allylacetic acid ($CH_2:CH\cdot CH_2\cdot CH_2\cdot CO\cdot OH$) has the following observed constants: $n_D^{7.5} = 1.4341$; $d_4^{7.5} = 0.9903$; molecular weight = 100.114. From this the calculated value of Mk is found as follows:

5 C	12.090
8 H	8.800
1 O′	1.525
1 O″	2.211
1 =	1.733
Mk_D	26.359

The observed value of Mk as calculated from the Lorentz and Lorenz formula is:

$$100.114 \times\frac{(1.4341)^2-1}{(1.4341)^2+2}\times\frac{1}{0.9903}=26.330$$

Optical exaltation is characteristic of substances containing conjugated double bonds (single bonds alternating with double bonds); *i.e.*, their molecular refractions (Mk) are usually one or more units in excess of the calculated values.

For Minerals.

Reference: Larsen and Berman, *The Microscopic Determination of the Nonopaque Minerals*, 2d edition (1934); U. S. Geological Survey, Bull. 848. Washington, D. C.

In the Gladstone and Dale formula:

$$\frac{n-1}{d}=K, \text{ and } K=k_1\frac{p_1}{100}+k_2\frac{p_2}{100}+etc.,$$

Table 10-73 (*Continued*)
INDEX OF REFRACTION CALCULATIONS

n is the mean index of refraction, d is the density, K is the specific refractive energy, k_1, k_2, etc. are the specific refractive energies of the components, and p_1, p_2, etc. the weight percentages of the components of a mineral. The mean index of refraction is $\dfrac{2\omega+\epsilon}{3}$ or $\dfrac{\alpha+\beta+\gamma}{3}$ unless the substance is highly birefringent when the mean index is better expressed by $\sqrt[3]{\omega^2\epsilon}$ or $\sqrt[3]{\alpha\beta\gamma}$. The specific refractive energies of the chief constituents of minerals are given in the following table:

	Molecular weight	k		Molecular weight	k
H_2O	18	$_a$0.3355, $_b$0.340,$_c$.354	As_2O_3	198	$_g$0.202,$_h$0.225
			Y_2O_3	226	.144
Li_2O	30	.31	Sb_2O_3	228.4	$_g$.209,$_i$.232
$(NH_4)_2O$	52	.503	La_2O_3	326	.149
Na_2O	62	.181	Ce_2O_3	328.5	.16
K_2O	94	.189	Bi_2O_3	464	.163
Cu_2O	143	.250	CO_2	44	.217
Rb_2O	187	.129	SiO_2	60	.207
Ag_2O	232	.154	TiO_2	80	.397
Cs_2O	282	.124	SeO_2	111	.147
Hg_2O	416	.189$_{Li}$	ZrO_2	122.5	.201
Tl_2O	424	.120	SnO_2	151	.145
BeO	25	.238	SbO_2	152	.198
MgO	40.4	.200	TeO_2	159.5	$_e$.200$_{Li}$
CaO	56	.225	ThO_2	264.5	.12
MnO	71	$_d$.191,$_e$.224	N_2O_5	108	.240
FeO	72	.187	P_2O_5	142	.190
NiO	75	.184	Cl_2O_5	151	.218
CoO	75	.184	V_2O_5	182.4	.43
CuO	79.6	$_d$.191,$_e$.253$_{Li}$	As_2O_5	230	.169
ZnO	81.4	$_d$.153,$_e$.183	Br_2O_5	240	.183
SrO	103.6	.143	Cb_2O_5	268	.295
CdO	128.4	.134	Sb_2O_5	320.4	.152,.222(?)
BaO	153.4	.127	I_2O_5	334	.177
HgO	216	.18	Ta_2O_5	446	.133
PbO	223	$_d$.137,$_e$.175$_{Li}$	SO_3	80	.177
B_2O_3	70	$_g$.220	CrO_3	100	.36
C_2O_3	72	.265	SeO_3	127	.165
Al_2O_3	102	.193,$_f$.214	MoO_3	144	.241$_{Li}$
Cr_2O_3	152	.27	TeO_3	175.6	.607
Mn_2O_3	158	$_d$.300,$_e$.304$_{Li}$	WO_3	235	.133
Fe_2O_3	160	$_d$.308,$_e$.36$_{Li}$	UO_3	286.5	.134
	Atomic weight	k		Atomic weight	k
H	1	1.256 or 1.44	S	32	$_k$0.502,$_m$1.00
C	12	.403	Cl	35.5	.303
O	16	.203	Br	80	.214
F	19	.043	I	127	.226

a Water and ice.
b Average.
c Alums, etc.
d Calculated from compounds containing the oxide.
e Calculated from the oxide.
f Calculated from feldspar, feldspathoids, etc.
g Isometric oxide.
h Monoclinic oxide.
i Orthorhombic oxide.
k Calculated from native sulfur.
m Calculated from sulfides; values vary.

Example: Sodium tetraborate, $Na_2B_4O_7 \cdot 5H_2O$, has the following observed constants: formula, $Na_2O \cdot 2B_2O_3 \cdot 5H_2O$; molecular weight, 292; contains, 21.2% Na_2O; 48% B_2O_3; 30.8% H_2O; d, 1.815; $\omega = 1.461$; $\epsilon = 1.474$. From the table k for Na_2O, 0.181; k for B_2O_3, 0.220; k for H_2O, 0.340. Therefore, $K = 0.181 \times 21.2\% + 0.220 \times 48\% + 0.340 \times 30.8\% = 0.24869$. Substituting in the Gladstone Dale formula:

$$\frac{n-1}{1.815} = 0.24869, \text{ or } n = 1.4514 \text{ by calculation.}$$

The observed average value of $n = \dfrac{2\omega+\epsilon}{3} = \dfrac{2(1.461)+1.474}{3} = 1.465$

Table 10-74
ISOTROPIC CRYSTALS FOR DETERMINING REFRACTIVE INDICES OF LIQUIDS BY IMMERSION METHODS

Reprinted by permission from *Handbook of Chemical Microscopy*, by Chamot and Mason, published by John Wiley & Sons, Inc.

Refr. index	Formula	Refr. index	Formula
1.326	NaF	1.617	$NaBrO_3$
1.340	K_2SiF_6	1.640	NH_4Cl
1.361	KF	1.641	$NaBr$
1.370	$(NH_4)_2SiF_6$	1.645	$CsCl$
1.439	$Na_2SO_4 \cdot Al_2(SO_4)_3 \cdot 24H_2O$	1.650	RbI
1.456	$K_2SO_4 \cdot Al_2(SO_4)_3 \cdot 24H_2O$	1.657	K_2SnCl_6
1.459	$(NH_4)_2SO_4 \cdot Al_2(SO_4)_3 \cdot 24H_2O$	1.667	KI
1.481	$K_2SO_4 \cdot Cr_2(SO_4)_3 \cdot 24H_2O$	1.678	$(NH_4)_2SnCl_6$
1.488	$(NH_4)_2SO_4 \cdot Fe_2(SO_4)_3 \cdot 24H_2O$	1.698	$CsBr$
1.490	KCl	1.703	NH_4I
1.494	$RbCl$	1.755	As_2O_3
1.504	$NaC_2H_3O_2 \cdot UO_2(C_2H_3O_2)_2$	1.774	NaI
1.515	$NaClO_3$	1.782	$Pb(NO_3)_2$
1.544	$NaCl$	1.787	CsI
1.553	$RbBr$	1.827	K_2PtCl_6
1.559	KBr	2.06	$AgCl$
1.571	$Ba(NO_3)_2$	2.25	$AgBr$
1.588	$Sr(NO_3)_2$		

Table 10-75
LIQUID MIXTURES FOR REFRACTIVE INDEX DETERMINATIONS

Reprinted by permission from *Handbook of Chemical Microscopy*, by Chamot and Mason, published by John Wiley & Sons, Inc.

Refr. Index (daylight, 20-22°C.)	Ethyl alcohol $n=1.362$	Ethylene glycol ethyl ether ("Cellosolve") $n=1.407$	Kerosene $n=1.443$
1.370	50 cc.	9 cc.	.....
1.380	50	30	.....
1.390	35	50	.....
1.400	10	50	.....
1.410	......	50	5 cc.
1.420	......	50	35
1.430	......	35	75
1.440	......	5	75

Refractive indices	Mixtures
1.450-1.475	Kerosene and turpentine.
1.480-1.535	Turpentine and ethylene bromide or clove oil.
1.540-1.635	Clove oil and α-bromonaphthalene.
1.640-1.655	α-Bromonaphthalene and α-chloronaphthalene.
1.650-1.740	α-Bromonaphthalene and methylene iodide.
1.740-1.790	Sulfur dissolved in methylene iodide.
1.790-1.960	Methylene iodide, antimony iodide, arsenic sulfide, antimony sulfide, and sulfur.

Table 10-76
REFRACTIVE INDICES (APPROXIMATE) OF VARIOUS SUBSTANCES

Reprinted by permission from *Handbook of Chemical Microscopy*, by Chamot and Mason, published by John Wiley & Sons, Inc.

Substance	ω	ϵ
"Acetate silk"	1.48	1.475
"Bakelite"	1.58—1.63	......
Celluloid	1.53	......
Cellulose fibers (cotton, flax, ramie, etc.)	1.53	1.59
"Collodion silk"	1.52	1.55
Gelatin, dry	1.54	......
Horn	1.56	......
Lacquer, dry ("Duco")	1.54	......
Linseed oil, dry	1.48—1.50	......
Rubber, pale crepe (vulcanized is higher)	1.51+	......
Shellac	1.54	......
Silk	1.54	1.59
Starch	1.53	......
Varnish, dry	1.51—1.54	......
"Viscose silk"	1.525	1.55
Wool	1.54+	1.55+

Table 10-77
LIQUID STANDARDS FOR REFRACTIVE INDEX DETERMINATIONS BY IMMERSION METHODS

The following compounds with the refractive indices as listed below can be obtained from the Organic Chemicals Department, Eastman Kodak Co., Rochester, N. Y. They are available in 25cc glass-stoppered bottles but only in lots of ten or more liquids.

$n^{20°}_D$	Liquid	$n^{20°}_D$	Liquid
1.3289	Methyl alcohol	1.5130	Ethyl iodide
1.3330	Water	1.5165	Anisole
1.3585	Acetone	1.5235	Trimethylene bromide
1.3725	Ethyl acetate	1.5246	Chlorobenzene
1.3875	n-Heptane	1.5300	Methyl iodide
1.3986	n-Butyl alcohol	1.5380	Ethylene bromide
1.4015	n-Butyl chloride	1.5465	o-Nitrotoluene
1.4221	Dioxan (1,4)	1.5520	Nitrobenzene
1.4225	Methylcyclohexane	1.5600	Tri-o-cresyl phosphate
1.4314	Ethylene glycol	1.5598	Bromobenzene
1.4420	Ethyl citrate	1.5720	o-Toluidine
1.4440	Ethylene chloride	1.5859	Aniline
1.4480	Trimethylene chloride	1.5960	Bromoform
1.4500	Cyclohexanone	1.6090	o-Iodotoluene
1.4663	Cyclohexanol	1.6200	Iodobenzene
1.4770	Diethanolamine	1.6235	Quinoline
1.4847	Triethanolamine	1.6377	s-Tetrabromoethane
1.4900	p-Cymene	1.6578	α-Bromonaphthalene
1.4940	s-Tetrachloroethane	1.6965	Phosphorus tribromide
1.4950	Toluene	1.7400	Methylene iodide
1.5005	Benzene	1.8400	Diiodophenylarsine

SURFACE TENSION

Table 10-78
SURFACE TENSIONS OF METALS AND FUSED SALTS

The table below gives the surface tension of various metals and fused salts in contact with an atmosphere as indicated; temperature °C.

Formula		Surface Tension in Dynes per Centimeter		
Ag	—air	$800^{970°}$	....	....
AgBr	—air	$121.4^{434°}$	....	....
AgCl	—air	$125.5^{452°}$	$121.6^{494°}$	$112.8^{573°}$
Au	—air	$(580-1000)^{1070°}$	....	....
BaCl$_2$	—air	$171^{962°}$	....	....
Bi	—H$_2$	$388^{300°}$	$354^{583°}$	....
BiCl$_3$	—N$_2$	$66.2^{271°}$	$58.1^{331°}$	$52.0^{382°}$
CaCl$_2$	—air	$152^{772°}$	....	....
Cd	—H$_2$	$621.5^{450°}$	$618.3^{500°}$	....
CsBr	—N$_2$	$81.8^{658°}$	$71.6^{808°}$	$62.7^{971°}$
CsCl	—N$_2$	$89.2^{664°}$	$73.7^{881°}$	$56.3^{1080°}$
CsF	—N$_2$	$104.5^{723°}$	$92.3^{877°}$	$78.9^{1100°}$
CsI	—N$_2$	$73.1^{654°}$	$62.5^{821°}$	$51.1^{1030°}$
CsNO$_3$	—N$_2$	$91.8^{426°}$	$83.7^{511°}$	$72.5^{686°}$
Cs$_2$SO$_4$	—N$_2$	$111.3^{1036°}$	$97.3^{1221°}$	$83.0^{1530°}$
Cu	—H$_2$	$1103^{1131°}$	....	....
Ga	—CO$_2$	$358^{300°}$	....	....
Hg	—Vac.	$480.3^{0°}$	$467.1^{60°}$	....
Hg	—air	$487^{15°}$	....	....
KBr	—N$_2$	$85.7^{775°}$	$82.0^{826°}$	$75.4^{920°}$
KCN	—air	$96.1^{634.5°}$	....	....
KCl	—N$_2$	$95.8^{800°}$	$82.2^{986°}$	$69.6^{1167°}$
K$_2$Cr$_2$O$_7$	—N$_2$	$140.1^{420°}$	$138.4^{480°}$	$135.0^{535°}$
KF	—N$_2$	$138.4^{913°}$	$119.9^{1147°}$	$104.9^{1310°}$
KI	—N$_2$	$75.2^{737°}$	$69.2^{812°}$	$66.5^{873°}$
KNO$_3$	—N$_2$	$110.4^{380°}$	$95.2^{578°}$	$80.2^{772°}$
KPO$_3$	—N$_2$	$155.5^{897°}$	$136.8^{1167°}$	$100.3^{1536°}$
K$_2$SO$_4$	—N$_2$	$143.7^{1070°}$	$126.2^{1347°}$	$106.8^{1656°}$
LiBO$_2$	—N$_2$	$261.8^{879°}$	$243.6^{1150°}$	$192.4^{1520°}$
LiCl	—N$_2$	$137.8^{614°}$	$123.2^{814°}$	$104.8^{1075°}$
LiF	—N$_2$	$249.5^{868.5°}$	$229.8^{1065°}$	$201.1^{1270°}$
LiNO$_3$	—N$_2$	$111.5^{359°}$	$106.0^{445°}$	$96.2^{609°}$
Li$_2$SO$_4$	—N$_2$	$223.8^{860°}$	$211.0^{1057°}$	$200.3^{1214°}$
Li$_2$SiO$_3$	—N$_2$	$374.6^{1254°}$	$356.2^{1421°}$	$346.6^{1601°}$
Na	—Vac.	$222^{100°}$	$211^{250°}$	....
NaBr	—N$_2$	$105.8^{761°}$	$92.9^{942°}$	$78.0^{1166°}$
NaCl	—N$_2$	$113.8^{803°}$	$106.4^{908°}$	$88.0^{1172°}$
NaF	—N$_2$	$199.5^{1010°}$	$180.5^{1189°}$	$162.9^{1357°}$
NaI	—N$_2$	$85.6^{706°}$	$80.5^{816°}$	$77.6^{861°}$
NaNO$_3$	—N$_2$	$119.7^{322°}$	$108.9^{513°}$	$93.7^{738°}$
NaPO$_3$	—N$_2$	$197.5^{827°}$	$181.6^{1099°}$	$147.5^{1517°}$
Na$_2$SO$_4$	—N$_2$	$194.8^{900°}$	$188.2^{990°}$	$184.7^{1077°}$
Pb	—H$_2$	$453^{350°}$	$423^{750°}$	....
PbCl$_2$	—air	$138^{490°}$	$131^{539°}$	$126^{614°}$
Pt	—air	$1819^{2000°}$	....	....
RbBr	—N$_2$	$87.7^{729°}$	$77.2^{884°}$	$60.6^{1121°}$
RbCl	—N$_2$	$95.7^{750°}$	$81.1^{923°}$	$61.4^{1150°}$
RbF	—N$_2$	$127.2^{803°}$	$113.0^{936°}$	$102.2^{1085°}$
RbI	—N$_2$	$79.4^{673°}$	$65.1^{869°}$	$55.4^{1016°}$
RbNO$_3$	—N$_2$	$107.5^{327°}$	$92.5^{527°}$	$77.7^{726°}$
Rb$_2$SO$_4$	—N$_2$	$132.5^{1086°}$	$118.9^{1289°}$	$108.9^{1545°}$
Sb	—H$_2$	$350^{640°}$	....	....
Sn	—H$_2$	$526^{253°}$	$508^{878°}$	....
SnCl$_2$	—N$_2$	$97^{307°}$	$89.0^{405°}$	$81.6^{480°}$
Zn	—H$_2$	$753^{477°}$	$747^{543°}$	....

Table 10-79
SURFACE TENSIONS OF VARIOUS LIQUIDS

The table below gives the surface tension of various liquids in dynes per centimeter. The nature of the gas in contact with the surface of the liquid is given in parenthesis following the name of the compound. The temperature in degrees centigrade is indicated by the superscript which accompanies the value in each case.

The value k_E (Eötvös constant) $= \dfrac{d\Gamma}{dt} = \dfrac{d(M/d)^{2/3}\gamma}{dt}$, where d is the density of the substance, M is the gram formula weight, t the temperature and γ is the surface tension in dynes per centimeter.

Name	Surface Tension in Dynes per Centimeter			k_E
Acetal (air)	$23.2^{5°}$	$21.65^{20°}$	$13.5^{100°}$	....
Acetaldehyde	$22.4^{10°}$	$21.2^{20°}$	$17.0^{50°}$	1.8
Acetaldoxime	$30.1^{35°}$	$25.1^{80°}$	$17.8^{145°}$	1.5
Acetamide	$39.3^{85°}$	$37.3^{105°}$	$35.7^{120°}$	1.2
Acetic acid (air)	$28.6^{10°}$	$27.6^{20°}$	$22.2^{75°}$	1.3
Acetic acid (vapor)	$28.8^{10°}$	$27.8^{20°}$	$22.3^{75°}$	....
Acetic anhydride	$33.3^{15°}$	$32.7^{20°}$	$18.6^{140°}$	2.2
Acetone (air or vapor)	$26.2^{0°}$	$23.7^{20°}$	$18.6^{60°}$	1.9
Acetonitrile (air)	$30.6^{10°}$	$29.3^{20°}$	$20.2^{90°}$	1.5
Acetonitrile (vapor)	$30.6^{10°}$	$29.3^{20°}$	$20.2^{90°}$	....
Acetophenone (air or vapor)	$39.8^{20°}$	$33.3^{75°}$	$22.8^{175°}$	2.2
Acetylene (vapor)	$18^{-77.4}$	$16^{-70.5}$	$14^{-62.4}$	2.4
Allyl alcohol (air or vapor)	$27.6^{0°}$	$25.8^{20°}$	$19.2^{95°}$	1.5
Anethole (air)	$37.6^{10°}$	$36.5^{20°}$	$19.6^{200°}$	2.3
Aniline (air)	$44.0^{10°}$	$42.9^{20°}$	$24.4^{180°}$	2.1
Anisole (air)	$36.4^{10°}$	$35.2^{20°}$	$28.0^{80°}$	2.3
Ammonia	$23^{11.1°}$	$18^{34.1°}$	$13^{59°}$	1.3
iso-Amyl acetate (air)	$26.6^{0°}$	$24.7^{20°}$	$17.1^{100°}$	2.3
iso-Amyl alcohol (air)	$25.3^{0°}$	$23.8^{20°}$	$17.7^{100°}$	1.90
Argon (vapor)	$13.2^{-188°}$	$11.9^{-183°}$	....	2.0
Benzene (air)	$31.6^{0°}$	$28.9^{20°}$	$21.3^{80°}$	2.22
Benzene (vapor)	$31.7^{0°}$	$29.02^{20°}$	$18.8^{100°}$	....
Benzonitrile (air or vapor)	$40.2^{10°}$	$39.1^{20°}$	$29.9^{100°}$	2.1
Benzylamine	$39.5^{20°}$	$36.5^{45°}$	$33.1^{75°}$	2.1
Bromine (air or vapor)	$45.0^{0°}$	$41.5^{20°}$	$36.2^{50°}$	2.0
Bromobenzene (air)	$37.7^{10°}$	$36.5^{20°}$	$27.2^{100°}$	2.19
p-Bromotoluene (air)	$32^{43°}$	$26^{100°}$	$20^{164°}$	....
iso-Butyl acetate (air)	$24.7^{5°}$	$23.3^{20°}$	$14.9^{100°}$	2.32
n-Butyl alcohol (air or vapor)	$26.2^{0°}$	$24.6^{20°}$	$17.8^{100°}$	1.91
iso-Butyl alcohol (air)	$24.4^{0°}$	$22.8^{20°}$	$20.5^{50°}$	1.94
n-Butyric acid (air)	$28.8^{0°}$	$26.8^{20°}$	$19.5^{100°}$	1.6
iso-Butyric acid (air or vapor)	$27.1^{0°}$	$25.2^{20°}$	$17.9^{100°}$	1.6
n-Butyronitrile (air)	$28.7^{10°}$	$27.6^{20°}$	$18.2^{110°}$	1.7
Carbon dioxide	$4.5^{0°}$	$1.16^{20°}$	$0.06^{30°}$	....
Carbon disulfide (air or vapor)	$35.3^{0°}$	$32.3^{20°}$	$26.5^{60°}$	2.1
Carbon monoxide (vapor)	$12.1^{-203°}$	$9.8^{-193°}$	$8.7^{-188°}$	2.0
Carbon tetrachloride (air)	$28.0^{10°}$	$26.8^{20°}$	$20.2^{75°}$	2.21
Carbon tetrachloride (vapor)	$28.2^{10°}$	$27.0^{20°}$	$17.3^{100°}$	....
Chloral	$30^{20°}$	$20^{97°}$	....	2.2
Chlorine	$31^{-60°}$	$18^{20°}$	$13^{50°}$	2.1
Chlorobenzene (air)	$34.4^{10°}$	$33.2^{20°}$	$24.0^{100°}$	2.21
Chlorobenzene (vapor)	$34.8^{10°}$	$33.6^{20°}$	$30.0^{50°}$	....
p-Chlorobromobenzene (air)	$33^{70°}$	$27^{132°}$	$21^{194°}$	....
Chloroform (air)	$28.5^{10°}$	$27.1^{20°}$	$21.7^{60°}$	2.1
p-Chlorotoluene (air or vapor)	$32^{25°}$	$26^{79°}$	$19^{151°}$	.....
o-Cresol (air or vapor)	$39.8^{10°}$	$30.4^{100°}$	$22.0^{180°}$	2.0
m-Cresol (air or vapor)	$38.4^{10°}$	$37.4^{20°}$	$22.2^{180°}$	1.8

Table 10-79 (*Continued*)
SURFACE TENSIONS OF VARIOUS LIQUIDS

Name	Surface Tension in Dynes per Centimeter			k_E
p-Cresol	$36.7^{20°}$	$34.0^{50°}$	$29.3^{100°}$	1.7
Cyclohexane (air)	$27^{10°}$	$25.3^{20°}$	$15.7^{80°}$	2.25
Dibenzylamine	$41.1^{20°}$	$38.4^{45°}$	$35.1^{75°}$	2.9
p-Dichlorobenzene (air)	$31^{68°}$	$25^{117°}$	$20^{170°}$	....
Diethyl ketone (air)	$26.9^{0°}$	$25.3^{15°}$	$22.2^{45°}$	....
Diethyl oxalate (air or vapor)	$33.2^{10°}$	$32.0^{20°}$	$26.6^{70°}$	....
Diethyl sulfate (air)	$34.6^{13°}$	$32.5^{32.5°}$	$28.6^{70°}$	....
Diethylaniline	$34.7^{15°}$	$34.2^{20°}$	$24.9^{110°}$	....
Dimethyl sulfate (air)	$40.1^{18°}$	$35.5^{55°}$	$31.0^{93°}$	....
Dimethylamine (N_2)	$25.2^{-78°}$	$20.2^{-23°}$	$17.7^{5°}$	....
Dimethylaniline (air)	$37.7^{10°}$	$36.56^{20°}$	$27.6^{100°}$	2.4
Ethyl acetate (air)	$26.5^{0°}$	$23.9^{20°}$	$17.4^{75°}$	2.3
Ethyl acetate (vapor)	$26.9^{0°}$	$24.3^{20°}$	$14.4^{100°}$	2.3
Ethyl acetoacetate (air or vapor)	$34.8^{0°}$	$32.5^{20°}$	$25.0^{90°}$	....
Ethyl alcohol (air)	$24.05^{0°}$	$22.27^{20°}$	$18.22^{70°}$	1.0
Ethyl alcohol (vapor)	$22.75^{20°}$	$20.14^{50°}$	$15.47^{100°}$	1.3
Ethyl benzoate	$37.5^{0°}$	$35.5^{20°}$	$30.0^{75°}$	....
Ethyl bromide (air or vapor)	$25.5^{10°}$	$24.2^{20°}$	$21.5^{40°}$	2.12
Ethyl n-butyrate (air)	$26.1^{5°}$	$24.54^{20°}$	$20.6^{60°}$	2.42
Ethyl iso-butyrate (air or vapor)	$24.8^{5°}$	$23.25^{20°}$	$13.85^{110°}$	2.39
Ethyl ether	$17.02^{0°}$	$13.55^{0°}$	$7.97^{100°}$	2.25
Ethyl formate (air or vapor)	$26.2^{0°}$	$23.6^{20°}$	$16.0^{80°}$	2.1
Ethyl iodide (air or vapor)	$30.6^{10°}$	$29.4^{20°}$	$22.4^{75°}$	2.2
Ethyl mercaptan (air or vapor)	$25.4^{0°}$	$24.0^{10°}$	$22.5^{20°}$	2.1
Ethyl propionate (air or vapor)	$25.9^{5°}$	$24.2^{20°}$	$15.5^{100°}$	2.3
Ethylamine (N_2)	$28.9^{-74°}$	$23.2^{-21.5°}$	$20.3^{9.9°}$	1.25
Ethylbenzene (air)	$31.4^{0°}$	$29.2^{20°}$	$24.9^{60°}$	....
Ethylene bromide (air or vapor)	$40.05^{10°}$	$38.75^{20°}$	$28.4^{100°}$	2.2
Ethylene chloride (air)	$33.6^{10°}$	$32.2^{20°}$	$24.0^{80°}$	....
Ethylene oxide (vapor)	$35.8^{-5°}$	$29.2^{-10°}$	$24.3^{20°}$	1.8
Formic acid (air)	$38.7^{10°}$	$37.6^{20°}$	$29.0^{100°}$	0.9
Furfural (air or vapor)	$43.5^{20°}$	$40.9^{40°}$	$25.4^{160°}$	....
Glycerol (air)	$63^{20°}$	$59^{90°}$	$52^{150°}$	....
Glycol (air or vapor)	$49^{0°}$	$47.7^{20°}$	$42.3^{80°}$	....
Guaiacol (air)	$38.66^{19.6°}$	$31.9^{78°}$	$21.3^{201°}$	2.2
Helium (vapor)	$0.353^{-271.6°}$	$0.239^{-270.1°}$	$0.098^{-268.9}$	1.0
n-Hexane (air)	$20.5^{0°}$	$18.4^{20°}$	$13.4^{68°}$	2.26
Hydrogen (vapor)	$2.88^{-258.4°}$	$2.32^{-255.1}$	$1.91^{-252.7}$	1.36
Hydrogen cyanide (air)	$19.1^{10°}$	$18.2^{17°}$	$17.2^{25°}$	1.1
Iodobenzene (air)	$40.3^{15°}$	$39.7^{20°}$	$30.6^{100°}$	2.18
Mesitylene (air)	$30.1^{5°}$	$28.5^{20°}$	$20.6^{100°}$	2.2
Methyl acetate (air or vapor)	$27.4^{0°}$	$24.6^{20°}$	$16.6^{80°}$	2.2
Methyl alcohol (air)	$24.5^{0°}$	$22.6^{20°}$	$15.7^{100°}$	1.0
Methyl n-butyrate (air or vapor)	$26.15^{10°}$	$25.0^{20°}$	$16.1^{100°}$	2.3
Methyl iso-butyrate (air or vapor)	$24.9^{10°}$	$23.8^{20°}$	$17.6^{75°}$	2.3
Methyl chloride (vapor)	$19.5^{0°}$	$17.8^{10°}$	$16.2^{20°}$	2.00
Methyl ether (vapor)	$21^{-40°}$	$18^{-20°}$	$16^{-10°}$	2.0
Methyl ethyl ketone (air or vapor)	$26.9^{0°}$	$24.6^{20°}$	$18.4^{75°}$	....
Methyl formate (air)	$28.0^{0°}$	$25.0^{20°}$	$23.5^{30°}$	2.09
Methyl formate (vapor)	$28.3^{0°}$	$25.1^{20°}$	$13.04^{100°}$	2.09
Methyl propionate (air or vapor)	$26.1^{10°}$	$24.9^{20°}$	$17.6^{80°}$	2.2
Methylamine (N_2)	$29.2^{-70°}$	$23.6^{-20°}$	$22.2^{-12°}$	1.2
Naphthalene (air or vapor)	$28.8^{127°}$	$24.0^{170°}$	$21.8^{190°}$	....
Neon (vapor)	$5.90^{-249°}$	$5.15^{-247°}$	$4.45^{-245°}$	2.0
Nitrobenzene (air or vapor)	$46.4^{0°}$	$43.9^{20°}$	$34.4^{100°}$	2.2
Nitroethane (air or vapor)	$33.4^{10°}$	$32.3^{20°}$	$22.5^{100°}$	1.7

Table 10-79 (*Continued*)
SURFACE TENSIONS OF VARIOUS LIQUIDS

Name	Surface Tension in Dynes per Centimeter			k_E
Nitrogen (vapor)	$10.5^{-203°}$	$8.3^{-193°}$	$6.2^{-183°}$	2.0
Nitromethane (air)	$39.8^{0°}$	$36.8^{20°}$	$26.1^{100°}$	....
Nitrosyl chloride (vapor)	$34.5^{-33°}$	$33^{-22°}$	$30^{-5.5°}$	1.46
Nitrous oxide	$10.1^{-25°}$	$3.4^{10°}$	$1.75^{20°}$	....
n-Octane (air or vapor)	$23.8^{0°}$	$21.8^{20°}$	$17.9^{60°}$	2.3
Oxygen (vapor)	$18.3^{-203°}$	$15.7^{-193°}$	$13.2^{-183°}$	1.92
Paraldehyde (air)	$27.5^{5°}$	$25.9^{20°}$	$14.5^{124°}$	2.6
Phenetole (air)	$32.7^{20°}$	$24.2^{100°}$	$16.8^{170°}$	2.4
Phenol (air or vapor)	$40.9^{20°}$	$34.4^{80°}$	$26.8^{150°}$	1.85
Phenyl isothiocyanate (air or vapor)	$42.5^{13°}$	$41.6^{20°}$	$32.4^{100°}$	2.4
Phosphorus oxychloride (air or vapor)	$33.4^{10°}$	$32.2^{20°}$	$24.1^{85°}$	2.2
Piperidine (air)	$32.65^{0°}$	$30.2^{20°}$	$20.4^{105°}$	2.1
Propionic acid (air or vapor)	$27.71^{0°}$	$26.7^{20°}$	$20.8^{80°}$	1.5
Propionitrile (air or vapor)	$28.3^{10°}$	$27.2^{20°}$	$19.4^{90°}$	1.6
n-Propyl acetate (air or vapor)	$26.6^{0°}$	$24.3^{20°}$	$15.6^{100°}$	2.3
n-Propyl alcohol (air)	$25.9^{-5°}$	$23.8^{20°}$	$20.5^{60°}$	1.3
iso-Propyl alcohol (air or vapor)	$22.8^{5°}$	$21.7^{20°}$	$17.0^{80°}$	1.1
n-Propyl formate (air or vapor)	$26.8^{0°}$	$24.5^{20°}$	$15.5^{100°}$	2.2
n-Propylamine (air)	$23.5^{10°}$	$22.4^{20°}$	$19.4^{45°}$	1.9
p-iso-Propyltoluene (air)	$29.5^{5°}$	$28.1^{20°}$	$20.7^{100°}$	2.3
Pyridine (air)	$40.8^{0°}$	$38.0^{20°}$	$26.4^{100°}$	2.3
Quinoline (air)	$45.0^{20°}$	$35.8^{100°}$	$25.1^{200°}$	2.4
Thiophene (air)	$36.2^{0°}$	$33.1^{20°}$	$24.3^{80°}$	2.04
Thiophenol (air)	$41.0^{10°}$	$39.8^{20°}$	$31.3^{90°}$	2.0
Toluene (air)	$30.74^{0°}$	$28.43^{20°}$	$19.39^{100°}$	2.2
Toluene (vapor)	$28.5^{20°}$	$25.0^{50°}$	$16.3^{130°}$	2.2
o-Toluidine (air or vapor)	$42.3^{0°}$	$40.0^{20°}$	$31.2^{100°}$	....
Triethylamine (air)	$22.9^{0°}$	$20.9^{20°}$	$13.7^{90°}$	2.37
Trimethylamine (N_2)	$24.8^{-73°}$	$20.2^{-32°}$	$17.4^{-4°}$	1.65
Tripalmitin (air)	$29.5^{55.7°}$	$27.2^{87.6°}$	$25.3^{115.3°}$	5.4
Triphenylmethane	$33.8^{108.7°}$	$24.5^{208.2°}$	$15.4^{335.5°}$	2.1to1.5
iso-Valeric acid (air or vapor)	$26.16^{15°}$	$25.3^{20°}$	$20.4^{80°}$	1.7
Water (see special table)				
m-Xylene (air or vapor)	$31.15^{0°}$	$28.90^{20°}$	$20.46^{100°}$	2.25
p-Xylene (air)	$29.92^{5°}$	$28.37^{20°}$	$24.2^{60°}$	....

Table 10-80
SURFACE TENSION OF WATER
Temperature in Degrees C.

Surface Tension in Dynes/cm.		Surface Tension in Dynes/cm.	
$76.96^{-8°}$ (air)	$73.05^{18°}$	$71.18^{30°}$	$52.84^{130°}$
$76.42^{-5°}$	$72.90^{19°}$	$70.38^{35°}$	k_E
$75.64^{0°}$	$72.75^{20°}$	$69.56^{40°}$	$1.03^{25°}$
$74.92^{5°}$	$72.59^{21°}$	$68.74^{45°}$	$1.07^{70°}$
$74.22^{10°}$	$72.44^{22°}$	$67.91^{50°}$	$1.18^{100°}$
$74.07^{11°}$	$72.28^{23°}$	$66.18^{60°}$	$1.27^{120°}$
$73.93^{12°}$	$72.13^{24°}$	$64.42^{70°}$	
$73.78^{13°}$	$71.97^{25°}$	$62.61^{80°}$	
$73.64^{14°}$	$71.82^{26°}$	$60.75^{90°}$	
$73.49^{15°}$	$71.66^{27°}$	$58.85^{100°}$	
$73.34^{16°}$	$71.50^{28°}$	$56.89^{110°}$ (vapor)	
$73.19^{17°}$	$71.35^{29°}$	$54.89^{120°}$	

Table 10-81
FORMULAS RELATING SURFACE TENSION AND TEMPERATURE
OF VARIOUS SUBSTANCES

In this table γ is the surface tension in dynes per centimeter, T is the temperature in degrees absolute, and T_c is the critical temperature in degrees absolute.

Benzene: γ (vapor) $= 71.926 \left(1 - \dfrac{T}{561.6}\right)^{1.23} \pm 0.2$, from 0°C. to T_c

p-Bromotoluene: γ (air) $= 66.46 \left(1 - \dfrac{T}{694}\right)^{1.2}$

Carbon dioxide: $\gamma = 75 \left(1 - \dfrac{T}{T_c}\right)^{1.25}$

Carbon tetrachloride: γ (vapor) $= 67.671 \left(1 - \dfrac{T}{556.25}\right)^{1.23} \pm 0.2$, from 10°C. to 270°C.

Chlorine: $\gamma = 72 \left(1 - \dfrac{T}{T_c}\right)^{1.13}$

p-Chloroaniline: γ (air) $= 81.37 \left(1 - \dfrac{T}{787}\right)^{1.2}$

Chlorobenzene: γ (vapor) $= 72.20 \left(1 - \dfrac{T}{632.3}\right)^{1.23} \pm 0.2$, from 10°C. to 333°C.

p-Chlorobromobenzene: γ (air) $= 71.83 \left(1 - \dfrac{T}{722}\right)^{1.2}$

p-Chloroiodobenzene: γ (air) $= 74.71 \left(1 - \dfrac{T}{767}\right)^{1.2}$

p-Chloronitrobenzene: γ (air) $= 78.68 \left(1 - \dfrac{T}{768}\right)^{1.2}$

p-Chlorotoluene: γ (air or vapor) $= 66.65 \left(1 - \dfrac{T}{653}\right)^{1.2}$

p-Dichlorobenzene: γ (air) $= 71.45 \left(1 - \dfrac{T}{675}\right)^{1.2}$

Ethyl alcohol: $\gamma = 230.564 \left(1 - \dfrac{T}{516.2}\right)^{1.45} - 304.339 \left(1 - \dfrac{T}{516.2}\right)^{2} + 139.756 \left(1 - \dfrac{T}{516.2}\right)^{3} \pm$ 0.3, from 10°C. to 240°C.

Ethyl ether: γ (vapor) $= 57.358 \left(1 - \dfrac{T}{466.9}\right)^{1.23} \pm 0.2$, from 20°C. to 193°C.

p-Iodotoluene: γ (air) $= 69.28 \left(1 - \dfrac{T}{734}\right)^{1.2}$

Methyl alcohol: $\gamma = 173.245 \left(1 - \dfrac{T}{513.1}\right)^{1.33} - 243.042 \left(1 - \dfrac{T}{513.1}\right)^{2} + 146.344 \left(1 - \dfrac{T}{513.1}\right)^{3} \pm$ 0.2, from 70°C. to 235°C.

Methyl chloride: $\gamma = 72 \left(1 - \dfrac{T}{T_c}\right)^{1.22}$

Methyl formate: $\gamma = 77.83 \left(1 - \dfrac{T}{487.1}\right)^{1.23} \pm 0.2$, from 50°C. to 214°C.

p-Nitrotoluene: γ (air) $= 74.06 \left(1 - \dfrac{T}{754}\right)^{1.2}$

Nitrous oxide: $\gamma = 86 \left(1 - \dfrac{T}{T_c}\right)^{1.33}$

Table 10-82
SURFACE TENSION OF AQUEOUS INORGANIC SOLUTIONS AGAINST AIR

The table below gives values for the surface tension of various inorganic aqueous solutions of different concentrations. The concentrations are shown in the super-

	Formula	Surface Tension in Dynes per Centimeter		
1	$AgNO_3$ (20°)	$73.39^{0.5M}$	$73.98^{1.0M}$	$75.06^{2.0M}$
2	$BaCl_2$ (20°)	$72.94^{0.05M}$	$73.08^{0.1M}$	$73.53^{0.25M}$
3	$CaCl_2$ (25°)	$72.32^{0.1M}$	$73.49^{0.5M}$	$75.17^{1.0M}$
4	$CdCl_2$ (20°)	$73.65^{0.5M}$	$74.45^{1.0M}$	$76.85^{2.5M}$
5	$CuSO_4$ (20°)	$73.12^{0.2M}$	$73.67^{0.5M}$	$74.58^{1.0M}$
6	HCl (20°)	$72.59^{0.5M}$	$72.47^{1.0M}$	$72.27^{2.0M}$
7	HNO_3 (20°)	$72.1^{0.7M}$	$71.6^{1.5M}$	$70.9^{2.8M}$
8	H_2O_2 (18°)	$73.37^{4.3M}$	$73.82^{9.5M}$	$74.28^{15.0M}$
9	KBr (20°)	$73.42^{0.5M}$	$74.07^{1.0M}$	$74.74^{1.5M}$
10	K_2CO_3 (20°)	$74.21^{0.5M}$	$75.73^{1.0M}$	$79.07^{2.0M}$
11	KI (20°)	$73.17^{0.5M}$	$73.60^{1.0M}$	$74.01^{1.5M}$
12	KNO_3 (20°)	$72.78^{0.02M}$	$72.86^{0.1M}$	$73.26^{0.5M}$
13	KOH (18°)	$73.94^{0.5M}$	$74.83^{1.0M}$	$75.72^{1.5M}$
14	$LiCl$ (20°)	$73.56^{0.5M}$	$74.38^{1.0M}$	$75.18^{1.5M}$
15	$MgCl_2$ (20°)	$72.92^{0.05M}$	$73.07^{0.1M}$	$73.53^{0.25M}$
16	$MgSO_4$ (20°)	$72.83^{0.025M}$	$72.90^{0.05M}$	$73.01^{0.1M}$
17	$MnCl_2$ (18°)	$74.54^{0.5M}$	$76.02^{1.0M}$	$78.97^{2.0M}$
18	NH_4Cl (20°)	$72.91^{0.1M}$	$73.46^{0.5M}$	$74.14^{1.0M}$
19	NH_4NO_3 (20°)	$73.25^{0.5M}$	$73.75^{1.0M}$	$74.65^{2.0M}$
20	NH_4OH (18°)	$71.8^{0.5M}$	$70.8^{1.0M}$	$67.9^{3.0M}$
21	$(NH_4)_2SO_4$ (20°)	$73.84^{0.5M}$	$74.92^{1.0M}$	$77.10^{2.0M}$
22	Na_2CO_3 (20°)	$73.40^{0.25M}$	$74.03^{0.5M}$	$75.40^{1.0M}$
23	$NaCl$ (20°)	$72.92^{0.1M}$	$73.57^{0.5M}$	$74.39^{1.0M}$
24	Na_2CrO_4 (30°)	$72.6^{0.51M}$	$77.3^{1.95M}$	$82.6^{3.31M}$
25	$NaNO_3$ (20°)	$72.87^{0.1M}$	$73.35^{0.5M}$	$73.95^{1.0M}$
26	$NaOH$ (18°)	$74.4^{0.7M}$	$75.9^{1.5M}$	$83.1^{5.0M}$
27	$Na_2S_2O_3$ (40°)	$71.1^{0.49M}$	$72.4^{1.0M}$	$78.9^{3.04M}$
28	$SrCl_2$ (20°)	$72.93^{0.05M}$	$73.10^{0.1M}$	$73.57^{0.25M}$
29	$Zn(NO_3)_2$ (40°)	$71.1^{0.6M}$	$72.6^{1.1M}$	$77.4^{2.8M}$
30	$ZnSO_4$ (20°)	$73.25^{0.25M}$	$73.73^{0.5M}$	$74.71^{1.0M}$

Table 10-82 (*Continued*)
SURFACE TENSION OF AQUEOUS INORGANIC SOLUTIONS AGAINST AIR

script and are expressed in moles per kilogram of water. The temperature in degrees Centigrade is given in parenthesis after the formula.

Surface Tension in Dynes per Centimeter					
$76.03^{3.0M}$	$77.70^{5.0M}$	$78.60^{6.2M}$	....	....	1
$74.26^{0.5M}$	$75.72^{1.0M}$	$77.77^{1.7M}$	....	....	2
$78.87^{2.0M}$	$86.92^{4.0M}$	$90.37^{5.0M}$	$97.07^{7.0M}$	$106.97^{11.2M}$	3
$79.15^{4.0M}$	$81.55^{5.8M}$	....	....	....	4
$74.95^{1.2M}$	....	....	....	....	5
$71.90^{4.0M}$	$71.50^{6.0M}$	$70.55^{9.0M}$	$65.95^{17.7M}$	....	6
$68.3^{8.5M}$	....	....	....	....	7
$74.88^{20.5M}$	$75.82^{30.6M}$	....	....	....	8
$76.72^{3.0M}$	$79.06^{4.8°}$	....	....	....	9
$82.85^{3.0M}$	$87.15^{4.0M}$	$96.95^{6.0M}$	$102.55^{7.0M}$	$110.75^{8.3M}$	10
$75.26^{3.0M}$	$76.10^{4.0M}$	$77.95^{6.2M}$	....	....	11
$73.78^{1.0M}$	$74.30^{1.5M}$	$74.80^{2.0M}$	$75.25^{2.5M}$	$76.45^{3.9M}$	12
$76.58^{2.0M}$	$79.75^{3.8M}$	....	....	....	13
$76.01^{2.0M}$	$79.45^{4.0M}$	$86.28^{8.0M}$	$93.52^{13.0M}$	$100.76^{19.4M}$	14
$74.27^{0.5M}$	$75.79^{1.0M}$	$79.10^{2.0M}$	$82.98^{3.0M}$	$85.70^{3.65M}$	15
$73.28^{0.25M}$	$73.78^{0.5M}$	$74.85^{1.0M}$	$77.32^{2.0M}$	$79.27^{2.7M}$	16
$81.95^{3.0M}$	$84.89^{4.0M}$	$87.25^{4.8M}$	....	....	17
$75.4^{2.0M}$	$76.5^{3.0M}$	$77.6^{4.0M}$	$78.6^{5.0M}$	$80.4^{6.7M}$	18
$76.33^{4.0M}$	$77.82^{6.0M}$	$79.15^{8.0M}$	$80.30^{10.0M}$	$82.12^{14.0M}$	19
$65.3^{6.0M}$	$63.0^{10.0M}$	$61.1^{15.0M}$	$59.7^{20.0M}$	$57.1^{34.0M}$	20
$79.25^{3.0M}$	$81.42^{4.0M}$	$83.13^{4.8M}$	....	....	21
$76.75^{1.5M}$	....	....	....	....	22
$76.03^{2.0M}$	$77.65^{3.0M}$	$79.29^{4.0M}$	$80.92^{5.0M}$	$82.55^{6.0M}$	23
$90.9^{5.10M}$	$96.5^{6.12M}$	$98.6^{6.99M}$	....	....	24
$75.12^{2.0M}$	$77.44^{0.0M}$	$80.37^{0.0M}$	$82.71^{0.0M}$	$84.11^{2.2M}$	25
$87.8^{7.0M}$	$95.9^{11.0M}$	$100.7^{14.0M}$	....	....	26
$85.9^{5.04M}$	$93.5^{8.34M}$	$96.6^{9.73M}$	$98.4^{11.18M}$	....	27
$74.37^{0.5M}$	$76.00^{1.0M}$	$77.75^{1.5M}$	$79.51^{2.0M}$	$81.62^{2.6M}$	28
$81.1^{4.2M}$	$83.3^{5.2M}$	$84.9^{6.3M}$	....	....	29
$77.1^{2.0M}$	$78.1^{2.4M}$	$79.1^{2.75M}$	$84.5^{5.25M}$	....	30

Table 10-83
SURFACE TENSION OF AQUEOUS ORGANIC SOLUTIONS AGAINST AIR

The table below gives values for the surface tension of various organic aqueous solutions of different concentrations. The concentration is indicated directly below

	Formula	Surface Tension in Dynes per Centimeter		
1	Acetic acid (30°C)	67.98 1.000%	60.11 5.001%	54.56 10.01%
2	Acetone (25°C)	55.45 5%	48.94 10%	41.11 20%
3	iso-Amyl alcohol (15°C)	57.0 0.249%	49.3 0.5%	41.1 0.99%
4	n-Butyl alcohol (20°C)	72.8 0.0244%	72.4 0.0487%	67.2 0.1954%
5	iso-Butyl alcohol (15°C)	65.6 0.249%	53.0 0.99%	45.4 1.96%
6	n-Butyric acid (25°C)	69 0.14%	65 0.31%	60 0.73%
7	iso-Butyric acid (25°C)	69.6 0.10%	62.6 0.43%	49.6 1.92%
8	n-Caproic acid (19°C)	70 0.00212ML	63 0.0064ML	56 0.0128ML
9	iso-Caproic acid (18°C)	71 0.0016ML	67 0.0036ML	60 0.0081ML
10	Chloroacetic acid (25°C)	66.40 3.35%	61.22 7.70%	58.91 10.09%
11	Ethyl alcohol (25°C)	60.79 2.72%	54.87 5.21%	46.03 11.10%
12	Formic acid (30°C)	70.07 1.00%	66.20 5.00%	62.78 10.00%
13	Glycerol (18°C)	72.90 5%	72.85 10%	72.40 20%
14	Glycol (15°C)	73.27 0.125ML	73.02 0.25ML	72.36 0.5ML
15	Maleic acid (20°C)	72.6 0.169%	72.4 0.338%	72.0 0.673%
16	Malic acid (20°C)	72.6 0.154%	72.4 0.320%	72.1 0.645%
17	Methyl alcohol (30°C)	68.44 1.011%	65.32 2.500%	54.60 9.994%
18	Methyl propionate (15°C)	71.72 0.0078ML	70.29 0.0156ML	67.75 0.0313ML
19	Octyl alcohol (20°C)	72.8 1.009%	71.8 1.998%	69.0 2.96%
20	Phenol (20°C)	72.6 0.024%	71.3 0.118%	66.5 0.471%
21	Propionic acid (25°C)	64.5 0.998%	59.8 1.914%	49.1 5.84%
22	n-Propyl alcohol (15°C)	70.76 0.0313ML	68.64 0.0625ML	65.08 0.125ML
23	iso-Propyl alcohol (15°C)	70.77 0.0313ML	68.56 0.0625ML	65.21 0.125ML
24	Propylamine (15°C)	70.39 0.03125ML	68.43 0.0625ML	64.09 0.125ML
25	Pyrocatechol (20°C)	72.1 0.03ML	71.7 0.05ML	66.8 0.20ML
26	Pyrogallol (20°C)	71.7 0.1ML	70.4 0.2ML	65.7 0.5ML
27	Resorcinol (20°C)	72.1 0.03ML	70.6 0.10ML	66.8 0.40ML
28	Succinic acid (20°C)	72.5 0.0922%	72.4 0.184%	72.2 0.369%
29	Sucrose (20°C)	73.30 10%	73.75 20%	74.15 30%
30	n-Valeric acid (25°C)	65.0 0.11%	58.7 0.22%	53.5 0.38%
31	iso-Valeric acid (25°C)	63.7 0.15%	56.9 0.34%	45.3 1.03%

Table 10-83 (*Continued*)
SURFACE TENSION OF AQUEOUS ORGANIC SOLUTIONS
AGAINST AIR

the corresponding surface tension. These concentrations are expressed in percents when so designated, or in moles per liter of solution in which case they are marked *ML*.

Surface Tension in Dynes per Centimeter				
47.72 20.09%	40.68 40.11%	34.26 69.91%	26.58 100%	1
35.98 30%	30.38 50%	26.75 75%	23.04 100%	2
36.0 1.48%	32.3 1.96%	23.0 100%	....	3
56.0 0.772%	48.1 1.538%	40.4 3.105%	28.6 5.927%	4
40.3 2.91%	34.0 4.76%	28.0 7.41%	22.9 100%	5
42 3.83%	33 8.63%	28 24.96%	27 79.38%	6
37.3 5.65%	29.8 11.69%	27.3 18.95%	26.5 69.93%	7
49 0.0212ML	40 0.0425ML	34 0.068ML	31 0.085ML	8
51 0.0183ML	41 0.0411ML	36 0.0616ML	31 0.0924ML	9
50.34 31.54%	47.68 49.97%	46.81 60.01%	45.09 72.11%	10
37.53 20.50%	29.63 40.00%	23.64 87.92%	22.03 100%	11
57.92 20.00%	51.97 40.00%	42.09 80.01%	36.51 100%	12
69.9 50%	65.8 85%	63.8 98%	63.4 100%	13
70.94 1.0ML	....	....	....	14
71.4 1.338%	70.5 2.641%	69.1 5.13%	66.8 9.67%	15
71.5 1.28%	70.7 2.54%	69.3 4.92%	67.1 9.29%	16
46.05 20.00%	36.09 39.98%	26.03 80.03%	21.76 100%	17
63.87 0.0625ML	58.39 0.125ML	51.28 0.25ML	41.61 0.5ML	18
67.1 3.89%	58.6 7.46%	50.5 13.6%	41.8 21.9%	19
61.1 0.941%	54.0 1.881%	46.0 3.755%	42.3 5.623%	20
43.9 9.82%	36.4 21.71%	29.7 73.92%	25.6 99.99%	21
59.34 0.25ML	51.94 0.5ML	43.59 1.0ML	....	22
60.37 0.25ML	53.66 0.5ML	46.58 1.0ML	....	23
59.55 0.25ML	53.03 0.5ML	46.68 0.1ML	....	24
62.4 0.40ML	59.7 0.60ML	55.7 1.0ML	51.6 2.0ML	25
62.4 0.75ML	60.0 1.00ML	56.1 1.50ML	53.8 2.00ML	26
64.4 0.75ML	61.1 1.5ML	58.6 3.0ML	57.1 6.0ML	27
71.7 0.732%	71.0 1.45%	69.7 2.86%	68.1 5.55%	28
74.85 40%	76.45 55%	....	....	29
42.0 0.83%	32.1 2.34%	25.4 87.04%	26.5 98.91%	30
34.6 2.54%	29.2 4.24%	24.6 89.37%	24.9 99.77%	31

Table 10-84
INTERFACIAL TENSIONS

The table below gives values for the interfacial tensions (γ_i) in dynes per centimeter, between various compounds and either water or mercury. The temperature in degrees Centigrade is indicated in each case by a superscript.

Name	$\gamma_i, \dfrac{dynes}{cm}$	Name	$\gamma_i, \dfrac{dynes}{cm}$
iso-Amyl alcohol—H_2O	$5.01^{8°}$	Ethylene bromide—H_2O	$36.54^{20°}$
iso-Amyl butyrate—H_2O	$23.0^{20°}$	Ethylene bromide—Hg	$326^{20°}$
iso-Amyl chloride—H_2O	$15.44^{20°}$	n-Hexane—H_2O	$51.1^{20°}$
Aniline—H_2O	$5.77^{20°}$	n-Hexane—Hg	$378^{20°}$
Aniline—Hg	$341^{20°}$	Iodobenzene—H_2O	$41.84^{20°}$
Anisole—H_2O	$25.82^{20°}$	Mercury—H_2O	$375^{20°}$
Benzaldehyde—H_2O	$15.51^{20°}$	Mesitylene—H_2O	$38.7^{20°}$
Benzene—H_2O	$35.00^{20°}$	Methyl iodide—Hg	$304^{20°}$
Benzene—Hg	$357.2^{20°}$	Methylene chloride—H_2O	$28.31^{20°}$
Benzyl alcohol—H_2O	$4.75^{22.5°}$	Methylene chloride—Hg	$342.5^{20°}$
Bromobenzene—H_2O	$39.82^{20°}$	Nitrobenzene—H_2O	$25.66^{20°}$
o-Bromotoluene—H_2O	$41.15^{20°}$	Nitrobenzene—Hg	$350.5^{20°}$
iso-Butyl alcohol—H_2O	$2.1^{18°}$	Nitroethane—Hg	$378^{20°}$
iso-Butyl alcohol—Hg	$342.7^{20°}$	Nitromethane—H_2O	$9.66^{20°}$
iso-Butyl chloride—H_2O	$24.43^{20°}$	o-Nitrotoluene—H_2O	$27.19^{20°}$
tert.-Butyl chloride—H_2O	$23.75^{20°}$	n-Octane—H_2O	$50.8^{20°}$
Butyronitrile—H_2O	$10.38^{20°}$	n-Octane—Hg	$374.7^{20°}$
Carbon disulfide—H_2O	$48.36^{20°}$	n-Octyl alcohol—H_2O	$8.5^{20°}$
Carbon disulfide—Hg	$336^{20°}$	sec.-Octyl alcohol—Hg	$359.0^{20°}$
Carbon tetrachloride—H_2O	$45^{20°}$	Oleic acid—H_2O	$15.6^{20°}$
Chlorobenzene—H_2O	$37.41^{20°}$	Oleic acid—Hg	$322^{20°}$
Chloroform—H_2O	$32.8^{20°}$	iso-Pentane—H_2O	$49.64^{20°}$
α-Chloronaphthalene—H_2O	$40.74^{20°}$	Phenetole—H_2O	$29.4^{20°}$
p-Cymene—H_2O	$34.61^{20°}$	n-Propyl alcohol—Hg	$368^{20°}$
Diethyl carbonate—H_2O	$12.86^{20°}$	Styrene—H_2O	$35.48^{19°}$
Dipropylamine—H_2O	$1.66^{20°}$	Toluene—H_2O	$36.1^{25°}$
Ethyl alcohol—Hg	$364^{20°}$	Toluene—Hg	$359^{20°}$
Ethyl bromide—H_2O	$31.20^{20°}$	iso-Valeric acid—H_2O	$2.73^{20°}$
Ethyl ether—H_2O	$10.7^{20°}$	o-Xylene—H_2O	$36.06^{20°}$
Ethyl ether—Hg	$379^{20°}$	p-Xylene—H_2O	$37.77^{20°}$
Ethyl mercaptan—H_2O	$26.12^{20°}$	o-Xylene—Hg	$359^{20°}$
Ethyl oleate—H_2O	$21.34^{20°}$	m-Xylene—Hg	$357^{20°}$
Ethyl propyl ketone—H_2O	$13.58^{20°}$	p-Xylene—Hg	$361^{20°}$
Ethylbenzene—H_2O	$31.35^{17.5°}$		

TRANSPORT PROPERTIES

THERMAL CONDUCTIVITY

In the following tables dealing with heat conductivity, the coefficient λ is given in g.-cal./(sec.)(sq. cm.)(°C./cm.). This is the quantity of heat in gram calories, transmitted per second through a plate of the material one centimeter thick and one square centimeter in area when the temperature difference between the two sides of the plate is one degree Centigrade. To express λ in English units, btu./(sec.)(sq. in.)-(°F./inch), multiply the values given below by 0.00560.

Table 10-85
THERMAL CONDUCTIVITY OF VARIOUS METALS AND ALLOYS

Metal or Alloy	Temp. °C.	λ	Metal or Alloy	Temp. °C.	λ
Aluminum, commercial	−188.2	0.454	89Cu+11Zn	18	0.275
" "	0	0.461	87Cu+13Zn	18	0.301
" 99%	18	0.504	82Cu+18Zn	18	0.313
" "	100	0.49	68Cu+32Zn	18	0.260
" "	400	0.76	German silver (52Cu,	0	0.0700
" "	600	1.01	26Zn, 22Ni)	100	0.0887
Antimony	−77	0.0628	62Cu+15Ni+22Zn	18	0.0595
	0	0.0538	Gold	0	0.744
	100	0.0515	"	97	0.7464
20Sb+80Bi	0	0.0152	90Au+10Pd	25	0.234
	100	0.0205	50Au+50Pd	25	0.086
50Sb+50Bi	0	0.0196	10Au+90Pd	25	0.124
	100	0.0229	40Au+60Pt	25	0.062
70Sb+30Bi	0	0.0234	10Au+90Pt	25	0.182
	100	0.0281	Iridium	17	0.141
33.3Sb+66.7Cd	0	0.0267	Iron, with 0.1%C+ }	18	0.1436
50Sb+50Cd	0	0.00519	0.1%Mn+0.2%Si }	100	0.1420
66.7Sb+33.3Cd	0	0.00299	99Fe+1C	18	0.1085
Bismuth	−77	0.0257		100	0.1076
	0	0.0177	Fe+1.5C+0.19Mn+	18	0.119
	100	0.0164	0.05Si+0.03Cu+		
25Bi+75Pb(Vol.)	44	0.0468	0.01P+0.025S		
96.5Bi+3.5Pb(Vol.)	44	0.0129	Bessemer steel	8	0.0985
90Bi+10Sn(Vol.)	44	0.0126	Lead	18	0.0827
50Bi+50Sn	12.5	0.056	"	100	0.0815
25Bi+75Sn	12.5	0.102	Lithium	0	0.17
Brass, red	0	0.246	"	101.3	0.18
" "	100	0.2827	Magnesium	0−100	0.376
Brass, yellow	0	0.2041	Manganin (84Cu+4Ni	18	0.05186
" "	100	0.254	+12Mn)	100	0.0631
Cadmium	0	0.2213	Mercury, solid	−269.3	0.40
"	100	0.2045	" "	−44.2	0.0664
Constantin, see			" liquid	−37.2	0.0218
60 Cu+40 Ni			" "	0	0.0248
Cobalt, with 0.24%C,+			" "	50.4	0.0298
1.4Fe+1.1 Ni+0.14Si	30	0.1653	" "	149.4	0.0385
Chromium steel, 5%Cr	30	0.073	Molybdenum	17	0.346
" " 10%Cr	30	0.052	Nickel, 99%	−160	0.129
" " 15%Cr	30	0.044		18	0.140
Copper	−183	1.111	Nickel +2 or 3% Co	300	0.126
"	0	0.920		500	0.104
"	100	0.92		950	0.065
60Cu+40Ni	18	0.05401		1200	0.058
	100	0.06405	Nickel steel, (30.4Ni+	29	0.029
54Cu+46Ni	18	0.0484	0.14Si+0.84Mn+	71	0.031
99.37Cu+0.63P	30	0.250	0.26C)		
98.02Cu+1.98P	30	0.125	Palladium	18	0.1683

Table 10-85 (*Continued*)
THERMAL CONDUCTIVITY OF VARIOUS METALS AND ALLOYS

Metal or Alloy	Temp. °C.	λ	Metal or Alloy	Temp. °C.	λ
Palladium	100	0.1817	Silver, 99.9%	−160	0.998
90Pd+10Pt	25	0.134	″ ″	0	1.096
50Pd+50Pt	25	0.088	″ ″	10−97	0.9628
10Pd+90Pt	25	0.103	″ 99.98%	18	1.006
90Pd+10Ag	25	0.114	″ ″	100	0.9919
50Pd+50Ag	25	0.076	Sodium	5.7	0.321
10Pd+90Ag	25	0.337	″	21.2	0.317
Platinum	−252.8	0.93	″	88.1	0.288
″	−183	0.182	Tantalum	17	0.130
″	0−200	0.167	″	1827	0.198
90Pt+10Ir	17	0.074	Tin	−170	0.195
90Pt+10Rh	17	0.072	″	0	0.1528
30Pt+70Ag	25	0.074	″	100	0.1423
10Pt+90Ag	25	0.234	30Sn+70Zn (Vol.)	44	0.224
Potassium	5.0	0.234	91.1Sn+8.9Zn (Vol.)	44	0.157
″	20.7	0.232	Tungsten	0	0.383
″	57.6	0.217	″	2227	0.354
62.9K+37.1Na	6.0	0.0549	Wood's metal	7	0.0319
	42.9	0.0619	Zinc	−170	0.280
Rhodium	17	0.210	″	18	0.2653
				100	0.2619

Table 10-86
THERMAL CONDUCTIVITY OF VARIOUS SOLIDS

Substance	Temp. °C.	λ ×10³	Substance	Temp. °C.	λ ×10³
Aluminum oxide (powder)	46.8	1.62	Cork, Sp.G. =0.204	30	0.128
″ ″ (fused)	650-1350	8.0	Cork meal	100	0.133
Asbestos board	20	1.78	Cotton, Sp.G. =0.081	0	0.136
Asbestos fabric	20	0.666	Diatomaceous earth	20	0.13
Asbestos fiber	0	0.267	Earth's crust, average	20	4.0
″ ″	100	0.284	Ebonite	0	0.378
Asbestos paper	20	0.345	Eiderdown	20	0.011
Asphalt	20	1.78	Feathers (with air)	9	0.0574
Basalt	20	5.2	Feldspar	20	5.6
Bauxite	600	1.33	Felt (dark gray)	40	0.149
Boiler scale	65.5	3.13	Ferric oxide	200	1.41
Brick, common	20	1.5	(pressed powder)		
Blotting paper	20	0.15	Ferrous oxide (pressed	49.8	1.33
Cadmium oxide	46.5	1.63	powder)		
(pressed powder)			Fiber, vegetable (with	9	0.0645
Carborundum*			air)		
Cardboard	20	0.5	″ , ″ (with-	9	1.42
Cement, Portland	89.5	0.71	out air)		
Chalk	20	2.2	Fire brick	20	1.1
Clay (fire-hardened)	360 to	2.09 to	Flannel	50	0.0355
	600	2.21	Flint	20	2.4
Coal	<0	0.405	Fluorite	0	24.68
″	1427	20.1	″	100	19.10
Cobalt oxide (pressed	48.5	1.00	Gas carbon	20	8.5
powder)			″ ″	100	9.5
Concrete	20	2.2	Glass, crown	12.5	1.63
Copper oxide (pressed	45.6	2.42	″ , flint	12.5	1.43
powder)			″ , Jena	22	2.27
Copper sulfide (pure)	0	1.06	″ , soda	20	1.7

*Trade name, see Silicon carbide.

Table 10-86 (*Continued*)
THERMAL CONDUCTIVITY OF VARIOUS SOLIDS

Substance	Temp. °C.	$\lambda \times 10^3$	Substance	Temp. °C.	$\lambda \times 10^3$
Glass, soda	100	1.8	Quartz glass	0	3.32
Granite	20	8.17	" "	100	4.57
Graphite, Sp.G. =1.58	50	105.5	Rock salt	0	16.67
" ⊥ to axis	142	42.6	" "	100	11.59
" " " "	555	279	Roofing paper	<0	0.453
Graphite powder, Sp.G. =0.7	40	2.85	Rubber, hard, gray	49	0.55
Gutta percha	20	0.48	" soft, "	49	0.44
Gypsum	0	3.1	" " red	49	0.34
Hair-cloth	<0	0.0402	Sand, dry	20	0.93
Horn	<0	0.087	Sandstone, Sp.G. =2.259	40	4.39
Horse hair, Sp.G. =0.172	20	0.122	Sawdust, Sp.G. =0.19	30	0.14
Ice	....	5.7	Serpentine	20	2.4
Infusorial earth	100	0.34	Silica, see Quartz		
" "	300	0.40	Silicon carbide	650–1350	37.2
Lampblack, Sp.G. =0.165	40	0.156	Silk, Sp.G. =0.101	0	0.122
Lava	16 to 99	2.01	" " "	50	0.136
Leather, cowhide	84	0.42	Silver bromide	0	2.46
" sole, Sp.G. =1.00	30	0.38	Silver chloride	0	2.6
Lime, clayey	20	7.8	Slate	20	4.70
Linen	20	0.21	Snow, fresh, Sp.G. =0.111	....	2.56
Linoleum, Sp.G. =1.183	20	0.445	" , old, Sp.G. =0.450	....	0.115
Magnesia (MgO) (pressed powder), Sp.G. =0.797	47.6	1.45	Sodium chlorate	0	2.665
			Soil, dry	20	0.33
Magnesia brick	50 to 1130	2.7 to 7.2	Sugar, cane	0	1.39
			Sulfur, rhombic	0	0.70
Magnesite	1000	3.98	" plastic	20 to 100	0.63
Marble, white	....	7.8	Sylvite	0	16.65
" black	30	6.85	"	100	11.76
Mica	41.3	0.860	Thymol	12	0.359
Mica, pressed plates	60	0.627	Wadding, Sp.G. =0.01	18	0.93
Naphthalene	0	0.90	Wax, bees'	20	0.207
α-Naphthol	35	0.76	Wood, maple∥to face	20	1.015
β-Naphthol	35	0.80	Wood, maple, ⊥ to face	50	0.434
Nickel oxide (pressed powder, Sp.G. =1.445)	46.2	2.24	Wood, oak, ⊥ to face Sp.G. =0.825	15	0.500
Onyx	30	5.56	Wood, oak,∥to face Sp.G. =0.819	15	0.834
Paper	20	0.3	Wood, pine,∥to face Sp.G. =0.551	20	0.834
Paraffin	0	0.688			
Plaster of Paris	20	0.70	Wood, pine, ⊥ to face Sp.G. =0.546	15	0.361
Porcelain	95	2.48			
Potassium chloride	0	16.6	Wood, teak,∥to face Sp.G. =0.604	15	0.903
Potassium iodide	0	12			
Quartz, ∥ to axis	0	32.5	Wood, teak, ⊥ to face Sp.G. =0.642	15	0.417
" , " " "	100	21.5			
" ⊥ to axis	0	17.31	Zinc oxide (pressed powder, Sp.G. =2.886)	49.7	1.42
" " " "	100	13.33			

Table 10-87
THERMAL CONDUCTIVITY OF VARIOUS LIQUIDS AND SOLUTIONS

Liquid or Solution	Temp. °C.	$\lambda \times 10^3$	Liquid or Solution	Temp. °C.	$\lambda \times 10^3$
Acetic acid	25	0.43	Octane	4	0.375
" " , 50%	25	0.85	Olive oil, Sp.G. =0.911	15.7	0.4515
Acetone	0	0.4228	Paraffin oil	17	0.346
Ammonia, 26%	18	1.09	Pentane	−185	0.3955
Amyl acetate	12	0.302	"	14	0.2856
Amyl alcohol	12	0.328	Petrolatum	20	0.222
Amyl chloride	12	0.283	Petroleum	13	0.355
Amyl iodide	12	0.203	Potassium bromide, 40%	32	1.176
Aniline	12	0.408	Potassium carbonate, 20%	32	1.373
Barium chloride, 21%	32	1.396	Potassium chlorate,	13	1.16
Benzene	12	0.333	Sp.G. =1.026		
Bromobenzene	12	0.265	Potassium chloride, 20%	32	1.334
iso-Butyl alcohol	12	0.340	Potassium hydroxide, 21%	32	1.385
iso-Butyl bromide	12	0.278	" , 42%	32	1.313
iso-Butyl chloride	12	0.278	Potassium nitrate, 10%	32	1.409
iso-Butyl iodide	12	0.208	" " , 20%	32	1.337
n-Butyric acid	12	0.360	Potassium sulfate, 10%	32	1.440
iso-Butyric acid	12	0.340	Propionic acid	12	0.390
Calcium chloride, 15%	32	1.383	Propyl acetate	12	0.327
" " , 30%	32	1.315	n-Propyl alcohol	12	0.373
iso-Caproic acid	12	0.298	iso-Propyl alcohol	0	0.3683
Carbon disulfide	12	0.343	Propyl bromide	12	0.257
Carbon tetrachloride	12	0.252	Propyl chloride	12	0.283
Castor oil	....	0.425	Propyl formate	12	0.357
Chlorobenzene	12	0.302	Propyl iodide	12	0.220
Chloroform	12	0.288	Sodium bromide, 20%	32	1.348
Copper sulfate, 18%	32	1.379	" " , 40%	32	1.289
Cylinder oil	81	0.290	Sodium carbonate, 10%	32	1.403
Cymene	12	0.272	Sodium chloride, 12.5%	32	1.403
Ethyl acetate	12	0.348	" " , 25%	32	1.141
Ethyl alcohol	5.2	0.487	Sodium nitrate, 20%	32	1.376
" "	51	0.369	" " , 44%	32	1.311
" " , 90%	15	0.4391	Sodium sulfate, 10%	32	1.447
" " , 70%	14.1	0.5711	Strontium nitrate, 20%	32	1.398
" " , 50%	12.9	0.7461	" " , 40%	32	1.346
" " , 30%	12.3	1.002	Sulfuric acid, 30%	32	1.244
" " , 10%	11.9	1.247	" " , 60%	32	1.047
Ethyl bromide	12	0.247	" " , 90%	32	0.846
Ethyl ether	12	0.303	" " , Sp.G. =1.054	20.5	1.26
Ethyl formate	12	0.378	" " , Sp.G. =1.18	21	1.30
Ethyl iodide	12	0.222	Thymol	13	0.313
Ethyl sulfide	12	0.328	Toluene	0	0.3492
Ethylene glycol	0	0.6353	"	12	0.307
Formic acid	12	0.648	Turpentine oil	12	0.260
Glycerol	12	0.670	n-Valeric acid	12	0.325
"	48	0.613	iso-Valeric acid	12	0.312
Heptane	4	0.337	Water	4.1	1.29
Hexane	4	0.364	"	12	1.36
Hydrochloric acid, 12.5%	32	1.262	"	40.8	1.555
" " , 25%	32	1.151	Wood tar	79.5	0.324
" " , 38%	32	1.052	o-Xylene	0	0.3443
Magnesium chloride, 11%	32	1.376	" "	33	0.2519
" " , 29%	32	1.238	m-Xylene	0	0.3429
Magnesium sulfate, 22%	32	1.414	Zinc chloride, 17.5%	32	1.327
Methyl acetate	12	0.385	" " , 35%	32	1.213
Methyl alcohol	12	0.495	Zinc sulfate, 16%	32	1.382
Methyl butyrato	12	0.335	" sulfate, 32%	32	1.327
Methyl valerate	12	0.315	" " , Sp.G. =1.134	4.5	1.18
Nitrobenzene	12.5	0.3801	" " , Sp.G. =1.382	45.2	1.44

Table 10-88
THERMAL CONDUCTIVITY OF VARIOUS GASES AND VAPORS

Gas or Vapor	Temp. °C.	$\lambda \times 10^5$	Gas or Vapor	Temp. °C.	$\lambda \times 10^5$
Acetone	0	2.301	Helium	100	39.85
"	100	3.96	n-Heptane	100	4.136
"	184	5.90	n-Hexane	20	2.854
Acetylene	0	4.40	Hexylene	100	4.396
Air	−191.1	1.80	Hydrogen	−252.2	3.22
"	−78.4	4.256	"	−78.4	30.65
"	0	5.572	"	0	39.60
"	100	7.197	"	100	49.94
"	531	15.95	Hydrogen sulfide	0	3.045
Ammonia	−57.6	3.82	Mercury vapor	203	1.846
"	0	5.135	Methane	−181.6	2.248
"	100	7.09	"	−75.6	4.940
Argon	−182.6	1.42	"	0	7.200
"	0	3.88	Methyl alcohol	0	3.357
"	100	5.087	" "	100	5.161
Benzene	0	2.094	Methyl bromide	4.6	1.74
"	100	4.144	Methyl chloride	0	2.216
"	212.5	7.08	" "	100	3.841
Butylamine	6.5	3.003	" "	212.5	6.113
Carbon dioxide	−78.5	2.546	Methyl dichloride	0	1.562
" "	0	3.393	" "	100	2.524
" "	100	5.06	" "	212.5	3.804
" "	546	14.20	Methyl iodide	0	1.098
Carbon disulfide	0	1.615	" "	100	1.804
Carbon monoxide	−191	1.650	Neon	−181.4	4.99
" "	0	5.425	"	−74.4	8.79
Carbon tetrachloride	46	1.666	"	0	10.87
" "	100	2.048	"	105.8	13.44
" "	184	2.599	Nitric oxide, NO	−71.4	4.160
Chlorine	0	1.829	" "	0	5.55
Chloroform	0	1.523	Nitrogen	−191.4	1.829
"	100	2.333	"	−78.4	4.305
"	184	3.103	"	0	5.68
Ethane	−70.4	2.727	"	100	7.18
"	0	4.306	Nitrogen dioxide	55	8.88
"	100	7.673	Nitrous oxide, N_2O	−71.8	2.710
Ethyl acetate	46	2.88	" "	0	3.515
" "	100	3.862	" "	100	5.06
" "	184	5.69	Oxygen	−191.4	1.721
Ethyl alcohol	20	3.583	"	−78.4	4.292
" "	100	4.98	"	0	5.70
Ethyl ether	0	3.101	"	100	7.427
" "	100	5.278	n-Pentane	20	3.267
" "	212.5	8.400	iso-Pentane	0	2.912
Ethylene	−71.1	2.572	" "	100	5.105
"	0	4.02	" "	184	7.52
"	100	6.36	Sulfur dioxide	0	1.950
Helium	−252.2	5.18	Water vapor	46	4.580
"	−191.7	14.84	" "	100	5.510
"	0	33.60			

VISCOSITY

The viscosity of a substance is the shearing resistance of a liquid film which separates two horizontal plates, one of which is being moved across the other. The absolute viscosity of a substance is the force in dynes which will move one square centimeter of a plane surface with a speed of one centimeter per second relative to another parallel plane surface from which it is separated by a layer of the substance one centimeter thick. This may be expressed by the following formula:

$$\text{viscosity} = \frac{\text{force} \times \text{film thickness}}{\text{area of plate} \times \text{velocity}}.$$

The unit of absolute viscosity is the poise which is equal to one dyne second per square centimeter. That is, when two plates have a shearing area of 1 square centimeter and a film thickness of 1 centimeter, if a force of one dyne is required to maintain a velocity of 1 centimeter per second, the fluid is said to have a viscosity of 1 poise. Since the poise (η) is such a large unit it is generally more convenient to use the millipoise (1/1000 poise) or, in the case of gases, the micropoise (1/1,000,000 poise). Frequently the viscosity of a liquid is expressed in terms of specific viscosity, which is its absolute viscosity compared with that of water at the same temperature. If the temperature in question is 20°C. the specific viscosity equals the viscosity in centipoises (1/100 poise) since the viscosity of water at 20°C. is approximately 1 centipoise.

The kinematic viscosity of a substance is the ratio of the absolute viscosity to the density of the substance at the temperature of measurement. The unit of kinematic viscosity is the stoke and is equal to that possessed by a fluid which has a viscosity of one poise and a density of one gram per cubic centimeter.

Of special importance in the field of lubrication are the so-called kinematic viscosimeters which measure the time in seconds for a given volume of liquid to flow through a definite orifice at some specified temperature. See special table for conversion of kinematic viscosity in centistokes to Saybolt, Redwood, and Engler viscosimeter values.

Table 10-89
VISCOSITY OF VARIOUS LIQUIDS IN MILLIPOISES

In the table below is given the viscosity of various liquids in millipoises (0.001η) at the temperature in degrees Centigrade as indicated by the superscript.

Values in millipoises may be converted to centistokes by dividing by $10d^t$, where d^t is the density of the substance at the same temperature t.

Name		Name	
Acetaldehyde	$2.751^{0\circ}$	Acetone	$4.013^{0\circ}$
	$2.521^{10\circ}$		$3.311^{20\circ}$
	$2.307^{20\circ}$		$2.561^{50\circ}$
Acetamide	$13.21^{05\circ}$	Allyl alcohol	$21.45^{30\circ}$
	$10.6^{120\circ}$		$13.632^{20\circ}$
Acetanilide	$22.2^{120\circ}$		$4.137^{90\circ}$
	$19.0^{130\circ}$	Allyl bromide	$6.258^{0\circ}$
Acetic acid	$12.22^{20\circ}$		$5.037^{20\circ}$
	$10.396^{30\circ}$		$4.192^{40\circ}$
	$7.956^{50\circ}$		$3.279^{70\circ}$
	$4.244^{110\circ}$	Allyl chloride	$4.127^{0\circ}$
Acetic anhydride	$12.448^{0\circ}$		$3.374^{20\circ}$
	$9.065^{20\circ}$		$2.825^{40\circ}$
	$6.991^{40\circ}$		
	$3.079^{130\circ}$		

Table 10-89 (*Continued*)
VISCOSITY OF VARIOUS LIQUIDS

Name		Name	
Allyl iodide	$9.358^{0°}$	*iso*-Butyl chloride	$5.878^{0°}$
	$7.338^{20°}$		$4.617^{20°}$
	$5.972^{40°}$		$3.729^{40°}$
	$3.652^{100°}$		$3.090^{60°}$
Ammonia	$2.66^{-33.5°}$	Butyl formate	$6.91^{20°}$
Amyl acetate	$8.055^{45°}$	*iso*-Butyl iodide	$11.664^{0°}$
Amyl alcohol	$89.22^{0°}$		$8.753^{20°}$
	$62.34^{10°}$		$6.970^{40°}$
	$40.04^{23°}$		$3.496^{120°}$
Aniline	$60.23^{12°}$	*n*-Butyric acid	$22.855^{0°}$
	$44.67^{20°}$		$15.402^{20°}$
	$29.3^{32°}$		$11.205^{40°}$
	$15.55^{60°}$		$3.230^{160°}$
Anisole	$10.89^{20°}$	*iso*-Butyric acid	$18.868^{0°}$
	$7.409^{45°}$		$13.175^{20°}$
Benzene	$9.00^{0°}$		$9.804^{40°}$
	$7.57^{10°}$		$3.258^{150°}$
	$6.47^{20°}$	Caproic acid	$32.01^{20°}$
	$5.61^{30°}$	Carbon dioxide	$0.925^{5°}$
	$4.36^{50°}$	(Under the pressure	$0.852^{10.2°}$
	$3.502^{70°}$	of its saturated vapor)	$0.784^{15°}$
Benzotrichloride	$30.71^{0°}$		$0.712^{20°}$
	$25.5^{17°}$		$0.625^{25°}$
Benzyl acetate	$13.99^{45°}$		$0.539^{29°}$
Benzyl alcohol	$55.82^{20°}$		$0.321^{31.1°}$
	$30.08^{45°}$		$4.36^{0.4°}$
Benzyl benzoate	$84.54^{25°}$	Carbon disulfide	$3.763^{19.94°}$
Bismuth	$16.20^{283°}$		$3.167^{45.96°}$
	$15.25^{330°}$		$3.36^{70.6°}$
	$14.55^{365°}$	Carbon tetrachloride	$9.578^{21.21°}$
Bromine	$12.541^{0.56°}$		$7.928^{35.21°}$
	$10.466^{16.16°}$		$4.056^{99.6°}$
	$9.46^{26°}$	Castor oil	$32950^{6.5°}$
	$7.211^{56.41°}$		$10272^{19.6°}$
Bromobenzene	$15.66^{-0.1°}$		$2245^{40.6°}$
	$11.23^{21.9°}$	Chlorobenzene	$10.50^{-0.3°}$
	$8.53^{43.6°}$		$8.03^{20.1°}$
	$3.85^{142.5°}$		$6.37^{40.2°}$
iso-Butyl acetate	$7.04^{20°}$		$3.20^{123.6°}$
	$3.66^{78.1°}$	Chloroform	$7.86^{-10°}$
	$2.163^{130.9°}$		$6.99^{0°}$
	$1.426^{183.8°}$		$5.63^{20°}$
n-Butyl alcohol	$51.85^{0°}$		$4.64^{40°}$
	$29.48^{20°}$		$3.89^{60°}$
	$17.816^{40°}$	*o*-Chlorophenol	$22.50^{45°}$
	$4.602^{110°}$	*m*-Chlorophenol	$47.22^{45°}$
iso-Butyl alcohol	$80.38^{0°}$	*p*-Chlorophenol	$60.18^{45°}$
	$39.06^{20°}$	Copal lac	$48002^{22.2°}$
	$21.223^{40°}$	*o*-Cresol	$35.06^{45°}$
	$5.270^{100°}$	*m*-Cresol	$184.23^{20°}$
tert.-Butyl alcohol	$58.88^{22.4°}$		$50.57^{45°}$
	$23.69^{37.2°}$	*p*-Cresol	$56.07^{45°}$
	$6.49^{77.1°}$	Cyclohexane	$10.3^{17°}$
iso-Butyl bromide	$8.277^{0°}$		$9.32^{20°}$
	$6.433^{20°}$		$8.62^{7°}$
	$5.179^{40°}$		$7.53^{5°}$
	$3.260^{90°}$		

Table 10-89 (Continued)
VISCOSITY OF VARIOUS LIQUIDS

Name		Name	
Decane	$7.75^{22.3°}$	Ethyl propyl ether	$4.007^{0.35°}$
Diallyl	$3.415^{0.37°}$		$3.231^{20.33°}$
	$2.752^{0°}$		$2.248^{60.18°}$
	$2.436^{36.06°}$	Ethyl salicylate	$17.72^{45°}$
	$2.059^{56.2°}$	Ethyl sulfide	$5.619^{0.21°}$
Diethyl ketone	$5.990^{0°}$		$4.292^{24.64°}$
	$4.705^{20°}$		$2.498^{88°}$
	$3.854^{40°}$	Ethyl valerate	$8.47^{20°}$
	$2.349^{100°}$	Ethylaniline	$20.42^{5°}$
Diethylamine	$8.236^{-33.5°}$		$10.81^{55°}$
	$3.672^{5°}$	Ethylene bromide	$24.38^{0°}$
Diethylaniline	$19.5^{25°}$		$17.21^{20°}$
Dimethylamine	$4.368^{-33.5°}$		$12.86^{40°}$
Dimethylaniline	$12.85^{25°}$		$5.309^{130°}$
Diphenylamine	$46.60^{55°}$	Ethylene chloride	$11.322^{0°}$
	$10.41^{30°}$		$8.385^{20°}$
Dipropyl ether	$5.40^{-5.9°}$		$6.523^{40°}$
	$4.25^{20°}$		$4.357^{80°}$
	$2.241^{88°}$	Ethylidene chloride	$6.282^{0°}$
Dipropyl ketone	$7.36^{20°}$		$4.983^{20°}$
Dodecane	$12.57^{23.3°}$		$3.731^{50°}$
Ethyl acetate	$5.825^{0°}$	Eugenol	$69.31^{25°}$
	$4.546^{20°}$	Fluorobenzene	$7.52^{0.2°}$
	$3.668^{40°}$		$5.851^{9.9°}$
	$2.786^{70°}$		$3.258^{80.9°}$
Ethyl alcohol	$17.716^{0°}$	Formamide	$7.68^{105°}$
	$14.510^{10°}$		$6.59^{120°}$
	$11.943^{20°}$	Formanilide	$16.5^{123°}$
	$8.309^{40°}$	Formic acid	$22.469^{10°}$
	$5.098^{70°}$		$17.844^{20°}$
Ethyl benzene	$8.769^{0°}$		$12.190^{40°}$
	$6.687^{20°}$		$5.492^{100°}$
	$5.311^{40°}$	Glycerol	$42200^{2.8°}$
	$2.512^{130°}$		$10690^{20°}$
Ethyl benzoate	$22.42^{0°}$		$7776^{20.9°}$
	$20.14^{25°}$	Glycol	$173.3^{25°}$
Ethyl bromide	$4.866^{0°}$	Heptane	$5.236^{0°}$
	$4.407^{10°}$		$4.163^{20°}$
	$4.020^{20°}$		$3.410^{40°}$
	$3.678^{30°}$		$2.239^{90°}$
Ethyl butyrate	$6.68^{20°}$	Heptyl alcohol	$70.14^{20°}$
Ethyl ether	$2.950^{0°}$	Heptylic acid	$43.56^{20°}$
	$2.681^{10°}$	Hexadecane	$35.91^{22.2°}$
	$2.448^{20°}$	Hexane	$4.012^{0°}$
	$2.230^{30°}$		$3.258^{20°}$
Ethyl formate	$5.103^{0°}$		$2.708^{40°}$
	$4.085^{20°}$		$2.288^{60°}$
	$3.356^{40°}$	Iodobenzene	$21.94^{-4.6°}$
	$3.077^{50°}$		$17.74^{17.4°}$
Ethyl iodide	$7.269^{0°}$		$4.88^{148.8°}$
	$5.925^{20°}$	Isoprene	$2.664^{0.35°}$
	$4.951^{40°}$		$2.234^{20.41°}$
	$3.914^{70°}$		$2.037^{32.02°}$
Ethyl propionate	$6.967^{0°}$	Lactic acid	$403.3^{25°}$
	$5.367^{20°}$	Lead bromide	$101.93^{72°}$
	$4.282^{40°}$		$69.70^{412°}$
	$2.705^{90°}$		$40.73^{492°}$

Table 10-89 (*Continued*)
VISCOSITY OF VARIOUS LIQUIDS

Name	
Lead chloride	$55.32^{498°}$
	$40.20^{538°}$
	$29.56^{608°}$
Lithium nitrate	$55.9^{259°}$
	$45.0^{284°}$
	$29.4^{344°}$
Menthol	$68.93^{4.9°}$
Mercury	$18.68^{-21.4°}$
	$16.84^{0°}$
	$15.47^{20°}$
	$14.83^{40°}$
	$13.60^{62.9°}$
	$12.99^{79.4°}$
	$12.60^{98°}$
	$11.71^{124°}$
	$10.92^{154°}$
	$10.18^{196.7°}$
	$9.86^{237.8°}$
	$9.615^{262.5°}$
	$9.183^{314.7°}$
Methyl acetate	$4.837^{0°}$
	$3.88^{20°}$
	$3.202^{40°}$
	$2.929^{50°}$
Methyl alcohol	$8.08^{0°}$
	$6.90^{10°}$
	$5.93^{20°}$
	$4.49^{40°}$
	$3.49^{60°}$
Methyl benzoate	$20.59^{20°}$
Methyl n-butyrate	$7.625^{0°}$
	$5.795^{20°}$
	$4.585^{40°}$
	$2.647^{100°}$
Methyl iso-butyrate	$6.764^{0°}$
	$5.228^{20°}$
	$4.190^{40°}$
	$2.648^{90°}$
Methyl ethyl ketone	$5.429^{0°}$
	$4.284^{20°}$
	$3.490^{40°}$
	$2.482^{80°}$
Methyl formate	$4.355^{0°}$
	$3.548^{20°}$
	$3.253^{30°}$
Methyl iodide	$6.055^{0°}$
	$5.001^{20°}$
	$4.240^{40°}$
Methyl propionate	$5.866^{0°}$
	$4.597^{20°}$
	$3.748^{40°}$
	$2.861^{70°}$
Methyl propyl ketone	$6.477^{0°}$
	$5.056^{20°}$
	$4.096^{40°}$
	$2.465^{100°}$

Name	
Methyl sulfide	$3.599^{0.27°}$
	$3.008^{20.19°}$
	$2.649^{35.81°}$
Methyl valerate	$7.13^{20°}$
Methylaniline	$20.2^{25°}$
	$10.84^{55°}$
Methylene chloride	$5.40^{0.46°}$
	$4.414^{20.53°}$
	$3.802^{37.51°}$
Nitric acid	$22.75^{0°}$
	$17.70^{10°}$
Nitrobenzene	$28.21^{0°}$
	$19.8^{20°}$
	$15.5^{35°}$
Nitrogen dioxide	$5.349^{0°}$
	$4.766^{10°}$
	$4.275^{20°}$
o-Nitrophenol	$23.43^{45°}$
Nonane	$6.192^{2.3°}$
Nonylic acid	$83.1^{20°}$
Octane	$7.060^{0°}$
	$6.159^{10°}$
	$5.419^{20°}$
	$4.828^{30°}$
	$4.328^{40°}$
	$3.551^{60°}$
	$2.971^{80°}$
	$2.160^{120°}$
Octyl alcohol	$89.47^{20°}$
Octylic acid	$57.49^{20°}$
Olive oil	$10081^{5.6°}$
	$3773^{7.9°}$
	$1546^{5.7°}$
	$70^{100°}$
Paraffin oil	$1018^{18°}$
Pentadecane	$28.14^{22°}$
Pentane	$2.894^{0°}$
	$2.395^{20°}$
	$2.200^{30°}$
Phenetole	$12.62^{20°}$
	$8.249^{45°}$
Phenol	$127.44^{18.3°}$
	$60.24^{35°}$
	$34.21^{50°}$
	$13.14^{91°}$
Phenyl acetate	$17.99^{45°}$
Phenyl ethane	$6.076^{25°}$
Potassium dichromate	$133.9^{397°}$
	$93.8^{457°}$
	$66.4^{507°}$
Potassium nitrate	$29.70^{333°}$
	$24.42^{373°}$
	$20.07^{413°}$
Propionic acid	$15.21^{40°}$
	$11.022^{20°}$
	$8.451^{40°}$
	$3.308^{140°}$

Table 10-89 (*Continued*)
VISCOSITY OF VARIOUS LIQUIDS

Name	
Propyl acetate	$7.734^{0°}$
	$5.854^{20°}$
	$4.604^{40°}$
	$2.587^{100°}$
n-Propyl alcohol	$38.827^{0°}$
	$22.563^{20°}$
	$14.050^{40°}$
	$5.310^{90°}$
iso-Propyl alcohol	$45.646^{0°}$
	$23.702^{20°}$
	$13.311^{40°}$
	$5.249^{80°}$
n-Propyl bromide	$6.509^{0°}$
	$5.241^{20°}$
	$4.334^{40°}$
	$3.379^{70°}$
iso-Propyl bromide	$6.106^{0°}$
	$4.894^{20°}$
	$4.028^{40°}$
	$3.680^{50°}$
Propyl *n*-butyrate	$8.31^{20°}$
Propyl *iso*-butyrate	$7.41^{20°}$
n-Propyl chloride	$4.416^{0°}$
	$3.589^{20°}$
	$2.990^{40°}$
iso-Propyl chloride	$4.080^{0°}$
	$3.292^{20°}$
	$2.993^{30°}$
n-Propyl formate	$6.720^{0°}$
	$5.210^{20°}$
	$4.171^{40°}$
	$2.883^{80°}$
iso-Propyl formate	$5.122^{20°}$
n-Propyl iodide	$9.435^{0°}$
	$7.438^{20°}$
	$6.067^{40°}$
	$3.714^{100°}$
iso-Propyl iodide	$8.841^{0°}$
	$6.971^{20°}$
	$5.676^{40°}$
	$4.005^{80°}$
Propyl propionate	$6.732^{20°}$
Propyl valerate	$10.53^{20°}$
n-Propylamine	$3.53^{25°}$
Pyridine	$8.775^{25°}$
Rape oil	$1118^{15.6°}$
	$422^{37.9°}$
	$177^{65.7°}$
	$80^{100°}$
Rosin	$(1.0\times10^{21})^{7.1°}$
	$(6\times10^{18})^{20°}$
Safrol	$22.94^{25°}$
iso-Safrol	$39.81^{25°}$
Salicylaldehyde	$16.69^{45°}$
Silver bromide	$18.636^{609°}$
	$14.876^{688°}$
	$11.928^{803°}$
Silver chloride	$16.06^{603°}$
	$13.72^{669°}$
	$11.86^{734°}$
Silver iodide	$30.26^{605°}$
	$21.23^{730°}$
	$15.56^{827°}$
Silver nitrate	$37.7^{244°}$
	$30.5^{275°}$
	$23.0^{342°}$
Sodium bromide	$14.2^{762°}$
	$13.5^{766°}$
	$12.8^{780°}$
Sodium chloride	$13.0^{841°}$
	$12.0^{850°}$
	$10.1^{896°}$
	$9.7^{924°}$
Sodium nitrate	$29.19^{308°}$
	$26.61^{328°}$
	$24.39^{348°}$
	$22.37^{368°}$
	$20.57^{388°}$
	$18.28^{418°}$
Sperm oil	$420^{15.6°}$
	$185^{37.9°}$
	$85^{65.7°}$
	$46^{100°}$
Succinonitrile	$27.6^{58.7°}$
	$23.6^{69.4°}$
	$18.1^{83.0°}$
Sulfur	$109.4^{123°}$
	$86.6^{135.5°}$
	$70.9^{149.5°}$
	$71.9^{156.3°}$
	$75.9^{158.2°}$
	$94.8^{159.2°}$
	$144.5^{159.5°}$
	$228.3^{160.0°}$
Sulfur	$773.2^{160.3°}$
	$5000^{165.0°}$
	$45000^{171.0°}$
	$215000^{200°}$
	$205000^{210°}$
	$186000^{222°}$
Sulfuric acid (100%)	$618^{0°}$
	$269.4^{15°}$
	$191.5^{25°}$
	$172^{33°}$
	$106^{50°}$
	$63.5^{70°}$
	$42.5^{90°}$
Tallow	$176^{65.7°}$
	$78^{100°}$
Tetrachloroethylene	$11.384^{0.43°}$
	$8.759^{22.3°}$
	$6.539^{52.68°}$
	$4.034^{117.09°}$

Table 10-89 (Continued)
VISCOSITY OF VARIOUS LIQUIDS

Name		Name	
Tetradecane	$21.31^{21.9°}$	Undecane	$9.47^{22.7°}$
Thiophene	$8.71^{20.24°}$	Urethane	$9.16^{105°}$
	$6.432^{22.5°}$		$7.15^{120°}$
	$3.528^{82.53°}$	n-Valeric acid	$22.36^{200°}$
Tin	$16.78^{280°}$		$12.55^{0°}$
	$16.64^{296°}$	n-Valeric acid	$9.79^{70°}$
	$14.21^{357°}$	iso-Valeric acid	$24.11^{200°}$
	$13.11^{389°}$	Water (see special table)	$10.02^{200°}$
Toluene	$7.71^{90°}$	Wool fat, neutral	$16726^{5.7°}$
	$5.90^{320°}$		$3141^{100°}$
	$4.71^{340°}$	o-Xylene	$11.049^{0°}$
	$3.87^{460°}$		$8.102^{200°}$
	$2.588^{110°}$		$6.270^{40°}$
Tridecane	$15.50^{23.3°}$		$4.168^{80°}$
Triethylamine	$7.726^{-33.5°}$		$2.626^{140°}$
	$3.63^{25°}$	m-Xylene	$8.05^{90°}$
Trimethylamine	$3.208^{-33.5°}$		$6.200^{200°}$
Turpentine	$22.48^{0°}$		$4.970^{40°}$
	$17.83^{100°}$		$3.455^{80°}$
	$14.87^{200°}$		$2.418^{130°}$
	$12.72^{30°}$	p-Xylene	$7.385^{10°}$
	$10.71^{40°}$		$6.475^{20°}$
	$8.21^{60°}$		$5.134^{40°}$
	$6.71^{80°}$		$3.519^{80°}$
			$2.424^{130°}$

Table 10-90
VISCOSITY OF GASES AND VAPORS IN MICROPOISES

In the table below is given the viscosity of various gases and vapors in micropoises ($\eta \times 10^{-6}$) at the temperature in degrees Centigrade as indicated in each case by the superscript.

Name		Name	
Acetone	$72.5^{0°}$	Air	$343.3^{450°}$
	$78.0^{18°}$		$358.3^{500°}$
	$94.27^{100°}$	Ammonia	$67.2^{-78.5°}$
	$125.7^{212.5°}$		$92.6^{0°}$
Acetylene	$94.3^{0°}$		$108.0^{20°}$
Air	$24^{-145°}$		$130.3^{100°}$
	$87^{-98°}$	Argon	$73.56^{-183.2°}$
	$147^{-49.7°}$		$138.0^{-104.4°}$
	$170.9^{0°}$		$169.7^{-60.2°}$
	$175.9^{100°}$		$210.4^{0°}$
	$180.8^{200°}$		$216.8^{12.3°}$
	$185.6^{300°}$		$221.0^{23°}$
	$190.4^{400°}$		$270.2^{99.6°}$
	$195.1^{500°}$		$323.1^{183°}$
	$199.7^{600°}$	Arsine	$145.8^{0°}$
	$208.8^{800°}$		$198.1^{100°}$
	$217.5^{1000°}$	Benzene	$70.9^{0°}$
	$238.5^{1500°}$		$75.9^{16.8°}$
	$258.2^{2000°}$		$100.77^{0.1°}$
	$277.0^{2500°}$		$117.6^{100°}$
	$294.6^{3000°}$		$124.7^{212.5°}$
	$311.3^{3500°}$	Bromine	$151.1^{12.8°}$
	$327.7^{4000°}$		$170.5^{65.7°}$

Table 10-90 (*Continued*)
VISCOSITY OF GASES AND VAPORS

Name		Name	
Bromine	$188.5^{99.7°}$	Chloroform	$130.7^{100°}$
	$207.9^{139.7°}$		$166^{212.5°}$
	$227.3^{179.7°}$	Cyanogen	$93.5^{0°}$
	$287.4^{302°}$		$107^{20°}$
n-Butane	$84.04^{14.7°}$		$128.1^{100°}$
	$84.13^{16.0°}$	Diethylamine	$92^{99.9°}$
	$109.2^{100°}$	Ethane	$63.4^{-78.5°}$
iso-Butane	$75.5^{23°}$		$84.8^{0°}$
iso-Butyl acetate	$76.4^{16.1°}$	Ethyl acetate	$68.4^{0°}$
	$112.0^{100°}$		$94.3^{100°}$
iso-Butyl formate	$83.0^{17.7°}$		$126^{212.5°}$
	$97.2^{63.6°}$	Ethyl alcohol	$82.7^{0°}$
	$114.2^{99.9°}$		$88.5^{16.8°}$
n-Butylamine	$82^{99.8°}$		$108.8^{100°}$
iso-Butylamine	$88^{99.8°}$		$141.7^{212.5°}$
Carbon dioxide	$103.3^{-78.4°}$	Ethyl chloride	$93.5^{0°}$
	$129.4^{-20.7°}$		$105.0^{20°}$
	$138.0^{0°}$		$144.0^{157.3°}$
	$143.6^{12°}$		$171.4^{240.6°}$
	$160.0^{20°}$	Ethyl ether	$68.4^{0°}$
	$197.2^{100°}$		$71.6^{100°}$
	$214.3^{162.4°}$		$73.5^{18.9°}$
	$238.5^{222°}$		$75.5^{25.8°}$
	$268.2^{302°}$		$79.3^{36.5°}$
pressure 1 atm.	$148^{20°}$		$95.5^{100°}$
	$153^{30°}$		$123.4^{212.5°}$
	$156^{35°}$	Ethyl formate	$92^{99.8°}$
	$157^{40°}$	Ethyl propionate	$75.0^{16.1°}$
pressure 20 atm.	$156^{20°}$		$105.4^{68.6°}$
	$159^{30°}$		$116.1^{99.9°}$
	$163^{35°}$	Ethylene	$69.93^{-75.7°}$
pressure 40 atm.	$166^{20°}$		$76.85^{-44.08°}$
	$168^{30°}$		$85.1^{-21.5°}$
	$176^{40°}$		$90.66^{-0.05°}$
pressure 80 atm.	$565^{30°}$		$101.6^{15°}$
	$528^{32°}$		$109.0^{20°}$
	$361^{35°}$		$127.8^{99.25°}$
	$218^{40°}$		$153.0^{182.4°}$
Carbon disulfide	$91.1^{0°}$		$182.6^{302.6°}$
	$96.4^{14.2°}$	Helium	$29.46^{-258.1°}$
Carbon monoxide	$56.5^{-191.6°}$		$81.54^{-198.00°}$
	$86.8^{-149.2°}$		$139.2^{-102.6°}$
	$128^{-78.4°}$		$158.7^{-60.9°}$
	$148^{-42.3°}$		$178.8^{-22.8°}$
	$167^{0°}$		$187.6^{0°}$
	$175^{11.4°}$		$191.4^{9.8°}$
	$184^{20°}$		$196.0^{15°}$
	$210^{100°}$		$199.4^{21.4°}$
Carbon oxysulfide	$119^{15°}$		$234.1^{100°}$
	$154^{100°}$		$269.9^{184.6°}$
Chlorine	$129.7^{12.7°}$	Hydrogen	$5.7^{-257.7°}$
	$147.0^{20°}$		$9.9^{-251.6°}$
	$168.8^{99.1°}$		$31.93^{-202.2°}$
	$189.7^{146°}$		$36.9^{-191.8°}$
Chloroform	$94.4^{0°}$		$60.93^{-102.9°}$
	$98.9^{14.2°}$		$71.03^{-60.2°}$
	$102.9^{17.4°}$		$81.9^{-20.6°}$

Table 10-90 (*Continued*)
VISCOSITY OF GASES AND VAPORS

Name		Name	
Hydrogen	$84.9^{0°}$	Nitric oxide	$186^{20°}$
	$92.3^{15°}$	Nitrogen (chemical)	$56^{-191.5°}$
	$93.1^{20°}$		$156.3^{-21.5°}$
	$97.6^{53.4°}$		$167.4^{0°}$
	$104.6^{100.5°}$		$173.7^{13.9°}$
	$121.5^{182.4°}$		$184^{20°}$
	$139.2^{302°}$		$189.4^{53.5°}$
Hydrogen bromide	$181.9^{18.7°}$		$212^{100°}$
	$234.4^{100.2°}$		$246^{182.7°}$
Hydrogen chloride	$137.9^{0°}$	Nitrogen (atm.)	$169.5^{0°}$
	$156.0^{20°}$		$174.4^{15°}$
	$182.2^{100°}$		$213.4^{101.1°}$
Hydrogen iodide	$185.7^{20.6°}$		$246.4^{183°}$
	$238.3^{100.2°}$	Nitrous oxide	$124.9^{-21.5°}$
Hydrogen sulfide	$116.6^{0°}$		$136.2^{0°}$
	$124.1^{17°}$		$149.8^{25°}$
	$158.7^{100°}$		$160.6^{53.6°}$
Iodine	$184.3^{124°}$		$173.9^{75.8^{d}}$
	$203.8^{170°}$		$182.9^{100.3°}$
	$219.8^{205.4°}$		$216.1^{183.1°}$
	$239.7^{247.1°}$		$261.0^{289.9°}$
Krypton	$233.4^{0°}$		$307.3^{413.6°}$
	$240.5^{10.6°}$	Oxygen	$65^{-191.1°}$
	$245.9^{16.3°}$		$112.8^{-129.8°}$
	$306.3^{100°}$		$145^{-78.7°}$
Mercury	$494^{273°}$		$169.3^{-39.5°}$
	$551^{313°}$		$192^{0°}$
	$641^{369°}$		$201.4^{15°}$
	$654^{380°}$		$206^{20°}$
Methane	$35.1^{-181.6°}$		$216^{53.5°}$
	$76.7^{-78.4°}$		$248.5^{99.74°}$
	$103.5^{0°}$		$288.5^{185.8°}$
	$120.1^{20°}$	*iso*-Pentane	$88.5^{100°}$
	$136.3^{100°}$		$116.4^{212.5°}$
Methyl acetate	$101.5^{100°}$	Phosphine	$106.1^{0°}$
	$135.9^{212.5°}$		$112.0^{15°}$
Methyl bromide	$103.6^{0°}$		$143.8^{100°}$
Methyl chloride	$93.6^{-15.3°}$	Propionic acid	$118^{139.8°}$
	$102.5^{0°}$	Propyl acetate	$74.3^{15°}$
	$116^{20°}$		$95.4^{77.8°}$
	$138.4^{99.1°}$		$109.6^{100°}$
	$170.6^{182.4°}$	*n*-Propyl alcohol	$93^{99.9°}$
	$213.9^{302°}$	*iso*-Propyl alcohol	$109^{99.8°}$
Methyl *iso*-butyrate	$75.4^{24°}$	*n*-Propyl ether	$78.89^{100.1°}$
	$99.9^{65.5°}$	*iso*-Propyl ether	$84.24^{100.2°}$
	$112.2^{100°}$	Propyl formate	$92^{99.9°}$
Methyl ether	$90.5^{0°}$	Radon (Niton)	$212.4^{0°}$
	$102^{20°}$	Silane (SiH$_4$)	$112.4^{15°}$
	$109.2^{99.85°}$		$142.4^{100°}$
Methyl ethyl ether	$102.9^{100°}$	Sulfur dioxide	$117^{0°}$
Methyl formate	$92.3^{20°}$		$124.2^{18°}$
	$135.2^{100°}$		$161.6^{100°}$
Methyl propionate	$94^{99.8°}$	Water vapor	$90.4^{0°}$
Neon	$298.1^{0°}$		$97.5^{20.6°}$
	$308^{13.8°}$		$100.6^{28.9°}$
	$365.2^{100°}$		$127^{100°}$
Nitric oxide	$178^{0°}$		$145^{151.2°}$

Table 10-90 (*Continued*)
VISCOSITY OF GASES AND VAPORS

Name		Name	
Water vapor	$168^{207.1°}$	Xenon	$218^{10.9°}$
	$190^{261.3°}$		$222.2^{15.3°}$
Xenon	$210.7^{0°}$		$282.7^{100.1°}$

Table 10-91
VISCOSITY OF AQUEOUS SOLUTIONS AT 20°C

Formula	Viscosity in millipoises				
	1.0 Molar	0.5 Molar	0.25 Molar	0.125 Molar	0.0625 Molar
$BaCl_2$	13.156	11.530	10.800	10.467	10.310
$Ba(NO_3)_2$	......	......	10.661	10.388	......
$CaCl_2$	13.320	11.660	10.843	10.500	......
$Ca(NO_3)_2$	13.062	11.536	10.786	10.402	......
KCl	10.095	10.116	10.135	10.138	10.140
KNO_3	10.056	10.104	10.127	10.130	......
$MgCl_2$	13.503	11.785	10.925	10.530	......
$Mg(NO_3)_2$	13.896	11.910	11.016	10.588	......
$NaCl$	11.225	10.620	10.385	10.261	10.201
$NaNO_3$	11.200	10.583	10.334	10.226	......
$SrCl_2$	13.300	11.628	10.831	10.481	......
$Sr(NO_3)_2$	12.876	11.389	10.740	10.400	......

SPECIFIC VISCOSITY OF AQUEOUS SOLUTIONS AT 25°C

In the table below are given the specific viscosities of aqueous solutions of different concentrations. To convert these values to absolute viscosities multiply by the viscosity of water at 25°C. See special table for viscosity of water.

Name	1.0 Molar	0.5 Molar	0.25 Molar	0.125 Molar
Acetic acid	1.1131	1.0596	1.0304	1.0171
Aluminum sulfate	1.4064	1.1782	1.0825	1.0381
Ammonia	1.0245	1.0105	1.0058	1.0030
Ammonium chloride	0.9884	0.9976	0.9990	0.9999
Ammonium nitrate	0.9722	0.9862	0.9908	0.9958
Ammonium sulfate	1.1114	1.0552	1.0302	1.0148
Arsenic acid, ortho	1.2707	1.1291	1.0595	1.0309
Barium chloride	1.1228	1.0572	1.0263	1.0128
Barium nitrate	1.0893	1.0437	1.0214	1.0084
Beryllium sulfate	1.3600	1.1620	1.0749	1.0151
n-Butyric acid	1.2803	1.1317	1.0637	1.0308
iso-Butyric acid	1.2728	1.1287	1.0661	1.0322
Cadmium chloride	1.1342	1.0631	1.0310	1.0202
Cadmium nitrate	1.1648	1.0742	1.0385	1.0177
Cadmium sulfate	1.3476	1.1574	1.0780	1.0335
Calcium chloride	1.1563	1.0764	1.0362	1.0172
Calcium nitrate	1.1172	1.0553	1.0218	1.0076
Cesium chloride	0.9775	......	......	......
Chloric acid	1.0520	1.0255	1.0145	1.0059
Cobalt chloride	1.2041	1.0975	1.0482	1.0232
Cobalt nitrate	1.1657	1.0754	1.0318	1.0180
Cobalt sulfate	1.3543	1.1598	1.0766	1.0402
Copper chloride	1.2050	1.0977	1.0470	1.0268
Copper nitrate	1.1792	1.0802	1.0400	1.0179
Copper sulfate	1.3580	1.1603	1.0802	1.0384
Dichloroacetic acid	1.2649	1.1318	1.0640	1.0287
Dimethylamine	1.3044	1.1440	1.0632	1.0300
Ferric chloride	1.2816	1.1334	1.0602	1.0302

Table 10-91 (*Continued*)
SPECIFIC VISCOSITY OF AQUEOUS SOLUTIONS AT 25°C

Name	1.0 Molar	0.5 Molar	0.25 Molar	0.125 Molar
Formic acid	1.0312	1.0169	1.0092	1.0049
Hydrogen bromide	1.0320	1.0164	1.0095	1.0068
Hydrogen chloride	1.0671	1.0338	1.0166	1.0095
Lactic acid	1.2499	1.1192	1.0585	1.0319
Lead nitrate	1.1010	1.0418	1.0174	1.0066
Lithium chloride	1.1423	1.0665	1.0314	1.0116
Lithium sulfate	1.2905	1.1372	1.0655	1.0320
Magnesium chloride	1.2015	1.0940	1.0445	1.0206
Magnesium nitrate	1.1706	1.0824	1.0396	1.0198
Magnesium sulfate	1.3673	1.1639	1.0784	1.0320
Manganese chloride	1.2089	1.0982	1.0481	1.0230
Manganese nitrate	1.1831	1.0867	1.0426	1.0235
Manganese sulfate	1.3640	1.1690	1.0761	1.0366
Mercuric chloride	1.0460	......	1.0116	1.0042
Methylamine	1.1554	1.0821	1.0340	1.0170
Nickel chloride	1.2055	1.0968	1.0443	1.0210
Nickel nitrate	1.1800	1.0840	1.0422	1.0195
Nickel sulfate	1.3615	1.1615	1.0751	1.0323
Nitric acid	1.0266	1.0115	1.0052	1.0027
Perchloric acid	1.0118	1.0032	0.9998	0.9992
Phosphoric acid, ortho	1.2871	1.1331	1.0656	1.0312
Potassium carbonate	1.1667	1.0784	1.0391	1.0192
Potassium chloride	0.9872	0.9874	0.9903	0.9928
Potassium chromate	1.1133	1.0528	1.0224	1.0116
Potassium dichromate	......	1.0061	1.0034	0.9999
Potassium ferricyanide	1.0610	1.0211	1.0108	1.0182
Potassium ferrocyanide	1.1124	1.0514	1.0228	1.0116
Potassium hydroxide	1.1294	1.0637	1.0313	1.0130
Potassium nitrate	0.9753	0.9822	0.9870	0.9921
Potassium sulfate	1.1051	1.0486	1.0206	1.0078
Propionic acid	1.1968	1.0991	1.0471	1.0264
Rubidium chloride	0.9846	......	......	......
Silver nitrate	1.1150	1.0491	1.0240	1.0114
Sodium acetate	1.3915	1.1806	1.0889	1.0439
Sodium benzoate	1.6498	1.2788	1.1303	1.0633
Sodium bromide	1.0639	1.0299	1.0148	1.0078
Sodium *n*-butyrate	1.6773	1.2933	1.1363	1.0659
Sodium *iso*-butyrate	1.6845	1.2997	1.1428	1.0707
Sodium carbonate	.....	1.2847	1.1367	1.0610
Sodium chlorate	1.0901	1.0421	1.0219	1.0117
Sodium chloride	1.0973	1.0471	1.0239	1.0126
Sodium citrate	1.3856	1.1730	1.0847	1.0470
Sodium formate	1.2069	1.0947	1.0447	1.0231
Sodium hydroxide	1.2355	1.1087	1.0560	1.0302
Sodium lactate	1.4988	1.2232	1.1043	1.0512
Sodium nitrate	1.0655	1.0259	1.0122	1.0069
Sodium oxalate	......	......	1.0573	1.0282
Sodium perchlorate	1.0462	1.0183	1.0096	1.0028
Sodium phthalate	1.4905	1.2246	1.1111	1.0614
Sodium propionate	1.5380	1.2352	1.1122	1.0548
Sodium succinate	1.3899	1.1792	1.0849	1.0448
Sodium sulfate	1.2291	1.1058	1.0522	1.0235
Sodium tartrate	1.3365	1.1502	1.0724	1.0300
Strontium chloride	1.1411	1.0674	1.0338	1.0141
Strontium nitrate	1.1150	1.0491	1.0240	1.0114
Sulfuric acid	1.0898	1.0433	1.0216	1.0082
Thallous nitrate	0.9471	......	0.9865	0.9932
Trimethylamine	1.6838	1.2992	1.1430	1.0677
iso-Valeric acid	......	......	1.0750	1.0375
Zinc chloride	1.1890	1.0959	1.0526	1.0238
Zinc nitrate	1.1642	1.0857	1.0390	1.0186
Zinc sulfate	1.3671	1.1726	1.0824	1.0358

Table 10-92
VISCOSITY OF AQUEOUS SUCROSE SOLUTIONS

The values are expressed in centipoises according to the tables of Bingham and Jackson, *Bur. Standards Bull.* 14, 59 (1918).

$t°$ C	Per cent by weight of sucrose			$t°$ C	Per cent by weight of sucrose		
	20	40	60		20	40	60
0	3.804	14.77	238	55	0.884	2.219	11.67
5	3.154	11.56	156	60	0.808	1.982	9.83
10	2.652	9.794	109.8	65	0.742	1.778	8.34
15	2.267	7.468	74.6	70	0.685	1.608	7.15
20	1.960	6.200	56.5	75	0.635	1.462	6.20
25	1.704	5.187	43.86	80	0.590	1.334	5.40
30	1.504	4.382	33.78	85	0.550	1.221	4.73
35	1.331	3.762	26.52	90	.....	1.123	4.15
40	1.193	3.249	21.28	95	.....	1.037	3.72
45	1.070	2.847	17.18	100	.....	0.960	3.34
50	0.970	2.497	14.01				

Table 10-93
VISCOSITY OF AQUEOUS ETHYL ALCOHOL SOLUTIONS

The values are expressed in centipoises according to the tables of Bingham and Jackson, *Bur. Standards Bull.* 14, 59 (1918).

% by wt. / % by vol.	10 / 12.36	20 / 24.09	30 / 35.23	40 / 45.83	50 / 55.93	60 / 65.56	70 / 74.80	80 / 83.59	90 / 92.01	100 / 100
$t°$ C.										
0	3.311	5.319	6.94	7.14	6.58	5.75	4.762	3.690	2.732	1.773
5	2.577	4.065	5.29	5.59	5.26	4.63	3.906	3.125	2.309	1.623
10	2.179	3.165	4.05	4.39	4.18	3.77	3.268	2.710	2.101	1.466
15	1.792	2.618	3.26	3.53	3.44	3.14	2.770	2.309	1.802	1.332
20	1.538	2.183	2.71	2.91	2.87	2.67	2.370	2.008	1.610	1.200
25	1.323	1.815	2.18	2.35	2.40	2.24	2.037	1.748	1.424	1.096
30	1.160	1.553	1.87	2.02	2.02	1.93	1.767	1.531	1.279	1.003
35	1.006	1.332	1.58	1.72	1.72	1.66	1.529	1.355	1.147	0.914
40	0.907	1.160	1.368	1.482	1.499	1.447	1.344	1.203	1.035	0.834
45	0.812	1.015	1.189	1.289	1.294	1.271	1.189	1.081	0.939	0.764
50	0.734	0.907	1.050	1.132	1.155	1.127	1.062	0.968	0.848	0.702
55	0.663	0.814	0.929	0.998	1.020	0.997	0.943	0.867	0.764	0.644
60	0.609	0.736	0.834	0.893	0.913	0.902	0.856	0.789	0.704	0.592
65	0.554	0.666	0.752	0.802	0.818	0.806	0.766	0.711	0.641	0.551
70	0.514	0.608	0.683	0.727	0.740	0.729	0.695	0.650	0.589	0.504
75	0.476	0.559	0.624	0.663	0.672	0.663	0.636	0.600	0.546	0.471
80	0.430	0.505	0.567	0.601	0.612	0.604				

Table 10-94
VISCOSITY OF WATER
IN CENTIPOISES FROM 0° TO 100° C.

Calculated by the formula: $\eta = 2.1482[(t - 8.435) + \sqrt{8078.4 + (t - 8.435)^2}] - 120$. Bingham and Jackson, *Bur. Standards Bull.* **14**, 75 (1918); Bingham, "Fluidity and Plasticity," p. 340, McGraw-Hill, New York (1922).

$t°$ C	η	$t°$ C	η	$t°$ C	η
0	1.7921	34	0.7371	69	0.4117
1	1.7313	35	0.7225	70	0.4061
2	1.6728	36	0.7085	71	0.4006
3	1.6191	37	0.6947	72	0.3952
4	1.5674	38	0.6814	73	0.3900
5	1.5188	39	0.6685	74	0.3849
6	1.4728	40	0.6560	75	0.3799
7	1.4284	41	0.6439	76	0.3750
8	1.3860	42	0.6321	77	0.3702
9	1.3462	43	0.6207	78	0.3655
10	1.3077	44	0.6097	79	0.3610
11	1.2713	45	0.5988	80	0.3565
12	1.2363	46	0.5883	81	0.3521
13	1.2028	47	0.5782	82	0.3478
14	1.1709	48	0.5683	83	0.3436
15	1.1404	49	0.5588	84	0.3395
16	1.1111	50	0.5494	85	0.3355
17	1.0828	51	0.5404	86	0.3315
18	1.0559	52	0.5315	87	0.3276
19	1.0299	53	0.5229	88	0.3239
20	1.0050	54	0.5146	89	0.3202
20.20	1.0000	55	0.5064	90	0.3165
21	0.9810	56	0.4985	91	0.3130
22	0.9579	57	0.4907	92	0.3095
23	0.9358	58	0.4832	93	0.3060
24	0.9142	59	0.4759	94	0.3027
25	0.8937	60	0.4688	95	0.2994
26	0.8737	61	0.4618	96	0.2962
27	0.8545	62	0.4550	97	0.2930
28	0.8360	63	0.4483	98	0.2899
29	0.8180	64	0.4418	99	0.2868
30	0.8007	65	0.4355	100	0.2838
31	0.7840	66	0.4293		
32	0.7679	67	0.4233		
33	0.7523	68	0.4174		

IN CENTIPOISES FROM −10° TO 0°C. AND FROM 100° TO 150°C.*

$t°$ C	η	$t°$ C	η	$t°$ C	η
−10	2.60	−2	1.91	120	0.232
−8	2.40	0	1.79	130	0.212
−6	2.22	+100	0.284	140	0.196
−4	2.05	+110	0.256	150	0.184

* At the pressure of the saturated vapor at the indicated temperature.

Table 10-95
ABSOLUTE VISCOSITIES OF AQUEOUS GLYCEROL SOLUTIONS

M. L. Sheeley, *Ind. Eng. Chem. 24, 1060 (1932)*.*

Aqueous solutions of glycerol are suitable liquids for use as standards in calibrating technical viscosimeters. By means of an accurate specific gravity determination the percentage of glycerol concentration can be determined within ±0.02 per cent. (See special table of density of aqueous glycerol solutions.)

% Glycerol by Weight	Absolute Viscosity in Centipoises			% Glycerol by Weight	Absolute Viscosity in Centipoises		
	20°C.	25°C.	30°C.		20°C.	25°C.	30°C.
0.00	1.005	0.893	0.800	55.00	7.997	6.582	5.494
1.00	1.029	0.912	0.817	56.00	8.482	6.963	5.816
2.00	1.055	0.935	0.836	57.00	9.018	7.394	6.148
3.00	1.083	0.959	0.856	58.00	9.586	7.830	6.495
4.00	1.112	0.984	0.877	59.00	10.25	8.312	6.870
5.00	1.143	1.010	0.900	60.00	10.96	8.823	7.312
6.00	1.175	1.037	0.924	61.00	11.71	9.428	7.740
7.00	1.207	1.064	0.948	62.00	12.52	10.11	8.260
8.00	1.239	1.092	0.972	63.00	13.43	10.83	8.812
9.00	1.274	1.121	0.997	64.00	14.42	11.57	9.386
10.00	1.311	1.153	1.024	65.00	15.54	12.36	10.02
11.00	1.350	1.186	1.052	66.00	16.73	13.22	10.68
12.00	1.390	1.221	1.082	67.00	17.96	14.18	11.45
13.00	1.431	1.256	1.112	68.00	19.40	15.33	12.33
14.00	1.473	1.292	1.143	69.00	21.07	16.62	13.27
15.00	1.517	1.331	1.174	70.00	22.94	17.96	14.32
16.00	1.565	1.370	1.207	71.00	25.17	19.53	15.56
17.00	1.614	1.411	1.244	72.00	27.56	21.29	16.88
18.00	1.664	1.453	1.281	73.00	30.21	23.28	18.34
19.00	1.715	1.495	1.320	74.00	33.04	25.46	19.93
20.00	1.769	1.542	1.360	75.00	36.46	27.73	21.68
21.00	1.829	1.592	1.403	76.00	40.19	30.56	23.60
22.00	1.892	1.644	1.447	77.00	44.53	33.58	25.90
23.00	1.957	1.699	1.494	78.00	49.57	37.18	28.68
24.00	2.025	1.754	1.541	79.00	55.47	41.16	31.62
25.00	2.095	1.810	1.590	80.00	62.0	45.86	34.92
26.00	2.167	1.870	1.641	81.00	69.3	51.02	38.56
27.00	2.242	1.934	1.695	82.00	77.9	56.90	42.92
28.00	2.324	2.008	1.752	83.00	87.9	64.2	47.90
29.00	2.410	2.082	1.812	84.00	99.6	72.2	53.63
30.00	2.501	2.157	1.876	85.00	112.9	81.5	60.05
31.00	2.597	2.235	1.942	86.00	129.6	92.6	68.1
32.00	2.700	2.318	2.012	87.00	150.4	106.1	77.5
33.00	2.809	2.407	2.088	88.00	174.5	122.6	88.8
34.00	2.921	2.502	2.167	89.00	201.4	141.8	101.1
35.00	3.040	2.600	2.249	90.00	234.6	163.6	115.3
36.00	3.169	2.706	2.335	90.50	255.0	175.6	124.3
37.00	3.300	2.817	2.427	91.00	278.4	189.3	134.4
38.00	3.440	2.932	2.523	91.50	302.8	204.0	145.0
39.00	3.593	3.052	2.624	92.00	328.4	221.8	156.5
40.00	3.750	3.181	2.731	92.50	356.2	241.2	169.3
41.00	3.917	3.319	2.845	93.00	387.7	262.9	182.8
42.00	4.106	3.466	2.966	93.50	421.3	285.7	196.2
43.00	4.307	3.624	3.094	94.00	457.7	308.7	212.0
44.00	4.509	3.787	3.231	94.50	498.5	335.6	229.0
45.00	4.715	3.967	3.380	95.00	545	366.0	248.8
46.00	4.952	4.165	3.540	95.50	601	397.8	271.4
47.00	5.206	4.367	3.706	96.00	661	435.0	296.7
48.00	5.465	4.571	3.873	96.50	731	476.8	324.3
49.00	5.730	4.787	4.051	97.00	805	522.9	354.0
50.00	6.050	5.041	4.247	97.50	885	571	387.4
51.00	6.396	5.319	4.467	98.00	974	629	424.0
52.00	6.764	5.597	4.709	98.50	1080	698	465.3
53.00	7.158	5.910	4.957	99.00	1337	856	564
54.00	7.562	6.230	5.210	100.00	1499	945	624

* See also Segur and Oberstar, *Ind. Eng. Chem.*, 43, 2117 (1951).

COMPRESSIBILITY AND THERMAL EXPANSION

Table 10-96
COMPRESSIBILITY OF WATER

In the table below are given the relative volumes of water at various temperatures and pressures. The volume at 0°C. and one normal atmosphere (760 mm of Hg) is taken as unity.

P, atm	−10°C.	0°C.	10°C.	20°C.	40°C.	60°C.	80°C.
1	1.0017	1.0000	1.0001	1.0016	1.0076	1.0168	1.0287
500	0.9788	0.9767	0.9778	0.9804	0.9867	0.9967	1.0071
1000	0.9581	0.9566	0.9591	0.9619	0.9689	0.9780	0.9884
1500	0.9399	0.9394	0.9424	0.9456	0.9529	0.9617	0.9717
2000	0.9223	0.9241	0.9277	0.9312	0.9386	0.9472	0.9568
2500	0.9083	0.9112	0.9147	0.9183	0.9257	0.9343	0.9437
3000	0.8962	0.8993	0.9028	0.9065	0.9139	0.9225	0.9315
3500	0.8852	0.8884	0.8919	0.8956	0.9030	0.9115	0.9203
4000	0.8751	0.8783	0.8818	0.8855	0.8931	0.9012	0.9097
4500	0.8658	0.8692	0.8725	0.8762	0.8838	0.8919	0.9001
5000	0.8573	0.8606	0.8639	0.8675	0.8752	0.8832	0.8913
6000	......	0.8452	0.8481	0.8517	0.8595	0.8674	0.8752
7000	......	......	0.8340	0.8374	0.8456	0.8534	0.8610
8000	......	......	......	0.8244	0.8330	0.8408	0.8483
9000	......	......	......	0.8128	0.8219	0.8297	0.8371
10000	......	......	......	0.8027	0.8119	0.8196	0.8268
11000	......	......	......	......	0.8023	0.8101	0.8172
12000	......	......	......	......	0.7931	0.8009	0.8080

Table 10-97
COEFFICIENTS OF COMPRESSIBILITY OF VARIOUS LIQUIDS AND AQUEOUS SOLUTIONS

In the table below are given values of β_t (the compressibility coefficient of a liquid at $t°_1$.) to be used with the formula

$$\beta_t = \frac{1}{V_1}\left(\frac{V_1 - V_2}{P_2 - P_1}\right)$$

where V_1 is the volume of a liquid under a pressure of P_1 in atmospheres at $t°C.$, and V_2 is the volume at some other pressure P_2, at the same temperature.

Liquid	t °C.	Pressure Range, atm.	$\beta_t \times 10^6$
Acetic acid	25	at 92.5	81.4
" "	25	at 218.5	72.6
" "	25	at 494	57.1
Acetone	14.2	8.90 to 36.51	111
"	0	100 to 500	82
"	0	500 to 1000	59
"	0	1000 to 1500	47
"	0	1500 to 2000	40
"	25	at 82.5	111.8
Allyl alcohol	9.6	1 to 500	69
" "	9.6	500 to 1000	51
" "	9.6	1000 to 1500	43
" "	9.6	1500 to 2000	36
Ammonia, 2.29%	0	up to 10	50.9
" , 17.58%	0	up to 10	41.6
" , 21.58%	15	up to 10	36.5
Amyl alcohol	17.75	at 8	83.5
" "	13.8	8.5 to 37.12	88.2
Aniline	25	at 85.5	43.2
"	25	at 181.5	40.5

Table 10-97 (*Continued*)
COEFFICIENTS OF COMPRESSIBILITY OF VARIOUS LIQUIDS

Liquid	t °C.	Pressure Range, atm.	$\beta t \times 10^6$
Aniline	25	at 281.5	38.3
"	25	at 390	36.1
Benzene	20	1 to 2	95.3
"	15.4	1 to 4	87
"	12.9	1 to 18.5	86.8
"	10	8.12 to 37.2	90
"	20	98.7 to 296	78.7
"	20	296 to 494	67.5
Bromine	20	0 to 98.7	63.5
"	20	98.7 to 197.4	58.4
"	20	197.4 to 296	54.6
"	20	296 to 395	52.1
"	20	395 to 494	49.9
Bromoform	20	0 to 98.7	51.0
"	20	98.7 to 197.4	47.5
"	20	197.4 to 296	44.0
"	20	296 to 395	42.0
"	20	395 to 494	41.0
n-Butyl alcohol	17.4	at 8	90
iso-Butyl alcohol	17.95	at 8	98
Butyl benzoate	10	1 to 5.25	59
Butyl butyrate	10	1 to 5.25	90
Butyl valerate	10	1 to 5.25	92
Calcium chloride, 5.8%	20	2 to 20	39.7
Calcium chloride, 9.9%	20	2 to 20	37.1
" , 17.8%	20	2 to 20	31.3
" , 40.9%	20	2 to 20	21.7
Caproic acid	30	20 to 400	68
" "	65	20 to 100	90
" "	100	20 to 200	109
Carbon dioxide	13	at 60	1740
" "	13	at 70	960
" "	13	at 80	660
" "	13	at 90	440
Carbon disulfide	20	1 to 2	80.95
" "	0	1 to 500	66
" "	0	500 to 1000	53
" "	0	1000 to 1500	43
" "	0	1500 to 2000	37
" "	0	2000 to 2500	33
Carbon tetrachloride	10	1 to 5.25	70
" "	20	0 to 98.7	91.6
" "	20	98.7 to 197.4	89.9
" "	20	197.4 to 296	83.5
" "	20	296 to 395	75.5
" "	20	395 to 494	69.9
" "	20	at 542.5	62.5
" "	20	at 664	55.0
Castor oil	14.8	1 to 10	47.2
Chlorine	20	9.9 to 98.7	118
"	20	98.7 to 197.4	110
"	20	197.4 to 296	102
"	20	296 to 395	90.7
"	20	395 to 494	84.5
Chlorobenzene	13.3	1 to 18.5	67.1
"	80	1 to 2	108.4

Table 10-97 (*Continued*)
COEFFICIENTS OF COMPRESSIBILITY OF VARIOUS LIQUIDS

Liquid	t °C.	Pressure Range, atm.	βt ×10⁶
Chloroform	20	0 to 98.7	94.9
"	20	98.7 to 197.4	89.8
"	20	197.4 to 296	80.1
"	20	296 to 395	72.9
"	20	395 to 494	67.8
"	0	1 to 2	87.27
"	60	1 to 2	139.13
Cyclohexanol	40	98.7 to 296	56.4
Decane	23	0 to 1	105
Diphenylamine	65	0 to 500	57
"	100	0 to 300	64
"	185	0 to 100	110
Ethyl acetate	13.3	8.12 to 37.45	104
" "	99.6	8.13 to 37.15	250
Ethyl alcohol	20	1 to 50	112
" "	20	50 to 100	102
" "	20	100 to 200	95
" "	20	200 to 300	86
" "	20	300 to 400	80
" "	20	400 to 500	73
" "	20	500 to 600	69
" "	0	1 to 50	96
" "	0	600 to 700	60
" "	0	900 to 1000	52
" "	100	900 to 1000	73
Ethyl bromide	10.1	1 to 500	90
" "	10.1	500 to 1000	63
" "	10.1	1000 to 1500	50
" "	10.1	2000 to 2500	36
" "	13.7	1 to 18.5	113.4
Ethyl chloride	0	1 to 500	103
" "	0	500 to 1000	69
" "	0	1000 to 1500	55
" "	0	2000 to 2500	39
" "	15.2	8.7 to 37.22	153
" "	99	12.77 to 34.47	495
Ethyl ether	8.1	1 to 8	163.8
" "	13.5	8.43 to 25.4	169
" "	185	100 to 200	741
" "	185	100 to 400	478
" "	35	1 to 2000	42.5
" "	0	at 0	159
" "	0	at 100	125.7
" "	0	at 200	112.2
" "	0	at 500	84.5
" "	0	at 1000	63.5
" "	−109.8	at 1000	34.5
Ethyl iodide	10.6	1 to 500	74
" "	10.6	500 to 1000	56
" "	10.6	1000 to 1500	46
" "	10.6	1500 to 2000	38
" "	10.6	2000 to 2500	34
" "	10.6	2500 to 3000	31
Ethylene bromide	10	1 to 5.25	55.8
" "	64	1 to 5.25	76.6
Ethylene chloride	10	1 to 5.25	67.7

Table 10-97 (*Continued*)

COEFFICIENTS OF COMPRESSIBILITY OF VARIOUS LIQUIDS

Liquid	t °C.	Pressure Range, atm.	$\beta t \times 10^6$
Ethylene chloride	75	1 to 5.25	111.1
Fluorobenzene	13.9	1 to 18.5	87.7
Glycerol	14.8	1 to 10	22.1
Heptane	23	0 to 1	134
Hexane	23	0 to 1	159
Mercury	22.8	1 to 500	3.0
"	22.8	500 to 1000	3.8
"	22.8	1000 to 1500	3.7
"	22.8	1500 to 2000	3.6
"	22.8	2000 to 2500	3.5
"	22.8	2500 to 3000	3.4
"	191.8	1 to 500	4.6
"	191.8	2500 to 3000	4.3
"	0	1 to 50	3.92
"	0	5810 to 6780	3.30
Mesitylene	20	98.7 to 296	68.4
"	20	296 to 494	59.1
Methyl acetate	14.3	8.10 to 37.53	97
Methyl alcohol	0	1 to 500	79
" "	0	500 to 1000	58
" "	0	1000 to 1500	47
" "	0	1500 to 2000	40
" "	0	2500 to 3000	29
" "	14.7	8.50 to 37.12	104
" "	18.1	at 8	120
Methyl butyrate	10	1 to 5.25	89
Methyl valerate	10	1 to 5.25	91
Nitrobenzene	25	at 86.5	46.1
"	25	at 192	43.0
"	25	at 303	40.1
"	25	at 419	38.1
Octane	23	0 to 1	121
Olive oil	20.5	1 to 10	63.3
" "	14.8	1 to 10	56.3
Palmitic acid	65	20 to 100	90
" "	185	20 to 300	134
" "	310	20 to 400	220
Paraffin, M. P. =55°	64	20 to 100	83
"	100	20 to 400	94
"	185	20 to 400	137
"	310	20 to 400	236
Paraffin oil	34	at 1	87
" "	34	at 2000	34
" "	34	at 4500	17
Pentane	0	1 to 29	174
"	20	1 to 29	242
"	0	at 0	175.9
"	0	at 484	94.6
"	0	at 967	68.9
Petroleum	1	1 to 15	67.91
"	16.1	1 to 15	76.77
"	35.1	1 to 15	82.83
"	52.2	1 to 15	92.21
"	72.1	1 to 15	100.16
"	94.0	1 to 15	108.8
Phosphorus trichloride	10.1	1 to 500	72

Table 10-97 (*Continued*)
COEFFICIENTS OF COMPRESSIBILITY OF VARIOUS LIQUIDS

Liquid	t °C.	Pressure Range, atm.	$\beta t \times 10^6$
Phosphorus trichloride	10.1	500 to 1000	54
" "	10.1	1000 to 1500	45
" "	10.1	1500 to 2000	38
" "	10.1	2000 to 2500	33
Potassium chloride, 2.49%	20	2 to 20	42.6
" " , 4.40%	20	2 to 20	41.2
" " , 8.28%	20	2 to 20	38.9
" " , 16.75%	20	2 to 20	34.1
" " , 24.31%	20	2 to 20	30.1
n-Propyl alcohol	0	1 to 500	69
" "	0	500 to 1000	52
" "	0	1000 to 1500	42
" "	0	2500 to 3000	27
" "	5.6	at 8	89.5
iso-Propyl alcohol	5.65	at 8	95
" "	17.85	at 8	103
n-Propyl benzene	20	98.7 to 296	70.7
" "	20	296 to 494	61.0
iso-Propyl benzene	20	98.7 to 296	71.3
"	20	296 to 494	61.2
Pseudocumene	20	98.7 to 296	65.2
"	20	296 to 494	56.9
Sodium chloride, 1.32%	15	at 10	45.3
" " , 13.53%	15	at 10	33.9
" " , 22.22%	15	at 10	28.2
" " , 26.21%	15	at 10	25.5
Sugar, cane, 0%	12.4	1 to 146	46.5
" , " , 10%	12.4	1 to 146	42.65
" , " , 20%	12.4	1 to 146	39.65
Sulfurous acid	0	1 to 16	302.5
Thymol	64	20 to 100	69
"	64	20 to 400	66
"	100	20 to 400	80
"	310	20 to 400	268
Toluene	10	1 to 5.25	79
"	66	1 to 5.25	114
"	20	1 to 2	91.47
"	25	at 114.5	80.6
"	25	at 230.5	70.3
"	25	at 355	62.1
p-Toluidine	45	1 to 150	51.2
"	28	20 to 100	56
"	100	20 to 400	77
"	185	20 to 400	112
"	310	20 to 400	243
Water, see special table			
Xylene	10	1 to 5.25	74
"	65	1 to 5.25	106
"	100	1 to 5.25	132

Table 10-98
COEFFICIENTS OF CUBICAL EXPANSION FOR VARIOUS LIQUIDS AND AQUEOUS SOLUTIONS

See also the table *Coefficients of Cubical Expansion For Various Solids.*

If V_0 is the volume at 0°C, then the volume at $t°C = V_t = V_0(1 + at + bt^2 + ct^3)$. Where the compound is marked with * the basic volume is not that at 0°C, but at some other temperature, t_0. The equation then reads $V_t = Vt_0(1 + a(t - t_0) + b(t - t_0)^2 + c(t - t_0)^3]$. In the case of aqueous solutions, the concentrations in percent are given after the name.

Liquid	Coef. at 20°C. $\times 10^3$	Range °C.	$a \times 10^3$	$b \times 10^6$	$c \times 10^8$
Acetic acid	1.071	16 to 107	1.063	−0.12636	1.0876
Acetone	1.487	0 to 54	1.324	3.809	−0.87983
Allyl alcohol	1.049	0 to 94	0.97019	1.8725	0.36452
Allyl bromide	1.241	0 to 69	1.2275	−0.44365	2.5843
Allyl chloride	1.475	9 to 44	1.3218	5.078	−4.1915
Allyl ether	1.346	0 to 88	1.2519	2.2401	0.35775
Allyl iodide	1.091	0 to 101	1.0539	0.63572	1.0036
Amyl acetate	1.162	0 to 124	1.1501	−0.09046	1.3015
Amyl alcohol	0.902	−15 to 80	0.89001	0.65729	1.18458
Amyl benzoate	0.848	0 to 198	0.81711	0.7377	0.10593
Amyl bromide	1.102	0 to 80	1.02321	1.90086	0.19756
Amyl chloride	1.208	0 to 100	1.17155	0.50077	1.35368
Amyl iodide	0.986	20 to 142	0.92658	1.4647	0.05962
Aniline	0.858	0 to 141	0.82349	0.8408	0.10741
Arsenous chloride	1.020	−15 to 130	0.97907	0.96695	0.17772
Benzene	1.237	11 to 81	1.17626	1.27755	0.80648
Benzoyl chloride	0.880	12 to 146	0.85893	0.44219	0.27139
Bromine	1.113	−7 to 60	1.03819	1.711138	0.54471
n-Butyl alcohol	0.950	6 to 108	0.83751	2.8634	−0.12415
n-Butyric acid	1.063	0 to 100	1.02573	0.83760	0.34694
iso-Butyric acid	1.068	16 to 118	0.97625	2.3976	−0.32145
Calcium chloride, 40.9%	0.458	17 to 24	0.42383	0.8571	....
Cane sugar, 43.2%	0.343	0 to 35	0.2536	2.247	....
n-Caproic acid	0.975	15 to 155	0.94413	0.68358	0.26586
Carbon disulfide	1.218	−34 to 68	1.1398	1.37065	1.91225
Carbon tetrachloride	1.236	0 to 76	1.18384	0.89881	1.35135
Chloral	0.934	13 to 51	0.9545	−2.2139	5.6392
Chloroform	1.273	0 to 63	1.10715	4.66473	−1.74328
o-Cresol	......	66 to 186	0.71072	1.1464	0.2242
m-Cresol	......	65 to 194	0.77526	0.27102	0.3868
p-Cresol	......	66 to 186	0.86476	0.53912	0.64418
Cymene	0.946	0 to 100	0.895	1.277	....
Diallyl	1.375	0 to 60	1.3423	−0.34339	3.8693
Diethyl ketone	1.233	0 to 95	1.15342	1.88396	0.32021
Dipropyl	1.381	0 to 66	1.2948	1.7471	1.2363
Ethyl acetate	1.389	−36 to 72	1.2585	2.95688	0.14922
Ethyl alcohol	......	0–80	1.04139	0.7836	1.7618
Ethyl alchol 50%	......	0–39	0.7450	1.85	0.730
Ethyl benzoate	0.900	0 to 159	0.86606	0.8229	0.12084
Ethyl benzene	0.961	24 to 131	0.86172	2.5344	−0.18319
Ethyl bromide	1.418	−32 to 54	1.33763	1.50135	1.6900
Ethyl chloride	1.706	−32 to 26	1.57458	2.81366	1.56987
Ethyl ether	1.656	−15 to 38	1.51324	2.35918	4.00512
Ethyl formate	1.417	0 to 63	1.36446	0.13538	3.9248
Ethyl iodide	1.179	10 to 65	1.1520	0.26032	1.4181
Ethyl nitrate	1.299	9 to 72	1.1290	4.7915	−1.8413
Ethyl oxalate	1.136	0 to 141	1.06031	1.0983	2.6657

Table 10-98 (*Continued*)
COEFFICIENTS OF CUBICAL EXPANSION FOR VARIOUS LIQUIDS AND AQUEOUS SOLUTIONS

Liquid	Coef. at 20°C. $\times 10^3$	Range °C.	$a \times 10^3$	$b \times 10^6$	$c \times 10^8$
Ethyl sulfide	1.278	0 to 90	1.19643	1.80653	0.78821
Ethylene chloride	1.161	−28 to 84	1.11893	1.0469	0.10342
Ethylene glycol	0.6375	11 to 136	0.5657	1.7074	0.293
Formic acid	1.025	5 to 104	0.99269	0.62514	0.5965
Glycerol	0.505	....	0.4853	0.4895	....
iso-Hexane	1.445	0 to 55	1.37022	0.97649	2.9819
Hydrochloric acid, 33.2%	0.455	0 to 33	0.4460	0.215	....
4.2%	0.239	0 to 33	0.0652	4.355	....
1.0%	0.211	0 to 32	0.0153	4.899	....
*3.4%, to =110°	......	110 to 140	0.620	4.5	....
Isoprene	1.567	0 to 33	1.4603	0.99793	5.60149
Mercury	......	0 to 100	0.18169041	0.002951266	0.0114562
	......	24 to 299	0.181163	0.01155	0.0021187
Methyl acetate	1.427	0 to 58	1.34982	0.87098	3.5562
Methyl alcohol	1.259	−38 to 70	1.18557	1.56493	0.91113
Methyl benzoate	0.895	0 to 162	0.8633	0.7414	0.15896
Methyl bromide	1.684	−35 to 28	1.41521	3.31528	11.3809
Methyl cyanide	1.301	6 to 66	1.2118	1.7780	1.5322
Methyl ethyl ketone	1.315	0 to 76	1.18654	3.37043	−0.53365
Methyl formate	1.563	0 to 10	1.35824	10.538	−1.8085
Methyl iodide	1.273	5 to 39	1.1440	4.0465	−2.7393
Methyl propionate	1.304	0 to 74	1.3049	−1.3275	4.6943
Methyl sulfide	1.082	0 to 111	1.01705	1.57606	0.19072
Nitrobenzene	......	144 to 164	0.8263	0.52249	0.13779
Olive oil	0.721	....	0.68215	1.14053	−0.539
n-Pentane	1.656	−190 to 30	1.50697	3.435	0.975
iso-Pentane	1.680	0 to 27	1.46834	5.09626	0.6979
Petroleum, sp. gr. 0.8467	0.955	24 to 120	0.8994	1.396	....
Petroleum ether	2.26	−190 to 0	1.46	1.60	....
Phenol	1.090	36 to 157	0.834	0.10732	0.4446
Phosphorus tribromide	0.868	0 to 100	0.8472	0.43672	0.25276
Phosphorus trichloride	1.154	−36 to 75	1.12862	0.87288	1.79236
Phosphorus oxychloride	1.116	0 to 107	1.06431	1.12666	0.5299
Potassium chloride, 24.3%	0.353	16 to 25	0.2695	2.080	....
Propionic acid	1.102	0 to 133	1.0396	1.5487	0.04301
n-Propyl alcohol	0.956	0 to 94	0.7743	4.9689	−1.4069
iso-Propyl alcohol	1.094	0 to 83	1.04345	0.44303	2.7274
n-Propyl chloride	1.447	0 to 42	1.3306	3.8313	−1.3859
iso-Propyl chloride	1.591	0 to 34	1.3696	5.5287	....
n-Propyl ether	1.354	0 to 88	1.2132	3.9318	−1.3644
iso-Propyl ether	1.452	0 to 67	1.2872	4.2923	−0.58573
Propyl iodide	1.102	10 to 98	1.0276	1.8658	−0.0051
Silicon tetrachloride	1.430	−32 to 59	1.29412	2.18414	4.08642
Sodium acid sulfate, 21%	0.555	0 to 34	0.5364	4.75	....
Sodium chloride, 20.6%	0.414	0 to 29	0.3640	1.237	....
Sodium sulfate,1.9%	0.235	0 to 40	0.0449	4.749	....
24%	0.410	11 to 40	0.3599	1.258	....
Stannic chloride	1.178	−19 to 113	1.1328	0.91171	0.75798
Sulfur chloride	0.968	12 to 111	0.9591	−0.03819	0.73186
Sulfuric acid, conc.	......	0–30	0.5758	−0.864	....
Sulfuric acid 10.9%	0.387	0 to 30	0.2835	2.580	....
5.4%	0.311	0 to 30	0.1450	4.143	....
1.4%	0.234	0 to 30	0.03335	5.025	....
*2.3%, to =100°	......	110 to 140	0.729	2.8	....

Table 10-98 (*Continued*)
COEFFICIENTS OF CUBICAL EXPANSION FOR VARIOUS LIQUIDS AND AQUEOUS SOLUTIONS

Liquid	Coef. at 20°C. $\times 10^3$	Range °C.	$a \times 10^3$	$b \times 10^6$	$c \times 10^8$
*4.5%, to =110°	......	110 to 140	0.648	4.2	....
Thymol	......	62 to 157	0.84369	0.26625	0.35997
Titanium tetrachloride	0.998	−22 to 134	0.94257	1.34579	0.0888
Toluene	1.099	0 to 100	1.028	1.779	....
o-Toluidine	0.847	0 to 141	0.82136	0.6046	0.14696
Trimethyl carbinol	1.023	20 to 77	1.31261	−8.8155	3.61209
n-Valeric acid	1.004	8 to 144	0.97557	0.61852	0.30378
Water (see also below)		−13 to 0	−0.09417	1.449	−59.85
o-Xylene	0.973	16 to 131	0.91734	1.3245	0.19586
m-Xylene	1.009	0 to 141	0.96396	1.0251	0.32753
p-Xylene	1.011	19 to 131	0.97013	0.8714	0.5287

WATER

To find the cubical expansion of water, substitute in the following formulas:

$0°$ to $33°$: $V_t = V_o (1 - 0.0_4 64268t + 0.0_5 850526t^2 - 0.0_7 678977t^3 + 0.0_9 401209t^4$

$0°$ to $80°$: $V_t = V_o (1 - 0.0_4 53255t + 0.0_5 761532t^2 - 0.0_7 437217t^3 + 0.0_9 164322t^4$

Table 10-99
CUBICAL EXPANSION OF SOLIDS

The coefficient of cubical expansion is the increase in volume per unit volume per degree C. rise in temperature. For ordinary work the assumption may be made that the coefficient of cubical expansion is about three times the coefficient of linear expansion.

Substance	Temp. °C.	Coef. $\times 10^7$	Substance	Temp. °C.	Coef. $\times 10^7$
Antimony	0 to 100	316.7	Paraffin	20	5880
Beryl	0 to 100	10.5	Platinum	0 to 100	265
Bismuth	0 to 100	394.8	Porcelain	0 to 100	108.0
Copper	0 to 100	499.8	'' , Berlin	20	81.4
Diamond	40	35.4	Potassium chloride	0 to 100	1094
Emerald	40	16.8	'' nitrate	0 to 100	1967
Galena	0 to 100	558	'' sulfate	20	1075.4
Glass, Corning 790	0 to 350	24	Quartz	0 to 100	384
'' , Corning 774	0 to 350	96	Rock salt	50 to 60	1212.0
'' , Corning 8800	0 to 350	183	Rubber	20	4870
'' , Jena 59 III	20 to 100	156	Silver	0 to 100	583.1
'' , soda lime tubing	0 to 300	276	Sodium	20	2136.4
Gold	0 to 100	441.1	Stearic acid	33.8 to 45.5	8100
Ice	−20 to −1	1125.0	Sulfur, native	13.2 to 50.3	2230
Iceland spar	50 to 60	144.7	Tin	0 to 100	688.9
Iron	0 to 100	355.0	Zinc	0 to 100	892.8
Lead	0 to 100	839.9			

Table 10-100
COEFFICIENT OF LINEAR EXPANSION

In the table below are given the values of C in the formula:—
$l_t = l_0 (1+Ct)$, where l_0 is the length of the material at 0°C., l_t is its length at t°C., and C is the coefficient of linear expansion. See also tables on properties of elements and properties of alloys.

Substance	Temp. °C.	$C \times 10^6$
Alundum	25–900	8.7
Bakelite	20–60	22
Bauxite brick	25–100	4.4
Caoutchouc	16.7–25.3	77.0
Condensite, No. 100	16–79	44.0
" , No. 128	18–56	20.0
Diamond	40	1.18
Ebonite	25.3–35.4	84.2
Emerald, ‖ to axis	0–85	−1.35
" , ⊥ to axis	0–85	1.00
Fire clay brick	25–100	8.1
Fluorspar	0–100	19.50
Formica	20–60	30.0
Gas carbon	40	5.40
Glass, American plate	0–100 *ca.*	9.0
" , American window	0–100 *ca.*	9.0
" , (standard flint) soda lime tubing	0–300	9.2
" , Kimble N-51A	0–300	5.0
" , Jena 16 III	0–100	8.1
" , Jena 59 III	0–100	5.8
" , Jena 59 III	−191 to +16	4.24
" , Corning 790	0–350	0.8
" , Corning 774	0–350	3.2
" , Corning 8800	0–350	6.1
" , Corning 8810	0–350	8.8
" , B & L BSC-2*	25–525	8.3
" , B & L LBC-2*	15–570	9.9
" , B & L DBC-1*	25–600	8.2
" , B & L C-1*	40–501	10.1
" , B & L BF-1*	25–505	9.8
" , B & L DF-2*	25–450	8.0
" , B & L EDF-1*	25–420	9.0
Graphite	40	7.86
Gutta percha	20	198.3
Ice	−20 to −1	51
Iceland spar, ‖to axis	0–80	26.31
" " , ⊥ to axis	0–80	5.44
Limestone	25–100	9
Magnesium oxide	25–100	9.7–11.4
Marble	15–100	11.7
Paraffin	0–16	106.62
"	16–38	130.30
"	38–49	477.07
Porcelain	20–790	4.13
" , Bayeux	0	2.5
" , Bayeux	1000–1400	5.53

Substance	Temp. °C.	$C \times 10^6$
Porcelain, Berlin	0–100	3.1
Quartz, ‖ to axis	−191 to +16	5.21
" , ‖ " "	0–80	7.97
" , ⊥ " "	0–80	13.37
Quartz glass	−190 to +16	−0.26
" "	16–500	0.57
" "	16–1000	0.58
Rock salt	40	40.4
Rubber, hard	0	69.1
" , "	−160	30.0
Sandstone	20	7–12
Slate	20	6–10
Topaz, ‖ to lesser horizontal axis	0–100	8.32
Topaz, ‖ to greater horizontal axis	0–100	8.36
Topaz, ‖ to vertical axis	0–100	4.72
Tourmaline, ‖ to longitudinal axis	0–100	9.37
Tourmaline, ‖ to horizontal axis	0–100	7.73
Vulcanite	0–18	63.60
Wedgwood ware	0–100	8.90
Wood, ‖ to fiber		
" , " Ash	0–100	9.51
" , " Beech	2–34	2.57
" , " Chestnut	2–34	6.49
" , " Elm	2–34	5.65
" , " Mahogany	2–34	3.61
" , " Maple	2–34	6.38
" , " Oak	2–34	4.92
" , " Pine	2–34	5.41
" , " Walnut	2–34	6.58
Wood, across fiber		
" , " Beech	2–34	61.4
" , " Chestnut	2–34	32.5
" , " Elm	2–34	44.3
" , " Mahogany	2–34	40.4
" , " Maple	2–34	48.4
" , " Oak	2–34	54.4
" , " Pine	2–34	34.1
" , " Walnut	2–34	48.4
Wax, white	10–26	230.0
" , "	26–31	312.0
" , "	31–43	486.0
" , "	43–57	1522.7

*Optical glass.

Section 11

MISCELLANEOUS

SIEVES AND SCREENS

The following terms are often employed in reporting the results of a screen analysis.

Cumulative Per Cent is the term employed to express the percentage of the total amount of a material which would remain on a testing sieve if only one sieve were used for testing the entire sample; to obtain the **cumulative weight,** it is necessary to add the weight of all of the material which remains on sieves coarser than the one in question to the amount remaining on the sieve in question. When such data are plotted as a curve, each point on the curve represents the *total* material that would be retained if only the one sieve represented by that particular point were used in the sieve analysis.

Effective Size is a term customarily employed in specifications for filtration sands; effective size is measured by that opening in millimeters which will just pass 10% of a representative sample of sand.

Uniformity Co-efficient is the numerical value obtained by dividing the sieve opening (in millimeters) which will pass 60% of the sample by the sieve opening (in millimeters) which will just pass 10% of the sample.

Table 11-1
U.S. SIEVE SERIES
ASTM E-11-61 Adopted 1961

Sieve Designation		Sieve Opening		Wire Diameter	
Standard	Alternate	mm	ca. inch	mm	ca. inch
26.9 mm	1.06 in.	26.9	1.06	3.90	0.1535
*22.6 mm	7/8 in.	22.6	0.875	3.50	0.1378
*16.0 mm	5/8 in.	16.0	0.625	3.00	0.1181
*11.2 mm	7/16 in.	11.2	0.438	2.45	0.0965
*8.00 mm	5/16 in.	8.00	0.312	2.07	0.0815
6.73 mm	0.265 in.	6.73	0.265	1.87	0.0736
*5.66 mm	No. 3½	5.66	0.223	1.68	0.0661
4.76 mm	No. 4	4.76	0.187	1.54	0.0606
*4.00 mm	No. 5	4.00	0.157	1.37	0.0539
3.36 mm	No. 6	3.36	0.132	1.23	0.0484
*2.83 mm	No. 7	2.83	0.111	1.10	0.0430
2.38 mm	No. 8	2.38	0.0937	1.00	0.0394
*2.00 mm	No. 10	2.00	0.0787	0.900	0.0354
1.68 mm	No. 12	1.68	0.0661	0.810	0.0319
*1.41 mm	No. 14	1.41	0.0555	0.725	0.0285
1.19 mm	No. 16	1.19	0.0469	0.650	0.0256
*1.00 mm	No. 18	1.00	0.0394	0.580	0.0228
841 μ	No. 20	0.841	0.0331	0.510	0.0201
*707 μ	No. 25	0.707	0.0278	0.450	0.0177
595 μ	No. 30	0.595	0.0231	0.390	0.0154
*500 μ	No. 35	0.500	0.0197	0.340	0.0134
420 μ	No. 40	0.420	0.0165	0.290	0.0114
*354 μ	No. 45	0.354	0.0139	0.247	0.0097
297 μ	No. 50	0.297	0.0117	0.215	0.0085
*250 μ	No. 60	0.250	0.0098	0.180	0.0071
210 μ	No. 70	0.210	0.0083	0.152	0.0060
*177 μ	No. 80	0.177	0.0070	0.131	0.0052
149 μ	No. 100	0.149	0.0059	0.110	0.0043
*125 μ	No. 120	0.125	0.0049	0.091	0.0036
105 μ	No. 140	0.105	0.0041	0.076	0.0030
*88 μ	No. 170	0.088	0.0035	0.064	0.0025
74 μ	No. 200	0.074	0.0029	0.053	0.0021
*63 μ	No. 230	0.063	0.0025	0.044	0.0017
53 μ	No. 270	0.053	0.0021	0.037	0.0015
44 μ	No. 325	0.044	0.0017	0.030	0.0012

* These sieves correspond to those proposed as an International (ISO) Standard. It is recommended that wherever possible these sieves be included in all sieve analysis data or reports intended for international publication.

STANDARD CALIBRATION TABLES FOR THERMOCOUPLES

The following tables represent arbitrary reference curves which show fairly well the relation between emf and temperature and consequently should be used with an appropriate correction curve which must be determined for any individual thermocouple by means of a calibration at three or more fixed temperature points. Such a correction curve is constructed by plotting ΔE the difference between the observed emf at a known temperature and the emf for this same temperature (as read from the table) against the standard emf given in the table. The corrected emf for any temperature is then obtained by subtracting algebraically ΔE as read from the correction curve from the observed emf, before converting the latter into a temperature as read from the table. The tables which follow are for a fixed (or cold) junction maintained at 0°C. If this fixed junction is not at 0°C. but at some other temperature which is very near to 0°C. a correction should be applied as follows: $E_h = E_o + E$ where E_o is the observed emf, E is the emf corresponding to the temperature of the fixed junction and E_h is the desired emf corrected for the temperature at which the fixed junction is being maintained. In the tables which follow the emf values are expressed in absolute millivolts ($=0.99967$ international millivolts, U.S.) and the temperatures in degrees Celsius (centigrade) on the International Temperature Scale of 1948. References: National Bureau Standards *Circular 561 (1955)*; *Jour. Res. Natl. Bur. Stds.*, **50**, 229 *(1953)*.

Table 11-2
PLATINUM *vs* PLATINUM (90%)-RHODIUM (10%)

°C	0°	10°	20°	30°	40°	50°	60°	70°	80°	90°
0°	0.000	0.056	0.113	0.173	0.235	0.299	0.364	0.431	0.500	0.571
100°	0.643	0.717	0.792	0.869	0.946	1.025	1.106	1.187	1.269	1.352
200°	1.436	1.521	1.607	1.693	1.780	1.868	1.956	2.045	2.135	2.225
300°	2.316	2.408	2.499	2.592	2.685	2.778	2.872	2.966	3.061	3.156
400°	3.251	3.347	3.442	3.539	3.635	3.732	3.829	3.926	4.024	4.122
500°	4.221	4.319	4.419	4.518	4.618	4.718	4.818	4.919	5.020	5.122
600°	5.224	5.326	5.429	5.532	5.635	5.738	5.842	5.946	6.050	6.155
700°	6.260	6.365	6.471	6.577	6.683	6.790	6.897	7.005	7.112	7.220
800°	7.329	7.438	7.547	7.656	7.766	7.876	7.987	8.098	8.209	8.320
900°	8.432	8.545	8.657	8.770	8.883	8.997	9.111	9.225	9.340	9.455
1000°	9.570	9.686	9.802	9.918	10.035	10.152	10.269	10.387	10.505	10.623
1100°	10.741	10.860	10.979	11.098	11.217	11.336	11.456	11.575	11.695	11.815
1200°	11.935	12.055	12.175	12.296	12.416	12.536	12.657	12.777	12.897	13.018
1300°	13.138	13.258	13.378	13.498	13.618	13.738	13.858	13.978	14.098	14.217
1400°	14.337	14.457	14.576	14.696	14.815	14.935	15.054	15.173	15.292	15.411
1500°	15.530	15.649	15.768	15.887	16.006	16.124	16.243	16.361	16.479	16.597
1600°	16.716	16.834	16.952	17.069	17.187	17.305	17.422	17.539	17.657	17.774
1700°	17.891	18.008	18.124	18.241	18.358	18.474	18.590	......	......	......

Table 11-3
PLATINUM *vs* PLATINUM (87%)-RHODIUM (13%)

°C	0°	10°	20°	30°	40°	50°	60°	70°	80°	90°
0°	0.000	0.055	0.112	0.172	0.234	0.298	0.363	0.431	0.500	0.572
100°	0.645	0.721	0.798	0.877	0.957	1.039	1.121	1.205	1.290	1.377
200°	1.465	1.553	1.643	1.734	1.826	1.918	2.012	2.107	2.202	2.298
300°	2.395	2.493	2.591	2.690	2.790	2.890	2.991	3.092	3.194	3.296
400°	3.399	3.502	3.607	3.712	3.817	3.923	4.029	4.134	4.241	4.348
500°	4.455	4.563	4.672	4.782	4.893	5.004	5.115	5.226	5.338	5.450
600°	5.563	5.677	5.792	5.907	6.022	6.137	6.252	6.368	6.485	6.602
700°	6.720	6.838	6.957	7.076	7.195	7.315	7.436	7.557	7.679	7.801
800°	7.924	8.047	8.170	8.294	8.419	8.544	8.669	8.795	8.921	9.047
900°	9.175	9.303	9.431	9.559	9.687	9.816	9.946	10.077	10.208	10.339
1000°	10.471	10.603	10.735	10.869	11.003	11.138	11.273	11.408	11.544	11.681

Table 11-3 (*Continued*)
PLATINUM *vs* PLATINUM (87%)-RHODIUM (13%)

°C	0°	10°	20°	30°	40°	50°	60°	70°	80°	90°
1100°	11.817	11.954	12.090	12.227	12.365	12.503	12.641	12.779	12.917	13.055
1200°	13.193	13.332	13.471	13.610	13.749	13.888	14.027	14.165	14.304	14.443
1300°	14.582	14.721	14.860	14.999	15.138	15.276	15.415	15.553	15.692	15.831
1400°	15.969	16.108	16.247	16.386	16.524	16.663	16.802	16.940	17.079	17.217
1500°	17.355	17.493	17.631	17.768	17.906	18.043	18.179	18.316	18.453	18.590
1600°	18.727	18.864	19.001	19.137	19.273	19.409	19.545	19.682	19.818	19.954

Table 11-4
CHROMEL-ALUMEL

°C	0°	10°	20°	30°	40°	50°	60°	70°	80°	90°
−190°	−5.60	−5.43	−5.24	−5.03	−4.81	−4.58	−4.32	−4.06	−3.78	−3.49
−90°	−3.19	−2.87	−2.54	−2.20	−1.86	−1.50	−1.14	−0.77	−0.39	0.00
0°	0.00	0.40	0.80	1.20	1.61	2.02	2.43	2.85	3.26	3.68
100°	4.10	4.51	4.92	5.33	5.73	6.13	6.53	6.93	7.33	7.73
200°	8.13	8.54	8.94	9.34	9.75	10.16	10.57	10.98	11.39	11.80
300°	12.21	12.63	13.04	13.46	13.88	14.29	14.71	15.13	15.55	15.98
400°	16.40	16.82	17.24	17.67	18.09	18.51	18.94	19.36	19.79	20.22
500°	20.65	21.07	21.50	21.92	22.35	22.78	23.20	23.63	24.06	24.49
600°	24.91	25.34	25.76	26.19	26.61	27.03	27.45	27.87	28.29	28.72
700°	29.14	29.56	29.97	30.39	30.81	31.23	31.65	32.06	32.48	32.89
800°	33.30	33.71	34.12	34.53	34.93	35.34	35.75	36.15	36.55	36.96
900°	37.36	37.76	38.16	38.56	38.95	39.35	39.75	40.14	40.53	40.92
1000°	41.31	41.70	42.09	42.48	42.87	43.25	43.63	44.02	44.40	44.78
1100°	45.16	45.54	45.92	46.29	46.67	47.04	47.41	47.78	48.15	48.52
1200°	48.89	49.25	49.62	49.98	50.34	50.69	51.05	51.41	51.76	52.11
1300°	52.46	52.81	53.16	53.51	53.85	54.20	54.54	54.88	.....	.....

Table 11-5
CHROMEL-CONSTANTAN

°C	0°	10°	20°	30°	40°	50°	60°	70°	80°	90°
−200°	−8.71	−8.45	−8.17	−7.87	−7.55	−7.20	−6.83	−6.44	−6.04	−5.62
−100°	−5.18	−4.73	−4.26	−3.78	−3.28	−2.77	−2.24	−1.70	−1.14	−0.58
0°	0.00	0.59	1.19	1.80	2.41	3.04	3.68	4.33	4.99	5.65
100°	6.32	7.00	7.69	8.38	9.08	9.79	10.51	11.23	11.95	12.68
200°	13.42	14.17	14.92	15.67	16.42	17.18	17.95	18.72	19.49	20.26
300°	21.04	21.82	22.60	22.39	24.18	24.97	25.76	26.56	27.35	28.15
400°	28.95	29.75	30.55	31.36	32.16	32.96	33.77	34.58	35.39	36.20
500°	37.01	37.82	38.62	39.43	40.24	41.05	41.86	42.67	43.48	44.29
600°	45.10	45.91	46.72	47.53	48.33	49.13	49.93	50.73	51.54	52.34
700°	53.14	53.94	54.74	55.53	56.33	57.12	57.92	58.71	59.50	60.29
800°	61.08	61.86	62.65	63.43	64.21	64.99	65.77	66.54	67.31	68.08
900°	68.85	69.62	70.39	71.15	71.92	72.68	73.44	74.20	74.95	75.70

Table 11-6
IRON-CONSTANTAN

°C	0°	10°	20°	30°	40°	50°	60°	70°	80°	90°
−190°	−7.66	−7.40	−7.12	−6.82	−6.50	−6.16	−5.80	−5.42	−5.03	−4.63
−90°	−4.21	−3.78	−3.34	−2.89	−2.43	−1.96	−1.48	−1.00	−0.50	0.00
0°	0.00	0.50	1.02	1.54	2.06	2.58	3.11	3.65	4.19	4.73
100°	5.27	5.81	6.36	6.90	7.45	8.00	8.56	9.11	9.67	10.22
200°	10.78	11.34	11.89	12.45	13.01	13.56	14.12	14.67	15.22	15.77
300°	16.33	16.88	17.43	17.98	18.54	19.09	19.64	20.20	20.75	21.30
400°	21.85	22.40	22.95	23.50	24.06	24.61	25.16	25.72	26.27	26.83
500°	27.39	27.95	28.52	29.08	29.65	30.22	30.80	31.37	31.95	32.53
600°	33.11	33.70	34.29	34.88	35.48	36.08	36.69	37.30	37.91	38.53
700°	39.15	39.78	40.41	41.05	41.68	42.32	42.96	43.60	44.25	44.89
800°	45.53	46.18	46.82	47.46	48.09	48.73	49.36	49.98	.....	.....

Table 11-7
COPPER-CONSTANTAN

°C	0°	10°	20°	30°	40°	50°	60°	70°	80°	90°
−190°	−5.379	−5.205	−5.018	−4.817	−4.603	−4.377	−4.138	−3.887	−3.624	−3.349
−90°	−3.062	−2.764	−2.455	−2.135	−1.804	−1.463	−1.112	−0.751	−0.380	0.000
0°	0.000	0.389	0.787	1.194	1.610	2.035	2.467	2.908	3.357	3.813
100°	4.277	4.749	5.227	5.712	6.204	6.703	7.208	7.719	8.236	8.759
200°	9.288	9.823	10.363	10.909	11.459	12.015	12.575	13.140	13.710	14.285
300°	14.864	15.447	16.035	16.626	17.222	17.821	18.425	19.032	19.642	20.257

FIXED POINTS FOR THERMOMETER CALIBRATION

INTERNATIONAL PRACTICAL TEMPERATURE SCALE OF 1948

Revision of 1960

The International Practical Temperature scale of 1948 is a text revision of the International Temperature Scale of 1948, the numerical values of temperature remaining the same. The adjective "Practical" was added to the name by the International Committee on Weights and Measures. The scale continues to be based upon six (primary) fixed and reproducible equilibrium temperatures to which values have been assigned, and upon the same interpolation formulas relating temperatures to the indications of specified measuring instruments. The specific measuring instruments are the standard platinum resistance thermometer, the standard platinum platinum-rhodium thermocouple and the optical pyrometer. For further definitions and recommendations as to apparatus and procedures the reader should consult H. F. Stimson, *Jour. Res. Natl. Bur. Stds.*, **42**, 209 (1949); *ibid.* **65A**, No. 3 (1961); *Natl. Bur. Stds., Monograph* **37** (1961).

The triple point of water, with the value 0.01°C replaces the former ice point as a defining fixed point of the scale. In addition to the six fundamental and primary fixed points certain other (secondary) fixed points are available and may be used for various purposes. The values of these secondary, fixed-point temperatures are not defined but have been derived from temperature determinations with standard instruments.

Table 11-8
PRIMARY AND SECONDARY FIXED POINTS

Under the pressure of 1 standard atmosphere (except for the triple points).

Temperature of	In °C (Int. 1948)	Pressure corrections and conditions
Equilibrium between liquid oxygen and its vapor. Oxygen point.	−182.97*	$+0.01254(P_{mm} - 760) - 0.000006(P_m - 760)^2$ †See footnote a.
Equilibrium between solid carbon dioxide and its vapor.	−78.5	$+0.01595(P_{mm} - 760) - 0.000011(P_{mm} - 760)^2$ † See footnote b.
Freezing (mp) mercury.	−38.87	....
Equilibrium between ice and air-saturated water.	0.000	Ice point.
Equilibrium between ice, water and water vapor. Triple point.	+0.01*	Equals 273.16°K exactly.
Triple point of phenoxy benzene.	26.88	Triple point of diphenyl ether.
Transition of sodium sulfate deca-hydrate.	32.38	....
Equilibrium between liquid water and its vapor.	100*	$+0.03686(P_{mm} - 760) - 0.000020(P_{mm} - 760)^2$ † See footnote c. Steam point.
Triple point of benzoic acid.	122.36	....
Freezing (mp) indium.	156.61	....
Equilibrium between liquid naphthalene and its vapor.	218.0	† See footnote d.
Freezing (mp) tin.	231.91	....
Equilibrium between liquid benzophenone and its vapor.	305.9	† See footnote e.
Freezing (mp) cadmium.	321.03	....
Freezing (mp) lead.	327.3	....
Equilibrium between liquid mercury and its vapor.	356.58	$+0.07309(P_{mm} - 760) - 0.000040(P_{mm} - 760)^2$ † See footnote f.
Freezing (mp) zinc.	419.505‡	....
Equilibrium between liquid sulfur and its vapor. S point.	444.6*	$+0.09080(P_{mm} - 760) - 0.000048(P_{mm} - 760)^2$ † See footnote g.
Freezing (mp) aluminum.	660.1	....
Freezing (mp) silver.	960.8*	Silver point.
Freezing (mp) gold.	1063*	Gold point.
Freezing (mp) copper.	1083	In a reducing atmosphere.
Freezing (mp) nickel.	1453	....
Freezing (mp) cobalt.	1492	....
Freezing (mp) paladium.	1552	....
Freezing (mp) platinum.	1769	....
Freezing (mp) rhodium.	1960	....
Freezing (mp) iridium.	2443	....
Melting tungsten.	3380	....

* Primary fixed point.
† a From 660 to 860 mm: $t_p = -182.97 + 9.530(p/p_0 - 1) - 3.72(p/p_0 - 1)^2 + 2.2(p/p_0 - 1)^3$
  b From 680 to 780 mm: $t_p = -78.5 + 12.12(p/p_0 - 1) - 6.4(p/p_0 - 1)^2$
  c From 660 to 860 mm: $t_p = 100 + 28.012(p/p_0 - 1) - 11.64(p/p_0 - 1)^2 + 7.1(p/p_0 - 1)^3$
  d From 680 to 780 mm: $t_p = 218.0 + 44.4(p/p_0 - 1) - 19(p/p_0 - 1)^2$
  e From 680 to 780 mm: $t_p = 305.9 + 48.8(p/p_0 - 1) - 21(p/p_0 - 1)^2$
  f From 680 to 780 mm: $t_p = 356.58 + 55.552(p/p_0 - 1) - 23.03(p/p_0 - 1)^2 + 14.0(p/p_0 - 1)^3$
  g From 660 to 800 mm: $t_p = 444.6 + 69.010(p/p_0 - 1) - 27.48(p/p_0 - 1)^2 + 19.14(p/p_0 - 1)^3$
‡ More reproducible than the sulfur point; recommended.

Table 11-9
TEMPERATURE REFERENCE POINTS FOR CALIBRATION OF LABORATORY INSTRUMENTS

The following compilation of fixed temperature points is based upon constants of certain organic compounds which have been measured with sufficient care to warrant their acceptance for *fairly close approximations* of temperature readings. The values are for *carefully* purified compounds and should *not* be applied except to material which has been carefully prepared and purified, to a constant set of physical properties, by repeated fractional distillations and/or freezings. The list was compiled by making a selection from the compounds listed in Timmermans' *PHYSICO-CHEMICAL CONSTANTS OF PURE ORGANIC COMPOUNDS*, Elsevier Publishing Co., Inc., New York (1950).

Abbreviations: B, boiling point at 760 mm of mercury pressure; F, freezing point; M, melting point.

Substance	Point	°C	Pressure corrections
iso-Pentane	F	−159.9	....
Methylcyclohexane	F	−126.6	....
Diethyl ether	M*	−116.3	....
iso-Propylbenzene (Cumene)	F	−96.0	....
Toluene	F	−95.0	....
n-Octane	F	−56.8	....
m-Xylene	F	−47.9	....
Chlorobenzene	F	−45.2	....
Bromobenzene	F	−30.6	....
Carbon tetrachloride	F	−22.9	....
Benzene	F	5.5	....
Ethylene dibromide, $(CH_2Br)_2$	F	9.9	....
Acetophenone	M	19.6	....
Diphenyl ether	M	26.9	....
iso-Pentane	B	27.9	$+0.038(P_{mm}-760)$
Diethyl ether	B	34.6	$+0.036(P_{mm}-760)$
p-Toluidine	F	43.7	....
Carbon disulfide	B	46.2	$+0.042(P_{mm}-760)$
Benzophenone	M	48.1	....
Acetone	B	56.2	$+0.040(P_{mm}-760)$
Carbon tetrachloride	B	76.7	$+0.044(P_{mm}-760)$
Benzene	B	80.1	$+0.043(P_{mm}-760)$
Naphthalene	M	80.3	....
Toluene	B	110.6	$+0.046(P_{mm}-760)$
Benzoic acid	M	122.4	....
n-Octane	B	125.6	$+0.047(P_{mm}-760)$
Ethylene dibromide, $(CH_2Br)_2$	B	131.7	$+0.048(P_{mm}-760)$
m-Xylene	B	139.1	$+0.049(P_{mm}-760)$
iso-Propylbenzene (Cumene)	B	152.4	$+0.051(P_{mm}-760)$
Bromobenzene	B	156.2	$+0.053(P_{mm}-760)$
p-Toluidine	B	200.6	$+0.054(P_{mm}-760)$
Acetophenone	B	202.0	$+0.055(P_{mm}-760)$
Naphthalene	B	218.0	$+0.055(P_{mm}-760)$
Quinoline	B	237.2	$+0.057(P_{mm}-760)$
Diphenyl ether	B	258.3	....
Benzophenone	B	305.9	$+0.064(P_{mm}-760)$

* Slow melting to allow for crystalline transition from a metastable crystalline form which melts at −123.4°C.

CHANGES IN CALIBRATION

Tests of a liquid-in-glass thermometer consist of comparisons of readings at an adequate number of points on its scale, usually at intervals from 40 to 100 divisions, with those on a standard instrument. Such a standard instrument may be either a standard platinum resistance thermometer or a suitable mercury-in-glass thermometer which has been standardized on the International Temperature Scale. Tests may also be made at the ice and steam points and a number of secondary fixed points if the equipment for attaining the temperatures of the fixed points is available.

Thermometers are subject to changes with time and use primarily as a result of changes in the volume of the bulb. For well-annealed thermometers, not subject to excessively high temperatures, bulb changes are small. After a thermometer has been heated, the bulb does not at once return to its original volume but remains somewhat larger thus temporarily lowering the readings; and if a thermometer is kept at a high temperature for long periods the ice point and the boiling point may be permanently lowered. All such changes may be determined by making a test at some reference point and changing all readings by the amount of the observed change.

Use of a thermometer at pressures differing appreciably from those prevailing during test will yield different readings. A greater pressure on the bulb will result in higher readings and a lesser pressure in lower readings. Such changes for cylindrical bulbs having a diameter of from 5 to 7 mm are of the order of 0.1°C (0.2°F) per atmosphere of pressure.

See also section on *Correction for Emergent Stem.*

CORRECTION FOR EMERGENT STEM OF LIQUID-IN-GLASS THERMOMETERS

When a thermometer which has been standardized for total immersion is used with a part of the liquid column at a temperature below that of the bulb, the reading is low and a correction must be applied. For this correction the following formula is employed:

$$T_c = T_o + f \times 1 \times (T_o - T_m)$$

where T_c = corrected temperature
T_o = observed temperature
l = length of column in degrees above the surface of the liquid the temperature of which is being taken
T_m = mean temperature of mercury (or other thermometer liquid) column; i.e., the temperature of the middle point of the mercury (or other liquid) column as read from another thermometer
f = correction factor as given in the table below. In calculating the emergent stem correction for thermometers containing organic liquids (alcohol, pentane, toluene) it is sufficient to use the approximate value, $f = 0.001$. In such thermometers the value of f is practically independent of the kind of glass.

T_m °C.	Values of f for various glasses				
	Corning 0041	Corning 8800	Corning 8810	Jena 16 III	Jena 59 III
50	0.000157	0.000166	0.000156	0.000158	0.000164
150	0.000159	0.000167	0.000157	0.000158	0.000165
250	0.000163	0.000168	0.000161	0.000161	0.000170
350	0.000168	0.000173	0.000166	....	0.000177
450	....	0.000180	0.000174	....	0.000187
500	....	....	....	....	0.000195

Table 11-10
PROPERTIES OF COMBUSTIBLE MIXTURES

Fuel	Flame temperature, °C (stoichiometric mixture)		Lowest ignition temperature, °C		Flammability limit, % by volume fuel (25°C, 760 mm)				Maximum burning velocity, cm/sec	
					Air		Oxygen			
	Air	Oxygen	Air	Oxygen	Lower	Upper	Lower	Upper	Air	Oxygen
Acetone	2105		700	568	2.5	13.0			46	
Acetylene	2400	3140	335	350	2.4	52.3	3.5	89.4	266	2480
Acrylonitrile	2444				4.6				43	
Ammonia					15.5	27.0	13.5	79.0		
Amyl alcohol			409	390	1.2					
Aniline			770	530						
Benzene	2290		740	662	1.2	9.1			40	
n-Butane	1895	2900	490	460	1.9	8.4			83	
1-Butene	1930				1.8	12.0			43	
Carbon disulfide					1.2	73.0			49	
Carbon monoxide	1960	2600	610	590	15.7	70.9	16.7	93.5	60	250
Cyanogen		4500			6.6	42.6				270
Cyclopropane	2310				2.4	10.4	2.5	63.1	50	
Decane	2265		463	202	0.6	5.4			37	
Diethyl ether	2235		343	178	1.8	36.5	2.1	82.0	40	
Ethane	1895		530	500	4.2	9.5	4.1	50.5	86	
Ethene	1975		540	485	4.0	28.6	2.9	79.9	67	
Ethyl alcohol			558	425	3.3	18.9				
Hexane	2220		487	268	1.3	8.6			40	
Hydrogen	2045	2660	530	450	9.5	65.2	9.4	91.6	440	1140
Hydrogen sulfide					4.3	45.5				
Isopropyl alcohol			590	512	1.7				34	
Methane	1875	2680	645	645	6.3	11.9	6.5	59.2	70	
Methyl alcohol				555	6.7	50			49	
Natural gas	1700–1910	2740	560	450	9.8	24.8	10.0	73.6	55	
Pentane	2230				1.3	4.9			40	
Propane	1925	2850	510	490	2.4	9.5			43	390
Propene	1935				2.2	12.1	2.1	52.8	43	
Propyl alcohol			505	445	2.2	13.5				
Toluene	2325		810	552	1.0	7.3	1.3	6.8	37	

	Fluorine									
Hydrogen	4030									

	Nitrous oxide									
Acetylene	2800								160	
Hydrogen	2690								390	

FORMULAS FOR CALCULATING MINERAL-FREE btu AND FIXED CARBON

$$FC = \% \text{ Fixed carbon} \qquad A = \% \text{ Ash}$$
$$M = \% \text{ H}_2\text{O} \qquad\qquad S = \% \text{ Sulfur}$$

In cases of litigation, coal samples containing more than 1% CO_2 as carbonate must be floated on a heavy liquid to reduce the amount to less than 1%.

The long method (or Parr) formula used only for referee purposes:

$$\% \text{ Dry, mineral-free } FC = \frac{FC - 0.15S}{100 - (M + 1.08A + 0.55S)} \times 100$$

$$\text{Moist, mineral-free } btu = \frac{btu - 50S}{100 - (1.08A + 0.55S)} \times 100$$

The short (or A. S. T. M.) approximation formula:

$$\% \text{ Dry, mineral-free } FC = \frac{FC}{100 - (M + 1.1A + 0.1S)} \times 100$$

$$\text{Moist, mineral-free } btu = \frac{btu}{100 - (1.1A + 0.1S)} \times 100$$

HEAT OF COMBUSTION OF COAL FROM ANALYSES

The usual formula employed in calculating the heat of combustion of coal from an ultimate analysis is that of Dulong:

$$BTU/pound = 14,544C + 62,028 (H - \frac{O}{8}) + 4050S$$

where C, H, O, and S are the decimal percents of carbon, hydrogen, oxygen, sulfur found on ultimate analysis made on the "as received" basis.

A proximate analysis of coal gives moisture, volatile matter, fixed carbon and ash. It is possible to compute the heat of combustion of coal from the values obtained in a proximate analysis by first finding the percent of carbon, hydrogen and oxygen by means of the formulas below and then substituting in Dulong's formula given above.

The method is as follows:—From the values of a proximate analysis find the per cent of volatile matter (V) and fixed carbon *on an ash- and moisture-free basis.* These two values should add up to a total of 100%. Then,

$$\%H = \%V (\frac{7.35}{\%V+10} - 0.013)$$

$\%N = 0.07V$ for anthracite and semi-anthracite coals.
$\%N = 2.10 - 0.012V$ for bituminous and lignite coals.
$\%C = $ Fixed $C + 0.02V^2$ for anthracite coals.
$\%C = $ Fixed $C + 0.9 (V-10)$ for semi-anthracite coals.
$\%C = $ Fixed $C + 0.9 (V-14)$ for bituminous coals.
$\%C = $ Fixed $C + 0.9 (V-18)$ for lignite.

Sulfur increases V, therefore the volatile carbon is too high by the extent of the sulfur in the coal.

Example.—A bituminous coal from Jefferson, Pa. gives a proximate analysis as follows: moisture, 2.59%; volatile matter, 30.41%; fixed carbon, 59.08%; ash, 7.9%. The %ash plus %moisture = 7.9 + 2.59 = 10.49 Therefore,

$$\frac{30.41}{100-10.49} \times 100 = 33.97\%$$ volatile matter on ash- and moisture-free basis.

$$\frac{59.08}{100-10.49} \times 100 = 66.00\%$$ fixed carbon on ash- and moisture-free basis.

$$H = 33.97 (\frac{7.35}{33.97+10} - 0.013) = 5.23\%H$$

$N = 2.10 - 0.012 \times 33.97 = 1.69\%N$
$C = 66.00 + 0.9 (33.97 - 14) = 83.97\%C$
$83.97 \times 0.8951 = 75.16\%C$ as received
$5.23 \times 0.8951 = 4.68\%H$ as received
$1.69 \times 0.8951 = 1.51\%N$ as received
7.9% ash
2.59% moisture
91.84%, therefore $100 - 91.94 = 8.16\%$ oxygen

Then by substitution of the values above in the Dulong formula

$$14,544 \times 0.7516 + 62,028 (0.0468 - \frac{0.0816}{8}) = 13,201 \text{ BTU/pound.}$$

The experimental value reported for this coal was 13,860 BTU/pound.

COMMON PHYSICAL CHEMISTRY EQUATIONS

By Frank Hovorka, Ph. D.

Gases

1. Molecular weight, volume, and temperature.

Perfect Gas Law: $PV = \dfrac{W}{M}RT$

Van der Waals' Equation: $\left[P + \dfrac{a}{V^2}\left(\dfrac{W}{M}\right)^2\right]\left[V - b\dfrac{W}{M}\right] = \dfrac{W}{M}RT$

Berthelot's Equation: $PV = \dfrac{W}{M}RT\left[1 + \dfrac{9}{128}\dfrac{P}{Pc}\dfrac{Tc}{T}\left(1 - 6\dfrac{Tc^2}{T^2}\right)\right]$

R is a constant independent of the nature of the gas and must be expressed in the same units as the pressure and the volume. It has the following values:

$R = 0.08205$ liter-atmospheres per degree per mole

82.05 cc-atmospheres per degree per mole

8.314 joules per degree per mole

1.987 calories per degree per mole

8.314×10^7 ergs per degree per mole

P, V, T, M, and W are the pressure, volume, absolute temperature, molecular weight, and the mass of the gas respectively. Pc and Tc are the critical pressure and the critical absolute temperature; a and b are individual constants that must be determined empirically for each substance. (See critical temperature and volume.) When V is in cubic centimeters then b must be expressed in cubic centimeters per molecular weight. When P is in atmospheres a must be in cc^2-atmospheres per molecular weight.

2. Critical temperature, pressure, and volume.

$$Pc = \frac{a}{3Vc^2} \qquad\qquad Tc = \frac{8PcVc}{3R}$$

$$Vc^3 = \frac{ab}{Pc} \qquad\qquad a = 3PcVc^2 \qquad\qquad b = \frac{Vc}{3}$$

Vc is the critical volume. For explanation of the other terms see part (1).

3. Effusion of Gases. $c = \sqrt{\dfrac{3PV}{mn}} = \sqrt{\dfrac{3RT}{M}} = \sqrt{\dfrac{3P}{d}}$

c is the root mean square velocity; m is the mass of each molecule; n is the number of molecules in a volume V at the pressure P or at the temperature T. When c is in centimeters per second, m is expressed in grams, V in cubic centimeters, P in dynes per square centimeter and R in ergs; d is the density; M is the molecular weight.

COMMON PHYSICAL CHEMISTRY EQUATIONS

Graham's Law:
$$\frac{c_1}{c_2} = \sqrt{\frac{d_2}{d_1}} = \sqrt{\frac{M_2}{M_1}}$$

c_1 and c_2, d_1 and d_2, M_1 and M_2 are the root mean square velocities, densities, and molecular weights of the two gases under consideration.

$$\frac{t_1}{t_2} = \frac{c_2}{c_1} = \sqrt{\frac{M_1}{M_2}}$$

t_1 and t_2 are the times of passage of each gas under a given pressure through an orifice. For an explanation of the other terms in the equation see the preceding equation.

$$\bar{n} = n\sqrt{\frac{RT}{2\pi M}}$$

$\bar{n}$ is the number of molecules striking a surface per square centimeter per second; n is the total number of molecules per cubic centimeter; M is the molecular weight; T is the absolute temperature; R is expressed in ergs per degree and is equal to 8.315×10^7 ergs.

4. Dissociation.
$$\alpha = \frac{d_1 - d_2}{d_2(n-1)} = \frac{M_1 - M_2}{M_2(n-1)} = \frac{V_2 - V_1}{V_1(n-1)}$$

d_1, M_1, and V_1 are respectively the density, the molecular weight, and the volume of the undissociated gas; d_2, M_2, and V_2 are respectively the density, the apparent molecular weight, and the volume of the dissociated gas at the same temperature; α is the degree of dissociation, and n is the number of particles into which a molecule dissociates.

5. Adsorption.

Freundlich's Isotherm:
$$\log \frac{x}{m} = k + \frac{1}{n} \log P$$

x is the mass of the gas absorbed; m is the mass of the absorbent; k and n are experimentally determined constants; P is the pressure of the gas.

Langmuir's Equation:
$$\frac{P}{x} = a + bP$$

x is the number of gram molecules adsorbed per square centimeter at a pressure P; a and b are experimentally determined constants.

Liquids

1. Viscosity.
$$\frac{n_1}{n_2} = \frac{d_1 t_1}{d_2 t_2}$$

n_1 and n_2, d_1 and d_2, t_1 and t_2 are respectively the absolute viscosities, densities, and times of flow in seconds of the two liquids through the same capillary at the same temperature and pressure; $\frac{n_1}{n_2}$ is the relative viscosity where n_2 usually refers to water.

COMMON PHYSICAL CHEMISTRY EQUATIONS

2. Surface tension. $\quad \gamma = \dfrac{1}{2}hdgr$

γ is the surface tension in dynes per centimeter, h is the height in centimeters to which a liquid of the density d rises in a capillary of the radius r (in centimeters); g is the gravitational constant expressed in centimeters per sec.[2]

Molecular Surface Energy: $\quad \gamma\left(\dfrac{M}{d}\right)^{2/3} = k(t_c - t - 6)$

M, d, and t_c are respectively the molecular weight, density, and the critical temperature of the liquid; t is the experimental temperature; k is the constant (Eötvos constant) equal to 2.12 for unassociated liquids.

Molecular Weight: $\quad M = \left[\dfrac{k(t_2 - t_1)}{\gamma_1 d_1^{-2/3} - \gamma_2 d_2^{-2/3}}\right]^{3/2}$

γ_1 and γ_2, d_1 and d_2 are respectively the surface tensions and densities of a liquid at the temperature t_1 and t_2 respectively.

Eötvos Constant: $\quad k = \dfrac{\gamma_1\left(\dfrac{M}{d_1}\right)^{2/3} - \gamma_2\left(\dfrac{M}{d_2}\right)^{2/3}}{t_2 - t_1}$

For explanation of the terms in this equation see equations on molecular surface energy and molecular weight.

Association: $\quad x = \left(\dfrac{2.12}{k}\right)^{3/2}$

x is the degree of association.

Gibbs Adsorption Equation: $\quad \mu = -\dfrac{C}{RT} \times \dfrac{d\gamma}{dC}$

μ is the excess of solute per unit surface or interface; C is the concentration of the solute, and $\dfrac{d\gamma}{dC}$ is the rate at which surface tension changes with concentration.

Sugden's Parachor: $\quad P = \dfrac{m}{D-d} \times \gamma^{1/4}$

P is the parachor of the liquid under consideration, m is its molecular weight, D its density, γ its surface tension, and d the density of its saturated vapor (d may be neglected at low pressures).

3. Boiling Point and the Heat of Vaporization.

The Law of Ramsay and Young: $\quad \dfrac{T_A}{T_B} = \dfrac{T_{A1}}{T_{B1}} + C(T_{A1} - T_A)$

T_A and T_B are the absolute boiling points of the substances A and B under the same pressure. T_{A1} and T_{B1} are the corresponding boiling points under another pressure. C is a constant. This relation holds very closely for related substances for which C is practically zero.

COMMON PHYSICAL CHEMISTRY EQUATIONS

Heat of Vaporization. *Trouton's Rule*: $\dfrac{\Delta H}{T} = 21$

Nernst-Bingham Rule: $\dfrac{\Delta H}{T} = 17 + 0.011T$

ΔH is the heat of vaporization per molecular weight and T its absolute boiling point under one atmosphere of pressure.
Both rules hold quite well for non-associated liquids.

4. Index of Refraction.

Lorenz and Lorentz Equation: $\dfrac{n^2 - 1}{n^2 + 2} \times \dfrac{1}{d} = r$

n is the index of refraction, d the density, and r the specific refraction.

Solutions

1. Vapor Pressure. Raoult's Law: $P = P_0 X_0$

P is the vapor pressure of the solvent or volatile solute over its solution; X_0 is the mole fraction and P_0 is the vapor pressure of the pure solvent or the volatile solute at the same temperature as the solution.

2. Molecular Weight of Solutes.

Rise of Boiling Point: $M = 1000 K_b \dfrac{g}{G \Delta T_b}$

Lowering of the Freezing Point: $M = 1000 K_f \dfrac{g}{G \Delta T_f}$

Lowering of the Vapor Pressure: $M = \dfrac{gm}{G} \times \dfrac{P}{P_0 - P}$

M is the molecular weight of the solute, g is the mass of the solute and G is the weight of the solvent. K_b and K_f are the molal boiling point and molal freezing point constants. T_b and T_f are the experimental boiling point raising and freezing point lowering respectively. m is the molecular weight of the solvent, P_o is the vapor pressure of pure solvent and P is the vapor pressure of the solvent from its solution at the same temperature.

3. Conductivity.

Resistance: $R = \dfrac{E}{I}$

R is the resistance in ohms, E the voltage and I the amperage.

Conductance: $L = \dfrac{1}{R}$

L is the conductivity in reciprocal ohms; R is the resistance in ohms.

COMMON PHYSICAL CHEMISTRY EQUATIONS

Specific Conductivity: $\quad \overline{L} = \dfrac{l}{A} \times L$

$\overline{L}$ is the specific conductivity, l is the distance between the electrodes, A the area of the electrodes, and L is the conductivity in reciprocal ohms.

Cell Constant: $\quad K = \dfrac{A}{l}$

K is the cell constant; l is the distance between the electrodes and A the area of the electrodes.

Equivalent Conductivity: $\quad \Lambda = \dfrac{1000\overline{L}}{C}$

Λ is the equivalent conductivity, $\overline{L}$ is the specific conductivity, and C is the normality of the solute.

Thermodynamics

1. First Law of Thermodynamics: $\quad \Delta E = q - W$

ΔE is the change of internal energy of a system, q is the heat absorbed by the system, and W is the work done by the system against the surroundings.

Work Done by an Ideal Gas During Expansion:

    (1)   Against constant external pressure: $\quad W = P(V_2 - V_1)$

    (2)   Against variable external pressure:

$$W = RT \ln \frac{V_2}{V_1} = RT \ln \frac{P_1}{P_2}$$

W is the work done, and P is the external pressure. P_1, V_1, P_2, V_2 are the initial and final pressures and volumes. If P is expressed in atmospheres and V is expressed in cubic centimeters, then W is in cc-atmospheres. One calorie is equal to 41.3 cc-atmospheres. When R is in calories then the term W (in 2) is in calories regardless of the units of V or P.

2. Effect of Temperature on Equilibrium.

In all of the following equations ΔH is positive when heat is absorbed.

Clapeyron Equation: $\quad \dfrac{dP}{dT} = \dfrac{\Delta H}{T(V_2 - V_1)}$

$\dfrac{dP}{dT}$ is the rate of change of vapor pressure of a pure substance with temperature. V_1 and V_2 are the initial and final volumes. ΔH is the amount of heat absorbed by the substance undergoing a change of phase. T is the absolute equilibrium temperature of the two phases. When P is in atmospheres and V in cubic centimeters, ΔH must be in cc-atmospheres. (One calorie = 41.3 cc-atmospheres.) Likewise when V_1 refers to one gram of a substance, V_2 and ΔH also must refer to one gram.

COMMON PHYSICAL CHEMISTRY EQUATIONS

Clausius-Clapeyron Equation: The following equation is applicable to vaporization and sublimation only:

$$ln \frac{P_1}{P_2} = \frac{\Delta H}{R}\left(\frac{1}{T_2} - \frac{1}{T_1}\right)$$

P_1 and P_2 are the vapor pressures at absolute temperatures T_1 and T_2 respectively. ΔH is the molecular heat of vaporization or sublimation. If R is in calories, ΔH will be in calories regardless of the units of pressure.

$$ln \frac{C_1}{C_2} = \frac{\Delta H}{R}\left(\frac{1}{T_2} - \frac{1}{T_1}\right)$$

C_1 and C_2 are the solubilities of a substance (in any units) at the absolute temperature T_1 and T_2. ΔH is the molecular heat of solution. When R is expressed in calories, then ΔH must be expressed in calories.

Van't Hoff's Isochore:

$$ln \frac{K_{c1}}{K_{c2}} = \frac{\Delta H}{R}\left(\frac{1}{T_2} - \frac{1}{T_1}\right)$$

$$ln \frac{K_{p1}}{K_{p2}} = \frac{\Delta H}{R}\left(\frac{1}{T_2} - \frac{1}{T_1}\right)$$

All K_c and K_p values are obtained by placing the products in the numerator and the reactants in the denominator. K_{c1} and K_{p1} are the equilibrium constants at absolute temperature T_1; K_{c2} and K_{p2} are the equilibrium constants at absolute temperature T_2. ΔH is the heat of reaction for that reaction on which K_c and K_p are based. When R is in calories ΔH must also be in calories.

Free Energy and Equilibrium Constant: $\quad \Delta F^0 = - RT \, ln \, K_p$

ΔF^0 is equal to the change in free energy for a reaction in which the reactants are in their standard states (1 atmosphere) and are converted into products at 1 atmosphere.

Nernst Heat Theorem: $\quad \log K_p = - \dfrac{\Delta H}{4.58T} + \Sigma n (1.75 \log T) + \Sigma (nc)$

ΔH is the heat of reaction at the absolute temperature T. Σn is equal to the number of moles of gaseous products minus the number of moles of gaseous reactants. $\Sigma (nc)$ is the sum of the Nernst constants; it is equal to the sum of individual constants (each constant must be multiplied by the number of moles of that substance) for gaseous products minus the sum of constants for gaseous reactants. For the values of Nernst constants see Taylor, *Treatise on Physical Chemistry*, p 586 (1931). In cases where the constant has not been determined for a particular substance, 3 to 3.2 may be used as a fair approximation.

Rates of Reaction

1. **Unimolecular Reaction:** $\quad k = \dfrac{1}{t} ln \left(\dfrac{a}{a - x}\right)$

$$t_{\frac{1}{2}} = \frac{0.693}{k}$$

COMMON PHYSICAL CHEMISTRY EQUATIONS

k is the velocity constant, a is the initial concentration, $(a - x)$ is the concentration remaining after the time t, x is the amount reacting in the time t, and $t_{\frac{1}{2}}$ is the period of half life of a unimolecular reaction.

2. Bimolecular Reaction.

(1) When equimolecular quantities (a and b) of the initial substance are used:

$$k = \frac{1}{t} \times \frac{x}{a(a - x)}$$

(2) When other than equimolecular quantities of the initial substances are used:

$$k = \frac{1}{t(a-b)} \ln \frac{b(a - x)}{a(b - x)}$$

3. Influence of Temperature on the Rate of Reaction.
Arrhenius Equations:

$$\frac{d \ln k}{dt} = \frac{E}{RT^2}$$

$$\ln \frac{k_1}{k_2} = \frac{E}{R}\left(\frac{1}{T_2} - \frac{1}{T_1}\right)$$

E represents the energy of activation per gram mole of a substance; k_1 and k_2 are the specific reaction rates at absolute temperatures T_1 and T_2 respectively; R is the gas constant.

Electromotive Force

1. Variation with Temperature: $E = T\left(\frac{dE}{dT}\right) - \frac{\Delta H}{nF}$

E is the electromotive force of a system; n is the change in valence of the ion causing the potential difference; F is equal to the charge carried by gram ion or is equal to 23,060 calories; $\frac{dE}{dT}$ is the rate of change of electromotive force with temperature; T is the experimental absolute temperature; ΔH is the heat of reaction.

2. Concentration Cell.

(1) *Without transference:* $E = \frac{RT}{nF} \ln \frac{C_1}{C_2}$

(2) *With transference:* $E = \left(\frac{2\mu}{\mu + v}\right)\frac{RT}{nF} \ln \frac{C_1}{C_2}$

μ and v are the velocities of the positive and negative ions respectively; C_1 and C_2 are the concentrations (in any units); the velocity of the faster ion is inserted in the numerator. See part 1 for the explanation of the other terms.

Table 11-11
COMPOSITION AND PHYSICAL PROPERTIES OF ALLOYS

The following table has the alloys arranged alphabetically according to the name of the element which predominates. The following abbreviations are employed: A, annealed; C, cold worked; AC, annealed and cold worked; H, hardened; HR, hot rolled. The ultimate strength is given in thousands of pounds per square inch.

Composition %	Trade Name	Sp. Gr.	M. P. °C.	Thermal Expansion ×10⁶	Ultimate Strength
Aluminum					
99.2Al	Aluminum 2S	2.71	660	23.94	12AC
95Al, 5Cu	Lynite, body alloy	.....	650	26	......
94Al, 4Cu, 0.5Mg, 0.5Mn, coated with					
99.7+Al	Alclad 17 ST	2.79	538–46	21.96	55
90Al, 10Mg	Magnalium	2.50	600	24	......
70Al, 30Mg	Magnalium	2.00	435	......	......
98Al, 1.25Mn	Aluminum Alloy 3S	2.74	640–55	23.04	16–31 AC
95Al, 5Si	Aluminum-silicon 43	2.58	577–630	21.96	19
Bismuth					
53Bi, 32Pb, 15Sn	Eutectic fusible alloy	......	96	......	......
52Bi, 40Pb, 8Cd	Eutectic fusible alloy	......	91.5	......	......
50Bi, 27Pb, 13Sn,10Cd	Lipowitz alloy	......	70–2	......	......
50Bi, 25Pb, 12.5Sn, 12.5Cd	Wood's metal	9.7	70–2	......	......
40Bi, 40Pb, 20Sn	Bismuth solder	......	111	......	......
Cerium					
70–3Ce, 17–24Zn, 1.6– 6Fe, 0–2.4Al, Mn	Ignition pin alloy	......	......	......	......
61Ce, 37Fe	Ignition pin alloy	......	......	......	......
Cobalt					
56Co, 34–40Cr, 9.2W, 1.2–2C, 0–1Fe	Stellite No. 2	......	......	......	......
55Co,20–23Cr, 15–20W, 3–5Fe, 1.5–4C	Stellite No. 3	......	......	......	......
40–80Co, 20–35Cr, 0–25W, 0.75–2.5C	Stellite	8.50	1150	14.99	......
Copper					
99.9+Cu	Deoxidized copper	8.50	1082	17.71	32–55
99.9Cu, 0.01P	Deoxidized copper	8.91	1082	17.71	35–55
90Cu, 10Al	Aluminum bronze	7.6	1050	16.5	......
90Cu, 9Al, 1Fe	Resistac	......	1066	......	75
80–90Cu, 8–10Al, 6–7Fe	Ampco Metal	7.2	649	......	60–100
88–96Cu, 2.3–10.5Al, Fe, Sn	Aluminum bronze	7.5–8.2	1038–71	......	......
47Cu, 33Au, 20Ag	8 Carat gold	......	......	......	......
40Cu, 31Au, 19Ag, 10Pd	Palladium gold	......	......	......	......
95Cu, 5Mn	Manganese bronze	8.8	1060	......	......
82Cu, 15Mn, 3Ni	Manganin	8.5	......	......	......
61Cu, 26Mn, 13Al	Magnetic alloy	......	......	......	......
88.5Cu, 5Ni, 5Sn, 1.5Si	Barberite	8.8	1070	......	60
75Cu, 25Ni	Nickel coinage, U.S.A.	......	......	......	

Table 11-11 (*Continued*)
COMPOSITION AND PHYSICAL PROPERTIES OF ALLOYS

Composition %	Trade Name	Sp. Gr.	M.P. °C.	Thermal Expansion ×10⁶	Ultimate Strength
Copper					
75Cu, 20Ni, 5Zn	Ambrac A	8.58	1150	16.40	50-130
65Cu, 18Ni, 17Zn	Nickel silver	8.75	1110	18.36	55-110
60Cu, 40Ni	Constantan	8.4	1280	......	......
55Cu, 18Ni, 27Zn	Nickel silver	8.7	1055	......	......
94.8-96Cu, 3-4Si, 1-1.2Mn	Everdur	8.46	1000	16.99	50-145
95.5Cu, 4.3Sn, 0.2P	Phosphor bronze 30	8.91	1050	18.90	45-115
91.6Cu, 8.25Sn, 0.15P	Phosphor bronze 47	8.91	......	......	51-130
90Cu, 10Sn, P trace	Phosphor bronze 209	9.00	......	......	Up to 135
90Cu, 10Sn	Gun metal bronze	8.80	......	......	......
82Cu, 16Sn, 2Zn	Bearing bronze	......	......	......	......
67Cu, 33Sn	Speculum metal	......	......	......	......
90Cu, 10Zn	Red brass	8.80	1050	18.18	37-100
89Cu, 9Zn, 2Pb	Hardware bronze	8.83	1050	18.18	35-60
85Cu, 15Zn	Red brass	8.75	1030	18.72	43-75
70Cu, 29Ni, 1Sn	Adnic	......	1205	16.29	113H
70Cu, 29Zn, 1Sn	Admiralty	8.17	935	20.16	50-95
67Cu, 33Zn	Yellow brass	8.40	940	18.50	......
60Cu, 40Zn	Muntz metal	......	840	......	......
60Cu, 25Zn, 15Ni	German silver	......	......	18.40	......
60Cu, 19Zn, 10Al, 6Fe, 5Mn	Hytensl bronze	......	980	......	105
55Cu, 25Zn, 20Ni	Usual German silver	......	......	......	......
Gold					
92Au, 8Cu	Great Britain standard	......	900	......	......
90Au, 10Cu	Coinage gold	17.17	940	......	......
67Au, 8-27Cu, 6-26Ag	16 Carat gold	......	......	......	......
62Au, 13Cu, 11Ag	15 Carat gold	......	......	......	......
58Au, 14-28Cu, 4-28Ag	14 Carat gold	......	...••.	......	......
50Au, 35Cu, 15Ag	12 Carat gold solder	......	......	......	......
42Au, 38-46Cu, 12-20Ag	10 Carat gold	......	......	......	......
90Au, 10Pd	Palladium or white gold	......	1265	......	......
80Au, 20Pd	Palau	......	1375	......	......
60Au, 40Pt	White, platinum gold	......	1500	......	......
92Au, 4.9Ag, 3.4Cu	22 Carat, dark dental	......	......	......	......
92Au, 4.2Ag, 3.8Cu	22 Carat gold	......	......	......	......
84Au, 8.3-11Ag, 6-8.3Cu	20 Carat gold	......	......	......	......
75Au, 10-20Ag, 5-15Cu	18 Carat gold	......	......	......	......
68-70Au, 25Ag, 5-7.5 Pt or Ni	Platinum substitute, electrical	......	......	......	......
50Au, 33Ag, 17Cu	14 Carat gold solder	......	......	......	......
41Au, 37Ag, 21Cu, 0.6 brass	10 Carat gold solder	......	......	......	......
40Au, 37Ag, 23Cu	8 Carat gold solder	......	......	......	......
Iridium					
95Ir, 5Pt	....	22.38	......	......	......
Iron					
99.8Fe, 0.11O, 0.025 S, 0.017Mn 0.012C, 0.005P	Armco ingot iron	7.86	1530	......	48

Table 11-11 (*Continued*)
COMPOSITION AND PHYSICAL PROPERTIES OF ALLOYS

Composition %	Trade Name	Sp. Gr.	M. P. °C.	Thermal Expansion $\times 10^6$	Ultimate Strength
Iron					
98.5Fe	Wrought iron	7.7	1510	......	48
80Fe, 20Al	Ferro-aluminum	6.30	1480	......	......
99Fe, 1C	Steel	7.83	1430	12.0	......
97Fe, 3C	White cast iron	7.60	1150	......	......
94Fe, 3.5C, 2.5Si	Gray cast iron	7.0	1230	11.2	......
Fe, 30-40Co, 5-9W, 1.5-3Cr, 0.4-0.8C	K. S. Magnet steel	......	......	......	
Fe, 0.45Cu, 0.07Mo, 0.03C	Molybdenum iron; Toncan copper	7.83	1525	11.99	......
Fe, +10Cr, 0.25C, <0.5Mn	Stainless steel	7.75	1510	......	98-250
90-92Fe, +Cr, 0.4Mn, <0.12C	Stainless steel	7.75	1450	11.00	250
90Fe, +Cr, 0.4Mn, <0.12C	Stainless iron	7.75	1450	11.00	80
88Fe, 16-17Cr, 0.4Mn, 0.1C max.	Stainless iron	......	......	9.99	60-80
86-88Fe, 12-14Cr, 0.3C	Carpenter stainless steel No. 2	7.75	1425	......	260
86-88Fe, 12-14Cr, <0.5 Mn, <0.1C, Ni trace	Defirust rustless iron	7.75	1480	......	80-160
86-88Fe, 12-14Cr, 0.1C	Carpenter stainless steel No. 1	7.78	1490	......	85A-180H
85-89Fe, 10-14Cr, <0.5Mn, <0.13C	Stainless iron	7.78	1490	10.19	80-200
82-86Fe, 12-16Cr, <0.5Ni, <0.05Si, <0.5Mn, <0.12C	Ascoloy 33	7.64	1495	10.89	85
86.7Fe, 12.5Cr, 0.35Mn, 0.35Ni, 0.12C	Sterling stainless steel T	7.75	1430	9.99	125H
86.4Fe, 13.5Cr, 0.1C	Stainless I	7.75	......	10.91	125
84-86Fe, 12.5-14.5Cr, 0.5Mn max., 0.5Si max., 0.5Ni min., 0.12-0.18C	Enduro S 15	7.86	1475	10.89	100A
84-86Fe, 12.5-14.5Cr 0.5 Mn max., 0.5Si max., 0.25Ni, 0.05-0.12C	Enduro S	7.86	1500	10.89	......
85.8Fe, 13.5Cr, 0.35 Mn, 0.35C	Sterling stainless steel A	7.75	1425	10.30	200H
85.6Fe, 14Cr, 0.35C	Stainless A	7.75	......	10.91	100
85Fe, 13-14Cr, 2Ni max., 0.3-0.6Mn, 0.12C max.	Enduro KM 1	7.75	1490	9.99	......
82-84Fe, 16-18Cr, <0.5Mn, C	Duraloy B	7.61	1510	......	100
84.3Fe, 12Cr, 2.15C, 0.75V, 0.75Co	Crocar	......	......	......	103
82-84Fe, 16-18Cr, <0.5Mn, <0.1C, Ni trace	Special defirust rustless iron	7.75	1480	......	80
81-83Fe, 16.5-18.5 Cr, 0.75Si, 0.1C max.	Enduro A	7.64	1510	11.00	75

Table 11-11 (*Continued*)
COMPOSITION AND PHYSICAL PROPERTIES OF ALLOYS

Composition %	Trade Name	Sp. Gr.	M. P. °C.	Thermal Expansion ×10⁶	Ultimate Strength
Iron					
82.8Fe, 16.5Cr, 0.65C	Stainless B	......	......	......	......
82.5Fe, 16.5Cr, 0.65C, 0.35Mn	Sterling stainless steel B	7.72	1425	10.91	200H
82Fe, 16-18Cr, <0.5 Mn, 0.5Ni, 0.35C	Sweetaloy 16	7.83	1495	11.00	75
79-82Fe, 16-19Cr, <0.5Mn, <0.5Ni, <0.5Si, <0.12C	Ascoloy 66	7.64	......	......	87
79-82Fe, 16.5-18.5Cr, 0.5 Mn max., 0.5-1.25Si max., 0.25Ni max., 0.1C max	Enduro A	7.86	1490	10.80	78A
79-81Fe, 16.5-18Cr, 1-1.1C, 0.75-1.0Si, 0.35-0.5Mn	Delhi hard	7.75	1500	9.99	125A
78.7Fe, 20Cr, 1Cu, 0.3C	Carpenter stainless steel 3	7.70	1475	......	100
71-76Fe, 17-19Cr, 7-10 Ni, 0.05Mn, 0.2C	Defistain rustless iron	7.83	1455	......	90
7i-75Fe, 17-19Cr, 8-9 Ni, 0.5Mn, 0.06-0.25C	Midvale V2A	7.89	1450	16.99	90-115
70-75Fe, 25-30Cr, <0.5Mn, 0.25C, Ni trace	Defiheat rustless iron	7.89	1595	......	75-100
Fe, 17-20Cr, 7-10Ni, <0.5Mn, <0.5Si, <0.2C	Allegheny metal	7.85	1430-70	17.30	90A
69-75Fe, 16.5-19.5Cr, 7-I0Ni, 0.75Si max., 0.5Mn max., 0.15C max.	Enduro KA2	7.86	1400	15.98	90
74Fe, 18Cr, 8Ni, 0.18C	Stainless N	......	......	18.00	90
71-74Fe, 17.5-19Cr, 8-9Ni, 0.5 Mn, 0.15C max.	Rezistal KA2	7.86	......	15.98	90
73.5Fe, 18Cr, 8Ni, 0.35 Mn, 0.15C	Sterling nirosta steel	7.92	1425	16.99	200C
73Fe, 18Cr, 9Ni, 0.5Mn, C	Duraloy 18-8	7.86	1475	14.99	105-115
70-73Fe, 27-30Cr, 0.5Mn, C	Duraloy A	7.61	1510	......	40-50
72.4Fe, 18Cr, 9.5Ni, 0.1C	Carpenter stainless steel No. 4	7.72	1400	......	89A-100HR
70-72Fe, 17.5-19Cr, 8-9Ni, 2-2.5Si, 0.1-0.2C	Rezistal 2C	7.86	......	15.98	108
68-72Fe, 26-30Cr, <1Mn, <0.6Ni, <0.6Si, <0.25C	Ascoloy 55	7.61	......	10.19	90
70Fe, 28Cr, 0.5Mn, 0.5 Ni, 0.5C	Sweetaloy 19	7.86	1495	11.00	50
70Fe, 19Cr, 9Ni, 1Cu, 1Mo, 0.2C	Stainless U	......	......	18.00	100

Table 11-11 (*Continued*)
COMPOSITION AND PHYSICAL PROPERTIES OF ALLOYS

Composition %	Trade Name	Sp. Gr.	M. P. °C.	Thermal Expansion ×10⁶	Ultimate Strength
Iron					
69Fe, 18-20Cr, 8-10Ni, 0.5Mn, 0.15C	Sweetaloy 17	7.86	1450	15.98	85
68.1Fe, 20Cr, 7Ni, 4W, 0.35C, 0.5Mn	Midvale HR	8.03	......	......	85-135
60-66Fe, 22-25Cr, 10-13 Ni, <1Mn, <0.5Si, <0.2C	Ascoloy 44	7.9	1400-25	16.20	100A
57-62Fe, 28-30Cr, 8-10 Ni, 1.5Si, 0.5-0.7C, 0.4-0.5Mn	Misco C	7.89	1540	......	82.6
60Fe, 28Cr, 10Ni, 0.5 Mn, 0.35C	Sweetaloy 22	7.97	1495	......	70
50Fe, 50Cr	Ferro-chromium	6.9	1460	......	......
50-54Fe, 25-26Cr, 19-21 Ni, 2-3Si, 0.2C	Rezistal 7	......	......	......	......
85-88Fe, 11-14Mn, 1-1.3C	Rol-Man, manganese steel	......	1290	......	160
86Fe, 13Mn, 1C	Manganese steel	7.81	1510	......	......
50Fe, 50Mn	Ferro-manganese	......	1325	......	......
96.5Fe, 3.5Ni	Nickel steel	.....	1530	......	......
95.1Fe, 3Ni, 1.5Cr, 0.4C	Nickel-chrome steel	......	......	......	......
79Fe, 15Ni, 2.5Cr, 3Si, 0.6C	Durimet D	......	1450	......	75
74.2Fe, 25Ni, 0.8C	Ferro-nickel	8.1	1500	18	......
70.9Fe, 20Ni, 8Cr, 0.75 Mn, 0.4C	Cyclops 17 Metal	8.00	1425-80	......	125
70Fe, 25Ni, 5Si, 0.25C	Durimet A	7.89	1500	......	......
67.8Fe, 32Ni, 0.2C	Ferro-nickel valve steel	8.0	1480	4	......
67Fe, 22Ni, 10Cr, 0.5 Mn, 0.2C	Sweetaloy 18	7.97	1450	18.90	79
63.8Fe, 36Ni, 0.2C	Invar	8.0	1495	0.8	......
57-61Fe, 24-26Ni, 10-12Cr, 4.5-5.5Si, 0.15C	Rezistal 255C	7.81	......	......	116
51-58Fe, 25-28Ni, 13-15Cr, 3-4W, 1-1.5 Mn, 0.4-0.5C	Midvale ATV 3	8.11	......	......	108
57Fe, 25Ni, 15Cr, 0.3C	Pyrasteel	7.89	1450	17.10	80
47-56Fe, 33-39Ni, 10-12 Cr, 1.1-1.8Mn, 0.25-0.35C	Midvale ATV 1	8.06	1450	......	85-100
53-56Fe, 24-26Ni, 17-18 Cr, 2.5Si, 0.15-0.25C	Rezistal 4	7.78	......	16.29	113
53.85Fe, 46Ni, 0.15C	Platinite	8.2	1470	7.5	......
47-52Fe, 34-36Ni, 10-12Cr, 4.5-5.5Si, 0.15C	Rezistal 355C	7.81	......	......	......
Fe, 35-37Ni, 15-17Cr, 1.4-1.6Si, 0.6-0.8Mn, 0.5-0.7C	Standard Misco	7.97	1540	13.50	75
50Fe, 35Ni, 15Cr	Chromax castings	7.81	1480	12.19	60
48Fe, 35Ni, 12Cr, 5Si, 0.25C	Durimet B	7.89	1500	......	100

Table 11-11 (*Continued*)
COMPOSITION AND PHYSICAL PROPERTIES OF ALLOYS

Composition %	Trade Name	Sp. Gr.	M. P. °C.	Thermal Expansion ×10⁶	Ultimate Strength
Iron					
45Fe, 36Ni, 18Cr, 0.5 Mn, 0.3C	Sweetaloy 20	7.97	1495	......	64
97.6Fe, 2Si, 0.4C	Silicon steel	......	......	......	......
73-97Fe, 1-24Si, 2-3C, 0.1P, 0.04-0.14S	Meehanite metal	......	......	......	50-65
84.86Fe, 13.5Si, 1C, 0.4 Mn, 0.18P, 0.05S	Tantiron	7.83	1315	......	......
84.3Fe, 14.5Si, 0.85C, 0.35Mn	Duriron	7.00	1265	15.59	14
94.5Fe, 5W, 0.5C	Tungsten steel	......	......	......	......
75Fe, 18W, 6Cr, 0.3V, 0.7C	High speed steel	......	......	......	......
66Fe, 17W, 10Cr 3.5C, 2.5Mo	Cristite 1	7.61	......	15.59	......
Lead					
99.8Pb, 0.2As	Lead for shot	......	......	......	......
94Pb, 6Sb	Battery plate	......	300	......	......
92-94Pb, 6-8Sb	Antimonial lead	11.0	245-90	27.00	3
90Pb, 10Sb	Magnolia	......	270	......	.. ..
82Pb, 15Sb, 3Sn	Type metal	......	......	......	......
75Pb, 19Sb, 5Sn, 1Cu	White metal	9.5	238	......	......
92Pb, 8Cd	Aluminum solder, U.S.P.1,333,666	......	310	......	......
99.93Pb, 0.08Cu	Chemical lead	11.35	327	28.98	1.74
67Pb, 33Sn	Plumber's solder	9.4	275	25.0	......
58Pb, 26Sn, 15Sb, 1Cu	Standard type metal	......	......	......	......
50Pb, 50Sn	Half and half solder	9.4	275	25.0	......
Magnesium					
98.5Mg, 1.5Mn	Dowmetal M	1.76	650	16	......
95.7Mg, 4Al, 0.3Mn	Dowmetal F	1.76	625	16	......
93.7Mg, 6Al, 0.3Mn	Dowmetal E	1.78	610	16	......
91.8Mg, 8Al, 0.2Mn	Dowmetal A	1.80	600	16	......
90Mg, 3-7Al, 2-5Zn, 0.5Mn	Electron	......	......	......	......
89.9Mg, 1.5 Mn	Dowmetal G	1.80	595	16	......
88Mg, 12Al	Dowmetal B	......	575	......	......
88Mg, 8.3Al, 2.0Cu, 1.0 Cd, 0.5Zn, 0.2Mn	Dowmetal D	......	......	......	......
		......	......	......	......
85Mg, 15Al	Dowmetal C	......	590	......	......
91.8Mg, 4Cu, 2Al, 2Cd, 0.2Mn	Dowmetal T	1.82	640	16	......
95Mg, 4-5Zn, 0-0.6Cu	Electron	......	......	......	......
Mercury					
80Hg, 20Bi	Bismuth amalgam	......	90	......	......
70Hg, 30Cu	Dentist amalgam	......	......	......	......
Nickel					
99-99.5Ni (+Co), 0.25-1.0C, 0.25-1.0Si, 0.3-1.0Mn, 0.6-1Fe, 0.25-1Cu	Nickel	8.86	1450	......	65-140

Table 11-11 (*Continued*)
COMPOSITION AND PHYSICAL PROPERTIES OF ALLOYS

Composition %	Trade Name	Sp. Gr.	M. P. °C.	Thermal Expansion ×10⁶	Ultimate Strength
Nickel					
Ni-Cr steel with high Si	Elcomet	8.03	......	......	......
80Ni, 20Cr	Chromel A	8.4	......	......	......
80Ni, 20Cr	Tophet A	8.50	1345	13.00	120
80Ni, 20Cr	Nichrome IV	8.50	1390	13.21	110
63Ni, 21Cr, 6.5Cu, 5Mo, 2W, 1Fe, 1Al, 1Mn	Illium	......	......	......	......
60Ni, 25Cr, 7Cu	Illium G	8.31	1300	13.50	73
60Ni, 20Cr, 10Fe, 1.75 Mn, 0.5C	Fireamor	8.00	1330	13.99	60
73Ni, 17.5Co, 6.5Fe, 2.5 Ti, 0.2Mn	Konel	8.61	1450-1500	10.66	100
90Ni, 3Cu, ±10Si, 1.5Al	Hastelloy	7.81	1160	11.59	......
60-70Ni, 25-35Cu, 1-3Fe, 0.25-2.0Mn, 0.02-1.5 Si, 0.3-0.5C	Monel metal	8.80	1330	......	65-170
60Ni, 33Cu, 6.5Fe	Monel metal	8.90	1360	14	......
75Ni, 12Fe, 11Cr, 2Mn	Nichrome wire	......	......	......	......
61Ni, 23Fe, 16Cr	Chromel C	8.24	......	......	......
60-62Ni, 23-26Fe, 10-11Cr, 2-2.5W, 1.2-1.5Mn, 0.3-0.35C	Midvale BTG	8.47	1450	......	
60Ni, 28Fe, 12Cr	Tophet C	8.19	1350	13.70	100
60Ni, 25Fe, 15Cr,0.7C	Nichrome castings	8.08	......	12.10	64
60Ni, 24Fe, 16Cr, 0.1C	Nichrome	8.17	1350	13.70	100
60 Ni, 20Fe, 20Mo	Hastelloy A	8.80	1300	10.71	100-120
35Ni, 17Fe, 15Cr, 1.75 Mn, 0.5C	Zorite	7.92	1300	......	50
Ni, Fe, Mo	Hastelloy C	8.91	1350	......	60
60Ni, 20Pt, 10Pd, 10V	Palau	......	......	......	......
Palladium					
67Pd, 33Ag	Palladium alloy	......	1415	......	......
Platinum					
90Pt, 10Ir	Platinum-iridium	21.61	......	8.8	......
80-100Pt, 0.-20Rh	Thermocouple rhodium	......	......	8.8	......
Silver					
92.5Ag, 7.5Cu	Standard silver	......	920	18	......
90Ag, 10Cu	U. S. silver coin	10.3	890	......	......
63Ag, 30Cu, 7.5Zn	Silver solder	......	......	......	......
40Ag, 40Sn, 14Cu, 6Zn	Bu. Stand. silver solder	......	......	......	......
Tantalum					
99.5Ta	Tantalum	16.6	2850	6.50	130
Tin					
78Sn, 9Al, 8Zn, 5Cd	Bu. Stand. SN1 aluminum solder	......	......	......	......

Table 11-11 (*Continued*)
COMPOSITION AND PHYSICAL PROPERTIES OF ALLOYS

Composition %	Trade Name	Sp. Gr.	M.P. °C.	Thermal Expansion ×10⁶	Ultimate Strength
Tin					
90Sn, 10Sb	Britannia metal	......	255	......	......
90Sn, 7Sb, 3Cu	Babbitt metal	......	......	......	......
75Sn, 12.5Sb, 12.5Cu	Antifriction	7.53	233	......	......
85Sn, 6.8Cu, 6Bi,1.7Sb	Pewter	......	....	....	......
50Sn, 32Pb, 18Cd	Eutectic alloy	......	145	......	......
86Sn, 9Zn, 5Al	Bu. Stand. SN3 aluminum solder	......	....	......	......
69Sn, 29Zn, 2.4Al, 2.4P	Bu. Stand. SN2 aluminum solder	......	....	......	......
41Sn, 28Zn, 3Cu, 0.6Mn, 0.1Al	Aluminum solder, U.S.P. 1,332,899	......	....	......	......
Tungsten					
W, 0.5-0.75ThO₂	Filament tungsten	......	....	......	......
Zinc					
90Zn, 6Al, 4Cu	Geophys. Lab., aluminum solder	......	....	......	......
75Zn, 20Cd, 5Al	Bu. Stand. ZN1, aluminum solder	......	....	......	......
67Zn, 33Cu	Easily fusible solder	......	795	20	......
60Zn, 40Cu	White solder	......	840	21	......
63 Zn,21Sn,12Pb, 3.2Cu	Battery plate	......	....	......	......

Table 11-12
MELTING POINT AND COMPOSITION OF FUSIBLE ALLOYS

Melting Point Deg. C.	Composition			
	% Bismuth	% Lead	% Tin	% Cadmium
46.9 + 18.1%In	41	22.1	10.6	8.2
66	50	25	12.5	12.5
70*	50	27	13	10
75	27.5	27.5	10	34.5
80	43.8	25	25	6.2
92	52	40	0	8
96	53	32	15	0
100	50	20	30	0
108	42.1	42.1	15.8	0
111	40	40	20	0
117	36.5	36.5	27	0
123	33.3	33.3	33.3	0
130	30.8	38.4	30.8	0
136	0	26.5	59.3	14.2
144.4	0	25	50	25
145	20	40	40	0
155	16	36	48	0
160	11.2	44.4	44.4	0
165	11.4	45.6	43	0
172	12.8	49	38.2	0
178	12.5	50	37.5	0
187	0	31	69	0
200	0	20	80	0
230	0	55.6	44.4	0
265	0	71.4	28.6	0
290	0	87.5	12.5	0

*This alloy (Lipowitz's alloy) and many others can be alloyed with mercury to further reduce the melting point; after melting the alloy and removing the molten mass from the source of heat, the mercury is introduced and the mixture stirred with an iron rod. **CAUTION:** mercury vapors are poisonous.

Table 11-13
VAN DER WAALS' CONSTANTS FOR GASES

Calculated from the values given in *Landolt-Bornstein Phys.-Chem. Tab.*, 5th Ed., p. 254 (1923); published by J. Springer, Berlin.

$$\left(P + \frac{a}{V^2}\right)\left(V - b\right) = RT \text{ for } one \text{ mole}$$

To use the values of a and b in the table below, P must be expressed in atmospheres and V in liters per mole; then $R = 0.08207$ liter atmospheres per mole per degree; T is degrees absolute.

$$\left(P + \frac{n^2a}{V^2}\right)\left(V - nb\right) = nRT \text{ for } n \text{ moles}$$

Name	Formula	a $\dfrac{(\text{liters})^2 \times \text{atm.}}{(\text{mole})^2}$	b $\dfrac{\text{liters}}{\text{mole}}$
Acetic acid	CH_3CO_2H	17.59	0.1068
Acetic anhydride	$(CH_3CO)_2O$	19.90	0.1263
Acetone	$(CH_3)_2CO$	13.91	0.0994
Acetonitrile	CH_3CN	17.58	0.1168
Acetylene	$HCCH$	4.390	0.05136
Ammonia	NH_3	4.170	0.03707
Amyl formate	$HCO_2C_5H_{11}$	27.58	0.1730
Amylene	C_5H_{10}	15.90	0.1207
Aniline	$C_6H_5NH_2$	26.50	0.1369
Argon	Ar	1.345	0.03219
Benzene	C_6H_6	18.00	0.1154
Benzonitrile	C_6H_5CN	33.39	0.1724
Bromobenzene	C_6H_5Br	28.56	0.1539
Butane n	C_4H_{10}	14.47	0.1226
Butyronitrile	C_3H_7CN	25.72	0.1596
Capronitrile	$C_5H_{11}CN$	34.16	0.1984
Carbon dioxide	CO_2	3.592	0.04267
Carbon disulfide	CS_2	11.62	0.07685
Carbon monoxide	CO	1.485	0.03985
Carbon oxysulfide	COS	3.933	0.05817
Carbon tetrachloride	CCl_4	20.39	0.1383
Chlorine	Cl_2	6.493	0.05622
Chlorobenzene	C_6H_5Cl	25.43	0.1453
Chloroform	$CHCl_3$	15.17	0.1022
Cresol m	$CH_3C_6H_4OH$	31.38	0.1607
Cyanogen	$(CN)_2$	7.667	0.06901
Cyclohexane	C_6H_{12}	22.81	0.1424
Cymene	$CH_3C_6H_4C_3H_7$	42.16	0.2336
Decane	$C_{10}H_{12}$	48.55	0.2905
Di-isobutyl	C_8H_{18}	34.97	0.2296
Diethylamine	$(C_2H_5)_2NH$	19.15	0.1392
Dimethylamine	$(CH_3)_2NH$	10.38	0.08570
Dimethylaniline	$C_6H_5N(CH_3)_2$	37.49	0.1970
Diphenyl	$(C_6H_5)_2$	52.79	0.2480
Diphenyl methane	$(C_6H_5)_2CH_2$	38.20	0.2240
Dipropylamine	$(C_3H_7)_2NH$	27.72	0.1820
Di-isopropyl	$(C_3H_7)_2$	23.13	0.1669
Durene	$C_{10}H_{14}$	45.32	0.2424
Ethane	C_2H_6	5.489	0.06380
Ethyl acetate	$CH_3CO_2C_2H_5$	20.45	0.1412
Ethyl alcohol	C_2H_5OH	12.02	0.08407
Ethylamine	$C_2H_5NH_2$	10.60	0.08409
Ethyl benzene	$C_2H_5C_6H_5$	28.60	0.1667
Ethyl butyrate	$C_3H_7CO_2C_2H_5$	30.07	0.1919

Table 11-13 (*Continued*)
VAN DER WAALS' CONSTANTS FOR GASES

Name	Formula	$\dfrac{a}{\text{(liters)}^2 \times \text{atm.}}$ $\dfrac{}{\text{(mole)}^2}$	$\dfrac{b}{\dfrac{\text{liters}}{\text{mole}}}$
Ethyl isobutyrate	$C_3H_7CO_2C_2H_5$	28.87	0.1994
Ethyl chloride	C_2H_5Cl	10.91	0.08651
Ethyl ether	$(C_2H_5)_2O$	17.38	0.1344
Ethyl formate	$HCO_2C_2H_5$	14.80	0.1056
Ethyl mercaptan	C_2H_5SH	11.24	0.08098
Ethyl propionate	$C_2H_5CO_2C_2H_5$	24.39	0.1615
Ethyl sulfide	$(C_2H_5)_2S$	18.75	0.1214
Ethylene	C_2H_4	4.471	0.05714
Ethylene bromide	$(CH_2Br)_2$	13.98	0.08664
Ethylene chloride	$(CH_2Cl)_2$	16.91	0.1086
Ethylidene chloride	CH_3CHCl_2	15.50	0.1073
Fluorobenzene	C_6H_5F	19.93	0.1286
Germanium tetrachloride	$GeCl_4$	22.60	0.1485
Helium	He	0.03412	0.02370
Heptane n	C_7H_{16}	31.51	0.2654
Hexane n	C_6H_{14}	24.39	0.1735
Hydrogen	H_2	0.2444	0.02661
Hydrogen bromide	HBr	4.451	0.04431
Hydrogen chloride	HCl	3.667	0.04081
Hydrogen selenide	H_2Se	5.268	0.04637
Hydrogen sulfide	H_2S	4.431	0.04287
Iodobenzene	C_6H_5I	33.08	0.1656
Isoamylene	C_5H_{10}	18.08	0.1405
Isobutane	C_4H_{10}	12.87	0.1142
Isobutyl acetate	$CH_3CO_2C_4H_9$	28.50	0.1833
Isobutyl alcohol	C_4H_9OH	17.03	0.1143
Isobutylbenzene	$C_6H_5C_4H_9$	38.59	0.2144
Isobutyl formate	$HCO_2C_4H_9$	22.54	0.1476
Isopentane	C_5H_{12}	18.05	0.1417
Isopropyl alcohol	C_3H_7OH	13.78	0.09804
Isopropylbenzene	$C_6H_5C_3H_7$	35.64	0.2025
Krypton	Kr	2.318	0.03978
Mercury	Hg	8.093	0.01696
Mesitylene	$(CH_3)_3C_6H_3$	34.32	0.1979
Methane	CH_4	2.253	0.04278
Methyl acetate	$CH_3CO_2CH_3$	15.29	0.1091
Methyl alcohol	CH_3OH	9.523	0.06702
Methylamine	CH_3NH_2	7.130	0.05992
Methyl butyrate	$C_3H_7CO_2CH_3$	23.94	0.1569
Methyl isobutyrate	$C_3H_7CO_2CH_3$	24.50	0.1637
Methyl chloride	CH_3Cl	7.471	0.06483
Methyl ether	$(CH_3)_2O$	8.073	0.07246
Methyl ethyl ether	$CH_3OC_2H_5$	11.95	0.09775
Methyl ethyl sulfide	$CH_3SC_2H_5$	19.23	0.1304
Methyl fluoride	CH_3F	4.631	0.05264
Methyl formate	HCO_2CH_3	10.84	0.08068
Methyl propionate	$C_2H_5CO_2CH_3$	19.91	0.1360
Methyl sulfide	$(CH_3)_2S$	12.87	0.09213
Methyl valerate	$C_4H_9CO_2CH_3$	28.96	0.1845
Naphthalene	$C_{10}H_8$	39.74	0.1937
Neon	Ne	0.2107	0.01709
Nitric oxide	NO	1.340	0.02789
Nitrogen	N_2	1.390	0.03913
Nitrogen dioxide	NO_2	5.284	0.04424
Nitrous oxide	N_2O	3.782	0.04415
Octane n	C_8H_{18}	37.32	0.2368

Table 11-13 (*Continued*)
VAN DER WAALS' CONSTANTS FOR GASES

Name	Formula	a $\dfrac{(\text{liters})^2 \times \text{atm.}}{(\text{mole})^2}$	b $\dfrac{\text{liters}}{\text{mole}}$
Oxygen (Ozone, cf. below)	O_2	1.360	0.03803
Pentane *n*	C_5H_{12}	19.01	0.1460
Phenetole	$C_6H_5OC_2H_5$	35.16	0.1963
Phosphine	PH_3	4.631	0.05156
Phosphonium chloride	PH_4Cl	4.054	0.04545
Phosphorus	P	52.94	0.1566
Propane	C_3H_8	8.664	0.08445
Propionic acid	$C_2H_5CO_2H$	20.11	0.1187
Propionitrile	C_2H_5CN	16.44	0.1064
Propyl acetate	$CH_3CO_2C_3H_7$	24.63	0.1619
Propyl alcohol	C_3H_7OH	14.92	0.1019
Propylamine	$C_3H_7NH_2$	14.99	0.1090
Propyl benzene	$C_6H_5C_3H_7$	35.85	0.2028
Propyl chloride	C_3H_7Cl	15.91	0.1141
Propyl formate	$HCO_2C_3H_7$	18.95	0.1280
Propylene	C_3H_6	8.379	0.08272
Pseudo-cumene	$C_6H_3(CH_3)_3$	36.61	0.2021
Silicon fluoride	SiF_4	4.195	0.05571
Silicon tetrahydride	SiH_4	4.320	0.05786
Stannic chloride	$SnCl_4$	26.91	0.1642
Sulfur dioxide	SO_2	6.714	0.05636
Thiophene	C_4H_4S	20.72	0.1270
Toluene	$C_6H_5CH_3$	24.06	0.1463
Triethylamine	$(C_2H_5)_3N$	27.17	0.1831
Trimethylamine	$(CH_3)_3N$	13.02	0.1084
Xenon	Xe	4.194	0.05105
Xylene *m*	$C_6H_4(CH_3)_2$	30.36	0.1772
Xylene *o*	$C_6H_4(CH_3)_2$	29.98	0.1755
Xylene *p*	$C_6H_4(CH_3)_2$	30.93	0.1809
Water	H_2O	5.464	0.03049

Ozone, O_3 $a = 3.545$; $b = 0.04903$

RELATION OF CONCENTRATION UNITS

$1 \ \mu g/ml = 1 \ mg/liter = 1 \ ppm \ (w/v)$

$1 \ meq/liter = (weight \ in \ grams \ per \ equivalent/1000) \ in \ 1 \ liter$

$1 \ mg \ atom \ per \ liter = (atomic \ weight \ in \ grams/1000) \ in \ 1 \ liter$

$1 \ \mu g/ml = (1 \ meq/liter)(1000/weight \ in \ grams \ per \ equivalent)$

Table 11-14
STANDARD STOCK SOLUTIONS*

Element	Procedure
Aluminum	Dissolve 1.000 g Al wire in minimum amount of 2 M HCl; dilute to volume.
Antimony	Dissolve 1.000 g Sb in (1) 10 ml HNO_3 plus 5 ml HCl, and dilute to volume when dissolution is complete; or (2) 18 ml HBr plus 2 ml liquid Br_2, when dissolution is complete add 10 ml $HClO_4$, heat in a well-ventilated hood while swirling until white fumes appear and continue for several minutes to expel all HBr, then cool and dilute to volume.
Arsenic	Dissolve 1.3203 g of As_2O_3 in 3 ml 8 M HCl and dilute to volume; or treat the oxide with 2 g NaOH and 20 ml water, after dissolution dilute to 200 ml, neutralize with HCl (pH meter), and dilute to volume.
Barium	(1) Dissolve 1.7787 g $BaCl_2 \cdot 2H_2O$ (fresh crystals) in water and dilute to volume. (2) Dissolve 1.516 g $BaCl_2$ (dried at 250°C for 2 hr) in water and dilute to volume. (3) Treat 1.4367 g $BaCO_3$ with 300 ml water, slowly add 10 ml of HCl and, after the CO_2 is released by swirling, dilute to volume.
Beryllium	(1) Dissolve 19.655 g $BeSO_4 \cdot 4H_2O$ in water, add 5 ml HCl (or HNO_3), and dilute to volume. (2) Dissolve 1.000 g Be in 25 ml 2 M HCl, then dilute to volume.
Bismuth	Dissolve 1.000 g Bi in 8 ml of 10 M HNO_3, boil gently to expel brown fumes, and dilute to volume.
Boron	Dissolve 5.720 g fresh crystals of H_3BO_3 and dilute to volume.
Bromine	Dissolve 1.489 g KBr (or 1.288 g NaBr) in water and dilute to volume.
Cadmium	(1) Dissolve 1.000 g Cd in 10 ml of 2 M HCl; dilute to volume. (2) Dissolve 2.282 g $3CdSO_4 \cdot 8H_2O$ in water; dilute to volume.
Calcium	Place 2.4973 g $CaCO_3$ in volumetric flask with 300 ml water, carefully add 10 ml HCl, after CO_2 is released by swirling, dilute to volume.
Cerium	(1) Dissolve 4.515 g $(NH_4)_4Ce(SO_4)_4 \cdot 2H_2O$ in 500 ml water to which 30 ml H_2SO_4 had been added, cool, and dilute to volume. Advisable to standardize against As_2O_3. (2) Dissolve 3.913 g $(NH_4)_2Ce(NO_3)_6$ in 10 ml H_2SO_4, stir 2 min, cautiously introduce 15 ml water and again stir 2 min. Repeat addition of water and stirring until all the salt has dissolved, then dilute to volume.
Cesium	Dissolve 1.267 g CsCl and dilute to volume. Standardize: Pipette 25 ml of final solution to Pt dish, add 1 drop H_2SO_4, evaporate to dryness, and heat to constant weight at $>$ 800°C. Cs (in $\mu g/ml$) = (40)(0.734)(wt of residue)
Chlorine	Dissolve 1.648 g NaCl and dilute to volume.
Chromium	(1) Dissolve 2.829 g $K_2Cr_2O_7$ in water and dilute to volume. (2) Dissolve 1.000 g Cr in 10 ml HCl, and dilute to volume.
Cobalt	Dissolve 1.000 g Co in 10 ml of 2 M HCl, and dilute to volume.
Copper	(1) Dissolve 3.929 g fresh crystals of $CuSO_4 \cdot 5H_2O$, and dilute to volume. (2) Dissolve 1.000 g Cu in 10 ml HCl plus 5 ml water to which HNO_3(or 30%H_2O_2) is added dropwise until dissolution is complete. Boil to expel oxides of nitrogen and chlorine, then dilute to volume.
Dysprosium	Dissolve 1.1477 g Dy_2O_3 in 50 ml of 2 M HCl; dilute to volume.

* 1000 $\mu g/ml$ as the element in a final volume of 1 liter unless stated otherwise.
From J. A. Dean and T. C. Rains, "Standard Solutions for Flame Spectrometry," in *Flame Emission and Atomic Absorption Spectrometry*, J. A. Dean and T. C. Rains (Eds.), Vol. 2, Chap. 13, Marcel Dekker, New York, 1971.

Table 11-14 (*Continued*)
STANDARD STOCK SOLUTIONS

Element	Procedure
Erbium	Dissolve 1.1436 g Er_2O_3 in 50 ml of 2 M HCl; dilute to volume.
Europium	Dissolve 1.1579 g Eu_2O_3 in 50 ml of 2 M HCl; dilute to volume.
Fluorine	Dissolve 2.210 g NaF in water and dilute to volume.
Gadolinium	Dissolve 1.152 g Gd_2O_3 in 50 ml of 2 M HCl; dilute to volume.
Gallium	Dissolve 1.000 g Ga in 50 ml of 2 M HCl; dilute to volume.
Germanium	Dissolve 1.4408 g GeO_2 with 50 g oxalic acid in 100 ml of water; dilute to volume.
Gold	Dissolve 1.000 g Au in 10 ml of hot HNO_3 by dropwise addition of HCl, boil to expel oxides of nitrogen and chlorine, and dilute to volume. Store in amber container away from light.
Hafnium	Transfer 1.000 g Hf to Pt dish, add 10 ml of 9 M H_2SO_4, and then slowly add HF dropwise until dissolution is complete. Dilute to volume with 10% H_2SO_4.
Holmium	Dissolve 1.1455 g Ho_2O_3 in 50 ml of 2 M HCl; dilute to volume.
Indium	Dissolve 1.000 g In in 50 ml of 2 M HCl; dilute to volume.
Iodine	Dissolve 1.308 g KI in water and dilute to volume.
Iridium	(1) Dissolve 2.465 g Na_3IrCl_6 in water and dilute to volume. (2) Transfer 1.000 g Ir sponge to a glass tube, add 20 ml of HCl and 1 ml of $HClO_4$. Seal the tube and place in an oven at 300°C for 24 hr. Cool, break open the tube, transfer the solution to a volumetric flask, and dilute to volume. Observe all safety precautions in opening the glass tube.
Iron	Dissolve 1.000 g Fe wire in 20 ml of 5 M HCl; dilute to volume.
Lanthanum	Dissolve 1.1717 g La_2O_3 (dried at 110°C) in 50 ml of 5 M HCl, and dilute to volume.
Lead	(1) Dissolve 1.5985 g $Pb(NO_3)_2$ in water plus 10 ml HNO_3, and dilute to volume. (2) Dissolve 1.000 g Pb in 10 ml HNO_3, and dilute to volume.
Lithium	Dissolve a slurry of 5.3228 g Li_2CO_3 in 300 ml of water by addition of 15 ml HCl, after release of CO_2 by swirling, dilute to volume.
Lutetium	Dissolve 1.6079 g $LuCl_3$ in water and dilute to volume.
Magnesium	Dissolve 1.000 g Mg in 50 ml of 1 M HCl and dilute to volume.
Manganese	(1) Dissolve 1.000 g Mn in 10 ml HCl plus 1 ml HNO_3, and dilute to volume. (2) Dissolve 3.0764 g $MnSO_4 \cdot H_2O$ (dried at 105°C for 4 hr) in water and dilute to volume. (3) Dissolve 1.5824 g MnO_2 in 10 HCl in a good hood, evaporate to gentle dryness, dissolve residue in water and dilute to volume.
Mercury	Dissolve 1.000 g Hg in 10 ml of 5 M HNO_3 and dilute to volume.
Molybdenum	(1) Dissolve 2.0425 g $(NH_4)_2MoO_4$ in water and dilute to volume. (2) Dissolve 1.5003 g MoO_3 in 100 ml of 2 M ammonia, and dilute to volume.
Neodymium	Dissolve 1.7373 g $NdCl_3$ in 100 ml 1 M HCl and dilute to volume.
Nickel	Dissolve 1.000 g Ni in 10 ml hot HNO_3, cool, and dilute to volume.
Niobium	Transfer 1.000 g Nb (or 1.4305 g Nb_2O_5) to Pt dish, add 20 ml HF, and heat gently to complete dissolution. Cool, add 40 ml H_2SO_4, and evaporate to fumes of SO_3. Cool and dilute to volume with 8 M H_2SO_4.
Osmium	Dissolve 1.3360 g OsO_4 in water and dilute to 100 ml. Prepare only as needed as solution loses strength on standing unless Os is reduced by SO_2 and water is replaced by 100 ml 0.1 M HCl.
Palladium	Dissolve 1.000 g Pd in 10 ml of HNO_3 by dropwise addition of HCl to hot solution; dilute to volume.
Phosphorus	Dissolve 4.260 g $(NH_4)_2HPO_4$ in water and dilute to volume.
Platinum	Dissolve 1.000 g Pt in 40 ml of hot aqua regia, evaporate to incipient dryness, add 10 ml HCl and again evaporate to moist residue. Add 10 ml HCl and dilute to volume.
Potassium	Dissolve 1.9067 g KCl (or 2.8415 g KNO_3) in water and dilute to volume.
Praseodymium	Dissolve 1.1703 g Pr_2O_3 in 50 ml of 2 M HCl; dilute to volume.
Rhenium	Dissolve 1.000 g Re in 10 ml of 8 M HNO_3 in an ice bath until initial reaction subsides, then dilute to volume.

Table 11-14 (*Continued*)
STANDARD STOCK SOLUTIONS

Element	Procedure
Rhodium	Dissolve 1.000 g Rh by the sealed-tube method described under iridium.
Rubidium	Dissolve 1.4148 g RbCl in water. Standardize as described under cesium. Rb (in $\mu g/ml$) = (40)(0.320)(wt of residue).
Ruthenium	Dissolve 1.317 g RuO_2 in 15 ml of HCl; dilute to volume.
Samarium	Dissolve 1.1596 g Sm_2O_3 in 50 ml of 2 *M* HCl; dilute to volume.
Scandium	Dissolve 1.5338 g Sc_2O_3 in 50 ml of 2 *M* HCl; dilute to volume.
Selenium	Dissolve 1.4050 g SeO_2 in water and dilute to volume or dissolve 1.000 g Se in 5 ml of HNO_3, then dilute to volume.
Silicon	Fuse 2.1393 g SiO_2 with 4.60 g Na_2CO_3, maintaining melt for 15 min in Pt crucible. Cool, dissolve in warm water, and dilute to volume. Solution contains also 2000 $\mu g/ml$ sodium.
Silver	(1) Dissolve 1.5748 g $AgNO_3$ in water and dilute to volume. (2) Dissolve 1.000 g Ag in 10 ml of HNO_3; dilute to volume. Store in amber glass container away from light.
Sodium	Dissolve 2.5421 g NaCl in water and dilute to volume.
Strontium	Dissolve a slurry of 1.6849 g $SrCO_3$ in 300 ml of water by careful addition of 10 ml of HCl, after release of CO_2 by swirling, dilute to volume.
Sulfur	Dissolve 4.122 g $(NH_4)_2SO_4$ in water and dilute to volume.
Tantalum	Transfer 1.000 g Ta (or 1.2210 g Ta_2O_5) to Pt dish, add 20 ml of HF, and heat gently to complete the dissolution. Cool, add 40 ml of H_2SO_4 and evaporate to heavy fumes of SO_3. Cool and dilute to volume with 50% H_2SO_4.
Tellurium	(1) Dissolve 1.2508 g TeO_2 in 10 ml of HCl; dilute to volume. (2) Dissolve 1.000 g Te in 10 ml of warm HCl with dropwise addition of HNO_3, then dilute to volume.
Terbium	Dissolve 1.6692 g of $TbCl_3$ in water, add 1 ml of HCl, and dilute to volume.
Thallium	Dissolve 1.3034 g $TlNO_3$ in water and dilute to volume.
Thorium	Dissolve 2.3794 g $Th(NO_3)_4 \cdot 4H_2O$ in water, add 5 ml HNO_3, and dilute to volume.
Thulium	Dissolve 1.142 g Tm_2O_3 in 50 ml of 2 *M* HCl; dilute to volume.
Tin	Dissolve 1.000 g Sn in 15 ml of warm HCl; dilute to volume.
Titanium	Dissolve 1.000 g Ti in 10 ml of H_2SO_4 with dropwise addition of HNO_3; dilute to volume with 5% H_2SO_4.
Tungsten	Dissolve 1.7941 g of $Na_2WO_4 \cdot 2H_2O$ in water and dilute to volume.
Uranium	Dissolve 2.1095 g $UO_2(NO_3)_2 \cdot 6H_2O$ (or 1.7734 g uranyl acetate dihydrate) in water and dilute to volume.
Vanadium	Dissolve 2.2963 g NH_4VO_3 in 100 ml of water plus 10 ml of HNO_3; dilute to volume.
Ytterbium	Dissolve 1.6147 g $YbCl_3$ in water and dilute to volume.
Yttrium	Dissolve 1.2692 g Y_2O_3 in 50 ml of 2 *M* HCl and dilute to volume.
Zinc	Dissolve 1.000 g Zn in 10 ml of HCl; dilute to volume.
Zirconium	Dissolve 3.533 g $ZrOCl_2 \cdot 8H_2O$ in 50 ml of 2 *M* HCl, and dilute to volume. Solution should be standardized.

CONCENTRATIONS OF COMMONLY USED ACIDS AND AQUEOUS AMMONIA

Freshly opened bottles of these reagents are uniformly of the concentrations indicated below. This may not be true of bottles long opened and this is especially true of ammonia which rapidly loses its strength. In preparing volumetric solutions, it is well to be on the safe side and take a little more than the calculated volume of the concentrated reagent, since it is much easier to dilute a concentrated solution than to strengthen one that is too weak. The 5N solutions present obvious advantages over the usual indefinite dilutions for use as shelf reagents; however, it is not necessary that these shelf reagents be highly accurate.

Reagent	Molarity	Reagent	Molarity
Acetic acid (99.5%)	17.4	HNO_3 (69%)	15.4
HCl (38%)	12.0	H_3PO_4 (85%)	14.7
$HClO_4$ (72%)	11.6	H_2SO_4 (94%)	17.6
HF (45%)	25.7	Ammonia (27%)	14.3

Routine Normal Solutions†

A concentrated C. P. acid usually comes to the laboratory in a bottle having a label which states: its molecular weight, M; its specific gravity, d; and its percentage assay, p. When such an acid is used in preparing an aqueous solution of desired normality, N, a convenient formula to employ is:

$$V = \frac{100 \; MN}{bpd}$$

wherein b is the basicity of the acid and V is the number of milliliters of concentrated acid required for one liter of the dilute solution.

Example. Acid sulfuric is dibasic and has the molecular weight 98.08. If the concentrated acid assays 95.5% and has the specific gravity 1.84, the volume required for one liter of a 0.2 normal solution is:

$$V = \frac{100 \times 98.08 \times 0.2}{2 \times 95.5 \times 1.84} = 5.58 \text{ ml}$$

Such a solution may often be used just as it stands. Where accuracy is required, it is most conveniently standardized by comparison with a high grade Potassium Biphthalate or other primary standard, *q.v.*

---

† Hague, *Chemist Analyst* **30**(1), 21 (1941); reprinted by permission of J. T. Baker Chemical Co.

INDEX

SUBJECT INDEX